WIN the complete library of RSMeans Online data!

Register your book below to receive your FREE quarterly updates, PLUS you will be entered into a quarterly drawing to win the COMPLETE RSMeans Online library of 2015 data!

Be sure to keep up-to-date in 2015!

Fill out the card below for RSMeans' free quarterly updates, as well as a chance to win the complete RSMeans Online library. Please provide your name, address, and email below and return this card by mail, or to register online:

http://info.thegordiangroup.com/RSMeans.html

Name _____

email _____ Title _____

Company _____

Street _____

City/Town _____ State/Prov. _____ Zip/Postal Code _____

RSMeans

WIN the complete library of RSMeans Online 2015 data!

See other side for details.

BUSINESS REPLY MAIL
FIRST-CLASS MAIL   PERMIT NO. 31   NORWELL, MA

POSTAGE WILL BE PAID BY ADDRESSEE

RSMEANS
700 LONGWATER DRIVE
NORWELL MA 02061-1624

NO POSTAGE
NECESSARY
IF MAILED
IN THE
UNITED STATES

How the Book Is Built: An Overview

The Construction Specifications Institute (CSI) and Construction Specifications Canada (CSC) have produced the 2014 edition of MasterFormat®, a system of titles and numbers used extensively to organize construction information.

All unit price data in the RSMeans cost data books is now arranged in the 50-division MasterFormat® 2014 system.

Powerful Construction Tool

You have in your hands one of the most powerful construction tools available today. A successful project is built on the foundation of an accurate and dependable estimate. This book will enable you to construct just such an estimate.

For the casual user the book is designed to be:

- quickly and easily understood so you can get right to your estimate.
- filled with valuable information so you can understand the necessary factors that go into the cost estimate.

For the regular user, the book is designed to be:

- a handy desk reference that can be quickly referred to for key costs.
- a comprehensive, fully reliable source of current construction costs and productivity rates so you'll be prepared to estimate any project.
- a source book for preliminary project cost, product selections, and alternate materials and methods.

To meet all of these requirements, we have organized the book into the following clearly defined sections.

Quick Start

See our "Quick Start" instructions on the following page to get started right away.

Estimating with RSMeans Unit Price Cost Data

Please refer to these steps for guidance on completing an estimate using RSMeans unit price cost data.

How to Use the Book: The Details

This section contains an in-depth explanation of how the book is arranged . . . and how you can use it to determine a reliable construction cost estimate. It includes information about how we develop our cost figures and how to completely prepare your estimate.

Unit Price Section

All cost data has been divided into the 50 divisions according to the MasterFormat system of classification and numbering. For a listing of these divisions and an outline of their subdivisions, see the Unit Price Section Table of Contents.

Estimating tips are included at the beginning of each division.

Assemblies Section

The cost data in this section has been organized in an "Assemblies" format. These assemblies are the functional elements of a building and are arranged according to the 7 elements of the UNIFORMAT II classification system. For a complete explanation of a typical "Assemblies" page, see "How RSMeans Assemblies Data Works."

Reference Section

This section includes information on Equipment Rental Costs, Crew Listings, Historical Cost Indexes, City Cost Indexes, Location Factors, Reference Tables, Change Orders, Square Foot Costs, and a listing of Abbreviations.

Equipment Rental Costs: This section contains the average costs to rent and operate hundreds of pieces of construction equipment.

Crew Listings: This section lists all of the crews referenced in the book. For the purposes of this book, a crew is composed of more than one trade classification and/or the addition of power equipment to any trade classification. Power equipment is included in the cost of the crew. Costs are shown both with bare labor rates and with the installing contractor's overhead and profit added. For each, the total crew cost per eight-hour day and the composite cost per labor-hour are listed.

Historical Cost Indexes: These indexes provide you with data to adjust construction costs over time.

City Cost Indexes: All costs in this book are U.S. national averages. Costs vary because of the regional economy. You can adjust costs by CSI Division to over 700 locations throughout the U.S. and Canada by using the data in this section.

Location Factors: You can adjust total project costs to over 900 locations throughout the U.S. and Canada by using the data in this section.

Reference Tables: At the beginning of selected major classifications in the Unit Price Section are reference numbers shown in a shaded box. These numbers refer you to related information in the Reference Section. In this section, you'll find reference tables, explanations, estimating information that support how we develop the unit price data, technical data, and estimating procedures.

Change Orders: This section includes information on the factors that influence the pricing of change orders.

Square Foot Costs: This section contains costs for 59 different building types that allow you to make a rough estimate for the overall cost of a project or its major components.

Abbreviations: A listing of abbreviations used throughout this book, along with the terms they represent, is included in this section.

Index

A comprehensive listing of all terms and subjects in this book will help you quickly find what you need when you are not sure where it occurs in MasterFormat.

The Scope of This Book

This book is designed to be as comprehensive and as easy to use as possible. To that end we have made certain assumptions and limited its scope in two key ways:

1. We have established material prices based on a national average.
2. We have computed labor costs based on a 30-city national average of union wage rates.

For a more detailed explanation of how the cost data is developed, see "How To Use the Book: The Details."

Project Size/Type

The material prices in RSMeans cost data books are "contractor's prices." They are the prices that contractors can expect to pay at the lumberyards, suppliers/distributers warehouses, etc. Small orders of speciality items would be higher than the costs shown, while very large orders, such as truckload lots, would be less. The variation would depend on the size, timing, and negotiating power of the contractor. The labor costs are primarily for new construction or major renovation rather than repairs or minor alterations.

With reasonable exercise of judgment, the figures can be used for any building work.

Absolute Essentials for a Quick Start

If you feel you are ready to use this book and don't think you will need the detailed instructions that begin on the following page, this Absolute Essentials for a Quick Start page is for you. These steps will allow you to get started estimating in a matter of minutes.

1 Scope
Think through the project that you will be estimating, and identify the many individual work tasks that will need to be covered in your estimate.

2 Quantify
Determine the number of units that will be required for each work task that you identified.

3 Pricing
Locate individual Unit Price line items that match the work tasks you identified. The Unit Price Section Table of Contents that begins on page 1 and the Index in the back of the book will help you find these line items.

4 Multiply
Multiply the Total Incl O&P cost for a Unit Price line item in the book by your quantity for that item. The price you calculate will be an estimate for a completed item of work performed by a subcontractor. Keep adding line items in this manner to build your estimate.

5 Project Overhead
Include project overhead items in your estimate. These items are needed to make the job run and are typically, but not always, provided by the General Contractor. They can be found in Division 1. An alternate method of estimating project overhead costs is to apply a percentage of the total project cost.

Include rented tools not included in crews, waste, rubbish handling, and cleanup.

6 Estimate Summary
Include General Contractor's markup on subcontractors, General Contractor's office overhead and profit, and sales tax on materials and equipment.

Adjust your estimate to the project's location by using the City Cost Indexes or Location Factors found in the Reference Section.

Editors' Note: We urge you to spend time reading and understanding the supporting material in the front of this book. An accurate estimate requires experience, knowledge, and careful calculation. The more you know about how we at RSMeans developed the data, the more accurate your estimate will be. In addition, it is important to take into consideration the reference material in the back of the book such as Equipment Listings, Crew Listings, City Cost Indexes, Location Factors, and Reference Tables.

Estimating with RSMeans Unit Price Cost Data

[Fol]lowing these steps will allow you to [com]plete an accurate estimate using [RS]Means Unit Price cost data.

Scope Out the Project

Think through the project and identify those CSI Divisions needed in your estimate.

Identify the individual work tasks that will need to be covered in your estimate.

The Unit Price data in this book has been divided into 50 Divisions according to the CSI MasterFormat® 2014—their titles are listed on the back cover of your book.

The Unit Price Section Table of Contents on page 1 may also be helpful when scoping out your project.

Experienced estimators find it helpful to begin with Division 2 and continue through completion. Division 1 can be estimated after the full project scope is known.

Quantify

Determine the number of units required for each work task that you identified.

Experienced estimators include an allowance for waste in their quantities. (Waste is not included in RSMeans Unit Price line items unless so stated.)

Price the Quantities

Use the Unit Price Table of Contents, and the Index, to locate individual Unit Price line items for your estimate.

Reference Numbers indicated within a Unit Price section refer to additional information that you may find useful.

The crew indicates who is performing the work for that task. Crew codes are expanded in the Crew Listings in the Reference Section to include all trades and equipment that comprise the crew.

- The Daily Output is the amount of work the crew is expected to do in one day.
- The Labor-Hours value is the amount of time it will take for the crew to install one unit of work.
- The abbreviated Unit designation indicates the unit of measure upon which the crew, productivity, and prices are based.
- Bare Costs are shown for materials, labor, and equipment needed to complete the Unit Price line item. Bare costs do not include waste, project overhead, payroll insurance, payroll taxes, main office overhead, or profit.
- The Total Incl O&P cost is the billing rate or invoice amount of the installing contractor or subcontractor who performs the work for the Unit Price line item.

4 Multiply

- Multiply the total number of units needed for your project by the Total Incl O&P cost for each Unit Price line item.
- Be careful that your take off unit of measure matches the unit of measure in the Unit column.
- The price you calculate is an estimate for a completed item of work.
- Keep scoping individual tasks, determining the number of units required for those tasks, matching each task with individual Unit Price line items in the book, and multiply quantities by Total Incl O&P costs.
- An estimate completed in this manner is priced as if a subcontractor, or set of subcontractors, is performing the work. The estimate does not yet include Project Overhead or Estimate Summary components such as General Contractor markups on subcontracted work, General Contractor office overhead and profit, contingency, and location factor.

5 Project Overhead

- Include project overhead items from Division 1 – General Requirements.
- These items are needed to make the job run. They are typically, but not always, provided by the General Contractor. Items include, but are not limited to, field personnel, insurance, performance bond, permits, testing, temporary utilities, field office and storage facilities, temporary scaffolding and platforms, equipment mobilization and demobilization, temporary roads and sidewalks, winter protection, temporary barricades and fencing, temporary security, temporary signs, field engineering and layout, final cleaning and commissioning.
- Each item should be quantified, and matched to individual Unit Price line items in Division 1, and then priced and added to your estimate.
- An alternate method of estimating project overhead costs is to apply a percentage of the total project cost, usually 5% to 15% with an average of 10% (see General Conditions, page ix).
- Include other project related expenses in your estimate such as:
    - Rented equipment not itemized in the Crew Listings
    - Rubbish handling throughout the project (see 02 41 19.19)

6 Estimate Summary

- Include sales tax as required by laws of your state or county.
- Include the General Contractor's markup on self-performed work, usually 5% to 15% with an average of 10%.
- Include the General Contractor's markup on subcontracted work, usually 5% to 15% with an average of 10%.
- Include General Contractor's main office overhead and profit:
  - RSMeans gives general guidelines on the General Contractor's main office overhead (see section 01 31 13.60 and Reference Number R013113-50).
  - RSMeans gives no guidance on the General Contractor's profit.
  - Markups will depend on the size of the General Contractor's operations, his projected annual revenue, the level of risk he is taking on, and on the level of competition in the local area and for this project in particular.
- Include a contingency, usually 3% to 5%, if appropriate.
- Adjust your estimate to the project's location by using the City Cost Indexes or the Location Factors in the Reference Section:
  - Look at the rules on the pages for "How to Use the City Cost Indexes" to see how to apply the Indexes for your location.
  - When the proper Index or Factor has been identified for the project's location, convert it to a multiplier by dividing it by 100, and then multiply that multiplier by your estimated total cost. The original estimated total cost will now be adjusted up or down from the national average to a total that is appropriate for your location.

Editors' Notes:
We urge you to spend time reading and understanding the supporting material in the front of this book. An accurate estimate requires experience, knowledge, and careful calculation. The more you know about how we at RSMeans developed the data, the more accurate your estimate will be. In addition, it is important to take into consideration the reference material in the back of the book such as Equipment Listings, Crew Listings, City Cost Indexes, Location Factors, and Reference Tables.

How to Use the Book: The Details

What's Behind the Numbers? The Development of Cost Data

The staff at RSMeans continually monitors developments in the construction industry in order to ensure reliable, thorough, and up-to-date cost information. While overall construction costs may vary relative to general economic conditions, price fluctuations within the industry are dependent upon many factors. Individual price variations may, in fact, be opposite to overall economic trends. Therefore, costs are constantly tracked and complete updates are published yearly. Also, new items are frequently added in response to changes in materials and methods.

Costs—$ (U.S.)

All costs represent U.S. national averages and are given in U.S. dollars. The RSMeans City Cost Indexes can be used to adjust costs to a particular location. The City Cost Indexes for Canada can be used to adjust U.S. national averages to local costs in Canadian dollars. No exchange rate conversion is necessary.

G The processes or products identified by the green symbol in our publications have been determined to be environmentally responsible and/or resource-efficient solely by the RSMeans engineering staff. The inclusion of the green symbol does not represent compliance with any specific industry association or standard.

Material Costs

The RSMeans staff contacts manufacturers, dealers, distributors, and contractors all across the U.S. and Canada to determine national average material costs. If you have access to current material costs for your specific location, you may wish to make adjustments to reflect differences from the national average. Included within material costs are fasteners for a normal installation. RSMeans engineers use manufacturers' recommendations, written specifications, and/or standard construction practice for size and spacing of fasteners. Adjustments to material costs may be required for your specific application or location. The manufacturer's warranty is assumed. Extended warranties are not included in the material costs. Material costs do not include sales tax.

Labor Costs

Labor costs are based on the average of wage rates from 30 major U.S. cities. Rates are determined from labor union agreements or prevailing wages for construction trades for the current year. Rates, along with overhead and profit markups, are listed on the inside back cover of this book.

- If wage rates in your area vary from those used in this book, or if rate increases are expected within a given year, labor costs should be adjusted accordingly.

Labor costs reflect productivity based on actual working conditions. In addition to actual installation, these figures include time spent during a normal weekday on tasks such as, material receiving and handling, mobilization at site, site movement, breaks, and cleanup.

Productivity data is developed over an extended period so as not to be influenced by abnormal variations and reflects a typical average.

Equipment Costs

Equipment costs include not only rental, but also operating costs for equipment under normal use. The operating costs include parts and labor for routine servicing such as repair and replacement of pumps, filters, and worn lines. Normal operating expendables, such as fuel, lubricants, tires, and electricity (where applicable), are also included. Extraordinary operating expendables with highly variable wear patterns, such as diamond bits and blades, are excluded. These costs are included under materials. Equipment rental rates are obtained from industry sources throughout North America—contractors, suppliers, dealers, manufacturers, and distributors.

Rental rates can also be treated as reimbursement costs for contractor-owned equipment. Owned equipment costs include depreciation, loan payments, interest, taxes, insurance, storage and major repairs.

Equipment costs do not include operators' wages; nor do they include the cost to move equipment to a job site (mobilization) or from a job site (demobilization).

Equipment Cost/Day—The cost of power equipment required for each crew is included in the Crew Listings in the Reference Section (small tools that are considered as essential everyday tools are not listed out separately). The Crew Listings itemize specialized tools and heavy equipment along with labor trades. The daily cost of itemized equipment included in a crew is based on dividing the weekly bare rental rate by 5 (number of working days per week) and then adding the hourly operating cost times 8 (the number of hours per day). This Equipment Cost/Day is shown in the last column of the Equipment Rental Cost pages in the Reference Section.

Mobilization/Demobilization—The cost to move construction equipment from an equipment yard or rental company to the job site and back again is not included in equipment costs. Mobilization (to the site) and demobilization (from the site) costs can be found in the Unit Price Section. If a piece of equipment is already at the job site, it is not appropriate to utilize mob./demob. costs again in an estimate.

Overhead and Profit

Total Cost including O&P for the *Installing Contractor* is shown in the last column on the Unit Price and/or the Assemblies pages of this book. This figure is the sum of the bare material cost plus 10% for profit, the bare labor cost plus total overhead and profit, and the bare equipment cost plus 10% for profit. Details for the calculation of Overhead and Profit on labor are shown on the inside back cover and in the Reference Section of this book. (See "How RSMeans Unit Price Data Works" for an example of this calculation.)

General Conditions

Cost data in this book is presented in two ways: Bare Costs and Total Cost including O&P (Overhead and Profit). General Conditions, or General Requirements, of the contract should also be added to the Total Cost including O&P when applicable. Costs for General Conditions are listed in Division 1 of the Unit Price Section and the Reference

Section of this book. General Conditions for the *Installing Contractor* may range from 0% to 10% of the Total Cost including O&P. For the *General* or *Prime Contractor*, costs for General Conditions may range from 5% to 15% of the Total Cost including O&P, with a figure of 10% as the most typical allowance. If applicable, the Assemblies and Models sections of this book use costs that include the installing contractor's overhead and profit (O&P).

Factors Affecting Costs

Costs can vary depending upon a number of variables. Here's how we have handled the main factors affecting costs.

Quality—The prices for materials and the workmanship upon which productivity is based represent sound construction work. They are also in line with U.S. government specifications.

Overtime—We have made no allowance for overtime. If you anticipate premium time or work beyond normal working hours, be sure to make an appropriate adjustment to your labor costs.

Productivity—The productivity, daily output, and labor-hour figures for each line item are based on working an eight-hour day in daylight hours in moderate temperatures. For work that extends beyond normal work hours or is performed under adverse conditions, productivity may decrease. (See "How RSMeans Unit Price Data Works" for more on productivity.)

Size of Project—The size, scope of work, and type of construction project will have a significant impact on cost. Economies of scale can reduce costs for large projects. Unit costs can often run higher for small projects.

Location—Material prices in this book are for metropolitan areas. However, in dense urban areas, traffic and site storage limitations may increase costs. Beyond a 20-mile radius of large cities, extra trucking or transportation charges may also increase the material costs slightly. On the other hand, lower wage rates may be in effect. Be sure to consider both of these factors when preparing an estimate, particularly if the job site is located in a central city or remote rural location. In addition, highly specialized subcontract items may require travel and per-diem expenses for mechanics.

Other Factors—
- season of year
- contractor management
- weather conditions
- local union restrictions
- building code requirements
- availability of:
  - adequate energy
  - skilled labor
  - building materials
- owner's special requirements/restrictions
- safety requirements
- environmental considerations

Unpredictable Factors—General business conditions influence "in-place" costs of all items. Substitute materials and construction methods may have to be employed. These may affect the installed cost and/or life cycle costs. Such factors may be difficult to evaluate and cannot necessarily be predicted on the basis of the job's location in a particular section of the country. Thus, where these factors apply, you may find significant but unavoidable cost variations for which you will have to apply a measure of judgment to your estimate.

Rounding of Costs

In general, all unit prices in excess of $5.0 have been rounded to make them easier to use and still maintain adequate precision of the results. The rounding rules we have chosen are in the following table.

Prices from . . .	Rounded to the nearest . . .
$.01 to $5.00	$.0
$5.01 to $20.00	$.0
$20.01 to $100.00	$.5
$100.01 to $300.00	$1.0
$300.01 to $1,000.00	$5.0
$1,000.01 to $10,000.00	$25.0
$10,000.01 to $50,000.00	$100.0
$50,000.01 and above	$500.0

How Subcontracted Items Affect Costs

A considerable portion of all large construction jobs is usually subcontracted. I fact, the percentage done by subcontractors is constantly increasing and may run over 90%. Since the workers employed by these companies do nothing else but install their particular product, they soon become expert in that line. The result is, installation by these firms is accomplished so efficiently that the total in-place cost, even adding the general contractor's overhead and profit, is no more, and often less, than if the principal contractor had handled the installation himself/herself. Companies that deal with construction specialties are anxious to have their product perform well and, consequently the installation will be the best possible.

Contingencies

The allowance for contingencies generally provides for unforeseen construction difficulties. On alterations or repair jobs, 20% is not too much. If drawings are final and only field contingencies are being considered, 2% or 3% is probably sufficient, and often nothing need be added. Contractually, changes in plans will be covered by extras. The contractor should consider inflationary price trends and possible material shortages during the course of the job. These escalation factors are dependent upon both economic conditions and the anticipated time between the estimate and actual construction. If drawings are not complete or approved, or a budget cost is wanted, it is wise to add 5% to 10%. Contingencies, then, are a matter of judgment. Additional allowances for contingencies are shown in Division 1.

Important Estimating Considerations

The productivity or daily output of each craftsman or crew assumes a well-managed job where tradesmen with the proper tools and equipment, along with the appropriate construction materials, are present. Included are daily mobilization and cleanup time, break time and plan layout time. Unless otherwise indicated, time for material movement on site (for items that can be transported by hand) of up to 200' into the building and to the first or second floor is also included. If material has to be transported by other means, over greater distances, or to higher floors, an additional allowance should be considered by the estimator.

While horizontal movement is typically a sole function of distances, vertical transport introduces other variables that can significantly impact productivity. In an occupied building, the use of elevators (assuming access, size, and required protective measures are acceptable) must be understood at the time of the estimate. For new construction, hoist wait and cycle times can easily be 15 minutes and may result in scheduled access extending beyond the normal work day. Finally, all vertical transport will impose strict weight limits likely to preclude the use of any motorized material handling.

The productivity, or daily output, also assumes installation that meets manufacturer/designer/standard specifications. A time allowance for quality control checks, minor adjustments and any task required to ensure the proper function or operation is also included. For items that require connections to services, time is included for positioning, leveling, securing the unit and for making all the necessary connections (and start up where applicable) ensuring a complete installation. Estimating of the services themselves (electrical, plumbing, water, steam, hydraulics, dust collection, etc.) is separate.

In some cases, the estimator must consider the use of a crane and an appropriate crew for the installation of large or heavy items. For those situations where a crane is not included in the assigned crew and as part of the line item cost, equipment rental costs, mobilization and demobilization costs, and operator and support personnel costs must be considered.

Labor-Hours

The labor-hours expressed in this publication are derived by dividing the total daily labor hours for the crew by the daily output. Based on average installation time and the assumptions listed above, the labor-hours include: direct labor, indirect labor, and nonproductive time. A typical day for a craftsman might include, but not be limited to:

- Direct Work
    - Measuring and layout
    - Preparing materials
    - Actual installation
    - Quality assurance/quality control
- Indirect Work
    - Reading plans or specifications
    - Preparing space
    - Receiving materials
    - Material movement
    - Giving or receiving instruction
    - Miscellaneous
- Non-Work
    - Chatting
    - Personal issues
    - Breaks
    - Interruptions (i.e., sickness, weather, material or equipment shortages, etc.)

If any of the items for a typical day do not apply to the particular work or project situation, the estimator should make any necessary adjustments.

Final Checklist

Estimating can be a straightforward process provided you remember the basics. Here's a checklist of some of the steps you should remember to complete before finalizing your estimate.

Did you remember to . . .

- factor in the City Cost Index for your locale?
- take into consideration which items have been marked up and by how much?
- mark up the entire estimate sufficiently for your purposes?
- read the background information on techniques and technical matters that could impact your project time span and cost?
- include all components of your project in the final estimate?
- double check your figures for accuracy?
- call RSMeans if you have any questions about your estimate or the data you've found in our publications? Remember, RSMeans stands behind its publications. If you have any questions about your estimate . . . about the costs you've used from our books . . . or even about the technical aspects of the job that may affect your estimate, feel free to call the RSMeans editors at 1-877-79-4977.

Free Quarterly Updates

Stay up-to-date throughout 2015 with RSMeans' free cost data updates four times a year. Sign up online to make sure you have access to the newest data. Every quarter we provide the city cost adjustment factors for hundreds of cities and key materials. Register at:
http://info.thegordiangroup.com/RSMeans.html.

Using Minimum Labor/Equipment Charges for Small Quantities

Estimating small construction or repair tasks often creates situations in which the quantity of work to be performed is very small. When this occurs, the labor and/or equipment costs to perform the work may be too low to allow for the crew to get to the job, receive instructions, find materials, get set up, perform the work, clean up, and get to the next job. In these situations, the estimator should compare the developed labor and/or equipment costs for performing the work (e.g., quantity × labor and/or equipment costs) with the *minimum labor/equipment charge* within that Unit Price section of the book. (These minimum labor/equipment charge line items appear only in *Facilities Construction Cost Data* and *Commercial Renovation Cost Data*.)

If the labor and/or equipment costs developed by the estimator are LOWER THAN the *minimum labor/equipment charge* listed at the bottom of specific sections of Unit Price costs, the estimator should adjust the developed costs upward to the *minimum labor/equipment charge*. The proper use of a *minimum labor/equipment charge* results in having enough money in the estimate to cover the contractor's higher cost of performing a very small amount of work during a partial workday.

A *minimum labor/equipment charge* should be used only when the task being estimated is the only task the crew will perform at the job site that day. **If, however, the crew will be able to perform other tasks at the job site that day, the use of a minimum labor/equipment charge is not appropriate.**

08 52 Wood Windows

08 52 10 – Plain Wood Windows

08 52 10.40 Casement Window		Crew	Daily Output	Labor-Hours	Unit	Material	2015 Bare Costs Labor	2015 Bare Costs Equipment	Total	Total Incl O&P
0010	**CASEMENT WINDOW**, including frame, screen and grilles									
0100	2'-0" x 3'-0" H, dbl. insulated glass [G]	1 Carp	10	.800	Ea.	268	37.50		305.50	355
0150	Low E glass [G]		10	.800		259	37.50		296.50	345
0200	2'-0" x 4'-6" high, double insulated glass [G]		9	.889		360	41.50		401.50	465
0250	Low E glass [G]		9	.889		370	41.50		411.50	475
9000	Minimum labor/equipment charge	1 Carp	3	2.667	Job		125		125	205

Example:
Establish the bid price to install two casement windows. Assume installation of 2'×4'-6" metal clad windows with insulating glass [Unit Price line number 08 52 10.40 0250], and that this is the only task this crew will perform at the job site that day.

Solution:
Step One — Develop the Bare Labor Cost for this task:
Bare Labor Cost = 2 windows @ $41.50/each = $83.00

Step Two — Evaluate the *minimum labor/equipment charge* for this Unit Price section against the developed Bare Labor Cost for this task:
Minimum labor/equipment charge = $125.00 (compare with $83.00)

Step Three — Choose to adjust the developed labor cost upward to the *minimum labor/equipment charge*.

Step Four — Develop the bid price for this task (including O&P):
Add together the marked-up Bare Material Cost for this task and the marked-up *minimum labor/equipment charge* for this Unit Price section.

2 × ($370.00 + 10%) + ($125.00 + 63.9%)

= 2 × ($370.00 + $37.00) + ($125.00 + $79.88)

= 2 × ($407.00) + $204.88

= $814.00 + $204.88

= $1,018.88

ANSWER: $1,018.88 is the correct bid price to use. This sum takes into consideration the Material Cost (with 10% for profit) for these two windows, plus the *minimum labor/equipment charge* (with O&P included) for this section of the Unit Price.

Unit Price Section

Table of Contents

Sect. No.		Page
	General Requirements	
01 11	Summary of Work	10
01 21	Allowances	10
01 31	Project Management and Coordination	11
01 41	Regulatory Requirements	13
01 45	Quality Control	13
01 51	Temporary Utilities	14
01 52	Construction Facilities	14
01 54	Construction Aids	14
01 55	Vehicular Access and Parking	18
01 56	Temporary Barriers and Enclosures	19
01 58	Project Identification	21
01 71	Examination and Preparation	21
01 74	Cleaning and Waste Management	21
01 76	Protecting Installed Construction	21
01 91	Commissioning	21
01 93	Facility Maintenance	22
	Existing Conditions	
02 21	Surveys	26
02 32	Geotechnical Investigations	26
02 41	Demolition	26
02 42	Removal and Salvage of Construction Materials	33
02 43	Structure Moving	35
02 58	Snow Control	36
02 65	Underground Storage Tank Removal	36
02 81	Transportation and Disposal of Hazardous Materials	37
02 82	Asbestos Remediation	37
02 83	Lead Remediation	41
02 85	Mold Remediation	43
	Concrete	
03 01	Maintenance of Concrete	46
03 05	Common Work Results for Concrete	47
03 11	Concrete Forming	49
03 15	Concrete Accessories	54
03 21	Reinforcement Bars	58
03 22	Fabric and Grid Reinforcing	59
03 24	Fibrous Reinforcing	60
03 30	Cast-In-Place Concrete	60
03 31	Structural Concrete	63
03 35	Concrete Finishing	65
03 37	Specialty Placed Concrete	67
03 39	Concrete Curing	68
03 41	Precast Structural Concrete	68
03 45	Precast Architectural Concrete	69
03 48	Precast Concrete Specialties	69
03 51	Cast Roof Decks	70
03 52	Lightweight Concrete Roof Insulation	70
03 53	Concrete Topping	70
03 54	Cast Underlayment	70
03 63	Epoxy Grouting	71
03 81	Concrete Cutting	71
03 82	Concrete Boring	72
	Masonry	
04 01	Maintenance of Masonry	76
04 05	Common Work Results for Masonry	77
04 21	Clay Unit Masonry	83
04 22	Concrete Unit Masonry	88
04 23	Glass Unit Masonry	92
04 27	Multiple-Wythe Unit Masonry	93
04 41	Dry-Placed Stone	94
04 43	Stone Masonry	94
04 54	Refractory Brick Masonry	96
04 57	Masonry Fireplaces	97
04 71	Manufactured Brick Masonry	97
04 72	Cast Stone Masonry	97
04 73	Manufactured Stone Masonry	98

Sect. No.		Page
	Metals	
05 05	Common Work Results for Metals	100
05 12	Structural Steel Framing	103
05 15	Wire Rope Assemblies	107
05 21	Steel Joist Framing	107
05 31	Steel Decking	108
05 35	Raceway Decking Assemblies	109
05 41	Structural Metal Stud Framing	109
05 42	Cold-Formed Metal Joist Framing	113
05 44	Cold-Formed Metal Trusses	119
05 51	Metal Stairs	120
05 52	Metal Railings	122
05 55	Metal Stair Treads and Nosings	123
05 56	Metal Castings	124
05 58	Formed Metal Fabrications	125
05 71	Decorative Metal Stairs	126
05 73	Decorative Metal Railings	126
05 75	Decorative Formed Metal	127
	Wood, Plastics & Composites	
06 05	Common Work Results for Wood, Plastics, and Composites	130
06 11	Wood Framing	137
06 12	Structural Panels	146
06 13	Heavy Timber Construction	147
06 15	Wood Decking	148
06 16	Sheathing	149
06 17	Shop-Fabricated Structural Wood	151
06 18	Glued-Laminated Construction	152
06 22	Millwork	153
06 25	Prefinished Paneling	164
06 26	Board Paneling	166
06 43	Wood Stairs and Railings	166
06 44	Ornamental Woodwork	168
06 48	Wood Frames	170
06 49	Wood Screens and Exterior Wood Shutters	171
06 51	Structural Plastic Shapes and Plates	172
06 52	Plastic Structural Assemblies	173
06 63	Plastic Railings	173
06 65	Plastic Trim	174
06 80	Composite Fabrications	175
06 81	Composite Railings	175
	Thermal & Moisture Protection	
07 01	Operation and Maint. of Thermal and Moisture Protection	178
07 05	Common Work Results for Thermal and Moisture Protection	179
07 11	Dampproofing	180
07 12	Built-up Bituminous Waterproofing	180
07 13	Sheet Waterproofing	181
07 16	Cementitious and Reactive Waterproofing	181
07 19	Water Repellents	182
07 21	Thermal Insulation	182
07 22	Roof and Deck Insulation	186
07 24	Exterior Insulation and Finish Systems	188
07 25	Weather Barriers	188
07 26	Vapor Retarders	188
07 27	Air Barriers	189
07 31	Shingles and Shakes	189
07 32	Roof Tiles	191
07 33	Natural Roof Coverings	192
07 41	Roof Panels	193
07 42	Wall Panels	194
07 46	Siding	195
07 51	Built-Up Bituminous Roofing	198
07 52	Modified Bituminous Membrane Roofing	200
07 53	Elastomeric Membrane Roofing	200
07 54	Thermoplastic Membrane Roofing	201
07 55	Protected Membrane Roofing	202
07 56	Fluid-Applied Roofing	202
07 57	Coated Foamed Roofing	203

Sect. No.		Page
07 58	Roll Roofing	203
07 61	Sheet Metal Roofing	203
07 62	Sheet Metal Flashing and Trim	204
07 65	Flexible Flashing	205
07 71	Roof Specialties	207
07 72	Roof Accessories	211
07 76	Roof Pavers	213
07 81	Applied Fireproofing	213
07 84	Firestopping	214
07 91	Preformed Joint Seals	216
07 92	Joint Sealants	217
07 95	Expansion Control	218
	Openings	
08 01	Operation and Maintenance of Openings	220
08 05	Common Work Results for Openings	221
08 11	Metal Doors and Frames	222
08 12	Metal Frames	223
08 13	Metal Doors	226
08 14	Wood Doors	229
08 16	Composite Doors	236
08 17	Integrated Door Opening Assemblies	237
08 31	Access Doors and Panels	238
08 32	Sliding Glass Doors	240
08 33	Coiling Doors and Grilles	241
08 34	Special Function Doors	242
08 36	Panel Doors	244
08 38	Traffic Doors	245
08 41	Entrances and Storefronts	246
08 42	Entrances	248
08 43	Storefronts	248
08 45	Translucent Wall and Roof Assemblies	249
08 51	Metal Windows	250
08 52	Wood Windows	251
08 53	Plastic Windows	258
08 54	Composite Windows	260
08 56	Special Function Windows	261
08 61	Roof Windows	261
08 62	Unit Skylights	262
08 63	Metal-Framed Skylights	262
08 71	Door Hardware	263
08 74	Access Control Hardware	273
08 75	Window Hardware	274
08 79	Hardware Accessories	274
08 81	Glass Glazing	275
08 83	Mirrors	277
08 84	Plastic Glazing	277
08 85	Glazing Accessories	278
08 87	Glazing Surface Films	278
08 88	Special Function Glazing	278
08 91	Louvers	280
08 95	Vents	281
	Finishes	
09 01	Maintenance of Finishes	284
09 05	Common Work Results for Finishes	284
09 21	Plaster and Gypsum Board Assemblies	286
09 22	Supports for Plaster and Gypsum Board	287
09 23	Gypsum Plastering	291
09 24	Cement Plastering	292
09 25	Other Plastering	292
09 26	Veneer Plastering	292
09 28	Backing Boards and Underlayments	293
09 29	Gypsum Board	293
09 30	Tiling	297
09 34	Waterproofing-Membrane Tiling	299
09 51	Acoustical Ceilings	299
09 53	Acoustical Ceiling Suspension Assemblies	301

Table of Contents (cont.)

Sect. No.		Page
09 54	Specialty Ceilings	301
09 61	Flooring Treatment	302
09 62	Specialty Flooring	302
09 63	Masonry Flooring	303
09 64	Wood Flooring	304
09 65	Resilient Flooring	305
09 66	Terrazzo Flooring	307
09 67	Fluid-Applied Flooring	308
09 68	Carpeting	308
09 69	Access Flooring	310
09 72	Wall Coverings	310
09 77	Special Wall Surfacing	311
09 81	Acoustic Insulation	312
09 84	Acoustic Room Components	313
09 91	Painting	313
09 93	Staining and Transparent Finishing	326
09 96	High-Performance Coatings	327
Specialties		
10 11	Visual Display Units	330
10 13	Directories	331
10 14	Signage	331
10 17	Telephone Specialties	332
10 21	Compartments and Cubicles	332
10 22	Partitions	336
10 26	Wall and Door Protection	338
10 28	Toilet, Bath, and Laundry Accessories	338
10 31	Manufactured Fireplaces	340
10 32	Fireplace Specialties	341
10 35	Stoves	342
10 43	Emergency Aid Specialties	342
10 44	Fire Protection Specialties	342
10 51	Lockers	343
10 55	Postal Specialties	344
10 56	Storage Assemblies	344
10 57	Wardrobe and Closet Specialties	344
10 73	Protective Covers	345
10 74	Manufactured Exterior Specialties	347
10 86	Security Mirrors and Domes	347
Equipment		
11 11	Vehicle Service Equipment	350
11 12	Parking Control Equipment	350
11 13	Loading Dock Equipment	351
11 21	Retail and Service Equipment	351
11 22	Banking Equipment	353
11 30	Residential Equipment	354
11 32	Unit Kitchens	355
11 41	Foodservice Storage Equipment	356
11 42	Food Preparation Equipment	357
11 43	Food Delivery Carts and Conveyors	358
11 44	Food Cooking Equipment	358
11 46	Food Dispensing Equipment	359
11 48	Foodservice Cleaning and Disposal Equipment	360
11 52	Audio-Visual Equipment	361
11 53	Laboratory Equipment	361
11 57	Vocational Shop Equipment	362
11 61	Broadcast, Theater, and Stage Equipment	362
11 66	Athletic Equipment	363
11 71	Medical Sterilizing Equipment	364
11 76	Operating Room Equipment	364
11 81	Facility Maintenance Equipment	364
11 82	Facility Solid Waste Handling Equipment	365
11 91	Religious Equipment	366
11 97	Security Equipment	366
11 98	Detention Equipment	366
Furnishings		
12 21	Window Blinds	368
12 23	Interior Shutters	368
12 24	Window Shades	368
12 32	Manufactured Wood Casework	369

Sect. No.		Page
12 34	Manufactured Plastic Casework	372
12 35	Specialty Casework	372
12 36	Countertops	373
12 48	Rugs and Mats	375
12 51	Office Furniture	375
12 54	Hospitality Furniture	376
12 55	Detention Furniture	376
12 56	Institutional Furniture	376
12 63	Stadium and Arena Seating	377
12 67	Pews and Benches	377
12 93	Interior Public Space Furnishings	377
Special Construction		
13 05	Common Work Results for Special Construction	380
13 12	Fountains	383
13 17	Tubs and Pools	383
13 21	Controlled Environment Rooms	384
13 24	Special Activity Rooms	384
13 28	Athletic and Recreational Special Construction	384
13 34	Fabricated Engineered Structures	385
13 42	Building Modules	386
13 48	Sound, Vibration, and Seismic Control	386
13 49	Radiation Protection	386
Conveying Equipment		
14 11	Manual Dumbwaiters	390
14 12	Electric Dumbwaiters	390
14 21	Electric Traction Elevators	390
14 24	Hydraulic Elevators	391
14 27	Custom Elevator Cabs and Doors	392
14 28	Elevator Equipment and Controls	393
14 42	Wheelchair Lifts	394
14 45	Vehicle Lifts	394
14 91	Facility Chutes	395
14 92	Pneumatic Tube Systems	395
Fire Suppression		
21 05	Common Work Results for Fire Suppression	398
21 11	Facility Fire-Suppression Water-Service Piping	398
21 12	Fire-Suppression Standpipes	398
21 13	Fire-Suppression Sprinkler Systems	399
Plumbing		
22 01	Operation and Maintenance of Plumbing	402
22 05	Common Work Results for Plumbing	402
22 07	Plumbing Insulation	405
22 11	Facility Water Distribution	406
22 13	Facility Sanitary Sewerage	419
22 14	Facility Storm Drainage	423
22 15	General Service Compressed-Air Systems	423
22 31	Domestic Water Softeners	424
22 33	Electric Domestic Water Heaters	424
22 34	Fuel-Fired Domestic Water Heaters	425
22 41	Residential Plumbing Fixtures	425
22 42	Commercial Plumbing Fixtures	429
22 43	Healthcare Plumbing Fixtures	431
22 45	Emergency Plumbing Fixtures	431
22 47	Drinking Fountains and Water Coolers	432
22 66	Chemical-Waste Systems for Lab. and Healthcare Facilities	432
Heating Ventilation Air Conditioning		
23 05	Common Work Results for HVAC	436
23 07	HVAC Insulation	437
23 09	Instrumentation and Control for HVAC	437
23 21	Hydronic Piping and Pumps	438
23 23	Refrigerant Piping	439
23 31	HVAC Ducts and Casings	440

Sect. No.		Page
23 33	Air Duct Accessories	442
23 34	HVAC Fans	444
23 36	Air Terminal Units	446
23 37	Air Outlets and Inlets	446
23 38	Ventilation Hoods	448
23 41	Particulate Air Filtration	448
23 42	Gas-Phase Air Filtration	448
23 43	Electronic Air Cleaners	448
23 51	Breechings, Chimneys, and Stacks	449
23 52	Heating Boilers	450
23 54	Furnaces	452
23 55	Fuel-Fired Heaters	453
23 56	Solar Energy Heating Equipment	454
23 61	Refrigerant Compressors	456
23 62	Packaged Compressor and Condenser Units	456
23 64	Packaged Water Chillers	456
23 65	Cooling Towers	457
23 72	Air-to-Air Energy Recovery Equipment	458
23 73	Indoor Central-Station Air-Handling Units	458
23 74	Packaged Outdoor HVAC Equipment	458
23 81	Decentralized Unitary HVAC Equipment	459
23 82	Convection Heating and Cooling Units	461
23 83	Radiant Heating Units	464
23 84	Humidity Control Equipment	465
23 91	Prefabricated Equipment Supports	465
Electrical		
26 01	Operation and Maintenance of Electrical Systems	468
26 05	Common Work Results for Electrical	468
26 22	Low-Voltage Transformers	483
26 24	Switchboards and Panelboards	484
26 27	Low-Voltage Distribution Equipment	485
26 28	Low-Voltage Circuit Protective Devices	487
26 29	Low-Voltage Controllers	488
26 32	Packaged Generator Assemblies	489
26 51	Interior Lighting	489
26 52	Emergency Lighting	491
26 53	Exit Signs	491
26 55	Special Purpose Lighting	491
26 56	Exterior Lighting	492
26 61	Lighting Systems and Accessories	492
Communications		
27 15	Communications Horizontal Cabling	496
27 41	Audio-Video Systems	496
27 51	Distributed Audio-Video Communications Systems	496
Electronic Safety & Security		
28 13	Access Control	500
28 16	Intrusion Detection	500
28 23	Video Surveillance	501
28 31	Fire Detection and Alarm	501
28 39	Mass Notification Systems	502
Earthwork		
31 05	Common Work Results for Earthwork	504
31 06	Schedules for Earthwork	504
31 14	Earth Stripping and Stockpiling	505
31 23	Excavation and Fill	505
31 25	Erosion and Sedimentation Controls	519
31 31	Soil Treatment	519
31 41	Shoring	519
31 43	Concrete Raising	520
31 46	Needle Beams	520

Table of Contents (cont.)

Sect. No.		Page
31 48	Underpinning	520
31 62	Driven Piles	521
31 63	Bored Piles	521
Exterior Improvements		
32 01	Operation and Maintenance of Exterior Improvements	524
32 06	Schedules for Exterior Improvements	530
32 11	Base Courses	531
32 12	Flexible Paving	531
32 13	Rigid Paving	532
32 14	Unit Paving	533
32 16	Curbs, Gutters, Sidewalks, and Driveways	534
32 17	Paving Specialties	535
32 31	Fences and Gates	536

Sect. No.		Page
32 32	Retaining Walls	539
32 33	Site Furnishings	539
32 84	Planting Irrigation	540
32 91	Planting Preparation	540
32 92	Turf and Grasses	542
32 93	Plants	543
32 94	Planting Accessories	544
32 96	Transplanting	545
Utilities		
33 01	Operation and Maintenance of Utilities	548
33 05	Common Work Results for Utilities	549
33 11	Water Utility Distribution Piping	550
33 12	Water Utility Distribution Equipment	550

Sect. No.		Page
33 21	Water Supply Wells	550
33 31	Sanitary Utility Sewerage Piping	551
33 36	Utility Septic Tanks	551
33 41	Storm Utility Drainage Piping	553
33 44	Storm Utility Water Drains	553
33 49	Storm Drainage Structures	553
33 51	Natural-Gas Distribution	554
33 52	Liquid Fuel Distribution	554
33 71	Electrical Utility Transmission and Distribution	555
Transportation		
34 71	Roadway Construction	558
Material Processing & Handling Equipment		
41 22	Cranes and Hoists	560

How RSMeans Unit Price Data Works

All RSMeans unit price data is organized in the same way.

It is important to understand the structure, so that you can find information easily and use it correctly.

RSMeans **Line Numbers** consist of 12 characters, which identify a unique location in the database for each task. The first 6 or 8 digits conform to the Construction Specifications Institute MasterFormat® 2014. The remainder of the digits are a further breakdown by RSMeans in order to arrange items in understandable groups of similar tasks. Line numbers are consistent across all RSMeans publications, so a line number in any RSMeans product will always refer to the same unit of work.

RSMeans engineers have created **reference** information to assist you in your estimate. If there is information that applies to a section, it will be indicated at the start of the section. In this case, R033105-10 provides information on the proportionate quantities of formwork, reinforcing, and concrete used in cast-in-place concrete items such as footings, slabs, beams, and columns. The Reference Section is located in the back of the book on the pages with a gray edge.

RSMeans **Descriptions** are shown in a hierarchical structure to make them readable. In order to read a complete description, read up through the indents to the top of the section. Include everything that is above and to the left that is not contradicted by information below. For instance, the complete description for line 03 30 53.40 3550 is "Concrete in place, including forms (4 uses), Grade 60 rebar, concrete (Portland cement Type 1), placement and finishing unless otherwise indicated; Equipment pad (3000 psi), 4' x 4' x 6" thick."

When using **RSMeans data**, it is important to read through an entire section to ensure that you use the data that most closely matches your work. Note that sometimes there is additional information shown in the section that may improve your price. There are frequently lines that further describe, add to, or adjust data for specific situations.

03 30 Cast-In-Place Concrete
03 30 53 – Miscellaneous Cast-In-Place Concrete

03 30 53.40	Concrete In Place	
0010	**CONCRETE IN PLACE**	R033105-10
0020	Including forms (4 uses), Grade 60 rebar, concrete (Portland cement	R033105-20
0050	Type I), placement and finishing unless otherwise indicated	R033105-70
0300	Beams (3500 psi), 5 kip per L.F., 10' span	R033105-85
0350	25' span	
0500	Chimney foundations (5000 psi), over 5 C.Y.	
0510	(3500 psi), under 5 C.Y.	
0700	Columns, square (4000 psi), 12" x 12", less than 2% reinforcing	
3100	Elevated slabs, flat plate, including finish, not	
3400	Cellular concrete, 1-5/8" fill, under 5000 S.F.	
3450	Over 10,000 S.F.	
3500	Add per floor for 3 to 6 stories high	
3520	For 7 to 20 stories high	
3540	Equipment pad (3000 psi), 3' x 3' x 6" thick	
3550	4' x 4' x 6" thick	
3560	5' x 5' x 8" thick	
3570	6' x 6' x 8" thick	

The data published in RSMeans print books represents a "national average" cost. This data should be modified to the project location using the **City Cost Indexes** or **Location Factors** tables found in the Reference Section (see pages 757–805). Use the location factors to adjust estimate totals if the project covers multiple trades. Use the city cost indexes (CCI) for single trade projects or projects where a more detailed analysis is required. All figures in the two tables are derived from the same research. The last row of data in the CCI, the weighted average, is the same as the numbers reported for each location in the location factor table.

Crews include labor or labor and equipment necessary to accomplish each task. In this case, Crew C-14H is used. RSMeans selects a crew to represent the workers and equipment that are typically used for that task. In this case, Crew C-14H consists of one carpenter foreman (outside), two carpenters, one rodman, one laborer, one cement finisher, and one gas engine vibrator. Details of all crews can be found in the reference section.

Crews

Crew No.	Bare Costs		Incl. Subs O&P		Cost Per Labor-Hour	
Crew C-14C	Hr.	Daily	Hr.	Daily	Bare Costs	Incl. O&P
1 Carpenter Foreman (outside)	$48.95	$391.60	$80.25	$642.00	$45.08	$73.66
6 Carpenters	46.95	2253.60	76.95	3693.60		
2 Rodmen (reinf.)	52.55	840.80	85.75	1372.00		
4 Laborers	37.60	1203.20	61.65	1972.80		
1 Cement Finisher	45.00	360.00	71.15	569.20		
1 Gas Engine Vibrator		31.20		34.32	.28	.31
112 L.H., Daily Totals		$5080.40		$8283.92	$45.36	$73.96

The **Daily Output** is the amount of work that the crew can do in a normal 8-hour workday, including mobilization, layout, movement of materials, and cleanup. In this case, crew C-14H can install thirty 4' x 4' x 6" thick concrete pads in a day. Daily output is variable, based on many factors, including the size of the job, location, and environmental conditions. RSMeans data represents work done in daylight (or adequate lighting) and temperate conditions.

Bare Costs are the costs of materials, labor, and equipment that the installing contractor pays. They represent the cost, in U.S. dollars, for one unit of work. They do not include any markups for profit or labor burden.

Crew	Daily Output	Labor-Hours	Unit	Material	2015 Bare Costs Labor	Equipment	Total	Total Incl O&P
C-14A	15.62	12.804	C.Y.	320	605	48	973	1,400
"	18.55	10.782		335	510	40.50	885.50	1,250
C-14C	32.22	3.476		150	157	.97	307.97	420
"	23.71	4.724		177	213	1.32	391.32	545
C-14A	11.96	16.722		365	790	63	1,218	1,775
	2000	.028		.99	1.17	.36	2.52	3.38
	2200	.025		.94	1.07	.33	2.34	3.12
	31800	.002			.07	.02	.09	.14
	21200	.003			.11	.03	.14	.22
C-14H	45	1.067	Ea.	47	49.50	.69	97.19	133
	30	1.600		69.50	74	1.04	144.54	198
	18	2.667		122	124	1.73	247.73	335
	14	3.429		164	159	2.23	325.23	440

The **Total Incl O&P** column is the total cost, including overhead and profit, that the installing contractor will charge the customer. This represents the cost of materials plus 10% profit, the cost of labor plus labor burden and 10% profit, and the cost of equipment plus 10% profit. It does not include the general contractor's overhead and profit. Note: See the inside back cover for details of how RSMeans calculates labor burden.

The **Total column** represents the total bare cost for the installing contractor, in U.S. dollars. In this case, the sum of $69.50 for material + $74.00 for labor + $1.04 for equipment is $144.54.

The figure in the **Labor Hours** column is the amount of labor required to perform one unit of work—in this case the amount of labor required to construct one 4' x 4' equipment pad. This figure is calculated by dividing the number of hours of labor in the crew by the daily output (48 labor hours divided by 30 pads = 1.6 hours of labor per pad). Multiply 1.600 times 60 to see the value in minutes: 60 x 1.6 = 96 minutes. Note: the labor hour figure is not dependent on the crew size. A change in crew size will result in a corresponding change in daily output, but the labor hours per unit of work will not change.

All RSMeans unit cost data includes the typical **Unit of Measure** used for estimating that item. For concrete-in-place the typical unit is cubic yards (C.Y.) or each (Ea.). For installing broadloom carpet it is square yard, and for gypsum board it is square foot. The estimator needs to take special care that the unit in the data matches the unit in the take-off. Unit conversions may be found in the Reference Section.

How RSMeans Unit Price Works (Continued)

Sample Estimate

This sample demonstrates the elements of an estimate, including a tally of the RSMeans data lines, and a summary of the markups on a contractor's work to arrive at a total cost to the owner. The RSMeans Location Factor is added at the bottom of the estimate to adjust the cost of the work to a specific location.

Work Performed: The body of the estimate shows the RSMeans data selected, including line number, a brief description of each item, its take-off unit and quantity, and the bare costs of materials, labor, and equipment. This estimate also includes a column titled "SubContract." This data is taken from the RSMeans column "Total Incl O&P," and represents the total that a subcontractor would charge a general contractor for the work, including the sub's markup for overhead and profit.

Division 1, General Requirements: This is the first division numerically, but the last division estimated. Division 1 includes project-wide needs provided by the general contractor. These requirements vary by project, but may include temporary facilities and utilities, security, testing, project cleanup, etc. For small projects a percentage can be used, typically between 5% and 15% of project cost. For large projects the costs may be itemized and priced individually.

Bonds: Bond costs should be added to the estimate. The figures here represent a typical performance bond, ensuring the owner that if the general contractor does not complete the obligations in the construction contract the bonding company will pay the cost for completion of the work.

Location Adjustment: RSMeans published data is based on national average costs. If necessary, adjust the total cost of the project using a location factor from the "Location Factor" table or the "City Cost Index" table. Use location factors if the work is general, covering multiple trades. If the work is by a single trade (e.g., masonry) use the more specific data found in the "City Cost Indexes."

This estimate is based on an interactive spreadsheet. A copy of this spreadsheet is located on the RSMeans website at **http://www.reedconstructiondata.com/rsmeans/extras/546011.** You are free to download it and adjust it to your methodology.

Project Name: Pre-Engineered Steel Building			Architect: As Shown	
Location:	Anywhere, USA			
Line Number	**Description**	**Qty**	**Unit**	**Material**
03 30 53.40 3940	Strip footing, 12" x 24", reinforced	34	C.Y.	$4,726.00
03 30 53.40 3950	Strip footing, 12" x 36", reinforced	15	C.Y.	$1,995.00
03 11 13.65 3000	Concrete slab edge forms	500	L.F.	$175.00
03 22 11.10 0200	Welded wire fabric reinforcing	150	C.S.F.	$2,580.00
03 31 13.35 0300	Ready mix concrete, 4000 psi for slab on grade	278	C.Y.	$29,746.00
03 31 13.70 4300	Place, strike off & consolidate concrete slab	278	C.Y.	$0.00
03 35 13.30 0250	Machine float & trowel concrete slab	15,000	S.F.	$0.00
03 15 16.20 0140	Cut control joints in concrete slab	950	L.F.	$57.00
03 39 23.13 0300	Sprayed concrete curing membrane	150	C.S.F.	$1,657.50
Division 03	**Subtotal**			**$40,936.50**
08 36 13.10 2650	Manual 10' x 10' steel sectional overhead door	8	Ea.	$8,800.00
08 36 13.10 2860	Insulation and steel back panel for OH door	800	S.F.	$3,800.00
Division 08	**Subtotal**			**$12,600.00**
13 34 19.50 1100	Pre-Engineered Steel Building, 100' x 150' x 24'	15,000	SF Flr.	$0.00
13 34 19.50 6050	Framing for PESB door opening, 3' x 7'	4	Opng.	$0.00
13 34 19.50 6100	Framing for PESB door opening, 10' x 10'	8	Opng.	$0.00
13 34 19.50 6200	Framing for PESB window opening, 4' x 3'	6	Opng.	$0.00
13 34 19.50 5750	PESB door, 3' x 7', single leaf	4	Opng.	$2,340.00
13 34 19.50 7750	PESB sliding window, 4' x 3' with screen	6	Opng.	$2,280.00
13 34 19.50 6550	PESB gutter, eave type, 26 ga., painted	300	L.F.	$2,085.00
13 34 19.50 8650	PESB roof vent, 12" wide x 10' long	15	Ea.	$540.00
13 34 19.50 6900	PESB insulation, vinyl faced, 4" thick	27,400	S.F.	$11,508.00
Division 13	**Subtotal**			**$18,753.00**
			Subtotal	$72,289.50
Division 01	**General Requirements @ 7%**			**5,060.27**
			Estimate Subtotal	$77,349.77
			Sales Tax @ 5%	3,867.49
			Subtotal A	81,217.25
			GC O & P	8,121.73
			Subtotal B	89,338.98
			Contingency @ 5%	
			Subtotal C	
			Bond @ $12/1000 +10% O&P	
			Subtotal D	
			Location Adjustment Factor	
			Grand Total	

This example shows the cost to construct a pre-engineered steel building. The foundation, doors, windows, and insulation will be installed by the general contractor. A subcontractor will install the structural steel, roofing, and siding.

Labor	Equipment	SubContract	01/01/15 Estimate Total	R+R	
$3,570.00	$22.10	$0.00			
$1,260.00	$7.80	$0.00			
$1,190.00	$0.00	$0.00			
$4,050.00	$0.00	$0.00			
$0.00	$0.00	$0.00			
$4,753.80	$158.46	$0.00			
$9,000.00	$450.00	$0.00			
$380.00	$85.50	$0.00			
$952.50	$0.00	$0.00			
$25,156.30	**$723.86**	**$0.00**	**$66,816.66**	**Division 03**	
$3,320.00	$0.00	$0.00			
$0.00	$0.00	$0.00			
$3,320.00	**$0.00**	**$0.00**	**$15,920.00**	**Division 08**	
$0.00	$0.00	$360,000.00			
$0.00	$0.00	$2,340.00			
$0.00	$0.00	$9,400.00			
$0.00	$0.00	$3,420.00			
$672.00	$0.00	$0.00			
$576.00	$67.20	$0.00			
$789.00	$0.00	$0.00			
$3,165.00	$0.00	$0.00			
$9,042.00	$0.00	$0.00			
$14,244.00	**$67.20**	**$375,160.00**	**$408,224.20**	**Division 13**	
$42,720.30	$791.06	$375,160.00	$490,960.86	Subtotal	
2,990.42	55.37	26,261.20		Gen. Requirements	
$45,710.72	$846.43	$401,421.20	$490,960.86	Estimate Subtotal	
	42.32	10,035.53		Sales tax	
45,710.72	888.76	411,456.73		Subtotal	
28,569.20	88.88	41,145.67		GC O&P	
74,279.92	977.63	452,602.40	$617,198.93	Subtotal	
			30,859.95	Contingency	
			$648,058.88	Subtotal	
			8,554.38	Bond	
			$656,613.26	Subtotal	
102.30			15,102.10	Location Adjustment	
			$671,715.36	**Grand Total**	

Sales Tax: If the work is subject to state or local sales taxes, the amount must be added to the estimate. Sales tax may be added to material costs, equipment costs, and subcontracted work. In this case, sales tax was added in all three categories. It was assumed that approximately half the subcontracted work would be material cost, so the tax was applied to 50% of the subcontract total.

GC O&P: This entry represents the general contractor's markup on material, labor, equipment, and subcontractor costs. RSMeans' standard markup on materials, equipment, and subcontracted work is 10%. In this estimate, the markup on the labor performed by the GC's workers uses "Skilled Workers Average" shown in Column F on the table "Installing Contractor's Overhead & Profit," which can be found on the inside-back cover of the book.

Contingency: A factor for contingency may be added to any estimate to represent the cost of unknowns that may occur between the time that the estimate is performed and the time the project is constructed. The amount of the allowance will depend on the stage of design at which the estimate is done, and the contractor's assessment of the risk involved. Refer to section 01 21 16.50 for contigency allowances.

Estimating Tips

01 20 00 Price and Payment Procedures

- Allowances that should be added to estimates to cover contingencies and job conditions that are not included in the national average material and labor costs are shown in section 01 21.
- When estimating historic preservation projects (depending on the condition of the existing structure and the owner's requirements), a 15%–20% contingency or allowance is recommended, regardless of the stage of the drawings.

01 30 00 Administrative Requirements

- Before determining a final cost estimate, it is a good practice to review all the items listed in Subdivisions 01 31 and 01 32 to make final adjustments for items that may need customizing to specific job conditions.
- Requirements for initial and periodic submittals can represent a significant cost to the General Requirements of a job. Thoroughly check the submittal specifications when estimating a project to determine any costs that should be included.

01 40 00 Quality Requirements

- All projects will require some degree of Quality Control. This cost is not included in the unit cost of construction listed in each division. Depending upon the terms of the contract, the various costs of inspection and testing can be the responsibility of either the owner or the contractor. Be sure to include the required costs in your estimate.

01 50 00 Temporary Facilities and Controls

- Barricades, access roads, safety nets, scaffolding, security, and many more requirements for the execution of a safe project are elements of direct cost. These costs can easily be overlooked when preparing an estimate. When looking through the major classifications of this subdivision, determine which items apply to each division in your estimate.
- Construction Equipment Rental Costs can be found in the Reference Section in section 01 54 33. Operators' wages are not included in equipment rental costs.
- Equipment mobilization and demobilization costs are not included in equipment rental costs and must be considered separately in section 01 54 36.50.

- The cost of small tools provided by the installing contractor for his workers is covered in the "Overhead" column on the "Installing Contractor's Overhead and Profit" table that lists labor trades, base rates and markups and, therefore, is included in the "Total Incl. O&P" cost of any Unit Price line item.

01 70 00 Execution and Closeout Requirements

- When preparing an estimate, thoroughly read the specifications to determine the requirements for Contract Closeout. Final cleaning, record documentation, operation and maintenance data, warranties and bonds, and spare parts and maintenance materials can all be elements of cost for the completion of a contract. Do not overlook these in your estimate.

Reference Numbers

Reference numbers are shown in shaded boxes at the beginning of some major classifications. These numbers refer to related items in the Reference Section. The reference information may be an estimating procedure, an alternate pricing method, or technical information.

Note: Not all subdivisions listed here necessarily appear in this publication. ■

No part of this publication may be reproduced, stored in a retrieval system, or transmitted in any form or by any means without prior written permission of RSMeans.

01 11 Summary of Work

01 11 31 – Professional Consultants

01 11 31.10 Architectural Fees	Crew	Daily Output	Labor-Hours	Unit	Material	2015 Bare Costs Labor	Equipment	Total	Total Incl O&P
0010 **ARCHITECTURAL FEES** R011110-10									
0020 For new construction									
0060 Minimum				Project				4.90%	4.90%
0090 Maximum								16%	16%
0100 For alteration work, to $500,000, add to new construction fee								50%	50%
0150 Over $500,000, add to new construction fee								25%	25%
01 11 31.20 Construction Management Fees									
0010 **CONSTRUCTION MANAGEMENT FEES**									
0060 For work to $100,000				Project				10%	10%
0070 To $250,000								9%	9%
0090 To $1,000,000								6%	6%
0100 To $5,000,000								5%	5%
0110 To $10,000,000								4%	4%
01 11 31.30 Engineering Fees									
0010 **ENGINEERING FEES** R011110-30									
0020 Educational planning consultant, minimum				Project				.50%	.50%
0100 Maximum				"				2.50%	2.50%
0400 Elevator & conveying systems, minimum				Contrct				2.50%	2.50%
0500 Maximum								5%	5%
1000 Mechanical (plumbing & HVAC), minimum								4.10%	4.10%
1100 Maximum								10.10%	10.10%
1200 Structural, minimum				Project				1%	1%
1300 Maximum				"				2.50%	2.50%
01 11 31.75 Renderings									
0010 **RENDERINGS** Color, matted, 20" x 30", eye level,									
0020 1 building, minimum				Ea.	2,050			2,050	2,275
0050 Average					2,875			2,875	3,175
0100 Maximum					4,625			4,625	5,100
1000 5 buildings, minimum					4,125			4,125	4,525
1100 Maximum					8,250			8,250	9,075

01 21 Allowances

01 21 16 – Contingency Allowances

01 21 16.50 Contingencies

0010 **CONTINGENCIES**, Add to estimate									
0020 Conceptual stage				Project				20%	20%
0050 Schematic stage								15%	15%
0100 Preliminary working drawing stage (Design Dev.)								10%	10%
0150 Final working drawing stage								3%	3%

01 21 53 – Factors Allowance

01 21 53.50 Factors

0010 **FACTORS** Cost adjustments R012153-10									
0100 Add to construction costs for particular job requirements									
0500 Cut & patch to match existing construction, add, minimum				Costs	2%	3%			
0550 Maximum					5%	9%			
0800 Dust protection, add, minimum					1%	2%			
0850 Maximum					4%	11%			
1100 Equipment usage curtailment, add, minimum					1%	1%			
1150 Maximum					3%	10%			
1400 Material handling & storage limitation, add, minimum					1%	1%			

01 21 Allowances

01 21 53 – Factors Allowance

01 21 53.50 Factors

		Crew	Daily Output	Labor-Hours	Unit	Material	2015 Bare Costs Labor	Equipment	Total	Total Incl O&P
1450	Maximum				Costs	6%	7%			
1700	Protection of existing work, add, minimum					2%	2%			
1750	Maximum					5%	7%			
2000	Shift work requirements, add, minimum						5%			
2050	Maximum						30%			
2300	Temporary shoring and bracing, add, minimum					2%	5%			
2350	Maximum					5%	12%			
2400	Work inside prisons and high security areas, add, minimum						30%			
2450	Maximum						50%			

01 21 55 – Job Conditions Allowance

01 21 55.50 Job Conditions

		Crew	Daily Output	Labor-Hours	Unit	Material	2015 Bare Costs Labor	Equipment	Total	Total Incl O&P
0010	**JOB CONDITIONS** Modifications to applicable									
0020	cost summaries									
0100	Economic conditions, favorable, deduct				Project				2%	2%
0200	Unfavorable, add								5%	5%
0300	Hoisting conditions, favorable, deduct								2%	2%
0400	Unfavorable, add								5%	5%
0500	General Contractor management, experienced, deduct								2%	2%
0600	Inexperienced, add								10%	10%
0700	Labor availability, surplus, deduct								1%	1%
0800	Shortage, add								10%	10%
0900	Material storage area, available, deduct								1%	1%
1000	Not available, add								2%	2%
1100	Subcontractor availability, surplus, deduct								5%	5%
1200	Shortage, add								12%	12%
1300	Work space, available, deduct								2%	2%
1400	Not available, add								5%	5%

01 21 63 – Taxes

01 21 63.10 Taxes

			Crew	Daily Output	Labor-Hours	Unit	Material	2015 Bare Costs Labor	Equipment	Total	Total Incl O&P
0010	**TAXES**	R012909-80									
0020	Sales tax, State, average					%	5.04%				
0050	Maximum	R012909-85					7.50%				
0200	Social Security, on first $117,000 of wages							7.65%			
0300	Unemployment, combined Federal and State, minimum							.60%			
0350	Average							7.80%			
0400	Maximum							12.87%			

01 31 Project Management and Coordination

01 31 13 – Project Coordination

01 31 13.20 Field Personnel

		Crew	Daily Output	Labor-Hours	Unit	Material	2015 Bare Costs Labor	Equipment	Total	Total Incl O&P
0010	**FIELD PERSONNEL**									
0020	Clerk, average				Week		450		450	730
0160	General purpose laborer, average						1,500		1,500	2,475
0180	Project manager, minimum						1,975		1,975	3,225
0200	Average						2,275		2,275	3,700
0220	Maximum						2,600		2,600	4,225
0240	Superintendent, minimum						1,925		1,925	3,125
0260	Average						2,125		2,125	3,450
0280	Maximum						2,400		2,400	3,925

01 31 Project Management and Coordination

01 31 13 – Project Coordination

01 31 13.30 Insurance

01 31 13.30 Insurance		Crew	Daily Output	Labor-Hours	Unit	Material	2015 Bare Costs Labor	Equipment	Total	Total Incl O&P
0010	**INSURANCE** R013113-40									
0020	Builders risk, standard, minimum				Job				.24%	.24%
0050	Maximum R013113-60								.64%	.64%
0200	All-risk type, minimum								.25%	.25%
0250	Maximum								.62%	.62%
0400	Contractor's equipment floater, minimum				Value				.50%	.50%
0450	Maximum				"				1.50%	1.50%
0600	Public liability, average				Job				2.02%	2.02%
0810	Workers' compensation & employer's liability									
2000	Range of 35 trades in 50 states, excl. wrecking, min.				Payroll		1.31%			
2100	Average						13.50%			
2200	Maximum						132%			

01 31 13.40 Main Office Expense

0010	**MAIN OFFICE EXPENSE** Average for General Contractors									
0020	As a percentage of their annual volume									
0030	Annual volume to $300,000, minimum				% Vol.				20%	
0040	Maximum								30%	
0060	To $500,000, minimum								17%	
0070	Maximum								22%	
0080	To $1,000,000, minimum								16%	
0090	Maximum								19%	
0110	To $3,000,000, minimum								14%	
0120	Maximum								16%	
0130	To $5,000,000, minimum								8%	
0140	Maximum								10%	

01 31 13.50 General Contractor's Mark-Up

0010	**GENERAL CONTRACTOR'S MARK-UP** on Change Orders									
0200	Extra work, by subcontractors, add				%				10%	10%
0250	By General Contractor, add								15%	15%
0400	Omitted work, by subcontractors, deduct all but								5%	5%
0450	By General Contractor, deduct all but								7.50%	7.50%
0600	Overtime work, by subcontractors, add								15%	15%
0650	By General Contractor, add								10%	10%

01 31 13.60 Installing Contractor's Main Office Overhead

0010	**INSTALLING CONTRACTOR'S MAIN OFFICE OVERHEAD** R013113-50									
0020	As percent of direct costs, minimum				%				5%	
0050	Average								13%	
0100	Maximum								30%	

01 31 13.80 Overhead and Profit

0010	**OVERHEAD & PROFIT** Allowance to add to items in this R013113-50									
0020	book that do not include Subs O&P, average				%				25%	
0100	Allowance to add to items in this book that									
0110	do include Subs O&P, minimum				%				5%	5%
0150	Average								10%	10%
0200	Maximum								15%	15%
0290	Typical, by size of project, under $50,000								40%	
0310	$50,000 to $100,000								35%	
0320	$100,000 to $500,000								25%	
0330	$500,000 to $1,000,000								20%	

01 31 Project Management and Coordination

01 31 13 – Project Coordination

01 31 13.90 Performance Bond	Crew	Daily Output	Labor-Hours	Unit	Material	2015 Bare Costs Labor	Equipment	Total	Total Incl O&P
0010 **PERFORMANCE BOND**									
0020 For buildings, minimum				Job				.60%	.60%
0100 Maximum				"				2.50%	2.50%

01 41 Regulatory Requirements

01 41 26 – Permit Requirements

01 41 26.50 Permits									
0010 **PERMITS**									
0020 Rule of thumb, most cities, minimum				Job				.50%	.50%
0100 Maximum				"				2%	2%

01 45 Quality Control

01 45 23 – Testing and Inspecting Services

01 45 23.50 Testing									
0010 **TESTING** and Inspecting Services									
1800 Compressive test, cylinder, delivered to lab, ASTM C 39				Ea.				12	13
1900 Picked up by lab, minimum								14	15
1950 Average								18	20
2000 Maximum								27	30
2200 Compressive strength, cores (not incl. drilling), ASTM C 42								36	40
2300 Patching core holes				↓				22	24
4730 Soil testing									
4735 Soil density, nuclear method, ASTM D2922				Ea.				35	38.50
4740 Sand cone method ASTM D1556								27	30
4750 Moisture content, ASTM D 2216								9	10
4780 Permeability test, double ring infiltrometer								500	550
4800 Permeability, var. or constant head, undist., ASTM D 2434								227	250
4850 Recompacted								250	275
4900 Proctor compaction, 4" standard mold, ASTM D 698								123	135
4950 6" modified mold								68	75
5100 Shear tests, triaxial, minimum								410	450
5150 Maximum								545	600
5550 Technician for inspection, per day, earthwork								320	350
5650 Bolting								400	440
5750 Roofing								480	530
5790 Welding				↓				480	530
5820 Non-destructive metal testing, dye penetrant				Day				310	340
5840 Magnetic particle								310	340
5860 Radiography								450	495
5880 Ultrasonic				↓				310	340
6000 Welding certification, minimum				Ea.				91	100
6100 Maximum				"				250	275
7000 Underground storage tank									
7500 Volumetric tightness test, <=12,000 gal.				Ea.				435	480
7510 <=30,000 gal.				"				615	675
7600 Vadose zone (soil gas) sampling, 10-40 samples, min.				Day				1,375	1,500
7610 Maximum				"				2,275	2,500
7700 Ground water monitoring incl. drilling 3 wells, min.				Total				4,550	5,000
7710 Maximum				"				6,375	7,000

01 45 Quality Control

01 45 23 – Testing and Inspecting Services

01 45 23.50 Testing	Crew	Daily Output	Labor-Hours	Unit	Material	2015 Bare Costs Labor	Equipment	Total	Total Incl O&P
8000 X-ray concrete slabs				Ea.				182	200

01 51 Temporary Utilities

01 51 13 – Temporary Electricity

01 51 13.80 Temporary Utilities

		Crew	Daily Output	Labor-Hours	Unit	Material	Labor	Equipment	Total	Total Incl O&P
0010	TEMPORARY UTILITIES									
0350	Lighting, lamps, wiring, outlets, 40,000 SF building, 8 strings	1 Elec	34	.235	CSF Flr	4.91	12.85		17.76	25.50
0360	16 strings	"	17	.471		9.80	25.50		35.30	51
0400	Power for temp lighting only, 6.6 KWH, per month								.92	1.01
0430	11.8 KWH, per month								1.65	1.82
0450	23.6 KWH, per month								3.30	3.63
0600	Power for job duration incl. elevator, etc., minimum								47	51.50
0650	Maximum								110	121

01 52 Construction Facilities

01 52 13 – Field Offices and Sheds

01 52 13.20 Office and Storage Space

		Crew	Daily Output	Labor-Hours	Unit	Material	Labor	Equipment	Total	Total Incl O&P
0010	OFFICE AND STORAGE SPACE									
0020	Office trailer, furnished, no hookups, 20' x 8', buy	2 Skwk	1	16	Ea.	10,400	780		11,180	12,700
0250	Rent per month					188			188	206
0300	32' x 8', buy	2 Skwk	.70	22.857		15,500	1,100		16,600	18,900
0350	Rent per month					239			239	262
0400	50' x 10', buy	2 Skwk	.60	26.667		24,100	1,300		25,400	28,600
0450	Rent per month					340			340	375
0500	50' x 12', buy	2 Skwk	.50	32		29,400	1,550		30,950	34,800
0550	Rent per month					410			410	455
0700	For air conditioning, rent per month, add					48.50			48.50	53.50
0800	For delivery, add per mile				Mile	11			11	12.10
0890	Delivery each way				Ea.	200			200	220
1200	Storage boxes, 20' x 8', buy	2 Skwk	1.80	8.889		2,775	430		3,205	3,750
1250	Rent per month					82.50			82.50	91
1300	40' x 8', buy	2 Skwk	1.40	11.429		3,850	555		4,405	5,125
1350	Rent per month					101			101	111

01 54 Construction Aids

01 54 16 – Temporary Hoists

01 54 16.50 Weekly Forklift Crew

		Crew	Daily Output	Labor-Hours	Unit	Material	Labor	Equipment	Total	Total Incl O&P
0010	WEEKLY FORKLIFT CREW									
0100	All-terrain forklift, 45' lift, 35' reach, 9000 lb. capacity	A-3P	.20	40	Week		1,950	2,625	4,575	5,975

01 54 19 – Temporary Cranes

01 54 19.50 Daily Crane Crews

		Crew	Daily Output	Labor-Hours	Unit	Material	Labor	Equipment	Total	Total Incl O&P
0010	DAILY CRANE CREWS for small jobs, portal to portal									
0100	12-ton truck-mounted hydraulic crane	A-3H	1	8	Day		415	860	1,275	1,600
0200	25-ton	A-3I	1	8			415	990	1,405	1,750
0300	40-ton	A-3J	1	8			415	1,225	1,640	2,000
0400	55-ton	A-3K	1	16			775	1,625	2,400	3,025
0500	80-ton	A-3L	1	16			775	2,350	3,125	3,800

01 54 Construction Aids

01 54 19 – Temporary Cranes

01 54 19.50 Daily Crane Crews

		Crew	Daily Output	Labor-Hours	Unit	Material	2015 Bare Costs Labor	2015 Bare Costs Equipment	Total	Total Incl O&P
0600	100-ton	A-3M	1	16	Day		775	2,325	3,100	3,800
0900	If crane is needed on a Saturday, Sunday or Holiday									
0910	At time-and-a-half, add				Day		50%			
0920	At double time, add				"		100%			

01 54 19.60 Monthly Tower Crane Crew

		Crew	Daily Output	Labor-Hours	Unit	Material	Labor	Equipment	Total	Total Incl O&P
0010	**MONTHLY TOWER CRANE CREW**, excludes concrete footing									
0100	Static tower crane, 130' high, 106' jib, 6200 lb. capacity	A-3N	.05	176	Month		9,100	24,800	33,900	41,700

01 54 23 – Temporary Scaffolding and Platforms

01 54 23.60 Pump Staging

		Crew	Daily Output	Labor-Hours	Unit	Material	Labor	Equipment	Total	Total Incl O&P
0010	**PUMP STAGING**, Aluminum R015423-20									
0200	24' long pole section, buy				Ea.	410			410	450
0300	18' long pole section, buy					320			320	350
0400	12' long pole section, buy					215			215	236
0500	6' long pole section, buy					113			113	124
0600	6' long splice joint section, buy					84			84	92.50
0700	Pump jack, buy					172			172	189
0900	Foldable brace, buy					67.50			67.50	74
1000	Workbench/back safety rail support, buy					91			91	100
1100	Scaffolding planks/workbench, 14" wide x 24' long, buy					775			775	850
1200	Plank end safety rail, buy					345			345	380
1250	Safety net, 22' long, buy					410			410	450
1300	System in place, 50' working height, per use based on 50 uses	2 Carp	84.80	.189	C.S.F.	6.95	8.85		15.80	22
1400	100 uses		84.80	.189		3.46	8.85		12.31	18.30
1500	150 uses		84.80	.189		2.32	8.85		11.17	17.05

01 54 23.70 Scaffolding

		Crew	Daily Output	Labor-Hours	Unit	Material	Labor	Equipment	Total	Total Incl O&P
0010	**SCAFFOLDING** R015423-10									
0015	Steel tube, regular, no plank, labor only to erect & dismantle									
0090	Building exterior, wall face, 1 to 5 stories, 6'-4" x 5' frames	3 Carp	8	3	C.S.F.		141		141	231
0200	6 to 12 stories	4 Carp	8	4			188		188	310
0301	13 to 20 stories	5 Clab	8	5			188		188	310
0460	Building interior, wall face area, up to 16' high	3 Carp	12	2			94		94	154
0560	16' to 40' high		10	2.400			113		113	185
0800	Building interior floor area, up to 30' high		150	.160	C.C.F.		7.50		7.50	12.30
0900	Over 30' high	4 Carp	160	.200	"		9.40		9.40	15.40
0906	Complete system for face of walls, no plank, material only rent/mo				C.S.F.	35.50			35.50	39.50
0908	Interior spaces, no plank, material only rent/mo				C.C.F.	4.33			4.33	4.76
0910	Steel tubular, heavy duty shoring, buy									
0920	Frames 5' high 2' wide				Ea.	81.50			81.50	90
0925	5' high 4' wide					93			93	102
0930	6' high 2' wide					93.50			93.50	103
0935	6' high 4' wide					109			109	120
0940	Accessories									
0945	Cross braces				Ea.	15.50			15.50	17.05
0950	U-head, 8" x 8"					19.10			19.10	21
0955	J-head, 4" x 8"					13.90			13.90	15.30
0960	Base plate, 8" x 8"					15.50			15.50	17.05
0965	Leveling jack					33.50			33.50	36.50
1000	Steel tubular, regular, buy									
1100	Frames 3' high 5' wide				Ea.	75			75	82.50
1150	5' high 5' wide					86.50			86.50	95
1200	6'-4" high 5' wide					118			118	129
1350	7'-6" high 6' wide					158			158	173

01 54 Construction Aids

01 54 23 – Temporary Scaffolding and Platforms

01 54 23.70 Scaffolding

		Crew	Daily Output	Labor-Hours	Unit	Material	2015 Bare Costs Labor	Equipment	Total	Total Incl O&P
1500	Accessories cross braces				Ea.	15.50			15.50	17.05
1550	Guardrail post					16.40			16.40	18.05
1600	Guardrail 7' section					8.70			8.70	9.60
1650	Screw jacks & plates					21.50			21.50	24
1700	Sidearm brackets					30.50			30.50	33.50
1750	8" casters					31			31	34
1800	Plank 2" x 10" x 16'-0"					57.50			57.50	63.50
1900	Stairway section					286			286	315
1910	Stairway starter bar					32			32	35
1920	Stairway inside handrail					58.50			58.50	64
1930	Stairway outside handrail					86.50			86.50	95
1940	Walk-thru frame guardrail					41.50			41.50	46
2000	Steel tubular, regular, rent/mo.									
2100	Frames 3' high 5' wide				Ea.	5			5	5.50
2150	5' high 5' wide					5			5	5.50
2200	6'-4" high 5' wide					5.50			5.50	6.05
2250	7'-6" high 6' wide					10			10	11
2500	Accessories, cross braces					1			1	1.10
2550	Guardrail post					1			1	1.10
2600	Guardrail 7' section					1			1	1.10
2650	Screw jacks & plates					2			2	2.20
2700	Sidearm brackets					2			2	2.20
2750	8" casters					8			8	8.80
2800	Outrigger for rolling tower					3			3	3.30
2850	Plank 2" x 10" x 16'-0"					10			10	11
2900	Stairway section					35			35	38.50
2940	Walk-thru frame guardrail					2.50			2.50	2.75
3000	Steel tubular, heavy duty shoring, rent/mo.									
3250	5' high 2' & 4' wide				Ea.	8.50			8.50	9.35
3300	6' high 2' & 4' wide					8.50			8.50	9.35
3500	Accessories, cross braces					1			1	1.10
3600	U - head, 8" x 8"					2.50			2.50	2.75
3650	J - head, 4" x 8"					2.50			2.50	2.75
3700	Base plate, 8" x 8"					1			1	1.10
3750	Leveling jack					2.50			2.50	2.75
5700	Planks, 2" x 10" x 16'-0", labor only to erect & remove to 50' H	3 Carp	72	.333			15.65		15.65	25.50
5800	Over 50' high	4 Carp	80	.400			18.80		18.80	31

01 54 23.75 Scaffolding Specialties

		Crew	Daily Output	Labor-Hours	Unit	Material	2015 Bare Costs Labor	Equipment	Total	Total Incl O&P
0010	**SCAFFOLDING SPECIALTIES**									
1200	Sidewalk bridge, heavy duty steel posts & beams, including									
1210	parapet protection & waterproofing (material cost is rent/month)									
1220	8' to 10' wide, 2 posts	3 Carp	15	1.600	L.F.	45	75		120	173
1230	3 posts	"	10	2.400	"	69.50	113		182.50	262
1500	Sidewalk bridge using tubular steel scaffold frames including									
1510	planking (material cost is rent/month)	3 Carp	45	.533	L.F.	8.55	25		33.55	50.50
1600	For 2 uses per month, deduct from all above					50%				
1700	For 1 use every 2 months, add to all above					100%				
1900	Catwalks, 20" wide, no guardrails, 7' span, buy				Ea.	166			166	183
2000	10' span, buy					233			233	256
3720	Putlog, standard, 8' span, with hangers, buy					162			162	178
3730	Rent per month					20			20	22
3750	12' span, buy					203			203	223

01 54 Construction Aids

01 54 23 – Temporary Scaffolding and Platforms

01 54 23.75 Scaffolding Specialties		Crew	Daily Output	Labor-Hours	Unit	Material	2015 Bare Costs Labor	Equipment	Total	Total Incl O&P
3755	Rent per month				Ea.	25			25	27.50
3760	Trussed type, 16' span, buy					365			365	400
3770	Rent per month					30			30	33
3790	22' span, buy					420			420	465
3795	Rent per month					40			40	44
4000	7 step					760			760	835
4100	Rolling towers, buy, 5' wide, 7' long, 10' high					1,200			1,200	1,325
4200	For 5' high added sections, to buy, add					204			204	224
4300	Complete incl. wheels, railings, outriggers,									
4350	21' high, to buy				Ea.	2,025			2,025	2,225
4400	Rent/month = 5% of purchase cost				"	200			200	220
5000	Motorized work platform, mast climber									
5050	Base unit, 50' W, less than 100' tall, rent/mo				Ea.	2,900			2,900	3,175
5100	Less than 200' tall, rent/mo					3,700			3,700	4,075
5150	Less than 300' tall, rent/mo					4,375			4,375	4,825
5200	Less than 400' tall, rent/mo					5,000			5,000	5,500
5250	Set up and demob, per unit, less than 100' tall	B-68F	16.60	1.446	C.S.F.		71.50	20	91.50	138
5300	Less than 200' tall		25	.960			47.50	13.30	60.80	91.50
5350	Less than 300' tall		30	.800			39.50	11.05	50.55	76
5400	Less than 400' tall		33	.727			36	10.05	46.05	69.50
5500	Mobilization (price includes freight in and out) per unit				Ea.				1,100	1,225

01 54 23.80 Staging Aids		Crew	Daily Output	Labor-Hours	Unit	Material	2015 Bare Costs Labor	Equipment	Total	Total Incl O&P
0010	**STAGING AIDS** and fall protection equipment									
0100	Sidewall staging bracket, tubular, buy				Ea.	55			55	60.50
0110	Cost each per day, based on 250 days use				Day	.22			.22	.24
0200	Guard post, buy				Ea.	50.50			50.50	55.50
0210	Cost each per day, based on 250 days use				Day	.20			.20	.22
0300	End guard chains, buy per pair				Pair	36			36	39.50
0310	Cost per set per day, based on 250 days use				Day	.22			.22	.24
1010	Cost each per day, based on 250 days use				"	.04			.04	.04
1100	Wood bracket, buy				Ea.	16.80			16.80	18.50
1110	Cost each per day, based on 250 days use				Day	.07			.07	.07
2010	Cost per pair per day, based on 250 days use				"	.47			.47	.52
2100	Steel siderail jack, buy per pair				Pair	97.50			97.50	107
2110	Cost per pair per day, based on 250 days use				Day	.39			.39	.43
3010	Cost each per day, based on 250 days use				"	.23			.23	.25
3100	Aluminum scaffolding plank, 20" wide x 24' long, buy				Ea.	755			755	830
3110	Cost each per day, based on 250 days use				Day	3.01			3.01	3.31
4000	Nylon full body harness, lanyard and rope grab				Ea.	172			172	189
4010	Cost each per day, based on 250 days use				Day	.69			.69	.76
4100	Rope for safety line, 5/8" x 100' nylon, buy				Ea.	56			56	61.50
4110	Cost each per day, based on 250 days use				Day	.22			.22	.25
4200	Permanent U-Bolt roof anchor, buy				Ea.	35.50			35.50	39.50
4300	Temporary (one use) roof ridge anchor, buy				"	32			32	35.50
5000	Installation (setup and removal) of staging aids									
5010	Sidewall staging bracket	2 Carp	64	.250	Ea.		11.75		11.75	19.25
5020	Guard post with 2 wood rails	"	64	.250			11.75		11.75	19.25
5030	End guard chains, set	1 Carp	64	.125			5.85		5.85	9.60
5100	Roof shingling bracket		96	.083			3.91		3.91	6.40
5200	Ladder jack		64	.125			5.85		5.85	9.60
5300	Wood plank, 2" x 10" x 16'	2 Carp	80	.200			9.40		9.40	15.40
5310	Aluminum scaffold plank, 20" x 24'	"	40	.400			18.80		18.80	31

01 54 Construction Aids

01 54 23 – Temporary Scaffolding and Platforms

01 54 23.80 Staging Aids

		Crew	Daily Output	Labor-Hours	Unit	Material	2015 Bare Costs Labor	Equipment	Total	Total Incl O&P
5410	Safety rope	1 Carp	40	.200	Ea.		9.40		9.40	15.40
5420	Permanent U-Bolt roof anchor (install only)	2 Carp	40	.400			18.80		18.80	31
5430	Temporary roof ridge anchor (install only)	1 Carp	64	.125	↓		5.85		5.85	9.60

01 54 26 – Temporary Swing Staging

01 54 26.50 Swing Staging

		Crew	Daily Output	Labor-Hours	Unit	Material	Labor	Equipment	Total	Total Incl O&P
0010	**SWING STAGING**, 500 lb. cap., 2' wide to 24' long, hand operated									
0020	steel cable type, with 60' cables, buy				Ea.	4,900			4,900	5,375
0030	Rent per month				"	490			490	540
0600	Lightweight (not for masons) 24' long for 150' height,									
0610	manual type, buy				Ea.	10,200			10,200	11,200
0620	Rent per month					1,025			1,025	1,125
0700	Powered, electric or air, to 150' high, buy					25,600			25,600	28,200
0710	Rent per month					1,800			1,800	1,975
0780	To 300' high, buy					26,000			26,000	28,600
0800	Rent per month					1,825			1,825	2,000
1000	Bosun's chair or work basket 3' x 3.5', to 300' high, electric, buy					10,400			10,400	11,400
1010	Rent per month				↓	730			730	800
2200	Move swing staging (setup and remove)	E-4	2	16	Move		850	73	923	1,600

01 54 36 – Equipment Mobilization

01 54 36.50 Mobilization

		Crew	Daily Output	Labor-Hours	Unit	Material	Labor	Equipment	Total	Total Incl O&P
0010	**MOBILIZATION** (Use line item again for demobilization) R015436-50									
0015	Up to 25 mi. haul dist. (50 mi. RT for mob/demob crew)									
1200	Small equipment, placed in rear of, or towed by pickup truck	A-3A	4	2	Ea.		97	39	136	197
1300	Equipment hauled on 3-ton capacity towed trailer	A-3Q	2.67	3			146	67	213	305
1400	20-ton capacity	B-34U	2	8			355	237	592	825
1500	40-ton capacity	B-34N	2	8			365	375	740	995
1600	50-ton capacity	B-34V	1	24			1,125	1,050	2,175	2,950
1700	Crane, truck-mounted, up to 75 ton (driver only)	1 Eqhv	4	2			103		103	164
1800	Over 75 ton (with chase vehicle)	A-3E	2.50	6.400			294	62.50	356.50	540
2400	Crane, large lattice boom, requiring assembly	B-34W	.50	144	↓		6,275	7,500	13,775	18,400
2500	For each additional 5 miles haul distance, add						10%	10%		
3000	For large pieces of equipment, allow for assembly/knockdown									
3100	For mob/demob of micro-tunneling equip, see Section 33 05 23.19									
3200	For mob/demob of pile driving equip, see Section 31 62 19.10									

01 55 Vehicular Access and Parking

01 55 23 – Temporary Roads

01 55 23.50 Roads and Sidewalks

		Crew	Daily Output	Labor-Hours	Unit	Material	Labor	Equipment	Total	Total Incl O&P
0010	**ROADS AND SIDEWALKS** Temporary									
0050	Roads, gravel fill, no surfacing, 4" gravel depth	B-14	715	.067	S.Y.	4.04	2.67	.51	7.22	9.35
0100	8" gravel depth	"	615	.078	"	8.10	3.10	.59	11.79	14.60

01 56 Temporary Barriers and Enclosures

01 56 13 – Temporary Air Barriers

01 56 13.60 Tarpaulins

		Crew	Daily Output	Labor-Hours	Unit	Material	2015 Bare Costs Labor	Equipment	Total	Total Incl O&P
0010	**TARPAULINS**									
0020	Cotton duck, 10 oz. to 13.13 oz. per S.Y., 6'x8'				S.F.	.83			.83	.91
0050	30'x30'					.59			.59	.65
0100	Polyvinyl coated nylon, 14 oz. to 18 oz., minimum					1.36			1.36	1.50
0150	Maximum					1.36			1.36	1.50
0200	Reinforced polyethylene 3 mils thick, white					.04			.04	.04
0300	4 mils thick, white, clear or black					.09			.09	.10
0400	5.5 mils thick, clear					.18			.18	.20
0500	White, fire retardant					.44			.44	.48
0600	12 mils, oil resistant, fire retardant					.39			.39	.43
0700	8.5 mils, black					.57			.57	.63
0720	Steel reinforced polyethylene, 20 mils thick					.52			.52	.57
0730	Polyester reinforced w/integral fastening system 11 mils thick					.21			.21	.23
0740	Mylar polyester, non-reinforced, 7 mils thick					1.17			1.17	1.29

01 56 13.90 Winter Protection

		Crew	Daily Output	Labor-Hours	Unit	Material	Labor	Equipment	Total	Total Incl O&P
0010	**WINTER PROTECTION**									
0100	Framing to close openings	2 Clab	500	.032	S.F.	.46	1.20		1.66	2.47
0200	Tarpaulins hung over scaffolding, 8 uses, not incl. scaffolding		1500	.011		.24	.40		.64	.92
0250	Tarpaulin polyester reinf. w/integral fastening system 11 mils thick		1600	.010		.21	.38		.59	.85
0300	Prefab fiberglass panels, steel frame, 8 uses		1200	.013		2.40	.50		2.90	3.46

01 56 16 – Temporary Dust Barriers

01 56 16.10 Dust Barriers, Temporary

		Crew	Daily Output	Labor-Hours	Unit	Material	Labor	Equipment	Total	Total Incl O&P
0010	**DUST BARRIERS, TEMPORARY**									
0020	Spring loaded telescoping pole & head, to 12', erect and dismantle	1 Clab	240	.033	Ea.		1.25		1.25	2.05
0025	Cost per day (based upon 250 days)				Day	.26			.26	.29
0030	To 21', erect and dismantle	1 Clab	240	.033	Ea.		1.25		1.25	2.05
0035	Cost per day (based upon 250 days)				Day	.44			.44	.48
0040	Accessories, caution tape reel, erect and dismantle	1 Clab	480	.017	Ea.		.63		.63	1.03
0045	Cost per day (based upon 250 days)				Day	.36			.36	.40
0060	Foam rail and connector, erect and dismantle	1 Clab	240	.033	Ea.		1.25		1.25	2.05
0065	Cost per day (based upon 250 days)				Day	.10			.10	.11
0070	Caution tape	1 Clab	384	.021	C.L.F.	2.70	.78		3.48	4.25
0080	Zipper, standard duty		60	.133	Ea.	8	5		13	17
0090	Heavy duty		48	.167	"	9.50	6.25		15.75	21
0100	Polyethylene sheet, 4 mil		37	.216	Sq.	2.90	8.15		11.05	16.55
0110	6 mil		37	.216	"	4.02	8.15		12.17	17.75
1000	Dust partition, 6 mil polyethylene, 1" x 3" frame	2 Carp	2000	.008	S.F.	.30	.38		.68	.94
1080	2" x 4" frame	"	2000	.008	"	.35	.38		.73	1.01

01 56 23 – Temporary Barricades

01 56 23.10 Barricades

		Crew	Daily Output	Labor-Hours	Unit	Material	Labor	Equipment	Total	Total Incl O&P
0010	**BARRICADES**									
0020	5' high, 3 rail @ 2" x 8", fixed	2 Carp	20	.800	L.F.	6	37.50		43.50	68
0150	Movable	"	30	.533	"	5	25		30	46.50
0300	Stock units, 6' high, 8' wide, plain, buy				Ea.	390			390	430
0350	With reflective tape, buy				"	405			405	445
0400	Break-a-way 3" PVC pipe barricade									
0410	with 3 ea. 1' x 4' reflectorized panels, buy				Ea.	106			106	117
0500	Plywood with steel legs, 24" wide					59			59	65
0600	Warning signal flag tree, 11' high, 2 flags, buy					257			257	283
0800	Traffic cones, PVC, 18" high					12.50			12.50	13.75
0850	28" high					19.50			19.50	21.50

01 56 Temporary Barriers and Enclosures

01 56 23 – Temporary Barricades

01 56 23.10 Barricades

		Crew	Daily Output	Labor-Hours	Unit	Material	2015 Bare Costs Labor	2015 Bare Costs Equipment	Total	Total Incl O&P
0900	Barrels, 55 gal., with flasher	1 Clab	96	.083	Ea.	91	3.13		94.13	105
1000	Guardrail, wooden, 3' high, 1" x 6", on 2" x 4" posts	2 Carp	200	.080	L.F.	1.27	3.76		5.03	7.55
1100	2" x 6", on 4" x 4" posts	"	165	.097		2.46	4.55		7.01	10.15
1200	Portable metal with base pads, buy					15.55			15.55	17.10
1250	Typical installation, assume 10 reuses	2 Carp	600	.027		2.55	1.25		3.80	4.86
1300	Barricade tape, polyethylene, 7 mil, 3" wide x 500' long roll				Ea.	25			25	27.50
3000	Detour signs, set up and remove									
3010	Reflective aluminum, MUTCD, 24" x 24", post mounted	1 Clab	20	.400	Ea.	2.46	15.05		17.51	27
4000	Roof edge portable barrier stands and warning flags, 50 uses	1 Rohe	9100	.001	L.F.	.07	.03		.10	.12
4010	100 uses	"	9100	.001	"	.03	.03		.06	.09

01 56 26 – Temporary Fencing

01 56 26.50 Temporary Fencing

		Crew	Daily Output	Labor-Hours	Unit	Material	Labor	Equipment	Total	Total Incl O&P
0010	**TEMPORARY FENCING**									
0020	Chain link, 11 ga., 4' high	2 Clab	400	.040	L.F.	2.95	1.50		4.45	5.70
0100	6' high		300	.053		2.95	2.01		4.96	6.55
0200	Rented chain link, 6' high, to 1000' (up to 12 mo.)		400	.040		4.29	1.50		5.79	7.20
0250	Over 1000' (up to 12 mo.)		300	.053		4.29	2.01		6.30	8
0350	Plywood, painted, 2" x 4" frame, 4' high	A-4	135	.178		6.20	7.95		14.15	19.80
0400	4" x 4" frame, 8' high	"	110	.218		12	9.75		21.75	29
0500	Wire mesh on 4" x 4" posts, 4' high	2 Carp	100	.160		9.85	7.50		17.35	23
0550	8' high	"	80	.200		14.90	9.40		24.30	32

01 56 29 – Temporary Protective Walkways

01 56 29.50 Protection

		Crew	Daily Output	Labor-Hours	Unit	Material	Labor	Equipment	Total	Total Incl O&P
0010	**PROTECTION**									
0020	Stair tread, 2" x 12" planks, 1 use	1 Carp	75	.107	Tread	5.10	5		10.10	13.80
0100	Exterior plywood, 1/2" thick, 1 use		65	.123		1.97	5.80		7.77	11.60
0200	3/4" thick, 1 use		60	.133		2.78	6.25		9.03	13.30
2200	Sidewalks, 2" x 12" planks, 2 uses		350	.023	S.F.	.85	1.07		1.92	2.70
2300	Exterior plywood, 2 uses, 1/2" thick		750	.011		.33	.50		.83	1.18
2400	5/8" thick		650	.012		.39	.58		.97	1.38
2500	3/4" thick		600	.013		.46	.63		1.09	1.54

01 56 32 – Temporary Security

01 56 32.50 Watchman

		Crew	Daily Output	Labor-Hours	Unit	Material	Labor	Equipment	Total	Total Incl O&P
0010	**WATCHMAN**									
0020	Service, monthly basis, uniformed person, minimum				Hr.				25	27.50
0100	Maximum								45.50	50
0200	Person and command dog, minimum								31	34
0300	Maximum								54.50	60
0500	Sentry dog, leased, with job patrol (yard dog), 1 dog				Week				290	320
0600	2 dogs				"				390	430
0800	Purchase, trained sentry dog, minimum				Ea.				1,375	1,500
0900	Maximum				"				2,725	3,000

01 58 Project Identification

01 58 13 – Temporary Project Signage

01 58 13.50 Signs	Crew	Daily Output	Labor-Hours	Unit	Material	2015 Bare Costs Labor	Equipment	Total	Total Incl O&P
0010 **SIGNS**									
0020 High intensity reflectorized, no posts, buy				Ea.	25			25	27.50

01 71 Examination and Preparation

01 71 23 – Field Engineering

01 71 23.13 Construction Layout

	Crew	Daily Output	Labor-Hours	Unit	Material	Labor	Equipment	Total	Total Incl O&P
0010 **CONSTRUCTION LAYOUT**									
1100 Crew for layout of building, trenching or pipe laying, 2 person crew	A-6	1	16	Day		750	55	805	1,275
1200 3 person crew	A-7	1	24			1,225	54.50	1,279.50	2,050
1400 Crew for roadway layout, 4 person crew	A-8	1	32	↓		1,600	54.50	1,654.50	2,625

01 74 Cleaning and Waste Management

01 74 13 – Progress Cleaning

01 74 13.20 Cleaning Up

	Crew	Daily Output	Labor-Hours	Unit	Material	Labor	Equipment	Total	Total Incl O&P
0010 **CLEANING UP**									
0020 After job completion, allow, minimum				Job				.30%	.30%
0040 Maximum				"				1%	1%
0052 Cleanup of floor area, continuous, per day, during const.	A-5	16	1.125	M.S.F.	2.18	42.50	3.84	48.52	76
0100 Final by GC at end of job	"	11.50	1.565	"	2.31	59	5.35	66.66	105

01 76 Protecting Installed Construction

01 76 13 – Temporary Protection of Installed Construction

01 76 13.20 Temporary Protection

	Crew	Daily Output	Labor-Hours	Unit	Material	Labor	Equipment	Total	Total Incl O&P
0010 **TEMPORARY PROTECTION**									
0020 Flooring, 1/8" tempered hardboard, taped seams	2 Carp	1500	.011	S.F.	.41	.50		.91	1.27
0030 Peel away carpet protection	1 Clab	3200	.003	"	.11	.09		.20	.27

01 91 Commissioning

01 91 13 – General Commissioning Requirements

01 91 13.50 Building Commissioning

	Crew	Daily Output	Labor-Hours	Unit	Material	Labor	Equipment	Total	Total Incl O&P
0010 **BUILDING COMMISSIONING**									
0100 Basic building commissioning, minimum				%				.25%	.25%
0150 Maximum								.50%	.50%
0200 Enhanced building commissioning, minimum								.50%	.50%
0250 Maximum				↓				1%	1%

01 93 Facility Maintenance

01 93 09 – Facility Equipment

01 93 09.50 Moving Equipment	Crew	Daily Output	Labor-Hours	Unit	Material	2015 Bare Costs Labor	2015 Bare Costs Equipment	Total	Total Incl O&P
0010 **MOVING EQUIPMENT**, Remove and reset, 100' distance,									
0020 No obstructions, no assembly or leveling unless noted									
0100 Annealing furnace, 24' overall	B-67	4	4	Ea.		196	78	274	395
0200 Annealing oven, small		14	1.143			56	22	78	113
0240 Very large		1	16			780	310	1,090	1,575
0400 Band saw, small		12	1.333			65	26	91	132
0440 Large		8	2			98	39	137	197
0500 Blue print copy machine		7	2.286			112	44.50	156.50	225
0600 Bonding mill, 6"		7	2.286			112	44.50	156.50	225
0620 12"		6	2.667			130	52	182	263
0640 18"		4	4			196	78	274	395
0660 24"		2	8			390	156	546	790
0700 Boring machine (jig)	B-68	7	3.429			168	44.50	212.50	315
0800 Bridgeport mill, standard	B-67	14	1.143			56	22	78	113
1000 Calibrator, 6 unit	"	14	1.143			56	22	78	113
1100 Comparitor, bench top	2 Clab	14	1.143			43		43	70.50
1140 Floor mounted	B-67	7	2.286			112	44.50	156.50	225
1200 Computer, desk top	2 Clab	25	.640			24		24	39.50
1300 Copy machine	"	25	.640			24		24	39.50
1500 Deflasher	B-67	14	1.143			56	22	78	113
1600 Degreaser, small		14	1.143			56	22	78	113
1640 Large 24' overall		1	16			780	310	1,090	1,575
1700 Desk with chair	2 Clab	25	.640			24		24	39.50
1725 Desk only	"	22	.727			27.50		27.50	45
1750 Chair only	1 Clab	300	.027			1		1	1.64
1800 Dial press	B-67	7	2.286			112	44.50	156.50	225
1900 Drafting table	2 Clab	14	1.143			43		43	70.50
2000 Drill press, bench top	"	14	1.143			43		43	70.50
2040 Floor mounted	B-67	14	1.143			56	22	78	113
2080 Industrial radial	"	7	2.286			112	44.50	156.50	225
2100 Dust collector, portable	2 Clab	25	.640			24		24	39.50
2140 Stationary, small	B-67	7	2.286			112	44.50	156.50	225
2180 Stationary, large	"	2	8			390	156	546	790
2300 Electric discharge machine	B-68	7	3.429			168	44.50	212.50	315
2400 Environmental chamber walls, including assembly	4 Clab	18	1.778	L.F.		67		67	110
2600 File cabinet	2 Clab	25	.640	Ea.		24		24	39.50
2800 Grinder/sander, pedestal mount	B-67	14	1.143			56	22	78	113
3000 Hack saw, power	2 Clab	24	.667			25		25	41
3100 Hydraulic press	B-67	14	1.143			56	22	78	113
3500 Laminar flow tables	"	14	1.143			56	22	78	113
3600 Lathe, bench	2 Clab	14	1.143			43		43	70.50
3640 6"	B-67	14	1.143			56	22	78	113
3680 10"		13	1.231			60	24	84	122
3720 12"		12	1.333			65	26	91	132
4000 Milling machine		8	2			98	39	137	197
4100 Molding press, 25 ton		5	3.200			156	62	218	315
4140 60 ton		4	4			196	78	274	395
4180 100 ton		2	8			390	156	546	790
4220 150 ton		1.50	10.667			520	207	727	1,050
4260 200 ton		1	16			780	310	1,090	1,575
4300 300 ton		.75	21.333			1,050	415	1,465	2,100
4700 Oil pot stand		14	1.143			56	22	78	113
5000 Press, 10 ton		14	1.143			56	22	78	113

01 93 Facility Maintenance

01 93 09 – Facility Equipment

01 93 09.50 Moving Equipment		Crew	Daily Output	Labor-Hours	Unit	Material	2015 Bare Costs Labor	Equipment	Total	Total Incl O&P
5040	15 ton	B-67	12	1.333	Ea.		65	26	91	132
5080	20 ton		10	1.600			78	31	109	158
5120	30 ton		8	2			98	39	137	197
5160	45 ton		6	2.667			130	52	182	263
5200	60 ton		4	4			196	78	274	395
5240	75 ton		2.50	6.400			315	124	439	630
5280	100 ton		2	8			390	156	546	790
5500	Raised floor, including assembly	2 Carp	250	.064	S.F.		3		3	4.92
5600	Rolling mill, 6"	B-67	7	2.286	Ea.		112	44.50	156.50	225
5640	9"		6	2.667			130	52	182	263
5680	12"		4	4			196	78	274	395
5720	13"		3.50	4.571			223	89	312	455
5760	18"		2	8			390	156	546	790
5800	25"		1	16			780	310	1,090	1,575
6000	Sander, floor stand		14	1.143			56	22	78	113
6100	Screw machine		7	2.286			112	44.50	156.50	225
6200	Shaper, 16"		14	1.143			56	22	78	113
6300	Shear, power assist		4	4			196	78	274	395
6400	Slitter, 6"		14	1.143			56	22	78	113
6440	8"		13	1.231			60	24	84	122
6480	10"		12	1.333			65	26	91	132
6520	12"		11	1.455			71	28.50	99.50	143
6560	16"		10	1.600			78	31	109	158
6600	20"		8	2			98	39	137	197
6640	24"		6	2.667			130	52	182	263
6800	Snag and tap machine		7	2.286			112	44.50	156.50	225
6900	Solder machine (auto)		7	2.286			112	44.50	156.50	225
7000	Storage cabinet metal, small	2 Clab	36	.444			16.70		16.70	27.50
7040	Large		25	.640			24		24	39.50
7100	Storage rack open, small		14	1.143			43		43	70.50
7140	Large		7	2.286			86		86	141
7200	Surface bench, small	B-67	14	1.143			56	22	78	113
7240	Large	"	5	3.200			156	62	218	315
7300	Surface grinder, large wet	B-68	5	4.800			235	62.50	297.50	440
7500	Time check machine	2 Clab	14	1.143			43		43	70.50
8000	Welder, 30 kVA (bench)		14	1.143			43		43	70.50
8100	Work bench with chair		25	.640			24		24	39.50
8110	For haul up to 200 feet, add				%		25%			

Division Notes

		CREW	DAILY OUTPUT	LABOR-HOURS	UNIT	BARE COSTS				TOTAL INCL O&P
						MAT.	LABOR	EQUIP.	TOTAL	

Estimating Tips

02 30 00 Subsurface Investigation

In preparing estimates on structures involving earthwork or foundations, all information concerning soil characteristics should be obtained. Look particularly for hazardous waste, evidence of prior dumping of debris, and previous stream beds.

02 40 00 Demolition and Structure Moving

The costs shown for selective demolition do not include rubbish handling or disposal. These items should be estimated separately using RSMeans data or other sources.

- Historic preservation often requires that the contractor remove materials from the existing structure, rehab them, and replace them. The estimator must be aware of any related measures and precautions that must be taken when doing selective demolition and cutting and patching. Requirements may include special handling and storage, as well as security.

- In addition to Subdivision 02 41 00, you can find selective demolition items in each division. Example: Roofing demolition is in Division 7.

02 40 00 Building Deconstruction

This section provides costs for the careful dismantling and recycling of most of low-rise building materials.

02 50 00 Containment of Hazardous Waste

This section addresses on-site hazardous waste disposal costs.

02 80 00 Hazardous Material Disposal/ Remediation

This subdivision includes information on hazardous waste handling, asbestos remediation, lead remediation, and mold remediation. See reference R028213-20 and R028319-60 for further guidance in using these unit price lines.

02 90 00 Monitoring Chemical Sampling, Testing Analysis

This section provides costs for on-site sampling and testing hazardous waste.

Reference Numbers

Reference numbers are shown in shaded boxes at the beginning of some major classifications. These numbers refer to related items in the Reference Section. The reference information may be an estimating procedure, an alternate pricing method, or technical information.

Note: Not all subdivisions listed here necessarily appear in this publication. ■

Did you know?
RSMeans Online gives you the same access to RSMeans' data with 24/7 access:
- Quickly locate costs in the searchable database.
- Build cost lists, estimates, and reports in minutes.
- Adjust costs to any location in the U.S. and Canada with the click of a button.

Start your free trial today at **www.rsmeansonline.com**

RSMeansOnline

No part of this publication may be reproduced, stored in a retrieval system, or transmitted in any form or by any means without prior written permission of RSMeans.

02 21 Surveys

02 21 13 – Site Surveys

02 21 13.09 Topographical Surveys

		Crew	Daily Output	Labor-Hours	Unit	Material	2015 Bare Costs Labor	2015 Bare Costs Equipment	Total	Total Incl O&P
0010	**TOPOGRAPHICAL SURVEYS**									
0020	Topographical surveying, conventional, minimum	A-7	3.30	7.273	Acre	20	375	16.60	411.60	640
0100	Maximum	A-8	.60	53.333	"	60	2,675	91	2,826	4,475

02 21 13.13 Boundary and Survey Markers

		Crew	Daily Output	Labor-Hours	Unit	Material	Labor	Equipment	Total	Total Incl O&P
0010	**BOUNDARY AND SURVEY MARKERS**									
0300	Lot location and lines, large quantities, minimum	A-7	2	12	Acre	35	620	27.50	682.50	1,075
0320	Average	"	1.25	19.200		55	990	44	1,089	1,700
0400	Small quantities, maximum	A-8	1	32	↓	75	1,600	54.50	1,729.50	2,725

02 32 Geotechnical Investigations

02 32 13 – Subsurface Drilling and Sampling

02 32 13.10 Boring and Exploratory Drilling

		Crew	Daily Output	Labor-Hours	Unit	Material	Labor	Equipment	Total	Total Incl O&P
0010	**BORING AND EXPLORATORY DRILLING**									
0020	Borings, initial field stake out & determination of elevations	A-6	1	16	Day		750	55	805	1,275
0100	Drawings showing boring details				Total		335		335	425
0200	Report and recommendations from P.E.						775		775	970
0300	Mobilization and demobilization	B-55	4	6	↓		229	271	500	675
0350	For over 100 miles, per added mile		450	.053	Mile		2.03	2.41	4.44	5.95
0600	Auger holes in earth, no samples, 2-1/2" diameter		78.60	.305	L.F.		11.65	13.75	25.40	34
0650	4" diameter		67.50	.356			13.55	16.05	29.60	39.50
0800	Cased borings in earth, with samples, 2-1/2" diameter		55.50	.432		14	16.50	19.50	50	64
0850	4" diameter	↓	32.60	.736		18	28	33	79	102
1000	Drilling in rock, "BX" core, no sampling	B-56	34.90	.458			19.75	45.50	65.25	82
1050	With casing & sampling		31.70	.505		14	22	50.50	86.50	106
1200	"NX" core, no sampling		25.92	.617			26.50	61.50	88	111
1250	With casing and sampling	↓	25	.640	↓	15	27.50	63.50	106	131
1400	Borings, earth, drill rig and crew with truck mounted auger	B-55	1	24	Day		915	1,075	1,990	2,700
1450	Rock using crawler type drill	B-56	1	16	"		690	1,600	2,290	2,850
1500	For inner city borings add, minimum								10%	10%
1510	Maximum								20%	20%

02 41 Demolition

02 41 13 – Selective Site Demolition

02 41 13.15 Hydrodemolition

		Crew	Daily Output	Labor-Hours	Unit	Material	Labor	Equipment	Total	Total Incl O&P
0010	**HYDRODEMOLITION**									
0015	Hydrodemolition, concrete pavement									
0120	20,000 PSI, Crew to include loader / vacuum truck as required									
0130	2" depth	B-5	1000	.056	S.F.		2.33	1.42	3.75	5.35
0410	4" depth		800	.070			2.91	1.78	4.69	6.65
0420	6" depth	↓	600	.093	↓		3.88	2.37	6.25	8.90

02 41 13.17 Demolish, Remove Pavement and Curb

		Crew	Daily Output	Labor-Hours	Unit	Material	Labor	Equipment	Total	Total Incl O&P
0010	**DEMOLISH, REMOVE PAVEMENT AND CURB** R024119-10									
5010	Pavement removal, bituminous roads, up to 3" thick	B-38	690	.058	S.Y.		2.48	1.85	4.33	6.05
5050	4" to 6" thick		420	.095			4.08	3.04	7.12	9.95
5100	Bituminous driveways		640	.063			2.68	1.99	4.67	6.50
5200	Concrete to 6" thick, hydraulic hammer, mesh reinforced		255	.157			6.70	5	11.70	16.35
5300	Rod reinforced		200	.200	↓		8.55	6.40	14.95	21
5400	Concrete, 7" to 24" thick, plain		33	1.212	C.Y.		52	38.50	90.50	127
5500	Reinforced	↓	24	1.667	"		71.50	53	124.50	174

02 41 Demolition

02 41 13 – Selective Site Demolition

02 41 13.17 Demolish, Remove Pavement and Curb

		Crew	Daily Output	Labor-Hours	Unit	Material	2015 Bare Costs Labor	2015 Bare Costs Equipment	Total	Total Incl O&P
5590	Minimum labor/equipment charge	B-38	6	6.667	Job		285	213	498	695
5600	With hand held air equipment, bituminous, to 6" thick	B-39	1900	.025	S.F.		1	.12	1.12	1.77
5700	Concrete to 6" thick, no reinforcing		1600	.030			1.19	.15	1.34	2.10
5800	Mesh reinforced		1400	.034			1.36	.17	1.53	2.40
5900	Rod reinforced		765	.063			2.50	.30	2.80	4.39
5990	Minimum labor/equipment charge	B-38	6	6.667	Job		285	213	498	695
6000	Curbs, concrete, plain	B-6	360	.067	L.F.		2.75	1.01	3.76	5.55
6100	Reinforced		275	.087			3.60	1.32	4.92	7.30
6200	Granite		360	.067			2.75	1.01	3.76	5.55
6300	Bituminous		528	.045			1.88	.69	2.57	3.80
6390	Minimum labor/equipment charge		6	4	Job		165	60.50	225.50	335

02 41 13.23 Utility Line Removal

		Crew	Daily Output	Labor-Hours	Unit	Material	Labor	Equipment	Total	Total Incl O&P
0010	**UTILITY LINE REMOVAL**									
0015	No hauling, abandon catch basin or manhole	B-6	7	3.429	Ea.		142	52	194	286
0020	Remove existing catch basin or manhole, masonry		4	6			248	91	339	500
0030	Catch basin or manhole frames and covers, stored		13	1.846			76	28	104	154
0040	Remove and reset		7	3.429			142	52	194	286
0900	Hydrants, fire, remove only	B-21A	5	8			375	94.50	469.50	700
0950	Remove and reset	"	2	20			940	237	1,177	1,750
0990	Minimum labor/equipment charge	2 Plum	2	8	Job		470		470	730
2900	Pipe removal, sewer/water, no excavation, 12" diameter	B-6	175	.137	L.F.		5.65	2.08	7.73	11.45
2930	15"-18" diameter	B-12Z	150	.160			6.75	10.70	17.45	22.50
2960	21"-24" diameter		120	.200			8.45	13.40	21.85	28.50
3000	27"-36" diameter		90	.267			11.30	17.85	29.15	38
3200	Steel, welded connections, 4" diameter	B-6	160	.150			6.20	2.28	8.48	12.50
3300	10" diameter		80	.300			12.40	4.55	16.95	25
3390	Minimum labor/equipment charge		3	8	Job		330	121	451	670

02 41 13.30 Minor Site Demolition

		Crew	Daily Output	Labor-Hours	Unit	Material	Labor	Equipment	Total	Total Incl O&P
0010	**MINOR SITE DEMOLITION** R024119-10									
0100	Roadside delineators, remove only	B-80	175	.183	Ea.		7.55	4.12	11.67	16.75
0110	Remove and reset	"	100	.320	"		13.20	7.20	20.40	29.50
0400	Minimum labor/equipment charge	B-6	4	6	Job		248	91	339	500
0800	Guiderail, corrugated steel, remove only	B-80A	100	.240	L.F.		9	3.03	12.03	18.15
0850	Remove and reset	"	40	.600	"		22.50	7.60	30.10	45.50
0860	Guide posts, remove only	B-80B	120	.267	Ea.		10.75	2.03	12.78	19.70
0870	Remove and reset	B-55	50	.480	"		18.30	21.50	39.80	54
0890	Minimum labor/equipment charge	2 Clab	4	4	Job		150		150	247
1000	Masonry walls, block, solid	B-5	1800	.031	C.F.		1.29	.79	2.08	2.97
1200	Brick, solid		900	.062			2.59	1.58	4.17	5.95
1400	Stone, with mortar		900	.062			2.59	1.58	4.17	5.95
1500	Dry set		1500	.037			1.55	.95	2.50	3.56
1600	Median barrier, precast concrete, remove and store	B-3	430	.112	L.F.		4.57	6	10.57	14
1610	Remove and reset	"	390	.123	"		5.05	6.60	11.65	15.40
1650	Minimum labor/equipment charge	A-1	4	2	Job		75	20.50	95.50	146
4000	Sidewalk removal, bituminous, 2" thick	B-6	350	.069	S.Y.		2.83	1.04	3.87	5.70
4010	2-1/2" thick		325	.074			3.05	1.12	4.17	6.15
4050	Brick, set in mortar		185	.130			5.35	1.97	7.32	10.80
4100	Concrete, plain, 4"		160	.150			6.20	2.28	8.48	12.50
4110	Plain, 5"		140	.171			7.05	2.60	9.65	14.30
4120	Plain, 6"		120	.200			8.25	3.04	11.29	16.70
4200	Mesh reinforced, concrete, 4"		150	.160			6.60	2.43	9.03	13.35
4210	5" thick		131	.183			7.55	2.78	10.33	15.30

02 41 Demolition

02 41 13 – Selective Site Demolition

02 41 13.30 Minor Site Demolition		Crew	Daily Output	Labor-Hours	Unit	Material	2015 Bare Costs Labor	2015 Bare Costs Equipment	Total	Total Incl O&P
4220	6" thick	B-6	112	.214	S.Y.		8.85	3.25	12.10	17.90
4290	Minimum labor/equipment charge	B-39	12	4	Job		159	19.40	178.40	281

02 41 13.33 Railtrack Removal

0010	**RAILTRACK REMOVAL**									
3500	Railroad track removal, ties and track	B-13	330	.170	L.F.		6.95	2.23	9.18	13.75
3600	Ballast	B-14	500	.096	C.Y.		3.82	.73	4.55	7
3700	Remove and re-install, ties & track using new bolts & spikes		50	.960	L.F.		38	7.30	45.30	70
3800	Turnouts using new bolts and spikes		1	48	Ea.		1,900	365	2,265	3,500
3890	Minimum labor/equipment charge		5	9.600	Job		380	73	453	700

02 41 13.60 Selective Demolition Fencing

0010	**SELECTIVE DEMOLITION FENCING** R024119-10									
1600	Fencing, barbed wire, 3 strand	2 Clab	430	.037	L.F.		1.40		1.40	2.29
1650	5 strand	"	280	.057			2.15		2.15	3.52
1700	Chain link, posts & fabric, 8' to 10' high, remove only	B-6	445	.054			2.23	.82	3.05	4.50
1750	Remove and reset	"	70	.343			14.15	5.20	19.35	28.50

02 41 16 – Structure Demolition

02 41 16.13 Building Demolition

0010	**BUILDING DEMOLITION** Large urban projects, incl. 20 mi. haul R024119-10									
0011	No foundation or dump fees, C.F. is vol. of building standing									
0020	Steel	B-8	21500	.003	C.F.		.13	.15	.28	.37
0050	Concrete		15300	.004			.18	.22	.40	.53
0080	Masonry		20100	.003			.14	.16	.30	.40
0100	Mixture of types		20100	.003			.14	.16	.30	.40
0500	Small bldgs, or single bldgs, no salvage included, steel	B-3	14800	.003			.13	.17	.30	.40
0600	Concrete		11300	.004			.17	.23	.40	.53
0650	Masonry		14800	.003			.13	.17	.30	.40
0700	Wood		14800	.003			.13	.17	.30	.40
0750	For buildings with no interior walls, deduct						50%			50%
1000	Demoliton single family house, one story, wood 1600 S.F.	B-3	1	48	Ea.		1,975	2,575	4,550	6,000
1020	3200 S.F.		.50	96			3,925	5,150	9,075	12,000
1200	Demoliton two family house, two story, wood 2400 S.F.		.67	71.964			2,950	3,850	6,800	9,025
1220	4200 S.F.		.38	128			5,250	6,850	12,100	16,000
1300	Demoliton three family house, three story, wood 3200 S.F.		.50	96			3,925	5,150	9,075	12,000
1320	5400 S.F.		.30	160			6,550	8,575	15,125	20,000
5000	For buildings with no interior walls, deduct						50%			50%

02 41 16.17 Building Demolition Footings and Foundations

0010	**BUILDING DEMOLITION FOOTINGS AND FOUNDATIONS** R024119-10									
0200	Floors, concrete slab on grade,									
0240	4" thick, plain concrete	B-13L	5000	.003	S.F.		.17	.41	.58	.71
0280	Reinforced, wire mesh		4000	.004			.21	.51	.72	.90
0300	Rods		4500	.004			.18	.46	.64	.79
0400	6" thick, plain concrete		4000	.004			.21	.51	.72	.90
0420	Reinforced, wire mesh		3200	.005			.26	.64	.90	1.12
0440	Rods		3600	.004			.23	.57	.80	.99
1000	Footings, concrete, 1' thick, 2' wide	B-5	300	.187	L.F.		7.75	4.74	12.49	17.80
1080	1'-6" thick, 2' wide		250	.224			9.30	5.70	15	21.50
1120	3' wide		200	.280			11.65	7.10	18.75	26.50
1140	2' thick, 3' wide		175	.320			13.30	8.10	21.40	30.50
1200	Average reinforcing, add								10%	10%
1220	Heavy reinforcing, add								20%	20%
2000	Walls, block, 4" thick	B-13L	8000	.002	S.F.		.10	.26	.36	.44

02 41 Demolition

02 41 16 – Structure Demolition

02 41 16.17 Building Demolition Footings and Foundations

		Crew	Daily Output	Labor-Hours	Unit	Material	2015 Bare Costs Labor	2015 Bare Costs Equipment	Total	Total Incl O&P
2040	6" thick	B-13L	6000	.003	S.F.		.14	.34	.48	.60
2080	8" thick		4000	.004			.21	.51	.72	.90
2100	12" thick		3000	.005			.28	.68	.96	1.19
2200	For horizontal reinforcing, add								10%	10%
2220	For vertical reinforcing, add								20%	20%
2400	Concrete, plain concrete, 6" thick	B-13L	4000	.004			.21	.51	.72	.90
2420	8" thick		3500	.005			.24	.59	.83	1.03
2440	10" thick		3000	.005			.28	.68	.96	1.19
2500	12" thick		2500	.006			.33	.82	1.15	1.43
2600	For average reinforcing, add								10%	10%
2620	For heavy reinforcing, add								20%	20%
9000	Minimum labor/equipment charge	A-1	2	4	Job		150	40.50	190.50	292

02 41 16.19 Minor Building Deconstruction

			Crew	Daily Output	Labor-Hours	Unit	Material	Labor	Equipment	Total	Total Incl O&P
0010	**MINOR BUILDING DECONSTRUCTION**	R024119-10									
0011	For salvage, avg house										
0225	2 story, pre 1970 house, 1400 S.F., labor for all salv mat, min	G	6 Clab	25	1.920	SF Flr.		72		72	118
0230	Maximum	G	"	15	3.200	"		120		120	197
0235	Salvage of carpet, tackless	G	2 Clab	2400	.007	S.F.		.25		.25	.41
0240	Wood floors, incl denailing and packaging	G	3 Clab	270	.089	"		3.34		3.34	5.50
0260	Wood doors and trim, standard	G	1 Clab	16	.500	Ea.		18.80		18.80	31
0280	Base or cove mouldings, incl denailing and packaging	G		500	.016	L.F.		.60		.60	.99
0320	Closet shelving and trim, incl denailing and packaging	G		18	.444	Set		16.70		16.70	27.50
0340	Kitchen cabinets, uppers and lowers, prefab type	G	2 Clab	24	.667	L.F.		25		25	41
0360	Kitchen cabinets, uppers and lowers, built-in	G		12	1.333	"		50		50	82
0370	Bath fixt, incl toilet, tub, vanity, and med cabinet	G		3	5.333	Set		201		201	330

02 41 19 – Selective Demolition

02 41 19.13 Selective Building Demolition

0010	**SELECTIVE BUILDING DEMOLITION**
0020	Costs related to selective demolition of specific building components
0025	are included under Common Work Results (XX 05)
0030	in the component's appropriate division.

02 41 19.16 Selective Demolition, Cutout

			Crew	Daily Output	Labor-Hours	Unit	Material	Labor	Equipment	Total	Total Incl O&P
0010	**SELECTIVE DEMOLITION, CUTOUT**	R024119-10									
0020	Concrete, elev. slab, light reinforcement, under 6 C.F.		B-9	65	.615	C.F.		23.50	3.58	27.08	42.50
0050	Light reinforcing, over 6 C.F.			75	.533	"		20.50	3.10	23.60	36.50
0200	Slab on grade to 6" thick, not reinforced, under 8 S.F.			85	.471	S.F.		17.90	2.74	20.64	32.50
0250	8 – 16 S.F.			175	.229	"		8.70	1.33	10.03	15.70
0255	For over 16 S.F. see Line 02 41 16.17 0400										
0600	Walls, not reinforced, under 6 C.F.		B-9	60	.667	C.F.		25.50	3.88	29.38	46
0650	6 – 12 C.F.		"	80	.500	"		19	2.91	21.91	34
0655	For over 12 C.F. see Line 02 41 16.17 2500										
1000	Concrete, elevated slab, bar reinforced, under 6 C.F.		B-9	45	.889	C.F.		34	5.15	39.15	61
1050	Bar reinforced, over 6 C.F.			50	.800	"		30.50	4.66	35.16	55
1200	Slab on grade to 6" thick, bar reinforced, under 8 S.F.			75	.533	S.F.		20.50	3.10	23.60	36.50
1250	8 – 16 S.F.			150	.267	"		10.15	1.55	11.70	18.30
1255	For over 16 S.F. see Line 02 41 16.17 0440										
1400	Walls, bar reinforced, under 6 C.F.		B-9	50	.800	C.F.		30.50	4.66	35.16	55
1450	6 – 12 C.F.		"	70	.571	"		21.50	3.33	24.83	39
1455	For over 12 C.F. see Lines 02 41 16.17 2500 and 2600										
2000	Brick, to 4 S.F. opening, not including toothing										
2040	4" thick		B-9	30	1.333	Ea.		50.50	7.75	58.25	91.50
2060	8" thick			18	2.222			84.50	12.95	97.45	152

02 41 Demolition

02 41 19 – Selective Demolition

02 41 19.16 Selective Demolition, Cutout

		Crew	Daily Output	Labor-Hours	Unit	Material	2015 Bare Costs Labor	Equipment	Total	Total Incl O&P
2080	12" thick	B-9	10	4	Ea.		152	23.50	175.50	275
2400	Concrete block, to 4 S.F. opening, 2" thick		35	1.143			43.50	6.65	50.15	78.50
2420	4" thick		30	1.333			50.50	7.75	58.25	91.50
2440	8" thick		27	1.481			56.50	8.60	65.10	102
2460	12" thick		24	1.667			63.50	9.70	73.20	115
2600	Gypsum block, to 4 S.F. opening, 2" thick		80	.500			19	2.91	21.91	34
2620	4" thick		70	.571			21.50	3.33	24.83	39
2640	8" thick		55	.727			27.50	4.23	31.73	50
2800	Terra cotta, to 4 S.F. opening, 4" thick		70	.571			21.50	3.33	24.83	39
2840	8" thick		65	.615			23.50	3.58	27.08	42.50
2880	12" thick		50	.800			30.50	4.66	35.16	55
4000	For toothing masonry, see Section 04 01 20.50									
6000	Walls, interior, not including re-framing,									
6010	openings to 5 S.F.									
6100	Drywall to 5/8" thick	1 Clab	24	.333	Ea.		12.55		12.55	20.50
6200	Paneling to 3/4" thick		20	.400			15.05		15.05	24.50
6300	Plaster, on gypsum lath		20	.400			15.05		15.05	24.50
6340	On wire lath		14	.571			21.50		21.50	35
7000	Wood frame, not including re-framing, openings to 5 S.F.									
7200	Floors, sheathing and flooring to 2" thick	1 Clab	5	1.600	Ea.		60		60	98.50
7310	Roofs, sheathing to 1" thick, not including roofing		6	1.333			50		50	82
7410	Walls, sheathing to 1" thick, not including siding		7	1.143			43		43	70.50
8500	Minimum labor/equipment charge		4	2	Job		75		75	123

02 41 19.18 Selective Demolition, Disposal Only

			Crew	Daily Output	Labor-Hours	Unit	Material	Labor	Equipment	Total	Total Incl O&P
0010	**SELECTIVE DEMOLITION, DISPOSAL ONLY**	R024119-10									
0015	Urban bldg w/salvage value allowed										
0020	Including loading and 5 mile haul to dump										
0200	Steel frame		B-3	430	.112	C.Y.		4.57	6	10.57	14
0300	Concrete frame			365	.132			5.40	7.05	12.45	16.45
0400	Masonry construction			445	.108			4.41	5.80	10.21	13.50
0500	Wood frame			247	.194			7.95	10.40	18.35	24.50

02 41 19.19 Selective Demolition

			Crew	Daily Output	Labor-Hours	Unit	Material	Labor	Equipment	Total	Total Incl O&P
0010	**SELECTIVE DEMOLITION,** Rubbish Handling	R024119-10									
0020	The following are to be added to the demolition prices										
0050	The following are components for a complete chute system										
0100	Top chute circular steel, 4' long, 18" diameter		B-1C	15	1.600	Ea.	266	61	29	356	425
0102	23" diameter			15	1.600		288	61	29	378	445
0104	27" diameter			15	1.600		310	61	29	400	470
0106	30" diameter			15	1.600		330	61	29	420	495
0108	33" diameter			15	1.600		355	61	29	445	520
0110	36" diameter			15	1.600		375	61	29	465	545
0112	Regular chute, 18" diameter			15	1.600		199	61	29	289	350
0114	23" diameter			15	1.600		222	61	29	312	375
0116	27" diameter			15	1.600		243	61	29	333	400
0118	30" diameter			15	1.600		266	61	29	356	425
0120	33" diameter			15	1.600		288	61	29	378	445
0122	36" diameter			15	1.600		310	61	29	400	470
0124	Control door chute, 18" diameter			15	1.600		375	61	29	465	545
0126	23" diameter			15	1.600		400	61	29	490	570
0128	27" diameter			15	1.600		420	61	29	510	595
0130	30" diameter			15	1.600		445	61	29	535	620
0132	33" diameter			15	1.600		465	61	29	555	640

02 41 Demolition

02 41 19 – Selective Demolition

02 41 19.19 Selective Demolition		Crew	Daily Output	Labor-Hours	Unit	Material	2015 Bare Costs Labor	Equipment	Total	Total Incl O&P
0134	36" diameter	B-1C	15	1.600	Ea.	490	61	29	580	665
0136	Chute liners, 14 ga., 18-30" diameter		15	1.600		223	61	29	313	380
0138	33-36" diameter		15	1.600		280	61	29	370	440
0140	17% thinner chute, 30" diameter		15	1.600		222	61	29	312	375
0142	33% thinner chute, 30" diameter		15	1.600		167	61	29	257	315
0144	Top chute cover	1 Clab	24	.333		153	12.55		165.55	189
0146	Door chute cover	"	24	.333		153	12.55		165.55	189
0148	Top chute trough	2 Clab	12	1.333		510	50		560	640
0150	Bolt down frame & counter weights, 250 lb.	B-1	4	6		4,250	230		4,480	5,025
0152	500 lb.		4	6		6,200	230		6,430	7,200
0154	750 lb.		4	6		9,100	230		9,330	10,400
0156	1000 lb.		2.67	8.989		10,200	345		10,545	11,800
0158	1500 lb.		2.67	8.989		13,000	345		13,345	14,900
0160	Chute warning light system, 5 stories	B-1C	4	6		8,975	230	109	9,314	10,400
0162	10 stories	"	2	12		14,300	460	218	14,978	16,700
0164	Dust control device for dumpsters	1 Clab	8	1		102	37.50		139.50	174
0166	Install or replace breakaway cord		8	1		27	37.50		64.50	91
0168	Install or replace warning sign		16	.500		11.45	18.80		30.25	43.50
0600	Dumpster, weekly rental, 1 dump/week, 6 C.Y. capacity (2 Tons)				Week	415			415	455
0700	10 C.Y. capacity (3 Tons)					480			480	530
0725	20 C.Y. capacity (5 Tons)					565			565	625
0800	30 C.Y. capacity (7 Tons)					730			730	800
0840	40 C.Y. capacity (10 Tons)					775			775	850
0900	Alternate pricing for dumpsters									
0910	Delivery, average for all sizes				Ea.	75			75	82.50
0920	Haul, average for all sizes					235			235	259
0930	Rent per day, average for all sizes					20			20	22
0940	Rent per month, average for all sizes					80			80	88
0950	Disposal fee per ton, average for all sizes				Ton	88			88	97
2000	Load, haul, dump and return, 0 – 50' haul, hand carried	2 Clab	24	.667	C.Y.		25		25	41
2005	Wheeled		37	.432			16.25		16.25	26.50
2040	0 – 100' haul, hand carried		16.50	.970			36.50		36.50	60
2045	Wheeled		25	.640			24		24	39.50
2080	Haul and return, add per each extra 100' haul, hand carried		35.50	.451			16.95		16.95	28
2085	Wheeled		54	.296			11.15		11.15	18.25
2120	For travel in elevators, up to 10 floors, add		140	.114			4.30		4.30	7.05
2130	0 – 50' haul, incl. up to 5 riser stair, hand carried		23	.696			26		26	43
2135	Wheeled		35	.457			17.20		17.20	28
2140	6 – 10 riser stairs, hand carried		22	.727			27.50		27.50	45
2145	Wheeled		34	.471			17.70		17.70	29
2150	11 – 20 riser stairs, hand carried		20	.800			30		30	49.50
2155	Wheeled		31	.516			19.40		19.40	32
2160	21 – 40 riser stairs, hand carried		16	1			37.50		37.50	61.50
2165	Wheeled		24	.667			25		25	41
2170	0 – 100' haul, incl. 5 riser stair, hand carried		15	1.067			40		40	66
2175	Wheeled		23	.696			26		26	43
2180	6 – 10 riser stair, hand carried		14	1.143			43		43	70.50
2185	Wheeled		21	.762			28.50		28.50	47
2190	11 – 20 riser stair, hand carried		12	1.333			50		50	82
2195	Wheeled		18	.889			33.50		33.50	55
2200	21 – 40 riser stair, hand carried		8	2			75		75	123
2205	Wheeled		12	1.333			50		50	82
2210	Haul and return, add per each extra 100' haul, hand carried		35.50	.451			16.95		16.95	28

02 41 Demolition

02 41 19 – Selective Demolition

02 41 19.19 Selective Demolition

	02 41 19.19 Selective Demolition	Crew	Daily Output	Labor-Hours	Unit	Material	2015 Bare Costs Labor	2015 Bare Costs Equipment	Total	Total Incl O&P
2215	Wheeled	2 Clab	54	.296	C.Y.		11.15		11.15	18.25
2220	For each additional flight of stairs, up to 5 risers, add		550	.029	Flight		1.09		1.09	1.79
2225	6 – 10 risers, add		275	.058			2.19		2.19	3.59
2230	11 – 20 risers, add		138	.116			4.36		4.36	7.15
2235	21 – 40 risers, add		69	.232			8.70		8.70	14.30
3000	Loading & trucking, including 2 mile haul, chute loaded	B-16	45	.711	C.Y.		27.50	15.35	42.85	62
3040	Hand loading truck, 50' haul	"	48	.667			26	14.40	40.40	58
3080	Machine loading truck	B-17	120	.267			10.90	6.45	17.35	25
3120	Wheeled 50' and ramp dump loaded	2 Clab	24	.667			25		25	41
5000	Haul, per mile, up to 8 C.Y. truck	B-34B	1165	.007			.28	.59	.87	1.09
5100	Over 8 C.Y. truck	"	1550	.005			.21	.45	.66	.82

02 41 19.20 Selective Demolition, Dump Charges

0010	**SELECTIVE DEMOLITION, DUMP CHARGES**	R024119-10								
0020	Dump charges, typical urban city, tipping fees only									
0100	Building construction materials				Ton	74			74	81
0200	Trees, brush, lumber					63			63	69.50
0300	Rubbish only					63			63	69.50
0500	Reclamation station, usual charge					74			74	81

02 41 19.21 Selective Demolition, Gutting

		Crew	Daily Output	Labor-Hours	Unit	Material	Labor	Equipment	Total	Total Incl O&P
0010	**SELECTIVE DEMOLITION, GUTTING**	R024119-10								
0020	Building interior, including disposal, dumpster fees not included									
0500	Residential building									
0560	Minimum	B-16	400	.080	SF Flr.		3.10	1.73	4.83	6.95
0580	Maximum	"	360	.089	"		3.44	1.92	5.36	7.70
0900	Commercial building									
1000	Minimum	B-16	350	.091	SF Flr.		3.54	1.97	5.51	7.95
1020	Maximum		250	.128	"		4.95	2.76	7.71	11.15
3000	Minimum labor/equipment charge		4	8	Job		310	173	483	695

02 41 19.25 Selective Demolition, Saw Cutting

		Crew	Daily Output	Labor-Hours	Unit	Material	Labor	Equipment	Total	Total Incl O&P
0010	**SELECTIVE DEMOLITION, SAW CUTTING**	R024119-10								
0015	Asphalt, up to 3" deep	B-89	1050	.015	L.F.	.12	.67	.46	1.25	1.71
0020	Each additional inch of depth	"	1800	.009		.04	.39	.27	.70	.96
1200	Masonry walls, hydraulic saw, brick, per inch of depth	B-89B	300	.053		.04	2.34	2.86	5.24	6.95
1220	Block walls, solid, per inch of depth	"	250	.064		.04	2.81	3.43	6.28	8.30
2000	Brick or masonry w/hand held saw, per inch of depth	A-1	125	.064		.05	2.41	.65	3.11	4.71
5000	Wood sheathing to 1" thick, on walls	1 Carp	200	.040			1.88		1.88	3.08
5020	On roof	"	250	.032			1.50		1.50	2.46
9000	Minimum labor/equipment charge	A-1	2	4	Job		150	40.50	190.50	292

02 41 19.27 Selective Demolition, Torch Cutting

		Crew	Daily Output	Labor-Hours	Unit	Material	Labor	Equipment	Total	Total Incl O&P
0010	**SELECTIVE DEMOLITION, TORCH CUTTING**	R024119-10								
0020	Steel, 1" thick plate	E-25	333	.024	L.F.	.84	1.31	.03	2.18	3.33
0040	1" diameter bar	"	600	.013	Ea.	.14	.73	.02	.89	1.49
1000	Oxygen lance cutting, reinforced concrete walls									
1040	12" to 16" thick walls	1 Clab	10	.800	L.F.		30		30	49.50
1080	24" thick walls	"	6	1.333	"		50		50	82
1090	Minimum labor/equipment charge	E-25	2	4	Job		219	5.70	224.70	400
1100	See Section 05 05 21.10									

02 42 Removal and Salvage of Construction Materials

02 42 10 – Building Deconstruction

02 42 10.10 Estimated Salvage Value or Savings

		Crew	Daily Output	Labor-Hours	Unit	Material	2015 Bare Costs Labor	2015 Bare Costs Equipment	Total	Total Incl O&P
0010	**ESTIMATED SALVAGE VALUE OR SAVINGS**									
0015	Excludes material handling, packaging, container costs and									
0020	transportation for salvage or disposal									
0050	All Items in Section 02 42 10.10 are credit deducts and not costs									
0100	Copper Wire Salvage Value	G			Lb.				1.60	1.60
0110	Disposal Savings	G							.04	.04
0200	Copper Pipe Salvage Value	G							2.50	2.50
0210	Disposal Savings	G							.05	.05
0300	Steel Pipe Salvage Value	G							.06	.06
0310	Disposal Savings	G							.03	.03
0400	Cast Iron Pipe Salvage Value	G							.03	.03
0410	Disposal Savings	G							.01	.01
0500	Steel Doors or Windows Salvage Value	G							.06	.06
0510	Aluminum	G							.55	.55
0520	Disposal Savings	G							.03	.03
0600	Aluminum Siding Salvage Value	G							.49	.49
0630	Disposal Savings	G							.03	.03
0640	Wood Siding (no lead or asbestos)	G			C.Y.				12	12
0800	Clean Concrete Disposal Savings	G			Ton				62	62
0850	Asphalt Shingles Disposal Savings	G			"				60	60
1000	Wood wall framing clean salvage value	G			M.B.F.				55	55
1010	Painted	G							44	44
1020	Floor framing	G							55	55
1030	Painted	G							44	44
1050	Roof framing	G							55	55
1060	Painted	G							44	44
1100	Wood beams salvage value	G							55	55
1200	Wood framing and beams disposal savings	G			Ton				66	66
1220	Wood sheating and sub-base flooring	G							72.50	72.50
1230	Wood wall paneling (1/4 inch thick)	G							66	66
1300	Wood panel 3/4-1 inch thick low salvage value	G			S.F.				.55	.55
1350	high salvage value	G			"				2.20	2.20
1400	Disposal savings	G			Ton				66	66
1500	Flooring tongue and groove 25/32 inch thick low salvage value	G			S.F.				.55	.55
1530	High salvage value	G			"				1.10	1.10
1560	Disposal savings	G			Ton				66	66
1600	Drywall or sheet rock salvage value	G							22	22
1650	Disposal savings	G							66	66

02 42 10.20 Deconstruction of Building Components

		Crew	Daily Output	Labor-Hours	Unit	Material	2015 Bare Costs Labor	2015 Bare Costs Equipment	Total	Total Incl O&P
0010	**DECONSTRUCTION OF BUILDING COMPONENTS**									
0012	Buildings one or two stories only									
0015	Excludes material handling, packaging, container costs and									
0020	transportation for salvage or disposal									
0050	Deconstruction of Plumbing Fixtures									
0100	Wall hung or countertop lavatory	G	2 Clab	16	1	Ea.		37.50		37.50	61.50
0110	Single or double compartment kitchen sink	G		14	1.143			43		43	70.50
0120	Wall hung urinal	G		14	1.143			43		43	70.50
0130	Floor mounted	G		8	2			75		75	123
0140	Floor mounted water closet	G		16	1			37.50		37.50	61.50
0150	Wall hung	G		14	1.143			43		43	70.50
0160	Water fountain, free standing	G		16	1			37.50		37.50	61.50
0170	Wall hung or deck mounted	G		12	1.333			50		50	82

02 42 Removal and Salvage of Construction Materials

02 42 10 – Building Deconstruction

02 42 10.20 Deconstruction of Building Components

		Crew	Daily Output	Labor-Hours	Unit	Material	2015 Bare Costs Labor	2015 Bare Costs Equipment	Total	Total Incl O&P
0180	Bathtub, steel or fiberglass	G 2 Clab	10	1.600	Ea.		60		60	98.50
0190	Cast iron	G	8	2			75		75	123
0200	Shower, single	G	6	2.667			100		100	164
0210	Group	G	7	2.286			86		86	141
0300	Deconstruction of Electrical Fixtures									
0310	Surface mount incandescent fixtures	G 2 Clab	48	.333	Ea.		12.55		12.55	20.50
0320	Fluorescent, 2 lamp	G	32	.500			18.80		18.80	31
0330	4 lamp	G	24	.667			25		25	41
0340	Strip Fluorescent, 1 lamp	G	40	.400			15.05		15.05	24.50
0350	2 lamp	G	32	.500			18.80		18.80	31
0400	Recessed drop-in fluorescent fixture, 2 lamp	G	27	.593			22.50		22.50	36.50
0410	4 lamp	G	18	.889			33.50		33.50	55
0500	Deconstruction of appliances									
0510	Cooking stoves	G 2 Clab	26	.615	Ea.		23		23	38
0520	Dishwashers	G "	26	.615	"		23		23	38
0600	Deconstruction of millwork and trim									
0610	Cabinets, wood	G 2 Carp	40	.400	L.F.		18.80		18.80	31
0620	Countertops	G	100	.160	"		7.50		7.50	12.30
0630	Wall paneling, 1 inch thick	G	500	.032	S.F.		1.50		1.50	2.46
0640	Ceiling trim	G	500	.032	L.F.		1.50		1.50	2.46
0650	Wainscoting	G	500	.032	S.F.		1.50		1.50	2.46
0660	Base, 3/4" to 1" thick	G	600	.027	L.F.		1.25		1.25	2.05
0700	Deconstruction of doors and windows									
0710	Doors, wrap, interior, wood, single, no closers	G 2 Carp	21	.762	Ea.	4.50	36		40.50	63.50
0720	Double	G	13	1.231		9	58		67	104
0730	Solid core, single, exterior or interior	G	10	1.600		4.50	75		79.50	128
0740	Double	G	8	2		9	94		103	164
0810	Windows, wrap, wood, single									
0812	with no casement or cladding	G 2 Carp	21	.762	Ea.	4.50	36		40.50	63.50
0820	with casement and/or cladding	G "	18	.889	"	4.50	41.50		46	73.50
0900	Deconstruction of interior finishes									
0910	Drywall for recycling	G 2 Clab	1775	.009	S.F.		.34		.34	.56
0920	Plaster wall, first floor	G	1775	.009			.34		.34	.56
0930	Second floor	G	1330	.012			.45		.45	.74
1000	Deconstruction of roofing and accessories									
1010	Built-up roofs	G 2 Clab	570	.028	S.F.		1.06		1.06	1.73
1020	Gutters, facia and rakes	G "	1140	.014	L.F.		.53		.53	.87
2000	Deconstruction of wood components									
2010	Roof sheeting	G 2 Clab	570	.028	S.F.		1.06		1.06	1.73
2020	Main roof framing	G	760	.021	L.F.		.79		.79	1.30
2030	Porch roof framing	G	445	.036			1.35		1.35	2.22
2040	Beams 4" x 8"	G B-1	375	.064			2.45		2.45	4.01
2050	4" x 10"	G	300	.080			3.06		3.06	5
2055	4" x 12"	G	250	.096			3.67		3.67	6
2060	6" x 8"	G	250	.096			3.67		3.67	6
2065	6" x 10"	G	200	.120			4.59		4.59	7.55
2070	6" x 12"	G	170	.141			5.40		5.40	8.85
2075	8" x 12"	G	126	.190			7.30		7.30	11.95
2080	10" x 12"	G	100	.240			9.20		9.20	15.05
2100	Ceiling joists	G 2 Clab	800	.020			.75		.75	1.23
2150	Wall framing, interior	G	1230	.013			.49		.49	.80
2160	Sub-floor	G	2000	.008	S.F.		.30		.30	.49
2170	Floor joists	G	2000	.008	L.F.		.30		.30	.49

02 42 Removal and Salvage of Construction Materials

02 42 10 – Building Deconstruction

02 42 10.20 Deconstruction of Building Components

		Crew	Daily Output	Labor-Hours	Unit	Material	2015 Bare Costs Labor	Equipment	Total	Total Incl O&P
2200	Wood siding (no lead or asbestos)	G 2 Clab	1300	.012	S.F.		.46		.46	.76
2300	Wall framing, exterior	G	1600	.010	L.F.		.38		.38	.62
2400	Stair risers	G	53	.302	Ea.		11.35		11.35	18.60
2500	Posts	G	800	.020	L.F.		.75		.75	1.23
3000	Deconstruction of exterior brick walls									
3010	Exterior brick walls, first floor	G 2 Clab	200	.080	S.F.		3.01		3.01	4.93
3020	Second floor	G	64	.250	"		9.40		9.40	15.40
3030	Brick chimney	G	100	.160	C.F.		6		6	9.85
4000	Deconstruction of concrete									
4010	Slab on grade, 4" thick, plain concrete	G B-9	500	.080	S.F.		3.04	.47	3.51	5.50
4020	Wire mesh reinforced	G	470	.085			3.23	.50	3.73	5.85
4030	Rod reinforced	G	400	.100			3.80	.58	4.38	6.90
4110	Foundation wall, 6" thick, plain concrete	G	160	.250			9.50	1.46	10.96	17.20
4120	8" thick	G	140	.286			10.85	1.66	12.51	19.65
4130	10" thick	G	120	.333			12.65	1.94	14.59	23
9000	Deconstruction process, support equipment as needed									
9010	Daily use, portal to portal, 12-ton truck-mounted hydraulic crane crew	G A-3H	1	8	Day		415	860	1,275	1,600
9020	Daily use, skid steer and operator	G A-3C	1	8			390	320	710	965
9030	Daily use, backhoe 48 H.P., operator and labor	G "	1	8			390	320	710	965

02 42 10.30 Deconstruction Material Handling

		Crew	Daily Output	Labor-Hours	Unit	Material	Labor	Equipment	Total	Total Incl O&P
0010	**DECONSTRUCTION MATERIAL HANDLING**									
0012	Buildings one or two stories only									
0100	Clean and stack brick on pallet	G 2 Clab	1200	.013	Ea.		.50		.50	.82
0200	Haul 50' and load rough lumber up to 2" x 8" size	G	2000	.008	"		.30		.30	.49
0210	Lumber larger than 2" x 8"	G	3200	.005	B.F.		.19		.19	.31
0300	Finish wood for recycling stack and wrap per pallet	G	8	2	Ea.	36	75		111	163
0350	Light fixtures		6	2.667		65	100		165	236
0375	Windows		6	2.667		61	100		161	232
0400	Miscellaneous materials		8	2		18	75		93	143
1000	See Section 02 41 19.19 for bulk material handling									

02 43 Structure Moving

02 43 13 – Structure Relocation

02 43 13.13 Building Relocation

		Crew	Daily Output	Labor-Hours	Unit	Material	Labor	Equipment	Total	Total Incl O&P
0010	**BUILDING RELOCATION**									
0011	One day move, up to 24' wide									
0020	Reset on existing foundation				Total				11,500	11,500
0040	Wood or steel frame bldg., based on ground floor area	G B-4	185	.259	S.F.		9.95	2.78	12.73	19.30
0060	Masonry bldg., based on ground floor area	G "	137	.350			13.45	3.75	17.20	26
0200	For 24' to 42' wide, add								15%	15%

02 58 Snow Control

02 58 13 – Snow Fencing

02 58 13.10 Snow Fencing System		Crew	Daily Output	Labor-Hours	Unit	Material	2015 Bare Costs Labor	Equipment	Total	Total Incl O&P
0010	**SNOW FENCING SYSTEM**									
7001	Snow fence on steel posts 10' O.C., 4' high	B-1	500	.048	L.F.	.93	1.84		2.77	4.03

02 65 Underground Storage Tank Removal

02 65 10 – Underground Tank and Contaminated Soil Removal

02 65 10.30 Removal of Underground Storage Tanks

			Crew	Daily Output	Labor-Hours	Unit	Material	Labor	Equipment	Total	Total Incl O&P
0010	**REMOVAL OF UNDERGROUND STORAGE TANKS** R026510-20										
0011	Petroleum storage tanks, non-leaking										
0100	Excavate & load onto trailer										
0110	3000 gal. to 5000 gal. tank	G	B-14	4	12	Ea.		475	91	566	875
0120	6000 gal. to 8000 gal. tank	G	B-3A	3	13.333			535	345	880	1,250
0130	9000 gal. to 12000 gal. tank	G	"	2	20	↓		805	515	1,320	1,875
0190	Known leaking tank, add					%				100%	100%
0200	Remove sludge, water and remaining product from tank bottom										
0201	of tank with vacuum truck										
0300	3000 gal. to 5000 gal. tank	G	A-13	5	1.600	Ea.		78	153	231	291
0310	6000 gal. to 8000 gal. tank	G		4	2			97	191	288	365
0320	9000 gal. to 12000 gal. tank	G	↓	3	2.667	↓		130	254	384	485
0390	Dispose of sludge off-site, average					Gal.				6.25	6.80
0400	Insert inert solid CO_2 "dry ice" into tank										
0401	For cleaning/transporting tanks (1.5 lb./100 gal. cap)	G	1 Clab	500	.016	Lb.	1.17	.60		1.77	2.28
0403	Insert solid carbon dioxide, 1.5 lb./100 gal.	G	"	400	.020	"	1.17	.75		1.92	2.52
0503	Disconnect and remove piping	G	1 Plum	160	.050	L.F.		2.94		2.94	4.58
0603	Transfer liquids, 10% of volume	G	"	1600	.005	Gal.		.29		.29	.46
0703	Cut accessway into underground storage tank	G	1 Clab	5.33	1.501	Ea.		56.50		56.50	92.50
0813	Remove sludge, wash and wipe tank, 500 gal.	G	1 Plum	8	1			58.50		58.50	91.50
0823	3,000 gal.	G		6.67	1.199			70.50		70.50	110
0833	5,000 gal.	G		6.15	1.301			76.50		76.50	119
0843	8,000 gal.	G		5.33	1.501			88		88	137
0853	10,000 gal.	G		4.57	1.751			103		103	160
0863	12,000 gal.	G	↓	4.21	1.900	↓		112		112	174
1020	Haul tank to certified salvage dump, 100 miles round trip										
1023	3000 gal. to 5000 gal. tank					Ea.				760	830
1026	6000 gal. to 8000 gal. tank									880	960
1029	9,000 gal. to 12,000 gal. tank					↓				1,050	1,150
1100	Disposal of contaminated soil to landfill										
1110	Minimum					C.Y.				145	160
1111	Maximum					"				400	440
1120	Disposal of contaminated soil to										
1121	bituminous concrete batch plant										
1130	Minimum					C.Y.				80	88
1131	Maximum					"				115	125
1203	Excavate, pull, & load tank, backfill hole, 8,000 gal. +	G	B-12C	.50	32	Ea.		1,425	2,350	3,775	4,875
1213	Haul tank to certified dump, 100 miles rt, 8,000 gal. +	G	B-34K	1	8			320	950	1,270	1,575
1223	Excavate, pull, & load tank, backfill hole, 500 gal.	G	B-11C	1	16			705	365	1,070	1,525
1233	Excavate, pull, & load tank, backfill hole, 3,000 – 5,000 gal.	G	B-11M	.50	32			1,400	785	2,185	3,125
1243	Haul tank to certified dump, 100 miles rt, 500 gal.	G	B-34L	1	8			390	245	635	885
1253	Haul tank to certified dump, 100 miles rt, 3,000 – 5,000 gal.	G	B-34M	1	8			390	305	695	950
2010	Decontamination of soil on site incl poly tarp on top/bottom										
2011	Soil containment berm, and chemical treatment										
2020	Minimum	G	B-11C	100	.160	C.Y.	7.80	7.05	3.64	18.49	24

02 65 Underground Storage Tank Removal

02 65 10 – Underground Tank and Contaminated Soil Removal

02 65 10.30 Removal of Underground Storage Tanks		Crew	Daily Output	Labor-Hours	Unit	Material	2015 Bare Costs Labor	Equipment	Total	Total Incl O&P
2021	Maximum	G B-11C	100	.160	C.Y.	10.10	7.05	3.64	20.79	26.50
2050	Disposal of decontaminated soil, minimum								135	150
2055	Maximum								400	440

02 81 Transportation and Disposal of Hazardous Materials

02 81 20 – Hazardous Waste Handling

02 81 20.10 Hazardous Waste Cleanup/Pickup/Disposal

		Crew	Daily Output	Labor-Hours	Unit	Material	2015 Bare Costs Labor	Equipment	Total	Total Incl O&P
0010	**HAZARDOUS WASTE CLEANUP/PICKUP/DISPOSAL**									
0100	For contractor rental equipment, i.e., Dozer,									
0110	Front end loader, Dump truck, etc., see 01 54 33 Reference Section									
1000	Solid pickup									
1100	55 gal. drums				Ea.				240	265
1120	Bulk material, minimum				Ton				190	210
1130	Maximum				"				595	655
1200	Transportation to disposal site									
1220	Truckload = 80 drums or 25 C.Y. or 18 tons									
1260	Minimum				Mile				3.95	4.45
1270	Maximum				"				7.25	7.35
3000	Liquid pickup, vacuum truck, stainless steel tank									
3100	Minimum charge, 4 hours									
3110	1 compartment, 2200 gallon				Hr.				140	155
3120	2 compartment, 5000 gallon				"				200	225
3400	Transportation in 6900 gallon bulk truck				Mile				7.95	8.75
3410	In teflon lined truck				"				10.20	11.25
5000	Heavy sludge or dry vacuumable material				Hr.				140	160
6000	Dumpsite disposal charge, minimum				Ton				140	155
6020	Maximum				"				415	455

02 82 Asbestos Remediation

02 82 13 – Asbestos Abatement

02 82 13.39 Asbestos Remediation Plans and Methods

		Crew	Daily Output	Labor-Hours	Unit	Material	2015 Bare Costs Labor	Equipment	Total	Total Incl O&P
0010	**ASBESTOS REMEDIATION PLANS AND METHODS**									
0100	Building Survey-Commercial Building				Ea.				2,200	2,400
0200	Asbestos Abatement Remediation Plan				"				1,350	1,475

02 82 13.41 Asbestos Abatement Equipment

		Crew	Daily Output	Labor-Hours	Unit	Material	2015 Bare Costs Labor	Equipment	Total	Total Incl O&P
0010	**ASBESTOS ABATEMENT EQUIPMENT** R028213-20									
0011	Equipment and supplies, buy									
0200	Air filtration device, 2000 CFM				Ea.	960			960	1,050
0250	Large volume air sampling pump, minimum					345			345	380
0260	Maximum					335			335	365
0300	Airless sprayer unit, 2 gun					4,500			4,500	4,950
0350	Light stand, 500 watt					49			49	54
0400	Personal respirators									
0410	Negative pressure, 1/2 face, dual operation, min.				Ea.	26.50			26.50	29
0420	Maximum					29.50			29.50	32.50
0450	P.A.P.R., full face, minimum					124			124	137
0460	Maximum					165			165	182
0470	Supplied air, full face, incl. air line, minimum					168			168	185
0480	Maximum					405			405	445

02 82 Asbestos Remediation

02 82 13 – Asbestos Abatement

02 82 13.41 Asbestos Abatement Equipment

		Crew	Daily Output	Labor-Hours	Unit	Material	2015 Bare Costs Labor	2015 Bare Costs Equipment	Total	Total Incl O&P
0500	Personnel sampling pump				Ea.	226			226	249
1500	Power panel, 20 unit, incl. GFI					600			600	660
1600	Shower unit, including pump and filters					1,275			1,275	1,425
1700	Supplied air system (type C)					3,425			3,425	3,750
1750	Vacuum cleaner, HEPA, 16 gal., stainless steel, wet/dry					435			435	475
1760	55 gallon					560			560	615
1800	Vacuum loader, 9 – 18 ton/hr.					94,500			94,500	104,000
1900	Water atomizer unit, including 55 gal. drum					281			281	310
2000	Worker protection, whole body, foot, head cover & gloves, plastic					8.65			8.65	9.55
2500	Respirator, single use					25			25	27.50
2550	Cartridge for respirator					5.75			5.75	6.30
2570	Glove bag, 7 mil, 50" x 64"					9.20			9.20	10.10
2580	10 mil, 44" x 60"					5.75			5.75	6.35
3000	HEPA vacuum for work area, minimum					305			305	335
3050	Maximum					815			815	900
6000	Disposable polyethylene bags, 6 mil, 3 C.F.					.82			.82	.90
6300	Disposable fiber drums, 3 C.F.					18.20			18.20	20
6400	Pressure sensitive caution labels, 3" x 5"					3.47			3.47	3.82
6450	11" x 17"					7.35			7.35	8.05
6500	Negative air machine, 1800 CFM					845			845	930

02 82 13.42 Preparation of Asbestos Containment Area

		Crew	Daily Output	Labor-Hours	Unit	Material	2015 Bare Costs Labor	2015 Bare Costs Equipment	Total	Total Incl O&P
0010	**PREPARATION OF ASBESTOS CONTAINMENT AREA**									
0100	Pre-cleaning, HEPA vacuum and wet wipe, flat surfaces	A-9	12000	.005	S.F.	.02	.28		.30	.47
0200	Protect carpeted area, 2 layers 6 mil poly on 3/4" plywood	"	1000	.064		2.04	3.35		5.39	7.65
0300	Separation barrier, 2" x 4" @ 16", 1/2" plywood ea. side, 8' high	2 Carp	400	.040		3.32	1.88		5.20	6.75
0310	12' high		320	.050		3.32	2.35		5.67	7.50
0320	16' high		200	.080		2.30	3.76		6.06	8.70
0400	Personnel decontam. chamber, 2" x 4" @ 16", 3/4" ply ea. side		280	.057		4.34	2.68		7.02	9.15
0450	Waste decontam. chamber, 2" x 4" studs @ 16", 3/4" ply ea. side		360	.044		4.34	2.09		6.43	8.20
0500	Cover surfaces with polyethylene sheeting									
0501	Including glue and tape									
0550	Floors, each layer, 6 mil	A-9	8000	.008	S.F.	.04	.42		.46	.72
0551	4 mil		9000	.007		.03	.37		.40	.63
0560	Walls, each layer, 6 mil		6000	.011		.04	.56		.60	.95
0561	4 mil		7000	.009		.03	.48		.51	.80
0570	For heights above 12', add						20%			
0575	For heights above 20', add						30%			
0580	For fire retardant poly, add					100%				
0590	For large open areas, deduct					10%	20%			
0600	Seal floor penetrations with foam firestop to 36 sq. in.	2 Carp	200	.080	Ea.	7.75	3.76		11.51	14.65
0610	36 sq. in. to 72 sq. in.		125	.128		15.50	6		21.50	27
0615	72 sq. in. to 144 sq. in.		80	.200		31	9.40		40.40	49.50
0620	Wall penetrations, to 36 square inches		180	.089		7.75	4.17		11.92	15.35
0630	36 sq. in. to 72 sq. in.		100	.160		15.50	7.50		23	29.50
0640	72 sq. in. to 144 sq. in.		60	.267		31	12.50		43.50	54.50
0800	Caulk seams with latex	1 Carp	230	.035	L.F.	.17	1.63		1.80	2.87
0900	Set up neg. air machine, 1-2k CFM/25 M.C.F. volume	1 Asbe	4.30	1.860	Ea.		97.50		97.50	157
0950	Set up and remove portable shower unit	2 Asbe	4	4	"		209		209	335

02 82 Asbestos Remediation

02 82 13 – Asbestos Abatement

02 82 13.43 Bulk Asbestos Removal

		Crew	Daily Output	Labor-Hours	Unit	Material	2015 Bare Costs Labor	2015 Bare Costs Equipment	Total	Total Incl O&P
0010	**BULK ASBESTOS REMOVAL**									
0020	Includes disposable tools and 2 suits and 1 respirator filter/day/worker									
0100	Beams, W 10 x 19	A-9	235	.272	L.F.	.79	14.25		15.04	24
0110	W 12 x 22		210	.305		.88	15.95		16.83	26.50
0120	W 14 x 26		180	.356		1.03	18.65		19.68	31
0130	W 16 x 31		160	.400		1.15	21		22.15	35
0140	W 18 x 40		140	.457		1.32	24		25.32	40
0150	W 24 x 55		110	.582		1.68	30.50		32.18	51
0160	W 30 x 108		85	.753		2.17	39.50		41.67	66
0170	W 36 x 150		72	.889		2.56	46.50		49.06	78
0200	Boiler insulation		480	.133	S.F.	.45	7		7.45	11.75
0210	With metal lath, add				%				50%	50%
0300	Boiler breeching or flue insulation	A-9	520	.123	S.F.	.36	6.45		6.81	10.75
0310	For active boiler, add				%				100%	100%
0400	Duct or AHU insulation	A-10B	440	.073	S.F.	.21	3.82		4.03	6.40
0500	Duct vibration isolation joints, up to 24 sq. in. duct	A-9	56	1.143	Ea.	3.30	60		63.30	100
0520	25 sq. in. to 48 sq. in. duct		48	1.333		3.85	70		73.85	116
0530	49 sq. in. to 76 sq. in. duct		40	1.600		4.62	84		88.62	140
0600	Pipe insulation, air cell type, up to 4" diameter pipe		900	.071	L.F.	.21	3.73		3.94	6.25
0610	4" to 8" diameter pipe		800	.080		.23	4.19		4.42	7
0620	10" to 12" diameter pipe		700	.091		.26	4.79		5.05	8
0630	14" to 16" diameter pipe		550	.116		.34	6.10		6.44	10.15
0650	Over 16" diameter pipe		650	.098	S.F.	.28	5.15		5.43	8.60
0700	With glove bag up to 3" diameter pipe		200	.320	L.F.	9.15	16.75		25.90	37
1000	Pipe fitting insulation up to 4" diameter pipe		320	.200	Ea.	.58	10.50		11.08	17.50
1100	6" to 8" diameter pipe		304	.211		.61	11.05		11.66	18.40
1110	10" to 12" diameter pipe		192	.333		.96	17.45		18.41	29
1120	14" to 16" diameter pipe		128	.500		1.44	26		27.44	43.50
1130	Over 16" diameter pipe		176	.364	S.F.	1.05	19.05		20.10	31.50
1200	With glove bag, up to 8" diameter pipe		75	.853	L.F.	6.25	44.50		50.75	79
2000	Scrape foam fireproofing from flat surface		2400	.027	S.F.	.08	1.40		1.48	2.33
2100	Irregular surfaces		1200	.053		.15	2.80		2.95	4.66
3000	Remove cementitious material from flat surface		1800	.036		.10	1.86		1.96	3.11
3100	Irregular surface		1000	.064		.13	3.35		3.48	5.55
4000	Scrape acoustical coating/fireproofing, from ceiling		3200	.020		.06	1.05		1.11	1.75
5000	Remove VAT and mastic from floor by hand		2400	.027		.08	1.40		1.48	2.33
5100	By machine	A-11	4800	.013		.04	.70	.01	.75	1.17
5150	For 2 layers, add				%				50%	50%
6000	Remove contaminated soil from crawl space by hand	A-9	400	.160	C.F.	.46	8.40		8.86	14
6100	With large production vacuum loader	A-12	700	.091	"	.26	4.79	1.09	6.14	9.20
7000	Radiator backing, not including radiator removal	A-9	1200	.053	S.F.	.15	2.80		2.95	4.66
8000	Cement-asbestos transite board and cement wall board	2 Asbe	1000	.016		.15	.84		.99	1.51
8100	Transite shingle siding	A-10B	750	.043		.21	2.24		2.45	3.83
8200	Shingle roofing	"	2000	.016		.08	.84		.92	1.44
8250	Built-up, no gravel, non-friable	B-2	1400	.029		.08	1.09		1.17	1.87
8260	Bituminous flashing	1 Rofc	300	.027		.08	1.07		1.15	2.02
8300	Asbestos millboard, flat board and VAT contaminated plywood	2 Asbe	1000	.016		.08	.84		.92	1.44
9000	For type B (supplied air) respirator equipment, add				%				10%	10%

02 82 13.44 Demolition In Asbestos Contaminated Area

		Crew	Daily Output	Labor-Hours	Unit	Material	Labor	Equipment	Total	Total Incl O&P
0010	**DEMOLITION IN ASBESTOS CONTAMINATED AREA**									
0200	Ceiling, including suspension system, plaster and lath	A-9	2100	.030	S.F.	.09	1.60		1.69	2.67
0210	Finished plaster, leaving wire lath		585	.109		.32	5.75		6.07	9.55

02 82 Asbestos Remediation

02 82 13 – Asbestos Abatement

02 82 13.44 Demolition In Asbestos Contaminated Area

		Crew	Daily Output	Labor-Hours	Unit	Material	2015 Bare Costs Labor	2015 Bare Costs Equipment	Total	Total Incl O&P
0220	Suspended acoustical tile	A-9	3500	.018	S.F.	.05	.96		1.01	1.60
0230	Concealed tile grid system		3000	.021		.06	1.12		1.18	1.87
0240	Metal pan grid system		1500	.043		.12	2.24		2.36	3.73
0250	Gypsum board		2500	.026		.07	1.34		1.41	2.24
0260	Lighting fixtures up to 2' x 4'		72	.889	Ea.	2.56	46.50		49.06	78
0400	Partitions, non load bearing									
0410	Plaster, lath, and studs	A-9	690	.093	S.F.	.88	4.86		5.74	8.75
0450	Gypsum board and studs	"	1390	.046	"	.13	2.41		2.54	4.03
9000	For type B (supplied air) respirator equipment, add				%				10%	10%

02 82 13.45 OSHA Testing

		Crew	Daily Output	Labor-Hours	Unit	Material	Labor	Equipment	Total	Total Incl O&P
0010	**OSHA TESTING**									
0100	Certified technician, minimum				Day				200	220
0110	Maximum								300	330
0120	Industrial hygienist, minimum								250	250
0130	Maximum								400	440
0200	Asbestos sampling and PCM analysis, NIOSH 7400, minimum	1 Asbe	8	1	Ea.	2.96	52.50		55.46	87.50
0210	Maximum		4	2		3.26	105		108.26	172
1000	Cleaned area samples		8	1		2.81	52.50		55.31	87
1100	PCM air sample analysis, NIOSH 7400, minimum		8	1		31.50	52.50		84	119
1110	Maximum		4	2		3.42	105		108.42	172
1200	TEM air sample analysis, NIOSH 7402, minimum								80	106
1210	Maximum								360	450

02 82 13.46 Decontamination of Asbestos Containment Area

		Crew	Daily Output	Labor-Hours	Unit	Material	Labor	Equipment	Total	Total Incl O&P
0010	**DECONTAMINATION OF ASBESTOS CONTAINMENT AREA**									
0100	Spray exposed substrate with surfactant (bridging)									
0200	Flat surfaces	A-9	6000	.011	S.F.	.36	.56		.92	1.30
0250	Irregular surfaces		4000	.016	"	.31	.84		1.15	1.69
0300	Pipes, beams, and columns		2000	.032	L.F.	.56	1.68		2.24	3.32
1000	Spray encapsulate polyethylene sheeting		8000	.008	S.F.	.34	.42		.76	1.04
1100	Roll down polyethylene sheeting		8000	.008	"		.42		.42	.67
1500	Bag polyethylene sheeting		400	.160	Ea.	.81	8.40		9.21	14.40
2000	Fine clean exposed substrate, with nylon brush		2400	.027	S.F.		1.40		1.40	2.25
2500	Wet wipe substrate		4800	.013			.70		.70	1.12
2600	Vacuum surfaces, fine brush		6400	.010			.52		.52	.84
3000	Structural demolition									
3100	Wood stud walls	A-9	2800	.023	S.F.		1.20		1.20	1.93
3500	Window manifolds, not incl. window replacement		4200	.015			.80		.80	1.28
3600	Plywood carpet protection		2000	.032			1.68		1.68	2.70
4000	Remove custom decontamination facility	A-10A	8	3	Ea.	15.40	158		173.40	270
4100	Remove portable decontamination facility	3 Asbe	12	2	"	13.05	105		118.05	182
5000	HEPA vacuum, shampoo carpeting	A-9	4800	.013	S.F.	.07	.70		.77	1.20
9000	Final cleaning of protected surfaces	A-10A	8000	.003	"		.16		.16	.25

02 82 13.47 Asbestos Waste Pkg., Handling, and Disp.

		Crew	Daily Output	Labor-Hours	Unit	Material	Labor	Equipment	Total	Total Incl O&P
0010	**ASBESTOS WASTE PACKAGING, HANDLING, AND DISPOSAL**									
0100	Collect and bag bulk material, 3 C.F. bags, by hand	A-9	400	.160	Ea.	.82	8.40		9.22	14.40
0200	Large production vacuum loader	A-12	880	.073		.86	3.81	.87	5.54	8.05
1000	Double bag and decontaminate	A-9	960	.067		.82	3.49		4.31	6.50
2000	Containerize bagged material in drums, per 3 C.F. drum	"	800	.080		18.20	4.19		22.39	27
3000	Cart bags 50' to dumpster	2 Asbe	400	.040			2.09		2.09	3.37
5000	Disposal charges, not including haul, minimum				C.Y.				61	67
5020	Maximum				"				355	395
9000	For type B (supplied air) respirator equipment, add				%				10%	10%

02 82 Asbestos Remediation

02 82 13 – Asbestos Abatement

02 82 13.48 Asbestos Encapsulation With Sealants

		Crew	Daily Output	Labor-Hours	Unit	Material	2015 Bare Costs Labor	Equipment	Total	Total Incl O&P
0010	**ASBESTOS ENCAPSULATION WITH SEALANTS**									
0100	Ceilings and walls, minimum	A-9	21000	.003	S.F.	.31	.16		.47	.60
0110	Maximum		10600	.006		.45	.32		.77	1.01
0200	Columns and beams, minimum		13300	.005		.31	.25		.56	.75
0210	Maximum		5325	.012		.51	.63		1.14	1.57
0300	Pipes to 12" diameter including minor repairs, minimum		800	.080	L.F.	.43	4.19		4.62	7.20
0310	Maximum		400	.160	"	1.12	8.40		9.52	14.75

02 83 Lead Remediation

02 83 19 – Lead-Based Paint Remediation

02 83 19.21 Lead Paint Remediation Plans and Methods

		Crew	Daily Output	Labor-Hours	Unit	Material	2015 Bare Costs Labor	Equipment	Total	Total Incl O&P
0010	**LEAD PAINT REMEDIATION PLANS AND METHODS**									
0100	Building Survey-Commercial Building				Ea.				2,050	2,250
0200	Lead Abatement Remediation Plan								1,225	1,350
0300	Lead Paint Testing, AAS Analysis								51	56
0400	Lead Paint Testing, X-Ray Fluorescence								51	56

02 83 19.22 Preparation of Lead Containment Area

		Crew	Daily Output	Labor-Hours	Unit	Material	Labor	Equipment	Total	Total Incl O&P
0010	**PREPARATION OF LEAD CONTAINMENT AREA**									
0020	Lead abatement work area, test kit, per swab	1 Skwk	16	.500	Ea.	4.21	24.50		28.71	44
0025	For dust barriers see Section 01 56 16.10									
0050	Caution sign	1 Skwk	48	.167	Ea.	8.35	8.10		16.45	22.50
0100	Pre-cleaning, HEPA vacuum and wet wipe, floor and wall surfaces	3 Skwk	5000	.005	S.F.	.01	.23		.24	.40
0105	Ceiling, 6' - 11' high		4100	.006		.08	.28		.36	.54
0108	12' - 15' high		3550	.007		.07	.33		.40	.61
0115	Over 15' high		3000	.008		.09	.39		.48	.73
0500	Cover surfaces with polyethylene sheeting									
0550	Floors, each layer, 6 mil	3 Skwk	8000	.003	S.F.	.04	.15		.19	.29
0560	Walls, each layer, 6 mil	"	6000	.004	"	.04	.19		.23	.37
0570	For heights above 12', add						20%			
2400	Post abatement cleaning of protective sheeting, HEPA vacuum & wet wipe	3 Skwk	5000	.005	S.F.	.01	.23		.24	.40
2450	Doff, bag and seal protective sheeting		12000	.002		.01	.10		.11	.17
2500	Post abatement cleaning, HEPA vacuum & wet wipe		5000	.005		.01	.23		.24	.40

02 83 19.23 Encapsulation of Lead-Based Paint

		Crew	Daily Output	Labor-Hours	Unit	Material	Labor	Equipment	Total	Total Incl O&P
0010	**ENCAPSULATION OF LEAD-BASED PAINT**									
0020	Interior, brushwork, trim, under 6"	1 Pord	240	.033	L.F.	2.30	1.34		3.64	4.69
0030	6" to 12" wide		180	.044		3.06	1.79		4.85	6.25
0040	Balustrades		300	.027		1.84	1.08		2.92	3.75
0050	Pipe to 4" diameter		500	.016		1.12	.65		1.77	2.27
0060	To 8" diameter		375	.021		1.48	.86		2.34	3.01
0070	To 12" diameter		250	.032		2.19	1.29		3.48	4.49
0080	To 16" diameter		170	.047		3.26	1.90		5.16	6.65
0090	Cabinets, ornate design		200	.040	S.F.	2.81	1.61		4.42	5.70
0100	Simple design		250	.032	"	2.24	1.29		3.53	4.54
0110	Doors, 3' x 7', both sides, incl. frame & trim									
0120	Flush	1 Pord	6	1.333	Ea.	28	54		82	118
0130	French, 10 – 15 lite		3	2.667		5.65	108		113.65	179
0140	Panel		4	2		34	80.50		114.50	168
0150	Louvered		2.75	2.909		31	117		148	224
0160	Windows, per interior side, per 15 S.F.									
0170	1 to 6 lite	1 Pord	14	.571	Ea.	19.40	23		42.40	58.50

02 83 Lead Remediation

02 83 19 – Lead-Based Paint Remediation

02 83 19.23 Encapsulation of Lead-Based Paint

		Crew	Daily Output	Labor-Hours	Unit	Material	2015 Bare Costs Labor	2015 Bare Costs Equipment	Total	Total Incl O&P
0180	7 to 10 lite	1 Pord	7.50	1.067	Ea.	21.50	43		64.50	93
0190	12 lite		5.75	1.391		29	56		85	122
0200	Radiators		8	1		69	40.50		109.50	141
0210	Grilles, vents		275	.029	S.F.	2.04	1.17		3.21	4.13
0220	Walls, roller, drywall or plaster		1000	.008		.56	.32		.88	1.14
0230	With spunbonded reinforcing fabric		720	.011		.63	.45		1.08	1.41
0240	Wood		800	.010		.69	.40		1.09	1.41
0250	Ceilings, roller, drywall or plaster		900	.009		.63	.36		.99	1.27
0260	Wood		700	.011		.78	.46		1.24	1.60
0270	Exterior, brushwork, gutters and downspouts		300	.027	L.F.	1.84	1.08		2.92	3.75
0280	Columns		400	.020	S.F.	1.38	.81		2.19	2.82
0290	Spray, siding		600	.013	"	.93	.54		1.47	1.88
0300	Miscellaneous									
0310	Electrical conduit, brushwork, to 2" diameter	1 Pord	500	.016	L.F.	1.12	.65		1.77	2.27
0320	Brick, block or concrete, spray		500	.016	S.F.	1.12	.65		1.77	2.27
0330	Steel, flat surfaces and tanks to 12"		500	.016		1.12	.65		1.77	2.27
0340	Beams, brushwork		400	.020		1.38	.81		2.19	2.82
0350	Trusses		400	.020		1.38	.81		2.19	2.82

02 83 19.26 Removal of Lead-Based Paint

		Crew	Daily Output	Labor-Hours	Unit	Material	Labor	Equipment	Total	Total Incl O&P
0010	**REMOVAL OF LEAD-BASED PAINT**	R028319-60								
0011	By chemicals, per application									
0050	Baseboard, to 6" wide	1 Pord	64	.125	L.F.	.72	5.05		5.77	8.90
0070	To 12" wide		32	.250	"	1.40	10.10		11.50	17.75
0200	Balustrades, one side		28	.286	S.F.	1.45	11.55		13	20
1400	Cabinets, simple design		32	.250		1.31	10.10		11.41	17.65
1420	Ornate design		25	.320		1.58	12.90		14.48	22.50
1600	Cornice, simple design		60	.133		1.50	5.40		6.90	10.30
1620	Ornate design		20	.400		5.35	16.15		21.50	32
2800	Doors, one side, flush		84	.095		1.88	3.84		5.72	8.25
2820	Two panel		80	.100		1.31	4.04		5.35	7.95
2840	Four panel		45	.178		1.40	7.15		8.55	13.10
2880	For trim, one side, add		64	.125	L.F.	.71	5.05		5.76	8.90
3000	Fence, picket, one side		30	.267	S.F.	1.31	10.75		12.06	18.75
3200	Grilles, one side, simple design		30	.267		1.33	10.75		12.08	18.75
3220	Ornate design		25	.320		1.43	12.90		14.33	22.50
3240	Handrails		90	.089	L.F.	1.33	3.59		4.92	7.20
4400	Pipes, to 4" diameter		90	.089		1.88	3.59		5.47	7.80
4420	To 8" diameter		50	.160		3.75	6.45		10.20	14.55
4440	To 12" diameter		36	.222		5.65	8.95		14.60	20.50
4460	To 16" diameter		20	.400		7.50	16.15		23.65	34.50
4500	For hangers, add		40	.200	Ea.	2.49	8.05		10.54	15.70
4800	Siding		90	.089	S.F.	1.24	3.59		4.83	7.10
5000	Trusses, open		55	.145	SF Face	1.96	5.85		7.81	11.60
6200	Windows, one side only, double hung, 1/1 light, 24" x 48" high		4	2	Ea.	23	80.50		103.50	156
6220	30" x 60" high		3	2.667		31	108		139	207
6240	36" x 72" high		2.50	3.200		37	129		166	249
6280	40" x 80" high		2	4		46.50	161		207.50	310
6400	Colonial window, 6/6 light, 24" x 48" high		2	4		46.50	161		207.50	310
6420	30" x 60" high		1.50	5.333		62	215		277	415
6440	36" x 72" high		1	8		93	325		418	620
6480	40" x 80" high		1	8		93	325		418	620
6600	8/8 light, 24" x 48" high		2	4		46.50	161		207.50	310

02 83 Lead Remediation

02 83 19 – Lead-Based Paint Remediation

02 83 19.26 Removal of Lead-Based Paint

		Crew	Daily Output	Labor-Hours	Unit	Material	2015 Bare Costs Labor	Equipment	Total	Total Incl O&P
6620	40" x 80" high	1 Pord	1	8	Ea.	93	325		418	620
6800	12/12 light, 24" x 48" high		1	8		93	325		418	620
6820	40" x 80" high	↓	.75	10.667	↓	124	430		554	825
6840	Window frame & trim items, included in pricing above									
7000	Hand scraping and HEPA vacuum, less than 4 S.F.	1 Pord	8	1	Ea.	.82	40.50		41.32	66
8000	Collect and bag bulk material, 3 C.F. bags, by hand		30	.267	"	.82	10.75		11.57	18.20
9000	Minimum labor/equipment charge	↓	3	2.667	Job		108		108	173

02 85 Mold Remediation

02 85 16 – Mold Remediation Preparation and Containment

02 85 16.40 Mold Remediation Plans and Methods

		Crew	Daily Output	Labor-Hours	Unit	Material	2015 Bare Costs Labor	Equipment	Total	Total Incl O&P
0010	**MOLD REMEDIATION PLANS AND METHODS**									
0020	Initial inspection, areas to 2500 S.F.				Total				268	295
0030	Areas to 5000 S.F.								440	485
0032	Areas to 10000 S.F.								440	485
0040	Testing, air sample each								126	139
0050	Swab sample								115	127
0060	Tape sample								125	140
0070	Post remediation air test								126	139
0080	Mold abatement plan, area to 2500 S.F.								1,225	1,325
0090	Areas to 5000 S.F.								1,625	1,775
0095	Areas to 10000 S.F.								2,550	2,800
0100	Packup & removal of contents, average 3 bedroom home, excl storage								8,150	8,975
0110	Average 5 bedroom home, excl storage				↓				15,300	16,800
0600	For demolition in mold contaminated areas, see Section 02 85 33.50									
0610	For personal protection equipment, see Section 02 82 13.41									

02 85 16.50 Preparation of Mold Containment Area

		Crew	Daily Output	Labor-Hours	Unit	Material	2015 Bare Costs Labor	Equipment	Total	Total Incl O&P
0010	**PREPARATION OF MOLD CONTAINMENT AREA**									
0100	Pre-cleaning, HEPA vacuum and wet wipe, flat surfaces	A-9	12000	.005	S.F.	.02	.28		.30	.47
0300	Separation barrier, 2" x 4" @ 16", 1/2" plywood ea. side, 8' high	2 Carp	400	.040		3.32	1.88		5.20	6.75
0310	12' high		320	.050		3.32	2.35		5.67	7.50
0320	16' high		200	.080		2.30	3.76		6.06	8.70
0400	Personnel decontam. chamber, 2" x 4" @ 16", 3/4" ply ea. side		280	.057		4.34	2.68		7.02	9.15
0450	Waste decontam. chamber, 2" x 4" studs @ 16", 3/4" ply each side	↓	360	.044	↓	4.34	2.09		6.43	8.20
0500	Cover surfaces with polyethylene sheeting									
0501	Including glue and tape									
0550	Floors, each layer, 6 mil	A-9	8000	.008	S.F.	.04	.42		.46	.72
0551	4 mil		9000	.007		.03	.37		.40	.63
0560	Walls, each layer, 6 mil		6000	.011		.04	.56		.60	.95
0561	4 mil	↓	7000	.009	↓	.03	.48		.51	.80
0570	For heights above 12', add						20%			
0575	For heights above 20', add						30%			
0580	For fire retardant poly, add					100%				
0590	For large open areas, deduct					10%	20%			
0600	Seal floor penetrations with foam firestop to 36 sq. in.	2 Carp	200	.080	Ea.	7.75	3.76		11.51	14.65
0610	36 sq. in. to 72 sq. in.		125	.128		15.50	6		21.50	27
0615	72 sq. in. to 144 sq. in.		80	.200		31	9.40		40.40	49.50
0620	Wall penetrations, to 36 square inches		180	.089		7.75	4.17		11.92	15.35
0630	36 sq. in. to 72 sq. in.		100	.160		15.50	7.50		23	29.50
0640	72 sq. in. to 144 sq. in.	↓	60	.267		31	12.50		43.50	54.50
0800	Caulk seams with latex caulk	1 Carp	230	.035	L.F.	.17	1.63		1.80	2.87

02 85 Mold Remediation

02 85 16 – Mold Remediation Preparation and Containment

02 85 16.50 Preparation of Mold Containment Area	Crew	Daily Output	Labor-Hours	Unit	Material	2015 Bare Costs Labor	2015 Bare Costs Equipment	Total	Total Incl O&P
0900 Set up neg. air machine, 1-2k CFM/25 M.C.F. volume	1 Asbe	4.30	1.860	Ea.		97.50		97.50	157

02 85 33 – Removal and Disposal of Materials with Mold

02 85 33.50 Demolition in Mold Contaminated Area

		Crew	Daily Output	Labor-Hours	Unit	Material	Labor	Equipment	Total	Total Incl O&P
0010	DEMOLITION IN MOLD CONTAMINATED AREA									
0200	Ceiling, including suspension system, plaster and lath	A-9	2100	.030	S.F.	.09	1.60		1.69	2.67
0210	Finished plaster, leaving wire lath		585	.109		.32	5.75		6.07	9.55
0220	Suspended acoustical tile		3500	.018		.05	.96		1.01	1.60
0230	Concealed tile grid system		3000	.021		.06	1.12		1.18	1.87
0240	Metal pan grid system		1500	.043		.12	2.24		2.36	3.73
0250	Gypsum board		2500	.026		.07	1.34		1.41	2.24
0255	Plywood		2500	.026		.07	1.34		1.41	2.24
0260	Lighting fixtures up to 2' x 4'		72	.889	Ea.	2.56	46.50		49.06	78
0400	Partitions, non load bearing									
0410	Plaster, lath, and studs	A-9	690	.093	S.F.	.88	4.86		5.74	8.75
0450	Gypsum board and studs		1390	.046		.13	2.41		2.54	4.03
0465	Carpet & pad		1390	.046		.13	2.41		2.54	4.03
0600	Pipe insulation, air cell type, up to 4" diameter pipe		900	.071	L.F.	.21	3.73		3.94	6.25
0610	4" to 8" diameter pipe		800	.080		.23	4.19		4.42	7
0620	10" to 12" diameter pipe		700	.091		.26	4.79		5.05	8
0630	14" to 16" diameter pipe		550	.116		.34	6.10		6.44	10.15
0650	Over 16" diameter pipe		650	.098	S.F.	.28	5.15		5.43	8.60
9000	For type B (supplied air) respirator equipment, add				%		10%			10%

Estimating Tips
General
- Carefully check all the plans and specifications. Concrete often appears on drawings other than structural drawings, including mechanical and electrical drawings for equipment pads. The cost of cutting and patching is often difficult to estimate. See Subdivision 03 81 for Concrete Cutting, Subdivision 02 41 19.16 for Cutout Demolition, Subdivision 03 05 05.10 for Concrete Demolition, and Subdivision 02 41 19.19 for Rubbish Handling (handling, loading and hauling of debris).
- Always obtain concrete prices from suppliers near the job site. A volume discount can often be negotiated, depending upon competition in the area. Remember to add for waste, particularly for slabs and footings on grade.

03 10 00 Concrete Forming and Accessories
- A primary cost for concrete construction is forming. Most jobs today are constructed with prefabricated forms. The selection of the forms best suited for the job and the total square feet of forms required for efficient concrete forming and placing are key elements in estimating concrete construction. Enough forms must be available for erection to make efficient use of the concrete placing equipment and crew.
- Concrete accessories for forming and placing depend upon the systems used. Study the plans and specifications to ensure that all special accessory requirements have been included in the cost estimate, such as anchor bolts, inserts, and hangers.
- Included within costs for forms-in-place are all necessary bracing and shoring.

03 20 00 Concrete Reinforcing
- Ascertain that the reinforcing steel supplier has included all accessories, cutting, bending, and an allowance for lapping, splicing, and waste. A good rule of thumb is 10% for lapping, splicing, and waste. Also, 10% waste should be allowed for welded wire fabric.
- The unit price items in the subdivisions for Reinforcing In Place, Glass Fiber Reinforcing, and Welded Wire Fabric include the labor to install accessories such as beam and slab bolsters, high chairs, and bar ties and tie wire. The material cost for these accessories is not included; they may be obtained from the Accessories Division.

03 30 00 Cast-In-Place Concrete
- When estimating structural concrete, pay particular attention to requirements for concrete additives, curing methods, and surface treatments. Special consideration for climate, hot or cold, must be included in your estimate. Be sure to include requirements for concrete placing equipment, and concrete finishing.
- For accurate concrete estimating, the estimator must consider each of the following major components individually: forms, reinforcing steel, ready-mix concrete, placement of the concrete, and finishing of the top surface. For faster estimating, Subdivision 03 30 53.40 for Concrete-In-Place can be used; here, various items of concrete work are presented that include the costs of all five major components (unless specifically stated otherwise).

03 40 00 Precast Concrete
03 50 00 Cast Decks and Underlayment
- The cost of hauling precast concrete structural members is often an important factor. For this reason, it is important to get a quote from the nearest supplier. It may become economically feasible to set up precasting beds on the site if the hauling costs are prohibitive.

Reference Numbers
Reference numbers are shown in shaded boxes at the beginning of some major classifications. These numbers refer to related items in the Reference Section. The reference information may be an estimating procedure, an alternate pricing method, or technical information.

Note: Not all subdivisions listed here necessarily appear in this publication. ■

No part of this publication may be reproduced, stored in a retrieval system, or transmitted in any form or by any means without prior written permission of RSMeans.

03 01 Maintenance of Concrete

03 01 30 – Maintenance of Cast-In-Place Concrete

03 01 30.62 Concrete Patching

		Crew	Daily Output	Labor-Hours	Unit	Material	2015 Bare Costs Labor	2015 Bare Costs Equipment	Total	Total Incl O&P
0010	**CONCRETE PATCHING**									
0100	Floors, 1/4" thick, small areas, regular grout	1 Cefi	170	.047	S.F.	1.46	2.12		3.58	4.95
0150	Epoxy grout	"	100	.080	"	8.05	3.60		11.65	14.60
2000	Walls, including chipping, cleaning and epoxy grout									
2100	1/4" deep	1 Cefi	65	.123	S.F.	7.65	5.55		13.20	17.20
2150	1/2" deep		50	.160		15.35	7.20		22.55	28.50
2200	3/4" deep		40	.200		23	9		32	40
9000	Minimum labor/equipment charge	↓	4.50	1.778	Job		80		80	126

03 01 30.71 Concrete Crack Repair

		Crew	Daily Output	Labor-Hours	Unit	Material	2015 Bare Costs Labor	2015 Bare Costs Equipment	Total	Total Incl O&P
0010	**CONCRETE CRACK REPAIR**									
1000	Structural repair of concrete cracks by epoxy injection (ACI RAP-1)									
1001	suitable for horizontal, vertical and overhead repairs									
1010	Clean/grind concrete surface(s) free of contaminants	1 Cefi	400	.020	L.F.		.90		.90	1.42
1015	Rout crack with v-notch crack chaser, if needed	C-32	600	.027		.03	1.10	.15	1.28	1.98
1020	Blow out crack with oil-free dry compressed air (1 pass)	C-28	3000	.003	↓		.12	.01	.13	.20
1030	Install surface-mounted entry ports (spacing = concrete depth)	1 Cefi	400	.020	Ea.	1.39	.90		2.29	2.95
1040	Cap crack at surface with epoxy gel (per side/face)		400	.020	L.F.	.34	.90		1.24	1.80
1050	Snap off ports, grind off epoxy cap residue after injection		200	.040	"		1.80		1.80	2.85
1100	Manual injection with 2-part epoxy cartridge, excludes prep									
1110	Up to 1/32" (0.03125") wide x 4" deep	1 Cefi	160	.050	L.F.	.17	2.25		2.42	3.75
1120	6" deep		120	.067		.26	3		3.26	5
1130	8" deep		107	.075		.34	3.36		3.70	5.70
1140	10" deep		100	.080		.43	3.60		4.03	6.15
1150	12" deep		96	.083		.51	3.75		4.26	6.50
1210	Up to 1/16" (0.0625") wide x 4" deep		160	.050		.34	2.25		2.59	3.94
1220	6" deep		120	.067		.51	3		3.51	5.30
1230	8" deep		107	.075		.69	3.36		4.05	6.05
1240	10" deep		100	.080		.86	3.60		4.46	6.65
1250	12" deep		96	.083		1.03	3.75		4.78	7.10
1310	Up to 3/32" (0.09375") wide x 4" deep		160	.050		.51	2.25		2.76	4.13
1320	6" deep		120	.067		.77	3		3.77	5.60
1330	8" deep		107	.075		1.03	3.36		4.39	6.45
1340	10" deep		100	.080		1.29	3.60		4.89	7.10
1350	12" deep		96	.083		1.54	3.75		5.29	7.65
1410	Up to 1/8" (0.125") wide x 4" deep		160	.050		.69	2.25		2.94	4.31
1420	6" deep		120	.067		1.03	3		4.03	5.85
1430	8" deep		107	.075		1.37	3.36		4.73	6.80
1440	10" deep		100	.080		1.72	3.60		5.32	7.60
1450	12" deep	↓	96	.083	↓	2.06	3.75		5.81	8.20
1500	Pneumatic injection with 2-part bulk epoxy, excludes prep									
1510	Up to 5/32" (0.15625") wide x 4" deep	C-31	240	.033	L.F.	.24	1.50	1.55	3.29	4.34
1520	6" deep		180	.044		.36	2	2.07	4.43	5.85
1530	8" deep		160	.050		.48	2.25	2.32	5.05	6.65
1540	10" deep		150	.053		.61	2.40	2.48	5.49	7.20
1550	12" deep		144	.056		.73	2.50	2.58	5.81	7.60
1610	Up to 3/16" (0.1875") wide x 4" deep		240	.033		.29	1.50	1.55	3.34	4.39
1620	6" deep		180	.044		.44	2	2.07	4.51	5.90
1630	8" deep		160	.050		.58	2.25	2.32	5.15	6.75
1640	10" deep		150	.053		.73	2.40	2.48	5.61	7.30
1650	12" deep		144	.056		.87	2.50	2.58	5.95	7.75
1710	Up to 1/4" (0.1875") wide x 4" deep		240	.033		.39	1.50	1.55	3.44	4.50
1720	6" deep	↓	180	.044		.58	2	2.07	4.65	6.05

03 01 Maintenance of Concrete

03 01 30 – Maintenance of Cast-In-Place Concrete

03 01 30.71 Concrete Crack Repair

		Crew	Daily Output	Labor-Hours	Unit	Material	2015 Bare Costs Labor	Equipment	Total	Total Incl O&P
1730	8" deep	C-31	160	.050	L.F.	.77	2.25	2.32	5.34	6.95
1740	10" deep		150	.053		.97	2.40	2.48	5.85	7.60
1750	12" deep	↓	144	.056	↓	1.16	2.50	2.58	6.24	8.05
2000	Non-structural filling of concrete cracks by gravity-fed resin (ACI RAP-2)									
2001	suitable for individual cracks in stable horizontal surfaces only									
2010	Clean/grind concrete surface(s) free of contaminants	1 Cefi	400	.020	L.F.		.90		.90	1.42
2020	Rout crack with v-notch crack chaser, if needed	C-32	600	.027		.03	1.10	.15	1.28	1.98
2030	Blow out crack with oil-free dry compressed air (1 pass)	C-28	3000	.003			.12	.01	.13	.20
2040	Cap crack with epoxy gel at underside of elevated slabs, if needed	1 Cefi	400	.020		.34	.90		1.24	1.80
2050	Insert backer rod into crack, if needed		400	.020		.02	.90		.92	1.44
2060	Partially fill crack with fine dry sand, if needed		800	.010		.02	.45		.47	.73
2070	Apply two beads of sealant alongside crack to form reservoir, if needed	↓	400	.020	↓	.25	.90		1.15	1.69
2100	Manual filling with squeeze bottle of 2-part epoxy resin									
2110	Full depth crack up to 1/16" (0.0625") wide x 4" deep	1 Cefi	300	.027	L.F.	.09	1.20		1.29	2
2120	6" deep		240	.033		.14	1.50		1.64	2.53
2130	8" deep		200	.040		.19	1.80		1.99	3.06
2140	10" deep		185	.043		.24	1.95		2.19	3.34
2150	12" deep		175	.046		.28	2.06		2.34	3.56
2210	Full depth crack up to 1/8" (0.125") wide x 4" deep		270	.030		.19	1.33		1.52	2.32
2220	6" deep		215	.037		.28	1.67		1.95	2.96
2230	8" deep		180	.044		.38	2		2.38	3.58
2240	10" deep		165	.048		.47	2.18		2.65	3.97
2250	12" deep		155	.052		.57	2.32		2.89	4.29
2310	Partial depth crack up to 3/16" (0.1875") wide x 1" deep		300	.027		.07	1.20		1.27	1.98
2410	Up to 1/4" (0.250") wide x 1" deep		270	.030		.09	1.33		1.42	2.21
2510	Up to 5/16" (0.3125") wide x 1" deep		250	.032		.12	1.44		1.56	2.41
2610	Up to 3/8" (0.375") wide x 1" deep	↓	240	.033	↓	.14	1.50		1.64	2.53

03 01 30.72 Concrete Surface Repairs

		Crew	Daily Output	Labor-Hours	Unit	Material	2015 Bare Costs Labor	Equipment	Total	Total Incl O&P
0010	**CONCRETE SURFACE REPAIRS**									
9000	Surface repair by Methacrylate flood coat (ACI RAP-13)									
9010	Suitable for healing and sealing horizontal surfaces only									
9020	Large cracks must previously have been repaired or filled									
9100	Shotblast entire surface to remove contaminants	A-1A	4000	.002	S.F.		.10	.05	.15	.22
9200	Blow off dust and debris with oil-free dry compressed air	C-28	16000	.001			.02		.02	.04
9300	Flood coat surface w/Methacrylate, distribute w/broom/squeegee, no prep.	3 Cefi	8000	.003		.98	.14		1.12	1.29
9400	Lightly broadcast even coat of dry silica sand while sealer coat is tacky	1 Cefi	8000	.001	↓	.01	.05		.06	.08

03 05 Common Work Results for Concrete

03 05 05 – Selective Demolition for Concrete

03 05 05.10 Selective Demolition, Concrete

		Crew	Daily Output	Labor-Hours	Unit	Material	2015 Bare Costs Labor	Equipment	Total	Total Incl O&P
0010	**SELECTIVE DEMOLITION, CONCRETE** R024119-10									
0012	Excludes saw cutting, torch cutting, loading or hauling									
0050	Break into small pieces, reinf. less than 1% of cross-sectional area	B-9	24	1.667	C.Y.		63.50	9.70	73.20	115
0060	Reinforcing 1% to 2% of cross-sectional area		16	2.500			95	14.55	109.55	172
0070	Reinforcing more than 2% of cross-sectional area	↓	8	5	↓		190	29	219	340
0150	Remove whole pieces, up to 2 tons per piece	E-18	36	1.111	Ea.		58.50	26.50	85	132
0160	2 – 5 tons per piece		30	1.333			70	31.50	101.50	159
0170	5 – 10 tons per piece		24	1.667			87.50	39.50	127	199
0180	10 – 15 tons per piece	↓	18	2.222			117	52.50	169.50	264
0250	Precast unit embedded in masonry, up to 1 C.F.	D-1	16	1			42		42	68.50
0260	1 – 2 C.F.		12	1.333	↓		56		56	91.50

03 05 Common Work Results for Concrete

03 05 05 – Selective Demolition for Concrete

03 05 05.10 Selective Demolition, Concrete		Crew	Daily Output	Labor-Hours	Unit	Material	2015 Bare Costs Labor	Equipment	Total	Total Incl O&P
0270	2 – 5 C.F.	D-1	10	1.600	Ea.		67.50		67.50	110
0280	5 – 10 C.F.	↓	8	2	↓		84		84	137
0990	For hydrodemolition see Section 02 41 13.15									
1910	Minimum labor/equipment charge	B-9	2	20	Job		760	116	876	1,375

03 05 13 – Basic Concrete Materials

03 05 13.20 Concrete Admixtures and Surface Treatments

		Crew	Daily Output	Labor-Hours	Unit	Material	2015 Bare Costs Labor	Equipment	Total	Total Incl O&P
0010	**CONCRETE ADMIXTURES AND SURFACE TREATMENTS**									
0040	Abrasives, aluminum oxide, over 20 tons				Lb.	1.88			1.88	2.07
0050	1 to 20 tons					2.02			2.02	2.22
0070	Under 1 ton					2.10			2.10	2.31
0100	Silicon carbide, black, over 20 tons					2.87			2.87	3.16
0110	1 to 20 tons					3.04			3.04	3.35
0120	Under 1 ton				↓	3.17			3.17	3.49
0200	Air entraining agent, .7 to 1.5 oz. per bag, 55 gallon drum				Gal.	14			14	15.40
0220	5 gallon pail					19.25			19.25	21
0300	Bonding agent, acrylic latex, 250 S.F. per gallon, 5 gallon pail					22.50			22.50	25
0320	Epoxy resin, 80 S.F. per gallon, 4 gallon case					61			61	67
0400	Calcium chloride, 50 lb. bags, TL lots				Ton	810			810	890
0420	Less than truckload lots				Bag	24			24	26.50
0500	Carbon black, liquid, 2 to 8 lb. per bag of cement				Lb.	9.10			9.10	10
0600	Colored pigments, integral, 2 to 10 lb. per bag of cement, subtle colors					2.40			2.40	2.64
0610	Standard colors					3.26			3.26	3.59
0620	Premium colors				↓	5.40			5.40	5.95
0920	Dustproofing compound, 250 S.F./gal., 5 gallon pail				Gal.	6.65			6.65	7.30
1010	Epoxy based, 125 S.F./gal., 5 gallon pail				"	55.50			55.50	61
1100	Hardeners, metallic, 55 lb. bags, natural (grey)				Lb.	.70			.70	.76
1200	Colors					2.27			2.27	2.50
1300	Non-metallic, 55 lb. bags, natural grey					.43			.43	.47
1320	Colors				↓	.94			.94	1.04
1550	Release agent, for tilt slabs, 5 gallon pail				Gal.	17.60			17.60	19.40
1570	For forms, 5 gallon pail					12.50			12.50	13.75
1590	Concrete release agent for forms, 100% biodegradeable, zero VOC, 5 gal pail G					19.95			19.95	22
1595	55 gallon drum G					16.80			16.80	18.50
1600	Sealer, hardener and dustproofer, epoxy-based, 125 S.F./gal., 5 gallon unit					55.50			55.50	61
1620	3 gallon unit					61			61	67.50
1630	Sealer, solvent-based, 250 S.F./gal., 55 gallon drum					24.50			24.50	27
1640	5 gallon pail					30.50			30.50	34
1650	Sealer, water based, 350 S.F., 55 gallon drum					22.50			22.50	25
1660	5 gallon pail					25.50			25.50	28.50
1900	Set retarder, 100 S.F./gal., 1 gallon pail				↓	22.50			22.50	24.50
2000	Waterproofing, integral 1 lb. per bag of cement				Lb.	2.12			2.12	2.33
2100	Powdered metallic, 40 lb. per 100 S.F., standard colors					2.52			2.52	2.77
2120	Premium colors				↓	3.53			3.53	3.88
3000	For integral colored pigments, 2500 psi (5 bag mix)									
3100	Standard colors, 1.8 lb. per bag, add				C.Y.	21.50			21.50	24
3200	9.4 lb. per bag, add					113			113	124
3400	Premium colors, 1.8 lb. per bag, add					29.50			29.50	32.50
3500	7.5 lb. per bag, add					122			122	134
3700	Ultra premium colors, 1.8 lb. per bag, add					49			49	53.50
3800	7.5 lb. per bag, add				↓	203			203	224

03 05 Common Work Results for Concrete

03 05 13 – Basic Concrete Materials

03 05 13.85 Winter Protection

		Crew	Daily Output	Labor-Hours	Unit	Material	2015 Bare Costs Labor	Equipment	Total	Total Incl O&P
0010	**WINTER PROTECTION**									
0012	For heated ready mix, add				C.Y.	4.05			4.05	4.46
0100	Temporary heat to protect concrete, 24 hours	2 Clab	50	.320	M.S.F.	535	12.05		547.05	605
0200	Temporary shelter for slab on grade, wood frame/polyethylene sheeting									
0201	Build or remove, light framing for short spans	2 Carp	10	1.600	M.S.F.	310	75		385	470
0210	Large framing for long spans	"	3	5.333	"	415	250		665	865
0710	Electrically, heated pads, 15 watts/S.F., 20 uses				S.F.	.52			.52	.57

03 11 Concrete Forming

03 11 13 – Structural Cast-In-Place Concrete Forming

03 11 13.20 Forms In Place, Beams and Girders

			Crew	Daily Output	Labor-Hours	Unit	Material	Labor	Equipment	Total	Total Incl O&P
0010	**FORMS IN PLACE, BEAMS AND GIRDERS**	R031113-40									
0500	Exterior spandrel, job-built plywood, 12" wide, 1 use	R031113-60	C-2	225	.213	SFCA	3.12	9.75		12.87	19.45
0650	4 use			310	.155		1.01	7.10		8.11	12.70
1000	18" wide, 1 use			250	.192		2.78	8.80		11.58	17.45
1150	4 use			315	.152		.91	6.95		7.86	12.40
1500	24" wide, 1 use			265	.181		2.54	8.30		10.84	16.40
1650	4 use			325	.148		.83	6.75		7.58	11.95
2000	Interior beam, job-built plywood, 12" wide, 1 use			300	.160		3.67	7.30		10.97	16.05
2150	4 use			377	.127		1.19	5.80		6.99	10.85
2500	24" wide, 1 use			320	.150		2.60	6.85		9.45	14.10
2650	4 use			395	.122		.84	5.55		6.39	10
3000	Encasing steel beam, hung, job-built plywood, 1 use			325	.148		3.20	6.75		9.95	14.55
3150	4 use			430	.112		1.04	5.10		6.14	9.50
9000	Minimum labor/equipment charge		2 Carp	2	8	Job		375		375	615

03 11 13.25 Forms In Place, Columns

			Crew	Daily Output	Labor-Hours	Unit	Material	Labor	Equipment	Total	Total Incl O&P
0010	**FORMS IN PLACE, COLUMNS**	R031113-40									
0500	Round fiberglass, 4 use per mo., rent, 12" diameter		C-1	160	.200	L.F.	8.95	8.90		17.85	24.50
0550	16" diameter	R031113-60		150	.213		10.65	9.50		20.15	27.50
0600	18" diameter			140	.229		11.95	10.20		22.15	30
0650	24" diameter			135	.237		14.85	10.55		25.40	33.50
0700	28" diameter			130	.246		16.60	11		27.60	36.50
0800	30" diameter			125	.256		17.35	11.40		28.75	38
0850	36" diameter			120	.267		23	11.90		34.90	45
1500	Round fiber tube, recycled paper, 1 use, 8" diameter	G		155	.206		1.77	9.20		10.97	17.05
1550	10" diameter	G		155	.206		2.40	9.20		11.60	17.75
1600	12" diameter	G		150	.213		2.83	9.50		12.33	18.70
1650	14" diameter	G		145	.221		3.91	9.85		13.76	20.50
1700	16" diameter	G		140	.229		4.88	10.20		15.08	22
1720	18" diameter	G		140	.229		5.70	10.20		15.90	23
1750	20" diameter	G		135	.237		7.60	10.55		18.15	26
1800	24" diameter	G		130	.246		9.70	11		20.70	28.50
1850	30" diameter	G		125	.256		14.50	11.40		25.90	34.50
1900	36" diameter	G		115	.278		17.10	12.40		29.50	39.50
1950	42" diameter	G		100	.320		44.50	14.30		58.80	72.50
2000	48" diameter	G		85	.376		51.50	16.80		68.30	84.50
2200	For seamless type, add						15%				
3000	Round, steel, 4 use per mo., rent, regular duty, 14" diameter	G	C-1	145	.221	L.F.	12.30	9.85		22.15	29.50
3050	16" diameter	G		125	.256		12.50	11.40		23.90	32.50
3100	Heavy duty, 20" diameter	G		105	.305		13.80	13.60		27.40	37.50

03 11 Concrete Forming

03 11 13 – Structural Cast-In-Place Concrete Forming

03 11 13.25 Forms In Place, Columns

		Crew	Daily Output	Labor-Hours	Unit	Material	2015 Bare Costs Labor	2015 Bare Costs Equipment	Total	Total Incl O&P
3150	24" diameter [G]	C-1	85	.376	L.F.	15.05	16.80		31.85	44
3200	30" diameter [G]		70	.457		17.30	20.50		37.80	52.50
3250	36" diameter [G]		60	.533		18.60	24		42.60	59.50
3300	48" diameter [G]		50	.640		27.50	28.50		56	77.50
3350	60" diameter [G]		45	.711		34	31.50		65.50	89.50
4500	For second and succeeding months, deduct					50%				
5000	Job-built plywood, 8" x 8" columns, 1 use	C-1	165	.194	SFCA	2.73	8.65		11.38	17.20
5050	2 use		195	.164		1.56	7.30		8.86	13.70
5100	3 use		210	.152		1.09	6.80		7.89	12.35
5150	4 use		215	.149		.90	6.65		7.55	11.90
5500	12" x 12" columns, 1 use		180	.178		2.62	7.95		10.57	15.90
5550	2 use		210	.152		1.44	6.80		8.24	12.75
5600	3 use		220	.145		1.05	6.50		7.55	11.80
5650	4 use		225	.142		.85	6.35		7.20	11.35
6000	16" x 16" columns, 1 use		185	.173		2.62	7.70		10.32	15.55
6050	2 use		215	.149		1.41	6.65		8.06	12.45
6100	3 use		230	.139		1.05	6.20		7.25	11.30
6150	4 use		235	.136		.86	6.05		6.91	10.90
6500	24" x 24" columns, 1 use		190	.168		2.97	7.50		10.47	15.55
6550	2 use		216	.148		1.63	6.60		8.23	12.65
6600	3 use		230	.139		1.19	6.20		7.39	11.45
6650	4 use		238	.134		.96	6		6.96	10.90
7000	36" x 36" columns, 1 use		200	.160		2.31	7.15		9.46	14.25
7050	2 use		230	.139		1.31	6.20		7.51	11.60
7100	3 use		245	.131		.92	5.85		6.77	10.55
7150	4 use		250	.128		.75	5.70		6.45	10.20
7400	Steel framed plywood, based on 50 uses of purchased									
7420	forms, and 4 uses of bracing lumber									
7500	8" x 8" column	C-1	340	.094	SFCA	2.14	4.20		6.34	9.25
7550	10" x 10"		350	.091		1.87	4.08		5.95	8.75
7600	12" x 12"		370	.086		1.59	3.86		5.45	8.10
7650	16" x 16"		400	.080		1.23	3.57		4.80	7.20
7700	20" x 20"		420	.076		1.09	3.40		4.49	6.75
7750	24" x 24"		440	.073		.79	3.24		4.03	6.15
7755	30" x 30"		440	.073		1	3.24		4.24	6.40
7760	36" x 36"		460	.070		.88	3.10		3.98	6.05
9000	Minimum labor/equipment charge	2 Carp	2	8	Job		375		375	615

03 11 13.35 Forms In Place, Elevated Slabs

		Crew	Daily Output	Labor-Hours	Unit	Material	Labor	Equipment	Total	Total Incl O&P
0010	**FORMS IN PLACE, ELEVATED SLABS** R031113-40									
1000	Flat plate, job-built plywood, to 15' high, 1 use	C-2	470	.102	S.F.	3.86	4.67		8.53	11.90
1150	4 use R031113-60		560	.086		1.26	3.92		5.18	7.80
1500	15' to 20' high ceilings, 4 use		495	.097		1.30	4.43		5.73	8.70
2000	Flat slab, drop panels, job-built plywood, to 15' high, 1 use		449	.107		4.42	4.89		9.31	12.85
2150	4 use		544	.088		1.43	4.04		5.47	8.20
2250	15' to 20' high ceilings, 4 use		480	.100		2.47	4.57		7.04	10.20
3500	Floor slab, with 1-way joist pans, 1 use		415	.116		6.35	5.30		11.65	15.65
3650	4 use		500	.096		3.37	4.39		7.76	10.90
4500	With 2-way waffle domes, 1 use		405	.119		6.50	5.40		11.90	16.05
4550	4 use		470	.102		3.51	4.67		8.18	11.50
5000	Box out for slab openings, over 16" deep, 1 use		190	.253	SFCA	4.57	11.55		16.12	24
5050	2 use		240	.200	"	2.51	9.15		11.66	17.75
5500	Shallow slab box outs, to 10 S.F.		42	1.143	Ea.	12.10	52.50		64.60	99

03 11 Concrete Forming

03 11 13 – Structural Cast-In-Place Concrete Forming

03 11 13.35 Forms In Place, Elevated Slabs

		Crew	Daily Output	Labor-Hours	Unit	Material	2015 Bare Costs Labor	Equipment	Total	Total Incl O&P
5550	Over 10 S.F. (use perimeter)	C-2	600	.080	L.F.	1.61	3.66		5.27	7.80
6000	Bulkhead forms for slab, with keyway, 1 use, 2 piece		500	.096		2.14	4.39		6.53	9.55
6100	3 piece (see also edge forms)		460	.104		2.31	4.77		7.08	10.35
6200	Slab bulkhead form, 4-1/2" high, exp metal, w/keyway & stakes G	C-1	1200	.027		.92	1.19		2.11	2.96
6210	5-1/2" high G		1100	.029		1.12	1.30		2.42	3.36
6215	7-1/2" high G		960	.033		1.31	1.49		2.80	3.88
6220	9-1/2" high G		840	.038		1.43	1.70		3.13	4.36
6500	Curb forms, wood, 6" to 12" high, on elevated slabs, 1 use		180	.178	SFCA	1.68	7.95		9.63	14.85
6550	2 use		205	.156		.93	6.95		7.88	12.40
6600	3 use		220	.145		.67	6.50		7.17	11.40
6650	4 use		225	.142		.55	6.35		6.90	11
7000	Edge forms to 6" high, on elevated slab, 4 use		500	.064	L.F.	.21	2.86		3.07	4.91
7070	7" to 12" high, 1 use		162	.198	SFCA	1.27	8.80		10.07	15.85
7080	2 use		198	.162		.70	7.20		7.90	12.55
7090	3 use		222	.144		.51	6.45		6.96	11.10
7101	4 use		350	.091		.21	4.08		4.29	6.95
7500	Depressed area forms to 12" high, 4 use		300	.107	L.F.	1.02	4.76		5.78	8.90
7550	12" to 24" high, 4 use		175	.183		1.38	8.15		9.53	14.85
8000	Perimeter deck and rail for elevated slabs, straight		90	.356		12.45	15.85		28.30	39.50
8050	Curved		65	.492		17.05	22		39.05	55
8500	Void forms, round plastic, 8" high x 3" diameter G		450	.071	Ea.	.71	3.17		3.88	6
8550	4" diameter G		425	.075		1.06	3.36		4.42	6.65
8600	6" diameter G		400	.080		1.75	3.57		5.32	7.80
8650	8" diameter G		375	.085		3.13	3.81		6.94	9.70
9000	Minimum labor/equipment charge	2 Carp	2	8	Job		375		375	615

03 11 13.40 Forms In Place, Equipment Foundations

			Crew	Daily Output	Labor-Hours	Unit	Material	Labor	Equipment	Total	Total Incl O&P
0010	**FORMS IN PLACE, EQUIPMENT FOUNDATIONS**	R031113-40									
0020	1 use		C-2	160	.300	SFCA	3.45	13.70		17.15	26.50
0050	2 use	R031113-60		190	.253		1.90	11.55		13.45	21
0100	3 use			200	.240		1.38	11		12.38	19.50
0150	4 use			205	.234		1.12	10.70		11.82	18.80
9000	Minimum labor/equipment charge		1 Carp	3	2.667	Job		125		125	205

03 11 13.45 Forms In Place, Footings

			Crew	Daily Output	Labor-Hours	Unit	Material	Labor	Equipment	Total	Total Incl O&P
0010	**FORMS IN PLACE, FOOTINGS**	R031113-40									
0020	Continuous wall, plywood, 1 use		C-1	375	.085	SFCA	6.80	3.81		10.61	13.75
0150	4 use	R031113-60	"	485	.066	"	2.22	2.94		5.16	7.25
1500	Keyway, 4 use, tapered wood, 2" x 4"		1 Carp	530	.015	L.F.	.22	.71		.93	1.41
1550	2" x 6"		"	500	.016	"	.33	.75		1.08	1.59
3000	Pile cap, square or rectangular, job-built plywood, 1 use		C-1	290	.110	SFCA	2.91	4.92		7.83	11.25
3150	4 use			383	.084		.95	3.73		4.68	7.15
5000	Spread footings, job-built lumber, 1 use			305	.105		2.28	4.68		6.96	10.15
5150	4 use			414	.077		.74	3.45		4.19	6.45
9000	Minimum labor/equipment charge		1 Carp	3	2.667	Job		125		125	205

03 11 13.47 Forms In Place, Gas Station Forms

			Crew	Daily Output	Labor-Hours	Unit	Material	Labor	Equipment	Total	Total Incl O&P
0010	**FORMS IN PLACE, GAS STATION FORMS**										
0050	Curb fascia, with template, 12 ga. steel, left in place, 9" high G	1 Carp	50	.160	L.F.	13.90	7.50		21.40	27.50
1000	Sign or light bases, 18" diameter, 9" high G		9	.889	Ea.	87.50	41.50		129	165
1050	30" diameter, 13" high G		8	1	"	139	47		186	230
1990	Minimum labor/equipment charge			2	4	Job		188		188	310
2000	Island forms, 10' long, 9" high, 3'-6" wide G	C-1	10	3.200	Ea.	390	143		533	660
2050	4' wide G		9	3.556		400	159		559	700
2500	20' long, 9" high, 4' wide G		6	5.333		645	238		883	1,100

For customer support on your Commercial Renovation Cost Data, call 877.791.4977.

03 11 Concrete Forming

03 11 13 – Structural Cast-In-Place Concrete Forming

03 11 13.47 Forms In Place, Gas Station Forms

		Crew	Daily Output	Labor-Hours	Unit	Material	2015 Bare Costs Labor	Equipment	Total	Total Incl O&P
2550	5' wide [G]	C-1	5	6.400	Ea.	670	286		956	1,200
9000	Minimum labor/equipment charge	1 Carp	3	2.667	Job		125		125	205

03 11 13.50 Forms In Place, Grade Beam

			Crew	Daily Output	Labor-Hours	Unit	Material	Labor	Equipment	Total	Total Incl O&P
0010	**FORMS IN PLACE, GRADE BEAM**	R031113-40									
0020	Job-built plywood, 1 use		C-2	530	.091	SFCA	3.14	4.14		7.28	10.25
0150	4 use	R031113-60	"	605	.079	"	1.02	3.63		4.65	7.05
9000	Minimum labor/equipment charge		2 Carp	2	8	Job		375		375	615

03 11 13.65 Forms In Place, Slab On Grade

			Crew	Daily Output	Labor-Hours	Unit	Material	Labor	Equipment	Total	Total Incl O&P
0010	**FORMS IN PLACE, SLAB ON GRADE**	R031113-40									
1000	Bulkhead forms w/keyway, wood, 6" high, 1 use	R031113-60	C-1	510	.063	L.F.	1.04	2.80		3.84	5.75
1400	Bulkhead form for slab, 4-1/2" high, exp metal, incl keyway & stakes [G]			1200	.027		.92	1.19		2.11	2.96
1410	5-1/2" high [G]			1100	.029		1.12	1.30		2.42	3.36
1420	7-1/2" high [G]			960	.033		1.31	1.49		2.80	3.88
1430	9-1/2" high [G]			840	.038	↓	1.43	1.70		3.13	4.36
2000	Curb forms, wood, 6" to 12" high, on grade, 1 use			215	.149	SFCA	2.44	6.65		9.09	13.60
2150	4 use			275	.116	"	.79	5.20		5.99	9.35
3000	Edge forms, wood, 4 use, on grade, to 6" high			600	.053	L.F.	.35	2.38		2.73	4.28
3050	7" to 12" high			435	.074	SFCA	.79	3.28		4.07	6.25
3060	Over 12"			350	.091	"	.91	4.08		4.99	7.70
3500	For depressed slabs, 4 use, to 12" high			300	.107	L.F.	.75	4.76		5.51	8.65
3550	To 24" high			175	.183		.99	8.15		9.14	14.45
4000	For slab blockouts, to 12" high, 1 use			200	.160		.80	7.15		7.95	12.55
4050	To 24" high, 1 use			120	.267		1.01	11.90		12.91	20.50
4100	Plastic (extruded), to 6" high, multiple use, on grade			800	.040	↓	8.20	1.78		9.98	11.95
5000	Screed, 24 ga. metal key joint, see Section 03 15 16.30										
5020	Wood, incl. wood stakes, 1" x 3"		C-1	900	.036	L.F.	.77	1.59		2.36	3.44
5050	2" x 4"			900	.036	"	.80	1.59		2.39	3.48
6000	Trench forms in floor, wood, 1 use			160	.200	SFCA	1.94	8.90		10.84	16.80
6150	4 use			185	.173	"	.63	7.70		8.33	13.35
8760	Void form, corrugated fiberboard, 4" x 12", 4' long [G]			3000	.011	S.F.	2.87	.48		3.35	3.93
8770	6" x 12", 4' long		↓	3000	.011		3.42	.48		3.90	4.54
8780	1/4" thick hardboard protective cover for void form		2 Carp	1500	.011	↓	.58	.50		1.08	1.46
9000	Minimum labor/equipment charge		1 Carp	2	4	Job		188		188	310

03 11 13.85 Forms In Place, Walls

			Crew	Daily Output	Labor-Hours	Unit	Material	Labor	Equipment	Total	Total Incl O&P
0010	**FORMS IN PLACE, WALLS**	R031113-10									
0100	Box out for wall openings, to 16" thick, to 10 S.F.		C-2	24	2	Ea.	27	91.50		118.50	180
0150	Over 10 S.F. (use perimeter)	R031113-40	"	280	.171	L.F.	2.31	7.85		10.16	15.40
0250	Brick shelf, 4" w, add to wall forms, use wall area above shelf										
0260	1 use	R031113-60	C-2	240	.200	SFCA	2.50	9.15		11.65	17.75
0350	4 use			300	.160	"	1	7.30		8.30	13.10
0500	Bulkhead, wood with keyway, 1 use, 2 piece		↓	265	.181	L.F.	2.12	8.30		10.42	15.95
0600	Bulkhead forms with keyway, 1 piece expanded metal, 8" wall [G]		C-1	1000	.032		1.31	1.43		2.74	3.78
0610	10" wall [G]			800	.040		1.43	1.78		3.21	4.50
0620	12" wall [G]		↓	525	.061	↓	1.72	2.72		4.44	6.35
0700	Buttress, to 8' high, 1 use		C-2	350	.137	SFCA	4.28	6.25		10.53	15
0850	4 use			480	.100	"	1.41	4.57		5.98	9.05
1000	Corbel or haunch, to 12" wide, add to wall forms, 1 use			150	.320	L.F.	2.37	14.65		17.02	26.50
1150	4 use			180	.267	"	.77	12.20		12.97	21
2000	Wall, job-built plywood, to 8' high, 1 use			370	.130	SFCA	2.74	5.95		8.69	12.70
2050	2 use			435	.110		1.75	5.05		6.80	10.15
2100	3 use			495	.097		1.27	4.43		5.70	8.65
2150	4 use		↓	505	.095	↓	1.03	4.35		5.38	8.25

03 11 Concrete Forming

03 11 13 – Structural Cast-In-Place Concrete Forming

03 11 13.85 Forms In Place, Walls		Crew	Daily Output	Labor-Hours	Unit	Material	2015 Bare Costs Labor	Equipment	Total	Total Incl O&P
2400	Over 8' to 16' high, 1 use	C-2	280	.171	SFCA	3.03	7.85		10.88	16.20
2450	2 use		345	.139		1.34	6.35		7.69	11.90
2500	3 use		375	.128		.95	5.85		6.80	10.65
2550	4 use		395	.122		.78	5.55		6.33	9.95
2700	Over 16' high, 1 use		235	.204		2.71	9.35		12.06	18.30
2750	2 use		290	.166		1.49	7.55		9.04	14.05
2800	3 use		315	.152		1.09	6.95		8.04	12.60
2850	4 use		330	.145		.88	6.65		7.53	11.85
4000	Radial, smooth curved, job-built plywood, 1 use		245	.196		2.55	8.95		11.50	17.50
4150	4 use		335	.143		.83	6.55		7.38	11.65
4200	Below grade, job-built plywood, 1 use		225	.213		2.77	9.75		12.52	19.05
4210	2 use		225	.213		1.53	9.75		11.28	17.70
4220	3 use		225	.213		1.27	9.75		11.02	17.40
4230	4 use		225	.213		.90	9.75		10.65	17
4600	Retaining wall, battered, job-built plyw'd, to 8' high, 1 use		300	.160		2.03	7.30		9.33	14.25
4750	4 use		390	.123		.66	5.65		6.31	9.95
4900	Over 8' to 16' high, 1 use		240	.200		2.22	9.15		11.37	17.45
5050	4 use		320	.150		.72	6.85		7.57	12.05
7500	Lintel or sill forms, 1 use	1 Carp	30	.267		3.26	12.50		15.76	24
7560	4 use	"	37	.216		1.06	10.15		11.21	17.80
7800	Modular prefabricated plywood, based on 20 uses of purchased									
7820	forms, and 4 uses of bracing lumber									
7860	To 8' high	C-2	800	.060	SFCA	.94	2.74		3.68	5.55
8060	Over 8' to 16' high		600	.080		.99	3.66		4.65	7.10
8600	Pilasters, 1 use		270	.178		3.19	8.15		11.34	16.80
8660	4 use		385	.125		1.04	5.70		6.74	10.50
9010	Steel framed plywood, based on 50 uses of purchased									
9020	forms, and 4 uses of bracing lumber									
9060	To 8' high	C-2	600	.080	SFCA	.74	3.66		4.40	6.80
9260	Over 8' to 16' high		450	.107		.74	4.88		5.62	8.80
9460	Over 16' to 20' high		400	.120		.74	5.50		6.24	9.80
9475	For elevated walls, add						10%			
9480	For battered walls, 1 side battered, add					10%	10%			
9485	For battered walls, 2 sides battered, add					15%	15%			
9900	Minimum labor/equipment charge	2 Carp	2	8	Job		375		375	615

03 11 16 – Architectural Cast-in Place Concrete Forming

03 11 16.13 Concrete Form Liners

0010	**CONCRETE FORM LINERS**									
5750	Liners for forms (add to wall forms), ABS plastic									
5800	Aged wood, 4" wide, 1 use	1 Carp	256	.031	SFCA	3.30	1.47		4.77	6.05
5820	2 use		256	.031		1.82	1.47		3.29	4.40
5830	3 use		256	.031		1.32	1.47		2.79	3.85
5840	4 use		256	.031		1.07	1.47		2.54	3.58
5900	Fractured rope rib, 1 use		192	.042		4.79	1.96		6.75	8.45
5925	2 use		192	.042		2.63	1.96		4.59	6.10
5950	3 use		192	.042		1.92	1.96		3.88	5.30
6000	4 use		192	.042		1.56	1.96		3.52	4.92
6100	Ribbed, 3/4" deep x 1-1/2" O.C., 1 use		224	.036		4.79	1.68		6.47	8
6125	2 use		224	.036		2.63	1.68		4.31	5.65
6150	3 use		224	.036		1.92	1.68		3.60	4.86
6200	4 use		224	.036		1.56	1.68		3.24	4.46
6300	Rustic brick pattern, 1 use		224	.036		3.30	1.68		4.98	6.40

03 11 Concrete Forming

03 11 16 – Architectural Cast-in Place Concrete Forming

03 11 16.13 Concrete Form Liners

		Crew	Daily Output	Labor-Hours	Unit	Material	2015 Bare Costs Labor	2015 Bare Costs Equipment	Total	Total Incl O&P
6325	2 use	1 Carp	224	.036	SFCA	1.82	1.68		3.50	4.75
6350	3 use		224	.036		1.32	1.68		3	4.20
6400	4 use		224	.036		1.07	1.68		2.75	3.93
6500	3/8" striated, random, 1 use		224	.036		3.30	1.68		4.98	6.40
6525	2 use		224	.036		1.82	1.68		3.50	4.75
6550	3 use		224	.036		1.32	1.68		3	4.20
6600	4 use		224	.036		1.07	1.68		2.75	3.93

03 15 Concrete Accessories

03 15 05 – Concrete Forming Accessories

03 15 05.70 Shores

			Crew	Daily Output	Labor-Hours	Unit	Material	Labor	Equipment	Total	Total Incl O&P
0010	**SHORES**										
0020	Erect and strip, by hand, horizontal members										
0500	Aluminum joists and stringers	G	2 Carp	60	.267	Ea.		12.50		12.50	20.50
0600	Steel, adjustable beams	G		45	.356			16.70		16.70	27.50
0700	Wood joists			50	.320			15		15	24.50
0800	Wood stringers			30	.533			25		25	41
1000	Vertical members to 10' high	G		55	.291			13.65		13.65	22.50
1050	To 13' high	G		50	.320			15		15	24.50
1100	To 16' high	G		45	.356			16.70		16.70	27.50
1500	Reshoring	G		1400	.011	S.F.	.58	.54		1.12	1.52
1600	Flying truss system	G	C-17D	9600	.009	SFCA		.43	.08	.51	.79
1760	Horizontal, aluminum joists, 6-1/4" high x 5' to 21' span, buy	G				L.F.	15.45			15.45	17
1770	Beams, 7-1/4" high x 4' to 30' span	G				"	18.10			18.10	19.90
1810	Horizontal, steel beam, W8x10, 7' span, buy	G				Ea.	57			57	63
1830	10' span	G					67.50			67.50	74.50
1920	15' span	G					117			117	128
1940	20' span	G					164			164	180
1970	Steel stringer, W8x10, 4' to 16' span, buy	G				L.F.	6.80			6.80	7.45
3000	Rent for job duration, aluminum joist @ 2' O.C., per mo.	G				SF Flr.	.39			.39	.43
3050	Steel W8x10	G					.17			.17	.19
3060	Steel adjustable	G					.17			.17	.19
3500	#1 post shore, steel, 5'-7" to 9'-6" high, 10000# cap., buy	G				Ea.	148			148	163
3550	#2 post shore, 7'-3" to 12'-10" high, 7800# capacity	G					171			171	188
3600	#3 post shore, 8'-10" to 16'-1" high, 3800# capacity	G					187			187	206
5010	Frame shoring systems, steel, 12000#/leg, buy										
5040	Frame, 2' wide x 6' high	G				Ea.	93.50			93.50	103
5250	X-brace	G					15.50			15.50	17.05
5550	Base plate	G					15.50			15.50	17.05
5600	Screw jack	G					33.50			33.50	36.50
5650	U-head, 8" x 8"	G					19.10			19.10	21

03 15 05.85 Stair Tread Inserts

		Crew	Daily Output	Labor-Hours	Unit	Material	Labor	Equipment	Total	Total Incl O&P
0010	**STAIR TREAD INSERTS**									
0105	Cast nosing insert, abrasive surface, pre-drilled, includes screws									
0110	Aluminum, 3" wide x 3' long	1 Cefi	32	.250	Ea.	55	11.25		66.25	78.50
0120	4' long		31	.258		70	11.60		81.60	95.50
0130	5' long		30	.267		87	12		99	114
0135	Extruded nosing insert, black abrasive strips, continuous anchor									
0140	Aluminum, 3" wide x 3' long	1 Cefi	64	.125	Ea.	35.50	5.65		41.15	48
0150	4' long		60	.133		55	6		61	70
0160	5' long		56	.143		62	6.45		68.45	78

03 15 Concrete Accessories

03 15 05 – Concrete Forming Accessories

03 15 05.85 Stair Tread Inserts

		Crew	Daily Output	Labor-Hours	Unit	Material	2015 Bare Costs Labor	2015 Bare Costs Equipment	Total	Total Incl O&P
0165	Extruded nosing insert, black abrasive strips, pre-drilled, incl. screws									
0170	Aluminum, 3" wide x 3' long	1 Cefi	32	.250	Ea.	49	11.25		60.25	72
0180	4' long		31	.258		63.50	11.60		75.10	88.50
0190	5' long	↓	30	.267	↓	74.50	12		86.50	101
9000	Minimum labor/equipment charge	1 Carp	4	2	Job		94		94	154

03 15 13 – Waterstops

03 15 13.50 Waterstops

		Crew	Daily Output	Labor-Hours	Unit	Material	Labor	Equipment	Total	Incl O&P
0010	**WATERSTOPS**, PVC and Rubber									
0020	PVC, ribbed 3/16" thick, 4" wide	1 Carp	155	.052	L.F.	1.35	2.42		3.77	5.45
0050	6" wide		145	.055		2.27	2.59		4.86	6.75
0500	With center bulb, 6" wide, 3/16" thick		135	.059		1.99	2.78		4.77	6.75
0550	3/8" thick		130	.062		3.67	2.89		6.56	8.80
0600	9" wide x 3/8" thick	↓	125	.064	↓	6.05	3		9.05	11.55

03 15 16 – Concrete Construction Joints

03 15 16.20 Control Joints, Saw Cut

		Crew	Daily Output	Labor-Hours	Unit	Material	Labor	Equipment	Total	Incl O&P
0010	**CONTROL JOINTS, SAW CUT**									
0100	Sawcut control joints in green concrete									
0120	1" depth	C-27	2000	.008	L.F.	.04	.36	.08	.48	.70
0140	1-1/2" depth		1800	.009		.06	.40	.09	.55	.79
0160	2" depth	↓	1600	.010	↓	.08	.45	.10	.63	.92
0180	Sawcut joint reservoir in cured concrete									
0182	3/8" wide x 3/4" deep, with single saw blade	C-27	1000	.016	L.F.	.06	.72	.17	.95	1.38
0184	1/2" wide x 1" deep, with double saw blades		900	.018		.11	.80	.19	1.10	1.60
0186	3/4" wide x 1-1/2" deep, with double saw blades	↓	800	.020		.23	.90	.21	1.34	1.91
0190	Water blast joint to wash away laitance, 2 passes	C-29	2500	.003			.12	.03	.15	.23
0200	Air blast joint to blow out debris and air dry, 2 passes	C-28	2000	.004	↓		.18	.01	.19	.29
0300	For backer rod, see Section 07 91 23.10									
0340	For joint sealant, see Sections 03 15 16.30 or 07 92 13.20									
0900	For replacement of joint sealant, see Section 07 01 90.81									

03 15 16.30 Expansion Joints

			Crew	Daily Output	Labor-Hours	Unit	Material	Labor	Equipment	Total	Incl O&P
0010	**EXPANSION JOINTS**										
0020	Keyed, cold, 24 ga., incl. stakes, 3-1/2" high	G	1 Carp	200	.040	L.F.	.81	1.88		2.69	3.97
0050	4-1/2" high	G		200	.040		.92	1.88		2.80	4.09
0100	5-1/2" high	G		195	.041		1.12	1.93		3.05	4.39
0150	7-1/2" high	G		190	.042		1.31	1.98		3.29	4.68
0160	9-1/2" high	G	↓	185	.043		1.43	2.03		3.46	4.90
0300	Poured asphalt, plain, 1/2" x 1"		1 Clab	450	.018		.69	.67		1.36	1.86
0350	1" x 2"			400	.020		2.76	.75		3.51	4.27
0500	Neoprene, liquid, cold applied, 1/2" x 1"			450	.018		2.88	.67		3.55	4.27
0550	1" x 2"			400	.020		11.50	.75		12.25	13.90
0700	Polyurethane, poured, 2 part, 1/2" x 1"			400	.020		1.28	.75		2.03	2.64
0750	1" x 2"			350	.023		5.10	.86		5.96	7.05
0900	Rubberized asphalt, hot or cold applied, 1/2" x 1"			450	.018		.40	.67		1.07	1.54
0950	1" x 2"			400	.020		1.60	.75		2.35	2.99
1100	Hot applied, fuel resistant, 1/2" x 1"			450	.018		.60	.67		1.27	1.76
1150	1" x 2"		↓	400	.020		2.40	.75		3.15	3.87
2000	Premolded, bituminous fiber, 1/2" x 6"		1 Carp	375	.021		.44	1		1.44	2.12
2050	1" x 12"			300	.027		1.99	1.25		3.24	4.24
2140	Concrete expansion joint, recycled paper and fiber, 1/2" x 6"	G		390	.021		.42	.96		1.38	2.04
2150	1/2" x 12"	G		360	.022		.83	1.04		1.87	2.63
2250	Cork with resin binder, 1/2" x 6"			375	.021		1.11	1		2.11	2.86

03 15 Concrete Accessories

03 15 16 – Concrete Construction Joints

03 15 16.30 Expansion Joints		Crew	Daily Output	Labor-Hours	Unit	Material	2015 Bare Costs Labor	Equipment	Total	Total Incl O&P
2300	1" x 12"	1 Carp	300	.027	L.F.	3.11	1.25		4.36	5.45
2500	Neoprene sponge, closed cell, 1/2" x 6"		375	.021		2.37	1		3.37	4.25
2550	1" x 12"		300	.027		8.25	1.25		9.50	11.15
2750	Polyethylene foam, 1/2" x 6"		375	.021		.68	1		1.68	2.39
2800	1" x 12"		300	.027		2.19	1.25		3.44	4.46
3000	Polyethylene backer rod, 3/8" diameter		460	.017		.03	.82		.85	1.37
3050	3/4" diameter		460	.017		.06	.82		.88	1.40
3100	1" diameter		460	.017		.09	.82		.91	1.44
3500	Polyurethane foam, with polybutylene, 1/2" x 1/2"		475	.017		1.14	.79		1.93	2.55
3550	1" x 1"		450	.018		2.88	.83		3.71	4.54
3750	Polyurethane foam, regular, closed cell, 1/2" x 6"		375	.021		.84	1		1.84	2.56
3800	1" x 12"		300	.027		3	1.25		4.25	5.35
4000	Polyvinyl chloride foam, closed cell, 1/2" x 6"		375	.021		2.28	1		3.28	4.15
4050	1" x 12"		300	.027		7.85	1.25		9.10	10.70
4250	Rubber, gray sponge, 1/2" x 6"		375	.021		1.92	1		2.92	3.75
4300	1" x 12"		300	.027		6.90	1.25		8.15	9.65
4400	Redwood heartwood, 1" x 4"		400	.020		1.16	.94		2.10	2.82
4450	1" x 6"		375	.021		1.75	1		2.75	3.57
5000	For installation in walls, add						75%			
5250	For installation in boxouts, add						25%			

03 15 19 – Cast-In Concrete Anchors

03 15 19.05 Anchor Bolt Accessories

		Crew	Daily Output	Labor-Hours	Unit	Material	Labor	Equipment	Total	Total Incl O&P
0010	**ANCHOR BOLT ACCESSORIES**									
0015	For anchor bolts set in fresh concrete, see Section 03 15 19.10									
8150	Anchor bolt sleeve, plastic, 1" diam. bolts	1 Carp	60	.133	Ea.	8.20	6.25		14.45	19.30
8500	1-1/2" diameter		28	.286		14.60	13.40		28	38
8600	2" diameter		24	.333		18.60	15.65		34.25	46
8650	3" diameter		20	.400		34	18.80		52.80	68.50

03 15 19.10 Anchor Bolts

			Crew	Daily Output	Labor-Hours	Unit	Material	Labor	Equipment	Total	Total Incl O&P
0010	**ANCHOR BOLTS**										
0015	Made from recycled materials										
0025	Single bolts installed in fresh concrete, no templates										
0030	Hooked w/nut and washer, 1/2" diameter, 8" long	G	1 Carp	132	.061	Ea.	1.33	2.85		4.18	6.15
0040	12" long	G		131	.061		1.48	2.87		4.35	6.35
0050	5/8" diameter, 8" long	G		129	.062		2.98	2.91		5.89	8.05
0060	12" long	G		127	.063		3.67	2.96		6.63	8.90
0070	3/4" diameter, 8" long	G		127	.063		3.67	2.96		6.63	8.90
0080	12" long	G		125	.064		4.59	3		7.59	9.95
0090	2-bolt pattern, including job-built 2-hole template, per set										
0100	J-type, incl. hex nut & washer, 1/2" diameter x 6" long	G	1 Carp	21	.381	Set	5.20	17.90		23.10	35.50
0110	12" long	G		21	.381		5.80	17.90		23.70	36
0120	18" long	G		21	.381		6.70	17.90		24.60	37
0130	3/4" diameter x 8" long	G		20	.400		10.20	18.80		29	42
0140	12" long	G		20	.400		12	18.80		30.80	44
0150	18" long	G		20	.400		14.75	18.80		33.55	47.50
0160	1" diameter x 12" long	G		19	.421		20.50	19.75		40.25	55
0170	18" long	G		19	.421		24	19.75		43.75	59
0180	24" long	G		19	.421		28.50	19.75		48.25	64
0190	36" long	G		18	.444		38	21		59	76
0200	1-1/2" diameter x 18" long	G		17	.471		37.50	22		59.50	77.50
0210	24" long	G		16	.500		44	23.50		67.50	87
0300	L-type, incl. hex nut & washer, 3/4" diameter x 12" long	G		20	.400		12.90	18.80		31.70	45

03 15 Concrete Accessories

03 15 19 – Cast-In Concrete Anchors

03 15 19.10 Anchor Bolts

			Crew	Daily Output	Labor-Hours	Unit	Material	2015 Bare Costs Labor	Equipment	Total	Total Incl O&P
0310	18" long	G	1 Carp	20	.400	Set	15.75	18.80		34.55	48.50
0320	24" long	G		20	.400		18.60	18.80		37.40	51.50
0330	30" long	G		20	.400		23	18.80		41.80	56
0340	36" long	G		20	.400		25.50	18.80		44.30	59.50
0350	1" diameter x 12" long	G		19	.421		19.55	19.75		39.30	54
0360	18" long	G		19	.421		23.50	19.75		43.25	58.50
0370	24" long	G		19	.421		28.50	19.75		48.25	64
0380	30" long	G		19	.421		33	19.75		52.75	69
0390	36" long	G		18	.444		37.50	21		58.50	75.50
0400	42" long	G		18	.444		45	21		66	83.50
0410	48" long	G		18	.444		50	21		71	89
0420	1-1/4" diameter x 18" long	G		18	.444		30	21		51	67
0430	24" long	G		18	.444		35	21		56	72.50
0440	30" long	G		17	.471		40	22		62	80
0450	36" long	G		17	.471		45	22		67	85.50
1000	4-bolt pattern, including job-built 4-hole template, per set										
1100	J-type, incl. hex nut & washer, 1/2" diameter x 6" long	G	1 Carp	19	.421	Set	7.55	19.75		27.30	41
1110	12" long	G		19	.421		8.75	19.75		28.50	42
1120	18" long	G		18	.444		10.55	21		31.55	45.50
1130	3/4" diameter x 8" long	G		17	.471		17.55	22		39.55	55.50
1140	12" long	G		17	.471		21	22		43	59.50
1150	18" long	G		17	.471		26.50	22		48.50	65.50
1160	1" diameter x 12" long	G		16	.500		38	23.50		61.50	80
1170	18" long	G		15	.533		45	25		70	90.50
1180	24" long	G		15	.533		54	25		79	101
1190	36" long	G		15	.533		73	25		98	122
1200	1-1/2" diameter x 18" long	G		13	.615		72	29		101	127
1210	24" long	G		12	.667		85.50	31.50		117	146
1300	L-type, incl. hex nut & washer, 3/4" diameter x 12" long	G		17	.471		23	22		45	61.50
1310	18" long	G		17	.471		28.50	22		50.50	67.50
1320	24" long	G		17	.471		34.50	22		56.50	74
1330	30" long	G		16	.500		43	23.50		66.50	85.50
1340	36" long	G		16	.500		48.50	23.50		72	92
1350	1" diameter x 12" long	G		16	.500		36	23.50		59.50	78.50
1360	18" long	G		15	.533		44.50	25		69.50	90
1370	24" long	G		15	.533		54	25		79	101
1380	30" long	G		15	.533		63.50	25		88.50	111
1390	36" long	G		15	.533		72	25		97	121
1400	42" long	G		14	.571		87	27		114	140
1410	48" long	G		14	.571		97.50	27		124.50	151
1420	1-1/4" diameter x 18" long	G		14	.571		57	27		84	107
1430	24" long	G		14	.571		67	27		94	118
1440	30" long	G		13	.615		77	29		106	132
1450	36" long	G		13	.615		87	29		116	144

03 15 19.45 Machinery Anchors

			Crew	Daily Output	Labor-Hours	Unit	Material	2015 Bare Costs Labor	Equipment	Total	Total Incl O&P
0010	**MACHINERY ANCHORS**, heavy duty, incl. sleeve, floating base nut,										
0020	lower stud & coupling nut, fiber plug, connecting stud, washer & nut.										
0030	For flush mounted embedment in poured concrete heavy equip. pads.										
0200	Stud & bolt, 1/2" diameter	G	E-16	40	.400	Ea.	72.50	21.50	3.65	97.65	123
0300	5/8" diameter	G		35	.457		80.50	24.50	4.17	109.17	138
0500	3/4" diameter	G		30	.533		93	28.50	4.86	126.36	159
0600	7/8" diameter	G		25	.640		101	34.50	5.85	141.35	179

03 15 Concrete Accessories

03 15 19 – Cast-In Concrete Anchors

03 15 19.45 Machinery Anchors		Crew	Daily Output	Labor-Hours	Unit	Material	2015 Bare Costs Labor	Equipment	Total	Total Incl O&P
0800	1" diameter	G E-16	20	.800	Ea.	117	43	7.30	167.30	215
0900	1-1/4" diameter	G	15	1.067		141	57	9.75	207.75	269

03 21 Reinforcement Bars

03 21 11 – Plain Steel Reinforcement Bars

03 21 11.60 Reinforcing In Place

			Crew	Daily Output	Labor-Hours	Unit	Material	Labor	Equipment	Total	Incl O&P
0010	**REINFORCING IN PLACE**, 50-60 ton lots, A615 Grade 60	R032110-10									
0020	Includes labor, but not material cost, to install accessories										
0030	Made from recycled materials										
0102	Beams & Girders, #3 to #7	G	4 Rodm	3200	.010	Lb.	.48	.53		1.01	1.39
0152	#8 to #18	G		5400	.006		.48	.31		.79	1.04
0202	Columns, #3 to #7	G		3000	.011		.48	.56		1.04	1.45
0252	#8 to #18	G		4600	.007		.48	.37		.85	1.13
0402	Elevated slabs, #4 to #7	G		5800	.006		.48	.29		.77	1
0502	Footings, #4 to #7	G		4200	.008		.48	.40		.88	1.18
0552	#8 to #18	G		7200	.004		.48	.23		.71	.91
0602	Slab on grade, #3 to #7	G		4200	.008		.48	.40		.88	1.18
0702	Walls, #3 to #7	G		6000	.005		.48	.28		.76	.99
0752	#8 to #18	G		8000	.004		.48	.21		.69	.87
0900	For other than 50 – 60 ton lots										
1000	Under 10 ton job, #3 to #7, add						25%	10%			
1010	#8 to #18, add						20%	10%			
1050	10 – 50 ton job, #3 to #7, add						10%				
1060	#8 to #18, add						5%				
1100	60 – 100 ton job, #3 to #7, deduct						5%				
1110	#8 to #18, deduct						10%				
1150	Over 100 ton job, #3 to #7, deduct						10%				
1160	#8 to #18, deduct						15%				
2000	Unloading & sorting, add to above		C-5	100	.560	Ton		29	7.35	36.35	55
2200	Crane cost for handling, 90 picks/day, up to 1.5 Tons/bundle, add to above			135	.415			21.50	5.45	26.95	40.50
2210	1.0 Ton/bundle			92	.609			31.50	8	39.50	60
2220	0.5 Ton/bundle			35	1.600			82.50	21	103.50	157
2400	Dowels, 2 feet long, deformed, #3	G	2 Rodm	520	.031	Ea.	.40	1.62		2.02	3.08
2410	#4	G		480	.033		.71	1.75		2.46	3.64
2420	#5	G		435	.037		1.11	1.93		3.04	4.37
2430	#6	G		360	.044		1.60	2.34		3.94	5.55
2450	Longer and heavier dowels, add	G		725	.022	Lb.	.53	1.16		1.69	2.48
2500	Smooth dowels, 12" long, 1/4" or 3/8" diameter	G		140	.114	Ea.	.72	6		6.72	10.60
2520	5/8" diameter	G		125	.128		1.26	6.75		8.01	12.40
2530	3/4" diameter	G		110	.145		1.57	7.65		9.22	14.15
2600	Dowel sleeves for CIP concrete, 2-part system										
2610	Sleeve base, plastic, for 5/8" smooth dowel sleeve, fasten to edge form		1 Rodm	200	.040	Ea.	.53	2.10		2.63	4.01
2615	Sleeve, plastic, 12" long, for 5/8" smooth dowel, snap onto base			400	.020		1.18	1.05		2.23	3.02
2620	Sleeve base, for 3/4" smooth dowel sleeve			175	.046		.53	2.40		2.93	4.50
2625	Sleeve, 12" long, for 3/4" smooth dowel			350	.023		1.33	1.20		2.53	3.42
2630	Sleeve base, for 1" smooth dowel sleeve			150	.053		.67	2.80		3.47	5.30
2635	Sleeve, 12" long, for 1" smooth dowel			300	.027		1.40	1.40		2.80	3.83
2700	Dowel caps, visual warning only, plastic, #3 to #8		2 Rodm	800	.020		.26	1.05		1.31	2.01
2720	#8 to #18			750	.021		.63	1.12		1.75	2.52
2750	Impalement protective, plastic, #4 to #9			800	.020		1.22	1.05		2.27	3.06
9000	Minimum labor/equipment charge		1 Rodm	4	2	Job		105		105	172

03 21 Reinforcement Bars

03 21 21 – Composite Reinforcement Bars

03 21 21.11 Glass Fiber-Reinforced Polymer Reinforcement Bars

	Crew	Daily Output	Labor-Hours	Unit	Material	2015 Bare Costs Labor	2015 Bare Costs Equipment	Total	Total Incl O&P
0010 **GLASS FIBER-REINFORCED POLYMER REINFORCEMENT BARS**									
0020 Includes labor, but not material cost, to install accessories									
0050 #2 bar, .043 lb./L.F.	4 Rodm	9500	.003	L.F.	.40	.18		.58	.73
0100 #3 bar, .092 lb./L.F.		9300	.003		.49	.18		.67	.84
0150 #4 bar, .160 lb./L.F.		9100	.004		.71	.19		.90	1.08
0200 #5 bar, .258 lb./L.F.		8700	.004		1.09	.19		1.28	1.52
0250 #6 bar, .372 lb./L.F.		8300	.004		1.41	.20		1.61	1.88
0300 #7 bar, .497 lb./L.F.		7900	.004		1.84	.21		2.05	2.37
0350 #8 bar, .620 lb./L.F.		7400	.004		2.38	.23		2.61	2.99
0400 #9 bar, .800 lb./L.F.		6800	.005		3.10	.25		3.35	3.81
0450 #10 bar, 1.08 lb./L.F.		5800	.006		3.75	.29		4.04	4.60
0500 For Bends, add per bend				Ea.	1.48			1.48	1.63

03 22 Fabric and Grid Reinforcing

03 22 11 – Plain Welded Wire Fabric Reinforcing

03 22 11.10 Plain Welded Wire Fabric

		Crew	Daily Output	Labor-Hours	Unit	Material	2015 Bare Costs Labor	2015 Bare Costs Equipment	Total	Total Incl O&P
0010 **PLAIN WELDED WIRE FABRIC** ASTM A185 R032205-30										
0020 Includes labor, but not material cost, to install accessories										
0030 Made from recycled materials										
0050 Sheets										
0100 6 x 6 - W1.4 x W1.4 (10 x 10) 21 lb. per C.S.F.	G	2 Rodm	35	.457	C.S.F.	14.50	24		38.50	55
0200 6 x 6 - W2.1 x W2.1 (8 x 8) 30 lb. per C.S.F.	G		31	.516		17.20	27		44.20	63.50
0300 6 x 6 - W2.9 x W2.9 (6 x 6) 42 lb. per C.S.F.	G		29	.552		22.50	29		51.50	72
0400 6 x 6 - W4 x W4 (4 x 4) 58 lb. per C.S.F.	G		27	.593		31.50	31		62.50	85.50
0500 4 x 4 - W1.4 x W1.4 (10 x 10) 31 lb. per C.S.F.	G		31	.516		20	27		47	66.50
0600 4 x 4 - W2.1 x W2.1 (8 x 8) 44 lb. per C.S.F.	G		29	.552		25	29		54	75
0650 4 x 4 - W2.9 x W2.9 (6 x 6) 61 lb. per C.S.F.	G		27	.593		40.50	31		71.50	95.50
0700 4 x 4 - W4 x W4 (4 x 4) 85 lb. per C.S.F.	G		25	.640		50.50	33.50		84	111
0750 Rolls										
0800 2 x 2 - #14 galv., 21 lb./C.S.F., beam & column wrap	G	2 Rodm	6.50	2.462	C.S.F.	41.50	129		170.50	257
0900 2 x 2 - #12 galv. for gunite reinforcing	G	"	6.50	2.462	"	62.50	129		191.50	280
9000 Minimum labor/equipment charge		1 Rodm	4	2	Job		105		105	172

03 22 13 – Galvanized Welded Wire Fabric Reinforcing

03 22 13.10 Galvanized Welded Wire Fabric

	Crew	Daily Output	Labor-Hours	Unit	Material	2015 Bare Costs Labor	2015 Bare Costs Equipment	Total	Total Incl O&P
0010 **GALVANIZED WELDED WIRE FABRIC**									
0100 Add to plain welded wire pricing for galvanized welded wire				Lb.	.23			.23	.25

03 22 16 – Epoxy-Coated Welded Wire Fabric Reinforcing

03 22 16.10 Epoxy-Coated Welded Wire Fabric

	Crew	Daily Output	Labor-Hours	Unit	Material	2015 Bare Costs Labor	2015 Bare Costs Equipment	Total	Total Incl O&P
0010 **EPOXY-COATED WELDED WIRE FABRIC**									
0100 Add to plain welded wire pricing for epoxy-coated welded wire				Lb.	.21			.21	.23

03 24 Fibrous Reinforcing

03 24 05 – Reinforcing Fibers

03 24 05.30 Synthetic Fibers

		Crew	Daily Output	Labor-Hours	Unit	Material	2015 Bare Costs Labor	2015 Bare Costs Equipment	Total	Total Incl O&P
0010	**SYNTHETIC FIBERS**									
0100	Synthetic fibers, add to concrete				Lb.	4.40			4.40	4.84
0110	1-1/2 lb. per C.Y.				C.Y.	6.80			6.80	7.50

03 24 05.70 Steel Fibers

			Crew	Daily Output	Labor-Hours	Unit	Material	Labor	Equipment	Total	Total Incl O&P
0010	**STEEL FIBERS**										
0140	ASTM A850, Type V, continuously deformed, 1-1/2" long x 0.045" diam.										
0150	Add to price of ready mix concrete	G				Lb.	1.13			1.13	1.24
0205	Alternate pricing, dosing at 5 lb. per C.Y., add to price of RMC	G				C.Y.	5.65			5.65	6.20
0210	10 lb. per C.Y.	G					11.30			11.30	12.45
0215	15 lb. per C.Y.	G					16.95			16.95	18.65
0220	20 lb. per C.Y.	G					22.50			22.50	25
0225	25 lb. per C.Y.	G					28.50			28.50	31
0230	30 lb. per C.Y.	G					34			34	37.50
0235	35 lb. per C.Y.	G					39.50			39.50	43.50
0240	40 lb. per C.Y.	G					45			45	49.50
0250	50 lb. per C.Y.	G					56.50			56.50	62
0275	75 lb. per C.Y.	G					85			85	93
0300	100 lb. per C.Y.	G					113			113	124

03 30 Cast-In-Place Concrete

03 30 53 – Miscellaneous Cast-In-Place Concrete

03 30 53.40 Concrete In Place

			Crew	Daily Output	Labor-Hours	Unit	Material	Labor	Equipment	Total	Total Incl O&P
0010	**CONCRETE IN PLACE**	R033105-10									
0020	Including forms (4 uses), Grade 60 rebar, concrete (Portland cement	R033105-20									
0050	Type I), placement and finishing unless otherwise indicated	R033105-70									
0300	Beams (3500 psi), 5 kip per L.F., 10' span	R033105-85	C-14A	15.62	12.804	C.Y.	320	605	48	973	1,400
0350	25' span		"	18.55	10.782		335	510	40.50	885.50	1,250
0500	Chimney foundations (5000 psi), over 5 C.Y.		C-14C	32.22	3.476		150	157	.97	307.97	420
0510	(3500 psi), under 5 C.Y.		"	23.71	4.724		177	213	1.32	391.32	545
0700	Columns, square (4000 psi), 12" x 12", less than 2% reinforcing		C-14A	11.96	16.722		365	790	63	1,218	1,775
0720	2% to 3% reinforcing			10.13	19.743		565	935	74	1,574	2,225
0740	Over 3% reinforcing			9.03	22.148		840	1,050	83.50	1,973.50	2,700
0800	16" x 16", less than 2% reinforcing			16.22	12.330		286	585	46.50	917.50	1,325
0820	2% to 3% reinforcing			12.57	15.911		480	750	60	1,290	1,825
0840	Over 3% reinforcing			10.25	19.512		735	920	73.50	1,728.50	2,375
0900	24" x 24", less than 2% reinforcing			23.66	8.453		241	400	32	673	950
0920	2% to 3% reinforcing			17.71	11.293		425	535	42.50	1,002.50	1,375
0940	Over 3% reinforcing			14.15	14.134		670	670	53	1,393	1,900
1000	36" x 36", less than 2% reinforcing			33.69	5.936		212	281	22.50	515.50	720
1020	2% to 3% reinforcing			23.32	8.576		370	405	32.50	807.50	1,100
1040	Over 3% reinforcing			17.82	11.223		625	530	42	1,197	1,600
1100	Columns, round (4000 psi), tied, 12" diameter, less than 2% reinforcing			20.97	9.537		315	450	36	801	1,125
1120	2% to 3% reinforcing			15.27	13.098		515	620	49.50	1,184.50	1,625
1140	Over 3% reinforcing			12.11	16.515		780	780	62	1,622	2,200
1200	16" diameter, less than 2% reinforcing			31.49	6.351		289	300	24	613	830
1220	2% to 3% reinforcing			19.12	10.460		490	495	39.50	1,024.50	1,400
1240	Over 3% reinforcing			13.77	14.524		735	685	54.50	1,474.50	2,000
1300	20" diameter, less than 2% reinforcing			41.04	4.873		289	230	18.30	537.30	710
1320	2% to 3% reinforcing			24.05	8.316		475	395	31.50	901.50	1,200
1340	Over 3% reinforcing			17.01	11.758		735	555	44	1,334	1,750
1400	24" diameter, less than 2% reinforcing			51.85	3.857		269	182	14.50	465.50	610

03 30 Cast-In-Place Concrete

03 30 53 – Miscellaneous Cast-In-Place Concrete

03 30 53.40 Concrete In Place		Crew	Daily Output	Labor-Hours	Unit	Material	2015 Bare Costs Labor	Equipment	Total	Total Incl O&P
1420	2% to 3% reinforcing	C-14A	27.06	7.391	C.Y.	470	350	28	848	1,125
1440	Over 3% reinforcing		18.29	10.935		715	515	41	1,271	1,675
1500	36" diameter, less than 2% reinforcing		75.04	2.665		266	126	10	402	510
1520	2% to 3% reinforcing		37.49	5.335		445	252	20	717	920
1540	Over 3% reinforcing		22.84	8.757		695	415	33	1,143	1,475
1900	Elevated slab (4000 psi), flat slab with drops, 125 psf Sup. Load, 20' span	C-14B	38.45	5.410		261	255	19.55	535.55	725
1950	30' span		50.99	4.079		276	192	14.75	482.75	635
2100	Flat plate, 125 psf Sup. Load, 15' span		30.24	6.878		240	325	25	590	820
2150	25' span		49.60	4.194		249	198	15.15	462.15	615
2300	Waffle const., 30" domes, 125 psf Sup. Load, 20' span		37.07	5.611		259	265	20.50	544.50	740
2350	30' span		44.07	4.720		241	223	17.05	481.05	650
2500	One way joists, 30" pans, 125 psf Sup. Load, 15' span		27.38	7.597		310	360	27.50	697.50	955
2550	25' span		31.15	6.677		292	315	24	631	860
2700	One way beam & slab, 125 psf Sup. Load, 15' span		20.59	10.102		259	475	36.50	770.50	1,100
2750	25' span		28.36	7.334		245	345	26.50	616.50	865
2900	Two way beam & slab, 125 psf Sup. Load, 15' span		24.04	8.652		250	410	31	691	975
2950	25' span		35.87	5.799		216	273	21	510	705
3100	Elevated slabs, flat plate, including finish, not									
3110	including forms or reinforcing									
3150	Regular concrete (4000 psi), 4" slab	C-8	2613	.021	S.F.	1.43	.90	.28	2.61	3.33
3200	6" slab		2585	.022		2.09	.91	.28	3.28	4.07
3250	2-1/2" thick floor fill		2685	.021		.94	.87	.27	2.08	2.74
3300	Lightweight, 110# per C.F., 2-1/2" thick floor fill		2585	.022		1.46	.91	.28	2.65	3.38
3400	Cellular concrete, 1-5/8" fill, under 5000 S.F.		2000	.028		.99	1.17	.36	2.52	3.38
3450	Over 10,000 S.F.		2200	.025		.94	1.07	.33	2.34	3.12
3500	Add per floor for 3 to 6 stories high		31800	.002			.07	.02	.09	.14
3520	For 7 to 20 stories high		21200	.003			.11	.03	.14	.22
3540	Equipment pad (3000 psi), 3' x 3' x 6" thick	C-14H	45	1.067	Ea.	47	49.50	.69	97.19	133
3550	4' x 4' x 6" thick		30	1.600		69.50	74	1.04	144.54	198
3560	5' x 5' x 8" thick		18	2.667		122	124	1.73	247.73	335
3570	6' x 6' x 8" thick		14	3.429		164	159	2.23	325.23	440
3580	8' x 8' x 10" thick		8	6		350	278	3.90	631.90	845
3590	10' x 10' x 12" thick		5	9.600		595	445	6.25	1,046.25	1,375
3800	Footings (3000 psi), spread under 1 C.Y.	C-14C	28	4	C.Y.	166	180	1.12	347.12	480
3825	1 C.Y. to 5 C.Y.		43	2.605		201	117	.73	318.73	415
3850	Over 5 C.Y.		75	1.493		185	67.50	.42	252.92	315
3900	Footings, strip (3000 psi), 18" x 9", unreinforced	C-14L	40	2.400		125	105	.79	230.79	310
3920	18" x 9", reinforced	C-14C	35	3.200		148	144	.90	292.90	400
3925	20" x 10", unreinforced	C-14L	45	2.133		122	93.50	.70	216.20	288
3930	20" x 10", reinforced	C-14C	40	2.800		140	126	.78	266.78	360
3935	24" x 12", unreinforced	C-14L	55	1.745		120	76.50	.58	197.08	258
3940	24" x 12", reinforced	C-14C	48	2.333		139	105	.65	244.65	325
3945	36" x 12", unreinforced	C-14L	70	1.371		116	60	.45	176.45	226
3950	36" x 12", reinforced	C-14C	60	1.867		133	84	.52	217.52	285
4000	Foundation mat (3000 psi), under 10 C.Y.		38.67	2.896		204	131	.81	335.81	440
4050	Over 20 C.Y.		56.40	1.986		178	89.50	.56	268.06	345
4200	Wall, free-standing (3000 psi), 8" thick, 8' high	C-14D	45.83	4.364		160	204	16.40	380.40	530
4250	14' high		27.26	7.337		190	345	27.50	562.50	800
4260	12" thick, 8' high		64.32	3.109		145	146	11.70	302.70	410
4270	14' high		40.01	4.999		154	234	18.80	406.80	570
4300	15" thick, 8' high		80.02	2.499		139	117	9.40	265.40	355
4350	12' high		51.26	3.902		139	183	14.65	336.65	465
4500	18' high		48.85	4.094		157	192	15.40	364.40	505

03 30 Cast-In-Place Concrete

03 30 53 – Miscellaneous Cast-In-Place Concrete

03 30 53.40 Concrete In Place		Crew	Daily Output	Labor-Hours	Unit	Material	2015 Bare Costs Labor	Equipment	Total	Total Incl O&P
4520	Handicap access ramp (4000 psi), railing both sides, 3' wide	C-14H	14.58	3.292	L.F.	320	153	2.14	475.14	600
4525	5' wide		12.22	3.928		330	182	2.55	514.55	665
4530	With 6" curb and rails both sides, 3' wide		8.55	5.614		330	260	3.65	593.65	790
4535	5' wide	↓	7.31	6.566	↓	335	305	4.27	644.27	870
4650	Slab on grade (3500 psi), not including finish, 4" thick	C-14E	60.75	1.449	C.Y.	124	67.50	.51	192.01	248
4700	6" thick	"	92	.957	"	119	44.50	.33	163.83	204
4701	Thickened slab edge (3500 psi), for slab on grade poured									
4702	monolithically with slab; depth is in addition to slab thickness;									
4703	formed vertical outside edge, earthen bottom and inside slope									
4705	8" deep x 8" wide bottom, unreinforced	C-14L	2190	.044	L.F.	3.47	1.92	.01	5.40	7
4710	8" x 8", reinforced	C-14C	1670	.067		5.75	3.02	.02	8.79	11.25
4715	12" deep x 12" wide bottom, unreinforced	C-14L	1800	.053		7.05	2.34	.02	9.41	11.60
4720	12" x 12", reinforced	C-14C	1310	.086		11.20	3.85	.02	15.07	18.65
4725	16" deep x 16" wide bottom, unreinforced	C-14L	1440	.067		11.85	2.92	.02	14.79	17.80
4730	16" x 16", reinforced	C-14C	1120	.100		16.80	4.51	.03	21.34	26
4735	20" deep x 20" wide bottom, unreinforced	C-14L	1150	.083		17.90	3.66	.03	21.59	25.50
4740	20" x 20", reinforced	C-14C	920	.122		24	5.50	.03	29.53	35.50
4745	24" deep x 24" wide bottom, unreinforced	C-14L	930	.103		25	4.53	.03	29.56	35.50
4750	24" x 24", reinforced	C-14C	740	.151	↓	33.50	6.80	.04	40.34	47.50
4751	Slab on grade (3500 psi), incl. troweled finish, not incl. forms									
4760	or reinforcing, over 10,000 S.F., 4" thick	C-14F	3425	.021	S.F.	1.35	.90	.01	2.26	2.94
4820	6" thick		3350	.021		1.98	.92	.01	2.91	3.66
4840	8" thick		3184	.023		2.71	.97	.01	3.69	4.54
4900	12" thick		2734	.026		4.06	1.13	.01	5.20	6.30
4950	15" thick	↓	2505	.029	↓	5.10	1.23	.01	6.34	7.55
5000	Slab on grade (3000 psi), incl. broom finish, not incl. forms									
5001	or reinforcing, 4" thick	C-14G	2873	.019	S.F.	1.32	.82	.01	2.15	2.78
5010	6" thick		2590	.022		2.07	.91	.01	2.99	3.74
5020	8" thick	↓	2320	.024	↓	2.70	1.02	.01	3.73	4.60
5200	Lift slab in place above the foundation, incl. forms, reinforcing,									
5210	concrete (4000 psi) and columns, over 20,000 S.F. per floor	C-14B	2113	.098	S.F.	6.90	4.64	.36	11.90	15.55
5250	10,000 S.F. to 20,000 S.F. per floor		1650	.126		7.55	5.95	.46	13.96	18.50
5300	Under 10,000 S.F. per floor	↓	1500	.139	↓	8.20	6.55	.50	15.25	20.50
5500	Lightweight, ready mix, including screed finish only,									
5510	not including forms or reinforcing									
5550	1:4 (2500 psi) for structural roof decks	C-14B	260	.800	C.Y.	166	37.50	2.89	206.39	248
5600	1:6 (3000 psi) for ground slab with radiant heat	C-14F	92	.783		168	33.50	.34	201.84	239
5650	1:3:2 (2000 psi) with sand aggregate, roof deck	C-14B	260	.800		164	37.50	2.89	204.39	246
5700	Ground slab (2000 psi)	C-14F	107	.673		164	29	.29	193.29	227
5900	Pile caps (3000 psi), incl. forms and reinf., sq. or rect., under 10 C.Y.	C-14C	54.14	2.069		168	93.50	.58	262.08	340
5950	Over 10 C.Y.		75	1.493		157	67.50	.42	224.92	283
6000	Triangular or hexagonal, under 10 C.Y.		53	2.113		123	95.50	.59	219.09	292
6050	Over 10 C.Y.	↓	85	1.318		138	59.50	.37	197.87	249
6200	Retaining walls (3000 psi), gravity, 4' high see Section 32 32	C-14D	66.20	3.021		140	141	11.35	292.35	400
6250	10' high		125	1.600		134	75	6	215	276
6300	Cantilever, level backfill loading, 8' high		70	2.857		150	134	10.75	294.75	395
6350	16' high	↓	91	2.198	↓	145	103	8.25	256.25	335
6800	Stairs (3500 psi), not including safety treads, free standing, 3'-6" wide	C-14H	83	.578	LF Nose	5.60	27	.38	32.98	50
6850	Cast on ground		125	.384	"	4.63	17.80	.25	22.68	34.50
7000	Stair landings, free standing		200	.240	S.F.	4.52	11.10	.16	15.78	23
7050	Cast on ground	↓	475	.101	"	3.52	4.68	.07	8.27	11.55
9000	Minimum labor/equipment charge	2 Carp	1	16	Job		750		750	1,225

03 31 Structural Concrete

03 31 13 – Heavyweight Structural Concrete

03 31 13.25 Concrete, Hand Mix

		Crew	Daily Output	Labor-Hours	Unit	Material	2015 Bare Costs Labor	Equipment	Total	Total Incl O&P
0010	**CONCRETE, HAND MIX** for small quantities or remote areas									
0050	Includes bulk local aggregate, bulk sand, bagged Portland									
0060	cement (Type I) and water, using gas powered cement mixer									
0125	2500 psi	C-30	135	.059	C.F.	3.29	2.23	1.28	6.80	8.65
0130	3000 psi		135	.059		3.52	2.23	1.28	7.03	8.95
0135	3500 psi		135	.059		3.66	2.23	1.28	7.17	9.10
0140	4000 psi		135	.059		3.82	2.23	1.28	7.33	9.25
0145	4500 psi		135	.059		4	2.23	1.28	7.51	9.45
0150	5000 psi	↓	135	.059	↓	4.26	2.23	1.28	7.77	9.75
0300	Using pre-bagged dry mix and wheelbarrow (80-lb. bag = 0.6 C.F.)									
0340	4000 psi	1 Clab	48	.167	C.F.	6.25	6.25		12.50	17.20

03 31 13.30 Concrete, Volumetric Site-Mixed

		Crew	Daily Output	Labor-Hours	Unit	Material	2015 Bare Costs Labor	Equipment	Total	Total Incl O&P
0010	**CONCRETE, VOLUMETRIC SITE-MIXED**									
0015	Mixed on-site in volumetric truck									
0020	Includes local aggregate, sand, Portland cement (Type I) and water									
0025	Excludes all additives and treatments									
0100	3000 psi, 1 C.Y. mixed and discharged				C.Y.	177			177	194
0110	2 C.Y.					138			138	151
0120	3 C.Y.					124			124	136
0130	4 C.Y.					113			113	124
0140	5 C.Y.				↓	106			106	117
0200	For truck holding/waiting time past first 2 on-site hours, add				Hr.	78			78	86
0210	For trip charge beyond first 20 miles, each way, add				Mile	3.50			3.50	3.85
0220	For each additional increase of 500 psi, add				Ea.	4.17			4.17	4.59

03 31 13.35 Heavyweight Concrete, Ready Mix

		Crew	Daily Output	Labor-Hours	Unit	Material	2015 Bare Costs Labor	Equipment	Total	Total Incl O&P
0010	**HEAVYWEIGHT CONCRETE, READY MIX**, delivered									
0012	Includes local aggregate, sand, Portland cement (Type I) and water									
0015	Excludes all additives and treatments R033105-20									
0020	2000 psi				C.Y.	97			97	107
0100	2500 psi					99.50			99.50	109
0150	3000 psi					102			102	112
0200	3500 psi					104			104	115
0300	4000 psi					107			107	118
0350	4500 psi					110			110	121
0400	5000 psi					113			113	125
0411	6000 psi					116			116	128
0412	8000 psi					123			123	135
0413	10,000 psi					129			129	142
0414	12,000 psi					135			135	149
1000	For high early strength (Portland cement Type III), add					10%				
1300	For winter concrete (hot water), add					4.05			4.05	4.46
1410	For mid-range water reducer, add					3.31			3.31	3.64
1420	For high-range water reducer/superplasticizer, add					5.65			5.65	6.20
1430	For retarder, add					3.23			3.23	3.55
1440	For non-Chloride accelerator, add					5.50			5.50	6.05
1450	For Chloride accelerator, per 1%, add					2.90			2.90	3.19
1460	For fiber reinforcing, synthetic (1 lb./C.Y.), add					6.65			6.65	7.30
1500	For Saturday delivery, add				↓	10.80			10.80	11.90
1510	For truck holding/waiting time past 1st hour per load, add				Hr.	94			94	103
1520	For short load (less than 4 C.Y.), add per load				Ea.	60.50			60.50	66.50
2000	For all lightweight aggregate, add				C.Y.	45%				

03 31 Structural Concrete
03 31 13 – Heavyweight Structural Concrete

03 31 13.70 Placing Concrete

		Crew	Daily Output	Labor-Hours	Unit	Material	2015 Bare Costs Labor	2015 Bare Costs Equipment	Total	Total Incl O&P
0010	**PLACING CONCRETE** R033105-70									
0020	Includes labor and equipment to place, level (strike off) and consolidate									
0050	Beams, elevated, small beams, pumped	C-20	60	1.067	C.Y.		43	13.05	56.05	84.50
0100	With crane and bucket	C-7	45	1.600			65.50	27	92.50	136
0200	Large beams, pumped	C-20	90	.711			28.50	8.70	37.20	56
0250	With crane and bucket	C-7	65	1.108			45.50	18.70	64.20	94
0400	Columns, square or round, 12" thick, pumped	C-20	60	1.067			43	13.05	56.05	84.50
0450	With crane and bucket	C-7	40	1.800			73.50	30.50	104	153
0600	18" thick, pumped	C-20	90	.711			28.50	8.70	37.20	56
0650	With crane and bucket	C-7	55	1.309			53.50	22	75.50	111
0800	24" thick, pumped	C-20	92	.696			28	8.50	36.50	55
0850	With crane and bucket	C-7	70	1.029			42	17.35	59.35	87
1000	36" thick, pumped	C-20	140	.457			18.45	5.60	24.05	36
1050	With crane and bucket	C-7	100	.720			29.50	12.15	41.65	61
1400	Elevated slabs, less than 6" thick, pumped	C-20	140	.457			18.45	5.60	24.05	36
1450	With crane and bucket	C-7	95	.758			31	12.80	43.80	64
1500	6" to 10" thick, pumped	C-20	160	.400			16.15	4.89	21.04	31.50
1550	With crane and bucket	C-7	110	.655			27	11.05	38.05	55.50
1600	Slabs over 10" thick, pumped	C-20	180	.356			14.35	4.35	18.70	28.50
1650	With crane and bucket	C-7	130	.554			22.50	9.35	31.85	47
1900	Footings, continuous, shallow, direct chute	C-6	120	.400			15.65	.52	16.17	26
1950	Pumped	C-20	150	.427			17.25	5.20	22.45	34
2000	With crane and bucket	C-7	90	.800			32.50	13.50	46	68
2100	Footings, continuous, deep, direct chute	C-6	140	.343			13.45	.45	13.90	22.50
2150	Pumped	C-20	160	.400			16.15	4.89	21.04	31.50
2200	With crane and bucket	C-7	110	.655			27	11.05	38.05	55.50
2400	Footings, spread, under 1 C.Y., direct chute	C-6	55	.873			34	1.13	35.13	57
2450	Pumped	C-20	65	.985			40	12.05	52.05	78
2500	With crane and bucket	C-7	45	1.600			65.50	27	92.50	136
2600	Over 5 C.Y., direct chute	C-6	120	.400			15.65	.52	16.17	26
2650	Pumped	C-20	150	.427			17.25	5.20	22.45	34
2700	With crane and bucket	C-7	100	.720			29.50	12.15	41.65	61
2900	Foundation mats, over 20 C.Y., direct chute	C-6	350	.137			5.35	.18	5.53	8.95
2950	Pumped	C-20	400	.160			6.45	1.96	8.41	12.65
3000	With crane and bucket	C-7	300	.240			9.80	4.05	13.85	20.50
3200	Grade beams, direct chute	C-6	150	.320			12.55	.42	12.97	21
3250	Pumped	C-20	180	.356			14.35	4.35	18.70	28.50
3300	With crane and bucket	C-7	120	.600			24.50	10.10	34.60	51
3500	High rise, for more than 5 stories, pumped, add per story	C-20	2100	.030			1.23	.37	1.60	2.41
3510	With crane and bucket, add per story	C-7	2100	.034			1.40	.58	1.98	2.91
3700	Pile caps, under 5 C.Y., direct chute	C-6	90	.533			21	.69	21.69	35
3750	Pumped	C-20	110	.582			23.50	7.10	30.60	46
3800	With crane and bucket	C-7	80	.900			37	15.15	52.15	76
3850	Pile cap, 5 C.Y. to 10 C.Y., direct chute	C-6	175	.274			10.75	.36	11.11	17.90
3900	Pumped	C-20	200	.320			12.95	3.91	16.86	25.50
3950	With crane and bucket	C-7	150	.480			19.65	8.10	27.75	41
4000	Over 10 C.Y., direct chute	C-6	215	.223			8.75	.29	9.04	14.55
4050	Pumped	C-20	240	.267			10.75	3.26	14.01	21
4100	With crane and bucket	C-7	185	.389			15.95	6.55	22.50	33
4300	Slab on grade, up to 6" thick, direct chute	C-6	110	.436			17.10	.57	17.67	28.50
4350	Pumped	C-20	130	.492			19.90	6	25.90	39
4400	With crane and bucket	C-7	110	.655			27	11.05	38.05	55.50

03 31 Structural Concrete

03 31 13 – Heavyweight Structural Concrete

03 31 13.70 Placing Concrete

		Crew	Daily Output	Labor-Hours	Unit	Material	2015 Bare Costs Labor	2015 Bare Costs Equipment	Total	Total Incl O&P
4600	Over 6" thick, direct chute	C-6	165	.291	C.Y.		11.40	.38	11.78	18.95
4650	Pumped	C-20	185	.346			14	4.23	18.23	27
4700	With crane and bucket	C-7	145	.497			20.50	8.35	28.85	42
4900	Walls, 8" thick, direct chute	C-6	90	.533			21	.69	21.69	35
4950	Pumped	C-20	100	.640			26	7.85	33.85	50.50
5000	With crane and bucket	C-7	80	.900			37	15.15	52.15	76
5050	12" thick, direct chute	C-6	100	.480			18.80	.62	19.42	31
5100	Pumped	C-20	110	.582			23.50	7.10	30.60	46
5200	With crane and bucket	C-7	90	.800			32.50	13.50	46	68
5300	15" thick, direct chute	C-6	105	.457			17.90	.59	18.49	29.50
5350	Pumped	C-20	120	.533			21.50	6.50	28	42
5400	With crane and bucket	C-7	95	.758			31	12.80	43.80	64
5600	Wheeled concrete dumping, add to placing costs above									
5610	Walking cart, 50' haul, add	C-18	32	.281	C.Y.		10.65	1.88	12.53	19.50
5620	150' haul, add		24	.375			14.20	2.50	16.70	26.50
5700	250' haul, add		18	.500			18.90	3.34	22.24	34.50
5800	Riding cart, 50' haul, add	C-19	80	.113			4.25	1.25	5.50	8.35
5810	150' haul, add		60	.150			5.65	1.66	7.31	11.15
5900	250' haul, add		45	.200			7.55	2.22	9.77	14.85
6000	Concrete in-fill for pan-type metal stairs and landings. Manual placement									
6010	includes up to 50' horizontal haul from point of concrete discharge.									
6100	Stair pan treads, 2" deep									
6110	Flights in 1st floor level up/down from discharge point	C-8A	3200	.015	S.F.		.61		.61	.98
6120	2nd floor level		2500	.019			.78		.78	1.25
6130	3rd floor level		2000	.024			.97		.97	1.57
6140	4th floor level		1800	.027			1.08		1.08	1.74
6200	Intermediate stair landings, pan-type 4" deep									
6210	Flights in 1st floor level up/down from discharge point	C-8A	2000	.024	S.F.		.97		.97	1.57
6220	2nd floor level		1500	.032			1.29		1.29	2.09
6230	3rd floor level		1200	.040			1.62		1.62	2.61
6240	4th floor level		1000	.048			1.94		1.94	3.14
9000	Minimum labor/equipment charge	C-6	2	24	Job		940	31	971	1,550

03 35 Concrete Finishing

03 35 13 – High-Tolerance Concrete Floor Finishing

03 35 13.30 Finishing Floors, High Tolerance

		Crew	Daily Output	Labor-Hours	Unit	Material	Labor	Equipment	Total	Total Incl O&P
0010	**FINISHING FLOORS, HIGH TOLERANCE**									
0012	Finishing of fresh concrete flatwork requires that concrete									
0013	first be placed, struck off & consolidated									
0015	Basic finishing for various unspecified flatwork									
0100	Bull float only	C-10	4000	.006	S.F.		.26		.26	.41
0125	Bull float & manual float		2000	.012			.51		.51	.82
0150	Bull float, manual float, & broom finish, w/edging & joints		1850	.013			.55		.55	.88
0200	Bull float, manual float & manual steel trowel		1265	.019			.81		.81	1.29
0210	For specified Random Access Floors in ACI Classes 1, 2, 3 and 4 to achieve									
0215	Composite Overall Floor Flatness and Levelness values up to FF35/FL25									
0250	Bull float, machine float & machine trowel (walk-behind)	C-10C	1715	.014	S.F.		.60	.03	.63	.98
0300	Power screed, bull float, machine float & trowel (walk-behind)	C-10D	2400	.010			.43	.05	.48	.73
0350	Power screed, bull float, machine float & trowel (ride-on)	C-10E	4000	.006			.26	.06	.32	.48
0352	For specified Random Access Floors in ACI Classes 5, 6, 7 and 8 to achieve									
0354	Composite Overall Floor Flatness and Levelness values up to FF50/FL50									

03 35 Concrete Finishing

03 35 13 – High-Tolerance Concrete Floor Finishing

03 35 13.30 Finishing Floors, High Tolerance		Crew	Daily Output	Labor-Hours	Unit	Material	2015 Bare Costs Labor	2015 Bare Costs Equipment	Total	Total Incl O&P
0356	Add for two-dimensional restraightening after power float	C-10	6000	.004	S.F.		.17		.17	.27
0358	For specified Random or Defined Access Floors in ACI Class 9 to achieve									
0360	Composite Overall Floor Flatness and Levelness values up to FF100/FL100									
0362	Add for two-dimensional restraightening after bull float & power float	C-10	3000	.008	S.F.		.34		.34	.54
0364	For specified Superflat Defined Access Floors in ACI Class 9 to achieve									
0366	Minimum Floor Flatness and Levelness values of FF100/FL100									
0368	Add for 2-dim'l restraightening after bull float, power float, power trowel	C-10	2000	.012	S.F.		.51		.51	.82
9100	Minimum labor/equipment charge	"	2	12	Job		510		510	815

03 35 29 – Tooled Concrete Finishing

03 35 29.60 Finishing Walls

		Crew	Daily Output	Labor-Hours	Unit	Material	Labor	Equipment	Total	Total Incl O&P
0010	**FINISHING WALLS**									
0020	Break ties and patch voids	1 Cefi	540	.015	S.F.	.04	.67		.71	1.09
0050	Burlap rub with grout	"	450	.018		.04	.80		.84	1.31
0300	Bush hammer, green concrete	B-39	1000	.048			1.91	.23	2.14	3.37
0350	Cured concrete	"	650	.074			2.94	.36	3.30	5.15
0500	Acid etch	1 Cefi	575	.014		.14	.63		.77	1.14
0700	Sandblast, light penetration	E-11	1100	.029		.46	1.23	.21	1.90	2.88
0750	Heavy penetration	"	375	.085		.92	3.61	.63	5.16	7.95
0850	Grind form fins flush	1 Clab	700	.011	L.F.		.43		.43	.70
9000	Minimum labor/equipment charge	C-10	2	12	Job		510		510	815

03 35 33 – Stamped Concrete Finishing

03 35 33.50 Slab Texture Stamping

		Crew	Daily Output	Labor-Hours	Unit	Material	Labor	Equipment	Total	Total Incl O&P
0010	**SLAB TEXTURE STAMPING**									
0050	Stamping requires that concrete first be placed, struck off, consolidated,									
0060	bull floated and free of bleed water. Decorative stamping tasks include:									
0100	Step 1 - first application of dry shake colored hardener	1 Cefi	6400	.001	S.F.	.43	.06		.49	.56
0110	Step 2 - bull float		6400	.001			.06		.06	.09
0130	Step 3 - second application of dry shake colored hardener		6400	.001		.21	.06		.27	.32
0140	Step 4 - bull float, manual float & steel trowel	3 Cefi	1280	.019			.84		.84	1.33
0150	Step 5 - application of dry shake colored release agent	1 Cefi	6400	.001		.10	.06		.16	.20
0160	Step 6 - place, tamp & remove mats	3 Cefi	2400	.010		1.43	.45		1.88	2.29
0170	Step 7 - touch up edges, mat joints & simulated grout lines	1 Cefi	1280	.006			.28		.28	.44
0300	Alternate stamping estimating method includes all tasks above	4 Cefi	800	.040		2.17	1.80		3.97	5.25
0400	Step 8 - pressure wash @ 3000 psi after 24 hours	1 Cefi	1600	.005			.23		.23	.36
0500	Step 9 - roll 2 coats cure/seal compound when dry	"	800	.010		.62	.45		1.07	1.39

03 35 43 – Polished Concrete Finishing

03 35 43.10 Polished Concrete Floors

		Crew	Daily Output	Labor-Hours	Unit	Material	Labor	Equipment	Total	Total Incl O&P
0010	**POLISHED CONCRETE FLOORS**									
0015	Processing of cured concrete to include grinding, honing,									
0020	and polishing of interior floors with 22" segmented diamond									
0025	planetary floor grinder (2 passes in different directions per grit)									
0100	Removal of pre-existing coatings, dry, with carbide discs using									
0105	dry vacuum pick-up system, final hand sweeping									
0110	Glue, adhesive or tar	J-4	1.60	15	M.S.F.	19.75	640	135	794.75	1,200
0120	Paint, epoxy, 1 coat		3.60	6.667		19.75	284	60	363.75	545
0130	2 coats		1.80	13.333		19.75	565	120	704.75	1,050
0200	Grinding and edging, wet, including wet vac pick-up and auto									
0205	scrubbing between grit changes									
0210	40-grit diamond/metal matrix	J-4A	1.60	20	M.S.F.	31.50	825	281	1,137.50	1,675
0220	80-grit diamond/metal matrix		2	16		31.50	660	225	916.50	1,325
0230	120-grit diamond/metal matrix		2.40	13.333		31.50	550	187	768.50	1,125

03 35 Concrete Finishing

03 35 43 – Polished Concrete Finishing

03 35 43.10 Polished Concrete Floors		Crew	Daily Output	Labor-Hours	Unit	Material	2015 Bare Costs Labor	Equipment	Total	Total Incl O&P
0240	200-grit diamond/metal matrix	J-4A	2.80	11.429	M.S.F.	31.50	470	161	662.50	970
0300	Spray on dye or stain (1 coat)	1 Cefi	16	.500		216	22.50		238.50	274
0400	Spray on densifier/hardener (2 coats)	"	8	1		350	45		395	455
0410	Auto scrubbing after 2nd coat, when dry	J-4B	16	.500	↓		18.80	14.60	33.40	47
0500	Honing and edging, wet, including wet vac pick-up and auto									
0505	scrubbing between grit changes									
0510	100-grit diamond/resin matrix	J-4A	2.80	11.429	M.S.F.	31.50	470	161	662.50	970
0520	200-grit diamond/resin matrix	"	2.80	11.429	"	31.50	470	161	662.50	970
0530	Dry, including dry vacuum pick-up system, final hand sweeping									
0540	400-grit diamond/resin matrix	J-4A	2.80	11.429	M.S.F.	31.50	470	161	662.50	970
0600	Polishing and edging, dry, including dry vac pick-up and hand									
0605	sweeping between grit changes									
0610	800-grit diamond/resin matrix	J-4A	2.80	11.429	M.S.F.	31.50	470	161	662.50	970
0620	1500-grit diamond/resin matrix		2.80	11.429		31.50	470	161	662.50	970
0630	3000-grit diamond/resin matrix	↓	2.80	11.429		31.50	470	161	662.50	970
0700	Auto scrubbing after final polishing step	J-4B	16	.500	↓		18.80	14.60	33.40	47

03 37 Specialty Placed Concrete

03 37 13 – Shotcrete

03 37 13.30 Gunite (Dry-Mix)

		Crew	Daily Output	Labor-Hours	Unit	Material	Labor	Equipment	Total	Total Incl O&P
0010	**GUNITE (DRY-MIX)**									
0020	Typical in place, 1" layers, no mesh included	C-16	2000	.028	S.F.	.35	1.17	.20	1.72	2.49
0300	Typical in place, including mesh, 2" thick, flat surfaces		1000	.056		1.32	2.34	.39	4.05	5.65
0350	Curved surfaces		500	.112		1.32	4.69	.79	6.80	9.85
0500	4" thick, flat surfaces		750	.075		2.02	3.13	.52	5.67	7.85
0550	Curved surfaces	↓	350	.160		2.02	6.70	1.12	9.84	14.25
0900	Prepare old walls, no scaffolding, good condition	C-10	1000	.024			1.02		1.02	1.63
0950	Poor condition	"	275	.087			3.71		3.71	5.95
1100	For high finish requirement or close tolerance, add						50%			
1150	Very high				↓		110%			
9000	Minimum labor/equipment charge	C-10	1	24	Job		1,025		1,025	1,625

03 37 13.60 Shotcrete (Wet-Mix)

		Crew	Daily Output	Labor-Hours	Unit	Material	Labor	Equipment	Total	Total Incl O&P
0010	**SHOTCRETE (WET-MIX)**									
0020	Wet mix, placed @ up to 12 C.Y. per hour, 3000 psi	C-8C	80	.600	C.Y.	113	25	5.70	143.70	170
0100	Up to 35 C.Y. per hour	C-8E	240	.200	"	101	8.20	2.21	111.41	128
1010	Fiber reinforced, 1" thick	C-8C	1740	.028	S.F.	.83	1.14	.26	2.23	3.05
1020	2" thick		900	.053		1.66	2.20	.51	4.37	5.95
1030	3" thick		825	.058		2.48	2.40	.55	5.43	7.25
1040	4" thick	↓	750	.064	↓	3.31	2.65	.61	6.57	8.60

03 39 Concrete Curing

03 39 13 – Water Concrete Curing

03 39 13.50 Water Curing

		Crew	Daily Output	Labor-Hours	Unit	Material	2015 Bare Costs Labor	2015 Bare Costs Equipment	Total	Total Incl O&P
0010	**WATER CURING**									
0015	With burlap, 4 uses assumed, 7.5 oz.	2 Clab	55	.291	C.S.F.	13.05	10.95		24	32.50
0100	10 oz.	"	55	.291	"	23.50	10.95		34.45	44
0400	Curing blankets, 1" to 2" thick, buy				S.F.	.19			.19	.21
9000	Minimum labor/equipment charge	1 Clab	5	1.600	Job		60		60	98.50

03 39 23 – Membrane Concrete Curing

03 39 23.13 Chemical Compound Membrane Concrete Curing

		Crew	Daily Output	Labor-Hours	Unit	Material	Labor	Equipment	Total	Total Incl O&P
0010	**CHEMICAL COMPOUND MEMBRANE CONCRETE CURING**									
0300	Sprayed membrane curing compound	2 Clab	95	.168	C.S.F.	11.05	6.35		17.40	22.50
0700	Curing compound, solvent based, 400 S.F./gal., 55 gallon lots				Gal.	24.50			24.50	27
0720	5 gallon lots					30.50			30.50	34
0800	Curing compound, water based, 250 S.F./gal., 55 gallon lots					22.50			22.50	25
0820	5 gallon lots					25.50			25.50	28.50

03 39 23.23 Sheet Membrane Concrete Curing

		Crew	Daily Output	Labor-Hours	Unit	Material	Labor	Equipment	Total	Total Incl O&P
0010	**SHEET MEMBRANE CONCRETE CURING**									
0200	Curing blanket, burlap/poly, 2-ply	2 Clab	70	.229	C.S.F.	23	8.60		31.60	39.50

03 41 Precast Structural Concrete

03 41 13 – Precast Concrete Hollow Core Planks

03 41 13.50 Precast Slab Planks

			Crew	Daily Output	Labor-Hours	Unit	Material	Labor	Equipment	Total	Total Incl O&P
0010	**PRECAST SLAB PLANKS**	R034105-30									
0020	Prestressed roof/floor members, grouted, solid, 4" thick		C-11	2400	.030	S.F.	5.85	1.56	.76	8.17	10
0050	6" thick			2800	.026		6.65	1.34	.65	8.64	10.40
0100	Hollow, 8" thick			3200	.023		7.10	1.17	.57	8.84	10.50
0150	10" thick			3600	.020		7.75	1.04	.50	9.29	10.90
0200	12" thick			4000	.018		7.90	.93	.45	9.28	10.80

03 41 23 – Precast Concrete Stairs

03 41 23.50 Precast Stairs

		Crew	Daily Output	Labor-Hours	Unit	Material	Labor	Equipment	Total	Total Incl O&P
0010	**PRECAST STAIRS**									
0020	Precast concrete treads on steel stringers, 3' wide	C-12	75	.640	Riser	135	30	8.70	173.70	206
0300	Front entrance, 5' wide with 48" platform, 2 risers		16	3	Flight	485	140	41	666	800
0350	5 risers		12	4		765	186	54.50	1,005.50	1,200
0500	6' wide, 2 risers		15	3.200		535	149	43.50	727.50	880
0550	5 risers		11	4.364		845	203	59.50	1,107.50	1,325
1200	Basement entrance stairwell, 6 steps, incl. steel bulkhead door	B-51	22	2.182		1,450	83.50	11.15	1,544.65	1,725
1250	14 steps	"	11	4.364		2,400	167	22.50	2,589.50	2,925

03 41 33 – Precast Structural Pretensioned Concrete

03 41 33.10 Precast Beams

			Crew	Daily Output	Labor-Hours	Unit	Material	Labor	Equipment	Total	Total Incl O&P
0010	**PRECAST BEAMS**	R034105-30									
0011	L-shaped, 20' span, 12" x 20"		C-11	32	2.250	Ea.	3,000	117	56.50	3,173.50	3,575
0060	18" x 36"			24	3		4,125	156	75.50	4,356.50	4,875
0100	24" x 44"			22	3.273		4,925	170	82.50	5,177.50	5,825
0150	30' span, 12" x 36"			24	3		5,550	156	75.50	5,781.50	6,475
0200	18" x 44"			20	3.600		6,775	187	90.50	7,052.50	7,875
0250	24" x 52"			16	4.500		8,125	234	113	8,472	9,450
0400	40' span, 12" x 52"			20	3.600		8,625	187	90.50	8,902.50	9,925
0450	18" x 52"			16	4.500		9,625	234	113	9,972	11,100
0500	24" x 52"			12	6		10,800	310	151	11,261	12,600
1200	Rectangular, 20' span, 12" x 20"			32	2.250		2,925	117	56.50	3,098.50	3,475

03 41 Precast Structural Concrete

03 41 33 – Precast Structural Pretensioned Concrete

03 41 33.10 Precast Beams

		Crew	Daily Output	Labor-Hours	Unit	Material	2015 Bare Costs Labor	Equipment	Total	Total Incl O&P
1250	18" x 36"	C-11	24	3	Ea.	3,600	156	75.50	3,831.50	4,325
1300	24" x 44"		22	3.273		4,325	170	82.50	4,577.50	5,150
1400	30' span, 12" x 36"		24	3		4,875	156	75.50	5,106.50	5,700
1450	18" x 44"		20	3.600		5,850	187	90.50	6,127.50	6,875
1500	24" x 52"		16	4.500		7,075	234	113	7,422	8,300
1600	40' span, 12" x 52"		20	3.600		7,225	187	90.50	7,502.50	8,375
1650	18" x 52"		16	4.500		8,225	234	113	8,572	9,575
1700	24" x 52"		12	6		9,425	310	151	9,886	11,100
2000	"T" shaped, 20' span, 12" x 20"		32	2.250		3,425	117	56.50	3,598.50	4,050
2050	18" x 36"		24	3		4,600	156	75.50	4,831.50	5,425
2100	24" x 44"		22	3.273		5,525	170	82.50	5,777.50	6,475
2200	30' span, 12" x 36"		24	3		6,325	156	75.50	6,556.50	7,300
2250	18" x 44"		20	3.600		7,675	187	90.50	7,952.50	8,850
2300	24" x 52"		16	4.500		9,175	234	113	9,522	10,600
2500	40' span, 12" x 52"		20	3.600		10,400	187	90.50	10,677.50	11,900
2550	18" x 52"		16	4.500		11,000	234	113	11,347	12,700
2600	24" x 52"		12	6		12,200	310	151	12,661	14,200

03 41 33.15 Precast Columns

			Crew	Daily Output	Labor-Hours	Unit	Material	Labor	Equipment	Total	Total Incl O&P
0010	**PRECAST COLUMNS**	R034105-30									
0020	Rectangular to 12' high, 16" x 16"		C-11	120	.600	L.F.	172	31	15.10	218.10	261
0050	24" x 24"		"	96	.750	"	234	39	18.90	291.90	345

03 45 Precast Architectural Concrete

03 45 13 – Faced Architectural Precast Concrete

03 45 13.50 Precast Wall Panels

		Crew	Daily Output	Labor-Hours	Unit	Material	Labor	Equipment	Total	Total Incl O&P
0010	**PRECAST WALL PANELS**									
2200	Fiberglass reinforced cement with urethane core									
2210	R20, 8' x 8', 5" plain finish	E-2	750	.075	S.F.	21	3.86	2.01	26.87	32.50
2220	Exposed aggregate or brick finish	"	600	.093	"	32	4.83	2.52	39.35	46

03 48 Precast Concrete Specialties

03 48 43 – Precast Concrete Trim

03 48 43.40 Precast Lintels

		Crew	Daily Output	Labor-Hours	Unit	Material	Labor	Equipment	Total	Total Incl O&P
0010	**PRECAST LINTELS**, smooth gray, prestressed, stock units only									
0800	4" wide, 8" high, x 4' long	D-10	28	1.143	Ea.	25	52.50	16.90	94.40	131
0850	8' long		24	1.333		60.50	61.50	19.70	141.70	187
1000	6" wide, 8" high, x 4' long		26	1.231		35.50	56.50	18.20	110.20	151
1050	10' long		22	1.455		91	67	21.50	179.50	232
1200	8" wide, 8" high, x 4' long		24	1.333		45	61.50	19.70	126.20	170
1250	12' long		20	1.600		146	73.50	23.50	243	305
1275	For custom sizes, types, colors, or finishes of precast lintels, add					150%				

03 48 43.90 Precast Window Sills

		Crew	Daily Output	Labor-Hours	Unit	Material	Labor	Equipment	Total	Total Incl O&P
0010	**PRECAST WINDOW SILLS**									
0600	Precast concrete, 4" tapers to 3", 9" wide	D-1	70	.229	L.F.	11.90	9.60		21.50	29
0650	11" wide		60	.267		15.75	11.25		27	35.50
0700	13" wide, 3 1/2" tapers to 2 1/2", 12" wall		50	.320		15.50	13.45		28.95	39

03 51 Cast Roof Decks

03 51 13 – Cementitious Wood Fiber Decks

03 51 13.50 Cementitious/Wood Fiber Planks	Crew	Daily Output	Labor-Hours	Unit	Material	2015 Bare Costs Labor	Equipment	Total	Total Incl O&P
0010 **CEMENTITIOUS/WOOD FIBER PLANKS**									
0050 Plank, beveled edge, 1" thick	2 Carp	1000	.016	S.F.	2.50	.75		3.25	3.98
0100 1-1/2" thick		975	.016		3.25	.77		4.02	4.84
0150 T & G, 2" thick		950	.017		2.75	.79		3.54	4.33
0200 2-1/2" thick		925	.017		3	.81		3.81	4.63
1000 Bulb tee, sub-purlin and grout, 6' span, add	E-1	5000	.005		2.16	.25	.03	2.44	2.83
1100 8' span	"	4200	.006		2.16	.30	.03	2.49	2.93

03 52 Lightweight Concrete Roof Insulation

03 52 16 – Lightweight Insulating Concrete

03 52 16.13 Lightweight Cellular Insulating Concrete

	Crew	Daily Output	Labor-Hours	Unit	Material	Labor	Equipment	Total	Total Incl O&P
0010 **LIGHTWEIGHT CELLULAR INSULATING CONCRETE** R035216-10									
0020 Portland cement and foaming agent G	C-8	50	1.120	C.Y.	123	47	14.40	184.40	226

03 52 16.16 Lightweight Aggregate Insulating Concrete

	Crew	Daily Output	Labor-Hours	Unit	Material	Labor	Equipment	Total	Total Incl O&P
0010 **LIGHTWEIGHT AGGREGATE INSULATING CONCRETE** R035216-10									
0100 Poured vermiculite or perlite, field mix,									
0110 1:6 field mix G	C-8	50	1.120	C.Y.	252	47	14.40	313.40	370
0200 Ready mix, 1:6 mix, roof fill, 2" thick G		10000	.006	S.F.	1.40	.23	.07	1.70	2
0250 3" thick G		7700	.007	"	2.10	.30	.09	2.49	2.90

03 53 Concrete Topping

03 53 16 – Iron-Aggregate Concrete Topping

03 53 16.50 Floor Topping

	Crew	Daily Output	Labor-Hours	Unit	Material	Labor	Equipment	Total	Total Incl O&P
0010 **FLOOR TOPPING**									
0400 Integral topping/finish, on fresh concrete, using 1:1:2 mix, 3/16" thick	C-10B	1000	.040	S.F.	.10	1.62	.27	1.99	3.03
0450 1/2" thick		950	.042		.28	1.71	.29	2.28	3.39
0500 3/4" thick		850	.047		.42	1.91	.32	2.65	3.89
0600 1" thick		750	.053		.56	2.16	.36	3.08	4.50
0800 Granolithic topping, on fresh or cured concrete, 1:1:1-1/2 mix, 1/2" thick		590	.068		.31	2.75	.46	3.52	5.30
0820 3/4" thick		580	.069		.46	2.80	.47	3.73	5.55
0850 1" thick		575	.070		.62	2.82	.47	3.91	5.75
0950 2" thick		500	.080		1.23	3.24	.55	5.02	7.20

03 54 Cast Underlayment

03 54 13 – Gypsum Cement Underlayment

03 54 13.50 Poured Gypsum Underlayment

	Crew	Daily Output	Labor-Hours	Unit	Material	Labor	Equipment	Total	Total Incl O&P
0010 **POURED GYPSUM UNDERLAYMENT**									
0400 Underlayment, gypsum based, self-leveling 2500 psi, pumped, 1/2" thick	C-8	24000	.002	S.F.	.36	.10	.03	.49	.59
0500 3/4" thick		20000	.003		.54	.12	.04	.70	.83
0600 1" thick		16000	.004		.72	.15	.05	.92	1.09
1400 Hand placed, 1/2" thick	C-18	450	.020		.36	.76	.13	1.25	1.79
1500 3/4" thick	"	300	.030		.54	1.13	.20	1.87	2.68

03 54 16 – Hydraulic Cement Underlayment

03 54 16.50 Cement Underlayment

	Crew	Daily Output	Labor-Hours	Unit	Material	Labor	Equipment	Total	Total Incl O&P
0010 **CEMENT UNDERLAYMENT**									
2510 Underlayment, P.C based, self-leveling, 4100 psi, pumped, 1/4" thick	C-8	20000	.003	S.F.	1.55	.12	.04	1.71	1.93
2520 1/2" thick		19000	.003		3.09	.12	.04	3.25	3.64

03 54 Cast Underlayment

03 54 16 – Hydraulic Cement Underlayment

03 54 16.50 Cement Underlayment		Crew	Daily Output	Labor-Hours	Unit	Material	2015 Bare Costs Labor	Equipment	Total	Total Incl O&P
2530	3/4" thick	C-8	18000	.003	S.F.	4.64	.13	.04	4.81	5.35
2540	1" thick		17000	.003		6.20	.14	.04	6.38	7.05
2550	1-1/2" thick		15000	.004		9.30	.16	.05	9.51	10.50
2560	Hand placed, 1/2" thick	C-18	450	.020		3.09	.76	.13	3.98	4.79
2610	Topping, P.C based, self-leveling, 6100 psi, pumped, 1/4" thick	C-8	20000	.003		2.26	.12	.04	2.42	2.72
2620	1/2" thick		19000	.003		4.52	.12	.04	4.68	5.20
2630	3/4" thick		18000	.003		6.80	.13	.04	6.97	7.70
2660	1" thick		17000	.003		9.05	.14	.04	9.23	10.20
2670	1-1/2" thick		15000	.004		13.55	.16	.05	13.76	15.20
2680	Hand placed, 1/2" thick	C-18	450	.020		4.52	.76	.13	5.41	6.35

03 63 Epoxy Grouting

03 63 05 – Grouting of Dowels and Fasteners

03 63 05.10 Epoxy Only		Crew	Daily Output	Labor-Hours	Unit	Material	Labor	Equipment	Total	Total Incl O&P
0010	**EPOXY ONLY**									
1500	Chemical anchoring, epoxy cartridge, excludes layout, drilling, fastener									
1530	For fastener 3/4" diam. x 6" embedment	2 Skwk	72	.222	Ea.	5.65	10.80		16.45	24
1535	1" diam. x 8" embedment		66	.242		8.45	11.80		20.25	28.50
1540	1-1/4" diam. x 10" embedment		60	.267		16.90	12.95		29.85	39.50
1545	1-3/4" diam. x 12" embedment		54	.296		28	14.40		42.40	54.50
1550	14" embedment		48	.333		34	16.20		50.20	63.50
1555	2" diam. x 12" embedment		42	.381		45	18.55		63.55	79.50
1560	18" embedment		32	.500		56.50	24.50		81	102

03 81 Concrete Cutting

03 81 13 – Flat Concrete Sawing

03 81 13.50 Concrete Floor/Slab Cutting		Crew	Daily Output	Labor-Hours	Unit	Material	Labor	Equipment	Total	Total Incl O&P
0010	**CONCRETE FLOOR/SLAB CUTTING**									
0050	Includes blade cost, layout and set-up time									
0300	Saw cut concrete slabs, plain, up to 3" deep	B-89	1060	.015	L.F.	.14	.66	.46	1.26	1.72
0320	Each additional inch of depth		3180	.005		.05	.22	.15	.42	.57
0400	Mesh reinforced, up to 3" deep		980	.016		.16	.72	.50	1.38	1.86
0420	Each additional inch of depth		2940	.005		.05	.24	.17	.46	.62
0500	Rod reinforced, up to 3" deep		800	.020		.19	.88	.61	1.68	2.28
0520	Each additional inch of depth		2400	.007		.06	.29	.20	.55	.76
0590	Minimum labor/equipment charge		2	8	Job		350	244	594	830

03 81 13.75 Concrete Saw Blades		Crew	Daily Output	Labor-Hours	Unit	Material	Labor	Equipment	Total	Total Incl O&P
0010	**CONCRETE SAW BLADES**									
3000	Blades for saw cutting, included in cutting line items									
3020	Diamond, 12" diameter				Ea.	241			241	265
3040	18" diameter					465			465	510
3080	24" diameter					770			770	845
3120	30" diameter					1,100			1,100	1,200
3160	36" diameter					1,425			1,425	1,550
3200	42" diameter					2,625			2,625	2,875

03 81 Concrete Cutting

03 81 16 – Track Mounted Concrete Wall Sawing

03 81 16.50 Concrete Wall Cutting		Crew	Daily Output	Labor-Hours	Unit	Material	2015 Bare Costs Labor	Equipment	Total	Total Incl O&P
0010	**CONCRETE WALL CUTTING**									
0750	Includes blade cost, layout and set-up time									
0800	Concrete walls, hydraulic saw, plain, per inch of depth	B-89B	250	.064	L.F.	.05	2.81	3.43	6.29	8.30
0820	Rod reinforcing, per inch of depth		150	.107	"	.06	4.68	5.70	10.44	13.85
0890	Minimum labor/equipment charge	▼	2	8	Job		350	430	780	1,025

03 82 Concrete Boring

03 82 13 – Concrete Core Drilling

03 82 13.10 Core Drilling		Crew	Daily Output	Labor-Hours	Unit	Material	2015 Bare Costs Labor	Equipment	Total	Total Incl O&P
0010	**CORE DRILLING**									
0015	Includes bit cost, layout and set-up time									
0020	Reinforced concrete slab, up to 6" thick									
0100	1" diameter core	B-89A	17	.941	Ea.	.18	40.50	6.80	47.48	73.50
0150	For each additional inch of slab thickness in same hole, add		1440	.011		.03	.48	.08	.59	.90
0200	2" diameter core		16.50	.970		.28	42	7	49.28	76
0250	For each additional inch of slab thickness in same hole, add		1080	.015		.05	.64	.11	.80	1.21
0300	3" diameter core		16	1		.38	43	7.25	50.63	79
0350	For each additional inch of slab thickness in same hole, add		720	.022		.06	.96	.16	1.18	1.81
0500	4" diameter core		15	1.067		.49	46	7.70	54.19	84
0550	For each additional inch of slab thickness in same hole, add		480	.033		.08	1.44	.24	1.76	2.70
0700	6" diameter core		14	1.143		.76	49.50	8.25	58.51	90.50
0750	For each additional inch of slab thickness in same hole, add		360	.044		.13	1.92	.32	2.37	3.62
0900	8" diameter core		13	1.231		1.05	53	8.90	62.95	97.50
0950	For each additional inch of slab thickness in same hole, add		288	.056		.17	2.40	.40	2.97	4.54
1100	10" diameter core		12	1.333		1.48	57.50	9.65	68.63	106
1150	For each additional inch of slab thickness in same hole, add		240	.067		.25	2.88	.48	3.61	5.50
1300	12" diameter core		11	1.455		1.78	62.50	10.50	74.78	116
1350	For each additional inch of slab thickness in same hole, add		206	.078		.30	3.35	.56	4.21	6.40
1500	14" diameter core		10	1.600		2.08	69	11.55	82.63	128
1550	For each additional inch of slab thickness in same hole, add		180	.089		.35	3.83	.64	4.82	7.35
1700	18" diameter core		9	1.778		2.82	76.50	12.85	92.17	142
1750	For each additional inch of slab thickness in same hole, add		144	.111		.47	4.79	.80	6.06	9.20
1754	24" diameter core		8	2		4.02	86.50	14.45	104.97	161
1756	For each additional inch of slab thickness in same hole, add	▼	120	.133		.67	5.75	.96	7.38	11.20
1760	For horizontal holes, add to above				▼		20%	20%		
1770	Prestressed hollow core plank, 8" thick									
1780	1" diameter core	B-89A	17.50	.914	Ea.	.24	39.50	6.60	46.34	72
1790	For each additional inch of plank thickness in same hole, add		3840	.004		.03	.18	.03	.24	.35
1794	2" diameter core		17.25	.928		.38	40	6.70	47.08	73.50
1796	For each additional inch of plank thickness in same hole, add		2880	.006		.05	.24	.04	.33	.48
1800	3" diameter core		17	.941		.51	40.50	6.80	47.81	74
1810	For each additional inch of plank thickness in same hole, add		1920	.008		.06	.36	.06	.48	.73
1820	4" diameter core		16.50	.970		.66	42	7	49.66	76.50
1830	For each additional inch of plank thickness in same hole, add		1280	.013		.08	.54	.09	.71	1.07
1840	6" diameter core		15.50	1.032		1.01	44.50	7.45	52.96	82
1850	For each additional inch of plank thickness in same hole, add		960	.017		.13	.72	.12	.97	1.44
1860	8" diameter core		15	1.067		1.40	46	7.70	55.10	85
1870	For each additional inch of plank thickness in same hole, add		768	.021		.17	.90	.15	1.22	1.83
1880	10" diameter core		14	1.143		1.98	49.50	8.25	59.73	92
1890	For each additional inch of plank thickness in same hole, add	▼	640	.025	▼	.25	1.08	.18	1.51	2.23

03 82 Concrete Boring

03 82 13 – Concrete Core Drilling

03 82 13.10 Core Drilling

		Crew	Daily Output	Labor-Hours	Unit	Material	2015 Bare Costs Labor	2015 Bare Costs Equipment	Total	Total Incl O&P
1900	12" diameter core	B-89A	13.50	1.185	Ea.	2.37	51	8.55	61.92	95.50
1910	For each additional inch of plank thickness in same hole, add		548	.029		.30	1.26	.21	1.77	2.61
1999	Drilling, core, minimum labor/equipment charge		2	8	Job		345	58	403	630
3000	Bits for core drilling, included in drilling line items									
3010	Diamond, premium, 1" diameter				Ea.	70.50			70.50	78
3020	2" diameter					113			113	124
3030	3" diameter					152			152	168
3040	4" diameter					197			197	217
3060	6" diameter					305			305	335
3080	8" diameter					420			420	460
3110	10" diameter					595			595	650
3120	12" diameter					710			710	780
3140	14" diameter					835			835	915
3180	18" diameter					1,125			1,125	1,250
3240	24" diameter					1,600			1,600	1,775

03 82 16 – Concrete Drilling

03 82 16.10 Concrete Impact Drilling

		Crew	Daily Output	Labor-Hours	Unit	Material	2015 Bare Costs Labor	2015 Bare Costs Equipment	Total	Total Incl O&P
0010	**CONCRETE IMPACT DRILLING**									
0020	Includes bit cost, layout and set-up time, no anchors									
0050	Up to 4" deep in concrete/brick floors/walls									
0100	Holes, 1/4" diameter	1 Carp	75	.107	Ea.	.07	5		5.07	8.25
0150	For each additional inch of depth in same hole, add		430	.019		.02	.87		.89	1.45
0200	3/8" diameter		63	.127		.06	5.95		6.01	9.80
0250	For each additional inch of depth in same hole, add		340	.024		.01	1.10		1.11	1.83
0300	1/2" diameter		50	.160		.06	7.50		7.56	12.35
0350	For each additional inch of depth in same hole, add		250	.032		.01	1.50		1.51	2.48
0400	5/8" diameter		48	.167		.09	7.85		7.94	12.95
0450	For each additional inch of depth in same hole, add		240	.033		.02	1.56		1.58	2.58
0500	3/4" diameter		45	.178		.12	8.35		8.47	13.85
0550	For each additional inch of depth in same hole, add		220	.036		.03	1.71		1.74	2.83
0600	7/8" diameter		43	.186		.16	8.75		8.91	14.45
0650	For each additional inch of depth in same hole, add		210	.038		.04	1.79		1.83	2.97
0700	1" diameter		40	.200		.17	9.40		9.57	15.60
0750	For each additional inch of depth in same hole, add		190	.042		.04	1.98		2.02	3.29
0800	1-1/4" diameter		38	.211		.26	9.90		10.16	16.50
0850	For each additional inch of depth in same hole, add		180	.044		.06	2.09		2.15	3.49
0900	1-1/2" diameter		35	.229		.39	10.75		11.14	18.05
0950	For each additional inch of depth in same hole, add		165	.048		.10	2.28		2.38	3.84
1000	For ceiling installations, add						40%			

Division Notes

	CREW	DAILY OUTPUT	LABOR-HOURS	UNIT	BARE COSTS				TOTAL INCL O&P
					MAT.	LABOR	EQUIP.	TOTAL	

Division 4 – Masonry

Estimating Tips
04 05 00 Common Work Results for Masonry

- The terms mortar and grout are often used interchangeably, and incorrectly. Mortar is used to bed masonry units, seal the entry of air and moisture, provide architectural appearance, and allow for size variations in the units. Grout is used primarily in reinforced masonry construction and is used to bond the masonry to the reinforcing steel. Common mortar types are M(2500 psi), S(1800 psi), N(750 psi), and O(350 psi), and conform to ASTM C270. Grout is either fine or coarse and conforms to ASTM C476, and in-place strengths generally exceed 2500 psi. Mortar and grout are different components of masonry construction and are placed by entirely different methods. An estimator should be aware of their unique uses and costs.
- Mortar is included in all assembled masonry line items. The mortar cost, part of the assembled masonry material cost, includes all ingredients, all labor, and all equipment required. Please see reference number R040513-10.
- Waste, specifically the loss/droppings of mortar and the breakage of brick and block, is included in all masonry assemblies in this division. A factor of 25% is added for mortar and 3% for brick and concrete masonry units.
- Scaffolding or staging is not included in any of the Division 4 costs. Refer to Subdivision 01 54 23 for scaffolding and staging costs.

04 20 00 Unit Masonry

- The most common types of unit masonry are brick and concrete masonry. The major classifications of brick are building brick (ASTM C62), facing brick (ASTM C216), glazed brick, fire brick, and pavers. Many varieties of texture and appearance can exist within these classifications, and the estimator would be wise to check local custom and availability within the project area. For repair and remodeling jobs, matching the existing brick may be the most important criteria.
- Brick and concrete block are priced by the piece and then converted into a price per square foot of wall. Openings less than two square feet are generally ignored by the estimator because any savings in units used is offset by the cutting and trimming required.
- It is often difficult and expensive to find and purchase small lots of historic brick. Costs can vary widely. Many design issues affect costs, selection of mortar mix, and repairs or replacement of masonry materials. Cleaning techniques must be reflected in the estimate.
- All masonry walls, whether interior or exterior, require bracing. The cost of bracing walls during construction should be included by the estimator, and this bracing must remain in place until permanent bracing is complete. Permanent bracing of masonry walls is accomplished by masonry itself, in the form of pilasters or abutting wall corners, or by anchoring the walls to the structural frame. Accessories in the form of anchors, anchor slots, and ties are used, but their supply and installation can be by different trades. For instance, anchor slots on spandrel beams and columns are supplied and welded in place by the steel fabricator, but the ties from the slots into the masonry are installed by the bricklayer. Regardless of the installation method, the estimator must be certain that these accessories are accounted for in pricing.

Reference Numbers

Reference numbers are shown in shaded boxes at the beginning of some major classifications. These numbers refer to related items in the Reference Section. The reference information may be an estimating procedure, an alternate pricing method, or technical information.

Note: Not all subdivisions listed here necessarily appear in this publication. ■

04 01 Maintenance of Masonry

04 01 20 – Maintenance of Unit Masonry

04 01 20.20 Pointing Masonry

		Crew	Daily Output	Labor-Hours	Unit	Material	2015 Bare Costs Labor	2015 Bare Costs Equipment	Total	Total Incl O&P
0010	**POINTING MASONRY**									
0300	Cut and repoint brick, hard mortar, running bond	1 Bric	80	.100	S.F.	.56	4.62		5.18	8.10
0320	Common bond		77	.104		.56	4.80		5.36	8.40
0360	Flemish bond		70	.114		.59	5.25		5.84	9.25
0400	English bond		65	.123		.59	5.70		6.29	9.90
0600	Soft old mortar, running bond		100	.080		.56	3.69		4.25	6.60
0620	Common bond		96	.083		.56	3.85		4.41	6.85
0640	Flemish bond		90	.089		.59	4.10		4.69	7.35
0680	English bond		82	.098		.59	4.50		5.09	8
0700	Stonework, hard mortar		140	.057	L.F.	.74	2.64		3.38	5.10
0720	Soft old mortar		160	.050	"	.74	2.31		3.05	4.58
1000	Repoint, mask and grout method, running bond		95	.084	S.F.	.74	3.89		4.63	7.10
1020	Common bond		90	.089		.74	4.10		4.84	7.50
1040	Flemish bond		86	.093		.78	4.29		5.07	7.85
1060	English bond		77	.104		.78	4.80		5.58	8.65
2000	Scrub coat, sand grout on walls, thin mix, brushed		120	.067		2.96	3.08		6.04	8.25
2020	Troweled		98	.082		4.12	3.77		7.89	10.70
9000	Minimum labor/equipment charge		3	2.667	Job		123		123	200

04 01 20.30 Pointing CMU

		Crew	Daily Output	Labor-Hours	Unit	Material	Labor	Equipment	Total	Total Incl O&P
0010	**POINTING CMU**									
0300	Cut and repoint block, hard mortar, running bond	1 Bric	190	.042	S.F.	.23	1.94		2.17	3.41
0310	Stacked bond		200	.040		.23	1.85		2.08	3.25
0600	Soft old mortar, running bond		230	.035		.23	1.61		1.84	2.86
0610	Stacked bond		245	.033		.23	1.51		1.74	2.70

04 01 20.41 Unit Masonry Stabilization

		Crew	Daily Output	Labor-Hours	Unit	Material	Labor	Equipment	Total	Total Incl O&P
0010	**UNIT MASONRY STABILIZATION**									
0100	Structural repointing method									
0110	Cut / grind mortar joint	1 Bric	240	.033	L.F.		1.54		1.54	2.50
0120	Clean and mask joint		2500	.003		.11	.15		.26	.36
0130	Epoxy paste and 1/4" FRP rod		240	.033		1.67	1.54		3.21	4.34
0132	3/8" FRP rod		160	.050		2.48	2.31		4.79	6.50
0134	1/4" CFRP rod		240	.033		14.50	1.54		16.04	18.45
0136	3/8" CFRP rod		160	.050		18.35	2.31		20.66	24
0140	Remove masking		14400	.001			.03		.03	.04
0300	Structural fabric method									
0310	Primer	1 Bric	600	.013	S.F.	.90	.62		1.52	1.99
0320	Apply filling/leveling paste		720	.011		.72	.51		1.23	1.62
0330	Epoxy, glass fiber fabric		720	.011		8.10	.51		8.61	9.75
0340	Carbon fiber fabric		720	.011		18.90	.51		19.41	22

04 01 20.50 Toothing Masonry

		Crew	Daily Output	Labor-Hours	Unit	Material	Labor	Equipment	Total	Total Incl O&P
0010	**TOOTHING MASONRY**									
0500	Brickwork, soft old mortar	1 Clab	40	.200	V.L.F.		7.50		7.50	12.35
0520	Hard mortar		30	.267			10.05		10.05	16.45
0700	Blockwork, soft old mortar		70	.114			4.30		4.30	7.05
0720	Hard mortar		50	.160			6		6	9.85
9000	Minimum labor/equipment charge		4	2	Job		75		75	123

04 01 30 – Unit Masonry Cleaning

04 01 30.20 Cleaning Masonry

		Crew	Daily Output	Labor-Hours	Unit	Material	Labor	Equipment	Total	Total Incl O&P
0010	**CLEANING MASONRY**									
0200	By chemical, brush and rinse, new work, light construction dust	D-1	1000	.016	S.F.	.06	.67		.73	1.16
0220	Medium construction dust		800	.020		.09	.84		.93	1.46

04 01 Maintenance of Masonry

04 01 30 – Unit Masonry Cleaning

04 01 30.20 Cleaning Masonry

		Crew	Daily Output	Labor-Hours	Unit	Material	2015 Bare Costs Labor	2015 Bare Costs Equipment	Total	Total Incl O&P
0240	Heavy construction dust, drips or stains	D-1	600	.027	S.F.	.11	1.12		1.23	1.96
0260	Low pressure wash and rinse, light restoration, light soil		800	.020		.13	.84		.97	1.51
0270	Average soil, biological staining		400	.040		.19	1.68		1.87	2.95
0280	Heavy soil, biological and mineral staining, paint		330	.048		.26	2.04		2.30	3.61
0300	High pressure wash and rinse, heavy restoration, light soil		600	.027		.12	1.12		1.24	1.96
0310	Average soil, biological staining		400	.040		.17	1.68		1.85	2.93
0320	Heavy soil, biological and mineral staining, paint	↓	250	.064		.23	2.69		2.92	4.63
0400	High pressure wash, water only, light soil	C-29	500	.016			.60	.14	.74	1.14
0420	Average soil, biological staining		375	.021			.80	.19	.99	1.53
0440	Heavy soil, biological and mineral staining, paint		250	.032			1.20	.28	1.48	2.28
0800	High pressure water and chemical, light soil		450	.018		.16	.67	.16	.99	1.44
0820	Average soil, biological staining		300	.027		.23	1	.23	1.46	2.16
0840	Heavy soil, biological and mineral staining, paint	↓	200	.040		.31	1.50	.35	2.16	3.20
1200	Sandblast, wet system, light soil	J-6	1750	.018		.31	.76	.13	1.20	1.72
1220	Average soil, biological staining		1100	.029		.46	1.21	.21	1.88	2.70
1240	Heavy soil, biological and mineral staining, paint		700	.046		.62	1.91	.34	2.87	4.12
1400	Dry system, light soil		2500	.013		.31	.53	.09	.93	1.30
1420	Average soil, biological staining		1750	.018		.46	.76	.13	1.35	1.89
1440	Heavy soil, biological and mineral staining, paint		1000	.032		.62	1.34	.24	2.20	3.09
1800	For walnut shells, add					.69			.69	.76
1820	For corn chips, add					.69			.69	.76
2000	Steam cleaning, light soil	A-1H	750	.011			.40	.10	.50	.77
2020	Average soil, biological staining		625	.013			.48	.12	.60	.92
2040	Heavy soil, biological and mineral staining	↓	375	.021			.80	.20	1	1.54
4000	Add for masking doors and windows	1 Clab	800	.010		.07	.38		.45	.70
4200	Add for pedestrian protection				Job				10%	10%
9000	Minimum labor/equipment charge	D-4	2	16	"		685	66.50	751.50	1,175

04 01 30.60 Brick Washing

		Crew	Daily Output	Labor-Hours	Unit	Material	Labor	Equipment	Total	Total Incl O&P
0010	**BRICK WASHING**									
0012	Acid cleanser, smooth brick surface	1 Bric	560	.014	S.F.	.04	.66		.70	1.12
0050	Rough brick		400	.020		.06	.92		.98	1.56
0060	Stone, acid wash	↓	600	.013	↓	.07	.62		.69	1.08
1000	Muriatic acid, price per gallon in 5 gallon lots				Gal.	8.70			8.70	9.55

04 05 Common Work Results for Masonry

04 05 05 – Selective Demolition for Masonry

04 05 05.10 Selective Demolition

		Crew	Daily Output	Labor-Hours	Unit	Material	Labor	Equipment	Total	Total Incl O&P
0010	**SELECTIVE DEMOLITION** R024119-10									
0200	Bond beams, 8" block with #4 bar	2 Clab	32	.500	L.F.		18.80		18.80	31
0300	Concrete block walls, unreinforced, 2" thick		1200	.013	S.F.		.50		.50	.82
0310	4" thick		1150	.014			.52		.52	.86
0320	6" thick		1100	.015			.55		.55	.90
0330	8" thick		1050	.015			.57		.57	.94
0340	10" thick		1000	.016			.60		.60	.99
0360	12" thick		950	.017			.63		.63	1.04
0380	Reinforced alternate courses, 2" thick		1130	.014			.53		.53	.87
0390	4" thick		1080	.015			.56		.56	.91
0400	6" thick		1035	.015			.58		.58	.95
0410	8" thick		990	.016			.61		.61	1
0420	10" thick		940	.017			.64		.64	1.05
0430	12" thick	↓	890	.018	↓		.68		.68	1.11

04 05 Common Work Results for Masonry

04 05 05 – Selective Demolition for Masonry

04 05 05.10 Selective Demolition		Crew	Daily Output	Labor-Hours	Unit	Material	2015 Bare Costs Labor	2015 Bare Costs Equipment	Total	Total Incl O&P
0440	Reinforced alternate courses & vertically 48" OC, 4" thick	2 Clab	900	.018	S.F.		.67		.67	1.10
0450	6" thick		850	.019			.71		.71	1.16
0460	8" thick		800	.020			.75		.75	1.23
0480	10" thick		750	.021			.80		.80	1.32
0490	12" thick		700	.023			.86		.86	1.41
1000	Chimney, 16" x 16", soft old mortar	1 Clab	55	.145	C.F.		5.45		5.45	8.95
1020	Hard mortar		40	.200			7.50		7.50	12.35
1030	16" x 20", soft old mortar		55	.145			5.45		5.45	8.95
1040	Hard mortar		40	.200			7.50		7.50	12.35
1050	16" x 24", soft old mortar		55	.145			5.45		5.45	8.95
1060	Hard mortar		40	.200			7.50		7.50	12.35
1080	20" x 20", soft old mortar		55	.145			5.45		5.45	8.95
1100	Hard mortar		40	.200			7.50		7.50	12.35
1110	20" x 24", soft old mortar		55	.145			5.45		5.45	8.95
1120	Hard mortar		40	.200			7.50		7.50	12.35
1140	20" x 32", soft old mortar		55	.145			5.45		5.45	8.95
1160	Hard mortar		40	.200			7.50		7.50	12.35
1200	48" x 48", soft old mortar		55	.145			5.45		5.45	8.95
1220	Hard mortar		40	.200			7.50		7.50	12.35
1250	Metal, high temp steel jacket, 24" diameter	E-2	130	.431	V.L.F.		22.50	11.60	34.10	52
1260	60" diameter	"	60	.933			48.50	25	73.50	112
1280	Flue lining, up to 12" x 12"	1 Clab	200	.040			1.50		1.50	2.47
1282	Up to 24" x 24"		150	.053			2.01		2.01	3.29
2000	Columns, 8" x 8", soft old mortar		48	.167			6.25		6.25	10.30
2020	Hard mortar		40	.200			7.50		7.50	12.35
2060	16" x 16", soft old mortar		16	.500			18.80		18.80	31
2100	Hard mortar		14	.571			21.50		21.50	35
2140	24" x 24", soft old mortar		8	1			37.50		37.50	61.50
2160	Hard mortar		6	1.333			50		50	82
2200	36" x 36", soft old mortar		4	2			75		75	123
2220	Hard mortar		3	2.667			100		100	164
2230	Alternate pricing method, soft old mortar		30	.267	C.F.		10.05		10.05	16.45
2240	Hard mortar		23	.348	"		13.10		13.10	21.50
3000	Copings, precast or masonry, to 8" wide									
3020	Soft old mortar	1 Clab	180	.044	L.F.		1.67		1.67	2.74
3040	Hard mortar	"	160	.050	"		1.88		1.88	3.08
3100	To 12" wide									
3120	Soft old mortar	1 Clab	160	.050	L.F.		1.88		1.88	3.08
3140	Hard mortar	"	140	.057	"		2.15		2.15	3.52
4000	Fireplace, brick, 30" x 24" opening									
4020	Soft old mortar	1 Clab	2	4	Ea.		150		150	247
4040	Hard mortar		1.25	6.400			241		241	395
4100	Stone, soft old mortar		1.50	5.333			201		201	330
4120	Hard mortar		1	8			300		300	495
5000	Veneers, brick, soft old mortar		140	.057	S.F.		2.15		2.15	3.52
5020	Hard mortar		125	.064			2.41		2.41	3.95
5050	Glass block, up to 4" thick		500	.016			.60		.60	.99
5100	Granite and marble, 2" thick		180	.044			1.67		1.67	2.74
5120	4" thick		170	.047			1.77		1.77	2.90
5140	Stone, 4" thick		180	.044			1.67		1.67	2.74
5160	8" thick		175	.046			1.72		1.72	2.82
5400	Alternate pricing method, stone, 4" thick		60	.133	C.F.		5		5	8.20
5420	8" thick		85	.094	"		3.54		3.54	5.80

04 05 Common Work Results for Masonry

04 05 05 – Selective Demolition for Masonry

04 05 05.10 Selective Demolition	Crew	Daily Output	Labor-Hours	Unit	Material	2015 Bare Costs Labor	Equipment	Total	Total Incl O&P
9000 Minimum labor/equipment charge	1 Clab	2	4	Job		150		150	247

04 05 13 – Masonry Mortaring

04 05 13.10 Cement

0010 **CEMENT**									
0100 Masonry, 70 lb. bag, T.L. lots				Bag	12.15			12.15	13.35
0150 L.T.L. lots					12.85			12.85	14.15
0200 White, 70 lb. bag, T.L. lots					15.85			15.85	17.45
0250 L.T.L. lots					16.70			16.70	18.40

04 05 13.20 Lime

0010 **LIME**									
0020 Masons, hydrated, 50 lb. bag, T.L. lots				Bag	10.10			10.10	11.10
0050 L.T.L. lots					11.10			11.10	12.20
0200 Finish, double hydrated, 50 lb. bag, T.L. lots					8.65			8.65	9.55
0250 L.T.L. lots					9.55			9.55	10.50

04 05 13.23 Surface Bonding Masonry Mortaring

0010 **SURFACE BONDING MASONRY MORTARING**									
0020 Gray or white colors, not incl. block work	1 Bric	540	.015	S.F.	.13	.68		.81	1.26

04 05 13.30 Mortar

0010 **MORTAR**									
0020 With masonry cement									
0100 Type M, 1:1:6 mix	1 Brhe	143	.056	C.F.	5.25	2.13		7.38	9.20
0200 Type N, 1:3 mix		143	.056		5.40	2.13		7.53	9.40
0300 Type O, 1:3 mix		143	.056		4.18	2.13		6.31	8.05
0400 Type PM, 1:1:6 mix, 2500 psi		143	.056		6	2.13		8.13	10.05
0500 Type S, 1/2:1:4 mix		143	.056		5.25	2.13		7.38	9.25
2000 With portland cement and lime									
2100 Type M, 1:1/4:3 mix	1 Brhe	143	.056	C.F.	8.95	2.13		11.08	13.25
2200 Type N, 1:1:6 mix, 750 psi		143	.056		7.15	2.13		9.28	11.35
2300 Type O, 1:2:9 mix (Pointing Mortar)		143	.056		8.35	2.13		10.48	12.60
2400 Type PL, 1:1/2:4 mix, 2500 psi		143	.056		6	2.13		8.13	10.05
2600 Type S, 1:1/2:4 mix, 1800 psi		143	.056		8.20	2.13		10.33	12.50
2650 Pre-mixed, type S or N					5.05			5.05	5.55
2700 Mortar for glass block	1 Brhe	143	.056		11.05	2.13		13.18	15.60
2900 Mortar for fire brick, dry mix, 10 lb. pail				Ea.	27			27	29.50

04 05 13.91 Masonry Restoration Mortaring

0010 **MASONRY RESTORATION MORTARING**									
0020 Masonry restoration mix				Lb.	.29			.29	.32
0050 White				"	.29			.29	.32

04 05 13.93 Mortar Pigments

0010 **MORTAR PIGMENTS**, 50 lb. bags (2 bags per M bricks)									
0020 Color admixture, range 2 to 10 lb. per bag of cement, light colors				Lb.	4.37			4.37	4.81
0050 Medium colors					6.40			6.40	7
0100 Dark colors					13.25			13.25	14.60

04 05 13.95 Sand

0010 **SAND**, screened and washed at pit									
0020 For mortar, per ton				Ton	19.35			19.35	21.50
0050 With 10 mile haul					33.50			33.50	37
0100 With 30 mile haul					65.50			65.50	72
0200 Screened and washed, at the pit				C.Y.	27			27	29.50
0250 With 10 mile haul					47			47	51.50

04 05 Common Work Results for Masonry

04 05 13 – Masonry Mortaring

04 05 13.95 Sand		Crew	Daily Output	Labor-Hours	Unit	Material	2015 Bare Costs Labor	Equipment	Total	Total Incl O&P
0300	With 30 mile haul				C.Y.	91			91	100

04 05 13.98 Mortar Admixtures

0010	**MORTAR ADMIXTURES**									
0020	Waterproofing admixture, per quart (1 qt. to 2 bags of masonry cement)				Qt.	3.22			3.22	3.54

04 05 16 – Masonry Grouting

04 05 16.30 Grouting

		Crew	Daily Output	Labor-Hours	Unit	Material	Labor	Equipment	Total	Total Incl O&P
0010	**GROUTING**									
0011	Bond beams & lintels, 8" deep, 6" thick, 0.15 C.F. per L.F.	D-4	1480	.022	L.F.	.66	.92	.09	1.67	2.32
0020	8" thick, 0.2 C.F. per L.F.		1400	.023		1.06	.98	.10	2.14	2.84
0050	10" thick, 0.25 C.F. per L.F.		1200	.027		1.11	1.14	.11	2.36	3.18
0060	12" thick, 0.3 C.F. per L.F.		1040	.031		1.33	1.31	.13	2.77	3.72
0200	Concrete block cores, solid, 4" thk., by hand, 0.067 C.F./S.F. of wall	D-8	1100	.036	S.F.	.30	1.56		1.86	2.87
0210	6" thick, pumped, 0.175 C.F. per S.F.	D-4	720	.044		.77	1.90	.19	2.86	4.12
0250	8" thick, pumped, 0.258 C.F. per S.F.		680	.047		1.14	2.01	.20	3.35	4.72
0300	10" thick, pumped, 0.340 C.F. per S.F.		660	.048		1.50	2.07	.20	3.77	5.20
0350	12" thick, pumped, 0.422 C.F. per S.F.		640	.050		1.87	2.14	.21	4.22	5.75
0500	Cavity walls, 2" space, pumped, 0.167 C.F./S.F. of wall		1700	.019		.74	.80	.08	1.62	2.20
0550	3" space, 0.250 C.F./S.F.		1200	.027		1.11	1.14	.11	2.36	3.18
0600	4" space, 0.333 C.F. per S.F.		1150	.028		1.47	1.19	.12	2.78	3.67
0700	6" space, 0.500 C.F. per S.F.		800	.040		2.21	1.71	.17	4.09	5.35
0800	Door frames, 3' x 7' opening, 2.5 C.F. per opening		60	.533	Opng.	11.05	23	2.22	36.27	51.50
0850	6' x 7' opening, 3.5 C.F. per opening		45	.711	"	15.45	30.50	2.97	48.92	69.50
2000	Grout, C476, for bond beams, lintels and CMU cores		350	.091	C.F.	4.42	3.91	.38	8.71	11.60
9000	Minimum labor/equipment charge	1 Bric	2	4	Job		185		185	300

04 05 19 – Masonry Anchorage and Reinforcing

04 05 19.05 Anchor Bolts

		Crew	Daily Output	Labor-Hours	Unit	Material	Labor	Equipment	Total	Total Incl O&P
0010	**ANCHOR BOLTS**									
0015	Installed in fresh grout in CMU bond beams or filled cores, no templates									
0020	Hooked, with nut and washer, 1/2" diam., 8" long	1 Bric	132	.061	Ea.	1.33	2.80		4.13	6
0030	12" long		131	.061		1.48	2.82		4.30	6.20
0040	5/8" diameter, 8" long		129	.062		2.98	2.86		5.84	7.95
0050	12" long		127	.063		3.67	2.91		6.58	8.75
0060	3/4" diameter, 8" long		127	.063		3.67	2.91		6.58	8.75
0070	12" long		125	.064		4.59	2.95		7.54	9.85

04 05 19.16 Masonry Anchors

		Crew	Daily Output	Labor-Hours	Unit	Material	Labor	Equipment	Total	Total Incl O&P
0010	**MASONRY ANCHORS**									
0020	For brick veneer, galv., corrugated, 7/8" x 7", 22 Ga.	1 Bric	10.50	.762	C	14.35	35		49.35	73
0100	24 Ga.		10.50	.762		9.85	35		44.85	68
0150	16 Ga.		10.50	.762		28	35		63	87.50
0200	Buck anchors, galv., corrugated, 16 ga., 2" bend, 8" x 2"		10.50	.762		56.50	35		91.50	119
0250	8" x 3"		10.50	.762		58.50	35		93.50	122
0300	Adjustable, rectangular, 4-1/8" wide									
0350	Anchor and tie, 3/16" wire, mill galv.									
0400	2-3/4" eye, 3-1/4" tie	1 Bric	1.05	7.619	M	450	350		800	1,075
0500	4-3/4" tie		1.05	7.619		485	350		835	1,100
0520	5-1/2" tie		1.05	7.619		525	350		875	1,150
0550	4-3/4" eye, 3-1/4" tie		1.05	7.619		495	350		845	1,125
0570	4-3/4" tie		1.05	7.619		535	350		885	1,150
0580	5-1/2" tie		1.05	7.619		585	350		935	1,225
0660	Cavity wall, Z-type, galvanized, 6" long, 1/8" diam.		10.50	.762	C	25	35		60	84.50
0670	3/16" diameter		10.50	.762		29	35		64	89

04 05 Common Work Results for Masonry

04 05 19 – Masonry Anchorage and Reinforcing

04 05 19.16 Masonry Anchors

		Crew	Daily Output	Labor-Hours	Unit	Material	2015 Bare Costs Labor	2015 Bare Costs Equipment	Total	Total Incl O&P
0680	1/4" diameter	1 Bric	10.50	.762	C	40.50	35		75.50	102
0850	8" long, 3/16" diameter		10.50	.762		25	35		60	84.50
0855	1/4" diameter		10.50	.762		49	35		84	111
1000	Rectangular type, galvanized, 1/4" diameter, 2" x 6"		10.50	.762		72.50	35		107.50	137
1050	4" x 6"		10.50	.762		87.50	35		122.50	154
1100	3/16" diameter, 2" x 6"		10.50	.762		45	35		80	107
1150	4" x 6"		10.50	.762		51.50	35		86.50	114
1200	Mesh wall tie, 1/2" mesh, hot dip galvanized									
1400	16 ga., 12" long, 3" wide	1 Bric	9	.889	C	88.50	41		129.50	164
1420	6" wide		9	.889		130	41		171	210
1440	12" wide		8.50	.941		208	43.50		251.50	300
1500	Rigid partition anchors, plain, 8" long, 1" x 1/8"		10.50	.762		235	35		270	315
1550	1" x 1/4"		10.50	.762		277	35		312	360
1580	1-1/2" x 1/8"		10.50	.762		259	35		294	340
1600	1-1/2" x 1/4"		10.50	.762		325	35		360	410
1650	2" x 1/8"		10.50	.762		305	35		340	390
1700	2" x 1/4"		10.50	.762		405	35		440	500
2000	Column flange ties, wire, galvanized									
2300	3/16" diameter, up to 3" wide	1 Bric	10.50	.762	C	87.50	35		122.50	153
2350	To 5" wide		10.50	.762		95	35		130	161
2400	To 7" wide		10.50	.762		102	35		137	170
2600	To 9" wide		10.50	.762		109	35		144	177
2650	1/4" diameter, up to 3" wide		10.50	.762		110	35		145	178
2700	To 5" wide		10.50	.762		119	35		154	188
2800	To 7" wide		10.50	.762		133	35		168	204
2850	To 9" wide		10.50	.762		144	35		179	215
2900	For hot dip galvanized, add					35%				
4000	Channel slots, 1-3/8" x 1/2" x 8"									
4100	12 ga., plain	1 Bric	10.50	.762	C	213	35		248	291
4150	16 ga., galvanized	"	10.50	.762	"	146	35		181	217
4200	Channel slot anchors									
4300	16 ga., galvanized, 1-1/4" x 3-1/2"				C	56.50			56.50	62.50
4350	1-1/4" x 5-1/2"					66			66	73
4400	1-1/4" x 7-1/2"					75			75	82.50
4500	1/8" plain, 1-1/4" x 3-1/2"					143			143	157
4550	1-1/4" x 5-1/2"					152			152	167
4600	1-1/4" x 7-1/2"					166			166	183
4700	For corrugation, add					76.50			76.50	84.50
4750	For hot dip galvanized, add					35%				
5000	Dowels									
5100	Plain, 1/4" diameter, 3" long				C	55.50			55.50	61
5150	4" long					61.50			61.50	67.50
5200	6" long					75			75	82.50
5300	3/8" diameter, 3" long					73			73	80.50
5350	4" long					91			91	100
5400	6" long					105			105	115
5500	1/2" diameter, 3" long					107			107	117
5550	4" long					127			127	139
5600	6" long					164			164	180
5700	5/8" diameter, 3" long					146			146	161
5750	4" long					180			180	198
5800	6" long					247			247	272
6000	3/4" diameter, 3" long					182			182	200

04 05 Common Work Results for Masonry

04 05 19 – Masonry Anchorage and Reinforcing

04 05 19.16 Masonry Anchors

		Crew	Daily Output	Labor-Hours	Unit	Material	2015 Bare Costs Labor	2015 Bare Costs Equipment	Total	Total Incl O&P
6100	4" long				C	231			231	254
6150	6" long					330			330	365
6300	For hot dip galvanized, add					35%				

04 05 19.26 Masonry Reinforcing Bars

		Crew	Daily Output	Labor-Hours	Unit	Material	Labor	Equipment	Total	Total Incl O&P
0010	**MASONRY REINFORCING BARS**									
0015	Steel bars A615, placed horiz., #3 & #4 bars	1 Bric	450	.018	Lb.	.48	.82		1.30	1.87
0020	#5 & #6 bars		800	.010		.48	.46		.94	1.28
0050	Placed vertical, #3 & #4 bars		350	.023		.48	1.06		1.54	2.25
0060	#5 & #6 bars		650	.012		.48	.57		1.05	1.45
0200	Joint reinforcing, regular truss, to 6" wide, mill std galvanized		30	.267	C.L.F.	22.50	12.30		34.80	44.50
0250	12" wide		20	.400		26	18.45		44.45	58.50
0400	Cavity truss with drip section, to 6" wide		30	.267		22.50	12.30		34.80	44.50
0450	12" wide		20	.400		26	18.45		44.45	58.50
0500	Joint reinforcing, ladder type, mill std galvanized									
0600	9 ga. sides, 9 ga. ties, 4" wall	1 Bric	30	.267	C.L.F.	19.85	12.30		32.15	42
0650	6" wall		30	.267		18.40	12.30		30.70	40.50
0700	8" wall		25	.320		19.70	14.75		34.45	45.50
0750	10" wall		20	.400		22	18.45		40.45	54
0800	12" wall		20	.400		23	18.45		41.45	55.50
1000	Truss type									
1100	9 ga. sides, 9 ga. ties, 4" wall	1 Bric	30	.267	C.L.F.	21.50	12.30		33.80	43.50
1150	6" wall		30	.267		22	12.30		34.30	44
1200	8" wall		25	.320		26	14.75		40.75	53
1250	10" wall		20	.400		21.50	18.45		39.95	53.50
1300	12" wall		20	.400		22.50	18.45		40.95	54.50
1500	3/16" sides, 9 ga. ties, 4" wall		30	.267		25	12.30		37.30	47.50
1550	6" wall		30	.267		31.50	12.30		43.80	54.50
1600	8" wall		25	.320		33	14.75		47.75	60
1650	10" wall		20	.400		33.50	18.45		51.95	67
1700	12" wall		20	.400		33.50	18.45		51.95	67
2000	3/16" sides, 3/16" ties, 4" wall		30	.267		35.50	12.30		47.80	59
2050	6" wall		30	.267		36.50	12.30		48.80	60
2100	8" wall		25	.320		37.50	14.75		52.25	65.50
2150	10" wall		20	.400		39.50	18.45		57.95	73.50
2200	12" wall		20	.400		41.50	18.45		59.95	76
2500	Cavity truss type, galvanized									
2600	9 ga. sides, 9 ga. ties, 4" wall	1 Bric	25	.320	C.L.F.	37	14.75		51.75	65
2650	6" wall		25	.320		35	14.75		49.75	62.50
2700	8" wall		20	.400		39	18.45		57.45	73
2750	10" wall		15	.533		41	24.50		65.50	85
2800	12" wall		15	.533		37.50	24.50		62	81.50
3000	3/16" sides, 9 ga. ties, 4" wall		25	.320		35.50	14.75		50.25	63
3050	6" wall		25	.320		32.50	14.75		47.25	60
3100	8" wall		20	.400		47	18.45		65.45	81.50
3150	10" wall		15	.533		33.50	24.50		58	77
3200	12" wall		15	.533		37.50	24.50		62	81
3500	For hot dip galvanizing, add				Ton	460			460	505

04 05 23 – Masonry Accessories

04 05 23.13 Masonry Control and Expansion Joints

		Crew	Daily Output	Labor-Hours	Unit	Material	Labor	Equipment	Total	Total Incl O&P
0010	**MASONRY CONTROL AND EXPANSION JOINTS**									
0020	Rubber, for double wythe 8" minimum wall (Brick/CMU)	1 Bric	400	.020	L.F.	2.07	.92		2.99	3.78
0025	"T" shaped		320	.025		1.23	1.15		2.38	3.23

04 05 Common Work Results for Masonry

04 05 23 – Masonry Accessories

04 05 23.13 Masonry Control and Expansion Joints		Crew	Daily Output	Labor-Hours	Unit	Material	2015 Bare Costs Labor	Equipment	Total	Total Incl O&P
0030	Cross-shaped for CMU units	1 Bric	280	.029	L.F.	1.45	1.32		2.77	3.75
0050	PVC, for double wythe 8" minimum wall (Brick/CMU)		400	.020		1.47	.92		2.39	3.12
0120	"T" shaped		320	.025		.75	1.15		1.90	2.71
0160	Cross-shaped for CMU units		280	.029		.92	1.32		2.24	3.16

04 05 23.19 Masonry Cavity Drainage, Weepholes, and Vents

		Crew	Daily Output	Labor-Hours	Unit	Material	Labor	Equipment	Total	Total Incl O&P
0010	**MASONRY CAVITY DRAINAGE, WEEPHOLES, AND VENTS**									
0020	Extruded aluminum, 4" deep, 2-3/8" x 8-1/8"	1 Bric	30	.267	Ea.	34	12.30		46.30	57.50
0050	5" x 8-1/8"		25	.320		45	14.75		59.75	73.50
0100	2-1/4" x 25"		25	.320		78	14.75		92.75	110
0150	5" x 16-1/2"		22	.364		62.50	16.80		79.30	96
0175	5" x 24"		22	.364		84	16.80		100.80	120
0200	6" x 16-1/2"		22	.364		90.50	16.80		107.30	127
0250	7-3/4" x 16-1/2"		20	.400		77	18.45		95.45	115
0400	For baked enamel finish, add					35%				
0500	For cast aluminum, painted, add					60%				
1000	Stainless steel ventilators, 6" x 6"	1 Bric	25	.320		210	14.75		224.75	256
1050	8" x 8"		24	.333		232	15.40		247.40	281
1100	12" x 12"		23	.348		266	16.05		282.05	320
1150	12" x 6"		24	.333		254	15.40		269.40	305
1200	Foundation block vent, galv., 1-1/4" thk, 8" high, 16" long, no damper		30	.267		16.20	12.30		28.50	38
1250	For damper, add					3.11			3.11	3.42

04 05 23.95 Wall Plugs

0010	**WALL PLUGS** (for nailing to brickwork)									
0020	25 ga., galvanized, plain	1 Bric	10.50	.762	C	26.50	35		61.50	86
0050	Wood filled	"	10.50	.762	"	68	35		103	132

04 21 Clay Unit Masonry

04 21 13 – Brick Masonry

04 21 13.13 Brick Veneer Masonry

			Crew	Daily Output	Labor-Hours	Unit	Material	Labor	Equipment	Total	Total Incl O&P
0010	**BRICK VENEER MASONRY**, T.L. lots, excl. scaff., grout & reinforcing	R042110-20									
0015	Material costs incl. 3% brick and 25% mortar waste										
2000	Standard, sel. common, 4" x 2-2/3" x 8", (6.75/S.F.)	R042110-50	D-8	230	.174	S.F.	4.16	7.45		11.61	16.75
2020	Red, 4" x 2-2/3" x 8", running bond			220	.182		3.72	7.80		11.52	16.80
2050	Full header every 6th course (7.88/S.F.)			185	.216		4.34	9.30		13.64	19.85
2100	English, full header every 2nd course (10.13/S.F.)			140	.286		5.55	12.25		17.80	26
2150	Flemish, alternate header every course (9.00/S.F.)			150	.267		4.94	11.45		16.39	24
2200	Flemish, alt. header every 6th course (7.13/S.F.)			205	.195		3.93	8.35		12.28	17.90
2250	Full headers throughout (13.50/S.F.)			105	.381		7.40	16.35		23.75	34.50
2300	Rowlock course (13.50/S.F.)			100	.400		7.40	17.15		24.55	36
2350	Rowlock stretcher (4.50/S.F.)			310	.129		2.51	5.55		8.06	11.75
2400	Soldier course (6.75/S.F.)			200	.200		3.72	8.60		12.32	18.05
2450	Sailor course (4.50/S.F.)			290	.138		2.51	5.90		8.41	12.40
2600	Buff or gray face, running bond, (6.75/S.F.)			220	.182		3.93	7.80		11.73	17
2700	Glazed face brick, running bond			210	.190		11.75	8.15		19.90	26.50
2750	Full header every 6th course (7.88/S.F.)			170	.235		13.70	10.10		23.80	31.50
3000	Jumbo, 6" x 4" x 12" running bond (3.00/S.F.)			435	.092		4.67	3.95		8.62	11.55
3050	Norman, 4" x 2-2/3" x 12" running bond, (4.5/S.F.)			320	.125		5.95	5.35		11.30	15.25
3100	Norwegian, 4" x 3-1/5" x 12" (3.75/S.F.)			375	.107		5.15	4.58		9.73	13.10
3150	Economy, 4" x 4" x 8" (4.50/S.F.)			310	.129		4.08	5.55		9.63	13.50
3200	Engineer, 4" x 3-1/5" x 8" (5.63/S.F.)			260	.154		3.61	6.60		10.21	14.70

04 21 Clay Unit Masonry

04 21 13 – Brick Masonry

04 21 13.13 Brick Veneer Masonry

		Crew	Daily Output	Labor-Hours	Unit	Material	2015 Bare Costs Labor	Equipment	Total	Total Incl O&P
3250	Roman, 4" x 2" x 12" (6.00/S.F.)	D-8	250	.160	S.F.	6.95	6.85		13.80	18.75
3300	SCR, 6" x 2-2/3" x 12" (4.50/S.F.)		310	.129		6	5.55		11.55	15.60
3350	Utility, 4" x 4" x 12" (3.00/S.F.)	↓	360	.111	↓	4.65	4.77		9.42	12.85
3360	For less than truck load lots, add				M	15%				
3400	For cavity wall construction, add						15%			
3450	For stacked bond, add						10%			
3500	For interior veneer construction, add						15%			
3550	For curved walls, add						30%			
9000	Minimum labor/equipment charge	D-1	2	8	Job		335		335	550

04 21 13.14 Thin Brick Veneer

		Crew	Daily Output	Labor-Hours	Unit	Material	Labor	Equipment	Total	Total Incl O&P
0010	**THIN BRICK VENEER**									
0015	Material costs incl. 3% brick and 25% mortar waste									
0020	On & incl. metal panel support sys, modular, 2-2/3" x 5/8" x 8", red	D-7	92	.174	S.F.	9	6.65		15.65	20.50
0100	Closure, 4" x 5/8" x 8"		110	.145		8.80	5.55		14.35	18.40
0110	Norman, 2-2/3" x 5/8" x 12"		110	.145		8.85	5.55		14.40	18.45
0120	Utility, 4" x 5/8" x 12"		125	.128		8.55	4.89		13.44	17.10
0130	Emperor, 4" x 3/4" x 16"		175	.091		9.65	3.49		13.14	16.15
0140	Super emperor, 8" x 3/4" x 16"	↓	195	.082		9.95	3.13		13.08	15.90
0150	For L shaped corners with 4" return, add				L.F.	9.25			9.25	10.20
0200	On masonry/plaster back-up, modular, 2-2/3" x 5/8" x 8", red	D-7	137	.117	S.F.	4.21	4.46		8.67	11.70
0210	Closure, 4" x 5/8" x 8"		165	.097		3.98	3.70		7.68	10.25
0220	Norman, 2-2/3" x 5/8" x 12"		165	.097		4.02	3.70		7.72	10.25
0230	Utility, 4" x 5/8" x 12"		185	.086		3.75	3.30		7.05	9.35
0240	Emperor, 4" x 3/4" x 16"		260	.062		4.85	2.35		7.20	9.05
0250	Super emperor, 8" x 3/4" x 16"	↓	285	.056	↓	5.15	2.14		7.29	9.05
0260	For L shaped corners with 4" return, add				L.F.	9.25			9.25	10.20
0270	For embedment into pre-cast concrete panels, add				S.F.	14.40			14.40	15.85

04 21 13.15 Chimney

		Crew	Daily Output	Labor-Hours	Unit	Material	Labor	Equipment	Total	Total Incl O&P
0010	**CHIMNEY**, excludes foundation, scaffolding, grout and reinforcing									
0100	Brick, 16" x 16", 8" flue	D-1	18.20	.879	V.L.F.	24	37		61	86.50
0150	16" x 20" with one 8" x 12" flue		16	1		37.50	42		79.50	110
0200	16" x 24" with two 8" x 8" flues		14	1.143		55	48		103	139
0250	20" x 20" with one 12" x 12" flue		13.70	1.168		45.50	49		94.50	130
0300	20" x 24" with two 8" x 12" flues		12	1.333		62.50	56		118.50	160
0350	20" x 32" with two 12" x 12" flues	↓	10	1.600	↓	80	67.50		147.50	198

04 21 13.18 Columns

		Crew	Daily Output	Labor-Hours	Unit	Material	Labor	Equipment	Total	Total Incl O&P
0010	**COLUMNS**, solid, excludes scaffolding, grout and reinforcing									
0050	Brick, 8" x 8", 9 brick per V.L.F.	D-1	56	.286	V.L.F.	4.76	12.05		16.81	25
0100	12" x 8", 13.5 brick per V.L.F.		37	.432		7.15	18.20		25.35	37.50
0200	12" x 12", 20 brick per V.L.F.		25	.640		10.60	27		37.60	55.50
0300	16" x 12", 27 brick per V.L.F.		19	.842		14.30	35.50		49.80	73
0400	16" x 16", 36 brick per V.L.F.		14	1.143		19.05	48		67.05	99.50
0500	20" x 16", 45 brick per V.L.F.		11	1.455		24	61		85	126
0600	20" x 20", 56 brick per V.L.F.		9	1.778		29.50	75		104.50	155
0700	24" x 20", 68 brick per V.L.F.		7	2.286		36	96		132	197
0800	24" x 24", 81 brick per V.L.F.		6	2.667		43	112		155	230
1000	36" x 36", 182 brick per V.L.F.	↓	3	5.333	↓	96.50	225		321.50	470
9000	Minimum labor/equipment charge		2	8	Job		335		335	550

04 21 13.30 Oversized Brick

		Crew	Daily Output	Labor-Hours	Unit	Material	Labor	Equipment	Total	Total Incl O&P
0010	**OVERSIZED BRICK**, excludes scaffolding, grout and reinforcing									
0100	Veneer, 4" x 2.25" x 16"	D-8	387	.103	S.F.	5.10	4.44		9.54	12.85
0102	8" x 2.25" x 16", multicell	↓	265	.151	↓	16.10	6.50		22.60	28.50

04 21 Clay Unit Masonry

04 21 13 – Brick Masonry

04 21 13.30 Oversized Brick

		Crew	Daily Output	Labor-Hours	Unit	Material	2015 Bare Costs Labor	Equipment	Total	Total Incl O&P
0105	4" x 2.75" x 16"	D-8	412	.097	S.F.	5.30	4.17		9.47	12.65
0107	8" x 2.75" x 16", multicell		295	.136		16.10	5.80		21.90	27
0110	4" x 4" x 16"		460	.087		3.48	3.73		7.21	9.90
0120	4" x 8" x 16"		533	.075		4.08	3.22		7.30	9.75
0122	4" x 8" x 16" multi cell		327	.122		15.10	5.25		20.35	25
0125	Loadbearing, 6" x 4" x 16", grouted and reinforced		387	.103		10.30	4.44		14.74	18.55
0130	8" x 4" x 16", grouted and reinforced		327	.122		11.30	5.25		16.55	21
0132	10" x 4" x 16", grouted and reinforced		327	.122		23	5.25		28.25	33.50
0135	6" x 8" x 16", grouted and reinforced		440	.091		13.40	3.90		17.30	21
0140	8" x 8" x 16", grouted and reinforced		400	.100		14.30	4.29		18.59	22.50
0145	Curtainwall/reinforced veneer, 6" x 4" x 16"		387	.103		14.65	4.44		19.09	23.50
0150	8" x 4" x 16"		327	.122		17.80	5.25		23.05	28
0152	10" x 4" x 16"		327	.122		25	5.25		30.25	36
0155	6" x 8" x 16"		440	.091		18.20	3.90		22.10	26.50
0160	8" x 8" x 16"		400	.100		25.50	4.29		29.79	35
0200	For 1 to 3 slots in face, add					15%				
0210	For 4 to 7 slots in face, add					25%				
0220	For bond beams, add					20%				
0230	For bullnose shapes, add					20%				
0240	For open end knockout, add					10%				
0250	For white or gray color group, add					10%				
0260	For 135 degree corner, add					250%				

04 21 13.35 Common Building Brick

0010	**COMMON BUILDING BRICK**, C62, TL lots, material only R042110-20									
0020	Standard				M	490			490	540
0050	Select				"	500			500	550

04 21 13.40 Structural Brick

0010	**STRUCTURAL BRICK** C652, Grade SW, incl. mortar, scaffolding not incl.									
0100	Standard unit, 4-5/8" x 2-3/4" x 9-5/8"	D-8	245	.163	S.F.	4.08	7		11.08	15.90
0120	Bond beam		225	.178		4.08	7.65		11.73	16.90
0140	V cut bond beam		225	.178		4.08	7.65		11.73	16.90
0160	Stretcher quoin, 5-5/8" x 2-3/4" x 9-5/8"		245	.163		7.55	7		14.55	19.70
0180	Corner quoin		245	.163		7.55	7		14.55	19.70
0200	Corner, 45 deg, 4-5/8" x 2-3/4" x 10-7/16"		235	.170		7.55	7.30		14.85	20

04 21 13.45 Face Brick

0010	**FACE BRICK** Material Only, C216, TL lots R042110-20									
0300	Standard modular, 4" x 2-2/3" x 8"				M	435			435	480
0450	Economy, 4" x 4" x 8"					775			775	855
0510	Economy, 4" x 4" x 12"					1,225			1,225	1,350
0550	Jumbo, 6" x 4" x 12"					1,400			1,400	1,550
0610	Jumbo, 8" x 4" x 12"					1,400			1,400	1,550
0650	Norwegian, 4" x 3-1/5" x 12"					1,225			1,225	1,350
0710	Norwegian, 6" x 3-1/5" x 12"					1,525			1,525	1,675
0850	Standard glazed, plain colors, 4" x 2-2/3" x 8"					1,600			1,600	1,750
1000	Deep trim shades, 4" x 2-2/3" x 8"					2,025			2,025	2,225
1080	Jumbo utility, 4" x 4" x 12"					1,400			1,400	1,525
1120	4" x 8" x 8"					1,775			1,775	1,950
1140	4" x 8" x 16"					5,225			5,225	5,750
1260	Engineer, 4" x 3-1/5" x 8"					525			525	575
1350	King, 4" x 2-3/4" x 10"					485			485	530
1770	Standard modular, double glazed, 4" x 2-2/3" x 8"					2,400			2,400	2,650
1850	Jumbo, colored glazed ceramic, 6" x 4" x 12"					2,525			2,525	2,775

For customer support on your Commercial Renovation Cost Data, call 877.791.4977.

04 21 Clay Unit Masonry

04 21 13 – Brick Masonry

04 21 13.45 Face Brick

		Crew	Daily Output	Labor-Hours	Unit	Material	2015 Bare Costs Labor	Equipment	Total	Total Incl O&P
2050	Jumbo utility, glazed, 4" x 4" x 12"				M	4,750			4,750	5,225
2100	4" x 8" x 8"					5,600			5,600	6,150
2150	4" x 16" x 8"					6,550			6,550	7,200
2170	For less than truck load lots, add					15			15	16.50
2180	For buff or gray brick, add					16			16	17.60
3050	Used brick					405			405	445
3150	Add for brick to match existing work, minimum					5%				
3200	Maximum					50%				

04 21 26 – Glazed Structural Clay Tile Masonry

04 21 26.10 Structural Facing Tile

		Crew	Daily Output	Labor-Hours	Unit	Material	Labor	Equipment	Total	Total Incl O&P
0010	**STRUCTURAL FACING TILE**, std. colors, excl. scaffolding, grout, reinforcing									
0020	6T series, 5-1/3" x 12", 2.3 pieces per S.F., glazed 1 side, 2" thick	D-8	225	.178	S.F.	8.95	7.65		16.60	22.50
0100	4" thick		220	.182		12.20	7.80		20	26
0150	Glazed 2 sides		195	.205		16	8.80		24.80	32
0250	6" thick		210	.190		18.20	8.15		26.35	33.50
0300	Glazed 2 sides		185	.216		21.50	9.30		30.80	39
0400	8" thick		180	.222		24	9.55		33.55	42
0500	Special shapes, group 1		400	.100	Ea.	7.70	4.29		11.99	15.45
0550	Group 2		375	.107		12.50	4.58		17.08	21
0600	Group 3		350	.114		16.05	4.90		20.95	25.50
0650	Group 4		325	.123		33.50	5.30		38.80	45
0700	Group 5		300	.133		39.50	5.70		45.20	53
0750	Group 6		275	.145		54	6.25		60.25	69.50
1000	Fire rated, 4" thick, 1 hr. rating		210	.190	S.F.	17.45	8.15		25.60	32.50
1050	2 hr. rating		170	.235		27	10.10		37.10	46.50
1300	Acoustic, 4" thick		210	.190		34	8.15		42.15	51
1310	6" thick		170	.235		35	10.10		45.10	55
1320	8" thick		145	.276		37	11.85		48.85	60.50
1350	8T acoustic, 4" thick		265	.151		16.95	6.50		23.45	29
1360	6" thick		256	.156		25	6.70		31.70	38.50
1370	8" thick		208	.192		26.50	8.25		34.75	43
2000	8W series, 8" x 16", 1.125 pieces per S.F.									
2050	2" thick, glazed 1 side	D-8	360	.111	S.F.	10.45	4.77		15.22	19.25
2100	4" thick, glazed 1 side		345	.116		14.35	4.98		19.33	24
2150	Glazed 2 sides		325	.123		17.15	5.30		22.45	27.50
2200	6" thick, glazed 1 side		330	.121		24.50	5.20		29.70	35.50
2210	Glazed 2 sides		250	.160		29.50	6.85		36.35	43.50
2250	8" thick, glazed 1 side		310	.129		25	5.55		30.55	36
2260	Glazed 2 sides		220	.182		20.50	7.80		28.30	35
2500	Special shapes, group 1		300	.133	Ea.	14.85	5.70		20.55	25.50
2550	Group 2		280	.143		19.85	6.15		26	32
2600	Group 3		260	.154		21	6.60		27.60	34
2650	Group 4		250	.160		43.50	6.85		50.35	59
2700	Group 5		240	.167		39.50	7.15		46.65	55
2750	Group 6		230	.174		85.50	7.45		92.95	106
2900	Fire rated, 2" thick, glazed 1 side		360	.111	S.F.	14.50	4.77		19.27	23.50
3000	4" thick, glazed 1 side		345	.116		13.95	4.98		18.93	23.50
3020	Gazed 2 sides		345	.116		20.50	4.98		25.48	30.50
3040	6" thick, glazed 1 side		330	.121		21.50	5.20		26.70	32
3045	Glazed 2 sides		250	.160		21.50	6.85		28.35	34.50
3047	8" thick, glazed 1 side		310	.129		35	5.55		40.55	47.50
3048	Glazed 2 sides		220	.182		37	7.80		44.80	53

04 21 Clay Unit Masonry

04 21 26 – Glazed Structural Clay Tile Masonry

04 21 26.10 Structural Facing Tile	Crew	Daily Output	Labor-Hours	Unit	Material	2015 Bare Costs Labor	Equipment	Total	Total Incl O&P
3050	Special shapes, group 1, fire rated	D-8	300	.133	S.F.	17.65	5.70		23.35	28.50
3060	Group 2		280	.143		23.50	6.15		29.65	36
3070	Group 3		260	.154		25	6.60		31.60	38.50
3080	Group 4		250	.160		48.50	6.85		55.35	64
3090	Group 5		240	.167		57	7.15		64.15	74.50
3100	Acoustic, 4" thick		345	.116		18.55	4.98		23.53	28.50
3120	4W series, 8" x 8", 2.25 pieces per S.F.									
3125	2" thick, glazed 1 side	D-8	360	.111	S.F.	9.60	4.77		14.37	18.30
3130	4" thick, glazed 1 side		345	.116		11.30	4.98		16.28	20.50
3135	Glazed 2 sides		325	.123		15.55	5.30		20.85	25.50
3140	6" thick, glazed 1 side		330	.121		16.05	5.20		21.25	26
3145	Glazed 2 sides		250	.160		19.35	6.85		26.20	32.50
3150	8" thick, glazed 1 side		310	.129		23.50	5.55		29.05	34.50
3155	Special shapes, group I		300	.133	Ea.	7.50	5.70		13.20	17.55
3160	Group II		280	.143	"	8.65	6.15		14.80	19.45
3200	For designer colors, add					25%				
3300	For epoxy mortar joints, add				S.F.	1.74			1.74	1.91
9000	Minimum labor/equipment charge	D-1	2	8	Job		335		335	550

04 21 29 – Terra Cotta Masonry

04 21 29.10 Terra Cotta Masonry Components

		Crew	Daily Output	Labor-Hours	Unit	Material	Labor	Equipment	Total	Total Incl O&P
0010	**TERRA COTTA MASONRY COMPONENTS**									
0020	Coping, split type, not glazed, 9" wide	D-1	90	.178	L.F.	13.15	7.50		20.65	26.50
0100	13" wide		80	.200		18.35	8.40		26.75	33.50
0200	Coping, split type, glazed, 9" wide		90	.178		11.95	7.50		19.45	25.50
0250	13" wide		80	.200		15.75	8.40		24.15	31
0500	Partition or back-up blocks, scored, in C.L. lots									
0700	Non-load bearing 12" x 12", 3" thick, special order	D-8	550	.073	S.F.	16.85	3.12		19.97	23.50
0750	4" thick, standard		500	.080		5.35	3.43		8.78	11.45
0800	6" thick		450	.089		7.20	3.81		11.01	14.10
0850	8" thick		400	.100		9.05	4.29		13.34	16.95
1000	Load bearing, 12" x 12", 4" thick, in walls		500	.080		5.45	3.43		8.88	11.60
1050	In floors		750	.053		5.45	2.29		7.74	9.70
1200	6" thick, in walls		450	.089		7.35	3.81		11.16	14.30
1250	In floors		675	.059		7.35	2.54		9.89	12.25
1400	8" thick, in walls		400	.100		9.15	4.29		13.44	17.05
1450	In floors		575	.070		9.15	2.99		12.14	14.90
1600	10" thick, in walls, special order		350	.114		24.50	4.90		29.40	35
1650	In floors, special order		500	.080		24.50	3.43		27.93	32.50
1800	12" thick, in walls, special order		300	.133		23	5.70		28.70	35
1850	In floors, special order		450	.089		23	3.81		26.81	31.50
2000	For reinforcing with steel rods, add to above					15%	5%			
2025	Placed vertical, #3 & #4 bars	1 Bric	350	.023	Lb.	.48	1.06		1.54	2.25
2100	For smooth tile instead of scored, add				S.F.	3.66			3.66	4.03
2200	For L.C.L. quantities, add				"	10%	10%			
9000	Minimum labor/equipment charge	D-1	2	8	Job		335		335	550

04 21 29.20 Terra Cotta Tile

		Crew	Daily Output	Labor-Hours	Unit	Material	Labor	Equipment	Total	Total Incl O&P
0010	**TERRA COTTA TILE**, on walls, dry set, 1/2" thick									
0100	Square, hexagonal or lattice shapes, unglazed	1 Tilf	135	.059	S.F.	4.62	2.54		7.16	9.10
0300	Glazed, plain colors		130	.062		7.10	2.63		9.73	11.95
0400	Intense colors		125	.064		8.30	2.74		11.04	13.50

04 22 Concrete Unit Masonry

04 22 10 – Concrete Masonry Units

04 22 10.11 Autoclave Aerated Concrete Block

			Crew	Daily Output	Labor-Hours	Unit	Material	2015 Bare Costs Labor	Equipment	Total	Total Incl O&P
0010	**AUTOCLAVE AERATED CONCRETE BLOCK**, excl. scaffolding, grout & reinforcing										
0050	Solid, 4" x 8" x 24", incl. mortar	G	D-8	600	.067	S.F.	1.48	2.86		4.34	6.30
0060	6" x 8" x 24"	G		600	.067		2.24	2.86		5.10	7.10
0070	8" x 8" x 24"	G		575	.070		2.98	2.99		5.97	8.15
0080	10" x 8" x 24"	G		575	.070		3.64	2.99		6.63	8.85
0090	12" x 8" x 24"	G		550	.073		4.47	3.12		7.59	10

04 22 10.14 Concrete Block, Back-Up

		Crew	Daily Output	Labor-Hours	Unit	Material	Labor	Equipment	Total	Total Incl O&P
0010	**CONCRETE BLOCK, BACK-UP**, C90, 2000 psi									
0020	Normal weight, 8" x 16" units, tooled joint 1 side									
0050	Not-reinforced, 2000 psi, 2" thick	D-8	475	.084	S.F.	1.50	3.61		5.11	7.55
0200	4" thick		460	.087		1.79	3.73		5.52	8
0300	6" thick		440	.091		2.31	3.90		6.21	8.90
0350	8" thick		400	.100		2.84	4.29		7.13	10.10
0400	10" thick		330	.121		2.96	5.20		8.16	11.70
0450	12" thick	D-9	310	.155		4.05	6.50		10.55	15.05
1000	Reinforced, alternate courses, 4" thick	D-8	450	.089		1.95	3.81		5.76	8.35
1100	6" thick		430	.093		2.48	3.99		6.47	9.25
1150	8" thick		395	.101		3.03	4.35		7.38	10.40
1200	10" thick		320	.125		3.12	5.35		8.47	12.20
1250	12" thick	D-9	300	.160		4.21	6.75		10.96	15.60
9000	Minimum labor/equipment charge	D-1	2	8	Job		335		335	550

04 22 10.16 Concrete Block, Bond Beam

		Crew	Daily Output	Labor-Hours	Unit	Material	Labor	Equipment	Total	Total Incl O&P
0010	**CONCRETE BLOCK, BOND BEAM**, C90, 2000 psi									
0020	Not including grout or reinforcing									
0125	Regular block, 6" thick	D-8	584	.068	L.F.	2.65	2.94		5.59	7.70
0130	8" high, 8" thick	"	565	.071		2.71	3.04		5.75	7.90
0150	12" thick	D-9	510	.094		4.02	3.96		7.98	10.85
0525	Lightweight, 6" thick	D-8	592	.068		2.73	2.90		5.63	7.70
0530	8" high, 8" thick	"	575	.070		3.29	2.99		6.28	8.50
0550	12" thick	D-9	520	.092		4.43	3.89		8.32	11.15
2000	Including grout and 2 #5 bars									
2100	Regular block, 8" high, 8" thick	D-8	300	.133	L.F.	4.78	5.70		10.48	14.55
2150	12" thick	D-9	250	.192		6.65	8.10		14.75	20.50
2500	Lightweight, 8" high, 8" thick	D-8	305	.131		5.35	5.65		11	15.05
2550	12" thick	D-9	255	.188		7.05	7.90		14.95	20.50
9000	Minimum labor/equipment charge	D-1	2	8	Job		335		335	550

04 22 10.19 Concrete Block, Insulation Inserts

		Crew	Daily Output	Labor-Hours	Unit	Material	Labor	Equipment	Total	Total Incl O&P
0010	**CONCRETE BLOCK, INSULATION INSERTS**									
0100	Styrofoam, plant installed, add to block prices									
0200	8" x 16" units, 6" thick				S.F.	1.20			1.20	1.32
0250	8" thick					1.35			1.35	1.49
0300	10" thick					1.40			1.40	1.54
0350	12" thick					1.55			1.55	1.71
0500	8" x 8" units, 8" thick					1.20			1.20	1.32
0550	12" thick					1.40			1.40	1.54

04 22 10.23 Concrete Block, Decorative

		Crew	Daily Output	Labor-Hours	Unit	Material	Labor	Equipment	Total	Total Incl O&P
0010	**CONCRETE BLOCK, DECORATIVE**, C90, 2000 psi									
0020	Embossed, simulated brick face									
0100	8" x 16" units, 4" thick	D-8	400	.100	S.F.	2.70	4.29		6.99	9.95
0200	8" thick		340	.118		2.95	5.05		8	11.45
0250	12" thick		300	.133		4.98	5.70		10.68	14.75
0400	Embossed both sides									

04 22 Concrete Unit Masonry

04 22 10 – Concrete Masonry Units

04 22 10.23 Concrete Block, Decorative		Crew	Daily Output	Labor-Hours	Unit	Material	2015 Bare Costs Labor	Equipment	Total	Total Incl O&P
0500	8" thick	D-8	300	.133	S.F.	3.97	5.70		9.67	13.65
0550	12" thick	"	275	.145	"	5.20	6.25		11.45	15.90
1000	Fluted high strength									
1100	8" x 16" x 4" thick, flutes 1 side,	D-8	345	.116	S.F.	3.83	4.98		8.81	12.30
1150	Flutes 2 sides		335	.119		4.40	5.10		9.50	13.20
1200	8" thick		300	.133		5.15	5.70		10.85	15
1250	For special colors, add					.62			.62	.68
1400	Deep grooved, smooth face									
1450	8" x 16" x 4" thick	D-8	345	.116	S.F.	2.46	4.98		7.44	10.80
1500	8" thick	"	300	.133	"	3.89	5.70		9.59	13.60
2000	Formblock, incl. inserts & reinforcing									
2100	8" x 16" x 8" thick	D-8	345	.116	S.F.	3.44	4.98		8.42	11.90
2150	12" thick	"	310	.129	"	4.55	5.55		10.10	14
2500	Ground face									
2600	8" x 16" x 4" thick	D-8	345	.116	S.F.	3.58	4.98		8.56	12.05
2650	6" thick		325	.123		3.98	5.30		9.28	13
2700	8" thick		300	.133		4.39	5.70		10.09	14.10
2725	10" thick		280	.143		5.10	6.15		11.25	15.55
2750	12" thick	D-9	265	.181		5.60	7.65		13.25	18.55
2900	For special colors, add, minimum					15%				
2950	For special colors, add, maximum					45%				
4000	Slump block									
4100	4" face height x 16" x 4" thick	D-1	165	.097	S.F.	3.89	4.08		7.97	10.95
4150	6" thick		160	.100		5.65	4.21		9.86	13.10
4200	8" thick		155	.103		5.60	4.35		9.95	13.20
4250	10" thick		140	.114		10.95	4.81		15.76	19.90
4300	12" thick		130	.123		11.85	5.20		17.05	21.50
4400	6" face height x 16" x 6" thick		155	.103		5.25	4.35		9.60	12.85
4450	8" thick		150	.107		7.95	4.49		12.44	16.05
4500	10" thick		130	.123		12.80	5.20		18	22.50
4550	12" thick		120	.133		13.15	5.60		18.75	23.50
5000	Split rib profile units, 1" deep ribs, 8 ribs									
5100	8" x 16" x 4" thick	D-8	345	.116	S.F.	3.84	4.98		8.82	12.35
5150	6" thick		325	.123		4.37	5.30		9.67	13.40
5200	8" thick		300	.133		4.94	5.70		10.64	14.75
5225	10" thick		300	.133		5.40	5.70		11.10	15.25
5250	12" thick	D-9	275	.175		5.80	7.35		13.15	18.30
5400	For special deeper colors, 4" thick, add					1.24			1.24	1.37
5450	12" thick, add					1.28			1.28	1.40
5600	For white, 4" thick, add					1.24			1.24	1.37
5650	6" thick, add					1.27			1.27	1.39
5700	8" thick, add					1.29			1.29	1.42
5750	12" thick, add					1.33			1.33	1.47
6000	Split face									
6100	8" x 16" x 4" thick	D-8	350	.114	S.F.	3.48	4.90		8.38	11.85
6150	6" thick		325	.123		3.98	5.30		9.28	12.95
6200	8" thick		300	.133		4.47	5.70		10.17	14.20
6225	10" thick		275	.145		5.05	6.25		11.30	15.70
6250	12" thick	D-9	270	.178		5.40	7.50		12.90	18.15
6300	For scored, add					.37			.37	.41
6400	For special deeper colors, 4" thick, add					.60			.60	.66
6450	6" thick, add					.71			.71	.78
6500	8" thick, add					.73			.73	.81

04 22 Concrete Unit Masonry

04 22 10 – Concrete Masonry Units

04 22 10.23 Concrete Block, Decorative

		Crew	Daily Output	Labor-Hours	Unit	Material	2015 Bare Costs Labor	Equipment	Total	Total Incl O&P
6550	12" thick, add				S.F.	.76			.76	.83
6650	For white, 4" thick, add					1.23			1.23	1.35
6700	6" thick, add					1.24			1.24	1.37
6750	8" thick, add					1.25			1.25	1.38
6800	12" thick, add				↓	1.28			1.28	1.40
7000	Scored ground face, 2 to 5 scores									
7100	8" x 16" x 4" thick	D-8	340	.118	S.F.	7.55	5.05		12.60	16.50
7150	6" thick		310	.129		8.40	5.55		13.95	18.20
7200	8" thick	↓	290	.138		9.45	5.90		15.35	20
7250	12" thick	D-9	265	.181	↓	12.70	7.65		20.35	26.50
8000	Hexagonal face profile units, 8" x 16" units									
8100	4" thick, hollow	D-8	340	.118	S.F.	3.63	5.05		8.68	12.20
8200	Solid		340	.118		4.65	5.05		9.70	13.30
8300	6" thick, hollow		310	.129		3.82	5.55		9.37	13.20
8350	8" thick, hollow	↓	290	.138	↓	4.36	5.90		10.26	14.45
8500	For stacked bond, add						26%			
8550	For high rise construction, add per story	D-8	67.80	.590	M.S.F.		25.50		25.50	41
8600	For scored block, add					10%				
8650	For honed or ground face, per face, add				Ea.	1.13			1.13	1.24
8700	For honed or ground end, per end, add				"	1.13			1.13	1.24
8750	For bullnose block, add					10%				
8800	For special color, add					13%				
9000	Minimum labor/equipment charge	D-1	2	8	Job		335		335	550

04 22 10.24 Concrete Block, Exterior

		Crew	Daily Output	Labor-Hours	Unit	Material	Labor	Equipment	Total	Total Incl O&P
0010	**CONCRETE BLOCK, EXTERIOR**, C90, 2000 psi									
0020	Reinforced alt courses, tooled joints 2 sides									
0100	Normal weight, 8" x 16" x 6" thick	D-8	395	.101	S.F.	2.32	4.35		6.67	9.60
0200	8" thick		360	.111		4	4.77		8.77	12.15
0250	10" thick	↓	290	.138		4.24	5.90		10.14	14.30
0300	12" thick	D-9	250	.192	↓	4.84	8.10		12.94	18.45
9000	Minimum labor/equipment charge	D-1	2	8	Job		335		335	550

04 22 10.26 Concrete Block Foundation Wall

		Crew	Daily Output	Labor-Hours	Unit	Material	Labor	Equipment	Total	Total Incl O&P
0010	**CONCRETE BLOCK FOUNDATION WALL**, C90/C145									
0050	Normal-weight, cut joints, horiz joint reinf, no vert reinf.									
0200	Hollow, 8" x 16" x 6" thick	D-8	455	.088	S.F.	2.89	3.77		6.66	9.30
0250	8" thick		425	.094		3.45	4.04		7.49	10.35
0300	10" thick	↓	350	.114		3.57	4.90		8.47	11.90
0350	12" thick	D-9	300	.160		4.68	6.75		11.43	16.10
0500	Solid, 8" x 16" block, 6" thick	D-8	440	.091		3.05	3.90		6.95	9.70
0550	8" thick	"	415	.096		4.15	4.14		8.29	11.30
0600	12" thick	D-9	350	.137	↓	5.95	5.75		11.70	15.95
1000	Reinforced, #4 vert @ 48"									
1100	Hollow, 8" x 16" block, 4" thick	D-8	455	.088	S.F.	2.88	3.77		6.65	9.30
1125	6" thick		445	.090		3.76	3.86		7.62	10.45
1150	8" thick		415	.096		4.69	4.14		8.83	11.90
1200	10" thick	↓	340	.118		5.15	5.05		10.20	13.90
1250	12" thick	D-9	290	.166		6.65	6.95		13.60	18.65
1500	Solid, 8" x 16" block, 6" thick	D-8	430	.093		3.08	3.99		7.07	9.90
1600	8" thick	"	405	.099		4.15	4.24		8.39	11.45
1650	12" thick	D-9	340	.141		5.95	5.95		11.90	16.20
9000	Minimum labor/equipment charge	D-1	2	8	Job		335		335	550

04 22 Concrete Unit Masonry

04 22 10 – Concrete Masonry Units

04 22 10.30 Concrete Block, Interlocking

		Crew	Daily Output	Labor-Hours	Unit	Material	2015 Bare Costs Labor	2015 Bare Costs Equipment	Total	Total Incl O&P
0010	**CONCRETE BLOCK, INTERLOCKING**									
0100	Not including grout or reinforcing									
0200	8" x 16" units, 2,000 psi, 8" thick	D-1	245	.065	S.F.	2.90	2.75		5.65	7.65
0300	12" thick	"	220	.073		4.30	3.06		7.36	9.70
0400	Including grout & reinforcing, 8" thick	D-4	245	.131		7.75	5.60	.54	13.89	18.10
0450	12" thick		220	.145		9.35	6.20	.61	16.16	21
0500	16" thick		185	.173		11.70	7.40	.72	19.82	25.50
9000	Minimum labor/equipment charge	D-1	2	8	Job		335		335	550

04 22 10.32 Concrete Block, Lintels

		Crew	Daily Output	Labor-Hours	Unit	Material	Labor	Equipment	Total	Total Incl O&P
0010	**CONCRETE BLOCK, LINTELS**, C90, normal weight									
0100	Including grout and horizontal reinforcing									
0200	8" x 8" x 8", 1 #4 bar	D-4	300	.107	L.F.	3.74	4.56	.44	8.74	11.95
0250	2 #4 bars		295	.108		3.96	4.63	.45	9.04	12.35
0400	8" x 16" x 8", 1 #4 bar		275	.116		3.83	4.97	.49	9.29	12.80
0450	2 #4 bars		270	.119		4.05	5.05	.49	9.59	13.20
1000	12" x 8" x 8", 1 #4 bar		275	.116		5.20	4.97	.49	10.66	14.30
1100	2 #4 bars		270	.119		5.40	5.05	.49	10.94	14.70
1150	2 #5 bars		270	.119		5.65	5.05	.49	11.19	14.95
1200	2 #6 bars		265	.121		5.95	5.15	.50	11.60	15.45
1500	12" x 16" x 8", 1 #4 bar		250	.128		5.80	5.45	.53	11.78	15.85
1600	2 #3 bars		245	.131		5.80	5.60	.54	11.94	16
1650	2 #4 bars		245	.131		6	5.60	.54	12.14	16.20
1700	2 #5 bars		240	.133		6.25	5.70	.56	12.51	16.70
9000	Minimum labor/equipment charge	D-1	2	8	Job		335		335	550

04 22 10.33 Lintel Block

		Crew	Daily Output	Labor-Hours	Unit	Material	Labor	Equipment	Total	Total Incl O&P
0010	**LINTEL BLOCK**									
3481	Lintel block 6" x 8" x 8"	D-1	300	.053	Ea.	1.28	2.25		3.53	5.05
3501	6" x 16" x 8"		275	.058		2	2.45		4.45	6.20
3521	8" x 8" x 8"		275	.058		1.15	2.45		3.60	5.25
3561	8" x 16" x 8"		250	.064		2.02	2.69		4.71	6.60

04 22 10.34 Concrete Block, Partitions

		Crew	Daily Output	Labor-Hours	Unit	Material	Labor	Equipment	Total	Total Incl O&P
0010	**CONCRETE BLOCK, PARTITIONS**, excludes scaffolding									
1000	Lightweight block, tooled joints, 2 sides, hollow									
1100	Not reinforced, 8" x 16" x 4" thick	D-8	440	.091	S.F.	1.80	3.90		5.70	8.35
1150	6" thick		410	.098		2.56	4.19		6.75	9.60
1200	8" thick		385	.104		3.13	4.46		7.59	10.70
1250	10" thick		370	.108		3.79	4.64		8.43	11.70
1300	12" thick	D-9	350	.137		4.01	5.75		9.76	13.80
1500	Reinforced alternate courses, 4" thick	D-8	435	.092		1.95	3.95		5.90	8.55
1600	6" thick		405	.099		2.70	4.24		6.94	9.85
1650	8" thick		380	.105		3.28	4.52		7.80	10.95
1700	10" thick		365	.110		3.95	4.70		8.65	12
1750	12" thick	D-9	345	.139		4.18	5.85		10.03	14.15
4000	Regular block, tooled joints, 2 sides, hollow									
4100	Not reinforced, 8" x 16" x 4" thick	D-8	430	.093	S.F.	1.70	3.99		5.69	8.35
4150	6" thick		400	.100		2.22	4.29		6.51	9.45
4200	8" thick		375	.107		2.74	4.58		7.32	10.45
4250	10" thick		360	.111		2.86	4.77		7.63	10.90
4300	12" thick	D-9	340	.141		3.95	5.95		9.90	14
4500	Reinforced alternate courses, 8" x 16" x 4" thick	D-8	425	.094		1.86	4.04		5.90	8.60
4550	6" thick		395	.101		2.39	4.35		6.74	9.70

04 22 Concrete Unit Masonry

04 22 10 – Concrete Masonry Units

04 22 10.34 Concrete Block, Partitions

		Crew	Daily Output	Labor-Hours	Unit	Material	2015 Bare Costs Labor	2015 Bare Costs Equipment	Total	Total Incl O&P
4600	8" thick	D-8	370	.108	S.F.	2.94	4.64		7.58	10.80
4650	10" thick	↓	355	.113		3.98	4.84		8.82	12.25
4700	12" thick	D-9	335	.143	↓	4.12	6.05		10.17	14.35
9000	Minimum labor/equipment charge	D-1	2	8	Job		335		335	550

04 22 10.38 Concrete Brick

0010	**CONCRETE BRICK**, C55, grade N, type 1									
0100	Regular, 4 x 2-1/4 x 8	D-8	660	.061	Ea.	.54	2.60		3.14	4.82
0125	Rusticated, 4 x 2-1/4 x 8		660	.061		.60	2.60		3.20	4.90
0150	Frog, 4 x 2-1/4 x 8		660	.061		.58	2.60		3.18	4.87
0200	Double, 4 x 4-7/8 x 8	↓	535	.075	↓	.95	3.21		4.16	6.25

04 22 10.44 Glazed Concrete Block

0010	**GLAZED CONCRETE BLOCK** C744									
0100	Single face, 8" x 16" units, 2" thick	D-8	360	.111	S.F.	9.60	4.77		14.37	18.35
0200	4" thick		345	.116		9.90	4.98		14.88	19
0250	6" thick		330	.121		10.70	5.20		15.90	20
0300	8" thick		310	.129		11.45	5.55		17	21.50
0350	10" thick	↓	295	.136		13.05	5.80		18.85	24
0400	12" thick	D-9	280	.171		13.95	7.20		21.15	27
0700	Double face, 8" x 16" units, 4" thick	D-8	340	.118		13.85	5.05		18.90	23.50
0750	6" thick		320	.125		17	5.35		22.35	27.50
0800	8" thick		300	.133	↓	17.80	5.70		23.50	29
1000	Jambs, bullnose or square, single face, 8" x 16", 2" thick		315	.127	Ea.	18.35	5.45		23.80	29
1050	4" thick		285	.140	"	19.30	6		25.30	31
1200	Caps, bullnose or square, 8" x 16", 2" thick		420	.095	L.F.	17.05	4.09		21.14	25.50
1250	4" thick		380	.105	"	18.50	4.52		23.02	28
1256	Corner, bullnose or square, 2" thick		280	.143	Ea.	19.95	6.15		26.10	32
1258	4" thick		270	.148		21.50	6.35		27.85	34
1260	6" thick		260	.154		27	6.60		33.60	40.50
1270	8" thick		250	.160		32.50	6.85		39.35	47
1280	10" thick		240	.167		33	7.15		40.15	48
1290	12" thick		230	.174	↓	35	7.45		42.45	50.50
1500	Cove base, 8" x 16", 2" thick		315	.127	L.F.	8.55	5.45		14	18.30
1550	4" thick		285	.140		8.65	6		14.65	19.35
1600	6" thick		265	.151		9.30	6.50		15.80	21
1650	8" thick	↓	245	.163	↓	9.75	7		16.75	22
9000	Minimum labor/equipment charge	D-1	2	8	Job		335		335	550

04 23 Glass Unit Masonry

04 23 13 – Vertical Glass Unit Masonry

04 23 13.10 Glass Block

0010	**GLASS BLOCK**									
0100	Plain, 4" thick, under 1,000 S.F., 6" x 6"	D-8	115	.348	S.F.	26.50	14.95		41.45	53.50
0150	8" x 8"		160	.250		15.35	10.75		26.10	34.50
0160	end block		160	.250		59	10.75		69.75	82.50
0170	90 deg corner		160	.250		53.50	10.75		64.25	76.50
0180	45 deg corner		160	.250		44	10.75		54.75	66
0200	12" x 12"		175	.229		22	9.80		31.80	40.50
0210	4" x 8"		160	.250		28	10.75		38.75	48.50
0220	6" x 8"	↓	160	.250		19.90	10.75		30.65	39.50
0300	1,000 to 5,000 S.F., 6" x 6"		135	.296		26	12.70		38.70	49

04 23 Glass Unit Masonry

04 23 13 – Vertical Glass Unit Masonry

04 23 13.10 Glass Block

		Crew	Daily Output	Labor-Hours	Unit	Material	2015 Bare Costs Labor	Equipment	Total	Total Incl O&P
0350	8" x 8"	D-8	190	.211	S.F.	15.05	9.05		24.10	31.50
0400	12" x 12"		215	.186		21.50	8		29.50	37
0410	4" x 8"		215	.186		27.50	8		35.50	43
0420	6" x 8"		215	.186		19.50	8		27.50	34.50
0500	Over 5,000 S.F., 6" x 6"		145	.276		25	11.85		36.85	47
0700	For solar reflective blocks, add					100%				
1000	Thinline, plain, 3-1/8" thick, under 1,000 S.F., 6" x 6"	D-8	115	.348	S.F.	21	14.95		35.95	47.50
1050	8" x 8"		160	.250		11.60	10.75		22.35	30
1250	8" x 8"		215	.186		11.05	8		19.05	25
1400	For cleaning block after installation (both sides), add		1000	.040		.16	1.72		1.88	2.97
9000	Minimum labor/equipment charge	D-1	2	8	Job		335		335	550

04 27 Multiple-Wythe Unit Masonry

04 27 10 – Multiple-Wythe Masonry

04 27 10.10 Cornices

		Crew	Daily Output	Labor-Hours	Unit	Material	Labor	Equipment	Total	Total Incl O&P
0010	**CORNICES**									
0110	Face bricks, 12 brick/S.F.	D-1	30	.533	SF Face	5.65	22.50		28.15	42.50
0150	15 brick/S.F.		23	.696	"	6.75	29.50		36.25	55
9000	Minimum labor/equipment charge		1.50	10.667	Job		450		450	730

04 27 10.30 Brick Walls

		Crew	Daily Output	Labor-Hours	Unit	Material	Labor	Equipment	Total	Total Incl O&P
0010	**BRICK WALLS**, including mortar, excludes scaffolding									
0020	Estimating by number of brick									
0140	Face brick, 4" thick wall, 6.75 brick/S.F.	D-8	1.45	27.586	M	540	1,175		1,715	2,525
0150	Common brick, 4" thick wall, 6.75 brick/S.F.		1.60	25		595	1,075		1,670	2,400
0204	8" thick, 13.50 bricks per S.F.		1.80	22.222		615	955		1,570	2,225
0250	12" thick, 20.25 bricks per S.F.		1.90	21.053		620	905		1,525	2,150
0304	16" thick, 27.00 bricks per S.F.		2	20		630	860		1,490	2,100
0500	Reinforced, face brick, 4" thick wall, 6.75 brick/S.F.		1.40	28.571		565	1,225		1,790	2,625
0520	Common brick, 4" thick wall, 6.75 brick/S.F.		1.55	25.806		620	1,100		1,720	2,475
0550	8" thick, 13.50 bricks per S.F.		1.75	22.857		640	980		1,620	2,300
0600	12" thick, 20.25 bricks per S.F.		1.85	21.622		645	930		1,575	2,200
0650	16" thick, 27.00 bricks per S.F.		1.95	20.513		655	880		1,535	2,150
0790	Alternate method of figuring by square foot									
0800	Face brick, 4" thick wall, 6.75 brick/S.F.	D-8	215	.186	S.F.	3.65	8		11.65	17
0850	Common brick, 4" thick wall, 6.75 brick/S.F. R042110-50		240	.167		4.03	7.15		11.18	16.10
0900	8" thick, 13.50 bricks per S.F.		135	.296		8.30	12.70		21	29.50
1000	12" thick, 20.25 bricks per S.F.		95	.421		12.55	18.05		30.60	43.50
1050	16" thick, 27.00 bricks per S.F.		75	.533		17	23		40	55.50
1200	Reinforced, face brick, 4" thick wall, 6.75 brick/S.F.		210	.190		3.82	8.15		11.97	17.50
1220	Common brick, 4" thick wall, 6.75 brick/S.F.		235	.170		4.20	7.30		11.50	16.50
1250	8" thick, 13.50 bricks per S.F.		130	.308		8.65	13.20		21.85	31
1300	12" thick, 20.25 bricks per S.F.		90	.444		13.05	19.05		32.10	45.50
1350	16" thick, 27.00 bricks per S.F.		70	.571		17.65	24.50		42.15	59.50
9000	Minimum labor/equipment charge	D-1	2	8	Job		335		335	550

04 27 10.40 Steps

		Crew	Daily Output	Labor-Hours	Unit	Material	Labor	Equipment	Total	Total Incl O&P
0010	**STEPS**									
0012	Entry steps, select common brick	D-1	.30	53.333	M	500	2,250		2,750	4,200

04 41 Dry-Placed Stone

04 41 10 – Dry Placed Stone

04 41 10.10 Rough Stone Wall		Crew	Daily Output	Labor-Hours	Unit	Material	2015 Bare Costs Labor	Equipment	Total	Total Incl O&P
0011	**ROUGH STONE WALL**, Dry									
0012	Dry laid (no mortar), under 18" thick	G	D-1	60	.267	C.F.	12.90	11.25		24.15	32.50
0100	Random fieldstone, under 18" thick	G	D-12	60	.533		12.90	22.50		35.40	51
0150	Over 18" thick	G	"	63	.508		15.50	21.50		37	52
0500	Field stone veneer	G	D-8	120	.333	S.F.	12.20	14.30		26.50	37
0510	Valley stone veneer	G		120	.333		12.20	14.30		26.50	37
0520	River stone veneer	G		120	.333		12.20	14.30		26.50	37
0600	Rubble stone walls, in mortar bed, up to 18" thick	G	D-11	75	.320	C.F.	15.55	14.10		29.65	40
9000	Minimum labor/equipment charge		D-1	2	8	Job		335		335	550

04 43 Stone Masonry

04 43 10 – Masonry with Natural and Processed Stone

04 43 10.05 Ashlar Veneer

		Crew	Daily Output	Labor-Hours	Unit	Material	Labor	Equipment	Total	Total Incl O&P
0011	**ASHLAR VENEER** 4" + or - thk, random or random rectangular									
0150	Sawn face, split joints, low priced stone	D-8	140	.286	S.F.	11.60	12.25		23.85	32.50
0200	Medium priced stone		130	.308		13.35	13.20		26.55	36
0300	High priced stone		120	.333		17.90	14.30		32.20	43
0600	Seam face, split joints, medium price stone		125	.320		19	13.75		32.75	43.50
0700	High price stone		120	.333		18.75	14.30		33.05	44
1000	Split or rock face, split joints, medium price stone		125	.320		11.40	13.75		25.15	35
1100	High price stone		120	.333		17.25	14.30		31.55	42.50

04 43 10.10 Bluestone

		Crew	Daily Output	Labor-Hours	Unit	Material	Labor	Equipment	Total	Total Incl O&P
0010	**BLUESTONE**, cut to size									
0500	Sills, natural cleft, 10" wide to 6' long, 1-1/2" thick	D-11	70	.343	L.F.	13.35	15.15		28.50	39
0550	2" thick		63	.381		14.60	16.80		31.40	43.50
0600	Smooth finish, 1-1/2" thick		70	.343		12	15.15		27.15	37.50
0650	2" thick		63	.381		13	16.80		29.80	42
0800	Thermal finish, 1-1/2" thick		70	.343		12	15.15		27.15	37.50
0850	2" thick		63	.381		13	16.80		29.80	42
1000	Stair treads, natural cleft, 12" wide, 6' long, 1-1/2" thick	D-10	115	.278		12	12.80	4.12	28.92	38
1050	2" thick		105	.305		13	14	4.51	31.51	42
1100	Smooth finish, 1-1/2" thick		115	.278		12	12.80	4.12	28.92	38
1150	2" thick		105	.305		13	14	4.51	31.51	42
1300	Thermal finish, 1-1/2" thick		115	.278		12	12.80	4.12	28.92	38
1350	2" thick		105	.305		13	14	4.51	31.51	42
9000	Minimum labor/equipment charge	D-1	2.50	6.400	Job		269		269	440

04 43 10.45 Granite

		Crew	Daily Output	Labor-Hours	Unit	Material	Labor	Equipment	Total	Total Incl O&P
0010	**GRANITE**, cut to size									
0050	Veneer, polished face, 3/4" to 1-1/2" thick									
0150	Low price, gray, light gray, etc.	D-10	130	.246	S.F.	26.50	11.35	3.64	41.49	52
0180	Medium price, pink, brown, etc.		130	.246		29.50	11.35	3.64	44.49	55
0220	High price, red, black, etc.		130	.246		42	11.35	3.64	56.99	69
0300	1-1/2" to 2-1/2" thick, veneer									
0350	Low price, gray, light gray, etc.	D-10	130	.246	S.F.	28.50	11.35	3.64	43.49	54
0500	Medium price, pink, brown, etc.	"	130	.246	"	33.50	11.35	3.64	48.49	59.50
2500	Steps, copings, etc., finished on more than one surface									
2550	Low price, gray, light gray, etc.	D-10	50	.640	C.F.	91	29.50	9.45	129.95	158
2575	Medium price, pink, brown, etc.		50	.640		118	29.50	9.45	156.95	188
2600	High price, red, black, etc.		50	.640		146	29.50	9.45	184.95	218
9000	Minimum labor/equipment charge	D-1	2	8	Job		335		335	550

04 43 Stone Masonry

04 43 10 – Masonry with Natural and Processed Stone

04 43 10.55 Limestone

		Crew	Daily Output	Labor-Hours	Unit	Material	2015 Bare Costs Labor	Equipment	Total	Total Incl O&P
0010	**LIMESTONE**, cut to size									
0020	Veneer facing panels									
0500	Texture finish, light stick, 4-1/2" thick, 5' x 12'	D-4	300	.107	S.F.	19.50	4.56	.44	24.50	29.50
0750	5" thick, 5' x 14' panels	D-10	275	.116		20.50	5.35	1.72	27.57	33
1000	Sugarcube finish, 2" Thick, 3' x 5' panels		275	.116		28.50	5.35	1.72	35.57	42
1050	3" Thick, 4' x 9' panels		275	.116		22.50	5.35	1.72	29.57	35.50
1200	4" Thick, 5' x 11' panels		275	.116		30	5.35	1.72	37.07	43.50
1400	Sugarcube, textured finish, 4-1/2" thick, 5' x 12'		275	.116		31	5.35	1.72	38.07	44.50
1450	5" thick, 5' x 14' panels		275	.116		32	5.35	1.72	39.07	46
2000	Coping, sugarcube finish, top & 2 sides		30	1.067	C.F.	67.50	49	15.80	132.30	171
2100	Sills, lintels, jambs, trim, stops, sugarcube finish, simple		20	1.600		67.50	73.50	23.50	164.50	219
2150	Detailed		20	1.600		67.50	73.50	23.50	164.50	219
2300	Steps, extra hard, 14" wide, 6" rise		50	.640	L.F.	25	29.50	9.45	63.95	85.50
3000	Quoins, plain finish, 6" x 12" x 12"	D-12	25	1.280	Ea.	37.50	54.50		92	130
3050	6" x 16" x 24"	"	25	1.280	"	50	54.50		104.50	144
9000	Minimum labor/equipment charge	D-1	2	8	Job		335		335	550

04 43 10.60 Marble

		Crew	Daily Output	Labor-Hours	Unit	Material	2015 Bare Costs Labor	Equipment	Total	Total Incl O&P
0011	**MARBLE**, ashlar, split face, 4" + or - thick, random									
0040	Lengths 1' to 4' & heights 2" to 7-1/2", average	D-8	175	.229	S.F.	17.45	9.80		27.25	35
0100	Base, polished, 3/4" or 7/8" thick, polished, 6" high	D-10	65	.492	L.F.	12	22.50	7.30	41.80	57.50
0300	Carvings or bas relief, from templates, simple design		80	.400	S.F.	144	18.40	5.90	168.30	194
0350	Intricate design		80	.400	"	335	18.40	5.90	359.30	405
1000	Facing, polished finish, cut to size, 3/4" to 7/8" thick									
1050	Carrara or equal	D-10	130	.246	S.F.	19	11.35	3.64	33.99	43.50
1100	Arabescato or equal		130	.246		39	11.35	3.64	53.99	65.50
1300	1-1/4" thick, Botticino Classico or equal		125	.256		24	11.80	3.79	39.59	49.50
1350	Statuarietto or equal		125	.256		41.50	11.80	3.79	57.09	68.50
1500	2" thick, Crema Marfil or equal		120	.267		42.50	12.25	3.94	58.69	71
1550	Cafe Pinta or equal		120	.267		70	12.25	3.94	86.19	101
1700	Rubbed finish, cut to size, 4" thick									
1740	Average	D-10	100	.320	S.F.	40	14.70	4.73	59.43	73
1780	Maximum	"	100	.320	"	69	14.70	4.73	88.43	105
2200	Window sills, 6" x 3/4" thick	D-1	85	.188	L.F.	10.90	7.90		18.80	25
2500	Flooring, polished tiles, 12" x 12" x 3/8" thick									
2510	Thin set, Giallo Solare or equal	D-11	90	.267	S.F.	14.75	11.75		26.50	35.50
2600	Sky Blue or equal		90	.267		16	11.75		27.75	37
2700	Mortar bed, Giallo Solare or equal		65	.369		14.75	16.30		31.05	43
2740	Sky Blue or equal		65	.369		16	16.30		32.30	44
2780	Travertine, 3/8" thick, Sierra or equal	D-10	130	.246		9.25	11.35	3.64	24.24	32.50
2790	Silver or equal	"	130	.246		25.50	11.35	3.64	40.49	50.50
2800	Patio tile, non-slip, 1/2" thick, flame finish	D-11	75	.320		11.90	14.10		26	36
2900	Shower or toilet partitions, 7/8" thick partitions									
3050	3/4" or 1-1/4" thick stiles, polished 2 sides, average	D-11	75	.320	S.F.	45.50	14.10		59.60	73
3201	Soffits, add to above prices				"	20%	100%			
3210	Stairs, risers, 7/8" thick x 6" high	D-10	115	.278	L.F.	15.25	12.80	4.12	32.17	42
3360	Treads, 12" wide x 1-1/4" thick	"	115	.278	"	43.50	12.80	4.12	60.42	73
3500	Thresholds, 3' long, 7/8" thick, 4" to 5" wide, plain	D-12	24	1.333	Ea.	35.50	57		92.50	132
3550	Beveled		24	1.333	"	70.50	57		127.50	170
3700	Window stools, polished, 7/8" thick, 5" wide		85	.376	L.F.	21.50	16.05		37.55	49.50
9000	Minimum labor/equipment charge	D-1	2	8	Job		335		335	550

04 43 Stone Masonry

04 43 10 – Masonry with Natural and Processed Stone

04 43 10.75 Sandstone or Brownstone

		Crew	Daily Output	Labor-Hours	Unit	Material	2015 Bare Costs Labor	2015 Bare Costs Equipment	Total	Total Incl O&P
0011	**SANDSTONE OR BROWNSTONE**									
0100	Sawed face veneer, 2-1/2" thick, to 2' x 4' panels	D-10	130	.246	S.F.	19	11.35	3.64	33.99	43.50
0150	4" thick, to 3'-6" x 8' panels		100	.320		19	14.70	4.73	38.43	50
0300	Split face, random sizes		100	.320		13.65	14.70	4.73	33.08	44
9000	Minimum labor/equipment charge	D-1	2.50	6.400	Job		269		269	440

04 43 10.80 Slate

		Crew	Daily Output	Labor-Hours	Unit	Material	Labor	Equipment	Total	Total Incl O&P
0010	**SLATE**									
0040	Pennsylvania - blue gray to black									
0050	Vermont - unfading green, mottled green & purple, gray & purple									
0100	Virginia - blue black									
3100	Stair landings, 1" thick, black, clear	D-1	65	.246	S.F.	21	10.35		31.35	40.50
3200	Ribbon	"	65	.246	"	23.50	10.35		33.85	42.50
3500	Stair treads, sand finish, 1" thick x 12" wide									
3550	Under 3 L.F.	D-10	85	.376	L.F.	23.50	17.30	5.55	46.35	59.50
3600	3 L.F. to 6 L.F.	"	120	.267	"	25	12.25	3.94	41.19	51.50
3700	Ribbon, sand finish, 1" thick x 12" wide									
3750	To 6 L.F.	D-10	120	.267	L.F.	21	12.25	3.94	37.19	47.50
4000	Stools or sills, sand finish, 1" thick, 6" wide	D-12	160	.200		12.20	8.50		20.70	27.50
4100	Honed finish		160	.200		11.65	8.50		20.15	26.50
4200	10" wide		90	.356		18.80	15.15		33.95	45
4250	Honed finish		90	.356		17.50	15.15		32.65	44
4400	2" thick, 6" wide		140	.229		19.60	9.75		29.35	37.50
4450	Honed finish		140	.229		18.65	9.75		28.40	36.50
4600	10" wide		90	.356		30.50	15.15		45.65	58.50
4650	Honed finish		90	.356		29	15.15		44.15	56.50
4800	For lengths over 3', add					25%				
9000	Minimum labor/equipment charge	D-1	2.50	6.400	Job		269		269	440

04 43 10.85 Window Sill

		Crew	Daily Output	Labor-Hours	Unit	Material	Labor	Equipment	Total	Total Incl O&P
0010	**WINDOW SILL**									
0020	Bluestone, thermal top, 10" wide, 1-1/2" thick	D-1	85	.188	S.F.	10	7.90		17.90	24
0050	2" thick		75	.213	"	10	9		19	25.50
0100	Cut stone, 5" x 8" plain		48	.333	L.F.	12.10	14.05		26.15	36.50
0200	Face brick on edge, brick, 8" wide		80	.200		5.05	8.40		13.45	19.25
0400	Marble, 9" wide, 1" thick		85	.188		8.65	7.90		16.55	22.50
0900	Slate, colored, unfading, honed, 12" wide, 1" thick		85	.188		8.50	7.90		16.40	22.50
0950	2" thick		70	.229		8.50	9.60		18.10	25
9000	Minimum labor/equipment charge	1 Bric	2	4	Job		185		185	300

04 54 Refractory Brick Masonry

04 54 10 – Refractory Brick Work

04 54 10.10 Fire Brick

		Crew	Daily Output	Labor-Hours	Unit	Material	Labor	Equipment	Total	Total Incl O&P
0010	**FIRE BRICK**									
0012	Low duty, 2000°F, 9" x 2-1/2" x 4-1/2"	D-1	.60	26.667	M	1,425	1,125		2,550	3,400
0050	High duty, 3000°F	"	.60	26.667	"	2,600	1,125		3,725	4,675

04 54 10.20 Fire Clay

		Crew	Daily Output	Labor-Hours	Unit	Material	Labor	Equipment	Total	Total Incl O&P
0010	**FIRE CLAY**									
0020	Gray, high duty, 100 lb. bag				Bag	33			33	36.50
0050	100 lb. drum, premixed (400 brick per drum)				Drum	41			41	45

04 57 Masonry Fireplaces

04 57 10 – Brick or Stone Fireplaces

04 57 10.10 Fireplace		Crew	Daily Output	Labor-Hours	Unit	Material	2015 Bare Costs Labor	2015 Bare Costs Equipment	Total	Total Incl O&P
0010	**FIREPLACE**									
0100	Brick fireplace, not incl. foundations or chimneys									
0110	30" x 29" opening, incl. chamber, plain brickwork	D-1	.40	40	Ea.	550	1,675		2,225	3,350
0200	Fireplace box only (110 brick)	"	2	8	"	157	335		492	720
0300	For elaborate brickwork and details, add					35%	35%			
0400	For hearth, brick & stone, add	D-1	2	8	Ea.	205	335		540	775
0410	For steel, damper, cleanouts, add		4	4		19.70	168		187.70	296
0600	Plain brickwork, incl. metal circulator		.50	32		995	1,350		2,345	3,300
0800	Face brick only, standard size, 8" x 2-2/3" x 4"		.30	53.333	M	550	2,250		2,800	4,250
0900	Stone fireplace, fieldstone, add				SF Face	8.50			8.50	9.35
1000	Cut stone, add				"	8			8	8.80
9000	Minimum labor/equipment charge	D-1	2	8	Job		335		335	550

04 71 Manufactured Brick Masonry

04 71 10 – Simulated or Manufactured Brick

04 71 10.10 Simulated Brick		Crew	Daily Output	Labor-Hours	Unit	Material	2015 Bare Costs Labor	2015 Bare Costs Equipment	Total	Total Incl O&P
0010	**SIMULATED BRICK**									
0020	Aluminum, baked on colors	1 Carp	200	.040	S.F.	4.31	1.88		6.19	7.80
0050	Fiberglass panels		200	.040		8.05	1.88		9.93	11.95
0100	Urethane pieces cemented in mastic		150	.053		8.25	2.50		10.75	13.15
0150	Vinyl siding panels		200	.040		10	1.88		11.88	14.10

04 72 Cast Stone Masonry

04 72 10 – Cast Stone Masonry Features

04 72 10.10 Coping		Crew	Daily Output	Labor-Hours	Unit	Material	2015 Bare Costs Labor	2015 Bare Costs Equipment	Total	Total Incl O&P
0010	**COPING**, stock units									
0050	Precast concrete, 10" wide, 4" tapers to 3-1/2", 8" wall	D-1	75	.213	L.F.	17.10	9		26.10	33.50
0100	12" wide, 3-1/2" tapers to 3", 10" wall		70	.229		18.45	9.60		28.05	36
0110	14" wide, 4" tapers to 3-1/2", 12" wall		65	.246		20.50	10.35		30.85	39.50
0150	16" wide, 4" tapers to 3-1/2", 14" wall		60	.267		22	11.25		33.25	43
0300	Limestone for 12" wall, 4" thick		90	.178		14.70	7.50		22.20	28.50
0350	6" thick		80	.200		22	8.40		30.40	38
0500	Marble, to 4" thick, no wash, 9" wide		90	.178		12.05	7.50		19.55	25.50
0550	12" wide		80	.200		16.15	8.40		24.55	31.50
0700	Terra cotta, 9" wide		90	.178		6.35	7.50		13.85	19.15
0750	12" wide		80	.200		8.65	8.40		17.05	23
0800	Aluminum, for 12" wall		80	.200		8.90	8.40		17.30	23.50
9000	Minimum labor/equipment charge		2	8	Job		335		335	550

04 72 20 – Cultured Stone Veneer

04 72 20.10 Cultured Stone Veneer Components		Crew	Daily Output	Labor-Hours	Unit	Material	2015 Bare Costs Labor	2015 Bare Costs Equipment	Total	Total Incl O&P
0010	**CULTURED STONE VENEER COMPONENTS**									
0110	On wood frame and sheathing substrate, random sized cobbles, corner stones	D-8	70	.571	V.L.F.	10.15	24.50		34.65	51
0120	Field stones		140	.286	S.F.	7.40	12.25		19.65	28
0130	Random sized flats, corner stones		70	.571	V.L.F.	10.05	24.50		34.55	51
0140	Field stones		140	.286	S.F.	8.70	12.25		20.95	29.50
0150	Horizontal lined ledgestones, corner stones		75	.533	V.L.F.	10.15	23		33.15	48
0160	Field stones		150	.267	S.F.	7.40	11.45		18.85	26.50
0170	Random shaped flats, corner stones		65	.615	V.L.F.	10.15	26.50		36.65	54
0180	Field stones		150	.267	S.F.	7.40	11.45		18.85	26.50

For customer support on your Commercial Renovation Cost Data, call 877.791.4977.

04 72 Cast Stone Masonry

04 72 20 – Cultured Stone Veneer

04 72 20.10 Cultured Stone Veneer Components		Crew	Daily Output	Labor-Hours	Unit	Material	2015 Bare Costs Labor	Equipment	Total	Total Incl O&P
0190	Random shaped/textured face, corner stones	D-8	65	.615	V.L.F.	10.15	26.50		36.65	54
0200	Field stones		130	.308	S.F.	7.40	13.20		20.60	29.50
0210	Random shaped river rock, corner stones		65	.615	V.L.F.	10.15	26.50		36.65	54
0220	Field stones		130	.308	S.F.	7.40	13.20		20.60	29.50
0240	On concrete or CMU substrate, random sized cobbles, corner stones		70	.571	V.L.F.	9.55	24.50		34.05	50.50
0250	Field stones		140	.286	S.F.	7.10	12.25		19.35	28
0260	Random sized flats, corner stones		70	.571	V.L.F.	9.45	24.50		33.95	50.50
0270	Field stones		140	.286	S.F.	8.40	12.25		20.65	29
0280	Horizontal lined ledgestones, corner stones		75	.533	V.L.F.	9.55	23		32.55	47.50
0290	Field stones		150	.267	S.F.	7.10	11.45		18.55	26.50
0300	Random shaped flats, corner stones		70	.571	V.L.F.	9.55	24.50		34.05	50.50
0310	Field stones		140	.286	S.F.	7.10	12.25		19.35	28
0320	Random shaped/textured face, corner stones		65	.615	V.L.F.	9.55	26.50		36.05	53.50
0330	Field stones		130	.308	S.F.	7.10	13.20		20.30	29.50
0340	Random shaped river rock, corner stones		65	.615	V.L.F.	9.55	26.50		36.05	53.50
0350	Field stones	▼	130	.308	S.F.	7.10	13.20		20.30	29.50
0360	Cultured stone veneer, #15 felt weather resistant barrier	1 Clab	3700	.002	Sq.	5.40	.08		5.48	6.10
0370	Expanded metal lath, diamond, 2.5 lb./S.Y., galvanized	1 Lath	85	.094	S.Y.	2.72	4.05		6.77	9.35
0390	Water table or window sill, 18" long	1 Bric	80	.100	Ea.	10.30	4.62		14.92	18.85

04 73 Manufactured Stone Masonry

04 73 20 – Simulated or Manufactured Stone

04 73 20.10 Simulated Stone		Crew	Daily Output	Labor-Hours	Unit	Material	2015 Bare Costs Labor	Equipment	Total	Total Incl O&P
0010	**SIMULATED STONE**									
0100	Insulated fiberglass panels, 5/8" ply backer	L-4	200	.120	S.F.	10.80	5.30		16.10	20.50

Estimating Tips

05 05 00 Common Work Results for Metals

- Nuts, bolts, washers, connection angles, and plates can add a significant amount to both the tonnage of a structural steel job and the estimated cost. As a rule of thumb, add 10% to the total weight to account for these accessories.
- Type 2 steel construction, commonly referred to as "simple construction," consists generally of field-bolted connections with lateral bracing supplied by other elements of the building, such as masonry walls or x-bracing. The estimator should be aware, however, that shop connections may be accomplished by welding or bolting. The method may be particular to the fabrication shop and may have an impact on the estimated cost.

05 10 00 Structural Steel

- Steel items can be obtained from two sources: a fabrication shop or a metals service center. Fabrication shops can fabricate items under more controlled conditions than can crews in the field. They are also more efficient and can produce items more economically. Metal service centers serve as a source of long mill shapes to both fabrication shops and contractors.
- Most line items in this structural steel subdivision, and most items in 05 50 00 Metal Fabrications, are indicated as being shop fabricated. The bare material cost for these shop fabricated items is the "Invoice Cost" from the shop and includes the mill base price of steel plus mill extras, transportation to the shop, shop drawings and detailing where warranted, shop fabrication and handling, sandblasting and a shop coat of primer paint, all necessary structural bolts, and delivery to the job site. The bare labor cost and bare equipment cost for these shop fabricated items is for field installation or erection.
- Line items in Subdivision 05 12 23.40 Lightweight Framing, and other items scattered in Division 5, are indicated as being field fabricated. The bare material cost for these field fabricated items is the "Invoice Cost" from the metals service center and includes the mill base price of steel plus mill extras, transportation to the metals service center, material handling, and delivery of long lengths of mill shapes to the job site. Material costs for structural bolts and welding rods should be added to the estimate. The bare labor cost and bare equipment cost for these items is for both field fabrication and field installation or erection, and include time for cutting, welding and drilling in the fabricated metal items. Drilling into concrete and fasteners to fasten field fabricated items to other work are not included and should be added to the estimate.

05 20 00 Steel Joist Framing

- In any given project the total weight of open web steel joists is determined by the loads to be supported and the design. However, economies can be realized in minimizing the amount of labor used to place the joists. This is done by maximizing the joist spacing, and therefore minimizing the number of joists required to be installed on the job. Certain spacings and locations may be required by the design, but in other cases maximizing the spacing and keeping it as uniform as possible will keep the costs down.

05 30 00 Steel Decking

- The takeoff and estimating of metal deck involves more than simply the area of the floor or roof and the type of deck specified or shown on the drawings. Many different sizes and types of openings may exist. Small openings for individual pipes or conduits may be drilled after the floor/roof is installed, but larger openings may require special deck lengths as well as reinforcing or structural support. The estimator should determine who will be supplying this reinforcing. Additionally, some deck terminations are part of the deck package, such as screed angles and pour stops, and others will be part of the steel contract, such as angles attached to structural members and cast-in-place angles and plates. The estimator must ensure that all pieces are accounted for in the complete estimate.

05 50 00 Metal Fabrications

- The most economical steel stairs are those that use common materials, standard details, and most importantly, a uniform and relatively simple method of field assembly. Commonly available A36 channels and plates are very good choices for the main stringers of the stairs, as are angles and tees for the carrier members. Risers and treads are usually made by specialty shops, and it is most economical to use a typical detail in as many places as possible. The stairs should be pre-assembled and shipped directly to the site. The field connections should be simple and straightforward to be accomplished efficiently, and with minimum equipment and labor.

Reference Numbers

Reference numbers are shown in shaded boxes at the beginning of some major classifications. These numbers refer to related items in the Reference Section. The reference information may be an estimating procedure, an alternate pricing method, or technical information.

Note: Not all subdivisions listed here necessarily appear in this publication. ∎

No part of this publication may be reproduced, stored in a retrieval system, or transmitted in any form or by any means without prior written permission of RSMeans.

05 05 Common Work Results for Metals

05 05 05 – Selective Demolition for Metals

05 05 05.10 Selective Demolition, Metals

		Crew	Daily Output	Labor-Hours	Unit	Material	2015 Bare Costs Labor	2015 Bare Costs Equipment	Total	Total Incl O&P
0010	**SELECTIVE DEMOLITION, METALS** R024119-10									
0015	Excludes shores, bracing, cutting, loading, hauling, dumping									
0020	Remove nuts only up to 3/4" diameter	1 Sswk	480	.017	Ea.		.88		.88	1.59
0030	7/8" to 1-1/4" diameter		240	.033			1.75		1.75	3.17
0040	1-3/8" to 2" diameter		160	.050			2.63		2.63	4.76
0060	Unbolt and remove structural bolts up to 3/4" diameter		240	.033			1.75		1.75	3.17
0070	7/8" to 2" diameter		160	.050			2.63		2.63	4.76
0140	Light weight framing members, remove whole or cut up, up to 20 lb.		240	.033			1.75		1.75	3.17
0150	21 – 40 lb.	2 Sswk	210	.076			4.01		4.01	7.25
0160	41 – 80 lb.	3 Sswk	180	.133			7		7	12.70
0170	81 – 120 lb.	4 Sswk	150	.213			11.25		11.25	20.50
0230	Structural members, remove whole or cut up, up to 500 lb.	E-19	48	.500			26	19.75	45.75	67
0240	1/4 – 2 tons	E-18	36	1.111			58.50	26.50	85	132
0250	2 – 5 tons	E-24	30	1.067			55.50	24.50	80	125
0260	5 – 10 tons	E-20	24	2.667			138	48.50	186.50	297
0270	10 – 15 tons	E-2	18	3.111			161	84	245	375
0340	Fabricated item, remove whole or cut up, up to 20 lb.	1 Sswk	96	.083			4.39		4.39	7.95
0350	21 – 40 lb.	2 Sswk	84	.190			10.05		10.05	18.10
0360	41 – 80 lb.	3 Sswk	72	.333			17.55		17.55	31.50
0370	81 – 120 lb.	4 Sswk	60	.533			28		28	51
0380	121 – 500 lb.	E-19	48	.500			26	19.75	45.75	67
0390	501 – 1000 lb.	"	36	.667			34.50	26.50	61	89
0500	Steel roof decking, uncovered, bare	B-2	5000	.008	S.F.		.30		.30	.50
2950	Minimum labor/equipment charge	E-19	2	12	Job		625	475	1,100	1,600

05 05 19 – Post-Installed Concrete Anchors

05 05 19.10 Chemical Anchors

		Crew	Daily Output	Labor-Hours	Unit	Material	Labor	Equipment	Total	Total Incl O&P
0010	**CHEMICAL ANCHORS**									
0020	Includes layout & drilling									
1430	Chemical anchor, w/rod & epoxy cartridge, 3/4" diam. x 9-1/2" long	B-89A	27	.593	Ea.	9.80	25.50	4.28	39.58	57
1435	1" diameter x 11-3/4" long		24	.667		17.55	29	4.82	51.37	71.50
1440	1-1/4" diameter x 14" long		21	.762		36.50	33	5.50	75	99.50
1445	1-3/4" diameter x 15" long		20	.800		65	34.50	5.80	105.30	134
1450	18" long		17	.941		78	40.50	6.80	125.30	160
1455	2" diameter x 18" long		16	1		103	43	7.25	153.25	191
1460	24" long		15	1.067		134	46	7.70	187.70	231

05 05 19.20 Expansion Anchors

			Crew	Daily Output	Labor-Hours	Unit	Material	Labor	Equipment	Total	Total Incl O&P
0010	**EXPANSION ANCHORS**										
0100	Anchors for concrete, brick or stone, no layout and drilling										
0200	Expansion shields, zinc, 1/4" diameter, 1-5/16" long, single	G	1 Carp	90	.089	Ea.	.43	4.17		4.60	7.30
0300	1-3/8" long, double	G		85	.094		.54	4.42		4.96	7.85
0400	3/8" diameter, 1-1/2" long, single	G		85	.094		.64	4.42		5.06	7.95
0500	2" long, double	G		80	.100		1.18	4.70		5.88	9
0600	1/2" diameter, 2-1/16" long, single	G		80	.100		1.18	4.70		5.88	9
0700	2-1/2" long, double	G		75	.107		1.91	5		6.91	10.30
0800	5/8" diameter, 2-5/8" long, single	G		75	.107		2.05	5		7.05	10.45
0900	2-3/4" long, double	G		70	.114		2.70	5.35		8.05	11.75
1000	3/4" diameter, 2-3/4" long, single	G		70	.114		3.09	5.35		8.44	12.20
1100	3-15/16" long, double	G		65	.123		5	5.80		10.80	14.95
2100	Hollow wall anchors for gypsum wall board, plaster or tile										
2300	1/8" diameter, short	G	1 Carp	160	.050	Ea.	.23	2.35		2.58	4.10
2400	Long	G		150	.053		.23	2.50		2.73	4.35
2500	3/16" diameter, short	G		150	.053		.40	2.50		2.90	4.54

05 05 Common Work Results for Metals

05 05 19 – Post-Installed Concrete Anchors

05 05 19.20 Expansion Anchors

		Crew	Daily Output	Labor-Hours	Unit	Material	2015 Bare Costs Labor	2015 Bare Costs Equipment	Total	Total Incl O&P
2600	Long	1 Carp G	140	.057	Ea.	.59	2.68		3.27	5.05
2700	1/4" diameter, short	G	140	.057		.59	2.68		3.27	5.05
2800	Long	G	130	.062		.65	2.89		3.54	5.45
3000	Toggle bolts, bright steel, 1/8" diameter, 2" long	G	85	.094		.19	4.42		4.61	7.45
3100	4" long	G	80	.100		.25	4.70		4.95	8
3200	3/16" diameter, 3" long	G	80	.100		.26	4.70		4.96	8
3300	6" long	G	75	.107		.36	5		5.36	8.60
3400	1/4" diameter, 3" long	G	75	.107		.34	5		5.34	8.55
3500	6" long	G	70	.114		.51	5.35		5.86	9.35
3600	3/8" diameter, 3" long	G	70	.114		.84	5.35		6.19	9.70
3700	6" long	G	60	.133		1.36	6.25		7.61	11.75
3800	1/2" diameter, 4" long	G	60	.133		1.88	6.25		8.13	12.30
3900	6" long	G	50	.160		2.45	7.50		9.95	15
4000	Nailing anchors									
4100	Nylon nailing anchor, 1/4" diameter, 1" long	1 Carp	3.20	2.500	C	21	117		138	215
4200	1-1/2" long		2.80	2.857		26	134		160	249
4300	2" long		2.40	3.333		30.50	157		187.50	291
4400	Metal nailing anchor, 1/4" diameter, 1" long	G	3.20	2.500		17.85	117		134.85	212
4500	1-1/2" long	G	2.80	2.857		23.50	134		157.50	246
4600	2" long	G	2.40	3.333		27.50	157		184.50	288
8000	Wedge anchors, not including layout or drilling									
8050	Carbon steel, 1/4" diameter, 1-3/4" long	G 1 Carp	150	.053	Ea.	.41	2.50		2.91	4.55
8100	3-1/4" long	G	140	.057		.54	2.68		3.22	4.99
8150	3/8" diameter, 2-1/4" long	G	145	.055		.50	2.59		3.09	4.80
8200	5" long	G	140	.057		.88	2.68		3.56	5.35
8250	1/2" diameter, 2-3/4" long	G	140	.057		.99	2.68		3.67	5.50
8300	7" long	G	125	.064		1.70	3		4.70	6.80
8350	5/8" diameter, 3-1/2" long	G	130	.062		1.86	2.89		4.75	6.80
8400	8-1/2" long	G	115	.070		3.95	3.27		7.22	9.70
8450	3/4" diameter, 4-1/4" long	G	115	.070		2.90	3.27		6.17	8.55
8500	10" long	G	95	.084		6.60	3.95		10.55	13.75
8550	1" diameter, 6" long	G	100	.080		9.30	3.76		13.06	16.40
8575	9" long	G	85	.094		12.10	4.42		16.52	20.50
8600	12" long	G	75	.107		13.10	5		18.10	22.50
8650	1-1/4" diameter, 9" long	G	70	.114		24.50	5.35		29.85	36
8700	12" long	G	60	.133		31.50	6.25		37.75	45
8750	For type 303 stainless steel, add					350%				
8800	For type 316 stainless steel, add					450%				
8950	Self-drilling concrete screw, hex washer head, 3/16" diam. x 1-3/4" long	G 1 Carp	300	.027	Ea.	.20	1.25		1.45	2.27
8960	2-1/4" long	G	250	.032		.24	1.50		1.74	2.72
8970	Phillips flat head, 3/16" diam. x 1-3/4" long	G	300	.027		.21	1.25		1.46	2.28
8980	2-1/4" long	G	250	.032		.23	1.50		1.73	2.71
9000	Minimum labor/equipment charge		4	2	Job		94		94	154

05 05 21 – Fastening Methods for Metal

05 05 21.10 Cutting Steel

		Crew	Daily Output	Labor-Hours	Unit	Material	Labor	Equipment	Total	Total Incl O&P
0010	**CUTTING STEEL**									
0020	Hand burning, incl. preparation, torch cutting & grinding, no staging									
0050	Steel to 1/4" thick	E-25	400	.020	L.F.	.19	1.09	.03	1.31	2.21
0100	1/2" thick		320	.025		.35	1.37	.04	1.76	2.90
0150	3/4" thick		260	.031		.58	1.68	.04	2.30	3.73
0200	1" thick		200	.040		.84	2.19	.06	3.09	4.93
9000	Minimum labor/equipment charge		2	4	Job		219	5.70	224.70	400

05 05 Common Work Results for Metals

05 05 21 – Fastening Methods for Metal

05 05 21.15 Drilling Steel

05 05 21.15 Drilling Steel	Crew	Daily Output	Labor-Hours	Unit	Material	2015 Bare Costs Labor	2015 Bare Costs Equipment	Total	Total Incl O&P
0010 **DRILLING STEEL**									
1910 Drilling & layout for steel, up to 1/4" deep, no anchor									
1920 Holes, 1/4" diameter	1 Sswk	112	.071	Ea.	.09	3.76		3.85	6.90
1925 For each additional 1/4" depth, add		336	.024		.09	1.25		1.34	2.37
1930 3/8" diameter		104	.077		.09	4.05		4.14	7.40
1935 For each additional 1/4" depth, add		312	.026		.09	1.35		1.44	2.54
1940 1/2" diameter		96	.083		.10	4.39		4.49	8.05
1945 For each additional 1/4" depth, add		288	.028		.10	1.46		1.56	2.75
1950 5/8" diameter		88	.091		.14	4.79		4.93	8.80
1955 For each additional 1/4" depth, add		264	.030		.14	1.60		1.74	3.04
1960 3/4" diameter		80	.100		.18	5.25		5.43	9.70
1965 For each additional 1/4" depth, add		240	.033		.18	1.75		1.93	3.37
1970 7/8" diameter		72	.111		.23	5.85		6.08	10.80
1975 For each additional 1/4" depth, add		216	.037		.23	1.95		2.18	3.78
1980 1" diameter		64	.125		.24	6.60		6.84	12.15
1985 For each additional 1/4" depth, add		192	.042		.24	2.19		2.43	4.22
1990 For drilling up, add						40%			
2000 Minimum labor/equipment charge	1 Carp	4	2	Job		94		94	154

05 05 21.90 Welding Steel

05 05 21.90 Welding Steel	Crew	Daily Output	Labor-Hours	Unit	Material	2015 Bare Costs Labor	2015 Bare Costs Equipment	Total	Total Incl O&P
0010 **WELDING STEEL**, Structural R050521-20									
0020 Field welding, 1/8" E6011, cost per welder, no operating engineer	E-14	8	1	Hr.	4.33	54.50	18.25	77.08	124
0200 With 1/2 operating engineer	E-13	8	1.500		4.33	79	18.25	101.58	162
0300 With 1 operating engineer	E-12	8	2		4.33	103	18.25	125.58	201
0500 With no operating engineer, 2# weld rod per ton	E-14	8	1	Ton	4.33	54.50	18.25	77.08	124
0600 8# E6011 per ton	"	2	4		17.30	219	73	309.30	495
0800 With one operating engineer per welder, 2# E6011 per ton	E-12	8	2		4.33	103	18.25	125.58	201
0900 8# E6011 per ton	"	2	8		17.30	415	73	505.30	805
1200 Continuous fillet, down welding									
1300 Single pass, 1/8" thick, 0.1#/L.F.	E-14	150	.053	L.F.	.22	2.91	.97	4.10	6.55
1400 3/16" thick, 0.2#/L.F.		75	.107		.43	5.85	1.94	8.22	13.15
1500 1/4" thick, 0.3#/L.F.		50	.160		.65	8.75	2.92	12.32	19.70
1610 5/16" thick, 0.4#/L.F.		38	.211		.87	11.50	3.84	16.21	26
1800 3 passes, 3/8" thick, 0.5#/L.F.		30	.267		1.08	14.55	4.86	20.49	33
2010 4 passes, 1/2" thick, 0.7#/L.F.		22	.364		1.52	19.85	6.65	28.02	45
2200 5 to 6 passes, 3/4" thick, 1.3#/L.F.		12	.667		2.81	36.50	12.15	51.46	82.50
2400 8 to 11 passes, 1" thick, 2.4#/L.F.		6	1.333		5.20	73	24.50	102.70	164
2600 For vertical joint welding, add						20%			
2700 Overhead joint welding, add						300%			
2900 For semi-automatic welding, obstructed joints, deduct						5%			
3000 Exposed joints, deduct						15%			
4000 Cleaning and welding plates, bars, or rods									
4010 to existing beams, columns, or trusses	E-14	12	.667	L.F.	1.08	36.50	12.15	49.73	80.50
9000 Minimum labor/equipment charge	"	4	2	Job		109	36.50	145.50	238

05 05 23 – Metal Fastenings

05 05 23.30 Lag Screws

05 05 23.30 Lag Screws		Crew	Daily Output	Labor-Hours	Unit	Material	2015 Bare Costs Labor	2015 Bare Costs Equipment	Total	Total Incl O&P
0010 **LAG SCREWS**										
0020 Steel, 1/4" diameter, 2" long	G	1 Carp	200	.040	Ea.	.09	1.88		1.97	3.18
0200 1/2" diameter, 3" long	G		130	.062		.64	2.89		3.53	5.45
0300 5/8" diameter, 3" long	G		120	.067		1.15	3.13		4.28	6.40

05 05 Common Work Results for Metals

05 05 23 – Metal Fastenings

05 05 23.50 Powder Actuated Tools and Fasteners

		Crew	Daily Output	Labor-Hours	Unit	Material	2015 Bare Costs Labor	2015 Bare Costs Equipment	Total	Total Incl O&P
0010	**POWDER ACTUATED TOOLS & FASTENERS**									
0020	Stud driver, .22 caliber, single shot					Ea.	148			148	163
0100	.27 caliber, semi automatic, strip					"	440			440	485
0300	Powder load, single shot, .22 cal, power level 2, brown					C	5.35			5.35	5.90
0400	Strip, .27 cal, power level 4, red						7.70			7.70	8.50
0600	Drive pin, .300 x 3/4" long	G	1 Carp	4.80	1.667		4.11	78.50		82.61	133
0700	.300 x 3" long with washer	G	"	4	2		12.65	94		106.65	168

05 05 23.70 Structural Blind Bolts

			Crew	Daily Output	Labor-Hours	Unit	Material	Labor	Equipment	Total	Total Incl O&P
0010	**STRUCTURAL BLIND BOLTS**										
0100	1/4" diameter x 1/4" grip	G	1 Sswk	240	.033	Ea.	1.24	1.75		2.99	4.53
0150	1/2" grip	G		216	.037		1.33	1.95		3.28	4.98
0200	3/8" diameter x 1/2" grip	G		232	.034		1.75	1.82		3.57	5.20
0250	3/4" grip	G		208	.038		1.84	2.02		3.86	5.70
0300	1/2" diameter x 1/2" grip	G		224	.036		3.99	1.88		5.87	7.80
0350	3/4" grip	G		200	.040		5.60	2.11		7.71	9.95
0400	5/8" diameter x 3/4" grip	G		216	.037		8.25	1.95		10.20	12.60
0450	1" grip	G		192	.042		9.50	2.19		11.69	14.40

05 05 23.87 Weld Studs

			Crew	Daily Output	Labor-Hours	Unit	Material	Labor	Equipment	Total	Total Incl O&P
0010	**WELD STUDS**										
0020	1/4" diameter, 2-11/16" long	G	E-10	1120	.014	Ea.	.35	.77	.40	1.52	2.22
0100	4-1/8" long	G		1080	.015		.33	.79	.42	1.54	2.26
0200	3/8" diameter, 4-1/8" long	G		1080	.015		.38	.79	.42	1.59	2.32
0300	6-1/8" long	G		1040	.015		.49	.83	.43	1.75	2.50
9000	Minimum labor/equipment charge		1 Sswk	2	4	Job		211		211	380

05 12 Structural Steel Framing

05 12 23 – Structural Steel for Buildings

05 12 23.05 Canopy Framing

			Crew	Daily Output	Labor-Hours	Unit	Material	Labor	Equipment	Total	Total Incl O&P
0010	**CANOPY FRAMING**										
0020	6" and 8" members, shop fabricated	G	E-4	3000	.011	Lb.	1.59	.57	.05	2.21	2.82
9000	Minimum labor/equipment charge		1 Sswk	1	8	Job		420		420	760

05 12 23.10 Ceiling Supports

			Crew	Daily Output	Labor-Hours	Unit	Material	Labor	Equipment	Total	Total Incl O&P
0010	**CEILING SUPPORTS**										
1000	Entrance door/folding partition supports, shop fabricated	G	E-4	60	.533	L.F.	26.50	28.50	2.43	57.43	82.50
1100	Linear accelerator door supports	G		14	2.286		121	121	10.40	252.40	365
1200	Lintels or shelf angles, hung, exterior hot dipped galv.	G		267	.120		18.10	6.35	.55	25	32
1250	Two coats primer paint instead of galv.	G		267	.120		15.65	6.35	.55	22.55	29.50
1400	Monitor support, ceiling hung, expansion bolted	G		4	8	Ea.	420	425	36.50	881.50	1,275
1450	Hung from pre-set inserts	G		6	5.333		450	283	24.50	757.50	1,025
1600	Motor supports for overhead doors	G		4	8		214	425	36.50	675.50	1,050
1700	Partition support for heavy folding partitions, without pocket	G		24	1.333	L.F.	60.50	71	6.10	137.60	201
1750	Supports at pocket only	G		12	2.667		121	142	12.15	275.15	400
2000	Rolling grilles & fire door supports	G		34	.941		51.50	50	4.29	105.79	152
2100	Spider-leg light supports, expansion bolted to ceiling slab	G		8	4	Ea.	172	213	18.25	403.25	595
2150	Hung from pre-set inserts	G		12	2.667	"	186	142	12.15	340.15	475
2400	Toilet partition support	G		36	.889	L.F.	60.50	47	4.05	111.55	156
2500	X-ray travel gantry support	G		12	2.667	"	207	142	12.15	361.15	495

05 12 Structural Steel Framing

05 12 23 – Structural Steel for Buildings

05 12 23.17 Columns, Structural

			Crew	Daily Output	Labor-Hours	Unit	Material	2015 Bare Costs Labor	2015 Bare Costs Equipment	Total	Total Incl O&P
0010	**COLUMNS, STRUCTURAL**	R051223-10									
0015	Made from recycled materials										
0020	Shop fab'd for 100-ton, 1-2 story project, bolted connections										
0800	Steel, concrete filled, extra strong pipe, 3-1/2" diameter		E-2	660	.085	L.F.	43.50	4.39	2.29	50.18	58
0830	4" diameter			780	.072		48.50	3.71	1.94	54.15	62
0890	5" diameter			1020	.055		58	2.84	1.48	62.32	70.50
0930	6" diameter			1200	.047		77	2.41	1.26	80.67	90
0940	8" diameter			1100	.051		77	2.63	1.37	81	90.50
1500	Steel pipe, extra strong, no concrete, 3" to 5" diameter	G		16000	.004	Lb.	1.33	.18	.09	1.60	1.88
3300	Structural tubing, square, A500GrB, 4" to 6" square, light section	G		11270	.005	"	1.33	.26	.13	1.72	2.06
8090	For projects 75 to 99 tons, add					All	10%				
8092	50 to 74 tons, add						20%				
8094	25 to 49 tons, add						30%	10%			
8096	10 to 24 tons, add						50%	25%			
8098	2 to 9 tons, add						75%	50%			
8099	Less than 2 tons, add						100%	100%			
9000	Minimum labor/equipment charge		1 Sswk	1	8	Job		420		420	760

05 12 23.20 Curb Edging

			Crew	Daily Output	Labor-Hours	Unit	Material	Labor	Equipment	Total	Total Incl O&P
0010	**CURB EDGING**										
0020	Steel angle w/anchors, shop fabricated, on forms, 1" x 1", 0.8#/L.F.	G	E-4	350	.091	L.F.	1.67	4.86	.42	6.95	11.10
0300	4" x 4" angles, 8.2#/L.F.	G		275	.116	"	13.85	6.20	.53	20.58	27
9000	Minimum labor/equipment charge			4	8	Job		425	36.50	461.50	810

05 12 23.40 Lightweight Framing

			Crew	Daily Output	Labor-Hours	Unit	Material	Labor	Equipment	Total	Total Incl O&P
0010	**LIGHTWEIGHT FRAMING**										
0015	Made from recycled materials										
0200	For load-bearing steel studs see Section 05 41 13.30										
0400	Angle framing, field fabricated, 4" and larger	G	E-3	440	.055	Lb.	.77	2.91	.33	4.01	6.45
0450	Less than 4" angles	G		265	.091		.80	4.83	.55	6.18	10.25
0600	Channel framing, field fabricated, 8" and larger	G		500	.048		.80	2.56	.29	3.65	5.80
0650	Less than 8" channels	G		335	.072		.80	3.82	.44	5.06	8.25
1000	Continuous slotted channel framing system, shop fab, simple framing	G	2 Sswk	2400	.007		4.11	.35		4.46	5.15
1200	Complex framing	G	"	1600	.010		4.64	.53		5.17	6.05
1250	Plate & bar stock for reinforcing beams and trusses	G					1.46			1.46	1.60
1300	Cross bracing, rods, shop fabricated, 3/4" diameter	G	E-3	700	.034		1.59	1.83	.21	3.63	5.30
1310	7/8" diameter	G		850	.028		1.59	1.51	.17	3.27	4.66
1320	1" diameter	G		1000	.024		1.59	1.28	.15	3.02	4.22
1330	Angle, 5" x 5" x 3/8"	G		2800	.009		1.59	.46	.05	2.10	2.64
1350	Hanging lintels, shop fabricated	G		850	.028		1.59	1.51	.17	3.27	4.66
1380	Roof frames, shop fabricated, 3'-0" square, 5' span	G	E-2	4200	.013		1.59	.69	.36	2.64	3.36
1400	Tie rod, not upset, 1-1/2" to 4" diameter, with turnbuckle	G	2 Sswk	800	.020		1.72	1.05		2.77	3.79
1420	No turnbuckle	G		700	.023		1.66	1.20		2.86	4
1500	Upset, 1-3/4" to 4" diameter, with turnbuckle	G		800	.020		1.72	1.05		2.77	3.79
1520	No turnbuckle	G		700	.023		1.66	1.20		2.86	4
9000	Minimum labor/equipment charge			2	8	Job		420		420	760

05 12 23.45 Lintels

			Crew	Daily Output	Labor-Hours	Unit	Material	Labor	Equipment	Total	Total Incl O&P
0010	**LINTELS**										
0015	Made from recycled materials										
0020	Plain steel angles, shop fabricated, under 500 lb.	G	1 Bric	550	.015	Lb.	1.02	.67		1.69	2.21
0100	500 to 1000 lb.	G		640	.013		.99	.58		1.57	2.03
0200	1,000 to 2,000 lb.	G		640	.013		.97	.58		1.55	2
0300	2,000 to 4,000 lb.	G		640	.013		.94	.58		1.52	1.97

05 12 Structural Steel Framing

05 12 23 – Structural Steel for Buildings

05 12 23.45 Lintels

		Crew	Daily Output	Labor-Hours	Unit	Material	2015 Bare Costs Labor	Equipment	Total	Total Incl O&P
0500	For built-up angles and plates, add to above	G			Lb.	1.33			1.33	1.46
0700	For engineering, add to above					.13			.13	.15
0900	For galvanizing, add to above, under 500 lb.					.30			.30	.33
0950	500 to 2,000 lb.					.28			.28	.30
1000	Over 2,000 lb.					.25			.25	.28
2000	Steel angles, 3-1/2" x 3", 1/4" thick, 2'-6" long	G	1 Bric	47	.170	Ea.	14.30	7.85		22.15	28.50
2100	4'-6" long	G		26	.308		26	14.20		40.20	51.50
2500	3-1/2" x 3-1/2" x 5/16", 5'-0" long	G		18	.444		38	20.50		58.50	75.50
2600	4" x 3-1/2", 1/4" thick, 5'-0" long	G		21	.381		33	17.60		50.60	64.50
2700	9'-0" long	G		12	.667		59	31		90	115
2800	4" x 3-1/2" x 5/16", 7'-0" long	G		12	.667		57	31		88	113
2900	5" x 3-1/2" x 5/16", 10'-0" long	G		8	1		92	46		138	176
9000	Minimum labor/equipment charge			4	2	Job		92.50		92.50	150

05 12 23.60 Pipe Support Framing

		Crew	Daily Output	Labor-Hours	Unit	Material	Labor	Equipment	Total	Total Incl O&P
0010	**PIPE SUPPORT FRAMING**									
0020	Under 10#/L.F., shop fabricated	G	E-4	3900	.008	Lb.	1.78	.44	.04	2.26	2.78
0200	10.1 to 15#/L.F.	G		4300	.007		1.75	.40	.03	2.18	2.67
0400	15.1 to 20#/L.F.	G		4800	.007		1.72	.35	.03	2.10	2.56
0600	Over 20#/L.F.	G		5400	.006		1.70	.32	.03	2.05	2.47

05 12 23.65 Plates

		Crew	Daily Output	Labor-Hours	Unit	Material	Labor	Equipment	Total	Total Incl O&P
0010	**PLATES**									
0015	Made from recycled materials									
0020	For connections & stiffener plates, shop fabricated									
0050	1/8" thick (5.1 lb./S.F.)	G				S.F.	6.75			6.75	7.45
0100	1/4" thick (10.2 lb./S.F.)	G					13.50			13.50	14.85
0300	3/8" thick (15.3 lb./S.F.)	G					20.50			20.50	22.50
0400	1/2" thick (20.4 lb./S.F.)	G					27			27	29.50
0450	3/4" thick (30.6 lb./S.F.)	G					40.50			40.50	44.50
0500	1" thick (40.8 lb./S.F.)	G					54			54	59.50
2000	Steel plate, warehouse prices, no shop fabrication										
2100	1/4" thick (10.2 lb./S.F.)	G				S.F.	7.30			7.30	8

05 12 23.75 Structural Steel Members

		Crew	Daily Output	Labor-Hours	Unit	Material	Labor	Equipment	Total	Total Incl O&P
0010	**STRUCTURAL STEEL MEMBERS** R051223-10									
0015	Made from recycled materials									
0020	Shop fab'd for 100-ton, 1-2 story project, bolted connections									
0102	Beam or girder, W 6 x 9 R051223-15	G	E-2	600	.093	L.F.	13.10	4.83	2.52	20.45	25.50
0302	W 8 x 10	G		600	.093		14.60	4.83	2.52	21.95	27.50
0502	x 31	G		550	.102		45	5.25	2.75	53	61.50
0702	W 10 x 22	G		600	.093		32	4.83	2.52	39.35	46.50
0902	x 49	G		550	.102		71.50	5.25	2.75	79.50	90.50
1102	W 12 x 16	G		880	.064		23.50	3.29	1.72	28.51	33
1302	x 22	G		880	.064		32	3.29	1.72	37.01	43
1502	x 26	G		880	.064		38	3.29	1.72	43.01	49
1902	W 14 x 26	G		990	.057		38	2.93	1.53	42.46	48.50
2102	x 30	G		900	.062		43.50	3.22	1.68	48.40	55.50
2702	W 16 x 26	G		1000	.056		38	2.90	1.51	42.41	48
2902	x 31	G		900	.062		45	3.22	1.68	49.90	57
8490	For projects 75 to 99 tons, add						10%				
8492	50 to 74 tons, add						20%				
8494	25 to 49 tons, add						30%	10%			
8496	10 to 24 tons, add						50%	25%			
8498	2 to 9 tons, add						75%	50%			

05 12 Structural Steel Framing

05 12 23 – Structural Steel for Buildings

05 12 23.75 Structural Steel Members	Crew	Daily Output	Labor-Hours	Unit	Material	2015 Bare Costs Labor	Equipment	Total	Total Incl O&P
8499 Less than 2 tons, add				L.F.	100%	100%			
9000 Minimum labor/equipment charge	E-2	2	28	Job		1,450	755	2,205	3,350

05 12 23.77 Structural Steel Projects

0010 **STRUCTURAL STEEL PROJECTS**									
0015 Made from recycled materials									
0020 Shop fab'd for 100-ton, 1-2 story project, bolted connections									
0700 Offices, hospitals, etc., steel bearing, 1 to 2 stories R050521-20 G	E-5	10.30	7.767	Ton	2,650	405	161	3,216	3,825
1300 Industrial bldgs., 1 story, beams & girders, steel bearing G		12.90	6.202		2,650	325	128	3,103	3,625
1400 Masonry bearing R051223-10 G		10	8		2,650	420	166	3,236	3,850
1600 1 story with roof trusses, steel bearing G		10.60	7.547		3,125	395	156	3,676	4,325
1700 Masonry bearing R051223-15 G		8.30	9.639		3,125	505	200	3,830	4,550
5390 For projects 75 to 99 tons, add					10%				
5392 50 to 74 tons, add					20%				
5394 25 to 49 tons, add					30%	10%			
5396 10 to 24 tons, add					50%	25%			
5398 2 to 9 tons, add					75%	50%			
5399 Less than 2 tons, add					100%	100%			

05 12 23.78 Structural Steel Secondary Members

0010 **STRUCTURAL STEEL SECONDARY MEMBERS**									
0015 Made from recycled materials									
0020 Shop fabricated for 20-ton girt/purlin framing package, materials only									
0100 Girts/purlins, C/Z-shapes, includes clips and bolts									
0110 6" x 2-1/2" x 2-1/2", 16 ga., 3.0 lb./L.F.				L.F.	3.58			3.58	3.94
0115 14 ga., 3.5 lb./L.F.					4.17			4.17	4.59
0120 8" x 2-3/4" x 2-3/4", 16 ga., 3.4 lb./L.F.					4.05			4.05	4.46
0125 14 ga., 4.1 lb./L.F.					4.89			4.89	5.40
0130 12 ga., 5.6 lb./L.F.					6.70			6.70	7.35
0135 10" x 3-1/2" x 3-1/2", 14 ga., 4.7 lb./L.F.					5.60			5.60	6.15
0140 12 ga., 6.7 lb./L.F.					8			8	8.80
0145 12" x 3-1/2" x 3-1/2", 14 ga., 5.3 lb./L.F.					6.30			6.30	6.95
0150 12 ga., 7.4 lb./L.F.					8.80			8.80	9.70
0200 Eave struts, C-shape, includes clips and bolts									
0210 6" x 4" x 3", 16 ga., 3.1 lb./L.F.				L.F.	3.70			3.70	4.07
0215 14 ga., 3.9 lb./L.F.					4.65			4.65	5.10
0220 8" x 4" x 3", 16 ga., 3.5 lb./L.F.					4.17			4.17	4.59
0225 14 ga., 4.4 lb./L.F.					5.25			5.25	5.75
0230 12 ga., 6.2 lb./L.F.					7.40			7.40	8.15
0235 10" x 5" x 3", 14 ga., 5.2 lb./L.F.					6.20			6.20	6.80
0240 12 ga., 7.3 lb./L.F.					8.70			8.70	9.60
0245 12" x 5" x 4", 14 ga., 6.0 lb./L.F.					7.15			7.15	7.85
0250 12 ga., 8.4 lb./L.F.					10			10	11
0300 Rake/base angle, excludes concrete drilling and expansion anchors									
0310 2" x 2", 14 ga., 1.0 lb./L.F.	2 Sswk	640	.025	L.F.	1.19	1.32		2.51	3.69
0315 3" x 2", 14 ga., 1.3 lb./L.F.		535	.030		1.55	1.57		3.12	4.56
0320 3" x 3", 14 ga., 1.6 lb./L.F.		500	.032		1.91	1.68		3.59	5.15
0325 4" x 3", 14 ga., 1.8 lb./L.F.		480	.033		2.15	1.75		3.90	5.55
0600 Installation of secondary members, erection only									
0610 Girts, purlins, eave struts, 16 ga., 6" deep	E-18	100	.400	Ea.		21	9.50	30.50	47.50
0615 8" deep		80	.500			26.50	11.85	38.35	59.50
0620 14 ga., 6" deep		80	.500			26.50	11.85	38.35	59.50
0625 8" deep		65	.615			32.50	14.60	47.10	73
0630 10" deep		55	.727			38.50	17.25	55.75	86.50

05 12 Structural Steel Framing

05 12 23 – Structural Steel for Buildings

05 12 23.78 Structural Steel Secondary Members		Crew	Daily Output	Labor-Hours	Unit	Material	2015 Bare Costs Labor	Equipment	Total	Total Incl O&P
0635	12" deep	E-18	50	.800	Ea.		42	19	61	95.50
0640	12 ga., 8" deep		50	.800			42	19	61	95.50
0645	10" deep		45	.889			47	21	68	106
0650	12" deep	↓	40	1	↓		52.50	23.50	76	119
0900	For less than 20-ton job lots									
0905	For 15 to 19 tons, add					10%				
0910	For 10 to 14 tons, add					25%				
0915	For 5 to 9 tons, add					50%	50%	50%		
0920	For 1 to 4 tons, add					75%	75%	75%		
0925	For less than 1 ton, add					100%	100%	100%		
0990	Minimum labor/equipment charge	E-18	1	40	Job		2,100	950	3,050	4,775

05 15 Wire Rope Assemblies

05 15 16 – Steel Wire Rope Assemblies

05 15 16.70 Temporary Cable Safety Railing

			Crew	Daily Output	Labor-Hours	Unit	Material	Labor	Equipment	Total	Total Incl O&P
0010	**TEMPORARY CABLE SAFETY RAILING**, Each 100' strand incl.										
0020	2 eyebolts, 1 turnbuckle, 100' cable, 2 thimbles, 6 clips										
0025	Made from recycled materials										
0100	One strand using 1/4" cable & accessories	G	2 Sswk	4	4	C.L.F.	209	211		420	610
0200	1/2" cable & accessories	G	"	2	8	"	440	420		860	1,250

05 21 Steel Joist Framing

05 21 13 – Deep Longspan Steel Joist Framing

05 21 13.50 Deep Longspan Joists

			Crew	Daily Output	Labor-Hours	Unit	Material	Labor	Equipment	Total	Total Incl O&P
0010	**DEEP LONGSPAN JOISTS**										
3010	DLH series, 40-ton job lots, bolted cross bridging, shop primer										
3015	Made from recycled materials										
3040	Spans to 144' (shipped in 2 pieces)	G	E-7	13	6.154	Ton	1,925	320	139	2,384	2,850
3500	For less than 40-ton job lots										
3502	For 30 to 39 tons, add						10%				
3504	20 to 29 tons, add						20%				
3506	10 to 19 tons, add						30%				
3507	5 to 9 tons, add						50%	25%			
3508	1 to 4 tons, add						75%	50%			
3509	Less than 1 ton, add						100%	100%			
4010	SLH series, 40-ton job lots, bolted cross bridging, shop primer										
4040	Spans to 200' (shipped in 3 pieces)	G	E-7	13	6.154	Ton	1,975	320	139	2,434	2,900
6100	For less than 40-ton job lots										
6102	For 30 to 39 tons, add						10%				
6104	20 to 29 tons, add						20%				
6106	10 to 19 tons, add						30%				
6107	5 to 9 tons, add						50%	25%			
6108	1 to 4 tons, add						75%	50%			
6109	Less than 1 ton, add						100%	100%			

05 21 Steel Joist Framing

05 21 16 – Longspan Steel Joist Framing

05 21 16.50 Longspan Joists

		Crew	Daily Output	Labor-Hours	Unit	Material	2015 Bare Costs Labor	Equipment	Total	Total Incl O&P
0010	**LONGSPAN JOISTS**									
2000	LH series, 40-ton job lots, bolted cross bridging, shop primer									
2015	Made from recycled materials									
2040	Longspan joists, LH series, up to 96' G	E-7	13	6.154	Ton	1,825	320	139	2,284	2,750
2600	For less than 40-ton job lots									
2602	For 30 to 39 tons, add					10%				
2604	20 to 29 tons, add					20%				
2606	10 to 19 tons, add					30%				
2607	5 to 9 tons, add					50%	25%			
2608	1 to 4 tons, add					75%	50%			
2609	Less than 1 ton, add					100%	100%			

05 21 19 – Open Web Steel Joist Framing

05 21 19.10 Open Web Joists

		Crew	Daily Output	Labor-Hours	Unit	Material	2015 Bare Costs Labor	Equipment	Total	Total Incl O&P
0010	**OPEN WEB JOISTS**									
0015	Made from recycled materials									
0050	K series, 40-ton lots, horiz. bridging, spans to 30', shop primer G	E-7	12	6.667	Ton	1,650	350	150	2,150	2,600
0440	K series, 30' to 50' spans G	"	17	4.706	"	1,625	246	106	1,977	2,350
0800	For less than 40-ton job lots									
0802	For 30 to 39 tons, add					10%				
0804	20 to 29 tons, add					20%				
0806	10 to 19 tons, add					30%				
0807	5 to 9 tons, add					50%	25%			
0808	1 to 4 tons, add					75%	50%			
0809	Less than 1 ton, add					100%	100%			
1010	CS series, 40-ton job lots, horizontal bridging, shop primer									
1040	Spans to 30' G	E-7	12	6.667	Ton	1,700	350	150	2,200	2,650
1500	For less than 40-ton job lots									
1502	For 30 to 39 tons, add					10%				
1504	20 to 29 tons, add					20%				
1506	10 to 19 tons, add					30%				
1507	5 to 9 tons, add					50%	25%			
1508	1 to 4 tons, add					75%	50%			
1509	Less than 1 ton, add					100%	100%			
6400	Individual steel bearing plate, 6" x 6" x 1/4" with J-hook G	1 Bric	160	.050	Ea.	7.95	2.31		10.26	12.50
9000	Minimum labor and equipment, bar joists	F-6	1	40	Job		1,775	655	2,430	3,600

05 31 Steel Decking

05 31 13 – Steel Floor Decking

05 31 13.50 Floor Decking

		Crew	Daily Output	Labor-Hours	Unit	Material	2015 Bare Costs Labor	Equipment	Total	Total Incl O&P
0010	**FLOOR DECKING** R053100-10									
0015	Made from recycled materials									
5100	Non-cellular composite decking, galvanized, 1-1/2" deep, 16 ga. G	E-4	3500	.009	S.F.	3.44	.49	.04	3.97	4.72
5120	18 ga. G		3650	.009		2.78	.47	.04	3.29	3.94
5140	20 ga. G		3800	.008		2.22	.45	.04	2.71	3.29
5200	2" deep, 22 ga. G		3860	.008		1.93	.44	.04	2.41	2.96
5300	20 ga. G		3600	.009		2.13	.47	.04	2.64	3.23
5400	18 ga. G		3380	.009		2.73	.50	.04	3.27	3.96
5500	16 ga. G		3200	.010		3.41	.53	.05	3.99	4.76
5700	3" deep, 22 ga. G		3200	.010		2.10	.53	.05	2.68	3.32
5800	20 ga. G		3000	.011		2.34	.57	.05	2.96	3.65

05 31 Steel Decking

05 31 13 – Steel Floor Decking

05 31 13.50 Floor Decking		Crew	Daily Output	Labor-Hours	Unit	Material	2015 Bare Costs Labor	Equipment	Total	Total Incl O&P
5900	18 ga.	G E-4	2850	.011	S.F.	2.90	.60	.05	3.55	4.33
6000	16 ga.	G	2700	.012		3.87	.63	.05	4.55	5.45
9000	Minimum labor/equipment charge	1 Sswk	1	8	Job		420		420	760

05 31 23 – Steel Roof Decking

05 31 23.50 Roof Decking

0010	**ROOF DECKING**									
0015	Made from recycled materials									
2100	Open type, 1-1/2" deep, Type B, wide rib, galv., 22 ga., under 50 sq.	G E-4	4500	.007	S.F.	2.05	.38	.03	2.46	2.97
2400	Over 500 squares	G "	5100	.006	"	1.47	.33	.03	1.83	2.25

05 31 33 – Steel Form Decking

05 31 33.50 Form Decking

0010	**FORM DECKING**									
0015	Made from recycled materials									
6100	Slab form, steel, 28 ga., 9/16" deep, Type UFS, uncoated	G E-4	4000	.008	S.F.	1.55	.43	.04	2.02	2.51
6200	Galvanized	G "	4000	.008	"	1.37	.43	.04	1.84	2.32

05 35 Raceway Decking Assemblies

05 35 13 – Steel Cellular Decking

05 35 13.50 Cellular Decking

0010	**CELLULAR DECKING**									
0015	Made from recycled materials									
0200	Cellular units, galv, 1-1/2" deep, Type BC, 20-20 ga., over 15 squares	G E-4	1460	.022	S.F.	8.10	1.17	.10	9.37	11.10
0400	3" deep, Type NC, galvanized, 20-20 ga.	G	1375	.023		8.90	1.24	.11	10.25	12.15
1000	4-1/2" deep, Type JC, galvanized, 18-20 ga.	G	1100	.029		12.40	1.55	.13	14.08	16.60
1500	For acoustical deck, add					15%				
1900	For multi-story or congested site, add						50%			

05 41 Structural Metal Stud Framing

05 41 13 – Load-Bearing Metal Stud Framing

05 41 13.05 Bracing

0010	**BRACING**, shear wall X-bracing, per 10' x 10' bay, one face									
0015	Made of recycled materials									
0120	Metal strap, 20 ga. x 4" wide	G 2 Carp	18	.889	Ea.	17.55	41.50		59.05	88
0130	6" wide	G	18	.889		29.50	41.50		71	101
0160	18 ga. x 4" wide	G	16	1		30.50	47		77.50	111
0170	6" wide	G	16	1		45	47		92	127
0410	Continuous strap bracing, per horizontal row on both faces									
0420	Metal strap, 20 ga. x 2" wide, studs 12" O.C.	G 1 Carp	7	1.143	C.L.F.	53	53.50		106.50	146
0430	16" O.C.	G	8	1		53	47		100	135
0440	24" O.C.	G	10	.800		53	37.50		90.50	120
0450	18 ga. x 2" wide, studs 12" O.C.	G	6	1.333		75	62.50		137.50	186
0460	16" O.C.	G	7	1.143		75	53.50		128.50	171
0470	24" O.C.	G	8	1		75	47		122	160

05 41 13.10 Bridging

0010	**BRIDGING**, solid between studs w/1-1/4" leg track, per stud bay									
0015	Made from recycled materials									
0200	Studs 12" O.C., 18 ga. x 2-1/2" wide	G 1 Carp	125	.064	Ea.	.89	3		3.89	5.90
0210	3-5/8" wide	G	120	.067		1.07	3.13		4.20	6.35

05 41 Structural Metal Stud Framing

05 41 13 – Load-Bearing Metal Stud Framing

05 41 13.10 Bridging

			Crew	Daily Output	Labor-Hours	Unit	Material	2015 Bare Costs Labor	Equipment	Total	Total Incl O&P
0220	4" wide	G	1 Carp	120	.067	Ea.	1.13	3.13		4.26	6.40
0230	6" wide	G		115	.070		1.48	3.27		4.75	7
0240	8" wide	G		110	.073		1.84	3.41		5.25	7.60
0300	16 ga. x 2-1/2" wide	G		115	.070		1.13	3.27		4.40	6.60
0310	3-5/8" wide	G		110	.073		1.38	3.41		4.79	7.10
0320	4" wide	G		110	.073		1.47	3.41		4.88	7.20
0330	6" wide	G		105	.076		1.87	3.58		5.45	7.90
0340	8" wide	G		100	.080		2.34	3.76		6.10	8.75
1200	Studs 16" O.C., 18 ga. x 2-1/2" wide	G		125	.064		1.14	3		4.14	6.15
1210	3-5/8" wide	G		120	.067		1.37	3.13		4.50	6.65
1220	4" wide	G		120	.067		1.45	3.13		4.58	6.75
1230	6" wide	G		115	.070		1.90	3.27		5.17	7.45
1240	8" wide	G		110	.073		2.36	3.41		5.77	8.20
1300	16 ga. x 2-1/2" wide	G		115	.070		1.45	3.27		4.72	6.95
1310	3-5/8" wide	G		110	.073		1.77	3.41		5.18	7.55
1320	4" wide	G		110	.073		1.88	3.41		5.29	7.65
1330	6" wide	G		105	.076		2.39	3.58		5.97	8.50
1340	8" wide	G		100	.080		3	3.76		6.76	9.45
2200	Studs 24" O.C., 18 ga. x 2-1/2" wide	G		125	.064		1.65	3		4.65	6.75
2210	3-5/8" wide	G		120	.067		1.98	3.13		5.11	7.35
2220	4" wide	G		120	.067		2.10	3.13		5.23	7.45
2230	6" wide	G		115	.070		2.75	3.27		6.02	8.35
2240	8" wide	G		110	.073		3.41	3.41		6.82	9.35
2300	16 ga. x 2-1/2" wide	G		115	.070		2.10	3.27		5.37	7.65
2310	3-5/8" wide	G		110	.073		2.55	3.41		5.96	8.40
2320	4" wide	G		110	.073		2.72	3.41		6.13	8.60
2330	6" wide	G		105	.076		3.46	3.58		7.04	9.65
2340	8" wide	G		100	.080		4.34	3.76		8.10	10.95
3000	Continuous bridging, per row										
3100	16 ga. x 1-1/2" channel thru studs 12" O.C.	G	1 Carp	6	1.333	C.L.F.	48	62.50		110.50	156
3110	16" O.C.	G		7	1.143		48	53.50		101.50	141
3120	24" O.C.	G		8.80	.909		48	42.50		90.50	123
4100	2" x 2" angle x 18 ga., studs 12" O.C.	G		7	1.143		75	53.50		128.50	171
4110	16" O.C.	G		9	.889		75	41.50		116.50	151
4120	24" O.C.	G		12	.667		75	31.50		106.50	134
4200	16 ga., studs 12" O.C.	G		5	1.600		94.50	75		169.50	227
4210	16" O.C.	G		7	1.143		94.50	53.50		148	192
4220	24" O.C.	G		10	.800		94.50	37.50		132	166

05 41 13.25 Framing, Boxed Headers/Beams

			Crew	Daily Output	Labor-Hours	Unit	Material	Labor	Equipment	Total	Total Incl O&P
0010	**FRAMING, BOXED HEADERS/BEAMS**										
0015	Made from recycled materials										
0200	Double, 18 ga. x 6" deep	G	2 Carp	220	.073	L.F.	5.10	3.41		8.51	11.20
0210	8" deep	G		210	.076		5.65	3.58		9.23	12.05
0220	10" deep	G		200	.080		6.90	3.76		10.66	13.75
0230	12" deep	G		190	.084		7.55	3.95		11.50	14.80
0300	16 ga. x 8" deep	G		180	.089		6.50	4.17		10.67	14
0310	10" deep	G		170	.094		7.90	4.42		12.32	15.90
0320	12" deep	G		160	.100		8.60	4.70		13.30	17.15
0400	14 ga. x 10" deep	G		140	.114		9.10	5.35		14.45	18.80
0410	12" deep	G		130	.123		10	5.80		15.80	20.50
1210	Triple, 18 ga. x 8" deep	G		170	.094		8.20	4.42		12.62	16.25
1220	10" deep	G		165	.097		9.85	4.55		14.40	18.30

05 41 Structural Metal Stud Framing

05 41 13 – Load-Bearing Metal Stud Framing

05 41 13.25 Framing, Boxed Headers/Beams		Crew	Daily Output	Labor-Hours	Unit	Material	2015 Bare Costs Labor	Equipment	Total	Total Incl O&P
1230	12" deep	2 Carp G	160	.100	L.F.	10.85	4.70		15.55	19.60
1300	16 ga. x 8" deep	G	145	.110		9.50	5.20		14.70	18.90
1310	10" deep	G	140	.114		11.35	5.35		16.70	21.50
1320	12" deep	G	135	.119		12.40	5.55		17.95	22.50
1400	14 ga. x 10" deep	G	115	.139		12.40	6.55		18.95	24.50
1410	12" deep	G	110	.145		13.70	6.85		20.55	26.50

05 41 13.30 Framing, Stud Walls

		Crew	Daily Output	Labor-Hours	Unit	Material	Labor	Equipment	Total	Total Incl O&P
0010	**FRAMING, STUD WALLS** w/top & bottom track, no openings,									
0020	Headers, beams, bridging or bracing									
0025	Made from recycled materials									
4100	8' high walls, 18 ga. x 2-1/2" wide, studs 12" O.C.	2 Carp G	54	.296	L.F.	8.50	13.90		22.40	32.50
4110	16" O.C.	G	77	.208		6.80	9.75		16.55	23.50
4120	24" O.C.	G	107	.150		5.10	7		12.10	17.10
4130	3-5/8" wide, studs 12" O.C.	G	53	.302		10.05	14.15		24.20	34
4140	16" O.C.	G	76	.211		8.05	9.90		17.95	25
4150	24" O.C.	G	105	.152		6.05	7.15		13.20	18.40
4160	4" wide, studs 12" O.C.	G	52	.308		10.55	14.45		25	35
4170	16" O.C.	G	74	.216		8.45	10.15		18.60	26
4180	24" O.C.	G	103	.155		6.35	7.30		13.65	18.95
4190	6" wide, studs 12" O.C.	G	51	.314		13.40	14.75		28.15	38.50
4200	16" O.C.	G	73	.219		10.75	10.30		21.05	28.50
4210	24" O.C.	G	101	.158		8.10	7.45		15.55	21
4220	8" wide, studs 12" O.C.	G	50	.320		16.30	15		31.30	42.50
4230	16" O.C.	G	72	.222		13.10	10.45		23.55	31.50
4240	24" O.C.	G	100	.160		9.90	7.50		17.40	23
4300	16 ga. x 2-1/2" wide, studs 12" O.C.	G	47	.340		10.10	16		26.10	37
4310	16" O.C.	G	68	.235		8	11.05		19.05	27
4320	24" O.C.	G	94	.170		5.90	8		13.90	19.60
4330	3-5/8" wide, studs 12" O.C.	G	46	.348		12.05	16.35		28.40	40.50
4340	16" O.C.	G	66	.242		9.55	11.40		20.95	29
4350	24" O.C.	G	92	.174		7.05	8.15		15.20	21
4360	4" wide, studs 12" O.C.	G	45	.356		12.65	16.70		29.35	41.50
4370	16" O.C.	G	65	.246		10	11.55		21.55	30
4380	24" O.C.	G	90	.178		7.40	8.35		15.75	22
4390	6" wide, studs 12" O.C.	G	44	.364		15.80	17.05		32.85	45.50
4400	16" O.C.	G	64	.250		12.55	11.75		24.30	33
4410	24" O.C.	G	88	.182		9.30	8.55		17.85	24.50
4420	8" wide, studs 12" O.C.	G	43	.372		19.50	17.45		36.95	50
4430	16" O.C.	G	63	.254		15.50	11.90		27.40	36.50
4440	24" O.C.	G	86	.186		11.50	8.75		20.25	27
5100	10' high walls, 18 ga. x 2-1/2" wide, studs 12" O.C.	G	54	.296		10.20	13.90		24.10	34
5110	16" O.C.	G	77	.208		8.05	9.75		17.80	25
5120	24" O.C.	G	107	.150		5.95	7		12.95	18.05
5130	3-5/8" wide, studs 12" O.C.	G	53	.302		12.05	14.15		26.20	36.50
5140	16" O.C.	G	76	.211		9.55	9.90		19.45	26.50
5150	24" O.C.	G	105	.152		7.05	7.15		14.20	19.50
5160	4" wide, studs 12" O.C.	G	52	.308		12.65	14.45		27.10	37.50
5170	16" O.C.	G	74	.216		10.05	10.15		20.20	27.50
5180	24" O.C.	G	103	.155		7.40	7.30		14.70	20
5190	6" wide, studs 12" O.C.	G	51	.314		16	14.75		30.75	41.50
5200	16" O.C.	G	73	.219		12.70	10.30		23	31
5210	24" O.C.	G	101	.158		9.40	7.45		16.85	22.50

05 41 Structural Metal Stud Framing

05 41 13 – Load-Bearing Metal Stud Framing

05 41 13.30 Framing, Stud Walls			Crew	Daily Output	Labor-Hours	Unit	Material	2015 Bare Costs Labor	Equipment	Total	Total Incl O&P
5220	8" wide, studs 12" O.C.	G	2 Carp	50	.320	L.F.	19.50	15		34.50	46
5230	16" O.C.	G		72	.222		15.50	10.45		25.95	34
5240	24" O.C.	G		100	.160		11.50	7.50		19	25
5300	16 ga. x 2-1/2" wide, studs 12" O.C.	G		47	.340		12.20	16		28.20	39.50
5310	16" O.C.	G		68	.235		9.55	11.05		20.60	28.50
5320	24" O.C.	G		94	.170		6.95	8		14.95	21
5330	3-5/8" wide, studs 12" O.C.	G		46	.348		14.55	16.35		30.90	43
5340	16" O.C.	G		66	.242		11.40	11.40		22.80	31
5350	24" O.C.	G		92	.174		8.30	8.15		16.45	22.50
5360	4" wide, studs 12" O.C.	G		45	.356		15.25	16.70		31.95	44.50
5370	16" O.C.	G		65	.246		12	11.55		23.55	32
5380	24" O.C.	G		90	.178		8.70	8.35		17.05	23.50
5390	6" wide, studs 12" O.C.	G		44	.364		19	17.05		36.05	49
5400	16" O.C.	G		64	.250		14.95	11.75		26.70	35.50
5410	24" O.C.	G		88	.182		10.90	8.55		19.45	26
5420	8" wide, studs 12" O.C.	G		43	.372		23.50	17.45		40.95	54.50
5430	16" O.C.	G		63	.254		18.50	11.90		30.40	40
5440	24" O.C.	G		86	.186		13.50	8.75		22.25	29
6190	12' high walls, 18 ga. x 6" wide, studs 12" O.C.	G		41	.390		18.65	18.30		36.95	50.50
6200	16" O.C.	G		58	.276		14.70	12.95		27.65	37
6210	24" O.C.	G		81	.198		10.75	9.25		20	27
6220	8" wide, studs 12" O.C.	G		40	.400		22.50	18.80		41.30	56
6230	16" O.C.	G		57	.281		17.90	13.20		31.10	41
6240	24" O.C.	G		80	.200		13.10	9.40		22.50	30
6390	16 ga. x 6" wide, studs 12" O.C.	G		35	.457		22.50	21.50		44	59.50
6400	16" O.C.	G		51	.314		17.40	14.75		32.15	43
6410	24" O.C.	G		70	.229		12.55	10.75		23.30	31.50
6420	8" wide, studs 12" O.C.	G		34	.471		27.50	22		49.50	66.50
6430	16" O.C.	G		50	.320		21.50	15		36.50	48
6440	24" O.C.	G		69	.232		15.50	10.90		26.40	35
6530	14 ga. x 3-5/8" wide, studs 12" O.C.	G		34	.471		21	22		43	59.50
6540	16" O.C.	G		48	.333		16.55	15.65		32.20	44
6550	24" O.C.	G		65	.246		11.90	11.55		23.45	32
6560	4" wide, studs 12" O.C.	G		33	.485		22.50	23		45.50	62
6570	16" O.C.	G		47	.340		17.55	16		33.55	45.50
6580	24" O.C.	G		64	.250		12.65	11.75		24.40	33
6730	12 ga. x 3-5/8" wide, studs 12" O.C.	G		31	.516		29.50	24		53.50	72
6740	16" O.C.	G		43	.372		23	17.45		40.45	53.50
6750	24" O.C.	G		59	.271		16.05	12.75		28.80	38.50
6760	4" wide, studs 12" O.C.	G		30	.533		31.50	25		56.50	75.50
6770	16" O.C.	G		42	.381		24.50	17.90		42.40	56
6780	24" O.C.	G		58	.276		17.15	12.95		30.10	40
7390	16' high walls, 16 ga. x 6" wide, studs 12" O.C.	G		33	.485		28.50	23		51.50	69
7400	16" O.C.	G		48	.333		22.50	15.65		38.15	50
7410	24" O.C.	G		67	.239		15.80	11.20		27	36
7420	8" wide, studs 12" O.C.	G		32	.500		35.50	23.50		59	77.50
7430	16" O.C.	G		47	.340		27.50	16		43.50	56.50
7440	24" O.C.	G		66	.242		19.50	11.40		30.90	40
7560	14 ga. x 4" wide, studs 12" O.C.	G		31	.516		29	24		53	71.50
7570	16" O.C.	G		45	.356		22.50	16.70		39.20	52
7580	24" O.C.	G		61	.262		15.90	12.30		28.20	37.50
7590	6" wide, studs 12" O.C.	G		30	.533		36.50	25		61.50	81
7600	16" O.C.	G		44	.364		28.50	17.05		45.55	59

05 41 Structural Metal Stud Framing

05 41 13 – Load-Bearing Metal Stud Framing

05 41 13.30 Framing, Stud Walls

			Crew	Daily Output	Labor-Hours	Unit	Material	2015 Bare Costs Labor	Equipment	Total	Total Incl O&P
7610	24" O.C.	G	2 Carp	60	.267	L.F.	20	12.50		32.50	42.50
7760	12 ga. x 4" wide, studs 12" O.C.	G		29	.552		41	26		67	87.50
7770	16" O.C.	G		40	.400		31.50	18.80		50.30	65.50
7780	24" O.C.	G		55	.291		22	13.65		35.65	46.50
7790	6" wide, studs 12" O.C.	G		28	.571		51.50	27		78.50	101
7800	16" O.C.	G		39	.410		39.50	19.25		58.75	75
7810	24" O.C.	G		54	.296		27.50	13.90		41.40	53.50
8590	20' high walls, 14 ga. x 6" wide, studs 12" O.C.	G		29	.552		45	26		71	91.50
8600	16" O.C.	G		42	.381		34.50	17.90		52.40	67.50
8610	24" O.C.	G		57	.281		24	13.20		37.20	48
8620	8" wide, studs 12" O.C.	G		28	.571		48.50	27		75.50	97.50
8630	16" O.C.	G		41	.390		37.50	18.30		55.80	71
8640	24" O.C.	G		56	.286		26.50	13.40		39.90	51
8790	12 ga. x 6" wide, studs 12" O.C.	G		27	.593		64	28		92	116
8800	16" O.C.	G		37	.432		48.50	20.50		69	87
8810	24" O.C.	G		51	.314		33.50	14.75		48.25	61
8820	8" wide, studs 12" O.C.	G		26	.615		77.50	29		106.50	133
8830	16" O.C.	G		36	.444		59	21		80	99
8840	24" O.C.	G		50	.320		41	15		56	69.50
9000	Minimum labor/equipment charge			4	4	Job		188		188	310

05 42 Cold-Formed Metal Joist Framing

05 42 13 – Cold-Formed Metal Floor Joist Framing

05 42 13.05 Bracing

			Crew	Daily Output	Labor-Hours	Unit	Material	Labor	Equipment	Total	Total Incl O&P
0010	**BRACING**, continuous, per row, top & bottom										
0015	Made from recycled materials										
0120	Flat strap, 20 ga. x 2" wide, joists at 12" O.C.	G	1 Carp	4.67	1.713	C.L.F.	55	80.50		135.50	193
0130	16" O.C.	G		5.33	1.501		53.50	70.50		124	175
0140	24" O.C.	G		6.66	1.201		51.50	56.50		108	149
0150	18 ga. x 2" wide, joists at 12" O.C.	G		4	2		74	94		168	236
0160	16" O.C.	G		4.67	1.713		73	80.50		153.50	212
0170	24" O.C.	G		5.33	1.501		71.50	70.50		142	195

05 42 13.10 Bridging

			Crew	Daily Output	Labor-Hours	Unit	Material	Labor	Equipment	Total	Total Incl O&P
0010	**BRIDGING**, solid between joists w/1-1/4" leg track, per joist bay										
0015	Made from recycled materials										
0230	Joists 12" O.C., 18 ga. track x 6" wide	G	1 Carp	80	.100	Ea.	1.48	4.70		6.18	9.35
0240	8" wide	G		75	.107		1.84	5		6.84	10.20
0250	10" wide	G		70	.114		2.29	5.35		7.64	11.30
0260	12" wide	G		65	.123		2.60	5.80		8.40	12.30
0330	16 ga. track x 6" wide	G		70	.114		1.87	5.35		7.22	10.85
0340	8" wide	G		65	.123		2.34	5.80		8.14	12.05
0350	10" wide	G		60	.133		2.91	6.25		9.16	13.45
0360	12" wide	G		55	.145		3.35	6.85		10.20	14.90
0440	14 ga. track x 8" wide	G		60	.133		2.93	6.25		9.18	13.50
0450	10" wide	G		55	.145		3.64	6.85		10.49	15.20
0460	12" wide	G		50	.160		4.20	7.50		11.70	16.90
0550	12 ga. track x 10" wide	G		45	.178		5.35	8.35		13.70	19.55
0560	12" wide	G		40	.200		5.45	9.40		14.85	21.50
1230	16" O.C., 18 ga. track x 6" wide	G		80	.100		1.90	4.70		6.60	9.80
1240	8" wide	G		75	.107		2.36	5		7.36	10.80
1250	10" wide	G		70	.114		2.94	5.35		8.29	12.05

05 42 Cold-Formed Metal Joist Framing

05 42 13 – Cold-Formed Metal Floor Joist Framing

05 42 13.10 Bridging

			Crew	Daily Output	Labor-Hours	Unit	Material	2015 Bare Costs Labor	Equipment	Total	Total Incl O&P
1260	12" wide	G	1 Carp	65	.123	Ea.	3.33	5.80		9.13	13.10
1330	16 ga. track x 6" wide	G		70	.114		2.39	5.35		7.74	11.45
1340	8" wide	G		65	.123		3	5.80		8.80	12.75
1350	10" wide	G		60	.133		3.73	6.25		9.98	14.35
1360	12" wide	G		55	.145		4.29	6.85		11.14	15.90
1440	14 ga. track x 8" wide	G		60	.133		3.76	6.25		10.01	14.40
1450	10" wide	G		55	.145		4.67	6.85		11.52	16.35
1460	12" wide	G		50	.160		5.40	7.50		12.90	18.20
1550	12 ga. track x 10" wide	G		45	.178		6.85	8.35		15.20	21
1560	12" wide	G		40	.200		7	9.40		16.40	23
2230	24" O.C., 18 ga. track x 6" wide	G		80	.100		2.75	4.70		7.45	10.70
2240	8" wide	G		75	.107		3.41	5		8.41	11.95
2250	10" wide	G		70	.114		4.25	5.35		9.60	13.45
2260	12" wide	G		65	.123		4.82	5.80		10.62	14.75
2330	16 ga. track x 6" wide	G		70	.114		3.46	5.35		8.81	12.60
2340	8" wide	G		65	.123		4.34	5.80		10.14	14.25
2350	10" wide	G		60	.133		5.40	6.25		11.65	16.20
2360	12" wide	G		55	.145		6.20	6.85		13.05	18.05
2440	14 ga. track x 8" wide	G		60	.133		5.45	6.25		11.70	16.25
2450	10" wide	G		55	.145		6.75	6.85		13.60	18.65
2460	12" wide	G		50	.160		7.80	7.50		15.30	21
2550	12 ga. track x 10" wide	G		45	.178		9.90	8.35		18.25	24.50
2560	12" wide	G		40	.200		10.10	9.40		19.50	26.50

05 42 13.25 Framing, Band Joist

			Crew	Daily Output	Labor-Hours	Unit	Material	Labor	Equipment	Total	Total Incl O&P
0010	**FRAMING, BAND JOIST** (track) fastened to bearing wall										
0015	Made from recycled materials										
0220	18 ga. track x 6" deep	G	2 Carp	1000	.016	L.F.	1.21	.75		1.96	2.56
0230	8" deep	G		920	.017		1.50	.82		2.32	2.99
0240	10" deep	G		860	.019		1.87	.87		2.74	3.49
0320	16 ga. track x 6" deep	G		900	.018		1.52	.83		2.35	3.04
0330	8" deep	G		840	.019		1.91	.89		2.80	3.57
0340	10" deep	G		780	.021		2.37	.96		3.33	4.19
0350	12" deep	G		740	.022		2.73	1.02		3.75	4.66
0430	14 ga. track x 8" deep	G		750	.021		2.39	1		3.39	4.27
0440	10" deep	G		720	.022		2.97	1.04		4.01	4.98
0450	12" deep	G		700	.023		3.42	1.07		4.49	5.55
0540	12 ga. track x 10" deep	G		670	.024		4.35	1.12		5.47	6.60
0550	12" deep	G		650	.025		4.45	1.16		5.61	6.80

05 42 13.30 Framing, Boxed Headers/Beams

			Crew	Daily Output	Labor-Hours	Unit	Material	Labor	Equipment	Total	Total Incl O&P
0010	**FRAMING, BOXED HEADERS/BEAMS**										
0015	Made from recycled materials										
0200	Double, 18 ga. x 6" deep	G	2 Carp	220	.073	L.F.	5.10	3.41		8.51	11.20
0210	8" deep	G		210	.076		5.65	3.58		9.23	12.05
0220	10" deep	G		200	.080		6.90	3.76		10.66	13.75
0230	12" deep	G		190	.084		7.55	3.95		11.50	14.80
0300	16 ga. x 8" deep	G		180	.089		6.50	4.17		10.67	14
0310	10" deep	G		170	.094		7.90	4.42		12.32	15.90
0320	12" deep	G		160	.100		8.60	4.70		13.30	17.15
0400	14 ga. x 10" deep	G		140	.114		9.10	5.35		14.45	18.80
0410	12" deep	G		130	.123		10	5.80		15.80	20.50
0500	12 ga. x 10" deep	G		110	.145		12	6.85		18.85	24.50
0510	12" deep	G		100	.160		13.25	7.50		20.75	27

05 42 Cold-Formed Metal Joist Framing

05 42 13 – Cold-Formed Metal Floor Joist Framing

05 42 13.30 Framing, Boxed Headers/Beams		Crew	Daily Output	Labor-Hours	Unit	Material	2015 Bare Costs Labor	Equipment	Total	Total Incl O&P
1210	Triple, 18 ga. x 8" deep	G 2 Carp	170	.094	L.F.	8.20	4.42		12.62	16.25
1220	10" deep	G	165	.097		9.85	4.55		14.40	18.30
1230	12" deep	G	160	.100		10.85	4.70		15.55	19.60
1300	16 ga. x 8" deep	G	145	.110		9.50	5.20		14.70	18.90
1310	10" deep	G	140	.114		11.35	5.35		16.70	21.50
1320	12" deep	G	135	.119		12.40	5.55		17.95	22.50
1400	14 ga. x 10" deep	G	115	.139		13.15	6.55		19.70	25
1410	12" deep	G	110	.145		14.50	6.85		21.35	27
1500	12 ga. x 10" deep	G	90	.178		17.50	8.35		25.85	33
1510	12" deep	G	85	.188		19.40	8.85		28.25	36

05 42 13.40 Framing, Joists		Crew	Daily Output	Labor-Hours	Unit	Material	Labor	Equipment	Total	Total Incl O&P
0010	**FRAMING, JOISTS**, no band joists (track), web stiffeners, headers,									
0020	Beams, bridging or bracing									
0025	Made from recycled materials									
0030	Joists (2" flange) and fasteners, materials only									
0220	18 ga. x 6" deep	G			L.F.	1.58			1.58	1.73
0230	8" deep	G				1.86			1.86	2.04
0240	10" deep	G				2.18			2.18	2.40
0320	16 ga. x 6" deep	G				1.93			1.93	2.13
0330	8" deep	G				2.31			2.31	2.54
0340	10" deep	G				2.70			2.70	2.97
0350	12" deep	G				3.07			3.07	3.37
0430	14 ga. x 8" deep	G				2.90			2.90	3.19
0440	10" deep	G				3.34			3.34	3.67
0450	12" deep	G				3.80			3.80	4.18
0540	12 ga. x 10" deep	G				4.86			4.86	5.35
0550	12" deep	G				5.50			5.50	6.10
1010	Installation of joists to band joists, beams & headers, labor only									
1220	18 ga. x 6" deep	2 Carp	110	.145	Ea.		6.85		6.85	11.20
1230	8" deep		90	.178			8.35		8.35	13.70
1240	10" deep		80	.200			9.40		9.40	15.40
1320	16 ga. x 6" deep		95	.168			7.90		7.90	12.95
1330	8" deep		70	.229			10.75		10.75	17.60
1340	10" deep		60	.267			12.50		12.50	20.50
1350	12" deep		55	.291			13.65		13.65	22.50
1430	14 ga. x 8" deep		65	.246			11.55		11.55	18.95
1440	10" deep		45	.356			16.70		16.70	27.50
1450	12" deep		35	.457			21.50		21.50	35
1540	12 ga. x 10" deep		40	.400			18.80		18.80	31
1550	12" deep		30	.533			25		25	41
9000	Minimum labor/equipment charge		4	4	Job		188		188	310

05 42 13.45 Framing, Web Stiffeners		Crew	Daily Output	Labor-Hours	Unit	Material	Labor	Equipment	Total	Total Incl O&P
0010	**FRAMING, WEB STIFFENERS** at joist bearing, fabricated from									
0020	Stud piece (1-5/8" flange) to stiffen joist (2" flange)									
0025	Made from recycled materials									
2120	For 6" deep joist, with 18 ga. x 2-1/2" stud	G 1 Carp	120	.067	Ea.	.85	3.13		3.98	6.10
2130	3-5/8" stud	G	110	.073		1	3.41		4.41	6.70
2140	4" stud	G	105	.076		1.05	3.58		4.63	7
2150	6" stud	G	100	.080		1.32	3.76		5.08	7.60
2160	8" stud	G	95	.084		1.60	3.95		5.55	8.25
2220	8" deep joist, with 2-1/2" stud	G	120	.067		1.14	3.13		4.27	6.40
2230	3-5/8" stud	G	110	.073		1.34	3.41		4.75	7.05

05 42 Cold-Formed Metal Joist Framing

05 42 13 – Cold-Formed Metal Floor Joist Framing

05 42 13.45 Framing, Web Stiffeners		Crew	Daily Output	Labor-Hours	Unit	Material	2015 Bare Costs Labor	Equipment	Total	Total Incl O&P
2240	4" stud	G 1 Carp	105	.076	Ea.	1.41	3.58		4.99	7.40
2250	6" stud	G	100	.080		1.77	3.76		5.53	8.10
2260	8" stud	G	95	.084		2.14	3.95		6.09	8.85
2320	10" deep joist, with 2-1/2" stud	G	110	.073		1.41	3.41		4.82	7.15
2330	3-5/8" stud	G	100	.080		1.66	3.76		5.42	8
2340	4" stud	G	95	.084		1.74	3.95		5.69	8.40
2350	6" stud	G	90	.089		2.19	4.17		6.36	9.25
2360	8" stud	G	85	.094		2.66	4.42		7.08	10.15
2420	12" deep joist, with 2-1/2" stud	G	110	.073		1.70	3.41		5.11	7.45
2430	3-5/8" stud	G	100	.080		2	3.76		5.76	8.35
2440	4" stud	G	95	.084		2.10	3.95		6.05	8.80
2450	6" stud	G	90	.089		2.64	4.17		6.81	9.75
2460	8" stud	G	85	.094		3.20	4.42		7.62	10.75
3130	For 6" deep joist, with 16 ga. x 3-5/8" stud	G	100	.080		1.25	3.76		5.01	7.55
3140	4" stud	G	95	.084		1.31	3.95		5.26	7.95
3150	6" stud	G	90	.089		1.62	4.17		5.79	8.65
3160	8" stud	G	85	.094		2	4.42		6.42	9.45
3230	8" deep joist, with 3-5/8" stud	G	100	.080		1.68	3.76		5.44	8
3240	4" stud	G	95	.084		1.76	3.95		5.71	8.45
3250	6" stud	G	90	.089		2.17	4.17		6.34	9.25
3260	8" stud	G	85	.094		2.68	4.42		7.10	10.20
3330	10" deep joist, with 3-5/8" stud	G	85	.094		2.08	4.42		6.50	9.55
3340	4" stud	G	80	.100		2.17	4.70		6.87	10.10
3350	6" stud	G	75	.107		2.69	5		7.69	11.15
3360	8" stud	G	70	.114		3.32	5.35		8.67	12.45
3430	12" deep joist, with 3-5/8" stud	G	85	.094		2.50	4.42		6.92	10
3440	4" stud	G	80	.100		2.62	4.70		7.32	10.60
3450	6" stud	G	75	.107		3.24	5		8.24	11.75
3460	8" stud	G	70	.114		4	5.35		9.35	13.20
4230	For 8" deep joist, with 14 ga. x 3-5/8" stud	G	90	.089		2.08	4.17		6.25	9.15
4240	4" stud	G	85	.094		2.20	4.42		6.62	9.65
4250	6" stud	G	80	.100		2.76	4.70		7.46	10.75
4260	8" stud	G	75	.107		2.95	5		7.95	11.45
4330	10" deep joist, with 3-5/8" stud	G	75	.107		2.57	5		7.57	11.05
4340	4" stud	G	70	.114		2.72	5.35		8.07	11.80
4350	6" stud	G	65	.123		3.42	5.80		9.22	13.20
4360	8" stud	G	60	.133		3.65	6.25		9.90	14.25
4430	12" deep joist, with 3-5/8" stud	G	75	.107		3.10	5		8.10	11.60
4440	4" stud	G	70	.114		3.28	5.35		8.63	12.40
4450	6" stud	G	65	.123		4.12	5.80		9.92	14
4460	8" stud	G	60	.133		4.40	6.25		10.65	15.10
5330	For 10" deep joist, with 12 ga. x 3-5/8" stud	G	65	.123		3.72	5.80		9.52	13.55
5340	4" stud	G	60	.133		3.97	6.25		10.22	14.60
5350	6" stud	G	55	.145		5	6.85		11.85	16.70
5360	8" stud	G	50	.160		6.05	7.50		13.55	18.95
5430	12" deep joist, with 3-5/8" stud	G	65	.123		4.48	5.80		10.28	14.40
5440	4" stud	G	60	.133		4.78	6.25		11.03	15.50
5450	6" stud	G	55	.145		6	6.85		12.85	17.80
5460	8" stud	G	50	.160		7.30	7.50		14.80	20.50

05 42 Cold-Formed Metal Joist Framing

05 42 23 – Cold-Formed Metal Roof Joist Framing

05 42 23.05 Framing, Bracing

		Crew	Daily Output	Labor-Hours	Unit	Material	2015 Bare Costs Labor	2015 Bare Costs Equipment	Total	Total Incl O&P
0010	**FRAMING, BRACING**									
0015	Made from recycled materials									
0020	Continuous bracing, per row									
0100	16 ga. x 1-1/2" channel thru rafters/trusses @ 16" O.C.	G 1 Carp	4.50	1.778	C.L.F.	48	83.50		131.50	190
0120	24" O.C.	G	6	1.333		48	62.50		110.50	156
0300	2" x 2" angle x 18 ga., rafters/trusses @ 16" O.C.	G	6	1.333		75	62.50		137.50	186
0320	24" O.C.	G	8	1		75	47		122	160
0400	16 ga., rafters/trusses @ 16" O.C.	G	4.50	1.778		94.50	83.50		178	241
0420	24" O.C.	G	6.50	1.231		94.50	58		152.50	199

05 42 23.10 Framing, Bridging

		Crew	Daily Output	Labor-Hours	Unit	Material	Labor	Equipment	Total	Total Incl O&P
0010	**FRAMING, BRIDGING**									
0015	Made from recycled materials									
0020	Solid, between rafters w/1-1/4" leg track, per rafter bay									
1200	Rafters 16" O.C., 18 ga. x 4" deep	G 1 Carp	60	.133	Ea.	1.45	6.25		7.70	11.85
1210	6" deep	G	57	.140		1.90	6.60		8.50	12.90
1220	8" deep	G	55	.145		2.36	6.85		9.21	13.80
1230	10" deep	G	52	.154		2.94	7.20		10.14	15.10
1240	12" deep	G	50	.160		3.33	7.50		10.83	15.95
2200	24" O.C., 18 ga. x 4" deep	G	60	.133		2.10	6.25		8.35	12.55
2210	6" deep	G	57	.140		2.75	6.60		9.35	13.80
2220	8" deep	G	55	.145		3.41	6.85		10.26	14.95
2230	10" deep	G	52	.154		4.25	7.20		11.45	16.50
2240	12" deep	G	50	.160		4.82	7.50		12.32	17.60

05 42 23.50 Framing, Parapets

		Crew	Daily Output	Labor-Hours	Unit	Material	Labor	Equipment	Total	Total Incl O&P
0010	**FRAMING, PARAPETS**									
0015	Made from recycled materials									
0100	3' high installed on 1st story, 18 ga. x 4" wide studs, 12" O.C.	G 2 Carp	100	.160	L.F.	5.30	7.50		12.80	18.15
0110	16" O.C.	G	150	.107		4.52	5		9.52	13.15
0120	24" O.C.	G	200	.080		3.73	3.76		7.49	10.25
0200	6" wide studs, 12" O.C.	G	100	.160		6.80	7.50		14.30	19.75
0210	16" O.C.	G	150	.107		5.80	5		10.80	14.55
0220	24" O.C.	G	200	.080		4.80	3.76		8.56	11.45
1100	Installed on 2nd story, 18 ga. x 4" wide studs, 12" O.C.	G	95	.168		5.30	7.90		13.20	18.80
1110	16" O.C.	G	145	.110		4.52	5.20		9.72	13.45
1120	24" O.C.	G	190	.084		3.73	3.95		7.68	10.60
1200	6" wide studs, 12" O.C.	G	95	.168		6.80	7.90		14.70	20.50
1210	16" O.C.	G	145	.110		5.80	5.20		11	14.85
1220	24" O.C.	G	190	.084		4.80	3.95		8.75	11.80
2100	Installed on gable, 18 ga. x 4" wide studs, 12" O.C.	G	85	.188		5.30	8.85		14.15	20.50
2110	16" O.C.	G	130	.123		4.52	5.80		10.32	14.40
2120	24" O.C.	G	170	.094		3.73	4.42		8.15	11.35
2200	6" wide studs, 12" O.C.	G	85	.188		6.80	8.85		15.65	22
2210	16" O.C.	G	130	.123		5.80	5.80		11.60	15.80
2220	24" O.C.	G	170	.094		4.80	4.42		9.22	12.55

05 42 23.60 Framing, Roof Rafters

		Crew	Daily Output	Labor-Hours	Unit	Material	Labor	Equipment	Total	Total Incl O&P
0010	**FRAMING, ROOF RAFTERS**									
0015	Made from recycled materials									
0100	Boxed ridge beam, double, 18 ga. x 6" deep	G 2 Carp	160	.100	L.F.	5.10	4.70		9.80	13.30
0110	8" deep	G	150	.107		5.65	5		10.65	14.40
0120	10" deep	G	140	.114		6.90	5.35		12.25	16.40
0130	12" deep	G	130	.123		7.55	5.80		13.35	17.75

05 42 Cold-Formed Metal Joist Framing

05 42 23 – Cold-Formed Metal Roof Joist Framing

05 42 23.60 Framing, Roof Rafters

			Crew	Daily Output	Labor-Hours	Unit	Material	2015 Bare Costs Labor	Equipment	Total	Total Incl O&P
0200		16 ga. x 6" deep	G	2 Carp	150	.107	L.F.	5.80	5		10.80	14.55
0210		8" deep	G		140	.114		6.50	5.35		11.85	15.95
0220		10" deep	G		130	.123		7.90	5.80		13.70	18.10
0230		12" deep	G		120	.133		8.60	6.25		14.85	19.70
1100	Rafters, 2" flange, material only, 18 ga. x 6" deep		G					1.58			1.58	1.73
1110		8" deep	G					1.86			1.86	2.04
1120		10" deep	G					2.18			2.18	2.40
1130		12" deep	G					2.52			2.52	2.77
1200		16 ga. x 6" deep	G					1.93			1.93	2.13
1210		8" deep	G					2.31			2.31	2.54
1220		10" deep	G					2.70			2.70	2.97
1230		12" deep	G					3.07			3.07	3.37
2100	Installation only, ordinary rafter to 4:12 pitch, 18 ga. x 6" deep			2 Carp	35	.457	Ea.		21.50		21.50	35
2110		8" deep			30	.533			25		25	41
2120		10" deep			25	.640			30		30	49.50
2130		12" deep			20	.800			37.50		37.50	61.50
2200		16 ga. x 6" deep			30	.533			25		25	41
2210		8" deep			25	.640			30		30	49.50
2220		10" deep			20	.800			37.50		37.50	61.50
2230		12" deep			15	1.067			50		50	82
8100	Add to labor, ordinary rafters on steep roofs								25%			
8110	Dormers & complex roofs								50%			
8200	Hip & valley rafters to 4:12 pitch								25%			
8210	Steep roofs								50%			
8220	Dormers & complex roofs								75%			
8300	Hip & valley jack rafters to 4:12 pitch								50%			
8310	Steep roofs								75%			
8320	Dormers & complex roofs								100%			
9000	Minimum labor/equipment charge			2 Carp	4	4	Job		188		188	310

05 42 23.70 Framing, Soffits and Canopies

			Crew	Daily Output	Labor-Hours	Unit	Material	2015 Bare Costs Labor	Equipment	Total	Total Incl O&P
0010	**FRAMING, SOFFITS & CANOPIES**											
0015	Made from recycled materials											
0130	Continuous ledger track @ wall, studs @ 16" O.C., 18 ga. x 4" wide		G	2 Carp	535	.030	L.F.	.97	1.40		2.37	3.36
0140		6" wide	G		500	.032		1.27	1.50		2.77	3.85
0150		8" wide	G		465	.034		1.57	1.62		3.19	4.38
0160		10" wide	G		430	.037		1.96	1.75		3.71	5
0230	Studs @ 24" O.C., 18 ga. x 4" wide		G		800	.020		.92	.94		1.86	2.56
0240		6" wide	G		750	.021		1.21	1		2.21	2.97
0250		8" wide	G		700	.023		1.50	1.07		2.57	3.41
0260		10" wide	G		650	.025		1.87	1.16		3.03	3.95
1000	Horizontal soffit and canopy members, material only											
1030	1-5/8" flange studs, 18 ga. x 4" deep		G				L.F.	1.26			1.26	1.39
1040		6" deep	G					1.58			1.58	1.74
1050		8" deep	G					1.92			1.92	2.11
1140	2" flange joists, 18 ga. x 6" deep		G					1.80			1.80	1.98
1150		8" deep	G					2.12			2.12	2.34
1160		10" deep	G					2.50			2.50	2.75
4030	Installation only, 18 ga., 1-5/8" flange x 4" deep			2 Carp	130	.123	Ea.		5.80		5.80	9.45
4040		6" deep			110	.145			6.85		6.85	11.20
4050		8" deep			90	.178			8.35		8.35	13.70
4140	2" flange, 18 ga. x 6" deep				110	.145			6.85		6.85	11.20
4150		8" deep			90	.178			8.35		8.35	13.70

05 42 Cold-Formed Metal Joist Framing

05 42 23 – Cold-Formed Metal Roof Joist Framing

05 42 23.70 Framing, Soffits and Canopies		Crew	Daily Output	Labor-Hours	Unit	Material	2015 Bare Costs Labor	Equipment	Total	Total Incl O&P
4160	10" deep	2 Carp	80	.200	Ea.		9.40		9.40	15.40
6010	Clips to attach facia to rafter tails, 2" x 2" x 18 ga. angle	G	1 Carp	120	.067		.88	3.13		4.01	6.10
6020	16 ga. angle	G	"	100	.080	↓	1.12	3.76		4.88	7.40
9000	Minimum labor/equipment charge		2 Carp	4	4	Job		188		188	310

05 44 Cold-Formed Metal Trusses

05 44 13 – Cold-Formed Metal Roof Trusses

05 44 13.60 Framing, Roof Trusses

			Crew	Daily Output	Labor-Hours	Unit	Material	Labor	Equipment	Total	Total Incl O&P
0010	**FRAMING, ROOF TRUSSES**										
0015	Made from recycled materials										
0020	Fabrication of trusses on ground, Fink (W) or King Post, to 4:12 pitch										
0120	18 ga. x 4" chords, 16' span	G	2 Carp	12	1.333	Ea.	59	62.50		121.50	168
0130	20' span	G		11	1.455		73.50	68.50		142	193
0140	24' span	G		11	1.455		88	68.50		156.50	209
0150	28' span	G		10	1.600		103	75		178	236
0160	32' span	G		10	1.600		118	75		193	252
0250	6" chords, 28' span	G		9	1.778		129	83.50		212.50	279
0260	32' span	G		9	1.778		148	83.50		231.50	300
0270	36' span	G		8	2		166	94		260	335
0280	40' span	G		8	2		185	94		279	355
1120	5:12 to 8:12 pitch, 18 ga. x 4" chords, 16' span	G		10	1.600		67	75		142	197
1130	20' span	G		9	1.778		84	83.50		167.50	230
1140	24' span	G		9	1.778		101	83.50		184.50	248
1150	28' span	G		8	2		118	94		212	283
1160	32' span	G		8	2		134	94		228	300
1250	6" chords, 28' span	G		7	2.286		148	107		255	340
1260	32' span	G		7	2.286		169	107		276	360
1270	36' span	G		6	2.667		190	125		315	415
1280	40' span	G		6	2.667		211	125		336	435
2120	9:12 to 12:12 pitch, 18 ga. x 4" chords, 16' span	G		8	2		84	94		178	247
2130	20' span	G		7	2.286		105	107		212	292
2140	24' span	G		7	2.286		126	107		233	315
2150	28' span	G		6	2.667		147	125		272	365
2160	32' span	G		6	2.667		168	125		293	390
2250	6" chords, 28' span	G		5	3.200		185	150		335	450
2260	32' span	G		5	3.200		211	150		361	480
2270	36' span	G		4	4		238	188		426	570
2280	40' span	G		4	4	↓	264	188		452	600
4900	Minimum labor/equipment charge		↓	4	4	Job		188		188	310
5120	Erection only of roof trusses, to 4:12 pitch, 16' span		F-6	48	.833	Ea.		37	13.60	50.60	75
5130	20' span			46	.870			38.50	14.20	52.70	78
5140	24' span			44	.909			40	14.85	54.85	82
5150	28' span			42	.952			42	15.55	57.55	85.50
5160	32' span			40	1			44	16.35	60.35	90
5170	36' span			38	1.053			46.50	17.20	63.70	94.50
5180	40' span			36	1.111			49	18.15	67.15	100
5220	5:12 to 8:12 pitch, 16' span			42	.952			42	15.55	57.55	85.50
5230	20' span			40	1			44	16.35	60.35	90
5240	24' span			38	1.053			46.50	17.20	63.70	94.50
5250	28' span			36	1.111			49	18.15	67.15	100
5260	32' span		↓	34	1.176	↓		52	19.25	71.25	106

05 44 Cold-Formed Metal Trusses

05 44 13 – Cold-Formed Metal Roof Trusses

05 44 13.60 Framing, Roof Trusses		Crew	Daily Output	Labor-Hours	Unit	Material	2015 Bare Costs Labor	Equipment	Total	Total Incl O&P
5270	36' span	F-6	32	1.250	Ea.		55	20.50	75.50	113
5280	40' span		30	1.333			59	22	81	120
5320	9:12 to 12:12 pitch, 16' span		36	1.111			49	18.15	67.15	100
5330	20' span		34	1.176			52	19.25	71.25	106
5340	24' span		32	1.250			55	20.50	75.50	113
5350	28' span		30	1.333			59	22	81	120
5360	32' span		28	1.429			63	23.50	86.50	129
5370	36' span		26	1.538			68	25	93	139
5380	40' span		24	1.667	↓		73.50	27.50	101	150
9000	Minimum labor/equipment charge	↓	2	20	Job		885	325	1,210	1,775

05 51 Metal Stairs

05 51 13 – Metal Pan Stairs

05 51 13.50 Pan Stairs

			Crew	Daily Output	Labor-Hours	Unit	Material	Labor	Equipment	Total	Total Incl O&P
0010	**PAN STAIRS**, shop fabricated, steel stringers										
0015	Made from recycled materials										
0200	Cement fill metal pan, picket rail, 3'-6" wide	G	E-4	35	.914	Riser	500	48.50	4.17	552.67	645
0300	4'-0" wide	G		30	1.067		560	56.50	4.86	621.36	720
0350	Wall rail, both sides, 3'-6" wide	G		53	.604	↓	380	32	2.75	414.75	480
1500	Landing, steel pan, conventional	G		160	.200	S.F.	66	10.65	.91	77.56	93

05 51 16 – Metal Floor Plate Stairs

05 51 16.50 Floor Plate Stairs

0010	**FLOOR PLATE STAIRS**, shop fabricated, steel stringers										
0015	Made from recycled materials										
0400	Cast iron tread and pipe rail, 3'-6" wide	G	E-4	35	.914	Riser	535	48.50	4.17	587.67	680
0450	4'-0" wide	G		30	1.067		560	56.50	4.86	621.36	720
0475	5'-0" wide	G		28	1.143		615	60.50	5.20	680.70	795
0500	Checkered plate tread, industrial, 3'-6" wide	G		28	1.143		330	60.50	5.20	395.70	475
0550	Circular, for tanks, 3'-0" wide	G		33	.970		370	51.50	4.42	425.92	505
0600	For isolated stairs, add							100%			
0800	Custom steel stairs, 3'-6" wide, economy	G	E-4	35	.914		500	48.50	4.17	552.67	645
0810	Medium priced	G		30	1.067		660	56.50	4.86	721.36	835
0900	Deluxe	G	↓	20	1.600		825	85	7.30	917.30	1,075
1100	For 4' wide stairs, add						5%	5%			
1300	For 5' wide stairs, add					↓	10%	10%			

05 51 19 – Metal Grating Stairs

05 51 19.50 Grating Stairs

0010	**GRATING STAIRS**, shop fabricated, steel stringers, safety nosing on treads										
0015	Made from recycled materials										
0020	Grating tread and pipe railing, 3'-6" wide	G	E-4	35	.914	Riser	330	48.50	4.17	382.67	455
0100	4'-0" wide	G		30	1.067	"	430	56.50	4.86	491.36	580
9000	Minimum labor/equipment charge		↓	2	16	Job		850	73	923	1,600

05 51 23 – Metal Fire Escapes

05 51 23.25 Fire Escapes

0010	**FIRE ESCAPES**, shop fabricated										
0200	2' wide balcony, 1" x 1/4" bars 1-1/2" O.C., with railing	G	2 Sswk	10	1.600	L.F.	58.50	84		142.50	217
0400	1st story cantilevered stair, standard, with railing	G		.50	32	Ea.	2,425	1,675		4,100	5,725
0500	Cable counterweighted, with railing	G		.40	40	"	2,250	2,100		4,350	6,275
0700	36" x 40" platform & fixed stair, with railing	G	↓	.40	40	Flight	1,075	2,100		3,175	4,975

05 51 Metal Stairs

05 51 23 – Metal Fire Escapes

05 51 23.25 Fire Escapes		Crew	Daily Output	Labor-Hours	Unit	Material	2015 Bare Costs Labor	Equipment	Total	Total Incl O&P
0900	For 3'-6" wide escapes, add to above					100%	150%			

05 51 23.50 Fire Escape Stairs

0010	**FIRE ESCAPE STAIRS**, portable									
0100	Portable ladder				Ea.	112			112	123

05 51 33 – Metal Ladders

05 51 33.13 Vertical Metal Ladders

0010	**VERTICAL METAL LADDERS**, shop fabricated									
0015	Made from recycled materials									
0020	Steel, 20" wide, bolted to concrete, with cage	G	E-4	50	.640	V.L.F.	64	34	2.92	100.92	135
0100	Without cage	G		85	.376		38	20	1.72	59.72	80
0300	Aluminum, bolted to concrete, with cage	G		50	.640		120	34	2.92	156.92	197
0400	Without cage	G		85	.376		51	20	1.72	72.72	94
9000	Minimum labor/equipment charge			2	16	Job		850	73	923	1,600

05 51 33.16 Inclined Metal Ladders

0010	**INCLINED METAL LADDERS**, shop fabricated									
0015	Made from recycled materials									
3900	Industrial ships ladder, steel, 24" W, grating treads, 2 line pipe rail	G	E-4	30	1.067	Riser	193	56.50	4.86	254.36	320
4000	Aluminum	G	"	30	1.067	"	270	56.50	4.86	331.36	405

05 51 33.23 Alternating Tread Ladders

0010	**ALTERNATING TREAD LADDERS**, shop fabricated									
0015	Made from recycled materials									
0800	Alternating tread ladders, 68-degree angle of inclline									
0810	8 foot vertical rise, steel, 149 lb., standard paint color		B-68G	3	5.333	Ea.	2,275	281	111	2,667	3,125
0820	Non-standard paint color			3	5.333		2,625	281	111	3,017	3,525
0830	Galvanized			3	5.333		2,650	281	111	3,042	3,525
0840	Stainless			3	5.333		3,850	281	111	4,242	4,850
0850	Aluminum, 87 lb.			3	5.333		2,800	281	111	3,192	3,700
1010	10 foot vertical rise, steel, 181 lb., standard paint color			2.75	5.818		2,750	305	121	3,176	3,725
1020	Non-standard paint color			2.75	5.818		3,150	305	121	3,576	4,175
1030	Galvanized			2.75	5.818		3,200	305	121	3,626	4,225
1040	Stainless			2.75	5.818		4,625	305	121	5,051	5,775
1050	Aluminum, 103 lb.			2.75	5.818		3,375	305	121	3,801	4,425
1210	12 foot vertical rise, steel, 245 lb., standard paint color			2.50	6.400		3,225	335	133	3,693	4,300
1220	Non-standard paint color			2.50	6.400		3,650	335	133	4,118	4,775
1230	Galvanized			2.50	6.400		3,750	335	133	4,218	4,875
1240	Stainless			2.50	6.400		5,400	335	133	5,868	6,700
1250	Aluminum, 103 lb.			2.50	6.400		5,200	335	133	5,668	6,475
1410	14 foot vertical rise, steel, 281 lb., standard paint color			2.25	7.111		3,700	375	147	4,222	4,900
1420	Non-standard paint color			2.25	7.111		4,175	375	147	4,697	5,425
1430	Galvanized			2.25	7.111		4,325	375	147	4,847	5,575
1440	Stainless			2.25	7.111		6,200	375	147	6,722	7,625
1450	Aluminum, 136 lb.			2.25	7.111		4,525	375	147	5,047	5,800
1610	16 foot vertical rise, steel, 317 lb., standard paint color			2	8		4,175	420	166	4,761	5,550
1620	Non-standard paint color			2	8		4,700	420	166	5,286	6,100
1630	Galvanized			2	8		4,875	420	166	5,461	6,300
1640	Stainless			2	8		6,975	420	166	7,561	8,625
1650	Aluminum, 153 lb.			2	8		5,100	420	166	5,686	6,575

05 52 Metal Railings

05 52 13 – Pipe and Tube Railings

05 52 13.50 Railings, Pipe		Crew	Daily Output	Labor-Hours	Unit	Material	2015 Bare Costs Labor	Equipment	Total	Total Incl O&P
0010	**RAILINGS, PIPE**, shop fab'd, 3'-6" high, posts @ 5' O.C.									
0015	Made from recycled materials									
0020	Aluminum, 2 rail, satin finish, 1-1/4" diameter G	E-4	160	.200	L.F.	36	10.65	.91	47.56	59.50
0030	Clear anodized G		160	.200		44	10.65	.91	55.56	68.50
0040	Dark anodized G		160	.200		49	10.65	.91	60.56	74
0080	1-1/2" diameter, satin finish G		160	.200		42.50	10.65	.91	54.06	66.50
0090	Clear anodized G		160	.200		47.50	10.65	.91	59.06	72
0100	Dark anodized G		160	.200		52.50	10.65	.91	64.06	77.50
0140	Aluminum, 3 rail, 1-1/4" diam., satin finish G		137	.234		54.50	12.40	1.07	67.97	83
0150	Clear anodized G		137	.234		67.50	12.40	1.07	80.97	98
0160	Dark anodized G		137	.234		75	12.40	1.07	88.47	106
0200	1-1/2" diameter, satin finish G		137	.234		64.50	12.40	1.07	77.97	94.50
0210	Clear anodized G		137	.234		74	12.40	1.07	87.47	105
0220	Dark anodized G		137	.234		80.50	12.40	1.07	93.97	112
0500	Steel, 2 rail, on stairs, primed, 1-1/4" diameter G		160	.200		25	10.65	.91	36.56	47.50
0520	1-1/2" diameter G		160	.200		27	10.65	.91	38.56	50
0540	Galvanized, 1-1/4" diameter G		160	.200		34	10.65	.91	45.56	57.50
0560	1-1/2" diameter G		160	.200		38.50	10.65	.91	50.06	62
0580	Steel, 3 rail, primed, 1-1/4" diameter G		137	.234		37.50	12.40	1.07	50.97	64.50
0600	1-1/2" diameter G		137	.234		39	12.40	1.07	52.47	66.50
0620	Galvanized, 1-1/4" diameter G		137	.234		52.50	12.40	1.07	65.97	81
0640	1-1/2" diameter G		137	.234		60.50	12.40	1.07	73.97	90
0700	Stainless steel, 2 rail, 1-1/4" diam. #4 finish G		137	.234		110	12.40	1.07	123.47	145
0720	High polish G		137	.234		178	12.40	1.07	191.47	220
0740	Mirror polish G		137	.234		223	12.40	1.07	236.47	269
0760	Stainless steel, 3 rail, 1-1/2" diam., #4 finish G		120	.267		166	14.15	1.22	181.37	210
0770	High polish G		120	.267		275	14.15	1.22	290.37	330
0780	Mirror finish G		120	.267		335	14.15	1.22	350.37	395
0900	Wall rail, alum. pipe, 1-1/4" diam., satin finish G		213	.150		20	8	.69	28.69	37
0905	Clear anodized G		213	.150		25	8	.69	33.69	42.50
0910	Dark anodized G		213	.150		29.50	8	.69	38.19	47.50
0915	1-1/2" diameter, satin finish G		213	.150		22.50	8	.69	31.19	40
0920	Clear anodized G		213	.150		28	8	.69	36.69	46
0925	Dark anodized G		213	.150		35	8	.69	43.69	53.50
0930	Steel pipe, 1-1/4" diameter, primed G		213	.150		15	8	.69	23.69	31.50
0935	Galvanized G		213	.150		22	8	.69	30.69	39
0940	1-1/2" diameter G		176	.182		15.50	9.65	.83	25.98	35.50
0945	Galvanized G		213	.150		22	8	.69	30.69	39
0955	Stainless steel pipe, 1-1/2" diam., #4 finish G		107	.299		88	15.90	1.36	105.26	127
0960	High polish G		107	.299		179	15.90	1.36	196.26	227
0965	Mirror polish G		107	.299		212	15.90	1.36	229.26	263
2000	2-line pipe rail (1-1/2" T&B) with 1/2" pickets @ 4-1/2" O.C.,									
2005	attached handrail on brackets									
2010	42" high aluminum, satin finish, straight & level G	E-4	120	.267	L.F.	198	14.15	1.22	213.37	245
2050	42" high steel, primed, straight & level G	"	120	.267		127	14.15	1.22	142.37	167
4000	For curved and level rails, add					10%	10%			
4100	For sloped rails for stairs, add					30%	30%			
9000	Minimum labor/equipment charge	1 Sswk	2	4	Job		211		211	380

05 52 Metal Railings

05 52 16 – Industrial Railings

05 52 16.50 Railings, Industrial

		Crew	Daily Output	Labor-Hours	Unit	Material	2015 Bare Costs Labor	2015 Bare Costs Equipment	Total	Total Incl O&P
0010	**RAILINGS, INDUSTRIAL**, shop fab'd, 3'-6" high, posts @ 5' O.C.									
0020	2 rail, 3'-6" high, 1-1/2" pipe [G]	E-4	255	.125	L.F.	27	6.65	.57	34.22	42.50
0200	For 4" high kick plate, 10 ga., add [G]					5.65			5.65	6.20
0500	For curved level rails, add					10%	10%			
0550	For sloped rails for stairs, add					30%	30%			
9000	Minimum labor/equipment charge	1 Sswk	2	4	Job		211		211	380

05 55 Metal Stair Treads and Nosings

05 55 13 – Metal Stair Treads

05 55 13.50 Stair Treads

		Crew	Daily Output	Labor-Hours	Unit	Material	2015 Bare Costs Labor	2015 Bare Costs Equipment	Total	Total Incl O&P
0010	**STAIR TREADS**, stringers and bolts not included									
3000	Diamond plate treads, steel, 1/8" thick									
3005	Open riser, black enamel									
3010	9" deep x 36" long [G]	2 Sswk	48	.333	Ea.	91.50	17.55		109.05	132
3020	42" long [G]		48	.333		96.50	17.55		114.05	138
3030	48" long [G]		48	.333		101	17.55		118.55	143
3040	11" deep x 36" long [G]		44	.364		96.50	19.15		115.65	141
3050	42" long [G]		44	.364		102	19.15		121.15	147
3060	48" long [G]		44	.364		108	19.15		127.15	154
3110	Galvanized, 9" deep x 36" long [G]		48	.333		144	17.55		161.55	191
3120	42" long [G]		48	.333		154	17.55		171.55	201
3130	48" long [G]		48	.333		163	17.55		180.55	211
3140	11" deep x 36" long [G]		44	.364		149	19.15		168.15	199
3150	42" long [G]		44	.364		163	19.15		182.15	214
3160	48" long [G]		44	.364		169	19.15		188.15	221
3200	Closed riser, black enamel									
3210	12" deep x 36" long [G]	2 Sswk	40	.400	Ea.	110	21		131	159
3220	42" long [G]		40	.400		119	21		140	169
3230	48" long [G]		40	.400		126	21		147	176
3240	Galvanized, 12" deep x 36" long [G]		40	.400		173	21		194	229
3250	42" long [G]		40	.400		189	21		210	246
3260	48" long [G]		40	.400		198	21		219	256
4000	Bar grating treads									
4005	Steel, 1-1/4" x 3/16" bars, anti-skid nosing, black enamel									
4010	8-5/8" deep x 30" long [G]	2 Sswk	48	.333	Ea.	54	17.55		71.55	91
4020	36" long [G]		48	.333		63.50	17.55		81.05	101
4030	48" long [G]		48	.333		97.50	17.55		115.05	139
4040	10-15/16" deep x 36" long [G]		44	.364		70	19.15		89.15	112
4050	48" long [G]		44	.364		100	19.15		119.15	145
4060	Galvanized, 8-5/8" deep x 30" long [G]		48	.333		62	17.55		79.55	100
4070	36" long [G]		48	.333		73.50	17.55		91.05	113
4080	48" long [G]		48	.333		108	17.55		125.55	151
4090	10-15/16" deep x 36" long [G]		44	.364		86.50	19.15		105.65	130
4100	48" long [G]		44	.364		112	19.15		131.15	158
4200	Aluminum, 1-1/4" x 3/16" bars, serrated, with nosing									
4210	7-5/8" deep x 18" long [G]	2 Sswk	52	.308	Ea.	50	16.20		66.20	84.50
4220	24" long [G]		52	.308		59	16.20		75.20	94.50
4230	30" long [G]		52	.308		68.50	16.20		84.70	105
4240	36" long [G]		52	.308		177	16.20		193.20	225
4250	8-13/16" deep x 18" long [G]		48	.333		67	17.55		84.55	106

05 55 Metal Stair Treads and Nosings

05 55 13 – Metal Stair Treads

05 55 13.50 Stair Treads

		Crew	Daily Output	Labor-Hours	Unit	Material	2015 Bare Costs Labor	2015 Bare Costs Equipment	Total	Total Incl O&P
4260	24" long	[G] 2 Sswk	48	.333	Ea.	99.50	17.55		117.05	141
4270	30" long	[G]	48	.333		115	17.55		132.55	158
4280	36" long	[G]	48	.333		194	17.55		211.55	246
4290	10" deep x 18" long	[G]	44	.364		130	19.15		149.15	178
4300	30" long	[G]	44	.364		173	19.15		192.15	225
4310	36" long	[G]	44	.364		210	19.15		229.15	266
5000	Channel grating treads									
5005	Steel, 14 ga., 2-12" thick, galvanized									
5010	9" deep x 36" long	[G] 2 Sswk	48	.333	Ea.	103	17.55		120.55	145
5020	48" long	[G]	48	.333	"	139	17.55		156.55	184
9000	Minimum labor/equipment charge		4	4	Job		211		211	380

05 55 19 – Metal Stair Tread Covers

05 55 19.50 Stair Tread Covers for Renovation

		Crew	Daily Output	Labor-Hours	Unit	Material	Labor	Equipment	Total	Total Incl O&P
0010	**STAIR TREAD COVERS FOR RENOVATION**									
0205	Extruded tread cover with nosing, pre-drilled, includes screws									
0210	Aluminum with black abrasive strips, 9" wide x 3' long	1 Carp	24	.333	Ea.	107	15.65		122.65	144
0220	4' long		22	.364		142	17.05		159.05	184
0230	5' long		20	.400		177	18.80		195.80	225
0240	11" wide x 3' long		24	.333		139	15.65		154.65	179
0250	4' long		22	.364		179	17.05		196.05	224
0260	5' long		20	.400		230	18.80		248.80	283
0305	Black abrasive strips with yellow front strips									
0310	Aluminum, 9" wide x 3' long	1 Carp	24	.333	Ea.	116	15.65		131.65	153
0320	4' long		22	.364		154	17.05		171.05	198
0330	5' long		20	.400		196	18.80		214.80	247
0340	11" wide x 3' long		24	.333		149	15.65		164.65	190
0350	4' long		22	.364		192	17.05		209.05	239
0360	5' long		20	.400		247	18.80		265.80	305
0405	Black abrasive strips with photoluminescent front strips									
0410	Aluminum, 9" wide x 3' long	1 Carp	24	.333	Ea.	156	15.65		171.65	197
0420	4' long		22	.364		172	17.05		189.05	217
0430	5' long		20	.400		215	18.80		233.80	267
0440	11" wide x 3' long		24	.333		153	15.65		168.65	194
0450	4' long		22	.364		204	17.05		221.05	252
0460	5' long		20	.400		255	18.80		273.80	310
9000	Minimum labor/equipment charge		4	2	Job		94		94	154

05 56 Metal Castings

05 56 13 – Metal Construction Castings

05 56 13.50 Construction Castings

		Crew	Daily Output	Labor-Hours	Unit	Material	Labor	Equipment	Total	Total Incl O&P
0010	**CONSTRUCTION CASTINGS**									
0020	Manhole covers and frames, see Section 33 44 13.13									
0100	Column bases, cast iron, 16" x 16", approx. 65 lb.	[G] E-4	46	.696	Ea.	142	37	3.17	182.17	226
0200	32" x 32", approx. 256 lb.	[G]	23	1.391	"	520	74	6.35	600.35	715
0600	Miscellaneous C.I. castings, light sections, less than 150 lb.	[G]	3200	.010	Lb.	8.95	.53	.05	9.53	10.85
1300	Special low volume items	[G]	3200	.010	"	11.25	.53	.05	11.83	13.35

05 58 Formed Metal Fabrications

05 58 13 – Column Covers

05 58 13.05 Column Covers

			Crew	Daily Output	Labor-Hours	Unit	Material	2015 Bare Costs Labor	Equipment	Total	Total Incl O&P
0010	**COLUMN COVERS**										
0015	Made from recycled materials										
0020	Excludes structural steel, light ga. metal framing, misc. metals, sealants										
0100	Round covers, 2 halves with 2 vertical joints for backer rod and sealant										
0110	Up to 12' high, no horizontal joints										
0120	12" diameter, 0.125" aluminum, anodized/painted finish	G	2 Sswk	32	.500	V.L.F.	30.50	26.50		57	81
0130	Type 304 stainless steel, 16 gauge, #4 brushed finish	G		32	.500		46.50	26.50		73	98.50
0140	Type 316 stainless steel, 16 gauge, #4 brushed finish	G		32	.500		54	26.50		80.50	107
0150	18" diameter, aluminum	G		32	.500		46	26.50		72.50	98
0160	Type 304 stainless steel	G		32	.500		69.50	26.50		96	124
0170	Type 316 stainless steel	G		32	.500		81	26.50		107.50	137
0180	24" diameter, aluminum	G		32	.500		61	26.50		87.50	115
0190	Type 304 stainless steel	G		32	.500		93	26.50		119.50	150
0200	Type 316 stainless steel	G		32	.500		108	26.50		134.50	167
0210	30" diameter, aluminum	G		30	.533		76.50	28		104.50	135
0220	Type 304 stainless steel	G		30	.533		116	28		144	179
0230	Type 316 stainless steel	G		30	.533		135	28		163	199
0240	36" diameter, aluminum	G		30	.533		91.50	28		119.50	152
0250	Type 304 stainless steel	G		30	.533		139	28		167	204
0260	Type 316 stainless steel	G		30	.533		162	28		190	229
0400	Up to 24' high, 2 stacked sections with 1 horizontal joint										
0410	18" diameter, aluminum	G	2 Sswk	28	.571	V.L.F.	48	30		78	108
0450	Type 304 stainless steel	G		28	.571		73	30		103	135
0460	Type 316 stainless steel	G		28	.571		85	30		115	148
0470	24" diameter, aluminum	G		28	.571		64	30		94	125
0480	Type 304 stainless steel	G		28	.571		97.50	30		127.50	162
0490	Type 316 stainless steel	G		28	.571		113	30		143	180
0500	30" diameter, aluminum	G		24	.667		80	35		115	152
0510	Type 304 stainless steel	G		24	.667		122	35		157	198
0520	Type 316 stainless steel	G		24	.667		142	35		177	220
0530	36" diameter, aluminum	G		24	.667		96.50	35		131.50	170
0540	Type 304 stainless steel	G		24	.667		146	35		181	225
0550	Type 316 stainless steel	G		24	.667		170	35		205	251

05 58 23 – Formed Metal Guards

05 58 23.90 Window Guards

			Crew	Daily Output	Labor-Hours	Unit	Material	Labor	Equipment	Total	Total Incl O&P
0010	**WINDOW GUARDS**, shop fabricated										
0015	Expanded metal, steel angle frame, permanent	G	E-4	350	.091	S.F.	23	4.86	.42	28.28	34.50
0025	Steel bars, 1/2" x 1/2", spaced 5" O.C.	G	"	290	.110	"	15.90	5.85	.50	22.25	28.50
0030	Hinge mounted, add	G				Opng.	46			46	50.50
0040	Removable type, add	G				"	29			29	32
0050	For galvanized guards, add					S.F.	35%				
0070	For pivoted or projected type, add						105%	40%			
0100	Mild steel, stock units, economy	G	E-4	405	.079		6.25	4.20	.36	10.81	14.90
0200	Deluxe	G		405	.079		12.70	4.20	.36	17.26	22
0400	Woven wire, stock units, 3/8" channel frame, 3' x 5' opening	G		40	.800	Opng.	169	42.50	3.65	215.15	267
0500	4' x 6' opening	G		38	.842		270	45	3.84	318.84	380
0800	Basket guards for above, add	G					233			233	256
1000	Swinging guards for above, add	G					79.50			79.50	87.50
9000	Minimum labor/equipment charge		1 Sswk	2	4	Job		211		211	380

05 58 Formed Metal Fabrications

05 58 25 – Formed Lamp Posts

05 58 25.40 Lamp Posts

		Crew	Daily Output	Labor-Hours	Unit	Material	2015 Bare Costs Labor	2015 Bare Costs Equipment	Total	Total Incl O&P
0010	**LAMP POSTS**									
0020	Aluminum, 7' high, stock units, post only	G 1 Carp	16	.500	Ea.	82	23.50		105.50	129
0100	Mild steel, plain	G	16	.500	"	73	23.50		96.50	119
9000	Minimum labor/equipment charge		4	2	Job		94		94	154

05 71 Decorative Metal Stairs

05 71 13 – Fabricated Metal Spiral Stairs

05 71 13.50 Spiral Stairs

		Crew	Daily Output	Labor-Hours	Unit	Material	Labor	Equipment	Total	Incl O&P
0010	**SPIRAL STAIRS**									
1805	Shop fabricated, custom ordered									
1810	Aluminum, 5'-0" diameter, plain units	G E-4	45	.711	Riser	575	38	3.24	616.24	705
1820	Fancy units	G	45	.711		1,100	38	3.24	1,141.24	1,275
1900	Cast iron, 4'-0" diameter, plain units	G	45	.711		540	38	3.24	581.24	665
1920	Fancy Units	G	25	1.280		735	68	5.85	808.85	940
2000	Steel, industrial checkered plate, 4' diameter	G	45	.711		540	38	3.24	581.24	665
2200	6' diameter	G	40	.800		650	42.50	3.65	696.15	795
3100	Spiral stair kits, 12 stacking risers to fit exact floor height									
3110	Steel, flat metal treads, primed, 3'-6" diameter	G 2 Carp	1.60	10	Flight	1,275	470		1,745	2,175
3120	4'-0" diameter	G	1.45	11.034		1,450	520		1,970	2,450
3130	4'-6" diameter	G	1.35	11.852		1,600	555		2,155	2,675
3140	5'-0" diameter	G	1.25	12.800		1,750	600		2,350	2,900
3210	Galvanized, 3'-6" diameter	G	1.60	10		1,525	470		1,995	2,450
3220	4'-0" diameter	G	1.45	11.034		1,750	520		2,270	2,775
3230	4'-6" diameter	G	1.35	11.852		1,925	555		2,480	3,025
3240	5'-0" diameter	G	1.25	12.800		2,100	600		2,700	3,275
3310	Checkered plate tread, primed, 3'-6" diameter	G	1.45	11.034		1,500	520		2,020	2,500
3320	4'-0" diameter	G	1.35	11.852		1,700	555		2,255	2,775
3330	4'-6" diameter	G	1.25	12.800		1,850	600		2,450	3,025
3340	5'-0" diameter	G	1.15	13.913		2,025	655		2,680	3,300
3410	Galvanized, 3'-6" diameter	G	1.45	11.034		1,800	520		2,320	2,850
3420	4'-0" diameter	G	1.35	11.852		2,050	555		2,605	3,150
3430	4'-6" diameter	G	1.25	12.800		2,225	600		2,825	3,425
3440	5'-0" diameter	G	1.15	13.913		2,425	655		3,080	3,750
3510	Red oak covers on flat metal treads, 3'-6" diameter		1.35	11.852		2,600	555		3,155	3,775
3520	4'-0" diameter		1.25	12.800		2,875	600		3,475	4,150
3530	4'-6" diameter		1.15	13.913		3,100	655		3,755	4,475
3540	5'-0" diameter		1.05	15.238		3,375	715		4,090	4,900

05 73 Decorative Metal Railings

05 73 16 – Wire Rope Decorative Metal Railings

05 73 16.10 Cable Railings

		Crew	Daily Output	Labor-Hours	Unit	Material	Labor	Equipment	Total	Incl O&P
0010	**CABLE RAILINGS**, with 316 stainless steel 1 x 19 cable, 3/16" diameter									
0015	Made from recycled materials									
0100	1-3/4" diameter stainless steel posts x 42" high, cables 4" OC	G 2 Sswk	25	.640	L.F.	33	33.50		66.50	97.50

05 73 Decorative Metal Railings

05 73 23 – Ornamental Railings

05 73 23.50 Railings, Ornamental		Crew	Daily Output	Labor-Hours	Unit	Material	2015 Bare Costs Labor	Equipment	Total	Total Incl O&P
0010	**RAILINGS, ORNAMENTAL**, 3'-6" high, posts @ 6' O.C.									
0020	Bronze or stainless, hand forged, plain	G	2 Sswk	24	.667	L.F.	122	35		157	198
0100	Fancy	G		18	.889		223	47		270	330
0200	Aluminum, panelized, plain	G		24	.667		13.60	35		48.60	78.50
0300	Fancy	G		18	.889		30	47		77	118
0400	Wrought iron, hand forged, plain	G		24	.667		81	35		116	153
0500	Fancy	G		18	.889		159	47		206	260
0550	Steel, panelized, plain	G		24	.667		20.50	35		55.50	86
0560	Fancy	G		18	.889		32	47		79	120
0600	Composite metal/wood/glass, plain			18	.889		119	47		166	216
0700	Fancy			12	1.333		238	70		308	390
9000	Minimum labor/equipment charge		1 Sswk	2	4	Job		211		211	380

05 75 Decorative Formed Metal

05 75 13 – Columns

05 75 13.10 Aluminum Columns

0010	**ALUMINUM COLUMNS**										
0015	Made from recycled materials										
0020	Aluminum, extruded, stock units, no cap or base, 6" diameter	G	E-4	240	.133	L.F.	10.50	7.10	.61	18.21	25
0100	8" diameter	G	"	170	.188	"	13.85	10	.86	24.71	34
0460	Caps, ornamental, plain	G				Set	340			340	370
0470	Fancy	G				"	1,675			1,675	1,850
0500	For square columns, add to column prices above					L.F.	50%				

Division Notes

		CREW	DAILY OUTPUT	LABOR-HOURS	UNIT	BARE COSTS				TOTAL INCL O&P
						MAT.	LABOR	EQUIP.	TOTAL	

Estimating Tips
06 05 00 Common Work Results for Wood, Plastics, and Composites

- Common to any wood-framed structure are the accessory connector items such as screws, nails, adhesives, hangers, connector plates, straps, angles, and hold-downs. For typical wood-framed buildings, such as residential projects, the aggregate total for these items can be significant, especially in areas where seismic loading is a concern. For floor and wall framing, the material cost is based on 10 to 25 lbs. per MBF. Hold-downs, hangers, and other connectors should be taken off by the piece.

  Included with material costs are fasteners for a normal installation. RSMeans engineers use manufacturer's recommendations, written specifications, and/or standard construction practice for size and spacing of fasteners. Prices for various fasteners are shown for informational purposes only. Adjustments should be made if unusual fastening conditions exist.

06 10 00 Carpentry

- Lumber is a traded commodity and therefore sensitive to supply and demand in the marketplace. Even in "budgetary" estimating of wood-framed projects, it is advisable to call local suppliers for the latest market pricing.

- Common quantity units for wood-framed projects are "thousand board feet" (MBF). A board foot is a volume of wood, 1" x 1' x 1', or 144 cubic inches. Board-foot quantities are generally calculated using nominal material dimensions— dressed sizes are ignored. Board foot per lineal foot of any stick of lumber can be calculated by dividing the nominal cross-sectional area by 12. As an example, 2,000 lineal feet of 2 x 12 equates to 4 MBF by dividing the nominal area, 2 x 12, by 12, which equals 2, and multiplying by 2,000 to give 4,000 board feet. This simple rule applies to all nominal dimensioned lumber.

- Waste is an issue of concern at the quantity takeoff for any area of construction. Framing lumber is sold in even foot lengths, i.e., 10', 12', 14', 16' and, depending on spans, wall heights, and the grade of lumber, waste is inevitable. A rule of thumb for lumber waste is 5%–10% depending on material quality and the complexity of the framing.

- Wood in various forms and shapes is used in many projects, even where the main structural framing is steel, concrete, or masonry. Plywood as a back-up partition material and 2x boards used as blocking and cant strips around roof edges are two common examples. The estimator should ensure that the costs of all wood materials are included in the final estimate.

06 20 00 Finish Carpentry

- It is necessary to consider the grade of workmanship when estimating labor costs for erecting millwork and interior finish. In practice, there are three grades: premium, custom, and economy. The RSMeans daily output for base and case moldings is in the range of 200 to 250 L.F. per carpenter per day. This is appropriate for most average custom-grade projects. For premium projects, an adjustment to productivity of 25%–50% should be made, depending on the complexity of the job.

Reference Numbers

Reference numbers are shown in shaded boxes at the beginning of some major classifications. These numbers refer to related items in the Reference Section. The reference information may be an estimating procedure, an alternate pricing method, or technical information.

Note: Not all subdivisions listed here necessarily appear in this publication. ■

No part of this publication may be reproduced, stored in a retrieval system, or transmitted in any form or by any means without prior written permission of RSMeans.

06 05 Common Work Results for Wood, Plastics, and Composites

06 05 05 – Selective Demolition for Wood, Plastics, and Composites

06 05 05.10 Selective Demolition Wood Framing		Crew	Daily Output	Labor-Hours	Unit	Material	2015 Bare Costs Labor	2015 Bare Costs Equipment	Total	Total Incl O&P
0010	**SELECTIVE DEMOLITION WOOD FRAMING** R024119-10									
0100	Timber connector, nailed, small	1 Clab	96	.083	Ea.		3.13		3.13	5.15
0110	Medium		60	.133			5		5	8.20
0120	Large		48	.167			6.25		6.25	10.30
0130	Bolted, small		48	.167			6.25		6.25	10.30
0140	Medium		32	.250			9.40		9.40	15.40
0150	Large		24	.333			12.55		12.55	20.50
2958	Beams, 2" x 6"	2 Clab	1100	.015	L.F.		.55		.55	.90
2960	2" x 8"		825	.019			.73		.73	1.20
2965	2" x 10"		665	.024			.90		.90	1.48
2970	2" x 12"		550	.029			1.09		1.09	1.79
2972	2" x 14"		470	.034			1.28		1.28	2.10
2975	4" x 8"	B-1	413	.058			2.22		2.22	3.65
2980	4" x 10"		330	.073			2.78		2.78	4.56
2985	4" x 12"		275	.087			3.34		3.34	5.45
3000	6" x 8"		275	.087			3.34		3.34	5.45
3040	6" x 10"		220	.109			4.17		4.17	6.85
3080	6" x 12"		185	.130			4.96		4.96	8.15
3120	8" x 12"		140	.171			6.55		6.55	10.75
3160	10" x 12"		110	.218			8.35		8.35	13.70
3162	Alternate pricing method		1.10	21.818	M.B.F.		835		835	1,375
3170	Blocking, in 16" OC wall framing, 2" x 4"	1 Clab	600	.013	L.F.		.50		.50	.82
3172	2" x 6"		400	.020			.75		.75	1.23
3174	In 24" OC wall framing, 2" x 4"		600	.013			.50		.50	.82
3176	2" x 6"		400	.020			.75		.75	1.23
3178	Alt method, wood blocking removal from wood framing		.40	20	M.B.F.		750		750	1,225
3179	Wood blocking removal from steel framing		.36	22.222	"		835		835	1,375
3180	Bracing, let in, 1" x 3", studs 16" OC		1050	.008	L.F.		.29		.29	.47
3181	Studs 24" OC		1080	.007			.28		.28	.46
3182	1" x 4", studs 16" OC		1050	.008			.29		.29	.47
3183	Studs 24" OC		1080	.007			.28		.28	.46
3184	1" x 6", studs 16" OC		1050	.008			.29		.29	.47
3185	Studs 24" OC		1080	.007			.28		.28	.46
3186	2" x 3", studs 16" OC		800	.010			.38		.38	.62
3187	Studs 24" OC		830	.010			.36		.36	.59
3188	2" x 4", studs 16" OC		800	.010			.38		.38	.62
3189	Studs 24" OC		830	.010			.36		.36	.59
3190	2" x 6", studs 16" OC		800	.010			.38		.38	.62
3191	Studs 24" OC		830	.010			.36		.36	.59
3192	2" x 8", studs 16" OC		800	.010			.38		.38	.62
3193	Studs 24" OC		830	.010			.36		.36	.59
3194	"T" shaped metal bracing, studs at 16" OC		1060	.008			.28		.28	.47
3195	Studs at 24" OC		1200	.007			.25		.25	.41
3196	Metal straps, studs at 16" OC		1200	.007			.25		.25	.41
3197	Studs at 24" OC		1240	.006			.24		.24	.40
3200	Columns, round, 8' to 14' tall		40	.200	Ea.		7.50		7.50	12.35
3202	Dimensional lumber sizes	2 Clab	1.10	14.545	M.B.F.		545		545	895
3250	Blocking, between joists	1 Clab	320	.025	Ea.		.94		.94	1.54
3252	Bridging, metal strap, between joists		320	.025	Pr.		.94		.94	1.54
3254	Wood, between joists		320	.025	"		.94		.94	1.54
3260	Door buck, studs, header & access., 8' high 2" x 4" wall, 3' wide		32	.250	Ea.		9.40		9.40	15.40
3261	4' wide		32	.250			9.40		9.40	15.40
3262	5' wide		32	.250			9.40		9.40	15.40

06 05 Common Work Results for Wood, Plastics, and Composites

06 05 05 – Selective Demolition for Wood, Plastics, and Composites

06 05 05.10 Selective Demolition Wood Framing		Crew	Daily Output	Labor-Hours	Unit	Material	2015 Bare Costs Labor	Equipment	Total	Total Incl O&P
3263	6' wide	1 Clab	32	.250	Ea.		9.40		9.40	15.40
3264	8' wide		30	.267			10.05		10.05	16.45
3265	10' wide		30	.267			10.05		10.05	16.45
3266	12' wide		30	.267			10.05		10.05	16.45
3267	2" x 6" wall, 3' wide		32	.250			9.40		9.40	15.40
3268	4' wide		32	.250			9.40		9.40	15.40
3269	5' wide		32	.250			9.40		9.40	15.40
3270	6' wide		32	.250			9.40		9.40	15.40
3271	8' wide		30	.267			10.05		10.05	16.45
3272	10' wide		30	.267			10.05		10.05	16.45
3273	12' wide		30	.267			10.05		10.05	16.45
3274	Window buck, studs, header & access, 8' high 2" x 4" wall, 2' wide		24	.333			12.55		12.55	20.50
3275	3' wide		24	.333			12.55		12.55	20.50
3276	4' wide		24	.333			12.55		12.55	20.50
3277	5' wide		24	.333			12.55		12.55	20.50
3278	6' wide		24	.333			12.55		12.55	20.50
3279	7' wide		24	.333			12.55		12.55	20.50
3280	8' wide		22	.364			13.65		13.65	22.50
3281	10' wide		22	.364			13.65		13.65	22.50
3282	12' wide		22	.364			13.65		13.65	22.50
3283	2" x 6" wall, 2' wide		24	.333			12.55		12.55	20.50
3284	3' wide		24	.333			12.55		12.55	20.50
3285	4' wide		24	.333			12.55		12.55	20.50
3286	5' wide		24	.333			12.55		12.55	20.50
3287	6' wide		24	.333			12.55		12.55	20.50
3288	7' wide		24	.333			12.55		12.55	20.50
3289	8' wide		22	.364			13.65		13.65	22.50
3290	10' wide		22	.364			13.65		13.65	22.50
3291	12' wide		22	.364			13.65		13.65	22.50
3360	Deck or porch decking		825	.010	L.F.		.36		.36	.60
3400	Fascia boards, 1" x 6"		500	.016			.60		.60	.99
3440	1" x 8"		450	.018			.67		.67	1.10
3480	1" x 10"		400	.020			.75		.75	1.23
3490	2" x 6"		450	.018			.67		.67	1.10
3500	2" x 8"		400	.020			.75		.75	1.23
3510	2" x 10"		350	.023			.86		.86	1.41
3610	Furring, on wood walls or ceiling		4000	.002	S.F.		.08		.08	.12
3620	On masonry or concrete walls or ceiling		1200	.007	"		.25		.25	.41
3800	Headers over openings, 2 @ 2" x 6"		110	.073	L.F.		2.73		2.73	4.48
3840	2 @ 2" x 8"		100	.080			3.01		3.01	4.93
3880	2 @ 2" x 10"		90	.089			3.34		3.34	5.50
3885	Alternate pricing method		.26	30.651	M.B.F.		1,150		1,150	1,900
3920	Joists, 1" x 4"		1250	.006	L.F.		.24		.24	.39
3930	1" x 6"		1135	.007			.27		.27	.43
3940	1" x 8"		1000	.008			.30		.30	.49
3950	1" x 10"		895	.009			.34		.34	.55
3960	1" x 12"		765	.010			.39		.39	.64
4200	2" x 4"	2 Clab	1000	.016			.60		.60	.99
4230	2" x 6"		970	.016			.62		.62	1.02
4240	2" x 8"		940	.017			.64		.64	1.05
4250	2" x 10"		910	.018			.66		.66	1.08
4280	2" x 12"		880	.018			.68		.68	1.12
4281	2" x 14"		850	.019			.71		.71	1.16

For customer support on your Commercial Renovation Cost Data, call 877.791.4977.

06 05 Common Work Results for Wood, Plastics, and Composites

06 05 05 – Selective Demolition for Wood, Plastics, and Composites

06 05 05.10 Selective Demolition Wood Framing		Crew	Daily Output	Labor-Hours	Unit	Material	2015 Bare Costs Labor	Equipment	Total	Total Incl O&P
4282	Composite joists, 9-1/2"	2 Clab	960	.017	L.F.		.63		.63	1.03
4283	11-7/8"		930	.017			.65		.65	1.06
4284	14"		897	.018			.67		.67	1.10
4285	16"		865	.019			.70		.70	1.14
4290	Wood joists, alternate pricing method		1.50	10.667	M.B.F.		400		400	660
4500	Open web joist, 12" deep		500	.032	L.F.		1.20		1.20	1.97
4505	14" deep		475	.034			1.27		1.27	2.08
4510	16" deep		450	.036			1.34		1.34	2.19
4520	18" deep		425	.038			1.42		1.42	2.32
4530	24" deep		400	.040			1.50		1.50	2.47
4550	Ledger strips, 1" x 2"	1 Clab	1200	.007			.25		.25	.41
4560	1" x 3"		1200	.007			.25		.25	.41
4570	1" x 4"		1200	.007			.25		.25	.41
4580	2" x 2"		1100	.007			.27		.27	.45
4590	2" x 4"		1000	.008			.30		.30	.49
4600	2" x 6"		1000	.008			.30		.30	.49
4601	2" x 8" or 2" x 10"		800	.010			.38		.38	.62
4602	4" x 6"		600	.013			.50		.50	.82
4604	4" x 8"		450	.018			.67		.67	1.10
5400	Posts, 4" x 4"	2 Clab	800	.020			.75		.75	1.23
5405	4" x 6"		550	.029			1.09		1.09	1.79
5410	4" x 8"		440	.036			1.37		1.37	2.24
5425	4" x 10"		390	.041			1.54		1.54	2.53
5430	4" x 12"		350	.046			1.72		1.72	2.82
5440	6" x 6"		400	.040			1.50		1.50	2.47
5445	6" x 8"		350	.046			1.72		1.72	2.82
5450	6" x 10"		320	.050			1.88		1.88	3.08
5455	6" x 12"		290	.055			2.07		2.07	3.40
5480	8" x 8"		300	.053			2.01		2.01	3.29
5500	10" x 10"		240	.067			2.51		2.51	4.11
5660	Tongue and groove floor planks		2	8	M.B.F.		300		300	495
5682	Rafters, ordinary, 16" OC, 2" x 4"		880	.018	S.F.		.68		.68	1.12
5683	2" x 6"		840	.019			.72		.72	1.17
5684	2" x 8"		820	.020			.73		.73	1.20
5685	2" x 10"		820	.020			.73		.73	1.20
5686	2" x 12"		810	.020			.74		.74	1.22
5687	24" OC, 2" x 4"		1170	.014			.51		.51	.84
5688	2" x 6"		1117	.014			.54		.54	.88
5689	2" x 8"		1091	.015			.55		.55	.90
5690	2" x 10"		1091	.015			.55		.55	.90
5691	2" x 12"		1077	.015			.56		.56	.92
5795	Rafters, ordinary, 2" x 4" (alternate method)		862	.019	L.F.		.70		.70	1.14
5800	2" x 6" (alternate method)		850	.019			.71		.71	1.16
5840	2" x 8" (alternate method)		837	.019			.72		.72	1.18
5855	2" x 10" (alternate method)		825	.019			.73		.73	1.20
5865	2" x 12" (alternate method)		812	.020			.74		.74	1.21
5870	Sill plate, 2" x 4"	1 Clab	1170	.007			.26		.26	.42
5871	2" x 6"		780	.010			.39		.39	.63
5872	2" x 8"		586	.014			.51		.51	.84
5873	Alternate pricing method		.78	10.256	M.B.F.		385		385	630
5885	Ridge board, 1" x 4"	2 Clab	900	.018	L.F.		.67		.67	1.10
5886	1" x 6"		875	.018			.69		.69	1.13
5887	1" x 8"		850	.019			.71		.71	1.16

06 05 Common Work Results for Wood, Plastics, and Composites

06 05 05 – Selective Demolition for Wood, Plastics, and Composites

06 05 05.10 Selective Demolition Wood Framing		Crew	Daily Output	Labor-Hours	Unit	Material	2015 Bare Costs Labor	Equipment	Total	Total Incl O&P
5888	1" x 10"	2 Clab	825	.019	L.F.		.73		.73	1.20
5889	1" x 12"		800	.020			.75		.75	1.23
5890	2" x 4"		900	.018			.67		.67	1.10
5892	2" x 6"		875	.018			.69		.69	1.13
5894	2" x 8"		850	.019			.71		.71	1.16
5896	2" x 10"		825	.019			.73		.73	1.20
5898	2" x 12"		800	.020			.75		.75	1.23
5900	Hip & valley rafters, 2" x 6"		500	.032			1.20		1.20	1.97
5940	2" x 8"		420	.038			1.43		1.43	2.35
6050	Rafter tie, 1" x 4"		1250	.013			.48		.48	.79
6052	1" x 6"		1135	.014			.53		.53	.87
6054	2" x 4"		1000	.016			.60		.60	.99
6056	2" x 6"		970	.016			.62		.62	1.02
6070	Sleepers, on concrete, 1" x 2"	1 Clab	4700	.002			.06		.06	.10
6075	1" x 3"		4000	.002			.08		.08	.12
6080	2" x 4"		3000	.003			.10		.10	.16
6085	2" x 6"		2600	.003			.12		.12	.19
6086	Sheathing from roof, 5/16"	2 Clab	1600	.010	S.F.		.38		.38	.62
6088	3/8"		1525	.010			.39		.39	.65
6090	1/2"		1400	.011			.43		.43	.70
6092	5/8"		1300	.012			.46		.46	.76
6094	3/4"		1200	.013			.50		.50	.82
6096	Board sheathing from roof		1400	.011			.43		.43	.70
6100	Sheathing, from walls, 1/4"		1200	.013			.50		.50	.82
6110	5/16"		1175	.014			.51		.51	.84
6120	3/8"		1150	.014			.52		.52	.86
6130	1/2"		1125	.014			.53		.53	.88
6140	5/8"		1100	.015			.55		.55	.90
6150	3/4"		1075	.015			.56		.56	.92
6152	Board sheathing from walls		1500	.011			.40		.40	.66
6158	Subfloor/roof deck, with boards		2200	.007			.27		.27	.45
6159	Subfloor/roof deck, with tongue & groove boards		2000	.008			.30		.30	.49
6160	Plywood, 1/2" thick		768	.021			.78		.78	1.28
6162	5/8" thick		760	.021			.79		.79	1.30
6164	3/4" thick		750	.021			.80		.80	1.32
6165	1-1/8" thick		720	.022			.84		.84	1.37
6166	Underlayment, particle board, 3/8" thick	1 Clab	780	.010			.39		.39	.63
6168	1/2" thick		768	.010			.39		.39	.64
6170	5/8" thick		760	.011			.40		.40	.65
6172	3/4" thick		750	.011			.40		.40	.66
6200	Stairs and stringers, straight run	2 Clab	40	.400	Riser		15.05		15.05	24.50
6240	With platforms, winders or curves	"	26	.615	"		23		23	38
6300	Components, tread	1 Clab	110	.073	Ea.		2.73		2.73	4.48
6320	Riser		80	.100	"		3.76		3.76	6.15
6390	Stringer, 2" x 10"		260	.031	L.F.		1.16		1.16	1.90
6400	2" x 12"		260	.031			1.16		1.16	1.90
6410	3" x 10"		250	.032			1.20		1.20	1.97
6420	3" x 12"		250	.032			1.20		1.20	1.97
6590	Wood studs, 2" x 3"	2 Clab	3076	.005			.20		.20	.32
6600	2" x 4"		2000	.008			.30		.30	.49
6640	2" x 6"		1600	.010			.38		.38	.62
6720	Wall framing, including studs, plates and blocking, 2" x 4"	1 Clab	600	.013	S.F.		.50		.50	.82
6740	2" x 6"		480	.017	"		.63		.63	1.03

06 05 Common Work Results for Wood, Plastics, and Composites

06 05 05 – Selective Demolition for Wood, Plastics, and Composites

06 05 05.10 Selective Demolition Wood Framing		Crew	Daily Output	Labor-Hours	Unit	Material	2015 Bare Costs Labor	2015 Bare Costs Equipment	Total	Total Incl O&P
6750	Headers, 2" x 4"	1 Clab	1125	.007	L.F.		.27		.27	.44
6755	2" x 6"		1125	.007			.27		.27	.44
6760	2" x 8"		1050	.008			.29		.29	.47
6765	2" x 10"		1050	.008			.29		.29	.47
6770	2" x 12"		1000	.008			.30		.30	.49
6780	4" x 10"		525	.015			.57		.57	.94
6785	4" x 12"		500	.016			.60		.60	.99
6790	6" x 8"		560	.014			.54		.54	.88
6795	6" x 10"		525	.015			.57		.57	.94
6797	6" x 12"		500	.016			.60		.60	.99
7000	Trusses									
7050	12' span	2 Clab	74	.216	Ea.		8.15		8.15	13.35
7150	24' span	F-3	66	.606			29	9.90	38.90	58.50
7200	26' span		64	.625			30	10.20	40.20	60
7250	28' span		62	.645			31	10.55	41.55	62
7300	30' span		58	.690			33	11.30	44.30	66.50
7350	32' span		56	.714			34	11.70	45.70	68.50
7400	34' span		54	.741			35.50	12.10	47.60	71.50
7450	36' span		52	.769			37	12.60	49.60	74
8000	Soffit, T & G wood	1 Clab	520	.015	S.F.		.58		.58	.95
8010	Hardboard, vinyl or aluminum	"	640	.013			.47		.47	.77
8030	Plywood	2 Carp	315	.051			2.38		2.38	3.91
9000	Minimum labor/equipment charge	1 Clab	4	2	Job		75		75	123
9500	See Section 02 41 19.19 for rubbish handling									

06 05 05.20 Selective Demolition Millwork and Trim		Crew	Daily Output	Labor-Hours	Unit	Material	Labor	Equipment	Total	Total Incl O&P
0010	**SELECTIVE DEMOLITION MILLWORK AND TRIM** R024119-10									
1000	Cabinets, wood, base cabinets, per L.F.	2 Clab	80	.200	L.F.		7.50		7.50	12.35
1020	Wall cabinets, per L.F.	"	80	.200	"		7.50		7.50	12.35
1060	Remove and reset, base cabinets	2 Carp	18	.889	Ea.		41.50		41.50	68.50
1070	Wall cabinets		20	.800			37.50		37.50	61.50
1072	Oven cabinet, 7' high		11	1.455			68.50		68.50	112
1074	Cabinet door, up to 2' high	1 Clab	66	.121			4.56		4.56	7.45
1076	2' - 4' high	"	46	.174			6.55		6.55	10.70
1100	Steel, painted, base cabinets	2 Clab	60	.267	L.F.		10.05		10.05	16.45
1120	Wall cabinets		60	.267	"		10.05		10.05	16.45
1200	Casework, large area		320	.050	S.F.		1.88		1.88	3.08
1220	Selective		200	.080	"		3.01		3.01	4.93
1500	Counter top, straight runs		200	.080	L.F.		3.01		3.01	4.93
1510	L, U or C shapes		120	.133			5		5	8.20
1550	Remove and reset, straight runs	2 Carp	50	.320			15		15	24.50
1560	L, U or C shape	"	40	.400			18.80		18.80	31
2000	Paneling, 4' x 8' sheets	2 Clab	2000	.008	S.F.		.30		.30	.49
2100	Boards, 1" x 4"		700	.023			.86		.86	1.41
2120	1" x 6"		750	.021			.80		.80	1.32
2140	1" x 8"		800	.020			.75		.75	1.23
3000	Trim, baseboard, to 6" wide		1200	.013	L.F.		.50		.50	.82
3040	Greater than 6" and up to 12" wide		1000	.016			.60		.60	.99
3080	Remove and reset, minimum	2 Carp	400	.040			1.88		1.88	3.08
3090	Maximum	"	300	.053			2.50		2.50	4.10
3100	Ceiling trim	2 Clab	1000	.016			.60		.60	.99
3120	Chair rail		1200	.013			.50		.50	.82
3140	Railings with balusters		240	.067			2.51		2.51	4.11

06 05 Common Work Results for Wood, Plastics, and Composites

06 05 05 – Selective Demolition for Wood, Plastics, and Composites

06 05 05.20 Selective Demolition Millwork and Trim		Crew	Daily Output	Labor-Hours	Unit	Material	2015 Bare Costs Labor	Equipment	Total	Total Incl O&P
3160	Wainscoting	2 Clab	700	.023	S.F.		.86		.86	1.41
9000	Minimum labor/equipment charge	1 Clab	4	2	Job		75		75	123

06 05 23 – Wood, Plastic, and Composite Fastenings

06 05 23.10 Nails

0010	**NAILS**, material only, based upon 50# box purchase									
0020	Copper nails, plain				Lb.	9.85			9.85	10.80
0400	Stainless steel, plain					9.20			9.20	10.15
0500	Box, 3d to 20d, bright					1.37			1.37	1.51
0520	Galvanized					1.68			1.68	1.85
0600	Common, 3d to 60d, plain					1.30			1.30	1.43
0700	Galvanized					1.97			1.97	2.17
0800	Aluminum					10.80			10.80	11.90
1000	Annular or spiral thread, 4d to 60d, plain					2.63			2.63	2.89
1200	Galvanized					3.15			3.15	3.47
1400	Drywall nails, plain					1.50			1.50	1.65
1600	Galvanized					2.20			2.20	2.42
1800	Finish nails, 4d to 10d, plain					1.25			1.25	1.38
2000	Galvanized					1.86			1.86	2.05
2100	Aluminum					5.85			5.85	6.45
2300	Flooring nails, hardened steel, 2d to 10d, plain					3.38			3.38	3.72
2400	Galvanized					4.04			4.04	4.44
2500	Gypsum lath nails, 1-1/8", 13 ga. flathead, blued					2			2	2.20
2600	Masonry nails, hardened steel, 3/4" to 3" long, plain					1.71			1.71	1.88
2700	Galvanized					3.27			3.27	3.60
2900	Roofing nails, threaded, galvanized					1.76			1.76	1.94
3100	Aluminum					4.85			4.85	5.35
3300	Compressed lead head, threaded, galvanized					2.49			2.49	2.74
3600	Siding nails, plain shank, galvanized					2.02			2.02	2.22
3800	Aluminum					5.85			5.85	6.45
5000	Add to prices above for cement coating					.11			.11	.12
5200	Zinc or tin plating					.14			.14	.15
5500	Vinyl coated sinkers, 8d to 16d					2.50			2.50	2.75

06 05 23.50 Wood Screws

0010	**WOOD SCREWS**									
0020	#8 x 1" long, steel				C	2.70			2.70	2.97
0100	Brass					12.20			12.20	13.40
0200	#8, 2" long, steel					4.16			4.16	4.58
0300	Brass					22			22	24
0400	#10, 1" long, steel					3.30			3.30	3.63
0500	Brass					15.60			15.60	17.15
0600	#10, 2" long, steel					5.30			5.30	5.85
0700	Brass					27			27	29.50
0800	#10, 3" long, steel					8.85			8.85	9.75
1000	#12, 2" long, steel					7.30			7.30	8.05
1100	Brass					34.50			34.50	38
1500	#12, 3" long, steel					11.15			11.15	12.30
2000	#12, 4" long, steel					18.75			18.75	20.50

06 05 23.60 Timber Connectors

0010	**TIMBER CONNECTORS**									
0020	Add up cost of each part for total cost of connection									
0100	Connector plates, steel, with bolts, straight	2 Carp	75	.213	Ea.	30.50	10		40.50	50
0110	Tee, 7 ga.		50	.320		35	15		50	63

06 05 Common Work Results for Wood, Plastics, and Composites

06 05 23 – Wood, Plastic, and Composite Fastenings

06 05 23.60 Timber Connectors

		Crew	Daily Output	Labor-Hours	Unit	Material	2015 Bare Costs Labor	Equipment	Total	Total Incl O&P
0120	T-Strap, 14 ga., 12" x 8" x 2"	2 Carp	50	.320	Ea.	35	15		50	63
0150	Anchor plates, 7 ga., 9" x 7"		75	.213		30.50	10		40.50	50
0200	Bolts, machine, sq. hd. with nut & washer, 1/2" diameter, 4" long	1 Carp	140	.057		.75	2.68		3.43	5.25
0300	7-1/2" long		130	.062		1.37	2.89		4.26	6.25
0500	3/4" diameter, 7-1/2" long		130	.062		3.20	2.89		6.09	8.25
0610	Machine bolts, w/nut, washer, 3/4" diam., 15" L, HD's & beam hangers		95	.084		5.95	3.95		9.90	13.05
0800	Drilling bolt holes in timber, 1/2" diameter		450	.018	Inch		.83		.83	1.37
0900	1" diameter		350	.023	"		1.07		1.07	1.76
1100	Framing anchor, angle, 3" x 3" x 1-1/2", 12 ga		175	.046	Ea.	2.61	2.15		4.76	6.40
1150	Framing anchors, 18 ga., 4-1/2" x 2-3/4"		175	.046		2.61	2.15		4.76	6.40
1160	Framing anchors, 18 ga., 4-1/2" x 3"		175	.046		2.61	2.15		4.76	6.40
1170	Clip anchors plates, 18 ga., 12" x 1-1/8"		175	.046		2.61	2.15		4.76	6.40
1250	Holdowns, 3 ga. base, 10 ga. body		8	1		25	47		72	105
1260	Holdowns, 7 ga. 11-1/16" x 3-1/4"		8	1		25	47		72	105
1270	Holdowns, 7 ga. 14-3/8" x 3-1/8"		8	1		25	47		72	105
1275	Holdowns, 12 ga. 8" x 2-1/2"		8	1		25	47		72	105
1300	Joist and beam hangers, 18 ga. galv., for 2" x 4" joist		175	.046		.73	2.15		2.88	4.32
1400	2" x 6" to 2" x 10" joist		165	.048		1.39	2.28		3.67	5.25
1600	16 ga. galv., 3" x 6" to 3" x 10" joist		160	.050		2.96	2.35		5.31	7.10
1700	3" x 10" to 3" x 14" joist		160	.050		4.92	2.35		7.27	9.25
1800	4" x 6" to 4" x 10" joist		155	.052		3.07	2.42		5.49	7.35
1900	4" x 10" to 4" x 14" joist		155	.052		5	2.42		7.42	9.45
2000	Two-2" x 6" to two-2" x 10" joists		150	.053		4.19	2.50		6.69	8.70
2100	Two-2" x 10" to two-2" x 14" joists		150	.053		4.68	2.50		7.18	9.25
2300	3/16" thick, 6" x 8" joist		145	.055		65	2.59		67.59	76
2400	6" x 10" joist		140	.057		67.50	2.68		70.18	79
2500	6" x 12" joist		135	.059		70.50	2.78		73.28	82
2700	1/4" thick, 6" x 14" joist		130	.062		73	2.89		75.89	85
2900	Plywood clips, extruded aluminum H clip, for 3/4" panels					.24			.24	.26
3000	Galvanized 18 ga. back-up clip					.18			.18	.20
3200	Post framing, 16 ga. galv. for 4" x 4" base, 2 piece	1 Carp	130	.062		16.40	2.89		19.29	23
3300	Cap		130	.062		23	2.89		25.89	29.50
3500	Rafter anchors, 18 ga. galv., 1-1/2" wide, 5-1/4" long		145	.055		.49	2.59		3.08	4.79
3600	10-3/4" long		145	.055		1.46	2.59		4.05	5.85
3800	Shear plates, 2-5/8" diameter		120	.067		2.38	3.13		5.51	7.75
3900	4" diameter		115	.070		5.65	3.27		8.92	11.55
4000	Sill anchors, embedded in concrete or block, 25-1/2" long		115	.070		12.70	3.27		15.97	19.30
4100	Spike grids, 3" x 6"		120	.067		.96	3.13		4.09	6.20
4400	Split rings, 2-1/2" diameter		120	.067		1.96	3.13		5.09	7.30
4500	4" diameter		110	.073		2.97	3.41		6.38	8.85
4550	Tie plate, 20 ga., 7" x 3 1/8"		110	.073		2.97	3.41		6.38	8.85
4560	Tie plate, 20 ga., 5" x 4 1/8"		110	.073		2.97	3.41		6.38	8.85
4575	Twist straps, 18 ga., 12" x 1 1/4"		110	.073		2.97	3.41		6.38	8.85
4580	Twist straps, 18 ga., 16" x 1 1/4"		110	.073		2.97	3.41		6.38	8.85
4600	Strap ties, 20 ga., 2-1/16" wide, 12 13/16" long		180	.044		.94	2.09		3.03	4.45
4700	Strap ties, 16 ga., 1-3/8" wide, 12" long		180	.044		.94	2.09		3.03	4.45
4800	21-5/8" x 1-1/4"		160	.050		2.96	2.35		5.31	7.10
5000	Toothed rings, 2-5/8" or 4" diameter		90	.089		1.78	4.17		5.95	8.80
5200	Truss plates, nailed, 20 ga., up to 32' span		17	.471	Truss	12.90	22		34.90	50
5400	Washers, 2" x 2" x 1/8"				Ea.	.40			.40	.44
5500	3" x 3" x 3/16"				"	1.06			1.06	1.17
9000	Minimum labor/equipment charge	1 Carp	4	2	Job		94		94	154

06 05 Common Work Results for Wood, Plastics, and Composites

06 05 23 – Wood, Plastic, and Composite Fastenings

06 05 23.80 Metal Bracing

		Crew	Daily Output	Labor-Hours	Unit	Material	2015 Bare Costs Labor	Equipment	Total	Total Incl O&P
0010	**METAL BRACING**									
0302	Let-in, "T" shaped, 22 ga. galv. steel, studs at 16" O.C.	1 Carp	580	.014	L.F.	.81	.65		1.46	1.95
0402	Studs at 24" O.C.		600	.013		.81	.63		1.44	1.92
0502	Steel straps, 16 ga. galv. steel, studs at 16" O.C.		600	.013		1.05	.63		1.68	2.19
0602	Studs at 24" O.C.		620	.013		1.05	.61		1.66	2.15

06 11 Wood Framing

06 11 10 – Framing with Dimensional, Engineered or Composite Lumber

06 11 10.01 Forest Stewardship Council Certification

		Crew	Daily Output	Labor-Hours	Unit	Material	2015 Bare Costs Labor	Equipment	Total	Total Incl O&P
0010	**FOREST STEWARDSHIP COUNCIL CERTIFICATION**									
0020	For Forest Stewardship Council (FSC) cert dimension lumber, add [G]					65%				

06 11 10.02 Blocking

		Crew	Daily Output	Labor-Hours	Unit	Material	Labor	Equipment	Total	Total Incl O&P
0010	**BLOCKING**									
2600	Miscellaneous, to wood construction									
2620	2" x 4"	1 Carp	.17	47.059	M.B.F.	635	2,200		2,835	4,325
2625	Pneumatic nailed		.21	38.095		640	1,800		2,440	3,625
2660	2" x 8"		.27	29.630		690	1,400		2,090	3,025
2665	Pneumatic nailed		.33	24.242		695	1,150		1,845	2,650
2720	To steel construction									
2740	2" x 4"	1 Carp	.14	57.143	M.B.F.	635	2,675		3,310	5,100
2780	2" x 8"		.21	38.095	"	690	1,800		2,490	3,675
9000	Minimum labor/equipment charge		4	2	Job		94		94	154

06 11 10.04 Wood Bracing

		Crew	Daily Output	Labor-Hours	Unit	Material	Labor	Equipment	Total	Total Incl O&P
0010	**WOOD BRACING**									
0012	Let-in, with 1" x 6" boards, studs @ 16" O.C.	1 Carp	150	.053	L.F.	.73	2.50		3.23	4.90
0202	Studs @ 24" O.C.	"	230	.035	"	.73	1.63		2.36	3.48

06 11 10.06 Bridging

		Crew	Daily Output	Labor-Hours	Unit	Material	Labor	Equipment	Total	Total Incl O&P
0010	**BRIDGING**									
0012	Wood, for joists 16" O.C., 1" x 3"	1 Carp	130	.062	Pr.	.63	2.89		3.52	5.45
0017	Pneumatic nailed		170	.047		.68	2.21		2.89	4.37
0102	2" x 3" bridging		130	.062		.69	2.89		3.58	5.50
0107	Pneumatic nailed		170	.047		.69	2.21		2.90	4.38
0302	Steel, galvanized, 18 ga., for 2" x 10" joists at 12" O.C.		130	.062		.92	2.89		3.81	5.75
0352	16" O.C.		135	.059		.93	2.78		3.71	5.60
0402	24" O.C.		140	.057		1.90	2.68		4.58	6.50
0602	For 2" x 14" joists at 16" O.C.		130	.062		1.40	2.89		4.29	6.30
0902	Compression type, 16" O.C., 2" x 8" joists		200	.040		.93	1.88		2.81	4.10
1002	2" x 12" joists		200	.040		.93	1.88		2.81	4.10

06 11 10.10 Beam and Girder Framing

		Crew	Daily Output	Labor-Hours	Unit	Material	Labor	Equipment	Total	Total Incl O&P
0010	**BEAM AND GIRDER FRAMING** R061110-30									
1000	Single, 2" x 6"	2 Carp	700	.023	L.F.	.67	1.07		1.74	2.49
1005	Pneumatic nailed		812	.020		.67	.92		1.59	2.26
1020	2" x 8"		650	.025		.92	1.16		2.08	2.90
1025	Pneumatic nailed		754	.021		.93	1		1.93	2.65
1040	2" x 10"		600	.027		1.34	1.25		2.59	3.52
1045	Pneumatic nailed		696	.023		1.35	1.08		2.43	3.25
1060	2" x 12"		550	.029		1.72	1.37		3.09	4.14
1065	Pneumatic nailed		638	.025		1.73	1.18		2.91	3.84
1080	2" x 14"		500	.032		2.01	1.50		3.51	4.67

06 11 Wood Framing

06 11 10 – Framing with Dimensional, Engineered or Composite Lumber

06 11 10.10 Beam and Girder Framing

		Crew	Daily Output	Labor-Hours	Unit	Material	2015 Bare Costs Labor	Equipment	Total	Total Incl O&P
1085	Pneumatic nailed	2 Carp	580	.028	L.F.	2.02	1.30		3.32	4.35
1100	3" x 8"		550	.029		2.55	1.37		3.92	5.05
1120	3" x 10"		500	.032		3.22	1.50		4.72	6
1140	3" x 12"		450	.036		3.86	1.67		5.53	7
1160	3" x 14"		400	.040		4.51	1.88		6.39	8.05
1170	4" x 6"	F-3	1100	.036		2.84	1.74	.59	5.17	6.60
1180	4" x 8"		1000	.040		2.89	1.92	.65	5.46	7
1200	4" x 10"		950	.042		4.28	2.02	.69	6.99	8.75
1220	4" x 12"		900	.044		5.15	2.13	.73	8.01	9.90
1240	4" x 14"		850	.047		6	2.25	.77	9.02	11.10
1250	6" x 8"		525	.076		7.40	3.65	1.25	12.30	15.45
1260	6" x 10"		500	.080		6.45	3.83	1.31	11.59	14.80
1290	8" x 12"		300	.133		15.20	6.40	2.18	23.78	29.50
2000	Double, 2" x 6"	2 Carp	625	.026		1.33	1.20		2.53	3.44
2005	Pneumatic nailed		725	.022		1.34	1.04		2.38	3.18
2020	2" x 8"		575	.028		1.84	1.31		3.15	4.17
2025	Pneumatic nailed		667	.024		1.86	1.13		2.99	3.89
2040	2" x 10"		550	.029		2.68	1.37		4.05	5.20
2045	Pneumatic nailed		638	.025		2.70	1.18		3.88	4.90
2060	2" x 12"		525	.030		3.45	1.43		4.88	6.15
2065	Pneumatic nailed		610	.026		3.47	1.23		4.70	5.85
2080	2" x 14"		475	.034		4.02	1.58		5.60	7
2085	Pneumatic nailed		551	.029		4.05	1.36		5.41	6.70
3000	Triple, 2" x 6"		550	.029		2	1.37		3.37	4.44
3005	Pneumatic nailed		638	.025		2.01	1.18		3.19	4.14
3020	2" x 8"		525	.030		2.77	1.43		4.20	5.40
3025	Pneumatic nailed		609	.026		2.78	1.23		4.01	5.10
3040	2" x 10"		500	.032		4.02	1.50		5.52	6.90
3045	Pneumatic nailed		580	.028		4.04	1.30		5.34	6.55
3060	2" x 12"		475	.034		5.15	1.58		6.73	8.30
3065	Pneumatic nailed		551	.029		5.20	1.36		6.56	7.95
3080	2" x 14"		450	.036		6.70	1.67		8.37	10.10
3085	Pneumatic nailed		522	.031		6.05	1.44		7.49	9.05
9000	Minimum labor/equipment charge	1 Carp	2	4	Job		188		188	310

06 11 10.12 Ceiling Framing

		Crew	Daily Output	Labor-Hours	Unit	Material	2015 Bare Costs Labor	Equipment	Total	Total Incl O&P
0010	**CEILING FRAMING**									
6000	Suspended, 2" x 3"	2 Carp	1000	.016	L.F.	.39	.75		1.14	1.66
6050	2" x 4"		900	.018		.42	.83		1.25	1.84
6100	2" x 6"		800	.020		.67	.94		1.61	2.27
6150	2" x 8"		650	.025		.92	1.16		2.08	2.90
9000	Minimum labor/equipment charge	1 Carp	4	2	Job		94		94	154

06 11 10.14 Posts and Columns

		Crew	Daily Output	Labor-Hours	Unit	Material	2015 Bare Costs Labor	Equipment	Total	Total Incl O&P
0010	**POSTS AND COLUMNS**									
0100	4" x 4"	2 Carp	390	.041	L.F.	1.73	1.93		3.66	5.05
0150	4" x 6"		275	.058		2.84	2.73		5.57	7.60
0200	4" x 8"		220	.073		2.89	3.41		6.30	8.80
0250	6" x 6"		215	.074		5.05	3.49		8.54	11.30
0300	6" x 8"		175	.091		7.40	4.29		11.69	15.20
0350	6" x 10"		150	.107		6.45	5		11.45	15.30
9000	Minimum labor/equipment charge	1 Carp	2	4	Job		188		188	310

06 11 Wood Framing

06 11 10 – Framing with Dimensional, Engineered or Composite Lumber

06 11 10.18 Joist Framing

		Crew	Daily Output	Labor-Hours	Unit	Material	2015 Bare Costs Labor	Equipment	Total	Total Incl O&P
0010	**JOIST FRAMING** R061110-30									
2000	Joists, 2" x 4"	2 Carp	1250	.013	L.F.	.42	.60		1.02	1.46
2005	Pneumatic nailed		1438	.011		.43	.52		.95	1.33
2100	2" x 6"		1250	.013		.67	.60		1.27	1.72
2105	Pneumatic nailed		1438	.011		.67	.52		1.19	1.60
2150	2" x 8"		1100	.015		.92	.68		1.60	2.13
2155	Pneumatic nailed		1265	.013		.93	.59		1.52	1.99
2200	2" x 10"		900	.018		1.34	.83		2.17	2.84
2205	Pneumatic nailed		1035	.015		1.35	.73		2.08	2.67
2250	2" x 12"		875	.018		1.72	.86		2.58	3.31
2255	Pneumatic nailed		1006	.016		1.73	.75		2.48	3.13
2300	2" x 14"		770	.021		2.01	.98		2.99	3.81
2305	Pneumatic nailed		886	.018		2.02	.85		2.87	3.62
2350	3" x 6"		925	.017		1.90	.81		2.71	3.41
2400	3" x 10"		780	.021		3.22	.96		4.18	5.10
2450	3" x 12"		600	.027		3.86	1.25		5.11	6.30
2500	4" x 6"		800	.020		2.84	.94		3.78	4.66
2550	4" x 10"		600	.027		4.28	1.25		5.53	6.75
2600	4" x 12"		450	.036		5.15	1.67		6.82	8.40
2605	Sister joist, 2" x 6"		800	.020		.67	.94		1.61	2.27
2606	Pneumatic nailed		960	.017		.67	.78		1.45	2.02
2610	2" x 8"		640	.025		.92	1.17		2.09	2.93
2611	Pneumatic nailed		768	.021		.93	.98		1.91	2.62
2615	2" x 10"		535	.030		1.34	1.40		2.74	3.77
2616	Pneumatic nailed		642	.025		1.35	1.17		2.52	3.40
2620	2" x 12"		455	.035		1.72	1.65		3.37	4.61
2625	Pneumatic nailed		546	.029		1.73	1.38		3.11	4.16
3000	Composite wood joist 9-1/2" deep		.90	17.778	M.L.F.	1,800	835		2,635	3,350
3010	11-1/2" deep		.88	18.182		1,975	855		2,830	3,575
3020	14" deep		.82	19.512		2,100	915		3,015	3,800
3030	16" deep		.78	20.513		3,475	965		4,440	5,400
4000	Open web joist 12" deep		.88	18.182		3,525	855		4,380	5,275
4002	Per linear foot		880	.018	L.F.	3.52	.85		4.37	5.25
4004	Treated, per linear foot		880	.018	"	4.39	.85		5.24	6.25
4010	14" deep		.82	19.512	M.L.F.	3,725	915		4,640	5,600
4012	Per linear foot		820	.020	L.F.	3.72	.92		4.64	5.60
4014	Treated, per linear foot		820	.020	"	4.74	.92		5.66	6.70
4020	16" deep		.78	20.513	M.L.F.	3,825	965		4,790	5,775
4022	Per linear foot		780	.021	L.F.	3.83	.96		4.79	5.80
4024	Treated, per linear foot		780	.021	"	4.99	.96		5.95	7.10
4030	18" deep		.74	21.622	M.L.F.	4,000	1,025		5,025	6,075
4032	Per linear foot		740	.022	L.F.	4	1.02		5.02	6.05
4034	Treated, per linear foot		740	.022	"	5.30	1.02		6.32	7.50
6000	Composite rim joist, 1-1/4" x 9-1/2"		.90	17.778	M.L.F.	1,775	835		2,610	3,325
6010	1-1/4" x 11-1/2"		.88	18.182		2,050	855		2,905	3,650
6020	1-1/4" x 14-1/2"		.82	19.512		2,750	915		3,665	4,525
6030	1-1/4" x 16-1/2"		.78	20.513		3,375	965		4,340	5,300
9000	Minimum labor/equipment charge	1 Carp	4	2	Job		94		94	154

06 11 10.24 Miscellaneous Framing

		Crew	Daily Output	Labor-Hours	Unit	Material	Labor	Equipment	Total	Total Incl O&P
0010	**MISCELLANEOUS FRAMING**									
2000	Firestops, 2" x 4"	2 Carp	780	.021	L.F.	.42	.96		1.38	2.05
2005	Pneumatic nailed		952	.017		.43	.79		1.22	1.76

06 11 Wood Framing

06 11 10 – Framing with Dimensional, Engineered or Composite Lumber

06 11 10.24 Miscellaneous Framing

		Crew	Daily Output	Labor-Hours	Unit	Material	2015 Bare Costs Labor	2015 Bare Costs Equipment	Total	Total Incl O&P
2100	2" x 6"	2 Carp	600	.027	L.F.	.67	1.25		1.92	2.78
2105	Pneumatic nailed		732	.022		.67	1.03		1.70	2.42
5000	Nailers, treated, wood construction, 2" x 4"		800	.020		.51	.94		1.45	2.10
5005	Pneumatic nailed		960	.017		.51	.78		1.29	1.84
5100	2" x 6"		750	.021		.78	1		1.78	2.50
5105	Pneumatic nailed		900	.018		.78	.83		1.61	2.23
5120	2" x 8"		700	.023		1.05	1.07		2.12	2.92
5125	Pneumatic nailed		840	.019		1.06	.89		1.95	2.63
5200	Steel construction, 2" x 4"		750	.021		.51	1		1.51	2.20
5220	2" x 6"		700	.023		.78	1.07		1.85	2.62
5240	2" x 8"		650	.025		1.05	1.16		2.21	3.05
7000	Rough bucks, treated, for doors or windows, 2" x 6"		400	.040		.78	1.88		2.66	3.94
7005	Pneumatic nailed		480	.033		.78	1.56		2.34	3.42
7100	2" x 8"		380	.042		1.05	1.98		3.03	4.40
7105	Pneumatic nailed		456	.035		1.06	1.65		2.71	3.86
8000	Stair stringers, 2" x 10"		130	.123		1.34	5.80		7.14	10.90
8100	2" x 12"		130	.123		1.72	5.80		7.52	11.35
8150	3" x 10"		125	.128		3.22	6		9.22	13.40
8200	3" x 12"		125	.128		3.86	6		9.86	14.10
8870	Laminated structural lumber, 1-1/4" x 11-1/2"		130	.123		2.05	5.80		7.85	11.70
8880	1-1/4" x 14-1/2"		130	.123		2.72	5.80		8.52	12.45
9000	Minimum labor/equipment charge	1 Carp	4	2	Job		94		94	154

06 11 10.26 Partitions

		Crew	Daily Output	Labor-Hours	Unit	Material	2015 Bare Costs Labor	2015 Bare Costs Equipment	Total	Total Incl O&P
0010	**PARTITIONS**									
0020	Single bottom and double top plate, no waste, std. & better lumber									
0180	2" x 4" studs, 8' high, studs 12" O.C.	2 Carp	80	.200	L.F.	4.65	9.40		14.05	20.50
0185	12" O.C., pneumatic nailed		96	.167		4.69	7.85		12.54	18
0200	16" O.C.		100	.160		3.81	7.50		11.31	16.50
0205	16" O.C., pneumatic nailed		120	.133		3.84	6.25		10.09	14.45
0300	24" O.C.		125	.128		2.96	6		8.96	13.10
0305	24" O.C., pneumatic nailed		150	.107		2.98	5		7.98	11.50
0380	10' high, studs 12" O.C.		80	.200		5.50	9.40		14.90	21.50
0385	12" O.C., pneumatic nailed		96	.167		5.55	7.85		13.40	18.95
0400	16" O.C.		100	.160		4.44	7.50		11.94	17.20
0405	16" O.C., pneumatic nailed		120	.133		4.47	6.25		10.72	15.15
0500	24" O.C.		125	.128		3.38	6		9.38	13.55
0505	24" O.C., pneumatic nailed		150	.107		3.41	5		8.41	11.95
0580	12' high, studs 12" O.C.		65	.246		6.35	11.55		17.90	26
0585	12" O.C., pneumatic nailed		78	.205		6.40	9.65		16.05	23
0600	16" O.C.		80	.200		5.10	9.40		14.50	21
0605	16" O.C., pneumatic nailed		96	.167		5.10	7.85		12.95	18.45
0700	24" O.C.		100	.160		3.81	7.50		11.31	16.50
0705	24" O.C., pneumatic nailed		120	.133		3.84	6.25		10.09	14.45
0780	2" x 6" studs, 8' high, studs 12" O.C.		70	.229		7.35	10.75		18.10	25.50
0785	12" O.C., pneumatic nailed		84	.190		7.40	8.95		16.35	23
0800	16" O.C.		90	.178		6	8.35		14.35	20.50
0805	16" O.C., pneumatic nailed		108	.148		6.05	6.95		13	18.05
0900	24" O.C.		115	.139		4.66	6.55		11.21	15.85
0905	24" O.C., pneumatic nailed		138	.116		4.70	5.45		10.15	14.05
0980	10' high, studs 12" O.C.		70	.229		8.65	10.75		19.40	27
0985	12" O.C., pneumatic nailed		84	.190		8.75	8.95		17.70	24.50
1000	16" O.C.		90	.178		7	8.35		15.35	21.50

06 11 Wood Framing

06 11 10 – Framing with Dimensional, Engineered or Composite Lumber

06 11 10.26 Partitions

		Crew	Daily Output	Labor-Hours	Unit	Material	2015 Bare Costs Labor	Equipment	Total	Total Incl O&P
1005	16" O.C., pneumatic nailed	2 Carp	108	.148	L.F.	7.05	6.95		14	19.15
1100	24" O.C.		115	.139		5.35	6.55		11.90	16.55
1105	24" O.C., pneumatic nailed		138	.116		5.35	5.45		10.80	14.80
1180	12' high, studs 12" O.C.		55	.291		10	13.65		23.65	33.50
1185	12" O.C., pneumatic nailed		66	.242		10.05	11.40		21.45	29.50
1200	16" O.C.		70	.229		8	10.75		18.75	26.50
1205	16" O.C., pneumatic nailed		84	.190		8.05	8.95		17	23.50
1300	24" O.C.		90	.178		6	8.35		14.35	20.50
1305	24" O.C., pneumatic nailed		108	.148		6.05	6.95		13	18.05
1400	For horizontal blocking, 2" x 4", add		600	.027		.42	1.25		1.67	2.52
1500	2" x 6", add		600	.027		.67	1.25		1.92	2.78
1600	For openings, add		250	.064			3		3	4.92
1702	Headers for above openings, material only, add				B.F.	.76			.76	.84
9000	Minimum labor/equipment charge	1 Carp	4	2	Job		94		94	154

06 11 10.28 Porch or Deck Framing

		Crew	Daily Output	Labor-Hours	Unit	Material	Labor	Equipment	Total	Total Incl O&P
0010	**PORCH OR DECK FRAMING**									
0100	Treated lumber, posts or columns, 4" x 4"	2 Carp	390	.041	L.F.	1.34	1.93		3.27	4.63
0110	4" x 6"		275	.058		1.91	2.73		4.64	6.60
0120	4" x 8"		220	.073		3.81	3.41		7.22	9.80
0130	Girder, single, 4" x 4"		675	.024		1.34	1.11		2.45	3.29
0140	4" x 6"		600	.027		1.91	1.25		3.16	4.15
0150	4" x 8"		525	.030		3.81	1.43		5.24	6.55
0160	Double, 2" x 4"		625	.026		1.05	1.20		2.25	3.12
0170	2" x 6"		600	.027		1.60	1.25		2.85	3.81
0180	2" x 8"		575	.028		2.15	1.31		3.46	4.51
0190	2" x 10"		550	.029		2.86	1.37		4.23	5.40
0200	2" x 12"		525	.030		3.95	1.43		5.38	6.70
0210	Triple, 2" x 4"		575	.028		1.57	1.31		2.88	3.86
0220	2" x 6"		550	.029		2.40	1.37		3.77	4.88
0230	2" x 8"		525	.030		3.23	1.43		4.66	5.90
0240	2" x 10"		500	.032		4.29	1.50		5.79	7.20
0250	2" x 12"		475	.034		5.95	1.58		7.53	9.10
0260	Ledger, bolted 4' O.C., 2" x 4"		400	.040		.66	1.88		2.54	3.81
0270	2" x 6"		395	.041		.93	1.90		2.83	4.14
0280	2" x 8"		390	.041		1.20	1.93		3.13	4.48
0290	2" x 10"		385	.042		1.54	1.95		3.49	4.90
0300	2" x 12"		380	.042		2.08	1.98		4.06	5.55
0310	Joists, 2" x 4"		1250	.013		.52	.60		1.12	1.56
0320	2" x 6"		1250	.013		.79	.60		1.39	1.86
0330	2" x 8"		1100	.015		1.07	.68		1.75	2.30
0340	2" x 10"		900	.018		1.42	.83		2.25	2.94
0350	2" x 12"		875	.018		1.71	.86		2.57	3.29
0360	Railings and trim, 1" x 4"	1 Carp	300	.027		.47	1.25		1.72	2.57
0370	2" x 2"		300	.027		.33	1.25		1.58	2.42
0380	2" x 4"		300	.027		.51	1.25		1.76	2.61
0390	2" x 6"		300	.027		.79	1.25		2.04	2.91
0400	Decking, 1" x 4"		275	.029	S.F.	2.16	1.37		3.53	4.62
0410	2" x 4"		300	.027		1.73	1.25		2.98	3.95
0420	2" x 6"		320	.025		1.69	1.17		2.86	3.78
0430	5/4" x 6"		320	.025		2.25	1.17		3.42	4.40
0440	Balusters, square, 2" x 2"	2 Carp	660	.024	L.F.	.34	1.14		1.48	2.24
0450	Turned, 2" x 2"		420	.038		.45	1.79		2.24	3.42

06 11 Wood Framing

06 11 10 – Framing with Dimensional, Engineered or Composite Lumber

06 11 10.28 Porch or Deck Framing

		Crew	Daily Output	Labor-Hours	Unit	Material	2015 Bare Costs Labor	Equipment	Total	Total Incl O&P
0460	Stair stringer, 2" x 10"	2 Carp	130	.123	L.F.	1.42	5.80		7.22	11
0470	2" x 12"		130	.123		1.71	5.80		7.51	11.35
0480	Stair treads, 1" x 4"		140	.114		2.17	5.35		7.52	11.20
0490	2" x 4"		140	.114		.52	5.35		5.87	9.35
0500	2" x 6"		160	.100		.78	4.70		5.48	8.55
0510	5/4" x 6"		160	.100		1.05	4.70		5.75	8.85
0520	Turned handrail post, 4" x 4"		64	.250	Ea.	31	11.75		42.75	53.50
0530	Lattice panel, 4' x 8', 1/2"		1600	.010	S.F.	.81	.47		1.28	1.66
0535	3/4"		1600	.010	"	1.21	.47		1.68	2.10
0540	Cedar, posts or columns, 4" x 4"		390	.041	L.F.	3.93	1.93		5.86	7.50
0550	4" x 6"		275	.058		7.35	2.73		10.08	12.60
0560	4" x 8"		220	.073		10	3.41		13.41	16.60
0800	Decking, 1" x 4"		550	.029		1.81	1.37		3.18	4.23
0810	2" x 4"		600	.027		3.62	1.25		4.87	6.05
0820	2" x 6"		640	.025		6.60	1.17		7.77	9.15
0830	5/4" x 6"		640	.025		4.45	1.17		5.62	6.80
0840	Railings and trim, 1" x 4"		600	.027		1.81	1.25		3.06	4.04
0860	2" x 4"		600	.027		3.62	1.25		4.87	6.05
0870	2" x 6"		600	.027		6.60	1.25		7.85	9.30
0920	Stair treads, 1" x 4"		140	.114		1.81	5.35		7.16	10.80
0930	2" x 4"		140	.114		3.62	5.35		8.97	12.80
0940	2" x 6"		160	.100		6.60	4.70		11.30	14.95
0950	5/4" x 6"		160	.100		4.45	4.70		9.15	12.60
0980	Redwood, posts or columns, 4" x 4"		390	.041		6.45	1.93		8.38	10.20
0990	4" x 6"		275	.058		12.60	2.73		15.33	18.35
1000	4" x 8"		220	.073		23.50	3.41		26.91	31.50
1240	Decking, 1" x 4"	1 Carp	275	.029	S.F.	3.97	1.37		5.34	6.60
1260	2" x 6"		340	.024		7.50	1.10		8.60	10.05
1270	5/4" x 6"		320	.025		4.77	1.17		5.94	7.15
1280	Railings and trim, 1" x 4"	2 Carp	600	.027	L.F.	1.17	1.25		2.42	3.34
1310	2" x 6"		600	.027		7.50	1.25		8.75	10.30
1420	Alternative decking, wood/plastic composite, 5/4" x 6" G		640	.025		3.14	1.17		4.31	5.35
1440	1" x 4" square edge fir		550	.029		2.18	1.37		3.55	4.64
1450	1" x 4" tongue and groove fir		450	.036		1.46	1.67		3.13	4.34
1460	1" x 4" mahogany		550	.029		2.01	1.37		3.38	4.45
1462	5/4" x 6" PVC		550	.029		3.73	1.37		5.10	6.35
1465	Framing, porch or deck, alt deck fastening, screws, add	1 Carp	240	.033	S.F.		1.56		1.56	2.56
1470	Accessories, joist hangers, 2" x 4"		160	.050	Ea.	.73	2.35		3.08	4.65
1480	2" x 6" through 2" x 12"		150	.053		1.39	2.50		3.89	5.65
1530	Post footing, incl excav, backfill, tube form & concrete, 4' deep, 8" dia	F-7	12	2.667		12.15	113		125.15	198
1540	10" diameter		11	2.909		17.75	123		140.75	222
1550	12" diameter		10	3.200		23.50	135		158.50	248

06 11 10.30 Roof Framing

		Crew	Daily Output	Labor-Hours	Unit	Material	Labor	Equipment	Total	Total Incl O&P
0010	**ROOF FRAMING** R061110-30									
1900	Rough fascia, 2" x 6"	2 Carp	250	.064	L.F.	.67	3		3.67	5.65
2000	2" x 8"		225	.071		.92	3.34		4.26	6.45
2100	2" x 10"		180	.089		1.34	4.17		5.51	8.30
5000	Rafters, to 4 in 12 pitch, 2" x 6", ordinary		1000	.016		.67	.75		1.42	1.96
5060	2" x 8", ordinary		950	.017		.92	.79		1.71	2.31
5120	2" x 10", ordinary		630	.025		1.34	1.19		2.53	3.42
5180	2" x 12", ordinary		575	.028		1.72	1.31		3.03	4.04
5250	Composite rafter, 9-1/2" deep		575	.028		1.79	1.31		3.10	4.11

06 11 Wood Framing

06 11 10 – Framing with Dimensional, Engineered or Composite Lumber

06 11 10.30 Roof Framing		Crew	Daily Output	Labor-Hours	Unit	Material	2015 Bare Costs Labor	Equipment	Total	Total Incl O&P
5260	11-1/2" deep	2 Carp	575	.028	L.F.	1.97	1.31		3.28	4.31
5300	Hip and valley rafters, 2" x 6", ordinary		760	.021		.67	.99		1.66	2.35
5360	2" x 8", ordinary		720	.022		.92	1.04		1.96	2.72
5420	2" x 10", ordinary		570	.028		1.34	1.32		2.66	3.63
5480	2" x 12", ordinary		525	.030		1.72	1.43		3.15	4.25
5540	Hip and valley jacks, 2" x 6", ordinary		600	.027		.67	1.25		1.92	2.78
5600	2" x 8", ordinary		490	.033		.92	1.53		2.45	3.52
5660	2" x 10", ordinary		450	.036		1.34	1.67		3.01	4.21
5720	2" x 12", ordinary		375	.043		1.72	2		3.72	5.20
5761	For slopes steeper than 4 in 12, add						30%			
5780	Rafter tie, 1" x 4", #3	2 Carp	800	.020	L.F.	.47	.94		1.41	2.06
5790	2" x 4", #3		800	.020		.42	.94		1.36	2.01
5800	Ridge board, #2 or better, 1" x 6"		600	.027		.73	1.25		1.98	2.85
5820	1" x 8"		550	.029		1.20	1.37		2.57	3.56
5840	1" x 10"		500	.032		1.56	1.50		3.06	4.17
5860	2" x 6"		500	.032		.67	1.50		2.17	3.19
5880	2" x 8"		450	.036		.92	1.67		2.59	3.75
5900	2" x 10"		400	.040		1.34	1.88		3.22	4.55
5920	Roof cants, split, 4" x 4"		650	.025		1.73	1.16		2.89	3.79
5940	6" x 6"		600	.027		5.05	1.25		6.30	7.60
5960	Roof curbs, untreated, 2" x 6"		520	.031		.67	1.44		2.11	3.10
5980	2" x 12"		400	.040		1.72	1.88		3.60	4.98
6000	Sister rafters, 2" x 6"		800	.020		.67	.94		1.61	2.27
6020	2" x 8"		640	.025		.92	1.17		2.09	2.93
6040	2" x 10"		535	.030		1.34	1.40		2.74	3.77
6060	2" x 12"		455	.035		1.72	1.65		3.37	4.61
9000	Minimum labor/equipment charge	1 Carp	4	2	Job		94		94	154
06 11 10.32 Sill and Ledger Framing										
0010	**SILL AND LEDGER FRAMING**									
2002	Ledgers, nailed, 2" x 4"	2 Carp	755	.021	L.F.	.42	.99		1.41	2.10
2052	2" x 6"		600	.027		.67	1.25		1.92	2.78
2102	Bolted, not including bolts, 3" x 6"		325	.049		1.88	2.31		4.19	5.85
2152	3" x 12"		233	.069		3.83	3.22		7.05	9.50
2602	Mud sills, redwood, construction grade, 2" x 4"		895	.018		2.25	.84		3.09	3.85
2622	2" x 6"		780	.021		3.40	.96		4.36	5.30
4002	Sills, 2" x 4"		600	.027		.42	1.25		1.67	2.51
4052	2" x 6"		550	.029		.66	1.37		2.03	2.96
4082	2" x 8"		500	.032		.91	1.50		2.41	3.46
4101	2" x 10"		450	.036		1.32	1.67		2.99	4.19
4121	2" x 12"		400	.040		1.70	1.88		3.58	4.95
4202	Treated, 2" x 4"		550	.029		.50	1.37		1.87	2.79
4222	2" x 6"		500	.032		.77	1.50		2.27	3.31
4242	2" x 8"		450	.036		1.04	1.67		2.71	3.88
4261	2" x 10"		400	.040		1.38	1.88		3.26	4.60
4281	2" x 12"		350	.046		1.92	2.15		4.07	5.65
4402	4" x 4"		450	.036		1.30	1.67		2.97	4.17
4422	4" x 6"		350	.046		1.85	2.15		4	5.55
4462	4" x 8"		300	.053		3.73	2.50		6.23	8.20
4481	4" x 10"		260	.062		4.67	2.89		7.56	9.90
9000	Minimum labor/equipment charge	1 Carp	4	2	Job		94		94	154

06 11 Wood Framing

06 11 10 – Framing with Dimensional, Engineered or Composite Lumber

06 11 10.34 Sleepers

		Crew	Daily Output	Labor-Hours	Unit	Material	2015 Bare Costs Labor	2015 Bare Costs Equipment	Total	Total Incl O&P
0010	**SLEEPERS**									
0100	On concrete, treated, 1" x 2"	2 Carp	2350	.007	L.F.	.27	.32		.59	.82
0150	1" x 3"		2000	.008		.45	.38		.83	1.12
0200	2" x 4"		1500	.011		.55	.50		1.05	1.42
0250	2" x 6"		1300	.012		.86	.58		1.44	1.90
9000	Minimum labor/equipment charge	1 Carp	4	2	Job		94		94	154

06 11 10.36 Soffit and Canopy Framing

		Crew	Daily Output	Labor-Hours	Unit	Material	Labor	Equipment	Total	Total Incl O&P
0010	**SOFFIT AND CANOPY FRAMING**									
1002	Canopy or soffit framing, 1" x 4"	2 Carp	900	.018	L.F.	.47	.83		1.30	1.89
1021	1" x 6"		850	.019		.73	.88		1.61	2.25
1042	1" x 8"		750	.021		1.20	1		2.20	2.96
1102	2" x 4"		620	.026		.42	1.21		1.63	2.46
1121	2" x 6"		560	.029		.67	1.34		2.01	2.93
1142	2" x 8"		500	.032		.92	1.50		2.42	3.47
1202	3" x 4"		500	.032		1.07	1.50		2.57	3.64
1221	3" x 6"		400	.040		1.90	1.88		3.78	5.15
1242	3" x 10"		300	.053		3.22	2.50		5.72	7.65
9000	Minimum labor/equipment charge	1 Carp	4	2	Job		94		94	154

06 11 10.38 Treated Lumber Framing Material

		Crew	Daily Output	Labor-Hours	Unit	Material	Labor	Equipment	Total	Total Incl O&P
0010	**TREATED LUMBER FRAMING MATERIAL**									
0100	2" x 4"				M.B.F.	755			755	830
0110	2" x 6"					770			770	845
0120	2" x 8"					780			780	855
0130	2" x 10"					830			830	910
0140	2" x 12"					960			960	1,050
0200	4" x 4"					975			975	1,075
0210	4" x 6"					925			925	1,025
0220	4" x 8"					1,400			1,400	1,550

06 11 10.40 Wall Framing

		Crew	Daily Output	Labor-Hours	Unit	Material	Labor	Equipment	Total	Total Incl O&P
0010	**WALL FRAMING** R061110-30									
0100	Door buck, studs, header, access, 8' H, 2" x 4" wall, 3' W	1 Carp	32	.250	Ea.	17.85	11.75		29.60	39
0110	4' wide		32	.250		19.15	11.75		30.90	40.50
0120	5' wide		32	.250		23	11.75		34.75	45
0130	6' wide		32	.250		25	11.75		36.75	47
0140	8' wide		30	.267		35.50	12.50		48	59.50
0150	10' wide		30	.267		48.50	12.50		61	74
0160	12' wide		30	.267		62.50	12.50		75	89
0170	2" x 6" wall, 3' wide		32	.250		25.50	11.75		37.25	47.50
0180	4' wide		32	.250		27	11.75		38.75	49
0190	5' wide		32	.250		31	11.75		42.75	53.50
0200	6' wide		32	.250		33	11.75		44.75	55.50
0210	8' wide		30	.267		43.50	12.50		56	68
0220	10' wide		30	.267		56.50	12.50		69	82.50
0230	12' wide		30	.267		70.50	12.50		83	98
0240	Window buck, studs, header & access, 8' high 2" x 4" wall, 2' wide		24	.333		18.80	15.65		34.45	46
0250	3' wide		24	.333		22	15.65		37.65	50
0260	4' wide		24	.333		24	15.65		39.65	52
0270	5' wide		24	.333		28	15.65		43.65	56.50
0280	6' wide		24	.333		31.50	15.65		47.15	60
0290	7' wide		24	.333		40	15.65		55.65	69.50
0300	8' wide		22	.364		44.50	17.05		61.55	76.50

06 11 Wood Framing

06 11 10 – Framing with Dimensional, Engineered or Composite Lumber

06 11 10.40 Wall Framing		Crew	Daily Output	Labor-Hours	Unit	Material	2015 Bare Costs Labor	Equipment	Total	Total Incl O&P
0310	10' wide	1 Carp	22	.364	Ea.	59	17.05		76.05	92.50
0320	12' wide		22	.364		75	17.05		92.05	111
0330	2" x 6" wall, 2' wide		24	.333		28.50	15.65		44.15	57
0340	3' wide		24	.333		32	15.65		47.65	60.50
0350	4' wide		24	.333		34.50	15.65		50.15	63.50
0360	5' wide		24	.333		39	15.65		54.65	68.50
0370	6' wide		24	.333		43	15.65		58.65	72.50
0380	7' wide		24	.333		52.50	15.65		68.15	83.50
0390	8' wide		22	.364		57.50	17.05		74.55	91
0400	10' wide		22	.364		72.50	17.05		89.55	108
0410	12' wide		22	.364		90.50	17.05		107.55	128
2002	Headers over openings, 2" x 6"	2 Carp	360	.044	L.F.	.67	2.09		2.76	4.15
2007	2" x 6", pneumatic nailed		432	.037		.67	1.74		2.41	3.59
2052	2" x 8"		340	.047		.92	2.21		3.13	4.63
2057	2" x 8", pneumatic nailed		408	.039		.93	1.84		2.77	4.04
2101	2" x 10"		320	.050		1.34	2.35		3.69	5.30
2106	2" x 10", pneumatic nailed		384	.042		1.35	1.96		3.31	4.69
2152	2" x 12"		300	.053		1.72	2.50		4.22	6
2157	2" x 12", pneumatic nailed		360	.044		1.73	2.09		3.82	5.35
2180	4" x 8"		260	.062		2.89	2.89		5.78	7.90
2185	4" x 8", pneumatic nailed		312	.051		2.90	2.41		5.31	7.15
2191	4" x 10"		240	.067		4.28	3.13		7.41	9.85
2196	4" x 10", pneumatic nailed		288	.056		4.30	2.61		6.91	9
2202	4" x 12"		190	.084		5.15	3.95		9.10	12.15
2207	4" x 12", pneumatic nailed		228	.070		5.15	3.30		8.45	11.10
2241	6" x 10"		165	.097		6.45	4.55		11	14.55
2246	6" x 10", pneumatic nailed		198	.081		6.50	3.79		10.29	13.35
2251	6" x 12"		140	.114		8.20	5.35		13.55	17.85
2256	6" x 12", pneumatic nailed		168	.095		8.25	4.47		12.72	16.40
5002	Plates, untreated, 2" x 3"		850	.019		.39	.88		1.27	1.88
5007	2" x 3", pneumatic nailed		1020	.016		.40	.74		1.14	1.64
5022	2" x 4"		800	.020		.42	.94		1.36	2.01
5027	2" x 4", pneumatic nailed		960	.017		.43	.78		1.21	1.75
5041	2" x 6"		750	.021		.67	1		1.67	2.37
5046	2" x 6", pneumatic nailed		900	.018		.67	.83		1.50	2.11
5122	Studs, 8' high wall, 2" x 3"		1200	.013		.39	.63		1.02	1.46
5127	2" x 3", pneumatic nailed		1440	.011		.40	.52		.92	1.28
5142	2" x 4"		1100	.015		.42	.68		1.10	1.58
5147	2" x 4", pneumatic nailed		1320	.012		.43	.57		1	1.40
5162	2" x 6"		1000	.016		.67	.75		1.42	1.96
5167	2" x 6", pneumatic nailed		1200	.013		.67	.63		1.30	1.77
5182	3" x 4"		800	.020		1.07	.94		2.01	2.72
5187	3" x 4", pneumatic nailed		960	.017		1.08	.78		1.86	2.46
8200	For 12' high walls, deduct						5%			
8220	For stub wall, 6' high, add						20%			
8240	3' high, add						40%			
8250	For second story & above, add						5%			
8300	For dormer & gable, add						15%			
9000	Minimum labor/equipment charge	1 Carp	4	2	Job		94		94	154

06 11 Wood Framing

06 11 10 – Framing with Dimensional, Engineered or Composite Lumber

06 11 10.42 Furring

		Crew	Daily Output	Labor-Hours	Unit	Material	2015 Bare Costs Labor	2015 Bare Costs Equipment	Total	Total Incl O&P
0010	**FURRING**									
0012	Wood strips, 1" x 2", on walls, on wood	1 Carp	550	.015	L.F.	.24	.68		.92	1.38
0015	On wood, pneumatic nailed		710	.011		.24	.53		.77	1.13
0300	On masonry		495	.016		.26	.76		1.02	1.53
0400	On concrete		260	.031		.26	1.44		1.70	2.66
0600	1" x 3", on walls, on wood		550	.015		.39	.68		1.07	1.54
0605	On wood, pneumatic nailed		710	.011		.39	.53		.92	1.29
0700	On masonry		495	.016		.42	.76		1.18	1.70
0800	On concrete		260	.031		.42	1.44		1.86	2.83
0850	On ceilings, on wood		350	.023		.39	1.07		1.46	2.18
0855	On wood, pneumatic nailed		450	.018		.39	.83		1.22	1.79
0900	On masonry		320	.025		.42	1.17		1.59	2.38
0950	On concrete		210	.038		.42	1.79		2.21	3.39
9000	Minimum labor/equipment charge		4	2	Job		94		94	154

06 11 10.44 Grounds

		Crew	Daily Output	Labor-Hours	Unit	Material	2015 Bare Costs Labor	2015 Bare Costs Equipment	Total	Total Incl O&P
0010	**GROUNDS**									
0020	For casework, 1" x 2" wood strips, on wood	1 Carp	330	.024	L.F.	.24	1.14		1.38	2.13
0100	On masonry		285	.028		.26	1.32		1.58	2.45
0200	On concrete		250	.032		.26	1.50		1.76	2.75
0400	For plaster, 3/4" deep, on wood		450	.018		.24	.83		1.07	1.63
0500	On masonry		225	.036		.26	1.67		1.93	3.03
0600	On concrete		175	.046		.26	2.15		2.41	3.81
0700	On metal lath		200	.040		.26	1.88		2.14	3.37
9000	Minimum labor/equipment charge		4	2	Job		94		94	154

06 12 Structural Panels

06 12 10 – Structural Insulated Panels

06 12 10.10 OSB Faced Panels

			Crew	Daily Output	Labor-Hours	Unit	Material	2015 Bare Costs Labor	2015 Bare Costs Equipment	Total	Total Incl O&P
0010	**OSB FACED PANELS**										
0100	Structural insul. panels, 7/16" OSB both faces, EPS insul, 3-5/8" T	G	F-3	2075	.019	S.F.	3.60	.92	.32	4.84	5.80
0110	5-5/8" thick	G		1725	.023		4.05	1.11	.38	5.54	6.70
0120	7-3/8" thick	G		1425	.028		4.40	1.34	.46	6.20	7.55
0130	9-3/8" thick	G		1125	.036		4.70	1.70	.58	6.98	8.55
0140	7/16" OSB one face, EPS insul, 3-5/8" thick	G		2175	.018		3.70	.88	.30	4.88	5.85
0150	5-5/8" thick	G		1825	.022		4.30	1.05	.36	5.71	6.85
0160	7-3/8" thick	G		1525	.026		4.80	1.26	.43	6.49	7.80
0170	9-3/8" thick	G		1225	.033		5.30	1.56	.53	7.39	9
0190	7/16" OSB - 1/2" GWB faces, EPS insul, 3-5/8" T	G		2075	.019		3.29	.92	.32	4.53	5.45
0200	5-5/8" thick	G		1725	.023		3.90	1.11	.38	5.39	6.50
0210	7-3/8" thick	G		1425	.028		4.42	1.34	.46	6.22	7.55
0220	9-3/8" thick	G		1125	.036		5	1.70	.58	7.28	8.90
0240	7/16" OSB - 1/2" MRGWB faces, EPS insul, 3-5/8" T	G		2075	.019		3.39	.92	.32	4.63	5.60
0250	5-5/8" thick	G		1725	.023		4	1.11	.38	5.49	6.65
0260	7-3/8" thick	G		1425	.028		4.52	1.34	.46	6.32	7.65
0270	9-3/8" thick	G		1125	.036		5.10	1.70	.58	7.38	9.05
0300	For 1/2" GWB added to OSB skin, add	G					1.28			1.28	1.41
0310	For 1/2" MRGWB added to OSB skin, add	G					1.28			1.28	1.41
0320	For one T1-11 skin, add to OSB-OSB	G					1.90			1.90	2.09
0330	For one 19/32" CDX skin, add to OSB-OSB	G					1.46			1.46	1.61
0500	Structural insulated panel, 7/16" OSB both sides, straw core										

06 12 Structural Panels

06 12 10 – Structural Insulated Panels

06 12 10.10 OSB Faced Panels

		Crew	Daily Output	Labor-Hours	Unit	Material	2015 Bare Costs Labor	Equipment	Total	Total Incl O&P
0510	4-3/8" T, walls (w/sill, splines, plates) G	F-6	2400	.017	S.F.	7.40	.74	.27	8.41	9.60
0520	Floors (w/splines) G		2400	.017		7.40	.74	.27	8.41	9.60
0530	Roof (w/splines) G		2400	.017		7.40	.74	.27	8.41	9.60
0550	7-7/8" T, walls (w/sill, splines, plates) G		2400	.017		11.20	.74	.27	12.21	13.80
0560	Floors (w/splines) G		2400	.017		11.20	.74	.27	12.21	13.80
0570	Roof (w/splines) G		2400	.017		11.20	.74	.27	12.21	13.80

06 12 19 – Composite Shearwall Panels

06 12 19.10 Steel and Wood Composite Shearwall Panels

		Crew	Daily Output	Labor-Hours	Unit	Material	Labor	Equipment	Total	Total Incl O&P
0010	**STEEL & WOOD COMPOSITE SHEARWALL PANELS**									
0020	Anchor bolts, 36" long (must be placed in wet concrete)	1 Carp	150	.053	Ea.	32	2.50		34.50	39.50
0030	On concrete, 2 x 4 & 2 x 6 walls, 7' - 10' high, 360 lb. shear, 12" wide	2 Carp	8	2		420	94		514	615
0040	715 lb. shear, 15" wide		8	2		500	94		594	705
0050	1860 lb. shear, 18" wide		8	2		520	94		614	725
0060	2780 lb. shear, 21" wide		8	2		620	94		714	835
0070	3790 lb. shear, 24" wide		8	2		710	94		804	935
0080	2 x 6 walls, 11' to 13' high, 1180 lb. shear, 18" wide		6	2.667		635	125		760	905
0090	1555 lb. shear, 21" wide		6	2.667		800	125		925	1,075
0100	2280 lb. shear, 24" wide		6	2.667		895	125		1,020	1,200
0110	For installing above on wood floor frame, add									
0120	Coupler nuts, threaded rods, bolts, shear transfer plate kit	1 Carp	16	.500	Ea.	63	23.50		86.50	108
0130	Framing anchors, angle (2 required)	"	96	.083	"	2.61	3.91		6.52	9.25
0140	For blocking see Section 06 11 10.02									
0150	For installing above, first floor to second floor, wood floor frame, add									
0160	Add stack option to first floor wall panel				Ea.	69.50			69.50	76
0170	Threaded rods, bolts, shear transfer plate kit	1 Carp	16	.500		76	23.50		99.50	122
0180	Framing anchors, angle (2 required)	"	96	.083		2.61	3.91		6.52	9.25
0190	For blocking see section 06 11 10.02									
0200	For installing stacked panels, balloon framing									
0210	Add stack option to first floor wall panel				Ea.	69.50			69.50	76
0220	Threaded rods, bolts kit	1 Carp	16	.500	"	44	23.50		67.50	87

06 13 Heavy Timber Construction

06 13 23 – Heavy Timber Framing

06 13 23.10 Heavy Framing

		Crew	Daily Output	Labor-Hours	Unit	Material	Labor	Equipment	Total	Total Incl O&P
0010	**HEAVY FRAMING**									
0020	Beams, single 6" x 10"	2 Carp	1.10	14.545	M.B.F.	1,600	685		2,285	2,875
0100	Single 8" x 16"		1.20	13.333	"	2,000	625		2,625	3,225
0202	Built from 2" lumber, multiple 2" x 14"		900	.018	B.F.	.85	.83		1.68	2.31
0212	Built from 3" lumber, multiple 3" x 6"		700	.023		1.25	1.07		2.32	3.14
0222	Multiple 3" x 8"		800	.020		1.27	.94		2.21	2.93
0232	Multiple 3" x 10"		900	.018		1.28	.83		2.11	2.77
0242	Multiple 3" x 12"		1000	.016		1.28	.75		2.03	2.63
0252	Built from 4" lumber, multiple 4" x 6"		800	.020		1.41	.94		2.35	3.09
0262	Multiple 4" x 8"		900	.018		1.07	.83		1.90	2.55
0272	Multiple 4" x 10"		1000	.016		1.28	.75		2.03	2.63
0282	Multiple 4" x 12"		1100	.015		1.28	.68		1.96	2.52
0292	Columns, structural grade, 1500f, 4" x 4"		450	.036	L.F.	1.70	1.67		3.37	4.61
0302	6" x 6"		225	.071		4.21	3.34		7.55	10.10
0402	8" x 8"		240	.067		7.20	3.13		10.33	13.05
0502	10" x 10"		90	.178		12.10	8.35		20.45	27

06 13 Heavy Timber Construction

06 13 23 – Heavy Timber Framing

06 13 23.10 Heavy Framing		Crew	Daily Output	Labor-Hours	Unit	Material	2015 Bare Costs Labor	Equipment	Total	Total Incl O&P
0602	12" x 12"	2 Carp	70	.229	L.F.	18.35	10.75		29.10	37.50
0802	Floor planks, 2" thick, T & G, 2" x 6"		1050	.015	B.F.	1.59	.72		2.31	2.92
0902	2" x 10"		1100	.015		1.59	.68		2.27	2.87
1102	3" thick, 3" x 6"		1050	.015		1.59	.72		2.31	2.92
1202	3" x 10"		1100	.015		1.59	.68		2.27	2.87
1402	Girders, structural grade, 12" x 12"		800	.020		1.53	.94		2.47	3.22
1502	10" x 16"		1000	.016		2.40	.75		3.15	3.87
2302	Roof purlins, 4" thick, structural grade	↓	1050	.015	↓	1.07	.72		1.79	2.35
9000	Minimum labor/equipment charge	1 Carp	2	4	Job		188		188	310

06 15 Wood Decking

06 15 16 – Wood Roof Decking

06 15 16.10 Solid Wood Roof Decking

		Crew	Daily Output	Labor-Hours	Unit	Material	Labor	Equipment	Total	Total Incl O&P
0010	**SOLID WOOD ROOF DECKING**									
0350	Cedar planks, 2" thick	2 Carp	350	.046	S.F.	6.25	2.15		8.40	10.35
0400	3" thick		320	.050		9.35	2.35		11.70	14.15
0500	4" thick		250	.064		12.45	3		15.45	18.60
0550	6" thick		200	.080		18.70	3.76		22.46	26.50
0650	Douglas fir, 2" thick		350	.046		2.69	2.15		4.84	6.50
0700	3" thick		320	.050		4.04	2.35		6.39	8.30
0800	4" thick		250	.064		5.40	3		8.40	10.80
0850	6" thick		200	.080		8.05	3.76		11.81	15.05
0950	Hemlock, 2" thick		350	.046		2.74	2.15		4.89	6.55
1000	3" thick		320	.050		4.11	2.35		6.46	8.35
1100	4" thick		250	.064		5.50	3		8.50	10.90
1150	6" thick		200	.080		8.20	3.76		11.96	15.20
1250	Western white spruce, 2" thick		350	.046		1.75	2.15		3.90	5.45
1300	3" thick		320	.050		2.62	2.35		4.97	6.75
1400	4" thick		250	.064		3.50	3		6.50	8.75
1450	6" thick	↓	200	.080	↓	5.25	3.76		9.01	11.90
9000	Minimum labor/equipment charge	1 Carp	2	4	Job		188		188	310

06 15 23 – Laminated Wood Decking

06 15 23.10 Laminated Roof Deck

		Crew	Daily Output	Labor-Hours	Unit	Material	Labor	Equipment	Total	Total Incl O&P
0010	**LAMINATED ROOF DECK**									
0020	Pine or hemlock, 3" thick	2 Carp	425	.038	S.F.	6.20	1.77		7.97	9.70
0100	4" thick		325	.049		8.10	2.31		10.41	12.70
0300	Cedar, 3" thick		425	.038		7	1.77		8.77	10.60
0400	4" thick		325	.049		9.40	2.31		11.71	14.15
0600	Fir, 3" thick		425	.038		5.40	1.77		7.17	8.85
0700	4" thick	↓	325	.049	↓	7.40	2.31		9.71	11.95
9000	Minimum labor/equipment charge	1 Carp	3	2.667	Job		125		125	205

06 16 Sheathing

06 16 13 – Insulating Sheathing

06 16 13.10 Insulating Sheathing		Crew	Daily Output	Labor-Hours	Unit	Material	2015 Bare Costs Labor	2015 Bare Costs Equipment	Total	Total Incl O&P
0010	**INSULATING SHEATHING**									
0020	Expanded polystyrene, 1#/C.F. density, 3/4" thick R2.89 G	2 Carp	1400	.011	S.F.	.33	.54		.87	1.24
0030	1" thick R3.85 G		1300	.012		.39	.58		.97	1.38
0040	2" thick R7.69 G		1200	.013		.64	.63		1.27	1.74
0050	Extruded polystyrene, 15 PSI compressive strength, 1" thick, R5 G		1300	.012		.69	.58		1.27	1.71
0060	2" thick, R10 G		1200	.013		.86	.63		1.49	1.97
0070	Polyisocyanurate, 2#/C.F. density, 3/4" thick G		1400	.011		.60	.54		1.14	1.54
0080	1" thick G		1300	.012		.62	.58		1.20	1.64
0090	1-1/2" thick G		1250	.013		.78	.60		1.38	1.85
0100	2" thick G		1200	.013		.97	.63		1.60	2.10

06 16 23 – Subflooring

06 16 23.10 Subfloor

		Crew	Daily Output	Labor-Hours	Unit	Material	Labor	Equipment	Total	Total Incl O&P
0010	**SUBFLOOR** R061636-20									
0011	Plywood, CDX, 1/2" thick	2 Carp	1500	.011	SF Flr.	.66	.50		1.16	1.54
0015	Pneumatic nailed		1860	.009		.66	.40		1.06	1.38
0102	5/8" thick		1350	.012		.78	.56		1.34	1.77
0107	Pneumatic nailed		1674	.010		.78	.45		1.23	1.60
0202	3/4" thick		1250	.013		.93	.60		1.53	2.01
0207	Pneumatic nailed		1550	.010		.93	.48		1.41	1.81
0302	1-1/8" thick, 2-4-1 including underlayment		1050	.015		2.03	.72		2.75	3.40
0440	With boards, 1" x 6", S4S, laid regular		900	.018		1.57	.83		2.40	3.10
0452	1" x 8", laid regular		1000	.016		1.90	.75		2.65	3.32
0462	Laid diagonal		850	.019		1.90	.88		2.78	3.54
0502	1" x 10", laid regular		1100	.015		1.95	.68		2.63	3.26
0602	Laid diagonal		900	.018		1.95	.83		2.78	3.51
8990	Subfloor adhesive, 3/8" bead	1 Carp	2300	.003	L.F.	.12	.16		.28	.40
9000	Minimum labor/equipment charge	"	4	2	Job		94		94	154

06 16 26 – Underlayment

06 16 26.10 Wood Product Underlayment

		Crew	Daily Output	Labor-Hours	Unit	Material	Labor	Equipment	Total	Total Incl O&P
0010	**WOOD PRODUCT UNDERLAYMENT** R061636-20									
0015	Plywood, underlayment grade, 1/4" thick	2 Carp	1500	.011	S.F.	.83	.50		1.33	1.73
0018	Pneumatic nailed		1860	.009		.83	.40		1.23	1.57
0030	3/8" thick		1500	.011		.92	.50		1.42	1.83
0070	Pneumatic nailed		1860	.009		.92	.40		1.32	1.67
0102	1/2" thick		1450	.011		1.10	.52		1.62	2.06
0107	Pneumatic nailed		1798	.009		1.10	.42		1.52	1.89
0202	5/8" thick		1400	.011		1.20	.54		1.74	2.20
0207	Pneumatic nailed		1736	.009		1.20	.43		1.63	2.03
0302	3/4" thick		1300	.012		1.43	.58		2.01	2.52
0306	Pneumatic nailed		1612	.010		1.43	.47		1.90	2.33
0502	Particle board, 3/8" thick		1500	.011		.40	.50		.90	1.26
0507	Pneumatic nailed		1860	.009		.40	.40		.80	1.10
0602	1/2" thick		1450	.011		.43	.52		.95	1.32
0607	Pneumatic nailed		1798	.009		.43	.42		.85	1.15
0802	5/8" thick		1400	.011		.52	.54		1.06	1.45
0807	Pneumatic nailed		1736	.009		.52	.43		.95	1.28
0902	3/4" thick		1300	.012		.63	.58		1.21	1.64
0907	Pneumatic nailed		1612	.010		.63	.47		1.10	1.45
1100	Hardboard, underlayment grade, 4' x 4', .215" thick G		1500	.011		.58	.50		1.08	1.46
9000	Minimum labor/equipment charge	1 Carp	4	2	Job		94		94	154

06 16 Sheathing

06 16 33 – Wood Board Sheathing

06 16 33.10 Board Sheathing		Crew	Daily Output	Labor-Hours	Unit	Material	2015 Bare Costs Labor	2015 Bare Costs Equipment	Total	Total Incl O&P
0009	**BOARD SHEATHING**									
0010	Roof, 1" x 6" boards, laid horizontal	2 Carp	725	.022	S.F.	1.57	1.04		2.61	3.43
0020	On steep roof		520	.031		1.57	1.44		3.01	4.10
0040	On dormers, hips, & valleys		480	.033		1.57	1.56		3.13	4.29
0050	Laid diagonal		650	.025		1.57	1.16		2.73	3.62
0070	1" x 8" boards, laid horizontal		875	.018		1.90	.86		2.76	3.50
0080	On steep roof		635	.025		1.90	1.18		3.08	4.03
0090	On dormers, hips, & valleys		580	.028		1.95	1.30		3.25	4.26
0100	Laid diagonal		725	.022		1.90	1.04		2.94	3.79
0110	Skip sheathing, 1" x 4", 7" OC	1 Carp	1200	.007		.58	.31		.89	1.15
0120	1" x 6", 9" OC		1450	.006		.72	.26		.98	1.21
0180	Tongue and groove sheathing/decking, 1" x 6"		1000	.008		1.80	.38		2.18	2.60
0190	2" x 6"		1000	.008		3.81	.38		4.19	4.81
0200	Walls, 1" x 6" boards, laid regular	2 Carp	650	.025		1.57	1.16		2.73	3.62
0210	Laid diagonal		585	.027		1.57	1.28		2.85	3.83
0220	1" x 8" boards, laid regular		765	.021		1.90	.98		2.88	3.70
0230	Laid diagonal		650	.025		1.90	1.16		3.06	3.98

06 16 36 – Wood Panel Product Sheathing

06 16 36.10 Sheathing			Crew	Daily Output	Labor-Hours	Unit	Material	2015 Bare Costs Labor	2015 Bare Costs Equipment	Total	Total Incl O&P
0010	**SHEATHING**	R061110-30									
0012	Plywood on roofs, CDX										
0032	5/16" thick		2 Carp	1600	.010	S.F.	.56	.47		1.03	1.38
0037	Pneumatic nailed	R061636-20		1952	.008		.56	.39		.95	1.24
0052	3/8" thick			1525	.010		.60	.49		1.09	1.47
0057	Pneumatic nailed			1860	.009		.60	.40		1	1.32
0102	1/2" thick			1400	.011		.66	.54		1.20	1.60
0103	Pneumatic nailed			1708	.009		.66	.44		1.10	1.44
0202	5/8" thick			1300	.012		.78	.58		1.36	1.81
0207	Pneumatic nailed			1586	.010		.78	.47		1.25	1.64
0302	3/4" thick			1200	.013		.93	.63		1.56	2.05
0307	Pneumatic nailed			1464	.011		.93	.51		1.44	1.86
0502	Plywood on walls, with exterior CDX, 3/8" thick			1200	.013		.60	.63		1.23	1.69
0507	Pneumatic nailed			1488	.011		.60	.50		1.10	1.49
0602	1/2" thick			1125	.014		.66	.67		1.33	1.81
0607	Pneumatic nailed			1395	.011		.66	.54		1.20	1.60
0702	5/8" thick			1050	.015		.78	.72		1.50	2.03
0707	Pneumatic nailed			1302	.012		.78	.58		1.36	1.81
0802	3/4" thick			975	.016		.93	.77		1.70	2.28
0807	Pneumatic nailed			1209	.013		.93	.62		1.55	2.04
1000	For shear wall construction, add							20%			
1200	For structural 1 exterior plywood, add					S.F.	10%				
3000	Wood fiber, regular, no vapor barrier, 1/2" thick		2 Carp	1200	.013		.61	.63		1.24	1.70
3100	5/8" thick			1200	.013		.77	.63		1.40	1.88
3300	No vapor barrier, in colors, 1/2" thick			1200	.013		.74	.63		1.37	1.84
3400	5/8" thick			1200	.013		.78	.63		1.41	1.89
3600	With vapor barrier one side, white, 1/2" thick			1200	.013		.60	.63		1.23	1.69
3700	Vapor barrier 2 sides, 1/2" thick			1200	.013		.83	.63		1.46	1.94
3800	Asphalt impregnated, 25/32" thick			1200	.013		.33	.63		.96	1.39
3850	Intermediate, 1/2" thick			1200	.013		.25	.63		.88	1.31
4500	Oriented strand board, on roof, 7/16" thick	G		1460	.011		.50	.51		1.01	1.39
4505	Pneumatic nailed	G		1780	.009		.50	.42		.92	1.24
4550	1/2" thick	G		1400	.011		.50	.54		1.04	1.43

06 16 Sheathing

06 16 36 – Wood Panel Product Sheathing

06 16 36.10 Sheathing

		Crew	Daily Output	Labor-Hours	Unit	Material	2015 Bare Costs Labor	Equipment	Total	Total Incl O&P
4555	Pneumatic nailed [G]	2 Carp	1736	.009	S.F.	.50	.43		.93	1.26
4600	5/8" thick [G]		1300	.012		.69	.58		1.27	1.71
4605	Pneumatic nailed [G]		1586	.010		.69	.47		1.16	1.54
4610	On walls, 7/16" thick		1200	.013		.50	.63		1.13	1.58
4615	Pneumatic nailed		1488	.011		.50	.50		1	1.38
4620	1/2" thick		1195	.013		.50	.63		1.13	1.58
4625	Pneumatic nailed		1325	.012		.50	.57		1.07	1.48
4630	5/8" thick		1050	.015		.69	.72		1.41	1.93
4635	Pneumatic nailed		1302	.012		.69	.58		1.27	1.71
4700	Oriented strand board, factory laminated W.R. barrier, on roof, 1/2" thick [G]		1400	.011		.78	.54		1.32	1.74
4705	Pneumatic nailed [G]		1736	.009		.78	.43		1.21	1.57
4720	5/8" thick [G]		1300	.012		.93	.58		1.51	1.97
4725	Pneumatic nailed [G]		1586	.010		.93	.47		1.40	1.80
4730	5/8" thick, T&G [G]		1150	.014		.93	.65		1.58	2.09
4735	Pneumatic nailed, T&G [G]		1400	.011		.93	.54		1.47	1.90
4740	On walls, 7/16" thick [G]		1200	.013		.66	.63		1.29	1.76
4745	Pneumatic nailed [G]		1488	.011		.66	.50		1.16	1.56
4750	1/2" thick [G]		1195	.013		.78	.63		1.41	1.89
4755	Pneumatic nailed [G]		1325	.012		.78	.57		1.35	1.79
4800	Joint sealant tape, 3-1/2"		7600	.002	L.F.	.31	.10		.41	.50
4810	Joint sealant tape, 6"		7600	.002	"	.44	.10		.54	.64
9000	Minimum labor/equipment charge	1 Carp	2	4	Job		188		188	310

06 16 43 – Gypsum Sheathing

06 16 43.10 Gypsum Sheathing

		Crew	Daily Output	Labor-Hours	Unit	Material	Labor	Equipment	Total	Total Incl O&P
0010	**GYPSUM SHEATHING**									
0020	Gypsum, weatherproof, 1/2" thick	2 Carp	1125	.014	S.F.	.45	.67		1.12	1.59
0040	With embedded glass mats	"	1100	.015	"	.72	.68		1.40	1.91

06 17 Shop-Fabricated Structural Wood

06 17 33 – Wood I-Joists

06 17 33.10 Wood and Composite I-Joists

		Crew	Daily Output	Labor-Hours	Unit	Material	Labor	Equipment	Total	Total Incl O&P
0010	**WOOD AND COMPOSITE I-JOISTS**									
0100	Plywood webs, incl. bridging & blocking, panels 24" O.C.									
1200	15' to 24' span, 50 psf live load	F-5	2400	.013	SF Flr.	2.02	.63		2.65	3.27
1300	55 psf live load		2250	.014		2.23	.67		2.90	3.56
1400	24' to 30' span, 45 psf live load		2600	.012		2.36	.58		2.94	3.55
1500	55 psf live load		2400	.013		3.93	.63		4.56	5.35

06 17 53 – Shop-Fabricated Wood Trusses

06 17 53.10 Roof Trusses

		Crew	Daily Output	Labor-Hours	Unit	Material	Labor	Equipment	Total	Total Incl O&P
0010	**ROOF TRUSSES**									
0100	Fink (W) or King post type, 2'-0" O.C.									
0200	Metal plate connected, 4 in 12 slope									
0210	24' to 29' span	F-3	3000	.013	SF Flr.	1.70	.64	.22	2.56	3.15
0300	30' to 43' span		3000	.013		2.16	.64	.22	3.02	3.65
0400	44' to 60' span		3000	.013		2.29	.64	.22	3.15	3.80
0700	Glued and nailed, add					50%				

06 18 Glued-Laminated Construction

06 18 13 – Glued-Laminated Beams

06 18 13.20 Laminated Framing		Crew	Daily Output	Labor-Hours	Unit	Material	2015 Bare Costs Labor	Equipment	Total	Total Incl O&P
0010	**LAMINATED FRAMING**									
0020	30 lb., short term live load, 15 lb. dead load									
0200	Straight roof beams, 20' clear span, beams 8' O.C.	F-3	2560	.016	SF Flr.	2.04	.75	.26	3.05	3.74
0300	Beams 16' O.C.		3200	.013		1.48	.60	.20	2.28	2.82
0500	40' clear span, beams 8' O.C.		3200	.013		3.90	.60	.20	4.70	5.50
0600	Beams 16' O.C.		3840	.010		3.20	.50	.17	3.87	4.52
0800	60' clear span, beams 8' O.C.	F-4	2880	.017		6.70	.79	.39	7.88	9.10
0900	Beams 16' O.C.	"	3840	.013		5	.59	.29	5.88	6.80
1100	Tudor arches, 30' to 40' clear span, frames 8' O.C.	F-3	1680	.024		8.75	1.14	.39	10.28	11.95
1200	Frames 16' O.C.	"	2240	.018		6.85	.86	.29	8	9.25
1400	50' to 60' clear span, frames 8' O.C.	F-4	2200	.022		9.45	1.04	.51	11	12.65
1500	Frames 16' O.C.		2640	.018		8.05	.86	.43	9.34	10.70
1700	Radial arches, 60' clear span, frames 8' O.C.		1920	.025		8.85	1.19	.59	10.63	12.25
1800	Frames 16' O.C.		2880	.017		6.80	.79	.39	7.98	9.15
2000	100' clear span, frames 8' O.C.		1600	.030		9.15	1.42	.71	11.28	13.15
2100	Frames 16' O.C.		2400	.020		8.05	.95	.47	9.47	10.90
2300	120' clear span, frames 8' O.C.		1440	.033		12.15	1.58	.78	14.51	16.75
2400	Frames 16' O.C.		1920	.025		11.10	1.19	.59	12.88	14.75
2600	Bowstring trusses, 20' O.C., 40' clear span	F-3	2400	.017		5.50	.80	.27	6.57	7.65
2700	60' clear span	F-4	3600	.013		4.92	.63	.31	5.86	6.75
2800	100' clear span		4000	.012		6.95	.57	.28	7.80	8.90
2900	120' clear span		3600	.013		7.45	.63	.31	8.39	9.55
3000	For less than 1000 B.F., add					20%				
3100	For premium appearance, add to S.F. prices					5%				
3300	For industrial type, deduct					15%				
3500	For stain and varnish, add					5%				
3900	For 3/4" laminations, add to straight					25%				
4100	Add to curved					15%				
4300	Alternate pricing method: (use nominal footage of									
4310	components). Straight beams, camber less than 6"	F-3	3.50	11.429	M.B.F.	2,925	545	187	3,657	4,300
4400	Columns, including hardware		2	20		3,125	960	325	4,410	5,350
4600	Curved members, radius over 32'		2.50	16		3,200	765	262	4,227	5,075
4700	Radius 10' to 32'		3	13.333		3,175	640	218	4,033	4,800
4900	For complicated shapes, add maximum					100%				
5100	For pressure treating, add to straight					35%				
5200	Add to curved					45%				
6000	Laminated veneer members, southern pine or western species									
6050	1-3/4" wide x 5-1/2" deep	2 Carp	480	.033	L.F.	3.44	1.56		5	6.35
6100	9-1/2" deep		480	.033		4.74	1.56		6.30	7.75
6150	14" deep		450	.036		7.85	1.67		9.52	11.40
6200	18" deep		450	.036		10.65	1.67		12.32	14.50
6300	Parallel strand members, southern pine or western species									
6350	1-3/4" wide x 9-1/4" deep	2 Carp	480	.033	L.F.	5.10	1.56		6.66	8.15
6400	11-1/4" deep		450	.036		5.75	1.67		7.42	9.10
6450	14" deep		400	.040		7.85	1.88		9.73	11.75
6500	3-1/2" wide x 9-1/4" deep		480	.033		16.30	1.56		17.86	20.50
6550	11-1/4" deep		450	.036		20.50	1.67		22.17	25
6600	14" deep		400	.040		24	1.88		25.88	29.50
6650	7" wide x 9-1/4" deep		450	.036		34.50	1.67		36.17	40.50
6700	11-1/4" deep		420	.038		43.50	1.79		45.29	50.50
6750	14" deep		400	.040		52	1.88		53.88	60
9000	Minimum labor/equipment charge	F-3	2.50	16	Job		765	262	1,027	1,550

06 22 Millwork

06 22 13 – Standard Pattern Wood Trim

06 22 13.15 Moldings, Base

		Crew	Daily Output	Labor-Hours	Unit	Material	2015 Bare Costs Labor	Equipment	Total	Total Incl O&P
0010	**MOLDINGS, BASE**									
5100	Classic profile, 5/8" x 5-1/2", finger jointed and primed	1 Carp	250	.032	L.F.	1.46	1.50		2.96	4.06
5105	Poplar		240	.033		1.54	1.56		3.10	4.25
5110	Red oak		220	.036		2.36	1.71		4.07	5.40
5115	Maple		220	.036		3.57	1.71		5.28	6.75
5120	Cherry		220	.036		4.24	1.71		5.95	7.45
5125	3/4 x 7-1/2", finger jointed and primed		250	.032		1.85	1.50		3.35	4.49
5130	Poplar		240	.033		2.46	1.56		4.02	5.25
5135	Red oak		220	.036		3.43	1.71		5.14	6.55
5140	Maple		220	.036		4.75	1.71		6.46	8
5145	Cherry		220	.036		5.65	1.71		7.36	9
5150	Modern profile, 5/8" x 3-1/2", finger jointed and primed		250	.032		.89	1.50		2.39	3.44
5155	Poplar		240	.033		.98	1.56		2.54	3.64
5160	Red oak		220	.036		1.62	1.71		3.33	4.58
5165	Maple		220	.036		2.50	1.71		4.21	5.55
5170	Cherry		220	.036		2.70	1.71		4.41	5.75
5175	Ogee profile, 7/16" x 3", finger jointed and primed		250	.032		.59	1.50		2.09	3.11
5180	Poplar		240	.033		.64	1.56		2.20	3.26
5185	Red oak		220	.036		.90	1.71		2.61	3.79
5200	9/16" x 3-1/2", finger jointed and primed		250	.032		.66	1.50		2.16	3.18
5205	Pine		240	.033		1.13	1.56		2.69	3.80
5210	Red oak		220	.036		2.38	1.71		4.09	5.40
5215	9/16" x 4-1/2", red oak		220	.036		3.56	1.71		5.27	6.70
5220	5/8" x 3-1/2", finger jointed and primed		250	.032		.89	1.50		2.39	3.44
5225	Poplar		240	.033		.98	1.56		2.54	3.64
5230	Red oak		220	.036		1.62	1.71		3.33	4.58
5235	Maple		220	.036		2.50	1.71		4.21	5.55
5240	Cherry		220	.036		2.70	1.71		4.41	5.75
5245	5/8" x 4", finger jointed and primed		250	.032		1.15	1.50		2.65	3.72
5250	Poplar		240	.033		1.26	1.56		2.82	3.94
5255	Red oak		220	.036		1.81	1.71		3.52	4.79
5260	Maple		220	.036		2.79	1.71		4.50	5.85
5265	Cherry		220	.036		2.93	1.71		4.64	6
5270	Rectangular profile, oak, 3/8" x 1-1/4"		260	.031		1.25	1.44		2.69	3.75
5275	1/2" x 2-1/2"		255	.031		1.94	1.47		3.41	4.54
5280	1/2" x 3-1/2"		250	.032		2.36	1.50		3.86	5.05
5285	1" x 6"		240	.033		4.09	1.56		5.65	7.05
5290	1" x 8"		240	.033		5.70	1.56		7.26	8.80
5295	Pine, 3/8" x 1-3/4"		260	.031		.46	1.44		1.90	2.87
5300	7/16" x 2-1/2"		255	.031		.74	1.47		2.21	3.22
5305	1" x 6"		240	.033		.83	1.56		2.39	3.47
5310	1" x 8"		240	.033		1.06	1.56		2.62	3.72
5315	Shoe, 1/2" x 3/4", primed		260	.031		.48	1.44		1.92	2.90
5320	Pine		240	.033		.30	1.56		1.86	2.89
5325	Poplar		240	.033		.38	1.56		1.94	2.98
5330	Red oak		220	.036		.52	1.71		2.23	3.37
5335	Maple		220	.036		.69	1.71		2.40	3.56
5340	Cherry		220	.036		.78	1.71		2.49	3.66
5345	11/16" x 1-1/2", pine		240	.033		.70	1.56		2.26	3.33
5350	Caps, 11/16" x 1-3/8", pine		240	.033		.55	1.56		2.11	3.16
5355	3/4" x 1-3/4", finger jointed and primed		260	.031		.70	1.44		2.14	3.14
5360	Poplar		240	.033		.99	1.56		2.55	3.65
5365	Red oak		220	.036		1.11	1.71		2.82	4.02

06 22 Millwork

06 22 13 – Standard Pattern Wood Trim

06 22 13.15 Moldings, Base

		Crew	Daily Output	Labor-Hours	Unit	Material	2015 Bare Costs Labor	Equipment	Total	Total Incl O&P
5370	Maple	1 Carp	220	.036	L.F.	1.55	1.71		3.26	4.50
5375	Cherry		220	.036		3.21	1.71		4.92	6.35
5380	Combination base & shoe, 9/16" x 3-1/2" & 1/2" x 3/4", pine		125	.064		1.43	3		4.43	6.50
5385	Three piece oak, 6" high		80	.100		5.70	4.70		10.40	14
5390	Including 3/4" x 1" base shoe		70	.114		5.90	5.35		11.25	15.30
5395	Flooring cant strip, 3/4" x 3/4", pre-finished pine		260	.031		.51	1.44		1.95	2.93
5400	For pre-finished, stain and clear coat, add					.51			.51	.56
5405	Clear coat only, add					.42			.42	.46
9000	Minimum labor/equipment charge	1 Carp	4	2	Job		94		94	154

06 22 13.30 Moldings, Casings

		Crew	Daily Output	Labor-Hours	Unit	Material	Labor	Equipment	Total	Total Incl O&P
0010	**MOLDINGS, CASINGS**									
0085	Apron, 9/16" x 2-1/2", pine	1 Carp	250	.032	L.F.	1.63	1.50		3.13	4.25
0090	5/8" x 2-1/2", pine		250	.032		1.63	1.50		3.13	4.25
0110	5/8" x 3-1/2", pine		220	.036		1.86	1.71		3.57	4.84
0300	Band, 11/16" x 1-1/8", pine		270	.030		.75	1.39		2.14	3.10
0310	11/16" x 1-1/2", finger jointed and primed		270	.030		.76	1.39		2.15	3.11
0320	Pine		270	.030		.96	1.39		2.35	3.33
0330	11/16" x 1-3/4", finger jointed and primed		270	.030		.95	1.39		2.34	3.32
0350	Pine		270	.030		1.16	1.39		2.55	3.55
0355	Beaded, 3/4" x 3-1/2", finger jointed and primed		220	.036		1.09	1.71		2.80	4
0360	Poplar		220	.036		1.09	1.71		2.80	4
0365	Red oak		220	.036		1.62	1.71		3.33	4.58
0370	Maple		220	.036		2.50	1.71		4.21	5.55
0375	Cherry		220	.036		3.02	1.71		4.73	6.10
0380	3/4" x 4", finger jointed and primed		220	.036		1.15	1.71		2.86	4.06
0385	Poplar		220	.036		1.41	1.71		3.12	4.35
0390	Red oak		220	.036		1.98	1.71		3.69	4.98
0395	Maple		220	.036		2.60	1.71		4.31	5.65
0400	Cherry		220	.036		3.29	1.71		5	6.40
0405	3/4" x 5-1/2", finger jointed and primed		200	.040		1.15	1.88		3.03	4.34
0410	Poplar		200	.040		1.89	1.88		3.77	5.15
0415	Red oak		200	.040		2.78	1.88		4.66	6.15
0420	Maple		200	.040		3.66	1.88		5.54	7.10
0425	Cherry		200	.040		4.22	1.88		6.10	7.70
0430	Classic profile, 3/4" x 2-3/4", finger jointed and primed		250	.032		.77	1.50		2.27	3.31
0435	Poplar		250	.032		.98	1.50		2.48	3.54
0440	Red oak		250	.032		1.43	1.50		2.93	4.03
0445	Maple		250	.032		2.05	1.50		3.55	4.71
0450	Cherry		250	.032		2.32	1.50		3.82	5
0455	Fluted, 3/4" x 3-1/2", poplar		220	.036		1.09	1.71		2.80	4
0460	Red oak		220	.036		1.62	1.71		3.33	4.58
0465	Maple		220	.036		2.50	1.71		4.21	5.55
0470	Cherry		220	.036		3.02	1.71		4.73	6.10
0475	3/4" x 4", poplar		220	.036		1.41	1.71		3.12	4.35
0480	Red oak		220	.036		1.98	1.71		3.69	4.98
0485	Maple		220	.036		2.60	1.71		4.31	5.65
0490	Cherry		220	.036		3.29	1.71		5	6.40
0495	3/4" x 5-1/2", poplar		200	.040		1.89	1.88		3.77	5.15
0500	Red oak		200	.040		2.78	1.88		4.66	6.15
0505	Maple		200	.040		3.66	1.88		5.54	7.10
0510	Cherry		200	.040		4.22	1.88		6.10	7.70
0515	3/4" x 7-1/2", poplar		190	.042		1.27	1.98		3.25	4.64

06 22 Millwork

06 22 13 – Standard Pattern Wood Trim

06 22 13.30 Moldings, Casings

		Crew	Daily Output	Labor-Hours	Unit	Material	2015 Bare Costs Labor	Equipment	Total	Total Incl O&P
0520	Red oak	1 Carp	190	.042	L.F.	3.85	1.98		5.83	7.45
0525	Maple		190	.042		5.55	1.98		7.53	9.35
0530	Cherry		190	.042		6.65	1.98		8.63	10.55
0535	3/4" x 9-1/2", poplar		180	.044		4.04	2.09		6.13	7.85
0540	Red oak		180	.044		6.45	2.09		8.54	10.45
0545	Maple		180	.044		8.80	2.09		10.89	13.10
0550	Cherry		180	.044		9.60	2.09		11.69	14
0555	Modern profile, 9/16" x 2-1/4", poplar		250	.032		.56	1.50		2.06	3.07
0560	Red oak		250	.032		.76	1.50		2.26	3.29
0565	11/16" x 2-1/2", finger jointed & primed		250	.032		.84	1.50		2.34	3.38
0570	Pine		250	.032		1.28	1.50		2.78	3.87
0575	3/4" x 2-1/2", poplar		250	.032		.86	1.50		2.36	3.40
0580	Red oak		250	.032		1.17	1.50		2.67	3.75
0585	Maple		250	.032		1.76	1.50		3.26	4.39
0590	Cherry		250	.032		2.23	1.50		3.73	4.91
0595	Mullion, 5/16" x 2", pine		270	.030		.86	1.39		2.25	3.22
0600	9/16" x 2-1/2", fingerjointed and primed		250	.032		.98	1.50		2.48	3.54
0605	Pine		250	.032		1.28	1.50		2.78	3.87
0610	Red oak		250	.032		2.86	1.50		4.36	5.60
0615	1-1/16" x 3-3/4", red oak		220	.036		6.95	1.71		8.66	10.40
0620	Ogee, 7/16" x 2-1/2", poplar		250	.032		.56	1.50		2.06	3.07
0625	Red oak		250	.032		.77	1.50		2.27	3.31
0630	9/16" x 2-1/4", finger jointed and primed		250	.032		.49	1.50		1.99	3
0635	Poplar		250	.032		.56	1.50		2.06	3.07
0640	Red oak		250	.032		.76	1.50		2.26	3.29
0645	11/16" x 2-1/2", finger jointed and primed		250	.032		.72	1.50		2.22	3.25
0700	Pine		250	.032		1.41	1.50		2.91	4.01
0701	Red oak		250	.032		2.70	1.50		4.20	5.45
0730	11/16" x 3-1/2", finger jointed and primed		220	.036		1.11	1.71		2.82	4.02
0750	Pine		220	.036		1.72	1.71		3.43	4.69
0755	3/4" x 2-1/2", finger jointed and primed		250	.032		.89	1.50		2.39	3.44
0760	Poplar		250	.032		.86	1.50		2.36	3.40
0765	Red oak		250	.032		1.17	1.50		2.67	3.75
0770	Maple		250	.032		1.76	1.50		3.26	4.39
0775	Cherry		250	.032		2.23	1.50		3.73	4.91
0780	3/4" x 3-1/2", finger jointed and primed		220	.036		.89	1.71		2.60	3.78
0785	Poplar		220	.036		1.09	1.71		2.80	4
0790	Red oak		220	.036		1.62	1.71		3.33	4.58
0795	Maple		220	.036		2.50	1.71		4.21	5.55
0800	Cherry		220	.036		3.02	1.71		4.73	6.10
4700	Square profile, 1" x 1", teak		215	.037		2.26	1.75		4.01	5.35
4800	Rectangular profile, 1" x 3", teak		200	.040		6.20	1.88		8.08	9.95
9000	Minimum labor/equipment charge		4	2	Job		94		94	154

06 22 13.35 Moldings, Ceilings

		Crew	Daily Output	Labor-Hours	Unit	Material	Labor	Equipment	Total	Total Incl O&P
0010	**MOLDINGS, CEILINGS**									
0600	Bed, 9/16" x 1-3/4", pine	1 Carp	270	.030	L.F.	1.06	1.39		2.45	3.44
0650	9/16" x 2", pine		270	.030		1.14	1.39		2.53	3.53
0710	9/16" x 1-3/4", oak		270	.030		1.86	1.39		3.25	4.32
1200	Cornice, 9/16" x 1-3/4", pine		270	.030		.96	1.39		2.35	3.33
1300	9/16" x 2-1/4", pine		265	.030		1.27	1.42		2.69	3.71
1350	Cove, 1/2" x 2-1/4", poplar		265	.030		1	1.42		2.42	3.42
1360	Red oak		265	.030		1.33	1.42		2.75	3.78

06 22 Millwork

06 22 13 – Standard Pattern Wood Trim

06 22 13.35 Moldings, Ceilings		Crew	Daily Output	Labor-Hours	Unit	Material	2015 Bare Costs Labor	Equipment	Total	Total Incl O&P
1370	Hard maple	1 Carp	265	.030	L.F.	1.83	1.42		3.25	4.33
1380	Cherry		265	.030		2.24	1.42		3.66	4.78
2400	9/16" x 1-3/4", pine		270	.030		.97	1.39		2.36	3.35
2500	11/16" x 2-3/4", pine		265	.030		1.80	1.42		3.22	4.30
2510	Crown, 5/8" x 5/8", poplar		300	.027		.43	1.25		1.68	2.52
2520	Red oak		300	.027		.51	1.25		1.76	2.61
2530	Hard maple		300	.027		.69	1.25		1.94	2.81
2540	Cherry		300	.027		.73	1.25		1.98	2.85
2600	9/16" x 3-5/8", pine		250	.032		2.02	1.50		3.52	4.68
2700	11/16" x 4-1/4", pine		250	.032		2.90	1.50		4.40	5.65
2705	Oak		250	.032		6.40	1.50		7.90	9.50
2710	3/4" x 1-3/4", poplar		270	.030		.70	1.39		2.09	3.05
2720	Red oak		270	.030		1.03	1.39		2.42	3.41
2730	Hard maple		270	.030		1.36	1.39		2.75	3.77
2740	Cherry		270	.030		1.68	1.39		3.07	4.13
2750	3/4" x 2", poplar		270	.030		.92	1.39		2.31	3.29
2760	Red oak		270	.030		1.24	1.39		2.63	3.64
2770	Hard maple		270	.030		1.65	1.39		3.04	4.09
2780	Cherry		270	.030		1.84	1.39		3.23	4.30
2790	3/4" x 2-3/4", poplar		265	.030		1	1.42		2.42	3.42
2800	Red oak		265	.030		1.54	1.42		2.96	4.01
2810	Hard maple		265	.030		2.51	1.42		3.93	5.10
2820	Cherry		265	.030		2.35	1.42		3.77	4.90
2830	3/4" x 3-1/2", poplar		250	.032		1.27	1.50		2.77	3.85
2840	Red oak		250	.032		1.92	1.50		3.42	4.57
2850	Hard maple		250	.032		2.51	1.50		4.01	5.20
2860	Cherry		250	.032		2.94	1.50		4.44	5.70
2870	FJP poplar		250	.032		.90	1.50		2.40	3.45
2880	3/4" x 5", poplar		245	.033		1.89	1.53		3.42	4.59
2890	Red oak		245	.033		2.79	1.53		4.32	5.60
2900	Hard maple		245	.033		3.67	1.53		5.20	6.55
2910	Cherry		245	.033		4.24	1.53		5.77	7.15
2920	FJP poplar		245	.033		1.29	1.53		2.82	3.93
2930	3/4" x 6-1/4", poplar		240	.033		2.26	1.56		3.82	5.05
2940	Red oak		240	.033		3.40	1.56		4.96	6.30
2950	Hard maple		240	.033		4.44	1.56		6	7.45
2960	Cherry		240	.033		5.25	1.56		6.81	8.35
2970	7/8" x 8-3/4", poplar		220	.036		3.51	1.71		5.22	6.65
2980	Red oak		220	.036		5.35	1.71		7.06	8.65
2990	Hard maple		220	.036		7.35	1.71		9.06	10.90
3000	Cherry		220	.036		8.40	1.71		10.11	12.05
3010	1" x 7-1/4", poplar		220	.036		3.71	1.71		5.42	6.90
3020	Red oak		220	.036		5.75	1.71		7.46	9.15
3030	Hard maple		220	.036		8	1.71		9.71	11.55
3040	Cherry		220	.036		9.30	1.71		11.01	13.05
3050	1-1/16" x 4-1/4", poplar		250	.032		2.25	1.50		3.75	4.93
3060	Red oak		250	.032		2.64	1.50		4.14	5.35
3070	Hard maple		250	.032		3.49	1.50		4.99	6.30
3080	Cherry		250	.032		4.98	1.50		6.48	7.90
3090	Dentil crown, 3/4" x 5", poplar		250	.032		1.89	1.50		3.39	4.54
3100	Red oak		250	.032		2.79	1.50		4.29	5.55
3110	Hard maple		250	.032		3.67	1.50		5.17	6.50
3120	Cherry		250	.032		4.24	1.50		5.74	7.10

06 22 Millwork

06 22 13 – Standard Pattern Wood Trim

06 22 13.35 Moldings, Ceilings

		Crew	Daily Output	Labor-Hours	Unit	Material	2015 Bare Costs Labor	2015 Bare Costs Equipment	Total	Total Incl O&P
3130	Dentil piece for above, 1/2" x 1/2", poplar	1 Carp	300	.027	L.F.	2.68	1.25		3.93	4.99
3140	Red oak		300	.027		3.08	1.25		4.33	5.45
3150	Hard maple		300	.027		3.50	1.25		4.75	5.90
3160	Cherry		300	.027		3.81	1.25		5.06	6.25
9000	Minimum labor/equipment charge		4	2	Job		94		94	154

06 22 13.40 Moldings, Exterior

		Crew	Daily Output	Labor-Hours	Unit	Material	2015 Bare Costs Labor	2015 Bare Costs Equipment	Total	Total Incl O&P
0010	**MOLDINGS, EXTERIOR**									
0100	Band board, cedar, rough sawn, 1" x 2"	1 Carp	300	.027	L.F.	.54	1.25		1.79	2.65
0110	1" x 3"		300	.027		.81	1.25		2.06	2.94
0120	1" x 4"		250	.032		1.07	1.50		2.57	3.64
0130	1" x 6"		250	.032		1.61	1.50		3.11	4.23
0140	1" x 8"		225	.036		2.15	1.67		3.82	5.10
0150	1" x 10"		225	.036		2.67	1.67		4.34	5.70
0160	1" x 12"		200	.040		3.21	1.88		5.09	6.60
0240	STK, 1" x 2"		300	.027		.51	1.25		1.76	2.61
0250	1" x 3"		300	.027		.56	1.25		1.81	2.67
0260	1" x 4"		250	.032		.65	1.50		2.15	3.17
0270	1" x 6"		250	.032		1.03	1.50		2.53	3.59
0280	1" x 8"		225	.036		1.60	1.67		3.27	4.50
0290	1" x 10"		225	.036		2.14	1.67		3.81	5.10
0300	1" x 12"		200	.040		3.54	1.88		5.42	6.95
0310	Pine, #2, 1" x 2"		300	.027		.25	1.25		1.50	2.33
0320	1" x 3"		300	.027		.40	1.25		1.65	2.49
0330	1" x 4"		250	.032		.49	1.50		1.99	3
0340	1" x 6"		250	.032		.75	1.50		2.25	3.28
0350	1" x 8"		225	.036		1.22	1.67		2.89	4.08
0360	1" x 10"		225	.036		1.59	1.67		3.26	4.49
0370	1" x 12"		200	.040		2.05	1.88		3.93	5.35
0380	D & better, 1" x 2"		300	.027		.41	1.25		1.66	2.50
0390	1" x 3"		300	.027		.61	1.25		1.86	2.72
0400	1" x 4"		250	.032		.82	1.50		2.32	3.36
0410	1" x 6"		250	.032		1.12	1.50		2.62	3.69
0420	1" x 8"		225	.036		1.62	1.67		3.29	4.52
0430	1" x 10"		225	.036		2.03	1.67		3.70	4.97
0440	1" x 12"		200	.040		2.63	1.88		4.51	5.95
0450	Redwood, clear all heart, 1" x 2"		300	.027		.63	1.25		1.88	2.75
0460	1" x 3"		300	.027		.94	1.25		2.19	3.09
0470	1" x 4"		250	.032		1.19	1.50		2.69	3.77
0480	1" x 6"		252	.032		1.78	1.49		3.27	4.39
0490	1" x 8"		225	.036		2.36	1.67		4.03	5.35
0500	1" x 10"		225	.036		3.96	1.67		5.63	7.10
0510	1" x 12"		200	.040		4.75	1.88		6.63	8.35
0530	Corner board, cedar, rough sawn, 1" x 2"		225	.036		.54	1.67		2.21	3.34
0540	1" x 3"		225	.036		.81	1.67		2.48	3.63
0550	1" x 4"		200	.040		1.07	1.88		2.95	4.26
0560	1" x 6"		200	.040		1.61	1.88		3.49	4.85
0570	1" x 8"		200	.040		2.15	1.88		4.03	5.45
0580	1" x 10"		175	.046		2.67	2.15		4.82	6.45
0590	1" x 12"		175	.046		3.21	2.15		5.36	7.05
0670	STK, 1" x 2"		225	.036		.51	1.67		2.18	3.30
0680	1" x 3"		225	.036		.56	1.67		2.23	3.36
0690	1" x 4"		200	.040		.63	1.88		2.51	3.78

06 22 Millwork

06 22 13 – Standard Pattern Wood Trim

06 22 13.40 Moldings, Exterior		Crew	Daily Output	Labor-Hours	Unit	Material	2015 Bare Costs Labor	Equipment	Total	Total Incl O&P
0700	1" x 6"	1 Carp	200	.040	L.F.	1.03	1.88		2.91	4.21
0710	1" x 8"		200	.040		1.60	1.88		3.48	4.84
0720	1" x 10"		175	.046		2.14	2.15		4.29	5.85
0730	1" x 12"		175	.046		3.54	2.15		5.69	7.40
0740	Pine, #2, 1" x 2"		225	.036		.25	1.67		1.92	3.02
0750	1" x 3"		225	.036		.40	1.67		2.07	3.18
0760	1" x 4"		200	.040		.49	1.88		2.37	3.62
0770	1" x 6"		200	.040		.75	1.88		2.63	3.90
0780	1" x 8"		200	.040		1.22	1.88		3.10	4.42
0790	1" x 10"		175	.046		1.59	2.15		3.74	5.25
0800	1" x 12"		175	.046		2.05	2.15		4.20	5.75
0810	D & better, 1" x 2"		225	.036		.41	1.67		2.08	3.19
0820	1" x 3"		225	.036		.61	1.67		2.28	3.41
0830	1" x 4"		200	.040		.82	1.88		2.70	3.98
0840	1" x 6"		200	.040		1.12	1.88		3	4.31
0850	1" x 8"		200	.040		1.62	1.88		3.50	4.86
0860	1" x 10"		175	.046		2.03	2.15		4.18	5.75
0870	1" x 12"		175	.046		2.63	2.15		4.78	6.40
0880	Redwood, clear all heart, 1" x 2"		225	.036		.63	1.67		2.30	3.44
0890	1" x 3"		225	.036		.94	1.67		2.61	3.78
0900	1" x 4"		200	.040		1.19	1.88		3.07	4.39
0910	1" x 6"		200	.040		1.78	1.88		3.66	5.05
0920	1" x 8"		200	.040		2.36	1.88		4.24	5.65
0930	1" x 10"		175	.046		3.96	2.15		6.11	7.90
0940	1" x 12"		175	.046		4.75	2.15		6.90	8.75
0950	Cornice board, cedar, rough sawn, 1" x 2"		330	.024		.54	1.14		1.68	2.47
0960	1" x 3"		290	.028		.81	1.30		2.11	3.01
0970	1" x 4"		250	.032		1.07	1.50		2.57	3.64
0980	1" x 6"		250	.032		1.61	1.50		3.11	4.23
0990	1" x 8"		200	.040		2.15	1.88		4.03	5.45
1000	1" x 10"		180	.044		2.67	2.09		4.76	6.35
1010	1" x 12"		180	.044		3.21	2.09		5.30	6.95
1020	STK, 1" x 2"		330	.024		.51	1.14		1.65	2.43
1030	1" x 3"		290	.028		.56	1.30		1.86	2.74
1040	1" x 4"		250	.032		.65	1.50		2.15	3.17
1050	1" x 6"		250	.032		1.03	1.50		2.53	3.59
1060	1" x 8"		200	.040		1.60	1.88		3.48	4.84
1070	1" x 10"		180	.044		2.14	2.09		4.23	5.75
1080	1" x 12"		180	.044		3.54	2.09		5.63	7.30
1500	Pine, #2, 1" x 2"		330	.024		.25	1.14		1.39	2.15
1510	1" x 3"		290	.028		.27	1.30		1.57	2.42
1600	1" x 4"		250	.032		.49	1.50		1.99	3
1700	1" x 6"		250	.032		.75	1.50		2.25	3.28
1800	1" x 8"		200	.040		1.22	1.88		3.10	4.42
1900	1" x 10"		180	.044		1.59	2.09		3.68	5.15
2000	1" x 12"		180	.044		2.05	2.09		4.14	5.65
2020	D & better, 1" x 2"		330	.024		.41	1.14		1.55	2.32
2030	1" x 3"		290	.028		.61	1.30		1.91	2.79
2040	1" x 4"		250	.032		.82	1.50		2.32	3.36
2050	1" x 6"		250	.032		1.12	1.50		2.62	3.69
2060	1" x 8"		200	.040		1.62	1.88		3.50	4.86
2070	1" x 10"		180	.044		2.03	2.09		4.12	5.65
2080	1" x 12"		180	.044		2.63	2.09		4.72	6.30

06 22 Millwork

06 22 13 – Standard Pattern Wood Trim

06 22 13.40 Moldings, Exterior

		Crew	Daily Output	Labor-Hours	Unit	Material	2015 Bare Costs Labor	2015 Bare Costs Equipment	Total	Total Incl O&P
2090	Redwood, clear all heart, 1" x 2"	1 Carp	330	.024	L.F.	.63	1.14		1.77	2.57
2100	1" x 3"		290	.028		.94	1.30		2.24	3.16
2110	1" x 4"		250	.032		1.19	1.50		2.69	3.77
2120	1" x 6"		250	.032		1.78	1.50		3.28	4.41
2130	1" x 8"		200	.040		2.36	1.88		4.24	5.65
2140	1" x 10"		180	.044		3.96	2.09		6.05	7.80
2150	1" x 12"		180	.044		4.75	2.09		6.84	8.65
2160	3 piece, 1" x 2", 1" x 4", 1" x 6", rough sawn cedar		80	.100		3.24	4.70		7.94	11.25
2180	STK cedar		80	.100		2.19	4.70		6.89	10.10
2200	#2 pine		80	.100		1.49	4.70		6.19	9.35
2210	D & better pine		80	.100		2.35	4.70		7.05	10.30
2220	Clear all heart redwood		80	.100		3.60	4.70		8.30	11.65
2230	1" x 8", 1" x 10", 1" x 12", rough sawn cedar		65	.123		8	5.80		13.80	18.25
2240	STK cedar		65	.123		7.25	5.80		13.05	17.45
2300	#2 pine		65	.123		4.82	5.80		10.62	14.75
2320	D & better pine		65	.123		6.25	5.80		12.05	16.35
2330	Clear all heart redwood		65	.123		11.05	5.80		16.85	21.50
2340	Door/window casing, cedar, rough sawn, 1" x 2"		275	.029		.54	1.37		1.91	2.84
2350	1" x 3"		275	.029		.81	1.37		2.18	3.13
2360	1" x 4"		250	.032		1.07	1.50		2.57	3.64
2370	1" x 6"		250	.032		1.61	1.50		3.11	4.23
2380	1" x 8"		230	.035		2.15	1.63		3.78	5.05
2390	1" x 10"		230	.035		2.67	1.63		4.30	5.60
2395	1" x 12"		210	.038		3.21	1.79		5	6.45
2410	STK, 1" x 2"		275	.029		.51	1.37		1.88	2.80
2420	1" x 3"		275	.029		.56	1.37		1.93	2.86
2430	1" x 4"		250	.032		.65	1.50		2.15	3.17
2440	1" x 6"		250	.032		1.03	1.50		2.53	3.59
2450	1" x 8"		230	.035		1.60	1.63		3.23	4.44
2460	1" x 10"		230	.035		2.14	1.63		3.77	5.05
2470	1" x 12"		210	.038		3.54	1.79		5.33	6.80
2550	Pine, #2, 1" x 2"		275	.029		.25	1.37		1.62	2.52
2560	1" x 3"		275	.029		.40	1.37		1.77	2.68
2570	1" x 4"		250	.032		.49	1.50		1.99	3
2580	1" x 6"		250	.032		.75	1.50		2.25	3.28
2590	1" x 8"		230	.035		1.22	1.63		2.85	4.02
2600	1" x 10"		230	.035		1.59	1.63		3.22	4.43
2610	1" x 12"		210	.038		2.05	1.79		3.84	5.20
2620	Pine, D & better, 1" x 2"		275	.029		.41	1.37		1.78	2.69
2630	1" x 3"		275	.029		.61	1.37		1.98	2.91
2640	1" x 4"		250	.032		.82	1.50		2.32	3.36
2650	1" x 6"		250	.032		1.12	1.50		2.62	3.69
2660	1" x 8"		230	.035		1.62	1.63		3.25	4.46
2670	1" x 10"		230	.035		2.03	1.63		3.66	4.91
2680	1" x 12"		210	.038		2.63	1.79		4.42	5.80
2690	Redwood, clear all heart, 1" x 2"		275	.029		.63	1.37		2	2.94
2695	1" x 3"		275	.029		.94	1.37		2.31	3.28
2710	1" x 4"		250	.032		1.19	1.50		2.69	3.77
2715	1" x 6"		250	.032		1.78	1.50		3.28	4.41
2730	1" x 8"		230	.035		2.36	1.63		3.99	5.25
2740	1" x 10"		230	.035		3.96	1.63		5.59	7.05
2750	1" x 12"		210	.038		4.75	1.79		6.54	8.20
3500	Bellyband, pine, 11/16" x 4-1/4"		250	.032		3.23	1.50		4.73	6

06 22 Millwork

06 22 13 – Standard Pattern Wood Trim

06 22 13.40 Moldings, Exterior

		Crew	Daily Output	Labor-Hours	Unit	Material	2015 Bare Costs Labor	Equipment	Total	Total Incl O&P
3610	Brickmold, pine, 1-1/4" x 2"	1 Carp	200	.040	L.F.	2.43	1.88		4.31	5.75
3620	FJP, 1-1/4" x 2"		200	.040		1.09	1.88		2.97	4.28
5100	Fascia, cedar, rough sawn, 1" x 2"		275	.029		.54	1.37		1.91	2.84
5110	1" x 3"		275	.029		.81	1.37		2.18	3.13
5120	1" x 4"		250	.032		1.07	1.50		2.57	3.64
5200	1" x 6"		250	.032		1.61	1.50		3.11	4.23
5300	1" x 8"		230	.035		2.15	1.63		3.78	5.05
5310	1" x 10"		230	.035		2.67	1.63		4.30	5.60
5320	1" x 12"		210	.038		3.21	1.79		5	6.45
5400	2" x 4"		220	.036		1.05	1.71		2.76	3.96
5500	2" x 6"		220	.036		1.58	1.71		3.29	4.54
5600	2" x 8"		200	.040		2.11	1.88		3.99	5.40
5700	2" x 10"		180	.044		2.62	2.09		4.71	6.30
5800	2" x 12"		170	.047		6.25	2.21		8.46	10.50
6120	STK, 1" x 2"		275	.029		.51	1.37		1.88	2.80
6130	1" x 3"		275	.029		.56	1.37		1.93	2.86
6140	1" x 4"		250	.032		.65	1.50		2.15	3.17
6150	1" x 6"		250	.032		1.03	1.50		2.53	3.59
6160	1" x 8"		230	.035		1.60	1.63		3.23	4.44
6170	1" x 10"		230	.035		2.14	1.63		3.77	5.05
6180	1" x 12"		210	.038		3.54	1.79		5.33	6.80
6185	2" x 2"		260	.031		.63	1.44		2.07	3.07
6190	Pine, #2, 1" x 2"		275	.029		.25	1.37		1.62	2.52
6200	1" x 3"		275	.029		.40	1.37		1.77	2.68
6210	1" x 4"		250	.032		.49	1.50		1.99	3
6220	1" x 6"		250	.032		.75	1.50		2.25	3.28
6230	1" x 8"		230	.035		1.22	1.63		2.85	4.02
6240	1" x 10"		230	.035		1.59	1.63		3.22	4.43
6250	1" x 12"		210	.038		2.05	1.79		3.84	5.20
6260	D & better, 1" x 2"		275	.029		.41	1.37		1.78	2.69
6270	1" x 3"		275	.029		.61	1.37		1.98	2.91
6280	1" x 4"		250	.032		.82	1.50		2.32	3.36
6290	1" x 6"		250	.032		1.12	1.50		2.62	3.69
6300	1" x 8"		230	.035		1.62	1.63		3.25	4.46
6310	1" x 10"		230	.035		2.03	1.63		3.66	4.91
6312	1" x 12"		210	.038		2.63	1.79		4.42	5.80
6330	Southern yellow, 1-1/4" x 5"		240	.033		2.27	1.56		3.83	5.05
6340	1-1/4" x 6"		240	.033		2.52	1.56		4.08	5.35
6350	1-1/4" x 8"		215	.037		3.43	1.75		5.18	6.65
6360	1-1/4" x 12"		190	.042		5.05	1.98		7.03	8.80
6370	Redwood, clear all heart, 1" x 2"		275	.029		.63	1.37		2	2.94
6380	1" x 3"		275	.029		1.19	1.37		2.56	3.55
6390	1" x 4"		250	.032		1.19	1.50		2.69	3.77
6400	1" x 6"		250	.032		1.78	1.50		3.28	4.41
6410	1" x 8"		230	.035		2.36	1.63		3.99	5.25
6420	1" x 10"		230	.035		3.96	1.63		5.59	7.05
6430	1" x 12"		210	.038		4.75	1.79		6.54	8.20
6440	1-1/4" x 5"		240	.033		1.86	1.56		3.42	4.60
6450	1-1/4" x 6"		240	.033		2.22	1.56		3.78	5
6460	1-1/4" x 8"		215	.037		3.53	1.75		5.28	6.75
6470	1-1/4" x 12"		190	.042		7.10	1.98		9.08	11.10
6580	Frieze, cedar, rough sawn, 1" x 2"		275	.029		.54	1.37		1.91	2.84
6590	1" x 3"		275	.029		.81	1.37		2.18	3.13

06 22 Millwork

06 22 13 – Standard Pattern Wood Trim

06 22 13.40 Moldings, Exterior		Crew	Daily Output	Labor-Hours	Unit	Material	2015 Bare Costs Labor	Equipment	Total	Total Incl O&P
6600	1" x 4"	1 Carp	250	.032	L.F.	1.07	1.50		2.57	3.64
6610	1" x 6"		250	.032		1.61	1.50		3.11	4.23
6620	1" x 8"		250	.032		2.15	1.50		3.65	4.83
6630	1" x 10"		225	.036		2.67	1.67		4.34	5.70
6640	1" x 12"		200	.040		3.18	1.88		5.06	6.60
6650	STK, 1" x 2"		275	.029		.51	1.37		1.88	2.80
6660	1" x 3"		275	.029		.56	1.37		1.93	2.86
6670	1" x 4"		250	.032		.65	1.50		2.15	3.17
6680	1" x 6"		250	.032		1.03	1.50		2.53	3.59
6690	1" x 8"		250	.032		1.60	1.50		3.10	4.22
6700	1" x 10"		225	.036		2.14	1.67		3.81	5.10
6710	1" x 12"		200	.040		3.54	1.88		5.42	6.95
6790	Pine, #2, 1" x 2"		275	.029		.25	1.37		1.62	2.52
6800	1" x 3"		275	.029		.40	1.37		1.77	2.68
6810	1" x 4"		250	.032		.49	1.50		1.99	3
6820	1" x 6"		250	.032		.75	1.50		2.25	3.28
6830	1" x 8"		250	.032		1.22	1.50		2.72	3.80
6840	1" x 10"		225	.036		1.59	1.67		3.26	4.49
6850	1" x 12"		200	.040		2.05	1.88		3.93	5.35
6860	D & better, 1" x 2"		275	.029		.41	1.37		1.78	2.69
6870	1" x 3"		275	.029		.61	1.37		1.98	2.91
6880	1" x 4"		250	.032		.82	1.50		2.32	3.36
6890	1" x 6"		250	.032		1.12	1.50		2.62	3.69
6900	1" x 8"		250	.032		1.62	1.50		3.12	4.24
6910	1" x 10"		225	.036		2.03	1.67		3.70	4.97
6920	1" x 12"		200	.040		2.63	1.88		4.51	5.95
6930	Redwood, clear all heart, 1" x 2"		275	.029		.63	1.37		2	2.94
6940	1" x 3"		275	.029		.94	1.37		2.31	3.28
6950	1" x 4"		250	.032		1.19	1.50		2.69	3.77
6960	1" x 6"		250	.032		1.78	1.50		3.28	4.41
6970	1" x 8"		250	.032		2.36	1.50		3.86	5.05
6980	1" x 10"		225	.036		3.96	1.67		5.63	7.10
6990	1" x 12"		200	.040		4.75	1.88		6.63	8.35
7000	Grounds, 1" x 1", cedar, rough sawn		300	.027		.28	1.25		1.53	2.35
7010	STK		300	.027		.31	1.25		1.56	2.39
7020	Pine, #2		300	.027		.16	1.25		1.41	2.22
7030	D & better		300	.027		.25	1.25		1.50	2.33
7050	Redwood		300	.027		.39	1.25		1.64	2.47
7060	Rake/verge board, cedar, rough sawn, 1" x 2"		225	.036		.54	1.67		2.21	3.34
7070	1" x 3"		225	.036		.81	1.67		2.48	3.63
7080	1" x 4"		200	.040		1.07	1.88		2.95	4.26
7090	1" x 6"		200	.040		1.61	1.88		3.49	4.85
7100	1" x 8"		190	.042		2.15	1.98		4.13	5.60
7110	1" x 10"		190	.042		2.67	1.98		4.65	6.20
7120	1" x 12"		180	.044		3.21	2.09		5.30	6.95
7130	STK, 1" x 2"		225	.036		.51	1.67		2.18	3.30
7140	1" x 3"		225	.036		.56	1.67		2.23	3.36
7150	1" x 4"		200	.040		.65	1.88		2.53	3.79
7160	1" x 6"		200	.040		1.03	1.88		2.91	4.21
7170	1" x 8"		190	.042		1.60	1.98		3.58	5
7180	1" x 10"		190	.042		2.14	1.98		4.12	5.60
7190	1" x 12"		180	.044		3.54	2.09		5.63	7.30
7200	Pine, #2, 1" x 2"		225	.036		.25	1.67		1.92	3.02

06 22 Millwork

06 22 13 – Standard Pattern Wood Trim

06 22 13.40 Moldings, Exterior		Crew	Daily Output	Labor-Hours	Unit	Material	2015 Bare Costs Labor	Equipment	Total	Total Incl O&P
7210	1" x 3"	1 Carp	225	.036	L.F.	.40	1.67		2.07	3.18
7220	1" x 4"		200	.040		.49	1.88		2.37	3.62
7230	1" x 6"		200	.040		.75	1.88		2.63	3.90
7240	1" x 8"		190	.042		1.22	1.98		3.20	4.58
7250	1" x 10"		190	.042		1.59	1.98		3.57	4.99
7260	1" x 12"		180	.044		2.05	2.09		4.14	5.65
7340	D & better, 1" x 2"		225	.036		.41	1.67		2.08	3.19
7350	1" x 3"		225	.036		.61	1.67		2.28	3.41
7360	1" x 4"		200	.040		.82	1.88		2.70	3.98
7370	1" x 6"		200	.040		1.12	1.88		3	4.31
7380	1" x 8"		190	.042		1.62	1.98		3.60	5
7390	1" x 10"		190	.042		2.03	1.98		4.01	5.45
7400	1" x 12"		180	.044		2.63	2.09		4.72	6.30
7410	Redwood, clear all heart, 1" x 2"		225	.036		.63	1.67		2.30	3.44
7420	1" x 3"		225	.036		.94	1.67		2.61	3.78
7430	1" x 4"		200	.040		1.19	1.88		3.07	4.39
7440	1" x 6"		200	.040		1.78	1.88		3.66	5.05
7450	1" x 8"		190	.042		2.36	1.98		4.34	5.85
7460	1" x 10"		190	.042		3.96	1.98		5.94	7.60
7470	1" x 12"		180	.044		4.75	2.09		6.84	8.65
7480	2" x 4"		200	.040		2.28	1.88		4.16	5.60
7490	2" x 6"		182	.044		3.43	2.06		5.49	7.15
7500	2" x 8"		165	.048		4.56	2.28		6.84	8.75
7630	Soffit, cedar, rough sawn, 1" x 2"	2 Carp	440	.036		.54	1.71		2.25	3.40
7640	1" x 3"		440	.036		.81	1.71		2.52	3.69
7650	1" x 4"		420	.038		1.07	1.79		2.86	4.11
7660	1" x 6"		420	.038		1.61	1.79		3.40	4.70
7670	1" x 8"		420	.038		2.15	1.79		3.94	5.30
7680	1" x 10"		400	.040		2.67	1.88		4.55	6
7690	1" x 12"		400	.040		3.21	1.88		5.09	6.60
7700	STK, 1" x 2"		440	.036		.51	1.71		2.22	3.36
7710	1" x 3"		440	.036		.56	1.71		2.27	3.42
7720	1" x 4"		420	.038		.65	1.79		2.44	3.64
7730	1" x 6"		420	.038		1.03	1.79		2.82	4.06
7740	1" x 8"		420	.038		1.60	1.79		3.39	4.69
7750	1" x 10"		400	.040		2.14	1.88		4.02	5.45
7760	1" x 12"		400	.040		3.54	1.88		5.42	6.95
7770	Pine, #2, 1" x 2"		440	.036		.25	1.71		1.96	3.08
7780	1" x 3"		440	.036		.40	1.71		2.11	3.24
7790	1" x 4"		420	.038		.49	1.79		2.28	3.47
7800	1" x 6"		420	.038		.75	1.79		2.54	3.75
7810	1" x 8"		420	.038		1.22	1.79		3.01	4.27
7820	1" x 10"		400	.040		1.59	1.88		3.47	4.83
7830	1" x 12"		400	.040		2.05	1.88		3.93	5.35
7840	D & better, 1" x 2"		440	.036		.41	1.71		2.12	3.25
7850	1" x 3"		440	.036		.61	1.71		2.32	3.47
7860	1" x 4"		420	.038		.82	1.79		2.61	3.83
7870	1" x 6"		420	.038		1.12	1.79		2.91	4.16
7880	1" x 8"		420	.038		1.62	1.79		3.41	4.71
7890	1" x 10"		400	.040		2.03	1.88		3.91	5.30
7900	1" x 12"		400	.040		2.63	1.88		4.51	5.95
7910	Redwood, clear all heart, 1" x 2"		440	.036		.63	1.71		2.34	3.50
7920	1" x 3"		440	.036		.94	1.71		2.65	3.84

06 22 Millwork

06 22 13 – Standard Pattern Wood Trim

06 22 13.40 Moldings, Exterior

		Crew	Daily Output	Labor-Hours	Unit	Material	2015 Bare Costs Labor	Equipment	Total	Total Incl O&P
7930	1" x 4"	2 Carp	420	.038	L.F.	1.19	1.79		2.98	4.24
7940	1" x 6"		420	.038		1.78	1.79		3.57	4.88
7950	1" x 8"		420	.038		2.36	1.79		4.15	5.50
7960	1" x 10"		400	.040		3.96	1.88		5.84	7.45
7970	1" x 12"		400	.040		4.75	1.88		6.63	8.35
8050	Trim, crown molding, pine, 11/16" x 4-1/4"	1 Carp	250	.032		4.54	1.50		6.04	7.45
8060	Back band, 11/16" x 1-1/16"		250	.032		.99	1.50		2.49	3.55
8070	Insect screen frame stock, 1-1/16" x 1-3/4"		395	.020		2.39	.95		3.34	4.19
8080	Dentils, 2-1/2" x 2-1/2" x 4", 6" O.C.		30	.267		1.22	12.50		13.72	22
8100	Fluted, 5-1/2"		165	.048		5.05	2.28		7.33	9.30
8110	Stucco bead, 1-3/8" x 1-5/8"		250	.032		2.50	1.50		4	5.20
9000	Minimum labor/equipment charge		4	2	Job		94		94	154

06 22 13.45 Moldings, Trim

		Crew	Daily Output	Labor-Hours	Unit	Material	Labor	Equipment	Total	Total Incl O&P
0010	**MOLDINGS, TRIM**									
0200	Astragal, stock pine, 11/16" x 1-3/4"	1 Carp	255	.031	L.F.	1.42	1.47		2.89	3.97
0250	1-5/16" x 2-3/16"		240	.033		2.10	1.56		3.66	4.87
0800	Chair rail, stock pine, 5/8" x 2-1/2"		270	.030		1.58	1.39		2.97	4.02
0900	5/8" x 3-1/2"		240	.033		2.40	1.56		3.96	5.20
1000	Closet pole, stock pine, 1-1/8" diameter		200	.040		1.16	1.88		3.04	4.36
1100	Fir, 1-5/8" diameter		200	.040		2.20	1.88		4.08	5.50
3300	Half round, stock pine, 1/4" x 1/2"		270	.030		.24	1.39		1.63	2.54
3350	1/2" x 1"		255	.031		.73	1.47		2.20	3.21
3400	Handrail, fir, single piece, stock, hardware not included									
3450	1-1/2" x 1-3/4"	1 Carp	80	.100	L.F.	2.56	4.70		7.26	10.50
3470	Pine, 1-1/2" x 1-3/4"		80	.100		2.40	4.70		7.10	10.35
3500	1-1/2" x 2-1/2"		76	.105		2.46	4.94		7.40	10.80
3600	Lattice, stock pine, 1/4" x 1-1/8"		270	.030		.35	1.39		1.74	2.67
3700	1/4" x 1-3/4"		250	.032		.92	1.50		2.42	3.47
3800	Miscellaneous, custom, pine, 1" x 1"		270	.030		.44	1.39		1.83	2.76
3850	1" x 2"		265	.030		.87	1.42		2.29	3.28
3900	1" x 3"		240	.033		1.31	1.56		2.87	4
4100	Birch or oak, nominal 1" x 1"		240	.033		.42	1.56		1.98	3.02
4200	Nominal 1" x 3"		215	.037		1.26	1.75		3.01	4.24
4400	Walnut, nominal 1" x 1"		215	.037		.66	1.75		2.41	3.59
4500	Nominal 1" x 3"		200	.040		1.99	1.88		3.87	5.25
4700	Teak, nominal 1" x 1"		215	.037		2.80	1.75		4.55	5.95
4800	Nominal 1" x 3"		200	.040		8.40	1.88		10.28	12.35
4900	Quarter round, stock pine, 1/4" x 1/4"		275	.029		.24	1.37		1.61	2.50
4950	3/4" x 3/4"		255	.031		.50	1.47		1.97	2.96
5600	Wainscot moldings, 1-1/8" x 9/16", 2' high, minimum		76	.105	S.F.	11.65	4.94		16.59	21
5700	Maximum		65	.123	"	15.75	5.80		21.55	27
9000	Minimum labor/equipment charge		4	2	Job		94		94	154

06 22 13.50 Moldings, Window and Door

		Crew	Daily Output	Labor-Hours	Unit	Material	Labor	Equipment	Total	Total Incl O&P
0010	**MOLDINGS, WINDOW AND DOOR**									
2800	Door moldings, stock, decorative, 1-1/8" wide, plain	1 Carp	17	.471	Set	47.50	22		69.50	88.50
2900	Detailed		17	.471	"	91.50	22		113.50	137
2960	Clear pine door jamb, no stops, 11/16" x 4-9/16"		240	.033	L.F.	5.20	1.56		6.76	8.25
3150	Door trim set, 1 head and 2 sides, pine, 2-1/2 wide		12	.667	Opng.	24	31.50		55.50	77.50
3170	3-1/2" wide		11	.727	"	29	34		63	88
3250	Glass beads, stock pine, 3/8" x 1/2"		275	.029	L.F.	.33	1.37		1.70	2.60
3270	3/8" x 7/8"		270	.030		.42	1.39		1.81	2.74
4850	Parting bead, stock pine, 3/8" x 3/4"		275	.029		.44	1.37		1.81	2.72

06 22 Millwork

06 22 13 – Standard Pattern Wood Trim

06 22 13.50 Moldings, Window and Door

		Crew	Daily Output	Labor-Hours	Unit	Material	2015 Bare Costs Labor	Equipment	Total	Total Incl O&P
4870	1/2" x 3/4"	1 Carp	255	.031	L.F.	.42	1.47		1.89	2.87
5000	Stool caps, stock pine, 11/16" x 3-1/2"		200	.040		2.07	1.88		3.95	5.35
5100	1-1/16" x 3-1/4"		150	.053	↓	3.24	2.50		5.74	7.65
5300	Threshold, oak, 3' long, inside, 5/8" x 3-5/8"		32	.250	Ea.	6.50	11.75		18.25	26.50
5400	Outside, 1-1/2" x 7-5/8"	↓	16	.500	"	45	23.50		68.50	88
5900	Window trim sets, including casings, header, stops,									
5910	stool and apron, 2-1/2" wide, FJP	1 Carp	13	.615	Opng.	31.50	29		60.50	82
5950	Pine		10	.800		37	37.50		74.50	103
6000	Oak		6	1.333	↓	64.50	62.50		127	174
9000	Minimum labor/equipment charge	↓	4	2	Job		94		94	154

06 22 13.60 Moldings, Soffits

			Crew	Daily Output	Labor-Hours	Unit	Material	Labor	Equipment	Total	Total Incl O&P
0010	**MOLDINGS, SOFFITS**	R061110-30									
0200	Soffits, pine, 1" x 4"		2 Carp	420	.038	L.F.	.47	1.79		2.26	3.44
0210	1" x 6"			420	.038		.72	1.79		2.51	3.72
0220	1" x 8"			420	.038		1.19	1.79		2.98	4.24
0230	1" x 10"			400	.040		1.55	1.88		3.43	4.78
0240	1" x 12"			400	.040		2.01	1.88		3.89	5.30
0250	STK cedar, 1" x 4"			420	.038		.62	1.79		2.41	3.61
0260	1" x 6"			420	.038		1	1.79		2.79	4.03
0270	1" x 8"			420	.038		1.57	1.79		3.36	4.66
0280	1" x 10"			400	.040		2.10	1.88		3.98	5.40
0290	1" x 12"			400	.040	↓	3.50	1.88		5.38	6.95
1000	Exterior AC plywood, 1/4" thick			400	.040	S.F.	.88	1.88		2.76	4.05
1050	3/8" thick			400	.040		.92	1.88		2.80	4.09
1100	1/2" thick		↓	400	.040		1.10	1.88		2.98	4.29
1150	Polyvinyl chloride, white, solid		1 Carp	230	.035		2.10	1.63		3.73	4.99
1160	Perforated		"	230	.035	↓	2.10	1.63		3.73	4.99
1170	Accessories, "J" channel 5/8"		2 Carp	700	.023	L.F.	.45	1.07		1.52	2.26
9000	Minimum labor/equipment charge		"	5	3.200	Job		150		150	246

06 25 Prefinished Paneling

06 25 13 – Prefinished Hardboard Paneling

06 25 13.10 Paneling, Hardboard

			Crew	Daily Output	Labor-Hours	Unit	Material	Labor	Equipment	Total	Total Incl O&P
0010	**PANELING, HARDBOARD**										
0050	Not incl. furring or trim, hardboard, tempered, 1/8" thick	G	2 Carp	500	.032	S.F.	.41	1.50		1.91	2.91
0100	1/4" thick	G		500	.032		.64	1.50		2.14	3.16
0300	Tempered pegboard, 1/8" thick	G		500	.032		.40	1.50		1.90	2.90
0400	1/4" thick	G		500	.032		.68	1.50		2.18	3.21
0600	Untempered hardboard, natural finish, 1/8" thick	G		500	.032		.42	1.50		1.92	2.92
0700	1/4" thick	G		500	.032		.51	1.50		2.01	3.02
0900	Untempered pegboard, 1/8" thick	G		500	.032		.44	1.50		1.94	2.94
1000	1/4" thick	G		500	.032		.48	1.50		1.98	2.99
1200	Plastic faced hardboard, 1/8" thick	G		500	.032		.66	1.50		2.16	3.19
1300	1/4" thick	G		500	.032		.89	1.50		2.39	3.44
1500	Plastic faced pegboard, 1/8" thick	G		500	.032		.67	1.50		2.17	3.20
1600	1/4" thick	G		500	.032		.85	1.50		2.35	3.40
1800	Wood grained, plain or grooved, 1/8" thick	G		500	.032		.69	1.50		2.19	3.22
1900	1/4" thick	G		425	.038	↓	1.40	1.77		3.17	4.44
2100	Moldings, wood grained MDF			500	.032	L.F.	.41	1.50		1.91	2.91
2200	Pine		↓	425	.038	"	1.40	1.77		3.17	4.44
9000	Minimum labor/equipment charge		1 Carp	2	4	Job		188		188	310

06 25 Prefinished Paneling

06 25 16 – Prefinished Plywood Paneling

06 25 16.10 Paneling, Plywood

		Crew	Daily Output	Labor-Hours	Unit	Material	2015 Bare Costs Labor	Equipment	Total	Total Incl O&P
0010	**PANELING, PLYWOOD** R061636-20									
2400	Plywood, prefinished, 1/4" thick, 4' x 8' sheets									
2410	with vertical grooves. Birch faced, economy	2 Carp	500	.032	S.F.	1.40	1.50		2.90	4
2420	Average		420	.038		1.20	1.79		2.99	4.25
2430	Custom		350	.046		1.25	2.15		3.40	4.90
2600	Mahogany, African		400	.040		2.75	1.88		4.63	6.10
2700	Philippine (Lauan)		500	.032		.65	1.50		2.15	3.18
2900	Oak		500	.032		1.40	1.50		2.90	4
3000	Cherry		400	.040		2.05	1.88		3.93	5.35
3200	Rosewood		320	.050		2.90	2.35		5.25	7.05
3400	Teak		400	.040		2.90	1.88		4.78	6.25
3600	Chestnut		375	.043		4.80	2		6.80	8.60
3800	Pecan		400	.040		2.50	1.88		4.38	5.85
3900	Walnut, average		500	.032		2.40	1.50		3.90	5.10
3950	Custom		400	.040		5.25	1.88		7.13	8.90
4000	Plywood, prefinished, 3/4" thick, stock grades, economy		320	.050		1.56	2.35		3.91	5.55
4100	Average		224	.071		4.68	3.35		8.03	10.65
4300	Architectural grade, custom		224	.071		5.20	3.35		8.55	11.20
4400	Luxury		160	.100		5.20	4.70		9.90	13.40
4600	Plywood, "A" face, birch, VC, 1/2" thick, natural		450	.036		2.05	1.67		3.72	5
4700	Select		450	.036		2.15	1.67		3.82	5.10
4900	Veneer core, 3/4" thick, natural		320	.050		2.24	2.35		4.59	6.30
5000	Select		320	.050		2.44	2.35		4.79	6.55
5200	Lumber core, 3/4" thick, natural		320	.050		3.05	2.35		5.40	7.20
5500	Plywood, knotty pine, 1/4" thick, A2 grade		450	.036		1.70	1.67		3.37	4.61
5600	A3 grade		450	.036		2.10	1.67		3.77	5.05
5800	3/4" thick, veneer core, A2 grade		320	.050		2.15	2.35		4.50	6.20
5900	A3 grade		320	.050		2.42	2.35		4.77	6.50
6100	Aromatic cedar, 1/4" thick, plywood		400	.040		2.20	1.88		4.08	5.50
6200	1/4" thick, particle board		400	.040		1.05	1.88		2.93	4.24
9000	Minimum labor/equipment charge	1 Carp	2	4	Job		188		188	310

06 25 26 – Panel System

06 25 26.10 Panel Systems

		Crew	Daily Output	Labor-Hours	Unit	Material	2015 Bare Costs Labor	Equipment	Total	Total Incl O&P
0010	**PANEL SYSTEMS**									
0100	Raised panel, eng. wood core w/wood veneer, std., paint grade	2 Carp	300	.053	S.F.	11.20	2.50		13.70	16.40
0110	Oak veneer		300	.053		25.50	2.50		28	32
0120	Maple veneer		300	.053		32.50	2.50		35	40
0130	Cherry veneer		300	.053		37	2.50		39.50	45
0300	Class I fire rated, paint grade		300	.053		13	2.50		15.50	18.40
0310	Oak veneer		300	.053		30	2.50		32.50	37
0320	Maple veneer		300	.053		40.50	2.50		43	48.50
0330	Cherry veneer		300	.053		49	2.50		51.50	57.50
0510	Beadboard, 5/8" MDF, standard, primed		300	.053		8.45	2.50		10.95	13.40
0520	Oak veneer, unfinished		300	.053		13.35	2.50		15.85	18.80
0530	Maple veneer, unfinished		300	.053		14.65	2.50		17.15	20
0610	Rustic paneling, 5/8" MDF, standard, maple veneer, unfinished		300	.053		18.15	2.50		20.65	24

06 26 Board Paneling

06 26 13 - Profile Board Paneling

06 26 13.10 Paneling, Boards

		Crew	Daily Output	Labor-Hours	Unit	Material	2015 Bare Costs Labor	2015 Bare Costs Equipment	Total	Total Incl O&P
0010	**PANELING, BOARDS**									
6400	Wood board paneling, 3/4" thick, knotty pine	2 Carp	300	.053	S.F.	1.91	2.50		4.41	6.20
6500	Rough sawn cedar		300	.053		3.20	2.50		5.70	7.60
6700	Redwood, clear, 1" x 4" boards		300	.053		4.96	2.50		7.46	9.55
6900	Aromatic cedar, closet lining, boards		275	.058		2.32	2.73		5.05	7.05
9000	Minimum labor/equipment charge	1 Carp	2	4	Job		188		188	310

06 43 Wood Stairs and Railings

06 43 13 - Wood Stairs

06 43 13.20 Prefabricated Wood Stairs

		Crew	Daily Output	Labor-Hours	Unit	Material	2015 Bare Costs Labor	2015 Bare Costs Equipment	Total	Total Incl O&P
0010	**PREFABRICATED WOOD STAIRS**									
0100	Box stairs, prefabricated, 3'-0" wide									
0110	Oak treads, up to 14 risers	2 Carp	39	.410	Riser	92.50	19.25		111.75	134
0600	With pine treads for carpet, up to 14 risers	"	39	.410	"	59.50	19.25		78.75	97
1100	For 4' wide stairs, add				Flight	25%				
1550	Stairs, prefabricated stair handrail with balusters	1 Carp	30	.267	L.F.	78.50	12.50		91	107
1700	Basement stairs, prefabricated, pine treads									
1710	Pine risers, 3' wide, up to 14 risers	2 Carp	52	.308	Riser	59.50	14.45		73.95	89
4000	Residential, wood, oak treads, prefabricated		1.50	10.667	Flight	1,200	500		1,700	2,150
4200	Built in place		.44	36.364	"	2,175	1,700		3,875	5,175
4400	Spiral, oak, 4'-6" diameter, unfinished, prefabricated,									
4500	incl. railing, 9' high	2 Carp	1.50	10.667	Flight	3,425	500		3,925	4,600
9000	Minimum labor/equipment charge	"	3	5.333	Job		250		250	410

06 43 13.40 Wood Stair Parts

		Crew	Daily Output	Labor-Hours	Unit	Material	2015 Bare Costs Labor	2015 Bare Costs Equipment	Total	Total Incl O&P
0010	**WOOD STAIR PARTS**									
0020	Pin top balusters, 1-1/4", oak, 34"	1 Carp	96	.083	Ea.	5.10	3.91		9.01	12
0030	38"		96	.083		5.85	3.91		9.76	12.85
0040	42"		96	.083		6.35	3.91		10.26	13.40
0050	Poplar, 34"		96	.083		3.05	3.91		6.96	9.75
0060	38"		96	.083		3.88	3.91		7.79	10.65
0070	42"		96	.083		8.85	3.91		12.76	16.10
0080	Maple, 34"		96	.083		4.90	3.91		8.81	11.80
0090	38"		96	.083		5.60	3.91		9.51	12.60
0100	42"		96	.083		6.50	3.91		10.41	13.55
0130	Primed, 34"		96	.083		3	3.91		6.91	9.70
0140	38"		96	.083		3.62	3.91		7.53	10.40
0150	42"		96	.083		4.42	3.91		8.33	11.25
0180	Box top balusters, 1-1/4", oak, 34"		60	.133		8.95	6.25		15.20	20
0190	38"		60	.133		9.95	6.25		16.20	21
0200	42"		60	.133		10.95	6.25		17.20	22.50
0210	Poplar, 34"		60	.133		6.25	6.25		12.50	17.15
0220	38"		60	.133		6.95	6.25		13.20	17.90
0230	42"		60	.133		7.50	6.25		13.75	18.50
0240	Maple, 34"		60	.133		8.25	6.25		14.50	19.35
0250	38"		60	.133		9	6.25		15.25	20
0260	42"		60	.133		10	6.25		16.25	21.50
0290	Primed, 34"		60	.133		7	6.25		13.25	17.95
0300	38"		60	.133		8	6.25		14.25	19.05
0310	42"		60	.133		8.35	6.25		14.60	19.45
0340	Square balusters, cut from lineal stock, pine, 1-1/16" x 1-1/16"		180	.044	L.F.	1.50	2.09		3.59	5.05
0350	1-5/16" x 1-5/16"		180	.044		2.10	2.09		4.19	5.75

06 43 Wood Stairs and Railings

06 43 13 – Wood Stairs

06 43 13.40 Wood Stair Parts		Crew	Daily Output	Labor-Hours	Unit	Material	2015 Bare Costs Labor	Equipment	Total	Total Incl O&P
0360	1-5/8" x 1-5/8"	1 Carp	180	.044	L.F.	3.30	2.09		5.39	7.05
0370	Turned newel, oak, 3-1/2" square, 48" high		8	1	Ea.	92	47		139	178
0380	62" high		8	1		92	47		139	178
0390	Poplar, 3-1/2" square, 48" high		8	1		54	47		101	137
0400	62" high		8	1		68	47		115	152
0410	Maple, 3-1/2" square, 48" high		8	1		72	47		119	156
0420	62" high		8	1		92	47		139	178
0430	Square newel, oak, 3-1/2" square, 48" high		8	1		54	47		101	137
0440	58" high		8	1		68	47		115	152
0450	Poplar, 3-1/2" square, 48" high		8	1		35	47		82	116
0460	58" high		8	1		42	47		89	123
0470	Maple, 3" square, 48" high		8	1		52	47		99	134
0480	58" high		8	1		64	47		111	148
0490	Railings, oak, economy		96	.083	L.F.	8.65	3.91		12.56	15.90
0500	Average		96	.083		13	3.91		16.91	20.50
0510	Custom		96	.083		16.50	3.91		20.41	24.50
0520	Maple, economy		96	.083		11	3.91		14.91	18.50
0530	Average		96	.083		13.50	3.91		17.41	21.50
0540	Custom		96	.083		16.95	3.91		20.86	25
0550	Oak, for bending rail, economy		48	.167		23.50	7.85		31.35	39
0560	Average		48	.167		26	7.85		33.85	41.50
0570	Custom		48	.167		29.50	7.85		37.35	45
0580	Maple, for bending rail, economy		48	.167		32	7.85		39.85	48
0590	Average		48	.167		32	7.85		39.85	48
0600	Custom		48	.167		32	7.85		39.85	48
0610	Risers, oak, 3/4" x 8", 36" long		80	.100	Ea.	13	4.70		17.70	22
0620	42" long		70	.114		15.15	5.35		20.50	25.50
0630	48" long		63	.127		17.30	5.95		23.25	29
0640	54" long		56	.143		19.50	6.70		26.20	32.50
0650	60" long		50	.160		21.50	7.50		29	36.50
0660	72" long		42	.190		26	8.95		34.95	43
0670	Poplar, 3/4" x 8", 36" long		80	.100		12.50	4.70		17.20	21.50
0680	42" long		71	.113		14.55	5.30		19.85	24.50
0690	48" long		63	.127		16.65	5.95		22.60	28
0700	54" long		56	.143		18.70	6.70		25.40	31.50
0710	60" long		50	.160		21	7.50		28.50	35.50
0720	72" long		42	.190		25	8.95		33.95	42
0730	Pine, 1" x 8", 36" long		80	.100		3.57	4.70		8.27	11.65
0740	42" long		70	.114		4.17	5.35		9.52	13.40
0750	48" long		63	.127		4.76	5.95		10.71	15
0760	54" long		56	.143		5.35	6.70		12.05	16.90
0770	60" long		50	.160		5.95	7.50		13.45	18.85
0780	72" long		42	.190		7.15	8.95		16.10	22.50
0790	Treads, oak, no returns, 1-1/32" x 11-1/2" x 36" long		32	.250		27	11.75		38.75	49
0800	42" long		32	.250		31.50	11.75		43.25	54
0810	48" long		32	.250		36	11.75		47.75	59
0820	54" long		32	.250		40.50	11.75		52.25	64
0830	60" long		32	.250		45	11.75		56.75	69
0840	72" long		32	.250		54	11.75		65.75	79
0850	Mitred return one end, 1-1/32" x 11-1/2" x 36" long		24	.333		36	15.65		51.65	65
0860	42" long		24	.333		42	15.65		57.65	71.50
0870	48" long		24	.333		48	15.65		63.65	78.50
0880	54" long		24	.333		54	15.65		69.65	85

06 43 Wood Stairs and Railings

06 43 13 – Wood Stairs

06 43 13.40 Wood Stair Parts

		Crew	Daily Output	Labor-Hours	Unit	Material	2015 Bare Costs Labor	Equipment	Total	Total Incl O&P
0890	60" long	1 Carp	24	.333	Ea.	60	15.65		75.65	91.50
0900	72" long		24	.333		72	15.65		87.65	105
0910	Mitred return two ends, 1-1/32" x 11-1/2" x 36" long		12	.667		46	31.50		77.50	102
0920	42" long		12	.667		53.50	31.50		85	111
0930	48" long		12	.667		61.50	31.50		93	119
0940	54" long		12	.667		69	31.50		100.50	128
0950	60" long		12	.667		76.50	31.50		108	136
0960	72" long		12	.667		92	31.50		123.50	153
0970	Starting step, oak, 48", bullnose		8	1		172	47		219	266
0980	Double end bullnose		8	1		254	47		301	355
1030	Skirt board, pine, 1" x 10"		55	.145	L.F.	1.55	6.85		8.40	12.90
1040	1" x 12"		52	.154	"	2.01	7.20		9.21	14.05
1050	Oak landing tread, 1-1/16" thick		54	.148	S.F.	9	6.95		15.95	21.50
1060	Oak cove molding		96	.083	L.F.	1	3.91		4.91	7.50
1070	Oak stringer molding		96	.083	"	4	3.91		7.91	10.80
1090	Rail bolt, 5/16" x 3-1/2"		48	.167	Ea.	2.75	7.85		10.60	15.90
1100	5/16" x 4-1/2"		48	.167		2.75	7.85		10.60	15.90
1120	Newel post anchor		16	.500		13	23.50		36.50	53
1130	Tapered plug, 1/2"		240	.033		1	1.56		2.56	3.66
1140	1"		240	.033		.99	1.56		2.55	3.65
9000	Minimum labor/equipment charge		3	2.667	Job		125		125	205

06 43 16 – Wood Railings

06 43 16.10 Wood Handrails and Railings

		Crew	Daily Output	Labor-Hours	Unit	Material	Labor	Equipment	Total	Total Incl O&P
0010	**WOOD HANDRAILS AND RAILINGS**									
0020	Custom design, architectural grade, hardwood, plain	1 Carp	38	.211	L.F.	12.05	9.90		21.95	29.50
0100	Shaped		30	.267		62.50	12.50		75	89.50
0300	Stock interior railing with spindles 4" O.C., 4' long		40	.200		38.50	9.40		47.90	58
0400	8' long		48	.167		38.50	7.85		46.35	55.50
9000	Minimum labor/equipment charge		3	2.667	Job		125		125	205

06 44 Ornamental Woodwork

06 44 19 – Wood Grilles

06 44 19.10 Grilles

		Crew	Daily Output	Labor-Hours	Unit	Material	Labor	Equipment	Total	Total Incl O&P
0010	**GRILLES** and panels, hardwood, sanded									
0020	2' x 4' to 4' x 8', custom designs, unfinished, economy	1 Carp	38	.211	S.F.	62	9.90		71.90	84.50
0050	Average		30	.267		69	12.50		81.50	96.50
0100	Custom		19	.421		72	19.75		91.75	112
9000	Minimum labor/equipment charge		2	4	Job		188		188	310

06 44 33 – Wood Mantels

06 44 33.10 Fireplace Mantels

		Crew	Daily Output	Labor-Hours	Unit	Material	Labor	Equipment	Total	Total Incl O&P
0010	**FIREPLACE MANTELS**									
0015	6" molding, 6' x 3'-6" opening, plain, paint grade	1 Carp	5	1.600	Opng.	440	75		515	605
0100	Ornate, oak		5	1.600		590	75		665	775
0300	Prefabricated pine, colonial type, stock, deluxe		2	4		1,500	188		1,688	1,950
0400	Economy		3	2.667		690	125		815	965
9000	Minimum labor/equipment charge		3	2.667	Job		125		125	205

06 44 33.20 Fireplace Mantel Beam

		Crew	Daily Output	Labor-Hours	Unit	Material	Labor	Equipment	Total	Total Incl O&P
0010	**FIREPLACE MANTEL BEAM**									
0020	Rough texture wood, 4" x 8"	1 Carp	36	.222	L.F.	8.30	10.45		18.75	26
0100	4" x 10"		35	.229	"	10.90	10.75		21.65	29.50

06 44 Ornamental Woodwork

06 44 33 – Wood Mantels

06 44 33.20 Fireplace Mantel Beam		Crew	Daily Output	Labor-Hours	Unit	Material	2015 Bare Costs Labor	Equipment	Total	Total Incl O&P
0300	Laminated hardwood, 2-1/4" x 10-1/2" wide, 6' long	1 Carp	5	1.600	Ea.	110	75		185	244
0400	8' long		5	1.600	"	150	75		225	288
0600	Brackets for above, rough sawn		12	.667	Pr.	10	31.50		41.50	62.50
0700	Laminated		12	.667	"	15	31.50		46.50	68
9000	Minimum labor/equipment charge		4	2	Job		94		94	154

06 44 39 – Wood Posts and Columns

06 44 39.10 Decorative Beams

		Crew	Daily Output	Labor-Hours	Unit	Material	Labor	Equipment	Total	Total Incl O&P
0010	**DECORATIVE BEAMS**									
0020	Rough sawn cedar, non-load bearing, 4" x 4"	2 Carp	180	.089	L.F.	1.19	4.17		5.36	8.15
0100	4" x 6"		170	.094		1.79	4.42		6.21	9.20
0200	4" x 8"		160	.100		2.39	4.70		7.09	10.35
0300	4" x 10"		150	.107		3.63	5		8.63	12.20
0400	4" x 12"		140	.114		4.61	5.35		9.96	13.85
0500	8" x 8"		130	.123		4.78	5.80		10.58	14.70
9000	Minimum labor/equipment charge	1 Carp	3	2.667	Job		125		125	205

06 44 39.20 Columns

		Crew	Daily Output	Labor-Hours	Unit	Material	Labor	Equipment	Total	Total Incl O&P
0010	**COLUMNS**									
0050	Aluminum, round colonial, 6" diameter	2 Carp	80	.200	V.L.F.	19	9.40		28.40	36.50
0100	8" diameter		62.25	.257		22	12.05		34.05	44
0200	10" diameter		55	.291		22.50	13.65		36.15	47.50
0250	Fir, stock units, hollow round, 6" diameter		80	.200		29.50	9.40		38.90	48
0300	8" diameter		80	.200		35.50	9.40		44.90	54.50
0350	10" diameter		70	.229		44.50	10.75		55.25	66.50
0360	12" diameter		65	.246		54.50	11.55		66.05	79
0400	Solid turned, to 8' high, 3-1/2" diameter		80	.200		9.60	9.40		19	26
0500	4-1/2" diameter		75	.213		11.90	10		21.90	29.50
0600	5-1/2" diameter		70	.229		16	10.75		26.75	35
0800	Square columns, built-up, 5" x 5"		65	.246		14.20	11.55		25.75	34.50
0900	Solid, 3-1/2" x 3-1/2"		130	.123		9.60	5.80		15.40	20
1600	Hemlock, tapered, T & G, 12" diam., 10' high		100	.160		39.50	7.50		47	56
1700	16' high		65	.246		73	11.55		84.55	99
1900	14" diameter, 10' high		100	.160		113	7.50		120.50	137
2000	18' high		65	.246		103	11.55		114.55	132
2200	18" diameter, 12' high		65	.246		165	11.55		176.55	201
2300	20' high		50	.320		118	15		133	155
2500	20" diameter, 14' high		40	.400		180	18.80		198.80	229
2600	20' high		35	.457		170	21.50		191.50	222
2800	For flat pilasters, deduct					33%				
3000	For splitting into halves, add				Ea.	106			106	117
4000	Rough sawn cedar posts, 4" x 4"	2 Carp	250	.064	V.L.F.	3.90	3		6.90	9.20
4100	4" x 6"		235	.068		6.80	3.20		10	12.75
4200	6" x 6"		220	.073		9.90	3.41		13.31	16.50
4300	8" x 8"		200	.080		19.20	3.76		22.96	27
9000	Minimum labor/equipment charge	1 Carp	3	2.667	Job		125		125	205

06 48 Wood Frames

06 48 13 – Exterior Wood Door Frames

06 48 13.10 Exterior Wood Door Frames and Accessories

		Crew	Daily Output	Labor-Hours	Unit	Material	2015 Bare Costs Labor	2015 Bare Costs Equipment	Total	Total Incl O&P
0010	**EXTERIOR WOOD DOOR FRAMES AND ACCESSORIES**									
0400	Exterior frame, incl. ext. trim, pine, 5/4 x 4-9/16" deep	2 Carp	375	.043	L.F.	6.60	2		8.60	10.60
0420	5-3/16" deep		375	.043		7.90	2		9.90	11.95
0440	6-9/16" deep		375	.043		8.80	2		10.80	12.95
0600	Oak, 5/4 x 4-9/16" deep		350	.046		19.75	2.15		21.90	25
0620	5-3/16" deep		350	.046		21.50	2.15		23.65	27.50
0640	6-9/16" deep		350	.046		19.50	2.15		21.65	25
1000	Sills, 8/4 x 8" deep, oak, no horns		100	.160		6.65	7.50		14.15	19.60
1020	2" horns		100	.160		20.50	7.50		28	35
1040	3" horns		100	.160		20.50	7.50		28	35
1100	8/4 x 10" deep, oak, no horns		90	.178		6.40	8.35		14.75	21
1120	2" horns		90	.178		26.50	8.35		34.85	42.50
1140	3" horns		90	.178		26.50	8.35		34.85	42.50
2000	Wood frame & trim, ext, colonial, 3' opng, fluted pilasters, flat head		22	.727	Ea.	505	34		539	610
2010	Dentil head		21	.762		580	36		616	695
2020	Ram's head		20	.800		695	37.50		732.50	820
2100	5'-4" opening, in-swing, fluted pilasters, flat head		17	.941		440	44		484	560
2120	Ram's head		15	1.067		1,400	50		1,450	1,600
2140	Out swing, fluted pilasters, flat head		17	.941		520	44		564	645
2160	Ram's head		15	1.067		1,475	50		1,525	1,700
2400	6'-0" opening, in-swing, fluted pilasters, flat head		16	1		520	47		567	645
2420	Ram's head		10	1.600		1,475	75		1,550	1,750
2460	Out-swing, fluted pilasters, flat head		16	1		520	47		567	645
2480	Ram's head		10	1.600		1,475	75		1,550	1,750
2600	For two sidelights, flat head, add		30	.533	Opng.	226	25		251	290
2620	Ram's head, add		20	.800	"	840	37.50		877.50	985
2700	Custom birch frame, 3'-0" opening		16	1	Ea.	240	47		287	340
2750	6'-0" opening		16	1		360	47		407	470
2900	Exterior, modern, plain trim, 3' opng., in-swing, FJP		26	.615		46.50	29		75.50	98.50
2920	Fir		24	.667		55	31.50		86.50	112
2940	Oak		22	.727		63	34		97	126

06 48 16 – Interior Wood Door Frames

06 48 16.10 Interior Wood Door Jamb and Frames

		Crew	Daily Output	Labor-Hours	Unit	Material	2015 Bare Costs Labor	2015 Bare Costs Equipment	Total	Total Incl O&P
0010	**INTERIOR WOOD DOOR JAMB AND FRAMES**									
3000	Interior frame, pine, 11/16" x 3-5/8" deep	2 Carp	375	.043	L.F.	4.44	2		6.44	8.15
3020	4-9/16" deep		375	.043		4.92	2		6.92	8.70
3200	Oak, 11/16" x 3-5/8" deep		350	.046		9.85	2.15		12	14.30
3220	4-9/16" deep		350	.046		9.95	2.15		12.10	14.45
3240	5-3/16" deep		350	.046		13.90	2.15		16.05	18.75
3400	Walnut, 11/16" x 3-5/8" deep		350	.046		9	2.15		11.15	13.40
3420	4-9/16" deep		350	.046		9.45	2.15		11.60	13.90
3440	5-3/16" deep		350	.046		9.55	2.15		11.70	14
3600	Pocket door frame		16	1	Ea.	81	47		128	166
3800	Threshold, oak, 5/8" x 3-5/8" deep		200	.080	L.F.	3.56	3.76		7.32	10.05
3820	4-5/8" deep		190	.084		4.13	3.95		8.08	11.05
3840	5-5/8" deep		180	.089		6.50	4.17		10.67	14
9000	Minimum labor/equipment charge	1 Carp	4	2	Job		94		94	154

06 49 Wood Screens and Exterior Wood Shutters

06 49 19 – Exterior Wood Shutters

	06 49 19.10 Shutters, Exterior	Crew	Daily Output	Labor-Hours	Unit	Material	2015 Bare Costs Labor	Equipment	Total	Total Incl O&P
0010	**SHUTTERS, EXTERIOR**									
0012	Aluminum, louvered, 1'-4" wide, 3'-0" long	1 Carp	10	.800	Pr.	200	37.50		237.50	282
0200	4'-0" long		10	.800		240	37.50		277.50	325
0300	5'-4" long		10	.800		281	37.50		318.50	370
0400	6'-8" long		9	.889		355	41.50		396.50	460
1000	Pine, louvered, primed, each 1'-2" wide, 3'-3" long		10	.800		216	37.50		253.50	300
1100	4'-7" long		10	.800		270	37.50		307.50	360
1250	Each 1'-4" wide, 3'-0" long		10	.800		223	37.50		260.50	305
1350	5'-3" long		10	.800		330	37.50		367.50	420
1500	Each 1'-6" wide, 3'-3" long		10	.800		244	37.50		281.50	330
1600	4'-7" long		10	.800		325	37.50		362.50	415
1620	Cedar, louvered, 1'-2" wide, 5'-7" long		10	.800		315	37.50		352.50	405
1630	Each 1'-4" wide, 2'-2" long		10	.800		175	37.50		212.50	255
1640	3'-0" long		10	.800		217	37.50		254.50	300
1650	3'-3" long		10	.800		228	37.50		265.50	315
1660	3'-11" long		10	.800		265	37.50		302.50	355
1670	4'-3" long		10	.800		275	37.50		312.50	365
1680	5'-3" long		10	.800		290	37.50		327.50	380
1690	5'-11" long		10	.800		350	37.50		387.50	445
1700	Door blinds, 6'-9" long, each 1'-3" wide		9	.889		380	41.50		421.50	485
1710	1'-6" wide		9	.889		435	41.50		476.50	550
1720	Cedar, solid raised panel, each 1'-4" wide, 3'-3" long		10	.800		315	37.50		352.50	405
1730	3'-11" long		10	.800		360	37.50		397.50	455
1740	4'-3" long		10	.800		365	37.50		402.50	460
1750	4'-7" long		10	.800		390	37.50		427.50	490
1760	4'-11" long		10	.800		415	37.50		452.50	520
1770	5'-11" long		10	.800		520	37.50		557.50	630
1800	Door blinds, 6'-9" long, each 1'-3" wide		9	.889		535	41.50		576.50	660
1900	1'-6" wide		9	.889		630	41.50		671.50	765
2500	Polystyrene, solid raised panel, each 1'-4" wide, 3'-3" long		10	.800		71.50	37.50		109	140
2600	3'-11" long		10	.800		93	37.50		130.50	164
2700	4'-7" long		10	.800		105	37.50		142.50	178
2800	5'-3" long		10	.800		120	37.50		157.50	194
2900	6'-8" long		9	.889		152	41.50		193.50	236
4500	Polystyrene, louvered, each 1'-2" wide, 3'-3" long		10	.800		36	37.50		73.50	101
4600	4'-7" long		10	.800		46	37.50		83.50	112
4750	5'-3" long		10	.800		59	37.50		96.50	127
4850	6'-8" long		9	.889		68	41.50		109.50	144
6000	Vinyl, louvered, each 1'-2" x 4'-7" long		10	.800		64	37.50		101.50	132
6200	Each 1'-4" x 6'-8" long		9	.889		76	41.50		117.50	152
9000	Minimum labor/equipment charge		4	2	Job		94		94	154

06 51 Structural Plastic Shapes and Plates

06 51 13 – Plastic Lumber

06 51 13.10 Recycled Plastic Lumber		Crew	Daily Output	Labor-Hours	Unit	Material	2015 Bare Costs Labor	Equipment	Total	Total Incl O&P
0010	**RECYCLED PLASTIC LUMBER**									
4000	Sheeting, recycled plastic, black or white, 4' x 8' x 1/8" G	2 Carp	1100	.015	S.F.	1.26	.68		1.94	2.51
4010	4' x 8' x 3/16" G		1100	.015		1.90	.68		2.58	3.21
4020	4' x 8' x 1/4" G		950	.017		2.25	.79		3.04	3.78
4030	4' x 8' x 3/8" G		950	.017		3.80	.79		4.59	5.50
4040	4' x 8' x 1/2" G		900	.018		5	.83		5.83	6.85
4050	4' x 8' x 5/8" G		900	.018		7.50	.83		8.33	9.60
4060	4' x 8' x 3/4" G		850	.019		9.40	.88		10.28	11.80
4070	Add for colors G				Ea.	5%				
8500	100% recycled plastic, var colors, NLB, 2" x 2" G				L.F.	1.76			1.76	1.94
8510	2" x 4" G					3.65			3.65	4.02
8520	2" x 6" G					5.75			5.75	6.35
8530	2" x 8" G					7.90			7.90	8.70
8540	2" x 10" G					11.50			11.50	12.65
8550	5/4" x 4" G					4.45			4.45	4.90
8560	5/4" x 6" G					5			5	5.50
8570	1" x 6" G					2.87			2.87	3.16
8580	1/2" x 8" G					3.05			3.05	3.36
8590	2" x 10" T & G G					11.50			11.50	12.65
8600	3" x 10" T & G G					15.30			15.30	16.85
8610	Add for premium colors G					20%				

06 51 13.12 Structural Plastic Lumber

		Crew	Daily Output	Labor-Hours	Unit	Material	Labor	Equipment	Total	Total Incl O&P
0010	**STRUCTURAL PLASTIC LUMBER**									
1320	Plastic lumber, posts or columns, 4" x 4"	2 Carp	390	.041	L.F.	9	1.93		10.93	13.05
1325	4" x 6"		275	.058		13.15	2.73		15.88	19
1330	4" x 8"		220	.073		19.20	3.41		22.61	26.50
1340	Girder, single, 4" x 4"		675	.024		9	1.11		10.11	11.70
1345	4" x 6"		600	.027		13.15	1.25		14.40	16.55
1350	4" x 8"		525	.030		19.20	1.43		20.63	23.50
1352	Double, 2" x 4"		625	.026		8.25	1.20		9.45	11
1354	2" x 6"		600	.027		12.65	1.25		13.90	16
1356	2" x 8"		575	.028		16.10	1.31		17.41	19.85
1358	2" x 10"		550	.029		20	1.37		21.37	24
1360	2" x 12"		525	.030		25	1.43		26.43	30
1362	Triple, 2" x 4"		575	.028		12.35	1.31		13.66	15.75
1364	2" x 6"		550	.029		19	1.37		20.37	23
1366	2" x 8"		525	.030		24	1.43		25.43	29
1368	2" x 10"		500	.032		30	1.50		31.50	35.50
1370	2" x 12"		475	.034		37	1.58		38.58	43.50
1372	Ledger, bolted 4' O.C., 2" x 4"		400	.040		4.26	1.88		6.14	7.75
1374	2" x 6"		550	.029		6.40	1.37		7.77	9.30
1376	2" x 8"		390	.041		8.15	1.93		10.08	12.15
1378	2" x 10"		385	.042		10.10	1.95		12.05	14.30
1380	2" x 12"		380	.042		12.50	1.98		14.48	17
1382	Joists, 2" x 4"		1250	.013		4.12	.60		4.72	5.50
1384	2" x 6"		1250	.013		6.35	.60		6.95	7.95
1386	2" x 8"		1100	.015		8.05	.68		8.73	9.95
1388	2" x 10"		500	.032		10.10	1.50		11.60	13.55
1390	2" x 12"		875	.018		12.40	.86		13.26	15.05
1392	Railings and trim, 5/4" x 4"	1 Carp	300	.027		4.25	1.25		5.50	6.75
1394	2" x 2"		300	.027		2.35	1.25		3.60	4.64
1396	2" x 4"		300	.027		4.10	1.25		5.35	6.55

06 51 Structural Plastic Shapes and Plates

06 51 13 – Plastic Lumber

06 51 13.12 Structural Plastic Lumber		Crew	Daily Output	Labor-Hours	Unit	Material	2015 Bare Costs Labor	Equipment	Total	Total Incl O&P
1398	2" x 6"	1 Carp	300	.027	L.F.	6.30	1.25		7.55	9

06 52 Plastic Structural Assemblies

06 52 10 – Fiberglass Structural Assemblies

06 52 10.30 Fiberglass Grating

		Crew	Daily Output	Labor-Hours	Unit	Material	2015 Bare Costs Labor	Equipment	Total	Total Incl O&P
0010	**FIBERGLASS GRATING**									
0100	Molded, green (for mod. corrosive environment)									
0140	1" x 4" mesh, 1" thick	2 Sswk	400	.040	S.F.	8.50	2.11		10.61	13.15
0180	1-1/2" square mesh, 1" thick		400	.040		18.70	2.11		20.81	24.50
0220	1-1/4" thick		400	.040		9.35	2.11		11.46	14.05
0260	1-1/2" thick		400	.040		23	2.11		25.11	29
0300	2" square mesh, 2" thick		320	.050		26	2.63		28.63	34
1000	Orange (for highly corrosive environment)									
1040	1" x 4" mesh, 1" thick	2 Sswk	400	.040	S.F.	16.50	2.11		18.61	22
1080	1-1/2" square mesh, 1" thick		400	.040		19.55	2.11		21.66	25.50
1120	1-1/4" thick		400	.040		20	2.11		22.11	26
1160	1-1/2" thick		400	.040		20.50	2.11		22.61	26.50
1200	2" square mesh, 2" thick		320	.050		20.50	2.63		23.13	27.50
3000	Pultruded, green (for mod. corrosive environment)									
3040	1" O.C. bar spacing, 1" thick	2 Sswk	400	.040	S.F.	14.85	2.11		16.96	20
3080	1-1/2" thick		320	.050		15.30	2.63		17.93	21.50
3120	1-1/2" O.C. bar spacing, 1" thick		400	.040		13.70	2.11		15.81	18.85
3160	1-1/2" thick		400	.040		18.05	2.11		20.16	23.50
4000	Grating support legs, fixed height, no base				Ea.	46.50			46.50	51.50
4040	With base					41			41	45
4080	Adjustable to 60"					61.50			61.50	67.50

06 52 10.40 Fiberglass Floor Grating

		Crew	Daily Output	Labor-Hours	Unit	Material	2015 Bare Costs Labor	Equipment	Total	Total Incl O&P
0010	**FIBERGLASS FLOOR GRATING**									
0100	Reinforced polyester, fire retardant, 1" x 4" grid, 1" thick	E-4	510	.063	S.F.	14	3.34	.29	17.63	22
0200	1-1/2" x 6" mesh, 1-1/2" thick		500	.064		16.50	3.40	.29	20.19	24.50
0300	With grit surface, 1-1/2" x 6" grid, 1-1/2" thick		500	.064		16.80	3.40	.29	20.49	25

06 63 Plastic Railings

06 63 10 – Plastic (PVC) Railings

06 63 10.10 Plastic Railings

		Crew	Daily Output	Labor-Hours	Unit	Material	2015 Bare Costs Labor	Equipment	Total	Total Incl O&P
0010	**PLASTIC RAILINGS**									
0100	Horizontal PVC handrail with balusters, 3-1/2" wide, 36" high	1 Carp	96	.083	L.F.	24.50	3.91		28.41	33.50
0150	42" high		96	.083		28	3.91		31.91	37.50
0200	Angled PVC handrail with balusters, 3-1/2" wide, 36" high		72	.111		28.50	5.20		33.70	39.50
0250	42" high		72	.111		31.50	5.20		36.70	43
0300	Post sleeve for 4 x 4 post		96	.083		12.80	3.91		16.71	20.50
0400	Post cap for 4 x 4 post, flat profile		48	.167	Ea.	12.80	7.85		20.65	27
0450	Newel post style profile		48	.167		23.50	7.85		31.35	39
0500	Raised corbeled profile		48	.167		35.50	7.85		43.35	52
0550	Post base trim for 4 x 4 post		96	.083		18.25	3.91		22.16	26.50

06 65 Plastic Trim

06 65 10 – PVC Trim

06 65 10.10 PVC Trim, Exterior		Crew	Daily Output	Labor-Hours	Unit	Material	2015 Bare Costs Labor	Equipment	Total	Total Incl O&P
0010	**PVC TRIM, EXTERIOR**									
0100	Cornerboards, 5/4" x 6" x 6"	1 Carp	240	.033	L.F.	8.45	1.56		10.01	11.85
0110	Door/window casing, 1" x 4"		200	.040		1.39	1.88		3.27	4.61
0120	1" x 6"		200	.040		2.11	1.88		3.99	5.40
0130	1" x 8"		195	.041		2.78	1.93		4.71	6.20
0140	1" x 10"		195	.041		3.50	1.93		5.43	7
0150	1" x 12"		190	.042		4.33	1.98		6.31	8
0160	5/4" x 4"		195	.041		1.69	1.93		3.62	5
0170	5/4" x 6"		195	.041		2.72	1.93		4.65	6.15
0180	5/4" x 8"		190	.042		3.56	1.98		5.54	7.15
0190	5/4" x 10"		190	.042		4.56	1.98		6.54	8.25
0200	5/4" x 12"		185	.043		5.25	2.03		7.28	9.15
0210	Fascia, 1" x 4"		250	.032		1.39	1.50		2.89	3.99
0220	1" x 6"		250	.032		2.11	1.50		3.61	4.78
0230	1" x 8"		225	.036		2.78	1.67		4.45	5.80
0240	1" x 10"		225	.036		3.50	1.67		5.17	6.60
0250	1" x 12"		200	.040		4.33	1.88		6.21	7.85
0260	5/4" x 4"		240	.033		1.69	1.56		3.25	4.42
0270	5/4" x 6"		240	.033		2.72	1.56		4.28	5.55
0280	5/4" x 8"		215	.037		3.56	1.75		5.31	6.80
0290	5/4" x 10"		215	.037		4.56	1.75		6.31	7.85
0300	5/4" x 12"		190	.042		5.25	1.98		7.23	9.05
0310	Frieze, 1" x 4"		250	.032		1.39	1.50		2.89	3.99
0320	1" x 6"		250	.032		2.11	1.50		3.61	4.78
0330	1" x 8"		225	.036		2.78	1.67		4.45	5.80
0340	1" x 10"		225	.036		3.50	1.67		5.17	6.60
0350	1" x 12"		200	.040		4.33	1.88		6.21	7.85
0360	5/4" x 4"		240	.033		1.69	1.56		3.25	4.42
0370	5/4" x 6"		240	.033		2.72	1.56		4.28	5.55
0380	5/4" x 8"		215	.037		3.56	1.75		5.31	6.80
0390	5/4" x 10"		215	.037		4.56	1.75		6.31	7.85
0400	5/4" x 12"		190	.042		5.25	1.98		7.23	9.05
0410	Rake, 1" x 4"		200	.040		1.39	1.88		3.27	4.61
0420	1" x 6"		200	.040		2.11	1.88		3.99	5.40
0430	1" x 8"		190	.042		2.78	1.98		4.76	6.30
0440	1" x 10"		190	.042		3.50	1.98		5.48	7.10
0450	1" x 12"		180	.044		4.33	2.09		6.42	8.20
0460	5/4" x 4"		195	.041		1.69	1.93		3.62	5
0470	5/4" x 6"		195	.041		2.72	1.93		4.65	6.15
0480	5/4" x 8"		185	.043		3.56	2.03		5.59	7.25
0490	5/4" x 10"		185	.043		4.56	2.03		6.59	8.35
0500	5/4" x 12"		175	.046		5.25	2.15		7.40	9.30
0510	Rake trim, 1" x 4"		225	.036		1.39	1.67		3.06	4.27
0520	1" x 6"		225	.036		2.11	1.67		3.78	5.05
0560	5/4" x 4"		220	.036		1.69	1.71		3.40	4.66
0570	5/4" x 6"		220	.036		2.72	1.71		4.43	5.80
0610	Soffit, 1" x 4"	2 Carp	420	.038		1.39	1.79		3.18	4.46
0620	1" x 6"		420	.038		2.11	1.79		3.90	5.25
0630	1" x 8"		420	.038		2.78	1.79		4.57	6
0640	1" x 10"		400	.040		3.50	1.88		5.38	6.95
0650	1" x 12"		400	.040		4.33	1.88		6.21	7.85
0660	5/4" x 4"		410	.039		1.69	1.83		3.52	4.86
0670	5/4" x 6"		410	.039		2.72	1.83		4.55	6

06 65 Plastic Trim

06 65 10 – PVC Trim

06 65 10.10 PVC Trim, Exterior

		Crew	Daily Output	Labor-Hours	Unit	Material	2015 Bare Costs Labor	2015 Bare Costs Equipment	Total	Total Incl O&P
0680	5/4" x 8"	2 Carp	410	.039	L.F.	3.56	1.83		5.39	6.90
0690	5/4" x 10"		390	.041		4.56	1.93		6.49	8.15
0700	5/4" x 12"		390	.041		5.25	1.93		7.18	8.95

06 80 Composite Fabrications

06 80 10 – Composite Decking

06 80 10.10 Woodgrained Composite Decking

		Crew	Daily Output	Labor-Hours	Unit	Material	Labor	Equipment	Total	Total Incl O&P
0010	**WOODGRAINED COMPOSITE DECKING**									
0100	Woodgrained composite decking, 1" x 6"	2 Carp	640	.025	L.F.	3.54	1.17		4.71	5.80
0110	Grooved edge		660	.024		3.68	1.14		4.82	5.90
0120	2" x 6"		640	.025		3.67	1.17		4.84	5.95
0130	Encased, 1" x 6"		640	.025		3.67	1.17		4.84	5.95
0140	Grooved edge		660	.024		3.81	1.14		4.95	6.05
0150	2" x 6"		640	.025		4.98	1.17		6.15	7.40

06 81 Composite Railings

06 81 10 – Encased Railings

06 81 10.10 Encased Composite Railings

		Crew	Daily Output	Labor-Hours	Unit	Material	Labor	Equipment	Total	Total Incl O&P
0010	**ENCASED COMPOSITE RAILINGS**									
0100	Encased composite railing, 6' long, 36" high, incl. balusters	1 Carp	16	.500	Ea.	156	23.50		179.50	210
0110	42" high, incl. balusters		16	.500		172	23.50		195.50	229
0120	8' long, 36" high, incl. balusters		12	.667		156	31.50		187.50	224
0130	42" high, incl. balusters		12	.667		172	31.50		203.50	242
0140	Accessories, post sleeve, 4" x 4", 39" long		32	.250		23	11.75		34.75	45
0150	96" long		24	.333		86.50	15.65		102.15	121
0160	6" x 6", 39" long		32	.250		47	11.75		58.75	71
0170	96" long		24	.333		137	15.65		152.65	176
0180	Accessories, post skirt, 4" x 4"		96	.083		4	3.91		7.91	10.80
0190	6" x 6"		96	.083		4.85	3.91		8.76	11.75
0200	Post cap, 4" x 4", flat		48	.167		6	7.85		13.85	19.45
0210	Pyramid		48	.167		6	7.85		13.85	19.45
0220	Post cap, 6" x 6", flat		48	.167		9.60	7.85		17.45	23.50
0230	Pyramid		48	.167		9.60	7.85		17.45	23.50

Division Notes

	CREW	DAILY OUTPUT	LABOR-HOURS	UNIT	BARE COSTS				TOTAL INCL O&P
					MAT.	LABOR	EQUIP.	TOTAL	

Estimating Tips
07 10 00 Dampproofing and Waterproofing
- Be sure of the job specifications before pricing this subdivision. The difference in cost between waterproofing and dampproofing can be great. Waterproofing will hold back standing water. Dampproofing prevents the transmission of water vapor. Also included in this section are vapor retarding membranes.

07 20 00 Thermal Protection
- Insulation and fireproofing products are measured by area, thickness, volume or R-value. Specifications may give only what the specific R-value should be in a certain situation. The estimator may need to choose the type of insulation to meet that R-value.

07 30 00 Steep Slope Roofing
07 40 00 Roofing and Siding Panels
- Many roofing and siding products are bought and sold by the square. One square is equal to an area that measures 100 square feet.

  This simple change in unit of measure could create a large error if the estimator is not observant. Accessories necessary for a complete installation must be figured into any calculations for both material and labor.

07 50 00 Membrane Roofing
07 60 00 Flashing and Sheet Metal
07 70 00 Roofing and Wall Specialties and Accessories
- The items in these subdivisions compose a roofing system. No one component completes the installation, and all must be estimated. Built-up or single-ply membrane roofing systems are made up of many products and installation trades. Wood blocking at roof perimeters or penetrations, parapet coverings, reglets, roof drains, gutters, downspouts, sheet metal flashing, skylights, smoke vents, and roof hatches all need to be considered along with the roofing material. Several different installation trades will need to work together on the roofing system. Inherent difficulties in the scheduling and coordination of various trades must be accounted for when estimating labor costs.

07 90 00 Joint Protection
- To complete the weather-tight shell, the sealants and caulkings must be estimated. Where different materials meet—at expansion joints, at flashing penetrations, and at hundreds of other locations throughout a construction project—they provide another line of defense against water penetration. Often, an entire system is based on the proper location and placement of caulking or sealants. The detailed drawings that are included as part of a set of architectural plans show typical locations for these materials. When caulking or sealants are shown at typical locations, this means the estimator must include them for all the locations where this detail is applicable. Be careful to keep different types of sealants separate, and remember to consider backer rods and primers if necessary.

Reference Numbers
Reference numbers are shown in shaded boxes at the beginning of some major classifications. These numbers refer to related items in the Reference Section. The reference information may be an estimating procedure, an alternate pricing method, or technical information.

Note: Not all subdivisions listed here necessarily appear in this publication. ■

Estimate with precision with RSMeans Online...
Quick, intuitive and easy-to-use, RSMeans Online gives you instant access to hundreds of thousands of material, labor, and equipment costs from RSMeans' comprehensive database, delivering the information you need to build competitive estimates every time.

Start your free trial today at **www.rsmeansonline.com**

RSMeansOnline

No part of this publication may be reproduced, stored in a retrieval system, or transmitted in any form or by any means without prior written permission of RSMeans.

07 01 Operation and Maint. of Thermal and Moisture Protection

07 01 50 – Maintenance of Membrane Roofing

07 01 50.10 Roof Coatings

		Crew	Daily Output	Labor-Hours	Unit	Material	2015 Bare Costs Labor	Equipment	Total	Total Incl O&P
0010	**ROOF COATINGS**									
0012	Asphalt, brush grade, material only				Gal.	8.95			8.95	9.80
0200	Asphalt base, fibered aluminum coating [G]					8.25			8.25	9.10
0300	Asphalt primer, 5 gallon					7.50			7.50	8.25
0600	Coal tar pitch, 200 lb. barrels				Ton	1,525			1,525	1,675
0700	Tar roof cement, 5 gal. lots				Gal.	14.05			14.05	15.45
0800	Glass fibered roof & patching cement, 5 gallon				"	8.25			8.25	9.10
0900	Reinforcing glass membrane, 450 S.F./roll				Ea.	59.50			59.50	65.50
1000	Neoprene roof coating, 5 gal., 2 gal./sq.				Gal.	30.50			30.50	33.50
1100	Roof patch & flashing cement, 5 gallon					8.70			8.70	9.55
1200	Roof resaturant, glass fibered, 3 gal./sq.					8.85			8.85	9.75
1600	Reflective roof coating, white, elastomeric, approx. 50 S.F. per gal. [G]					18.90			18.90	21

07 01 90 – Maintenance of Joint Protection

07 01 90.81 Joint Sealant Replacement

		Crew	Daily Output	Labor-Hours	Unit	Material	2015 Bare Costs Labor	Equipment	Total	Total Incl O&P
0010	**JOINT SEALANT REPLACEMENT**									
0050	Control joints in concrete floors/slabs									
0100	Option 1 for joints with hard dry sealant									
0110	Step 1: Sawcut to remove 95% of old sealant									
0112	1/4" wide x 1/2" deep, with single saw blade	C-27	4800	.003	L.F.	.02	.15	.03	.20	.30
0114	3/8" wide x 3/4" deep, with single saw blade		4000	.004		.03	.18	.04	.25	.37
0116	1/2" wide x 1" deep, with double saw blades		3600	.004		.06	.20	.05	.31	.44
0118	3/4" wide x 1-1/2" deep, with double saw blades		3200	.005		.13	.23	.05	.41	.56
0120	Step 2: Water blast joint faces and edges	C-29	2500	.003			.12	.03	.15	.23
0130	Step 3: Air blast joint faces and edges	C-28	2000	.004			.18	.01	.19	.29
0140	Step 4: Sand blast joint faces and edges	E-11	2000	.016			.68	.12	.80	1.30
0150	Step 5: Air blast joint faces and edges	C-28	2000	.004			.18	.01	.19	.29
0200	Option 2 for joints with soft pliable sealant									
0210	Step 1: Plow joint with rectangular blade	B-62	2600	.009	L.F.		.38	.07	.45	.69
0220	Step 2: Sawcut to re-face joint faces									
0222	1/4" wide x 1/2" deep, with single saw blade	C-27	2400	.007	L.F.	.02	.30	.07	.39	.58
0224	3/8" wide x 3/4" deep, with single saw blade		2000	.008		.04	.36	.08	.48	.71
0226	1/2" wide x 1" deep, with double saw blades		1800	.009		.09	.40	.09	.58	.82
0228	3/4" wide x 1-1/2" deep, with double saw blades		1600	.010		.17	.45	.10	.72	1.02
0230	Step 3: Water blast joint faces and edges	C-29	2500	.003			.12	.03	.15	.23
0240	Step 4: Air blast joint faces and edges	C-28	2000	.004			.18	.01	.19	.29
0250	Step 5: Sand blast joint faces and edges	E-11	2000	.016			.68	.12	.80	1.30
0260	Step 6: Air blast joint faces and edges	C-28	2000	.004			.18	.01	.19	.29
0290	For saw cutting new control joints, see Section 03 15 16.20									
8910	For backer rod, see Section 07 91 23.10									
8920	For joint sealant, see Sections 03 15 16.30 or 07 92 13.20									

07 05 Common Work Results for Thermal and Moisture Protection

07 05 05 – Selective Demolition for Thermal and Moisture Protection

07 05 05.10 Selective Demo., Thermal and Moist. Protection		Crew	Daily Output	Labor-Hours	Unit	Material	2015 Bare Costs Labor	Equipment	Total	Total Incl O&P
0010	**SELECTIVE DEMO., THERMAL AND MOISTURE PROTECTION**									
0020	Caulking/sealant, to 1" x 1" joint R024119-10	1 Clab	600	.013	L.F.		.50		.50	.82
0120	Downspouts, including hangers		350	.023	"		.86		.86	1.41
0220	Flashing, sheet metal		290	.028	S.F.		1.04		1.04	1.70
0420	Gutters, aluminum or wood, edge hung		240	.033	L.F.		1.25		1.25	2.05
0520	Built-in		100	.080	"		3.01		3.01	4.93
0620	Insulation, air/vapor barrier		3500	.002	S.F.		.09		.09	.14
0670	Batts or blankets		1400	.006	C.F.		.21		.21	.35
0720	Foamed or sprayed in place	2 Clab	1000	.016	B.F.		.60		.60	.99
0770	Loose fitting	1 Clab	3000	.003	C.F.		.10		.10	.16
0870	Rigid board		3450	.002	B.F.		.09		.09	.14
1120	Roll roofing, cold adhesive		12	.667	Sq.		25		25	41
1170	Roof accessories, adjustable metal chimney flashing		9	.889	Ea.		33.50		33.50	55
1325	Plumbing vent flashing		32	.250	"		9.40		9.40	15.40
1375	Ridge vent strip, aluminum		310	.026	L.F.		.97		.97	1.59
1620	Skylight to 10 S.F.		8	1	Ea.		37.50		37.50	61.50
2120	Roof edge, aluminum soffit and fascia		570	.014	L.F.		.53		.53	.87
2170	Concrete coping, up to 12" wide	2 Clab	160	.100			3.76		3.76	6.15
2220	Drip edge	1 Clab	1000	.008			.30		.30	.49
2270	Gravel stop		950	.008			.32		.32	.52
2370	Sheet metal coping, up to 12" wide		240	.033			1.25		1.25	2.05
2470	Roof insulation board, over 2" thick	B-2	7800	.005	B.F.		.19		.19	.32
2520	Up to 2" thick	"	3900	.010	S.F.		.39		.39	.64
2620	Roof ventilation, louvered gable vent	1 Clab	16	.500	Ea.		18.80		18.80	31
2670	Remove, roof hatch	G-3	15	2.133			100		100	160
2675	Rafter vents	1 Clab	960	.008			.31		.31	.51
2720	Soffit vent and/or fascia vent		575	.014	L.F.		.52		.52	.86
2775	Soffit vent strip, aluminum, 3" to 4" wide		160	.050			1.88		1.88	3.08
2820	Roofing accessories, shingle moulding, to 1" x 4"		1600	.005			.19		.19	.31
2870	Cant strip	B-2	2000	.020			.76		.76	1.25
2920	Concrete block walkway	1 Clab	230	.035			1.31		1.31	2.14
3070	Roofing, felt paper, 15#		70	.114	Sq.		4.30		4.30	7.05
3125	#30 felt		30	.267	"		10.05		10.05	16.45
3170	Asphalt shingles, 1 layer	B-2	3500	.011	S.F.		.43		.43	.71
3180	2 layers		1750	.023	"		.87		.87	1.42
3370	Modified bitumen		26	1.538	Sq.		58.50		58.50	96
3420	Built-up, no gravel, 3 ply		25	1.600			61		61	99.50
3470	4 ply		21	1.905			72.50		72.50	119
3620	5 ply		1600	.025	S.F.		.95		.95	1.56
3720	5 ply, with gravel		890	.045			1.71		1.71	2.80
3725	Loose gravel removal		5000	.008			.30		.30	.50
3730	Embedded gravel removal		2000	.020			.76		.76	1.25
3870	Fiberglass sheet		1200	.033			1.27		1.27	2.08
4120	Slate shingles		1900	.021			.80		.80	1.31
4170	Ridge shingles, clay or slate		2000	.020	L.F.		.76		.76	1.25
4320	Single ply membrane, attached at seams		52	.769	Sq.		29		29	48
4370	Ballasted		75	.533			20.50		20.50	33
4420	Fully adhered		39	1.026			39		39	64
4550	Roof hatch, 2'-6" x 3'-0"	1 Clab	10	.800	Ea.		30		30	49.50
4670	Wood shingles	B-2	2200	.018	S.F.		.69		.69	1.13
4820	Sheet metal roofing	"	2150	.019			.71		.71	1.16
4970	Siding, horizontal wood clapboards	1 Clab	380	.021			.79		.79	1.30
5025	Exterior insulation finish system	"	120	.067			2.51		2.51	4.11

07 05 Common Work Results for Thermal and Moisture Protection

07 05 05 – Selective Demolition for Thermal and Moisture Protection

07 05 05.10 Selective Demo., Thermal and Moist. Protection		Crew	Daily Output	Labor-Hours	Unit	Material	2015 Bare Costs Labor	Equipment	Total	Total Incl O&P
5070	Tempered hardboard, remove and reset	1 Carp	380	.021	S.F.		.99		.99	1.62
5120	Tempered hardboard sheet siding	"	375	.021	↓		1		1	1.64
5170	Metal, corner strips	1 Clab	850	.009	L.F.		.35		.35	.58
5225	Horizontal strips		444	.018	S.F.		.68		.68	1.11
5320	Vertical strips		400	.020			.75		.75	1.23
5520	Wood shingles		350	.023			.86		.86	1.41
5620	Stucco siding		360	.022			.84		.84	1.37
5670	Textured plywood		725	.011			.41		.41	.68
5720	Vinyl siding		510	.016	↓		.59		.59	.97
5770	Corner strips		900	.009	L.F.		.33		.33	.55
5870	Wood, boards, vertical	↓	400	.020	S.F.		.75		.75	1.23
5920	Waterproofing, protection/drain board	2 Clab	3900	.004	B.F.		.15		.15	.25
5970	Over 1/2" thick		1750	.009	S.F.		.34		.34	.56
6020	To 1/2" thick	↓	2000	.008	"		.30		.30	.49
9000	Minimum labor/equipment charge	1 Clab	2	4	Job		150		150	247

07 11 Dampproofing

07 11 13 – Bituminous Dampproofing

07 11 13.10 Bituminous Asphalt Coating

0010	BITUMINOUS ASPHALT COATING									
0030	Brushed on, below grade, 1 coat	1 Rofc	665	.012	S.F.	.22	.48		.70	1.12
0100	2 coat		500	.016		.45	.64		1.09	1.65
0300	Sprayed on, below grade, 1 coat		830	.010		.22	.39		.61	.95
0400	2 coat	↓	500	.016	↓	.44	.64		1.08	1.64
0500	Asphalt coating, with fibers				Gal.	8.25			8.25	9.10
0600	Troweled on, asphalt with fibers, 1/16" thick	1 Rofc	500	.016	S.F.	.36	.64		1	1.56
0700	1/8" thick		400	.020		.64	.80		1.44	2.15
1000	1/2" thick		350	.023		2.07	.92		2.99	3.93
9000	Minimum labor/equipment charge	↓	3	2.667	Job		107		107	193

07 11 16 – Cementitious Dampproofing

07 11 16.20 Cementitious Parging

0010	CEMENTITIOUS PARGING									
0020	Portland cement, 2 coats, 1/2" thick	D-1	250	.064	S.F.	.33	2.69		3.02	4.74
0100	Waterproofed Portland cement, 1/2" thick, 2 coats	"	250	.064	"	3.93	2.69		6.62	8.70

07 12 Built-up Bituminous Waterproofing

07 12 13 – Built-Up Asphalt Waterproofing

07 12 13.20 Membrane Waterproofing

0010	MEMBRANE WATERPROOFING									
0012	On slabs, 1 ply, felt, mopped	G-1	3000	.019	S.F.	.42	.70	.19	1.31	1.94
0015	On walls, 1 ply, felt, mopped		3000	.019		.42	.70	.19	1.31	1.94
0100	On slabs, 1 ply, glass fiber fabric, mopped		2100	.027		.46	1	.27	1.73	2.61
0105	On walls, 1 ply, glass fiber fabric, mopped		2100	.027		.46	1	.27	1.73	2.61
0300	On slabs, 2 ply, felt, mopped		2500	.022		.83	.84	.23	1.90	2.68
0305	On walls, 2 ply, felt, mopped		2500	.022		.83	.84	.23	1.90	2.68
0400	On slabs, 2 ply, glass fiber fabric, mopped		1650	.034		1.01	1.27	.34	2.62	3.79
0405	On walls, 2 ply, glass fiber fabric, mopped		1650	.034		1.01	1.27	.34	2.62	3.79
0600	On slabs, 3 ply, felt, mopped		2100	.027		1.25	1	.27	2.52	3.48
0605	On walls, 3 ply, felt, mopped		2100	.027		1.25	1	.27	2.52	3.48

07 12 Built-up Bituminous Waterproofing

07 12 13 – Built-Up Asphalt Waterproofing

07 12 13.20 Membrane Waterproofing

		Crew	Daily Output	Labor-Hours	Unit	Material	2015 Bare Costs Labor	Equipment	Total	Total Incl O&P
0700	On slabs, 3 ply, glass fiber fabric, mopped	G-1	1550	.036	S.F.	1.37	1.36	.37	3.10	4.36
0705	On walls, 3 ply, glass fiber fabric, mopped	↓	1550	.036		1.37	1.36	.37	3.10	4.36
0710	Asphaltic hardboard protection board, 1/8" thick	2 Rofc	500	.032		.61	1.28		1.89	2.99
0715	1/4" thick		450	.036		1.70	1.43		3.13	4.45
1000	EPS membrane protection board, 1/4"		3500	.005		.33	.18		.51	.70
1050	3/8" thick		3500	.005		.36	.18		.54	.73
1060	1/2" thick		3500	.005	↓	.40	.18		.58	.76
1070	Fiberglass fabric, black, 20/10 mesh		116	.138	Sq.	15.40	5.55		20.95	27
9000	Minimum labor/equipment charge	↓	2	8	Job		320		320	580

07 13 Sheet Waterproofing

07 13 53 – Elastomeric Sheet Waterproofing

07 13 53.10 Elastomeric Sheet Waterproofing and Access.

		Crew	Daily Output	Labor-Hours	Unit	Material	Labor	Equipment	Total	Total Incl O&P
0010	**ELASTOMERIC SHEET WATERPROOFING AND ACCESS.**									
0090	EPDM, plain, 45 mils thick	2 Rofc	580	.028	S.F.	1.41	1.11		2.52	3.55
0100	60 mils thick		570	.028		1.47	1.13		2.60	3.65
0300	Nylon reinforced sheets, 45 mils thick		580	.028		1.55	1.11		2.66	3.70
0400	60 mils thick	↓	570	.028	↓	1.64	1.13		2.77	3.84
0600	Vulcanizing splicing tape for above, 2" wide				C.L.F.	60.50			60.50	66.50
0700	4" wide				"	121			121	133
0900	Adhesive, bonding, 60 S.F. per gal.				Gal.	27			27	29.50
1000	Splicing, 75 S.F. per gal.				"	41			41	45
1200	Neoprene sheets, plain, 45 mils thick	2 Rofc	580	.028	S.F.	1.70	1.11		2.81	3.87
1300	60 mils thick		570	.028		2.17	1.13		3.30	4.42
1500	Nylon reinforced, 45 mils thick		580	.028		1.96	1.11		3.07	4.16
1600	60 mils thick		570	.028		3.05	1.13		4.18	5.40
1800	120 mils thick	↓	500	.032	↓	6.05	1.28		7.33	8.95
1900	Adhesive, splicing, 150 S.F. per gal. per coat				Gal.	41			41	45
2100	Fiberglass reinforced, fluid applied, 1/8" thick	2 Rofc	500	.032	S.F.	1.63	1.28		2.91	4.11
2200	Polyethylene and rubberized asphalt sheets, 60 mils thick		550	.029		.92	1.17		2.09	3.12
2400	Polyvinyl chloride sheets, plain, 10 mils thick		580	.028		.15	1.11		1.26	2.17
2500	20 mils thick		570	.028		.19	1.13		1.32	2.24
2700	30 mils thick		560	.029	↓	.24	1.15		1.39	2.33
3000	Adhesives, trowel grade, 40-100 S.F. per gal.				Gal.	24			24	26.50
3100	Brush grade, 100-250 S.F. per gal.				"	21.50			21.50	24
3300	Bitumen modified polyurethane, fluid applied, 55 mils thick	2 Rofc	665	.024	S.F.	.93	.96		1.89	2.76
9000	Minimum labor/equipment charge	"	2	8	Job		320		320	580

07 16 Cementitious and Reactive Waterproofing

07 16 16 – Crystalline Waterproofing

07 16 16.20 Cementitious Waterproofing

		Crew	Daily Output	Labor-Hours	Unit	Material	Labor	Equipment	Total	Total Incl O&P
0010	**CEMENTITIOUS WATERPROOFING**									
0020	1/8" application, sprayed on	G-2A	1000	.024	S.F.	.73	.86	.72	2.31	3.10
0050	4 coat cementitious metallic slurry	1 Cefi	1.20	6.667	C.S.F.	32	300		332	510

07 19 Water Repellents

07 19 19 – Silicone Water Repellents

07 19 19.10 Silicone Based Water Repellents

		Crew	Daily Output	Labor-Hours	Unit	Material	2015 Bare Costs Labor	Equipment	Total	Total Incl O&P
0010	**SILICONE BASED WATER REPELLENTS**									
0020	Water base liquid, roller applied	2 Rofc	7000	.002	S.F.	.53	.09		.62	.75
0200	Silicone or stearate, sprayed on CMU, 1 coat	1 Rofc	4000	.002		.34	.08		.42	.51
0300	2 coats		3000	.003	↓	.68	.11		.79	.94
9000	Minimum labor/equipment charge	↓	3	2.667	Job		107		107	193

07 21 Thermal Insulation

07 21 13 – Board Insulation

07 21 13.10 Rigid Insulation

			Crew	Daily Output	Labor-Hours	Unit	Material	Labor	Equipment	Total	Total Incl O&P
0010	**RIGID INSULATION**, for walls										
0040	Fiberglass, 1.5#/C.F., unfaced, 1" thick, R4.1	G	1 Carp	1000	.008	S.F.	.27	.38		.65	.92
0060	1-1/2" thick, R6.2	G		1000	.008		.40	.38		.78	1.06
0080	2" thick, R8.3	G		1000	.008		.45	.38		.83	1.12
0120	3" thick, R12.4	G		800	.010		.56	.47		1.03	1.39
0370	3#/C.F., unfaced, 1" thick, R4.3	G		1000	.008		.52	.38		.90	1.19
0390	1-1/2" thick, R6.5	G		1000	.008		.78	.38		1.16	1.48
0400	2" thick, R8.7	G		890	.009		1.05	.42		1.47	1.85
0420	2-1/2" thick, R10.9	G		800	.010		1.10	.47		1.57	1.98
0440	3" thick, R13	G		800	.010		1.59	.47		2.06	2.52
0520	Foil faced, 1" thick, R4.3	G		1000	.008		.90	.38		1.28	1.61
0540	1-1/2" thick, R6.5	G		1000	.008		1.35	.38		1.73	2.11
0560	2" thick, R8.7	G		890	.009		1.69	.42		2.11	2.55
0580	2-1/2" thick, R10.9	G		800	.010		1.98	.47		2.45	2.95
0600	3" thick, R13	G	↓	800	.010	↓	2.18	.47		2.65	3.17
1600	Isocyanurate, 4' x 8' sheet, foil faced, both sides										
1610	1/2" thick	G	1 Carp	800	.010	S.F.	.31	.47		.78	1.11
1620	5/8" thick	G		800	.010		.33	.47		.80	1.13
1630	3/4" thick	G		800	.010		.36	.47		.83	1.17
1640	1" thick	G		800	.010		.52	.47		.99	1.34
1650	1-1/2" thick	G		730	.011		.63	.51		1.14	1.53
1660	2" thick	G		730	.011		.80	.51		1.31	1.72
1670	3" thick	G		730	.011		1.80	.51		2.31	2.82
1680	4" thick	G		730	.011		2.05	.51		2.56	3.10
1700	Perlite, 1" thick, R2.77	G		800	.010		.42	.47		.89	1.23
1750	2" thick, R5.55	G		730	.011		.75	.51		1.26	1.67
1900	Extruded polystyrene, 25 PSI compressive strength, 1" thick, R5	G		800	.010		.53	.47		1	1.35
1940	2" thick R10	G		730	.011		1.04	.51		1.55	1.98
1960	3" thick, R15	G		730	.011		1.50	.51		2.01	2.49
2100	Expanded polystyrene, 1" thick, R3.85	G		800	.010		.25	.47		.72	1.05
2120	2" thick, R7.69	G		730	.011		.50	.51		1.01	1.39
2140	3" thick, R11.49	G	↓	730	.011	↓	.75	.51		1.26	1.67
9000	Minimum labor/equipment charge			4	2	Job		94		94	154

07 21 13.13 Foam Board Insulation

			Crew	Daily Output	Labor-Hours	Unit	Material	Labor	Equipment	Total	Total Incl O&P
0010	**FOAM BOARD INSULATION**										
0600	Polystyrene, expanded, 1" thick, R4	G	1 Carp	680	.012	S.F.	.25	.55		.80	1.18
0700	2" thick, R8	G	↓	675	.012	"	.50	.56		1.06	1.46
9000	Minimum labor/equipment charge			4	2	Job		94		94	154

07 21 Thermal Insulation

07 21 16 – Blanket Insulation

07 21 16.10 Blanket Insulation for Floors/Ceilings

		Crew	Daily Output	Labor-Hours	Unit	Material	2015 Bare Costs Labor	2015 Bare Costs Equipment	Total	Total Incl O&P
0010	**BLANKET INSULATION FOR FLOORS/CEILINGS**									
0020	Including spring type wire fasteners									
2000	Fiberglass, blankets or batts, paper or foil backing									
2100	3-1/2" thick, R13	G	1 Carp	700	.011	S.F.	.37	.54		.91	1.29
2150	6-1/4" thick, R19	G		600	.013		.49	.63		1.12	1.57
2210	9-1/2" thick, R30	G		500	.016		.71	.75		1.46	2.01
2220	12" thick, R38	G		475	.017		.96	.79		1.75	2.36
3000	Unfaced, 3-1/2" thick, R13	G		600	.013		.31	.63		.94	1.37
3010	6-1/4" thick, R19	G		500	.016		.36	.75		1.11	1.63
3020	9-1/2" thick, R30	G		450	.018		.58	.83		1.41	2.01
3030	12" thick, R38	G		425	.019		.74	.88		1.62	2.26
9000	Minimum labor/equipment charge			4	2	Job		94		94	154

07 21 16.20 Blanket Insulation for Walls

		Crew	Daily Output	Labor-Hours	Unit	Material	2015 Bare Costs Labor	2015 Bare Costs Equipment	Total	Total Incl O&P
0010	**BLANKET INSULATION FOR WALLS**									
0020	Kraft faced fiberglass, 3-1/2" thick, R11, 15" wide	G	1 Carp	1350	.006	S.F.	.27	.28		.55	.76
0030	23" wide	G		1600	.005		.27	.23		.50	.68
0060	R13, 11" wide	G		1150	.007		.30	.33		.63	.87
0080	15" wide	G		1350	.006		.30	.28		.58	.79
0100	23" wide	G		1600	.005		.30	.23		.53	.71
0110	R15, 11" wide	G		1150	.007		.45	.33		.78	1.04
0120	15" wide	G		1350	.006		.45	.28		.73	.96
0130	23" wide	G		1600	.005		.45	.23		.68	.88
0140	6" thick, R19, 11" wide	G		1150	.007		.42	.33		.75	1
0160	15" wide	G		1350	.006		.42	.28		.70	.92
0180	23" wide	G		1600	.005		.42	.23		.65	.84
0182	R21, 11" wide	G		1150	.007		.61	.33		.94	1.21
0184	15" wide	G		1350	.006		.61	.28		.89	1.13
0186	23" wide	G		1600	.005		.61	.23		.84	1.05
0188	9" thick, R30, 11" wide	G		985	.008		.71	.38		1.09	1.40
0200	15" wide	G		1150	.007		.71	.33		1.04	1.32
0220	23" wide	G		1350	.006		.71	.28		.99	1.24
0230	12" thick, R38, 11" wide	G		985	.008		.96	.38		1.34	1.68
0240	15" wide	G		1150	.007		.96	.33		1.29	1.60
0260	23" wide	G		1350	.006		.96	.28		1.24	1.52
0410	Foil faced fiberglass, 3-1/2" thick, R13, 11" wide	G		1150	.007		.45	.33		.78	1.04
0420	15" wide	G		1350	.006		.45	.28		.73	.96
0440	23" wide	G		1600	.005		.45	.23		.68	.88
0442	R15, 11" wide	G		1150	.007		.47	.33		.80	1.06
0444	15" wide	G		1350	.006		.47	.28		.75	.98
0446	23" wide	G		1600	.005		.47	.23		.70	.90
0448	6" thick, R19, 11" wide	G		1150	.007		.60	.33		.93	1.20
0460	15" wide	G		1350	.006		.60	.28		.88	1.12
0480	23" wide	G		1600	.005		.60	.23		.83	1.04
0482	R21, 11" wide	G		1150	.007		.62	.33		.95	1.22
0484	15" wide	G		1350	.006		.62	.28		.90	1.14
0486	23" wide	G		1600	.005		.62	.23		.85	1.06
0488	9" thick, R30, 11" wide	G		985	.008		.90	.38		1.28	1.61
0500	15" wide	G		1150	.007		.90	.33		1.23	1.53
0550	23" wide	G		1350	.006		.90	.28		1.18	1.45
0560	12" thick, R38, 11" wide	G		985	.008		1.05	.38		1.43	1.78
0570	15" wide	G		1150	.007		1.05	.33		1.38	1.70
0580	23" wide	G		1350	.006		1.05	.28		1.33	1.62

07 21 Thermal Insulation

07 21 16 – Blanket Insulation

07 21 16.20 Blanket Insulation for Walls		Crew	Daily Output	Labor-Hours	Unit	Material	2015 Bare Costs Labor	Equipment	Total	Total Incl O&P
0620	Unfaced fiberglass, 3-1/2" thick, R13, 11" wide	G	1 Carp	1150	.007	S.F.	.31	.33		.64	.88
0820	15" wide	G		1350	.006		.31	.28		.59	.80
0830	23" wide	G		1600	.005		.31	.23		.54	.72
0832	R15, 11" wide	G		1150	.007		.42	.33		.75	1
0836	23" wide	G		1600	.005		.42	.23		.65	.84
0838	6" thick, R19, 11" wide	G		1150	.007		.36	.33		.69	.94
0860	15" wide	G		1150	.007		.36	.33		.69	.94
0880	23" wide	G		1350	.006		.36	.28		.64	.86
0882	R21, 11" wide	G		1150	.007		.54	.33		.87	1.13
0886	15" wide	G		1350	.006		.54	.28		.82	1.05
0888	23" wide	G		1600	.005		.54	.23		.77	.97
0890	9" thick, R30, 11" wide	G		985	.008		.58	.38		.96	1.26
0900	15" wide	G		1150	.007		.58	.33		.91	1.18
0920	23" wide	G		1350	.006		.58	.28		.86	1.10
0930	12" thick, R38, 11" wide	G		985	.008		.74	.38		1.12	1.43
0940	15" wide	G		1000	.008		.74	.38		1.12	1.43
0960	23" wide	G		1150	.007		.74	.33		1.07	1.35
1300	Wall or ceiling insulation, mineral wool batts										
1320	3-1/2" thick, R15	G	1 Carp	1600	.005	S.F.	.60	.23		.83	1.04
1340	5-1/2" thick, R23	G		1600	.005		.94	.23		1.17	1.42
1380	7-1/4" thick, R30	G		1350	.006		1.24	.28		1.52	1.83
1700	Non-rigid insul, recycled blue cotton fiber, unfaced batts, R13, 16" wide	G		1600	.005		1.01	.23		1.24	1.49
1710	R19, 16" wide	G		1600	.005		1.38	.23		1.61	1.90
1850	Friction fit wire insulation supports, 16" O.C.			960	.008	Ea.	.07	.39		.46	.72
9000	Minimum labor/equipment charge			4	2	Job		94		94	154

07 21 19 – Foamed In Place Insulation

07 21 19.10 Masonry Foamed In Place Insulation		Crew	Daily Output	Labor-Hours	Unit	Material	Labor	Equipment	Total	Total Incl O&P
0010	**MASONRY FOAMED IN PLACE INSULATION**										
0100	Amino-plast foam, injected into block core, 6" block	G	G-2A	6000	.004	Ea.	.15	.14	.12	.41	.55
0110	8" block	G		5000	.005		.19	.17	.14	.50	.66
0120	10" block	G		4000	.006		.23	.22	.18	.63	.84
0130	12" block	G		3000	.008		.31	.29	.24	.84	1.10
0140	Injected into cavity wall	G		13000	.002	B.F.	.05	.07	.06	.18	.24
0150	Preparation, drill holes into mortar joint every 4 VLF, 5/8" dia		1 Clab	960	.008	Ea.		.31		.31	.51
0160	7/8" dia			680	.012			.44		.44	.73
0170	Patch drilled holes, 5/8" diameter			1800	.004		.03	.17		.20	.31
0180	7/8" diameter			1200	.007		.05	.25		.30	.46
9000	Minimum labor/equipment charge		G-2A	2	12	Job		430	360	790	1,150
9010	Minimum labor/equipment charge		1 Clab	4	2	"		75		75	123

07 21 23 – Loose-Fill Insulation

07 21 23.10 Poured Loose-Fill Insulation		Crew	Daily Output	Labor-Hours	Unit	Material	Labor	Equipment	Total	Total Incl O&P
0010	**POURED LOOSE-FILL INSULATION**										
0020	Cellulose fiber, R3.8 per inch	G	1 Carp	200	.040	C.F.	.69	1.88		2.57	3.84
0021	4" thick	G		1000	.008	S.F.	.17	.38		.55	.80
0022	6" thick	G		800	.010	"	.28	.47		.75	1.08
0080	Fiberglass wool, R4 per inch	G		200	.040	C.F.	.55	1.88		2.43	3.68
0081	4" thick	G		600	.013	S.F.	.19	.63		.82	1.24
0082	6" thick	G		400	.020	"	.26	.94		1.20	1.83
0100	Mineral wool, R3 per inch	G		200	.040	C.F.	.41	1.88		2.29	3.53
0101	4" thick	G		600	.013	S.F.	.14	.63		.77	1.18
0102	6" thick	G		400	.020	"	.21	.94		1.15	1.77
0300	Polystyrene, R4 per inch	G		200	.040	C.F.	1.60	1.88		3.48	4.84

07 21 Thermal Insulation

07 21 23 – Loose-Fill Insulation

07 21 23.10 Poured Loose-Fill Insulation

		Crew	Daily Output	Labor-Hours	Unit	Material	2015 Bare Costs Labor	Equipment	Total	Total Incl O&P
0301	4" thick	1 Carp [G]	600	.013	S.F.	.53	.63		1.16	1.61
0302	6" thick	[G]	400	.020	"	.80	.94		1.74	2.42
0400	Perlite, R2.78 per inch	[G]	200	.040	C.F.	5.20	1.88		7.08	8.85
0401	4" thick	[G]	1000	.008	S.F.	1.73	.38		2.11	2.53
0402	6" thick	[G]	800	.010	"	2.61	.47		3.08	3.64
9000	Minimum labor/equipment charge		4	2	Job		94		94	154

07 21 23.20 Masonry Loose-Fill Insulation

		Crew	Daily Output	Labor-Hours	Unit	Material	Labor	Equipment	Total	Total Incl O&P
0010	**MASONRY LOOSE-FILL INSULATION**, vermiculite or perlite									
0100	In cores of concrete block, 4" thick wall, .115 C.F./S.F.	D-1 [G]	4800	.003	S.F.	.60	.14		.74	.89
0200	6" thick wall, .175 C.F./S.F.	[G]	3000	.005		.91	.22		1.13	1.37
0300	8" thick wall, .258 C.F./S.F.	[G]	2400	.007		1.34	.28		1.62	1.94
0400	10" thick wall, .340 C.F./S.F.	[G]	1850	.009		1.77	.36		2.13	2.54
0500	12" thick wall, .422 C.F./S.F.	[G]	1200	.013		2.20	.56		2.76	3.33
0600	Poured cavity wall, vermiculite or perlite, water repellant	[G]	250	.064	C.F.	5.20	2.69		7.89	10.15
0700	Foamed in place, urethane in 2-5/8" cavity	G-2A [G]	1035	.023	S.F.	1.38	.83	.69	2.90	3.74
0800	For each 1" added thickness, add	[G] "	2372	.010	"	.53	.36	.30	1.19	1.55

07 21 26 – Blown Insulation

07 21 26.10 Blown Insulation

		Crew	Daily Output	Labor-Hours	Unit	Material	Labor	Equipment	Total	Total Incl O&P
0010	**BLOWN INSULATION** Ceilings, with open access									
0020	Cellulose, 3-1/2" thick, R13	G-4 [G]	5000	.005	S.F.	.24	.18	.08	.50	.65
0030	5-3/16" thick, R19	[G]	3800	.006		.35	.24	.11	.70	.91
0050	6-1/2" thick, R22	[G]	3000	.008		.45	.31	.13	.89	1.15
0100	8-11/16" thick, R30	[G]	2600	.009		.61	.35	.15	1.11	1.42
0120	10-7/8" thick, R38	[G]	1800	.013		.78	.51	.22	1.51	1.94
1000	Fiberglass, 5.5" thick, R11	[G]	3800	.006		.19	.24	.11	.54	.72
1050	6" thick, R12	[G]	3000	.008		.26	.31	.13	.70	.94
1100	8.8" thick, R19	[G]	2200	.011		.33	.42	.18	.93	1.24
1200	10" thick, R22	[G]	1800	.013		.38	.51	.22	1.11	1.51
1300	11.5" thick, R26	[G]	1500	.016		.46	.61	.27	1.34	1.79
1350	13" thick, R30	[G]	1400	.017		.53	.66	.29	1.48	1.98
1450	16" thick, R38	[G]	1145	.021		.68	.80	.35	1.83	2.44
1500	20" thick, R49	[G]	920	.026		.89	1	.44	2.33	3.10
9000	Minimum labor/equipment charge		4	6	Job		230	100	330	485

07 21 27 – Reflective Insulation

07 21 27.10 Reflective Insulation Options

		Crew	Daily Output	Labor-Hours	Unit	Material	Labor	Equipment	Total	Total Incl O&P
0010	**REFLECTIVE INSULATION OPTIONS**									
0020	Aluminum foil on reinforced scrim	1 Carp [G]	19	.421	C.S.F.	14.20	19.75		33.95	48
0100	Reinforced with woven polyolefin	[G]	19	.421		22	19.75		41.75	56.50
0500	With single bubble air space, R8.8	[G]	15	.533		28	25		53	72
0600	With double bubble air space, R9.8	[G]	15	.533		32	25		57	76
9000	Minimum labor/equipment charge		4	2	Job		94		94	154

07 21 29 – Sprayed Insulation

07 21 29.10 Sprayed-On Insulation

		Crew	Daily Output	Labor-Hours	Unit	Material	Labor	Equipment	Total	Total Incl O&P
0010	**SPRAYED-ON INSULATION**									
0020	Fibrous/cementitious, finished wall, 1" thick, R3.7	G-2 [G]	2050	.012	S.F.	.31	.46	.07	.84	1.16
0100	Attic, 5.2" thick, R19	[G]	1550	.015	"	.42	.61	.09	1.12	1.54
0200	Fiberglass, R4 per inch, vertical	[G]	1600	.015	B.F.	.18	.59	.08	.85	1.25
0210	Horizontal	[G]	1200	.020	"	.18	.79	.11	1.08	1.60
0300	Closed cell, spray polyurethane foam, 2 pounds per cubic foot density									
0310	1" thick	G-2A [G]	6000	.004	S.F.	.53	.14	.12	.79	.96
0320	2" thick	[G]	3000	.008		1.05	.29	.24	1.58	1.92

07 21 Thermal Insulation

07 21 29 – Sprayed Insulation

07 21 29.10 Sprayed-On Insulation		Crew	Daily Output	Labor-Hours	Unit	Material	2015 Bare Costs Labor	Equipment	Total	Total Incl O&P
0330	3" thick G	G-2A	2000	.012	S.F.	1.58	.43	.36	2.37	2.87
0335	3-1/2" thick G		1715	.014		1.84	.50	.42	2.76	3.36
0340	4" thick G		1500	.016		2.10	.57	.48	3.15	3.84
0350	5" thick G		1200	.020		2.63	.72	.60	3.95	4.81
0355	5-1/2" thick G		1090	.022		2.89	.79	.66	4.34	5.30
0360	6" thick G		1000	.024		3.15	.86	.72	4.73	5.75
9000	Minimum labor/equipment charge	G-2	2	12	Job		475	66.50	541.50	840

07 22 Roof and Deck Insulation

07 22 16 – Roof Board Insulation

07 22 16.10 Roof Deck Insulation

		Crew	Daily Output	Labor-Hours	Unit	Material	Labor	Equipment	Total	Total Incl O&P
0010	**ROOF DECK INSULATION**, fastening excluded									
0016	Asphaltic cover board, fiberglass lined, 1/8" thick	1 Rofc	1400	.006	S.F.	.47	.23		.70	.93
0018	1/4" thick		1400	.006		.94	.23		1.17	1.44
0020	Fiberboard low density, 1/2" thick R1.39 G		1300	.006		.30	.25		.55	.78
0030	1" thick R2.78 G		1040	.008		.52	.31		.83	1.13
0080	1-1/2" thick R4.17 G		1040	.008		.80	.31		1.11	1.44
0100	2" thick R5.56 G		1040	.008		1.06	.31		1.37	1.73
0110	Fiberboard high density, 1/2" thick R1.3 G		1300	.006		.30	.25		.55	.78
0120	1" thick R2.5 G		1040	.008		.58	.31		.89	1.20
0130	1-1/2" thick R3.8 G		1040	.008		.88	.31		1.19	1.53
0200	Fiberglass, 3/4" thick R2.78 G		1300	.006		.61	.25		.86	1.12
0400	15/16" thick R3.70 G		1300	.006		.81	.25		1.06	1.34
0460	1-1/16" thick R4.17 G		1300	.006		1.01	.25		1.26	1.56
0600	1-5/16" thick R5.26 G		1300	.006		1.37	.25		1.62	1.96
0650	2-1/16" thick R8.33 G		1040	.008		1.45	.31		1.76	2.16
0700	2-7/16" thick R10 G		1040	.008		1.68	.31		1.99	2.41
0800	Gypsum cover board, fiberglass mat facer, 1/4" thick		1400	.006		.47	.23		.70	.93
0810	1/2" thick		1300	.006		.57	.25		.82	1.08
0820	5/8" thick		1200	.007		.61	.27		.88	1.15
0830	Primed fiberglass mat facer, 1/4" thick		1400	.006		.51	.23		.74	.97
0840	1/2" thick		1300	.006		.62	.25		.87	1.13
0850	5/8" thick		1200	.007		.65	.27		.92	1.20
1650	Perlite, 1/2" thick R1.32 G		1365	.006		.28	.24		.52	.73
1655	3/4" thick R2.08 G		1040	.008		.34	.31		.65	.93
1660	1" thick R2.78 G		1040	.008		.50	.31		.81	1.11
1670	1-1/2" thick R4.17 G		1040	.008		.73	.31		1.04	1.36
1680	2" thick R5.56 G		910	.009		1	.35		1.35	1.74
1685	2-1/2" thick R6.67 G		910	.009		1.30	.35		1.65	2.07
1690	Tapered for drainage G		1040	.008	B.F.	1.01	.31		1.32	1.67
1700	Polyisocyanurate, 2#/C.F. density, 3/4" thick G		1950	.004	S.F.	.46	.16		.62	.81
1705	1" thick G		1820	.004		.48	.18		.66	.85
1715	1-1/2" thick G		1625	.005		.64	.20		.84	1.06
1725	2" thick G		1430	.006		.83	.22		1.05	1.32
1735	2-1/2" thick G		1365	.006		1.06	.24		1.30	1.59
1745	3" thick G		1300	.006		1.27	.25		1.52	1.85
1755	3-1/2" thick G		1300	.006		1.95	.25		2.20	2.60
1765	Tapered for drainage G		1820	.004	B.F.	.64	.18		.82	1.02
1900	Extruded Polystyrene									
1910	15 PSI compressive strength, 1" thick, R5 G	1 Rofc	1950	.004	S.F.	.55	.16		.71	.91
1920	2" thick, R10 G		1625	.005		.72	.20		.92	1.15

07 22 Roof and Deck Insulation

07 22 16 – Roof Board Insulation

	07 22 16.10 Roof Deck Insulation		Crew	Daily Output	Labor-Hours	Unit	Material	2015 Bare Costs Labor	Equipment	Total	Total Incl O&P
1930	3" thick, R15	G	1 Rofc	1300	.006	S.F.	1.43	.25		1.68	2.02
1932	4" thick, R20	G		1300	.006		1.93	.25		2.18	2.57
1934	Tapered for drainage	G		1950	.004	B.F.	.59	.16		.75	.95
1940	25 PSI compressive strength, 1" thick, R5	G		1950	.004	S.F.	.69	.16		.85	1.06
1942	2" thick, R10	G		1625	.005		1.31	.20		1.51	1.80
1944	3" thick, R15	G		1300	.006		2	.25		2.25	2.65
1946	4" thick, R20	G		1300	.006		2.76	.25		3.01	3.49
1948	Tapered for drainage	G		1950	.004	B.F.	.61	.16		.77	.97
1950	40 psi compressive strength, 1" thick, R5	G		1950	.004	S.F.	.53	.16		.69	.88
1952	2" thick, R10	G		1625	.005		1.01	.20		1.21	1.47
1954	3" thick, R15	G		1300	.006		1.46	.25		1.71	2.05
1956	4" thick, R20	G		1300	.006		1.91	.25		2.16	2.55
1958	Tapered for drainage	G		1820	.004	B.F.	.76	.18		.94	1.16
1960	60 PSI compressive strength, 1" thick, R5	G		1885	.004	S.F.	.74	.17		.91	1.12
1962	2" thick, R10	G		1560	.005		1.41	.21		1.62	1.92
1964	3" thick, R15	G		1270	.006		2.29	.25		2.54	2.98
1966	4" thick, R20	G		1235	.006		2.85	.26		3.11	3.60
1968	Tapered for drainage	G		1820	.004	B.F.	.96	.18		1.14	1.38
2010	Expanded polystyrene, 1#/C.F. density, 3/4" thick, R2.89	G		1950	.004	S.F.	.19	.16		.35	.51
2020	1" thick, R3.85	G		1950	.004		.25	.16		.41	.58
2100	2" thick, R7.69	G		1625	.005		.50	.20		.70	.91
2110	3" thick, R11.49	G		1625	.005		.75	.20		.95	1.19
2120	4" thick, R15.38	G		1625	.005		1	.20		1.20	1.46
2130	5" thick, R19.23	G		1495	.005		1.25	.21		1.46	1.77
2140	6" thick, R23.26	G		1495	.005		1.50	.21		1.71	2.04
2150	Tapered for drainage	G		1950	.004	B.F.	.53	.16		.69	.88
2400	Composites with 2" EPS										
2410	1" fiberboard	G	1 Rofc	1325	.006	S.F.	1.40	.24		1.64	1.98
2420	7/16" oriented strand board	G		1040	.008		1.13	.31		1.44	1.80
2430	1/2" plywood	G		1040	.008		1.39	.31		1.70	2.09
2440	1" perlite	G		1040	.008		1.14	.31		1.45	1.81
2450	Composites with 1-1/2" polyisocyanurate										
2460	1" fiberboard	G	1 Rofc	1040	.008	S.F.	1.19	.31		1.50	1.87
2470	1" perlite	G		1105	.007		1.06	.29		1.35	1.69
2480	7/16" oriented strand board	G		1040	.008		.93	.31		1.24	1.58
3000	Fastening alternatives, coated screws, 2" long			3744	.002	Ea.	.05	.09		.14	.22
3010	4" long			3120	.003		.10	.10		.20	.30
3020	6" long			2675	.003		.17	.12		.29	.41
3030	8" long			2340	.003		.25	.14		.39	.53
3040	10" long			1872	.004		.43	.17		.60	.78
3050	Pre-drill and drive wedge spike, 2-1/2"			1248	.006		.37	.26		.63	.87
3060	3-1/2"			1101	.007		.48	.29		.77	1.06
3070	4-1/2"			936	.009		.60	.34		.94	1.28
3075	3" galvanized deck plates			7488	.001		.07	.04		.11	.16
3080	Spot mop asphalt		G-1	295	.190	Sq.	5.55	7.10	1.93	14.58	21
3090	Full mop asphalt		"	192	.292		11.10	10.95	2.96	25.01	35
3110	Low-rise polyurethane adhesive, from 5 gallon kit, 12" OC beads		1 Rofc	45	.178		33	7.15		40.15	49
3120	6" OC beads			32	.250		65.50	10.05		75.55	90
3130	4" OC beads			30	.267		98.50	10.70		109.20	127
9000	Minimum labor/equipment charge			3.25	2.462	Job		98.50		98.50	178

07 24 Exterior Insulation and Finish Systems

07 24 13 – Polymer-Based Exterior Insulation and Finish System

07 24 13.10 Exterior Insulation and Finish Systems

		Crew	Daily Output	Labor-Hours	Unit	Material	2015 Bare Costs Labor	Equipment	Total	Total Incl O&P
0010	EXTERIOR INSULATION AND FINISH SYSTEMS									
0095	Field applied, 1" EPS insulation G	J-1	390	.103	S.F.	1.91	4.21	.36	6.48	9.25
0100	With 1/2" cement board sheathing G		268	.149		2.65	6.10	.52	9.27	13.30
0105	2" EPS insulation G		390	.103		2.16	4.21	.36	6.73	9.55
0110	With 1/2" cement board sheathing G		268	.149		2.90	6.10	.52	9.52	13.55
0115	3" EPS insulation G		390	.103		2.41	4.21	.36	6.98	9.80
0120	With 1/2" cement board sheathing G		268	.149		3.15	6.10	.52	9.77	13.85
0125	4" EPS insulation G		390	.103		2.66	4.21	.36	7.23	10.10
0130	With 1/2" cement board sheathing G		268	.149		4.14	6.10	.52	10.76	14.95
0140	Premium finish add		1265	.032		.33	1.30	.11	1.74	2.55
0150	Heavy duty reinforcement add		914	.044		.83	1.79	.15	2.77	3.95
0160	2.5#/S.Y. metal lath substrate add	1 Lath	75	.107	S.Y.	2.72	4.59		7.31	10.20
0170	3.4#/S.Y. metal lath substrate add	"	75	.107	"	4.13	4.59		8.72	11.75
0180	Color or texture change,	J-1	1265	.032	S.F.	.77	1.30	.11	2.18	3.04
0190	With substrate leveling base coat	1 Plas	530	.015		.77	.65		1.42	1.89
0210	With substrate sealing base coat	1 Pord	1224	.007		.10	.26		.36	.53
0370	V groove shape in panel face				L.F.	.62			.62	.68
0380	U groove shape in panel face				"	.80			.80	.88
0433	Crack repair, acrylic rubber, fluid applied, 20 mils thick	1 Plas	350	.023	S.F.	1.45	.98		2.43	3.17
0437	50 mils thick, reinforced	"	200	.040	"	2.65	1.72		4.37	5.65
0440	For higher than one story, add						25%			

07 25 Weather Barriers

07 25 10 – Weather Barriers or Wraps

07 25 10.10 Weather Barriers

		Crew	Daily Output	Labor-Hours	Unit	Material	Labor	Equipment	Total	Total Incl O&P
0010	WEATHER BARRIERS									
0400	Asphalt felt paper, 15#	1 Carp	37	.216	Sq.	5.40	10.15		15.55	22.50
0401	Per square foot	"	3700	.002	S.F.	.05	.10		.15	.23
0450	Housewrap, exterior, spun bonded polypropylene									
0470	Small roll	1 Carp	3800	.002	S.F.	.15	.10		.25	.32
0480	Large roll	"	4000	.002	"	.14	.09		.23	.30
2100	Asphalt felt roof deck vapor barrier, class 1 metal decks	1 Rofc	37	.216	Sq.	22	8.65		30.65	39.50
2200	For all other decks	"	37	.216		16.50	8.65		25.15	34
2800	Asphalt felt, 50% recycled content, 15 lb., 4 sq. per roll	1 Carp	36	.222		5.40	10.45		15.85	23
2810	30 lb., 2 sq. per roll	"	36	.222		10.80	10.45		21.25	29
3000	Building wrap, spunbonded polyethylene	2 Carp	8000	.002	S.F.	.15	.09		.24	.32
9960	Minimum labor/equipment charge	1 Carp	2	4	Job		188		188	310

07 26 Vapor Retarders

07 26 10 – Above-Grade Vapor Retarders

07 26 10.10 Vapor Retarders

		Crew	Daily Output	Labor-Hours	Unit	Material	Labor	Equipment	Total	Total Incl O&P
0010	VAPOR RETARDERS									
0020	Aluminum and kraft laminated, foil 1 side G	1 Carp	37	.216	Sq.	12.50	10.15		22.65	30.50
0100	Foil 2 sides G		37	.216		14	10.15		24.15	32
0600	Polyethylene vapor barrier, standard, 2 mil G		37	.216		1.50	10.15		11.65	18.30
0700	4 mil G		37	.216		2.90	10.15		13.05	19.85
0900	6 mil G		37	.216		4.02	10.15		14.17	21
1200	10 mil G		37	.216		8.85	10.15		19	26.50
1300	Clear reinforced, fire retardant, 8 mil G		37	.216		10.85	10.15		21	28.50

07 26 Vapor Retarders

07 26 10 – Above-Grade Vapor Retarders

07 26 10.10 Vapor Retarders

		Crew	Daily Output	Labor-Hours	Unit	Material	2015 Bare Costs Labor	Equipment	Total	Total Incl O&P
1350	Cross laminated type, 3 mil [G]	1 Carp	37	.216	Sq.	7.60	10.15		17.75	25
1400	4 mil [G]		37	.216		7.95	10.15		18.10	25.50
1800	Reinf. waterproof, 2 mil polyethylene backing, 1 side		37	.216		6.05	10.15		16.20	23.50
1900	2 sides		37	.216		7.95	10.15		18.10	25.50
2400	Waterproofed kraft with sisal or fiberglass fibers		37	.216		12.75	10.15		22.90	30.50
9950	Minimum labor/equipment charge		4	2	Job		94		94	154

07 27 Air Barriers

07 27 26 – Fluid-Applied Membrane Air Barriers

07 27 26.10 Fluid Applied Membrane Air Barrier

		Crew	Daily Output	Labor-Hours	Unit	Material	Labor	Equipment	Total	Total Incl O&P
0010	**FLUID APPLIED MEMBRANE AIR BARRIER**									
0100	Spray applied vapor barrier, 25 S.F./gallon	1 Pord	1375	.006	S.F.	.01	.23		.24	.40

07 31 Shingles and Shakes

07 31 13 – Asphalt Shingles

07 31 13.10 Asphalt Roof Shingles

		Crew	Daily Output	Labor-Hours	Unit	Material	Labor	Equipment	Total	Total Incl O&P
0010	**ASPHALT ROOF SHINGLES**									
0100	Standard strip shingles									
0150	Inorganic, class A, 25 year	1 Rofc	5.50	1.455	Sq.	79.50	58.50		138	193
0155	Pneumatic nailed		7	1.143		79.50	46		125.50	171
0200	30 year		5	1.600		94	64		158	219
0205	Pneumatic nailed		6.25	1.280		94	51.50		145.50	196
0250	Standard laminated multi-layered shingles									
0300	Class A, 240-260 lb./square	1 Rofc	4.50	1.778	Sq.	110	71.50		181.50	250
0305	Pneumatic nailed		5.63	1.422		110	57		167	224
0350	Class A, 250-270 lb./square		4	2		110	80		190	266
0355	Pneumatic nailed		5	1.600		110	64		174	237
0400	Premium, laminated multi-layered shingles									
0450	Class A, 260-300 lb./square	1 Rofc	3.50	2.286	Sq.	153	91.50		244.50	335
0455	Pneumatic nailed		4.37	1.831		153	73.50		226.50	300
0500	Class A, 300-385 lb./square		3	2.667		230	107		337	445
0505	Pneumatic nailed		3.75	2.133		230	85.50		315.50	410
0800	#15 felt underlayment		64	.125		5.40	5		10.40	15
0825	#30 felt underlayment		58	.138		10.60	5.55		16.15	21.50
0850	Self adhering polyethylene and rubberized asphalt underlayment		22	.364		75	14.60		89.60	109
0900	Ridge shingles		330	.024	L.F.	2.10	.97		3.07	4.07
0905	Pneumatic nailed		412.50	.019	"	2.10	.78		2.88	3.71
1000	For steep roofs (7 to 12 pitch or greater), add						50%			
9000	Minimum labor/equipment charge	1 Rofc	3	2.667	Job		107		107	193

07 31 16 – Metal Shingles

07 31 16.10 Aluminum Shingles

		Crew	Daily Output	Labor-Hours	Unit	Material	Labor	Equipment	Total	Total Incl O&P
0010	**ALUMINUM SHINGLES**									
0020	Mill finish, .019 thick	1 Carp	5	1.600	Sq.	221	75		296	365
0100	.020" thick	"	5	1.600		224	75		299	370
0300	For colors, add					21			21	23
0600	Ridge cap, .024" thick	1 Carp	170	.047	L.F.	3.63	2.21		5.84	7.60
0700	End wall flashing, .024" thick		170	.047		2.10	2.21		4.31	5.95
0900	Valley section, .024" thick		170	.047		3.57	2.21		5.78	7.55
1000	Starter strip, .024" thick		400	.020		1.67	.94		2.61	3.38

07 31 Shingles and Shakes

07 31 16 – Metal Shingles

07 31 16.10 Aluminum Shingles

		Crew	Daily Output	Labor-Hours	Unit	Material	2015 Bare Costs Labor	Equipment	Total	Total Incl O&P
1200	Side wall flashing, .024" thick	1 Carp	170	.047	L.F.	2.03	2.21		4.24	5.85
1500	Gable flashing, .024" thick		400	.020		1.63	.94		2.57	3.33
9000	Minimum labor/equipment charge		3	2.667	Job		125		125	205

07 31 16.20 Steel Shingles

		Crew	Daily Output	Labor-Hours	Unit	Material	Labor	Equipment	Total	Total Incl O&P
0010	**STEEL SHINGLES**									
0012	Galvanized, 26 ga.	1 Rots	2.20	3.636	Sq.	335	146		481	635
0200	24 ga.	"	2.20	3.636		335	146		481	635
0300	For colored galvanized shingles, add					56			56	61.50
9000	Minimum labor/equipment charge	1 Rots	3	2.667	Job		107		107	193

07 31 26 – Slate Shingles

07 31 26.10 Slate Roof Shingles

			Crew	Daily Output	Labor-Hours	Unit	Material	Labor	Equipment	Total	Total Incl O&P
0010	**SLATE ROOF SHINGLES**										
0100	Buckingham Virginia black, 3/16" - 1/4" thick	G	1 Rots	1.75	4.571	Sq.	470	183		653	845
0200	1/4" thick	G		1.75	4.571		470	183		653	845
0900	Pennsylvania black, Bangor, #1 clear	G		1.75	4.571		490	183		673	870
1200	Vermont, unfading, green, mottled green	G		1.75	4.571		475	183		658	855
1300	Semi-weathering green & gray	G		1.75	4.571		345	183		528	710
1400	Purple	G		1.75	4.571		425	183		608	795
1500	Black or gray	G		1.75	4.571		455	183		638	830
2500	Slate roof repair, extensive replacement			1	8		600	320		920	1,250
2600	Repair individual pieces, scattered			19	.421	Ea.	7	16.90		23.90	38
2700	Ridge shingles, slate			200	.040	L.F.	10	1.60		11.60	13.90
9000	Minimum labor/equipment charge			3	2.667	Job		107		107	193

07 31 29 – Wood Shingles and Shakes

07 31 29.13 Wood Shingles

			Crew	Daily Output	Labor-Hours	Unit	Material	Labor	Equipment	Total	Total Incl O&P
0010	**WOOD SHINGLES**	R061110-30									
0012	16" No. 1 red cedar shingles, 5" exposure, on roof		1 Carp	2.50	3.200	Sq.	287	150		437	560
0015	Pneumatic nailed			3.25	2.462		287	116		403	505
0200	7-1/2" exposure, on walls			2.05	3.902		191	183		374	510
0205	Pneumatic nailed			2.67	2.996		191	141		332	440
0300	18" No. 1 red cedar perfections, 5-1/2" exposure, on roof			2.75	2.909		246	137		383	495
0305	Pneumatic nailed			3.57	2.241		246	105		351	445
0500	7-1/2" exposure, on walls			2.25	3.556		181	167		348	475
0505	Pneumatic nailed			2.92	2.740		181	129		310	410
0600	Resquared, and rebutted, 5-1/2" exposure, on roof			3	2.667		279	125		404	510
0605	Pneumatic nailed			3.90	2.051		279	96.50		375.50	465
0900	7-1/2" exposure, on walls			2.45	3.265		205	153		358	475
0905	Pneumatic nailed			3.18	2.516		205	118		323	420
1000	Add to above for fire retardant shingles						56			56	61.50
1060	Preformed ridge shingles		1 Carp	400	.020	L.F.	5	.94		5.94	7.05
2000	White cedar shingles, 16" long, extras, 5" exposure, on roof			2.40	3.333	Sq.	192	157		349	470
2005	Pneumatic nailed			3.12	2.564		192	120		312	410
2050	5" exposure on walls			2	4		192	188		380	520
2055	Pneumatic nailed			2.60	3.077		192	144		336	450
2100	7-1/2" exposure, on walls			2	4		137	188		325	460
2105	Pneumatic nailed			2.60	3.077		137	144		281	390
2150	"B" grade, 5" exposure on walls			2	4		165	188		353	490
2155	Pneumatic nailed			2.60	3.077		165	144		309	420
2300	For 15# organic felt underlayment on roof, 1 layer, add			64	.125		5.40	5.85		11.25	15.55
2400	2 layers, add			32	.250		10.80	11.75		22.55	31
2600	For steep roofs (7/12 pitch or greater), add to above							50%			

07 31 Shingles and Shakes

07 31 29 – Wood Shingles and Shakes

07 31 29.13 Wood Shingles

		Crew	Daily Output	Labor-Hours	Unit	Material	2015 Bare Costs Labor	2015 Bare Costs Equipment	Total	Total Incl O&P
2700	Panelized systems, No.1 cedar shingles on 5/16" CDX plywood									
2800	On walls, 8' strips, 7" or 14" exposure	2 Carp	700	.023	S.F.	6	1.07		7.07	8.35
3500	On roofs, 8' strips, 7" or 14" exposure	1 Carp	3	2.667	Sq.	600	125		725	865
3505	Pneumatic nailed		4	2	"	600	94		694	815
9000	Minimum labor/equipment charge	↓	3	2.667	Job		125		125	205

07 31 29.16 Wood Shakes

		Crew	Daily Output	Labor-Hours	Unit	Material	Labor	Equipment	Total	Total Incl O&P
0010	**WOOD SHAKES**									
1100	Hand-split red cedar shakes, 1/2" thick x 24" long, 10" exp. on roof	1 Carp	2.50	3.200	Sq.	272	150		422	545
1105	Pneumatic nailed		3.25	2.462		272	116		388	490
1110	3/4" thick x 24" long, 10" exp. on roof		2.25	3.556		272	167		439	575
1115	Pneumatic nailed		2.92	2.740		272	129		401	510
1200	1/2" thick, 18" long, 8-1/2" exp. on roof		2	4		180	188		368	510
1205	Pneumatic nailed		2.60	3.077		180	144		324	435
1210	3/4" thick x 18" long, 8 1/2" exp. on roof		1.80	4.444		180	209		389	540
1215	Pneumatic nailed		2.34	3.419		180	161		341	460
1255	10" exp. on walls		2	4		174	188		362	500
1260	10" exposure on walls, pneumatic nailed	↓	2.60	3.077		174	144		318	430
1700	Add to above for fire retardant shakes, 24" long					56			56	61.50
1800	18" long				↓	56			56	61.50
1810	Ridge shakes	1 Carp	350	.023	L.F.	5	1.07		6.07	7.25

07 32 Roof Tiles

07 32 13 – Clay Roof Tiles

07 32 13.10 Clay Tiles

		Crew	Daily Output	Labor-Hours	Unit	Material	Labor	Equipment	Total	Total Incl O&P
0010	**CLAY TILES**, including accessories									
0300	Flat shingle, interlocking, 15", 166 pcs/sq, fireflashed blend	3 Rots	6	4	Sq.	505	160		665	845
0500	Terra cotta red		6	4		510	160		670	850
0600	Roman pan and top, 18", 102 pcs/sq, fireflashed blend	↓	5.50	4.364		555	175		730	925
0640	Terra cotta red	1 Rots	2.40	3.333		555	134		689	850
1100	Barrel mission tile, 18", 166 pcs/sq, fireflashed blend	3 Rots	5.50	4.364		410	175		585	765
1140	Terra cotta red		5.50	4.364		410	175		585	765
1700	Scalloped edge flat shingle, 14", 145 pcs/sq, fireflashed blend		6	4		1,125	160		1,285	1,550
1800	Terra cotta red	↓	6	4		1,100	160		1,260	1,525
3010	#15 felt underlayment	1 Rofc	64	.125		5.40	5		10.40	15
3020	#30 felt underlayment		58	.138		10.60	5.55		16.15	21.50
3040	Polyethylene and rubberized asph. underlayment	↓	22	.364	↓	75	14.60		89.60	109
9010	Minimum labor/equipment charge	1 Rots	2	4	Job		160		160	290

07 32 16 – Concrete Roof Tiles

07 32 16.10 Concrete Tiles

		Crew	Daily Output	Labor-Hours	Unit	Material	Labor	Equipment	Total	Total Incl O&P
0010	**CONCRETE TILES**									
0020	Corrugated, 13" x 16-1/2", 90 per sq., 950 lb. per sq.									
0050	Earthtone colors, nailed to wood deck	1 Rots	1.35	5.926	Sq.	106	238		344	545
0150	Blues		1.35	5.926		106	238		344	545
0200	Greens		1.35	5.926		106	238		344	545
0250	Premium colors	↓	1.35	5.926	↓	106	238		344	545
0500	Shakes, 13" x 16-1/2", 90 per sq., 950 lb. per sq.									
0600	All colors, nailed to wood deck	1 Rots	1.50	5.333	Sq.	108	214		322	505
1500	Accessory pieces, ridge & hip, 10" x 16-1/2", 8 lb. each	"	120	.067	Ea.	3.20	2.67		5.87	8.35
1700	Rake, 6-1/2" x 16-3/4", 9 lb. each					3.20			3.20	3.52
1800	Mansard hip, 10" x 16-1/2", 9.2 lb. each					3.20			3.20	3.52

07 32 Roof Tiles

07 32 16 – Concrete Roof Tiles

07 32 16.10 Concrete Tiles		Crew	Daily Output	Labor-Hours	Unit	Material	2015 Bare Costs Labor	2015 Bare Costs Equipment	Total	Total Incl O&P
1900	Hip starter, 10" x 16-1/2", 10.5 lb. each				Ea.	9.90			9.90	10.90
2000	3 or 4 way apex, 10" each side, 11.5 lb. each					11.40			11.40	12.55
9000	Minimum labor/equipment charge	1 Rots	3	2.667	Job		107		107	193

07 32 19 – Metal Roof Tiles

07 32 19.10 Metal Roof Tiles

		Crew	Daily Output	Labor-Hours	Unit	Material	Labor	Equipment	Total	Total Incl O&P
0010	**METAL ROOF TILES**									
0020	Accessories included, .032" thick aluminum, mission tile	1 Carp	2.50	3.200	Sq.	825	150		975	1,150
0200	Spanish tiles		3	2.667	"	570	125		695	830
9000	Minimum labor/equipment charge		3	2.667	Job		125		125	205

07 33 Natural Roof Coverings

07 33 63 – Vegetated Roofing

07 33 63.10 Green Roof Systems

			Crew	Daily Output	Labor-Hours	Unit	Material	Labor	Equipment	Total	Total Incl O&P
0010	**GREEN ROOF SYSTEMS**										
0020	Soil mixture for green roof 30% sand, 55% gravel, 15% soil										
0100	Hoist and spread soil mixture 4 inch depth up to five stories tall roof	G	B-13B	4000	.014	S.F.	.25	.57	.28	1.10	1.52
0150	6 inch depth	G		2667	.021		.38	.86	.42	1.66	2.29
0200	8 inch depth	G		2000	.028		.50	1.15	.56	2.21	3.03
0250	10 inch depth	G		1600	.035		.63	1.43	.70	2.76	3.80
0300	12 inch depth	G		1335	.042		.76	1.72	.84	3.32	4.55
0310	Alt. man-made soil mix, hoist & spread, 4" deep up to 5 stories tall roof	G		4000	.014		1.86	.57	.28	2.71	3.29
0350	Mobilization 55 ton crane to site	G	1 Eqhv	3.60	2.222	Ea.		115		115	182
0355	Hoisting cost to five stories per day (Avg. 28 picks per day)	G	B-13B	1	56	Day		2,300	1,125	3,425	4,975
0360	Mobilization or demobilization, 100 ton crane to site driver & escort	G	A-3E	2.50	6.400	Ea.		294	62.50	356.50	540
0365	Hoisting cost six to ten stories per day (Avg. 21 picks per day)	G	B-13C	1	56	Day		2,300	1,675	3,975	5,550
0370	Hoist and spread soil mixture 4 inch depth six to ten stories tall roof	G		4000	.014	S.F.	.25	.57	.42	1.24	1.67
0375	6 inch depth	G		2667	.021		.38	.86	.63	1.87	2.51
0380	8 inch depth	G		2000	.028		.50	1.15	.83	2.48	3.33
0385	10 inch depth	G		1600	.035		.63	1.43	1.04	3.10	4.17
0390	12 inch depth	G		1335	.042		.76	1.72	1.25	3.73	4.99
0400	Green roof edging treated lumber 4" x 4" no hoisting included	G	2 Carp	400	.040	L.F.	1.34	1.88		3.22	4.55
0410	4" x 6"	G		400	.040		1.91	1.88		3.79	5.20
0420	4" x 8"	G		360	.044		3.81	2.09		5.90	7.60
0430	4" x 6" double stacked	G		300	.053		3.83	2.50		6.33	8.30
0500	Green roof edging redwood lumber 4" x 4" no hoisting included	G		400	.040		6.45	1.88		8.33	10.15
0510	4" x 6"	G		400	.040		12.60	1.88		14.48	16.95
0520	4" x 8"	G		360	.044		23.50	2.09		25.59	29.50
0530	4" x 6" double stacked	G		300	.053		25	2.50		27.50	31.50
0550	Components, not including membrane or insulation:										
0560	Fluid applied rubber membrane, reinforced, 215 mil thick	G	G-5	350	.114	S.F.	.28	4.17	.55	5	8.45
0570	Root barrier	G	2 Rofc	775	.021		.60	.83		1.43	2.16
0580	Moisture retention barrier and reservoir	G	"	900	.018		2.45	.71		3.16	3.99
0600	Planting sedum, light soil, potted, 2-1/4" diameter, two per S.F.	G	1 Clab	420	.019		4.90	.72		5.62	6.55
0610	one per S.F.	G	"	840	.010		2.45	.36		2.81	3.29
0630	Planting sedum mat per S.F. including shipping (4000 S.F. min)	G	4 Clab	4000	.008		5.90	.30		6.20	6.95
0640	Installation sedum mat system (no soil required) per S.F. (4000 S.F. min)	G	"	4000	.008		8.20	.30		8.50	9.50
0645	Note: pricing of sedum mats shipped in full truck loads (4000-5000 S.F.)										

07 41 Roof Panels

07 41 13 – Metal Roof Panels

07 41 13.10 Aluminum Roof Panels

		Crew	Daily Output	Labor-Hours	Unit	Material	2015 Bare Costs Labor	Equipment	Total	Total Incl O&P
0010	**ALUMINUM ROOF PANELS**									
0020	Corrugated or ribbed, .0155" thick, natural	G-3	1200	.027	S.F.	.99	1.25		2.24	3.09
0300	Painted		1200	.027		1.44	1.25		2.69	3.58
0400	Corrugated, .018" thick, on steel frame, natural finish		1200	.027		1.30	1.25		2.55	3.43
0600	Painted		1200	.027		1.60	1.25		2.85	3.76
0700	Corrugated, on steel frame, natural, .024" thick		1200	.027		1.85	1.25		3.10	4.04
0800	Painted		1200	.027		2.25	1.25		3.50	4.48
0900	.032" thick, natural		1200	.027		2.39	1.25		3.64	4.63
1200	Painted		1200	.027		3.06	1.25		4.31	5.35
9000	Minimum labor/equipment charge	1 Rofc	3	2.667	Job		107		107	193

07 41 13.20 Steel Roofing Panels

		Crew	Daily Output	Labor-Hours	Unit	Material	2015 Bare Costs Labor	Equipment	Total	Total Incl O&P
0010	**STEEL ROOFING PANELS**									
0012	Corrugated or ribbed, on steel framing, 30 ga. galv	G-3	1100	.029	S.F.	1.60	1.36		2.96	3.94
0100	28 ga.		1050	.030		1.65	1.43		3.08	4.11
0300	26 ga.		1000	.032		1.68	1.50		3.18	4.25
0400	24 ga.		950	.034		2.90	1.58		4.48	5.70
0600	Colored, 28 ga.		1050	.030		1.67	1.43		3.10	4.13
0700	26 ga.		1000	.032		1.98	1.50		3.48	4.58
0710	Flat profile, 1-3/4" standing seams, 10" wide, standard finish, 26 ga.		1000	.032		3.75	1.50		5.25	6.55
0715	24 ga.		950	.034		4.35	1.58		5.93	7.30
0720	22 ga.		900	.036		5.40	1.66		7.06	8.60
0725	Zinc aluminum alloy finish, 26 ga.		1000	.032		3	1.50		4.50	5.70
0730	24 ga.		950	.034		3.55	1.58		5.13	6.45
0735	22 ga.		900	.036		4.05	1.66		5.71	7.15
0740	12" wide, standard finish, 26 ga.		1000	.032		3.75	1.50		5.25	6.55
0745	24 ga.		950	.034		4.90	1.58		6.48	7.95
0750	Zinc aluminum alloy finish, 26 ga.		1000	.032		4.26	1.50		5.76	7.10
0755	24 ga.		950	.034		3.50	1.58		5.08	6.40
0840	Flat profile, 1" x 3/8" batten, 12" wide, standard finish, 26 ga.		1000	.032		3.30	1.50		4.80	6.05
0845	24 ga.		950	.034		3.90	1.58		5.48	6.80
0850	22 ga.		900	.036		4.65	1.66		6.31	7.75
0855	Zinc aluminum alloy finish, 26 ga.		1000	.032		3.20	1.50		4.70	5.90
0860	24 ga.		950	.034		3.60	1.58		5.18	6.50
0865	22 ga.		900	.036		4.15	1.66		5.81	7.25
0870	16-1/2" wide, standard finish, 24 ga.		950	.034		3.85	1.58		5.43	6.75
0875	22 ga.		900	.036		4.30	1.66		5.96	7.40
0880	Zinc aluminum alloy finish, 24 ga.		950	.034		3.35	1.58		4.93	6.20
0885	22 ga.		900	.036		3.75	1.66		5.41	6.80
0890	Flat profile, 2" x 2" batten, 12" wide, standard finish, 26 ga.		1000	.032		3.85	1.50		5.35	6.65
0895	24 ga.		950	.034		4.55	1.58		6.13	7.55
0900	22 ga.		900	.036		5.55	1.66		7.21	8.75
0905	Zinc aluminum alloy finish, 26 ga.		1000	.032		3.55	1.50		5.05	6.30
0910	24 ga.		950	.034		4.05	1.58		5.63	7
0915	22 ga.		900	.036		4.70	1.66		6.36	7.80
0920	16-1/2" wide, standard finish, 24 ga.		950	.034		4.20	1.58		5.78	7.15
0925	22 ga.		900	.036		4.90	1.66		6.56	8.05
0930	Zinc aluminum alloy finish, 24 ga.		950	.034		3.80	1.58		5.38	6.70
0935	22 ga.		900	.036		4.30	1.66		5.96	7.40
9000	Minimum labor/equipment charge	1 Rofc	2	4	Job		160		160	290

07 41 Roof Panels

07 41 33 – Plastic Roof Panels

07 41 33.10 Fiberglass Panels

		Crew	Daily Output	Labor-Hours	Unit	Material	2015 Bare Costs Labor	Equipment	Total	Total Incl O&P
0010	**FIBERGLASS PANELS**									
0012	Corrugated panels, roofing, 8 oz. per S.F.	G-3	1000	.032	S.F.	2.10	1.50		3.60	4.71
0100	12 oz. per S.F.		1000	.032		4.18	1.50		5.68	7
0300	Corrugated siding, 6 oz. per S.F.		880	.036		2.10	1.70		3.80	5.05
0400	8 oz. per S.F.		880	.036		2.10	1.70		3.80	5.05
0500	Fire retardant		880	.036		3.80	1.70		5.50	6.90
0600	12 oz. siding, textured		880	.036		3.75	1.70		5.45	6.85
0700	Fire retardant		880	.036		4.35	1.70		6.05	7.50
0900	Flat panels, 6 oz. per S.F., clear or colors		880	.036		2.40	1.70		4.10	5.35
1100	Fire retardant, class A		880	.036		3.45	1.70		5.15	6.55
1300	8 oz. per S.F., clear or colors		880	.036		2.52	1.70		4.22	5.50
9000	Minimum labor/equipment charge	1 Rofc	2	4	Job		160		160	290

07 42 Wall Panels

07 42 13 – Metal Wall Panels

07 42 13.10 Mansard Panels

		Crew	Daily Output	Labor-Hours	Unit	Material	2015 Bare Costs Labor	Equipment	Total	Total Incl O&P
0010	**MANSARD PANELS**									
0600	Aluminum, stock units, straight surfaces	1 Shee	115	.070	S.F.	3.90	3.89		7.79	10.45
0700	Concave or convex surfaces		75	.107	"	2.06	5.95		8.01	11.65
0800	For framing, to 5' high, add		115	.070	L.F.	3.60	3.89		7.49	10.10
0900	Soffits, to 1' wide		125	.064	S.F.	2.20	3.58		5.78	8.05
9000	Minimum labor/equipment charge		2.50	3.200	Job		179		179	283

07 42 13.20 Aluminum Siding Panels

		Crew	Daily Output	Labor-Hours	Unit	Material	2015 Bare Costs Labor	Equipment	Total	Total Incl O&P
0010	**ALUMINUM SIDING PANELS**									
0012	Corrugated, on steel framing, .019 thick, natural finish	G-3	775	.041	S.F.	1.52	1.93		3.45	4.77
0100	Painted		775	.041		1.66	1.93		3.59	4.93
0400	Farm type, .021" thick on steel frame, natural		775	.041		1.55	1.93		3.48	4.81
0600	Painted		775	.041		1.65	1.93		3.58	4.92
0700	Industrial type, corrugated, on steel, .024" thick, mill		775	.041		2.15	1.93		4.08	5.45
0900	Painted		775	.041		2.31	1.93		4.24	5.65
1000	.032" thick, mill		775	.041		2.45	1.93		4.38	5.80
1200	Painted		775	.041		3	1.93		4.93	6.40
1300	V-Beam, on steel frame, .032" thick, mill		775	.041		2.80	1.93		4.73	6.20
1500	Painted		775	.041		3.20	1.93		5.13	6.60
1600	.040" thick, mill		775	.041		3.38	1.93		5.31	6.80
1800	Painted		775	.041		3.94	1.93		5.87	7.45
3800	Horizontal, colored clapboard, 8" wide, plain	2 Carp	515	.031		2.45	1.46		3.91	5.10
3810	Insulated		515	.031		2.67	1.46		4.13	5.35
3830	8" embossed, painted		515	.031		2.65	1.46		4.11	5.30
3840	Insulated		515	.031		2.84	1.46		4.30	5.50
3860	12" painted, smooth		600	.027		2.50	1.25		3.75	4.80
3870	Insulated		600	.027		2.67	1.25		3.92	4.99
3890	12" embossed, painted		600	.027		2.65	1.25		3.90	4.97
3900	Insulated		515	.031		2.55	1.46		4.01	5.20
4000	Vertical board & batten, colored, non-insulated		515	.031		1.90	1.46		3.36	4.48
4200	For simulated wood design, add					.12			.12	.13
4300	Corners for above, outside	2 Carp	515	.031	V.L.F.	3.36	1.46		4.82	6.10
4500	Inside corners	"	515	.031	"	1.61	1.46		3.07	4.16
9000	Minimum labor/equipment charge	1 Carp	3	2.667	Job		125		125	205

07 42 Wall Panels

07 42 13 – Metal Wall Panels

07 42 13.30 Steel Siding

		Daily Output	Labor-Hours	Unit	Material	2015 Bare Costs Labor	2015 Bare Costs Equipment	Total	Total Incl O&P
		Crew							
0010	**STEEL SIDING**								
0020	Beveled, vinyl coated, 8" wide	1 Carp	265	.030	S.F.	1.85	1.42		3.27	4.36
0050	10" wide	"	275	.029		1.88	1.37		3.25	4.31
0080	Galv, corrugated or ribbed, on steel frame, 30 ga.	G-3	800	.040		1.20	1.87		3.07	4.32
0100	28 ga.		795	.040		1.28	1.88		3.16	4.43
0300	26 ga.		790	.041		1.80	1.90		3.70	5
0400	24 ga.		785	.041		1.85	1.91		3.76	5.10
0600	22 ga.		770	.042		2.15	1.94		4.09	5.50
0700	Colored, corrugated/ribbed, on steel frame, 10 yr. finish, 28 ga.		800	.040		1.95	1.87		3.82	5.15
0900	26 ga.		795	.040		2.09	1.88		3.97	5.30
1000	24 ga.		790	.041		2.37	1.90		4.27	5.65
1020	20 ga.		785	.041		3	1.91		4.91	6.35
9000	Minimum labor/equipment charge	1 Carp	3	2.667	Job		125		125	205

07 46 Siding

07 46 23 – Wood Siding

07 46 23.10 Wood Board Siding

		Crew	Daily Output	Labor-Hours	Unit	Material	Labor	Equipment	Total	Total Incl O&P
0010	**WOOD BOARD SIDING**									
3200	Wood, cedar bevel, A grade, 1/2" x 6"	1 Carp	295	.027	S.F.	4.14	1.27		5.41	6.65
3300	1/2" x 8"		330	.024		5.90	1.14		7.04	8.35
3500	3/4" x 10", clear grade		375	.021		6.30	1		7.30	8.55
3600	"B" grade		375	.021		3.79	1		4.79	5.80
3800	Cedar, rough sawn, 1" x 4", A grade, natural		220	.036		6.55	1.71		8.26	10.05
3900	Stained		220	.036		6.70	1.71		8.41	10.15
4100	1" x 12", board & batten, #3 & Btr., natural		420	.019		4.48	.89		5.37	6.40
4200	Stained		420	.019		4.81	.89		5.70	6.75
4400	1" x 8" channel siding, #3 & Btr., natural		330	.024		4.57	1.14		5.71	6.85
4500	Stained		330	.024		4.92	1.14		6.06	7.25
4700	Redwood, clear, beveled, vertical grain, 1/2" x 4"		220	.036		4.11	1.71		5.82	7.30
4750	1/2" x 6"		295	.027		4.49	1.27		5.76	7.05
4800	1/2" x 8"		330	.024		4.89	1.14		6.03	7.25
5000	3/4" x 10"		375	.021		4.50	1		5.50	6.60
5200	Channel siding, 1" x 10", B grade		375	.021		4.12	1		5.12	6.15
5250	Redwood, T&G boards, B grade, 1" x 4"		220	.036		2.54	1.71		4.25	5.60
5270	1" x 8"		330	.024		4.10	1.14		5.24	6.40
5400	White pine, rough sawn, 1" x 8", natural		330	.024		2.19	1.14		3.33	4.28
5500	Stained		330	.024		2.19	1.14		3.33	4.28
9000	Minimum labor/equipment charge		2	4	Job		188		188	310

07 46 29 – Plywood Siding

07 46 29.10 Plywood Siding Options

		Crew	Daily Output	Labor-Hours	Unit	Material	Labor	Equipment	Total	Total Incl O&P
0010	**PLYWOOD SIDING OPTIONS**									
0900	Plywood, medium density overlaid, 3/8" thick	2 Carp	750	.021	S.F.	1.25	1		2.25	3.02
1000	1/2" thick		700	.023		1.44	1.07		2.51	3.34
1100	3/4" thick		650	.025		1.85	1.16		3.01	3.93
1600	Texture 1-11, cedar, 5/8" thick, natural		675	.024		2.57	1.11		3.68	4.65
1700	Factory stained		675	.024		2.80	1.11		3.91	4.90
1900	Texture 1-11, fir, 5/8" thick, natural		675	.024		1.57	1.11		2.68	3.55
2000	Factory stained		675	.024		1.80	1.11		2.91	3.80
2050	Texture 1-11, S.Y.P., 5/8" thick, natural		675	.024		1.37	1.11		2.48	3.33
2100	Factory stained		675	.024		1.44	1.11		2.55	3.40

07 46 Siding

07 46 29 – Plywood Siding

07 46 29.10 Plywood Siding Options		Crew	Daily Output	Labor-Hours	Unit	Material	2015 Bare Costs Labor	Equipment	Total	Total Incl O&P
2200	Rough sawn cedar, 3/8" thick, natural	2 Carp	675	.024	S.F.	1.25	1.11		2.36	3.20
2300	Factory stained		675	.024		1.50	1.11		2.61	3.47
2500	Rough sawn fir, 3/8" thick, natural		675	.024		.84	1.11		1.95	2.74
2600	Factory stained		675	.024		1.04	1.11		2.15	2.96
2800	Redwood, textured siding, 5/8" thick		675	.024		1.92	1.11		3.03	3.93
3000	Polyvinyl chloride coated, 3/8" thick		750	.021		1.11	1		2.11	2.86
9000	Minimum labor/equipment charge	1 Carp	2	4	Job		188		188	310

07 46 33 – Plastic Siding

07 46 33.10 Vinyl Siding

		Crew	Daily Output	Labor-Hours	Unit	Material	Labor	Equipment	Total	Total Incl O&P
0010	**VINYL SIDING**									
3995	Clapboard profile, woodgrain texture, .048 thick, double 4	2 Carp	495	.032	S.F.	1.04	1.52		2.56	3.64
4000	Double 5		550	.029		1.04	1.37		2.41	3.39
4005	Single 8		495	.032		1.26	1.52		2.78	3.88
4010	Single 10		550	.029		1.51	1.37		2.88	3.91
4015	.044 thick, double 4		495	.032		.99	1.52		2.51	3.58
4020	Double 5		550	.029		.99	1.37		2.36	3.33
4025	.042 thick, double 4		495	.032		.92	1.52		2.44	3.51
4030	Double 5		550	.029		.92	1.37		2.29	3.26
4035	Cross sawn texture, .040 thick, double 4		495	.032		.67	1.52		2.19	3.23
4040	Double 5		550	.029		.67	1.37		2.04	2.98
4045	Smooth texture, .042 thick, double 4		495	.032		.75	1.52		2.27	3.32
4050	Double 5		550	.029		.75	1.37		2.12	3.07
4055	Single 8		495	.032		.75	1.52		2.27	3.32
4060	Cedar texture, .044 thick, double 4		495	.032		1.01	1.52		2.53	3.61
4065	Double 6		600	.027		1.01	1.25		2.26	3.17
4070	Dutch lap profile, woodgrain texture, .048 thick, double 5		550	.029		1.04	1.37		2.41	3.39
4075	.044 thick, double 4.5		525	.030		1.01	1.43		2.44	3.47
4080	.042 thick, double 4.5		525	.030		.84	1.43		2.27	3.28
4085	.040 thick, double 4.5		525	.030		.67	1.43		2.10	3.09
4100	Shake profile, 10" wide		400	.040		3.45	1.88		5.33	6.90
4105	Vertical pattern, .046 thick, double 5		550	.029		1.42	1.37		2.79	3.81
4110	.044 thick, triple 3		550	.029		1.56	1.37		2.93	3.96
4115	.040 thick, triple 4		550	.029		1.50	1.37		2.87	3.90
4120	.040 thick, triple 2.66		550	.029		1.81	1.37		3.18	4.24
4125	Insulation, fan folded extruded polystyrene, 1/4"		2000	.008		.28	.38		.66	.93
4130	3/8"		2000	.008		.31	.38		.69	.96
4135	Accessories, J channel, 5/8" pocket		700	.023	L.F.	.46	1.07		1.53	2.26
4140	3/4" pocket		695	.023		.50	1.08		1.58	2.32
4145	1-1/4" pocket		680	.024		.77	1.10		1.87	2.65
4150	Flexible, 3/4" pocket		600	.027		2.33	1.25		3.58	4.61
4155	Under sill finish trim		500	.032		.50	1.50		2	3.01
4160	Vinyl starter strip		700	.023		.58	1.07		1.65	2.40
4165	Aluminum starter strip		700	.023		.28	1.07		1.35	2.07
4170	Window casing, 2-1/2" wide, 3/4" pocket		510	.031		1.57	1.47		3.04	4.13
4175	Outside corner, woodgrain finish, 4" face, 3/4" pocket		700	.023		1.92	1.07		2.99	3.88
4180	5/8" pocket		700	.023		1.92	1.07		2.99	3.88
4185	Smooth finish, 4" face, 3/4" pocket		700	.023		1.92	1.07		2.99	3.88
4190	7/8" pocket		690	.023		1.89	1.09		2.98	3.86
4195	1-1/4" pocket		700	.023		1.33	1.07		2.40	3.23
4200	Soffit and fascia, 1' overhang, solid		120	.133		4.29	6.25		10.54	14.95
4205	Vented		120	.133		4.29	6.25		10.54	14.95
4207	18" overhang, solid		110	.145		5	6.85		11.85	16.70

07 46 Siding

07 46 33 – Plastic Siding

07 46 33.10 Vinyl Siding

		Crew	Daily Output	Labor-Hours	Unit	Material	2015 Bare Costs Labor	2015 Bare Costs Equipment	Total	Total Incl O&P
4208	Vented	2 Carp	110	.145	L.F.	5	6.85		11.85	16.70
4210	2' overhang, solid		100	.160		5.70	7.50		13.20	18.60
4215	Vented		100	.160		5.70	7.50		13.20	18.60
4217	3' overhang, solid		100	.160		7.10	7.50		14.60	20
4218	Vented		100	.160		7.10	7.50		14.60	20
4220	Colors for siding and soffits, add				S.F.	.14			.14	.15
4225	Colors for accessories and trim, add				L.F.	.30			.30	.33
9000	Minimum labor/equipment charge	1 Carp	3	2.667	Job		125		125	205

07 46 33.20 Polypropylene Siding

		Crew	Daily Output	Labor-Hours	Unit	Material	Labor	Equipment	Total	Total Incl O&P
0010	**POLYPROPYLENE SIDING**									
4090	Shingle profile, random grooves, double 7	2 Carp	400	.040	S.F.	2.99	1.88		4.87	6.35
4092	Cornerpost for above	1 Carp	365	.022	L.F.	12	1.03		13.03	14.90
4095	Triple 5	2 Carp	400	.040	S.F.	2.99	1.88		4.87	6.35
4097	Cornerpost for above	1 Carp	365	.022	L.F.	11.05	1.03		12.08	13.85
5000	Staggered butt, double 7"	2 Carp	400	.040	S.F.	3.45	1.88		5.33	6.90
5002	Cornerpost for above	1 Carp	365	.022	L.F.	12	1.03		13.03	14.90
5010	Half round, double 6-1/4"	2 Carp	360	.044	S.F.	3.40	2.09		5.49	7.15
5020	Shake profile, staggered butt, double 9"	"	510	.031	"	3.45	1.47		4.92	6.20
5022	Cornerpost for above	1 Carp	365	.022	L.F.	9	1.03		10.03	11.60
5030	Straight butt, double 7"	2 Carp	400	.040	S.F.	3.40	1.88		5.28	6.85
5032	Cornerpost for above	1 Carp	365	.022	L.F.	11.90	1.03		12.93	14.75
6000	Accessories, J channel, 5/8" pocket	2 Carp	700	.023		.46	1.07		1.53	2.26
6010	3/4" pocket		695	.023		.50	1.08		1.58	2.32
6020	1-1/4" pocket		680	.024		.77	1.10		1.87	2.65
6030	Aluminum starter strip		700	.023		.28	1.07		1.35	2.07

07 46 46 – Fiber Cement Siding

07 46 46.10 Fiber Cement Siding

		Crew	Daily Output	Labor-Hours	Unit	Material	Labor	Equipment	Total	Total Incl O&P
0010	**FIBER CEMENT SIDING**									
0020	Lap siding, 5/16" thick, 6" wide, 4-3/4" exposure, smooth texture	2 Carp	415	.039	S.F.	1.18	1.81		2.99	4.27
0025	Woodgrain texture		415	.039		1.18	1.81		2.99	4.27
0030	7-1/2" wide, 6-1/4" exposure, smooth texture		425	.038		1.46	1.77		3.23	4.51
0035	Woodgrain texture		425	.038		1.46	1.77		3.23	4.51
0040	8" wide, 6-3/4" exposure, smooth texture		425	.038		1.42	1.77		3.19	4.46
0045	Roughsawn texture		425	.038		1.42	1.77		3.19	4.46
0050	9-1/2" wide, 8-1/4" exposure, smooth texture		440	.036		1.22	1.71		2.93	4.14
0055	Woodgrain texture		440	.036		1.22	1.71		2.93	4.14
0060	12" wide, 10-3/8" exposure, smooth texture		455	.035		2.01	1.65		3.66	4.92
0065	Woodgrain texture		455	.035		2.01	1.65		3.66	4.92
0070	Panel siding, 5/16" thick, smooth texture		750	.021		1.25	1		2.25	3.02
0075	Stucco texture		750	.021		1.25	1		2.25	3.02
0080	Grooved woodgrain texture		750	.021		1.25	1		2.25	3.02
0085	V - grooved woodgrain texture		750	.021		1.25	1		2.25	3.02
0088	Shingle siding, 48" x 15-1/4" panels, 7" exposure		700	.023		3.94	1.07		5.01	6.10
0090	Wood starter strip		400	.040	L.F.	.42	1.88		2.30	3.54

07 46 73 – Soffit

07 46 73.10 Soffit Options

		Crew	Daily Output	Labor-Hours	Unit	Material	Labor	Equipment	Total	Total Incl O&P
0010	**SOFFIT OPTIONS**									
0012	Aluminum, residential, .020" thick	1 Carp	210	.038	S.F.	2	1.79		3.79	5.15
0100	Baked enamel on steel, 16 or 18 ga.		105	.076		5.90	3.58		9.48	12.35
0300	Polyvinyl chloride, white, solid		230	.035		2.10	1.63		3.73	4.99
0400	Perforated		230	.035		2.10	1.63		3.73	4.99

07 46 Siding

07 46 73 – Soffit

07 46 73.10 Soffit Options	Crew	Daily Output	Labor-Hours	Unit	Material	2015 Bare Costs Labor	2015 Bare Costs Equipment	Total	Total Incl O&P
0500 For colors, add				S.F.	.15			.15	.17
9000 Minimum labor/equipment charge	1 Carp	3	2.667	Job		125		125	205

07 51 Built-Up Bituminous Roofing

07 51 13 – Built-Up Asphalt Roofing

07 51 13.10 Built-Up Roofing Components

	Crew	Daily Output	Labor-Hours	Unit	Material	Labor	Equipment	Total	Total Incl O&P
0010 **BUILT-UP ROOFING COMPONENTS**									
0012 Asphalt saturated felt, #30, 2 square per roll	1 Rofc	58	.138	Sq.	10.60	5.55		16.15	21.50
0200 #15, 4 sq. per roll, plain or perforated, not mopped		58	.138		5.40	5.55		10.95	15.95
0300 Roll roofing, smooth, #65		15	.533		9.65	21.50		31.15	49
0500 #90		12	.667		37	26.50		63.50	89
0520 Mineralized		12	.667		35	26.50		61.50	87
0540 D.C. (Double coverage), 19" selvage edge		10	.800		52.50	32		84.50	116
0580 Adhesive (lap cement)				Gal.	8.85			8.85	9.75
0800 Steep, flat or dead level asphalt, 10 ton lots, packaged				Ton	925			925	1,025
9000 Minimum labor/equipment charge	1 Rofc	4	2	Job		80		80	145

07 51 13.13 Cold-Applied Built-Up Asphalt Roofing

	Crew	Daily Output	Labor-Hours	Unit	Material	Labor	Equipment	Total	Total Incl O&P
0010 **COLD-APPLIED BUILT-UP ASPHALT ROOFING**									
0020 3 ply system, installation only (components listed below)	G-5	50	.800	Sq.		29	3.86	32.86	57
0100 Spunbond poly. fabric, 1.35 oz./S.Y., 36"W, 10.8 sq./roll				Ea.	135			135	149
0500 Base & finish coat, 3 gal./sq., 5 gal./can				Gal.	7.75			7.75	8.55
0600 Coating, ceramic granules, 1/2 sq./bag				Ea.	21.50			21.50	23.50
0700 Aluminum, 2 gal./sq.				Gal.	13.50			13.50	14.85
0800 Emulsion, fibered or non-fibered, 4 gal./sq.				"	6.50			6.50	7.15

07 51 13.20 Built-Up Roofing Systems

	Crew	Daily Output	Labor-Hours	Unit	Material	Labor	Equipment	Total	Total Incl O&P
0010 **BUILT-UP ROOFING SYSTEMS**									
0120 Asphalt flood coat with gravel/slag surfacing, not including									
0140 Insulation, flashing or wood nailers									
0200 Asphalt base sheet, 3 plies #15 asphalt felt, mopped	G-1	22	2.545	Sq.	100	95.50	26	221.50	310
0350 On nailable decks		21	2.667		102	100	27	229	325
0500 4 plies #15 asphalt felt, mopped		20	2.800		136	105	28.50	269.50	370
0550 On nailable decks		19	2.947		120	111	30	261	365
0700 Coated glass base sheet, 2 plies glass (type IV), mopped		22	2.545		107	95.50	26	228.50	320
0850 3 plies glass, mopped		20	2.800		130	105	28.50	263.50	365
0950 On nailable decks		19	2.947		122	111	30	263	365
1100 4 plies glass fiber felt (type IV), mopped		20	2.800		160	105	28.50	293.50	400
1150 On nailable decks		19	2.947		145	111	30	286	390
1200 Coated & saturated base sheet, 3 plies #15 asph. felt, mopped		20	2.800		112	105	28.50	245.50	345
1250 On nailable decks		19	2.947		104	111	30	245	345
1300 4 plies #15 asphalt felt, mopped		22	2.545		130	95.50	26	251.50	345
2000 Asphalt flood coat, smooth surface									
2200 Asphalt base sheet & 3 plies #15 asphalt felt, mopped	G-1	24	2.333	Sq.	105	87.50	23.50	216	299
2400 On nailable decks		23	2.435		96.50	91.50	24.50	212.50	298
2600 4 plies #15 asphalt felt, mopped		24	2.333		123	87.50	23.50	234	320
2700 On nailable decks		23	2.435		115	91.50	24.50	231	320
2900 Coated glass fiber base sheet, mopped, and 2 plies of									
2910 glass fiber felt (type IV)	G-1	25	2.240	Sq.	102	84	23	209	290
3100 On nailable decks		24	2.333		96.50	87.50	23.50	207.50	290
3200 3 plies, mopped		23	2.435		125	91.50	24.50	241	330
3300 On nailable decks		22	2.545		117	95.50	26	238.50	330

07 51 Built-Up Bituminous Roofing

07 51 13 – Built-Up Asphalt Roofing

07 51 13.20 Built-Up Roofing Systems

		Crew	Daily Output	Labor-Hours	Unit	Material	2015 Bare Costs Labor	Equipment	Total	Total Incl O&P
3800	4 plies glass fiber felt (type IV), mopped	G-1	23	2.435	Sq.	147	91.50	24.50	263	355
3900	On nailable decks		22	2.545		140	95.50	26	261.50	355
4000	Coated & saturated base sheet, 3 plies #15 asph. felt, mopped		24	2.333		107	87.50	23.50	218	300
4200	On nailable decks		23	2.435		99	91.50	24.50	215	300
4300	4 plies #15 organic felt, mopped		22	2.545		125	95.50	26	246.50	340
4500	Coal tar pitch with gravel/slag surfacing									
4600	4 plies #15 tarred felt, mopped	G-1	21	2.667	Sq.	233	100	27	360	470
4800	3 plies glass fiber felt (type IV), mopped	"	19	2.947	"	193	111	30	334	445
5000	Coated glass fiber base sheet, and 2 plies of									
5010	glass fiber felt, (type IV), mopped	G-1	19	2.947	Sq.	196	111	30	337	450
5300	On nailable decks		18	3.111		170	117	31.50	318.50	435
5600	4 plies glass fiber felt (type IV), mopped		21	2.667		264	100	27	391	500
5800	On nailable decks		20	2.800		237	105	28.50	370.50	485

07 51 13.30 Cants

		Crew	Daily Output	Labor-Hours	Unit	Material	Labor	Equipment	Total	Total Incl O&P
0010	**CANTS**									
0012	Lumber, treated, 4" x 4" cut diagonally	1 Rofc	325	.025	L.F.	1.75	.99		2.74	3.71
0300	Mineral or fiber, trapezoidal, 1" x 4" x 48"		325	.025		.25	.99		1.24	2.06
0400	1-1/2" x 5-5/8" x 48"		325	.025		.44	.99		1.43	2.26
9000	Minimum labor/equipment charge		4	2	Job		80		80	145

07 51 13.40 Felts

		Crew	Daily Output	Labor-Hours	Unit	Material	Labor	Equipment	Total	Total Incl O&P
0010	**FELTS**									
0012	Glass fibered roofing felt, #15, not mopped	1 Rofc	58	.138	Sq.	9.65	5.55		15.20	20.50
0300	Base sheet, #80, channel vented		58	.138		42.50	5.55		48.05	57
0400	#70, coated		58	.138		17.20	5.55		22.75	29
0500	Cap, #87, mineral surfaced		58	.138		80	5.55		85.55	98
0600	Flashing membrane, #65		16	.500		9.65	20		29.65	46.50
0800	Coal tar fibered, #15, no mopping		58	.138		18.10	5.55		23.65	30
0900	Asphalt felt, #15, 4 sq. per roll, no mopping		58	.138		5.40	5.55		10.95	15.95
1100	#30, 2 sq. per roll		58	.138		10.60	5.55		16.15	21.50
1200	Double coated, #33		58	.138		11	5.55		16.55	22
1400	#40, base sheet		58	.138		9.65	5.55		15.20	20.50
1450	Coated and saturated		58	.138		12	5.55		17.55	23
1500	Tarred felt, organic, #15, 4 sq. rolls		58	.138		13.95	5.55		19.50	25.50
1550	#30, 2 sq. roll		58	.138		28	5.55		33.55	40.50
1700	Add for mopping above felts, per ply, asphalt, 24 lb. per sq.	G-1	192	.292		11.10	10.95	2.96	25.01	35
1800	Coal tar mopping, 30 lb. per sq.		186	.301		23	11.30	3.06	37.36	49
1900	Flood coat, with asphalt, 60 lb. per sq.		60	.933		27.50	35	9.50	72	104
2000	With coal tar, 75 lb. per sq.		56	1		57	37.50	10.15	104.65	142
9000	Minimum labor/equipment charge	1 Rofc	4	2	Job		80		80	145

07 51 13.50 Walkways for Built-Up Roofs

		Crew	Daily Output	Labor-Hours	Unit	Material	Labor	Equipment	Total	Total Incl O&P
0010	**WALKWAYS FOR BUILT-UP ROOFS**									
0020	Asphalt impregnated, 3' x 6' x 1/2" thick	1 Rofc	400	.020	S.F.	1.56	.80		2.36	3.16
0100	3' x 3' x 3/4" thick		400	.020	"	4.48	.80		5.28	6.40
0600	100% recycled rubber, 3' x 4' x 3/8" G		400	.020	L.F.	6.35	.80		7.15	8.40
0610	3' x 4' x 1/2" G		400	.020		6.80	.80		7.60	8.90
0620	3' x 4' x 3/4" G		400	.020		8.45	.80		9.25	10.75
9000	Minimum labor/equipment charge		2.75	2.909	Job		117		117	211

07 52 Modified Bituminous Membrane Roofing

07 52 13 – Atactic-Polypropylene-Modified Bituminous Membrane Roofing

07 52 13.10 APP Modified Bituminous Membrane

		Crew	Daily Output	Labor-Hours	Unit	Material	2015 Bare Costs Labor	2015 Bare Costs Equipment	Total	Total Incl O&P
0010	APP MODIFIED BITUMINOUS MEMBRANE R075213-30									
0020	Base sheet, #15 glass fiber felt, nailed to deck	1 Rofc	58	.138	Sq.	10.65	5.55		16.20	21.50
0030	Spot mopped to deck	G-1	295	.190		15.20	7.10	1.93	24.23	31.50
0040	Fully mopped to deck	"	192	.292		20.50	10.95	2.96	34.41	46
0050	#15 organic felt, nailed to deck	1 Rofc	58	.138		6.40	5.55		11.95	17.05
0060	Spot mopped to deck	G-1	295	.190		10.95	7.10	1.93	19.98	27
0070	Fully mopped to deck	"	192	.292		16.50	10.95	2.96	30.41	41
2100	APP mod., smooth surf. cap sheet, poly. reinf., torched, 160 mils	G-5	2100	.019	S.F.	.68	.69	.09	1.46	2.11
2150	170 mils		2100	.019		.70	.69	.09	1.48	2.13
2200	Granule surface cap sheet, poly. reinf., torched, 180 mils		2000	.020		.89	.73	.10	1.72	2.41
2250	Smooth surface flashing, torched, 160 mils		1260	.032		.68	1.16	.15	1.99	3.01
2300	170 mils		1260	.032		.70	1.16	.15	2.01	3.03
2350	Granule surface flashing, torched, 180 mils		1260	.032		.89	1.16	.15	2.20	3.24
2400	Fibrated aluminum coating	1 Rofc	3800	.002		.08	.08		.16	.24
2450	Seam heat welding	"	205	.039	L.F.	.08	1.56		1.64	2.92

07 52 16 – Styrene-Butadiene-Styrene Modified Bituminous Membrane Roofing

07 52 16.10 SBS Modified Bituminous Membrane

		Crew	Daily Output	Labor-Hours	Unit	Material	Labor	Equipment	Total	Total Incl O&P
0010	SBS MODIFIED BITUMINOUS MEMBRANE									
0080	Mod bit rfng, SBS mod, gran surf cap sheet, poly reinf									
1150	For reflective granules, add				S.F.	.71	1.02	.28	2.01	2.93
1600	Smooth surface cap sheet, mopped, 145 mils	G-1	2100	.027		.78	1	.27	2.05	2.97
1620	Lightweight base sheet, fiberglass reinforced, 35 to 47 mil		2100	.027		.27	1	.27	1.54	2.41
1625	Heavyweight base/ply sheet, reinforced, 87 to 120 mil thick		2100	.027		.89	1	.27	2.16	3.09
1650	Granulated walkpad, 180 – 220 mils	1 Rofc	400	.020		1.82	.80		2.62	3.45
1700	Smooth surface flashing, 145 mils	G-1	1260	.044		.78	1.67	.45	2.90	4.37
1800	150 mils		1260	.044		.49	1.67	.45	2.61	4.05
1900	Granular surface flashing, 150 mils		1260	.044		.65	1.67	.45	2.77	4.23
2000	160 mils		1260	.044		.71	1.67	.45	2.83	4.29
2010	Elastomeric asphalt primer	1 Rofc	2600	.003		.17	.12		.29	.41
2015	Roofing asphalt, 30 lb. per square	G-1	19000	.003		.14	.11	.03	.28	.38
2020	Cold process adhesive, 20 – 30 mils thick	1 Rofc	750	.011		.24	.43		.67	1.03
2025	Self adhering vapor retarder, 30 to 45 mils thick	G-5	2150	.019		1.02	.68	.09	1.79	2.45
2050	Seam heat welding	1 Rofc	205	.039	L.F.	.08	1.56		1.64	2.92

07 53 Elastomeric Membrane Roofing

07 53 16 – Chlorosulfonate-Polyethylene Roofing

07 53 16.10 Chlorosulfonated Polyethylene Roofing

		Crew	Daily Output	Labor-Hours	Unit	Material	Labor	Equipment	Total	Total Incl O&P
0010	CHLOROSULFONATED POLYETHYLENE ROOFING									
0800	Chlorosulfonated polyethylene (CSPE)									
0900	45 mils, heat welded seams, plate attachment	G-5	35	1.143	Sq.	310	41.50	5.50	357	420
1100	Heat welded seams, plate attachment and ballasted		26	1.538		320	56	7.45	383.45	465
1200	60 mils, heat welded seams, plate attachment		35	1.143		410	41.50	5.50	457	530
1300	Heat welded seams, plate attachment and ballasted		26	1.538		420	56	7.45	483.45	570

07 53 23 – Ethylene-Propylene-Diene-Monomer Roofing

07 53 23.20 Ethylene-Propylene-Diene-Monomer Roofing

		Crew	Daily Output	Labor-Hours	Unit	Material	Labor	Equipment	Total	Total Incl O&P
0010	ETHYLENE-PROPYLENE-DIENE-MONOMER ROOFING (EPDM)									
3500	Ethylene-propylene-diene-monomer (EPDM), 45 mils, 0.28 psf									
3600	Loose-laid & ballasted with stone (10 psf)	G-5	51	.784	Sq.	85	28.50	3.79	117.29	149
3700	Mechanically attached		35	1.143		79	41.50	5.50	126	168
3800	Fully adhered with adhesive		26	1.538		111	56	7.45	174.45	231

07 53 Elastomeric Membrane Roofing

07 53 23 – Ethylene-Propylene-Diene-Monomer Roofing

07 53 23.20 Ethylene-Propylene-Diene-Monomer Roofing	Crew	Daily Output	Labor-Hours	Unit	Material	2015 Bare Costs Labor	Equipment	Total	Total Incl O&P
4500	60 mils, 0.40 psf									
4600	Loose-laid & ballasted with stone (10 psf)	G-5	51	.784	Sq.	100	28.50	3.79	132.29	166
4700	Mechanically attached		35	1.143		93	41.50	5.50	140	184
4800	Fully adhered with adhesive	↓	26	1.538		125	56	7.45	188.45	247
4810	45 mil, .28 psf, membrane only					50.50			50.50	55.50
4820	60 mil, .40 psf, membrane only				↓	63			63	69.50
4850	Seam tape for membrane, 3" x 100' roll				Ea.	40			40	43.50
4900	Batten strips, 10' sections					3.43			3.43	3.77
4910	Cover tape for batten strips, 6" x 100' roll				↓	183			183	202
4930	Plate anchors				M	78			78	86
4970	Adhesive for fully adhered systems, 60 S.F./gal.				Gal.	19.65			19.65	21.50

07 53 29 – Polyisobutylene Roofing

07 53 29.10 Polyisobutylene Roofing	Crew	Daily Output	Labor-Hours	Unit	Material	2015 Bare Costs Labor	Equipment	Total	Total Incl O&P
0010	POLYISOBUTYLENE ROOFING									
7500	Polyisobutylene (PIB), 100 mils, 0.57 psf									
7600	Loose-laid & ballasted with stone/gravel (10 psf)	G-5	51	.784	Sq.	200	28.50	3.79	232.29	276
7700	Partially adhered with adhesive		35	1.143		240	41.50	5.50	287	345
7800	Hot asphalt attachment		35	1.143		230	41.50	5.50	277	335
7900	Fully adhered with contact cement	↓	26	1.538	↓	250	56	7.45	313.45	385

07 54 Thermoplastic Membrane Roofing

07 54 19 – Polyvinyl-Chloride Roofing

07 54 19.10 Polyvinyl-Chloride Roofing (PVC)	Crew	Daily Output	Labor-Hours	Unit	Material	2015 Bare Costs Labor	Equipment	Total	Total Incl O&P
0010	POLYVINYL-CHLORIDE ROOFING (PVC)									
8200	Heat welded seams									
8700	Reinforced, 48 mils, 0.33 psf									
8750	Loose-laid & ballasted with stone/gravel (12 psf)	G-5	51	.784	Sq.	118	28.50	3.79	150.29	186
8800	Mechanically attached		35	1.143		111	41.50	5.50	158	204
8850	Fully adhered with adhesive	↓	26	1.538	↓	153	56	7.45	216.45	277
8860	Reinforced, 60 mils, .40 psf									
8870	Loose-laid & ballasted with stone/gravel (12 psf)	G-5	51	.784	Sq.	120	28.50	3.79	152.29	188
8880	Mechanically attached		35	1.143		112	41.50	5.50	159	206
8890	Fully adhered with adhesive	↓	26	1.538		154	56	7.45	217.45	279

07 54 23 – Thermoplastic-Polyolefin Roofing

07 54 23.10 Thermoplastic Polyolefin Roofing (T.P.O)	Crew	Daily Output	Labor-Hours	Unit	Material	2015 Bare Costs Labor	Equipment	Total	Total Incl O&P
0010	THERMOPLASTIC POLYOLEFIN ROOFING (T.P.O.)									
0100	45 mil, loose laid & ballasted with stone (1/2 ton/sq.)	G-5	51	.784	Sq.	87.50	28.50	3.79	119.79	152
0120	Fully adhered		25	1.600		78	58.50	7.75	144.25	199
0140	Mechanically attached		34	1.176		77	43	5.70	125.70	169
0160	Self adhered		35	1.143		87	41.50	5.50	134	178
0180	60 mil membrane, heat welded seams, ballasted		50	.800		100	29	3.86	132.86	167
0200	Fully adhered		25	1.600		90	58.50	7.75	156.25	213
0220	Mechanically attached		34	1.176		93.50	43	5.70	142.20	187
0240	Self adhered	↓	35	1.143	↓	107	41.50	5.50	154	200

07 54 30 – Ketone Ethylene Ester Roofing

07 54 30.10 Ketone Ethylene Ester Roofing	Crew	Daily Output	Labor-Hours	Unit	Material	2015 Bare Costs Labor	Equipment	Total	Total Incl O&P
0010	KETONE ETHYLENE ESTER ROOFING									
0100	Ketone ethylene ester roofing, 50 mil, fully adhered	G-5	26	1.538	Sq.	215	56	7.45	278.45	345
0120	Mechanically attached	↓	35	1.143		141	41.50	5.50	188	237

07 54 Thermoplastic Membrane Roofing

07 54 30 – Ketone Ethylene Ester Roofing

07 54 30.10 Ketone Ethylene Ester Roofing		Crew	Daily Output	Labor-Hours	Unit	Material	2015 Bare Costs Labor	Equipment	Total	Total Incl O&P
0140	Ballasted with stone	G-5	51	.784	Sq.	149	28.50	3.79	181.29	219
0160	50 mil, fleece backed, adhered w/hot asphalt	G-1	26	2.154		174	81	22	277	360
0180	Accessories, pipe boot	1 Rofc	32	.250	Ea.	24.50	10.05		34.55	44.50
0200	Pre-formed corners		32	.250	"	8.95	10.05		19	28
0220	Ketone clad metal, including up to 4 bends		330	.024	S.F.	3.95	.97		4.92	6.10
0240	Walkway pad	2 Rofc	800	.020	"	4.27	.80		5.07	6.15
0260	Stripping material	1 Rofc	310	.026	L.F.	.95	1.04		1.99	2.92

07 55 Protected Membrane Roofing

07 55 10 – Protected Membrane Roofing Components

07 55 10.10 Protected Membrane Roofing Components		Crew	Daily Output	Labor-Hours	Unit	Material	Labor	Equipment	Total	Total Incl O&P
0010	**PROTECTED MEMBRANE ROOFING COMPONENTS**									
0100	Choose roofing membrane from 07 50									
0120	Then choose roof deck insulation from 07 22									
0130	Filter fabric	2 Rofc	10000	.002	S.F.	.12	.06		.18	.26
0140	Ballast, 3/8" - 1/2" in place	G-1	36	1.556	Ton	20	58.50	15.80	94.30	144
0150	3/4" - 1-1/2" in place	"	36	1.556	"	20	58.50	15.80	94.30	144
0200	2" concrete blocks, natural	1 Clab	115	.070	S.F.	3.37	2.62		5.99	8
0210	Colors	"	115	.070	"	3.73	2.62		6.35	8.40

07 56 Fluid-Applied Roofing

07 56 10 – Fluid-Applied Roofing Elastomers

07 56 10.10 Elastomeric Roofing		Crew	Daily Output	Labor-Hours	Unit	Material	Labor	Equipment	Total	Total Incl O&P
0010	**ELASTOMERIC ROOFING**									
0020	Acrylic, 44% solids, 2 coats, on corrugated metal	2 Rofc	2400	.007	S.F.	.58	.27		.85	1.11
0025	On smooth metal		3000	.005		.46	.21		.67	.90
0030	On foam or modified bitumen		1500	.011		.92	.43		1.35	1.78
0035	On concrete		1500	.011		.92	.43		1.35	1.78
0040	On tar and gravel		1500	.011		.92	.43		1.35	1.78
0045	36% solids, 2 coats, on corrugated metal		2400	.007		.56	.27		.83	1.10
0050	On smooth metal		3000	.005		.45	.21		.66	.88
0055	On foam or modified bitumen		1500	.011		.90	.43		1.33	1.76
0060	On concrete		1500	.011		.90	.43		1.33	1.76
0065	On tar and gravel		1500	.011		.90	.43		1.33	1.76
0070	Primer if required; 2 coats on corrugated metal		2400	.007		.52	.27		.79	1.05
0075	On smooth metal		3000	.005		.52	.21		.73	.96
0080	On foam or modified bitumen		1500	.011		.75	.43		1.18	1.60
0085	On concrete		1500	.011		.56	.43		.99	1.39
0090	On tar & gravel/rolled roof		1500	.011		.75	.43		1.18	1.60
0110	Acrylic rubber, fluid applied, 20 mils thick	G-5	2000	.020		2	.73	.10	2.83	3.63
0120	50 mils, reinforced		1200	.033		3	1.22	.16	4.38	5.70
0130	For walking surface, add		900	.044		1.05	1.62	.21	2.88	4.33
0300	Neoprene, fluid applied, 20 mil thick, not-reinforced	G-1	1135	.049		1.32	1.85	.50	3.67	5.35
0600	Non-woven polyester, reinforced		960	.058		1.45	2.19	.59	4.23	6.20
0700	5 coat neoprene deck, 60 mil thick, under 10,000 S.F.		325	.172		4.40	6.45	1.75	12.60	18.45
0900	Over 10,000 S.F.		625	.090		4.40	3.36	.91	8.67	11.90
9000	Minimum labor/equipment charge	1 Rofc	2	4	Job		160		160	290

07 57 Coated Foamed Roofing

07 57 13 – Sprayed Polyurethane Foam Roofing

07 57 13.10 Sprayed Polyurethane Foam Roofing (S.P.F.)	Crew	Daily Output	Labor-Hours	Unit	Material	2015 Bare Costs Labor	Equipment	Total	Total Incl O&P
0010 **SPRAYED POLYURETHANE FOAM ROOFING (S.P.F.)**									
0100 Primer for metal substrate (when required)	G-2A	3000	.008	S.F.	.44	.29	.24	.97	1.24
0200 Primer for non-metal substrate (when required)		3000	.008		.17	.29	.24	.70	.95
0300 Closed cell spray, polyurethane foam, 3 lb. per C.F. density, 1", R6.7		15000	.002		.66	.06	.05	.77	.88
0400 2", R13.4		13125	.002		1.32	.07	.05	1.44	1.62
0500 3", R18.6		11485	.002		1.98	.08	.06	2.12	2.38
0550 4", R24.8		10080	.002		2.64	.09	.07	2.80	3.14
0700 Spray-on silicone coating		2500	.010		1.13	.34	.29	1.76	2.16
0800 Warranty 5-20 year manufacturer's								.15	.15
0900 Warranty 20 year, no dollar limit								.20	.20

07 58 Roll Roofing

07 58 10 – Asphalt Roll Roofing

07 58 10.10 Roll Roofing

	Crew	Daily Output	Labor-Hours	Unit	Material	Labor	Equipment	Total	Total Incl O&P
0010 **ROLL ROOFING**									
0100 Asphalt, mineral surface									
0200 1 ply #15 organic felt, 1 ply mineral surfaced									
0300 Selvage roofing, lap 19", nailed & mopped	G-1	27	2.074	Sq.	74.50	78	21	173.50	246
0400 3 plies glass fiber felt (type IV), 1 ply mineral surfaced									
0500 Selvage roofing, lapped 19", mopped	G-1	25	2.240	Sq.	126	84	23	233	315
0600 Coated glass fiber base sheet, 2 plies of glass fiber									
0700 Felt (type IV), 1 ply mineral surfaced selvage									
0800 Roofing, lapped 19", mopped	G-1	25	2.240	Sq.	133	84	23	240	325
0900 On nailable decks	"	24	2.333	"	122	87.50	23.50	233	320
1000 3 plies glass fiber felt (type III), 1 ply mineral surfaced									
1100 Selvage roofing, lapped 19", mopped	G-1	25	2.240	Sq.	126	84	23	233	315

07 61 Sheet Metal Roofing

07 61 13 – Standing Seam Sheet Metal Roofing

07 61 13.10 Standing Seam Sheet Metal Roofing, Field Fab.

	Crew	Daily Output	Labor-Hours	Unit	Material	Labor	Equipment	Total	Total Incl O&P
0010 **STANDING SEAM SHEET METAL ROOFING, FIELD FABRICATED**									
0400 Copper standing seam roofing, over 10 squares, 16 oz., 125 lb. per sq.	1 Shee	1.30	6.154	Sq.	960	345		1,305	1,600
0600 18 oz., 140 lb. per sq.		1.20	6.667		1,075	375		1,450	1,775
0700 20 oz., 150 lb. per sq.		1.10	7.273		1,175	405		1,580	1,950
1200 For abnormal conditions or small areas, add					25%	100%			
1300 For lead-coated copper, add					25%				
1400 Lead standing seam roofing, 5 lb. per S.F.	1 Shee	1.30	6.154		1,300	345		1,645	1,975
1500 Zinc standing seam roofing, .020" thick		1.20	6.667		1,175	375		1,550	1,900
1510 .027" thick		1.15	6.957		1,475	390		1,865	2,250
1520 .032" thick		1.10	7.273		1,900	405		2,305	2,725
1530 .040" thick		1.05	7.619		2,375	425		2,800	3,275
9000 Minimum labor/equipment charge		2	4	Job		224		224	355

07 61 16 – Batten Seam Sheet Metal Roofing

07 61 16.10 Batten Seam Sheet Metal Roofing, Field Fabricated

	Crew	Daily Output	Labor-Hours	Unit	Material	Labor	Equipment	Total	Total Incl O&P
0010 **BATTEN SEAM SHEET METAL ROOFING, FIELD FABRICATED**									
0012 Copper batten seam roofing, over 10 sq., 16 oz., 130 lb. per sq.	1 Shee	1.10	7.273	Sq.	1,225	405		1,630	1,975
0020 Lead batten seam roofing, 5 lb. per S.F.		1.20	6.667		1,350	375		1,725	2,100
0100 Zinc/copper alloy batten seam roofing, .020 thick		1.20	6.667		1,225	375		1,600	1,950
0200 Copper roofing, batten seam, over 10 sq., 18 oz., 145 lb. per sq.		1	8		1,350	450		1,800	2,200

07 61 Sheet Metal Roofing

07 61 16 – Batten Seam Sheet Metal Roofing

07 61 16.10 Batten Seam Sheet Metal Roofing, Field Fabricated

		Crew	Daily Output	Labor-Hours	Unit	Material	2015 Bare Costs Labor	2015 Bare Costs Equipment	Total	Total Incl O&P
0300	20 oz., 160 lb. per sq.	1 Shee	1	8	Sq.	1,500	450		1,950	2,325
0500	Stainless steel batten seam roofing, type 304, 28 ga.		1.20	6.667		585	375		960	1,225
0600	26 ga.		1.15	6.957		545	390		935	1,225
0800	Zinc, copper alloy roofing, batten seam, .027" thick		1.15	6.957		1,550	390		1,940	2,325
0900	.032" thick		1.10	7.273		1,975	405		2,380	2,825
1000	.040" thick		1.05	7.619		2,475	425		2,900	3,400
9000	Minimum labor/equipment charge		2	4	Job		224		224	355

07 61 19 – Flat Seam Sheet Metal Roofing

07 61 19.10 Flat Seam Sheet Metal Roofing, Field Fabricated

		Crew	Daily Output	Labor-Hours	Unit	Material	Labor	Equipment	Total	Total Incl O&P
0010	**FLAT SEAM SHEET METAL ROOFING, FIELD FABRICATED**									
0900	Copper flat seam roofing, over 10 squares, 16 oz., 115 lb./sq.	1 Shee	1.20	6.667	Sq.	890	375		1,265	1,575
0950	18 oz., 130 lb./sq.		1.15	6.957		995	390		1,385	1,725
1000	20 oz., 145 lb./sq.		1.10	7.273		1,100	405		1,505	1,850
1008	Zinc flat seam roofing, .020" thick		1.20	6.667		1,050	375		1,425	1,750
1010	.027" thick		1.15	6.957		1,325	390		1,715	2,075
1020	.032" thick		1.12	7.143		1,675	400		2,075	2,475
1030	.040" thick		1.05	7.619		2,100	425		2,525	3,000
1100	Lead flat seam roofing, 5 lb. per S.F.		1.30	6.154		1,150	345		1,495	1,825
9000	Minimum labor/equipment charge		2.75	2.909	Job		163		163	257

07 62 Sheet Metal Flashing and Trim

07 62 10 – Sheet Metal Trim

07 62 10.10 Sheet Metal Cladding

		Crew	Daily Output	Labor-Hours	Unit	Material	Labor	Equipment	Total	Total Incl O&P
0010	**SHEET METAL CLADDING**									
0100	Aluminum, up to 6 bends, .032" thick, window casing	1 Carp	180	.044	S.F.	1.40	2.09		3.49	4.96
0200	Window sill		72	.111	L.F.	1.40	5.20		6.60	10.10
0300	Door casing		180	.044	S.F.	1.40	2.09		3.49	4.96
0400	Fascia		250	.032		1.40	1.50		2.90	4
0500	Rake trim		225	.036		1.40	1.67		3.07	4.28
0700	.024" thick, window casing		180	.044		1.22	2.09		3.31	4.76
0800	Window sill		72	.111	L.F.	1.22	5.20		6.42	9.90
0900	Door casing		180	.044	S.F.	1.22	2.09		3.31	4.76
1000	Fascia		250	.032		1.22	1.50		2.72	3.80
1100	Rake trim		225	.036		1.22	1.67		2.89	4.08
1200	Vinyl coated aluminum, up to 6 bends, window casing		180	.044		1.35	2.09		3.44	4.91
1300	Window sill		72	.111	L.F.	1.35	5.20		6.55	10.05
1400	Door casing		180	.044	S.F.	1.35	2.09		3.44	4.91
1500	Fascia		250	.032		1.35	1.50		2.85	3.95
1600	Rake trim		225	.036		1.35	1.67		3.02	4.23

07 65 Flexible Flashing

07 65 10 – Sheet Metal Flashing

07 65 10.10 Sheet Metal Flashing and Counter Flashing	Crew	Daily Output	Labor-Hours	Unit	Material	2015 Bare Costs Labor	Equipment	Total	Total Incl O&P
0010 **SHEET METAL FLASHING AND COUNTER FLASHING**									
0011 Including up to 4 bends									
0020 Aluminum, mill finish, .013" thick	1 Rofc	145	.055	S.F.	.79	2.21		3	4.87
0030 .016" thick		145	.055		.91	2.21		3.12	5
0060 .019" thick		145	.055		1.15	2.21		3.36	5.25
0100 .032" thick		145	.055		1.35	2.21		3.56	5.50
0200 .040" thick		145	.055		2.31	2.21		4.52	6.55
0300 .050" thick		145	.055		2.43	2.21		4.64	6.65
0325 Mill finish 5" x 7" step flashing, .016" thick		1920	.004	Ea.	.15	.17		.32	.47
0350 Mill finish 12" x 12" step flashing, .016" thick		1600	.005	"	.50	.20		.70	.91
0400 Painted finish, add				S.F.	.29			.29	.32
1000 Mastic-coated 2 sides, .005" thick	1 Rofc	330	.024		1.82	.97		2.79	3.76
1100 .016" thick		330	.024		2.01	.97		2.98	3.97
1600 Copper, 16 oz., sheets, under 1000 lb.		115	.070		7.65	2.79		10.44	13.50
1700 Over 4000 lb.		155	.052		7.65	2.07		9.72	12.20
1900 20 oz. sheets, under 1000 lb.		110	.073		9.40	2.92		12.32	15.60
2000 Over 4000 lb.		145	.055		8.95	2.21		11.16	13.80
2200 24 oz. sheets, under 1000 lb.		105	.076		13.30	3.06		16.36	20
2300 Over 4000 lb.		135	.059		12.65	2.38		15.03	18.20
2500 32 oz. sheets, under 1000 lb.		100	.080		17.80	3.21		21.01	25.50
2600 Over 4000 lb.		130	.062		16.90	2.47		19.37	23
2700 W shape for valleys, 16 oz., 24" wide		100	.080	L.F.	15.40	3.21		18.61	23
5800 Lead, 2.5 lb. per S.F., up to 12" wide		135	.059	S.F.	4.99	2.38		7.37	9.80
5900 Over 12" wide		135	.059		3.99	2.38		6.37	8.70
8900 Stainless steel sheets, 32 ga.,		155	.052		3	2.07		5.07	7.05
9000 28 ga.		155	.052		3.99	2.07		6.06	8.15
9100 26 ga.		155	.052		4.25	2.07		6.32	8.40
9200 24 ga.		155	.052		4.75	2.07		6.82	9
9290 For mechanically keyed flashing, add					40%				
9320 Steel sheets, galvanized, 20 ga.	1 Rofc	130	.062	S.F.	1.25	2.47		3.72	5.85
9322 22 ga.		135	.059		1.21	2.38		3.59	5.60
9324 24 ga.		140	.057		.92	2.29		3.21	5.15
9326 26 ga.		148	.054		.80	2.17		2.97	4.80
9328 28 ga.		155	.052		.69	2.07		2.76	4.50
9340 30 ga.		160	.050		.58	2.01		2.59	4.26
9400 Terne coated stainless steel, .015" thick, 28 ga		155	.052		7.35	2.07		9.42	11.85
9500 .018" thick, 26 ga		155	.052		8.30	2.07		10.37	12.85
9600 Zinc and copper alloy (brass), .020" thick		155	.052		9.30	2.07		11.37	14
9700 .027" thick		155	.052		11	2.07		13.07	15.85
9800 .032" thick		155	.052		14	2.07		16.07	19.15
9900 .040" thick		155	.052		18.50	2.07		20.57	24
9950 Minimum labor/equipment charge		3	2.667	Job		107		107	193

07 65 12 – Fabric and Mastic Flashings

07 65 12.10 Fabric and Mastic Flashing and Counter Flashing	Crew	Daily Output	Labor-Hours	Unit	Material	2015 Bare Costs Labor	Equipment	Total	Total Incl O&P
0010 **FABRIC AND MASTIC FLASHING AND COUNTER FLASHING**									
1300 Asphalt flashing cement, 5 gallon				Gal.	9.20			9.20	10.10
4900 Fabric, asphalt-saturated cotton, specification grade	1 Rofc	35	.229	S.Y.	2.96	9.15		12.11	19.80
5000 Utility grade		35	.229		1.42	9.15		10.57	18.10
5300 Close-mesh fabric, saturated, 17 oz. per S.Y.		35	.229		2.15	9.15		11.30	18.90
5500 Fiberglass, resin-coated		35	.229		.97	9.15		10.12	17.60
8500 Shower pan, bituminous membrane, 7 oz.		155	.052	S.F.	1.62	2.07		3.69	5.50

07 65 Flexible Flashing

07 65 13 – Laminated Sheet Flashing

07 65 13.10 Laminated Sheet Flashing	Crew	Daily Output	Labor-Hours	Unit	Material	2015 Bare Costs Labor	Equipment	Total	Total Incl O&P
0010 **LAMINATED SHEET FLASHING**, Including up to 4 bends									
0500 Aluminum, fabric-backed 2 sides, mill finish, .004" thick	1 Rofc	330	.024	S.F.	1.54	.97		2.51	3.45
0700 .005" thick		330	.024		1.82	.97		2.79	3.76
0750 Mastic-backed, self adhesive		460	.017		3.51	.70		4.21	5.10
0800 Mastic-coated 2 sides, .004" thick		330	.024		1.54	.97		2.51	3.45
2800 Copper, paperbacked 1 side, 2 oz.		330	.024		1.84	.97		2.81	3.78
2900 3 oz.		330	.024		2.15	.97		3.12	4.13
3100 Paperbacked 2 sides, 2 oz.		330	.024		1.97	.97		2.94	3.93
3150 3 oz.		330	.024		2.25	.97		3.22	4.24
3200 5 oz.		330	.024		3.03	.97		4	5.10
3250 7 oz.		330	.024		5.30	.97		6.27	7.60
3400 Mastic-backed 2 sides, copper, 2 oz.		330	.024		1.82	.97		2.79	3.76
3500 3 oz.		330	.024		2.20	.97		3.17	4.18
3700 5 oz.		330	.024		3.30	.97		4.27	5.40
3800 Fabric-backed 2 sides, copper, 2 oz.		330	.024		2.10	.97		3.07	4.07
4000 3 oz.		330	.024		2.40	.97		3.37	4.40
4100 5 oz.		330	.024		3.40	.97		4.37	5.50
4300 Copper-clad stainless steel, .015" thick, under 500 lb.		115	.070		6.40	2.79		9.19	12.10
4400 Over 2000 lb.		155	.052		6.10	2.07		8.17	10.45
4600 .018" thick, under 500 lb.		100	.080		7.20	3.21		10.41	13.70
4700 Over 2000 lb.		145	.055		6.80	2.21		9.01	11.50
8550 Shower pan, 3 ply copper and fabric, 3 oz.		155	.052		3.50	2.07		5.57	7.60
8600 7 oz.		155	.052		4.20	2.07		6.27	8.35
9300 Stainless steel, paperbacked 2 sides, .005" thick		330	.024		3.62	.97		4.59	5.75

07 65 19 – Plastic Sheet Flashing

07 65 19.10 Plastic Sheet Flashing and Counter Flashing

	Crew	Daily Output	Labor-Hours	Unit	Material	Labor	Equipment	Total	Total Incl O&P
0010 **PLASTIC SHEET FLASHING AND COUNTER FLASHING**									
7300 Polyvinyl chloride, black, 10 mil	1 Rofc	285	.028	S.F.	.21	1.13		1.34	2.26
7400 20 mil		285	.028		.28	1.13		1.41	2.34
7600 30 mil		285	.028		.35	1.13		1.48	2.42
7700 60 mil		285	.028		.85	1.13		1.98	2.97
7900 Black or white for exposed roofs, 60 mil		285	.028		1.80	1.13		2.93	4.01
8060 PVC tape, 5" x 45 mils, for joint covers, 100 L.F./roll				Ea.	170			170	187
8850 Polyvinyl chloride, 30 mil	1 Rofc	160	.050	S.F.	1.30	2.01		3.31	5.05

07 65 23 – Rubber Sheet Flashing

07 65 23.10 Rubber Sheet Flashing and Counterflashing

	Crew	Daily Output	Labor-Hours	Unit	Material	Labor	Equipment	Total	Total Incl O&P
0010 **RUBBER SHEET FLASHING AND COUNTERFLASHING**									
4810 EPDM 90 mils, 1" diameter pipe flashing	1 Rofc	32	.250	Ea.	19.35	10.05		29.40	39.50
4820 2" diameter		30	.267		19.25	10.70		29.95	40.50
4830 3" diameter		28	.286		20.50	11.45		31.95	43.50
4840 4" diameter		24	.333		23.50	13.35		36.85	50
4850 6" diameter		22	.364		23.50	14.60		38.10	52.50
8100 Rubber, butyl, 1/32" thick		285	.028	S.F.	1.92	1.13		3.05	4.14
8200 1/16" thick		285	.028		2.60	1.13		3.73	4.89
8300 Neoprene, cured, 1/16" thick		285	.028		2.23	1.13		3.36	4.48
8400 1/8" thick		285	.028		6.05	1.13		7.18	8.70

07 71 Roof Specialties

07 71 16 – Manufactured Counterflashing Systems

07 71 16.10 Roof Drain Boot		Crew	Daily Output	Labor-Hours	Unit	Material	2015 Bare Costs Labor	Equipment	Total	Total Incl O&P
0010	**ROOF DRAIN BOOT**									
0100	Cast iron, 4" diameter	1 Shee	125	.064	L.F.	50	3.58		53.58	61
0300	4" x 3"		125	.064		57.50	3.58		61.08	69
0400	5" x 4"		125	.064		117	3.58		120.58	135

07 71 19 – Manufactured Gravel Stops and Fasciae

07 71 19.10 Gravel Stop

		Crew	Daily Output	Labor-Hours	Unit	Material	Labor	Equipment	Total	Total Incl O&P
0010	**GRAVEL STOP**									
0020	Aluminum, .050" thick, 4" face height, mill finish	1 Shee	145	.055	L.F.	6.30	3.09		9.39	11.75
0080	Duranodic finish		145	.055		6.75	3.09		9.84	12.30
0100	Painted		145	.055		6.95	3.09		10.04	12.50
0300	6" face height		135	.059		6.45	3.32		9.77	12.35
0350	Duranodic finish		135	.059		7.25	3.32		10.57	13.25
0400	Painted		135	.059		8.20	3.32		11.52	14.25
0600	8" face height		125	.064		7.35	3.58		10.93	13.75
0650	Duranodic finish		125	.064		8.40	3.58		11.98	14.90
0700	Painted		125	.064		8.75	3.58		12.33	15.30
0900	12" face height, .080 thick, 2 piece		100	.080		9.95	4.48		14.43	17.95
0950	Duranodic finish		100	.080		10.25	4.48		14.73	18.35
1000	Painted		100	.080		12	4.48		16.48	20.50
1200	Copper, 16 oz., 3" face height		145	.055		24	3.09		27.09	31.50
1300	6" face height		135	.059		33	3.32		36.32	42
1350	Galv steel, 24 ga., 4" leg, plain, with continuous cleat, 4" face		145	.055		6.25	3.09		9.34	11.75
1360	6" face height		145	.055		6.50	3.09		9.59	12
1500	Polyvinyl chloride, 6" face height		135	.059		5.20	3.32		8.52	10.95
1600	9" face height		125	.064		6	3.58		9.58	12.25
1800	Stainless steel, 24 ga., 6" face height		135	.059		15.30	3.32		18.62	22
1900	12" face height		100	.080		22.50	4.48		26.98	32
2100	20 ga., 6" face height		135	.059		17.80	3.32		21.12	25
2200	12" face height		100	.080		27	4.48		31.48	36.50
9000	Minimum labor/equipment charge		3.50	2.286	Job		128		128	202

07 71 19.30 Fascia

		Crew	Daily Output	Labor-Hours	Unit	Material	Labor	Equipment	Total	Total Incl O&P
0010	**FASCIA**									
0100	Aluminum, reverse board and batten, .032" thick, colored, no furring incl	1 Shee	145	.055	S.F.	6.75	3.09		9.84	12.30
0200	Residential type, aluminum	1 Carp	200	.040	L.F.	1.95	1.88		3.83	5.25
0220	Vinyl	"	200	.040	"	2.10	1.88		3.98	5.40
0300	Steel, galv and enameled, stock, no furring, long panels	1 Shee	145	.055	S.F.	4.25	3.09		7.34	9.55
0600	Short panels		115	.070	"	5.80	3.89		9.69	12.55
9000	Minimum labor/equipment charge		4	2	Job		112		112	177

07 71 23 – Manufactured Gutters and Downspouts

07 71 23.10 Downspouts

		Crew	Daily Output	Labor-Hours	Unit	Material	Labor	Equipment	Total	Total Incl O&P
0010	**DOWNSPOUTS**									
0020	Aluminum, embossed, .020" thick, 2" x 3"	1 Shee	190	.042	L.F.	.98	2.36		3.34	4.80
0100	Enameled		190	.042		1.37	2.36		3.73	5.25
0300	.024" thick, 2' x 3"		180	.044		2.05	2.49		4.54	6.20
0400	3" x 4"		140	.057		2.33	3.20		5.53	7.60
0600	Round, corrugated aluminum, 3" diameter, .020" thick		190	.042		1.95	2.36		4.31	5.85
0700	4" diameter, .025" thick		140	.057		2.53	3.20		5.73	7.85
0900	Wire strainer, round, 2" diameter		155	.052	Ea.	2.67	2.89		5.56	7.50
1000	4" diameter		155	.052		3.67	2.89		6.56	8.60
1200	Rectangular, perforated, 2" x 3"		145	.055		2.30	3.09		5.39	7.40
1300	3" x 4"		145	.055		3.32	3.09		6.41	8.50

07 71 Roof Specialties

07 71 23 – Manufactured Gutters and Downspouts

07 71 23.10 Downspouts

		Crew	Daily Output	Labor-Hours	Unit	Material	2015 Bare Costs Labor	Equipment	Total	Total Incl O&P
1500	Copper, round, 16 oz., stock, 2" diameter	1 Shee	190	.042	L.F.	7.60	2.36		9.96	12.05
1600	3" diameter		190	.042		7.50	2.36		9.86	11.95
1800	4" diameter		145	.055		9.70	3.09		12.79	15.50
1900	5" diameter		130	.062		17.05	3.44		20.49	24.50
2100	Rectangular, corrugated copper, stock, 2" x 3"		190	.042		8.55	2.36		10.91	13.10
2200	3" x 4"		145	.055		10.40	3.09		13.49	16.30
2400	Rectangular, plain copper, stock, 2" x 3"		190	.042		11.05	2.36		13.41	15.85
2500	3" x 4"		145	.055		13.95	3.09		17.04	20
2700	Wire strainers, rectangular, 2" x 3"		145	.055	Ea.	17.05	3.09		20.14	23.50
2800	3" x 4"		145	.055		17.80	3.09		20.89	24.50
3000	Round, 2" diameter		145	.055		6.40	3.09		9.49	11.90
3100	3" diameter		145	.055		7.10	3.09		10.19	12.65
3300	4" diameter		145	.055		12.10	3.09		15.19	18.15
3400	5" diameter		115	.070		22.50	3.89		26.39	30.50
3600	Lead-coated copper, round, stock, 2" diameter		190	.042	L.F.	22	2.36		24.36	28
3700	3" diameter		190	.042		22	2.36		24.36	28
3900	4" diameter		145	.055		23	3.09		26.09	30.50
4000	5" diameter, corrugated		130	.062		22	3.44		25.44	30
4200	6" diameter, corrugated		105	.076		30	4.26		34.26	40
4300	Rectangular, corrugated, stock, 2" x 3"		190	.042		15.40	2.36		17.76	20.50
4500	Plain, stock, 2" x 3"		190	.042		24	2.36		26.36	30
4600	3" x 4"		145	.055		33	3.09		36.09	41.50
4800	Steel, galvanized, round, corrugated, 2" or 3" diameter, 28 ga.		190	.042		2.08	2.36		4.44	6
4900	4" diameter, 28 ga.		145	.055		2.52	3.09		5.61	7.65
5100	5" diameter, 26 ga.		130	.062		4	3.44		7.44	9.85
5400	6" diameter, 28 ga.		105	.076		4.20	4.26		8.46	11.35
5500	26 ga.		105	.076		4	4.26		8.26	11.15
5700	Rectangular, corrugated, 28 ga., 2" x 3"		190	.042		1.94	2.36		4.30	5.85
5800	3" x 4"		145	.055		1.82	3.09		4.91	6.85
6000	Rectangular, plain, 28 ga., galvanized, 2" x 3"		190	.042		3.65	2.36		6.01	7.75
6100	3" x 4"		145	.055		4.06	3.09		7.15	9.35
6300	Epoxy painted, 24 ga., corrugated, 2" x 3"		190	.042		2.25	2.36		4.61	6.20
6400	3" x 4"		145	.055		2.80	3.09		5.89	7.95
6600	Wire strainers, rectangular, 2" x 3"		145	.055	Ea.	17.05	3.09		20.14	23.50
6700	3" x 4"		145	.055		17.80	3.09		20.89	24.50
6900	Round strainers, 2" or 3" diameter		145	.055		3.86	3.09		6.95	9.10
7000	4" diameter		145	.055		5.80	3.09		8.89	11.20
9000	Minimum labor/equipment charge		4	2	Job		112		112	177

07 71 23.20 Downspout Elbows

		Crew	Daily Output	Labor-Hours	Unit	Material	2015 Bare Costs Labor	Equipment	Total	Total Incl O&P
0010	**DOWNSPOUT ELBOWS**									
0020	Aluminum, embossed, 2" x 3", .020" thick	1 Shee	100	.080	Ea.	.95	4.48		5.43	8.10
0100	Enameled		100	.080		1.75	4.48		6.23	9
0200	Embossed, 3" x 4", .025" thick		100	.080		4.51	4.48		8.99	12
0300	Enameled		100	.080		3.75	4.48		8.23	11.20
0400	Embossed, corrugated, 3" diameter, .020" thick		100	.080		2.74	4.48		7.22	10.05
0500	4" diameter, .025" thick		100	.080		5.80	4.48		10.28	13.40
0600	Copper, 16 oz., 2" diameter		100	.080		9.45	4.48		13.93	17.45
0700	3" diameter		100	.080		9	4.48		13.48	16.95
0800	4" diameter		100	.080		13.90	4.48		18.38	22.50
1000	2" x 3" corrugated		100	.080		9.35	4.48		13.83	17.35
1100	3" x 4" corrugated		100	.080		13.25	4.48		17.73	21.50
1300	Vinyl, 2-1/2" diameter, 45 or 75 degree bend		100	.080		3.88	4.48		8.36	11.30

07 71 Roof Specialties

07 71 23 – Manufactured Gutters and Downspouts

07 71 23.20 Downspout Elbows		Crew	Daily Output	Labor-Hours	Unit	Material	2015 Bare Costs Labor	Equipment	Total	Total Incl O&P
1400	Tee Y junction	1 Shee	75	.107	Ea.	12.90	5.95		18.85	23.50
9000	Minimum labor/equipment charge	↓	4	2	Job		112		112	177

07 71 23.30 Gutters

		Crew	Daily Output	Labor-Hours	Unit	Material	Labor	Equipment	Total	Total Incl O&P
0010	**GUTTERS**									
0012	Aluminum, stock units, 5" K type, .027" thick, plain	1 Shee	125	.064	L.F.	2.72	3.58		6.30	8.65
0100	Enameled		125	.064		2.66	3.58		6.24	8.55
0300	5" K type type, .032" thick, plain		125	.064		3.23	3.58		6.81	9.20
0400	Enameled		125	.064		3.02	3.58		6.60	8.95
0700	Copper, half round, 16 oz., stock units, 4" wide		125	.064		8.90	3.58		12.48	15.45
0900	5" wide		125	.064		7.25	3.58		10.83	13.60
1000	6" wide		118	.068		11.85	3.79		15.64	19
1200	K type, 16 oz., stock, 5" wide		125	.064		8.05	3.58		11.63	14.50
1300	6" wide		125	.064		8.50	3.58		12.08	15
1500	Lead coated copper, 16 oz., half round, stock, 4" wide		125	.064		13.80	3.58		17.38	21
1600	6" wide		118	.068		21.50	3.79		25.29	30
1800	K type, stock, 5" wide		125	.064		18.35	3.58		21.93	25.50
1900	6" wide		125	.064		17.90	3.58		21.48	25.50
2100	Copper clad stainless steel, K type, 5" wide		125	.064		7.50	3.58		11.08	13.90
2200	6" wide		125	.064		9.45	3.58		13.03	16.05
2400	Steel, galv, half round or box, 28 ga., 5" wide, plain		125	.064		2.10	3.58		5.68	7.95
2500	Enameled		125	.064		2.10	3.58		5.68	7.95
2700	26 ga., stock, 5" wide		125	.064		2.18	3.58		5.76	8.05
2800	6" wide		125	.064		2.71	3.58		6.29	8.65
3000	Vinyl, O.G., 4" wide	1 Carp	115	.070		1.25	3.27		4.52	6.75
3100	5" wide		115	.070		1.55	3.27		4.82	7.05
3200	4" half round, stock units	↓	115	.070	↓	1.30	3.27		4.57	6.80
3250	Joint connectors				Ea.	2.90			2.90	3.19
3300	Wood, clear treated cedar, fir or hemlock, 3" x 4"	1 Carp	100	.080	L.F.	10	3.76		13.76	17.15
3400	4" x 5"	"	100	.080	"	16.75	3.76		20.51	24.50
5000	Accessories, end cap, K type, aluminum 5"	1 Shee	625	.013	Ea.	.66	.72		1.38	1.86
5010	6"		625	.013		1.44	.72		2.16	2.71
5020	Copper, 5"		625	.013		3.22	.72		3.94	4.67
5030	6"		625	.013		3.54	.72		4.26	5
5040	Lead coated copper, 5"		625	.013		12.50	.72		13.22	14.90
5050	6"		625	.013		13.35	.72		14.07	15.85
5060	Copper clad stainless steel, 5"		625	.013		3	.72		3.72	4.43
5070	6"		625	.013		3.60	.72		4.32	5.10
5080	Galvanized steel, 5"		625	.013		1.28	.72		2	2.54
5090	6"		625	.013		2.18	.72		2.90	3.53
5100	Vinyl, 4"	1 Carp	625	.013		13.15	.60		13.75	15.50
5110	5"	"	625	.013		13.15	.60		13.75	15.50
5120	Half round, copper, 4"	1 Shee	625	.013		4.53	.72		5.25	6.10
5130	5"		625	.013		4.53	.72		5.25	6.10
5140	6"		625	.013		7.65	.72		8.37	9.60
5150	Lead coated copper, 5"		625	.013		14.15	.72		14.87	16.70
5160	6"		625	.013		21	.72		21.72	24.50
5170	Copper clad stainless steel, 5"		625	.013		4.50	.72		5.22	6.10
5180	6"		625	.013		4.50	.72		5.22	6.10
5190	Galvanized steel, 5"		625	.013		2.25	.72		2.97	3.61
5200	6"		625	.013		2.80	.72		3.52	4.21
5210	Outlet, aluminum, 2" x 3"		420	.019		.62	1.07		1.69	2.36
5220	3" x 4"		420	.019		1.05	1.07		2.12	2.84

07 71 Roof Specialties

07 71 23 – Manufactured Gutters and Downspouts

07 71 23.30 Gutters

		Crew	Daily Output	Labor-Hours	Unit	Material	2015 Bare Costs Labor	2015 Bare Costs Equipment	Total	Total Incl O&P
5230	2-3/8" round	1 Shee	420	.019	Ea.	.56	1.07		1.63	2.30
5240	Copper, 2" x 3"		420	.019		6.50	1.07		7.57	8.85
5250	3" x 4"		420	.019		7.85	1.07		8.92	10.35
5260	2-3/8" round		420	.019		4.55	1.07		5.62	6.70
5270	Lead coated copper, 2" x 3"		420	.019		25.50	1.07		26.57	29.50
5280	3" x 4"		420	.019		28.50	1.07		29.57	33
5290	2-3/8" round		420	.019		25.50	1.07		26.57	29.50
5300	Copper clad stainless steel, 2" x 3"		420	.019		6.50	1.07		7.57	8.85
5310	3" x 4"		420	.019		7.85	1.07		8.92	10.35
5320	2-3/8" round		420	.019		4.55	1.07		5.62	6.70
5330	Galvanized steel, 2" x 3"		420	.019		3.18	1.07		4.25	5.20
5340	3" x 4"		420	.019		4.90	1.07		5.97	7.10
5350	2-3/8" round		420	.019		3.98	1.07		5.05	6.05
5360	K type mitres, aluminum		65	.123		3	6.90		9.90	14.15
5370	Copper		65	.123		17.40	6.90		24.30	30
5380	Lead coated copper		65	.123		54.50	6.90		61.40	71
5390	Copper clad stainless steel		65	.123		27.50	6.90		34.40	41.50
5400	Galvanized steel		65	.123		22.50	6.90		29.40	36
5420	Half round mitres, copper		65	.123		63.50	6.90		70.40	81
5430	Lead coated copper		65	.123		89.50	6.90		96.40	109
5440	Copper clad stainless steel		65	.123		56	6.90		62.90	73
5450	Galvanized steel		65	.123		25.50	6.90		32.40	39
5460	Vinyl mitres and outlets		65	.123		9.75	6.90		16.65	21.50
5470	Sealant		940	.009	L.F.	.01	.48		.49	.76
5480	Soldering		96	.083	"	.19	4.66		4.85	7.55
9000	Minimum labor/equipment charge		3.75	2.133	Job		119		119	188

07 71 23.35 Gutter Guard

		Crew	Daily Output	Labor-Hours	Unit	Material	2015 Bare Costs Labor	2015 Bare Costs Equipment	Total	Total Incl O&P
0010	**GUTTER GUARD**									
0020	6" wide strip, aluminum mesh	1 Carp	500	.016	L.F.	2.42	.75		3.17	3.89
0100	Vinyl mesh		500	.016	"	2.61	.75		3.36	4.10
9000	Minimum labor/equipment charge		4	2	Job		94		94	154

07 71 26 – Reglets

07 71 26.10 Reglets and Accessories

		Crew	Daily Output	Labor-Hours	Unit	Material	2015 Bare Costs Labor	2015 Bare Costs Equipment	Total	Total Incl O&P
0010	**REGLETS AND ACCESSORIES**									
0020	Reglet, aluminum, .025" thick, in parapet	1 Carp	225	.036	L.F.	1.93	1.67		3.60	4.86
0300	16 oz. copper		225	.036		6.40	1.67		8.07	9.80
0400	Galvanized steel, 24 ga.		225	.036		1.34	1.67		3.01	4.21
0600	Stainless steel, .020" thick		225	.036		3.75	1.67		5.42	6.85
0900	Counter flashing for above, 12" wide, .032" aluminum	1 Shee	150	.053		2.40	2.98		5.38	7.35
1200	16 oz. copper		150	.053		6.10	2.98		9.08	11.40
1300	Galvanized steel, 26 ga.		150	.053		1.42	2.98		4.40	6.25
1500	Stainless steel, .020" thick		150	.053		6.15	2.98		9.13	11.45
9000	Minimum labor/equipment charge	1 Carp	3	2.667	Job		125		125	205

07 71 29 – Manufactured Roof Expansion Joints

07 71 29.10 Expansion Joints

		Crew	Daily Output	Labor-Hours	Unit	Material	2015 Bare Costs Labor	2015 Bare Costs Equipment	Total	Total Incl O&P
0010	**EXPANSION JOINTS**									
0300	Butyl or neoprene center with foam insulation, metal flanges									
0400	Aluminum, .032" thick for openings to 2-1/2"	1 Rofc	165	.048	L.F.	11.30	1.94		13.24	15.95
0600	For joint openings to 3-1/2"		165	.048		11.30	1.94		13.24	15.95
0610	For joint openings to 5"		165	.048		13.10	1.94		15.04	17.90
0620	For joint openings to 8"		165	.048		15.80	1.94		17.74	21

07 71 Roof Specialties

07 71 29 – Manufactured Roof Expansion Joints

07 71 29.10 Expansion Joints		Crew	Daily Output	Labor-Hours	Unit	Material	2015 Bare Costs Labor	Equipment	Total	Total Incl O&P
0700	Copper, 16 oz. for openings to 2-1/2"	1 Rofc	165	.048	L.F.	16.70	1.94		18.64	22
0900	For joint openings to 3-1/2"		165	.048		16.70	1.94		18.64	22
0910	For joint openings to 5"		165	.048		18.90	1.94		20.84	24.50
0920	For joint openings to 8"		165	.048		22	1.94		23.94	27.50
1000	Galvanized steel, 26 ga. for openings to 2-1/2"		165	.048		9.80	1.94		11.74	14.30
1200	For joint openings to 3-1/2"		165	.048		9.80	1.94		11.74	14.30
1210	For joint openings to 5"		165	.048		11.30	1.94		13.24	15.95
1220	For joint openings to 8"		165	.048		14.80	1.94		16.74	19.80
1300	Lead-coated copper, 16 oz. for openings to 2-1/2"		165	.048		32	1.94		33.94	38.50
1500	For joint openings to 3-1/2"		165	.048		32	1.94		33.94	38.50
1600	Stainless steel, .018", for openings to 2-1/2"		165	.048		15.40	1.94		17.34	20.50
1800	For joint openings to 3-1/2"		165	.048		15.40	1.94		17.34	20.50
1810	For joint openings to 5"		165	.048		17.20	1.94		19.14	22.50
1820	For joint openings to 8"		165	.048		21	1.94		22.94	26.50
1900	Neoprene, double-seal type with thick center, 4-1/2" wide		125	.064		14.50	2.57		17.07	20.50
1950	Polyethylene bellows, with galv steel flat flanges		100	.080		6.20	3.21		9.41	12.60
1960	With galvanized angle flanges		100	.080		6.40	3.21		9.61	12.85
2000	Roof joint with extruded aluminum cover, 2"	1 Shee	115	.070		28	3.89		31.89	37
2100	Roof joint, plastic curbs, foam center, standard	1 Rofc	100	.080		13	3.21		16.21	20
2200	Large	"	100	.080		17	3.21		20.21	24.50
2500	Roof to wall joint with extruded aluminum cover	1 Shee	115	.070		28	3.89		31.89	37
2700	Wall joint, closed cell foam on PVC cover, 9" wide	1 Rofc	125	.064		5	2.57		7.57	10.15
2800	12" wide	"	115	.070		6	2.79		8.79	11.65
9000	Minimum labor/equipment charge	1 Shee	3	2.667	Job		149		149	235

07 71 43 – Drip Edge

07 71 43.10 Drip Edge, Rake Edge, Ice Belts

		Crew	Daily Output	Labor-Hours	Unit	Material	Labor	Equipment	Total	Total Incl O&P
0010	**DRIP EDGE, RAKE EDGE, ICE BELTS**									
0020	Aluminum, .016" thick, 5" wide, mill finish	1 Carp	400	.020	L.F.	.55	.94		1.49	2.15
0100	White finish		400	.020		.63	.94		1.57	2.23
0200	8" wide, mill finish		400	.020		1.45	.94		2.39	3.14
0300	Ice belt, 28" wide, mill finish		100	.080		7.65	3.76		11.41	14.60
0310	Vented, mill finish		400	.020		1.99	.94		2.93	3.73
0320	Painted finish		400	.020		2.24	.94		3.18	4
0400	Galvanized, 5" wide		400	.020		.53	.94		1.47	2.12
0500	8" wide, mill finish		400	.020		.80	.94		1.74	2.42
0510	Rake edge, aluminum, 1-1/2" x 1-1/2"		400	.020		.30	.94		1.24	1.87
0520	3-1/2" x 1-1/2"		400	.020		.42	.94		1.36	2
9000	Minimum labor/equipment charge		4	2	Job		94		94	154

07 72 Roof Accessories

07 72 23 – Relief Vents

07 72 23.10 Roof Vents

		Crew	Daily Output	Labor-Hours	Unit	Material	Labor	Equipment	Total	Total Incl O&P
0010	**ROOF VENTS**									
0020	Mushroom shape, for built-up roofs, aluminum	1 Rofc	30	.267	Ea.	65	10.70		75.70	91
0100	PVC, 6" high		30	.267	"	21	10.70		31.70	42.50
9000	Minimum labor/equipment charge		2.75	2.909	Job		117		117	211

07 72 23.20 Vents

		Crew	Daily Output	Labor-Hours	Unit	Material	Labor	Equipment	Total	Total Incl O&P
0010	**VENTS**									
0100	Soffit or eave, aluminum, mill finish, strips, 2-1/2" wide	1 Carp	200	.040	L.F.	.43	1.88		2.31	3.55
0200	3" wide		200	.040		.41	1.88		2.29	3.53

07 72 Roof Accessories

07 72 23 – Relief Vents

07 72 23.20 Vents

		Crew	Daily Output	Labor-Hours	Unit	Material	2015 Bare Costs Labor	Equipment	Total	Total Incl O&P
0300	Enamel finish, 3" wide	1 Carp	200	.040	L.F.	.50	1.88		2.38	3.63
0400	Mill finish, rectangular, 4" x 16"		72	.111	Ea.	1.45	5.20		6.65	10.15
0500	8" x 16"	↓	72	.111		2.26	5.20		7.46	11.05
2420	Roof ventilator	Q-9	16	1		52	50.50		102.50	137
2500	Vent, roof vent	1 Rofc	24	.333	↓	26	13.35		39.35	52.50

07 72 26 – Ridge Vents

07 72 26.10 Ridge Vents and Accessories

		Crew	Daily Output	Labor-Hours	Unit	Material	Labor	Equipment	Total	Total Incl O&P
0010	**RIDGE VENTS AND ACCESSORIES**									
0100	Aluminum strips, mill finish	1 Rofc	160	.050	L.F.	2.75	2.01		4.76	6.65
0150	Painted finish		160	.050	"	4.05	2.01		6.06	8.10
0200	Connectors		48	.167	Ea.	4.77	6.70		11.47	17.35
0300	End caps		48	.167	"	2.03	6.70		8.73	14.35
0400	Galvanized strips		160	.050	L.F.	3.58	2.01		5.59	7.55
0430	Molded polyethylene, shingles not included		160	.050	"	2.74	2.01		4.75	6.65
0440	End plugs		48	.167	Ea.	2.03	6.70		8.73	14.35
0450	Flexible roll, shingles not included	↓	160	.050	L.F.	2.38	2.01		4.39	6.25
2300	Ridge vent strip, mill finish	1 Shee	155	.052	"	3.85	2.89		6.74	8.80

07 72 33 – Roof Hatches

07 72 33.10 Roof Hatch Options

		Crew	Daily Output	Labor-Hours	Unit	Material	Labor	Equipment	Total	Total Incl O&P
0010	**ROOF HATCH OPTIONS**									
0500	2'-6" x 3', aluminum curb and cover	G-3	10	3.200	Ea.	1,025	150		1,175	1,375
0520	Galvanized steel curb and aluminum cover		10	3.200		640	150		790	945
0540	Galvanized steel curb and cover		10	3.200		715	150		865	1,025
0600	2'-6" x 4'-6", aluminum curb and cover		9	3.556		1,375	166		1,541	1,800
0800	Galvanized steel curb and aluminum cover		9	3.556		885	166		1,051	1,250
0900	Galvanized steel curb and cover		9	3.556		965	166		1,131	1,325
1100	4' x 4' aluminum curb and cover		8	4		1,675	187		1,862	2,150
1120	Galvanized steel curb and aluminum cover		8	4		1,725	187		1,912	2,200
1140	Galvanized steel curb and cover		8	4		1,200	187		1,387	1,600
1200	2'-6" x 8'-0", aluminum curb and cover		6.60	4.848		1,900	227		2,127	2,475
1400	Galvanized steel curb and aluminum cover		6.60	4.848		1,800	227		2,027	2,350
1500	Galvanized steel curb and cover	↓	6.60	4.848		1,525	227		1,752	2,050
1800	For plexiglass panels, 2'-6" x 3'-0", add to above				↓	440			440	485
9000	Minimum labor/equipment charge	2 Carp	2	8	Job		375		375	615

07 72 36 – Smoke Vents

07 72 36.10 Smoke Hatches

		Crew	Daily Output	Labor-Hours	Unit	Material	Labor	Equipment	Total	Total Incl O&P
0010	**SMOKE HATCHES**									
0200	For 3'-0" long, add to roof hatches from Section 07 72 33.10				Ea.	25%	5%			
0250	For 4'-0" long, add to roof hatches from Section 07 72 33.10					20%	5%			
0300	For 8'-0" long, add to roof hatches from Section 07 72 33.10				↓	10%	5%			

07 72 36.20 Smoke Vent Options

		Crew	Daily Output	Labor-Hours	Unit	Material	Labor	Equipment	Total	Total Incl O&P
0010	**SMOKE VENT OPTIONS**									
0100	4' x 4' aluminum cover and frame	G-3	13	2.462	Ea.	2,050	115		2,165	2,425
0200	Galvanized steel cover and frame		13	2.462		1,800	115		1,915	2,150
0300	4' x 8' aluminum cover and frame		8	4		2,800	187		2,987	3,375
0400	Galvanized steel cover and frame	↓	8	4	↓	2,375	187		2,562	2,925
9000	Minimum labor/equipment charge	2 Carp	2	8	Job		375		375	615

07 72 Roof Accessories

07 72 53 – Snow Guards

07 72 53.10 Snow Guard Options

		Crew	Daily Output	Labor-Hours	Unit	Material	2015 Bare Costs Labor	Equipment	Total	Total Incl O&P
0010	**SNOW GUARD OPTIONS**									
0100	Slate & asphalt shingle roofs, fastened with nails	1 Rofc	160	.050	Ea.	12.55	2.01		14.56	17.40
0200	Standing seam metal roofs, fastened with set screws		48	.167		16.95	6.70		23.65	31
0300	Surface mount for metal roofs, fastened with solder		48	.167	↓	6.05	6.70		12.75	18.75
0400	Double rail pipe type, including pipe	↓	130	.062	L.F.	32	2.47		34.47	39.50

07 72 73 – Pitch Pockets

07 72 73.10 Pitch Pockets, Variable Sizes

0010	**PITCH POCKETS, VARIABLE SIZES**									
0100	Adjustable, 4" to 7", welded corners, 4" deep	1 Rofc	48	.167	Ea.	16.25	6.70		22.95	30
0200	Side extenders, 6"	"	240	.033	"	2.50	1.34		3.84	5.15

07 72 80 – Vents

07 72 80.30 Vent Options

0010	**VENT OPTIONS**									
0020	Plastic, for insulated decks, 1 per M.S.F.	1 Rofc	40	.200	Ea.	21	8		29	37.50
0100	Heavy duty		20	.400		38	16.05		54.05	71
0300	Aluminum	↓	30	.267		22	10.70		32.70	43.50
0800	Polystyrene baffles, 12" wide for 16" O.C. rafter spacing	1 Carp	90	.089		.42	4.17		4.59	7.30
0900	For 24" O.C. rafter spacing		110	.073	↓	.77	3.41		4.18	6.45
9000	Minimum labor/equipment charge	↓	3	2.667	Job		125		125	205

07 76 Roof Pavers

07 76 16 – Roof Decking Pavers

07 76 16.10 Roof Pavers and Supports

0010	**ROOF PAVERS AND SUPPORTS**									
1000	Roof decking pavers, concrete blocks, 2" thick, natural	1 Clab	115	.070	S.F.	3.37	2.62		5.99	8
1100	Colors		115	.070	"	3.73	2.62		6.35	8.40
1200	Support pedestal, bottom cap		960	.008	Ea.	3	.31		3.31	3.81
1300	Top cap		960	.008		4.80	.31		5.11	5.80
1400	Leveling shims, 1/16"		1920	.004		1.20	.16		1.36	1.58
1500	1/8"		1920	.004		1.20	.16		1.36	1.58
1600	Buffer pad		960	.008	↓	2.50	.31		2.81	3.26
1700	PVC legs (4" SDR 35)		2880	.003	Inch	.12	.10		.22	.30
2000	Alternate pricing method, system in place	↓	101	.079	S.F.	7	2.98		9.98	12.60

07 81 Applied Fireproofing

07 81 16 – Cementitious Fireproofing

07 81 16.10 Sprayed Cementitious Fireproofing

0010	**SPRAYED CEMENTITIOUS FIREPROOFING**									
0050	Not including canvas protection, normal density									
0100	Per 1" thick, on flat plate steel	G-2	3000	.008	S.F.	.53	.32	.04	.89	1.14
0200	Flat decking		2400	.010		.53	.40	.06	.99	1.28
0400	Beams		1500	.016		.53	.63	.09	1.25	1.70
0500	Corrugated or fluted decks		1250	.019		.79	.76	.11	1.66	2.21
0700	Columns, 1-1/8" thick		1100	.022		.59	.86	.12	1.57	2.17
0800	2-3/16" thick		700	.034		1.25	1.36	.19	2.80	3.78
0900	For canvas protection, add	↓	5000	.005	↓	.08	.19	.03	.30	.43
1000	Not including canvas protection, high density									
1100	Per 1" thick, on flat plate steel	G-2	3000	.008	S.F.	1.85	.32	.04	2.21	2.60

07 81 Applied Fireproofing

07 81 16 – Cementitious Fireproofing

07 81 16.10 Sprayed Cementitious Fireproofing

		Crew	Daily Output	Labor-Hours	Unit	Material	2015 Bare Costs Labor	Equipment	Total	Total Incl O&P
1110	On flat decking	G-2	2400	.010	S.F.	1.85	.40	.06	2.31	2.74
1120	On beams		1500	.016		1.85	.63	.09	2.57	3.16
1130	Corrugated or fluted decks		1250	.019		1.85	.76	.11	2.72	3.38
1140	Columns, 1-1/8" thick		1100	.022		2.08	.86	.12	3.06	3.81
1150	2-3/16" thick		1100	.022		4.16	.86	.12	5.14	6.10
1170	For canvas protection, add		5000	.005		.08	.19	.03	.30	.43
1200	Not including canvas protection, retrofitting									
1210	Per 1" thick, on flat plate steel	G-2	1500	.016	S.F.	.38	.63	.09	1.10	1.54
1220	On flat decking		1200	.020		.38	.79	.11	1.28	1.82
1230	On beams		750	.032		.38	1.27	.18	1.83	2.66
1240	Corrugated or fluted decks		625	.038		.58	1.52	.21	2.31	3.31
1250	Columns, 1-1/8" thick		550	.044		.43	1.73	.24	2.40	3.53
1260	2-3/16" thick		500	.048		.86	1.90	.27	3.03	4.30
1400	Accessories, preliminary spattered texture coat		4500	.005		.02	.21	.03	.26	.39
1410	Bonding agent	1 Plas	1000	.008		.09	.34		.43	.65
9000	Minimum labor/equipment charge	G-2	3	8	Job		315	44.50	359.50	560

07 84 Firestopping

07 84 13 – Penetration Firestopping

07 84 13.10 Firestopping

		Crew	Daily Output	Labor-Hours	Unit	Material	2015 Bare Costs Labor	Equipment	Total	Total Incl O&P
0010	**FIRESTOPPING** R078413-30									
0100	Metallic piping, non insulated									
0110	Through walls, 2" diameter	1 Carp	16	.500	Ea.	17.20	23.50		40.70	57.50
0120	4" diameter		14	.571		26	27		53	73
0130	6" diameter		12	.667		35.50	31.50		67	90.50
0140	12" diameter		10	.800		62.50	37.50		100	130
0150	Through floors, 2" diameter		32	.250		9.80	11.75		21.55	30
0160	4" diameter		28	.286		14.20	13.40		27.60	37.50
0170	6" diameter		24	.333		18.50	15.65		34.15	46
0180	12" diameter		20	.400		31.50	18.80		50.30	65.50
0190	Metallic piping, insulated									
0200	Through walls, 2" diameter	1 Carp	16	.500	Ea.	24	23.50		47.50	65
0210	4" diameter		14	.571		33	27		60	80
0220	6" diameter		12	.667		42	31.50		73.50	97.50
0230	12" diameter		10	.800		68.50	37.50		106	137
0240	Through floors, 2" diameter		32	.250		16.60	11.75		28.35	37.50
0250	4" diameter		28	.286		21	13.40		34.40	45
0260	6" diameter		24	.333		25.50	15.65		41.15	53.50
0270	12" diameter		20	.400		31.50	18.80		50.30	65.50
0280	Non metallic piping, non insulated									
0290	Through walls, 2" diameter	1 Carp	12	.667	Ea.	68.50	31.50		100	127
0300	4" diameter		10	.800		86	37.50		123.50	157
0310	6" diameter		8	1		120	47		167	209
0330	Through floors, 2" diameter		16	.500		54	23.50		77.50	98
0340	4" diameter		6	1.333		66.50	62.50		129	177
0350	6" diameter		6	1.333		80.50	62.50		143	192
0370	Ductwork, insulated & non insulated, round									
0380	Through walls, 6" diameter	1 Carp	12	.667	Ea.	35	31.50		66.50	90
0390	12" diameter		10	.800		69.50	37.50		107	138
0400	18" diameter		8	1		113	47		160	202
0410	Through floors, 6" diameter		16	.500		18.60	23.50		42.10	59

07 84 Firestopping

07 84 13 – Penetration Firestopping

07 84 13.10 Firestopping		Crew	Daily Output	Labor-Hours	Unit	Material	2015 Bare Costs Labor	Equipment	Total	Total Incl O&P
0420	12" diameter	1 Carp	14	.571	Ea.	34	27		61	81.50
0430	18" diameter	↓	12	.667	↓	59	31.50		90.50	117
0440	Ductwork, insulated & non insulated, rectangular									
0450	With stiffener/closure angle, through walls, 6" x 12"	1 Carp	8	1	Ea.	28	47		75	108
0460	12" x 24"		6	1.333		37.50	62.50		100	144
0470	24" x 48"		4	2		107	94		201	271
0480	With stiffener/closure angle, through floors, 6" x 12"		10	.800		15.40	37.50		52.90	78.50
0490	12" x 24"		8	1		28	47		75	108
0500	24" x 48"	↓	6	1.333		54	62.50		116.50	163
0510	Multi trade openings									
0520	Through walls, 6" x 12"	1 Carp	2	4	Ea.	59	188		247	375
0530	12" x 24"	"	1	8		238	375		613	875
0540	24" x 48"	2 Carp	1	16		950	750		1,700	2,275
0550	48" x 96"	"	.75	21.333		3,700	1,000		4,700	5,725
0560	Through floors, 6" x 12"	1 Carp	2	4		39	188		227	355
0570	12" x 24"	"	1	8		157	375		532	790
0580	24" x 48"	2 Carp	.75	21.333		625	1,000		1,625	2,350
0590	48" x 96"	"	.50	32	↓	2,450	1,500		3,950	5,125
0600	Structural penetrations, through walls									
0610	Steel beams, W8 x 10	1 Carp	8	1	Ea.	37.50	47		84.50	118
0620	W12 x 14		6	1.333		59	62.50		121.50	168
0630	W21 x 44		5	1.600		118	75		193	253
0640	W36 x 135		3	2.667		287	125		412	520
0650	Bar joists, 18" deep		6	1.333		54.50	62.50		117	163
0660	24" deep		6	1.333		67.50	62.50		130	177
0670	36" deep		5	1.600		101	75		176	235
0680	48" deep	↓	4	2	↓	118	94		212	284
0690	Construction joints, floor slab at exterior wall									
0700	Precast, brick, block or drywall exterior									
0710	2" wide joint	1 Carp	125	.064	L.F.	8.50	3		11.50	14.25
0720	4" wide joint	"	75	.107	"	17	5		22	27
0730	Metal panel, glass or curtain wall exterior									
0740	2" wide joint	1 Carp	40	.200	L.F.	20	9.40		29.40	37.50
0750	4" wide joint	"	25	.320	"	28	15		43	55
0760	Floor slab to drywall partition									
0770	Flat joint	1 Carp	100	.080	L.F.	8.40	3.76		12.16	15.40
0780	Fluted joint		50	.160		17.40	7.50		24.90	31.50
0790	Etched fluted joint	↓	75	.107	↓	11.10	5		16.10	20.50
0800	Floor slab to concrete/masonry partition									
0810	Flat joint	1 Carp	75	.107	L.F.	18.75	5		23.75	28.50
0820	Fluted joint	"	50	.160	"	22.50	7.50		30	37
0830	Concrete/CMU wall joints									
0840	1" wide	1 Carp	100	.080	L.F.	10.25	3.76		14.01	17.45
0850	2" wide		75	.107		18.75	5		23.75	28.50
0860	4" wide	↓	50	.160	↓	38	7.50		45.50	54
0870	Concrete/CMU floor joints									
0880	1" wide	1 Carp	200	.040	L.F.	5.15	1.88		7.03	8.75
0890	2" wide		150	.053		9.40	2.50		11.90	14.45
0900	4" wide	↓	100	.080	↓	17.90	3.76		21.66	26

07 91 Preformed Joint Seals

07 91 13 – Compression Seals

07 91 13.10 Compression Seals		Crew	Daily Output	Labor-Hours	Unit	Material	2015 Bare Costs Labor	Equipment	Total	Total Incl O&P
0010	**COMPRESSION SEALS**									
4900	O-ring type cord, 1/4"	1 Bric	472	.017	L.F.	.37	.78		1.15	1.68
4910	1/2"		440	.018		.95	.84		1.79	2.42
4920	3/4"		424	.019		1.80	.87		2.67	3.40
4930	1"		408	.020		3.40	.91		4.31	5.20
4940	1-1/4"		384	.021		6.30	.96		7.26	8.50
4950	1-1/2"		368	.022		7.95	1		8.95	10.40
4960	1-3/4"		352	.023		13.25	1.05		14.30	16.30
4970	2"	▼	344	.023	▼	18.50	1.07		19.57	22.50

07 91 16 – Joint Gaskets

07 91 16.10 Joint Gaskets		Crew	Daily Output	Labor-Hours	Unit	Material	Labor	Equipment	Total	Total Incl O&P
0010	**JOINT GASKETS**									
4400	Joint gaskets, neoprene, closed cell w/adh, 1/8" x 3/8"	1 Bric	240	.033	L.F.	.28	1.54		1.82	2.81
4500	1/4" x 3/4"		215	.037		.59	1.72		2.31	3.44
4700	1/2" x 1"		200	.040		1.30	1.85		3.15	4.43
4800	3/4" x 1-1/2"	▼	165	.048	▼	1.45	2.24		3.69	5.25

07 91 23 – Backer Rods

07 91 23.10 Backer Rods		Crew	Daily Output	Labor-Hours	Unit	Material	Labor	Equipment	Total	Total Incl O&P
0010	**BACKER RODS**									
0032	Backer rod, polyethylene, 1/4" diameter	1 Bric	460	.017	L.F.	.02	.80		.82	1.33
0052	1/2" diameter		460	.017		.03	.80		.83	1.35
0072	3/4" diameter		460	.017		.06	.80		.86	1.37
0092	1" diameter	▼	460	.017	▼	.09	.80		.89	1.41

07 91 26 – Joint Fillers

07 91 26.10 Joint Fillers		Crew	Daily Output	Labor-Hours	Unit	Material	Labor	Equipment	Total	Total Incl O&P
0010	**JOINT FILLERS**									
4360	Butyl rubber filler, 1/4" x 1/4"	1 Bric	290	.028	L.F.	.22	1.27		1.49	2.32
4365	1/2" x 1/2"		250	.032		.89	1.48		2.37	3.38
4370	1/2" x 3/4"		210	.038		1.34	1.76		3.10	4.33
4375	3/4" x 3/4"		230	.035		2.01	1.61		3.62	4.82
4380	1" x 1"	▼	180	.044	▼	2.68	2.05		4.73	6.30
4390	For coloring, add					12%				
4980	Polyethylene joint backing, 1/4" x 2"	1 Bric	2.08	3.846	C.L.F.	12	178		190	300
4990	1/4" x 6"		1.28	6.250	"	28	288		316	500
5600	Silicone, room temp vulcanizing foam seal, 1/4" x 1/2"		1312	.006	L.F.	.33	.28		.61	.83
5610	1/2" x 1/2"		656	.012		.67	.56		1.23	1.65
5620	1/2" x 3/4"		442	.018		1	.84		1.84	2.46
5630	3/4" x 3/4"		328	.024		1.50	1.13		2.63	3.48
5640	1/8" x 1"		1312	.006		.33	.28		.61	.83
5650	1/8" x 3"		442	.018		1	.84		1.84	2.46
5670	1/4" x 3"		295	.027		2	1.25		3.25	4.24
5680	1/4" x 6"		148	.054		3.99	2.49		6.48	8.45
5690	1/2" x 6"		82	.098		8	4.50		12.50	16.15
5700	1/2" x 9"		52.50	.152		12	7.05		19.05	24.50
5710	1/2" x 12"	▼	33	.242	▼	15.95	11.20		27.15	36

07 92 Joint Sealants

07 92 13 – Elastomeric Joint Sealants

07 92 13.10 Masonry Joint Sealants

		Crew	Daily Output	Labor-Hours	Unit	Material	2015 Bare Costs Labor	Equipment	Total	Total Incl O&P
0010	**MASONRY JOINT SEALANTS**, 1/2" x 1/2" joint									
0050	Re-caulk only, oil base	1 Bric	225	.036	L.F.	.52	1.64		2.16	3.24
0100	Acrylic latex		205	.039		.23	1.80		2.03	3.18
0200	Polyurethane		200	.040		4.10	1.85		5.95	7.50
0300	Silicone		195	.041		4.25	1.89		6.14	7.75
1000	Cut out and re-caulk, oil base		145	.055		.52	2.55		3.07	4.71
1050	Acrylic latex		130	.062		.23	2.84		3.07	4.87
1100	Polyurethane		125	.064		4.10	2.95		7.05	9.30
1150	Silicone		120	.067		4.25	3.08		7.33	9.70
9000	Minimum labor/equipment charge		4	2	Job		92.50		92.50	150

07 92 13.20 Caulking and Sealant Options

		Crew	Daily Output	Labor-Hours	Unit	Material	2015 Bare Costs Labor	Equipment	Total	Total Incl O&P
0010	**CAULKING AND SEALANT OPTIONS**									
0050	Latex acrylic based, bulk				Gal.	26.50			26.50	29.50
0055	Bulk in place 1/4" x 1/4" bead	1 Bric	300	.027	L.F.	.08	1.23		1.31	2.09
0060	1/4" x 3/8"		294	.027		.14	1.26		1.40	2.19
0065	1/4" x 1/2"		288	.028		.19	1.28		1.47	2.30
0075	3/8" x 3/8"		284	.028		.21	1.30		1.51	2.35
0080	3/8" x 1/2"		280	.029		.28	1.32		1.60	2.46
0085	3/8" x 5/8"		276	.029		.35	1.34		1.69	2.57
0095	3/8" x 3/4"		272	.029		.42	1.36		1.78	2.67
0100	1/2" x 1/2"		275	.029		.37	1.34		1.71	2.59
0105	1/2" x 5/8"		269	.030		.47	1.37		1.84	2.74
0110	1/2" x 3/4"		263	.030		.56	1.40		1.96	2.90
0115	1/2" x 7/8"		256	.031		.65	1.44		2.09	3.07
0120	1/2" x 1"		250	.032		.75	1.48		2.23	3.22
0125	3/4" x 3/4"		244	.033		.84	1.51		2.35	3.39
0130	3/4" x 1"		225	.036		1.12	1.64		2.76	3.90
0135	1" x 1"		200	.040		1.50	1.85		3.35	4.65
0190	Cartridges				Gal.	32			32	35.50
0200	11 fl. oz. cartridge				Ea.	2.76			2.76	3.04
0500	1/4" x 1/2"	1 Bric	288	.028	L.F.	.23	1.28		1.51	2.34
0600	1/2" x 1/2"		275	.029		.45	1.34		1.79	2.68
0800	3/4" x 3/4"		244	.033		1.01	1.51		2.52	3.58
0900	3/4" x 1"		225	.036		1.35	1.64		2.99	4.16
1000	1" x 1"		200	.040		1.69	1.85		3.54	4.86
1400	Butyl based, bulk				Gal.	35			35	38.50
1500	Cartridges				"	38.50			38.50	42
1700	1/4" x 1/2", 154 L.F./gal.	1 Bric	288	.028	L.F.	.23	1.28		1.51	2.34
1800	1/2" x 1/2", 77 L.F./gal.	"	275	.029	"	.45	1.34		1.79	2.68
2300	Polysulfide compounds, 1 component, bulk				Gal.	72			72	79
2400	Cartridges				"	116			116	128
2600	1 or 2 component, in place, 1/4" x 1/4", 308 L.F./gal.	1 Bric	300	.027	L.F.	.23	1.23		1.46	2.26
2700	1/2" x 1/4", 154 L.F./gal.		288	.028		.47	1.28		1.75	2.60
2900	3/4" x 3/8", 68 L.F./gal.		272	.029		1.06	1.36		2.42	3.37
3000	1" x 1/2", 38 L.F./gal.		250	.032		1.89	1.48		3.37	4.48
3200	Polyurethane, 1 or 2 component				Gal.	49			49	54
3300	Cartridges				"	64			64	70.50
3500	Bulk, in place, 1/4" x 1/4"	1 Bric	300	.027	L.F.	.16	1.23		1.39	2.18
3655	1/2" x 1/4"		288	.028		.32	1.28		1.60	2.44
3800	3/4" x 3/8"		272	.029		.72	1.36		2.08	3.01
3900	1" x 1/2"		250	.032		1.28	1.48		2.76	3.81
4100	Silicone rubber, bulk				Gal.	49			49	54

07 92 Joint Sealants

07 92 13 – Elastomeric Joint Sealants

07 92 13.20 Caulking and Sealant Options	Crew	Daily Output	Labor-Hours	Unit	Material	2015 Bare Costs Labor	2015 Bare Costs Equipment	Total	Total Incl O&P
4200 Cartridges				Gal.	51			51	56
4300 Bulk in place, 1/4" x 1/2", 154 L.F./gal.	1 Bric	235	.034	L.F.	.32	1.57		1.89	2.91

07 92 16 – Rigid Joint Sealants

07 92 16.10 Rigid Joint Sealants

0010 **RIGID JOINT SEALANTS**									
5802 Tapes, sealant, PVC foam adhesive, 1/16" x 1/4"				L.F.	.10			.10	.11
5902 1/16" x 1/2"					.10			.10	.11
5952 1/16" x 1"					.16			.16	.18
6002 1/8" x 1/2"					.10			.10	.10

07 92 19 – Acoustical Joint Sealants

07 92 19.10 Acoustical Sealant

0010 **ACOUSTICAL SEALANT**									
0020 Acoustical sealant, elastomeric, cartridges				Ea.	8.50			8.50	9.35
0025 In place, 1/4" x 1/4"	1 Bric	300	.027	L.F.	.35	1.23		1.58	2.38
0030 1/4" x 1/2"		288	.028		.69	1.28		1.97	2.85
0035 1/2" x 1/2"		275	.029		1.39	1.34		2.73	3.70
0040 1/2" x 3/4"		263	.030		2.08	1.40		3.48	4.57
0045 3/4" x 3/4"		244	.033		3.12	1.51		4.63	5.90
0050 1" x 1"		200	.040		5.55	1.85		7.40	9.10

07 95 Expansion Control

07 95 13 – Expansion Joint Cover Assemblies

07 95 13.50 Expansion Joint Assemblies

	Crew	Daily Output	Labor-Hours	Unit	Material	Labor	Equipment	Total	Total Incl O&P
0010 **EXPANSION JOINT ASSEMBLIES**									
0200 Floor cover assemblies, 1" space, aluminum	1 Sswk	38	.211	L.F.	16.85	11.10		27.95	38.50
0300 Bronze		38	.211		54.50	11.10		65.60	80
0500 2" space, aluminum		38	.211		16.85	11.10		27.95	38.50
0600 Bronze		38	.211		54.50	11.10		65.60	80
0800 Wall and ceiling assemblies, 1" space, aluminum		38	.211		14.85	11.10		25.95	36.50
0900 Bronze		38	.211		48.50	11.10		59.60	73.50
1100 2" space, aluminum		38	.211		15.15	11.10		26.25	36.50
1200 Bronze		38	.211		48.50	11.10		59.60	73.50
1400 Floor to wall assemblies, 1" space, aluminum		38	.211		18	11.10		29.10	40
1500 Bronze or stainless		38	.211		59	11.10		70.10	85
1700 Gym floor angle covers, aluminum, 3" x 3" angle		46	.174		18	9.15		27.15	36.50
1800 3" x 4" angle		46	.174		21	9.15		30.15	39.50
2000 Roof closures, aluminum, flat roof, low profile, 1" space		57	.140		22.50	7.40		29.90	38.50
2100 High profile		57	.140		24	7.40		31.40	40
2300 Roof to wall, low profile, 1" space		57	.140		21.50	7.40		28.90	37
2400 High profile		57	.140		24	7.40		31.40	40
9000 Minimum labor/equipment charge		2	4	Job		211		211	380

Division 8 – Openings

Estimating Tips
08 10 00 Doors and Frames
All exterior doors should be addressed for their energy conservation (insulation and seals).

- Most metal doors and frames look alike, but there may be significant differences among them. When estimating these items, be sure to choose the line item that most closely compares to the specification or door schedule requirements regarding:
  - type of metal
  - metal gauge
  - door core material
  - fire rating
  - finish

- Wood and plastic doors vary considerably in price. The primary determinant is the veneer material. Lauan, birch, and oak are the most common veneers. Other variables include the following:
  - hollow or solid core
  - fire rating
  - flush or raised panel
  - finish

- Door pricing includes bore for cylindrical lockset and mortise for hinges.

08 30 00 Specialty Doors and Frames
- There are many varieties of special doors, and they are usually priced per each. Add frames, hardware, or operators required for a complete installation.

08 40 00 Entrances, Storefronts, and Curtain Walls
- Glazed curtain walls consist of the metal tube framing and the glazing material. The cost data in this subdivision is presented for the metal tube framing alone or the composite wall. If your estimate requires a detailed takeoff of the framing, be sure to add the glazing cost and any tints.

08 50 00 Windows
- Most metal windows are delivered preglazed. However, some metal windows are priced without glass. Refer to 08 80 00 Glazing for glass pricing. The grade C indicates commercial grade windows, usually ASTM C-35.

- All wood windows and vinyl are priced preglazed. The glazing is insulating glass. Add the cost of screens and grills if required, and not already included.

08 70 00 Hardware
- Hardware costs add considerably to the cost of a door. The most efficient method to determine the hardware requirements for a project is to review the door and hardware schedule together. One type of door may have different hardware, depending on the door usage.

- Door hinges are priced by the pair, with most doors requiring 1-1/2 pairs per door. The hinge prices do not include installation labor, because it is included in door installation. Hinges are classified according to the frequency of use, base material, and finish.

08 80 00 Glazing
- Different openings require different types of glass. The most common types are:
  - float
  - tempered
  - insulating
  - impact-resistant
  - ballistic-resistant

- Most exterior windows are glazed with insulating glass. Entrance doors and window walls, where the glass is less than 18" from the floor, are generally glazed with tempered glass. Interior windows and some residential windows are glazed with float glass.

- Coastal communities require the use of impact-resistant glass, dependant on wind speed.

- The insulation or 'u' value is a strong consideration, along with solar heat gain, to determine total energy efficiency.

Reference Numbers
Reference numbers are shown in shaded boxes at the beginning of some major classifications. These numbers refer to related items in the Reference Section. The reference information may be an estimating procedure, an alternate pricing method, or technical information.

Note: Not all subdivisions listed here necessarily appear in this publication. ∎

No part of this publication may be reproduced, stored in a retrieval system, or transmitted in any form or by any means without prior written permission of RSMeans.

08 01 Operation and Maintenance of Openings

08 01 11 – Operation and Maintenance of Metal Doors and Frames

08 01 11.10 Door and Window Maintenance

		Crew	Daily Output	Labor-Hours	Unit	Material	2015 Bare Costs Labor	Equipment	Total	Total Incl O&P
0010	**DOOR & WINDOW MAINTENANCE**									
1130	Install door	2 Carp	13.07	1.224	Ea.		57.50		57.50	94
1140	Remove deadbolt	1 Carp	13.85	.578			27		27	44.50
1150	Remove panic bar		7.69	1.040			49		49	80
1420	Remove door, plane to fit, rail, reinstall door		9.60	.833			39		39	64

08 01 14 – Operation and Maintenance of Wood Doors

08 01 14.15 Door and Window Maintenance

		Crew	Daily Output	Labor-Hours	Unit	Material	2015 Bare Costs Labor	Equipment	Total	Total Incl O&P
0010	**DOOR & WINDOW MAINTENANCE**									
0050	Rehang single door	1 Carp	32	.250	Ea.		11.75		11.75	19.25
0100	Rehang, double door		16	.500	Pr.		23.50		23.50	38.50
0350	Install window sash cord		32	.250	Ea.	3.80	11.75		15.55	23.50

08 01 53 – Operation and Maintenance of Plastic Windows

08 01 53.81 Solid Vinyl Replacement Windows

			Crew	Daily Output	Labor-Hours	Unit	Material	2015 Bare Costs Labor	Equipment	Total	Total Incl O&P
0010	**SOLID VINYL REPLACEMENT WINDOWS**	R085313-20									
0020	Double hung, insulated glass, up to 83 united inches	G	2 Carp	8	2	Ea.	340	94		434	530
0040	84 to 93	G		8	2		380	94		474	570
0060	94 to 101	G		6	2.667		380	125		505	620
0080	102 to 111	G		6	2.667		395	125		520	640
0100	112 to 120	G		6	2.667		430	125		555	680
0120	For each united inch over 120, add	G		800	.020	Inch	5.50	.94		6.44	7.60
0140	Casement windows, one operating sash, 42 to 60 united inches	G		8	2	Ea.	230	94		324	405
0160	61 to 70	G		8	2		260	94		354	440
0180	71 to 80	G		8	2		280	94		374	465
0200	81 to 96	G		8	2		300	94		394	485
0220	Two operating sash, 58 to 78 united inches	G		8	2		455	94		549	655
0240	79 to 88	G		8	2		485	94		579	690
0260	89 to 98	G		8	2		525	94		619	735
0280	99 to 108	G		6	2.667		550	125		675	810
0300	109 to 121	G		6	2.667		590	125		715	855
0320	Two operating, one fixed sash, 73 to 108 united inches	G		8	2		715	94		809	945
0340	109 to 118	G		8	2		755	94		849	985
0360	119 to 128	G		6	2.667		775	125		900	1,050
0380	129 to 138	G		6	2.667		870	125		995	1,150
0400	139 to 156	G		6	2.667		945	125		1,070	1,250
0420	Four operating sash, 98 to 118 united inches	G		8	2		1,025	94		1,119	1,275
0440	119 to 128	G		8	2		1,100	94		1,194	1,350
0460	129 to 138	G		6	2.667		1,150	125		1,275	1,475
0480	139 to 148	G		6	2.667		1,225	125		1,350	1,525
0500	149 to 168	G		6	2.667		1,300	125		1,425	1,625
0520	169 to 178	G		6	2.667		1,400	125		1,525	1,725
0560	Fixed picture window, up to 63 united inches	G		8	2		180	94		274	350
0580	64 to 83	G		8	2		206	94		300	380
0600	84 to 101	G		8	2		257	94		351	435
0620	For each united inch over 101, add	G		900	.018	Inch	3.10	.83		3.93	4.78
0800	Cellulose fiber insulation, poured into sash balance cavity	G	1 Carp	36	.222	C.F.	.69	10.45		11.14	17.85
0820	Silicone caulking at perimeter	G	"	800	.010	L.F.	.17	.47		.64	.95

08 05 Common Work Results for Openings

08 05 05 – Selective Demolition for Openings

08 05 05.10 Selective Demolition Doors

		Crew	Daily Output	Labor-Hours	Unit	Material	2015 Bare Costs Labor	2015 Bare Costs Equipment	Total	Total Incl O&P
0010	**SELECTIVE DEMOLITION DOORS** R024119-10									
0200	Doors, exterior, 1-3/4" thick, single, 3' x 7' high	1 Clab	16	.500	Ea.		18.80		18.80	31
0202	Doors, exterior, 3' - 6' wide x 7' high		12	.667			25		25	41
0210	3' x 8' high		10	.800			30		30	49.50
0215	Double, 3' x 8' high		6	1.333			50		50	82
0220	Double, 6' x 7' high		12	.667			25		25	41
0500	Interior, 1-3/8" thick, single, 3' x 7' high		20	.400			15.05		15.05	24.50
0520	Double, 6' x 7' high		16	.500			18.80		18.80	31
0700	Bi-folding, 3' x 6'-8" high		20	.400			15.05		15.05	24.50
0720	6' x 6'-8" high		18	.444			16.70		16.70	27.50
0900	Bi-passing, 3' x 6'-8" high		16	.500			18.80		18.80	31
0940	6' x 6'-8" high		14	.571			21.50		21.50	35
1500	Remove and reset, hollow core	1 Carp	8	1			47		47	77
1520	Solid		6	1.333			62.50		62.50	103
2000	Frames, including trim, metal		8	1			47		47	77
2200	Wood	2 Carp	32	.500			23.50		23.50	38.50
2201	Alternate pricing method	1 Carp	200	.040	L.F.		1.88		1.88	3.08
2205	Remove door hardware	"	45.70	.175	Ea.		8.20		8.20	13.45
3000	Special doors, counter doors	2 Carp	6	2.667			125		125	205
3100	Double acting		10	1.600			75		75	123
3200	Floor door (trap type), or access type		8	2			94		94	154
3300	Glass, sliding, including frames		12	1.333			62.50		62.50	103
3400	Overhead, commercial, 12' x 12' high		4	4			188		188	310
3440	up to 20' x 16' high		3	5.333			250		250	410
3445	up to 35' x 30' high		1	16			750		750	1,225
3500	Residential, 9' x 7' high		8	2			94		94	154
3540	16' x 7' high		7	2.286			107		107	176
3600	Remove and reset, small		4	4			188		188	310
3620	Large		2.50	6.400			300		300	490
3700	Roll-up grille		5	3.200			150		150	246
3800	Revolving door		2	8			375		375	615
3900	Storefront swing door		3	5.333			250		250	410
3902	Cafe/bar swing door	2 Clab	8	2			75		75	123
4000	Residential lockset, exterior	1 Carp	28	.286			13.40		13.40	22
4224	Pocket door		8	1			47		47	77
4300	Remove and reset deadbolt		8	1			47		47	77
5585	Remove panic device	1 Clab	10	.800			30		30	49.50
5590	Remove mail slot		45	.178			6.70		6.70	10.95
5595	Remove peep hole		45	.178			6.70		6.70	10.95
5604	Remove 9' x 7' garage door	2 Carp	9.50	1.684			79		79	130
5624	16' x 7' garage door		8	2			94		94	154
5626	9' x 7' swing up garage door		7	2.286			107		107	176
5628	16' x 7' swing up garage door		7	2.286			107		107	176
5630	Garage door demolition, remove garage door track	1 Carp	6	1.333			62.50		62.50	103
5644	Remove overhead door opener	1 Clab	5	1.600			60		60	98.50
5774	Remove heavy gage sectional door, 20 ga., 8' x 8'	2 Carp	7	2.286			107		107	176
5784	10' x 10'		6	2.667			125		125	205
5794	12' x 12'		5	3.200			150		150	246
5805	14' x 14'		4	4			188		188	310
6384	Remove French door unit	1 Clab	8	1			37.50		37.50	61.50
7100	Remove double swing pneumatic doors, openers and sensors	2 Skwk	.50	32	Opng.		1,550		1,550	2,525
7570	Remove shock absorbing door	2 Sswk	1.90	8.421	"		445		445	800
9000	Minimum labor/equipment charge	1 Carp	4	2	Job		94		94	154

08 05 Common Work Results for Openings

08 05 05 – Selective Demolition for Openings

08 05 05.20 Selective Demolition of Windows

		Crew	Daily Output	Labor-Hours	Unit	Material	2015 Bare Costs Labor	2015 Bare Costs Equipment	Total	Total Incl O&P
0010	**SELECTIVE DEMOLITION OF WINDOWS** R024119-10									
0200	Aluminum, including trim, to 12 S.F.	1 Clab	16	.500	Ea.		18.80		18.80	31
0240	To 25 S.F.		11	.727			27.50		27.50	45
0280	To 50 S.F.		5	1.600			60		60	98.50
0320	Storm windows/screens, to 12 S.F.		27	.296			11.15		11.15	18.25
0360	To 25 S.F.		21	.381			14.30		14.30	23.50
0400	To 50 S.F.		16	.500			18.80		18.80	31
0600	Glass, up to 10 SF per window		200	.040	S.F.		1.50		1.50	2.47
0620	Over 10 SF per window		150	.053	"		2.01		2.01	3.29
1000	Steel, including trim, to 12 S.F.		13	.615	Ea.		23		23	38
1020	To 25 S.F.		9	.889			33.50		33.50	55
1040	To 50 S.F.		4	2			75		75	123
2000	Wood, including trim, to 12 S.F.		22	.364			13.65		13.65	22.50
2020	To 25 S.F.		18	.444			16.70		16.70	27.50
2060	To 50 S.F.		13	.615			23		23	38
2065	To 180 S.F.		8	1			37.50		37.50	61.50
4400	Remove skylight, prefabricated glass block with metal frame	G-3	180	.178	S.F.		8.30		8.30	13.35
4410	Remove skylight, plstc domes, flush/curb mtd	"	395	.081	"		3.79		3.79	6.05
5020	Remove and reset window, up to a 2'x2' widow	1 Carp	6	1.333	Ea.		62.50		62.50	103
5040	Up to a 3'x3' window		4	2			94		94	154
5080	Up to a 4'x5' window		2	4			188		188	310
6000	Screening only	1 Clab	4000	.002	S.F.		.08		.08	.12
9000	Minimum labor/equipment charge	"	4	2	Job		75		75	123

08 11 Metal Doors and Frames

08 11 16 – Aluminum Doors and Frames

08 11 16.10 Entrance Doors

		Crew	Daily Output	Labor-Hours	Unit	Material	2015 Bare Costs Labor	2015 Bare Costs Equipment	Total	Total Incl O&P
0010	**ENTRANCE DOORS** and frame, Aluminum, narrow stile									
0011	Including standard hardware, clear finish, no glass									
0012	Top and bottom offset pivots, 1/4" beveled glass stops, threshold									
0013	Dead bolt lock with inside thumb screw, standard push pull									
0020	3'-0" x 7'-0" opening	2 Sswk	2	8	Ea.	915	420		1,335	1,750
0025	Anodizing aluminum entr. door & frame, add					104			104	114
0030	3'-6" x 7'-0" opening	2 Sswk	2	8		840	420		1,260	1,675
0100	3'-0" x 10'-0" opening, 3' high transom		1.80	8.889		1,300	470		1,770	2,275
0200	3'-6" x 10'-0" opening, 3' high transom		1.80	8.889		1,350	470		1,820	2,350
0280	5'-0" x 7'-0" opening		2	8		1,425	420		1,845	2,325
0300	6'-0" x 7'-0" opening		1.30	12.308		1,200	650		1,850	2,500
0301	6'-0" x 7'-0" opening		1.30	12.308	Pr.	1,200	650		1,850	2,500
0400	6'-0" x 10'-0" opening, 3' high transom		1.10	14.545	"	1,675	765		2,440	3,200
0520	3'-0" x 7'-0" opening, wide stile		2	8	Ea.	1,000	420		1,420	1,850
0540	3'-6" x 7'-0" opening		2	8		1,200	420		1,620	2,075
0560	5'-0" x 7'-0" opening		2	8		1,525	420		1,945	2,425
0580	6'-0" x 7'-0" opening		1.30	12.308	Pr.	1,575	650		2,225	2,925
0600	7'-0" x 7'-0" opening		1	16	"	1,725	840		2,565	3,400
1200	For non-standard size, add				Leaf	80%				
1250	For installation of non-standard size, add						20%			
1300	Light bronze finish, add				Leaf	36%				
1400	Dark bronze finish, add					25%				
1500	For black finish, add					40%				

08 11 Metal Doors and Frames

08 11 16 – Aluminum Doors and Frames

08 11 16.10 Entrance Doors

		Crew	Daily Output	Labor-Hours	Unit	Material	2015 Bare Costs Labor	Equipment	Total	Total Incl O&P
1600	Concealed panic device, add				Leaf	940			940	1,025
1700	Electric striker release, add				Opng.	280			280	310
1800	Floor check, add				Leaf	650			650	710
1900	Concealed closer, add				"	530			530	580
9000	Minimum labor/equipment charge	2 Carp	4	4	Job		188		188	310

08 11 63 – Metal Screen and Storm Doors and Frames

08 11 63.23 Aluminum Screen and Storm Doors and Frames

		Crew	Daily Output	Labor-Hours	Unit	Material	Labor	Equipment	Total	Total Incl O&P
0010	**ALUMINUM SCREEN AND STORM DOORS AND FRAMES**									
0020	Combination storm and screen									
0420	Clear anodic coating, 2'-8" wide	2 Carp	14	1.143	Ea.	205	53.50		258.50	315
0440	3'-0" wide	"	14	1.143	"	178	53.50		231.50	283
0500	For 7' door height, add					8%				
1020	Mill finish, 2'-8" wide	2 Carp	14	1.143	Ea.	235	53.50		288.50	345
1040	3'-0" wide	"	14	1.143		258	53.50		311.50	370
1100	For 7'-0" door, add					8%				
1520	White painted, 2'-8" wide	2 Carp	14	1.143		286	53.50		339.50	405
1540	3'-0" wide	"	14	1.143		310	53.50		363.50	430
1600	For 7'-0" door, add					8%				
2000	Wood door & screen, see Section 08 14 33.20									
9000	Minimum labor/equipment charge	1 Carp	4	2	Job		94		94	154

08 12 Metal Frames

08 12 13 – Hollow Metal Frames

08 12 13.13 Standard Hollow Metal Frames

			Crew	Daily Output	Labor-Hours	Unit	Material	Labor	Equipment	Total	Total Incl O&P
0010	**STANDARD HOLLOW METAL FRAMES**										
0020	16 ga., up to 5-3/4" jamb depth										
0025	3'-0" x 6'-8" single	G	2 Carp	16	1	Ea.	146	47		193	237
0028	3'-6" wide, single	G		16	1		153	47		200	246
0030	4'-0" wide, single	G		16	1		152	47		199	244
0040	6'-0" wide, double	G		14	1.143		204	53.50		257.50	310
0045	8'-0" wide, double	G		14	1.143		213	53.50		266.50	320
0100	3'-0" x 7'-0" single	G		16	1		151	47		198	243
0110	3'-6" wide, single			16	1		159	47		206	251
0112	4'-0" wide, single			16	1		159	47		206	251
0140	6'-0" wide, double			14	1.143		194	53.50		247.50	300
0145	8'-0" wide, double			14	1.143		229	53.50		282.50	340
1000	16 ga., up to 4-7/8" deep, 3'-0" x 7'-0" single	G		16	1		166	47		213	260
1140	6'-0" wide, double	G		14	1.143		188	53.50		241.50	295
1200	16 ga., 8-3/4" deep, 3'-0" x 7'-0" single	G		16	1		199	47		246	296
1240	6'-0" wide, double	G		14	1.143		234	53.50		287.50	345
2800	14 ga., up to 3-7/8" deep, 3'-0" x 7'-0" single	G		16	1		181	47		228	276
2840	6'-0" wide, double	G		14	1.143		217	53.50		270.50	325
3000	14 ga., up to 5-3/4" deep, 3'-0" x 6'-8" single	G		16	1		155	47		202	248
3002	3'-6" wide, single	G		16	1		198	47		245	294
3005	4'-0" wide, single	G		16	1		203	47		250	300
3600	up to 5-3/4" jamb depth, 4'-0" x 7'-0" single			15	1.067		185	50		235	285
3620	6'-0" wide, double			12	1.333		235	62.50		297.50	360
3640	8'-0" wide, double			12	1.333		246	62.50		308.50	375
3700	8'-0" high, 4'-0" wide, single			15	1.067		235	50		285	340
3740	8'-0" wide, double			12	1.333		290	62.50		352.50	425

08 12 Metal Frames

08 12 13 – Hollow Metal Frames

08 12 13.13 Standard Hollow Metal Frames

		Crew	Daily Output	Labor-Hours	Unit	Material	2015 Bare Costs Labor	Equipment	Total	Total Incl O&P
4000	6-3/4" deep, 4'-0" x 7'-0" single G	2 Carp	15	1.067	Ea.	222	50		272	325
4020	6'-0" wide, double G		12	1.333		276	62.50		338.50	410
4040	8'-0", wide double G		12	1.333		284	62.50		346.50	415
4100	8'-0" high, 4'-0" wide, single G		15	1.067		276	50		326	385
4140	8'-0" wide, double G		12	1.333		320	62.50		382.50	460
4400	8-3/4" deep, 4'-0" x 7'-0", single		15	1.067		252	50		302	360
4440	8'-0" wide, double		12	1.333		315	62.50		377.50	450
4500	4'-0" x 8'-0", single		15	1.067		283	50		333	390
4540	8'-0" wide, double		12	1.333		350	62.50		412.50	490
4900	For welded frames, add					63			63	69.50
5380	Steel frames, KD, 14 ga., "B" label, to 5-3/4" throat, to 3'-0" x 7'-0"	2 Carp	15	1.067		215	50		265	320
5400	14 ga., "B" label, up to 5-3/4" deep, 4'-0" x 7'-0" single G		15	1.067		210	50		260	315
5440	8'-0" wide, double G		12	1.333		272	62.50		334.50	400
5800	6-3/4" deep, 7'-0" high, 4'-0" wide, single G		15	1.067		218	50		268	320
5840	8'-0" wide, double G		12	1.333		385	62.50		447.50	530
6200	8-3/4" deep, 4'-0" x 7'-0" single G		15	1.067		291	50		341	400
6240	8'-0" wide, double G		12	1.333		380	62.50		442.50	520
6300	For "A" label use same price as "B" label									
6400	For baked enamel finish, add					30%	15%			
6500	For galvanizing, add					20%				
6600	For hospital stop, add				Ea.	295			295	325
6620	For hospital stop, stainless steel add				"	380			380	420
7900	Transom lite frames, fixed, add	2 Carp	155	.103	S.F.	54	4.85		58.85	67.50
8000	Movable, add	"	130	.123	"	68	5.80		73.80	84.50
9000	Minimum labor/equipment charge	1 Carp	4	2	Job		94		94	154

08 12 13.25 Channel Metal Frames

		Crew	Daily Output	Labor-Hours	Unit	Material	Labor	Equipment	Total	Total Incl O&P
0010	**CHANNEL METAL FRAMES**									
0020	Steel channels with anchors and bar stops									
0100	6" channel @ 8.2#/L.F., 3' x 7' door, weighs 150# G	E-4	13	2.462	Ea.	239	131	11.20	381.20	510
0200	8" channel @ 11.5#/L.F., 6' x 8' door, weighs 275# G		9	3.556		435	189	16.20	640.20	840
0300	8' x 12' door, weighs 400# G		6.50	4.923		635	262	22.50	919.50	1,200
0400	10" channel @ 15.3#/L.F., 10' x 10' door, weighs 500# G		6	5.333		795	283	24.50	1,102.50	1,400
0500	12' x 12' door, weighs 600# G		5.50	5.818		955	310	26.50	1,291.50	1,650
0600	12" channel @ 20.7#/L.F., 12' x 12' door, weighs 825# G		4.50	7.111		1,300	380	32.50	1,712.50	2,175
0700	12' x 16' door, weighs 1000# G		4	8		1,600	425	36.50	2,061.50	2,550
0800	For frames without bar stops, light sections, deduct					15%				
0900	Heavy sections, deduct					10%				
9000	Minimum labor/equipment charge	E-4	4	8	Job		425	36.50	461.50	810

08 12 13.53 Borrowed Lites

		Crew	Daily Output	Labor-Hours	Unit	Material	Labor	Equipment	Total	Total Incl O&P
0010	**BORROWED LITES**									
0100	Hollow metal section 20 ga., 3 1/2" with glass stop	2 Carp	100	.160	L.F.	8.15	7.50		15.65	21.50
0110	3-3/4" with glass stop		100	.160		8.30	7.50		15.80	21.50
0120	4" with glass stop		100	.160		8.45	7.50		15.95	21.50
0130	4-5/8" with glass stop		100	.160		8.75	7.50		16.25	22
0140	4-7/8" with glass stop		100	.160		8.90	7.50		16.40	22
0150	5" with glass stop		100	.160		9	7.50		16.50	22
0160	5-3/8" with glass stop		100	.160		9.25	7.50		16.75	22.50
0300	Hollow metal section 18 ga., 3-1/2" with glass stop		80	.200		9	9.40		18.40	25.50
0310	3-3/4" with glass stop		80	.200		9.20	9.40		18.60	25.50
0320	4" with glass stop		80	.200		9.30	9.40		18.70	25.50
0330	4-5/8" with glass stop		80	.200		9.75	9.40		19.15	26
0340	4-7/8" with glass stop		80	.200		10	9.40		19.40	26.50

08 12 Metal Frames

08 12 13 – Hollow Metal Frames

08 12 13.53 Borrowed Lites

		Crew	Daily Output	Labor-Hours	Unit	Material	2015 Bare Costs Labor	Equipment	Total	Total Incl O&P
0350	5" with glass stop	2 Carp	80	.200	L.F.	10.10	9.40		19.50	26.50
0360	5-3/8" with glass stop		80	.200		10.35	9.40		19.75	27
0370	6-5/8" with glass stop		80	.200		11.25	9.40		20.65	28
0380	6-7/8" with glass stop		80	.200		11.40	9.40		20.80	28
0390	7-1/4" with glass stop		80	.200		11.90	9.40		21.30	28.50
0500	Mullion section 18 ga., 3-1/2" with double stop		50	.320		21	15		36	47.50
0510	3-3/4" with double stop		50	.320		21.50	15		36.50	48
0530	4-5/8" with double stop		50	.320		23	15		38	49.50
0540	4-7/8" with double stop		50	.320		23.50	15		38.50	50
0550	5" with double stop		50	.320		23.50	15		38.50	50
0560	5-3/8" with double stop		50	.320		24	15		39	51
0570	6-5/8" with double stop		50	.320		26	15		41	53.50
0580	6-7/8" with double stop		50	.320		26.50	15		41.50	54
0590	7-1/4" with double stop		50	.320		27	15		42	54.50
1000	Assembled frame 2'-0" x 2'-0" 16 ga., 3-3/4" with glass stop		10	1.600	Ea.	180	75		255	320
1010	3'-0" x 2'-0"		10	1.600		180	75		255	320
1020	4'-0" x 2'-0"		10	1.600		180	75		255	320
1030	5'-0" x 2'-0"		10	1.600		180	75		255	320
1040	6'-0" x 2'-0"		8	2		270	94		364	450
1050	7'-0" x 2'-0"		8	2		270	94		364	450
1060	8'-0" x 2'-0"		8	2		270	94		364	450
1100	2'-0" x 7'-0"		8	2		270	94		364	450
1110	3'-0" x 7'-0"		8	2		270	94		364	450
1120	4'-0" x 7'-0"		8	2		270	94		364	450
1130	5'-0" x 7'-0"		8	2		270	94		364	450
1140	6'-0" x 7'-0"		6	2.667		295	125		420	530
1150	7'-0" x 7'-0"		6	2.667		295	125		420	530
1160	8'-0" x 7'-0"		6	2.667		295	125		420	530
1200	Assembled frame 2'-0" x 2'-0" 16 ga., 4-7/8" with glass stop		10	1.600		180	75		255	320
1210	3'-0" x 2'-0"		10	1.600		180	75		255	320
1220	4'-0" x 2'-0"		8	2		180	94		274	350
1230	5'-0" x 2'-0"		8	2		180	94		274	350
1240	6'-0" x 2'-0"		8	2		270	94		364	450
1250	7'-0" x 2'-0"		8	2		270	94		364	450
1260	8'-0" x 2'-0"		8	2		270	94		364	450
1300	2'-0" x 7'-0"		8	2		270	94		364	450
1310	3'-0" x 7'-0"		8	2		270	94		364	450
1320	4'-0" x 7'-0"		8	2		270	94		364	450
1330	5'-0" x 7'-0"		6	2.667		270	125		395	500
1340	6'-0" x 7'-0"		6	2.667		295	125		420	530
1350	7'-0" x 7'-0"		6	2.667		270	125		395	500
1360	8'-0" x 7'-0"		6	2.667		295	125		420	530
1400	Assembled frame 2'-0" x 3'-0" 16 ga., 6-1/8" with glass stop		6	2.667		180	125		305	405
1410	3'-0" x 3'-0"		8	2		180	94		274	350
1420	4'-0" x 3'-0"		8	2		180	94		274	350
1430	5'-0" x 3'-0"		6	2.667		180	125		305	405
1440	6'-0" x 3'-0"		6	2.667		270	125		395	500
1450	7'-0" x 3'-0"		6	2.667		270	125		395	500
1460	8'-0" x 3'-0"		10	1.600		270	75		345	420

08 13 Metal Doors

08 13 13 – Hollow Metal Doors

08 13 13.13 Standard Hollow Metal Doors		Crew	Daily Output	Labor-Hours	Unit	Material	2015 Bare Costs Labor	2015 Bare Costs Equipment	Total	Total Incl O&P
0010	STANDARD HOLLOW METAL DOORS R081313-20									
0015	Flush, full panel, hollow core									
0017	When noted doors are prepared but do not include glass or louvers									
0020	1-3/8" thick, 20 ga., 2'-0" x 6'-8" G	2 Carp	20	.800	Ea.	335	37.50		372.50	430
0040	2'-8" x 6'-8" G		18	.889		350	41.50		391.50	455
0060	3'-0" x 6'-8" G		17	.941		350	44		394	460
0100	3'-0" x 7'-0" G		17	.941		360	44		404	470
0120	For vision lite, add					94.50			94.50	104
0140	For narrow lite, add					102			102	113
0320	Half glass, 20 ga., 2'-0" x 6'-8" G	2 Carp	20	.800		490	37.50		527.50	600
0340	2'-8" x 6'-8" G		18	.889		515	41.50		556.50	635
0360	3'-0" x 6'-8" G		17	.941		510	44		554	635
0400	3'-0" x 7'-0" G		17	.941		625	44		669	765
0410	1-3/8" thick, 18 ga., 2'-0" x 6'-8" G		20	.800		400	37.50		437.50	500
0420	3'-0" x 6'-8" G		17	.941		405	44		449	520
0425	3'-0" x 7'-0" G		17	.941		415	44		459	535
0450	For vision lite, add					94.50			94.50	104
0452	For narrow lite, add					102			102	113
0460	Half glass, 18 ga., 2'-0" x 6'-8" G	2 Carp	20	.800		555	37.50		592.50	675
0465	2'-8" x 6'-8" G		18	.889		575	41.50		616.50	705
0470	3'-0" x 6'-8" G		17	.941		565	44		609	695
0475	3'-0" x 7'-0" G		17	.941		565	44		609	700
0500	Hollow core, 1-3/4" thick, full panel, 20 ga., 2'-8" x 6'-8" G		18	.889		420	41.50		461.50	530
0520	3'-0" x 6'-8" G		17	.941		420	44		464	535
0640	3'-0" x 7'-0" G		17	.941		435	44		479	555
0680	4'-0" x 7'-0" G		15	1.067		630	50		680	775
0700	4'-0" x 8'-0" G		13	1.231		735	58		793	905
1000	18 ga., 2'-8" x 6'-8" G		17	.941		490	44		534	615
1020	3'-0" x 6'-8" G		16	1		475	47		522	600
1120	3'-0" x 7'-0" G		17	.941		515	44		559	640
1150	3'-6" x 7'-0"	1 Carp	14	.571		570	27		597	675
1180	4'-0" x 7'-0" G	2 Carp	14	1.143		630	53.50		683.50	785
1200	4'-0" x 8'-0" G	"	17	.941		735	44		779	885
1212	For vision lite, add					94.50			94.50	104
1214	For narrow lite, add					102			102	113
1230	Half glass, 20 ga., 2'-8" x 6'-8" G	2 Carp	20	.800		575	37.50		612.50	695
1240	3'-0" x 6'-8" G		18	.889		575	41.50		616.50	705
1260	3'-0" x 7'-0" G		18	.889		595	41.50		636.50	720
1280	Embossed panel, 1-3/4" thick, poly core, 20 ga., 3'-0" x 7'-0" G		18	.889		415	41.50		456.50	525
1290	Half glass, 1-3/4" thick, poly core, 20 ga., 3'-0" x 7'-0" G		18	.889		610	41.50		651.50	740
1320	18 ga., 2'-8" x 6'-8" G		18	.889		640	41.50		681.50	775
1340	3'-0" x 6'-8" G		17	.941		635	44		679	770
1360	3'-0" x 7'-0" G		17	.941		645	44		689	785
1380	4'-0" x 7'-0" G		15	1.067		785	50		835	945
1400	4'-0" x 8'-0" G		14	1.143		890	53.50		943.50	1,075
1500	Flush full panel, 16 ga., steel hollow core									
1520	2'-0" x 6'-8" G	2 Carp	20	.800	Ea.	555	37.50		592.50	670
1530	2'-8" x 6'-8" G		20	.800		555	37.50		592.50	675
1540	3'-0" x 6'-8" G		20	.800		550	37.50		587.50	665
1560	2'-8" x 7'-0" G		18	.889		575	41.50		616.50	700
1570	3'-0" x 7'-0" G		18	.889		560	41.50		601.50	685
1580	3'-6" x 7'-0" G		18	.889		655	41.50		696.50	790
1590	4'-0" x 7'-0" G		18	.889		725	41.50		766.50	865

08 13 Metal Doors

08 13 13 – Hollow Metal Doors

08 13 13.13 Standard Hollow Metal Doors

		Crew	Daily Output	Labor-Hours	Unit	Material	2015 Bare Costs Labor	Equipment	Total	Total Incl O&P
1600	2'-8" x 8'-0" [G]	2 Carp	18	.889	Ea.	700	41.50		741.50	840
1620	3'-0" x 8'-0" [G]		18	.889		720	41.50		761.50	860
1630	3'-6" x 8'-0" [G]		18	.889		800	41.50		841.50	950
1640	4'-0" x 8'-0" [G]		18	.889		850	41.50		891.50	1,000
1650	1-13/16", 14 Ga., 2'-8" x 7'-0" [G]		10	1.600		1,125	75		1,200	1,375
1670	3'-0" x 7'-0" [G]		10	1.600		1,075	75		1,150	1,325
1690	3'-6" x 7'-0" [G]		10	1.600		1,200	75		1,275	1,425
1700	4'-0" x 7'-0" [G]		10	1.600		1,250	75		1,325	1,500
1720	Insulated, 1-3/4" thick, full panel, 18 ga., 3'-0" x 6'-8" [G]		15	1.067		475	50		525	605
1740	2'-8" x 7'-0" [G]		16	1		505	47		552	630
1760	3'-0" x 7'-0" [G]		15	1.067		490	50		540	620
1800	4'-0" x 8'-0" [G]		13	1.231		740	58		798	905
1820	Half glass, 18 ga., 3'-0" x 6'-8" [G]		16	1		635	47		682	770
1840	2'-8" x 7'-0" [G]		17	.941		660	44		704	800
1860	3'-0" x 7'-0" [G]		16	1		685	47		732	830
1900	4'-0" x 8'-0" [G]		14	1.143		670	53.50		723.50	825
2000	For vision lite, add					94.50			94.50	104
2010	For narrow lite, add					102			102	113
8100	For bottom louver, add					282			282	310
8110	For baked enamel finish, add					30%	15%			
8120	For galvanizing, add					20%				
8170	For soundproofing STC 40, add				Ea.	1,725			1,725	1,900
8190	For 3 hour door, add					370			370	405
8270	For dutch door with shelf, add to standard door					300			300	330
9000	Minimum labor/equipment charge	1 Carp	4	2	Job		94		94	154

08 13 13.15 Metal Fire Doors

		Crew	Daily Output	Labor-Hours	Unit	Material	Labor	Equipment	Total	Total Incl O&P
0010	**METAL FIRE DOORS** R081313-20									
0015	Steel, flush, "B" label, 90 minute									
0020	Full panel, 20 ga., 2'-0" x 6'-8"	2 Carp	20	.800	Ea.	420	37.50		457.50	520
0040	2'-8" x 6'-8"		18	.889		435	41.50		476.50	550
0060	3'-0" x 6'-8"		17	.941		435	44		479	555
0080	3'-0" x 7'-0"		17	.941		455	44		499	575
0140	18 ga., 3'-0" x 6'-8"		16	1		495	47		542	620
0160	2'-8" x 7'-0"		17	.941		520	44		564	645
0180	3'-0" x 7'-0"		16	1		505	47		552	630
0200	4'-0" x 7'-0"		15	1.067		650	50		700	795
0220	For "A" label, 3 hour, 18 ga., use same price as "B" label									
0240	For vision lite, add				Ea.	156			156	172
0300	Full panel, 16 ga., 2'-0" x 6'-8"	2 Carp	20	.800		545	37.50		582.50	660
0310	2'-8" x 6'-8"		18	.889		545	41.50		586.50	670
0320	3'-0" x 6'-8"		17	.941		540	44		584	670
0350	2'-8" x 7'-0"		17	.941		565	44		609	695
0360	3'-0" x 7'-0"		16	1		545	47		592	675
0370	4'-0" x 7'-0"		15	1.067		705	50		755	860
0520	Flush, "B" label 90 min., egress core, 20 ga., 2'-0" x 6'-8"		18	.889		655	41.50		696.50	790
0540	2'-8" x 6'-8"		17	.941		665	44		709	805
0560	3'-0" x 6'-8"		16	1		665	47		712	805
0580	3'-0" x 7'-0"		16	1		685	47		732	825
0640	Flush, "A" label 3 hour, egress core, 18 ga., 3'-0" x 6'-8"		15	1.067		720	50		770	875
0660	2'-8" x 7'-0"		16	1		750	47		797	895
0680	3'-0" x 7'-0"		15	1.067		740	50		790	895
0700	4'-0" x 7'-0"		14	1.143		885	53.50		938.50	1,050

08 13 Metal Doors

08 13 13 – Hollow Metal Doors

08 13 13.15 Metal Fire Doors

			Crew	Daily Output	Labor-Hours	Unit	Material	2015 Bare Costs Labor	2015 Bare Costs Equipment	Total	Total Incl O&P
9000	Minimum labor/equipment charge		1 Carp	4	2	Job		94		94	154

08 13 13.20 Residential Steel Doors

			Crew	Daily Output	Labor-Hours	Unit	Material	Labor	Equipment	Total	Total Incl O&P
0010	**RESIDENTIAL STEEL DOORS**										
0020	Prehung, insulated, exterior										
0030	Embossed, full panel, 2'-8" x 6'-8"	G	2 Carp	17	.941	Ea.	315	44		359	420
0040	3'-0" x 6'-8"	G		15	1.067		270	50		320	380
0060	3'-0" x 7'-0"	G		15	1.067		330	50		380	445
0070	5'-4" x 6'-8", double	G		8	2		630	94		724	850
0220	Half glass, 2'-8" x 6'-8"	G		17	.941		320	44		364	425
0240	3'-0" x 6'-8"	G		16	1		320	47		367	425
0260	3'-0" x 7'-0"	G		16	1		370	47		417	480
0270	5'-4" x 6'-8", double	G		8	2		650	94		744	870
1320	Flush face, full panel, 2'-8" x 6'-8"	G		16	1		264	47		311	365
1340	3'-0" x 6'-8"	G		15	1.067		264	50		314	370
1360	3'-0" x 7'-0"	G		15	1.067		296	50		346	405
1380	5'-4" x 6'-8", double	G		8	2		560	94		654	770
1420	Half glass, 2'-8" x 6'-8"	G		17	.941		325	44		369	430
1440	3'-0" x 6'-8"	G		16	1		325	47		372	430
1460	3'-0" x 7'-0"	G		16	1		360	47		407	470
1480	5'-4" x 6'-8", double	G		8	2		640	94		734	860
1500	Sidelight, full lite, 1'-0" x 6'-8" with grille	G					252			252	277
1510	1'-0" x 6'-8", low e	G					268			268	295
1520	1'-0" x 6'-8", half lite	G					280			280	310
1530	1'-0" x 6'-8", half lite, low e	G					284			284	310
2300	Interior, residential, closet, bi-fold, 2'-0" x 6'-8"	G	2 Carp	16	1		161	47		208	254
2330	3'-0" wide	G		16	1		200	47		247	297
2360	4'-0" wide	G		15	1.067		260	50		310	370
2400	5'-0" wide	G		14	1.143		315	53.50		368.50	440
2420	6'-0" wide	G		13	1.231		335	58		393	465
9000	Minimum labor/equipment charge		1 Carp	4	2	Job		94		94	154

08 13 13.25 Doors Hollow Metal

			Crew	Daily Output	Labor-Hours	Unit	Material	Labor	Equipment	Total	Total Incl O&P
0010	**DOORS HOLLOW METAL**										
0500	Exterior, commercial, flush, 20 ga., 1-3/4" x 7'-0" x 2'-6" wide	G	2 Carp	15	1.067	Ea.	395	50		445	510
0530	2'-8" wide	G		15	1.067		435	50		485	555
0560	3'-0" wide	G		14	1.143		435	53.50		488.50	565
0590	3'-6" wide	G		14	1.143		440	53.50		493.50	575
1000	18 ga., 1-3/4" x 7'-0" x 2'-6" wide	G		15	1.067		495	50		545	625
1030	2'-8" wide	G		15	1.067		500	50		550	630
1060	3'-0" wide	G		14	1.143		485	53.50		538.50	625
1500	16 ga., 1-3/4 x 7'-0" x 2'-6" wide	G		15	1.067		560	50		610	695
1530	2'-8" wide	G		15	1.067		565	50		615	705
1560	3'-0" wide	G		14	1.143		555	53.50		608.50	700
1590	3'-6" wide	G		14	1.143		645	53.50		698.50	800
2900	Fire door, "A" label, 18 gauge, 1-3/4" x 2'-6" x 7'-0"	G		15	1.067		645	50		695	790
2930	2'-8" wide	G		15	1.067		665	50		715	810
2960	3'-0" wide	G		14	1.143		640	53.50		693.50	795
2990	3'-6" wide	G		14	1.143		735	53.50		788.50	895
3100	"B" label, 2'-6" wide	G		15	1.067		585	50		635	725
3130	2'-8" wide	G		15	1.067		590	50		640	730
3160	3'-0" wide	G		14	1.143		580	53.50		633.50	730

08 13 Metal Doors

08 13 16 – Aluminum Doors

08 13 16.10 Commercial Aluminum Doors

		Crew	Daily Output	Labor-Hours	Unit	Material	2015 Bare Costs Labor	Equipment	Total	Total Incl O&P
0010	**COMMERCIAL ALUMINUM DOORS**, flush, no glazing									
5000	Flush panel doors, pair of 2'-6" x 7'-0"	2 Sswk	2	8	Pr.	1,475	420		1,895	2,350
5050	3'-0" x 7'-0", single		2.50	6.400	Ea.	825	335		1,160	1,525
5100	Pair of 3'-0" x 7'-0"		2	8	Pr.	1,550	420		1,970	2,475
5150	3'-6" x 7'-0", single		2.50	6.400	Ea.	985	335		1,320	1,675

08 14 Wood Doors

08 14 13 – Carved Wood Doors

08 14 13.10 Types of Wood Doors, Carved

		Crew	Daily Output	Labor-Hours	Unit	Material	2015 Bare Costs Labor	Equipment	Total	Total Incl O&P
0010	**TYPES OF WOOD DOORS, CARVED**									
3000	Solid wood, 1-3/4" thick stile and rail									
3020	Mahogany, 3'-0" x 7'-0", six panel	2 Carp	14	1.143	Ea.	1,125	53.50		1,178.50	1,350
3030	With two lites		10	1.600		1,850	75		1,925	2,150
3040	3'-6" x 8'-0", six panel		10	1.600		1,500	75		1,575	1,775
3050	With two lites		8	2		2,500	94		2,594	2,900
3100	Pine, 3'-0" x 7'-0", six panel		14	1.143		525	53.50		578.50	670
3110	With two lites		10	1.600		825	75		900	1,025
3120	3'-6" x 8'-0", six panel		10	1.600		920	75		995	1,125
3130	With two lites		8	2		1,825	94		1,919	2,175
3200	Red oak, 3'-0" x 7'-0", six panel		14	1.143		1,750	53.50		1,803.50	2,000
3210	With two lites		10	1.600		2,300	75		2,375	2,650
3220	3'-6" x 8'-0", six panel		10	1.600		2,600	75		2,675	2,975
3230	With two lites		8	2		3,300	94		3,394	3,775
4000	Hand carved door, mahogany									
4020	3'-0" x 7'-0", simple design	2 Carp	14	1.143	Ea.	1,750	53.50		1,803.50	2,025
4030	Intricate design		11	1.455		3,700	68.50		3,768.50	4,175
4040	3'-6" x 8'-0", simple design		10	1.600		3,000	75		3,075	3,425
4050	Intricate design		8	2		3,700	94		3,794	4,225
4400	For custom finish, add					475			475	525
4600	Side light, mahogany, 7'-0" x 1'-6" wide, 4 lites	2 Carp	18	.889		1,100	41.50		1,141.50	1,275
4610	6 lites		14	1.143		2,625	53.50		2,678.50	3,000
4620	8'-0" x 1'-6" wide, 4 lites		14	1.143		1,800	53.50		1,853.50	2,075
4630	6 lites		10	1.600		2,100	75		2,175	2,425
4640	Side light, oak, 7'-0" x 1'-6" wide, 4 lites		18	.889		1,200	41.50		1,241.50	1,400
4650	6 lites		14	1.143		2,100	53.50		2,153.50	2,400
4660	8'-0" x 1'-6" wide, 4 lites		14	1.143		1,100	53.50		1,153.50	1,300
4670	6 lites		10	1.600		2,100	75		2,175	2,425

08 14 16 – Flush Wood Doors

08 14 16.09 Smooth Wood Doors

		Crew	Daily Output	Labor-Hours	Unit	Material	2015 Bare Costs Labor	Equipment	Total	Total Incl O&P
0010	**SMOOTH WOOD DOORS**									
0015	Flush, interior, hollow core									
0025	Lauan face, 1-3/8", 3'-0" x 6'-8"	2 Carp	17	.941	Ea.	54.50	44		98.50	133
0030	4'-0" x 6'-8"		16	1		126	47		173	216
0080	1-3/4", 2'-0" x 6'-8"		17	.941		39	44		83	116
0085	3'-0" x 6'-8"	1 Carp	16	.500		54.50	23.50		78	98.50
0108	3'-0" x 7'-0"	2 Carp	16	1		145	47		192	237
0112	Pair of 3'-0" x 7'-0"		9	1.778	Pr.	104	83.50		187.50	252
0140	Birch face, 1-3/8", 2'-6" x 6'-8"		17	.941	Ea.	85	44		129	166
0180	3'-0" x 6'-8"		17	.941		95.50	44		139.50	178
0200	4'-0" x 6'-8"		16	1		158	47		205	251

08 14 Wood Doors

08 14 16 – Flush Wood Doors

08 14 16.09 Smooth Wood Doors		Crew	Daily Output	Labor-Hours	Unit	Material	2015 Bare Costs Labor	Equipment	Total	Total Incl O&P
0202	1-3/4", 2'-0" x 6'-8"	2 Carp	17	.941	Ea.	50.50	44		94.50	128
0204	2'-4" x 7'-0"		16	1		96.50	47		143.50	183
0206	2'-6" x 7'-0"		16	1		100	47		147	187
0210	3'-0" x 7'-0"		16	1		110	47		157	198
0214	Pair of 3'-0" x 7'-0"		9	1.778	Pr.	213	83.50		296.50	370
0220	Oak face, 1-3/8", 2'-0" x 6'-8"		17	.941	Ea.	104	44		148	187
0280	3'-0" x 6'-8"		17	.941		115	44		159	200
0300	4'-0" x 6'-8"		16	1		139	47		186	230
0305	1-3/4", 2'-6" x 6'-8"		17	.941		110	44		154	194
0310	3'-0" x 7'-0"		16	1		211	47		258	310
0320	Walnut face, 1-3/8", 2'-0" x 6'-8"		17	.941		182	44		226	274
0340	2'-6" x 6'-8"		17	.941		189	44		233	281
0380	3'-0" x 6'-8"		17	.941		197	44		241	289
0400	4'-0" x 6'-8"		16	1		216	47		263	315
0430	For 7'-0" high, add					26			26	28.50
0440	For 8'-0" high, add					36			36	39.50
0480	For prefinishing, clear, add					46			46	50.50
0500	For prefinishing, stain, add					57			57	62.50
1320	M.D. overlay on hardboard, 1-3/8", 2'-0" x 6'-8"	2 Carp	17	.941		115	44		159	200
1340	2'-6" x 6'-8"		17	.941		115	44		159	200
1380	3'-0" x 6'-8"		17	.941		126	44		170	212
1400	4'-0" x 6'-8"		16	1		178	47		225	273
1420	For 7'-0" high, add					15.75			15.75	17.35
1440	For 8'-0" high, add					31.50			31.50	34.50
1720	H.P. plastic laminate, 1-3/8", 2'-0" x 6'-8"	2 Carp	16	1		260	47		307	365
1740	2'-6" x 6'-8"		16	1		260	47		307	365
1780	3'-0" x 6'-8"		15	1.067		290	50		340	400
1800	4'-0" x 6'-8"		14	1.143		385	53.50		438.50	510
1820	For 7'-0" high, add					15.75			15.75	17.35
1840	For 8'-0" high, add					31.50			31.50	34.50
2020	Particle core, lauan face, 1-3/8", 2'-6" x 6'-8"	2 Carp	15	1.067		92	50		142	183
2040	3'-0" x 6'-8"		14	1.143		94	53.50		147.50	191
2080	3'-0" x 7'-0"		13	1.231		101	58		159	206
2085	4'-0" x 7'-0"		12	1.333		123	62.50		185.50	238
2110	1-3/4", 3'-0" x 7'-0"		13	1.231		145	58		203	255
2120	Birch face, 1-3/8", 2'-6" x 6'-8"		15	1.067		103	50		153	195
2140	3'-0" x 6'-8"		14	1.143		113	53.50		166.50	212
2180	3'-0" x 7'-0"		13	1.231		123	58		181	230
2200	4'-0" x 7'-0"		12	1.333		139	62.50		201.50	256
2205	1-3/4", 3'-0" x 7'-0"		13	1.231		123	58		181	230
2220	Oak face, 1-3/8", 2'-6" x 6'-8"		15	1.067		116	50		166	210
2240	3'-0" x 6'-8"		14	1.143		128	53.50		181.50	229
2280	3'-0" x 7'-0"		13	1.231		133	58		191	241
2300	4'-0" x 7'-0"		12	1.333		156	62.50		218.50	275
2305	1-3/4" 3'-0" x 7'-0"		13	1.231		190	58		248	305
2320	Walnut face, 1-3/8", 2'-0" x 6'-8"		15	1.067		123	50		173	217
2340	2'-6" x 6'-8"		14	1.143		139	53.50		192.50	241
2380	3'-0" x 6'-8"		13	1.231		156	58		214	267
2400	4'-0" x 6'-8"		12	1.333		206	62.50		268.50	330
2440	For 8'-0" high, add					41			41	45
2460	For 8'-0" high walnut, add					36			36	39.50
2720	For prefinishing, clear, add					36			36	39.50
2740	For prefinishing, stain, add					53			53	58.50

08 14 Wood Doors

08 14 16 – Flush Wood Doors

08 14 16.09 Smooth Wood Doors

		Crew	Daily Output	Labor-Hours	Unit	Material	2015 Bare Costs Labor	Equipment	Total	Total Incl O&P
3320	M.D. overlay on hardboard, 1-3/8", 2'-6" x 6'-8"	2 Carp	14	1.143	Ea.	106	53.50		159.50	205
3340	3'-0" x 6'-8"		13	1.231		115	58		173	222
3380	3'-0" x 7'-0"		12	1.333		117	62.50		179.50	232
3400	4'-0" x 7'-0"		10	1.600		155	75		230	294
3440	For 8'-0" height, add					37			37	40.50
3460	For solid wood core, add					42			42	46
3720	H.P. plastic laminate, 1-3/8", 2'-6" x 6'-8"	2 Carp	13	1.231		160	58		218	271
3740	3'-0" x 6'-8"		12	1.333		185	62.50		247.50	305
3780	3'-0" x 7'-0"		11	1.455		190	68.50		258.50	320
3800	4'-0" x 7'-0"		8	2		225	94		319	400
3840	For 8'-0" height, add					37			37	40.50
3860	For solid wood core, add					42			42	46
4000	Exterior, flush, solid core, birch, 1-3/4" x 2'-6" x 7'-0"	2 Carp	15	1.067		173	50		223	272
4020	2'-8" wide		15	1.067		159	50		209	257
4040	3'-0" wide		14	1.143		209	53.50		262.50	320
4045	3'-0" x 8'-0"	1 Carp	8	1		430	47		477	550
4100	Oak faced 1-3/4" x 2'-6" x 7'-0"	2 Carp	15	1.067		215	50		265	320
4120	2'-8" wide		15	1.067		225	50		275	330
4140	3'-0" wide		14	1.143		230	53.50		283.50	340
4160	Walnut faced, 1-3/4" x 3'-0" x 6'-8"	1 Carp	17	.471		295	22		317	360
4180	3'-6" wide	"	17	.471		365	22		387	435
4200	Walnut faced, 1-3/4" x 2'-6" x 7'-0"	2 Carp	15	1.067		310	50		360	420
4220	2'-8" wide		15	1.067		315	50		365	425
4240	3'-0" wide		14	1.143		320	53.50		373.50	440
4300	For 6'-8" high door, deduct from 7'-0" door					16.80			16.80	18.50
5000	Wood doors, for vision lite, add					94.50			94.50	104
5010	Wood doors, for narrow lite, add					102			102	113
5015	Wood doors, for bottom (or top) louver, add					282			282	310
9000	Minimum labor/equipment charge	1 Carp	4	2	Job		94		94	154

08 14 16.10 Wood Doors Decorator

		Crew	Daily Output	Labor-Hours	Unit	Material	Labor	Equipment	Total	Total Incl O&P
0010	**WOOD DOORS DECORATOR**									
1800	Exterior, flush, solid wood core, birch 1-3/4" x 2'-6" x 7'-0"	2 Carp	15	1.067	Ea.	320	50		370	430
1820	2'-8" wide		15	1.067		325	50		375	440
1840	3'-0" wide		14	1.143		335	53.50		388.50	460
1900	Oak faced, 1-3/4" x 2'-6" x 7'-0"		15	1.067		320	50		370	430
1920	2'-8" wide		15	1.067		330	50		380	445
1940	3'-0" wide		14	1.143		340	53.50		393.50	465
2100	Walnut faced, 1-3/4" x 2'-6" x 7'-0"		15	1.067		385	50		435	505
2120	2'-8" wide		15	1.067		400	50		450	520
2140	3'-0" wide		14	1.143		430	53.50		483.50	565

08 14 16.20 Wood Fire Doors

		Crew	Daily Output	Labor-Hours	Unit	Material	Labor	Equipment	Total	Total Incl O&P
0010	**WOOD FIRE DOORS**									
0020	Particle core, 7 face plys, "B" label,									
0040	1 hour, birch face, 1-3/4" x 2'-6" x 6'-8"	2 Carp	14	1.143	Ea.	420	53.50		473.50	550
0080	3'-0" x 6'-8"		13	1.231		400	58		458	535
0090	3'-0" x 7'-0"		12	1.333		435	62.50		497.50	585
0100	4'-0" x 7'-0"		12	1.333		540	62.50		602.50	700
0140	Oak face, 2'-6" x 6'-8"		14	1.143		450	53.50		503.50	585
0180	3'-0" x 6'-8"		13	1.231		455	58		513	595
0190	3'-0" x 7'-0"		12	1.333		460	62.50		522.50	610
0200	4'-0" x 7'-0"		12	1.333		590	62.50		652.50	755
0240	Walnut face, 2'-6" x 6'-8"		14	1.143		470	53.50		523.50	605

08 14 Wood Doors

08 14 16 – Flush Wood Doors

08 14 16.20 Wood Fire Doors

		Crew	Daily Output	Labor-Hours	Unit	Material	2015 Bare Costs Labor	Equipment	Total	Total Incl O&P
0280	3'-0" x 6'-8"	2 Carp	13	1.231	Ea.	490	58		548	635
0290	3'-0" x 7'-0"		12	1.333		510	62.50		572.50	665
0300	4'-0" x 7'-0"		12	1.333		630	62.50		692.50	800
0440	M.D. overlay on hardboard, 2'-6" x 6'-8"		15	1.067		315	50		365	425
0480	3'-0" x 6'-8"		14	1.143		375	53.50		428.50	505
0490	3'-0" x 7'-0"		13	1.231		395	58		453	530
0500	4'-0" x 7'-0"		12	1.333		415	62.50		477.50	560
0740	90 minutes, birch face, 1-3/4" x 2'-6" x 6'-8"		14	1.143		325	53.50		378.50	445
0780	3'-0" x 6'-8"		13	1.231		320	58		378	445
0790	3'-0" x 7'-0"		12	1.333		390	62.50		452.50	535
0800	4'-0" x 7'-0"		12	1.333		545	62.50		607.50	700
0840	Oak face, 2'-6" x 6'-8"		14	1.143		430	53.50		483.50	565
0880	3'-0" x 6'-8"		13	1.231		440	58		498	580
0890	3'-0" x 7'-0"		12	1.333		455	62.50		517.50	605
0900	4'-0" x 7'-0"		12	1.333		590	62.50		652.50	755
0940	Walnut face, 2'-6" x 6'-8"		14	1.143		400	53.50		453.50	530
0980	3'-0" x 6'-8"		13	1.231		410	58		468	545
0990	3'-0" x 7'-0"		12	1.333		470	62.50		532.50	620
1000	4'-0" x 7'-0"		12	1.333		620	62.50		682.50	785
1140	M.D. overlay on hardboard, 2'-6" x 6'-8"		15	1.067		355	50		405	470
1180	3'-0" x 6'-8"		14	1.143		375	53.50		428.50	505
1190	3'-0" x 7'-0"		13	1.231		405	58		463	540
1200	4'-0" x 7'-0"		12	1.333		455	62.50		517.50	605
1240	For 8'-0" height, add					75			75	82.50
1260	For 8'-0" height walnut, add					90			90	99
2200	Custom architectural "B" label, flush, 1-3/4" thick, birch,									
2210	Solid core									
2220	2'-6" x 7'-0"	2 Carp	15	1.067	Ea.	305	50		355	415
2260	3'-0" x 7'-0"		14	1.143		315	53.50		368.50	435
2300	4'-0" x 7'-0"		13	1.231		400	58		458	535
2420	4'-0" x 8'-0"		11	1.455		430	68.50		498.50	585
2480	For oak veneer, add					50%				
2500	For walnut veneer, add					75%				
9000	Minimum labor/equipment charge	1 Carp	4	2	Job		94		94	154

08 14 33 – Stile and Rail Wood Doors

08 14 33.10 Wood Doors Paneled

0010	**WOOD DOORS PANELED**									
0020	Interior, six panel, hollow core, 1-3/8" thick									
0040	Molded hardboard, 2'-0" x 6'-8"	2 Carp	17	.941	Ea.	62	44		106	141
0060	2'-6" x 6'-8"		17	.941		64	44		108	143
0070	2'-8" x 6'-8"		17	.941		67	44		111	146
0080	3'-0" x 6'-8"		17	.941		72	44		116	152
0140	Embossed print, molded hardboard, 2'-0" x 6'-8"		17	.941		64	44		108	143
0160	2'-6" x 6'-8"		17	.941		64	44		108	143
0180	3'-0" x 6'-8"		17	.941		72	44		116	152
0540	Six panel, solid, 1-3/8" thick, pine, 2'-0" x 6'-8"		15	1.067		155	50		205	253
0560	2'-6" x 6'-8"		14	1.143		170	53.50		223.50	275
0580	3'-0" x 6'-8"		13	1.231		145	58		203	255
1020	Two panel, bored rail, solid, 1-3/8" thick, pine, 1'-6" x 6'-8"		16	1		270	47		317	375
1040	2'-0" x 6'-8"		15	1.067		355	50		405	470
1060	2'-6" x 6'-8"		14	1.143		400	53.50		453.50	530
1340	Two panel, solid, 1-3/8" thick, fir, 2'-0" x 6'-8"		15	1.067		160	50		210	258

08 14 Wood Doors

08 14 33 – Stile and Rail Wood Doors

08 14 33.10 Wood Doors Paneled

		Crew	Daily Output	Labor-Hours	Unit	Material	2015 Bare Costs Labor	Equipment	Total	Total Incl O&P
1360	2'-6" x 6'-8"	2 Carp	14	1.143	Ea.	210	53.50		263.50	320
1380	3'-0" x 6'-8"		13	1.231		415	58		473	550
1740	Five panel, solid, 1-3/8" thick, fir, 2'-0" x 6'-8"		15	1.067		280	50		330	390
1760	2'-6" x 6'-8"		14	1.143		420	53.50		473.50	550
1780	3'-0" x 6'-8"		13	1.231		420	58		478	555
4190	Exterior, Knotty pine, paneled, 1-3/4", 3'-0" x 6'-8"		16	1		780	47		827	935
4195	Double 1-3/4", 3'-0" x 6'-8"		16	1		1,550	47		1,597	1,800
4200	Ash, paneled, 1-3/4", 3'-0" x 6'-8"		16	1		890	47		937	1,050
4205	Double 1-3/4", 3'-0" x 6'-8"		16	1		1,775	47		1,822	2,025
4210	Cherry, paneled, 1-3/4", 3'-0" x 6'-8"		16	1		890	47		937	1,050
4215	Double 1-3/4", 3'-0" x 6'-8"		16	1		1,775	47		1,822	2,025
4230	Ash, paneled, 1-3/4", 3'-0" x 8'-0"		16	1		950	47		997	1,125
4235	Double 1-3/4", 3'-0" x 8'-0"		16	1		1,900	47		1,947	2,175
4240	Hard Maple, paneled, 1-3/4", 3'-0" x 8'-0"		16	1		950	47		997	1,125
4245	Double 1-3/4", 3'-0" x 8'-0"		16	1		1,900	47		1,947	2,175
4250	Cherry, paneled, 1-3/4", 3'-0" x 8'-0"		16	1		1,000	47		1,047	1,175
4255	Double 1-3/4", 3'-0" x 8'-0"		16	1		2,000	47		2,047	2,275
9000	Minimum labor/equipment charge	1 Carp	4	2	Job		94		94	154

08 14 33.20 Wood Doors Residential

		Crew	Daily Output	Labor-Hours	Unit	Material	2015 Bare Costs Labor	Equipment	Total	Total Incl O&P
0010	**WOOD DOORS RESIDENTIAL**									
0200	Exterior, combination storm & screen, pine									
0260	2'-8" wide	2 Carp	10	1.600	Ea.	300	75		375	455
0280	3'-0" wide		9	1.778		310	83.50		393.50	475
0300	7'-1" x 3'-0" wide		9	1.778		335	83.50		418.50	505
0400	Full lite, 6'-9" x 2'-6" wide		11	1.455		320	68.50		388.50	460
0420	2'-8" wide		10	1.600		320	75		395	475
0440	3'-0" wide		9	1.778		325	83.50		408.50	490
0500	7'-1" x 3'-0" wide		9	1.778		355	83.50		438.50	525
0700	Dutch door, pine, 1-3/4" x 2'-8" x 6'-8", 6 panel		12	1.333		790	62.50		852.50	970
0720	Half glass		10	1.600		900	75		975	1,125
0800	3'-0" wide, 6 panel		12	1.333		790	62.50		852.50	970
0820	Half glass		10	1.600		900	75		975	1,125
1000	Entrance door, colonial, 1-3/4" x 6'-8" x 2'-8" wide		16	1		500	47		547	625
1020	6 panel pine, 3'-0" wide		15	1.067		445	50		495	570
1100	8 panel pine, 2'-8" wide		16	1		580	47		627	710
1120	3'-0" wide		15	1.067		570	50		620	705
1200	For tempered safety glass lites, (min of 2) add					79			79	87
1300	Flush, birch, solid core, 1-3/4" x 6'-8" x 2'-8" wide	2 Carp	16	1		113	47		160	202
1320	3'-0" wide		15	1.067		119	50		169	212
1350	7'-0" x 2'-8" wide		16	1		122	47		169	211
1360	3'-0" wide		15	1.067		143	50		193	239
1380	For tempered safety glass lites, add					105			105	116
1420	6'-8" x 3'-0" wide, fir	2 Carp	16	1		470	47		517	590
1720	Carved mahogany 3'-0" x 6'-8"		15	1.067		1,375	50		1,425	1,600
1800	Lauan, solid core, 1-3/4" x 7'-0" x 2'-4" wide		16	1		110	47		157	199
1810	2'-6" wide		15	1.067		114	50		164	207
1820	2'-8" wide		9	1.778		118	83.50		201.50	267
1830	3'-0" wide		16	1		130	47		177	220
1840	3'-4" wide		16	1		205	47		252	305
1850	Pair of 3'-0" wide		15	1.067	Pr.	259	50		309	365
2700	Interior, closet, bi-fold, w/hardware, no frame or trim incl.									
2720	Flush, birch, 2'-6" x 6'-8"	2 Carp	13	1.231	Ea.	67.50	58		125.50	169

08 14 Wood Doors

08 14 33 – Stile and Rail Wood Doors

08 14 33.20 Wood Doors Residential		Crew	Daily Output	Labor-Hours	Unit	Material	2015 Bare Costs Labor	Equipment	Total	Total Incl O&P
2740	3'-0" wide	2 Carp	13	1.231	Ea.	71.50	58		129.50	173
2760	4'-0" wide		12	1.333		109	62.50		171.50	223
2780	5'-0" wide		11	1.455		105	68.50		173.50	228
2800	6'-0" wide		10	1.600		128	75		203	264
2820	Flush, hardboard, primed, 6'-8" x 2'-6" wide		13	1.231		56.50	58		114.50	157
2840	3'-0" wide		13	1.231		64	58		122	165
2860	4'-0" wide		12	1.333		94	62.50		156.50	206
2880	5'-0" wide		11	1.455		101	68.50		169.50	223
2900	6'-0" wide		10	1.600		139	75		214	276
3000	Raised panel pine, 6'-6" or 6'-8" x 2'-6" wide		13	1.231		198	58		256	315
3020	3'-0" wide		13	1.231		278	58		336	400
3040	4'-0" wide		12	1.333		305	62.50		367.50	440
3060	5'-0" wide		11	1.455		365	68.50		433.50	510
3080	6'-0" wide		10	1.600		400	75		475	565
3200	Louvered, pine 6'-6" or 6'-8" x 2'-6" wide		13	1.231		144	58		202	253
3220	3'-0" wide		13	1.231		208	58		266	325
3240	4'-0" wide		12	1.333		235	62.50		297.50	360
3260	5'-0" wide		11	1.455		262	68.50		330.50	400
3280	6'-0" wide		10	1.600		289	75		364	445
4400	Bi-passing closet, incl. hardware and frame, no trim incl.									
4420	Flush, lauan, 6'-8" x 4'-0" wide	2 Carp	12	1.333	Opng.	176	62.50		238.50	297
4440	5'-0" wide		11	1.455		187	68.50		255.50	315
4460	6'-0" wide		10	1.600		209	75		284	355
4600	Flush, birch, 6'-8" x 4'-0" wide		12	1.333		223	62.50		285.50	350
4620	5'-0" wide		11	1.455		212	68.50		280.50	345
4640	6'-0" wide		10	1.600		260	75		335	410
4800	Louvered, pine, 6'-8" x 4'-0" wide		12	1.333		410	62.50		472.50	560
4820	5'-0" wide		11	1.455		395	68.50		463.50	545
4840	6'-0" wide		10	1.600		535	75		610	710
5000	Paneled, pine, 6'-8" x 4'-0" wide		12	1.333		515	62.50		577.50	670
5020	5'-0" wide		11	1.455		410	68.50		478.50	560
5040	6'-0" wide		10	1.600		610	75		685	795
5042	8'-0" wide		12	1.333		1,025	62.50		1,087.50	1,225
6100	Folding accordion, closet, including track and frame									
6120	Vinyl, 2 layer, stock	2 Carp	10	1.600	Ea.	66	75		141	196
6200	Rigid PVC	"	10	1.600	"	55.50	75		130.50	184
7310	Passage doors, flush, no frame included									
7320	Hardboard, hollow core, 1-3/8" x 6'-8" x 1'-6" wide	2 Carp	18	.889	Ea.	40.50	41.50		82	113
7330	2'-0" wide		18	.889		43	41.50		84.50	116
7340	2'-6" wide		18	.889		47.50	41.50		89	121
7350	2'-8" wide		18	.889		49	41.50		90.50	123
7360	3'-0" wide		17	.941		52	44		96	130
7420	Lauan, hollow core, 1-3/8" x 6'-8" x 1'-6" wide		18	.889		35	41.50		76.50	107
7440	2'-0" wide		18	.889		35	41.50		76.50	107
7450	2'-4" wide		18	.889		38.50	41.50		80	111
7460	2'-6" wide		18	.889		38.50	41.50		80	111
7480	2'-8" wide		18	.889		40	41.50		81.50	113
7500	3'-0" wide		17	.941		42.50	44		86.50	119
7700	Birch, hollow core, 1-3/8" x 6'-8" x 1'-6" wide		18	.889		41	41.50		82.50	114
7720	2'-0" wide		18	.889		42.50	41.50		84	115
7740	2'-6" wide		18	.889		50	41.50		91.50	124
7760	2'-8" wide		18	.889		50.50	41.50		92	124
7780	3'-0" wide		17	.941		52	44		96	130

08 14 Wood Doors

08 14 33 – Stile and Rail Wood Doors

08 14 33.20 Wood Doors Residential

		Crew	Daily Output	Labor-Hours	Unit	Material	2015 Bare Costs Labor	Equipment	Total	Total Incl O&P
8000	Pine louvered, 1-3/8" x 6'-8" x 1'-6" wide	2 Carp	19	.842	Ea.	120	39.50		159.50	197
8020	2'-0" wide		18	.889		131	41.50		172.50	213
8040	2'-6" wide		18	.889		150	41.50		191.50	234
8060	2'-8" wide		18	.889		161	41.50		202.50	246
8080	3'-0" wide		17	.941		175	44		219	265
8300	Pine paneled, 1-3/8" x 6'-8" x 1'-6" wide		19	.842		128	39.50		167.50	206
8320	2'-0" wide		18	.889		153	41.50		194.50	237
8330	2'-4" wide		18	.889		182	41.50		223.50	269
8340	2'-6" wide		18	.889		182	41.50		223.50	269
8360	2'-8" wide		18	.889		185	41.50		226.50	273
8380	3'-0" wide		17	.941		204	44		248	297
8490	3'-0" wide		17	.941		276	44		320	380
8804	Pocket door, 6 panel pine, 2'-6" x 6'-8" with frame		10.50	1.524		315	71.50		386.50	460
8814	2'-8" x 6'-8"		10.50	1.524		325	71.50		396.50	470
8824	3'-0" x 6'-8"		10.50	1.524		355	71.50		426.50	505
9000	Passage doors, flush, no frame, birch, solid core, 1-3/8" x 2'-4" x 7'-0"		16	1		118	47		165	207
9020	2'-8" wide		16	1		125	47		172	215
9040	3'-0" wide		16	1		135	47		182	226
9060	3'-4" wide		15	1.067		156	50		206	254
9080	Pair of 3'-0" wide		9	1.778	Pr.	267	83.50		350.50	430
9100	Lauan, solid core, 1-3/8" x 7'-0" x 2'-4" wide		16	1	Ea.	108	47		155	196
9120	2'-8" wide		16	1		116	47		163	204
9140	3'-0" wide		16	1		123	47		170	212
9160	3'-4" wide		15	1.067		129	50		179	224
9180	Pair of 3'-0" wide		9	1.778	Pr.	223	83.50		306.50	380
9200	Hardboard, solid core, 1-3/8" x 7'-0" x 2'-4" wide		16	1	Ea.	134	47		181	225
9220	2'-8" wide		16	1		139	47		186	230
9240	3'-0" wide		16	1		143	47		190	234
9260	3'-4" wide		15	1.067		158	50		208	256
9900	Minimum labor/equipment charge	1 Carp	4	2	Job		94		94	154

08 14 35 – Torrified Doors

08 14 35.10 Torrified Exterior Doors

		Crew	Daily Output	Labor-Hours	Unit	Material	Labor	Equipment	Total	Total Incl O&P
0010	**TORRIFIED EXTERIOR DOORS**									
0020	Wood doors made from torrified wood, exterior.									
0030	All doors require a finish be applied, all glass is insulated									
0040	All doors require pilot holes for all fasteners									
0100	6 panel, Paint grade poplar, 1-3/4" x 3'-0"x 6'-8"	2 Carp	12	1.333	Ea.	1,075	62.50		1,137.50	1,275
0120	Half glass 3'-0"x 6'-8"	"	12	1.333		1,175	62.50		1,237.50	1,400
0200	Side lite, full glass, 1-3/4" x 1'-2" x 6'-8"					905			905	995
0220	Side lite, half glass, 1-3/4" x 1'-2" x 6'-8"					905			905	995
0300	Raised Face, 2 Panel, Paint grade poplar, 1-3/4" x 3'-0"x 7'-0"	2 Carp	12	1.333		1,275	62.50		1,337.50	1,500
0320	Side lite, raised face, half glass, 1-3/4" x 1'-2" x 7'-0"					1,050			1,050	1,175
0500	6 panel, Fir, 1-3/4" x 3'-0"x 6'-8"	2 Carp	12	1.333		1,550	62.50		1,612.50	1,800
0520	Half glass 3'-0"x 6'-8"	"	12	1.333		1,650	62.50		1,712.50	1,925
0600	Side lite, full glass, 1-3/4" x 1'-2" x 6'-8"					1,150			1,150	1,275
0620	Side lite, half glass, 1-3/4" x 1'-2" x 6'-8"					1,450			1,450	1,600
0700	6 panel, Mahogany, 1-3/4" x 3'-0"x 6'-8"	2 Carp	12	1.333		1,625	62.50		1,687.50	1,875
0800	Side lite, full glass, 1-3/4" x 1'-2" x 6'-8"					1,225			1,225	1,350
0820	Side lite, half glass, 1-3/4" x 1'-2" x 6'-8"					1,200			1,200	1,325

08 14 Wood Doors

08 14 40 – Interior Cafe Doors

08 14 40.10 Cafe Style Doors

		Crew	Daily Output	Labor-Hours	Unit	Material	2015 Bare Costs Labor	2015 Bare Costs Equipment	Total	Total Incl O&P
0010	**CAFE STYLE DOORS**									
6520	Interior cafe doors, 2'-6" opening, stock, panel pine	2 Carp	16	1	Ea.	218	47		265	315
6540	3'-0" opening	"	16	1	"	240	47		287	340
6550	Louvered pine									
6560	2'-6" opening	2 Carp	16	1	Ea.	180	47		227	275
8000	3'-0" opening		16	1		193	47		240	289
8010	2'-6" opening, hardwood		16	1		264	47		311	365
8020	3'-0" opening		16	1		281	47		328	385
9000	Minimum labor/equipment charge	1 Carp	4	2	Job		94		94	154

08 16 Composite Doors

08 16 13 – Fiberglass Doors

08 16 13.10 Entrance Doors, Fiberous Glass

			Crew	Daily Output	Labor-Hours	Unit	Material	2015 Bare Costs Labor	2015 Bare Costs Equipment	Total	Total Incl O&P
0010	**ENTRANCE DOORS, FIBEROUS GLASS**										
0020	Exterior, fiberglass, door, 2'-8" wide x 6'-8" high	G	2 Carp	15	1.067	Ea.	270	50		320	380
0040	3'-0" wide x 6'-8" high	G		15	1.067		270	50		320	380
0060	3'-0" wide x 7'-0" high	G		15	1.067		460	50		510	585
0080	3'-0" wide x 6'-8" high, with two lites	G		15	1.067		315	50		365	425
0100	3'-0" wide x 8'-0" high, with two lites	G		15	1.067		525	50		575	660
0110	Half glass, 3'-0" wide x 6'-8" high	G		15	1.067		435	50		485	560
0120	3'-0" wide x 6'-8" high, low e	G		15	1.067		465	50		515	590
0130	3'-0" wide x 8'-0" high	G		15	1.067		585	50		635	725
0140	3'-0" wide x 8'-0" high, low e	G		15	1.067		655	50		705	800
0150	Side lights, 1'-0" wide x 6'-8" high	G					266			266	293
0160	1'-0" wide x 6'-8" high, low e	G					281			281	310
0180	1'-0" wide x 6'-8" high, full glass	G					310			310	340
0190	1'-0" wide x 6'-8" high, low e	G					340			340	370

08 16 14 – French Doors

08 16 14.10 Exterior Doors With Glass Lites

		Crew	Daily Output	Labor-Hours	Unit	Material	2015 Bare Costs Labor	2015 Bare Costs Equipment	Total	Total Incl O&P
0010	**EXTERIOR DOORS WITH GLASS LITES**									
0020	French, Fir, 1-3/4", 3'-0"wide x 6'-8" high	2 Carp	12	1.333	Ea.	600	62.50		662.50	765
0025	Double		12	1.333		1,200	62.50		1,262.50	1,425
0030	Maple, 1-3/4", 3'-0"wide x 6'-8" high		12	1.333		675	62.50		737.50	850
0035	Double		12	1.333		1,350	62.50		1,412.50	1,575
0040	Cherry, 1-3/4", 3'-0"wide x 6'-8" high		12	1.333		790	62.50		852.50	970
0045	Double		12	1.333		1,575	62.50		1,637.50	1,825
0100	Mahogany, 1-3/4", 3'-0"wide x 8'-0" high		10	1.600		800	75		875	1,000
0105	Double		10	1.600		1,600	75		1,675	1,875
0110	Fir, 1-3/4", 3'-0"wide x 8'-0" high		10	1.600		1,200	75		1,275	1,450
0115	Double		10	1.600		2,400	75		2,475	2,775
0120	Oak, 1-3/4", 3'-0"wide x 8'-0" high		10	1.600		1,825	75		1,900	2,125
0125	Double		10	1.600		3,650	75		3,725	4,125

08 17 Integrated Door Opening Assemblies

08 17 13 – Integrated Metal Door Opening Assemblies

08 17 13.10 Hollow Metal Doors and Frames

	08 17 13.10 Hollow Metal Doors and Frames		Crew	Daily Output	Labor-Hours	Unit	Material	2015 Bare Costs Labor	Equipment	Total	Total Incl O&P
0010	**HOLLOW METAL DOORS AND FRAMES**										
0100	Prehung, flush, 18 ga., 1-3/4" x 6'-8" x 2'-8" wide		1 Carp	16	.500	Ea.	665	23.50		688.50	770
0120	3'-0" wide			15	.533		620	25		645	725
0130	3'-6" wide			13	.615		710	29		739	830
0140	4'-0" wide			10	.800		770	37.50		807.50	905
0300	Double, 1-3/4" x 3'-0" x 6'-8"			5	1.600		700	75		775	895
0350	3'-0" x 8'-0"		↓	4	2	↓	1,550	94		1,644	1,850

08 17 13.20 Stainless Steel Doors and Frames

	08 17 13.20 Stainless Steel Doors and Frames		Crew	Daily Output	Labor-Hours	Unit	Material	Labor	Equipment	Total	Total Incl O&P
0010	**STAINLESS STEEL DOORS AND FRAMES**										
0020	Stainless Steel(304) prehung 24 g 2'-6" x 6'-8" door w/16 g frame	G	2 Carp	6	2.667	Ea.	1,800	125		1,925	2,175
0025	2'-8" x 6'-8"	G		6	2.667		1,825	125		1,950	2,225
0030	3'-0" x 6'-8"	G		6	2.667		1,825	125		1,950	2,200
0040	3'-0" x 7'-0"	G		6	2.667		1,850	125		1,975	2,225
0050	4'-0" x 7'-0"	G		5	3.200		2,400	150		2,550	2,900
0100	Stainless Steel(316) prehung 24 g 2'-6" x 6'-8" door w/16 g frame	G		6	2.667		2,700	125		2,825	3,175
0110	2'-8" x 6'-8"	G		6	2.667		2,725	125		2,850	3,175
0120	3'-0" x 6'-8"	G		6	2.667		2,500	125		2,625	2,950
0150	3'-0" x 7'-0"	G		6	2.667		2,675	125		2,800	3,150
0160	4'-0" x 7'-0"	G		6	2.667		3,300	125		3,425	3,825
0300	Stainless Steel(304) prehung 18 g 2'-6" x 6'-8" door w/16 g frame	G		6	2.667		3,075	125		3,200	3,575
0310	2'-8" x 6'-8"	G		6	2.667		3,100	125		3,225	3,600
0320	3'-0" x 6'-8"	G		6	2.667		3,150	125		3,275	3,675
0350	3'-0" x 7'-0"	G		6	2.667		3,200	125		3,325	3,725
0360	4'-0" x 7'-0"	G		5	3.200		3,400	150		3,550	3,975
0500	Stainless steel, prehung door, foam core, 14 ga, 3'-0" x 7'-0"	G		5	3.200		3,400	150		3,550	4,000
0600	Stainless steel, prehung double door, foam core, 14 ga, 3'-0" x 7'-0"	G	↓	4	4	↓	6,700	188		6,888	7,675

08 17 23 – Integrated Wood Door Opening Assemblies

08 17 23.10 Pre-Hung Doors

	08 17 23.10 Pre-Hung Doors	Crew	Daily Output	Labor-Hours	Unit	Material	Labor	Equipment	Total	Total Incl O&P
0010	**PRE-HUNG DOORS**									
0300	Exterior, wood, comb. storm & screen, 6'-9" x 2'-6" wide	2 Carp	15	1.067	Ea.	296	50		346	405
0320	2'-8" wide		15	1.067		296	50		346	405
0340	3'-0" wide	↓	15	1.067		305	50		355	415
0360	For 7'-0" high door, add					30			30	33
1600	Entrance door, flush, birch, solid core									
1620	4-5/8" solid jamb, 1-3/4" x 6'-8" x 2'-8" wide	2 Carp	16	1	Ea.	289	47		336	395
1640	3'-0" wide		16	1		375	47		422	490
1642	5-5/8" jamb	↓	16	1		325	47		372	435
1680	For 7'-0" high door, add					25			25	27.50
2000	Entrance door, colonial, 6 panel pine									
2020	4-5/8" solid jamb, 1-3/4" x 6'-8" x 2'-8" wide	2 Carp	16	1	Ea.	640	47		687	780
2040	3'-0" wide	"	16	1		675	47		722	815
2060	For 7'-0" high door, add					54			54	59
2200	For 5-5/8" solid jamb, add					41.50			41.50	46
2230	French style, exterior, 1 lite, 1-3/4" x 3'-0" x 6'-8"	1 Carp	14	.571		590	27		617	695
2235	9 lites	"	14	.571		660	27		687	770
2245	15 lites	2 Carp	14	1.143	↓	640	53.50		693.50	795
2250	Double, 15 lites, 2'-0" x 6'-8", 4'-0" opening		7	2.286	Pr.	1,200	107		1,307	1,500
2260	2'-6" x 6'-8", 5'-0" opening		7	2.286		1,325	107		1,432	1,625
2280	3'-0" x 6'-8", 6'-0" opening		7	2.286	↓	1,350	107		1,457	1,650
2430	3'-0" x 7'-0", 15 lites		14	1.143	Ea.	950	53.50		1,003.50	1,150
2432	Two 3'-0" x 7'-0"		7	2.286	Pr.	1,975	107		2,082	2,325
2435	3'-0" x 8'-0"	↓	14	1.143	Ea.	1,000	53.50		1,053.50	1,200

08 17 Integrated Door Opening Assemblies

08 17 23 – Integrated Wood Door Opening Assemblies

08 17 23.10 Pre-Hung Doors

		Crew	Daily Output	Labor-Hours	Unit	Material	2015 Bare Costs Labor	Equipment	Total	Total Incl O&P
2437	Two, 3'-0" x 8'-0"	2 Carp	7	2.286	Pr.	2,075	107		2,182	2,475
4000	Interior, passage door, 4-5/8" solid jamb									
4350	Paneled, primed, hollow core, 2'-8" wide	2 Carp	17	.941	Ea.	116	44		160	201
4360	3'-0" wide		17	.941		180	44		224	271
4400	Lauan, flush, solid core, 1-3/8" x 6'-8" x 2'-6" wide		17	.941		186	44		230	278
4420	2'-8" wide		17	.941		186	44		230	278
4440	3'-0" wide		16	1		201	47		248	298
4600	Hollow core, 1-3/8" x 6'-8" x 2'-6" wide		17	.941		125	44		169	211
4620	2'-8" wide		17	.941		125	44		169	210
4640	3'-0" wide		16	1		140	47		187	231
4700	For 7'-0" high door, add					35.50			35.50	39
5000	Birch, flush, solid core, 1-3/8" x 6'-8" x 2'-6" wide	2 Carp	17	.941		275	44		319	380
5020	2'-8" wide		17	.941		201	44		245	294
5040	3'-0" wide		16	1		300	47		347	410
5200	Hollow core, 1-3/8" x 6'-8" x 2'-6" wide		17	.941		222	44		266	315
5220	2'-8" wide		17	.941		266	44		310	365
5240	3'-0" wide		16	1		230	47		277	330
5280	For 7'-0" high door, add					30.50			30.50	33.50
5500	Hardboard paneled, 1-3/8" x 6'-8" x 2'-6" wide	2 Carp	17	.941		145	44		189	232
5520	2'-8" wide		17	.941		153	44		197	241
5540	3'-0" wide		16	1		150	47		197	242
6000	Pine paneled, 1-3/8" x 6'-8" x 2'-6" wide		17	.941		255	44		299	355
6020	2'-8" wide		17	.941		274	44		318	375
6040	3'-0" wide		16	1		282	47		329	385
8200	Birch, flush, solid core, 1-3/4" x 6'-8" x 2'-4" wide	1 Carp	17	.471		229	22		251	288
8220	2'-6" wide		17	.471		227	22		249	286
8240	2'-8" wide		17	.471		221	22		243	279
8260	3'-0" wide		16	.500		229	23.50		252.50	291
8280	3'-6" wide		15	.533		360	25		385	435
8500	Pocket door frame with lauan, flush, hollow core, 1-3/8" x 3'-0" x 6'-8"		17	.471		223	22		245	282
9000	Minimum labor/equipment charge		4	2	Job		94		94	154

08 31 Access Doors and Panels

08 31 13 – Access Doors and Frames

08 31 13.10 Types of Framed Access Doors

		Crew	Daily Output	Labor-Hours	Unit	Material	2015 Bare Costs Labor	Equipment	Total	Total Incl O&P
0010	**TYPES OF FRAMED ACCESS DOORS**									
1000	Fire rated door with lock									
1100	Metal, 12" x 12"	1 Carp	10	.800	Ea.	165	37.50		202.50	244
1150	18" x 18"		9	.889		220	41.50		261.50	310
1200	24" x 24"		9	.889		325	41.50		366.50	430
1250	24" x 36"		8	1		335	47		382	440
1300	24" x 48"		8	1		430	47		477	550
1350	36" x 36"		7.50	1.067		495	50		545	625
1400	48" x 48"		7.50	1.067		635	50		685	775
1600	Stainless steel, 12" x 12"		10	.800		282	37.50		319.50	370
1650	18" x 18"		9	.889		410	41.50		451.50	520
1700	24" x 24"		9	.889		500	41.50		541.50	620
1750	24" x 36"		8	1		640	47		687	780
2000	Flush door for finishing									
2100	Metal 8" x 8"	1 Carp	10	.800	Ea.	38	37.50		75.50	104
2150	12" x 12"	"	10	.800	"	45	37.50		82.50	111

08 31 Access Doors and Panels

08 31 13 – Access Doors and Frames

08 31 13.10 Types of Framed Access Doors

		Crew	Daily Output	Labor-Hours	Unit	Material	2015 Bare Costs Labor	2015 Bare Costs Equipment	Total	Total Incl O&P
3000	Recessed door for acoustic tile									
3100	Metal, 12" x 12"	1 Carp	4.50	1.778	Ea.	82	83.50		165.50	227
3150	12" x 24"		4.50	1.778		100	83.50		183.50	247
3200	24" x 24"		4	2		130	94		224	297
3250	24" x 36"		4	2		180	94		274	350
4000	Recessed door for drywall									
4100	Metal 12" x 12"	1 Carp	6	1.333	Ea.	80	62.50		142.50	191
4150	12" x 24"		5.50	1.455		114	68.50		182.50	237
4200	24" x 36"		5	1.600		182	75		257	325
6000	Standard door									
6100	Metal, 8" x 8"	1 Carp	10	.800	Ea.	45	37.50		82.50	111
6150	12" x 12"		10	.800		50	37.50		87.50	117
6200	18" x 18"		9	.889		70	41.50		111.50	146
6250	24" x 24"		9	.889		85	41.50		126.50	162
6300	24" x 36"		8	1		125	47		172	215
6350	36" x 36"		8	1		145	47		192	237
6500	Stainless steel, 8" x 8"		10	.800		85	37.50		122.50	155
6550	12" x 12"		10	.800		110	37.50		147.50	183
6600	18" x 18"		9	.889		205	41.50		246.50	295
6650	24" x 24"		9	.889		265	41.50		306.50	360
9000	Minimum labor/equipment charge		4	2	Job		94		94	154

08 31 13.20 Bulkhead/Cellar Doors

		Crew	Daily Output	Labor-Hours	Unit	Material	Labor	Equipment	Total	Total Incl O&P
0010	**BULKHEAD/CELLAR DOORS**									
0020	Steel, not incl. sides, 44" x 62"	1 Carp	5.50	1.455	Ea.	560	68.50		628.50	725
0100	52" x 73"		5.10	1.569		790	73.50		863.50	990
0500	With sides and foundation plates, 57" x 45" x 24"		4.70	1.702		845	80		925	1,050
0600	42" x 49" x 51"		4.30	1.860		915	87.50		1,002.50	1,150
9000	Minimum labor/equipment charge		2	4	Job		188		188	310

08 31 13.30 Commercial Floor Doors

		Crew	Daily Output	Labor-Hours	Unit	Material	Labor	Equipment	Total	Total Incl O&P
0010	**COMMERCIAL FLOOR DOORS**									
0020	Aluminum tile, steel frame, one leaf, 2' x 2' opng.	2 Sswk	3.50	4.571	Opng.	880	241		1,121	1,400
0050	3'-6" x 3'-6" opening		3.50	4.571		1,600	241		1,841	2,175
0500	Double leaf, 4' x 4' opening		3	5.333		1,750	281		2,031	2,425
0550	5' x 5' opening		3	5.333		3,100	281		3,381	3,900
9000	Minimum labor/equipment charge		2	8	Job		420		420	760

08 31 13.35 Industrial Floor Doors

		Crew	Daily Output	Labor-Hours	Unit	Material	Labor	Equipment	Total	Total Incl O&P
0010	**INDUSTRIAL FLOOR DOORS**									
0020	Steel 300 psf L.L., single leaf, 2' x 2', 175#	2 Sswk	6	2.667	Opng.	760	140		900	1,100
0050	3' x 3' opening, 300#		5.50	2.909		1,075	153		1,228	1,450
0300	Double leaf, 4' x 4' opening, 455#		5	3.200		2,275	168		2,443	2,800
0350	5' x 5' opening, 645#		4.50	3.556		3,300	187		3,487	3,975
1000	Aluminum, 300 psf L.L., single leaf, 2' x 2', 60#		6	2.667		800	140		940	1,125
1050	3' x 3' opening, 100#		5.50	2.909		1,300	153		1,453	1,700
1500	Double leaf, 4' x 4' opening, 160#		5	3.200		2,100	168		2,268	2,600
1550	5' x 5' opening, 235#		4.50	3.556		2,800	187		2,987	3,425
9000	Minimum labor/equipment charge		2	8	Job		420		420	760

08 32 Sliding Glass Doors

08 32 13 – Sliding Aluminum-Framed Glass Doors

08 32 13.10 Sliding Aluminum Doors

		Crew	Daily Output	Labor-Hours	Unit	Material	2015 Bare Costs Labor	2015 Bare Costs Equipment	Total	Total Incl O&P
0010	**SLIDING ALUMINUM DOORS**									
0350	Aluminum, 5/8" tempered insulated glass, 6' wide									
0400	Premium	2 Carp	4	4	Ea.	1,575	188		1,763	2,025
0450	Economy		4	4		815	188		1,003	1,200
0500	8' wide, premium		3	5.333		1,650	250		1,900	2,225
0550	Economy		3	5.333		1,425	250		1,675	1,950
0600	12' wide, premium		2.50	6.400		2,975	300		3,275	3,775
0650	Economy		2.50	6.400		1,550	300		1,850	2,200
4000	Aluminum, baked on enamel, temp glass, 6'-8" x 10'-0" wide		4	4		1,075	188		1,263	1,475
4020	Insulating glass, 6'-8" x 6'-0" wide		4	4		935	188		1,123	1,325
4040	8'-0" wide		3	5.333		1,100	250		1,350	1,600
4060	10'-0" wide		2	8		1,350	375		1,725	2,100
4080	Anodized, temp glass, 6'-8" x 6'-0" wide		4	4		455	188		643	810
4100	8'-0" wide		3	5.333		575	250		825	1,050
4120	10'-0" wide		2	8		645	375		1,020	1,325
5000	Aluminum sliding glass door system									
5010	Sliding door 4' wide opening single side	2 Carp	2	8	Ea.	5,400	375		5,775	6,575
5015	8' wide opening single side		2	8		8,100	375		8,475	9,525
5020	Telescoping glass door system, 4' wide opening biparting		2	8		4,500	375		4,875	5,575
5025	8' wide opening biparting		2	8		5,400	375		5,775	6,575
5030	Folding glass door, 4' wide opening biparting		2	8		7,200	375		7,575	8,550
5035	8' wide opening biparting		2	8		9,000	375		9,375	10,500
5040	ICU-CCU Sliding telescoping glass door, 4' x 7', single side opening		2	8		2,650	375		3,025	3,550
5045	8' x 7', single side opening		2	8		4,125	375		4,500	5,150
7000	Electric swing door operator and control, single door w/sensors	1 Carp	4	2		2,525	94		2,619	2,925
7005	Double door w/sensors		2	4		4,950	188		5,138	5,750
7010	Electric folding door operator and control, single door w/sensors		4	2		5,850	94		5,944	6,575
7015	Bi-folding door		4	2		7,025	94		7,119	7,875
7020	Electric swing door operator and control, single door		4	2		2,700	94		2,794	3,125

08 32 19 – Sliding Wood-Framed Glass Doors

08 32 19.10 Sliding Wood Doors

		Crew	Daily Output	Labor-Hours	Unit	Material	Labor	Equipment	Total	Total Incl O&P
0010	**SLIDING WOOD DOORS**									
0020	Wood, tempered insul. glass, 6' wide, premium	2 Carp	4	4	Ea.	1,425	188		1,613	1,875
0100	Economy		4	4		1,175	188		1,363	1,575
0150	8' wide, wood, premium		3	5.333		1,825	250		2,075	2,400
0200	Economy		3	5.333		1,500	250		1,750	2,050
0235	10' wide, wood, premium		2.50	6.400		2,600	300		2,900	3,350
0240	Economy		2.50	6.400		2,200	300		2,500	2,925
0250	12' wide, wood, premium		2.50	6.400		3,000	300		3,300	3,800
0300	Economy		2.50	6.400		2,450	300		2,750	3,200
9000	Minimum labor/equipment charge	1 Carp	2	4	Job		188		188	310

08 32 19.15 Sliding Glass Vinyl-Clad Wood Doors

			Crew	Daily Output	Labor-Hours	Unit	Material	Labor	Equipment	Total	Total Incl O&P
0010	**SLIDING GLASS VINYL-CLAD WOOD DOORS**										
0020	Glass, sliding vinyl clad, insul. glass, 6'-0" x 6'-8"	G	2 Carp	4	4	Opng.	1,500	188		1,688	1,975
0025	6'-0" x 6'-10" high	G		4	4		1,650	188		1,838	2,125
0030	6'-0" x 8'-0" high	G		4	4		1,975	188		2,163	2,475
0050	5'-0" x 6'-8" high	G		4	4		1,500	188		1,688	1,950
0100	8'-0" x 6'-10" high	G		4	4		2,025	188		2,213	2,525
0150	8'-0" x 8'-0" high	G		4	4		2,225	188		2,413	2,750
0500	4 leaf, 9'-0" x 6'-10" high	G		3	5.333		3,250	250		3,500	3,975
0550	9'-0" x 8'-0" high	G		3	5.333		3,775	250		4,025	4,550
0600	12'-0" x 6'-10" high	G		3	5.333		3,875	250		4,125	4,675

08 32 Sliding Glass Doors

08 32 19 – Sliding Wood-Framed Glass Doors

08 32 19.15 Sliding Glass Vinyl-Clad Wood Doors

		Crew	Daily Output	Labor-Hours	Unit	Material	2015 Bare Costs Labor	Equipment	Total	Total Incl O&P
0650	12'-0" x 8'-0" high [G]	2 Carp	3	5.333	Opng.	3,875	250		4,125	4,675
9000	Minimum labor/equipment charge	1 Carp	4	2	Job		94		94	154

08 33 Coiling Doors and Grilles

08 33 13 – Coiling Counter Doors

08 33 13.10 Counter Doors, Coiling Type

		Crew	Daily Output	Labor-Hours	Unit	Material	2015 Bare Costs Labor	Equipment	Total	Total Incl O&P
0010	**COUNTER DOORS, COILING TYPE**									
0020	Manual, incl. frame and hardware, galv. stl., 4' roll-up, 6' long	2 Carp	2	8	Opng.	1,225	375		1,600	1,975
0300	Galvanized steel, UL label		1.80	8.889		1,225	415		1,640	2,025
0600	Stainless steel, 4' high roll-up, 6' long		2	8		2,150	375		2,525	2,975
0700	10' long		1.80	8.889		2,500	415		2,915	3,425
2000	Aluminum, 4' high, 4' long		2.20	7.273		1,525	340		1,865	2,225
2020	6' long		2	8		1,800	375		2,175	2,625
2040	8' long		1.90	8.421		2,050	395		2,445	2,900
2060	10' long		1.80	8.889		2,100	415		2,515	2,975
2080	14' long		1.40	11.429		2,675	535		3,210	3,825
2100	6' high, 4' long		2	8		1,850	375		2,225	2,650
2120	6' long		1.60	10		1,675	470		2,145	2,600
2140	10' long		1.40	11.429		2,200	535		2,735	3,300
9000	Minimum labor/equipment charge	1 Carp	2	4	Job		188		188	310

08 33 16 – Coiling Counter Grilles

08 33 16.10 Coiling Grilles

		Crew	Daily Output	Labor-Hours	Unit	Material	2015 Bare Costs Labor	Equipment	Total	Total Incl O&P
0010	**COILING GRILLES**									
2020	Aluminum, manual operated, mill finish	2 Sswk	82	.195	S.F.	28	10.25		38.25	49.50
2040	Bronze anodized		82	.195	"	44.50	10.25		54.75	67.50
2060	Steel, manual operated, 10' x 10' high		1	16	Opng.	2,500	840		3,340	4,275
2080	15' x 8' high		.80	20	"	2,900	1,050		3,950	5,075
3000	For safety edge bottom bar, electric, add				L.F.	49			49	54
8000	For motor operation, add	2 Sswk	5	3.200	Opng.	1,250	168		1,418	1,675
9000	Minimum labor/equipment charge	"	1	16	Job		840		840	1,525

08 33 23 – Overhead Coiling Doors

08 33 23.10 Coiling Service Doors

		Crew	Daily Output	Labor-Hours	Unit	Material	2015 Bare Costs Labor	Equipment	Total	Total Incl O&P
0010	**COILING SERVICE DOORS** Steel, manual, 20 ga., incl. hardware									
0050	8' x 8' high	2 Sswk	1.60	10	Ea.	1,150	525		1,675	2,225
0130	12' x 12' high, standard		1.20	13.333		2,175	700		2,875	3,675
0160	10' x 20' high, standard		.50	32		2,625	1,675		4,300	5,950
2000	Class A fire doors, manual, 20 ga., 8' x 8' high		1.40	11.429		1,600	600		2,200	2,825
2100	10' x 10' high		1.10	14.545		2,175	765		2,940	3,775
2200	20' x 10' high		.80	20		4,450	1,050		5,500	6,800
2300	12' x 12' high		1	16		3,425	840		4,265	5,300
2400	20' x 12' high		.80	20		4,825	1,050		5,875	7,225
2500	14' x 14' high		.60	26.667		3,750	1,400		5,150	6,650
2600	20' x 16' high		.50	32		5,900	1,675		7,575	9,550
2700	10' x 20' high		.40	40		4,675	2,100		6,775	8,925
3000	For 18 ga. doors, add				S.F.	1.65			1.65	1.82
3300	For enamel finish, add				"	1.90			1.90	2.09
3600	For safety edge bottom bar, pneumatic, add				L.F.	23			23	25
4000	For weatherstripping, extruded rubber, jambs, add					14.40			14.40	15.85
4100	Hood, add					8.20			8.20	9
4200	Sill, add					5.15			5.15	5.65

08 33 Coiling Doors and Grilles

08 33 23 – Overhead Coiling Doors

08 33 23.10 Coiling Service Doors		Crew	Daily Output	Labor-Hours	Unit	Material	2015 Bare Costs Labor	2015 Bare Costs Equipment	Total	Total Incl O&P
4500	Motor operators, to 14' x 14' opening	2 Sswk	5	3.200	Ea.	1,150	168		1,318	1,575
4700	For fire door, additional fusible link, add				"	28			28	31
9000	Minimum labor/equipment charge	2 Sswk	1	16	Job		840		840	1,525

08 34 Special Function Doors

08 34 13 – Cold Storage Doors

08 34 13.10 Doors for Cold Area Storage

		Crew	Daily Output	Labor-Hours	Unit	Material	Labor	Equipment	Total	Total Incl O&P
0010	**DOORS FOR COLD AREA STORAGE**									
0020	Single, 20 ga. galvanized steel									
0300	Horizontal sliding, 5' x 7', manual operation, 3.5" thick	2 Carp	2	8	Ea.	3,250	375		3,625	4,200
0400	4" thick		2	8		3,250	375		3,625	4,200
0500	6" thick		2	8		3,425	375		3,800	4,400
0800	5' x 7', power operation, 2" thick		1.90	8.421		5,600	395		5,995	6,800
0900	4" thick		1.90	8.421		5,700	395		6,095	6,925
1000	6" thick		1.90	8.421		6,500	395		6,895	7,800
1300	9' x 10', manual operation, 2" insulation		1.70	9.412		4,550	440		4,990	5,725
1400	4" insulation		1.70	9.412		4,675	440		5,115	5,875
1500	6" insulation		1.70	9.412		5,575	440		6,015	6,875
1800	Power operation, 2" insulation		1.60	10		7,675	470		8,145	9,225
1900	4" insulation		1.60	10		7,900	470		8,370	9,450
2000	6" insulation		1.70	9.412		8,850	440		9,290	10,500
2300	For stainless steel face, add					25%				
3000	Hinged, lightweight, 3' x 7'-0", 2" thick	2 Carp	2	8	Ea.	1,425	375		1,800	2,175
3050	4" thick		1.90	8.421		1,775	395		2,170	2,600
3300	Polymer doors, 3' x 7'-0"		1.90	8.421		1,350	395		1,745	2,125
3350	6" thick		1.40	11.429		2,350	535		2,885	3,450
3600	Stainless steel, 3' x 7'-0", 4" thick		1.90	8.421		1,725	395		2,120	2,550
3650	6" thick		1.40	11.429		2,950	535		3,485	4,100
3900	Painted, 3' x 7'-0", 4" thick		1.90	8.421		1,275	395		1,670	2,050
3950	6" thick		1.40	11.429		2,350	535		2,885	3,450
5000	Bi-parting, electric operated									
5010	6' x 8' opening, galv. faces, 4" thick for cooler	2 Carp	.80	20	Opng.	7,225	940		8,165	9,475
5050	For freezer, 4" thick		.80	20		7,950	940		8,890	10,300
5300	For door buck framing and door protection, add		2.50	6.400		640	300		940	1,200
6000	Galvanized batten door, galvanized hinges, 4' x 7'		2	8		1,825	375		2,200	2,650
6050	6' x 8'		1.80	8.889		2,475	415		2,890	3,400
6500	Fire door, 3 hr., 6' x 8', single slide		.80	20		8,275	940		9,215	10,700
6550	Double, bi-parting		.70	22.857		12,200	1,075		13,275	15,200
9000	Minimum labor/equipment charge	1 Carp	2	4	Job		188		188	310

08 34 36 – Darkroom Doors

08 34 36.10 Various Types of Darkroom Doors

		Crew	Daily Output	Labor-Hours	Unit	Material	Labor	Equipment	Total	Total Incl O&P
0010	**VARIOUS TYPES OF DARKROOM DOORS**									
0015	Revolving, standard, 2 way, 36" diameter	2 Carp	3.10	5.161	Opng.	2,900	242		3,142	3,575
0020	41" diameter		3.10	5.161		3,000	242		3,242	3,725
0050	3 way, 51" diameter		1.40	11.429		3,825	535		4,360	5,075
1000	4 way, 49" diameter		1.40	11.429		3,950	535		4,485	5,225
2000	Hinged safety, 2 way, 41" diameter		2.30	6.957		3,750	325		4,075	4,625
2500	3 way, 51" diameter		1.40	11.429		4,050	535		4,585	5,350
3000	Pop out safety, 2 way, 41" diameter		3.10	5.161		4,025	242		4,267	4,825
4000	3 way, 51" diameter		1.40	11.429		4,875	535		5,410	6,250

08 34 Special Function Doors

08 34 36 – Darkroom Doors

08 34 36.10 Various Types of Darkroom Doors

		Crew	Daily Output	Labor-Hours	Unit	Material	2015 Bare Costs Labor	Equipment	Total	Total Incl O&P
5000	Wheelchair-type, pop out, 51" diameter	2 Carp	1.40	11.429	Opng	5,500	535		6,035	6,925
5020	72" diameter		.90	17.778		8,950	835		9,785	11,200

08 34 53 – Security Doors and Frames

08 34 53.20 Steel Door

		Crew	Daily Output	Labor-Hours	Unit	Material	Labor	Equipment	Total	Total Incl O&P
0010	**STEEL DOOR** with ballistic core and welded frame both 14 ga.									
0050	Flush, UL 752 Level 3, 1-3/4", 3'-0" x 6'-8"	2 Carp	1.50	10.667	Opng.	2,325	500		2,825	3,400
0055	1-3/4", 3'-6" x 6'-8"		1.50	10.667		2,425	500		2,925	3,500
0060	1-3/4", 4'-0" x 6'-8"		1.20	13.333		2,650	625		3,275	3,950
0100	UL 752 Level 8, 1-3/4", 3'-0" x 6'-8"		1.50	10.667		10,200	500		10,700	12,000
0105	1-3/4", 3'-6" x 6'-8"		1.50	10.667		10,300	500		10,800	12,100
0110	1-3/4", 4'-0" x 6'-8"		1.20	13.333		11,700	625		12,325	13,900
0120	UL 752 Level 3, 1-3/4", 3'-0" x 7'-0"		1.50	10.667		2,350	500		2,850	3,425
0125	1-3/4", 3'-6" x 7'-0"		1.50	10.667		2,475	500		2,975	3,550
0130	1-3/4", 4'-0" x 7'-0"		1.20	13.333		2,675	625		3,300	3,975
0150	UL 752 Level 8, 1-3/4", 3'-0" x 7'-0"		1.50	10.667		10,200	500		10,700	12,000
0155	1-3/4", 3'-6" x 7'-0"		1.50	10.667		10,300	500		10,800	12,100
0160	1-3/4", 4'-0" x 7'-0"		1.20	13.333		11,700	625		12,325	13,900
1000	Safe Room sliding door and hardware, 1-3/4", 3'-0" x 7'-0" UL 752 Level 3		.50	32		23,500	1,500		25,000	28,400
1050	Safe Room swinging door and hardware, 1-3/4", 3'-0" x 7'-0" UL 752 Level 3		.50	32		28,100	1,500		29,600	33,400

08 34 53.30 Wood Ballistic Doors

		Crew	Daily Output	Labor-Hours	Unit	Material	Labor	Equipment	Total	Total Incl O&P
0010	**WOOD BALLISTIC DOORS** with frames and hardware									
0050	Wood, 1-3/4", 3'-0" x 7'-0" UL 752 Level 3	2 Carp	1.50	10.667	Opng.	2,250	500		2,750	3,300

08 34 56 – Security Gates

08 34 56.10 Gates

		Crew	Daily Output	Labor-Hours	Unit	Material	Labor	Equipment	Total	Total Incl O&P
0010	**GATES**									
0015	Driveway Gates include mounting hardware									
0500	Wood, security gate, driveway, dual, 10' wide	H-4	.80	25	Opng.	3,900	1,100		5,000	6,075
0505	12' wide		.80	25		4,300	1,100		5,400	6,500
0510	15' wide		.80	25		5,100	1,100		6,200	7,375
0600	Steel, security gate, driveway, single, 10' wide		.80	25		2,150	1,100		3,250	4,125
0605	12' wide		.80	25		2,225	1,100		3,325	4,200
0620	Steel, security gate, driveway, dual, 12' wide		.80	25		2,050	1,100		3,150	4,050
0625	14' wide		.80	25		2,200	1,100		3,300	4,200
0630	16' wide		.80	25		2,525	1,100		3,625	4,550
0700	Aluminum, security gate, driveway, dual, 10' wide		.80	25		3,000	1,100		4,100	5,075
0705	12' wide		.80	25		3,200	1,100		4,300	5,300
0710	16' wide		.80	25		3,700	1,100		4,800	5,850
1000	Security gate, driveway, opener 12 VDC				Ea.	800			800	880
1010	Wireless					1,800			1,800	1,975
1020	Security gate, driveway, opener 24 VDC					1,300			1,300	1,425
1030	Wireless					1,450			1,450	1,600
1040	Security gate, driveway, opener 12 VDC, solar panel 10 watt	1 Elec	2	4		300	219		519	670
1050	20 watt	"	2	4		450	219		669	835

08 34 73 – Sound Control Door Assemblies

08 34 73.10 Acoustical Doors

		Crew	Daily Output	Labor-Hours	Unit	Material	Labor	Equipment	Total	Total Incl O&P
0010	**ACOUSTICAL DOORS**									
0020	Including framed seals, 3' x 7', wood, 40 STC rating	2 Carp	1.50	10.667	Ea.	1,300	500		1,800	2,250
0100	Steel, 41 STC rating		1.50	10.667		3,275	500		3,775	4,425
0200	45 STC rating		1.50	10.667		3,675	500		4,175	4,875
0300	48 STC rating		1.50	10.667		4,275	500		4,775	5,550
0400	52 STC rating		1.50	10.667		4,900	500		5,400	6,225

08 34 Special Function Doors

08 34 73 – Sound Control Door Assemblies

08 34 73.10 Acoustical Doors	Crew	Daily Output	Labor-Hours	Unit	Material	2015 Bare Costs Labor	Equipment	Total	Total Incl O&P
9000 Minimum labor/equipment charge	1 Carp	4	2	Job		94		94	154

08 36 Panel Doors

08 36 13 – Sectional Doors

08 36 13.10 Overhead Commercial Doors

		Crew	Daily Output	Labor-Hours	Unit	Material	Labor	Equipment	Total	Total Incl O&P
0010	OVERHEAD COMMERCIAL DOORS									
1000	Stock, sectional, heavy duty, wood, 1-3/4" thick, 8' x 8' high	2 Carp	2	8	Ea.	1,050	375		1,425	1,775
1200	12' x 12' high		1.50	10.667		2,050	500		2,550	3,075
1300	Chain hoist, 14' x 14' high		1.30	12.308		3,300	580		3,880	4,575
1600	20' x 16' high		.65	24.615		5,475	1,150		6,625	7,925
2100	For medium duty custom door, deduct					5%	5%			
2150	For medium duty stock doors, deduct					10%	5%			
2300	Fiberglass and aluminum, heavy duty, sectional, 12' x 12' high	2 Carp	1.50	10.667	Ea.	2,700	500		3,200	3,800
2450	Chain hoist, 20' x 20' high	"	.50	32		6,675	1,500		8,175	9,800
2900	For electric trolley operator, 1/3 H.P., to 12' x 12', add	1 Carp	2	4		950	188		1,138	1,350
2950	Over 12' x 12', 1/2 H.P., add		1	8		1,125	375		1,500	1,875
2980	Overhead, for row of clear lites add		1	8		110	375		485	735
9000	Minimum labor/equipment charge	2 Carp	1.50	10.667	Job		500		500	820

08 36 13.20 Residential Garage Doors

		Crew	Daily Output	Labor-Hours	Unit	Material	Labor	Equipment	Total	Total Incl O&P
0010	RESIDENTIAL GARAGE DOORS									
0050	Hinged, wood, custom, double door, 9' x 7'	2 Carp	4	4	Ea.	800	188		988	1,200
0070	16' x 7'		3	5.333		1,200	250		1,450	1,725
0200	Overhead, sectional, incl. hardware, fiberglass, 9' x 7', standard		5.28	3.030		945	142		1,087	1,275
0220	Deluxe		5.28	3.030		1,150	142		1,292	1,475
0300	16' x 7', standard		6	2.667		1,575	125		1,700	1,925
0320	Deluxe		6	2.667		2,150	125		2,275	2,550
0500	Hardboard, 9' x 7', standard		8	2		625	94		719	845
0520	Deluxe		8	2		815	94		909	1,050
0600	16' x 7', standard		6	2.667		1,225	125		1,350	1,550
0620	Deluxe		6	2.667		1,425	125		1,550	1,775
0700	Metal, 9' x 7', standard		5.28	3.030		740	142		882	1,050
0720	Deluxe		8	2		915	94		1,009	1,150
0800	16' x 7', standard		3	5.333		940	250		1,190	1,425
0820	Deluxe		6	2.667		1,400	125		1,525	1,750
0900	Wood, 9' x 7', standard		8	2		970	94		1,064	1,225
0920	Deluxe		8	2		2,150	94		2,244	2,500
1000	16' x 7', standard		6	2.667		1,625	125		1,750	2,000
1020	Deluxe		6	2.667		3,025	125		3,150	3,525
1800	Door hardware, sectional	1 Carp	4	2		350	94		444	540
1810	Door tracks only		4	2		163	94		257	335
1820	One side only		7	1.143		120	53.50		173.50	220
3000	Swing-up, including hardware, fiberglass, 9' x 7', standard	2 Carp	8	2		1,000	94		1,094	1,250
3020	Deluxe		8	2		1,100	94		1,194	1,350
3100	16' x 7', standard		6	2.667		1,250	125		1,375	1,575
3120	Deluxe		6	2.667		1,600	125		1,725	1,950
3200	Hardboard, 9' x 7', standard		8	2		550	94		644	760
3220	Deluxe		8	2		650	94		744	870
3300	16' x 7', standard		6	2.667		670	125		795	940
3320	Deluxe		6	2.667		850	125		975	1,150
3400	Metal, 9' x 7', standard		8	2		600	94		694	815
3420	Deluxe		8	2		965	94		1,059	1,200

08 36 Panel Doors

08 36 13 – Sectional Doors

08 36 13.20 Residential Garage Doors

		Crew	Daily Output	Labor-Hours	Unit	Material	2015 Bare Costs Labor	Equipment	Total	Total Incl O&P
3500	16' x 7', standard	2 Carp	6	2.667	Ea.	800	125		925	1,075
3520	Deluxe		6	2.667		1,100	125		1,225	1,400
3600	Wood, 9' x 7', standard		8	2		700	94		794	925
3620	Deluxe		8	2		1,125	94		1,219	1,375
3700	16' x 7', standard		6	2.667		900	125		1,025	1,200
3720	Deluxe		6	2.667		2,100	125		2,225	2,500
3900	Door hardware only, swing up	1 Carp	4	2		168	94		262	340
3920	One side only		7	1.143		90	53.50		143.50	187
4000	For electric operator, economy, add		8	1		425	47		472	545
4100	Deluxe, including remote control		8	1		610	47		657	750
4500	For transmitter/receiver control, add to operator				Total	110			110	121
4600	Transmitters, additional				"	60			60	66
6000	Replace section, on sectional door, fiberglass, 9' x 7'	1 Carp	4	2	Ea.	640	94		734	855
6020	16' x 7'		3.50	2.286		745	107		852	995
6200	Hardboard, 9' x 7'		4	2		115	94		209	280
6220	16' x 7'		3.50	2.286		218	107		325	415
6300	Metal, 9' x 7'		4	2		184	94		278	355
6320	16' x 7'		3.50	2.286		300	107		407	505
6500	Wood, 9' x 7'		4	2		116	94		210	282
6520	16' x 7'		3.50	2.286		225	107		332	425
7010	Garage doors, row of lites					115			115	127
9000	Minimum labor/equipment charge	1 Carp	2.50	3.200	Job		150		150	246

08 36 19 – Multi-Leaf Vertical Lift Doors

08 36 19.10 Sectional Vertical Lift Doors

		Crew	Daily Output	Labor-Hours	Unit	Material	Labor	Equipment	Total	Total Incl O&P
0010	**SECTIONAL VERTICAL LIFT DOORS**									
0020	Motorized, 14 ga. steel, incl. frame and control panel									
0050	16' x 16' high	L-10	.50	48	Ea.	21,500	2,550	1,300	25,350	29,600
0100	10' x 20' high		1.30	18.462		34,800	980	505	36,285	40,600
0120	15' x 20' high		1.30	18.462		42,600	980	505	44,085	49,200
0140	20' x 20' high		1	24		49,900	1,275	655	51,830	58,000
0160	25' x 20' high		1	24		56,000	1,275	655	57,930	64,500
0170	32' x 24' high		.75	32		48,700	1,700	870	51,270	57,500
0180	20' x 25' high		1	24		57,500	1,275	655	59,430	66,000
0200	25' x 25' high		.70	34.286		66,000	1,825	935	68,760	77,000
0220	25' x 30' high		.70	34.286		71,000	1,825	935	73,760	82,500
0240	30' x 30' high		.70	34.286		82,500	1,825	935	85,260	94,500
0260	35' x 30' high		.70	34.286		92,500	1,825	935	95,260	106,000

08 38 Traffic Doors

08 38 19 – Rigid Traffic Doors

08 38 19.20 Double Acting Swing Doors

		Crew	Daily Output	Labor-Hours	Unit	Material	Labor	Equipment	Total	Total Incl O&P
0010	**DOUBLE ACTING SWING DOORS**									
0020	Including frame, closer, hardware and vision panel									
1000	Polymer, 7'-0" high, 4'-0" wide	2 Carp	4.20	3.810	Pr.	2,100	179		2,279	2,600
1025	6'-0" wide		4	4		2,200	188		2,388	2,725
1050	6'-8" wide		4	4		2,400	188		2,588	2,950
2000	3/4" thick, stainless steel									
2010	Stainless steel, 7' high opening, 4' wide	2 Carp	4	4	Pr.	2,600	188		2,788	3,150
2050	7' wide		3.80	4.211	"	2,800	198		2,998	3,400
9000	Minimum labor/equipment charge		2	8	Job		375		375	615

08 38 Traffic Doors

08 38 19 – Rigid Traffic Doors

08 38 19.30 Shock Absorbing Doors

		Crew	Daily Output	Labor-Hours	Unit	Material	2015 Bare Costs Labor	2015 Bare Costs Equipment	Total	Total Incl O&P
0010	**SHOCK ABSORBING DOORS**									
0020	Rigid, no frame, 1-1/2" thick, 5' x 7'	2 Sswk	1.90	8.421	Opng.	1,525	445		1,970	2,475
0100	8' x 8'		1.80	8.889		2,000	470		2,470	3,050
0500	Flexible, no frame, insulated, .16" thick, economy, 5' x 7'		2	8		1,750	420		2,170	2,675
0600	Deluxe		1.90	8.421		2,625	445		3,070	3,675
1000	8' x 8' opening, economy		2	8		2,750	420		3,170	3,775
1100	Deluxe		1.90	8.421		3,500	445		3,945	4,650
9000	Minimum labor/equipment charge		2	8	Job		420		420	760

08 41 Entrances and Storefronts

08 41 13 – Aluminum-Framed Entrances and Storefronts

08 41 13.20 Tube Framing

		Crew	Daily Output	Labor-Hours	Unit	Material	2015 Bare Costs Labor	2015 Bare Costs Equipment	Total	Total Incl O&P
0010	**TUBE FRAMING**, For window walls and store fronts, aluminum stock									
0050	Plain tube frame, mill finish, 1-3/4" x 1-3/4"	2 Glaz	103	.155	L.F.	9.90	7		16.90	22.50
0150	1-3/4" x 4"		98	.163		13.35	7.35		20.70	26.50
0200	1-3/4" x 4-1/2"		95	.168		16	7.60		23.60	30
0250	2" x 6"		89	.180		23.50	8.10		31.60	39
0350	4" x 4"		87	.184		26.50	8.30		34.80	43
0400	4-1/2" x 4-1/2"		85	.188		28	8.50		36.50	45
0450	Glass bead		240	.067		3.07	3.01		6.08	8.25
1000	Flush tube frame, mill finish, 1/4" glass, 1-3/4" x 4", open header		80	.200		13.25	9		22.25	29.50
1050	Open sill		82	.195		10.85	8.80		19.65	26
1100	Closed back header		83	.193		19	8.70		27.70	35
1150	Closed back sill		85	.188		18.15	8.50		26.65	34
1160	Tube fmg., spandrel cover both sides, alum 1" wide	1 Sswk	85	.094	S.F.	98	4.96		102.96	117
1170	Tube fmg., spandrel cover both sides, alum 2" wide	"	85	.094	"	38.50	4.96		43.46	51
1200	Vertical mullion, one piece	2 Glaz	75	.213	L.F.	19.90	9.60		29.50	37.50
1250	Two piece		73	.219		21	9.90		30.90	39.50
1300	90° or 180° vertical corner post		75	.213		32.50	9.60		42.10	51
1400	1-3/4" x 4-1/2", open header		80	.200		16	9		25	32.50
1450	Open sill		82	.195		13.55	8.80		22.35	29
1500	Closed back header		83	.193		19.10	8.70		27.80	35
1550	Closed back sill		85	.188		18.95	8.50		27.45	35
1600	Vertical mullion, one piece		75	.213		21.50	9.60		31.10	39
1650	Two piece		73	.219		22.50	9.90		32.40	40.50
1700	90° or 180° vertical corner post		75	.213		23	9.60		32.60	41
2000	Flush tube frame, mil fin., ins. glass w/thml brk, 2" x 4-1/2", open header		75	.213		16.30	9.60		25.90	33.50
2050	Open sill		77	.208		13.70	9.35		23.05	30.50
2100	Closed back header		78	.205		15.50	9.25		24.75	32
2150	Closed back sill		80	.200		14.95	9		23.95	31
2200	Vertical mullion, one piece		70	.229		17.20	10.30		27.50	35.50
2250	Two piece		68	.235		18.60	10.60		29.20	37.50
2300	90° or 180° vertical corner post		70	.229		17.70	10.30		28	36.50
5000	Flush tube frame, mill fin., thermal brk., 2-1/4" x 4-1/2", open header		74	.216		17.15	9.75		26.90	34.50
5050	Open sill		75	.213		15.10	9.60		24.70	32
5100	Vertical mullion, one piece		69	.232		17.75	10.45		28.20	36.50
5150	Two piece		67	.239		21.50	10.75		32.25	41
5200	90° or 180° vertical corner post		69	.232		19.10	10.45		29.55	38
5295	Flush tube frame, mill finish, 4-1/2" x 4-1/2"		85	.188		28	8.50		36.50	45
5300	Plain tube frame, mill finish, 2" x 3" jamb		115	.139		10.90	6.25		17.15	22

08 41 Entrances and Storefronts

08 41 13 – Aluminum-Framed Entrances and Storefronts

08 41 13.20 Tube Framing

		Crew	Daily Output	Labor-Hours	Unit	Material	2015 Bare Costs Labor	Equipment	Total	Total Incl O&P
5310	2" x 4"	2 Glaz	115	.139	L.F.	16.60	6.25		22.85	28.50
5320	3" x 5-1/2"		107	.150		26.50	6.75		33.25	40.50
5330	Head, 2" x 3"		200	.080		19.95	3.61		23.56	28
5340	Mullion, 2" x 3"		200	.080		20.50	3.61		24.11	28.50
5350	2" x 4"		115	.139		24	6.25		30.25	36.50
5360	3" x 5-1/2"		107	.150		31.50	6.75		38.25	46
5370	4" corner mullion		90	.178		29.50	8		37.50	45.50
5380	Horizontal, 2" x 3"		115	.139		23.50	6.25		29.75	36
5390	3" x 5-1/2"		107	.150		29	6.75		35.75	43
5430	Sill section, 1/8" x 6"		133	.120		36	5.45		41.45	48.50
5440	1/8" x 7"		133	.120		43	5.45		48.45	56.50
5450	1/8" x 8-1/2"		133	.120		50.50	5.45		55.95	64.50
5460	Column covers, aluminum, 1/8" x 26"		57	.281		65	12.65		77.65	92
5470	1/8" x 34"		53	.302		83.50	13.60		97.10	114
5480	1/8" x 38"		53	.302		93	13.60		106.60	124
5700	Vertical mullions, clear finish, 1/4" thick glass		528	.030		12.45	1.37		13.82	15.90
5720	3/8" thick		445	.036		12.45	1.62		14.07	16.35
5740	1/2" thick		400	.040		12.90	1.80		14.70	17.10
5760	3/4" thick		344	.047		13.35	2.10		15.45	18.10
5780	1" thick		304	.053		13.50	2.37		15.87	18.70
6980	Door stop (snap in)		380	.042		3.40	1.90		5.30	6.80
7000	For joints, 90°, clip type, add				Ea.	25.50			25.50	28.50
7050	Screw spline joint, add					24			24	26
7100	For joint other than 90°, add					49.50			49.50	54.50
8000	For bronze anodized aluminum, add					15%				
8020	For black finish, add					30%				
8050	For stainless steel materials, add					350%				
8100	For monumental grade, add					53%				
8150	For steel stiffener, add	2 Glaz	200	.080	L.F.	11.40	3.61		15.01	18.40
8200	For 2 to 5 stories, add per story				Story		8%			
9000	Minimum labor/equipment charge	2 Glaz	2	8	Job		360		360	585

08 41 19 – Stainless-Steel-Framed Entrances and Storefronts

08 41 19.10 Stainless-Steel and Glass Entrance Unit

		Crew	Daily Output	Labor-Hours	Unit	Material	2015 Bare Costs Labor	Equipment	Total	Total Incl O&P
0010	**STAINLESS-STEEL AND GLASS ENTRANCE UNIT**, narrow stiles									
0020	3' x 7' opening, including hardware, minimum	2 Sswk	1.60	10	Opng.	6,800	525		7,325	8,425
0050	Average		1.40	11.429		7,250	600		7,850	9,050
0100	Maximum		1.20	13.333		7,825	700		8,525	9,900
1000	For solid bronze entrance units, statuary finish, add					64%				
1100	Without statuary finish, add					45%				
2000	Balanced doors, 3' x 7', economy	2 Sswk	.90	17.778	Ea.	9,200	935		10,135	11,800
2100	Premium		.70	22.857	"	15,500	1,200		16,700	19,300
9000	Minimum labor/equipment charge		2	8	Job		420		420	760

08 41 26 – All-Glass Entrances and Storefronts

08 41 26.10 Window Walls Aluminum, Stock

		Crew	Daily Output	Labor-Hours	Unit	Material	2015 Bare Costs Labor	Equipment	Total	Total Incl O&P
0010	**WINDOW WALLS ALUMINUM, STOCK**, including glazing									
0020	Minimum	H-2	160	.150	S.F.	46.50	6.40		52.90	62
0050	Average		140	.171		64	7.30		71.30	82.50
0100	Maximum		110	.218		172	9.30		181.30	204
0500	For translucent sandwich wall systems, see Section 07 41 33.10									
0850	Cost of the above walls depends on material,									
0860	finish, repetition, and size of units.									
0870	The larger the opening, the lower the S.F. cost									

08 41 Entrances and Storefronts

08 41 26 – All-Glass Entrances and Storefronts

08 41 26.10 Window Walls Aluminum, Stock

		Crew	Daily Output	Labor-Hours	Unit	Material	2015 Bare Costs Labor	2015 Bare Costs Equipment	Total	Total Incl O&P
1200	Double glazed acoustical window wall for airports,									

08 42 Entrances

08 42 26 – All-Glass Entrances

08 42 26.10 Swinging Glass Doors

		Crew	Daily Output	Labor-Hours	Unit	Material	Labor	Equipment	Total	Total Incl O&P
0010	**SWINGING GLASS DOORS**									
0020	Including hardware, 1/2" thick, tempered, 3' x 7' opening	2 Glaz	2	8	Opng.	2,200	360		2,560	3,000
0100	6' x 7' opening		1.40	11.429	"	4,400	515		4,915	5,650
9000	Minimum labor/equipment charge		2	8	Job		360		360	585

08 42 36 – Balanced Door Entrances

08 42 36.10 Balanced Entrance Doors

0010	**BALANCED ENTRANCE DOORS**									
0020	Hardware & frame, alum. & glass, 3' x 7', econ.	2 Sswk	.90	17.778	Ea.	6,600	935		7,535	8,975
0150	Premium		.70	22.857	"	7,925	1,200		9,125	10,900
9000	Minimum labor/equipment charge		1	16	Job		840		840	1,525

08 43 Storefronts

08 43 13 – Aluminum-Framed Storefronts

08 43 13.10 Aluminum-Framed Entrance Doors and Frames

		Crew	Daily Output	Labor-Hours	Unit	Material	Labor	Equipment	Total	Total Incl O&P
0010	**ALUMINUM-FRAMED ENTRANCE DOORS AND FRAMES**									
0015	Standard hardware and glass stops but no glass									
0020	Entrance door, 3' x 7' opening, clear anodized finish	2 Sswk	7	2.286	Opng.	525	120		645	795
0040	Bronze finish		7	2.286		530	120		650	800
0060	Black finish		7	2.286		575	120		695	845
0200	3'-6" x 7'-0", mill finish		7	2.286		635	120		755	910
0220	Bronze finish		7	2.286		655	120		775	935
0240	Black finish		7	2.286		750	120		870	1,050
0500	6' x 7' opening, clear finish		6	2.667		840	140		980	1,175
0520	Bronze finish		6	2.667		910	140		1,050	1,250
0600	Door Frame for above doors 3'-0" x 7'-0", mill finish		6	2.667		405	140		545	700
0620	Bronze finish		6	2.667		465	140		605	765
0640	Black finish		6	2.667		505	140		645	810
0700	3'-6" x 7'-0", mill finish		6	2.667		330	140		470	620
0720	Bronze finish		6	2.667		330	140		470	620
0740	Black finish		6	2.667		330	140		470	620
0800	6'-0" x 7'-0", mill finish		6	2.667		330	140		470	620
0820	Bronze finish		6	2.667		335	140		475	625
0840	Black finish		6	2.667		360	140		500	650
1000	With 3' high transom above, 3' x 7' opening, clear finish		5.50	2.909		490	153		643	815
1050	Bronze finish		5.50	2.909		510	153		663	835
1100	Black finish		5.50	2.909		535	153		688	860
1300	3'-6" x 7'-0" opening, clear finish		5.50	2.909		340	153		493	650
1320	Bronze finish		5.50	2.909		355	153		508	670
1340	Black finish		5.50	2.909		365	153		518	675
1500	6' x 7' opening, clear finish		5.50	2.909		595	153		748	930
1550	Bronze finish		5.50	2.909		615	153		768	950
1600	Black finish		5.50	2.909		675	153		828	1,025
8000	For 8' high doors add				Ea.	200			200	220
9000	Minimum labor/equipment charge	2 Sswk	4	4	Job		211		211	380

08 43 Storefronts

08 43 13 – Aluminum-Framed Storefronts

08 43 13.20 Storefront Systems

		Crew	Daily Output	Labor-Hours	Unit	Material	2015 Bare Costs Labor	2015 Bare Costs Equipment	Total	Total Incl O&P
0010	**STOREFRONT SYSTEMS**, aluminum frame clear 3/8" plate glass									
0020	incl. 3' x 7' door with hardware (400 sq. ft. max. wall)									
0500	Wall height to 12' high, commercial grade	2 Glaz	150	.107	S.F.	22.50	4.81		27.31	33
0600	Institutional grade		130	.123		28	5.55		33.55	40
0700	Monumental grade		115	.139		40.50	6.25		46.75	54.50
1000	6' x 7' door with hardware, commercial grade		135	.119		30	5.35		35.35	41.50
1100	Institutional grade		115	.139		28.50	6.25		34.75	41.50
1200	Monumental grade		100	.160		54.50	7.20		61.70	71.50
1500	For bronze anodized finish, add					15%				
1600	For black anodized finish, add					36%				
1700	For stainless steel framing, add to monumental					78%				
9000	Minimum labor/equipment charge	2 Glaz	1	16	Job		720		720	1,175

08 43 29 – Sliding Storefronts

08 43 29.10 Sliding Panels

		Crew	Daily Output	Labor-Hours	Unit	Material	2015 Bare Costs Labor	2015 Bare Costs Equipment	Total	Total Incl O&P
0010	**SLIDING PANELS**									
0020	Mall fronts, aluminum & glass, 15' x 9' high	2 Glaz	1.30	12.308	Opng.	3,475	555		4,030	4,725
0100	24' x 9' high		.70	22.857		4,925	1,025		5,950	7,100
0200	48' x 9' high, with fixed panels		.90	17.778		9,100	800		9,900	11,300
0500	For bronze finish, add					17%				
9000	Minimum labor/equipment charge	2 Glaz	1	16	Job		720		720	1,175

08 45 Translucent Wall and Roof Assemblies

08 45 10 – Translucent Roof Assemblies

08 45 10.10 Skyroofs

		Crew	Daily Output	Labor-Hours	Unit	Material	2015 Bare Costs Labor	2015 Bare Costs Equipment	Total	Total Incl O&P
0010	**SKYROOFS**,									
1200	Skylights, circular, clear, double glazed acrylic									
1230	30" diameter	2 Carp	3	5.333	Ea.	3,000	250		3,250	3,700
1250	60" diameter		3	5.333		4,000	250		4,250	4,800
1290	96" diameter		2	8		5,000	375		5,375	6,125
1300	Skylight Barrel Vault, clear, double glazed, acrylic									
1330	3'-0" X 12'-0"	G-3	3	10.667	Ea.	5,000	500		5,500	6,300
1350	4'-0" X 12'-0"		3	10.667		5,500	500		6,000	6,850
1390	5'-0" X 12'-0"		2	16		6,000	750		6,750	7,800
1400	Skylight Pyramid, Aluminum frame, clear low-E laminated glass									
1410	The glass is installed in the frame except where noted									
1430	Square, 3' X 3'	G-3	3	10.667	Ea.	6,000	500		6,500	7,400
1440	4' X 4'		3	10.667		7,000	500		7,500	8,500
1450	5' X 5', glass must be field installed		3	10.667		8,000	500		8,500	9,600
1460	6' X 6', glass must be field installed		2	16		10,000	750		10,750	12,200
1550	Install pre-cut laminated glass in aluminum frame on a flat roof	2 Glaz	55	.291	SF Surf		13.10		13.10	21.50
1560	Install pre-cut laminated glass in aluminum frame on a sloped roof	"	40	.400	"		18.05		18.05	29.50
9000	Minimum labor/equipment charge	2 Carp	8	2	Job		94		94	154

08 51 Metal Windows

08 51 13 – Aluminum Windows

08 51 13.10 Aluminum Sash

		Crew	Daily Output	Labor-Hours	Unit	Material	2015 Bare Costs Labor	2015 Bare Costs Equipment	Total	Total Incl O&P
0010	**ALUMINUM SASH**									
0020	Stock, grade C, glaze & trim not incl., casement	2 Sswk	200	.080	S.F.	39	4.21		43.21	50.50
0050	Double hung		200	.080		39.50	4.21		43.71	51
0100	Fixed casement		200	.080		17.35	4.21		21.56	26.50
0150	Picture window		200	.080		18.50	4.21		22.71	28
0200	Projected window		200	.080		35.50	4.21		39.71	46.50
0250	Single hung		200	.080		16.55	4.21		20.76	26
0300	Sliding		200	.080		21.50	4.21		25.71	31
1000	Mullions for above, tubular		240	.067	L.F.	6.15	3.51		9.66	13.15
9000	Minimum labor/equipment charge	1 Sswk	2	4	Job		211		211	380

08 51 13.20 Aluminum Windows

		Crew	Daily Output	Labor-Hours	Unit	Material	2015 Bare Costs Labor	2015 Bare Costs Equipment	Total	Total Incl O&P
0010	**ALUMINUM WINDOWS**, incl. frame and glazing, commercial grade									
1000	Stock units, casement, 3'-1" x 3'-2" opening	2 Sswk	10	1.600	Ea.	375	84		459	560
1050	Add for storms					120			120	132
1600	Projected, with screen, 3'-1" x 3'-2" opening	2 Sswk	10	1.600		355	84		439	540
1700	Add for storms					117			117	129
2000	4'-5" x 5'-3" opening	2 Sswk	8	2		400	105		505	630
2100	Add for storms					126			126	139
2500	Enamel finish windows, 3'-1" x 3'-2"	2 Sswk	10	1.600		360	84		444	545
2600	4'-5" x 5'-3"		8	2		405	105		510	635
3000	Single hung, 2' x 3' opening, enameled, standard glazed		10	1.600		208	84		292	380
3100	Insulating glass		10	1.600		252	84		336	430
3300	2'-8" x 6'-8" opening, standard glazed		8	2		365	105		470	590
3400	Insulating glass		8	2		475	105		580	710
3700	3'-4" x 5'-0" opening, standard glazed		9	1.778		300	93.50		393.50	500
3800	Insulating glass		9	1.778		335	93.50		428.50	535
4000	Sliding aluminum, 3' x 2' opening, standard glazed		10	1.600		217	84		301	390
4100	Insulating glass		10	1.600		232	84		316	405
4300	5' x 3' opening, standard glazed		9	1.778		330	93.50		423.50	535
4400	Insulating glass		9	1.778		385	93.50		478.50	595
4600	8' x 4' opening, standard glazed		6	2.667		350	140		490	640
4700	Insulating glass		6	2.667		565	140		705	875
5000	9' x 5' opening, standard glazed		4	4		530	211		741	965
5100	Insulating glass		4	4		850	211		1,061	1,325
5500	Sliding, with thermal barrier and screen, 6' x 4', 2 track		8	2		725	105		830	985
5700	4 track		8	2		910	105		1,015	1,200
6000	For above units with bronze finish, add					15%				
6200	For installation in concrete openings, add					8%				
7000	Double hung, insulating glass, 2'-0" x 2'-0"	2 Sswk	11	1.455		115	76.50		191.50	265
7020	2'-0" x 2'-6"		11	1.455		124	76.50		200.50	274
7040	3'-0" x 1'-6"		10	1.600		270	84		354	450
7060	3'-0" x 2'-0"		10	1.600		315	84		399	495
7080	3'-0" x 2'-6"		10	1.600		380	84		464	570
7100	3'-0" x 3'-0"		10	1.600		425	84		509	615
7120	3'-0" x 3'-6"		10	1.600		445	84		529	640
7140	3'-0" x 4'-0"		10	1.600		480	84		564	680
7160	3'-0" x 5'-0"		10	1.600		530	84		614	730
7180	3'-0" x 6'-0"		9	1.778		560	93.50		653.50	785

08 51 Metal Windows

08 51 23 – Steel Windows

08 51 23.10 Steel Sash

		Crew	Daily Output	Labor-Hours	Unit	Material	2015 Bare Costs Labor	Equipment	Total	Total Incl O&P
0010	**STEEL SASH** Custom units, glazing and trim not included									
0100	Casement, 100% vented	2 Sswk	200	.080	S.F.	66.50	4.21		70.71	80.50
0200	50% vented		200	.080		54.50	4.21		58.71	67
0300	Fixed		200	.080		29	4.21		33.21	39.50
1000	Projected, commercial, 40% vented		200	.080		51.50	4.21		55.71	64
1100	Intermediate, 50% vented		200	.080		58.50	4.21		62.71	71.50
1500	Industrial, horizontally pivoted		200	.080		53	4.21		57.21	65.50
1600	Fixed		200	.080		31	4.21		35.21	41.50
2000	Industrial security sash, 50% vented		200	.080		57.50	4.21		61.71	70.50
2100	Fixed		200	.080		47	4.21		51.21	59
2500	Picture window		200	.080		30	4.21		34.21	40.50
3000	Double hung		200	.080		59.50	4.21		63.71	73
5000	Mullions for above, open interior face		240	.067	L.F.	10.45	3.51		13.96	17.85
5100	With interior cover		240	.067	"	17.30	3.51		20.81	25.50
5200	Single glazing for above, add	2 Glaz	200	.080	S.F.	6.90	3.61		10.51	13.45
6000	Double glazing for above, add		200	.080		13.10	3.61		16.71	20.50
6100	Triple glazing for above, add		85	.188		12.50	8.50		21	27.50
9000	Minimum labor/equipment charge	1 Sswk	2	4	Job		211		211	380

08 51 23.20 Steel Windows

		Crew	Daily Output	Labor-Hours	Unit	Material	Labor	Equipment	Total	Total Incl O&P
0010	**STEEL WINDOWS** Stock, including frame, trim and insul. glass R085123-10									
1000	Custom units, double hung, 2'-8" x 4'-6" opening	2 Sswk	12	1.333	Ea.	705	70		775	905
1100	2'-4" x 3'-9" opening		12	1.333		585	70		655	765
1500	Commercial projected, 3'-9" x 5'-5" opening		10	1.600		1,225	84		1,309	1,500
1600	6'-9" x 4'-1" opening		7	2.286		1,625	120		1,745	2,000
2000	Intermediate projected, 2'-9" x 4'-1" opening		12	1.333		685	70		755	880
2100	4'-1" x 5'-5" opening		10	1.600		1,400	84		1,484	1,700
9000	Minimum labor/equipment charge	1 Sswk	3	2.667	Job		140		140	254

08 51 66 – Metal Window Screens

08 51 66.10 Screens

		Crew	Daily Output	Labor-Hours	Unit	Material	Labor	Equipment	Total	Total Incl O&P
0010	**SCREENS**									
0020	For metal sash, aluminum or bronze mesh, flat screen	2 Sswk	1200	.013	S.F.	4.30	.70		5	6
0500	Wicket screen, inside window		1000	.016		6.65	.84		7.49	8.80
0800	Security screen, aluminum frame with stainless steel cloth		1200	.013		24	.70		24.70	27.50
0900	Steel grate, painted, on steel frame		1600	.010		13.10	.53		13.63	15.35
1000	Screens for solar louvers		160	.100		24.50	5.25		29.75	36.50
4000	See Section 05 58 23.90									

08 52 Wood Windows

08 52 10 – Plain Wood Windows

08 52 10.20 Awning Window

		Crew	Daily Output	Labor-Hours	Unit	Material	Labor	Equipment	Total	Total Incl O&P
0010	**AWNING WINDOW**, Including frame, screens and grilles									
0100	34" x 22", insulated glass	1 Carp	10	.800	Ea.	270	37.50		307.50	360
0200	Low E glass		10	.800		271	37.50		308.50	360
0300	40" x 28", insulated glass		9	.889		315	41.50		356.50	415
0400	Low E Glass		9	.889		340	41.50		381.50	445
0500	48" x 36", insulated glass		8	1		465	47		512	590
0600	Low E glass		8	1		490	47		537	615
4000	Impact windows, minimum, add					60%				
4010	Impact windows, maximum, add					160%				
9000	Minimum labor/equipment charge	1 Carp	4	2	Job		94		94	154

08 52 Wood Windows

08 52 10 – Plain Wood Windows

08 52 10.30 Wood Windows

			Crew	Daily Output	Labor-Hours	Unit	Material	2015 Bare Costs Labor	Equipment	Total	Total Incl O&P
0010	**WOOD WINDOWS**, double hung										
0020	Including frame, double insulated glass, screens and grilles										
0040	Double hung, 2'-2" x 3'-4" high		2 Carp	15	1.067	Ea.	210	50		260	315
0060	2'-2" x 4'-4"			14	1.143		229	53.50		282.50	340
0080	2'-6" x 3'-4"			13	1.231		219	58		277	335
0100	2'-6" x 4'-0"			12	1.333		229	62.50		291.50	355
0120	2'-6" x 4'-8"			12	1.333		250	62.50		312.50	380
0140	2'-10" x 3'-4"			10	1.600		224	75		299	370
0160	2'-10" x 4'-0"			10	1.600		247	75		322	395
0180	3'-7" x 3'-4"			9	1.778		254	83.50		337.50	415
0200	3'-7" x 5'-4"			9	1.778		284	83.50		367.50	445
0220	3'-10" x 5'-4"			8	2		510	94		604	715
3800	Triple glazing for above, add						25%				

08 52 10.40 Casement Window

			Crew	Daily Output	Labor-Hours	Unit	Material	Labor	Equipment	Total	Total Incl O&P
0010	**CASEMENT WINDOW**, including frame, screen and grilles										
0100	2'-0" x 3'-0" H, dbl. insulated glass	G	1 Carp	10	.800	Ea.	268	37.50		305.50	355
0150	Low E glass	G		10	.800		259	37.50		296.50	345
0200	2'-0" x 4'-6" high, double insulated glass	G		9	.889		360	41.50		401.50	465
0250	Low E glass	G		9	.889		370	41.50		411.50	475
0260	Casement 4'-2" x 4'-2" double insulated glass	G		11	.727		875	34		909	1,025
0270	4'-0" x 4'-0" Low E glass	G		11	.727		535	34		569	645
0290	6'-4" x 5'-7" Low E glass	G		9	.889		1,125	41.50		1,166.50	1,300
0300	2'-4" x 6'-0" high, double insulated glass	G		8	1		440	47		487	560
0350	Low E glass	G		8	1		475	47		522	600
0522	Vinyl clad, premium, double insulated glass, 2'-0" x 3'-0"	G		10	.800		271	37.50		308.50	360
0524	2'-0" x 4'-0"	G		9	.889		315	41.50		356.50	420
0525	2'-0" x 5'-0"	G		8	1		360	47		407	475
0528	2'-0" x 6'-0"	G		8	1		380	47		427	495
0600	3'-0" x 5'-0"	G		8	1		665	47		712	805
0700	4'-0" x 3'-0"	G		8	1		730	47		777	880
0710	4'-0" x 4'-0"	G		8	1		625	47		672	760
0720	4'-8" x 4'-0"	G		8	1		690	47		737	835
0730	4'-8" x 5'-0"	G		6	1.333		790	62.50		852.50	970
0740	4'-8" x 6'-0"	G		6	1.333		880	62.50		942.50	1,075
0750	6'-0" x 4'-0"	G		6	1.333		805	62.50		867.50	990
0800	6'-0" x 5'-0"	G		6	1.333		895	62.50		957.50	1,100
0900	5'-6" x 5'-6"	G	2 Carp	15	1.067		1,425	50		1,475	1,650
8190	For installation, add per leaf							15%			
8200	For multiple leaf units, deduct for stationary sash										
8220	2' high					Ea.	23			23	25.50
8240	4'-6" high						26			26	28.50
8260	6' high						34.50			34.50	38
8300	Impact windows, minimum, add						60%				
8310	Impact windows, maximum, add						160%				
9000	Minimum labor/equipment charge		1 Carp	3	2.667	Job		125		125	205

08 52 10.50 Double Hung

			Crew	Daily Output	Labor-Hours	Unit	Material	Labor	Equipment	Total	Total Incl O&P
0010	**DOUBLE HUNG**, Including frame, screens and grilles										
0100	2'-0" x 3'-0" high, low E insul. glass	G	1 Carp	10	.800	Ea.	210	37.50		247.50	293
0200	3'-0" x 4'-0" high, double insulated glass	G		9	.889		279	41.50		320.50	375
0300	4'-0" x 4'-6" high, low E insulated glass	G		8	1		320	47		367	430
8000	Impact windows, minimum, add						60%				
8010	Impact windows, maximum, add						160%				

08 52 Wood Windows

08 52 10 – Plain Wood Windows

08 52 10.50 Double Hung

		Crew	Daily Output	Labor-Hours	Unit	Material	2015 Bare Costs Labor	2015 Bare Costs Equipment	Total	Total Incl O&P
9000	Minimum labor/equipment charge	1 Carp	3	2.667	Job		125		125	205

08 52 10.55 Picture Window

		Crew	Daily Output	Labor-Hours	Unit	Material	Labor	Equipment	Total	Incl O&P
0010	**PICTURE WINDOW**, Including frame and grilles									
0100	3'-6" x 4'-0" high, dbl. insulated glass	2 Carp	12	1.333	Ea.	420	62.50		482.50	570
0150	Low E glass		12	1.333		435	62.50		497.50	580
0200	4'-0" x 4'-6" high, double insulated glass		11	1.455		550	68.50		618.50	715
0250	Low E glass		11	1.455		530	68.50		598.50	695
0300	5'-0" x 4'-0" high, double insulated glass		11	1.455		580	68.50		648.50	750
0350	Low E glass		11	1.455		605	68.50		673.50	775
0400	6'-0" x 4'-6" high, double insulated glass		10	1.600		625	75		700	815
0450	Low E glass		10	1.600		635	75		710	825

08 52 10.65 Wood Sash

		Crew	Daily Output	Labor-Hours	Unit	Material	Labor	Equipment	Total	Incl O&P
0010	**WOOD SASH**, Including glazing but not trim									
0050	Custom, 5'-0" x 4'-0", 1" dbl. glazed, 3/16" thick lites	2 Carp	3.20	5	Ea.	230	235		465	640
0100	1/4" thick lites		5	3.200		245	150		395	515
0200	1" thick, triple glazed		5	3.200		415	150		565	700
0300	7'-0" x 4'-6" high, 1" double glazed, 3/16" thick lites		4.30	3.721		420	175		595	745
0400	1/4" thick lites		4.30	3.721		475	175		650	805
0500	1" thick, triple glazed		4.30	3.721		540	175		715	880
0600	8'-6" x 5'-0" high, 1" double glazed, 3/16" thick lites		3.50	4.571		565	215		780	975
0700	1/4" thick lites		3.50	4.571		620	215		835	1,025
0800	1" thick, triple glazed		3.50	4.571		625	215		840	1,025
0900	Window frames only, based on perimeter length				L.F.	4.02			4.02	4.42
1200	Window sill, stock, per lineal foot					8.50			8.50	9.35
1250	Casing, stock					3.30			3.30	3.63

08 52 10.70 Sliding Windows

			Crew	Daily Output	Labor-Hours	Unit	Material	Labor	Equipment	Total	Incl O&P
0010	**SLIDING WINDOWS**										
0100	3'-0" x 3'-0" high, double insulated	G	1 Carp	10	.800	Ea.	284	37.50		321.50	370
0120	Low E glass	G		10	.800		310	37.50		347.50	400
0200	4'-0" x 3'-6" high, double insulated	G		9	.889		355	41.50		396.50	460
0220	Low E glass	G		9	.889		360	41.50		401.50	470
0300	6'-0" x 5'-0" high, double insulated	G		8	1		490	47		537	610
0320	Low E glass	G		8	1		530	47		577	655
9000	Minimum labor/equipment charge			3	2.667	Job		125		125	205

08 52 13 – Metal-Clad Wood Windows

08 52 13.10 Awning Windows, Metal-Clad

		Crew	Daily Output	Labor-Hours	Unit	Material	Labor	Equipment	Total	Incl O&P
0010	**AWNING WINDOWS, METAL-CLAD**									
2000	Metal clad, awning deluxe, double insulated glass, 34" x 22"	1 Carp	9	.889	Ea.	247	41.50		288.50	340
2050	36" x 25"		9	.889		272	41.50		313.50	370
2100	40" x 22"		9	.889		291	41.50		332.50	390
2150	40" x 30"		9	.889		340	41.50		381.50	445
2200	48" x 28"		8	1		345	47		392	455
2250	60" x 36"		8	1		370	47		417	485

08 52 13.20 Casement Windows, Metal-Clad

			Crew	Daily Output	Labor-Hours	Unit	Material	Labor	Equipment	Total	Incl O&P
0010	**CASEMENT WINDOWS, METAL-CLAD**										
0100	Metal clad, deluxe, dbl. insul. glass, 2'-0" x 3'-0" high	G	1 Carp	10	.800	Ea.	279	37.50		316.50	365
0120	2'-0" x 4'-0" high	G		9	.889		315	41.50		356.50	415
0130	2'-0" x 5'-0" high	G		8	1		325	47		372	435
0140	2'-0" x 6'-0" high	G		8	1		365	47		412	475
0300	Metal clad, casement, bldrs mdl, 6'-0" x 4'-0", dbl. insltd gls, 3 panels		2 Carp	10	1.600		1,200	75		1,275	1,450
0310	9'-0" x 4'-0", 4 panels			8	2		1,550	94		1,644	1,850

08 52 Wood Windows

08 52 13 – Metal-Clad Wood Windows

08 52 13.20 Casement Windows, Metal-Clad

		Crew	Daily Output	Labor-Hours	Unit	Material	2015 Bare Costs Labor	Equipment	Total	Total Incl O&P
0320	10'-0" x 5'-0", 5 panels	2 Carp	7	2.286	Ea.	2,100	107		2,207	2,500
0330	12'-0" x 6'-0", 6 panels		6	2.667		2,700	125		2,825	3,150

08 52 13.30 Double-Hung Windows, Metal-clad

			Crew	Daily Output	Labor-Hours	Unit	Material	Labor	Equipment	Total	Total Incl O&P
0010	**DOUBLE-HUNG WINDOWS, METAL-CLAD**										
0100	Metal clad, deluxe, dbl. insul. glass, 2'-6" x 3'-0" high	G	1 Carp	10	.800	Ea.	272	37.50		309.50	360
0120	3'-0" x 3'-6" high	G		10	.800		315	37.50		352.50	405
0140	3'-0" x 4'-0" high	G		9	.889		325	41.50		366.50	430
0160	3'-0" x 4'-6" high	G		9	.889		345	41.50		386.50	450
0180	3'-0" x 5'-0" high	G		8	1		370	47		417	485
0200	3'-6" x 6'-0" high	G		8	1		450	47		497	570

08 52 13.35 Picture and Sliding Windows Metal-Clad

			Crew	Daily Output	Labor-Hours	Unit	Material	Labor	Equipment	Total	Total Incl O&P
0010	**PICTURE AND SLIDING WINDOWS METAL-CLAD**										
2000	Metal clad, dlx picture, dbl. insul. glass, 4'-0" x 4'-0" high		2 Carp	12	1.333	Ea.	375	62.50		437.50	515
2100	4'-0" x 6'-0" high			11	1.455		545	68.50		613.50	710
2200	5'-0" x 6'-0" high			10	1.600		610	75		685	795
2300	6'-0" x 6'-0" high			10	1.600		695	75		770	890
2400	Metal clad, dlx sliding, double insulated glass, 3'-0" x 3'-0" high	G	1 Carp	10	.800		330	37.50		367.50	420
2420	4'-0" x 3'-6" high	G		9	.889		400	41.50		441.50	510
2440	5'-0" x 4'-0" high	G		9	.889		480	41.50		521.50	595
2460	6'-0" x 5'-0" high	G		8	1		730	47		777	875
9000	Minimum labor/equipment charge		2 Carp	2.75	5.818	Job		273		273	450

08 52 13.40 Bow and Bay Windows, Metal-Clad

		Crew	Daily Output	Labor-Hours	Unit	Material	Labor	Equipment	Total	Total Incl O&P
0010	**BOW AND BAY WINDOWS, METAL-CLAD**									
0100	Metal clad, deluxe, dbl. insul. glass, 8'-0" x 5'-0" high, 4 panels	2 Carp	10	1.600	Ea.	1,675	75		1,750	1,950
0120	10'-0" x 5'-0" high, 5 panels		8	2		1,800	94		1,894	2,125
0140	10'-0" x 6'-0" high, 5 panels		7	2.286		2,100	107		2,207	2,500
0160	12'-0" x 6'-0" high, 6 panels		6	2.667		2,925	125		3,050	3,425
0400	Double hung, bldrs. model, bay, 8' x 4' high, dbl. insulated glass		10	1.600		1,325	75		1,400	1,575
0440	Low E glass		10	1.600		1,425	75		1,500	1,700
0460	9'-0" x 5'-0" high, double insulated glass		6	2.667		1,425	125		1,550	1,775
0480	Low E glass		6	2.667		1,500	125		1,625	1,850
0500	Metal clad, deluxe, dbl. insul. glass, 7'-0" x 4'-0" high		10	1.600		1,275	75		1,350	1,525
0520	8'-0" x 4'-0" high		8	2		1,300	94		1,394	1,600
0540	8'-0" x 5'-0" high		7	2.286		1,350	107		1,457	1,675
0560	9'-0" x 5'-0" high		6	2.667		1,450	125		1,575	1,775

08 52 16 – Plastic-Clad Wood Windows

08 52 16.10 Bow Window

		Crew	Daily Output	Labor-Hours	Unit	Material	Labor	Equipment	Total	Total Incl O&P
0010	**BOW WINDOW** Including frames, screens, and grilles									
0020	End panels operable									
1000	Bow type, casement, wood, bldrs. mdl., 8' x 5' dbl. insltd glass, 4 panel	2 Carp	10	1.600	Ea.	1,500	75		1,575	1,775
1050	Low E glass		10	1.600		1,325	75		1,400	1,575
1100	10'-0" x 5'-0", double insulated glass, 6 panels		6	2.667		1,350	125		1,475	1,700
1200	Low E glass, 6 panels		6	2.667		1,450	125		1,575	1,800
1300	Vinyl clad, bldrs. model, double insulated glass, 6'-0" x 4'-0", 3 panel		10	1.600		1,025	75		1,100	1,250
1340	9'-0" x 4'-0", 4 panel		8	2		1,350	94		1,444	1,650
1380	10'-0" x 6'-0", 5 panels		7	2.286		2,250	107		2,357	2,650
1420	12'-0" x 6'-0", 6 panels		6	2.667		2,925	125		3,050	3,425
2000	Bay window, 8' x 5', dbl. insul glass		10	1.600		1,875	75		1,950	2,200
2050	Low E glass		10	1.600		2,275	75		2,350	2,625
2100	12'-0" x 6'-0", double insulated glass, 6 panels		6	2.667		2,350	125		2,475	2,775
2200	Low E glass		6	2.667		2,375	125		2,500	2,825

08 52 Wood Windows

08 52 16 – Plastic-Clad Wood Windows

08 52 16.10 Bow Window

		Crew	Daily Output	Labor-Hours	Unit	Material	2015 Bare Costs Labor	Equipment	Total	Total Incl O&P
2300	Vinyl clad, premium, double insulated glass, 8'-0" x 5'-0"	2 Carp	10	1.600	Ea.	1,750	75		1,825	2,050
2340	10'-0" x 5'-0"		8	2		2,300	94		2,394	2,675
2380	10'-0" x 6'-0"		7	2.286		2,625	107		2,732	3,050
2420	12'-0" x 6'-0"		6	2.667		3,200	125		3,325	3,725
3300	Vinyl clad, premium, double insulated glass, 7'-0" x 4'-6"		10	1.600		1,375	75		1,450	1,625
3340	8'-0" x 4'-6"		8	2		1,400	94		1,494	1,675
3380	8'-0" x 5'-0"		7	2.286		1,450	107		1,557	1,775
3420	9'-0" x 5'-0"		6	2.667		1,500	125		1,625	1,850
9000	Minimum labor/equipment charge		2.50	6.400	Job		300		300	490

08 52 16.15 Awning Window Vinyl-Clad

		Crew	Daily Output	Labor-Hours	Unit	Material	Labor	Equipment	Total	Total Incl O&P
0010	**AWNING WINDOW VINYL-CLAD** Including frames, screens, and grilles									
0240	Vinyl clad, 34" x 22"	1 Carp	10	.800	Ea.	264	37.50		301.50	350
0280	36" x 28"		9	.889		305	41.50		346.50	405
0300	36" x 36"		9	.889		340	41.50		381.50	445
0340	40" x 22"		10	.800		288	37.50		325.50	375
0360	48" x 28"		8	1		370	47		417	480
0380	60" x 36"		8	1		500	47		547	625

08 52 16.20 Half Round, Vinyl Clad

		Crew	Daily Output	Labor-Hours	Unit	Material	Labor	Equipment	Total	Total Incl O&P
0010	**HALF ROUND, VINYL CLAD**, double insulated glass, incl. grille									
0800	14" height x 24" base	2 Carp	9	1.778	Ea.	415	83.50		498.50	590
1040	15" height x 25" base		8	2		385	94		479	575
1060	16" height x 28" base		7	2.286		405	107		512	620
1080	17" height x 29" base		7	2.286		420	107		527	635
2000	19" height x 33" base	1 Carp	6	1.333		465	62.50		527.50	615
2100	20" height x 35" base		6	1.333		465	62.50		527.50	615
2200	21" height x 37" base		6	1.333		480	62.50		542.50	635
2250	23" height x 41" base	2 Carp	6	2.667		520	125		645	775
2300	26" height x 48" base		6	2.667		535	125		660	790
2350	30" height x 56" base		6	2.667		615	125		740	880
3000	36" height x 67" base	1 Carp	4	2		1,050	94		1,144	1,300
3040	38" height x 71" base	2 Carp	5	3.200		975	150		1,125	1,325
3050	40" height x 75" base	"	5	3.200		1,275	150		1,425	1,650
5000	Elliptical, 71" x 16"	1 Carp	11	.727		975	34		1,009	1,125
5100	95" x 21"	"	10	.800		1,375	37.50		1,412.50	1,575

08 52 16.35 Double-Hung Window

			Crew	Daily Output	Labor-Hours	Unit	Material	Labor	Equipment	Total	Total Incl O&P
0010	**DOUBLE-HUNG WINDOW** Including frames, screens, and grilles										
0300	Vinyl clad, premium, double insulated glass, 2'-6" x 3'-0"	G	1 Carp	10	.800	Ea.	310	37.50		347.50	400
0305	2'-6" x 4'-0"	G		10	.800		360	37.50		397.50	455
0400	3'-0" x 3'-6"	G		10	.800		335	37.50		372.50	430
0500	3'-0" x 4'-0"	G		9	.889		395	41.50		436.50	505
0600	3'-0" x 4'-6"	G		9	.889		410	41.50		451.50	520
0700	3'-0" x 5'-0"	G		8	1		445	47		492	560
0790	3'-4" x 5'-0"	G		8	1		455	47		502	575
0800	3'-6" x 6'-0"	G		8	1		490	47		537	615
0820	4'-0" x 5'-0"	G		7	1.143		560	53.50		613.50	705
0830	4'-0" x 6'-0"	G		7	1.143		700	53.50		753.50	855

08 52 16.40 Transom Windows

		Crew	Daily Output	Labor-Hours	Unit	Material	Labor	Equipment	Total	Total Incl O&P
0010	**TRANSOM WINDOWS**									
0050	Vinyl clad, premium, double insulated glass, 32" x 8"	1 Carp	16	.500	Ea.	188	23.50		211.50	245
0100	36" x 8"		16	.500		199	23.50		222.50	258
0110	36" x 12"		16	.500		212	23.50		235.50	272
2000	Custom sizes, up to 350 sq. in.		12	.667		256	31.50		287.50	335

08 52 Wood Windows

08 52 16 – Plastic-Clad Wood Windows

08 52 16.40 Transom Windows

		Crew	Daily Output	Labor-Hours	Unit	Material	2015 Bare Costs Labor	Equipment	Total	Total Incl O&P
2100	351 to 750 sq. in.	1 Carp	12	.667	Ea.	325	31.50		356.50	405
2200	751 to 1150 sq. in.		11	.727		400	34		434	495
2300	1151 to 1450 sq. in.		11	.727		460	34		494	565
2400	1451 to 1850 sq. in.	2 Carp	12	1.333		530	62.50		592.50	690
2500	1851 to 2250 sq. in.		12	1.333		605	62.50		667.50	770
2600	2251 to 2650 sq. in.		11	1.455		680	68.50		748.50	860
2700	2651 to 3050 sq. in.		11	1.455		685	68.50		753.50	865
2800	3051 to 3450 sq. in.		11	1.455		755	68.50		823.50	940
2900	3451 to 3850 sq. in.		10	1.600		810	75		885	1,025
3000	3851 to 4250 sq. in.		10	1.600		870	75		945	1,075
3100	4251 to 4650 sq. in.		10	1.600		890	75		965	1,100
3200	4651 to 5050 sq. in.		10	1.600		935	75		1,010	1,150
3300	5051 to 5450 sq. in.		9	1.778		1,000	83.50		1,083.50	1,225
3400	5451 to 5850 sq. in.		9	1.778		975	83.50		1,058.50	1,200
3600	6251 to 6650 sq. in.		8	2		1,175	94		1,269	1,425
3700	6651 to 7050 sq. in.		8	2		1,250	94		1,344	1,525

08 52 16.45 Trapezoid Windows

		Crew	Daily Output	Labor-Hours	Unit	Material	Labor	Equipment	Total	Total Incl O&P
0010	**TRAPEZOID WINDOWS**									
0900	20" base x 44" leg x 53" leg	2 Carp	13	1.231	Ea.	390	58		448	525
1000	24" base x 90" leg x 102" leg		8	2		670	94		764	895
3000	36" base x 40" leg x 22" leg		12	1.333		430	62.50		492.50	575
3010	36" base x 44" leg x 25" leg		13	1.231		450	58		508	590
3050	36" base x 26" leg x 48" leg		9	1.778		455	83.50		538.50	635
3100	36" base x 42" legs, 50" peak		9	1.778		545	83.50		628.50	735
3200	36" base x 60" leg x 81" leg		11	1.455		710	68.50		778.50	895
4320	44" base x 23" leg x 56" leg		11	1.455		560	68.50		628.50	730
4350	44" base x 59" leg x 92" leg		10	1.600		840	75		915	1,050
4500	46" base x 15" leg x 46" leg		8	2		425	94		519	620
4550	46" base x 16" leg x 48" leg		8	2		450	94		544	650
4600	46" base x 50" leg x 80" leg		7	2.286		680	107		787	925
6600	66" base x 12" leg x 42" leg		8	2		575	94		669	790
6650	66" base x 12" legs, 28" peak		9	1.778		455	83.50		538.50	635
6700	68" base x 23" legs, 31" peak		8	2		575	94		669	790

08 52 16.70 Vinyl Clad, Premium, Dbl. Insulated Glass

			Crew	Daily Output	Labor-Hours	Unit	Material	Labor	Equipment	Total	Total Incl O&P
0010	**VINYL CLAD, PREMIUM, DBL. INSULATED GLASS**										
1000	Sliding, 3'-0" x 3'-0"	G	1 Carp	10	.800	Ea.	605	37.50		642.50	725
1050	4'-0" x 3'-6"	G		9	.889		685	41.50		726.50	825
1100	5'-0" x 4'-0"	G		9	.889		900	41.50		941.50	1,050
1150	6'-0" x 5'-0"	G		8	1		1,125	47		1,172	1,325

08 52 50 – Window Accessories

08 52 50.10 Window Grille or Muntin

		Crew	Daily Output	Labor-Hours	Unit	Material	Labor	Equipment	Total	Total Incl O&P
0010	**WINDOW GRILLE OR MUNTIN**, snap in type									
0020	Standard pattern interior grilles									
2000	Wood, awning window, glass size 28" x 16" high	1 Carp	30	.267	Ea.	28	12.50		40.50	51.50
2060	44" x 24" high		32	.250		40	11.75		51.75	63.50
2100	Casement, glass size, 20" x 36" high		30	.267		32	12.50		44.50	55.50
2180	20" x 56" high		32	.250		43	11.75		54.75	67
2200	Double hung, glass size, 16" x 24" high		24	.333	Set	51	15.65		66.65	82
2280	32" x 32" high		34	.235	"	131	11.05		142.05	162
2500	Picture, glass size, 48" x 48" high		30	.267	Ea.	120	12.50		132.50	153
2580	60" x 68" high		28	.286	"	183	13.40		196.40	224
2600	Sliding, glass size, 14" x 36" high		24	.333	Set	35.50	15.65		51.15	64.50

08 52 Wood Windows

08 52 50 – Window Accessories

08 52 50.10 Window Grille or Muntin		Crew	Daily Output	Labor-Hours	Unit	Material	2015 Bare Costs Labor	Equipment	Total	Total Incl O&P
2680	36" x 36" high	1 Carp	22	.364	Set	43.50	17.05		60.55	75.50
9000	Minimum labor/equipment charge		5	1.600	Job		75		75	123

08 52 66 – Wood Window Screens

08 52 66.10 Wood Screens

		Crew	Daily Output	Labor-Hours	Unit	Material	Labor	Equipment	Total	Total Incl O&P
0010	**WOOD SCREENS**									
0020	Over 3 S.F., 3/4" frames	2 Carp	375	.043	S.F.	4.89	2		6.89	8.70
0100	1-1/8" frames		375	.043		8.20	2		10.20	12.30
0200	Rescreen wood frame		500	.032		1.61	1.50		3.11	4.23
9000	Minimum labor/equipment charge	1 Carp	4	2	Job		94		94	154

08 52 69 – Wood Storm Windows

08 52 69.10 Storm Windows

			Crew	Daily Output	Labor-Hours	Unit	Material	Labor	Equipment	Total	Total Incl O&P
0010	**STORM WINDOWS**, aluminum residential										
0300	Basement, mill finish, incl. fiberglass screen										
0320	1'-10" x 1'-0" high	G	2 Carp	30	.533	Ea.	35	25		60	79.50
0340	2'-9" x 1'-6" high	G		30	.533		38	25		63	83
0360	3'-4" x 2'-0" high	G		30	.533		45	25		70	90.50
1600	Double-hung, combination, storm & screen										
1700	Custom, clear anodic coating, 2'-0" x 3'-5" high		2 Carp	30	.533	Ea.	95	25		120	146
1720	2'-6" x 5'-0" high			28	.571		115	27		142	171
1740	4'-0" x 6'-0" high			25	.640		230	30		260	305
1800	White painted, 2'-0" x 3'-5" high			30	.533		110	25		135	162
1820	2'-6" x 5'-0" high			28	.571		170	27		197	231
1840	4'-0" x 6'-0" high			25	.640		280	30		310	360
2000	Clear anodic coating, 2'-0" x 3'-5" high	G		30	.533		95	25		120	146
2020	2'-6" x 5'-0" high	G		28	.571		117	27		144	172
2040	4'-0" x 6'-0" high	G		25	.640		130	30		160	193
2400	White painted, 2'-0" x 3'-5" high	G		30	.533		90	25		115	140
2420	2'-6" x 5'-0" high	G		28	.571		95	27		122	149
2440	4'-0" x 6'-0" high	G		25	.640		110	30		140	171
2600	Mill finish, 2'-0" x 3'-5" high	G		30	.533		85	25		110	135
2620	2'-6" x 5'-0" high	G		28	.571		90	27		117	143
2640	4'-0" x 6'-8" high	G		25	.640		110	30		140	171
4000	Picture window, storm, 1 lite, white or bronze finish										
4020	4'-6" x 4'-6" high		2 Carp	25	.640	Ea.	130	30		160	193
4040	5'-8" x 4'-6" high			20	.800		148	37.50		185.50	225
4400	Mill finish, 4'-6" x 4'-6" high			25	.640		130	30		160	193
4420	5'-8" x 4'-6" high			20	.800		150	37.50		187.50	227
4600	3 lite, white or bronze finish										
4620	4'-6" x 4'-6" high		2 Carp	25	.640	Ea.	160	30		190	226
4640	5'-8" x 4'-6" high			20	.800		170	37.50		207.50	249
4800	Mill finish, 4'-6" x 4'-6" high			25	.640		150	30		180	215
4820	5'-8" x 4'-6" high			20	.800		160	37.50		197.50	238
6000	Sliding window, storm, 2 lite, white or bronze finish										
6020	3'-4" x 2'-7" high		2 Carp	28	.571	Ea.	135	27		162	193
6040	4'-4" x 3'-3" high			25	.640		150	30		180	215
6060	5'-4" x 6'-0" high			20	.800		240	37.50		277.50	325
8000	PVC framed										
8100	Double-hung, combination, storm & screen										
8120	2'-6" x 3'-5"		2 Carp	15	1.067	Ea.	135	50		185	231
8140	4' x 6'		"	12.50	1.280	"	171	60		231	287
8600	Single lite picture storm										
8620	4'-6" x 4'-6"		2 Carp	12.50	1.280	Ea.	171	60		231	287

08 52 Wood Windows

08 52 69 – Wood Storm Windows

08 52 69.10 Storm Windows		Crew	Daily Output	Labor-Hours	Unit	Material	2015 Bare Costs Labor	Equipment	Total	Total Incl O&P
8640	5'-8" x 4'-6"	2 Carp	10	1.600	Ea.	187	75		262	330
9000	Magnetic interior storm window									
9100	3/16" plate glass	1 Glaz	107	.075	S.F.	6.10	3.37		9.47	12.15
9410	Minimum labor/equipment charge	1 Carp	4	2	Job		94		94	154

08 53 Plastic Windows

08 53 13 – Vinyl Windows

08 53 13.20 Vinyl Single Hung Windows

			Crew	Daily Output	Labor-Hours	Unit	Material	Labor	Equipment	Total	Total Incl O&P
0010	**VINYL SINGLE HUNG WINDOWS**, insulated glass										
0100	Grids, low E, J fin, ext. jambs, 21" x 53"	G	2 Carp	18	.889	Ea.	198	41.50		239.50	287
0110	21" x 57"	G		17	.941		200	44		244	293
0120	21" x 65"	G		16	1		205	47		252	305
0130	25" x 41"	G		20	.800		190	37.50		227.50	271
0140	25" x 49"	G		18	.889		200	41.50		241.50	289
0150	25" x 57"	G		17	.941		205	44		249	299
0160	25" x 65"	G		16	1		240	47		287	340
0170	29" x 41"	G		18	.889		195	41.50		236.50	284
0180	29" x 53"	G		18	.889		205	41.50		246.50	295
0190	29" x 57"	G		17	.941		210	44		254	305
0200	29" x 65"	G		16	1		215	47		262	315
0210	33" x 41"	G		20	.800		200	37.50		237.50	282
0220	33" x 53"	G		18	.889		215	41.50		256.50	305
0230	33" x 57"	G		17	.941		215	44		259	310
0240	33" x 65"	G		16	1		220	47		267	320
0250	37" x 41"	G		20	.800		225	37.50		262.50	310
0260	37" x 53"	G		18	.889		235	41.50		276.50	330
0270	37" x 57"	G		17	.941		240	44		284	335
0280	37" x 65"	G		16	1		256	47		303	360

08 53 13.30 Vinyl Double Hung Windows

			Crew	Daily Output	Labor-Hours	Unit	Material	Labor	Equipment	Total	Total Incl O&P
0010	**VINYL DOUBLE HUNG WINDOWS**, insulated glass										
0100	Grids, low E, J fin, ext. jambs, 21" x 53"	G	2 Carp	18	.889	Ea.	215	41.50		256.50	305
0102	21" x 37"	G		18	.889		223	41.50		264.50	315
0104	21" x 41"	G		18	.889		232	41.50		273.50	325
0106	21" x 49"	G		18	.889		247	41.50		288.50	340
0110	21" x 57"	G		17	.941		263	44		307	360
0120	21" x 65"	G		16	1		279	47		326	380
0128	25" x 37"	G		20	.800		234	37.50		271.50	320
0130	25" x 41"	G		20	.800		242	37.50		279.50	330
0140	25" x 49"	G		18	.889		258	41.50		299.50	355
0145	25" x 53"	G		18	.889		265	41.50		306.50	360
0150	25" x 57"	G		17	.941		274	44		318	375
0160	25" x 65"	G		16	1		290	47		337	395
0162	25" x 69"	G		16	1		297	47		344	400
0164	25" x 77"	G		16	1		325	47		372	430
0168	29" x 37"	G		18	.889		242	41.50		283.50	335
0170	29" x 41"	G		18	.889		250	41.50		291.50	345
0172	29" x 49"	G		18	.889		267	41.50		308.50	365
0180	29" x 53"	G		18	.889		271	41.50		312.50	365
0190	29" x 57"	G		17	.941		280	44		324	385
0200	29" x 65"	G		16	1		300	47		347	405
0202	29" x 69"	G		16	1		310	47		357	415

08 53 Plastic Windows

08 53 13 – Vinyl Windows

08 53 13.30 Vinyl Double Hung Windows

			Crew	Daily Output	Labor-Hours	Unit	Material	2015 Bare Costs Labor	2015 Bare Costs Equipment	Total	Total Incl O&P
0205	29" x 77"	G	2 Carp	16	1	Ea.	335	47		382	440
0208	33" x 37"	G		20	.800		255	37.50		292.50	345
0210	33" x 41"	G		20	.800		263	37.50		300.50	350
0215	33" x 49"	G		20	.800		280	37.50		317.50	370
0220	33" x 53"	G		18	.889		285	41.50		326.50	385
0230	33" x 57"	G		17	.941		297	44		341	400
0240	33" x 65"	G		16	1		315	47		362	420
0242	33" x 69"	G		16	1		330	47		377	435
0246	33" x 77"	G		16	1		350	47		397	460
0250	37" x 41"	G		20	.800		289	37.50		326.50	380
0255	37" x 49"	G		20	.800		282	37.50		319.50	370
0260	37" x 53"	G		18	.889		315	41.50		356.50	415
0270	37" x 57"	G		17	.941		325	44		369	435
0280	37" x 65"	G		16	1		350	47		397	460
0282	37" x 69"	G		16	1		360	47		407	470
0286	37" x 77"	G		16	1		380	47		427	490
0300	Solid vinyl, average quality, double insulated glass, 2'-0" x 3'-0"	G	1 Carp	10	.800		291	37.50		328.50	380
0310	3'-0" x 4'-0"	G		9	.889		207	41.50		248.50	296
0320	4'-0" x 4'-6"	G		8	1		335	47		382	440
0330	Premium, double insulated glass, 2'-6" x 3'-0"	G		10	.800		271	37.50		308.50	360
0340	3'-0" x 3'-6"	G		9	.889		292	41.50		333.50	390
0350	3'-0" x 4'-0"	G		9	.889		315	41.50		356.50	420
0360	3'-0" x 4'-6"	G		9	.889		330	41.50		371.50	435
0370	3'-0" x 5'-0"	G		8	1		355	47		402	465
0380	3'-6" x 6'-0"	G		8	1		380	47		427	495

08 53 13.40 Vinyl Casement Windows

			Crew	Daily Output	Labor-Hours	Unit	Material	2015 Bare Costs Labor	2015 Bare Costs Equipment	Total	Total Incl O&P
0010	**VINYL CASEMENT WINDOWS**, insulated glass										
0015	Grids, low E, J fin, extension jambs, screens										
0100	One lite, 21" x 41"	G	2 Carp	20	.800	Ea.	295	37.50		332.50	385
0110	21" x 47"	G		20	.800		320	37.50		357.50	415
0120	21" x 53"	G		20	.800		345	37.50		382.50	440
0128	24" x 35"	G		19	.842		283	39.50		322.50	375
0130	24" x 41"	G		19	.842		310	39.50		349.50	405
0140	24" x 47"	G		19	.842		335	39.50		374.50	430
0150	24" x 53"	G		19	.842		360	39.50		399.50	460
0158	28" x 35"	G		19	.842		300	39.50		339.50	395
0160	28" x 41"	G		19	.842		330	39.50		369.50	430
0170	28" x 47"	G		19	.842		350	39.50		389.50	450
0180	28" x 53"	G		19	.842		385	39.50		424.50	490
0184	28" x 59"	G		19	.842		395	39.50		434.50	500
0188	Two lites, 33" x 35"	G		18	.889		470	41.50		511.50	585
0190	33" x 41"	G		18	.889		500	41.50		541.50	620
0200	33" x 47"	G		18	.889		540	41.50		581.50	660
0210	33" x 53"	G		18	.889		575	41.50		616.50	700
0212	33" x 59"	G		18	.889		610	41.50		651.50	740
0215	33" x 72"	G		18	.889		635	41.50		676.50	765
0220	41" x 41"	G		18	.889		545	41.50		586.50	670
0230	41" x 47"	G		18	.889		585	41.50		626.50	710
0240	41" x 53"	G		17	.941		620	44		664	755
0242	41" x 59"	G		17	.941		650	44		694	790
0246	41" x 72"	G		17	.941		685	44		729	830
0250	47" x 41"	G		17	.941		555	44		599	685

For customer support on your Commercial Renovation Cost Data, call 877.791.4977.

08 53 Plastic Windows

08 53 13 – Vinyl Windows

08 53 13.40 Vinyl Casement Windows

			Crew	Daily Output	Labor-Hours	Unit	Material	2015 Bare Costs Labor	Equipment	Total	Total Incl O&P
0260	47" x 47"	G	2 Carp	17	.941	Ea.	590	44		634	725
0270	47" x 53"	G		17	.941		625	44		669	765
0272	47" x 59"	G		17	.941		680	44		724	825
0280	56" x 41"	G		15	1.067		595	50		645	735
0290	56" x 47"	G		15	1.067		625	50		675	770
0300	56" x 53"	G		15	1.067		680	50		730	830
0302	56" x 59"	G		15	1.067		710	50		760	860
0310	56" x 72"	G		15	1.067		770	50		820	930
0340	Solid vinyl, premium, double insulated glass, 2'-0" x 3'-0" high	G	1 Carp	10	.800		270	37.50		307.50	360
0360	2'-0" x 4'-0" high	G		9	.889		299	41.50		340.50	400
0380	2'-0" x 5'-0" high	G		8	1		335	47		382	445

08 53 13.50 Vinyl Picture Windows

		Crew	Daily Output	Labor-Hours	Unit	Material	Labor	Equipment	Total	Total Incl O&P
0010	**VINYL PICTURE WINDOWS**, insulated glass									
0100	Grids, low E, J fin, ext. jambs, 33" x 47"	2 Carp	12	1.333	Ea.	280	62.50		342.50	415
0110	35" x 71"		12	1.333		375	62.50		437.50	520
0120	47" x 35"		12	1.333		300	62.50		362.50	435
0130	47" x 41"		12	1.333		390	62.50		452.50	535
0140	47" x 47"		12	1.333		345	62.50		407.50	480
0150	47" x 53"		11	1.455		370	68.50		438.50	515
0160	71" x 35"		11	1.455		390	68.50		458.50	540
0170	71" x 41"		11	1.455		410	68.50		478.50	560
0180	71" x 47"		11	1.455		440	68.50		508.50	595

08 53 13.60 Vinyl Half Round Windows

		Crew	Daily Output	Labor-Hours	Unit	Material	Labor	Equipment	Total	Total Incl O&P
0010	**VINYL HALF ROUND WINDOWS**, Including grille, j fin low E, ext. jambs									
0100	10" height x 20" base	2 Carp	8	2	Ea.	475	94		569	675
0110	15" height x 30" base		8	2		440	94		534	640
0120	17" height x 34" base		7	2.286		320	107		427	525
0130	19" height x 38" base		7	2.286		345	107		452	555
0140	19" height x 33" base		7	2.286		555	107		662	790
0150	24" height x 48" base	1 Carp	6	1.333		365	62.50		427.50	505
0160	25" height x 50" base	"	6	1.333		755	62.50		817.50	935
0170	30" height x 60" base	2 Carp	6	2.667		660	125		785	930

08 54 Composite Windows

08 54 13 – Fiberglass Windows

08 54 13.10 Fiberglass Single Hung Windows

			Crew	Daily Output	Labor-Hours	Unit	Material	Labor	Equipment	Total	Total Incl O&P
0010	**FIBERGLASS SINGLE HUNG WINDOWS**										
0100	Grids, low E, 18" x 24"	G	2 Carp	18	.889	Ea.	335	41.50		376.50	440
0110	18" x 40"	G		17	.941		340	44		384	450
0130	24" x 40"	G		20	.800		360	37.50		397.50	455
0230	36" x 36"	G		17	.941		370	44		414	480
0250	36" x 48"	G		20	.800		405	37.50		442.50	505
0260	36" x 60"	G		18	.889		445	41.50		486.50	560
0280	36" x 72"	G		16	1		470	47		517	590
0290	48" x 40"	G		16	1		470	47		517	590

08 54 13.30 Fiberglass Slider Windows

			Crew	Daily Output	Labor-Hours	Unit	Material	Labor	Equipment	Total	Total Incl O&P
0010	**FIBERGLASS SLIDER WINDOWS**										
0100	Grids, low E, 36" x 24"	G	2 Carp	20	.800	Ea.	340	37.50		377.50	435
0110	36" x 36"	G	"	20	.800	"	390	37.50		427.50	490

08 54 Composite Windows

08 54 13 – Fiberglass Windows

08 54 13.50 Fiberglass Bay Windows

		Crew	Daily Output	Labor-Hours	Unit	Material	2015 Bare Costs Labor	2015 Bare Costs Equipment	Total	Total Incl O&P
0010	**FIBERGLASS BAY WINDOWS**									
0150	48" x 36"	[G] 2 Carp	11	1.455	Ea.	1,100	68.50		1,168.50	1,300

08 56 Special Function Windows

08 56 46 – Radio-Frequency-Interference Shielding Windows

08 56 46.10 Radio-Frequency-interference Mesh

		Crew	Daily Output	Labor-Hours	Unit	Material	Labor	Equipment	Total	Total Incl O&P
0010	**RADIO-FREQUENCY-INTERFERENCE MESH**									
0100	16 mesh copper 0.011" wire				S.F.	4.79			4.79	5.25
0150	22 mesh copper 0.015" wire					6.10			6.10	6.70
0200	100 mesh copper 0.022" wire					8.65			8.65	9.50
0250	100 mesh stainless steel 0.0012" wire					14			14	15.40

08 56 63 – Detention Windows

08 56 63.13 Visitor Cubicle Windows

		Crew	Daily Output	Labor-Hours	Unit	Material	Labor	Equipment	Total	Total Incl O&P
0010	**VISITOR CUBICLE WINDOWS**									
4000	Visitor cubicle, vision panel, no intercom	E-4	2	16	Ea.	3,200	850	73	4,123	5,125

08 61 Roof Windows

08 61 13 – Metal Roof Windows

08 61 13.10 Roof Windows

		Crew	Daily Output	Labor-Hours	Unit	Material	Labor	Equipment	Total	Total Incl O&P
0010	**ROOF WINDOWS**, fixed high perf tmpd glazing, metallic framed									
0020	46" x 21-1/2", Flashed for shingled roof	1 Carp	8	1	Ea.	270	47		317	375
0100	46" x 28"		8	1		300	47		347	405
0125	57" x 44"		6	1.333		370	62.50		432.50	510
0130	72" x 28"		7	1.143		370	53.50		423.50	495
0150	Fixed, laminated tempered glazing, 46" x 21-1/2"		8	1		455	47		502	575
0175	46" x 28"		8	1		505	47		552	630
0200	57" x 44"		6	1.333		475	62.50		537.50	625
0500	Vented flashing set for shingled roof, 46" x 21-1/2"		7	1.143		455	53.50		508.50	590
0525	46" x 28"		6	1.333		505	62.50		567.50	660
0550	57" x 44"		5	1.600		635	75		710	820
0560	72" x 28"		5	1.600		635	75		710	820
0575	Flashing set for low pitched roof, 46" x 21-1/2"		7	1.143		525	53.50		578.50	670
0600	46" x 28"		7	1.143		575	53.50		628.50	725
0625	57" x 44"		5	1.600		715	75		790	910
0650	Flashing set for curb 46" x 21-1/2"		7	1.143		605	53.50		658.50	755
0675	46" x 28"		7	1.143		660	53.50		713.50	815
0700	57" x 44"		5	1.600		810	75		885	1,025

08 62 Unit Skylights

08 62 13 – Domed Unit Skylights

08 62 13.10 Domed Skylights

			Crew	Daily Output	Labor-Hours	Unit	Material	2015 Bare Costs Labor	2015 Bare Costs Equipment	Total	Total Incl O&P
0010	**DOMED SKYLIGHTS**										
0020	Skylight, fixed dome type, 22" x 22"	G	G-3	12	2.667	Ea.	216	125		341	440
0030	22" x 46"	G		10	3.200		269	150		419	535
0040	30" x 30"	G		12	2.667		286	125		411	515
0050	30" x 46"	G		10	3.200		380	150		530	660
0110	Fixed, double glazed, 22" x 27"	G		12	2.667		275	125		400	505
0120	22" x 46"	G		10	3.200		293	150		443	560
0130	44" x 46"	G		10	3.200		430	150		580	715
0210	Operable, double glazed, 22" x 27"	G		12	2.667		405	125		530	645
0220	22" x 46"	G		10	3.200		465	150		615	755
0230	44" x 46"	G		10	3.200		875	150		1,025	1,200
9000	Minimum labor/equipment charge			2	16	Job		750		750	1,200

08 62 13.20 Skylights

			Crew	Daily Output	Labor-Hours	Unit	Material	Labor	Equipment	Total	Total Incl O&P
0010	**SKYLIGHTS**, flush or curb mounted										
2120	Ventilating insulated plexiglass dome with										
2130	curb mounting, 36" x 36"	G	G-3	12	2.667	Ea.	480	125		605	730
2150	52" x 52"	G		12	2.667		670	125		795	935
2160	28" x 52"	G		10	3.200		490	150		640	780
2170	36" x 52"	G		10	3.200		545	150		695	835
2180	For electric opening system, add	G					315			315	345
2210	Operating skylight, with thermopane glass, 24" x 48"	G	G-3	10	3.200		585	150		735	885
2220	32" x 48"	G		9	3.556		615	166		781	940
2300	Insulated safety glass with aluminum frame	G		160	.200	S.F.	94	9.35		103.35	118
4000	Skylight, solar tube kit, incl dome, flashing, diffuser, 1 pipe, 10" dia.	G	1 Carp	2	4	Ea.	235	188		423	570
4010	14" diam.	G		2	4		325	188		513	670
4020	21" diam.	G		2	4		410	188		598	760
4030	Accessories for, 1' long x 9" dia pipe	G		24	.333		47	15.65		62.65	77
4040	2' long x 9" dia pipe	G		24	.333		37	15.65		52.65	66
4050	4' long x 9" dia pipe	G		20	.400		70	18.80		88.80	108
4060	1' long x 13" dia pipe	G		24	.333		65	15.65		80.65	97
4070	2' long x 13" dia pipe	G		24	.333		51	15.65		66.65	81.50
4080	4' long x 13" dia pipe	G		20	.400		97	18.80		115.80	138
4090	6.5" turret ext for 21" dia pipe	G		16	.500		101	23.50		124.50	150
4100	12' long x 21" dia flexible pipe	G		12	.667		82	31.50		113.50	142
4110	45 degree elbow, 10"	G		16	.500		144	23.50		167.50	197
4120	14"	G		16	.500		76	23.50		99.50	122
4130	Interior decorative ring, 9"	G		20	.400		35	18.80		53.80	69.50
4140	13"	G		20	.400		55	18.80		73.80	91.50

08 63 Metal-Framed Skylights

08 63 13 – Domed Metal-Framed Skylights

08 63 13.20 Skylight Rigid Metal-Framed

		Crew	Daily Output	Labor-Hours	Unit	Material	Labor	Equipment	Total	Total Incl O&P
0010	**SKYLIGHT RIGID METAL-FRAMED** Skylght framing is aluminum									
0050	Fixed acrylic double domes, curb mount, 25-1/2" x 25-1/2"	G-3	10	3.200	Ea.	200	150		350	460
0060	25-1/2" x 33-1/2"		10	3.200		230	150		380	495
0070	25-1/2" x 49-1/2"		6	5.333		265	249		514	690
0080	33-1/2" x 33-1/2"		8	4		290	187		477	620
0090	37-1/2" x 25-1/2"		8	4		240	187		427	565
0100	37-1/2" x 37-1/2"		8	4		235	187		422	560
0110	37-1/2" x 49-1/2"		6	5.333		400	249		649	840
0120	49-1/2" x 33-1/2"		6	5.333		355	249		604	790

08 63 Metal-Framed Skylights

08 63 13 – Domed Metal-Framed Skylights

08 63 13.20 Skylight Rigid Metal-Framed

		Crew	Daily Output	Labor-Hours	Unit	Material	2015 Bare Costs Labor	Equipment	Total	Total Incl O&P
0130	49-1/2" x 49-1/2"	G-3	6	5.333	Ea.	445	249		694	890
1000	Fixed tempered glass, curb mount, 17-1/2" x 33-1/2"		10	3.200		170	150		320	425
1020	17-1/2" x 49-1/2"		6	5.333		190	249		439	610
1030	25-1/2" x 25-1/2"		10	3.200		170	150		320	425
1040	25-1/2" x 33-1/2"		10	3.200		200	150		350	460
1050	25-1/2" x 37-1/2"		8	4		210	187		397	530
1060	25-1/2" x 49-1/2"		6	5.333		222	249		471	645
1070	25-1/2" x 73-1/2"		6	5.333		375	249		624	810
1080	33-1/2" x 33-1/2"		8	4		235	187		422	560
2000	Manual vent tempered glass & screen, curb, 25-1/2" x 25-1/2"		10	3.200		410	150		560	690
2020	25-1/2" x 37-1/2"		10	3.200		465	150		615	750
2030	25-1/2" x 49-1/2"		8	4		505	187		692	855
2040	33-1/2" x 33-1/2"		8	4		535	187		722	890
2050	33-1/2" x 49-1/2"		6	5.333		675	249		924	1,150
2060	37-1/2" x 37-1/2"		6	5.333		615	249		864	1,075
2070	49-1/2" x 49-1/2"		6	5.333		790	249		1,039	1,275
3000	Electric vent tempered glass, curb mount, 25-1/2" x 25-1/2"		10	3.200		960	150		1,110	1,300
3020	25-1/2" x 37-1/2"		10	3.200		1,050	150		1,200	1,400
3030	25-1/2" x 49-1/2"		8	4		1,125	187		1,312	1,525
3040	33-1/2" x 33-1/2"		8	4		1,125	187		1,312	1,550
3050	33-1/2" x 49-1/2"		6	5.333		1,225	249		1,474	1,725
3060	37-1/2" x 37-1/2"		6	5.333		1,200	249		1,449	1,700
3070	49-1/2" x 49-1/2"		6	5.333		1,325	249		1,574	1,850

08 71 Door Hardware

08 71 13 – Automatic Door Operators

08 71 13.10 Automatic Openers Commercial

		Crew	Daily Output	Labor-Hours	Unit	Material	2015 Bare Costs Labor	Equipment	Total	Total Incl O&P
0010	**AUTOMATIC OPENERS COMMERCIAL**									
0020	Pneumatic, incl opener, motion sens, control box, tubing, compressor									
0050	For single swing door, per opening	2 Skwk	.80	20	Ea.	4,550	975		5,525	6,575
0100	Pair, per opening		.50	32	Opng.	7,375	1,550		8,925	10,700
1000	For single sliding door, per opening		.60	26.667		4,975	1,300		6,275	7,575
1300	Bi-parting pair		.50	32		7,475	1,550		9,025	10,800
1420	Electronic door opener incl motion sens, 12 V control box, motor									
1450	For single swing door, per opening	2 Skwk	.80	20	Opng.	3,600	975		4,575	5,550
1500	Pair, per opening		.50	32		6,650	1,550		8,200	9,850
1600	For single sliding door, per opening		.60	26.667		4,400	1,300		5,700	6,950
1700	Bi-parting pair		.50	32		5,350	1,550		6,900	8,425
1750	Handicap actuator buttons, 2, including 12 V DC wiring, add	1 Carp	1.50	5.333	Pr.	470	250		720	925
2000	Electric Panic Button for door	2 Skwk	.50	32	Opng.	213	1,550		1,763	2,750

08 71 20 – Hardware

08 71 20.15 Hardware

		Crew	Daily Output	Labor-Hours	Unit	Material	2015 Bare Costs Labor	Equipment	Total	Total Incl O&P
0010	**HARDWARE**									
0020	Average percentage for hardware, total job cost									
0500	Total hardware for building, average distribution				Job	85%	15%			
1000	Door hardware, apartment, interior	1 Carp	4	2	Door	455	94		549	655
1300	Average, door hardware, motel/hotel interior, with access card		4	2		580	94		674	795
1500	Hospital bedroom, average quality		4	2		640	94		734	860
2000	High quality		3	2.667		740	125		865	1,025
2250	School, single exterior, incl. lever, incl. panic device		3	2.667		1,300	125		1,425	1,650

08 71 Door Hardware

08 71 20 – Hardware

08 71 20.15 Hardware		Crew	Daily Output	Labor-Hours	Unit	Material	2015 Bare Costs Labor	Equipment	Total	Total Incl O&P
2500	Single interior, regular use, lever included	1 Carp	3	2.667	Door	635	125		760	905
2600	Average, door hdwe.set, school, classroom, ANSI F88, incl. lever		3	2.667		905	125		1,030	1,200
2850	Stairway, single interior		3	2.667		590	125		715	850
3100	Double exterior, with panic device		2	4	Pr.	2,425	188		2,613	2,975
6020	Add for fire alarm door holder, electro-magnetic	1 Elec	4	2	Ea.	103	109		212	282

08 71 20.30 Door Closers

		Crew	Daily Output	Labor-Hours	Unit	Material	Labor	Equipment	Total	Incl O&P
0010	**DOOR CLOSERS** Adjustable backcheck, multiple mounting									
0015	and rack and pinion									
0020	Standard Regular Arm	1 Carp	6	1.333	Ea.	192	62.50		254.50	315
0040	Hold open arm		6	1.333		200	62.50		262.50	325
0100	Fusible link		6.50	1.231		165	58		223	277
1500	Concealed closers, normal use, head, pivot hung, interior		5.50	1.455		345	68.50		413.50	490
1510	Exterior		5.50	1.455		400	68.50		468.50	550
1520	Overhead concealed, all sizes, regular arm		5.50	1.455		200	68.50		268.50	330
1525	Concealed arm		5	1.600		315	75		390	470
1530	Concealed in door, all sizes, regular arm		5.50	1.455		325	68.50		393.50	470
1535	Concealed arm		5	1.600		255	75		330	405
1560	Floor concealed, all sizes, single acting		2.20	3.636		500	171		671	830
1565	Double acting		2.20	3.636		475	171		646	805
1570	Interior, floor, offset pivot, single acting		3.50	2.286		830	107		937	1,100
1590	Exterior		3.50	2.286		905	107		1,012	1,175
1610	Hold open arm		6	1.333		420	62.50		482.50	565
1620	Double acting, standard arm		6	1.333		760	62.50		822.50	940
1630	Hold open arm		6	1.333		765	62.50		827.50	945
1640	Floor, center hung, single acting, bottom arm		6	1.333		410	62.50		472.50	555
1650	Double acting		6	1.333		465	62.50		527.50	615
1660	Offset hung, single acting, bottom arm		6	1.333		550	62.50		612.50	710
2000	Backcheck and adjustable power, hinge face mount									
5000	For cast aluminum cylinder, deduct				Ea.	35			35	38.50
5040	For delayed action, add					46			46	50.50
5080	For fusible link arm, add					35			35	38.50
5120	For shock absorbing arm, add					50			50	55
5160	For spring power adjustment, add					40			40	44
6000	Closer-holder, hinge face mount, all sizes, exposed arm	1 Carp	6.50	1.231		205	58		263	320
6500	Electro magnetic closer/holder									
6510	Single point, no detector	1 Carp	4	2	Ea.	505	94		599	715
6515	Including detector		4	2		665	94		759	890
6520	Multi-point, no detector		4	2		885	94		979	1,125
6524	Including detector		4	2		1,325	94		1,419	1,600
6550	Electric automatic operators									
6555	Operator	1 Carp	4	2	Ea.	1,975	94		2,069	2,325
6570	Wall plate actuator		4	2		217	94		311	395
7000	Electronic closer-holder, hinge facemount, concealed arm		5	1.600		420	75		495	585
7400	With built-in detector		5	1.600		605	75		680	790
9000	Minimum labor/equipment charge		4	2	Job		94		94	154

08 71 20.35 Panic Devices

		Crew	Daily Output	Labor-Hours	Unit	Material	Labor	Equipment	Total	Incl O&P
0010	**PANIC DEVICES**									
0015	For rim locks, single door exit only	1 Carp	6	1.333	Ea.	460	62.50		522.50	610
0020	Outside key and pull		5	1.600		570	75		645	755
0200	Bar and vertical rod, exit only		5	1.600		865	75		940	1,075
0210	Outside key and pull		4	2		975	94		1,069	1,225
0400	Bar and concealed rod		4	2		745	94		839	975

08 71 Door Hardware

08 71 20 – Hardware

08 71 20.35 Panic Devices

		Crew	Daily Output	Labor-Hours	Unit	Material	2015 Bare Costs Labor	Equipment	Total	Total Incl O&P
0600	Touch bar, exit only	1 Carp	6	1.333	Ea.	555	62.50		617.50	715
0610	Outside key and pull		5	1.600		685	75		760	875
0700	Touch bar and vertical rod, exit only		5	1.600		845	75		920	1,050
0710	Outside key and pull		4	2		970	94		1,064	1,225
1000	Mortise, bar, exit only		4	2		725	94		819	955
1600	Touch bar, exit only		4	2		760	94		854	990
2000	Narrow stile, rim mounted, bar, exit only		6	1.333		610	62.50		672.50	775
2010	Outside key and pull		5	1.600		890	75		965	1,100
2200	Bar and vertical rod, exit only		5	1.600		920	75		995	1,125
2210	Outside key and pull		4	2		920	94		1,014	1,150
2400	Bar and concealed rod, exit only		3	2.667		1,075	125		1,200	1,375
3000	Mortise, bar, exit only		4	2		600	94		694	815
3600	Touch bar, exit only		4	2		770	94		864	1,000
4000	Double doors, exit only		2	4	Pr.	1,425	188		1,613	1,850
4500	Exit & entrance		2	4	"	1,975	188		2,163	2,475
9000	Minimum labor/equipment charge		2.50	3.200	Job		150		150	246

08 71 20.40 Lockset

		Crew	Daily Output	Labor-Hours	Unit	Material	2015 Bare Costs Labor	Equipment	Total	Total Incl O&P
0010	**LOCKSET,** Standard duty									
0020	Non-keyed, passage, w/sect. trim	1 Carp	12	.667	Ea.	70	31.50		101.50	129
0100	Privacy		12	.667		75	31.50		106.50	134
0400	Keyed, single cylinder function		10	.800		108	37.50		145.50	180
0420	Hotel (see also Section 08 71 20.15)		8	1		200	47		247	297
0500	Lever handled, keyed, single cylinder function		10	.800		128	37.50		165.50	203
1000	Heavy duty with sectional trim, non-keyed, passages		12	.667		125	31.50		156.50	190
1100	Privacy		12	.667		155	31.50		186.50	223
1400	Keyed, single cylinder function		10	.800		185	37.50		222.50	266
1420	Hotel		8	1		500	47		547	625
1600	Communicating		10	.800		300	37.50		337.50	390
1690	For re-core cylinder, add					50			50	55
3800	Cipher lockset w/key pad (security item)	1 Carp	13	.615		920	29		949	1,075
3900	Cipher lockset with dial for swinging doors (security item)		13	.615		1,800	29		1,829	2,050
3920	with dial for swinging doors & drill resistant plate (security item)		12	.667		2,200	31.50		2,231.50	2,475
3950	Cipher lockset with dial for safe/vault door (security item)		12	.667		1,500	31.50		1,531.50	1,700
3980	Keyless, pushbutton type									
4000	Residential/light commercial, deadbolt, standard	1 Carp	9	.889	Ea.	140	41.50		181.50	223
4010	Heavy duty		9	.889		230	41.50		271.50	320
4020	Industrial, heavy duty, with deadbolt		9	.889		380	41.50		421.50	485
4030	Key override		9	.889		390	41.50		431.50	500
4040	Lever activated handle		9	.889		415	41.50		456.50	525
4050	Key override		9	.889		440	41.50		481.50	555
4060	Double sided pushbutton type		8	1		755	47		802	905
4070	Key override		8	1		790	47		837	945
9000	Minimum labor/equipment charge		6	1.333	Job		62.50		62.50	103

08 71 20.41 Dead Locks

		Crew	Daily Output	Labor-Hours	Unit	Material	2015 Bare Costs Labor	Equipment	Total	Total Incl O&P
0010	**DEAD LOCKS**									
0011	Mortise heavy duty outside key (security item)	1 Carp	9	.889	Ea.	175	41.50		216.50	262
0020	Double cylinder		9	.889		175	41.50		216.50	262
0100	Medium duty, outside key		10	.800		110	37.50		147.50	183
0110	Double cylinder		10	.800		120	37.50		157.50	194
1000	Tubular, standard duty, outside key		10	.800		45	37.50		82.50	111
1010	Double cylinder		10	.800		60	37.50		97.50	128
1200	Night latch, outside key		10	.800		45	37.50		82.50	111

08 71 Door Hardware

08 71 20 – Hardware

08 71 20.41 Dead Locks		Crew	Daily Output	Labor-Hours	Unit	Material	2015 Bare Costs Labor	Equipment	Total	Total Incl O&P
1203	Deadlock night latch	1 Carp	7.70	1.039	Ea.	45	49		94	130
1420	Deadbolt lock, single cylinder		10	.800		48.50	37.50		86	115
1440	Double cylinder		10	.800		63	37.50		100.50	131
08 71 20.42 Mortise Locksets										
0010	**MORTISE LOCKSETS**, Comm., wrought knobs & full escutcheon trim									
0015	Assumes mortise is cut									
0020	Non-keyed, passage, grade 3	1 Carp	9	.889	Ea.	149	41.50		190.50	233
0030	Grade 1		8	1		405	47		452	520
0040	Privacy set Grade 3		9	.889		169	41.50		210.50	255
0050	Grade 1		8	1		455	47		502	575
0100	Keyed, office/entrance/apartment, Grade 2		8	1		197	47		244	294
0110	Grade 1		7	1.143		515	53.50		568.50	660
0120	Single cylinder, typical, Grade 3		8	1		189	47		236	285
0130	Grade 1		7	1.143		500	53.50		553.50	640
0200	Hotel, room, Grade 3		7	1.143		190	53.50		243.50	297
0210	Grade 1 (see also Section 08 71 20.15)		6	1.333		510	62.50		572.50	665
0300	Double cylinder, Grade 3		8	1		225	47		272	325
0310	Grade 1		7	1.143		515	53.50		568.50	660
1000	Wrought knobs and sectional trim, non-keyed, passage, Grade 3		10	.800		130	37.50		167.50	205
1010	Grade 1		9	.889		405	41.50		446.50	515
1040	Privacy, Grade 3		10	.800		145	37.50		182.50	222
1050	Grade 1		9	.889		455	41.50		496.50	570
1100	Keyed, entrance, office/apartment, Grade 3		9	.889		220	41.50		261.50	310
1103	Install lockset		6.92	1.156		220	54.50		274.50	330
1110	Grade 1		8	1		520	47		567	645
1120	Single cylinder, Grade 3		9	.889		225	41.50		266.50	315
1130	Grade 1		8	1		500	47		547	625
2000	Cast knobs and full escutcheon trim									
2010	Non-keyed, passage, Grade 3	1 Carp	9	.889	Ea.	275	41.50		316.50	375
2020	Grade 1		8	1		380	47		427	495
2040	Privacy, Grade 3		9	.889		320	41.50		361.50	420
2050	Grade 1		8	1		440	47		487	560
2120	Keyed, single cylinder, Grade 3		8	1		330	47		377	440
2123	Mortise lock		6.15	1.301		330	61		391	465
2130	Grade 1		7	1.143		525	53.50		578.50	670
3000	Cast knob and sectional trim, non-keyed, passage, Grade 3		10	.800		210	37.50		247.50	293
3010	Grade 1		10	.800		380	37.50		417.50	480
3040	Privacy, Grade 3		10	.800		225	37.50		262.50	310
3050	Grade 1		10	.800		440	37.50		477.50	545
3100	Keyed, office/entrance/apartment, Grade 3		9	.889		255	41.50		296.50	350
3110	Grade 1		9	.889		580	41.50		621.50	710
3120	Single cylinder, Grade 3		9	.889		260	41.50		301.50	355
3130	Grade 1		9	.889		525	41.50		566.50	650
3190	For re-core cylinder, add					75			75	82.50
08 71 20.45 Peepholes										
0010	**PEEPHOLES**									
2010	Peephole	1 Carp	32	.250	Ea.	15.80	11.75		27.55	36.50
2020	Peephole, wide view	"	32	.250	"	16.50	11.75		28.25	37.50
08 71 20.50 Door Stops										
0010	**DOOR STOPS**									
0020	Holder & bumper, floor or wall	1 Carp	32	.250	Ea.	34.50	11.75		46.25	57
1300	Wall bumper, 4" diameter, with rubber pad, aluminum		32	.250		11.70	11.75		23.45	32

08 71 Door Hardware

08 71 20 – Hardware

08 71 20.50 Door Stops

		Crew	Daily Output	Labor-Hours	Unit	Material	2015 Bare Costs Labor	2015 Bare Costs Equipment	Total	Total Incl O&P
1600	Door bumper, floor type, aluminum	1 Carp	32	.250	Ea.	8.10	11.75		19.85	28
1620	Brass		32	.250		10	11.75		21.75	30.50
1630	Bronze		32	.250		18.20	11.75		29.95	39.50
1900	Plunger type, door mounted		32	.250		28	11.75		39.75	50
2500	Holder, floor type, aluminum		32	.250		33.50	11.75		45.25	56.50
2520	Wall type, aluminum		32	.250		35	11.75		46.75	58
2530	Overhead type, bronze		32	.250		102	11.75		113.75	132
2540	Plunger type, aluminum		32	.250		28	11.75		39.75	50
2560	Brass		32	.250		43	11.75		54.75	66.50
3000	Electro-magnetic, wall mounted, US3		3	2.667		258	125		383	490
3020	Floor mounted, US3		3	2.667		315	125		440	555
4000	Doorstop, ceiling mounted		3	2.667		20	125		145	227
4030	Doorstop, header		3	2.667		45	125		170	255
9000	Minimum labor/equipment charge		6	1.333	Job		62.50		62.50	103

08 71 20.55 Push-Pull Plates

		Crew	Daily Output	Labor-Hours	Unit	Material	Labor	Equipment	Total	Total Incl O&P
0010	**PUSH-PULL PLATES**									
0090	Push plate, 0.050 thick, 3" x 12", aluminum	1 Carp	12	.667	Ea.	6.25	31.50		37.75	58.50
0100	4" x 16"		12	.667		12.80	31.50		44.30	65.50
0110	6" x 16"		12	.667		8.50	31.50		40	61
0120	8" x 16"		12	.667		9.90	31.50		41.40	62.50
0200	Push plate, 0.050 thick, 3" x 12", brass		12	.667		14	31.50		45.50	67
0210	4" x 16"		12	.667		17.50	31.50		49	71
0220	6" x 16"		12	.667		27	31.50		58.50	81
0230	8" x 16"		12	.667		35	31.50		66.50	90
0250	Push plate, 0.050 thick, 3" x 12", satin brass		12	.667		14.25	31.50		45.75	67
0260	4" x 16"		12	.667		17.70	31.50		49.20	71
0270	6" x 16"		12	.667		27	31.50		58.50	81
0280	8" x 16"		12	.667		35	31.50		66.50	90
0490	Push plate, 0.050 thick, 3" x 12", bronze		12	.667		17	31.50		48.50	70
0500	4" x 16"		12	.667		25	31.50		56.50	79
0510	6" x 16"		12	.667		32	31.50		63.50	86.50
0520	8" x 16"		12	.667		38.50	31.50		70	94
0600	Push plate, antimicrobial copper alloy finish, 3.5" x 15"		13	.615		19.80	29		48.80	69.50
0610	4" x 16"		13	.615		23.50	29		52.50	73
0620	6" x 16"		13	.615		26	29		55	76
0630	6" x 20"		13	.615		26	29		55	76
0740	Push plate, 0.050 thick, 3" x 12", stainless steel		12	.667		12	31.50		43.50	64.50
0750	4" x 16"		12	.667		27.50	31.50		59	81.50
0760	6" x 16"		12	.667		18.65	31.50		50.15	72
0780	8" x 16"		12	.667		23.50	31.50		55	77.50
0790	Push plate, 0.050 thick, 3" x 12", satin stainless steel		12	.667		7	31.50		38.50	59
0810	4" x 16"		12	.667		8.50	31.50		40	61
0820	6" x 16"		12	.667		11.70	31.50		43.20	64.50
0830	8" x 16"		12	.667		16	31.50		47.50	69
0980	Pull plate, 0.050 thick, 3" x 12", aluminum		12	.667		25.50	31.50		57	79.50
1000	4" x 16"		12	.667		25.50	31.50		57	79.50
1050	Pull plate, 0.050 thick, 3" x 12", brass		12	.667		39.50	31.50		71	95
1060	4" x 16"		12	.667		34.50	31.50		66	89.50
1080	Pull plate, 0.050 thick, 3" x 12", bronze		12	.667		50	31.50		81.50	107
1100	4" x 16"		12	.667		56	31.50		87.50	113
1180	Pull plate, 0.050 thick, 3" x 12", stainless steel		12	.667		45.50	31.50		77	102
1200	4" x 16"		12	.667		58.50	31.50		90	116

08 71 Door Hardware

08 71 20 – Hardware

08 71 20.55 Push-Pull Plates

		Crew	Daily Output	Labor-Hours	Unit	Material	2015 Bare Costs Labor	Equipment	Total	Total Incl O&P
1250	Pull plate, 0.050 thick, 3" x 12", chrome	1 Carp	12	.667	Ea.	44.50	31.50		76	101
1270	4" x 16"		12	.667		46.50	31.50		78	103
1500	Pull handle and push bar, aluminum		11	.727		118	34		152	186
2000	Bronze		10	.800		159	37.50		196.50	237
9800	Minimum labor/equipment charge		5	1.600	Job		75		75	123

08 71 20.60 Entrance Locks

0010	**ENTRANCE LOCKS**									
0015	Cylinder, grip handle deadlocking latch	1 Carp	9	.889	Ea.	175	41.50		216.50	262
0020	Deadbolt		8	1		175	47		222	270
0100	Push and pull plate, dead bolt		8	1		225	47		272	325
0900	For handicapped lever, add					150			150	165

08 71 20.65 Thresholds

0010	**THRESHOLDS**									
0011	Threshold 3' long saddles aluminum	1 Carp	48	.167	L.F.	9.40	7.85		17.25	23
0100	Aluminum, 8" wide, 1/2" thick		12	.667	Ea.	49	31.50		80.50	105
0500	Bronze		60	.133	L.F.	42	6.25		48.25	57
0600	Bronze, panic threshold, 5" wide, 1/2" thick		12	.667	Ea.	158	31.50		189.50	225
0700	Rubber, 1/2" thick, 5-1/2" wide		20	.400		40	18.80		58.80	75
0800	2-3/4" wide		20	.400		45	18.80		63.80	80.50
1950	ADA Compliant Thresholds									
2000	Threshold, wood oak 3-1/2" wide x 24" long	1 Carp	12	.667	Ea.	9	31.50		40.50	61.50
2010	3-1/2" wide x 36" long		12	.667		13.50	31.50		45	66.50
2020	3-1/2" wide x 48" long		12	.667		18	31.50		49.50	71.50
2030	4-1/2" wide x 24" long		12	.667		12	31.50		43.50	64.50
2040	4-1/2" wide x 36" long		12	.667		17	31.50		48.50	70
2050	4-1/2" wide x 48" long		12	.667		23	31.50		54.50	77
2060	6-1/2" wide x 24" long		12	.667		16.50	31.50		48	69.50
2070	6-1/2" wide x 36" long		12	.667		25	31.50		56.50	79
2080	6-1/2" wide x 48" long		12	.667		33	31.50		64.50	88
2090	Threshold, wood cherry 3-1/2" wide x 24" long		12	.667		13.50	31.50		45	66.50
2100	3-1/2" wide x 36" long		12	.667		21	31.50		52.50	74.50
2110	3-1/2" wide x 48" long		12	.667		27	31.50		58.50	81
2120	4-1/2" wide x 24" long		12	.667		17.50	31.50		49	71
2130	4-1/2" wide x 36" long		12	.667		25.50	31.50		57	79.50
2140	4-1/2" wide x 48" long		12	.667		34	31.50		65.50	89
2150	6-1/2" wide x 24" long		12	.667		25.50	31.50		57	79.50
2160	6-1/2" wide x 36" long		12	.667		36.50	31.50		68	91.50
2170	6-1/2" wide x 48" long		12	.667		49	31.50		80.50	106
2180	Threshold, wood walnut 3-1/2" wide x 24" long		12	.667		15.50	31.50		47	68.50
2190	3-1/2" wide x 36" long		12	.667		23	31.50		54.50	77
2200	3-1/2" wide x 48" long		12	.667		30	31.50		61.50	84.50
2210	4-1/2" wide x 24" long		12	.667		19.50	31.50		51	73
2220	4-1/2" wide x 36" long		12	.667		28.50	31.50		60	83
2230	4-1/2" wide x 48" long		12	.667		38	31.50		69.50	93.50
2240	6-1/2" wide x 24" long		12	.667		28	31.50		59.50	82.50
2250	6-1/2" wide x 36" long		12	.667		41.50	31.50		73	97
2260	6-1/2" wide x 48" long		12	.667		55	31.50		86.50	112
2300	Threshold, aluminum 4" wide x 36" long		12	.667		33	31.50		64.50	88
2310	4" wide x 48" long		12	.667		41	31.50		72.50	96.50
2320	4" wide x 72" long		12	.667		66	31.50		97.50	124
2330	5" wide x 36" long		12	.667		45	31.50		76.50	101
2340	5" wide x 48" long		12	.667		56	31.50		87.50	113

08 71 Door Hardware

08 71 20 – Hardware

08 71 20.65 Thresholds

		Crew	Daily Output	Labor-Hours	Unit	Material	2015 Bare Costs Labor	2015 Bare Costs Equipment	Total	Total Incl O&P
2350	5" wide x 72" long	1 Carp	12	.667	Ea.	89	31.50		120.50	150
2360	6" wide x 36" long		12	.667		53	31.50		84.50	110
2370	6" wide x 48" long		12	.667		68	31.50		99.50	127
2380	6" wide x 72" long		12	.667		106	31.50		137.50	169
2390	7" wide x 36" long		12	.667		69	31.50		100.50	128
2400	7" wide x 48" long		12	.667		90	31.50		121.50	151
2410	7" wide x 72" long		12	.667		137	31.50		168.50	203
2500	Threshold, ramp, aluminum or rubber 24" x 24"		12	.667		190	31.50		221.50	261
9000	Minimum labor/equipment charge		4	2	Job		94		94	154

08 71 20.75 Door Hardware Accessories

		Crew	Daily Output	Labor-Hours	Unit	Material	Labor	Equipment	Total	Total Incl O&P
0010	**DOOR HARDWARE ACCESSORIES**									
0050	Door closing coordinator, 36" (for paired openings up to 56")	1 Carp	8	1	Ea.	98	47		145	185
0060	48" (for paired openings up to 84")		8	1		105	47		152	193
0070	56" (for paired openings up to 96")		8	1		116	47		163	204
1000	Knockers, brass, standard		16	.500		46.50	23.50		70	89.50
1100	Deluxe		10	.800		121	37.50		158.50	195
4500	Rubber door silencers		540	.015		.44	.70		1.14	1.62
9000	Minimum labor/equipment charge		6	1.333	Job		62.50		62.50	103

08 71 20.80 Hasps

		Crew	Daily Output	Labor-Hours	Unit	Material	Labor	Equipment	Total	Total Incl O&P
0010	**HASPS**, steel assembly									
0015	3"	1 Carp	26	.308	Ea.	4.90	14.45		19.35	29
0020	4-1/2"		13	.615		6.60	29		35.60	55
0040	6"		12.50	.640		9.35	30		39.35	60

08 71 20.90 Hinges

		Crew	Daily Output	Labor-Hours	Unit	Material	Labor	Equipment	Total	Total Incl O&P
0010	**HINGES**									
0012	Full mortise, avg. freq., steel base, USP, 4-1/2" x 4-1/2"				Pr.	36			36	40
0100	5" x 5", USP					57.50			57.50	63
0200	6" x 6", USP					119			119	131
0400	Brass base, 4-1/2" x 4-1/2", US10					58.50			58.50	64.50
0500	5" x 5", US10					86			86	94.50
0600	6" x 6", US10					162			162	178
0800	Stainless steel base, 4-1/2" x 4-1/2", US32					74			74	81.50
0900	For non removable pin, add (security item)				Ea.	4.88			4.88	5.35
0910	For floating pin, driven tips, add					3.30			3.30	3.63
0930	For hospital type tip on pin, add					13.95			13.95	15.35
0940	For steeple type tip on pin, add					18.95			18.95	21
0950	Full mortise, high frequency, steel base, 3-1/2" x 3-1/2", US26D				Pr.	30			30	33
1000	4-1/2" x 4-1/2", USP					63			63	69
1100	5" x 5", USP					50			50	55
1200	6" x 6", USP					134			134	148
1400	Brass base, 3-1/2" x 3-1/2", US4					51			51	56
1430	4-1/2" x 4-1/2", US10					73.50			73.50	81
1500	5" x 5", US10					122			122	134
1600	6" x 6", US10					165			165	181
1800	Stainless steel base, 4-1/2" x 4-1/2", US32					103			103	114
1810	5" x 4-1/2", US32					137			137	151
1930	For hospital type tip on pin, add				Ea.	13.15			13.15	14.45
1950	Full mortise, low frequency, steel base, 3-1/2" x 3-1/2", US26D				Pr.	23.50			23.50	26
2000	4-1/2" x 4-1/2", USP					22			22	24.50
2100	5" x 5", USP					47.50			47.50	52
2200	6" x 6", USP					89			89	98
2300	4-1/2" x 4-1/2", US3					16.85			16.85	18.50

08 71 Door Hardware

08 71 20 – Hardware

08 71 20.90 Hinges

		Crew	Daily Output	Labor-Hours	Unit	Material	2015 Bare Costs Labor	2015 Bare Costs Equipment	Total	Total Incl O&P
2310	5" x 5", US3				Pr.	41			41	45
2400	Brass base, 4-1/2" x 4-1/2", US10					53			53	58.50
2500	5" x 5", US10					76.50			76.50	84
2800	Stainless steel base, 4-1/2" x 4-1/2", US32					74.50			74.50	82
8000	Install hinge	1 Carp	34	.235			11.05		11.05	18.10

08 71 20.91 Special Hinges

		Crew	Daily Output	Labor-Hours	Unit	Material	Labor	Equipment	Total	Total Incl O&P
0010	**SPECIAL HINGES**									
0015	Paumelle, high frequency									
0020	Steel base, 6" x 4-1/2", US10				Pr.	145			145	160
0100	Brass base, 5" x 4-1/2", US10				Ea.	248			248	273
0200	Paumelle, average frequency, steel base, 4-1/2" x 3-1/2", US10				Pr.	98.50			98.50	108
0400	Olive knuckle, low frequency, brass base, 6" x 4-1/2", US10				Ea.	144			144	159
1000	Electric hinge with concealed conductor, average frequency									
1010	Steel base, 4-1/2" x 4-1/2", US26D				Pr.	310			310	340
1100	Bronze base, 4-1/2" x 4-1/2", US26D				"	320			320	350
1200	Electric hinge with concealed conductor, high frequency									
1210	Steel base, 4-1/2" x 4-1/2", US26D				Pr.	262			262	288
1600	Double weight, 800 lb., steel base, removable pin, 5" x 6", USP					410			410	450
1700	Steel base-welded pin, 5" x 6", USP					166			166	183
1800	Triple weight, 2000 lb., steel base, welded pin, 5" x 6", USP					535			535	590
2000	Pivot reinf., high frequency, steel base, 7-3/4" door plate, USP					149			149	164
2200	Bronze base, 7-3/4" door plate, US10					232			232	255
3000	Swing clear, full mortise, full or half surface, high frequency,									
3010	Steel base, 5" high, USP				Pr.	143			143	157
3200	Swing clear, full mortise, average frequency									
3210	Steel base, 4-1/2" high, USP				Pr.	128			128	141
4000	Wide throw, average frequency, steel base, 4-1/2" x 6", USP					94.50			94.50	104
4200	High frequency, steel base, 4-1/2" x 6", USP					112			112	124
4440	US26D					89.50			89.50	98
4600	Spring hinge, single acting, 6" flange, steel				Ea.	50			50	55
4700	Brass					94.50			94.50	104
4900	Double acting, 6" flange, steel					80			80	88
4950	Brass					133			133	146
5000	T-strap, galvanized, 4"				Pr.	17.25			17.25	19
5010	6"					29.50			29.50	32.50
5020	8"					52.50			52.50	58
8000	Continuous hinges									
8010	Steel, piano, 2" x 72"	1 Carp	20	.400	Ea.	22	18.80		40.80	55
8020	Brass, piano, 1-1/16" x 30"		30	.267		8	12.50		20.50	29.50
8030	Acrylic, piano, 1-3/4" x 12"		40	.200		15	9.40		24.40	32
8040	Aluminum, door, standard duty, 7'		3	2.667		135	125		260	355
8050	Heavy duty, 7'		3	2.667		142	125		267	360
8060	8'		3	2.667		160	125		285	380
8070	Steel, door, heavy duty, 7'		3	2.667		200	125		325	425
8080	8'		3	2.667		250	125		375	480
8090	Stainless steel, door, heavy duty, 7'		3	2.667		260	125		385	490
8100	8'		3	2.667		280	125		405	515
9000	Continuous hinge, steel, full mortise, heavy duty, 96 inch		2	4		460	188		648	815
9200	Continuous geared hinge, aluminum, full mortise, standard duty, 83 inch		3	2.667		120	125		245	335
9250	Continuous geared hinge, aluminum, full mortise, heavy duty, 83 inch		3	2.667		200	125		325	425

08 71 Door Hardware

08 71 20 – Hardware

08 71 20.95 Kick Plates

		Crew	Daily Output	Labor-Hours	Unit	Material	2015 Bare Costs Labor	2015 Bare Costs Equipment	Total	Total Incl O&P
0010	**KICK PLATES**									
0020	Stainless steel, .050, 16 ga., 8" x 28", US32	1 Carp	15	.533	Ea.	38	25		63	83
0030	8" x 30"		15	.533		41	25		66	86
0040	8" x 34"		15	.533		46	25		71	91.50
0050	10" x 28"		15	.533		76	25		101	125
0060	10" x 30"		15	.533		82	25		107	131
0070	10" x 34"		15	.533		92	25		117	142
0080	Mop/Kick, 4" x 28"		15	.533		34	25		59	78.50
0090	4" x 30"		15	.533		36	25		61	80.50
0100	4" x 34"		15	.533		41	25		66	86
0110	6" x 28"		15	.533		43	25		68	88.50
0120	6" x 30"		15	.533		47	25		72	92.50
0130	6" x 34"		15	.533		53	25		78	99.50
0500	Bronze, .050", 8" x 28"		15	.533		66.50	25		91.50	115
0510	8" x 30"		15	.533		65	25		90	113
0520	8" x 34"		15	.533		73	25		98	122
0530	10" x 28"		15	.533		75	25		100	124
0540	10" x 30"		15	.533		80	25		105	129
0550	10" x 34"		15	.533		91	25		116	141
0560	Mop/Kick, 4" x 28"		15	.533		33	25		58	77.50
0570	4" x 30"		15	.533		36	25		61	80.50
0580	4" x 34"		15	.533		37	25		62	81.50
0590	6" x 28"		15	.533		46	25		71	91.50
0600	6" x 30"		15	.533		52	25		77	98
0610	6" x 34"		15	.533		56	25		81	103
1000	Acrylic, .125", 8" x 26"		15	.533		28	25		53	72
1010	8" x 36"		15	.533		38	25		63	83
1020	8" x 42"		15	.533		45	25		70	90.50
1030	10" x 26"		15	.533		35	25		60	79.50
1040	10" x 36"		15	.533		48	25		73	94
1050	10" x 42"		15	.533		68.50	25		93.50	117
1060	Mop/Kick, 4" x 26"		15	.533		17	25		42	59.50
1070	4" x 36"		15	.533		24	25		49	67.50
1080	4" x 42"		15	.533		27	25		52	70.50
1090	6" x 26"		15	.533		23	25		48	66.50
1100	6" x 36"		15	.533		34	25		59	78.50
1110	6" x 42"		15	.533		39	25		64	84
1220	Brass, .050", 8" x 26"		15	.533		56	25		81	103
1230	8" x 36"		15	.533		75	25		100	124
1240	8" x 42"		15	.533		86	25		111	136
1250	10" x 26"		15	.533		71	25		96	119
1260	10" x 36"		15	.533		91	25		116	141
1270	10" x 42"		15	.533		105	25		130	157
1320	Mop/Kick, 4" x 26"		15	.533		28	25		53	72
1330	4" x 36"		15	.533		39	25		64	84
1340	4" x 42"		15	.533		44	25		69	89.50
1350	6" x 26"		15	.533		38	25		63	83
1360	6" x 36"		15	.533		48	25		73	94
1370	6" x 42"		15	.533		55	25		80	102
1800	Aluminum, .050", 8" x 26"		15	.533		35	25		60	79.50
1810	8" x 36"		15	.533		40	25		65	85
1820	8" x 42"		15	.533		47	25		72	92.50

08 71 Door Hardware

08 71 20 – Hardware

08 71 20.95 Kick Plates

		Crew	Daily Output	Labor-Hours	Unit	Material	2015 Bare Costs Labor	2015 Bare Costs Equipment	Total	Total Incl O&P
1830	10" x 26"	1 Carp	15	.533	Ea.	36	25		61	80.50
1840	10" x 36"		15	.533		50	25		75	96
1850	10" x 42"		15	.533		59	25		84	106
1860	Mop/Kick, 4" x 26"		15	.533		15	25		40	57.50
1870	4" x 36"		15	.533		20	25		45	63
1880	4" x 42"		15	.533		24	25		49	67.50
1890	6" x 26"		15	.533		22	25		47	65
1900	6" x 36"		15	.533		30	25		55	74
1910	6" x 42"		15	.533		35	25		60	79.50
9000	Minimum labor/equipment charge		6	1.333	Job		62.50		62.50	103

08 71 21 – Astragals

08 71 21.10 Exterior Mouldings, Astragals

		Crew	Daily Output	Labor-Hours	Unit	Material	Labor	Equipment	Total	Total Incl O&P
0010	**EXTERIOR MOULDINGS, ASTRAGALS**									
0400	One piece, overlapping cadmium plated steel, flat, 3/16" x 2"	1 Carp	90	.089	L.F.	4	4.17		8.17	11.25
0600	Prime coated steel, flat, 1/8" x 3"		90	.089		5.75	4.17		9.92	13.20
0800	Stainless steel, flat, 3/32" x 1-5/8"		90	.089		16	4.17		20.17	24.50
1000	Aluminum, flat, 1/8" x 2"		90	.089		4.10	4.17		8.27	11.35
1200	Nail on, "T" extrusion		120	.067		1.90	3.13		5.03	7.25
1300	Vinyl bulb insert		105	.076		2.50	3.58		6.08	8.60
1600	Screw on, "T" extrusion		90	.089		3.75	4.17		7.92	11
1700	Vinyl insert		75	.107		4.50	5		9.50	13.15
2000	"L" extrusion, neoprene bulbs		75	.107		4.10	5		9.10	12.70
2100	Neoprene sponge insert		75	.107		6.90	5		11.90	15.80
2200	Magnetic		75	.107		10.60	5		15.60	19.85
2400	Spring hinged security seal, with cam		75	.107		6.80	5		11.80	15.70
2600	Spring loaded locking bolt, vinyl insert		45	.178		9.20	8.35		17.55	24
2800	Neoprene sponge strip, "Z" shaped, aluminum		60	.133		8.30	6.25		14.55	19.40
2900	Solid neoprene strip, nail on aluminum strip		90	.089		4.05	4.17		8.22	11.30
3000	One piece stile protection									
3020	Neoprene fabric loop, nail on aluminum strips	1 Carp	60	.133	L.F.	1.10	6.25		7.35	11.45
3110	Flush mounted aluminum extrusion, 1/2" x 1-1/4"		60	.133		6.90	6.25		13.15	17.85
3140	3/4" x 1-3/8"		60	.133		4.10	6.25		10.35	14.75
3160	1-1/8" x 1-3/4"		60	.133		4.70	6.25		10.95	15.40
3300	Mortise, 9/16" x 3/4"		60	.133		4.10	6.25		10.35	14.75
3320	13/16" x 1-3/8"		60	.133		4.30	6.25		10.55	15
3600	Spring bronze strip, nail on type		105	.076		1.85	3.58		5.43	7.90
3620	Screw on, with retainer		75	.107		2.70	5		7.70	11.15
3800	Flexible stainless steel housing, pile insert, 1/2" door		105	.076		7.25	3.58		10.83	13.85
3820	3/4" door		105	.076		8.10	3.58		11.68	14.75
4000	Extruded aluminum retainer, flush mount, pile insert		105	.076		2.25	3.58		5.83	8.35
4080	Mortise, felt insert		90	.089		4.55	4.17		8.72	11.85
4160	Mortise with spring, pile insert		90	.089		3.40	4.17		7.57	10.60
4400	Rigid vinyl retainer, mortise, pile insert		105	.076		2.70	3.58		6.28	8.80
4600	Wool pile filler strip, aluminum backing		105	.076		2.70	3.58		6.28	8.80
5000	Two piece overlapping astragal, extruded aluminum retainer									
5010	Pile insert	1 Carp	60	.133	L.F.	3.35	6.25		9.60	13.95
5020	Vinyl bulb insert		60	.133		1.85	6.25		8.10	12.30
5040	Vinyl flap insert		60	.133		3.65	6.25		9.90	14.25
5060	Solid neoprene flap insert		60	.133		6.55	6.25		12.80	17.45
5080	Hypalon rubber flap insert		60	.133		6.65	6.25		12.90	17.55
5090	Snap on cover, pile insert		60	.133		9.45	6.25		15.70	20.50
5400	Magnetic aluminum, surface mounted		60	.133		23	6.25		29.25	35.50

08 71 Door Hardware

08 71 21 – Astragals

08 71 21.10 Exterior Mouldings, Astragals	Crew	Daily Output	Labor-Hours	Unit	Material	2015 Bare Costs Labor	Equipment	Total	Total Incl O&P
5500 Interlocking aluminum, 5/8" x 1" neoprene bulb insert	1 Carp	45	.178	L.F.	5.70	8.35		14.05	19.95
5600 Adjustable aluminum, 9/16" x 21/32", pile insert		45	.178		17.55	8.35		25.90	33
5800 Magnetic, adjustable, 9/16" x 21/32"	↓	45	.178	↓	23	8.35		31.35	38.50
6000 Two piece stile protection									
6010 Cloth backed rubber loop, 1" gap, nail on aluminum strips	1 Carp	45	.178	L.F.	4.35	8.35		12.70	18.50
6040 Screw on aluminum strips		45	.178		6.55	8.35		14.90	21
6100 1-1/2" gap, screw on aluminum extrusion		45	.178		5.85	8.35		14.20	20
6240 Vinyl fabric loop, slotted aluminum extrusion, 1" gap		45	.178		2.20	8.35		10.55	16.10
6300 1-1/4" gap	↓	45	.178		6.20	8.35		14.55	20.50

08 71 25 – Weatherstripping

08 71 25.10 Mechanical Seals, Weatherstripping	Crew	Daily Output	Labor-Hours	Unit	Material	2015 Bare Costs Labor	Equipment	Total	Total Incl O&P
0010 **MECHANICAL SEALS, WEATHERSTRIPPING**									
1000 Doors, wood frame, interlocking, for 3' x 7' door, zinc	1 Carp	3	2.667	Opng.	44	125		169	254
1100 Bronze		3	2.667		56	125		181	267
1300 6' x 7' opening, zinc		2	4		54	188		242	370
1400 Bronze	↓	2	4	↓	65	188		253	380
1700 Wood frame, spring type, bronze									
1800 3' x 7' door	1 Carp	7.60	1.053	Opng.	23	49.50		72.50	107
1900 6' x 7' door	"	7	1.143	"	29.50	53.50		83	121
2200 Metal frame, spring type, bronze									
2300 3' x 7' door	1 Carp	3	2.667	Opng.	46.50	125		171.50	256
2400 6' x 7' door	"	2.50	3.200	"	52	150		202	305
2500 For stainless steel, spring type, add					133%				
2700 Metal frame, extruded sections, 3' x 7' door, aluminum	1 Carp	3	2.667	Opng.	28	125		153	236
2800 Bronze		3	2.667		82	125		207	295
3100 6' x 7' door, aluminum		1.50	5.333		35	250		285	450
3200 Bronze	↓	1.50	5.333	↓	137	250		387	560
3500 Threshold weatherstripping									
3650 Door sweep, flush mounted, aluminum	1 Carp	25	.320	Ea.	19	15		34	45.50
3700 Vinyl		25	.320		18	15		33	44.50
5000 Garage door bottom weatherstrip, 12' aluminum, clear		14	.571		25	27		52	71.50
5010 Bronze		14	.571		90	27		117	143
5050 Bottom protection, Rubber		14	.571		37	27		64	84.50
5100 Threshold		14	.571	↓	72	27		99	123
9000 Minimum labor/equipment charge	↓	3	2.667	Job		125		125	205

08 74 Access Control Hardware

08 74 13 – Card Key Access Control Hardware

08 74 13.50 Card Key Access	Crew	Daily Output	Labor-Hours	Unit	Material	2015 Bare Costs Labor	Equipment	Total	Total Incl O&P
0010 **CARD KEY ACCESS**									
0020 Computerized system, processor, proximity reader and cards									
0030 Does not include door hardware, lockset or wiring									
0040 Card key system for 1 door				Ea.	1,225			1,225	1,350
0060 Card key system for 2 doors					2,125			2,125	2,350
0080 Card key system for 4 doors					2,650			2,650	2,900
0100 Processor for card key access system					850			850	935
0160 Magnetic lock for electric access, 600 Pound holding force					190			190	209
0170 Magnetic lock for electric access, 1200 Pound holding force					190			190	209
0200 Proximity card reader				↓	130			130	143

08 74 Access Control Hardware

08 74 19 – Biometric Identity Access Control Hardware

08 74 19.50 Biometric Identity Access	Crew	Daily Output	Labor-Hours	Unit	Material	2015 Bare Costs Labor	2015 Bare Costs Equipment	Total	Total Incl O&P
0010 **BIOMETRIC IDENTITY ACCESS**									
0220 Hand geometry scanner, mem of 512 users, excl striker/power	1 Elec	3	2.667	Ea.	2,100	146		2,246	2,525
0230 Memory upgrade for, adds 9,700 user profiles		8	1		300	54.50		354.50	415
0240 Adds 32,500 user profiles		8	1		600	54.50		654.50	745
0250 Prison type, memory of 256 users, excl striker, power		3	2.667		2,600	146		2,746	3,075
0260 Memory upgrade for, adds 3,300 user profiles		8	1		250	54.50		304.50	360
0270 Adds 9,700 user profiles		8	1		460	54.50		514.50	590
0280 Adds 27,900 user profiles		8	1		610	54.50		664.50	755
0290 All weather, mem of 512 users, excl striker/power		3	2.667		3,900	146		4,046	4,525
0300 Facial & fingerprint scanner, combination unit, excl striker/power		3	2.667		4,300	146		4,446	4,950
0310 Access for, for initial setup, excl striker/power	↓	3	2.667	↓	1,100	146		1,246	1,425

08 75 Window Hardware

08 75 10 – Window Handles and Latches

08 75 10.10 Handles and Latches

	Crew	Daily Output	Labor-Hours	Unit	Material	Labor	Equipment	Total	Total Incl O&P
0010 **HANDLES AND LATCHES**									
1000 Handles, surface mounted, aluminum	1 Carp	24	.333	Ea.	4.40	15.65		20.05	30.50
1020 Brass		24	.333		5.70	15.65		21.35	32
1040 Chrome		24	.333		7	15.65		22.65	33
1500 Recessed, aluminum		12	.667		2.51	31.50		34.01	54.50
1520 Brass		12	.667		3.95	31.50		35.45	56
1540 Chrome		12	.667		3.51	31.50		35.01	55.50
2000 Latches, aluminum		20	.400		3.36	18.80		22.16	34.50
2020 Brass		20	.400		4.55	18.80		23.35	36
2040 Chrome		20	.400	↓	3.28	18.80		22.08	34.50
9000 Minimum labor/equipment charge	↓	6	1.333	Job		62.50		62.50	103

08 75 30 – Weatherstripping

08 75 30.10 Mechanical Weather Seals

	Crew	Daily Output	Labor-Hours	Unit	Material	Labor	Equipment	Total	Total Incl O&P
0010 **MECHANICAL WEATHER SEALS**, Window, double hung, 3' X 5'									
0020 Zinc	1 Carp	7.20	1.111	Opng.	20	52		72	108
0100 Bronze		7.20	1.111		40	52		92	130
0200 Vinyl V strip		7	1.143		9.50	53.50		63	98.50
0500 As above but heavy duty, zinc		4.60	1.739		20	81.50		101.50	156
0600 Bronze	↓	4.60	1.739	↓	70	81.50		151.50	211
9000 Minimum labor/equipment charge	1 Clab	4.60	1.739	Job		65.50		65.50	107

08 79 Hardware Accessories

08 79 20 – Door Accessories

08 79 20.10 Door Hardware Accessories

	Crew	Daily Output	Labor-Hours	Unit	Material	Labor	Equipment	Total	Total Incl O&P
0010 **DOOR HARDWARE ACCESSORIES**									
0140 Door bolt, surface, 4"	1 Carp	32	.250	Ea.	11.70	11.75		23.45	32
0160 Door latch	"	12	.667	"	8.55	31.50		40.05	61
0200 Sliding closet door									
0220 Track and hanger, single	1 Carp	10	.800	Ea.	58.50	37.50		96	126
0240 Double		8	1		80	47		127	165
0260 Door guide, single		48	.167		30	7.85		37.85	46
0280 Double		48	.167		40	7.85		47.85	57
0600 Deadbolt and lock cover plate, brass or stainless steel	↓	30	.267	↓	28	12.50		40.50	51.50

08 79 Hardware Accessories

08 79 20 – Door Accessories

08 79 20.10 Door Hardware Accessories	Crew	Daily Output	Labor-Hours	Unit	Material	2015 Bare Costs Labor	Equipment	Total	Total Incl O&P
0620 Hole cover plate, brass or chrome	1 Carp	35	.229	Ea.	8	10.75		18.75	26.50
2240 Mortise lockset, passage, lever handle		9	.889		160	41.50		201.50	245
4000 Security chain, standard		18	.444		10	21		31	45
4100 Deluxe	↓	18	.444	↓	45	21		66	83.50

08 81 Glass Glazing

08 81 10 – Float Glass

08 81 10.10 Various Types and Thickness of Float Glass

	Crew	Daily Output	Labor-Hours	Unit	Material	Labor	Equipment	Total	Total Incl O&P
0010 **VARIOUS TYPES AND THICKNESS OF FLOAT GLASS**									
0020 3/16" Plain	2 Glaz	130	.123	S.F.	5.05	5.55		10.60	14.55
0200 Tempered, clear		130	.123		6.95	5.55		12.50	16.65
0300 Tinted		130	.123		8	5.55		13.55	17.80
0600 1/4" thick, clear, plain		120	.133		5.95	6		11.95	16.30
0700 Tinted		120	.133		9.10	6		15.10	19.75
0800 Tempered, clear		120	.133		8.85	6		14.85	19.50
0900 Tinted		120	.133		10.90	6		16.90	22
1600 3/8" thick, clear, plain		75	.213		10.20	9.60		19.80	27
1700 Tinted		75	.213		15.70	9.60		25.30	33
1800 Tempered, clear		75	.213		16.75	9.60		26.35	34
1900 Tinted		75	.213		18.85	9.60		28.45	36
2200 1/2" thick, clear, plain		55	.291		17.30	13.10		30.40	40.50
2300 Tinted		55	.291		27.50	13.10		40.60	51.50
2400 Tempered, clear		55	.291		25	13.10		38.10	49
2500 Tinted		55	.291		26	13.10		39.10	50
2800 5/8" thick, clear, plain		45	.356		27.50	16.05		43.55	56
2900 Tempered, clear		45	.356		31.50	16.05		47.55	60.50
3200 3/4" thick, clear, plain		35	.457		35.50	20.50		56	72.50
3300 Tempered, clear		35	.457		41	20.50		61.50	79
3600 1" thick, clear, plain	↓	30	.533		59	24		83	104
8900 For low emissivity coating for 3/16" & 1/4" only, add to above				↓	18%				
9000 Minimum labor/equipment charge	1 Glaz	2	4	Job		180		180	293

08 81 17 – Fire Glass

08 81 17.10 Fire Resistant Glass

	Crew	Daily Output	Labor-Hours	Unit	Material	Labor	Equipment	Total	Total Incl O&P
0010 **FIRE RESISTANT GLASS**									
0020 Fire Glass Minimum	2 Glaz	40	.400	S.F.	35	18.05		53.05	68
0030 Mid Range		40	.400		76	18.05		94.05	113
0050 High End	↓	40	.400	↓	340	18.05		358.05	405

08 81 20 – Vision Panels

08 81 20.10 Full Vision

	Crew	Daily Output	Labor-Hours	Unit	Material	Labor	Equipment	Total	Total Incl O&P
0010 **FULL VISION**, window system with 3/4" glass mullions									
0020 Up to 10' high	H-2	130	.185	S.F.	63.50	7.85		71.35	82.50
0100 10' to 20' high, minimum		110	.218		67.50	9.30		76.80	89.50
0150 Average		100	.240		73	10.20		83.20	96.50
0200 Maximum	↓	80	.300	↓	81.50	12.80		94.30	111
9000 Minimum labor/equipment charge	1 Glaz	2	4	Job		180		180	293

08 81 Glass Glazing

08 81 25 – Glazing Variables

08 81 25.10 Applications of Glazing

		Crew	Daily Output	Labor-Hours	Unit	Material	2015 Bare Costs Labor	Equipment	Total	Total Incl O&P
0010	APPLICATIONS OF GLAZING R088110-10									
0600	For glass replacement, add				S.F.		100%			
0700	For gasket settings, add				L.F.	5.75			5.75	6.35
0900	For sloped glazing, add				S.F.		26%			
2000	Fabrication, polished edges, 1/4" thick				Inch	.55			.55	.61
2100	1/2" thick					1.30			1.30	1.43
2500	Mitered edges, 1/4" thick					1.30			1.30	1.43
2600	1/2" thick					2.15			2.15	2.37

08 81 30 – Insulating Glass

08 81 30.10 Reduce Heat Transfer Glass

		Crew	Daily Output	Labor-Hours	Unit	Material	Labor	Equipment	Total	Total Incl O&P
0010	REDUCE HEAT TRANSFER GLASS									
0015	2 lites 1/8" float, 1/2" thk under 15 S.F.									
0020	Clear R088110-10 G	2 Glaz	95	.168	S.F.	10.05	7.60		17.65	23.50
0100	Tinted G		95	.168		14.05	7.60		21.65	28
0200	2 lites 3/16" float, for 5/8" thk unit, 15 to 30 S.F., clear G		90	.178		13.70	8		21.70	28
0400	1" thk, dbl. glazed, 1/4" float, 30-70 S.F., clear G		75	.213		16.80	9.60		26.40	34
0500	Tinted G		75	.213		23.50	9.60		33.10	41.50
2000	Both lites, light & heat reflective G		85	.188		31.50	8.50		40	48.50
2500	Heat reflective, film inside, 1" thick unit, clear G		85	.188		27.50	8.50		36	44.50
2600	Tinted G		85	.188		28.50	8.50		37	45.50
3000	Film on weatherside, clear, 1/2" thick unit G		95	.168		19.60	7.60		27.20	34
3100	5/8" thick unit G		90	.178		20	8		28	35
3200	1" thick unit G		85	.188		27	8.50		35.50	44
3350	Clear heat reflective film on inside G	1 Glaz	50	.160		12.90	7.20		20.10	26
3360	Heat reflective film on inside, tinted G		25	.320		13.70	14.45		28.15	38.50
3370	Heat reflective film on the inside metalized G		20	.400		14	18.05		32.05	45
5000	Spectrally selective film, on ext, blocks solar gain/allows 70% of light G	2 Glaz	95	.168		14.25	7.60		21.85	28
9000	Minimum labor/equipment charge	1 Glaz	2	4	Job		180		180	293

08 81 40 – Plate Glass

08 81 40.10 Plate Glass

		Crew	Daily Output	Labor-Hours	Unit	Material	Labor	Equipment	Total	Total Incl O&P
0010	PLATE GLASS Twin ground, polished,									
0020	3/16" thick	2 Glaz	100	.160	S.F.	5.30	7.20		12.50	17.50
0100	1/4" thick		94	.170		7.25	7.70		14.95	20.50
0200	3/8" thick		60	.267		12.55	12.05		24.60	33.50
0300	1/2" thick		40	.400		24.50	18.05		42.55	56.50

08 81 55 – Window Glass

08 81 55.10 Sheet Glass

		Crew	Daily Output	Labor-Hours	Unit	Material	Labor	Equipment	Total	Total Incl O&P
0010	SHEET GLASS (window), clear float, stops, putty bed									
0015	1/8" thick, clear float	2 Glaz	480	.033	S.F.	3.65	1.50		5.15	6.45
0500	3/16" thick, clear		480	.033		5.90	1.50		7.40	8.90
0600	Tinted		480	.033		7.45	1.50		8.95	10.65
0700	Tempered		480	.033		9.20	1.50		10.70	12.55
9000	Minimum labor/equipment charge		5	3.200	Job		144		144	234

08 81 65 – Wire Glass

08 81 65.10 Glass Reinforced With Wire

		Crew	Daily Output	Labor-Hours	Unit	Material	Labor	Equipment	Total	Total Incl O&P
0010	GLASS REINFORCED WITH WIRE									
0012	1/4" thick rough obscure	2 Glaz	135	.119	S.F.	23.50	5.35		28.85	34.50
1000	Polished wire, 1/4" thick, diamond, clear		135	.119		28	5.35		33.35	39
1500	Pinstripe, obscure		135	.119		41	5.35		46.35	53.50

08 83 Mirrors

08 83 13 – Mirrored Glass Glazing

08 83 13.10 Mirrors

		Crew	Daily Output	Labor-Hours	Unit	Material	2015 Bare Costs Labor	2015 Bare Costs Equipment	Total	Total Incl O&P
0010	**MIRRORS**, No frames, wall type, 1/4" plate glass, polished edge									
0100	Up to 5 S.F.	2 Glaz	125	.128	S.F.	9.50	5.75		15.25	19.80
0200	Over 5 S.F.		160	.100		9.25	4.51		13.76	17.50
0500	Door type, 1/4" plate glass, up to 12 S.F.		160	.100		8.70	4.51		13.21	16.90
1000	Float glass, up to 10 S.F., 1/8" thick		160	.100		5.90	4.51		10.41	13.80
1100	3/16" thick		150	.107		7.30	4.81		12.11	15.85
1500	12" x 12" wall tiles, square edge, clear		195	.082		2.22	3.70		5.92	8.45
1600	Veined		195	.082		5.85	3.70		9.55	12.40
2000	1/4" thick, stock sizes, one way transparent		125	.128		19.60	5.75		25.35	31
2010	Bathroom, unframed, laminated		160	.100		13.90	4.51		18.41	22.50
2500	Tempered		160	.100		17.80	4.51		22.31	27

08 83 13.15 Reflective Glass

			Crew	Daily Output	Labor-Hours	Unit	Material	Labor	Equipment	Total	Total Incl O&P
0010	**REFLECTIVE GLASS**										
0100	1/4" float with fused metallic oxide fixed	G	2 Glaz	115	.139	S.F.	16.95	6.25		23.20	29
0500	1/4" float glass with reflective applied coating	G	"	115	.139	"	13.70	6.25		19.95	25.50

08 84 Plastic Glazing

08 84 10 – Plexiglass Glazing

08 84 10.10 Plexiglass Acrylic

		Crew	Daily Output	Labor-Hours	Unit	Material	Labor	Equipment	Total	Total Incl O&P
0010	**PLEXIGLASS ACRYLIC**, clear, masked,									
0020	1/8" thick, cut sheets	2 Glaz	170	.094	S.F.	12	4.24		16.24	20
0200	Full sheets		195	.082		5	3.70		8.70	11.50
0500	1/4" thick, cut sheets		165	.097		14	4.37		18.37	22.50
0600	Full sheets		185	.086		9	3.90		12.90	16.25
0900	3/8" thick, cut sheets		155	.103		20	4.66		24.66	29.50
1000	Full sheets		180	.089		15	4.01		19.01	23
1300	1/2" thick, cut sheets		135	.119		28	5.35		33.35	39.50
1400	Full sheets		150	.107		20	4.81		24.81	30
1700	3/4" thick, cut sheets		115	.139		69	6.25		75.25	86
1800	Full sheets		130	.123		40	5.55		45.55	53
2100	1" thick, cut sheets		105	.152		77.50	6.85		84.35	96.50
2200	Full sheets		125	.128		48	5.75		53.75	62.50
3000	Colored, 1/8" thick, cut sheets		170	.094		18	4.24		22.24	26.50
3200	Full sheets		195	.082		11	3.70		14.70	18.10
3500	1/4" thick, cut sheets		165	.097		20	4.37		24.37	29
3600	Full sheets		185	.086		14	3.90		17.90	22
4000	Mirrors, untinted, cut sheets, 1/8" thick		185	.086		12	3.90		15.90	19.55
4200	1/4" thick		180	.089		16	4.01		20.01	24

08 84 20 – Polycarbonate

08 84 20.10 Thermoplastic

		Crew	Daily Output	Labor-Hours	Unit	Material	Labor	Equipment	Total	Total Incl O&P
0010	**THERMOPLASTIC**, clear, masked, cut sheets									
0020	1/8" thick	2 Glaz	170	.094	S.F.	14	4.24		18.24	22.50
0500	3/16" thick		165	.097		16	4.37		20.37	24.50
1000	1/4" thick		155	.103		17	4.66		21.66	26.50
1500	3/8" thick		150	.107		26	4.81		30.81	36.50
9000	Minimum labor/equipment charge	1 Glaz	2	4	Job		180		180	293

08 85 Glazing Accessories

08 85 10 – Miscellaneous Glazing Accessories

08 85 10.10 Glazing Gaskets

		Crew	Daily Output	Labor-Hours	Unit	Material	2015 Bare Costs Labor	Equipment	Total	Total Incl O&P
0010	**GLAZING GASKETS**, Neoprene for glass tongued mullion									
0015	1/4" glass	2 Glaz	200	.080	L.F.	.73	3.61		4.34	6.65
0020	3/8"		200	.080		.91	3.61		4.52	6.85
0040	1/2"		200	.080		1.03	3.61		4.64	7
0060	3/4"		180	.089		1.50	4.01		5.51	8.15
0080	1"		180	.089		1.22	4.01		5.23	7.85
1000	Glazing compound, wood	1 Glaz	58	.138		1.10	6.20		7.30	11.30
1005	Glazing compound, per window, up to 30 L.F.					1.10			1.10	1.21
1006	Glazing compound, per window, up to 30 L.F.				Ea.	33			33	36.50
1020	Metal	1 Glaz	58	.138	L.F.	1.10	6.20		7.30	11.30

08 87 Glazing Surface Films

08 87 13 – Solar Control Films

08 87 13.10 Solar Films On Glass

			Crew	Daily Output	Labor-Hours	Unit	Material	Labor	Equipment	Total	Total Incl O&P
0010	**SOLAR FILMS ON GLASS** (glass not included)										
2000	Minimum	G	2 Glaz	180	.089	S.F.	6.80	4.01		10.81	14
2050	Maximum	G	"	225	.071	"	15.30	3.21		18.51	22

08 87 23 – Safety and Security Films

08 87 23.13 Safety Films

		Crew	Daily Output	Labor-Hours	Unit	Material	Labor	Equipment	Total	Total Incl O&P
0010	**SAFETY FILMS**									
2010	Window protection film, controls blast damage	2 Glaz	80	.200	S.F.	6.80	9		15.80	22

08 87 23.16 Security Films

		Crew	Daily Output	Labor-Hours	Unit	Material	Labor	Equipment	Total	Total Incl O&P
0010	**SECURITY FILMS**, clear, 32000 psi tensile strength, adhered to glass R088110-10									
0100	.002" thick, daylight installation	H-2	950	.025	S.F.	2.70	1.08		3.78	4.72
0150	.004" thick, daylight installation		800	.030		3.35	1.28		4.63	5.75
0200	.006" thick, daylight installation		700	.034		3.60	1.46		5.06	6.35
0210	Install for anchorage		600	.040		4	1.70		5.70	7.15
0400	.007" thick, daylight installation		600	.040		4.45	1.70		6.15	7.65
0410	Install for anchorage		500	.048		4.94	2.04		6.98	8.80
0500	.008" thick, daylight installation		500	.048		5	2.04		7.04	8.85
0510	Install for anchorage		500	.048		5.55	2.04		7.59	9.45
0600	.015" thick, daylight installation		400	.060		8.80	2.56		11.36	13.85
0610	Install for anchorage		400	.060		5.55	2.56		8.11	10.25
0900	Security film anchorage, mechanical attachment and cover plate	H-3	370	.043	L.F.	9.90	1.74		11.64	13.75
0950	Security film anchorage, wet glaze structural caulking	1 Glaz	225	.036	"	.99	1.60		2.59	3.69
1000	Adhered security film removal	1 Clab	275	.029	S.F.		1.09		1.09	1.79

08 88 Special Function Glazing

08 88 52 – Prefabricated Glass Block Windows

08 88 52.10 Prefabricated Glass Block Windows

		Crew	Daily Output	Labor-Hours	Unit	Material	Labor	Equipment	Total	Total Incl O&P
0010	**PREFABRICATED GLASS BLOCK WINDOWS**									
0015	Includes frame and silicone seal									
0020	Glass Block Window 16" x 8"	2 Glaz	6	2.667	Ea.	185	120		305	400
0025	16" x 16"		6	2.667		223	120		343	440
0030	16" x 24"		6	2.667		265	120		385	485
0050	16" x 48"		4	4		380	180		560	715
0100	Glass Block Window 24" x 8"		6	2.667		211	120		331	425
0110	24" x 16"		6	2.667		265	120		385	485

08 88 Special Function Glazing

08 88 52 – Prefabricated Glass Block Windows

08 88 52.10 Prefabricated Glass Block Windows

		Crew	Daily Output	Labor-Hours	Unit	Material	2015 Bare Costs Labor	Equipment	Total	Total Incl O&P
0120	24" x 24"	2 Glaz	6	2.667	Ea.	315	120		435	540
0150	24" x 48"		6	2.667		470	120		590	710
0200	Glass Block Window 32" x 8"		6	2.667		211	120		331	425
0210	32" x 16"		6	2.667		300	120		420	525
0220	32" x 24"		4	4		365	180		545	695
0250	32" x 48"		4	4		555	180		735	905
0300	Glass Block Window 40" x 8"		6	2.667		271	120		391	495
0310	40" x 16"		6	2.667		340	120		460	570
0320	40" x 24"		6	2.667		410	120		530	650
0350	40" x 48"		6	2.667		645	120		765	905
0400	Glass Block Window 48" x 8"		6	2.667		300	120		420	525
0410	48" x 16"		6	2.667		380	120		500	610
0420	48" x 24"		6	2.667		470	120		590	710
0450	48" x 48"		6	2.667		730	120		850	1,000
0500	Glass Block Window 56" x 8"		6	2.667		330	120		450	560
0510	56" x 16"		4	4		415	180		595	755
0520	56" x 24"		4	4		520	180		700	865
0550	56" x 48"		4	4		825	180		1,005	1,200

08 88 52.20 Prefabricated Glass Block Windows Hurricane Resistant

		Crew	Daily Output	Labor-Hours	Unit	Material	Labor	Equipment	Total	Total Incl O&P
0010	**PREFABRICATED GLASS BLOCK WINDOWS HURRICANE RESISTANT**									
0015	Includes frame and silicone seal									
3020	Glass Block Window 16" x 16"	2 Glaz	6	2.667	Ea.	500	120		620	745
3030	16" x 24"		6	2.667		580	120		700	835
3040	16" x 32"		4	4		640	180		820	1,000
3050	16" x 40"		4	4		690	180		870	1,050
3060	16" x 48"		4	4		700	180		880	1,075
3070	16" x 56"		4	4		880	180		1,060	1,275
3080	Glass Block Window 24" x 24"		6	2.667		640	120		760	900
3090	24" x 32"		4	4		700	180		880	1,075
3100	24" x 40"		4	4		880	180		1,060	1,275
3110	24" x 48"		4	4		910	180		1,090	1,300
3120	24" x 56"		4	4		950	180		1,130	1,350
3130	Glass Block Window 32" x 32"		4	4		800	180		980	1,175
3140	32" x 40"		4	4		900	180		1,080	1,275
3150	32" x 48"		4	4		1,050	180		1,230	1,450
3160	32" x 56"		4	4		1,200	180		1,380	1,625
3200	Glass Block Window 40" x 40"		4	4		1,025	180		1,205	1,425
3210	40" x 48"		3	5.333		1,250	241		1,491	1,775
3220	40" x 56"		3	5.333		1,400	241		1,641	1,950
3260	Glass Block Window 48" x 48"		3	5.333		1,450	241		1,691	2,000
3270	48" x 56"		3	5.333		1,625	241		1,866	2,175
3310	Glass Block Window 56" x 56"		3	5.333		1,775	241		2,016	2,350

08 88 56 – Ballistics-Resistant Glazing

08 88 56.10 Laminated Glass

		Crew	Daily Output	Labor-Hours	Unit	Material	Labor	Equipment	Total	Total Incl O&P
0010	**LAMINATED GLASS**									
0020	Clear float .03" vinyl 1/4"	2 Glaz	90	.178	S.F.	12.40	8		20.40	26.50
0100	3/8" thick		78	.205		22.50	9.25		31.75	39.50
0200	.06" vinyl, 1/2" thick		65	.246		25.50	11.10		36.60	46
1000	5/8" thick		90	.178		29.50	8		37.50	45.50
2000	Bullet-resisting, 1-3/16" thick, to 15 S.F.		16	1		105	45		150	189
2100	Over 15 S.F.		16	1		119	45		164	204
2200	2" thick, to 15 S.F.		10	1.600		140	72		212	271

08 88 Special Function Glazing

08 88 56 – Ballistics-Resistant Glazing

08 88 56.10 Laminated Glass		Crew	Daily Output	Labor-Hours	Unit	Material	2015 Bare Costs Labor	Equipment	Total	Total Incl O&P
2300	Over 15 S.F.	2 Glaz	10	1.600	S.F.	120	72		192	249
2500	2-1/4" thick, to 15 S.F.		12	1.333		179	60		239	295
2600	Over 15 S.F.		12	1.333		168	60		228	283
2700	Level 2 (.357 magnum), NIJ and UL		12	1.333		77	60		137	182
2750	Level 3A (.44 magnum) NIJ, UL 3		12	1.333		82	60		142	188
2800	Level 4 (AK-47) NIJ, UL 7 & 8		12	1.333		113	60		173	222
2850	Level 5 (M-16) UL		12	1.333		116	60		176	226
2900	Level 3 (7.62 Armor Piercing) NIJ, UL 4 & 5		12	1.333		137	60		197	249

08 91 Louvers

08 91 16 – Operable Louvers

08 91 16.10 Movable Blade Louvers

		Crew	Daily Output	Labor-Hours	Unit	Material	Labor	Equipment	Total	Total Incl O&P
0010	**MOVABLE BLADE LOUVERS**									
0100	PVC, commercial grade, 12" x 12"	1 Shee	20	.400	Ea.	82	22.50		104.50	126
0110	16" x 16"		14	.571		96	32		128	157
0120	18" x 18"		14	.571		99	32		131	160
0130	24" x 24"		14	.571		140	32		172	205
0140	30" x 30"		12	.667		168	37.50		205.50	244
0150	36" x 36"		12	.667		197	37.50		234.50	276
0300	Stainless steel, commercial grade, 12" x 12"		20	.400		219	22.50		241.50	277
0310	16" x 16"		14	.571		265	32		297	345
0320	18" x 18"		14	.571		282	32		314	360
0330	20" x 20"		14	.571		325	32		357	405
0340	24" x 24"		14	.571		385	32		417	475
0350	30" x 30"		12	.667		430	37.50		467.50	535
0360	36" x 36"		12	.667		520	37.50		557.50	635

08 91 19 – Fixed Louvers

08 91 19.10 Aluminum Louvers

		Crew	Daily Output	Labor-Hours	Unit	Material	Labor	Equipment	Total	Total Incl O&P
0010	**ALUMINUM LOUVERS**									
0020	Aluminum with screen, residential, 8" x 8"	1 Carp	38	.211	Ea.	20	9.90		29.90	38
0100	12" x 12"		38	.211		16	9.90		25.90	34
0200	12" x 18"		35	.229		20	10.75		30.75	39.50
0250	14" x 24"		30	.267		29	12.50		41.50	52.50
0300	18" x 24"		27	.296		32	13.90		45.90	58
0500	24" x 30"		24	.333		60	15.65		75.65	91.50
0700	Triangle, adjustable, small		20	.400		55	18.80		73.80	91.50
0800	Large		15	.533		76	25		101	125
2100	Midget, aluminum, 3/4" deep, 1" diameter		85	.094		.75	4.42		5.17	8.10
2150	3" diameter		60	.133		2.47	6.25		8.72	12.95
2200	4" diameter		50	.160		4.95	7.50		12.45	17.75
2250	6" diameter		30	.267		4.10	12.50		16.60	25
3000	PVC, commercial grade, 12" x 12"	1 Shee	20	.400		97	22.50		119.50	143
3010	12" x 18"		20	.400		109	22.50		131.50	156
3020	12" x 24"		20	.400		146	22.50		160.50	197
3030	14" x 24"		20	.400		152	22.50		174.50	203
3100	Aluminum, commercial grade, 12" x 12"		20	.400		171	22.50		193.50	224
3110	24" x 24"		16	.500		249	28		277	320
3120	24" x 36"		16	.500		300	28		328	375
3130	36" x 24"		16	.500		310	28		338	385
3140	36" x 36"		14	.571		390	32		422	480

08 91 Louvers

08 91 19 – Fixed Louvers

08 91 19.10 Aluminum Louvers

		Crew	Daily Output	Labor-Hours	Unit	Material	2015 Bare Costs Labor	Equipment	Total	Total Incl O&P
3150	36" x 48"	1 Shee	10	.800	Ea.	400	45		445	510
3160	48" x 36"		10	.800		390	45		435	500
3170	48" x 48"		10	.800		435	45		480	545
3180	60" x 48"		8	1		485	56		541	625
3190	60" x 60"		8	1		600	56		656	750

08 91 19.20 Steel Louvers

		Crew	Daily Output	Labor-Hours	Unit	Material	Labor	Equipment	Total	Total Incl O&P
0010	**STEEL LOUVERS**									
3300	Galvanized Steel, fixed blades, commercial grade, 18" x 18"	1 Shee	20	.400	Ea.	197	22.50		219.50	253
3310	24" x 24"		20	.400		232	22.50		254.50	291
3320	24" x 36"		16	.500		300	28		328	375
3330	36" x 24"		16	.500		298	28		326	375
3340	36" x 36"		14	.571		380	32		412	470
3350	36" x 48"		10	.800		385	45		430	495
3360	48" x 36"		10	.800		385	45		430	495
3370	48" x 48"		10	.800		425	45		470	535
3380	60" x 48"		10	.800		490	45		535	610
3390	60" x 60"		10	.800		590	45		635	715

08 91 26 – Door Louvers

08 91 26.10 Steel Louvers, 18 Gauge, Fixed Blade

		Crew	Daily Output	Labor-Hours	Unit	Material	Labor	Equipment	Total	Total Incl O&P
0010	**STEEL LOUVERS, 18 GAUGE, FIXED BLADE**									
0050	12" x 12", with enamel or powder coat	1 Carp	20	.400	Ea.	82	18.80		100.80	121
0055	18" x 12"		20	.400		87.50	18.80		106.30	127
0060	18" x 18"		20	.400		103	18.80		121.80	145
0065	24" x 12"		20	.400		108	18.80		126.80	150
0070	24" x 18"		20	.400		117	18.80		135.80	160
0075	24" x 24"		20	.400		144	18.80		162.80	189
0100	12" x 12", galvanized		20	.400		72.50	18.80		91.30	111
0105	18" x 12"		20	.400		85.50	18.80		104.30	125
0115	24" x 12"		20	.400		102	18.80		120.80	143
0125	24" x 24"		20	.400		143	18.80		161.80	188

08 95 Vents

08 95 13 – Soffit Vents

08 95 13.10 Wall Louvers

		Crew	Daily Output	Labor-Hours	Unit	Material	Labor	Equipment	Total	Total Incl O&P
0010	**WALL LOUVERS**									
2330	Soffit vent, continuous, 3" wide, aluminum, mill finish	1 Carp	200	.040	L.F.	.67	1.88		2.55	3.82
2340	Baked enamel finish		200	.040	"	5.50	1.88		7.38	9.15
2400	Under eaves vent, aluminum, mill finish, 16" x 4"		48	.167	Ea.	1.90	7.85		9.75	14.95
2500	16" x 8"		48	.167	"	2.18	7.85		10.03	15.25

08 95 16 – Wall Vents

08 95 16.10 Louvers

		Crew	Daily Output	Labor-Hours	Unit	Material	Labor	Equipment	Total	Total Incl O&P
0010	**LOUVERS**									
0020	Redwood, 2'-0" diameter, full circle	1 Carp	16	.500	Ea.	190	23.50		213.50	248
0100	Half circle		16	.500		180	23.50		203.50	237
0200	Octagonal		16	.500		142	23.50		165.50	195
0300	Triangular, 5/12 pitch, 5'-0" at base		16	.500		200	23.50		223.50	259
1000	Rectangular, 1'-4" x 1'-3"		16	.500		19.50	23.50		43	60
1100	Rectangular, 1'-4" x 1'-8"		16	.500		28	23.50		51.50	69.50
1200	1'-4" x 2'-2"		15	.533		31	25		56	75.50
1300	1'-9" x 2'-2"		15	.533		37.50	25		62.50	82.50

08 95 Vents

08 95 16 – Wall Vents

08 95 16.10 Louvers		Crew	Daily Output	Labor-Hours	Unit	Material	2015 Bare Costs Labor	2015 Bare Costs Equipment	Total	Total Incl O&P
1400	2'-3" x 2'-2"	1 Carp	14	.571	Ea.	49.50	27		76.50	98.50
1700	2'-4" x 2'-11"		13	.615		49.50	29		78.50	102
2000	Aluminum, 12" x 16"		25	.320		20	15		35	46.50
2010	16" x 20"		25	.320		27.50	15		42.50	55
2020	24" x 30"		25	.320		59	15		74	89.50
2100	6' triangle		12	.667		168	31.50		199.50	237
3100	Round, 2'-2" diameter		16	.500		124	23.50		147.50	175
7000	Vinyl gable vent, 8" x 8"		38	.211		14	9.90		23.90	31.50
7020	12" x 12"		38	.211		27	9.90		36.90	45.50
7080	12" x 18"		35	.229		35	10.75		45.75	56
7200	18" x 24"		30	.267		45	12.50		57.50	70
9000	Minimum labor/equipment charge		3.50	2.286	Job		107		107	176

Estimating Tips
General
- Room Finish Schedule: A complete set of plans should contain a room finish schedule. If one is not available, it would be well worth the time and effort to obtain one.

09 20 00 Plaster and Gypsum Board
- Lath is estimated by the square yard plus a 5% allowance for waste. Furring, channels, and accessories are measured by the linear foot. An extra foot should be allowed for each accessory miter or stop.
- Plaster is also estimated by the square yard. Deductions for openings vary by preference, from zero deduction to 50% of all openings over 2 feet in width. The estimator should allow one extra square foot for each linear foot of horizontal interior or exterior angle located below the ceiling level. Also, double the areas of small radius work.
- Drywall accessories, studs, track, and acoustical caulking are all measured by the linear foot. Drywall taping is figured by the square foot. Gypsum wallboard is estimated by the square foot. No material deductions should be made for door or window openings under 32 S.F.

09 60 00 Flooring
- Tile and terrazzo areas are taken off on a square foot basis. Trim and base materials are measured by the linear foot. Accent tiles are listed per each. Two basic methods of installation are used. Mud set is approximately 30% more expensive than thin set. In terrazzo work, be sure to include the linear footage of embedded decorative strips, grounds, machine rubbing, and power cleanup.
- Wood flooring is available in strip, parquet, or block configuration. The latter two types are set in adhesives with quantities estimated by the square foot. The laying pattern will influence labor costs and material waste. In addition to the material and labor for laying wood floors, the estimator must make allowances for sanding and finishing these areas, unless the flooring is prefinished.
- Sheet flooring is measured by the square yard. Roll widths vary, so consideration should be given to use the most economical width, as waste must be figured into the total quantity. Consider also the installation methods available, direct glue down or stretched.

09 70 00 Wall Finishes
- Wall coverings are estimated by the square foot. The area to be covered is measured, length by height of wall above baseboards, to calculate the square footage of each wall. This figure is divided by the number of square feet in the single roll which is being used. Deduct, in full, the areas of openings such as doors and windows. Where a pattern match is required allow 25%–30% waste.

09 80 00 Acoustic Treatment
- Acoustical systems fall into several categories. The takeoff of these materials should be by the square foot of area with a 5% allowance for waste. Do not forget about scaffolding, if applicable, when estimating these systems.

09 90 00 Painting and Coating
- A major portion of the work in painting involves surface preparation. Be sure to include cleaning, sanding, filling, and masking costs in the estimate.
- Protection of adjacent surfaces is not included in painting costs. When considering the method of paint application, an important factor is the amount of protection and masking required. These must be estimated separately and may be the determining factor in choosing the method of application.

Reference Numbers
Reference numbers are shown in shaded boxes at the beginning of some major classifications. These numbers refer to related items in the Reference Section. The reference information may be an estimating procedure, an alternate pricing method, or technical information.

Note: Not all subdivisions listed here necessarily appear in this publication. ∎

09 01 Maintenance of Finishes

09 01 70 – Maintenance of Wall Finishes

09 01 70.10 Gypsum Wallboard Repairs

		Crew	Daily Output	Labor-Hours	Unit	Material	2015 Bare Costs Labor	2015 Bare Costs Equipment	Total	Total Incl O&P
0010	**GYPSUM WALLBOARD REPAIRS**									
0100	Fill and sand, pin/nail holes	1 Carp	960	.008	Ea.		.39		.39	.64
0110	Screw head pops		480	.017			.78		.78	1.28
0120	Dents, up to 2" square		48	.167		.01	7.85		7.86	12.85
0130	2" to 4" square		24	.333		.03	15.65		15.68	25.50
0140	Cut square, patch, sand and finish, holes, up to 2" square		12	.667		.03	31.50		31.53	51.50
0150	2" to 4" square		11	.727		.09	34		34.09	56
0160	4" to 8" square		10	.800		.23	37.50		37.73	62
0170	8" to 12" square		8	1		.46	47		47.46	77.50
0180	12" to 32" square		6	1.333		1.55	62.50		64.05	105
0210	16" by 48"		5	1.600		2.65	75		77.65	126
0220	32" by 48"		4	2		4.09	94		98.09	159
0230	48" square		3.50	2.286		5.75	107		112.75	182
0240	60" square		3.20	2.500	↓	9.50	117		126.50	202
0500	Skim coat surface with joint compound		1600	.005	S.F.	.03	.23		.26	.42
0510	Prepare, retape and refinish joints		60	.133	L.F.	.64	6.25		6.89	10.95
9000	Minimum labor/equipment charge	↓	2	4	Job		188		188	310

09 05 Common Work Results for Finishes

09 05 05 – Selective Demolition for Finishes

09 05 05.10 Selective Demolition, Ceilings

		Crew	Daily Output	Labor-Hours	Unit	Material	2015 Bare Costs Labor	2015 Bare Costs Equipment	Total	Total Incl O&P
0010	**SELECTIVE DEMOLITION, CEILINGS** R024119-10									
0200	Ceiling, drywall, furred and nailed or screwed	2 Clab	800	.020	S.F.		.75		.75	1.23
0220	On metal frame		760	.021			.79		.79	1.30
0240	On suspension system, including system		720	.022			.84		.84	1.37
1000	Plaster, lime and horse hair, on wood lath, incl. lath		700	.023			.86		.86	1.41
1020	On metal lath		570	.028			1.06		1.06	1.73
1100	Gypsum, on gypsum lath		720	.022			.84		.84	1.37
1120	On metal lath		500	.032			1.20		1.20	1.97
1200	Suspended ceiling, mineral fiber, 2' x 2' or 2' x 4'		1500	.011			.40		.40	.66
1250	On suspension system, incl. system		1200	.013			.50		.50	.82
1500	Tile, wood fiber, 12" x 12", glued		900	.018			.67		.67	1.10
1540	Stapled		1500	.011			.40		.40	.66
1580	On suspension system, incl. system		760	.021			.79		.79	1.30
2000	Wood, tongue and groove, 1" x 4"		1000	.016			.60		.60	.99
2040	1" x 8"		1100	.015			.55		.55	.90
2400	Plywood or wood fiberboard, 4' x 8' sheets	↓	1200	.013	↓		.50		.50	.82
9000	Minimum labor/equipment charge	1 Clab	2	4	Job		150		150	247

09 05 05.20 Selective Demolition, Flooring

		Crew	Daily Output	Labor-Hours	Unit	Material	2015 Bare Costs Labor	2015 Bare Costs Equipment	Total	Total Incl O&P
0010	**SELECTIVE DEMOLITION, FLOORING** R024119-10									
0200	Brick with mortar	2 Clab	475	.034	S.F.		1.27		1.27	2.08
0400	Carpet, bonded, including surface scraping		2000	.008			.30		.30	.49
0480	Tackless		9000	.002			.07		.07	.11
0550	Carpet tile, releasable adhesive		5000	.003			.12		.12	.20
0560	Permanent adhesive	↓	1850	.009			.33		.33	.53
0600	Composition, acrylic or epoxy		400	.040			1.50		1.50	2.47
0700	Concrete, scarify skin	A-1A	225	.036			1.73	.95	2.68	3.86
0800	Resilient, sheet goods	2 Clab	1400	.011			.43		.43	.70
0820	For gym floors	"	900	.018	↓		.67		.67	1.10
0850	Vinyl or rubber cove base	1 Clab	1000	.008	L.F.		.30		.30	.49
0860	Vinyl or rubber cove base, molded corner	"	1000	.008	Ea.		.30		.30	.49

09 05 Common Work Results for Finishes

09 05 05 – Selective Demolition for Finishes

09 05 05.20 Selective Demolition, Flooring

		Crew	Daily Output	Labor-Hours	Unit	Material	2015 Bare Costs Labor	Equipment	Total	Total Incl O&P
0870	For glued and caulked installation, add to labor					50%				
0900	Vinyl composition tile, 12" x 12"	2 Clab	1000	.016	S.F.		.60		.60	.99
2000	Tile, ceramic, thin set		675	.024			.89		.89	1.46
2020	Mud set		625	.026			.96		.96	1.58
2200	Marble, slate, thin set		675	.024			.89		.89	1.46
2220	Mud set		625	.026			.96		.96	1.58
2600	Terrazzo, thin set		450	.036			1.34		1.34	2.19
2620	Mud set		425	.038			1.42		1.42	2.32
2640	Terrazzo, cast in place	▼	300	.053			2.01		2.01	3.29
3000	Wood, block, on end	1 Carp	400	.020			.94		.94	1.54
3200	Parquet		450	.018			.83		.83	1.37
3400	Strip flooring, interior, 2-1/4" x 25/32" thick		325	.025			1.16		1.16	1.89
3500	Exterior, porch flooring, 1" x 4"		220	.036			1.71		1.71	2.80
3800	Subfloor, tongue and groove, 1" x 6"		325	.025			1.16		1.16	1.89
3820	1" x 8"		430	.019			.87		.87	1.43
3840	1" x 10"		520	.015			.72		.72	1.18
4000	Plywood, nailed		600	.013			.63		.63	1.03
4100	Glued and nailed		400	.020			.94		.94	1.54
4200	Hardboard, 1/4" thick	▼	760	.011			.49		.49	.81
8000	Remove flooring, bead blast, simple floor plan	A-1A	1000	.008			.39	.21	.60	.87
8100	complex floor plan		400	.020			.97	.54	1.51	2.17
8150	Mastic only		1500	.005	▼		.26	.14	.40	.58
9000	Minimum labor/equipment charge	1 Clab	4	2	Job		75		75	123

09 05 05.30 Selective Demolition, Walls and Partitions

		Crew	Daily Output	Labor-Hours	Unit	Material	2015 Bare Costs Labor	Equipment	Total	Total Incl O&P
0010	**SELECTIVE DEMOLITION, WALLS AND PARTITIONS** R024119-10									
0020	Walls, concrete, reinforced	B-39	120	.400	C.F.		15.90	1.94	17.84	28
0025	Plain	"	160	.300			11.95	1.46	13.41	21
0100	Brick, 4" to 12" thick	B-9	220	.182	▼		6.90	1.06	7.96	12.50
0200	Concrete block, 4" thick		1150	.035	S.F.		1.32	.20	1.52	2.39
0280	8" thick		1050	.038			1.45	.22	1.67	2.61
0300	Exterior stucco 1" thick over mesh	▼	3200	.013			.48	.07	.55	.86
1000	Drywall, nailed or screwed	1 Clab	1000	.008			.30		.30	.49
1010	2 layers		400	.020			.75		.75	1.23
1020	Glued and nailed		900	.009			.33		.33	.55
1500	Fiberboard, nailed		900	.009			.33		.33	.55
1520	Glued and nailed		800	.010			.38		.38	.62
1568	Plenum barrier, sheet lead		300	.027			1		1	1.64
1600	Glass block		65	.123			4.63		4.63	7.60
2000	Movable walls, metal, 5' high		300	.027			1		1	1.64
2020	8' high		400	.020			.75		.75	1.23
2200	Metal or wood studs, finish 2 sides, fiberboard	B-1	520	.046			1.77		1.77	2.90
2250	Lath and plaster		260	.092			3.53		3.53	5.80
2300	Plasterboard (drywall)		520	.046			1.77		1.77	2.90
2350	Plywood	▼	450	.053			2.04		2.04	3.35
2800	Paneling, 4' x 8' sheets	1 Clab	475	.017			.63		.63	1.04
3000	Plaster, lime and horsehair, on wood lath		400	.020			.75		.75	1.23
3020	On metal lath		335	.024			.90		.90	1.47
3400	Gypsum or perlite, on gypsum lath		410	.020			.73		.73	1.20
3420	On metal lath		300	.027	▼		1		1	1.64
3450	Plaster, interior gypsum, acoustic, or cement		60	.133	S.Y.		5		5	8.20
3500	Stucco, on masonry		145	.055			2.07		2.07	3.40
3510	Commercial 3-coat		80	.100			3.76		3.76	6.15

09 05 Common Work Results for Finishes

09 05 05 – Selective Demolition for Finishes

09 05 05.30 Selective Demolition, Walls and Partitions

		Crew	Daily Output	Labor-Hours	Unit	Material	2015 Bare Costs Labor	2015 Bare Costs Equipment	Total	Total Incl O&P
3520	Interior stucco	1 Clab	25	.320	S.Y.		12.05		12.05	19.75
3753	Remove damaged glass block	B-1	134.62	.178	S.F.		6.80		6.80	11.20
3760	Tile, ceramic, on walls, thin set	1 Clab	300	.027			1		1	1.64
3765	Mud set		250	.032			1.20		1.20	1.97
3800	Toilet partitions, slate or marble		5	1.600	Ea.		60		60	98.50
3820	Metal or plastic		8	1	"		37.50		37.50	61.50
5000	Wallcovering, vinyl	1 Pape	700	.011	S.F.		.47		.47	.75
5010	With release agent		1500	.005			.22		.22	.35
5025	Wallpaper, 2 layers or less, by hand		250	.032			1.31		1.31	2.10
5035	3 layers or more		165	.048			1.98		1.98	3.19
5040	Designer		480	.017			.68		.68	1.10
9000	Minimum labor/equipment charge	1 Clab	4	2	Job		75		75	123

09 05 71 – Acoustic Underlayment

09 05 71.10 Acoustical Underlayment

		Crew	Daily Output	Labor-Hours	Unit	Material	2015 Bare Costs Labor	2015 Bare Costs Equipment	Total	Total Incl O&P
0010	**ACOUSTICAL UNDERLAYMENT**									
4000	Nylon matting 0.4" thick, with carbon black spinerette									
4010	plus polyester fabric, on floor	D-7	1600	.010	S.F.	1.36	.38		1.74	2.10
4200	Fiberglass reinf. backer board underlayment, 7/16" thick, on floor	"	1500	.011	"	2.77	.41		3.18	3.69

09 21 Plaster and Gypsum Board Assemblies

09 21 16 – Gypsum Board Assemblies

09 21 16.23 Gypsum Board Shaft Wall Assemblies

		Crew	Daily Output	Labor-Hours	Unit	Material	2015 Bare Costs Labor	2015 Bare Costs Equipment	Total	Total Incl O&P
0010	**GYPSUM BOARD SHAFT WALL ASSEMBLIES**									
0020	Cavity type on 25 ga. J-track & C-H studs, 24" O.C.									
0030	1" thick coreboard wall liner on shaft side									
0040	2-hour assembly with double layer									
0060	5/8" fire rated gypsum board on room side	2 Carp	220	.073	S.F.	2.10	3.41		5.51	7.90
0100	3-hour assembly with triple layer									
0300	5/8" fire rated gypsum board on room side	2 Carp	180	.089	S.F.	1.77	4.17		5.94	8.80
0400	4-hour assembly, 1" coreboard, 5/8" fire rated gypsum board									
0600	and 3/4" galv. metal furring channels, 24" O.C., with									
0700	Double layer 5/8" fire rated gypsum board on room side	2 Carp	110	.145	S.F.	1.67	6.85		8.52	13.05
0900	For taping & finishing, add per side	1 Carp	1050	.008	"	.05	.36		.41	.64
1000	For insulation, see Section 07 21									
5200	For work over 8' high, add	2 Carp	3060	.005	S.F.		.25		.25	.40
5300	For distribution cost over 3 stories high, add per story	"	6100	.003	"		.12		.12	.20

09 21 16.33 Partition Wall

		Crew	Daily Output	Labor-Hours	Unit	Material	2015 Bare Costs Labor	2015 Bare Costs Equipment	Total	Total Incl O&P
0010	**PARTITION WALL** Stud wall, 8' to 12' high									
0050	1/2", interior, gypsum board, std, tape & finish 2 sides									
0500	Installed on and incl., 2" x 4" wood studs, 16" O.C.	2 Carp	310	.052	S.F.	1.16	2.42		3.58	5.25
1000	Metal studs, NLB, 25 ga., 16" O.C., 3-5/8" wide		350	.046		1.06	2.15		3.21	4.69
1200	6" wide		330	.048		1.19	2.28		3.47	5.05
1400	Water resistant, on 2" x 4" wood studs, 16" O.C.		310	.052		1.34	2.42		3.76	5.45
1600	Metal studs, NLB, 25 ga., 16" O.C., 3-5/8" wide		350	.046		1.24	2.15		3.39	4.89
1800	6" wide		330	.048		1.37	2.28		3.65	5.25
2000	Fire res., 2 layers, 1-1/2 hr., on 2" x 4" wood studs, 16" O.C.		210	.076		1.96	3.58		5.54	8
2200	Metal studs, NLB, 25 ga., 16" O.C., 3-5/8" wide		250	.064		1.86	3		4.86	6.95
2400	6" wide		230	.070		1.99	3.27		5.26	7.55
2600	Fire & water res., 2 layers, 1-1/2 hr., 2" x 4" studs, 16" O.C.		210	.076		1.96	3.58		5.54	8
2800	Metal studs, NLB, 25 ga., 16" O.C., 3-5/8" wide		250	.064		1.86	3		4.86	6.95

09 21 Plaster and Gypsum Board Assemblies

09 21 16 – Gypsum Board Assemblies

09 21 16.33 Partition Wall

		Crew	Daily Output	Labor-Hours	Unit	Material	2015 Bare Costs Labor	2015 Bare Costs Equipment	Total	Total Incl O&P
3000	6" wide	2 Carp	230	.070	S.F.	1.99	3.27		5.26	7.55
3200	5/8", interior, gypsum board, standard, tape & finish 2 sides									
3400	Installed on and including 2" x 4" wood studs, 16" O.C.	2 Carp	300	.053	S.F.	1.22	2.50		3.72	5.45
3600	24" O.C.		330	.048		1.12	2.28		3.40	4.96
3800	Metal studs, NLB, 25 ga., 16" O.C., 3-5/8" wide		340	.047		1.12	2.21		3.33	4.85
4000	6" wide		320	.050		1.25	2.35		3.60	5.20
4200	24" O.C., 3-5/8" wide		360	.044		1.02	2.09		3.11	4.55
4400	6" wide		340	.047		1.12	2.21		3.33	4.85
4800	Water resistant, on 2" x 4" wood studs, 16" O.C.		300	.053		1.40	2.50		3.90	5.65
5000	24" O.C.		330	.048		1.30	2.28		3.58	5.15
5200	Metal studs, NLB, 25 ga. 16" O.C., 3-5/8" wide		340	.047		1.30	2.21		3.51	5.05
5400	6" wide		320	.050		1.43	2.35		3.78	5.40
5600	24" O.C., 3-5/8" wide		360	.044		1.20	2.09		3.29	4.75
5800	6" wide		340	.047		1.30	2.21		3.51	5.05
6000	Fire resistant, 2 layers, 2 hr., on 2" x 4" wood studs, 16" O.C.		205	.078		1.82	3.66		5.48	8
6200	24" O.C.		235	.068		1.82	3.20		5.02	7.25
6400	Metal studs, NLB, 25 ga., 16" O.C., 3-5/8" wide		245	.065		1.85	3.07		4.92	7.10
6600	6" wide		225	.071		1.95	3.34		5.29	7.60
6800	24" O.C., 3-5/8" wide		265	.060		1.72	2.83		4.55	6.55
7000	6" wide		245	.065		1.82	3.07		4.89	7.05
7200	Fire & water resistant, 2 layers, 2 hr., 2" x 4" studs, 16" O.C.		205	.078		1.92	3.66		5.58	8.10
7400	24" O.C.		235	.068		1.82	3.20		5.02	7.25
7600	Metal studs, NLB, 25 ga., 16" O.C., 3-5/8" wide		245	.065		1.82	3.07		4.89	7.05
7800	6" wide		225	.071		1.95	3.34		5.29	7.60
8000	24" O.C., 3-5/8" wide		265	.060		1.72	2.83		4.55	6.55
8200	6" wide		245	.065		1.82	3.07		4.89	7.05
8600	1/2" blueboard, mesh tape both sides									
8620	Installed on and including 2" x 4" wood studs, 16" O.C.	2 Carp	300	.053	S.F.	1.22	2.50		3.72	5.45
8640	Metal studs, NLB, 25 ga., 16" O.C., 3-5/8" wide		340	.047		1.12	2.21		3.33	4.85
8660	6" wide		320	.050		1.25	2.35		3.60	5.20
8800	Hospital security partition, 5/8" fiber reinf. high abuse gyp. bd.									
8810	Mtl. studs, NLB, 20 ga., 16" O.C., 3-5/8" wide, w/sec. mesh, gyp. bd.	2 Carp	208	.077	S.F.	4.18	3.61		7.79	10.50
9000	Exterior, 1/2" gypsum sheathing, 1/2" gypsum finished, interior,									
9100	including foil faced insulation, metal studs, 20 ga.									
9200	16" O.C., 3-5/8" wide	2 Carp	290	.055	S.F.	1.70	2.59		4.29	6.10
9400	6" wide		270	.059		1.88	2.78		4.66	6.60
9600	Partitions, for work over 8' high, add		1530	.010			.49		.49	.80

09 22 Supports for Plaster and Gypsum Board

09 22 03 – Fastening Methods for Finishes

09 22 03.20 Drilling Plaster/Drywall

		Crew	Daily Output	Labor-Hours	Unit	Material	2015 Bare Costs Labor	2015 Bare Costs Equipment	Total	Total Incl O&P
0010	**DRILLING PLASTER/DRYWALL**									
1100	Drilling & layout for drywall/plaster walls, up to 1" deep, no anchor									
1200	Holes, 1/4" diameter	1 Carp	150	.053	Ea.	.01	2.50		2.51	4.11
1300	3/8" diameter		140	.057		.01	2.68		2.69	4.41
1400	1/2" diameter		130	.062		.01	2.89		2.90	4.75
1500	3/4" diameter		120	.067		.01	3.13		3.14	5.15
1600	1" diameter		110	.073		.02	3.41		3.43	5.60
1700	1-1/4" diameter		100	.080		.03	3.76		3.79	6.20
1800	1-1/2" diameter		90	.089		.05	4.17		4.22	6.90
1900	For ceiling installations, add						40%			

09 22 Supports for Plaster and Gypsum Board

09 22 13 – Metal Furring

09 22 13.13 Metal Channel Furring

	09 22 13.13 Metal Channel Furring	Crew	Daily Output	Labor-Hours	Unit	Material	2015 Bare Costs Labor	Equipment	Total	Total Incl O&P
0010	**METAL CHANNEL FURRING**									
0030	Beams and columns, 7/8" channels, galvanized, 12" O.C.	1 Lath	155	.052	S.F.	.40	2.22		2.62	3.94
0050	16" O.C.		170	.047		.33	2.02		2.35	3.55
0070	24" O.C.		185	.043		.22	1.86		2.08	3.17
0100	Ceilings, on steel, 7/8" channels, galvanized, 12" O.C.		210	.038		.37	1.64		2.01	2.98
0300	16" O.C.		290	.028		.33	1.19		1.52	2.23
0400	24" O.C.		420	.019		.22	.82		1.04	1.53
0600	1-5/8" channels, galvanized, 12" O.C.		190	.042		.49	1.81		2.30	3.39
0700	16" O.C.		260	.031		.44	1.32		1.76	2.56
0900	24" O.C.		390	.021		.29	.88		1.17	1.71
0930	7/8" channels with sound isolation clips, 12" O.C.		120	.067		1.74	2.87		4.61	6.45
0940	16" O.C.		100	.080		1.31	3.44		4.75	6.85
0950	24" O.C.		165	.048		.87	2.08		2.95	4.24
0960	1-5/8" channels, galvanized, 12" O.C.		110	.073		1.86	3.13		4.99	6.95
0970	16" O.C.		100	.080		1.40	3.44		4.84	6.95
0980	24" O.C.		155	.052		.93	2.22		3.15	4.52
1000	Walls, 7/8" channels, galvanized, 12" O.C.		235	.034		.37	1.46		1.83	2.70
1200	16" O.C.		265	.030		.33	1.30		1.63	2.40
1300	24" O.C.		350	.023		.22	.98		1.20	1.79
1500	1-5/8" channels, galvanized, 12" O.C.		210	.038		.49	1.64		2.13	3.12
1600	16" O.C.		240	.033		.44	1.43		1.87	2.74
1800	24" O.C.		305	.026		.29	1.13		1.42	2.10
1920	7/8" channels with sound isolation clips, 12" O.C.		125	.064		1.74	2.75		4.49	6.25
1940	16" O.C.		100	.080		1.31	3.44		4.75	6.85
1950	24" O.C.		150	.053		.87	2.29		3.16	4.57
1960	1-5/8" channels, galvanized, 12" O.C.		115	.070		1.86	2.99		4.85	6.75
1970	16" O.C.		95	.084		1.40	3.62		5.02	7.25
1980	24" O.C.		140	.057		.93	2.46		3.39	4.90
9000	Minimum labor/equipment charge		4	2	Job		86		86	135

09 22 16 – Non-Structural Metal Framing

09 22 16.13 Non-Structural Metal Stud Framing

		Crew	Daily Output	Labor-Hours	Unit	Material	Labor	Equipment	Total	Total Incl O&P
0010	**NON-STRUCTURAL METAL STUD FRAMING**									
1600	Non-load bearing, galv., 8' high, 25 ga. 1-5/8" wide, 16" O.C.	1 Carp	619	.013	S.F.	.27	.61		.88	1.28
1610	24" O.C.		950	.008		.20	.40		.60	.87
1620	2-1/2" wide, 16" O.C.		613	.013		.33	.61		.94	1.36
1630	24" O.C.		938	.009		.24	.40		.64	.93
1640	3-5/8" wide, 16" O.C.		600	.013		.39	.63		1.02	1.46
1650	24" O.C.		925	.009		.29	.41		.70	.99
1660	4" wide, 16" O.C.		594	.013		.43	.63		1.06	1.51
1670	24" O.C.		925	.009		.32	.41		.73	1.03
1680	6" wide, 16" O.C.		588	.014		.52	.64		1.16	1.62
1690	24" O.C.		906	.009		.39	.41		.80	1.11
1700	20 ga. studs, 1-5/8" wide, 16" O.C.		494	.016		.34	.76		1.10	1.62
1710	24" O.C.		763	.010		.25	.49		.74	1.09
1720	2-1/2" wide, 16" O.C.		488	.016		.42	.77		1.19	1.72
1730	24" O.C.		750	.011		.31	.50		.81	1.17
1740	3-5/8" wide, 16" O.C.		481	.017		.48	.78		1.26	1.81
1750	24" O.C.		738	.011		.36	.51		.87	1.23
1760	4" wide, 16" O.C.		475	.017		.57	.79		1.36	1.93
1770	24" O.C.		738	.011		.43	.51		.94	1.30
1780	6" wide, 16" O.C.		469	.017		.66	.80		1.46	2.04
1790	24" O.C.		725	.011		.50	.52		1.02	1.40

09 22 Supports for Plaster and Gypsum Board

09 22 16 – Non-Structural Metal Framing

09 22 16.13 Non-Structural Metal Stud Framing

		Crew	Daily Output	Labor-Hours	Unit	Material	2015 Bare Costs Labor	Equipment	Total	Total Incl O&P
2000	Non-load bearing, galv., 10' high, 25 ga. 1-5/8" wide, 16" O.C.	1 Carp	495	.016	S.F.	.25	.76		1.01	1.52
2100	24" O.C.		760	.011		.19	.49		.68	1.01
2200	2-1/2" wide, 16" O.C.		490	.016		.31	.77		1.08	1.60
2250	24" O.C.		750	.011		.23	.50		.73	1.07
2300	3-5/8" wide, 16" O.C.		480	.017		.37	.78		1.15	1.68
2350	24" O.C.		740	.011		.27	.51		.78	1.13
2400	4" wide, 16" O.C.		475	.017		.41	.79		1.20	1.75
2450	24" O.C.		740	.011		.30	.51		.81	1.16
2500	6" wide, 16" O.C.		470	.017		.49	.80		1.29	1.85
2550	24" O.C.		725	.011		.36	.52		.88	1.25
2600	20 ga. studs, 1-5/8" wide, 16" O.C.		395	.020		.32	.95		1.27	1.91
2650	24" O.C.		610	.013		.23	.62		.85	1.27
2700	2-1/2" wide, 16" O.C.		390	.021		.39	.96		1.35	2.01
2750	24" O.C.		600	.013		.29	.63		.92	1.35
2800	3-5/8" wide, 16" OC		385	.021		.46	.98		1.44	2.10
2850	24" O.C.		590	.014		.34	.64		.98	1.41
2900	4" wide, 16" O.C.		380	.021		.54	.99		1.53	2.21
2950	24" O.C.		590	.014		.40	.64		1.04	1.48
3000	6" wide, 16" O.C.		375	.021		.63	1		1.63	2.33
3050	24" O.C.		580	.014		.46	.65		1.11	1.57
3060	Non-load bearing, galv., 12' high, 25 ga. 1-5/8" wide, 16" O.C.		413	.019		.24	.91		1.15	1.76
3070	24" O.C.		633	.013		.18	.59		.77	1.16
3080	2-1/2" wide, 16" O.C.		408	.020		.30	.92		1.22	1.84
3090	24" O.C.		625	.013		.22	.60		.82	1.23
3100	3-5/8" wide, 16" O.C.		400	.020		.35	.94		1.29	1.93
3110	24" O.C.		617	.013		.26	.61		.87	1.28
3120	4" wide, 16" O.C.		396	.020		.39	.95		1.34	1.98
3130	24" O.C.		617	.013		.28	.61		.89	1.31
3140	6" wide, 16" O.C.		392	.020		.47	.96		1.43	2.09
3150	24" O.C.		604	.013		.34	.62		.96	1.40
3160	20 ga. studs, 1-5/8" wide, 16" O.C.		329	.024		.30	1.14		1.44	2.20
3170	24" O.C.		508	.016		.22	.74		.96	1.45
3180	2-1/2" wide, 16" O.C.		325	.025		.38	1.16		1.54	2.30
3190	24" O.C.		500	.016		.27	.75		1.02	1.53
3200	3-5/8" wide, 16" O.C.		321	.025		.44	1.17		1.61	2.40
3210	24" O.C.		492	.016		.32	.76		1.08	1.60
3220	4" wide, 16" O.C.		317	.025		.52	1.19		1.71	2.51
3230	24" O.C.		492	.016		.38	.76		1.14	1.66
3240	6" wide, 16" O.C.		313	.026		.60	1.20		1.80	2.63
3250	24" O.C.		483	.017		.44	.78		1.22	1.75
5000	Load bearing studs, see Section 05 41 13.30									
9000	Minimum labor/equipment charge	1 Carp	4	2	Job		94		94	154

09 22 26 – Suspension Systems

09 22 26.13 Ceiling Suspension Systems

		Crew	Daily Output	Labor-Hours	Unit	Material	Labor	Equipment	Total	Total Incl O&P
0010	**CEILING SUSPENSION SYSTEMS** for gypsum board or plaster									
8000	Suspended ceilings, including carriers									
8200	1-1/2" carriers, 24" O.C. with:									
8300	7/8" channels, 16" O.C.	1 Lath	275	.029	S.F.	.54	1.25		1.79	2.56
8320	24" O.C.		310	.026		.42	1.11		1.53	2.22
8400	1-5/8" channels, 16" O.C.		205	.039		.64	1.68		2.32	3.35
8420	24" O.C.		250	.032		.50	1.38		1.88	2.72
8600	2" carriers, 24" O.C. with:									

09 22 Supports for Plaster and Gypsum Board

09 22 26 – Suspension Systems

09 22 26.13 Ceiling Suspension Systems		Crew	Daily Output	Labor-Hours	Unit	Material	2015 Bare Costs Labor	Equipment	Total	Total Incl O&P
8700	7/8" channels, 16" O.C.	1 Lath	250	.032	S.F.	.59	1.38		1.97	2.82
8720	24" O.C.		285	.028		.48	1.21		1.69	2.43
8800	1-5/8" channels, 16" O.C.		190	.042		.70	1.81		2.51	3.62
8820	24" O.C.		225	.036		.55	1.53		2.08	3.02

09 22 36 – Lath

09 22 36.13 Gypsum Lath

		Crew	Daily Output	Labor-Hours	Unit	Material	Labor	Equipment	Total	Total Incl O&P
0010	**GYPSUM LATH** R092000-50									
0020	Plain or perforated, nailed, 3/8" thick	1 Lath	85	.094	S.Y.	3.06	4.05		7.11	9.70
0100	1/2" thick		80	.100		2.43	4.30		6.73	9.40
0300	Clipped to steel studs, 3/8" thick		75	.107		3.06	4.59		7.65	10.55
0400	1/2" thick		70	.114		2.43	4.91		7.34	10.40
1500	For ceiling installations, add		216	.037			1.59		1.59	2.51
1600	For columns and beams, add		170	.047			2.02		2.02	3.19
9000	Minimum labor/equipment charge		4.25	1.882	Job		81		81	127

09 22 36.23 Metal Lath

		Crew	Daily Output	Labor-Hours	Unit	Material	Labor	Equipment	Total	Total Incl O&P
0010	**METAL LATH** R092000-50									
0020	Diamond, expanded, 2.5 lb. per S.Y., painted				S.Y.	3.61			3.61	3.97
0100	Galvanized					2.72			2.72	2.99
0300	3.4 lb. per S.Y., painted					4.08			4.08	4.49
0400	Galvanized					4.13			4.13	4.54
0600	For 15# asphalt sheathing paper, add					.49			.49	.53
0900	Flat rib, 1/8" high, 2.75 lb., painted					3.40			3.40	3.74
1000	Foil backed					3.58			3.58	3.94
1200	3.4 lb. per S.Y., painted					4.24			4.24	4.66
1300	Galvanized					4.35			4.35	4.79
1500	For 15# asphalt sheathing paper, add					.49			.49	.53
1800	High rib, 3/8" high, 3.4 lb. per S.Y., painted					4.12			4.12	4.53
1900	Galvanized					3.70			3.70	4.07
2400	3/4" high, painted, .60 lb. per S.F.				S.F.	.62			.62	.68
2500	.75 lb. per S.F.				"	1.33			1.33	1.46
2800	Stucco mesh, painted, 3.6 lb.				S.Y.	3.69			3.69	4.06
3000	K-lath, perforated, absorbent paper, regular					4.42			4.42	4.86
3100	Heavy duty					5.20			5.20	5.75
3300	Waterproof, heavy duty, grade B backing					5.10			5.10	5.60
3400	Fire resistant backing					5.65			5.65	6.20
3600	2.5 lb. diamond painted, on wood framing, on walls	1 Lath	85	.094		3.61	4.05		7.66	10.30
3700	On ceilings		75	.107		3.61	4.59		8.20	11.15
3900	3.4 lb. diamond painted, on wood framing, on walls		80	.100		4.24	4.30		8.54	11.40
4000	On ceilings		70	.114		4.24	4.91		9.15	12.40
4200	3.4 lb. diamond painted, wired to steel framing		75	.107		4.24	4.59		8.83	11.85
4300	On ceilings		60	.133		4.24	5.75		9.99	13.70
4600	Cornices, wired to steel		35	.229		4.24	9.85		14.09	20
4800	Screwed to steel studs, 2.5 lb.		80	.100		3.61	4.30		7.91	10.70
4900	3.4 lb.		75	.107		4.08	4.59		8.67	11.70
5100	Rib lath, painted, wired to steel, on walls, 2.5 lb.		75	.107		3.40	4.59		7.99	10.95
5200	3.4 lb.		70	.114		4.12	4.91		9.03	12.30
5400	4.0 lb.		65	.123		5.65	5.30		10.95	14.55
5500	For self-furring lath, add					.11			.11	.12
5700	Suspended ceiling system, incl. 3.4 lb. diamond lath, painted	1 Lath	15	.533		4.31	23		27.31	40.50
5800	Galvanized	"	15	.533		4.24	23		27.24	40.50
6000	Hollow metal stud partitions, 3.4 lb. painted lath both sides									
6010	Non-load bearing, 25 ga., w/rib lath 2-1/2" studs, 12" O.C.	1 Lath	20.30	.394	S.Y.	11.75	16.95		28.70	39.50

09 22 Supports for Plaster and Gypsum Board

09 22 36 – Lath

09 22 36.23 Metal Lath

		Crew	Daily Output	Labor-Hours	Unit	Material	2015 Bare Costs Labor	Equipment	Total	Total Incl O&P
6300	16" O.C.	1 Lath	21.10	.379	S.Y.	11	16.30		27.30	37.50
6350	24" O.C.		22.70	.352		10.30	15.15		25.45	35.50
6400	3-5/8" studs, 16" O.C.		19.50	.410		11.55	17.65		29.20	40.50
6600	24" O.C.		20.40	.392		10.65	16.85		27.50	38.50
6700	4" studs, 16" O.C.		20.40	.392		11.90	16.85		28.75	39.50
6900	24" O.C.		21.60	.370		10.95	15.95		26.90	37
7000	6" studs, 16" O.C.		19.50	.410		12.65	17.65		30.30	42
7100	24" O.C.		21.10	.379		11.50	16.30		27.80	38
7200	L.B. partitions, 16 ga., w/rib lath, 2-1/2" studs, 16" O.C.		20	.400		12.45	17.20		29.65	40.50
7300	3-5/8" studs, 16 ga.		19.70	.406		14.15	17.45		31.60	43
7500	4" studs, 16 ga.		19.50	.410		14.65	17.65		32.30	44
7600	6" studs, 16 ga.		18.70	.428		17.30	18.40		35.70	48
9000	Minimum labor/equipment charge		4.25	1.882	Job		81		81	127

09 23 Gypsum Plastering

09 23 13 – Acoustical Gypsum Plastering

09 23 13.10 Perlite or Vermiculite Plaster

			Crew	Daily Output	Labor-Hours	Unit	Material	Labor	Equipment	Total	Total Incl O&P
0010	**PERLITE OR VERMICULITE PLASTER**	R092000-50									
0020	In 100 lb. bags, under 200 bags					Bag	17.55			17.55	19.35
0300	2 coats, no lath included, on walls		J-1	92	.435	S.Y.	5.85	17.85	1.53	25.23	36.50
0400	On ceilings			79	.506		5.85	21	1.78	28.63	41.50
0900	3 coats, no lath included, on walls			74	.541		6.40	22	1.90	30.30	44.50
1000	On ceilings			63	.635		6.40	26	2.23	34.63	51
1700	For irregular or curved surfaces, add to above							30%			
1800	For columns and beams, add to above							50%			
1900	For soffits, add to ceiling prices							40%			
9000	Minimum labor/equipment charge		1 Plas	1	8	Job		345		345	550

09 23 20 – Gypsum Plaster

09 23 20.10 Gypsum Plaster On Walls and Ceilings

			Crew	Daily Output	Labor-Hours	Unit	Material	Labor	Equipment	Total	Total Incl O&P
0010	**GYPSUM PLASTER ON WALLS AND CEILINGS**	R092000-50									
0020	80# bag, less than 1 ton					Bag	15.95			15.95	17.55
0300	2 coats, no lath included, on walls		J-1	105	.381	S.Y.	3.61	15.60	1.34	20.55	30.50
0400	On ceilings			92	.435		3.61	17.85	1.53	22.99	34
0900	3 coats, no lath included, on walls			87	.460		5.20	18.85	1.61	25.66	37.50
1000	On ceilings			78	.513		5.20	21	1.80	28	41
1600	For irregular or curved surfaces, add							30%			
1800	For columns & beams, add							50%			
9000	Minimum labor/equipment charge		1 Plas	1	8	Job		345		345	550

09 24 Cement Plastering

09 24 23 – Cement Stucco

09 24 23.40 Stucco

		Crew	Daily Output	Labor-Hours	Unit	Material	2015 Bare Costs Labor	Equipment	Total	Total Incl O&P
0010	**STUCCO** R092000-50									
0015	3 coats 1" thick, float finish, with mesh, on wood frame	J-2	63	.762	S.Y.	6.25	31.50	2.22	39.97	59.50
0100	On masonry construction, no mesh incl.	J-1	67	.597		2.55	24.50	2.10	29.15	44
0300	For trowel finish, add	1 Plas	170	.047			2.02		2.02	3.23
0400	For 3/4" thick, on masonry, deduct	J-1	880	.045		.63	1.86	.16	2.65	3.85
0600	For coloring add		685	.058		.40	2.39	.20	2.99	4.50
0700	For special texture add	↓	200	.200		1.41	8.20	.70	10.31	15.40
0900	For soffits, add	J-2	155	.310		2.18	12.80	.90	15.88	24
1000	Exterior stucco, with bonding agent, 3 coats, on walls, no mesh incl.	J-1	200	.200		3.65	8.20	.70	12.55	17.90
1200	Ceilings		180	.222		3.65	9.10	.78	13.53	19.45
1300	Beams		80	.500		3.65	20.50	1.76	25.91	39
1500	Columns		100	.400		3.65	16.40	1.40	21.45	31.50
1550	Minimum labor/equipment charge	1 Plas	1	8	Job		345		345	550
1600	Mesh, painted, nailed to wood, 1.8 lb.	1 Lath	60	.133	S.Y.	6.20	5.75		11.95	15.85
1800	3.6 lb.		55	.145		3.69	6.25		9.94	13.90
1900	Wired to steel, painted, 1.8 lb.		53	.151		6.20	6.50		12.70	17
2100	3.6 lb.	↓	50	.160	↓	3.69	6.90		10.59	14.90
9000	Minimum labor/equipment charge		4	2	Job		86		86	135

09 25 Other Plastering

09 25 23 – Lime Based Plastering

09 25 23.10 Venetian Plaster

		Crew	Daily Output	Labor-Hours	Unit	Material	2015 Bare Costs Labor	Equipment	Total	Total Incl O&P
0010	**VENETIAN PLASTER**									
0100	Walls, 1 coat primer, roller applied	1 Plas	950	.008	S.F.	.16	.36		.52	.76
0200	Plaster, 3 coats, incl. sanding	2 Plas	700	.023	"	.46	.98		1.44	2.07
0210	For pigment, light colors add per ea.				Ea.	3			3	3.30
0220	For pigment, dark colors add per ea.				"	9			9	9.90
0300	For sealer/wax coat incl. burnishing, add	1 Plas	300	.027	S.F.	.41	1.15		1.56	2.28

09 26 Veneer Plastering

09 26 13 – Gypsum Veneer Plastering

09 26 13.20 Blueboard

		Crew	Daily Output	Labor-Hours	Unit	Material	2015 Bare Costs Labor	Equipment	Total	Total Incl O&P
0010	**BLUEBOARD** For use with thin coat									
0100	plaster application see Section 09 26 13.80									
1000	3/8" thick, on walls or ceilings, standard, no finish included	2 Carp	1900	.008	S.F.	.34	.40		.74	1.02
1100	With thin coat plaster finish		875	.018		.45	.86		1.31	1.91
1400	On beams, columns, or soffits, standard, no finish included		675	.024		.39	1.11		1.50	2.25
1450	With thin coat plaster finish		475	.034		.50	1.58		2.08	3.14
3000	1/2" thick, on walls or ceilings, standard, no finish included		1900	.008		.33	.40		.73	1.01
3100	With thin coat plaster finish		875	.018		.44	.86		1.30	1.90
3300	Fire resistant, no finish included		1900	.008		.33	.40		.73	1.01
3400	With thin coat plaster finish		875	.018		.44	.86		1.30	1.90
3450	On beams, columns, or soffits, standard, no finish included		675	.024		.38	1.11		1.49	2.24
3500	With thin coat plaster finish		475	.034		.49	1.58		2.07	3.13
3700	Fire resistant, no finish included		675	.024		.38	1.11		1.49	2.24
3800	With thin coat plaster finish		475	.034		.49	1.58		2.07	3.13
5000	5/8" thick, on walls or ceilings, fire resistant, no finish included		1900	.008		.34	.40		.74	1.02
5100	With thin coat plaster finish		875	.018		.45	.86		1.31	1.91
5500	On beams, columns, or soffits, no finish included		675	.024		.39	1.11		1.50	2.25

09 26 Veneer Plastering

09 26 13 – Gypsum Veneer Plastering

09 26 13.20 Blueboard

		Crew	Daily Output	Labor-Hours	Unit	Material	2015 Bare Costs Labor	2015 Bare Costs Equipment	Total	Total Incl O&P
5600	With thin coat plaster finish	2 Carp	475	.034	S.F.	.50	1.58		2.08	3.14
6000	For high ceilings, over 8' high, add		3060	.005			.25		.25	.40
6500	For over 3 stories high, add per story		6100	.003			.12		.12	.20
9000	Minimum labor/equipment charge	1 Carp	2	4	Job		188		188	310

09 26 13.80 Thin Coat Plaster

		Crew	Daily Output	Labor-Hours	Unit	Material	Labor	Equipment	Total	Total Incl O&P
0010	**THIN COAT PLASTER** R092000-50									
0012	1 coat veneer, not incl. lath	J-1	3600	.011	S.F.	.11	.46	.04	.61	.89
1000	In 50 lb. bags				Bag	15.25			15.25	16.80

09 28 Backing Boards and Underlayments

09 28 13 – Cementitious Backing Boards

09 28 13.10 Cementitious Backerboard

		Crew	Daily Output	Labor-Hours	Unit	Material	Labor	Equipment	Total	Total Incl O&P
0010	**CEMENTITIOUS BACKERBOARD**									
0070	Cementitious backerboard, on floor, 3' x 4' x 1/2" sheets	2 Carp	525	.030	S.F.	.78	1.43		2.21	3.21
0080	3' x 5' x 1/2" sheets		525	.030		.76	1.43		2.19	3.19
0090	3' x 6' x 1/2" sheets		525	.030		.74	1.43		2.17	3.16
0100	3' x 4' x 5/8" sheets		525	.030		.99	1.43		2.42	3.44
0110	3' x 5' x 5/8" sheets		525	.030		.99	1.43		2.42	3.44
0120	3' x 6' x 5/8" sheets		525	.030		.96	1.43		2.39	3.41
0150	On wall, 3' x 4' x 1/2" sheets		350	.046		.78	2.15		2.93	4.38
0160	3' x 5' x 1/2" sheets		350	.046		.76	2.15		2.91	4.36
0170	3' x 6' x 1/2" sheets		350	.046		.74	2.15		2.89	4.33
0180	3' x 4' x 5/8" sheets		350	.046		.99	2.15		3.14	4.61
0190	3' x 5' x 5/8" sheets		350	.046		.99	2.15		3.14	4.61
0200	3' x 6' x 5/8" sheets		350	.046		.96	2.15		3.11	4.58
0250	On counter, 3' x 4' x 1/2" sheets		180	.089		.78	4.17		4.95	7.70
0260	3' x 5' x 1/2" sheets		180	.089		.76	4.17		4.93	7.70
0270	3' x 6' x 1/2" sheets		180	.089		.74	4.17		4.91	7.65
0300	3' x 4' x 5/8" sheets		180	.089		.99	4.17		5.16	7.95
0310	3' x 5' x 5/8" sheets		180	.089		.99	4.17		5.16	7.95
0320	3' x 6' x 5/8" sheets		180	.089		.96	4.17		5.13	7.90

09 29 Gypsum Board

09 29 10 – Gypsum Board Panels

09 29 10.30 Gypsum Board

		Crew	Daily Output	Labor-Hours	Unit	Material	Labor	Equipment	Total	Total Incl O&P
0010	**GYPSUM BOARD** on walls & ceilings R092910-10									
0100	Nailed or screwed to studs unless otherwise noted									
0110	1/4" thick, on walls or ceilings, standard, no finish included	2 Carp	1330	.012	S.F.	.35	.56		.91	1.32
0115	1/4" thick, on walls or ceilings, flexible, no finish included		1050	.015		.50	.72		1.22	1.72
0117	1/4" thick, on columns or soffits, flexible, no finish included		1050	.015		.50	.72		1.22	1.72
0130	1/4" thick, standard, no finish included, less than 800 S.F.		510	.031		.35	1.47		1.82	2.80
0150	3/8" thick, on walls, standard, no finish included		2000	.008		.34	.38		.72	.99
0200	On ceilings, standard, no finish included		1800	.009		.34	.42		.76	1.05
0250	On beams, columns, or soffits, no finish included		675	.024		.34	1.11		1.45	2.19
0300	1/2" thick, on walls, standard, no finish included		2000	.008		.30	.38		.68	.95
0350	Taped and finished (level 4 finish)		965	.017		.35	.78		1.13	1.66
0390	With compound skim coat (level 5 finish)		775	.021		.40	.97		1.37	2.03
0400	Fire resistant, no finish included		2000	.008		.35	.38		.73	1.01
0450	Taped and finished (level 4 finish)		965	.017		.40	.78		1.18	1.72

09 29 Gypsum Board

09 29 10 – Gypsum Board Panels

09 29 10.30 Gypsum Board

		Crew	Daily Output	Labor-Hours	Unit	Material	2015 Bare Costs Labor	Equipment	Total	Total Incl O&P
0490	With compound skim coat (level 5 finish)	2 Carp	775	.021	S.F.	.45	.97		1.42	2.09
0500	Water resistant, no finish included		2000	.008		.39	.38		.77	1.05
0550	Taped and finished (level 4 finish)		965	.017		.44	.78		1.22	1.76
0590	With compound skim coat (level 5 finish)		775	.021		.49	.97		1.46	2.13
0600	Prefinished, vinyl, clipped to studs		900	.018		.48	.83		1.31	1.90
0700	Mold resistant, no finish included		2000	.008		.44	.38		.82	1.10
0710	Taped and finished (level 4 finish)		965	.017		.49	.78		1.27	1.82
0720	With compound skim coat (level 5 finish)		775	.021		.54	.97		1.51	2.18
1000	On ceilings, standard, no finish included		1800	.009		.30	.42		.72	1.01
1050	Taped and finished (level 4 finish)		765	.021		.35	.98		1.33	1.99
1090	With compound skim coat (level 5 finish)		610	.026		.40	1.23		1.63	2.46
1100	Fire resistant, no finish included		1800	.009		.35	.42		.77	1.07
1150	Taped and finished (level 4 finish)		765	.021		.40	.98		1.38	2.05
1195	With compound skim coat (level 5 finish)		610	.026		.45	1.23		1.68	2.52
1200	Water resistant, no finish included		1800	.009		.39	.42		.81	1.11
1250	Taped and finished (level 4 finish)		765	.021		.44	.98		1.42	2.09
1290	With compound skim coat (level 5 finish)		610	.026		.49	1.23		1.72	2.56
1310	Mold resistant, no finish included		1800	.009		.44	.42		.86	1.16
1320	Taped and finished (level 4 finish)		765	.021		.49	.98		1.47	2.15
1330	With compound skim coat (level 5 finish)		610	.026		.54	1.23		1.77	2.61
1350	Sag resistant, no finish included		1600	.010		.34	.47		.81	1.14
1360	Taped and finished (level 4 finish)		765	.021		.39	.98		1.37	2.04
1370	With compound skim coat (level 5 finish)		610	.026		.44	1.23		1.67	2.50
1500	On beams, columns, or soffits, standard, no finish included		675	.024		.35	1.11		1.46	2.20
1550	Taped and finished (level 4 finish)		540	.030		.35	1.39		1.74	2.66
1590	With compound skim coat (level 5 finish)		475	.034		.40	1.58		1.98	3.03
1600	Fire resistant, no finish included		675	.024		.35	1.11		1.46	2.21
1650	Taped and finished (level 4 finish)		540	.030		.40	1.39		1.79	2.72
1690	With compound skim coat (level 5 finish)		475	.034		.45	1.58		2.03	3.09
1700	Water resistant, no finish included		675	.024		.45	1.11		1.56	2.31
1750	Taped and finished (level 4 finish)		540	.030		.44	1.39		1.83	2.76
1790	With compound skim coat (level 5 finish)		475	.034		.49	1.58		2.07	3.13
1800	Mold resistant, no finish included		675	.024		.51	1.11		1.62	2.38
1810	Taped and finished (level 4 finish)		540	.030		.49	1.39		1.88	2.82
1820	With compound skim coat (level 5 finish)		475	.034		.54	1.58		2.12	3.18
1850	Sag resistant, no finish included		675	.024		.39	1.11		1.50	2.25
1860	Taped and finished (level 4 finish)		540	.030		.39	1.39		1.78	2.71
1870	With compound skim coat (level 5 finish)		475	.034		.44	1.58		2.02	3.07
2000	5/8" thick, on walls, standard, no finish included		2000	.008		.33	.38		.71	.98
2050	Taped and finished (level 4 finish)		965	.017		.38	.78		1.16	1.69
2090	With compound skim coat (level 5 finish)		775	.021		.43	.97		1.40	2.06
2100	Fire resistant, no finish included		2000	.008		.34	.38		.72	.99
2150	Taped and finished (level 4 finish)		965	.017		.39	.78		1.17	1.71
2195	With compound skim coat (level 5 finish)		775	.021		.44	.97		1.41	2.07
2200	Water resistant, no finish included		2000	.008		.42	.38		.80	1.08
2250	Taped and finished (level 4 finish)		965	.017		.47	.78		1.25	1.79
2290	With compound skim coat (level 5 finish)		775	.021		.52	.97		1.49	2.16
2300	Prefinished, vinyl, clipped to studs		900	.018		.76	.83		1.59	2.21
2510	Mold resistant, no finish included		2000	.008		.46	.38		.84	1.13
2520	Taped and finished (level 4 finish)		965	.017		.51	.78		1.29	1.84
2530	With compound skim coat (level 5 finish)		775	.021		.56	.97		1.53	2.21
3000	On ceilings, standard, no finish included		1800	.009		.33	.42		.75	1.04
3050	Taped and finished (level 4 finish)		765	.021		.38	.98		1.36	2.02

09 29 Gypsum Board

09 29 10 – Gypsum Board Panels

09 29 10.30 Gypsum Board

		Crew	Daily Output	Labor-Hours	Unit	Material	2015 Bare Costs Labor	Equipment	Total	Total Incl O&P
3090	With compound skim coat (level 5 finish)	2 Carp	615	.026	S.F.	.43	1.22		1.65	2.47
3100	Fire resistant, no finish included		1800	.009		.34	.42		.76	1.05
3150	Taped and finished (level 4 finish)		765	.021		.39	.98		1.37	2.04
3190	With compound skim coat (level 5 finish)		615	.026		.44	1.22		1.66	2.48
3200	Water resistant, no finish included		1800	.009		.42	.42		.84	1.14
3250	Taped and finished (level 4 finish)		765	.021		.47	.98		1.45	2.12
3290	With compound skim coat (level 5 finish)		615	.026		.52	1.22		1.74	2.57
3300	Mold resistant, no finish included		1800	.009		.46	.42		.88	1.19
3310	Taped and finished (level 4 finish)		765	.021		.51	.98		1.49	2.17
3320	With compound skim coat (level 5 finish)		615	.026		.56	1.22		1.78	2.62
3500	On beams, columns, or soffits, no finish included		675	.024		.38	1.11		1.49	2.24
3550	Taped and finished (level 4 finish)		475	.034		.43	1.58		2.01	3.07
3590	With compound skim coat (level 5 finish)		380	.042		.49	1.98		2.47	3.78
3600	Fire resistant, no finish included		675	.024		.39	1.11		1.50	2.25
3650	Taped and finished (level 4 finish)		475	.034		.45	1.58		2.03	3.08
3690	With compound skim coat (level 5 finish)		380	.042		.44	1.98		2.42	3.72
3700	Water resistant, no finish included		675	.024		.48	1.11		1.59	2.35
3750	Taped and finished (level 4 finish)		475	.034		.52	1.58		2.10	3.16
3790	With compound skim coat (level 5 finish)		380	.042		.54	1.98		2.52	3.83
3800	Mold resistant, no finish included		675	.024		.53	1.11		1.64	2.40
3810	Taped and finished (level 4 finish)		475	.034		.56	1.58		2.14	3.21
3820	With compound skim coat (level 5 finish)		380	.042		.58	1.98		2.56	3.88
4000	Fireproofing, beams or columns, 2 layers, 1/2" thick, incl finish		330	.048		.79	2.28		3.07	4.60
4010	Mold resistant		330	.048		.97	2.28		3.25	4.80
4050	5/8" thick		300	.053		.77	2.50		3.27	4.95
4060	Mold resistant		300	.053		1.01	2.50		3.51	5.20
4100	3 layers, 1/2" thick		225	.071		1.19	3.34		4.53	6.75
4110	Mold resistant		225	.071		1.46	3.34		4.80	7.05
4150	5/8" thick		210	.076		1.16	3.58		4.74	7.15
4160	Mold resistant		210	.076		1.52	3.58		5.10	7.50
5050	For 1" thick coreboard on columns		480	.033		.78	1.56		2.34	3.42
5100	For foil-backed board, add					.15			.15	.17
5200	For work over 8' high, add	2 Carp	3060	.005			.25		.25	.40
5270	For textured spray, add	2 Lath	1600	.010		.04	.43		.47	.72
5300	For distribution cost over 3 stories high, add per story	2 Carp	6100	.003			.12		.12	.20
5350	For finishing inner corners, add		950	.017	L.F.	.10	.79		.89	1.41
5355	For finishing outer corners, add		1250	.013		.22	.60		.82	1.24
5500	For acoustical sealant, add per bead	1 Carp	500	.016		.04	.75		.79	1.28
5550	Sealant, 1 quart tube				Ea.	7.05			7.05	7.80
6000	Gypsum sound dampening panels									
6010	1/2" thick on walls, multi-layer, light weight, no finish included	2 Carp	1500	.011	S.F.	1.87	.50		2.37	2.88
6015	Taped and finished (level 4 finish)		725	.022		1.92	1.04		2.96	3.81
6020	With compound skim coat (level 5 finish)		580	.028		1.97	1.30		3.27	4.29
6025	5/8" thick on walls, for wood studs, no finish included		1500	.011		2.17	.50		2.67	3.21
6030	Taped and finished (level 4 finish)		725	.022		2.22	1.04		3.26	4.14
6035	With compound skim coat (level 5 finish)		580	.028		2.27	1.30		3.57	4.62
6040	For metal stud, no finish included		1500	.011		2.06	.50		2.56	3.09
6045	Taped and finished (level 4 finish)		725	.022		2.11	1.04		3.15	4.02
6050	With compound skim coat (level 5 finish)		580	.028		2.16	1.30		3.46	4.50
6055	Abuse resist, no finish included		1500	.011		3.75	.50		4.25	4.95
6060	Taped and finished (level 4 finish)		725	.022		3.80	1.04		4.84	5.90
6065	With compound skim coat (level 5 finish)		580	.028		3.85	1.30		5.15	6.35
6070	Shear rated, no finish included		1500	.011		4.30	.50		4.80	5.55

09 29 Gypsum Board

09 29 10 – Gypsum Board Panels

09 29 10.30 Gypsum Board

		Crew	Daily Output	Labor-Hours	Unit	Material	2015 Bare Costs Labor	Equipment	Total	Total Incl O&P
6075	Taped and finished (level 4 finish)	2 Carp	725	.022	S.F.	4.35	1.04		5.39	6.50
6080	With compound skim coat (level 5 finish)		580	.028		4.40	1.30		5.70	6.95
6085	For SCIF applications, no finish included		1500	.011		4.72	.50		5.22	6
6090	Taped and finished (level 4 finish)		725	.022		4.77	1.04		5.81	6.95
6095	With compound skim coat (level 5 finish)		580	.028		4.82	1.30		6.12	7.40
6100	1-3/8" thick on walls, THX Certified, no finish included		1500	.011		8.40	.50		8.90	10.05
6105	Taped and finished (level 4 finish)		725	.022		8.45	1.04		9.49	11
6110	With compound skim coat (level 5 finish)		580	.028		8.50	1.30		9.80	11.45
6115	5/8" thick on walls, score & snap installation, no finish included		2000	.008		1.69	.38		2.07	2.48
6120	Taped and finished (level 4 finish)		965	.017		1.74	.78		2.52	3.19
6125	With compound skim coat (level 5 finish)		775	.021		1.79	.97		2.76	3.56
7020	5/8" thick on ceilings, for wood joists, no finish included		1200	.013		2.17	.63		2.80	3.42
7025	Taped and finished (level 4 finish)		510	.031		2.22	1.47		3.69	4.85
7030	With compound skim coat (level 5 finish)		410	.039		2.27	1.83		4.10	5.50
7035	For metal joists, no finish included		1200	.013		2.06	.63		2.69	3.30
7040	Taped and finished (level 4 finish)		510	.031		2.11	1.47		3.58	4.73
7045	With compound skim coat (level 5 finish)		410	.039		2.16	1.83		3.99	5.40
7050	Abuse resist, no finish included		1200	.013		3.75	.63		4.38	5.15
7055	Taped and finished (level 4 finish)		510	.031		3.80	1.47		5.27	6.60
7060	With compound skim coat (level 5 finish)		410	.039		3.85	1.83		5.68	7.25
7065	Shear rated, no finish included		1200	.013		4.30	.63		4.93	5.75
7070	Taped and finished (level 4 finish)		510	.031		4.35	1.47		5.82	7.20
7075	With compound skim coat (level 5 finish)		410	.039		4.40	1.83		6.23	7.85
7080	For SCIF applications, no finish included		1200	.013		4.72	.63		5.35	6.25
7085	Taped and finished (level 4 finish)		510	.031		4.77	1.47		6.24	7.65
7090	With compound skim coat (level 5 finish)		410	.039		4.82	1.83		6.65	8.30
8010	5/8" thick on ceilings, score & snap installation, no finish included		1600	.010		1.69	.47		2.16	2.63
8015	Taped and finished (level 4 finish)		680	.024		1.74	1.10		2.84	3.72
8020	With compound skim coat (level 5 finish)		545	.029		1.79	1.38		3.17	4.23
9000	Minimum labor/equipment charge	1 Carp	2	4	Job		188		188	310

09 29 15 – Gypsum Board Accessories

09 29 15.10 Accessories, Gypsum Board

		Crew	Daily Output	Labor-Hours	Unit	Material	2015 Bare Costs Labor	Equipment	Total	Total Incl O&P
0010	**ACCESSORIES, GYPSUM BOARD**									
0020	Casing bead, galvanized steel	1 Carp	2.90	2.759	C.L.F.	24	130		154	239
0100	Vinyl		3	2.667		22	125		147	230
0300	Corner bead, galvanized steel, 1" x 1"		4	2		14.70	94		108.70	170
0400	1-1/4" x 1-1/4"		3.50	2.286		16.15	107		123.15	194
0600	Vinyl		4	2		20	94		114	176
0900	Furring channel, galv. steel, 7/8" deep, standard		2.60	3.077		33.50	144		177.50	274
1000	Resilient		2.55	3.137		25.50	147		172.50	269
1100	J trim, galvanized steel, 1/2" wide		3	2.667		22	125		147	229
1120	5/8" wide		2.95	2.712		31	127		158	243
1500	Z stud, galvanized steel, 1-1/2" wide		2.60	3.077		38.50	144		182.50	280
9000	Minimum labor/equipment charge		3	2.667	Job		125		125	205

09 30 Tiling

09 30 13 – Ceramic Tiling

09 30 13.10 Ceramic Tile	Crew	Daily Output	Labor-Hours	Unit	Material	2015 Bare Costs Labor	Equipment	Total	Total Incl O&P
0010 **CERAMIC TILE**									
0600 Cove base, 4-1/4" x 4-1/4" high, mud set	D-7	91	.176	L.F.	3.99	6.70		10.69	15
0700 Thin set		128	.125		3.87	4.77		8.64	11.80
0900 6" x 4-1/4" high, mud set		100	.160		4.54	6.10		10.64	14.65
1000 Thin set		137	.117		4.42	4.46		8.88	11.90
1200 Sanitary cove base, 6" x 4-1/4" high, mud set		93	.172		4.36	6.55		10.91	15.15
1300 Thin set		124	.129		4.24	4.93		9.17	12.45
1500 6" x 6" high, mud set		84	.190		5.35	7.25		12.60	17.40
1600 Thin set		117	.137		5.20	5.20		10.40	14
2400 Bullnose trim, 4-1/4" x 4-1/4", mud set		82	.195		3.92	7.45		11.37	16.05
2500 Thin set		128	.125		3.84	4.77		8.61	11.75
2700 2" x 6" bullnose trim, mud set		84	.190		4.05	7.25		11.30	15.95
2800 Thin set		124	.129		3.99	4.93		8.92	12.20
3000 Floors, natural clay, random or uniform, thin set, color group 1		183	.087	S.F.	4.15	3.34		7.49	9.80
3100 Color group 2		183	.087		5.85	3.34		9.19	11.70
3300 Porcelain type, 1 color, color group 2, 1" x 1"		183	.087		5.20	3.34		8.54	10.95
3310 2" x 2" or 2" x 1", thin set		190	.084		6.10	3.22		9.32	11.80
3350 For random blend, 2 colors, add					1			1	1.10
3360 4 colors, add					1.50			1.50	1.65
4300 Specialty tile, 4-1/4" x 4-1/4" x 1/2", decorator finish	D-7	183	.087		10.40	3.34		13.74	16.70
4500 Add for epoxy grout, 1/16" joint, 1" x 1" tile		800	.020		.67	.76		1.43	1.95
4600 2" x 2" tile		820	.020		.62	.74		1.36	1.86
4610 Add for epoxy grout, 1/8" joint, 8" x 8" x 3/8" tile, add		900	.018		1.44	.68		2.12	2.66
4800 Pregrouted sheets, walls, 4-1/4" x 4-1/4", 6" x 4-1/4"									
4810 and 8-1/2" x 4-1/4", 4 S.F. sheets, silicone grout	D-7	240	.067	S.F.	5.10	2.55		7.65	9.65
5100 Floors, unglazed, 2 S.F. sheets,									
5110 Urethane adhesive	D-7	180	.089	S.F.	5.10	3.39		8.49	10.95
5400 Walls, interior, thin set, 4-1/4" x 4-1/4" tile		190	.084		2.26	3.22		5.48	7.60
5500 6" x 4-1/4" tile		190	.084		2.92	3.22		6.14	8.30
5700 8-1/2" x 4-1/4" tile		190	.084		4.86	3.22		8.08	10.45
5800 6" x 6" tile		175	.091		3.28	3.49		6.77	9.10
5810 8" x 8" tile		170	.094		4.44	3.59		8.03	10.60
5820 12" x 12" tile		160	.100		4.35	3.82		8.17	10.85
5830 16" x 16" tile		150	.107		4.77	4.07		8.84	11.70
6000 Decorated wall tile, 4-1/4" x 4-1/4", color group 1		270	.059		3.18	2.26		5.44	7.05
6100 Color group 4		180	.089		49.50	3.39		52.89	60
6600 Crystalline glazed, 4-1/4" x 4-1/4", mud set, plain		100	.160		4.36	6.10		10.46	14.45
6700 4-1/4" x 4-1/4", scored tile		100	.160		5.80	6.10		11.90	16.05
6900 6" x 6" plain		93	.172		6.70	6.55		13.25	17.70
7000 For epoxy grout, 1/16" joints, 4-1/4" tile, add		800	.020		.41	.76		1.17	1.66
7200 For tile set in dry mortar, add		1735	.009			.35		.35	.56
7300 For tile set in Portland cement mortar, add		290	.055		.16	2.11		2.27	3.51
9300 Ceramic tiles, recycled glass, standard colors, 2" x 2" thru 6" x 6" [G]		190	.084		21	3.22		24.22	28.50
9310 6" x 6" [G]		175	.091		21.50	3.49		24.99	29
9320 8" x 8" [G]		170	.094		22.50	3.59		26.09	30.50
9330 12" x 12" [G]		160	.100		22.50	3.82		26.32	31
9340 Earthtones, 2" x 2" to 4" x 8" [G]		190	.084		25	3.22		28.22	32.50
9350 6" x 6" [G]		175	.091		25	3.49		28.49	33
9360 8" x 8" [G]		170	.094		26	3.59		29.59	34
9370 12" x 12" [G]		160	.100		26	3.82		29.82	34.50
9380 Deep colors, 2" x 2" to 4" x 8" [G]		190	.084		29.50	3.22		32.72	37.50
9390 6" x 6" [G]		175	.091		29.50	3.49		32.99	38
9400 8" x 8" [G]		170	.094		31	3.59		34.59	39.50

For customer support on your Commercial Renovation Cost Data, call 877.791.4977.

09 30 Tiling

09 30 13 – Ceramic Tiling

09 30 13.10 Ceramic Tile

		Crew	Daily Output	Labor-Hours	Unit	Material	2015 Bare Costs Labor	Equipment	Total	Total Incl O&P
9410	12" x 12" G	D-7	160	.100	S.F.	31	3.82		34.82	40
9500	Minimum labor/equipment charge		3.25	4.923	Job		188		188	297

09 30 13.20 Ceramic Tile Repairs

		Crew	Daily Output	Labor-Hours	Unit	Material	Labor	Equipment	Total	Total Incl O&P
0010	**CERAMIC TILE REPAIRS**									
1000	Grout removal, carbide tipped, rotary grinder	1 Clab	240	.033	L.F.		1.25		1.25	2.05

09 30 13.45 Ceramic Tile Accessories

		Crew	Daily Output	Labor-Hours	Unit	Material	Labor	Equipment	Total	Total Incl O&P
0010	**CERAMIC TILE ACCESSORIES**									
0100	Spacers, 1/8"				C	1.98			1.98	2.18
1310	Sealer for natural stone tile, installed	1 Tilf	650	.012	S.F.	.05	.53		.58	.89

09 30 16 – Quarry Tiling

09 30 16.10 Quarry Tile

		Crew	Daily Output	Labor-Hours	Unit	Material	Labor	Equipment	Total	Total Incl O&P
0010	**QUARRY TILE**									
0100	Base, cove or sanitary, mud set, to 5" high, 1/2" thick	D-7	110	.145	L.F.	5.35	5.55		10.90	14.65
0300	Bullnose trim, red, mud set, 6" x 6" x 1/2" thick		120	.133		4.39	5.10		9.49	12.90
0400	4" x 4" x 1/2" thick		110	.145		4.50	5.55		10.05	13.70
0600	4" x 8" x 1/2" thick, using 8" as edge		130	.123		4.50	4.70		9.20	12.35
0700	Floors, mud set, 1,000 S.F. lots, red, 4" x 4" x 1/2" thick		120	.133	S.F.	8.05	5.10		13.15	16.90
0900	6" x 6" x 1/2" thick		140	.114		7.55	4.36		11.91	15.20
1000	4" x 8" x 1/2" thick		130	.123		5.55	4.70		10.25	13.50
1300	For waxed coating, add					.75			.75	.83
1500	For non-standard colors, add					.46			.46	.51
1600	For abrasive surface, add					.52			.52	.57
1800	Brown tile, imported, 6" x 6" x 3/4"	D-7	120	.133		7.95	5.10		13.05	16.80
1900	8" x 8" x 1"		110	.145		8.70	5.55		14.25	18.30
2100	For thin set mortar application, deduct		700	.023			.87		.87	1.38
2700	Stair tread, 6" x 6" x 3/4", plain		50	.320		7.20	12.20		19.40	27
2800	Abrasive		47	.340		6.05	13		19.05	27
3000	Wainscot, 6" x 6" x 1/2", thin set, red		105	.152		4.68	5.80		10.48	14.35
3100	Non-standard colors		105	.152		5.20	5.80		11	14.90
3300	Window sill, 6" wide, 3/4" thick		90	.178	L.F.	9.20	6.80		16	21
3400	Corners		80	.200	Ea.	6.65	7.65		14.30	19.35
9000	Minimum labor/equipment charge		3.25	4.923	Job		188		188	297

09 30 23 – Glass Mosaic Tiling

09 30 23.10 Glass Mosaics

		Crew	Daily Output	Labor-Hours	Unit	Material	Labor	Equipment	Total	Total Incl O&P
0010	**GLASS MOSAICS** 3/4" tile on 12" sheets, standard grout									
1020	1" tile on 12" sheets, opalescent finish	D-7	73	.219	S.F.	16.85	8.35		25.20	32
1040	1" x 2" tile on 12" sheet, blend		73	.219		18.15	8.35		26.50	33
1060	2" tile on 12" sheet, blend		73	.219		16.50	8.35		24.85	31.50
1080	5/8" x random tile, linear, on 12" sheet, blend		73	.219		26	8.35		34.35	41.50
1600	Dots on 12" sheet		73	.219		26	8.35		34.35	41.50
1700	For glass mosaic tiles set in dry mortar, add		290	.055		.45	2.11		2.56	3.83
1720	For glass mosaic tile set in Portland cement mortar, add		290	.055		.01	2.11		2.12	3.34
1730	For polyblend sanded tile grout		96.15	.166	Lb.	2.19	6.35		8.54	12.45

09 30 29 – Metal Tiling

09 30 29.10 Metal Tile

		Crew	Daily Output	Labor-Hours	Unit	Material	Labor	Equipment	Total	Total Incl O&P
0010	**METAL TILE** 4' x 4' sheet, 24 ga., tile pattern, nailed									
0200	Stainless steel	2 Carp	512	.031	S.F.	28	1.47		29.47	33.50
0400	Aluminized steel	"	512	.031	"	15.10	1.47		16.57	19
9000	Minimum labor/equipment charge	1 Carp	4	2	Job		94		94	154

09 34 Waterproofing-Membrane Tiling

09 34 13 – Waterproofing-Membrane Ceramic Tiling

09 34 13.10 Ceramic Tile Waterproofing Membrane	Crew	Daily Output	Labor-Hours	Unit	Material	2015 Bare Costs Labor	Equipment	Total	Total Incl O&P
0010 **CERAMIC TILE WATERPROOFING MEMBRANE**									
0020 On floors, including thinset									
0030 Fleece laminated polyethylene grid, 1/8" thick	D-7	250	.064	S.F.	2.26	2.44		4.70	6.35
0040 5/16" thick	"	250	.064	"	2.58	2.44		5.02	6.70
0050 On walls, including thinset									
0060 Fleece laminated polyethylene sheet, 8 mil thick	D-7	480	.033	S.F.	2.26	1.27		3.53	4.50
0070 Accessories, including thinset									
0080 Joint and corner sheet, 4 mils thick, 5" wide	1 Tilf	240	.033	L.F.	1.33	1.43		2.76	3.72
0090 7-1/4" wide		180	.044		1.69	1.90		3.59	4.86
0100 10" wide		120	.067		2.06	2.85		4.91	6.80
0110 Pre-formed corners, inside		32	.250	Ea.	6.90	10.70		17.60	24.50
0120 Outside		32	.250		7.65	10.70		18.35	25.50
0130 2" flanged floor drain with 6" stainless steel grate		16	.500		370	21.50		391.50	445
0140 EPS, sloped shower floor		480	.017	S.F.	4.95	.71		5.66	6.60
0150 Curb		32	.250	L.F.	14	10.70		24.70	32.50

09 51 Acoustical Ceilings

09 51 23 – Acoustical Tile Ceilings

09 51 23.10 Suspended Acoustic Ceiling Tiles

	Crew	Daily Output	Labor-Hours	Unit	Material	Labor	Equipment	Total	Total Incl O&P
0010 **SUSPENDED ACOUSTIC CEILING TILES**, not including									
0100 suspension system									
0300 Fiberglass boards, film faced, 2' x 2' or 2' x 4', 5/8" thick	1 Carp	625	.013	S.F.	1.24	.60		1.84	2.35
0400 3/4" thick		600	.013		2.63	.63		3.26	3.92
0500 3" thick, thermal, R11		450	.018		2.42	.83		3.25	4.03
0600 Glass cloth faced fiberglass, 3/4" thick		500	.016		2.65	.75		3.40	4.15
0700 1" thick		485	.016		3.13	.77		3.90	4.71
0820 1-1/2" thick, nubby face		475	.017		2.53	.79		3.32	4.08
1110 Mineral fiber tile, lay-in, 2' x 2' or 2' x 4', 5/8" thick, fine texture		625	.013		.91	.60		1.51	1.99
1115 Rough textured		625	.013		.85	.60		1.45	1.93
1125 3/4" thick, fine textured		600	.013		1.94	.63		2.57	3.16
1130 Rough textured		600	.013		1.56	.63		2.19	2.75
1135 Fissured		600	.013		1.96	.63		2.59	3.19
1150 Tegular, 5/8" thick, fine textured		470	.017		1.03	.80		1.83	2.44
1155 Rough textured		470	.017		1.14	.80		1.94	2.56
1165 3/4" thick, fine textured		450	.018		2.14	.83		2.97	3.72
1170 Rough textured		450	.018		1.43	.83		2.26	2.94
1175 Fissured		450	.018		2.16	.83		2.99	3.75
1185 For plastic film face, add					.75			.75	.83
1190 For fire rating, add					.44			.44	.48
1300 Metal panel, lay-in, 2' x 2', sq. edge	1 Carp	500	.016		9.55	.75		10.30	11.75
1350 Tegular edge		500	.016		13.20	.75		13.95	15.75
1400 2' x 4', sq. edge		500	.016		12.90	.75		13.65	15.45
1450 Tegular edge		500	.016		13.20	.75		13.95	15.75
1500 Perforated alum. clip-in, 2' x 2'		500	.016		13.45	.75		14.20	16
1550 2' x 4'		500	.016		10.85	.75		11.60	13.20
1600 Solid alum. planks, 3-1/4"x12', open reveal		500	.016		2.35	.75		3.10	3.82
1650 Closed reveal		500	.016		3	.75		3.75	4.53
1700 7-1/4"x12', open reveal		500	.016		4	.75		4.75	5.65
1750 Closed reveal		500	.016		5.10	.75		5.85	6.85
1775 Metal, open cell, 2'x2', 6" cell		500	.016		8	.75		8.75	10.05
1800 8" cell		500	.016		8.85	.75		9.60	11

09 51 Acoustical Ceilings

09 51 23 – Acoustical Tile Ceilings

09 51 23.10 Suspended Acoustic Ceiling Tiles

		Crew	Daily Output	Labor-Hours	Unit	Material	2015 Bare Costs Labor	Equipment	Total	Total Incl O&P
1825	2'x4', 6" cell	1 Carp	500	.016	S.F.	5.10	.75		5.85	6.85
1850	8" cell		500	.016		5.10	.75		5.85	6.85
1870	Translucent lay-in panels, 2'x2'		500	.016		23	.75		23.75	26
1890	2'x6'		500	.016		17.20	.75		17.95	20
3720	Mineral fiber, 24" x 24" or 48", reveal edge, painted, 5/8" thick		600	.013		1.15	.63		1.78	2.30
3740	3/4" thick		575	.014		1.52	.65		2.17	2.74
5020	66 – 78% recycled content, 3/4" thick G		600	.013		1.93	.63		2.56	3.15
5040	Mylar, 42% recycled content, 3/4" thick G		600	.013		4.54	.63		5.17	6
6000	Remove and replace ceiling tiles, min fiber, 2x2 or 2x4, 5/8"thk.		335	.024		.91	1.12		2.03	2.84
9000	Minimum labor/equipment charge		4	2	Job		94		94	154

09 51 23.30 Suspended Ceilings, Complete

		Crew	Daily Output	Labor-Hours	Unit	Material	Labor	Equipment	Total	Total Incl O&P
0010	**SUSPENDED CEILINGS, COMPLETE**, including standard									
0100	suspension system but not incl. 1-1/2" carrier channels									
0600	Fiberglass ceiling board, 2' x 4' x 5/8", plain faced	1 Carp	500	.016	S.F.	1.97	.75		2.72	3.40
0700	Offices, 2' x 4' x 3/4"		380	.021		3.36	.99		4.35	5.30
0800	Mineral fiber, on 15/16" T bar susp. 2' x 2' x 3/4" lay-in board		345	.023		2.89	1.09		3.98	4.96
0810	2' x 4' x 5/8" tile		380	.021		2.31	.99		3.30	4.17
0820	Tegular, 2' x 2' x 5/8" tile on 9/16" grid		250	.032		2.40	1.50		3.90	5.10
0830	2' x 4' x 3/4" tile		275	.029		2.60	1.37		3.97	5.10
0900	Luminous panels, prismatic, acrylic		255	.031		3.37	1.47		4.84	6.10
1200	Metal pan with acoustic pad, steel		75	.107		4.51	5		9.51	13.15
1300	Painted aluminum		75	.107		3.07	5		8.07	11.60
1500	Aluminum, degreased finish		75	.107		5.20	5		10.20	13.90
1600	Stainless steel		75	.107		9.75	5		14.75	18.90
1800	Tile, Z bar suspension, 5/8" mineral fiber tile		150	.053		2.32	2.50		4.82	6.65
1900	3/4" mineral fiber tile		150	.053		2.48	2.50		4.98	6.85
2402	For strip lighting, see Section 26 51 13.50									
2500	For rooms under 500 S.F., add				S.F.		25%			
9000	Minimum labor/equipment charge	1 Carp	2	4	Job		188		188	310

09 51 53 – Direct-Applied Acoustical Ceilings

09 51 53.10 Ceiling Tile

		Crew	Daily Output	Labor-Hours	Unit	Material	Labor	Equipment	Total	Total Incl O&P
0010	**CEILING TILE**, stapled or cemented									
0100	12" x 12" or 12" x 24", not including furring									
0600	Mineral fiber, vinyl coated, 5/8" thick	1 Carp	300	.027	S.F.	2.18	1.25		3.43	4.45
0700	3/4" thick		300	.027		2.40	1.25		3.65	4.69
0900	Fire rated, 3/4" thick, plain faced		300	.027		1.27	1.25		2.52	3.45
1000	Plastic coated face		300	.027		1.84	1.25		3.09	4.07
1200	Aluminum faced, 5/8" thick, plain		300	.027		1.66	1.25		2.91	3.88
3000	Wood fiber tile, 1/2" thick		400	.020		1.12	.94		2.06	2.77
3100	3/4" thick		400	.020		1.23	.94		2.17	2.89
3300	For flameproofing, add					.10			.10	.11
3400	For sculptured 3 dimensional, add					.31			.31	.34
3900	For ceiling primer, add					.13			.13	.14
4000	For ceiling cement, add					.39			.39	.43
9000	Minimum labor/equipment charge	1 Carp	4	2	Job		94		94	154

09 53 Acoustical Ceiling Suspension Assemblies

09 53 23 – Metal Acoustical Ceiling Suspension Assemblies

09 53 23.30 Ceiling Suspension Systems		Crew	Daily Output	Labor-Hours	Unit	Material	2015 Bare Costs Labor	Equipment	Total	Total Incl O&P
0010	**CEILING SUSPENSION SYSTEMS** for boards and tile									
0050	Class A suspension system, 15/16" T bar, 2' x 4' grid	1 Carp	800	.010	S.F.	.73	.47		1.20	1.58
0300	2' x 2' grid		650	.012		.95	.58		1.53	1.99
0310	25% recycled steel, 2' x 4' grid [G]		800	.010		.77	.47		1.24	1.62
0320	2' x 2' grid [G]		650	.012		.96	.58		1.54	2.01
0350	For 9/16" grid, add					.16			.16	.18
0360	For fire rated grid, add					.09			.09	.10
0370	For colored grid, add					.21			.21	.23
0400	Concealed Z bar suspension system, 12" module	1 Carp	520	.015		.84	.72		1.56	2.10
0600	1-1/2" carrier channels, 4' O.C., add	"	470	.017		.11	.80		.91	1.43
0700	Carrier channels for ceilings with									
0900	recessed lighting fixtures, add	1 Carp	460	.017	S.F.	.20	.82		1.02	1.56
3000	Seismic ceiling bracing, IBC Site Class D, Occupancy Category II									
3050	For ceilings less than 2500 S.F.									
3060	Seismic clips at attached walls	1 Carp	180	.044	Ea.	1.12	2.09		3.21	4.65
3100	For ceilings greater than 2500 S.F., add									
3120	Seismic clips, joints at cross tees	1 Carp	120	.067	Ea.	3.29	3.13		6.42	8.75
3140	At cross tees and mains, mains field cut	"	60	.133	"	3.29	6.25		9.54	13.85
3200	Compression posts, telescopic, attached to structure above									
3210	To 30" high	1 Carp	26	.308	Ea.	39	14.45		53.45	66.50
3220	30" to 48" high		25.50	.314		43.50	14.75		58.25	72
3230	48" to 84" high		25	.320		52.50	15		67.50	82
3240	84" to 102" high		24.50	.327		60	15.35		75.35	91
3250	102" to 120" high		24	.333		85.50	15.65		101.15	120
3260	120" to 144" high		24	.333		95	15.65		110.65	131
3300	Stabilizer bars									
3310	12" long	1 Carp	240	.033	Ea.	.97	1.56		2.53	3.63
3320	24" long		235	.034		.92	1.60		2.52	3.63
3330	36" long		230	.035		.89	1.63		2.52	3.66
3340	48" long		220	.036		.73	1.71		2.44	3.61
3400	Wire support for light fixtures, per L.F. height to structure above									
3410	Less than 10 lb.	1 Carp	400	.020	L.F.	.28	.94		1.22	1.85
3420	10 lb. to 56 lb.	"	240	.033	"	.56	1.56		2.12	3.17
3500	Retrofit existing suspended ceiling grid to current code									
3510	Less than 2500 S.F. (using clips @ perimeter)	1 Carp	1455	.006	S.F.	.07	.26		.33	.49
3520	Greater than 2500 S.F. (using compression posts and clips)		550	.015		.44	.68		1.12	1.61
4000	Remove and replace ceiling grid, class A, 15/16" T bar, 2' x 2' grid		435	.018		.95	.86		1.81	2.46
4010	2' x 4' grid		535	.015		.73	.70		1.43	1.96

09 54 Specialty Ceilings

09 54 26 – Suspended Wood Ceilings

09 54 26.10 Wood Ceilings

		Crew	Daily Output	Labor-Hours	Unit	Material	Labor	Equipment	Total	Total Incl O&P
0010	**WOOD CEILINGS**									
1000	4" - 6" wood slats on heavy duty 15/16" T-bar grid	2 Carp	250	.064	S.F.	24	3		27	31.50

09 54 33 – Decorative Panel Ceilings

09 54 33.20 Metal Panel Ceilings

		Crew	Daily Output	Labor-Hours	Unit	Material	Labor	Equipment	Total	Total Incl O&P
0010	**METAL PANEL CEILINGS**									
0020	Lay-in or screwed to furring, not including grid									
0100	Tin ceilings, 2' x 2' or 2' x 4', bare steel finish	2 Carp	300	.053	S.F.	2.46	2.50		4.96	6.80
0120	Painted white finish		300	.053	"	3.71	2.50		6.21	8.20

09 54 Specialty Ceilings

09 54 33 – Decorative Panel Ceilings

09 54 33.20 Metal Panel Ceilings		Crew	Daily Output	Labor-Hours	Unit	Material	2015 Bare Costs Labor	Equipment	Total	Total Incl O&P
0140	Copper, chrome or brass finish	2 Carp	300	.053	L.F.	6.55	2.50		9.05	11.30
0200	Cornice molding, 2-1/2" to 3-1/2" wide, 4' long, bare steel finish		200	.080	S.F.	2.21	3.76		5.97	8.60
0220	Painted white finish		200	.080		2.75	3.76		6.51	9.15
0240	Copper, chrome or brass finish		200	.080		3.91	3.76		7.67	10.45
0320	5" to 6-1/2" wide, 4' long, bare steel finish		150	.107		3.19	5		8.19	11.70
0340	Painted white finish		150	.107		4.05	5		9.05	12.65
0360	Copper, chrome or brass finish		150	.107		6.45	5		11.45	15.30
0420	Flat molding, 3-1/2" to 5" wide, 4' long, bare steel finish		250	.064		3.71	3		6.71	9
0440	Painted white finish		250	.064		3.96	3		6.96	9.30
0460	Copper, chrome or brass finish		250	.064		7.50	3		10.50	13.15

09 61 Flooring Treatment

09 61 19 – Concrete Floor Staining

09 61 19.40 Floors, Interior

		Crew	Daily Output	Labor-Hours	Unit	Material	Labor	Equipment	Total	Total Incl O&P
0010	**FLOORS, INTERIOR**									
0300	Acid stain and sealer									
0310	Stain, one coat	1 Pord	650	.012	S.F.	.12	.50		.62	.93
0320	Two coats		570	.014		.23	.57		.80	1.17
0330	Acrylic sealer, one coat		2600	.003		.23	.12		.35	.45
0340	Two coats		1400	.006		.46	.23		.69	.88

09 62 Specialty Flooring

09 62 19 – Laminate Flooring

09 62 19.10 Floating Floor

		Crew	Daily Output	Labor-Hours	Unit	Material	Labor	Equipment	Total	Total Incl O&P
0010	**FLOATING FLOOR**									
8300	Floating floor, laminate, wood pattern strip, complete	1 Clab	133	.060	S.F.	4.35	2.26		6.61	8.50
8310	Components, T & G wood composite strips					3.91			3.91	4.30
8320	Film					.14			.14	.15
8330	Foam					.25			.25	.28
8340	Adhesive					.65			.65	.72
8350	Installation kit					.17			.17	.19
8360	Trim, 2" wide x 3' long				L.F.	4.30			4.30	4.73
8370	Reducer moulding				"	5.70			5.70	6.25

09 62 23 – Bamboo Flooring

09 62 23.10 Flooring, Bamboo

			Crew	Daily Output	Labor-Hours	Unit	Material	Labor	Equipment	Total	Total Incl O&P
0010	**FLOORING, BAMBOO**										
8600	Flooring, wood, bamboo strips, unfinished, 5/8" x 4" x 3'	G	1 Carp	255	.031	S.F.	4.60	1.47		6.07	7.45
8610	5/8" x 4" x 4'	G		275	.029		4.77	1.37		6.14	7.50
8620	5/8" x 4" x 6'	G		295	.027		5.25	1.27		6.52	7.85
8630	Finished, 5/8" x 4" x 3'	G		255	.031		5.05	1.47		6.52	7.95
8640	5/8" x 4" x 4'	G		275	.029		5.30	1.37		6.67	8.10
8650	5/8" x 4" x 6'	G		295	.027		4.61	1.27		5.88	7.15
8660	Stair treads, unfinished, 1-1/16" x 11-1/2" x 4'	G		18	.444	Ea.	44	21		65	82.50
8670	Finished, 1-1/16" x 11-1/2" x 4'	G		18	.444		78	21		99	120
8680	Stair risers, unfinished, 5/8" x 7-1/2" x 4'	G		18	.444		16.30	21		37.30	52
8690	Finished, 5/8" x 7-1/2" x 4'	G		18	.444		31	21		52	68
8700	Stair nosing, unfinished, 6' long	G		16	.500		36	23.50		59.50	78
8710	Finished, 6' long	G		16	.500		42.50	23.50		66	85.50

09 63 Masonry Flooring

09 63 13 – Brick Flooring

09 63 13.10 Miscellaneous Brick Flooring

		Crew	Daily Output	Labor-Hours	Unit	Material	2015 Bare Costs Labor	Equipment	Total	Total Incl O&P
0010	**MISCELLANEOUS BRICK FLOORING**									
0020	Acid-proof shales, red, 8" x 3-3/4" x 1-1/4" thick	D-7	.43	37.209	M	695	1,425		2,120	3,025
0050	2-1/4" thick	D-1	.40	40		965	1,675		2,640	3,800
0200	Acid-proof clay brick, 8" x 3-3/4" x 2-1/4" thick [G]		.40	40		935	1,675		2,610	3,775
0250	9" x 4-1/2" x 3" [G]		95	.168	S.F.	4.14	7.10		11.24	16.10
0260	Cast ceramic, pressed, 4" x 8" x 1/2", unglazed	D-7	100	.160		6.35	6.10		12.45	16.60
0270	Glazed		100	.160		8.45	6.10		14.55	18.95
0280	Hand molded flooring, 4" x 8" x 3/4", unglazed		95	.168		8.35	6.45		14.80	19.35
0290	Glazed		95	.168		10.50	6.45		16.95	21.50
0300	8" hexagonal, 3/4" thick, unglazed		85	.188		9.20	7.20		16.40	21.50
0310	Glazed		85	.188		16.60	7.20		23.80	29.50
0400	Heavy duty industrial, cement mortar bed, 2" thick, not incl. brick	D-1	80	.200		1.06	8.40		9.46	14.85
0450	Acid-proof joints, 1/4" wide	"	65	.246		1.46	10.35		11.81	18.45
0500	Pavers, 8" x 4", 1" to 1-1/4" thick, red	D-7	95	.168		3.69	6.45		10.14	14.20
0510	Ironspot	"	95	.168		5.20	6.45		11.65	15.90
0540	1-3/8" to 1-3/4" thick, red	D-1	95	.168		3.56	7.10		10.66	15.45
0560	Ironspot		95	.168		5.15	7.10		12.25	17.25
0580	2-1/4" thick, red		90	.178		3.62	7.50		11.12	16.20
0590	Ironspot		90	.178		5.60	7.50		13.10	18.40
0700	Paver, adobe brick, 6" x 12", 1/2" joint [G]		42	.381		1.30	16.05		17.35	27.50
0710	Mexican red, 12" x 12" [G]	1 Tilf	48	.167		1.66	7.15		8.81	13.10
0720	Saltillo, 12" x 12" [G]	"	48	.167		1.40	7.15		8.55	12.80
0800	For sidewalks and patios with pavers, see Section 32 14 16.10									
0870	For epoxy joints, add	D-1	600	.027	S.F.	2.81	1.12		3.93	4.92
0880	For Furan underlayment, add	"	600	.027		2.33	1.12		3.45	4.39
0890	For waxed surface, steam cleaned, add	A-1H	1000	.008		.20	.30	.08	.58	.79
9000	Minimum labor/equipment charge	1 Bric	2	4	Job		185		185	300

09 63 40 – Stone Flooring

09 63 40.10 Marble

		Crew	Daily Output	Labor-Hours	Unit	Material	2015 Bare Costs Labor	Equipment	Total	Total Incl O&P
0010	**MARBLE**									
0020	Thin gauge tile, 12" x 6", 3/8", white Carara	D-7	60	.267	S.F.	14.40	10.20		24.60	32
0100	Travertine		60	.267		11.60	10.20		21.80	29
0200	12" x 12" x 3/8", thin set, floors		60	.267		10.15	10.20		20.35	27.50
0300	On walls		52	.308		9.85	11.75		21.60	29.50
1000	Marble threshold, 4" wide x 36" long x 5/8" thick, white		60	.267	Ea.	10.15	10.20		20.35	27.50
9000	Minimum labor/equipment charge		3	5.333	Job		204		204	320

09 63 40.20 Slate Tile

		Crew	Daily Output	Labor-Hours	Unit	Material	2015 Bare Costs Labor	Equipment	Total	Total Incl O&P
0010	**SLATE TILE**									
0020	Vermont, 6" x 6" x 1/4" thick, thin set	D-7	180	.089	S.F.	7.50	3.39		10.89	13.60
9000	Minimum labor/equipment charge	"	3	5.333	Job		204		204	320

09 64 Wood Flooring

09 64 16 – Wood Block Flooring

09 64 16.10 End Grain Block Flooring

		Crew	Daily Output	Labor-Hours	Unit	Material	2015 Bare Costs Labor	2015 Bare Costs Equipment	Total	Total Incl O&P
0010	**END GRAIN BLOCK FLOORING**									
0020	End grain flooring, coated, 2" thick	1 Carp	295	.027	S.F.	3.57	1.27		4.84	6
0400	Natural finish, 1" thick, fir		125	.064		3.69	3		6.69	9
0600	1-1/2" thick, pine		125	.064		3.62	3		6.62	8.90
0700	2" thick, pine		125	.064		4.44	3		7.44	9.80
9000	Minimum labor/equipment charge		2	4	Job		188		188	310

09 64 23 – Wood Parquet Flooring

09 64 23.10 Wood Parquet

		Crew	Daily Output	Labor-Hours	Unit	Material	2015 Bare Costs Labor	2015 Bare Costs Equipment	Total	Total Incl O&P
0010	**WOOD PARQUET** flooring									
5200	Parquetry, 5/16" thk, no finish, oak, plain pattern	1 Carp	160	.050	S.F.	5.25	2.35		7.60	9.60
5300	Intricate pattern		100	.080		9.60	3.76		13.36	16.70
5500	Teak, plain pattern		160	.050		5.85	2.35		8.20	10.30
5600	Intricate pattern		100	.080		10.05	3.76		13.81	17.20
5650	13/16" thick, select grade oak, plain pattern		160	.050		9.75	2.35		12.10	14.55
5700	Intricate pattern		100	.080		16.40	3.76		20.16	24
5800	Custom parquetry, including finish, plain pattern		100	.080		16.80	3.76		20.56	24.50
5900	Intricate pattern		50	.160		24	7.50		31.50	38.50
6700	Parquetry, prefinished white oak, 5/16" thick, plain pattern		160	.050		7.90	2.35		10.25	12.55
6800	Intricate pattern		100	.080		8.45	3.76		12.21	15.40
7000	Walnut or teak, parquetry, plain pattern		160	.050		8	2.35		10.35	12.65
7100	Intricate pattern		100	.080		11.50	3.76		15.26	18.80
7200	Acrylic wood parquet blocks, 12" x 12" x 5/16",									
7210	Irradiated, set in epoxy	1 Carp	160	.050	S.F.	10	2.35		12.35	14.85

09 64 29 – Wood Strip and Plank Flooring

09 64 29.10 Wood

		Crew	Daily Output	Labor-Hours	Unit	Material	2015 Bare Costs Labor	2015 Bare Costs Equipment	Total	Total Incl O&P
0010	**WOOD** R061110-30									
0020	Fir, vertical grain, 1" x 4", not incl. finish, grade B & better	1 Carp	255	.031	S.F.	2.79	1.47		4.26	5.50
0100	C grade & better		255	.031		2.63	1.47		4.10	5.30
0300	Flat grain, 1" x 4", not incl. finish, B & better		255	.031		3.19	1.47		4.66	5.90
0400	C & better		255	.031		3.07	1.47		4.54	5.80
4000	Maple, strip, 25/32" x 2-1/4", not incl. finish, select		170	.047		4.95	2.21		7.16	9.05
4100	#2 & better		170	.047		4.29	2.21		6.50	8.35
4300	33/32" x 3-1/4", not incl. finish, #1 grade		170	.047		4.56	2.21		6.77	8.60
4400	#2 & better		170	.047		4.06	2.21		6.27	8.10
4600	Oak, white or red, 25/32" x 2-1/4", not incl. finish									
4700	#1 common	1 Carp	170	.047	S.F.	3.19	2.21		5.40	7.15
4900	Select quartered, 2-1/4" wide		170	.047		3.89	2.21		6.10	7.90
5000	Clear		170	.047		4.01	2.21		6.22	8.05
6100	Prefinished, white oak, prime grade, 2-1/4" wide		170	.047		4.69	2.21		6.90	8.75
6200	3-1/4" wide		185	.043		5.10	2.03		7.13	8.95
6400	Ranch plank		145	.055		7.15	2.59		9.74	12.10
6500	Hardwood blocks, 9" x 9", 25/32" thick		160	.050		6	2.35		8.35	10.45
7400	Yellow pine, 3/4" x 3-1/8", T & G, C & better, not incl. finish		200	.040		1.49	1.88		3.37	4.72
7500	Refinish wood floor, sand, 2 coats poly, wax, soft wood	1 Clab	400	.020		.90	.75		1.65	2.22
7600	Hard wood		130	.062		1.34	2.31		3.65	5.25
7800	Sanding and finishing, 2 coats polyurethane		295	.027		.90	1.02		1.92	2.66
7900	Subfloor and underlayment, see Section 06 16									
8015	Transition molding, 2 1/4" wide, 5' long	1 Carp	19.20	.417	Ea.	10.85	19.55		30.40	44
9000	Minimum labor/equipment charge	"	2	4	Job		188		188	310

09 65 Resilient Flooring

09 65 10 – Resilient Tile Underlayment

09 65 10.10 Latex Underlayment

		Crew	Daily Output	Labor-Hours	Unit	Material	2015 Bare Costs Labor	Equipment	Total	Total Incl O&P
0010	**LATEX UNDERLAYMENT**									
3600	Latex underlayment, 1/8" thk., cementitious for resilient flooring	1 Tilf	160	.050	S.F.	1.19	2.14		3.33	4.69
4000	Liquid, fortified				Gal.	33			33	36

09 65 13 – Resilient Base and Accessories

09 65 13.13 Resilient Base

		Crew	Daily Output	Labor-Hours	Unit	Material	Labor	Equipment	Total	Total Incl O&P
0010	**RESILIENT BASE**									
0690	1/8" vinyl base, 2 1/2" H, straight or cove, standard colors	1 Tilf	315	.025	L.F.	.67	1.09		1.76	2.46
0700	4" high		315	.025		1.32	1.09		2.41	3.17
0710	6" high		315	.025		1.40	1.09		2.49	3.26
0720	Corners, 2 1/2" high		315	.025	Ea.	2.02	1.09		3.11	3.94
0730	4" high		315	.025		2.14	1.09		3.23	4.07
0740	6" high		315	.025		2.45	1.09		3.54	4.42
0800	1/8" rubber base, 2 1/2" H, straight or cove, standard colors		315	.025	L.F.	1.13	1.09		2.22	2.96
1100	4" high		315	.025		1.02	1.09		2.11	2.84
1110	6" high		315	.025		1.77	1.09		2.86	3.67
1150	Corners, 2 1/2" high		315	.025	Ea.	2	1.09		3.09	3.92
1153	4" high		315	.025		2.05	1.09		3.14	3.98
1155	6" high		315	.025		2.51	1.09		3.60	4.48
1450	For premium color/finish add					50%				
1500	Millwork profile	1 Tilf	315	.025	L.F.	5.85	1.09		6.94	8.15

09 65 13.23 Resilient Stair Treads and Risers

		Crew	Daily Output	Labor-Hours	Unit	Material	Labor	Equipment	Total	Total Incl O&P
0010	**RESILIENT STAIR TREADS AND RISERS**									
0300	Rubber, molded tread, 12" wide, 5/16" thick, black	1 Tilf	115	.070	L.F.	14.75	2.98		17.73	21
0400	Colors		115	.070		15.35	2.98		18.33	21.50
0600	1/4" thick, black		115	.070		13.35	2.98		16.33	19.40
0700	Colors		115	.070		14.95	2.98		17.93	21
0900	Grip strip safety tread, colors, 5/16" thick		115	.070		20.50	2.98		23.48	27
1000	3/16" thick		120	.067		15.25	2.85		18.10	21.50
1200	Landings, smooth sheet rubber, 1/8" thick		120	.067	S.F.	7.80	2.85		10.65	13.10
1300	3/16" thick		120	.067	"	8.30	2.85		11.15	13.65
1500	Nosings, 3" wide, 3/16" thick, black		140	.057	L.F.	4.23	2.45		6.68	8.50
1600	Colors		140	.057		4.88	2.45		7.33	9.20
1800	Risers, 7" high, 1/8" thick, flat		250	.032		7.70	1.37		9.07	10.60
1900	Coved		250	.032		8.55	1.37		9.92	11.55
2100	Vinyl, molded tread, 12" wide, colors, 1/8" thick		115	.070		5.35	2.98		8.33	10.55
2200	1/4" thick		115	.070		6.70	2.98		9.68	12.10
2300	Landing material, 1/8" thick		200	.040	S.F.	6.05	1.71		7.76	9.35
2400	Riser, 7" high, 1/8" thick, coved		175	.046	L.F.	2.70	1.96		4.66	6.05
2500	Tread and riser combined, 1/8" thick		80	.100	"	9.95	4.28		14.23	17.65
9000	Minimum labor/equipment charge		3	2.667	Job		114		114	180

09 65 16 – Resilient Sheet Flooring

09 65 16.10 Rubber and Vinyl Sheet Flooring

			Crew	Daily Output	Labor-Hours	Unit	Material	Labor	Equipment	Total	Total Incl O&P
0010	**RUBBER AND VINYL SHEET FLOORING**										
5500	Linoleum, sheet goods	G	1 Tilf	360	.022	S.F.	3.59	.95		4.54	5.45
5900	Rubber, sheet goods, 36" wide, 1/8" thick			120	.067		7.75	2.85		10.60	13.05
5950	3/16" thick			100	.080		10.50	3.42		13.92	16.95
6000	1/4" thick			90	.089		12.45	3.80		16.25	19.65
8000	Vinyl sheet goods, backed, .065" thick, plain pattern/colors			250	.032		4.09	1.37		5.46	6.65
8050	Intricate pattern/colors			200	.040		4.57	1.71		6.28	7.75
8100	.080" thick, plain pattern/colors			230	.035		4.10	1.49		5.59	6.85
8150	Intricate pattern/colors			200	.040		5.90	1.71		7.61	9.20

09 65 Resilient Flooring

09 65 16 – Resilient Sheet Flooring

09 65 16.10 Rubber and Vinyl Sheet Flooring

		Crew	Daily Output	Labor-Hours	Unit	Material	2015 Bare Costs Labor	Equipment	Total	Total Incl O&P
8200	.125" thick, plain pattern/colors	1 Tilf	230	.035	S.F.	4.25	1.49		5.74	7.05
8250	intricate pattern/colors		200	.040		7.50	1.71		9.21	10.95
8400	For welding seams, add		100	.080	L.F.	.25	3.42		3.67	5.70
8450	For integral cove base, add		175	.046	"	.75	1.96		2.71	3.92
8700	Adhesive cement, 1 gallon per 200 to 300 S.F.				Gal.	27			27	29.50
8800	Asphalt primer, 1 gallon per 300 S.F.					14			14	15.40
8900	Emulsion, 1 gallon per 140 S.F.					18			18	19.80

09 65 19 – Resilient Tile Flooring

09 65 19.10 Miscellaneous Resilient Tile Flooring

			Crew	Daily Output	Labor-Hours	Unit	Material	Labor	Equipment	Total	Total Incl O&P
0010	**MISCELLANEOUS RESILIENT TILE FLOORING**										
2200	Cork tile, standard finish, 1/8" thick	G	1 Tilf	315	.025	S.F.	6.95	1.09		8.04	9.35
2250	3/16" thick	G		315	.025		6.65	1.09		7.74	9
2300	5/16" thick	G		315	.025		8.30	1.09		9.39	10.80
2350	1/2" thick	G		315	.025		10.95	1.09		12.04	13.70
2500	Urethane finish, 1/8" thick	G		315	.025		8.10	1.09		9.19	10.60
2550	3/16" thick	G		315	.025		8.25	1.09		9.34	10.75
2600	5/16" thick	G		315	.025		8.85	1.09		9.94	11.45
2650	1/2" thick	G		315	.025		12.15	1.09		13.24	15.05

09 65 19.19 Vinyl Composition Tile Flooring

		Crew	Daily Output	Labor-Hours	Unit	Material	Labor	Equipment	Total	Total Incl O&P
0010	**VINYL COMPOSITION TILE FLOORING**									
7000	Vinyl composition tile, 12" x 12", 1/16" thick	1 Tilf	500	.016	S.F.	1.18	.68		1.86	2.38
7050	Embossed		500	.016		2.21	.68		2.89	3.51
7100	Marbleized		500	.016		2.21	.68		2.89	3.51
7150	Solid		500	.016		2.85	.68		3.53	4.22
7200	3/32" thick, embossed		500	.016		1.51	.68		2.19	2.74
7250	Marbleized		500	.016		2.54	.68		3.22	3.87
7300	Solid		500	.016		2.36	.68		3.04	3.68
7350	1/8" thick, marbleized		500	.016		2.40	.68		3.08	3.72
7400	Solid		500	.016		1.53	.68		2.21	2.76
7450	Conductive		500	.016		6.05	.68		6.73	7.75

09 65 19.23 Vinyl Tile Flooring

		Crew	Daily Output	Labor-Hours	Unit	Material	Labor	Equipment	Total	Total Incl O&P
0010	**VINYL TILE FLOORING**									
7500	Vinyl tile, 12" x 12", 3/32" thick,	1 Tilf	500	.016	S.F.	3.59	.68		4.27	5.05
7550	3/32" thick, premium colors/patterns		500	.016		7.25	.68		7.93	9.10
7600	1/8" thick, standard colors/patterns		500	.016		5.65	.68		6.33	7.30
7650	Solid colors		500	.016		3.25	.68		3.93	4.66
7700	Marbleized or Travertine pattern		500	.016		5.90	.68		6.58	7.60
7750	Florentine pattern		500	.016		6.30	.68		6.98	8.05
7800	Premium colors/patterns		500	.016		6.20	.68		6.88	7.95
9500	Minimum labor/equipment charge		4	2	Job		85.50		85.50	135

09 65 19.33 Rubber Tile Flooring

		Crew	Daily Output	Labor-Hours	Unit	Material	Labor	Equipment	Total	Total Incl O&P
0010	**RUBBER TILE FLOORING**									
6050	Rubber tile, marbleized colors, 12" x 12", 1/8" thick	1 Tilf	400	.020	S.F.	5.70	.86		6.56	7.60
6100	3/16" thick		400	.020		9.25	.86		10.11	11.55
6300	Special tile, plain colors, 1/8" thick		400	.020		7.70	.86		8.56	9.85
6350	3/16" thick		400	.020		9.05	.86		9.91	11.30

09 65 33 – Conductive Resilient Flooring

09 65 33.10 Conductive Rubber and Vinyl Flooring

		Crew	Daily Output	Labor-Hours	Unit	Material	Labor	Equipment	Total	Total Incl O&P
0010	**CONDUCTIVE RUBBER AND VINYL FLOORING**									
1700	Conductive flooring, rubber tile, 1/8" thick	1 Tilf	315	.025	S.F.	6.95	1.09		8.04	9.35
1800	Homogeneous vinyl tile, 1/8" thick	"	315	.025	"	6.40	1.09		7.49	8.75

09 66 Terrazzo Flooring

09 66 13 – Portland Cement Terrazzo Flooring

09 66 13.10 Portland Cement Terrazzo

		Crew	Daily Output	Labor-Hours	Unit	Material	2015 Bare Costs Labor	Equipment	Total	Total Incl O&P
0010	**PORTLAND CEMENT TERRAZZO**, cast-in-place									
0020	Cove base, 6" high, 16 ga. zinc	1 Mstz	20	.400	L.F.	3.28	17.20		20.48	30.50
0100	Curb, 6" high and 6" wide		6	1.333		5.65	57.50		63.15	96.50
0300	Divider strip for floors, 14 ga., 1-1/4" deep, zinc		375	.021		1.42	.92		2.34	3.01
0400	Brass		375	.021		2.46	.92		3.38	4.16
0600	Heavy top strip 1/4" thick, 1-1/4" deep, zinc		300	.027		2.17	1.15		3.32	4.20
1200	For thin set floors, 16 ga., 1/2" x 1/2", zinc		350	.023		1.12	.98		2.10	2.78
1500	Floor, bonded to concrete, 1-3/4" thick, gray cement	J-3	75	.213	S.F.	3.35	8.35	4.05	15.75	21.50
1600	White cement, mud set		75	.213		3.73	8.35	4.05	16.13	22
1800	Not bonded, 3" total thickness, gray cement		70	.229		4.17	8.95	4.34	17.46	23.50
1900	White cement, mud set		70	.229		4.86	8.95	4.34	18.15	24.50
9000	Minimum labor/equipment charge	1 Mstz	1	8	Job		345		345	540

09 66 16 – Terrazzo Floor Tile

09 66 16.10 Tile or Terrazzo Base

		Crew	Daily Output	Labor-Hours	Unit	Material	Labor	Equipment	Total	Total Incl O&P
0010	**TILE OR TERRAZZO BASE**									
0020	Scratch coat only	1 Mstz	150	.053	S.F.	.43	2.29		2.72	4.09
0500	Scratch and brown coat only		75	.107	"	.82	4.58		5.40	8.15
9000	Minimum labor/equipment charge		1	8	Job		345		345	540

09 66 16.13 Portland Cement Terrazzo Floor Tile

		Crew	Daily Output	Labor-Hours	Unit	Material	Labor	Equipment	Total	Total Incl O&P
0010	**PORTLAND CEMENT TERRAZZO FLOOR TILE**									
1200	Floor tiles, non-slip, 1" thick, 12" x 12"	D-1	60	.267	S.F.	20	11.25		31.25	40.50
1300	1-1/4" thick, 12" x 12"		60	.267		20.50	11.25		31.75	41.50
1500	16" x 16"		50	.320		22.50	13.45		35.95	46.50
1600	1-1/2" thick, 16" x 16"		45	.356		20.50	14.95		35.45	47
1800	For Venetian terrazzo, add					6.15			6.15	6.80
1900	For white cement, add					.58			.58	.64

09 66 16.16 Plastic Matrix Terrazzo Floor Tile

		Crew	Daily Output	Labor-Hours	Unit	Material	Labor	Equipment	Total	Total Incl O&P
0010	**PLASTIC MATRIX TERRAZZO FLOOR TILE**									

09 66 16.30 Terrazzo, Precast

		Crew	Daily Output	Labor-Hours	Unit	Material	Labor	Equipment	Total	Total Incl O&P
0010	**TERRAZZO, PRECAST**									
0020	Base, 6" high, straight	1 Mstz	70	.114	L.F.	11.95	4.91		16.86	21
0100	Cove		60	.133		12.80	5.75		18.55	23
0300	8" high, straight		60	.133		11.45	5.75		17.20	21.50
0400	Cove		50	.160		16.85	6.85		23.70	29.50
0600	For white cement, add					.45			.45	.50
0700	For 16 ga. zinc toe strip, add					1.72			1.72	1.89
0900	Curbs, 4" x 4" high	1 Mstz	40	.200		32.50	8.60		41.10	49
1000	8" x 8" high	"	30	.267		38	11.45		49.45	60
2400	Stair treads, 1-1/2" thick, non-slip, three line pattern	2 Mstz	70	.229		41	9.80		50.80	60.50
2500	Nosing and two lines		70	.229		41	9.80		50.80	60.50
2700	2" thick treads, straight		60	.267		45.50	11.45		56.95	68
2800	Curved		50	.320		59	13.75		72.75	86.50
3000	Stair risers, 1" thick, to 6" high, straight sections		60	.267		10.80	11.45		22.25	30
3100	Cove		50	.320		15.65	13.75		29.40	38.50
3300	Curved, 1" thick, to 6" high, vertical		48	.333		21.50	14.30		35.80	46
3400	Cove		38	.421		40.50	18.10		58.60	73
3600	Stair tread and riser, single piece, straight, smooth surface		60	.267		52.50	11.45		63.95	76
3700	Non skid surface		40	.400		68	17.20		85.20	102
3900	Curved tread and riser, smooth surface		40	.400		74	17.20		91.20	108
4000	Non skid surface		32	.500		93	21.50		114.50	136
4200	Stair stringers, notched, 1" thick		25	.640		30	27.50		57.50	76.50

09 66 Terrazzo Flooring

09 66 16 – Terrazzo Floor Tile

09 66 16.30 Terrazzo, Precast		Crew	Daily Output	Labor-Hours	Unit	Material	2015 Bare Costs Labor	2015 Bare Costs Equipment	Total	Total Incl O&P
4300	2" thick	2 Mstz	22	.727	L.F.	35.50	31		66.50	88.50
4500	Stair landings, structural, non-slip, 1-1/2" thick		85	.188	S.F.	33	8.10		41.10	49
4600	3" thick	↓	75	.213		46.50	9.15		55.65	66
4800	Wainscot, 12" x 12" x 1" tiles	1 Mstz	12	.667		6.85	28.50		35.35	52.50
4900	16" x 16" x 1-1/2" tiles	"	8	1		14.25	43		57.25	83.50
9500	Minimum labor/equipment charge	1 Tilf	2	4	Job		171		171	270

09 66 23 – Resinous Matrix Terrazzo Flooring

09 66 23.16 Epoxy-Resin Terrazzo Flooring

		Crew	Daily Output	Labor-Hours	Unit	Material	Labor	Equipment	Total	Total Incl O&P
0010	**EPOXY-RESIN TERRAZZO FLOORING**									
1800	Epoxy terrazzo, 1/4" thick, chemical resistant, granite chips	J-3	200	.080	S.F.	5.95	3.14	1.52	10.61	13.15
1900	Recycled porcelain	"	150	.107	"	9.05	4.18	2.02	15.25	18.80

09 67 Fluid-Applied Flooring

09 67 13 – Elastomeric Liquid Flooring

09 67 13.13 Elastomeric Liquid Flooring

		Crew	Daily Output	Labor-Hours	Unit	Material	Labor	Equipment	Total	Total Incl O&P
0010	**ELASTOMERIC LIQUID FLOORING**									
0020	Cementitious acrylic, 1/4" thick	C-6	520	.092	S.F.	1.66	3.62	.12	5.40	7.85
0200	Methyl methachrylate, 1/4" thick	C-8A	3000	.016	"	6.50	.65		7.15	8.20

09 67 26 – Quartz Flooring

09 67 26.26 Quartz Flooring

		Crew	Daily Output	Labor-Hours	Unit	Material	Labor	Equipment	Total	Total Incl O&P
0010	**QUARTZ FLOORING**									
0600	Epoxy, with colored quartz chips, broadcast, 3/8" thick	C-6	675	.071	S.F.	2.81	2.79	.09	5.69	7.75
0700	1/2" thick		490	.098		4.05	3.84	.13	8.02	10.85
0900	Troweled, minimum		560	.086		3.57	3.36	.11	7.04	9.50
1000	Maximum	↓	480	.100	↓	5.40	3.92	.13	9.45	12.45
1200	Heavy duty epoxy topping, 1/4" thick,									
1300	500 to 1,000 S.F.	C-6	420	.114	S.F.	5.55	4.48	.15	10.18	13.60
1500	1,000 to 2,000 S.F.		450	.107		5.05	4.18	.14	9.37	12.50
1600	Over 10,000 S.F.	↓	480	.100	↓	4.66	3.92	.13	8.71	11.70

09 68 Carpeting

09 68 05 – Carpet Accessories

09 68 05.11 Flooring Transition Strip

		Crew	Daily Output	Labor-Hours	Unit	Material	Labor	Equipment	Total	Total Incl O&P
0010	**FLOORING TRANSITION STRIP**									
0107	Clamp down brass divider, 12' strip, vinyl to carpet	1 Tilf	31.25	.256	Ea.	13.30	10.95		24.25	32
0117	Vinyl to hard surface	"	31.25	.256	"	13.30	10.95		24.25	32

09 68 10 – Carpet Pad

09 68 10.10 Commercial Grade Carpet Pad

		Crew	Daily Output	Labor-Hours	Unit	Material	Labor	Equipment	Total	Total Incl O&P
0010	**COMMERCIAL GRADE CARPET PAD**									
9000	Sponge rubber pad, 20 oz./sq. yd.	1 Tilf	150	.053	S.Y.	4.51	2.28		6.79	8.55
9100	40-62 oz./sq. yd.		150	.053		8.75	2.28		11.03	13.25
9200	Felt pad, 20 oz./sq. yd.		150	.053		5.30	2.28		7.58	9.45
9300	32 to 56 oz./sq. yd.		150	.053		8.95	2.28		11.23	13.45
9400	Bonded urethane pad, 2.7 density		150	.053		5.95	2.28		8.23	10.15
9500	13.0 Density		150	.053		8	2.28		10.28	12.40
9600	Prime urethane pad, 2.7 density		150	.053		3.15	2.28		5.43	7.10
9700	13.0 density	↓	150	.053	↓	4.95	2.28		7.23	9.05

09 68 Carpeting

09 68 13 – Tile Carpeting

09 68 13.10 Carpet Tile

		Crew	Daily Output	Labor-Hours	Unit	Material	2015 Bare Costs Labor	2015 Bare Costs Equipment	Total	Total Incl O&P
0010	**CARPET TILE**									
0100	Tufted nylon, 18" x 18", hard back, 20 oz.	1 Tilf	80	.100	S.Y.	24	4.28		28.28	33.50
0110	26 oz.		80	.100		32	4.28		36.28	42
0200	Cushion back, 20 oz.		80	.100		28	4.28		32.28	37.50
0210	26 oz.		80	.100		43	4.28		47.28	54
1100	Tufted, 24" x 24", hard back, 24 oz. nylon		80	.100		30	4.28		34.28	40
1180	35 oz.		80	.100		35	4.28		39.28	45.50
5060	42 oz.		80	.100		46	4.28		50.28	58

09 68 16 – Sheet Carpeting

09 68 16.10 Sheet Carpet

		Crew	Daily Output	Labor-Hours	Unit	Material	2015 Bare Costs Labor	2015 Bare Costs Equipment	Total	Total Incl O&P
0010	**SHEET CARPET**									
0700	Nylon, level loop, 26 oz., light to medium traffic	1 Tilf	75	.107	S.Y.	25	4.57		29.57	34.50
0720	28 oz., light to medium traffic		75	.107		32	4.57		36.57	42
0900	32 oz., medium traffic		75	.107		42.50	4.57		47.07	54
1100	40 oz., medium to heavy traffic		75	.107		59.50	4.57		64.07	72.50
2920	Nylon plush, 30 oz., medium traffic		57	.140		29.50	6		35.50	42
3000	36 oz., medium traffic		75	.107		34.50	4.57		39.07	45
3100	42 oz., medium to heavy traffic		70	.114		45	4.89		49.89	57.50
3200	46 oz., medium to heavy traffic		70	.114		49	4.89		53.89	62
3300	54 oz., heavy traffic		70	.114		55	4.89		59.89	68.50
4500	50 oz., medium to heavy traffic		75	.107		100	4.57		104.57	117
4700	Patterned, 32 oz., medium to heavy traffic		70	.114		91	4.89		95.89	108
4900	48 oz., heavy traffic		70	.114		100	4.89		104.89	118
5000	For less than full roll (approx. 1500 S.F.), add					25%				
5100	For small rooms, less than 12' wide, add						25%			
5200	For large open areas (no cuts), deduct						25%			
5600	For bound carpet baseboard, add	1 Tilf	300	.027	L.F.	3	1.14		4.14	5.10
5610	For stairs, not incl. price of carpet, add	"	30	.267	Riser		11.40		11.40	18.05
5620	For borders and patterns, add to labor						18%			
8950	For tackless, stretched installation, add padding from 09 68 10.10 to above									
9850	For brand-named specific fiber, add				S.Y.	25%				
9900	Carpet cleaning machine, rent				Day				38	42
9910	Minimum labor/equipment charge	1 Tilf	3	2.667	Job		114		114	180

09 68 20 – Athletic Carpet

09 68 20.10 Indoor Athletic Carpet

		Crew	Daily Output	Labor-Hours	Unit	Material	2015 Bare Costs Labor	2015 Bare Costs Equipment	Total	Total Incl O&P
0010	**INDOOR ATHLETIC CARPET**									
3700	Polyethylene, in rolls, no base incl., landscape surfaces	1 Tilf	275	.029	S.F.	4.09	1.25		5.34	6.45
3800	Nylon action surface, 1/8" thick		275	.029		3.76	1.25		5.01	6.10
3900	1/4" thick		275	.029		5.45	1.25		6.70	7.90
4000	3/8" thick		275	.029		6.80	1.25		8.05	9.45
5500	Polyvinyl chloride, sheet goods for gyms, 1/4" thick		80	.100		7.90	4.28		12.18	15.45
5600	3/8" thick		60	.133		11.80	5.70		17.50	22

09 69 Access Flooring

09 69 13 – Rigid-Grid Access Flooring

09 69 13.10 Access Floors

		Crew	Daily Output	Labor-Hours	Unit	Material	2015 Bare Costs Labor	Equipment	Total	Total Incl O&P
0010	**ACCESS FLOORS**									
0015	Access floor package including panel, pedestal, stringers & laminate cover									
0100	Computer room, greater than 6,000 S.F.	4 Carp	750	.043	S.F.	7.50	2		9.50	11.55
0110	Less than 6,000 S.F.	2 Carp	375	.043		8.15	2		10.15	12.30
0120	Office, greater than 6,000 S.F.	4 Carp	1050	.030		5.35	1.43		6.78	8.25
0250	Panels, particle board or steel, 1250# load, no covering, under 6,000 S.F.	2 Carp	600	.027		3.75	1.25		5	6.20
0300	Over 6,000 S.F.		640	.025		3.21	1.17		4.38	5.45
0400	Aluminum, 24" panels		500	.032		33	1.50		34.50	38.50
0600	For carpet covering, add					8.75			8.75	9.65
0700	For vinyl floor covering, add					9.05			9.05	9.95
0900	For high pressure laminate covering, add					7.50			7.50	8.25
0910	For snap on stringer system, add	2 Carp	1000	.016		1.56	.75		2.31	2.95
0950	Office applications, steel or concrete panels,									
0960	no covering, over 6,000 S.F.	2 Carp	960	.017	S.F.	10.55	.78		11.33	12.90
1050	Pedestals, 6" to 12"	"	85	.188	Ea.	8.40	8.85		17.25	24

09 72 Wall Coverings

09 72 19 – Textile Wall Covering

09 72 19.10 Textile Wall Covering

		Crew	Daily Output	Labor-Hours	Unit	Material	Labor	Equipment	Total	Total Incl O&P
0010	**TEXTILE WALL COVERING**, including sizing; add 10-30% waste @ takeoff									
0020	Silk	1 Pape	640	.013	S.F.	4.15	.51		4.66	5.40
0030	Cotton		640	.013		6.65	.51		7.16	8.10
0040	Linen		640	.013		1.85	.51		2.36	2.86
0050	Blend		640	.013		2.97	.51		3.48	4.09

09 72 20 – Natural Fiber Wall Covering

09 72 20.10 Natural Fiber Wall Covering

		Crew	Daily Output	Labor-Hours	Unit	Material	Labor	Equipment	Total	Total Incl O&P
0010	**NATURAL FIBER WALL COVERING**, including sizing; add 10-30% waste @ takeoff									
0015	Bamboo	1 Pape	640	.013	S.F.	2.26	.51		2.77	3.31
0030	Burlap		640	.013		2.03	.51		2.54	3.05
0045	Jute		640	.013		1.29	.51		1.80	2.24
0060	Sisal		640	.013		1.46	.51		1.97	2.43

09 72 23 – Wallpapering

09 72 23.10 Wallpaper

		Crew	Daily Output	Labor-Hours	Unit	Material	Labor	Equipment	Total	Total Incl O&P
0010	**WALLPAPER** including sizing; add 10-30 percent waste @ takeoff									
0050	Aluminum foil	1 Pape	275	.029	S.F.	1.04	1.19		2.23	3.05
0100	Copper sheets, .025" thick, vinyl backing		240	.033		5.55	1.36		6.91	8.30
0300	Phenolic backing		240	.033		7.20	1.36		8.56	10.15
0600	Cork tiles, light or dark, 12" x 12" x 3/16"		240	.033		4.51	1.36		5.87	7.15
0700	5/16" thick		235	.034		3.13	1.39		4.52	5.70
0900	1/4" basketweave		240	.033		3.50	1.36		4.86	6.05
1000	1/2" natural, non-directional pattern		240	.033		6.90	1.36		8.26	9.80
1100	3/4" natural, non-directional pattern		240	.033		11.35	1.36		12.71	14.70
1200	Granular surface, 12" x 36", 1/2" thick		385	.021		1.31	.85		2.16	2.81
1300	1" thick		370	.022		1.68	.88		2.56	3.27
1500	Polyurethane coated, 12" x 12" x 3/16" thick		240	.033		4.08	1.36		5.44	6.70
1600	5/16" thick		235	.034		6.60	1.39		7.99	9.50
1800	Cork wallpaper, paperbacked, natural		480	.017		2.06	.68		2.74	3.37
1900	Colors		480	.017		2.87	.68		3.55	4.26
2100	Flexible wood veneer, 1/32" thick, plain woods		100	.080		2.44	3.27		5.71	7.95
2200	Exotic woods		95	.084		3.69	3.44		7.13	9.60

09 72 Wall Coverings

09 72 23 – Wallpapering

09 72 23.10 Wallpaper

		Crew	Daily Output	Labor-Hours	Unit	Material	2015 Bare Costs Labor	Equipment	Total	Total Incl O&P
2400	Gypsum-based, fabric-backed, fire resistant									
2500	for masonry walls, 21 oz./S.Y.	1 Pape	800	.010	S.F.	.85	.41		1.26	1.60
2600	Average		720	.011		1.27	.45		1.72	2.13
2700	Small quantities		640	.013		.76	.51		1.27	1.66
2750	Acrylic, modified, semi-rigid PVC, .028" thick	2 Carp	330	.048		1.37	2.28		3.65	5.25
2800	.040" thick	"	320	.050		1.81	2.35		4.16	5.85
3000	Vinyl wall covering, fabric-backed, lightweight, type 1 (12-15 oz./S.Y.)	1 Pape	640	.013		.97	.51		1.48	1.89
3300	Medium weight, type 2 (20-24 oz./S.Y.)		480	.017		.90	.68		1.58	2.09
3400	Heavy weight, type 3 (28 oz./S.Y.)		435	.018		1.37	.75		2.12	2.72
3600	Adhesive, 5 gal. lots (18 S.Y./gal.)				Gal.	14.20			14.20	15.60
3700	Wallpaper, average workmanship, solid pattern, low cost paper	1 Pape	640	.013	S.F.	.61	.51		1.12	1.49
3900	basic patterns (matching required), avg. cost paper		535	.015		1.11	.61		1.72	2.20
4000	Paper at $85 per double roll, quality workmanship		435	.018		2.20	.75		2.95	3.63
4190	Grass cloth, natural fabric	G	400	.020		2.01	.82		2.83	3.53
4200	Grass cloths with lining paper	G	400	.020		.91	.82		1.73	2.32
4300	Premium texture/color	G	350	.023		2.91	.94		3.85	4.70
5990	Wallpaper removal, 1 layer		800	.010		.03	.41		.44	.69
6000	Wallpaper removal, 3 layer		400	.020		.08	.82		.90	1.41
9000	Minimum labor/equipment charge		2	4	Job		164		164	263

09 77 Special Wall Surfacing

09 77 30 – Fiberglass Reinforced Panels

09 77 30.10 Fiberglass Reinforced Plastic Panels

		Crew	Daily Output	Labor-Hours	Unit	Material	Labor	Equipment	Total	Total Incl O&P
0010	**FIBERGLASS REINFORCED PLASTIC PANELS**, .090" thick									
0020	On walls, adhesive mounted, embossed surface	2 Carp	640	.025	S.F.	1.11	1.17		2.28	3.14
0030	Smooth surface		640	.025		1.37	1.17		2.54	3.43
0040	Fire rated, embossed surface		640	.025		1.99	1.17		3.16	4.11
0050	Nylon rivet mounted, on drywall, embossed surface		480	.033		1.11	1.56		2.67	3.78
0060	Smooth surface		480	.033		1.37	1.56		2.93	4.07
0070	Fire rated, embossed surface		480	.033		1.99	1.56		3.55	4.75
0080	On masonry, embossed surface		320	.050		1.11	2.35		3.46	5.05
0090	Smooth surface		320	.050		1.37	2.35		3.72	5.35
0100	Fire rated, embossed surface		320	.050		1.99	2.35		4.34	6.05
0110	Nylon rivet and adhesive mounted, on drywall, embossed surface		240	.067		1.26	3.13		4.39	6.55
0120	Smooth surface		240	.067		1.26	3.13		4.39	6.55
0130	Fire rated, embossed surface		240	.067		2.19	3.13		5.32	7.55
0140	On masonry, embossed surface		190	.084		1.26	3.95		5.21	7.90
0150	Smooth surface		190	.084		1.26	3.95		5.21	7.90
0160	Fire rated, embossed surface		190	.084		2.19	3.95		6.14	8.90
0170	For moldings add	1 Carp	250	.032	L.F.	.26	1.50		1.76	2.75
0180	On ceilings, for lay in grid system, embossed surface		400	.020	S.F.	1.11	.94		2.05	2.76
0190	Smooth surface		400	.020		1.37	.94		2.31	3.05
0200	Fire rated, embossed surface		400	.020		1.99	.94		2.93	3.73

09 77 43 – Panel Systems

09 77 43.20 Slatwall Panels and Accessories

		Crew	Daily Output	Labor-Hours	Unit	Material	Labor	Equipment	Total	Total Incl O&P
0010	**SLATWALL PANELS AND ACCESSORIES**									
0100	Slatwall panel, 4' x 8' x 3/4" T, MDF, paint grade	1 Carp	500	.016	S.F.	1.41	.75		2.16	2.78
0110	Melamine finish		500	.016		2.12	.75		2.87	3.56
0120	High pressure plastic laminate finish		500	.016		3.72	.75		4.47	5.30
0125	Wood veneer		500	.016		3.41	.75		4.16	4.98

09 77 Special Wall Surfacing

09 77 43 – Panel Systems

09 77 43.20 Slatwall Panels and Accessories

		Crew	Daily Output	Labor-Hours	Unit	Material	2015 Bare Costs Labor	Equipment	Total	Total Incl O&P
0130	Aluminum channel inserts, add				S.F.	2.43			2.43	2.67
0200	Accessories, corner forms, 8' L				L.F.	5.80			5.80	6.40
0210	T-connector, 8' L					8.40			8.40	9.25
0220	J-mold, 8' L					1.26			1.26	1.39
0230	Edge cap, 8' L					1.33			1.33	1.46
0240	Finish end cap, 8' L					3.39			3.39	3.73
0300	Display hook, metal, 4" L				Ea.	.43			.43	.47
0310	6" L					.48			.48	.53
0320	8" L					.52			.52	.57
0330	10" L					.59			.59	.65
0340	12" L					.64			.64	.70
0350	Acrylic, 4" L					.92			.92	1.01
0360	6" L					1.06			1.06	1.17
0370	8" L					1.11			1.11	1.22
0380	10" L					1.25			1.25	1.38
0400	Waterfall hanger, metal, 12" - 16"					3.84			3.84	4.22
0410	Acrylic					11.30			11.30	12.40
0500	Shelf bracket, metal, 8"					1.79			1.79	1.97
0510	10"					1.97			1.97	2.17
0520	12"					2.17			2.17	2.39
0530	14"					2.39			2.39	2.63
0540	16"					2.81			2.81	3.09
0550	Acrylic, 8"					3.79			3.79	4.17
0560	10"					4.03			4.03	4.43
0570	12"					4.26			4.26	4.69
0580	14"					4.48			4.48	4.93
0600	Shelf, acrylic, 12" x 16" x 1/4"					17.30			17.30	19
0610	12" x 24" x 1/4"					24			24	26

09 81 Acoustic Insulation

09 81 16 – Acoustic Blanket Insulation

09 81 16.10 Sound Attenuation Blanket

		Crew	Daily Output	Labor-Hours	Unit	Material	2015 Bare Costs Labor	Equipment	Total	Total Incl O&P
0010	**SOUND ATTENUATION BLANKET**									
0020	Blanket, 1" thick	1 Carp	925	.009	S.F.	.25	.41		.66	.95
0500	1-1/2" thick		920	.009		.25	.41		.66	.95
1000	2" thick		915	.009		.36	.41		.77	1.07
1500	3" thick		910	.009		.52	.41		.93	1.25
2000	Wall hung, STC 18 – 21, 1" thick, 4' x 20'	2 Carp	22	.727	Ea.	385	34		419	475
2010	10' x 20'	"	19	.842		960	39.50		999.50	1,125
2020	Wall hung, STC 27 – 28, 3" thick, 4' x 20'	3 Carp	12	2		545	94		639	755
2030	10' x 20'	"	9	2.667		1,375	125		1,500	1,700
3400	Urethane plastic foam, open cell, on wall, 2" thick	2 Carp	2050	.008	S.F.	3.11	.37		3.48	4.02
3500	3" thick		1550	.010		4.13	.48		4.61	5.35
3600	4" thick		1050	.015		5.80	.72		6.52	7.50
3700	On ceiling, 2" thick		1700	.009		3.10	.44		3.54	4.13
3800	3" thick		1300	.012		4.13	.58		4.71	5.50
3900	4" thick		900	.018		5.80	.83		6.63	7.70
9000	Minimum labor/equipment charge	1 Carp	5	1.600	Job		75		75	123

09 84 Acoustic Room Components

09 84 13 – Fixed Sound-Absorptive Panels

09 84 13.10 Fixed Panels		Crew	Daily Output	Labor-Hours	Unit	Material	2015 Bare Costs Labor	Equipment	Total	Total Incl O&P
0010	**FIXED PANELS** Perforated steel facing, painted with									
0100	Fiberglass or mineral filler, no backs, 2-1/4" thick, modular									
0200	space units, ceiling or wall hung, white or colored	1 Carp	100	.080	S.F.	8.80	3.76		12.56	15.80
0300	Fiberboard sound deadening panels, 1/2" thick	"	600	.013	"	.33	.63		.96	1.39
0500	Fiberglass panels, 4' x 8' x 1" thick, with									
0600	glass cloth face for walls, cemented	1 Carp	155	.052	S.F.	8.35	2.42		10.77	13.15
0700	1-1/2" thick, dacron covered, inner aluminum frame,									
0710	wall mounted	1 Carp	300	.027	S.F.	8.90	1.25		10.15	11.85
0900	Mineral fiberboard panels, fabric covered, 30" x 108",									
1000	3/4" thick, concealed spline, wall mounted	1 Carp	150	.053	S.F.	6.40	2.50		8.90	11.15
9000	Minimum labor/equipment charge	"	4	2	Job		94		94	154

09 91 Painting

09 91 03 – Paint Restoration

09 91 03.20 Sanding

		Crew	Daily Output	Labor-Hours	Unit	Material	Labor	Equipment	Total	Total Incl O&P
0010	**SANDING** and puttying interior trim, compared to									
0100	Painting 1 coat, on quality work				L.F.		100%			
0300	Medium work						50%			
0400	Industrial grade						25%			
0500	Surface protection, placement and removal									
0510	Basic drop cloths	1 Pord	6400	.001	S.F.		.05		.05	.08
0520	Masking with paper		800	.010		.07	.40		.47	.73
0530	Volume cover up (using plastic sheathing, or building paper)		16000	.001			.02		.02	.03

09 91 03.30 Exterior Surface Preparation

		Crew	Daily Output	Labor-Hours	Unit	Material	Labor	Equipment	Total	Total Incl O&P
0010	**EXTERIOR SURFACE PREPARATION**									
0015	Doors, per side, not incl. frames or trim									
0020	Scrape & sand									
0030	Wood, flush	1 Pord	616	.013	S.F.		.52		.52	.84
0040	Wood, detail		496	.016			.65		.65	1.05
0050	Wood, louvered		280	.029			1.15		1.15	1.85
0060	Wood, overhead		616	.013			.52		.52	.84
0070	Wire brush									
0080	Metal, flush	1 Pord	640	.013	S.F.		.50		.50	.81
0090	Metal, detail		520	.015			.62		.62	1
0100	Metal, louvered		360	.022			.90		.90	1.44
0110	Metal or fibr., overhead		640	.013			.50		.50	.81
0120	Metal, roll up		560	.014			.58		.58	.93
0130	Metal, bulkhead		640	.013			.50		.50	.81
0140	Power wash, based on 2500 lb. operating pressure									
0150	Metal, flush	A-1H	2240	.004	S.F.		.13	.03	.16	.26
0160	Metal, detail		2120	.004			.14	.04	.18	.27
0170	Metal, louvered		2000	.004			.15	.04	.19	.29
0180	Metal or fibr., overhead		2400	.003			.13	.03	.16	.24
0190	Metal, roll up		2400	.003			.13	.03	.16	.24
0200	Metal, bulkhead		2200	.004			.14	.03	.17	.26
0400	Windows, per side, not incl. trim									
0410	Scrape & sand									
0420	Wood, 1-2 lite	1 Pord	320	.025	S.F.		1.01		1.01	1.62
0430	Wood, 3-6 lite		280	.029			1.15		1.15	1.85
0440	Wood, 7-10 lite		240	.033			1.34		1.34	2.16
0450	Wood, 12 lite		200	.040			1.61		1.61	2.59

09 91 Painting

09 91 03 – Paint Restoration

09 91 03.30 Exterior Surface Preparation

		Crew	Daily Output	Labor-Hours	Unit	Material	2015 Bare Costs Labor	2015 Bare Costs Equipment	Total	Total Incl O&P
0460	Wood, Bay/Bow	1 Pord	320	.025	S.F.		1.01		1.01	1.62
0470	Wire brush									
0480	Metal, 1-2 lite	1 Pord	480	.017	S.F.		.67		.67	1.08
0490	Metal, 3-6 lite		400	.020			.81		.81	1.30
0500	Metal, Bay/Bow	↓	480	.017	↓		.67		.67	1.08
0510	Power wash, based on 2500 lb. operating pressure									
0520	1-2 lite	A-1H	4400	.002	S.F.		.07	.02	.09	.13
0530	3-6 lite		4320	.002			.07	.02	.09	.13
0540	7-10 lite		4240	.002			.07	.02	.09	.14
0550	12 lite		4160	.002			.07	.02	.09	.14
0560	Bay/Bow	↓	4400	.002	↓		.07	.02	.09	.13
0600	Siding, scrape and sand, light=10-30%, med.=30-70%									
0610	Heavy=70-100% of surface to sand									
0650	Texture 1-11, light	1 Pord	480	.017	S.F.		.67		.67	1.08
0660	Med.		440	.018			.73		.73	1.18
0670	Heavy		360	.022			.90		.90	1.44
0680	Wood shingles, shakes, light		440	.018			.73		.73	1.18
0690	Med.		360	.022			.90		.90	1.44
0700	Heavy		280	.029			1.15		1.15	1.85
0710	Clapboard, light		520	.015			.62		.62	1
0720	Med.		480	.017			.67		.67	1.08
0730	Heavy	↓	400	.020	↓		.81		.81	1.30
0740	Wire brush									
0750	Aluminum, light	1 Pord	600	.013	S.F.		.54		.54	.86
0760	Med.		520	.015			.62		.62	1
0770	Heavy	↓	440	.018	↓		.73		.73	1.18
0780	Pressure wash, based on 2500 lb. operating pressure									
0790	Stucco	A-1H	3080	.003	S.F.		.10	.02	.12	.19
0800	Aluminum or vinyl		3200	.003			.09	.02	.11	.18
0810	Siding, masonry, brick & block	↓	2400	.003	↓		.13	.03	.16	.24
1300	Miscellaneous, wire brush									
1310	Metal, pedestrian gate	1 Pord	100	.080	S.F.		3.23		3.23	5.20
1320	Aluminum chain link, both sides		250	.032			1.29		1.29	2.08
1400	Existing galvanized surface, clean and prime, prep for painting	↓	380	.021	↓	.11	.85		.96	1.50
8000	For chemical washing, see Section 04 01 30									
8010	For steam cleaning, see Section 04 01 30.20									

09 91 03.40 Interior Surface Preparation

		Crew	Daily Output	Labor-Hours	Unit	Material	Labor	Equipment	Total	Total Incl O&P
0010	**INTERIOR SURFACE PREPARATION**									
0020	Doors, per side, not incl. frames or trim									
0030	Scrape & sand									
0040	Wood, flush	1 Pord	616	.013	S.F.		.52		.52	.84
0050	Wood, detail		496	.016			.65		.65	1.05
0060	Wood, louvered	↓	280	.029	↓		1.15		1.15	1.85
0070	Wire brush									
0080	Metal, flush	1 Pord	640	.013	S.F.		.50		.50	.81
0090	Metal, detail		520	.015			.62		.62	1
0100	Metal, louvered	↓	360	.022	↓		.90		.90	1.44
0110	Hand wash									
0120	Wood, flush	1 Pord	2160	.004	S.F.		.15		.15	.24
0130	Wood, detailed		2000	.004			.16		.16	.26
0140	Wood, louvered		1360	.006			.24		.24	.38
0150	Metal, flush	↓	2160	.004	↓		.15		.15	.24

09 91 Painting

09 91 03 – Paint Restoration

09 91 03.40 Interior Surface Preparation		Crew	Daily Output	Labor-Hours	Unit	Material	2015 Bare Costs Labor	Equipment	Total	Total Incl O&P
0160	Metal, detail	1 Pord	2000	.004	S.F.		.16		.16	.26
0170	Metal, louvered		1360	.006			.24		.24	.38
0400	Windows, per side, not incl. trim									
0410	Scrape & sand									
0420	Wood, 1-2 lite	1 Pord	360	.022	S.F.		.90		.90	1.44
0430	Wood, 3-6 lite		320	.025			1.01		1.01	1.62
0440	Wood, 7-10 lite		280	.029			1.15		1.15	1.85
0450	Wood, 12 lite		240	.033			1.34		1.34	2.16
0460	Wood, Bay/Bow		360	.022			.90		.90	1.44
0470	Wire brush									
0480	Metal, 1-2 lite	1 Pord	520	.015	S.F.		.62		.62	1
0490	Metal, 3-6 lite		440	.018			.73		.73	1.18
0500	Metal, Bay/Bow		520	.015			.62		.62	1
0600	Walls, sanding, light=10-30%, medium - 30-70%,									
0610	heavy=70-100% of surface to sand									
0650	Walls, sand									
0660	Gypsum board or plaster, light	1 Pord	3077	.003	S.F.		.10		.10	.17
0670	Gypsum board or plaster, medium		2160	.004			.15		.15	.24
0680	Gypsum board or plaster, heavy		923	.009			.35		.35	.56
0690	Wood, T&G, light		2400	.003			.13		.13	.22
0700	Wood, T&G, med.		1600	.005			.20		.20	.32
0710	Wood, T&G, heavy		800	.010			.40		.40	.65
0720	Walls, wash									
0730	Gypsum board or plaster	1 Pord	3200	.003	S.F.		.10		.10	.16
0740	Wood, T&G		3200	.003			.10		.10	.16
0750	Masonry, brick & block, smooth		2800	.003			.12		.12	.19
0760	Masonry, brick & block, coarse		2000	.004			.16		.16	.26
8000	For chemical washing, see Section 04 01 30									
8010	For steam cleaning, see Section 04 01 30.20									
9010	Minimum labor/equipment charge	1 Pord	3	2.667	Job		108		108	173

09 91 03.41 Scrape After Fire Damage

		Crew	Daily Output	Labor-Hours	Unit	Material	Labor	Equipment	Total	Total Incl O&P
0010	**SCRAPE AFTER FIRE DAMAGE**									
0050	Boards, 1" x 4"	1 Pord	336	.024	L.F.		.96		.96	1.54
0060	1" x 6"		260	.031			1.24		1.24	2
0070	1" x 8"		207	.039			1.56		1.56	2.51
0080	1" x 10"		174	.046			1.86		1.86	2.98
0500	Framing, 2" x 4"		265	.030			1.22		1.22	1.96
0510	2" x 6"		221	.036			1.46		1.46	2.35
0520	2" x 8"		190	.042			1.70		1.70	2.73
0530	2" x 10"		165	.048			1.96		1.96	3.14
0540	2" x 12"		144	.056			2.24		2.24	3.60
1000	Heavy framing, 3" x 4"		226	.035			1.43		1.43	2.30
1010	4" x 4"		210	.038			1.54		1.54	2.47
1020	4" x 6"		191	.042			1.69		1.69	2.72
1030	4" x 8"		165	.048			1.96		1.96	3.14
1040	4" x 10"		144	.056			2.24		2.24	3.60
1060	4" x 12"		131	.061			2.46		2.46	3.96
2900	For sealing, light damage		825	.010	S.F.	.14	.39		.53	.78
2920	Heavy damage		460	.017	"	.31	.70		1.01	1.47
3000	For sandblasting, see Section 04 01 30.20									
9000	Minimum labor/equipment charge	1 Pord	3	2.667	Job		108		108	173

09 91 Painting

09 91 13 – Exterior Painting

09 91 13.30 Fences

		Crew	Daily Output	Labor-Hours	Unit	Material	2015 Bare Costs Labor	2015 Bare Costs Equipment	Total	Total Incl O&P
0010	**FENCES**									
0100	Chain link or wire metal, one side, water base									
0110	Roll & brush, first coat	1 Pord	960	.008	S.F.	.08	.34		.42	.63
0120	Second coat		1280	.006		.07	.25		.32	.49
0130	Spray, first coat		2275	.004		.08	.14		.22	.32
0140	Second coat	↓	2600	.003	↓	.08	.12		.20	.29
0150	Picket, water base									
0160	Roll & brush, first coat	1 Pord	865	.009	S.F.	.08	.37		.45	.69
0170	Second coat		1050	.008		.08	.31		.39	.58
0180	Spray, first coat		2275	.004		.08	.14		.22	.32
0190	Second coat	↓	2600	.003	↓	.08	.12		.20	.29
0200	Stockade, water base									
0210	Roll & brush, first coat	1 Pord	1040	.008	S.F.	.08	.31		.39	.59
0220	Second coat		1200	.007		.08	.27		.35	.52
0230	Spray, first coat		2275	.004		.08	.14		.22	.32
0240	Second coat	↓	2600	.003	↓	.08	.12		.20	.29
9000	Minimum labor/equipment charge		2	4	Job		161		161	259

09 91 13.42 Miscellaneous, Exterior

		Crew	Daily Output	Labor-Hours	Unit	Material	2015 Bare Costs Labor	2015 Bare Costs Equipment	Total	Total Incl O&P
0010	**MISCELLANEOUS, EXTERIOR**									
0100	Railing, ext., decorative wood, incl. cap & baluster									
0110	Newels & spindles @ 12" O.C.									
0120	Brushwork, stain, sand, seal & varnish									
0130	First coat	1 Pord	90	.089	L.F.	.81	3.59		4.40	6.65
0140	Second coat	"	120	.067	"	.81	2.69		3.50	5.20
0150	Rough sawn wood, 42" high, 2" x 2" verticals, 6" O.C.									
0160	Brushwork, stain, each coat	1 Pord	90	.089	L.F.	.26	3.59		3.85	6.05
0170	Wrought iron, 1" rail, 1/2" sq. verticals									
0180	Brushwork, zinc chromate, 60" high, bars 6" O.C.									
0190	Primer	1 Pord	130	.062	L.F.	.86	2.48		3.34	4.94
0200	Finish coat		130	.062		1.13	2.48		3.61	5.25
0210	Additional coat	↓	190	.042	↓	1.32	1.70		3.02	4.18
0220	Shutters or blinds, single panel, 2' x 4', paint all sides									
0230	Brushwork, primer	1 Pord	20	.400	Ea.	.66	16.15		16.81	26.50
0240	Finish coat, exterior latex		20	.400		.62	16.15		16.77	26.50
0250	Primer & 1 coat, exterior latex		13	.615		1.14	25		26.14	41.50
0260	Spray, primer		35	.229		.96	9.20		10.16	15.85
0270	Finish coat, exterior latex		35	.229		1.33	9.20		10.53	16.25
0280	Primer & 1 coat, exterior latex	↓	20	.400	↓	1.04	16.15		17.19	27
0290	For louvered shutters, add				S.F.	10%				
0300	Stair stringers, exterior, metal									
0310	Roll & brush, zinc chromate, to 14", each coat	1 Pord	320	.025	L.F.	.38	1.01		1.39	2.03
0320	Rough sawn wood, 4" x 12"									
0330	Roll & brush, exterior latex, each coat	1 Pord	215	.037	L.F.	.09	1.50		1.59	2.51
0340	Trellis/lattice, 2" x 2" @ 3" O.C. with 2" x 8" supports									
0350	Spray, latex, per side, each coat	1 Pord	475	.017	S.F.	.09	.68		.77	1.19
0450	Decking, ext., sealer, alkyd, brushwork, sealer coat		1140	.007		.10	.28		.38	.57
0460	1st coat		1140	.007		.10	.28		.38	.57
0470	2nd coat		1300	.006		.07	.25		.32	.48
0500	Paint, alkyd, brushwork, primer coat		1140	.007		.11	.28		.39	.59
0510	1st coat		1140	.007		.13	.28		.41	.60
0520	2nd coat		1300	.006		.09	.25		.34	.50
0600	Sand paint, alkyd, brushwork, 1 coat		150	.053		.14	2.15		2.29	3.61

09 91 Painting

09 91 13 – Exterior Painting

09 91 13.42 Miscellaneous, Exterior

		Crew	Daily Output	Labor-Hours	Unit	Material	2015 Bare Costs Labor	2015 Bare Costs Equipment	Total	Total Incl O&P
9000	Minimum labor/equipment charge	1 Pord	2	4	Job		161		161	259

09 91 13.60 Siding Exterior

		Crew	Daily Output	Labor-Hours	Unit	Material	Labor	Equipment	Total	Total Incl O&P
0010	**SIDING EXTERIOR**, Alkyd (oil base)									
0450	Steel siding, oil base, paint 1 coat, brushwork	2 Pord	2015	.008	S.F.	.10	.32		.42	.62
0500	Spray		4550	.004		.15	.14		.29	.40
0800	Paint 2 coats, brushwork		1300	.012		.20	.50		.70	1.02
1000	Spray		2750	.006		.17	.23		.40	.57
1200	Stucco, rough, oil base, paint 2 coats, brushwork		1300	.012		.20	.50		.70	1.02
1400	Roller		1625	.010		.21	.40		.61	.87
1600	Spray		2925	.005		.22	.22		.44	.60
1800	Texture 1-11 or clapboard, oil base, primer coat, brushwork		1300	.012		.15	.50		.65	.96
2000	Spray		4550	.004		.15	.14		.29	.39
2400	Paint 2 coats, brushwork		810	.020		.30	.80		1.10	1.60
2600	Spray		2600	.006		.33	.25		.58	.76
3400	Stain 2 coats, brushwork		950	.017		.17	.68		.85	1.28
4000	Spray		3050	.005		.19	.21		.40	.55
4200	Wood shingles, oil base primer coat, brushwork		1300	.012		.14	.50		.64	.95
4400	Spray		3900	.004		.13	.17		.30	.41
5000	Paint 2 coats, brushwork		810	.020		.25	.80		1.05	1.55
5200	Spray		2275	.007		.23	.28		.51	.72
6500	Stain 2 coats, brushwork		950	.017		.17	.68		.85	1.28
7000	Spray		2660	.006		.24	.24		.48	.65
8000	For latex paint, deduct					10%				
8100	For work over 12' H, from pipe scaffolding, add						15%			
8200	For work over 12' H, from extension ladder, add						25%			
8300	For work over 12' H, from swing staging, add						35%			
9000	Minimum labor/equipment charge	1 Pord	2	4	Job		161		161	259

09 91 13.62 Siding, Misc.

		Crew	Daily Output	Labor-Hours	Unit	Material	Labor	Equipment	Total	Total Incl O&P
0010	**SIDING, MISC.**, latex paint									
0100	Aluminum siding									
0110	Brushwork, primer	2 Pord	2275	.007	S.F.	.06	.28		.34	.53
0120	Finish coat, exterior latex		2275	.007		.06	.28		.34	.52
0130	Primer & 1 coat exterior latex		1300	.012		.13	.50		.63	.94
0140	Primer & 2 coats exterior latex		975	.016		.19	.66		.85	1.26
0150	Mineral fiber shingles									
0160	Brushwork, primer	2 Pord	1495	.011	S.F.	.15	.43		.58	.85
0170	Finish coat, industrial enamel		1495	.011		.18	.43		.61	.89
0180	Primer & 1 coat enamel		810	.020		.33	.80		1.13	1.64
0190	Primer & 2 coats enamel		540	.030		.51	1.20		1.71	2.49
0200	Roll, primer		1625	.010		.17	.40		.57	.82
0210	Finish coat, industrial enamel		1625	.010		.20	.40		.60	.86
0220	Primer & 1 coat enamel		975	.016		.36	.66		1.02	1.46
0230	Primer & 2 coats enamel		650	.025		.56	.99		1.55	2.22
0240	Spray, primer		3900	.004		.13	.17		.30	.41
0250	Finish coat, industrial enamel		3900	.004		.16	.17		.33	.45
0260	Primer & 1 coat enamel		2275	.007		.29	.28		.57	.78
0270	Primer & 2 coats enamel		1625	.010		.46	.40		.86	1.14
0280	Waterproof sealer, first coat		4485	.004		.09	.14		.23	.32
0290	Second coat		5235	.003		.08	.12		.20	.29
0300	Rough wood incl. shingles, shakes or rough sawn siding									
0310	Brushwork, primer	2 Pord	1280	.013	S.F.	.13	.50		.63	.96
0320	Finish coat, exterior latex		1280	.013		.10	.50		.60	.92

09 91 Painting

09 91 13 – Exterior Painting

09 91 13.62 Siding, Misc.

		Crew	Daily Output	Labor-Hours	Unit	Material	2015 Bare Costs Labor	Equipment	Total	Total Incl O&P
0330	Primer & 1 coat exterior latex	2 Pord	960	.017	S.F.	.24	.67		.91	1.34
0340	Primer & 2 coats exterior latex		700	.023		.34	.92		1.26	1.85
0350	Roll, primer		2925	.005		.18	.22		.40	.55
0360	Finish coat, exterior latex		2925	.005		.12	.22		.34	.48
0370	Primer & 1 coat exterior latex		1790	.009		.30	.36		.66	.91
0380	Primer & 2 coats exterior latex		1300	.012		.42	.50		.92	1.26
0390	Spray, primer		3900	.004		.15	.17		.32	.43
0400	Finish coat, exterior latex		3900	.004		.09	.17		.26	.37
0410	Primer & 1 coat exterior latex		2600	.006		.24	.25		.49	.67
0420	Primer & 2 coats exterior latex		2080	.008		.34	.31		.65	.87
0430	Waterproof sealer, first coat		4485	.004		.16	.14		.30	.40
0440	Second coat		4485	.004		.09	.14		.23	.32
0450	Smooth wood incl. butt, T&G, beveled, drop or B&B siding									
0460	Brushwork, primer	2 Pord	2325	.007	S.F.	.10	.28		.38	.56
0470	Finish coat, exterior latex		1280	.013		.10	.50		.60	.92
0480	Primer & 1 coat exterior latex		800	.020		.20	.81		1.01	1.52
0490	Primer & 2 coats exterior latex		630	.025		.30	1.02		1.32	1.98
0500	Roll, primer		2275	.007		.11	.28		.39	.58
0510	Finish coat, exterior latex		2275	.007		.11	.28		.39	.58
0520	Primer & 1 coat exterior latex		1300	.012		.22	.50		.72	1.04
0530	Primer & 2 coats exterior latex		975	.016		.33	.66		.99	1.42
0540	Spray, primer		4550	.004		.08	.14		.22	.32
0550	Finish coat, exterior latex		4550	.004		.09	.14		.23	.33
0560	Primer & 1 coat exterior latex		2600	.006		.18	.25		.43	.60
0570	Primer & 2 coats exterior latex		1950	.008		.27	.33		.60	.83
0580	Waterproof sealer, first coat		5230	.003		.09	.12		.21	.29
0590	Second coat		5980	.003		.09	.11		.20	.26
0600	For oil base paint, add					10%				
9000	Minimum labor/equipment charge	1 Pord	2	4	Job		161		161	259

09 91 13.70 Doors and Windows, Exterior

		Crew	Daily Output	Labor-Hours	Unit	Material	Labor	Equipment	Total	Total Incl O&P
0010	**DOORS AND WINDOWS, EXTERIOR**									
0100	Door frames & trim, only									
0110	Brushwork, primer	1 Pord	512	.016	L.F.	.06	.63		.69	1.08
0120	Finish coat, exterior latex		512	.016		.08	.63		.71	1.10
0130	Primer & 1 coat, exterior latex		300	.027		.14	1.08		1.22	1.88
0135	2 coats, exterior latex, both sides		15	.533	Ea.	6.95	21.50		28.45	42
0140	Primer & 2 coats, exterior latex		265	.030	L.F.	.22	1.22		1.44	2.20
0150	Doors, flush, both sides, incl. frame & trim									
0160	Roll & brush, primer	1 Pord	10	.800	Ea.	4.55	32.50		37.05	57
0170	Finish coat, exterior latex		10	.800		5.95	32.50		38.45	58.50
0180	Primer & 1 coat, exterior latex		7	1.143		10.50	46		56.50	85.50
0190	Primer & 2 coats, exterior latex		5	1.600		16.40	64.50		80.90	122
0200	Brushwork, stain, sealer & 2 coats polyurethane		4	2		28.50	80.50		109	162
0210	Doors, French, both sides, 10-15 lite, incl. frame & trim									
0220	Brushwork, primer	1 Pord	6	1.333	Ea.	2.27	54		56.27	89
0230	Finish coat, exterior latex		6	1.333		2.97	54		56.97	90
0240	Primer & 1 coat, exterior latex		3	2.667		5.25	108		113.25	179
0250	Primer & 2 coats, exterior latex		2	4		8.05	161		169.05	268
0260	Brushwork, stain, sealer & 2 coats polyurethane		2.50	3.200		10.30	129		139.30	219
0270	Doors, louvered, both sides, incl. frame & trim									
0280	Brushwork, primer	1 Pord	7	1.143	Ea.	4.55	46		50.55	79
0290	Finish coat, exterior latex		7	1.143		5.95	46		51.95	80.50

09 91 Painting

09 91 13 – Exterior Painting

09 91 13.70 Doors and Windows, Exterior

		Crew	Daily Output	Labor-Hours	Unit	Material	2015 Bare Costs Labor	Equipment	Total	Total Incl O&P
0300	Primer & 1 coat, exterior latex	1 Pord	4	2	Ea.	10.50	80.50		91	142
0310	Primer & 2 coats, exterior latex		3	2.667		16.10	108		124.10	191
0320	Brushwork, stain, sealer & 2 coats polyurethane		4.50	1.778		28.50	71.50		100	147
0330	Doors, panel, both sides, incl. frame & trim									
0340	Roll & brush, primer	1 Pord	6	1.333	Ea.	4.55	54		58.55	91.50
0350	Finish coat, exterior latex		6	1.333		5.95	54		59.95	93
0360	Primer & 1 coat, exterior latex		3	2.667		10.50	108		118.50	185
0370	Primer & 2 coats, exterior latex		2.50	3.200		16.10	129		145.10	226
0380	Brushwork, stain, sealer & 2 coats polyurethane		3	2.667		28.50	108		136.50	205
0400	Windows, per ext. side, based on 15 S.F.									
0410	1 to 6 lite									
0420	Brushwork, primer	1 Pord	13	.615	Ea.	.90	25		25.90	41
0430	Finish coat, exterior latex		13	.615		1.17	25		26.17	41.50
0440	Primer & 1 coat, exterior latex		8	1		2.07	40.50		42.57	67.50
0450	Primer & 2 coats, exterior latex		6	1.333		3.17	54		57.17	90
0460	Stain, sealer & 1 coat varnish		7	1.143		4.06	46		50.06	78.50
0470	7 to 10 lite									
0480	Brushwork, primer	1 Pord	11	.727	Ea.	.90	29.50		30.40	48
0490	Finish coat, exterior latex		11	.727		1.17	29.50		30.67	48.50
0500	Primer & 1 coat, exterior latex		7	1.143		2.07	46		48.07	76.50
0510	Primer & 2 coats, exterior latex		5	1.600		3.17	64.50		67.67	107
0520	Stain, sealer & 1 coat varnish		6	1.333		4.06	54		58.06	91
0530	12 lite									
0540	Brushwork, primer	1 Pord	10	.800	Ea.	.90	32.50		33.40	53
0550	Finish coat, exterior latex		10	.800		1.17	32.50		33.67	53.50
0560	Primer & 1 coat, exterior latex		6	1.333		2.07	54		56.07	89
0570	Primer & 2 coats, exterior latex		5	1.600		3.17	64.50		67.67	107
0580	Stain, sealer & 1 coat varnish		6	1.333		4.15	54		58.15	91
0590	For oil base paint, add					10%				
9000	Minimum labor/equipment charge	1 Pord	2	4	Job		161		161	259

09 91 13.80 Trim, Exterior

		Crew	Daily Output	Labor-Hours	Unit	Material	Labor	Equipment	Total	Total Incl O&P
0010	**TRIM, EXTERIOR**									
0100	Door frames & trim (see Doors, interior or exterior)									
0110	Fascia, latex paint, one coat coverage									
0120	1" x 4", brushwork	1 Pord	640	.013	L.F.	.02	.50		.52	.84
0130	Roll		1280	.006		.03	.25		.28	.44
0140	Spray		2080	.004		.02	.16		.18	.27
0150	1" x 6" to 1" x 10", brushwork		640	.013		.08	.50		.58	.90
0160	Roll		1230	.007		.09	.26		.35	.52
0170	Spray		2100	.004		.07	.15		.22	.32
0180	1" x 12", brushwork		640	.013		.08	.50		.58	.90
0190	Roll		1050	.008		.09	.31		.40	.59
0200	Spray		2200	.004		.07	.15		.22	.31
0210	Gutters & downspouts, metal, zinc chromate paint									
0220	Brushwork, gutters, 5", first coat	1 Pord	640	.013	L.F.	.40	.50		.90	1.25
0230	Second coat		960	.008		.38	.34		.72	.95
0240	Third coat		1280	.006		.30	.25		.55	.75
0250	Downspouts, 4", first coat		640	.013		.40	.50		.90	1.25
0260	Second coat		960	.008		.38	.34		.72	.95
0270	Third coat		1280	.006		.30	.25		.55	.75
0280	Gutters & downspouts, wood									
0290	Brushwork, gutters, 5", primer	1 Pord	640	.013	L.F.	.06	.50		.56	.88

09 91 Painting

09 91 13 – Exterior Painting

09 91 13.80 Trim, Exterior

		Crew	Daily Output	Labor-Hours	Unit	Material	2015 Bare Costs Labor	Equipment	Total	Total Incl O&P
0300	Finish coat, exterior latex	1 Pord	640	.013	L.F.	.07	.50		.57	.89
0310	Primer & 1 coat exterior latex		400	.020		.14	.81		.95	1.45
0320	Primer & 2 coats exterior latex		325	.025		.22	.99		1.21	1.84
0330	Downspouts, 4", primer		640	.013		.06	.50		.56	.88
0340	Finish coat, exterior latex		640	.013		.07	.50		.57	.89
0350	Primer & 1 coat exterior latex		400	.020		.14	.81		.95	1.45
0360	Primer & 2 coats exterior latex		325	.025		.11	.99		1.10	1.72
0370	Molding, exterior, up to 14" wide									
0380	Brushwork, primer	1 Pord	640	.013	L.F.	.07	.50		.57	.89
0390	Finish coat, exterior latex		640	.013		.08	.50		.58	.90
0400	Primer & 1 coat exterior latex		400	.020		.17	.81		.98	1.48
0410	Primer & 2 coats exterior latex		315	.025		.17	1.02		1.19	1.83
0420	Stain & fill		1050	.008		.10	.31		.41	.60
0430	Shellac		1850	.004		.13	.17		.30	.42
0440	Varnish		1275	.006		.11	.25		.36	.53
9000	Minimum labor/equipment charge		2	4	Job		161		161	259

09 91 13.90 Walls, Masonry (CMU), Exterior

		Crew	Daily Output	Labor-Hours	Unit	Material	Labor	Equipment	Total	Total Incl O&P
0010	**WALLS, MASONRY (CMU), EXTERIOR**									
0360	Concrete masonry units (CMU), smooth surface									
0370	Brushwork, latex, first coat	1 Pord	640	.013	S.F.	.07	.50		.57	.89
0380	Second coat		960	.008		.06	.34		.40	.60
0390	Waterproof sealer, first coat		736	.011		.25	.44		.69	.97
0400	Second coat		1104	.007		.25	.29		.54	.74
0410	Roll, latex, paint, first coat		1465	.005		.09	.22		.31	.44
0420	Second coat		1790	.004		.07	.18		.25	.36
0430	Waterproof sealer, first coat		1680	.005		.25	.19		.44	.58
0440	Second coat		2060	.004		.25	.16		.41	.52
0450	Spray, latex, paint, first coat		1950	.004		.07	.17		.24	.34
0460	Second coat		2600	.003		.05	.12		.17	.26
0470	Waterproof sealer, first coat		2245	.004		.25	.14		.39	.50
0480	Second coat		2990	.003		.25	.11		.36	.44
0490	Concrete masonry unit (CMU), porous									
0500	Brushwork, latex, first coat	1 Pord	640	.013	S.F.	.14	.50		.64	.97
0510	Second coat		960	.008		.07	.34		.41	.62
0520	Waterproof sealer, first coat		736	.011		.25	.44		.69	.97
0530	Second coat		1104	.007		.25	.29		.54	.74
0540	Roll latex, first coat		1465	.005		.11	.22		.33	.47
0550	Second coat		1790	.004		.07	.18		.25	.37
0560	Waterproof sealer, first coat		1680	.005		.25	.19		.44	.58
0570	Second coat		2060	.004		.25	.16		.41	.52
0580	Spray latex, first coat		1950	.004		.08	.17		.25	.36
0590	Second coat		2600	.003		.05	.12		.17	.26
0600	Waterproof sealer, first coat		2245	.004		.25	.14		.39	.50
0610	Second coat		2990	.003		.25	.11		.36	.44
9000	Minimum labor/equipment charge		2	4	Job		161		161	259

09 91 23 – Interior Painting

09 91 23.20 Cabinets and Casework

		Crew	Daily Output	Labor-Hours	Unit	Material	Labor	Equipment	Total	Total Incl O&P
0010	**CABINETS AND CASEWORK**									
1000	Primer coat, oil base, brushwork	1 Pord	650	.012	S.F.	.07	.50		.57	.88
2000	Paint, oil base, brushwork, 1 coat		650	.012		.11	.50		.61	.92
3000	Stain, brushwork, wipe off		650	.012		.09	.50		.59	.89
4000	Shellac, 1 coat, brushwork		650	.012		.11	.50		.61	.92

09 91 Painting

09 91 23 – Interior Painting

09 91 23.20 Cabinets and Casework

		Crew	Daily Output	Labor-Hours	Unit	Material	2015 Bare Costs Labor	2015 Bare Costs Equipment	Total	Total Incl O&P
4500	Varnish, 3 coats, brushwork, sand after 1st coat	1 Pord	325	.025	S.F.	.28	.99		1.27	1.91
5000	For latex paint, deduct					10%				
6300	Strip, prep and refinish wood furniture									
6310	Remove paint using chemicals, wood furniture	1 Pord	28	.286	S.F.	1.58	11.55		13.13	20.50
6320	Prep for painting, sanding		75	.107		.23	4.30		4.53	7.15
6350	Stain and wipe, brushwork		600	.013		.09	.54		.63	.95
6355	Spray applied		900	.009		.08	.36		.44	.66
6360	Sealer or varnish, brushwork		1080	.007		.09	.30		.39	.58
6365	Spray applied		2100	.004		.09	.15		.24	.35
6370	Paint, primer, brushwork		720	.011		.07	.45		.52	.80
6375	Spray applied		2100	.004		.07	.15		.22	.32
6380	Finish coat, brushwork		810	.010		.11	.40		.51	.76
6385	Spray applied		2100	.004		.10	.15		.25	.36
9010	Minimum labor/equipment charge		2	4	Job		161		161	259

09 91 23.33 Doors and Windows, Interior Alkyd (Oil Base)

		Crew	Daily Output	Labor-Hours	Unit	Material	2015 Bare Costs Labor	2015 Bare Costs Equipment	Total	Total Incl O&P
0010	**DOORS AND WINDOWS, INTERIOR ALKYD (OIL BASE)**									
0500	Flush door & frame, 3' x 7', oil, primer, brushwork	1 Pord	10	.800	Ea.	3.45	32.50		35.95	56
1000	Paint, 1 coat		10	.800		4.14	32.50		36.64	56.50
1400	Stain, brushwork, wipe off		18	.444		1.81	17.95		19.76	31
1600	Shellac, 1 coat, brushwork		25	.320		2.25	12.90		15.15	23.50
1800	Varnish, 3 coats, brushwork, sand after 1st coat		9	.889		5.85	36		41.85	64
2000	Panel door & frame, 3' x 7', oil, primer, brushwork		6	1.333		2.66	54		56.66	89.50
2200	Paint, 1 coat		6	1.333		4.14	54		58.14	91
2600	Stain, brushwork, panel door, 3' x 7', not incl. frame		16	.500		1.81	20		21.81	34.50
2800	Shellac, 1 coat, brushwork		22	.364		2.25	14.65		16.90	26
3000	Varnish, 3 coats, brushwork, sand after 1st coat		7.50	1.067		5.85	43		48.85	75.50
4400	Windows, including frame and trim, per side									
4600	Colonial type, 6/6 lites, 2' x 3', oil, primer, brushwork	1 Pord	14	.571	Ea.	.42	23		23.42	37.50
5800	Paint, 1 coat		14	.571		.65	23		23.65	37.50
6200	3' x 5' opening, 6/6 lites, primer coat, brushwork		12	.667		1.05	27		28.05	44
6400	Paint, 1 coat		12	.667		1.64	27		28.64	45
6800	4' x 8' opening, 6/6 lites, primer coat, brushwork		8	1		2.24	40.50		42.74	67.50
7000	Paint, 1 coat		8	1		3.49	40.50		43.99	69
8000	Single lite type, 2' x 3', oil base, primer coat, brushwork		33	.242		.42	9.80		10.22	16.15
8200	Paint, 1 coat		33	.242		.65	9.80		10.45	16.40
8600	3' x 5' opening, primer coat, brushwork		20	.400		1.05	16.15		17.20	27
8800	Paint, 1 coat		20	.400		1.64	16.15		17.79	28
9200	4' x 8' opening, primer coat, brushwork		14	.571		2.24	23		25.24	39.50
9400	Paint, 1 coat		14	.571		3.49	23		26.49	41
9900	Minimum labor/equipment charge		2	4	Job		161		161	259

09 91 23.35 Doors and Windows, Interior Latex

		Crew	Daily Output	Labor-Hours	Unit	Material	2015 Bare Costs Labor	2015 Bare Costs Equipment	Total	Total Incl O&P
0010	**DOORS & WINDOWS, INTERIOR LATEX**									
0100	Doors, flush, both sides, incl. frame & trim									
0110	Roll & brush, primer	1 Pord	10	.800	Ea.	4.13	32.50		36.63	56.50
0120	Finish coat, latex		10	.800		5.30	32.50		37.80	58
0130	Primer & 1 coat latex		7	1.143		9.40	46		55.40	84.50
0140	Primer & 2 coats latex		5	1.600		14.40	64.50		78.90	120
0160	Spray, both sides, primer		20	.400		4.35	16.15		20.50	31
0170	Finish coat, latex		20	.400		5.55	16.15		21.70	32
0180	Primer & 1 coat latex		11	.727		9.95	29.50		39.45	58
0190	Primer & 2 coats latex		8	1		15.25	40.50		55.75	82
0200	Doors, French, both sides, 10-15 lite, incl. frame & trim									

09 91 Painting

09 91 23 – Interior Painting

09 91 23.35 Doors and Windows, Interior Latex

		Crew	Daily Output	Labor-Hours	Unit	Material	2015 Bare Costs Labor	2015 Bare Costs Equipment	Total	Total Incl O&P
0210	Roll & brush, primer	1 Pord	6	1.333	Ea.	2.06	54		56.06	89
0220	Finish coat, latex		6	1.333		2.65	54		56.65	89.50
0230	Primer & 1 coat latex		3	2.667		4.71	108		112.71	178
0240	Primer & 2 coats latex		2	4		7.20	161		168.20	267
0260	Doors, louvered, both sides, incl. frame & trim									
0270	Roll & brush, primer	1 Pord	7	1.143	Ea.	4.13	46		50.13	78.50
0280	Finish coat, latex		7	1.143		5.30	46		51.30	80
0290	Primer & 1 coat, latex		4	2		9.20	80.50		89.70	140
0300	Primer & 2 coats, latex		3	2.667		14.70	108		122.70	189
0320	Spray, both sides, primer		20	.400		4.35	16.15		20.50	31
0330	Finish coat, latex		20	.400		5.55	16.15		21.70	32
0340	Primer & 1 coat, latex		11	.727		9.95	29.50		39.45	58
0350	Primer & 2 coats, latex		8	1		15.60	40.50		56.10	82
0360	Doors, panel, both sides, incl. frame & trim									
0370	Roll & brush, primer	1 Pord	6	1.333	Ea.	4.35	54		58.35	91.50
0380	Finish coat, latex		6	1.333		5.30	54		59.30	92.50
0390	Primer & 1 coat, latex		3	2.667		9.40	108		117.40	183
0400	Primer & 2 coats, latex		2.50	3.200		14.70	129		143.70	224
0420	Spray, both sides, primer		10	.800		4.35	32.50		36.85	57
0430	Finish coat, latex		10	.800		5.55	32.50		38.05	58
0440	Primer & 1 coat, latex		5	1.600		9.95	64.50		74.45	115
0450	Primer & 2 coats, latex		4	2		15.60	80.50		96.10	147
0460	Windows, per interior side, based on 15 S.F.									
0470	1 to 6 lite									
0480	Brushwork, primer	1 Pord	13	.615	Ea.	.81	25		25.81	41
0490	Finish coat, enamel		13	.615		1.04	25		26.04	41
0500	Primer & 1 coat enamel		8	1		1.86	40.50		42.36	67
0510	Primer & 2 coats enamel		6	1.333		2.90	54		56.90	89.50
0530	7 to 10 lite									
0540	Brushwork, primer	1 Pord	11	.727	Ea.	.81	29.50		30.31	48
0550	Finish coat, enamel		11	.727		1.04	29.50		30.54	48
0560	Primer & 1 coat enamel		7	1.143		1.86	46		47.86	76
0570	Primer & 2 coats enamel		5	1.600		2.90	64.50		67.40	107
0590	12 lite									
0600	Brushwork, primer	1 Pord	10	.800	Ea.	.81	32.50		33.31	53
0610	Finish coat, enamel		10	.800		1.04	32.50		33.54	53
0620	Primer & 1 coat enamel		6	1.333		1.86	54		55.86	88.50
0630	Primer & 2 coats enamel		5	1.600		2.90	64.50		67.40	107
0650	For oil base paint, add					10%				
9000	Minimum labor/equipment charge	1 Pord	2	4	Job		161		161	259

09 91 23.39 Doors and Windows, Interior Latex, Zero Voc

			Crew	Daily Output	Labor-Hours	Unit	Material	2015 Bare Costs Labor	2015 Bare Costs Equipment	Total	Total Incl O&P
0010	**DOORS & WINDOWS, INTERIOR LATEX, ZERO VOC**										
0100	Doors flush, both sides, incl. frame & trim										
0110	Roll & brush, primer	G	1 Pord	10	.800	Ea.	4.90	32.50		37.40	57.50
0120	Finish coat, latex	G		10	.800		5.70	32.50		38.20	58.50
0130	Primer & 1 coat latex	G		7	1.143		10.60	46		56.60	85.50
0140	Primer & 2 coats latex	G		5	1.600		16	64.50		80.50	122
0160	Spray, both sides, primer	G		20	.400		5.15	16.15		21.30	31.50
0170	Finish coat, latex	G		20	.400		6	16.15		22.15	32.50
0180	Primer & 1 coat latex	G		11	.727		11.25	29.50		40.75	59.50
0190	Primer & 2 coats latex	G		8	1		16.95	40.50		57.45	83.50
0200	Doors, French, both sides, 10-15 lite, incl. frame & trim										

09 91 Painting

09 91 23 – Interior Painting

09 91 23.39 Doors and Windows, Interior Latex, Zero Voc

		Crew	Daily Output	Labor-Hours	Unit	Material	2015 Bare Costs Labor	Equipment	Total	Total Incl O&P
0210	Roll & brush, primer	1 Pord [G]	6	1.333	Ea.	2.45	54		56.45	89
0220	Finish coat, latex	[G]	6	1.333		2.86	54		56.86	89.50
0230	Primer & 1 coat latex	[G]	3	2.667		5.30	108		113.30	179
0240	Primer & 2 coats latex	[G]	2	4		8	161		169	268
0360	Doors, panel, both sides, incl. frame & trim									
0370	Roll & brush, primer	1 Pord [G]	6	1.333	Ea.	5.15	54		59.15	92
0380	Finish coat, latex	[G]	6	1.333		5.70	54		59.70	93
0390	Primer & 1 coat, latex	[G]	3	2.667		10.60	108		118.60	185
0400	Primer & 2 coats, latex	[G]	2.50	3.200		16.35	129		145.35	226
0420	Spray, both sides, primer	[G]	10	.800		5.15	32.50		37.65	57.50
0430	Finish coat, latex	[G]	10	.800		6	32.50		38.50	58.50
0440	Primer & 1 coat, latex	[G]	5	1.600		11.25	64.50		75.75	116
0450	Primer & 2 coats, latex	[G]	4	2		17.30	80.50		97.80	149
0460	Windows, per interior side, based on 15 S.F.									
0470	1 to 6 lite									
0480	Brushwork, primer	1 Pord [G]	13	.615	Ea.	.97	25		25.97	41
0490	Finish coat, enamel	[G]	13	.615		1.13	25		26.13	41
0500	Primer & 1 coat enamel	[G]	8	1		2.09	40.50		42.59	67.50
0510	Primer & 2 coats enamel	[G]	6	1.333		3.22	54		57.22	90
9000	Minimum labor/equipment charge		2	4	Job		161		161	259

09 91 23.40 Floors, Interior

		Crew	Daily Output	Labor-Hours	Unit	Material	Labor	Equipment	Total	Total Incl O&P
0010	**FLOORS, INTERIOR**									
0100	Concrete paint, latex									
0110	Brushwork									
0120	1st coat	1 Pord	975	.008	S.F.	.15	.33		.48	.69
0130	2nd coat		1150	.007		.10	.28		.38	.56
0140	3rd coat		1300	.006		.08	.25		.33	.49
0150	Roll									
0160	1st coat	1 Pord	2600	.003	S.F.	.20	.12		.32	.42
0170	2nd coat		3250	.002		.12	.10		.22	.29
0180	3rd coat		3900	.002		.09	.08		.17	.23
0190	Spray									
0200	1st coat	1 Pord	2600	.003	S.F.	.17	.12		.29	.39
0210	2nd coat		3250	.002		.09	.10		.19	.26
0220	3rd coat		3900	.002		.07	.08		.15	.21

09 91 23.44 Anti-Slip Floor Treatments

		Crew	Daily Output	Labor-Hours	Unit	Material	Labor	Equipment	Total	Total Incl O&P
0010	**ANTI-SLIP FLOOR TREATMENTS**									
1000	Walking surface treatment, ADA compliant, mop on and rinse									
1100	For tile, terrazzo, stone or smooth concrete	1 Pord	4000	.002	S.F.	.32	.08		.40	.49
1110	For marble		4000	.002		.21	.08		.29	.36
1120	For wood		4000	.002		.20	.08		.28	.35
1130	For baths and showers		500	.016		.37	.65		1.02	1.45
2000	Granular additive for paint or sealer, add to paint cost					.02			.02	.02

09 91 23.52 Miscellaneous, Interior

		Crew	Daily Output	Labor-Hours	Unit	Material	Labor	Equipment	Total	Total Incl O&P
0010	**MISCELLANEOUS, INTERIOR**									
2400	Floors, conc./wood, oil base, primer/sealer coat, brushwork	2 Pord	1950	.008	S.F.	.09	.33		.42	.63
2450	Roller		5200	.003		.10	.12		.22	.31
2600	Spray		6000	.003		.10	.11		.21	.28
2650	Paint 1 coat, brushwork		1950	.008		.10	.33		.43	.64
2800	Roller		5200	.003		.10	.12		.22	.31
2850	Spray		6000	.003		.11	.11		.22	.29
3000	Stain, wood floor, brushwork, 1 coat		4550	.004		.09	.14		.23	.32

09 91 Painting

09 91 23 – Interior Painting

09 91 23.52 Miscellaneous, Interior

		Crew	Daily Output	Labor-Hours	Unit	Material	2015 Bare Costs Labor	Equipment	Total	Total Incl O&P
3200	Roller	2 Pord	5200	.003	S.F.	.09	.12		.21	.30
3250	Spray		6000	.003		.09	.11		.20	.27
3400	Varnish, wood floor, brushwork		4550	.004		.09	.14		.23	.33
3450	Roller		5200	.003		.10	.12		.22	.31
3600	Spray	↓	6000	.003		.10	.11		.21	.28
3800	Grilles, per side, oil base, primer coat, brushwork	1 Pord	520	.015		.14	.62		.76	1.15
3850	Spray		1140	.007		.15	.28		.43	.62
3920	Paint 2 coats, brushwork		325	.025		.42	.99		1.41	2.07
3940	Spray	↓	650	.012	↓	.48	.50		.98	1.33
4600	Miscellaneous surfaces, metallic paint, spray applied									
4610	Water based, non-tintable, warm silver	1 Pord	1140	.007	S.F.	.75	.28		1.03	1.29
4620	Rusted iron		1140	.007		.97	.28		1.25	1.53
4630	Low VOC, tintable	↓	1140	.007		.33	.28		.61	.82
5000	Pipe, 1" - 4" diameter, primer or sealer coat, oil base, brushwork	2 Pord	1250	.013	L.F.	.10	.52		.62	.94
5100	Spray		2165	.007		.09	.30		.39	.58
5350	Paint 2 coats, brushwork		775	.021		.19	.83		1.02	1.55
5400	Spray		1240	.013		.22	.52		.74	1.08
6300	13" - 16" diameter, primer or sealer coat, brushwork		310	.052		.39	2.08		2.47	3.78
6450	Spray		540	.030		.44	1.20		1.64	2.41
6500	Paint 2 coats, brushwork		195	.082		.78	3.31		4.09	6.15
6550	Spray	↓	310	.052	↓	.86	2.08		2.94	4.30
7000	Trim, wood, incl. puttying, under 6" wide									
7200	Primer coat, oil base, brushwork	1 Pord	650	.012	L.F.	.03	.50		.53	.84
7250	Paint, 1 coat, brushwork		650	.012		.05	.50		.55	.86
7450	3 coats		325	.025		.16	.99		1.15	1.77
7500	Over 6" wide, primer coat, brushwork		650	.012		.07	.50		.57	.88
7550	Paint, 1 coat, brushwork		650	.012		.11	.50		.61	.92
7650	3 coats		325	.025	↓	.31	.99		1.30	1.95
8000	Cornice, simple design, primer coat, oil base, brushwork		650	.012	S.F.	.07	.50		.57	.88
8250	Paint, 1 coat		650	.012		.11	.50		.61	.92
8350	Ornate design, primer coat		350	.023		.07	.92		.99	1.56
8400	Paint, 1 coat		350	.023		.11	.92		1.03	1.60
8600	Balustrades, primer coat, oil base, brushwork		520	.015		.07	.62		.69	1.08
8900	Trusses and wood frames, primer coat, oil base, brushwork		800	.010		.07	.40		.47	.73
8950	Spray		1200	.007		.07	.27		.34	.51
9220	Paint 2 coats, brushwork		500	.016		.21	.65		.86	1.27
9240	Spray		600	.013		.24	.54		.78	1.12
9260	Stain, brushwork, wipe off		600	.013		.09	.54		.63	.95
9280	Varnish, 3 coats, brushwork	↓	275	.029	↓	.28	1.17		1.45	2.20
9350	For latex paint, deduct					10%				
9900	Minimum labor/equipment charge	1 Pord	2	4	Job		161		161	259

09 91 23.62 Electrostatic Painting

		Crew	Daily Output	Labor-Hours	Unit	Material	2015 Bare Costs Labor	Equipment	Total	Total Incl O&P
0010	**ELECTROSTATIC PAINTING**									
0100	In shop									
0200	Flat surfaces (lockers, casework, elevator doors. etc.)									
0300	One coat	1 Pord	200	.040	S.F.	.53	1.61		2.14	3.17
0400	Two coats	"	120	.067	"	.77	2.69		3.46	5.15
0500	Irregular surfaces (furniture, door frames, etc.)									
0600	One coat	1 Pord	150	.053	S.F.	.53	2.15		2.68	4.04
0700	Two coats	"	100	.080	"	.77	3.23		4	6.05
0800	On site									
0900	Flat surfaces (lockers, casework, elevator doors, etc.)									

09 91 Painting

09 91 23 – Interior Painting

09 91 23.62 Electrostatic Painting

		Crew	Daily Output	Labor-Hours	Unit	Material	2015 Bare Costs Labor	2015 Bare Costs Equipment	Total	Total Incl O&P
1000	One coat	1 Pord	150	.053	S.F.	.53	2.15		2.68	4.04
1100	Two coats	"	100	.080	"	.77	3.23		4	6.05
1200	Irregular surfaces (furniture, door frames, etc)									
1300	One coat	1 Pord	115	.070	S.F.	.53	2.81		3.34	5.10
1400	Two coats		70	.114		.77	4.61		5.38	8.25
2000	Anti-microbial coating, hospital application	↓	150	.053	↓	.06	2.15		2.21	3.53

09 91 23.72 Walls and Ceilings, Interior

		Crew	Daily Output	Labor-Hours	Unit	Material	2015 Bare Costs Labor	2015 Bare Costs Equipment	Total	Total Incl O&P
0010	**WALLS AND CEILINGS, INTERIOR**									
0100	Concrete, drywall or plaster, latex, primer or sealer coat									
0200	Smooth finish, brushwork	1 Pord	1150	.007	S.F.	.06	.28		.34	.52
0240	Roller		1350	.006		.06	.24		.30	.45
0280	Spray		2750	.003		.05	.12		.17	.25
0300	Sand finish, brushwork		975	.008		.06	.33		.39	.60
0340	Roller		1150	.007		.06	.28		.34	.52
0380	Spray		2275	.004		.05	.14		.19	.29
0800	Paint 2 coats, smooth finish, brushwork		680	.012		.13	.47		.60	.91
0840	Roller		800	.010		.13	.40		.53	.80
0880	Spray		1625	.005		.12	.20		.32	.46
0900	Sand finish, brushwork		605	.013		.13	.53		.66	1.01
0940	Roller		1020	.008		.13	.32		.45	.66
0980	Spray		1700	.005		.12	.19		.31	.45
1600	Glaze coating, 2 coats, spray, clear		1200	.007		.49	.27		.76	.97
1640	Multicolor	↓	1200	.007	↓	.99	.27		1.26	1.52
1660	Painting walls, complete, including surface prep, primer &									
1670	2 coats finish, on drywall or plaster, with roller	1 Pord	325	.025	S.F.	.20	.99		1.19	1.82
1700	For oil base paint, add					10%				
1800	For ceiling installations, add						25%			
2000	Masonry or concrete block, primer/sealer, latex paint									
2100	Primer, smooth finish, brushwork	1 Pord	1000	.008	S.F.	.11	.32		.43	.64
2110	Roller		1150	.007		.11	.28		.39	.57
2180	Spray		2400	.003		.10	.13		.23	.33
2200	Sand finish, brushwork		850	.009		.11	.38		.49	.73
2210	Roller		975	.008		.11	.33		.44	.65
2280	Spray		2050	.004		.10	.16		.26	.36
2800	Primer plus one finish coat, smooth brush		525	.015		.30	.61		.91	1.32
2810	Roller		615	.013		.19	.53		.72	1.05
2880	Spray		1200	.007		.17	.27		.44	.61
2900	Sand finish, brushwork		450	.018		.19	.72		.91	1.36
2910	Roller		515	.016		.19	.63		.82	1.22
2980	Spray		1025	.008		.17	.31		.48	.69
3600	Glaze coating, 3 coats, spray, clear		900	.009		.70	.36		1.06	1.35
3620	Multicolor		900	.009		1.15	.36		1.51	1.85
4000	Block filler, 1 coat, brushwork		425	.019		.12	.76		.88	1.35
4100	Silicone, water repellent, 2 coats, spray	↓	2000	.004		.33	.16		.49	.63
4120	For oil base paint, add					10%				
8200	For work 8' - 15' H, add						10%			
8300	For work over 15' H, add						20%			
8400	For light textured surfaces, add						10%			
8410	Heavy textured, add						25%			
9900	Minimum labor/equipment charge	1 Pord	2	4	Job		161		161	259

09 91 Painting

09 91 23 – Interior Painting

09 91 23.74 Walls and Ceilings, Interior, Zero VOC Latex

		Crew	Daily Output	Labor-Hours	Unit	Material	2015 Bare Costs Labor	2015 Bare Costs Equipment	Total	Total Incl O&P
0010	**WALLS AND CEILINGS, INTERIOR, ZERO VOC LATEX**									
0100	Concrete, dry wall or plaster, latex, primer or sealer coat									
0200	Smooth finish, brushwork	1 Pord G	1150	.007	S.F.	.06	.28		.34	.52
0240	Roller	G	1350	.006		.06	.24		.30	.45
0280	Spray	G	2750	.003		.05	.12		.17	.24
0300	Sand finish, brushwork	G	975	.008		.06	.33		.39	.60
0340	Roller	G	1150	.007		.07	.28		.35	.52
0380	Spray	G	2275	.004		.05	.14		.19	.29
0800	Paint 2 coats, smooth finish, brushwork	G	680	.012		.15	.47		.62	.93
0840	Roller	G	800	.010		.16	.40		.56	.82
0880	Spray	G	1625	.005		.13	.20		.33	.47
0900	Sand finish, brushwork	G	605	.013		.15	.53		.68	1.02
0940	Roller	G	1020	.008		.16	.32		.48	.68
0980	Spray	G	1700	.005		.13	.19		.32	.46
1800	For ceiling installations, add						25%			
8200	For work 8' - 15' H, add						10%			
8300	For work over 15' H, add						20%			
9900	Minimum labor/equipment charge	1 Pord	2	4	Job		161		161	259

09 91 23.75 Dry Fall Painting

		Crew	Daily Output	Labor-Hours	Unit	Material	Labor	Equipment	Total	Total Incl O&P
0010	**DRY FALL PAINTING**									
0100	Sprayed on walls, gypsum board or plaster									
0220	One coat	1 Pord	2600	.003	S.F.	.06	.12		.18	.26
0250	Two coats		1560	.005		.11	.21		.32	.45
0280	Concrete or textured plaster, one coat		1560	.005		.06	.21		.27	.39
0310	Two coats		1300	.006		.11	.25		.36	.52
0340	Concrete block, one coat		1560	.005		.06	.21		.27	.39
0370	Two coats		1300	.006		.11	.25		.36	.52
0400	Wood, one coat		877	.009		.06	.37		.43	.65
0430	Two coats		650	.012		.11	.50		.61	.92
0440	On ceilings, gypsum board or plaster									
0470	One coat	1 Pord	1560	.005	S.F.	.06	.21		.27	.39
0500	Two coats		1300	.006		.11	.25		.36	.52
0530	Concrete or textured plaster, one coat		1560	.005		.06	.21		.27	.39
0560	Two coats		1300	.006		.11	.25		.36	.52
0570	Structural steel, bar joists or metal deck, one coat		1560	.005		.06	.21		.27	.39
0580	Two coats		1040	.008		.11	.31		.42	.62
9900	Minimum labor/equipment charge		2	4	Job		161		161	259

09 93 Staining and Transparent Finishing

09 93 23 – Interior Staining and Finishing

09 93 23.10 Varnish

		Crew	Daily Output	Labor-Hours	Unit	Material	Labor	Equipment	Total	Total Incl O&P
0010	**VARNISH**									
0012	1 coat + sealer, on wood trim, brush, no sanding included	1 Pord	400	.020	S.F.	.07	.81		.88	1.38
0020	1 coat + sealer, on wood trim, brush, no sanding included, no VOC		400	.020		.21	.81		1.02	1.53
0100	Hardwood floors, 2 coats, no sanding included, roller		1890	.004		.15	.17		.32	.44
9000	Minimum labor/equipment charge		4	2	Job		80.50		80.50	130

09 96 High-Performance Coatings

09 96 23 – Graffiti-Resistant Coatings

09 96 23.10 Graffiti Resistant Treatments

		Crew	Daily Output	Labor-Hours	Unit	Material	2015 Bare Costs Labor	2015 Bare Costs Equipment	Total	Total Incl O&P
0010	**GRAFFITI RESISTANT TREATMENTS**, sprayed on walls									
0100	Non-sacrificial, permanent non-stick coating, clear, on metals	1 Pord	2000	.004	S.F.	2.06	.16		2.22	2.53
0200	Concrete		2000	.004		2.35	.16		2.51	2.84
0300	Concrete block		2000	.004		3.03	.16		3.19	3.60
0400	Brick		2000	.004		3.44	.16		3.60	4.04
0500	Stone		2000	.004		3.44	.16		3.60	4.04
0600	Unpainted wood		2000	.004		3.97	.16		4.13	4.63
2000	Semi-permanent cross linking polymer primer, on metals		2000	.004		.70	.16		.86	1.03
2100	Concrete		2000	.004		.84	.16		1	1.18
2200	Concrete block		2000	.004		1.05	.16		1.21	1.41
2300	Brick		2000	.004		.84	.16		1	1.18
2400	Stone		2000	.004		.84	.16		1	1.18
2500	Unpainted wood		2000	.004		1.16	.16		1.32	1.54
3000	Top coat, on metals		2000	.004		.55	.16		.71	.86
3100	Concrete		2000	.004		.62	.16		.78	.95
3200	Concrete block		2000	.004		.87	.16		1.03	1.22
3300	Brick		2000	.004		.73	.16		.89	1.06
3400	Stone		2000	.004		.73	.16		.89	1.06
3500	Unpainted wood		2000	.004		.87	.16		1.03	1.22
5000	Sacrificial, water based, on metal		2000	.004		.32	.16		.48	.62
5100	Concrete		2000	.004		.32	.16		.48	.62
5200	Concrete block		2000	.004		.32	.16		.48	.62
5300	Brick		2000	.004		.32	.16		.48	.62
5400	Stone		2000	.004		.32	.16		.48	.62
5500	Unpainted wood		2000	.004		.32	.16		.48	.62
8000	Cleaner for use after treatment									
8100	Towels or wipes, per package of 30				Ea.	.63			.63	.70
8200	Aerosol spray, 24 oz. can				"	18			18	19.80

09 96 46 – Intumescent Coatings

09 96 46.10 Coatings, Intumescent

		Crew	Daily Output	Labor-Hours	Unit	Material	Labor	Equipment	Total	Total Incl O&P
0010	**COATINGS, INTUMESCENT**, spray applied									
0100	On exterior structural steel, 0.25" d.f.t.	1 Pord	475	.017	S.F.	.41	.68		1.09	1.54
0150	0.51" d.f.t.		350	.023		.41	.92		1.33	1.93
0200	0.98" d.f.t.		280	.029		.41	1.15		1.56	2.30
0300	On interior structural steel, 0.108" d.f.t.		300	.027		.41	1.08		1.49	2.18
0350	0.310" d.f.t.		150	.053		.41	2.15		2.56	3.91
0400	0.670" d.f.t.		100	.080		.41	3.23		3.64	5.65

09 96 53 – Elastomeric Coatings

09 96 53.10 Coatings, Elastomeric

		Crew	Daily Output	Labor-Hours	Unit	Material	Labor	Equipment	Total	Total Incl O&P
0010	**COATINGS, ELASTOMERIC**									
0020	High build, water proof, one coat system									
0100	Concrete, brush	1 Pord	650	.012	S.F.	.27	.50		.77	1.10

09 96 56 – Epoxy Coatings

09 96 56.20 Wall Coatings

		Crew	Daily Output	Labor-Hours	Unit	Material	Labor	Equipment	Total	Total Incl O&P
0010	**WALL COATINGS**									
0100	Acrylic glazed coatings, matte	1 Pord	525	.015	S.F.	.31	.61		.92	1.33
0200	Gloss		305	.026		.65	1.06		1.71	2.42
0300	Epoxy coatings, solvent based		525	.015		.40	.61		1.01	1.43
0400	Water based		170	.047		1.20	1.90		3.10	4.37
2400	Sprayed perlite or vermiculite, 1/16" thick, solvent based		2935	.003		.27	.11		.38	.48
2500	Water based		640	.013		.74	.50		1.24	1.62

09 96 High-Performance Coatings

09 96 56 – Epoxy Coatings

09 96 56.20 Wall Coatings		Crew	Daily Output	Labor-Hours	Unit	Material	2015 Bare Costs Labor	2015 Bare Costs Equipment	Total	Total Incl O&P
2700	Vinyl plastic wall coating, solvent based	1 Pord	735	.011	S.F.	.33	.44		.77	1.07
2800	Water based		240	.033		.82	1.34		2.16	3.06
3000	Urethane on smooth surface, 2 coats, solvent based		1135	.007		.27	.28		.55	.76
3100	Water based		665	.012		.59	.49		1.08	1.43

Division 10 – Specialties

Estimating Tips

General

- The items in this division are usually priced per square foot or each.
- Many items in Division 10 require some type of support system or special anchors that are not usually furnished with the item. The required anchors must be added to the estimate in the appropriate division.
- Some items in Division 10, such as lockers, may require assembly before installation. Verify the amount of assembly required. Assembly can often exceed installation time.

10 20 00 Interior Specialties

- Support angles and blocking are not included in the installation of toilet compartments, shower/dressing compartments, or cubicles. Appropriate line items from Divisions 5 or 6 may need to be added to support the installations.
- Toilet partitions are priced by the stall. A stall consists of a side wall, pilaster, and door with hardware. Toilet tissue holders and grab bars are extra.
- The required acoustical rating of a folding partition can have a significant impact on costs. Verify the sound transmission coefficient rating of the panel priced to the specification requirements.
- Grab bar installation does not include supplemental blocking or backing to support the required load. When grab bars are installed at an existing facility, provisions must be made to attach the grab bars to solid structure.

Reference Numbers

Reference numbers are shown in shaded boxes at the beginning of some major classifications. These numbers refer to related items in the Reference Section. The reference information may be an estimating procedure, an alternate pricing method, or technical information.

Note: Not all subdivisions listed here necessarily appear in this publication. ∎

Did you know?
RSMeans Online gives you the same access to RSMeans' data with 24/7 access:
- Quickly locate costs in the searchable database.
- Build cost lists, estimates, and reports in minutes.
- Adjust costs to any location in the U.S. and Canada with the click of a button.

Start your free trial today at **www.rsmeansonline.com**

RSMeansOnline

No part of this publication may be reproduced, stored in a retrieval system, or transmitted in any form or by any means without prior written permission of RSMeans.

10 11 Visual Display Units

10 11 13 – Chalkboards

10 11 13.13 Fixed Chalkboards

		Crew	Daily Output	Labor-Hours	Unit	Material	2015 Bare Costs Labor	2015 Bare Costs Equipment	Total	Total Incl O&P
0010	**FIXED CHALKBOARDS** Porcelain enamel steel									
3900	Wall hung									
4000	Aluminum frame and chalktrough									
4300	3' x 5'	2 Carp	15	1.067	Ea.	320	50		370	430
4600	4' x 12'	"	13	1.231	"	600	58		658	755
4700	Wood frame and chalktrough									
4800	3' x 4'	2 Carp	16	1	Ea.	194	47		241	291
5300	4' x 8'	"	13	1.231	"	345	58		403	475
5400	Liquid chalk, white porcelain enamel, wall hung									
5420	Deluxe units, aluminum trim and chalktrough									
5450	4' x 4'	2 Carp	16	1	Ea.	253	47		300	355
5550	4' x 12'	"	12	1.333	"	535	62.50		597.50	695
5700	Wood trim and chalktrough									
5900	4' x 4'	2 Carp	16	1	Ea.	715	47		762	860
6200	4' x 8'		14	1.143	"	970	53.50		1,023.50	1,175
9000	Minimum labor/equipment charge	↓	3	5.333	Job		250		250	410

10 11 13.43 Portable Chalkboards

		Crew	Daily Output	Labor-Hours	Unit	Material	Labor	Equipment	Total	Total Incl O&P
0010	**PORTABLE CHALKBOARDS**									
0100	Freestanding, reversible									
0120	Economy, wood frame, 4' x 6'									
0140	Chalkboard both sides				Ea.	610			610	670
0160	Chalkboard one side, cork other side				"	575			575	630
0200	Standard, lightweight satin finished aluminum, 4' x 6'									
0220	Chalkboard both sides				Ea.	635			635	695
0240	Chalkboard one side, cork other side				"	640			640	705
0300	Deluxe, heavy duty extruded aluminum, 4' x 6'									
0320	Chalkboard both sides				Ea.	1,050			1,050	1,150
0340	Chalkboard one side, cork other side				"	970			970	1,075

10 11 16 – Markerboards

10 11 16.53 Electronic Markerboards

		Crew	Daily Output	Labor-Hours	Unit	Material	Labor	Equipment	Total	Total Incl O&P
0010	**ELECTRONIC MARKERBOARDS**									
0100	Wall hung or free standing, 3' x 4' to 4' x 6'	2 Carp	8	2	S.F.	87.50	94		181.50	250
0150	5' x 6' to 4' x 8'		8	2	"	61	94		155	221
0500	Interactive projection module for existing whiteboards	↓	8	2	Ea.	1,300	94		1,394	1,575

10 11 23 – Tackboards

10 11 23.10 Fixed Tackboards

		Crew	Daily Output	Labor-Hours	Unit	Material	Labor	Equipment	Total	Total Incl O&P
0010	**FIXED TACKBOARDS**									
2120	Prefabricated, 1/4" cork, 3' x 5' with aluminum frame	2 Carp	16	1	Ea.	132	47		179	222
2140	Wood frame	"	16	1	"	156	47		203	248
2300	Glass enclosed cabinets, alum., cork panel, hinged doors									
2600	4' x 7', 3 door	2 Carp	10	1.600	Ea.	1,775	75		1,850	2,075
9000	Minimum labor/equipment charge	"	4	4	Job		188		188	310

10 11 23.20 Control Boards

		Crew	Daily Output	Labor-Hours	Unit	Material	Labor	Equipment	Total	Total Incl O&P
0010	**CONTROL BOARDS**									
1000	Hospital patient display board, 4-color custom design									
1010	Porcelain steel dry erase board, 36" x 24"	2 Carp	7.50	2.133	Ea.	246	100		346	435

10 13 Directories

10 13 10 – Building Directories

10 13 10.10 Directory Boards		Crew	Daily Output	Labor-Hours	Unit	Material	2015 Bare Costs Labor	Equipment	Total	Total Incl O&P
0010	**DIRECTORY BOARDS**									
0050	Plastic, glass covered, 30" x 20"	2 Carp	3	5.333	Ea.	198	250		448	625
0100	36" x 48"		2	8		845	375		1,220	1,550
0900	Outdoor, weatherproof, black plastic, 36" x 24"		2	8		760	375		1,135	1,450
1000	36" x 36"	↓	1.50	10.667	↓	880	500		1,380	1,800
9000	Minimum labor/equipment charge	1 Carp	1	8	Job		375		375	615

10 14 Signage

10 14 19 – Dimensional Letter Signage

10 14 19.10 Exterior Signs

		Crew	Daily Output	Labor-Hours	Unit	Material	Labor	Equipment	Total	Total Incl O&P
0010	**EXTERIOR SIGNS**									
3900	Plaques, custom, 20" x 30", for up to 450 letters, cast aluminum	2 Carp	4	4	Ea.	1,850	188		2,038	2,325
4000	Cast bronze		4	4		1,750	188		1,938	2,200
4200	30" x 36", up to 900 letters cast aluminum		3	5.333		2,625	250		2,875	3,300
4300	Cast bronze	↓	3	5.333		4,025	250		4,275	4,825
5100	Exit signs, 24 ga. alum., 14" x 12" surface mounted	1 Carp	30	.267		47.50	12.50		60	73
5200	10" x 7"	"	20	.400		25.50	18.80		44.30	59
6400	Replacement sign faces, 6" or 8"	1 Clab	50	.160	↓	62.50	6		68.50	78.50
9000	Minimum labor/equipment charge	1 Carp	4	2	Job		94		94	154

10 14 23 – Panel Signage

10 14 23.13 Engraved Panel Signage

		Crew	Daily Output	Labor-Hours	Unit	Material	Labor	Equipment	Total	Total Incl O&P
0010	**ENGRAVED PANEL SIGNAGE**, interior									
1010	Flexible door sign, adhesive back, w/Braille, 5/8" letters, 4" x 4"	1 Clab	32	.250	Ea.	33	9.40		42.40	52
1050	6" x 6"		32	.250		48.50	9.40		57.90	69
1100	8" x 2"		32	.250		33	9.40		42.40	52
1150	8" x 4"		32	.250		43.50	9.40		52.90	63
1200	8" x 8"		32	.250		53	9.40		62.40	74
1250	12" x 2"		32	.250		36	9.40		45.40	55
1300	12" x 6"		32	.250		39	9.40		48.40	58.50
1350	12" x 12"		32	.250		150	9.40		159.40	180
1500	Graphic symbols, 2" x 2"		32	.250		12	9.40		21.40	28.50
1550	6" x 6"		32	.250		31	9.40		40.40	49.50
1600	8" x 8"	↓	32	.250		39	9.40		48.40	58
2010	Corridor, stock acrylic, 2-sided, with mounting bracket, 2" x 8"	1 Carp	24	.333		24.50	15.65		40.15	52.50
2020	2" x 10"		24	.333		35.50	15.65		51.15	64.50
2050	3" x 8"		24	.333		28.50	15.65		44.15	57
2060	3" x 10"		24	.333		40	15.65		55.65	70
2070	3" x 12"		24	.333		37	15.65		52.65	66
2100	4" x 8"		24	.333		21	15.65		36.65	48.50
2110	4" x 10"		24	.333		38	15.65		53.65	67.50
2120	4" x 12"		24	.333	↓	51	15.65		66.65	82

10 14 53 – Traffic Signage

10 14 53.20 Traffic Signs

		Crew	Daily Output	Labor-Hours	Unit	Material	Labor	Equipment	Total	Total Incl O&P
0010	**TRAFFIC SIGNS**									
0012	Stock, 24" x 24", no posts, .080" alum. reflectorized	B-80	70	.457	Ea.	85	18.85	10.30	114.15	135
0100	High intensity		70	.457		97.50	18.85	10.30	126.65	150
0300	30" x 30", reflectorized		70	.457		123	18.85	10.30	152.15	177
0400	High intensity		70	.457		135	18.85	10.30	164.15	190
0600	Guide and directional signs, 12" x 18", reflectorized		70	.457		34.50	18.85	10.30	63.65	79.50
0700	High intensity	↓	70	.457	↓	52	18.85	10.30	81.15	99

10 14 Signage

10 14 53 – Traffic Signage

10 14 53.20 Traffic Signs

		Crew	Daily Output	Labor-Hours	Unit	Material	2015 Bare Costs Labor	2015 Bare Costs Equipment	Total	Total Incl O&P
0900	18" x 24", stock signs, reflectorized	B-80	70	.457	Ea.	47	18.85	10.30	76.15	93.50
1000	High intensity		70	.457		52	18.85	10.30	81.15	99
1200	24" x 24", stock signs, reflectorized		70	.457		57	18.85	10.30	86.15	104
1300	High intensity		70	.457		62	18.85	10.30	91.15	110
1500	Add to above for steel posts, galvanized, 10'-0" upright, bolted		200	.160		32.50	6.60	3.60	42.70	50.50
1600	12'-0" upright, bolted		140	.229		39	9.40	5.15	53.55	64
1800	Highway road signs, aluminum, over 20 S.F., reflectorized		350	.091	S.F.	33.50	3.77	2.06	39.33	45.50
2000	High intensity		350	.091		33.50	3.77	2.06	39.33	45.50
2200	Highway, suspended over road, 80 S.F. min., reflectorized		165	.194		32	8	4.37	44.37	53.50
2300	High intensity		165	.194		30.50	8	4.37	42.87	52

10 17 Telephone Specialties

10 17 16 – Telephone Enclosures

10 17 16.10 Commercial Telephone Enclosures

		Crew	Daily Output	Labor-Hours	Unit	Material	Labor	Equipment	Total	Total Incl O&P
0010	**COMMERCIAL TELEPHONE ENCLOSURES**									
0300	Shelf type, wall hung, recessed	2 Carp	5	3.200	Ea.	745	150		895	1,075
0400	Surface mount		5	3.200	"	1,625	150		1,775	2,025
9000	Minimum labor/equipment charge		4	4	Job		188		188	310

10 21 Compartments and Cubicles

10 21 13 – Toilet Compartments

10 21 13.13 Metal Toilet Compartments

		Crew	Daily Output	Labor-Hours	Unit	Material	Labor	Equipment	Total	Total Incl O&P
0010	**METAL TOILET COMPARTMENTS**									
0110	Cubicles, ceiling hung									
0200	Powder coated steel	2 Carp	4	4	Ea.	525	188		713	885
0500	Stainless steel	"	4	4		1,075	188		1,263	1,475
0600	For handicap units, incl. 52" grab bars, add					450			450	495
0900	Floor and ceiling anchored									
1000	Powder coated steel	2 Carp	5	3.200	Ea.	590	150		740	890
1300	Stainless steel	"	5	3.200		1,275	150		1,425	1,650
1400	For handicap units, incl. 52" grab bars, add					315			315	345
1610	Floor anchored									
1700	Powder coated steel	2 Carp	7	2.286	Ea.	590	107		697	825
2000	Stainless steel	"	7	2.286		1,400	107		1,507	1,725
2100	For handicap units, incl. 52" grab bars, add					310			310	345
2200	For juvenile units, deduct					41.50			41.50	45.50
2450	Floor anchored, headrail braced									
2500	Powder coated steel	2 Carp	6	2.667	Ea.	390	125		515	635
2804	Stainless steel	"	4.60	3.478		1,025	163		1,188	1,400
2900	For handicap units, incl. 52" grab bars, add					370			370	410
3000	Wall hung partitions, powder coated steel	2 Carp	7	2.286		635	107		742	875
3300	Stainless steel	"	7	2.286		1,650	107		1,757	2,000
3400	For handicap units, incl. 52" grab bars, add					370			370	410
4000	Screens, entrance, floor mounted, 58" high, 48" wide									
4200	Powder coated steel	2 Carp	15	1.067	Ea.	242	50		292	350
4500	Stainless steel	"	15	1.067	"	910	50		960	1,075
4650	Urinal screen, 18" wide									
4704	Powder coated steel	2 Carp	6.15	2.602	Ea.	211	122		333	430
5004	Stainless steel	"	6.15	2.602	"	580	122		702	840

10 21 Compartments and Cubicles

10 21 13 – Toilet Compartments

10 21 13.13 Metal Toilet Compartments

		Crew	Daily Output	Labor-Hours	Unit	Material	2015 Bare Costs Labor	2015 Bare Costs Equipment	Total	Total Incl O&P
5100	Floor mounted, head rail braced									
5300	Powder coated steel	2 Carp	8	2	Ea.	230	94		324	405
5600	Stainless steel	"	8	2	"	570	94		664	780
5750	Pilaster, flush									
5800	Powder coated steel	2 Carp	10	1.600	Ea.	278	75		353	430
6100	Stainless steel		10	1.600		625	75		700	815
6300	Post braced, powder coated steel		10	1.600		163	75		238	300
6600	Stainless steel		10	1.600		450	75		525	620
6700	Wall hung, bracket supported									
6800	Powder coated steel	2 Carp	10	1.600	Ea.	163	75		238	300
7100	Stainless steel		10	1.600		278	75		353	430
7400	Flange supported, powder coated steel		10	1.600		106	75		181	239
7700	Stainless steel		10	1.600		310	75		385	465
7800	Wedge type, powder coated steel		10	1.600		134	75		209	271
8100	Stainless steel		10	1.600		575	75		650	760
9000	Minimum labor/equipment charge	1 Carp	2.50	3.200	Job		150		150	246

10 21 13.14 Metal Toilet Compartment Components

		Crew	Daily Output	Labor-Hours	Unit	Material	2015 Bare Costs Labor	2015 Bare Costs Equipment	Total	Total Incl O&P
0010	**METAL TOILET COMPARTMENT COMPONENTS**									
0100	Pilasters									
0110	Overhead braced, powder coated steel, 7" wide x 82" high	2 Carp	22.20	.721	Ea.	73.50	34		107.50	137
0120	Stainless steel		22.20	.721		125	34		159	193
0130	Floor braced, powder coated steel, 7" wide x 70" high		23.30	.687		128	32		160	194
0140	Stainless steel		23.30	.687		245	32		277	325
0150	Ceiling hung, powder coated steel, 7" wide x 83" high		13.30	1.203		136	56.50		192.50	242
0160	Stainless steel		13.30	1.203		274	56.50		330.50	395
0170	Wall hung, powder coated steel, 3" wide x 58" high		18.90	.847		133	40		173	211
0180	Stainless steel		18.90	.847		193	40		233	278
0200	Panels									
0210	Powder coated steel, 31" wide x 58" high	2 Carp	18.90	.847	Ea.	137	40		177	215
0220	Stainless steel		18.90	.847		365	40		405	465
0230	Powder coated steel, 53" wide x 58" high		18.90	.847		169	40		209	251
0240	Stainless steel		18.90	.847		475	40		515	585
0250	Powder coated steel, 63" wide x 58" high		18.90	.847		205	40		245	291
0260	Stainless steel		18.90	.847		520	40		560	640
0300	Doors									
0310	Powder coated steel, 24" wide x 58" high	2 Carp	14.10	1.135	Ea.	139	53.50		192.50	241
0320	Stainless steel		14.10	1.135		290	53.50		343.50	410
0330	Powder coated steel, 26" wide x 58" high		14.10	1.135		141	53.50		194.50	244
0340	Stainless steel		14.10	1.135		300	53.50		353.50	420
0350	Powder coated steel, 28" wide x 58" high		14.10	1.135		162	53.50		215.50	266
0360	Stainless steel		14.10	1.135		335	53.50		388.50	455
0370	Powder coated steel, 36" wide x 58" high		14.10	1.135		174	53.50		227.50	280
0380	Stainless steel		14.10	1.135		375	53.50		428.50	500
0400	Headrails									
0410	For powder coated steel, 62" long	2 Carp	65	.246	Ea.	22	11.55		33.55	43.50
0420	Stainless steel		65	.246		22	11.55		33.55	43.50
0430	For powder coated steel, 84" long		50	.320		31.50	15		46.50	59.50
0440	Stainless steel		50	.320		31.50	15		46.50	59.50
0450	For powder coated steel, 120" long		30	.533		43	25		68	88.50
0460	Stainless steel		30	.533		42.50	25		67.50	88
9000	Minimum labor/equipment charge	1 Carp	4	2	Job		94		94	154

10 21 Compartments and Cubicles

10 21 13 – Toilet Compartments

10 21 13.16 Plastic-Laminate-Clad Toilet Compartments

		Crew	Daily Output	Labor-Hours	Unit	Material	2015 Bare Costs Labor	Equipment	Total	Total Incl O&P
0010	**PLASTIC-LAMINATE-CLAD TOILET COMPARTMENTS**									
0110	Cubicles, ceiling hung									
0300	Plastic laminate on particle board	2 Carp	4	4	Ea.	520	188		708	880
0600	For handicap units, incl. 52" grab bars, add				"	450			450	495
0900	Floor and ceiling anchored									
1100	Plastic laminate on particle board	2 Carp	5	3.200	Ea.	790	150		940	1,100
1400	For handicap units, incl. 52" grab bars, add				"	315			315	345
1610	Floor mounted									
1800	Plastic laminate on particle board	2 Carp	7	2.286	Ea.	535	107		642	760
2450	Floor mounted, headrail braced									
2600	Plastic laminate on particle board	2 Carp	6	2.667	Ea.	750	125		875	1,025
3400	For handicap units, incl. 52" grab bars, add					370			370	410
4300	Entrance screen, floor mtd., plas. lam., 58" high, 48" wide	2 Carp	15	1.067		610	50		660	750
4800	Urinal screen, 18" wide, ceiling braced, plastic laminate		8	2		194	94		288	365
5400	Floor mounted, headrail braced		8	2		200	94		294	375
5900	Pilaster, flush, plastic laminate		10	1.600		505	75		580	685
6400	Post braced, plastic laminate		10	1.600		305	75		380	460
6700	Wall hung, bracket supported									
6900	Plastic laminate on particle board	2 Carp	10	1.600	Ea.	94.50	75		169.50	227
7450	Flange supported									
7500	Plastic laminate on particle board	2 Carp	10	1.600	Ea.	230	75		305	375
9000	Minimum labor/equipment charge	1 Carp	2.50	3.200	Job		150		150	246

10 21 13.17 Plastic-Lam. Clad Toilet Compart. Components

		Crew	Daily Output	Labor-Hours	Unit	Material	Labor	Equipment	Total	Total Incl O&P
0010	**PLASTIC-LAMINATE CLAD TOILET COMPARTMENT COMPONENTS**									
0100	Pilasters									
0110	Overhead braced, 7" wide x 82" high	2 Carp	22.20	.721	Ea.	99.50	34		133.50	165
0130	Floor anchored, 7" wide x 70" high		23.30	.687		99.50	32		131.50	162
0150	Ceiling hung, 7" wide x 83" high		13.30	1.203		104	56.50		160.50	207
0180	Wall hung, 3" wide x 58" high		18.90	.847		93	40		133	167
0200	Panels									
0210	31" wide x 58" high	2 Carp	18.90	.847	Ea.	148	40		188	227
0230	51" wide x 58" high		18.90	.847		202	40		242	287
0250	63" wide x 58" high		18.90	.847		235	40		275	325
0300	Doors									
0310	24" wide x 58" high	2 Carp	14.10	1.135	Ea.	142	53.50		195.50	244
0330	26" wide x 58" high		14.10	1.135		147	53.50		200.50	250
0350	28" wide x 58" high		14.10	1.135		152	53.50		205.50	255
0370	36" wide x 58" high		14.10	1.135		182	53.50		235.50	288
0400	Headrails									
0410	62" long	2 Carp	65	.246	Ea.	22.50	11.55		34.05	44
0430	84" long		60	.267		31	12.50		43.50	54.50
0450	120" long		30	.533		42	25		67	87
9000	Minimum labor/equipment charge	1 Carp	4	2	Job		94		94	154

10 21 13.19 Plastic Toilet Compartments

		Crew	Daily Output	Labor-Hours	Unit	Material	Labor	Equipment	Total	Total Incl O&P
0010	**PLASTIC TOILET COMPARTMENTS**									
0110	Cubicles, ceiling hung									
0250	Phenolic	2 Carp	4	4	Ea.	865	188		1,053	1,250
0600	For handicap units, incl. 52" grab bars, add				"	450			450	495
0900	Floor and ceiling anchored									
1050	Phenolic	2 Carp	5	3.200	Ea.	810	150		960	1,125
1400	For handicap units, incl. 52" grab bars, add				"	315			315	345
1610	Floor mounted									

10 21 Compartments and Cubicles

10 21 13 – Toilet Compartments

10 21 13.19 Plastic Toilet Compartments

		Crew	Daily Output	Labor-Hours	Unit	Material	2015 Bare Costs Labor	Equipment	Total	Total Incl O&P
1750	Phenolic	2 Carp	7	2.286	Ea.	750	107		857	1,000
2100	For handicap units, incl. 52" grab bars, add					310			310	345
2200	For juvenile units, deduct					41.50			41.50	45.50
2450	Floor mounted, headrail braced									
2550	Phenolic	2 Carp	6	2.667	Ea.	750	125		875	1,025
9000	Minimum labor/equipment charge	1 Carp	2.50	3.200	Job		150		150	246

10 21 13.20 Plastic Toilet Compartment Components

		Crew	Daily Output	Labor-Hours	Unit	Material	2015 Bare Costs Labor	Equipment	Total	Total Incl O&P
0010	**PLASTIC TOILET COMPARTMENT COMPONENTS**									
0100	Pilasters									
0110	Overhead braced, polymer plastic, 7" wide x 82" high	2 Carp	22.20	.721	Ea.	121	34		155	189
0120	Phenolic		22.20	.721		151	34		185	222
0130	Floor braced, polymer plastic, 7" wide x 70" high		23.30	.687		171	32		203	242
0140	Phenolic		23.30	.687		142	32		174	209
0150	Ceiling hung, polymer plastic, 7" wide x 83" high		13.30	1.203		171	56.50		227.50	281
0160	Phenolic		13.30	1.203		161	56.50		217.50	270
0180	Wall hung, phenolic, 3" wide x 58" high		18.90	.847		96	40		136	170
0200	Panels									
0203	Polymer plastic, 18" high x 55" high	2 Carp	18.90	.847	Ea.	252	40		292	340
0206	Phenolic, 18" wide x 58" high		18.90	.847		222	40		262	310
0210	Polymer plastic, 31" high x 55" high		18.90	.847		305	40		345	400
0220	Phenolic, 31" wide x 58" high		18.90	.847		263	40		303	355
0223	Polymer plastic, 48" high x 55" high		18.90	.847		470	40		510	580
0226	Phenolic, 48" wide x 58" high		18.90	.847		430	40		470	535
0230	Polymer plastic, 51" wide x 55" high		18.90	.847		410	40		450	515
0240	Phenolic, 51" wide x 58" high		18.90	.847		400	40		440	505
0250	Polymer plastic, 63" wide x 55" high		18.90	.847		560	40		600	680
0260	Phenolic, 63" wide x 58" high		18.90	.847		420	40		460	525
0300	Doors									
0310	Polymer plastic, 24" wide x 55" high	2 Carp	14.10	1.135	Ea.	211	53.50		264.50	320
0320	Phenolic, 24" wide x 58" high		14.10	1.135		296	53.50		349.50	415
0330	Polymer plastic, 26" high x 55" high		14.10	1.135		225	53.50		278.50	335
0340	Phenolic, 26" wide x 58" high		14.10	1.135		315	53.50		368.50	435
0350	Polymer plastic, 28" wide x 55" high		14.10	1.135		250	53.50		303.50	365
0360	Phenolic, 28" wide x 58" high		14.10	1.135		330	53.50		383.50	455
0370	Polymer plastic, 36" wide x 55" high		14.10	1.135		291	53.50		344.50	410
0380	Phenolic, 36" wide x 58" high		14.10	1.135		415	53.50		468.50	545
0400	Headrails									
0410	For polymer plastic, 62" long	2 Carp	65	.246	Ea.	22.50	11.55		34.05	44
0420	Phenolic		65	.246		22.50	11.55		34.05	44
0430	For polymer plastic, 84" long		50	.320		32	15		47	60
0440	Phenolic		50	.320		32	15		47	60
0450	For polymer plastic, 120" long		30	.533		43	25		68	88.50
0460	Phenolic		30	.533		43	25		68	88.50
9000	Minimum labor/equipment charge	1 Carp	4	2	Job		94		94	154

10 21 13.40 Stone Toilet Compartments

		Crew	Daily Output	Labor-Hours	Unit	Material	2015 Bare Costs Labor	Equipment	Total	Total Incl O&P
0010	**STONE TOILET COMPARTMENTS**									
0100	Cubicles, ceiling hung, marble	2 Marb	2	8	Ea.	1,800	350		2,150	2,550
0600	For handicap units, incl. 52" grab bars, add					450			450	495
0800	Floor & ceiling anchored, marble	2 Marb	2.50	6.400		1,975	278		2,253	2,625
1400	For handicap units, incl. 52" grab bars, add					315			315	345
1600	Floor mounted, marble	2 Marb	3	5.333		1,225	232		1,457	1,725
2400	Floor mounted, headrail braced, marble	"	3	5.333		1,150	232		1,382	1,650

10 21 Compartments and Cubicles

10 21 13 – Toilet Compartments

10 21 13.40 Stone Toilet Compartments

		Crew	Daily Output	Labor-Hours	Unit	Material	2015 Bare Costs Labor	Equipment	Total	Total Incl O&P
2900	For handicap units, incl. 52" grab bars, add				Ea.	370			370	410
4100	Entrance screen, floor mounted marble, 58" high, 48" wide	2 Marb	9	1.778		795	77.50		872.50	995
4600	Urinal screen, 18" wide, ceiling braced, marble	D-1	6	2.667		755	112		867	1,025
5100	Floor mounted, head rail braced									
5200	Marble	D-1	6	2.667	Ea.	645	112		757	895
5700	Pilaster, flush, marble		9	1.778		840	75		915	1,050
6200	Post braced, marble		9	1.778		825	75		900	1,025
9000	Minimum labor/equipment charge	1 Carp	2.50	3.200	Job		150		150	246

10 21 23 – Cubicle Curtains and Track

10 21 23.16 Cubicle Track and Hardware

0010	**CUBICLE TRACK AND HARDWARE**									
0550	Polyester, antimicrobial, 9' ceiling height	1 Carp	425	.019	L.F.	19.65	.88		20.53	23
0560	8' ceiling height	"	425	.019	"	17.50	.88		18.38	20.50

10 22 Partitions

10 22 16 – Folding Gates

10 22 16.10 Security Gates

0010	**SECURITY GATES** for roll up type, see Section 08 33 13.10									
0300	Scissors type folding gate, ptd. steel, single, 6-1/2' high, 5-1/2' wide	2 Sswk	4	4	Opng.	226	211		437	630
0350	6-1/2' wide		4	4		246	211		457	650
0400	7-1/2' wide		4	4		237	211		448	640
0600	Double gate, 8' high, 8' wide		2.50	6.400		390	335		725	1,050
0650	10' wide		2.50	6.400		425	335		760	1,075
0700	12' wide		2	8		620	420		1,040	1,450
0750	14' wide		2	8		630	420		1,050	1,450
0900	Door gate, folding steel, 4' wide, 61" high		4	4		139	211		350	535
1000	71" high		4	4		169	211		380	565
1200	81" high		4	4		195	211		406	595
1300	Window gates, 2' to 4' wide, 31" high		4	4		80	211		291	470
1500	55" high		3.75	4.267		122	225		347	540
1600	79" high		3.50	4.571		144	241		385	595

10 22 19 – Demountable Partitions

10 22 19.43 Demountable Composite Partitions

0010	**DEMOUNTABLE COMPOSITE PARTITIONS**, add for doors									
0100	Do not deduct door openings from total L.F.									
0900	Demountable gypsum system on 2" to 2-1/2"									
1000	steel studs, 9' high, 3" to 3-3/4" thick									
1200	Vinyl clad gypsum	2 Carp	48	.333	L.F.	60	15.65		75.65	91.50
1300	Fabric clad gypsum		44	.364		150	17.05		167.05	193
1500	Steel clad gypsum		40	.400		167	18.80		185.80	215
1600	1.75 system, aluminum framing, vinyl clad hardboard,									
1800	paper honeycomb core panel, 1-3/4" to 2-1/2" thick									
1900	9' high	2 Carp	48	.333	L.F.	101	15.65		116.65	137
2100	7' high		60	.267		90.50	12.50		103	120
2200	5' high		80	.200		76.50	9.40		85.90	99.50
2250	Unitized gypsum system									
2300	Unitized panel, 9' high, 2" to 2-1/2" thick									
2350	Vinyl clad gypsum	2 Carp	48	.333	L.F.	130	15.65		145.65	169
2400	Fabric clad gypsum	"	44	.364	"	214	17.05		231.05	263
2500	Unitized mineral fiber system									

10 22 Partitions

10 22 19 – Demountable Partitions

10 22 19.43 Demountable Composite Partitions

		Crew	Daily Output	Labor-Hours	Unit	Material	2015 Bare Costs Labor	Equipment	Total	Total Incl O&P
2510	Unitized panel, 9' high, 2-1/4" thick, aluminum frame									
2550	Vinyl clad mineral fiber	2 Carp	48	.333	L.F.	129	15.65		144.65	168
2600	Fabric clad mineral fiber	"	44	.364	"	193	17.05		210.05	240
2800	Movable steel walls, modular system									
2900	Unitized panels, 9' high, 48" wide									
3100	Baked enamel, pre-finished	2 Carp	60	.267	L.F.	146	12.50		158.50	182
3200	Fabric clad steel	"	56	.286	"	212	13.40		225.40	255
5500	For acoustical partitions, add, unsealed				S.F.	2.36			2.36	2.60
5550	Sealed				"	11			11	12.10
5700	For doors, see Sections 08 11 & 08 16									
5800	For door hardware, see Section 08 71									
6100	In-plant modular office system, w/prehung hollow core door									
6200	3" thick polystyrene core panels									
6250	12' x 12', 2 wall	2 Clab	3.80	4.211	Ea.	4,025	158		4,183	4,700
6300	4 wall		1.90	8.421		6,150	315		6,465	7,275
6350	16' x 16', 2 wall		3.60	4.444		6,125	167		6,292	7,025
6400	4 wall		1.80	8.889		8,325	335		8,660	9,700
9000	Minimum labor/equipment charge	2 Carp	3	5.333	Job		250		250	410

10 22 23 – Portable Partitions, Screens, and Panels

10 22 23.13 Wall Screens

		Crew	Daily Output	Labor-Hours	Unit	Material	2015 Bare Costs Labor	Equipment	Total	Total Incl O&P
0010	**WALL SCREENS**, divider panels, free standing, fiber core									
0020	Fabric face straight									
0100	3'-0" long, 4'-0" high	2 Carp	100	.160	L.F.	123	7.50		130.50	147
0200	5'-0" high		90	.178		103	8.35		111.35	127
0500	6'-0" high		75	.213		104	10		114	131
0900	5'-0" long, 4'-0" high		175	.091		67.50	4.29		71.79	81.50
1000	5'-0" high		150	.107		75.50	5		80.50	91
1500	6"-0" high		125	.128		90.50	6		96.50	109
1600	6'-0" long, 5'-0" high		162	.099		75.50	4.64		80.14	90.50
3200	Economical panels, fabric face, 4'-0" long, 5'-0" high		132	.121		50	5.70		55.70	64.50
3250	6'-0" high		112	.143		55.50	6.70		62.20	72
3300	5'-0" long, 5'-0" high		150	.107		54.50	5		59.50	67.50
3350	6'-0" high		125	.128		50	6		56	65
3450	Acoustical panels, 60 to 90 NRC, 3'-0" long, 5'-0" high		90	.178		76.50	8.35		84.85	97.50
3550	6'-0" high		75	.213		89.50	10		99.50	115
3600	5'-0" long, 5'-0" high		150	.107		61	5		66	75
3650	6'-0" high		125	.128		69	6		75	85.50
3700	6'-0" long, 5'-0" high		162	.099		53	4.64		57.64	66
3750	6'-0" high		138	.116		82	5.45		87.45	99
3800	Economy acoustical panels, 40 N.R.C., 4'-0" long, 5'-0" high		132	.121		50	5.70		55.70	64.50
3850	6'-0" high		112	.143		55.50	6.70		62.20	72
3900	5'-0" long, 6'-0" high		125	.128		50	6		56	65
3950	6'-0" long, 5'-0" high		162	.099		47	4.64		51.64	59.50
9000	Minimum labor/equipment charge		3	5.333	Job		250		250	410

10 22 33 – Accordion Folding Partitions

10 22 33.10 Partitions, Accordion Folding

		Crew	Daily Output	Labor-Hours	Unit	Material	2015 Bare Costs Labor	Equipment	Total	Total Incl O&P
0010	**PARTITIONS, ACCORDION FOLDING**									
0100	Vinyl covered, over 150 S.F., frame not included									
0300	Residential, 1.25 lb. per S.F., 8' maximum height	2 Carp	300	.053	S.F.	25	2.50		27.50	31.50
0400	Commercial, 1.75 lb. per S.F., 8' maximum height		225	.071		28.50	3.34		31.84	37
0900	Acoustical, 3 lb. per S.F., 17' maximum height		100	.160		31.50	7.50		39	47.50
1200	5 lb. per S.F., 20' maximum height		95	.168		44	7.90		51.90	61.50

10 22 Partitions

10 22 33 – Accordion Folding Partitions

	10 22 33.10 Partitions, Accordion Folding	Crew	Daily Output	Labor-Hours	Unit	Material	2015 Bare Costs Labor	2015 Bare Costs Equipment	Total	Total Incl O&P
1500	Vinyl clad wood or steel, electric operation, 5.0 psf	2 Carp	160	.100	S.F.	63	4.70		67.70	76.50
1900	Wood, non-acoustic, birch or mahogany, to 10' high		300	.053		34	2.50		36.50	41
9000	Minimum labor/equipment charge		4	4	Job		188		188	310

10 22 39 – Folding Panel Partitions

10 22 39.10 Partitions, Folding Panel

		Crew	Daily Output	Labor-Hours	Unit	Material	Labor	Equipment	Total	Total Incl O&P
0010	**PARTITIONS, FOLDING PANEL**, acoustic, wood									
0100	Vinyl faced, to 18' high, 6 psf, economy trim	2 Carp	60	.267	S.F.	57	12.50		69.50	83
0150	Standard trim		45	.356		68	16.70		84.70	102
0200	Premium trim		30	.533		87.50	25		112.50	138
0400	Plastic laminate or hardwood finish, standard trim		60	.267		58.50	12.50		71	85
0500	Premium trim		30	.533		62.50	25		87.50	110
0600	Wood, low acoustical type, 4.5 psf, to 14' high		50	.320		42.50	15		57.50	71.50
9000	Minimum labor/equipment charge		4	4	Job		188		188	310

10 26 Wall and Door Protection

10 26 13 – Corner Guards

10 26 13.10 Metal Corner Guards

		Crew	Daily Output	Labor-Hours	Unit	Material	Labor	Equipment	Total	Total Incl O&P
0010	**METAL CORNER GUARDS**									
0020	Steel angle w/anchors, 1" x 1" x 1/4", 1.5#/L.F.	2 Carp	160	.100	L.F.	7.10	4.70		11.80	15.50
0100	2" x 2" x 1/4" angles, 3.2#/L.F.	"	150	.107	"	10.20	5		15.20	19.40
1000	Wall protection, stainless steel, 16 ga, 48" x 36" tall, screwed to studs	2 Skwk	500	.032	S.F.	8.25	1.56		9.81	11.65
1050	Wall end guard, stainless steel, 16 ga, 36" tall, screwed to studs	1 Skwk	30	.267	Ea.	25	12.95		37.95	48.50
9000	Minimum labor/equipment charge	1 Carp	2	4	Job		188		188	310

10 28 Toilet, Bath, and Laundry Accessories

10 28 13 – Toilet Accessories

10 28 13.13 Commercial Toilet Accessories

		Crew	Daily Output	Labor-Hours	Unit	Material	Labor	Equipment	Total	Total Incl O&P
0010	**COMMERCIAL TOILET ACCESSORIES**									
0200	Curtain rod, stainless steel, 5' long, 1" diameter	1 Carp	13	.615	Ea.	28.50	29		57.50	78.50
0300	1-1/4" diameter		13	.615		29	29		58	79
0350	Chrome, 1" diameter		13	.615		32.50	29		61.50	83
0360	For vinyl curtain, add		1950	.004	S.F.	.91	.19		1.10	1.32
0400	Diaper changing station, horizontal, wall mounted, plastic		10	.800	Ea.	229	37.50		266.50	315
0420	Vertical		10	.800		229	37.50		266.50	315
0430	Oval shaped		10	.800		225	37.50		262.50	310
0440	Recessed, with stainless steel flange		6	1.333		565	62.50		627.50	730
0500	Dispenser units, combined soap & towel dispensers,									
0510	mirror and shelf, flush mounted	1 Carp	10	.800	Ea.	310	37.50		347.50	400
0600	Towel dispenser and waste receptacle,									
0610	18 gallon capacity	1 Carp	10	.800	Ea.	300	37.50		337.50	390
0800	Grab bar, straight, 1-1/4" diameter, stainless steel, 18" long		24	.333		29	15.65		44.65	57.50
0900	24" long		23	.348		29	16.35		45.35	58.50
1000	30" long		22	.364		31.50	17.05		48.55	62.50
1100	36" long		20	.400		38.50	18.80		57.30	73.50
1105	42" long		20	.400		46	18.80		64.80	81.50
1120	Corner, 36" long		20	.400		85.50	18.80		104.30	125
1200	1-1/2" diameter, 24" long		23	.348		31	16.35		47.35	61
1300	36" long		20	.400		33.50	18.80		52.30	67.50
1310	42" long		18	.444		38	21		59	75.50

10 28 Toilet, Bath, and Laundry Accessories

10 28 13 – Toilet Accessories

10 28 13.13 Commercial Toilet Accessories		Crew	Daily Output	Labor-Hours	Unit	Material	2015 Bare Costs Labor	Equipment	Total	Total Incl O&P
1500	Tub bar, 1-1/4" diameter, 24" x 36"	1 Carp	14	.571	Ea.	92.50	27		119.50	146
1600	Plus vertical arm		12	.667		97.50	31.50		129	159
1900	End tub bar, 1" diameter, 90° angle, 16" x 32"		12	.667		109	31.50		140.50	172
2010	Tub/shower/toilet, 2-wall, 36" x 24"		12	.667		91	31.50		122.50	152
2110	Antimicrobial copper alloy finish, straight, 18" long		24	.333		68.50	15.65		84.15	101
2120	24" long		23	.348		75.50	16.35		91.85	110
2130	36" long		20	.400		90	18.80		108.80	130
2140	48" long		19	.421		103	19.75		122.75	146
2300	Hand dryer, surface mounted, electric, 115 volt, 20 amp		4	2		445	94		539	640
2400	230 volt, 10 amp		4	2		745	94		839	975
2450	Hand dryer, touch free, 1400 watt, 81,000 rpm		4	2		1,025	94		1,119	1,275
2600	Hat and coat strip, stainless steel, 4 hook, 36" long		24	.333		68	15.65		83.65	101
2700	6 hook, 60" long		20	.400		124	18.80		142.80	168
3000	Mirror, with stainless steel 3/4" square frame, 18" x 24"		20	.400		49	18.80		67.80	85
3100	36" x 24"		15	.533		109	25		134	161
3200	48" x 24"		10	.800		150	37.50		187.50	227
3300	72" x 24"		6	1.333		289	62.50		351.50	425
3500	With 5" stainless steel shelf, 18" x 24"		20	.400		194	18.80		212.80	244
3600	36" x 24"		15	.533		236	25		261	300
3700	48" x 24"		10	.800		245	37.50		282.50	330
3800	72" x 24"		6	1.333		300	62.50		362.50	435
4100	Mop holder strip, stainless steel, 5 holders, 48" long		20	.400		89	18.80		107.80	129
4200	Napkin/tampon dispenser, recessed		15	.533		590	25		615	690
4220	Semi-recessed		6.50	1.231		310	58		368	435
4250	Napkin receptacle, recessed		6.50	1.231		165	58		223	277
4300	Robe hook, single, regular		36	.222		18.95	10.45		29.40	38
4400	Heavy duty, concealed mounting		36	.222		19.60	10.45		30.05	38.50
4600	Soap dispenser, chrome, surface mounted, liquid		20	.400		46.50	18.80		65.30	82
4700	Powder		20	.400		56.50	18.80		75.30	93.50
5000	Recessed stainless steel, liquid		10	.800		154	37.50		191.50	231
5600	Shelf, stainless steel, 5" wide, 18 ga., 24" long		24	.333		80.50	15.65		96.15	114
5700	48" long		16	.500		157	23.50		180.50	211
5800	8" wide shelf, 18 ga., 24" long		22	.364		69	17.05		86.05	104
5900	48" long		14	.571		121	27		148	177
6000	Toilet seat cover dispenser, stainless steel, recessed		20	.400		167	18.80		185.80	214
6050	Surface mounted		15	.533		34.50	25		59.50	79
6100	Toilet tissue dispenser, surface mounted, SS, single roll		30	.267		17.80	12.50		30.30	40
6200	Double roll		24	.333		23.50	15.65		39.15	51.50
6240	Plastic, twin/jumbo dbl. roll		24	.333		28.50	15.65		44.15	56.50
6290	Toilet seat	1 Plum	40	.200		30	11.75		41.75	51.50
6400	Towel bar, stainless steel, 18" long	1 Carp	23	.348		41.50	16.35		57.85	73
6500	30" long	"	21	.381		111	17.90		128.90	153
6610	Antimicrobial copper alloy finish, 3/4" round, straight, w/o mounting				L.F.	10.95			10.95	12.05
6620	Antimicrobial copper alloy finish, 1" round, straight, w/o mounting				"	13.15			13.15	14.50
6630	24" long, including mounting	1 Carp	23	.348	Ea.	93.50	16.35		109.85	130
6700	Towel dispenser, stainless steel, surface mounted		16	.500		44.50	23.50		68	87.50
6800	Flush mounted, recessed		10	.800		257	37.50		294.50	345
6900	Plastic, touchless, battery operated		16	.500		88	23.50		111.50	136
7000	Towel holder, hotel type, 2 guest size		20	.400		52	18.80		70.80	88.50
7200	Towel shelf, stainless steel, 24" long, 8" wide		20	.400		60	18.80		78.80	97
7400	Tumbler holder, for tumbler only		30	.267		17.80	12.50		30.30	40
7410	Tumbler holder, recessed		20	.400		9.80	18.80		28.60	42
7500	Soap, tumbler & toothbrush		30	.267		19.60	12.50		32.10	42

10 28 Toilet, Bath, and Laundry Accessories

10 28 13 – Toilet Accessories

10 28 13.13 Commercial Toilet Accessories

		Crew	Daily Output	Labor-Hours	Unit	Material	2015 Bare Costs Labor	Equipment	Total	Total Incl O&P
7510	Tumbler & toothbrush holder	1 Carp	20	.400	Ea.	13.60	18.80		32.40	46
7700	Wall urn ash receiver, surface mount, 11" long		12	.667		95	31.50		126.50	156
8000	Waste receptacles, stainless steel, with top, 13 gallon		10	.800		295	37.50		332.50	385
8100	36 gallon		8	1		405	47		452	520
9000	Minimum labor/equipment charge		5	1.600	Job		75		75	123

10 28 16 – Bath Accessories

10 28 16.20 Medicine Cabinets

		Crew	Daily Output	Labor-Hours	Unit	Material	Labor	Equipment	Total	Total Incl O&P
0010	**MEDICINE CABINETS**									
0020	With mirror, sst frame, 16" x 22", unlighted	1 Carp	14	.571	Ea.	98	27		125	152
0100	Wood frame		14	.571		128	27		155	185
0300	Sliding mirror doors, 20" x 16" x 4-3/4", unlighted		7	1.143		124	53.50		177.50	224
0400	24" x 19" x 8-1/2", lighted		5	1.600		179	75		254	320
0600	Triple door, 30" x 32", unlighted, plywood body		7	1.143		325	53.50		378.50	450
0700	Steel body		7	1.143		375	53.50		428.50	500
0900	Oak door, wood body, beveled mirror, single door		7	1.143		199	53.50		252.50	305
1000	Double door		6	1.333		380	62.50		442.50	525
1200	Hotel cabinets, stainless, with lower shelf, unlighted		10	.800		200	37.50		237.50	281
1300	Lighted		5	1.600		305	75		380	460
9000	Minimum labor/equipment charge		4	2	Job		94		94	154

10 28 19 – Tub and Shower Enclosures

10 28 19.10 Partitions, Shower

		Crew	Daily Output	Labor-Hours	Unit	Material	Labor	Equipment	Total	Total Incl O&P
0010	**PARTITIONS, SHOWER** floor mounted, no plumbing									
0400	Cabinet, one piece, fiberglass, 32" x 32"	2 Carp	5	3.200	Ea.	530	150		680	830
0420	36" x 36"		5	3.200		570	150		720	870
0440	36" x 48"		5	3.200		1,375	150		1,525	1,750
0460	Acrylic, 32" x 32"		5	3.200		310	150		460	585
0480	36" x 36"		5	3.200		1,025	150		1,175	1,375
0500	36" x 48"		5	3.200		1,500	150		1,650	1,925
0520	Shower door for above, clear plastic, 24" wide	1 Carp	8	1		186	47		233	281
0540	28" wide		8	1		206	47		253	305
0560	Tempered glass, 24" wide		8	1		200	47		247	297
0580	28" wide		8	1		226	47		273	325
4100	Shower doors, economy plastic, 24" wide	1 Shee	9	.889		144	49.50		193.50	237
4200	Tempered glass door, economy		8	1		258	56		314	375
4700	Deluxe, tempered glass, chrome on brass frame, 42" to 44"		8	1		385	56		441	510
4800	39" to 48" wide		1	8		625	450		1,075	1,400
9990	Minimum labor/equipment charge		2.50	3.200	Job		179		179	283

10 31 Manufactured Fireplaces

10 31 13 – Manufactured Fireplace Chimneys

10 31 13.10 Fireplace Chimneys

		Crew	Daily Output	Labor-Hours	Unit	Material	Labor	Equipment	Total	Total Incl O&P
0010	**FIREPLACE CHIMNEYS**									
0500	Chimney dbl. wall, all stainless, over 8'-6", 7" diam., add to fireplace	1 Carp	33	.242	V.L.F.	83.50	11.40		94.90	110
0600	10" diameter, add to fireplace		32	.250		132	11.75		143.75	164
0700	12" diameter, add to fireplace		31	.258		158	12.10		170.10	194
0800	14" diameter, add to fireplace		30	.267		227	12.50		239.50	271
1000	Simulated brick chimney top, 4' high, 16" x 16"		10	.800	Ea.	425	37.50		462.50	530
1100	24" x 24"		7	1.143	"	540	53.50		593.50	685

10 31 Manufactured Fireplaces

10 31 13 – Manufactured Fireplace Chimneys

10 31 13.20 Chimney Accessories

		Crew	Daily Output	Labor-Hours	Unit	Material	2015 Bare Costs Labor	Equipment	Total	Total Incl O&P
0010	**CHIMNEY ACCESSORIES**									
0020	Chimney screens, galv., 13" x 13" flue	1 Bric	8	1	Ea.	58.50	46		104.50	139
0050	24" x 24" flue		5	1.600		124	74		198	257
0200	Stainless steel, 13" x 13" flue		8	1		97.50	46		143.50	182
0250	20" x 20" flue		5	1.600		152	74		226	287
2400	Squirrel and bird screens, galvanized, 8" x 8" flue		16	.500		53	23		76	96
2450	13" x 13" flue		12	.667		55	31		86	111
9000	Minimum labor/equipment charge		3.50	2.286	Job		105		105	172

10 31 16 – Manufactured Fireplace Forms

10 31 16.10 Fireplace Forms

		Crew	Daily Output	Labor-Hours	Unit	Material	Labor	Equipment	Total	Total Incl O&P
0010	**FIREPLACE FORMS**									
1800	Fireplace forms, no accessories, 32" opening	1 Bric	3	2.667	Ea.	670	123		793	940
1900	36" opening		2.50	3.200		855	148		1,003	1,175
2000	40" opening		2	4		1,125	185		1,310	1,550
2100	78" opening		1.50	5.333		1,650	246		1,896	2,225

10 31 23 – Prefabricated Fireplaces

10 31 23.10 Fireplace, Prefabricated

		Crew	Daily Output	Labor-Hours	Unit	Material	Labor	Equipment	Total	Total Incl O&P
0010	**FIREPLACE, PREFABRICATED**, free standing or wall hung									
0100	With hood & screen, painted	1 Carp	1.30	6.154	Ea.	1,375	289		1,664	2,000
0150	Average		1	8		1,625	375		2,000	2,400
0200	Stainless steel		.90	8.889		3,050	415		3,465	4,050
1500	Simulated logs, gas fired, 40,000 BTU, 2' long, manual safety pilot		7	1.143	Set	425	53.50		478.50	555
1600	Adjustable flame remote pilot		6	1.333		1,100	62.50		1,162.50	1,325
1700	Electric, 1,500 BTU, 1'-6" long, incandescent flame		7	1.143		197	53.50		250.50	305
1800	1,500 BTU, LED flame		6	1.333		300	62.50		362.50	440
2000	Fireplace, built-in, 36" hearth, radiant		1.30	6.154	Ea.	660	289		949	1,200
2100	Recirculating, small fan		1	8		880	375		1,255	1,575
2150	Large fan		.90	8.889		1,900	415		2,315	2,775
2200	42" hearth, radiant		1.20	6.667		890	315		1,205	1,500
2300	Recirculating, small fan		.90	8.889		1,175	415		1,590	1,975
2350	Large fan		.80	10		1,300	470		1,770	2,200
2400	48" hearth, radiant		1.10	7.273		2,075	340		2,415	2,825
2500	Recirculating, small fan		.80	10		2,350	470		2,820	3,375
2550	Large fan		.70	11.429		2,375	535		2,910	3,475
3000	See through, including doors		.80	10		2,225	470		2,695	3,225
3200	Corner (2 wall)		1	8		3,250	375		3,625	4,200

10 32 Fireplace Specialties

10 32 13 – Fireplace Dampers

10 32 13.10 Dampers

		Crew	Daily Output	Labor-Hours	Unit	Material	Labor	Equipment	Total	Total Incl O&P
0010	**DAMPERS**									
0800	Damper, rotary control, steel, 30" opening	1 Bric	6	1.333	Ea.	119	61.50		180.50	231
0850	Cast iron, 30" opening		6	1.333		125	61.50		186.50	237
0880	36" opening		6	1.333		127	61.50		188.50	240
0900	48" opening		6	1.333		167	61.50		228.50	284
0920	60" opening		6	1.333		355	61.50		416.50	490
1000	84" opening, special order		5	1.600		910	74		984	1,125
1050	96" opening, special order		4	2		925	92.50		1,017.50	1,175
1200	Steel plate, poker control, 60" opening		8	1		320	46		366	430

10 32 Fireplace Specialties

10 32 13 – Fireplace Dampers

10 32 13.10 Dampers

		Crew	Daily Output	Labor-Hours	Unit	Material	2015 Bare Costs Labor	2015 Bare Costs Equipment	Total	Total Incl O&P
1250	84" opening, special order	1 Bric	5	1.600	Ea.	585	74		659	765
1400	"Universal" type, chain operated, 32" x 20" opening		8	1		250	46		296	350
1450	48" x 24" opening		5	1.600		375	74		449	530

10 32 23 – Fireplace Doors

10 32 23.10 Doors

		Crew	Daily Output	Labor-Hours	Unit	Material	Labor	Equipment	Total	Total Incl O&P
0010	**DOORS**									
0400	Cleanout doors and frames, cast iron, 8" x 8"	1 Bric	12	.667	Ea.	41	31		72	95
0450	12" x 12"		10	.800		108	37		145	179
0500	18" x 24"		8	1		150	46		196	240
0550	Cast iron frame, steel door, 24" x 30"		5	1.600		315	74		389	465
1600	Dutch Oven door and frame, cast iron, 12" x 15" opening		13	.615		131	28.50		159.50	191
1650	Copper plated, 12" x 15" opening		13	.615		257	28.50		285.50	330

10 35 Stoves

10 35 13 – Heating Stoves

10 35 13.10 Woodburning Stoves

		Crew	Daily Output	Labor-Hours	Unit	Material	Labor	Equipment	Total	Total Incl O&P
0010	**WOODBURNING STOVES**									
0015	Cast iron, °1500sf	2 Carp	1.30	12.308	Ea.	1,175	580		1,755	2,250
0020	1500-2000sf		1	16		2,025	750		2,775	3,450
0030	2,000 sf		.80	20		2,775	940		3,715	4,600
0050	For gas log lighter, add					45			45	49.50

10 43 Emergency Aid Specialties

10 43 13 – Defibrillator Cabinets

10 43 13.05 Defibrillator Cabinets

		Crew	Daily Output	Labor-Hours	Unit	Material	Labor	Equipment	Total	Total Incl O&P
0010	**DEFIBRILLATOR CABINETS**, not equipped, stainless steel									
0050	Defibrillator cabinet, stainless steel with strobe & alarm 12" x 27"	1 Carp	10	.800	Ea.	430	37.50		467.50	530
0100	Automatic External Defibrillator	"	30	.267	"	1,350	12.50		1,362.50	1,500

10 44 Fire Protection Specialties

10 44 13 – Fire Protection Cabinets

10 44 13.53 Fire Equipment Cabinets

		Crew	Daily Output	Labor-Hours	Unit	Material	Labor	Equipment	Total	Total Incl O&P
0010	**FIRE EQUIPMENT CABINETS**, not equipped, 20 ga. steel box									
0040	recessed, D.S. glass in door, box size given									
1000	Portable extinguisher, single, 8" x 12" x 27", alum. door & frame	Q-12	8	2	Ea.	155	101		256	330
1100	Steel door and frame	"	8	2	"	116	101		217	286
3000	Hose rack assy., 1-1/2" valve & 100' hose, 24" x 40" x 5-1/2"									
3200	Steel door and frame	Q-12	6	2.667	Ea.	246	135		381	480
4000	Hose rack assy., 2-1/2" x 1-1/2" valve, 100' hose, 24" x 40" x 8"									
4200	Steel door and frame	Q-12	6	2.667	Ea.	253	135		388	490
5000	Hose rack assy., 2-1/2" x 1-1/2" valve, 100' hose									
5010	and extinguisher, 30" x 40" x 8"									
5200	Steel door and frame	Q-12	5	3.200	Ea.	264	162		426	545

10 44 Fire Protection Specialties

10 44 16 – Fire Extinguishers

10 44 16.13 Portable Fire Extinguishers	Crew	Daily Output	Labor-Hours	Unit	Material	2015 Bare Costs Labor	Equipment	Total	Total Incl O&P
0010 **PORTABLE FIRE EXTINGUISHERS**									
0140 CO_2, with hose and "H" horn, 10 lb.				Ea.	275			275	305
1000 Dry chemical, pressurized									
1040 Standard type, portable, painted, 2-1/2 lb.				Ea.	37.50			37.50	41
1080 10 lb.					81			81	89.50
1100 20 lb.					136			136	150
1120 30 lb.					425			425	470
2000 ABC all purpose type, portable, 2-1/2 lb.					21			21	23.50
2080 9-1/2 lb.					44			44	48
3500 Halotron 1, 2-1/2 lb.					126			126	139
3600 5 lb.					198			198	218
3700 11 lb.					390			390	430

10 51 Lockers

10 51 13 – Metal Lockers

10 51 13.10 Lockers

		Crew	Daily Output	Labor-Hours	Unit	Material	Labor	Equipment	Total	Total Incl O&P
0011	**LOCKERS** steel, baked enamel, pre-assembled									
0110	Single tier box locker, 12" x 15" x 72"	1 Shee	20	.400	Ea.	223	22.50		245.50	282
0120	18" x 15" x 72"		20	.400		240	22.50		262.50	300
0130	12" x 18" x 72"		20	.400		228	22.50		250.50	287
0140	18" x 18" x 72"		20	.400		283	22.50		305.50	345
0410	Double tier, 12" x 15" x 36"		30	.267		232	14.90		246.90	279
0420	18" x 15" x 36"		30	.267		235	14.90		249.90	283
0430	12" x 18" x 36"		30	.267		264	14.90		278.90	315
0440	18" x 18" x 36"		30	.267		241	14.90		255.90	289
0500	Two person, 18" x 15" x 72"		20	.400		291	22.50		313.50	355
0510	18" x 18" x 72"		20	.400		325	22.50		347.50	390
0520	Duplex, 15" x 15" x 72"		20	.400		325	22.50		347.50	390
0530	15" x 21" x 72"		20	.400		365	22.50		387.50	435
1100	Wire meshed wardrobe, floor. mtd., open front varsity type		7.50	1.067		272	59.50		331.50	395
2400	16-person locker unit with clothing rack									
2500	72 wide x 15" deep x 72" high	1 Shee	15	.533	Ea.	455	30		485	545
2550	18" deep		15	.533	"	605	30		635	710
3250	Rack w/24 wire mesh baskets		1.50	5.333	Set	400	298		698	910
3260	30 baskets		1.25	6.400		350	360		710	950
3270	36 baskets		.95	8.421		510	470		980	1,300
3280	42 baskets		.80	10		555	560		1,115	1,500
3600	For hanger rods, add				Ea.	1.90			1.90	2.09
3650	For number plate kit, 100 plates #1 - #100, add	1 Shee	4	2		82	112		194	267
3700	For locker base, closed front panel		90	.089		7.25	4.97		12.22	15.85
3710	End panel, bolted		36	.222		9	12.45		21.45	29.50
3800	For sloping top, 12" wide		24	.333		31.50	18.65		50.15	64
3810	15" wide		24	.333		34	18.65		52.65	67
3820	18" wide		24	.333		35.50	18.65		54.15	69
3850	Sloping top end panel, 12" deep		72	.111		12.20	6.20		18.40	23
3860	15" deep		72	.111		12.20	6.20		18.40	23
3870	18" deep		72	.111		12.20	6.20		18.40	23
3900	For finish end panels, steel, 60" high, 15" deep		12	.667		36	37.50		73.50	98.50
3910	72" high, 12" deep		12	.667		29	37.50		66.50	91
3920	18" deep		12	.667		43	37.50		80.50	106

10 51 Lockers

10 51 13 – Metal Lockers

10 51 13.10 Lockers

		Crew	Daily Output	Labor-Hours	Unit	Material	2015 Bare Costs Labor	2015 Bare Costs Equipment	Total	Total Incl O&P
5000	For 'ready to assemble' lockers,									
5010	Add to labor						75%			
5020	Deduct from material					20%				
6000	Heavy duty for detention facility, tamper proof, 14 ga. welded steel, solid									
6100	24" W x 24" D x 74" H, single tier	1 Shee	18	.444	Ea.	515	25		540	610
6110	Double tier		18	.444		505	25		530	595
6120	Triple tier		18	.444		520	25		545	615
9000	Minimum labor/equipment charge		2.50	3.200	Job		179		179	283

10 55 Postal Specialties

10 55 23 – Mail Boxes

10 55 23.10 Commercial Mail Boxes

		Crew	Daily Output	Labor-Hours	Unit	Material	Labor	Equipment	Total	Total Incl O&P
0010	**COMMERCIAL MAIL BOXES**									
0020	Horiz., key lock, 5"H x 6"W x 15"D, alum., rear load	1 Carp	34	.235	Ea.	40	11.05		51.05	62
0100	Front loading		34	.235		40	11.05		51.05	62
0200	Double, 5"H x 12"W x 15"D, rear loading		26	.308		66.50	14.45		80.95	97
0300	Front loading		26	.308		70	14.45		84.45	101
0500	Quadruple, 10"H x 12"W x 15"D, rear loading		20	.400		106	18.80		124.80	148
0600	Front loading		20	.400		88.50	18.80		107.30	128
1600	Vault type, horizontal, for apartments, 4" x 5"		34	.235		24	11.05		35.05	44
1700	Alphabetical directories, 120 names		10	.800		125	37.50		162.50	200
1900	Letter slot, residential		20	.400		80	18.80		98.80	119
2000	Post office type		8	1		104	47		151	191
9000	Minimum labor/equipment charge		5	1.600	Job		75		75	123

10 56 Storage Assemblies

10 56 13 – Metal Storage Shelving

10 56 13.10 Shelving

		Crew	Daily Output	Labor-Hours	Unit	Material	Labor	Equipment	Total	Total Incl O&P
0010	**SHELVING**									
0020	Metal, industrial, cross-braced, 3' wide, 12" deep	1 Sswk	175	.046	SF Shlf	7.65	2.41		10.06	12.75
0100	24" deep		330	.024		5.05	1.28		6.33	7.85
2200	Wide span, 1600 lb. capacity per shelf, 6' wide, 24" deep		380	.021		7	1.11		8.11	9.70
2400	36" deep		440	.018		6.85	.96		7.81	9.25
9000	Minimum labor/equipment charge	1 Carp	4	2	Job		94		94	154

10 57 Wardrobe and Closet Specialties

10 57 23 – Closet and Utility Shelving

10 57 23.19 Wood Closet and Utility Shelving

		Crew	Daily Output	Labor-Hours	Unit	Material	Labor	Equipment	Total	Total Incl O&P
0010	**WOOD CLOSET AND UTILITY SHELVING**									
0020	Pine, clear grade, no edge band, 1" x 8"	1 Carp	115	.070	L.F.	3.49	3.27		6.76	9.20
0100	1" x 10"		110	.073		4.34	3.41		7.75	10.35
0200	1" x 12"		105	.076		5.25	3.58		8.83	11.60
0600	Plywood, 3/4" thick with lumber edge, 12" wide		75	.107		1.87	5		6.87	10.25
0700	24" wide		70	.114		3.30	5.35		8.65	12.45
0900	Bookcase, clear grade pine, shelves 12" O.C., 8" deep, per S.F. shelf		70	.114	S.F.	11.35	5.35		16.70	21.50
1000	12" deep shelves		65	.123	"	17	5.80		22.80	28
1200	Adjustable closet rod and shelf, 12" wide, 3' long		20	.400	Ea.	11.15	18.80		29.95	43.50

10 57 Wardrobe and Closet Specialties

10 57 23 – Closet and Utility Shelving

	10 57 23.19 Wood Closet and Utility Shelving	Crew	Daily Output	Labor-Hours	Unit	Material	2015 Bare Costs Labor	Equipment	Total	Total Incl O&P
1300	8' long	1 Carp	15	.533	Ea.	21.50	25		46.50	64.50
1500	Prefinished shelves with supports, stock, 8" wide		75	.107	L.F.	4.58	5		9.58	13.25
1600	10" wide		70	.114	"	6.50	5.35		11.85	15.95
9000	Minimum labor/equipment charge		4	2	Job		94		94	154

10 73 Protective Covers

10 73 13 – Awnings

10 73 13.10 Awnings, Fabric

		Crew	Daily Output	Labor-Hours	Unit	Material	Labor	Equipment	Total	Total Incl O&P
0010	**AWNINGS, FABRIC**									
0020	Including acrylic canvas and frame, standard design									
0100	Door and window, slope, 3' high, 4' wide	1 Carp	4.50	1.778	Ea.	720	83.50		803.50	925
0110	6' wide		3.50	2.286		925	107		1,032	1,200
0120	8' wide		3	2.667		1,125	125		1,250	1,450
0200	Quarter round convex, 4' wide		3	2.667		1,125	125		1,250	1,425
0210	6' wide		2.25	3.556		1,450	167		1,617	1,875
0220	8' wide		1.80	4.444		1,775	209		1,984	2,300
0300	Dome, 4' wide		7.50	1.067		430	50		480	555
0310	6' wide		3.50	2.286		970	107		1,077	1,250
0320	8' wide		2	4		1,725	188		1,913	2,200
0350	Elongated dome, 4' wide		1.33	6.015		1,625	282		1,907	2,250
0360	6' wide		1.11	7.207		1,925	340		2,265	2,675
0370	8' wide		1	8		2,275	375		2,650	3,125
1000	Entry or walkway, peak, 12' long, 4' wide	2 Carp	.90	17.778		5,225	835		6,060	7,125
1010	6' wide		.60	26.667		8,050	1,250		9,300	10,900
1020	8' wide		.40	40		11,100	1,875		12,975	15,300
1100	Radius with dome end, 4' wide		1.10	14.545		3,950	685		4,635	5,475
1110	6' wide		.70	22.857		6,350	1,075		7,425	8,750
1120	8' wide		.50	32		9,050	1,500		10,550	12,400
2000	Retractable lateral arm awning, manual									
2010	To 12' wide, 8' - 6" projection	2 Carp	1.70	9.412	Ea.	1,175	440		1,615	2,025
2020	To 14' wide, 8' - 6" projection		1.10	14.545		1,375	685		2,060	2,650
2030	To 19' wide, 8' - 6" projection		.85	18.824		1,875	885		2,760	3,500
2040	To 24' wide, 8' - 6" projection		.67	23.881		2,350	1,125		3,475	4,450
2050	Motor for above, add	1 Carp	2.67	3		1,000	141		1,141	1,350
3000	Patio/deck canopy with frame									
3010	12' wide, 12' projection	2 Carp	2	8	Ea.	1,675	375		2,050	2,450
3020	16' wide, 14' projection	"	1.20	13.333		2,600	625		3,225	3,875
9000	For fire retardant canvas, add					7%				
9010	For lettering or graphics, add					35%				
9020	For painted or coated acrylic canvas, deduct					8%				
9030	For translucent or opaque vinyl canvas, add					10%				
9040	For 6 or more units, deduct					20%	15%			

10 73 16 – Canopies

10 73 16.10 Canopies, Residential

		Crew	Daily Output	Labor-Hours	Unit	Material	Labor	Equipment	Total	Total Incl O&P
0010	**CANOPIES, RESIDENTIAL** Prefabricated									
0500	Carport, free standing, baked enamel, alum., .032", 40 psf									
0520	16' x 8', 4 posts	2 Carp	3	5.333	Ea.	4,725	250		4,975	5,600
0600	20' x 10', 6 posts		2	8		4,825	375		5,200	5,950
0605	30' x 10', 8 posts		2	8		7,250	375		7,625	8,600
1000	Door canopies, extruded alum., .032", 42" projection, 4' wide	1 Carp	8	1		201	47		248	298

10 73 Protective Covers

10 73 16 – Canopies

10 73 16.10 Canopies, Residential

		Crew	Daily Output	Labor-Hours	Unit	Material	2015 Bare Costs Labor	Equipment	Total	Total Incl O&P
1020	6' wide	1 Carp	6	1.333	Ea.	284	62.50		346.50	415
1040	8' wide	2 Carp	9	1.778		370	83.50		453.50	545
1060	10' wide		7	2.286		390	107		497	605
1080	12' wide		5	3.200		620	150		770	925
1200	54" projection, 4' wide	1 Carp	8	1		248	47		295	350
1220	6' wide	"	6	1.333		290	62.50		352.50	425
1240	8' wide	2 Carp	9	1.778		350	83.50		433.50	520
1260	10' wide		7	2.286		515	107		622	745
1280	12' wide		5	3.200		555	150		705	860
1300	Painted, add					20%				
1310	Bronze anodized, add					50%				
3000	Window awnings, aluminum, window 3' high, 4' wide	1 Carp	10	.800		126	37.50		163.50	201
3020	6' wide	"	8	1		180	47		227	275
3040	9' wide	2 Carp	9	1.778		305	83.50		388.50	470
3060	12' wide	"	5	3.200		305	150		455	580
3100	Window, 4' high, 4' wide	1 Carp	10	.800		159	37.50		196.50	237
3120	6' wide	"	8	1		229	47		276	330
3140	9' wide	2 Carp	9	1.778		375	83.50		458.50	545
3160	12' wide	"	5	3.200		375	150		525	655
3200	Window, 6' high, 4' wide	1 Carp	10	.800		295	37.50		332.50	385
3220	6' wide	"	8	1		340	47		387	450
3240	9' wide	2 Carp	9	1.778		900	83.50		983.50	1,125
3260	12' wide	"	5	3.200		1,375	150		1,525	1,775
3400	Roll-up aluminum, 2'-6" wide	1 Carp	14	.571		190	27		217	253
3420	3' wide		12	.667		195	31.50		226.50	266
3440	4' wide		10	.800		244	37.50		281.50	330
3460	6' wide		8	1		290	47		337	395
3480	9' wide	2 Carp	9	1.778		400	83.50		483.50	575
3500	12' wide	"	5	3.200		535	150		685	835
3600	Window awnings, canvas, 24" drop, 3' wide	1 Carp	30	.267	L.F.	50	12.50		62.50	75.50
3620	4' wide		40	.200		43	9.40		52.40	63
3700	30" drop, 3' wide		30	.267		51.50	12.50		64	77.50
3720	4' wide		40	.200		44.50	9.40		53.90	64.50
3740	5' wide		45	.178		40	8.35		48.35	57.50
3760	6' wide		48	.167		37	7.85		44.85	53.50
3780	8' wide		48	.167		34.50	7.85		42.35	51
3800	10' wide		50	.160		42	7.50		49.50	58.50
3900	Repair canvas, minimum		8	1	Ea.		47		47	77
3920	Maximum		4	2	"		94		94	154
9000	Minimum labor/equipment charge		3	2.667	Job		125		125	205

10 73 16.20 Metal Canopies

		Crew	Daily Output	Labor-Hours	Unit	Material	Labor	Equipment	Total	Total Incl O&P
0010	**METAL CANOPIES**									
0020	Wall hung, .032", aluminum, prefinished, 8' x 10'	K-2	1.30	18.462	Ea.	2,150	900	233	3,283	4,200
0300	8' x 20'		1.10	21.818		3,650	1,075	276	5,001	6,175
1000	12' x 20'		1	24		5,250	1,175	305	6,730	8,150
2300	Aluminum entrance canopies, flat soffit, .032"									
2500	3'-6" x 4'-0", clear anodized	2 Carp	4	4	Ea.	915	188		1,103	1,300
9000	Minimum labor/equipment charge	"	2	8	Job		375		375	615

10 74 Manufactured Exterior Specialties

10 74 23 – Cupolas

10 74 23.10 Wood Cupolas

		Crew	Daily Output	Labor-Hours	Unit	Material	2015 Bare Costs Labor	2015 Bare Costs Equipment	Total	Total Incl O&P
0010	**WOOD CUPOLAS**									
0020	Stock units, pine, painted, 18" sq., 28" high, alum. roof	1 Carp	4.10	1.951	Ea.	184	91.50		275.50	350
0100	Copper roof		3.80	2.105		255	99		354	445
0300	23" square, 33" high, aluminum roof		3.70	2.162		360	102		462	560
0400	Copper roof		3.30	2.424		470	114		584	700
0600	30" square, 37" high, aluminum roof		3.70	2.162		560	102		662	780
0700	Copper roof		3.30	2.424		670	114		784	920
0900	Hexagonal, 31" wide, 46" high, copper roof		4	2		930	94		1,024	1,175
1000	36" wide, 50" high, copper roof		3.50	2.286		1,400	107		1,507	1,725
1200	For deluxe stock units, add to above					25%				
1400	For custom built units, add to above					50%	50%			
9000	Minimum labor/equipment charge	1 Carp	2.75	2.909	Job		137		137	224

10 74 29 – Steeples

10 74 29.10 Prefabricated Steeples

		Crew	Daily Output	Labor-Hours	Unit	Material	Labor	Equipment	Total	Total Incl O&P
0010	**PREFABRICATED STEEPLES**									
4000	Steeples, translucent fiberglass, 30" square, 15' high	F-3	2	20	Ea.	8,375	960	325	9,660	11,100
4150	25' high	"	1.80	22.222		9,750	1,075	365	11,190	12,800
4600	Aluminum, baked finish, 16" square, 14' high					5,700			5,700	6,275
4640	35' high, 8' base					36,700			36,700	40,400
4680	152' high, custom					632,000			632,000	695,000

10 74 33 – Weathervanes

10 74 33.10 Residential Weathervanes

		Crew	Daily Output	Labor-Hours	Unit	Material	Labor	Equipment	Total	Total Incl O&P
0010	**RESIDENTIAL WEATHERVANES**									
0020	Residential types, 18" to 24"	1 Carp	8	1	Ea.	151	47		198	243
0100	24" to 48"		2	4	"	1,700	188		1,888	2,175
9000	Minimum labor/equipment charge		4	2	Job		94		94	154

10 74 46 – Window Wells

10 74 46.10 Area Window Wells

		Crew	Daily Output	Labor-Hours	Unit	Material	Labor	Equipment	Total	Total Incl O&P
0010	**AREA WINDOW WELLS**, Galvanized steel									
0020	20 ga., 3'-2" wide, 1' deep	1 Sswk	29	.276	Ea.	17.45	14.50		31.95	45.50
0100	2' deep		23	.348		31	18.30		49.30	67.50
0300	16 ga., 3'-2" wide, 1' deep		29	.276		23	14.50		37.50	52
0400	3' deep		23	.348		47	18.30		65.30	84.50
0600	Welded grating for above, 15 lb., painted		45	.178		87.50	9.35		96.85	113
0700	Galvanized		45	.178		118	9.35		127.35	147
0900	Translucent plastic cap for above		60	.133		19	7		26	33.50

10 86 Security Mirrors and Domes

10 86 10 – Security Mirrors

10 86 10.10 Exterior Traffic Control Mirrors

		Crew	Daily Output	Labor-Hours	Unit	Material	Labor	Equipment	Total	Total Incl O&P
0010	**EXTERIOR TRAFFIC CONTROL MIRRORS**									
0100	Convex, stainless steel, 20 ga., 26" diameter	1 Carp	12	.667	Ea.	213	31.50		244.50	286

10 86 20 – Security Domes

10 86 20.10 Domes

		Crew	Daily Output	Labor-Hours	Unit	Material	Labor	Equipment	Total	Total Incl O&P
0010	**DOMES** for security cameras (CCTV)									
0100	Ceiling mounted, 10" diameter	1 Carp	30	.267	Ea.	11.60	12.50		24.10	33.50
0110	12" diameter	"	30	.267	"	7.55	12.50		20.05	29

Division Notes

		CREW	DAILY OUTPUT	LABOR-HOURS	UNIT	BARE COSTS				TOTAL INCL O&P
						MAT.	LABOR	EQUIP.	TOTAL	

Division 11 – Equipment

Estimating Tips
General
- The items in this division are usually priced per square foot or each. Many of these items are purchased by the owner for installation by the contractor. Check the specifications for responsibilities and include time for receiving, storage, installation, and mechanical and electrical hookups in the appropriate divisions.
- Many items in Division 11 require some type of support system that is not usually furnished with the item. Examples of these systems include blocking for the attachment of casework and support angles for ceiling-hung projection screens. The required blocking or supports must be added to the estimate in the appropriate division.
- Some items in Division 11 may require assembly or electrical hookups. Verify the amount of assembly required or the need for a hard electrical connection and add the appropriate costs.

Reference Numbers
Reference numbers are shown in shaded boxes at the beginning of some major classifications. These numbers refer to related items in the Reference Section. The reference information may be an estimating procedure, an alternate pricing method, or technical information.

Note: Not all subdivisions listed here necessarily appear in this publication. ■

Estimate with precision with RSMeans Online...
Quick, intuitive and easy-to-use, RSMeans Online gives you instant access to hundreds of thousands of material, labor, and equipment costs from RSMeans' comprehensive database, delivering the information you need to build competitive estimates every time.

Start your free trial today at **www.rsmeansonline.com**

RSMeansOnline

No part of this publication may be reproduced, stored in a retrieval system, or transmitted in any form or by any means without prior written permission of RSMeans.

11 11 Vehicle Service Equipment

11 11 13 – Compressed-Air Vehicle Service Equipment

11 11 13.10 Compressed Air Equipment		Crew	Daily Output	Labor-Hours	Unit	Material	2015 Bare Costs Labor	2015 Bare Costs Equipment	Total	Total Incl O&P
0010	COMPRESSED AIR EQUIPMENT									
0030	Compressors, electric, 1-1/2 H.P., standard controls	L-4	1.50	16	Ea.	455	710		1,165	1,650
0550	Dual controls		1.50	16		810	710		1,520	2,050
0600	5 H.P., 115/230 volt, standard controls		1	24		2,575	1,050		3,625	4,550
0650	Dual controls		1	24		3,400	1,050		4,450	5,450

11 11 19 – Vehicle Lubrication Equipment

11 11 19.10 Lubrication Equipment

0010	LUBRICATION EQUIPMENT									
3000	Lube equipment, 3 reel type, with pumps, not including piping	L-4	.50	48	Set	8,800	2,125		10,925	13,200
3100	Hose reel, including hose, oil/lube, 1000 PSI	2 Sswk	2	8	Ea.	740	420		1,160	1,575
3200	Grease, 5000 PSI		2	8		800	420		1,220	1,650
3300	Air, 50 feet, 160 PSI		2	8		875	420		1,295	1,725
3350	25 feet, 160 PSI		2	8		525	420		945	1,325

11 11 33 – Vehicle Spray Painting Equipment

11 11 33.10 Spray Painting Equipment

0010	SPRAY PAINTING EQUIPMENT									
4000	Spray painting booth, 26' long, complete	L-4	.40	60	Ea.	16,400	2,650		19,050	22,300

11 12 Parking Control Equipment

11 12 13 – Parking Key and Card Control Units

11 12 13.10 Parking Control Units

0010	PARKING CONTROL UNITS									
5100	Card reader	1 Elec	2	4	Ea.	1,950	219		2,169	2,500
5120	Proximity with customer display	2 Elec	1	16		5,575	875		6,450	7,475
6000	Parking control software, basic functionality	1 Elec	.50	16		23,900	875		24,775	27,600
6020	multi-function	"	.20	40		104,500	2,200		106,700	118,500

11 12 16 – Parking Ticket Dispensers

11 12 16.10 Ticket Dispensers

0010	TICKET DISPENSERS									
5900	Ticket spitter with time/date stamp, standard	2 Elec	2	8	Ea.	6,200	440		6,640	7,500
5920	Mag stripe encoding	"	2	8	"	18,800	440		19,240	21,400

11 12 26 – Parking Fee Collection Equipment

11 12 26.13 Parking Fee Coin Collection Equipment

0010	PARKING FEE COIN COLLECTION EQUIPMENT									
5200	Cashier booth, average	B-22	1	30	Ea.	10,400	1,325	207	11,932	13,900
5300	Collector station, pay on foot	2 Elec	.20	80		111,000	4,375		115,375	129,000
5320	Credit card only	"	.50	32		20,500	1,750		22,250	25,200

11 12 26.23 Fee Equipment

0010	FEE EQUIPMENT									
5600	Fee computer	1 Elec	1.50	5.333	Ea.	14,200	292		14,492	16,100

11 12 33 – Parking Gates

11 12 33.13 Lift Arm Parking Gates

0010	LIFT ARM PARKING GATES									
5000	Barrier gate with programmable controller	2 Elec	3	5.333	Ea.	3,400	292		3,692	4,175
5020	Industrial		3	5.333		5,050	292		5,342	6,000
5050	Non-programmable, with reader and 12' arm		3	5.333		1,875	292		2,167	2,500
5500	Exit verifier		1	16		17,600	875		18,475	20,800
5700	Full sign, 4" letters	1 Elec	2	4		1,225	219		1,444	1,700

11 12 Parking Control Equipment

11 12 33 – Parking Gates

11 12 33.13 Lift Arm Parking Gates

		Crew	Daily Output	Labor-Hours	Unit	Material	2015 Bare Costs Labor	2015 Bare Costs Equipment	Total	Total Incl O&P
5800	Inductive loop	2 Elec	4	4	Ea.	171	219		390	530
5950	Vehicle detector, microprocessor based	1 Elec	3	2.667		425	146		571	695
7100	Traffic spike unit, flush mount, spring loaded, 72" L	B-89	4	4		1,250	175	122	1,547	1,800
7200	Surface mount, 72" L	2 Skwk	10	1.600		2,275	78		2,353	2,625

11 13 Loading Dock Equipment

11 13 13 – Loading Dock Bumpers

11 13 13.10 Dock Bumpers

		Crew	Daily Output	Labor-Hours	Unit	Material	Labor	Equipment	Total	Total Incl O&P
0010	**DOCK BUMPERS** Bolts not included									
0012	2" x 6" to 4" x 8", average	1 Carp	300	.027	B.F.	.70	1.25		1.95	2.82
0050	Bumpers, rubber blocks 4-1/2" thick, 10" high, 14" long		26	.308	Ea.	64	14.45		78.45	93.50
0200	24" long		22	.364		96	17.05		113.05	134
0300	36" long		17	.471		78	22		100	122
0500	12" high, 14" long		25	.320		101	15		116	137

11 13 16 – Loading Dock Seals and Shelters

11 13 16.10 Dock Seals and Shelters

		Crew	Daily Output	Labor-Hours	Unit	Material	Labor	Equipment	Total	Total Incl O&P
0010	**DOCK SEALS AND SHELTERS**									
6200	Shelters, fabric, for truck or train, scissor arms, minimum	1 Carp	1	8	Ea.	1,775	375		2,150	2,575
6300	Maximum	"	.50	16	"	2,425	750		3,175	3,900

11 13 19 – Stationary Loading Dock Equipment

11 13 19.10 Dock Equipment

		Crew	Daily Output	Labor-Hours	Unit	Material	Labor	Equipment	Total	Total Incl O&P
0010	**DOCK EQUIPMENT**									
2200	Dock boards, heavy duty, 60" x 60", aluminum, 5,000 lb. capacity				Ea.	1,450			1,450	1,600
4200	Platform lifter, 6' x 6', portable, 3,000 lb. capacity				"	9,250			9,250	10,200
6000	Dock leveler, 15 ton capacity									
6100	I beam construction, mechanical, 6' x 8'	E-16	.50	32	Ea.	4,400	1,725	292	6,417	8,250
6150	Hydraulic		.50	32		5,225	1,725	292	7,242	9,175
6200	Formed beam deck construction, mechanical, 6' x 8'		.50	32		3,250	1,725	292	5,267	7,000
6250	Hydraulic		.50	32		4,450	1,725	292	6,467	8,325
6300	Edge of dock leveler, mechanical, 15 ton capacity		2	8		970	430	73	1,473	1,925
7000	22.5 ton capacity									
7100	Vertical storing dock leveler, hydraulic, 6' x 6'	E-16	.40	40	Ea.	7,200	2,150	365	9,715	12,200

11 21 Retail and Service Equipment

11 21 13 – Cash Registers and Checking Equipment

11 21 13.10 Checkout Counter

		Crew	Daily Output	Labor-Hours	Unit	Material	Labor	Equipment	Total	Total Incl O&P
0010	**CHECKOUT COUNTER**									
0020	Supermarket conveyor, single belt	2 Clab	10	1.600	Ea.	3,175	60		3,235	3,600
0100	Double belt, power take-away		9	1.778		4,575	67		4,642	5,125
0400	Double belt, power take-away, incl. side scanning		7	2.286		5,375	86		5,461	6,050
0800	Warehouse or bulk type		6	2.667		6,350	100		6,450	7,150
1000	Scanning system, 2 lanes, w/registers, scan gun & memory				System	16,600			16,600	18,300
1100	10 lanes, single processor, full scan, with scales				"	158,000			158,000	173,500
2000	Register, restaurant, minimum				Ea.	740			740	815
2100	Maximum					3,125			3,125	3,425
2150	Store, minimum					740			740	815
2200	Maximum					3,125			3,125	3,425

11 21 Retail and Service Equipment

11 21 33 – Checkroom Equipment

11 21 33.10 Clothes Check Equipment

		Crew	Daily Output	Labor-Hours	Unit	Material	2015 Bare Costs Labor	Equipment	Total	Total Incl O&P
0010	**CLOTHES CHECK EQUIPMENT**									
0030	Clothes check rack, free standing, st. stl., 2-tier, 90 bag capacity				Ea.	1,525			1,525	1,700
0050	Wall mounted, 45 bag capacity	L-2	8	2		750	82.50		832.50	960
0100	Garment checking bag, green mesh fabric, 21" H x 17" W with 4.5" hook					19.10			19.10	21

11 21 53 – Barber and Beauty Shop Equipment

11 21 53.10 Barber Equipment

		Crew	Daily Output	Labor-Hours	Unit	Material	Labor	Equipment	Total	Total Incl O&P
0010	**BARBER EQUIPMENT**									
0020	Chair, hydraulic, movable, minimum	1 Carp	24	.333	Ea.	560	15.65		575.65	640
0050	Maximum	"	16	.500		3,475	23.50		3,498.50	3,875
0200	Wall hung styling station with mirrors, minimum	L-2	8	2		460	82.50		542.50	640
0300	Maximum	"	4	4		2,325	165		2,490	2,825
0500	Sink, hair washing basin, rough plumbing not incl.	1 Plum	8	1		495	58.50		553.50	635
1000	Sterilizer, liquid solution for tools					161			161	177
1100	Total equipment, rule of thumb, per chair, minimum	L-8	1	20		1,925	985		2,910	3,725
1150	Maximum	"	1	20		5,150	985		6,135	7,275

11 21 73 – Commercial Laundry and Dry Cleaning Equipment

11 21 73.13 Dry Cleaning Equipment

		Crew	Daily Output	Labor-Hours	Unit	Material	Labor	Equipment	Total	Total Incl O&P
0010	**DRY CLEANING EQUIPMENT**									
2000	Dry cleaners, electric, 20 lb. capacity, not incl. rough-in	L-1	.20	80	Ea.	33,700	4,525		38,225	44,100
2050	25 lb. capacity		.17	94.118		48,600	5,325		53,925	62,000
2100	30 lb. capacity		.15	106		51,000	6,050		57,050	66,000
2150	60 lb. capacity		.09	177		79,000	10,100		89,100	102,500

11 21 73.16 Drying and Conditioning Equipment

		Crew	Daily Output	Labor-Hours	Unit	Material	Labor	Equipment	Total	Total Incl O&P
0010	**DRYING AND CONDITIONING EQUIPMENT**									
0100	Dryers, Not including rough-in									
1500	Industrial, 30 lb. capacity	1 Plum	2	4	Ea.	3,175	235		3,410	3,850
1600	50 lb. capacity	"	1.70	4.706		3,400	276		3,676	4,175
4700	Lint collector, ductwork not included, 8,000 to 10,000 CFM	Q-10	.30	80		9,200	4,175		13,375	16,700

11 21 73.19 Finishing Equipment

		Crew	Daily Output	Labor-Hours	Unit	Material	Labor	Equipment	Total	Total Incl O&P
0010	**FINISHING EQUIPMENT**									
3500	Folders, blankets & sheets, minimum	1 Elec	.17	47.059	Ea.	33,100	2,575		35,675	40,400
3700	King size with automatic stacker		.10	80		60,000	4,375		64,375	73,000
3800	For conveyor delivery, add		.45	17.778		14,500	970		15,470	17,400

11 21 73.23 Commercial Ironing Equipment

		Crew	Daily Output	Labor-Hours	Unit	Material	Labor	Equipment	Total	Total Incl O&P
0010	**COMMERCIAL IRONING EQUIPMENT**									
4500	Ironers, institutional, 110", single roll	1 Elec	.20	40	Ea.	32,100	2,200		34,300	38,700

11 21 73.26 Commercial Washers and Extractors

		Crew	Daily Output	Labor-Hours	Unit	Material	Labor	Equipment	Total	Total Incl O&P
0010	**COMMERCIAL WASHERS AND EXTRACTORS**, not including rough-in									
6000	Combination washer/extractor, 20 lb. capacity	L-6	1.50	8	Ea.	5,950	460		6,410	7,250
6100	30 lb. capacity		.80	15		9,275	860		10,135	11,600
6200	50 lb. capacity		.68	17.647		10,900	1,000		11,900	13,600
6300	75 lb. capacity		.30	40		20,400	2,300		22,700	26,100
6350	125 lb. capacity		.16	75		27,700	4,300		32,000	37,100

11 21 73.33 Coin-Operated Laundry Equipment

		Crew	Daily Output	Labor-Hours	Unit	Material	Labor	Equipment	Total	Total Incl O&P
0010	**COIN-OPERATED LAUNDRY EQUIPMENT**									
0990	Dryer, gas fired									
1000	Commercial, 30 lb. capacity, coin operated, single	1 Plum	3	2.667	Ea.	3,300	157		3,457	3,875
1100	Double stacked	"	2	4	"	7,275	235		7,510	8,375
5290	Clothes washer									
5300	Commercial, coin operated, average	1 Plum	3	2.667	Ea.	1,250	157		1,407	1,625

11 21 Retail and Service Equipment

11 21 83 – Photo Processing Equipment

11 21 83.13 Darkroom Equipment

		Crew	Daily Output	Labor-Hours	Unit	Material	2015 Bare Costs Labor	2015 Bare Costs Equipment	Total	Total Incl O&P
0010	**DARKROOM EQUIPMENT**									
0020	Developing sink, 5" deep, 24" x 48"	Q-1	2	8	Ea.	4,150	425		4,575	5,225
0050	48" x 52"		1.70	9.412		4,425	495		4,920	5,650
0200	10" deep, 24" x 48"		1.70	9.412		1,550	495		2,045	2,475
0250	24" x 108"	↓	1.50	10.667		3,650	565		4,215	4,900
0500	Dryers, dehumidified filtered air, 36" x 25" x 68" high	L-7	6	4.667		4,300	212		4,512	5,075
0550	48" x 25" x 68" high		5	5.600		10,100	254		10,354	11,500
2000	Processors, automatic, color print, minimum		4	7		15,900	320		16,220	18,000
2050	Maximum		.60	46.667		25,000	2,125		27,125	31,000
2300	Black and white print, minimum		2	14		12,100	635		12,735	14,300
2350	Maximum		.80	35		62,000	1,600		63,600	70,500
2600	Manual processor, 16" x 20" maximum print size		2	14		9,575	635		10,210	11,500
2650	20" x 24" maximum print size		1	28		9,100	1,275		10,375	12,100
3000	Viewing lights, 20" x 24"		6	4.667		288	212		500	660
3100	20" x 24" with color correction	↓	6	4.667		345	212		557	725
3500	Washers, round, minimum sheet 11" x 14"	Q-1	2	8		3,275	425		3,700	4,250
3550	Maximum sheet 20" x 24"		1	16		3,600	845		4,445	5,300
3800	Square, minimum sheet 20" x 24"		1	16		3,150	845		3,995	4,800
3900	Maximum sheet 50" x 56"	↓	.80	20	↓	4,900	1,050		5,950	7,050
4500	Combination tank sink, tray sink, washers, with									
4510	Dry side tables, average	Q-1	.45	35.556	Ea.	10,800	1,875		12,675	14,700

11 22 Banking Equipment

11 22 13 – Vault Equipment

11 22 13.16 Safes

		Crew	Daily Output	Labor-Hours	Unit	Material	Labor	Equipment	Total	Total Incl O&P
0010	**SAFES**									
0800	Money, "B" label, 9" x 14" x 14"				Ea.	540			540	595
0900	Tool resistive, 24" x 24" x 20"					4,150			4,150	4,550
1050	Tool and torch resistive, 24" x 24" x 20"				↓	7,875			7,875	8,675

11 22 16 – Teller and Service Equipment

11 22 16.13 Teller Equipment Systems

		Crew	Daily Output	Labor-Hours	Unit	Material	Labor	Equipment	Total	Total Incl O&P
0010	**TELLER EQUIPMENT SYSTEMS**									
0020	Alarm system, police	2 Elec	1.60	10	Ea.	4,975	545		5,520	6,325
0400	Bullet resistant teller window, 44" x 60"	1 Glaz	.60	13.333		3,850	600		4,450	5,225
0500	48" x 60"	"	.60	13.333	↓	5,600	600		6,200	7,125
3000	Counters for banks, frontal only	2 Carp	1	16	Station	1,850	750		2,600	3,250
3100	Complete with steel undercounter		.50	32	"	3,625	1,500		5,125	6,425
5400	Partitions, bullet-resistant, 1-3/16" glass, 8' high	↓	10	1.600	L.F.	201	75		276	345
5600	Pass thru, bullet-res. window, painted steel, 24" x 36"	2 Sswk	1.60	10	Ea.	2,650	525		3,175	3,875

11 30 Residential Equipment

11 30 13 – Residential Appliances

11 30 13.15 Cooking Equipment

		Crew	Daily Output	Labor-Hours	Unit	Material	2015 Bare Costs Labor	2015 Bare Costs Equipment	Total	Total Incl O&P
0010	**COOKING EQUIPMENT**									
0020	Cooking range, 30" free standing, 1 oven, minimum	2 Clab	10	1.600	Ea.	440	60		500	585
0050	Maximum	"	4	4		2,025	150		2,175	2,475
0350	Built-in, 30" wide, 1 oven, minimum	1 Elec	6	1.333		680	73		753	865
0400	Maximum	2 Carp	2	8		1,350	375		1,725	2,125
0900	Countertop cooktops, 4 burner, standard, minimum	1 Elec	6	1.333		296	73		369	440
0950	Maximum		3	2.667		1,300	146		1,446	1,675
1250	Microwave oven, minimum		4	2		120	109		229	300
1300	Maximum		2	4		465	219		684	850
5380	Oven, built in, standard		4	2		635	109		744	865
5390	Deluxe		2	4		2,275	219		2,494	2,850

11 30 13.16 Refrigeration Equipment

		Crew	Daily Output	Labor-Hours	Unit	Material	Labor	Equipment	Total	Total Incl O&P
0010	**REFRIGERATION EQUIPMENT**									
5450	Refrigerator, no frost, 6 C.F.	2 Clab	15	1.067	Ea.	244	40		284	335
5500	Refrigerator, no frost, 10 C.F. to 12 C.F., minimum		10	1.600		425	60		485	565
5600	Maximum		6	2.667		450	100		550	660
6790	Energy-star qualified, 18 C.F., minimum G	2 Carp	4	4		510	188		698	870
6795	Maximum G		2	8		1,175	375		1,550	1,900
6797	21.7 C.F., minimum G		4	4		1,025	188		1,213	1,425
6799	Maximum G		4	4		2,100	188		2,288	2,625

11 30 13.17 Kitchen Cleaning Equipment

		Crew	Daily Output	Labor-Hours	Unit	Material	Labor	Equipment	Total	Total Incl O&P
0010	**KITCHEN CLEANING EQUIPMENT**									
2750	Dishwasher, built-in, 2 cycles, minimum	L-1	4	4	Ea.	238	227		465	615
2800	Maximum		2	8		435	455		890	1,175
3100	Energy-star qualified, minimum G		4	4		370	227		597	760
3110	Maximum G		2	8		1,500	455		1,955	2,350

11 30 13.18 Waste Disposal Equipment

		Crew	Daily Output	Labor-Hours	Unit	Material	Labor	Equipment	Total	Total Incl O&P
0010	**WASTE DISPOSAL EQUIPMENT**									
1750	Compactor, residential size, 4 to 1 compaction, minimum	1 Carp	5	1.600	Ea.	645	75		720	835
1800	Maximum	"	3	2.667		1,025	125		1,150	1,325
3300	Garbage disposal, sink type, minimum	L-1	10	1.600		87.50	90.50		178	237
3350	Maximum	"	10	1.600		220	90.50		310.50	385

11 30 13.19 Kitchen Ventilation Equipment

		Crew	Daily Output	Labor-Hours	Unit	Material	Labor	Equipment	Total	Total Incl O&P
0010	**KITCHEN VENTILATION EQUIPMENT**									
4150	Hood for range, 2 speed, vented, 30" wide, minimum	L-3	5	3.200	Ea.	70.50	164		234.50	340
4200	Maximum		3	5.333		780	273		1,053	1,300
4300	42" wide, minimum		5	3.200		168	164		332	445
4330	Custom		5	3.200		1,650	164		1,814	2,075
4350	Maximum		3	5.333		2,025	273		2,298	2,650
4400	Ventless hood, 2 speed, 30" wide, minimum	1 Elec	8	1		86	54.50		140.50	179
4450	Maximum		7	1.143		590	62.50		652.50	745
4510	36" wide, minimum		7	1.143		76	62.50		138.50	181
4550	Maximum		6	1.333		570	73		643	745
4580	42" wide, minimum		6	1.333		188	73		261	320
4600	Maximum		5	1.600		690	87.50		777.50	895
4650	For vented 1 speed, deduct from maximum					50			50	55

11 30 13.24 Washers

		Crew	Daily Output	Labor-Hours	Unit	Material	Labor	Equipment	Total	Total Incl O&P
0010	**WASHERS**									
5000	Residential, 4 cycle, average	1 Plum	3	2.667	Ea.	875	157		1,032	1,200
6750	Energy star qualified, front loading, minimum G		3	2.667		655	157		812	965
6760	Maximum G		1	8		1,600	470		2,070	2,475

11 30 Residential Equipment

11 30 13 – Residential Appliances

11 30 13.24 Washers

		Crew	Daily Output	Labor-Hours	Unit	Material	2015 Bare Costs Labor	Equipment	Total	Total Incl O&P
6764	Top loading, minimum [G]	1 Plum	3	2.667	Ea.	450	157		607	740
6766	Maximum [G]	↓	3	2.667	↓	1,250	157		1,407	1,625

11 30 13.25 Dryers

		Crew	Daily Output	Labor-Hours	Unit	Material	Labor	Equipment	Total	Total Incl O&P
0010	**DRYERS**									
0500	Gas fired residential, 16 lb. capacity, average	1 Plum	3	2.667	Ea.	675	157		832	985
6770	Electric, front loading, energy-star qualified, minimum [G]	L-2	3	5.333		385	220		605	785
6780	Maximum [G]	"	2	8	↓	1,925	330		2,255	2,650

11 30 15 – Miscellaneous Residential Appliances

11 30 15.13 Sump Pumps

		Crew	Daily Output	Labor-Hours	Unit	Material	Labor	Equipment	Total	Total Incl O&P
0010	**SUMP PUMPS**									
6400	Cellar drainer, pedestal, 1/3 H.P., molded PVC base	1 Plum	3	2.667	Ea.	135	157		292	395
6450	Solid brass	"	2	4	"	289	235		524	685
6460	Sump pump, see also Section 22 14 29.16									

11 30 33 – Retractable Stairs

11 30 33.10 Disappearing Stairway

		Crew	Daily Output	Labor-Hours	Unit	Material	Labor	Equipment	Total	Total Incl O&P
0010	**DISAPPEARING STAIRWAY** No trim included									
0020	One piece, yellow pine, 8'-0" ceiling	2 Carp	4	4	Ea.	257	188		445	590
0030	9'-0" ceiling		4	4		263	188		451	600
0040	10'-0" ceiling		3	5.333		274	250		524	710
0050	11'-0" ceiling		3	5.333		430	250		680	885
0060	12'-0" ceiling	↓	3	5.333		495	250		745	955
0100	Custom grade, pine, 8'-6" ceiling, minimum	1 Carp	4	2		177	94		271	350
0150	Average		3.50	2.286		253	107		360	455
0200	Maximum		3	2.667		325	125		450	565
0500	Heavy duty, pivoted, from 7'-7" to 12'-10" floor to floor		3	2.667		740	125		865	1,025
0600	16'-0" ceiling		2	4		1,525	188		1,713	1,975
0800	Economy folding, pine, 8'-6" ceiling		4	2		176	94		270	350
0900	9'-6" ceiling	↓	4	2		196	94		290	370
1100	Automatic electric, aluminum, floor to floor height, 8' to 9'	2 Carp	1	16		8,550	750		9,300	10,600
1400	11' to 12'		.90	17.778		9,075	835		9,910	11,400
1700	14' to 15'	↓	.70	22.857	↓	9,775	1,075		10,850	12,600
9000	Minimum labor/equipment charge	1 Carp	2	4	Job		188		188	310

11 32 Unit Kitchens

11 32 13 – Metal Unit Kitchens

11 32 13.10 Commercial Unit Kitchens

		Crew	Daily Output	Labor-Hours	Unit	Material	Labor	Equipment	Total	Total Incl O&P
0010	**COMMERCIAL UNIT KITCHENS**									
1500	Combination range, refrigerator and sink, 30" wide, minimum	L-1	2	8	Ea.	1,100	455		1,555	1,925
1550	Maximum	"	1	16	"	1,475	905		2,380	3,025
1640	Combination range, refrigerator, sink, microwave									
1660	Oven and ice maker	L-1	.80	20	Ea.	4,550	1,125		5,675	6,775

11 41 Foodservice Storage Equipment

11 41 13 – Refrigerated Food Storage Cases

11 41 13.10 Refrigerated Food Cases

		Crew	Daily Output	Labor-Hours	Unit	Material	2015 Bare Costs Labor	2015 Bare Costs Equipment	Total	Total Incl O&P
0010	REFRIGERATED FOOD CASES									
0030	Dairy, multi-deck, 12' long	Q-5	3	5.333	Ea.	11,300	287		11,587	12,800
0100	For rear sliding doors, add					1,800			1,800	2,000
0200	Delicatessen case, service deli, 12' long, single deck	Q-5	3.90	4.103		7,750	221		7,971	8,875
0300	Multi-deck, 18 S.F. shelf display		3	5.333		7,150	287		7,437	8,325
0400	Freezer, self-contained, chest-type, 30 C.F.		3.90	4.103		8,525	221		8,746	9,750
0500	Glass door, upright, 78 C.F.		3.30	4.848		10,200	261		10,461	11,600
0600	Frozen food, chest type, 12' long		3.30	4.848		8,225	261		8,486	9,450
0700	Glass door, reach-in, 5 door		3	5.333		14,100	287		14,387	15,900
0800	Island case, 12' long, single deck		3.30	4.848		7,275	261		7,536	8,425
0900	Multi-deck		3	5.333		8,500	287		8,787	9,800
1000	Meat case, 12' long, single deck		3.30	4.848		7,275	261		7,536	8,425
1050	Multi-deck		3.10	5.161		10,400	278		10,678	11,800
1100	Produce, 12' long, single deck		3.30	4.848		6,875	261		7,136	7,950
1200	Multi-deck		3.10	5.161		8,725	278		9,003	10,000

11 41 13.20 Refrigerated Food Storage Equipment

		Crew	Daily Output	Labor-Hours	Unit	Material	Labor	Equipment	Total	Total Incl O&P
0010	REFRIGERATED FOOD STORAGE EQUIPMENT									
2350	Cooler, reach-in, beverage, 6' long	Q-1	6	2.667	Ea.	3,600	141		3,741	4,175
4300	Freezers, reach-in, 44 C.F.		4	4		4,425	211		4,636	5,175
4500	68 C.F.		3	5.333		4,825	282		5,107	5,750
4600	Freezer, pre-fab, 8' x 8' w/refrigeration	2 Carp	.45	35.556		11,100	1,675		12,775	14,900
4620	8' x 12'		.35	45.714		11,200	2,150		13,350	15,800
4640	8' x 16'		.25	64		14,300	3,000		17,300	20,600
4660	8' x 20'		.17	94.118		19,400	4,425		23,825	28,600
4680	Reach-in, 1 compartment	Q-1	4	4		2,500	211		2,711	3,100
4685	Energy star rated G	R-18	7.80	3.333		2,575	143		2,718	3,050
4700	2 compartment	Q-1	3	5.333		4,100	282		4,382	4,975
4705	Energy star rated G	R-18	6.20	4.194		3,100	180		3,280	3,700
4710	3 compartment	Q-1	3	5.333		5,300	282		5,582	6,275
4715	Energy star rated G	R-18	5.60	4.643		4,175	199		4,374	4,925
8320	Refrigerator, reach-in, 1 compartment		7.80	3.333		2,425	143		2,568	2,900
8325	Energy star rated G		7.80	3.333		2,575	143		2,718	3,050
8330	2 compartment		6.20	4.194		3,800	180		3,980	4,475
8335	Energy star rated G		6.20	4.194		3,100	180		3,280	3,700
8340	3 compartment		5.60	4.643		4,775	199		4,974	5,575
8345	Energy star rated G		5.60	4.643		4,175	199		4,374	4,925
8350	Pre-fab, with refrigeration, 8' x 8'	2 Carp	.45	35.556		7,050	1,675		8,725	10,500
8360	8' x 12'		.35	45.714		8,000	2,150		10,150	12,300
8370	8' x 16'		.25	64		12,300	3,000		15,300	18,500
8380	8' x 20'		.17	94.118		15,700	4,425		20,125	24,500
8390	Pass-thru/roll-in, 1 compartment	R-18	7.80	3.333		4,500	143		4,643	5,175
8400	2 compartment		6.24	4.167		6,325	179		6,504	7,225
8410	3 compartment		5.60	4.643		8,625	199		8,824	9,825
8420	Walk-in, alum, door & floor only, no refrig, 6' x 6' x 7'-6"	2 Carp	1.40	11.429		8,775	535		9,310	10,500
8430	10' x 6' x 7'-6"		.55	29.091		12,700	1,375		14,075	16,300
8440	12' x 14' x 7'-6"		.25	64		17,500	3,000		20,500	24,100
8450	12' x 20' x 7'-6"		.17	94.118		19,300	4,425		23,725	28,500
8460	Refrigerated cabinets, mobile					3,925			3,925	4,325
8470	Refrigerator/freezer, reach-in, 1 compartment	R-18	5.60	4.643		5,800	199		5,999	6,700
8480	2 compartment	"	4.80	5.417		7,575	232		7,807	8,700

11 41 Foodservice Storage Equipment

11 41 13 – Refrigerated Food Storage Cases

11 41 13.30 Wine Cellar		Crew	Daily Output	Labor-Hours	Unit	Material	2015 Bare Costs Labor	2015 Bare Costs Equipment	Total	Total Incl O&P
0010	**WINE CELLAR**, refrigerated, Redwood interior, carpeted, walk-in type									
0020	6'-8" high, including racks									
0200	80" W x 48" D for 900 bottles	2 Carp	1.50	10.667	Ea.	4,300	500		4,800	5,550
0250	80" W x 72" D for 1300 bottles		1.33	12.030		5,225	565		5,790	6,675
0300	80" W x 94" D for 1900 bottles		1.17	13.675		6,350	640		6,990	8,025
0400	80" W x 124" D for 2500 bottles	↓	1	16	↓	7,450	750		8,200	9,425

11 41 33 – Foodservice Shelving

11 41 33.20 Metal Food Storage Shelving

		Crew	Daily Output	Labor-Hours	Unit	Material	Labor	Equipment	Total	Total Incl O&P
0010	**METAL FOOD STORAGE SHELVING**									
8600	Stainless steel shelving, louvered 4-tier, 20" x 3'	1 Clab	6	1.333	Ea.	1,400	50		1,450	1,625
8605	20" x 4'		6	1.333		1,550	50		1,600	1,775
8610	20" x 6'		6	1.333		2,200	50		2,250	2,500
8615	24" x 3'		6	1.333		1,975	50		2,025	2,250
8620	24" x 4'		6	1.333		2,350	50		2,400	2,675
8625	24" x 6'		6	1.333		3,275	50		3,325	3,675
8630	Flat 4-tier, 20" x 3'		6	1.333		1,150	50		1,200	1,350
8635	20" x 4'		6	1.333		1,400	50		1,450	1,625
8640	20" x 5'		6	1.333		1,625	50		1,675	1,850
8645	24" x 3'		6	1.333		1,275	50		1,325	1,475
8650	24" x 4'		6	1.333		2,225	50		2,275	2,525
8655	24" x 6'		6	1.333		2,675	50		2,725	3,000
8700	Galvanized shelving, louvered 4-tier, 20" x 3'		6	1.333		760	50		810	915
8705	20" x 4'		6	1.333		860	50		910	1,025
8710	20" x 6'		6	1.333		1,000	50		1,050	1,175
8715	24" x 3'		6	1.333		705	50		755	860
8720	24" x 4'		6	1.333		995	50		1,045	1,175
8725	24" x 6'		6	1.333		1,325	50		1,375	1,525
8730	Flat 4-tier, 20" x 3'		6	1.333		700	50		750	850
8735	20" x 4'		6	1.333		695	50		745	845
8740	20" x 6'		6	1.333		905	50		955	1,075
8745	24" x 3'		6	1.333		670	50		720	815
8750	24" x 4'		6	1.333		755	50		805	910
8755	24" x 6'		6	1.333		950	50		1,000	1,125
8760	Stainless steel dunnage rack, 24" x 3'		8	1		330	37.50		367.50	425
8765	24" x 4'		8	1		425	37.50		462.50	525
8770	Galvanized dunnage rack, 24" x 3'		8	1		168	37.50		205.50	247
8775	24" x 4'	↓	8	1	↓	190	37.50		227.50	271

11 42 Food Preparation Equipment

11 42 10 – Commercial Food Preparation Equipment

11 42 10.10 Choppers, Mixers and Misc. Equipment

		Crew	Daily Output	Labor-Hours	Unit	Material	Labor	Equipment	Total	Total Incl O&P
0010	**CHOPPERS, MIXERS AND MISC. EQUIPMENT**									
1700	Choppers, 5 pounds	R-18	7	3.714	Ea.	2,150	159		2,309	2,625
1720	16 pounds		5	5.200		2,200	223		2,423	2,775
1740	35 to 40 pounds	↓	4	6.500		3,550	279		3,829	4,350
1840	Coffee brewer, 5 burners	1 Plum	3	2.667		1,325	157		1,482	1,700
1850	Coffee urn, twin 6 gallon urns		2	4		2,400	235		2,635	3,025
1860	Single, 3 gallon	↓	3	2.667		1,825	157		1,982	2,250
3000	Fast food equipment, total package, minimum	6 Skwk	.08	600		203,500	29,200		232,700	271,500
3100	Maximum	"	.07	685	↓	277,500	33,400		310,900	359,000

11 42 Food Preparation Equipment

11 42 10 – Commercial Food Preparation Equipment

11 42 10.10 Choppers, Mixers and Misc. Equipment		Crew	Daily Output	Labor-Hours	Unit	Material	2015 Bare Costs Labor	Equipment	Total	Total Incl O&P
3800	Food mixers, bench type, 20 quarts	L-7	7	4	Ea.	2,775	182		2,957	3,350
3850	40 quarts		5.40	5.185		6,725	235		6,960	7,775
3900	60 quarts		5	5.600		11,300	254		11,554	12,900
4040	80 quarts		3.90	7.179		12,700	325		13,025	14,500
4100	Floor type, 20 quarts		15	1.867		3,225	84.50		3,309.50	3,700
4120	60 quarts		14	2		9,950	91		10,041	11,000
4140	80 quarts		12	2.333		15,400	106		15,506	17,200
4160	140 quarts	▼	8.60	3.256		25,600	148		25,748	28,400
6700	Peelers, small	R-18	8	3.250		1,900	139		2,039	2,300
6720	Large	"	6	4.333		4,825	186		5,011	5,600
6800	Pulper/extractor, close coupled, 5 HP	1 Plum	1.90	4.211		3,525	247		3,772	4,250
8580	Slicer with table	R-18	9	2.889	▼	4,675	124		4,799	5,350

11 43 Food Delivery Carts and Conveyors

11 43 13 – Food Delivery Carts

11 43 13.10 Mobile Carts, Racks and Trays

0010	**MOBILE CARTS, RACKS AND TRAYS**									
1650	Cabinet, heated, 1 compartment, reach-in	R-18	5.60	4.643	Ea.	3,300	199		3,499	3,950
1655	Pass-thru roll-in		5.60	4.643		3,825	199		4,024	4,525
1660	2 compartment, reach-in	▼	4.80	5.417		9,050	232		9,282	10,300
1670	Mobile					3,525			3,525	3,875
2000	Hospital food cart, hot and cold service, 20 tray capacity					15,300			15,300	16,800
6850	Mobile rack w/pan slide					1,400			1,400	1,525
9180	Tray and silver dispenser, mobile	1 Clab	16	.500	▼	915	18.80		933.80	1,025

11 44 Food Cooking Equipment

11 44 13 – Commercial Ranges

11 44 13.10 Cooking Equipment

0010	**COOKING EQUIPMENT**									
0020	Bake oven, gas, one section	Q-1	8	2	Ea.	5,450	106		5,556	6,175
0300	Two sections		7	2.286		9,075	121		9,196	10,200
0600	Three sections	▼	6	2.667		11,200	141		11,341	12,500
0900	Electric convection, single deck	L-7	4	7		6,225	320		6,545	7,375
1300	Broiler, without oven, standard	Q-1	8	2		3,550	106		3,656	4,075
1550	Infrared	L-7	4	7		7,425	320		7,745	8,675
4750	Fryer, with twin baskets, modular model	Q-1	7	2.286		1,250	121		1,371	1,575
5000	Floor model, on 6" legs	"	5	3.200		2,425	169		2,594	2,950
5100	Extra single basket, large					100			100	110
5170	Energy star rated, 50 lb. capacity [G]	R-18	4	6.500		4,650	279		4,929	5,550
5175	85 lb. capacity [G]	"	4	6.500		8,525	279		8,804	9,825
5300	Griddle, SS, 24" plate, w/4" legs, elec, 208 V, 3 phase, 3' long	Q-1	7	2.286		1,400	121		1,521	1,750
5550	4' long	"	6	2.667		2,250	141		2,391	2,700
6200	Iced tea brewer	1 Plum	3.44	2.326		750	137		887	1,050
6350	Kettle, w/steam jacket, tilting, w/positive lock, SS, 20 gallons	L-7	7	4		8,225	182		8,407	9,350
6600	60 gallons	"	6	4.667		11,000	212		11,212	12,400
6900	Range, restaurant type, 6 burners and 1 standard oven, 36" wide	Q-1	7	2.286		2,500	121		2,621	2,950
6950	Convection		7	2.286		4,450	121		4,571	5,100
7150	2 standard ovens, 24" griddle, 60" wide		6	2.667		4,575	141		4,716	5,250
7200	1 standard, 1 convection oven	▼	6	2.667	▼	9,475	141		9,616	10,600

11 44 Food Cooking Equipment

11 44 13 – Commercial Ranges

11 44 13.10 Cooking Equipment

		Crew	Daily Output	Labor-Hours	Unit	Material	2015 Bare Costs Labor	Equipment	Total	Total Incl O&P
7450	Heavy duty, single 34" standard oven, open top	Q-1	5	3.200	Ea.	5,050	169		5,219	5,825
7500	Convection oven		5	3.200		5,450	169		5,619	6,275
7700	Griddle top		6	2.667		2,850	141		2,991	3,375
7750	Convection oven		6	2.667		7,775	141		7,916	8,775
8850	Steamer, electric 27 KW	L-7	7	4		10,700	182		10,882	12,100
9100	Electric, 10 KW or gas 100,000 BTU	"	5	5.600		6,325	254		6,579	7,375
9150	Toaster, conveyor type, 16-22 slices per minute					1,075			1,075	1,200
9160	Pop-up, 2 slot					615			615	680
9200	For deluxe models of above equipment, add					75%				
9400	Rule of thumb: Equipment cost based									
9410	on kitchen work area									
9420	Office buildings, minimum	L-7	77	.364	S.F.	90.50	16.50		107	127
9450	Maximum		58	.483		153	22		175	204
9550	Public eating facilities, minimum		77	.364		119	16.50		135.50	158
9600	Maximum		46	.609		193	27.50		220.50	257
9750	Hospitals, minimum		58	.483		122	22		144	170
9800	Maximum		39	.718		225	32.50		257.50	300

11 46 Food Dispensing Equipment

11 46 16 – Service Line Equipment

11 46 16.10 Commercial Food Dispensing Equipment

		Crew	Daily Output	Labor-Hours	Unit	Material	Labor	Equipment	Total	Total Incl O&P
0010	**COMMERCIAL FOOD DISPENSING EQUIPMENT**									
1050	Butter pat dispenser	1 Clab	13	.615	Ea.	1,000	23		1,023	1,150
1100	Bread dispenser, counter top		13	.615		890	23		913	1,025
1900	Cup and glass dispenser, drop in		4	2		1,075	75		1,150	1,300
1920	Disposable cup, drop in		16	.500		495	18.80		513.80	575
2650	Dish dispenser, drop in, 12"		11	.727		2,250	27.50		2,277.50	2,525
2660	Mobile		10	.800		2,625	30		2,655	2,950
3300	Food warmer, counter, 1.2 KW					665			665	735
3550	1.6 KW					2,150			2,150	2,350
3600	Well, hot food, built-in, rectangular, 12" x 20"	R-30	10	2.600		815	115		930	1,075
3610	Circular, 7 qt.		10	2.600		420	115		535	645
3620	Refrigerated, 2 compartments		10	2.600		3,075	115		3,190	3,550
3630	3 compartments		9	2.889		3,700	128		3,828	4,275
3640	4 compartments		8	3.250		4,350	144		4,494	5,025
4720	Frost cold plate		9	2.889		18,400	128		18,528	20,500
5700	Hot chocolate dispenser	1 Plum	4	2		1,175	117		1,292	1,450
5750	Ice dispenser 567 pound	Q-1	6	2.667		5,200	141		5,341	5,950
6250	Jet spray dispenser	R-18	4.50	5.778		3,175	248		3,423	3,900
6300	Juice dispenser, concentrate	"	4.50	5.778		1,900	248		2,148	2,475
6690	Milk dispenser, bulk, 2 flavor	R-30	8	3.250		1,850	144		1,994	2,250
6695	3 flavor	"	8	3.250		2,475	144		2,619	2,950
8800	Serving counter, straight	1 Carp	40	.200	L.F.	925	9.40		934.40	1,050
8820	Curved section	"	30	.267	"	1,125	12.50		1,137.50	1,275
8825	Solid surface, see Section 12 36 61.16									
8860	Sneeze guard with lights, 60" L	1 Clab	16	.500	Ea.	395	18.80		413.80	465
8900	Sneeze guard, stainless steel and glass, single sided									
8910	Portable, 48" W				Ea.	305			305	335
8920	Portable, 72" W					325			325	360
8930	Adjustable, 36" W	1 Carp	24	.333		253	15.65		268.65	305
8940	Adjustable, 48" W	"	20	.400		320	18.80		338.80	385

11 46 Food Dispensing Equipment

11 46 16 – Service Line Equipment

11 46 16.10 Commercial Food Dispensing Equipment		Crew	Daily Output	Labor-Hours	Unit	Material	2015 Bare Costs Labor	Equipment	Total	Total Incl O&P
9100	Soft serve ice cream machine, medium	R-18	11	2.364	Ea.	12,100	101		12,201	13,500
9110	Large	"	9	2.889		21,800	124		21,924	24,100

11 46 83 – Ice Machines

11 46 83.10 Commercial Ice Equipment

		Crew	Daily Output	Labor-Hours	Unit	Material	Labor	Equipment	Total	Total Incl O&P
0010	**COMMERCIAL ICE EQUIPMENT**									
5800	Ice cube maker, 50 pounds per day	Q-1	6	2.667	Ea.	1,600	141		1,741	2,000
5810	65 pounds per day, energy star rated		6	2.667		1,475	141		1,616	1,850
5900	250 pounds per day		1.20	13.333		2,675	705		3,380	4,050
5950	300 pounds per day, remote condensing		1.20	13.333		2,350	705		3,055	3,700
6050	500 pounds per day		4	4		2,775	211		2,986	3,375
6060	With bin		1.20	13.333		3,400	705		4,105	4,825
6070	Modular, with bin and condenser		1.20	13.333		3,875	705		4,580	5,375
6090	1000 pounds per day, with bin		1	16		4,950	845		5,795	6,750
6100	Ice flakers, 300 pounds per day		1.60	10		2,800	530		3,330	3,900
6120	600 pounds per day		.95	16.842		3,950	890		4,840	5,725
6130	1000 pounds per day		.75	21.333		4,750	1,125		5,875	6,975
6140	2000 pounds per day		.65	24.615		21,600	1,300		22,900	25,700
6160	Ice storage bin, 500 pound capacity	Q-5	1	16		1,050	860		1,910	2,500
6180	1000 pound	"	.56	28.571		2,525	1,525		4,050	5,175

11 48 Foodservice Cleaning and Disposal Equipment

11 48 13 – Commercial Dishwashers

11 48 13.10 Dishwashers

		Crew	Daily Output	Labor-Hours	Unit	Material	Labor	Equipment	Total	Total Incl O&P
0010	**DISHWASHERS**									
2700	Dishwasher, commercial, rack type									
2720	10 to 12 racks per hour	Q-1	3.20	5	Ea.	3,525	264		3,789	4,275
2730	Energy star rated, 35 to 40 racks/hour **G**		1.30	12.308		4,700	650		5,350	6,200
2740	50 to 60 racks/hour **G**		1.30	12.308		10,200	650		10,850	12,200
2800	Automatic, 190 to 230 racks per hour	L-6	.35	34.286		13,900	1,975		15,875	18,400
2820	235 to 275 racks per hour		.25	48		31,500	2,750		34,250	38,900
2840	8,750 to 12,500 dishes per hour		.10	120		55,500	6,875		62,375	71,500
2950	Dishwasher hood, canopy type	L-3A	10	1.200	L.F.	865	61.50		926.50	1,050
2960	Pant leg type	"	2.50	4.800	Ea.	8,700	246		8,946	9,975
5200	Garbage disposal 1.5 HP, 100 GPH	L-1	4.80	3.333		2,175	189		2,364	2,700
5210	3 HP, 120 GPH		4.60	3.478		2,625	197		2,822	3,200
5220	5 HP, 250 GPH		4.50	3.556		3,600	202		3,802	4,300
6750	Pot sink, 3 compartment	1 Plum	7.25	1.103	L.F.	980	65		1,045	1,175
6760	Pot washer, low temp wash/rinse		1.60	5	Ea.	4,050	294		4,344	4,900
6770	High pressure wash, high temperature rinse		1.20	6.667		36,900	390		37,290	41,200
9170	Trash compactor, small, up to 125 lb. compacted weight	L-4	4	6		24,200	266		24,466	27,000
9175	Large, up to 175 lb. compacted weight	"	3	8		29,500	355		29,855	33,100

11 52 Audio-Visual Equipment

11 52 13 – Projection Screens

11 52 13.10 Projection Screens, Wall or Ceiling Hung	Crew	Daily Output	Labor-Hours	Unit	Material	2015 Bare Costs Labor	Equipment	Total	Total Incl O&P
0010	**PROJECTION SCREENS, WALL OR CEILING HUNG**, matte white									
0100	Manually operated, economy	2 Carp	500	.032	S.F.	5.90	1.50		7.40	8.95
0300	Intermediate		450	.036		6.90	1.67		8.57	10.35
0400	Deluxe		400	.040		9.55	1.88		11.43	13.60
9000	Minimum labor/equipment charge		3	5.333	Job		250		250	410

11 52 16 – Projectors

11 52 16.10 Movie Equipment

		Crew	Daily Output	Labor-Hours	Unit	Material	Labor	Equipment	Total	Total Incl O&P
0011	**MOVIE EQUIPMENT**									
0020	Changeover, minimum				Ea.	470			470	520
3000	Projection screens, rigid, in wall, acrylic, 1/4" thick	2 Glaz	195	.082	S.F.	42.50	3.70		46.20	52.50
3100	1/2" thick	"	130	.123	"	49	5.55		54.55	63
3700	Sound systems, incl. amplifier, mono, minimum	1 Elec	.90	8.889	Ea.	3,350	485		3,835	4,425
3800	Dolby/Super Sound, maximum		.40	20		18,300	1,100		19,400	21,800
4100	Dual system, 2 channel, front surround, minimum		.70	11.429		4,675	625		5,300	6,125
4200	Dolby/Super Sound, 4 channel, maximum		.40	20		16,700	1,100		17,800	20,100
5700	Seating, painted steel, upholstered, minimum	2 Carp	35	.457		133	21.50		154.50	182
5800	Maximum	"	28	.571		425	27		452	515

11 53 Laboratory Equipment

11 53 33 – Emergency Safety Appliances

11 53 33.13 Emergency Equipment

		Crew	Daily Output	Labor-Hours	Unit	Material	Labor	Equipment	Total	Total Incl O&P
0010	**EMERGENCY EQUIPMENT**									
1400	Safety equipment, eye wash, hand held				Ea.	410			410	450
1450	Deluge shower				"	770			770	850

11 53 43 – Service Fittings and Accessories

11 53 43.13 Fittings

		Crew	Daily Output	Labor-Hours	Unit	Material	Labor	Equipment	Total	Total Incl O&P
0010	**FITTINGS**									
1600	Sink, one piece plastic, flask wash, hose, free standing	1 Plum	1.60	5	Ea.	1,950	294		2,244	2,600
1610	Epoxy resin sink, 25" x 16" x 10"	"	2	4	"	221	235		456	610
1630	Steel table, open underneath				L.F.	500			500	545
1640	Doors underneath				"	780			780	860
1900	Utility cabinet, stainless steel				Ea.	935			935	1,025
1950	Utility table, acid resistant top with drawers	2 Carp	30	.533	L.F.	164	25		189	221
8000	Alternate pricing method: as percent of lab furniture									
8050	Installation, not incl. plumbing & duct work				% Furn.				22%	22%
8100	Plumbing, final connections, simple system								10%	10%
8110	Moderately complex system								15%	15%
8120	Complex system								20%	20%
8150	Electrical, simple system								10%	10%
8160	Moderately complex system								20%	20%
8170	Complex system								35%	35%

11 53 53 – Biological Safety Cabinets

11 53 53.10 Pharmacy Cabinets

		Crew	Daily Output	Labor-Hours	Unit	Material	Labor	Equipment	Total	Total Incl O&P
0010	**PHARMACY CABINETS**, vertical flow									
0100	Class II, type B2, 6' L	2 Carp	1.50	10.667	Ea.	13,200	500		13,700	15,300

11 57 Vocational Shop Equipment

11 57 10 – Shop Equipment

11 57 10.10 Vocational School Shop Equipment	Crew	Daily Output	Labor-Hours	Unit	Material	2015 Bare Costs Labor	2015 Bare Costs Equipment	Total	Total Incl O&P
0010 **VOCATIONAL SCHOOL SHOP EQUIPMENT**									
0020 Benches, work, wood, average	2 Carp	5	3.200	Ea.	635	150		785	945
0100 Metal, average		5	3.200		550	150		700	850
0400 Combination belt & disc sander, 6"		4	4		1,650	188		1,838	2,125
0700 Drill press, floor mounted, 12", 1/2 H.P.		4	4		415	188		603	765
0800 Dust collector, not incl. ductwork, 6" diameter	1 Shee	1.10	7.273		4,650	405		5,055	5,775
0810 Dust collector bag, 20" diameter	"	5	1.600		440	89.50		529.50	625
1000 Grinders, double wheel, 1/2 H.P.	2 Carp	5	3.200		217	150		367	485
1300 Jointer, 4", 3/4 H.P.		4	4		1,375	188		1,563	1,800
1600 Kilns, 16 C.F., to 2000°		4	4		1,475	188		1,663	1,925
1900 Lathe, woodworking, 10", 1/2 H.P.		4	4		545	188		733	910
2200 Planer, 13" x 6"		4	4		1,100	188		1,288	1,500
2500 Potter's wheel, motorized		4	4		1,125	188		1,313	1,550
2800 Saws, band, 14", 3/4 H.P.		4	4		930	188		1,118	1,325
3100 Metal cutting band saw, 14"		4	4		2,525	188		2,713	3,075
3400 Radial arm saw, 10", 2 H.P.		4	4		1,375	188		1,563	1,800
3700 Scroll saw, 24"		4	4		585	188		773	955
4000 Table saw, 10", 3 H.P.		4	4		2,725	188		2,913	3,300
4300 Welder AC arc, 30 amp capacity		4	4		3,175	188		3,363	3,800

11 61 Broadcast, Theater, and Stage Equipment

11 61 23 – Folding and Portable Stages

11 61 23.10 Portable Stages

	Crew	Daily Output	Labor-Hours	Unit	Material	Labor	Equipment	Total	Total Incl O&P
0010 **PORTABLE STAGES**									
5000 Stages, portable with steps, folding legs, stock, 8" high				SF Stg.	32.50			32.50	36
5100 16" high					49.50			49.50	54.50
5200 32" high					53.50			53.50	59
5300 40" high					60			60	66

11 61 33 – Rigging Systems and Controls

11 61 33.10 Controls

	Crew	Daily Output	Labor-Hours	Unit	Material	Labor	Equipment	Total	Total Incl O&P
0010 **CONTROLS**									
0050 Control boards with dimmers and breakers, minimum	1 Elec	1	8	Ea.	12,600	440		13,040	14,600
0150 Maximum	"	.20	40	"	129,000	2,200		131,200	145,500

11 61 43 – Stage Curtains

11 61 43.10 Curtains

	Crew	Daily Output	Labor-Hours	Unit	Material	Labor	Equipment	Total	Total Incl O&P
0010 **CURTAINS**									
0500 Curtain track, straight, light duty	2 Carp	20	.800	L.F.	27.50	37.50		65	92
0700 Curved sections		12	1.333	"	177	62.50		239.50	298
1000 Curtains, velour, medium weight		600	.027	S.F.	8	1.25		9.25	10.85

11 66 Athletic Equipment

11 66 13 – Exercise Equipment

11 66 13.10 Physical Training Equipment

		Crew	Daily Output	Labor-Hours	Unit	Material	2015 Bare Costs Labor	Equipment	Total	Total Incl O&P
0010	**PHYSICAL TRAINING EQUIPMENT**									
0020	Abdominal rack, 2 board capacity				Ea.	490			490	535
0050	Abdominal board, upholstered					665			665	730
0200	Bicycle trainer, minimum					485			485	535
0300	Deluxe, electric					4,350			4,350	4,775
0400	Barbell set, chrome plated steel, 25 lb.					252			252	277
0420	100 lb.					465			465	510
0450	200 lb.					705			705	780
0500	Weight plates, cast iron, per lb.				Lb.	5.25			5.25	5.80
0520	Storage rack, 10 station				Ea.	910			910	1,000
0600	Circuit training apparatus, 12 machines minimum	2 Clab	1.25	12.800	Set	29,000	480		29,480	32,700
0700	Average		1	16		35,600	600		36,200	40,200
0800	Maximum		.75	21.333		42,200	800		43,000	47,700
0820	Dumbbell set, cast iron, with rack and 5 pair					625			625	690
0900	Squat racks	2 Clab	5	3.200	Ea.	900	120		1,020	1,175
4150	Exercise equipment, bicycle trainer					760			760	835
4180	Chinning bar, adjustable, wall mounted	1 Carp	5	1.600		212	75		287	355
4200	Exercise ladder, 16' x 1'-7", suspended	L-2	3	5.333		1,375	220		1,595	1,850
4210	High bar, floor plate attached	1 Carp	4	2		2,250	94		2,344	2,625
4240	Parallel bars, adjustable		4	2		1,700	94		1,794	2,025
4270	Uneven parallel bars, adjustable		4	2		3,225	94		3,319	3,700
4280	Wall mounted, adjustable	L-2	1.50	10.667	Set	865	440		1,305	1,675
4300	Rope, ceiling mounted, 18' long	1 Carp	3.66	2.186	Ea.	190	103		293	375
4330	Side horse, vaulting		5	1.600		1,375	75		1,450	1,625
4360	Treadmill, motorized, deluxe, training type		5	1.600		3,850	75		3,925	4,350
4390	Weight lifting multi-station, minimum	2 Clab	1	16		335	600		935	1,350

11 66 23 – Gymnasium Equipment

11 66 23.13 Basketball Equipment

		Crew	Daily Output	Labor-Hours	Unit	Material	2015 Bare Costs Labor	Equipment	Total	Total Incl O&P
0010	**BASKETBALL EQUIPMENT**									
1000	Backstops, wall mtd., 6' extended, fixed, minimum	L-2	1	16	Ea.	1,375	660		2,035	2,600
1100	Maximum		1	16		1,900	660		2,560	3,175
1200	Swing up, minimum		1	16		1,450	660		2,110	2,650
1250	Maximum		1	16		2,850	660		3,510	4,200
1300	Portable, manual, heavy duty, spring operated		1.90	8.421		12,700	345		13,045	14,600
1400	Ceiling suspended, stationary, minimum		.78	20.513		4,050	845		4,895	5,850
1450	Fold up, with accessories, maximum		.40	40		6,075	1,650		7,725	9,375
1600	For electrically operated, add	1 Elec	1	8		2,350	440		2,790	3,250
5800	Wall pads, 1-1/2" thick, standard (not fire rated)	2 Carp	640	.025	S.F.	6.25	1.17		7.42	8.80

11 66 23.19 Boxing Ring

		Crew	Daily Output	Labor-Hours	Unit	Material	2015 Bare Costs Labor	Equipment	Total	Total Incl O&P
0010	**BOXING RING**									
4100	Elevated, 22' x 22'	L-4	.10	240	Ea.	7,075	10,600		17,675	25,100
4110	For cellular plastic foam padding, add		.10	240		1,025	10,600		11,625	18,500
4120	Floor level, including posts and ropes only, 20' x 20'		.80	30		4,550	1,325		5,875	7,175
4130	Canvas, 30' x 30'		5	4.800		1,275	212		1,487	1,750

11 66 23.47 Gym Mats

		Crew	Daily Output	Labor-Hours	Unit	Material	2015 Bare Costs Labor	Equipment	Total	Total Incl O&P
0010	**GYM MATS**									
5500	2" thick, naugahyde covered				S.F.	3.71			3.71	4.08
5600	Vinyl/nylon covered					8.05			8.05	8.85
6000	Wrestling mats, 1" thick, heavy duty					5.60			5.60	6.15

11 66 Athletic Equipment

11 66 43 – Interior Scoreboards

11 66 43.10 Scoreboards

		Crew	Daily Output	Labor-Hours	Unit	Material	2015 Bare Costs Labor	2015 Bare Costs Equipment	Total	Total Incl O&P
0010	**SCOREBOARDS**									
7000	Baseball, minimum	R-3	1.30	15.385	Ea.	4,125	835	106	5,066	5,975
7200	Maximum		.05	400		18,100	21,700	2,750	42,550	56,500
7300	Football, minimum		.86	23.256		5,325	1,275	160	6,760	8,025
7400	Maximum		.20	100		15,900	5,425	690	22,015	26,700
7500	Basketball (one side), minimum		2.07	9.662		2,400	525	66.50	2,991.50	3,550
7600	Maximum		.30	66.667		3,525	3,625	460	7,610	10,000
7700	Hockey-basketball (four sides), minimum		.25	80		5,675	4,350	550	10,575	13,600
7800	Maximum		.15	133		5,750	7,250	920	13,920	18,600

11 66 53 – Gymnasium Dividers

11 66 53.10 Divider Curtains

		Crew	Daily Output	Labor-Hours	Unit	Material	Labor	Equipment	Total	Total Incl O&P
0010	**DIVIDER CURTAINS**									
4500	Gym divider curtain, mesh top, vinyl bottom, manual	L-4	500	.048	S.F.	9	2.12		11.12	13.35
4700	Electric roll up	L-7	400	.070	"	12.05	3.18		15.23	18.40

11 71 Medical Sterilizing Equipment

11 71 10 – Medical Sterilizers & Distillers

11 71 10.10 Sterilizers and Distillers

		Crew	Daily Output	Labor-Hours	Unit	Material	Labor	Equipment	Total	Total Incl O&P
0010	**STERILIZERS AND DISTILLERS**									
3010	Portable, top loading, 105 – 135 degree C, 3 to 30 psi, 50 L chamber				Ea.	10,000			10,000	11,000
3020	Stainless steel basket, 10.7" diam. x 11.8" H					223			223	245
3025	Stainless steel pail, 10.7" diam. x 10.7" H					243			243	267
3050	85 L chamber					14,300			14,300	15,700
3060	Stainless steel basket, 15.3" diam. x 11.5" H					305			305	335
3065	Stainless steel pail, 15.3" diam. x 11" H					425			425	470

11 76 Operating Room Equipment

11 76 10 – Operating Room Equipment

11 76 10.10 Surgical Equipment

		Crew	Daily Output	Labor-Hours	Unit	Material	Labor	Equipment	Total	Total Incl O&P
0010	**SURGICAL EQUIPMENT**									
6600	Hydraulic, hand-held control, general surgery	1 Sswk	.60	13.333	Ea.	30,400	700		31,100	34,700
6650	Stationary, universal	"	.50	16		39,200	840		40,040	44,600
6900	Ceiling mount articulation, single arm	2 Elec	1	16		3,800	875		4,675	5,525

11 81 Facility Maintenance Equipment

11 81 19 – Vacuum Cleaning Systems

11 81 19.10 Vacuum Cleaning

		Crew	Daily Output	Labor-Hours	Unit	Material	Labor	Equipment	Total	Total Incl O&P
0010	**VACUUM CLEANING**									
0020	Central, 3 inlet, residential	1 Skwk	.90	8.889	Total	1,075	430		1,505	1,875
0200	Commercial		.70	11.429		1,225	555		1,780	2,250
0400	5 inlet system, residential		.50	16		1,500	780		2,280	2,925
0600	7 inlet system, commercial		.40	20		1,700	975		2,675	3,450
4010	Rule of thumb: First 1200 S.F., installed								1,425	1,575
4020	For each additional S.F., add				S.F.				.26	.26

11 82 Facility Solid Waste Handling Equipment

11 82 26 – Facility Waste Compactors

11 82 26.10 Compactors

		Crew	Daily Output	Labor-Hours	Unit	Material	2015 Bare Costs Labor	Equipment	Total	Total Incl O&P
0010	**COMPACTORS**									
0020	Compactors, 115 volt, 250#/hr., chute fed	L-4	1	24	Ea.	12,100	1,050		13,150	15,000
0100	Hand fed		2.40	10		15,100	445		15,545	17,300
1000	Heavy duty industrial compactor, 0.5 C.Y. capacity		1	24		10,100	1,050		11,150	12,800
1050	1.0 C.Y. capacity	▼	1	24		15,200	1,050		16,250	18,400
1400	For handling hazardous waste materials, 55 gallon drum packer, std.					19,700			19,700	21,700
1410	55 gallon drum packer w/HEPA filter					24,600			24,600	27,100
1420	55 gallon drum packer w/charcoal & HEPA filter					32,800			32,800	36,100
1430	All of the above made explosion proof, add				▼	1,450			1,450	1,575

11 82 39 – Medical Waste Disposal Systems

11 82 39.10 Off-Site Disposal

		Crew	Daily Output	Labor-Hours	Unit	Material	Labor	Equipment	Total	Total Incl O&P
0010	**OFF-SITE DISPOSAL**									
0100	Medical waste disposal, Red Bag system, pick up & treat, 200 lb. per week				Week	177			177	195
0110	Per month				Month	700			700	770
0150	Red bags, 7-10 gal., 1.2 mil, pkg of 500				Ea.	65.50			65.50	72.50
0200	15 gal., package of 250					61.50			61.50	68
0250	33 gal., package of 250					64			64	70
0300	45 gal., package of 100				▼	54.50			54.50	60

11 82 39.20 Disposal Carts

		Crew	Daily Output	Labor-Hours	Unit	Material	Labor	Equipment	Total	Total Incl O&P
0010	**DISPOSAL CARTS**									
2010	Medical waste disposal cart, HDPE, w/lid, 28 gal. capacity				Ea.	261			261	287
2020	96 gal. capacity					305			305	335
2030	150 gal. capacity, low profile					470			470	515
2040	200 gal. capacity				▼	910			910	1,000

11 82 39.30 Medical Waste Sanitizers

		Crew	Daily Output	Labor-Hours	Unit	Material	Labor	Equipment	Total	Total Incl O&P
0010	**MEDICAL WASTE SANITIZERS**									
2010	Small, hand loaded, 1.5 C.Y., 225 lb. capacity				Ea.	78,000			78,000	86,000
2020	Medium, cart loaded, 6.25 C.Y., 938 lb. capacity					104,000			104,000	114,500
2030	Large, cart loaded, 15 C.Y., 2250 lb. capacity					130,500			130,500	143,500
3010	Cart, aluminum, 75 lb. capacity					2,150			2,150	2,375
3020	95 lb. capacity					2,350			2,350	2,575
4010	Stainless steel, 173 lb. capacity					3,075			3,075	3,375
4020	232 lb. capacity					3,450			3,450	3,775
4030	Cart lift, hydraulic scissor type					6,100			6,100	6,700
4040	Portable aluminum ramp					1,725			1,725	1,900
4050	Fold-down steel tracks					1,375			1,375	1,525
4060	Pull-out drawer, small					6,575			6,575	7,225
4070	Medium					9,475			9,475	10,400
4080	Large				▼	13,400			13,400	14,800
5000	Medical waste treatment, sanitize, on-site									
5010	Less than 15,000 lb. per month				Lb.	.20			.20	.22
5020	Over 15,000 lb. per month				"	.16			.16	.18

11 91 Religious Equipment

11 91 13 – Baptisteries

11 91 13.10 Baptistry		Crew	Daily Output	Labor-Hours	Unit	Material	2015 Bare Costs Labor	Equipment	Total	Total Incl O&P
0010	**BAPTISTRY**									
0150	Fiberglass, 3'-6" deep, x 13'-7" long,									
0160	steps at both ends, incl. plumbing, minimum	L-8	1	20	Ea.	5,725	985		6,710	7,900
0200	Maximum	"	.70	28.571		9,325	1,400		10,725	12,600
0250	Add for filter, heater and lights					1,850			1,850	2,050

11 91 23 – Sanctuary Equipment

11 91 23.10 Sanctuary Furnishings

		Crew	Daily Output	Labor-Hours	Unit	Material	Labor	Equipment	Total	Total Incl O&P
0010	**SANCTUARY FURNISHINGS**									
0020	Altar, wood, custom design, plain	1 Carp	1.40	5.714	Ea.	2,550	268		2,818	3,275
0050	Deluxe		.20	40	"	12,400	1,875		14,275	16,700
5000	Wall cross, aluminum, extruded, 2" x 2" section		34	.235	L.F.	218	11.05		229.05	258
5150	4" x 4" section		29	.276		315	12.95		327.95	365
5300	Bronze, extruded, 1" x 2" section		31	.258		430	12.10		442.10	490
5350	2-1/2" x 2-1/2" section		34	.235		650	11.05		661.05	735

11 97 Security Equipment

11 97 30 – Security Drawers

11 97 30.10 Pass Through Drawer

		Crew	Daily Output	Labor-Hours	Unit	Material	Labor	Equipment	Total	Total Incl O&P
0010	**PASS THROUGH DRAWER**									
0100	Pass-thru drawer for personal items, 18" x 15" x 24"	1 Skwk	2	4	Ea.	2,800	195		2,995	3,400
0110	Including speakers	"	1.50	5.333	"	3,150	259		3,409	3,900

11 98 Detention Equipment

11 98 30 – Detention Cell Equipment

11 98 30.10 Cell Equipment

		Crew	Daily Output	Labor-Hours	Unit	Material	Labor	Equipment	Total	Total Incl O&P
0010	**CELL EQUIPMENT**									
3000	Toilet apparatus including wash basin, average	L-8	1.50	13.333	Ea.	3,400	655		4,055	4,825

Estimating Tips
General

- The items in this division are usually priced per square foot or each. Most of these items are purchased by the owner and installed by the contractor. Do not assume the items in Division 12 will be purchased and installed by the contractor. Check the specifications for responsibilities and include receiving, storage, installation, and mechanical and electrical hookups in the appropriate divisions.

- Some items in this division require some type of support system that is not usually furnished with the item. Examples of these systems include blocking for the attachment of casework and heavy drapery rods. The required blocking must be added to the estimate in the appropriate division.

Reference Numbers
Reference numbers are shown in shaded boxes at the beginning of some major classifications. These numbers refer to related items in the Reference Section. The reference information may be an estimating procedure, an alternate pricing method, or technical information.

Note: Not all subdivisions listed here necessarily appear in this publication. ■

Division 12 – Furnishings

Did you know?
RSMeans Online gives you the same access to RSMeans' data with 24/7 access:
- Quickly locate costs in the searchable database.
- Build cost lists, estimates, and reports in minutes.
- Adjust costs to any location in the U.S. and Canada with the click of a button.

Start your free trial today at **www.rsmeansonline.com**

RSMeansOnline

No part of this publication may be reproduced, stored in a retrieval system, or transmitted in any form or by any means without prior written permission of RSMeans.

12 21 Window Blinds

12 21 13 – Horizontal Louver Blinds

12 21 13.13 Metal Horizontal Louver Blinds

		Crew	Daily Output	Labor-Hours	Unit	Material	2015 Bare Costs Labor	Equipment	Total	Total Incl O&P
0010	METAL HORIZONTAL LOUVER BLINDS									
0020	Horizontal, 1" aluminum slats, solid color, stock	1 Carp	590	.014	S.F.	4.90	.64		5.54	6.45
0070	Horizontal, 1" aluminum slats, custom color	"	590	.014	"	5.40	.64		6.04	7

12 21 13.33 Vinyl Horizontal Louver Blinds

		Crew	Daily Output	Labor-Hours	Unit	Material	Labor	Equipment	Total	Total Incl O&P
0010	VINYL HORIZONTAL LOUVER BLINDS									
0100	2" composite, 48" wide, 48" high	1 Carp	30	.267	Ea.	92	12.50		104.50	122
0120	72" high		29	.276		131	12.95		143.95	165
0140	96" high		28	.286		180	13.40		193.40	220
0200	60" wide, 60" high		27	.296		99.50	13.90		113.40	132
0220	72" high		25	.320		114	15		129	151
0240	96" high		24	.333		182	15.65		197.65	226
0300	72" wide, 72" high		25	.320		194	15		209	239
0320	96" high		23	.348		271	16.35		287.35	325
0400	96" wide, 96" high		20	.400		315	18.80		333.80	375
1000	2" faux wood, 48" wide, 48" high		30	.267		59	12.50		71.50	85.50
1020	72" high		29	.276		81	12.95		93.95	110
1040	96" high		28	.286		100	13.40		113.40	132
1300	72" wide, 72" high		25	.320		125	15		140	162
1320	96" high		23	.348		196	16.35		212.35	242
1400	96" wide, 96" high		20	.400		217	18.80		235.80	269

12 23 Interior Shutters

12 23 10 – Wood Interior Shutters

12 23 10.10 Wood Interior Shutters

		Crew	Daily Output	Labor-Hours	Unit	Material	Labor	Equipment	Total	Total Incl O&P
0010	WOOD INTERIOR SHUTTERS, louvered									
0200	Two panel, 27" wide, 36" high	1 Carp	5	1.600	Set	150	75		225	288
0300	33" wide, 36" high		5	1.600		194	75		269	335
0500	47" wide, 36" high		5	1.600		260	75		335	410
1000	Four panel, 27" wide, 36" high		5	1.600		220	75		295	365
1100	33" wide, 36" high		5	1.600		282	75		357	435
1300	47" wide, 36" high		5	1.600		375	75		450	540

12 23 10.13 Wood Panels

		Crew	Daily Output	Labor-Hours	Unit	Material	Labor	Equipment	Total	Total Incl O&P
0010	WOOD PANELS									
3000	Wood folding panels with movable louvers, 7" x 20" each	1 Carp	17	.471	Pr.	79.50	22		101.50	124
4000	Fixed louver type, stock units, 8" x 20" each		17	.471		94	22		116	139
4450	18" x 40" each		17	.471		134	22		156	184

12 24 Window Shades

12 24 13 – Roller Window Shades

12 24 13.10 Shades

		Crew	Daily Output	Labor-Hours	Unit	Material	Labor	Equipment	Total	Total Incl O&P
0010	SHADES									
0020	Basswood, roll-up, stain finish, 3/8" slats	1 Carp	300	.027	S.F.	14.95	1.25		16.20	18.45
0030	Double layered, heat reflective		685	.012		9.35	.55		9.90	11.20
0200	7/8" slats		300	.027		14.10	1.25		15.35	17.55
0900	Mylar, single layer, non-heat reflective		685	.012		5.05	.55		5.60	6.45
0910	Mylar, single layer, heat reflective		685	.012		5.45	.55		6	6.90
1000	Double layered, heat reflective		685	.012		5.85	.55		6.40	7.35
1100	Triple layered, heat reflective		685	.012		6.40	.55		6.95	7.90
2000	Polyester, room darkening, with continuous cord, GEI									

12 24 Window Shades

12 24 13 – Roller Window Shades

12 24 13.10 Shades

		Crew	Daily Output	Labor-Hours	Unit	Material	2015 Bare Costs Labor	Equipment	Total	Total Incl O&P
2010	36" x 72"	1 Carp	38	.211	Ea.	224	9.90		233.90	263
2020	48" x 72"		28	.286		293	13.40		306.40	340
2030	60" x 72"		23	.348		345	16.35		361.35	405
2040	72" x 72"		19	.421		395	19.75		414.75	470
5000	Thermal, roll up, R4		44	.182	S.F.	12.05	8.55		20.60	27.50
5030	R10.7		44	.182	"	13.85	8.55		22.40	29.50
5050	Magnetic clips, set of 20				Set	26			26	28.50

12 32 Manufactured Wood Casework

12 32 16 – Manufactured Plastic-Laminate-Clad Casework

12 32 16.20 Plastic Laminate Casework Doors

		Crew	Daily Output	Labor-Hours	Unit	Material	Labor	Equipment	Total	Total Incl O&P
0010	**PLASTIC LAMINATE CASEWORK DOORS**									
1000	For casework frames, see Section 12 32 23.15									
1100	For casework hardware, see Section 12 32 23.35									
6000	Plastic laminate on particle board									
6100	12" wide, 18" high	1 Carp	25	.320	Ea.	24	15		39	51
6140	30" high		23	.348		40	16.35		56.35	71
6500	18" wide, 18" high		24	.333		36	15.65		51.65	65
6600	30" high		22	.364		60	17.05		77.05	94

12 32 16.25 Plastic Laminate Drawer Fronts

		Crew	Daily Output	Labor-Hours	Unit	Material	Labor	Equipment	Total	Total Incl O&P
0010	**PLASTIC LAMINATE DRAWER FRONTS**									
2800	Plastic laminate on particle board front									
3000	4" high, 12" wide	1 Carp	17	.471	Ea.	4.51	22		26.51	41
3200	18" wide	"	16	.500	"	6.75	23.50		30.25	46

12 32 23 – Hardwood Casework

12 32 23.10 Manufactured Wood Casework, Stock Units

		Crew	Daily Output	Labor-Hours	Unit	Material	Labor	Equipment	Total	Total Incl O&P
0010	**MANUFACTURED WOOD CASEWORK, STOCK UNITS**									
0300	Built-in drawer units, pine, 18" deep, 32" high, unfinished									
0400	Minimum	2 Carp	53	.302	L.F.	120	14.15		134.15	155
0500	Maximum	"	40	.400	"	141	18.80		159.80	186
0700	Kitchen base cabinets, hardwood, not incl. counter tops,									
0710	24" deep, 35" high, prefinished									
0800	One top drawer, one door below, 12" wide	2 Carp	24.80	.645	Ea.	265	30.50		295.50	340
0820	15" wide		24	.667		276	31.50		307.50	355
0840	18" wide		23.30	.687		300	32		332	385
0860	21" wide		22.70	.705		305	33		338	395
0880	24" wide		22.30	.717		365	33.50		398.50	455
1000	Four drawers, 12" wide		24.80	.645		279	30.50		309.50	355
1020	15" wide		24	.667		283	31.50		314.50	360
1040	18" wide		23.30	.687		310	32		342	400
1060	24" wide		22.30	.717		345	33.50		378.50	435
1200	Two top drawers, two doors below, 27" wide		22	.727		390	34		424	485
1220	30" wide		21.40	.748		425	35		460	530
1240	33" wide		20.90	.766		440	36		476	545
1260	36" wide		20.30	.788		455	37		492	560
1280	42" wide		19.80	.808		480	38		518	590
1300	48" wide		18.90	.847		515	40		555	635
1500	Range or sink base, two doors below, 30" wide		21.40	.748		350	35		385	445
1520	33" wide		20.90	.766		375	36		411	475
1540	36" wide		20.30	.788		395	37		432	495

12 32 Manufactured Wood Casework

12 32 23 – Hardwood Casework

12 32 23.10 Manufactured Wood Casework, Stock Units	Crew	Daily Output	Labor-Hours	Unit	Material	2015 Bare Costs Labor	Equipment	Total	Total Incl O&P
1560	42" wide	2 Carp	19.80	.808	Ea.	415	38		453	515
1580	48" wide		18.90	.847		435	40		475	545
1800	For sink front units, deduct					161			161	177
2000	Corner base cabinets, 36" wide, standard	2 Carp	18	.889		625	41.50		666.50	760
2100	Lazy Susan with revolving door	"	16.50	.970		840	45.50		885.50	1,000
4000	Kitchen wall cabinets, hardwood, 12" deep with two doors									
4050	12" high, 30" wide	2 Carp	24.80	.645	Ea.	237	30.50		267.50	310
4100	36" wide		24	.667		282	31.50		313.50	360
4400	15" high, 30" wide		24	.667		241	31.50		272.50	315
4420	33" wide		23.30	.687		298	32		330	385
4440	36" wide		22.70	.705		290	33		323	375
4450	42" wide		22.70	.705		325	33		358	415
4700	24" high, 30" wide		23.30	.687		325	32		357	410
4720	36" wide		22.70	.705		355	33		388	445
4740	42" wide		22.30	.717		405	33.50		438.50	505
5000	30" high, one door, 12" wide		22	.727		216	34		250	294
5020	15" wide		21.40	.748		241	35		276	325
5040	18" wide		20.90	.766		265	36		301	350
5060	24" wide		20.30	.788		310	37		347	400
5300	Two doors, 27" wide		19.80	.808		340	38		378	435
5320	30" wide		19.30	.829		355	39		394	460
5340	36" wide		18.80	.851		405	40		445	515
5360	42" wide		18.50	.865		445	40.50		485.50	555
5380	48" wide		18.40	.870		500	41		541	615
6000	Corner wall, 30" high, 24" wide		18	.889		355	41.50		396.50	460
6050	30" wide		17.20	.930		380	43.50		423.50	485
6100	36" wide		16.50	.970		430	45.50		475.50	545
6500	Revolving Lazy Susan		15.20	1.053		480	49.50		529.50	605
7000	Broom cabinet, 84" high, 24" deep, 18" wide		10	1.600		650	75		725	840
7500	Oven cabinets, 84" high, 24" deep, 27" wide		8	2		1,000	94		1,094	1,250
7750	Valance board trim		396	.040	L.F.	13	1.90		14.90	17.40
7780	Toe kick trim	1 Carp	256	.031	"	2.79	1.47		4.26	5.45
7790	Base cabinet corner filler		16	.500	Ea.	41.50	23.50		65	84
7800	Cabinet filler, 3" x 24"		20	.400		18.05	18.80		36.85	51
7810	3" x 30"		20	.400		22.50	18.80		41.30	56
7820	3" x 42"		18	.444		31.50	21		52.50	68.50
7830	3" x 80"		16	.500		60	23.50		83.50	105
7850	Cabinet panel		50	.160	S.F.	8.65	7.50		16.15	22
9000	For deluxe models of all cabinets, add					40%				
9500	For custom built in place, add					25%	10%			
9558	Rule of thumb, kitchen cabinets not including									
9560	appliances & counter top, minimum	2 Carp	30	.533	L.F.	176	25		201	234
9600	Maximum	"	25	.640	"	395	30		425	485
9700	Minimum labor/equipment charge	1 Carp	3	2.667	Job		125		125	205

12 32 23.15 Manufactured Wood Casework Frames

		Crew	Daily Output	Labor-Hours	Unit	Material	Labor	Equipment	Total	Total Incl O&P
0010	**MANUFACTURED WOOD CASEWORK FRAMES**									
0050	Base cabinets, counter storage, 36" high									
0100	One bay, 18" wide	1 Carp	2.70	2.963	Ea.	171	139		310	415
0400	Two bay, 36" wide		2.20	3.636		261	171		432	565
1100	Three bay, 54" wide		1.50	5.333		310	250		560	750
2800	Bookcases, one bay, 7' high, 18" wide		2.40	3.333		201	157		358	480
3500	Two bay, 36" wide		1.60	5		292	235		527	705

12 32 Manufactured Wood Casework

12 32 23 – Hardwood Casework

12 32 23.15 Manufactured Wood Casework Frames

		Crew	Daily Output	Labor-Hours	Unit	Material	2015 Bare Costs Labor	Equipment	Total	Total Incl O&P
4100	Three bay, 54" wide	1 Carp	1.20	6.667	Ea.	485	315		800	1,050
5100	Coat racks, one bay, 7' high, 24" wide		4.50	1.778		201	83.50		284.50	360
5300	Two bay, 48" wide		2.75	2.909		279	137		416	530
5800	Three bay, 72" wide		2.10	3.810		410	179		589	745
6100	Wall mounted cabinet, one bay, 24" high, 18" wide		3.60	2.222		110	104		214	292
6800	Two bay, 36" wide		2.20	3.636		161	171		332	455
7400	Three bay, 54" wide		1.70	4.706		201	221		422	580
8400	30" high, one bay, 18" wide		3.60	2.222		120	104		224	305
9000	Two bay, 36" wide		2.15	3.721		160	175		335	460
9400	Three bay, 54" wide		1.60	5		199	235		434	605
9800	Wardrobe, 7' high, single, 24" wide		2.70	2.963		222	139		361	470
9880	Partition & adjustable shelves, 48" wide		1.70	4.706		282	221		503	670
9950	Partition, adjustable shelves & drawers, 48" wide		1.40	5.714		425	268		693	905
9970	Minimum labor/equipment charge		4	2	Job		94		94	154

12 32 23.20 Manufactured Hardwood Casework Doors

		Crew	Daily Output	Labor-Hours	Unit	Material	Labor	Equipment	Total	Total Incl O&P
0010	**MANUFACTURED HARDWOOD CASEWORK DOORS**									
2000	Glass panel, hardwood frame									
2200	12" wide, 18" high	1 Carp	34	.235	Ea.	27	11.05		38.05	47.50
2600	30" high		32	.250		45	11.75		56.75	69
4450	18" wide, 18" high		32	.250		40.50	11.75		52.25	64
4550	30" high		29	.276		67.50	12.95		80.45	95.50
5000	Hardwood, raised panel									
5100	12" wide, 18" high	1 Carp	16	.500	Ea.	28.50	23.50		52	70
5200	30" high		15	.533		47.50	25		72.50	93.50
5500	18" wide, 18" high		15	.533		43	25		68	88
5600	30" high		14	.571		71.50	27		98.50	123
9000	Minimum labor/equipment charge		4	2	Job		94		94	154

12 32 23.25 Manufactured Wood Casework Drawer Fronts

		Crew	Daily Output	Labor-Hours	Unit	Material	Labor	Equipment	Total	Total Incl O&P
0010	**MANUFACTURED WOOD CASEWORK DRAWER FRONTS**									
0100	Solid hardwood front									
1000	4" high, 12" wide	1 Carp	17	.471	Ea.	4.33	22		26.33	41
1200	18" wide		16	.500	"	6.50	23.50		30	45.50
9000	Minimum labor/equipment charge		4	2	Job		94		94	154

12 32 23.30 Manufactured Wood Casework Vanities

		Crew	Daily Output	Labor-Hours	Unit	Material	Labor	Equipment	Total	Total Incl O&P
0010	**MANUFACTURED WOOD CASEWORK VANITIES**									
8000	Vanity bases, 2 doors, 30" high, 21" deep, 24" wide	2 Carp	20	.800	Ea.	310	37.50		347.50	400
8050	30" wide		16	1		370	47		417	480
8100	36" wide		13.33	1.200		360	56.50		416.50	490
8150	48" wide		11.43	1.400		470	65.50		535.50	630
9000	For deluxe models of all vanities, add to above					40%				
9500	For custom built in place, add to above					25%	10%			

12 32 23.35 Manufactured Wood Casework Hardware

		Crew	Daily Output	Labor-Hours	Unit	Material	Labor	Equipment	Total	Total Incl O&P
0010	**MANUFACTURED WOOD CASEWORK HARDWARE**									
1000	Catches, minimum	1 Carp	235	.034	Ea.	1.22	1.60		2.82	3.96
1040	Maximum	"	80	.100	"	7.55	4.70		12.25	16
2000	Door/drawer pulls, handles									
2200	Handles and pulls, projecting, metal, minimum	1 Carp	48	.167	Ea.	5	7.85		12.85	18.35
2240	Maximum		36	.222		10.60	10.45		21.05	29
2300	Wood, minimum		48	.167		5.25	7.85		13.10	18.65
2340	Maximum		36	.222		9.65	10.45		20.10	28
2400	Drawer pulls, antimicrobial copper alloy finish		50	.160		18.75	7.50		26.25	33
2600	Flush, metal, minimum		48	.167		5.25	7.85		13.10	18.65

12 32 Manufactured Wood Casework

12 32 23 – Hardwood Casework

12 32 23.35 Manufactured Wood Casework Hardware		Crew	Daily Output	Labor-Hours	Unit	Material	2015 Bare Costs Labor	Equipment	Total	Total Incl O&P
2640	Maximum	1 Carp	36	.222	Ea.	9.65	10.45		20.10	28
2900	Drawer knobs, antimicrobial copper alloy finish		50	.160	↓	12.50	7.50		20	26
3000	Drawer tracks/glides, minimum		48	.167	Pr.	8.95	7.85		16.80	22.50
3040	Maximum		24	.333		26	15.65		41.65	54
4000	Cabinet hinges, minimum		160	.050		3.02	2.35		5.37	7.15
4040	Maximum		68	.118	↓	11.45	5.50		16.95	21.50
7000	Appliance pulls, antimicrobial copper alloy finish	↓	50	.160	L.F.	62.50	7.50		70	81.50

12 34 Manufactured Plastic Casework

12 34 16 – Manufactured Solid-Plastic Casework

12 34 16.10 Outdoor Casework

		Crew	Daily Output	Labor-Hours	Unit	Material	2015 Bare Costs Labor	Equipment	Total	Total Incl O&P
0010	**OUTDOOR CASEWORK**									
0020	Cabinet, base, sink/range, 36"	2 Carp	20.30	.788	Ea.	1,875	37		1,912	2,100
0100	Base, 36"		20.30	.788		2,100	37		2,137	2,350
0200	Filler strip, 1" x 30"		158	.101		32.50	4.75		37.25	43.50
0210	Filler strip, 2" x 30"	↓	158	.101	↓	43	4.75		47.75	55

12 35 Specialty Casework

12 35 53 – Laboratory Casework

12 35 53.13 Metal Laboratory Casework

		Crew	Daily Output	Labor-Hours	Unit	Material	2015 Bare Costs Labor	Equipment	Total	Total Incl O&P
0010	**METAL LABORATORY CASEWORK**									
0020	Cabinets, base, door units, metal	2 Carp	18	.889	L.F.	231	41.50		272.50	325
0300	Drawer units		18	.889		515	41.50		556.50	635
0700	Tall storage cabinets, open, 7' high		20	.800		495	37.50		532.50	605
0900	With glazed doors		20	.800		740	37.50		777.50	875
1300	Wall cabinets, metal, 12-1/2" deep, open		20	.800		166	37.50		203.50	245
1500	With doors	↓	20	.800	↓	345	37.50		382.50	440

12 35 70 – Healthcare Casework

12 35 70.13 Hospital Casework

		Crew	Daily Output	Labor-Hours	Unit	Material	2015 Bare Costs Labor	Equipment	Total	Total Incl O&P
0010	**HOSPITAL CASEWORK**									
3000	Hospital cabinets, stainless steel with glass door(s), lockable									
3010	One door, 24" W x 18" D x 60" H	2 Clab	18	.889	Ea.	2,500	33.50		2,533.50	2,775
3020	Two doors, 36" W x 18" D x 60" H		15	1.067		2,825	40		2,865	3,175
3030	36" W x 24" D x 67" H		15	1.067		4,250	40		4,290	4,750
3040	48" W x 24" D x 66" H		12	1.333		4,000	50		4,050	4,475
3050	48" W x 24" D x 72" H		12	1.333		5,050	50		5,100	5,625
3060	60" W x 24" D x 72" H	↓	9	1.778	↓	5,500	67		5,567	6,150

12 36 Countertops

12 36 16 – Metal Countertops

12 36 16.10 Stainless Steel Countertops

		Crew	Daily Output	Labor-Hours	Unit	Material	2015 Bare Costs Labor	Equipment	Total	Total Incl O&P
0010	**STAINLESS STEEL COUNTERTOPS**									
3200	Stainless steel, custom	1 Carp	24	.333	S.F.	153	15.65		168.65	194

12 36 19 – Wood Countertops

12 36 19.10 Maple Countertops

		Crew	Daily Output	Labor-Hours	Unit	Material	Labor	Equipment	Total	Total Incl O&P
0010	**MAPLE COUNTERTOPS**									
2900	Solid, laminated, 1-1/2" thick, no splash	1 Carp	28	.286	L.F.	75.50	13.40		88.90	105
3000	With square splash		28	.286	"	90	13.40		103.40	121
3400	Recessed cutting block with trim, 16" x 20" x 1"		8	1	Ea.	92	47		139	178

12 36 23 – Plastic Countertops

12 36 23.13 Plastic-Laminate-Clad Countertops

		Crew	Daily Output	Labor-Hours	Unit	Material	Labor	Equipment	Total	Total Incl O&P
0010	**PLASTIC-LAMINATE-CLAD COUNTERTOPS**									
0020	Stock, 24" wide w/backsplash, minimum	1 Carp	30	.267	L.F.	17	12.50		29.50	39
0100	Maximum		25	.320		34.50	15		49.50	62.50
0300	Custom plastic, 7/8" thick, aluminum molding, no splash		30	.267		30	12.50		42.50	53.50
0400	Cove splash		30	.267		29	12.50		41.50	52.50
0600	1-1/4" thick, no splash		28	.286		35.50	13.40		48.90	61
0700	Square splash		28	.286		42.50	13.40		55.90	68.50
0900	Square edge, plastic face, 7/8" thick, no splash		30	.267		33	12.50		45.50	57
1000	With splash		30	.267		39.50	12.50		52	63.50
1200	For stainless channel edge, 7/8" thick, add					3.12			3.12	3.43
1300	1-1/4" thick, add					3.72			3.72	4.09
1500	For solid color suede finish, add					4.08			4.08	4.49
1700	For end splash, add				Ea.	18.35			18.35	20
1900	For cut outs, standard, add, minimum	1 Carp	32	.250		12.25	11.75		24	32.50
2000	Maximum		8	1		6.10	47		53.10	84
2100	Postformed, including backsplash and front edge		30	.267	L.F.	10.20	12.50		22.70	31.50
2110	Mitred, add		12	.667	Ea.		31.50		31.50	51.50
2200	Built-in place, 25" wide, plastic laminate		25	.320	L.F.	41.50	15		56.50	70
9000	Minimum labor/equipment charge		3.75	2.133	Job		100		100	164

12 36 33 – Tile Countertops

12 36 33.10 Ceramic Tile Countertops

		Crew	Daily Output	Labor-Hours	Unit	Material	Labor	Equipment	Total	Total Incl O&P
0010	**CERAMIC TILE COUNTERTOPS**									
2300	Ceramic tile mosaic	1 Carp	25	.320	L.F.	33.50	15		48.50	61.50

12 36 40 – Stone Countertops

12 36 40.10 Natural Stone Countertops

		Crew	Daily Output	Labor-Hours	Unit	Material	Labor	Equipment	Total	Total Incl O&P
0010	**NATURAL STONE COUNTERTOPS**									
2500	Marble, stock, with splash, 1/2" thick, minimum	1 Bric	17	.471	L.F.	42	21.50		63.50	81.50
2700	3/4" thick, maximum	"	13	.615	"	105	28.50		133.50	162

12 36 53 – Laboratory Countertops

12 36 53.10 Laboratory Countertops and Sinks

		Crew	Daily Output	Labor-Hours	Unit	Material	Labor	Equipment	Total	Total Incl O&P
0010	**LABORATORY COUNTERTOPS AND SINKS**									
0020	Countertops, epoxy resin, not incl. base cabinets, acid-proof, minimum	2 Carp	82	.195	S.F.	40.50	9.15		49.65	59.50
0030	Maximum		70	.229		50	10.75		60.75	72.50
0040	Stainless steel		82	.195		131	9.15		140.15	159

12 36 Countertops

12 36 61 – Simulated Stone Countertops

12 36 61.16 Solid Surface Countertops

		Crew	Daily Output	Labor-Hours	Unit	Material	2015 Bare Costs Labor	Equipment	Total	Total Incl O&P
0010	**SOLID SURFACE COUNTERTOPS**, Acrylic polymer									
0020	Pricing for orders of 100 L.F. or greater									
0100	25" wide, solid colors	2 Carp	28	.571	L.F.	54.50	27		81.50	104
0200	Patterned colors		28	.571		69	27		96	120
0300	Premium patterned colors		28	.571		86.50	27		113.50	139
0400	With silicone attached 4" backsplash, solid colors		27	.593		60	28		88	112
0500	Patterned colors		27	.593		76	28		104	129
0600	Premium patterned colors		27	.593		94.50	28		122.50	150
0700	With hard seam attached 4" backsplash, solid colors		23	.696		60	32.50		92.50	120
0800	Patterned colors		23	.696		76	32.50		108.50	137
0900	Premium patterned colors	↓	23	.696	↓	94.50	32.50		127	158
1000	Pricing for order of 51 – 99 L.F.									
1100	25" wide, solid colors	2 Carp	24	.667	L.F.	63	31.50		94.50	121
1200	Patterned colors		24	.667		79.50	31.50		111	139
1300	Premium patterned colors		24	.667		99.50	31.50		131	161
1400	With silicone attached 4" backsplash, solid colors		23	.696		69	32.50		101.50	130
1500	Patterned colors		23	.696		87.50	32.50		120	150
1600	Premium patterned colors		23	.696		109	32.50		141.50	174
1700	With hard seam attached 4" backsplash, solid colors		20	.800		69	37.50		106.50	138
1800	Patterned colors		20	.800		87.50	37.50		125	158
1900	Premium patterned colors	↓	20	.800	↓	109	37.50		146.50	182
2000	Pricing for order of 1 – 50 L.F.									
2100	25" wide, solid colors	2 Carp	20	.800	L.F.	73.50	37.50		111	143
2200	Patterned colors		20	.800		93.50	37.50		131	165
2300	Premium patterned colors		20	.800		117	37.50		154.50	191
2400	With silicone attached 4" backsplash, solid colors		19	.842		81	39.50		120.50	154
2500	Patterned colors		19	.842		102	39.50		141.50	178
2600	Premium patterned colors		19	.842		128	39.50		167.50	205
2700	With hard seam attached 4" backsplash, solid colors		15	1.067		81	50		131	171
2800	Patterned colors		15	1.067		102	50		152	195
2900	Premium patterned colors	↓	15	1.067	↓	128	50		178	222
3000	Sinks, pricing for order of 100 or greater units									
3100	Single bowl, hard seamed, solid colors, 13" x 17"	1 Carp	3	2.667	Ea.	370	125		495	610
3200	10" x 15"		7	1.143		170	53.50		223.50	275
3300	Cutouts for sinks	↓	8	1	↓		47		47	77
3400	Sinks, pricing for order of 51 – 99 units									
3500	Single bowl, hard seamed, solid colors, 13" x 17"	1 Carp	2.55	3.137	Ea.	425	147		572	705
3600	10" x 15"		6	1.333		196	62.50		258.50	320
3700	Cutouts for sinks	↓	7	1.143	↓		53.50		53.50	88
3800	Sinks, pricing for order of 1 – 50 units									
3900	Single bowl, hard seamed, solid colors, 13" x 17"	1 Carp	2	4	Ea.	500	188		688	860
4000	10" x 15"		4.55	1.758		230	82.50		312.50	390
4100	Cutouts for sinks		5.25	1.524			71.50		71.50	117
4200	Cooktop cutouts, pricing for 100 or greater units		4	2		27	94		121	184
4300	51 – 99 units		3.40	2.353		31.50	110		141.50	216
4400	1 – 50 units	↓	3	2.667	↓	36.50	125		161.50	246

12 36 61.17 Solid Surface Vanity Tops

		Crew	Daily Output	Labor-Hours	Unit	Material	Labor	Equipment	Total	Total Incl O&P
0010	**SOLID SURFACE VANITY TOPS**									
0015	Solid surface, center bowl, 17" x 19"	1 Carp	12	.667	Ea.	190	31.50		221.50	261
0020	19" x 25"		12	.667		194	31.50		225.50	266
0030	19" x 31"		12	.667		227	31.50		258.50	300
0040	19" x 37"	↓	12	.667	↓	264	31.50		295.50	345

12 36 Countertops

12 36 61 – Simulated Stone Countertops

12 36 61.17 Solid Surface Vanity Tops

		Crew	Daily Output	Labor-Hours	Unit	Material	2015 Bare Costs Labor	Equipment	Total	Total Incl O&P
0050	22" x 25"	1 Carp	10	.800	Ea.	345	37.50		382.50	440
0060	22" x 31"		10	.800		405	37.50		442.50	505
0070	22" x 37"		10	.800		470	37.50		507.50	580
0080	22" x 43"		10	.800		535	37.50		572.50	650
0090	22" x 49"		10	.800		595	37.50		632.50	715
0110	22" x 55"		8	1		675	47		722	820
0120	22" x 61"		8	1		770	47		817	925
0220	Double bowl, 22" x 61"		8	1		870	47		917	1,025
0230	Double bowl, 22" x 73"		8	1		950	47		997	1,125
0240	For aggregate colors, add					35%				
0250	For faucets and fittings, see Section 22 41 39.10									

12 36 61.19 Quartz Agglomerate Countertops

		Crew	Daily Output	Labor-Hours	Unit	Material	Labor	Equipment	Total	Total Incl O&P
0010	**QUARTZ AGGLOMERATE COUNTERTOPS**									
0100	25" wide, 4" backsplash, color group A, minimum	2 Carp	15	1.067	L.F.	64.50	50		114.50	153
0110	Maximum		15	1.067		90	50		140	181
0120	Color group B, minimum		15	1.067		66.50	50		116.50	156
0130	Maximum		15	1.067		94.50	50		144.50	186
0140	Color group C, minimum		15	1.067		78	50		128	168
0150	Maximum		15	1.067		107	50		157	199
0160	Color group D, minimum		15	1.067		84.50	50		134.50	175
0170	Maximum		15	1.067		115	50		165	208

12 48 Rugs and Mats

12 48 13 – Entrance Floor Mats and Frames

12 48 13.13 Entrance Floor Mats

			Crew	Daily Output	Labor-Hours	Unit	Material	Labor	Equipment	Total	Total Incl O&P
0010	**ENTRANCE FLOOR MATS**										
2000	Recycled rubber tire tile, 12" x 12" x 3/8" thick	G	1 Clab	125	.064	S.F.	10.10	2.41		12.51	15.05
2510	Natural cocoa fiber, 1/2" thick	G		125	.064		8.35	2.41		10.76	13.15
2520	3/4" thick	G		125	.064		6.90	2.41		9.31	11.55
2530	1" thick	G		125	.064		9.60	2.41		12.01	14.50
3000	Hospital tacky mats, package of 30 with frame					Ea.	56			56	61.50
3010	4 packages of 30					"	86.50			86.50	95.50

12 51 Office Furniture

12 51 16 – Case Goods

12 51 16.13 Metal Case Goods

		Crew	Daily Output	Labor-Hours	Unit	Material	Labor	Equipment	Total	Total Incl O&P
0010	**METAL CASE GOODS**									
0020	Desks, 29" high, double pedestal, 30" x 60", metal, minimum				Ea.	595			595	655
0030	Maximum				"	1,550			1,550	1,700

12 54 Hospitality Furniture

12 54 13 – Hotel and Motel Furniture

12 54 13.10 Hotel Furniture	Crew	Daily Output	Labor-Hours	Unit	Material	2015 Bare Costs Labor	2015 Bare Costs Equipment	Total	Total Incl O&P
0010 **HOTEL FURNITURE**									
0020 Standard quality set, minimum				Room	2,400			2,400	2,650
0200 Maximum				"	8,600			8,600	9,450

12 54 16 – Restaurant Furniture

12 54 16.20 Furniture, Restaurant

	Crew	Daily Output	Labor-Hours	Unit	Material	Labor	Equipment	Total	Total Incl O&P
0010 **FURNITURE, RESTAURANT**									
0020 Bars, built-in, front bar	1 Carp	5	1.600	L.F.	280	75		355	435
0200 Back bar		5	1.600	"	203	75		278	345
0500 Booth unit, molded plastic, stub wall and 2 seats, minimum		2	4	Set	370	188		558	715
0600 Maximum		1.50	5.333	"	1,450	250		1,700	2,000
0800 Booth seat, upholstered, foursome, single (end) minimum		5	1.600	Ea.	780	75		855	985
0900 Maximum		4	2		1,050	94		1,144	1,300
1000 Foursome, double, minimum		4	2		1,275	94		1,369	1,550
1100 Maximum		3	2.667		1,750	125		1,875	2,125
1300 Circle booth, upholstered, 1/4 circle, minimum		3	2.667		1,400	125		1,525	1,750
1400 Maximum		2	4		1,950	188		2,138	2,425
1500 3/4 circle, minimum		1.50	5.333		5,525	250		5,775	6,475
1600 Maximum		1	8		6,700	375		7,075	8,000

12 55 Detention Furniture

12 55 13 – Detention Bunks

12 55 13.13 Cots

	Crew	Daily Output	Labor-Hours	Unit	Material	Labor	Equipment	Total	Total Incl O&P
0010 **COTS**									
2500 Bolted, single, painted steel	E-4	20	1.600	Ea.	335	85	7.30	427.30	530
2700 Stainless steel	"	20	1.600	"	975	85	7.30	1,067.30	1,225

12 56 Institutional Furniture

12 56 43 – Dormitory Furniture

12 56 43.10 Dormitory Furnishings

	Crew	Daily Output	Labor-Hours	Unit	Material	Labor	Equipment	Total	Total Incl O&P
0010 **DORMITORY FURNISHINGS**									
0300 Bunkable bed, twin, minimum				Ea.	375			375	415
0320 Maximum				"	550			550	605

12 56 51 – Library Furniture

12 56 51.10 Library Furnishings

	Crew	Daily Output	Labor-Hours	Unit	Material	Labor	Equipment	Total	Total Incl O&P
0010 **LIBRARY FURNISHINGS**									
1710 Carrels, hardwood, 36" x 24", minimum	1 Carp	5	1.600	Ea.	785	75		860	985
1720 Maximum		4	2		2,000	94		2,094	2,350
1730 Metal, minimum		5	1.600		260	75		335	410
1740 Maximum		4	2		455	94		549	660
6010 Bookshelf, metal, 90" high, 10" shelf, double face		11.50	.696	L.F.	150	32.50		182.50	219
6020 Single face		12	.667	"	124	31.50		155.50	189
6050 For 8" shelving, subtract from above					10%				
6070 For 42" high with countertop, subtract from above					20%				

12 56 70 – Healthcare Furniture

12 56 70.10 Furniture, Hospital

	Crew	Daily Output	Labor-Hours	Unit	Material	Labor	Equipment	Total	Total Incl O&P
0010 **FURNITURE, HOSPITAL**									
0020 Beds, manual, minimum				Ea.	800			800	880
0100 Maximum				"	2,550			2,550	2,800

12 56 Institutional Furniture

12 56 70 – Healthcare Furniture

12 56 70.10 Furniture, Hospital		Crew	Daily Output	Labor-Hours	Unit	Material	2015 Bare Costs Labor	Equipment	Total	Total Incl O&P
1100	Patient wall systems, not incl. plumbing, minimum				Room	1,425			1,425	1,575
1200	Maximum				"	1,925			1,925	2,125

12 63 Stadium and Arena Seating

12 63 13 – Stadium and Arena Bench Seating

12 63 13.13 Bleachers

		Crew	Daily Output	Labor-Hours	Unit	Material	Labor	Equipment	Total	Total Incl O&P
0010	**BLEACHERS**									
3000	Telescoping, manual to 15 tier, minimum	F-5	65	.492	Seat	91.50	23.50		115	140
3100	Maximum		60	.533		137	25.50		162.50	193
3300	16 to 20 tier, minimum		60	.533		220	25.50		245.50	284
3400	Maximum		55	.582		275	27.50		302.50	345
3600	21 to 30 tier, minimum		50	.640		229	30.50		259.50	300
3700	Maximum		40	.800		300	38		338	390
3900	For integral power operation, add, minimum	2 Elec	300	.053		46	2.92		48.92	55
4000	Maximum	"	250	.064		73.50	3.50		77	86
5000	Benches, folding, in wall, 14' table, 2 benches	L-4	2	12	Set	775	530		1,305	1,725

12 67 Pews and Benches

12 67 13 – Pews

12 67 13.13 Sanctuary Pews

		Crew	Daily Output	Labor-Hours	Unit	Material	Labor	Equipment	Total	Total Incl O&P
0010	**SANCTUARY PEWS**									
1500	Bench type, hardwood, minimum	1 Carp	20	.400	L.F.	94.50	18.80		113.30	135
1550	Maximum	"	15	.533		187	25		212	247
1570	For kneeler, add					22.50			22.50	24.50

12 93 Interior Public Space Furnishings

12 93 23 – Trash and Litter Receptacles

12 93 23.10 Trash Receptacles

			Crew	Daily Output	Labor-Hours	Unit	Material	Labor	Equipment	Total	Total Incl O&P
0010	**TRASH RECEPTACLES**										
0500	Recycled plastic, var colors, round, 32 gal., 28" x 38" H	G	2 Clab	5	3.200	Ea.	510	120		630	760
0510	32 gal., 31" x 32" H	G		5	3.200		585	120		705	840
9110	Plastic, with dome lid, 32 gal. capacity			35	.457		58	17.20		75.20	92
9120	Recycled plastic slats, plastic dome lid, 32 gal. capacity			35	.457		284	17.20		301.20	340

Division Notes

		CREW	DAILY OUTPUT	LABOR-HOURS	UNIT	BARE COSTS				TOTAL INCL O&P
						MAT.	LABOR	EQUIP.	TOTAL	

Estimating Tips
General
- The items and systems in this division are usually estimated, purchased, supplied, and installed as a unit by one or more subcontractors. The estimator must ensure that all parties are operating from the same set of specifications and assumptions, and that all necessary items are estimated and will be provided. Many times the complex items and systems are covered, but the more common ones, such as excavation or a crane, are overlooked for the very reason that everyone assumes nobody could miss them. The estimator should be the central focus and be able to ensure that all systems are complete.
- Another area where problems can develop in this division is at the interface between systems. The estimator must ensure, for instance, that anchor bolts, nuts, and washers are estimated and included for the air-supported structures and pre-engineered buildings to be bolted to their foundations. Utility supply is a common area where essential items or pieces of equipment can be missed or overlooked, because each subcontractor may feel it is another's responsibility. The estimator should also be aware of certain items which may be supplied as part of a package but installed by others, and ensure that the installing contractor's estimate includes the cost of installation. Conversely, the estimator must also ensure that items are not costed by two different subcontractors, resulting in an inflated overall estimate.

13 30 00 Special Structures
- The foundations and floor slab, as well as rough mechanical and electrical, should be estimated, as this work is required for the assembly and erection of the structure. Generally, as noted in the book, the pre-engineered building comes as a shell. Pricing is based on the size and structural design parameters stated in the reference section. Additional features, such as windows and doors with their related structural framing, must also be included by the estimator. Here again, the estimator must have a clear understanding of the scope of each portion of the work and all the necessary interfaces.

Reference Numbers
Reference numbers are shown in shaded boxes at the beginning of some major classifications. These numbers refer to related items in the Reference Section. The reference information may be an estimating procedure, an alternate pricing method, or technical information.

Note: Not all subdivisions listed here necessarily appear in this publication. ■

Did you know?
RSMeans Online gives you the same access to RSMeans' data with 24/7 access:
- Quickly locate costs in the searchable database.
- Build cost lists, estimates, and reports in minutes.
- Adjust costs to any location in the U.S. and Canada with the click of a button.

Start your free trial today at **www.rsmeansonline.com**

RSMeansOnline

13 05 Common Work Results for Special Construction

13 05 05 – Selective Demolition for Special Construction

13 05 05.10 Selective Demolition, Air Supported Structures

		Crew	Daily Output	Labor-Hours	Unit	Material	2015 Bare Costs Labor	2015 Bare Costs Equipment	Total	Total Incl O&P
0010	SELECTIVE DEMOLITION, AIR SUPPORTED STRUCTURES									
0020	Tank covers, scrim, dbl. layer, vinyl poly w/hdwe., blower & controls									
0050	Round and rectangular	R024119-10	B-2	9000	.004	S.F.		.17		.17	.28
0100	Warehouse structures									
0120	Poly/vinyl fabric, 28 oz., incl. tension cables & inflation system		4 Clab	9000	.004	SF Flr.		.13		.13	.22
0150	Reinforced vinyl, 12 oz., 3000 S.F.		"	5000	.006			.24		.24	.39
0200	12,000 to 24,000 S.F.		8 Clab	20000	.003			.12		.12	.20
0250	Tedlar vinyl fabric, 28 oz. w/liner, to 3000 S.F.		4 Clab	5000	.006			.24		.24	.39
0300	12,000 to 24,000 S.F.		8 Clab	20000	.003			.12		.12	.20
0350	Greenhouse/shelter, woven polyethylene with liner									
0400	3000 S.F.		4 Clab	5000	.006	SF Flr.		.24		.24	.39
0450	12,000 to 24,000 S.F.		8 Clab	20000	.003			.12		.12	.20
0500	Tennis/gymnasium, poly/vinyl fabric, 28 oz., incl. thermal liner		4 Clab	9000	.004			.13		.13	.22
0600	Stadium/convention center, teflon coated fiberglass, incl. thermal liner		9 Clab	40000	.002			.07		.07	.11
0700	Doors, air lock, 15' long, 10' x 10'		2 Carp	1.50	10.667	Ea.		500		500	820
0720	15' x 15'			.80	20			940		940	1,550
0750	Revolving personnel door, 6' diam. x 6'-6" high			1.50	10.667			500		500	820

13 05 05.20 Selective Demolition, Garden Houses

		Crew	Daily Output	Labor-Hours	Unit	Material	Labor	Equipment	Total	Total Incl O&P
0010	SELECTIVE DEMOLITION, GARDEN HOUSES									
0020	Prefab, wood, excl foundation, average	2 Clab	400	.040	SF Flr.		1.50		1.50	2.47

13 05 05.25 Selective Demolition, Geodesic Domes

		Crew	Daily Output	Labor-Hours	Unit	Material	Labor	Equipment	Total	Total Incl O&P
0010	SELECTIVE DEMOLITION, GEODESIC DOMES									
0050	Shell only, interlocking plywood panels, 30' diameter	F-5	3.20	10	Ea.		475		475	780
0060	34' diameter		2.30	13.913			660		660	1,075
0070	39' diameter		2	16			760		760	1,250
0080	45' diameter	F-3	2.20	18.182			870	297	1,167	1,750
0090	55' diameter		2	20			960	325	1,285	1,900
0100	60' diameter		2	20			960	325	1,285	1,900
0110	65' diameter		1.60	25			1,200	410	1,610	2,400

13 05 05.30 Selective Demolition, Greenhouses

		Crew	Daily Output	Labor-Hours	Unit	Material	Labor	Equipment	Total	Total Incl O&P
0010	SELECTIVE DEMOLITION, GREENHOUSES									
0020	Resi-type, free standing, excl. foundations, 9' long x 8' wide	2 Clab	160	.100	SF Flr.		3.76		3.76	6.15
0030	9' long x 11' wide		170	.094			3.54		3.54	5.80
0040	9' long x 14' wide		220	.073			2.73		2.73	4.48
0050	9' long x 17' wide		320	.050			1.88		1.88	3.08
0060	Lean-to type, 4' wide		64	.250			9.40		9.40	15.40
0070	7' wide		120	.133			5		5	8.20
0080	Geodesic hemisphere, 1/8" plexiglass glazing, 8' diam.		4	4	Ea.		150		150	247
0090	24' diam.		.80	20			750		750	1,225
0100	48' diam.		.40	40			1,500		1,500	2,475

13 05 05.35 Selective Demolition, Hangars

		Crew	Daily Output	Labor-Hours	Unit	Material	Labor	Equipment	Total	Total Incl O&P
0010	SELECTIVE DEMOLITION, HANGARS									
0020	T type hangars, prefab, steel, galv roof & walls, incl doors, excl fndtn	E-2	2550	.022	SF Flr.		1.14	.59	1.73	2.64
0030	Circular type, prefab, steel frame, plastic skin, incl foundation, 80' diam	"	.50	112	Total		5,800	3,025	8,825	13,400

13 05 05.45 Selective Demolition, Lightning Protection

		Crew	Daily Output	Labor-Hours	Unit	Material	Labor	Equipment	Total	Total Incl O&P
0010	SELECTIVE DEMOLITION, LIGHTNING PROTECTION									
0020	Air terminal & base, copper, 3/8" diam. x 10", to 75' h	1 Clab	16	.500	Ea.		18.80		18.80	31
0030	1/2" diam. x 12", over 75' h		16	.500			18.80		18.80	31
0050	Aluminum, 1/2" diam. x 12", to 75' h		16	.500			18.80		18.80	31
0060	5/8" diam. x 12", over 75' h		16	.500			18.80		18.80	31
0070	Cable, copper, 220 lb. per thousand feet, to 75' high		640	.013	L.F.		.47		.47	.77

13 05 Common Work Results for Special Construction

13 05 05 – Selective Demolition for Special Construction

13 05 05.45 Selective Demolition, Lightning Protection	Crew	Daily Output	Labor-Hours	Unit	Material	2015 Bare Costs Labor	Equipment	Total	Total Incl O&P
0080 375 lb. per thousand feet, over 75' high	1 Clab	460	.017	L.F.		.65		.65	1.07
0090 Aluminum, 101 lb. per thousand feet, to 75' high		560	.014			.54		.54	.88
0100 199 lb. per thousand feet, over 75' high		480	.017			.63		.63	1.03
0110 Arrester, 175 V AC, to ground		16	.500	Ea.		18.80		18.80	31
0120 650 V AC, to ground		13	.615	"		23		23	38

13 05 05.50 Selective Demolition, Pre-Engineered Steel Buildings

	Crew	Daily Output	Labor-Hours	Unit	Material	Labor	Equipment	Total	Total Incl O&P
0010 **SELECTIVE DEMOLITION, PRE-ENGINEERED STEEL BUILDINGS**									
0500 Pre-engd. steel bldgs., rigid frame, clear span & multi post, excl. salvage									
0550 3,500 to 7,500 S.F.	L-10	1000	.024	SF Flr.		1.27	.65	1.92	2.93
0600 7,501 to 12,500 S.F.		1500	.016			.85	.44	1.29	1.95
0650 12,500 S.F. or greater		1650	.015			.77	.40	1.17	1.78
0700 Pre-engd. steel building components									
0710 Entrance canopy, including frame 4' x 4'	E-24	8	4	Ea.		209	92	301	465
0720 4' x 8'	"	7	4.571			238	105	343	535
0730 HM doors, self framing, single leaf	2 Skwk	8	2			97.50		97.50	158
0740 Double leaf		5	3.200			156		156	253
0760 Gutter, eave type		600	.027	L.F.		1.30		1.30	2.11
0770 Sash, single slide, double slide or fixed		24	.667	Ea.		32.50		32.50	52.50
0780 Skylight, fiberglass, to 30 S.F.		16	1			48.50		48.50	79
0785 Roof vents, circular, 12" to 24" diameter		12	1.333			65		65	105
0790 Continuous, 10' long		8	2			97.50		97.50	158
0900 Shelters, aluminum frame									
0910 Acrylic glazing, 3' x 9' x 8' high	2 Skwk	2	8	Ea.		390		390	630
0920 9' x 12' x 8' high	"	1.50	10.667	"		520		520	845

13 05 05.60 Selective Demolition, Silos

	Crew	Daily Output	Labor-Hours	Unit	Material	Labor	Equipment	Total	Total Incl O&P
0010 **SELECTIVE DEMOLITION, SILOS**									
0020 Conc stave, indstrl, conical/sloping bott, excl fndtn, 12' diam., 35' h	E-24	.18	177	Ea.		9,275	4,100	13,375	20,800
0030 16' diam., 45' h		.12	266			13,900	6,150	20,050	31,200
0040 25' diam., 75' h		.08	400			20,900	9,200	30,100	46,700
0050 Steel, factory fabricated, 30,000 gal. cap, painted or epoxy lined	L-5	2	28			1,475	370	1,845	3,025

13 05 05.65 Selective Demolition, Sound Control

	Crew	Daily Output	Labor-Hours	Unit	Material	Labor	Equipment	Total	Total Incl O&P
0010 **SELECTIVE DEMOLITION, SOUND CONTROL**									
0120 Acoustical enclosure, 4" thick walls & ceiling panels, 8 lb./S.F.	3 Carp	144	.167	SF Surf		7.85		7.85	12.85
0130 10.5 lb./S.F.		128	.188			8.80		8.80	14.45
0140 Reverb chamber, parallel walls, 4" thick		120	.200			9.40		9.40	15.40
0150 Skewed walls, parallel roof, 4" thick		110	.218			10.25		10.25	16.80
0160 Skewed walls/roof, 4" layer/air space		96	.250			11.75		11.75	19.25
0170 Sound-absorbing panels, painted metal, 2'-6" x 8', under 1,000 S.F.		430	.056			2.62		2.62	4.29
0180 Over 1,000 S.F.		480	.050			2.35		2.35	3.85
0190 Flexible transparent curtain, clear	3 Shee	430	.056			3.12		3.12	4.93
0192 50% clear, 50% foam		430	.056			3.12		3.12	4.93
0194 25% clear, 75% foam		430	.056			3.12		3.12	4.93
0196 100% foam		430	.056			3.12		3.12	4.93
0200 Audio-masking sys., incl. speakers, amplfr., signal gnrtr.									
0205 Ceiling mounted, 5,000 S.F.	2 Elec	4800	.003	S.F.		.18		.18	.28
0210 10,000 S.F.		5600	.003			.16		.16	.24
0220 Plenum mounted, 5,000 S.F.		7600	.002			.12		.12	.18
0230 10,000 S.F.		8800	.002			.10		.10	.15

13 05 05.70 Selective Demolition, Special Purpose Rooms

	Crew	Daily Output	Labor-Hours	Unit	Material	Labor	Equipment	Total	Total Incl O&P
0010 **SELECTIVE DEMOLITION, SPECIAL PURPOSE ROOMS**									
0100 Audiometric rooms, under 500 S.F. surface	4 Carp	200	.160	SF Surf		7.50		7.50	12.30
0110 Over 500 S.F. surface	"	240	.133	"		6.25		6.25	10.25

13 05 Common Work Results for Special Construction

13 05 05 – Selective Demolition for Special Construction

13 05 05.70 Selective Demolition, Special Purpose Rooms

		Crew	Daily Output	Labor-Hours	Unit	Material	2015 Bare Costs Labor	2015 Bare Costs Equipment	Total	Total Incl O&P
0200	Clean rooms, 12' x 12' soft wall, class 100	1 Carp	.30	26.667	Ea.		1,250		1,250	2,050
0210	Class 1000		.30	26.667			1,250		1,250	2,050
0220	Class 10,000		.35	22.857			1,075		1,075	1,750
0230	Class 100,000		.35	22.857			1,075		1,075	1,750
0300	Darkrooms, shell complete, 8' high	2 Carp	220	.073	SF Flr.		3.41		3.41	5.60
0310	12' high		110	.145	"		6.85		6.85	11.20
0350	Darkrooms doors, mini-cylindrical, revolving		4	4	Ea.		188		188	310
0400	Music room, practice modular		140	.114	SF Surf		5.35		5.35	8.80
0500	Refrigeration structures and finishes									
0510	Wall finish, 2 coat portland cement plaster, 1/2" thick	1 Clab	200	.040	S.F.		1.50		1.50	2.47
0520	Fiberglass panels, 1/8" thick		400	.020			.75		.75	1.23
0530	Ceiling finish, polystyrene plastic, 1" to 2" thick		500	.016			.60		.60	.99
0540	4" thick		450	.018			.67		.67	1.10
0550	Refrigerator, prefab aluminum walk-in, 7'-6" high, 6' x 6' OD	2 Carp	100	.160	SF Flr.		7.50		7.50	12.30
0560	10' x 10' OD		160	.100			4.70		4.70	7.70
0570	Over 150 S.F.		200	.080			3.76		3.76	6.15
0600	Sauna, prefabricated, including heater & controls, 7' high, to 30 S.F.		120	.133			6.25		6.25	10.25
0610	To 40 S.F.		140	.114			5.35		5.35	8.80
0620	To 60 S.F.		175	.091			4.29		4.29	7.05
0630	To 100 S.F.		220	.073			3.41		3.41	5.60
0640	To 130 S.F.		250	.064			3		3	4.92
0650	Steam bath, heater, timer, head, single, to 140 C.F.	1 Plum	2.20	3.636	Ea.		213		213	335
0660	To 300 C.F.		2.20	3.636			213		213	335
0670	Steam bath, comm. size, w/blow-down assembly, to 800 C.F.		1.80	4.444			261		261	405
0680	To 2500 C.F.		1.60	5			294		294	460
0690	Steam bath, comm. size, multiple, for motels, apts, 500 C.F., 2 baths		2	4			235		235	365
0700	1,000 C.F., 4 baths		1.40	5.714			335		335	525

13 05 05.75 Selective Demolition, Storage Tanks

		Crew	Daily Output	Labor-Hours	Unit	Material	Labor	Equipment	Total	Total Incl O&P
0010	**SELECTIVE DEMOLITION, STORAGE TANKS**									
0500	Steel tank, single wall, above ground, not incl. fdn., pumps or piping									
0510	Single wall, 275 gallon	Q-1	3	5.333	Ea.		282		282	440
0520	550 thru 2,000 gallon	B-34P	2	12			600	335	935	1,300
0530	5,000 thru 10,000 gallon	B-34Q	2	12			600	630	1,230	1,650
0540	15,000 thru 30,000 gallon	B-34S	2	16			845	1,775	2,620	3,275
0600	Steel tank, double wall, above ground not incl. fdn., pumps & piping									
0620	500 thru 2,000 gallon	B-34P	2	12	Ea.		600	335	935	1,300

13 05 05.85 Selective Demolition, Swimming Pool Equip

		Crew	Daily Output	Labor-Hours	Unit	Material	Labor	Equipment	Total	Total Incl O&P
0010	**SELECTIVE DEMOLITION, SWIMMING POOL EQUIP**									
0020	Diving stand, stainless steel, 3 meter	2 Clab	3	5.333	Ea.		201		201	330
0030	1 meter		5	3.200			120		120	197
0040	Diving board, 16' long, aluminum		5.40	2.963			111		111	183
0050	Fiberglass		5.40	2.963			111		111	183
0070	Ladders, heavy duty, stainless steel, 2 tread		14	1.143			43		43	70.50
0080	4 tread		12	1.333			50		50	82
0090	Lifeguard chair, stainless steel, fixed		5	3.200			120		120	197
0100	Slide, tubular, fiberglass, aluminum handrails & ladder, 5', straight		4	4			150		150	247
0110	8', curved		6	2.667			100		100	164
0120	10', curved		3	5.333			201		201	330
0130	12' straight, with platform		2.50	6.400			241		241	395
0140	Removable access ramp, stainless steel		4	4			150		150	247
0150	Removable stairs, stainless steel, collapsible		4	4			150		150	247

13 05 Common Work Results for Special Construction

13 05 05 – Selective Demolition for Special Construction

13 05 05.90 Selective Demolition, Tension Structures

		Crew	Daily Output	Labor-Hours	Unit	Material	2015 Bare Costs Labor	2015 Bare Costs Equipment	Total	Total Incl O&P
0010	**SELECTIVE DEMOLITION, TENSION STRUCTURES**									
0020	Steel/alum. frame, fabric shell, 60' clear span, 6,000 S.F.	B-41	2000	.022	SF Flr.		.86	.14	1	1.56
0030	12,000 S.F.		2200	.020			.78	.13	.91	1.41
0040	80' clear span, 20,800 S.F.		2440	.018			.70	.12	.82	1.28
0050	100' clear span, 10,000 S.F.	L-5	4350	.013			.68	.17	.85	1.40
0060	26,000 S.F.		4600	.012			.64	.16	.80	1.32
0070	36,000 S.F.		5000	.011			.59	.15	.74	1.21

13 05 05.95 Selective Demo, X-Ray/Radio Freq Protection

		Crew	Daily Output	Labor-Hours	Unit	Material	2015 Bare Costs Labor	2015 Bare Costs Equipment	Total	Total Incl O&P
0010	**SELECTIVE DEMO, X-RAY/RADIO FREQ PROTECTION**									
0020	Shielding lead, lined door frame, excl. hdwe., 1/16" thick	1 Clab	4.80	1.667	Ea.		62.50		62.50	103
0030	Lead sheets, 1/16" thick	2 Clab	270	.059	S.F.		2.23		2.23	3.65
0040	1/8" thick		240	.067			2.51		2.51	4.11
0050	Lead shielding, 1/4" thick		270	.059			2.23		2.23	3.65
0060	1/2" thick		240	.067			2.51		2.51	4.11
0070	Lead glass, 1/4" thick, 2.0 mm LE, 12" x 16"	2 Glaz	16	1	Ea.		45		45	73
0080	24" x 36"		8	2			90		90	146
0090	36" x 60"		4	4			180		180	293
0100	Lead glass window frame, with 1/16" lead & voice passage, 36" x 60"		4	4			180		180	293
0110	Lead glass window frame, 24" x 36"		8	2			90		90	146
0120	Lead gypsum board, 5/8" thick with 1/16" lead	2 Clab	320	.050	S.F.		1.88		1.88	3.08
0130	1/8" lead		280	.057			2.15		2.15	3.52
0140	1/32" lead		400	.040			1.50		1.50	2.47
0150	Butt joints, 1/8" lead or thicker, 2" x 7' long batten strip		480	.033	Ea.		1.25		1.25	2.05
0160	X-ray protection, average radiography room, up to 300 S.F., 1/16" lead, min		.50	32	Total		1,200		1,200	1,975
0170	Maximum		.30	53.333			2,000		2,000	3,300
0180	Deep therapy X-ray room, 250 kV cap, up to 300 S.F., 1/4" lead, min		.20	80			3,000		3,000	4,925
0190	Maximum		.12	133			5,025		5,025	8,225
0880	Radio frequency shielding, prefab or screen-type copper or steel, minimum		360	.044	SF Surf		1.67		1.67	2.74
0890	Average		310	.052			1.94		1.94	3.18
0895	Maximum		290	.055			2.07		2.07	3.40

13 12 Fountains

13 12 13 – Exterior Fountains

13 12 13.10 Yard Fountains

		Crew	Daily Output	Labor-Hours	Unit	Material	2015 Bare Costs Labor	2015 Bare Costs Equipment	Total	Total Incl O&P
0010	**YARD FOUNTAINS**									
0100	Outdoor fountain, 48" high with bowl and figures	2 Clab	2	8	Ea.	570	300		870	1,125

13 17 Tubs and Pools

13 17 13 – Hot Tubs

13 17 13.10 Redwood Hot Tub System

		Crew	Daily Output	Labor-Hours	Unit	Material	2015 Bare Costs Labor	2015 Bare Costs Equipment	Total	Total Incl O&P
0010	**REDWOOD HOT TUB SYSTEM**									
7050	4' diameter x 4' deep	Q-1	1	16	Ea.	3,225	845		4,070	4,875
7150	6' diameter x 4' deep		.80	20		4,950	1,050		6,000	7,075
7200	8' diameter x 4' deep		.80	20		7,250	1,050		8,300	9,625

13 17 Tubs and Pools

13 17 33 – Whirlpool Tubs

13 17 33.10 Whirlpool Bath

		Crew	Daily Output	Labor-Hours	Unit	Material	2015 Bare Costs Labor	2015 Bare Costs Equipment	Total	Total Incl O&P
0010	**WHIRLPOOL BATH**									
6000	Whirlpool, bath with vented overflow, molded fiberglass									
6100	66" x 36" x 24"	Q-1	1	16	Ea.	3,475	845		4,320	5,150
6400	72" x 36" x 21"		1	16		1,800	845		2,645	3,325
6500	60" x 34" x 21"		1	16		1,800	845		2,645	3,300
6600	72" x 42" x 23"		1	16		2,175	845		3,020	3,725
6710	For color add					10%				
6711	For designer colors and trim add					25%				

13 21 Controlled Environment Rooms

13 21 26 – Cold Storage Rooms

13 21 26.50 Refrigeration

		Crew	Daily Output	Labor-Hours	Unit	Material	2015 Bare Costs Labor	2015 Bare Costs Equipment	Total	Total Incl O&P
0010	**REFRIGERATION**									
0020	Curbs, 12" high, 4" thick, concrete	2 Carp	58	.276	L.F.	5.15	12.95		18.10	26.50
6300	Rule of thumb for complete units, w/o doors & refrigeration, cooler		146	.110	SF Flr.	149	5.15		154.15	172
6400	Freezer		109.60	.146	"	176	6.85		182.85	205

13 24 Special Activity Rooms

13 24 16 – Saunas

13 24 16.50 Saunas and Heaters

		Crew	Daily Output	Labor-Hours	Unit	Material	2015 Bare Costs Labor	2015 Bare Costs Equipment	Total	Total Incl O&P
0010	**SAUNAS AND HEATERS**									
0020	Prefabricated, incl. heater & controls, 7' high, 6' x 4', C/C	L-7	2.20	12.727	Ea.	5,175	580		5,755	6,650
1700	Door only, cedar, 2'x6', with 1'x4' tempered insulated glass window	2 Carp	3.40	4.706		750	221		971	1,175
1800	Prehung, incl. jambs, pulls & hardware	"	12	1.333		745	62.50		807.50	925
2500	Heaters only (incl. above), wall mounted, to 200 C.F.					685			685	755
4480	For additional equipment, see Section 11 66 13.10									

13 24 26 – Steam Baths

13 24 26.50 Steam Baths and Components

		Crew	Daily Output	Labor-Hours	Unit	Material	2015 Bare Costs Labor	2015 Bare Costs Equipment	Total	Total Incl O&P
0010	**STEAM BATHS AND COMPONENTS**									
0020	Heater, timer & head, single, to 140 C.F.	1 Plum	1.20	6.667	Ea.	2,200	390		2,590	3,000
0500	To 300 C.F.	"	1.10	7.273		2,425	425		2,850	3,350
2000	Multiple, motels, apts., 2 baths, w/blow-down assm., 500 C.F.	Q-1	1.30	12.308		6,325	650		6,975	8,000
2500	4 baths	"	.70	22.857		10,200	1,200		11,400	13,100

13 28 Athletic and Recreational Special Construction

13 28 33 – Athletic and Recreational Court Walls

13 28 33.50 Sport Court

		Crew	Daily Output	Labor-Hours	Unit	Material	2015 Bare Costs Labor	2015 Bare Costs Equipment	Total	Total Incl O&P
0010	**SPORT COURT**									
0020	Floors, No. 2 & better maple, 25/32" thick				SF Flr.				6.05	6.65
0300	Squash, regulation court in existing building, minimum				Court	36,800			36,800	40,400
0400	Maximum				"	41,000			41,000	45,000
0450	Rule of thumb for components:									
0470	Walls	3 Carp	.15	160	Court	11,000	7,500		18,500	24,400
0500	Floor	"	.25	96		8,725	4,500		13,225	17,000
0550	Lighting	2 Elec	.60	26.667		2,100	1,450		3,550	4,550

13 34 Fabricated Engineered Structures

13 34 13 – Glazed Structures

13 34 13.13 Greenhouses

		Crew	Daily Output	Labor-Hours	Unit	Material	2015 Bare Costs Labor	Equipment	Total	Total Incl O&P
0010	**GREENHOUSES**, Shell only, stock units, not incl. 2' stub walls,									
0020	foundation, floors, heat or compartments									
0300	Residential type, free standing, 8'-6" long x 7'-6" wide	2 Carp	59	.271	SF Flr.	20	12.75		32.75	43
0400	10'-6" wide		85	.188		37	8.85		45.85	55
0600	13'-6" wide		108	.148		39	6.95		45.95	54
0700	17'-0" wide		160	.100		43.50	4.70		48.20	55.50
0900	Lean-to type, 3'-10" wide		34	.471		41.50	22		63.50	82
1000	6'-10" wide		58	.276		50	12.95		62.95	76
1050	8'-0" wide		60	.267		57.50	12.50		70	83.50
1100	Wall mounted to existing window, 3' x 3'	1 Carp	4	2	Ea.	1,575	94		1,669	1,900
1120	4' x 5'	"	3	2.667		2,025	125		2,150	2,425
3900	Cooling, 1200 CFM exhaust fan, add					310			310	340
4000	7850 CFM					1,050			1,050	1,150
4200	For heaters, 10 MBH, add					215			215	237
4300	60 MBH, add					780			780	855
4500	For benches, 2' x 8', add					160			160	176
4600	4' x 10', add					195			195	214
4800	For ventilation & humidity control w/ 4 integrated outlets, add				Total	240			240	264
4900	For environmental controls and automation, 8 outputs, 9 stages, add				"	765			765	840
5100	For humidification equipment, add				Ea.	299			299	330
5200	For vinyl shading, add				S.F.	.24			.24	.26

13 34 13.19 Swimming Pool Enclosures

		Crew	Daily Output	Labor-Hours	Unit	Material	Labor	Equipment	Total	Total Incl O&P
0010	**SWIMMING POOL ENCLOSURES** Translucent, free standing									
0020	not including foundations, heat or light									
0200	Economy	2 Carp	200	.080	SF Hor.	37	3.76		40.76	47
0600	Deluxe	"	70	.229	"	93	10.75		103.75	120

13 34 23 – Fabricated Structures

13 34 23.16 Fabricated Control Booths

		Crew	Daily Output	Labor-Hours	Unit	Material	Labor	Equipment	Total	Total Incl O&P
0010	**FABRICATED CONTROL BOOTHS**									
0100	Guard House, prefab conc. w/bullet resistant doors & windows, roof & wiring									
0110	8' x 8', Level III	L-10	1	24	Ea.	44,200	1,275	655	46,130	51,500
0120	8' x 8', Level IV	"	1	24	"	50,500	1,275	655	52,430	58,500

13 34 23.45 Kiosks

		Crew	Daily Output	Labor-Hours	Unit	Material	Labor	Equipment	Total	Total Incl O&P
0010	**KIOSKS**									
0020	Round, advertising type, 5' diameter, 7' high, aluminum wall, illuminated				Ea.	23,500			23,500	25,800
0100	Aluminum wall, non-illuminated					22,500			22,500	24,700
0500	Rectangular, 5' x 9', 7'-6" high, aluminum wall, illuminated					25,500			25,500	28,000
0600	Aluminum wall, non-illuminated					24,000			24,000	26,400

13 34 63 – Natural Fiber Construction

13 34 63.50 Straw Bale Construction

		Crew	Daily Output	Labor-Hours	Unit	Material	Labor	Equipment	Total	Total Incl O&P
0010	**STRAW BALE CONSTRUCTION**									
2020	Straw bales in walls w/modified post and beam frame [G]	2 Carp	320	.050	S.F.	6.15	2.35		8.50	10.60

13 42 Building Modules

13 42 63 – Detention Cell Modules

13 42 63.16 Steel Detention Cell Modules		Crew	Daily Output	Labor-Hours	Unit	Material	2015 Bare Costs Labor	2015 Bare Costs Equipment	Total	Total Incl O&P
0010	**STEEL DETENTION CELL MODULES**									
2000	Cells, prefab., 5' to 6' wide, 7' to 8' high, 7' to 8' deep,									
2010	bar front, cot, not incl. plumbing	E-4	1.50	21.333	Ea.	9,775	1,125	97.50	10,997.50	12,900

13 48 Sound, Vibration, and Seismic Control

13 48 13 – Manufactured Sound and Vibration Control Components

13 48 13.50 Audio Masking		Crew	Daily Output	Labor-Hours	Unit	Material	Labor	Equipment	Total	Total Incl O&P
0010	**AUDIO MASKING**, acoustical enclosure, 4" thick wall and ceiling									
0020	8# per S.F., up to 12' span	3 Carp	72	.333	SF Surf	33	15.65		48.65	61.50
0300	Better quality panels, 10.5# per S.F.	↓	64	.375		37	17.60		54.60	70
0400	Reverb-chamber, 4" thick, parallel walls	↓	60	.400	↓	46.50	18.80		65.30	82

13 49 Radiation Protection

13 49 13 – Integrated X-Ray Shielding Assemblies

13 49 13.50 Lead Sheets		Crew	Daily Output	Labor-Hours	Unit	Material	Labor	Equipment	Total	Total Incl O&P
0010	**LEAD SHEETS**									
0300	Lead sheets, 1/16" thick	2 Lath	135	.119	S.F.	10.50	5.10		15.60	19.55
0400	1/8" thick		120	.133		31	5.75		36.75	43
0500	Lead shielding, 1/4" thick		135	.119		41.50	5.10		46.60	54
0550	1/2" thick	↓	120	.133	↓	73.50	5.75		79.25	90
1200	X-ray protection, average radiography or fluoroscopy									
1210	room, up to 300 S.F. floor, 1/16" lead, economy	2 Lath	.25	64	Total	10,100	2,750		12,850	15,400
1500	7'-0" walls, deluxe	"	.15	106	"	12,200	4,575		16,775	20,600
1600	Deep therapy X-ray room, 250 kV capacity,									
1800	up to 300 S.F. floor, 1/4" lead, economy	2 Lath	.08	200	Total	28,300	8,600		36,900	44,600
1900	7'-0" walls, deluxe		.06	266	"	34,900	11,500		46,400	56,500
1999	Minimum labor/equipment charge	↓	4.50	3.556	Job		153		153	241

13 49 21 – Lead Glazing

13 49 21.50 Lead Glazing		Crew	Daily Output	Labor-Hours	Unit	Material	Labor	Equipment	Total	Total Incl O&P
0010	**LEAD GLAZING**									
0600	Lead glass, 1/4" thick, 2.0 mm LE, 12" x 16"	2 Glaz	13	1.231	Ea.	380	55.50		435.50	505
0700	24" x 36"	"	8	2	"	1,325	90		1,415	1,600
2000	X-ray viewing panels, clear lead plastic									
2010	7 mm thick, 0.3 mm LE, 2.3 lb./S.F.	H-3	139	.115	S.F.	241	4.64		245.64	273
2020	12 mm thick, 0.5 mm LE, 3.9 lb./S.F.		82	.195		355	7.85		362.85	405
2030	18 mm thick, 0.8 mm LE, 5.9 lb./S.F.		54	.296		405	11.95		416.95	465
2040	22 mm thick, 1.0 mm LE, 7.2 lb./S.F.		44	.364		530	14.65		544.65	610
2050	35 mm thick, 1.5 mm LE, 11.5 lb./S.F.		28	.571		815	23		838	935
2060	46 mm thick, 2.0 mm LE, 15.0 lb./S.F.	↓	21	.762	↓	1,050	30.50		1,080.50	1,225
2090	For panels 12 S.F. to 48 S.F., add crating charge				Ea.				50	50

13 49 Radiation Protection

13 49 23 – Integrated RFI/EMI Shielding Assemblies

13 49 23.50 Modular Shielding Partitions

		Crew	Daily Output	Labor-Hours	Unit	Material	2015 Bare Costs Labor	Equipment	Total	Total Incl O&P
0010	**MODULAR SHIELDING PARTITIONS**									
4000	X-ray barriers, modular, panels mounted within framework for									
4002	attaching to floor, wall or ceiling, upper portion is clear lead									
4005	plastic window panels 48"H, lower portion is opaque leaded									
4008	steel panels 36"H, structural supports not incl.									
4010	1-section barrier, 36"W x 84"H overall									
4020	0.5 mm LE panels	H-3	6.40	2.500	Ea.	8,100	101		8,201	9,075
4030	0.8 mm LE panels		6.40	2.500		8,725	101		8,826	9,775
4040	1.0 mm LE panels		5.33	3.002		10,200	121		10,321	11,400
4050	1.5 mm LE panels		5.33	3.002		13,600	121		13,721	15,200
4060	2-section barrier, 72"W x 84"H overall									
4070	0.5 mm LE panels	H-3	4	4	Ea.	11,800	161		11,961	13,300
4080	0.8 mm LE panels		4	4		13,100	161		13,261	14,700
4090	1.0 mm LE panels		3.56	4.494		16,000	181		16,181	17,900
5000	1.5 mm LE panels		3.20	5		22,800	201		23,001	25,400
5010	3-section barrier, 108"W x 84"H overall									
5020	0.5 mm LE panels	H-3	3.20	5	Ea.	17,700	201		17,901	19,800
5030	0.8 mm LE panels		3.20	5		19,600	201		19,801	21,800
5040	1.0 mm LE panels		2.67	5.993		24,000	241		24,241	26,800
5050	1.5 mm LE panels		2.46	6.504		34,200	262		34,462	38,000
7000	X-ray barriers, mobile, mounted within framework w/casters on									
7005	bottom, clear lead plastic window panels on upper portion,									
7010	opaque on lower, 30"W x 75"H overall, incl. framework									
7020	24"H upper w/0.5 mm LE, 48"H lower w/0.8 mm LE	1 Carp	16	.500	Ea.	3,800	23.50		3,823.50	4,225
7030	48"W x 75"H overall, incl. framework									
7040	36"H upper w/0.5 mm LE, 36"H lower w/0.8 mm LE	1 Carp	16	.500	Ea.	6,175	23.50		6,198.50	6,850
7050	36"H upper w/1.0 mm LE, 36"H lower w/1.5 mm LE	"	16	.500	"	7,300	23.50		7,323.50	8,100
7060	72"W x 75"H overall, incl. framework									
7070	36"H upper w/0.5 mm LE, 36"H lower w/0.8 mm LE	1 Carp	16	.500	Ea.	7,300	23.50		7,323.50	8,100
7080	36"H upper w/1.0 mm LE, 36"H lower w/1.5 mm LE	"	16	.500	"	9,150	23.50		9,173.50	10,100

Division Notes

Estimating Tips
General
- Many products in Division 14 will require some type of support or blocking for installation not included with the item itself. Examples are supports for conveyors or tube systems, attachment points for lifts, and footings for hoists or cranes. Add these supports in the appropriate division.

14 10 00 Dumbwaiters
14 20 00 Elevators
- Dumbwaiters and elevators are estimated and purchased in a method similar to buying a car. The manufacturer has a base unit with standard features. Added to this base unit price will be whatever options the owner or specifications require. Increased load capacity, additional vertical travel, additional stops, higher speed, and cab finish options are items to be considered. When developing an estimate for dumbwaiters and elevators, remember that some items needed by the installers may have to be included as part of the general contract.

Examples are:
- shaftway
- rail support brackets
- machine room
- electrical supply
- sill angles
- electrical connections
- pits
- roof penthouses
- pit ladders

Check the job specifications and drawings before pricing.
- Installation of elevators and handicapped lifts in historic structures can require significant additional costs. The associated structural requirements may involve cutting into and repairing finishes, moldings, flooring, etc. The estimator must account for these special conditions.

14 30 00 Escalators and Moving Walks
- Escalators and moving walks are specialty items installed by specialty contractors. There are numerous options associated with these items. For specific options, contact a manufacturer or contractor. In a method similar to estimating dumbwaiters and elevators, you should verify the extent of general contract work and add items as necessary.

14 40 00 Lifts
14 90 00 Other Conveying Equipment
- Products such as correspondence lifts, chutes, and pneumatic tube systems, as well as other items specified in this subdivision, may require trained installers. The general contractor might not have any choice as to who will perform the installation, or when it will be performed. Long lead times are often required for these products, making early decisions in scheduling necessary.

Reference Numbers
Reference numbers are shown in shaded boxes at the beginning of some major classifications. These numbers refer to related items in the Reference Section. The reference information may be an estimating procedure, an alternate pricing method, or technical information.

Note: Not all subdivisions listed here necessarily appear in this publication. ∎

Estimate with precision with RSMeans Online…
Quick, intuitive and easy-to-use, RSMeans Online gives you instant access to hundreds of thousands of material, labor, and equipment costs from RSMeans' comprehensive database, delivering the information you need to build competitive estimates every time.

Start your free trial today at **www.rsmeansonline.com**

RSMeansOnline

14 11 Manual Dumbwaiters
14 11 10 – Hand Operated Dumbwaiters

14 11 10.20 Manual Dumbwaiters		Crew	Daily Output	Labor-Hours	Unit	Material	2015 Bare Costs Labor	Equipment	Total	Total Incl O&P
0010	MANUAL DUMBWAITERS									
0020	2 stop, hand powered, up to 75 lb. capacity	2 Elev	.75	21.333	Ea.	3,000	1,625		4,625	5,825
0100	76 lb capacity and up		.50	32	"	6,700	2,450		9,150	11,200
0300	For each additional stop, add		.75	21.333	Stop	1,100	1,625		2,725	3,725

14 12 Electric Dumbwaiters
14 12 10 – Dumbwaiters
14 12 10.10 Electric Dumbwaiters

		Crew	Daily Output	Labor-Hours	Unit	Material	Labor	Equipment	Total	Total Incl O&P
0010	ELECTRIC DUMBWAITERS									
0020	2 stop, up to 75 lb capacity	2 Elev	.13	123	Ea.	7,400	9,425		16,825	22,700
0100	76 lb capacity and up		.11	145	"	22,300	11,100		33,400	41,700
0600	For each additional stop, add		.54	29.630	Stop	3,300	2,275		5,575	7,125

14 21 Electric Traction Elevators
14 21 13 – Electric Traction Freight Elevators
14 21 13.10 Electric Traction Freight Elevators and Options

			Crew	Daily Output	Labor-Hours	Unit	Material	Labor	Equipment	Total	Total Incl O&P
0010	ELECTRIC TRACTION FREIGHT ELEVATORS AND OPTIONS	R142000-10									
0425	Electric freight, base unit, 4000 lb., 200 fpm, 4 stop, std. fin.		2 Elev	.05	320	Ea.	110,000	24,500		134,500	159,500
0450	For 5000 lb. capacity, add	R142000-40					5,850			5,850	6,425
0500	For 6000 lb. capacity, add						14,500			14,500	15,900
0525	For 7000 lb. capacity, add						18,100			18,100	19,900
0550	For 8000 lb. capacity, add						22,300			22,300	24,500
0575	For 10000 lb. capacity, add						30,500			30,500	33,600
0600	For 12000 lb. capacity, add						37,500			37,500	41,300
0625	For 16000 lb. capacity, add						45,000			45,000	49,500
0650	For 20000 lb. capacity, add						50,000			50,000	55,000
0675	For increased speed, 250 fpm, add						17,000			17,000	18,700
0700	300 fpm, geared electric, add						20,600			20,600	22,700
0725	350 fpm, geared electric, add						25,100			25,100	27,600
0750	400 fpm, geared electric, add						29,400			29,400	32,300
0775	500 fpm, gearless electric, add						36,400			36,400	40,100
0800	600 fpm, gearless electric, add						43,600			43,600	47,900
0825	700 fpm, gearless electric, add						52,000			52,000	57,000
0850	800 fpm, gearless electric, add						59,500			59,500	65,500
0875	For class "B" loading, add						4,800			4,800	5,275
0900	For class "C-1" loading, add						6,950			6,950	7,650
0925	For class "C-2" loading, add						8,000			8,000	8,800
0950	For class "C-3" loading, add						10,700			10,700	11,700
0975	For travel over 40 V.L.F., add		2 Elev	7.25	2.207	V.L.F.	645	169		814	970
1000	For number of stops over 4, add		"	.27	59.259	Stop	4,225	4,525		8,750	11,700

14 21 23 – Electric Traction Passenger Elevators
14 21 23.10 Electric Traction Passenger Elevators and Options

		Crew	Daily Output	Labor-Hours	Unit	Material	Labor	Equipment	Total	Total Incl O&P
0010	ELECTRIC TRACTION PASSENGER ELEVATORS AND OPTIONS									
1625	Electric pass., base unit, 2000 lb., 200 fpm, 4 stop, std. fin.	2 Elev	.05	320	Ea.	92,000	24,500		116,500	139,000
1650	For 2500 lb. capacity, add					3,800			3,800	4,175
1675	For 3000 lb. capacity, add					4,400			4,400	4,825
1700	For 3500 lb. capacity, add					5,675			5,675	6,250
1725	For 4000 lb. capacity, add					6,825			6,825	7,500
1750	For 4500 lb. capacity, add					9,150			9,150	10,100

14 21 Electric Traction Elevators

14 21 23 – Electric Traction Passenger Elevators

14 21 23.10 Electric Traction Passenger Elevators and Options

		Crew	Daily Output	Labor-Hours	Unit	Material	2015 Bare Costs Labor	Equipment	Total	Total Incl O&P
1775	For 5000 lb. capacity, add				Ea.	11,400			11,400	12,600
1800	For increased speed, 250 fpm, geared electric, add					4,800			4,800	5,275
1825	300 fpm, geared electric, add					6,575			6,575	7,225
1850	350 fpm, geared electric, add					8,775			8,775	9,650
1875	400 fpm, geared electric, add					11,500			11,500	12,600
1900	500 fpm, gearless electric, add					28,000			28,000	30,800
1925	600 fpm, gearless electric, add					43,700			43,700	48,000
1950	700 fpm, gearless electric, add					50,000			50,000	55,500
1975	800 fpm, gearless electric, add					55,500			55,500	61,000
2000	For travel over 40 V.L.F., add	2 Elev	7.25	2.207	V.L.F.	695	169		864	1,025
2025	For number of stops over 4, add		.27	59.259	Stop	3,075	4,525		7,600	10,400
2400	Electric hospital, base unit, 4000 lb., 200 fpm, 4 stop, std fin.		.05	320	Ea.	79,500	24,500		104,000	125,500
2425	For 4500 lb. capacity, add					5,475			5,475	6,025
2450	For 5000 lb. capacity, add					7,175			7,175	7,900
2475	For increased speed, 250 fpm, geared electric, add					4,925			4,925	5,425
2500	300 fpm, geared electric, add					7,600			7,600	8,375
2525	350 fpm, geared electric, add					8,775			8,775	9,650
2550	400 fpm, geared electric, add					11,500			11,500	12,700
2575	500 fpm, gearless electric, add					32,600			32,600	35,900
2600	600 fpm, gearless electric, add					47,600			47,600	52,500
2625	700 fpm, gearless electric, add					52,500			52,500	58,000
2650	800 fpm, gearless electric, add					59,000			59,000	65,000
2675	For travel over 40 V.L.F., add	2 Elev	7.25	2.207	V.L.F.	435	169		604	740
2700	For number of stops over 4, add	"	.27	59.259	Stop	4,325	4,525		8,850	11,800

14 21 33 – Electric Traction Residential Elevators

14 21 33.20 Residential Elevators

		Crew	Daily Output	Labor-Hours	Unit	Material	Labor	Equipment	Total	Total Incl O&P
0010	**RESIDENTIAL ELEVATORS**									
7000	Residential, cab type, 1 floor, 2 stop, economy model	2 Elev	.20	80	Ea.	11,400	6,125		17,525	22,100
7100	Custom model		.10	160		19,300	12,200		31,500	40,100
7200	2 floor, 3 stop, economy model		.12	133		17,000	10,200		27,200	34,400
7300	Custom model		.06	266		27,700	20,400		48,100	62,000

14 24 Hydraulic Elevators

14 24 13 – Hydraulic Freight Elevators

14 24 13.10 Hydraulic Freight Elevators and Options

		Crew	Daily Output	Labor-Hours	Unit	Material	Labor	Equipment	Total	Total Incl O&P
0010	**HYDRAULIC FREIGHT ELEVATORS AND OPTIONS**									
1025	Hydraulic freight, base unit, 2000 lb., 50 fpm, 2 stop, std. fin.	2 Elev	.10	160	Ea.	79,500	12,200		91,700	106,500
1050	For 2500 lb. capacity, add					3,875			3,875	4,250
1075	For 3000 lb. capacity, add					5,250			5,250	5,775
1100	For 3500 lb. capacity, add					8,350			8,350	9,200
1125	For 4000 lb. capacity, add					9,750			9,750	10,700
1150	For 4500 lb. capacity, add					14,300			14,300	15,700
1175	For 5000 lb. capacity, add					15,000			15,000	16,500
1200	For 6000 lb. capacity, add					15,800			15,800	17,400
1225	For 7000 lb. capacity, add					23,500			23,500	25,800
1250	For 8000 lb. capacity, add					30,900			30,900	33,900
1275	For 10000 lb. capacity, add					37,800			37,800	41,600
1300	For 12000 lb. capacity, add					49,100			49,100	54,000
1325	For 16000 lb. capacity, add					70,000			70,000	77,000
1350	For 20000 lb. capacity, add					80,500			80,500	89,000

14 24 Hydraulic Elevators

14 24 13 – Hydraulic Freight Elevators

14 24 13.10 Hydraulic Freight Elevators and Options		Crew	Daily Output	Labor-Hours	Unit	Material	2015 Bare Costs Labor	Equipment	Total	Total Incl O&P
1375	For increased speed, 100 fpm, add				Ea.	1,550			1,550	1,700
1400	125 fpm, add					3,425			3,425	3,775
1425	150 fpm, add					4,675			4,675	5,150
1450	175 fpm, add					7,150			7,150	7,875
1475	For class "B" loading, add					4,525			4,525	4,975
1500	For class "C-1" loading, add					6,900			6,900	7,600
1525	For class "C-2" loading, add					7,950			7,950	8,750
1550	For class "C-3" loading, add					10,700			10,700	11,800
1575	For travel over 20 V.L.F., add	2 Elev	7.25	2.207	V.L.F.	845	169		1,014	1,200
1600	For number of stops over 2, add	"	.27	59.259	Stop	2,150	4,525		6,675	9,350

14 24 23 – Hydraulic Passenger Elevators

14 24 23.10 Hydraulic Passenger Elevators and Options

		Crew	Daily Output	Labor-Hours	Unit	Material	Labor	Equipment	Total	Total Incl O&P
0010	**HYDRAULIC PASSENGER ELEVATORS AND OPTIONS**									
2050	Hyd. pass., base unit, 1500 lb., 100 fpm, 2 stop, std. fin.	2 Elev	.10	160	Ea.	37,700	12,200		49,900	60,500
2075	For 2000 lb. capacity, add					810			810	890
2100	For 2500 lb. capacity, add					2,800			2,800	3,100
2125	For 3000 lb. capacity, add					3,975			3,975	4,375
2150	For 3500 lb. capacity, add					6,875			6,875	7,550
2175	For 4000 lb. capacity, add					8,225			8,225	9,050
2200	For 4500 lb. capacity, add					10,800			10,800	11,900
2225	For 5000 lb. capacity, add					15,200			15,200	16,700
2250	For increased speed, 125 fpm, add					1,975			1,975	2,175
2275	150 fpm, add					2,550			2,550	2,825
2300	175 fpm, add					4,850			4,850	5,325
2325	200 fpm, add					9,275			9,275	10,200
2350	For travel over 12 V.L.F., add	2 Elev	7.25	2.207	V.L.F.	690	169		859	1,025
2375	For number of stops over 2, add		.27	59.259	Stop	960	4,525		5,485	8,050
2725	Hydraulic hospital, base unit, 4000 lb., 100 fpm, 2 stop, std. fin.		.10	160	Ea.	59,500	12,200		71,700	84,500
2775	For 4500 lb. capacity, add					6,700			6,700	7,350
2800	For 5000 lb. capacity, add					9,800			9,800	10,800
2825	For increased speed, 125 fpm, add					2,450			2,450	2,700
2850	150 fpm, add					3,450			3,450	3,800
2875	175 fpm, add					5,775			5,775	6,350
2900	200 fpm, add					8,450			8,450	9,300
2925	For travel over 12 V.L.F., add	2 Elev	7.25	2.207	V.L.F.	530	169		699	840
2950	For number of stops over 2, add	"	.27	59.259	Stop	4,175	4,525		8,700	11,600

14 27 Custom Elevator Cabs and Doors

14 27 13 – Custom Elevator Cab Finishes

14 27 13.10 Cab Finishes

		Crew	Daily Output	Labor-Hours	Unit	Material	Labor	Equipment	Total	Total Incl O&P
0010	**CAB FINISHES**									
3325	Passenger elevator cab finishes (based on 3500 lb. cab size)									
3350	Acrylic panel ceiling				Ea.	755			755	830
3375	Aluminum eggcrate ceiling					865			865	950
3400	Stainless steel doors					3,950			3,950	4,350
3425	Carpet flooring					610			610	670
3450	Epoxy flooring					465			465	510
3475	Quarry tile flooring					875			875	960
3500	Slate flooring					1,600			1,600	1,750
3525	Textured rubber flooring					650			650	715

14 27 Custom Elevator Cabs and Doors

14 27 13 – Custom Elevator Cab Finishes

14 27 13.10 Cab Finishes	Crew	Daily Output	Labor-Hours	Unit	Material	2015 Bare Costs Labor	Equipment	Total	Total Incl O&P
3550	Stainless steel walls				Ea.	4,100			4,100	4,500
3575	Stainless steel returns at door					1,175			1,175	1,275
4450	Hospital elevator cab finishes (based on 3500 lb. cab size)									
4475	Aluminum eggcrate ceiling				Ea.	900			900	990
4500	Stainless steel doors					3,950			3,950	4,350
4525	Epoxy flooring					465			465	510
4550	Quarry tile flooring					875			875	960
4575	Textured rubber flooring					650			650	715
4600	Stainless steel walls					4,550			4,550	5,000
4625	Stainless steel returns at door					960			960	1,050

14 28 Elevator Equipment and Controls

14 28 10 – Elevator Equipment and Control Options

14 28 10.10 Elevator Controls and Doors

		Crew	Daily Output	Labor-Hours	Unit	Material	Labor	Equipment	Total	Total Incl O&P
0010	**ELEVATOR CONTROLS AND DOORS**									
2975	Passenger elevator options									
3000	2 car group automatic controls	2 Elev	.66	24.242	Ea.	4,625	1,850		6,475	7,925
3025	3 car group automatic controls		.44	36.364		8,825	2,775		11,600	14,000
3050	4 car group automatic controls		.33	48.485		17,600	3,700		21,300	25,100
3075	5 car group automatic controls		.26	61.538		31,800	4,700		36,500	42,200
3100	6 car group automatic controls		.22	72.727		64,500	5,575		70,075	79,000
3125	Intercom service		3	5.333		955	410		1,365	1,675
3150	Duplex car selective collective		.66	24.242		8,100	1,850		9,950	11,800
3175	Center opening 1 speed doors		2	8		1,950	610		2,560	3,100
3200	Center opening 2 speed doors		2	8		2,750	610		3,360	3,975
3225	Rear opening doors (opposite front)		2	8		4,225	610		4,835	5,600
3250	Side opening 2 speed doors		2	8		4,350	610		4,960	5,725
3275	Automatic emergency power switching		.66	24.242		1,200	1,850		3,050	4,150
3300	Manual emergency power switching		8	2		515	153		668	800
3625	Hall finishes, stainless steel doors					1,425			1,425	1,575
3650	Stainless steel frames					1,450			1,450	1,600
3675	12 month maintenance contract								3,600	3,950
3700	Signal devices, hall lanterns	2 Elev	8	2		520	153		673	805
3725	Position indicators, up to 3		9.40	1.702		475	130		605	725
3750	Position indicators, per each over 3		32	.500		420	38.50		458.50	520
3775	High speed heavy duty door opener					3,250			3,250	3,575
3800	Variable voltage, O.H. gearless machine, min.	2 Elev	.16	100		34,000	7,650		41,650	49,200
3815	Maximum		.07	228		81,000	17,500		98,500	116,000
3825	Basement installed geared machine		.33	48.485		49,200	3,700		52,900	59,500
3850	Freight elevator options									
3875	Doors, bi-parting	2 Elev	.66	24.242	Ea.	7,550	1,850		9,400	11,200
3900	Power operated door and gate	"	.66	24.242		24,300	1,850		26,150	29,600
3925	Finishes, steel plate floor					1,825			1,825	2,000
3950	14 ga. 1/4" x 4' steel plate walls					2,000			2,000	2,200
3975	12 month maintenance contract								3,600	3,950
4000	Signal devices, hall lanterns	2 Elev	8	2		505	153		658	790
4025	Position indicators, up to 3		9.40	1.702		490	130		620	735
4050	Position indicators, per each over 3		32	.500		425	38.50		463.50	530
4075	Variable voltage basement installed geared machine		.66	24.242		20,000	1,850		21,850	24,900
4100	Hospital elevator options									
4125	2 car group automatic controls	2 Elev	.66	24.242	Ea.	4,850	1,850		6,700	8,200

14 28 Elevator Equipment and Controls

14 28 10 – Elevator Equipment and Control Options

14 28 10.10 Elevator Controls and Doors		Crew	Daily Output	Labor-Hours	Unit	Material	2015 Bare Costs Labor	Equipment	Total	Total Incl O&P
4150	3 car group automatic controls	2 Elev	.44	36.364	Ea.	9,275	2,775		12,050	14,500
4175	4 car group automatic controls		.33	48.485		12,400	3,700		16,100	19,400
4200	5 car group automatic controls		.26	61.538		33,500	4,700		38,200	44,200
4225	6 car group automatic controls		.22	72.727		67,500	5,575		73,075	83,000
4250	Intercom service		3	5.333		1,000	410		1,410	1,725
4275	Duplex car selective collective		.66	24.242		8,100	1,850		9,950	11,800
4300	Center opening 1 speed doors		2	8		1,950	610		2,560	3,100
4325	Center opening 2 speed doors		2	8		2,575	610		3,185	3,800
4350	Rear opening doors (opposite front)		2	8		4,250	610		4,860	5,625
4375	Side opening 2 speed doors		2	8		6,350	610		6,960	7,925
4400	Automatic emergency power switching		.66	24.242		1,175	1,850		3,025	4,125
4425	Manual emergency power switching		8	2		505	153		658	790
4675	Hall finishes, stainless steel doors					1,600			1,600	1,750
4700	Stainless steel frames					1,475			1,475	1,625
4725	12 month maintenance contract								3,600	3,950
4750	Signal devices, hall lanterns	2 Elev	8	2		490	153		643	770
4775	Position indicators, up to 3		9.40	1.702		475	130		605	720
4800	Position indicators, per each over 3		32	.500		415	38.50		453.50	520
4825	High speed heavy duty door opener					3,250			3,250	3,575
4850	Variable voltage, O.H. gearless machine, min.	2 Elev	.16	100		51,500	7,650		59,150	68,500
4865	Maximum		.07	228		79,500	17,500		97,000	114,500
4875	Basement installed geared machine		.33	48.485		20,300	3,700		24,000	28,100
5000	Drilling for piston, casing included, 18" diameter	B-48	80	.700	V.L.F.	56.50	30	36.50	123	150

14 42 Wheelchair Lifts

14 42 13 – Inclined Wheelchair Lifts

14 42 13.10 Inclined Wheelchair Lifts and Stairclimbers

		Crew	Daily Output	Labor-Hours	Unit	Material	Labor	Equipment	Total	Total Incl O&P
0010	**INCLINED WHEELCHAIR LIFTS AND STAIRCLIMBERS**									
7700	Stair climber (chair lift), single seat, minimum	2 Elev	1	16	Ea.	5,325	1,225		6,550	7,775
7800	Maximum		.20	80		7,350	6,125		13,475	17,500
8700	Stair lift, minimum		1	16		14,500	1,225		15,725	17,800
8900	Maximum		.20	80		22,900	6,125		29,025	34,700

14 42 16 – Vertical Wheelchair Lifts

14 42 16.10 Wheelchair Lifts

		Crew	Daily Output	Labor-Hours	Unit	Material	Labor	Equipment	Total	Total Incl O&P
0010	**WHEELCHAIR LIFTS**									
8000	Wheelchair lift, minimum	2 Elev	1	16	Ea.	7,325	1,225		8,550	9,950
8500	Maximum	"	.50	32	"	17,300	2,450		19,750	22,800

14 45 Vehicle Lifts

14 45 10 – Hydraulic Vehicle Lifts

14 45 10.10 Hydraulic Lifts

		Crew	Daily Output	Labor-Hours	Unit	Material	Labor	Equipment	Total	Total Incl O&P
0010	**HYDRAULIC LIFTS**									
2200	Single post, 8000 lb. capacity	L-4	.40	60	Ea.	5,550	2,650		8,200	10,500
2810	Double post, 6000 lb. capacity		2.67	8.989		7,825	400		8,225	9,275
2815	9000 lb. capacity		2.29	10.480		18,600	465		19,065	21,300
2820	15,000 lb. capacity		2	12		21,000	530		21,530	24,000
2822	Four post, 26,000 lb. capacity		1.80	13.333		14,200	590		14,790	16,600
2825	30,000 lb. capacity		1.60	15		46,400	665		47,065	52,000
2830	Ramp style, 4 post, 25,000 lb. capacity		2	12		18,200	530		18,730	20,900

14 45 Vehicle Lifts

14 45 10 – Hydraulic Vehicle Lifts

14 45 10.10 Hydraulic Lifts

		Crew	Daily Output	Labor-Hours	Unit	Material	2015 Bare Costs Labor	Equipment	Total	Total Incl O&P
2835	35,000 lb. capacity	L-4	1	24	Ea.	84,500	1,050		85,550	94,500
2840	50,000 lb. capacity		1	24		94,500	1,050		95,550	105,500
2845	75,000 lb. capacity	↓	1	24		110,000	1,050		111,050	122,500
2850	For drive thru tracks, add, minimum					1,175			1,175	1,275
2855	Maximum					2,000			2,000	2,200
2860	Ramp extensions, 3' (set of 2)					960			960	1,050
2865	Rolling jack platform					3,325			3,325	3,675
2870	Electric/hydraulic jacking beam					8,925			8,925	9,825
2880	Scissor lift, portable, 6000 lb. capacity				↓	8,750			8,750	9,625

14 91 Facility Chutes

14 91 33 – Laundry and Linen Chutes

14 91 33.10 Chutes

0011	**CHUTES**, linen, trash or refuse									
0050	Aluminized steel, 16 ga., 18" diameter	2 Shee	3.50	4.571	Floor	1,700	256		1,956	2,275
0100	24" diameter		3.20	5		1,775	280		2,055	2,400
0400	Galvanized steel, 16 ga., 18" diameter		3.50	4.571		1,000	256		1,256	1,500
0500	24" diameter		3.20	5		1,125	280		1,405	1,700
0800	Stainless steel, 18" diameter		3.50	4.571		3,000	256		3,256	3,700
0900	24" diameter	↓	3.20	5	↓	3,150	280		3,430	3,900

14 91 82 – Trash Chutes

14 91 82.10 Trash Chutes and Accessories

0010	**TRASH CHUTES AND ACCESSORIES**									
9000	Minimum labor/equipment charge	1 Shee	1	8	Job		450		450	705

14 92 Pneumatic Tube Systems

14 92 10 – Conventional, Automatic and Computer Controlled Pneumatic Tube Systems

14 92 10.10 Pneumatic Tube Systems

0010	**PNEUMATIC TUBE SYSTEMS**									
0020	100' long, single tube, 2 stations, stock									
0100	3" diameter	2 Stpi	.12	133	Total	3,375	7,975		11,350	16,100
0300	4" diameter	"	.09	177	"	4,275	10,600		14,875	21,300
0400	Twin tube, two stations or more, conventional system									
0700	3" round	2 Stpi	46	.348	L.F.	38	21		59	74.50
1050	Add for blower		2	8	System	5,200	480		5,680	6,475
1110	Plus for each round station, add		7.50	2.133	Ea.	1,350	127		1,477	1,700
1200	Alternate pricing method: base cost, economy model		.75	21.333	Total	5,750	1,275		7,025	8,325
1300	Custom model		.25	64	"	11,500	3,825		15,325	18,700
1500	Plus total system length, add, for economy model		93.40	.171	L.F.	8.25	10.25		18.50	25
1600	For custom model	↓	37.60	.426	"	25	25.50		50.50	67

Division Notes

	CREW	DAILY OUTPUT	LABOR-HOURS	UNIT	BARE COSTS				TOTAL INCL O&P
					MAT.	LABOR	EQUIP.	TOTAL	

Estimating Tips

Pipe for fire protection and all uses is located in Subdivisions 21 11 13 and 22 11 13.

The labor adjustment factors listed in Subdivision 22 01 02.20 also apply to Division 21.

Many, but not all, areas in the U.S. require backflow protection in the fire system. It is advisable to check local building codes for specific requirements.

For your reference, the following is a list of the most applicable Fire Codes and Standards which may be purchased from the NFPA, 1 Batterymarch Park, Quincy, MA 02169-7471.

- NFPA 1: Uniform Fire Code
- NFPA 10: Portable Fire Extinguishers
- NFPA 11: Low-, Medium-, and High-Expansion Foam
- NFPA 12: Carbon Dioxide Extinguishing Systems (Also companion 12A)
- NFPA 13: Installation of Sprinkler Systems (Also companion 13D, 13E, and 13R)
- NFPA 14: Installation of Standpipe and Hose Systems
- NFPA 15: Water Spray Fixed Systems for Fire Protection
- NFPA 16: Installation of Foam-Water Sprinkler and Foam-Water Spray Systems
- NFPA 17: Dry Chemical Extinguishing Systems (Also companion 17A)
- NFPA 18: Wetting Agents
- NFPA 20: Installation of Stationary Pumps for Fire Protection
- NFPA 22: Water Tanks for Private Fire Protection
- NFPA 24: Installation of Private Fire Service Mains and their Appurtenances
- NFPA 25: Inspection, Testing and Maintenance of Water-Based Fire Protection

Reference Numbers

Reference numbers are shown in shaded boxes at the beginning of some major classifications. These numbers refer to related items in the Reference Section. The reference information may be an estimating procedure, an alternate pricing method, or technical information.

Note: Not all subdivisions listed here necessarily appear in this publication. ■

Did you know?

RSMeans Online gives you the same access to RSMeans' data with 24/7 access:

- Quickly locate costs in the searchable database.
- Build cost lists, estimates, and reports in minutes.
- Adjust costs to any location in the U.S. and Canada with the click of a button.

Start your free trial today at **www.rsmeansonline.com**

RSMeansOnline

21 05 Common Work Results for Fire Suppression

21 05 23 – General-Duty Valves for Water-Based Fire-Suppression Piping

21 05 23.50 General-Duty Valves		Crew	Daily Output	Labor-Hours	Unit	Material	2015 Bare Costs Labor	2015 Bare Costs Equipment	Total	Total Incl O&P
0010	**GENERAL-DUTY VALVES**, for water-based fire suppression									
6500	Check, swing, C.I. body, brass fittings, auto. ball drip									
6520	4" size	Q-12	3	5.333	Ea.	350	269		619	805
9990	Minimum labor/equipment charge	1 Spri	3	2.667	Job		150		150	234

21 11 Facility Fire-Suppression Water-Service Piping

21 11 16 – Facility Fire Hydrants

21 11 16.50 Fire Hydrants for Buildings

		Crew	Daily Output	Labor-Hours	Unit	Material	Labor	Equipment	Total	Total Incl O&P
0010	**FIRE HYDRANTS FOR BUILDINGS**									
3750	Hydrants, wall, w/caps, single, flush, polished brass									
3800	2-1/2" x 2-1/2"	Q-12	5	3.200	Ea.	213	162		375	490
4350	Double, projecting, polished brass									
4400	2-1/2" x 2-1/2" x 4"	Q-12	5	3.200	Ea.	254	162		416	530

21 11 19 – Fire-Department Connections

21 11 19.50 Connections for the Fire-Department

		Crew	Daily Output	Labor-Hours	Unit	Material	Labor	Equipment	Total	Total Incl O&P
0010	**CONNECTIONS FOR THE FIRE-DEPARTMENT**									
7140	Standpipe connections, wall, w/plugs & chains									
7280	Double, flush, polished brass									
7300	2-1/2" x 2-1/2" x 4"	Q-12	5	3.200	Ea.	520	162		682	825
9990	Minimum labor/equipment charge	1 Spri	4	2	Job		112		112	176

21 12 Fire-Suppression Standpipes

21 12 13 – Fire-Suppression Hoses and Nozzles

21 12 13.50 Fire Hoses and Nozzles

		Crew	Daily Output	Labor-Hours	Unit	Material	Labor	Equipment	Total	Total Incl O&P
0010	**FIRE HOSES AND NOZZLES**									
0200	Adapters, rough brass, straight hose threads									
0220	One piece, female to male, rocker lugs									
0240	1" x 1"				Ea.	48			48	53
0320	2" x 2"					42			42	46
0380	2-1/2" x 2-1/2"					19			19	21
2200	Hose, less couplings									
2260	Synthetic jacket, lined, 300 lb. test, 1-1/2" diameter	Q-12	2600	.006	L.F.	3.24	.31		3.55	4.05
2280	2-1/2" diameter		2200	.007		5.60	.37		5.97	6.70
2360	High strength, 500 lb. test, 1-1/2" diameter		2600	.006		3.35	.31		3.66	4.18
2380	2-1/2" diameter		2200	.007		5.90	.37		6.27	7.05
5600	Nozzles, brass									
5620	Adjustable fog, 3/4" booster line				Ea.	111			111	122
5640	1-1/2" leader line					101			101	112
5660	2-1/2" direct connection					151			151	166
5680	2-1/2" playpipe nozzle					223			223	246
5780	For chrome plated, add					8%				
9900	Minimum labor/equipment charge	1 Plum	2	4	Job		235		235	365

21 12 19 – Fire-Suppression Hose Racks

21 12 19.50 Fire Hose Racks

		Crew	Daily Output	Labor-Hours	Unit	Material	Labor	Equipment	Total	Total Incl O&P
0010	**FIRE HOSE RACKS**									
2600	Hose rack, swinging, for 1-1/2" diameter hose,									
2620	Enameled steel, 50' & 75' lengths of hose	Q-12	20	.800	Ea.	61	40.50		101.50	130

21 12 Fire-Suppression Standpipes

21 12 23 – Fire-Suppression Hose Valves

21 12 23.70 Fire Hose Valves		Crew	Daily Output	Labor-Hours	Unit	Material	2015 Bare Costs Labor	Equipment	Total	Total Incl O&P
0010	**FIRE HOSE VALVES**									
3000	Gate, hose, wheel handle, N.R.S., rough brass, 1-1/2"	1 Spri	12	.667	Ea.	135	37.50		172.50	207
3040	2-1/2", 300 lb.	"	7	1.143		189	64		253	310
3080	For polished brass, add					40%				
3090	For polished chrome, add					50%				

21 13 Fire-Suppression Sprinkler Systems

21 13 13 – Wet-Pipe Sprinkler Systems

21 13 13.50 Wet-Pipe Sprinkler System Components

		Crew	Daily Output	Labor-Hours	Unit	Material	Labor	Equipment	Total	Total Incl O&P
0010	**WET-PIPE SPRINKLER SYSTEM COMPONENTS**									
1220	Water motor, complete with gong	1 Spri	4	2	Ea.	415	112		527	630
1800	Firecycle system, controls, includes panel,									
1820	batteries, solenoid valves and pressure switches	Q-13	1	32	Ea.	20,900	1,700		22,600	25,700
1860	Detector	1 Spri	16	.500	"	715	28		743	830
1900	Flexible sprinkler head connectors									
1910	Braided stainless steel hose with mounting bracket									
1920	1/2" and 3/4" outlet size									
1940	40" length	1 Spri	30	.267	Ea.	70.50	14.95		85.45	101
1960	60" length	"	22	.364	"	81	20.50		101.50	121
1982	May replace hard-pipe armovers									
1984	For wet, pre-action, deluge or dry pipe systems									
2600	Sprinkler heads, not including supply piping									
3700	Standard spray, pendent or upright, brass, 135°F to 286°F									
3740	1/2" NPT, 1/2" orifice	1 Spri	16	.500	Ea.	9.90	28		37.90	55
3790	For riser & feeder piping, see Section 22 11 13.00									

21 13 16 – Dry-Pipe Sprinkler Systems

21 13 16.50 Dry-Pipe Sprinkler System Components

		Crew	Daily Output	Labor-Hours	Unit	Material	Labor	Equipment	Total	Total Incl O&P
0010	**DRY-PIPE SPRINKLER SYSTEM COMPONENTS**									
0600	Accelerator	1 Spri	8	1	Ea.	755	56		811	920
0800	Air compressor for dry pipe system, automatic, complete									
0820	30 gal. system capacity, 3/4 HP	1 Spri	1.30	6.154	Ea.	765	345		1,110	1,375
2600	Sprinkler heads, not including supply piping									
2640	Dry, pendent, 1/2" orifice, 3/4" or 1" NPT									
2660	1/2" to 6" length	1 Spri	14	.571	Ea.	124	32		156	186
2700	15-1/4" to 18" length	"	14	.571	"	145	32		177	210
6330	Valves and components									
8200	Dry pipe valve, incl. trim and gauges, 3" size	Q-12	2	8	Ea.	2,575	405		2,980	3,475
8220	4" size	"	1	16	"	2,850	810		3,660	4,425

21 13 39 – Foam-Water Systems

21 13 39.50 Foam-Water System Components

		Crew	Daily Output	Labor-Hours	Unit	Material	Labor	Equipment	Total	Total Incl O&P
0010	**FOAM-WATER SYSTEM COMPONENTS**									
2600	Sprinkler heads, not including supply piping									
3600	Foam-water, pendent or upright, 1/2" NPT	1 Spri	12	.667	Ea.	210	37.50		247.50	289

Division Notes

	CREW	DAILY OUTPUT	LABOR-HOURS	UNIT	BARE COSTS				TOTAL INCL O&P
					MAT.	LABOR	EQUIP.	TOTAL	

Division 22 – Plumbing

Estimating Tips
22 10 00 Plumbing Piping and Pumps

This subdivision is primarily basic pipe and related materials. The pipe may be used by any of the mechanical disciplines, i.e., plumbing, fire protection, heating, and air conditioning.

Note: CPVC plastic piping approved for fire protection is located in 21 11 13.

- The labor adjustment factors listed in Subdivision 22 01 02.20 apply throughout Divisions 21, 22, and 23. CAUTION: the correct percentage may vary for the same items. For example, the percentage add for the basic pipe installation should be based on the maximum height that the craftsman must install for that particular section. If the pipe is to be located 14' above the floor but it is suspended on threaded rod from beams, the bottom flange of which is 18' high (4' rods), then the height is actually 18' and the add is 20%. The pipe coverer, however, does not have to go above the 14', and so the add should be 10%.
- Most pipe is priced first as straight pipe with a joint (coupling, weld, etc.) every 10' and a hanger usually every 10'. There are exceptions with hanger spacing such as for cast iron pipe (5') and plastic pipe (3 per 10'). Following each type of pipe there are several lines listing sizes and the amount to be subtracted to delete couplings and hangers. This is for pipe that is to be buried or supported together on trapeze hangers. The reason that the couplings are deleted is that these runs are usually long, and frequently longer lengths of pipe are used. By deleting the couplings, the estimator is expected to look up and add back the correct reduced number of couplings.
- When preparing an estimate, it may be necessary to approximate the fittings. Fittings usually run between 25% and 50% of the cost of the pipe. The lower percentage is for simpler runs, and the higher number is for complex areas, such as mechanical rooms.
- For historic restoration projects, the systems must be as invisible as possible, and pathways must be sought for pipes, conduit, and ductwork. While installations in accessible spaces (such as basements and attics) are relatively straightforward to estimate, labor costs may be more difficult to determine when delivery systems must be concealed.

22 40 00 Plumbing Fixtures

- Plumbing fixture costs usually require two lines: the fixture itself and its "rough-in, supply, and waste."
- In the Assemblies Section (Plumbing D2010) for the desired fixture, the System Components Group at the center of the page shows the fixture on the first line. The rest of the list (fittings, pipe, tubing, etc.) will total up to what we refer to in the Unit Price section as "Rough-in, supply, waste, and vent." Note that for most fixtures we allow a nominal 5' of tubing to reach from the fixture to a main or riser.
- Remember that gas- and oil-fired units need venting.

Reference Numbers

Reference numbers are shown in shaded boxes at the beginning of some major classifications. These numbers refer to related items in the Reference Section. The reference information may be an estimating procedure, an alternate pricing method, or technical information.

Note: Not all subdivisions listed here necessarily appear in this publication. ■

Estimate with precision with RSMeans Online...
Quick, intuitive and easy-to-use, RSMeans Online gives you instant access to hundreds of thousands of material, labor, and equipment costs from RSMeans' comprehensive database, delivering the information you need to build competitive estimates every time.

Start your free trial today at **www.rsmeansonline.com**

RSMeansOnline

No part of this publication may be reproduced, stored in a retrieval system, or transmitted in any form or by any means without prior written permission of RSMeans.

22 01 Operation and Maintenance of Plumbing

22 01 02 – Labor Adjustments

22 01 02.20 Labor Adjustment Factors

			Crew	Daily Output	Labor-Hours	Unit	Material	2015 Bare Costs Labor	2015 Bare Costs Equipment	Total	Total Incl O&P
0010	**LABOR ADJUSTMENT FACTORS**, (For Div. 21, 22 and 23)	R220102-20									
0100	Labor factors, The below are reasonable suggestions, however										
0110	each project must be evaluated for its own peculiarities, and										
0120	the adjustments be increased or decreased depending on the										
0130	severity of the special conditions.										
1000	Add to labor for elevated installation (Above floor level)										
1080	10' to 14.5' high							10%			
1100	15' to 19.5' high							20%			
1120	20' to 24.5' high							25%			
1140	25' to 29.5' high							35%			
1160	30' to 34.5' high							40%			
1180	35' to 39.5' high							50%			
1200	40' and higher							55%			
2000	Add to labor for crawl space										
2100	3' high							40%			
2140	4' high							30%			
3000	Add to labor for multi-story building										
3100	Add per floor for floors 3 thru 19							2%			
3140	Add per floor for floors 20 and up							4%			
4000	Add to labor for working in existing occupied buildings										
4100	Hospital							35%			
4140	Office building							25%			
4180	School							20%			
4220	Factory or warehouse							15%			
4260	Multi dwelling							15%			
5000	Add to labor, miscellaneous										
5100	Cramped shaft							35%			
5140	Congested area							15%			
5180	Excessive heat or cold							30%			
9000	Labor factors, The above are reasonable suggestions, however										
9010	each project should be evaluated for its own peculiarities.										
9100	Other factors to be considered are:										
9140	Movement of material and equipment through finished areas										
9180	Equipment room										
9220	Attic space										
9260	No service road										
9300	Poor unloading/storage area										
9340	Congested site area/heavy traffic										

22 05 Common Work Results for Plumbing

22 05 05 – Selective Demolition for Plumbing

22 05 05.10 Plumbing Demolition

			Crew	Daily Output	Labor-Hours	Unit	Material	Labor	Equipment	Total	Total Incl O&P
0010	**PLUMBING DEMOLITION**	R220105-10									
1020	Fixtures, including 10' piping										
1100	Bathtubs, cast iron	R024119-10	1 Plum	4	2	Ea.		117		117	183
1120	Fiberglass			6	1.333			78.50		78.50	122
1140	Steel			5	1.600			94		94	146
1200	Lavatory, wall hung			10	.800			47		47	73
1220	Counter top			8	1			58.50		58.50	91.50
1300	Sink, single compartment			8	1			58.50		58.50	91.50
1320	Double compartment			7	1.143			67		67	105

22 05 Common Work Results for Plumbing

22 05 05 – Selective Demolition for Plumbing

22 05 05.10 Plumbing Demolition

		Crew	Daily Output	Labor-Hours	Unit	Material	2015 Bare Costs Labor	Equipment	Total	Total Incl O&P
1400	Water closet, floor mounted	1 Plum	8	1	Ea.		58.50		58.50	91.50
1420	Wall mounted		7	1.143			67		67	105
1500	Urinal, floor mounted		4	2			117		117	183
1520	Wall mounted		7	1.143			67		67	105
1600	Water fountains, free standing		8	1			58.50		58.50	91.50
1620	Wall or deck mounted		6	1.333	↓		78.50		78.50	122
2000	Piping, metal, up thru 1-1/2" diameter		200	.040	L.F.		2.35		2.35	3.66
2050	2" thru 3-1/2" diameter		150	.053			3.13		3.13	4.88
2100	4" thru 6" diameter	2 Plum	100	.160			9.40		9.40	14.65
2150	8" thru 14" diameter	"	60	.267			15.65		15.65	24.50
2153	16" thru 20" diameter	Q-18	70	.343			19.10	.82	19.92	31
2155	24" thru 26" diameter		55	.436			24.50	1.05	25.55	39
2156	30" thru 36" diameter	↓	40	.600			33.50	1.44	34.94	53.50
2160	Plastic pipe with fittings, up thru 1-1/2" diameter	1 Plum	250	.032			1.88		1.88	2.93
2162	2" thru 3" diameter	"	200	.040			2.35		2.35	3.66
2164	4" thru 6" diameter	Q-1	200	.080			4.23		4.23	6.60
2166	8" thru 14" diameter		150	.107			5.65		5.65	8.80
2168	16" diameter	↓	100	.160	↓		8.45		8.45	13.20
2250	Water heater, 40 gal.	1 Plum	6	1.333	Ea.		78.50		78.50	122
6000	Remove and reset fixtures, easy access		6	1.333			78.50		78.50	122
6100	Difficult access		4	2	↓		117		117	183
9000	Minimum labor/equipment charge	↓	2	4	Job		235		235	365

22 05 23 – General-Duty Valves for Plumbing Piping

22 05 23.20 Valves, Bronze

		Crew	Daily Output	Labor-Hours	Unit	Material	Labor	Equipment	Total	Total Incl O&P
0010	**VALVES, BRONZE**									
1020	Angle, 150 lb., rising stem, threaded									
1070	3/4"	1 Plum	20	.400	Ea.	193	23.50		216.50	250
1080	1"		19	.421		279	24.50		303.50	345
1100	1-1/2"		13	.615		470	36		506	570
1110	2"	↓	11	.727	↓	755	42.50		797.50	895
1300	Ball									
1398	Threaded, 150 psi									
1460	3/4"	1 Plum	20	.400	Ea.	23.50	23.50		47	62
1750	Check, swing, class 150, regrinding disc, threaded									
1850	1/2"	1 Plum	24	.333	Ea.	75.50	19.55		95.05	114
1860	3/4"		20	.400		100	23.50		123.50	147
1870	1"		19	.421		144	24.50		168.50	197
1890	1-1/2"		13	.615		242	36		278	325
1900	2"	↓	11	.727		355	42.50		397.50	455
2850	Gate, N.R.S., soldered, 125 psi									
2920	1/2"	1 Plum	24	.333	Ea.	54	19.55		73.55	90
2940	3/4"		20	.400		63.50	23.50		87	107
2950	1"		19	.421		76.50	24.50		101	123
2970	1-1/2"		13	.615		142	36		178	214
2980	2"	↓	11	.727	↓	185	42.50		227.50	271
4850	Globe, class 150, rising stem, threaded									
4950	1/2"	1 Plum	24	.333	Ea.	95.50	19.55		115.05	136
4960	3/4"	"	20	.400	"	99	23.50		122.50	146
5600	Relief, pressure & temperature, self-closing, ASME, threaded									
5640	3/4"	1 Plum	28	.286	Ea.	253	16.75		269.75	305
5650	1"		24	.333		405	19.55		424.55	475
5660	1-1/4"	↓	20	.400	↓	895	23.50		918.50	1,025

22 05 Common Work Results for Plumbing

22 05 23 – General-Duty Valves for Plumbing Piping

22 05 23.20 Valves, Bronze

		Crew	Daily Output	Labor-Hours	Unit	Material	2015 Bare Costs Labor	Equipment	Total	Total Incl O&P
5670	1-1/2"	1 Plum	18	.444	Ea.	1,250	26		1,276	1,425
5680	2"		16	.500		1,350	29.50		1,379.50	1,550
6400	Pressure, water, ASME, threaded									
6440	3/4"	1 Plum	28	.286	Ea.	116	16.75		132.75	153
6450	1"		24	.333		260	19.55		279.55	315
6470	1-1/2"		18	.444		570	26		596	665
6900	Reducing, water pressure									
6920	300 psi to 25-75 psi, threaded or sweat									
6940	1/2"	1 Plum	24	.333	Ea.	395	19.55		414.55	465
6950	3/4"		20	.400		405	23.50		428.50	485
6960	1"		19	.421		630	24.50		654.50	730
8350	Tempering, water, sweat connections									
8400	1/2"	1 Plum	24	.333	Ea.	98	19.55		117.55	139
8440	3/4"	"	20	.400	"	126	23.50		149.50	176
8650	Threaded connections									
8700	1/2"	1 Plum	24	.333	Ea.	126	19.55		145.55	170
8740	3/4"		20	.400	"	770	23.50		793.50	880
9000	Minimum labor/equipment charge		4	2	Job		117		117	183

22 05 23.60 Valves, Plastic

		Crew	Daily Output	Labor-Hours	Unit	Material	2015 Bare Costs Labor	Equipment	Total	Total Incl O&P
0010	**VALVES, PLASTIC**									
1150	Ball, PVC, socket or threaded, true union									
1230	1/2"	1 Plum	26	.308	Ea.	40.50	18.05		58.55	72.50
1240	3/4"		25	.320		40.50	18.80		59.30	74
1250	1"		23	.348		48	20.50		68.50	85
1260	1-1/4"		21	.381		83.50	22.50		106	127
1270	1-1/2"		20	.400		83.50	23.50		107	129
1280	2"		17	.471		110	27.50		137.50	164
1650	CPVC, socket or threaded, single union									
1700	1/2"	1 Plum	26	.308	Ea.	58	18.05		76.05	91.50
1720	3/4"		25	.320		77	18.80		95.80	114
1730	1"		23	.348		87.50	20.50		108	128
1750	1-1/4"		21	.381		140	22.50		162.50	189
1760	1-1/2"		20	.400		140	23.50		163.50	191
1770	2"		17	.471		192	27.50		219.50	255
3150	Ball check, PVC, socket or threaded									
3290	2"	1 Plum	17	.471	Ea.	146	27.50		173.50	204
9000	Minimum labor/equipment charge	"	3.50	2.286	Job		134		134	209

22 05 76 – Facility Drainage Piping Cleanouts

22 05 76.10 Cleanouts

		Crew	Daily Output	Labor-Hours	Unit	Material	2015 Bare Costs Labor	Equipment	Total	Total Incl O&P
0010	**CLEANOUTS**									
0060	Floor type									
0080	Round or square, scoriated nickel bronze top									
0100	2" pipe size	1 Plum	10	.800	Ea.	197	47		244	290
0120	3" pipe size		8	1		295	58.50		353.50	415
0140	4" pipe size		6	1.333		295	78.50		373.50	445
0980	Round top, recessed for terrazzo									
1000	2" pipe size	1 Plum	9	.889	Ea.	197	52		249	299
1080	3" pipe size		6	1.333		295	78.50		373.50	445
1100	4" pipe size		4	2		295	117		412	510
1120	5" pipe size	Q-1	6	2.667		375	141		516	635
9000	Minimum labor/equipment charge	1 Plum	3	2.667	Job		157		157	244

22 05 Common Work Results for Plumbing

22 05 76 – Facility Drainage Piping Cleanouts

22 05 76.20 Cleanout Tees

		Crew	Daily Output	Labor-Hours	Unit	Material	2015 Bare Costs Labor	2015 Bare Costs Equipment	Total	Total Incl O&P
0010	**CLEANOUT TEES**									
0100	Cast iron, B&S, with countersunk plug									
0220	3" pipe size	1 Plum	3.60	2.222	Ea.	292	130		422	525
0240	4" pipe size	"	3.30	2.424	"	365	142		507	620
0500	For round smooth access cover, same price									
4000	Plastic, tees and adapters. Add plugs									
4010	ABS, DWV									
4020	Cleanout tee, 1-1/2" pipe size	1 Plum	15	.533	Ea.	23	31.50		54.50	74
9000	Minimum labor/equipment charge	"	2.75	2.909	Job		171		171	266

22 07 Plumbing Insulation

22 07 16 – Plumbing Equipment Insulation

22 07 16.10 Insulation for Plumbing Equipment

			Crew	Daily Output	Labor-Hours	Unit	Material	Labor	Equipment	Total	Total Incl O&P
0010	**INSULATION FOR PLUMBING EQUIPMENT**										
2900	Domestic water heater wrap kit										
2920	1-1/2" with vinyl jacket, 20-60 gal.	G	1 Plum	8	1	Ea.	18.60	58.50		77.10	112
2925	50 to 80 gallons	G	"	8	1	"	23.50	58.50		82	118

22 07 19 – Plumbing Piping Insulation

22 07 19.10 Piping Insulation

			Crew	Daily Output	Labor-Hours	Unit	Material	Labor	Equipment	Total	Total Incl O&P
0010	**PIPING INSULATION**										
0100	Rule of thumb, as a percentage of total mechanical costs					Job				10%	10%
0110	Insulation req'd. is based on the surface size/area to be covered										
0230	Insulated protectors, (ADA)										
0235	For exposed piping under sinks or lavatories										
0240	Vinyl coated foam, velcro tabs										
0245	P Trap, 1-1/4" or 1-1/2"		1 Plum	32	.250	Ea.	22.50	14.70		37.20	47.50
0260	Valve and supply cover										
0265	1/2", 3/8", and 7/16" pipe size		1 Plum	32	.250	Ea.	22	14.70		36.70	47
0280	Tailpiece offset (wheelchair)										
0285	1-1/4" pipe size		1 Plum	32	.250	Ea.	13.65	14.70		28.35	38
0600	Pipe covering (price copper tube one size less than IPS)										
6120	2" wall, 1/2" iron pipe size	G	Q-14	145	.110	L.F.	6.20	5.20		11.40	15.15
6210	4" iron pipe size	G	"	125	.128	"	11.25	6.05		17.30	22
6600	Fiberglass, with all service jacket										
6840	1" wall, 1/2" iron pipe size	G	Q-14	240	.067	L.F.	.83	3.14		3.97	5.95
6860	3/4" iron pipe size	G		230	.070		.90	3.28		4.18	6.25
6870	1" iron pipe size	G		220	.073		.97	3.43		4.40	6.55
6880	1-1/4" iron pipe size	G		210	.076		1.05	3.59		4.64	6.90
6890	1-1/2" iron pipe size	G		210	.076		1.13	3.59		4.72	7
6900	2" iron pipe size	G		200	.080		1.22	3.77		4.99	7.40
7080	1-1/2" wall, 1/2" iron pipe size	G		230	.070		1.57	3.28		4.85	7
7140	2" iron pipe size	G		190	.084		2.11	3.97		6.08	8.70
7160	3" iron pipe size	G		170	.094		2.37	4.44		6.81	9.75
7879	Rubber tubing, flexible closed cell foam										
8100	1/2" wall, 1/4" iron pipe size	G	1 Asbe	90	.089	L.F.	.54	4.65		5.19	8.10
8120	3/8" iron pipe size	G		90	.089		.60	4.65		5.25	8.15
8130	1/2" iron pipe size	G		89	.090		.67	4.71		5.38	8.30
8140	3/4" iron pipe size	G		89	.090		.75	4.71		5.46	8.40
8150	1" iron pipe size	G		88	.091		.82	4.76		5.58	8.55
8170	1-1/2" iron pipe size	G		87	.092		1.15	4.81		5.96	9

22 07 Plumbing Insulation

22 07 19 – Plumbing Piping Insulation

22 07 19.10 Piping Insulation

		Crew	Daily Output	Labor-Hours	Unit	Material	2015 Bare Costs Labor	Equipment	Total	Total Incl O&P
8180	2" iron pipe size	G 1 Asbe	86	.093	L.F.	1.47	4.87		6.34	9.45
8200	3" iron pipe size	G	85	.094		2.06	4.93		6.99	10.15
8220	4" iron pipe size	G	80	.100		3.06	5.25		8.31	11.75
8300	3/4" wall, 1/4" iron pipe size	G	90	.089		.85	4.65		5.50	8.45
8330	1/2" iron pipe size	G	89	.090		1.10	4.71		5.81	8.75
8340	3/4" iron pipe size	G	89	.090		1.35	4.71		6.06	9.05
8350	1" iron pipe size	G	88	.091		1.54	4.76		6.30	9.35
8360	1-1/4" iron pipe size	G	87	.092		2.05	4.81		6.86	10
8370	1-1/2" iron pipe size	G	87	.092		2.32	4.81		7.13	10.30
8380	2" iron pipe size	G	86	.093		2.68	4.87		7.55	10.80
8444	1" wall, 1/2" iron pipe size	G	86	.093		2.05	4.87		6.92	10.10
8445	3/4" iron pipe size	G	84	.095		2.48	4.99		7.47	10.75
8446	1" iron pipe size	G	84	.095		2.89	4.99		7.88	11.20
8447	1-1/4" iron pipe size	G	82	.098		3.23	5.10		8.33	11.75
8448	1-1/2" iron pipe size	G	82	.098		3.76	5.10		8.86	12.35
8449	2" iron pipe size	G	80	.100		4.95	5.25		10.20	13.85
8450	2-1/2" iron pipe size	G	80	.100		6.45	5.25		11.70	15.50
8456	Rubber insulation tape, 1/8" x 2" x 30'	G			Ea.	12.05			12.05	13.25
9600	Minimum labor/equipment charge	1 Plum	4	2	Job		117		117	183

22 11 Facility Water Distribution

22 11 13 – Facility Water Distribution Piping

22 11 13.14 Pipe, Brass

0010	**PIPE, BRASS**, Plain end									
0900	Field threaded, coupling & clevis hanger assembly 10' O.C.									
0920	Regular weight									
1120	1/2" diameter	1 Plum	48	.167	L.F.	7.15	9.80		16.95	23
1140	3/4" diameter		46	.174		9.30	10.20		19.50	26
1160	1" diameter		43	.186		13.30	10.90		24.20	31.50
1180	1-1/4" diameter	Q-1	72	.222		19.90	11.75		31.65	40.50
1200	1-1/2" diameter		65	.246		23.50	13		36.50	46.50
1220	2" diameter		53	.302		33	15.95		48.95	61.50
9000	Minimum labor/equipment charge	1 Plum	4	2	Job		117		117	183

22 11 13.16 Pipe Fittings, Brass

0010	**PIPE FITTINGS, BRASS**, Rough bronze, threaded.									
1000	Standard wt., 90° Elbow									
1100	1/2"	1 Plum	12	.667	Ea.	18	39		57	81
1120	3/4"		11	.727		24	42.50		66.50	93
1140	1"		10	.800		39	47		86	116
1160	1-1/4"	Q-1	17	.941		63.50	49.50		113	147
1180	1-1/2"	"	16	1		78.50	53		131.50	169
1500	45° Elbow, 1/8"	1 Plum	13	.615		22	36		58	81
1580	1/2"		12	.667		22	39		61	85.50
1600	3/4"		11	.727		31.50	42.50		74	101
1620	1"		10	.800		53.50	47		100.50	132
1640	1-1/4"	Q-1	17	.941		85	49.50		134.50	171
1660	1-1/2"	"	16	1		107	53		160	201
2000	Tee, 1/8"	1 Plum	9	.889		21	52		73	105
2080	1/2"		8	1		21	58.50		79.50	115
2100	3/4"		7	1.143		30	67		97	138
2120	1"		6	1.333		54.50	78.50		133	182

22 11 Facility Water Distribution

22 11 13 – Facility Water Distribution Piping

22 11 13.16 Pipe Fittings, Brass

		Crew	Daily Output	Labor-Hours	Unit	Material	2015 Bare Costs Labor	Equipment	Total	Total Incl O&P
2140	1-1/4"	Q-1	10	1.600	Ea.	93.50	84.50		178	235
2160	1-1/2"	"	9	1.778		105	94		199	262
2500	Coupling, 1/8"	1 Plum	26	.308		15.05	18.05		33.10	44.50
2580	1/2"		15	.533		15.05	31.50		46.55	65.50
2600	3/4"		14	.571		21	33.50		54.50	76
2620	1"		13	.615		36	36		72	96
2640	1-1/4"	Q-1	22	.727		60	38.50		98.50	127
2660	1-1/2"		20	.800		78.50	42.50		121	152
2680	2"		18	.889		130	47		177	215
9000	Minimum labor/equipment charge	1 Plum	4	2	Job		117		117	183

22 11 13.23 Pipe/Tube, Copper

		Crew	Daily Output	Labor-Hours	Unit	Material	2015 Bare Costs Labor	Equipment	Total	Total Incl O&P
0010	**PIPE/TUBE, COPPER**, Solder joints									
1000	Type K tubing, couplings & clevis hanger assemblies 10' O.C.									
1100	1/4" diameter	1 Plum	84	.095	L.F.	3.71	5.60		9.31	12.80
1200	1" diameter		66	.121		10.35	7.10		17.45	22.50
1260	2" diameter		40	.200		23.50	11.75		35.25	44
2000	Type L tubing, couplings & clevis hanger assemblies 10' O.C.									
2100	1/4" diameter	1 Plum	88	.091	L.F.	2.77	5.35		8.12	11.35
2120	3/8" diameter		84	.095		3.50	5.60		9.10	12.55
2140	1/2" diameter		81	.099		3.70	5.80		9.50	13.10
2160	5/8" diameter		79	.101		5.35	5.95		11.30	15.10
2180	3/4" diameter		76	.105		5.20	6.20		11.40	15.35
2200	1" diameter		68	.118		7.85	6.90		14.75	19.35
2220	1-1/4" diameter		58	.138		10.35	8.10		18.45	24
2240	1-1/2" diameter		52	.154		12.95	9.05		22	28.50
2260	2" diameter		42	.190		18.80	11.20		30	38
2280	2-1/2" diameter	Q-1	62	.258		29.50	13.65		43.15	54
2300	3" diameter		56	.286		37	15.10		52.10	64
2320	3-1/2" diameter		43	.372		53	19.65		72.65	88.50
2340	4" diameter		39	.410		65.50	21.50		87	106
2360	5" diameter		34	.471		120	25		145	171
2380	6" diameter	Q-2	40	.600		166	33		199	234
2410	For other than full hard temper, add					21%				
2590	For silver solder, add						15%			
3000	Type M tubing, couplings & clevis hanger assemblies 10' O.C.									
3140	1/2" diameter	1 Plum	84	.095	L.F.	3.05	5.60		8.65	12.05
3180	3/4" diameter		78	.103		4.13	6		10.13	13.95
3200	1" diameter		70	.114		6.60	6.70		13.30	17.70
4000	Type DWV tubing, couplings & clevis hanger assemblies 10' O.C.									
4100	1-1/4" diameter	1 Plum	60	.133	L.F.	9.15	7.85		17	22.50
4120	1-1/2" diameter		54	.148		11.15	8.70		19.85	26
4140	2" diameter		44	.182		14.60	10.65		25.25	33
4160	3" diameter	Q-1	58	.276		26.50	14.55		41.05	51.50
4180	4" diameter		40	.400		44	21		65	81
4200	5" diameter		36	.444		107	23.50		130.50	155
4220	6" diameter	Q-2	42	.571		153	31.50		184.50	217
9000	Minimum labor/equipment charge	1 Plum	4	2	Job		117		117	183

22 11 13.25 Pipe/Tube Fittings, Copper

		Crew	Daily Output	Labor-Hours	Unit	Material	2015 Bare Costs Labor	Equipment	Total	Total Incl O&P
0010	**PIPE/TUBE FITTINGS, COPPER**, Wrought unless otherwise noted									
0020	For silver solder, add						15%			
0040	Solder joints, copper x copper									
0070	90° elbow, 1/4"	1 Plum	22	.364	Ea.	3.28	21.50		24.78	37

22 11 Facility Water Distribution

22 11 13 – Facility Water Distribution Piping

22 11 13.25 Pipe/Tube Fittings, Copper		Crew	Daily Output	Labor-Hours	Unit	Material	2015 Bare Costs Labor	Equipment	Total	Total Incl O&P
0100	1/2"	1 Plum	20	.400	Ea.	1.10	23.50		24.60	37.50
0120	3/4"		19	.421		2.46	24.50		26.96	41
0130	1"		16	.500		6.05	29.50		35.55	52.50
0250	45° elbow, 1/4"		22	.364		5.90	21.50		27.40	40
0270	3/8"		22	.364		5	21.50		26.50	39
0280	1/2"		20	.400		2.01	23.50		25.51	38.50
0290	5/8"		19	.421		9.55	24.50		34.05	49
0300	3/4"		19	.421		3.52	24.50		28.02	42.50
0310	1"		16	.500		8.85	29.50		38.35	55.50
0320	1-1/4"		15	.533		11.85	31.50		43.35	62
0330	1-1/2"		13	.615		14.25	36		50.25	72
0340	2"		11	.727		24	42.50		66.50	92.50
0350	2-1/2"	Q-1	13	1.231		50.50	65		115.50	157
0360	3"		13	1.231		70.50	65		135.50	179
0370	3-1/2"		10	1.600		125	84.50		209.50	270
0380	4"		9	1.778		151	94		245	310
0390	5"		6	2.667		545	141		686	820
0400	6"	Q-2	9	2.667		860	146		1,006	1,175
0450	Tee, 1/4"	1 Plum	14	.571		6.55	33.50		40.05	59.50
0470	3/8"		14	.571		5.25	33.50		38.75	58.50
0480	1/2"		13	.615		1.87	36		37.87	58.50
0490	5/8"		12	.667		12.05	39		51.05	74.50
0500	3/4"		12	.667		4.51	39		43.51	66
0510	1"		10	.800		13.95	47		60.95	88.50
0520	1-1/4"		9	.889		18.85	52		70.85	102
0530	1-1/2"		8	1		29	58.50		87.50	124
0540	2"		7	1.143		45.50	67		112.50	155
0550	2-1/2"	Q-1	8	2		91.50	106		197.50	266
0560	3"		7	2.286		131	121		252	330
0570	3-1/2"		6	2.667		380	141		521	640
0580	4"		5	3.200		320	169		489	615
0590	5"		4	4		975	211		1,186	1,400
0600	6"	Q-2	6	4		1,325	219		1,544	1,825
0612	Tee, reducing on the outlet, 1/4"	1 Plum	15	.533		16.45	31.50		47.95	67
0613	3/8"		15	.533		15.40	31.50		46.90	66
0614	1/2"		14	.571		10.05	33.50		43.55	63.50
0615	5/8"		13	.615		32	36		68	92
0616	3/4"		12	.667		6.45	39		45.45	68
0617	1"		11	.727		18.50	42.50		61	87
0618	1-1/4"		10	.800		20.50	47		67.50	95.50
0619	1-1/2"		9	.889		21.50	52		73.50	106
0620	2"		8	1		39	58.50		97.50	135
0621	2-1/2"	Q-1	9	1.778		107	94		201	263
0622	3"		8	2		110	106		216	286
0623	4"		6	2.667		203	141		344	445
0624	5"		5	3.200		1,125	169		1,294	1,525
0625	6"	Q-2	7	3.429		1,550	188		1,738	2,000
0626	8"	"	6	4		5,950	219		6,169	6,900
0630	Tee, reducing on the run, 1/4"	1 Plum	15	.533		13.90	31.50		45.40	64.50
0631	3/8"		15	.533		19.60	31.50		51.10	70.50
0632	1/2"		14	.571		11.35	33.50		44.85	65
0633	5/8"		13	.615		19.40	36		55.40	78
0634	3/4"		12	.667		15.15	39		54.15	77.50

22 11 Facility Water Distribution

22 11 13 – Facility Water Distribution Piping

22 11 13.25 Pipe/Tube Fittings, Copper

		Crew	Daily Output	Labor-Hours	Unit	Material	2015 Bare Costs Labor	Equipment	Total	Total Incl O&P
0635	1"	1 Plum	11	.727	Ea.	16.65	42.50		59.15	85
0636	1-1/4"		10	.800		26	47		73	102
0637	1-1/2"		9	.889		46	52		98	133
0638	2"	↓	8	1		59	58.50		117.50	157
0639	2-1/2"	Q-1	9	1.778		152	94		246	315
0640	3"		8	2		202	106		308	385
0641	4"		6	2.667		425	141		566	685
0642	5"	↓	5	3.200		1,075	169		1,244	1,450
0643	6"	Q-2	7	3.429		1,625	188		1,813	2,100
0644	8"	"	6	4		5,950	219		6,169	6,900
0650	Coupling, 1/4"	1 Plum	24	.333		.75	19.55		20.30	31.50
0670	3/8"		24	.333		1	19.55		20.55	31.50
0680	1/2"		22	.364		.83	21.50		22.33	34.50
0690	5/8"		21	.381		2.50	22.50		25	38
0700	3/4"		21	.381		1.67	22.50		24.17	37
0710	1"		18	.444		5.55	26		31.55	46.50
0715	1-1/4"		17	.471		5.85	27.50		33.35	49.50
0716	1-1/2"		15	.533		7.70	31.50		39.20	57.50
0718	2"	↓	13	.615		12.90	36		48.90	70.50
0721	2-1/2"	Q-1	15	1.067		27.50	56.50		84	118
0722	3"		13	1.231		38.50	65		103.50	144
0724	3-1/2"		8	2		74	106		180	247
0726	4"		7	2.286		81.50	121		202.50	278
0728	5"	↓	6	2.667		187	141		328	425
0731	6"	Q-2	8	3	↓	310	164		474	595
2000	DWV, solder joints, copper x copper									
2030	90° Elbow, 1-1/4"	1 Plum	13	.615	Ea.	10.45	36		46.45	68
2050	1-1/2"		12	.667		14.55	39		53.55	77
2070	2"	↓	10	.800		28	47		75	104
2090	3"	Q-1	10	1.600		56.50	84.50		141	195
2100	4"	"	9	1.778		335	94		429	510
2250	Tee, Sanitary, 1-1/4"	1 Plum	9	.889		20.50	52		72.50	105
2270	1-1/2"		8	1		26	58.50		84.50	120
2290	2"		7	1.143		30	67		97	138
2310	3"	Q-1	7	2.286		138	121		259	340
2330	4"	"	6	2.667		330	141		471	580
2400	Coupling, 1-1/4"	1 Plum	14	.571		4.91	33.50		38.41	58
2420	1-1/2"		13	.615		6.10	36		42.10	63
2440	2"	↓	11	.727		8.45	42.50		50.95	76
2460	3"	Q-1	11	1.455		15.40	77		92.40	137
2480	4"	"	10	1.600	↓	49	84.50		133.50	186
9000	Minimum labor/equipment charge	1 Plum	4	2	Job		117		117	183

22 11 13.44 Pipe, Steel

		Crew	Daily Output	Labor-Hours	Unit	Material	Labor	Equipment	Total	Total Incl O&P
0010	**PIPE, STEEL** R221113-50									
0020	All pipe sizes are to Spec. A-53 unless noted otherwise									
0050	Schedule 40, threaded, with couplings, and clevis hanger									
0060	assemblies sized for covering, 10' O.C.									
0540	Black, 1/4" diameter	1 Plum	66	.121	L.F.	6.55	7.10		13.65	18.30
0560	1/2" diameter		63	.127		3.32	7.45		10.77	15.30
0570	3/4" diameter		61	.131		3.91	7.70		11.61	16.30
0580	1" diameter	↓	53	.151		5.05	8.85		13.90	19.35
0590	1-1/4" diameter	Q-1	89	.180	↓	6.20	9.50		15.70	21.50

22 11 Facility Water Distribution

22 11 13 – Facility Water Distribution Piping

22 11 13.44 Pipe, Steel

		Crew	Daily Output	Labor-Hours	Unit	Material	2015 Bare Costs Labor	Equipment	Total	Total Incl O&P
0600	1-1/2" diameter	Q-1	80	.200	L.F.	7.10	10.55		17.65	24.50
0610	2" diameter		64	.250		9	13.20		22.20	30.50
0620	2-1/2" diameter		50	.320		14	16.90		30.90	42
0630	3" diameter		43	.372		17.80	19.65		37.45	50
0640	3-1/2" diameter		40	.400		24.50	21		45.50	60
0650	4" diameter		36	.444		27	23.50		50.50	66
1280	All pipe sizes are to Spec. A-53 unless noted otherwise									
1281	Schedule 40, threaded, with couplings and clevis hanger									
1282	assemblies sized for covering, 10' O.C.									
1290	Galvanized, 1/4" diameter	1 Plum	66	.121	L.F.	9.10	7.10		16.20	21
1310	1/2" diameter		63	.127		3.69	7.45		11.14	15.70
1320	3/4" diameter		61	.131		4.24	7.70		11.94	16.65
1330	1" diameter		53	.151		5.75	8.85		14.60	20
1350	1-1/2" diameter	Q-1	80	.200		8.05	10.55		18.60	25.50
1360	2" diameter		64	.250		10.35	13.20		23.55	32
1370	2-1/2" diameter		50	.320		16.30	16.90		33.20	44.50
1380	3" diameter		43	.372		20.50	19.65		40.15	53.50
1400	4" diameter		36	.444		30	23.50		53.50	69.50
2000	Welded, sch. 40, on yoke & roll hanger assy's, sized for covering, 10' O.C.									
2040	Black, 1" diameter	Q-15	93	.172	L.F.	4.72	9.10	.62	14.44	20
2070	2" diameter		61	.262		7.85	13.85	.95	22.65	31
2090	3" diameter		43	.372		15.45	19.65	1.34	36.44	49
2110	4" diameter		37	.432		21.50	23	1.56	46.06	61
2120	5" diameter		32	.500		34.50	26.50	1.81	62.81	81
2130	6" diameter	Q-16	36	.667		43	36.50	1.60	81.10	106
9990	Minimum labor/equipment charge	1 Plum	3	2.667	Job		157		157	244

22 11 13.45 Pipe Fittings, Steel, Threaded

		Crew	Daily Output	Labor-Hours	Unit	Material	2015 Bare Costs Labor	Equipment	Total	Total Incl O&P
0010	**PIPE FITTINGS, STEEL, THREADED** R221113-50									
0020	Cast Iron									
0040	Standard weight, black									
0060	90° Elbow, straight									
0070	1/4"	1 Plum	16	.500	Ea.	9.70	29.50		39.20	56.50
0080	3/8"		16	.500		14.05	29.50		43.55	61.50
0090	1/2"		15	.533		6.15	31.50		37.65	56
0100	3/4"		14	.571		6.40	33.50		39.90	59.50
0110	1"		13	.615		7.60	36		43.60	65
0130	1-1/2"	Q-1	20	.800		14.90	42.50		57.40	82.50
0140	2"		18	.889		23.50	47		70.50	98.50
0150	2-1/2"		14	1.143		56	60.50		116.50	156
0160	3"		10	1.600		91.50	84.50		176	233
0180	4"		6	2.667		169	141		310	405
0500	Tee, straight									
0510	1/4"	1 Plum	10	.800	Ea.	15.20	47		62.20	89.50
0520	3/8"		10	.800		14.75	47		61.75	89.50
0540	3/4"		9	.889		11.20	52		63.20	94
0550	1"		8	1		9.95	58.50		68.45	102
0570	1-1/2"	Q-1	13	1.231		23.50	65		88.50	127
0580	2"		11	1.455		33	77		110	156
0590	2-1/2"		9	1.778		85.50	94		179.50	240
0600	3"		6	2.667		131	141		272	365
0620	4"		4	4		264	211		475	620
0700	Standard weight, galvanized cast iron									

22 11 Facility Water Distribution

22 11 13 – Facility Water Distribution Piping

22 11 13.45 Pipe Fittings, Steel, Threaded		Crew	Daily Output	Labor-Hours	Unit	Material	2015 Bare Costs Labor	Equipment	Total	Total Incl O&P
0720	90° Elbow, straight									
0730	1/4"	1 Plum	16	.500	Ea.	15.95	29.50		45.45	63.50
0740	3/8"		16	.500		15.95	29.50		45.45	63.50
0750	1/2"		15	.533		18.20	31.50		49.70	69
0760	3/4"		14	.571		17.85	33.50		51.35	72
0770	1"		13	.615		20.50	36		56.50	79
0790	1-1/2"	Q-1	20	.800		44	42.50		86.50	115
0800	2"		18	.889		65	47		112	145
0810	2-1/2"		14	1.143		134	60.50		194.50	241
0820	3"		10	1.600		203	84.50		287.50	355
0840	4"		6	2.667		370	141		511	630
1100	Tee, straight									
1110	1/4"	1 Plum	10	.800	Ea.	16.75	47		63.75	91.50
1120	3/8"		10	.800		19.30	47		66.30	94
1140	3/4"		9	.889		25.50	52		77.50	110
1150	1"		8	1		28	58.50		86.50	122
1170	1-1/2"	Q-1	13	1.231		64.50	65		129.50	172
1180	2"		11	1.455		80.50	77		157.50	209
1190	2-1/2"		9	1.778		164	94		258	325
1200	3"		6	2.667		365	141		506	620
1220	4"		4	4		515	211		726	900
5000	Malleable iron, 150 lb.									
5020	Black									
5040	90° elbow, straight									
5060	1/4"	1 Plum	16	.500	Ea.	4.78	29.50		34.28	51.50
5070	3/8"		16	.500		4.78	29.50		34.28	51.50
5090	3/4"		14	.571		3.99	33.50		37.49	57
5100	1"		13	.615		6.95	36		42.95	64
5120	1-1/2"	Q-1	20	.800		15.05	42.50		57.55	82.50
5130	2"		18	.889		26	47		73	102
5140	2-1/2"		14	1.143		58	60.50		118.50	158
5150	3"		10	1.600		84.50	84.50		169	225
5170	4"		6	2.667		182	141		323	420
5450	Tee, straight									
5470	1/4"	1 Plum	10	.800	Ea.	6.95	47		53.95	80.50
5480	3/8"		10	.800		6.95	47		53.95	80.50
5500	3/4"		9	.889		6.35	52		58.35	88.50
5510	1"		8	1		10.85	58.50		69.35	103
5520	1-1/4"	Q-1	14	1.143		17.60	60.50		78.10	113
5530	1-1/2"		13	1.231		22	65		87	125
5540	2"		11	1.455		37.50	77		114.50	161
5550	2-1/2"		9	1.778		80.50	94		174.50	235
5560	3"		6	2.667		119	141		260	350
5570	3-1/2"		5	3.200		275	169		444	565
5580	4"		4	4		286	211		497	645
5650	Coupling									
5670	1/4"	1 Plum	19	.421	Ea.	5.95	24.50		30.45	45
5680	3/8"		19	.421		5.95	24.50		30.45	45
5690	1/2"		19	.421		4.58	24.50		29.08	43.50
5700	3/4"		18	.444		5.35	26		31.35	46.50
5710	1"		15	.533		8	31.50		39.50	58
5720	1-1/4"	Q-1	26	.615		10.40	32.50		42.90	62
5730	1-1/2"		24	.667		14.05	35		49.05	70.50

22 11 Facility Water Distribution

22 11 13 – Facility Water Distribution Piping

22 11 13.45 Pipe Fittings, Steel, Threaded

		Crew	Daily Output	Labor-Hours	Unit	Material	2015 Bare Costs Labor	2015 Bare Costs Equipment	Total	Total Incl O&P
5740	2"	Q-1	21	.762	Ea.	21	40.50		61.50	86
5750	2-1/2"		18	.889		57.50	47		104.50	136
5760	3"		14	1.143		77.50	60.50		138	180
5780	4"	▼	10	1.600		156	84.50		240.50	305
6000	For galvanized elbows, tees, and couplings add					20%				
9990	Minimum labor/equipment charge	1 Plum	4	2	Job		117		117	183

22 11 13.47 Pipe Fittings, Steel

		Crew	Daily Output	Labor-Hours	Unit	Material	Labor	Equipment	Total	Total Incl O&P
0010	**PIPE FITTINGS, STEEL**, Flanged, Welded & Special									
3000	Weld joint, butt, carbon steel, standard weight									
3040	90° elbow, long radius									
3050	1/2" pipe size	Q-15	16	1	Ea.	45	53	3.61	101.61	136
3060	3/4" pipe size		16	1		45	53	3.61	101.61	136
3070	1" pipe size		16	1		21	53	3.61	77.61	109
3080	1-1/4" pipe size		14	1.143		21	60.50	4.13	85.63	122
3090	1-1/2" pipe size		13	1.231		21	65	4.44	90.44	129
3100	2" pipe size		10	1.600		22.50	84.50	5.80	112.80	163
3110	2-1/2" pipe size		8	2		27.50	106	7.20	140.70	203
3120	3" pipe size		7	2.286		29	121	8.25	158.25	229
3130	4" pipe size		5	3.200		48	169	11.55	228.55	330
3136	5" pipe size	▼	4	4		103	211	14.45	328.45	460
3140	6" pipe size	Q-16	5	4.800	▼	106	263	11.50	380.50	540
3350	Tee, straight									
3360	1/2" pipe size	Q-15	10	1.600	Ea.	110	84.50	5.80	200.30	259
3370	3/4" pipe size		10	1.600		110	84.50	5.80	200.30	259
3380	1" pipe size		10	1.600		54.50	84.50	5.80	144.80	198
3390	1-1/4" pipe size		9	1.778		68.50	94	6.40	168.90	229
3400	1-1/2" pipe size		8	2		68.50	106	7.20	181.70	248
3410	2" pipe size		6	2.667		54.50	141	9.65	205.15	291
3420	2-1/2" pipe size		5	3.200		75.50	169	11.55	256.05	360
3430	3" pipe size		4	4		84	211	14.45	309.45	440
3440	4" pipe size		3	5.333		118	282	19.25	419.25	590
3446	5" pipe size	▼	2.50	6.400		195	340	23	558	765
3450	6" pipe size	Q-16	3	8	▼	203	440	19.20	662.20	930
9990	Minimum labor/equipment charge	Q-15	3	5.333	Job		282	19.25	301.25	460

22 11 13.48 Pipe, Fittings and Valves, Steel, Grooved-Joint

		Crew	Daily Output	Labor-Hours	Unit	Material	Labor	Equipment	Total	Total Incl O&P
0010	**PIPE, FITTINGS AND VALVES, STEEL, GROOVED-JOINT**									
0012	Fittings are ductile iron. Steel fittings noted.									
0020	Pipe includes coupling & clevis type hanger assemblies, 10' O.C.									
1000	Schedule 40, black									
1040	3/4" diameter	1 Plum	71	.113	L.F.	5.50	6.60		12.10	16.30
1050	1" diameter		63	.127		5.30	7.45		12.75	17.45
1060	1-1/4" diameter		58	.138		6.60	8.10		14.70	19.90
1070	1-1/2" diameter		51	.157		7.35	9.20		16.55	22.50
1080	2" diameter	▼	40	.200		8.70	11.75		20.45	28
1090	2-1/2" diameter	Q-1	57	.281		13.65	14.85		28.50	38
1100	3" diameter		50	.320		16.75	16.90		33.65	45
1110	4" diameter		45	.356		23.50	18.80		42.30	55.50
1120	5" diameter	▼	37	.432	▼	38	23		61	77.50
3990	Fittings: coupling material required at joints not incl. in fitting price.									
3994	Add 1 selected coupling, material only, per joint for installed price.									
4000	Elbow, 90° or 45°, painted									
4030	3/4" diameter	1 Plum	50	.160	Ea.	60	9.40		69.40	80.50

22 11 Facility Water Distribution

22 11 13 – Facility Water Distribution Piping

22 11 13.48 Pipe, Fittings and Valves, Steel, Grooved-Joint

		Crew	Daily Output	Labor-Hours	Unit	Material	2015 Bare Costs Labor	Equipment	Total	Total Incl O&P
4040	1" diameter	1 Plum	50	.160	Ea.	32	9.40		41.40	49.50
4050	1-1/4" diameter		40	.200		32	11.75		43.75	53.50
4060	1-1/2" diameter		33	.242		32	14.25		46.25	57
4070	2" diameter		25	.320		32	18.80		50.80	64.50
4080	2-1/2" diameter	Q-1	40	.400		32	21		53	68
4090	3" diameter		33	.485		56.50	25.50		82	103
4100	4" diameter		25	.640		61.50	34		95.50	120
4110	5" diameter		20	.800		146	42.50		188.50	227
4250	For galvanized elbows, add					26%				
4690	Tee, painted									
4700	3/4" diameter	1 Plum	38	.211	Ea.	64.50	12.35		76.85	90.50
4740	1" diameter		33	.242		50	14.25		64.25	77
4750	1-1/4" diameter		27	.296		50	17.40		67.40	82
4760	1-1/2" diameter		22	.364		50	21.50		71.50	88.50
4770	2" diameter		17	.471		50	27.50		77.50	98
4780	2-1/2" diameter	Q-1	27	.593		50	31.50		81.50	104
4790	3" diameter		22	.727		68	38.50		106.50	135
4800	4" diameter		17	.941		103	49.50		152.50	192
4810	5" diameter		13	1.231		241	65		306	365
4900	For galvanized tees, add					24%				
4933	20" diameter	Q-2	16	1.500		565	82		647	750
4940	Flexible, standard, painted									
4950	3/4" diameter	1 Plum	100	.080	Ea.	17.80	4.70		22.50	27
4960	1" diameter		100	.080		17.80	4.70		22.50	27
4970	1-1/4" diameter		80	.100		23.50	5.85		29.35	34.50
4980	1-1/2" diameter		67	.119		25.50	7		32.50	39
4990	2" diameter		50	.160		27	9.40		36.40	44.50
5000	2-1/2" diameter	Q-1	80	.200		31.50	10.55		42.05	51.50
5010	3" diameter		67	.239		35	12.60		47.60	58
5020	3-1/2" diameter		57	.281		50	14.85		64.85	78
5030	4" diameter		50	.320		50.50	16.90		67.40	82
5040	5" diameter		40	.400		76.50	21		97.50	118
5050	6" diameter	Q-2	50	.480		90.50	26.50		117	141
5200	For galvanized couplings, add					33%				
9990	Minimum labor/equipment charge	1 Plum	4	2	Job		117		117	183

22 11 13.74 Pipe, Plastic

		Crew	Daily Output	Labor-Hours	Unit	Material	2015 Bare Costs Labor	Equipment	Total	Total Incl O&P
0010	**PIPE, PLASTIC**									
1800	PVC, couplings 10' O.C., clevis hanger assemblies, 3 per 10'									
1820	Schedule 40									
1860	1/2" diameter	1 Plum	54	.148	L.F.	4.88	8.70		13.58	18.90
1870	3/4" diameter		51	.157		5.20	9.20		14.40	20
1880	1" diameter		46	.174		5.85	10.20		16.05	22.50
1890	1-1/4" diameter		42	.190		6.55	11.20		17.75	24.50
1900	1-1/2" diameter		36	.222		6.85	13.05		19.90	28
1910	2" diameter	Q-1	59	.271		8.05	14.35		22.40	31.50
1920	2-1/2" diameter		56	.286		10.35	15.10		25.45	35
1930	3" diameter		53	.302		12.60	15.95		28.55	39
1940	4" diameter		48	.333		16.10	17.60		33.70	45
1950	5" diameter		43	.372		26.50	19.65		46.15	59.50
1960	6" diameter		39	.410		27.50	21.50		49	64
4100	DWV type, schedule 40, couplings 10' O.C., clevis hanger assy's, 3 per 10'									
4210	ABS, schedule 40, foam core type									

22 11 Facility Water Distribution

22 11 13 – Facility Water Distribution Piping

22 11 13.74 Pipe, Plastic

		Crew	Daily Output	Labor-Hours	Unit	Material	2015 Bare Costs Labor	2015 Bare Costs Equipment	Total	Total Incl O&P
4212	Plain end black									
4214	1-1/2" diameter	1 Plum	39	.205	L.F.	5.20	12.05		17.25	24.50
4216	2" diameter	Q-1	62	.258		5.60	13.65		19.25	27.50
4218	3" diameter		56	.286		8	15.10		23.10	32.50
4220	4" diameter		51	.314		10.20	16.55		26.75	37
4222	6" diameter	↓	42	.381	↓	18	20		38	51.50
4240	To delete coupling & hangers, subtract									
4244	1-1/2" diam. to 6" diam.					43%	48%			
4400	PVC									
4410	1-1/4" diameter	1 Plum	42	.190	L.F.	5.35	11.20		16.55	23.50
4460	2" diameter	Q-1	59	.271		5.65	14.35		20	28.50
4470	3" diameter		53	.302		8.05	15.95		24	34
4480	4" diameter		48	.333		10	17.60		27.60	38.50
4490	6" diameter	↓	39	.410	↓	16.40	21.50		37.90	52
5300	CPVC, socket joint, couplings 10' O.C., clevis hanger assemblies, 3 per 10'									
5302	Schedule 40									
5304	1/2" diameter	1 Plum	54	.148	L.F.	5.90	8.70		14.60	20
5305	3/4" diameter		51	.157		6.75	9.20		15.95	22
5306	1" diameter		46	.174		8.10	10.20		18.30	25
5307	1-1/4" diameter		42	.190		9.65	11.20		20.85	28
5308	1-1/2" diameter	↓	36	.222		10.85	13.05		23.90	32.50
5309	2" diameter	Q-1	59	.271		12.85	14.35		27.20	36.50
5310	2-1/2" diameter		56	.286		19.90	15.10		35	45.50
5311	3" diameter	↓	53	.302	↓	23.50	15.95		39.45	51
5360	CPVC, threaded, couplings 10' O.C., clevis hanger assemblies, 3 per 10'									
5380	Schedule 40									
5460	1/2" diameter	1 Plum	54	.148	L.F.	6.70	8.70		15.40	21
5470	3/4" diameter		51	.157		8.15	9.20		17.35	23.50
5480	1" diameter		46	.174		9.55	10.20		19.75	26.50
5490	1-1/4" diameter		42	.190		10.80	11.20		22	29.50
5500	1-1/2" diameter	↓	36	.222		11.80	13.05		24.85	33.50
5510	2" diameter	Q-1	59	.271		14	14.35		28.35	38
5520	2-1/2" diameter		56	.286		21	15.10		36.10	46.50
5530	3" diameter	↓	53	.302	↓	25.50	15.95		41.45	53
9900	Minimum labor/equipment charge	1 Plum	4	2	Job		117		117	183

22 11 13.76 Pipe Fittings, Plastic

		Crew	Daily Output	Labor-Hours	Unit	Material	2015 Bare Costs Labor	2015 Bare Costs Equipment	Total	Total Incl O&P
0010	**PIPE FITTINGS, PLASTIC**									
2700	PVC (white), schedule 40, socket joints									
2760	90° elbow, 1/2"	1 Plum	33.30	.240	Ea.	.48	14.10		14.58	22.50
2810	2"	Q-1	36.40	.440	"	2.89	23		25.89	39
4500	DWV, ABS, non pressure, socket joints									
4540	1/4 Bend, 1-1/4"	1 Plum	20.20	.396	Ea.	2.57	23.50		26.07	39.50
4560	1-1/2"	"	18.20	.440		1.92	26		27.92	42
4570	2"	Q-1	33.10	.483	↓	3.05	25.50		28.55	43.50
4650	1/8 Bend, same as 1/4 Bend									
4800	Tee, sanitary									
4820	1-1/4"	1 Plum	13.50	.593	Ea.	3.42	35		38.42	58.50
4830	1-1/2"	"	12.10	.661		2.92	39		41.92	63.50
4840	2"	Q-1	20	.800	↓	4.52	42.50		47.02	71
5000	DWV, PVC, schedule 40, socket joints									
5040	1/4 bend, 1-1/4"	1 Plum	20.20	.396	Ea.	5.20	23.50		28.70	42
5060	1-1/2"	"	18.20	.440	↓	1.48	26		27.48	41.50

22 11 Facility Water Distribution
22 11 13 – Facility Water Distribution Piping

22 11 13.76 Pipe Fittings, Plastic		Crew	Daily Output	Labor-Hours	Unit	Material	2015 Bare Costs Labor	Equipment	Total	Total Incl O&P
5070	2"	Q-1	33.10	.483	Ea.	2.33	25.50		27.83	42.50
5080	3"		20.80	.769		6.85	40.50		47.35	71
5090	4"		16.50	.970		13.05	51		64.05	94.50
5100	6"	↓	10.10	1.584		46	83.50		129.50	182
5105	8"	Q-2	9.30	2.581		65	141		206	293
5110	1/4 bend, long sweep, 1-1/2"	1 Plum	18.20	.440		3.42	26		29.42	44
5112	2"	Q-1	33.10	.483		3.83	25.50		29.33	44
5114	3"		20.80	.769		8.75	40.50		49.25	73
5116	4"	↓	16.50	.970		16.65	51		67.65	98.50
5150	1/8 bend, 1-1/4"	1 Plum	20.20	.396		3.40	23.50		26.90	40
5215	8"	Q-2	9.30	2.581		158	141		299	395
5250	Tee, sanitary 1-1/4"	1 Plum	13.50	.593		5.20	35		40.20	60.50
5254	1-1/2"	"	12.10	.661		2.58	39		41.58	63.50
5255	2"	Q-1	20	.800		3.80	42.50		46.30	70
5256	3"		13.90	1.151		10	61		71	106
5257	4"		11	1.455		17.75	77		94.75	140
5259	6"	↓	6.70	2.388		74	126		200	279
5261	8"	Q-2	6.20	3.871		173	212		385	520
5264	2" x 1-1/2"	Q-1	22	.727		3.35	38.50		41.85	63.50
5266	3" x 1-1/2"		15.50	1.032		7.05	54.50		61.55	93
5268	4" x 3"		12.10	1.322		21.50	70		91.50	133
5271	6" x 4"	↓	6.90	2.319		71.50	123		194.50	270
5314	Combination Y & 1/8 bend, 1-1/2"	1 Plum	12.10	.661		6.25	39		45.25	67.50
5315	2"	Q-1	20	.800		6.70	42.50		49.20	73.50
5317	3"		13.90	1.151		16.60	61		77.60	113
5318	4"	↓	11	1.455	↓	33	77		110	156
5324	Combination Y & 1/8 bend, reducing									
5325	2" x 2" x 1-1/2"	Q-1	22	.727	Ea.	8.80	38.50		47.30	69.50
5327	3" x 3" x 1-1/2"		15.50	1.032		15.65	54.50		70.15	102
5328	3" x 3" x 2"		15.30	1.046		11.35	55.50		66.85	98.50
5329	4" x 4" x 2"		12.20	1.311		17.85	69.50		87.35	128
5331	Wye, 1-1/4"	1 Plum	13.50	.593		6.65	35		41.65	62
5332	1-1/2"	"	12.10	.661		4.87	39		43.87	66
5333	2"	Q-1	20	.800		4.63	42.50		47.13	71
5334	3"		13.90	1.151		12.50	61		73.50	109
5335	4"		11	1.455		22.50	77		99.50	145
5336	6"	↓	6.70	2.388		68.50	126		194.50	273
5337	8"	Q-2	6.20	3.871		124	212		336	465
5341	2" x 1-1/2"	Q-1	22	.727		5.70	38.50		44.20	66.50
5342	3" x 1-1/2"		15.50	1.032		8.20	54.50		62.70	94
5343	4" x 3"		12.10	1.322		18.40	70		88.40	129
5344	6" x 4"	↓	6.90	2.319		50.50	123		173.50	247
5345	8" x 6"	Q-2	6.40	3.750		93.50	205		298.50	425
5347	Double wye, 1-1/2"	1 Plum	9.10	.879		10.55	51.50		62.05	92
5348	2"	Q-1	16.60	.964		12.20	51		63.20	93
5349	3"		10.40	1.538		25	81.50		106.50	155
5350	4"	↓	8.25	1.939	↓	50.50	102		152.50	216
5353	Double wye, reducing									
5354	2" x 2" x 1-1/2" x 1-1/2"	Q-1	16.80	.952	Ea.	10.75	50.50		61.25	90.50
5355	3" x 3" x 2" x 2"		10.60	1.509		18.45	79.50		97.95	145
5356	4" x 4" x 3" x 3"		8.45	1.893		40	100		140	200
5357	6" x 6" x 4" x 4"	↓	7.25	2.207		138	117		255	335
5374	Coupling, 1-1/4"	1 Plum	20.20	.396	↓	3.15	23.50		26.65	40

22 11 Facility Water Distribution

22 11 13 – Facility Water Distribution Piping

22 11 13.76 Pipe Fittings, Plastic

		Crew	Daily Output	Labor-Hours	Unit	Material	2015 Bare Costs Labor	Equipment	Total	Total Incl O&P
5376	1-1/2"	1 Plum	18.20	.440	Ea.	.72	26		26.72	41
5378	2"	Q-1	33.10	.483		.95	25.50		26.45	41
5380	3"		20.80	.769		3.29	40.50		43.79	67
5390	4"		16.50	.970		5.65	51		56.65	86
5410	Reducer bushing, 2" x 1-1/4"		36.50	.438		2.82	23		25.82	39
5412	3" x 1-1/2"		27.30	.586		5.85	31		36.85	55
5414	4" x 2"		18.20	.879		10.20	46.50		56.70	83.50
5416	6" x 4"		11.10	1.441		27	76		103	149
5418	8" x 6"	Q-2	10.20	2.353		46	129		175	252
5500	CPVC, Schedule 80, threaded joints									
5540	90° Elbow, 1/4"	1 Plum	32	.250	Ea.	12.15	14.70		26.85	36.50
5560	1/2"		30.30	.264		7.05	15.50		22.55	32
5570	3/4"		26	.308		10.55	18.05		28.60	39.50
5580	1"		22.70	.352		14.80	20.50		35.30	49
5590	1-1/4"		20.20	.396		28.50	23.50		52	68
5600	1-1/2"		18.20	.440		30.50	26		56.50	74
5610	2"	Q-1	33.10	.483		41	25.50		66.50	85
5620	2-1/2"		24.20	.661		128	35		163	195
5630	3"		20.80	.769		137	40.50		177.50	214
5730	Coupling, 1/4"	1 Plum	32	.250		15.45	14.70		30.15	40
5732	1/2"		30.30	.264		12.75	15.50		28.25	38
5734	3/4"		26	.308		20.50	18.05		38.55	50.50
5736	1"		22.70	.352		23.50	20.50		44	58.50
5738	1-1/4"		20.20	.396		25	23.50		48.50	63.50
5740	1-1/2"		18.20	.440		26.50	26		52.50	69.50
5742	2"	Q-1	33.10	.483		31.50	25.50		57	74.50
5744	2-1/2"		24.20	.661		56.50	35		91.50	117
5746	3"		20.80	.769		65.50	40.50		106	136
5900	CPVC, Schedule 80, socket joints									
5904	90° Elbow, 1/4"	1 Plum	32	.250	Ea.	11.55	14.70		26.25	35.50
5906	1/2"		30.30	.264		4.53	15.50		20.03	29
5908	3/4"		26	.308		5.80	18.05		23.85	34.50
5910	1"		22.70	.352		9.20	20.50		29.70	42.50
5912	1-1/4"		20.20	.396		19.85	23.50		43.35	58.50
5914	1-1/2"		18.20	.440		22	26		48	64.50
5916	2"	Q-1	33.10	.483		26.50	25.50		52	69.50
5918	2-1/2"		24.20	.661		61.50	35		96.50	122
5920	3"		20.80	.769		69.50	40.50		110	140
5930	45° Elbow, 1/4"	1 Plum	32	.250		17.20	14.70		31.90	42
5932	1/2"		30.30	.264		5.55	15.50		21.05	30
5934	3/4"		26	.308		7.95	18.05		26	37
5936	1"		22.70	.352		12.75	20.50		33.25	46.50
5938	1-1/4"		20.20	.396		25	23.50		48.50	64
5940	1-1/2"		18.20	.440		25.50	26		51.50	68
5942	2"	Q-1	33.10	.483		29	25.50		54.50	72
5944	2-1/2"		24.20	.661		59	35		94	119
5946	3"		20.80	.769		75.50	40.50		116	147
5990	Coupling, 1/4"	1 Plum	32	.250		12.30	14.70		27	36.50
5992	1/2"		30.30	.264		4.79	15.50		20.29	29.50
5994	3/4"		26	.308		6.70	18.05		24.75	35.50
5996	1"		22.70	.352		9	20.50		29.50	42.50
5998	1-1/4"		20.20	.396		13.50	23.50		37	51.50
6000	1-1/2"		18.20	.440		17	26		43	58.50

22 11 Facility Water Distribution

22 11 13 – Facility Water Distribution Piping

22 11 13.76 Pipe Fittings, Plastic

		Crew	Daily Output	Labor-Hours	Unit	Material	2015 Bare Costs Labor	Equipment	Total	Total Incl O&P
6002	2"	Q-1	33.10	.483	Ea.	19.75	25.50		45.25	61.50
6004	2-1/2"		24.20	.661		44	35		79	103
6006	3"	↓	20.80	.769	↓	47.50	40.50		88	116
9990	Minimum labor/equipment charge	1 Plum	4	2	Job		117		117	183

22 11 19 – Domestic Water Piping Specialties

22 11 19.10 Flexible Connectors

		Crew	Daily Output	Labor-Hours	Unit	Material	Labor	Equipment	Total	Total Incl O&P
0010	**FLEXIBLE CONNECTORS**, Corrugated, 7/8" O.D., 1/2" I.D.									
0050	Gas, seamless brass, steel fittings									
0200	12" long	1 Plum	36	.222	Ea.	17.40	13.05		30.45	39.50
0220	18" long		36	.222		21.50	13.05		34.55	44.50
0240	24" long		34	.235		25.50	13.80		39.30	49.50
0260	30" long	↓	34	.235	↓	27.50	13.80		41.30	52
9000	Minimum labor/equipment charge	↓	4	2	Job		117		117	183

22 11 19.18 Mixing Valve

		Crew	Daily Output	Labor-Hours	Unit	Material	Labor	Equipment	Total	Total Incl O&P
0010	**MIXING VALVE**, Automatic, water tempering.									
0040	1/2" size	1 Stpi	19	.421	Ea.	555	25		580	650
0050	3/4" size		18	.444	"	555	26.50		581.50	650
9000	Minimum labor/equipment charge	↓	5	1.600	Job		95.50		95.50	149

22 11 19.38 Water Supply Meters

		Crew	Daily Output	Labor-Hours	Unit	Material	Labor	Equipment	Total	Total Incl O&P
0010	**WATER SUPPLY METERS**									
2000	Domestic/commercial, bronze									
2020	Threaded									
2060	5/8" diameter, to 20 GPM	1 Plum	16	.500	Ea.	50	29.50		79.50	101
2080	3/4" diameter, to 30 GPM		14	.571		91	33.50		124.50	153
2100	1" diameter, to 50 GPM	↓	12	.667	↓	138	39		177	213
2300	Threaded/flanged									
2340	1-1/2" diameter, to 100 GPM	1 Plum	8	1	Ea.	340	58.50		398.50	460
9000	Minimum labor/equipment charge	"	3.25	2.462	Job		144		144	225

22 11 19.42 Backflow Preventers

		Crew	Daily Output	Labor-Hours	Unit	Material	Labor	Equipment	Total	Total Incl O&P
0010	**BACKFLOW PREVENTERS**, Includes valves									
0020	and four test cocks, corrosion resistant, automatic operation									
1000	Double check principle									
1010	Threaded, with ball valves									
1020	3/4" pipe size	1 Plum	16	.500	Ea.	231	29.50		260.50	300
1030	1" pipe size		14	.571		264	33.50		297.50	345
1040	1-1/2" pipe size		10	.800		560	47		607	695
1050	2" pipe size	↓	7	1.143	↓	655	67		722	825
1080	Threaded, with gate valves									
1100	3/4" pipe size	1 Plum	16	.500	Ea.	1,075	29.50		1,104.50	1,250
1120	1" pipe size		14	.571		1,100	33.50		1,133.50	1,250
1140	1-1/2" pipe size		10	.800		1,400	47		1,447	1,625
1160	2" pipe size	↓	7	1.143	↓	1,725	67		1,792	2,000
4000	Reduced pressure principle									
4100	Threaded, bronze, valves are ball									
4120	3/4" pipe size	1 Plum	16	.500	Ea.	445	29.50		474.50	535
4140	1" pipe size		14	.571		480	33.50		513.50	580
4150	1-1/4" pipe size		12	.667		845	39		884	990
4160	1-1/2" pipe size		10	.800		960	47		1,007	1,125
4180	2" pipe size		7	1.143		1,075	67		1,142	1,300
5000	Flanged, bronze, valves are OS&Y									
5060	2-1/2" pipe size	Q-1	5	3.200	Ea.	4,025	169		4,194	4,700

22 11 Facility Water Distribution

22 11 19 – Domestic Water Piping Specialties

22 11 19.42 Backflow Preventers

		Crew	Daily Output	Labor-Hours	Unit	Material	2015 Bare Costs Labor	2015 Bare Costs Equipment	Total	Total Incl O&P
5080	3" pipe size	Q-1	4.50	3.556	Ea.	4,650	188		4,838	5,400
5100	4" pipe size	↓	3	5.333		5,500	282		5,782	6,500
5120	6" pipe size	Q-2	3	8	↓	8,750	440		9,190	10,300
5600	Flanged, iron, valves are OS&Y									
5660	2-1/2" pipe size	Q-1	5	3.200	Ea.	3,025	169		3,194	3,600
5680	3" pipe size		4.50	3.556		3,175	188		3,363	3,800
5700	4" pipe size	↓	3	5.333		3,975	282		4,257	4,825
5720	6" pipe size	Q-2	3	8		5,775	440		6,215	7,025
5800	Rebuild 4" diameter reduced pressure	1 Plum	4	2		355	117		472	575
5810	6" diameter		2.66	3.008		375	177		552	690
5820	8" diameter		2	4		465	235		700	875
5830	10" diameter		1.60	5	↓	545	294		839	1,050
9010	Minimum labor/equipment charge	↓	2	4	Job		235		235	365

22 11 19.50 Vacuum Breakers

		Crew	Daily Output	Labor-Hours	Unit	Material	Labor	Equipment	Total	Total Incl O&P
0010	**VACUUM BREAKERS**									
0013	See also backflow preventers Section 22 11 19.42									
1000	Anti-siphon continuous pressure type									
1010	Max. 150 PSI - 210°F									
1020	Bronze body									
1030	1/2" size	1 Stpi	24	.333	Ea.	187	19.90		206.90	237
1040	3/4" size		20	.400		187	24		211	244
1050	1" size		19	.421		194	25		219	252
1060	1-1/4" size		15	.533		380	32		412	470
1070	1-1/2" size		13	.615		470	37		507	575
1080	2" size	↓	11	.727	↓	485	43.50		528.50	600
1200	Max. 125 PSI with atmospheric vent									
1210	Brass, in-line construction									
1220	1/4" size	1 Stpi	24	.333	Ea.	117	19.90		136.90	160
1230	3/8" size	"	24	.333		117	19.90		136.90	160
1260	For polished chrome finish, add				↓	13%				
2000	Anti-siphon, non-continuous pressure type									
2010	Hot or cold water 125 PSI - 210°F									
2020	Bronze body									
2030	1/4" size	1 Stpi	24	.333	Ea.	64	19.90		83.90	102
2040	3/8" size		24	.333		64	19.90		83.90	102
2050	1/2" size		24	.333		72.50	19.90		92.40	111
2060	3/4" size		20	.400		86.50	24		110.50	133
2070	1" size		19	.421		134	25		159	186
2080	1-1/4" size		15	.533		235	32		267	310
2090	1-1/2" size		13	.615		276	37		313	365
2100	2" size		11	.727		430	43.50		473.50	540
2110	2-1/2" size		8	1		1,225	60		1,285	1,450
2120	3" size	↓	6	1.333	↓	1,625	79.50		1,704.50	1,925
2150	For polished chrome finish, add					50%				
3000	Air gap fitting									
3020	1/2" NPT size	1 Plum	19	.421	Ea.	50	24.50		74.50	93.50
3030	1" NPT size	"	15	.533		57.50	31.50		89	113
3040	2" NPT size	Q-1	21	.762		113	40.50		153.50	187
3050	3" NPT size		14	1.143		223	60.50		283.50	340
3060	4" NPT size	↓	10	1.600		223	84.50		307.50	375

22 11 Facility Water Distribution

22 11 19 – Domestic Water Piping Specialties

22 11 19.54 Water Hammer Arresters/Shock Absorbers

		Crew	Daily Output	Labor-Hours	Unit	Material	2015 Bare Costs Labor	2015 Bare Costs Equipment	Total	Total Incl O&P
0010	**WATER HAMMER ARRESTERS/SHOCK ABSORBERS**									
0490	Copper									
0500	3/4" male I.P.S. For 1 to 11 fixtures	1 Plum	12	.667	Ea.	28	39		67	92
0600	1" male I.P.S. For 12 to 32 fixtures		8	1		45.50	58.50		104	142
0700	1-1/4" male I.P.S. For 33 to 60 fixtures		8	1		47	58.50		105.50	143
0800	1-1/2" male I.P.S. For 61 to 113 fixtures		8	1		67.50	58.50		126	166
0900	2" male I.P.S. For 114 to 154 fixtures		8	1		98.50	58.50		157	200
1000	2-1/2" male I.P.S. For 155 to 330 fixtures		4	2		305	117		422	520
9000	Minimum labor/equipment charge		3.50	2.286	Job		134		134	209

22 11 19.64 Hydrants

		Crew	Daily Output	Labor-Hours	Unit	Material	Labor	Equipment	Total	Total Incl O&P
0010	**HYDRANTS**									
0050	Wall type, moderate climate, bronze, encased									
0200	3/4" IPS connection	1 Plum	16	.500	Ea.	745	29.50		774.50	860
1000	Non-freeze, bronze, exposed									
1100	3/4" IPS connection, 4" to 9" thick wall	1 Plum	14	.571	Ea.	500	33.50		533.50	605
1120	10" to 14" thick wall	"	12	.667		545	39		584	660
1280	For anti-siphon type, add					132			132	145
9000	Minimum labor/equipment charge	1 Plum	3	2.667	Job		157		157	244

22 13 Facility Sanitary Sewerage

22 13 16 – Sanitary Waste and Vent Piping

22 13 16.20 Pipe, Cast Iron

			Crew	Daily Output	Labor-Hours	Unit	Material	Labor	Equipment	Total	Total Incl O&P
0010	**PIPE, CAST IRON**, Soil, on clevis hanger assemblies, 5' O.C.	R221113-50									
0020	Single hub, service wt., lead & oakum joints 10' O.C.										
2120	2" diameter		Q-1	63	.254	L.F.	11.35	13.40		24.75	33.50
2140	3" diameter			60	.267		15	14.10		29.10	38.50
2160	4" diameter			55	.291		18.55	15.35		33.90	44.50
2180	5" diameter		Q-2	76	.316		25.50	17.30		42.80	55
2200	6" diameter		"	73	.329		31	18		49	62.50
2220	8" diameter		Q-3	59	.542		47	30.50		77.50	99.50
4000	No hub, couplings 10' O.C.										
4100	1-1/2" diameter		Q-1	71	.225	L.F.	11.05	11.90		22.95	30.50
4120	2" diameter			67	.239		11.25	12.60		23.85	32
4140	3" diameter			64	.250		15.20	13.20		28.40	37
4160	4" diameter			58	.276		19.10	14.55		33.65	43.50
4180	5" diameter		Q-2	83	.289		25.50	15.85		41.35	52.50
9000	Minimum labor/equipment charge		1 Plum	4	2	Job		117		117	183

22 13 16.30 Pipe Fittings, Cast Iron

			Crew	Daily Output	Labor-Hours	Unit	Material	Labor	Equipment	Total	Total Incl O&P
0010	**PIPE FITTINGS, CAST IRON**, Soil	R221113-50									
0040	Hub and spigot, service weight, lead & oakum joints										
0080	1/4 bend, 2"		Q-1	16	1	Ea.	19.90	53		72.90	105
0120	3"			14	1.143		26.50	60.50		87	123
0140	4"			13	1.231		41.50	65		106.50	147
0160	5"		Q-2	18	1.333		58	73		131	178
0180	6"		"	17	1.412		72	77.50		149.50	201
0200	8"		Q-3	11	2.909		217	163		380	495
0340	1/8 bend, 2"		Q-1	16	1		14.15	53		67.15	98
0350	3"			14	1.143		22	60.50		82.50	119
0360	4"			13	1.231		33.50	65		98.50	138
0380	5"		Q-2	18	1.333		45.50	73		118.50	164

22 13 Facility Sanitary Sewerage

22 13 16 – Sanitary Waste and Vent Piping

22 13 16.30 Pipe Fittings, Cast Iron		Crew	Daily Output	Labor-Hours	Unit	Material	2015 Bare Costs Labor	Equipment	Total	Total Incl O&P
0400	6"	Q-2	17	1.412	Ea.	54.50	77.50		132	181
0420	8"	Q-3	11	2.909		164	163		327	435
0500	Sanitary tee, 2"	Q-1	10	1.600		27.50	84.50		112	163
0540	3"		9	1.778		45	94		139	196
0620	4"		8	2		55	106		161	226
0700	5"	Q-2	12	2		109	110		219	291
0800	6"	"	11	2.182		126	120		246	325
0880	8"	Q-3	7	4.571		330	256		586	760
5990	No hub									
6000	Cplg. & labor required at joints not incl. in fitting									
6010	price. Add 1 coupling per joint for installed price									
6020	1/4 Bend, 1-1/2"				Ea.	10.20			10.20	11.25
6060	2"					11.15			11.15	12.25
6080	3"					16.25			16.25	17.85
6120	4"					23			23	25.50
6184	1/4 Bend, long sweep, 1-1/2"					25.50			25.50	28
6186	2"					24.50			24.50	26.50
6188	3"					29.50			29.50	32.50
6189	4"					47			47	51.50
6190	5"					90.50			90.50	99.50
6191	6"					103			103	114
6192	8"					282			282	310
6193	10"					570			570	625
6200	1/8 Bend, 1-1/2"					8.60			8.60	9.45
6210	2"					9.60			9.60	10.55
6212	3"					12.80			12.80	14.10
6214	4"					16.80			16.80	18.50
6380	Sanitary Tee, tapped, 1-1/2"					20.50			20.50	22.50
6382	2" x 1-1/2"					17.95			17.95	19.70
6384	2"					19.25			19.25	21
6386	3" x 2"					28.50			28.50	31.50
6388	3"					49.50			49.50	54.50
6390	4" x 1-1/2"					25.50			25.50	28
6392	4" x 2"					29			29	32
6393	4"					29			29	32
6394	6" x 1-1/2"					66			66	73
6396	6" x 2"					67.50			67.50	74
6459	Sanitary Tee, 1-1/2"					14.35			14.35	15.80
6460	2"					15.35			15.35	16.85
6470	3"					18.90			18.90	21
6472	4"					36			36	39.50
8000	Coupling, standard (by CISPI Mfrs.)									
8020	1-1/2"	Q-1	48	.333	Ea.	13.95	17.60		31.55	43
8040	2"		44	.364		15.25	19.20		34.45	47
8080	3"		38	.421		17.10	22		39.10	53.50
8120	4"		33	.485		20	25.50		45.50	62
8300	Coupling, cast iron clamp & neoprene gasket (by MG)									
8310	1-1/2"	Q-1	48	.333	Ea.	6.40	17.60		24	34.50
8320	2"		44	.364		8.35	19.20		27.55	39
8330	3"		38	.421		12.65	22		34.65	48.50
8340	4"		33	.485		14.60	25.50		40.10	56
8600	Coupling, Stainless steel, heavy duty									
8620	1-1/2"	Q-1	48	.333	Ea.	10	17.60		27.60	38.50

22 13 Facility Sanitary Sewerage

22 13 16 – Sanitary Waste and Vent Piping

22 13 16.30 Pipe Fittings, Cast Iron		Crew	Daily Output	Labor-Hours	Unit	Material	2015 Bare Costs Labor	Equipment	Total	Total Incl O&P
8630	2"	Q-1	44	.364	Ea.	10.40	19.20		29.60	41.50
8640	2" x 1-1/2"		44	.364		12.20	19.20		31.40	43.50
8650	3"		38	.421		11.30	22		33.30	47
8660	4"		33	.485		12.75	25.50		38.25	54
9000	Minimum labor/equipment charge	1 Plum	4	2	Job		117		117	183

22 13 16.60 Traps

		Crew	Daily Output	Labor-Hours	Unit	Material	Labor	Equipment	Total	Total Incl O&P
0010	**TRAPS**									
0030	Cast iron, service weight									
0050	Running P trap, without vent									
1100	2"	Q-1	16	1	Ea.	143	53		196	240
1150	4"	"	13	1.231		143	65		208	258
1160	6"	Q-2	17	1.412		635	77.50		712.50	815
3000	P trap, B&S, 2" pipe size	Q-1	16	1		34	53		87	120
3040	3" pipe size	"	14	1.143		50.50	60.50		111	150
4700	Copper, drainage, drum trap									
4840	3" x 6" swivel, 1-1/2" pipe size	1 Plum	16	.500	Ea.	160	29.50		189.50	222
5100	P trap, standard pattern									
5200	1-1/4" pipe size	1 Plum	18	.444	Ea.	78.50	26		104.50	127
5240	1-1/2" pipe size		17	.471		72	27.50		99.50	122
5260	2" pipe size		15	.533		111	31.50		142.50	171
5280	3" pipe size		11	.727		281	42.50		323.50	375
9000	Minimum labor/equipment charge		3	2.667	Job		157		157	244

22 13 16.80 Vent Flashing and Caps

		Crew	Daily Output	Labor-Hours	Unit	Material	Labor	Equipment	Total	Total Incl O&P
0010	**VENT FLASHING AND CAPS**									
0120	Vent caps									
0140	Cast iron									
0180	2-1/2" - 3-5/8" pipe	1 Plum	21	.381	Ea.	45	22.50		67.50	84.50
0190	4" - 4-1/8" pipe	"	19	.421	"	55	24.50		79.50	99
0900	Vent flashing									
1000	Aluminum with lead ring									
1050	3" pipe	1 Plum	17	.471	Ea.	9.55	27.50		37.05	53.50
1060	4" pipe	"	16	.500	"	11.50	29.50		41	58.50
1350	Copper with neoprene ring									
1440	2" pipe	1 Plum	18	.444	Ea.	61.50	26		87.50	108
1450	3" pipe		17	.471		74.50	27.50		102	125
1460	4" pipe		16	.500		74.50	29.50		104	128
9000	Minimum labor/equipment charge		4	2	Job		117		117	183

22 13 19 – Sanitary Waste Piping Specialties

22 13 19.13 Sanitary Drains

		Crew	Daily Output	Labor-Hours	Unit	Material	Labor	Equipment	Total	Total Incl O&P
0010	**SANITARY DRAINS**									
0400	Deck, auto park, C.I., 13" top									
0440	3", 4", 5", and 6" pipe size	Q-1	8	2	Ea.	1,450	106		1,556	1,750
0480	For galvanized body, add				"	780			780	855
2000	Floor, medium duty, C.I., deep flange, 7" diam. top									
2040	2" and 3" pipe size	Q-1	12	1.333	Ea.	208	70.50		278.50	340
2080	For galvanized body, add					96.50			96.50	106
2120	With polished bronze top					315			315	345
2500	Heavy duty, cleanout & trap w/bucket, C.I., 15" top									
2540	2", 3", and 4" pipe size	Q-1	6	2.667	Ea.	6,575	141		6,716	7,450
2560	For galvanized body, add					1,675			1,675	1,850
2580	With polished bronze top					7,300			7,300	8,025

22 13 Facility Sanitary Sewerage

22 13 19 – Sanitary Waste Piping Specialties

22 13 19.14 Floor Receptors		Crew	Daily Output	Labor-Hours	Unit	Material	2015 Bare Costs Labor	Equipment	Total	Total Incl O&P
0010	**FLOOR RECEPTORS**, For connection to 2", 3" & 4" diameter pipe									
0200	12-1/2" square top, 25 sq. in. open area	Q-1	10	1.600	Ea.	970	84.50		1,054.50	1,200
0300	For grate with 4" diam. x 3-3/4" high funnel, add					174			174	191
0400	For grate with 6" diameter x 6" high funnel, add					223			223	246
0700	For acid-resisting bucket, add					274			274	300
0900	For stainless steel mesh bucket liner, add					216			216	238
2000	12-5/8" diameter top, 40 sq. in. open area	Q-1	10	1.600		770	84.50		854.50	975
2100	For options, add same prices as square top									
3000	8" x 4" rectangular top, 7.5 sq. in. open area	Q-1	14	1.143	Ea.	745	60.50		805.50	915
3100	For trap primer connections, add				"	88			88	97
9000	Minimum labor/equipment charge	Q-1	3	5.333	Job		282		282	440

22 13 19.15 Sink Waste Treatment

		Crew	Daily Output	Labor-Hours	Unit	Material	Labor	Equipment	Total	Total Incl O&P
0010	**SINK WASTE TREATMENT**, System for commercial kitchens									
0100	includes clock timer, & fittings									
0200	System less chemical, wall mounted cabinet	1 Plum	16	.500	Ea.	550	29.50		579.50	650
2000	Chemical, 1 gallon, add					68.50			68.50	75.50
2100	6 gallons, add					310			310	340
2200	15 gallons, add					840			840	925
2300	30 gallons, add					1,575			1,575	1,725
2400	55 gallons, add					2,675			2,675	2,950

22 13 23 – Sanitary Waste Interceptors

22 13 23.10 Interceptors

		Crew	Daily Output	Labor-Hours	Unit	Material	Labor	Equipment	Total	Total Incl O&P
0010	**INTERCEPTORS**									
0150	Grease, fabricated steel, 4 GPM, 8 lb. fat capacity	1 Plum	4	2	Ea.	1,150	117		1,267	1,450
0200	7 GPM, 14 lb. fat capacity		4	2		1,600	117		1,717	1,950
1000	10 GPM, 20 lb. fat capacity		4	2		1,875	117		1,992	2,250
1120	50 GPM, 100 lb. fat capacity	Q-1	2	8		6,300	425		6,725	7,575
1160	100 GPM, 200 lb. fat capacity		2	8		15,100	425		15,525	17,300
1240	300 GPM, 600 lb. fat capacity		1	16		31,500	845		32,345	36,000
9000	Minimum labor/equipment charge	1 Plum	3	2.667	Job		157		157	244

22 13 29 – Sanitary Sewerage Pumps

22 13 29.14 Sewage Ejector Pumps

		Crew	Daily Output	Labor-Hours	Unit	Material	Labor	Equipment	Total	Total Incl O&P
0010	**SEWAGE EJECTOR PUMPS**, With operating and level controls									
0100	Simplex system incl. tank, cover, pump 15' head									
0500	37 gal. PE tank, 12 GPM, 1/2 HP, 2" discharge	Q-1	3.20	5	Ea.	480	264		744	935
0510	3" discharge		3.10	5.161		520	273		793	995
0530	87 GPM, .7 HP, 2" discharge		3.20	5		735	264		999	1,225
0540	3" discharge		3.10	5.161		795	273		1,068	1,300
0600	45 gal. coated stl. tank, 12 GPM, 1/2 HP, 2" discharge		3	5.333		855	282		1,137	1,375
0610	3" discharge		2.90	5.517		890	291		1,181	1,425
0630	87 GPM, .7 HP, 2" discharge		3	5.333		1,100	282		1,382	1,650
0640	3" discharge		2.90	5.517		1,150	291		1,441	1,725
0660	134 GPM, 1 HP, 2" discharge		2.80	5.714		1,175	300		1,475	1,775
0680	3" discharge		2.70	5.926		1,250	315		1,565	1,875
0700	70 gal. PE tank, 12 GPM, 1/2 HP, 2" discharge		2.60	6.154		920	325		1,245	1,525
0710	3" discharge		2.40	6.667		980	350		1,330	1,625
0730	87 GPM, 0.7 HP, 2" discharge		2.50	6.400		1,200	340		1,540	1,825
0740	3" discharge		2.30	6.957		1,275	370		1,645	1,975
0760	134 GPM, 1 HP, 2" discharge		2.20	7.273		1,300	385		1,685	2,025
0770	3" discharge		2	8		1,375	425		1,800	2,175
9000	Minimum labor/equipment charge		2.50	6.400	Job		340		340	525

22 14 Facility Storm Drainage

22 14 26 – Facility Storm Drains

22 14 26.13 Roof Drains

		Crew	Daily Output	Labor-Hours	Unit	Material	2015 Bare Costs Labor	Equipment	Total	Total Incl O&P
0010	**ROOF DRAINS**									
0140	Cornice, C.I., 45° or 90° outlet									
0200	3" and 4" pipe size	Q-1	12	1.333	Ea.	330	70.50		400.50	475
0260	For galvanized body, add					75.50			75.50	83
0280	For polished bronze dome, add					85.50			85.50	94
3860	Roof, flat metal deck, C.I. body, 12" C.I. dome									
3890	3" pipe size	Q-1	14	1.143	Ea.	395	60.50		455.50	530
3900	4" pipe size	"	13	1.231	"	395	65		460	535
4620	Main, all aluminum, 12" low profile dome									
4640	2", 3" and 4" pipe size	Q-1	14	1.143	Ea.	435	60.50		495.50	575
9000	Minimum labor/equipment charge	1 Plum	4	2	Job		117		117	183

22 14 29 – Sump Pumps

22 14 29.16 Submersible Sump Pumps

		Crew	Daily Output	Labor-Hours	Unit	Material	2015 Bare Costs Labor	Equipment	Total	Total Incl O&P
0010	**SUBMERSIBLE SUMP PUMPS**									
1000	Elevator sump pumps, automatic									
1010	Complete systems, pump, oil detector, controls and alarm									
1020	1-1/2" discharge, does not include the sump pit/tank.									
1040	1/3 HP, 115 V	1 Plum	4.40	1.818	Ea.	1,875	107		1,982	2,225
1050	1/2 HP, 115 V		4	2		1,925	117		2,042	2,300
1060	1/2 HP, 230 V		4	2		1,975	117		2,092	2,350
1070	3/4 HP, 115 V		3.60	2.222		2,025	130		2,155	2,425
1080	3/4 HP, 230 V		3.60	2.222		2,075	130		2,205	2,475
1100	Sump pump only									
1110	1/3 HP, 115 V	1 Plum	6.40	1.250	Ea.	170	73.50		243.50	300
1120	1/2 HP, 115 V		5.80	1.379		243	81		324	395
1130	1/2 HP, 230 V		5.80	1.379		287	81		368	440
1140	3/4 HP, 115 V		5.40	1.481		325	87		412	490
1150	3/4 HP, 230 V		5.40	1.481		370	87		457	545
1200	Oil detector, control and alarm only									
1210	115 V	1 Plum	8	1	Ea.	1,700	58.50		1,758.50	1,975
1220	230 V	"	8	1	"	1,700	58.50		1,758.50	1,975
7000	Sump pump, automatic									
7100	Plastic, 1-1/4" discharge, 1/4 HP	1 Plum	6.40	1.250	Ea.	138	73.50		211.50	266
7140	1/3 HP		6	1.333		200	78.50		278.50	340
7180	1-1/2" discharge, 1/2 HP		5.20	1.538		281	90.50		371.50	450
7500	Cast iron, 1-1/4" discharge, 1/4 HP		6	1.333		194	78.50		272.50	335
7540	1/3 HP		6	1.333		229	78.50		307.50	375
7560	1/2 HP		5	1.600		277	94		371	450
9000	Minimum labor/equipment charge		4	2	Job		117		117	183

22 15 General Service Compressed-Air Systems

22 15 19 – General Service Packaged Air Compressors and Receivers

22 15 19.10 Air Compressors

		Crew	Daily Output	Labor-Hours	Unit	Material	2015 Bare Costs Labor	Equipment	Total	Total Incl O&P
0010	**AIR COMPRESSORS**									
5250	Air, reciprocating air cooled, splash lubricated, tank mounted									
5300	Single stage, 1 phase, 140 psi									
5303	1/2 HP, 30 gal. tank	1 Stpi	3	2.667	Ea.	1,950	159		2,109	2,400
5305	3/4 HP, 30 gal. tank		2.60	3.077		1,950	184		2,134	2,425
5307	1 HP, 30 gal. tank		2.20	3.636		2,450	217		2,667	3,050

22 31 Domestic Water Softeners

22 31 13 – Residential Domestic Water Softeners

22 31 13.10 Residential Water Softeners

		Crew	Daily Output	Labor-Hours	Unit	Material	2015 Bare Costs Labor	2015 Bare Costs Equipment	Total	Total Incl O&P
0010	**RESIDENTIAL WATER SOFTENERS**									
7350	Water softener, automatic, to 30 grains per gallon	2 Plum	5	3.200	Ea.	405	188		593	740
7400	To 100 grains per gallon	"	4	4	"	660	235		895	1,100

22 33 Electric Domestic Water Heaters

22 33 13 – Instantaneous Electric Domestic Water Heaters

22 33 13.20 Instantaneous Electric Point-Of-Use Water Heaters

			Crew	Daily Output	Labor-Hours	Unit	Material	Labor	Equipment	Total	Total Incl O&P
0010	**INSTANTANEOUS ELECTRIC POINT-OF-USE WATER HEATERS**										
8965	Point of use, electric, glass lined										
8969	Energy saver										
8970	2.5 gal. single element	G	1 Plum	2.80	2.857	Ea.	245	168		413	530
8971	4 gal. single element	G		2.80	2.857		253	168		421	540
8974	6 gal. single element	G		2.50	3.200		261	188		449	580
8975	10 gal. single element	G		2.50	3.200		330	188		518	655
8976	15 gal. single element	G		2.40	3.333		370	196		566	710
8977	20 gal. single element	G		2.40	3.333		410	196		606	755
8978	30 gal. single element	G		2.30	3.478		480	204		684	850
8979	40 gal. single element	G		2.20	3.636		800	213		1,013	1,225
8988	Commercial (ASHRAE energy std. 90)										
8989	6 gallon	G	1 Plum	2.50	3.200	Ea.	665	188		853	1,025
8990	10 gallon	G		2.50	3.200		710	188		898	1,075
8991	15 gallon	G		2.40	3.333		750	196		946	1,125
8992	20 gallon	G		2.40	3.333		785	196		981	1,175
8993	30 gallon	G		2.30	3.478		1,700	204		1,904	2,200
8995	Under the sink, copper, w/bracket										
8996	2.5 gallon	G	1 Plum	4	2	Ea.	485	117		602	720
9000	Minimum labor/equipment charge		"	1.75	4.571	Job		268		268	420

22 33 30 – Residential, Electric Domestic Water Heaters

22 33 30.13 Residential, Small-Capacity Elec. Water Heaters

		Crew	Daily Output	Labor-Hours	Unit	Material	Labor	Equipment	Total	Total Incl O&P
0010	**RESIDENTIAL, SMALL-CAPACITY ELECTRIC DOMESTIC WATER HEATERS**									
1000	Residential, electric, glass lined tank, 5 yr., 10 gal., single element	1 Plum	2.30	3.478	Ea.	330	204		534	680
1060	30 gallon, double element		2.20	3.636		475	213		688	860
1080	40 gallon, double element		2	4		800	235		1,035	1,250
1100	52 gallon, double element		2	4		895	235		1,130	1,350
1120	66 gallon, double element		1.80	4.444		1,200	261		1,461	1,725
1140	80 gallon, double element		1.60	5		1,350	294		1,644	1,925

22 33 33 – Light-Commercial Electric Domestic Water Heaters

22 33 33.10 Commercial Electric Water Heaters

		Crew	Daily Output	Labor-Hours	Unit	Material	Labor	Equipment	Total	Total Incl O&P
0010	**COMMERCIAL ELECTRIC WATER HEATERS**									
4000	Commercial, 100° rise. NOTE: for each size tank, a range of									
4010	heaters between the ones shown are available									
4020	Electric									
4100	5 gal., 3 kW, 12 GPH, 208 volt	1 Plum	2	4	Ea.	2,850	235		3,085	3,525
4160	50 gal., 36 kW, 148 GPH, 208 volt	"	1.80	4.444		6,650	261		6,911	7,700
4480	400 gal., 210 kW, 860 GPH, 480 volt	Q-1	1	16		40,000	845		40,845	45,300

22 34 Fuel-Fired Domestic Water Heaters

22 34 13 – Instantaneous, Tankless, Gas Domestic Water Heaters

22 34 13.10 Instantaneous, Tankless, Gas Water Heaters		Crew	Daily Output	Labor-Hours	Unit	Material	2015 Bare Costs Labor	Equipment	Total	Total Incl O&P
0010	**INSTANTANEOUS, TANKLESS, GAS WATER HEATERS**									
9410	Natural gas/propane, 3.2 GPM	G	1 Plum	2	4	Ea.	370	235		605	770
9420	6.4 GPM	G		1.90	4.211		625	247		872	1,075
9430	8.4 GPM	G		1.80	4.444		730	261		991	1,200
9440	9.5 GPM	G		1.60	5		930	294		1,224	1,475

22 34 30 – Residential Gas Domestic Water Heaters

22 34 30.13 Residential, Atmos, Gas Domestic Wtr Heaters

		Crew	Daily Output	Labor-Hours	Unit	Material	Labor	Equipment	Total	Incl O&P
0010	**RESIDENTIAL, ATMOSPHERIC, GAS DOMESTIC WATER HEATERS**									
2000	Gas fired, foam lined tank, 10 yr., vent not incl.									
2040	30 gallon	1 Plum	2	4	Ea.	895	235		1,130	1,350
2060	40 gallon		1.90	4.211		895	247		1,142	1,375
2100	75 gallon		1.50	5.333		1,350	315		1,665	2,000

22 34 36 – Commercial Gas Domestic Water Heaters

22 34 36.13 Commercial, Atmos., Gas Domestic Water Htrs.

		Crew	Daily Output	Labor-Hours	Unit	Material	Labor	Equipment	Total	Incl O&P
0010	**COMMERCIAL, ATMOSPHERIC, GAS DOMESTIC WATER HEATERS**									
6000	Gas fired, flush jacket, std. controls, vent not incl.									
6040	75 MBH input, 73 GPH	1 Plum	1.40	5.714	Ea.	3,500	335		3,835	4,375
6060	98 MBH input, 95 GPH		1.40	5.714		5,300	335		5,635	6,350
6180	200 MBH input, 192 GPH		.60	13.333		8,925	785		9,710	11,100
6900	For low water cutoff, add		8	1		350	58.50		408.50	475
6960	For bronze body hot water circulator, add		4	2		1,925	117		2,042	2,300

22 34 46 – Oil-Fired Domestic Water Heaters

22 34 46.10 Residential Oil-Fired Water Heaters

		Crew	Daily Output	Labor-Hours	Unit	Material	Labor	Equipment	Total	Incl O&P
0010	**RESIDENTIAL OIL-FIRED WATER HEATERS**									
3000	Oil fired, glass lined tank, 5 yr., vent not included, 30 gallon	1 Plum	2	4	Ea.	1,175	235		1,410	1,650
3040	50 gallon	"	1.80	4.444	"	1,375	261		1,636	1,925

22 34 46.20 Commercial Oil-Fired Water Heaters

		Crew	Daily Output	Labor-Hours	Unit	Material	Labor	Equipment	Total	Incl O&P
0010	**COMMERCIAL OIL-FIRED WATER HEATERS**									
8000	Oil fired, glass lined, UL listed, std. controls, vent not incl.									
8060	140 gal., 140 MBH input, 134 GPH	Q-1	2.13	7.512	Ea.	19,900	395		20,295	22,500
8080	140 gal., 199 MBH input, 191 GPH		2	8		20,600	425		21,025	23,400
8160	140 gal., 540 MBH input, 519 GPH		.96	16.667		28,100	880		28,980	32,300
8280	201 gal., 1250 MBH input, 1200 GPH	Q-2	1.22	19.672		43,900	1,075		44,975	50,000

22 41 Residential Plumbing Fixtures

22 41 13 – Residential Water Closets, Urinals, and Bidets

22 41 13.13 Water Closets

		Crew	Daily Output	Labor-Hours	Unit	Material	Labor	Equipment	Total	Incl O&P
0010	**WATER CLOSETS**									
0032	For automatic flush, see Line 22 42 39.10 0972									
0150	Tank type, vitreous china, incl. seat, supply pipe w/stop, 1.6 gpf or noted									
0200	Wall hung									
0400	Two piece, close coupled	Q-1	5.30	3.019	Ea.	630	159		789	945
0960	For rough-in, supply, waste, vent and carrier	"	2.73	5.861	"	1,025	310		1,335	1,600
0999	Floor mounted									
1100	Two piece, close coupled	Q-1	5.30	3.019	Ea.	237	159		396	510
1102	Economy		5.30	3.019		132	159		291	395
1110	Two piece, close coupled, dual flush		5.30	3.019		310	159		469	595
1140	Two piece, close coupled, 1.28 gpf, ADA	G	5.30	3.019		310	159		469	595
1960	For color, add					30%				

22 41 Residential Plumbing Fixtures

22 41 13 – Residential Water Closets, Urinals, and Bidets

22 41 13.13 Water Closets

		Crew	Daily Output	Labor-Hours	Unit	Material	2015 Bare Costs Labor	2015 Bare Costs Equipment	Total	Total Incl O&P
1980	For rough-in, supply, waste and vent	Q-1	3.05	5.246	Ea.	330	277		607	795

22 41 16 – Residential Lavatories and Sinks

22 41 16.13 Lavatories

		Crew	Daily Output	Labor-Hours	Unit	Material	Labor	Equipment	Total	Total Incl O&P
0010	**LAVATORIES**, With trim, white unless noted otherwise									
0500	Vanity top, porcelain enamel on cast iron									
0600	20" x 18"	Q-1	6.40	2.500	Ea.	335	132		467	575
0640	33" x 19" oval	"	6.40	2.500	"	475	132		607	730
0860	For color, add					25%				
1000	Cultured marble, 19" x 17", single bowl	Q-1	6.40	2.500	Ea.	175	132		307	400
1120	25" x 22", single bowl	"	6.40	2.500	"	215	132		347	440
1580	For color, same price									
1900	Stainless steel, self-rimming, 25" x 22", single bowl, ledge	Q-1	6.40	2.500	Ea.	365	132		497	610
1960	17" x 22", single bowl		6.40	2.500		355	132		487	595
2600	Steel, enameled, 20" x 17", single bowl		5.80	2.759		161	146		307	405
2900	Vitreous china, 20" x 16", single bowl		5.40	2.963		260	157		417	530
2960	20" x 17", single bowl		5.40	2.963		176	157		333	440
3580	Rough-in, supply, waste and vent for all above lavatories		2.30	6.957		231	370		601	830
4000	Wall hung									
4040	Porcelain enamel on cast iron, 16" x 14", single bowl	Q-1	8	2	Ea.	520	106		626	735
4180	20" x 18", single bowl	"	8	2	"	277	106		383	470
4580	For color, add					30%				
6000	Vitreous china, 18" x 15", single bowl with backsplash	Q-1	7	2.286	Ea.	231	121		352	440
6960	Rough-in, supply, waste and vent for above lavatories	"	1.66	9.639	"	455	510		965	1,300
7000	Pedestal type									
7600	Vitreous china, 27" x 21", white	Q-1	6.60	2.424	Ea.	700	128		828	970
7610	27" x 21", colored		6.60	2.424		880	128		1,008	1,175
7620	27" x 21", premium color		6.60	2.424		995	128		1,123	1,300
7660	26" x 20", white		6.60	2.424		795	128		923	1,075
7670	26" x 20", colored		6.60	2.424		1,000	128		1,128	1,300
7680	26" x 20", premium color		6.60	2.424		1,125	128		1,253	1,450
7700	24" x 20", white		6.60	2.424		405	128		533	645
7710	24" x 20", colored		6.60	2.424		475	128		603	725
7720	24" x 20", premium color		6.60	2.424		495	128		623	745
7760	21" x 18", white		6.60	2.424		291	128		419	520
7770	21" x 18", colored		6.60	2.424		291	128		419	520
7990	Rough-in, supply, waste and vent for pedestal lavatories		1.66	9.639		455	510		965	1,300
9000	Minimum labor/equipment charge	1 Plum	3	2.667	Job		157		157	244

22 41 16.16 Sinks

		Crew	Daily Output	Labor-Hours	Unit	Material	Labor	Equipment	Total	Total Incl O&P
0010	**SINKS**, With faucets and drain									
2000	Kitchen, counter top style, P.E. on C.I., 24" x 21" single bowl	Q-1	5.60	2.857	Ea.	285	151		436	550
2100	31" x 22" single bowl		5.60	2.857		615	151		766	915
2200	32" x 21" double bowl		4.80	3.333		355	176		531	665
3000	Stainless steel, self rimming, 19" x 18" single bowl		5.60	2.857		590	151		741	885
3100	25" x 22" single bowl		5.60	2.857		660	151		811	960
3200	33" x 22" double bowl		4.80	3.333		960	176		1,136	1,325
3300	43" x 22" double bowl		4.80	3.333		1,125	176		1,301	1,500
4000	Steel, enameled, with ledge, 24" x 21" single bowl		5.60	2.857		510	151		661	795
4100	32" x 21" double bowl		4.80	3.333		495	176		671	820
4960	For color sinks except stainless steel, add					10%				
4980	For rough-in, supply, waste and vent, counter top sinks	Q-1	2.14	7.477		260	395		655	900
5790	For rough-in, supply, waste & vent, sinks	"	1.85	8.649		260	455		715	1,000

22 41 Residential Plumbing Fixtures

22 41 19 – Residential Bathtubs

22 41 19.10 Baths

		Crew	Daily Output	Labor-Hours	Unit	Material	2015 Bare Costs Labor	2015 Bare Costs Equipment	Total	Total Incl O&P
0010	**BATHS**									
0100	Tubs, recessed porcelain enamel on cast iron, with trim									
0180	48" x 42"	Q-1	4	4	Ea.	2,625	211		2,836	3,225
0220	72" x 36"	"	3	5.333	"	2,725	282		3,007	3,450
0300	Mat bottom									
0380	5' long	Q-1	4.40	3.636	Ea.	1,125	192		1,317	1,550
0560	Corner 48" x 44"		4.40	3.636		2,625	192		2,817	3,200
2000	Enameled formed steel, 4'-6" long		5.80	2.759		495	146		641	770
4600	Module tub & showerwall surround, molded fiberglass									
4610	5' long x 34" wide x 76" high	Q-1	4	4	Ea.	840	211		1,051	1,250
4750	Handicap with 1-1/2" OD grab bar, antiskid bottom									
4760	60" x 32-3/4" x 72" high	Q-1	4	4	Ea.	840	211		1,051	1,250
4770	60" x 30" x 71" high with molded seat		3.50	4.571		875	242		1,117	1,350
9600	Rough-in, supply, waste and vent, for all above tubs, add		2.07	7.729		345	410		755	1,025
9900	Minimum labor/equipment charge		3	5.333	Job		282		282	440

22 41 23 – Residential Showers

22 41 23.20 Showers

		Crew	Daily Output	Labor-Hours	Unit	Material	Labor	Equipment	Total	Total Incl O&P
0011	**SHOWERS**, Stall, with drain only									
1520	32" square	Q-1	5	3.200	Ea.	1,175	169		1,344	1,550
1530	36" square		4.80	3.333		2,925	176		3,101	3,475
1540	Terrazzo receptor, 32" square		5	3.200		1,350	169		1,519	1,750
1580	36" corner angle		4.80	3.333		1,725	176		1,901	2,175
3000	Fiberglass, one piece, with 3 walls, 32" x 32" square		5.50	2.909		490	154		644	780
3100	36" x 36" square		5.50	2.909		505	154		659	795
4200	Rough-in, supply, waste and vent for above showers		2.05	7.805		345	410		755	1,025

22 41 23.40 Shower System Components

		Crew	Daily Output	Labor-Hours	Unit	Material	Labor	Equipment	Total	Total Incl O&P
0010	**SHOWER SYSTEM COMPONENTS**									
5500	Head, water economizer, 1.6 GPM [G]	1 Plum	24	.333	Ea.	45	19.55		64.55	80

22 41 36 – Residential Laundry Trays

22 41 36.10 Laundry Sinks

		Crew	Daily Output	Labor-Hours	Unit	Material	Labor	Equipment	Total	Total Incl O&P
0010	**LAUNDRY SINKS**, With trim									
0020	Porcelain enamel on cast iron, black iron frame									
0050	24" x 21", single compartment	Q-1	6	2.667	Ea.	580	141		721	855
0100	26" x 21", single compartment	"	6	2.667	"	610	141		751	890
2000	Molded stone, on wall hanger or legs									
2020	22" x 23", single compartment	Q-1	6	2.667	Ea.	167	141		308	405
2100	45" x 21", double compartment	"	5	3.200	"	330	169		499	630
3000	Plastic, on wall hanger or legs									
3020	18" x 23", single compartment	Q-1	6.50	2.462	Ea.	135	130		265	350
3100	20" x 24", single compartment		6.50	2.462		154	130		284	375
3200	36" x 23", double compartment		5.50	2.909		185	154		339	445
3300	40" x 24", double compartment		5.50	2.909		278	154		432	545
5000	Stainless steel, counter top, 22" x 17" single compartment		6	2.667		64	141		205	291
5200	33" x 22", double compartment		5	3.200		79	169		248	350
9600	Rough-in, supply, waste and vent, for all laundry sinks		2.14	7.477		260	395		655	900
9810	Minimum labor/equipment charge	1 Plum	3	2.667	Job		157		157	244

22 41 Residential Plumbing Fixtures

22 41 39 – Residential Faucets, Supplies and Trim

22 41 39.10 Faucets and Fittings		Crew	Daily Output	Labor-Hours	Unit	Material	2015 Bare Costs Labor	2015 Bare Costs Equipment	Total	Total Incl O&P
0010	**FAUCETS AND FITTINGS**									
0150	Bath, faucets, diverter spout combination, sweat	1 Plum	8	1	Ea.	86.50	58.50		145	187
0200	For integral stops, IPS unions, add					109			109	120
0420	Bath, press-bal mix valve w/diverter, spout, shower head, arm/flange	1 Plum	8	1		168	58.50		226.50	277
0500	Drain, central lift, 1-1/2" IPS male		20	.400		71	23.50		94.50	115
0600	Trip lever, 1-1/2" IPS male		20	.400		45	23.50		68.50	86
0810	Bidet									
0812	Fitting, over the rim, swivel spray/pop-up drain	1 Plum	8	1	Ea.	206	58.50		264.50	320
1000	Kitchen sink faucets, top mount, cast spout		10	.800		61.50	47		108.50	141
1100	For spray, add		24	.333		16.15	19.55		35.70	48.50
1200	Wall type, swing tube spout		10	.800		74.50	47		121.50	155
1300	Single control lever handle									
1310	With pull out spray									
1320	Polished chrome	1 Plum	10	.800	Ea.	196	47		243	288
2000	Laundry faucets, shelf type, IPS or copper unions		12	.667		49.50	39		88.50	116
2100	Lavatory faucet, centerset, without drain		10	.800		44.50	47		91.50	122
2120	With pop-up drain		6.66	1.201		62.50	70.50		133	179
2210	Porcelain cross handles and pop-up drain									
2220	Polished chrome	1 Plum	6.66	1.201	Ea.	189	70.50		259.50	320
2230	Polished brass	"	6.66	1.201	"	298	70.50		368.50	440
2260	Single lever handle and pop-up drain									
2280	Satin nickel	1 Plum	6.66	1.201	Ea.	277	70.50		347.50	415
2290	Polished chrome		6.66	1.201		198	70.50		268.50	330
2800	Self-closing, center set		10	.800		131	47		178	217
2810	Automatic sensor and operator, with faucet head G		6.15	1.301		450	76.50		526.50	615
4000	Shower by-pass valve with union		18	.444		68.50	26		94.50	116
4100	Shower arm with flange and head		22	.364		19.25	21.50		40.75	54.50
4140	Shower, hand held, pin mount, massage action, chrome		22	.364		76	21.50		97.50	117
4142	Polished brass		22	.364		142	21.50		163.50	190
4144	Shower, hand held, wall mtd, adj. spray, 2 wall mounts, chrome		20	.400		102	23.50		125.50	149
4146	Polished brass		20	.400		192	23.50		215.50	249
4148	Shower, hand held head, bar mounted 24", adj. spray, chrome		20	.400		164	23.50		187.50	217
4150	Polished brass		20	.400		340	23.50		363.50	405
4200	Shower thermostatic mixing valve, concealed, with shower head trim kit		8	1		345	58.50		403.50	470
4220	Shower pressure balancing mixing valve,									
4230	With shower head, arm, flange and diverter tub spout									
4240	Chrome	1 Plum	6.14	1.303	Ea.	360	76.50		436.50	515
4250	Satin nickel		6.14	1.303		555	76.50		631.50	730
4260	Polished graphite		6.14	1.303		555	76.50		631.50	730
5000	Sillcock, compact, brass, IPS or copper to hose		24	.333		9.70	19.55		29.25	41
6000	Stop and waste valves, bronze									
6100	Angle, solder end 1/2"	1 Plum	24	.333	Ea.	12.05	19.55		31.60	44
6110	3/4"		20	.400		14.15	23.50		37.65	52
6300	Straightway, solder end 3/8"		24	.333		12.35	19.55		31.90	44
6310	1/2"		24	.333		12.35	19.55		31.90	44
6410	Straightway, threaded 1/2"		24	.333		15.95	19.55		35.50	48
6420	3/4"		20	.400		18.30	23.50		41.80	56.50
6430	1"		19	.421		13.40	24.50		37.90	53.50
7800	Water closet, wax gasket		96	.083		1.38	4.89		6.27	9.15
7820	Gasket toilet tank to bowl		32	.250		2.38	14.70		17.08	25.50
7830	Replacement diaphragm washer assy for ballcock valve		12	.667		2.38	39		41.38	63.50
7850	Dual flush valve		12	.667		20	39		59	83

22 41 Residential Plumbing Fixtures

22 41 39 – Residential Faucets, Supplies and Trim

22 41 39.10 Faucets and Fittings		Crew	Daily Output	Labor-Hours	Unit	Material	2015 Bare Costs Labor	Equipment	Total	Total Incl O&P
8000	Water supply stops, polished chrome plate									
8200	Angle, 3/8"	1 Plum	24	.333	Ea.	12.90	19.55		32.45	44.50
8300	1/2"		22	.364		12.90	21.50		34.40	47.50
8400	Straight, 3/8"		26	.308		13.55	18.05		31.60	43
8500	1/2"		24	.333		13.55	19.55		33.10	45.50
8600	Water closet, angle, w/flex riser, 3/8"		24	.333		44	19.55		63.55	79
9000	Minimum labor/equipment charge		4	2	Job		117		117	183

22 42 Commercial Plumbing Fixtures

22 42 13 – Commercial Water Closets, Urinals, and Bidets

22 42 13.13 Water Closets

		Crew	Daily Output	Labor-Hours	Unit	Material	Labor	Equipment	Total	Total Incl O&P
0010	**WATER CLOSETS**									
3000	Bowl only, with flush valve, seat, 1.6 gpf unless noted									
3100	Wall hung	Q-1	5.80	2.759	Ea.	945	146		1,091	1,275
3200	For rough-in, supply, waste and vent, single WC		2.56	6.250		1,075	330		1,405	1,700
3360	With floor outlet, 1.28 gpf G		5.80	2.759		555	146		701	840
3362	With floor outlet, 1.28 gpf, ADA G		5.80	2.759		580	146		726	860
3370	For rough-in, supply, waste and vent, single WC		2.84	5.634		370	298		668	870
3390	Floor mounted children's size, 10-3/4" high									
3392	With automatic flush sensor, 1.6 gpf	Q-1	6.20	2.581	Ea.	615	136		751	895
3396	With automatic flush sensor, 1.28 gpf		6.20	2.581	"	620	136		756	895
9000	Minimum labor/equipment charge		4	4	Job		211		211	330

22 42 13.16 Urinals

		Crew	Daily Output	Labor-Hours	Unit	Material	Labor	Equipment	Total	Total Incl O&P
0010	**URINALS**									
3000	Wall hung, vitreous china, with self-closing valve									
3100	Siphon jet type	Q-1	3	5.333	Ea.	282	282		564	750
3120	Blowout type		3	5.333		465	282		747	950
3140	Water saving .5 gpf G		3	5.333		550	282		832	1,050
3300	Rough-in, supply, waste & vent		2.83	5.654		595	299		894	1,125
5000	Stall type, vitreous china, includes valve		2.50	6.400		740	340		1,080	1,325
6980	Rough-in, supply, waste and vent		1.99	8.040		365	425		790	1,075
9000	Minimum labor/equipment charge		4	4	Job		211		211	330

22 42 16 – Commercial Lavatories and Sinks

22 42 16.13 Lavatories

0010	**LAVATORIES**, With trim, white unless noted otherwise
0020	Commercial lavatories same as residential. See Section 22 41 16

22 42 16.40 Service Sinks

		Crew	Daily Output	Labor-Hours	Unit	Material	Labor	Equipment	Total	Total Incl O&P
0010	**SERVICE SINKS**									
6650	Service, floor, corner, P.E. on C.I., 28" x 28"	Q-1	4.40	3.636	Ea.	1,000	192		1,192	1,400
6790	For rough-in, supply, waste & vent, floor service sinks		1.64	9.756	"	705	515		1,220	1,575
9000	Minimum labor/equipment charge		4	4	Job		211		211	330

22 42 23 – Commercial Showers

22 42 23.30 Group Showers

		Crew	Daily Output	Labor-Hours	Unit	Material	Labor	Equipment	Total	Total Incl O&P
0010	**GROUP SHOWERS**									
6000	Group, w/pressure balancing valve, rough-in and rigging not included									
6800	Column, 6 heads, no receptors, less partitions	Q-1	3	5.333	Ea.	9,200	282		9,482	10,500
6900	With stainless steel partitions		1	16		11,900	845		12,745	14,400
7600	5 heads, no receptors, less partitions		3	5.333		6,350	282		6,632	7,425
7620	4 heads (1 handicap) no receptors, less partitions		3	5.333		5,650	282		5,932	6,650

22 42 Commercial Plumbing Fixtures

22 42 23 – Commercial Showers

22 42 23.30 Group Showers

		Crew	Daily Output	Labor-Hours	Unit	Material	2015 Bare Costs Labor	2015 Bare Costs Equipment	Total	Total Incl O&P
7700	With stainless steel partitions	Q-1	1	16	Ea.	5,650	845		6,495	7,525
9000	Minimum labor/equipment charge	↓	4	4	Job		211		211	330

22 42 33 – Wash Fountains

22 42 33.20 Commercial Wash Fountains

		Crew	Daily Output	Labor-Hours	Unit	Material	Labor	Equipment	Total	Total Incl O&P
0010	**COMMERCIAL WASH FOUNTAINS**									
1900	Group, foot control									
2000	Precast terrazzo, circular, 36" diam., 5 or 6 persons	Q-2	3	8	Ea.	7,275	440		7,715	8,675
2100	54" diameter for 8 or 10 persons		2.50	9.600		9,050	525		9,575	10,800
2400	Semi-circular, 36" diam. for 3 persons		3	8		6,375	440		6,815	7,700
2500	54" diam. for 4 or 5 persons	↓	2.50	9.600		8,575	525		9,100	10,200
5610	Group, infrared control, barrier free									
5614	Precast terrazzo									
5620	Semi-circular 36" diam. for 3 persons	Q-2	3	8	Ea.	7,250	440		7,690	8,650
5630	46" diam. for 4 persons		2.80	8.571		7,825	470		8,295	9,325
5640	Circular, 54" diam. for 8 persons, button control	↓	2.50	9.600		9,475	525		10,000	11,200
5700	Rough-in, supply, waste and vent for above wash fountains	Q-1	1.82	8.791	↓	370	465		835	1,125
9000	Minimum labor/equipment charge	Q-2	3	8	Job		440		440	685

22 42 39 – Commercial Faucets, Supplies, and Trim

22 42 39.10 Faucets and Fittings

		Crew	Daily Output	Labor-Hours	Unit	Material	Labor	Equipment	Total	Total Incl O&P
0010	**FAUCETS AND FITTINGS**									
0840	Flush valves, with vacuum breaker									
0850	Water closet									
0860	Exposed, rear spud	1 Plum	8	1	Ea.	148	58.50		206.50	255
0870	Top spud		8	1		149	58.50		207.50	255
0880	Concealed, rear spud		8	1		197	58.50		255.50	310
0890	Top spud		8	1		159	58.50		217.50	267
0900	Wall hung		8	1		177	58.50		235.50	287
0910	Dual flush flushometer		12	.667		197	39		236	277
0912	Flushometer retrofit kit	↓	18	.444	↓	16.30	26		42.30	58.50
0920	Urinal									
0930	Exposed, stall	1 Plum	8	1	Ea.	149	58.50		207.50	256
0940	Wall, (washout)		8	1		149	58.50		207.50	255
0950	Pedestal, top spud		8	1		143	58.50		201.50	249
0960	Concealed, stall		8	1		156	58.50		214.50	264
0970	Wall (washout)	↓	8	1	↓	168	58.50		226.50	277
0971	Automatic flush sensor and operator for									
0972	urinals or water closets, standard [G]	1 Plum	8	1	Ea.	450	58.50		508.50	585
0980	High efficiency water saving									
0984	Water closets, 1.28 gpf [G]	1 Plum	8	1	Ea.	415	58.50		473.50	550
0988	Urinals, .5 gpf [G]	"	8	1	"	415	58.50		473.50	550
2790	Faucets for lavatories									
2800	Self-closing, center set	1 Plum	10	.800	Ea.	131	47		178	217
2810	Automatic sensor and operator, with faucet head		6.15	1.301		450	76.50		526.50	615
3000	Service sink faucet, cast spout, pail hook, hose end	↓	14	.571	↓	80	33.50		113.50	141

22 42 39.30 Carriers and Supports

		Crew	Daily Output	Labor-Hours	Unit	Material	Labor	Equipment	Total	Total Incl O&P
0010	**CARRIERS AND SUPPORTS**, For plumbing fixtures									
3000	Lavatory, concealed arm									
3050	Floor mounted, single									
3100	High back fixture	1 Plum	6	1.333	Ea.	525	78.50		603.50	700
3200	Flat slab fixture		6	1.333		455	78.50		533.50	620
3220	Paraplegic	↓	6	1.333	↓	590	78.50		668.50	770

22 42 Commercial Plumbing Fixtures

22 42 39 – Commercial Faucets, Supplies, and Trim

22 42 39.30 Carriers and Supports

		Crew	Daily Output	Labor-Hours	Unit	Material	2015 Bare Costs Labor	Equipment	Total	Total Incl O&P
6980	Water closet, siphon jet									
7000	Horizontal, adjustable, caulk ♿									
7040	Single, 4" pipe size	1 Plum	5.33	1.501	Ea.	890	88		978	1,100
7060	5" pipe size		5.33	1.501		965	88		1,053	1,175
7100	Double, 4" pipe size		5	1.600		1,575	94		1,669	1,875
7120	5" pipe size		5	1.600		1,700	94		1,794	2,025
8200	Water closet, residential									
8220	Vertical centerline, floor mount									
8240	Single, 3" caulk, 2" or 3" vent	1 Plum	6	1.333	Ea.	610	78.50		688.50	790
8260	4" caulk, 2" or 4" vent		6	1.333	"	785	78.50		863.50	980
9990	Minimum labor/equipment charge		3.50	2.286	Job		134		134	209

22 43 Healthcare Plumbing Fixtures

22 43 39 – Healthcare Faucets

22 43 39.10 Faucets and Fittings

		Crew	Daily Output	Labor-Hours	Unit	Material	2015 Bare Costs Labor	Equipment	Total	Total Incl O&P
0010	**FAUCETS AND FITTINGS**									
2850	Medical, bedpan cleanser, with pedal valve,	1 Plum	12	.667	Ea.	760	39		799	895
2860	With screwdriver stop valve		12	.667		395	39		434	490
2870	With self-closing spray valve		12	.667		242	39		281	325
2900	Faucet, gooseneck spout, wrist handles, grid drain		10	.800		194	47		241	286
2940	Mixing valve, knee action, screwdriver stops		4	2		425	117		542	650

22 45 Emergency Plumbing Fixtures

22 45 13 – Emergency Showers

22 45 13.10 Emergency Showers

		Crew	Daily Output	Labor-Hours	Unit	Material	2015 Bare Costs Labor	Equipment	Total	Total Incl O&P
0010	**EMERGENCY SHOWERS**, Rough-in not included									
5000	Shower, single head, drench, ball valve, pull, freestanding	Q-1	4	4	Ea.	380	211		591	745
5200	Horizontal or vertical supply		4	4		555	211		766	940
6000	Multi-nozzle, eye/face wash combination		4	4		660	211		871	1,050
6400	Multi-nozzle, 12 spray, shower only		4	4		2,000	211		2,211	2,525
6600	For freeze-proof, add		6	2.667		465	141		606	735
9000	Minimum labor/equipment charge		3	5.333	Job		282		282	440

22 45 16 – Eyewash Equipment

22 45 16.10 Eyewash Safety Equipment

		Crew	Daily Output	Labor-Hours	Unit	Material	2015 Bare Costs Labor	Equipment	Total	Total Incl O&P
0010	**EYEWASH SAFETY EQUIPMENT**, Rough-in not included									
1000	Eye wash fountain									
1400	Plastic bowl, pedestal mounted	Q-1	4	4	Ea.	282	211		493	640

22 47 Drinking Fountains and Water Coolers

22 47 13 – Drinking Fountains

22 47 13.10 Drinking Water Fountains

		Crew	Daily Output	Labor-Hours	Unit	Material	2015 Bare Costs Labor	2015 Bare Costs Equipment	Total	Total Incl O&P
0010	**DRINKING WATER FOUNTAINS**, For connection to cold water supply									
1000	Wall mounted, non-recessed									
2700	Stainless steel, single bubbler, no back	1 Plum	4	2	Ea.	1,050	117		1,167	1,325
2740	With back		4	2		565	117		682	805
2780	Dual handle & wheelchair projection type		4	2		745	117		862	1,000
2820	Dual level for handicapped type		3.20	2.500		1,575	147		1,722	1,950
3980	For rough-in, supply and waste, add		2.21	3.620		174	212		386	520
4000	Wall mounted, semi-recessed									
4200	Poly-marble, single bubbler	1 Plum	4	2	Ea.	955	117		1,072	1,225
4600	Stainless steel, satin finish, single bubbler	"	4	2	"	1,150	117		1,267	1,450
6000	Wall mounted, fully recessed									
6400	Poly-marble, single bubbler	1 Plum	4	2	Ea.	1,600	117		1,717	1,925
6800	Stainless steel, single bubbler		4	2		1,550	117		1,667	1,875
7580	For rough-in, supply and waste, add		1.83	4.372		174	257		431	590
7600	Floor mounted, pedestal type									
8600	Enameled iron, heavy duty service, 2 bubblers	1 Plum	2	4	Ea.	2,575	235		2,810	3,225
8880	For freeze-proof valve system, add		2	4		705	235		940	1,150
8900	For rough-in, supply and waste, add		1.83	4.372		174	257		431	590
9000	Minimum labor/equipment charge		2	4	Job		235		235	365

22 47 16 – Pressure Water Coolers

22 47 16.10 Electric Water Coolers

		Crew	Daily Output	Labor-Hours	Unit	Material	2015 Bare Costs Labor	2015 Bare Costs Equipment	Total	Total Incl O&P
0010	**ELECTRIC WATER COOLERS**									
0100	Wall mounted, non-recessed									
0140	4 GPH	Q-1	4	4	Ea.	680	211		891	1,075
0160	8 GPH, barrier free, sensor operated		4	4		1,025	211		1,236	1,475
1000	Dual height, 8.2 GPH		3.80	4.211		2,075	222		2,297	2,625
1040	14.3 GPH		3.80	4.211		975	222		1,197	1,425
3300	Semi-recessed, 8.1 GPH		4	4		750	211		961	1,150
4600	Floor mounted, flush-to-wall									
4640	4 GPH	1 Plum	3	2.667	Ea.	740	157		897	1,050
4980	For stainless steel cabinet, add					134			134	148
5000	Dual height, 8.2 GPH	1 Plum	2	4		1,175	235		1,410	1,650
9000	Minimum labor/equipment charge	"	2	4	Job		235		235	365

22 66 Chemical-Waste Systems for Lab. and Healthcare Facilities

22 66 53 – Laboratory Chemical-Waste and Vent Piping

22 66 53.30 Glass Pipe

		Crew	Daily Output	Labor-Hours	Unit	Material	2015 Bare Costs Labor	2015 Bare Costs Equipment	Total	Total Incl O&P
0010	**GLASS PIPE**, Borosilicate, couplings & clevis hanger assemblies, 10' O.C.									
0020	Drainage									
1100	1-1/2" diameter	Q-1	52	.308	L.F.	11.40	16.25		27.65	38
1120	2" diameter		44	.364		14.65	19.20		33.85	46
1140	3" diameter		39	.410		19.50	21.50		41	55.50
1160	4" diameter		30	.533		34	28		62	81.50
1180	6" diameter		26	.615		58.50	32.50		91	115
9000	Minimum labor/equipment charge	1 Plum	4	2	Job		117		117	183

22 66 Chemical-Waste Systems for Lab. and Healthcare Facilities

22 66 53 – Laboratory Chemical-Waste and Vent Piping

22 66 53.40 Pipe Fittings, Glass		Crew	Daily Output	Labor-Hours	Unit	Material	2015 Bare Costs Labor	Equipment	Total	Total Incl O&P
0010	**PIPE FITTINGS, GLASS**									
0020	Drainage, beaded ends									
0040	Coupling & labor required at joints not incl. in fitting									
0050	price. Add 1 per joint for installed price									
0070	90° Bend or sweep, 1-1/2"				Ea.	31.50			31.50	35
0090	2"					40			40	44
0100	3"					66			66	72.50
0110	4"					105			105	116
0120	6" (sweep only)					330			330	360
0350	Tee, single sanitary, 1-1/2"					51			51	56.50
0370	2"					51			51	56.50
0380	3"					76.50			76.50	84
0390	4"					137			137	150
0400	6"					365			365	405
0500	Coupling, stainless steel, TFE seal ring									
0520	1-1/2"	Q-1	32	.500	Ea.	23.50	26.50		50	66.50
0530	2"		30	.533		29.50	28		57.50	76
0540	3"		25	.640		39.50	34		73.50	96
0550	4"		23	.696		67.50	37		104.50	132
0560	6"		20	.800		152	42.50		194.50	233
9000	Minimum labor/equipment charge	1 Plum	4	2	Job		117		117	183

22 66 53.60 Corrosion Resistant Pipe

		Crew	Daily Output	Labor-Hours	Unit	Material	Labor	Equipment	Total	Total Incl O&P
0010	**CORROSION RESISTANT PIPE**, No couplings or hangers									
0020	Iron alloy, drain, mechanical joint									
1000	1-1/2" diameter	Q-1	70	.229	L.F.	46	12.10		58.10	69.50
1100	2" diameter		66	.242		47	12.80		59.80	71.50
1120	3" diameter		60	.267		60.50	14.10		74.60	88.50
1140	4" diameter		52	.308		77.50	16.25		93.75	111
2980	Plastic, epoxy, fiberglass filament wound, B&S joint									
3000	2" diameter	Q-1	62	.258	L.F.	12	13.65		25.65	34.50
3100	3" diameter		51	.314		14	16.55		30.55	41.50
3120	4" diameter		45	.356		20	18.80		38.80	51.50
3160	8" diameter	Q-2	38	.632		44	34.50		78.50	103
3200	12" diameter	"	28	.857		72	47		119	153
9800	Minimum labor/equipment charge	1 Plum	4	2	Job		117		117	183

22 66 53.70 Pipe Fittings, Corrosion Resistant

		Crew	Daily Output	Labor-Hours	Unit	Material	Labor	Equipment	Total	Total Incl O&P
0010	**PIPE FITTINGS, CORROSION RESISTANT**									
0030	Iron alloy									
0050	Mechanical joint									
0060	1/4 Bend, 1-1/2"	Q-1	12	1.333	Ea.	77	70.50		147.50	195
0080	2"		10	1.600		126	84.50		210.50	271
0090	3"		9	1.778		151	94		245	310
0100	4"		8	2		174	106		280	355
0160	Tee and Y, sanitary, straight									
0170	1-1/2"	Q-1	8	2	Ea.	84	106		190	258
0180	2"		7	2.286		112	121		233	310
0190	3"		6	2.667		174	141		315	410
0200	4"		5	3.200		320	169		489	615
0360	Coupling, 1-1/2"		14	1.143		47.50	60.50		108	147
0380	2"		12	1.333		54	70.50		124.50	170
0390	3"		11	1.455		56.50	77		133.50	183
0400	4"		10	1.600		64.50	84.50		149	203

22 66 Chemical-Waste Systems for Lab. and Healthcare Facilities

22 66 53 – Laboratory Chemical-Waste and Vent Piping

22 66 53.70 Pipe Fittings, Corrosion Resistant		Crew	Daily Output	Labor-Hours	Unit	Material	2015 Bare Costs Labor	Equipment	Total	Total Incl O&P
3000	Epoxy, filament wound									
3030	Quick-lock joint									
3040	90° Elbow, 2"	Q-1	28	.571	Ea.	97.50	30		127.50	154
3060	3"		16	1		112	53		165	206
3070	4"		13	1.231		153	65		218	269
3190	Tee, 2"		19	.842		233	44.50		277.50	325
3200	3"		11	1.455		280	77		357	430
3210	4"		9	1.778		335	94		429	515
9000	Minimum labor/equipment charge	1 Plum	4	2	Job		117		117	183

Estimating Tips
The labor adjustment factors listed in Subdivision 22 01 02.20 also apply to Division 23.

23 10 00 Facility Fuel Systems
- The prices in this subdivision for above- and below-ground storage tanks do not include foundations or hold-down slabs, unless noted. The estimator should refer to Divisions 3 and 31 for foundation system pricing. In addition to the foundations, required tank accessories, such as tank gauges, leak detection devices, and additional manholes and piping, must be added to the tank prices.

23 50 00 Central Heating Equipment
- When estimating the cost of an HVAC system, check to see who is responsible for providing and installing the temperature control system. It is possible to overlook controls, assuming that they would be included in the electrical estimate.
- When looking up a boiler, be careful on specified capacity. Some manufacturers rate their products on output while others use input.
- Include HVAC insulation for pipe, boiler, and duct (wrap and liner).
- Be careful when looking up mechanical items to get the correct pressure rating and connection type (thread, weld, flange).

23 70 00 Central HVAC Equipment
- Combination heating and cooling units are sized by the air conditioning requirements. (See Reference No. R236000-20 for preliminary sizing guide.)
- A ton of air conditioning is nominally 400 CFM.
- Rectangular duct is taken off by the linear foot for each size, but its cost is usually estimated by the pound. Remember that SMACNA standards now base duct on internal pressure.
- Prefabricated duct is estimated and purchased like pipe: straight sections and fittings.
- Note that cranes or other lifting equipment are not included on any lines in Division 23. For example, if a crane is required to lift a heavy piece of pipe into place high above a gym floor, or to put a rooftop unit on the roof of a four-story building, etc., it must be added. Due to the potential for extreme variation—from nothing additional required to a major crane or helicopter—we feel that including a nominal amount for "lifting contingency" would be useless and detract from the accuracy of the estimate. When using equipment rental cost data from RSMeans, do not forget to include the cost of the operator(s).

Reference Numbers
Reference numbers are shown in shaded boxes at the beginning of some major classifications. These numbers refer to related items in the Reference Section. The reference information may be an estimating procedure, an alternate pricing method, or technical information.

Note: Not all subdivisions listed here necessarily appear in this publication. ■

---

*Note: **Trade Service**, in part, has been used as a reference source for some of the material prices used in Division 23.*

Did you know?
RSMeans Online gives you the same access to RSMeans' data with 24/7 access:
- Quickly locate costs in the searchable database.
- Build cost lists, estimates, and reports in minutes.
- Adjust costs to any location in the U.S. and Canada with the click of a button.

Start your free trial today at www.rsmeansonline.com

RSMeansOnline

23 05 Common Work Results for HVAC

23 05 05 – Selective Demolition for HVAC

23 05 05.10 HVAC Demolition

		Crew	Daily Output	Labor-Hours	Unit	Material	2015 Bare Costs Labor	2015 Bare Costs Equipment	Total	Total Incl O&P
0010	**HVAC DEMOLITION** R220105-10									
0100	Air conditioner, split unit, 3 ton	Q-5	2	8	Ea.		430		430	670
0150	Package unit, 3 ton R024119-10	Q-6	3	8	"		445		445	695
0298	Boilers									
0300	Electric, up thru 148 kW	Q-19	2	12	Ea.		650		650	1,000
0310	150 thru 518 kW	"	1	24			1,300		1,300	2,025
0320	550 thru 2000 kW	Q-21	.40	80			4,450		4,450	6,925
0330	2070 kW and up	"	.30	106			5,925		5,925	9,225
0340	Gas and/or oil, up thru 150 MBH	Q-7	2.20	14.545			825		825	1,300
0350	160 thru 2000 MBH		.80	40			2,275		2,275	3,550
0360	2100 thru 4500 MBH		.50	64			3,650		3,650	5,675
0370	4600 thru 7000 MBH		.30	106			6,075		6,075	9,475
0380	7100 thru 12,000 MBH		.16	200			11,400		11,400	17,700
0390	12,200 thru 25,000 MBH		.12	266			15,200		15,200	23,700
1000	Ductwork, 4" high, 8" wide	1 Clab	200	.040	L.F.		1.50		1.50	2.47
1020	10" wide		190	.042			1.58		1.58	2.60
1040	14" wide		180	.044			1.67		1.67	2.74
1100	6" high, 8" wide		165	.048			1.82		1.82	2.99
1120	12" wide		150	.053			2.01		2.01	3.29
1140	18" wide		135	.059			2.23		2.23	3.65
1200	10" high, 12" wide		125	.064			2.41		2.41	3.95
1220	18" wide		115	.070			2.62		2.62	4.29
1240	24" wide		110	.073			2.73		2.73	4.48
1300	12"-14" high, 16"-18" wide		85	.094			3.54		3.54	5.80
1320	24" wide		75	.107			4.01		4.01	6.60
1340	48" wide		71	.113			4.24		4.24	6.95
1400	18" high, 24" wide		67	.119			4.49		4.49	7.35
1420	36" wide		63	.127			4.77		4.77	7.85
1440	48" wide		59	.136			5.10		5.10	8.35
1500	30" high, 36" wide		56	.143			5.35		5.35	8.80
1520	48" wide		53	.151			5.70		5.70	9.30
1540	72" wide		50	.160			6		6	9.85
1550	Duct heater, electric strip	1 Elec	8	1	Ea.		54.50		54.50	84.50
1850	Minimum labor/equipment charge	1 Clab	3	2.667	Job		100		100	164
2200	Furnace, electric	Q-20	2	10	Ea.		510		510	805
2300	Gas or oil, under 120 MBH	Q-9	4	4			201		201	320
2340	Over 120 MBH	"	3	5.333			269		269	425
2800	Heat pump, package unit, 3 ton	Q-5	2.40	6.667			360		360	560
2840	Split unit, 3 ton		2	8			430		430	670
3000	Mechanical equipment, light items. Unit is weight, not cooling.		.90	17.778	Ton		955		955	1,500
3600	Heavy items		1.10	14.545	"		780		780	1,225
5090	Remove refrigerant from system	1 Stpi	40	.200	Lb.		11.95		11.95	18.65
9000	Minimum labor/equipment charge	Q-6	3	8	Job		445		445	695

23 05 23 – General-Duty Valves for HVAC Piping

23 05 23.30 Valves, Iron Body

		Crew	Daily Output	Labor-Hours	Unit	Material	Labor	Equipment	Total	Total Incl O&P
0010	**VALVES, IRON BODY**									
1020	Butterfly, wafer type, gear actuator, 200 lb.									
1030	2"	1 Plum	14	.571	Ea.	96	33.50		129.50	158
1060	4"	Q-1	5	3.200	"	113	169		282	390
1650	Gate, 125 lb., N.R.S.									
2150	Flanged									
2240	2-1/2"	Q-1	5	3.200	Ea.	735	169		904	1,075

23 05 Common Work Results for HVAC

23 05 23 – General-Duty Valves for HVAC Piping

23 05 23.30 Valves, Iron Body		Crew	Daily Output	Labor-Hours	Unit	Material	2015 Bare Costs Labor	Equipment	Total	Total Incl O&P
2260	3"	Q-1	4.50	3.556	Ea.	825	188		1,013	1,200
2280	4"	↓	3	5.333	↓	1,175	282		1,457	1,750
3550	OS&Y, 125 lb., flanged									
3680	4"	Q-1	3	5.333	Ea.	770	282		1,052	1,300
3690	5"	Q-2	3.40	7.059		1,250	385		1,635	1,975
3700	6"	"	3	8	↓	1,250	440		1,690	2,050
9000	Minimum labor/equipment charge	1 Plum	3	2.667	Job		157		157	244

23 07 HVAC Insulation

23 07 13 – Duct Insulation

23 07 13.10 Duct Thermal Insulation

			Crew	Daily Output	Labor-Hours	Unit	Material	Labor	Equipment	Total	Total Incl O&P
0010	**DUCT THERMAL INSULATION**										
0100	Rule of thumb, as a percentage of total mechanical costs					Job				10%	10%
0110	Insulation req'd. is based on the surface size/area to be covered										
3000	Ductwork										
3020	Blanket type, fiberglass, flexible										
3030	Fire rated for grease and hazardous exhaust ducts										
3060	1-1/2" thick		Q-14	300	.053	S.F.	4.54	2.51		7.05	9.05
3090	Fire rated for plenums										
3100	1/2" x24" x 25'		Q-14	7.20	2.222	Roll	167	105		272	350
3110	1/2" x24" x 25'			360	.044	S.F.	3.35	2.09		5.44	7.05
3120	1/2" x 48" x 25'			3.80	4.211	Roll	335	198		533	690
3126	1/2" x 48" x 25'		↓	380	.042	S.F.	3.35	1.98		5.33	6.85
3140	FSK vapor barrier wrap, .75 lb. density										
3160	1" thick	G	Q-14	350	.046	S.F.	.18	2.15		2.33	3.66
3170	1-1/2" thick	G		320	.050		.22	2.36		2.58	4.03
3180	2" thick	G		300	.053		.26	2.51		2.77	4.33
3190	3" thick	G		260	.062		.36	2.90		3.26	5.05
3200	4" thick	G	↓	242	.066	↓	.51	3.12		3.63	5.55
3210	Vinyl jacket, same as FSK										
9600	Minimum labor/equipment charge		1 Stpi	4	2	Job		120		120	186

23 07 16 – HVAC Equipment Insulation

23 07 16.10 HVAC Equipment Thermal Insulation

		Crew	Daily Output	Labor-Hours	Unit	Material	Labor	Equipment	Total	Total Incl O&P
0010	**HVAC EQUIPMENT THERMAL INSULATION**									
0100	Rule of thumb, as a percentage of total mechanical costs				Job				10%	10%
0110	Insulation req'd. is based on the surface size/area to be covered									
9610	Minimum labor/equipment charge	1 Stpi	4	2	Job		120		120	186

23 09 Instrumentation and Control for HVAC

23 09 53 – Pneumatic and Electric Control System for HVAC

23 09 53.10 Control Components

			Crew	Daily Output	Labor-Hours	Unit	Material	Labor	Equipment	Total	Total Incl O&P
0010	**CONTROL COMPONENTS**										
5000	Thermostats										
5030	Manual		1 Stpi	8	1	Ea.	51	60		111	149
5040	1 set back, electric, timed	G		8	1		37	60		97	134
5050	2 set back, electric, timed	G	↓	8	1		186	60		246	298
5200	24 hour, automatic, clock	G	1 Shee	8	1		165	56		221	271
6000	Valves, motorized zone										
6100	Sweat connections, 1/2" C x C		1 Stpi	20	.400	Ea.	168	24		192	222

23 09 Instrumentation and Control for HVAC

23 09 53 – Pneumatic and Electric Control System for HVAC

23 09 53.10 Control Components	Crew	Daily Output	Labor-Hours	Unit	Material	2015 Bare Costs Labor	Equipment	Total	Total Incl O&P
6110 3/4" C x C	1 Stpi	20	.400	Ea.	168	24		192	223
6120 1" C x C	↓	19	.421	↓	219	25		244	280
9000 Minimum labor/equipment charge	1 Plum	4	2	Job		117		117	183

23 21 Hydronic Piping and Pumps

23 21 20 – Hydronic HVAC Piping Specialties

23 21 20.34 Dielectric Unions

0010 **DIELECTRIC UNIONS**, Standard gaskets for water and air									
0020 250 psi maximum pressure									
0280 Female IPT to sweat, straight									
0340 3/4" pipe size	1 Plum	20	.400	Ea.	6.20	23.50		29.70	43.50
0780 Female IPT to female IPT, straight									
0800 1/2" pipe size	1 Plum	24	.333	Ea.	13.85	19.55		33.40	45.50
0840 3/4" pipe size		20	.400	"	15.60	23.50		39.10	53.50
9000 Minimum labor/equipment charge	↓	4	2	Job		117		117	183

23 21 20.46 Expansion Tanks

0010 **EXPANSION TANKS**									
2000 Steel, liquid expansion, ASME, painted, 15 gallon capacity	Q-5	17	.941	Ea.	640	50.50		690.50	780
2040 30 gallon capacity		12	1.333		715	71.50		786.50	895
2080 60 gallon capacity		8	2		1,000	108		1,108	1,275
2120 100 gallon capacity		6	2.667		1,450	143		1,593	1,825
3000 Steel ASME expansion, rubber diaphragm, 19 gal. cap. accept.		12	1.333		2,425	71.50		2,496.50	2,775
3020 31 gallon capacity		8	2		2,700	108		2,808	3,150
3040 61 gallon capacity		6	2.667		3,800	143		3,943	4,400
3080 119 gallon capacity		4	4	↓	4,100	215		4,315	4,825
9000 Minimum labor/equipment charge	↓	4	4	Job		215		215	335

23 21 20.70 Steam Traps

0010 **STEAM TRAPS**									
0030 Cast iron body, threaded									
0040 Inverted bucket									
0050 1/2" pipe size	1 Stpi	12	.667	Ea.	157	40		197	234
0100 1" pipe size		9	.889		420	53		473	550
0120 1-1/4" pipe size	↓	8	1	↓	635	60		695	795
1000 Float & thermostatic, 15 psi									
1010 3/4" pipe size	1 Stpi	16	.500	Ea.	141	30		171	202
1020 1" pipe size		15	.533		169	32		201	236
1040 1-1/2" pipe size		9	.889		298	53		351	415
1060 2" pipe size	↓	6	1.333	↓	590	79.50		669.50	775
9000 Minimum labor/equipment charge		4	2	Job		120		120	186

23 21 20.76 Strainers, Y Type, Bronze Body

0010 **STRAINERS, Y TYPE, BRONZE BODY**									
0050 Screwed, 125 lb., 1/4" pipe size	1 Stpi	24	.333	Ea.	23	19.90		42.90	56.50
0100 1/2" pipe size		20	.400		27.50	24		51.50	68
0120 3/4" pipe size		19	.421		34	25		59	76.50
0140 1" pipe size		17	.471		50	28		78	99
0160 1-1/2" pipe size		14	.571		108	34		142	173
0180 2" pipe size		13	.615		144	37		181	216
0182 3" pipe size	↓	12	.667		845	40		885	990
0200 300 lb., 2-1/2" pipe size	Q-5	17	.941		510	50.50		560.50	640
0220 3" pipe size	↓	16	1	↓	1,000	54		1,054	1,175

23 21 Hydronic Piping and Pumps

23 21 20 – Hydronic HVAC Piping Specialties

23 21 20.76 Strainers, Y Type, Bronze Body

		Crew	Daily Output	Labor-Hours	Unit	Material	2015 Bare Costs Labor	Equipment	Total	Total Incl O&P
0240	4" pipe size	Q-5	15	1.067	Ea.	2,300	57.50		2,357.50	2,625
1000	Flanged, 150 lb., 1-1/2" pipe size	1 Stpi	11	.727		520	43.50		563.50	645
1020	2" pipe size	"	8	1		705	60		765	870
1030	2-1/2" pipe size	Q-5	5	3.200		950	172		1,122	1,325
1040	3" pipe size		4.50	3.556		1,175	191		1,366	1,575
1060	4" pipe size	▼	3	5.333		1,775	287		2,062	2,400
1100	6" pipe size	Q-6	3	8		3,400	445		3,845	4,425
1106	8" pipe size	"	2.60	9.231	▼	3,725	515		4,240	4,900
1500	For 300 lb. rating, add					40%				
9000	Minimum labor/equipment charge	1 Stpi	3.75	2.133	Job		127		127	199

23 21 23 – Hydronic Pumps

23 21 23.13 In-Line Centrifugal Hydronic Pumps

		Crew	Daily Output	Labor-Hours	Unit	Material	Labor	Equipment	Total	Total Incl O&P
0010	**IN-LINE CENTRIFUGAL HYDRONIC PUMPS**									
0600	Bronze, sweat connections, 1/40 HP, in line									
0640	3/4" size	Q-1	16	1	Ea.	218	53		271	325
1000	Flange connection, 3/4" to 1-1/2" size									
1040	1/12 HP	Q-1	6	2.667	Ea.	565	141		706	840
1060	1/8 HP		6	2.667		970	141		1,111	1,300
1100	1/3 HP		6	2.667		1,075	141		1,216	1,425
1140	2" size, 1/6 HP		5	3.200		1,400	169		1,569	1,800
1180	2-1/2" size, 1/4 HP		5	3.200		1,775	169		1,944	2,225
1220	3" size, 1/4 HP		4	4	▼	1,850	211		2,061	2,375
9000	Minimum labor/equipment charge	▼	3.25	4.923	Job		260		260	405

23 23 Refrigerant Piping

23 23 16 – Refrigerant Piping Specialties

23 23 16.16 Refrigerant Line Sets

		Crew	Daily Output	Labor-Hours	Unit	Material	Labor	Equipment	Total	Total Incl O&P
0010	**REFRIGERANT LINE SETS**									
0100	Copper tube									
0110	1/2" insulation, both tubes									
0120	Combination 1/4" and 1/2" tubes									
0130	10' set	Q-5	42	.381	Ea.	70.50	20.50		91	110
0140	20' set		40	.400		110	21.50		131.50	155
0150	30' set		37	.432		148	23.50		171.50	199
0160	40' set		35	.457		193	24.50		217.50	252
0170	50' set		32	.500		229	27		256	294
0180	100' set	▼	22	.727	▼	520	39		559	635
0300	Combination 1/4" and 3/4" tubes									
0310	10' set	Q-5	40	.400	Ea.	78.50	21.50		100	120
0320	20' set		38	.421		139	22.50		161.50	189
0330	30' set		35	.457		208	24.50		232.50	268
0340	40' set		33	.485		272	26		298	340
0350	50' set		30	.533		345	28.50		373.50	420
0380	100' set	▼	20	.800	▼	780	43		823	925
0500	Combination 3/8" & 3/4" tubes									
0510	10' set	Q-5	28	.571	Ea.	90.50	30.50		121	148
0520	20' set		36	.444		141	24		165	193
0530	30' set		34	.471		191	25.50		216.50	251
0540	40' set		31	.516		252	28		280	320
0550	50' set	▼	28	.571	▼	296	30.50		326.50	375

23 23 Refrigerant Piping

23 23 16 – Refrigerant Piping Specialties

23 23 16.16 Refrigerant Line Sets		Crew	Daily Output	Labor-Hours	Unit	Material	2015 Bare Costs Labor	Equipment	Total	Total Incl O&P
0580	100' set	Q-5	18	.889	Ea.	875	48		923	1,025
0700	Combination 3/8" & 1-1/8" tubes									
0710	10' set	Q-5	36	.444	Ea.	158	24		182	212
0720	20' set		33	.485		265	26		291	330
0730	30' set		31	.516		355	28		383	435
0740	40' set		28	.571		535	30.50		565.50	635
0750	50' set	▼	26	.615	▼	615	33		648	730
0900	Combination 1/2" & 3/4" tubes									
0910	10' set	Q-5	37	.432	Ea.	95.50	23.50		119	142
0920	20' set		35	.457		169	24.50		193.50	225
0930	30' set		33	.485		255	26		281	320
0940	40' set		30	.533		335	28.50		363.50	415
0950	50' set		27	.593		420	32		452	515
0980	100' set	▼	17	.941	▼	950	50.50		1,000.50	1,125
2100	Combination 1/2" & 1-1/8" tubes									
2110	10' set	Q-5	35	.457	Ea.	163	24.50		187.50	218
2120	20' set		31	.516		279	28		307	350
2130	30' set		29	.552		420	29.50		449.50	505
2140	40' set		25	.640		565	34.50		599.50	675
2150	50' set	▼	14	1.143	▼	625	61.50		686.50	780
2300	For 1" thick insulation add					30%	15%			

23 31 HVAC Ducts and Casings

23 31 13 – Metal Ducts

23 31 13.13 Rectangular Metal Ducts

0010	**RECTANGULAR METAL DUCTS**									
0020	Fabricated rectangular, includes fittings, joints, supports,									
0021	allowance for flexible connections and field sketches.									
0030	Does not include "as-built dwgs." or insulation.									
0031	NOTE: Fabrication and installation are combined									
0040	as LABOR cost. Approx. 25% fittings assumed.									
0042	Fabrication/Inst. is to commercial quality standards									
0043	(SMACNA or equiv.) for structure, sealing, leak testing, etc.									
0100	Aluminum, alloy 3003-H14, under 100 lb.	Q-10	75	.320	Lb.	3.20	16.70		19.90	30
0110	100 to 500 lb.		80	.300		1.88	15.65		17.53	26.50
0120	500 to 1,000 lb.		95	.253		1.82	13.20		15.02	23
0160	Over 5,000 lb.		145	.166		1.77	8.65		10.42	15.60
0500	Galvanized steel, under 200 lb.		235	.102		.65	5.35		6	9.10
0520	200 to 500 lb.		245	.098		.64	5.10		5.74	8.75
0540	500 to 1,000 lb.		255	.094		.62	4.91		5.53	8.45
0560	1,000 to 2,000 lb.		265	.091		.61	4.73		5.34	8.10
0580	Over 5,000 lb.		285	.084		.61	4.40		5.01	7.60
1000	Stainless steel, type 304, under 100 lb.		165	.145		6.35	7.60		13.95	18.95
1020	100 to 500 lb.		175	.137		4.05	7.15		11.20	15.75
1030	500 to 1,000 lb.	▼	190	.126	▼	2.95	6.60		9.55	13.65

23 31 13.16 Round and Flat-Oval Spiral Ducts

0010	**ROUND AND FLAT-OVAL SPIRAL DUCTS**									
5400	Spiral preformed, steel, galv., straight lengths, Max 10" spwg.									
5410	4" diameter, 26 ga.	Q-9	360	.044	L.F.	1.70	2.24		3.94	5.40
5416	5" diameter, 26 ga.		320	.050		1.82	2.52		4.34	5.95
5420	6" diameter, 26 ga.	▼	280	.057		1.82	2.88		4.70	6.55

23 31 HVAC Ducts and Casings

23 31 13 – Metal Ducts

23 31 13.16 Round and Flat-Oval Spiral Ducts		Crew	Daily Output	Labor-Hours	Unit	Material	2015 Bare Costs Labor	Equipment	Total	Total Incl O&P
5425	7" diameter, 26 ga.	Q-9	240	.067	L.F.	2.12	3.36		5.48	7.65
5430	8" diameter, 26 ga.		200	.080		2.42	4.03		6.45	9
5440	10" diameter, 26 ga.		160	.100		3.02	5.05		8.07	11.25
9990	Minimum labor/equipment charge	1 Shee	3	2.667	Job		149		149	235

23 31 16 – Nonmetal Ducts

23 31 16.13 Fibrous-Glass Ducts

		Crew	Daily Output	Labor-Hours	Unit	Material	Labor	Equipment	Total	Total Incl O&P
0010	**FIBROUS-GLASS DUCTS**									
3490	Rigid fiberglass duct board, foil reinf. kraft facing									
3500	Rectangular, 1" thick, alum. faced, (FRK), std. weight	Q-10	350	.069	SF Surf	.79	3.58		4.37	6.50
9990	Minimum labor/equipment charge	1 Shee	3	2.667	Job		149		149	235

23 31 16.19 PVC Ducts

		Crew	Daily Output	Labor-Hours	Unit	Material	Labor	Equipment	Total	Total Incl O&P
0010	**PVC DUCTS**									
4000	Rigid plastic, corrosive fume resistant PVC									
4020	Straight, 6" diameter	Q-9	220	.073	L.F.	11.55	3.66		15.21	18.50
4040	8" diameter		160	.100		16.75	5.05		21.80	26.50
4060	10" diameter		120	.133		21	6.70		27.70	33.50
4070	12" diameter		100	.160		24.50	8.05		32.55	39.50
4080	14" diameter		70	.229		30.50	11.50		42	51.50
4090	16" diameter		62	.258		35.50	13		48.50	59.50
4100	18" diameter		58	.276		48.50	13.90		62.40	75.50
4110	20" diameter	Q-10	75	.320		59.50	16.70		76.20	92
4130	24" diameter	"	55	.436		62	23		85	105
4250	Coupling, 6" diameter	Q-9	88	.182	Ea.	23.50	9.15		32.65	40.50
4270	8" diameter		78	.205		30	10.35		40.35	49.50
4290	10" diameter		70	.229		35.50	11.50		47	57
4300	12" diameter		55	.291		38.50	14.65		53.15	65.50
4310	14" diameter		44	.364		50	18.30		68.30	84
4320	16" diameter		40	.400		52.50	20		72.50	90
4330	18" diameter		38	.421		68	21		89	109
4340	20" diameter	Q-10	55	.436		82	23		105	126
4360	24" diameter	"	45	.533		103	28		131	157
4470	Elbow, 90°, 6" diameter	Q-9	44	.364		108	18.30		126.30	147
4490	8" diameter		28	.571		121	29		150	179
4510	10" diameter		18	.889		147	45		192	232
4520	12" diameter		15	1.067		167	53.50		220.50	268
4540	16" diameter		10	1.600		251	80.50		331.50	405
4550	18" diameter	Q-10	15	1.600		350	83.50		433.50	510
4560	20" diameter		14	1.714		520	89.50		609.50	710
4580	24" diameter		12	2		640	104		744	865
4750	Elbow 45°, use 90° and deduct					25%				
9990	Minimum labor/equipment charge	1 Shee	3	2.667	Job		149		149	235

23 33 Air Duct Accessories

23 33 13 – Dampers

23 33 13.13 Volume-Control Dampers		Crew	Daily Output	Labor-Hours	Unit	Material	2015 Bare Costs Labor	Equipment	Total	Total Incl O&P
0010	**VOLUME-CONTROL DAMPERS**									
5990	Multi-blade dampers, opposed blade, 8" x 6"	1 Shee	24	.333	Ea.	21.50	18.65		40.15	53.50
5994	8" x 8"		22	.364		22.50	20.50		43	56.50
5996	10" x 10"		21	.381		26.50	21.50		48	63
6000	12" x 12"		21	.381		29.50	21.50		51	66
6020	12" x 18"		18	.444		39.50	25		64.50	82.50
6030	14" x 10"		20	.400		29	22.50		51.50	67
6031	14" x 14"		17	.471		35.50	26.50		62	80.50
6033	16" x 12"		17	.471		35.50	26.50		62	80.50
6035	16" x 16"		16	.500		44	28		72	92.50
6037	18" x 16"		15	.533		48	30		78	100
6038	18" x 18"		15	.533		52	30		82	104
6070	20" x 16"		14	.571		52	32		84	108
6072	20" x 20"		13	.615		62.50	34.50		97	124
6074	22" x 18"		14	.571		62.50	32		94.50	120
6076	24" x 16"		11	.727		61	40.50		101.50	132
6078	24" x 20"		8	1		72.50	56		128.50	169
6080	24" x 24"		8	1		85	56		141	182
6110	26" x 26"		6	1.333		95	74.50		169.50	223
6133	30" x 30"	Q-9	6.60	2.424		137	122		259	345
6135	32" x 32"		6.40	2.500		154	126		280	370
6180	48" x 36"		5.60	2.857		253	144		397	505
7500	Variable volume modulating motorized damper, incl. elect. mtr.									
7504	8" x 6"	1 Shee	15	.533	Ea.	118	30		148	177
7506	10" x 6"		14	.571		118	32		150	181
7510	10" x 10"		13	.615		122	34.50		156.50	189
7520	12" x 12"		12	.667		127	37.50		164.50	198
7522	12" x 16"		11	.727		127	40.50		167.50	204
7524	16" x 10"		12	.667		125	37.50		162.50	196
7526	16" x 14"		10	.800		130	45		175	214
7528	16" x 18"		9	.889		135	49.50		184.50	227
7542	18" x 18"		8	1		138	56		194	240
7544	20" x 14"		8	1		141	56		197	244
7546	20" x 18"		7	1.143		147	64		211	263
7560	24" x 12"		8	1		143	56		199	246
7562	24" x 18"		7	1.143		154	64		218	270
7564	24" x 24"		6	1.333		160	74.50		234.50	294
7568	28" x 10"		7	1.143		143	64		207	258
7590	30" x 14"		5	1.600		151	89.50		240.50	305
7600	30" x 18"		4	2		198	112		310	395
7610	30" x 24"		3.80	2.105		264	118		382	475
7700	For thermostat, add		8	1		31	56		87	123
8000	Multi-blade dampers, parallel blade									
8100	8" x 8"	1 Shee	24	.333	Ea.	82	18.65		100.65	120
8140	16" x 10"		20	.400		103	22.50		125.50	150
8160	18" x 12"		18	.444		113	25		138	163
8220	28" x 16"		10	.800		150	45		195	236

23 33 13.16 Fire Dampers

		Crew	Daily Output	Labor-Hours	Unit	Material	Labor	Equipment	Total	Total Incl O&P
0010	**FIRE DAMPERS**									
3000	Fire damper, curtain type, 1-1/2 hr. rated, vertical, 6" x 6"	1 Shee	24	.333	Ea.	24	18.65		42.65	56
3020	8" x 6"		22	.364		24	20.50		44.50	58.50
3240	16" x 14"		18	.444		44	25		69	87.50

23 33 Air Duct Accessories

23 33 13 – Dampers

23 33 13.16 Fire Dampers		Crew	Daily Output	Labor-Hours	Unit	Material	2015 Bare Costs Labor	Equipment	Total	Total Incl O&P
3400	24" x 20"	1 Shee	8	1	Ea.	54	56		110	148

23 33 19 – Duct Silencers

23 33 19.10 Duct Silencers

		Crew	Daily Output	Labor-Hours	Unit	Material	Labor	Equipment	Total	Total Incl O&P
0010	**DUCT SILENCERS**									
9000	Silencers, noise control for air flow, duct				MCFM	57			57	63

23 33 23 – Turning Vanes

23 33 23.13 Air Turning Vanes

		Crew	Daily Output	Labor-Hours	Unit	Material	Labor	Equipment	Total	Total Incl O&P
0010	**AIR TURNING VANES**									
9400	Turning vane components									
9410	Turning vane rail	1 Shee	160	.050	L.F.	.75	2.80		3.55	5.25
9420	Double thick, factory fab. vane		300	.027		1.17	1.49		2.66	3.65
9428	12" high set		170	.047		1.68	2.63		4.31	6
9432	14" high set		160	.050		1.96	2.80		4.76	6.60
9900	Minimum labor/equipment charge		4	2	Job		112		112	177

23 33 46 – Flexible Ducts

23 33 46.10 Flexible Air Ducts

		Crew	Daily Output	Labor-Hours	Unit	Material	Labor	Equipment	Total	Total Incl O&P
0010	**FLEXIBLE AIR DUCTS**									
1280	Add to labor for elevated installation									
1282	of prefabricated (purchased) ductwork									
1283	10' to 15' high						10%			
1284	15' to 20' high						20%			
1285	20' to 25' high						25%			
1286	25' to 30' high						35%			
1287	30' to 35' high						40%			
1288	35' to 40' high						50%			
1289	Over 40' high						55%			
1300	Flexible, coated fiberglass fabric on corr. resist. metal helix									
1400	pressure to 12" (WG) UL-181									
1500	Noninsulated, 3" diameter	Q-9	400	.040	L.F.	1.12	2.01		3.13	4.41
1540	5" diameter		320	.050		1.30	2.52		3.82	5.40
1560	6" diameter		280	.057		1.50	2.88		4.38	6.20
1580	7" diameter		240	.067		1.53	3.36		4.89	7
1600	8" diameter		200	.080		1.91	4.03		5.94	8.45
1640	10" diameter		160	.100		2.46	5.05		7.51	10.65
1660	12" diameter		120	.133		2.94	6.70		9.64	13.85
1900	Insulated, 1" thick, PE jacket, 3" diameter [G]		380	.042		2.60	2.12		4.72	6.20
1910	4" diameter [G]		340	.047		2.60	2.37		4.97	6.60
1920	5" diameter [G]		300	.053		2.60	2.69		5.29	7.10
1940	6" diameter [G]		260	.062		2.94	3.10		6.04	8.10
1960	7" diameter [G]		220	.073		3.20	3.66		6.86	9.30
1980	8" diameter [G]		180	.089		3.49	4.48		7.97	10.90
2020	10" diameter [G]		140	.114		4.25	5.75		10	13.80
2040	12" diameter [G]		100	.160		4.90	8.05		12.95	18.10
2060	14" diameter [G]		80	.200		6.05	10.05		16.10	22.50
9990	Minimum labor/equipment charge	1 Shee	3	2.667	Job		149		149	235

For customer support on your Commercial Renovation Cost Data, call 877.791.4977.

23 33 Air Duct Accessories

23 33 53 – Duct Liners

23 33 53.10 Duct Liner Board

			Crew	Daily Output	Labor-Hours	Unit	Material	2015 Bare Costs Labor	Equipment	Total	Total Incl O&P
0010	**DUCT LINER BOARD**										
3490	Board type, fiberglass liner, 3 lb. density										
3680	No finish										
3700	1" thick	G	Q-14	170	.094	S.F.	.44	4.44		4.88	7.65
3710	1-1/2" thick	G		140	.114		.66	5.40		6.06	9.40
3720	2" thick	G		130	.123		.88	5.80		6.68	10.25
3940	Board type, non-fibrous foam										
3950	Temperature, bacteria and fungi resistant										
3960	1" thick	G	Q-14	150	.107	S.F.	2.38	5.05		7.43	10.70
3970	1-1/2" thick	G		130	.123		3.28	5.80		9.08	12.90
3980	2" thick	G		120	.133		3.98	6.30		10.28	14.50

23 34 HVAC Fans

23 34 13 – Axial HVAC Fans

23 34 13.10 Axial Flow HVAC Fans

		Crew	Daily Output	Labor-Hours	Unit	Material	Labor	Equipment	Total	Total Incl O&P
0010	**AXIAL FLOW HVAC FANS**									
0020	Air conditioning and process air handling									

23 34 14 – Blower HVAC Fans

23 34 14.10 Blower Type HVAC Fans

		Crew	Daily Output	Labor-Hours	Unit	Material	Labor	Equipment	Total	Total Incl O&P
0010	**BLOWER TYPE HVAC FANS**									
2500	Ceiling fan, right angle, extra quiet, 0.10" S.P.									
2540	210 CFM	Q-20	19	1.053	Ea.	355	54		409	475
2580	885 CFM		16	1.250		890	64		954	1,075
2620	2,960 CFM		11	1.818		1,650	93		1,743	1,950
2640	For wall or roof cap, add	1 Shee	16	.500		300	28		328	375

23 34 16 – Centrifugal HVAC Fans

23 34 16.10 Centrifugal Type HVAC Fans

		Crew	Daily Output	Labor-Hours	Unit	Material	Labor	Equipment	Total	Total Incl O&P
0010	**CENTRIFUGAL TYPE HVAC FANS**									
0200	In-line centrifugal, supply/exhaust booster									
0220	aluminum wheel/hub, disconnect switch, 1/4" S.P.									
0240	500 CFM, 10" diameter connection	Q-20	3	6.667	Ea.	1,300	340		1,640	1,950
0280	1,520 CFM, 16" diameter connection		2	10		1,500	510		2,010	2,450
0320	3,480 CFM, 20" diameter connection		.80	25		1,925	1,275		3,200	4,150
0326	5,080 CFM, 20" diameter connection		.75	26.667		2,100	1,375		3,475	4,475
5000	Utility set, centrifugal, V belt drive, motor									
5020	1/4" S.P., 1200 CFM, 1/4 HP	Q-20	6	3.333	Ea.	1,750	171		1,921	2,200
5040	1520 CFM, 1/3 HP		5	4		2,225	205		2,430	2,775
5060	1850 CFM, 1/2 HP		4	5		2,200	256		2,456	2,825
5080	2180 CFM, 3/4 HP		3	6.667		2,600	340		2,940	3,400
7000	Roof exhauster, centrifugal, aluminum housing, 12" galvanized									
7020	curb, bird screen, back draft damper, 1/4" S.P.									
7100	Direct drive, 320 CFM, 11" sq. damper	Q-20	7	2.857	Ea.	705	146		851	1,000
7120	600 CFM, 11" sq. damper	"	6	3.333	"	900	171		1,071	1,250
9900	Minimum labor/equipment charge	1 Elec	4	2	Job		109		109	169

23 34 HVAC Fans

23 34 23 – HVAC Power Ventilators

23 34 23.10 HVAC Power Circulators and Ventilators

		Crew	Daily Output	Labor-Hours	Unit	Material	2015 Bare Costs Labor	2015 Bare Costs Equipment	Total	Total Incl O&P
0010	**HVAC POWER CIRCULATORS AND VENTILATORS**									
6650	Residential, bath exhaust, grille, back draft damper									
6660	50 CFM	Q-20	24	.833	Ea.	63.50	42.50		106	137
6670	110 CFM		22	.909		98	46.50		144.50	181
6680	Light combination, squirrel cage, 100 watt, 70 CFM		24	.833		112	42.50		154.50	190
6700	Light/heater combination, ceiling mounted									
6710	70 CFM, 1450 watt	Q-20	24	.833	Ea.	162	42.50		204.50	245
6800	Heater combination, recessed, 70 CFM		24	.833		67.50	42.50		110	141
6820	With 2 infrared bulbs		23	.870		105	44.50		149.50	186
6900	Kitchen exhaust, grille, complete, 160 CFM		22	.909		108	46.50		154.50	192
6910	180 CFM		20	1		90.50	51		141.50	180
6920	270 CFM		18	1.111		171	57		228	278
6940	Residential roof jacks and wall caps									
6944	Wall cap with back draft damper									
6946	3" & 4" diam. round duct	1 Shee	11	.727	Ea.	26	40.50		66.50	92.50
6948	6" diam. round duct	"	11	.727	"	64	40.50		104.50	135
6958	Roof jack with bird screen and back draft damper									
6960	3" & 4" diam. round duct	1 Shee	11	.727	Ea.	26	40.50		66.50	92.50
6962	3-1/4" x 10" rectangular duct	"	10	.800	"	48.50	45		93.50	124
6980	Transition									
6982	3-1/4" x 10" to 6" diam. round	1 Shee	20	.400	Ea.	32	22.50		54.50	71
8020	Attic, roof type									
8030	Aluminum dome, damper & curb									
8080	12" diameter, 1000 CFM (gravity)	1 Elec	10	.800	Ea.	575	44		619	700
8090	16" diameter, 1500 CFM (gravity)		9	.889		690	48.50		738.50	835
8100	20" diameter, 2500 CFM (gravity)		8	1		850	54.50		904.50	1,025
8110	26" diameter, 4000 CFM (gravity)		7	1.143		1,025	62.50		1,087.50	1,225
8120	32" diameter, 6500 CFM (gravity)		6	1.333		1,400	73		1,473	1,675
8130	38" diameter, 8000 CFM (gravity)		5	1.600		2,100	87.50		2,187.50	2,425
8140	50" diameter, 13,000 CFM (gravity)		4	2		3,025	109		3,134	3,500
8160	Plastic, ABS dome									
8180	1050 CFM	1 Elec	14	.571	Ea.	168	31.50		199.50	234
8200	1600 CFM	"	12	.667	"	252	36.50		288.50	335
8240	Attic, wall type, with shutter, one speed									
8250	12" diameter, 1000 CFM	1 Elec	14	.571	Ea.	395	31.50		426.50	485
8260	14" diameter, 1500 CFM		12	.667		430	36.50		466.50	530
8270	16" diameter, 2000 CFM		9	.889		485	48.50		533.50	610
8290	Whole house, wall type, with shutter, one speed									
8300	30" diameter, 4800 CFM	1 Elec	7	1.143	Ea.	1,050	62.50		1,112.50	1,250
8310	36" diameter, 7000 CFM		6	1.333		1,125	73		1,198	1,375
8320	42" diameter, 10,000 CFM		5	1.600		1,275	87.50		1,362.50	1,525
8330	48" diameter, 16,000 CFM		4	2		1,575	109		1,684	1,925
8340	For two speed, add					95			95	105
8350	Whole house, lay-down type, with shutter, one speed									
8360	30" diameter, 4500 CFM	1 Elec	8	1	Ea.	1,100	54.50		1,154.50	1,300
8370	36" diameter, 6500 CFM		7	1.143		1,200	62.50		1,262.50	1,400
8380	42" diameter, 9000 CFM		6	1.333		1,300	73		1,373	1,575
8390	48" diameter, 12,000 CFM		5	1.600		1,500	87.50		1,587.50	1,750
8440	For two speed, add					71.50			71.50	78.50
8450	For 12 hour timer switch, add	1 Elec	32	.250		71.50	13.70		85.20	99.50

23 36 Air Terminal Units

23 36 13 – Constant-Air-Volume Units

23 36 13.10 Constant Volume Mixing Boxes	Crew	Daily Output	Labor-Hours	Unit	Material	2015 Bare Costs Labor	Equipment	Total	Total Incl O&P
0010 **CONSTANT VOLUME MIXING BOXES**									
5180 Mixing box, includes electric or pneumatic motor									
5200 Constant volume, 150 to 270 CFM	Q-9	12	1.333	Ea.	700	67		767	875
5210 270 to 600 CFM	"	11	1.455	"	720	73		793	910

23 36 16 – Variable-Air-Volume Units

23 36 16.10 Variable Volume Mixing Boxes

	Crew	Daily Output	Labor-Hours	Unit	Material	Labor	Equipment	Total	Total Incl O&P
0010 **VARIABLE VOLUME MIXING BOXES**									
5180 Mixing box, includes electric or pneumatic motor									
5500 VAV Cool only, pneumatic, pressure independent 300 to 600 CFM	Q-9	11	1.455	Ea.	730	73		803	920
5510 500 to 1000 CFM		9	1.778		750	89.50		839.50	965
5520 800 to 1600 CFM		9	1.778		775	89.50		864.50	995
5530 1100 to 2000 CFM		8	2		795	101		896	1,025
5540 1500 to 3000 CFM		7	2.286		845	115		960	1,100
5550 2000 to 4000 CFM	▼	6	2.667		855	134		989	1,150
5560 For electric, w/thermostat, pressure dependent, add				▼	23			23	25.50

23 37 Air Outlets and Inlets

23 37 13 – Diffusers, Registers, and Grilles

23 37 13.10 Diffusers

	Crew	Daily Output	Labor-Hours	Unit	Material	Labor	Equipment	Total	Total Incl O&P
0010 **DIFFUSERS**, Aluminum, opposed blade damper unless noted									
0100 Ceiling, linear, also for sidewall									
0120 2" wide	1 Shee	32	.250	L.F.	16.20	14		30.20	40
0160 4" wide		26	.308		21.50	17.20		38.70	50.50
0180 6" wide		24	.333		26.50	18.65		45.15	59
0200 8" wide		22	.364	▼	31	20.50		51.50	66.50
0500 Perforated, 24" x 24" lay-in panel size, 6" x 6"		16	.500	Ea.	151	28		179	210
0520 8" x 8"		15	.533		159	30		189	222
0530 9" x 9"		14	.571		161	32		193	228
0540 10" x 10"		14	.571		162	32		194	229
0560 12" x 12"		12	.667		168	37.50		205.50	244
0590 16" x 16"		11	.727		189	40.50		229.50	271
0600 18" x 18"		10	.800		202	45		247	293
0610 20" x 20"		10	.800		218	45		263	310
0620 24" x 24"		9	.889		239	49.50		288.50	340
1000 Rectangular, 1 to 4 way blow, 6" x 6"		16	.500		50	28		78	99
1010 8" x 8"		15	.533		58	30		88	111
1014 9" x 9"		15	.533		66	30		96	120
1016 10" x 10"		15	.533		80	30		110	135
1020 12" x 6"		15	.533		71.50	30		101.50	126
1040 12" x 9"		14	.571		75.50	32		107.50	134
1060 12" x 12"		12	.667		84.50	37.50		122	152
1070 14" x 6"		13	.615		77.50	34.50		112	140
1074 14" x 14"		12	.667		128	37.50		165.50	199
1150 18" x 18"		9	.889		138	49.50		187.50	230
1170 24" x 12"		10	.800		163	45		208	251
2000 T bar mounting, 24" x 24" lay-in frame, 6" x 6"		16	.500		61	28		89	111
2020 8" x 8"		14	.571		62	32		94	119
2040 12" x 12"		12	.667		74	37.50		111.50	140
2060 16" x 16"		11	.727		94.50	40.50		135	168
2080 18" x 18"	▼	10	.800		106	45		151	188

23 37 Air Outlets and Inlets

23 37 13 – Diffusers, Registers, and Grilles

23 37 13.10 Diffusers

		Crew	Daily Output	Labor-Hours	Unit	Material	2015 Bare Costs Labor	2015 Bare Costs Equipment	Total	Total Incl O&P
6000	For steel diffusers instead of aluminum, deduct				Ea.	10%				
9000	Minimum labor/equipment charge	1 Shee	4	2	Job		112		112	177

23 37 13.30 Grilles

		Crew	Daily Output	Labor-Hours	Unit	Material	Labor	Equipment	Total	Total Incl O&P
0010	**GRILLES**									
0020	Aluminum, unless noted otherwise									
1000	Air return, steel, 6" x 6"	1 Shee	26	.308	Ea.	18.70	17.20		35.90	47.50
1020	10" x 6"		24	.333		18.70	18.65		37.35	50
1080	16" x 8"		22	.364		26.50	20.50		47	61
1100	12" x 12"		22	.364		26.50	20.50		47	61
1120	24" x 12"		18	.444		36	25		61	78.50
1300	48" x 24"		12	.667		128	37.50		165.50	199
9000	Minimum labor/equipment charge		4	2	Job		112		112	177

23 37 13.60 Registers

		Crew	Daily Output	Labor-Hours	Unit	Material	Labor	Equipment	Total	Total Incl O&P
0010	**REGISTERS**									
0980	Air supply									
1000	Ceiling/wall, O.B. damper, anodized aluminum									
1010	One or two way deflection, adj. curved face bars									
1120	12" x 12"	1 Shee	18	.444	Ea.	21	25		46	62.50
1140	14" x 8"		17	.471		18.40	26.50		44.90	62
1240	20" x 6"		18	.444		18.90	25		43.90	60
1340	24" x 8"		13	.615		25.50	34.50		60	82.50
1350	24" x 18"		12	.667		47.50	37.50		85	111
2700	Above registers in steel instead of aluminum, deduct					10%				
3000	Baseboard, hand adj. damper, enameled steel									
3012	8" x 6"	1 Shee	26	.308	Ea.	4.85	17.20		22.05	32.50
3020	10" x 6"		24	.333		5.30	18.65		23.95	35.50
3040	12" x 5"		23	.348		5.60	19.45		25.05	36.50
3060	12" x 6"		23	.348		5.75	19.45		25.20	37
3080	12" x 8"		22	.364		8.30	20.50		28.80	41
3100	14" x 6"		20	.400		6.20	22.50		28.70	42.50
9000	Minimum labor/equipment charge		4	2	Job		112		112	177

23 37 23 – HVAC Gravity Ventilators

23 37 23.10 HVAC Gravity Air Ventilators

		Crew	Daily Output	Labor-Hours	Unit	Material	Labor	Equipment	Total	Total Incl O&P
0010	**HVAC GRAVITY AIR VENTILATORS**, Includes base									
1280	Rotary ventilators, wind driven, galvanized									
1300	4" neck diameter	Q-9	20	.800	Ea.	64.50	40.50		105	135
1400	12" neck diameter		10	1.600		91	80.50		171.50	227
1500	24" neck diameter		8	2		390	101		491	585
1600	For aluminum, add					300%				
9000	Minimum labor/equipment charge	1 Plum	2	4	Job		235		235	365

23 38 Ventilation Hoods

23 38 13 – Commercial-Kitchen Hoods

23 38 13.10 Hood and Ventilation Equipment

	Crew	Daily Output	Labor-Hours	Unit	Material	2015 Bare Costs Labor	Equipment	Total	Total Incl O&P
0010 HOOD AND VENTILATION EQUIPMENT									
2970 Exhaust hood, sst, gutter on all sides, 4' x 4' x 2'	1 Carp	1.80	4.444	Ea.	4,725	209		4,934	5,550
2980 4' x 4' x 7'	"	1.60	5	"	7,525	235		7,760	8,650

23 41 Particulate Air Filtration

23 41 13 – Panel Air Filters

23 41 13.10 Panel Type Air Filters

	Crew	Daily Output	Labor-Hours	Unit	Material	2015 Bare Costs Labor	Equipment	Total	Total Incl O&P
0010 PANEL TYPE AIR FILTERS									
2950 Mechanical media filtration units									
3000 High efficiency type, with frame, non-supported G				MCFM	35			35	38.50
3100 Supported type G				"	45			45	49.50
5500 Throwaway glass or paper media type				Ea.	3.42			3.42	3.76

23 41 16 – Renewable-Media Air Filters

23 41 16.10 Disposable Media Air Filters

	Crew	Daily Output	Labor-Hours	Unit	Material	Labor	Equipment	Total	Total Incl O&P
0010 DISPOSABLE MEDIA AIR FILTERS									
5000 Renewable disposable roll				C.S.F.	1.54			1.54	1.70

23 41 19 – Washable Air Filters

23 41 19.10 Permanent Air Filters

	Crew	Daily Output	Labor-Hours	Unit	Material	Labor	Equipment	Total	Total Incl O&P
0010 PERMANENT AIR FILTERS									
4500 Permanent washable G				MCFM	20			20	22

23 41 23 – Extended Surface Filters

23 41 23.10 Expanded Surface Filters

	Crew	Daily Output	Labor-Hours	Unit	Material	Labor	Equipment	Total	Total Incl O&P
0010 EXPANDED SURFACE FILTERS									
4000 Medium efficiency, extended surface G				MCFM	5.50			5.50	6.05

23 42 Gas-Phase Air Filtration

23 42 13 – Activated-Carbon Air Filtration

23 42 13.10 Charcoal Type Air Filtration

	Crew	Daily Output	Labor-Hours	Unit	Material	Labor	Equipment	Total	Total Incl O&P
0010 CHARCOAL TYPE AIR FILTRATION									
0050 Activated charcoal type, full flow				MCFM	600			600	660
0060 Full flow, impregnated media 12" deep					225			225	248
0070 HEPA filter & frame for field erection					350			350	385
0080 HEPA filter-diffuser, ceiling install.				↓	300			300	330

23 43 Electronic Air Cleaners

23 43 13 – Washable Electronic Air Cleaners

23 43 13.10 Electronic Air Cleaners

	Crew	Daily Output	Labor-Hours	Unit	Material	Labor	Equipment	Total	Total Incl O&P
0010 ELECTRONIC AIR CLEANERS									
2000 Electronic air cleaner, duct mounted									
2150 1000 CFM	1 Shee	4	2	Ea.	420	112		532	640
2200 1200 CFM	↓	3.80	2.105		505	118		623	740
2250 1400 CFM		3.60	2.222	↓	520	124		644	770

23 51 Breechings, Chimneys, and Stacks

23 51 13 – Draft Control Devices

23 51 13.16 Vent Dampers

		Crew	Daily Output	Labor-Hours	Unit	Material	2015 Bare Costs Labor	Equipment	Total	Total Incl O&P
0010	**VENT DAMPERS**									
5000	Vent damper, bi-metal, gas, 3" diameter	Q-9	24	.667	Ea.	44	33.50		77.50	102
5010	4" diameter		24	.667		44	33.50		77.50	102
5020	5" diameter		23	.696		44	35		79	104
5030	6" diameter		22	.727		44	36.50		80.50	107
5040	7" diameter		21	.762		45.50	38.50		84	111
5050	8" diameter		20	.800		49	40.50		89.50	118
9000	Minimum labor/equipment charge	1 Shee	4	2	Job		112		112	177

23 51 13.19 Barometric Dampers

		Crew	Daily Output	Labor-Hours	Unit	Material	2015 Bare Costs Labor	Equipment	Total	Total Incl O&P
0010	**BAROMETRIC DAMPERS**									
1000	Barometric, gas fired system only, 6" size for 5" and 6" pipes	1 Shee	20	.400	Ea.	105	22.50		127.50	151
1040	8" size, for 7" and 8" pipes	"	18	.444	"	144	25		169	198
2000	All fuel, oil, oil/gas, coal									
2020	10" for 9" and 10" pipes	1 Shee	15	.533	Ea.	242	30		272	315
3260	For thermal switch for above, add	"	24	.333	"	100	18.65		118.65	140

23 51 23 – Gas Vents

23 51 23.10 Gas Chimney Vents

		Crew	Daily Output	Labor-Hours	Unit	Material	2015 Bare Costs Labor	Equipment	Total	Total Incl O&P
0010	**GAS CHIMNEY VENTS**, Prefab metal, U.L. listed									
0020	Gas, double wall, galvanized steel									
0080	3" diameter	Q-9	72	.222	V.L.F.	5.55	11.20		16.75	24
0100	4" diameter		68	.235		7.20	11.85		19.05	26.50
0120	5" diameter		64	.250		7.80	12.60		20.40	28.50
0140	6" diameter		60	.267		9.45	13.45		22.90	31.50
0160	7" diameter		56	.286		15.10	14.40		29.50	39
0180	8" diameter		52	.308		17.40	15.50		32.90	43.50
0200	10" diameter		48	.333		36	16.80		52.80	66.50
0220	12" diameter		44	.364		43	18.30		61.30	76.50
0260	16" diameter		40	.400		103	20		123	146
0300	20" diameter	Q-10	36	.667		150	35		185	220
0340	24" diameter	"	32	.750		236	39		275	320

23 51 26 – All-Fuel Vent Chimneys

23 51 26.30 All-Fuel Vent Chimneys, Double Wall, St. Stl.

		Crew	Daily Output	Labor-Hours	Unit	Material	2015 Bare Costs Labor	Equipment	Total	Total Incl O&P
0010	**ALL-FUEL VENT CHIMNEYS, DOUBLE WALL, STAINLESS STEEL**									
7780	All fuel, pressure tight, double wall, 4" insulation, U.L. listed, 1400°F.									
7790	304 stainless steel liner, aluminized steel outer jacket									
7800	6" diameter	Q-9	60	.267	V.L.F.	64	13.45		77.45	91
7804	8" diameter		52	.308		73	15.50		88.50	105
7806	10" diameter		48	.333		81.50	16.80		98.30	117
7808	12" diameter		44	.364		93.50	18.30		111.80	132
7810	14" diameter		42	.381		105	19.20		124.20	146
7880	For 316 stainless steel liner add				L.F.	30%				
8000	All fuel, double wall, stainless steel fittings									
8010	Roof support 6" diameter	Q-9	30	.533	Ea.	113	27		140	167
8030	8" diameter		26	.615		132	31		163	194
8040	10" diameter		24	.667		139	33.50		172.50	206
8050	12" diameter		22	.727		148	36.50		184.50	220
8060	14" diameter		21	.762		155	38.50		193.50	232
8100	Elbow 45°, 6" diameter		30	.533		241	27		268	310
8140	8" diameter		26	.615		271	31		302	345
8160	10" diameter		24	.667		310	33.50		343.50	395
8180	12" diameter		22	.727		350	36.50		386.50	445

23 51 Breechings, Chimneys, and Stacks

23 51 26 – All-Fuel Vent Chimneys

23 51 26.30 All-Fuel Vent Chimneys, Double Wall, St. Stl.

		Crew	Daily Output	Labor-Hours	Unit	Material	2015 Bare Costs Labor	Equipment	Total	Total Incl O&P
8200	14" diameter	Q-9	21	.762	Ea.	395	38.50		433.50	495
8300	Insulated tee, 6" diameter		30	.533		283	27		310	355
8360	8" diameter		26	.615		310	31		341	390
8380	10" diameter		24	.667		350	33.50		383.50	440
8400	12" diameter		22	.727		395	36.50		431.50	495
8420	14" diameter		20	.800		460	40.50		500.50	575
8500	Boot tee, 6" diameter		28	.571		540	29		569	640
8520	8" diameter		24	.667		590	33.50		623.50	700
8530	10" diameter		22	.727		685	36.50		721.50	815
8540	12" diameter		20	.800		810	40.50		850.50	955
8550	14" diameter		18	.889		925	45		970	1,100
8600	Rain cap with bird screen, 6" diameter		30	.533		290	27		317	365
8640	8" diameter		26	.615		335	31		366	420
8660	10" diameter		24	.667		390	33.50		423.50	485
8680	12" diameter		22	.727		450	36.50		486.50	555
8700	14" diameter		21	.762		510	38.50		548.50	625
8800	Flat roof flashing, 6" diameter		30	.533		106	27		133	160
8840	8" diameter		26	.615		116	31		147	176
8860	10" diameter		24	.667		124	33.50		157.50	189
8880	12" diameter		22	.727		136	36.50		172.50	208
8900	14" diameter		21	.762		139	38.50		177.50	214

23 52 Heating Boilers

23 52 13 – Electric Boilers

23 52 13.10 Electric Boilers, ASME

		Crew	Daily Output	Labor-Hours	Unit	Material	Labor	Equipment	Total	Total Incl O&P
0010	**ELECTRIC BOILERS, ASME**, Standard controls and trim									
1000	Steam, 6 KW, 20.5 MBH	Q-19	1.20	20	Ea.	3,950	1,075		5,025	6,025
1160	60 KW, 205 MBH		1	24		6,650	1,300		7,950	9,350
1300	296 KW, 1010 MBH		.45	53.333		25,000	2,875		27,875	32,100
1400	592 KW, 2020 MBH	Q-21	.34	94.118		35,400	5,225		40,625	47,000
1540	1110 KW, 3788 MBH	"	.19	168		49,600	9,350		58,950	69,000
2000	Hot water, 7.5 KW, 25.6 MBH	Q-19	1.30	18.462		4,975	1,000		5,975	7,025
2040	30 KW, 102 MBH		1.20	20		5,325	1,075		6,400	7,525
2060	45 KW, 164 MBH		1.20	20		5,425	1,075		6,500	7,625
2070	60 KW, 205 MBH		1.20	20		5,525	1,075		6,600	7,750
2080	75 KW, 256 MBH		1.10	21.818		5,900	1,175		7,075	8,325
2100	90 KW, 307 MBH		1.10	21.818		6,000	1,175		7,175	8,425
9000	Minimum labor/equipment charge	Q-20	1	20	Job		1,025		1,025	1,600

23 52 23 – Cast-Iron Boilers

23 52 23.20 Gas-Fired Boilers

		Crew	Daily Output	Labor-Hours	Unit	Material	Labor	Equipment	Total	Total Incl O&P
0010	**GAS-FIRED BOILERS**, Natural or propane, standard controls, packaged									
1000	Cast iron, with insulated jacket									
2000	Steam, gross output, 81 MBH	Q-7	1.40	22.857	Ea.	2,450	1,300		3,750	4,725
2060	163 MBH		.90	35.556		3,275	2,025		5,300	6,750
2200	440 MBH		.51	62.500		6,350	3,550		9,900	12,600
2320	1,875 MBH		.30	106		25,200	6,075		31,275	37,200
2400	3570 MBH		.18	181		35,800	10,300		46,100	55,500
2540	6,970 MBH		.10	320		98,500	18,200		116,700	137,000
3000	Hot water, gross output, 80 MBH		1.46	21.918		1,975	1,250		3,225	4,125
3020	100 MBH		1.35	23.704		2,550	1,350		3,900	4,925

23 52 Heating Boilers

23 52 23 – Cast-Iron Boilers

23 52 23.20 Gas-Fired Boilers

		Crew	Daily Output	Labor-Hours	Unit	Material	2015 Bare Costs Labor	Equipment	Total	Total Incl O&P
3040	122 MBH	Q-7	1.10	29.091	Ea.	2,675	1,650		4,325	5,500
3060	163 MBH		1	32		3,150	1,825		4,975	6,325
3200	440 MBH		.58	54.983		5,950	3,125		9,075	11,400
3320	2,000 MBH		.26	125		21,400	7,100		28,500	34,600
3540	6,970 MBH		.09	359		118,500	20,500		139,000	162,500
7000	For tankless water heater, add					10%				
7050	For additional zone valves up to 312 MBH add					194			194	214
9900	Minimum labor/equipment charge	Q-6	1	24	Job		1,350		1,350	2,100

23 52 23.30 Gas/Oil Fired Boilers

		Crew	Daily Output	Labor-Hours	Unit	Material	Labor	Equipment	Total	Total Incl O&P
0010	**GAS/OIL FIRED BOILERS**, Combination with burners and controls, packaged									
1000	Cast iron with insulated jacket									
2000	Steam, gross output, 720 MBH	Q-7	.43	74.074	Ea.	14,700	4,225		18,925	22,800
2140	2,700 MBH		.19	165		28,100	9,425		37,525	45,600
2380	6,970 MBH		.09	372		102,500	21,200		123,700	146,000
2900	Hot water, gross output									
2910	200 MBH	Q-6	.62	39.024	Ea.	10,500	2,175		12,675	15,000
2920	300 MBH		.49	49.080		10,500	2,725		13,225	15,900
2930	400 MBH		.41	57.971		12,300	3,225		15,525	18,600
2940	500 MBH		.36	67.039		13,300	3,750		17,050	20,400
3000	584 MBH	Q-7	.44	72.072		14,700	4,100		18,800	22,500
3060	1,460 MBH		.28	113		36,400	6,425		42,825	50,000
3300	13,500 MBH, 403.3 BHP		.04	727		190,500	41,400		231,900	274,000

23 52 23.40 Oil-Fired Boilers

		Crew	Daily Output	Labor-Hours	Unit	Material	Labor	Equipment	Total	Total Incl O&P
0010	**OIL-FIRED BOILERS**, Standard controls, flame retention burner, packaged									
1000	Cast iron, with insulated flush jacket									
2000	Steam, gross output, 109 MBH	Q-7	1.20	26.667	Ea.	2,325	1,525		3,850	4,925
2060	207 MBH		.90	35.556		3,175	2,025		5,200	6,650
2180	1,084 MBH		.38	85.106		10,900	4,850		15,750	19,600
2200	1,360 MBH		.33	98.160		12,500	5,575		18,075	22,400
2240	2,175 MBH		.24	133		18,100	7,625		25,725	31,900
2340	4,360 MBH		.15	214		42,800	12,200		55,000	66,000
2460	6,970 MBH		.09	363		101,000	20,700		121,700	143,500
3000	Hot water, same price as steam									
4000	For tankless coil in smaller sizes, add				Ea.	15%				

23 52 26 – Steel Boilers

23 52 26.40 Oil-Fired Boilers

		Crew	Daily Output	Labor-Hours	Unit	Material	Labor	Equipment	Total	Total Incl O&P
0010	**OIL-FIRED BOILERS**, Standard controls, flame retention burner									
5000	Steel, with insulated flush jacket									
7000	Hot water, gross output, 103 MBH	Q-6	1.60	15	Ea.	1,775	835		2,610	3,250
7020	122 MBH		1.45	16.506		1,875	920		2,795	3,500
7040	137 MBH		1.36	17.595		2,000	980		2,980	3,725
7060	168 MBH		1.30	18.405		2,100	1,025		3,125	3,925
7080	225 MBH		1.22	19.704		2,575	1,100		3,675	4,550
7100	315 MBH		.96	25.105		6,550	1,400		7,950	9,400
7120	420 MBH		.70	34.483		6,900	1,925		8,825	10,600
9000	Minimum labor/equipment charge		1.75	13.714	Job		765		765	1,200

23 52 Heating Boilers

23 52 88 – Burners

23 52 88.10 Replacement Type Burners

		Crew	Daily Output	Labor-Hours	Unit	Material	2015 Bare Costs Labor	Equipment	Total	Total Incl O&P
0010	**REPLACEMENT TYPE BURNERS**									
0990	Residential, conversion, gas fired, LP or natural									
1000	Gun type, atmospheric input 50 to 225 MBH	Q-1	2.50	6.400	Ea.	720	340		1,060	1,325
1020	100 to 400 MBH		2	8		1,200	425		1,625	1,975
1040	300 to 1000 MBH	↓	1.70	9.412	↓	3,625	495		4,120	4,750
3000	Flame retention oil fired assembly, input									
3020	.50 to 2.25 GPH	Q-1	2.40	6.667	Ea.	284	350		634	860
3040	2.0 to 5.0 GPH		2	8		315	425		740	1,000
3060	3.0 to 7.0 GPH		1.80	8.889		570	470		1,040	1,350
3080	6.0 to 12.0 GPH	↓	1.60	10	↓	915	530		1,445	1,825

23 54 Furnaces

23 54 13 – Electric-Resistance Furnaces

23 54 13.10 Electric Furnaces

		Crew	Daily Output	Labor-Hours	Unit	Material	Labor	Equipment	Total	Total Incl O&P
0010	**ELECTRIC FURNACES**, Hot air, blowers, std. controls									
0011	not including gas, oil or flue piping									
1000	Electric, UL listed									
1100	34.1 MBH	Q-20	4.40	4.545	Ea.	455	233		688	865

23 54 16 – Fuel-Fired Furnaces

23 54 16.13 Gas-Fired Furnaces

		Crew	Daily Output	Labor-Hours	Unit	Material	Labor	Equipment	Total	Total Incl O&P
0010	**GAS-FIRED FURNACES**									
3000	Gas, AGA certified, upflow, direct drive models									
3020	45 MBH input	Q-9	4	4	Ea.	535	201		736	910
3040	60 MBH input		3.80	4.211		535	212		747	925
3060	75 MBH input		3.60	4.444		575	224		799	990
3100	100 MBH input		3.20	5		625	252		877	1,075
3120	125 MBH input		3	5.333		655	269		924	1,150
3130	150 MBH input		2.80	5.714		670	288		958	1,200
3140	200 MBH input	↓	2.60	6.154	↓	2,800	310		3,110	3,575

23 54 16.16 Oil-Fired Furnaces

		Crew	Daily Output	Labor-Hours	Unit	Material	Labor	Equipment	Total	Total Incl O&P
0010	**OIL-FIRED FURNACES**									
6000	Oil, UL listed, atomizing gun type burner									
6020	56 MBH output	Q-9	3.60	4.444	Ea.	1,725	224		1,949	2,250
6040	95 MBH output		3.40	4.706		1,875	237		2,112	2,450
6060	134 MBH output		3.20	5		2,175	252		2,427	2,800
6080	151 MBH output		3	5.333		2,250	269		2,519	2,900
6100	200 MBH input		2.60	6.154	↓	2,450	310		2,760	3,200
9000	Minimum labor/equipment charge	↓	2.75	5.818	Job		293		293	460

23 54 24 – Furnace Components for Cooling

23 54 24.10 Furnace Components and Combinations

		Crew	Daily Output	Labor-Hours	Unit	Material	Labor	Equipment	Total	Total Incl O&P
0010	**FURNACE COMPONENTS AND COMBINATIONS**									
0080	Coils, A.C. evaporator, for gas or oil furnaces									
0090	Add-on, with holding charge									
0100	Upflow									
0120	1-1/2 ton cooling	Q-5	4	4	Ea.	220	215		435	575
0130	2 ton cooling		3.70	4.324		241	233		474	630
0140	3 ton cooling		3.30	4.848		375	261		636	820
0150	4 ton cooling		3	5.333		465	287		752	960
0160	5 ton cooling	↓	2.70	5.926		505	320		825	1,050

23 54 Furnaces

23 54 24 – Furnace Components for Cooling

23 54 24.10 Furnace Components and Combinations

		Crew	Daily Output	Labor-Hours	Unit	Material	2015 Bare Costs Labor	Equipment	Total	Total Incl O&P
0300	Downflow									
0330	2-1/2 ton cooling	Q-5	3	5.333	Ea.	295	287		582	770
0340	3-1/2 ton cooling		2.60	6.154		435	330		765	990
0350	5 ton cooling	↓	2.20	7.273	↓	505	390		895	1,175
0600	Horizontal									
0630	2 ton cooling	Q-5	3.90	4.103	Ea.	320	221		541	700
0640	3 ton cooling		3.50	4.571		370	246		616	790
0650	4 ton cooling		3.20	5		460	269		729	925
0660	5 ton cooling	↓	2.90	5.517	↓	460	297		757	970
2000	Cased evaporator coils for air handlers									
2100	1-1/2 ton cooling	Q-5	4.40	3.636	Ea.	267	196		463	600
2110	2 ton cooling		4.10	3.902		300	210		510	655
2120	2-1/2 ton cooling		3.90	4.103		320	221		541	695
2130	3 ton cooling		3.70	4.324		360	233		593	760
2140	3-1/2 ton cooling		3.50	4.571		485	246		731	920
2150	4 ton cooling		3.20	5		515	269		784	990
2160	5 ton cooling	↓	2.90	5.517	↓	525	297		822	1,050
3010	Air handler, modular									
3100	With cased evaporator cooling coil									
3120	1-1/2 ton cooling	Q-5	3.80	4.211	Ea.	850	226		1,076	1,300
3130	2 ton cooling		3.50	4.571		940	246		1,186	1,400
3140	2-1/2 ton cooling		3.30	4.848		995	261		1,256	1,500
3150	3 ton cooling		3.10	5.161		1,150	278		1,428	1,675
3160	3-1/2 ton cooling		2.90	5.517		1,325	297		1,622	1,925
3170	4 ton cooling		2.50	6.400		1,425	345		1,770	2,075
3180	5 ton cooling	↓	2.10	7.619	↓	1,700	410		2,110	2,500
3500	With no cooling coil									
3520	1-1/2 ton coil size	Q-5	12	1.333	Ea.	620	71.50		691.50	790
3530	2 ton coil size		10	1.600		620	86		706	820
3540	2-1/2 ton coil size		10	1.600		725	86		811	930
3554	3 ton coil size		9	1.778		825	95.50		920.50	1,050
3560	3-1/2 ton coil size		9	1.778		845	95.50		940.50	1,075
3570	4 ton coil size		8.50	1.882		1,050	101		1,151	1,325
3580	5 ton coil size	↓	8	2	↓	1,075	108		1,183	1,350
4000	With heater									
4120	5 kW, 17.1 MBH	Q-5	16	1	Ea.	670	54		724	820
4130	7.5 kW, 25.6 MBH		15.60	1.026		680	55		735	830
4140	10 kW, 34.2 MBH		15.20	1.053		845	56.50		901.50	1,025
4150	12.5 KW, 42.7 MBH	↓	14.80	1.081	↓	950	58		1,008	1,150

23 55 Fuel-Fired Heaters

23 55 13 – Fuel-Fired Duct Heaters

23 55 13.16 Gas-Fired Duct Heaters

		Crew	Daily Output	Labor-Hours	Unit	Material	Labor	Equipment	Total	Total Incl O&P
0010	**GAS-FIRED DUCT HEATERS**, Includes burner, controls, stainless steel									
0020	heat exchanger. Gas fired, electric ignition									
1000	Outdoor installation, with power venter									
1120	225 MBH output	Q-5	2.50	6.400	Ea.	5,000	345		5,345	6,025
1160	375 MBH output		1.60	10		8,150	540		8,690	9,825
1180	450 MBH output	↓	1.40	11.429	↓	9,350	615		9,965	11,300

23 55 Fuel-Fired Heaters

23 55 33 – Fuel-Fired Unit Heaters

23 55 33.16 Gas-Fired Unit Heaters		Crew	Daily Output	Labor-Hours	Unit	Material	2015 Bare Costs Labor	Equipment	Total	Total Incl O&P
0010	**GAS-FIRED UNIT HEATERS**, Cabinet, grilles, fan, ctrls., burner, no piping									
0022	thermostat, no piping. For flue see Section 23 51 23.10									
1000	Gas fired, floor mounted									
1100	60 MBH output	Q-5	10	1.600	Ea.	870	86		956	1,100
1180	180 MBH output		6	2.667		1,375	143		1,518	1,725
2000	Suspension mounted, propeller fan, 20 MBH output		8.50	1.882		1,125	101		1,226	1,400
2040	60 MBH output		7	2.286		1,700	123		1,823	2,075
2100	130 MBH output		5	3.200		2,225	172		2,397	2,725
2240	320 MBH output		2	8		4,100	430		4,530	5,200
2500	For powered venter and adapter, add					490			490	540
5000	Wall furnace, 17.5 MBH output	Q-5	6	2.667		745	143		888	1,050
5020	24 MBH output		5	3.200		745	172		917	1,100
5040	35 MBH output		4	4		790	215		1,005	1,200
9000	Minimum labor/equipment charge		3.50	4.571	Job		246		246	385

23 56 Solar Energy Heating Equipment

23 56 16 – Packaged Solar Heating Equipment

23 56 16.40 Solar Heating Systems

			Crew	Daily Output	Labor-Hours	Unit	Material	Labor	Equipment	Total	Total Incl O&P
0010	**SOLAR HEATING SYSTEMS**										
0020	System/Package prices, not including connecting										
0030	pipe, insulation, or special heating/plumbing fixtures										
0500	Hot water, standard package, low temperature										
0580	2 collectors, circulator, fittings, 120 gal. tank	G	Q-1	.40	40	Ea.	5,050	2,125		7,175	8,850

23 56 19 – Solar Heating Components

23 56 19.50 Solar Heating Ancillary

			Crew	Daily Output	Labor-Hours	Unit	Material	Labor	Equipment	Total	Total Incl O&P
0010	**SOLAR HEATING ANCILLARY**										
2300	Circulators, air										
2310	Blowers										
2550	Booster fan 6" diameter, 120 CFM	G	Q-9	16	1	Ea.	36.50	50.50		87	120
2580	8" diameter, 150 CFM	G		16	1		41	50.50		91.50	125
2660	Shutter/damper	G		12	1.333		56	67		123	168
2670	Shutter motor	G		16	1		139	50.50		189.50	233
2800	Circulators, liquid, 1/25 HP, 5.3 GPM	G	Q-1	14	1.143		114	60.50		174.50	220
2820	1/20 HP, 17 GPM	G	"	12	1.333		161	70.50		231.50	287
3000	Collector panels, air with aluminum absorber plate										
3010	Wall or roof mount										
3040	Flat black, plastic glazing										
3080	4' x 8'	G	Q-9	6	2.667	Ea.	660	134		794	940
3300	Collector panels, liquid with copper absorber plate										
3320	Black chrome, tempered glass glazing										
3330	Alum. frame, 4' x 8', 5/32" single glazing	G	Q-1	9.50	1.684	Ea.	995	89		1,084	1,250
3450	Flat black, alum. frame, 3.5' x 7.5'	G		9	1.778		880	94		974	1,100
3500	4' x 8'	G		5.50	2.909		1,050	154		1,204	1,400
3600	Liquid, full wetted, plastic, alum. frame, 4' x 10'	G		5	3.200		320	169		489	620
3650	Collector panel mounting, flat roof or ground rack	G		7	2.286		244	121		365	455
3670	Roof clamps	G		70	.229	Set	2.80	12.10		14.90	22
3700	Roof strap, teflon	G	1 Plum	205	.039	L.F.	23.50	2.29		25.79	29.50
3900	Differential controller with two sensors										
3930	Thermostat, hard wired	G	1 Plum	8	1	Ea.	101	58.50		159.50	203
4050	Pool valve system	G		2.50	3.200		161	188		349	470

23 56 Solar Energy Heating Equipment

23 56 19 – Solar Heating Components

23 56 19.50 Solar Heating Ancillary

		Crew	Daily Output	Labor-Hours	Unit	Material	2015 Bare Costs Labor	Equipment	Total	Total Incl O&P
4070	With 12 VAC actuator	1 Plum [G]	2	4	Ea.	325	235		560	720
4150	Sensors									
4220	Freeze prevention	1 Plum [G]	32	.250	Ea.	22.50	14.70		37.20	48
4300	Heat exchanger									
4315	includes coil, blower, circulator									
4316	and controller for DHW and space hot air									
4330	Fluid to air coil, up flow, 45 MBH	Q-1 [G]	4	4	Ea.	315	211		526	680
4380	70 MBH	[G]	3.50	4.571		355	242		597	770
4400	80 MBH	[G]	3	5.333		475	282		757	965
4580	Fluid to fluid package includes two circulating pumps									
4590	expansion tank, check valve, relief valve									
4600	controller, high temperature cutoff and sensors	Q-1 [G]	2.50	6.400	Ea.	800	340		1,140	1,400
4650	Heat transfer fluid									
4700	Propylene glycol, inhibited anti-freeze	1 Plum [G]	28	.286	Gal.	15.55	16.75		32.30	43
4800	Solar storage tanks, knocked down									
5190	6'-3" high, 7' x 7' = 306 C.F./2000 gallons	Q-10 [G]	1.20	20	Ea.	14,200	1,050		15,250	17,400
5210	7' x 14' = 613 C.F./4000 gallons	[G]	.60	40		21,400	2,100		23,500	26,800
5230	10'-6" x 14' = 919 C.F./6000 gallons	[G]	.40	60		24,900	3,125		28,025	32,400
5250	14' x 17'-6" = 1531 C.F./10,000 gallons	Q-11 [G]	.30	106		32,100	5,675		37,775	44,300
7000	Solar control valves and vents									
7070	Air eliminator, automatic 3/4" size	1 Plum [G]	32	.250	Ea.	30	14.70		44.70	56
7090	Air vent, automatic, 1/8" fitting	[G]	32	.250		17.25	14.70		31.95	42
7100	Manual, 1/8" NPT	[G]	32	.250		2.44	14.70		17.14	25.50
7120	Backflow preventer, 1/2" pipe size	[G]	16	.500		75.50	29.50		105	129
7130	3/4" pipe size	[G]	16	.500		77	29.50		106.50	131
7150	Balancing valve, 3/4" pipe size	[G]	20	.400		63	23.50		86.50	106
7180	Draindown valve, 1/2" copper tube	[G]	9	.889		212	52		264	315
7200	Flow control valve, 1/2" pipe size	[G]	22	.364		120	21.50		141.50	166
7220	Expansion tank, up to 5 gal.	[G]	32	.250		68	14.70		82.70	98
7400	Pressure gauge, 2" dial	[G]	32	.250		24	14.70		38.70	49.50
7450	Relief valve, temp. and pressure 3/4" pipe size	[G]	30	.267		21	15.65		36.65	47.50
7500	Solenoid valve, normally closed									
7520	Brass, 3/4" NPT, 24V	1 Plum [G]	9	.889	Ea.	123	52		175	217
7530	1" NPT, 24V	[G]	9	.889		193	52		245	294
7750	Vacuum relief valve, 3/4" pipe size	[G]	32	.250		29.50	14.70		44.20	55.50
7800	Thermometers									
7820	Digital temperature monitoring, 4 locations	1 Plum [G]	2.50	3.200	Ea.	137	188		325	445
7900	Upright, 1/2" NPT	" [G]	8	1	"	39.50	58.50		98	135
8250	Water storage tank with heat exchanger and electric element									
8270	66 gal. with 2" x 2 lb. density insulation	1 Plum [G]	1.60	5	Ea.	1,600	294		1,894	2,225
8300	80 gal. with 2" x 2 lb. density insulation	" [G]	1.60	5	"	1,600	294		1,894	2,225
8500	Water storage module, plastic									
8600	Tubular, 12" diameter, 4' high	1 Carp [G]	48	.167	Ea.	114	7.85		121.85	138
8610	12" diameter, 8' high	" [G]	40	.200		175	9.40		184.40	208
8650	Cap, 12" diameter	[G]				22			22	24
9000	Minimum labor/equipment charge	1 Plum	2	4	Job		235		235	365

23 61 Refrigerant Compressors

23 61 16 – Reciprocating Refrigerant Compressors

23 61 16.10 Reciprocating Compressors	Crew	Daily Output	Labor-Hours	Unit	Material	2015 Bare Costs Labor	Equipment	Total	Total Incl O&P
0010 **RECIPROCATING COMPRESSORS**									
0990 Refrigeration, recip. hermetic, switches & protective devices									
1000 10 ton	Q-5	1	16	Ea.	12,800	860		13,660	15,500
1100 20 ton	Q-6	.72	33.333		27,200	1,850		29,050	32,800
1400 50 ton	"	.20	120		32,100	6,700		38,800	45,700
1600 130 ton	Q-7	.21	152		40,500	8,675		49,175	58,000

23 62 Packaged Compressor and Condenser Units

23 62 13 – Packaged Air-Cooled Refrigerant Compressor and Condenser Units

23 62 13.10 Packaged Air-Cooled Refrig. Condensing Units

	Crew	Daily Output	Labor-Hours	Unit	Material	Labor	Equipment	Total	Total Incl O&P
0010 **PACKAGED AIR-COOLED REFRIGERANT CONDENSING UNITS**									
0020 Condensing unit									
0030 Air cooled, compressor, standard controls									
0050 1.5 ton	Q-5	2.50	6.400	Ea.	1,250	345		1,595	1,900
0200 2.5 ton		1.70	9.412		1,425	505		1,930	2,375
0300 3 ton		1.30	12.308		1,500	660		2,160	2,675
0400 4 ton		.90	17.778		1,975	955		2,930	3,675
0500 5 ton		.60	26.667		2,375	1,425		3,800	4,825

23 64 Packaged Water Chillers

23 64 19 – Reciprocating Water Chillers

23 64 19.10 Reciprocating Type Water Chillers

	Crew	Daily Output	Labor-Hours	Unit	Material	Labor	Equipment	Total	Total Incl O&P
0010 **RECIPROCATING TYPE WATER CHILLERS**, With standard controls									
0494 Water chillers, integral air cooled condenser									
0980 Water cooled, multiple compressor, semi-hermetic, tower not incl.									
1090 45 ton cooling	Q-7	.29	111	Ea.	28,200	6,325		34,525	40,900
1100 50 ton cooling		.28	113		35,800	6,475		42,275	49,500
1160 100 ton cooling		.18	179		60,500	10,200		70,700	82,500
1200 145 ton cooling		.16	202		71,500	11,500		83,000	96,500

23 64 23 – Scroll Water Chillers

23 64 23.10 Scroll Water Chillers

	Crew	Daily Output	Labor-Hours	Unit	Material	Labor	Equipment	Total	Total Incl O&P
0010 **SCROLL WATER CHILLERS**, With standard controls									
0480 Packaged w/integral air cooled condenser									
0482 10 ton cooling	Q-7	.34	94.118	Ea.	20,300	5,350		25,650	30,800
0490 15 ton cooling		.37	86.486		20,800	4,925		25,725	30,600
0500 20 ton cooling		.34	94.118		21,500	5,350		26,850	32,000
0520 40 ton cooling		.30	108		32,900	6,150		39,050	45,800
0680 Scroll water cooled, single compressor, hermetic, tower not incl.									
0760 10 ton cooling	Q-6	.36	67.039	Ea.	6,625	3,750		10,375	13,100

23 64 26 – Rotary-Screw Water Chillers

23 64 26.10 Rotary-Screw Type Water Chillers

	Crew	Daily Output	Labor-Hours	Unit	Material	Labor	Equipment	Total	Total Incl O&P
0010 **ROTARY-SCREW TYPE WATER CHILLERS**, With standard controls									
0110 Screw, liquid chiller, air cooled, insulated evaporator									
0120 130 ton	Q-7	.14	228	Ea.	91,500	13,000		104,500	121,000
0124 160 ton	"	.13	246	"	112,500	14,000		126,500	145,500

23 64 Packaged Water Chillers

23 64 33 – Direct Expansion Water Chillers

23 64 33.10 Direct Expansion Type Water Chillers		Crew	Daily Output	Labor-Hours	Unit	Material	2015 Bare Costs Labor	Equipment	Total	Total Incl O&P
0010	**DIRECT EXPANSION TYPE WATER CHILLERS**, With standard controls									
8000	Direct expansion, shell and tube type, for built up systems									
8020	1 ton	Q-5	2	8	Ea.	7,300	430		7,730	8,700
8030	5 ton		1.90	8.421		12,100	455		12,555	14,000
8040	10 ton	↓	1.70	9.412	↓	15,000	505		15,505	17,300
9000	Minimum labor/equipment charge	Q-6	1	24	Job		1,350		1,350	2,100

23 65 Cooling Towers

23 65 13 – Forced-Draft Cooling Towers

23 65 13.10 Forced-Draft Type Cooling Towers

		Crew	Daily Output	Labor-Hours	Unit	Material	Labor	Equipment	Total	Total Incl O&P
0010	**FORCED-DRAFT TYPE COOLING TOWERS**, Packaged units									
0070	Galvanized steel									
0080	Induced draft, crossflow									
0100	Vertical, belt drive, 61 tons	Q-6	90	.267	TonAC	215	14.85		229.85	260
0150	100 ton		100	.240		208	13.40		221.40	249
0200	115 ton		109	.220		180	12.30		192.30	218
0250	131 ton		120	.200		217	11.15		228.15	255
0260	162 ton	↓	132	.182	↓	175	10.15		185.15	209
1000	For higher capacities, use multiples									
1500	Induced air, double flow									
1900	Vertical, gear drive, 167 ton	Q-6	126	.190	TonAC	170	10.60		180.60	204
2000	297 ton		129	.186		105	10.40		115.40	131
2100	582 ton		132	.182		58.50	10.15		68.65	80
2200	1016 ton	↓	150	.160	↓	79	8.90		87.90	101
3000	For higher capacities, use multiples									
3500	For pumps and piping, add	Q-6	38	.632	TonAC	108	35		143	173
4000	For absorption systems, add				"	75%	75%			
4100	Cooling water chemical feeder	Q-5	3	5.333	Ea.	365	287		652	845
5000	Fiberglass tower on galvanized steel support structure									
5010	Draw thru									
5100	100 ton	Q-6	1.40	17.143	Ea.	13,900	955		14,855	16,800
5120	120 ton		1.20	20		16,200	1,125		17,325	19,600
5140	140 ton		1	24		17,500	1,350		18,850	21,300
5160	160 ton		.80	30		19,400	1,675		21,075	24,000
5180	180 ton		.65	36.923		22,100	2,050		24,150	27,500
5200	200 ton	↓	.48	50	↓	25,000	2,800		27,800	31,900
5300	For stainless steel support structure, add					30%				
5360	For higher capacities, use multiples of each size									
6000	Stainless steel									
6010	Induced draft, crossflow, horizontal, belt drive									
6100	57 ton	Q-6	1.50	16	Ea.	27,100	890		27,990	31,200
6120	91 ton		.99	24.242		32,700	1,350		34,050	38,100
6140	111 ton		.43	55.814		41,600	3,125		44,725	50,500
6160	126 ton	↓	.22	109	↓	41,600	6,075		47,675	55,500
6170	Induced draft, crossflow, vertical, gear drive									
6172	167 ton	Q-6	.75	32	Ea.	53,000	1,775		54,775	61,500
6174	297 ton		.43	55.814		59,000	3,125		62,125	70,000
6176	582 ton		.23	104		93,500	5,825		99,325	112,000
6180	1016 ton	↓	.15	160	↓	159,000	8,925		167,925	188,500
9000	Minimum labor/equipment charge	↓	1	24	Job		1,350		1,350	2,100

23 72 Air-to-Air Energy Recovery Equipment

23 72 13 – Heat-Wheel Air-to-Air Energy-Recovery Equipment

23 72 13.10 Heat-Wheel Air-to-Air Energy Recovery Equip.	Crew	Daily Output	Labor-Hours	Unit	Material	2015 Bare Costs Labor	Equipment	Total	Total Incl O&P
0010 **HEAT-WHEEL AIR-TO-AIR ENERGY RECOVERY EQUIPMENT**									
0100 Air to air									
9000 Minimum labor/equipment charge	1 Shee	4	2	Job		112		112	177

23 73 Indoor Central-Station Air-Handling Units

23 73 39 – Indoor, Direct Gas-Fired Heating and Ventilating Units

23 73 39.10 Make-Up Air Unit

0010 **MAKE-UP AIR UNIT**									
0020 Indoor suspension, natural/LP gas, direct fired,									
0032 standard control. For flue see Section 23 51 23.10									
0040 70°F temperature rise, MBH is input									
0100 75 MBH input	Q-6	3.60	6.667	Ea.	4,925	370		5,295	6,000
0220 225 MBH input	↓	2.40	10		6,750	560		7,310	8,275
0300 400 MBH input	↓	1.60	15	↓	15,400	835		16,235	18,200

23 74 Packaged Outdoor HVAC Equipment

23 74 23 – Packaged, Outdoor, Heating-Only Makeup-Air Units

23 74 23.16 Packaged, Ind.-Fired, In/Outdoor

	Crew	Daily Output	Labor-Hours	Unit	Material	Labor	Equipment	Total	Total Incl O&P
0010 **PACKAGED, IND.-FIRED, IN/OUTDOOR**, Heat-Only Makeup-Air Units									
1000 Rooftop unit, natural gas, gravity vent, S.S. exchanger									
1010 70°F temperature rise, MBH is input									
1020 250 MBH	Q-6	4	6	Ea.	15,400	335		15,735	17,400
1040 400 MBH		3.60	6.667		17,300	370		17,670	19,600
1060 550 MBH	↓	3.30	7.273		17,800	405		18,205	20,200
1600 For cleanable filters, add					5%				
1700 For electric modulating gas control, add				↓	10%				
9000 Minimum labor/equipment charge	Q-6	2.75	8.727	Job		485		485	760

23 74 33 – Dedicated Outdoor-Air Units

23 74 33.10 Rooftop Air Conditioners

	Crew	Daily Output	Labor-Hours	Unit	Material	Labor	Equipment	Total	Total Incl O&P
0010 **ROOFTOP AIR CONDITIONERS**, Standard controls, curb, economizer									
1000 Single zone, electric cool, gas heat									
1100 3 ton cooling, 60 MBH heating	Q-5	.70	22.857	Ea.	3,925	1,225		5,150	6,250
1120 4 ton cooling, 95 MBH heating		.61	26.403		4,575	1,425		6,000	7,275
1150 7.5 ton cooling, 170 MBH heating		.50	32.258		6,825	1,725		8,550	10,200
1156 8.5 ton cooling, 170 MBH heating	↓	.46	34.783		8,125	1,875		10,000	11,900
1160 10 ton cooling, 200 MBH heating	Q-6	.67	35.982		9,425	2,000		11,425	13,500
1220 30 ton cooling, 540 MBH heating	Q-7	.47	68.376		30,800	3,900		34,700	40,000
1240 40 ton cooling, 675 MBH heating	"	.35	91.168	↓	40,200	5,175		45,375	52,500
2000 Multizone, electric cool, gas heat, economizer									
2120 20 ton cooling, 360 MBH heating	Q-7	.53	60.038	Ea.	70,000	3,425		73,425	82,500
2180 30 ton cooling, 540 MBH heating		.37	85.562		107,000	4,875		111,875	125,000
2210 50 ton cooling, 540 MBH heating		.23	142		152,000	8,100		160,100	180,000
2220 70 ton cooling, 1500 MBH heating		.16	198		164,500	11,300		175,800	198,500
2240 80 ton cooling, 1500 MBH heating		.14	228		188,000	13,000		201,000	227,500
2280 105 ton cooling, 1500 MBH heating	↓	.11	290		217,500	16,500		234,000	265,000
2400 For hot water heat coil, deduct					5%				
2500 For steam heat coil, deduct					2%				
2600 For electric heat, deduct				↓	3%		5%		
9000 Minimum labor/equipment charge	Q-5	1.50	10.667	Job		575		575	895

23 81 Decentralized Unitary HVAC Equipment

23 81 19 – Self-Contained Air-Conditioners

23 81 19.10 Window Unit Air Conditioners

		Crew	Daily Output	Labor-Hours	Unit	Material	2015 Bare Costs Labor	Equipment	Total	Total Incl O&P
0010	**WINDOW UNIT AIR CONDITIONERS**									
4000	Portable/window, 15 amp 125 V grounded receptacle required									
4500	10,000 BTUH	1 Carp	6	1.333	Ea.	590	62.50		652.50	755
4520	12,000 BTUH	L-2	8	2	"	725	82.50		807.50	930
9000	Minimum labor/equipment charge	1 Carp	2	4	Job		188		188	310

23 81 19.20 Self-Contained Single Package

		Crew	Daily Output	Labor-Hours	Unit	Material	Labor	Equipment	Total	Total Incl O&P
0010	**SELF-CONTAINED SINGLE PACKAGE**									
0100	Air cooled, for free blow or duct, not incl. remote condenser									
0110	Constant volume									
0200	3 ton cooling	Q-5	1	16	Ea.	3,750	860		4,610	5,450
0210	4 ton cooling	"	.80	20		4,075	1,075		5,150	6,150
0240	10 ton cooling	Q-7	1	32		7,125	1,825		8,950	10,700
0280	30 ton cooling	"	.80	40		25,300	2,275		27,575	31,400
1000	Water cooled for free blow or duct, not including tower									
1010	Constant volume									
1100	3 ton cooling	Q-6	1	24	Ea.	3,725	1,350		5,075	6,200
1120	5 ton cooling	"	1	24		4,850	1,350		6,200	7,425
1140	10 ton cooling	Q-7	.90	35.556		9,450	2,025		11,475	13,600
1180	30 ton cooling	"	.70	45.714		37,900	2,600		40,500	45,800
1240	60 ton cooling	Q-8	.30	106		69,500	6,075	192	75,767	86,000
1300	For hot water or steam heat coils, add					12%	10%			
9000	Minimum labor/equipment charge	Q-5	1	16	Job		860		860	1,350

23 81 23 – Computer-Room Air-Conditioners

23 81 23.10 Computer Room Units

		Crew	Daily Output	Labor-Hours	Unit	Material	Labor	Equipment	Total	Total Incl O&P
0010	**COMPUTER ROOM UNITS**									
1000	Air cooled, includes remote condenser but not									
1020	interconnecting tubing or refrigerant									
1160	6 ton	Q-5	.30	53.333	Ea.	37,000	2,875		39,875	45,200
1240	10 ton		.25	64		39,100	3,450		42,550	48,500
1260	12 ton		.24	66.667		40,500	3,575		44,075	50,000
1300	20 ton	Q-6	.26	92.308		51,500	5,150		56,650	64,500
1320	22 ton	"	.24	100		52,000	5,575		57,575	66,000

23 81 26 – Split-System Air-Conditioners

23 81 26.10 Split Ductless Systems

		Crew	Daily Output	Labor-Hours	Unit	Material	Labor	Equipment	Total	Total Incl O&P
0010	**SPLIT DUCTLESS SYSTEMS**									
0100	Cooling only, single zone									
0110	Wall mount									
0120	3/4 ton cooling	Q-5	2	8	Ea.	1,100	430		1,530	1,900
0130	1 ton cooling		1.80	8.889		1,225	480		1,705	2,100
0140	1-1/2 ton cooling		1.60	10		1,975	540		2,515	3,025
0150	2 ton cooling		1.40	11.429		2,250	615		2,865	3,425
1000	Ceiling mount									
1020	2 ton cooling	Q-5	1.40	11.429	Ea.	2,000	615		2,615	3,150
1030	3 ton cooling	"	1.20	13.333	"	2,550	715		3,265	3,925
3000	Multizone									
3010	Wall mount									
3020	2 @ 3/4 ton cooling	Q-5	1.80	8.889	Ea.	3,375	480		3,855	4,450
5000	Cooling/Heating									
5010	Wall mount									
5110	1 ton cooling	Q-5	1.70	9.412	Ea.	1,300	505		1,805	2,225
5120	1-1/2 ton cooling	"	1.50	10.667	"	1,975	575		2,550	3,075

23 81 Decentralized Unitary HVAC Equipment

23 81 26 – Split-System Air-Conditioners

23 81 26.10 Split Ductless Systems		Crew	Daily Output	Labor-Hours	Unit	Material	2015 Bare Costs Labor	2015 Bare Costs Equipment	Total	Total Incl O&P
7000	Accessories for all split ductless systems									
7010	Add for ambient frost control	Q-5	8	2	Ea.	126	108		234	305
7020	Add for tube/wiring kit, (Line sets)									
7030	15' kit	Q-5	32	.500	Ea.	103	27		130	155
7036	25' kit		28	.571		166	30.50		196.50	230
7040	35' kit		24	.667		174	36		210	247
7050	50' kit		20	.800		261	43		304	355

23 81 29 – Variable Refrigerant Flow HVAC Systems

23 81 29.10 Heat Pump, Gas Driven		Crew	Daily Output	Labor-Hours	Unit	Material	2015 Bare Costs Labor	2015 Bare Costs Equipment	Total	Total Incl O&P
0010	**HEAT PUMP, GAS DRIVEN**, Variable refrigerant volume (VRV) type									
0020	Not including interconnecting tubing or multi-zone controls									
1000	For indoor fan VRV type AHU see 23 82 19.40									
1010	Multi-zone split									
1020	Outdoor unit									
1100	8 tons cooling, for up to 17 zones	Q-5	1.30	12.308	Ea.	29,100	660		29,760	33,000
1110	Isolation rails		2.60	6.154	Pair	1,025	330		1,355	1,650
1160	15 tons cooling, for up to 33 zones		1	16	Ea.	38,600	860		39,460	43,800
1170	Isolation rails		2.60	6.154	Pair	1,200	330		1,530	1,850
2000	Packaged unit									
2020	Outdoor unit									
2200	11 tons cooling	Q-5	1.30	12.308	Ea.	37,200	660		37,860	41,900
2210	Roof curb adapter	"	2.60	6.154		645	330		975	1,225
2220	Thermostat	1 Stpi	1.30	6.154		385	370		755	1,000

23 81 43 – Air-Source Unitary Heat Pumps

23 81 43.10 Air-Source Heat Pumps		Crew	Daily Output	Labor-Hours	Unit	Material	2015 Bare Costs Labor	2015 Bare Costs Equipment	Total	Total Incl O&P
0010	**AIR-SOURCE HEAT PUMPS**, Not including interconnecting tubing									
1000	Air to air, split system, not including curbs, pads, fan coil and ductwork									
1012	Outside condensing unit only, for fan coil see Section 23 82 19.10									
1020	2 ton cooling, 8.5 MBH heat @ 0°F	Q-5	2	8	Ea.	2,400	430		2,830	3,325
1040	3 ton cooling, 13 MBH heat @ 0°F		1.20	13.333		2,775	715		3,490	4,175
1080	7.5 ton cooling, 33 MBH heat @ 0°F		.45	35.556		6,425	1,900		8,325	10,100
1120	15 ton cooling, 64 MBH heat @ 0°F	Q-6	.50	48		11,800	2,675		14,475	17,200
1130	20 ton cooling, 85 MBH heat @ 0°F		.35	68.571		17,000	3,825		20,825	24,700
1140	25 ton cooling, 119 MBH heat @ 0°F		.25	96		20,400	5,350		25,750	30,900
1500	Single package, not including curbs, pads, or plenums									
1520	2 ton cooling, 6.5 MBH heat @ 0°F	Q-5	1.50	10.667	Ea.	3,100	575		3,675	4,325
1560	3 ton cooling, 10 MBH heat @ 0°F	"	1.20	13.333	"	3,575	715		4,290	5,050
6000	Air to water, single package, excluding. storage tank and ductwork									
6010	Includes circulating water pump, air duct connections, digital temperature									
6020	controller with remote tank temp. probe and sensor for storage tank.									
6040	Water heating - air cooling capacity									
6110	35.5 MBH heat water, 2.3 Ton cool air	Q-5	1.60	10	Ea.	11,400	540		11,940	13,400
6120	58 MBH heat water, 3.8 Ton cool air		1.10	14.545		13,000	780		13,780	15,500
6130	76 MBH heat water, 4.9 Ton cool air		.87	18.391		15,500	990		16,490	18,700
6140	98 MBH heat water, 6.5 Ton cool air		.62	25.806		20,100	1,400		21,500	24,300
6150	113 MBH heat water, 7.4 Ton cool air		.59	27.119		22,000	1,450		23,450	26,500
6160	142 MBH heat water, 9.2 Ton cool air		.52	30.769		26,700	1,650		28,350	32,000
6170	171 MBH heat water, 11.1 Ton cool air		.49	32.653		31,400	1,750		33,150	37,300

23 81 Decentralized Unitary HVAC Equipment

23 81 46 – Water-Source Unitary Heat Pumps

23 81 46.10 Water Source Heat Pumps

		Crew	Daily Output	Labor-Hours	Unit	Material	2015 Bare Costs Labor	Equipment	Total	Total Incl O&P
0010	**WATER SOURCE HEAT PUMPS**, Not incl. connecting tubing or water source									
2000	Water source to air, single package									
2100	1 ton cooling, 13 MBH heat @ 75°F	Q-5	2	8	Ea.	2,075	430		2,505	2,975
2140	2 ton cooling, 19 MBH heat @ 75°F		1.70	9.412		2,575	505		3,080	3,625
2220	5 ton cooling, 29 MBH heat @ 75°F		.90	17.778		3,825	955		4,780	5,700
9000	Minimum labor/equipment charge		1.75	9.143	Job		490		490	765

23 82 Convection Heating and Cooling Units

23 82 16 – Air Coils

23 82 16.10 Flanged Coils

		Crew	Daily Output	Labor-Hours	Unit	Material	Labor	Equipment	Total	Total Incl O&P
0010	**FLANGED COILS**									
1500	Hot water heating, 1 row, 24" x 48"	Q-5	4	4	Ea.	1,625	215		1,840	2,125
9000	Minimum labor/equipment charge	1 Plum	1	8	Job		470		470	730

23 82 16.20 Duct Heaters

		Crew	Daily Output	Labor-Hours	Unit	Material	Labor	Equipment	Total	Total Incl O&P
0010	**DUCT HEATERS**, Electric, 480 V, 3 Ph.									
0020	Finned tubular insert, 500°F									
0100	8" wide x 6" high, 4.0 kW	Q-20	16	1.250	Ea.	765	64		829	945
0120	12" high, 8.0 kW		15	1.333		1,275	68.50		1,343.50	1,500
0140	18" high, 12.0 kW		14	1.429		1,775	73		1,848	2,075
0160	24" high, 16.0 kW		13	1.538		2,300	79		2,379	2,650
0180	30" high, 20.0 kW		12	1.667		2,800	85.50		2,885.50	3,200
0300	12" wide x 6" high, 6.7 kW		15	1.333		815	68.50		883.50	1,000
0320	12" high, 13.3 kW		14	1.429		1,325	73		1,398	1,575
0340	18" high, 20.0 kW		13	1.538		1,850	79		1,929	2,150
8000	To obtain BTU multiply kW by 3413									
9900	Minimum labor/equipment charge	1 Shee	4	2	Job		112		112	177

23 82 19 – Fan Coil Units

23 82 19.10 Fan Coil Air Conditioning

		Crew	Daily Output	Labor-Hours	Unit	Material	Labor	Equipment	Total	Total Incl O&P
0010	**FAN COIL AIR CONDITIONING**									
0030	Fan coil AC, cabinet mounted, filters and controls									
0100	Chilled water, 1/2 ton cooling	Q-5	8	2	Ea.	815	108		923	1,075
0120	1 ton cooling		6	2.667		950	143		1,093	1,275
0160	2.5 ton cooling		5	3.200		1,975	172		2,147	2,450
0180	3 ton cooling		4	4		2,175	215		2,390	2,725
0262	For hot water coil, add					40%	10%			
0940	Direct expansion, for use w/air cooled condensing unit, 1.5 ton cooling	Q-5	5	3.200	Ea.	680	172		852	1,025
1000	5 ton cooling	"	3	5.333		1,300	287		1,587	1,900
1040	10 ton cooling	Q-6	2.60	9.231		2,800	515		3,315	3,875
1060	20 ton cooling	"	.70	34.286		5,400	1,900		7,300	8,925
1500	For hot water coil, add					40%	10%			
1512	For condensing unit add see Section 23 62									
3000	Chilled water, horizontal unit, housing, 2 pipe, fan control, no valves									
3100	1/2 ton cooling	Q-5	8	2	Ea.	1,275	108		1,383	1,575
3110	1 ton cooling		6	2.667		1,475	143		1,618	1,850
3120	1.5 ton cooling		5.50	2.909		1,700	156		1,856	2,125
3130	2 ton cooling		5.25	3.048		2,050	164		2,214	2,500
3140	3 ton cooling		4	4		2,425	215		2,640	3,000
3150	3.5 ton cooling		3.80	4.211		2,675	226		2,901	3,300
3160	4 ton cooling		3.80	4.211		2,675	226		2,901	3,300
4000	With electric heat, 2 pipe									

23 82 Convection Heating and Cooling Units

23 82 19 – Fan Coil Units

23 82 19.10 Fan Coil Air Conditioning		Crew	Daily Output	Labor-Hours	Unit	Material	2015 Bare Costs Labor	Equipment	Total	Total Incl O&P
4100	1/2 ton cooling	Q-5	8	2	Ea.	1,475	108		1,583	1,800
4105	3/4 ton cooling		7	2.286		1,625	123		1,748	2,000
4110	1 ton cooling		6	2.667		1,800	143		1,943	2,200
4120	1.5 ton cooling		5.50	2.909		2,000	156		2,156	2,450
4130	2 ton cooling		5.25	3.048		2,525	164		2,689	3,025
4135	2.5 ton cooling		4.80	3.333		3,475	179		3,654	4,100
4140	3 ton cooling		4	4		4,725	215		4,940	5,525
4150	3.5 ton cooling		3.80	4.211		5,650	226		5,876	6,550
4160	4 ton cooling		3.60	4.444		6,525	239		6,764	7,550
9000	Minimum labor/equipment charge		3.75	4.267	Job		229		229	360

23 82 19.40 Fan Coil Air Conditioning

		Crew	Daily Output	Labor-Hours	Unit	Material	Labor	Equipment	Total	Total Incl O&P
0010	**FAN COIL AIR CONDITIONING**, Variable refrigerant volume (VRV) type									
0020	Not including interconnecting tubing or multi-zone controls									
0030	For VRV condensing unit see section 23 81 29.10									
0050	Indoor type, ducted									
0100	Vertical concealed									
0130	1 ton cooling G	Q-5	2.97	5.387	Ea.	2,475	290		2,765	3,175
0140	1.5 ton cooling G		2.77	5.776		2,525	310		2,835	3,250
0150	2 ton cooling G		2.70	5.926		2,725	320		3,045	3,500
0160	2.5 ton cooling G		2.60	6.154		2,875	330		3,205	3,675
0170	3 ton cooling G		2.50	6.400		2,925	345		3,270	3,725
0180	3.5 ton cooling G		2.30	6.957		3,075	375		3,450	3,950
0190	4 ton cooling G		2.19	7.306		3,075	395		3,470	4,025
0200	4.5 ton cooling G		2.08	7.692		3,400	415		3,815	4,375
0230	Outside air connection possible									
1100	Ceiling concealed									
1130	.6 ton cooling G	Q-5	2.60	6.154	Ea.	1,625	330		1,955	2,300
1140	.75 ton cooling G		2.60	6.154		1,675	330		2,005	2,375
1150	1 ton cooling G		2.60	6.154		1,775	330		2,105	2,475
1152	Screening door G	1 Stpi	5.20	1.538		98	92		190	251
1160	1.5 ton cooling G	Q-5	2.60	6.154		1,825	330		2,155	2,550
1162	Screening door G	1 Stpi	5.20	1.538		113	92		205	268
1170	2 ton cooling G	Q-5	2.60	6.154		2,150	330		2,480	2,875
1172	Screening door G	1 Stpi	5.20	1.538		98	92		190	251
1180	2.5 ton cooling G	Q-5	2.60	6.154		2,500	330		2,830	3,275
1182	Screening door G	1 Stpi	5.20	1.538		170	92		262	330
1190	3 ton cooling G	Q-5	2.60	6.154		2,650	330		2,980	3,450
1192	Screening door G	1 Stpi	5.20	1.538		170	92		262	330
1200	4 ton cooling G	Q-5	2.30	6.957		2,725	375		3,100	3,550
1202	Screening door G	1 Stpi	5.20	1.538		170	92		262	330
1220	6 ton cooling G	Q-5	2.08	7.692		4,625	415		5,040	5,725
1230	8 ton cooling G	"	1.89	8.466		5,200	455		5,655	6,425
1260	Outside air connection possible									
4050	Indoor type, duct-free									
4100	Ceiling mounted cassette									
4130	.75 ton cooling G	Q-5	3.46	4.624	Ea.	1,875	249		2,124	2,475
4140	1 ton cooling G		2.97	5.387		2,000	290		2,290	2,650
4150	1.5 ton cooling G		2.77	5.776		2,050	310		2,360	2,725
4160	2 ton cooling G		2.60	6.154		2,200	330		2,530	2,950
4170	2.5 ton cooling G		2.50	6.400		2,250	345		2,595	3,000
4180	3 ton cooling G		2.30	6.957		2,425	375		2,800	3,225
4190	4 ton cooling G		2.08	7.692		2,700	415		3,115	3,625

23 82 Convection Heating and Cooling Units

23 82 19 – Fan Coil Units

23 82 19.40 Fan Coil Air Conditioning		Crew	Daily Output	Labor-Hours	Unit	Material	2015 Bare Costs Labor	Equipment	Total	Total Incl O&P
4198	For ceiling cassette decoration panel, add	G 1 Stpi	5.20	1.538	Ea.	294	92		386	470
4230	Outside air connection possible									
4500	Wall mounted									
4530	.6 ton cooling	G Q-5	5.20	3.077	Ea.	995	165		1,160	1,350
4540	.75 ton cooling	G	5.20	3.077		1,050	165		1,215	1,400
4550	1 ton cooling	G	4.16	3.846		1,200	207		1,407	1,625
4560	1.5 ton cooling	G	3.77	4.244		1,300	228		1,528	1,800
4570	2 ton cooling	G	3.46	4.624		1,400	249		1,649	1,950
4590	Condensate pump for the above air handlers	G 1 Stpi	6.46	1.238		211	74		285	345
4700	Floor standing unit									
4730	1 ton cooling	G Q-5	2.60	6.154	Ea.	1,675	330		2,005	2,375
4740	1.5 ton cooling	G	2.60	6.154		1,900	330		2,230	2,625
4750	2 ton cooling	G	2.60	6.154		2,050	330		2,380	2,775
4800	Floor standing unit, concealed									
4830	1 ton cooling	G Q-5	2.60	6.154	Ea.	1,600	330		1,930	2,300
4840	1.5 ton cooling	G	2.60	6.154		1,825	330		2,155	2,525
4850	2 ton cooling	G	2.60	6.154		1,950	330		2,280	2,675
4880	Outside air connection possible									
8000	Accessories									
8100	Branch divergence pipe fitting									
8110	Capacity under 76 MBH	1 Stpi	2.50	3.200	Ea.	151	191		342	465
8120	76 MBH to 112 MBH		2.50	3.200		178	191		369	495
8130	112 MBH to 234 MBH		2.50	3.200		505	191		696	855
8200	Header pipe fitting									
8210	Max 4 branches									
8220	Capacity under 76 MBH	1 Stpi	1.70	4.706	Ea.	221	281		502	685
8260	Max 8 branches									
8270	76 MBH to 112 MBH	1 Stpi	1.70	4.706	Ea.	450	281		731	935
8280	112 MBH to 234 MBH	"	1.30	6.154	"	720	370		1,090	1,375

23 82 29 – Radiators

23 82 29.10 Hydronic Heating

0010	HYDRONIC HEATING, Terminal units, not incl. main supply pipe									
1000	Radiation									
1100	Panel, baseboard, C.I., including supports, no covers	Q-5	46	.348	L.F.	38	18.70		56.70	70.50
3000	Radiators, cast iron									
3100	Free standing or wall hung, 6 tube, 25" high	Q-5	96	.167	Section	44	8.95		52.95	62.50
9000	Minimum labor/equipment charge	"	3.20	5	Job		269		269	420

23 82 36 – Finned-Tube Radiation Heaters

23 82 36.10 Finned Tube Radiation

0010	FINNED TUBE RADIATION, Terminal units, not incl. main supply pipe									
1150	Fin tube, wall hung, 14" slope top cover, with damper									
1200	1-1/4" copper tube, 4-1/4" alum. fin	Q-5	38	.421	L.F.	43	22.50		65.50	82.50
1250	1-1/4" steel tube, 4-1/4" steel fin		36	.444		39	24		63	80.50
1255	2" steel tube, 4-1/4" steel fin		32	.500		43	27		70	89
1310	Baseboard, pkgd, 1/2" copper tube, alum. fin, 7" high		60	.267		11.20	14.35		25.55	35
1320	3/4" copper tube, alum. fin, 7" high		58	.276		7.35	14.85		22.20	31
1340	1" copper tube, alum. fin, 8-7/8" high		56	.286		18.85	15.35		34.20	44.50
1360	1-1/4" copper tube, alum. fin, 8-7/8" high		54	.296		28	15.95		43.95	55.50
1500	Note: fin tube may also require corners, caps, etc.									

23 82 Convection Heating and Cooling Units

23 82 39 – Unit Heaters

23 82 39.16 Propeller Unit Heaters

		Crew	Daily Output	Labor-Hours	Unit	Material	2015 Bare Costs Labor	2015 Bare Costs Equipment	Total	Total Incl O&P
0010	**PROPELLER UNIT HEATERS**									
3950	Unit heaters, propeller, 115 V 2 psi steam, 60°F entering air									
4000	Horizontal, 12 MBH	Q-5	12	1.333	Ea.	360	71.50		431.50	505
4060	43.9 MBH		8	2		545	108		653	770
4240	286.9 MBH		2	8		1,650	430		2,080	2,500
4260	364 MBH		1.80	8.889		2,075	480		2,555	3,025
4270	404 MBH		1.60	10		2,150	540		2,690	3,225
4300	Vertical diffuser same price									
4310	Vertical flow, 40 MBH	Q-5	11	1.455	Ea.	540	78		618	715
4326	131.0 MBH	"	4	4		845	215		1,060	1,275
4354	420 MBH, (460 V)	Q-6	1.80	13.333		2,150	745		2,895	3,525
4358	500 MBH, (460 V)		1.71	14.035		2,875	785		3,660	4,375
4362	570 MBH, (460 V)		1.40	17.143		3,925	955		4,880	5,825
4366	620 MBH, (460 V)		1.30	18.462		3,925	1,025		4,950	5,925
4370	960 MBH, (460 V)		1.10	21.818		7,450	1,225		8,675	10,100

23 83 Radiant Heating Units

23 83 33 – Electric Radiant Heaters

23 83 33.10 Electric Heating

		Crew	Daily Output	Labor-Hours	Unit	Material	2015 Bare Costs Labor	2015 Bare Costs Equipment	Total	Total Incl O&P
0010	**ELECTRIC HEATING**, not incl. conduit or feed wiring									
1100	Rule of thumb: Baseboard units, including control	1 Elec	4.40	1.818	kW	104	99.50		203.50	268
1300	Baseboard heaters, 2' long, 350 watt		8	1	Ea.	28.50	54.50		83	116
1400	3' long, 750 watt		8	1		33	54.50		87.50	121
1600	4' long, 1000 watt		6.70	1.194		39	65.50		104.50	144
1800	5' long, 935 watt		5.70	1.404		48	77		125	172
2000	6' long, 1500 watt		5	1.600		54.50	87.50		142	196
2400	8' long, 2000 watt		4	2		66	109		175	242
2600	9' long, 1680 watt		3.60	2.222		102	122		224	300
2800	10' long, 1875 watt		3.30	2.424		190	133		323	415
2950	Wall heaters with fan, 120 to 277 volt									
3170	1000 watt	1 Elec	6	1.333	Ea.	134	73		207	260
3180	1250 watt		5	1.600		134	87.50		221.50	283
3190	1500 watt		4	2		134	109		243	315
3230	2500 watt		3.50	2.286		335	125		460	565
3250	4000 watt		2.70	2.963		495	162		657	795
3600	Thermostats, integral		16	.500		30	27.50		57.50	75.50
3800	Line voltage, 1 pole		8	1		19.20	54.50		73.70	106
3810	2 pole		8	1		29.50	54.50		84	117
9990	Minimum labor/equipment charge		4	2	Job		109		109	169

23 84 Humidity Control Equipment

23 84 13 – Humidifiers

23 84 13.10 Humidifier Units	Crew	Daily Output	Labor-Hours	Unit	Material	2015 Bare Costs Labor	2015 Bare Costs Equipment	Total	Total Incl O&P
0010 **HUMIDIFIER UNITS**									
0520 Steam, room or duct, filter, regulators, auto. controls, 220 V									
0560 22 lb. per hour	Q-5	5	3.200	Ea.	2,975	172		3,147	3,550
0580 33 lb. per hour		4	4		3,050	215		3,265	3,700
0620 100 lb. per hour		3	5.333		4,475	287		4,762	5,375
9000 Minimum labor/equipment charge		3.50	4.571	Job		246		246	385

23 91 Prefabricated Equipment Supports

23 91 10 – Prefabricated Curbs, Pads and Stands

23 91 10.10 Prefabricated Pads and Stands

	Crew	Daily Output	Labor-Hours	Unit	Material	Labor	Equipment	Total	Total Incl O&P
0010 **PREFABRICATED PADS AND STANDS**									
6000 Pad, fiberglass reinforced concrete with polystyrene foam core									
6050 Condenser, 2" thick, 20" x 38"	1 Shee	8	1	Ea.	27.50	56		83.50	119
6090 24" x 36"	"	12	.667		35	37.50		72.50	97.50
6280 36" x 36"	Q-9	8	2		50.50	101		151.50	215
6300 36" x 40"		7	2.286		57.50	115		172.50	246
6320 36" x 48"		7	2.286		67.50	115		182.50	256

Division Notes

		CREW	DAILY OUTPUT	LABOR-HOURS	UNIT	BARE COSTS				TOTAL INCL O&P
						MAT.	LABOR	EQUIP.	TOTAL	

Estimating Tips
26 05 00 Common Work Results for Electrical

- Conduit should be taken off in three main categories—power distribution, branch power, and branch lighting—so the estimator can concentrate on systems and components, therefore making it easier to ensure all items have been accounted for.
- For cost modifications for elevated conduit installation, add the percentages to labor according to the height of installation, and only to the quantities exceeding the different height levels, not to the total conduit quantities.
- Remember that aluminum wiring of equal ampacity is larger in diameter than copper and may require larger conduit.
- If more than three wires at a time are being pulled, deduct percentages from the labor hours of that grouping of wires.
- When taking off grounding systems, identify separately the type and size of wire, and list each unique type of ground connection.
- The estimator should take the weights of materials into consideration when completing a takeoff. Topics to consider include: How will the materials be supported? What methods of support are available? How high will the support structure have to reach? Will the final support structure be able to withstand the total burden? Is the support material included or separate from the fixture, equipment, and material specified?
- Do not overlook the costs for equipment used in the installation. If scaffolding or highlifts are available in the field, contractors may use them in lieu of the proposed ladders and rolling staging.

26 20 00 Low-Voltage Electrical Transmission

- Supports and concrete pads may be shown on drawings for the larger equipment, or the support system may be only a piece of plywood for the back of a panelboard. In either case, it must be included in the costs.

26 40 00 Electrical and Cathodic Protection

- When taking off cathodic protections systems, identify the type and size of cable, and list each unique type of anode connection.

26 50 00 Lighting

- Fixtures should be taken off room by room, using the fixture schedule, specifications, and the ceiling plan. For large concentrations of lighting fixtures in the same area, deduct the percentages from labor hours.

Reference Numbers

Reference numbers are shown in shaded boxes at the beginning of some major classifications. These numbers refer to related items in the Reference Section. The reference information may be an estimating procedure, an alternate pricing method, or technical information.

Note: Not all subdivisions listed here necessarily appear in this publication. ■

*Note: **Trade Service**, in part, has been used as a reference source for some of the material prices used in Division 26.*

Did you know?
RSMeans Online gives you the same access to RSMeans' data with 24/7 access:
- Quickly locate costs in the searchable database.
- Build cost lists, estimates, and reports in minutes.
- Adjust costs to any location in the U.S. and Canada with the click of a button.

Start your free trial today at **www.rsmeansonline.com**

RSMeansOnline

26 01 Operation and Maintenance of Electrical Systems

26 01 50 – Operation and Maintenance of Lighting

26 01 50.81 Luminaire Replacement	Crew	Daily Output	Labor-Hours	Unit	Material	2015 Bare Costs Labor	Equipment	Total	Total Incl O&P
0010 **LUMINAIRE REPLACEMENT**									
0962 Remove and install new replacement lens, 2' x 2' troffer	1 Elec	16	.500	Ea.	15.90	27.50		43.40	60
0964 2' x 4' troffer		16	.500		14.90	27.50		42.40	59
3200 Remove and replace (reinstall), lighting fixture	↓	4	2	↓		109		109	169

26 05 Common Work Results for Electrical

26 05 05 – Selective Demolition for Electrical

26 05 05.10 Electrical Demolition

			Crew	Daily Output	Labor-Hours	Unit	Material	2015 Bare Costs Labor	Equipment	Total	Total Incl O&P
0010	**ELECTRICAL DEMOLITION**	R260105-30									
0020	Conduit to 15' high, including fittings & hangers										
0100	Rigid galvanized steel, 1/2" to 1" diameter	R024119-10	1 Elec	242	.033	L.F.		1.81		1.81	2.80
0120	1-1/4" to 2"		"	200	.040			2.19		2.19	3.39
0140	2-1/2" to 3-1/2"		2 Elec	302	.053			2.90		2.90	4.49
0200	Electric metallic tubing (EMT), 1/2" to 1"		1 Elec	394	.020			1.11		1.11	1.72
0220	1-1/4" to 1-1/2"			326	.025			1.34		1.34	2.08
0240	2" to 3"		↓	236	.034	↓		1.85		1.85	2.87
0400	Wiremold raceway, including fittings & hangers										
0420	No. 3000		1 Elec	250	.032	L.F.		1.75		1.75	2.71
0440	No. 4000		"	217	.037	"		2.02		2.02	3.12
0500	Channels, steel, including fittings & hangers										
0520	3/4" x 1-1/2"		1 Elec	308	.026	L.F.		1.42		1.42	2.20
0540	1-1/2" x 1-1/2"			269	.030			1.63		1.63	2.52
0560	1-1/2" x 1-7/8"		↓	229	.035	↓		1.91		1.91	2.96
0600	Copper bus duct, indoor, 3 phase										
0610	Including hangers & supports										
0620	225 amp		2 Elec	135	.119	L.F.		6.50		6.50	10.05
0640	400 amp			106	.151			8.25		8.25	12.80
0660	600 amp			86	.186			10.20		10.20	15.75
0680	1000 amp			60	.267			14.60		14.60	22.50
0700	1600 amp			40	.400			22		22	34
0720	3000 amp		↓	10	1.600	↓		87.50		87.50	136
0800	Plug-in switches, 600V 3 ph., incl. disconnecting										
0820	wire, conduit terminations, 30 amp		1 Elec	15.50	.516	Ea.		28		28	43.50
0840	60 amp			13.90	.576			31.50		31.50	49
0850	100 amp			10.40	.769			42		42	65
0860	200 amp		↓	6.20	1.290			70.50		70.50	109
0880	400 amp		2 Elec	5.40	2.963			162		162	251
0900	600 amp			3.40	4.706			257		257	400
0920	800 amp			2.60	6.154			335		335	520
0940	1200 amp			2	8			440		440	680
0960	1600 amp		↓	1.70	9.412	↓		515		515	795
1010	Safety switches, 250 or 600V, incl. disconnection										
1050	of wire & conduit terminations										
1100	30 amp		1 Elec	12.30	.650	Ea.		35.50		35.50	55
1120	60 amp			8.80	.909			49.50		49.50	77
1140	100 amp			7.30	1.096			60		60	93
1160	200 amp		↓	5	1.600	↓		87.50		87.50	136
1210	Panel boards, incl. removal of all breakers,										
1220	conduit terminations & wire connections										
1230	3 wire, 120/240 V, 100A, to 20 circuits		1 Elec	2.60	3.077	Ea.		168		168	261
1240	200 amps, to 42 circuits		2 Elec	2.60	6.154	↓		335		335	520

26 05 Common Work Results for Electrical

26 05 05 – Selective Demolition for Electrical

26 05 05.10 Electrical Demolition	Crew	Daily Output	Labor-Hours	Unit	Material	2015 Bare Costs Labor	Equipment	Total	Total Incl O&P
1260	4 wire, 120/208 V, 125A, to 20 circuits	1 Elec	2.40	3.333	Ea.		182		182	282
1270	200 amps, to 42 circuits	2 Elec	2.40	6.667			365		365	565
1300	Transformer, dry type, 1 phase, incl. removal of									
1320	supports, wire & conduit terminations									
1340	1 kVA	1 Elec	7.70	1.039	Ea.		57		57	88
1360	5 kVA	"	4.70	1.702			93		93	144
1420	75 kVA	2 Elec	2.50	6.400			350		350	540
1440	3 phase to 600V, primary									
1460	3 kVA	1 Elec	3.87	2.067	Ea.		113		113	175
1480	15 kVA	2 Elec	3.67	4.360			238		238	370
1490	25 kVA		3.51	4.558			249		249	385
1500	30 kVA		3.42	4.678			256		256	395
1530	112.5 kVA	R-3	2.90	6.897			375	47.50	422.50	635
1560	500 kVA		1.40	14.286			775	98.50	873.50	1,300
1570	750 kVA		1.10	18.182			985	125	1,110	1,675
1600	Pull boxes & cabinets, sheet metal, incl. removal									
1620	of supports and conduit terminations									
1640	6" x 6" x 4"	1 Elec	31.10	.257	Ea.		14.05		14.05	22
1660	12" x 12" x 4"		23.30	.343			18.80		18.80	29
1720	Junction boxes, 4" sq. & oct.		80	.100			5.45		5.45	8.45
1740	Handy box		107	.075			4.09		4.09	6.35
1760	Switch box		107	.075			4.09		4.09	6.35
1780	Receptacle & switch plates		257	.031			1.70		1.70	2.64
1800	Wire, THW-THWN-THHN, removed from									
1810	in place conduit, to 15' high									
1830	#14	1 Elec	65	.123	C.L.F.		6.75		6.75	10.40
1840	#12		55	.145			7.95		7.95	12.30
1850	#10		45.50	.176			9.60		9.60	14.90
1880	#4	2 Elec	53	.302			16.50		16.50	25.50
1890	#3		50	.320			17.50		17.50	27
1910	1/0		33.20	.482			26.50		26.50	41
1920	2/0		29.20	.548			30		30	46.50
1930	3/0		25	.640			35		35	54
1980	400 kcmil		17	.941			51.50		51.50	79.50
1990	500 kcmil		16.20	.988			54		54	83.50
2000	Interior fluorescent fixtures, incl. supports									
2010	& whips, to 15' high									
2100	Recessed drop-in 2' x 2', 2 lamp	2 Elec	35	.457	Ea.		25		25	38.50
2120	2' x 4', 2 lamp		33	.485			26.50		26.50	41
2140	2' x 4', 4 lamp		30	.533			29		29	45
2160	4' x 4', 4 lamp		20	.800			44		44	68
2180	Surface mount, acrylic lens & hinged frame									
2200	1' x 4', 2 lamp	2 Elec	44	.364	Ea.		19.90		19.90	31
2220	2' x 2', 2 lamp		44	.364			19.90		19.90	31
2260	2' x 4', 4 lamp		33	.485			26.50		26.50	41
2280	4' x 4', 4 lamp		23	.696			38		38	59
2300	Strip fixtures, surface mount									
2320	4' long, 1 lamp	2 Elec	53	.302	Ea.		16.50		16.50	25.50
2340	4' long, 2 lamp		50	.320			17.50		17.50	27
2360	8' long, 1 lamp		42	.381			21		21	32.50
2380	8' long, 2 lamp		40	.400			22		22	34
2400	Pendant mount, industrial, incl. removal									
2410	of chain or rod hangers, to 15' high									

26 05 Common Work Results for Electrical

26 05 05 – Selective Demolition for Electrical

26 05 05.10 Electrical Demolition		Crew	Daily Output	Labor-Hours	Unit	Material	2015 Bare Costs Labor	Equipment	Total	Total Incl O&P
2420	4' long, 2 lamp	2 Elec	35	.457	Ea.		25		25	38.50
2440	8' long, 2 lamp	"	27	.593	"		32.50		32.50	50
2460	Interior incandescent, surface, ceiling									
2470	or wall mount, to 12' high									
2480	Metal cylinder type, 75 Watt	2 Elec	62	.258	Ea.		14.10		14.10	22
2500	150 Watt	"	62	.258	"		14.10		14.10	22
2520	Metal halide, high bay									
2540	400 Watt	2 Elec	15	1.067	Ea.		58.50		58.50	90.50
2560	1000 Watt		12	1.333			73		73	113
2580	150 Watt, low bay		20	.800			44		44	68
2600	Exterior fixtures, incandescent, wall mount									
2620	100 Watt	2 Elec	50	.320	Ea.		17.50		17.50	27
2640	Quartz, 500 Watt		33	.485			26.50		26.50	41
2660	1500 Watt		27	.593			32.50		32.50	50
2680	Wall pack, mercury vapor									
2700	175 Watt	2 Elec	25	.640	Ea.		35		35	54
2720	250 Watt	"	25	.640	"		35		35	54
9000	Minimum labor/equipment charge	1 Elec	4	2	Job		109		109	169
9900	Add to labor for higher elevated installation									
9910	15' to 20' high, add						10%			
9920	20' to 25' high, add						20%			
9930	25' to 30' high, add						25%			
9940	30' to 35' high, add						30%			
9950	35' to 40' high, add						35%			
9960	Over 40' high, add						40%			

26 05 19 – Low-Voltage Electrical Power Conductors and Cables

26 05 19.20 Armored Cable

		Crew	Daily Output	Labor-Hours	Unit	Material	Labor	Equipment	Total	Incl O&P
0010	**ARMORED CABLE**									
0050	600 volt, copper (BX), #14, 2 conductor, solid	1 Elec	2.40	3.333	C.L.F.	46	182		228	335
0100	3 conductor, solid		2.20	3.636		71	199		270	390
0150	#12, 2 conductor, solid		2.30	3.478		46	190		236	345
0200	3 conductor, solid		2	4		75.50	219		294.50	425
0250	#10, 2 conductor, solid		2	4		85	219		304	435
0300	3 conductor, solid		1.60	5		118	274		392	555
0340	#8, 2 conductor, stranded		1.50	5.333		222	292		514	695
0350	3 conductor, stranded		1.30	6.154		222	335		557	765
9010	600 volt, copper (MC) steel clad, #14, 2 wire		2.40	3.333		46.50	182		228.50	335
9020	3 wire		2.20	3.636		72	199		271	390
9030	4 wire		2	4		100	219		319	450
9040	#12, 2 wire		2.30	3.478		47	190		237	345
9050	3 wire		2	4		79.50	219		298.50	430
9070	#10, 2 wire		2	4		98.50	219		317.50	450
9080	3 wire		1.60	5		138	274		412	575
9100	#8, 2 wire, stranded		1.80	4.444		190	243		433	585
9110	3 wire, stranded		1.30	6.154		267	335		602	815
9900	Minimum labor/equipment charge		4	2	Job		109		109	169

26 05 19.50 Mineral Insulated Cable

		Crew	Daily Output	Labor-Hours	Unit	Material	Labor	Equipment	Total	Incl O&P
0010	**MINERAL INSULATED CABLE** 600 volt									
0100	1 conductor, #12	1 Elec	1.60	5	C.L.F.	365	274		639	825
0200	#10		1.60	5		470	274		744	940
0400	#8		1.50	5.333		520	292		812	1,025
0500	#6		1.40	5.714		620	315		935	1,175

26 05 Common Work Results for Electrical

26 05 19 – Low-Voltage Electrical Power Conductors and Cables

26 05 19.50 Mineral Insulated Cable		Crew	Daily Output	Labor-Hours	Unit	Material	2015 Bare Costs Labor	Equipment	Total	Total Incl O&P
0600	#4	2 Elec	2.40	6.667	C.L.F.	835	365		1,200	1,475
0800	#2		2.20	7.273		1,175	400		1,575	1,925
0900	#1		2.10	7.619		1,350	415		1,765	2,150
1000	1/0		2	8		1,600	440		2,040	2,425
1100	2/0		1.90	8.421		1,900	460		2,360	2,825
1200	3/0		1.80	8.889		2,275	485		2,760	3,250
1400	4/0		1.60	10		2,625	545		3,170	3,750
1410	250 kcmil	3 Elec	2.40	10		2,975	545		3,520	4,125
1420	350 kcmil		1.95	12.308		3,400	675		4,075	4,775
1430	500 kcmil		1.95	12.308		4,375	675		5,050	5,850
1500	2 conductor, #12	1 Elec	1.40	5.714		760	315		1,075	1,325
1600	#10		1.20	6.667		935	365		1,300	1,600
1800	#8		1.10	7.273		1,075	400		1,475	1,800
2000	#6		1.05	7.619		1,475	415		1,890	2,275
2100	#4	2 Elec	2	8		2,000	440		2,440	2,875
2200	3 conductor, #12	1 Elec	1.20	6.667		950	365		1,315	1,625
2400	#10		1.10	7.273		1,075	400		1,475	1,800
2600	#8		1.05	7.619		1,350	415		1,765	2,125
2800	#6		1	8		1,775	440		2,215	2,625
3000	#4	2 Elec	1.80	8.889		2,225	485		2,710	3,200
3100	4 conductor, #12	1 Elec	1.20	6.667		1,000	365		1,365	1,675
3200	#10		1.10	7.273		1,200	400		1,600	1,950
3400	#8		1	8		1,575	440		2,015	2,400
3600	#6		.90	8.889		2,050	485		2,535	3,000
3620	7 conductor, #12		1.10	7.273		1,275	400		1,675	2,025
3640	#10		1	8		1,675	440		2,115	2,525
3800	Terminations, 600 volt, 1 conductor, #12		8	1	Ea.	16.05	54.50		70.55	102
5500	2 conductor, #12		6.70	1.194		16.05	65.50		81.55	119
5600	#10		6.40	1.250		24	68.50		92.50	133
5800	#8		6.20	1.290		24	70.50		94.50	136
6000	#6		5.70	1.404		24	77		101	146
6200	#4		5.30	1.509		53.50	82.50		136	187
6400	3 conductor, #12		5.70	1.404		24	77		101	146
6500	#10		5.50	1.455		24	79.50		103.50	150
6600	#8		5.20	1.538		24	84		108	157
6800	#6		4.80	1.667		24	91		115	168
7200	#4		4.60	1.739		53.50	95		148.50	206
7400	4 conductor, #12		4.60	1.739		27	95		122	177
7500	#10		4.40	1.818		27	99.50		126.50	184
7600	#8		4.20	1.905		27	104		131	191
8400	#6		4	2		56.50	109		165.50	231
8500	7 conductor, #12		3.50	2.286		27	125		152	224
8600	#10		3	2.667		58	146		204	290
8800	Crimping tool, plier type					60			60	66
9000	Stripping tool					227			227	249
9200	Hand vise					67.50			67.50	74.50
9500	Minimum labor/equipment charge	1 Elec	4	2	Job		109		109	169

26 05 19.55 Non-Metallic Sheathed Cable

		Crew	Daily Output	Labor-Hours	Unit	Material	Labor	Equipment	Total	Total Incl O&P
0010	**NON-METALLIC SHEATHED CABLE** 600 volt									
0100	Copper with ground wire, (Romex)									
0150	#14, 2 conductor	1 Elec	2.70	2.963	C.L.F.	24	162		186	278
0200	3 conductor		2.40	3.333		34	182		216	320

26 05 Common Work Results for Electrical

26 05 19 – Low-Voltage Electrical Power Conductors and Cables

26 05 19.55 Non-Metallic Sheathed Cable

		Crew	Daily Output	Labor-Hours	Unit	Material	2015 Bare Costs Labor	Equipment	Total	Total Incl O&P
0220	4 conductor	1 Elec	2.20	3.636	C.L.F.	49	199		248	365
0250	#12, 2 conductor		2.50	3.200		36.50	175		211.50	310
0300	3 conductor		2.20	3.636		52	199		251	370
0320	4 conductor		2	4		76.50	219		295.50	425
0350	#10, 2 conductor		2.20	3.636		56.50	199		255.50	370
0400	3 conductor		1.80	4.444		82.50	243		325.50	465
0430	#8, 2 conductor		1.60	5		88.50	274		362.50	525
0450	3 conductor		1.50	5.333		133	292		425	595
0500	#6, 3 conductor	↓	1.40	5.714	↓	215	315		530	720
0550	SE type SER aluminum cable, 3 RHW and									
0600	1 bare neutral, 3 #8 & 1 #8	1 Elec	1.60	5	C.L.F.	158	274		432	600
0650	3 #6 & 1 #6	"	1.40	5.714		179	315		494	680
0700	3 #4 & 1 #6	2 Elec	2.40	6.667		165	365		530	745
0750	3 #2 & 1 #4		2.20	7.273		296	400		696	940
0800	3 #1/0 & 1 #2		2	8		450	440		890	1,175
0850	3 #2/0 & 1 #1		1.80	8.889		530	485		1,015	1,325
0900	3 #4/0 & 1 #2/0	↓	1.60	10	↓	755	545		1,300	1,675
6500	Service entrance cap for copper SEU									
6700	150 amp	1 Elec	10	.800	Ea.	16.10	44		60.10	85.50
6800	200 amp		8	1	"	24.50	54.50		79	112
9000	Minimum labor/equipment charge	↓	4	2	Job		109		109	169

26 05 19.90 Wire

		Crew	Daily Output	Labor-Hours	Unit	Material	Labor	Equipment	Total	Total Incl O&P
0010	**WIRE** R260533-22									
0020	600 volt, copper type THW, solid, #14	1 Elec	13	.615	C.L.F.	7.80	33.50		41.30	60.50
0030	#12		11	.727		11.95	40		51.95	74.50
0040	#10		10	.800		18.65	44		62.65	88.50
0140	#8		8	1		37.50	54.50		92	126
0160	#6	↓	6.50	1.231		63.50	67.50		131	174
0180	#4	2 Elec	10.60	1.509		100	82.50		182.50	238
0200	#3		10	1.600		126	87.50		213.50	274
0220	#2		9	1.778		158	97		255	325
0240	#1		8	2		200	109		309	390
0260	1/0		6.60	2.424		250	133		383	480
0280	2/0		5.80	2.759		315	151		466	580
0300	3/0		5	3.200		395	175		570	705
0350	4/0	↓	4.40	3.636		500	199		699	860
0400	250 kcmil	3 Elec	6	4		585	219		804	985
0420	300 kcmil		5.70	4.211		700	230		930	1,125
0450	350 kcmil		5.40	4.444		855	243		1,098	1,325
0480	400 kcmil		5.10	4.706		985	257		1,242	1,475
0490	500 kcmil		4.80	5		1,150	274		1,424	1,700
0510	750 kcmil	↓	3.30	7.273		1,975	400		2,375	2,775
0540	600 volt, aluminum type THHN, stranded, #6	1 Elec	8	1		43	54.50		97.50	132
0560	#4	2 Elec	13	1.231		53.50	67.50		121	163
0580	#2		10.60	1.509		72.50	82.50		155	208
0600	#1		9	1.778		106	97		203	267
0620	1/0		8	2		127	109		236	310
0640	2/0		7.20	2.222		150	122		272	355
0680	3/0		6.60	2.424		186	133		319	410
0700	4/0	↓	6.20	2.581		207	141		348	445
0720	250 kcmil	3 Elec	8.70	2.759		253	151		404	510
0740	300 kcmil		8.10	2.963		350	162		512	635

26 05 Common Work Results for Electrical

26 05 19 – Low-Voltage Electrical Power Conductors and Cables

26 05 19.90 Wire

		Crew	Daily Output	Labor-Hours	Unit	Material	2015 Bare Costs Labor	Equipment	Total	Total Incl O&P
0760	350 kcmil	3 Elec	7.50	3.200	C.L.F.	355	175		530	660
0780	400 kcmil		6.90	3.478		415	190		605	750
0800	500 kcmil		6	4		460	219		679	845
0850	600 kcmil		5.70	4.211		580	230		810	995
0880	700 kcmil		5.10	4.706		670	257		927	1,125
9000	Minimum labor/equipment charge	1 Elec	4	2	Job		109		109	169

26 05 26 – Grounding and Bonding for Electrical Systems

26 05 26.80 Grounding

		Crew	Daily Output	Labor-Hours	Unit	Material	Labor	Equipment	Total	Total Incl O&P
0010	**GROUNDING**									
0030	Rod, copper clad, 8' long, 1/2" diameter	1 Elec	5.50	1.455	Ea.	20	79.50		99.50	145
0040	5/8" diameter		5.50	1.455		19.20	79.50		98.70	144
0050	3/4" diameter		5.30	1.509		33.50	82.50		116	165
0080	10' long, 1/2" diameter		4.80	1.667		22	91		113	166
0090	5/8" diameter		4.60	1.739		24	95		119	174
0100	3/4" diameter		4.40	1.818		38.50	99.50		138	197
0130	15' long, 3/4" diameter		4	2		54	109		163	229
0260	Wire ground bare armored, #8-1 conductor		2	4	C.L.F.	76	219		295	425
0280	#4-1 conductor		1.60	5		128	274		402	565
0390	Bare copper wire, stranded, #8		11	.727		29.50	40		69.50	94
0400	#6		10	.800		52	44		96	126
0600	#2	2 Elec	10	1.600		138	87.50		225.50	288
0800	3/0		6.60	2.424		310	133		443	545
1000	4/0		5.70	2.807		395	154		549	675
1200	250 kcmil	3 Elec	7.20	3.333		465	182		647	790
1800	Water pipe ground clamps, heavy duty									
2000	Bronze, 1/2" to 1" diameter	1 Elec	8	1	Ea.	24	54.50		78.50	111
2100	1-1/4" to 2" diameter		8	1		33	54.50		87.50	121
2200	2-1/2" to 3" diameter		6	1.333		44.50	73		117.50	162
2800	Brazed connections, #6 wire		12	.667		16.15	36.50		52.65	74.50
3000	#2 wire		10	.800		21.50	44		65.50	92
3100	3/0 wire		8	1		32.50	54.50		87	120
3200	4/0 wire		7	1.143		37	62.50		99.50	138
3400	250 kcmil wire		5	1.600		43.50	87.50		131	184
3600	500 kcmil wire		4	2		53.50	109		162.50	228
9000	Minimum labor/equipment charge		4	2	Job		109		109	169

26 05 33 – Raceway and Boxes for Electrical Systems

26 05 33.13 Conduit

			Crew	Daily Output	Labor-Hours	Unit	Material	Labor	Equipment	Total	Total Incl O&P
0010	**CONDUIT** To 15' high, includes 2 terminations, 2 elbows,	R260533-22									
0020	11 beam clamps, and 11 couplings per 100 L.F.										
1161	Field bends, 45° to 90°, 1/2" diameter		1 Elec	53	.151	Ea.		8.25		8.25	12.80
1162	3/4" diameter			47	.170			9.30		9.30	14.40
1163	1" diameter			44	.182			9.95		9.95	15.40
1164	1-1/4" diameter			23	.348			19.05		19.05	29.50
1165	1-1/2" diameter			21	.381			21		21	32.50
1166	2" diameter			16	.500			27.50		27.50	42.50
1991	Field bends, 45° to 90°, 1/2" diameter			44	.182			9.95		9.95	15.40
1992	3/4" diameter			40	.200			10.95		10.95	16.95
1993	1" diameter			36	.222			12.15		12.15	18.80
1994	1-1/4" diameter			19	.421			23		23	35.50
1995	1-1/2" diameter			18	.444			24.50		24.50	37.50
1996	2" diameter			13	.615			33.50		33.50	52
2500	Steel, intermediate conduit (IMC), 1/2" diameter			100	.080	L.F.	1.79	4.38		6.17	8.75

26 05 Common Work Results for Electrical

26 05 33 – Raceway and Boxes for Electrical Systems

26 05 33.13 Conduit

		Crew	Daily Output	Labor-Hours	Unit	Material	2015 Bare Costs Labor	Equipment	Total	Total Incl O&P
2530	3/4" diameter	1 Elec	90	.089	L.F.	2.22	4.86		7.08	10
2550	1" diameter		70	.114		3.28	6.25		9.53	13.30
2570	1-1/4" diameter		65	.123		4.01	6.75		10.76	14.80
2600	1-1/2" diameter		60	.133		5.35	7.30		12.65	17.20
2630	2" diameter		50	.160		6.40	8.75		15.15	20.50
2650	2-1/2" diameter		40	.200		11.35	10.95		22.30	29.50
2670	3" diameter	2 Elec	60	.267		15.40	14.60		30	39.50
2700	3-1/2" diameter		54	.296		20.50	16.20		36.70	47.50
2730	4" diameter		50	.320		22	17.50		39.50	51.50
2731	Field bends, 45° to 90°, 1/2" diameter	1 Elec	44	.182	Ea.		9.95		9.95	15.40
2732	3/4" diameter		40	.200			10.95		10.95	16.95
2733	1" diameter		36	.222			12.15		12.15	18.80
2734	1-1/4" diameter		19	.421			23		23	35.50
2735	1-1/2" diameter		18	.444			24.50		24.50	37.50
2736	2" diameter		13	.615			33.50		33.50	52
5000	Electric metallic tubing (EMT), 1/2" diameter		170	.047	L.F.	.67	2.57		3.24	4.73
5020	3/4" diameter		130	.062		.94	3.37		4.31	6.25
5040	1" diameter		115	.070		1.60	3.81		5.41	7.65
5060	1-1/4" diameter		100	.080		2.61	4.38		6.99	9.65
5080	1-1/2" diameter		90	.089		3.35	4.86		8.21	11.25
5100	2" diameter		80	.100		4.20	5.45		9.65	13.05
5120	2-1/2" diameter		60	.133		9.05	7.30		16.35	21.50
5140	3" diameter	2 Elec	100	.160		10.60	8.75		19.35	25.50
5160	3-1/2" diameter		90	.178		13.25	9.70		22.95	29.50
5180	4" diameter		80	.200		14.50	10.95		25.45	33
5200	Field bends, 45° to 90°, 1/2" diameter	1 Elec	89	.090	Ea.		4.92		4.92	7.60
5220	3/4" diameter		80	.100			5.45		5.45	8.45
5240	1" diameter		73	.110			6		6	9.30
5260	1-1/4" diameter		38	.211			11.50		11.50	17.85
5280	1-1/2" diameter		36	.222			12.15		12.15	18.80
5300	2" diameter		26	.308			16.85		16.85	26
5320	Offsets, 1/2" diameter		65	.123			6.75		6.75	10.40
5340	3/4" diameter		62	.129			7.05		7.05	10.95
5360	1" diameter		53	.151			8.25		8.25	12.80
5380	1-1/4" diameter		30	.267			14.60		14.60	22.50
5400	1-1/2" diameter		28	.286			15.65		15.65	24
7600	EMT, "T" fittings with covers, 1/2" diameter, set screw		16	.500		7.90	27.50		35.40	51
9990	Minimum labor/equipment charge		4	2	Job		109		109	169

26 05 33.16 Outlet Boxes

		Crew	Daily Output	Labor-Hours	Unit	Material	2015 Bare Costs Labor	Equipment	Total	Total Incl O&P
0010	**OUTLET BOXES**									
0020	Pressed steel, octagon, 4"	1 Elec	20	.400	Ea.	2.61	22		24.61	37
0060	Covers, blank		64	.125		1.10	6.85		7.95	11.80
0100	Extension rings		40	.200		4.33	10.95		15.28	21.50
0150	Square, 4"		20	.400		2.42	22		24.42	36.50
0200	Extension rings		40	.200		4.37	10.95		15.32	22
0250	Covers, blank		64	.125		1.18	6.85		8.03	11.90
0300	Plaster rings		64	.125		2.40	6.85		9.25	13.25
0652	Switchbox		24	.333		4.18	18.25		22.43	32.50
1100	Concrete, floor, 1 gang		5.30	1.509		89.50	82.50		172	227
9000	Minimum labor/equipment charge		4	2	Job		109		109	169

26 05 Common Work Results for Electrical

26 05 33 – Raceway and Boxes for Electrical Systems

26 05 33.17 Outlet Boxes, Plastic

		Crew	Daily Output	Labor-Hours	Unit	Material	2015 Bare Costs Labor	Equipment	Total	Total Incl O&P
0010	**OUTLET BOXES, PLASTIC**									
0050	4" diameter, round with 2 mounting nails	1 Elec	25	.320	Ea.	2.39	17.50		19.89	29.50
0100	Bar hanger mounted		25	.320		5.10	17.50		22.60	32.50
0200	4", square with 2 mounting nails		25	.320		4.63	17.50		22.13	32
0300	Plaster ring		64	.125		1.98	6.85		8.83	12.80
0400	Switch box with 2 mounting nails, 1 gang		30	.267		4.39	14.60		18.99	27.50
0500	2 gang		25	.320		3.33	17.50		20.83	30.50
0600	3 gang		20	.400		4.93	22		26.93	39.50
0700	Old work box		30	.267		5.70	14.60		20.30	29
9000	Minimum labor/equipment charge		4	2	Job		109		109	169

26 05 33.18 Pull Boxes

		Crew	Daily Output	Labor-Hours	Unit	Material	Labor	Equipment	Total	Total Incl O&P
0010	**PULL BOXES**									
0100	Steel, pull box, NEMA 1, type SC, 6" W x 6" H x 4" D	1 Elec	8	1	Ea.	9.65	54.50		64.15	95
0200	8" W x 8" H x 4" D		8	1		14.50	54.50		69	100
0300	10" W x 12" H x 6" D		5.30	1.509		24.50	82.50		107	155
0400	16" W x 20" H x 8" D		4	2		84.50	109		193.50	262
0500	20" W x 24" H x 8" D		3.20	2.500		98	137		235	320
0600	24" W x 36" H x 8" D		2.70	2.963		151	162		313	415
0650	Pull box, hinged, NEMA 1, 6" W x 6" H x 4" D		8	1		11.60	54.50		66.10	97.50
0800	12" W x 16" H x 6" D		4.70	1.702		49.50	93		142.50	199
1000	20" W x 20" H x 6" D		3.60	2.222		86	122		208	283
1200	20" W x 20" H x 8" D		3.20	2.500		149	137		286	375
1400	24" W x 36" H x 8" D		2.70	2.963		237	162		399	510
1600	24" W x 42" H x 8" D		2	4		350	219		569	725

26 05 33.23 Wireway

		Crew	Daily Output	Labor-Hours	Unit	Material	Labor	Equipment	Total	Total Incl O&P
0010	**WIREWAY** to 15' high									
0100	NEMA 1, Screw cover w/fittings and supports, 2-1/2" x 2-1/2"	1 Elec	45	.178	L.F.	10.85	9.70		20.55	27
0200	4" x 4"	"	40	.200		11.35	10.95		22.30	29.50
0400	6" x 6"	2 Elec	60	.267		19.60	14.60		34.20	44
0600	8" x 8"	"	40	.400		33.50	22		55.50	71
4475	NEMA 3R, Screw cover w/fittings and supports, 4" x 4"	1 Elec	36	.222		16.50	12.15		28.65	37
4480	6" x 6"	2 Elec	55	.291		22	15.90		37.90	49
4485	8" x 8"		36	.444		34	24.50		58.50	75
4490	12" x 12"		18	.889		54.50	48.50		103	136

26 05 33.95 Cutting and Drilling

		Crew	Daily Output	Labor-Hours	Unit	Material	Labor	Equipment	Total	Total Incl O&P
0010	**CUTTING AND DRILLING**									
0100	Hole drilling to 10' high, concrete wall									
0110	8" thick, 1/2" pipe size	R-31	12	.667	Ea.	.24	36.50	4.59	41.33	62
0120	3/4" pipe size		12	.667		.24	36.50	4.59	41.33	62
0130	1" pipe size		9.50	.842		.38	46	5.80	52.18	78.50
0140	1-1/4" pipe size		9.50	.842		.38	46	5.80	52.18	78.50
0150	1-1/2" pipe size		9.50	.842		.38	46	5.80	52.18	78.50
0160	2" pipe size		4.40	1.818		.51	99.50	12.55	112.56	168
0170	2-1/2" pipe size		4.40	1.818		.51	99.50	12.55	112.56	168
0180	3" pipe size		4.40	1.818		.51	99.50	12.55	112.56	168
0190	3-1/2" pipe size		3.30	2.424		.66	133	16.70	150.36	224
0200	4" pipe size		3.30	2.424		.66	133	16.70	150.36	224
0500	12" thick, 1/2" pipe size		9.40	.851		.35	46.50	5.85	52.70	79
0520	3/4" pipe size		9.40	.851		.35	46.50	5.85	52.70	79
0540	1" pipe size		7.30	1.096		.56	60	7.55	68.11	102
0560	1-1/4" pipe size		7.30	1.096		.56	60	7.55	68.11	102

26 05 Common Work Results for Electrical

26 05 33 – Raceway and Boxes for Electrical Systems

26 05 33.95 Cutting and Drilling		Crew	Daily Output	Labor-Hours	Unit	Material	2015 Bare Costs Labor	Equipment	Total	Total Incl O&P
0570	1-1/2" pipe size	R-31	7.30	1.096	Ea.	.56	60	7.55	68.11	102
0580	2" pipe size		3.60	2.222		.76	122	15.30	138.06	206
0590	2-1/2" pipe size		3.60	2.222		.76	122	15.30	138.06	206
0600	3" pipe size		3.60	2.222		.76	122	15.30	138.06	206
0610	3-1/2" pipe size		2.80	2.857		.99	156	19.70	176.69	265
0630	4" pipe size		2.50	3.200		.99	175	22	197.99	297
0650	16" thick, 1/2" pipe size		7.60	1.053		.47	57.50	7.25	65.22	97.50
0670	3/4" pipe size		7	1.143		.47	62.50	7.85	70.82	106
0690	1" pipe size		6	1.333		.75	73	9.20	82.95	124
0710	1-1/4" pipe size		5.50	1.455		.75	79.50	10	90.25	135
0730	1-1/2" pipe size		5.50	1.455		.75	79.50	10	90.25	135
0750	2" pipe size		3	2.667		1.02	146	18.35	165.37	247
0770	2-1/2" pipe size		2.70	2.963		1.02	162	20.50	183.52	275
0790	3" pipe size		2.50	3.200		1.02	175	22	198.02	297
0810	3-1/2" pipe size		2.30	3.478		1.32	190	24	215.32	325
0830	4" pipe size		2	4		1.32	219	27.50	247.82	370
0850	20" thick, 1/2" pipe size		6.40	1.250		.59	68.50	8.60	77.69	116
0870	3/4" pipe size		6	1.333		.59	73	9.20	82.79	124
0890	1" pipe size		5	1.600		.94	87.50	11	99.44	149
0910	1-1/4" pipe size		4.80	1.667		.94	91	11.50	103.44	155
0930	1-1/2" pipe size		4.60	1.739		.94	95	12	107.94	161
0950	2" pipe size		2.70	2.963		1.27	162	20.50	183.77	275
0970	2-1/2" pipe size		2.40	3.333		1.27	182	23	206.27	310
0990	3" pipe size		2.20	3.636		1.27	199	25	225.27	340
1010	3-1/2" pipe size		2	4		1.64	219	27.50	248.14	370
1030	4" pipe size		1.70	4.706		1.64	257	32.50	291.14	435
1050	24" thick, 1/2" pipe size		5.50	1.455		.71	79.50	10	90.21	135
1070	3/4" pipe size		5.10	1.569		.71	86	10.80	97.51	146
1090	1" pipe size		4.30	1.860		1.13	102	12.80	115.93	173
1110	1-1/4" pipe size		4	2		1.13	109	13.80	123.93	185
1130	1-1/2" pipe size		4	2		1.13	109	13.80	123.93	185
1150	2" pipe size		2.40	3.333		1.52	182	23	206.52	310
1170	2-1/2" pipe size		2.20	3.636		1.52	199	25	225.52	340
1190	3" pipe size		2	4		1.52	219	27.50	248.02	370
1210	3-1/2" pipe size		1.80	4.444		1.97	243	30.50	275.47	410
1230	4" pipe size		1.50	5.333		1.97	292	37	330.97	495
1500	Brick wall, 8" thick, 1/2" pipe size		18	.444		.24	24.50	3.06	27.80	41
1520	3/4" pipe size		18	.444		.24	24.50	3.06	27.80	41
1540	1" pipe size		13.30	.602		.38	33	4.14	37.52	56
1560	1-1/4" pipe size		13.30	.602		.38	33	4.14	37.52	56
1580	1-1/2" pipe size		13.30	.602		.38	33	4.14	37.52	56
1600	2" pipe size		5.70	1.404		.51	77	9.65	87.16	130
1620	2-1/2" pipe size		5.70	1.404		.51	77	9.65	87.16	130
1640	3" pipe size		5.70	1.404		.51	77	9.65	87.16	130
1660	3-1/2" pipe size		4.40	1.818		.66	99.50	12.55	112.71	169
1680	4" pipe size		4	2		.66	109	13.80	123.46	185
1700	12" thick, 1/2" pipe size		14.50	.552		.35	30	3.80	34.15	51
1720	3/4" pipe size		14.50	.552		.35	30	3.80	34.15	51
1740	1" pipe size		11	.727		.56	40	5	45.56	67.50
1760	1-1/4" pipe size		11	.727		.56	40	5	45.56	67.50
1780	1-1/2" pipe size		11	.727		.56	40	5	45.56	67.50
1800	2" pipe size		5	1.600		.76	87.50	11	99.26	149
1820	2-1/2" pipe size		5	1.600		.76	87.50	11	99.26	149

26 05 Common Work Results for Electrical

26 05 33 – Raceway and Boxes for Electrical Systems

26 05 33.95 Cutting and Drilling		Crew	Daily Output	Labor-Hours	Unit	Material	2015 Bare Costs Labor	Equipment	Total	Total Incl O&P
1840	3" pipe size	R-31	5	1.600	Ea.	.76	87.50	11	99.26	149
1860	3-1/2" pipe size		3.80	2.105		.99	115	14.50	130.49	195
1880	4" pipe size		3.30	2.424		.99	133	16.70	150.69	224
1900	16" thick, 1/2" pipe size		12.30	.650		.47	35.50	4.48	40.45	60.50
1920	3/4" pipe size		12.30	.650		.47	35.50	4.48	40.45	60.50
1940	1" pipe size		9.30	.860		.75	47	5.95	53.70	80.50
1960	1-1/4" pipe size		9.30	.860		.75	47	5.95	53.70	80.50
1980	1-1/2" pipe size		9.30	.860		.75	47	5.95	53.70	80.50
2000	2" pipe size		4.40	1.818		1.02	99.50	12.55	113.07	169
2010	2-1/2" pipe size		4.40	1.818		1.02	99.50	12.55	113.07	169
2030	3" pipe size		4.40	1.818		1.02	99.50	12.55	113.07	169
2050	3-1/2" pipe size		3.30	2.424		1.32	133	16.70	151.02	225
2070	4" pipe size		3	2.667		1.32	146	18.35	165.67	247
2090	20" thick, 1/2" pipe size		10.70	.748		.59	41	5.15	46.74	70
2110	3/4" pipe size		10.70	.748		.59	41	5.15	46.74	70
2130	1" pipe size		8	1		.94	54.50	6.90	62.34	93
2150	1-1/4" pipe size		8	1		.94	54.50	6.90	62.34	93
2170	1-1/2" pipe size		8	1		.94	54.50	6.90	62.34	93
2190	2" pipe size		4	2		1.27	109	13.80	124.07	186
2210	2-1/2" pipe size		4	2		1.27	109	13.80	124.07	186
2230	3" pipe size		4	2		1.27	109	13.80	124.07	186
2250	3-1/2" pipe size		3	2.667		1.64	146	18.35	165.99	248
2270	4" pipe size		2.70	2.963		1.64	162	20.50	184.14	275
2290	24" thick, 1/2" pipe size		9.40	.851		.71	46.50	5.85	53.06	79
2310	3/4" pipe size		9.40	.851		.71	46.50	5.85	53.06	79
2330	1" pipe size		7.10	1.127		1.13	61.50	7.75	70.38	105
2350	1-1/4" pipe size		7.10	1.127		1.13	61.50	7.75	70.38	105
2370	1-1/2" pipe size		7.10	1.127		1.13	61.50	7.75	70.38	105
2390	2" pipe size		3.60	2.222		1.52	122	15.30	138.82	207
2410	2-1/2" pipe size		3.60	2.222		1.52	122	15.30	138.82	207
2430	3" pipe size		3.60	2.222		1.52	122	15.30	138.82	207
2450	3-1/2" pipe size		2.80	2.857		1.97	156	19.70	177.67	266
2470	4" pipe size		2.50	3.200		1.97	175	22	198.97	298
3000	Knockouts to 8' high, metal boxes & enclosures									
3020	With hole saw, 1/2" pipe size	1 Elec	53	.151	Ea.		8.25		8.25	12.80
3040	3/4" pipe size		47	.170			9.30		9.30	14.40
3050	1" pipe size		40	.200			10.95		10.95	16.95
3060	1-1/4" pipe size		36	.222			12.15		12.15	18.80
3070	1-1/2" pipe size		32	.250			13.70		13.70	21
3080	2" pipe size		27	.296			16.20		16.20	25
3090	2-1/2" pipe size		20	.400			22		22	34
4010	3" pipe size		16	.500			27.50		27.50	42.50
4030	3-1/2" pipe size		13	.615			33.50		33.50	52
4050	4" pipe size		11	.727			40		40	61.50

26 05 39 – Underfloor Raceways for Electrical Systems

26 05 39.30 Conduit In Concrete Slab

		Crew	Daily Output	Labor-Hours	Unit	Material	Labor	Equipment	Total	Total Incl O&P
0010	**CONDUIT IN CONCRETE SLAB** Including terminations,									
0020	fittings and supports									
3230	PVC, schedule 40, 1/2" diameter	1 Elec	270	.030	L.F.	.57	1.62		2.19	3.14
3250	3/4" diameter		230	.035		.66	1.90		2.56	3.67
3270	1" diameter		200	.040		.87	2.19		3.06	4.34
3300	1-1/4" diameter		170	.047		1.17	2.57		3.74	5.30

26 05 Common Work Results for Electrical

26 05 39 – Underfloor Raceways for Electrical Systems

26 05 39.30 Conduit In Concrete Slab		Crew	Daily Output	Labor-Hours	Unit	Material	2015 Bare Costs Labor	Equipment	Total	Total Incl O&P
3330	1-1/2" diameter	1 Elec	140	.057	L.F.	1.42	3.13		4.55	6.40
3350	2" diameter		120	.067		1.78	3.65		5.43	7.60
4350	Rigid galvanized steel, 1/2" diameter		200	.040		2.48	2.19		4.67	6.10
4400	3/4" diameter		170	.047		2.62	2.57		5.19	6.85
4450	1" diameter		130	.062		3.60	3.37		6.97	9.15
4500	1-1/4" diameter		110	.073		4.94	3.98		8.92	11.60
4600	1-1/2" diameter		100	.080		5.50	4.38		9.88	12.85
4800	2" diameter		90	.089		6.80	4.86		11.66	15.05
9000	Minimum labor/equipment charge		4	2	Job		109		109	169

26 05 39.40 Conduit In Trench

		Crew	Daily Output	Labor-Hours	Unit	Material	Labor	Equipment	Total	Total Incl O&P
0010	**CONDUIT IN TRENCH** Includes terminations and fittings									
0020	Does not include excavation or backfill, see Section 31 23 16.00									
0200	Rigid galvanized steel, 2" diameter	1 Elec	150	.053	L.F.	6.40	2.92		9.32	11.55
0400	2-1/2" diameter	"	100	.080		12.45	4.38		16.83	20.50
0600	3" diameter	2 Elec	160	.100		14.55	5.45		20	24.50
0800	3-1/2" diameter		140	.114		19.20	6.25		25.45	30.50
1000	4" diameter		100	.160		21	8.75		29.75	37
1200	5" diameter		80	.200		44	10.95		54.95	65
1400	6" diameter		60	.267		61	14.60		75.60	90
9000	Minimum labor/equipment charge	1 Elec	4	2	Job		109		109	169

26 05 80 – Wiring Connections

26 05 80.10 Motor Connections

		Crew	Daily Output	Labor-Hours	Unit	Material	Labor	Equipment	Total	Total Incl O&P
0010	**MOTOR CONNECTIONS**									
0020	Flexible conduit and fittings, 115 volt, 1 phase, up to 1 HP motor	1 Elec	8	1	Ea.	6.35	54.50		60.85	91.50
9000	Minimum labor/equipment charge	"	4	2	Job		109		109	169

26 05 90 – Residential Applications

26 05 90.10 Residential Wiring

		Crew	Daily Output	Labor-Hours	Unit	Material	Labor	Equipment	Total	Total Incl O&P
0010	**RESIDENTIAL WIRING**									
0020	20' avg. runs and #14/2 wiring incl. unless otherwise noted									
1000	Service & panel, includes 24' SE-AL cable, service eye, meter,									
1010	Socket, panel board, main bkr., ground rod, 15 or 20 amp									
1020	1-pole circuit breakers, and misc. hardware									
1100	100 amp, with 10 branch breakers	1 Elec	1.19	6.723	Ea.	560	370		930	1,175
1110	With PVC conduit and wire		.92	8.696		605	475		1,080	1,400
1120	With RGS conduit and wire		.73	10.959		765	600		1,365	1,775
1150	150 amp, with 14 branch breakers		1.03	7.767		860	425		1,285	1,600
1170	With PVC conduit and wire		.82	9.756		955	535		1,490	1,875
1180	With RGS conduit and wire		.67	11.940		1,275	655		1,930	2,400
1200	200 amp, with 18 branch breakers	2 Elec	1.80	8.889		1,150	485		1,635	2,025
1220	With PVC conduit and wire		1.46	10.959		1,250	600		1,850	2,300
1230	With RGS conduit and wire		1.24	12.903		1,650	705		2,355	2,900
1800	Lightning surge suppressor	1 Elec	32	.250		50.50	13.70		64.20	77
2000	Switch devices									
2100	Single pole, 15 amp, Ivory, with a 1-gang box, cover plate,									
2110	Type NM (Romex) cable	1 Elec	17.10	.468	Ea.	12.35	25.50		37.85	53
2120	Type MC (BX) cable		14.30	.559		21.50	30.50		52	71
2130	EMT & wire		5.71	1.401		30	76.50		106.50	152
2150	3-way, #14/3, type NM cable		14.55	.550		14.85	30		44.85	63
2170	Type MC cable		12.31	.650		27	35.50		62.50	84.50
2180	EMT & wire		5	1.600		32	87.50		119.50	171
2200	4-way, #14/3, type NM cable		14.55	.550		21	30		51	69.50

26 05 Common Work Results for Electrical

26 05 90 – Residential Applications

26 05 90.10 Residential Wiring		Crew	Daily Output	Labor-Hours	Unit	Material	2015 Bare Costs Labor	Equipment	Total	Total Incl O&P
2220	Type MC cable	1 Elec	12.31	.650	Ea.	33	35.50		68.50	91.50
2230	EMT & wire		5	1.600		38.50	87.50		126	178
2250	S.P., 20 amp, #12/2, type NM cable		13.33	.600		22	33		55	75
2270	Type MC cable		11.43	.700		28.50	38.50		67	91
2280	EMT & wire		4.85	1.649		41	90		131	185
2290	S.P. rotary dimmer, 600W, no wiring		17	.471		29.50	25.50		55	72.50
2300	S.P. rotary dimmer, 600W, type NM cable		14.55	.550		34.50	30		64.50	84.50
2320	Type MC cable		12.31	.650		43.50	35.50		79	103
2330	EMT & wire		5	1.600		53.50	87.50		141	195
2350	3-way rotary dimmer, type NM cable		13.33	.600		28.50	33		61.50	82.50
2370	Type MC cable		11.43	.700		37.50	38.50		76	101
2380	EMT & wire		4.85	1.649		47.50	90		137.50	193
2400	Interval timer wall switch, 20 amp, 1-30 min., #12/2									
2410	Type NM cable	1 Elec	14.55	.550	Ea.	56	30		86	108
2420	Type MC cable		12.31	.650		60	35.50		95.50	121
2430	EMT & wire		5	1.600		75	87.50		162.50	219
2500	Decorator style									
2510	S.P., 15 amp, type NM cable	1 Elec	17.10	.468	Ea.	16.25	25.50		41.75	57.50
2520	Type MC cable		14.30	.559		25	30.50		55.50	75
2530	EMT & wire		5.71	1.401		34	76.50		110.50	156
2550	3-way, #14/3, type NM cable		14.55	.550		18.75	30		48.75	67
2570	Type MC cable		12.31	.650		30.50	35.50		66	88.50
2580	EMT & wire		5	1.600		36	87.50		123.50	176
2600	4-way, #14/3, type NM cable		14.55	.550		25	30		55	74
2620	Type MC cable		12.31	.650		37	35.50		72.50	95.50
2630	EMT & wire		5	1.600		42	87.50		129.50	183
2650	S.P., 20 amp, #12/2, type NM cable		13.33	.600		26	33		59	79.50
2670	Type MC cable		11.43	.700		32.50	38.50		71	95
2680	EMT & wire		4.85	1.649		45	90		135	190
2700	S.P., slide dimmer, type NM cable		17.10	.468		28.50	25.50		54	71
2720	Type MC cable		14.30	.559		37.50	30.50		68	89
2730	EMT & wire		5.71	1.401		47.50	76.50		124	172
2750	S.P., touch dimmer, type NM cable		17.10	.468		32	25.50		57.50	75
2770	Type MC cable		14.30	.559		41	30.50		71.50	93
2780	EMT & wire		5.71	1.401		51.50	76.50		128	176
2800	3-way touch dimmer, type NM cable		13.33	.600		52	33		85	108
2820	Type MC cable		11.43	.700		60.50	38.50		99	127
2830	EMT & wire		4.85	1.649		71	90		161	218
3000	Combination devices									
3100	S.P. switch/15 amp recpt., Ivory, 1-gang box, plate									
3110	Type NM cable	1 Elec	11.43	.700	Ea.	24.50	38.50		63	86.50
3120	Type MC cable		10	.800		33.50	44		77.50	105
3130	EMT & wire		4.40	1.818		43.50	99.50		143	202
3150	S.P. switch/pilot light, type NM cable		11.43	.700		24.50	38.50		63	86.50
3170	Type MC cable		10	.800		33.50	44		77.50	105
3180	EMT & wire		4.43	1.806		43.50	99		142.50	201
3190	2-S.P. switches, 2-#14/2, no wiring		14	.571		6.90	31.50		38.40	56
3200	2-S.P. switches, 2-#14/2, type NM cables		10	.800		27.50	44		71.50	98
3220	Type MC cable		8.89	.900		41	49		90	121
3230	EMT & wire		4.10	1.951		46.50	107		153.50	216
3250	3-way switch/15 amp recpt., #14/3, type NM cable		10	.800		32	44		76	103
3270	Type MC cable		8.89	.900		44	49		93	124
3280	EMT & wire		4.10	1.951		49	107		156	219

26 05 Common Work Results for Electrical

26 05 90 – Residential Applications

26 05 90.10 Residential Wiring		Crew	Daily Output	Labor-Hours	Unit	Material	2015 Bare Costs Labor	Equipment	Total	Total Incl O&P
3300	2-3 way switches, 2-#14/3, type NM cables	1 Elec	8.89	.900	Ea.	41.50	49		90.50	122
3320	Type MC cable		8	1		60.50	54.50		115	151
3330	EMT & wire		4	2		56.50	109		165.50	231
3350	S.P. switch/20 amp recpt., #12/2, type NM cable		10	.800		32	44		76	103
3370	Type MC cable		8.89	.900		36	49		85	116
3380	EMT & wire		4.10	1.951		51	107		158	221
3400	Decorator style									
3410	S.P. switch/15 amp recpt., type NM cable	1 Elec	11.43	.700	Ea.	28.50	38.50		67	90.50
3420	Type MC cable		10	.800		37	44		81	109
3430	EMT & wire		4.40	1.818		47.50	99.50		147	206
3450	S.P. switch/pilot light, type NM cable		11.43	.700		28.50	38.50		67	90.50
3470	Type MC cable		10	.800		37.50	44		81.50	109
3480	EMT & wire		4.40	1.818		47.50	99.50		147	207
3500	2-S.P. switches, 2-#14/2, type NM cables		10	.800		31.50	44		75.50	103
3520	Type MC cable		8.89	.900		44.50	49		93.50	125
3530	EMT & wire		4.10	1.951		50.50	107		157.50	221
3550	3-way/15 amp recpt., #14/3, type NM cable		10	.800		36	44		80	108
3570	Type MC cable		8.89	.900		47.50	49		96.50	129
3580	EMT & wire		4.10	1.951		53	107		160	223
3650	2-3 way switches, 2-#14/3, type NM cables		8.89	.900		45.50	49		94.50	126
3670	Type MC cable		8	1		64.50	54.50		119	156
3680	EMT & wire		4	2		60.50	109		169.50	236
3700	S.P. switch/20 amp recpt., #12/2, type NM cable		10	.800		35.50	44		79.50	108
3720	Type MC cable		8.89	.900		39.50	49		88.50	120
3730	EMT & wire		4.10	1.951		55	107		162	226
4000	Receptacle devices									
4010	Duplex outlet, 15 amp recpt., Ivory, 1-gang box, plate									
4015	Type NM cable	1 Elec	14.55	.550	Ea.	10.80	30		40.80	58.50
4020	Type MC cable		12.31	.650		19.75	35.50		55.25	76.50
4030	EMT & wire		5.33	1.501		28.50	82		110.50	159
4050	With #12/2, type NM cable		12.31	.650		13.30	35.50		48.80	69.50
4070	Type MC cable		10.67	.750		19.85	41		60.85	85.50
4080	EMT & wire		4.71	1.699		32.50	93		125.50	180
4100	20 amp recpt., #12/2, type NM cable		12.31	.650		19.95	35.50		55.45	77
4120	Type MC cable		10.67	.750		26.50	41		67.50	92.50
4130	EMT & wire		4.71	1.699		39	93		132	187
4140	For GFI see Section 26 05 90.10 line 4300 below									
4150	Decorator style, 15 amp recpt., type NM cable	1 Elec	14.55	.550	Ea.	14.70	30		44.70	62.50
4170	Type MC cable		12.31	.650		23.50	35.50		59	81
4180	EMT & wire		5.33	1.501		32.50	82		114.50	163
4200	With #12/2, type NM cable		12.31	.650		17.20	35.50		52.70	74
4220	Type MC cable		10.67	.750		23.50	41		64.50	89.50
4230	EMT & wire		4.71	1.699		36.50	93		129.50	184
4250	20 amp recpt. #12/2, type NM cable		12.31	.650		24	35.50		59.50	81
4270	Type MC cable		10.67	.750		30.50	41		71.50	97
4280	EMT & wire		4.71	1.699		43	93		136	192
4300	GFI, 15 amp recpt., type NM cable		12.31	.650		41.50	35.50		77	101
4320	Type MC cable		10.67	.750		50.50	41		91.50	119
4330	EMT & wire		4.71	1.699		59	93		152	209
4350	GFI with #12/2, type NM cable		10.67	.750		44	41		85	112
4370	Type MC cable		9.20	.870		50.50	47.50		98	129
4380	EMT & wire		4.21	1.900		63	104		167	231
4400	20 amp recpt., #12/2 type NM cable		10.67	.750		52	41		93	121

26 05 Common Work Results for Electrical

26 05 90 – Residential Applications

26 05 90.10 Residential Wiring		Crew	Daily Output	Labor-Hours	Unit	Material	2015 Bare Costs Labor	Equipment	Total	Total Incl O&P
4420	Type MC cable	1 Elec	9.20	.870	Ea.	58.50	47.50		106	138
4430	EMT & wire		4.21	1.900		71.50	104		175.50	240
4500	Weather-proof cover for above receptacles, add	↓	32	.250	↓	3.46	13.70		17.16	25
4550	Air conditioner outlet, 20 amp-240 volt recpt.									
4560	30' of #12/2, 2 pole circuit breaker									
4570	Type NM cable	1 Elec	10	.800	Ea.	63	44		107	138
4580	Type MC cable		9	.889		70.50	48.50		119	153
4590	EMT & wire		4	2		82.50	109		191.50	260
4600	Decorator style, type NM cable		10	.800		68	44		112	143
4620	Type MC cable		9	.889		75.50	48.50		124	159
4630	EMT & wire	↓	4	2	↓	87	109		196	265
4650	Dryer outlet, 30 amp-240 volt recpt., 20' of #10/3									
4660	2 pole circuit breaker									
4670	Type NM cable	1 Elec	6.41	1.248	Ea.	61.50	68.50		130	174
4680	Type MC cable		5.71	1.401		64.50	76.50		141	190
4690	EMT & wire	↓	3.48	2.299	↓	76	126		202	279
4700	Range outlet, 50 amp-240 volt recpt., 30' of #8/3									
4710	Type NM cable	1 Elec	4.21	1.900	Ea.	90.50	104		194.50	261
4720	Type MC cable		4	2		124	109		233	305
4730	EMT & wire		2.96	2.703		108	148		256	350
4750	Central vacuum outlet, Type NM cable		6.40	1.250		58.50	68.50		127	171
4770	Type MC cable		5.71	1.401		69.50	76.50		146	196
4780	EMT & wire	↓	3.48	2.299	↓	83.50	126		209.50	287
4800	30 amp-110 volt locking recpt., #10/2 circ. bkr.									
4810	Type NM cable	1 Elec	6.20	1.290	Ea.	67	70.50		137.50	183
4820	Type MC cable		5.40	1.481		80	81		161	213
4830	EMT & wire	↓	3.20	2.500	↓	94	137		231	315
4900	Low voltage outlets									
4910	Telephone recpt., 20' of 4/C phone wire	1 Elec	26	.308	Ea.	9.30	16.85		26.15	36.50
4920	TV recpt., 20' of RG59U coax wire, F type connector	"	16	.500	"	18.75	27.50		46.25	63
4950	Door bell chime, transformer, 2 buttons, 60' of bellwire									
4970	Economy model	1 Elec	11.50	.696	Ea.	51.50	38		89.50	116
4980	Custom model		11.50	.696		99	38		137	168
4990	Luxury model, 3 buttons	↓	9.50	.842	↓	281	46		327	380
6000	Lighting outlets									
6050	Wire only (for fixture), type NM cable	1 Elec	32	.250	Ea.	6.90	13.70		20.60	28.50
6070	Type MC cable		24	.333		12.30	18.25		30.55	41.50
6080	EMT & wire		10	.800		20	44		64	90
6100	Box (4"), and wire (for fixture), type NM cable		25	.320		14.45	17.50		31.95	43
6120	Type MC cable		20	.400		19.90	22		41.90	56
6130	EMT & wire	↓	11	.727	↓	27.50	40		67.50	92
6200	Fixtures (use with lines 6050 or 6100 above)									
6210	Canopy style, economy grade	1 Elec	40	.200	Ea.	32	10.95		42.95	52
6220	Custom grade		40	.200		53.50	10.95		64.45	76
6250	Dining room chandelier, economy grade		19	.421		79.50	23		102.50	123
6270	Luxury grade		15	.533		715	29		744	830
6310	Kitchen fixture (fluorescent), economy grade		30	.267		71.50	14.60		86.10	102
6320	Custom grade		25	.320		219	17.50		236.50	268
6350	Outdoor, wall mounted, economy grade		30	.267		30	14.60		44.60	55.50
6360	Custom grade		30	.267		120	14.60		134.60	155
6370	Luxury grade		25	.320		248	17.50		265.50	300
6410	Outdoor PAR floodlights, 1 lamp, 150 watt		20	.400		32.50	22		54.50	69.50
6420	2 lamp, 150 watt each	↓	20	.400	↓	53.50	22		75.50	93

For customer support on your Commercial Renovation Cost Data, call 877.791.4977.

26 05 Common Work Results for Electrical

26 05 90 – Residential Applications

26 05 90.10 Residential Wiring

		Crew	Daily Output	Labor-Hours	Unit	Material	2015 Bare Costs Labor	Equipment	Total	Total Incl O&P
6425	Motion sensing, 2 lamp, 150 watt each	1 Elec	20	.400	Ea.	87	22		109	130
6430	For infrared security sensor, add		32	.250		132	13.70		145.70	166
6450	Outdoor, quartz-halogen, 300 watt flood		20	.400		39	22		61	76.50
6600	Recessed downlight, round, pre-wired, 50 or 75 watt trim		30	.267		80	14.60		94.60	111
6610	With shower light trim		30	.267		89	14.60		103.60	121
6620	With wall washer trim		28	.286		99.50	15.65		115.15	133
6630	With eye-ball trim		28	.286		99.50	15.65		115.15	133
6700	Porcelain lamp holder		40	.200		2.96	10.95		13.91	20
6710	With pull switch		40	.200		6.55	10.95		17.50	24
6750	Fluorescent strip, 2-20 watt tube, wrap around diffuser, 24"		24	.333		49.50	18.25		67.75	82.50
6760	1-34 watt tube, 48"		24	.333		87	18.25		105.25	124
6770	2-34 watt tubes, 48"		20	.400		103	22		125	147
6800	Bathroom heat lamp, 1-250 watt		28	.286		44	15.65		59.65	72
6810	2-250 watt lamps		28	.286		70	15.65		85.65	101
6820	For timer switch, see Section 26 05 90.10 line 2400									
6900	Outdoor post lamp, incl. post, fixture, 35' of #14/2									
6910	Type NMC cable	1 Elec	3.50	2.286	Ea.	258	125		383	480
6920	Photo-eye, add		27	.296		32	16.20		48.20	60.50
6950	Clock dial time switch, 24 hr., w/enclosure, type NM cable		11.43	.700		73.50	38.50		112	141
6970	Type MC cable		11	.727		82.50	40		122.50	152
6980	EMT & wire		4.85	1.649		91	90		181	240
7000	Alarm systems									
7050	Smoke detectors, box, #14/3, type NM cable	1 Elec	14.55	.550	Ea.	34.50	30		64.50	84.50
7070	Type MC cable		12.31	.650		44	35.50		79.50	104
7080	EMT & wire		5	1.600		49.50	87.50		137	191
7090	For relay output to security system, add					12			12	13.20
8000	Residential equipment									
8050	Disposal hook-up, incl. switch, outlet box, 3' of flex									
8060	20 amp-1 pole circ. bkr., and 25' of #12/2									
8070	Type NM cable	1 Elec	10	.800	Ea.	32.50	44		76.50	104
8080	Type MC cable		8	1		39.50	54.50		94	128
8090	EMT & wire		5	1.600		54.50	87.50		142	196
8100	Trash compactor or dishwasher hook-up, incl. outlet box,									
8110	3' of flex, 15 amp-1 pole circ. bkr., and 25' of #14/2									
8120	Type NM cable	1 Elec	10	.800	Ea.	24	44		68	94
8130	Type MC cable		8	1		34	54.50		88.50	122
8140	EMT & wire		5	1.600		46	87.50		133.50	187
8150	Hot water sink dispensor hook-up, use line 8100									
8200	Vent/exhaust fan hook-up, type NM cable	1 Elec	32	.250	Ea.	6.90	13.70		20.60	28.50
8220	Type MC cable		24	.333		12.30	18.25		30.55	41.50
8230	EMT & wire		10	.800		20	44		64	90
8250	Bathroom vent fan, 50 CFM (use with above hook-up)									
8260	Economy model	1 Elec	15	.533	Ea.	23	29		52	70.50
8270	Low noise model		15	.533		41	29		70	90
8280	Custom model		12	.667		127	36.50		163.50	197
8300	Bathroom or kitchen vent fan, 110 CFM									
8310	Economy model	1 Elec	15	.533	Ea.	67.50	29		96.50	120
8320	Low noise model	"	15	.533	"	92	29		121	146
8350	Paddle fan, variable speed (w/o lights)									
8360	Economy model (AC motor)	1 Elec	10	.800	Ea.	109	44		153	188
8362	With light kit		10	.800		150	44		194	233
8370	Custom model (AC motor)		10	.800		227	44		271	320
8372	With light kit		10	.800		268	44		312	365

26 05 Common Work Results for Electrical

26 05 90 – Residential Applications

26 05 90.10 Residential Wiring

		Crew	Daily Output	Labor-Hours	Unit	Material	2015 Bare Costs Labor	Equipment	Total	Total Incl O&P
8380	Luxury model (DC motor)	1 Elec	8	1	Ea.	330	54.50		384.50	450
8382	With light kit		8	1		375	54.50		429.50	495
8390	Remote speed switch for above, add	↓	12	.667	↓	37.50	36.50		74	97.50
8500	Whole house exhaust fan, ceiling mount, 36", variable speed									
8510	Remote switch, incl. shutters, 20 amp-1 pole circ. bkr.									
8520	30' of #12/2, type NM cable	1 Elec	4	2	Ea.	1,300	109		1,409	1,600
8530	Type MC cable		3.50	2.286		1,300	125		1,425	1,625
8540	EMT & wire	↓	3	2.667	↓	1,325	146		1,471	1,675
8600	Whirlpool tub hook-up, incl. timer switch, outlet box									
8610	3' of flex, 20 amp-1 pole GFI circ. bkr.									
8620	30' of #12/2, type NM cable	1 Elec	5	1.600	Ea.	142	87.50		229.50	292
8630	Type MC cable		4.20	1.905		145	104		249	320
8640	EMT & wire	↓	3.40	2.353		158	129		287	370
8650	Hot water heater hook-up, incl. 1-2 pole circ. bkr., box;									
8660	3' of flex, 20' of #10/2, type NM cable	1 Elec	5	1.600	Ea.	33	87.50		120.50	173
8670	Type MC cable		4.20	1.905		43	104		147	209
8680	EMT & wire	↓	3.40	2.353		48.50	129		177.50	252
9000	Heating/air conditioning									
9050	Furnace/boiler hook-up, incl. firestat, local on-off switch									
9060	Emergency switch, and 40' of type NM cable	1 Elec	4	2	Ea.	52.50	109		161.50	227
9070	Type MC cable		3.50	2.286		65.50	125		190.50	267
9080	EMT & wire	↓	1.50	5.333	↓	83	292		375	540
9100	Air conditioner hook-up, incl. local 60 amp disc. switch									
9110	3' sealtite, 40 amp, 2 pole circuit breaker									
9130	40' of #8/2, type NM cable	1 Elec	3.50	2.286	Ea.	163	125		288	375
9140	Type MC cable		3	2.667		216	146		362	465
9150	EMT & wire	↓	1.30	6.154		204	335		539	745
9200	Heat pump hook-up, 1-40 & 1-100 amp 2 pole circ. bkr.									
9210	Local disconnect switch, 3' sealtite									
9220	40' of #8/2 & 30' of #3/2									
9230	Type NM cable	1 Elec	1.30	6.154	Ea.	520	335		855	1,100
9240	Type MC cable		1.08	7.407		535	405		940	1,200
9250	EMT & wire	↓	.94	8.511	↓	585	465		1,050	1,375
9500	Thermostat hook-up, using low voltage wire									
9520	Heating only, 25' of #18-3	1 Elec	24	.333	Ea.	8.80	18.25		27.05	37.50
9530	Heating/cooling, 25' of #18-4	"	20	.400	"	11.45	22		33.45	46.50

26 22 Low-Voltage Transformers

26 22 13 – Low-Voltage Distribution Transformers

26 22 13.10 Transformer, Dry-Type

		Crew	Daily Output	Labor-Hours	Unit	Material	2015 Bare Costs Labor	Equipment	Total	Total Incl O&P
0010	**TRANSFORMER, DRY-TYPE**									
0050	Single phase, 240/480 volt primary, 120/240 volt secondary									
0100	1 kVA	1 Elec	2	4	Ea.	330	219		549	700
0300	2 kVA		1.60	5		490	274		764	965
0500	3 kVA		1.40	5.714		610	315		925	1,150
0700	5 kVA	↓	1.20	6.667		835	365		1,200	1,475
0900	7.5 kVA	2 Elec	2.20	7.273		1,175	400		1,575	1,900
1100	10 kVA		1.60	10		1,450	545		1,995	2,450
1300	15 kVA	↓	1.20	13.333	↓	1,700	730		2,430	3,000
2300	3 phase, 480 volt primary 120/208 volt secondary									
2310	Ventilated, 3 kVA	1 Elec	1	8	Ea.	910	440		1,350	1,675

26 22 Low-Voltage Transformers

26 22 13 – Low-Voltage Distribution Transformers

26 22 13.10 Transformer, Dry-Type		Crew	Daily Output	Labor-Hours	Unit	Material	2015 Bare Costs Labor	Equipment	Total	Total Incl O&P
2700	6 kVA	1 Elec	.80	10	Ea.	1,025	545		1,570	1,975
2900	9 kVA	↓	.70	11.429		1,075	625		1,700	2,175
3100	15 kVA	2 Elec	1.10	14.545		1,300	795		2,095	2,650
3300	30 kVA		.90	17.778		1,425	970		2,395	3,075
3500	45 kVA	↓	.80	20	↓	1,700	1,100		2,800	3,575
9000	Minimum labor/equipment charge	1 Elec	1	8	Job		440		440	680

26 24 Switchboards and Panelboards

26 24 16 – Panelboards

26 24 16.20 Panelboard and Load Center Circuit Breakers

		Crew	Daily Output	Labor-Hours	Unit	Material	Labor	Equipment	Total	Total Incl O&P
0010	**PANELBOARD AND LOAD CENTER CIRCUIT BREAKERS**									
0050	Bolt-on, 10,000 amp I.C., 120 volt, 1 pole									
0100	15 to 50 amp	1 Elec	10	.800	Ea.	16.65	44		60.65	86.50
0200	60 amp		8	1		19.20	54.50		73.70	106
0300	70 amp	↓	8	1	↓	28	54.50		82.50	116
0350	240 volt, 2 pole									
0400	15 to 50 amp	1 Elec	8	1	Ea.	36	54.50		90.50	124
0500	60 amp		7.50	1.067		49	58.50		107.50	144
0600	80 to 100 amp		5	1.600		93.50	87.50		181	239
0700	3 pole, 15 to 60 amp		6.20	1.290		115	70.50		185.50	236
0800	70 amp		5	1.600		146	87.50		233.50	297
0900	80 to 100 amp		3.60	2.222		166	122		288	370
1000	22,000 amp I.C., 240 volt, 2 pole, 70 – 225 amp		2.70	2.963		630	162		792	945
1100	3 pole, 70 – 225 amp		2.30	3.478		700	190		890	1,075
1200	14,000 amp I.C., 277 volts, 1 pole, 15 – 30 amp		8	1		44	54.50		98.50	133
1300	22,000 amp I.C., 480 volts, 2 pole, 70 – 225 amp		2.70	2.963		630	162		792	945
1400	3 pole, 70 – 225 amp		2.30	3.478		780	190		970	1,150
2060	Plug-in tandem, 120/240 V, 2-15 A, 1 pole		11	.727		24.50	40		64.50	88.50
2070	1-15 A & 1-20 A		11	.727		24.50	40		64.50	88.50
2080	2-20 A		11	.727		24.50	40		64.50	88.50
2082	Arc fault circuit interrupter, 120/240 V, 1-15 A & 1-20 A, 1 pole		11	.727	↓	65	40		105	133
9000	Minimum labor/equipment charge	↓	3	2.667	Job		146		146	226

26 24 16.30 Panelboards Commercial Applications

		Crew	Daily Output	Labor-Hours	Unit	Material	Labor	Equipment	Total	Total Incl O&P
0010	**PANELBOARDS COMMERCIAL APPLICATIONS**									
0050	NQOD, w/20 amp 1 pole bolt-on circuit breakers									
0100	3 wire, 120/240 volts, 100 amp main lugs									
0150	10 circuits	1 Elec	1	8	Ea.	545	440		985	1,275
0200	14 circuits		.88	9.091		660	495		1,155	1,500
0250	18 circuits		.75	10.667		720	585		1,305	1,700
0300	20 circuits		.65	12.308		805	675		1,480	1,925
0600	4 wire, 120/208 volts, 100 amp main lugs, 12 circuits		1	8		640	440		1,080	1,375
0650	16 circuits		.75	10.667		730	585		1,315	1,700
0700	20 circuits		.65	12.308		845	675		1,520	1,975
0750	24 circuits		.60	13.333		875	730		1,605	2,100
0800	30 circuits	↓	.53	15.094		1,050	825		1,875	2,425
0850	225 amp main lugs, 32 circuits	2 Elec	.90	17.778		1,175	970		2,145	2,800
0900	34 circuits		.84	19.048		1,200	1,050		2,250	2,950
0950	36 circuits		.80	20		1,225	1,100		2,325	3,050
1000	42 circuits	↓	.68	23.529	↓	1,375	1,275		2,650	3,500
1200	NEHB, w/20 amp, 1 pole bolt-on circuit breakers									
1250	4 wire, 277/480 volts, 100 amp main lugs, 12 circuits	1 Elec	.88	9.091	Ea.	1,225	495		1,720	2,125

26 24 Switchboards and Panelboards

26 24 16 – Panelboards

26 24 16.30 Panelboards Commercial Applications		Crew	Daily Output	Labor-Hours	Unit	Material	2015 Bare Costs Labor	Equipment	Total	Total Incl O&P
1300	20 circuits	1 Elec	.60	13.333	Ea.	1,825	730		2,555	3,125
1350	225 amp main lugs, 24 circuits	2 Elec	.90	17.778		2,075	970		3,045	3,775
1400	30 circuits		.80	20		2,475	1,100		3,575	4,425
1450	36 circuits		.72	22.222		2,875	1,225		4,100	5,050
1600	NQOD panel, w/20 amp, 1 pole, circuit breakers									
2000	4 wire, 120/208 volts with main circuit breaker									
2050	100 amp main, 24 circuits	1 Elec	.47	17.021	Ea.	1,175	930		2,105	2,750
2100	30 circuits	"	.40	20		1,325	1,100		2,425	3,150
2200	225 amp main, 32 circuits	2 Elec	.72	22.222		2,225	1,225		3,450	4,300
2250	42 circuits		.56	28.571		2,425	1,575		4,000	5,100
2300	400 amp main, 42 circuits		.48	33.333		3,250	1,825		5,075	6,400
2350	600 amp main, 42 circuits		.40	40		4,825	2,200		7,025	8,700
2400	NEHB, with 20 amp, 1 pole circuit breaker									
2450	4 wire, 277/480 volts with main circuit breaker									
2500	100 amp main, 24 circuits	1 Elec	.42	19.048	Ea.	2,375	1,050		3,425	4,250
2550	30 circuits	"	.38	21.053		2,775	1,150		3,925	4,850
2600	225 amp main, 30 circuits	2 Elec	.72	22.222		3,500	1,225		4,725	5,725
2650	42 circuits	"	.56	28.571		4,300	1,575		5,875	7,175
9000	Minimum labor/equipment charge	1 Elec	1	8	Job		440		440	680

26 27 Low-Voltage Distribution Equipment

26 27 13 – Electricity Metering

26 27 13.10 Meter Centers and Sockets

		Crew	Daily Output	Labor-Hours	Unit	Material	Labor	Equipment	Total	Total Incl O&P
0010	**METER CENTERS AND SOCKETS**									
0100	Sockets, single position, 4 terminal, 100 amp	1 Elec	3.20	2.500	Ea.	43	137		180	260
0200	150 amp		2.30	3.478		48	190		238	350
0300	200 amp		1.90	4.211		92	230		322	455
0400	Transformer rated, 20 amp		3.20	2.500		152	137		289	380
0500	Double position, 4 terminal, 100 amp		2.80	2.857		204	156		360	465
0600	150 amp		2.10	3.810		240	208		448	590
0700	200 amp		1.70	4.706		450	257		707	895
1100	Meter centers and sockets, three phase, single pos, 7 terminal, 100 amp		2.80	2.857		164	156		320	425
1200	200 amp		2.10	3.810		295	208		503	650
1400	400 amp		1.70	4.706		720	257		977	1,200
9000	Minimum labor/equipment charge		3	2.667	Job		146		146	226

26 27 16 – Electrical Cabinets and Enclosures

26 27 16.10 Cabinets

		Crew	Daily Output	Labor-Hours	Unit	Material	Labor	Equipment	Total	Total Incl O&P
0010	**CABINETS**									
7000	Cabinets, current transformer									
7050	Single door, 24" H x 24" W x 10" D	1 Elec	1.60	5	Ea.	152	274		426	590
7100	30" H x 24" W x 10" D		1.30	6.154		165	335		500	700
7150	36" H x 24" W x 10" D		1.10	7.273		177	400		577	810
7200	30" H x 30" W x 10" D		1	8		223	440		663	925
7250	36" H x 30" W x 10" D		.90	8.889		262	485		747	1,050
7300	36" H x 36" W x 10" D		.80	10		270	545		815	1,150
7500	Double door, 48" H x 36" W x 10" D		.60	13.333		590	730		1,320	1,775
7550	24" H x 24" W x 12" D		1	8		173	440		613	870
9990	Minimum labor/equipment charge		2	4	Job		219		219	340

26 27 Low-Voltage Distribution Equipment

26 27 23 – Indoor Service Poles

26 27 23.40 Surface Raceway

		Crew	Daily Output	Labor-Hours	Unit	Material	2015 Bare Costs Labor	Equipment	Total	Total Incl O&P
0010	**SURFACE RACEWAY**									
0090	Metal, straight section									
0100	No. 500	1 Elec	100	.080	L.F.	1.04	4.38		5.42	7.95
0110	No. 700		100	.080		1.17	4.38		5.55	8.10
0400	No. 1500, small pancake		90	.089		2.15	4.86		7.01	9.90
0600	No. 2000, base & cover, blank		90	.089		2.19	4.86		7.05	9.95
0800	No. 3000, base & cover, blank		75	.107		4.18	5.85		10.03	13.65
1000	No. 4000, base & cover, blank		65	.123		6.80	6.75		13.55	17.90
1200	No. 6000, base & cover, blank		50	.160	↓	11.40	8.75		20.15	26
2400	Fittings, elbows, No. 500		40	.200	Ea.	1.90	10.95		12.85	19.05
2800	Elbow cover, No. 2000		40	.200		3.57	10.95		14.52	21
2880	Tee, No. 500		42	.190		3.66	10.40		14.06	20
2900	No. 2000		27	.296		11.85	16.20		28.05	38
3000	Switch box, No. 500		16	.500		11	27.50		38.50	54.50
3400	Telephone outlet, No. 1500		16	.500		14.10	27.50		41.60	58
3600	Junction box, No. 1500	↓	16	.500	↓	9.65	27.50		37.15	53
3800	Plugmold wired sections, No. 2000									
4000	1 circuit, 6 outlets, 3 ft. long	1 Elec	8	1	Ea.	36	54.50		90.50	124
4100	2 circuits, 8 outlets, 6 ft. long	"	5.30	1.509	"	53	82.50		135.50	187
9300	Non-metallic, straight section									
9310	7/16" x 7/8", base & cover, blank	1 Elec	160	.050	L.F.	1.72	2.74		4.46	6.15
9320	Base & cover w/adhesive		160	.050		1.42	2.74		4.16	5.80
9340	7/16" x 1-5/16", base & cover, blank		145	.055		1.88	3.02		4.90	6.75
9350	Base & cover w/adhesive		145	.055		2.17	3.02		5.19	7.05
9370	11/16" x 2-1/4", base & cover, blank		130	.062		2.66	3.37		6.03	8.15
9380	Base & cover w/adhesive		130	.062		3.01	3.37		6.38	8.50
9385	1-11/16" x 5-1/4", two compartment base & cover w/screws		80	.100	↓	7.50	5.45		12.95	16.70
9400	Fittings, elbows, 7/16" x 7/8"		50	.160	Ea.	1.88	8.75		10.63	15.60
9410	7/16" x 1-5/16"		45	.178		1.95	9.70		11.65	17.20
9420	11/16" x 2-1/4"		40	.200		2.11	10.95		13.06	19.25
9425	1-11/16" x 5-1/4"		28	.286		11.05	15.65		26.70	36
9430	Tees, 7/16" x 7/8"		35	.229		2.42	12.50		14.92	22
9440	7/16" x 1-5/16"		32	.250		2.49	13.70		16.19	23.50
9450	11/16" x 2-1/4"		30	.267		2.56	14.60		17.16	25.50
9455	1-11/16" x 5-1/4"		24	.333		17.75	18.25		36	47.50
9460	Cover clip, 7/16" x 7/8"		80	.100		.49	5.45		5.94	9
9470	7/16" x 1-5/16"		72	.111		.44	6.10		6.54	9.90
9480	11/16" x 2-1/4"		64	.125		.75	6.85		7.60	11.45
9484	1-11/16" x 5-1/4"		42	.190		2.61	10.40		13.01	19
9486	Wire clip, 1-11/16" x 5-1/4"		68	.118		.45	6.45		6.90	10.45
9490	Blank end, 7/16" x 7/8"		50	.160		.70	8.75		9.45	14.30
9500	7/16" x 1-5/16"		45	.178		.78	9.70		10.48	15.90
9510	11/16" x 2-1/4"		40	.200		1.18	10.95		12.13	18.25
9515	1-11/16" x 5-1/4"		38	.211		5.45	11.50		16.95	24
9520	Round fixture box, 5.5" dia x 1"		25	.320		11.10	17.50		28.60	39
9530	Device box, 1 gang		30	.267		5	14.60		19.60	28
9540	2 gang		25	.320	↓	7.30	17.50		24.80	35
9990	Minimum labor/equipment charge	↓	5	1.600	Job		87.50		87.50	136

26 27 Low-Voltage Distribution Equipment

26 27 26 – Wiring Devices

26 27 26.20 Wiring Devices Elements

		Crew	Daily Output	Labor-Hours	Unit	Material	2015 Bare Costs Labor	2015 Bare Costs Equipment	Total	Total Incl O&P
0010	**WIRING DEVICES ELEMENTS**									
0200	Toggle switch, quiet type, single pole, 15 amp	1 Elec	40	.200	Ea.	6.55	10.95		17.50	24
0500	20 amp		27	.296		9.85	16.20		26.05	36
0550	Rocker, 15 amp		40	.200		4.10	10.95		15.05	21.50
0560	20 amp		27	.296		9.50	16.20		25.70	35.50
0600	3 way, 15 amp		23	.348		5.05	19.05		24.10	35
0850	Rocker, 15 amp		23	.348		5.75	19.05		24.80	36
0860	20 amp		18	.444		13.65	24.50		38.15	52.50
0900	4 way, 15 amp		15	.533		9.10	29		38.10	55
1030	Rocker, 15 amp		15	.533		19.15	29		48.15	66
1040	20 amp		11	.727		39	40		79	105
1650	Dimmer switch, 120 volt, incandescent, 600 watt, 1 pole [G]		16	.500		21	27.50		48.50	65.50
2460	Receptacle, duplex, 120 volt, grounded, 15 amp		40	.200		1.26	10.95		12.21	18.35
2470	20 amp		27	.296		7.90	16.20		24.10	33.50
2480	Ground fault interrupting, 15 amp		27	.296		32	16.20		48.20	60
2482	20 amp		27	.296		40	16.20		56.20	69
2490	Dryer, 30 amp		15	.533		4.39	29		33.39	50
2500	Range, 50 amp		11	.727		12.15	40		52.15	75
2600	Wall plates, stainless steel, 1 gang		80	.100		2.56	5.45		8.01	11.25
2800	2 gang		53	.151		4.33	8.25		12.58	17.55
3200	Lampholder, keyless		26	.308		12.65	16.85		29.50	40
3400	Pullchain with receptacle		22	.364		20.50	19.90		40.40	53.50
9000	Minimum labor/equipment charge		4	2	Job		109		109	169

26 27 73 – Door Chimes

26 27 73.10 Doorbell System

		Crew	Daily Output	Labor-Hours	Unit	Material	Labor	Equipment	Total	Total Incl O&P
0010	**DOORBELL SYSTEM**, incl. transformer, button & signal									
1000	Door chimes, 2 notes	1 Elec	16	.500	Ea.	22.50	27.50		50	67.50
1020	with ambient light		12	.667		109	36.50		145.50	177
1100	Tube type, 3 tube system		12	.667		187	36.50		223.50	263
1180	4 tube system		10	.800		385	44		429	490
1900	For transformer & button, add		5	1.600		15.85	87.50		103.35	153
3000	For push button only		24	.333		2.58	18.25		20.83	31
9000	Minimum labor/equipment charge		4	2	Job		109		109	169

26 28 Low-Voltage Circuit Protective Devices

26 28 16 – Enclosed Switches and Circuit Breakers

26 28 16.10 Circuit Breakers

		Crew	Daily Output	Labor-Hours	Unit	Material	Labor	Equipment	Total	Total Incl O&P
0010	**CIRCUIT BREAKERS** (in enclosure)									
0100	Enclosed (NEMA 1), 600 volt, 3 pole, 30 amp	1 Elec	3.20	2.500	Ea.	500	137		637	760
0200	60 amp		2.80	2.857		615	156		771	920
0400	100 amp		2.30	3.478		705	190		895	1,075
0500	200 amp		1.50	5.333		1,475	292		1,767	2,075
0600	225 amp		1.50	5.333		1,625	292		1,917	2,250
0700	400 amp	2 Elec	1.60	10		2,775	545		3,320	3,925
9000	Minimum labor/equipment charge	1 Elec	4	2	Job		109		109	169

26 28 16.20 Safety Switches

		Crew	Daily Output	Labor-Hours	Unit	Material	Labor	Equipment	Total	Total Incl O&P
0010	**SAFETY SWITCHES**									
0100	General duty 240 volt, 3 pole NEMA 1, fusible, 30 amp	1 Elec	3.20	2.500	Ea.	73.50	137		210.50	293
0200	60 amp		2.30	3.478		124	190		314	430
0300	100 amp		1.90	4.211		213	230		443	590

26 28 Low-Voltage Circuit Protective Devices

26 28 16 – Enclosed Switches and Circuit Breakers

26 28 16.20 Safety Switches		Crew	Daily Output	Labor-Hours	Unit	Material	2015 Bare Costs Labor	Equipment	Total	Total Incl O&P
0400	200 amp	1 Elec	1.30	6.154	Ea.	455	335		790	1,025
0500	400 amp	2 Elec	1.80	8.889	↓	1,150	485		1,635	2,025
9990	Minimum labor/equipment charge	1 Elec	3	2.667	Job		146		146	226

26 28 16.40 Time Switches

0010	**TIME SWITCHES**									
0100	Single pole, single throw, 24 hour dial	1 Elec	4	2	Ea.	145	109		254	330
0200	24 hour dial with reserve power		3.60	2.222		615	122		737	865
0300	Astronomic dial		3.60	2.222		209	122		331	420
0400	Astronomic dial with reserve power		3.30	2.424		665	133		798	935
0500	7 day calendar dial		3.30	2.424		150	133		283	370
0600	7 day calendar dial with reserve power		3.20	2.500		335	137		472	580
0700	Photo cell 2000 watt		8	1		28.50	54.50		83	116
1080	Load management device, 4 loads		2	4		995	219		1,214	1,450
1100	8 loads		1	8	↓	2,325	440		2,765	3,225
9000	Minimum labor/equipment charge		3.50	2.286	Job		125		125	194

26 29 Low-Voltage Controllers

26 29 13 – Enclosed Controllers

26 29 13.40 Relays

0010	**RELAYS** Enclosed (NEMA 1)									
0050	600 volt AC, 1 pole, 12 amp	1 Elec	5.30	1.509	Ea.	88.50	82.50		171	226
0100	2 pole, 12 amp		5	1.600		88.50	87.50		176	234
0200	4 pole, 10 amp	↓	4.50	1.778	↓	118	97		215	281

26 29 23 – Variable-Frequency Motor Controllers

26 29 23.10 Variable Frequency Drives/Adj. Frequency Drives

0010	**VARIABLE FREQUENCY DRIVES/ADJ. FREQUENCY DRIVES**										
0100	Enclosed (NEMA 1), 460 volt, for 3 HP motor size	G	1 Elec	.80	10	Ea.	1,700	545		2,245	2,725
0110	5 HP motor size	G		.80	10		1,925	545		2,470	2,950
0120	7.5 HP motor size	G		.67	11.940		2,300	655		2,955	3,525
0130	10 HP motor size	G	↓	.67	11.940		2,675	655		3,330	3,925
0140	15 HP motor size	G	2 Elec	.89	17.978		3,250	985		4,235	5,100
0150	20 HP motor size	G		.89	17.978		3,800	985		4,785	5,725
0160	25 HP motor size	G		.67	23.881		4,700	1,300		6,000	7,175
0170	30 HP motor size	G		.67	23.881		5,900	1,300		7,200	8,500
0180	40 HP motor size	G		.67	23.881		7,025	1,300		8,325	9,775
0190	50 HP motor size	G	↓	.53	30.189		8,675	1,650		10,325	12,100
0200	60 HP motor size	G	R-3	.56	35.714		10,500	1,950	246	12,696	14,800
0210	75 HP motor size	G		.56	35.714		12,100	1,950	246	14,296	16,600
0220	100 HP motor size	G		.50	40		14,100	2,175	276	16,551	19,200
0230	125 HP motor size	G		.50	40		15,700	2,175	276	18,151	21,000
0240	150 HP motor size	G		.50	40		17,800	2,175	276	20,251	23,300
0250	200 HP motor size	G	↓	.42	47.619		23,400	2,575	330	26,305	30,200
1100	Custom-engineered, 460 volt, for 3 HP motor size	G	1 Elec	.56	14.286		2,725	780		3,505	4,200
1110	5 HP motor size	G		.56	14.286		2,725	780		3,505	4,200
1120	7.5 HP motor size	G		.47	17.021		2,875	930		3,805	4,600
1130	10 HP motor size	G	↓	.47	17.021		3,000	930		3,930	4,750
1140	15 HP motor size	G	2 Elec	.62	25.806		3,750	1,400		5,150	6,300
1150	20 HP motor size	G		.62	25.806		4,225	1,400		5,625	6,825
1160	25 HP motor size	G		.47	34.043		4,950	1,850		6,800	8,300
1170	30 HP motor size	G	↓	.47	34.043	↓	6,175	1,850		8,025	9,675

26 29 Low-Voltage Controllers

26 29 23 – Variable-Frequency Motor Controllers

26 29 23.10 Variable Frequency Drives/Adj. Frequency Drives		Crew	Daily Output	Labor-Hours	Unit	Material	2015 Bare Costs Labor	Equipment	Total	Total Incl O&P
1180	40 HP motor size	2 Elec G	.47	34.043	Ea.	7,300	1,850		9,150	10,900
1190	50 HP motor size	G	.37	43.243		8,375	2,375		10,750	12,900
1200	60 HP motor size	R-3 G	.39	51.282		12,600	2,775	355	15,730	18,600
1210	75 HP motor size	G	.39	51.282		13,400	2,775	355	16,530	19,500
1220	100 HP motor size	G	.35	57.143		14,700	3,100	395	18,195	21,500
1230	125 HP motor size	G	.35	57.143		15,700	3,100	395	19,195	22,600
1240	150 HP motor size	G	.35	57.143		18,000	3,100	395	21,495	25,100
1250	200 HP motor size	G	.29	68.966		23,500	3,750	475	27,725	32,200
2000	For complex & special design systems to meet specific									
2010	requirements, obtain quote from vendor.									

26 32 Packaged Generator Assemblies

26 32 13 – Engine Generators

26 32 13.16 Gas-Engine-Driven Generator Sets

		Crew	Daily Output	Labor-Hours	Unit	Material	Labor	Equipment	Total	Total Incl O&P
0010	**GAS-ENGINE-DRIVEN GENERATOR SETS**									
0020	Gas or gasoline operated, includes battery,									
0050	charger, & muffler									
0200	3 phase 4 wire, 277/480 volt, 7.5 kW	R-3	.83	24.096	Ea.	7,350	1,300	166	8,816	10,300
0300	11.5 kW		.71	28.169		10,400	1,525	194	12,119	14,100
0400	20 kW		.63	31.746		12,300	1,725	219	14,244	16,400
0500	35 kW		.55	36.364		14,600	1,975	251	16,826	19,500

26 51 Interior Lighting

26 51 13 – Interior Lighting Fixtures, Lamps, and Ballasts

26 51 13.50 Interior Lighting Fixtures

		Crew	Daily Output	Labor-Hours	Unit	Material	Labor	Equipment	Total	Total Incl O&P
0010	**INTERIOR LIGHTING FIXTURES** Including lamps, mounting									
0030	hardware and connections									
0100	Fluorescent, C.W. lamps, troffer, recess mounted in grid, RS									
0130	Grid ceiling mount									
0200	Acrylic lens, 1'W x 4'L, two 40 watt	1 Elec	5.70	1.404	Ea.	47.50	77		124.50	171
0300	2'W x 2'L, two U40 watt		5.70	1.404		51	77		128	176
0600	2'W x 4'L, four 40 watt		4.70	1.702		57.50	93		150.50	208
0910	Acrylic lens, 1'W x 4'L, two 32 watt T8	G	5.70	1.404		59.50	77		136.50	185
0930	2'W x 2'L, two U32 watt T8	G	5.70	1.404		81	77		158	208
0940	2'W x 4'L, two 32 watt T8	G	5.30	1.509		67	82.50		149.50	202
0950	2'W x 4'L, three 32 watt T8	G	5	1.600		68	87.50		155.50	211
0960	2'W x 4'L, four 32 watt T8	G	4.70	1.702		71	93		164	222
1000	Surface mounted, RS									
1030	Acrylic lens with hinged & latched door frame									
1100	1'W x 4'L, two 40 watt	1 Elec	7	1.143	Ea.	64	62.50		126.50	168
1200	2'W x 2'L, two U40 watt		7	1.143		68.50	62.50		131	173
1500	2'W x 4'L, four 40 watt		5.30	1.509		81	82.50		163.50	217
2100	Strip fixture									
2130	Surface mounted									
2200	4' long, one 40 watt, RS	1 Elec	8.50	.941	Ea.	27.50	51.50		79	110
2300	4' long, two 40 watt, RS	"	8	1		38.50	54.50		93	127
2600	8' long, one 75 watt, SL	2 Elec	13.40	1.194		48	65.50		113.50	154
2700	8' long, two 75 watt, SL	"	12.40	1.290		58	70.50		128.50	173
3000	Strip, pendent mounted, industrial, white porcelain enamel									

26 51 Interior Lighting

26 51 13 – Interior Lighting Fixtures, Lamps, and Ballasts

26 51 13.50 Interior Lighting Fixtures		Crew	Daily Output	Labor-Hours	Unit	Material	2015 Bare Costs Labor	Equipment	Total	Total Incl O&P
3100	4' long, two 40 watt, RS	1 Elec	5.70	1.404	Ea.	52.50	77		129.50	177
3200	4' long, two 60 watt, HO	"	5	1.600		82.50	87.50		170	227
3300	8' long, two 75 watt, SL	2 Elec	8.80	1.818	↓	98	99.50		197.50	262
4220	Metal halide, integral ballast, ceiling, recess mounted									
4230	prismatic glass lens, floating door									
4240	2'W x 2'L, 250 watt	1 Elec	3.20	2.500	Ea.	315	137		452	560
4250	2'W x 2'L, 400 watt	2 Elec	5.80	2.759		365	151		516	635
4260	Surface mounted, 2'W x 2'L, 250 watt	1 Elec	2.70	2.963		345	162		507	630
4270	400 watt	2 Elec	4.80	3.333	↓	405	182		587	725
4280	High bay, aluminum reflector,									
4290	Single unit, 400 watt	2 Elec	4.60	3.478	Ea.	410	190		600	745
4300	Single unit, 1000 watt		4	4		590	219		809	990
4310	Twin unit, 400 watt	↓	3.20	5		820	274		1,094	1,325
4320	Low bay, aluminum reflector, 250W DX lamp	1 Elec	3.20	2.500	↓	360	137		497	610
4340	High pressure sodium integral ballast ceiling, recess mounted									
4350	prismatic glass lens, floating door									
4360	2'W x 2'L, 150 watt lamp	1 Elec	3.20	2.500	Ea.	385	137		522	635
4370	2'W x 2'L, 400 watt lamp	2 Elec	5.80	2.759		460	151		611	740
4380	Surface mounted, 2'W x 2'L, 150 watt lamp	1 Elec	2.70	2.963		470	162		632	770
4390	400 watt lamp	2 Elec	4.80	3.333	↓	525	182		707	860
4400	High bay, aluminum reflector,									
4410	Single unit, 400 watt lamp	2 Elec	4.60	3.478	Ea.	380	190		570	710
4430	Single unit, 1000 watt lamp	"	4	4		545	219		764	940
4440	Low bay, aluminum reflector, 150 watt lamp	1 Elec	3.20	2.500	↓	325	137		462	570
4450	Incandescent, high hat can, round alzak reflector, prewired									
4470	100 watt	1 Elec	8	1	Ea.	71.50	54.50		126	163
4480	150 watt		8	1		104	54.50		158.50	199
4500	300 watt	↓	6.70	1.194	↓	241	65.50		306.50	365
4600	Square glass lens with metal trim, prewired									
4630	100 watt	1 Elec	6.70	1.194	Ea.	55	65.50		120.50	162
4700	200 watt		6.70	1.194		97	65.50		162.50	208
6010	Vapor tight, incandescent, ceiling mounted, 200 watt		6.20	1.290		82.50	70.50		153	200
6100	Fluorescent, surface mounted, 2 lamps, 4'L, RS, 40 watt		3.20	2.500		115	137		252	340
6850	Vandalproof, surface mounted, fluorescent, two 32 watt T8	G	3.20	2.500		252	137		389	490
6860	Incandescent, one 150 watt		8	1		98.50	54.50		153	193
6900	Mirror light, fluorescent, RS, acrylic enclosure, two 40 watt		8	1		128	54.50		182.50	225
6910	One 40 watt		8	1		106	54.50		160.50	201
6920	One 20 watt	↓	12	.667	↓	81.50	36.50		118	147
7500	Ballast replacement, by weight of ballast, to 15' high									
7520	Indoor fluorescent, less than 2 lb.	1 Elec	10	.800	Ea.	26	44		70	96.50
7540	Two 40W, watt reducer, 2 to 5 lb.		9.40	.851		41	46.50		87.50	118
7560	Two F96 slimline, over 5 lb.		8	1		77	54.50		131.50	169
7580	Vaportite ballast, less than 2 lb.		9.40	.851		26	46.50		72.50	101
7600	2 lb. to 5 lb.		8.90	.899		41	49		90	122
7620	Over 5 lb.		7.60	1.053		77	57.50		134.50	174
7630	Electronic ballast for two tubes		8	1		37	54.50		91.50	126
7640	Dimmable ballast one lamp	G	8	1		106	54.50		160.50	202
7650	Dimmable ballast two-lamp	G	7.60	1.053		104	57.50		161.50	203
9000	Minimum labor/equipment charge	↓	3	2.667	Job		146		146	226

26 51 Interior Lighting

26 51 13 – Interior Lighting Fixtures, Lamps, and Ballasts

26 51 13.70 Residential Fixtures

	26 51 13.70 Residential Fixtures	Crew	Daily Output	Labor-Hours	Unit	Material	2015 Bare Costs Labor	2015 Bare Costs Equipment	Total	Total Incl O&P
0010	**RESIDENTIAL FIXTURES**									
0400	Fluorescent, interior, surface, circline, 32 watt & 40 watt	1 Elec	20	.400	Ea.	130	22		152	177
0500	2' x 2', two U-tube 32 watt T8		8	1		165	54.50		219.50	266
0700	Shallow under cabinet, two 20 watt		16	.500		62	27.50		89.50	111
0900	Wall mounted, 4'L, two 32 watt T8, with baffle		10	.800		133	44		177	214
2000	Incandescent, exterior lantern, wall mounted, 60 watt		16	.500		55.50	27.50		83	104
2100	Post light, 150W, with 7' post		4	2		243	109		352	435
2500	Lamp holder, weatherproof with 150W PAR		16	.500		32	27.50		59.50	77.50
2550	With reflector and guard		12	.667		56	36.50		92.50	119
2600	Interior pendent, globe with shade, 150 watt		20	.400		201	22		223	255
9000	Minimum labor/equipment charge		4	2	Job		109		109	169

26 52 Emergency Lighting

26 52 13 – Emergency Lighting Equipments

26 52 13.10 Emergency Lighting and Battery Units

		Crew	Daily Output	Labor-Hours	Unit	Material	Labor	Equipment	Total	Total Incl O&P
0010	**EMERGENCY LIGHTING AND BATTERY UNITS**									
0300	Emergency light units, battery operated									
0350	Twin sealed beam light, 25 watt, 6 volt each									
0500	Lead battery operated	1 Elec	4	2	Ea.	153	109		262	335
0700	Nickel cadmium battery operated		4	2	"	560	109		669	785
9000	Minimum labor/equipment charge		4	2	Job		109		109	169

26 53 Exit Signs

26 53 13 – Exit Lighting

26 53 13.10 Exit Lighting Fixtures

		Crew	Daily Output	Labor-Hours	Unit	Material	Labor	Equipment	Total	Total Incl O&P
0010	**EXIT LIGHTING FIXTURES**									
0080	Exit light ceiling or wall mount, incandescent, single face	1 Elec	8	1	Ea.	39	54.50		93.50	128
0100	Double face		6.70	1.194	"	39	65.50		104.50	144
9000	Minimum labor/equipment charge		4	2	Job		109		109	169

26 55 Special Purpose Lighting

26 55 59 – Display Lighting

26 55 59.10 Track Lighting

		Crew	Daily Output	Labor-Hours	Unit	Material	Labor	Equipment	Total	Total Incl O&P
0010	**TRACK LIGHTING**									
0080	Track, 1 circuit, 4' section	1 Elec	6.70	1.194	Ea.	44.50	65.50		110	150
0100	8' section	2 Elec	10.60	1.509		69	82.50		151.50	204
0300	3 circuits, 4' section	1 Elec	6.70	1.194		83	65.50		148.50	193
0400	8' section	2 Elec	10.60	1.509		109	82.50		191.50	248
9000	Minimum labor/equipment charge	1 Elec	3	2.667	Job		146		146	226

26 56 Exterior Lighting

26 56 23 – Area Lighting

26 56 23.10 Exterior Fixtures

		Crew	Daily Output	Labor-Hours	Unit	Material	2015 Bare Costs Labor	Equipment	Total	Total Incl O&P
0010	**EXTERIOR FIXTURES** With lamps									
0200	Wall mounted, incandescent, 100 watt	1 Elec	8	1	Ea.	35	54.50		89.50	123
0400	Quartz, 500 watt		5.30	1.509		50	82.50		132.50	183
1100	Wall pack, low pressure sodium, 35 watt		4	2		227	109		336	420
1150	55 watt		4	2		270	109		379	465
1160	High pressure sodium, 70 watt		4	2		224	109		333	415
1170	150 watt		4	2		244	109		353	435
1180	Metal Halide, 175 watt		4	2		248	109		357	440
1190	250 watt		4	2		282	109		391	480
1195	400 watt		4	2		335	109		444	535
1250	Induction lamp, 40 watt		4	2		340	109		449	545
1260	80 watt		4	2		580	109		689	810
1278	LED, poly lens, 26 watt		4	2		330	109		439	530
1280	110 watt		4	2		1,300	109		1,409	1,625
1500	LED, glass lens, 13 watt		4	2		330	109		439	530

26 56 33 – Walkway Lighting

26 56 33.10 Walkway Luminaire

		Crew	Daily Output	Labor-Hours	Unit	Material	Labor	Equipment	Total	Total Incl O&P
0010	**WALKWAY LUMINAIRE**									
9000	Minimum labor/equipment charge	1 Elec	3.75	2.133	Job		117		117	181

26 56 36 – Flood Lighting

26 56 36.20 Floodlights

		Crew	Daily Output	Labor-Hours	Unit	Material	Labor	Equipment	Total	Total Incl O&P
0010	**FLOODLIGHTS** with ballast and lamp,									
1290	floor mtd, mount with swivel bracket									
1400	Pole mounted, pole not included									
2250	Low pressure sodium, 55 watt	1 Elec	2.70	2.963	Ea.	585	162		747	895
2270	90 watt	"	2	4	"	645	219		864	1,050

26 61 Lighting Systems and Accessories

26 61 23 – Lamps Applications

26 61 23.10 Lamps

			Crew	Daily Output	Labor-Hours	Unit	Material	Labor	Equipment	Total	Total Incl O&P
0010	**LAMPS**										
0080	Fluorescent, rapid start, cool white, 2' long, 20 watt		1 Elec	1	8	C	310	440		750	1,025
0100	4' long, 40 watt			.90	8.889		236	485		721	1,025
0170	4' long, 34 watt energy saver	G		.90	8.889		430	485		915	1,225
0176	2' long, T8, 17 W energy saver	G		1	8		390	440		830	1,100
0178	3' long, T8, 25 W energy saver	G		.90	8.889		415	485		900	1,225
0180	4' long, T8, 32 watt energy saver	G		.90	8.889		305	485		790	1,100
0560	Twin tube compact lamp	G		.90	8.889		360	485		845	1,150
0570	Double twin tube compact lamp	G		.80	10		805	545		1,350	1,725
0600	Mercury vapor, mogul base, deluxe white, 100 watt			.30	26.667		3,650	1,450		5,100	6,275
0800	400 watt			.30	26.667		4,350	1,450		5,800	7,025
1000	Metal halide, mogul base, 175 watt			.30	26.667		2,550	1,450		4,000	5,075
1100	250 watt			.30	26.667		2,900	1,450		4,350	5,425
1350	High pressure sodium, 70 watt			.30	26.667		3,125	1,450		4,575	5,675
1370	150 watt			.30	26.667		3,325	1,450		4,775	5,925
1500	Low pressure sodium, 35 watt			.30	26.667		12,000	1,450		13,450	15,500
1600	90 watt			.30	26.667		13,400	1,450		14,850	17,100
1800	Incandescent, interior, A21, 100 watt			1.60	5		233	274		507	680
1900	A21, 150 watt			1.60	5		166	274		440	610
2300	R30, 75 watt			1.30	6.154		620	335		955	1,200

26 61 Lighting Systems and Accessories

26 61 23 – Lamps Applications

26 61 23.10 Lamps		Crew	Daily Output	Labor-Hours	Unit	Material	2015 Bare Costs Labor	Equipment	Total	Total Incl O&P
2500	Exterior, PAR 38, 75 watt	1 Elec	1.30	6.154	C	1,850	335		2,185	2,575
2600	PAR 38, 150 watt	↓	1.30	6.154	↓	2,000	335		2,335	2,725
9000	Minimum labor/equipment charge		4	2	Job		109		109	169

Division Notes

		CREW	DAILY OUTPUT	LABOR-HOURS	UNIT	BARE COSTS				TOTAL INCL O&P
						MAT.	LABOR	EQUIP.	TOTAL	

Estimating Tips
27 20 00 Data Communications
27 30 00 Voice Communications
27 40 00 Audio-Video Communications

When estimating material costs for special systems, it is always prudent to obtain manufacturers' quotations for equipment prices and special installation requirements which will affect the total costs.

Reference Numbers
Reference numbers are shown in shaded boxes at the beginning of some major classifications. These numbers refer to related items in the Reference Section. The reference information may be an estimating procedure, an alternate pricing method, or technical information.

Note: Not all subdivisions listed here necessarily appear in this publication. ∎

*Note: **Trade Service**, in part, has been used as a reference source for some of the material prices used in Division 27.*

Estimate with precision with RSMeans Online...
Quick, intuitive and easy-to-use, RSMeans Online gives you instant access to hundreds of thousands of material, labor, and equipment costs from RSMeans' comprehensive database, delivering the information you need to build competitive estimates every time.

Start your free trial today at **www.rsmeansonline.com**

RSMeansOnline

27 15 Communications Horizontal Cabling

27 15 10 – Special Communications Cabling

27 15 10.23 Sound and Video Cables and Fittings

		Crew	Daily Output	Labor-Hours	Unit	Material	2015 Bare Costs Labor	2015 Bare Costs Equipment	Total	Total Incl O&P
0010	**SOUND AND VIDEO CABLES & FITTINGS**									
1250	Nonshielded, #22-2 conductor	1 Elec	10	.800	C.L.F.	14.25	44		58.25	83.50

27 15 13 – Communications Copper Horizontal Cabling

27 15 13.13 Communication Cables

		Crew	Daily Output	Labor-Hours	Unit	Material	Labor	Equipment	Total	Total Incl O&P
0010	**COMMUNICATION CABLES**									
5000	High performance unshielded twisted pair (UTP)									
7000	Category 5, #24, 4 pair solid, PVC jacket	1 Elec	7	1.143	C.L.F.	14.95	62.50		77.45	113
7100	4 pair solid, plenum		7	1.143		45.50	62.50		108	147
7200	4 pair stranded, PVC jacket		7	1.143		22.50	62.50		85	122
7210	Category 5e, #24, 4 pair solid, PVC jacket		7	1.143		14.30	62.50		76.80	113
7212	4 pair solid, plenum		7	1.143		23.50	62.50		86	123
7214	4 pair stranded, PVC jacket		7	1.143		25.50	62.50		88	126
7240	Category 6, #24, 4 pair solid, PVC jacket		7	1.143		20.50	62.50		83	120
7242	4 pair solid, plenum		7	1.143		24.50	62.50		87	124
7244	4 pair stranded, PVC jacket		7	1.143		25.50	62.50		88	126
7300	Connector, RJ-45, category 5		80	.100	Ea.	1.33	5.45		6.78	9.90
7302	Shielded RJ-45, category 5		72	.111		3.39	6.10		9.49	13.15
7310	Jack, UTP RJ-45, category 3		72	.111		2.35	6.10		8.45	12
7312	Category 5		65	.123		5.05	6.75		11.80	15.95
7314	Category 5e		65	.123		3.29	6.75		10.04	14
7316	Category 6		65	.123		3.34	6.75		10.09	14.05

27 41 Audio-Video Systems

27 41 33 – Master Antenna Television Systems

27 41 33.10 T.V. Systems

		Crew	Daily Output	Labor-Hours	Unit	Material	Labor	Equipment	Total	Total Incl O&P
0010	**T.V. SYSTEMS**, not including rough-in wires, cables & conduits									
5000	Antenna, small	1 Elec	6	1.333	Ea.	47	73		120	165
5100	Large	"	4	2	"	198	109		307	385

27 51 Distributed Audio-Video Communications Systems

27 51 16 – Public Address and Mass Notification Systems

27 51 16.10 Public Address System

		Crew	Daily Output	Labor-Hours	Unit	Material	Labor	Equipment	Total	Total Incl O&P
0010	**PUBLIC ADDRESS SYSTEM**									
0100	Conventional, office	1 Elec	5.33	1.501	Speaker	135	82		217	276
0200	Industrial		2.70	2.963	"	261	162		423	540
9000	Minimum labor/equipment charge		3.50	2.286	Job		125		125	194

27 51 19 – Sound Masking Systems

27 51 19.10 Sound System

		Crew	Daily Output	Labor-Hours	Unit	Material	Labor	Equipment	Total	Total Incl O&P
0010	**SOUND SYSTEM**, not including rough-in wires, cables & conduits									
2000	Intercom, 30 station capacity, master station	2 Elec	2	8	Ea.	2,350	440		2,790	3,275
2020	10 station capacity	"	4	4		1,300	219		1,519	1,800
3600	House telephone, talking station	1 Elec	1.60	5		475	274		749	945
3800	Press to talk, release to listen	"	5.30	1.509		110	82.50		192.50	249
4000	System-on button					66			66	72.50
4200	Door release	1 Elec	4	2		118	109		227	299
4400	Combination speaker and microphone		8	1		201	54.50		255.50	305
4600	Termination box		3.20	2.500		63	137		200	282
4800	Amplifier or power supply		5.30	1.509		725	82.50		807.50	930

27 51 Distributed Audio-Video Communications Systems

27 51 19 – Sound Masking Systems

27 51 19.10 Sound System		Crew	Daily Output	Labor-Hours	Unit	Material	2015 Bare Costs Labor	Equipment	Total	Total Incl O&P
5000	Vestibule door unit	1 Elec	16	.500	Name	133	27.50		160.50	190
5200	Strip cabinet		27	.296	Ea.	252	16.20		268.20	300
5400	Directory		16	.500	"	119	27.50		146.50	174
9000	Minimum labor/equipment charge	↓	3.50	2.286	Job		125		125	194

Division Notes

		CREW	DAILY OUTPUT	LABOR-HOURS	UNIT	BARE COSTS				TOTAL INCL O&P
						MAT.	LABOR	EQUIP.	TOTAL	

Estimating Tips

- When estimating material costs for electronic safety and security systems, it is always prudent to obtain manufacturers' quotations for equipment prices and special installation requirements that affect the total cost.

- Fire alarm systems consist of control panels, annunciator panels, battery with rack, charger, and fire alarm actuating and indicating devices. Some fire alarm systems include speakers, telephone lines, door closer controls, and other components. Be careful not to overlook the costs related to installation for these items. Also be aware of costs for integrated automation instrumentation and terminal devices, control equipment, control wiring, and programming.

- Security equipment includes items such as CCTV, access control, and other detection and identification systems to perform alert and alarm functions. Be sure to consider the costs related to installation for this security equipment, such as for integrated automation instrumentation and terminal devices, control equipment, control wiring, and programming.

Reference Numbers

Reference numbers are shown in shaded boxes at the beginning of some major classifications. These numbers refer to related items in the Reference Section. The reference information may be an estimating procedure, an alternate pricing method, or technical information.

Note: Not all subdivisions listed here necessarily appear in this publication. ■

Did you know?

RSMeans Online gives you the same access to RSMeans' data with 24/7 access:
- Quickly locate costs in the searchable database.
- Build cost lists, estimates, and reports in minutes.
- Adjust costs to any location in the U.S. and Canada with the click of a button.

Start your free trial today at **www.rsmeansonline.com**

RSMeansOnline

28 13 Access Control

28 13 53 – Security Access Detection

28 13 53.13 Security Access Metal Detectors

		Crew	Daily Output	Labor-Hours	Unit	Material	2015 Bare Costs Labor	Equipment	Total	Total Incl O&P
0010	**SECURITY ACCESS METAL DETECTORS**									
0240	Metal detector, hand-held, wand type, unit only				Ea.	89			89	98
0250	Metal detector, walk through portal type, single zone	1 Elec	2	4		3,700	219		3,919	4,425
0260	Multi-zone	"	2	4		4,700	219		4,919	5,525

28 13 53.16 Security Access X-Ray Equipment

		Crew	Daily Output	Labor-Hours	Unit	Material	Labor	Equipment	Total	Total Incl O&P
0010	**SECURITY ACCESS X-RAY EQUIPMENT**									
0290	X-ray machine, desk top, for mail/small packages/letters	1 Elec	4	2	Ea.	3,425	109		3,534	3,950
0300	Conveyor type, incl monitor		2	4		16,000	219		16,219	17,900
0310	Includes additional features		2	4		28,600	219		28,819	31,700
0320	X-ray machine, large unit, for airports, incl monitor	2 Elec	1	16		40,000	875		40,875	45,400
0330	Full console	"	.50	32		68,500	1,750		70,250	78,000

28 13 53.23 Security Access Explosive Detection Equipment

		Crew	Daily Output	Labor-Hours	Unit	Material	Labor	Equipment	Total	Total Incl O&P
0010	**SECURITY ACCESS EXPLOSIVE DETECTION EQUIPMENT**									
0270	Explosives detector, walk through portal type	1 Elec	2	4	Ea.	44,200	219		44,419	48,900
0280	Hand-held, battery operated				"	25,500				28,100

28 16 Intrusion Detection

28 16 16 – Intrusion Detection Systems Infrastructure

28 16 16.50 Intrusion Detection

		Crew	Daily Output	Labor-Hours	Unit	Material	Labor	Equipment	Total	Total Incl O&P
0010	**INTRUSION DETECTION**, not including wires & conduits									
0100	Burglar alarm, battery operated, mechanical trigger	1 Elec	4	2	Ea.	278	109		387	475
0200	Electrical trigger		4	2		330	109		439	535
0400	For outside key control, add		8	1		84	54.50		138.50	177
0600	For remote signaling circuitry, add		8	1		125	54.50		179.50	222
0800	Card reader, flush type, standard		2.70	2.963		930	162		1,092	1,275
1000	Multi-code		2.70	2.963		1,200	162		1,362	1,575
1200	Door switches, hinge switch		5.30	1.509		58.50	82.50		141	193
1400	Magnetic switch		5.30	1.509		69	82.50		151.50	204
1600	Exit control locks, horn alarm		4	2		277	109		386	475
1800	Flashing light alarm		4	2		305	109		414	505
2000	Indicating panels, 1 channel		2.70	2.963		370	162		532	655
2200	10 channel	2 Elec	3.20	5		1,050	274		1,324	1,575
2400	20 channel		2	8		2,450	440		2,890	3,375
2600	40 channel		1.14	14.035		4,450	770		5,220	6,100
2800	Ultrasonic motion detector, 12 volt	1 Elec	2.30	3.478		230	190		420	550
3000	Infrared photoelectric detector		4	2		189	109		298	375
3200	Passive infrared detector		4	2		283	109		392	480
3400	Glass break alarm switch		8	1		93	54.50		147.50	187
3420	Switchmats, 30" x 5'		5.30	1.509		84.50	82.50		167	221
3440	30" x 25'		4	2		203	109		312	390
3460	Police connect panel		4	2		244	109		353	440
3480	Telephone dialer		5.30	1.509		385	82.50		467.50	555
3500	Alarm bell		4	2		104	109		213	284
3520	Siren		4	2		146	109		255	330
3540	Microwave detector, 10' to 200'		2	4		670	219		889	1,075
3560	10' to 350'		2	4		1,950	219		2,169	2,500

28 23 Video Surveillance

28 23 23 – Video Surveillance Systems Infrastructure

28 23 23.50 Video Surveillance Equipments

		Crew	Daily Output	Labor-Hours	Unit	Material	2015 Bare Costs Labor	2015 Bare Costs Equipment	Total	Total Incl O&P
0010	**VIDEO SURVEILLANCE EQUIPMENTS**									
0200	Video cameras, wireless, hidden in exit signs, clocks, etc., incl. receiver	1 Elec	3	2.667	Ea.	108	146		254	345
0210	Accessories for video recorder, single camera		3	2.667		183	146		329	425
0220	For multiple cameras		3	2.667		1,750	146		1,896	2,150
0230	Video cameras, wireless, for under vehicle searching, complete	↓	2	4	↓	10,200	219		10,419	11,500

28 31 Fire Detection and Alarm

28 31 23 – Fire Detection and Alarm Annunciation Panels and Fire Stations

28 31 23.50 Alarm Panels and Devices

		Crew	Daily Output	Labor-Hours	Unit	Material	2015 Bare Costs Labor	2015 Bare Costs Equipment	Total	Total Incl O&P
0010	**ALARM PANELS AND DEVICES**, not including wires & conduits									
3594	Fire, alarm control panel									
3600	4 zone	2 Elec	2	8	Ea.	400	440		840	1,125
3800	8 zone		1	16		780	875		1,655	2,200
4000	12 zone	↓	.67	23.988		2,400	1,300		3,700	4,650
4200	Battery and rack	1 Elec	4	2		410	109		519	625
4400	Automatic charger		8	1		585	54.50		639.50	725
4600	Signal bell		8	1		78.50	54.50		133	171
4800	Trouble buzzer or manual station		8	1		83.50	54.50		138	176
5600	Strobe and horn		5.30	1.509		152	82.50		234.50	295
5800	Fire alarm horn		6.70	1.194		61	65.50		126.50	168
6000	Door holder, electro-magnetic		4	2		103	109		212	282
6200	Combination holder and closer		3.20	2.500		123	137		260	350
6600	Drill switch		8	1		370	54.50		424.50	495
6800	Master box		2.70	2.963		6,400	162		6,562	7,300
7000	Break glass station		8	1		55.50	54.50		110	146
7800	Remote annunciator, 8 zone lamp	↓	1.80	4.444		209	243		452	605
8000	12 zone lamp	2 Elec	2.60	6.154		335	335		670	890
8200	16 zone lamp	"	2.20	7.273	↓	420	400		820	1,075

28 31 43 – Fire Detection Sensors

28 31 43.50 Fire and Heat Detectors

		Crew	Daily Output	Labor-Hours	Unit	Material	2015 Bare Costs Labor	2015 Bare Costs Equipment	Total	Total Incl O&P
0010	**FIRE & HEAT DETECTORS**									
5000	Detector, rate of rise	1 Elec	8	1	Ea.	51	54.50		105.50	141
5100	Fixed temperature	"	8	1	"	51	54.50		105.50	141

28 31 46 – Smoke Detection Sensors

28 31 46.50 Smoke Detectors

		Crew	Daily Output	Labor-Hours	Unit	Material	2015 Bare Costs Labor	2015 Bare Costs Equipment	Total	Total Incl O&P
0010	**SMOKE DETECTORS**									
5200	Smoke detector, ceiling type	1 Elec	6.20	1.290	Ea.	110	70.50		180.50	230
5400	Duct type	"	3.20	2.500	"	325	137		462	570

28 39 Mass Notification Systems

28 39 10 − Notification Systems

28 39 10.10 Mass Notification System	Crew	Daily Output	Labor-Hours	Unit	Material	2015 Bare Costs Labor	2015 Bare Costs Equipment	Total	Total Incl O&P
0010 **MASS NOTIFICATION SYSTEM**									
0100 Wireless command center, 10,000 devices	2 Elec	1.33	12.030	Ea.	3,600	660		4,260	4,975
0200 Option, email notification					1,750			1,750	1,925
0210 Remote device supervision & monitor					2,450			2,450	2,675
0300 Antenna VHF or UHF, for medium range	1 Elec	4	2		129	109		238	310
0310 For high-power transmitter		2	4		670	219		889	1,075
0400 Transmitter, 25 watt		4	2		2,025	109		2,134	2,400
0410 40 watt		2.66	3.008		2,700	165		2,865	3,225
0420 100 watt		1.33	6.015		6,775	330		7,105	7,950
0500 Wireless receiver/control module for speaker		8	1		265	54.50		319.50	375
0600 Desktop paging controller, stand alone		4	2		370	109		479	580

Estimating Tips
31 05 00 Common Work Results for Earthwork

- Estimating the actual cost of performing earthwork requires careful consideration of the variables involved. This includes items such as type of soil, whether water will be encountered, dewatering, whether banks need bracing, disposal of excavated earth, and length of haul to fill or spoil sites, etc. If the project has large quantities of cut or fill, consider raising or lowering the site to reduce costs, while paying close attention to the effect on site drainage and utilities.
- If the project has large quantities of fill, creating a borrow pit on the site can significantly lower the costs.
- It is very important to consider what time of year the project is scheduled for completion. Bad weather can create large cost overruns from dewatering, site repair, and lost productivity from cold weather.

Reference Numbers
Reference numbers are shown in shaded boxes at the beginning of some major classifications. These numbers refer to related items in the Reference Section. The reference information may be an estimating procedure, an alternate pricing method, or technical information.

Note: Not all subdivisions listed here necessarily appear in this publication. ■

Estimate with precision with RSMeans Online...
Quick, intuitive and easy-to-use, RSMeans Online gives you instant access to hundreds of thousands of material, labor, and equipment costs from RSMeans' comprehensive database, delivering the information you need to build competitive estimates every time.

Start your free trial today at **www.rsmeansonline.com**

RSMeansOnline

31 05 Common Work Results for Earthwork

31 05 13 – Soils for Earthwork

31 05 13.10 Borrow

		Crew	Daily Output	Labor-Hours	Unit	Material	2015 Bare Costs Labor	Equipment	Total	Total Incl O&P
0010	**BORROW**									
0020	Spread, 200 H.P. dozer, no compaction, 2 mi. RT haul									
0200	Common borrow	B-15	600	.047	C.Y.	12.40	1.99	4.62	19.01	22
0700	Screened loam		600	.047		27.50	1.99	4.62	34.11	39
0800	Topsoil, weed free	↓	600	.047		24.50	1.99	4.62	31.11	35.50
0900	For 5 mile haul, add	B-34B	200	.040	↓		1.60	3.46	5.06	6.40

31 05 16 – Aggregates for Earthwork

31 05 16.10 Borrow

		Crew	Daily Output	Labor-Hours	Unit	Material	2015 Bare Costs Labor	Equipment	Total	Total Incl O&P
0010	**BORROW**									
0020	Spread, with 200 H.P. dozer, no compaction, 2 mi. RT haul									
0100	Bank run gravel	B-15	600	.047	L.C.Y.	22	1.99	4.62	28.61	33
0300	Crushed stone (1.40 tons per CY), 1-1/2"		600	.047		23.50	1.99	4.62	30.11	34.50
0320	3/4"		600	.047		23.50	1.99	4.62	30.11	34.50
0340	1/2"		600	.047		26.50	1.99	4.62	33.11	37.50
0360	3/8"		600	.047		27.50	1.99	4.62	34.11	39
0400	Sand, washed, concrete		600	.047		41	1.99	4.62	47.61	53.50
0500	Dead or bank sand		600	.047		17.85	1.99	4.62	24.46	28
0600	Select structural fill	↓	600	.047		21	1.99	4.62	27.61	31.50
0900	For 5 mile haul, add	B-34B	200	.040	↓		1.60	3.46	5.06	6.40

31 06 Schedules for Earthwork

31 06 60 – Schedules for Special Foundations and Load Bearing Elements

31 06 60.14 Piling Special Costs

		Crew	Daily Output	Labor-Hours	Unit	Material	2015 Bare Costs Labor	Equipment	Total	Total Incl O&P
0010	**PILING SPECIAL COSTS**									
0011	Piling special costs, pile caps, see Section 03 30 53.40									
0500	Cutoffs, concrete piles, plain	1 Pile	5.50	1.455	Ea.		67		67	110
0600	With steel thin shell, add		38	.211			9.70		9.70	15.85
0700	Steel pile or "H" piles		19	.421			19.40		19.40	31.50
0800	Wood piles	↓	38	.211	↓		9.70		9.70	15.85
1000	Testing, any type piles, test load is twice the design load									
1050	50 ton design load, 100 ton test				Ea.				14,000	15,500
1100	100 ton design load, 200 ton test								20,000	22,000
1200	200 ton design load, 400 ton test				↓				28,000	31,000

31 06 60.15 Mobilization

		Crew	Daily Output	Labor-Hours	Unit	Material	2015 Bare Costs Labor	Equipment	Total	Total Incl O&P
0010	**MOBILIZATION**									
0020	Set up & remove, air compressor, 600 CFM	A-5	3.30	5.455	Ea.		206	18.60	224.60	355
0100	1200 CFM	"	2.20	8.182			310	28	338	535
0200	Crane, with pile leads and pile hammer, 75 ton	B-19	.60	106			5,075	2,900	7,975	11,400
0300	150 ton	"	.36	177	↓		8,475	4,825	13,300	19,000

31 14 Earth Stripping and Stockpiling

31 14 13 – Soil Stripping and Stockpiling

31 14 13.23 Topsoil Stripping and Stockpiling	Crew	Daily Output	Labor-Hours	Unit	Material	2015 Bare Costs Labor	2015 Bare Costs Equipment	Total	Total Incl O&P
0010 **TOPSOIL STRIPPING AND STOCKPILING**									
1500 Loam or topsoil, remove/stockpile on site									
1510 By hand, 6" deep, 50' haul, less than 100 S.Y.	B-1	100	.240	S.Y.		9.20		9.20	15.05
1520 By skid steer, 6" deep, 100' haul, 101-500 S.Y.	B-62	500	.048			1.98	.35	2.33	3.59
1530 100' haul, 501-900 S.Y.	"	900	.027			1.10	.19	1.29	1.99
1540 200' haul, 901-1100 S.Y.	B-63	1000	.040			1.59	.17	1.76	2.78
1550 By dozer, 200' haul, 1101-4000 S.Y.	B-10B	4000	.003			.14	.35	.49	.60

31 23 Excavation and Fill

31 23 16 – Excavation

31 23 16.13 Excavating, Trench

	Crew	Daily Output	Labor-Hours	Unit	Material	Labor	Equipment	Total	Total Incl O&P
0010 **EXCAVATING, TRENCH**									
0011 Or continuous footing									
0020 Common earth with no sheeting or dewatering included									
1400 By hand with pick and shovel 2' to 6' deep, light soil	1 Clab	8	1	B.C.Y.		37.50		37.50	61.50
1500 Heavy soil	"	4	2	"		75		75	123
1700 For tamping backfilled trenches, air tamp, add	A-1G	100	.080	E.C.Y.		3.01	.56	3.57	5.55
1900 Vibrating plate, add	B-18	180	.133	"		5.10	.26	5.36	8.65
2100 Trim sides and bottom for concrete pours, common earth		1500	.016	S.F.		.61	.03	.64	1.03
2300 Hardpan		600	.040	"		1.53	.08	1.61	2.60
9000 Minimum labor/equipment charge	1 Clab	4	2	Job		75		75	123

31 23 16.14 Excavating, Utility Trench

	Crew	Daily Output	Labor-Hours	Unit	Material	Labor	Equipment	Total	Total Incl O&P
0010 **EXCAVATING, UTILITY TRENCH**									
0011 Common earth									
0050 Trenching with chain trencher, 12 H.P., operator walking									
0100 4" wide trench, 12" deep	B-53	800	.010	L.F.		.49	.09	.58	.86
0150 18" deep		750	.011			.52	.09	.61	.92
0200 24" deep		700	.011			.56	.10	.66	.99
0300 6" wide trench, 12" deep		650	.012			.60	.10	.70	1.07
0350 18" deep		600	.013			.65	.11	.76	1.15
0400 24" deep		550	.015			.71	.12	.83	1.26
0450 36" deep		450	.018			.86	.15	1.01	1.54
0600 8" wide trench, 12" deep		475	.017			.82	.14	.96	1.46
0650 18" deep		400	.020			.97	.17	1.14	1.73
0700 24" deep		350	.023			1.11	.19	1.30	1.97
0750 36" deep		300	.027			1.30	.23	1.53	2.31
0900 Minimum labor/equipment charge		2	4	Job		194	34	228	350
1000 Backfill by hand including compaction, add									
1050 4" wide trench, 12" deep	A-1G	800	.010	L.F.		.38	.07	.45	.70
1100 18" deep		530	.015			.57	.11	.68	1.05
1150 24" deep		400	.020			.75	.14	.89	1.39
1300 6" wide trench, 12" deep		540	.015			.56	.10	.66	1.02
1350 18" deep		405	.020			.74	.14	.88	1.37
1400 24" deep		270	.030			1.11	.21	1.32	2.06
1450 36" deep		180	.044			1.67	.31	1.98	3.08
1600 8" wide trench, 12" deep		400	.020			.75	.14	.89	1.39
1650 18" deep		265	.030			1.14	.21	1.35	2.09
1700 24" deep		200	.040			1.50	.28	1.78	2.78
1750 36" deep		135	.059			2.23	.42	2.65	4.11
2000 Chain trencher, 40 H.P. operator riding									
2050 6" wide trench and backfill, 12" deep	B-54	1200	.007	L.F.		.32	.28	.60	.82

31 23 Excavation and Fill

31 23 16 – Excavation

31 23 16.14 Excavating, Utility Trench

		Crew	Daily Output	Labor-Hours	Unit	Material	2015 Bare Costs Labor	Equipment	Total	Total Incl O&P
2100	18" deep	B-54	1000	.008	L.F.		.39	.34	.73	.99
2150	24" deep		975	.008			.40	.34	.74	1.01
2200	36" deep		900	.009			.43	.37	.80	1.10
2250	48" deep		750	.011			.52	.45	.97	1.31
2300	60" deep		650	.012			.60	.52	1.12	1.52
2400	8" wide trench and backfill, 12" deep		1000	.008			.39	.34	.73	.99
2450	18" deep		950	.008			.41	.35	.76	1.04
2500	24" deep		900	.009			.43	.37	.80	1.10
2550	36" deep		800	.010			.49	.42	.91	1.23
2600	48" deep		650	.012			.60	.52	1.12	1.52
2700	12" wide trench and backfill, 12" deep		975	.008			.40	.34	.74	1.01
2750	18" deep		860	.009			.45	.39	.84	1.15
2800	24" deep		800	.010			.49	.42	.91	1.23
2850	36" deep		725	.011			.54	.46	1	1.36
3000	16" wide trench and backfill, 12" deep		835	.010			.47	.40	.87	1.18
3050	18" deep		750	.011			.52	.45	.97	1.31
3100	24" deep		700	.011			.56	.48	1.04	1.41
3200	Compaction with vibratory plate, add								35%	35%
5100	Hand excavate and trim for pipe bells after trench excavation									
5200	8" pipe	1 Clab	155	.052	L.F.		1.94		1.94	3.18
5300	18" pipe	"	130	.062	"		2.31		2.31	3.79
9000	Minimum labor/equipment charge	A-1G	4	2	Job		75	14.10	89.10	139

31 23 16.16 Structural Excavation for Minor Structures

		Crew	Daily Output	Labor-Hours	Unit	Material	2015 Bare Costs Labor	Equipment	Total	Total Incl O&P
0010	**STRUCTURAL EXCAVATION FOR MINOR STRUCTURES**									
0015	Hand, pits to 6' deep, sandy soil	1 Clab	8	1	B.C.Y.		37.50		37.50	61.50
0100	Heavy soil or clay		4	2			75		75	123
0300	Pits 6' to 12' deep, sandy soil		5	1.600			60		60	98.50
0500	Heavy soil or clay		3	2.667			100		100	164
0700	Pits 12' to 18' deep, sandy soil		4	2			75		75	123
0900	Heavy soil or clay		2	4			150		150	247
1100	Hand loading trucks from stock pile, sandy soil		12	.667			25		25	41
1300	Heavy soil or clay		8	1			37.50		37.50	61.50
1500	For wet or muck hand excavation, add to above								50%	50%
6000	Machine excavation, for spread and mat footings, elevator pits,									
6001	and small building foundations									
6030	Common earth, hydraulic backhoe, 1/2 C.Y. bucket	B-12E	55	.291	B.C.Y.		13	8.15	21.15	30
6035	3/4 C.Y. bucket	B-12F	90	.178			7.95	7.30	15.25	21
6040	1 C.Y. bucket	B-12A	108	.148			6.60	7.50	14.10	18.95
6050	1-1/2 C.Y. bucket	B-12B	144	.111			4.96	7.15	12.11	15.85
6060	2 C.Y. bucket	B-12C	200	.080			3.57	5.90	9.47	12.20
6070	Sand and gravel, 3/4 C.Y. bucket	B-12F	100	.160			7.15	6.55	13.70	18.70
6080	1 C.Y. bucket	B-12A	120	.133			5.95	6.75	12.70	17.05
6090	1-1/2 C.Y. bucket	B-12B	160	.100			4.47	6.45	10.92	14.30
6100	2 C.Y. bucket	B-12C	220	.073			3.25	5.35	8.60	11.15
6110	Clay, till, or blasted rock, 3/4 C.Y. bucket	B-12F	80	.200			8.95	8.20	17.15	23.50
6120	1 C.Y. bucket	B-12A	95	.168			7.50	8.55	16.05	21.50
6130	1-1/2 C.Y. bucket	B-12B	130	.123			5.50	7.90	13.40	17.55
6140	2 C.Y. bucket	B-12C	175	.091			4.08	6.70	10.78	13.95
6230	Sandy clay & loam, hydraulic backhoe, 1/2 C.Y. bucket	B-12E	60	.267			11.90	7.45	19.35	27.50
6235	3/4 C.Y. bucket	B-12F	98	.163			7.30	6.70	14	19.10
6240	1 C.Y. bucket	B-12A	116	.138			6.15	7	13.15	17.60
6250	1-1/2 C.Y. bucket	B-12B	156	.103			4.58	6.60	11.18	14.60

31 23 Excavation and Fill

31 23 16 – Excavation

31 23 16.16 Structural Excavation for Minor Structures

		Crew	Daily Output	Labor-Hours	Unit	Material	2015 Bare Costs Labor	Equipment	Total	Total Incl O&P
9000	Minimum labor/equipment charge	1 Clab	4	2	Job		75		75	123

31 23 16.42 Excavating, Bulk Bank Measure

		Crew	Daily Output	Labor-Hours	Unit	Material	Labor	Equipment	Total	Total Incl O&P
0010	**EXCAVATING, BULK BANK MEASURE** R312316-40									
0011	Common earth piled									
0020	For loading onto trucks, add									15%	15%
0200	Excavator, hydraulic, crawler mtd., 1 C.Y. cap. = 100 C.Y./hr. R312316-45	B-12A	800	.020	B.C.Y.		.89	1.02	1.91	2.56
1200	Front end loader, track mtd., 1-1/2 C.Y. cap. = 70 C.Y./hr.	B-10N	560	.021			.99	.93	1.92	2.62
1500	Wheel mounted, 3/4 C.Y. cap. = 45 C.Y./hr.	B-10R	360	.033			1.54	.83	2.37	3.38
5000	Excavating, bulk bank measure, sandy clay & loam piled									
5020	For loading onto trucks, add									15%	15%
5100	Excavator, hydraulic, crawler mtd., 1 C.Y. cap. = 120 C.Y./hr.	B-12A	960	.017	B.C.Y.		.74	.85	1.59	2.13
9000	Minimum labor/equipment charge	B-10L	2	6	Job		278	236	514	705

31 23 16.46 Excavating, Bulk, Dozer

		Crew	Daily Output	Labor-Hours	Unit	Material	Labor	Equipment	Total	Total Incl O&P
0010	**EXCAVATING, BULK, DOZER**									
0011	Open site									
2000	80 H.P., 50' haul, sand & gravel	B-10L	460	.026	B.C.Y.		1.21	1.03	2.24	3.06
2200	150' haul, sand & gravel		230	.052			2.41	2.05	4.46	6.10
2400	300' haul, sand & gravel		120	.100			4.63	3.94	8.57	11.75
3000	105 H.P., 50' haul, sand & gravel	B-10W	700	.017			.79	.86	1.65	2.22
3200	150' haul, sand & gravel		310	.039			1.79	1.94	3.73	5
3300	300' haul, sand & gravel		140	.086			3.97	4.31	8.28	11.10
4000	200 H.P., 50' haul, sand & gravel	B-10B	1400	.009			.40	.99	1.39	1.72
4200	150' haul, sand & gravel		595	.020			.93	2.33	3.26	4.05
4400	300' haul, sand & gravel		310	.039			1.79	4.47	6.26	7.80
5040	Clay	B-10M	1025	.012			.54	1.85	2.39	2.91
5400	300' haul, sand & gravel	"	470	.026			1.18	4.04	5.22	6.35

31 23 23 – Fill

31 23 23.13 Backfill

		Crew	Daily Output	Labor-Hours	Unit	Material	Labor	Equipment	Total	Total Incl O&P
0010	**BACKFILL** R312323-30									
0015	By hand, no compaction, light soil	1 Clab	14	.571	L.C.Y.		21.50		21.50	35
0100	Heavy soil		11	.727	"		27.50		27.50	45
0300	Compaction in 6" layers, hand tamp, add to above		20.60	.388	E.C.Y.		14.60		14.60	24
0400	Roller compaction operator walking, add	B-10A	100	.120			5.55	1.80	7.35	10.90
0500	Air tamp, add	B-9D	190	.211			8	1.40	9.40	14.65
0600	Vibrating plate, add	A-1D	60	.133			5	.60	5.60	8.85
0800	Compaction in 12" layers, hand tamp, add to above	1 Clab	34	.235			8.85		8.85	14.50
1000	Air tamp, add	B-9	285	.140			5.35	.82	6.17	9.65
1100	Vibrating plate, add	A-1E	90	.089			3.34	.52	3.86	6.05
1200	Trench, dozer, no compaction, 60 HP	B-10L	425	.028	L.C.Y.		1.31	1.11	2.42	3.31
1300	Dozer backfilling, bulk, up to 300' haul, no compaction	B-10B	1200	.010	"		.46	1.16	1.62	2.01
1400	Air tamped, add	B-11B	80	.200	E.C.Y.		8.60	3.77	12.37	18.05
1900	Dozer backfilling, trench, up to 300' haul, no compaction	B-10B	900	.013	L.C.Y.		.62	1.54	2.16	2.68
2000	Air tamped, add	B-11B	80	.200	E.C.Y.		8.60	3.77	12.37	18.05
2350	Spreading in 8" layers, small dozer	B-10B	1060	.011	L.C.Y.		.52	1.31	1.83	2.28
2450	Compacting with vibrating plate, 8" lifts	A-1D	73	.110	E.C.Y.		4.12	.50	4.62	7.30

31 23 23.14 Backfill, Structural

		Crew	Daily Output	Labor-Hours	Unit	Material	Labor	Equipment	Total	Total Incl O&P
0010	**BACKFILL, STRUCTURAL**									
0011	Dozer or F.E. loader									
0020	From existing stockpile, no compaction									
2000	80 H.P., 50' haul, sand & gravel	B-10L	1100	.011	L.C.Y.		.50	.43	.93	1.28
2010	Sandy clay & loam		1070	.011			.52	.44	.96	1.32

31 23 Excavation and Fill

31 23 23 – Fill

31 23 23.14 Backfill, Structural

		Crew	Daily Output	Labor-Hours	Unit	Material	2015 Bare Costs Labor	Equipment	Total	Total Incl O&P
2020	Common earth	B-10L	975	.012	L.C.Y.		.57	.48	1.05	1.44
2040	Clay		850	.014			.65	.56	1.21	1.66
2400	300' haul, sand & gravel		370	.032			1.50	1.28	2.78	3.80
2410	Sandy clay & loam		360	.033			1.54	1.31	2.85	3.91
2420	Common earth		330	.036			1.68	1.43	3.11	4.26
2440	Clay		290	.041			1.91	1.63	3.54	4.86
3000	105 H.P., 50' haul, sand & gravel	B-10W	1350	.009			.41	.45	.86	1.15
3010	Sandy clay & loam		1325	.009			.42	.46	.88	1.17
3020	Common earth		1225	.010			.45	.49	.94	1.27
3040	Clay		1100	.011			.50	.55	1.05	1.41
3300	300' haul, sand & gravel		465	.026			1.19	1.30	2.49	3.34
3310	Sandy clay & loam		455	.026			1.22	1.32	2.54	3.41
3320	Common earth		415	.029			1.34	1.45	2.79	3.74
3340	Clay		370	.032			1.50	1.63	3.13	4.19

31 23 23.16 Fill By Borrow and Utility Bedding

		Crew	Daily Output	Labor-Hours	Unit	Material	Labor	Equipment	Total	Total Incl O&P
0010	**FILL BY BORROW AND UTILITY BEDDING**									
0049	Utility bedding, for pipe & conduit, not incl. compaction									
0050	Crushed or screened bank run gravel	B-6	150	.160	L.C.Y.	25.50	6.60	2.43	34.53	41.50
0100	Crushed stone 3/4" to 1/2"		150	.160		23.50	6.60	2.43	32.53	39.50
0200	Sand, dead or bank		150	.160		17.85	6.60	2.43	26.88	33
0500	Compacting bedding in trench	A-1D	90	.089	E.C.Y.		3.34	.40	3.74	5.95
0600	If material source exceeds 2 miles, add for extra mileage.									

31 23 23.17 General Fill

		Crew	Daily Output	Labor-Hours	Unit	Material	Labor	Equipment	Total	Total Incl O&P
0010	**GENERAL FILL**									
0011	Spread dumped material, no compaction									
0020	By dozer, no compaction	B-10B	1000	.012	L.C.Y.		.56	1.39	1.95	2.42
0100	By hand	1 Clab	12	.667	"		25		25	41
9000	Minimum labor/equipment charge	"	4	2	Job		75		75	123

31 23 23.20 Hauling

		Crew	Daily Output	Labor-Hours	Unit	Material	Labor	Equipment	Total	Total Incl O&P
0010	**HAULING**									
0011	Excavated or borrow, loose cubic yards									
0012	no loading equipment, including hauling, waiting, loading/dumping									
0013	time per cycle (wait, load, travel, unload or dump & return)									
0014	8 C.Y. truck, 15 MPH ave, cycle 0.5 miles, 10 min. wait/Ld./Uld.	B-34A	320	.025	L.C.Y.		1	1.29	2.29	3.02
0016	cycle 1 mile		272	.029			1.18	1.51	2.69	3.56
0018	cycle 2 miles		208	.038			1.54	1.98	3.52	4.65
0020	cycle 4 miles		144	.056			2.23	2.86	5.09	6.75
0022	cycle 6 miles		112	.071			2.86	3.67	6.53	8.65
0024	cycle 8 miles		88	.091			3.64	4.67	8.31	11
0026	20 MPH ave, cycle 0.5 mile		336	.024			.95	1.22	2.17	2.89
0028	cycle 1 mile		296	.027			1.08	1.39	2.47	3.27
0030	cycle 2 miles		240	.033			1.33	1.71	3.04	4.03
0032	cycle 4 miles		176	.045			1.82	2.34	4.16	5.50
0034	cycle 6 miles		136	.059			2.36	3.02	5.38	7.15
0036	cycle 8 miles		112	.071			2.86	3.67	6.53	8.65
0044	25 MPH ave, cycle 4 miles		192	.042			1.67	2.14	3.81	5.05
0046	cycle 6 miles		160	.050			2	2.57	4.57	6.05
0048	cycle 8 miles		128	.063			2.50	3.21	5.71	7.55
0050	30 MPH ave, cycle 4 miles		216	.037			1.48	1.90	3.38	4.48
0052	cycle 6 miles		176	.045			1.82	2.34	4.16	5.50
0054	cycle 8 miles		144	.056			2.23	2.86	5.09	6.75
0114	15 MPH ave, cycle 0.5 mile, 15 min. wait/Ld./Uld.		224	.036			1.43	1.84	3.27	4.33

31 23 Excavation and Fill

31 23 23 – Fill

31 23 23.20 Hauling

		Crew	Daily Output	Labor-Hours	Unit	Material	2015 Bare Costs Labor	Equipment	Total	Total Incl O&P
0116	cycle 1 mile	B-34A	200	.040	L.C.Y.		1.60	2.06	3.66	4.84
0118	cycle 2 miles		168	.048			1.91	2.45	4.36	5.75
0120	cycle 4 miles		120	.067			2.67	3.43	6.10	8.05
0122	cycle 6 miles		96	.083			3.34	4.28	7.62	10.10
0124	cycle 8 miles		80	.100			4.01	5.15	9.16	12.10
0126	20 MPH ave, cycle 0.5 mile		232	.034			1.38	1.77	3.15	4.18
0128	cycle 1 mile		208	.038			1.54	1.98	3.52	4.65
0130	cycle 2 miles		184	.043			1.74	2.23	3.97	5.25
0132	cycle 4 miles		144	.056			2.23	2.86	5.09	6.75
0134	cycle 6 miles		112	.071			2.86	3.67	6.53	8.65
0136	cycle 8 miles		96	.083			3.34	4.28	7.62	10.10
0144	25 MPH ave, cycle 4 miles		152	.053			2.11	2.71	4.82	6.40
0146	cycle 6 miles		128	.063			2.50	3.21	5.71	7.55
0148	cycle 8 miles		112	.071			2.86	3.67	6.53	8.65
0150	30 MPH ave, cycle 4 miles		168	.048			1.91	2.45	4.36	5.75
0152	cycle 6 miles		144	.056			2.23	2.86	5.09	6.75
0154	cycle 8 miles		120	.067			2.67	3.43	6.10	8.05
0214	15 MPH ave, cycle 0.5 mile, 20 min wait/Ld./Uld.		176	.045			1.82	2.34	4.16	5.50
0216	cycle 1 mile		160	.050			2	2.57	4.57	6.05
0218	cycle 2 miles		136	.059			2.36	3.02	5.38	7.15
0220	cycle 4 miles		104	.077			3.08	3.95	7.03	9.30
0222	cycle 6 miles		88	.091			3.64	4.67	8.31	11
0224	cycle 8 miles		72	.111			4.45	5.70	10.15	13.45
0226	20 MPH ave, cycle 0.5 mile		176	.045			1.82	2.34	4.16	5.50
0228	cycle 1 mile		168	.048			1.91	2.45	4.36	5.75
0230	cycle 2 miles		144	.056			2.23	2.86	5.09	6.75
0232	cycle 4 miles		120	.067			2.67	3.43	6.10	8.05
0234	cycle 6 miles		96	.083			3.34	4.28	7.62	10.10
0236	cycle 8 miles		88	.091			3.64	4.67	8.31	11
0244	25 MPH ave, cycle 4 miles		128	.063			2.50	3.21	5.71	7.55
0246	cycle 6 miles		112	.071			2.86	3.67	6.53	8.65
0248	cycle 8 miles		96	.083			3.34	4.28	7.62	10.10
0250	30 MPH ave, cycle 4 miles		136	.059			2.36	3.02	5.38	7.15
0252	cycle 6 miles		120	.067			2.67	3.43	6.10	8.05
0254	cycle 8 miles		104	.077			3.08	3.95	7.03	9.30
0314	15 MPH ave, cycle 0.5 mile, 25 min wait/Ld./Uld.		144	.056			2.23	2.86	5.09	6.75
0316	cycle 1 mile		128	.063			2.50	3.21	5.71	7.55
0318	cycle 2 miles		112	.071			2.86	3.67	6.53	8.65
0320	cycle 4 miles		96	.083			3.34	4.28	7.62	10.10
0322	cycle 6 miles		80	.100			4.01	5.15	9.16	12.10
0324	cycle 8 miles		64	.125			5	6.45	11.45	15.10
0326	20 MPH ave, cycle 0.5 mile		144	.056			2.23	2.86	5.09	6.75
0328	cycle 1 mile		136	.059			2.36	3.02	5.38	7.15
0330	cycle 2 miles		120	.067			2.67	3.43	6.10	8.05
0332	cycle 4 miles		104	.077			3.08	3.95	7.03	9.30
0334	cycle 6 miles		88	.091			3.64	4.67	8.31	11
0336	cycle 8 miles		80	.100			4.01	5.15	9.16	12.10
0344	25 MPH ave, cycle 4 miles		112	.071			2.86	3.67	6.53	8.65
0346	cycle 6 miles		96	.083			3.34	4.28	7.62	10.10
0348	cycle 8 miles		88	.091			3.64	4.67	8.31	11
0350	30 MPH ave, cycle 4 miles		112	.071			2.86	3.67	6.53	8.65
0352	cycle 6 miles		104	.077			3.08	3.95	7.03	9.30
0354	cycle 8 miles		96	.083			3.34	4.28	7.62	10.10

31 23 Excavation and Fill

31 23 23 – Fill

31 23 23.20 Hauling

		Crew	Daily Output	Labor-Hours	Unit	Material	2015 Bare Costs Labor	Equipment	Total	Total Incl O&P
0414	15 MPH ave, cycle 0.5 mile, 30 min wait/Ld./Uld.	B-34A	120	.067	L.C.Y.		2.67	3.43	6.10	8.05
0416	cycle 1 mile		112	.071			2.86	3.67	6.53	8.65
0418	cycle 2 miles		96	.083			3.34	4.28	7.62	10.10
0420	cycle 4 miles		80	.100			4.01	5.15	9.16	12.10
0422	cycle 6 miles		72	.111			4.45	5.70	10.15	13.45
0424	cycle 8 miles		64	.125			5	6.45	11.45	15.10
0426	20 MPH ave, cycle 0.5 mile		120	.067			2.67	3.43	6.10	8.05
0428	cycle 1 mile		112	.071			2.86	3.67	6.53	8.65
0430	cycle 2 miles		104	.077			3.08	3.95	7.03	9.30
0432	cycle 4 miles		88	.091			3.64	4.67	8.31	11
0434	cycle 6 miles		80	.100			4.01	5.15	9.16	12.10
0436	cycle 8 miles		72	.111			4.45	5.70	10.15	13.45
0444	25 MPH ave, cycle 4 miles		96	.083			3.34	4.28	7.62	10.10
0446	cycle 6 miles		88	.091			3.64	4.67	8.31	11
0448	cycle 8 miles		80	.100			4.01	5.15	9.16	12.10
0450	30 MPH ave, cycle 4 miles		96	.083			3.34	4.28	7.62	10.10
0452	cycle 6 miles		88	.091			3.64	4.67	8.31	11
0454	cycle 8 miles		80	.100			4.01	5.15	9.16	12.10
0514	15 MPH ave, cycle 0.5 mile, 35 min wait/Ld./Uld.		104	.077			3.08	3.95	7.03	9.30
0516	cycle 1 mile		96	.083			3.34	4.28	7.62	10.10
0518	cycle 2 miles		88	.091			3.64	4.67	8.31	11
0520	cycle 4 miles		72	.111			4.45	5.70	10.15	13.45
0522	cycle 6 miles		64	.125			5	6.45	11.45	15.10
0524	cycle 8 miles		56	.143			5.70	7.35	13.05	17.30
0526	20 MPH ave, cycle 0.5 mile		104	.077			3.08	3.95	7.03	9.30
0528	cycle 1 mile		96	.083			3.34	4.28	7.62	10.10
0530	cycle 2 miles		96	.083			3.34	4.28	7.62	10.10
0532	cycle 4 miles		80	.100			4.01	5.15	9.16	12.10
0534	cycle 6 miles		72	.111			4.45	5.70	10.15	13.45
0536	cycle 8 miles		64	.125			5	6.45	11.45	15.10
0544	25 MPH ave, cycle 4 miles		88	.091			3.64	4.67	8.31	11
0546	cycle 6 miles		80	.100			4.01	5.15	9.16	12.10
0548	cycle 8 miles		72	.111			4.45	5.70	10.15	13.45
0550	30 MPH ave, cycle 4 miles		88	.091			3.64	4.67	8.31	11
0552	cycle 6 miles		80	.100			4.01	5.15	9.16	12.10
0554	cycle 8 miles		72	.111			4.45	5.70	10.15	13.45
1014	12 C.Y. truck, cycle 0.5 mile, 15 MPH ave, 15 min. wait/Ld./Uld.	B-34B	336	.024			.95	2.06	3.01	3.80
1016	cycle 1 mile		300	.027			1.07	2.30	3.37	4.25
1018	cycle 2 miles		252	.032			1.27	2.74	4.01	5.05
1020	cycle 4 miles		180	.044			1.78	3.84	5.62	7.10
1022	cycle 6 miles		144	.056			2.23	4.80	7.03	8.90
1024	cycle 8 miles		120	.067			2.67	5.75	8.42	10.65
1025	cycle 10 miles		96	.083			3.34	7.20	10.54	13.30
1026	20 MPH ave, cycle 0.5 mile		348	.023			.92	1.99	2.91	3.66
1028	cycle 1 mile		312	.026			1.03	2.21	3.24	4.10
1030	cycle 2 miles		276	.029			1.16	2.50	3.66	4.62
1032	cycle 4 miles		216	.037			1.48	3.20	4.68	5.90
1034	cycle 6 miles		168	.048			1.91	4.11	6.02	7.60
1036	cycle 8 miles		144	.056			2.23	4.80	7.03	8.90
1038	cycle 10 miles		120	.067			2.67	5.75	8.42	10.65
1040	25 MPH ave, cycle 4 miles		228	.035			1.41	3.03	4.44	5.60
1042	cycle 6 miles		192	.042			1.67	3.60	5.27	6.65
1044	cycle 8 miles		168	.048			1.91	4.11	6.02	7.60

31 23 Excavation and Fill

31 23 23 – Fill

31 23 23.20 Hauling

		Crew	Daily Output	Labor-Hours	Unit	Material	2015 Bare Costs Labor	Equipment	Total	Total Incl O&P
1046	cycle 10 miles	B-34B	144	.056	L.C.Y.		2.23	4.80	7.03	8.90
1050	30 MPH ave, cycle 4 miles		252	.032			1.27	2.74	4.01	5.05
1052	cycle 6 miles		216	.037			1.48	3.20	4.68	5.90
1054	cycle 8 miles		180	.044			1.78	3.84	5.62	7.10
1056	cycle 10 miles		156	.051			2.05	4.43	6.48	8.20
1060	35 MPH ave, cycle 4 miles		264	.030			1.21	2.62	3.83	4.84
1062	cycle 6 miles		228	.035			1.41	3.03	4.44	5.60
1064	cycle 8 miles		204	.039			1.57	3.39	4.96	6.25
1066	cycle 10 miles		180	.044			1.78	3.84	5.62	7.10
1068	cycle 20 miles		120	.067			2.67	5.75	8.42	10.65
1069	cycle 30 miles		84	.095			3.81	8.25	12.06	15.20
1070	cycle 40 miles		72	.111			4.45	9.60	14.05	17.70
1072	40 MPH ave, cycle 6 miles		240	.033			1.33	2.88	4.21	5.30
1074	cycle 8 miles		216	.037			1.48	3.20	4.68	5.90
1076	cycle 10 miles		192	.042			1.67	3.60	5.27	6.65
1078	cycle 20 miles		120	.067			2.67	5.75	8.42	10.65
1080	cycle 30 miles		96	.083			3.34	7.20	10.54	13.30
1082	cycle 40 miles		72	.111			4.45	9.60	14.05	17.70
1084	cycle 50 miles		60	.133			5.35	11.50	16.85	21.50
1094	45 MPH ave, cycle 8 miles		216	.037			1.48	3.20	4.68	5.90
1096	cycle 10 miles		204	.039			1.57	3.39	4.96	6.25
1098	cycle 20 miles		132	.061			2.43	5.25	7.68	9.65
1100	cycle 30 miles		108	.074			2.97	6.40	9.37	11.85
1102	cycle 40 miles		84	.095			3.81	8.25	12.06	15.20
1104	cycle 50 miles		72	.111			4.45	9.60	14.05	17.70
1106	50 MPH ave, cycle 10 miles		216	.037			1.48	3.20	4.68	5.90
1108	cycle 20 miles		144	.056			2.23	4.80	7.03	8.90
1110	cycle 30 miles		108	.074			2.97	6.40	9.37	11.85
1112	cycle 40 miles		84	.095			3.81	8.25	12.06	15.20
1114	cycle 50 miles		72	.111			4.45	9.60	14.05	17.70
1214	15 MPH ave, cycle 0.5 mile, 20 min. wait/Ld./Uld.		264	.030			1.21	2.62	3.83	4.84
1216	cycle 1 mile		240	.033			1.33	2.88	4.21	5.30
1218	cycle 2 miles		204	.039			1.57	3.39	4.96	6.25
1220	cycle 4 miles		156	.051			2.05	4.43	6.48	8.20
1222	cycle 6 miles		132	.061			2.43	5.25	7.68	9.65
1224	cycle 8 miles		108	.074			2.97	6.40	9.37	11.85
1225	cycle 10 miles		96	.083			3.34	7.20	10.54	13.30
1226	20 MPH ave, cycle 0.5 mile		264	.030			1.21	2.62	3.83	4.84
1228	cycle 1 mile		252	.032			1.27	2.74	4.01	5.05
1230	cycle 2 miles		216	.037			1.48	3.20	4.68	5.90
1232	cycle 4 miles		180	.044			1.78	3.84	5.62	7.10
1234	cycle 6 miles		144	.056			2.23	4.80	7.03	8.90
1236	cycle 8 miles		132	.061			2.43	5.25	7.68	9.65
1238	cycle 10 miles		108	.074			2.97	6.40	9.37	11.85
1240	25 MPH ave, cycle 4 miles		192	.042			1.67	3.60	5.27	6.65
1242	cycle 6 miles		168	.048			1.91	4.11	6.02	7.60
1244	cycle 8 miles		144	.056			2.23	4.80	7.03	8.90
1246	cycle 10 miles		132	.061			2.43	5.25	7.68	9.65
1250	30 MPH ave, cycle 4 miles		204	.039			1.57	3.39	4.96	6.25
1252	cycle 6 miles		180	.044			1.78	3.84	5.62	7.10
1254	cycle 8 miles		156	.051			2.05	4.43	6.48	8.20
1256	cycle 10 miles		144	.056			2.23	4.80	7.03	8.90
1260	35 MPH ave, cycle 4 miles		216	.037			1.48	3.20	4.68	5.90

31 23 Excavation and Fill

31 23 23 – Fill

31 23 23.20 Hauling		Crew	Daily Output	Labor-Hours	Unit	Material	2015 Bare Costs Labor	Equipment	Total	Total Incl O&P
1262	cycle 6 miles	B-34B	192	.042	L.C.Y.		1.67	3.60	5.27	6.65
1264	cycle 8 miles		168	.048			1.91	4.11	6.02	7.60
1266	cycle 10 miles		156	.051			2.05	4.43	6.48	8.20
1268	cycle 20 miles		108	.074			2.97	6.40	9.37	11.85
1269	cycle 30 miles		72	.111			4.45	9.60	14.05	17.70
1270	cycle 40 miles		60	.133			5.35	11.50	16.85	21.50
1272	40 MPH ave, cycle 6 miles		192	.042			1.67	3.60	5.27	6.65
1274	cycle 8 miles		180	.044			1.78	3.84	5.62	7.10
1276	cycle 10 miles		156	.051			2.05	4.43	6.48	8.20
1278	cycle 20 miles		108	.074			2.97	6.40	9.37	11.85
1280	cycle 30 miles		84	.095			3.81	8.25	12.06	15.20
1282	cycle 40 miles		72	.111			4.45	9.60	14.05	17.70
1284	cycle 50 miles		60	.133			5.35	11.50	16.85	21.50
1294	45 MPH ave, cycle 8 miles		180	.044			1.78	3.84	5.62	7.10
1296	cycle 10 miles		168	.048			1.91	4.11	6.02	7.60
1298	cycle 20 miles		120	.067			2.67	5.75	8.42	10.65
1300	cycle 30 miles		96	.083			3.34	7.20	10.54	13.30
1302	cycle 40 miles		72	.111			4.45	9.60	14.05	17.70
1304	cycle 50 miles		60	.133			5.35	11.50	16.85	21.50
1306	50 MPH ave, cycle 10 miles		180	.044			1.78	3.84	5.62	7.10
1308	cycle 20 miles		132	.061			2.43	5.25	7.68	9.65
1310	cycle 30 miles		96	.083			3.34	7.20	10.54	13.30
1312	cycle 40 miles		84	.095			3.81	8.25	12.06	15.20
1314	cycle 50 miles		72	.111			4.45	9.60	14.05	17.70
1414	15 MPH ave, cycle 0.5 mile, 25 min. wait/Ld./Uld.		204	.039			1.57	3.39	4.96	6.25
1416	cycle 1 mile		192	.042			1.67	3.60	5.27	6.65
1418	cycle 2 miles		168	.048			1.91	4.11	6.02	7.60
1420	cycle 4 miles		132	.061			2.43	5.25	7.68	9.65
1422	cycle 6 miles		120	.067			2.67	5.75	8.42	10.65
1424	cycle 8 miles		96	.083			3.34	7.20	10.54	13.30
1425	cycle 10 miles		84	.095			3.81	8.25	12.06	15.20
1426	20 MPH ave, cycle 0.5 mile		216	.037			1.48	3.20	4.68	5.90
1428	cycle 1 mile		204	.039			1.57	3.39	4.96	6.25
1430	cycle 2 miles		180	.044			1.78	3.84	5.62	7.10
1432	cycle 4 miles		156	.051			2.05	4.43	6.48	8.20
1434	cycle 6 miles		132	.061			2.43	5.25	7.68	9.65
1436	cycle 8 miles		120	.067			2.67	5.75	8.42	10.65
1438	cycle 10 miles		96	.083			3.34	7.20	10.54	13.30
1440	25 MPH ave, cycle 4 miles		168	.048			1.91	4.11	6.02	7.60
1442	cycle 6 miles		144	.056			2.23	4.80	7.03	8.90
1444	cycle 8 miles		132	.061			2.43	5.25	7.68	9.65
1446	cycle 10 miles		108	.074			2.97	6.40	9.37	11.85
1450	30 MPH ave, cycle 4 miles		168	.048			1.91	4.11	6.02	7.60
1452	cycle 6 miles		156	.051			2.05	4.43	6.48	8.20
1454	cycle 8 miles		132	.061			2.43	5.25	7.68	9.65
1456	cycle 10 miles		120	.067			2.67	5.75	8.42	10.65
1460	35 MPH ave, cycle 4 miles		180	.044			1.78	3.84	5.62	7.10
1462	cycle 6 miles		156	.051			2.05	4.43	6.48	8.20
1464	cycle 8 miles		144	.056			2.23	4.80	7.03	8.90
1466	cycle 10 miles		132	.061			2.43	5.25	7.68	9.65
1468	cycle 20 miles		96	.083			3.34	7.20	10.54	13.30
1469	cycle 30 miles		72	.111			4.45	9.60	14.05	17.70
1470	cycle 40 miles		60	.133			5.35	11.50	16.85	21.50

31 23 Excavation and Fill

31 23 23 – Fill

31 23 23.20 Hauling		Crew	Daily Output	Labor-Hours	Unit	Material	2015 Bare Costs Labor	Equipment	Total	Total Incl O&P
1472	40 MPH ave, cycle 6 miles	B-34B	168	.048	L.C.Y.		1.91	4.11	6.02	7.60
1474	cycle 8 miles		156	.051			2.05	4.43	6.48	8.20
1476	cycle 10 miles		144	.056			2.23	4.80	7.03	8.90
1478	cycle 20 miles		96	.083			3.34	7.20	10.54	13.30
1480	cycle 30 miles		84	.095			3.81	8.25	12.06	15.20
1482	cycle 40 miles		60	.133			5.35	11.50	16.85	21.50
1484	cycle 50 miles		60	.133			5.35	11.50	16.85	21.50
1494	45 MPH ave, cycle 8 miles		156	.051			2.05	4.43	6.48	8.20
1496	cycle 10 miles		144	.056			2.23	4.80	7.03	8.90
1498	cycle 20 miles		108	.074			2.97	6.40	9.37	11.85
1500	cycle 30 miles		84	.095			3.81	8.25	12.06	15.20
1502	cycle 40 miles		72	.111			4.45	9.60	14.05	17.70
1504	cycle 50 miles		60	.133			5.35	11.50	16.85	21.50
1506	50 MPH ave, cycle 10 miles		156	.051			2.05	4.43	6.48	8.20
1508	cycle 20 miles		120	.067			2.67	5.75	8.42	10.65
1510	cycle 30 miles		96	.083			3.34	7.20	10.54	13.30
1512	cycle 40 miles		72	.111			4.45	9.60	14.05	17.70
1514	cycle 50 miles		60	.133			5.35	11.50	16.85	21.50
1614	15 MPH, cycle 0.5 mile, 30 min. wait/Ld./Uld.		180	.044			1.78	3.84	5.62	7.10
1616	cycle 1 mile		168	.048			1.91	4.11	6.02	7.60
1618	cycle 2 miles		144	.056			2.23	4.80	7.03	8.90
1620	cycle 4 miles		120	.067			2.67	5.75	8.42	10.65
1622	cycle 6 miles		108	.074			2.97	6.40	9.37	11.85
1624	cycle 8 miles		84	.095			3.81	8.25	12.06	15.20
1625	cycle 10 miles		84	.095			3.81	8.25	12.06	15.20
1626	20 MPH ave, cycle 0.5 mile		180	.044			1.78	3.84	5.62	7.10
1628	cycle 1 mile		168	.048			1.91	4.11	6.02	7.60
1630	cycle 2 miles		156	.051			2.05	4.43	6.48	8.20
1632	cycle 4 miles		132	.061			2.43	5.25	7.68	9.65
1634	cycle 6 miles		120	.067			2.67	5.75	8.42	10.65
1636	cycle 8 miles		108	.074			2.97	6.40	9.37	11.85
1638	cycle 10 miles		96	.083			3.34	7.20	10.54	13.30
1640	25 MPH ave, cycle 4 miles		144	.056			2.23	4.80	7.03	8.90
1642	cycle 6 miles		132	.061			2.43	5.25	7.68	9.65
1644	cycle 8 miles		108	.074			2.97	6.40	9.37	11.85
1646	cycle 10 miles		108	.074			2.97	6.40	9.37	11.85
1650	30 MPH ave, cycle 4 miles		144	.056			2.23	4.80	7.03	8.90
1652	cycle 6 miles		132	.061			2.43	5.25	7.68	9.65
1654	cycle 8 miles		120	.067			2.67	5.75	8.42	10.65
1656	cycle 10 miles		108	.074			2.97	6.40	9.37	11.85
1660	35 MPH ave, cycle 4 miles		156	.051			2.05	4.43	6.48	8.20
1662	cycle 6 miles		144	.056			2.23	4.80	7.03	8.90
1664	cycle 8 miles		132	.061			2.43	5.25	7.68	9.65
1666	cycle 10 miles		120	.067			2.67	5.75	8.42	10.65
1668	cycle 20 miles		84	.095			3.81	8.25	12.06	15.20
1669	cycle 30 miles		72	.111			4.45	9.60	14.05	17.70
1670	cycle 40 miles		60	.133			5.35	11.50	16.85	21.50
1672	40 MPH, cycle 6 miles		144	.056			2.23	4.80	7.03	8.90
1674	cycle 8 miles		132	.061			2.43	5.25	7.68	9.65
1676	cycle 10 miles		120	.067			2.67	5.75	8.42	10.65
1678	cycle 20 miles		96	.083			3.34	7.20	10.54	13.30
1680	cycle 30 miles		72	.111			4.45	9.60	14.05	17.70
1682	cycle 40 miles		60	.133			5.35	11.50	16.85	21.50

31 23 Excavation and Fill

31 23 23 – Fill

31 23 23.20 Hauling

		Crew	Daily Output	Labor-Hours	Unit	Material	2015 Bare Costs Labor	2015 Bare Costs Equipment	Total	Total Incl O&P
1684	cycle 50 miles	B-34B	48	.167	L.C.Y.		6.70	14.40	21.10	26.50
1694	45 MPH ave, cycle 8 miles		144	.056			2.23	4.80	7.03	8.90
1696	cycle 10 miles		132	.061			2.43	5.25	7.68	9.65
1698	cycle 20 miles		96	.083			3.34	7.20	10.54	13.30
1700	cycle 30 miles		84	.095			3.81	8.25	12.06	15.20
1702	cycle 40 miles		60	.133			5.35	11.50	16.85	21.50
1704	cycle 50 miles		60	.133			5.35	11.50	16.85	21.50
1706	50 MPH ave, cycle 10 miles		132	.061			2.43	5.25	7.68	9.65
1708	cycle 20 miles		108	.074			2.97	6.40	9.37	11.85
1710	cycle 30 miles		84	.095			3.81	8.25	12.06	15.20
1712	cycle 40 miles		72	.111			4.45	9.60	14.05	17.70
1714	cycle 50 miles		60	.133			5.35	11.50	16.85	21.50
2000	Hauling, 8 C.Y. truck, small project cost per hour	B-34A	8	1	Hr.		40	51.50	91.50	121
2100	12 C.Y. Truck	B-34B	8	1			40	86.50	126.50	160
2150	16.5 C.Y. Truck	B-34C	8	1			40	91	131	165
2175	18 C.Y. 8 wheel Truck	B-34I	8	1			40	109	149	184
2200	20 C.Y. Truck	B-34D	8	1			40	92.50	132.50	167
2300	Grading at dump, or embankment if required, by dozer	B-10B	1000	.012	L.C.Y.		.56	1.39	1.95	2.42
2310	Spotter at fill or cut, if required	1 Clab	8	1	Hr.		37.50		37.50	61.50
9014	18 C.Y. truck, 8 wheels,15 min. wait/Ld./Uld.,15 MPH, cycle 0.5 mi.	B-34I	504	.016	L.C.Y.		.64	1.72	2.36	2.92
9016	cycle 1 mile		450	.018			.71	1.93	2.64	3.27
9018	cycle 2 miles		378	.021			.85	2.30	3.15	3.90
9020	cycle 4 miles		270	.030			1.19	3.22	4.41	5.45
9022	cycle 6 miles		216	.037			1.48	4.02	5.50	6.80
9024	cycle 8 miles		180	.044			1.78	4.83	6.61	8.15
9025	cycle 10 miles		144	.056			2.23	6.05	8.28	10.25
9026	20 MPH ave, cycle 0.5 mile		522	.015			.61	1.67	2.28	2.82
9028	cycle 1 mile		468	.017			.68	1.86	2.54	3.14
9030	cycle 2 miles		414	.019			.77	2.10	2.87	3.56
9032	cycle 4 miles		324	.025			.99	2.68	3.67	4.54
9034	cycle 6 miles		252	.032			1.27	3.45	4.72	5.85
9036	cycle 8 miles		216	.037			1.48	4.02	5.50	6.80
9038	cycle 10 miles		180	.044			1.78	4.83	6.61	8.15
9040	25 MPH ave, cycle 4 miles		342	.023			.94	2.54	3.48	4.30
9042	cycle 6 miles		288	.028			1.11	3.02	4.13	5.10
9044	cycle 8 miles		252	.032			1.27	3.45	4.72	5.85
9046	cycle 10 miles		216	.037			1.48	4.02	5.50	6.80
9050	30 MPH ave, cycle 4 miles		378	.021			.85	2.30	3.15	3.90
9052	cycle 6 miles		324	.025			.99	2.68	3.67	4.54
9054	cycle 8 miles		270	.030			1.19	3.22	4.41	5.45
9056	cycle 10 miles		234	.034			1.37	3.71	5.08	6.30
9060	35 MPH ave, cycle 4 miles		396	.020			.81	2.19	3	3.71
9062	cycle 6 miles		342	.023			.94	2.54	3.48	4.30
9064	cycle 8 miles		288	.028			1.11	3.02	4.13	5.10
9066	cycle 10 miles		270	.030			1.19	3.22	4.41	5.45
9068	cycle 20 miles		162	.049			1.98	5.35	7.33	9.10
9070	cycle 30 miles		126	.063			2.54	6.90	9.44	11.70
9072	cycle 40 miles		90	.089			3.56	9.65	13.21	16.35
9074	40 MPH ave, cycle 6 miles		360	.022			.89	2.41	3.30	4.09
9076	cycle 8 miles		324	.025			.99	2.68	3.67	4.54
9078	cycle 10 miles		288	.028			1.11	3.02	4.13	5.10
9080	cycle 20 miles		180	.044			1.78	4.83	6.61	8.15
9082	cycle 30 miles		144	.056			2.23	6.05	8.28	10.25

31 23 Excavation and Fill

31 23 23 – Fill

31 23 23.20 Hauling		Crew	Daily Output	Labor-Hours	Unit	Material	2015 Bare Costs Labor	Equipment	Total	Total Incl O&P
9084	cycle 40 miles	B-34I	108	.074	L.C.Y.		2.97	8.05	11.02	13.65
9086	cycle 50 miles		90	.089			3.56	9.65	13.21	16.35
9094	45 MPH ave, cycle 8 miles		324	.025			.99	2.68	3.67	4.54
9096	cycle 10 miles		306	.026			1.05	2.84	3.89	4.81
9098	cycle 20 miles		198	.040			1.62	4.39	6.01	7.45
9100	cycle 30 miles		144	.056			2.23	6.05	8.28	10.25
9102	cycle 40 miles		126	.063			2.54	6.90	9.44	11.70
9104	cycle 50 miles		108	.074			2.97	8.05	11.02	13.65
9106	50 MPH ave, cycle 10 miles		324	.025			.99	2.68	3.67	4.54
9108	cycle 20 miles		216	.037			1.48	4.02	5.50	6.80
9110	cycle 30 miles		162	.049			1.98	5.35	7.33	9.10
9112	cycle 40 miles		126	.063			2.54	6.90	9.44	11.70
9114	cycle 50 miles		108	.074			2.97	8.05	11.02	13.65
9214	20 min. wait/Ld./Uld.,15 MPH, cycle 0.5 mi.		396	.020			.81	2.19	3	3.71
9216	cycle 1 mile		360	.022			.89	2.41	3.30	4.09
9218	cycle 2 miles		306	.026			1.05	2.84	3.89	4.81
9220	cycle 4 miles		234	.034			1.37	3.71	5.08	6.30
9222	cycle 6 miles		198	.040			1.62	4.39	6.01	7.45
9224	cycle 8 miles		162	.049			1.98	5.35	7.33	9.10
9225	cycle 10 miles		144	.056			2.23	6.05	8.28	10.25
9226	20 MPH ave, cycle 0.5 mile		396	.020			.81	2.19	3	3.71
9228	cycle 1 mile		378	.021			.85	2.30	3.15	3.90
9230	cycle 2 miles		324	.025			.99	2.68	3.67	4.54
9232	cycle 4 miles		270	.030			1.19	3.22	4.41	5.45
9234	cycle 6 miles		216	.037			1.48	4.02	5.50	6.80
9236	cycle 8 miles		198	.040			1.62	4.39	6.01	7.45
9238	cycle 10 miles		162	.049			1.98	5.35	7.33	9.10
9240	25 MPH ave, cycle 4 miles		288	.028			1.11	3.02	4.13	5.10
9242	cycle 6 miles		252	.032			1.27	3.45	4.72	5.85
9244	cycle 8 miles		216	.037			1.48	4.02	5.50	6.80
9246	cycle 10 miles		198	.040			1.62	4.39	6.01	7.45
9250	30 MPH ave, cycle 4 miles		306	.026			1.05	2.84	3.89	4.81
9252	cycle 6 miles		270	.030			1.19	3.22	4.41	5.45
9254	cycle 8 miles		234	.034			1.37	3.71	5.08	6.30
9256	cycle 10 miles		216	.037			1.48	4.02	5.50	6.80
9260	35 MPH ave, cycle 4 miles		324	.025			.99	2.68	3.67	4.54
9262	cycle 6 miles		288	.028			1.11	3.02	4.13	5.10
9264	cycle 8 miles		252	.032			1.27	3.45	4.72	5.85
9266	cycle 10 miles		234	.034			1.37	3.71	5.08	6.30
9268	cycle 20 miles		162	.049			1.98	5.35	7.33	9.10
9270	cycle 30 miles		108	.074			2.97	8.05	11.02	13.65
9272	cycle 40 miles		90	.089			3.56	9.65	13.21	16.35
9274	40 MPH ave, cycle 6 miles		288	.028			1.11	3.02	4.13	5.10
9276	cycle 8 miles		270	.030			1.19	3.22	4.41	5.45
9278	cycle 10 miles		234	.034			1.37	3.71	5.08	6.30
9280	cycle 20 miles		162	.049			1.98	5.35	7.33	9.10
9282	cycle 30 miles		126	.063			2.54	6.90	9.44	11.70
9284	cycle 40 miles		108	.074			2.97	8.05	11.02	13.65
9286	cycle 50 miles		90	.089			3.56	9.65	13.21	16.35
9294	45 MPH ave, cycle 8 miles		270	.030			1.19	3.22	4.41	5.45
9296	cycle 10 miles		252	.032			1.27	3.45	4.72	5.85
9298	cycle 20 miles		180	.044			1.78	4.83	6.61	8.15
9300	cycle 30 miles		144	.056			2.23	6.05	8.28	10.25

31 23 Excavation and Fill

31 23 23 – Fill

31 23 23.20 Hauling

		Crew	Daily Output	Labor-Hours	Unit	Material	2015 Bare Costs Labor	2015 Bare Costs Equipment	Total	Total Incl O&P
9302	cycle 40 miles	B-34I	108	.074	L.C.Y.		2.97	8.05	11.02	13.65
9304	cycle 50 miles		90	.089			3.56	9.65	13.21	16.35
9306	50 MPH ave, cycle 10 miles		270	.030			1.19	3.22	4.41	5.45
9308	cycle 20 miles		198	.040			1.62	4.39	6.01	7.45
9310	cycle 30 miles		144	.056			2.23	6.05	8.28	10.25
9312	cycle 40 miles		126	.063			2.54	6.90	9.44	11.70
9314	cycle 50 miles		108	.074			2.97	8.05	11.02	13.65
9414	25 min. wait/Ld./Uld.,15 MPH, cycle 0.5 mi.		306	.026			1.05	2.84	3.89	4.81
9416	cycle 1 mile		288	.028			1.11	3.02	4.13	5.10
9418	cycle 2 miles		252	.032			1.27	3.45	4.72	5.85
9420	cycle 4 miles		198	.040			1.62	4.39	6.01	7.45
9422	cycle 6 miles		180	.044			1.78	4.83	6.61	8.15
9424	cycle 8 miles		144	.056			2.23	6.05	8.28	10.25
9425	cycle 10 miles		126	.063			2.54	6.90	9.44	11.70
9426	20 MPH ave, cycle 0.5 mile		324	.025			.99	2.68	3.67	4.54
9428	cycle 1 mile		306	.026			1.05	2.84	3.89	4.81
9430	cycle 2 miles		270	.030			1.19	3.22	4.41	5.45
9432	cycle 4 miles		234	.034			1.37	3.71	5.08	6.30
9434	cycle 6 miles		198	.040			1.62	4.39	6.01	7.45
9436	cycle 8 miles		180	.044			1.78	4.83	6.61	8.15
9438	cycle 10 miles		144	.056			2.23	6.05	8.28	10.25
9440	25 MPH ave, cycle 4 miles		252	.032			1.27	3.45	4.72	5.85
9442	cycle 6 miles		216	.037			1.48	4.02	5.50	6.80
9444	cycle 8 miles		198	.040			1.62	4.39	6.01	7.45
9446	cycle 10 miles		180	.044			1.78	4.83	6.61	8.15
9450	30 MPH ave, cycle 4 miles		252	.032			1.27	3.45	4.72	5.85
9452	cycle 6 miles		234	.034			1.37	3.71	5.08	6.30
9454	cycle 8 miles		198	.040			1.62	4.39	6.01	7.45
9456	cycle 10 miles		180	.044			1.78	4.83	6.61	8.15
9460	35 MPH ave, cycle 4 miles		270	.030			1.19	3.22	4.41	5.45
9462	cycle 6 miles		234	.034			1.37	3.71	5.08	6.30
9464	cycle 8 miles		216	.037			1.48	4.02	5.50	6.80
9466	cycle 10 miles		198	.040			1.62	4.39	6.01	7.45
9468	cycle 20 miles		144	.056			2.23	6.05	8.28	10.25
9470	cycle 30 miles		108	.074			2.97	8.05	11.02	13.65
9472	cycle 40 miles		90	.089			3.56	9.65	13.21	16.35
9474	40 MPH ave, cycle 6 miles		252	.032			1.27	3.45	4.72	5.85
9476	cycle 8 miles		234	.034			1.37	3.71	5.08	6.30
9478	cycle 10 miles		216	.037			1.48	4.02	5.50	6.80
9480	cycle 20 miles		144	.056			2.23	6.05	8.28	10.25
9482	cycle 30 miles		126	.063			2.54	6.90	9.44	11.70
9484	cycle 40 miles		90	.089			3.56	9.65	13.21	16.35
9486	cycle 50 miles		90	.089			3.56	9.65	13.21	16.35
9494	45 MPH ave, cycle 8 miles		234	.034			1.37	3.71	5.08	6.30
9496	cycle 10 miles		216	.037			1.48	4.02	5.50	6.80
9498	cycle 20 miles		162	.049			1.98	5.35	7.33	9.10
9500	cycle 30 miles		126	.063			2.54	6.90	9.44	11.70
9502	cycle 40 miles		108	.074			2.97	8.05	11.02	13.65
9504	cycle 50 miles		90	.089			3.56	9.65	13.21	16.35
9506	50 MPH ave, cycle 10 miles		234	.034			1.37	3.71	5.08	6.30
9508	cycle 20 miles		180	.044			1.78	4.83	6.61	8.15
9510	cycle 30 miles		144	.056			2.23	6.05	8.28	10.25
9512	cycle 40 miles		108	.074			2.97	8.05	11.02	13.65

31 23 Excavation and Fill

31 23 23 – Fill

31 23 23.20 Hauling

		Crew	Daily Output	Labor-Hours	Unit	Material	2015 Bare Costs Labor	2015 Bare Costs Equipment	Total	Total Incl O&P
9514	cycle 50 miles	B-34I	90	.089	L.C.Y.		3.56	9.65	13.21	16.35
9614	30 min. wait/Ld./Uld.,15 MPH, cycle 0.5 mi.		270	.030			1.19	3.22	4.41	5.45
9616	cycle 1 mile		252	.032			1.27	3.45	4.72	5.85
9618	cycle 2 miles		216	.037			1.48	4.02	5.50	6.80
9620	cycle 4 miles		180	.044			1.78	4.83	6.61	8.15
9622	cycle 6 miles		162	.049			1.98	5.35	7.33	9.10
9624	cycle 8 miles		126	.063			2.54	6.90	9.44	11.70
9625	cycle 10 miles		126	.063			2.54	6.90	9.44	11.70
9626	20 MPH ave, cycle 0.5 mile		270	.030			1.19	3.22	4.41	5.45
9628	cycle 1 mile		252	.032			1.27	3.45	4.72	5.85
9630	cycle 2 miles		234	.034			1.37	3.71	5.08	6.30
9632	cycle 4 miles		198	.040			1.62	4.39	6.01	7.45
9634	cycle 6 miles		180	.044			1.78	4.83	6.61	8.15
9636	cycle 8 miles		162	.049			1.98	5.35	7.33	9.10
9638	cycle 10 miles		144	.056			2.23	6.05	8.28	10.25
9640	25 MPH ave, cycle 4 miles		216	.037			1.48	4.02	5.50	6.80
9642	cycle 6 miles		198	.040			1.62	4.39	6.01	7.45
9644	cycle 8 miles		180	.044			1.78	4.83	6.61	8.15
9646	cycle 10 miles		162	.049			1.98	5.35	7.33	9.10
9650	30 MPH ave, cycle 4 miles		216	.037			1.48	4.02	5.50	6.80
9652	cycle 6 miles		198	.040			1.62	4.39	6.01	7.45
9654	cycle 8 miles		180	.044			1.78	4.83	6.61	8.15
9656	cycle 10 miles		162	.049			1.98	5.35	7.33	9.10
9660	35 MPH ave, cycle 4 miles		234	.034			1.37	3.71	5.08	6.30
9662	cycle 6 miles		216	.037			1.48	4.02	5.50	6.80
9664	cycle 8 miles		198	.040			1.62	4.39	6.01	7.45
9666	cycle 10 miles		180	.044			1.78	4.83	6.61	8.15
9668	cycle 20 miles		126	.063			2.54	6.90	9.44	11.70
9670	cycle 30 miles		108	.074			2.97	8.05	11.02	13.65
9672	cycle 40 miles		90	.089			3.56	9.65	13.21	16.35
9674	40 MPH ave, cycle 6 miles		216	.037			1.48	4.02	5.50	6.80
9676	cycle 8 miles		198	.040			1.62	4.39	6.01	7.45
9678	cycle 10 miles		180	.044			1.78	4.83	6.61	8.15
9680	cycle 20 miles		144	.056			2.23	6.05	8.28	10.25
9682	cycle 30 miles		108	.074			2.97	8.05	11.02	13.65
9684	cycle 40 miles		90	.089			3.56	9.65	13.21	16.35
9686	cycle 50 miles		72	.111			4.45	12.05	16.50	20.50
9694	45 MPH ave, cycle 8 miles		216	.037			1.48	4.02	5.50	6.80
9696	cycle 10 miles		198	.040			1.62	4.39	6.01	7.45
9698	cycle 20 miles		144	.056			2.23	6.05	8.28	10.25
9700	cycle 30 miles		126	.063			2.54	6.90	9.44	11.70
9702	cycle 40 miles		108	.074			2.97	8.05	11.02	13.65
9704	cycle 50 miles		90	.089			3.56	9.65	13.21	16.35
9706	50 MPH ave, cycle 10 miles		198	.040			1.62	4.39	6.01	7.45
9708	cycle 20 miles		162	.049			1.98	5.35	7.33	9.10
9710	cycle 30 miles		126	.063			2.54	6.90	9.44	11.70
9712	cycle 40 miles		108	.074			2.97	8.05	11.02	13.65
9714	cycle 50 miles		90	.089			3.56	9.65	13.21	16.35

31 23 23.23 Compaction

		Crew	Daily Output	Labor-Hours	Unit	Material	Labor	Equipment	Total	Total Incl O&P
0010	**COMPACTION**									
5000	Riding, vibrating roller, 6" lifts, 2 passes	B-10Y	3000	.004	E.C.Y.		.19	.18	.37	.50
5020	3 passes		2300	.005			.24	.24	.48	.65

31 23 Excavation and Fill

31 23 23 – Fill

31 23 23.23 Compaction

		Crew	Daily Output	Labor-Hours	Unit	Material	2015 Bare Costs Labor	Equipment	Total	Total Incl O&P
5040	4 passes	B-10Y	1900	.006	E.C.Y.		.29	.29	.58	.79
5050	8" lifts, 2 passes		4100	.003			.14	.13	.27	.37
5060	12" lifts, 2 passes		5200	.002			.11	.11	.22	.29
5080	3 passes		3500	.003			.16	.16	.32	.42
5100	4 passes		2600	.005			.21	.21	.42	.57
5600	Sheepsfoot or wobbly wheel roller, 6" lifts, 2 passes	B-10G	2400	.005			.23	.50	.73	.92
5620	3 passes		1735	.007			.32	.70	1.02	1.28
5640	4 passes		1300	.009			.43	.93	1.36	1.70
5680	12" lifts, 2 passes		5200	.002			.11	.23	.34	.43
5700	3 passes		3500	.003			.16	.34	.50	.63
5720	4 passes		2600	.005			.21	.46	.67	.85
7000	Walk behind, vibrating plate 18" wide, 6" lifts, 2 passes	A-1D	200	.040			1.50	.18	1.68	2.67
7020	3 passes		185	.043			1.63	.20	1.83	2.89
7040	4 passes		140	.057			2.15	.26	2.41	3.80
7200	12" lifts, 2 passes, 21" wide	A-1E	560	.014			.54	.08	.62	.97
7220	3 passes		375	.021			.80	.12	.92	1.46
7240	4 passes		280	.029			1.07	.17	1.24	1.94
7500	Vibrating roller 24" wide, 6" lifts, 2 passes	B-10A	420	.029			1.32	.43	1.75	2.59
7520	3 passes		280	.043			1.98	.64	2.62	3.89
7540	4 passes		210	.057			2.64	.86	3.50	5.15
7600	12" lifts, 2 passes		840	.014			.66	.21	.87	1.30
7620	3 passes		560	.021			.99	.32	1.31	1.94
7640	4 passes		420	.029			1.32	.43	1.75	2.59
8000	Rammer tamper, 6" to 11", 4" lifts, 2 passes	A-1F	130	.062			2.31	.38	2.69	4.21
8050	3 passes		97	.082			3.10	.52	3.62	5.65
8100	4 passes		65	.123			4.63	.77	5.40	8.45
8200	8" lifts, 2 passes		260	.031			1.16	.19	1.35	2.11
8250	3 passes		195	.041			1.54	.26	1.80	2.81
8300	4 passes		130	.062			2.31	.38	2.69	4.21
8400	13" to 18", 4" lifts, 2 passes	A-1G	390	.021			.77	.14	.91	1.42
8450	3 passes		290	.028			1.04	.19	1.23	1.91
8500	4 passes		195	.041			1.54	.29	1.83	2.85
8600	8" lifts, 2 passes		780	.010			.39	.07	.46	.71
8650	3 passes		585	.014			.51	.10	.61	.95
8700	4 passes		390	.021			.77	.14	.91	1.42
9000	Water, 3000 gal. truck, 3 mile haul	B-45	1888	.008		1.22	.38	.48	2.08	2.48
9010	6 mile haul		1444	.011		1.22	.50	.63	2.35	2.83
9020	12 mile haul		1000	.016		1.22	.73	.91	2.86	3.50
9030	6000 gal. wagon, 3 mile haul	B-59	2000	.004		1.22	.16	.25	1.63	1.88
9040	6 mile haul	"	1600	.005		1.22	.20	.31	1.73	2

31 23 23.24 Compaction, Structural

		Crew	Daily Output	Labor-Hours	Unit	Material	Labor	Equipment	Total	Total Incl O&P
0010	**COMPACTION, STRUCTURAL** R312323-30									
0020	Steel wheel tandem roller, 5 tons	B-10E	8	1.500	Hr.		69.50	19.70	89.20	133
0050	Air tamp, 6" to 8" lifts, common fill	B-9	250	.160	E.C.Y.		6.10	.93	7.03	10.95
0060	Select fill	"	300	.133			5.05	.78	5.83	9.15
0600	Vibratory plate, 8" lifts, common fill	A-1D	200	.040			1.50	.18	1.68	2.67
0700	Select fill	"	216	.037			1.39	.17	1.56	2.46
9000	Minimum labor/equipment charge	1 Clab	4	2	Job		75		75	123

31 25 Erosion and Sedimentation Controls

31 25 14 – Stabilization Measures for Erosion and Sedimentation Control

31 25 14.16 Rolled Erosion Control Mats and Blankets		Crew	Daily Output	Labor-Hours	Unit	Material	2015 Bare Costs Labor	Equipment	Total	Total Incl O&P
0010	**ROLLED EROSION CONTROL MATS AND BLANKETS**									
0020	Jute mesh, 100 S.Y. per roll, 4' wide, stapled	G	B-80A	2400	.010	S.Y.	1.03	.38	.13	1.54	1.89
0100	Plastic netting, stapled, 2" x 1" mesh, 20 mil	G	B-1	2500	.010		.24	.37		.61	.86
0200	Polypropylene mesh, stapled, 6.5 oz./S.Y.	G		2500	.010		2.30	.37		2.67	3.13
0300	Tobacco netting, or jute mesh #2, stapled	G	↓	2500	.010	↓	.19	.37		.56	.81
1000	Silt fence, install and maintain, remove	G	B-62	1300	.018	L.F.	.72	.76	.13	1.61	2.17
1100	Allow 25% per month for maintenance; 6-month max life										
1200	Place and remove hay bales	G	A-2	3	8	Ton	246	305	82	633	855
1250	Hay bales, staked	G	"	2500	.010	L.F.	9.85	.37	.10	10.32	11.50

31 31 Soil Treatment

31 31 16 – Termite Control

31 31 16.13 Chemical Termite Control

		Crew	Daily Output	Labor-Hours	Unit	Material	Labor	Equipment	Total	Total Incl O&P
0010	**CHEMICAL TERMITE CONTROL**									
0020	Slab and walls, residential	1 Skwk	1200	.007	SF Flr.	.32	.32		.64	.88
0100	Commercial, minimum		2496	.003		.33	.16		.49	.61
0200	Maximum		1645	.005	↓	.50	.24		.74	.92
0390	Minimum labor/equipment charge		4	2	Job		97.50		97.50	158
0400	Insecticides for termite control, minimum		14.20	.563	Gal.	68.50	27.50		96	120
0500	Maximum	↓	11	.727	"	117	35.50		152.50	187

31 41 Shoring

31 41 13 – Timber Shoring

31 41 13.10 Building Shoring

		Crew	Daily Output	Labor-Hours	Unit	Material	Labor	Equipment	Total	Total Incl O&P
0010	**BUILDING SHORING**									
0020	Shoring, existing building, with timber, no salvage allowance	B-51	2.20	21.818	M.B.F.	850	835	111	1,796	2,400
1000	On cribbing with 35 ton screw jacks, per box and jack		3.60	13.333	Jack	68.50	510	68	646.50	980
1090	Minimum labor/equipment charge	↓	2	24	Ea.		915	123	1,038	1,625
1100	Masonry openings in walls, see Section 02 41 19.16									

31 41 16 – Sheet Piling

31 41 16.10 Sheet Piling Systems

		Crew	Daily Output	Labor-Hours	Unit	Material	Labor	Equipment	Total	Total Incl O&P
0010	**SHEET PILING SYSTEMS**									
0020	Sheet piling steel, not incl. wales, 22 psf, 15' excav., left in place	B-40	10.81	5.920	Ton	1,600	282	345	2,227	2,575
0100	Drive, extract & salvage R314116-40		6	10.667	"	510	510	625	1,645	2,075
1200	15' deep excavation, 22 psf, left in place		983	.065	S.F.	18.60	3.10	3.81	25.51	29.50
1300	Drive, extract & salvage	↓	545	.117	"	5.70	5.60	6.85	18.15	23
2100	Rent steel sheet piling and wales, first month				Ton	310			310	340
2200	Per added month				"	31			31	34
3900	Wood, solid sheeting, incl. wales, braces and spacers,									
3910	drive, extract & salvage, 8' deep excavation	B-31	330	.121	S.F.	1.89	4.83	.66	7.38	10.70
4520	Left in place, 8' deep, 55 S.F./hr.		440	.091	"	3.41	3.62	.50	7.53	10.25
4990	Minimum labor/equipment charge	↓	2	20	Job		795	109	904	1,425
5000	For treated lumber add cost of treatment to lumber									

31 43 Concrete Raising
31 43 13 – Pressure Grouting

31 43 13.13 Concrete Pressure Grouting		Crew	Daily Output	Labor-Hours	Unit	Material	2015 Bare Costs Labor	2015 Bare Costs Equipment	Total	Total Incl O&P
0010	**CONCRETE PRESSURE GROUTING**									
0020	Grouting, pressure, cement & sand, 1:1 mix, minimum	B-61	124	.323	Bag	12.10	12.95	2.83	27.88	37.50
0100	Maximum		51	.784	"	12.10	31.50	6.90	50.50	72.50
0200	Cement and sand, 1:1 mix, minimum		250	.160	C.F.	24	6.45	1.40	31.85	38.50
0300	Maximum		100	.400		36.50	16.10	3.51	56.11	70
0400	Epoxy cement grout, minimum		137	.292		700	11.75	2.56	714.31	790
0500	Maximum		57	.702		700	28	6.15	734.15	825
0700	Alternate pricing method: (Add for materials)									
0710	5 person crew and equipment	B-61	1	40	Day		1,600	350	1,950	3,000

31 46 Needle Beams
31 46 13 – Cantilever Needle Beams
31 46 13.10 Needle Beams

0010	**NEEDLE BEAMS**									
0011	Incl. wood shoring 10' x 10' opening									
0400	Block, concrete, 8" thick	B-9	7.10	5.634	Ea.	50	214	33	297	440
0420	12" thick		6.70	5.970		60	227	35	322	475
0800	Brick, 4" thick with 8" backup block		5.70	7.018		60	267	41	368	545
1000	Brick, solid, 8" thick		6.20	6.452		50	245	37.50	332.50	495
1040	12" thick		4.90	8.163		60	310	47.50	417.50	630
1080	16" thick		4.50	8.889		80.50	340	51.50	472	700
2000	Add for additional floors of shoring	B-1	6	4		50	153		203	305
9000	Minimum labor/equipment charge	"	2	12	Job		460		460	755

31 48 Underpinning
31 48 13 – Underpinning Piers
31 48 13.10 Underpinning Foundations

0010	**UNDERPINNING FOUNDATIONS**									
0011	Including excavation,									
0020	forming, reinforcing, concrete and equipment									
0100	5' to 16' below grade, 100 to 500 C.Y.	B-52	2.30	24.348	C.Y.	283	1,050	258	1,591	2,325
0200	Over 500 C.Y.		2.50	22.400		255	975	238	1,468	2,150
0400	16' to 25' below grade, 100 to 500 C.Y.		2	28		310	1,225	297	1,832	2,650
0500	Over 500 C.Y.		2.10	26.667		295	1,175	283	1,753	2,525
0700	26' to 40' below grade, 100 to 500 C.Y.		1.60	35		340	1,525	370	2,235	3,250
0800	Over 500 C.Y.		1.80	31.111		310	1,350	330	1,990	2,900
0900	For under 50 C.Y., add					10%	40%			

31 62 Driven Piles

31 62 13 – Concrete Piles

31 62 13.23 Prestressed Concrete Piles

		Crew	Daily Output	Labor-Hours	Unit	Material	2015 Bare Costs Labor	Equipment	Total	Total Incl O&P
0010	**PRESTRESSED CONCRETE PILES**, 200 piles									
0020	Unless specified otherwise, not incl. pile caps or mobilization									
3100	Precast, prestressed, 40' long, 10" thick, square	B-19	700	.091	V.L.F.	18.35	4.36	2.48	25.19	30
3200	12" thick, square		680	.094		22.50	4.48	2.55	29.53	34.50
3400	14" thick, square	↓	600	.107		25.50	5.10	2.90	33.50	39.50
4000	18" thick, square	B-19A	520	.123	↓	47.50	5.85	4.16	57.51	66

31 62 16 – Steel Piles

31 62 16.13 Steel Piles

		Crew	Daily Output	Labor-Hours	Unit	Material	Labor	Equipment	Total	Total Incl O&P
0010	**STEEL PILES**									
0100	Step tapered, round, concrete filled									
0110	8" tip, 60 ton capacity, 30' depth	B-19	760	.084	V.L.F.	11.45	4.01	2.29	17.75	21.50
0120	60' depth		740	.086		12.95	4.12	2.35	19.42	23.50
0250	"H" Sections, 50' long, HP8 x 36	↓	640	.100		16.05	4.76	2.71	23.52	28.50
1300	HP14 X 102	B-19A	510	.125	↓	47	6	4.24	57.24	66

31 62 19 – Timber Piles

31 62 19.10 Wood Piles

		Crew	Daily Output	Labor-Hours	Unit	Material	Labor	Equipment	Total	Total Incl O&P
0010	**WOOD PILES**									
0011	Friction or end bearing, not including									
0050	mobilization or demobilization									
0100	Untreated piles, up to 30' long, 12" butts, 8" points	B-19	625	.102	V.L.F.	16.85	4.88	2.78	24.51	29.50
0200	30' to 39' long, 12" butts, 8" points	"	700	.091	"	16.85	4.36	2.48	23.69	28.50
0800	Treated piles, 12 lb. per C.F.,									
0810	friction or end bearing, ASTM class B									
1000	Up to 30' long, 12" butts, 8" points	B-19	625	.102	V.L.F.	15.85	4.88	2.78	23.51	28.50
1100	30' to 39' long, 12" butts, 8" points		700	.091		17.80	4.36	2.48	24.64	29.50
2700	Mobilization for 10,000 L.F. pile job, add		3300	.019			.92	.53	1.45	2.07
2800	25,000 L.F. pile job, add	↓	8500	.008	↓		.36	.20	.56	.80

31 62 23 – Composite Piles

31 62 23.13 Concrete-Filled Steel Piles

		Crew	Daily Output	Labor-Hours	Unit	Material	Labor	Equipment	Total	Total Incl O&P
0010	**CONCRETE-FILLED STEEL PILES** no mobilization or demobilization									
2600	Pipe piles, 50' lg. 8" diam., 29 lb. per L.F., no concrete	B-19	500	.128	V.L.F.	20.50	6.10	3.47	30.07	36
2700	Concrete filled		460	.139		21.50	6.65	3.78	31.93	38.50
3500	14" diameter, 46 lb. per L.F., no concrete		430	.149		35.50	7.10	4.04	46.64	55.50
3600	Concrete filled		355	.180		41.50	8.60	4.89	54.99	65
4100	18" diameter, 59 lb. per L.F., no concrete		355	.180		48.50	8.60	4.89	61.99	73
4200	Concrete filled	↓	310	.206	↓	54.50	9.85	5.60	69.95	82

31 63 Bored Piles

31 63 26 – Drilled Caissons

31 63 26.13 Fixed End Caisson Piles

		Crew	Daily Output	Labor-Hours	Unit	Material	Labor	Equipment	Total	Total Incl O&P
0010	**FIXED END CAISSON PILES**									
0015	Including excavation, concrete, 50 lb. reinforcing									
0020	per C.Y., not incl. mobilization, boulder removal, disposal									
0100	Open style, machine drilled, to 50' deep, in stable ground, no R316326-60									
0110	casings or ground water, 18" diam., 0.065 C.Y./L.F.	B-43	200	.240	V.L.F.	8.15	9.95	12.70	30.80	39
0200	24" diameter, 0.116 C.Y./L.F.		190	.253	"	14.60	10.50	13.40	38.50	48
0210	4' bell diameter, add		20	2.400	Ea.	251	99.50	127	477.50	575
0500	48" diameter, 0.465 C.Y./L.F.		100	.480	V.L.F.	58.50	19.95	25.50	103.95	125
0510	9' bell diameter, add	↓	2	24	Ea.	1,475	995	1,275	3,745	4,650

31 63 Bored Piles

31 63 26 – Drilled Caissons

31 63 26.13 Fixed End Caisson Piles

		Crew	Daily Output	Labor-Hours	Unit	Material	2015 Bare Costs Labor	2015 Bare Costs Equipment	Total	Total Incl O&P
1200	Open style, machine drilled, to 50' deep, in wet ground, pulled									
1220	pulled casing and pumping									
1400	24" diameter, 0.116 C.Y./L.F.	B-48	125	.448	V.L.F.	14.60	19.05	23.50	57.15	72.50
1410	4' bell diameter, add	"	19.80	2.828	Ea.	251	120	147	518	630
1700	48" diameter, 0.465 C.Y./L.F.	B-49	55	1.600	V.L.F.	58.50	71	66	195.50	252
1710	9' bell diameter, add	"	3.30	26.667	Ea.	1,475	1,175	1,100	3,750	4,750
2300	Open style, machine drilled, to 50' deep, in soft rocks and									
2320	medium hard shales									
2500	24" diameter, 0.116 C.Y./L.F.	B-49	30	2.933	V.L.F.	14.60	130	121	265.60	360
2510	4' bell diameter, add		10.90	8.073	Ea.	251	355	335	941	1,225
2800	48" diameter, 0.465 C.Y./L.F.		10	8.800	V.L.F.	58.50	390	365	813.50	1,100
2810	9' bell diameter, add		1.10	80	Ea.	1,125	3,550	3,300	7,975	10,600
3600	For rock excavation, sockets, add, minimum		120	.733	C.F.		32.50	30.50	63	86
3650	Average		95	.926			41	38.50	79.50	108
3700	Maximum		48	1.833			81	76	157	215
3900	For 50' to 100' deep, add				V.L.F.				7%	7%
4000	For 100' to 150' deep, add								25%	25%
4100	For 150' to 200' deep, add								30%	30%
4200	For casings left in place, add				Lb.	1.19			1.19	1.31
4300	For other than 50 lb. reinf. per C.Y., add or deduct				"	1.16			1.16	1.28
4400	For steel "I" beam cores, add	B-49	8.30	10.602	Ton	2,125	470	440	3,035	3,575
4500	Load and haul excess excavation, 2 miles	B-34B	178	.045	L.C.Y.		1.80	3.88	5.68	7.15
4600	For mobilization, 50 mile radius, rig to 36"	B-43	2	24	Ea.		995	1,275	2,270	3,025
4650	Rig to 84"	B-48	1.75	32			1,350	1,650	3,000	4,025
4700	For low headroom, add								50%	50%
4750	For difficult access, add								25%	25%

31 63 29 – Drilled Concrete Piers and Shafts

31 63 29.13 Uncased Drilled Concrete Piers

		Crew	Daily Output	Labor-Hours	Unit	Material	2015 Bare Costs Labor	2015 Bare Costs Equipment	Total	Total Incl O&P
0010	**UNCASED DRILLED CONCRETE PIERS**									
0020	Unless specified otherwise, not incl. pile caps or mobilization									
0800	Cast in place friction pile, 50' long, fluted,									
0810	tapered steel, 4000 psi concrete, no reinforcing									
0900	12" diameter, 7 ga.	B-19	600	.107	V.L.F.	29	5.10	2.90	37	43.50
1200	18" diameter, 7 ga.	"	480	.133	"	43.50	6.35	3.62	53.47	62
1300	End bearing, fluted, constant diameter,									
1320	4000 psi concrete, no reinforcing									
1340	12" diameter, 7 ga.	B-19	600	.107	V.L.F.	30.50	5.10	2.90	38.50	45
1400	18" diameter, 7 ga.	"	480	.133	"	48.50	6.35	3.62	58.47	67.50

Estimating Tips

32 01 00 Operations and Maintenance of Exterior Improvements

- Recycling of asphalt pavement is becoming very popular and is an alternative to removal and replacement. It can be a good value engineering proposal if removed pavement can be recycled, either at the project site or at another site that is reasonably close to the project site. Sections on repair of flexible and rigid pavement are included.

32 10 00 Bases, Ballasts, and Paving

- When estimating paving, keep in mind the project schedule. Also note that prices for asphalt and concrete are generally higher in the cold seasons. Lines for pavement markings, including tactile warning systems and fence lines, are included.

32 90 00 Planting

- The timing of planting and guarantee specifications often dictate the costs for establishing tree and shrub growth and a stand of grass or ground cover. Establish the work performance schedule to coincide with the local planting season. Maintenance and growth guarantees can add from 20%–100% to the total landscaping cost and can be contractually cumbersome. The cost to replace trees and shrubs can be as high as 5% of the total cost, depending on the planting zone, soil conditions, and time of year.

Reference Numbers

Reference numbers are shown in shaded boxes at the beginning of some major classifications. These numbers refer to related items in the Reference Section. The reference information may be an estimating procedure, an alternate pricing method, or technical information.

Note: Not all subdivisions listed here necessarily appear in this publication. ∎

Did you know?
RSMeans Online gives you the same access to RSMeans' data with 24/7 access:
- Quickly locate costs in the searchable database.
- Build cost lists, estimates, and reports in minutes.
- Adjust costs to any location in the U.S. and Canada with the click of a button.

Start your free trial today at **www.rsmeansonline.com**

RSMeansOnline

No part of this publication may be reproduced, stored in a retrieval system, or transmitted in any form or by any means without prior written permission of RSMeans.

32 01 Operation and Maintenance of Exterior Improvements

32 01 13 – Flexible Paving Surface Treatment

32 01 13.61 Slurry Seal (Latex Modified)

		Crew	Daily Output	Labor-Hours	Unit	Material	2015 Bare Costs Labor	Equipment	Total	Total Incl O&P
0010	**SLURRY SEAL (LATEX MODIFIED)**									
3600	Waterproofing, membrane, tar and fabric, small area	B-63	233	.172	S.Y.	14.05	6.85	.75	21.65	27.50
3640	Large area		1435	.028		12.95	1.11	.12	14.18	16.20
3680	Preformed rubberized asphalt, small area		100	.400		19.15	15.90	1.74	36.79	49
3720	Large area	↓	367	.109		17.45	4.34	.47	22.26	26.50
3780	Rubberized asphalt (latex) seal	B-45	5000	.003	↓	2.81	.15	.18	3.14	3.52
9000	Minimum labor/equipment charge	1 Clab	2	4	Job		150		150	247

32 01 13.62 Asphalt Surface Treatment

		Crew	Daily Output	Labor-Hours	Unit	Material	Labor	Equipment	Total	Total Incl O&P
0010	**ASPHALT SURFACE TREATMENT**									
3000	Pavement overlay, polypropylene									
3040	6 oz. per S.Y., ideal conditions	B-63	10000	.004	S.Y.	.98	.16	.02	1.16	1.36
3080	Adverse conditions		1000	.040		1.31	1.59	.17	3.07	4.22
3120	4 oz. per S.Y., ideal conditions		10000	.004		.80	.16	.02	.98	1.16
3160	Adverse conditions	↓	1000	.040		1.03	1.59	.17	2.79	3.92
3200	Tack coat, emulsion, .05 gal. per S.Y., 1000 S.Y.	B-45	2500	.006		.29	.29	.36	.94	1.18
3240	10,000 S.Y.		10000	.002		.23	.07	.09	.39	.48
3270	.10 gal. per S.Y., 1000 S.Y.		2500	.006		.54	.29	.36	1.19	1.46
3275	10,000 S.Y.		10000	.002		.44	.07	.09	.60	.70
3280	.15 gal. per S.Y., 1000 S.Y.		2500	.006		.80	.29	.36	1.45	1.74
3320	10,000 S.Y.	↓	10000	.002	↓	.64	.07	.09	.80	.92

32 01 13.64 Sand Seal

		Crew	Daily Output	Labor-Hours	Unit	Material	Labor	Equipment	Total	Total Incl O&P
0010	**SAND SEAL**									
2080	Sand sealing, sharp sand, asphalt emulsion, small area	B-91	10000	.006	S.Y.	1.48	.29	.23	2	2.35
2120	Roadway or large area	"	18000	.004	"	1.27	.16	.13	1.56	1.80
3000	Sealing random cracks, min 1/2" wide, to 1-1/2", 1,000 L.F.	B-77	2800	.014	L.F.	1.53	.55	.20	2.28	2.79
3040	10,000 L.F.		4000	.010	"	1.06	.38	.14	1.58	1.96
3080	Alternate method, 1,000 L.F.		200	.200	Gal.	39.50	7.65	2.83	49.98	59
3120	10,000 L.F.	↓	325	.123	"	32	4.71	1.74	38.45	44.50
3200	Multi-cracks (flooding), 1 coat, small area	B-92	460	.070	S.Y.	2.34	2.65	1.38	6.37	8.45
3240	Large area		2850	.011		2.08	.43	.22	2.73	3.23
3280	2 coat, small area		230	.139		14.50	5.30	2.76	22.56	27.50
3320	Large area	↓	1425	.022	↓	13.20	.86	.45	14.51	16.40
3360	Alternate method, small area		115	.278	Gal.	14.50	10.60	5.50	30.60	39.50
3400	Large area	↓	715	.045	"	13.65	1.71	.89	16.25	18.80

32 01 13.66 Fog Seal

		Crew	Daily Output	Labor-Hours	Unit	Material	Labor	Equipment	Total	Total Incl O&P
0010	**FOG SEAL**									
0012	Sealcoating, 2 coat coal tar pitch emulsion over 10,000 S.Y.	B-45	5000	.003	S.Y.	.90	.15	.18	1.23	1.42
0030	1000 to 10,000 S.Y.	"	3000	.005		.90	.24	.30	1.44	1.71
0100	Under 1000 S.Y.	B-1	1050	.023		.90	.87		1.77	2.42
0300	Petroleum resistant, over 10,000 S.Y.	B-45	5000	.003		1.30	.15	.18	1.63	1.86
0320	1000 to 10,000 S.Y.	"	3000	.005		1.30	.24	.30	1.84	2.15
0400	Under 1000 S.Y.	B-1	1050	.023		1.30	.87		2.17	2.86
0600	Non-skid pavement renewal, over 10,000 S.Y.	B-45	5000	.003		1.37	.15	.18	1.70	1.94
0620	1000 to 10,000 S.Y.	"	3000	.005		1.37	.24	.30	1.91	2.23
0700	Under 1000 S.Y.	B-1	1050	.023		1.37	.87		2.24	2.94
0800	Prepare and clean surface for above	A-2	8545	.003	↓		.11	.03	.14	.20
1000	Hand seal asphalt curbing	B-1	4420	.005	L.F.	.62	.21		.83	1.02
1900	Asphalt surface treatment, single course, small area									
1901	0.30 gal/S.Y. asphalt material, 20#/S.Y. aggregate	B-91	5000	.013	S.Y.	1.30	.57	.46	2.33	2.86
1910	Roadway or large area		10000	.006		1.20	.29	.23	1.72	2.04
1950	Asphalt surface treatment, dbl. course for small area		3000	.021		2.94	.95	.77	4.66	5.60
1960	Roadway or large area	↓	6000	.011		2.65	.48	.39	3.52	4.11

32 01 Operation and Maintenance of Exterior Improvements

32 01 13 – Flexible Paving Surface Treatment

32 01 13.66 Fog Seal

		Crew	Daily Output	Labor-Hours	Unit	Material	2015 Bare Costs Labor	Equipment	Total	Total Incl O&P
1980	Asphalt surface treatment, single course, for shoulders	B-91	7500	.009	S.Y.	1.46	.38	.31	2.15	2.56

32 01 13.68 Slurry Seal

		Crew	Daily Output	Labor-Hours	Unit	Material	Labor	Equipment	Total	Total Incl O&P
0010	**SLURRY SEAL**									
0100	Slurry seal, type I, 8 lb. agg./S.Y., 1 coat, small or irregular area	B-90	2800	.023	S.Y.	1.86	.94	.79	3.59	4.44
0150	Roadway or large area		10000	.006		1.86	.26	.22	2.34	2.72
0200	Type II, 12 lb. aggregate/S.Y., 2 coats, small or irregular area		2000	.032		3.72	1.32	1.10	6.14	7.45
0250	Roadway or large area		8000	.008		3.72	.33	.28	4.33	4.92
0300	Type III, 20 lb. aggregate/S.Y., 2 coats, small or irregular area		1800	.036		4.39	1.47	1.23	7.09	8.55
0350	Roadway or large area		6000	.011		4.39	.44	.37	5.20	5.95
0400	Slurry seal, thermoplastic coal-tar, type I, small or irregular area		2400	.027		3.76	1.10	.92	5.78	6.95
0450	Roadway or large area		8000	.008		3.76	.33	.28	4.37	4.97
0500	Type II, small or irregular area		2400	.027		4.80	1.10	.92	6.82	8.10
0550	Roadway or large area		7800	.008		4.80	.34	.28	5.42	6.15

32 01 16 – Flexible Paving Rehabilitation

32 01 16.71 Cold Milling Asphalt Paving

		Crew	Daily Output	Labor-Hours	Unit	Material	Labor	Equipment	Total	Total Incl O&P
0010	**COLD MILLING ASPHALT PAVING**									
5200	Cold planing & cleaning, 1" to 3" asphalt pavmt., over 25,000 S.Y.	B-71	6000	.009	S.Y.		.41	1.14	1.55	1.90
5280	5,000 S.Y. to 10,000 S.Y.	"	4000	.014	"		.61	1.71	2.32	2.86
5300	Asphalt pavement removal from conc. base, no haul									
5320	Rip, load & sweep 1" to 3"	B-70	8000	.007	S.Y.		.30	.23	.53	.75
5330	3" to 6" deep	"	5000	.011			.49	.37	.86	1.20
5340	Profile grooving, asphalt pavement load & sweep, 1" deep	B-71	12500	.004			.19	.55	.74	.91
5350	3" deep		9000	.006			.27	.76	1.03	1.28
5360	6" deep		5000	.011			.49	1.37	1.86	2.30

32 01 16.73 In Place Cold Reused Asphalt Paving

		Crew	Daily Output	Labor-Hours	Unit	Material	Labor	Equipment	Total	Total Incl O&P
0010	**IN PLACE COLD REUSED ASPHALT PAVING**									
5000	Reclamation, pulverizing and blending with existing base									
5040	Aggregate base, 4" thick pavement, over 15,000 S.Y.	B-73	2400	.027	S.Y.		1.23	2.18	3.41	4.36
5080	5,000 S.Y. to 15,000 S.Y.		2200	.029			1.34	2.37	3.71	4.75
5120	8" thick pavement, over 15,000 S.Y.		2200	.029			1.34	2.37	3.71	4.75
5160	5,000 S.Y. to 15,000 S.Y.		2000	.032			1.47	2.61	4.08	5.25

32 01 16.74 In Place Hot Reused Asphalt Paving

			Crew	Daily Output	Labor-Hours	Unit	Material	Labor	Equipment	Total	Total Incl O&P
0010	**IN PLACE HOT REUSED ASPHALT PAVING**										
5500	Recycle asphalt pavement at site										
5520	Remove, rejuvenate and spread 4" deep	G	B-72	2500	.026	S.Y.	4.73	1.14	4.56	10.43	12.05
5521	6" deep	G	"	2000	.032	"	6.95	1.42	5.70	14.07	16.15

32 01 17 – Flexible Paving Repair

32 01 17.10 Repair of Asphalt Pavement Holes

		Crew	Daily Output	Labor-Hours	Unit	Material	Labor	Equipment	Total	Total Incl O&P
0010	**REPAIR OF ASPHALT PAVEMENT HOLES** (cold patch)									
0100	Flexible pavement repair holes, roadway, light traffic, 1 C.F. size	B-37A	24	1	Ea.	9	38	15.75	62.75	89
0150	Group of two, 1 C.F. size each		16	1.500	Set	18	57	23.50	98.50	139
0200	Group of three, 1 C.F. size each		12	2	"	27	76	31.50	134.50	188
0300	Medium traffic, 1 C.F. size each	B-37B	24	1.333	Ea.	9	50.50	15.70	75.20	110
0350	Group of two, 1 C.F. size each		16	2	Set	18	76	23.50	117.50	170
0400	Group of three, 1 C.F. size each		12	2.667	"	27	101	31.50	159.50	229
0500	Highway/heavy traffic, 1 C.F. size each	B-37C	24	1.333	Ea.	9	51	26	86	121
0550	Group of two, 1 C.F. size each		16	2	Set	18	76.50	39	133.50	188
0600	Group of three, 1 C.F. size each		12	2.667	"	27	102	52	181	253
0700	Add police officer and car for traffic control				Hr.	45.50			45.50	50
1000	Flexible pavement repair holes, parking lot, bag material, 1 C.F. size each	B-37D	18	.889	Ea.	64.50	34	8	106.50	135
1010	Economy bag material, 1 C.F. each		18	.889	"	42	34	8	84	111

32 01 Operation and Maintenance of Exterior Improvements

32 01 17 – Flexible Paving Repair

32 01 17.10 Repair of Asphalt Pavement Holes		Crew	Daily Output	Labor-Hours	Unit	Material	2015 Bare Costs Labor	Equipment	Total	Total Incl O&P
1100	Group of two, 1 C.F. size each	B-37D	12	1.333	Set	129	51	12	192	238
1110	Group of two, economy bag, 1 C.F. size each		12	1.333		84	51	12	147	189
1200	Group of three 1 C.F. size each		10	1.600		193	61.50	14.40	268.90	330
1210	Economy bag, group of three, 1 C.F. size each		10	1.600		126	61.50	14.40	201.90	255
1300	Flexible pavement repair holes, parking lot, 1 C.F. single hole	A-3A	4	2	Ea.	64.50	97	39	200.50	268
1310	Economy material, 1 C.F. single hole		4	2	"	42	97	39	178	244
1400	Flexible pavement repair holes, parking lot, 1 C.F. four holes		2	4	Set	258	194	78	530	680
1410	Economy material, 1 C.F. four holes		2	4	"	168	194	78	440	580
1500	Flexible pavement repair holes, large parking lot, bulk matl, 1 C.F. size	B-37A	32	.750	Ea.	9	28.50	11.80	49.30	69.50

32 01 17.20 Repair of Asphalt Pavement Patches

		Crew	Daily Output	Labor-Hours	Unit	Material	Labor	Equipment	Total	Total Incl O&P
0010	**REPAIR OF ASPHALT PAVEMENT PATCHES**									
0100	Flexible pavement patches, roadway, light traffic, sawcut 10-25 S.F.	B-89	12	1.333	Ea.		58.50	40.50	99	138
0150	Sawcut 26-60 S.F.		10	1.600			70	49	119	166
0200	Sawcut 61-100 S.F.		8	2			87.50	61	148.50	207
0210	Sawcut groups of small size patches		24	.667			29	20.50	49.50	69
0220	Large size patches		16	1			44	30.50	74.50	104
0300	Flexible pavement patches, roadway, light traffic, digout 10-25 S.F.	B-6	16	1.500			62	23	85	125
0350	Digout 26-60 S.F.		12	2			82.50	30.50	113	168
0400	Digout 61-100 S.F.		8	3			124	45.50	169.50	250
0450	Add 8 C.Y. truck, small project debris haulaway	B-34A	8	1	Hr.		40	51.50	91.50	121
0460	Add 12 C.Y. truck, small project debris haulaway	B-34B	8	1			40	86.50	126.50	160
0480	Add flagger for non-intersection medium traffic	1 Clab	8	1			37.50		37.50	61.50
0490	Add flasher truck for intersection medium traffic or heavy traffic	A-2B	8	1			39	30.50	69.50	96.50
0500	Flexible pavement patches, roadway, repave, cold, 15 S.F., 4" D	B-37	8	6	Ea.	43	239	19.70	301.70	460
0510	6" depth		8	6		65	239	19.70	323.70	485
0520	Repave, cold, 20 S.F., 4" depth		8	6		57	239	19.70	315.70	475
0530	6" depth		8	6		86.50	239	19.70	345.20	505
0540	Repave, cold, 25 S.F., 4" depth		8	6		71.50	239	19.70	330.20	490
0550	6" depth		8	6		108	239	19.70	366.70	530
0600	Repave, cold, 30 S.F., 4" depth		8	6		85.50	239	19.70	344.20	505
0610	6" depth		8	6		130	239	19.70	388.70	555
0640	Repave, cold, 40 S.F., 4" depth		8	6		114	239	19.70	372.70	535
0650	6" depth		8	6		173	239	19.70	431.70	600
0680	Repave, cold, 50 S.F., 4" depth		8	6		143	239	19.70	401.70	570
0690	6" depth		8	6		216	239	19.70	474.70	650
0720	Repave, cold, 60 S.F., 4" depth		8	6		171	239	19.70	429.70	600
0730	6" depth		8	6		259	239	19.70	517.70	695
0800	Repave, cold, 70 S.F., 4" depth		7	6.857		199	273	22.50	494.50	690
0810	6" depth		7	6.857		300	273	22.50	595.50	800
0820	Repave, cold, 75 S.F., 4" depth		7	6.857		213	273	22.50	508.50	705
0830	6" depth		7	6.857		325	273	22.50	620.50	825
0900	Repave, cold, 80 S.F., 4" depth		6	8		228	320	26	574	800
0910	6" depth		6	8		345	320	26	691	930
0940	Repave, cold, 90 S.F., 4" depth		6	8		256	320	26	602	830
0950	6" depth		6	8		390	320	26	736	980
0980	Repave, cold, 100 S.F., 4" depth		6	8		285	320	26	631	865
0990	6" depth		6	8		430	320	26	776	1,025
1000	Add flasher truck for paving operations in medium or heavy traffic	A-2B	8	1	Hr.		39	30.50	69.50	96.50
1100	Prime coat for repair 15-40 S.F., 25% overspray	B-37A	48	.500	Ea.	7.25	19.05	7.85	34.15	47.50
1150	41-60 S.F.		40	.600		14.50	23	9.45	46.95	64
1175	61-80 S.F.		32	.750		19.65	28.50	11.80	59.95	81
1200	81-100 S.F.		32	.750		24	28.50	11.80	64.30	86

32 01 Operation and Maintenance of Exterior Improvements

32 01 17 – Flexible Paving Repair

32 01 17.20 Repair of Asphalt Pavement Patches

		Crew	Daily Output	Labor-Hours	Unit	Material	2015 Bare Costs Labor	2015 Bare Costs Equipment	Total	Total Incl O&P
1210	Groups of patches w/25% overspray	B-37A	3600	.007	S.F.	.24	.25	.10	.59	.80
1300	Flexible pavement repair patches, street repave, hot, 60 S.F., 4" D	B-37	8	6	Ea.	101	239	19.70	359.70	525
1310	6" depth		8	6		153	239	19.70	411.70	580
1320	Repave, hot, 70 S.F., 4" depth		7	6.857		118	273	22.50	413.50	600
1330	6" depth		7	6.857		179	273	22.50	474.50	665
1340	Repave, hot, 75 S.F., 4" depth		7	6.857		126	273	22.50	421.50	610
1350	6" depth		7	6.857		192	273	22.50	487.50	680
1360	Repave, hot, 80 S.F., 4" depth		6	8		135	320	26	481	695
1370	6" depth		6	8		205	320	26	551	775
1380	Repave, hot, 90 S.F., 4" depth		6	8		152	320	26	498	715
1390	6" depth		6	8		230	320	26	576	805
1400	Repave, hot, 100 S.F., 4" depth		6	8		169	320	26	515	735
1410	6" depth		6	8		256	320	26	602	830
1420	Pave hot groups of patches, 4" depth		900	.053	S.F.	1.68	2.12	.17	3.97	5.50
1430	6" depth		900	.053	"	2.56	2.12	.17	4.85	6.45
1500	Add 8 C.Y. truck for hot asphalt paving operations	B-34A	1	8	Day		320	410	730	965
1550	Add flasher truck for hot paving operations in med/heavy traffic	A-2B	8	1	Hr.		39	30.50	69.50	96.50
2000	Add police officer and car for traffic control				"	45.50			45.50	50

32 01 17.61 Sealing Cracks In Asphalt Paving

		Crew	Daily Output	Labor-Hours	Unit	Material	Labor	Equipment	Total	Total Incl O&P
0010	**SEALING CRACKS IN ASPHALT PAVING**									
0100	Sealing cracks in asphalt paving, 1/8" wide x 1/2" depth, slow set	B-37A	2000	.012	L.F.	.17	.46	.19	.82	1.15
0110	1/4" wide x 1/2" depth		2000	.012		.20	.46	.19	.85	1.18
0130	3/8" wide x 1/2" depth		2000	.012		.23	.46	.19	.88	1.21
0140	1/2" wide x 1/2" depth		1800	.013		.26	.51	.21	.98	1.34
0150	3/4" wide x 1/2" depth		1600	.015		.31	.57	.24	1.12	1.53
0160	1" wide x 1/2" depth		1600	.015		.37	.57	.24	1.18	1.59
0165	1/8" wide x 1" depth, rapid set	B-37F	2000	.016		.25	.61	.29	1.15	1.58
0170	1/4" wide x 1" depth		2000	.016		.48	.61	.29	1.38	1.84
0175	3/8" wide x 1" depth		1800	.018		.72	.68	.32	1.72	2.25
0180	1/2" wide x 1" depth		1800	.018		.97	.68	.32	1.97	2.51
0185	5/8" wide x 1" depth		1800	.018		1.21	.68	.32	2.21	2.78
0190	3/4" wide x 1" depth		1600	.020		1.45	.76	.36	2.57	3.23
0195	1" wide x 1" depth		1600	.020		1.93	.76	.36	3.05	3.76
0200	1/8" wide x 2" depth, rapid set		2000	.016		.48	.61	.29	1.38	1.84
0210	1/4" wide x 2" depth		2000	.016		.97	.61	.29	1.87	2.37
0220	3/8" wide x 2" depth		1800	.018		1.45	.68	.32	2.45	3.04
0230	1/2" wide x 2" depth		1800	.018		1.93	.68	.32	2.93	3.57
0240	5/8" wide x 2" depth		1800	.018		2.41	.68	.32	3.41	4.10
0250	3/4" wide x 2" depth		1600	.020		2.89	.76	.36	4.01	4.82
0260	1" wide x 2" depth		1600	.020		3.86	.76	.36	4.98	5.90
0300	Add flagger for non-intersection medium traffic	1 Clab	8	1	Hr.		37.50		37.50	61.50
0400	Add flasher truck for intersection medium traffic or heavy traffic	A-2B	8	1	"		39	30.50	69.50	96.50

32 01 29 – Rigid Paving Repair

32 01 29.61 Partial Depth Patching of Rigid Pavement

		Crew	Daily Output	Labor-Hours	Unit	Material	Labor	Equipment	Total	Total Incl O&P
0010	**PARTIAL DEPTH PATCHING OF RIGID PAVEMENT**									
0100	Rigid pavement repair, roadway, light traffic, 25% pitting 15 S.F.	B-37F	16	2	Ea.	27	76	36	139	194
0110	Pitting 20 S.F.		16	2		36	76	36	148	204
0120	Pitting 25 S.F.		16	2		45	76	36	157	214
0130	Pitting 30 S.F.		16	2		54	76	36	166	224
0140	Pitting 40 S.F.		16	2		72	76	36	184	244
0150	Pitting 50 S.F.		12	2.667		90	101	48.50	239.50	320
0160	Pitting 60 S.F.		12	2.667		108	101	48.50	257.50	335

32 01 Operation and Maintenance of Exterior Improvements

32 01 29 – Rigid Paving Repair

32 01 29.61 Partial Depth Patching of Rigid Pavement		Crew	Daily Output	Labor-Hours	Unit	Material	2015 Bare Costs Labor	Equipment	Total	Total Incl O&P
0170	Pitting 70 S.F.	B-37F	12	2.667	Ea.	126	101	48.50	275.50	355
0180	Pitting 75 S.F.		12	2.667		135	101	48.50	284.50	365
0190	Pitting 80 S.F.		8	4		144	152	72.50	368.50	485
0200	Pitting 90 S.F.		8	4		162	152	72.50	386.50	505
0210	Pitting 100 S.F.		8	4		180	152	72.50	404.50	525
0300	Roadway, light traffic, 50% pitting 15 S.F.		12	2.667		54	101	48.50	203.50	278
0310	Pitting 20 S.F.		12	2.667		72	101	48.50	221.50	298
0320	Pitting 25 S.F.		12	2.667		90	101	48.50	239.50	320
0330	Pitting 30 S.F.		12	2.667		108	101	48.50	257.50	335
0340	Pitting 40 S.F.		12	2.667		144	101	48.50	293.50	375
0350	Pitting 50 S.F.		8	4		180	152	72.50	404.50	525
0360	Pitting 60 S.F.		8	4		217	152	72.50	441.50	565
0370	Pitting 70 S.F.		8	4		253	152	72.50	477.50	605
0380	Pitting 75 S.F.		8	4		271	152	72.50	495.50	625
0390	Pitting 80 S.F.		6	5.333		289	203	96.50	588.50	755
0400	Pitting 90 S.F.		6	5.333		325	203	96.50	624.50	790
0410	Pitting 100 S.F.		6	5.333		360	203	96.50	659.50	830
1000	Rigid pavement repair, light traffic, surface patch, 2" deep, 15 S.F.		8	4		108	152	72.50	332.50	445
1010	Surface patch, 2" deep, 20 S.F.		8	4		144	152	72.50	368.50	485
1020	Surface patch, 2" deep, 25 S.F.		8	4		180	152	72.50	404.50	525
1030	Surface patch, 2" deep, 30 S.F.		8	4		217	152	72.50	441.50	565
1040	Surface patch, 2" deep, 40 S.F.		8	4		289	152	72.50	513.50	650
1050	Surface patch, 2" deep, 50 S.F.		6	5.333		360	203	96.50	659.50	830
1060	Surface patch, 2" deep, 60 S.F.		6	5.333		435	203	96.50	734.50	910
1070	Surface patch, 2" deep, 70 S.F.		6	5.333		505	203	96.50	804.50	990
1080	Surface patch, 2" deep, 75 S.F.		6	5.333		540	203	96.50	839.50	1,025
1090	Surface patch, 2" deep, 80 S.F.		5	6.400		580	243	116	939	1,150
1100	Surface patch, 2" deep, 90 S.F.		5	6.400		650	243	116	1,009	1,250
1200	Surface patch, 2" deep, 100 S.F.		5	6.400		720	243	116	1,079	1,325
2000	Add flagger for non-intersection medium traffic	1 Clab	8	1	Hr.		37.50		37.50	61.50
2100	Add flasher truck for intersection medium traffic or heavy traffic	A-2B	8	1			39	30.50	69.50	96.50
2200	Add police officer and car for traffic control					45.50			45.50	50

32 01 29.70 Full Depth Patching of Rigid Pavement

		Crew	Daily Output	Labor-Hours	Unit	Material	Labor	Equipment	Total	Total Incl O&P
0010	**FULL DEPTH PATCHING OF RIGID PAVEMENT**									
0015	Pavement preparation includes sawcut, remove pavement and replace									
0020	6 inches of subbase and one layer of reinforcement									
0030	Trucking and haulaway of debris excluded									
0100	Rigid pavement replace, light traffic, prep, 15 S.F., 6" depth	B-37E	16	3.500	Ea.	19.65	145	64	228.65	325
0110	Replacement preparation 20 S.F., 6" depth		16	3.500		26	145	64	235	335
0115	25 S.F., 6" depth		16	3.500		32.50	145	64	241.50	340
0120	30 S.F., 6" depth		14	4		39.50	166	73	278.50	390
0125	35 S.F., 6" depth		14	4		46	166	73	285	400
0130	40 S.F., 6" depth		14	4		52.50	166	73	291.50	405
0135	45 S.F., 6" depth		14	4		59	166	73	298	415
0140	50 S.F., 6" depth		12	4.667		65.50	193	85	343.50	475
0145	55 S.F., 6" depth		12	4.667		72	193	85	350	485
0150	60 S.F., 6" depth		10	5.600		78.50	232	102	412.50	575
0155	65 S.F., 6" depth		10	5.600		85	232	102	419	580
0160	70 S.F., 6" depth		10	5.600		91.50	232	102	425.50	590
0165	75 S.F., 6" depth		10	5.600		98	232	102	432	595
0170	80 S.F., 6" depth		8	7		105	290	128	523	725
0175	85 S.F., 6" depth		8	7		111	290	128	529	735

32 01 Operation and Maintenance of Exterior Improvements

32 01 29 – Rigid Paving Repair

32 01 29.70 Full Depth Patching of Rigid Pavement		Crew	Daily Output	Labor-Hours	Unit	Material	2015 Bare Costs Labor	Equipment	Total	Total Incl O&P
0180	90 S.F., 6" depth	B-37E	8	7	Ea.	118	290	128	536	740
0185	95 S.F., 6" depth		8	7		124	290	128	542	750
0190	100 S.F., 6" depth		8	7		131	290	128	549	755
0200	Pavement preparation for 8 inch rigid paving same as 6 inch									
0290	Pavement preparation for 9 inch rigid paving same as 10 inch									
0300	Rigid pavement replace, light traffic,prep, 15 S.F.,10" depth	B-37E	16	3.500	Ea.	32.50	145	64	241.50	340
0302	Pavement preparation 10 inch includes sawcut, remove pavement and									
0304	replace 6 inches of subbase and two layers of reinforcement									
0306	Trucking and haulaway of debris excluded									
0310	Replacement preparation 20 S.F., 10" depth	B-37E	16	3.500	Ea.	43.50	145	64	252.50	355
0315	25 S.F., 10" depth		16	3.500		54.50	145	64	263.50	365
0320	30 S.F., 10" depth		14	4		65.50	166	73	304.50	420
0325	35 S.F., 10" depth		14	4		50	166	73	289	405
0330	40 S.F., 10" depth		14	4		87	166	73	326	445
0335	45 S.F., 10" depth		12	4.667		98	193	85	376	510
0340	50 S.F., 10" depth		12	4.667		109	193	85	387	525
0345	55 S.F., 10" depth		12	4.667		120	193	85	398	535
0350	60 S.F., 10" depth		10	5.600		131	232	102	465	630
0355	65 S.F., 10" depth		10	5.600		142	232	102	476	645
0360	70 S.F., 10" depth		10	5.600		153	232	102	487	655
0365	75 S.F., 10" depth		8	7		163	290	128	581	790
0370	80 S.F., 10" depth		8	7		174	290	128	592	805
0375	85 S.F., 10" depth		8	7		185	290	128	603	815
0380	90 S.F., 10" depth		8	7		196	290	128	614	825
0385	95 S.F., 10" depth		8	7		207	290	128	625	840
0390	100 S.F., 10" depth		8	7		218	290	128	636	850
0500	Add 8 C.Y. truck, small project debris haulaway	B-34A	8	1	Hr.		40	51.50	91.50	121
0550	Add 12 C.Y. truck, small project debris haulaway	B-34B	8	1			40	86.50	126.50	160
0600	Add flagger for non-intersection medium traffic	1 Clab	8	1			37.50		37.50	61.50
0650	Add flasher truck for intersection medium traffic or heavy traffic	A-2B	8	1			39	30.50	69.50	96.50
0700	Add police officer and car for traffic control					45.50			45.50	50
0990	Concrete will replaced using quick set mix with aggregate									
1000	Rigid pavement replace, light traffic,repour, 15 S.F.,6" depth	B-37F	18	1.778	Ea.	161	67.50	32	260.50	325
1010	20 S.F., 6" depth		18	1.778		214	67.50	32	313.50	380
1015	25 S.F., 6" depth		18	1.778		268	67.50	32	367.50	440
1020	30 S.F., 6" depth		16	2		320	76	36	432	520
1025	35 S.F., 6" depth		16	2		375	76	36	487	580
1030	40 S.F., 6" depth		16	2		430	76	36	542	635
1035	45 S.F., 6" depth		16	2		480	76	36	592	695
1040	50 S.F., 6" depth		16	2		535	76	36	647	755
1045	55 S.F., 6" depth		12	2.667		590	101	48.50	739.50	870
1050	60 S.F., 6" depth		12	2.667		645	101	48.50	794.50	925
1055	65 S.F., 6" depth		12	2.667		695	101	48.50	844.50	985
1060	70 S.F., 6" depth		12	2.667		750	101	48.50	899.50	1,050
1065	75 S.F., 6" depth		10	3.200		805	122	58	985	1,150
1070	80 S.F., 6" depth		10	3.200		855	122	58	1,035	1,200
1075	85 S.F., 6" depth		8	4		910	152	72.50	1,134.50	1,325
1080	90 S.F., 6" depth		8	4		965	152	72.50	1,189.50	1,375
1085	95 S.F., 6" depth		8	4		1,025	152	72.50	1,249.50	1,450
1090	100 S.F., 6" depth		8	4		1,075	152	72.50	1,299.50	1,500
1099	Concrete will be replaced using 4500 PSI Concrete ready mix									
1100	Rigid pavement replace, light traffic, repour, 15 S.F., 6" depth	A-2	12	2	Ea.	278	76	20.50	374.50	450
1110	20-50 S.F., 6" depth		12	2		278	76	20.50	374.50	450

32 01 Operation and Maintenance of Exterior Improvements

32 01 29 – Rigid Paving Repair

32 01 29.70 Full Depth Patching of Rigid Pavement		Crew	Daily Output	Labor-Hours	Unit	Material	2015 Bare Costs Labor	Equipment	Total	Total Incl O&P
1120	55-65 S.F., 6" depth	A-2	10	2.400	Ea.	310	91.50	24.50	426	515
1130	70-80 S.F., 6" depth		10	2.400		340	91.50	24.50	456	545
1140	85-100 S.F., 6" depth		8	3		395	114	30.50	539.50	655
1200	Repour 15-40 S.F., 8" depth		12	2		278	76	20.50	374.50	450
1210	45-50 S.F., 8" depth		12	2		310	76	20.50	406.50	485
1220	55-60 S.F., 8" depth		10	2.400		340	91.50	24.50	456	545
1230	65-80 S.F., 8" depth		10	2.400		395	91.50	24.50	511	610
1240	85-90 S.F., 8" depth		8	3		430	114	30.50	574.50	690
1250	95-100 S.F., 8" depth		8	3		455	114	30.50	599.50	720
1300	Repour 15-30 S.F., 10" depth		12	2		278	76	20.50	374.50	450
1310	35-40 S.F., 10" depth		12	2		310	76	20.50	406.50	485
1320	45 S.F., 10" depth		12	2		340	76	20.50	436.50	515
1330	50-65 S.F., 10" depth		10	2.400		395	91.50	24.50	511	610
1340	70 S.F., 10" depth		10	2.400		430	91.50	24.50	546	645
1350	75-80 S.F., 10" depth		10	2.400		455	91.50	24.50	571	675
1360	85-95 S.F., 10" depth		8	3		515	114	30.50	659.50	790
1370	100 S.F., 10" depth		8	3		550	114	30.50	694.50	825

32 06 Schedules for Exterior Improvements

32 06 10 – Schedules for Bases, Ballasts, and Paving

32 06 10.10 Sidewalks, Driveways and Patios

		Crew	Daily Output	Labor-Hours	Unit	Material	Labor	Equipment	Total	Total Incl O&P
0010	**SIDEWALKS, DRIVEWAYS AND PATIOS** No base									
0020	Asphaltic concrete, 2" thick	B-37	720	.067	S.Y.	7.40	2.65	.22	10.27	12.70
0100	2-1/2" thick	"	660	.073	"	9.40	2.89	.24	12.53	15.30
0110	Bedding for brick or stone, mortar, 1" thick	D-1	300	.053	S.F.	.89	2.25		3.14	4.62
0120	2" thick	"	200	.080		2.21	3.37		5.58	7.95
0130	Sand, 2" thick	B-18	8000	.003		.28	.11	.01	.40	.51
0140	4" thick	"	4000	.006		.56	.23	.01	.80	1.01
0300	Concrete, 3000 psi, CIP, 6 x 6 - W1.4 x W1.4 mesh,									
0310	broomed finish, no base, 4" thick	B-24	600	.040	S.F.	1.69	1.73		3.42	4.66
0350	5" thick		545	.044		2.26	1.90		4.16	5.55
0400	6" thick		510	.047		2.64	2.03		4.67	6.20
0450	For bank run gravel base, 4" thick, add	B-18	2500	.010		.52	.37	.02	.91	1.19
0520	8" thick, add	"	1600	.015		1.04	.57	.03	1.64	2.12
0550	Exposed aggregate finish, add to above, minimum	B-24	1875	.013		.11	.55		.66	1.02
0600	Maximum	"	455	.053		.36	2.28		2.64	4.09
0950	Concrete tree grate, 5' square	B-6	25	.960	Ea.	420	39.50	14.55	474.05	545
0955	Tree well & cover, concrete, 3' square		25	.960		320	39.50	14.55	374.05	430
0960	Cast iron tree grate with frame, 2 piece, round, 5' diameter		25	.960		1,100	39.50	14.55	1,154.05	1,275
0980	Square, 5' side		25	.960		1,100	39.50	14.55	1,154.05	1,275
1700	Redwood, prefabricated, 4' x 4' sections	2 Carp	316	.051	S.F.	4.77	2.38		7.15	9.15
1750	Redwood planks, 1" thick, on sleepers	"	240	.067	"	4.77	3.13		7.90	10.40
2250	Stone dust, 4" thick	B-62	900	.027	S.Y.	3.33	1.10	.19	4.62	5.65
9000	Minimum labor/equipment charge	D-1	2	8	Job		335		335	550

32 06 10.20 Steps

		Crew	Daily Output	Labor-Hours	Unit	Material	Labor	Equipment	Total	Total Incl O&P
0010	**STEPS**									
0011	Incl. excav., borrow & concrete base as required									
0100	Brick steps	B-24	35	.686	LF Riser	15.35	29.50		44.85	65
0200	Railroad ties	2 Clab	25	.640		3.59	24		27.59	43.50
0300	Bluestone treads, 12" x 2" or 12" x 1-1/2"	B-24	30	.800		33.50	34.50		68	92.50
0490	Minimum labor/equipment charge	D-1	2	8	Job		335		335	550

32 06 Schedules for Exterior Improvements

32 06 10 – Schedules for Bases, Ballasts, and Paving

32 06 10.20 Steps		Crew	Daily Output	Labor-Hours	Unit	Material	2015 Bare Costs Labor	Equipment	Total	Total Incl O&P
0500	Concrete, cast in place, see Section 03 30 53.40									
0600	Precast concrete, see Section 03 41 23.50									

32 11 Base Courses

32 11 23 – Aggregate Base Courses

32 11 23.23 Base Course Drainage Layers

		Crew	Daily Output	Labor-Hours	Unit	Material	Labor	Equipment	Total	Total Incl O&P
0010	**BASE COURSE DRAINAGE LAYERS**									
0011	For roadways and large areas									
0050	Crushed 3/4" stone base, compacted, 3" deep	B-36C	5200	.008	S.Y.	2.30	.36	.81	3.47	3.99
0100	6" deep		5000	.008		4.61	.37	.84	5.82	6.55
0200	9" deep		4600	.009		6.90	.40	.91	8.21	9.25
0300	12" deep		4200	.010		9.20	.44	1	10.64	11.95
0301	Crushed 1-1/2" stone base, compacted to 4" deep	B-36B	6000	.011		4.46	.48	.80	5.74	6.55
0302	6" deep		5400	.012		6.70	.53	.89	8.12	9.20
0303	8" deep		4500	.014		8.90	.64	1.07	10.61	12
0304	12" deep		3800	.017		13.35	.75	1.27	15.37	17.30
0310	Minimum labor/equipment charge	B-36	1	40	Job		1,725	1,675	3,400	4,625
0350	Bank run gravel, spread and compacted									
0370	6" deep	B-32	6000	.005	S.Y.	4.33	.25	.39	4.97	5.60
0390	9" deep		4900	.007		6.50	.31	.48	7.29	8.15
0400	12" deep		4200	.008		8.65	.36	.56	9.57	10.70
6900	For small and irregular areas, add						50%	50%		
6990	Minimum labor/equipment charge	1 Clab	3	2.667	Job		100		100	164
7000	Prepare and roll sub-base, small areas to 2500 S.Y.	B-32A	1500	.016	S.Y.		.74	.95	1.69	2.23
8000	Large areas over 2500 S.Y.	"	3500	.007			.32	.41	.73	.96
8050	For roadways	B-32	4000	.008			.38	.59	.97	1.26
9000	Minimum labor/equipment charge	1 Clab	4	2	Job		75		75	123

32 12 Flexible Paving

32 12 16 – Asphalt Paving

32 12 16.13 Plant-Mix Asphalt Paving

		Crew	Daily Output	Labor-Hours	Unit	Material	Labor	Equipment	Total	Total Incl O&P
0010	**PLANT-MIX ASPHALT PAVING**									
0080	Binder course, 1-1/2" thick	B-25	7725	.011	S.Y.	5.70	.47	.35	6.52	7.45
0120	2" thick		6345	.014		7.60	.57	.43	8.60	9.75
0130	2-1/2" thick		5620	.016		9.50	.65	.48	10.63	12.05
0160	3" thick		4905	.018		11.40	.74	.55	12.69	14.35
0170	3-1/2" thick		4520	.019		13.30	.80	.60	14.70	16.60
0300	Wearing course, 1" thick	B-25B	10575	.009		3.78	.38	.28	4.44	5.10
0340	1-1/2" thick		7725	.012		6.35	.52	.38	7.25	8.25
0380	2" thick		6345	.015		8.50	.64	.46	9.60	10.95
0420	2-1/2" thick		5480	.018		10.50	.74	.54	11.78	13.35
0460	3" thick		4900	.020		12.55	.82	.60	13.97	15.80
0470	3-1/2" thick		4520	.021		14.70	.89	.65	16.24	18.35
0480	4" thick		4140	.023		16.80	.98	.71	18.49	21
0500	Open graded friction course	B-25C	5000	.010		2.13	.41	.47	3.01	3.52
0800	Alternate method of figuring paving costs									
0810	Binder course, 1-1/2" thick	B-25	630	.140	Ton	70	5.75	4.31	80.06	91
0811	2" thick		690	.128		70	5.25	3.93	79.18	90
0812	3" thick		800	.110		70	4.55	3.39	77.94	88

32 12 Flexible Paving

32 12 16 – Asphalt Paving

32 12 16.13 Plant-Mix Asphalt Paving

		Crew	Daily Output	Labor-Hours	Unit	Material	2015 Bare Costs Labor	Equipment	Total	Total Incl O&P
0813	4" thick	B-25	900	.098	Ton	70	4.04	3.01	77.05	87
0850	Wearing course, 1" thick	B-25B	575	.167		69	7.05	5.15	81.20	93
0851	1-1/2" thick		630	.152		69	6.40	4.68	80.08	91.50
0852	2" thick		690	.139		69	5.85	4.28	79.13	90
0853	2-1/2" thick		765	.125		69	5.30	3.86	78.16	89
0854	3" thick		800	.120		69	5.05	3.69	77.74	88.50
1000	Pavement replacement over trench, 2" thick	B-37	90	.533	S.Y.	7.85	21	1.75	30.60	45
1050	4" thick		70	.686		15.55	27.50	2.25	45.30	64
1080	6" thick		55	.873		25	34.50	2.86	62.36	86.50
3000	Prime coat, emulsion, .30 gal. per S.Y., 1000 S.Y.	B-45	2500	.006		1.74	.29	.36	2.39	2.77
3100	Tack coat, emulsion, .10 gal. per S.Y., 1000 S.Y.	"	2500	.006		.58	.29	.36	1.23	1.50

32 12 16.14 Asphaltic Concrete Paving

		Crew	Daily Output	Labor-Hours	Unit	Material	Labor	Equipment	Total	Total Incl O&P
0011	**ASPHALTIC CONCRETE PAVING**, parking lots & driveways									
0015	No asphalt hauling included									
0020	6" stone base, 2" binder course, 1" topping	B-25C	9000	.005	S.F.	1.86	.23	.26	2.35	2.70
0025	2" binder course, 2" topping		9000	.005		2.28	.23	.26	2.77	3.16
0030	3" binder course, 2" topping		9000	.005		2.71	.23	.26	3.20	3.63
0035	4" binder course, 2" topping		9000	.005		3.13	.23	.26	3.62	4.09
0040	1.5" binder course, 1" topping		9000	.005		1.65	.23	.26	2.14	2.47
0042	3" binder course, 1" topping		9000	.005		2.29	.23	.26	2.78	3.17
0045	3" binder course, 3" topping		9000	.005		3.13	.23	.26	3.62	4.09
0050	4" binder course, 3" topping		9000	.005		3.55	.23	.26	4.04	4.55
0055	4" binder course, 4" topping		9000	.005		3.96	.23	.26	4.45	5
0300	Binder course, 1-1/2" thick		35000	.001		.64	.06	.07	.77	.86
0400	2" thick		25000	.002		.82	.08	.09	.99	1.14
0500	3" thick		15000	.003		1.27	.14	.16	1.57	1.79
0600	4" thick		10800	.004		1.67	.19	.22	2.08	2.37
0800	Sand finish course, 3/4" thick		41000	.001		.32	.05	.06	.43	.49
0900	1" thick		34000	.001		.40	.06	.07	.53	.62
1000	Fill pot holes, hot mix, 2" thick	B-16	4200	.008		.85	.30	.16	1.31	1.60
1100	4" thick		3500	.009		1.25	.35	.20	1.80	2.17
1120	6" thick		3100	.010		1.67	.40	.22	2.29	2.74
1140	Cold patch, 2" thick	B-51	3000	.016		.99	.61	.08	1.68	2.18
1160	4" thick		2700	.018		1.89	.68	.09	2.66	3.29
1180	6" thick		1900	.025		2.94	.96	.13	4.03	4.96
3000	Prime coat, emulsion, .30 gal. per S.Y., 1000 S.Y.	B-45	2500	.006	S.Y.	1.74	.29	.36	2.39	2.77
3100	Tack coat, emulsion, .10 gal. per S.Y., 1000 S.Y.	"	2500	.006	"	.58	.29	.36	1.23	1.50

32 13 Rigid Paving

32 13 13 – Concrete Paving

32 13 13.23 Concrete Paving Surface Treatment

		Crew	Daily Output	Labor-Hours	Unit	Material	Labor	Equipment	Total	Total Incl O&P
0010	**CONCRETE PAVING SURFACE TREATMENT**									
0015	Including joints, finishing and curing									
0020	Fixed form, 12' pass, unreinforced, 6" thick	B-26	3000	.029	S.Y.	22	1.24	1.18	24.42	27.50
0100	8" thick		2750	.032		30	1.35	1.28	32.63	36.50
0400	12" thick		1800	.049		43	2.06	1.96	47.02	53
0410	Conc. pavement, w/jt.,fnsh.&curing,fix form,24' pass,unreinforced,6"T		6000	.015		21	.62	.59	22.21	24.50
0430	8" thick		5500	.016		28.50	.67	.64	29.81	33.50
0460	12" thick		3600	.024		41.50	1.03	.98	43.51	48.50
0520	Welded wire fabric, sheets for rigid paving 2.33 lb./S.Y.	2 Rodm	389	.041		1.31	2.16		3.47	4.97
0530	Reinforcing steel for rigid paving 12 lb./S.Y.		666	.024		6.10	1.26		7.36	8.75

32 13 Rigid Paving

32 13 13 - Concrete Paving

32 13 13.23 Concrete Paving Surface Treatment

		Crew	Daily Output	Labor-Hours	Unit	Material	2015 Bare Costs Labor	Equipment	Total	Total Incl O&P
0540	Reinforcing steel for rigid paving 18 lb./S.Y.	2 Rodm	444	.036	S.Y.	9.15	1.89		11.04	13.15
0620	Slip form, 12' pass, unreinforced, 6" thick	B-26A	5600	.016		21	.66	.66	22.32	25
0624	8" thick		5300	.017		29	.70	.70	30.40	34
0630	12" thick		3470	.025		42	1.07	1.07	44.14	49
0632	15" thick		2890	.030		53	1.28	1.28	55.56	62
0640	Slip form, 24' pass, unreinforced, 6" thick		11200	.008		20.50	.33	.33	21.16	24
0644	8" thick		10600	.008		28	.35	.35	28.70	31.50
0650	12" thick		6940	.013		41	.53	.53	42.06	46.50
0652	15" thick		5780	.015		51.50	.64	.64	52.78	58
0700	Finishing, broom finish small areas	2 Cefi	120	.133			6		6	9.50
1000	Curing, with sprayed membrane by hand	2 Clab	1500	.011		1	.40		1.40	1.75

32 14 Unit Paving

32 14 13 - Precast Concrete Unit Paving

32 14 13.18 Precast Concrete Plantable Pavers

		Crew	Daily Output	Labor-Hours	Unit	Material	2015 Bare Costs Labor	Equipment	Total	Total Incl O&P
0010	**PRECAST CONCRETE PLANTABLE PAVERS** (50% grass)									
0300	3/4" crushed stone base for plantable pavers, 6 inch depth	B-62	1000	.024	S.Y.	4.61	.99	.17	5.77	6.85
0400	8 inch depth		900	.027		6.15	1.10	.19	7.44	8.80
0500	10 inch depth		800	.030		7.70	1.24	.22	9.16	10.70
0600	12 inch depth		700	.034		9.20	1.42	.25	10.87	12.70
0700	Hydro seeding plantable pavers	B-81A	20	.800	M.S.F.	12.45	30.50	21.50	64.45	87
0800	Apply fertilizer and seed to plantable pavers	1 Clab	8	1	"	50.50	37.50		88	117

32 14 16 - Brick Unit Paving

32 14 16.10 Brick Paving

		Crew	Daily Output	Labor-Hours	Unit	Material	2015 Bare Costs Labor	Equipment	Total	Total Incl O&P
0010	**BRICK PAVING**									
0012	4" x 8" x 1-1/2", without joints (4.5 brick/S.F.)	D-1	110	.145	S.F.	2.21	6.10		8.31	12.40
0100	Grouted, 3/8" joint (3.9 brick/S.F.)		90	.178		1.95	7.50		9.45	14.35
0200	4" x 8" x 2-1/4", without joints (4.5 bricks/S.F.)		110	.145		2.25	6.10		8.35	12.45
0300	Grouted, 3/8" joint (3.9 brick/S.F.)		90	.178		1.95	7.50		9.45	14.35
0455	Pervious brick paving, 4" x 8" x 3-1/4", without joints (4.5 bricks/S.F.)		110	.145		3.38	6.10		9.48	13.65
0500	Bedding, asphalt, 3/4" thick	B-25	5130	.017		.70	.71	.53	1.94	2.50
0540	Course washed sand bed, 1" thick	B-18	5000	.005		.32	.18	.01	.51	.66
0580	Mortar, 1" thick	D-1	300	.053		.74	2.25		2.99	4.46
0620	2" thick		200	.080		1.48	3.37		4.85	7.10
1500	Brick on 1" thick sand bed laid flat, 4.5 per S.F.		100	.160		3.18	6.75		9.93	14.45
2000	Brick pavers, laid on edge, 7.2 per S.F.		70	.229		3.86	9.60		13.46	19.90
2500	For 4" thick concrete bed and joints, add		595	.027		1.19	1.13		2.32	3.15
2800	For steam cleaning, add	A-1H	950	.008		.09	.32	.08	.49	.71
9000	Minimum labor/equipment charge	1 Bric	2	4	Job		185		185	300

32 14 23 - Asphalt Unit Paving

32 14 23.10 Asphalt Blocks

		Crew	Daily Output	Labor-Hours	Unit	Material	2015 Bare Costs Labor	Equipment	Total	Total Incl O&P
0010	**ASPHALT BLOCKS**									
0020	Rectangular, 6" x 12" x 1-1/4", w/bed & neopr. adhesive	D-1	135	.119	S.F.	9.30	4.99		14.29	18.30
0100	3" thick		130	.123		13	5.20		18.20	23
0300	Hexagonal tile, 8" wide, 1-1/4" thick		135	.119		9.30	4.99		14.29	18.30
0400	2" thick		130	.123		13	5.20		18.20	23
0500	Square, 8" x 8", 1-1/4" thick		135	.119		9.30	4.99		14.29	18.30
0600	2" thick		130	.123		13	5.20		18.20	23
9000	Minimum labor/equipment charge	1 Bric	2	4	Job		185		185	300

32 14 Unit Paving

32 14 40 – Stone Paving

32 14 40.10 Stone Pavers

		Crew	Daily Output	Labor-Hours	Unit	Material	2015 Bare Costs Labor	Equipment	Total	Total Incl O&P
0010	**STONE PAVERS**									
1300	Slate, natural cleft, irregular, 3/4" thick	D-1	92	.174	S.F.	9.20	7.30		16.50	22
1350	Random rectangular, gauged, 1/2" thick		105	.152		19.85	6.40		26.25	32.50
1400	Random rectangular, butt joint, gauged, 1/4" thick		150	.107		21.50	4.49		25.99	31
1450	For sand rubbed finish, add					9.30			9.30	10.20
1550	Granite blocks, 3-1/2" x 3-1/2" x 3-1/2"	D-1	92	.174		12	7.30		19.30	25
1600	4" to 12" long, 3" to 5" wide, 3" to 5" thick	"	98	.163		10	6.85		16.85	22.50

32 16 Curbs, Gutters, Sidewalks, and Driveways

32 16 13 – Curbs and Gutters

32 16 13.13 Cast-in-Place Concrete Curbs and Gutters

		Crew	Daily Output	Labor-Hours	Unit	Material	Labor	Equipment	Total	Total Incl O&P
0010	**CAST-IN-PLACE CONCRETE CURBS AND GUTTERS**									
0290	Forms only, no concrete									
0300	Concrete, wood forms, 6" x 18", straight	C-2	500	.096	L.F.	2.99	4.39		7.38	10.50
0400	6" x 18", radius	"	200	.240		3.11	11		14.11	21.50
0404	Concrete, wood forms, 6" x 18", straight & concrete	C-2A	500	.096		5.85	4.36		10.21	13.50
0406	6" x 18", radius	"	200	.240		5.95	10.90		16.85	24.50
0415	Machine formed, 6" x 18", straight	B-69A	2000	.024		3.52	.99	.50	5.01	6.05
0416	6" x 18", radius	"	900	.053		3.55	2.20	1.11	6.86	8.70

32 16 13.23 Precast Concrete Curbs and Gutters

		Crew	Daily Output	Labor-Hours	Unit	Material	Labor	Equipment	Total	Total Incl O&P
0010	**PRECAST CONCRETE CURBS AND GUTTERS**									
0550	Precast, 6" x 18", straight	B-29	700	.080	L.F.	10.30	3.28	1.26	14.84	18.05
0600	6" x 18", radius	"	325	.172	"	10.80	7.05	2.71	20.56	26.50

32 16 13.33 Asphalt Curbs

		Crew	Daily Output	Labor-Hours	Unit	Material	Labor	Equipment	Total	Total Incl O&P
0010	**ASPHALT CURBS**									
0012	Curbs, asphaltic, machine formed, 8" wide, 6" high, 40 L.F./ton	B-27	1000	.032	L.F.	1.82	1.22	.30	3.34	4.33
0100	8" wide, 8" high, 30 L.F. per ton		900	.036		2.43	1.35	.33	4.11	5.25
0150	Asphaltic berm, 12" W, 3"-6" H, 35 L.F./ton, before pavement		700	.046		.04	1.74	.42	2.20	3.37
0200	12" W, 1-1/2" to 4" H, 60 L.F. per ton, laid with pavement	B-2	1050	.038		.02	1.45		1.47	2.40

32 16 13.43 Stone Curbs

		Crew	Daily Output	Labor-Hours	Unit	Material	Labor	Equipment	Total	Total Incl O&P
0010	**STONE CURBS**									
1000	Granite, split face, straight, 5" x 16"	D-13	275	.175	L.F.	11.65	7.85	1.72	21.22	27.50
1100	6" x 18"	"	250	.192		15.35	8.60	1.89	25.84	33
1300	Radius curbing, 6" x 18", over 10' radius	B-29	260	.215		18.75	8.85	3.39	30.99	38.50
1400	Corners, 2' radius	"	80	.700	Ea.	63	28.50	11.05	102.55	128
1600	Edging, 4-1/2" x 12", straight	d-13	300	.160	L.F.	5.85	7.15	1.58	14.58	19.80
1800	Curb inlets, (guttermouth) straight	B-29	41	1.366	Ea.	140	56	21.50	217.50	269
2000	Indian granite (belgian block)									
2100	Jumbo, 10-1/2" x 7-1/2" x 4", grey	D-1	150	.107	L.F.	6.15	4.49		10.64	14.05
2150	Pink		150	.107		7.75	4.49		12.24	15.80
2200	Regular, 9" x 4-1/2" x 4-1/2", grey		160	.100		4.67	4.21		8.88	12
2250	Pink		160	.100		6.15	4.21		10.36	13.60
2300	Cubes, 4" x 4" x 4", grey		175	.091		3.96	3.85		7.81	10.60
2350	Pink		175	.091		3.81	3.85		7.66	10.45
2400	6" x 6" x 6", pink		155	.103		12.60	4.35		16.95	21
2500	Alternate pricing method for indian granite									
2550	Jumbo, 10-1/2" x 7-1/2" x 4" (30 lb.), grey				Ton	350			350	385
2600	Pink					450			450	495
2650	Regular, 9" x 4-1/2" x 4-1/2" (20 lb.), grey					330			330	365
2700	Pink					430			430	475

32 16 Curbs, Gutters, Sidewalks, and Driveways

32 16 13 – Curbs and Gutters

32 16 13.43 Stone Curbs		Crew	Daily Output	Labor-Hours	Unit	Material	2015 Bare Costs Labor	Equipment	Total	Total Incl O&P
2750	Cubes, 4" x 4" x 4" (5 lb.), grey				Ton	480			480	530
2800	Pink					490			490	540
2850	6" x 6" x 6" (25 lb.), pink					490			490	540
2900	For pallets, add				↓	22			22	24

32 17 Paving Specialties

32 17 13 – Parking Bumpers

32 17 13.13 Metal Parking Bumpers

		Crew	Daily Output	Labor-Hours	Unit	Material	Labor	Equipment	Total	Incl O&P
0010	**METAL PARKING BUMPERS**									
0015	Bumper rails for garages, 12 Ga. rail, 6" wide, with steel									
0020	posts 12'-6" O.C., minimum	E-4	190	.168	L.F.	18.25	8.95	.77	27.97	37
0030	Average		165	.194		23	10.30	.88	34.18	44.50
0100	Maximum	↓	140	.229	↓	27.50	12.15	1.04	40.69	53
1300	Pipe bollards, conc. filled/paint, 8' L x 4' D hole, 6" diam.	B-6	20	1.200	Ea.	600	49.50	18.20	667.70	760
1400	8" diam.		15	1.600		685	66	24.50	775.50	885
1500	12" diam.	↓	12	2		965	82.50	30.50	1,078	1,225
8900	Security bollards, SS, lighted, hyd., incl. controls, group of 3	L-7	.06	509		48,400	23,100		71,500	91,000
8910	Group of 5	"	.04	682	↓	65,000	31,000		96,000	122,000
9000	Minimum labor/equipment charge	E-4	2	16	Job		850	73	923	1,600

32 17 13.16 Plastic Parking Bumpers

			Crew	Daily Output	Labor-Hours	Unit	Material	Labor	Equipment	Total	Incl O&P
0010	**PLASTIC PARKING BUMPERS**										
1600	Bollards, recycled plastic, 5" x 5" x 5' L	G	B-6	20	1.200	Ea.	98	49.50	18.20	165.70	208
1610	Wheel stops, recycled plastic, yellow, 4" x 6" x 6' L	G		20	1.200		70	49.50	18.20	137.70	177
1620	Speed bumps, recycled plastic, yellow, 3"H x 10" W x 6' L	G		20	1.200		144	49.50	18.20	211.70	258
1630	3"H x 10" W x 9' L	G	↓	20	1.200	↓	230	49.50	18.20	297.70	355

32 17 13.19 Precast Concrete Parking Bumpers

		Crew	Daily Output	Labor-Hours	Unit	Material	Labor	Equipment	Total	Incl O&P
0010	**PRECAST CONCRETE PARKING BUMPERS**									
1000	Wheel stops, precast concrete incl. dowels, 6" x 10" x 6'-0"	B-2	120	.333	Ea.	39.50	12.65		52.15	64.50
1100	8" x 13" x 6'-0"	"	120	.333		46	12.65		58.65	71.50
1540	Bollards, precast concrete, free standing, round, 15" Dia x 36" H	B-6	14	1.714		310	71	26	407	485
1550	16" Dia x 30" H		14	1.714		310	71	26	407	485
1560	18" Dia x 34" H		14	1.714		405	71	26	502	590
1570	18" Dia x 72" H		14	1.714		480	71	26	577	675
1572	26" Dia x 18" H		12	2		365	82.50	30.50	478	575
1574	26" Dia x 30" H		12	2		605	82.50	30.50	718	835
1578	26" Dia x 50" H		12	2		850	82.50	30.50	963	1,100
1580	Square, 22" x 32" H		12	2		520	82.50	30.50	633	740
1585	22" x 41" H		12	2		535	82.50	30.50	648	760
1590	Hexagon, 26" x 41" H	↓	10	2.400		550	99	36.50	685.50	805

32 17 13.26 Wood Parking Bumpers

		Crew	Daily Output	Labor-Hours	Unit	Material	Labor	Equipment	Total	Incl O&P
0010	**WOOD PARKING BUMPERS**									
0020	Parking barriers, timber w/saddles, treated type									
0100	4" x 4" for cars	B-2	520	.077	L.F.	2.92	2.92		5.84	8
0200	6" x 6" for trucks	"	520	.077	"	6.10	2.92		9.02	11.50

32 17 23 – Pavement Markings

32 17 23.13 Painted Pavement Markings

		Crew	Daily Output	Labor-Hours	Unit	Material	Labor	Equipment	Total	Incl O&P
0010	**PAINTED PAVEMENT MARKINGS**									
0020	Acrylic waterborne, white or yellow, 4" wide, less than 3000 L.F.	B-78	20000	.002	L.F.	.15	.09	.03	.27	.35
0200	6" wide, less than 3000 L.F.		11000	.004		.23	.17	.05	.45	.58
0500	8" wide, less than 3000 L.F.	↓	10000	.005	↓	.31	.18	.06	.55	.71

32 17 Paving Specialties

32 17 23 – Pavement Markings

32 17 23.13 Painted Pavement Markings

		Crew	Daily Output	Labor-Hours	Unit	Material	2015 Bare Costs Labor	Equipment	Total	Total Incl O&P
0600	12" wide, less than 3000 L.F.	B-78	4000	.012	L.F.	.46	.46	.15	1.07	1.42
0620	Arrows or gore lines		2300	.021	S.F.	.22	.80	.26	1.28	1.83
0640	Temporary paint, white or yellow, less than 3000 L.F.		15000	.003	L.F.	.08	.12	.04	.24	.33
0660	Removal	1 Clab	300	.027			1		1	1.64
0680	Temporary tape	2 Clab	1500	.011		.48	.40		.88	1.19
0710	Thermoplastic, white or yellow, 4" wide, less than 6000 L.F.	B-79	15000	.003		.31	.10	.10	.51	.62
0730	6" wide, less than 6000 L.F.		14000	.003		.47	.11	.11	.69	.81
0740	8" wide, less than 6000 L.F.		12000	.003		.62	.13	.12	.87	1.03
0750	12" wide, less than 6000 L.F.		6000	.007		.91	.26	.25	1.42	1.70
0760	Arrows		660	.061	S.F.	.61	2.32	2.24	5.17	6.90
0770	Gore lines		2500	.016		.61	.61	.59	1.81	2.32
0780	Letters		660	.061		.61	2.32	2.24	5.17	6.90

32 17 23.14 Pavement Parking Markings

		Crew	Daily Output	Labor-Hours	Unit	Material	Labor	Equipment	Total	Total Incl O&P
0010	**PAVEMENT PARKING MARKINGS**									
0790	Layout of pavement marking	A-2	25000	.001	L.F.		.04	.01	.05	.07
0800	Lines on pvmt., parking stall, paint, white, 4" wide	B-78B	400	.045	Stall	4.44	1.75	.94	7.13	8.75
0825	Parking stall, small quantities	2 Pord	80	.200		8.90	8.05		16.95	22.50
0830	Lines on pvmt., parking stall, thermoplastic, white, 4" wide	B-79	300	.133		13.50	5.10	4.92	23.52	28.50
1000	Street letters and numbers	B-78B	1600	.011	S.F.	.67	.44	.23	1.34	1.71

32 31 Fences and Gates

32 31 13 – Chain Link Fences and Gates

32 31 13.20 Fence, Chain Link Industrial

		Crew	Daily Output	Labor-Hours	Unit	Material	Labor	Equipment	Total	Total Incl O&P
0010	**FENCE, CHAIN LINK INDUSTRIAL**									
0011	Schedule 40, including concrete									
0020	3 strands barb wire, 2" post @ 10' O.C., set in concrete, 6' H									
0200	9 ga. wire, galv. steel, in concrete	B-80C	240	.100	L.F.	19.10	3.81	1.06	23.97	28.50
0248	Fence, add for vinyl coated fabric				S.F.	.66			.66	.73
0300	Aluminized steel	B-80C	240	.100	L.F.	20.50	3.81	1.06	25.37	30
0500	6 ga. wire, galv. steel		240	.100		21	3.81	1.06	25.87	30.50
0600	Aluminized steel		240	.100		30	3.81	1.06	34.87	40.50
0800	6 ga. wire, 6' high but omit barbed wire, galv. steel		250	.096		19.55	3.66	1.02	24.23	28.50
0900	Aluminized steel, in concrete		250	.096		23.50	3.66	1.02	28.18	33
0920	8' H, 6 ga. wire, 2-1/2" line post, galv. steel, in concrete		180	.133		31	5.10	1.41	37.51	44
0940	Aluminized steel, in concrete		180	.133		38	5.10	1.41	44.51	51.50
1100	Add for corner posts, 3" diam., galv. steel, in concrete		40	.600	Ea.	86.50	23	6.35	115.85	140
1200	Aluminized steel, in concrete		40	.600		86.50	23	6.35	115.85	140
1300	Add for braces, galv. steel		80	.300		35.50	11.45	3.17	50.12	61
1350	Aluminized steel		80	.300		46	11.45	3.17	60.62	72.50
1400	Gate for 6' high fence, 1-5/8" frame, 3' wide, galv. steel		10	2.400		203	91.50	25.50	320	400
1500	Aluminized steel, in concrete		10	2.400		205	91.50	25.50	322	400
2000	5'-0" high fence, 9 ga., no barbed wire, 2" line post, in concrete									
2010	10' O.C., 1-5/8" top rail, in concrete									
2100	Galvanized steel, in concrete	B-80C	300	.080	L.F.	18.45	3.05	.85	22.35	26.50
2200	Aluminized steel, in concrete		300	.080	"	18.75	3.05	.85	22.65	26.50
2400	Gate, 4' wide, 5' high, 2" frame, galv. steel, in concrete		10	2.400	Ea.	188	91.50	25.50	305	385
2500	Aluminized steel, in concrete		10	2.400	"	200	91.50	25.50	317	395
3100	Overhead slide gate, chain link, 6' high, to 18' wide, in concrete		38	.632	L.F.	97	24	6.70	127.70	153
3110	Cantilever type, in concrete	B-80	48	.667		129	27.50	15	171.50	203
3120	8' high, in concrete		24	1.333		155	55	30	240	293
3130	10' high, in concrete		18	1.778		190	73.50	40	303.50	370

32 31 Fences and Gates

32 31 13 – Chain Link Fences and Gates

32 31 13.20 Fence, Chain Link Industrial		Crew	Daily Output	Labor-Hours	Unit	Material	2015 Bare Costs Labor	Equipment	Total	Total Incl O&P
9000	Minimum labor/equipment charge	B-80	2	16	Job		660	360	1,020	1,475

32 31 13.25 Fence, Chain Link Residential		Crew	Daily Output	Labor-Hours	Unit	Material	Labor	Equipment	Total	Total Incl O&P
0010	**FENCE, CHAIN LINK RESIDENTIAL**									
0011	Schedule 20, 11 ga. wire, 1-5/8" post									
3300	Residential, 11 ga. wire, 1-5/8" line post @ 10' O.C.									
3310	1-3/8" top rail									
3350	4' high, galvanized steel	B-80C	475	.051	L.F.	12.40	1.93	.53	14.86	17.35
3400	Aluminized		475	.051	"	15.75	1.93	.53	18.21	21
3600	Gate, 3' wide, 1-3/8" frame, galv. steel		10	2.400	Ea.	100	91.50	25.50	217	287
3700	Aluminized		10	2.400	"	108	91.50	25.50	225	296
3900	3' high, galvanized steel		620	.039	L.F.	91.50	1.47	.41	93.38	104
4000	Aluminized		620	.039	"	94.50	1.47	.41	96.38	107
4200	Gate, 3' wide, 1-3/8" frame, galv. steel		12	2	Ea.	96.50	76	21	193.50	254
4300	Aluminized		12	2	"	110	76	21	207	269
9000	Minimum labor/equipment charge	B-1	2	12	Job		460		460	755

32 31 13.26 Tennis Court Fences and Gates		Crew	Daily Output	Labor-Hours	Unit	Material	Labor	Equipment	Total	Total Incl O&P
0010	**TENNIS COURT FENCES AND GATES**									
3150	Tennis courts, 11 ga. wire, 1-3/4" mesh, 2-1/2" line posts									
3170	1-5/8" top rail, 3" corner and gate posts									
3190	10' high, galvanized steel	B-80	190	.168	L.F.	26	6.95	3.79	36.74	44
3210	Vinyl covered 9 ga. wire		190	.168	"	32.50	6.95	3.79	43.24	51.50
3240	Corner posts for above, 3" diameter, 10' high		30	1.067	Ea.	84.50	44	24	152.50	191

32 31 13.40 Fence, Fabric and Accessories		Crew	Daily Output	Labor-Hours	Unit	Material	Labor	Equipment	Total	Total Incl O&P
0010	**FENCE, FABRIC & ACCESSORIES**									
8000	Components only, fabric, galvanized, 4' high	B-80C	585	.041	L.F.	3.44	1.56	.43	5.43	6.80
8040	6' high		430	.056	"	5.20	2.13	.59	7.92	9.80
8200	Posts, line post, 4' high		34	.706	Ea.	12.70	27	7.45	47.15	66
8240	6' high		26	.923	"	15.95	35	9.75	60.70	86
8400	Top rails, 1-3/8" O.D.		1000	.024	L.F.	1.45	.91	.25	2.61	3.37
8440	1-5/8" O.D.		1000	.024	"	2.29	.91	.25	3.45	4.29

32 31 19 – Decorative Metal Fences and Gates

32 31 19.10 Decorative Fence		Crew	Daily Output	Labor-Hours	Unit	Material	Labor	Equipment	Total	Total Incl O&P
0010	**DECORATIVE FENCE**									
5300	Tubular picket, steel, 6' sections, 1-9/16" posts, 4' high	B-80C	300	.080	L.F.	31	3.05	.85	34.90	40
5400	2" posts, 5' high		240	.100		35	3.81	1.06	39.87	46
5600	2" posts, 6' high		200	.120		42	4.57	1.27	47.84	55.50
5700	Staggered picket 1-9/16" posts, 4' high		300	.080		31	3.05	.85	34.90	40
5800	2" posts, 5' high		240	.100		35	3.81	1.06	39.87	46
5900	2" posts, 6' high		200	.120		42	4.57	1.27	47.84	55.50
6200	Gates, 4' high, 3' wide	B-1	10	2.400	Ea.	282	92		374	460
6300	5' high, 3' wide		10	2.400		350	92		442	535
6400	6' high, 3' wide		10	2.400		415	92		507	605
6500	4' wide		10	2.400		420	92		512	615

32 31 23 – Plastic Fences and Gates

32 31 23.20 Fence, Recycled Plastic			Crew	Daily Output	Labor-Hours	Unit	Material	Labor	Equipment	Total	Total Incl O&P
0010	**FENCE, RECYCLED PLASTIC**										
9015	Fence rail, made from recycled plastic, various colors, 2 rail	G	B-1	150	.160	L.F.	10.10	6.10		16.20	21
9018	3 rail	G		150	.160		12.85	6.10		18.95	24
9020	4 rail	G		150	.160		15.60	6.10		21.70	27
9030	Fence pole, made from recycled plastic, various colors, 7'	G		96	.250	Ea.	49.50	9.55		59.05	70
9040	Stockade fence, made from recycled plastic, various colors, 4' high	G	B-80C	160	.150	L.F.	34.50	5.70	1.59	41.79	49

32 31 Fences and Gates

32 31 23 – Plastic Fences and Gates

32 31 23.20 Fence, Recycled Plastic

			Crew	Daily Output	Labor-Hours	Unit	Material	2015 Bare Costs Labor	Equipment	Total	Total Incl O&P
9050	6' high	G	B-80C	160	.150	L.F.	36.50	5.70	1.59	43.79	51.50
9060	6' pole	G		96	.250	Ea.	39	9.55	2.65	51.20	61.50
9070	9' pole	G		96	.250	"	44.50	9.55	2.65	56.70	67.50
9080	Picket fence, made from recycled plastic, various colors, 3' high	G		160	.150	L.F.	23	5.70	1.59	30.29	36.50
9090	4' high	G		160	.150	"	34	5.70	1.59	41.29	48
9100	3' high gate	G		8	3	Ea.	23	114	31.50	168.50	246
9110	4' high gate	G		8	3		29.50	114	31.50	175	254
9120	5' high pole	G		96	.250		37	9.55	2.65	49.20	59.50
9130	6' high pole	G	↓	96	.250		43	9.55	2.65	55.20	66
9140	Pole cap only	G					5.30			5.30	5.85
9150	Keeper pins only	G			↓		.42			.42	.46

32 31 26 – Wire Fences and Gates

32 31 26.10 Fences, Misc. Metal

		Crew	Daily Output	Labor-Hours	Unit	Material	Labor	Equipment	Total	Total Incl O&P
0010	**FENCES, MISC. METAL**									
0012	Chicken wire, posts @ 4', 1" mesh, 4' high	B-80C	410	.059	L.F.	3.30	2.23	.62	6.15	7.95
0100	2" mesh, 6' high		350	.069		3.82	2.61	.73	7.16	9.25
0200	Galv. steel, 12 ga., 2" x 4" mesh, posts 5' O.C., 3' high		300	.080		2.78	3.05	.85	6.68	8.95
0300	5' high		300	.080		3.38	3.05	.85	7.28	9.60
0400	14 ga., 1" x 2" mesh, 3' high		300	.080		3.42	3.05	.85	7.32	9.65
0500	5' high	↓	300	.080	↓	4.57	3.05	.85	8.47	10.95
1000	Kennel fencing, 1-1/2" mesh, 6' long, 3'-6" wide, 6'-2" high	2 Clab	4	4	Ea.	510	150		660	805
1050	12' long		4	4		720	150		870	1,025
1200	Top covers, 1-1/2" mesh, 6' long		15	1.067		136	40		176	216
1250	12' long	↓	12	1.333	↓	191	50		241	292
4500	Security fence, prison grade, set in concrete, 12' high	B-80	25	1.280	L.F.	61.50	53	29	143.50	185
4600	16' high	"	20	1.600	"	79	66	36	181	234

32 31 29 – Wood Fences and Gates

32 31 29.20 Fence, Wood Rail

		Crew	Daily Output	Labor-Hours	Unit	Material	Labor	Equipment	Total	Total Incl O&P
0010	**FENCE, WOOD RAIL**									
0012	Picket, No. 2 cedar, Gothic, 2 rail, 3' high	B-1	160	.150	L.F.	7.60	5.75		13.35	17.75
0050	Gate, 3'-6" wide	B-80C	9	2.667	Ea.	77	102	28	207	282
0600	Open rail, rustic, No. 1 cedar, 2 rail, 3' high		160	.150	L.F.	5.90	5.70	1.59	13.19	17.55
0650	Gate, 3' wide		9	2.667	Ea.	82	102	28	212	287
1200	Stockade, No. 2 cedar, treated wood rails, 6' high		160	.150	L.F.	8.90	5.70	1.59	16.19	21
1250	Gate, 3' wide		9	2.667	Ea.	96	102	28	226	305
3300	Board, shadow box, 1" x 6", treated pine, 6' high		160	.150	L.F.	12.45	5.70	1.59	19.74	24.50
3400	No. 1 cedar, 6' high		150	.160		24.50	6.10	1.69	32.29	39
3900	Basket weave, No. 1 cedar, 6' high		160	.150		34	5.70	1.59	41.29	48
3950	Gate, 3'-6" wide	B-1	8	3	Ea.	232	115		347	445
4000	Treated pine, 6' high		150	.160	L.F.	16	6.10		22.10	27.50
4200	Gate, 3'-6" wide		9	2.667	Ea.	176	102		278	360
8000	Posts only, 4' high		40	.600		11.20	23		34.20	50
8040	6' high	↓	30	.800	↓	17.60	30.50		48.10	69.50
9000	Minimum labor/equipment charge	1 Clab	2	4	Job		150		150	247

32 32 Retaining Walls

32 32 29 – Timber Retaining Walls

32 32 29.10 Landscape Timber Retaining Walls

	32 32 29.10 Landscape Timber Retaining Walls	Crew	Daily Output	Labor-Hours	Unit	Material	2015 Bare Costs Labor	Equipment	Total	Total Incl O&P
0010	**LANDSCAPE TIMBER RETAINING WALLS**									
0100	Treated timbers, 6" x 6"	1 Clab	265	.030	L.F.	2.01	1.14		3.15	4.07
0110	6" x 8"	"	200	.040	"	2.65	1.50		4.15	5.40
0120	Drilling holes in timbers for fastening, 1/2"	1 Carp	450	.018	Inch		.83		.83	1.37
0130	5/8"	"	450	.018	"		.83		.83	1.37
0140	Reinforcing rods for fastening, 1/2"	1 Clab	312	.026	L.F.	.36	.96		1.32	1.97
0150	5/8"	"	312	.026	"	.56	.96		1.52	2.19
0160	Reinforcing fabric	2 Clab	2500	.006	S.Y.	1.90	.24		2.14	2.48
0170	Gravel backfill		28	.571	C.Y.	20	21.50		41.50	57
0180	Perforated pipe, 4" diameter with silt sock		1200	.013	L.F.	1.25	.50		1.75	2.20
0190	Galvanized 60d common nails	1 Clab	625	.013	Ea.	.16	.48		.64	.97
0200	20d common nails	"	3800	.002	"	.04	.08		.12	.17

32 32 53 – Stone Retaining Walls

32 32 53.10 Retaining Walls, Stone

	32 32 53.10 Retaining Walls, Stone	Crew	Daily Output	Labor-Hours	Unit	Material	Labor	Equipment	Total	Total Incl O&P
0010	**RETAINING WALLS, STONE**									
0015	Including excavation, concrete footing and									
0020	stone 3' below grade. Price is exposed face area.									
0200	Decorative random stone, to 6' high, 1'-6" thick, dry set	D-1	35	.457	S.F.	58.50	19.25		77.75	96
0300	Mortar set		40	.400		60.50	16.85		77.35	94
0500	Cut stone, to 6' high, 1'-6" thick, dry set		35	.457		60.50	19.25		79.75	98.50
0600	Mortar set		40	.400		61.50	16.85		78.35	95
0800	Random stone, 6' to 10' high, 2' thick, dry set		45	.356		66.50	14.95		81.45	97.50
0900	Mortar set		50	.320		69	13.45		82.45	97.50
1100	Cut stone, 6' to 10' high, 2' thick, dry set		45	.356		67	14.95		81.95	98
1200	Mortar set		50	.320		69	13.45		82.45	98
9000	Minimum labor/equipment charge		2	8	Job		335		335	550

32 33 Site Furnishings

32 33 33 – Site Manufactured Planters

32 33 33.10 Planters

	32 33 33.10 Planters	Crew	Daily Output	Labor-Hours	Unit	Material	Labor	Equipment	Total	Total Incl O&P
0010	**PLANTERS**									
0012	Concrete, sandblasted, precast, 48" diameter, 24" high	2 Clab	15	1.067	Ea.	630	40		670	760
0100	Fluted, precast, 7' diameter, 36" high		10	1.600		1,575	60		1,635	1,825
0300	Fiberglass, circular, 36" diameter, 24" high		15	1.067		725	40		765	860
0400	60" diameter, 24" high		10	1.600		1,125	60		1,185	1,325
9000	Minimum labor/equipment charge	1 Clab	2	4	Job		150		150	247

32 33 43 – Site Seating and Tables

32 33 43.13 Site Seating

	32 33 43.13 Site Seating	Crew	Daily Output	Labor-Hours	Unit	Material	Labor	Equipment	Total	Total Incl O&P
0010	**SITE SEATING**									
0012	Seating, benches, park, precast conc., w/backs, wood rails, 4' long	2 Clab	5	3.200	Ea.	595	120		715	850
0100	8' long		4	4		955	150		1,105	1,300
0300	Fiberglass, without back, one piece, 4' long		10	1.600		620	60		680	780
0400	8' long		7	2.286		815	86		901	1,050
0500	Steel barstock pedestals w/backs, 2" x 3" wood rails, 4' long		10	1.600		1,125	60		1,185	1,325
0510	8' long		7	2.286		1,425	86		1,511	1,725
0515	Powder coated steel, 4" x 4" plastic slats, 6' L		8	2		490	75		565	660
0520	3" x 8" wood plank, 4' long		10	1.600		1,200	60		1,260	1,425
0530	8' long		7	2.286		1,500	86		1,586	1,800
0540	Backless, 4" x 4" wood plank, 4' square		10	1.600		940	60		1,000	1,125
0550	8' long		7	2.286		1,000	86		1,086	1,250

32 33 Site Furnishings

32 33 43 – Site Seating and Tables

32 33 43.13 Site Seating

		Crew	Daily Output	Labor-Hours	Unit	Material	2015 Bare Costs Labor	Equipment	Total	Total Incl O&P
0560	Powder coated steel, with back and 2 anti-vagrant dividers, 6' long	2 Clab	8	2	Ea.	1,075	75		1,150	1,325
0600	Aluminum pedestals, with backs, aluminum slats, 8' long		8	2		480	75		555	650
0610	15' long		5	3.200		975	120		1,095	1,275
0620	Portable, aluminum slats, 8' long		8	2		465	75		540	635
0630	15' long		5	3.200		570	120		690	825
0800	Cast iron pedestals, back & arms, wood slats, 4' long		8	2		385	75		460	545
0820	8' long		5	3.200		1,075	120		1,195	1,375
0840	Backless, wood slats, 4' long		8	2		590	75		665	770
0860	8' long		5	3.200		1,150	120		1,270	1,475
1700	Steel frame, fir seat, 10' long		10	1.600		355	60		415	495
2000	Benches, park, with back, galv. stl. frame, 4" x 4" plastic slats, 6' L		7	2.286		490	86		576	675
9000	Minimum labor/equipment charge		2	8	Job		300		300	495

32 84 Planting Irrigation

32 84 23 – Underground Sprinklers

32 84 23.10 Sprinkler Irrigation System

		Crew	Daily Output	Labor-Hours	Unit	Material	Labor	Equipment	Total	Total Incl O&P
0010	**SPRINKLER IRRIGATION SYSTEM**									
0011	For lawns									
0100	Golf course with fully automatic system	C-17	.05	1600	9 holes	100,000	78,500		178,500	237,500
0200	24' diam. head at 15' O.C. incl. piping, auto oper., minimum	B-20	70	.343	Head	27	14.40		41.40	53
0300	Maximum		40	.600		45	25		70	90.50
0600	Sprinkler irrigation sys, golf course, auto sys, 60' dia HD		23	1.043		150	44		194	237
0800	Residential system, custom, 1" supply		2000	.012	S.F.	.26	.50		.76	1.11
0900	1-1/2" supply		1800	.013	"	.49	.56		1.05	1.45

32 91 Planting Preparation

32 91 13 – Soil Preparation

32 91 13.23 Structural Soil Mixing

		Crew	Daily Output	Labor-Hours	Unit	Material	Labor	Equipment	Total	Total Incl O&P
0010	**STRUCTURAL SOIL MIXING**									
0100	Rake topsoil, site material, harley rock rake, ideal	B-6	33	.727	M.S.F.		30	11.05	41.05	60.50
0200	Adverse	"	7	3.429			142	52	194	286
0300	Screened loam, york rake and finish, ideal	B-62	24	1			41.50	7.25	48.75	75
0400	Adverse	"	20	1.200			49.50	8.70	58.20	89.50
1000	Remove topsoil & stock pile on site, 75 HP dozer, 6" deep, 50' haul	B-10L	30	.400			18.50	15.75	34.25	47
1050	300' haul		6.10	1.967			91	77.50	168.50	231
1100	12" deep, 50' haul		15.50	.774			36	30.50	66.50	91
1150	300' haul		3.10	3.871			179	152	331	455
1200	200 HP dozer, 6" deep, 50' haul	B-10B	125	.096			4.44	11.10	15.54	19.30
1250	300' haul		30.70	.391			18.10	45	63.10	78.50
1300	12" deep, 50' haul		62	.194			8.95	22.50	31.45	39
1350	300' haul		15.40	.779			36	90	126	157
1400	Alternate method, 75 HP dozer, 50' haul	B-10L	860	.014	C.Y.		.65	.55	1.20	1.63
1450	300' haul	"	114	.105			4.87	4.14	9.01	12.35
1500	200 HP dozer, 50' haul	B-10B	2660	.005			.21	.52	.73	.90
1600	300' haul	"	570	.021			.97	2.43	3.40	4.24
1800	Rolling topsoil, hand push roller	1 Clab	3200	.003	S.F.		.09		.09	.15
1850	Tractor drawn roller	B-66	10666	.001	"		.04	.02	.06	.09
2000	Root raking and loading, residential, no boulders	B-6	53.30	.450	M.S.F.		18.60	6.85	25.45	37.50
2100	With boulders		32	.750			31	11.40	42.40	62.50

32 91 Planting Preparation

32 91 13 – Soil Preparation

32 91 13.23 Structural Soil Mixing

		Crew	Daily Output	Labor-Hours	Unit	Material	2015 Bare Costs Labor	2015 Bare Costs Equipment	Total	Total Incl O&P
2200	Municipal, no boulders	B-6	200	.120	M.S.F.		4.95	1.82	6.77	10
2300	With boulders	"	120	.200			8.25	3.04	11.29	16.70
2400	Large commercial, no boulders	B-10B	400	.030			1.39	3.47	4.86	6.05
2500	With boulders	"	240	.050			2.31	5.80	8.11	10.05
3000	Scarify subsoil, residential, skid steer loader w/scarifiers, 50 HP	B-66	32	.250			12.15	8.25	20.40	28.50
3050	Municipal, skid steer loader w/scarifiers, 50 HP	"	120	.067			3.24	2.19	5.43	7.55
3100	Large commercial, 75 HP, dozer w/scarifier	B-10L	240	.050			2.31	1.97	4.28	5.85
3500	Screen topsoil from stockpile, vibrating screen, wet material (organic)	B-10P	200	.060	C.Y.		2.78	5.95	8.73	11
3550	Dry material	"	300	.040			1.85	3.96	5.81	7.30
3600	Mixing with conditioners, manure and peat	B-10R	550	.022			1.01	.54	1.55	2.22
3650	Mobilization add for 2 days or less operation	B-34K	3	2.667	Job		107	315	422	520
3800	Spread conditioned topsoil, 6" deep, by hand	B-1	360	.067	S.Y.	5.45	2.55		8	10.20
3850	300 HP dozer	B-10M	27	.444	M.S.F.	590	20.50	70.50	681	755
4000	Spread soil conditioners, alum. sulfate, 1#/S.Y., hand push spreader	1 Clab	17500	.001	S.Y.	20	.02		20.02	22
4050	Tractor spreader	B-66	700	.011	M.S.F.	2,225	.56	.38	2,225.94	2,450
4100	Fertilizer, 0.2#/S.Y., push spreader	1 Clab	17500	.001	S.Y.	.12	.02		.14	.16
4150	Tractor spreader	B-66	700	.011	M.S.F.	13.35	.56	.38	14.29	15.95
4200	Ground limestone, 1#/S.Y., push spreader	1 Clab	17500	.001	S.Y.	.11	.02		.13	.15
4250	Tractor spreader	B-66	700	.011	M.S.F.	12.20	.56	.38	13.14	14.75
4400	Manure, 18#/S.Y., push spreader	1 Clab	2500	.003	S.Y.	8.15	.12		8.27	9.15
4450	Tractor spreader	B-66	280	.029	M.S.F.	905	1.39	.94	907.33	1,000
4500	Perlite, 1" deep, push spreader	1 Clab	17500	.001	S.Y.	10.35	.02		10.37	11.45
4550	Tractor spreader	B-66	700	.011	M.S.F.	1,150	.56	.38	1,150.94	1,275
4600	Vermiculite, push spreader	1 Clab	17500	.001	S.Y.	8.45	.02		8.47	9.35
4650	Tractor spreader	B-66	700	.011	M.S.F.	940	.56	.38	940.94	1,025
5000	Spread topsoil, skid steer loader and hand dress	B-62	270	.089	C.Y.	24.50	3.67	.64	28.81	33.50
5100	Articulated loader and hand dress	B-100	320	.038		24.50	1.74	2.99	29.23	33
5200	Articulated loader and 75HP dozer	B-10M	500	.024		24.50	1.11	3.79	29.40	33
5300	Road grader and hand dress	B-11L	1000	.016		24.50	.71	.74	25.95	29
6000	Tilling topsoil, 20 HP tractor, disk harrow, 2" deep	B-66	450	.018	M.S.F.		.86	.59	1.45	2.01
6050	4" deep		360	.022			1.08	.73	1.81	2.51
6100	6" deep		270	.030			1.44	.97	2.41	3.36
6150	26" rototiller, 2" deep	A-1J	1250	.006	S.Y.		.24	.04	.28	.43
6200	4" deep		1000	.008			.30	.05	.35	.54
6250	6" deep		750	.011			.40	.06	.46	.73

32 91 13.26 Planting Beds

		Crew	Daily Output	Labor-Hours	Unit	Material	2015 Bare Costs Labor	2015 Bare Costs Equipment	Total	Total Incl O&P
0010	**PLANTING BEDS**									
0100	Backfill planting pit, by hand, on site topsoil	2 Clab	18	.889	C.Y.		33.50		33.50	55
0200	Prepared planting mix, by hand	"	24	.667			25		25	41
0300	Skid steer loader, on site topsoil	B-62	340	.071			2.91	.51	3.42	5.30
0400	Prepared planting mix	"	410	.059			2.42	.42	2.84	4.38
1000	Excavate planting pit, by hand, sandy soil	2 Clab	16	1			37.50		37.50	61.50
1100	Heavy soil or clay	"	8	2			75		75	123
1200	1/2 C.Y. backhoe, sandy soil	B-11C	150	.107			4.70	2.43	7.13	10.20
1300	Heavy soil or clay	"	115	.139			6.15	3.17	9.32	13.40
2000	Mix planting soil, incl. loam, manure, peat, by hand	2 Clab	60	.267		43	10.05		53.05	63.50
2100	Skid steer loader	B-62	150	.160		43	6.60	1.16	50.76	59
3000	Pile sod, skid steer loader	"	2800	.009	S.Y.		.35	.06	.41	.64
3100	By hand	2 Clab	400	.040			1.50		1.50	2.47
4000	Remove sod, F.E. loader	B-10S	2000	.006			.28	.19	.47	.65
4100	Sod cutter	B-12K	3200	.005			.22	.31	.53	.70
4200	By hand	2 Clab	240	.067			2.51		2.51	4.11

32 91 Planting Preparation

32 91 19 – Landscape Grading

32 91 19.13 Topsoil Placement and Grading		Crew	Daily Output	Labor-Hours	Unit	Material	2015 Bare Costs Labor	Equipment	Total	Total Incl O&P
0010	**TOPSOIL PLACEMENT AND GRADING**									
0700	Furnish and place, truck dumped, screened, 4" deep	B-10S	1300	.009	S.Y.	3.44	.43	.29	4.16	4.78
0800	6" deep		820	.015	"	4.40	.68	.46	5.54	6.45
0810	Minimum labor/equipment charge		2	6	Ea.		278	189	467	655
0900	Fine grading and seeding, incl. lime, fertilizer & seed,									
1000	With equipment	B-14	1000	.048	S.Y.	.46	1.91	.36	2.73	4.01
2000	Minimum labor/equipment charge	1 Clab	4	2	Job		75		75	123

32 92 Turf and Grasses

32 92 19 – Seeding

32 92 19.13 Mechanical Seeding

		Crew	Daily Output	Labor-Hours	Unit	Material	Labor	Equipment	Total	Total Incl O&P
0010	**MECHANICAL SEEDING**									
0020	Mechanical seeding, 215 lb./acre	B-66	1.50	5.333	Acre	560	259	175	994	1,225
0100	44 lb./M.S.Y.	"	2500	.003	S.Y.	.18	.16	.11	.45	.57
0101	$2.00/lb., 44 lb./M.S.Y.	1 Clab	13950	.001	S.F.	.02	.02		.04	.06
0300	Fine grading and seeding incl. lime, fertilizer & seed,									
0310	with equipment	B-14	1000	.048	S.Y.	.44	1.91	.36	2.71	3.99
0400	Fertilizer hand push spreader, 35 lb. per M.S.F.	1 Clab	200	.040	M.S.F.	9.80	1.50		11.30	13.25
0600	Limestone hand push spreader, 50 lb. per M.S.F.		180	.044		5.40	1.67		7.07	8.65
0800	Grass seed hand push spreader, 4.5 lb. per M.S.F.		180	.044		20	1.67		21.67	24.50
1000	Hydro or air seeding for large areas, incl. seed and fertilizer	B-81	8900	.003	S.Y.	.43	.12	.08	.63	.75
1100	With wood fiber mulch added	"	8900	.003	"	1.76	.12	.08	1.96	2.22
1300	Seed only, over 100 lb., field seed, minimum				Lb.	1.75			1.75	1.93
1400	Maximum					1.75			1.75	1.93
1500	Lawn seed, minimum					1.41			1.41	1.55
1600	Maximum					2.46			2.46	2.71
1800	Aerial operations, seeding only, field seed	B-58	50	.480	Acre	570	19.80	63	652.80	725
1900	Lawn seed		50	.480		460	19.80	63	542.80	605
2100	Seed and liquid fertilizer, field seed		50	.480		690	19.80	63	772.80	860
2200	Lawn seed		50	.480		580	19.80	63	662.80	735
9000	Minimum labor/equipment charge	1 Clab	4	2	Job		75		75	123

32 92 23 – Sodding

32 92 23.10 Sodding Systems

		Crew	Daily Output	Labor-Hours	Unit	Material	Labor	Equipment	Total	Total Incl O&P
0010	**SODDING SYSTEMS**									
0020	Sodding, 1" deep, bluegrass sod, on level ground, over 8 M.S.F.	B-63	22	1.818	M.S.F.	243	72.50	7.90	323.40	395
0200	4 M.S.F.		17	2.353		254	93.50	10.20	357.70	440
0300	1000 S.F.		13.50	2.963		295	118	12.85	425.85	530
0500	Sloped ground, over 8 M.S.F.		6	6.667		243	265	29	537	730
0600	4 M.S.F.		5	8		254	320	34.50	608.50	835
0700	1000 S.F.		4	10		295	400	43.50	738.50	1,025
1000	Bent grass sod, on level ground, over 6 M.S.F.		20	2		252	79.50	8.70	340.20	420
1100	3 M.S.F.		18	2.222		261	88.50	9.65	359.15	440
1200	Sodding 1000 S.F. or less		14	2.857		286	114	12.40	412.40	515
1500	Sloped ground, over 6 M.S.F.		15	2.667		252	106	11.55	369.55	465
1600	3 M.S.F.		13.50	2.963		261	118	12.85	391.85	495
1700	1000 S.F.		12	3.333		286	133	14.45	433.45	545

32 93 Plants

32 93 13 – Ground Covers

32 93 13.10 Ground Cover Plants	Crew	Daily Output	Labor-Hours	Unit	Material	2015 Bare Costs Labor	Equipment	Total	Total Incl O&P
0010 **GROUND COVER PLANTS**									
0012 Plants, pachysandra, in prepared beds	B-1	15	1.600	C	62.50	61		123.50	169
0200 Vinca minor, 1 yr., bare root, in prepared beds		12	2	"	111	76.50		187.50	247
0600 Stone chips, in 50 lb. bags, Georgia marble		520	.046	Bag	3.76	1.77		5.53	7.05
0700 Onyx gemstone		260	.092		16.90	3.53		20.43	24.50
0800 Quartz		260	.092		16.80	3.53		20.33	24.50
0900 Pea gravel, truckload lots		28	.857	Ton	25.50	33		58.50	82

32 93 33 – Shrubs

32 93 33.10 Shrubs and Trees

	Crew	Daily Output	Labor-Hours	Unit	Material	Labor	Equipment	Total	Total Incl O&P
0010 **SHRUBS AND TREES**									
0011 Evergreen, in prepared beds, B & B									
0100 Arborvitae pyramidal, 4'-5'	B-17	30	1.067	Ea.	100	43.50	26	169.50	210
0150 Globe, 12"-15"	B-1	96	.250		21.50	9.55		31.05	39
0300 Cedar, blue, 8'-10'	B-17	18	1.778		231	73	43	347	420
0500 Hemlock, Canadian, 2-1/2'-3'	B-1	36	.667		31	25.50		56.50	76.50
0550 Holly, Savannah, 8' - 10' H		9.68	2.479		257	95		352	440
0600 Juniper, andorra, 18"-24"		80	.300		36.50	11.50		48	59
0620 Wiltoni, 15"-18"		80	.300		26	11.50		37.50	47.50
0640 Skyrocket, 4-1/2'-5'	B-17	55	.582		107	24	14.10	145.10	171
0660 Blue pfitzer, 2'-2-1/2'	B-1	44	.545		38.50	21		59.50	76.50
0680 Ketleerie, 2-1/2'-3'		50	.480		52	18.35		70.35	87
0700 Pine, black, 2-1/2'-3'		50	.480		59.50	18.35		77.85	95
0720 Mugo, 18"-24"		60	.400		59	15.30		74.30	90
0740 White, 4'-5'	B-17	75	.427		51	17.50	10.35	78.85	96
0800 Spruce, blue, 18"-24"	B-1	60	.400		66.50	15.30		81.80	98.50
0840 Norway, 4'-5'	B-17	75	.427		83	17.50	10.35	110.85	131
0900 Yew, denisforma, 12"-15"	B-1	60	.400		35	15.30		50.30	63.50
1000 Capitata, 18"-24"		30	.800		32.50	30.50		63	86
1100 Hicksi, 2'-2-1/2'		30	.800		78.50	30.50		109	137

32 93 33.20 Shrubs

	Crew	Daily Output	Labor-Hours	Unit	Material	Labor	Equipment	Total	Total Incl O&P
0010 **SHRUBS**									
0011 Broadleaf Evergreen, planted in prepared beds									
0100 Andromeda, 15"-18", container	B-1	96	.250	Ea.	32.50	9.55		42.05	51.50
0200 Azalea, 15" - 18", container		96	.250		29.50	9.55		39.05	48
0300 Barberry, 9"-12", container		130	.185		17.60	7.05		24.65	31
0400 Boxwood, 15"-18", B&B		96	.250		43	9.55		52.55	62.50
0500 Euonymus, emerald gaiety, 12" to 15", container		115	.209		24	8		32	39
0600 Holly, 15"-18", B & B		96	.250		36	9.55		45.55	55.50
0900 Mount laurel, 18" - 24", B & B		80	.300		70.50	11.50		82	96.50
1000 Paxistema, 9 – 12" high		130	.185		21	7.05		28.05	35
1100 Rhododendron, 18"-24", container		48	.500		37.50	19.15		56.65	73
1200 Rosemary, 1 gal. container		600	.040		17.70	1.53		19.23	22
2000 Deciduous, planted in prepared beds, amelanchier, 2'-3', B & B		57	.421		119	16.10		135.10	158
2100 Azalea, 15"-18", B & B		96	.250		29	9.55		38.55	47.50
2300 Bayberry, 2'-3', B & B		57	.421		24	16.10		40.10	53
2600 Cotoneaster, 15"-18", B & B		80	.300		26.50	11.50		38	48
2800 Dogwood, 3'-4', B & B	B-17	40	.800		32	33	19.40	84.40	110
2900 Euonymus, alatus compacta, 15" to 18", container	B-1	80	.300		26	11.50		37.50	48
3200 Forsythia, 2'-3', container	"	60	.400		18.45	15.30		33.75	45.50
3300 Hibiscus, 3'-4', B & B	B-17	75	.427		48	17.50	10.35	75.85	93
3400 Honeysuckle, 3'-4', B & B	B-1	60	.400		26.50	15.30		41.80	54
3500 Hydrangea, 2'-3', B & B	"	57	.421		30	16.10		46.10	59.50

32 93 Plants

32 93 33 – Shrubs

32 93 33.20 Shrubs

		Crew	Daily Output	Labor-Hours	Unit	Material	2015 Bare Costs Labor	Equipment	Total	Total Incl O&P
3600	Lilac, 3'-4', B & B	B-17	40	.800	Ea.	28	33	19.40	80.40	105
3900	Privet, bare root, 18"-24"	B-1	80	.300		14.50	11.50		26	35
4100	Quince, 2'-3', B & B	"	57	.421		28	16.10		44.10	57.50
4200	Russian olive, 3'-4', B & B	B-17	75	.427		29	17.50	10.35	56.85	71.50
4400	Spirea, 3'-4', B & B	B-1	70	.343		20.50	13.10		33.60	44
4500	Viburnum, 3'-4', B & B	B-17	40	.800		25	33	19.40	77.40	102

32 93 43 – Trees

32 93 43.20 Trees

			Crew	Daily Output	Labor-Hours	Unit	Material	2015 Bare Costs Labor	Equipment	Total	Total Incl O&P
0010	**TREES**										
0011	Deciduous, in prep. beds, balled & burlapped (B&B)										
0100	Ash, 2" caliper	G	B-17	8	4	Ea.	186	164	97	447	575
0200	Beech, 5'-6'	G		50	.640		187	26	15.50	228.50	266
0300	Birch, 6'-8', 3 stems	G		20	1.600		168	65.50	39	272.50	335
0500	Crabapple, 6'-8'	G		20	1.600		139	65.50	39	243.50	300
0600	Dogwood, 4'-5'	G		40	.800		131	33	19.40	183.40	219
0700	Eastern redbud 4'-5'	G		40	.800		149	33	19.40	201.40	239
0800	Elm, 8'-10'	G		20	1.600		258	65.50	39	362.50	435
0900	Ginkgo, 6'-7'	G		24	1.333		148	54.50	32.50	235	287
1000	Hawthorn, 8'-10', 1" caliper	G		20	1.600		162	65.50	39	266.50	325
1100	Honeylocust, 10'-12', 1-1/2" caliper	G		10	3.200		206	131	77.50	414.50	525
1300	Larch, 8'	G		32	1		127	41	24	192	233
1400	Linden, 8'-10', 1" caliper	G		20	1.600		144	65.50	39	248.50	305
1500	Magnolia, 4'-5'	G		20	1.600		101	65.50	39	205.50	260
1600	Maple, red, 8'-10', 1-1/2" caliper	G		10	3.200		202	131	77.50	410.50	520
1700	Mountain ash, 8'-10', 1" caliper	G		16	2		176	82	48.50	306.50	380
1800	Oak, 2-1/2"-3" caliper	G		6	5.333		325	218	129	672	855
2100	Planetree, 9'-11', 1-1/4" caliper	G		10	3.200		236	131	77.50	444.50	560
2200	Plum, 6'-8', 1" caliper	G		20	1.600		82	65.50	39	186.50	239
2300	Poplar, 9'-11', 1-1/4" caliper	G		10	3.200		144	131	77.50	352.50	455
2500	Sumac, 2'-3'	G		75	.427		44	17.50	10.35	71.85	88.50
2700	Tulip, 5'-6'	G		40	.800		47	33	19.40	99.40	126
2800	Willow, 6'-8', 1" caliper	G		20	1.600		96	65.50	39	200.50	255
9000	Minimum labor/equipment charge		1 Clab	4	2	Job		75		75	123

32 94 Planting Accessories

32 94 13 – Landscape Edging

32 94 13.20 Edging

		Crew	Daily Output	Labor-Hours	Unit	Material	2015 Bare Costs Labor	Equipment	Total	Total Incl O&P
0010	**EDGING**									
0050	Aluminum alloy, including stakes, 1/8" x 4", mill finish	B-1	390	.062	L.F.	2.30	2.36		4.66	6.40
0051	Black paint		390	.062		2.67	2.36		5.03	6.80
0052	Black anodized		390	.062		3.08	2.36		5.44	7.25
0100	Brick, set horizontally, 1-1/2 bricks per L.F.	D-1	370	.043		1.22	1.82		3.04	4.31
0150	Set vertically, 3 bricks per L.F.	"	135	.119		3.68	4.99		8.67	12.15
0200	Corrugated aluminum, roll, 4" wide	1 Carp	650	.012		1.93	.58		2.51	3.07
0250	6" wide	"	550	.015		2.41	.68		3.09	3.77
0600	Railroad ties, 6" x 8"	2 Carp	170	.094		2.79	4.42		7.21	10.30
0650	7" x 9"		136	.118		3.10	5.50		8.60	12.45
0750	Redwood 2" x 4"		330	.048		2.25	2.28		4.53	6.20
0800	Steel edge strips, incl. stakes, 1/4" x 5"	B-1	390	.062		3.92	2.36		6.28	8.15
0850	3/16" x 4"	"	390	.062		3.10	2.36		5.46	7.25

32 94 Planting Accessories

32 94 13 – Landscape Edging

32 94 13.20 Edging

		Crew	Daily Output	Labor-Hours	Unit	Material	2015 Bare Costs Labor	Equipment	Total	Total Incl O&P
9000	Minimum labor/equipment charge	1 Carp	4	2	Job		94		94	154

32 94 50 – Tree Guying

32 94 50.10 Tree Guying Systems

		Crew	Daily Output	Labor-Hours	Unit	Material	Labor	Equipment	Total	Total Incl O&P
0010	**TREE GUYING SYSTEMS**									
0015	Tree guying Including stakes, guy wire and wrap									
0100	Less than 3" caliper, 2 stakes	2 Clab	35	.457	Ea.	14.25	17.20		31.45	43.50
0200	3" to 4" caliper, 3 stakes	"	21	.762	"	19.25	28.50		47.75	68
1000	Including arrowhead anchor, cable, turnbuckles and wrap									
1100	Less than 3" caliper, 3" anchors	2 Clab	20	.800	Ea.	44.50	30		74.50	98.50
1200	3" to 6" caliper, 4" anchors		15	1.067		38.50	40		78.50	108
1300	6" caliper, 6" anchors		12	1.333		44.50	50		94.50	131
1400	8" caliper, 8" anchors		9	1.778		129	67		196	252

32 96 Transplanting

32 96 23 – Plant and Bulb Transplanting

32 96 23.23 Planting

		Crew	Daily Output	Labor-Hours	Unit	Material	Labor	Equipment	Total	Total Incl O&P
0010	**PLANTING**									
0012	Moving shrubs on site, 12" ball	B-62	28	.857	Ea.		35.50	6.20	41.70	64.50
0100	24" ball	"	22	1.091	"		45	7.90	52.90	81.50

32 96 23.43 Moving Trees

		Crew	Daily Output	Labor-Hours	Unit	Material	Labor	Equipment	Total	Total Incl O&P
0010	**MOVING TREES**, On site									
0300	Moving trees on site, 36" ball	B-6	3.75	6.400	Ea.		264	97	361	535
0400	60" ball	"	1	24	"		990	365	1,355	2,000

32 96 43 – Tree Transplanting

32 96 43.20 Tree Removal

		Crew	Daily Output	Labor-Hours	Unit	Material	Labor	Equipment	Total	Total Incl O&P
0010	**TREE REMOVAL**									
0100	Dig & lace, shrubs, broadleaf evergreen, 18"-24" high	B-1	55	.436	Ea.		16.70		16.70	27.50
0200	2'-3'	"	35	.686			26		26	43
0300	3'-4'	B-6	30	.800			33	12.15	45.15	67
0400	4'-5'	"	20	1.200			49.50	18.20	67.70	100
1000	Deciduous, 12"-15"	B-1	110	.218			8.35		8.35	13.70
1100	18"-24"		65	.369			14.15		14.15	23
1200	2'-3'		55	.436			16.70		16.70	27.50
1300	3'-4'	B-6	50	.480			19.80	7.30	27.10	40
2000	Evergreen, 18"-24"	B-1	55	.436			16.70		16.70	27.50
2100	2'-0" to 2'-6"		50	.480			18.35		18.35	30
2200	2'-6" to 3'-0"		35	.686			26		26	43
2300	3'-0" to 3'-6"		20	1.200			46		46	75.50
3000	Trees, deciduous, small, 2'-3'		55	.436			16.70		16.70	27.50
3100	3'-4'	B-6	50	.480			19.80	7.30	27.10	40
3200	4'-5'		35	.686			28.50	10.40	38.90	57.50
3300	5'-6'		30	.800			33	12.15	45.15	67
4000	Shade, 5'-6'		50	.480			19.80	7.30	27.10	40
4100	6'-8'		35	.686			28.50	10.40	38.90	57.50
4200	8'-10'		25	.960			39.50	14.55	54.05	80
4300	2" caliper		12	2			82.50	30.50	113	168
5000	Evergreen, 4'-5'		35	.686			28.50	10.40	38.90	57.50

32 96 Transplanting

32 96 43 – Tree Transplanting

32 96 43.20 Tree Removal		Crew	Daily Output	Labor-Hours	Unit	Material	2015 Bare Costs Labor	2015 Bare Costs Equipment	Total	Total Incl O&P
5100	5'-6'	B-6	25	.960	Ea.		39.50	14.55	54.05	80
5200	6'-7'	↓	19	1.263			52	19.15	71.15	106
5300	7'-8'		15	1.600			66	24.50	90.50	134
5400	8'-10'	↓	11	2.182	↓		90	33	123	183

Estimating Tips
33 10 00 Water Utilities
33 30 00 Sanitary Sewerage Utilities
33 40 00 Storm Drainage Utilities

- Never assume that the water, sewer, and drainage lines will go in at the early stages of the project. Consider the site access needs before dividing the site in half with open trenches, loose pipe, and machinery obstructions. Always inspect the site to establish that the site drawings are complete. Check off all existing utilities on your drawings as you locate them. Be especially careful with underground utilities because appurtenances are sometimes buried during regrading or repaving operations. If you find any discrepancies, mark up the site plan for further research. Differing site conditions can be very costly if discovered later in the project.

- See also Section 33 01 00 for restoration of pipe where removal/replacement may be undesirable. Use of new types of piping materials can reduce the overall project cost. Owners/design engineers should consider the installing contractor as a valuable source of current information on utility products and local conditions that could lead to significant cost savings.

Reference Numbers
Reference numbers are shown in shaded boxes at the beginning of some major classifications. These numbers refer to related items in the Reference Section. The reference information may be an estimating procedure, an alternate pricing method, or technical information.

Note: Not all subdivisions listed here necessarily appear in this publication. ■

*Note: **Trade Service**, in part, has been used as a reference source for some of the material prices used in Division 33.*

Estimate with precision with RSMeans Online...

Quick, intuitive and easy-to-use, RSMeans Online gives you instant access to hundreds of thousands of material, labor and equipment costs from RSMeans' comprehensive database, delivering the information you need to build competitive estimates every time.

Start your free trial today at www.rsmeansonline.com

RSMeansOnline

33 01 Operation and Maintenance of Utilities

33 01 30 – Operation and Maintenance of Sewer Utilities

33 01 30.16 TV Inspection of Sewer Pipelines	Crew	Daily Output	Labor-Hours	Unit	Material	2015 Bare Costs Labor	Equipment	Total	Total Incl O&P
0010 **TV INSPECTION OF SEWER PIPELINES**									
0100 Pipe internal cleaning & inspection, cleaning, pressure pipe systems									
0120 Pig method, lengths 1000' to 10,000'									
0140 4" diameter thru 24" diameter, minimum				L.F.				3.60	4.14
0160 Maximum				"				18	21
6000 Sewage/sanitary systems									
6100 Power rodder with header & cutters									
6110 Mobilization charge, minimum				Total				695	800
6120 Mobilization charge, maximum				"				9,125	10,600
6140 Cleaning 4"-12" diameter				L.F.				3.39	3.90
6190 14"-24" diameter								3.98	4.59
6240 30" diameter								5.80	6.65
6250 36" diameter								6.75	7.75
6260 48" diameter								7.70	8.85
6270 60" diameter								8.70	9.95
6280 72" diameter								9.65	11.10
9000 Inspection, television camera with video									
9060 up to 500 linear feet				Total				715	820

33 01 30.72 Relining Sewers	Crew	Daily Output	Labor-Hours	Unit	Material	2015 Bare Costs Labor	Equipment	Total	Total Incl O&P
0010 **RELINING SEWERS**									
0011 With cement incl. bypass & cleaning									
0020 Less than 10,000 L.F., urban, 6" to 10"	C-17E	130	.615	L.F.	9.55	30	.74	40.29	60.50
0050 10" to 12"		125	.640		11.70	31.50	.77	43.97	65
0070 12" to 16"		115	.696		12.05	34	.84	46.89	69.50
0100 16" to 20"		95	.842		14.15	41.50	1.02	56.67	83.50
0200 24" to 36"		90	.889		15.25	43.50	1.08	59.83	89
0300 48" to 72"		80	1		24.50	49	1.21	74.71	107
0500 Rural, 6" to 10"		180	.444		9.55	22	.54	32.09	46.50
0550 10" to 12"		175	.457		11.70	22.50	.55	34.75	50
0570 12" to 16"		160	.500		12.05	24.50	.61	37.16	54
0600 16" to 20"		135	.593		14.15	29	.72	43.87	63.50
0700 24" to 36"		125	.640		15.25	31.50	.77	47.52	68.50
0800 48" to 72"		100	.800		24.50	39	.97	64.47	91.50
1000 Greater than 10,000 L.F., urban, 6" to 10"		160	.500		9.55	24.50	.61	34.66	51
1050 10" to 12"		155	.516		11.70	25.50	.62	37.82	54.50
1070 12" to 16"		140	.571		12.05	28	.69	40.74	59.50
1100 16" to 20"		120	.667		14.15	32.50	.81	47.46	69.50
1200 24" to 36"		115	.696		15.25	34	.84	50.09	73
1300 48" to 72"		95	.842		24.50	41.50	1.02	67.02	94.50
1500 Rural, 6" to 10"		215	.372		9.55	18.25	.45	28.25	40.50
1550 10" to 12"		210	.381		11.70	18.70	.46	30.86	44
1570 12" to 16"		185	.432		12.05	21	.52	33.57	48.50
1600 16" to 20"		150	.533		14.15	26	.65	40.80	59
1700 24" to 36"		140	.571		15.25	28	.69	43.94	63
1800 48" to 72"		120	.667		24.50	32.50	.81	57.81	80.50
2000 Cured in place pipe, non-pressure, flexible felt resin, 400' runs									
2100 6" diameter				L.F.				25.50	28
2200 8" diameter								26.50	29
2300 10" diameter								29	32
2400 12" diameter								34.50	38
2500 15" diameter								59	65
2600 18" diameter								82	90

33 01 Operation and Maintenance of Utilities

33 01 30 – Operation and Maintenance of Sewer Utilities

33 01 30.72 Relining Sewers		Crew	Daily Output	Labor-Hours	Unit	Material	2015 Bare Costs Labor	Equipment	Total	Total Incl O&P
2700	21" diameter				L.F.				105	115
2800	24" diameter								177	195
2900	30" diameter								195	215
3000	36" diameter								205	225
3100	48" diameter								218	240

33 05 Common Work Results for Utilities

33 05 16 – Utility Structures

33 05 16.13 Precast Concrete Utility Boxes

		Crew	Daily Output	Labor-Hours	Unit	Material	Labor	Equipment	Total	Total Incl O&P
0010	**PRECAST CONCRETE UTILITY BOXES**, 6" thick									
0050	5' x 10' x 6' high, I.D.	B-13	2	28	Ea.	3,750	1,150	370	5,270	6,375
0350	Hand hole, precast concrete, 1-1/2" thick									
0400	1'-0" x 2'-0" x 1'-9", I.D., light duty	B-1	4	6	Ea.	400	230		630	815
0450	4'-6" x 3'-2" x 2'-0", O.D., heavy duty	B-6	3	8	"	1,475	330	121	1,926	2,275

33 05 23 – Trenchless Utility Installation

33 05 23.19 Microtunneling

		Crew	Daily Output	Labor-Hours	Unit	Material	Labor	Equipment	Total	Total Incl O&P
0010	**MICROTUNNELING**									
0011	Not including excavation, backfill, shoring,									
0020	or dewatering, average 50'/day, slurry method									
0100	24" to 48" outside diameter, minimum				L.F.				875	965
0110	Adverse conditions, add				%				50%	50%
1000	Rent microtunneling machine, average monthly lease				Month				97,500	107,000
1010	Operating technician				Day				630	705
1100	Mobilization and demobilization, minimum				Job				41,200	45,900
1110	Maximum				"				445,500	490,500

33 05 23.20 Horizontal Boring

		Crew	Daily Output	Labor-Hours	Unit	Material	Labor	Equipment	Total	Total Incl O&P
0010	**HORIZONTAL BORING**									
0011	Casing only, 100' minimum,									
0020	not incl. jacking pits or dewatering									
0100	Roadwork, 1/2" thick wall, 24" diameter casing	B-42	20	3.200	L.F.	121	136	68	325	430
0200	36" diameter		16	4		223	170	85	478	620
0300	48" diameter		15	4.267		310	181	90.50	581.50	740
0500	Railroad work, 24" diameter		15	4.267		121	181	90.50	392.50	530
0600	36" diameter		14	4.571		223	194	97	514	670
0700	48" diameter		12	5.333		310	226	113	649	840
0900	For ledge, add								20%	20%
1000	Small diameter boring, 3", sandy soil	B-82	900	.018		22	.77	.09	22.86	25.50
1040	Rocky soil	"	500	.032		22	1.38	.17	23.55	26.50

33 05 26 – Utility Identification

33 05 26.05 Utility Connection

		Crew	Daily Output	Labor-Hours	Unit	Material	Labor	Equipment	Total	Total Incl O&P
0010	**UTILITY CONNECTION**									
0020	Water, sanitary, stormwater, gas, single connection	B-14	1	48	Ea.	3,225	1,900	365	5,490	7,050
0030	Telecommunication	"	1	48	"	395	1,900	365	2,660	3,925

33 05 26.10 Utility Accessories

		Crew	Daily Output	Labor-Hours	Unit	Material	Labor	Equipment	Total	Total Incl O&P
0010	**UTILITY ACCESSORIES** R312316-40									
0400	Underground tape, detectable, reinforced, alum. foil core, 2"	1 Clab	150	.053	C.L.F.	6.50	2.01		8.51	10.45
0500	6"		140	.057	"	27.50	2.15		29.65	33.50
9000	Minimum labor/equipment charge		4	2	Job		75		75	123

33 11 Water Utility Distribution Piping

33 11 13 – Public Water Utility Distribution Piping

33 11 13.15 Water Supply, Ductile Iron Pipe		Crew	Daily Output	Labor-Hours	Unit	Material	2015 Bare Costs Labor	Equipment	Total	Total Incl O&P
0010	**WATER SUPPLY, DUCTILE IRON PIPE**	R331113-80								
0020	Not including excavation or backfill									
2000	Pipe, class 50 water piping, 18' lengths									
2020	Mechanical joint, 4" diameter		B-21A	200	.200	L.F.	30.50	9.40	2.37	42.27	51
3000	Push-on joint, 4" diameter			400	.100		21	4.69	1.18	26.87	32
3020	6" diameter		↓	333.33	.120	↓	17	5.65	1.42	24.07	29
8000	Piping, fittings, mechanical joint, AWWA C110										
8006	90° bend, 4" diameter		B-20A	16	2	Ea.	155	91.50		246.50	315
8020	6" diameter			12.80	2.500		229	114		343	435
8200	Wye or tee, 4" diameter			10.67	2.999		355	137		492	610
8220	6" diameter			8.53	3.751		535	171		706	865
8398	45° bends, 4" diameter			16	2		255	91.50		346.50	425
8400	6" diameter			12.80	2.500		199	114		313	400
8405	8" diameter			10.67	2.999		291	137		428	540
8450	Decreaser, 6" x 4" diameter			14.22	2.250		207	103		310	390
8460	8" x 6" diameter		↓	11.64	2.749		310	126		436	540
8550	Piping, butterfly valves, cast iron										
8560	4" diameter		B-20	6	4	Ea.	470	168		638	790
9600	Steel sleeve with tap, 4" diameter			3	8		435	335		770	1,025
9620	6" diameter		↓	2	12	↓	475	505		980	1,350

33 12 Water Utility Distribution Equipment

33 12 13 – Water Service Connections

33 12 13.15 Tapping, Crosses and Sleeves

		Crew	Daily Output	Labor-Hours	Unit	Material	Labor	Equipment	Total	Total Incl O&P
0010	**TAPPING, CROSSES AND SLEEVES**									
4000	Drill and tap pressurized main (labor only)									
4100	6" main, 1" to 2" service	Q-1	3	5.333	Ea.		282		282	440
4150	8" main, 1" to 2" service	"	2.75	5.818	"		305		305	480
4500	Tap and insert gate valve									
4600	8" main, 4" branch	B-21	3.20	8.750	Ea.		380	43	423	665
4650	6" branch		2.70	10.370			450	51	501	785
4800	12" main, 6" branch	↓	2.35	11.915	↓		515	58.50	573.50	905

33 21 Water Supply Wells

33 21 13 – Public Water Supply Wells

33 21 13.10 Wells and Accessories

			Crew	Daily Output	Labor-Hours	Unit	Material	Labor	Equipment	Total	Total Incl O&P
0010	**WELLS & ACCESSORIES**	R221113-50									
0011	Domestic										
0100	Drilled, 4" to 6" diameter		B-23	120	.333	L.F.		12.65	23.50	36.15	47
0200	8" diameter		"	95.20	.420	"		15.95	30	45.95	59
1400	Remove & reset pump, minimum		B-21	4	7	Ea.		305	34.50	339.50	535
1420	Maximum		"	2	14	"		605	69	674	1,050
1500	Pumps, installed in wells to 100' deep, 4" submersible										
1510	1/2 H.P.		Q-1	3.22	4.969	Ea.	350	263		613	795
1520	3/4 H.P.			2.66	6.015		400	320		720	935
1600	1 H.P.		↓	2.29	6.987		425	370		795	1,050
1700	1-1/2 H.P.		Q-22	1.60	10		800	530	410	1,740	2,150
1800	2 H.P.			1.33	12.030		830	635	490	1,955	2,450
1900	3 H.P.			1.14	14.035		1,150	740	575	2,465	3,050

33 21 Water Supply Wells

33 21 13 – Public Water Supply Wells

33 21 13.10 Wells and Accessories	Crew	Daily Output	Labor-Hours	Unit	Material	2015 Bare Costs Labor	Equipment	Total	Total Incl O&P
2000	5 H.P.	Q-22	1.14	14.035	Ea.	1,600	740	575	2,915	3,525
2050	Remove and install motor only, 4 H.P.	▼	1.14	14.035		1,100	740	575	2,415	2,975
5000	Wells to 180 ft. deep, 4" submersible, 1 HP	B-21	1.10	25.455		2,350	1,100	125	3,575	4,525
5500	2 HP		1.10	25.455		3,275	1,100	125	4,500	5,550
6000	3 HP		1	28		4,375	1,225	138	5,738	6,950
7000	5 HP		.90	31.111	▼	6,025	1,350	153	7,528	9,000
9000	Minimum labor/equipment charge	▼	1.80	15.556	Job		675	76.50	751.50	1,175

33 31 Sanitary Utility Sewerage Piping

33 31 13 – Public Sanitary Utility Sewerage Piping

33 31 13.15 Sewage Collection, Concrete Pipe

0010	**SEWAGE COLLECTION, CONCRETE PIPE**									
0020	See Section 33 41 13.60 for sewage/drainage collection, concrete pipe									

33 31 13.25 Sewage Collection, Polyvinyl Chloride Pipe

0010	**SEWAGE COLLECTION, POLYVINYL CHLORIDE PIPE**									
0020	Not including excavation or backfill									
2000	20' lengths, SDR 35, B&S, 4" diameter	B-20	375	.064	L.F.	1.45	2.68		4.13	6
2040	6" diameter		350	.069		3.28	2.88		6.16	8.30
2080	13' lengths, SDR 35, B&S, 8" diameter	▼	335	.072		6.90	3.01		9.91	12.50
2120	10" diameter	B-21	330	.085		11.40	3.68	.42	15.50	18.95
4000	Piping, DWV PVC, no exc./bkfill., 10' L, Sch 40, 4" diameter	B-20	375	.064		3.84	2.68		6.52	8.60
4010	6" diameter		350	.069		7.65	2.88		10.53	13.10
4020	8" diameter	▼	335	.072	▼	12	3.01		15.01	18.10

33 36 Utility Septic Tanks

33 36 13 – Utility Septic Tank and Effluent Wet Wells

33 36 13.13 Concrete Utility Septic Tank

0010	**CONCRETE UTILITY SEPTIC TANK**									
0011	Not including excavation or piping									
0015	Septic tanks, precast, 1,000 gallon	B-21	8	3.500	Ea.	1,000	152	17.20	1,169.20	1,375
0020	1,250 gallon		8	3.500		1,050	152	17.20	1,219.20	1,425
0060	1,500 gallon		7	4		1,325	173	19.70	1,517.70	1,775
0100	2,000 gallon		5	5.600		1,925	243	27.50	2,195.50	2,525
0140	2,500 gallon		5	5.600		2,025	243	27.50	2,295.50	2,650
0180	4,000 gallon	▼	4	7		5,550	305	34.50	5,889.50	6,625
0220	5,000 gal., 4 piece	B-13	3	18.667		9,575	765	245	10,585	12,000
0300	15,000 gallon, 4 piece	B-13B	1.70	32.941		20,000	1,350	665	22,015	24,900
0400	25,000 gallon, 4 piece		1.10	50.909		38,800	2,075	1,025	41,900	47,200
0500	40,000 gallon, 4 piece	▼	.80	70		49,700	2,875	1,400	53,975	60,500
0520	50,000 gallon, 5 piece	B-13C	.60	93.333		57,000	3,825	2,775	63,600	72,500
0640	75,000 gallon, cast in place	C-14C	.25	448		69,500	20,200	125	89,825	109,500
0660	100,000 gallon	"	.15	746		86,000	33,700	209	119,909	149,500
1150	Leaching field chambers, 13' x 3'-7" x 1'-4", standard	B-13	16	3.500		485	143	46	674	815
1200	Heavy duty, 8' x 4' x 1'-6"		14	4		284	164	52.50	500.50	635
1300	13' x 3'-9" x 1'-6"		12	4.667		1,125	191	61.50	1,377.50	1,600
1350	20' x 4' x 1'-6"	▼	5	11.200		1,175	460	147	1,782	2,175
1420	Leaching pit, precast concrete, 6', dia, 3' deep	B-21	4.70	5.957		885	258	29.50	1,172.50	1,425
1600	Leaching pit, 6'-6" diameter, 6' deep		5	5.600		1,025	243	27.50	1,295.50	1,550
1620	8' deep	▼	4	7		1,200	305	34.50	1,539.50	1,850

33 36 Utility Septic Tanks

33 36 13 – Utility Septic Tank and Effluent Wet Wells

33 36 13.13 Concrete Utility Septic Tank

		Crew	Daily Output	Labor-Hours	Unit	Material	2015 Bare Costs Labor	Equipment	Total	Total Incl O&P
1700	8' diameter, H-20 load, 6' deep	B-21	4	7	Ea.	1,500	305	34.50	1,839.50	2,175
1720	8' deep		3	9.333		2,500	405	46	2,951	3,450
2000	Velocity reducing pit, precast conc., 6' diameter, 3' deep		4.70	5.957		1,600	258	29.50	1,887.50	2,200

33 36 13.19 Polyethylene Utility Septic Tank

		Crew	Daily Output	Labor-Hours	Unit	Material	Labor	Equipment	Total	Total Incl O&P
0010	**POLYETHYLENE UTILITY SEPTIC TANK**									
0015	High density polyethylene, 1,000 gallon	B-21	8	3.500	Ea.	1,550	152	17.20	1,719.20	1,975
0020	1,250 gallon		8	3.500		1,200	152	17.20	1,369.20	1,600
0025	1,500 gallon		7	4		1,475	173	19.70	1,667.70	1,925

33 36 19 – Utility Septic Tank Effluent Filter

33 36 19.13 Utility Septic Tank Effluent Tube Filter

		Crew	Daily Output	Labor-Hours	Unit	Material	Labor	Equipment	Total	Total Incl O&P
0010	**UTILITY SEPTIC TANK EFFLUENT TUBE FILTER**									
3000	Effluent filter, 4" diameter	1 Skwk	8	1	Ea.	42	48.50		90.50	125
3020	6" diameter		7	1.143		249	55.50		304.50	365
3030	8" diameter		7	1.143		225	55.50		280.50	340
3040	8" diameter, very fine		7	1.143		460	55.50		515.50	595
3050	10" diameter, very fine		6	1.333		235	65		300	365
3060	10" diameter		6	1.333		280	65		345	415
3080	12" diameter		6	1.333		670	65		735	845
3090	15" diameter		5	1.600		1,150	78		1,228	1,375

33 36 33 – Utility Septic Tank Drainage Field

33 36 33.13 Utility Septic Tank Tile Drainage Field

		Crew	Daily Output	Labor-Hours	Unit	Material	Labor	Equipment	Total	Total Incl O&P
0010	**UTILITY SEPTIC TANK TILE DRAINAGE FIELD**									
0015	Distribution box, concrete, 5 outlets	2 Clab	20	.800	Ea.	80.50	30		110.50	138
0020	7 outlets		16	1		80.50	37.50		118	150
0025	9 outlets		8	2		490	75		565	665
0115	Distribution boxes, HDPE, 5 outlets		20	.800		64.50	30		94.50	121
0117	6 outlets		15	1.067		65	40		105	138
0118	7 outlets		15	1.067		65	40		105	138
0120	8 outlets		10	1.600		68	60		128	174
0240	Distribution boxes, Outlet Flow Leveler	1 Clab	50	.160		2.25	6		8.25	12.35
0300	Precast concrete, galley, 4' x 4' x 4'	B-21	16	1.750		250	76	8.60	334.60	405
0350	HDPE infiltration chamber 12" H X 15" W	2 Clab	300	.053	L.F.	7.30	2.01		9.31	11.35
0351	12" H X 15" W End Cap	1 Clab	32	.250	Ea.	19.75	9.40		29.15	37
0355	chamber 12" H X 22" W	2 Clab	300	.053	L.F.	6.70	2.01		8.71	10.65
0356	12" H X 22" W End Cap	1 Clab	32	.250	Ea.	16.85	9.40		26.25	34
0360	chamber 13" H X 34" W	2 Clab	300	.053	L.F.	15	2.01		17.01	19.80
0361	13" H X 34" W End Cap	1 Clab	32	.250	Ea.	53.50	9.40		62.90	74
0365	chamber 16" H X 34" W	2 Clab	300	.053	L.F.	10.70	2.01		12.71	15.10
0366	16" H X 34" W End Cap	1 Clab	32	.250	Ea.	8	9.40		17.40	24
0370	chamber 8" H X 16" W	2 Clab	300	.053	L.F.	9.65	2.01		11.66	13.90
0371	8" H X 16" W End Cap	1 Clab	32	.250	Ea.	10.80	9.40		20.20	27.50

33 36 50 – Drainage Field Systems

33 36 50.10 Drainage Field Excavation and Fill

		Crew	Daily Output	Labor-Hours	Unit	Material	Labor	Equipment	Total	Total Incl O&P
0010	**DRAINAGE FIELD EXCAVATION AND FILL**									
2200	Septic tank & drainage field excavation with 3/4 c.y backhoe	B-12F	145	.110	C.Y.		4.93	4.52	9.45	12.90
2400	4' trench for disposal field, 3/4 C.Y. backhoe	"	335	.048	L.F.		2.13	1.95	4.08	5.60
2600	Gravel fill, run of bank	B-6	150	.160	C.Y.	20	6.60	2.43	29.03	35.50
2800	Crushed stone, 3/4"	"	150	.160	"	36.50	6.60	2.43	45.53	53.50

33 41 Storm Utility Drainage Piping

33 41 13 – Public Storm Utility Drainage Piping

33 41 13.40 Piping, Storm Drainage, Corrugated Metal

		Crew	Daily Output	Labor-Hours	Unit	Material	2015 Bare Costs Labor	Equipment	Total	Total Incl O&P
0010	**PIPING, STORM DRAINAGE, CORRUGATED METAL** R221113-50									
0020	Not including excavation or backfill									
2000	Corrugated metal pipe, galvanized									
2020	Bituminous coated with paved invert, 20' lengths									
2040	8" diameter, 16 ga.	B-14	330	.145	L.F.	8.65	5.80	1.10	15.55	20
2080	12" diameter, 16 ga.	"	210	.229	"	11.05	9.10	1.73	21.88	29

33 41 13.60 Sewage/Drainage Collection, Concrete Pipe

		Crew	Daily Output	Labor-Hours	Unit	Material	Labor	Equipment	Total	Total Incl O&P
0010	**SEWAGE/DRAINAGE COLLECTION, CONCRETE PIPE**									
0020	Not including excavation or backfill									
1000	Non-reinforced pipe, extra strength, B&S or T&G joints									
1010	6" diameter	B-14	265.04	.181	L.F.	6	7.20	1.37	14.57	19.85
1020	8" diameter		224	.214		6.60	8.50	1.63	16.73	23
1040	12" diameter		200	.240		8.20	9.55	1.82	19.57	26.50
2000	Reinforced culvert, class 3, no gaskets									
2010	12" diameter	B-14	150	.320	L.F.	11	12.75	2.43	26.18	35.50
2020	15" diameter	"	150	.320	"	14	12.75	2.43	29.18	38.50

33 44 Storm Utility Water Drains

33 44 13 – Utility Area Drains

33 44 13.13 Catchbasins

		Crew	Daily Output	Labor-Hours	Unit	Material	Labor	Equipment	Total	Total Incl O&P
0010	**CATCHBASINS**									
0011	Not including footing & excavation									
1600	Frames & grates, C.I., 24" square, 500 lb.	B-6	7.80	3.077	Ea.	340	127	46.50	513.50	635
1700	26" D shape, 600 lb.	"	7	3.429	"	505	142	52	699	840
3320	Frames and covers, existing, raised for paving, 2", including									
3340	row of brick, concrete collar, up to 12" wide frame	B-6	18	1.333	Ea.	45	55	20	120	161

33 44 13.15 Remove and Replace Catch Basin Cover

		Crew	Daily Output	Labor-Hours	Unit	Material	Labor	Equipment	Total	Total Incl O&P
0010	**REMOVE AND REPLACE CATCH BASIN COVER**									
0023	Remove catch basin cover	1 Clab	80	.100	Ea.		3.76		3.76	6.15
0033	Replace catch basin cover	"	80	.100	"		3.76		3.76	6.15

33 49 Storm Drainage Structures

33 49 13 – Storm Drainage Manholes, Frames, and Covers

33 49 13.10 Storm Drainage Manholes, Frames and Covers

		Crew	Daily Output	Labor-Hours	Unit	Material	Labor	Equipment	Total	Total Incl O&P
0010	**STORM DRAINAGE MANHOLES, FRAMES & COVERS**									
0020	Excludes footing, excavation, backfill (See line items for frame & cover)									
0050	Brick, 4' inside diameter, 4' deep	D-1	1	16	Ea.	520	675		1,195	1,675
0100	6' deep		.70	22.857		740	960		1,700	2,400
0150	8' deep		.50	32		955	1,350		2,305	3,250
0200	For depths over 8', add		4	4	V.L.F.	83.50	168		251.50	365
1110	Precast, 4' I.D., 4' deep	B-22	4.10	7.317	Ea.	725	320	50.50	1,095.50	1,375
1120	6' deep		3	10		925	440	69	1,434	1,800
1130	8' deep		2	15		1,075	660	103	1,838	2,375
1140	For depths over 8', add		16	1.875	V.L.F.	127	82.50	12.90	222.40	288

33 51 Natural-Gas Distribution

33 51 13 – Natural-Gas Piping

33 51 13.10 Piping, Gas Service and Distribution, P.E.	Crew	Daily Output	Labor-Hours	Unit	Material	2015 Bare Costs Labor	2015 Bare Costs Equipment	Total	Total Incl O&P
0010 **PIPING, GAS SERVICE AND DISTRIBUTION, POLYETHYLENE**									
0020 Not including excavation or backfill									
1000 60 psi coils, compression coupling @ 100', 1/2" diameter, SDR 11	B-20A	608	.053	L.F.	.45	2.41		2.86	4.33
1010 1" diameter, SDR 11		544	.059		1.05	2.69		3.74	5.45
1040 1-1/4" diameter, SDR 11		544	.059		1.62	2.69		4.31	6.05
1100 2" diameter, SDR 11		488	.066		2.82	3		5.82	7.90
1160 3" diameter, SDR 11		408	.078		6.45	3.59		10.04	12.80
1500 60 PSI 40' joints with coupling, 3" diameter, SDR 11	B-21A	408	.098		10.05	4.60	1.16	15.81	19.65
1540 4" diameter, SDR 11		352	.114		14.25	5.35	1.35	20.95	25.50
1600 6" diameter, SDR 11		328	.122		33.50	5.70	1.44	40.64	47.50
1640 8" diameter, SDR 11		272	.147		51.50	6.90	1.74	60.14	69.50
9000 Minimum labor/equipment charge	B-20	2	12	Job		505		505	820

33 52 Liquid Fuel Distribution

33 52 16 – Gasoline Distribution

33 52 16.13 Gasoline Piping

	Crew	Daily Output	Labor-Hours	Unit	Material	2015 Bare Costs Labor	2015 Bare Costs Equipment	Total	Total Incl O&P
0010 **GASOLINE PIPING**									
0020 Primary containment pipe, fiberglass-reinforced									
0030 Plastic pipe 15' & 30' lengths									
0040 2" diameter	Q-6	425	.056	L.F.	5.95	3.15		9.10	11.40
0050 3" diameter		400	.060		10.35	3.35		13.70	16.60
0060 4" diameter		375	.064		13.65	3.57		17.22	20.50
0100 Fittings									
0110 Elbows, 90° & 45°, bell-ends, 2"	Q-6	24	1	Ea.	43.50	56		99.50	135
0120 3" diameter		22	1.091		55	61		116	156
0130 4" diameter		20	1.200		70	67		137	181
0200 Tees, bell ends, 2"		21	1.143		60.50	63.50		124	166
0210 3" diameter		18	1.333		64	74.50		138.50	187
0220 4" diameter		15	1.600		84	89		173	232
0230 Flanges bell ends, 2"		24	1		33.50	56		89.50	124
0240 3" diameter		22	1.091		39	61		100	138
0250 4" diameter		20	1.200		45	67		112	154
0260 Sleeve couplings, 2"		21	1.143		12.40	63.50		75.90	113
0270 3" diameter		18	1.333		17.80	74.50		92.30	136
0280 4" diameter		15	1.600		23	89		112	164
0290 Threaded adapters 2"		21	1.143		17.95	63.50		81.45	119
0300 3" diameter		18	1.333		34	74.50		108.50	154
0310 4" diameter		15	1.600		38	89		127	181
0320 Reducers, 2"		27	.889		27.50	49.50		77	108
0330 3" diameter		22	1.091		27.50	61		88.50	125
0340 4" diameter		20	1.200		37	67		104	145
1010 Gas station product line for secondary containment (double wall)									
1100 Fiberglass reinforced plastic pipe 25' lengths									
1120 Pipe, plain end, 3" diameter	Q-6	375	.064	L.F.	26.50	3.57		30.07	34.50
1130 4" diameter		350	.069		32	3.82		35.82	41.50
1140 5" diameter		325	.074		35.50	4.12		39.62	45.50
1150 6" diameter		300	.080		39	4.46		43.46	50
1200 Fittings									
1230 Elbows, 90° & 45°, 3" diameter	Q-6	18	1.333	Ea.	136	74.50		210.50	266
1240 4" diameter		16	1.500		167	83.50		250.50	315
1250 5" diameter		14	1.714		184	95.50		279.50	350

33 52 Liquid Fuel Distribution

33 52 16 – Gasoline Distribution

33 52 16.13 Gasoline Piping		Crew	Daily Output	Labor-Hours	Unit	Material	2015 Bare Costs Labor	Equipment	Total	Total Incl O&P
1260	6" diameter	Q-6	12	2	Ea.	204	112		316	400
1270	Tees, 3" diameter		15	1.600		166	89		255	320
1280	4" diameter		12	2		202	112		314	395
1290	5" diameter		9	2.667		315	149		464	580
1300	6" diameter		6	4		375	223		598	765
1310	Couplings, 3" diameter		18	1.333		55	74.50		129.50	177
1320	4" diameter		16	1.500		119	83.50		202.50	261
1330	5" diameter		14	1.714		214	95.50		309.50	385
1340	6" diameter		12	2		315	112		427	525
1350	Cross-over nipples, 3" diameter		18	1.333		10.60	74.50		85.10	128
1360	4" diameter		16	1.500		12.70	83.50		96.20	144
1370	5" diameter		14	1.714		15.90	95.50		111.40	167
1380	6" diameter		12	2		19.05	112		131.05	195
1400	Telescoping, reducers, concentric 4" x 3"		18	1.333		48	74.50		122.50	169
1410	5" x 4"		17	1.412		95.50	78.50		174	228
1420	6" x 5"		16	1.500		234	83.50		317.50	385

33 71 Electrical Utility Transmission and Distribution

33 71 19 – Electrical Underground Ducts and Manholes

33 71 19.17 Electric and Telephone Underground

		Crew	Daily Output	Labor-Hours	Unit	Material	2015 Bare Costs Labor	Equipment	Total	Total Incl O&P
0010	**ELECTRIC AND TELEPHONE UNDERGROUND**									
0011	Not including excavation									
0200	backfill and cast in place concrete									
4200	Underground duct, banks ready for concrete fill, min. of 7.5"									
4400	between conduits, center to center									
4600	2 @ 2" diameter	2 Elec	240	.067	L.F.	1.41	3.65		5.06	7.20
4800	4 @ 2" diameter		120	.133		2.82	7.30		10.12	14.40
5600	4 @ 4" diameter		80	.200		6.10	10.95		17.05	23.50
6200	Rigid galvanized steel, 2 @ 2" diameter		180	.089		12.90	4.86		17.76	22
6400	4 @ 2" diameter		90	.178		26	9.70		35.70	43.50
7400	4 @ 4" diameter		34	.471		83	25.50		108.50	131
9990	Minimum labor/equipment charge	1 Elec	3.50	2.286	Job		125		125	194

Division Notes

		CREW	DAILY OUTPUT	LABOR-HOURS	UNIT	BARE COSTS				TOTAL INCL O&P
						MAT.	LABOR	EQUIP.	TOTAL	

Estimating Tips
34 11 00 Rail Tracks
This subdivision includes items that may involve either repair of existing, or construction of new, railroad tracks. Additional preparation work, such as the roadbed earthwork, would be found in Division 31. Additional new construction siding and turnouts are found in Subdivision 34 72. Maintenance of railroads is found under 34 01 23 Operation and Maintenance of Railways.

34 40 00 Traffic Signals
This subdivision includes traffic signal systems. Other traffic control devices such as traffic signs are found in Subdivision 10 14 53 Traffic Signage.

34 70 00 Vehicle Barriers
This subdivision includes security vehicle barriers, guide and guard rails, crash barriers, and delineators. The actual maintenance and construction of concrete and asphalt pavement is found in Division 32.

Reference Numbers
Reference numbers are shown in shaded boxes at the beginning of some major classifications. These numbers refer to related items in the Reference Section. The reference information may be an estimating procedure, an alternate pricing method, or technical information.

Note: Not all subdivisions listed here necessarily appear in this publication. ■

Did you know?
RSMeans Online gives you the same access to RSMeans' data with 24/7 access:
- Quickly locate costs in the searchable database.
- Build cost lists, estimates, and reports in minutes.
- Adjust costs to any location in the U.S. and Canada with the click of a button.

Start your free trial today at **www.rsmeansonline.com**

RSMeansOnline

34 71 Roadway Construction

34 71 13 – Vehicle Barriers

34 71 13.26 Vehicle Guide Rails		Crew	Daily Output	Labor-Hours	Unit	Material	2015 Bare Costs Labor	Equipment	Total	Total Incl O&P
0010	**VEHICLE GUIDE RAILS**									
0012	Corrugated stl., galv. stl. posts, 6'-3" O.C.	B-80	850	.038	L.F.	23	1.55	.85	25.40	28.50
0200	End sections, galvanized, flared		50	.640	Ea.	98.50	26.50	14.40	139.40	166
0300	Wrap around end		50	.640	"	141	26.50	14.40	181.90	213
0400	Timber guide rail, 4" x 8" with 6" x 8" wood posts, treated	↓	960	.033	L.F.	12.05	1.37	.75	14.17	16.30

Estimating Tips

Products such as conveyors, material handling cranes and hoists, as well as other items specified in this division, require trained installers. The general contractor may not have any choice as to who will perform the installation or when it will be performed. Long lead times are often required for these products, making early decisions in purchasing and scheduling necessary. The installation of this type of equipment may require the embedment of mounting hardware during construction of floors, structural walls, or interior walls/partitions. Electrical connections will require coordination with the electrical contractor.

Reference Numbers

Reference numbers are shown in shaded boxes at the beginning of some major classifications. These numbers refer to related items in the Reference Section. The reference information may be an estimating procedure, an alternate pricing method, or technical information.

Note: Not all subdivisions listed here necessarily appear in this publication. ■

Did you know?

RSMeans Online gives you the same access to RSMeans' data with 24/7 access:
- Quickly locate costs in the searchable database.
- Build cost lists, estimates, and reports in minutes.
- Adjust costs to any location in the U.S. and Canada with the click of a button.

Start your free trial today at **www.rsmeansonline.com**

RSMeansOnline

No part of this publication may be reproduced, stored in a retrieval system, or transmitted in any form or by any means without prior written permission of RSMeans.

Division 41 – Mat'l. Processing & Handling Equip.

41 22 Cranes and Hoists

41 22 13 – Cranes

41 22 13.10 Crane Rail

		Crew	Daily Output	Labor-Hours	Unit	Material	2015 Bare Costs Labor	2015 Bare Costs Equipment	Total	Total Incl O&P
0010	**CRANE RAIL**									
0020	Box beam bridge, no equipment included	E-4	3400	.009	Lb.	1.33	.50	.04	1.87	2.41
0210	Running track only, 104 lb. per yard, 20' piece	"	160	.200	L.F.	23	10.65	.91	34.56	45.50

41 22 13.13 Bridge Cranes

		Crew	Daily Output	Labor-Hours	Unit	Material	Labor	Equipment	Total	Total Incl O&P
0010	**BRIDGE CRANES**									
0100	1 girder, 20' span, 3 ton	M-3	1	34	Ea.	25,400	1,950	139	27,489	31,200
0125	5 ton		1	34		27,900	1,950	139	29,989	33,900
0150	7.5 ton		1	34		33,100	1,950	139	35,189	39,600
0175	10 ton		.80	42.500		43,900	2,450	173	46,523	52,500
0200	15 ton		.80	42.500		56,500	2,450	173	59,123	66,000
0225	30' span, 3 ton		1	34		26,500	1,950	139	28,589	32,300
0250	5 ton		1	34		29,100	1,950	139	31,189	35,200
0275	7.5 ton		1	34		34,800	1,950	139	36,889	41,500
0300	10 ton		.80	42.500		45,400	2,450	173	48,023	54,000
0325	15 ton		.80	42.500		59,000	2,450	173	61,623	69,000
0350	2 girder, 40' span, 3 ton	M-4	.50	72		43,600	4,100	350	48,050	54,500
0375	5 ton		.50	72		45,600	4,100	350	50,050	57,000
0400	7.5 ton		.50	72		50,000	4,100	350	54,450	62,000
0425	10 ton		.40	90		58,500	5,125	440	64,065	73,000
0450	15 ton		.40	90		80,000	5,125	440	85,565	96,000
0475	25 ton		.30	120		94,000	6,825	585	101,410	115,000
0500	50' span, 3 ton		.50	72		49,700	4,100	350	54,150	61,500
0525	5 ton		.50	72		51,500	4,100	350	55,950	64,000
0550	7.5 ton		.50	72		55,500	4,100	350	59,950	68,000
0575	10 ton		.40	90		64,000	5,125	440	69,565	78,500
0600	15 ton		.40	90		83,500	5,125	440	89,065	100,500
0625	25 ton		.30	120		98,500	6,825	585	105,910	120,000

41 22 13.19 Jib Cranes

		Crew	Daily Output	Labor-Hours	Unit	Material	Labor	Equipment	Total	Total Incl O&P
0010	**JIB CRANES**									
0020	Jib crane, wall cantilever, 500 lb. capacity, 8' span	2 Mill	1	16	Ea.	1,400	785		2,185	2,750
0040	12' span		1	16		1,525	785		2,310	2,900
0060	16' span		1	16		1,700	785		2,485	3,100
0080	20' span		1	16		2,150	785		2,935	3,600
0100	1000 lb. capacity, 8' span		1	16		1,500	785		2,285	2,875
0120	12' span		1	16		1,625	785		2,410	3,025
0130	16' span		1	16		2,150	785		2,935	3,600
0150	20' span		1	16		2,525	785		3,310	4,000

Assemblies Section

Table of Contents

Table No.	Page

A SUBSTRUCTURE

A1010 Standard Foundations
- A1010 120 Strip Footing 570
- A1010 230 Spread Footing 571
- A1010 310 Foundation Underdrain 572

A1030 Slab on Grade
- A1030 110 Interior Slab on Grade 573

A2010 Basement Excavation
- A2010 130 Excavation, Footings or Trench 574
- A2010 140 Excavation, Foundation 575

A2020 Basement Walls
- A2020 120 Concrete Wall 576
- A2020 140 Concrete Block Wall 577

B SHELL

B1010 Floor Construction
- B1010 225 Floor - Ceiling, Concrete Slab 580
- B1010 232 Floor - Ceiling, Conc. Panel 581
- B1010 233 Floor - Ceiling, Conc. Plank 582
- B1010 243 Floor - Ceiling, Struc. Steel 583
- B1010 245 Floor - Ceiling, Steel Joists 584
- B1010 262 Open Web Wood Joists 585
- B1010 263 Floor-Ceiling, Wood Joists 586

B1020 Roof Construction
- B1020 106 Steel Joist Roof & Ceiling 587
- B1020 202 Wood Frame Roof & Ceiling 588
- B1020 210 Roof Truss, Wood 589

B2010 Exterior Walls
- B2010 108 Masonry Wall, Concrete Block 591
- B2010 131 Masonry Wall, Brick - Stone 592
- B2010 142 Parapet Wall 593
- B2010 143 Masonry Restoration - Cleaning 594
- B2010 150 Wood Frame Exterior Wall 595
- B2010 190 Selective Price Sheet 596

B2020 Exterior Windows
- B2020 108 Vinyl Clad Windows 598
- B2020 110 Windows - Aluminum 599
- B2020 112 Windows - Wood 600
- B2020 116 Storm Windows & Doors 601
- B2020 124 Aluminum Frame, Window Wall 602

B2030 Exterior Doors
- B2030 125 Doors, Metal - Commercial 603
- B2030 150 Doors, Sliding - Patio 604
- B2030 240 Doors, Residential - Exterior 605
- B2030 410 Doors, Overhead 606
- B2030 810 Selective Price Sheet 607

B3010 Roof Coverings
- B3010 160 Selective Price Sheet 609

B3020 Roof Openings
- B3020 230 Roof Hatches, Skylights 611
- B3020 240 Selective Price Sheet 612

C INTERIORS

C1010 Partitions
- C1010 106 Partitions, Concrete Block 616
- C1010 132 Partitions, Wood Stud 617
- C1010 134 Partitions, Metal Stud, NLB 618
- C1010 136 Partitions, Drywall 619
- C1010 138 Partitions, Metal Stud, LB 620
- C1010 148 Partitions, Plaster & Lath 621
- C1010 170 Selective Price Sheet 622
- C1010 180 Selective Price Sheet, Drywall & Plaster 624
- C1010 210 Partitions, Movable Office 625

C1020 Interior Doors
- C1020 106 Doors, Interior Flush, Wood 626
- C1020 108 Doors, Interior Solid & Louvered 627
- C1020 110 Doors, Interior Flush, Metal 628
- C1020 112 Doors, Closet 629

C2010 Stair Construction
- C2010 130 Stairs 630

C3010 Wall Finishes
- C3010 210 Selective Price Sheet 631

C3020 Floor Finishes
- C3020 430 Selective Price Sheet 633

C3030 Ceiling Finishes
- C3030 210 Ceiling, Suspended Acoustical 635
- C3030 220 Ceilings, Suspended Gypsum Board 636
- C3030 230 Ceiling, Suspended Plaster 637
- C3030 250 Selective Price Sheet 638

D SERVICES

D1010 Elevators and Lifts
- D1010 130 Oil Hydraulic Elevators 642

D2010 Plumbing Fixtures
- D2010 956 Plumbing - Public Restroom 643
- D2010 957 Plumbing - Public Restroom 645
- D2010 958 Plumbing - Two Fixture Bathroom 647
- D2010 959 Plumbing - Three Fixture Bathroom 648
- D2010 960 Plumbing - Three Fixture Bathroom 649
- D2010 961 Plumbing - Three Fixture Bathroom 650
- D2010 962 Plumbing - Four Fixture Bathroom 651
- D2010 963 Plumbing - Five Fixture Bathroom 652

Table of Contents (cont.)

Table No.	Page
D3010 Energy Supply	
D3010 540 Heating - Oil Fired Hot Water	654
D3010 550 Heating - Gas Fired Hot Water	655
D3010 991 Wood Burning Stoves	656
D3010 992 Masonry Fireplace	657
D3020 Heat Generating Systems	
D3020 122 Heating-Cooling, Oil, Forced Air	658
D3020 124 Heating-Cooling, Gas, Forced Air	659
D3020 130 Boiler, Cast Iron, Hot Water, Gas	660
D3020 134 Boiler, Cast Iron, Steam, Gas	661
D3020 136 Boiler, Cast Iron, Hot Water, Gas/Oil	661
D3020 138 Boiler, Cast Iron, Steam, Gas/Oil	661
D3020 330 Circulating Pump Systems, End Suction	662
D3040 Distribution Systems	
D3040 118 Fan Coil A/C Unit, Two Pipe	664
D3040 120 Fan Coil A/C Unit, Two Pipe, Electric Heat	664
D3040 122 Fan Coil A/C Unit, Four Pipe	665
D3040 124 Fan Coil A/C, Horizontal, Duct Mount, 2 Pipe	665
D3040 126 Fan Coil A/C, Horiz., Duct Mount, 2 Pipe, Elec. Ht.	665
D3040 610 Heat Exchanger, Plate Type	666
D3040 620 Heat Exchanger, Shell & Tube	666
D3050 Terminal & Package Units	
D3050 120 Unit Heaters, Gas	668
D3050 130 Unit Heaters, Hydronic	668
D3050 140 Cabinet Unit Heaters, Hydronic	669
D3050 175 Rooftop Air Conditioner, Const. Volume	670
D3050 180 Rooftop Air Conditioner, Variable Air Volume	670
D3050 210 AC Unit, Package, Elec. Ht., Water Cooled	672
D3050 215 AC Unit, Package, Elec. Ht., Water Cooled, VAV	672
D3050 220 AC Unit, Package, Hot Water Coil, Water Cooled	673
D3050 225 AC Unit, Package, HW Coil, Water Cooled, VAV	673
D3050 255 Thru-Wall A/C Unit	674
D3050 260 Thru-Wall Heat Pump	674
D4010 Sprinklers	
D4010 320 Fire Sprinkler Systems, Dry	675
D4010 420 Fire Sprinkler Systems, Wet	677
D5010 Electrical Service/Distribution	
D5010 210 Commercial Service - 3 Phase	678
D5010 220 Residential Service - Single Phase	679

Table No.	Page
D5020 Lighting and Branch Wiring	
D5020 180 Selective Price Sheet	680
D5020 248 Lighting, Fluorescent	682
D5020 250 Lighting, Incandescent	683
D5020 252 Lighting, High Intensity	684
D5090 Other Electrical Systems	
D5090 530 Heat, Baseboard	685
E EQUIPMENT & FURNISHINGS	
E1090 Other Equipment	
E1090 310 Kitchens	688
E1090 315 Selective Price Sheet	689
G BUILDING SITEWORK	
G1030 Site Earthwork	
G1030 810 Excavation, Utility Trench	692
G2010 Roadways	
G2010 240 Driveways	693
G2020 Parking Lots	
G2020 230 Parking Lots, Asphalt	694
G2020 240 Parking Lots, Concrete	695
G2030 Pedestrian Paving	
G2030 105 Sidewalks	696
G2050 Landscaping	
G2050 420 Landscaping - Lawn Establishment	697
G2050 510 Selective Price Sheet	698
G3020 Sanitary Sewer	
G3020 610 Septic Systems	700
G3060 Fuel Distribution	
G3060 310 Fiberglass Fuel Tank, Double Wall	701
G3060 320 Steel Fuel Tank, Double Wall	703
G3060 325 Steel Tank, Above Ground, Single Wall	704
G3060 330 Steel Tank, Above Ground, Double Wall	704
G4020 Site Lighting	
G4020 210 Light Pole (Installed)	705

How RSMeans Assemblies Data Works

Assemblies estimating provides a fast and reasonably accurate way to develop construction costs. An assembly is the grouping of individual work items, with appropriate quantities, to provide a cost for a major construction component in a convenient unit of measure.

An assemblies estimate is often used during early stages of design development to compare the cost impact of various design alternatives on total building cost.

Assemblies estimates are also used as an efficient tool to verify construction estimates.

Assemblies estimates do not require a completed design or detailed drawings. Instead, they are based on the general size of the structure and other known parameters of the project. The degree of accuracy of an assemblies estimate is generally within +/- 15%.

A10 Foundations

A1010 Standard Foundations

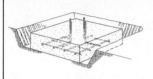

This page illustrates and describes a spread footing system including concrete, forms, reinforcing and anchor bolts. Lines within System Components give the unit price and total price on a cost each basis for this system. Prices for spread footing systems are on Line Items A1010 230 1200 thru 2300. Both material quantities and labor costs have been adjusted for the system listed.

Factors: To adjust for job conditions other than normal working situations use Lines A1010 230 2900 thru 4000.

Example: You are to install the system in an existing occupied building. Access to the site and protection to the building are mandatory. Go to Lines A1010 230 3600 and 3800 and apply these percentages to the appropriate MAT. and INST. costs.

RSMeans assemblies are identified by a **unique 12-character identifier**. The assemblies are numbered using UNIFORMAT II, ASTM Standard E1557. The first 6 characters represent this system to Level 4. The last 6 characters represent further breakdown by RSMeans in order to arrange items in understandable groups of similar tasks. Line numbers are consistent across all RSMeans publications, so a line number in any RSMeans assemblies data set will always refer to the same work.

System Components	QUANTITY	UNIT	COST EACH MAT.	COST EACH INST.	COST EACH TOTAL
Interior column footing, 3' square 1' thick, 2000 psi concrete, Including forms, reinforcing, and anchor bolts.					
Concrete, 2000 psi	.330	C.Y.	35.31		35.31
Placing concrete	.330	C.Y.		18.73	18.73
Forms, footing, 4 uses	12.000	SFCA	9.84	67.80	77.64
Reinforcing	11.000	Lb.	5.83	7.15	12.98
Anchor bolts, 3/4" diameter	2.000	Ea.	2.94	9.10	12.04
TOTAL			53.92	102.78	156.70

A1010 230	Spread Footing	MAT.	INST.	TOTAL
1200	3' square x 1' thick, 2000 psi concrete	54	102	156
1300	3000 psi concrete	55.50	102	157.50
1400	4000 psi concrete	57.50	102	159.50
1600	For alternate footing systems:			
1700	4' square x 1' thick, 2000 psi concrete	89.50	146	235.50
1800	3000 psi concrete	92.50	146	238.50
1900	4000 psi concrete	96	146	242
2100	5' square x 1'-3" thick, 2000 psi concrete	168	240	408
2200	3000 psi concrete	174	240	414
2300	4000 psi concrete	180	240	420
2600				
2700				
2900	Cut & patch to match existing construction, add, minimum	2%	3%	
3000	Maximum	5%	9%	
3100	Dust protection, add, minimum	1%	2%	
3200	Maximum	4%	11%	
3300	Equipment usage curtailment, add, minimum	1%	1%	
3400	Maximum	3%	10%	
3500	Material handling & storage limitation, add, minimum	1%	1%	
3600	Maximum	6%	7%	
3700	Protection of existing work, add, minimum	2%	2%	
3800	Maximum	5%	7%	
3900	Shift work requirements, add, minimum		5%	
4000	Maximum		30%	

RSMeans assemblies **Narrative descriptions** are shown in a hierarchical structure to make them readable. In order to read a complete description, read up through the indents to the top of the section. Include everything that is above and to the left that is not contradicted by information below.

For supplemental customizable square foot estimating forms, visit: http://www.reedconstructiondata.com/rsmeans/assemblies

Most assemblies consist of three major elements: a graphic, the system components, and the cost data itself. The **Graphic** is a visual representation showing the typical appearance of the assembly in question, frequently accompanied by additional explanatory technical information describing the class of items. The **System Components** is a listing of the individual tasks that make up the assembly, including the quantity and unit of measure for each item, along with the cost of material and installation. The **Assemblies Data** below lists prices for other similar systems with dimensional and/or size variations.

All RSMeans assemblies costs represent the cost for the installing contractor. An allowance for profit has been added to all material, labor, and equipment rental costs. A markup for labor burdens, including workers' compensation, fixed overhead, and business overhead, is included with installation costs.

The data published in RSMeans print books represents a "national average" cost. This data should be modified to the project location using the **City Cost Indexes** or **Location Factors** tables found in the Reference Section in the back of the book.

All RSMeans assemblies data includes a typical **Unit of Measure** used for estimating that item. For instance, while the unit of measure for the graphic shown here is Each, some A/C systems are estimated by the square foot (S.F.). The estimator needs to take special care that the unit in the data matches the unit in the takeoff. Abbreviations can be found in the Reference Section.

System components are listed separately to detail what is included in the development of the total system price.

A10 Foundations

A1010 Standard Foundations

This page illustrates and describes a spread footing system including concrete, forms, reinforcing and anchor bolts. Lines within System Components give the unit price and total price on a cost each basis for this system. Prices for spread footing systems are on Line Items A1010 230 1200 thru 2300. Both material quantities and labor costs have been adjusted for the system listed.

Factors: To adjust for job conditions other than normal working situations use Lines A1010 230 2900 thru 4000.

Example: You are to install the system in an existing occupied building. Access to the site and protection to the building are mandatory. Go to Lines A1010 230 3600 and 3800 and apply these percentages to the appropriate MAT. and INST. costs.

System Components	QUANTITY	UNIT	COST EACH		
			MAT.	INST.	TOTAL
Interior column footing, 3' square 1' thick, 2000 psi concrete, Including forms, reinforcing, and anchor bolts.					
Concrete, 2000 psi	.330	C.Y.	35.31		35.31
Placing concrete	.330	C.Y.		18.73	18.73
Forms, footing, 4 uses	12.000	SFCA	9.84	67.80	77.64
Reinforcing	11.000	Lb.	5.83	7.15	12.98
Anchor bolts, 3/4" diameter	2.000	Ea.	2.94	9.10	12.04
TOTAL			53.92	102.78	156.70

A1010 230	Spread Footing	COST EACH		
		MAT.	INST.	TOTAL
1200	3' square x 1' thick, 2000 psi concrete	54	102	156
1300	3000 psi concrete	55.50	102	157.50
1400	4000 psi concrete	57.50	102	159.50
1600	For alternate footing systems:			
1700	4' square x 1' thick, 2000 psi concrete	89.50	146	235.50
1800	3000 psi concrete	92.50	146	238.50
1900	4000 psi concrete	96	146	242
2100	5' square x 1'-3" thick, 2000 psi concrete	168	240	408
2200	3000 psi concrete	174	240	414
2300	4000 psi concrete	180	240	420
2600				
2700				
2900	Cut & patch to match existing construction, add, minimum	2%	3%	
3000	Maximum	5%	9%	
3100	Dust protection, add, minimum	1%	2%	
3200	Maximum	4%	11%	
3300	Equipment usage curtailment, add, minimum	1%	1%	
3400	Maximum	3%	10%	
3500	Material handling & storage limitation, add, minimum	1%	1%	
3600	Maximum	6%	7%	
3700	Protection of existing work, add, minimum	2%	2%	
3800	Maximum	5%	7%	
3900	Shift work requirements, add, minimum		5%	
4000	Maximum		30%	

How RSMeans Assemblies Data Works (Continued)

Sample Estimate
This sample demonstrates the elements of an estimate, including a tally of the RSMeans data lines. Published assemblies costs include all markups for labor burden and profit for the installing contractor. This estimate adds a summary of the markups applied by a general contractor on the installing contractors' work. These figures represent the total cost to the owner. The RSMeans location factor is added at the bottom of the estimate to adjust the cost of the work to a specific location

Work performed: The body of the estimate shows the RSMeans data selected, including line numbers, a brief description of each item, its takeoff quantity and unit, and the total installed cost, including the installing contractor's overhead and profit.

Location Factor: RSMeans published data is based on national average costs. If necessary, adjust the total cost of the project using a location factor from the "Location Factor" table or the "City Cost Indexes" table, found in the Reference Section. Use location factors if the work is general, covering the work of multiple trades. If the work is by a single trade (e.g. masonry) use the more specific data found in the City Cost Indexes.

To adjust costs by location factors, multiply the base cost by the factor and divide by 100.

Contingency: A factor for contingency may be added to any estimate to represent the cost of unknowns that may occur between the time that the estimate is performed and the time the project is constructed. The amount of the allowance will depend on the stage of design at which the estimate is done, and the contractor's assessment of the risk invloved.

Project Name:	Interior Fit-out, ABC Office
Location:	Anywhere, USA
Assembly Number	**Description**
C1010 124 1200	Wood partition, 2 x 4 @ 16" OC w/5/8" FR gypsum board
C1020 114 1800	Metal door & frame, flush hollow core, 3'-0" x 7'-0"
C3010 230 0080	Painting, brushwork, primer & 2 coats
C3020 410 0140	Carpet, tufted, nylon, roll goods, 12' wide, 26 oz
C3030 210 6000	Acoustic ceilings, 24" x 48" tile, tee grid suspension
D5020 125 0560	Receptacles incl plate, box, conduit, wire, 20 A duplex
D5020 125 0720	Light switch incl plate, box, conduit, wire, 20 A single pole
D5020 210 0560	Fluorescent fixtures, recess mounted, 20 per 1000 SF
Assembly Subtotal	
	Sales Tax @
	General Requirements @
Subtotal A	
	GC Overhead @
Subtotal B	
	GC Profit @
Subtotal C	
Adjusted by Location Factor	
	Architects Fee @
	Contingency @
Project Total Cost	

This estimate is based on an interactive spreadsheet.
A copy of this spreadsheet is located on the RSMeans website at
http://www.reedconstructiondata.com/rsmeans/extras/546011.
You are free to download it and adjust it to your methodology.

Date: 1/1/2015		R+R
Qty.	Unit	Subtotal
560.000	S.F.	$2,822.40
2.000	Ea.	$2,500.00
1,120.000	S.F.	$1,422.40
240.000	S.F.	$1,022.40
200.000	S.F.	$1,164.00
8.000	Ea.	$2,212.00
2.000	Ea.	$557.00
200.000	S.F.	$2,200.00
		$13,900.20
5 %		$ 347.51
7 %		$ 973.01
		$15,220.72
5 %		$ 761.04
		$15,981.75
5 %		$ 799.09
		$16,780.84
118.1		$ 19,818.18
8 %		$ 1,585.45
15 %		$ 2,972.73
		$ 24,376.36

Sales Tax: If the work is subject to state or local sales taxes, the amount must be added to the estimate. In a conceptual estimate it can be assumed that one half of the total represents material costs. Therefore, apply the sales tax rate to 50% of the assembly subtotal.

General Requirements: This item covers project-wide needs provided by the General Contractor. These items vary by project, but may include temporary facilities and utilities, security, testing, project cleanup, etc. In assemblies estimates a percentage is used, typically between 5% and 15% of project cost.

General Contractor Overhead: This entry represents the General Contractor's markup on all work to cover project administration costs.

General Contractor Profit: This entry represents the GC's profit on all work performed. The value included here can vary widely by project, and is influenced by the GC's perception of the project's financial risk and market conditions.

Architects Fee: If appropriate, add the design cost to the project estimate. These fees vary based on project complexity and size. Typical design and engineering fees can be found in the reference section.

A Substructure

No part of this publication may be reproduced, stored in a retrieval system, or transmitted in any form or by any means without prior written permission of RSMeans.

A10 Foundations

A1010 Standard Foundations

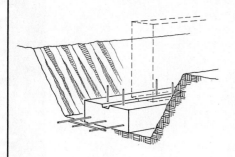

This page illustrates and describes a strip footing system including concrete, forms, reinforcing, keyway and dowels. Lines within System Components give the unit price and total price per linear foot for this system. Prices for strip footing systems are on Line Items A1010 120 1400 thru 2500. Both material quantities and labor costs have been adjusted for the system listed.

Factors: To adjust for job conditions other than normal working situations use Lines A1010 120 2900 thru 4000.

Example: You are to install this footing, and due to a lack of accessibility, only hand tools can be used. Material handling is also a problem. Go to Lines A1010 120 3400 and 3600 and apply these percentages to the appropriate MAT. and INST. costs.

System Components	QUANTITY	UNIT	COST PER L.F.		
			MAT.	INST.	TOTAL
Strip footing, 2'-0" wide x 1'-0" thick, 2000 psi concrete including forms Reinforcing, keyway, and dowels.					
Concrete, 2000 psi	.074	C.Y.	7.92		7.92
Placing concrete	.074	C.Y.		1.93	1.93
Forms, footing, 4 uses	2.000	S.F.	4.88	9.66	14.54
Reinforcing	3.170	Lb.	1.68	2.06	3.74
Keyway, 2" x 4", 4 uses	1.000	L.F.	.25	1.16	1.41
Dowels, #4 bars, 2' long, 24" O.C.	.500	Ea.	.39	1.43	1.82
TOTAL			15.12	16.24	31.36

A1010 120	Strip Footing	COST PER L.F.		
		MAT.	INST.	TOTAL
1400	2'-0" wide x 1' thick, 2000 psi concrete	15.10	16.25	31.35
1500	3000 psi concrete	15.50	16.25	31.75
1600	4000 psi concrete	15.95	16.25	32.20
1800	For alternate footing systems:			
1900	2'-6" wide x 1' thick, 2000 psi concrete	17.60	17.25	34.85
2000	3000 psi concrete	18.05	17.25	35.30
2100	4000 psi concrete	18.60	17.25	35.85
2300	3'-0" wide x 1' thick, 2000 psi concrete	19.85	18.25	38.10
2400	3000 psi concrete	20.50	18.25	38.75
2500	4000 psi concrete	21	18.25	39.25
2700				
2800				
2900	Cut & patch to match existing construction, add, minimum	2%	3%	
3000	Maximum	5%	9%	
3100	Dust protection, add, minimum	1%	2%	
3200	Maximum	4%	11%	
3300	Equipment usage curtailment, add, minimum	1%	1%	
3400	Maximum	3%	10%	
3500	Material handling & storage limitation, add, minimum	1%	1%	
3600	Maximum	6%	7%	
3700	Protection of existing work, add, minimum	2%	2%	
3800	Maximum	5%	7%	
3900	Shift work requirements, add, minimum		5%	
4000	Maximum		30%	

A10 Foundations

A1010 Standard Foundations

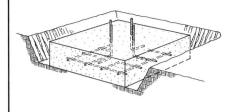

This page illustrates and describes a spread footing system including concrete, forms, reinforcing and anchor bolts. Lines within System Components give the unit price and total price on a cost each basis for this system. Prices for spread footing systems are on Line Items A1010 230 1200 thru 2300. Both material quantities and labor costs have been adjusted for the system listed.

Factors: To adjust for job conditions other than normal working situations use Lines A1010 230 2900 thru 4000.

Example: You are to install the system in an existing occupied building. Access to the site and protection to the building are mandatory. Go to Lines A1010 230 3600 and 3800 and apply these percentages to the appropriate MAT. and INST. costs.

System Components	QUANTITY	UNIT	COST EACH		
			MAT.	INST.	TOTAL
Interior column footing, 3' square 1' thick, 2000 psi concrete,					
Including forms, reinforcing, and anchor bolts.					
Concrete, 2000 psi	.330	C.Y.	35.31		35.31
Placing concrete	.330	C.Y.		18.73	18.73
Forms, footing, 4 uses	12.000	SFCA	9.84	67.80	77.64
Reinforcing	11.000	Lb.	5.83	7.15	12.98
Anchor bolts, 3/4" diameter	2.000	Ea.	2.94	9.10	12.04
TOTAL			53.92	102.78	156.70

A1010 230	Spread Footing	COST EACH		
		MAT.	INST.	TOTAL
1200	3' square x 1' thick, 2000 psi concrete	54	102	156
1300	3000 psi concrete	55.50	102	157.50
1400	4000 psi concrete	57.50	102	159.50
1600	For alternate footing systems:			
1700	4' square x 1' thick, 2000 psi concrete	89.50	146	235.50
1800	3000 psi concrete	92.50	146	238.50
1900	4000 psi concrete	96	146	242
2100	5' square x 1'-3" thick, 2000 psi concrete	168	240	408
2200	3000 psi concrete	174	240	414
2300	4000 psi concrete	180	240	420
2600				
2700				
2900	Cut & patch to match existing construction, add, minimum	2%	3%	
3000	Maximum	5%	9%	
3100	Dust protection, add, minimum	1%	2%	
3200	Maximum	4%	11%	
3300	Equipment usage curtailment, add, minimum	1%	1%	
3400	Maximum	3%	10%	
3500	Material handling & storage limitation, add, minimum	1%	1%	
3600	Maximum	6%	7%	
3700	Protection of existing work, add, minimum	2%	2%	
3800	Maximum	5%	7%	
3900	Shift work requirements, add, minimum		5%	
4000	Maximum		30%	

A10 Foundations

A1010 Standard Foundations

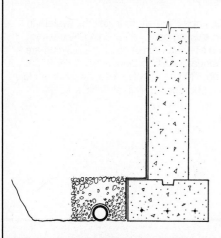

General: Footing drains can be placed either inside or outside of foundation walls depending upon the source of water to be intercepted. If the source of subsurface water is principally from grade or a subsurface stream above the bottom of the footing, outside drains should be used. For high water tables, use inside drains or both inside and outside.

The effectiveness of underdrains depends on good waterproofing. This must be carefully installed and protected during construction.

Costs below include the labor and materials for the pipe and 6" of crushed stone around pipe. Excavation and backfill are not included.

System Components	QUANTITY	UNIT	COST PER L.F.		
			MAT.	INST.	TOTAL
Foundation underdrain, outside, PVC 4" diameter.					
PVC pipe 4" diam. S.D.R. 35	1.000	L.F.	1.60	4.39	5.99
Crushed stone 3/4" to 1/2"	.070	C.Y.	1.82	.94	2.76
TOTAL			3.42	5.33	8.75

A1010 310	Foundation Underdrain	COST PER L.F.		
		MAT.	INST.	TOTAL
1000	Foundation underdrain, outside only, PVC, 4" diameter	3.42	5.35	8.77
1100	6" diameter	5.95	5.90	11.85
1400	Perforated HDPE, 6" diameter	3.91	2.30	6.21
1450	8" diameter	5.85	2.88	8.73
1500	12" diameter	11.20	7	18.20
1600	Corrugated metal, 16 ga. asphalt coated, 6" diameter	9.50	5.55	15.05
1650	8" diameter	12.25	5.90	18.15
1700	10" diameter	15.15	7.65	22.80
3000	Outside and inside, PVC 4" diameter	6.85	10.65	17.50
3100	6" diameter	11.90	11.85	23.75
3400	Perforated HDPE, 6" diameter	7.80	4.61	12.41
3450	8" diameter	11.65	5.75	17.40
3500	12" diameter	22.50	13.95	36.45
3600	Corrugated metal, 16 ga., asphalt coated, 6" diameter	19	11.10	30.10
3650	8" diameter	24.50	11.85	36.35
3700	10" diameter	30.50	15.35	45.85
4700	Cut & patch to match existing construction, add, minimum	2%	3%	
4800	Maximum	5%	9%	
4900	Dust protection, add, minimum	1%	2%	
5000	Maximum	4%	11%	
5100	Equipment usage curtailment, add, minimum	1%	1%	
5200	Maximum	3%	10%	
5300	Material handling & storage limitation, add, minimum	1%	1%	
5400	Maximum	6%	7%	
5500	Protection of existing work, add, minimum	2%	2%	
5600	Maximum	5%	7%	
5700	Shift work requirements, add, minimum		5%	
5800	Maximum		30%	
5900	Temporary shoring and bracing, add, minimum	2%	5%	
6000	Maximum	5%	12%	

A10 Foundations

A1030 Slab on Grade

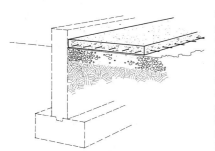

This page illustrates and describes a slab on grade system including slab, bank run gravel, bulkhead forms, placing concrete, welded wire fabric, vapor barrier, steel trowel finish and curing paper. Lines within System Components give the unit price and total price per square foot for this system. Prices for slab on grade systems are on Line Items A1030 110 1400 thru 2600. Both material quantities and labor costs have been adjusted for the system listed.

Factors: To adjust for job conditions other than normal working situations use Lines A1030 110 2900 thru 4000.

Example: You are to install the system at a site where protection of the existing building is required. Go to Line A1030 110 3800 and apply these percentages to the appropriate MAT. and INST. costs.

System Components	QUANTITY	UNIT	COST PER S.F.		
			MAT.	INST.	TOTAL
Ground slab, 4″ thick, 3000 psi concrete, 4″ granular base, vapor barrier					
Welded wire fabric, screed and steel trowel finish.					
Concrete, 4″ thick, 3000 psi concrete	.012	C.Y.	1.34		1.34
Bank run gravel, 4″ deep	.074	C.Y.	.29	.10	.39
Polyethylene vapor barrier, 10 mil.	.011	C.S.F.	.11	.18	.29
Bulkhead forms, expansion material	.100	L.F.	.04	.39	.43
Welded wire fabric, 6 x 6 - #10/10	.011	C.S.F.	.18	.43	.61
Place concrete	.012	C.Y.		.35	.35
Screed & steel trowel finish	1.000	S.F.		.98	.98
TOTAL			1.96	2.43	4.39

A1030 110	Interior Slab on Grade	COST PER S.F.		
		MAT.	INST.	TOTAL
1400	4″ thick slab, 3000 PSI concrete, 4″ deep bank run gravel	1.96	2.43	4.39
1500	6″ deep bank run gravel	2.19	2.42	4.61
1600	12″ deep bank run gravel	2.72	2.46	5.18
1700				
1800				
1900				
2000	For alternate slab systems:			
2100	5″ thick slab, 3000 psi concrete, 6″ deep bank run gravel	2.53	2.50	5.03
2200	12″ deep bank run gravel	3.06	2.54	5.60
2300				
2400				
2500	6″ thick slab, 3000 psi concrete, 6″ deep bank run gravel	2.98	2.61	5.59
2600	12″ deep bank run gravel	3.51	2.65	6.16
2700				
2900	Cut & patch to match existing construction, add, minimum	2%	3%	
3000	Maximum	5%	9%	
3100	Dust protection, add, minimum	1%	2%	
3200	Maximum	4%	11%	
3300	Equipment usage curtailment, add, minimum	1%	1%	
3400	Maximum	3%	10%	
3500	Material handling & storage limitation, add, minimum	1%	1%	
3600	Maximum	6%	7%	
3700	Protection of existing work, add, minimum	2%	2%	
3800	Maximum	5%	7%	
3900	Shift work requirements, add, minimum		5%	
4000	Maximum		30%	

A20 Basement Construction

A2010 Basement Excavation

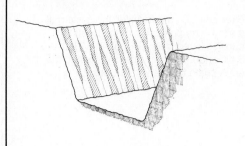

This page illustrates and describes continuous footing and trench excavation systems including a wheel mounted backhoe, operator, equipment rental, fuel, oil, mobilization, no hauling or backfill. Lines within System Components give the unit price and total price per linear foot for this system. Prices for alternate continuous footing and trench excavation systems are on Line Items A2010 130 1200 thru 2700. Both material quantities and labor costs have been adjusted for the system listed.

Factors: To adjust for job conditions other than normal working situations use Lines A2010 130 2900 thru 4000.

Example: You are to install the system and supply dust protection. Go to Line A2010 130 3000 and apply this percentage to the appropriate TOTAL costs.

System Components			COST PER L.F.		
	QUANTITY	UNIT	EQUIP.	LABOR	TOTAL
Continuous footing or trench excav. w/ 3/4 C.Y. wheel mntd backhoe Including operator, equipment rental, fuel, oil and mobilization. Trench Is 4' wide at bottom, 3' deep w/sides sloped 1 to 2 and 200' long. Costs Are based on a production rate of 240 C.Y./day. No hauling/backfill incl.					
Equipment operator	4.000	Hr.		320	320
3/4 C.Y. backhoe, wheel mounted	.500	Day	97.35		97.35
Operating expense (fuel, oil)	4.000	Hr.	102.96		102.96
Mobilization	1.000	Ea.	1,150	1,800	2,950
TOTAL			1,350.31	2,120	3,470.31
COST PER L.F.		L.F.	6.75	10.60	17.35

A2010 130	Excavation, Footings or Trench	COST PER L.F.		
		EQUIP.	LABOR	TOTAL
1200	For alternate trench sizes, 4' bottom with sloped sides:			
1300	2' deep, 50' long	7.55	17.70	25.25
1400	100' long	3.96	8.85	12.81
1500	300' long	4.47	7.05	11.52
1600	4' deep, 50' long	7.90	17.70	25.60
1700	100' long	4.34	8.85	13.19
1800	300' long	2.27	4.15	6.42
1900	6' deep, 50' long	8.50	17.70	26.20
2000	100' long	4.55	10.05	14.60
2100	300' long	3.54	6.15	9.69
2200	8' deep, 50' long	9.20	17.70	26.90
2300	100' long	6.60	12.05	18.65
2400	300' long	4.88	6.15	11.03
2500	10' deep, 50' long	11.20	21	32.20
2600	100' long	8.35	14.85	23.20
2700	300' long	6.45	10.80	17.25
2800				
2900	Dust protection, add, minimum		2%	2%
3000	Maximum		11%	11%
3100	Equipment usage curtailment, add, minimum		1%	1%
3200	Maximum		10%	10%
3300	Material handling & storage limitation, add, minimum		1%	1%
3400	Maximum		7%	7%
3500	Protection of existing work, add, minimum		2%	2%
3600	Maximum		7%	7%
3700	Shift work requirements, add, minimum		5%	5%
3800	Maximum		30%	30%
3900	Temporary shoring and bracing, add, minimum		5%	5%
4000	Maximum		12%	12%

A20 Basement Construction

A2010 Basement Excavation

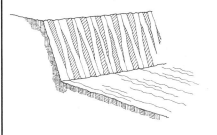

This page illustrates and describes foundation excavation systems including a backhoe-loader, operator, equipment rental, fuel, oil, mobilization, hauling material, no backfilling. Lines within System Components give the unit price and total price per cubic yard for this system. Prices for foundation excavation systems are on Line Items A2010 140 1600 thru 2200. Both material quantities and labor costs have been adjusted for the system listed.

Factors: To adjust for job conditions other than normal working situations use Lines A2010 140 2900 thru 4000.

Example: You are to install the system with the use of temporary shoring and bracing. Go to Line A2010 140 3900 and apply these percentages to the appropriate TOTAL costs.

System Components	QUANTITY	UNIT	COST PER C.Y.		
			EQUIP.	LABOR	TOTAL
Foundation excav w/ 3/4 C.Y. backhoe-loader, incl. operator, equip Rental, fuel, oil, and mobilization. Hauling of excavated material is Included. Prices based on one day production of 360 C.Y. in medium soil Without backfilling.					
Equipment operator	8.000	Hr.		640	640
Backhoe-loader, 3/4 C.Y.	1.000	Day	194.70		194.70
Operating expense (fuel, oil)	8.000	Hr.	205.92		205.92
Hauling, 12 C.Y. trucks, 1 mile round trip	360.000	L.C.Y.	1,087.20	738	1,825.20
Mobilization	1.000	Ea.	260	565	825
TOTAL			1,747.82	1,943	3,690.82
COST PER C.Y.		C.Y.	4.84	5.38	10.25

A2010 140	Excavation, Foundation		COST PER C.Y.		
			EQUIP.	LABOR	TOTAL
1600		100 C.Y.	6.20	11.30	17.50
1700		200 C.Y.	5.45	6.70	12.15
1800		300 C.Y.	5.05	5.70	10.75
1850		360 C.Y.	4.86	5.42	10.28
1900		400 C.Y.	4.78	5.20	9.98
2000		500 C.Y.	4.67	4.97	9.64
2100		600 C.Y.	4.54	4.76	9.30
2200		700 C.Y.	4.49	4.60	9.09
2300					
2400					
2900	Dust protection, add, minimum			2%	2%
3000	Maximum			11%	11%
3100	Equipment usage curtailment, add, minimum			1%	1%
3200	Maximum			10%	10%
3300	Material handling & storage limitation, add, minimum			1%	1%
3400	Maximum			7%	7%
3500	Protection of existing work, add, minimum			2%	2%
3600	Maximum			7%	7%
3700	Shift work requirements, add, minimum			5%	5%
3800	Maximum			30%	30%
3900	Temporary shoring and bracing, add, minimum			5%	5%
4000	Maximum			12%	12%

A20 Basement Construction

A2020 Basement Walls

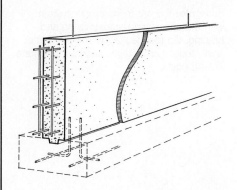

This page illustrates and describes a concrete wall system including concrete, placing concrete, forms, reinforcing, insulation, waterproofing and anchor bolts. Lines within System Components give the unit price and total price per linear foot for this system. Prices for concrete wall systems are on Line Items A2020 120 1400 thru 2600. Both material quantities and labor costs have been adjusted for the system listed.

Factors: To adjust for job conditions other than normal working situations use Lines A2020 120 2900 thru 4000.

Example: You are to install this wall system where delivery of material is difficult. Go to Line A2020 120 3600 and apply these percentages to the appropriate MAT. and INST. costs.

System Components	QUANTITY	UNIT	COST PER L.F. MAT.	COST PER L.F. INST.	COST PER L.F. TOTAL
Cast in place concrete foundation wall, 8" thick, 3' high, 2500 psi					
Concrete including forms, reinforcing, waterproofing, and anchor bolts.					
Concrete, 2500 psi, 8" thick, 3' high	.070	C.Y.	7.63		7.63
Forms, wall, 4 uses	6.000	S.F.	6.84	42.60	49.44
Reinforcing	6.000	Lb.	3.18	2.76	5.94
Placing concrete	.070	C.Y.		3.54	3.54
Waterproofing	3.000	S.F.	1.47	3.48	4.95
Rigid insulaton, 1" polystyrene	3.000	S.F.	.84	2.70	3.54
Anchor bolts, 1/2" diameter, 4' O.C.	.250	Ea.	.41	1.15	1.56
TOTAL			20.37	56.23	76.60

A2020 120	Concrete Wall	MAT.	INST.	TOTAL
1400	8" thick, 2500 psi concrete, 3' high	20.50	56	76.50
1500	4' high	28	75	103
1600	6' high	41.50	112	153.50
1700	8' high	55	149	204
1800	3500 psi concrete, 4' high	28.50	75	103.50
1900	6' high	42.50	112	154.50
2000	8' high	56.50	149	205.50
2100	12" thick, 2500 psi concrete, 4' high	35.50	79.50	115
2200	6' high	51	117	168
2300	8' high	68	156	224
2400	3500 psi concrete, 4' high	36	79.50	115.50
2500	8' high	70	156	226
2600	10' high	86.50	194	280.50
2700				
2900	Cut & patch to match existing construction, add, minimum	2%	3%	
3000	Maximum	5%	9%	
3100	Dust protection, add, minimum	1%	2%	
3200	Maximum	4%	11%	
3300	Equipment usage curtailment, add, minimum	1%	1%	
3400	Maximum	3%	10%	
3500	Material handling & storage limitation, add, minimum	1%	1%	
3600	Maximum	6%	7%	
3700	Protection of existing work, add, minimum	2%	2%	
3800	Maximum	5%	7%	
3900	Shift work requirements, add, minimum		5%	
4000	Maximum		30%	

A20 Basement Construction

A2020 Basement Walls

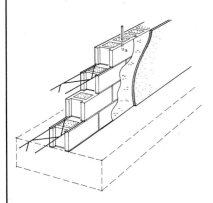

This page illustrates and describes a concrete block wall system including concrete block, masonry reinforcing, parging, waterproofing insulation and anchor bolts. Lines within System Components give the unit price and total price per linear foot for this system. Prices for concrete block wall systems are on Line Items A2020 140 1200 thru 2600. Both material quantities and labor costs have been adjusted for the system listed.

Factors: To adjust for job conditions other than normal working situations use Lines A2020 140 2900 thru 4000.

Example: You are to install the system to match an existing foundation wall. Go to Line A2020 140 3000 and apply these percentages to the appropriate MAT. and INST. costs.

System Components		QUANTITY	UNIT	COST PER L.F.		
				MAT.	INST.	TOTAL
Concrete block, 8" thick, masonry reinforcing, parged and Waterproofed, insulation and anchor bolts, wall 2'-8" high.						
Concrete block, 8" x 8" x 16"		2.670	S.F.	10.15	17.49	27.64
Masonry reinforcing		2.000	L.F.	.50	.40	.90
Parging		.193	S.Y.	.78	2.68	3.46
Waterproofing		2.670	S.F.	1.31	3.10	4.41
Insulation, 1" rigid polystyrene		2.670	S.F.	.75	2.40	3.15
Anchor bolts, 1/2" diameter, 4' O.C.		.250	Ea.	.41	1.15	1.56
	TOTAL			13.90	27.22	41.12

A2020 140	Concrete Block Wall	COST PER L.F.		
		MAT.	INST.	TOTAL
1200	8" thick block, 2'-8" high	13.90	27	40.90
1300	4' high	20.50	40	60.50
1400	6' high	30.50	60	90.50
1500	8' high	41	79.50	120.50
1600	Grouted solid, 4' high	25.50	54	79.50
1700	6' high	38	80.50	118.50
1800	8' high	51	107	158
2100	12" thick block, 4' high	26	58	84
2200	6' high	38.50	86.50	125
2300	8' high	51	115	166
2400	Grouted solid, 4' high	34	72.50	106.50
2500	6' high	50.50	109	159.50
2600	8' high	67.50	144	211.50
2700				
2900	Cut & patch to match existing construction, add, minimum	2%	3%	
3000	Maximum	5%	9%	
3100	Dust protection, add, minimum	1%	2%	
3200	Maximum	4%	11%	
3300	Equipment usage curtailment, add, minimum	1%	1%	
3400	Maximum	3%	10%	
3500	Material handling & storage limitation, add, minimum	1%	1%	
3600	Maximum	6%	7%	
3700	Protection of existing work, add, minimum	2%	2%	
3800	Maximum	5%	7%	
3900	Shift work requirements, add, minimum		5%	
4000	Maximum		30%	

No part of this publication may be reproduced, stored in a retrieval system, or transmitted in any form or by any means without prior written permission of RSMeans.

B Shell

B10 Superstructure

B1010 Floor Construction

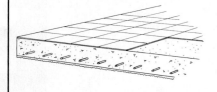

This page illustrates and describes a reinforced concrete slab system including concrete, placing concrete, formwork, reinforcing, steel trowel finish, V.A. floor tile and acoustical spray ceiling finish. Lines within System Components give the unit price per square foot for this system. Prices for reinforced concrete slab systems are on Line Items B1010 225 1400 thru 1800. Both material quantities and labor costs have been adjusted for the system listed.

Factors: To adjust for job conditions other than normal working situations use Lines B1010 225 2700 thru 4000.

Example: You are to install the system to match an existing floor system. Go to Line B1010 225 2800 and apply these percentages to the appropriate MAT. and INST. costs.

System Components	QUANTITY	UNIT	COST PER S.F.		
			MAT.	INST.	TOTAL
Flat slab system, reinforced concrete with vinyl tile floor, sprayed Acoustical ceiling finish, not including columns.					
Concrete, 4000 psi, 6" thick	.020	C.Y.	2.36		2.36
Placing concrete	.020	C.Y.		.72	.72
Formwork, 4 use	1.000	S.F.	1.38	6.40	7.78
Edge form	.120	S.F.	.07	1.25	1.32
Reinforcing steel	3.000	Lb.	1.59	1.95	3.54
Steel trowel finish	1.000	S.F.		1.29	1.29
Vinyl floor tile, 1/8" thick, standard colors/patterns	1.000	S.F.	6.20	1.08	7.28
Acoustical spray ceiling finish, solvent based	1.000	S.F.	.30	.18	.48
TOTAL			11.90	12.87	24.77

B1010 225	Floor - Ceiling, Concrete Slab	COST PER S.F.		
		MAT.	INST.	TOTAL
1400	Concrete, 4000 psi, 6" thick	11.90	12.85	24.75
1500	7" thick	12.35	13.45	25.80
1600	8" thick	13.05	14.10	27.15
1700	9" thick	13.60	14.65	28.25
1800	10" thick	14.30	15.40	29.70
1900				
2000				
2100				
2200				
2300				
2400				
2500				
2700	Cut & patch to match existing construction, add, minimum	2%	3%	
2800	Maximum	5%	9%	
2900	Dust protection, add, minimum	1%	2%	
3000	Maximum	4%	11%	
3100	Equipment usage curtailment, add, minimum	1%	1%	
3200	Maximum	3%	10%	
3300	Material handling & storage limitation, add, minimum	1%	1%	
3400	Maximum	6%	7%	
3500	Protection of existing work, add, minimum	2%	2%	
3600	Maximum	5%	7%	
3700	Shift work requirements, add, minimum		5%	
3800	Maximum		30%	
3900	Temporary shoring and bracing, add, minimum	2%	5%	
4000	Maximum	5%	12%	

B10 Superstructure

B1010 Floor Construction

This page illustrates and describes a hollow core prestressed concrete panel system including hollow core slab, grout, carpet, carpet padding, and sprayed ceiling. Lines within System Components give the unit price and total price per square foot for this system. Prices for hollow core prestressed concrete panel systems are on Line Items B1010 232 1200 thru 1600. Both material and labor costs have been adjusted for the system listed.

Factors: To adjust for job conditions other than normal working situations use Lines B1010 232 2700 thru 3800.

Example: You are to install the system where dust control is a major concern. Go to Line B1010 232 2800 and apply these percentages to the appropriate MAT. and INST. costs.

System Components	QUANTITY	UNIT	COST PER S.F.		
			MAT.	INST.	TOTAL
Precast hollow core plank with carpeted floors, padding					
And sprayed textured ceiling.					
Hollow core plank, 4" thick with grout topping	1.000	S.F.	6.45	3.57	10.02
Nylon carpet, 26 oz. medium traffic	.110	S.Y.	5.17	.79	5.96
Carpet padding, 20 oz./sq. yd.	.110	S.Y.	.55	.40	.95
Sprayed texture ceiling	1.000	S.F.	.04	.68	.72
TOTAL			12.21	5.44	17.65

B1010 232	Floor - Ceiling, Conc. Panel	COST PER S.F.		
		MAT.	INST.	TOTAL
1200	Hollow core concrete plank, with grout, 4" thick	12.20	5.45	17.65
1300	6" thick	13.10	4.93	18.03
1400	8" thick	13.55	4.55	18.10
1500	10" thick	14.25	4.25	18.50
1600	12" thick	14.40	4.02	18.42
1700				
1800				
1900				
2000				
2100				
2200				
2300				
2400				
2500				
2700	Dust protection, add, minimum	1%	2%	
2800	Maximum	4%	11%	
2900	Equipment usage curtailment, add, minimum	1%	1%	
3000	Maximum	3%	10%	
3100	Material handling & storage limitation, add, minimum	1%	1%	
3200	Maximum	6%	7%	
3300	Protection of existing work, add, minimum	2%	2%	
3400	Maximum	5%	7%	
3500	Shift work requirements, add, minimum		5%	
3600	Maximum		30%	
3700	Temporary shoring and bracing, add, minimum	2%	5%	
3800	Maximum	5%	12%	

B10 Superstructure

B1010 Floor Construction

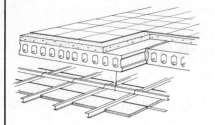

This page illustrates and describes a hollow core plank system including precast hollow core plank, concrete topping, V.A. tile, concealed suspension system and ceiling tile. Lines within System Components give the unit price and total price per square foot for this system. Prices for hollow core plank systems are on Line Items B1010 233 1400 thru 1800. Both material quantities and labor costs have been adjusted for the system listed.

Factors: To adjust for job conditions other than normal working situations use Lines B1010 233 2900 thru 4000.

Example: You are to install the system where equipment usage is a problem. Go to Line B1010 233 3200 and apply these percentages to the appropriate MAT. and INST. costs.

System Components	QUANTITY	UNIT	COST PER S.F.		
			MAT.	INST.	TOTAL
Precast hollow core plank with 2" topping, vinyl composition floor tile And suspended acoustical tile ceiling.					
Hollow core concrete plank, 4" thick	1.000	S.F.	6.45	3.57	10.02
Finishing floors, granolithic topping, 1:1:1-1/2 mix, 2" thick	1.000	S.F.	1.35	5.85	7.20
Vinyl composition tile, .08" thick	1.000	S.F.	1.66	1.08	2.74
Concealed Z bar suspension system, 12" module	1.000	S.F.	.92	1.18	2.10
Ceiling tile, mineral fiber, 3/4" thick	1.000	S.F.	2.64	2.05	4.69
TOTAL			13.02	13.73	26.75

B1010 233	Floor - Ceiling, Conc. Plank	COST PER S.F.		
		MAT.	INST.	TOTAL
1400	Hollow core concrete plank, 4" thick	13	13.75	26.75
1500	6" thick	13.90	13.20	27.10
1600	8" thick	14.35	12.80	27.15
1700	10" thick	15.05	12.55	27.60
1800	12" thick	15.20	12.30	27.50
1900				
2000				
2100				
2200				
2300				
2400				
2500				
2600				
2700				
2900	Dust protection, add, minimum	1%	2%	
3000	Maximum	4%	11%	
3100	Equipment usage curtailment, add, minimum	1%	1%	
3200	Maximum	3%	10%	
3300	Material handling & storage limitation, add, minimum	1%	1%	
3400	Maximum	6%	7%	
3500	Protection of existing work, add, minimum	2%	2%	
3600	Maximum	5%	7%	
3700	Shift work requirements, add, minimum		5%	
3800	Maximum		30%	
3900	Temporary shoring and bracing, add, minimum	2%	5%	
4000	Maximum	5%	12%	

B10 Superstructure

B1010 Floor Construction

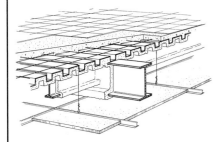

This page illustrates and describes a structural steel w/metal decking and concrete system including steel beams, steel decking, shear studs, concrete, placing concrete, edge form, steel trowel finish, curing, wire fabric, fireproofing, beams and decking, tile floor, and suspended ceiling. Lines within System Components give the unit price and total price per square foot for this system. Prices for alternate structural steel w/metal decking and concrete systems are on Line Items B1010 243 1900 thru 2400. Both material quantities and labor costs have been adjusted for the system listed.

Factors: To adjust for job conditions other than normal working situations use Lines B1010 243 2900 thru 4000.

Example: You are to install the system where material handling and storage are a problem. Go to Line B1010 243 3400 and apply these percentages to the appropriate MAT. and INST. costs.

System Components	QUANTITY	UNIT	COST PER S.F.		
			MAT.	INST.	TOTAL
Composite structural beams, 20 ga. 3" deep steel decking, shear studs, 3000 psi, concrete, placing concrete, edge forms, steel trowel finish, Welded wire fabric fireproofing, tile floor and suspended ceiling.					
Steel deck, 20 gage, 3" deep, galvanized	1.000	S.F.	2.58	1.07	3.65
Structural steel framing	.004	Ton	11.70	2.84	14.54
Shear studs, 3/4"	.250	Ea.	.14	.49	.63
Concrete, 3000 psi, 5" thick	.015	C.Y.	1.68		1.68
Place concrete	.015	C.Y.		.54	.54
Edge form	.140	L.F.	.03	.66	.69
Steel trowel finish	1.000	S.F.		1.29	1.29
Welded wire fabric 6 x 6 - #10/10	.011	C.S.F.	.18	.43	.61
Fireproofing, sprayed	1.880	S.F.	1.09	2.11	3.20
Vinyl composition floor tile	1.000	S.F.	1.30	1.08	2.38
Suspended acoustical ceiling	1.000	S.F.	3.70	1.62	5.32
TOTAL			22.40	12.13	34.53

B1010 243	Floor - Ceiling, Struc. Steel	COST PER S.F.		
		MAT.	INST.	TOTAL
1900	Composite deck, galvanized, 3" deep, 22 ga.	22	12.05	34.05
1950	20 ga.	22.50	12.10	34.60
2000	18 ga.	23	12.20	35.20
2100	16 ga.	24	12.25	36.25
2200	Composite deck, galvanized, 2" deep, 22 ga.	22	11.90	33.90
2300	20 ga.	22	11.95	33.95
2400	16 ga.	23.50	12.05	35.55
2500				
2900	Dust protection, add, minimum	1%	2%	
3000	Maximum	4%	11%	
3100	Equipment usage curtailment, add, minimum	1%	1%	
3200	Maximum	3%	10%	
3300	Material handling & storage limitation, add, minimum	1%	1%	
3400	Maximum	6%	7%	
3500	Protection of existing work, add, minimum	2%	2%	
3600	Maximum	5%	7%	
3700	Shift work requirements, add, minimum		5%	
3800	Maximum		30%	
3900	Temporary shoring and bracing, add, minimum	2%	5%	
4000	Maximum	5%	12%	

B10 Superstructure

B1010 Floor Construction

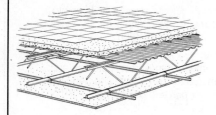

This page illustrates and describes an open web joist and steel slab-form system including open web steel joist, slab form, concrete, placing concrete, wire fabric, steel trowel finish, tile floor and plasterboard. Lines within System Components give the unit price and total price per square foot for this system. Prices for open web joists and steel slab-form systems are on Line Items B1010 245 1800 thru 2200. Both material quantities and labor costs have been adjusted for the systems listed.

Factors: To adjust for job conditions other than normal working situations use Lines B1010 245 2900 thru 4000.

Example: You are to install the system in a congested commercial area and most work will be done at night. Go to Line B1010 245 3800 and apply this percentage to the appropriate INST. cost.

System Components	QUANTITY	UNIT	COST PER S.F.		
			MAT.	INST.	TOTAL
Open web steel joists, 24" O.C., slab form, 3000 psi concrete, welded					
Wire fabric, steel trowel finish, floor tile, 5/8" drywall ceiling.					
Open web steel joists, 12" deep, 5.2#/L.F., 24" O.C.	2.630	Lb.	2.58	1.23	3.81
Slab form, 28 gage, 9/16" deep, galvanized	1.000	S.F.	1.51	.81	2.32
Concrete, 3000 psi, 2-1/2" thick	.008	C.Y.	.90		.90
Placing concrete	.008	C.Y.		.29	.29
Welded wire fabric 6 x 6 - #10/10	.011	C.S.F.	.18	.43	.61
Steel trowel finish	1.000	S.F.		1.29	1.29
Vinyl composition floor tile	1.000	S.F.	1.30	1.08	2.38
Ceiling furring, 3/4" channels, 24" O.C.	1.000	S.F.	.24	1.29	1.53
Gypsum drywall, 5/8" thick, finished	1.000	S.F.	.41	1.61	2.02
Paint ceiling	1.000	S.F.	.21	.68	.89
TOTAL			7.33	8.71	16.04

B1010 245		Floor - Ceiling, Steel Joists	COST PER S.F.		
			MAT.	INST.	TOTAL
1800	Open web joists, 12" deep, 5.2#/L.F.		7.35	8.70	16.05
1900		16" deep, 6.6#/L.F.	8	9	17
2000		20" deep, 8.4# L.F.	8.85	9.45	18.30
2100		24" deep, 11.5# L.F.	10.40	10.15	20.55
2200		26" deep, 12.8# L.F.	11	10.50	21.50
2300					
2400					
2500					
2600					
2700					
2900	Dust protection, add, minimum		1%	2%	
3000		Maximum	4%	11%	
3100	Equipment usage curtailment, add, minimum		1%	1%	
3200		Maximum	3%	10%	
3300	Material handling & storage limitation, add, minimum		1%	1%	
3400		Maximum	6%	7%	
3500	Protection of existing work, add, minimum		2%	2%	
3600		Maximum	5%	7%	
3700	Shift work requirements, add, minimum			5%	
3800		Maximum		30%	
3900	Temporary shoring and bracing, add, minimum		2%	5%	
4000		Maximum	5%	12%	

B10 Superstructure

B1010 Floor Construction

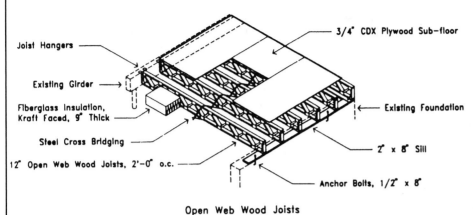

Open Web Wood Joists

This page illustrates and describes floor framing systems including sills, joist hangers, joists, bridging, subfloor and insulation. Lines within systems components give the unit price and total price per square foot for this system. Prices for alternate joist types are on line items B1010 262 0100 through 0700. Both material quantities and labor costs have been adjusted for the system listed.

Factors: To adjust for job conditions other than normal working situations use lines B1010 262 2700 through 4000.

Example: You are to install the system where delivery of material is difficult. Go to line B1010 262 3400 and apply these percentages to the appropriate MAT. and INST. costs.

System Components	QUANTITY	UNIT	COST PER S.F. MAT.	INST.	TOTAL
Open web wood joists, sill, butt to girder, 5/8" subfloor, insul.					
Anchor bolt, hook type, with nut and washer, 1/2" x 8" long	.250	Ea.	.04	.11	.15
Joist & beam hanger, 18 ga. galvanized	.750	Ea.	.11	.28	.39
Bridging, steel, compression, for 2" x 12" joists	.010	C.Pr.	.01	.03	.04
Framing, open web wood joists, 12" deep	.001	M.L.F.	3.88	1.40	5.28
Framing, sills, 2" x 8"	.001	M.B.F.	.75	1.85	2.60
Sub-floor, plywood, CDX, 3/4" thick	1.000	S.F.	.86	.91	1.77
Insulation, fiberglass, kraft face, 9" thick x 23" wide	1.000	S.F.	.78	.46	1.24
TOTAL			6.43	5.04	11.47

B1010 262	Open Web Wood Joists	MAT.	INST.	TOTAL
0100	12" open web joists, sill, butt to girder, 5/8" sub-floor	6.45	5.05	11.50
0200	14" joists	6.65	5.15	11.80
0300	16" joists	6.75	5.20	11.95
0400	18" joists	6.95	5.30	12.25
0450	Wood struc." I" joists, 24" O.C., to 24' span, 50 PSF LL, butt to girder	4.78	4.68	9.46
0500	55 PSF LL	5	4.75	9.75
0600	to 30' span, 45 PSF LL	5.15	4.60	9.75
0700	55 PSF LL	6.90	4.68	11.58
2700	Cut & patch to match existing construction, add, minimum	2%	3%	
2800	Maximum	5%	9%	
2900	Dust protection, add, minimum	1%	2%	
3000	Maximum	4%	11%	
3100	Equipment usage curtailment, add, minimum	1%	1%	
3200	Maximum	3%	10%	
3300	Material handling & storage limitation, add, minimum	1%	1%	
3400	Maximum	6%	7%	
3500	Protection of existing work, add, minimum	2%	2%	
3600	Maximum	5%	7%	
3700	Shift work requirements, add, minimum		5%	
3800	Maximum		30%	
3900	Temporary shoring and bracing, add, minimum	2%	5%	
4000	Maximum	5%	12%	

B10 Superstructure

B1010 Floor Construction

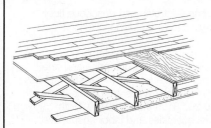

This page illustrates and describes a wood joist floor system including wood joist, oak floor, sub-floor, bridging, sand and finish floor, furring, plasterboard, taped, finished and painted ceiling. Lines within System Components give the unit price and total price per square foot for this system. Prices for wood joist floor systems are on Line Items B1010 263 1700 thru 2300. Both material quantities and labor costs have been adjusted for the system listed.

Factors: To adjust for job conditions other than normal working situations use Lines B1010 263 2700 thru 4000.

Example: You are to install the system during off peak hours, 6 P.M. to 2 A.M. Go to Line B1010 263 3800 and apply this percentage to the appropriate INST. costs.

System Components	QUANTITY	UNIT	COST PER S.F.		
			MAT.	INST.	TOTAL
Wood joists, 2" x 8", 16" O.C., oak floor (sanded & finished), 1/2" sub Floor, 1" x 3" bridging, furring, 5/8" drywall, taped, finished and painted.					
Wood joists, 2" x 8", 16" O.C.	1.000	L.F.	1.01	1.12	2.13
Subfloor plywood CDX 1/2"	1.000	SF Flr.	.72	.82	1.54
Bridging 1" x 3"	.150	Pr.	.10	.71	.81
Oak flooring, No. 1 common	1.000	S.F.	4.41	3.62	8.03
Sand and finish floor	1.000	S.F.	.99	1.67	2.66
Wood furring, 1" x 3", 16" O.C.	1.000	L.F.	.42	1.76	2.18
Gypsum drywall, 5/8" thick	1.000	S.F.	.36	.68	1.04
Tape and finishing	1.000	S.F.	.05	.62	.67
Paint ceiling	1.000	S.F.	.21	.68	.89
TOTAL			8.27	11.68	19.95

B1010 263	Floor-Ceiling, Wood Joists	COST PER S.F.		
		MAT.	INST.	TOTAL
1700	16" on center, 2" x 6" joists	8	11.55	19.55
1750	2" x 8"	8.25	11.70	19.95
1800	2" x 10"	8.75	11.95	20.70
1900	2" x 12"	9.15	11.95	21.10
2000	2" x 14"	9.45	12.15	21.60
2100	12" on center, 2" x 10" joists	9.10	12.25	21.35
2200	2" x 12"	9.65	12.30	21.95
2300	2" x 14"	10	12.55	22.55
2400				
2500				
2700	Cut & patch to match existing construction, add, minimum	2%	3%	
2800	Maximum	5%	9%	
2900	Dust protection, add, minimum	1%	2%	
3000	Maximum	4%	11%	
3100	Equipment usage curtailment, add, minimum	1%	1%	
3200	Maximum	3%	10%	
3300	Material handling & storage limitation, add, minimum	1%	1%	
3400	Maximum	6%	7%	
3500	Protection of existing work, add, minimum	2%	2%	
3600	Maximum	5%	7%	
3700	Shift work requirements, add, minimum		5%	
3800	Maximum		30%	
3900	Temporary shoring and bracing, add, minimum	2%	5%	
4000	Maximum	5%	12%	

B10 Superstructure

B1020 Roof Construction

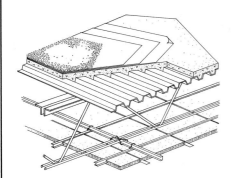

This page illustrates and describes a flat roof bar joist system including asphalt and gravel roof, lightweight concrete, metal decking, open web joists, and suspended acoustic ceiling. Lines within System Components give the unit price and total price per square foot for this system. Prices for flat roof bar joist systems are on Line Items B1020 106 1200 thru 1600. Both material quantities and labor costs have been adjusted for the system listed.

Factors: To adjust for job conditions other than normal working situations use Lines B1020 106 2700 thru 4000.

Example: You are to install the system where material handling and storage present some problem. Go to Line B1020 106 3300 and apply these percentages to the appropriate MAT. and INST. costs.

System Components	QUANTITY	UNIT	COST PER S.F.		
			MAT.	INST.	TOTAL
Three ply asphalt and gravel roof on lightweight concrete, metal decking on					
Web joist 4' O.C., with suspended acoustic ceiling.					
Open web joist, 10" deep, 4' O.C.	1.250	Lb.	1.23	.59	1.82
Metal decking, 1-1/2" deep, 22 Ga., galvanized	1.000	S.F.	2.25	.72	2.97
Gravel roofing, 3 ply asphalt	.010	C.S.F.	1.43	2.02	3.45
Perlite roof fill, 3" thick	1.000	S.F.	2.31	.59	2.90
Suspended ceiling, 3/4" mineral fiber on "Z" bar suspension	1.000	S.F.	2.73	4.10	6.83
TOTAL			9.95	8.02	17.97

B1020 106	Steel Joist Roof & Ceiling	COST PER S.F.		
		MAT.	INST.	TOTAL
1200	Open web bar joist, 10" deep, 5#/L.F.	9.95	8	17.95
1300	16" deep, 5#/L.F.	10.35	8.20	18.55
1400	18" deep, 6.6#/L.F.	10.70	8.40	19.10
1500	20" deep, 9.6#/L.F.	11.05	8.55	19.60
1600	24" deep, 11.52#/L.F.	11.55	8.75	20.30
1700				
1800				
1900				
2000				
2100				
2200				
2300				
2400				
2500				
2700	Cut & patch to match existing construction, add, minimum	2%	3%	
2800	Maximum	5%	9%	
2900	Dust protection, add, minimum	1%	2%	
3000	Maximum	4%	11%	
3100	Equipment usage curtailment, add, minimum	1%	1%	
3200	Maximum	3%	10%	
3300	Material handling & storage limitation, add, minimum	1%	1%	
3400	Maximum	6%	7%	
3500	Protection of existing work, add, minimum	2%	2%	
3600	Maximum	5%	7%	
3700	Shift work requirements, add, minimum		5%	
3800	Maximum		30%	
3900	Temporary shoring and bracing, add, minimum	2%	5%	
4000	Maximum	5%	12%	

B10 Superstructure

B1020 Roof Construction

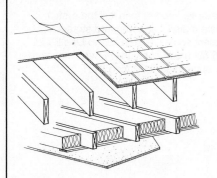

This page illustrates and describes a wood frame roof system including rafters, ceiling joists, sheathing, building paper, asphalt shingles, roof trim, furring, insulation, plaster and paint. Lines within System Components give the unit price and total price per square foot for this system. Prices for wood frame roof systems are on Line Items B1020 202 1800 thru 2700. Both material quantities and labor costs have been adjusted for the system listed.

Factors: To adjust for job conditions other than normal working situations use Lines B1020 202 3300 thru 4000.

Example: You are to install the system while protecting existing work. Go to Line B1020 202 3800 and apply these percentages to the appropriate MAT. and INST. costs.

System Components	QUANTITY	UNIT	COST PER S.F. MAT.	INST.	TOTAL
Wood frame roof system, 4 in 12 pitch, including rafters, sheathing, Shingles, insulation, drywall, thin coat plaster, and painting.					
Rafters, 2" x 6", 16" O.C., 4 in 12 pitch	1.080	L.F.	.79	1.33	2.12
Ceiling joists, 2" x 6", 16 O.C.	1.000	L.F.	.73	.99	1.72
Sheathing, 1/2" CDX	1.080	S.F.	.78	.95	1.73
Building paper, 15# felt	.011	C.S.F.	.07	.18	.25
Asphalt, standard strip shingles, inorganic, class A, 30 year	.011	C.S.F.	1.13	1.28	2.41
Roof trim	.100	L.F.	.18	.27	.45
Furring, 1" x 3", 16" O.C.	1.000	L.F.	.42	1.76	2.18
Fiberglass insulation, 6" batts	1.000	S.F.	.66	.46	1.12
Gypsum board, 1/2" thick	1.000	S.F.	.33	.68	1.01
Thin coat plaster	1.000	S.F.	.12	.77	.89
Paint, roller, 2 coats	1.000	S.F.	.21	.68	.89
TOTAL			5.42	9.35	14.77

B1020 202	Wood Frame Roof & Ceiling	COST PER S.F. MAT.	INST.	TOTAL
1800	Rafters 16" O.C., 2" x 6"	5.40	9.35	14.75
1900	2" x 8"	5.70	9.45	15.15
2000	2" x 10"	6.20	10.15	16.35
2100	2" x 12"	6.70	10.35	17.05
2200	Rafters 24" O.C., 2" x 6"	5.05	8.75	13.80
2300	2" x 8"	5.25	8.80	14.05
2400	2" x 10"	5.65	9.35	15
2500	2" x 12"	5.95	9.50	15.45
2600	Roof pitch, 6 in 12, add	3%	10%	
2700	8 in 12, add	5%	12%	
2800				
2900				
3000				
3100				
3300	Cut & patch to match existing construction, add, minimum	2%	3%	
3400	Maximum	5%	9%	
3500	Material handling & storage limitation, add, minimum	1%	1%	
3600	Maximum	6%	7%	
3700	Protection of existing work, add, minimum	2%	2%	
3800	Maximum	5%	7%	
3900	Shift work requirements, add, minimum		5%	
4000	Maximum		30%	

B10 Superstructure

B1020 Roof Construction

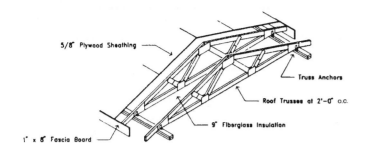

Wood Fabricated Roof Truss System

This page illustrates and describes a sloped roof truss framing system including trusses, truss anchors, sheathing, fascia and insulation. Lines B1020 210 1500 through 1800 are for a flat roof framing system including flat roof trusses, truss anchors, sheathing, fascia and insulation. Lines within system components give the unit price and total price per square foot for this system. Prices for alternate sizes and systems are given in line items B1020 210 0100 through 1800. Both material quantities and labor costs have been adjusted for the system listed.

Factors: To adjust for job conditions other than normal working situations use lines B1020 210 2700 through 4000.

Example: You are to install the system where crane placement and movement will be difficult. Go to line B1020 210 3200 and apply these percentages to the appropriate MAT. and INST. costs.

System Components	QUANTITY	UNIT	COST PER S.F.		
			MAT.	INST.	TOTAL
Roof truss, 2' O.C., 4/12 slope, 12' span, 5/8 sheath., fascia, insul.					
Timber connectors, rafter anchors, galv., 1 1/2" x 5 1/4"	.084	Ea.	.05	.36	.41
Plywood sheathing, CDX, 5/8" thick	1.054	S.F.	.91	1	1.91
Truss, 2' O.C., metal plate connectors, 12' span	.042	Ea.	1.91	1.91	3.82
Molding, fascia trim, 1" x 8"	.167	L.F.	.30	.45	.75
Insulation, fiberglass, kraft face, 9" thick x 23" wide	1.000	S.F.	.78	.46	1.24
TOTAL			3.95	4.18	8.13

B1020 210	Roof Truss, Wood	COST PER S.F.		
		MAT.	INST.	TOTAL
0100	Roof truss, 2'O.C., 4/12p, 1'ovhg, 12'span, 5/8"sheath, fascia, insul.	3.95	4.18	8.13
0200	20' span	3.80	3.54	7.34
0300	24' span	3.66	3.31	6.97
0400	26' span	5.20	4.43	9.63
0500	28' span	3.72	3.22	6.94
0600	30' span	3.73	3.20	6.93
0700	32' span	3.82	3.14	6.96
0800	34' span	3.73	3.11	6.84
0890	60' span	5.15	3.53	8.68
0900	8/12 pitch, 20' span	4.10	3.81	7.91
1000	24' span	4.22	3.57	7.79
1100	26' span	4.09	3.48	7.57
1200	28' span	4.24	3.46	7.70
1300	32' span	4.26	3.41	7.67
1400	36' span	4.42	3.34	7.76
1500	Flat roof frame, fab.strl.joists, 2' O.C., 15'-24' span, 50 PSF LL	4.08	2.83	6.91
1600	55 PSF LL	4.30	2.90	7.20
1700	24'-30'span, 45 PSF LL	4.41	2.71	7.12
1800	55 PSF LL	6.15	2.79	8.94
1900				
2700	Cut & patch to match existing construction, add, minimum	2%	3%	
2800	Maximum	5%	9%	
2900	Dust protection, add, minimum	1%	2%	
3000	Maximum	4%	11%	
3100	Equipment usage curtailment, add, minimum	1%	1%	
3200	Maximum	3%	10%	
3300	Material handling & storage limitation, add, minimum	1%	1%	
3400	Maximum	6%	7%	

B10 Superstructure

B1020 Roof Construction

B1020 210	Roof Truss, Wood	COST PER S.F.		
		MAT.	INST.	TOTAL
3500	Protection of existing work, add, minimum	2%	2%	
3600	Maximum	5%	7%	
3700	Shift work requirements, add, minimum		5%	
3800	Maximum		30%	
3900	Temporary shoring and bracing, add, minimum	2%	5%	
4000	Maximum	5%	12%	

B20 Exterior Enclosure

B2010 Exterior Walls

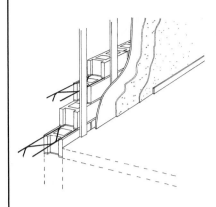

This page illustrates and describes a masonry concrete block system including concrete block wall, pointed, reinforcing, waterproofing, gypsum plaster on gypsum lath, and furring. Lines within System Components give the unit price and total price per square foot for this system. Prices for masonry concrete block wall systems are on Line Items B2010 108 1400 thru 2200. Both material quantities and labor costs have been adjusted for the system listed.

Factors: To adjust for job conditions other than normal working situations use Lines B2010 108 2700 thru 4000.

Example: You are to install the system and match existing construction at several locations. Go to Line B2010 108 2800 and apply these percentages to the appropriate MAT. and INST. costs.

System Components	QUANTITY	UNIT	COST PER S.F.		
			MAT.	INST.	TOTAL
Concrete block, 8" thick, reinforced every 2 courses, waterproofing gypsum Plaster over gypsum lath on 1" x 3" furring, interior painting & baseboard.					
Concrete block, 8" x 8" x 16", reinforced	1.000	S.F.	5.15	6.75	11.90
Silicone waterproofing, 2 coats	1.000	S.F.	.75	.19	.94
Bituminous coating, 1/16" thick	1.000	S.F.	.40	1.16	1.56
On concrete	1.000	L.F.	.46	2.37	2.83
Gypsum lath, 3/8" thick	.110	S.Y.	.37	.70	1.07
Gypsum plaster, 2 coats	.110	S.Y.	.44	2.91	3.35
Painting, 2 coats	1.000	S.F.	.14	.53	.67
Baseboard wood, 9/16" x 2-5/8"	.100	L.F.	.12	.26	.38
Paint baseboard, primer + 1 coat enamel	.100	L.F.	.01	.07	.08
TOTAL			7.84	14.94	22.78

B2010 108	Masonry Wall, Concrete Block	COST PER S.F.		
		MAT.	INST.	TOTAL
1400	8" thick block, regular	7.85	14.95	22.80
1500	Fluted 2 sides	8.40	17.50	25.90
1600	Deep grooved	6.95	17.50	24.45
1700	Slump block	8.85	15.25	24.10
1800	Split rib	8.15	17.50	25.65
1900				
2000	12" thick block, regular	7.05	17.85	24.90
2100	Slump block	15.75	16.65	32.40
2200	Split rib	9.05	20	29.05
2300				
2700	Cut & patch to match existing construction, add, minimum	2%	3%	
2800	Maximum	5%	9%	
2900	Dust protection, add, minimum	1%	2%	
3000	Maximum	4%	11%	
3100	Equipment usage curtailment, add, minimum	1%	1%	
3200	Maximum	3%	10%	
3300	Material handling & storage limitation, add, minimum	1%	1%	
3400	Maximum	6%	7%	
3500	Protection of existing work, add, minimum	2%	2%	
3600	Maximum	5%	7%	
3700	Shift work requirements, add, minimum		5%	
3800	Maximum		30%	
3900	Temporary shoring and bracing, add, minimum	2%	5%	
4000	Maximum	5%	12%	

B20 Exterior Enclosure

B2010 Exterior Walls

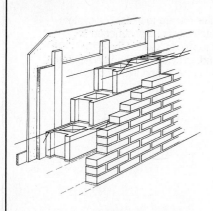

This page illustrates and describes a masonry wall, brick-stone system including brick, concrete block, durawall, insulation, plasterboard, taped and finished, furring, baseboard and painting interior. Lines within System Components give the unit price and total price per square foot for this system. Prices for masonry wall, brick-stone systems are on Line Item B2010 129 1400 thru 2500. Both material quantities and labor costs have been adjusted for the system listed.

Factors: To adjust for job conditions other than normal working situations use Lines B2010 129 3100 thru 4200.

Example: You are to install the system without damaging the existing work. Go to Line B2010 129 3900 and apply these percentages to the appropriate MAT. and INST. costs.

System Components	QUANTITY	UNIT	COST PER S.F.		
			MAT.	INST.	TOTAL
Face brick, 4"thick, concrete block back-up, reinforce every second course, 3/4"insulation, furring, 1/2"drywall, taped, finish, and painted, baseboard					
Face brick, 4" brick	1.000	S.F.	4.10	12.70	16.80
Concrete back-up block, reinforced 8" thick	1.000	S.F.	3.34	7.05	10.39
3/4" rigid polystyrene insulation	1.000	S.F.	.58	.77	1.35
Furring, 1" x 3", wood, 16" O.C.	1.000	L.F.	.46	1.24	1.70
Drywall, 1/2" thick	1.000	S.F.	.33	.62	.95
Taping & finishing	1.000	S.F.	.05	.62	.67
Painting, 2 coats	1.000	S.F.	.21	.68	.89
Baseboard, wood, 9/16" x 2-5/8"	.100	L.F.	.12	.26	.38
Paint baseboard, primer + 1 coat enamel	.100	L.F.	.01	.07	.08
TOTAL			9.20	24.01	33.21

B2010 131	Masonry Wall, Brick - Stone	COST PER S.F.		
		MAT.	INST.	TOTAL
1400	Face brick, standard, red, 4" x 2-2/3" x 8" (6.75/S.F.)	9.20	24	33.20
1500	Norman, 4" x 2-2/3" x 12" (4.5 per S.F.)	11.60	20	31.60
1600	Roman, 4" x 2" x 12" (6.0 per S.F.)	12.70	22.50	35.20
1700	Engineer, 4" x 3-1/5" x 8" (5.63 per S.F.)	9.05	22	31.05
1800	S.C.R., 6" x 2-2/3" x 12" (4.5 per S.F.)	11.70	20.50	32.20
1900	Jumbo, 6" x 4" x 12" (3.0 per S.F.)	10.25	17.70	27.95
2000	Norwegian, 6" x 3-1/5" x 12" (3.75 per S.F.)	10.75	18.75	29.50
2100				
2200				
2300	Stone, veneer, fieldstone, 6" thick	12.20	20.50	32.70
2400	Marble, 2" thick	52	35.50	87.50
2500	Limestone, 2" thick	36.50	22	58.50
3100	Cut & patch to match existing construction, add, minimum	2%	3%	
3200	Maximum	5%	9%	
3300	Dust protection, add, minimum	1%	2%	
3400	Maximum	4%	11%	
3500	Equipment usage curtailment, add, minimum	1%	1%	
3600	Maximum	3%	10%	
3700	Material handling & storage limitation, add, minimum	1%	1%	
3800	Maximum	6%	7%	
3900	Protection of existing work, add, minimum	2%	2%	
4000	Maximum	5%	7%	
4100	Shift work requirements, add, minimum		5%	
4200	Maximum		30%	

B20 Exterior Enclosure

B2010 Exterior Walls

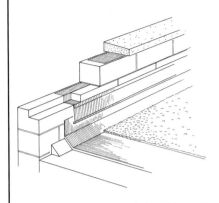

This page illustrates and describes a parapet wall system including wall, coping, flashing and cant strip. Lines within System Components give the unit price and cost per lineal foot for this system. Prices for parapet wall systems are on Lines B2010 142 1300 thru 2400. Both material quantities and labor costs have been adjusted for the system listed.

Factors: To adjust for job conditions other than normal working situations, use Lines B2010 142 2900 thru 4000.

Example: You are to install the system without damaging the adjacent property. Go to Line B2010 142 3600 and apply these percentages to the appropriate MAT. and INST. costs.

System Components	QUANTITY	UNIT	COST PER L.F.		
			MAT.	INST.	TOTAL
Concrete block parapet, incl. reinf., coping, flashing & cant strip, 2'high					
8" concrete block	2.000	S.F.	10.30	13.50	23.80
Masonry reinforcing	1.000	L.F.	.25	.20	.45
Coping, precast	1.000	L.F.	20.50	15.65	36.15
Roof cant	1.000	L.F.	.41	1.78	2.19
16 oz.	1.000	L.F.	7.05	2.74	9.79
Cap flashing	1.000	L.F.	8.45	5.05	13.50
TOTAL			46.96	38.92	85.88

B2010 142	Parapet Wall	COST PER L.F.		
		MAT.	INST.	TOTAL
1300	Concrete block, 8" thick	47	39	86
1400	12" thick	55	51	106
1500	Split rib block, 8" thick	47.50	44	91.50
1600	12" thick	53.50	52	105.50
1700	Brick, common brick, 8" wall	54.50	66	120.50
1800	12" wall	64	84	148
1900	4" brick, 4" backup block	49	64	113
2000	8" backup block	51.50	65.50	117
2100	Stucco on masonry	47.50	43.50	91
2200	On wood frame	20.50	35	55.50
2300	Wood, T 1-11 siding	32.50	42.50	75
2400	Boards, 1" x 6" cedar	39.50	47	86.50
2500				
2600				
2900	Dust protection, add, minimum	1%	2%	
3000	Maximum	4%	11%	
3100	Equipment usage curtailment, add, minimum	1%	1%	
3200	Maximum	3%	10%	
3300	Material handling & storage limitation, add, minimum	1%	1%	
3400	Maximum	6%	7%	
3500	Protection of existing work, add, minimum	2%	2%	
3600	Maximum	5%	7%	
3700	Shift work requirements, add, minimum		5%	
3800	Maximum		30%	
3900	Temporary shoring and bracing, add, minimum	2%	5%	
4000	Maximum	5%	12%	

B20 Exterior Enclosure

B2010 Exterior Walls

This page illustrates and describes a masonry cleaning and restoration system, including staging, cleaning, repointing. Lines within System Components give the unit price and cost per square foot for this system. Prices for systems are on Lines B2010 143 1300 thru 2500. Both material quantities and labor costs have been adjusted for the system listed.

Factors: To adjust for conditions other than normal working situations, use Lines B2010 143 2900 thru 4200.

Example: You are to clean a wall and be concerned about dust control. Go to line B2010 143 3100 and apply these percentages to the appropriate MAT. and INST. costs.

System Components	QUANTITY	UNIT	COST PER S.F.		
			MAT.	INST.	TOTAL
Repoint existing building, brick, common bond, high pressure cleaning, Water only, soft old mortar.					
Scaffolding, steel tubular, bldg ext wall face, labor only, 1 to 5 stories	.010	C.S.F.		2.31	2.31
Cleaning masonry, high pressure wash, water only, avg soil, bio staining	1.100	S.F.		1.68	1.68
Repoint, brick common bond	1.000	S.F.	.61	6.25	6.86
TOTAL			.61	10.24	10.85

B2010 143	Masonry Restoration - Cleaning	COST PER S.F.		
		MAT.	INST.	TOTAL
1300	Brick, common bond	.61	10.25	10.86
1400	Flemish bond	.65	10.70	11.35
1500	English bond	.65	11.35	12
1700	Stone, 2' x 2' blocks	.82	7.75	8.57
1800	2' x 4' blocks	.62	6.85	7.47
2000	Add to above prices for alternate cleaning systems:			
2100	Chemical brush and wash	.05	.69	.74
2200	High pressure chemical and water	.08	.57	.65
2300	Sandblasting, wet system	.38	1.64	2.02
2400	Dry system	.36	.97	1.33
2500	Steam cleaning		.47	.47
2600				
2900	Cut & patch to match existing construction, add, minimum	2%	3%	
3000	Maximum	5%	9%	
3100	Dust protection, add, minimum	1%	2%	
3200	Maximum	4%	11%	
3300	Equipment usage curtailment, add, minimum	1%	1%	
3400	Maximum	3%	10%	
3500	Material handling & storage limitation, add, minimum	1%	1%	
3600	Maximum	6%	7%	
3900	Protection of existing work, add, minimum	2%	2%	
4000	Maximum	5%	7%	
4100	Shift work requirements, add, minimum		5%	
4200	Maximum		30%	

B20 Exterior Enclosure

B2010 Exterior Walls

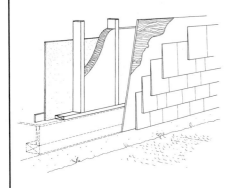

This page illustrates and describes a wood frame exterior wall system including wood studs, sheathing, felt, insulation, plasterboard, taped and finished, baseboard and painted interior. Lines within System Components give the unit price and total price per square foot for this system. Prices for wood frame exterior wall systems are on Line Items B2010 150 1600 thru 2700. Both material quantities and labor costs have been adjusted for the system listed.

Factors: To adjust for job conditions other than normal working situations use Lines B2010 150 3100 thru 4200.

Example: You are to install the system with need for complete temporary bracing. Go to Line B2010 150 4200 and apply these percentages to the appropriate MAT. and INST. costs.

System Components	QUANTITY	UNIT	COST PER S.F.		
			MAT.	INST.	TOTAL
Wood stud wall, cedar shingle siding, building paper, plywood sheathing, Insulation, 5/8" drywall, taped, finished and painted, baseboard.					
2" x 4" wood studs, 16" O.C.	.100	L.F.	.49	1.23	1.72
1/2" CDX sheathing	1.000	S.F.	.72	.88	1.60
18" No. 1 red cedar shingles, 7-1/2" exposure	.008	C.S.F.	1.59	2.19	3.78
15# felt paper	.010	C.S.F.	.06	.17	.23
3-1/2" fiberglass insulation	1.000	S.F.	.50	.46	.96
5/8" drywall	1.000	S.F.	.36	.62	.98
Drywall finishing adder	1.000	S.F.	.05	.62	.67
Baseboard trim, stock pine, 9/16" x 3-1/2", painted	.100	L.F.	.12	.26	.38
Paint baseboard, primer + 1 coat enamel	.100	L.F.	.01	.07	.08
Paint, 2 coats, interior	1.000	S.F.	.21	.68	.89
TOTAL			4.11	7.18	11.29

B2010 150	Wood Frame Exterior Wall	COST PER S.F.		
		MAT.	INST.	TOTAL
1600	Cedar shingle siding, painted	4.11	7.20	11.31
1700	Aluminum siding, horizontal clapboard	5.20	7.40	12.60
1800	Cedar bevel siding, 1/2" x 6", vertical, painted	7.05	7.10	14.15
1900	Redwood siding 1" x 4" to 1" x 6" vertical, T & G	7.05	7.80	14.85
2000	Board and batten	7.45	6.45	13.90
2100	Ship lap siding	7.50	6.85	14.35
2200	Plywood, grooved (T1-11) fir	4.25	6.80	11.05
2300	Redwood	5.35	6.80	12.15
2400	Southern yellow pine	4.03	6.80	10.83
2500	Masonry on stud wall, stucco, wire and plaster	3.27	10.75	14.02
2600	Stone veneer	9.60	14.10	23.70
2700	Brick veneer, brick $275 per M	6.60	17.70	24.30
3100	Cut & patch to match existing construction, add, minimum	2%	3%	
3200	Maximum	5%	9%	
3300	Dust protection, add, minimum	1%	2%	
3400	Maximum	4%	11%	
3500	Material handling & storage limitation, add, minimum	1%	1%	
3600	Maximum	6%	7%	
3700	Protection of existing work, add, minimum	2%	2%	
3800	Maximum	5%	7%	
3900	Shift work requirements, add, minimum		5%	
4000	Maximum		30%	
4100	Temporary shoring and bracing, add, minimum	2%	5%	
4200	Maximum	5%	12%	

B20 Exterior Enclosure

B2010 Exterior Walls

B2010 190	Selective Price Sheet	COST PER S.F.		
		MAT.	INST.	TOTAL
0100	Exterior surface, masonry, concrete block, standard 4" thick	1.87	6.50	8.37
0200	6" thick	2.45	7	9.45
0300	8" thick	3.02	7.45	10.47
0400	12" thick	4.35	9.65	14
0500	Split rib, 4" thick	4.23	8.10	12.33
0600	8" thick	5.45	9.30	14.75
0700	Brick running bond, standard size, 6.75/S.F.	4.10	12.70	16.80
0800	Buff, 6.75/S.F.	4.32	12.70	17.02
0900	Stucco, on frame	.76	5.80	6.56
1000	On masonry	.31	4.59	4.90
1100	Metal, aluminum, horizontal, plain	2.70	2.39	5.09
1200	Insulated	2.94	2.39	5.33
1300	Vertical, plain	2.09	2.39	4.48
1400	Insulated	2.18	2.49	4.67
1500	Wood, beveled siding, "A" grade cedar, 1/2" x 6"	4.55	2.09	6.64
1600	1/2" x 8"	6.50	1.87	8.37
1700	Shingles, 16" #1 red, 7-1/2" exposure	2.10	3	5.10
1800	18" perfections, 7-1/2" exposure	2.26	2.51	4.77
1900	Handsplit, 10" exposure	3	2.46	5.46
2000	White cedar, 7-1/2" exposure	1.51	3.10	4.61
2100	Vertical, board & batten, redwood	4.52	2.80	7.32
2200	White pine	2.41	1.87	4.28
2300	T. & G. boards, redwood, 1" x 4"	2.79	2.80	5.59
2400	1' x 8"	4.51	1.87	6.38
2500				
2600	Interior surface, drywall, taped & finished, standard, 1/2"	.38	1.28	1.66
2700	5/8" thick	.41	1.28	1.69
2800	Fire resistant, 1/2" thick	.44	1.28	1.72
2900	5/8" thick	.43	1.28	1.71
3000	Moisture resistant, 1/2" thick	.48	1.28	1.76
3100	5/8" thick	.51	1.28	1.79
3200	Core board, 1" thick	.86	2.56	3.42
3300	Plaster, gypsum, 2 coats	.44	2.94	3.38
3400	3 coats	.63	3.53	4.16
3500	Perlite or vermiculite, 2 coats	.72	3.35	4.07
3600	3 coats	.78	4.17	4.95
3700	Gypsum lath, standard, 3/8" thick	.37	.70	1.07
3800	1/2" thick	.30	.75	1.05
3900	Fire resistant, 3/8" thick	.30	.86	1.16
4000	1/2" thick	.34	.93	1.27
4100	Metal lath, diamond, 2.5 lb.	.44	.70	1.14
4200	Rib, 3.4 lb.	.50	.86	1.36
4300	Framing metal studs including top and bottom			
4400	Runners, walls 10' high			
4500	24" O.C., non load bearing 20 gauge, 2-1/2" wide	.32	1.03	1.35
4600	3-5/8" wide	.37	1.04	1.41
4700	4" wide	.44	1.04	1.48
4800	6" wide	.51	1.06	1.57
4900	Load bearing 18 gauge, 2-1/2" wide	.66	1.15	1.81
5000	3-5/8" wide	.78	1.18	1.96
5100	4" wide	.88	1.49	2.37
5200	6" wide	1.04	1.22	2.26
5300	16" O.C., non load bearing 20 gauge, 2-1/2" wide	.43	1.58	2.01
5400	3-5/8" wide	.50	1.60	2.10
5500	4" wide	.59	1.62	2.21
5600	6" wide	.69	1.64	2.33
5700	Load bearing 18 gauge, 2-1/2" wide	.89	1.60	2.49
5800	3-5/8" wide	1.05	1.62	2.67

B20 Exterior Enclosure

B2010 Exterior Walls

B2010 190	Selective Price Sheet	COST PER S.F.		
		MAT.	INST.	TOTAL
5900	4" wide	1.11	1.67	2.78
6000	6" wide	1.40	1.69	3.09
6100	Framing wood studs incl. double top plate and			
6200	Single bottom plate, walls 10' high			
6300	24" O.C., 2" x 4"	.37	.99	1.36
6400	2" x 6"	.59	1.07	1.66
6500	16" O.C., 2" x 4"	.49	1.23	1.72
6600	2" x 6"	.77	1.37	2.14
6700	Sheathing, boards, 1" x 6"	1.73	1.89	3.62
6800	1" x 8"	2.09	1.61	3.70
6900	Plywood, 3/8" thick	.66	1.03	1.69
7000	1/2" thick	.72	1.09	1.81
7100	5/8" thick	.86	1.17	2.03
7200	3/4" thick	1.02	1.26	2.28
7300	Wood fiber, 5/8" thick	.85	1.03	1.88
7400	Gypsum weatherproof, 1/2" thick	.50	1.09	1.59
7500	Insulation, fiberglass batts, 3-1/2" thick, R13	.41	.88	1.29
7600	6" thick, R19	.54	1.03	1.57
7700	Poured 4" thick, fiberglass wool, R4/inch	.60	3.08	3.68
7800	Mineral wool, R3/inch	.45	3.08	3.53
7900	Polystyrene, R4/inch	1.76	3.08	4.84
8000	Perlite or vermiculite, R2.7/inch	5.75	3.08	8.83
8100	Rigid, fiberglass, R4.3/inch, 1" thick	.57	.62	1.19
8200	R8.7/inch, 2" thick	1.16	.69	1.85

B20 Exterior Enclosure

B2020 Exterior Windows

B2020 108 Vinyl Clad Windows

	MATERIAL	TYPE	GLAZING	SIZE	DETAIL	COST PER UNIT		
						MAT.	INST.	TOTAL
1000	Vinyl clad	casement	insul. glass	1'-4" x 4'-0"		420	142	562
1010				2'-0" x 3'-0"		360	142	502
1020				2'-0" x 4'-0"		410	149	559
1030				2'-0" x 5'-0"		460	157	617
1040				2'-0" x 6'-0"		480	157	637
1050				2'-4" x 4'-0"		460	157	617
1060				2'-6" x 5'-0"		565	153	718
1070				3'-0" x 5'-0"		790	157	947
1080				4'-0" x 3'-0"		865	157	1,022
1090				4'-0" x 4'-0"		745	157	902
1100				4'-8" x 4'-0"		820	157	977
1110				4'-8" x 5'-0"		925	183	1,108
1120				4'-8" x 6'-0"		1,025	183	1,208
1130				6'-0" x 4'-0"		945	183	1,128
1140				6'-0" x 5'-0"		1,050	183	1,233
3000		double-hung	insul. glass	2'-0" x 4'-0"		440	136	576
3050				2'-0" x 5'-0"		500	136	636
3100				2'-4" x 4'-0"		455	142	597
3150				2'-4" x 4'-8"		455	142	597
3200				2'-4" x 6'-0"		515	142	657
3250				2'-6" x 4'-0"		455	142	597
3300				2'-8" x 4'-0"		680	183	863
3350				2'-8" x 5'-0"		535	157	692
3375				2'-8" x 6'-0"		545	157	702
3400				3'-0 x 3'-6"		430	142	572
3450				3'-0 x 4'-0"		495	149	644
3500				3'-0 x 4'-8"		530	149	679
3550				3'-0 x 5'-0"		545	157	702
3600				3'-0 x 6'-0"		490	157	647
3700				4'-0 x 5'-0"		680	168	848
3800				4'-0 x 6'-0"		830	168	998

B20 Exterior Enclosure

B2020 Exterior Windows

This page illustrates and describes an aluminum window system including double hung aluminum window, exterior and interior trim, hardware and insulating glass. Lines within System Components give the unit price and total price on a cost each basis for this system. Prices for aluminum window systems are on Line Items B2020 110 1000 thru 2400. Both material quantities and labor costs have been adjusted for the system listed.

Factors: To adjust for job conditions other than normal working situations use Lines B2020 110 3100 thru 4000.

Example: You are to install the system and cut and patch to match existing construction. Go to Line B2020 110 3200 and apply these percentages to the appropriate MAT. and INST. costs.

System Components	QUANTITY	UNIT	COST EACH		
			MAT.	INST.	TOTAL
Double hung aluminum window, 2'-4" x 2'-6" exterior and interior trim, Hardware glazed with insulating glass.					
Double hung aluminum window, 2'-4" x 2' 6"	1.000	Ea.	253.61	44.31	297.92
Interior Trim	1.000	Set	34.50	47.50	82
Hardware	1.000	Set	4.84	25.50	30.34
TOTAL			292.95	117.31	410.26

B2020 110	Windows - Aluminum	COST EACH		
		MAT.	INST.	TOTAL
1000	Aluminum, double hung, 2'-4" x 2'-6"	293	117	410
1100	2'-8" x 4'-6"	570	178	748
1200	3'-4" x 5'-6"	875	268	1,143
1300	Casement, 3'-6" x 2'-4"	390	135	525
1400	4'-6" x 2'-4"	495	167	662
1500	5'-6" x 2'-4"	625	226	851
1600	Projected window, 2'-1" x 3'-0"	283	121	404
1700	3'-5" x 3'-0"	445	165	610
1800	4'-0" x 4'-0"	700	250	950
1900	Horizontal sliding 3'-0" x 3'-0"	251	141	392
2000	3'-6" x 4'-0"	375	193	568
2100	5'-0" x 6'-0"	780	355	1,135
2200	Picture window, 3'-8" x 3'-1"	271	159	430
2300	4'-4" x 4'-5"	440	232	672
2400	5'-8" x 4'-9"	625	335	960
2500				
2600				
2700				
2800				
2900				
3100	Cut & patch to match existing construction, add, minimum	2%	3%	
3200	Maximum	5%	9%	
3300	Dust protection, add, minimum	1%	2%	
3400	Maximum	4%	11%	
3500	Material handling & storage limitation, add, minimum	1%	1%	
3600	Maximum	6%	7%	
3700	Protection of existing work, add, minimum	2%	2%	
3800	Maximum	5%	7%	
3900	Shift work requirements, add, minimum		5%	
4000	Maximum		30%	

B20 Exterior Enclosure

B2020 Exterior Windows

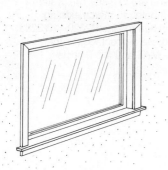

This page illustrates and describes a wood window system including wood picture window, exterior and interior trim, hardware and insulating glass. Lines within System Components give the unit price and total price on a cost each basis for this system. Prices for wood window systems are on Line Items B2020 112 1200 thru 2900. Both material quantities and labor costs have been adjusted for the system listed.

Factors: To adjust for job conditions other than normal working situations use Lines B2020 112 3100 thru 4000.

Example: You are to install the above system where dust control is vital. Go to Line B2020 112 3400 and apply these percentages to the appropriate MAT. and INST. costs.

System Components	QUANTITY	UNIT	COST EACH		
			MAT.	INST.	TOTAL
Wood picture window 4'-0" x 4'-6", exterior and interior trim, hardware, Glazed with insulating glass.					
4'-0" x 4'-6" wood picture window with insulating glass	1.000	Ea.	585	112	697
Oak interior trim	1.000	Set	71	103	174
Hardware	1.000	Set	4.84	25.50	30.34
TOTAL			660.84	240.50	901.34

B2020 112	Windows - Wood		COST EACH		
			MAT.	INST.	TOTAL
1200	Picture window, 4'-0" x 4'-6"		660	241	901
1300	5'-0" x 4'-0"		740	241	981
1400	6'-0" x 4'-6"		775	252	1,027
1500	Bow, bay window, 8'-0" x 5'-0", standard		2,575	252	2,827
1600	Deluxe		1,900	252	2,152
1700	Bow, bay window, 12'-0" x 6'-0" standard		3,300	335	3,635
1800	Deluxe		2,650	335	2,985
1900	Double hung, 3'-0" x 4'-0"		365	156	521
2000	4'-0" x 4'-6"		455	206	661
2100	Casement 2'-0" x 3'-0"		325	135	460
2200	2 leaf, 4'-0" x 4'-0"		855	224	1,079
2300	3 leaf, 6'-0" x 6'-0"		1,650	360	2,010
2400	Awning, 2'-10" x 1'-10"		335	135	470
2500	3'-6" x 2'-4"		420	156	576
2600	4'-0" x 3'-0"		615	206	821
2700	Horizontal sliding 3'-0" x 2'-0"		380	135	515
2800	4'-0" x 3'-6"		445	156	601
2900	6'-0" x 5'-0"		655	206	861
3100	Cut & patch to match existing construction, add, minimum		2%	3%	
3200	Maximum		5%	9%	
3300	Dust protection, add, minimum		1%	2%	
3400	Maximum		4%	11%	
3500	Material handling & storage limitation, add, minimum		1%	1%	
3600	Maximum		6%	7%	
3700	Protection of existing work, add, minimum		2%	2%	
3800	Maximum		5%	7%	
3900	Shift work requirements, add, minimum			5%	
4000	Maximum			30%	

B20 Exterior Enclosure

B2020 Exterior Windows

This page illustrates and describes storm window and door systems based on a cost each price. Prices for storm window and door systems are on Line Items B2020 116 0400 thru 2400. Both material quantities and labor costs have been adjusted for the system listed.

Factors: To adjust for job conditions other than normal working situations use Lines B2020 116 3100 thru 4000 and apply these percentages to the appropriate MAT and INST. costs.

Example: You are to install the system and protect all existing construction. Go to Line B2020 116 3800 and apply these percentages to the appropriate MAT. and INST. costs.

B2020 116	Storm Windows & Doors	MAT.	INST.	TOTAL
0350	Storm Window and door systems, single glazed			
0400	Window, custom aluminum anodized, 2'-0" x 3'-5"	105	41	146
0500	2'-6" x 5'-0"	127	44	171
0600	4'-0" x 6'-0"	253	49.50	302.50
0700	White painted aluminum, 2'-0" x 3'-5"	121	41	162
0800	2'-6" x 5'-0"	187	44	231
0900	4'-0" x 6'-0"	310	49.50	359.50
1000	Average quality aluminum, anodized, 2'-0" x 3'-5"	105	41	146
1100	2'-6" x 5'-0"	128	44	172
1200	4'-0" x 6'-0"	143	49.50	192.50
1300	White painted aluminum, 2'-0" x 3'-5"	99	41	140
1400	2'-6" x 5'-0"	105	44	149
1500	4'-0" x 6'-0"	121	49.50	170.50
1600	Mill finish, 2'-0" x 3'-5"	93.50	41	134.50
1700	2'-6" x 5'-0"	99	44	143
1800	4'-0" x 6'-0"	121	49.50	170.50
1900				
2000	Door, aluminum anodized 3'-0" x 6'-8"	195	88	283
2100	White painted aluminum, 3'-0" x 6'-8"	340	88	428
2200	Mill finish, 3'-0" x 6'-8"	284	88	372
2300				
2400	Wood, storm and screen, painted	340	137	477
2500				
2600				
2700				
2800				
3100	Cut & patch to match existing construction, add, minimum	2%	3%	
3200	Maximum	5%	9%	
3300	Dust protection, add, minimum	1%	2%	
3400	Maximum	4%	11%	
3500	Material handling & storage limitation, add, minimum	1%	1%	
3600	Maximum	6%	7%	
3700	Protection of existing work, add, minimum	2%	2%	
3800	Maximum	5%	7%	
3900	Shift work requirements, add, minimum		5%	
4000	Maximum		30%	

B20 Exterior Enclosure

B2020 Exterior Windows

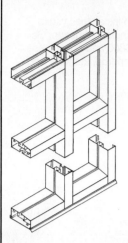

This page illustrates and describes a window wall system including aluminum tube framing, caulking, and glass. Lines within System Components give the unit price and total price per square foot for this system. Prices for window wall systems are on Line Items B2020 124 1300 thru 2600. Both material quantities and labor costs have been adjusted for the system listed.

Factors: To adjust for job conditions other than normal working situations use Lines B2020 124 3100 thru 4000.

Example: You are to install the system with need for complete dust protection. Go to Line B2020 124 3400 and apply these percentages to the appropriate MAT. and INST. costs.

System Components			COST PER S.F.		
	QUANTITY	UNIT	MAT.	INST.	TOTAL
Window wall, including aluminum header, sill, mullions, caulking And glass					
Header, mill finish, 1-3/4" x 4-1/2" deep	.167	L.F.	3.51	2.35	5.86
Sill, mill finish, 1-3/4" x 4-1/2" deep	.167	L.F.	3.51	2.30	5.81
Vertical mullion, 1-3/4" x 4-1/2" deep, 6' O.C.	.191	L.F.	8.77	5.75	14.52
Caulking	.381	L.F.	.10	.80	.90
Glass, 1/4" plate	1.000	S.F.	6.55	9.75	16.30
TOTAL			22.44	20.95	43.39

B2020 124	Aluminum Frame, Window Wall	COST PER S.F.		
		MAT.	INST.	TOTAL
1300	Mill finish, 1-3/4" x 4-1/2" deep, 1/4" plate glass	22.50	21	43.50
1400	2" x 4-1/2" deep, insulating glass	24	24.50	48.50
1500	Thermo-break, 2-1/4" x 4-1/2" deep, insulating glass	25.50	24.50	50
1600	Bronze finish, 1-3/4" x 4-1/2" deep, 1/4" plate glass	25.50	23	48.50
1700	2" x 4-1/2" deep, insulating glass	26.50	26.50	53
1800	Thermo-break, 2-1/4" x 4-1/2" deep, insulating glass	28	26.50	54.50
2000	Black finish, 1-3/4" x 4-1/2" deep, 1/4" plate glass	26.50	24	50.50
2100	2" x 4-1/2" deep, insulating glass	27.50	27.50	55
2200	Thermo-break, 2-1/4" x 4-1/2" deep, insulating glass	29.50	28	57.50
2300				
2400	Stainless steel, 1-3/4" x 4-1/2" deep, 1/4" plate glass	34.50	29	63.50
2500	2" x 4-1/2" deep, insulating glass	34	32.50	66.50
2600	Thermo-break, 2-1/4" x 4-1/2" deep, insulating glass	36.50	33.50	70
2700				
3100	Cut & patch to match existing construction, add, minimum	2%	3%	
3200	Maximum	5%	9%	
3300	Dust protection, add, minimum	1%	2%	
3400	Maximum	4%	11%	
3500	Material handling, & storage limitation, add, minimum	1%	1%	
3600	Maximum	6%	7%	
3700	Protection of existing work, add, minimum	2%	2%	
3800	Maximum	5%	7%	
3900	Shift work requirements, add, minimum		5%	
4000	Maximum		30%	

B20 Exterior Enclosure

B2030 Exterior Doors

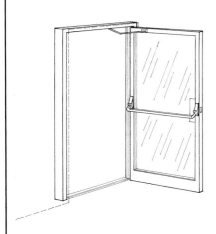

This page illustrates and describes a commercial metal door system, including a single aluminum and glass door, narrow stiles, jamb, hardware weatherstripping, panic hardware and closer. Lines within System Components give the unit price and total price on a cost each basis for this system. Prices for commercial metal door systems are on Line Items B2030 125 1200 thru 2500. Both material quantities and labor costs have been adjusted for the system listed.

Factors: To adjust for job conditions other than normal working situations, use Lines B2030 125 3100 thru 4000.

Example: You are to install the system and cut and patch to match existing construction. Go to Line B2030 125 3200 and apply these percentages to the appropriate MAT. and INST. costs.

System Components	QUANTITY	UNIT	COST EACH		
			MAT.	INST.	TOTAL
Single aluminum and glass door, 3'-0"x7'-0", with narrow stiles, ext. jamb. Weatherstripping, 1/2" tempered insul. glass, panic hardware, and closer.					
Aluminum door, 3'-0" x 7'-0" x 1-3/4", narrow stiles	1.000	Ea.	1,000	760	1,760
Tempered insulating glass, 1/2" thick	20.000	S.F.	550	430	980
Panic hardware	1.000	Set	505	103	608
Surface mount regular arm	1.000	Ea.	168	94.50	262.50
TOTAL			2,223	1,387.50	3,610.50

B2030 125	Doors, Metal - Commercial	COST EACH		
		MAT.	INST.	TOTAL
1200	Single aluminum and glass, 3'-0" x 7'-0"	2,225	1,400	3,625
1300	With transom, 3'-0" x 10'-0"	2,875	1,650	4,525
1400	Anodized aluminum and glass, 3'-0" x 7'-0"	2,575	1,650	4,225
1500	With transom, 3'-0" x 10'-0"	3,375	1,950	5,325
1600	Steel, deluxe, hollow metal 3'-0" x 7'-0"	1,675	530	2,205
1700	With transom 3'-0" x 10'-0"	2,200	600	2,800
1800	Fire door, "A" label, 3'-0" x 7'-0"	1,825	530	2,355
1900	Double, aluminum and glass, 6'-0" x 7'-0"	3,550	2,375	5,925
2000	With transom, 6'-0" x 10'-0"	4,500	2,925	7,425
2100	Anodized aluminum and glass 6'-0" x 7'-0"	4,025	2,800	6,825
2200	With transom, 6'-0" x 10'-0"	5,150	3,400	8,550
2300	Steel, deluxe, hollow metal, 6'-0" x 7'-0"	3,000	920	3,920
2400	With transom, 6'-0" x 10'-0"	4,075	1,075	5,150
2500	Fire door, "A" label, 6'-0" x 7'-0"	3,200	910	4,110
2800				
2900				
3100	Cut & patch to match existing construction, add, minimum	2%	3%	
3200	Maximum	5%	9%	
3300	Dust protection, add, minimum	1%	2%	
3400	Maximum	4%	11%	
3500	Material handling & storage limitation, add, minimum	1%	1%	
3600	Maximum	6%	7%	
3700	Protection of existing work, add, minimum	2%	2%	
3800	Maximum	5%	7%	
3900	Shift work requirements, add, minimum		5%	
4000	Maximum		30%	

B20 Exterior Enclosure

B2030 Exterior Doors

This page illustrates and describes sliding door systems including a sliding door, frame, interior and exterior trim with exterior staining. Lines within System Components give the unit price and total price on a cost each basis for this system. Prices for sliding door systems are on Line Items B2030 150 1000 thru 2400. Both material quantities and labor costs have been adjusted for the system listed.

Factors: To adjust for job conditions other than normal working situations use Lines B2030 150 2700 thru 4000.

Example: You are to install the system with temporary shoring and bracing. Go to Line B2030 150 3900 and apply these percentages to the appropriate MAT. and INST. costs.

System Components	QUANTITY	UNIT	COST EACH		
			MAT.	INST.	TOTAL
Sliding wood door, 6'-0" x 6'-8", with wood frame, interior and exterior Trim and exterior staining.					
Sliding wood door, standard, 6'-0" x 6'-8", insulated glass	1.000	Ea.	1,275	310	1,585
Interior & exterior trim	1.000	Set	31	49.20	80.20
Stain door	1.000	Ea.	2.43	30.50	32.93
Stain trim	1.000	Ea.	2.80	13	15.80
TOTAL			1,311.23	402.70	1,713.93

B2030 150	Doors, Sliding - Patio	COST EACH		
		MAT.	INST.	TOTAL
1000	Wood, standard, 6'-0" x 6'-8", insulated glass	1,300	405	1,705
1100	8'-0" x 6'-8"	1,700	510	2,210
1200	12'-0" x 6'-8"	2,750	615	3,365
1300	Vinyl coated, 6'-0" x 6'-8"	1,600	405	2,005
1400	8'-0" x 6'-8"	2,050	510	2,560
1500	12'-0" x 6'-8"	3,350	615	3,965
1700	Aluminum, standard, 6'-0" x 6'-8", insulated glass	955	435	1,390
1800	8'-0" x 6'-8"	1,625	545	2,170
1900	12'-0" x 6'-8"	1,775	650	2,425
2000	Anodized, 6'-0" x 6'-8"	1,775	435	2,210
2100	8'-0" x 6'-8"	1,900	545	2,445
2200	12'-0" x 6'-8"	3,350	650	4,000
2300				
2400	Deduct for single glazing	81		81
2500				
2700	Cut & patch to match existing construction, add, minimum	2%	3%	
2800	Maximum	5%	9%	
2900	Dust protection, add, minimum	1%	2%	
3000	Maximum	4%	11%	
3100	Equipment usage curtailment, add, minimum	1%	1%	
3200	Maximum	3%	10%	
3300	Material handling & storage limitation, add, minimum	1%	1%	
3400	Maximum	6%	7%	
3500	Protection of existing, work, add, minimum	2%	2%	
3600	Maximum	5%	7%	
3700	Shift work requirements, add, minimum		5%	
3800	Maximum		30%	
3900	Temporary shoring and bracing, add, minimum	2%	5%	
4000	Maximum	5%	12%	

B20 Exterior Enclosure

B2030 Exterior Doors

This page illustrates and describes residential door systems including a door, frame, trim, hardware, weatherstripping, stained and finished. Lines within System Components give the unit price and total price on a cost each basis for this system. Prices for residential door systems are on Line Items B2030 240 1500 thru 2200. Both material quantities and labor costs have been adjusted for the system listed.

Factors: To adjust for job conditions other than normal working situations use Lines B2030 240 3100 thru 4000.

Example: You are to install the system with a material handling and storage limitation. Go to Line B2030 240 3600 and apply these percentages to the appropriate MAT. and INST. costs.

System Components	QUANTITY	UNIT	COST EACH		
			MAT.	INST.	TOTAL
Single solid wood colonial door, 3' x 6'-8" with wood frame and trim,					
Stained and finished, including hardware and weatherstripping.					
Solid wood colonial door, fir 3' x 6'-8" x 1-3/4", hinges	1.000	Ea.	490	82	572
Hinges, ball bearing	1.000	Ea.	38.50		38.50
Exterior frame, with trim	1.000	Set	164.05	55.76	219.81
Interior trim	1.000	Set	26	51.50	77.50
Average quality	1.000	Set	55	44	99
Sill, oak 8" deep	3.500	Ea.	78.75	43.05	121.80
Weatherstripping	1.000	Set	25.50	81	106.50
Stained and finished	1.000	Ea.	5.20	65	70.20
TOTAL			883	422.31	1,305.31

B2030 240	Doors, Residential - Exterior	COST EACH		
		MAT.	INST.	TOTAL
1400	Single doors, 3' x 6'-8"			
1500	Solid colonial door, fir	885	420	1,305
1600	Hollow metal exterior door, plain	855	395	1,250
1700	Solid core wood door, plain	525	405	930
1800				
1900	Double doors, 6' x 6'-8"			
2000	Solid colonial double doors, fir	1,475	560	2,035
2100	Hollow metal exterior doors, plain	1,425	525	1,950
2200	Hollow core wood doors, plain	755	545	1,300
2300				
2800				
2900				
3100	Cut & patch to match existing construction, add, minimum	2%	3%	
3200	Maximum	5%	9%	
3300	Dust protection, add, minimum	1%	2%	
3400	Maximum	4%	11%	
3500	Material handling & storage limitation, add, minimum	1%	1%	
3600	Maximum	6%	7%	
3700	Protection of existing work, add, minimum	2%	2%	
3800	Maximum	5%	7%	
3900	Shift work requirements, add, minimum		5%	
4000	Maximum		30%	

B20 Exterior Enclosure

B2030 Exterior Doors

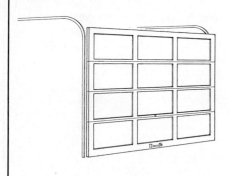

This page illustrates and describes overhead door systems including an overhead door, track, hardware, trim, and electric door opener. Lines within System Components give the unit price and total price on a cost each basis for this system. Prices for overhead door systems are on Line Items B2030 410 1200 thru 2300. Both material quantities and labor costs have been adjusted for the system listed.

Factors: To adjust for job conditions other than normal working situations use Lines B2030 410 3100 thru 4000.

Example: You are to install the system and match existing construction. Go to Line B2030 410 3200 and apply these percentages to the appropriate MAT. and INST. costs.

System Components			COST EACH		
	QUANTITY	UNIT	MAT.	INST.	TOTAL
Wood overhead door, commercial sectional, including track, door, hardware And trim, electrically operated.					
Commercial, heavy duty wood door, 8' x 8' x 1-3/4" thick	1.000	Ea.	1,150	615	1,765
Frame, 2 x 8, pressure treated	1.000	Set	27.84	77.76	105.60
Wood trim	1.000	Set	29.76	61.44	91.20
Painting, two coats	1.000	Ea.	21.60	207.36	228.96
Electric trolley operator	1.000	Ea.	1,050	310	1,360
TOTAL			2,279.20	1,271.56	3,550.76

B2030 410	Doors, Overhead	COST EACH		
		MAT.	INST.	TOTAL
1200	Commercial, wood, 1-3/4" thick, 8' x 8'	2,275	1,275	3,550
1300	12' x 12'	3,425	1,800	5,225
1400	14' x 14'	4,850	2,100	6,950
1500	Fiberglass & aluminum 12' x 12'	4,150	1,825	5,975
1600	20' x 20'	8,850	4,625	13,475
1700	Residential, wood, 9' x 7'	1,150	480	1,630
1800	16' x 7'	1,900	720	2,620
1900	Hardboard faced, 9' x 7'	760	480	1,240
2000	16' x 7'	1,450	720	2,170
2100	Fiberglass & aluminum, 9' x 7'	1,125	560	1,685
2200	16' x 7'	1,825	720	2,545
2300	For residential electric opener, add	675	77	752
2400				
2500				
2600				
2700				
2800				
2900				
3100	Cut & patch to match existing construction, add, minimum	2%	3%	
3200	Maximum	5%	9%	
3300	Dust protection, add, minimum	1%	2%	
3400	Maximum	4%	11%	
3500	Material handling & storage limitation, add, minimum	1%	1%	
3600	Maximum	6%	7%	
3700	Protection of existing work, add, minimum	2%	2%	
3800	Maximum	5%	7%	
3900	Shift work requirements, add, minimum		5%	
4000	Maximum		30%	

B20 Exterior Enclosure

B2030 Exterior Doors

B2030 810	Selective Price Sheet	COST EACH		
		MAT.	INST.	TOTAL
0100	Door closer, rack and pinion	211	103	314
0200	Backcheck and adjustable power	211	103	314
0300	Regular, hinge face mount, all sizes, regular arm	211	103	314
0400	Hold open arm	220	103	323
0500	Top jamb mount, all sizes, regular arm	211	103	314
0600	Hold open arm	220	103	323
0700	Stop face mount, all sizes, regular arm	211	103	314
0800	Hold open arm	220	103	323
0900	Fusible link, hinge face mount, all sizes, regular arm	250	103	353
1000	Hold open arm	259	103	362
1100	Top jamb mount, all sizes, regular arm	250	103	353
1200	Hold open arm	259	103	362
1300	Stop face mount, all sizes, regular arm	250	103	353
1400	Hold open arm	259	103	362
1500				
1600				
1700	Door stops			
1800				
1900	Holder & bumper, floor or wall	37.50	19.25	56.75
2000	Wall bumper	12.85	19.25	32.10
2100	Floor bumper	8.90	19.25	28.15
2200	Plunger type, door mounted	30.50	19.25	49.75
2300	Hinges, full mortise, material only, per pair			
2400	Low frequency, 4-1/2" x 4-1/2", steel base, USP	24.50		24.50
2500	Brass base, US10	58.50		58.50
2600	Stainless steel base, US32	82		82
2700	Average frequency, 4-1/2" x 4-1/2", steel base, USP	40		40
2800	Brass base, US10	64.50		64.50
2900	Stainless steel base, US32	81.50		81.50
3000	High frequency, 4-1/2" x 4-1/2", steel base, USP	69		69
3100	Brass base, US10	81		81
3200	Stainless steel base, US32	114		114
3300	Kick plate			
3400				
3500	6" high, for 3'-0" door, aluminum	31.50	41	72.50
3600	Bronze	61.50	41	102.50
3700	Panic device			
3800				
3900	For rim locks, single door, exit	505	103	608
4000	Outside key and pull	630	123	753
4100	Bar and vertical rod, exit only	955	123	1,078
4200	Outside key and pull	1,075	154	1,229
4300	Lockset			
4400				
4500	Heavy duty, cylindrical, passage doors	138	51.50	189.50
4600	Classroom	550	77	627
4700	Bedroom, bathroom, and inner office doors	171	51.50	222.50
4800	Apartment, office, and corridor doors	330	61.50	391.50
4900	Standard duty, cylindrical, exit doors	118	61.50	179.50
5100	Passage doors	82.50	51.50	134
5200	Public restroom, classroom, & office doors	220	77	297
5300	Deadlock, mortise, heavy duty	193	68.50	261.50
5400	Double cylinder	193	68.50	261.50
5500	Entrance lock, cylinder, deadlocking latch	193	68.50	261.50
5600	Deadbolt	193	77	270
5700	Commercial, mortise, wrought knob, keyed, minimum	208	77	285
5800	Maximum	550	88	638
5900	Cast knob, keyed, minimum	286	68.50	354.50

B20 Exterior Enclosure

B2030 Exterior Doors

B2030 810	Selective Price Sheet	COST EACH		
		MAT.	INST.	TOTAL
6000	Maximum	580	68.50	648.50
6100	Push-pull			
6200				
6300	Aluminum	14.05	51.50	65.55
6400	Bronze	27.50	51.50	79
6500	Door pull, designer style, minimum	28	51.50	79.50
6600	Maximum	260	112	372
6700	Threshold			
6800				
6900	3'-0" long door saddles, aluminum, minimum	10.35	12.85	23.20
7000	Maximum	53.50	51.50	105
7100	Bronze, minimum	46.50	10.25	56.75
7200	Maximum	173	51.50	224.50
7300	Rubber, 1/2" thick, 5-1/2" wide	44	31	75
7400	2-3/4" wide	49.50	31	80.50
7500	Weatherstripping, per set			
7600				
7700	Doors, wood frame, interlocking for 3' x 7' door, zinc	48.50	205	253.50
7800	Bronze	61.50	205	266.50
7900	Wood frame, spring type for 3' x 7' door, bronze	25.50	81	106.50
8000	Metal frame, spring type for 3' x 7' door, bronze	51	205	256
8100	For stainless steel, spring type, add	133%		
8200				
8300	Metal frame, extruded sections, 3' x 7' door, aluminum	31	205	236
8400	Bronze	90	205	295

B30 Roofing

B3010 Roof Coverings

B3010 160	Selective Price Sheet	COST PER S.F.		
		MAT.	INST.	TOTAL
0100	Roofing, built-up, asphalt roll roof, 3 ply organic/mineral surface	.82	1.64	2.46
0200	3 plies glass fiber felt type iv, 1 ply mineral surface	1.38	1.77	3.15
0300	Cold applied, 3 ply		.57	.57
0400	Coal tar pitch, 4 ply tarred felt	2.57	2.11	4.68
0500	Mopped, 3 ply glass fiber	2.12	2.33	4.45
0600	4 ply organic felt	2.57	2.11	4.68
0700	Elastomeric, hypalon, neoprene unreinforced	1.45	3.89	5.34
0800	Polyester reinforced	1.59	4.60	6.19
0900	Neoprene, 5 coats 60 mils	4.84	13.65	18.49
1000	Over 10,000 S.F.	4.84	7.05	11.89
1600				
1700	Asphalt, strip, 210-235#/sq.	.88	1.05	1.93
1800	235-240#/sq.	1.03	1.16	2.19
1900	Class A laminated	1.21	1.29	2.50
2000	Class C laminated	1.21	1.45	2.66
2100	Slate, buckingham, 3/16" thick	5.15	3.30	8.45
2200	Black, 1/4" thick	5.15	3.30	8.45
2300	Wood, shingles, 16" no. 1, 5" exp.	3.15	2.46	5.61
2400	Red cedar, 18" perfections	2.71	2.24	4.95
2500	Shakes, 24", 10" exposure	3	2.46	5.46
2600	18", 8-1/2" exposure	1.98	3.10	5.08
2700	Insulation, ceiling batts, fiberglass, 3-1/2" thick, R13	.33	.46	.79
2800	6" thick, R19	.46	.46	.92
2900	9" thick, R30	.78	.54	1.32
3000	12" thick, R38	1.06	.54	1.60
3100	Mineral fiber, 3-1/2" thick, R13	.66	.38	1.04
3200	6" thick, R19	1.04	.38	1.42
3300	Roof deck, fiberboard, 1" thick, R2.78	.68	.75	1.43
3400	Mineral fiber, 2" thick, R5.26	1.28	.75	2.03
3500	Perlite boards, 3/4" thick, R2.08	.48	.75	1.23
3600	2" thick, R5.26	1.21	.83	2.04
3700	Polystyrene extruded, 15 PSI, 1" thick, R5.26	.72	.49	1.21
3800	2" thick, R10	.90	.55	1.45
3900	40 PSI, 1" thick, R5	.69	.49	1.18
4000	Tapered for drainage	.95	.51	1.46
4300	Ceiling, plaster, gypsum, 2 coats	.44	3.35	3.79
4400	3 coats	.63	3.94	4.57
4500	Perlite or vermiculite, 2 coats	.72	3.88	4.60
4600	3 coats	.78	4.88	5.66
4700	Gypsum lath, plain 3/8" thick	.37	.70	1.07
4800	1/2" thick	.30	.75	1.05
4900	Firestop, 3/8" thick	.30	.86	1.16
5000	1/2" thick	.34	.93	1.27
5100	Metal lath, rib, 2.75 lb.	.42	.80	1.22
5200	3.40 lb.	.50	.86	1.36
5300	Diamond, 2.50 lb.	.44	.80	1.24
5400	3.40 lb.	.52	1	1.52
5500	Drywall, taped and finished standard, 1/2" thick	.38	1.61	1.99
5600	5/8" thick	.41	1.61	2.02
5700	Fire resistant, 1/2" thick	.44	1.61	2.05
5800	5/8" thick	.43	1.61	2.04
5900	Water resist., 1/2" thick	.48	1.61	2.09
6000	5/8" thick	.51	1.61	2.12
6100	Finish, instead of taping			
6200	For thin coat plaster, add	.12	.77	.89
6300	Finish textured spray, add	.04	.68	.72
6500	Tile, stapled glued, mineral fiber plastic coated, 5/8" thick	2.40	2.05	4.45
6600	3/4" thick	2.64	2.05	4.69

B30 Roofing

B3010 Roof Coverings

B3010 160	Selective Price Sheet	COST PER S.F.		
		MAT.	INST.	TOTAL
6700	Wood fiber, 1/2" thick	1.23	1.54	2.77
6800	3/4" thick	1.35	1.54	2.89
6900	Suspended, fiberglass film faced, 5/8" thick	1.36	.99	2.35
7000	3" thick	2.66	1.37	4.03
7100	Mineral fiber 5/8" thick, standard face	.75	.91	1.66
7200	Aluminum faced	3.98	1.03	5.01
7500	Ceiling suspension systems, for tile, "T" bar, class "A", 2' x 4' grid	.81	.77	1.58
7600	2' x 2' grid	1.04	.95	1.99
7700	Concealed "Z" bar, 12" module	.92	1.18	2.10
7800				
7900	Plaster/drywall, 3/4" channels, steel furring, 16" O.C.	.36	1.87	2.23
8000	24" O.C.	.24	1.29	1.53
8100	1-1/2" channels, 16" O.C.	.48	2.08	2.56
8200	24" O.C.	.32	1.39	1.71
8300	Ceiling framing, 2" x 4" studs, 16" O.C.	.35	1.03	1.38
8400	24" O.C.	.24	.69	.93

B30 Roofing

B3020 Roof Openings

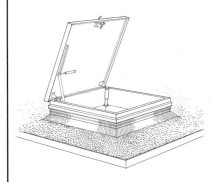

This page illustrates and describes a roof hatch system. Lines within System Components give the unit price and total cost each for this system. Prices for systems are on Lines B3020 230 1400 thru 2400. Both material quantities and labor costs have been adjusted for the system listed.

Factors: To adjust for job conditions other than normal working situtations use Lines B3020 230 2900 thru 4000.

Example: You are to install the system and cut and patch to match existing construction. Use line B3020 230 3000 and apply these percentages to the appropriate MAT. and INST. costs.

System Components	QUANTITY	UNIT	COST EACH		
			MAT.	INST.	TOTAL
Roof hatch, 2'-6" x 3'-0", aluminum, curb and cover included. Through steel construction.					
Roof hatch, 2'-6" x 3'-0", aluminum	1.000	Ea.	1,125	240	1,365
Cutout decking	11.000	L.F.	4.29	27.61	31.90
Frame opening	44.000	L.F.	38.28	411.84	450.12
Flashing, 16 oz. copper	16.000	S.F.	135.20	80.80	216
Cant strip, 4 x 4, mineral fiber	12.000	L.F.	4.92	21.36	26.28
TOTAL			1,307.69	781.61	2,089.30

B3020 230	Roof Hatches, Skylights	COST EACH		
		MAT.	INST.	TOTAL
1400	Roof hatch, aluminum, curb and cover, 2'-6" x 3'-0"	1,300	785	2,085
1500	2'-6" x 4'-6"	1,750	955	2,705
1600	2'-6" x 8'-0"	2,425	1,375	3,800
1700	Skylight, plexiglass dome, with curb mounting, 30" x 32"	715	745	1,460
1800	30" x 45"	770	925	1,695
1900	40" x 45"	915	1,250	2,165
2000	Smoke hatch, unlabeled, 2'-6" x 3'-0"	1,600	845	2,445
2100	2'-6" x 4'-6"	2,125	1,025	3,150
2200	2'-6" x 8'-0"	2,625	1,425	4,050
2300				
2400	For steel ladder, add per vertical linear foot	42	38	80
2500				
2600				
2700				
2900	Cut & patch to match existing construction, add, minimum	2%	3%	
3000	Maximum	5%	9%	
3100	Dust protection, add, minimum	1%	2%	
3200	Maximum	4%	11%	
3300	Equipment usage curtailment, add, minimum	1%	1%	
3400	Maximum	3%	10%	
3500	Material handling & storage limitation, add, minimum	1%	1%	
3600	Maximum	6%	7%	
3700	Protection of existing work, add, minimum	2%	2%	
3800	Maximum	5%	7%	
3900	Shift work requirements, add, minimum		5%	
4000	Maximum		30%	

B30 Roofing

B3020 Roof Openings

B3020 240	Selective Price Sheet	COST PER L.F.		
		MAT.	INST.	TOTAL
0100	Downspouts per L.F., aluminum, enameled .024" thick, 2" x 3"	2.26	3.92	6.18
0200	3" x 4"	2.56	5.05	7.61
0300	Round .025" thick, 3" diam.	2.15	3.72	5.87
0400	4" diam.	2.78	5.05	7.83
0500	Copper, round 16 oz. stock, 2" diam.	8.35	3.72	12.07
0600	3" diam.	8.25	3.72	11.97
0700	4" diam.	10.65	4.87	15.52
0800	5" diam.	18.80	5.45	24.25
0900	Rectangular, 2" x 3"	12.15	3.72	15.87
1000	3" x 4"	15.35	4.87	20.22
1100	Lead coated copper, round, 2" diam.	24.50	3.72	28.22
1200	3" diam.	24.50	3.72	28.22
1300	4" diam.	25.50	4.87	30.37
1400	5" diam.	24.50	5.45	29.95
1500	Rectangular, 2" x 3"	26.50	3.72	30.22
1600	3" x 4"	36.50	4.87	41.37
1700	Steel galvanized, round 28 gauge, 3" diam.	2.29	3.72	6.01
1800	4" diam.	2.77	4.87	7.64
1900	5" diam.	4.40	5.45	9.85
2000	6" diam.	4.62	6.75	11.37
2100	Rectangular, 2" x 3"	4.02	3.72	7.74
2200	3" x 4"	4.47	4.87	9.34
2300	Elbows, aluminum, round, 3" diam.	3.01	7.05	10.06
2400	4" diam.	6.35	7.05	13.40
2500	Rectangular, 2" x 3"	1.05	7.05	8.10
2600	3" x 4"	4.96	7.05	12.01
2700	Copper, round 16 oz., 2" diam.	10.40	7.05	17.45
2800	3" diam.	9.90	7.05	16.95
2900	4" diam.	15.30	7.05	22.35
3100	Rectangular, 2" x 3"	10.30	7.05	17.35
3200	3" x 4"	14.55	7.05	21.60
3300	Drip edge per L.F., aluminum, 5" wide	.61	1.54	2.15
3400	8" wide	1.60	1.54	3.14
3500	28" wide	8.45	6.15	14.60
3600				
3700	Steel galvanized, 5" wide	.58	1.54	2.12
3800	8" wide	.88	1.54	2.42
3900				
4000				
4100				
4200				
4300	Flashing 12" wide per S.F., aluminum, mill finish, .013" thick	.87	4	4.87
4400	.019" thick	1.27	4	5.27
4500	.040" thick	2.54	4	6.54
4600	.050" thick	2.67	4	6.67
4700	Copper, mill finish, 16 oz.	8.45	5.05	13.50
4800	20 oz.	10.35	5.25	15.60
4900	24 oz.	14.65	5.50	20.15
5000	32 oz.	19.60	5.80	25.40
5100	Lead, 2.5 lb./S.F., 12" wide	5.50	4.29	9.79
5200	Over 12" wide	4.39	4.29	8.68
5900	Polyvinyl chloride, black, .010" thick	.23	2.03	2.26
6000	.020" thick	.31	2.03	2.34
6100	.030" thick	.39	2.03	2.42
6200	.056" thick	.94	2.03	2.97
6300	Steel, galvanized, 20 gauge	1.38	4.46	5.84
6400	30 gauge	.64	3.62	4.26
6500	Stainless, 32 gauge, .010" thick	3.30	3.74	7.04

B30 Roofing

B3020 Roof Openings

B3020 240	Selective Price Sheet	COST PER L.F.		
		MAT.	INST.	TOTAL
6600	28 gauge, .015" thick	4.39	3.74	8.13
6700	26 gauge, .018" thick	4.68	3.74	8.42
6800	24 gauge, .025" thick	5.25	3.74	8.99
6900	Gutters per L.F., aluminum, 5" box, .027" thick	3.09	5.85	8.94
7000	.032" thick	3.55	5.65	9.20
7100	Copper, half round, 4" wide	9.80	5.65	15.45
7200	6" wide	13	6	19
7300	Steel, 26 gauge galvanized, 5" wide	2.40	5.65	8.05
7400	6" wide	2.98	5.65	8.63
7500	Wood, treated hem-fir, 3" x 4"	11	6.15	17.15
7600	4" x 5"	18.45	6.15	24.60
7700	Reglet per L.F., aluminum, .025" thick	2.12	2.74	4.86
7800	Copper, 10 oz.	7.05	2.74	9.79
7900	Steel, galvanized, 24 gauge	1.47	2.74	4.21
8000	Stainless, .020" thick	4.13	2.74	6.87
8100	Counter flashing 12" wide per L.F., aluminum, .032" thick	2.64	4.71	7.35
8200	Aluminum, .025" thick	2.64	4.71	7.35
8300	Steel, galvanized, 24 gauge	1.56	4.71	6.27
8400	Stainless, .020" thick	6.75	4.71	11.46

C Interiors

C10 Interior Construction

C1010 Partitions

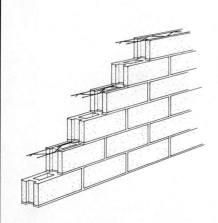

This page illustrates and describes a concrete block wall system including concrete block, horizontal reinforcing alternate courses, mortar, and tooled joints both sides. Lines within System Components give the unit price and total price per square foot for this system. Prices for concrete block wall systems are on Line Items C1010 106 0800 thru 2100. Both material quantities and labor costs have been adjusted for the system listed.

Factors: To adjust for job conditions other than normal working situations use Lines C1010 106 2900 thru 4000.

Example: You are to install the system and protect all existing work. Go to Line C1010 106 3600 and apply these percentages to the appropriate MAT. and INST. costs.

System Components	QUANTITY	UNIT	COST PER S.F. MAT.	INST.	TOTAL
Concrete block partition, including horizontal reinforcing every second Course, mortar, tooled joints both sides.					
8" x 16" concrete block, normal weight, 4" thick	1.000	S.F.	1.87	6.50	8.37
Horizontal reinforcing every second course	.750	L.F.	.19	.15	.34
TOTAL			2.06	6.65	8.71

C1010 106	Partitions, Concrete Block	MAT.	INST.	TOTAL
0800	8" x 16" concrete block, normal weight, 4" thick	2.06	6.65	8.71
0900	6" thick	2.64	7.15	9.79
1000	8" thick	3.21	7.60	10.81
1100	10" thick	3.34	7.90	11.24
1200	12" thick	4.54	9.80	14.34
1300	8" x 16" concrete block, lightweight, 4" thick	2.17	6.50	8.67
1400	6" thick	3.01	6.95	9.96
1500	8" thick	3.63	7.40	11.03
1600	10" thick	4.36	7.70	12.06
1700	12" thick	4.60	9.55	14.15
1800	8" x 16" glazed concrete block, 4" thick	11.10	8.25	19.35
1900	8" thick	12.80	9.15	21.95
2000	Structural facing tile, 6T series, glazed 2 sides, 4" thick	17.80	14.45	32.25
2100	6" thick	24	15.25	39.25
2200				
2300				
2400				
2500				
2600				
2900	Cut & patch to match existing construction, add, minimum	2%	3%	
3000	Maximum	5%	9%	
3100	Dust protection, add, minimum	1%	2%	
3200	Maximum	4%	11%	
3300	Material handling & storage limitation, add, minimum	1%	1%	
3400	Maximum	6%	7%	
3500	Protection of existing work, add, minimum	2%	2%	
3600	Maximum	5%	7%	
3700	Shift work requirements, add, minimum		5%	
3800	Maximum		30%	
3900	Temporary shoring and bracing, add, minimum	2%	5%	
4000	Maximum	5%	12%	

C10 Interior Construction

C1010 Partitions

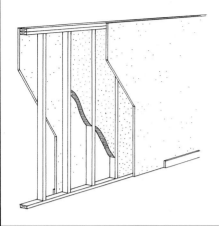

This page illustrates and describes a wood stud partition system including wood studs with plates, gypsum plasterboard – taped and finished, insulation, baseboard and painting. Lines within System Components give the unit price and total price per square foot for this system. Prices for wood stud partition systems are on Line Items C1010 132 1300 thru 2700. Both material quantities and labor costs have been adjusted for the system listed.

Factors: To adjust for job conditions other than normal working situations use Lines C1010 132 2900 thru 4000.

Example: You are to install the system where material handling and storage present a serious problem. Go to Line C1010 132 3400 and apply these percentages to the appropriate MAT. and INST. costs.

System Components	QUANTITY	UNIT	COST PER S.F.		
			MAT.	INST.	TOTAL
Wood stud wall, 2"x4", 16" O.C., dbl. top plate, sngl bot. plate, 5/8" dwl. taped, finished and painted on 2 faces, insulation, baseboard, wall 8' high.					
Wood studs, 2" x 4", 16" O.C., 8' high	1.000	S.F.	.52	1.54	2.06
Gypsum drywall, 5/8" thick	2.000	S.F.	.72	1.24	1.96
Taping and finishing	2.000	S.F.	.10	1.24	1.34
Insulation, 3-1/2" fiberglass batts	1.000	S.F.	.50	.46	.96
Baseboard	.200	L.F.	.31	.64	.95
Paint baseboard, primer + 2 coats	.200	L.F.	.05	.38	.43
Painting, roller, 2 coats	2.000	S.F.	.42	1.36	1.78
TOTAL			2.62	6.86	9.48

C1010 132	Partitions, Wood Stud	COST PER S.F.		
		MAT.	INST.	TOTAL
1300	2" x 3" studs, 8' high, 16" O.C.	2.59	6.75	9.34
1400	24" O.C.	2.48	6.50	8.98
1500	10' high, 16" O.C.	2.56	6.50	9.06
1600	24" O.C.	2.45	6.25	8.70
1650	2" x 4" studs, 8' high, 16" O.C.	2.62	6.85	9.47
1700	24" O.C.	2.51	6.55	9.06
1800	10' high, 16" O.C.	2.59	6.55	9.14
1900	24" O.C.	2.47	6.30	8.77
2000	12' high, 16" O.C.	2.55	6.55	9.10
2100	24" O.C.	2.44	6.30	8.74
2200	2" x 6" studs, 8' high, 16" O.C.	2.93	7.05	9.98
2300	24" O.C.	2.74	6.65	9.39
2400	10' high, 16" O.C.	2.87	6.70	9.57
2500	24" O.C.	2.69	6.40	9.09
2600	12' high, 16" O.C.	2.80	6.75	9.55
2700	24" O.C.	2.63	6.40	9.03
2900	Cut & patch to match existing construction, add, minimum	2%	3%	
3000	Maximum	5%	9%	
3100	Dust protection, add, minimum	1%	2%	
3200	Maximum	4%	11%	
3300	Material handling & storage limitation, add, minimum	1%	1%	
3400	Maximum	6%	7%	
3500	Protection of existing work, add, minimum	2%	2%	
3600	Maximum	5%	7%	
3700	Shift work requirements, add, minimum		5%	
3800	Maximum		30%	
3900	Temporary shoring and bracing, add, minimum	2%	5%	
4000	Maximum	5%	12%	

C10 Interior Construction

C1010 Partitions

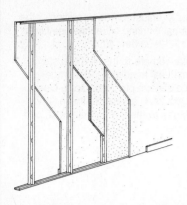

This page illustrates and describes a non-load bearing metal stud partition system including metal studs with runners, gypsum plasterboard, taped and finished, insulation, baseboard and painting. Lines within System Components give the unit price and total price per square foot for this system. Prices for non-load bearing metal stud partition systems are on Line Items C1010 134 1300 thru 2300. Both material quantities and labor costs have been adjusted for the system listed.

Factors: To adjust for job conditions other than normal working situations use Lines C1010 134 2900 thru 4000.

Example: You are to install the system and cut and patch to match existing construction. Go to Line C1010 134 3000 and apply these percentages to the appropriate MAT. and INST. costs.

System Components	QUANTITY	UNIT	COST PER S.F. MAT.	INST.	TOTAL
Non-load bearing metal studs, including top & bottom runners, 5/8" drywall, Taped, finished and painted 2 faces, insulation, painted baseboard.					
Metal studs, 25 ga., 3-5/8" wide, 24" O.C.	1.000	S.F.	.30	.83	1.13
Gypsum drywall, 5/8" thick	2.000	S.F.	.72	1.24	1.96
Taping & finishing	2.000	S.F.	.10	1.24	1.34
Insulation, 3-1/2" fiberglass batts	1.000	S.F.	.50	.46	.96
Baseboard	.200	L.F.	.25	.51	.76
Paint baseboard, primer + 2 coats	.200	L.F.	.04	.31	.35
Painting, roller work, 2 coats	2.000	S.F.	.42	1.36	1.78
TOTAL			2.33	5.95	8.28

C1010 134	Partitions, Metal Stud, NLB	MAT.	INST.	TOTAL
1300	Non-load bearing, 25 ga., 24" O.C., 2-1/2" wide	2.28	5.95	8.23
1350	3-5/8" wide	2.33	5.95	8.28
1400	6" wide	2.43	5.95	8.38
1500	16" O.C., 2-1/2" wide	2.34	6.15	8.49
1600	3-5/8" wide	2.41	6.15	8.56
1700	6" wide	2.53	6.20	8.73
1800	20 ga., 24" O.C., 2-1/2" wide	2.35	6.15	8.50
1900	3-5/8" wide	2.47	6.15	8.62
2000	6" wide	2.54	6.20	8.74
2100	16" O.C., 2-1/2" wide	2.43	6.40	8.83
2200	3-5/8" wide	2.58	6.40	8.98
2300	6" wide	2.67	6.45	9.12
2400				
2500				
2600				
2700				
2900	Cut & patch to match existing construction, add, minimum	2%	3%	
3000	Maximum	5%	9%	
3100	Dust protection, add, minimum	1%	2%	
3200	Maximum	4%	11%	
3300	Material handling & storage limitation, add, minimum	1%	1%	
3400	Maximum	6%	7%	
3500	Protection of existing work, add, minimum	2%	2%	
3600	Maximum	5%	7%	
3700	Shift work requirements, add, minimum		5%	
3800	Maximum		30%	
3900	Temporary shoring and bracing, add, minimum	2%	5%	
4000	Maximum	5%	12%	

C10 Interior Construction

C1010 Partitions

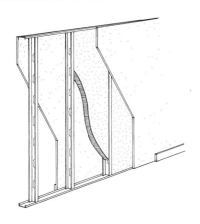

This page illustrates and describes a drywall system including gypsum plasterboard, taped and finished, metal studs with runners, insulation, baseboard and painting. Lines within System Components give the unit price and total price per square foot for this system. Prices for drywall systems are on Line Items C1010 136 1200 thru 1900. Both material quantities and labor costs have been adjusted for the system listed.

Factors: To adjust for job conditions other than normal working situations use Lines C1010 136 2900 thru 4000.

Example: You are to install the system and control dust in the work area. Go to Line C1010 136 3100 and apply these percentages to the appropriate MAT. and INST. costs.

System Components	QUANTITY	UNIT	COST PER S.F.		
			MAT.	INST.	TOTAL
Gypsum drywall, taped, finished and painted 2 faces, galvanized metal studs					
Including top & bottom runners, insulation, painted baseboard, wall 10'high					
Gypsum drywall, 5/8" thick, standard	2.000	S.F.	.72	1.24	1.96
Taping and finishing	2.000	S.F.	.10	1.24	1.34
Metal studs, 20 ga., 3-5/8" wide, 24" O.C.	1.000	S.F.	.37	1.04	1.41
Insulation, 3-1/2" fiberglass batts	1.000	S.F.	.50	.46	.96
Baseboard	.200	L.F.	.25	.51	.76
Paint baseboard, primer + 2 coats	.200	L.F.	.03	.13	.16
Painting, roller 2 coats	2.000	S.F.	.30	1.30	1.60
TOTAL			2.27	5.92	8.19

C1010 136	Partitions, Drywall	COST PER S.F.		
		MAT.	INST.	TOTAL
1200	Gypsum drywall, 5/8" thick, standard	2.27	5.90	8.17
1300	Fire resistant	2.22	5.70	7.92
1400	Water resistant	2.40	5.70	8.10
1500	1/2" thick, standard	2.14	5.70	7.84
1600	Fire resistant	2.26	5.70	7.96
1700	Water resistant	2.34	5.70	8.04
1800	3/8" thick, vinyl faced, standard	2.54	7.20	9.74
1900	5/8" thick, vinyl faced, fire resistant	3.16	7.20	10.36
2000				
2100				
2200				
2300				
2400				
2500				
2600				
2700				
2900	Cut & patch to match existing construction, add, minimum	2%	3%	
3000	Maximum	5%	9%	
3100	Dust protection, add, minimum	1%	2%	
3200	Maximum	4%	11%	
3300	Material handling & storage limitation, add, minimum	1%	1%	
3400	Maximum	6%	7%	
3500	Protection of existing work, add, minimum	2%	2%	
3600	Maximum	5%	7%	
3700	Shift work requirements, add, minimum		5%	
3800	Maximum		30%	
3900	Temporary shoring and bracing, add, minimum	2%	5%	
4000	Maximum	5%	12%	

C10 Interior Construction

C1010 Partitions

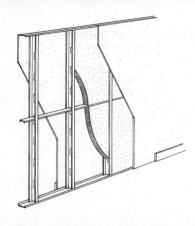

This page illustrates and describes a load bearing metal stud wall system including metal studs, sheetrock–taped and finished, insulation, baseboard and painting. Lines within System Components give the unit price and total price per square foot for this system. Prices for load bearing metal stud wall systems are on Line Items C1010 138 1500 thru 2500. Both material quantities and labor costs have been adjusted for the system listed.

Factors: To adjust for job conditions other than normal working situations use Lines C1010 138 2900 thru 4000.

Example: You are to install the system using temporary shoring and bracing. Go to Line C1010 138 3900 and apply these percentages to the appropriate MAT. and INST. costs.

System Components	QUANTITY	UNIT	COST PER S.F.		
			MAT.	INST.	TOTAL
Load bearing, 18 ga., 3-5/8", galvanized metal studs, 24" O.C., including Top and bottom runners, 1/2" drywall, taped, finished and painted 2 Faces, 3" insulation, and painted baseboard, wall 10' high.					
Metal studs, 24" O.C., 18 ga., 3-5/8" wide, galvanized	1.000	S.F.	.78	1.18	1.96
Gypsum drywall 1/2" thick	2.000	S.F.	.66	1.24	1.90
Taping and finishing	2.000	S.F.	.10	1.24	1.34
Insulation, 3-1/2" fiberglass batts	1.000	S.F.	.50	.46	.96
Baseboard	.200	L.F.	.25	.51	.76
Paint baseboard, primer + 2 coats	.200	L.F.	.04	.31	.35
Painting, roller 2 coats	2.000	S.F.	.42	1.36	1.78
TOTAL			2.75	6.30	9.05

C1010 138	Partitions, Metal Stud, LB	COST PER S.F.		
		MAT.	INST.	TOTAL
1500	Load bearing, 18 ga., 24" O.C., 2-1/2" wide	2.63	6.25	8.88
1550	3-5/8" wide	2.75	6.30	9.05
1600	6" wide	3.01	6.35	9.36
1700	16" O.C. 2-1/2" wide	2.79	6.55	9.34
1800	3-5/8" wide	2.94	6.60	9.54
1900	6" wide	3.26	6.65	9.91
2000	16 ga., 24" O.C., 2-1/2" wide	2.74	6.45	9.19
2100	3-5/8" wide	2.88	6.45	9.33
2200	6" wide	3.17	6.50	9.67
2300	16" O.C., 2-1/2" wide	2.93	6.75	9.68
2400	3-5/8" wide	3.11	6.80	9.91
2500	6" wide	3.47	6.85	10.32
2600				
2700				
2900	Cut & patch to match existing construction, add, minimum	2%	3%	
3000	Maximum	5%	9%	
3100	Dust protection, add, minumum	1%	2%	
3200	Maximum	4%	11%	
3300	Material handling & storage limitation, add, minimum	1%	1%	
3400	Maximum	6%	7%	
3500	Protection of existing work, add, minimum	2%	2%	
3600	Maximum	5%	7%	
3700	Shift work requirements, add, minimum		5%	
3800	Maximum		30%	
3900	Temporary shoring and bracing, add, minimum	2%	5%	
4000	Maximum	5%	12%	

C10 Interior Construction

C1010 Partitions

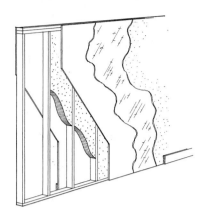

This page illustrates and describes a plaster and lath system including gypsum plaster, gypsum lath, wood studs with plates, insulation, baseboard and painting. Lines within System Components give the unit price and price per square foot for this system. Prices for plaster and lath systems are on Line Items C1010 148 1400 thru 2500. Both material quantities and labor costs have been adjusted for the system listed.

Factors: To adjust for job conditions other than normal working situations use Lines C1010 148 2900 thru 4000.

Example: You are to install the system during evening hours only. Go to Line C1010 148 3800 and apply this percentage to the appropriate INST. costs.

System Components	QUANTITY	UNIT	COST PER S.F. MAT.	COST PER S.F. INST.	COST PER S.F. TOTAL
Gypsum plaster, 2 coats, over 3/8" lath, 2 faces, 2" x 4" wood Stud partition, 24"O.C. incl. double top plate, single bottom plate, 3-1/2" Insulation, baseboard and painting.					
Gypsum plaster, 2 coats	.220	S.Y.	.87	5.82	6.69
Lath, gypsum, 3/8" thick	.220	S.Y.	.74	1.40	2.14
Wood studs, 2" x 4", 24" O.C.	1.000	S.F.	.37	.99	1.36
Insulation, 3-1/2" fiberglass batts	1.000	S.F.	.50	.46	.96
Baseboard, 9/16" x 3-1/2"	.200	L.F.	.25	.51	.76
Paint baseboard, primer + 2 coats	.200	L.F.	.03	.13	.16
Paint, 2 coats	2.000	S.F.	.42	1.36	1.78
TOTAL			3.18	10.67	13.85

C1010 148	Partitions, Plaster & Lath	MAT.	INST.	TOTAL
1400	Gypsum plaster, 2 coats	3.18	10.65	13.83
1500	3 coats	3.56	11.85	15.41
1600	Perlite plaster, 2 coats	3.73	11.45	15.18
1700	3 coats	3.86	13.10	16.96
1800				
1900				
2000	For alternate lath systems:			
2100	Gypsum lath, 1/2" thick	3.58	11.55	15.13
2200	Foil back, 3/8" thick	3.82	11.65	15.47
2300	1/2" thick	3.95	11.80	15.75
2400	Metal lath, 2.5 Lb. diamond	3.86	11.45	15.31
2500	3.4 Lb. diamond	4.02	11.65	15.67
2600				
2700				
2900	Cut & patch to match existing construction, add, minimum	2%	3%	
3000	Maximum	5%	9%	
3100	Dust protection, add, minimum	1%	2%	
3200	Maximum	4%	11%	
3300	Material handling & storage limitation, add, minimum	1%	1%	
3400	Maximum	6%	7%	
3500	Protection of existing work, add, minimum	2%	2%	
3600	Maximum	5%	7%	
3700	Shift work requirements, add, minimum		5%	
3800	Maximum		30%	
3900	Temporary shoring and bracing, add, minimum	2%	5%	
4000	Maximum	5%	12%	

C10 Interior Construction

C1010 Partitions

C1010 170	Selective Price Sheet	COST PER S.F.		
		MAT.	INST.	TOTAL
0100	Studs			
0200				
0300	24" O.C. metal, 10' high wall, including			
0400	Top and bottom runners			
0500	Non load bearing, galvanized 25 Ga., 1-5/8" wide	.20	.81	1.01
0600	2-1/2" wide	.25	.82	1.07
0700	3-5/8" wide	.30	.83	1.13
0800	4" wide	.33	.83	1.16
0900	6" wide	.40	.85	1.25
1000				
1100	Galvanized 20 Ga., 2-1/2" wide	.32	1.03	1.35
1200	3-5/8" wide	.37	1.04	1.41
1300	4" wide	.44	1.04	1.48
1400	6" wide	.51	1.06	1.57
1500	Load bearing, painted 18 Ga., 2-1/2" wide	.66	1.15	1.81
1600	3-5/8" wide	.78	1.18	1.96
1700	4" wide	.70	1.20	1.90
1800	6" wide	1.04	1.22	2.26
1900	Galvanized 18 Ga., 2-1/2" wide	.66	1.15	1.81
2000	3-5/8" wide	.78	1.18	1.96
2100	4" wide	.70	1.20	1.90
2200	6" wide	1.04	1.22	2.26
2300	Galvanized 16 Ga., 2-1/2" wide	.77	1.31	2.08
2400	3-5/8" wide	.91	1.34	2.25
2500	4" wide	.96	1.37	2.33
2600	6" wide	1.20	1.40	2.60
2700				
2800				
2900	24" O.C. wood, 10' high wall, including			
3000	Double top plate and shoe			
3100	2" x 4"	.44	1.16	1.60
3200	2" x 6"	.68	1.24	1.92
3300				
3400				
3500	Furring, 24" O.C. 10' high wall, metal, 3/4" channels	.24	1.29	1.53
3600	1-1/2" channels	.32	1.39	1.71
3700	Wood, on wood, 1" x 2" strips	.13	.56	.69
3800	1" x 3" strips	.21	.56	.77
3900	On masonry, 1" x 2" strips	.15	.62	.77
4000	1" x 3" strips	.23	.62	.85
4100	On concrete, 1" x 2" strips	.15	1.19	1.34
4200	1" x 3" strips	.23	1.19	1.42
4300	Studs			
4400				
4500	16" O.C., metal, 10' high wall, including			
4600	Top and bottom runners			
4700	Non load bearing, galvanized 25 Ga., 1-5/8" wide	.28	1.24	1.52
4800	2-1/2" wide	.34	1.26	1.60
4900	3-5/8" wide	.40	1.28	1.68
5000	4" wide	.45	1.30	1.75
5100	6" wide	.54	1.31	1.85
5200				
5300	Galvanized 20 Ga., 2-1/2" wide	.43	1.58	2.01
5400	3-5/8" wide	.50	1.60	2.10
5500	4" wide	.59	1.62	2.21
5600	6" wide	.69	1.64	2.33

C10 Interior Construction

C1010 Partitions

C1010 170	Selective Price Sheet	COST PER S.F.		
		MAT.	INST.	TOTAL
5700	Load bearing, painted 18 Ga., 2-1/2" wide	.89	1.60	2.49
5800	3-5/8" wide	1.05	1.62	2.67
5900	4" wide	1.11	1.67	2.78
6000	6" wide	1.40	1.69	3.09
6100	Galvanized 18 Ga., 2-1/2" wide	.89	1.60	2.49
6200	3-5/8" wide	1.05	1.62	2.67
6300	4" wide	1.11	1.67	2.78
6400	6" wide	1.40	1.69	3.09
6500	Galvanized 16 Ga., 2-1/2" wide	1.05	1.81	2.86
6600	3-5/8" wide	1.26	1.87	3.13
6700	4" wide	1.32	1.90	3.22
6800	6" wide	1.65	1.93	3.58
6900				
7000				
7100	16" O.C., wood, 10' high wall, including			
7200	Double top plate and shoe			
7300	2" x 4"	.53	1.34	1.87
7400	2" x 6"	.68	1.21	1.89
7500				
7600				
7700	Furring, 16" O.C. 10' high wall, metal, 3/4" channels	.36	1.87	2.23
7800	1-1/2" channels	.48	2.08	2.56
7900	Wood, on wood, 1" x 2" strips	.20	.84	1.04
8000	1" x 3" strips	.32	.84	1.16
8100	On masonry, 1" x 2" strips	.22	.93	1.15
8200	1" x 3" strips	.35	.93	1.28
8300	On concrete, 1" x 2" strips	.22	1.78	2
8400	1" x 3" strips	.35	1.78	2.13

C10 Interior Construction

C1010 Partitions

C1010 180	Selective Price Sheet, Drywall & Plaster	COST PER S.F.		
		MAT.	INST.	TOTAL
0100	Lath, gypsum perforated			
0200				
0300	Lath, gypsum, perforated, regular, 3/8" thick	.37	.70	1.07
0400	1/2" thick	.30	.86	1.16
0500	Fire resistant, 3/8" thick	.30	.86	1.16
0600	1/2" thick	.34	.93	1.27
0700	Foil back, 3/8" thick	.42	.80	1.22
0800	1/2" thick	.48	.86	1.34
0900				
1000	Metal lath			
1100	Metal lath, diamond, painted, 2.5 lb.	.44	.70	1.14
1200	3.4 lb.	.52	.80	1.32
1300	Rib painted, 2.75 lb	.42	.80	1.22
1400	3.40 lb	.50	.86	1.36
1500				
1600				
1700	Plaster, gypsum, 2 coats	.44	2.94	3.38
1800	3 coats	.63	3.53	4.16
1900	Perlite/vermiculite, 2 coats	.72	3.35	4.07
2000	3 coats	.78	4.17	4.95
2100	Bondcrete, 1 coat	.45	1.54	1.99
2200				
2500				
2600				
2700	Drywall, standard, 3/8" thick, no finish included	.37	.62	.99
2800	1/2" thick, no finish included	.33	.62	.95
2900	Taped and finished	.38	1.28	1.66
3000	5/8" thick, no finish included	.36	.62	.98
3100	Taped and finished	.41	1.28	1.69
3200	Fire resistant, 1/2" thick, no finish included	.39	.62	1.01
3300	Taped and finished	.44	1.28	1.72
3400	5/8" thick, no finish included	.37	.62	.99
3500	Taped and finished	.43	1.28	1.71
3600	Water resistant, 1/2" thick, no finish included	.43	.62	1.05
3700	Taped and finished	.48	1.28	1.76
3800	5/8" thick, no finish included	.46	.62	1.08
3900	Taped and finished	.51	1.28	1.79
4000	Finish, instead of taping			
4100	For thin coat plaster, add	.12	.77	.89
4200	Finish, textured spray, add	.04	.68	.72

C10 Interior Construction

C1010 Partitions

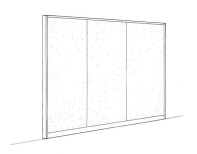

This page illustrates and describes a movable office partition system including demountable office partitions of various styles and sizes based on a cost per square foot basis. Prices for movable office partition systems are on Line Items C1010 210 0300 thru 3600.

C1010 210	Partitions, Movable Office	COST PER S.F.		
		MAT.	INST.	TOTAL
0250	Office partition, demountable, no deduction for door opening, add for doors			
0300	Air wall, cork finish, semi acoustic, 1-5/8" thick, minimum	42.50	3.79	46.29
0400	Maximum	47	6.50	53.50
0500	Acoustic, 2" thick, minimum	40	4.04	44.04
0600	Maximum	61.50	5.45	66.95
0700				
0800				
0900	In-plant modular office system, w/prehung steel door			
1000	3" thick honeycomb core panels			
1100	12' x 12', 2 wall	8.90	.52	9.42
1200	4 wall	20.50	1.56	22.06
1300				
1400				
1500	Gypsum, demountable, 3" to 3-3/4" thick x 9' high, vinyl clad	7.35	2.83	10.18
1600	Fabric clad	18.30	3.11	21.41
1700	1.75 system, vinyl clad hardboard, paper honeycomb core panel			
1800	1-3/4" to 2-1/2" thick x 9' high	12.30	2.83	15.13
1900	Unitized gypsum panel, 2" to 2-1/2" thick x 9' high, vinyl clad	15.85	2.83	18.68
2000	Fabric clad	26	3.11	29.11
2100				
2200				
2300	Unitized mineral fiber panel system, 2-1/4" thick x 9' high			
2400	Vinyl clad mineral fiber	15.75	2.83	18.58
2500	Fabric clad mineral fiber	23.50	3.11	26.61
2600				
2700	Movable steel walls, modular system			
2800	Unitized panels, 48" wide x 9' high			
2900	Baked enamel, pre-finished	17.85	2.28	20.13
3000	Fabric clad	26	2.44	28.44
3100	For acoustical partitions, add, unsealed	2.60		2.60
3200	Sealed	12.10		12.10
3300				
3400				
3500	Note: For door prices, see Div. 08			
3600	For door hardware prices, see Div. 08			
3700				
3800				
3900				
4000				

C10 Interior Construction

C1020 Interior Doors

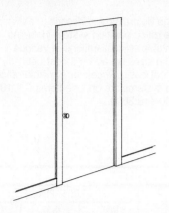

This page illustrates and describes flush interior door systems including hollow core door, jamb, header and trim with hardware. Lines within System Components give the unit price and total price on a cost each basis for this system. Prices for flush interior door systems are on Line items C1020 106 1000 thru 2400. Both material quantities and labor costs have been adjusted for the system listed.

Factors: To adjust for job conditions other than normal working situations use Lines C1020 106 2700 thru 3800.

Example: You are to install the system in an area where dust protection is vital. Go to Line C1020 106 3000 and apply these percentages to the appropriate MAT. and INST. costs.

System Components	QUANTITY	UNIT	COST EACH		
			MAT.	INST.	TOTAL
Single hollow core door, include jamb, header, trim and hardware, painted.					
Hollow core Lauan, 1-3/8" thick, 2'-0" x 6'-8", painted	1.000	Ea.	38.50	68.50	107
Paint door and frame, 1 coat	1.000	Ea.	2.43	30.50	32.93
Wood jamb, 4-9/16" deep	1.000	Set	86.40	52.48	138.88
Trim, casing	1.000	Set	49.60	78.72	128.32
Hardware, hinges	1.000	Set	24.50		24.50
Hardware, lockset	1.000	Set	33	38.50	71.50
TOTAL			234.43	268.70	503.13

C1020 106	Doors, Interior Flush, Wood	COST EACH		
		MAT.	INST.	TOTAL
1000	Lauan (Mahogany) hollow core, 1-3/8" x 2'-0" x 6'-8"	234	269	503
1100	2'-6" x 6'-8"	243	273	516
1200	2'-8" x 6'-8"	247	276	523
1300	3'-0" x 6'-8"	254	286	540
1500	Birch, hollow core, 1-3/8" x 2'-0" x 6'-8"	242	269	511
1600	2'-6" x 6'-8"	255	273	528
1700	2'-8" x 6'-8"	259	276	535
1800	3'-0" x 6'-8"	265	286	551
1900	Solid core, pre-hung, 1-3/8" x 2'-6" x 6'-8"	505	277	782
2000	2'-8" x 6'-8"	425	280	705
2100	3'-0" x 6'-8"	545	290	835
2200				
2400	For metal frame instead of wood, add	50%	20%	
2600				
2700	Cut & patch to match existing construction, add, minimum	2%	3%	
2800	Maximum	5%	9%	
2900	Dust protection, add, minimum	1%	2%	
3000	Maximum	4%	11%	
3100	Equipment usage curtailment, add, minimum	1%	1%	
3200	Maximum	3%	10%	
3300	Material handling & storage limitation, add, minimum	1%	1%	
3400	Maximum	6%	7%	
3500	Protection of existing work, add, minimum	2%	2%	
3600	Maximum	5%	7%	
3700	Shift work requirements, add, minimum		5%	
3800	Maximum		30%	
3900				

C10 Interior Construction

C1020 Interior Doors

This page illustrates and describes interior, solid and louvered door systems including a pine panel door, wood jambs, header, and trim with hardware. Lines within System Components give the unit price and total price on a cost each basis for this system. Prices for interior, solid and louvered systems are on Line Items C1020 108 1100 thru 2400. Both material quantities and labor costs have been adjusted for the system listed.

Factors: To adjust for job conditions other than normal working situations use Lines C1020 108 2900 thru 4000.

Example: You are to install the system during night hours only. Go to Line C1020 108 4000 and apply these percentages to the appropriate INST. costs.

System Components	QUANTITY	UNIT	COST EACH		
			MAT.	INST.	TOTAL
Single interior door, including jamb, header, trim and hardware, painted.					
Solid pine panel door, 1-3/8" thick, 2'-0" x 6'-8"	1.000	Ea.	168	68.50	236.50
Paint door and frame, 1 coat	1.000	Ea.	2.43	30.50	32.93
Wooden jamb, 4-5/8" deep	1.000	Set	86.40	52.48	138.88
Trim, casing	1.000	Set	49.60	78.72	128.32
Hardware, hinges	1.000	Set	24.50		24.50
Hardware, lockset	1.000	Set	33	38.50	71.50
TOTAL			363.93	268.70	632.63

C1020 108	Doors, Interior Solid & Louvered	COST EACH		
		MAT.	INST.	TOTAL
1000				
1100	Solid pine, painted raised panel, 1-3/8" x 2'-0" x 6'-8"	365	269	634
1200	2'-6" x 6'-8"	400	273	673
1300	2'-8" x 6'-8"	405	276	681
1400	3'-0" x 6'-8"	430	286	716
1500				
1600	Louvered pine, painted 1'-6" x 6'-8"	330	265	595
1700	2'-0" x 6'-8"	345	273	618
1800	2'-6" x 6'-8"	370	276	646
1900	3'-0" x 6'-8"	400	286	686
2200	For prehung door, deduct	5%	30%	
2300				
2400	For metal frame instead of wood, add	50%	20%	
2500				
2900	Cut & patch to match existing construction, add, minimum	2%	3%	
3000	Maximum	5%	9%	
3100	Dust protection, add, minimum	1%	2%	
3200	Maximum	4%	11%	
3300	Equipment usage curtailment, add, minimum	1%	1%	
3400	Maximum	3%	10%	
3500	Material handling & storage limitation, add, minimum	1%	1%	
3600	Maximum	6%	7%	
3700	Protection of existing work, add, minimum	2%	2%	
3800	Maximum	5%	7%	
3900	Shift work requirements, add, minimum		5%	
4000	Maximum		30%	

C10 Interior Construction

C1020 Interior Doors

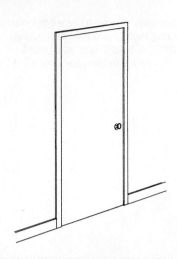

This page illustrates and describes interior metal door systems including a metal door, metal frame and hardware. Lines within System Components give the unit price and total price on a cost each for this system. Prices for interior metal door systems are on Line Items C1020 110 1000 thru 2000. Both material quantities and labor costs have been adjusted for the system listed.

Factors: To adjust for job conditions other than normal working situations use Lines C1020 110 2900 thru 4000.

Example: You are to install the system while protecting existing construction. Go to Line C1020 110 3700 and apply these percentages to the appropriate MAT. and INST. costs.

System Components	QUANTITY	UNIT	COST EACH		
			MAT.	INST.	TOTAL
Single metal door, including frame and hardware.					
Hollow metal door, 1-3/8" thick, 2'-6" x 6'-8", painted	1.000	Ea.	545	82	627
Metal frame, 6-1/4" deep	1.000	Set	186	77	263
Paint door and frame, 1 coat	1.000	Ea.	2.43	30.50	32.93
Hardware, hinges	1.000	Set	94.50		94.50
Hardware, passage lockset	1.000	Set	33	38.50	71.50
TOTAL			860.93	228	1,088.93

C1020 110	Doors, Interior Flush, Metal	COST EACH		
		MAT.	INST.	TOTAL
1000	Hollow metal doors, 1-3/8" thick, 2'-6" x 6'-8"	860	228	1,088
1100	1-3/8" thick, 2'-8" x 6'-8"	865	228	1,093
1200	3'-0" x 7'-0"	850	239	1,089
1300				
1400	Interior fire door, 1-3/8" thick, 2'-6" x 6'-8"	1,075	228	1,303
1500	2'-8" x 6'-8"	1,075	228	1,303
1600	3'-0" x 7'-0"	1,050	239	1,289
1700				
1800	Add to fire doors:			
1900	Baked enamel finish	30%	15%	
2000	Galvanizing	20%		
2200				
2300				
2400				
2900	Cut & patch to match existing construction, add, minimum	2%	3%	
3000	Maximum	5%	9%	
3100	Dust protection, add, minimum	1%	2%	
3200	Maximum	4%	11%	
3300	Equipment usage curtailment, add, minimum	1%	1%	
3400	Maximum	3%	10%	
3500	Material handling & storage limitation, add, minimum	1%	1%	
3600	Maximum	6%	7%	
3700	Protection of existing work, add, minimum	2%	2%	
3800	Maximum	5%	7%	
3900	Shift work requirements, add, minimum		5%	
4000	Maximum		30%	

C10 Interior Construction

C1020 Interior Doors

This page illustrates and describes an interior closet door system including an interior closet door, painted, with trim and hardware. Prices for interior closet door systems are on Line Items C1020 112 0500 thru 2200. Both material quantities and labor costs have been adjusted for the system listed.

Factors: To adjust for job conditions other than normal working situations use Lines C1020 112 2900 thru 4000.

Example: You are to install the system and match the existing construction. Go to Line C1020 112 2900 and apply these percentages to the appropriate MAT. and INST. costs.

C1020 112	Doors, Closet	COST PER SET		
		MAT.	INST.	TOTAL
0350	Interior closet door painted, including frame, trim and hardware, prehung.			
0400	Bi-fold doors			
0500	Pine paneled, 3'-0" x 6'-8"	415	223	638
0600	6'-0" x 6'-8"	575	299	874
0700	Birch, hollow core, 3'-0" x 6'-8"	206	241	447
0800	6'-0" x 6'-8"	305	325	630
0900	Lauan, hollow core, 3'-0" x 6'-8"	190	223	413
1000	6'-0" x 6'-8"	274	299	573
1100	Louvered pine, 3'-0" x 6'-8"	340	223	563
1200	6'-0" x 6'-8"	455	299	754
1300				
1400	Sliding, bi-passing closet doors			
1500	Pine paneled, 4'-0" x 6'-8"	685	237	922
1600	6'-0" x 6'-8"	805	299	1,104
1700	Birch, hollow core, 4'-0" x 6'-8"	365	237	602
1800	6'-0" x 6'-8"	420	299	719
1900	Lauan, hollow core, 4'-0" x 6'-8"	310	237	547
2000	6'-0" x 6'-8"	365	299	664
2100	Louvered pine, 4'-0" x 6'-8"	575	237	812
2200	6'-0" x 6'-8"	720	299	1,019
2300				
2400				
2500				
2600				
2700				
2800				
2900	Cut & patch to match existing construction, add, minimum	2%	3%	
3000	Maximum	5%	9%	
3100	Dust protection, add, minimum	1%	2%	
3200	Maximum	4%	11%	
3300	Equipment usage curtailment, add, minimum	1%	1%	
3400	Maximum	3%	10%	
3500	Material handling & storage limitation, add, minimum	1%	1%	
3600	Maximum	6%	7%	
3700	Protection of existing work, add, minimum	2%	2%	
3800	Maximum	5%	7%	
3900	Shift work requirements, add, minimum		5%	
4000	Maximum		30%	

C20 Stairs

C2010 Stair Construction

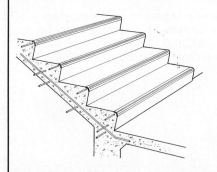

This page illustrates and describes a stair system based on a cost per flight price. Prices for various stair systems are on Line Items C2010 130 0700 thru 3200. Both material quantities and labor costs have been adjusted for the system listed.

Factors: To adjust for job conditions other than normal working situations use Lines C2010 130 3500 thru 4200.

Example: You are to install the system during evenings only. Go to Line C2010 130 4200 and apply this percentage to the appropriate MAT. and INST. costs.

System Components	QUANTITY	UNIT	COST PER FLIGHT		
			MAT.	INST.	TOTAL
Below are various stair systems based on cost per flight of stairs, no side Walls. Stairs are 4'-0" wide, railings are included unless otherwise noted.					

C2010 130	Stairs	COST PER FLIGHT		
		MAT.	INST.	TOTAL
0700	Concrete, cast in place, no nosings, no railings, 12 risers	298	2,125	2,423
0800	24 risers	595	4,225	4,820
0900	Add for 1 intermediate landing	159	585	744
1000	Concrete, cast in place, with nosings, no railings, 12 risers	1,225	2,325	3,550
1100	24 risers	2,450	4,675	7,125
1200	Add for 1 intermediate landing	236	605	841
1300	Steel, grating tread, safety nosing, 12 risers	4,750	1,200	5,950
1400	24 risers	9,500	2,450	11,950
1500	Add for intermediate landing	1,175	320	1,495
1600	Steel, cement fill pan tread, 12 risers	7,250	1,200	8,450
1700	24 risers	14,500	2,450	16,950
1800	Add for intermediate landing	1,175	320	1,495
1900	Spiral, industrial, 4' - 6" diameter, 12 risers	7,150	865	8,015
2000	24 risers	14,300	1,725	16,025
2100	Wood, box stairs, oak treads, 12 risers	2,375	695	3,070
2200	24 risers	4,750	1,400	6,150
2300	Add for 1 intermediate landing	193	157	350
2400	Wood, basement stairs, no risers, 12 steps	785	282	1,067
2500	24 steps	1,575	565	2,140
2600	Add for 1 intermediate landing	42.50	30	72.50
2700	Wood, open, rough sawn cedar, 12 steps	2,700	565	3,265
2800	24 steps	5,400	1,125	6,525
2900	Add for 1 intermediate landing	22.50	31	53.50
3000	Wood, residential, oak treads, 12 risers	2,625	3,075	5,700
3100	24 risers	5,225	6,150	11,375
3200	Add for 1 intermediate landing	172	143	315
3500	Dust protection, add, minimum	1%	2%	
3600	Maximum	4%	11%	
3700	Material handling & storage limitation, add, minimum	1%	1%	
3800	Maximum	6%	7%	
3900	Protection of existing work, add, minimum	2%	2%	
4000	Maximum	5%	7%	
4100	Shift work requirements, add, minimum		5%	
4200	Maximum		30%	

C30 Interior Finishes

C3010 Wall Finishes

C3010 210	Selective Price Sheet	COST PER S.F.		
		MAT.	INST.	TOTAL
0100	Painting, on plaster or drywall, brushwork, primer and 1 ct.	.15	.76	.91
0200	Primer and 2 ct.	.22	1.21	1.43
0300	Rollerwork, primer and 1 ct.	.15	.65	.80
0400	Primer and 2 ct.	.22	1.03	1.25
0500	Woodwork incl. puttying, brushwork, primer and 1 ct.	.14	1.15	1.29
0600	Primer and 2 ct.	.21	1.53	1.74
0700	Wood trim to 6" wide, enamel, primer and 1 ct.	.14	.65	.79
0800	Primer and 2 ct.	.22	.82	1.04
0900	Cabinets and casework, enamel, primer and 1 ct.	.14	1.30	1.44
1000	Primer and 2 ct.	.22	1.60	1.82
1100	On masonry or concrete, latex, brushwork, primer and 1 ct.	.31	1.08	1.39
1200	Primer and 2 ct.	.41	1.55	1.96
1300	For block filler, add	.24	1.33	1.57
1400				
1500	Varnish, wood trim, sealer 1 ct., sanding, puttying, quality work	.14	2.34	2.48
1600	Medium work	.11	1.82	1.93
1700	Without sanding	.17	.27	.44
1800				
1900	Wall coverings, wall paper, at low price per double roll, avg workmanship	.67	.82	1.49
2000	At medium price per double roll, average workmanship	1.22	.98	2.20
2100	At high price per double roll, quality workmanship	2.42	1.21	3.63
2200				
2300	Grass cloths with lining paper	1	1.32	2.32
2400	Premium texture/color	3.20	1.50	4.70
2500	Vinyl, fabric backed, light weight	1.07	.82	1.89
2600	Medium weight	.99	1.10	2.09
2700	Heavy weight	1.51	1.21	2.72
2800				
2900	Cork tiles, 12" x 12", 3/16" thick	4.96	2.19	7.15
3000	5/16" thick	3.44	2.24	5.68
3100	Granular surface, 12" x 36", 1/2" thick	1.44	1.37	2.81
3200	1" thick	1.85	1.42	3.27
3300	Aluminum foil	1.14	1.91	3.05
3400				
3500	Tile, ceramic, adhesive set, 4-1/4" x 4-1/4"	2.48	5.10	7.58
3600	6" x 6"	3.61	5.50	9.11
3700	Decorated, 4-1/4" x 4-1/4", minimum	3.50	3.57	7.07
3800	Color group 4	54.50	5.35	59.85
3900	For epoxy grout, add	.45	1.21	1.66
4000	Pregrouted sheets	5.65	4.02	9.67
4100	Glass mosaics, 3/4" tile on 12" sheets, minimum	18.90	13.20	32.10
4200	Color group 8	45.50	15.10	60.60
4300	Metal, tile pattern, 4' x 4' sheet, 24 ga., nailed			
4400	Stainless steel	31	2.40	33.40
4500	Aluminized steel	16.60	2.40	19
4600				
4700	Brick, interior veneer, 4" face brick, running bond, minimum	4.02	13	17.02
4800	Maximum	4.02	13	17.02
4900	Simulated, urethane pieces, set in mastic	9.05	4.10	13.15
5000	Fiberglass panels	8.85	3.08	11.93
5100	Wall coating, on drywall, thin coat, plain	.12	.77	.89
5200	Stipple	.12	.77	.89
5300	Textured spray	.04	.68	.72
5400				
5500	Paneling not incl. furring or trim, hardboard, tempered, 1/8" thick	.45	2.46	2.91
5600	1/4" thick	.70	2.46	3.16
5700	Plastic faced, 1/8" thick	.73	2.46	3.19
5800	1/4" thick	.98	2.46	3.44

C30 Interior Finishes

C3010 Wall Finishes

C3010 210	Selective Price Sheet	COST PER S.F.		
		MAT.	INST.	TOTAL
5900	Woodgrained, 1/4" thick, minimum	.76	2.46	3.22
6000	Maximum	1.54	2.90	4.44
6100	Plywood, 4' x 8' shts. 1/4" thick, prefin., birch faced, min.	1.54	2.46	4
6200	Maximum	1.38	3.52	4.90
6300	Walnut, minimum	2.64	2.46	5.10
6400	Maximum	5.80	3.08	8.88
6500	Mahogany, African	3.03	3.08	6.11
6600	Philippine	.72	2.46	3.18
6700	Chestnut	5.30	3.28	8.58
6800	Pecan	2.75	3.08	5.83
6900	Rosewood	3.19	3.85	7.04
7000	Teak	3.19	3.08	6.27
7100	Aromatic cedar, plywood	2.42	3.08	5.50
7200	Particle board	1.16	3.08	4.24
7300	Wood board, 3/4" thick, knotty pine	2.10	4.10	6.20
7400	Rough sawn cedar	3.52	4.10	7.62
7500	Redwood, clear	5.45	4.10	9.55
7600	Aromatic cedar	2.55	4.48	7.03

C30 Interior Finishes

C3020 Floor Finishes

C3020 430	Selective Price Sheet	COST PER S.F.		
		MAT.	INST.	TOTAL
0100	Flooring, carpet, acrylic, 26 oz. light traffic	3.05	.80	3.85
0200	35 oz. heavy traffic	7.25	.80	8.05
0300	Nylon anti-static, 15 oz. light traffic	2.10	1.05	3.15
0400	22 oz. medium traffic	4.22	.80	5.02
0500	26 oz. heavy traffic	6	.86	6.86
0600	28 oz. heavy traffic	6.70	.86	7.56
0700	Tile, foamed back, needle punch	4.75	.95	5.70
0800	Tufted loop	3.41	.95	4.36
0900	Wool, 36 oz. medium traffic	11.10	.86	11.96
1000	42 oz. heavy traffic	12.20	.86	13.06
1100	Composition, epoxy, with colored chips, minimum	3.09	4.64	7.73
1200	Maximum	4.46	6.40	10.86
1300	Trowelled, minimum	3.93	5.55	9.48
1400	Maximum	5.90	6.55	12.45
1500	Terrazzo, 1/4" thick, chemical resistant, minimum	6.55	6.60	13.15
1600	Maximum	9.95	8.85	18.80
1700	Resilient, asphalt tile, 1/8" thick	1.31	1.35	2.66
1800	Conductive flooring, rubber, 1/8" thick	7.65	1.72	9.37
1900	Cork tile 1/8" thick, standard finish	7.65	1.72	9.37
2000	Urethane finish	8.90	1.72	10.62
2100	PVC sheet goods for gyms, 1/4" thick	8.70	6.75	15.45
2200	3/8" thick	13	9	22
2300	Vinyl composition 12" x 12" tile, plain, 1/16" thick	1.30	1.08	2.38
2400	1/8" thick	1.68	1.08	2.76
2500	Vinyl tile, 12" x 12" x 1/8" thick, minimum	6.20	1.08	7.28
2600	Premium colors/pattern	6.85	1.08	7.93
2700	Vinyl sheet goods, backed, .080" thick, plain pattern/colors	4.51	2.35	6.86
2900	Slate, random rectangular, 1/4" thick	23.50	7.30	30.80
3000	1/2" thick	22	10.45	32.45
3100	Natural cleft, irregular, 3/4" thick	10.10	11.90	22
3200	For sand rubbed finish, add	10.20		10.20
3300	Terrazzo, cast in place, bonded 1-3/4" thick, gray cement	3.69	17.65	21.34
3400	White cement	4.10	17.65	21.75
3500	Not bonded 3" thick, gray cement	4.59	18.90	23.49
3600	White cement	5.35	18.90	24.25
3700	Precast, 12" x 12" x 1" thick	22	18.25	40.25
3800	1-1/4" thick	23	18.25	41.25
3900	16" x 16" x 1-1/4" thick	24.50	22	46.50
4000	1-1/2" thick	22.50	24.50	47
4100	Marble travertine, standard, 12" x 12" x 3/4" thick	12.75	16.10	28.85
4200				
4300	Tile, ceramic, natural clay, thin set	4.56	5.25	9.81
4400	Porcelain, thin set	6.70	5.10	11.80
4500	Specialty, decorator finish	11.45	5.25	16.70
4600				
4700	Quarry, red, mud set, 4" x 4" x 1/2" thick	8.85	8.05	16.90
4800	6" x 6" x 1/2" thick	8.30	6.90	15.20
4900	Brown, imported, 6" x 6" x 3/4" thick	8.75	8.05	16.80
5000	8" x 8" x 1" thick	9.55	8.75	18.30
5100	Slate, Vermont, thin set, 6" x 6" x 1/4" thick	8.25	5.35	13.60
5200				
5300	Wood, maple strip, 25/32" x 2-1/4", finished, select	5.60	3.89	9.49
5400	2nd and better	4.89	3.89	8.78
5500	Oak, 25/32" x 2-1/4" finished, clear	3.68	3.89	7.57
5600	No. 1 common	4.58	3.89	8.47
5700	Parquet, standard, 5/16", finished, minimum	5.90	4.12	10.02
5800	Maximum	10.70	6.40	17.10
5900	Custom, finished, plain pattern	18.50	6.15	24.65

C30 Interior Finishes

C3020 Floor Finishes

C3020 430	Selective Price Sheet	COST PER S.F.		
		MAT.	INST.	TOTAL
6000	Intricate pattern	26	12.30	38.30
6100	Prefinished, oak, 2-1/4" wide	5.15	3.62	8.77
6200	Ranch plank	7.85	4.25	12.10
6300	Sleepers on concrete, treated, 24" O.C., 1" x 2"	.23	.40	.63
6400	1" x 3"	.42	.52	.94
6500	2" x 4"	.53	.72	1.25
6600	2" x 6"	.84	.84	1.68
6700	Refinish old floors, soft wood	.99	1.23	2.22
6800	Hard wood	1.47	3.79	5.26
6900	Subflooring, plywood, CDX, 1/2" thick	.72	.82	1.54
7000	5/8" thick	.86	.91	1.77
7100	3/4" thick	1.02	.99	2.01
7200				
7300	1" x 10" boards, S4S, laid regular	2.14	1.12	3.26
7400	Laid diagonal	2.14	1.37	3.51
7500	1" x 8" boards, S4S, laid regular	2.09	1.23	3.32
7600	Laid diagonal	2.09	1.45	3.54
7700	Underlayment, plywood, underlayment grade, 3/8" thick	1.01	.82	1.83
7800	1/2" thick	1.21	.85	2.06
7900	5/8" thick	1.32	.88	2.20
8000	3/4" thick	1.57	.95	2.52
8100	Particle board, 3/8" thick	.44	.82	1.26
8200	1/2" thick	.47	.85	1.32
8300	5/8" thick	.57	.88	1.45
8400	3/4" thick	.69	.95	1.64

C30 Interior Finishes

C3030 Ceiling Finishes

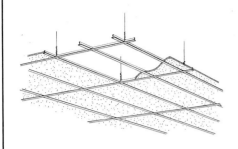

This page illustrates suspended acoustical board systems including acoustic ceiling board, hangers, and T bar suspension. Lines within System Components give the unit price and total price per square foot for this system. Prices for suspended acoustical board systems are on Line Items C3030 210 1000 thru 2400. Both material quantities and labor costs have been adjusted for the system listed.

Factors: To adjust for job conditions other than normal working situations use Lines C3030 210 2900 thru 4000.

Example: You are to install the system and protect existing construction. Go to Line C3030 210 3800 and apply these percentages to the appropriate MAT. and INST. costs.

System Components	QUANTITY	UNIT	COST PER S.F. MAT.	INST.	TOTAL
Suspended acoustical ceiling board installed on exposed grid system.					
Fiberglass boards, film faced, 2' x 4', 5/8" thick	1.000	S.F.	1.36	.99	2.35
Hangers, #12 wire	1.000	S.F.	.03	.59	.62
T bar suspension system, 2' x 4' grid	1.000	S.F.	.81	.77	1.58
TOTAL			2.20	2.35	4.55

C3030 210	Ceiling, Suspended Acoustical	COST PER S.F. MAT.	INST.	TOTAL
1000	2' x 4' grid, fiberglass board, film faced, 5/8" thick	2.20	2.35	4.55
1100	Mineral fiber board, aluminum faced	4.82	2.31	7.13
1200	Standard faced	1.59	2.27	3.86
1300	Plastic faced	3.40	2.90	6.30
1400	Fiberglass, film faced, 3" thick, R11	3.50	2.73	6.23
1500	Grass cloth faced, 3/4" thick	3.76	2.59	6.35
1600	1" thick	4.28	2.63	6.91
1700	1-1/2" thick, nubby face	3.62	2.66	6.28
2200				
2300				
2400	Add for 2' x 2' grid system	.21	.19	.40
2500				
2600				
2700				
2900	Cut & patch to match existing construction, add, minimum	2%	3%	
3000	Maximum	5%	9%	
3100	Dust protection, add, minimum	1%	2%	
3200	Maximum	4%	11%	
3300	Equipment usage curtailment, add, minimum	1%	1%	
3400	Maximum	3%	10%	
3500	Material handling & storage limitation, add, minimum	1%	1%	
3600	Maximum	6%	7%	
3700	Protection of existing work, add, minimum	2%	2%	
3800	Maximum	5%	7%	
3900	Shift work requirements, add, minimum		5%	
4000	Maximum		30%	

C30 Interior Finishes

C3030 Ceiling Finishes

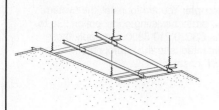

This page illustrates and describes suspended gypsum board systems including gypsum board, metal furring, taping, finished and painted. Lines within System Components give the unit price and total price per square foot for this system. Prices for suspended gypsum board systems are on Line Items C3030 220 1400 thru 1700. Both material quantities and labor costs have been adjusted for the system listed.

Factors: To adjust for job conditions other than normal working situations use Lines C3030 220 2900 thru 4000.

Example: You are to install the system and control dust in the working area. Go to Line C3030 220 3200 and apply these percentages to the appropriate MAT. and INST. costs.

System Components	QUANTITY	UNIT	COST PER S.F.		
			MAT.	INST.	TOTAL
Suspended ceiling gypsum board, 4' x 8' x 5/8" thick, On metal furring, taped, finished and painted.					
Gypsum drywall, 4' x 8', 5/8" thick, screwed	1.000	S.F.	.36	.68	1.04
Main runners, 1-1/2" C.R.C., 4' O.C.	.500	S.F.	.16	.70	.86
25 ga., channels, 2' O.C.	1.000	S.F.	.24	1.29	1.53
Taped and finished	1.000	S.F.	.05	.62	.67
Paint, 2 coats, roller work	1.000	S.F.	.21	.68	.89
TOTAL			1.02	3.97	4.99

C3030 220	Ceilings, Suspended Gypsum Board	COST PER S.F.		
		MAT.	INST.	TOTAL
1400	4' x 8' x 5/8", gypsum drywall, 2 coats paint	1.02	3.97	4.99
1500	Thin coat plaster, 2 coats paint	.88	3.44	4.32
1600	Spray-on sand finish, no paint	.80	3.35	4.15
1700	12" x 12" x 3/4" acoustical wood fiber tile	2.11	4.21	6.32
1800				
1900				
2000				
2100				
2200				
2300				
2400				
2500				
2600				
2700				
2900	Cut & patch to match existing construction, add, minimum	2%	3%	
3000	Maximum	5%	9%	
3100	Dust protection, add, minimum	1%	2%	
3200	Maximum	4%	11%	
3300	Equipment usage curtailment, add, minimum	1%	1%	
3400	Maximum	3%	10%	
3500	Material handling & storage limitation, add, minimum	1%	1%	
3600	Maximum	6%	7%	
3700	Protection of existing work, add, minimum	2%	2%	
3800	Maximum	5%	7%	
3900	Shift work requirements, add, minimum		5%	
4000	Maximum		30%	

C30 Interior Finishes

C3030 Ceiling Finishes

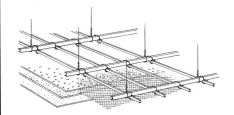

This page illustrates and describes suspended plaster and lath systems including gypsum plaster, lath, furring and runners with ceiling painted. Lines within System Components give the unit price and total price per square foot for this system. Prices for suspended plaster and lath systems are on Line Items C3030 230 1200 thru 2300. Both material quantities and labor costs have been adjusted for the system listed.

Factors: To adjust for job conditions other than normal working situations use Lines C3030 230 2900 thru 4000.

Example: You are to install the system to match existing construction. Go to Line C3030 230 2900 and apply these percentages to the appropriate MAT. and INST. costs.

System Components	QUANTITY	UNIT	COST PER S.F.		
			MAT.	INST.	TOTAL
Gypsum plaster, 3 coats, on 3.4# rib lath, on 3/4" C.R.C. furring on 1-1/2" main runners, ceiling painted.					
Gypsum plaster, 3 coats	.110	S.Y.	.63	3.91	4.54
3.4# rib lath	.110	S.Y.	.50	.85	1.35
Main runners, 1-1/2" C.R.C. 24" O.C.	.333	S.F.	.32	1.39	1.71
Furring 3/4" C.R.C. 16" O.C.	1.000	S.F.	.36	1.87	2.23
Painting, 2 coats, roller work	1.000	S.F.	.21	.68	.89
TOTAL			2.02	8.70	10.72

C3030 230	Ceiling, Suspended Plaster	COST PER S.F.		
		MAT.	INST.	TOTAL
1200	Gypsum plaster, 3 coats, on 3.4# rib lath	2.02	8.70	10.72
1300	On 2.5# diamond lath	1.96	8.60	10.56
1400	On 3/8" gypsum lath	1.89	8.60	10.49
1500	2 coats, on 3.4# rib lath	1.81	8.70	10.51
1600	On 2.5# diamond lath	1.77	8.05	9.82
1700	On 3/8" gypsum lath	1.70	8.05	9.75
1800	Perlite plaster, 3 coats, on 3.4# rib lath	2.17	9.60	11.77
1900	On 2.5# diamond lath	2.11	9.55	11.66
2000	On 3/8" gypsum lath	2.04	9.55	11.59
2100	2 coats, on 3.4# rib lath	2.10	8.60	10.70
2200	On 2.5# diamond lath	2.04	8.55	10.59
2300	On 3/8" gypsum lath	1.97	8.55	10.52
2400				
2500				
2600				
2700				
2900	Cut & patch to match existing construction, add, minimum	2%	3%	
3000	Maximum	5%	9%	
3100	Dust protection, add, minimum	1%	2%	
3200	Maximum	4%	11%	
3300	Equipment usage curtailment, add, minimum	1%	1%	
3400	Maximum	3%	10%	
3500	Material handling & storage limitation, add, minimum	1%	1%	
3600	Maximum	6%	7%	
3700	Protection of existing work, add, minimum	2%	2%	
3800	Maximum	5%	7%	
3900	Shift work requirements, add, minimum		5%	
4000	Maximum		30%	

C30 Interior Finishes

C3030 Ceiling Finishes

C3030 250	Selective Price Sheet	COST PER S.F.		
		MAT.	INST.	TOTAL
0100	Ceiling, plaster, gypsum, 2 coats	.44	3.35	3.79
0200	3 coats	.63	3.94	4.57
0300	Perlite or vermiculite, 2 coats	.72	3.88	4.60
0400	3 coats	.78	4.88	5.66
0500	Gypsum lath, plain, 3/8" thick	.37	.70	1.07
0600	1/2" thick	.30	.75	1.05
0700	Firestop, 3/8" thick	.30	.86	1.16
0800	1/2" thick	.34	.93	1.27
0900	Metal lath, rib, 2.75 lb.	.42	.80	1.22
1000	3.40 lb.	.50	.86	1.36
1100	Diamond, 2.50 lb.	.44	.80	1.24
1200	3.40 lb.	.52	1	1.52
1500	Drywall, standard, 1/2" thick, no finish included	.33	.68	1.01
1600	Taped and finished	.38	1.61	1.99
1700	5/8" thick, no finish included	.36	.68	1.04
1800	Taped and finished	.41	1.61	2.02
1900	Fire resistant, 1/2" thick, no finish included	.39	.68	1.07
2000	Taped and finished	.44	1.61	2.05
2100	5/8" thick, no finish included	.37	.68	1.05
2200	Taped and finished	.43	1.61	2.04
2300	Water resistant, 1/2" thick, no finish included	.43	.68	1.11
2400	Taped and finished	.48	1.61	2.09
2500	5/8" thick, no finish included	.46	.68	1.14
2600	Taped and finished	.51	1.61	2.12
2700	Finish, instead of taping			
2800	For thin coat plaster, add	.12	.77	.89
2900	Finish, textured spray, add	.04	.68	.72
3000				
3100				
3200				
3300	Tile, stapled or glued, mineral fiber plastic coated, 5/8" thick	2.40	2.05	4.45
3400	3/4" thick	2.64	2.05	4.69
3500	Wood fiber, 1/2" thick	1.23	1.54	2.77
3600	3/4" thick	1.35	1.54	2.89
3700	Suspended, fiberglass, film faced, 5/8" thick	1.36	.99	2.35
3800	3" thick	2.66	1.37	4.03
3900	Mineral fiber, 5/8" thick, standard	.75	.91	1.66
4000	Aluminum	3.98	1.03	5.01
4100	Wood fiber, reveal edge, 1" thick			
4200	3" thick			
4300	Framing, metal furring, 3/4" channels, 12" O.C.	.40	2.58	2.98
4400	16" O.C.	.36	1.87	2.23
4500	24" O.C.	.24	1.29	1.53
4600				
4700	1-1/2" channels, 12" O.C.	.54	2.85	3.39
4800	16" O.C.	.48	2.08	2.56
4900	24" O.C.	.32	1.39	1.71
5000				
5100				
5200				
5300	Ceiling suspension systems, for tile,			
5400	Concealed "Z" bar, 12" module	.92	1.18	2.10
5500	Class A, "T" bar 2'-0" x 4'-0" grid	.81	.77	1.58
5600	2'-0" x 2'-0" grid	1.04	.95	1.99
5700	Carrier channels for lighting fixtures, add	.22	1.34	1.56
5800				
5900				
6000				

C30 Interior Finishes

C3030 Ceiling Finishes

C3030 250	Selective Price Sheet	COST PER S.F.		
		MAT.	INST.	TOTAL
6100	Wood, furring 1" x 3", on wood, 12" O.C.	.42	1.76	2.18
6200	16" O.C.	.32	1.32	1.64
6300	24" O.C.	.21	.88	1.09
6400				
6500	On concrete, 12" O.C.	.46	2.93	3.39
6600	16" O.C.	.35	2.20	2.55
6700	24" O.C.	.23	1.47	1.70
6800				
6900	Joists, 2" x 4", 12" O.C.	.61	1.28	1.89
7000	16" O.C.	.51	1.07	1.58
7100	24" O.C.	.41	.87	1.28
7200	32" O.C.	.31	.66	.97
7300	2" x 6", 12" O.C.	.95	1.29	2.24
7400	16" O.C.	.79	1.07	1.86
7500	24" O.C.	.63	.85	1.48
7600	32" O.C.	.47	.63	1.10

No part of this publication may be reproduced, stored in a retrieval system, or transmitted in any form or by any means without prior written permission of RSMeans.

D10 Conveying

D1010 Elevators and Lifts

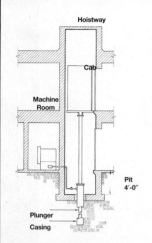

This page illustrates and describes oil hydraulic elevator systems. Prices for various oil hydraulic elevator systems are on Line Items D1010 130 0400 thru 2000. Both material quantities and labor costs have been adjusted for the system listed.

Factors: To adjust for job conditions other than normal working situations use Lines D1010 130 3100 thru 4000.

Example: You are to install the system with limited equipment usage. Go to Line D1010 130 3400 and apply this percentage to the appropriate TOTAL costs.

D1010 130	Oil Hydraulic Elevators	COST EACH		
		MAT.	INST.	TOTAL
0310	Oil-hydraulic elevator systems			
0320	Including piston and piston shaft			
0330				
0400	1500 Lb. passenger, 2 floors	41,400	18,900	60,300
0500	3 floors	51,500	29,200	80,700
0600	4 floors	81,500	41,100	122,600
0700	5 floors	91,500	51,000	142,500
0800	6 floors	100,500	61,000	161,500
0900				
1000	2500 Lb. passenger, 2 floors	44,500	18,900	63,400
1100	3 floors	55,000	29,200	84,200
1200	4 floors	80,000	41,100	121,100
1300	5 floors	90,000	51,000	141,000
1400	6 floors	103,500	61,000	164,500
1500				
1600	4000 Lb. passenger, 2 floors	50,500	18,900	69,400
1700	3 floors	63,000	29,200	92,200
1800	4 floors	90,500	41,100	131,600
1900	5 floors	100,500	51,000	151,500
2000	6 floors	109,500	61,000	170,500
2100				
2200				
2300				
2400				
2500				
2600				
2700				
2800				
2900				
3000				
3100	Dust protection, add, minimum	1%	2%	
3200	Maximum	4%	11%	
3300	Equipment usage curtailment, add, minimum	1%	1%	
3400	Maximum	3%	10%	
3500	Material handling & storage limitation, add, minimum	1%	1%	
3600	Maximum	6%	7%	
3700	Protection of existing work, add, minimum	2%	2%	
3800	Maximum	5%	7%	
3900	Shift work requirements, add, minimum		5%	
4000	Maximum		30%	

D20 Plumbing

D2010 Plumbing Fixtures

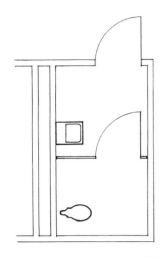

This page illustrates and describes a women's public restroom system including a water closet, lavatory, accessories, and service piping. Lines within System Components give the unit price and total price on a cost each basis for this system. Prices for women's public restroom systems are on Line Items D2010 956 1300 thru 1700. Both material quantities and labor costs have been adjusted for the system listed.

Factors: To adjust for job conditions other than normal working situations use Lines D2010 956 2900 thru 4000.

Example: You are to install the system and protect surrounding area from dust. Go to Line D2010 956 3100 and apply these percentages to the MAT. and INST. costs.

System Components	QUANTITY	UNIT	COST EACH MAT.	INST.	TOTAL
Public women's restroom incl. water closet, lavatory, accessories and Necessary service piping to install this system in one wall.					
Water closet, wall mounted, one piece	1.000	Ea.	1,050	227	1,277
Rough-in waste and vent for water closet	1.000	Set	1,175	515	1,690
Lavatory, 20" x 18" P.E. cast iron with accessories	1.000	Ea.	305	165	470
Rough-in waste and vent for lavatory	1.000	Set	500	795	1,295
Toilet partition, painted metal between walls, floor mounted	1.000	Ea.	430	205	635
For handicap unit, add	1.000	Ea.	410		410
Grab bar, 36" long	1.000	Ea.	85	62	147
Mirror, 18" x 24", with stainless steel shelf	1.000	Ea.	213	31	244
Napkin/tampon dispenser, recessed	1.000	Ea.	650	41	691
Soap Dispenser, chrome, surface mounted, liquid	1.000	Ea.	51	31	82
Toilet tissue dispenser, surface mounted, stainless steel	1.000	Ea.	19.60	20.50	40.10
Towel dispenser, surface mounted, stainless steel	1.000	Ea.	49	38.50	87.50
TOTAL			4,937.60	2,131	7,068.60

D2010 956	Plumbing - Public Restroom	COST EACH MAT.	INST.	TOTAL
1200				
1300	Public women's restroom, one water closet, one lavatory	4,950	2,125	7,075
1400	Two water closets, two lavatories	9,125	4,075	13,200
1500				
1600	For each additional water closet over 2, add	2,900	940	3,840
1700	For each additional lavatory over 2, add	1,075	1,025	2,100
1800				
1900				
2400	NOTE: PLUMBING APPROXIMATIONS			
2500	WATER CONTROL: water meter, backflow preventer,			
2600	Shock absorbers, vacuum breakers, mixer....10 to 15% of fixtures			
2700	PIPE AND FITTINGS: .... 30 to 60% of fixtures			
2800				
2900	Cut & patch to match existing construction, add, minimum	2%	3%	
3000	Maximum	5%	9%	
3100	Dust protection, add, minimum	1%	2%	
3200	Maximum	4%	11%	
3300	Equipment usage curtailment, add, minimum	1%	1%	
3400	Maximum	3%	10%	
3500	Material handling & storage limitation, add, minimum	1%	1%	

D20 Plumbing

D2010 Plumbing Fixtures

D2010 956	Plumbing - Public Restroom	COST EACH		
		MAT.	INST.	TOTAL
3600	Maximum	6%	7%	
3700	Protection of existing work, add, minimum	2%	2%	
3800	Maximum	5%	7%	
3900	Shift work requirements, add, minimum		5%	
4000	Maximum		30%	

D20 Plumbing

D2010 Plumbing Fixtures

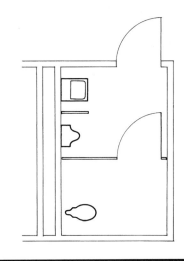

This page illustrates and describes a men's public restroom system including a water closet, urinal, lavatory, accessories and service piping. Lines within System Components give the unit price and total price on a cost each basis for this system. Prices for men's public restroom systems are on Line Items D2010 957 1700 thru 2200. Both material quantities and labor costs have been adjusted for the system listed.

Factors: To adjust for job conditions other than normal working situations use Lines D2010 957 2900 thru 4000.

Example: You are to install the system and match existing construction. Go to Line D2010 957 3000 and apply these percentages to the appropriate MAT. and INST. costs.

System Components	QUANTITY	UNIT	COST EACH MAT.	INST.	TOTAL
Public men's restroom incl. water closet, urinal, lavatory, accessories, And necessary service piping to install this system in one wall.					
Water closet, wall mounted, one piece	1.000	Ea.	1,050	227	1,277
Rough-in waste & vent for water closet	1.000	Set	1,175	515	1,690
Urinal, wall hung	1.000	Ea.	310	440	750
Rough-in waste & vent for urinal	1.000	Set	655	465	1,120
Lavatory, 20" x 18", P.E. cast iron with accessories	1.000	Ea.	305	165	470
Rough-in waste & vent for lavatory	1.000	Set	500	795	1,295
Partition, painted mtl., between walls, floor mntd.	1.000	Ea.	430	205	635
For handicap unit, add	1.000	Ea.	410		410
Grab bars	1.000	Ea.	85	62	147
Urinal screen, powder coated steel, wall mounted	1.000	Ea.	179	123	302
Mirror, 18" x 24" with stainless steel shelf	1.000	Ea.	213	31	244
Soap dispenser, surface mounted, liquid	1.000	Ea.	51	31	82
Toilet tissue dispenser, surface mounted, stainless steel	1.000	Ea.	19.60	20.50	40.10
Towel dispenser, surface mounted, stainless steel	1.000	Ea.	49	38.50	87.50
TOTAL			5,431.60	3,118	8,549.60

D2010 957	Plumbing - Public Restroom	COST EACH MAT.	INST.	TOTAL
1600				
1700	Public men's restroom, one water closet, one urinal, one lavatory	5,425	3,125	8,550
1800	Two water closets, two urinals, two lavatories	10,800	6,075	16,875
1900				
2000	For each additional water closet over 2, add	2,900	940	3,840
2100	For each additional urinal over 2, add	1,150	1,025	2,175
2200	For each additional lavatory over 2, add	1,075	1,025	2,100
2300				
2400	NOTE: PLUMBING APPROXIMATIONS			
2500	WATER CONTROL: water meter, backflow preventer,			
2600	Shock absorbers, vacuum breakers, mixer....10 to 15% of fixtures			
2700	PIPE AND FITTINGS: .... 30 to 60% of fixtures			
2800				
2900	Cut & patch to match existing construction, add, minimum	2%	3%	
3000	Maximum	5%	9%	
3100	Dust protection, add, minimum	1%	2%	
3200	Maximum	4%	11%	
3300	Equipment usage curtailment, add, minimum	1%	1%	

D20 Plumbing

D2010 Plumbing Fixtures

D2010 957	Plumbing - Public Restroom	COST EACH		
		MAT.	INST.	TOTAL
3400	Maximum	3%	10%	
3500	Material handling, & storage limitation, add, minimum	1%	1%	
3600	Maximum	6%	7%	
3700	Protection of existing work, add, minimum	2%	2%	
3800	Maximum	5%	7%	
3900	Shift work requirements, add, minimum		5%	
4000	Maximum		30%	

D20 Plumbing

D2010 Plumbing Fixtures

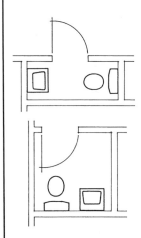

This page illustrates and describes a two fixture lavatory system including a water closet, lavatory, accessories and all service piping. Lines within System Components give the unit price and total price on a cost each basis for this system. Prices for two fixture lavatory systems are on Line Items D2010 958 1800 and 1900. Both material quantities and labor costs have been adjusted for the system listed.

Factors: To adjust for job conditions other than normal working situations use Lines D2010 958 2900 thru 4000.

Example: You are to install the system while controlling dust in the work area. Go to Line D2010 958 3200 and apply these percentages to the appropriate MAT. and INST. costs.

System Components	QUANTITY	UNIT	COST EACH		
			MAT.	INST.	TOTAL
Two fixture bathroom incl. water closet, lavatory, accessories and Necessary service piping to install this system in 2 walls.					
Water closet, floor mounted, 2 piece, close coupled	1.000	Ea.	261	249	510
Rough in waste & vent for water closet	1.000	Set	365	430	795
Lavatory, 20" x 18", P.E. cast iron with accessories	1.000	Ea.	305	165	470
Rough in waste & vent for lavatory	1.000	Set	500	795	1,295
Additional service piping					
1/2" copper pipe with sweat solder joints	10.000	L.F.	40.70	90.50	131.20
2" black schedule 40 steel pipe with threaded couplings	12.000	L.F.	118.80	246	364.80
4" cast iron soil pipe with lead and oakum joints	7.000	L.F.	143.50	168	311.50
Accessories					
Toilet tissue dispenser, chrome, single roll	1.000	Ea.	19.60	20.50	40.10
18" long stainless steel towel bar	1.000	Ea.	46	27	73
Medicine cabinet with mirror, 20" x 16", unlighted	1.000	Ea.	108	44	152
TOTAL			1,907.60	2,235	4,142.60

D2010 958	Plumbing - Two Fixture Bathroom	COST EACH		
		MAT.	INST.	TOTAL
1800	Water closet, lavatory & accessories, in 2 walls with all service piping	1,900	2,225	4,125
1900	Above system installed in 1 wall	1,775	1,950	3,725
2000				
2400	NOTE: PLUMBING APPROXIMATIONS			
2500	WATER CONTROL: water meter, backflow preventer,			
2600	Shock absorbers, vacuum breakers, mixer....10 to 15% of fixtures			
2700	PIPE AND FITTINGS: .... 30 to 60% of fixtures			
2800				
2900	Cut & patch to match existing construction, add, minimum	2%	3%	
3000	Maximum	5%	9%	
3100	Dust protection, add, minimum	1%	2%	
3200	Maximum	4%	11%	
3300	Equipment usage curtailment, add, minimum	1%	1%	
3400	Maximum	3%	10%	
3500	Material handling & storage limitation, add, minimum	1%	1%	
3600	Maximum	6%	7%	
3700	Protection of existing work, add, minimum	2%	2%	
3800	Maximum	5%	7%	
3900	Shift work requirements, add, minimum		5%	
4000	Maximum		30%	

D20 Plumbing

D2010 Plumbing Fixtures

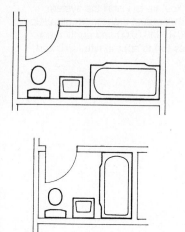

This page illustrates and describes a three fixture bathroom system including a water closet, tub, lavatory, accessories and service piping. Lines within System Components give the unit price and total price on a cost each basis for this system. Prices for a three fixture bathroom system is on Line Item D2010 959 1700. Both material quantities and labor costs have been adjusted for the system listed.

Factors: To adjust for job conditions other than normal working situations use Lines D2010 959 2900 thru 4000.

Example: You are to install the system and protect all existing work. Go to Line D2010 959 3800 and apply these percentages to the appropriate MAT. and INST. costs.

System Components	QUANTITY	UNIT	COST EACH		
			MAT.	INST.	TOTAL
Three fixture bathroom incl. water closet, bathtub, lavatory, accessories, And necessary service piping to install this system in 1 wall.					
Water closet, tank vit china flr mtd close cpld, 2 pc. w/seat supply & stop	1.000	Ea.	261	249	510
For rough in, supply, waste and vent	.930	Set	339.45	399.90	739.35
Bath tub, P.E., cast iron, 5' long, with accessories	1.000	Ea.	1,250	300	1,550
Rough-in, supply, waste and vent, for all above tubs, add	.950	Set	361	603.25	964.25
Lavatory, 20" x 18" P.E. cast iron with accessories	1.000	Ea.	305	165	470
Rough-in, supply, waste and vent for above lavatories	1.000	Set	500	795	1,295
Accessories					
Toilet tissue dispenser, surface mounted, SS, single roll	1.000	Ea.	19.60	20.50	40.10
Towel bar, stainless steel, 18" long	2.000	Ea.	92	54	146
Medicine cabinets with mirror, sst. frame, 16" x 22", unlighted	1.000	Ea.	108	44	152
TOTAL			3,236.05	2,630.65	5,866.70

D2010 959	Plumbing - Three Fixture Bathroom	COST EACH		
		MAT.	INST.	TOTAL
1600				
1700	Water closet, lavatory, bathtub, in 1 wall with all service piping	3,225	2,625	5,850
2400	NOTE: PLUMBING APPROXIMATIONS			
2500	WATER CONTROL: water meter, backflow preventer,			
2600	Shock absorbers, vacuum breakers, mixer....10 to 15% of fixtures			
2700	PIPE AND FITTINGS: .... 30 to 60% of fixtures			
2800				
2900	Cut & patch to match existing construction, add, minimum	2%	3%	
3000	Maximum	5%	9%	
3100	Dust protection, add, minimum	1%	2%	
3200	Maximum	4%	11%	
3300	Equipment usage curtailment, add, minimum	1%	1%	
3400	Maximum	3%	10%	
3500	Material handling & storage limitation, add, minimum	1%	1%	
3600	Maximum	6%	7%	
3700	Protection of existing work, add, minimum	2%	2%	
3800	Maximum	5%	7%	
3900	Shift work requirements, add, minimum		5%	
4000	Maximum		30%	

D20 Plumbing

D2010 Plumbing Fixtures

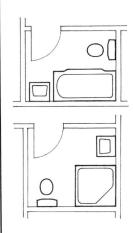

This page illustrates and describes a three fixture bathroom system including a water closet, tub, lavatory, accessories, and service piping. Lines within System Components give the unit price and total price on a cost each basis for this system. Prices for a three fixture bathroom system are on Line Item D2010 960 1900 thru 2000. Both material quantities and labor costs have been adjusted for the system listed.

Factors: To adjust for job conditions other than normal working situations use Lines D2010 960 2900 thru 4000.

Example: You are to install the system and protect the surrounding area from dust. Go to Line D2010 960 3100 and apply these percentages to the appropriate MAT. and INST. costs.

System Components	QUANTITY	UNIT	COST EACH		
			MAT.	INST.	TOTAL
Three fixture bathroom incl. water closet, bathtub, lavatory, accessories,					
And necessary service piping to install this system in 2 walls.					
Water closet, floor mounted, 2 piece, close coupled	1.000	Ea.	261	249	510
Rough-in waste & vent for water closet	1.000	Set	365	430	795
Bathtub, P.E. cast iron, 5' long with accessories	1.000	Ea.	1,250	300	1,550
Rough-in waste & vent for bathtub	1.000	Set	380	635	1,015
Lavatory, 20" x 18" P.E. cast iron with accessories	1.000	Ea.	305	165	470
Rough-in waste & vent for lavatory	1.000	Set	500	795	1,295
Additional service piping					
1-1/4" copper DWV type tubing, sweat solder joints	6.000	L.F.	60.30	73.20	133.50
2" black schedule 40 steel pipe with threaded couplings	12.000	L.F.	118.80	246	364.80
Accessories					
Toilet tissue dispenser, chrome, single roll	1.000	Ea.	19.60	20.50	40.10
18" long stainless steel towel bar	2.000	Ea.	92	54	146
Medicine cabinet with mirror, 20" x 16", unlighted	1.000	Ea.	108	44	152
TOTAL			3,459.70	3,011.70	6,471.40

D2010 960	Plumbing - Three Fixture Bathroom	COST EACH		
		MAT.	INST.	TOTAL
1900	Water closet, lavatory, bathtub, in 2 walls with all service piping	3,450	3,000	6,450
2000	Above system with corner contour tub, P.E. cast iron	5,100	3,000	8,100
2300				
2400	NOTE: PLUMBING APPROXIMATIONS			
2500	WATER CONTROL: water meter, backflow preventer,			
2600	Shock absorbers, vacuum breakers, mixer....10 to 15% of fixtures			
2700	PIPE AND FITTINGS: .... 30 to 60% of fixtures			
2800				
2900	Cut & patch to match existing construction, add, minimum	2%	3%	
3000	Maximum	5%	9%	
3100	Dust protection, add, minimum	1%	2%	
3200	Maximum	4%	11%	
3300	Equipment usage curtailment, add, minimum	1%	1%	
3400	Maximum	3%	10%	
3500	Material handling & storage limitation, add, minimum	1%	1%	
3600	Maximum	6%	7%	
3700	Protection of existing work, add, minimum	2%	2%	
3800	Maximum	5%	7%	
3900	Shift work requirements, add, minimum		5%	
4000	Maximum		30%	

D20 Plumbing

D2010 Plumbing Fixtures

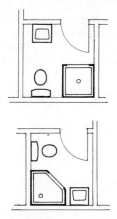

This page illustrates and describes a three fixture bathroom system including a water closet, shower, lavatory, accessories, and service piping. Lines within System Components give the unit price and total price on a cost each basis for this system. Prices for three fixture bathroom systems are on Line Items D2010 961 2100 thru 2200. Both material quantities and labor costs have been adjusted for the system listed.

Factors: To adjust for job conditions other than normal working situations use Lines D2010 961 2900 thru 4000.

Example: You are to install the system and protect existing construction. Go to Line D2010 961 3700 and apply these percentages to the appropriate MAT. and INST. costs.

System Components	QUANTITY	UNIT	COST EACH		
			MAT.	INST.	TOTAL
Three fixture bathroom incl. water closet, shower, lavatory, accessories, And necessary service piping to install this system in 2 walls.					
Water closet, floor mounted, 2 piece, close coupled	1.000	Ea.	261	249	510
Rough-in waste & vent for water closet	1.000	Set	365	430	795
32" shower, enameled stall, molded stone receptor	1.000	Ea.	1,275	264	1,539
Rough-in waste & vent for shower	1.000	Set	375	645	1,020
Lavatory, 20" x 18" P.E. cast iron with accessories	1.000	Ea.	305	165	470
Rough-in waste & vent for lavatory	1.000	Set	500	795	1,295
Additional service piping					
1-1/4" Copper DWV type tubing, sweat solder joints	4.000	L.F.	40.20	48.80	89
2" black schedule 40 steel pipe with threaded couplings	6.000	L.F.	59.40	123	182.40
4" cast iron soil with lead and oakum joints	7.000	L.F.	143.50	168	311.50
Accessories					
Toilet tissue dispenser, chrome, single roll	1.000	Ea.	19.60	20.50	40.10
18" long stainless steel towel bar	2.000	Ea.	92	54	146
Medicine cabinet with mirror, 20" x 16", unlighted	1.000	Ea.	108	44	152
TOTAL			3,543.70	3,006.30	6,550

D2010 961	Plumbing - Three Fixture Bathroom	COST EACH		
		MAT.	INST.	TOTAL
2100	Water closet, 32" shower, lavavatory, in 2 walls with all service piping	3,550	3,000	6,550
2200	Above system with 36" corner angle shower	4,175	3,025	7,200
2300				
2400	NOTE: PLUMBING APPROXIMATIONS			
2500	WATER CONTROL: water meter, backflow preventer,			
2600	Shock absorbers, vacuum breakers, mixer....10 to 15% of fixtures			
2700	PIPE AND FITTINGS: .... 30 to 60% of fixtures			
2800				
2900	Cut & patch to match existing construction, add, minimum	2%	3%	
3000	Maximum	5%	9%	
3100	Dust protection, add, minimum	1%	2%	
3200	Maximum	4%	11%	
3300	Equipment usage curtailment, add, minimum	1%	1%	
3400	Maximum	3%	10%	
3500	Material handling & storage limitation, add, minimum	1%	1%	
3600	Maximum	6%	7%	
3700	Protection of existing work, add, minimum	2%	2%	
3800	Maximum	5%	7%	
3900	Shift work requirements, add, minimum		5%	
4000	Maximum		30%	

D20 Plumbing

D2010 Plumbing Fixtures

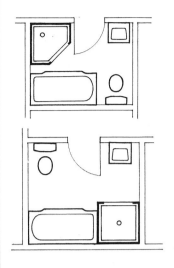

This page illustrates and describes a four fixture bathroom system including a water closet, shower, bathtub, lavatory, accessories, and service piping. Lines within System Components give the unit price and total price on a cost each basis for this system. Prices for four fixture bathroom systems are on Line Items D2010 962 1800 thru 1900. Both material quantities and labor costs have been adjusted for the system listed.

Factors: To adjust for job conditions other than normal working situations use Lines D2010 962 2900 thru 4000.

Example: You are to install the system during weekends and evenings. Go to Line D2010 962 4000 and apply these percentages to the appropriate MAT. and INST. costs.

System Components	QUANTITY	UNIT	COST EACH		
			MAT.	INST.	TOTAL
Four fixture bathroom incl. water closet, shower, bathtub, lavatory					
Accessories and necessary service piping to install this system in 2 walls.					
Water closet, floor mounted, 2 piece, close coupled	1.000	Ea.	261	249	510
Rough-in waste & vent for water closet	1.000	Set	365	430	795
32" shower, enameled steel stall, molded stone receptor	1.000	Ea.	1,275	264	1,539
Rough-in waste & vent for shower	1.000	Set	375	645	1,020
Bathtub, P.E. cast iron, 5' long with accessories	1.000	Ea.	1,250	300	1,550
Rough-in waste & vent for bathtub	1.000	Set	380	635	1,015
Lavatory, 20" x 18" P.E. cast iron with accessories	1.000	Ea.	305	165	470
Rough-in waste & vent for lavatory	1.000	Set	500	795	1,295
Accessories					
Toilet tissue dispenser, chrome, single roll	1.000	Ea.	19.60	20.50	40.10
18" long stainless steel towel bar	2.000	Ea.	92	54	146
Medicine cabinet with mirror, 20" x 16" unlighted	1.000	Ea.	108	44	152
TOTAL			4,930.60	3,601.50	8,532.10

D2010 962	Plumbing - Four Fixture Bathroom	COST EACH		
		MAT.	INST.	TOTAL
1800	Water closet, 32" shower, bathtub, lavavatory, in 2 walls	4,925	3,600	8,525
1900	Above system with 36" corner angle shower, plumbing in 3 walls	5,725	3,950	9,675
2000				
2300				
2400	NOTE: PLUMBING APPROXIMATIONS			
2500	WATER CONTROL: water meter, backflow preventer,			
2600	Shock absorbers, vacuum breakers, mixer....10 to 15% of fixtures			
2700	PIPE AND FITTINGS: .... 30 to 60% of fixtures			
2800				
2900	Cut & patch to match existing construction, add, minimum	2%	3%	
3000	Maximum	5%	9%	
3100	Dust protection, add, minimum	1%	2%	
3200	Maximum	4%	11%	
3300	Equipment usage curtailment, add, minimum	1%	1%	
3400	Maximum	3%	10%	
3500	Material handling & storage limitation, add, minimum	1%	1%	
3600	Maximum	6%	7%	
3700	Protection of existing work, add, minimum	2%	2%	
3800	Maximum	5%	7%	
3900	Shift work requirements, add, minimum		5%	
4000	Maximum		30%	

D20 Plumbing

D2010 Plumbing Fixtures

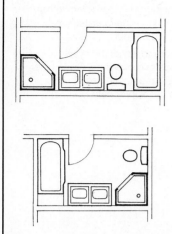

This page illustrates and describes a five fixture bathroom system including a water closet, shower, bathtub, lavatories, accessories, and service piping. Lines within System Components give the unit price and total price on a cost each basis for this system. Prices for five fixture bathroom systems are on Line Items D2010 963 1800 thru 1900. Both material quantities and labor costs have been adjusted for the system listed.

Factors: To adjust for job conditions other than normal working situations use Lines D2010 963 2900 thru 4000.

Example: You are to install and match any existing construction. Go to Line D2010 963 2900 and apply these percentages to the appropriate MAT. and INST. costs.

System Components	QUANTITY	UNIT	COST EACH MAT.	COST EACH INST.	COST EACH TOTAL
Five fixture bathroom incl. water closet, shower, bathtub, 2 lavatories, Accessories and necessary service piping to install this system in 1 wall.					
Water closet, floor mounted, 2 piece, close coupled	1.000	Ea.	261	249	510
Rough-in waste & vent for water closet	1.000	Set	365	430	795
36" corner angle shower, enameled steel stall, molded stone receptor	1.000	Ea.	1,900	275	2,175
Rough-in waste & vent for shower	1.000	Set	375	645	1,020
Bathtub, P.E. cast iron, 5' long with accessories	1.000	Ea.	1,250	300	1,550
Rough-in waste & vent for bathtub	1.000	Set	380	635	1,015
Countertop, plastic laminate, with backsplash	2.000	Ea.	172	82	254
Vanity base, 2 doors, 24" wide	2.000	Ea.	340	112	452
Lavatories, 20" x 18" cabinet mntd, PECI	2.000	Ea.	740	412	1,152
Rough-in waste & vent for lavatories	1.600	Set	406.40	920	1,326.40
Accessories					
Toilet tissue dispenser, chrome, single roll	1.000	Ea.	19.60	20.50	40.10
18" long stainless steel towel bars	2.000	Ea.	92	54	146
Medicine cabinet with mirror, 20" x 16", unlighted	2.000	Ea.	216	88	304
TOTAL			6,517	4,222.50	10,739.50

D2010 963	Plumbing - Five Fixture Bathroom	MAT.	INST.	TOTAL
1800	Water closet, shower, bathtub, 2 lavatories, on 1 wall	6,525	4,225	10,750
1900	Above system installed in 2 walls with all necessary service piping	6,575	4,375	10,950
2000				
2100				
2200				
2300				
2400	NOTE: PLUMBING APPROXIMATIONS			
2500	WATER CONTROL: water meter, backflow preventer,			
2600	Shock absorbers, vacuum breakers, mixer....10 to 15% of fixtures			
2700	PIPE AND FITTINGS: .... 30 to 60% of fixtures			
2800				
2900	Cut & patch to match existing construction, add, minimum	2%	3%	
3000	Maximum	5%	9%	
3100	Dust protection, add, minimum	1%	2%	
3200	Maximum	4%	11%	
3300	Equipment usage curtailment, add, minimum	1%	1%	
3400	Maximum	3%	10%	
3500	Material handling & storage limitation, add, minimum	1%	1%	

D20 Plumbing

D2010 Plumbing Fixtures

D2010 963	Plumbing - Five Fixture Bathroom	COST EACH		
		MAT.	INST.	TOTAL
3600	Maximum	6%	7%	
3700	Protection of existing work, add, minimum	2%	2%	
3800	Maximum	5%	7%	
3900	Shift work requirements, add, minimum		5%	
4000	Maximum		30%	

D30 HVAC

D3010 Energy Supply

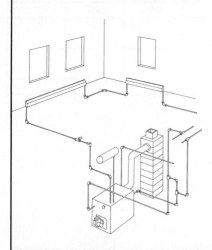

This page illustrates and describes an oil fired hot water baseboard system including an oil fired boiler, fin tube radiation and all fittings and piping. Lines within System Components give the unit price and total price per square foot for this system. Prices for oil fired hot water baseboard systems are on Line Items D3010 540 1600 thru 2100. Both material quantities and labor costs have been adjusted for the system listed.

Factors: To adjust for job conditions other than normal working situations use Lines D3010 540 2900 thru 4000.

Example: You are to install the system while protecting all existing work. Go to Line D3010 540 3700 and apply these percentages to the appropriate MAT. and INST. costs.

System Components	QUANTITY	UNIT	COST PER S.F. MAT.	INST.	TOTAL
Oil fired hot water baseboard system including boiler, fin tube radiation And all necessary fittings and piping. Area to 800 S.F.					
Boiler, cast iron, w/oil piping, 97 MBH	1.000	Ea.	2,437.50	1,625	4,062.50
Copper piping	130.000	L.F.	1,475.50	1,644.50	3,120
Fin tube radiation	68.000	L.F.	550.80	1,564	2,114.80
Circulator	1.000	Ea.	620	220	840
Oil tank	1.000	Ea.	540	268	808
Expansion tank, ASME	1.000	Ea.	785	96	881
TOTAL			6,408.80	5,417.50	11,826.30
COST PER S.F.		S.F.	8.01	6.77	14.78

D3010 540	Heating - Oil Fired Hot Water	MAT.	INST.	TOTAL
1600	Cast iron boiler, area to 800 S.F.	8	6.78	14.78
1700	To 1000 S.F.	6.95	5.85	12.80
1800	To 1200 S.F.	6	5.30	11.30
1900	To 1600 S.F.	7.15	5.30	12.45
2000	To 2000 S.F.	5.95	4.68	10.63
2100	To 3000 S.F.	4.39	3.95	8.34
2200				
2300				
2700				
2800				
2900	Cut & patch to match existing construction, add, minimum	2%	3%	
3000	Maximum	5%	9%	
3100	Dust protection, add, minimum	1%	2%	
3200	Maximum	4%	11%	
3300	Equipment usage curtailment, add, minimum	1%	1%	
3400	Maximum	3%	10%	
3500	Material handling & storage limitation, add, minimum	1%	1%	
3600	Maximum	6%	7%	
3700	Protection of existing work, add, minimum	2%	2%	
3800	Maximum	5%	7%	
3900	Shift work requirements, add, minimum		5%	
4000	Maximum		30%	

D30 HVAC

D3010 Energy Supply

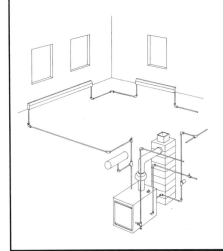

This page illustrates and describes a gas fired hot water baseboard system including a gas fired boiler, fin tube radiation, fittings and piping. Lines within System Components give the unit price and total price per square foot for this system. Prices for gas fired hot water baseboard systems are on Line Items D3010 550 1400 thru 1900. Both material quantities and labor costs have been adjusted for the system listed.

Factors: To adjust for job conditions other than normal working situations use Lines D3010 550 2900 thru 4000.

Example: You are to install the system with minimal equipment usage. Go to Line D3010 550 3400 and apply these percentages to the appropriate MAT. and INST. costs.

System Components	QUANTITY	UNIT	COST PER S.F. MAT.	INST.	TOTAL
Gas fired hot water baseboard system including boiler, fin tube radiation, All necessary fittings and piping. Area to 800 S.F.					
Cast iron boiler, insulating jacket, gas piping, 80 MBH	1.000	Ea.	2,718.75	2,437.50	5,156.25
Copper piping	130.000	L.F.	1,475.50	1,644.50	3,120
Fin tube radiation	68.000	L.F.	550.80	1,564	2,114.80
Circulator, flange connection	1.000	Ea.	620	220	840
Expansion tank, ASME	1.000	Ea.	785	96	881
TOTAL			6,150.05	5,962	12,112.05
COST PER S.F.		S.F.	7.69	7.45	15.14

D3010 550	Heating - Gas Fired Hot Water	MAT.	INST.	TOTAL
1400	Cast iron boiler, area to 800 S.F.	7.69	7.44	15.13
1500	To 1000 S.F.	6.45	6.55	13
1600	To 1200 S.F.	5.60	5.90	11.50
1700	To 1600 S.F.	6.80	5.75	12.55
1800	To 2000 S.F.	6	5.10	11.10
1900	To 3000 S.F.	4.66	4.49	9.15
2000				
2100				
2900	Cut & patch to match existing construction, add, minimum	2%	3%	
3000	Maximum	5%	9%	
3100	Dust protection, add, minimum	1%	2%	
3200	Maximum	4%	11%	
3300	Equipment usage curtailment, add, minimum	1%	1%	
3400	Maximum	3%	10%	
3500	Material handling & storage limitation, add, minimum	1%	1%	
3600	Maximum	6%	7%	
3700	Protection of existing work, add, minimum	2%	2%	
3800	Maximum	5%	7%	
3900	Shift work requirements, add, minimum		5%	
4000	Maximum		30%	

D30 HVAC

D3010 Energy Supply

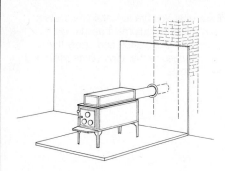

This page illustrates and describes a wood burning stove system including a free standing stove, preformed hearth, masonry chimney, and necessary piping and fittings. Lines within System Components give the unit price and total price on a cost each basis for this system. Prices for wood burning stove systems are on Line Items D3010 991 1400 thru 2000. Both material quantities and labor costs have been adjusted for the system listed.

Factors: To adjust for job conditions other than normal working situations use Lines D3010 991 2900 thru 4000.

Example: You are to install the system and protect existing construction. Go to Line D3010 991 3500 and apply these percentages to the appropriate MAT., and INST. costs.

System Components	QUANTITY	UNIT	COST EACH		
			MAT.	INST.	TOTAL
Cast iron, free standing, wood burning stove with preformed hearth, masonry Chimney, and all necessary piping and fittings to install in chimney.					
Cast iron wood burning stove, stove pipe	1.000	Ea.	1,300	945	2,245
Wall panel or hearth, non-combustible	1.000	Ea.	1,045	173.85	1,218.85
16" x 16" brick chimney, 8" x 8" flue	20.000	V.L.F.	662.50	1,500	2,162.50
Concrete in place, chimney foundations	.500	C.Y.	82.50	128.54	211.04
TOTAL			3,090	2,747.39	5,837.39

D3010 991	Wood Burning Stoves	COST EACH		
		MAT.	INST.	TOTAL
1400	Cast iron, free standing with preformed hearth	3,100	2,750	5,850
1500				
1600	Installed in existing fireplace	1,850	1,025	2,875
1700				
1800	Installed with insulated metal chimney			
1900	System including ceiling package,			
2000	Metal chimney to 10 L.F.	3,650	1,475	5,125
2100				
2200				
2300				
2400				
2500				
2600				
2700				
2900	Dust protection, add, minimum	1%	2%	
3000	Maximum	4%	11%	
3100	Equipment usage curtailment, add, minimum	1%	1%	
3200	Maximum	3%	10%	
3300	Material handling & storage limitation, add, minimum	1%	1%	
3400	Maximum	6%	7%	
3500	Protection of existing work, add, minimum	2%	2%	
3600	Maximum	5%	7%	
3700	Shift work requirements, add, minimum		5%	
3800	Maximum		30%	
3900	Temporary shoring and bracing, add, minimum	2%	5%	
4000	Maximum	5%	12%	

D30 HVAC

D3010 Energy Supply

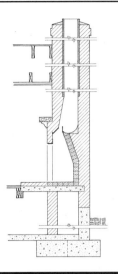

This page illustrates and describes masonry fireplace systems including a brick fireplace, footing, foundation, hearth, firebox, chimney and flue.
Lines within System Components give the unit price and total price on a cost each basis for this system. Prices for masonry fireplace systems are on Line Items D3010 992 1200 thru 1500. Both material quantities and labor costs have been adjusted for the system listed.

Factors: To adjust for job conditions other than normal working situations use Lines D3010 992 2700 thru 4000.

Example: You are to install the system with some temporary shoring and bracing. Go to Line D3010 992 3900 and apply these percentages to the appropriate MAT. and INST. costs.

System Components	QUANTITY	UNIT	COST EACH		
			MAT.	INST.	TOTAL
Brick masonry fireplace, including footing, foundation, hearth, firebox, Chimney and flue, chimney 12' above firebox.					
Footing, 4' x 7' x 12" thick, 3000 psi concrete	1.040	C.Y.	190.32	308.09	498.41
Foundation, 12" concrete block	180.000	S.F.	927	1,971	2,898
Fireplace, brick faced, 6'-0" wide x 5'-0" high	1.000	Ea.	605	2,750	3,355
Hearth	1.000	Ea.	225	550	775
Chimney, 20" x 20", one 12" x 12" flue, 12 V.L.F.	12.000	V.L.F.	600	960	1,560
Mantle, wood	7.000	L.F.	63.70	119.70	183.40
TOTAL			2,611.02	6,658.79	9,269.81

D3010 992	Masonry Fireplace	COST EACH		
		MAT.	INST.	TOTAL
1200	Chimney, 20" x 20", one 12" x 12" flue, 12 V.L.F.	2,600	6,650	9,250
1300	18 V.L.F.	2,900	7,150	10,050
1400	24 V.L.F.	3,200	7,625	10,825
1500	Fieldstone face instead of brick	2,900	6,650	9,550
1600				
1700				
1800				
1900				
2000				
2100				
2200				
2300				
2700	Cut & patch to match existing construction, add, minimum	2%	3%	
2800	Maximum	5%	9%	
2900	Dust protection, add, minimum	1%	2%	
3000	Maximum	4%	11%	
3100	Equipment usage curtailment, add, minimum	1%	1%	
3200	Maximum	3%	10%	
3300	Material handling & storage limitation, add, minimum	1%	1%	
3400	Maximum	6%	7%	
3500	Protection of existing work, add, minimum	2%	2%	
3600	Maximum	5%	7%	
3700	Shift work requirements, add, minimum		5%	
3800	Maximum		30%	
3900	Temporary shoring and bracing, add, minimum	2%	5%	
4000	Maximum	5%	12%	

D30 HVAC

D3020 Heat Generating Systems

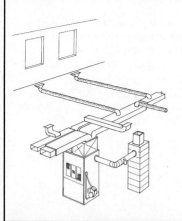

This page illustrates and describes an oil fired forced air system including an oil fired furnace, ductwork, registers and hookups. Lines within System Components give the unit price and total price per square foot for this system. Prices for oil fired forced air systems are on Line Items D3020 122 1600 thru 2800. Both material quantities and labor costs have been adjusted for the system listed.

Factors: To adjust for job conditions other than normal working situations use Lines D3020 122 3100 thru 4200.

Example: You are to install the system during evenings and weekends. Go to Line D3020 122 4200 and apply this percentage to the appropriate INST. cost.

System Components			COST PER S.F.		
	QUANTITY	UNIT	MAT.	INST.	TOTAL
Oil fired hot air heating system including furnace, ductwork, registers And all necessary hookups.					
Area to 800 S.F., heat only					
Furnace, oil, atomizing gun type burner, w/oil piping	1.000	Ea.	2,375	443.75	2,818.75
Oil tank, steel, 275 gallon	1.000	Ea.	540	268	808
Duct, galvanized, steel	312.000	Lb.	224.64	2,620.80	2,845.44
Insulation, blanket type, duct work	270.000	S.F.	1,347.30	1,090.80	2,438.10
Flexible duct, 6" diameter, insulated	100.000	L.F.	323	489	812
Registers, baseboard, gravity, 12" x 6"	8.000	Ea.	50.40	244	294.40
Return, damper, 36" x 18"	1.000	Ea.	80	70.50	150.50
TOTAL			4,940.34	5,226.85	10,167.19
COST PER S.F.		S.F.	6.18	6.53	12.71

D3020 122	Heating-Cooling, Oil, Forced Air	COST PER S.F.		
		MAT.	INST.	TOTAL
1600	Oil fired, area to 800 S.F.	6.19	6.53	12.72
1700	To 1000 S.F.	4.95	5.30	10.25
1800	To 1200 S.F.	4.27	4.66	8.93
1900	To 1600 S.F.	3.41	3.88	7.29
2000	To 2000 S.F.	3.73	5	8.73
2100	To 3000 S.F.	2.93	3.97	6.90
2200	For combined heating and cooling systems:			
2300	Oil fired, heating and cooling, area to 800 S.F.	8.40	7.75	16.15
2400	To 1000 S.F.	6.70	6.30	13
2500	To 1200 S.F.	5.75	5.50	11.25
2600	To 1600 S.F.	4.60	4.55	9.15
2700	To 2000 S.F.	4.77	5.60	10.37
2800	To 3000 S.F.	3.65	4.40	8.05
3100	Cut & patch to match existing construction, add, minimum	2%	3%	
3200	Maximum	5%	9%	
3300	Dust protection, add, minimum	1%	2%	
3400	Maximum	4%	11%	
3500	Equipment usage curtailment, add, minimum	1%	1%	
3600	Maximum	3%	10%	
3700	Material handling & storage limitation, add, minimum	1%	1%	
3800	Maximum	6%	7%	
3900	Protection of existing work, add, minimum	2%	2%	
4000	Maximum	5%	7%	
4100	Shift work requirements, add, minimum		5%	
4200	Maximum		30%	

D30 HVAC

D3020 Heat Generating Systems

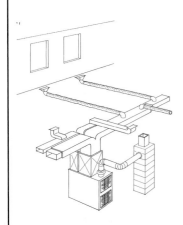

This page illustrates and describes a gas fired forced air system including a gas fired furnace, ductwork, registers and hookups. Lines within System Components give the unit price and total price per square foot for this system. Prices for gas fired forced air systems are on Line Items D3020 124 1400 thru 2600. Both material quantities and labor costs have been adjusted for the system listed.

Factors: To adjust for job conditions other than normal working situations use Lines D3020 124 2900 thru 4000.

Example: You are to install the system with material handling and storage limitations. Go to Line D3020 124 3500 and apply these percentages to the appropriate MAT. and INST. costs.

System Components	QUANTITY	UNIT	COST PER S.F.		
			MAT.	INST.	TOTAL
Gas fired hot air heating system including furnace, ductwork, registers And all necessary hookups.					
Area to 800 S.F., heat only					
Furnace, gas, AGA certified, direct drive, w/gas piping, 44 MBH	1.000	Ea.	737.50	400	1,137.50
Duct, galvanized steel	312.000	Lb.	224.64	2,620.80	2,845.44
Insulation, blanket type, ductwork	270.000	S.F.	1,347.30	1,090.80	2,438.10
Flexible duct, 6" diameter, insulated	100.000	L.F.	323	489	812
Registers, baseboard, gravity, 12" x 6"	8.000	Ea.	50.40	244	294.40
Return, damper, 36" x 18"	1.000	Ea.	80	70.50	150.50
TOTAL			2,762.84	4,915.10	7,677.94
COST PER S.F.		S.F.	3.45	6.14	9.59

D3020 124	Heating-Cooling, Gas, Forced Air	COST PER S.F.		
		MAT.	INST.	TOTAL
1400	Gas fired, area to 800 S.F.	3.47	6.16	9.63
1500	To 1000 S.F.	2.77	5	7.77
1600	To 1200 S.F.	2.45	4.42	6.87
1700	To 1600 S.F.	2.28	4.13	6.41
1800	To 2000 S.F.	2.60	4.89	7.49
1900	To 3000 S.F.	2.05	3.88	5.93
2000	For combined heating and cooling systems:			
2100	Gas fired, heating and cooling, area to 800 S.F.	5.65	7.35	13
2200	To 1000 S.F.	4.54	6	10.54
2300	To 1200 S.F.	3.93	5.25	9.18
2400	To 1600 S.F.	3.47	4.80	8.27
2500	To 2000 S.F.	3.63	5.50	9.13
2600	To 3000 S.F.	2.77	4.31	7.08
2900	Cut & patch to match existing construction, add, minimum	2%	3%	
3000	Maximum	5%	9%	
3100	Dust protection, add, minimum	1%	2%	
3200	Maximum	4%	11%	
3300	Equipment usage curtailment, add, minimum	1%	1%	
3400	Maximum	3%	10%	
3500	Material handling & storage limitation, add, minimum	1%	1%	
3600	Maximum	6%	7%	
3700	Protection of existing work, add, minimum	2%	2%	
3800	Maximum	5%	7%	
3900	Shift work requirements, add, minimum		5%	
4000	Maximum		30%	

D30 HVAC

D3020 Heat Generating Systems

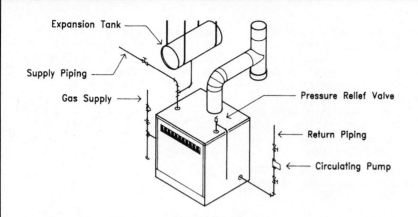

Cast Iron Boiler, Hot Water, Gas Fired

This page illustrates and describes boilers including expansion tank, circulating pump and all service piping. Lines within Systems Components give the material and installation price on a cost each basis for the components. Prices for boiler systems are on Line Items D3020 130 1010 thru D3020 138 1040. Material quantities and labor costs have been adjusted for the system listed.

Factors: To adjust for job conditions other than normal working situations use Line D3020 138 2700 thru 4000.

System Components	QUANTITY	UNIT	COST EACH		
			MAT.	INST.	TOTAL
Cast iron boiler, including piping, chimney and accessories					
Boilers, gas fired, std controls, CI, insulated, HW, gross output 100 MBH	1.000	Ea.	2,825	2,100	4,925
Pipe, black steel, Sch 40, threaded, W/coupling & hangers, 10' OC, 3/4" dia	20.000	L.F.	96.75	270	366.75
Pipe, black steel, Sch 40, threaded, W/coupling & hangers, 10' OC, 1" dia.	20.000	L.F.	119.33	296.70	416.03
Elbow, 90°, black, straight, 3/4" dia.	9.000	Ea.	39.51	472.50	512.01
Elbow, 90°, black, straight, 1" dia.	6.000	Ea.	45.60	339	384.60
Tee, black, straight, 3/4" dia.	2.000	Ea.	13.90	163	176.90
Tee, black, straight, 1" dia.	2.000	Ea.	23.90	183	206.90
Tee, black, reducing, 1" dia.	2.000	Ea.	38.40	183	221.40
Pipe cap, black, 3/4" dia.	1.000	Ea.	4.98	23	27.98
Union, black with brass seat, 3/4" dia.	2.000	Ea.	37	113	150
Union, black with brass seat, 1" dia.	2.000	Ea.	48	122	170
Pipe nipples, black, 3/4" dia.	5.000	Ea.	10.75	30	40.75
Pipe nipples, black, 1" dia.	3.000	Ea.	8.33	20.70	29.03
Valves, bronze, gate, N.R.S., threaded, class 150, 3/4" size	2.000	Ea.	179	73	252
Valves, bronze, gate, N.R.S., threaded, class 150, 1" size	2.000	Ea.	224	77	301
Gas cock, brass, 3/4" size	1.000	Ea.	17.05	33.50	50.55
Thermometer, stem type, 9" case, 8" stem, 3/4" NPT	2.000	Ea.	478	53	531
Tank, steel, liquid expansion, ASME, painted, 15 gallon capacity	1.000	Ea.	700	79	779
Pump, circulating, bronze, flange connection, 3/4" to 1-1/2" size, 1/8 HP	1.000	Ea.	1,075	220	1,295
Vent chimney, all fuel, pressure tight, double wall, SS, 6" dia.	20.000	L.F.	1,190	420	1,610
Vent chimney, elbow, 90° fixed, 6" dia.	2.000	Ea.	890	85	975
Vent chimney, Tee, 6" dia.	2.000	Ea.	522	106	628
Vent chimney, ventilated roof thimble, 6" dia.	1.000	Ea.	272	49	321
Vent chimney, flat roof flashing, 6" dia.	1.000	Ea.	117	42.50	159.50
Vent chimney, stack cap, 6" diameter	1.000	Ea.	320	27.50	347.50
Insulation, fiberglass pipe covering, 1" wall, 1" IPS	20.000	L.F.	21.40	110	131.40
TOTAL			9,316.90	5,691.40	15,008.30

D3020 130	Boiler, Cast Iron, Hot Water, Gas		COST EACH		
			MAT.	INST.	TOTAL
1000	For alternate boilers:				
1010	Boiler, cast iron, gas, hot water, 100 MBH		9,325	5,700	15,025
1020	200 MBH		10,700	6,525	17,225
1030	320 MBH		10,800	7,100	17,900
1040	440 MBH		14,600	9,150	23,750
1050	544 MBH		24,800	14,300	39,100
1060	765 MBH		30,000	15,400	45,400
1070	1088 MBH		32,000	16,400	48,400

D30 HVAC

D3020 Heat Generating Systems

D3020 130	Boiler, Cast Iron, Hot Water, Gas	COST EACH		
		MAT.	INST.	TOTAL
1080	1530 MBH	39,000	21,700	60,700
1090	2312 MBH	45,900	25,500	71,400

D3020 134	Boiler, Cast Iron, Steam, Gas	COST EACH		
		MAT.	INST.	TOTAL
1010	Boiler, cast iron, gas, steam, 100 MBH	7,800	5,500	13,300
1020	200 MBH	15,600	10,400	26,000
1030	320 MBH	17,400	11,600	29,000
1040	544 MBH	26,400	18,200	44,600
1050	765 MBH	30,000	19,000	49,000
1060	1275 MBH	35,700	20,800	56,500
1070	2675 MBH	51,000	30,400	81,400

D3020 136	Boiler, Cast Iron, Hot Water, Gas/Oil	COST EACH		
		MAT.	INST.	TOTAL
1010	Boiler, cast iron, gas & oil, hot water, 584 MBH	30,300	15,900	46,200
1020	876 MBH	38,600	17,000	55,600
1030	1168 MBH	48,600	19,300	67,900
1040	1460 MBH	59,000	23,100	82,100
1050	2044 MBH	64,500	23,900	88,400
1060	2628 MBH	77,500	29,000	106,500

D3020 138	Boiler, Cast Iron, Steam, Gas/Oil	COST EACH		
		MAT.	INST.	TOTAL
1010	Boiler, cast iron, gas & oil, steam, 810 MBH	33,100	20,600	53,700
1020	1360 MBH	38,400	22,000	60,400
1030	2040 MBH	47,600	28,200	75,800
1040	2700 MBH	51,000	31,400	82,400
2700	Cut & patch to match existing construction, add, minimum	2%	3%	
2800	Maximum	5%	9%	
2900	Dust protection, add, minimum	1%	2%	
3000	Maximum	4%	11%	
3100	Equipment usage curtailment, add, minimum	1%	1%	
3200	Maximum	3%	10%	
3300	Material handling & storage limitation, add, minimum	1%	1%	
3400	Maximum	6%	7%	
3500	Protection of existing work, add, minimum	2%	2%	
3600	Maximum	5%	7%	
3700	Shift work requirements, add, minimum		5%	
3800	Maximum		30%	
3900	Temporary shoring and bracing, add, minimum	2%	5%	
4000	Maximum	5%	12%	

D30 HVAC

D3020 Heat Generating Systems

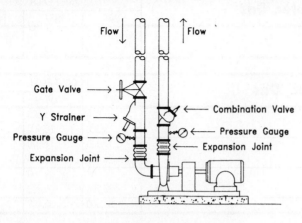

Base Mounted End−Suction Pump

This page illustrates and describes end suction base mounted circulating pumps, including strainer and piping. Lines within system components give the material and installation price on a cost each basis for the components. Prices for circulating pumps are on Line items D3020 330 1010 thru 1050. Material quantities and labor costs have been adjusting for the system listed.

Factors: To adjust for job conditions other than normal working situations uses Lines D3020 330 2700 thru 4000.

System Components	QUANTITY	UNIT	COST EACH MAT.	COST EACH INST.	COST EACH TOTAL
Circulator pump, end suction, including pipe, Valves, fittings and guages.					
Pump, circulating, CI, base mounted, 2-1/2" size, 3 HP, to 150 GPM	1.000	Ea.	8,675	730	9,405
Pipe, black steel, Sch. 40, on yoke & roll hangers, 10' O.C., 2-1/2" dia	12.000	L.F.	168.60	352.20	520.80
Elbow, 90°, weld joint, steel, 2-1/2" pipe size	1.000	Ea.	30.50	172.95	203.45
Flange, weld neck, 150 LB, 2-1/2" pipe size	8.000	Ea.	280	691.76	971.76
Valve, iron body, gate, 125 lb., N.R.S., flanged, 2-1/2" size	1.000	Ea.	810	264	1,074
Strainer, Y type, iron body, flanged, 125 lb., 2-1/2" pipe size	1.000	Ea.	127	268	395
Multipurpose valve, CI body, 2" size	1.000	Ea.	505	93	598
Expansion joint, flanged spool, 6" F to F, 2-1/2" dia	2.000	Ea.	588	216	804
T-O-L, weld joint, socket, 1/4" pipe size, nozzle	2.000	Ea.	15.60	120.52	136.12
T-O-L, weld joint, socket, 1/2" pipe size, nozzle	2.000	Ea.	15.60	125.78	141.38
Control gauges, pressure or vacuum, 3-1/2" diameter dial	2.000	Ea.	22.80	47	69.80
Pressure/temperature relief plug, 316 SS, 3/4" OD, 7-1/2" insertion	2.000	Ea.	153	47	200
Insulation, fiberglass pipe covering, 1-1/2" wall, 2-1/2" IPS	12.000	L.F.	30	81	111
Water flow	1.000	Ea.	1,325	1,100	2,425
Pump balancing	1.000	Ea.		330	330
TOTAL			12,746.10	4,639.21	17,385.31

D3020 330	Circulating Pump Systems, End Suction	COST EACH MAT.	COST EACH INST.	COST EACH TOTAL
1000	For alternate pump sizes:			
1010	Pump, base mtd with motor, end-suction, 2-1/2" size, 3 HP, to 150 GPM	12,700	4,650	17,350
1020	3" size, 5 HP, to 225 GPM	15,300	5,250	20,550
1030	4" size, 7-1/2 HP, to 350 GPM	18,100	6,250	24,350
1040	5" size, 15 HP, to 1000 GPM	26,500	9,225	35,725
1050	6" size, 25 HP, to 1550 GPM	35,100	11,200	46,300
2700	Cut & patch to match existing construction, add, minimum	2%	3%	
2800	Maximum	5%	9%	
2900	Dust protection, add, minimum	1%	2%	
3000	Maximum	4%	11%	
3100	Equipment usage curtailment, add, minimum	1%	1%	
3200	Maximum	3%	10%	
3300	Material handling & storage limitation, add, minimum	1%	1%	
3400	Maximum	6%	7%	
3500	Protection of existing work, add, minimum	2%	2%	
3600	Maximum	5%	7%	

D3020 Heat Generating Systems

D3020 330	Circulating Pump Systems, End Suction	COST EACH		
		MAT.	INST.	TOTAL
3700	Shift work requirements, add, minimum		5%	
3800	Maximum		30%	
3900	Temporary shoring and bracing, add, minimum	2%	5%	
4000	Maximum	5%	12%	

D30 HVAC

D3040 Distribution Systems

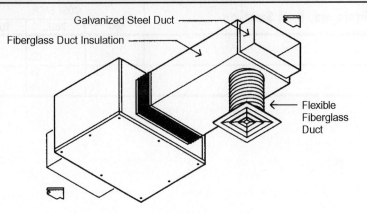

Horizontal Fan Coil Air Conditioning System

This page illustrates and describes fan coil air conditioners including duct work, duct installation, piping and diffusers. Lines within Systems Components give the material and installation price on a cost each basis for the components. Prices for fan coil A/C unit systems are on Line Items D3040 118 1010 thru D3040 126 1120. Material quantities and labor costs have been adjusted for the system listed.

Factors: To adjust for job conditions other than normal working situations use Lines D3040 126 2700 thru 4000.

System Components	QUANTITY	UNIT	COST EACH MAT.	COST EACH INST.	COST EACH TOTAL
Fan coil A/C system, horizontal, with housing, controls, 2 pipe, 1/2 ton					
Fan coil A/C, horizontal housing, filters, chilled water, 1/2 ton cooling	1.000	Ea.	1,400	168	1,568
Pipe, black steel, Sch 40, threaded, W/cplgs & hangers, 10' OC, 3/4" dia.	20.000	L.F.	92.45	258	350.45
Elbow, 90°, black, straight, 3/4" dia.	6.000	Ea.	26.34	315	341.34
Tee, black, straight, 3/4" dia.	2.000	Ea.	13.90	163	176.90
Union, black with brass seat, 3/4" dia.	2.000	Ea.	37	113	150
Pipe nipples, 3/4" diam	3.000	Ea.	6.45	18	24.45
Valves, bronze, gate, N.R.S., threaded, class 150, 3/4" size	2.000	Ea.	179	73	252
Circuit setter, balance valve, bronze body, threaded, 3/4" pipe size	1.000	Ea.	87	37.50	124.50
Insulation, fiberglass pipe covering, 1" wall, 3/4" IPS	20.000	L.F.	19.80	105	124.80
Ductwork, 12" x 8" fabricated, galvanized steel, 12 LF	55.000	Lb.	39.60	462	501.60
Insulation, ductwork, blanket type, fiberglass, 1" thk, 1-1/2 LB density	40.000	S.F.	199.60	161.60	361.20
Diffusers, aluminum, OB damper, ceiling, perf, 24"x24" panel size, 6"x6"	2.000	Ea.	332	88	420
Ductwork, flexible, fiberglass fabric, insulated, 1"thk, PE jacket, 6" dia	16.000	L.F.	51.68	78.24	129.92
Round volume control damper 6" dia.	2.000	Ea.	55	64	119
Fan coil unit balancing	1.000	Ea.		91.50	91.50
Re-heat coil balancing	1.000	Ea.		138	138
Diffuser/register, high, balancing	2.000	Ea.		254	254
TOTAL			2,539.82	2,587.84	5,127.66

D3040 118	Fan Coil A/C Unit, Two Pipe	MAT.	INST.	TOTAL
1000	For alternate A/C units:			
1010	Fan coil A/C system, cabinet mounted, controls, 2 pipe, 1/2 Ton	1,350	1,375	2,725
1020	1 Ton	1,650	1,550	3,200
1030	1-1/2 Ton	1,750	1,575	3,325
1040	2 Ton	2,850	1,900	4,750
1050	3 Ton	3,775	1,975	5,750

D3040 120	Fan Coil A/C Unit, Two Pipe, Electric Heat	MAT.	INST.	TOTAL
1010	Fan coil A/C system, cabinet mntd, elect. ht, controls, 2 pipe, 1/2 Ton	1,875	1,375	3,250
1020	1 Ton	2,300	1,550	3,850
1030	1-1/2 Ton	2,675	1,575	4,250
1040	2 Ton	4,125	1,950	6,075
1050	3 Ton	6,800	1,975	8,775

D30 HVAC

D3040 Distribution Systems

D3040 122	Fan Coil A/C Unit, Four Pipe	COST EACH		
		MAT.	INST.	TOTAL
1010	Fan coil A/C system, cabinet mounted, controls, 4 pipe, 1/2 Ton	2,200	2,675	4,875
1020	1 Ton	2,525	2,850	5,375
1030	1-1/2 Ton	2,675	2,875	5,550
1040	2 Ton	3,350	2,975	6,325
1050	3 Ton	4,800	3,200	8,000

D3040 124	Fan Coil A/C, Horizontal, Duct Mount, 2 Pipe	COST EACH		
		MAT.	INST.	TOTAL
1010	Fan coil A/C system, horizontal w/housing, controls, 2 pipe, 1/2 Ton	2,550	2,600	5,150
1020	1 Ton	3,425	3,775	7,200
1030	1-1/2 Ton	4,275	4,900	9,175
1040	2 Ton	5,725	5,900	11,625
1050	3 Ton	7,275	7,850	15,125
1060	3-1/2 Ton	7,550	8,125	15,675
1070	4 Ton	7,975	9,225	17,200
1080	5 Ton	9,650	11,700	21,350
1090	6 Ton	9,875	12,600	22,475
1100	7 Ton	9,850	13,000	22,850
1110	8 Ton	9,850	13,300	23,150
1120	10 Ton	10,400	13,800	24,200

D3040 126	Fan Coil A/C, Horiz., Duct Mount, 2 Pipe, Elec. Ht.	COST EACH		
		MAT.	INST.	TOTAL
1010	Fan coil A/C system, horiz. hsng, elect. ht, ctrls, 2 pipe, 1/2 Ton	2,775	2,600	5,375
1020	1 Ton	3,775	3,775	7,550
1030	1-1/2 Ton	4,600	4,900	9,500
1040	2 Ton	6,250	5,900	12,150
1050	3 Ton	9,800	8,100	17,900
1060	3-1/2 Ton	10,800	8,125	18,925
1070	4 Ton	12,200	9,250	21,450
1080	5 Ton	15,000	11,700	26,700
1090	6 Ton	15,200	12,600	27,800
1100	7 Ton	15,600	13,300	28,900
1110	8 Ton	15,800	13,900	29,700
1120	10 Ton	16,900	13,000	29,900
2700	Cut & patch to match existing construction, add, minimum	2%	3%	
2800	Maximum	5%	9%	
2900	Dust protection, add, minimum	1%	2%	
3000	Maximum	4%	11%	
3100	Equipment usage curtailment, add, minimum	1%	1%	
3200	Maximum	3%	10%	
3300	Material handling & storage limitation, add, minimum	1%	1%	
3400	Maximum	6%	7%	
3500	Protection of existing work, add, minimum	2%	2%	
3600	Maximum	5%	7%	
3700	Shift work requirements, add, minimum		5%	
3800	Maximum		30%	
3900	Temporary shoring and bracing, add, minimum	2%	5%	
4000	Maximum	5%	12%	

D30 HVAC

D3040 Distribution Systems

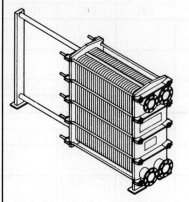

Plate Heat Exchanger

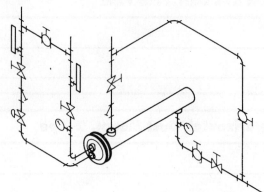

Shell and Tube Heat Exchanger

This page illustrates and describes plate and shell and tube type heat exchangers including pressure gauge and exchanger control system. Lines within Systems Components give the material and installation price on a cost each basis for the components. Prices for heat exchanger systems are on Line Items D3040 610 1010 thru D3040 620 1040. Material quantities and labor costs have been adjusted for the system listed.

Factors To adjust for job conditions other then normal working situations use Lines D3040 620 2700 thru 4000.

System Components	QUANTITY	UNIT	COST EACH MAT.	COST EACH INST.	COST EACH TOTAL
Shell and tube type heat exchanger with related valves and piping.					
Heat exchanger, 4 pass, 3/4" O.D. copper tubes, by steam at 10 psi, 40 GPM	1.000	Ea.	5,525	335	5,860
Pipe, black steel, Sch 40, threaded, W/couplings & hangers, 10' OC, 2" dia.	40.000	L.F.	499.95	1,035.25	1,535.20
Elbow, 90°, straight, 2" dia.	16.000	Ea.	456	1,168	1,624
Reducer, black steel, concentric, 2" dia.	2.000	Ea.	64	126	190
Valves, bronze, gate, rising stem, threaded, class 150, 2" size	6.000	Ea.	1,500	399	1,899
Valves, bronze, globe, class 150, rising stem, threaded, 2" size	2.000	Ea.	1,030	133	1,163
Strainers, Y type, bronze body, screwed, 125 lb, 2" pipe size	2.000	Ea.	316	115	431
Valve, electric motor actuated, brass, 2 way, screwed, 1-1/2" pipe size	1.000	Ea.	730	56.50	786.50
Union, black with brass seat, 2" dia.	8.000	Ea.	400	620	1,020
Tee, black, straight, 2" dia.	9.000	Ea.	324	1,080	1,404
Tee, black, reducing run and outlet, 2" dia.	4.000	Ea.	264	480	744
Pipe nipple, black, 2" dia.	21.000	Ea.	4.95	10.25	15.20
Thermometers, stem type, 9" case, 8" stem, 3/4" NPT	2.000	Ea.	478	53	531
Gauges, pressure or vacuum, 3-1/2" diameter dial	2.000	Ea.	22.80	47	69.80
Insulation, fiberglass pipe covering, 1" wall, 2" IPS	40.000	L.F.	53.60	242	295.60
Heat exchanger control system	1.000	Ea.	2,950	2,350	5,300
Coil balancing	1.000	Ea.		138	138
TOTAL			14,618.30	8,388	23,006.30

D3040 610	Heat Exchanger, Plate Type	COST EACH MAT.	COST EACH INST.	COST EACH TOTAL
1000	For alternate heat exchangers:			
1010	Plate heat exchanger, 400 GPM	57,000	16,900	73,900
1020	800 GPM	91,000	21,500	112,500
1030	1200 GPM	137,500	29,200	166,700
1040	1800 GPM	186,000	36,300	222,300

D3040 620	Heat Exchanger, Shell & Tube	COST EACH MAT.	COST EACH INST.	COST EACH TOTAL
1010	Shell & tube heat exchanger, 40 GPM	14,600	8,400	23,000
1020	96 GPM	26,800	14,900	41,700
1030	240 GPM	48,800	20,200	69,000
1040	600 GPM	94,500	28,500	123,000
2700	Cut & patch to match existing construction, add, minimum	2%	3%	
2800	Maximum	5%	9%	

D30 HVAC

D3040 Distribution Systems

D3040 620	Heat Exchanger, Shell & Tube	COST EACH		
		MAT.	INST.	TOTAL
2900	Dust protection, add, minimum	1%	2%	
3000	Maximum	4%	11%	
3100	Equipment usage curtailment, add, minimum	1%	1%	
3200	Maximum	3%	10%	
3300	Material handling & storage limitation, add, minimum	1%	1%	
3400	Maximum	6%	7%	
3500	Protection of existing work, add, minimum	2%	2%	
3600	Maximum	5%	7%	
3700	Shift work requirements, add, minimum		5%	
3800	Maximum		30%	
3900	Temporary shoring and bracing, add, minimum	2%	5%	
4000	Maximum	5%	12%	

D30 HVAC

D3050 Terminal & Package Units

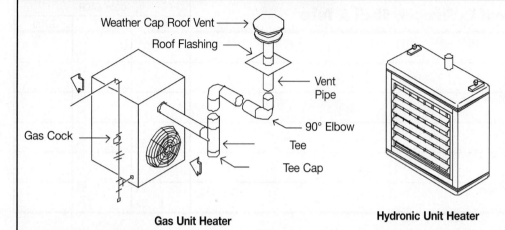

Gas Unit Heater | Hydronic Unit Heater

This page illustrates and describes unit heaters including piping and vents. Lines within Systems Components give the material and installation price on a cost each basis for this system. Prices for unit heater systems are on Line items D3050 120 1010 thru D3050 140 1040. Material quantities and labor costs have been adjusted for the system listed.

Factors: To adjust for job conditions other than normal working situations use Lines D3050 140 2700 thru 4000.

System Components	QUANTITY	UNIT	COST EACH		
			MAT.	INST.	TOTAL
Unit heaters complete with piping, thermostats and chimney as required					
Space heater, propeller fan, 20 MBH output	1.000	Ea.	1,250	158	1,408
Pipe, black steel, Sch 40, threaded, W/coupling & hangers, 10' OC, 3/4" dia	20.000	L.F.	90.30	252	342.30
Elbow, 90°, black steel, straight, 1/2" dia.	3.000	Ea.	10.89	147	157.89
Elbow, 90°, black steel, straight, 3/4" dia.	3.000	Ea.	13.17	157.50	170.67
Tee, black steel, reducing, 3/4" dia.	1.000	Ea.	12	81.50	93.50
Union, black with brass seat, 3/4" dia.	1.000	Ea.	18.50	56.50	75
Pipe nipples, black, 1/2" dia	2.000	Ea.	7.32	23.30	30.62
Pipe nipples, black, 3/4" dia	2.000	Ea.	4.30	12	16.30
Pipe cap, black, 3/4" dia.	1.000	Ea.	4.98	23	27.98
Gas cock, brass, 3/4" size	1.000	Ea.	17.05	33.50	50.55
Thermostat, 1 set back, electric, timed	1.000	Ea.	41	93	134
Wiring, thermostat hook-up, 25' of #18-3	1.000	Ea.	9.70	28	37.70
Vent chimney, prefab metal, U.L. listed, gas, double wall, galv. st, 4" dia	12.000	L.F.	95.40	224.40	319.80
Vent chimney, gas, double wall, galv. steel, elbow 90°, 4" dia.	2.000	Ea.	59	75	134
Vent chimney, gas, double wall, galv. steel, Tee, 4" dia.	1.000	Ea.	42.50	49	91.50
Vent chimney, gas, double wall, galv. steel, T cap, 4" dia.	1.000	Ea.	2.79	30.50	33.29
Vent chimney, gas, double wall, galv. steel, roof flashing, 4" dia.	1.000	Ea.	10.90	37.50	48.40
Vent chimney, gas, double wall, galv. steel, top, 4" dia.	1.000	Ea.	15.10	29	44.10
TOTAL			1,704.90	1,510.70	3,215.60

D3050 120	Unit Heaters, Gas	COST EACH		
		MAT.	INST.	TOTAL
1000	For alternate unit heaters:			
1010	Space heater, suspended, gas fired, propeller fan, 20 MBH	1,700	1,500	3,200
1020	60 MBH	2,375	1,575	3,950
1030	100 MBH	2,800	1,650	4,450
1040	160 MBH	3,300	1,825	5,125
1050	200 MBH	4,025	2,025	6,050
1060	280 MBH	4,700	2,175	6,875
1070	320 MBH	6,325	2,450	8,775

D3050 130	Unit Heaters, Hydronic	COST EACH		
		MAT.	INST.	TOTAL
1010	Space heater, suspended, horiz. mount, HW, prop. fan, 20 MBH	2,125	1,625	3,750
1020	60 MBH	2,875	1,825	4,700
1030	100 MBH	3,450	2,000	5,450
1040	150 MBH	3,900	2,200	6,100
1050	200 MBH	4,425	2,475	6,900
1060	300 MBH	5,225	2,900	8,125

D30 HVAC

D3050 Terminal & Package Units

D3050 140	Cabinet Unit Heaters, Hydronic		COST EACH		
			MAT.	INST.	TOTAL
1010	Unit heater, cabinet type, horizontal blower, hot water, 20 MBH		2,500	1,575	4,075
1020	60 MBH		3,625	1,750	5,375
1030	100 MBH		4,275	1,925	6,200
1040	120 MBH		4,375	2,000	6,375
2700	Cut & patch to match existing construction, add, minimum		2%	3%	
2800	Maximum		5%	9%	
2900	Dust protection, add, minimum		1%	2%	
3000	Maximum		4%	11%	
3100	Equipment usage curtailment, add, minimum		1%	1%	
3200	Maximum		3%	10%	
3300	Material handling & storage limitation, add, minimum		1%	1%	
3400	Maximum		6%	7%	
3500	Protection of existing work, add, minimum		2%	2%	
3600	Maximum		5%	7%	
3700	Shift work requirements, add, minimum			5%	
3800	Maximum			30%	
3900	Temporary shoring and bracing, add, minimum		2%	5%	
4000	Maximum		5%	12%	

D30 HVAC

D3050 Terminal & Package Units

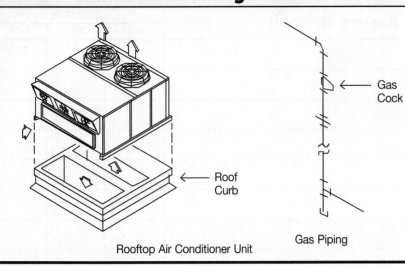

Rooftop Air Conditioner Unit

This page illustrates and describes packaged air conditioners including rooftop DX unit, pipes and fittings. Lines within Systems Components give the material and installation price on a cost each basis for the components. Prices for A/C systems are on Line Items D3050 175 1010 thru D3050 180 1060. Material quantities and labor costs have been adjusted for the system listed.

Factors: To adjust for job conditions other than normal working situations use Lines D3050 180 2700 thru 4000.

System Components	QUANTITY	UNIT	COST EACH		
			MAT.	INST.	TOTAL
Packaged rooftop DX unit, gas, with related piping and valves.					
Roof top A/C, curb, economizer, sgl zone, elec cool, gas ht, 5 ton, 112 MBH	1.000	Ea.	5,625	2,400	8,025
Pipe, black steel, Sch 40, threaded, W/cplgs & hangers, 10' OC, 1" dia	20.000	L.F.	116.55	289.80	406.35
Elbow, 90°, black, straight, 3/4" dia.	3.000	Ea.	13.17	157.50	170.67
Elbow, 90°, black, straight, 1" dia.	3.000	Ea.	22.80	169.50	192.30
Tee, black, reducing, 1" dia.	1.000	Ea.	19.20	91.50	110.70
Union, black with brass seat, 1" dia.	1.000	Ea.	24	61	85
Pipe nipple, black, 3/4" dia	2.000	Ea.	8.60	24	32.60
Pipe nipple, black, 1" dia	2.000	Ea.	5.55	13.80	19.35
Cap, black, 1" dia.	1.000	Ea.	6.05	24.50	30.55
Gas cock, brass, 1" size	1.000	Ea.	34	38.50	72.50
Control system, pneumatic, Rooftop A/C unit	1.000	Ea.	3,050	2,625	5,675
Rooftop unit heat/cool balancing	1.000	Ea.		495	495
TOTAL			8,924.92	6,390.10	15,315.02

D3050 175	Rooftop Air Conditioner, Const. Volume	COST EACH		
		MAT.	INST.	TOTAL
1000	For alternate A/C systems:			
1010	A/C, Rooftop, DX cool, gas heat, curb, economizer, fltrs, 5 Ton	8,925	6,400	15,325
1020	7-1/2 Ton	10,800	6,650	17,450
1030	12-1/2 Ton	16,500	7,300	23,800
1040	18 Ton	21,200	8,000	29,200
1050	25 Ton	33,200	9,100	42,300
1060	40 Ton	47,700	12,200	59,900

D3050 180	Rooftop Air Conditioner, Variable Air Volume	COST EACH		
		MAT.	INST.	TOTAL
1010	A/C, Rooftop, DX cool, gas heat, curb, ecmizr, fltrs, VAV, 12-1/2 Ton	19,400	7,800	27,200
1020	18 Ton	24,700	8,600	33,300
1030	25 Ton	37,400	9,875	47,275
1040	40 Ton	55,500	13,400	68,900
1050	60 Ton	79,500	18,300	97,800
1060	80 Ton	115,500	22,900	138,400
2700	Cut & patch to match existing construction, add, minimum		2%	3%
2800	Maximum		5%	9%
2900	Dust protection, add, minimum		1%	2%
3000	Maximum		4%	11%

D30 HVAC

D3050 Terminal & Package Units

D3050 180	Rooftop Air Conditioner, Variable Air Volume	COST EACH		
		MAT.	INST.	TOTAL
3100	Equipment usage curtailment, add, minimum	1%	1%	
3200	Maximum	3%	10%	
3300	Material handling & storage limitation, add, minimum	1%	1%	
3400	Maximum	6%	7%	
3500	Protection of existing work, add, minimum	2%	2%	
3600	Maximum	5%	7%	
3700	Shift work requirements, add, minimum		5%	
3800	Maximum		30%	
3900	Temporary shoring and bracing, add, minimum	2%	5%	
4000	Maximum	5%	12%	

D30 HVAC

D3050 Terminal & Package Units

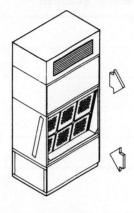
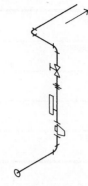

Condenser Supply Water Piping — Condenser Return Water Piping

Self-Contained Air Conditioner Unit, Water Cooled

This page illustrates and describes packaged electrical air conditioning units including multizone control and related piping. Lines within Systems Components give the material and installation price on a cost each basis for the components. Prices for packaged electrical A/C unit systems are on Line Items D3050 210 1010 thru D3050 225 1060. Material quantities and labor costs have been adjusted for the system listed.

Factors: To adjust for job conditions other than normal working situations use Lines D3050 225 2700 thru 4000.

System Components	QUANTITY	UNIT	COST EACH MAT.	COST EACH INST.	COST EACH TOTAL
Packaged water cooled electric air conditioning unit with related piping and valves.					
Self-contained, water cooled, elect. heat, not inc tower, 5 ton, const vol	1.000	Ea.	9,175	1,750	10,925
Pipe, black steel, Sch 40, threaded, W/cplg & hangers, 10' OC, 1-1/4" dia.	20.000	L.F.	146.20	318.20	464.40
Elbow, 90°, black, straight, 1-1/4" dia.	6.000	Ea.	75.30	360	435.30
Tee, black, straight, 1-1/4" dia.	2.000	Ea.	38.70	188	226.70
Tee, black, reducing, 1-1/4" dia.	2.000	Ea.	68	188	256
Thermometers, stem type, 9" case, 8" stem, 3/4" NPT	2.000	Ea.	478	53	531
Union, black with brass seat, 1-1/4" dia.	2.000	Ea.	69	126	195
Pipe nipple, black, 1-1/4" dia	3.000	Ea.	10.20	22.20	32.40
Valves, bronze, gate, N.R.S., threaded, class 150, 1-1/4" size	2.000	Ea.	304	98	402
Circuit setter, bal valve, bronze body, threaded, 1-1/4" pipe size	1.000	Ea.	165	49.50	214.50
Insulation, fiberglass pipe covering, 1" wall, 1-1/4" IPS	20.000	L.F.	23.20	115	138.20
Control system, pneumatic, A/C Unit with heat	1.000	Ea.	3,050	2,625	5,675
Re-heat coil balancing	1.000	Ea.		138	138
Rooftop unit heat/cool balancing	1.000	Ea.		495	495
TOTAL			13,602.60	6,525.90	20,128.50

D3050 210	AC Unit, Package, Elec. Ht., Water Cooled	MAT.	INST.	TOTAL
1000	For alternate A/C systems:			
1010	A/C, Self contained, single pkg., water cooled, elect. heat, 5 Ton	13,600	6,525	20,125
1020	10 Ton	21,300	8,025	29,325
1030	20 Ton	28,900	10,100	39,000
1040	30 Ton	47,500	11,100	58,600
1050	40 Ton	60,500	12,300	72,800
1060	50 Ton	79,000	14,700	93,700

D3050 215	AC Unit, Package, Elec. Ht., Water Cooled, VAV	MAT.	INST.	TOTAL
1010	A/C, Self contained, single pkg., water cooled, elect. ht, VAV, 10 Ton	23,800	8,025	31,825
1020	20 Ton	33,100	10,100	43,200
1030	30 Ton	54,000	11,100	65,100
1040	40 Ton	68,000	12,300	80,300
1050	50 Ton	88,500	14,700	103,200
1060	60 Ton	101,500	17,800	119,300

D30 HVAC

D3050 Terminal & Package Units

D3050 220	AC Unit, Package, Hot Water Coil, Water Cooled	COST EACH		
		MAT.	INST.	TOTAL
1010	A/C, Self contn'd, single pkg., water cool, H/W ht, const. vol, 5 Ton	12,200	9,050	21,250
1020	10 Ton	18,100	10,800	28,900
1030	20 Ton	43,300	13,400	56,700
1040	30 Ton	56,000	17,000	73,000
1050	40 Ton	69,000	20,600	89,600
1060	50 Ton	84,500	25,400	109,900

D3050 225	AC Unit, Package, HW Coil, Water Cooled, VAV	COST EACH		
		MAT.	INST.	TOTAL
1010	A/C, Self contn'd, single pkg., water cool, H/W ht, VAV, 10 Ton	20,600	10,100	30,700
1020	20 Ton	29,400	13,100	42,500
1030	30 Ton	47,300	16,200	63,500
1040	40 Ton	62,500	17,700	80,200
1050	50 Ton	79,000	20,900	99,900
1060	60 Ton	93,500	22,700	116,200
2700	Cut & patch to match existing construction, add, minimum	2%	3%	
2800	Maximum	5%	9%	
2900	Dust protection, add, minimum	1%	2%	
3000	Maximum	4%	11%	
3100	Equipment usage curtailment, add, minimum	1%	1%	
3200	Maximum	3%	10%	
3300	Material handling & storage limitation, add, minimum	1%	1%	
3400	Maximum	6%	7%	
3500	Protection of existing work, add, minimum	2%	2%	
3600	Maximum	5%	7%	
3700	Shift work requirements, add, minimum		5%	
3800	Maximum		30%	
3900	Temporary shoring and bracing, add, minimum	2%	5%	
4000	Maximum	5%	12%	

D30 HVAC

D3050 Terminal & Package Units

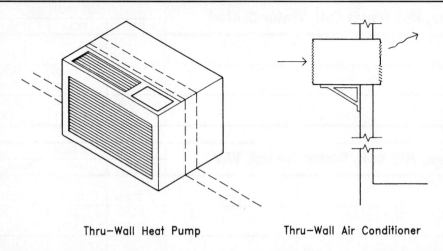

Thru-Wall Heat Pump   Thru-Wall Air Conditioner

This page illustrates and describes thru-wall air conditioning and heat pump units. Lines within Systems Components give the material and installation price on a cost each basis for the components. Prices for thru-wall unit systems are on Line Items D3050 255 1010 thru D3050 260 1050. Material quantities and labor costs have been adjusted for the system listed.

Factors: To adjust for job conditions other than normal working conditions use Lines D3050 260 2700 thru 4000.

System Components			COST EACH		
	QUANTITY	UNIT	MAT.	INST.	TOTAL
Electric thru-wall air conditioning unit.					
A/C unit, thru-wall, electric heat., cabinet, louver, 1/2 ton	1.000	Ea.	850	224	1,074
TOTAL			850	224	1,074

D3050 255	Thru-Wall A/C Unit	COST EACH		
		MAT.	INST.	TOTAL
1000	For alternate A/C units:			
1010	A/C Unit, thru-the-wall, supplemental elect. heat, cabinet, louver, 1/2 Ton	850	224	1,074
1020	3/4 Ton	1,325	268	1,593
1030	1 Ton	1,500	335	1,835
1040	1-1/2 Ton	2,100	560	2,660
1050	2 Ton	2,125	705	2,830

D3050 260	Thru-Wall Heat Pump	COST EACH		
		MAT.	INST.	TOTAL
1010	Heat pump, thru-the-wall, cabinet, louver, 1/2 Ton	2,700	168	2,868
1020	3/4 Ton	2,825	224	3,049
1030	1 Ton	3,275	335	3,610
1040	Supplemental. elect. heat, 1-1/2 Ton	3,350	865	4,215
1050	2 Ton	3,425	895	4,320
2700	Cut & patch to match existing construction, add, minimum	2%	3%	
2800	Maximum	5%	9%	
2900	Dust protection, add, minimum	1%	2%	
3000	Maximum	4%	11%	
3100	Equipment usage curtailment, add, minimum	1%	1%	
3200	Maximum	3%	10%	
3300	Material handling & storage limitation, add, minimum	1%	1%	
3400	Maximum	6%	7%	
3500	Protection of existing work, add, minimum	2%	2%	
3600	Maximum	5%	7%	
3700	Shift work requirements, add, minimum		5%	
3800	Maximum		30%	
3900	Temporary shoring and bracing, add, minimum	2%	5%	
4000	Maximum	5%	12%	

D40 Fire Protection

D4010 Sprinklers

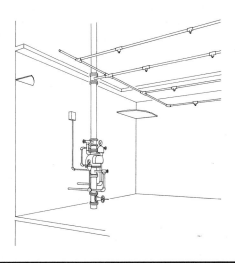

This page illustrates and describes a dry type fire sprinkler system. Lines within System Components give the unit cost of a system for a 2,000 square foot building. Lines D4010 320 1700 thru 2400 give the square foot costs for alternate systems. Both material quantities and labor costs have been adjusted for the system listed.

Factors: To adjust for conditions other than normal working conditions, use Lines D4010 320 2900 thru 4000.

System Components	QUANTITY	UNIT	COST EACH		
			MAT.	INST.	TOTAL
Pre-action fire sprinkler system, ordinary hazard, open area to 2000 S.F.					
On one floor.					
Air compressor	1.000	Ea.	840	540	1,380
4" OS & Y valve	1.000	Ea.	850	440	1,290
Pipe riser 4" diameter	10.000	L.F.	240	372.20	612.20
Sprinkler head supply piping, 1"	163.000	L.F.	416.25	1,035	1,451.25
Sprinkler head supply piping, 1-1/4"	163.000	L.F.	374	814	1,188
Sprinkler head supply piping, 2-1/2"	163.000	L.F.	585.20	1,007	1,592.20
Pipe fittings, 1"	8.000	Ea.	95.60	732	827.60
Pipe fittings, 1-1/4"	8.000	Ea.	154.80	752	906.80
Pipe fittings, 2-1/2"	4.000	Ea.	354	584	938
Pipe fittings, 4"	2.000	Ea.	630	660	1,290
Detectors	2.000	Ea.	1,570	88	1,658
Sprinkler heads	16.000	Ea.	174.40	704	878.40
Fire department connection	1.000	Ea.	570	253	823
TOTAL			6,854.25	7,981.20	14,835.45

D4010 320	Fire Sprinkler Systems, Dry	COST PER S.F.		
		MAT.	INST.	TOTAL
1700	Ordinary hazard, one floor, area to 2000 S.F./floor	3.46	4	7.46
1800	For each additional floor, add per floor	1.53	3.34	4.87
1900	Area to 3200 S.F./ floor	2.75	3.97	6.72
2000	For each additional floor, add per floor	1.55	3.55	5.10
2100	Area to 5000 S.F./ floor	3.09	4.05	7.14
2200	For each additional floor, add per floor	2.01	3.76	5.77
2300	Area to 8000 S.F./ floor	2.38	3.40	5.78
2400	For each additional floor, add per floor	1.60	3.21	4.81
2500				
2600				
2700				
2800				
2900	Cut & patch to match existing construction, add, minimum	2%	3%	
3000	Maximum	5%	9%	
3100	Dust protection, add, minimum	1%	2%	
3200	Maximum	4%	11%	
3300	Equipment usage curtailment, add, minimum	1%	1%	
3400	Maximum	3%	10%	

D40 Fire Protection

D4010 Sprinklers

D4010 320	Fire Sprinkler Systems, Dry	COST PER S.F.		
		MAT.	INST.	TOTAL
3500	Material handling & storage limitation, add, minimum	1%	1%	
3600	Maximum	6%	7%	
3700	Protection of existing work, add, minimum	2%	2%	
3800	Maximum	5%	7%	
3900	Shift work requirements, add, minimum		5%	
4000	Maximum		30%	

D40 Fire Protection

D4010 Sprinklers

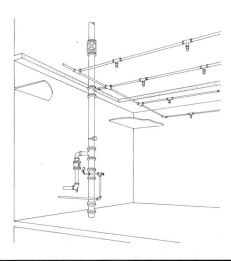

This page illustrates and describes a wet type fire sprinkler system. Lines within System Components give the unit cost of a system for a 2,000 square foot building. Lines D4010 420 1900 thru 2600 give the square foot costs for alternate systems. Both material quantities and labor costs have been adjusted for the system listed.

Factors: To adjust for conditions other than normal working conditions, use Lines D4010 420 2900 thru 3600.

System Components	QUANTITY	UNIT	COST EACH MAT.	COST EACH INST.	COST EACH TOTAL
Wet pipe fire sprinkler system, ordinary hazard, open area to 2000 S.F.					
On one floor.					
4" OS & Y valve	1.000	Ea.	850	440	1,290
Wet pipe alarm valve	1.000	Ea.	1,875	630	2,505
Water motor alarm	1.000	Ea.	455	176	631
3" check valve	1.000	Ea.	385	420	805
Pipe riser, 4" diameter	10.000	L.F.	240	372.20	612.20
Water gauges and trim	1.000	Set	3,150	1,275	4,425
Electric fire horn	1.000	Ea.	67	101	168
Sprinkler head supply piping, 1" diameter	168.000	L.F.	416.25	1,035	1,451.25
Sprinkler head supply piping, 1-1/2" diameter	168.000	L.F.	429	907.50	1,336.50
Sprinkler head supply piping, 2-1/2" diameter	168.000	L.F.	585.20	1,007	1,592.20
Pipe fittings, 1"	8.000	Ea.	95.60	732	827.60
Pipe fittings, 1-1/2"	8.000	Ea.	192	808	1,000
Pipe fittings, 2-1/2"	4.000	Ea.	354	584	938
Pipe fittings, 4"	2.000	Ea.	630	660	1,290
Sprinkler heads	16.000	Ea.	174.40	704	878.40
Fire department connection	1.000	Ea.	570	253	823
TOTAL			10,468.45	10,104.70	20,573.15

D4010 420	Fire Sprinkler Systems, Wet	COST PER S.F. MAT.	COST PER S.F. INST.	COST PER S.F. TOTAL
1900	Ordinary hazard, one floor, area to 2000 S.F./floor	5.25	5.05	10.30
2000	For each additional floor, add per floor	1.57	3.40	4.97
2100	Area to 3200 S.F./floor	3.85	4.59	8.44
2200	For each additional floor, add per floor	1.55	3.55	5.10
2300	Area to 5000 S.F./ floor	3.48	4.43	7.91
2400	For each additional floor, add per floor	2.01	3.76	5.77
2500	Area to 8000 S.F./ floor	2.52	3.62	6.14
2600	For each additional floor, add per floor	1.71	3.41	5.12
2700				
2900	Cut & patch to match existing construction, add minimum	2%	3%	
3000	Maximum	5%	9%	
3100	Dust protection, add, minimum	1%	2%	
3200	Maximum	4%	11%	
3300	Equipment usage curtailment, add, minimum	1%	1%	
3400	Maximum	3%	10%	
3500	Material handling & storage limitation, add, minimum	1%	1%	
3600	Maximum	6%	7%	

D50 Electrical

D5010 Electrical Service/Distribution

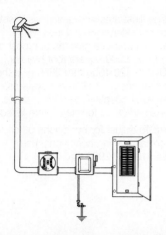

This page illustrates and describes commercial service systems including a meter socket, service head and cable, entrance switch, steel conduit, copper wire, panel board, ground rod, wire, and conduit. Lines within System Components give the unit price and total price on a cost each basis for this system. Prices for commercial service systems are on Line Items D5010 210 1500 thru 1700. Both material quantities and labor costs have been adjusted for the system listed.

Factors: To adjust for job conditions other than normal working situations use Lines D5010 210 2900 thru 4000.

Example: You are to install the system with maximum equipment usage curtailment. Go to Line D5010 210 3400 and apply these percentages to the appropriate MAT. and INST. costs.

System Components	QUANTITY	UNIT	COST EACH		
			MAT.	INST.	TOTAL
Commercial electric service including service breakers, metering 120/208 Volt, 3 phase, 4 wire, feeder, and panel board.					
100 Amp Service					
Meter socket	1.000	Ea.	47.50	212	259.50
Service head	.200	C.L.F.	65	136	201
Service entrance switch	1.000	Ea.	234	355	589
Rigid steel conduit	20.000	L.F.	88.20	208	296.20
600 volt copper wire #3	1.000	C.L.F.	138	136	274
Panel board, 24 circuits, 20 Amp breakers	1.000	Ea.	965	1,125	2,090
Ground rod	1.000	Ea.	22	123	145
TOTAL			1,559.70	2,295	3,854.70

D5010 210	Commercial Service - 3 Phase	COST EACH		
		MAT.	INST.	TOTAL
1500	120/208 Volt, 3 phase, 4 wire service, 60 Amp	1,075	1,775	2,850
1550	100 Amp	1,550	2,300	3,850
1600	200 Amp	3,200	4,050	7,250
1700	400 Amp	6,725	5,075	11,800
1800				
1900				
2000				
2100				
2200				
2300				
2400				
2500				
2900	Cut & patch to match existing construction, add, minimum	2%	3%	
3000	Maximum	5%	9%	
3100	Dust protection, add, minimum	1%	2%	
3200	Maximum	4%	11%	
3300	Equipment usage curtailment, add, minimum	1%	1%	
3400	Maximum	3%	10%	
3500	Material handling & storage limitation, add, minimum	1%	1%	
3600	Maximum	6%	7%	
3700	Protection of existing work, add, minimum	2%	2%	
3800	Maximum	5%	7%	
3900	Shift work requirements, add, minimum		5%	
4000	Maximum		30%	

D50 Electrical

D5010 Electrical Service/Distribution

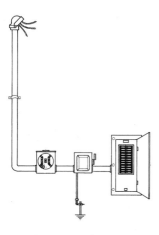

This page illustrates and describes a residential, single phase system including a weather cap, service entrance cable, meter socket, entrance switch, ground rod, ground cable, EMT, and panelboard. Lines within System Components give the unit price and total price on a cost each basis for this system. Prices for a residential, single phase system are also given. Both material quantities and labor costs have been adjusted for the system listed.

Factors: To adjust for job conditions other than normal working situations use Lines D5010 220 2900 thru 4000.

Example: You are to install the system with a minimum equipment usage curtailment. Go to Line D5010 220 3300 and apply these percentages to the appropriate MAT. and INST. costs.

System Components	QUANTITY	UNIT	COST EACH		
			MAT.	INST.	TOTAL
100 Amp Service, single phase					
Weathercap	1.000	Ea.	9.65	56.50	66.15
Service entrance cable	.200	C.L.F.	65	136	201
Meter socket	1.000	Ea.	47.50	212	259.50
Entrance disconnect switch	1.000	Ea.	234	355	589
Ground rod, with clamp	1.000	Ea.	24.50	141	165.50
Ground cable	.100	C.L.F.	14.10	42.50	56.60
Panelboard, 12 circuit	1.000	Ea.	273	565	838
TOTAL			667.75	1,508	2,175.75
200 Amp Service, single phase					
Weathercap	1.000	Ea.	27	84.50	111.50
Service entrance cable	.200	C.L.F.	166	194	360
Meter socket	1.000	Ea.	101	355	456
Entrance disconnect switch	1.000	Ea.	500	520	1,020
Ground rod, with clamp	1.000	Ea.	42.50	154	196.50
Ground cable	.100	C.L.F.	34.50	23.40	57.90
3/4" EMT	10.000	L.F.	10.40	52	62.40
Panelboard, 24 circuit	1.000	Ea.	600	880	1,480
TOTAL			1,481.40	2,262.90	3,744.30

D5010 220	Residential Service - Single Phase	COST EACH		
		MAT.	INST.	TOTAL
2800				
2810	100 Amp Service, single phase	670	1,500	2,170
2820	200 Amp Service, single phase	1,475	2,275	3,750
2900	Cut & patch to match existing construction, add, minimum	2%	3%	
3000	Maximum	5%	9%	
3100	Dust protection, add, minimum	1%	2%	
3200	Maximum	4%	11%	
3300	Equipment usage curtailment, add, minimum	1%	1%	
3400	Maximum	3%	10%	
3500	Material handling & storage limitation, add, minimum	1%	1%	
3600	Maximum	6%	7%	
3700	Protection of existing work, add, minimum	2%	2%	
3800	Maximum	5%	7%	
3900	Shift work requirements, add, minimum		5%	
4000	Maximum		30%	

D50 Electrical

D5020 Lighting and Branch Wiring

D5020 180	Selective Price Sheet	COST EACH		
		MAT.	INST.	TOTAL
0100	Using non-metallic sheathed, cable, air conditioning receptacle	22	68	90
0200	Disposal wiring	19.70	75.50	95.20
0300	Dryer circuit	39.50	123	162.50
0400	Duplex receptacle	22	52	74
0500	Fire alarm or smoke detector	100	68	168
0600	Furnace circuit & switch	30	113	143
0700	Ground fault receptacle	61	84.50	145.50
0800	Heater circuit	24	84.50	108.50
0900	Lighting wiring	24.50	42.50	67
1000	Range circuit	85	169	254
1100	Switches single pole	24	42.50	66.50
1200	3-way	27	56.50	83.50
1300	Water heater circuit	32	136	168
1400	Weatherproof receptacle	154	113	267
1500	Using BX cable, air conditioning receptacle	29	81.50	110.50
1600	Disposal wiring	26	90.50	116.50
1700	Dryer circuit	42.50	147	189.50
1800	Duplex receptacle	29	62.50	91.50
1900	Fire alarm or smoke detector	100	68	168
2000	Furnace circuit & switch	40	136	176
2100	Ground fault receptacle	68	103	171
2200	Heater circuit	28.50	103	131.50
2300	Lighting wiring	29	51	80
2400	Range circuit	112	205	317
2500	Switches, single pole	31.50	51	82.50
2600	3-way	32	68	100
2700	Water heater circuit	43	161	204
2800	Weatherproof receptacle	159	136	295
2900	Using EMT conduit, air conditioning receptacle	43	101	144
3000	Disposal wiring	42	113	155
3100	Dryer circuit	55	183	238
3200	Duplex receptacle	43	78	121
3300	Fire alarm or smoke detector	118	101	219
3400	Furnace circuit & switch	50	169	219
3500	Ground fault receptacle	96	125	221
3600	Heater circuit	41.50	125	166.50
3700	Lighting wiring	40	63.50	103.50
3800	Range circuit	98.50	251	349.50
3900	Switches, single pole	45	63.50	108.50
4000	3-way	42	84.50	126.50
4100	Water heater circuit	48.50	199	247.50
4200	Weatherproof receptacle	173	169	342
4300	Using aluminum conduit, air conditioning receptacle	47	136	183
4400	Disposal wiring	46	151	197
4500	Dryer circuit	68	242	310
4600	Duplex receptacle	44.50	104	148.50
4700	Fire alarm or smoke detector	127	136	263
4800	Furnace circuit & switch	57	226	283
4900	Ground fault receptacle	94.50	169	263.50
5000	Heater circuit	46	169	215
5100	Lighting wiring	52.50	84.50	137
5200	Range circuit	117	340	457
5300	Switches, single pole	56	84.50	140.50
5400	3-way	53	113	166
5500	Water heater circuit	60	271	331
5600	Weatherproof receptacle	191	226	417
5700	Using galvanized steel conduit	51.50	144	195.50
5800	Disposal wiring	45.50	161	206.50

D50 Electrical

D5020 Lighting and Branch Wiring

D5020 180	Selective Price Sheet	COST EACH		
		MAT.	INST.	TOTAL
5900	Dryer circuit	77.50	261	338.50
6000	Duplex receptacle	51.50	111	162.50
6100	Fire alarm or smoke detector	126	144	270
6200	Furnace circuit & switch	56.50	242	298.50
6300	Ground fault receptacle	102	178	280
6400	Heater circuit	49.50	178	227.50
6500	Lighting wiring	54.50	90.50	145
6600	Range circuit	124	355	479
6700	Switches, single pole	63	90.50	153.50
6800	3-way	60	117	177
6900	Water heater circuit	64.50	282	346.50
7000	Weatherproof receptacle	191	242	433
7100				
7200				
7300				
7400				
7500				
7600				
7700				
7800				
7900				
8000				
8100				
8200				
8300				
8400				

D50 Electrical

D5020 Lighting and Branch Wiring

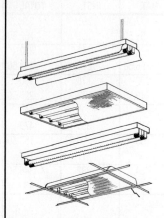

This page illustrates and describes fluorescent lighting systems including a fixture, lamp, outlet box and wiring. Lines within System Components give the unit price and total price on a cost each basis for this system. Prices for fluorescent lighting systems are on Line Items D5020 248 1200 thru 1500. Both material quantities and labor costs have been adjusted for the system listed.

Factors: To adjust for job conditions other than normal working situations use Lines D5020 248 2900 thru 4000.

Example: You are to install the system during evening hours. Go to Line D5020 248 3900 and apply this percentage to the appropriate INST. cost.

System Components	QUANTITY	UNIT	COST EACH		
			MAT.	INST.	TOTAL
Fluorescent lighting, including fixture, lamp, outlet box and wiring.					
Recessed lighting fixture, on suspended system	1.000	Ea.	63.50	144	207.50
Outlet box	1.000	Ea.	2.66	37.50	40.16
#12 wire	.660	C.L.F.	8.65	40.59	49.24
Conduit, EMT, 1/2" conduit	20.000	L.F.	14.80	79.80	94.60
TOTAL			89.61	301.89	391.50

D5020 248	Lighting, Fluorescent	COST EACH		
		MAT.	INST.	TOTAL
1200	Recessed lighting fixture, on suspended system	89.50	300	389.50
1300	Surface mounted, 2' x 4', acrylic prismatic diffuser	116	320	436
1400	Strip fixture, 8' long, two 8' lamps	90.50	299	389.50
1500	Pendant mounted, industrial, 8' long, with reflectors	135	345	480
1600				
1700				
1800				
1900				
2000				
2100				
2200				
2300				
2900	Cut & patch to match existing construction, add, minimum	2%	3%	
3000	Maximum	5%	9%	
3100	Dust protection, add, minimum	1%	2%	
3200	Maximum	4%	11%	
3300	Equipment usage curtailment, add, minimum	1%	1%	
3400	Maximum	3%	10%	
3500	Material handling & storage limitation, add, minimum	1%	1%	
3600	Maximum	6%	7%	
3700	Protection of existing work, add, minimum	2%	2%	
3800	Maximum	5%	7%	
3900	Shift work requirements, add, minimum		5%	
4000	Maximum		30%	

D50 Electrical

D5020 Lighting and Branch Wiring

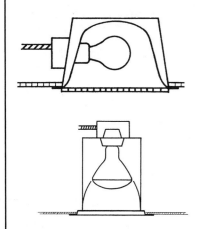

This page illustrates and describes incandescent lighting systems including a fixture, lamp, outlet box, conduit and wiring. Lines within System Components give the unit price and total price on a cost each basis for this system. Prices for an alternate incandescent lighting system are also given. Both material quantities and labor costs have been adjusted for the system listed.

Factors: To adjust for conditions other than normal working situations use Lines D5020 250 2900 thru 4000.

Example: You are to install the system and cut and match existing construction. Go to Line D5020 250 3000 and apply this percentage to the appropriate INST. costs.

System Components	QUANTITY	UNIT	COST EACH		
			MAT.	INST.	TOTAL
Incandescent light fixture, including lamp, outlet box, conduit and wiring.					
Recessed wide reflector with flat glass lens	1.000	Ea.	107	101	208
Outlet box	1.000	Ea.	2.66	37.50	40.16
Armored cable, 3 wire	.200	C.L.F.	15.60	62	77.60
TOTAL			125.26	200.50	325.76
Recessed flood light fixture, including lamp, outlet box, conduit & wire					
Recessed, R-40 flood lamp with refractor	1.000	Ea.	114	84.50	198.50
150 watt R-40 flood lamp	.010	Ea.	6.80	5.20	12
Outlet box	1.000	Ea.	2.66	37.50	40.16
Outlet box cover	1.000	Ea.	1.30	10.60	11.90
Romex, 12-2 with ground	.200	C.L.F.	8	62	70
Conduit, 1/2" EMT	20.000	L.F.	14.80	79.80	94.60
TOTAL			147.56	279.60	427.16

D5020 250	Lighting, Incandescent	COST EACH		
		MAT.	INST.	TOTAL
2810	Incandescent light fixture, including lamp, outlet box, conduit and wiring.	125	201	326
2820	Recessed flood light fixture, including lamp, outlet box, conduit & wire	148	280	428
2900	Cut & patch to match existing construction, add, minimum	2%	3%	
3000	Maximum	5%	9%	
3100	Dust protection, add, minimum	1%	2%	
3200	Maximum	4%	11%	
3300	Equipment usage curtailment, add, minimum	1%	1%	
3400	Maximum	3%	10%	
3500	Material handling & storage limitation, add, minimum	1%	1%	
3600	Maximum	6%	7%	
3700	Protection of existing work, add, minimum	2%	2%	
3800	Maximum	5%	7%	
3900	Shift work requirements, add, minimum		5%	
4000	Maximum		30%	

D50 Electrical

D5020 Lighting and Branch Wiring

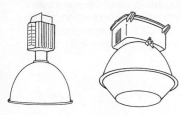

This page illustrates and describes high intensity lighting systems including a lamp, EMT conduit, EMT "T" fitting with cover, and wire. Lines within System Components give the unit price and total price on a cost each basis for this system. Prices for high intensity lighting systems are on Line Items D5020 252 1100 thru 1800. Both material quantities and labor costs have been adjusted for the system listed.

Factors: To adjust for job conditions other than normal working situations use Lines D5020 252 2900 thru 4000.

Example: You are to install the system and protect existing construction. Go to Line D5020 252 3700 and apply these percentages to the appropriate MAT. and INST. costs.

System Components			COST EACH		
	QUANTITY	UNIT	MAT.	INST.	TOTAL
High intensity lighting system consisting of 400 watt metal halide fixture And lamp with 1/2" EMT conduit and fittings using #12 wire, high bay.					
Single unit, 400 watt	1.000	Ea.	450	295	745
Electric metallic tubing (EMT), 1/2" diameter	30.000	Ea.	22.20	119.70	141.90
1/2" EMT "T" fitting with cover	1.000	Ea.	8.70	42.50	51.20
Wire, 600 volt, type THWN-THHN, copper, solid, #12	.600	Ea.	7.86	36.90	44.76
TOTAL			488.76	494.10	982.86

D5020 252	Lighting, High Intensity	COST EACH		
		MAT.	INST.	TOTAL
1100	High bay: 400 watt, metal halide fixture and lamp	490	495	985
1200	High pressure sodium and lamp	455	495	950
1400	1000 watt, metal halide	690	540	1,230
1500	High pressure sodium	640	540	1,180
1700	Low bay: 250 watt, metal halide	440	410	850
1800	150 watt, high pressure sodium	400	410	810
1900				
2000				
2100				
2200				
2300				
2400				
2500				
2600				
2700				
2900	Cut & patch to match existing construction, add, minimum	2%	3%	
3000	Maximum	5%	9%	
3100	Dust protection, add, minimum	1%	2%	
3200	Maximum	4%	11%	
3300	Equipment usage curtailment, add, minimum	1%	1%	
3400	Maximum	3%	10%	
3500	Material handling & storage limitation, add, minimum	1%	1%	
3600	Maximum	6%	7%	
3700	Protection of existing work, add, minimum	2%	2%	
3800	Maximum	5%	7%	
3900	Shift work requirements, add, minimum		5%	
4000	Maximum		30%	

D50 Electrical

D5090 Other Electrical Systems

This page illustrates and describes baseboard heat systems including a thermostat, outlet box, breaker, and feed. Lines within System Components give the unit price and total price on a cost each basis for this system. Prices for baseboard heat systems are on Line Items D5090 530 1500 thru 1800. Both material quantities and labor costs have been adjusted for the system listed.

Factors: To adjust for job conditions other than normal working situations use Lines D5090 530 2900 thru 4000.

Example: You are to install the system during evenings and weekends only. Go to Line D5090 530 4000 and apply this percentage to the appropriate INST. cost.

System Components	QUANTITY	UNIT	COST EACH		
			MAT.	INST.	TOTAL
Baseboard heat including thermostat, outlet box, breaker and feed.					
Electric baseboard heater, 4' long	1.000	Ea.	43	101	144
Thermostat, integral	1.000	Ea.	33	42.50	75.50
Romex, 12-3 with ground	.400	C.L.F.	23	136	159
Panel board breaker, 20 Amp	1.000	Ea.	39.50	84.50	124
Outlet box	1.000	Ea.	2.66	37.50	40.16
TOTAL			141.16	401.50	542.66

D5090 530	Heat, Baseboard	COST EACH		
		MAT.	INST.	TOTAL
1400	Heat, electric baseboard			
1500	2' long	130	385	515
1550	4' long	141	400	541
1600	6' long	158	435	593
1700	8' long	171	470	641
1800	10' long	305	505	810
1900				
2000				
2100				
2200				
2300				
2400				
2500				
2600				
2700				
2900	Cut & patch to match existing construction, add, minimum	2%	3%	
3000	Maximum	5%	9%	
3100	Dust protection, add, minimum	1%	2%	
3200	Maximum	4%	11%	
3300	Equipment usage curtailment, add, minimum	1%	1%	
3400	Maximum	3%	10%	
3500	Material handling & storage limitation, add, minimum	1%	1%	
3600	Maximum	6%	7%	
3700	Protection of existing work, add, minimum	2%	2%	
3800	Maximum	5%	7%	
3900	Shift work requirements, add, minimum		5%	
4000	Maximum		30%	

E Equipment & Furnishings

No part of this publication may be reproduced, stored in a retrieval system, or transmitted in any form or by any means without prior written permission of RSMeans.

E10 Equipment

E1090 Other Equipment

This page illustrates and describes kitchen systems including top and bottom cabinets, custom laminated plastic top, single bowl sink, and appliances. Lines within System Components give the unit price and total price on a cost each basis for this system. Prices for kitchen systems are on Line Items E1090 310 1400 thru 1600. Both material quantities and labor costs have been adjusted for the system listed.

Factors: To adjust for job conditions other than normal working situations use Lines E1090 310 2900 thru 4000.

Example: You are to install the system and protect the work area from dust. Go to Line E1090 310 3200 and apply these percentages to the appropriate MAT. and INST. costs.

System Components	QUANTITY	UNIT	COST EACH		
			MAT.	INST.	TOTAL
Kitchen cabinets including wall and base cabinets, custom laminated Plastic top, sink & appliances, no plumbing or electrical rough-in included.					
Prefinished wood cabinets, average quality, wall and base	20.000	L.F.	3,860	820	4,680
Custom laminated plastic counter top	20.000	L.F.	374	410	784
Stainless steel sink, 22" x 25"	1.000	Ea.	725	235	960
Faucet, top mount	1.000	Ea.	67.50	73	140.50
Dishwasher, built-in	1.000	Ea.	261	355	616
Compactor, built-in	1.000	Ea.	710	123	833
Range hood, vented, 30" wide	1.000	Ea.	77.50	262	339.50
TOTAL			6,075	2,278	8,353

E1090 310	Kitchens	COST EACH		
		MAT.	INST.	TOTAL
1400	Prefinished wood cabinets, average quality	6,075	2,275	8,350
1500	Prefinished wood cabinets, high quality	11,400	2,625	14,025
1600	Custom cabinets, built in place, high quality	15,000	3,000	18,000
1700				
1800				
1900				
2000				
2100				
2200	NOTE: No plumbing or electric rough-ins are included in the above			
2300	Prices, for plumbing see Division 22, for electric see Division 26.			
2400				
2500				
2600				
2700				
2900	Cut & patch to match existing construction, add, minimum	2%	3%	
3000	Maximum	5%	9%	
3100	Dust protection, add, minimum	1%	2%	
3200	Maximum	4%	11%	
3300	Equipment usage curtailment, add, minimum	1%	1%	
3400	Maximum	3%	10%	
3500	Material handling & storage limitation, add, minimum	1%	1%	
3600	Maximum	6%	7%	
3700	Protection of existing work, add, minimum	2%	2%	
3800	Maximum	5%	7%	
3900	Shift work requirements, add, minimum		5%	
4000	Maximum		30%	

E10 Equipment

E1090 Other Equipment

E1090 315	Selective Price Sheet	MAT.	INST.	TOTAL
0100	Cabinets standard wood, base, one drawer one door, 12" wide	292	49.50	341.50
0200	15" wide	305	51.50	356.50
0300	18" wide	330	53	383
0400	21" wide	340	54	394
0500	24" wide	400	55	455
0600				
0700	Two drawers two doors, 27" wide	430	56	486
0800	30" wide	470	57.50	527.50
0900	33" wide	485	59	544
1000	36" wide	500	60.50	560.50
1100	42" wide	530	62	592
1200	48" wide	570	65	635
1300	Drawer base (4 drawers), 12" wide	305	49.50	354.50
1400	15" wide	310	51.50	361.50
1500	18" wide	345	53	398
1600	24" wide	380	55	435
1700	Sink or range base, 30" wide	385	57.50	442.50
1800	33" wide	415	59	474
1900	36" wide	435	60.50	495.50
2000	42" wide	455	62	517
2100	Corner base, 36" wide	690	68.50	758.50
2200	Lazy susan with revolving door	925	74.50	999.50
2300	Cabinets standard wood, wall two doors, 12" high, 30" wide	261	49.50	310.50
2400	36" wide	310	51.50	361.50
2500	15" high, 30" high	265	51.50	316.50
2600	36" wide	320	54	374
2700	24" high, 30" wide	355	53	408
2800	36" wide	390	54	444
2900	30" high, 30" wide	395	64	459
3000	36" wide	450	65.50	515.50
3100	42" wide	490	66.50	556.50
3200	48" wide	550	67	617
3300	One door, 30" high, 12" wide	238	56	294
3400	15" wide	265	57.50	322.50
3500	18" wide	292	59	351
3600	24" wide	340	60.50	400.50
3700	Corner, 30" high, 24" wide	390	68.50	458.50
3800	36" wide	470	74.50	544.50
3900	Broom, 84" high, 24" deep, 18" wide	715	123	838
4000	Oven, 84" high, 24" deep, 27" wide	1,100	154	1,254
4100	Valance board, 4' long	57	12.45	69.45
4200	6' long	86	18.65	104.65
4300	Counter tops, laminated plastic, stock 25" wide w/backsplash, min.	18.70	20.50	39.20
4400	Maximum	38	24.50	62.50
4500	Custom, 7/8" thick, no splash	33	20.50	53.50
4600	Cove splash	32	20.50	52.50
4700	1-1/4" thick, no splash	39	22	61
4800	Square splash	46.50	22	68.50
4900	Post formed	11.20	20.50	31.70
5000				
5100	Maple laminated 1-1/2" thick, no splash	83	22	105
5200	Square splash	98.50	22	120.50
5300				
5400				
5500	Appliances, range, free standing, minimum	485	98.50	583.50
5600	Maximum	2,225	247	2,472
5700	Built-in, minimum	750	113	863
5800	Maximum	1,500	615	2,115

E10 Equipment

E1090 Other Equipment

E1090 315	Selective Price Sheet	COST EACH		
		MAT.	INST.	TOTAL
5900	Counter top range 4 burner, maximum	325	113	438
6000	Maximum	1,450	226	1,676
6100	Compactor, built-in, minimum	710	123	833
6200	Maximum	1,125	205	1,330
6300	Dishwasher, built-in, minimum	261	355	616
6400	Maximum	480	705	1,185
6500	Garbage disposer, sink-pipe, minimum	96	141	237
6600	Maximum	242	141	383
6700	Range hood, 30" wide, 2 speed, minimum	77.50	262	339.50
6800	Maximum	855	435	1,290
6900	Refrigerator, no frost, 12 cu. ft.	465	98.50	563.50
7000	20 cu. ft.	495	164	659
7100	Plumb. not incl. rough-ins, sinks porc. C.I., single bowl, 21" x 24"	315	235	550
7200	21" x 30"	680	235	915
7300	Double bowl, 20" x 32"	390	275	665
7400				
7500	Stainless steel, single bowl, 19" x 18"	650	235	885
7600	22" x 25"	725	235	960
7700				
7800				
7900				
8000				
8100				
8200				
8300				
8400				

G Building Sitework

No part of this publication may be reproduced, stored in a retrieval system, or transmitted in any form or by any means without prior written permission of RSMeans.

G10 Site Preparation

G1030 Site Earthwork

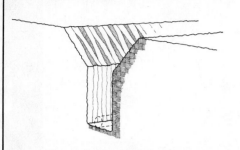

This page illustrates and describes utility trench excavation with backfill systems including a trench, backfill, excavated material, utility pipe or ductwork not included. Lines within System Components give the unit price and total price on a linear foot basis for this system. Prices for utility trench excavation with backfill systems are on Line Items G1030 810 1100 thru 2500. Both material quantities and labor costs have been adjusted for the system listed.

Factors: To adjust for job conditions other than normal working situations use Lines G1030 810 3100 thru 4000.

Example: You are to install the system and protect existing construction. Go to Line G1030 810 3600 and apply these percentages to the appropriate TOTAL cost.

System Components	QUANTITY	UNIT	COST PER L.F.		
			MAT.	INST.	TOTAL
Cont. utility trench excav. with 3/8 C.Y. wheel mtd. backhoe, Backfilling with excav. mat. and comp. in 12" lifts. Trench is 2' wide by 4' dp. Cost of utility piping or ductwork is not incl. Cost based on an Excavation production rate of 150 C.Y. daily no hauling included.					
Machine excavate trench, 2' wide by 4' deep	.296	C.Y.	.79	2.23	3.02
Dozer backfill with excavated material	.296	C.Y.	.36	.62	.98
Compact in 12" lifts, vibrating plate	.296	C.Y.	.09	2.47	2.56
TOTAL			1.24	5.32	6.56

G1030 810	Excavation, Utility Trench	COST PER L.F.		
		MAT.	INST.	TOTAL
1100	2' wide x 2' deep	.62	2.67	3.29
1200	3' deep	.92	3.99	4.91
1250	4' deep	1.24	5.30	6.55
1300	With sloping sides, 2' wide x 5' deep	3.48	15	18.50
1400	6' deep	4.65	20	24.50
1500	7' deep	5.95	25.50	31.50
1600	8' deep	7.45	32	39.50
1700	9' deep	9.05	39	48
1800	10' deep	10.85	46.50	57.50
2000	For hauling excavated material up to 2 miles & backfilling w/ gravel			
2100	Gravel, 2' wide by 2' deep, add	4.14	3.77	7.91
2200	4' deep, add	8.30	7.55	15.85
2300	6' deep, add	31	28.50	59.50
2400	8' deep, add	50	45.50	95.50
2500	10' deep, add	72.50	66	138.50
2600				
2700				
2800	For shallow, hand excavated trenches, no backfill, to 6' dp, light soil		61.50	61.50
2900	Heavy soil		123	123
3000				
3100	Equipment usage curtailment, add, minimum	1%	1%	
3200	Maximum	3%	10%	
3300	Material handling & storage limitation, add, minimum	1%	1%	
3400	Maximum	6%	7%	
3500	Protection of existing work, add, minimum	2%	2%	
3600	Maximum	5%	7%	
3700	Shift work requirements, add, minimum		5%	
3800	Maximum		30%	
3900	Temporary shoring and bracing, add, minimum	2%	5%	
4000	Maximum	5%	12%	

G20 Site Improvements

G2010 Roadways

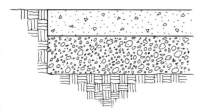

This page illustrates and describes driveway systems including concrete slab, gravel base, broom finish, compaction, and joists. Lines within System Components give the unit price and total price on a cost each basis for this system. Prices for driveway systems are on Line Items G2010 240 1200 thru 2700. Both material quantities and labor costs have been adjusted for the system listed.

Factors: To adjust for job conditions other than normal working situations use Lines G2010 240 2900 thru 4000.

Example: You are to install the system and match existing construction. Go to Line G2010 240 3000 and apply these percentages to the appropriate MAT. and INST. costs.

System Components			COST EACH		
	QUANTITY	UNIT	MAT.	INST.	TOTAL
Complete driveway, 10' x 30', 4" concrete slab, 3000 psi, on 6" crushed Stone base, concrete broom finished and cured with 10' x 10' joints.					
Grade and compact subgrade	33.000	S.Y.		73.59	73.59
Place and compact 6" crushed stone base	33.000	S.Y.	166.65	49.83	216.48
Place and remove edgeforms (4 use)	80.000	L.F.	30.40	312	342.40
Concrete for slab, 3000 psi	4.000	C.Y.	448		448
Place concrete, 4" thick slab, direct chute	4.000	C.Y.		114.48	114.48
Screed, float and broom finish slab	4.000	C.Y.		264	264
Spray on membrane curing compound	3.000	S.F.	36.45	31.20	67.65
Saw cut 3" deep joints	60.000	L.F.	10.20	101.40	111.60
TOTAL			691.70	946.50	1,638.20

G2010 240	Driveways	COST EACH		
		MAT.	INST.	TOTAL
1200	Concrete: 10' x 30' with 4" concrete on 6" crushed stone	690	945	1,635
1300	With 6" concrete	915	1,000	1,915
1400	20' x 30' with 4" concrete	1,375	1,825	3,200
1500	With 6" concrete	1,825	1,925	3,750
1600	10' wide, for each additional 10' length over 30', 4" thick, add	221	305	526
1700	6" thick, add	305	325	630
1800	20' wide, for each additional 10' length over 30', 4" thick, add	440	580	1,020
1900	6" thick, add	580	615	1,195
2000	Asphalt: 10' x 30' with 2" binder, 1" topping on 6" cr. st., sealed	625	1,075	1,700
2100	3" binder, 1" topping	765	1,100	1,865
2200	20' x 30' with 2" binder, 1" topping	1,250	1,350	2,600
2300	3" binding, 1" topping	1,525	1,375	2,900
2400	10' wide for each add'l. 10' length over 30', 2" b + 1" t, add	209	86.50	295.50
2500	3" binder, 1" topping	255	91	346
2600	20' wide for each add'l. 10' length over 30', 2" b + 1" t, add	420	173	593
2700	3" binder, 1" topping	510	182	692
2800				
2900	Cut & patch to match existing construction, add, minimum	2%	3%	
3000	Maximum	5%	9%	
3100	Dust protection, add, minimum	1%	2%	
3200	Maximum	4%	11%	
3300	Equipment usage curtailment, add, minimum	1%	1%	
3400	Maximum	3%	10%	
3500	Material handling & storage limitation, add, minimum	1%	1%	
3600	Maximum	6%	7%	
3700	Protection of existing work, add, minimum	2%	2%	
3800	Maximum	5%	7%	
3900	Shift work requirements, add, minimum		5%	
4000	Maximum		30%	

G20 Site Improvements

G2020 Parking Lots

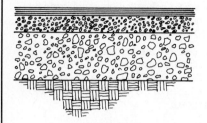

This page illustrates and describes asphalt parking lot systems including asphalt binder, topping, crushed stone base, painted parking stripes and concrete parking blocks. Lines within System Components give the unit price and total price per square yard for this system. Prices for asphalt parking lot systems are on Line Items G2020 230 1400 thru 2500. Both material quantities and labor costs have been adjusted for the system listed.

Factors: To adjust for job conditions other than normal working situations use Lines G2020 230 2900 thru 4000.

Example: You are to install the system and match existing construction. Go to Line G2020 230 2900 and apply these percentages to the appropriate MAT. and INST. costs.

System Components	QUANTITY	UNIT	COST PER S.Y.		
			MAT.	INST.	TOTAL
Parking lot consisting of 2" asphalt binder and 1" topping on 6"					
Crushed stone base with painted parking stripes and concrete parking blocks.					
Fine grade and compact subgrade	1.000	S.Y.		2.23	2.23
6" crushed stone base, stone	.320	Ton	5.05	1.51	6.56
2" asphalt binder	1.000	S.Y.	8.35	1.40	9.75
1" asphalt topping	1.000	S.Y.	4.15	.93	5.08
Paint parking stripes	.500	L.F.	.09	.10	.19
6" x 10" x 6' precast concrete parking blocks	.020	Ea.	.87	.42	1.29
Mobilization of equipment	.005	Ea.		4.13	4.13
TOTAL			18.51	10.72	29.23

G2020 230	Parking Lots, Asphalt	COST PER S.Y.		
		MAT.	INST.	TOTAL
1400	2" binder plus 1" topping on 6" crushed stone	18.50	10.70	29.20
1500	9" crushed stone	21	10.85	31.85
1600	12" crushed stone	23.50	11.05	34.55
1700	On bank run gravel, 6" deep	18.20	10	28.20
1800	9" deep	20.50	10.20	30.70
1900	12" deep	23	10.40	33.40
2000	3" binder plus 1" topping on 6" crushed stone	22.50	11.15	33.65
2100	9" deep crushed stone	25.50	11.30	36.80
2200	12" deep crushed stone	28	11.45	39.45
2300	On bank run gravel, 6" deep	22.50	10.45	32.95
2400	9" deep	25	10.65	35.65
2500	12" deep	27	10.80	37.80
2600				
2700				
2900	Cut & patch to match existing construction, add, minimum	2%	3%	
3000	Maximum	5%	9%	
3100	Dust protection, add, minimum	1%	2%	
3200	Maximum	4%	11%	
3300	Equipment usage curtailment, add, minimum	1%	1%	
3400	Maximum	3%	10%	
3500	Material handling & storage limitation, add, minimum	1%	1%	
3600	Maximum	6%	7%	
3700	Protection of existing work, add, minimum	2%	2%	
3800	Maximum	5%	7%	
3900	Shift work requirements, add, minimum		5%	
4000	Maximum		30%	

G20 Site Improvements

G2020 Parking Lots

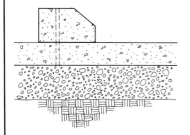

This page illustrates and describes concrete parking lot systems including a concrete slab, unreinforced, compacted gravel base, broom finished, cured, joints, painted parking stripes, and parking blocks. Lines within System Components give the unit price and total price per square yard for this system. Prices for concrete parking lot systems are on Line Items G2020 240 1500 thru 2500. All materials have been adjusted according to the system listed.

Factors: To adjust for job conditions other than normal working situations use Lines G2020 240 2900 thru 4000.

Example: You are to install the system and protect existing construction. Go to Line G2020 240 3800 and apply these percentages to the appropriate MAT. and INST. costs.

System Components	QUANTITY	UNIT	COST PER S.Y.		
			MAT.	INST.	TOTAL
Parking lot, 4" slab, 3000 psi concrete on 6" crushed stone base					
With 12' x 12' joints, painted parking strips, conc. parking blocks.					
Grade and compact subgrade	1.000	S.Y.		2.23	2.23
6" crushed stone base, stone	.320	Ton	5.05	1.51	6.56
Place and remove edge forms, 4 use	.250	L.F.	.10	.98	1.08
Concrete for slab, 3000 psi	1.000	S.F.	12.32		12.32
Place concrete, 4" thick slab, direct chute	1.000	S.F.		3.15	3.15
Spray on membrane curing compound	1.000	S.F.	1.09	.66	1.75
Saw cut 3" deep joints	2.250	L.F.	.38	3.81	4.19
Paint parking stripes	.500	L.F.	.09	.10	.19
6" x 10" x 6' precast concrete parking blocks	.020	Ea.	.87	.42	1.29
TOTAL			19.90	12.86	32.76

G2020 240	Parking Lots, Concrete	COST PER S.Y.		
		MAT.	INST.	TOTAL
1500	4" concrete slab, 3000 psi on 3" crushed stone	17.40	12.80	30.20
1550	6" crushed stone	19.90	12.85	32.75
1600	9" crushed stone	22.50	12.95	35.45
1700	On bank run gravel, 6" deep	19.60	12.20	31.80
1800	9" deep	22	12.35	34.35
1900	On compacted subgrade only	14.85	11.35	26.20
2000	6" concrete slab, 3000 psi on 3" crushed stone	23.50	14.35	37.85
2100	6" crushed stone	26	14.35	40.35
2200	9" crushed stone	28.50	14.50	43
2300	On bank run gravel, 6" deep	25.50	13.70	39.20
2400	9" deep	28	13.90	41.90
2500	On compacted subgrade only	26	14.35	40.35
2600				
2700				
2900	Cut & patch to match existing construction, add, minimum	2%	3%	
3000	Maximum	5%	9%	
3100	Dust protection, add, minimum	1%	2%	
3200	Maximum	4%	11%	
3300	Equipment usage curtailment, add, minimum	1%	1%	
3400	Maximum	3%	10%	
3500	Material handling & storage limitation, add, minimum	1%	1%	
3600	Maximum	6%	7%	
3700	Protection of existing work, add, minimum	2%	2%	
3800	Maximum	5%	7%	
3900	Shift work requirements, add, minimum		5%	
4000	Maximum		30%	

G20 Site Improvements

G2030 Pedestrian Paving

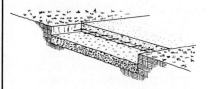

This page illustrates and describes sidewalk systems including concrete, welded wire and broom finish. Lines within System Components give unit price and total price per square foot for this system. Prices for sidewalk systems are on Line Items G2030 105 1700 thru 1900. Both material quantities and labor costs have been adjusted for the system listed.

Factors: To adjust for job conditions other than normal working situations use Lines G2030 105 2900 thru 4000.

Example: You are to install the system and match existing construction. Go to Line G2030 105 2900 and apply these percentages to the appropriate MAT. and INST. costs.

System Components			COST PER S.F.		
	QUANTITY	UNIT	MAT.	INST.	TOTAL
4" thick concrete sidewalk with welded wire fabric 3000 psi air entrained concrete, broom finish.					
Gravel fill, 4" deep	.012	C.Y.	.29	.10	.39
Compact fill	.012	C.Y.		.03	.03
Hand grade	1.000	S.F.		2.22	2.22
Edge form	.250	L.F.	.10	.98	1.08
Welded wire fabric	.011	S.F.	.18	.43	.61
Concrete, 3000 psi air entrained	.012	C.Y.	1.34		1.34
Place concrete	.012	C.Y.		.35	.35
Broom finish	1.000	S.F.		.88	.88
TOTAL			1.91	4.99	6.90

G2030 105	Sidewalks	COST PER S.F.		
		MAT.	INST.	TOTAL
1700	Asphalt (bituminous), 2" thick	1.19	2.86	4.05
1750	4" thick	1.91	4.99	6.90
1800	Brick, on sand, bed, 4.5 brick per S.F.	3.50	10.95	14.45
1900	Flagstone, slate, 1" thick, rectangular	10.70	11.90	22.60
2000				
2100				
2200				
2300				
2400				
2500				
2600				
2700				
2900	Cut & patch to match existing construction, add, minimum	2%	3%	
3000	Maximum	5%	9%	
3100	Dust protection, add, minimum	1%	2%	
3200	Maximum	4%	11%	
3300	Equipment usage curtailment, add, minimum	1%	1%	
3400	Maximum	3%	10%	
3500	Material handling & storage limitation, add, minimum	1%	1%	
3600	Maximum	6%	7%	
3700	Protection of existing work, add, minimum	2%	2%	
3800	Maximum	5%	7%	
3900	Shift work requirements, add, minimum		5%	
4000	Maximum		30%	

For customer support on your Commercial Renovation Cost Data, call 877.791.4977.

G20 Site Improvements

G2050 Landscaping

This page describes landscaping—lawn establishment systems including loam, lime, fertilizer, side and top mulching. Lines within system components give the unit price and total price per square yard for this system. Prices for landscaping—lawn establishment systems are on Line Items G2050 420 1000 thru 1200. Both material quantities and labor costs have been adjusted for the system listed.

Factors: To adjust for job conditions other than normal working situations use Lines G2050 420 2900 thru 4000.

Example: You are to install the system and provide dust protection. Go to Line G2050 420 3200 and apply these percentages to the appropriate MAT. and INST. costs.

System Components	QUANTITY	UNIT	COST PER S.Y. MAT.	INST.	TOTAL
Establishing lawns with loam, lime, fertilizer, seed and top mulching					
On rough graded areas.					
Furnish and place loam 4" deep	.110	C.Y.	3.78	1	4.78
Fine grade, lime, fertilize and seed	1.000	S.Y.	.50	3.51	4.01
Hay mulch, 1 bale/M.S.F.	1.000	S.Y.	.21	.60	.81
Rolling with hand roller	1.000	S.Y.		1.09	1.09
TOTAL			4.49	6.20	10.69

G2050 420	Landscaping - Lawn Establishment	MAT.	INST.	TOTAL
1000	Establishing lawns with loam, lime, fertilizer, seed and hay mulch	4.49	6.20	10.69
1100	Above system with jute mesh in place of hay mulch	5.40	6.35	11.75
1200	Above system with sod in place of seed	6.35	2.18	8.53
1300				
1400				
1500				
1600				
1700				
1800				
1900				
2000				
2100				
2200				
2300				
2400				
2500				
2600				
2700				
2900	Cut & patch to match existing construction, add, minimum	2%	3%	
3000	Maximum	5%	9%	
3100	Dust protection, add, minimum	1%	2%	
3200	Maximum	4%	11%	
3300	Equipment usage curtailment, add, minimum	1%	1%	
3400	Maximum	3%	10%	
3500	Material handling & storage limitation, add, minimum	1%	1%	
3600	Maximum	6%	7%	
3700	Protection of existing work, add, minimum	2%	2%	
3800	Maximum	5%	7%	
3900	Shift work requirements, add, minimum		5%	
4000	Maximum		30%	

G20 Site Improvements

G2050 Landscaping

G2050 510	Selective Price Sheet	COST PER UNIT		
		MAT.	INST.	TOTAL
0100	Shrubs and trees, Evergreen, in prepared beds			
0200				
0300	Arborvitae pyramidal, 4'-5'	111	99	210
0400	Globe, 12"-15"	23.50	15.70	39.20
0500	Cedar, blue, 8'-10'	254	166	420
0600	Hemlock, Canadian, 2-1/2'-3'	34.50	42	76.50
0700	Juniper, andora, 18"-24"	40	18.80	58.80
0800	Wiltoni, 15"-18"	28.50	18.80	47.30
0900	Skyrocket, 4-1/2'-5'	117	54	171
1000	Blue pfitzer, 2'-2-1/2'	42.50	34	76.50
1100	Ketleerie, 2-1/2'-3'	57	30	87
1200	Pine, black, 2-1/2'-3'	65	30	95
1300	Mugo, 18"-24"	65	25	90
1400	White, 4'-5'	56	40	96
1500	Spruce, blue, 18"-24"	73.50	25	98.50
1600	Norway, 4'-5'	91	40	131
1700	Yew, denisforma, 12"-15"	38.50	25	63.50
1800	Capitata, 18"-24"	36	50	86
1900	Hicksi, 2'-2-1/2'	86.50	50	136.50
2000				
2100	Trees, Deciduous, in prepared beds			
2200				
2300	Beech, 5'-6'	206	59.50	265.50
2400	Dogwood, 4'-5'	144	74.50	218.50
2500	Elm, 8'-10'	284	149	433
2600	Magnolia, 4'-5'	111	149	260
2700	Maple, red, 8'-10', 1-1/2" caliper	222	298	520
2800	Oak, 2-1/2"-3" caliper	360	495	855
2900	Willow, 6'-8', 1" caliper	106	149	255
3000				
3100	Shrubs, Broadleaf Evergreen, in prepared beds			
3200				
3300	Andromeda, 15"-18", container	36	15.70	51.70
3400	Azalea, 15"-18", container	32.50	15.70	48.20
3500	Barberry, 9"-12", container	19.35	11.60	30.95
3600	Boxwood, 15"-18", B & B	47	15.70	62.70
3700	Euonymus, emerald gaiety, 12"-15", container	26	13.10	39.10
3800	Holly, 15"-18", B & B	40	15.70	55.70
3900	Mount laurel, 18"-24", B & B	77.50	18.80	96.30
4000	Privet, 18"-24", B & B	23.50	11.60	35.10
4100	Rhododendron, 18"-24", container	41.50	31.50	73
4200	Rosemary, 1 gal. container	19.45	2.51	21.96
4300	Deciduous, amalanchier, 2'-3', B & B	131	26.50	157.50
4400	Azalea, 15"-18", B & B	32	15.70	47.70
4500	Bayberry, 2'-3', B & B	26.50	26.50	53
4600	Cotoneaster, 15"-18", B & B	29	18.80	47.80
4700	Dogwood, 3'-4', B & B	35	74.50	109.50
4800	Euonymus, alatus compacta, 15"-18", container	29	18.80	47.80
4900	Forsythia, 2'-3', container	20.50	25	45.50
5000	Honeysuckle, 3'-4', B & B	29	25	54
5100	Hydrangea, 2'-3', B & B	33	26.50	59.50
5200	Lilac, 3'-4', B & B	30.50	74.50	105
5300	Quince, 2'-3', B & B	31	26.50	57.50
5400				
5500	Ground Cover			
5600				
5700	Plants, Pachysandra, prepared beds, per hundred	69	100	169
5800	Vinca minor or English ivy, per hundred	122	125	247

G20 Site Improvements

G2050 Landscaping

G2050 510	Selective Price Sheet	COST PER UNIT		
		MAT.	INST.	TOTAL
5900	Plant bed prep. 18" dp., mach., per square foot	2.63	.45	3.08
6000	By hand, per square foot	2.63	3.18	5.81
6100	Stone chips, Georgia marble, 50# bags, per bag	4.14	2.90	7.04
6200	Onyx gemstone, per bag	18.60	5.80	24.40
6300	Quartz, per bag	18.50	5.80	24.30
6400	Pea gravel, truck load lots, per cubic yard	28	54	82
6500	Mulch, polyethylene mulch, per square yard	.51	.49	1
6600	Wood chips, 2" deep, per square yard	1.76	2.24	4
6700	Peat moss, 1" deep, per square yard	3.08	.55	3.63
6800	Erosion control per square yard, Jute mesh stapled	1.13	.76	1.89
6900	Plastic netting, stapled	.26	.60	.86
7000	Polypropylene mesh, stapled	2.53	.60	3.13
7100	Tobacco netting, stapled	.21	.60	.81
7200	Lawns per square yard, seeding incl. fine grade, limestone, fert. and seed	.50	3.51	4.01
7300	Sodding, incl. fine grade, level ground	2.41	1.14	3.55
7500	Edging, per linear foot, Redwood, untreated, 1" x 4"	1.28	2.46	3.74
7600	2" x 4"	2.47	3.73	6.20
7700	Stl. edge strips, 1/4" x 5" inc. stakes	4.40	3.73	8.13
7800	3/16" x 4"	2.64	3.73	6.37
7900	Brick edging, set on edge	2.69	8.10	10.79
8000	Set flat	1.35	2.96	4.31
8100				
8200				
8300				
8400				

G30 Site Mechanical Utilities

G3020 Sanitary Sewer

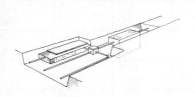

This page illustrates and describes drainage and utilities – septic system including a tank, distribution box, excavation, piping, crushed stone and backfill. Lines within System Components give the unit price and total price on a cost each basis for this system. Prices for drainage and utilities – septic systems are on Line Items G3020 610 1600 thru 2000. Both material quantities and labor costs have been adjusted for the system listed.

Factors: To adjust for job conditions other than normal working situations use Lines G3020 610 3100 thru 4000.

Example: You are to install the system with a material handling limitation. Go to Line G3020 610 3500 and apply these percentages to the appropriate MAT. and INST. costs.

System Components			COST EACH		
	QUANTITY	UNIT	MAT.	INST.	TOTAL
Septic system including tank, distribution box, excavation, piping, crushed Stone and backfill for a 1000 S.F. leaching field.					
Precast concrete septic tank, 1000 gal. capacity	1.000	Ea.	1,100	265.95	1,365.95
Concrete distribution box	1.000	Ea.	88.50	61.50	150
Sch. 40 sewer pipe, PVC, 4″ diameter	25.000	L.F.	272	746.30	1,018.30
Sch. 40 sewer pipe fittings, PVC, 4″ diameter	8.000	Ea.	118.40	632	750.40
Excavation for septic tank	120.000	C.Y.		232.56	232.56
Excavation for disposal field	120.000	C.Y.		563.58	563.58
Crushed stone backfill	76.000	C.Y.	1,672	1,016.12	2,688.12
Backfill with excavated material	26.000	C.Y.		52.26	52.26
Asphalt felt paper, 15#	6.000	Sq.	35.70	99.90	135.60
TOTAL			3,286.60	3,670.17	6,956.77

G3020 610	Septic Systems	COST EACH		
		MAT.	INST.	TOTAL
1600	1000 gal. tank with 1000 S.F. field	3,275	3,650	6,925
1700	1000 gal. tank with 2000 S.F. field	4,675	5,425	10,100
1800	With leaching pits	3,250	2,125	5,375
1900	2000 gal. tank with 2000 S.F. field	5,675	5,600	11,275
2000	With leaching pits	4,350	2,750	7,100
2100				
2200				
2300				
2400				
2500				
2600				
2700				
2800				
2900				
3100	Dust protection, add, minimum		1%	2%
3200	Maximum		4%	11%
3300	Equipment usage curtailment, add, minimum		1%	1%
3400	Maximum		3%	10%
3500	Material handling & storage limitation, add, minimum		1%	1%
3600	Maximum		6%	7%
3700	Protection of existing work, add, minimum		2%	2%
3800	Maximum		5%	7%
3900	Shift work requirements, add, minimum			5%
4000	Maximum			30%

G30 Site Mechanical Utilities
G3060 Fuel Distribution

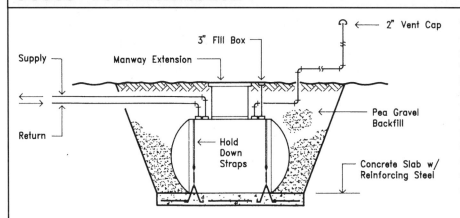

Fiberglass Underground Fuel Storage Tank, Single Wall

This page illustrates and describes fiberglass tanks including hold down slab, reinforcing steel and peastone gravel. Lines within Systems Components give the material and installation price on a cost each basis for the components. Prices for fiberglass tank systems are on Line Items G3060 305 1010 thru G3060 310 1010. Material quantities and labor costs have been adjusted for the system listed.

Factors: To adjust for job conditions other than normal working situations use Lines G3060 310 2700 thru 4000.

System Components	QUANTITY	UNIT	COST EACH MAT.	COST EACH INST.	COST EACH TOTAL
Double wall underground fiberglass storage tanks including, excavation, hold down slab, backfill, manway extension and related piping.					
Excavating trench, no sheeting or dewatering, 6'-10 ' D, 3/4 CY hyd backhoe	18.000	Ea.		149.40	149.40
Corrosion resistance wrap & coat, small diam. pipe, 1" diam., add	120.000	Ea.	134.40		134.40
Fiberglass-reinforced elbows 90 & 45 degree, 2 inch	7.000	Ea.	332.50	609	941.50
Fiberglass-reinforced threaded adapters, 2 inch	2.000	Ea.	39.50	199	238.50
Remote tank gauging system, 30', 5" pointer travel	1.000	Ea.	5,625	375	6,000
Corrosion resistance wrap & coat, small diam. pipe. 4 " diam. add	30.000	Ea.	72.90		72.90
Pea gravel	9.000	Ea.	1,062	486	1,548
Reinforcing in hold down pad, #3 to #7	120.000	Ea.	63.60	78	141.60
Tank leak detector system, 8 channel, external monitoring	1.000	Ea.	1,000		1,000
Tubbing copper, type L, 3/4" diam.	120.000	Ea.	684	1,158	1,842
Elbow 90 degree ,wrought cu x cu 3/4 " diam.	4.000	Ea.	10.84	154	164.84
Pipe, black steel welded, sch 40, 3" diam.	30.000	Ea.	510	959.40	1,469.40
Elbow 90 degree, steel welded butt joint, 3 " diam.	3.000	Ea.	96	591.15	687.15
Fiberglass-reinforced plastic pipe, 2 inch	156.000	Ea.	1,014	765.96	1,779.96
Fiberglass-reinforced plastic pipe, 3 inch	36.000	Ea.	1,044	199.80	1,243.80
Fiberglass-reinforced elbows 90 & 45 degree, 3 inch	3.000	Ea.	450	348	798
Tank leak detection system, probes, well monitoring, liquid phase detection	2.000	Ea.	1,050		1,050
Double wall fiberglass annular space	1.000	Ea.	355		355
Flange, weld neck, 150 lb, 3 " pipe size	1.000	Ea.	38.50	98.54	137.04
Vent protector/breather, 2" diam.	1.000	Ea.	33.50	23.50	57
Corrosion resistance wrap & coat, small diam. pipe, 2" diam., add	36.000	Ea.	54.72		54.72
Concrete hold down pad, 6' x 5' x 8" thick	30.000	Ea.	88.80	49.20	138
600 gallon capacity, underground storage tank, double wall fiberglass	1.000	Ea.	7,925	555	8,480
Tank, for hold-downs 500-4000 gal., add	1.000	Ea.	490	177	667
Foot valve, single poppet, 3/4" diam.	1.000	Ea.	78.50	41.50	120
Fuel fill box, locking inner cover, 3" diam.	1.000	Ea.	133	149	282
Fuel oil specialties, valve,ball chk, globe type fusible, 3/4" diam.	2.000	Ea.	154	75	229
TOTAL			22,539.76	7,241.45	29,781.21

G3060 310	Fiberglass Fuel Tank, Double Wall	COST EACH MAT.	COST EACH INST.	COST EACH TOTAL
1010	Storage tank, fuel, underground, double wall fiberglass, 600 Gal.	22,500	7,225	29,725
1020	2500 Gal.	35,700	10,000	45,700
1030	4000 Gal.	41,800	12,400	54,200
1040	6000 Gal.	48,800	14,000	62,800
1050	8000 Gal.	56,500	16,900	73,400
1060	10,000 Gal.	64,000	19,400	83,400

G30 Site Mechanical Utilities

G3060 Fuel Distribution

G3060 310	Fiberglass Fuel Tank, Double Wall	COST EACH		
		MAT.	INST.	TOTAL
1070	15,000 Gal.	87,500	22,400	109,900
1080	20,000 Gal.	110,500	27,100	137,600
1090	25,000 Gal.	140,500	31,500	172,000
1100	30,000 Gal.	160,500	33,300	193,800
2700	Cut & patch to match existing construction, add, minimum	2%	3%	
2800	Maximum	5%	9%	
2900	Dust protection, add, minimum	1%	2%	
3000	Maximum	4%	11%	
3100	Equipment usage curtailment, add, minimum	1%	1%	
3200	Maximum	3%	10%	
3300	Material handling & storage limitation, add, minimum	1%	1%	
3400	Maximum	6%	7%	
3500	Protection of existing work, add, minimum	2%	2%	
3600	Maximum	5%	7%	
3700	Shift work requirements, add, minimum		5%	
3800	Maximum		30%	
3900	Temporary shoring and bracing, add, minimum	2%	5%	
4000	Maximum	5%	12%	

G30 Site Mechanical Utilities

G3060 Fuel Distribution

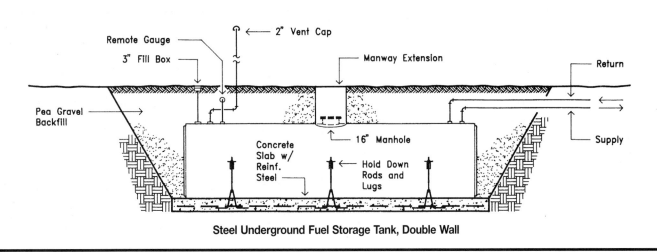

Steel Underground Fuel Storage Tank, Double Wall

System Components	QUANTITY	UNIT	COST EACH MAT.	COST EACH INST.	COST EACH TOTAL
Double wall underground steel storage tank including, excavation, hold down slab, backfill, manway extension and related piping.					
Fiberglass-reinforced plastic pipe, 2 inch	156.000	Ea.	1,014	765.96	1,779.96
Fiberglass-reinforced plastic pipe, 3 inch	36.000	Ea.	1,044	199.80	1,243.80
Fiberglass-reinforced elbows 90 & 45 degree, 3 inch	3.000	Ea.	450	348	798
Tank leak detection system, probes, well monitoring, liquid phase detection	2.000	Ea.	1,050		1,050
Flange, weld neck, 150 lb, 3" pipe size	1.000	Ea.	38.50	98.54	137.04
Vent protector/breather, 2" diam.	1.000	Ea.	33.50	23.50	57
Concrete hold down pad, 6' x 5' x 8" thick	30.000	Ea.	88.80	49.20	138
Tanks, for hold-downs, 500-2000 gal, add	1.000	Ea.	201	177	378
Double wall steel tank annular space	1.000	Ea.	370		370
Foot valve, single poppet, 3/4" diam.	1.000	Ea.	78.50	41.50	120
Fuel fill box, locking inner cover, 3" diam.	1.000	Ea.	133	149	282
Fuel oil specialties, valve, ball chk, globe type fusible, 3/4" diam.	2.000	Ea.	154	75	229
Pea gravel	9.000	Ea.	1,062	486	1,548
Reinforcing in hold down pad, #3 to #7	120.000	Ea.	63.60	78	141.60
Tank, for manways, add	1.000	Ea.	2,000		2,000
Tank leak detector system, 8 channel, external monitoring	1.000	Ea.	1,000		1,000
Tubbing copper, type L, 3/4" diam.	120.000	Ea.	684	1,158	1,842
Elbow 90 degree, wrought cu x cu 3/4" diam.	8.000	Ea.	21.68	308	329.68
Pipe, black steel welded, sch 40, 3" diam.	30.000	Ea.	510	959.40	1,469.40
Elbow 90 degree, steel welded butt joint, 3" diam.	3.000	Ea.	96	591.15	687.15
Excavating trench, no sheeting or dewatering, 6'-10' D, 3/4 CY hyd backhoe	18.000	Ea.		149.40	149.40
Fiberglass-reinforced elbows 90 & 45 degree, 2 inch	7.000	Ea.	332.50	609	941.50
Fiberglass-reinforced threaded adapters, 2 inch	2.000	Ea.	39.50	199	238.50
500 gallon capacity, underground storage tank, double wall steel	1.000	Ea.	6,150	560	6,710
Remote tank gauging system, 30', 5" pointer travel	1.000	Ea.	5,625	375	6,000
TOTAL			22,239.58	7,400.45	29,640.03

G3060 320		Steel Fuel Tank, Double Wall	MAT.	INST.	TOTAL
1010	Storage tank, fuel, underground, double wall steel, 500 Gal.		22,200	7,400	29,600
1020		2000 Gal.	29,900	9,700	39,600
1030		4000 Gal.	40,700	12,500	53,200
1040		6000 Gal.	55,000	13,800	68,800
1050		8000 Gal.	59,000	17,200	76,200
1060		10,000 Gal.	66,000	19,700	85,700

G30 Site Mechanical Utilities

G3060 Fuel Distribution

G3060 320	Steel Fuel Tank, Double Wall	COST EACH		
		MAT.	INST.	TOTAL
1070	15,000 Gal.	75,000	22,900	97,900
1080	20,000 Gal.	87,000	27,500	114,500
1090	25,000 Gal.	135,500	31,500	167,000
1100	30,000 Gal.	154,000	33,500	187,500
1110	40,000 Gal.	194,000	42,100	236,100
1120	50,000 Gal.	234,500	50,500	285,000

G3060 325	Steel Tank, Above Ground, Single Wall	COST EACH		
		MAT.	INST.	TOTAL
1010	Storage tank, fuel, above ground, single wall steel, 550 Gal.	14,200	5,150	19,350
1020	2000 Gal.	21,800	5,275	27,075
1030	5000 Gal.	32,900	8,075	40,975
1040	10,000 Gal.	50,500	8,600	59,100
1050	15,000 Gal.	61,500	8,850	70,350
1060	20,000 Gal.	76,000	9,125	85,125
1070	25,000 Gal.	87,000	9,350	96,350
1080	30,000 Gal.	102,000	9,750	111,750

G3060 330	Steel Tank, Above Ground, Double Wall	COST EACH		
		MAT.	INST.	TOTAL
1010	Storage tank, fuel, above ground, double wall steel, 500 Gal.	14,300	5,225	19,525
1020	2000 Gal.	22,500	5,350	27,850
1030	4000 Gal.	31,400	5,450	36,850
1040	6000 Gal.	36,700	8,350	45,050
1050	8000 Gal.	43,700	8,600	52,300
1060	10,000 Gal.	47,200	8,750	55,950
1070	15,000 Gal.	64,500	9,075	73,575
1080	20,000 Gal.	71,500	9,350	80,850
1090	25,000 Gal.	84,000	9,650	93,650
1100	30,000 Gal.	90,500	10,000	100,500
2700	Cut & patch to match existing construction, add, minimum	2%	3%	
2800	Maximum	5%	9%	
2900	Dust protection, add, minimum	1%	2%	
3000	Maximum	4%	11%	
3100	Equipment usage curtailment, add, minimum	1%	1%	
3200	Maximum	3%	10%	
3300	Material handling & storage limitation, add, minimum	1%	1%	
3400	Maximum	6%	7%	
3500	Protection of existing work, add, minimum	2%	2%	
3600	Maximum	5%	7%	
3700	Shift work requirements, add, minimum		5%	
3800	Maximum		30%	
3900	Temporary shoring and bracing, add, minimum	2%	5%	
4000	Maximum	5%	12%	

G40 Site Electrical Utilities

G4020 Site Lighting

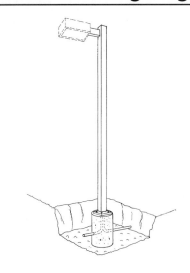

This page illustrates and describes light poles for parking or walkway area lighting. Included are aluminum or steel light poles, single or multiple fixture bracket arms, excavation, concrete footing, backfill and compaction. Lines within system components give the unit price and total price on a cost per each basis. Prices for systems are shown on lines G4020 210 0200 through 1440. Both material quantities and labor costs have been adjusted for the system listed.

Factors: To adjust for job conditions other than normal working situations use lines G4020 210 2700 through 3600.

Example: You are to install the system where you must be careful not to damage existing walks or parking areas. Go to line G4020 210 3300 and apply these percentages to the appropriate MAT. and INST. costs.

System Components	QUANTITY	UNIT	COST EACH		
			MAT.	INST.	TOTAL
Light poles, aluminum, 20' high, 1 arm bracket					
Aluminum light pole, 20', no concrete base	1.000	Ea.	1,025	637	1,662
Bracket arm for Aluminum light pole	1.000	Ea.	133	84.50	217.50
Excavation by hand, pits to 6' deep, heavy soil or clay	2.368	C.Y.		291.26	291.26
Footing, concrete incl forms, reinforcing, spread, under 1 C.Y.	.465	C.Y.	85.10	137.76	222.86
Backfill by hand	1.903	C.Y.		85.64	85.64
Compaction vibrating plate	1.903	C.Y.		11.55	11.55
TOTAL			1,243.10	1,247.71	2,490.81

G4020 210	Light Pole (Installed)	COST EACH		
		MAT.	INST.	TOTAL
0200	Light pole, aluminum, 20' high, 1 arm bracket	1,250	1,250	2,500
0240	2 arm brackets	1,375	1,250	2,625
0280	3 arm brackets	1,500	1,300	2,800
0320	4 arm brackets	1,650	1,300	2,950
0360	30' high, 1 arm bracket	2,200	1,575	3,775
0400	2 arm brackets	2,325	1,575	3,900
0440	3 arm brackets	2,450	1,600	4,050
0480	4 arm brackets	2,600	1,600	4,200
0680	40' high, 1 arm bracket	2,675	2,100	4,775
0720	2 arm brackets	2,825	2,100	4,925
0760	3 arm brackets	2,950	2,150	5,100
0800	4 arm brackets	3,100	2,150	5,250
0840	Steel, 20' high, 1 arm bracket	1,475	1,300	2,775
0880	2 arm brackets	1,525	1,300	2,825
0920	3 arm brackets	1,575	1,350	2,925
0960	4 arm brackets	1,625	1,350	2,975
1000	30' high, 1 arm bracket	1,725	1,675	3,400
1040	2 arm brackets	1,775	1,675	3,450
1080	3 arm brackets	1,800	1,725	3,525
1120	4 arm brackets	1,875	1,725	3,600
1320	40' high, 1 arm bracket	2,275	2,275	4,550
1360	2 arm brackets	2,325	2,275	4,600
1400	3 arm brackets	2,350	2,325	4,675
1440	4 arm brackets	2,425	2,325	4,750
2700	Cut & patch to match existing construction, add, minimum	2%	3%	
2800	Maximum	5%	9%	

G40 Site Electrical Utilities

G4020 Site Lighting

G4020 210	Light Pole (Installed)	COST EACH		
		MAT.	INST.	TOTAL
2900	Equipment usage curtailment, add, minimum	1%	1%	
3000	Maximum	3%	10%	
3100	Material handling & storage limitation, add, minimum	1%	1%	
3200	Maximum	6%	7%	
3300	Protection of existing work, add, minimum	2%	2%	
3400	Maximum	5%	7%	
3500	Shift work requirements, add, minimum		5%	
3600	Maximum		30%	

Reference Section

All the reference information is in one section, making it easy to find what you need to know . . . and easy to use the book on a daily basis. This section is visually identified by a vertical black bar on the page edges.

In this Reference Section, we've included Equipment Rental Costs, a listing of rental and operating costs; Crew Listings, a full listing of all crews and equipment, and their costs; Historical Cost Indexes for cost comparisons over time; City Cost Indexes and Location Factors for adjusting costs to the region you are in; Reference Tables, where you will find explanations, estimating information and procedures, or technical data; Change Orders, information on pricing changes to contract documents; Square Foot Costs that allow you to make a rough estimate for the overall cost of a project; and an explanation of all the Abbreviations in the book.

Table of Contents

Construction Equipment Rental Costs	709
Crew Listings	721
Historical Cost Indexes	756
City Cost Indexes	757
Location Factors	800
Reference Tables	806
R01 General Requirements	806
R02 Existing Conditions	817
R03 Concrete	820
R04 Masonry	830
R05 Metals	831
R06 Wood, Plastics & Composites	833
R07 Thermal & Moisture Protection	833
R08 Openings	835
R09 Finishes	838
R14 Conveying Equipment	840

Reference Tables (cont.)

R22 Plumbing	841
R26 Electrical	842
R31 Earthwork	843
R32 Exterior Improvements	845
R33 Utilities	846
Change Orders	847
Square Foot Costs	851
Abbreviations	861

Equipment Rental Costs

Estimating Tips
- This section contains the average costs to rent and operate hundreds of pieces of construction equipment. This is useful information when estimating the time and material requirements of any particular operation in order to establish a unit or total cost. Equipment costs include not only rental, but also operating costs for equipment under normal use.

Rental Costs
- Equipment rental rates are obtained from the following industry sources throughout North America: contractors, suppliers, dealers, manufacturers, and distributors.
- Rental rates vary throughout the country, with larger cities generally having lower rates. Lease plans for new equipment are available for periods in excess of six months, with a percentage of payments applying toward purchase.
- Monthly rental rates vary from 2% to 5% of the purchase price of the equipment depending on the anticipated life of the equipment and its wearing parts.
- Weekly rental rates are about 1/3 the monthly rates, and daily rental rates are about 1/3 the weekly rate.
- Rental rates can also be treated as reimbursement costs for contractor-owned equipment. Owned equipment costs include depreciation, loan payments, interest, taxes, insurance, storage, and major repairs.

Operating Costs
- The operating costs include parts and labor for routine servicing, such as repair and replacement of pumps, filters and worn lines. Normal operating expendables, such as fuel, lubricants, tires and electricity (where applicable), are also included.
- Extraordinary operating expendables with highly variable wear patterns, such as diamond bits and blades, are excluded. These costs can be found as material costs in the Unit Price section.
- The hourly operating costs listed do not include the operator's wages.

Equipment Cost/Day
- Any power equipment required by a crew is shown in the Crew Listings with a daily cost.
- The daily cost of equipment needed by a crew is based on dividing the weekly rental rate by 5 (number of working days in the week), and then adding the hourly operating cost times 8 (the number of hours in a day). This "Equipment Cost/Day" is shown in the far right column of the Equipment Rental pages.
- If equipment is needed for only one or two days, it is best to develop your own cost by including components for daily rent and hourly operating cost. This is important when the listed Crew for a task does not contain the equipment needed, such as a crane for lifting mechanical heating/cooling equipment up onto a roof.
- If the quantity of work is less than the crew's Daily Output shown for a Unit Price line item that includes a bare unit equipment cost, it is recommended to estimate one day's rental cost and operating cost for equipment shown in the Crew Listing for that line item.

Mobilization/ Demobilization
- The cost to move construction equipment from an equipment yard or rental company to the job site and back again is not included in equipment rental costs listed in the Reference Section, nor in the bare equipment cost of any unit price line item, nor in any equipment costs shown in the Crew listings.
- Mobilization (to the site) and demobilization (from the site) costs can be found in the Unit Price Section.
- If a piece of equipment is already at the job site, it is not appropriate to utilize mobilization/demobilization costs again in an estimate.

01 54 | Construction Aids

01 54 33 | Equipment Rental

			UNIT	HOURLY OPER. COST	RENT PER DAY	RENT PER WEEK	RENT PER MONTH	EQUIPMENT COST/DAY	
10	0010	**CONCRETE EQUIPMENT RENTAL** without operators	R015433 -10						10
	0200	Bucket, concrete lightweight, 1/2 C.Y.		Ea.	.80	23.50	70	210	20.40
	0300	1 C.Y.			.90	27.50	83	249	23.80
	0400	1-1/2 C.Y.			1.10	36.50	110	330	30.80
	0500	2 C.Y.			1.25	45	135	405	37
	0580	8 C.Y.			6.10	257	770	2,300	202.80
	0600	Cart, concrete, self-propelled, operator walking, 10 C.F.			3.25	56.50	170	510	60
	0700	Operator riding, 18 C.F.			5.35	95	285	855	99.80
	0800	Conveyer for concrete, portable, gas, 16" wide, 26' long			12.60	123	370	1,100	174.80
	0900	46' long			13.00	148	445	1,325	193
	1000	56' long			13.10	157	470	1,400	198.80
	1100	Core drill, electric, 2-1/2 H.P., 1" to 8" bit diameter			1.77	68.50	205	615	55.15
	1150	11 H.P., 8" to 18" cores			5.95	113	340	1,025	115.60
	1200	Finisher, concrete floor, gas, riding trowel, 96" wide			12.75	145	435	1,300	189
	1300	Gas, walk-behind, 3 blade, 36" trowel			2.15	20	60	180	29.20
	1400	4 blade, 48" trowel			4.20	27.50	83	249	50.20
	1500	Float, hand-operated (Bull float) 48" wide			.08	13.35	40	120	8.65
	1570	Curb builder, 14 H.P., gas, single screw			14.65	248	745	2,225	266.20
	1590	Double screw			15.35	290	870	2,600	296.80
	1600	Floor grinder, concrete and terrazzo, electric, 22" path			2.57	158	475	1,425	115.55
	1700	Edger, concrete, electric, 7" path			1.04	51.50	155	465	39.30
	1750	Vacuum pick-up system for floor grinders, wet/dry			1.49	81.50	245	735	60.90
	1800	Mixer, powered, mortar and concrete, gas, 6 C.F., 18 H.P.			8.65	118	355	1,075	140.20
	1900	10 C.F., 25 H.P.			10.80	143	430	1,300	172.40
	2000	16 C.F.			11.15	165	495	1,475	188.20
	2100	Concrete, stationary, tilt drum, 2 C.Y.			6.90	232	695	2,075	194.20
	2120	Pump, concrete, truck mounted 4" line 80' boom			25.00	865	2,600	7,800	720
	2140	5" line, 110' boom			32.40	1,150	3,435	10,300	946.20
	2160	Mud jack, 50 C.F. per hr.			7.30	125	375	1,125	133.40
	2180	225 C.F. per hr.			9.50	145	435	1,300	163
	2190	Shotcrete pump rig, 12 C.Y./hr.			14.85	220	660	1,975	250.80
	2200	35 C.Y./hr.			17.50	235	705	2,125	281
	2600	Saw, concrete, manual, gas, 18 H.P.			6.75	45	135	405	81
	2650	Self-propelled, gas, 30 H.P.			13.30	102	305	915	167.40
	2675	V-groove crack chaser, manual, gas, 6 H.P.			2.35	17.35	52	156	29.20
	2700	Vibrators, concrete, electric, 60 cycle, 2 H.P.			.46	8.65	26	78	8.90
	2800	3 H.P.			.60	11.65	35	105	11.80
	2900	Gas engine, 5 H.P.			2.00	16	48	144	25.60
	3000	8 H.P.			2.75	15.35	46	138	31.20
	3050	Vibrating screed, gas engine, 8 H.P.			2.91	71.50	215	645	66.30
	3120	Concrete transit mixer, 6 x 4, 250 H.P., 8 C.Y., rear discharge			61.55	570	1,715	5,150	835.40
	3200	Front discharge			72.45	700	2,095	6,275	998.60
	3300	6 x 6, 285 H.P., 12 C.Y., rear discharge			71.50	660	1,980	5,950	968
	3400	Front discharge			74.00	705	2,120	6,350	1,016
20	0010	**EARTHWORK EQUIPMENT RENTAL** without operators	R015433 -10						20
	0040	Aggregate spreader, push type 8' to 12' wide		Ea.	3.05	25.50	76	228	39.60
	0045	Tailgate type, 8' wide			2.90	32.50	98	294	42.80
	0055	Earth auger, truck-mounted, for fence & sign posts, utility poles			19.15	440	1,320	3,950	417.20
	0060	For borings and monitoring wells			46.65	675	2,030	6,100	779.20
	0070	Portable, trailer mounted			3.30	32.50	98	294	46
	0075	Truck-mounted, for caissons, water wells			100.15	2,900	8,705	26,100	2,542
	0080	Horizontal boring machine, 12" to 36" diameter, 45 H.P.			24.20	192	575	1,725	308.60
	0090	12" to 48" diameter, 65 H.P.			34.35	335	1,000	3,000	474.80
	0095	Auger, for fence posts, gas engine, hand held			.60	6	18	54	8.40
	0100	Excavator, diesel hydraulic, crawler mounted, 1/2 C.Y. cap.			24.65	420	1,255	3,775	448.20
	0120	5/8 C.Y. capacity			32.55	550	1,645	4,925	589.40
	0140	3/4 C.Y. capacity			35.60	615	1,850	5,550	654.80
	0150	1 C.Y. capacity			49.70	690	2,075	6,225	812.60

01 54 | Construction Aids

01 54 33 | Equipment Rental

		UNIT	HOURLY OPER. COST	RENT PER DAY	RENT PER WEEK	RENT PER MONTH	EQUIPMENT COST/DAY
0200	1-1/2 C.Y. capacity	Ea.	59.80	920	2,760	8,275	1,030
0300	2 C.Y. capacity		67.35	1,050	3,180	9,550	1,175
0320	2-1/2 C.Y. capacity		102.55	1,300	3,925	11,800	1,605
0325	3-1/2 C.Y. capacity		144.45	2,150	6,420	19,300	2,440
0330	4-1/2 C.Y. capacity		173.40	2,625	7,880	23,600	2,963
0335	6 C.Y. capacity		219.80	2,900	8,680	26,000	3,494
0340	7 C.Y. capacity		222.05	3,025	9,070	27,200	3,590
0342	Excavator attachments, bucket thumbs		3.20	245	735	2,200	172.60
0345	Grapples		2.75	193	580	1,750	138
0346	Hydraulic hammer for boom mounting, 4000 ft lb.		12.40	350	1,045	3,125	308.20
0347	5000 ft lb.		14.40	425	1,280	3,850	371.20
0348	8000 ft lb.		21.30	625	1,875	5,625	545.40
0349	12,000 ft lb.		23.35	750	2,245	6,725	635.80
0350	Gradall type, truck mounted, 3 ton @ 15' radius, 5/8 C.Y.		43.65	890	2,665	8,000	882.20
0370	1 C.Y. capacity		47.85	1,025	3,095	9,275	1,002
0400	Backhoe-loader, 40 to 45 H.P., 5/8 C.Y. capacity		14.90	240	720	2,150	263.20
0450	45 H.P. to 60 H.P., 3/4 C.Y. capacity		23.40	295	885	2,650	364.20
0460	80 H.P., 1-1/4 C.Y. capacity		25.60	310	935	2,800	391.80
0470	112 H.P., 1-1/2 C.Y. capacity		40.95	645	1,930	5,800	713.60
0482	Backhoe-loader attachment, compactor, 20,000 lb.		5.80	142	425	1,275	131.40
0485	Hydraulic hammer, 750 ft lb.		3.30	96.50	290	870	84.40
0486	Hydraulic hammer, 1200 ft lb.		6.30	217	650	1,950	180.40
0500	Brush chipper, gas engine, 6" cutter head, 35 H.P.		11.10	105	315	945	151.80
0550	Diesel engine, 12" cutter head, 130 H.P.		27.55	285	855	2,575	391.40
0600	15" cutter head, 165 H.P.		33.10	335	1,005	3,025	465.80
0750	Bucket, clamshell, general purpose, 3/8 C.Y.		1.30	38.50	115	345	33.40
0800	1/2 C.Y.		1.40	45	135	405	38.20
0850	3/4 C.Y.		1.55	55	165	495	45.40
0900	1 C.Y.		1.60	58.50	175	525	47.80
0950	1-1/2 C.Y.		2.55	81.50	245	735	69.40
1000	2 C.Y.		2.70	90	270	810	75.60
1010	Bucket, dragline, medium duty, 1/2 C.Y.		.75	23.50	70	210	20
1020	3/4 C.Y.		.75	24.50	73	219	20.60
1030	1 C.Y.		.80	26	78	234	22
1040	1-1/2 C.Y.		1.20	40	120	360	33.60
1050	2 C.Y.		1.30	45	135	405	37.40
1070	3 C.Y.		1.95	61.50	185	555	52.60
1200	Compactor, manually guided 2-drum vibratory smooth roller, 7.5 H.P.		7.25	203	610	1,825	180
1250	Rammer/tamper, gas, 8"		2.75	46.50	140	420	50
1260	15"		3.05	53.50	160	480	56.40
1300	Vibratory plate, gas, 18" plate, 3000 lb. blow		2.70	24.50	73	219	36.20
1350	21" plate, 5000 lb. blow		3.40	32.50	98	294	46.80
1370	Curb builder/extruder, 14 H.P., gas, single screw		14.65	248	745	2,225	266.20
1390	Double screw		15.35	290	870	2,600	296.80
1500	Disc harrow attachment, for tractor		.44	73.50	221	665	47.70
1810	Feller buncher, shearing & accumulating trees, 100 H.P.		48.30	755	2,260	6,775	838.40
1860	Grader, self-propelled, 25,000 lb.		39.50	640	1,925	5,775	701
1910	30,000 lb.		43.15	650	1,955	5,875	736.20
1920	40,000 lb.		64.90	1,100	3,275	9,825	1,174
1930	55,000 lb.		84.15	1,700	5,130	15,400	1,699
1950	Hammer, pavement breaker, self-propelled, diesel, 1000 to 1250 lb.		29.90	350	1,055	3,175	450.20
2000	1300 to 1500 lb.		44.85	705	2,110	6,325	780.80
2050	Pile driving hammer, steam or air, 4150 ft lb. @ 225 bpm		10.40	475	1,420	4,250	367.20
2100	8750 ft lb. @ 145 bpm		12.40	660	1,975	5,925	494.20
2150	15,000 ft lb. @ 60 bpm		14.00	790	2,370	7,100	586
2200	24,450 ft lb. @ 111 bpm		15.00	875	2,630	7,900	646
2250	Leads, 60' high for pile driving hammers up to 20,000 ft lb.		3.30	80.50	242	725	74.80
2300	90' high for hammers over 20,000 ft lb.		4.95	141	424	1,275	124.40

For customer support on your Commercial Renovation Cost Data, call 877.791.4977.

01 54 | Construction Aids

01 54 33 | Equipment Rental

		UNIT	HOURLY OPER. COST	RENT PER DAY	RENT PER WEEK	RENT PER MONTH	EQUIPMENT COST/DAY
2350	Diesel type hammer, 22,400 ft lb.	Ea.	22.55	460	1,380	4,150	456.40
2400	41,300 ft lb.		32.00	540	1,625	4,875	581
2450	141,000 ft lb.		51.65	960	2,875	8,625	988.20
2500	Vib. elec. hammer/extractor, 200 kW diesel generator, 34 H.P.		54.90	635	1,900	5,700	819.20
2550	80 H.P.		100.65	925	2,775	8,325	1,360
2600	150 H.P.		190.90	1,775	5,300	15,900	2,587
2800	Log chipper, up to 22" diameter, 600 H.P.		63.30	640	1,915	5,750	889.40
2850	Logger, for skidding & stacking logs, 150 H.P.		52.75	815	2,445	7,325	911
2860	Mulcher, diesel powered, trailer mounted		24.05	212	635	1,900	319.40
2900	Rake, spring tooth, with tractor		18.23	345	1,042	3,125	354.25
3000	Roller, vibratory, tandem, smooth drum, 20 H.P.		8.80	145	435	1,300	157.40
3050	35 H.P.		11.10	247	740	2,225	236.80
3100	Towed type vibratory compactor, smooth drum, 50 H.P.		26.00	315	940	2,825	396
3150	Sheepsfoot, 50 H.P.		27.35	345	1,035	3,100	425.80
3170	Landfill compactor, 220 H.P.		91.00	1,475	4,440	13,300	1,616
3200	Pneumatic tire roller, 80 H.P.		16.80	350	1,050	3,150	344.40
3250	120 H.P.		25.20	600	1,800	5,400	561.60
3300	Sheepsfoot vibratory roller, 240 H.P.		68.95	1,100	3,270	9,800	1,206
3320	340 H.P.		100.85	1,575	4,730	14,200	1,753
3350	Smooth drum vibratory roller, 75 H.P.		24.85	585	1,755	5,275	549.80
3400	125 H.P.		32.55	710	2,135	6,400	687.40
3410	Rotary mower, brush, 60", with tractor		22.60	315	950	2,850	370.80
3420	Rototiller, walk-behind, gas, 5 H.P.		1.96	51.50	155	465	46.70
3422	8 H.P.		3.08	80	240	720	72.65
3440	Scrapers, towed type, 7 C.Y. capacity		5.80	113	340	1,025	114.40
3450	10 C.Y. capacity		6.60	157	470	1,400	146.80
3500	15 C.Y. capacity		7.10	180	540	1,625	164.80
3525	Self-propelled, single engine, 14 C.Y. capacity		114.76	1,600	4,830	14,500	1,884
3550	Dual engine, 21 C.Y. capacity		173.80	2,175	6,490	19,500	2,688
3600	31 C.Y. capacity		231.70	3,075	9,195	27,600	3,693
3640	44 C.Y. capacity		287.80	3,950	11,885	35,700	4,679
3650	Elevating type, single engine, 11 C.Y. capacity		71.45	985	2,955	8,875	1,163
3700	22 C.Y. capacity		139.55	2,350	7,025	21,100	2,521
3710	Screening plant 110 H.P. w/5' x 10' screen		25.35	400	1,200	3,600	442.80
3720	5' x 16' screen		29.65	505	1,515	4,550	540.20
3850	Shovel, crawler-mounted, front-loading, 7 C.Y. capacity		258.85	3,200	9,610	28,800	3,993
3855	12 C.Y. capacity		386.45	4,400	13,220	39,700	5,736
3860	Shovel/backhoe bucket, 1/2 C.Y.		2.50	66.50	200	600	60
3870	3/4 C.Y.		2.55	75	225	675	65.40
3880	1 C.Y.		2.65	83.50	250	750	71.20
3890	1-1/2 C.Y.		2.75	96.50	290	870	80
3910	3 C.Y.		3.15	130	390	1,175	103.20
3950	Stump chipper, 18" deep, 30 H.P.		7.82	205	615	1,850	185.55
4110	Dozer, crawler, torque converter, diesel 80 H.P.		29.05	400	1,200	3,600	472.40
4150	105 H.P.		35.60	530	1,590	4,775	602.80
4200	140 H.P.		51.25	795	2,380	7,150	886
4260	200 H.P.		77.25	1,275	3,845	11,500	1,387
4310	300 H.P.		100.65	1,825	5,460	16,400	1,897
4360	410 H.P.		132.60	2,250	6,720	20,200	2,405
4370	500 H.P.		171.15	3,225	9,650	29,000	3,299
4380	700 H.P.		280.90	4,550	13,660	41,000	4,979
4400	Loader, crawler, torque conv., diesel, 1-1/2 C.Y., 80 H.P.		31.25	455	1,360	4,075	522
4450	1-1/2 to 1-3/4 C.Y., 95 H.P.		34.60	590	1,765	5,300	629.80
4510	1-3/4 to 2-1/4 C.Y., 130 H.P.		53.80	875	2,630	7,900	956.40
4530	2-1/2 to 3-1/4 C.Y., 190 H.P.		65.95	1,100	3,300	9,900	1,188
4560	3-1/2 to 5 C.Y., 275 H.P.		87.80	1,425	4,295	12,900	1,561
4610	Front end loader, 4WD, articulated frame, diesel, 1 to 1-1/4 C.Y., 70 H.P.		19.75	235	705	2,125	299
4620	1-1/2 to 1-3/4 C.Y., 95 H.P.		24.80	300	900	2,700	378.40

01 54 | Construction Aids

01 54 33 | Equipment Rental

			UNIT	HOURLY OPER. COST	RENT PER DAY	RENT PER WEEK	RENT PER MONTH	EQUIPMENT COST/DAY	
20	4650	1-3/4 to 2 C.Y., 130 H.P.	Ea.	29.15	380	1,140	3,425	461.20	20
	4710	2-1/2 to 3-1/2 C.Y., 145 H.P.		33.10	425	1,275	3,825	519.80	
	4730	3 to 4-1/2 C.Y., 185 H.P.		42.40	550	1,650	4,950	669.20	
	4760	5-1/4 to 5-3/4 C.Y., 270 H.P.		67.50	905	2,710	8,125	1,082	
	4810	7 to 9 C.Y., 475 H.P.		114.05	1,700	5,125	15,400	1,937	
	4870	9 - 11 C.Y., 620 H.P.		160.55	2,775	8,290	24,900	2,942	
	4880	Skid steer loader, wheeled, 10 C.F., 30 H.P. gas		10.45	150	450	1,350	173.60	
	4890	1 C.Y., 78 H.P., diesel		20.20	262	785	2,350	318.60	
	4892	Skid-steer attachment, auger		.48	80.50	242	725	52.25	
	4893	Backhoe		.66	111	332	995	71.70	
	4894	Broom		.74	124	372	1,125	80.30	
	4895	Forks		.22	36	108	325	23.35	
	4896	Grapple		.56	92.50	278	835	60.10	
	4897	Concrete hammer		1.06	177	531	1,600	114.70	
	4898	Tree spade		1.03	172	515	1,550	111.25	
	4899	Trencher		.54	90	270	810	58.30	
	4900	Trencher, chain, boom type, gas, operator walking, 12 H.P.		5.00	46.50	140	420	68	
	4910	Operator riding, 40 H.P.		20.15	290	870	2,600	335.20	
	5000	Wheel type, diesel, 4' deep, 12" wide		85.95	845	2,535	7,600	1,195	
	5100	6' deep, 20" wide		92.75	1,925	5,740	17,200	1,890	
	5150	Chain type, diesel, 5' deep, 8" wide		38.35	560	1,675	5,025	641.80	
	5200	Diesel, 8' deep, 16" wide		155.95	3,600	10,770	32,300	3,402	
	5202	Rock trencher, wheel type, 6" wide x 18" deep		21.50	335	1,000	3,000	372	
	5206	Chain type, 18" wide x 7' deep		112.00	2,800	8,395	25,200	2,575	
	5210	Tree spade, self-propelled		14.38	267	800	2,400	275.05	
	5250	Truck, dump, 2-axle, 12 ton, 8 C.Y. payload, 220 H.P.		34.40	227	680	2,050	411.20	
	5300	Three axle dump, 16 ton, 12 C.Y. payload, 400 H.P.		61.00	340	1,015	3,050	691	
	5310	Four axle dump, 25 ton, 18 C.Y. payload, 450 H.P.		71.75	490	1,475	4,425	869	
	5350	Dump trailer only, rear dump, 16-1/2 C.Y.		5.45	138	415	1,250	126.60	
	5400	20 C.Y.		5.90	157	470	1,400	141.20	
	5450	Flatbed, single axle, 1-1/2 ton rating		25.55	68.50	205	615	245.40	
	5500	3 ton rating		30.65	96.50	290	870	303.20	
	5550	Off highway rear dump, 25 ton capacity		73.90	1,275	3,800	11,400	1,351	
	5600	35 ton capacity		82.50	1,425	4,260	12,800	1,512	
	5610	50 ton capacity		102.65	1,700	5,080	15,200	1,837	
	5620	65 ton capacity		105.65	1,700	5,090	15,300	1,863	
	5630	100 ton capacity		151.40	2,850	8,570	25,700	2,925	
	6000	Vibratory plow, 25 H.P., walking		8.45	60	180	540	103.60	
40	0010	**GENERAL EQUIPMENT RENTAL** without operators R015433-10							40
	0150	Aerial lift, scissor type, to 15' high, 1000 lb. cap., electric	Ea.	3.05	51.50	155	465	55.40	
	0160	To 25' high, 2000 lb. capacity		3.45	66.50	200	600	67.60	
	0170	Telescoping boom to 40' high, 500 lb. capacity, diesel		13.55	320	965	2,900	301.40	
	0180	To 45' high, 500 lb. capacity		14.75	350	1,055	3,175	329	
	0190	To 60' high, 600 lb. capacity		17.20	495	1,490	4,475	435.60	
	0195	Air compressor, portable, 6.5 CFM, electric		.67	12.65	38	114	12.95	
	0196	Gasoline		.81	19	57	171	17.90	
	0200	Towed type, gas engine, 60 CFM		13.55	48.50	145	435	137.40	
	0300	160 CFM		15.80	50	150	450	156.40	
	0400	Diesel engine, rotary screw, 250 CFM		17.05	108	325	975	201.40	
	0500	365 CFM		23.10	132	395	1,175	263.80	
	0550	450 CFM		29.35	165	495	1,475	333.80	
	0600	600 CFM		51.70	228	685	2,050	550.60	
	0700	750 CFM		51.90	237	710	2,125	557.20	
	0800	For silenced models, small sizes, add to rent		3%	5%	5%	5%		
	0900	Large sizes, add to rent		5%	7%	7%	7%		
	0930	Air tools, breaker, pavement, 60 lb.	Ea.	.50	9.65	29	87	9.80	
	0940	80 lb.		.50	10.35	31	93	10.20	
	0950	Drills, hand (jackhammer) 65 lb.		.60	17.35	52	156	15.20	

01 54 | Construction Aids

01 54 33 | Equipment Rental

		UNIT	HOURLY OPER. COST	RENT PER DAY	RENT PER WEEK	RENT PER MONTH	EQUIPMENT COST/DAY
0960	Track or wagon, swing boom, 4" drifter	Ea.	62.50	880	2,640	7,925	1,028
0970	5" drifter		74.45	1,075	3,235	9,700	1,243
0975	Track mounted quarry drill, 6" diameter drill		128.10	1,600	4,765	14,300	1,978
0980	Dust control per drill		1.02	23.50	71	213	22.35
0990	Hammer, chipping, 12 lb.		.55	26	78	234	20
1000	Hose, air with couplings, 50' long, 3/4" diameter		.03	5	15	45	3.25
1100	1" diameter		.04	6.35	19	57	4.10
1200	1-1/2" diameter		.05	9	27	81	5.80
1300	2" diameter		.07	12	36	108	7.75
1400	2-1/2" diameter		.11	19	57	171	12.30
1410	3" diameter		.14	23	69	207	14.90
1450	Drill, steel, 7/8" x 2'		.05	8.65	26	78	5.60
1460	7/8" x 6'		.05	9	27	81	5.80
1520	Moil points		.02	3.33	10	30	2.15
1525	Pneumatic nailer w/accessories		.58	38.50	115	345	27.65
1530	Sheeting driver for 60 lb. breaker		.04	6	18	54	3.90
1540	For 90 lb. breaker		.12	8	24	72	5.75
1550	Spade, 25 lb.		.45	6.65	20	60	7.60
1560	Tamper, single, 35 lb.		.55	36.50	109	325	26.20
1570	Triple, 140 lb.		.82	54.50	164	490	39.35
1580	Wrenches, impact, air powered, up to 3/4" bolt		.40	12.65	38	114	10.80
1590	Up to 1-1/4" bolt		.50	23.50	70	210	18
1600	Barricades, barrels, reflectorized, 1 to 99 barrels		.03	4.60	13.80	41.50	3
1610	100 to 200 barrels		.02	3.53	10.60	32	2.30
1620	Barrels with flashers, 1 to 99 barrels		.03	5.25	15.80	47.50	3.40
1630	100 to 200 barrels		.03	4.20	12.60	38	2.75
1640	Barrels with steady burn type C lights		.04	7	21	63	4.50
1650	Illuminated board, trailer mounted, with generator		3.50	130	390	1,175	106
1670	Portable barricade, stock, with flashers, 1 to 6 units		.03	5.25	15.80	47.50	3.40
1680	25 to 50 units		.03	4.90	14.70	44	3.20
1685	Butt fusion machine, wheeled, 1.5 HP electric, 2" - 8" diameter pipe		2.63	167	500	1,500	121.05
1690	Tracked, 20 HP diesel, 4"-12" diameter pipe		11.14	490	1,465	4,400	382.10
1695	83 HP diesel, 8" - 24" diameter pipe		30.46	975	2,930	8,800	829.70
1700	Carts, brick, hand powered, 1000 lb. capacity		.50	83.50	251	755	54.20
1800	Gas engine, 1500 lb., 7-1/2' lift		4.17	115	345	1,025	102.35
1822	Dehumidifier, medium, 6 lb./hr., 150 CFM		1.00	62	186	560	45.20
1824	Large, 18 lb./hr., 600 CFM		2.02	126	378	1,125	91.75
1830	Distributor, asphalt, trailer mounted, 2000 gal., 38 H.P. diesel		9.95	325	980	2,950	275.60
1840	3000 gal., 38 H.P. diesel		11.45	355	1,070	3,200	305.60
1850	Drill, rotary hammer, electric		1.02	26	78	234	23.75
1860	Carbide bit, 1-1/2" diameter, add to electric rotary hammer		.02	3.61	10.84	32.50	2.35
1865	Rotary, crawler, 250 H.P.		148.50	2,050	6,185	18,600	2,425
1870	Emulsion sprayer, 65 gal., 5 H.P. gas engine		2.91	97	291	875	81.50
1880	200 gal., 5 H.P. engine		7.85	162	485	1,450	159.80
1900	Floor auto-scrubbing machine, walk-behind, 28" path		4.98	325	970	2,900	233.85
1930	Floodlight, mercury vapor, or quartz, on tripod, 1000 watt		.44	20.50	62	186	15.90
1940	2000 watt		.83	41	123	370	31.25
1950	Floodlights, trailer mounted with generator, 1 - 300 watt light		3.65	71.50	215	645	72.20
1960	2 - 1000 watt lights		4.85	96.50	290	870	96.80
2000	4 - 300 watt lights		4.55	93.50	280	840	92.40
2005	Foam spray rig, incl. box trailer, compressor, generator, proportioner		35.03	490	1,465	4,400	573.25
2020	Forklift, straight mast, 12' lift, 5000 lb., 2 wheel drive, gas		26.35	202	605	1,825	331.80
2040	21' lift, 5000 lb., 4 wheel drive, diesel		20.90	240	720	2,150	311.20
2050	For rough terrain, 42' lift, 35' reach, 9000 lb., 110 H.P.		29.55	485	1,450	4,350	526.40
2060	For plant, 4 ton capacity, 80 H.P., 2 wheel drive, gas		16.15	93.50	280	840	185.20
2080	10 ton capacity, 120 H.P., 2 wheel drive, diesel		24.10	162	485	1,450	289.80
2100	Generator, electric, gas engine, 1.5 kW to 3 kW		3.75	11.35	34	102	36.80
2200	5 kW		4.80	15.35	46	138	47.60

01 54 | Construction Aids

01 54 33 | Equipment Rental

		UNIT	HOURLY OPER. COST	RENT PER DAY	RENT PER WEEK	RENT PER MONTH	EQUIPMENT COST/DAY	
40	2300	10 kW	Ea.	9.10	36.50	110	330	94.80	40
	2400	25 kW		10.60	85	255	765	135.80	
	2500	Diesel engine, 20 kW		12.10	66.50	200	600	136.80	
	2600	50 kW		23.35	103	310	930	248.80	
	2700	100 kW		43.30	128	385	1,150	423.40	
	2800	250 kW		85.15	235	705	2,125	822.20	
	2850	Hammer, hydraulic, for mounting on boom, to 500 ft lb.		2.55	75	225	675	65.40	
	2860	1000 ft lb.		4.35	127	380	1,150	110.80	
	2900	Heaters, space, oil or electric, 50 MBH		2.01	7.65	23	69	20.70	
	3000	100 MBH		3.76	10.65	32	96	36.50	
	3100	300 MBH		11.03	38.50	115	345	111.25	
	3150	500 MBH		18.15	45	135	405	172.20	
	3200	Hose, water, suction with coupling, 20' long, 2" diameter		.02	3	9	27	1.95	
	3210	3" diameter		.03	4.33	13	39	2.85	
	3220	4" diameter		.03	5	15	45	3.25	
	3230	6" diameter		.11	17.65	53	159	11.50	
	3240	8" diameter		.27	44.50	133	400	28.75	
	3250	Discharge hose with coupling, 50' long, 2" diameter		.01	1.33	4	12	.90	
	3260	3" diameter		.01	2.33	7	21	1.50	
	3270	4" diameter		.02	3.67	11	33	2.35	
	3280	6" diameter		.06	9.35	28	84	6.10	
	3290	8" diameter		.18	30	90	270	19.45	
	3295	Insulation blower		.78	6	18	54	9.85	
	3300	Ladders, extension type, 16' to 36' long		.14	22.50	68	204	14.70	
	3400	40' to 60' long		.19	31	93	279	20.10	
	3405	Lance for cutting concrete		2.35	71.50	215	645	61.80	
	3407	Lawn mower, rotary, 22", 5 H.P.		1.80	49	147	440	43.80	
	3408	48" self propelled		2.98	75	225	675	68.85	
	3410	Level, electronic, automatic, with tripod and leveling rod		1.20	75.50	226	680	54.80	
	3430	Laser type, for pipe and sewer line and grade		.73	48.50	145	435	34.85	
	3440	Rotating beam for interior control		.78	52	156	470	37.45	
	3460	Builder's optical transit, with tripod and rod		.09	14.65	44	132	9.50	
	3500	Light towers, towable, with diesel generator, 2000 watt		4.55	93.50	280	840	92.40	
	3600	4000 watt		4.85	96.50	290	870	96.80	
	3700	Mixer, powered, plaster and mortar, 6 C.F., 7 H.P.		2.70	20	60	180	33.60	
	3800	10 C.F., 9 H.P.		2.80	31.50	94	282	41.20	
	3850	Nailer, pneumatic		.58	38.50	115	345	27.65	
	3900	Paint sprayers complete, 8 CFM		1.04	69.50	208	625	49.90	
	4000	17 CFM		1.90	127	380	1,150	91.20	
	4020	Pavers, bituminous, rubber tires, 8' wide, 50 H.P., diesel		31.15	490	1,465	4,400	542.20	
	4030	10' wide, 150 H.P.		106.10	1,800	5,370	16,100	1,923	
	4050	Crawler, 8' wide, 100 H.P., diesel		88.65	1,800	5,385	16,200	1,786	
	4060	10' wide, 150 H.P.		113.15	2,200	6,610	19,800	2,227	
	4070	Concrete paver, 12' to 24' wide, 250 H.P.		101.40	1,550	4,675	14,000	1,746	
	4080	Placer-spreader-trimmer, 24' wide, 300 H.P.		150.25	2,650	7,960	23,900	2,794	
	4100	Pump, centrifugal gas pump, 1-1/2" diam., 65 GPM		3.90	48.50	145	435	60.20	
	4200	2" diameter, 130 GPM		5.35	53.50	160	480	74.80	
	4300	3" diameter, 250 GPM		5.65	55	165	495	78.20	
	4400	6" diameter, 1500 GPM		30.45	172	515	1,550	346.60	
	4500	Submersible electric pump, 1-1/4" diameter, 55 GPM		.39	16.35	49	147	12.90	
	4600	1-1/2" diameter, 83 GPM		.43	18.65	56	168	14.65	
	4700	2" diameter, 120 GPM		1.50	23.50	70	210	26	
	4800	3" diameter, 300 GPM		2.69	41.50	125	375	46.50	
	4900	4" diameter, 560 GPM		12.34	158	475	1,425	193.70	
	5000	6" diameter, 1590 GPM		18.40	212	635	1,900	274.20	
	5100	Diaphragm pump, gas, single, 1-1/2" diameter		1.18	49.50	148	445	39.05	
	5200	2" diameter		4.25	61.50	185	555	71	
	5300	3" diameter		4.25	61.50	185	555	71	

For customer support on your Commercial Renovation Cost Data, call 877.791.4977.

01 54 | Construction Aids

01 54 33 | Equipment Rental

		UNIT	HOURLY OPER. COST	RENT PER DAY	RENT PER WEEK	RENT PER MONTH	EQUIPMENT COST/DAY
5400	Double, 4" diameter	Ea.	6.40	105	315	945	114.20
5450	Pressure washer 5 GPM, 3000 psi		4.90	51.50	155	465	70.20
5460	7 GPM, 3000 psi		6.45	60	180	540	87.60
5500	Trash pump, self-priming, gas, 2" diameter		4.60	21	63	189	49.40
5600	Diesel, 4" diameter		9.15	88.50	265	795	126.20
5650	Diesel, 6" diameter		25.25	147	440	1,325	290
5655	Grout Pump		26.35	268	805	2,425	371.80
5700	Salamanders, L.P. gas fired, 100,000 Btu		3.96	13.65	41	123	39.90
5705	50,000 Btu		2.21	10.35	31	93	23.90
5720	Sandblaster, portable, open top, 3 C.F. capacity		.55	26.50	80	240	20.40
5730	6 C.F. capacity		.95	40	120	360	31.60
5740	Accessories for above		.13	21.50	65	195	14.05
5750	Sander, floor		.70	14.35	43	129	14.20
5760	Edger		.50	14.35	43	129	12.60
5800	Saw, chain, gas engine, 18" long		2.30	21.50	64	192	31.20
5900	Hydraulic powered, 36" long		.75	65	195	585	45
5950	60" long		.75	66.50	200	600	46
6000	Masonry, table mounted, 14" diameter, 5 H.P.		1.32	56.50	170	510	44.55
6050	Portable cut-off, 8 H.P.		2.50	33.50	100	300	40
6100	Circular, hand held, electric, 7-1/4" diameter		.23	4.67	14	42	4.65
6200	12" diameter		.23	8	24	72	6.65
6250	Wall saw, w/hydraulic power, 10 H.P.		9.70	61.50	185	555	114.60
6275	Shot blaster, walk-behind, 20" wide		4.75	293	880	2,650	214
6280	Sidewalk broom, walk-behind		2.52	78.50	235	705	67.15
6300	Steam cleaner, 100 gallons per hour		3.70	76.50	230	690	75.60
6310	200 gallons per hour		5.35	95	285	855	99.80
6340	Tar Kettle/Pot, 400 gallons		15.60	75	225	675	169.80
6350	Torch, cutting, acetylene-oxygen, 150' hose, excludes gases		.30	15	45	135	11.40
6360	Hourly operating cost includes tips and gas		19.00				152
6410	Toilet, portable chemical		.13	21	63	189	13.65
6420	Recycle flush type		.15	25	75	225	16.20
6430	Toilet, fresh water flush, garden hose,		.18	30.50	91	273	19.65
6440	Hoisted, non-flush, for high rise		.15	24.50	74	222	16
6465	Tractor, farm with attachment		21.50	297	890	2,675	350
6480	Trailers, platform, flush deck, 2 axle, 3 ton capacity		1.45	20	60	180	23.60
6500	25 ton capacity		5.45	117	350	1,050	113.60
6600	40 ton capacity		7.00	163	490	1,475	154
6700	3 axle, 50 ton capacity		7.55	180	540	1,625	168.40
6800	75 ton capacity		9.40	235	705	2,125	216.20
6810	Trailer mounted cable reel for high voltage line work		5.45	260	779	2,325	199.40
6820	Trailer mounted cable tensioning rig		10.85	515	1,550	4,650	396.80
6830	Cable pulling rig		73.05	2,900	8,680	26,000	2,320
6900	Water tank trailer, engine driven discharge, 5000 gallons		7.00	142	425	1,275	141
6925	10,000 gallons		9.50	197	590	1,775	194
6950	Water truck, off highway, 6000 gallons		88.15	770	2,315	6,950	1,168
7010	Tram car for high voltage line work, powered, 2 conductor		6.60	141	423	1,275	137.40
7020	Transit (builder's level) with tripod		.09	14.65	44	132	9.50
7030	Trench box, 3000 lb., 6' x 8'		.56	93.50	280	840	60.50
7040	7200 lb., 6' x 20'		.75	125	375	1,125	81
7050	8000 lb., 8' x 16'		1.08	180	540	1,625	116.65
7060	9500 lb., 8' x 20'		1.21	201	603	1,800	130.30
7065	11,000 lb., 8' x 24'		1.27	211	633	1,900	136.75
7070	12,000 lb., 10' x 20'		1.50	249	748	2,250	161.60
7100	Truck, pickup, 3/4 ton, 2 wheel drive		13.65	58.50	175	525	144.20
7200	4 wheel drive		13.95	73.50	220	660	155.60
7250	Crew carrier, 9 passenger		19.30	86.50	260	780	206.40
7290	Flat bed truck, 20,000 lb. GVW		21.70	125	375	1,125	248.60
7300	Tractor, 4 x 2, 220 H.P.		30.20	197	590	1,775	359.60

01 54 | Construction Aids

01 54 33 | Equipment Rental

			UNIT	HOURLY OPER. COST	RENT PER DAY	RENT PER WEEK	RENT PER MONTH	EQUIPMENT COST/DAY	
40	7410	330 H.P.	Ea.	44.75	270	810	2,425	520	40
	7500	6 x 4, 380 H.P.		51.35	315	945	2,825	599.80	
	7600	450 H.P.		62.30	380	1,145	3,425	727.40	
	7610	Tractor, with A frame, boom and winch, 225 H.P.		33.00	272	815	2,450	427	
	7620	Vacuum truck, hazardous material, 2500 gallons		12.05	290	870	2,600	270.40	
	7625	5,000 gallons		14.26	405	1,220	3,650	358.10	
	7650	Vacuum, HEPA, 16 gallon, wet/dry		.85	18	54	162	17.60	
	7655	55 gallon, wet/dry		.80	27	81	243	22.60	
	7660	Water tank, portable		.16	26.50	80	240	17.30	
	7690	Sewer/catch basin vacuum, 14 C.Y., 1500 gallons		18.53	610	1,830	5,500	514.25	
	7700	Welder, electric, 200 amp		3.88	16.35	49	147	40.85	
	7800	300 amp		5.71	20	60	180	57.70	
	7900	Gas engine, 200 amp		14.65	23.50	70	210	131.20	
	8000	300 amp		16.38	24.50	74	222	145.85	
	8100	Wheelbarrow, any size		.08	13	39	117	8.45	
	8200	Wrecking ball, 4000 lb.		2.35	68.50	205	615	59.80	
50	0010	**HIGHWAY EQUIPMENT RENTAL** without operators	R015433 -10						50
	0050	Asphalt batch plant, portable drum mixer, 100 ton/hr.	Ea.	78.06	1,425	4,270	12,800	1,478	
	0060	200 ton/hr.		89.18	1,500	4,530	13,600	1,619	
	0070	300 ton/hr.		106.34	1,775	5,325	16,000	1,916	
	0100	Backhoe attachment, long stick, up to 185 H.P., 10.5' long		.35	23.50	70	210	16.80	
	0140	Up to 250 H.P., 12' long		.38	25	75	225	18.05	
	0180	Over 250 H.P., 15' long		.53	35	105	315	25.25	
	0200	Special dipper arm, up to 100 H.P., 32' long		1.08	71.50	215	645	51.65	
	0240	Over 100 H.P., 33' long		1.34	89.50	268	805	64.30	
	0280	Catch basin/sewer cleaning truck, 3 ton, 9 C.Y., 1000 gal.		42.70	385	1,160	3,475	573.60	
	0300	Concrete batch plant, portable, electric, 200 C.Y./hr.		22.87	515	1,550	4,650	492.95	
	0520	Grader/dozer attachment, ripper/scarifier, rear mounted, up to 135 H.P.		2.90	56.50	170	510	57.20	
	0540	Up to 180 H.P.		3.90	88.50	265	795	84.20	
	0580	Up to 250 H.P.		5.00	117	350	1,050	110	
	0700	Pvmt. removal bucket, for hyd. excavator, up to 90 H.P.		1.90	53.50	160	480	47.20	
	0740	Up to 200 H.P.		2.15	73.50	220	660	61.20	
	0780	Over 200 H.P.		2.25	85	255	765	69	
	0900	Aggregate spreader, self-propelled, 187 H.P.		53.00	685	2,050	6,150	834	
	1000	Chemical spreader, 3 C.Y.		3.30	43.50	130	390	52.40	
	1900	Hammermill, traveling, 250 H.P.		78.83	2,075	6,240	18,700	1,879	
	2000	Horizontal borer, 3" diameter, 13 H.P. gas driven		6.35	55	165	495	83.80	
	2150	Horizontal directional drill, 20,000 lb. thrust, 78 H.P. diesel		31.15	675	2,025	6,075	654.20	
	2160	30,000 lb. thrust, 115 H.P.		38.90	1,025	3,100	9,300	931.20	
	2170	50,000 lb. thrust, 170 H.P.		55.65	1,325	3,960	11,900	1,237	
	2190	Mud trailer for HDD, 1500 gallons, 175 H.P., gas		31.90	152	455	1,375	346.20	
	2200	Hydromulcher, diesel, 3000 gallon, for truck mounting		23.40	245	735	2,200	334.20	
	2300	Gas, 600 gallon		8.60	98.50	295	885	127.80	
	2400	Joint & crack cleaner, walk behind, 25 H.P.		3.95	50	150	450	61.60	
	2500	Filler, trailer mounted, 400 gallons, 20 H.P.		9.50	210	630	1,900	202	
	3000	Paint striper, self-propelled, 40 gallon, 22 H.P.		7.30	155	465	1,400	151.40	
	3100	120 gallon, 120 H.P.		23.00	395	1,185	3,550	421	
	3200	Post drivers, 6" I-Beam frame, for truck mounting		19.05	380	1,135	3,400	379.40	
	3400	Road sweeper, self-propelled, 8' wide, 90 H.P.		39.80	575	1,720	5,150	662.40	
	3450	Road sweeper, vacuum assisted, 4 C.Y., 220 gallons		75.75	635	1,900	5,700	986	
	4000	Road mixer, self-propelled, 130 H.P.		46.85	765	2,295	6,875	833.80	
	4100	310 H.P.		83.25	2,100	6,315	18,900	1,929	
	4220	Cold mix paver, incl. pug mill and bitumen tank, 165 H.P.		97.00	2,275	6,855	20,600	2,147	
	4240	Pavement brush, towed		3.20	93.50	280	840	81.60	
	4250	Paver, asphalt, wheel or crawler, 130 H.P., diesel		96.65	2,275	6,795	20,400	2,132	
	4300	Paver, road widener, gas 1' to 6', 67 H.P.		48.50	895	2,680	8,050	924	
	4400	Diesel, 2' to 14', 88 H.P.		62.70	1,050	3,135	9,400	1,129	
	4600	Slipform pavers, curb and gutter, 2 track, 75 H.P.		57.20	900	2,700	8,100	997.60	

01 54 | Construction Aids

01 54 33 | Equipment Rental

			UNIT	HOURLY OPER. COST	RENT PER DAY	RENT PER WEEK	RENT PER MONTH	EQUIPMENT COST/DAY	
50	4700	4 track, 165 H.P.	Ea.	41.50	740	2,220	6,650	776	50
	4800	Median barrier, 215 H.P.		60.50	1,050	3,150	9,450	1,114	
	4901	Trailer, low bed, 75 ton capacity		10.10	235	705	2,125	221.80	
	5000	Road planer, walk behind, 10" cutting width, 10 H.P.		3.70	33.50	100	300	49.60	
	5100	Self-propelled, 12" cutting width, 64 H.P.		10.40	113	340	1,025	151.20	
	5120	Traffic line remover, metal ball blaster, truck mounted, 115 H.P.		47.95	730	2,190	6,575	821.60	
	5140	Grinder, truck mounted, 115 H.P.		54.60	790	2,375	7,125	911.80	
	5160	Walk-behind, 11 H.P.		4.20	53.50	160	480	65.60	
	5200	Pavement profiler, 4' to 6' wide, 450 H.P.		254.50	3,325	9,960	29,900	4,028	
	5300	8' to 10' wide, 750 H.P.		399.20	4,350	13,055	39,200	5,805	
	5400	Roadway plate, steel, 1" x 8' x 20'		.08	13.35	40	120	8.65	
	5600	Stabilizer, self-propelled, 150 H.P.		49.30	600	1,795	5,375	753.40	
	5700	310 H.P.		99.05	1,675	5,030	15,100	1,798	
	5800	Striper, truck mounted, 120 gallon paint, 460 H.P.		64.20	485	1,455	4,375	804.60	
	5900	Thermal paint heating kettle, 115 gallons		8.20	25.50	77	231	81	
	6000	Tar kettle, 330 gallon, trailer mounted		12.25	56.50	170	510	132	
	7000	Tunnel locomotive, diesel, 8 to 12 ton		31.85	585	1,750	5,250	604.80	
	7005	Electric, 10 ton		26.00	665	1,995	5,975	607	
	7010	Muck cars, 1/2 C.Y. capacity		2.00	24.50	73	219	30.60	
	7020	1 C.Y. capacity		2.25	32.50	97	291	37.40	
	7030	2 C.Y. capacity		2.35	36.50	110	330	40.80	
	7040	Side dump, 2 C.Y. capacity		2.60	45	135	405	47.80	
	7050	3 C.Y. capacity		3.45	50	150	450	57.60	
	7060	5 C.Y. capacity		4.95	63.50	190	570	77.60	
	7100	Ventilating blower for tunnel, 7-1/2 H.P.		2.05	51.50	155	465	47.40	
	7110	10 H.P.		2.28	51.50	155	465	49.25	
	7120	20 H.P.		3.48	67.50	202	605	68.25	
	7140	40 H.P.		5.76	96.50	290	870	104.10	
	7160	60 H.P.		8.77	152	455	1,375	161.15	
	7175	75 H.P.		11.66	207	620	1,850	217.30	
	7180	200 H.P.		23.65	305	910	2,725	371.20	
	7800	Windrow loader, elevating		55.00	1,300	3,935	11,800	1,227	
60	0010	**LIFTING AND HOISTING EQUIPMENT RENTAL** without operators R015433-10							60
	0120	Aerial lift truck, 2 person, to 80'	Ea.	26.30	705	2,120	6,350	634.40	
	0140	Boom work platform, 40' snorkel		12.00	275	825	2,475	261	
	0150	Crane, flatbed mounted, 3 ton capacity R015433-15		16.35	188	565	1,700	243.80	
	0200	Crane, climbing, 106' jib, 6000 lb. capacity, 410 fpm R312316-45		39.35	1,650	4,960	14,900	1,307	
	0300	101' jib, 10,250 lb. capacity, 270 fpm		45.95	2,100	6,280	18,800	1,624	
	0500	Tower, static, 130' high, 106' jib, 6200 lb. capacity at 400 fpm		43.20	1,900	5,730	17,200	1,492	
	0600	Crawler mounted, lattice boom, 1/2 C.Y., 15 tons at 12' radius		36.98	625	1,870	5,600	669.85	
	0700	3/4 C.Y., 20 tons at 12' radius		49.31	780	2,340	7,025	862.50	
	0800	1 C.Y., 25 tons at 12' radius		65.75	1,050	3,120	9,350	1,150	
	0900	1-1/2 C.Y., 40 tons at 12' radius		65.75	1,050	3,150	9,450	1,156	
	1000	2 C.Y., 50 tons at 12' radius		69.70	1,225	3,680	11,000	1,294	
	1100	3 C.Y., 75 tons at 12' radius		74.70	1,425	4,305	12,900	1,459	
	1200	100 ton capacity, 60' boom		84.60	1,650	4,950	14,900	1,667	
	1300	165 ton capacity, 60' boom		107.75	1,925	5,785	17,400	2,019	
	1400	200 ton capacity, 70' boom		130.55	2,400	7,210	21,600	2,486	
	1500	350 ton capacity, 80' boom		182.90	3,625	10,845	32,500	3,632	
	1600	Truck mounted, lattice boom, 6 x 4, 20 tons at 10' radius		37.11	1,075	3,260	9,775	948.90	
	1700	25 tons at 10' radius		40.09	1,175	3,550	10,700	1,031	
	1800	8 x 4, 30 tons at 10' radius		43.61	1,250	3,780	11,300	1,105	
	1900	40 tons at 12' radius		46.70	1,325	3,950	11,900	1,164	
	2000	60 tons at 15' radius		53.07	1,400	4,180	12,500	1,261	
	2050	82 tons at 15' radius		59.80	1,475	4,460	13,400	1,370	
	2100	90 tons at 15' radius		67.36	1,625	4,860	14,600	1,511	
	2200	115 tons at 15' radius		76.11	1,800	5,430	16,300	1,695	
	2300	150 tons at 18' radius		83.65	1,900	5,720	17,200	1,813	

01 54 | Construction Aids

01 54 33 | Equipment Rental

			UNIT	HOURLY OPER. COST	RENT PER DAY	RENT PER WEEK	RENT PER MONTH	EQUIPMENT COST/DAY	
60	2350	165 tons at 18' radius	Ea.	90.05	2,025	6,060	18,200	1,932	60
	2400	Truck mounted, hydraulic, 12 ton capacity		42.60	520	1,565	4,700	653.80	
	2500	25 ton capacity		44.95	630	1,885	5,650	736.60	
	2550	33 ton capacity		45.50	645	1,930	5,800	750	
	2560	40 ton capacity		58.85	750	2,250	6,750	920.80	
	2600	55 ton capacity		76.80	855	2,570	7,700	1,128	
	2700	80 ton capacity		100.45	1,375	4,105	12,300	1,625	
	2720	100 ton capacity		94.20	1,425	4,250	12,800	1,604	
	2740	120 ton capacity		100.05	1,525	4,575	13,700	1,715	
	2760	150 ton capacity		127.60	2,000	5,995	18,000	2,220	
	2800	Self-propelled, 4 x 4, with telescoping boom, 5 ton		17.55	225	675	2,025	275.40	
	2900	12-1/2 ton capacity		32.30	360	1,075	3,225	473.40	
	3000	15 ton capacity		33.00	375	1,130	3,400	490	
	3050	20 ton capacity		35.95	445	1,335	4,000	554.60	
	3100	25 ton capacity		37.55	500	1,495	4,475	599.40	
	3150	40 ton capacity		46.00	560	1,675	5,025	703	
	3200	Derricks, guy, 20 ton capacity, 60' boom, 75' mast		27.55	405	1,214	3,650	463.20	
	3300	100' boom, 115' mast		43.21	695	2,090	6,275	763.70	
	3400	Stiffleg, 20 ton capacity, 70' boom, 37' mast		30.11	525	1,580	4,750	556.90	
	3500	100' boom, 47' mast		46.29	845	2,530	7,600	876.30	
	3550	Helicopter, small, lift to 1250 lb. maximum, w/pilot		103.70	3,275	9,800	29,400	2,790	
	3600	Hoists, chain type, overhead, manual, 3/4 ton		.10	.33	1	3	1	
	3900	10 ton		.70	6	18	54	9.20	
	4000	Hoist and tower, 5000 lb. cap., portable electric, 40' high		4.95	233	699	2,100	179.40	
	4100	For each added 10' section, add		.11	18.35	55	165	11.90	
	4200	Hoist and single tubular tower, 5000 lb. electric, 100' high		6.70	325	976	2,925	248.80	
	4300	For each added 6'-6" section, add		.19	31.50	95	285	20.50	
	4400	Hoist and double tubular tower, 5000 lb., 100' high		7.19	360	1,075	3,225	272.50	
	4500	For each added 6'-6" section, add		.21	35	105	315	22.70	
	4550	Hoist and tower, mast type, 6000 lb., 100' high		7.76	370	1,115	3,350	285.10	
	4570	For each added 10' section, add		.13	21.50	65	195	14.05	
	4600	Hoist and tower, personnel, electric, 2000 lb., 100' @ 125 fpm		16.31	990	2,970	8,900	724.50	
	4700	3000 lb., 100' @ 200 fpm		18.62	1,125	3,360	10,100	820.95	
	4800	3000 lb., 150' @ 300 fpm		20.62	1,250	3,760	11,300	916.95	
	4900	4000 lb., 100' @ 300 fpm		21.38	1,275	3,840	11,500	939.05	
	5000	6000 lb., 100' @ 275 fpm		23.06	1,350	4,030	12,100	990.50	
	5100	For added heights up to 500', add	L.F.	.01	1.67	5	15	1.10	
	5200	Jacks, hydraulic, 20 ton	Ea.	.05	2	6	18	1.60	
	5500	100 ton		.40	11.65	35	105	10.20	
	6100	Jacks, hydraulic, climbing w/50' jackrods, control console, 30 ton cap.		2.01	134	402	1,200	96.50	
	6150	For each added 10' jackrod section, add		.05	3.33	10	30	2.40	
	6300	50 ton capacity		3.23	215	646	1,950	155.05	
	6350	For each added 10' jackrod section, add		.06	4	12	36	2.90	
	6500	125 ton capacity		8.45	565	1,690	5,075	405.60	
	6550	For each added 10' jackrod section, add		.58	38.50	115	345	27.65	
	6600	Cable jack, 10 ton capacity with 200' cable		1.68	112	336	1,000	80.65	
	6650	For each added 50' of cable, add		.20	13.35	40	120	9.60	
70	0010	**WELLPOINT EQUIPMENT RENTAL** without operators	R015433 -10						70
	0020	Based on 2 months rental							
	0100	Combination jetting & wellpoint pump, 60 H.P. diesel	Ea.	18.40	330	996	3,000	346.40	
	0200	High pressure gas jet pump, 200 H.P., 300 psi	"	44.06	284	851	2,550	522.70	
	0300	Discharge pipe, 8" diameter	L.F.	.01	.54	1.62	4.86	.40	
	0350	12" diameter		.01	.79	2.38	7.15	.55	
	0400	Header pipe, flows up to 150 GPM, 4" diameter		.01	.49	1.47	4.41	.35	
	0500	400 GPM, 6" diameter		.01	.57	1.72	5.15	.40	
	0600	800 GPM, 8" diameter		.01	.79	2.38	7.15	.55	
	0700	1500 GPM, 10" diameter		.01	.84	2.51	7.55	.60	

01 54 | Construction Aids

01 54 33 | Equipment Rental

			UNIT	HOURLY OPER. COST	RENT PER DAY	RENT PER WEEK	RENT PER MONTH	EQUIPMENT COST/DAY	
70	0800	2500 GPM, 12" diameter	L.F.	.02	1.58	4.74	14.20	1.10	70
	0900	4500 GPM, 16" diameter		.03	2.02	6.07	18.20	1.45	
	0950	For quick coupling aluminum and plastic pipe, add	▼	.03	2.09	6.28	18.85	1.50	
	1100	Wellpoint, 25' long, with fittings & riser pipe, 1-1/2" or 2" diameter	Ea.	.06	4.18	12.54	37.50	3	
	1200	Wellpoint pump, diesel powered, 4" suction, 20 H.P.		7.83	191	574	1,725	177.45	
	1300	6" suction, 30 H.P.		10.70	237	712	2,125	228	
	1400	8" suction, 40 H.P.		14.45	325	976	2,925	310.80	
	1500	10" suction, 75 H.P.		22.27	380	1,141	3,425	406.35	
	1600	12" suction, 100 H.P.		31.72	605	1,810	5,425	615.75	
	1700	12" suction, 175 H.P.	▼	47.41	670	2,010	6,025	781.30	
80	0010	**MARINE EQUIPMENT RENTAL** without operators							80
	0200	Barge, 400 Ton, 30' wide x 90' long	Ea.	17.50	1,050	3,180	9,550	776	
	0240	800 Ton, 45' wide x 90' long		21.25	1,300	3,870	11,600	944	
	2000	Tugboat, diesel, 100 H.P.		38.20	217	650	1,950	435.60	
	2040	250 H.P.		79.25	395	1,180	3,550	870	
	2080	380 H.P.		156.80	1,175	3,535	10,600	1,961	
	3000	Small work boat, gas, 16-foot, 50 H.P.		17.15	60	180	540	173.20	
	4000	Large, diesel, 48-foot, 200 H.P.	▼	89.40	1,250	3,740	11,200	1,463	

Reference: R015433-10

Crews

Crew No.	Bare Costs Hr.	Bare Costs Daily	Incl. Subs O&P Hr.	Incl. Subs O&P Daily	Cost Per Labor-Hour Bare Costs	Cost Per Labor-Hour Incl. O&P
Crew A-1	Hr.	Daily	Hr.	Daily	Bare Costs	Incl. O&P
1 Building Laborer	$37.60	$300.80	$61.65	$493.20	$37.60	$61.65
1 Concrete Saw, Gas Manual		81.00		89.10	10.13	11.14
8 L.H., Daily Totals		$381.80		$582.30	$47.73	$72.79
Crew A-1A	Hr.	Daily	Hr.	Daily	Bare Costs	Incl. O&P
1 Skilled Worker	$48.65	$389.20	$79.05	$632.40	$48.65	$79.05
1 Shot Blaster, 20"		214.00		235.40	26.75	29.43
8 L.H., Daily Totals		$603.20		$867.80	$75.40	$108.47
Crew A-1B	Hr.	Daily	Hr.	Daily	Bare Costs	Incl. O&P
1 Building Laborer	$37.60	$300.80	$61.65	$493.20	$37.60	$61.65
1 Concrete Saw		167.40		184.14	20.93	23.02
8 L.H., Daily Totals		$468.20		$677.34	$58.52	$84.67
Crew A-1C	Hr.	Daily	Hr.	Daily	Bare Costs	Incl. O&P
1 Building Laborer	$37.60	$300.80	$61.65	$493.20	$37.60	$61.65
1 Chain Saw, Gas, 18"		31.20		34.32	3.90	4.29
8 L.H., Daily Totals		$332.00		$527.52	$41.50	$65.94
Crew A-1D	Hr.	Daily	Hr.	Daily	Bare Costs	Incl. O&P
1 Building Laborer	$37.60	$300.80	$61.65	$493.20	$37.60	$61.65
1 Vibrating Plate, Gas, 18"		36.20		39.82	4.53	4.98
8 L.H., Daily Totals		$337.00		$533.02	$42.13	$66.63
Crew A-1E	Hr.	Daily	Hr.	Daily	Bare Costs	Incl. O&P
1 Building Laborer	$37.60	$300.80	$61.65	$493.20	$37.60	$61.65
1 Vibrating Plate, Gas, 21"		46.80		51.48	5.85	6.43
8 L.H., Daily Totals		$347.60		$544.68	$43.45	$68.08
Crew A-1F	Hr.	Daily	Hr.	Daily	Bare Costs	Incl. O&P
1 Building Laborer	$37.60	$300.80	$61.65	$493.20	$37.60	$61.65
1 Rammer/Tamper, Gas, 8"		50.00		55.00	6.25	6.88
8 L.H., Daily Totals		$350.80		$548.20	$43.85	$68.53
Crew A-1G	Hr.	Daily	Hr.	Daily	Bare Costs	Incl. O&P
1 Building Laborer	$37.60	$300.80	$61.65	$493.20	$37.60	$61.65
1 Rammer/Tamper, Gas, 15"		56.40		62.04	7.05	7.75
8 L.H., Daily Totals		$357.20		$555.24	$44.65	$69.41
Crew A-1H	Hr.	Daily	Hr.	Daily	Bare Costs	Incl. O&P
1 Building Laborer	$37.60	$300.80	$61.65	$493.20	$37.60	$61.65
1 Exterior Steam Cleaner		75.60		83.16	9.45	10.40
8 L.H., Daily Totals		$376.40		$576.36	$47.05	$72.05
Crew A-1J	Hr.	Daily	Hr.	Daily	Bare Costs	Incl. O&P
1 Building Laborer	$37.60	$300.80	$61.65	$493.20	$37.60	$61.65
1 Cultivator, Walk-Behind, 5 H.P.		46.70		51.37	5.84	6.42
8 L.H., Daily Totals		$347.50		$544.57	$43.44	$68.07
Crew A-1K	Hr.	Daily	Hr.	Daily	Bare Costs	Incl. O&P
1 Building Laborer	$37.60	$300.80	$61.65	$493.20	$37.60	$61.65
1 Cultivator, Walk-Behind, 8 H.P.		72.65		79.92	9.08	9.99
8 L.H., Daily Totals		$373.45		$573.12	$46.68	$71.64
Crew A-1M	Hr.	Daily	Hr.	Daily	Bare Costs	Incl. O&P
1 Building Laborer	$37.60	$300.80	$61.65	$493.20	$37.60	$61.65
1 Snow Blower, Walk-Behind		67.15		73.86	8.39	9.23
8 L.H., Daily Totals		$367.95		$567.07	$45.99	$70.88
Crew A-2	Hr.	Daily	Hr.	Daily	Bare Costs	Incl. O&P
2 Laborers	$37.60	$601.60	$61.65	$986.40	$38.10	$62.12
1 Truck Driver (light)	39.10	312.80	63.05	504.40		
1 Flatbed Truck, Gas, 1.5 Ton		245.40		269.94	10.23	11.25
24 L.H., Daily Totals		$1159.80		$1760.74	$48.33	$73.36
Crew A-2A	Hr.	Daily	Hr.	Daily	Bare Costs	Incl. O&P
2 Laborers	$37.60	$601.60	$61.65	$986.40	$38.10	$62.12
1 Truck Driver (light)	39.10	312.80	63.05	504.40		
1 Flatbed Truck, Gas, 1.5 Ton		245.40		269.94		
1 Concrete Saw		167.40		184.14	17.20	18.92
24 L.H., Daily Totals		$1327.20		$1944.88	$55.30	$81.04
Crew A-2B	Hr.	Daily	Hr.	Daily	Bare Costs	Incl. O&P
1 Truck Driver (light)	$39.10	$312.80	$63.05	$504.40	$39.10	$63.05
1 Flatbed Truck, Gas, 1.5 Ton		245.40		269.94	30.68	33.74
8 L.H., Daily Totals		$558.20		$774.34	$69.78	$96.79
Crew A-3A	Hr.	Daily	Hr.	Daily	Bare Costs	Incl. O&P
1 Equip. Oper. (light)	$48.60	$388.80	$77.15	$617.20	$48.60	$77.15
1 Pickup Truck, 4 x 4, 3/4 Ton		155.60		171.16	19.45	21.40
8 L.H., Daily Totals		$544.40		$788.36	$68.05	$98.55
Crew A-3B	Hr.	Daily	Hr.	Daily	Bare Costs	Incl. O&P
1 Equip. Oper. (medium)	$50.60	$404.80	$80.30	$642.40	$45.33	$72.42
1 Truck Driver (heavy)	40.05	320.40	64.55	516.40		
1 Dump Truck, 12 C.Y., 400 H.P.		691.00		760.10		
1 F.E. Loader, W.M., 2.5 C.Y.		519.80		571.78	75.67	83.24
16 L.H., Daily Totals		$1936.00		$2490.68	$121.00	$155.67
Crew A-3C	Hr.	Daily	Hr.	Daily	Bare Costs	Incl. O&P
1 Equip. Oper. (light)	$48.60	$388.80	$77.15	$617.20	$48.60	$77.15
1 Loader, Skid Steer, 78 H.P.		318.60		350.46	39.83	43.81
8 L.H., Daily Totals		$707.40		$967.66	$88.42	$120.96
Crew A-3D	Hr.	Daily	Hr.	Daily	Bare Costs	Incl. O&P
1 Truck Driver (light)	$39.10	$312.80	$63.05	$504.40	$39.10	$63.05
1 Pickup Truck, 4 x 4, 3/4 Ton		155.60		171.16		
1 Flatbed Trailer, 25 Ton		113.60		124.96	33.65	37.02
8 L.H., Daily Totals		$582.00		$800.52	$72.75	$100.07
Crew A-3E	Hr.	Daily	Hr.	Daily	Bare Costs	Incl. O&P
1 Equip. Oper. (crane)	$51.70	$413.60	$82.05	$656.40	$45.88	$73.30
1 Truck Driver (heavy)	40.05	320.40	64.55	516.40		
1 Pickup Truck, 4 x 4, 3/4 Ton		155.60		171.16	9.72	10.70
16 L.H., Daily Totals		$889.60		$1343.96	$55.60	$84.00
Crew A-3F	Hr.	Daily	Hr.	Daily	Bare Costs	Incl. O&P
1 Equip. Oper. (crane)	$51.70	$413.60	$82.05	$656.40	$45.88	$73.30
1 Truck Driver (heavy)	40.05	320.40	64.55	516.40		
1 Pickup Truck, 4 x 4, 3/4 Ton		155.60		171.16		
1 Truck Tractor, 6x4, 380 H.P.		599.80		659.78		
1 Lowbed Trailer, 75 Ton		221.80		243.98	61.08	67.18
16 L.H., Daily Totals		$1711.20		$2247.72	$106.95	$140.48

Crews

Crew No.	Bare Costs		Incl. Subs O&P		Cost Per Labor-Hour	
Crew A-3G	**Hr.**	**Daily**	**Hr.**	**Daily**	**Bare Costs**	**Incl. O&P**
1 Equip. Oper. (crane)	$51.70	$413.60	$82.05	$656.40	$45.88	$73.30
1 Truck Driver (heavy)	40.05	320.40	64.55	516.40		
1 Pickup Truck, 4 x 4, 3/4 Ton		155.60		171.16		
1 Truck Tractor, 6x4, 450 H.P.		727.40		800.14		
1 Lowbed Trailer, 75 Ton		221.80		243.98	69.05	75.95
16 L.H., Daily Totals		$1838.80		$2388.08	$114.93	$149.26
Crew A-3H	**Hr.**	**Daily**	**Hr.**	**Daily**	**Bare Costs**	**Incl. O&P**
1 Equip. Oper. (crane)	$51.70	$413.60	$82.05	$656.40	$51.70	$82.05
1 Hyd. Crane, 12 Ton (Daily)		860.80		946.88	107.60	118.36
8 L.H., Daily Totals		$1274.40		$1603.28	$159.30	$200.41
Crew A-3I	**Hr.**	**Daily**	**Hr.**	**Daily**	**Bare Costs**	**Incl. O&P**
1 Equip. Oper. (crane)	$51.70	$413.60	$82.05	$656.40	$51.70	$82.05
1 Hyd. Crane, 25 Ton (Daily)		989.60		1088.56	123.70	136.07
8 L.H., Daily Totals		$1403.20		$1744.96	$175.40	$218.12
Crew A-3J	**Hr.**	**Daily**	**Hr.**	**Daily**	**Bare Costs**	**Incl. O&P**
1 Equip. Oper. (crane)	$51.70	$413.60	$82.05	$656.40	$51.70	$82.05
1 Hyd. Crane, 40 Ton (Daily)		1221.00		1343.10	152.63	167.89
8 L.H., Daily Totals		$1634.60		$1999.50	$204.32	$249.94
Crew A-3K	**Hr.**	**Daily**	**Hr.**	**Daily**	**Bare Costs**	**Incl. O&P**
1 Equip. Oper. (crane)	$51.70	$413.60	$82.05	$656.40	$48.45	$76.90
1 Equip. Oper. (oiler)	45.20	361.60	71.75	574.00		
1 Hyd. Crane, 55 Ton (Daily)		1469.00		1615.90		
1 P/U Truck, 3/4 Ton (Daily)		167.20		183.92	102.26	112.49
16 L.H., Daily Totals		$2411.40		$3030.22	$150.71	$189.39
Crew A-3L	**Hr.**	**Daily**	**Hr.**	**Daily**	**Bare Costs**	**Incl. O&P**
1 Equip. Oper. (crane)	$51.70	$413.60	$82.05	$656.40	$48.45	$76.90
1 Equip. Oper. (oiler)	45.20	361.60	71.75	574.00		
1 Hyd. Crane, 80 Ton (Daily)		2174.00		2391.40		
1 P/U Truck, 3/4 Ton (Daily)		167.20		183.92	146.32	160.96
16 L.H., Daily Totals		$3116.40		$3805.72	$194.78	$237.86
Crew A-3M	**Hr.**	**Daily**	**Hr.**	**Daily**	**Bare Costs**	**Incl. O&P**
1 Equip. Oper. (crane)	$51.70	$413.60	$82.05	$656.40	$48.45	$76.90
1 Equip. Oper. (oiler)	45.20	361.60	71.75	574.00		
1 Hyd. Crane, 100 Ton (Daily)		2169.00		2385.90		
1 P/U Truck, 3/4 Ton (Daily)		167.20		183.92	146.01	160.61
16 L.H., Daily Totals		$3111.40		$3800.22	$194.46	$237.51
Crew A-3N	**Hr.**	**Daily**	**Hr.**	**Daily**	**Bare Costs**	**Incl. O&P**
1 Equip. Oper. (crane)	$51.70	$413.60	$82.05	$656.40	$51.70	$82.05
1 Tower Cane (monthly)		1128.00		1240.80	141.00	155.10
8 L.H., Daily Totals		$1541.60		$1897.20	$192.70	$237.15
Crew A-3P	**Hr.**	**Daily**	**Hr.**	**Daily**	**Bare Costs**	**Incl. O&P**
1 Equip. Oper. (light)	$48.60	$388.80	$77.15	$617.20	$48.60	$77.15
1 A.T. Forklift, 42' lift		526.40		579.04	65.80	72.38
8 L.H., Daily Totals		$915.20		$1196.24	$114.40	$149.53
Crew A-3Q	**Hr.**	**Daily**	**Hr.**	**Daily**	**Bare Costs**	**Incl. O&P**
1 Equip. Oper. (light)	$48.60	$388.80	$77.15	$617.20	$48.60	$77.15
1 Pickup Truck, 4 x 4, 3/4 Ton		155.60		171.16		
1 Flatbed trailer, 3 Ton		23.60		25.96	22.40	24.64
8 L.H., Daily Totals		$568.00		$814.32	$71.00	$101.79

Crew No.	Bare Costs		Incl. Subs O&P		Cost Per Labor-Hour	
Crew A-4	**Hr.**	**Daily**	**Hr.**	**Daily**	**Bare Costs**	**Incl. O&P**
2 Carpenters	$46.95	$751.20	$76.95	$1231.20	$44.75	$72.92
1 Painter, Ordinary	40.35	322.80	64.85	518.80		
24 L.H., Daily Totals		$1074.00		$1750.00	$44.75	$72.92
Crew A-5	**Hr.**	**Daily**	**Hr.**	**Daily**	**Bare Costs**	**Incl. O&P**
2 Laborers	$37.60	$601.60	$61.65	$986.40	$37.77	$61.81
.25 Truck Driver (light)	39.10	78.20	63.05	126.10		
.25 Flatbed Truck, Gas, 1.5 Ton		61.35		67.48	3.41	3.75
18 L.H., Daily Totals		$741.15		$1179.98	$41.17	$65.55
Crew A-6	**Hr.**	**Daily**	**Hr.**	**Daily**	**Bare Costs**	**Incl. O&P**
1 Instrument Man	$48.65	$389.20	$79.05	$632.40	$46.88	$76.13
1 Rodman/Chainman	45.10	360.80	73.20	585.60		
1 Level, Electronic		54.80		60.28	3.42	3.77
16 L.H., Daily Totals		$804.80		$1278.28	$50.30	$79.89
Crew A-7	**Hr.**	**Daily**	**Hr.**	**Daily**	**Bare Costs**	**Incl. O&P**
1 Chief of Party	$60.90	$487.20	$96.20	$769.60	$51.55	$82.82
1 Instrument Man	48.65	389.20	79.05	632.40		
1 Rodman/Chainman	45.10	360.80	73.20	585.60		
1 Level, Electronic		54.80		60.28	2.28	2.51
24 L.H., Daily Totals		$1292.00		$2047.88	$53.83	$85.33
Crew A-8	**Hr.**	**Daily**	**Hr.**	**Daily**	**Bare Costs**	**Incl. O&P**
1 Chief of Party	$60.90	$487.20	$96.20	$769.60	$49.94	$80.41
1 Instrument Man	48.65	389.20	79.05	632.40		
2 Rodmen/Chainmen	45.10	721.60	73.20	1171.20		
1 Level, Electronic		54.80		60.28	1.71	1.88
32 L.H., Daily Totals		$1652.80		$2633.48	$51.65	$82.30
Crew A-9	**Hr.**	**Daily**	**Hr.**	**Daily**	**Bare Costs**	**Incl. O&P**
1 Asbestos Foreman	$52.85	$422.80	$84.95	$679.60	$52.41	$84.25
7 Asbestos Workers	52.35	2931.60	84.15	4712.40	52.41	84.25
64 L.H., Daily Totals		$3354.40		$5392.00	$52.41	$84.25
Crew A-10A	**Hr.**	**Daily**	**Hr.**	**Daily**	**Bare Costs**	**Incl. O&P**
1 Asbestos Foreman	$52.85	$422.80	$84.95	$679.60	$52.52	$84.42
2 Asbestos Workers	52.35	837.60	84.15	1346.40		
24 L.H., Daily Totals		$1260.40		$2026.00	$52.52	$84.42
Crew A-10B	**Hr.**	**Daily**	**Hr.**	**Daily**	**Bare Costs**	**Incl. O&P**
1 Asbestos Foreman	$52.85	$422.80	$84.95	$679.60	$52.48	$84.35
3 Asbestos Workers	52.35	1256.40	84.15	2019.60		
32 L.H., Daily Totals		$1679.20		$2699.20	$52.48	$84.35
Crew A-10C	**Hr.**	**Daily**	**Hr.**	**Daily**	**Bare Costs**	**Incl. O&P**
3 Asbestos Workers	$52.35	$1256.40	$84.15	$2019.60	$52.35	$84.15
1 Flatbed Truck, Gas, 1.5 Ton		245.40		269.94	10.23	11.25
24 L.H., Daily Totals		$1501.80		$2289.54	$62.58	$95.40
Crew A-10D	**Hr.**	**Daily**	**Hr.**	**Daily**	**Bare Costs**	**Incl. O&P**
2 Asbestos Workers	$52.35	$837.60	$84.15	$1346.40	$50.40	$80.53
1 Equip. Oper. (crane)	51.70	413.60	82.05	656.40		
1 Equip. Oper. (oiler)	45.20	361.60	71.75	574.00		
1 Hydraulic Crane, 33 Ton		750.00		825.00	23.44	25.78
32 L.H., Daily Totals		$2362.80		$3401.80	$73.84	$106.31

Crews

Crew No.		Bare Costs		Incl. Subs O&P		Cost Per Labor-Hour	
						Bare Costs	Incl. O&P
Crew A-11	Hr.	Daily	Hr.	Daily			
1 Asbestos Foreman	$52.85	$422.80	$84.95	$679.60		$52.41	$84.25
7 Asbestos Workers	52.35	2931.60	84.15	4712.40			
2 Chip. Hammers, 12 Lb., Elec.		40.00		44.00		.63	.69
64 L.H., Daily Totals		$3394.40		$5436.00		$53.04	$84.94
Crew A-12	Hr.	Daily	Hr.	Daily		Bare Costs	Incl. O&P
1 Asbestos Foreman	$52.85	$422.80	$84.95	$679.60		$52.41	$84.25
7 Asbestos Workers	52.35	2931.60	84.15	4712.40			
1 Trk-Mtd Vac, 14 CY, 1500 Gal.		514.25		565.67			
1 Flatbed Truck, 20,000 GVW		248.60		273.46		11.92	13.11
64 L.H., Daily Totals		$4117.25		$6231.14		$64.33	$97.36
Crew A-13	Hr.	Daily	Hr.	Daily		Bare Costs	Incl. O&P
1 Equip. Oper. (light)	$48.60	$388.80	$77.15	$617.20		$48.60	$77.15
1 Trk-Mtd Vac, 14 CY, 1500 Gal.		514.25		565.67			
1 Flatbed Truck, 20,000 GVW		248.60		273.46		95.36	104.89
8 L.H., Daily Totals		$1151.65		$1456.34		$143.96	$182.04
Crew B-1	Hr.	Daily	Hr.	Daily		Bare Costs	Incl. O&P
1 Labor Foreman (outside)	$39.60	$316.80	$64.90	$519.20		$38.27	$62.73
2 Laborers	37.60	601.60	61.65	986.40			
24 L.H., Daily Totals		$918.40		$1505.60		$38.27	$62.73
Crew B-1A	Hr.	Daily	Hr.	Daily		Bare Costs	Incl. O&P
1 Labor Foreman (outside)	$39.60	$316.80	$64.90	$519.20		$38.27	$62.73
2 Laborers	37.60	601.60	61.65	986.40			
2 Cutting Torches		22.80		25.08			
2 Sets of Gases		304.00		334.40		13.62	14.98
24 L.H., Daily Totals		$1245.20		$1865.08		$51.88	$77.71
Crew B-1B	Hr.	Daily	Hr.	Daily		Bare Costs	Incl. O&P
1 Labor Foreman (outside)	$39.60	$316.80	$64.90	$519.20		$41.63	$67.56
2 Laborers	37.60	601.60	61.65	986.40			
1 Equip. Oper. (crane)	51.70	413.60	82.05	656.40			
2 Cutting Torches		22.80		25.08			
2 Sets of Gases		304.00		334.40			
1 Hyd. Crane, 12 Ton		653.80		719.18		30.64	33.71
32 L.H., Daily Totals		$2312.60		$3240.66		$72.27	$101.27
Crew B-1C	Hr.	Daily	Hr.	Daily		Bare Costs	Incl. O&P
1 Labor Foreman (outside)	$39.60	$316.80	$64.90	$519.20		$38.27	$62.73
2 Laborers	37.60	601.60	61.65	986.40			
1 Aerial Lift Truck, 60' Boom		435.60		479.16		18.15	19.97
24 L.H., Daily Totals		$1354.00		$1984.76		$56.42	$82.70
Crew B-1D	Hr.	Daily	Hr.	Daily		Bare Costs	Incl. O&P
2 Laborers	$37.60	$601.60	$61.65	$986.40		$37.60	$61.65
1 Small Work Boat, Gas, 50 H.P.		173.20		190.52			
1 Pressure Washer, 7 GPM		87.60		96.36		16.30	17.93
16 L.H., Daily Totals		$862.40		$1273.28		$53.90	$79.58
Crew B-1E	Hr.	Daily	Hr.	Daily		Bare Costs	Incl. O&P
1 Labor Foreman (outside)	$39.60	$316.80	$64.90	$519.20		$38.10	$62.46
3 Laborers	37.60	902.40	61.65	1479.60			
1 Work Boat, Diesel, 200 H.P.		1463.00		1609.30			
2 Pressure Washer, 7 GPM		175.20		192.72		51.19	56.31
32 L.H., Daily Totals		$2857.40		$3800.82		$89.29	$118.78
Crew B-1F	Hr.	Daily	Hr.	Daily		Bare Costs	Incl. O&P
2 Skilled Workers	$48.65	$778.40	$79.05	$1264.80		$44.97	$73.25
1 Laborer	37.60	300.80	61.65	493.20			
1 Small Work Boat, Gas, 50 H.P.		173.20		190.52			
1 Pressure Washer, 7 GPM		87.60		96.36		10.87	11.95
24 L.H., Daily Totals		$1340.00		$2044.88		$55.83	$85.20
Crew B-1G	Hr.	Daily	Hr.	Daily		Bare Costs	Incl. O&P
2 Laborers	$37.60	$601.60	$61.65	$986.40		$37.60	$61.65
1 Small Work Boat, Gas, 50 H.P.		173.20		190.52		10.82	11.91
16 L.H., Daily Totals		$774.80		$1176.92		$48.42	$73.56
Crew B-1H	Hr.	Daily	Hr.	Daily		Bare Costs	Incl. O&P
2 Skilled Workers	$48.65	$778.40	$79.05	$1264.80		$44.97	$73.25
1 Laborer	37.60	300.80	61.65	493.20			
1 Small Work Boat, Gas, 50 H.P.		173.20		190.52		7.22	7.94
24 L.H., Daily Totals		$1252.40		$1948.52		$52.18	$81.19
Crew B-1J	Hr.	Daily	Hr.	Daily		Bare Costs	Incl. O&P
1 Labor Foreman (inside)	$38.10	$304.80	$62.45	$499.60		$37.85	$62.05
1 Laborer	37.60	300.80	61.65	493.20			
16 L.H., Daily Totals		$605.60		$992.80		$37.85	$62.05
Crew B-1K	Hr.	Daily	Hr.	Daily		Bare Costs	Incl. O&P
1 Carpenter Foreman (inside)	$47.45	$379.60	$77.75	$622.00		$47.20	$77.35
1 Carpenter	46.95	375.60	76.95	615.60			
16 L.H., Daily Totals		$755.20		$1237.60		$47.20	$77.35
Crew B-2	Hr.	Daily	Hr.	Daily		Bare Costs	Incl. O&P
1 Labor Foreman (outside)	$39.60	$316.80	$64.90	$519.20		$38.00	$62.30
4 Laborers	37.60	1203.20	61.65	1972.80			
40 L.H., Daily Totals		$1520.00		$2492.00		$38.00	$62.30
Crew B-2A	Hr.	Daily	Hr.	Daily		Bare Costs	Incl. O&P
1 Labor Foreman (outside)	$39.60	$316.80	$64.90	$519.20		$38.27	$62.73
2 Laborers	37.60	601.60	61.65	986.40			
1 Aerial Lift Truck, 60' Boom		435.60		479.16		18.15	19.97
24 L.H., Daily Totals		$1354.00		$1984.76		$56.42	$82.70
Crew B-3	Hr.	Daily	Hr.	Daily		Bare Costs	Incl. O&P
1 Labor Foreman (outside)	$39.60	$316.80	$64.90	$519.20		$40.92	$66.27
2 Laborers	37.60	601.60	61.65	986.40			
1 Equip. Oper. (medium)	50.60	404.80	80.30	642.40			
2 Truck Drivers (heavy)	40.05	640.80	64.55	1032.80			
1 Crawler Loader, 3 C.Y.		1188.00		1306.80			
2 Dump Trucks, 12 C.Y., 400 H.P.		1382.00		1520.20		53.54	58.90
48 L.H., Daily Totals		$4534.00		$6007.80		$94.46	$125.16
Crew B-3A	Hr.	Daily	Hr.	Daily		Bare Costs	Incl. O&P
4 Laborers	$37.60	$1203.20	$61.65	$1972.80		$40.20	$65.38
1 Equip. Oper. (medium)	50.60	404.80	80.30	642.40			
1 Hyd. Excavator, 1.5 C.Y.		1030.00		1133.00		25.75	28.32
40 L.H., Daily Totals		$2638.00		$3748.20		$65.95	$93.70
Crew B-3B	Hr.	Daily	Hr.	Daily		Bare Costs	Incl. O&P
2 Laborers	$37.60	$601.60	$61.65	$986.40		$41.46	$67.04
1 Equip. Oper. (medium)	50.60	404.80	80.30	642.40			
1 Truck Driver (heavy)	40.05	320.40	64.55	516.40			
1 Backhoe Loader, 80 H.P.		391.80		430.98			
1 Dump Truck, 12 C.Y., 400 H.P.		691.00		760.10		33.84	37.22
32 L.H., Daily Totals		$2409.60		$3336.28		$75.30	$104.26

Crews

Crew No.	Bare Costs		Incl. Subs O&P		Cost Per Labor-Hour	
Crew B-3C	**Hr.**	**Daily**	**Hr.**	**Daily**	**Bare Costs**	**Incl. O&P**
3 Laborers	$37.60	$902.40	$61.65	$1479.60	$40.85	$66.31
1 Equip. Oper. (medium)	50.60	404.80	80.30	642.40		
1 Crawler Loader, 4 C.Y.		1561.00		1717.10	48.78	53.66
32 L.H., Daily Totals		$2868.20		$3839.10	$89.63	$119.97
Crew B-4	**Hr.**	**Daily**	**Hr.**	**Daily**	**Bare Costs**	**Incl. O&P**
1 Labor Foreman (outside)	$39.60	$316.80	$64.90	$519.20	$38.34	$62.67
4 Laborers	37.60	1203.20	61.65	1972.80		
1 Truck Driver (heavy)	40.05	320.40	64.55	516.40		
1 Truck Tractor, 220 H.P.		359.60		395.56		
1 Flatbed Trailer, 40 Ton		154.00		169.40	10.70	11.77
48 L.H., Daily Totals		$2354.00		$3573.36	$49.04	$74.44
Crew B-5	**Hr.**	**Daily**	**Hr.**	**Daily**	**Bare Costs**	**Incl. O&P**
1 Labor Foreman (outside)	$39.60	$316.80	$64.90	$519.20	$41.60	$67.44
4 Laborers	37.60	1203.20	61.65	1972.80		
2 Equip. Oper. (medium)	50.60	809.60	80.30	1284.80		
1 Air Compressor, 250 cfm		201.40		221.54		
2 Breakers, Pavement, 60 lb.		19.60		21.56		
2 -50' Air Hoses, 1.5"		11.60		12.76		
1 Crawler Loader, 3 C.Y.		1188.00		1306.80	25.37	27.90
56 L.H., Daily Totals		$3750.20		$5339.46	$66.97	$95.35
Crew B-5D	**Hr.**	**Daily**	**Hr.**	**Daily**	**Bare Costs**	**Incl. O&P**
1 Labor Foreman (outside)	$39.60	$316.80	$64.90	$519.20	$41.41	$67.08
4 Laborers	37.60	1203.20	61.65	1972.80		
2 Equip. Oper. (medium)	50.60	809.60	80.30	1284.80		
1 Truck Driver (heavy)	40.05	320.40	64.55	516.40		
1 Air Compressor, 250 cfm		201.40		221.54		
2 Breakers, Pavement, 60 lb.		19.60		21.56		
2 -50' Air Hoses, 1.5"		11.60		12.76		
1 Crawler Loader, 3 C.Y.		1188.00		1306.80		
1 Dump Truck, 12 C.Y., 400 H.P.		691.00		760.10	32.99	36.29
64 L.H., Daily Totals		$4761.60		$6615.96	$74.40	$103.37
Crew B-6	**Hr.**	**Daily**	**Hr.**	**Daily**	**Bare Costs**	**Incl. O&P**
2 Laborers	$37.60	$601.60	$61.65	$986.40	$41.27	$66.82
1 Equip. Oper. (light)	48.60	388.80	77.15	617.20		
1 Backhoe Loader, 48 H.P.		364.20		400.62	15.18	16.69
24 L.H., Daily Totals		$1354.60		$2004.22	$56.44	$83.51
Crew B-6B	**Hr.**	**Daily**	**Hr.**	**Daily**	**Bare Costs**	**Incl. O&P**
2 Labor Foremen (outside)	$39.60	$633.60	$64.90	$1038.40	$38.27	$62.73
4 Laborers	37.60	1203.20	61.65	1972.80		
1 S.P. Crane, 4x4, 5 Ton		275.40		302.94		
1 Flatbed Truck, Gas, 1.5 Ton		245.40		269.94		
1 Butt Fusion Mach., 4"-12" diam.		382.10		420.31	18.81	20.69
48 L.H., Daily Totals		$2739.70		$4004.39	$57.08	$83.42
Crew B-6C	**Hr.**	**Daily**	**Hr.**	**Daily**	**Bare Costs**	**Incl. O&P**
2 Labor Foremen (outside)	$39.60	$633.60	$64.90	$1038.40	$38.27	$62.73
4 Laborers	37.60	1203.20	61.65	1972.80		
1 S.P. Crane, 4x4, 12 Ton		473.40		520.74		
1 Flatbed Truck, Gas, 3 Ton		303.20		333.52		
1 Butt Fusion Mach., 8"-24" diam.		829.70		912.67	33.46	36.81
48 L.H., Daily Totals		$3443.10		$4778.13	$71.73	$99.54
Crew B-7	**Hr.**	**Daily**	**Hr.**	**Daily**	**Bare Costs**	**Incl. O&P**
1 Labor Foreman (outside)	$39.60	$316.80	$64.90	$519.20	$40.10	$65.30
4 Laborers	37.60	1203.20	61.65	1972.80		
1 Equip. Oper. (medium)	50.60	404.80	80.30	642.40		
1 Brush Chipper, 12", 130 H.P.		391.40		430.54		
1 Crawler Loader, 3 C.Y.		1188.00		1306.80		
2 Chain Saws, Gas, 36" Long		90.00		99.00	34.78	38.26
48 L.H., Daily Totals		$3594.20		$4970.74	$74.88	$103.56
Crew B-7A	**Hr.**	**Daily**	**Hr.**	**Daily**	**Bare Costs**	**Incl. O&P**
2 Laborers	$37.60	$601.60	$61.65	$986.40	$41.27	$66.82
1 Equip. Oper. (light)	48.60	388.80	77.15	617.20		
1 Rake w/Tractor		354.25		389.68		
2 Chain Saw, Gas, 18"		62.40		68.64	17.36	19.10
24 L.H., Daily Totals		$1407.05		$2061.92	$58.63	$85.91
Crew B-7B	**Hr.**	**Daily**	**Hr.**	**Daily**	**Bare Costs**	**Incl. O&P**
1 Labor Foreman (outside)	$39.60	$316.80	$64.90	$519.20	$40.09	$65.19
4 Laborers	37.60	1203.20	61.65	1972.80		
1 Equip. Oper. (medium)	50.60	404.80	80.30	642.40		
1 Truck Driver (heavy)	40.05	320.40	64.55	516.40		
1 Brush Chipper, 12", 130 H.P.		391.40		430.54		
1 Crawler Loader, 3 C.Y.		1188.00		1306.80		
2 Chain Saws, Gas, 36" Long		90.00		99.00		
1 Dump Truck, 8 C.Y., 220 H.P.		411.20		452.32	37.15	40.87
56 L.H., Daily Totals		$4325.80		$5939.46	$77.25	$106.06
Crew B-7C	**Hr.**	**Daily**	**Hr.**	**Daily**	**Bare Costs**	**Incl. O&P**
1 Labor Foreman (outside)	$39.60	$316.80	$64.90	$519.20	$40.09	$65.19
4 Laborers	37.60	1203.20	61.65	1972.80		
1 Equip. Oper. (medium)	50.60	404.80	80.30	642.40		
1 Truck Driver (heavy)	40.05	320.40	64.55	516.40		
1 Brush Chipper, 12", 130 H.P.		391.40		430.54		
1 Crawler Loader, 3 C.Y.		1188.00		1306.80		
2 Chain Saws, Gas, 36" Long		90.00		99.00		
1 Dump Truck, 12 C.Y., 400 H.P.		691.00		760.10	42.15	46.37
56 L.H., Daily Totals		$4605.60		$6247.24	$82.24	$111.56
Crew B-8	**Hr.**	**Daily**	**Hr.**	**Daily**	**Bare Costs**	**Incl. O&P**
1 Labor Foreman (outside)	$39.60	$316.80	$64.90	$519.20	$42.66	$68.71
2 Laborers	37.60	601.60	61.65	986.40		
2 Equip. Oper. (medium)	50.60	809.60	80.30	1284.80		
1 Equip. Oper. (oiler)	45.20	361.60	71.75	574.00		
2 Truck Drivers (heavy)	40.05	640.80	64.55	1032.80		
1 Hyd. Crane, 25 Ton		736.60		810.26		
1 Crawler Loader, 3 C.Y.		1188.00		1306.80		
2 Dump Trucks, 12 C.Y., 400 H.P.		1382.00		1520.20	51.67	56.83
64 L.H., Daily Totals		$6037.00		$8034.46	$94.33	$125.54
Crew B-9	**Hr.**	**Daily**	**Hr.**	**Daily**	**Bare Costs**	**Incl. O&P**
1 Labor Foreman (outside)	$39.60	$316.80	$64.90	$519.20	$38.00	$62.30
4 Laborers	37.60	1203.20	61.65	1972.80		
1 Air Compressor, 250 cfm		201.40		221.54		
2 Breakers, Pavement, 60 lb.		19.60		21.56		
2 -50' Air Hoses, 1.5"		11.60		12.76	5.82	6.40
40 L.H., Daily Totals		$1752.60		$2747.86	$43.81	$68.70

Crews

Crew No.		Bare Costs		Incl. Subs O&P		Cost Per Labor-Hour	
Crew B-9A	Hr.	Daily	Hr.	Daily	Bare Costs	Incl. O&P
2 Laborers	$37.60	$601.60	$61.65	$986.40	$38.42	$62.62
1 Truck Driver (heavy)	40.05	320.40	64.55	516.40		
1 Water Tank Trailer, 5000 Gal.		141.00		155.10		
1 Truck Tractor, 220 H.P.		359.60		395.56		
2 -50' Discharge Hoses, 3"		3.00		3.30	20.98	23.08
24 L.H., Daily Totals		$1425.60		$2056.76	$59.40	$85.70
Crew B-9B	Hr.	Daily	Hr.	Daily	Bare Costs	Incl. O&P
2 Laborers	$37.60	$601.60	$61.65	$986.40	$38.42	$62.62
1 Truck Driver (heavy)	40.05	320.40	64.55	516.40		
2 -50' Discharge Hoses, 3"		3.00		3.30		
1 Water Tank Trailer, 5000 Gal.		141.00		155.10		
1 Truck Tractor, 220 H.P.		359.60		395.56		
1 Pressure Washer		70.20		77.22	23.91	26.30
24 L.H., Daily Totals		$1495.80		$2133.98	$62.33	$88.92
Crew B-9D	Hr.	Daily	Hr.	Daily	Bare Costs	Incl. O&P
1 Labor Foreman (outside)	$39.60	$316.80	$64.90	$519.20	$38.00	$62.30
4 Common Laborers	37.60	1203.20	61.65	1972.80		
1 Air Compressor, 250 cfm		201.40		221.54		
2 -50' Air Hoses, 1.5"		11.60		12.76		
2 Air Powered Tampers		52.40		57.64	6.63	7.30
40 L.H., Daily Totals		$1785.40		$2783.94	$44.63	$69.60
Crew B-10	Hr.	Daily	Hr.	Daily	Bare Costs	Incl. O&P
1 Equip. Oper. (medium)	$50.60	$404.80	$80.30	$642.40	$46.27	$74.08
.5 Laborer	37.60	150.40	61.65	246.60		
12 L.H., Daily Totals		$555.20		$889.00	$46.27	$74.08
Crew B-10A	Hr.	Daily	Hr.	Daily	Bare Costs	Incl. O&P
1 Equip. Oper. (medium)	$50.60	$404.80	$80.30	$642.40	$46.27	$74.08
.5 Laborer	37.60	150.40	61.65	246.60		
1 Roller, 2-Drum, W.B., 7.5 H.P.		180.00		198.00	15.00	16.50
12 L.H., Daily Totals		$735.20		$1087.00	$61.27	$90.58
Crew B-10B	Hr.	Daily	Hr.	Daily	Bare Costs	Incl. O&P
1 Equip. Oper. (medium)	$50.60	$404.80	$80.30	$642.40	$46.27	$74.08
.5 Laborer	37.60	150.40	61.65	246.60		
1 Dozer, 200 H.P.		1387.00		1525.70	115.58	127.14
12 L.H., Daily Totals		$1942.20		$2414.70	$161.85	$201.22
Crew B-10C	Hr.	Daily	Hr.	Daily	Bare Costs	Incl. O&P
1 Equip. Oper. (medium)	$50.60	$404.80	$80.30	$642.40	$46.27	$74.08
.5 Laborer	37.60	150.40	61.65	246.60		
1 Dozer, 200 H.P.		1387.00		1525.70		
1 Vibratory Roller, Towed, 23 Ton		396.00		435.60	148.58	163.44
12 L.H., Daily Totals		$2338.20		$2850.30	$194.85	$237.53
Crew B-10D	Hr.	Daily	Hr.	Daily	Bare Costs	Incl. O&P
1 Equip. Oper. (medium)	$50.60	$404.80	$80.30	$642.40	$46.27	$74.08
.5 Laborer	37.60	150.40	61.65	246.60		
1 Dozer, 200 H.P.		1387.00		1525.70		
1 Sheepsft. Roller, Towed		425.80		468.38	151.07	166.17
12 L.H., Daily Totals		$2368.00		$2883.08	$197.33	$240.26
Crew B-10E	Hr.	Daily	Hr.	Daily	Bare Costs	Incl. O&P
1 Equip. Oper. (medium)	$50.60	$404.80	$80.30	$642.40	$46.27	$74.08
.5 Laborer	37.60	150.40	61.65	246.60		
1 Tandem Roller, 5 Ton		157.40		173.14	13.12	14.43
12 L.H., Daily Totals		$712.60		$1062.14	$59.38	$88.51

Crew No.		Bare Costs		Incl. Subs O&P		Cost Per Labor-Hour	
Crew B-10F	Hr.	Daily	Hr.	Daily	Bare Costs	Incl. O&P
1 Equip. Oper. (medium)	$50.60	$404.80	$80.30	$642.40	$46.27	$74.08
.5 Laborer	37.60	150.40	61.65	246.60		
1 Tandem Roller, 10 Ton		236.80		260.48	19.73	21.71
12 L.H., Daily Totals		$792.00		$1149.48	$66.00	$95.79
Crew B-10G	Hr.	Daily	Hr.	Daily	Bare Costs	Incl. O&P
1 Equip. Oper. (medium)	$50.60	$404.80	$80.30	$642.40	$46.27	$74.08
.5 Laborer	37.60	150.40	61.65	246.60		
1 Sheepsfoot Roller, 240 H.P.		1206.00		1326.60	100.50	110.55
12 L.H., Daily Totals		$1761.20		$2215.60	$146.77	$184.63
Crew B-10H	Hr.	Daily	Hr.	Daily	Bare Costs	Incl. O&P
1 Equip. Oper. (medium)	$50.60	$404.80	$80.30	$642.40	$46.27	$74.08
.5 Laborer	37.60	150.40	61.65	246.60		
1 Diaphragm Water Pump, 2"		71.00		78.10		
1 -20' Suction Hose, 2"		1.95		2.15		
2 -50' Discharge Hoses, 2"		1.80		1.98	6.23	6.85
12 L.H., Daily Totals		$629.95		$971.23	$52.50	$80.94
Crew B-10I	Hr.	Daily	Hr.	Daily	Bare Costs	Incl. O&P
1 Equip. Oper. (medium)	$50.60	$404.80	$80.30	$642.40	$46.27	$74.08
.5 Laborer	37.60	150.40	61.65	246.60		
1 Diaphragm Water Pump, 4"		114.20		125.62		
1 -20' Suction Hose, 4"		3.25		3.58		
2 -50' Discharge Hoses, 4"		4.70		5.17	10.18	11.20
12 L.H., Daily Totals		$677.35		$1023.37	$56.45	$85.28
Crew B-10J	Hr.	Daily	Hr.	Daily	Bare Costs	Incl. O&P
1 Equip. Oper. (medium)	$50.60	$404.80	$80.30	$642.40	$46.27	$74.08
.5 Laborer	37.60	150.40	61.65	246.60		
1 Centrifugal Water Pump, 3"		78.20		86.02		
1 -20' Suction Hose, 3"		2.85		3.13		
2 -50' Discharge Hoses, 3"		3.00		3.30	7.00	7.70
12 L.H., Daily Totals		$639.25		$981.46	$53.27	$81.79
Crew B-10K	Hr.	Daily	Hr.	Daily	Bare Costs	Incl. O&P
1 Equip. Oper. (medium)	$50.60	$404.80	$80.30	$642.40	$46.27	$74.08
.5 Laborer	37.60	150.40	61.65	246.60		
1 Centr. Water Pump, 6"		346.60		381.26		
1 -20' Suction Hose, 6"		11.50		12.65		
2 -50' Discharge Hoses, 6"		12.20		13.42	30.86	33.94
12 L.H., Daily Totals		$925.50		$1296.33	$77.13	$108.03
Crew B-10L	Hr.	Daily	Hr.	Daily	Bare Costs	Incl. O&P
1 Equip. Oper. (medium)	$50.60	$404.80	$80.30	$642.40	$46.27	$74.08
.5 Laborer	37.60	150.40	61.65	246.60		
1 Dozer, 80 H.P.		472.40		519.64	39.37	43.30
12 L.H., Daily Totals		$1027.60		$1408.64	$85.63	$117.39
Crew B-10M	Hr.	Daily	Hr.	Daily	Bare Costs	Incl. O&P
1 Equip. Oper. (medium)	$50.60	$404.80	$80.30	$642.40	$46.27	$74.08
.5 Laborer	37.60	150.40	61.65	246.60		
1 Dozer, 300 H.P.		1897.00		2086.70	158.08	173.89
12 L.H., Daily Totals		$2452.20		$2975.70	$204.35	$247.97
Crew B-10N	Hr.	Daily	Hr.	Daily	Bare Costs	Incl. O&P
1 Equip. Oper. (medium)	$50.60	$404.80	$80.30	$642.40	$46.27	$74.08
.5 Laborer	37.60	150.40	61.65	246.60		
1 F.E. Loader, T.M., 1.5 C.Y		522.00		574.20	43.50	47.85
12 L.H., Daily Totals		$1077.20		$1463.20	$89.77	$121.93

Crews

Crew No.	Bare Costs		Incl. Subs O&P		Cost Per Labor-Hour	
Crew B-10O	Hr.	Daily	Hr.	Daily	Bare Costs	Incl. O&P
1 Equip. Oper. (medium)	$50.60	$404.80	$80.30	$642.40	$46.27	$74.08
.5 Laborer	37.60	150.40	61.65	246.60		
1 F.E. Loader, T.M., 2.25 C.Y.		956.40		1052.04	79.70	87.67
12 L.H., Daily Totals		$1511.60		$1941.04	$125.97	$161.75
Crew B-10P	Hr.	Daily	Hr.	Daily	Bare Costs	Incl. O&P
1 Equip. Oper. (medium)	$50.60	$404.80	$80.30	$642.40	$46.27	$74.08
.5 Laborer	37.60	150.40	61.65	246.60		
1 Crawler Loader, 3 C.Y.		1188.00		1306.80	99.00	108.90
12 L.H., Daily Totals		$1743.20		$2195.80	$145.27	$182.98
Crew B-10Q	Hr.	Daily	Hr.	Daily	Bare Costs	Incl. O&P
1 Equip. Oper. (medium)	$50.60	$404.80	$80.30	$642.40	$46.27	$74.08
.5 Laborer	37.60	150.40	61.65	246.60		
1 Crawler Loader, 4 C.Y.		1561.00		1717.10	130.08	143.09
12 L.H., Daily Totals		$2116.20		$2606.10	$176.35	$217.18
Crew B-10R	Hr.	Daily	Hr.	Daily	Bare Costs	Incl. O&P
1 Equip. Oper. (medium)	$50.60	$404.80	$80.30	$642.40	$46.27	$74.08
.5 Laborer	37.60	150.40	61.65	246.60		
1 F.E. Loader, W.M., 1 C.Y.		299.00		328.90	24.92	27.41
12 L.H., Daily Totals		$854.20		$1217.90	$71.18	$101.49
Crew B-10S	Hr.	Daily	Hr.	Daily	Bare Costs	Incl. O&P
1 Equip. Oper. (medium)	$50.60	$404.80	$80.30	$642.40	$46.27	$74.08
.5 Laborer	37.60	150.40	61.65	246.60		
1 F.E. Loader, W.M., 1.5 C.Y.		378.40		416.24	31.53	34.69
12 L.H., Daily Totals		$933.60		$1305.24	$77.80	$108.77
Crew B-10T	Hr.	Daily	Hr.	Daily	Bare Costs	Incl. O&P
1 Equip. Oper. (medium)	$50.60	$404.80	$80.30	$642.40	$46.27	$74.08
.5 Laborer	37.60	150.40	61.65	246.60		
1 F.E. Loader, W.M., 2.5 C.Y.		519.80		571.78	43.32	47.65
12 L.H., Daily Totals		$1075.00		$1460.78	$89.58	$121.73
Crew B-10U	Hr.	Daily	Hr.	Daily	Bare Costs	Incl. O&P
1 Equip. Oper. (medium)	$50.60	$404.80	$80.30	$642.40	$46.27	$74.08
.5 Laborer	37.60	150.40	61.65	246.60		
1 F.E. Loader, W.M., 5.5 C.Y.		1082.00		1190.20	90.17	99.18
12 L.H., Daily Totals		$1637.20		$2079.20	$136.43	$173.27
Crew B-10V	Hr.	Daily	Hr.	Daily	Bare Costs	Incl. O&P
1 Equip. Oper. (medium)	$50.60	$404.80	$80.30	$642.40	$46.27	$74.08
.5 Laborer	37.60	150.40	61.65	246.60		
1 Dozer, 700 H.P.		4979.00		5476.90	414.92	456.41
12 L.H., Daily Totals		$5534.20		$6365.90	$461.18	$530.49
Crew B-10W	Hr.	Daily	Hr.	Daily	Bare Costs	Incl. O&P
1 Equip. Oper. (medium)	$50.60	$404.80	$80.30	$642.40	$46.27	$74.08
.5 Laborer	37.60	150.40	61.65	246.60		
1 Dozer, 105 H.P.		602.80		663.08	50.23	55.26
12 L.H., Daily Totals		$1158.00		$1552.08	$96.50	$129.34
Crew B-10X	Hr.	Daily	Hr.	Daily	Bare Costs	Incl. O&P
1 Equip. Oper. (medium)	$50.60	$404.80	$80.30	$642.40	$46.27	$74.08
.5 Laborer	37.60	150.40	61.65	246.60		
1 Dozer, 410 H.P.		2405.00		2645.50	200.42	220.46
12 L.H., Daily Totals		$2960.20		$3534.50	$246.68	$294.54

Crew No.	Bare Costs		Incl. Subs O&P		Cost Per Labor-Hour	
Crew B-10Y	Hr.	Daily	Hr.	Daily	Bare Costs	Incl. O&P
1 Equip. Oper. (medium)	$50.60	$404.80	$80.30	$642.40	$46.27	$74.08
.5 Laborer	37.60	150.40	61.65	246.60		
1 Vibr. Roller, Towed, 12 Ton		549.80		604.78	45.82	50.40
12 L.H., Daily Totals		$1105.00		$1493.78	$92.08	$124.48
Crew B-11A	Hr.	Daily	Hr.	Daily	Bare Costs	Incl. O&P
1 Equipment Oper. (med.)	$50.60	$404.80	$80.30	$642.40	$44.10	$70.97
1 Laborer	37.60	300.80	61.65	493.20		
1 Dozer, 200 H.P.		1387.00		1525.70	86.69	95.36
16 L.H., Daily Totals		$2092.60		$2661.30	$130.79	$166.33
Crew B-11B	Hr.	Daily	Hr.	Daily	Bare Costs	Incl. O&P
1 Equipment Oper. (light)	$48.60	$388.80	$77.15	$617.20	$43.10	$69.40
1 Laborer	37.60	300.80	61.65	493.20		
1 Air Powered Tamper		26.20		28.82		
1 Air Compressor, 365 cfm		263.80		290.18		
2 -50' Air Hoses, 1.5"		11.60		12.76	18.85	20.73
16 L.H., Daily Totals		$991.20		$1442.16	$61.95	$90.14
Crew B-11C	Hr.	Daily	Hr.	Daily	Bare Costs	Incl. O&P
1 Equipment Oper. (med.)	$50.60	$404.80	$80.30	$642.40	$44.10	$70.97
1 Laborer	37.60	300.80	61.65	493.20		
1 Backhoe Loader, 48 H.P.		364.20		400.62	22.76	25.04
16 L.H., Daily Totals		$1069.80		$1536.22	$66.86	$96.01
Crew B-11K	Hr.	Daily	Hr.	Daily	Bare Costs	Incl. O&P
1 Equipment Oper. (med.)	$50.60	$404.80	$80.30	$642.40	$44.10	$70.97
1 Laborer	37.60	300.80	61.65	493.20		
1 Trencher, Chain Type, 8' D		3402.00		3742.20	212.63	233.89
16 L.H., Daily Totals		$4107.60		$4877.80	$256.73	$304.86
Crew B-11L	Hr.	Daily	Hr.	Daily	Bare Costs	Incl. O&P
1 Equipment Oper. (med.)	$50.60	$404.80	$80.30	$642.40	$44.10	$70.97
1 Laborer	37.60	300.80	61.65	493.20		
1 Grader, 30,000 Lbs.		736.20		809.82	46.01	50.61
16 L.H., Daily Totals		$1441.80		$1945.42	$90.11	$121.59
Crew B-11M	Hr.	Daily	Hr.	Daily	Bare Costs	Incl. O&P
1 Equipment Oper. (med.)	$50.60	$404.80	$80.30	$642.40	$44.10	$70.97
1 Laborer	37.60	300.80	61.65	493.20		
1 Backhoe Loader, 80 H.P.		391.80		430.98	24.49	26.94
16 L.H., Daily Totals		$1097.40		$1566.58	$68.59	$97.91
Crew B-11S	Hr.	Daily	Hr.	Daily	Bare Costs	Incl. O&P
1 Equipment Operator (med.)	$50.60	$404.80	$80.30	$642.40	$46.27	$74.08
.5 Laborer	37.60	150.40	61.65	246.60		
1 Dozer, 300 H.P.		1897.00		2086.70		
1 Ripper, Beam & 1 Shank		84.20		92.62	165.10	181.61
12 L.H., Daily Totals		$2536.40		$3068.32	$211.37	$255.69
Crew B-11T	Hr.	Daily	Hr.	Daily	Bare Costs	Incl. O&P
1 Equipment Operator (med.)	$50.60	$404.80	$80.30	$642.40	$46.27	$74.08
.5 Laborer	37.60	150.40	61.65	246.60		
1 Dozer, 410 H.P.		2405.00		2645.50		
1 Ripper, Beam & 2 Shanks		110.00		121.00	209.58	230.54
12 L.H., Daily Totals		$3070.20		$3655.50	$255.85	$304.63

Crews

Crew No.	Bare Costs		Incl. Subs O&P		Cost Per Labor-Hour	
Crew B-11W	Hr.	Daily	Hr.	Daily	Bare Costs	Incl. O&P
1 Equipment Operator (med.)	$50.60	$404.80	$80.30	$642.40	$40.73	$65.62
1 Common Laborer	37.60	300.80	61.65	493.20		
10 Truck Drivers (heavy)	40.05	3204.00	64.55	5164.00		
1 Dozer, 200 H.P.		1387.00		1525.70		
1 Vibratory Roller, Towed, 23 Ton		396.00		435.60		
10 Dump Trucks, 8 C.Y., 220 H.P.		4112.00		4523.20	61.41	67.55
96 L.H., Daily Totals		$9804.60		$12784.10	$102.13	$133.17
Crew B-11Y	Hr.	Daily	Hr.	Daily	Bare Costs	Incl. O&P
1 Labor Foreman (outside)	$39.60	$316.80	$64.90	$519.20	$42.16	$68.23
5 Common Laborers	37.60	1504.00	61.65	2466.00		
3 Equipment Operators (med.)	50.60	1214.40	80.30	1927.20		
1 Dozer, 80 H.P.		472.40		519.64		
2 Roller, 2-Drum, W.B., 7.5 H.P.		360.00		396.00		
4 Vibrating Plate, Gas, 21"		187.20		205.92	14.16	15.58
72 L.H., Daily Totals		$4054.80		$6033.96	$56.32	$83.81
Crew B-12A	Hr.	Daily	Hr.	Daily	Bare Costs	Incl. O&P
1 Equip. Oper. (crane)	$51.70	$413.60	$82.05	$656.40	$44.65	$71.85
1 Laborer	37.60	300.80	61.65	493.20		
1 Hyd. Excavator, 1 C.Y.		812.60		893.86	50.79	55.87
16 L.H., Daily Totals		$1527.00		$2043.46	$95.44	$127.72
Crew B-12B	Hr.	Daily	Hr.	Daily	Bare Costs	Incl. O&P
1 Equip. Oper. (crane)	$51.70	$413.60	$82.05	$656.40	$44.65	$71.85
1 Laborer	37.60	300.80	61.65	493.20		
1 Hyd. Excavator, 1.5 C.Y.		1030.00		1133.00	64.38	70.81
16 L.H., Daily Totals		$1744.40		$2282.60	$109.03	$142.66
Crew B-12C	Hr.	Daily	Hr.	Daily	Bare Costs	Incl. O&P
1 Equip. Oper. (crane)	$51.70	$413.60	$82.05	$656.40	$44.65	$71.85
1 Laborer	37.60	300.80	61.65	493.20		
1 Hyd. Excavator, 2 C.Y.		1175.00		1292.50	73.44	80.78
16 L.H., Daily Totals		$1889.40		$2442.10	$118.09	$152.63
Crew B-12D	Hr.	Daily	Hr.	Daily	Bare Costs	Incl. O&P
1 Equip. Oper. (crane)	$51.70	$413.60	$82.05	$656.40	$44.65	$71.85
1 Laborer	37.60	300.80	61.65	493.20		
1 Hyd. Excavator, 3.5 C.Y.		2440.00		2684.00	152.50	167.75
16 L.H., Daily Totals		$3154.40		$3833.60	$197.15	$239.60
Crew B-12E	Hr.	Daily	Hr.	Daily	Bare Costs	Incl. O&P
1 Equip. Oper. (crane)	$51.70	$413.60	$82.05	$656.40	$44.65	$71.85
1 Laborer	37.60	300.80	61.65	493.20		
1 Hyd. Excavator, .5 C.Y.		448.20		493.02	28.01	30.81
16 L.H., Daily Totals		$1162.60		$1642.62	$72.66	$102.66
Crew B-12F	Hr.	Daily	Hr.	Daily	Bare Costs	Incl. O&P
1 Equip. Oper. (crane)	$51.70	$413.60	$82.05	$656.40	$44.65	$71.85
1 Laborer	37.60	300.80	61.65	493.20		
1 Hyd. Excavator, .75 C.Y.		654.80		720.28	40.92	45.02
16 L.H., Daily Totals		$1369.20		$1869.88	$85.58	$116.87
Crew B-12G	Hr.	Daily	Hr.	Daily	Bare Costs	Incl. O&P
1 Equip. Oper. (crane)	$51.70	$413.60	$82.05	$656.40	$44.65	$71.85
1 Laborer	37.60	300.80	61.65	493.20		
1 Crawler Crane, 15 Ton		669.85		736.84		
1 Clamshell Bucket, .5 C.Y.		38.20		42.02	44.25	48.68
16 L.H., Daily Totals		$1422.45		$1928.45	$88.90	$120.53
Crew B-12H	Hr.	Daily	Hr.	Daily	Bare Costs	Incl. O&P
1 Equip. Oper. (crane)	$51.70	$413.60	$82.05	$656.40	$44.65	$71.85
1 Laborer	37.60	300.80	61.65	493.20		
1 Crawler Crane, 25 Ton		1150.00		1265.00		
1 Clamshell Bucket, 1 C.Y.		47.80		52.58	74.86	82.35
16 L.H., Daily Totals		$1912.20		$2467.18	$119.51	$154.20
Crew B-12I	Hr.	Daily	Hr.	Daily	Bare Costs	Incl. O&P
1 Equip. Oper. (crane)	$51.70	$413.60	$82.05	$656.40	$44.65	$71.85
1 Laborer	37.60	300.80	61.65	493.20		
1 Crawler Crane, 20 Ton		862.50		948.75		
1 Dragline Bucket, .75 C.Y.		20.60		22.66	55.19	60.71
16 L.H., Daily Totals		$1597.50		$2121.01	$99.84	$132.56
Crew B-12J	Hr.	Daily	Hr.	Daily	Bare Costs	Incl. O&P
1 Equip. Oper. (crane)	$51.70	$413.60	$82.05	$656.40	$44.65	$71.85
1 Laborer	37.60	300.80	61.65	493.20		
1 Gradall, 5/8 C.Y.		882.20		970.42	55.14	60.65
16 L.H., Daily Totals		$1596.60		$2120.02	$99.79	$132.50
Crew B-12K	Hr.	Daily	Hr.	Daily	Bare Costs	Incl. O&P
1 Equip. Oper. (crane)	$51.70	$413.60	$82.05	$656.40	$44.65	$71.85
1 Laborer	37.60	300.80	61.65	493.20		
1 Gradall, 3 Ton, 1 C.Y.		1002.00		1102.20	62.63	68.89
16 L.H., Daily Totals		$1716.40		$2251.80	$107.28	$140.74
Crew B-12L	Hr.	Daily	Hr.	Daily	Bare Costs	Incl. O&P
1 Equip. Oper. (crane)	$51.70	$413.60	$82.05	$656.40	$44.65	$71.85
1 Laborer	37.60	300.80	61.65	493.20		
1 Crawler Crane, 15 Ton		669.85		736.84		
1 F.E. Attachment, .5 C.Y.		60.00		66.00	45.62	50.18
16 L.H., Daily Totals		$1444.25		$1952.43	$90.27	$122.03
Crew B-12M	Hr.	Daily	Hr.	Daily	Bare Costs	Incl. O&P
1 Equip. Oper. (crane)	$51.70	$413.60	$82.05	$656.40	$44.65	$71.85
1 Laborer	37.60	300.80	61.65	493.20		
1 Crawler Crane, 20 Ton		862.50		948.75		
1 F.E. Attachment, .75 C.Y.		65.40		71.94	57.99	63.79
16 L.H., Daily Totals		$1642.30		$2170.29	$102.64	$135.64
Crew B-12N	Hr.	Daily	Hr.	Daily	Bare Costs	Incl. O&P
1 Equip. Oper. (crane)	$51.70	$413.60	$82.05	$656.40	$44.65	$71.85
1 Laborer	37.60	300.80	61.65	493.20		
1 Crawler Crane, 25 Ton		1150.00		1265.00		
1 F.E. Attachment, 1 C.Y.		71.20		78.32	76.33	83.96
16 L.H., Daily Totals		$1935.60		$2492.92	$120.97	$155.81
Crew B-12O	Hr.	Daily	Hr.	Daily	Bare Costs	Incl. O&P
1 Equip. Oper. (crane)	$51.70	$413.60	$82.05	$656.40	$44.65	$71.85
1 Laborer	37.60	300.80	61.65	493.20		
1 Crawler Crane, 40 Ton		1156.00		1271.60		
1 F.E. Attachment, 1.5 C.Y.		80.00		88.00	77.25	84.97
16 L.H., Daily Totals		$1950.40		$2509.20	$121.90	$156.82
Crew B-12P	Hr.	Daily	Hr.	Daily	Bare Costs	Incl. O&P
1 Equip. Oper. (crane)	$51.70	$413.60	$82.05	$656.40	$44.65	$71.85
1 Laborer	37.60	300.80	61.65	493.20		
1 Crawler Crane, 40 Ton		1156.00		1271.60		
1 Dragline Bucket, 1.5 C.Y.		33.60		36.96	74.35	81.78
16 L.H., Daily Totals		$1904.00		$2458.16	$119.00	$153.63

Crews

Crew No.	Bare Costs		Incl. Subs O&P		Cost Per Labor-Hour	
Crew B-12Q	Hr.	Daily	Hr.	Daily	Bare Costs	Incl. O&P
1 Equip. Oper. (crane)	$51.70	$413.60	$82.05	$656.40	$44.65	$71.85
1 Laborer	37.60	300.80	61.65	493.20		
1 Hyd. Excavator, 5/8 C.Y.		589.40		648.34	36.84	40.52
16 L.H., Daily Totals		$1303.80		$1797.94	$81.49	$112.37
Crew B-12S	Hr.	Daily	Hr.	Daily	Bare Costs	Incl. O&P
1 Equip. Oper. (crane)	$51.70	$413.60	$82.05	$656.40	$44.65	$71.85
1 Laborer	37.60	300.80	61.65	493.20		
1 Hyd. Excavator, 2.5 C.Y.		1605.00		1765.50	100.31	110.34
16 L.H., Daily Totals		$2319.40		$2915.10	$144.96	$182.19
Crew B-12T	Hr.	Daily	Hr.	Daily	Bare Costs	Incl. O&P
1 Equip. Oper. (crane)	$51.70	$413.60	$82.05	$656.40	$44.65	$71.85
1 Laborer	37.60	300.80	61.65	493.20		
1 Crawler Crane, 75 Ton		1459.00		1604.90		
1 F.E. Attachment, 3 C.Y.		103.20		113.52	97.64	107.40
16 L.H., Daily Totals		$2276.60		$2868.02	$142.29	$179.25
Crew B-12V	Hr.	Daily	Hr.	Daily	Bare Costs	Incl. O&P
1 Equip. Oper. (crane)	$51.70	$413.60	$82.05	$656.40	$44.65	$71.85
1 Laborer	37.60	300.80	61.65	493.20		
1 Crawler Crane, 75 Ton		1459.00		1604.90		
1 Dragline Bucket, 3 C.Y.		52.60		57.86	94.47	103.92
16 L.H., Daily Totals		$2226.00		$2812.36	$139.13	$175.77
Crew B-12Y	Hr.	Daily	Hr.	Daily	Bare Costs	Incl. O&P
1 Equip. Oper. (crane)	$51.70	$413.60	$82.05	$656.40	$42.30	$68.45
2 Laborers	37.60	601.60	61.65	986.40		
1 Hyd. Excavator, 3.5 C.Y.		2440.00		2684.00	101.67	111.83
24 L.H., Daily Totals		$3455.20		$4326.80	$143.97	$180.28
Crew B-12Z	Hr.	Daily	Hr.	Daily	Bare Costs	Incl. O&P
1 Equip. Oper. (crane)	$51.70	$413.60	$82.05	$656.40	$42.30	$68.45
2 Laborers	37.60	601.60	61.65	986.40		
1 Hyd. Excavator, 2.5 C.Y.		1605.00		1765.50	66.88	73.56
24 L.H., Daily Totals		$2620.20		$3408.30	$109.18	$142.01
Crew B-13	Hr.	Daily	Hr.	Daily	Bare Costs	Incl. O&P
1 Labor Foreman (outside)	$39.60	$316.80	$64.90	$519.20	$40.99	$66.47
4 Laborers	37.60	1203.20	61.65	1972.80		
1 Equip. Oper. (crane)	51.70	413.60	82.05	656.40		
1 Equip. Oper. (oiler)	45.20	361.60	71.75	574.00		
1 Hyd. Crane, 25 Ton		736.60		810.26	13.15	14.47
56 L.H., Daily Totals		$3031.80		$4532.66	$54.14	$80.94
Crew B-13A	Hr.	Daily	Hr.	Daily	Bare Costs	Incl. O&P
1 Labor Foreman (outside)	$39.60	$316.80	$64.90	$519.20	$42.30	$68.27
2 Laborers	37.60	601.60	61.65	986.40		
2 Equipment Operators (med.)	50.60	809.60	80.30	1284.80		
2 Truck Drivers (heavy)	40.05	640.80	64.55	1032.80		
1 Crawler Crane, 75 Ton		1459.00		1604.90		
1 Crawler Loader, 4 C.Y.		1561.00		1717.10		
2 Dump Trucks, 8 C.Y., 220 H.P.		822.40		904.64	68.61	75.48
56 L.H., Daily Totals		$6211.20		$8049.84	$110.91	$143.75

Crew No.	Bare Costs		Incl. Subs O&P		Cost Per Labor-Hour	
Crew B-13B	Hr.	Daily	Hr.	Daily	Bare Costs	Incl. O&P
1 Labor Foreman (outside)	$39.60	$316.80	$64.90	$519.20	$40.99	$66.47
4 Laborers	37.60	1203.20	61.65	1972.80		
1 Equip. Oper. (crane)	51.70	413.60	82.05	656.40		
1 Equip. Oper. (oiler)	45.20	361.60	71.75	574.00		
1 Hyd. Crane, 55 Ton		1128.00		1240.80	20.14	22.16
56 L.H., Daily Totals		$3423.20		$4963.20	$61.13	$88.63
Crew B-13C	Hr.	Daily	Hr.	Daily	Bare Costs	Incl. O&P
1 Labor Foreman (outside)	$39.60	$316.80	$64.90	$519.20	$40.99	$66.47
4 Laborers	37.60	1203.20	61.65	1972.80		
1 Equip. Oper. (crane)	51.70	413.60	82.05	656.40		
1 Equip. Oper. (oiler)	45.20	361.60	71.75	574.00		
1 Crawler Crane, 100 Ton		1667.00		1833.70	29.77	32.74
56 L.H., Daily Totals		$3962.20		$5556.10	$70.75	$99.22
Crew B-13D	Hr.	Daily	Hr.	Daily	Bare Costs	Incl. O&P
1 Laborer	$37.60	$300.80	$61.65	$493.20	$44.65	$71.85
1 Equip. Oper. (crane)	51.70	413.60	82.05	656.40		
1 Hyd. Excavator, 1 C.Y.		812.60		893.86		
1 Trench Box		81.00		89.10	55.85	61.44
16 L.H., Daily Totals		$1608.00		$2132.56	$100.50	$133.29
Crew B-13E	Hr.	Daily	Hr.	Daily	Bare Costs	Incl. O&P
1 Laborer	$37.60	$300.80	$61.65	$493.20	$44.65	$71.85
1 Equip. Oper. (crane)	51.70	413.60	82.05	656.40		
1 Hyd. Excavator, 1.5 C.Y.		1030.00		1133.00		
1 Trench Box		81.00		89.10	69.44	76.38
16 L.H., Daily Totals		$1825.40		$2371.70	$114.09	$148.23
Crew B-13F	Hr.	Daily	Hr.	Daily	Bare Costs	Incl. O&P
1 Laborer	$37.60	$300.80	$61.65	$493.20	$44.65	$71.85
1 Equip. Oper. (crane)	51.70	413.60	82.05	656.40		
1 Hyd. Excavator, 3.5 C.Y.		2440.00		2684.00		
1 Trench Box		81.00		89.10	157.56	173.32
16 L.H., Daily Totals		$3235.40		$3922.70	$202.21	$245.17
Crew B-13G	Hr.	Daily	Hr.	Daily	Bare Costs	Incl. O&P
1 Laborer	$37.60	$300.80	$61.65	$493.20	$44.65	$71.85
1 Equip. Oper. (crane)	51.70	413.60	82.05	656.40		
1 Hyd. Excavator, .75 C.Y.		654.80		720.28		
1 Trench Box		81.00		89.10	45.99	50.59
16 L.H., Daily Totals		$1450.20		$1958.98	$90.64	$122.44
Crew B-13H	Hr.	Daily	Hr.	Daily	Bare Costs	Incl. O&P
1 Laborer	$37.60	$300.80	$61.65	$493.20	$44.65	$71.85
1 Equip. Oper. (crane)	51.70	413.60	82.05	656.40		
1 Gradall, 5/8 C.Y.		882.20		970.42		
1 Trench Box		81.00		89.10	60.20	66.22
16 L.H., Daily Totals		$1677.60		$2209.12	$104.85	$138.07
Crew B-13I	Hr.	Daily	Hr.	Daily	Bare Costs	Incl. O&P
1 Laborer	$37.60	$300.80	$61.65	$493.20	$44.65	$71.85
1 Equip. Oper. (crane)	51.70	413.60	82.05	656.40		
1 Gradall, 3 Ton, 1 C.Y.		1002.00		1102.20		
1 Trench Box		81.00		89.10	67.69	74.46
16 L.H., Daily Totals		$1797.40		$2340.90	$112.34	$146.31

Crews

Crew No.		Bare Costs		Incl. Subs O&P		Cost Per Labor-Hour	
Crew B-13J	Hr.	Daily	Hr.	Daily	Bare Costs	Incl. O&P	
1 Laborer	$37.60	$300.80	$61.65	$493.20	$44.65	$71.85	
1 Equip. Oper. (crane)	51.70	413.60	82.05	656.40			
1 Hyd. Excavator, 2.5 C.Y.		1605.00		1765.50			
1 Trench Box		81.00		89.10	105.38	115.91	
16 L.H., Daily Totals		$2400.40		$3004.20	$150.03	$187.76	

Crew B-13K	Hr.	Daily	Hr.	Daily	Bare Costs	Incl. O&P
2 Equip. Opers. (crane)	$51.70	$827.20	$82.05	$1312.80	$51.70	$82.05
1 Hyd. Excavator, .75 C.Y.		654.80		720.28		
1 Hyd. Hammer, 4000 ft-lb		308.20		339.02		
1 Hyd. Excavator, .75 C.Y.		654.80		720.28	101.11	111.22
16 L.H., Daily Totals		$2445.00		$3092.38	$152.81	$193.27

Crew B-13L	Hr.	Daily	Hr.	Daily	Bare Costs	Incl. O&P
2 Equip. Opers. (crane)	$51.70	$827.20	$82.05	$1312.80	$51.70	$82.05
1 Hyd. Excavator, 1.5 C.Y.		1030.00		1133.00		
1 Hyd. Hammer, 5000 ft-lb		371.20		408.32		
1 Hyd. Excavator, .75 C.Y.		654.80		720.28	128.50	141.35
16 L.H., Daily Totals		$2883.20		$3574.40	$180.20	$223.40

Crew B-13M	Hr.	Daily	Hr.	Daily	Bare Costs	Incl. O&P
2 Equip. Opers. (crane)	$51.70	$827.20	$82.05	$1312.80	$51.70	$82.05
1 Hyd. Excavator, 2.5 C.Y.		1605.00		1765.50		
1 Hyd. Hammer, 8000 ft-lb		545.40		599.94		
1 Hyd. Excavator, 1.5 C.Y.		1030.00		1133.00	198.78	218.65
16 L.H., Daily Totals		$4007.60		$4811.24	$250.47	$300.70

Crew B-13N	Hr.	Daily	Hr.	Daily	Bare Costs	Incl. O&P
2 Equip. Opers. (crane)	$51.70	$827.20	$82.05	$1312.80	$51.70	$82.05
1 Hyd. Excavator, 3.5 C.Y.		2440.00		2684.00		
1 Hyd. Hammer, 12,000 ft-lb		635.80		699.38		
1 Hyd. Excavator, 1.5 C.Y.		1030.00		1133.00	256.61	282.27
16 L.H., Daily Totals		$4933.00		$5829.18	$308.31	$364.32

Crew B-14	Hr.	Daily	Hr.	Daily	Bare Costs	Incl. O&P
1 Labor Foreman (outside)	$39.60	$316.80	$64.90	$519.20	$39.77	$64.78
4 Laborers	37.60	1203.20	61.65	1972.80		
1 Equip. Oper. (light)	48.60	388.80	77.15	617.20		
1 Backhoe Loader, 48 H.P.		364.20		400.62	7.59	8.35
48 L.H., Daily Totals		$2273.00		$3509.82	$47.35	$73.12

Crew B-14A	Hr.	Daily	Hr.	Daily	Bare Costs	Incl. O&P
1 Equip. Oper. (crane)	$51.70	$413.60	$82.05	$656.40	$47.00	$75.25
.5 Laborer	37.60	150.40	61.65	246.60		
1 Hyd. Excavator, 4.5 C.Y.		2963.00		3259.30	246.92	271.61
12 L.H., Daily Totals		$3527.00		$4162.30	$293.92	$346.86

Crew B-14B	Hr.	Daily	Hr.	Daily	Bare Costs	Incl. O&P
1 Equip. Oper. (crane)	$51.70	$413.60	$82.05	$656.40	$47.00	$75.25
.5 Laborer	37.60	150.40	61.65	246.60		
1 Hyd. Excavator, 6 C.Y.		3494.00		3843.40	291.17	320.28
12 L.H., Daily Totals		$4058.00		$4746.40	$338.17	$395.53

Crew B-14C	Hr.	Daily	Hr.	Daily	Bare Costs	Incl. O&P
1 Equip. Oper. (crane)	$51.70	$413.60	$82.05	$656.40	$47.00	$75.25
.5 Laborer	37.60	150.40	61.65	246.60		
1 Hyd. Excavator, 7 C.Y.		3590.00		3949.00	299.17	329.08
12 L.H., Daily Totals		$4154.00		$4852.00	$346.17	$404.33

Crew B-14F	Hr.	Daily	Hr.	Daily	Bare Costs	Incl. O&P
1 Equip. Oper. (crane)	$51.70	$413.60	$82.05	$656.40	$47.00	$75.25
.5 Laborer	37.60	150.40	61.65	246.60		
1 Hyd. Shovel, 7 C.Y.		3993.00		4392.30	332.75	366.02
12 L.H., Daily Totals		$4557.00		$5295.30	$379.75	$441.27

Crew B-14G	Hr.	Daily	Hr.	Daily	Bare Costs	Incl. O&P
1 Equip. Oper. (crane)	$51.70	$413.60	$82.05	$656.40	$47.00	$75.25
.5 Laborer	37.60	150.40	61.65	246.60		
1 Hyd. Shovel, 12 C.Y.		5736.00		6309.60	478.00	525.80
12 L.H., Daily Totals		$6300.00		$7212.60	$525.00	$601.05

Crew B-14J	Hr.	Daily	Hr.	Daily	Bare Costs	Incl. O&P
1 Equip. Oper. (medium)	$50.60	$404.80	$80.30	$642.40	$46.27	$74.08
.5 Laborer	37.60	150.40	61.65	246.60		
1 F.E. Loader, 8 C.Y.		1937.00		2130.70	161.42	177.56
12 L.H., Daily Totals		$2492.20		$3019.70	$207.68	$251.64

Crew B-14K	Hr.	Daily	Hr.	Daily	Bare Costs	Incl. O&P
1 Equip. Oper. (medium)	$50.60	$404.80	$80.30	$642.40	$46.27	$74.08
.5 Laborer	37.60	150.40	61.65	246.60		
1 F.E. Loader, 10 C.Y.		2942.00		3236.20	245.17	269.68
12 L.H., Daily Totals		$3497.20		$4125.20	$291.43	$343.77

Crew B-15	Hr.	Daily	Hr.	Daily	Bare Costs	Incl. O&P
1 Equipment Oper. (med.)	$50.60	$404.80	$80.30	$642.40	$42.71	$68.64
.5 Laborer	37.60	150.40	61.65	246.60		
2 Truck Drivers (heavy)	40.05	640.80	64.55	1032.80		
2 Dump Trucks, 12 C.Y., 400 H.P.		1382.00		1520.20		
1 Dozer, 200 H.P.		1387.00		1525.70	98.89	108.78
28 L.H., Daily Totals		$3965.00		$4967.70	$141.61	$177.42

Crew B-16	Hr.	Daily	Hr.	Daily	Bare Costs	Incl. O&P
1 Labor Foreman (outside)	$39.60	$316.80	$64.90	$519.20	$38.71	$63.19
2 Laborers	37.60	601.60	61.65	986.40		
1 Truck Driver (heavy)	40.05	320.40	64.55	516.40		
1 Dump Truck, 12 C.Y., 400 H.P.		691.00		760.10	21.59	23.75
32 L.H., Daily Totals		$1929.80		$2782.10	$60.31	$86.94

Crew B-17	Hr.	Daily	Hr.	Daily	Bare Costs	Incl. O&P
2 Laborers	$37.60	$601.60	$61.65	$986.40	$40.96	$66.25
1 Equip. Oper. (light)	48.60	388.80	77.15	617.20		
1 Truck Driver (heavy)	40.05	320.40	64.55	516.40		
1 Backhoe Loader, 48 H.P.		364.20		400.62		
1 Dump Truck, 8 C.Y., 220 H.P.		411.20		452.32	24.23	26.65
32 L.H., Daily Totals		$2086.20		$2972.94	$65.19	$92.90

Crew B-17A	Hr.	Daily	Hr.	Daily	Bare Costs	Incl. O&P
2 Labor Foremen (outside)	$39.60	$633.60	$64.90	$1038.40	$40.41	$66.11
6 Laborers	37.60	1804.80	61.65	2959.20		
1 Skilled Worker Foreman (out)	50.65	405.20	82.30	658.40		
1 Skilled Worker	48.65	389.20	79.05	632.40		
80 L.H., Daily Totals		$3232.80		$5288.40	$40.41	$66.11

Crew B-17B	Hr.	Daily	Hr.	Daily	Bare Costs	Incl. O&P
2 Laborers	$37.60	$601.60	$61.65	$986.40	$40.96	$66.25
1 Equip. Oper. (light)	48.60	388.80	77.15	617.20		
1 Truck Driver (heavy)	40.05	320.40	64.55	516.40		
1 Backhoe Loader, 48 H.P.		364.20		400.62		
1 Dump Truck, 12 C.Y., 400 H.P.		691.00		760.10	32.98	36.27
32 L.H., Daily Totals		$2366.00		$3280.72	$73.94	$102.52

Crews

Crew No.	Bare Costs		Incl. Subs O&P		Cost Per Labor-Hour	
Crew B-18	Hr.	Daily	Hr.	Daily	Bare Costs	Incl. O&P
1 Labor Foreman (outside)	$39.60	$316.80	$64.90	$519.20	$38.27	$62.73
2 Laborers	37.60	601.60	61.65	986.40		
1 Vibrating Plate, Gas, 21"		46.80		51.48	1.95	2.15
24 L.H., Daily Totals		$965.20		$1557.08	$40.22	$64.88
Crew B-19	Hr.	Daily	Hr.	Daily	Bare Costs	Incl. O&P
1 Pile Driver Foreman (outside)	$48.10	$384.80	$78.55	$628.40	$47.64	$76.95
4 Pile Drivers	46.10	1475.20	75.30	2409.60		
2 Equip. Oper. (crane)	51.70	827.20	82.05	1312.80		
1 Equip. Oper. (oiler)	45.20	361.60	71.75	574.00		
1 Crawler Crane, 40 Ton		1156.00		1271.60		
1 Lead, 90' High		124.40		136.84		
1 Hammer, Diesel, 22k ft-lb		456.40		502.04	27.14	29.85
64 L.H., Daily Totals		$4785.60		$6835.28	$74.78	$106.80
Crew B-19A	Hr.	Daily	Hr.	Daily	Bare Costs	Incl. O&P
1 Pile Driver Foreman (outside)	$48.10	$384.80	$78.55	$628.40	$47.64	$76.95
4 Pile Drivers	46.10	1475.20	75.30	2409.60		
2 Equip. Oper. (crane)	51.70	827.20	82.05	1312.80		
1 Equip. Oper. (oiler)	45.20	361.60	71.75	574.00		
1 Crawler Crane, 75 Ton		1459.00		1604.90		
1 Lead, 90' high		124.40		136.84		
1 Hammer, Diesel, 41k ft-lb		581.00		639.10	33.82	37.20
64 L.H., Daily Totals		$5213.20		$7305.64	$81.46	$114.15
Crew B-19B	Hr.	Daily	Hr.	Daily	Bare Costs	Incl. O&P
1 Pile Driver Foreman (outside)	$48.10	$384.80	$78.55	$628.40	$47.64	$76.95
4 Pile Drivers	46.10	1475.20	75.30	2409.60		
2 Equip. Oper. (crane)	51.70	827.20	82.05	1312.80		
1 Equip. Oper. (oiler)	45.20	361.60	71.75	574.00		
1 Crawler Crane, 40 Ton		1156.00		1271.60		
1 Lead, 90' High		124.40		136.84		
1 Hammer, Diesel, 22k ft-lb		456.40		502.04		
1 Barge, 400 Ton		776.00		853.60	39.26	43.19
64 L.H., Daily Totals		$5561.60		$7688.88	$86.90	$120.14
Crew B-19C	Hr.	Daily	Hr.	Daily	Bare Costs	Incl. O&P
1 Pile Driver Foreman (outside)	$48.10	$384.80	$78.55	$628.40	$47.64	$76.95
4 Pile Drivers	46.10	1475.20	75.30	2409.60		
2 Equip. Oper. (crane)	51.70	827.20	82.05	1312.80		
1 Equip. Oper. (oiler)	45.20	361.60	71.75	574.00		
1 Crawler Crane, 75 Ton		1459.00		1604.90		
1 Lead, 90' High		124.40		136.84		
1 Hammer, Diesel, 41k ft-lb		581.00		639.10		
1 Barge, 400 Ton		776.00		853.60	45.94	50.54
64 L.H., Daily Totals		$5989.20		$8159.24	$93.58	$127.49
Crew B-20	Hr.	Daily	Hr.	Daily	Bare Costs	Incl. O&P
1 Labor Foreman (outside)	$39.60	$316.80	$64.90	$519.20	$41.95	$68.53
1 Skilled Worker	48.65	389.20	79.05	632.40		
1 Laborer	37.60	300.80	61.65	493.20		
24 L.H., Daily Totals		$1006.80		$1644.80	$41.95	$68.53
Crew B-20A	Hr.	Daily	Hr.	Daily	Bare Costs	Incl. O&P
1 Labor Foreman (outside)	$39.60	$316.80	$64.90	$519.20	$45.71	$72.84
1 Laborer	37.60	300.80	61.65	493.20		
1 Plumber	58.70	469.60	91.55	732.40		
1 Plumber Apprentice	46.95	375.60	73.25	586.00		
32 L.H., Daily Totals		$1462.80		$2330.80	$45.71	$72.84

Crew No.	Bare Costs		Incl. Subs O&P		Cost Per Labor-Hour	
Crew B-21	Hr.	Daily	Hr.	Daily	Bare Costs	Incl. O&P
1 Labor Foreman (outside)	$39.60	$316.80	$64.90	$519.20	$43.34	$70.46
1 Skilled Worker	48.65	389.20	79.05	632.40		
1 Laborer	37.60	300.80	61.65	493.20		
.5 Equip. Oper. (crane)	51.70	206.80	82.05	328.20		
.5 S.P. Crane, 4x4, 5 Ton		137.70		151.47	4.92	5.41
28 L.H., Daily Totals		$1351.30		$2124.47	$48.26	$75.87
Crew B-21A	Hr.	Daily	Hr.	Daily	Bare Costs	Incl. O&P
1 Labor Foreman (outside)	$39.60	$316.80	$64.90	$519.20	$46.91	$74.68
1 Laborer	37.60	300.80	61.65	493.20		
1 Plumber	58.70	469.60	91.55	732.40		
1 Plumber Apprentice	46.95	375.60	73.25	586.00		
1 Equip. Oper. (crane)	51.70	413.60	82.05	656.40		
1 S.P. Crane, 4x4, 12 Ton		473.40		520.74	11.84	13.02
40 L.H., Daily Totals		$2349.80		$3507.94	$58.74	$87.70
Crew B-21B	Hr.	Daily	Hr.	Daily	Bare Costs	Incl. O&P
1 Labor Foreman (outside)	$39.60	$316.80	$64.90	$519.20	$40.82	$66.38
3 Laborers	37.60	902.40	61.65	1479.60		
1 Equip. Oper. (crane)	51.70	413.60	82.05	656.40		
1 Hyd. Crane, 12 Ton		653.80		719.18	16.34	17.98
40 L.H., Daily Totals		$2286.60		$3374.38	$57.16	$84.36
Crew B-21C	Hr.	Daily	Hr.	Daily	Bare Costs	Incl. O&P
1 Labor Foreman (outside)	$39.60	$316.80	$64.90	$519.20	$40.99	$66.47
4 Laborers	37.60	1203.20	61.65	1972.80		
1 Equip. Oper. (crane)	51.70	413.60	82.05	656.40		
1 Equip. Oper. (oiler)	45.20	361.60	71.75	574.00		
2 Cutting Torches		22.80		25.08		
2 Sets of Gases		304.00		334.40		
1 Lattice Boom Crane, 90 Ton		1511.00		1662.10	32.82	36.10
56 L.H., Daily Totals		$4133.00		$5743.98	$73.80	$102.57
Crew B-22	Hr.	Daily	Hr.	Daily	Bare Costs	Incl. O&P
1 Labor Foreman (outside)	$39.60	$316.80	$64.90	$519.20	$43.90	$71.24
1 Skilled Worker	48.65	389.20	79.05	632.40		
1 Laborer	37.60	300.80	61.65	493.20		
.75 Equip. Oper. (crane)	51.70	310.20	82.05	492.30		
.75 S.P. Crane, 4x4, 5 Ton		206.55		227.21	6.88	7.57
30 L.H., Daily Totals		$1523.55		$2364.30	$50.78	$78.81
Crew B-22A	Hr.	Daily	Hr.	Daily	Bare Costs	Incl. O&P
1 Labor Foreman (outside)	$39.60	$316.80	$64.90	$519.20	$43.03	$69.86
1 Skilled Worker	48.65	389.20	79.05	632.40		
2 Laborers	37.60	601.60	61.65	986.40		
1 Equipment Operator, Crane	51.70	413.60	82.05	656.40		
1 S.P. Crane, 4x4, 5 Ton		275.40		302.94		
1 Butt Fusion Mach., 4"-12" diam.		382.10		420.31	16.44	18.08
40 L.H., Daily Totals		$2378.70		$3517.65	$59.47	$87.94
Crew B-22B	Hr.	Daily	Hr.	Daily	Bare Costs	Incl. O&P
1 Labor Foreman (outside)	$39.60	$316.80	$64.90	$519.20	$43.03	$69.86
1 Skilled Worker	48.65	389.20	79.05	632.40		
2 Laborers	37.60	601.60	61.65	986.40		
1 Equip. Oper. (crane)	51.70	413.60	82.05	656.40		
1 S.P. Crane, 4x4, 5 Ton		275.40		302.94		
1 Butt Fusion Mach., 8"-24" diam.		829.70		912.67	27.63	30.39
40 L.H., Daily Totals		$2826.30		$4010.01	$70.66	$100.25

Crews

Crew No.	Bare Costs		Incl. Subs O&P		Cost Per Labor-Hour	
Crew B-22C	Hr.	Daily	Hr.	Daily	Bare Costs	Incl. O&P
1 Skilled Worker	$48.65	$389.20	$79.05	$632.40	$43.13	$70.35
1 Laborer	37.60	300.80	61.65	493.20		
1 Butt Fusion Mach., 2"-8" diam.		121.05		133.16	7.57	8.32
16 L.H., Daily Totals		$811.05		$1258.76	$50.69	$78.67
Crew B-23	Hr.	Daily	Hr.	Daily	Bare Costs	Incl. O&P
1 Labor Foreman (outside)	$39.60	$316.80	$64.90	$519.20	$38.00	$62.30
4 Laborers	37.60	1203.20	61.65	1972.80		
1 Drill Rig, Truck-Mounted		2542.00		2796.20		
1 Flatbed Truck, Gas, 3 Ton		303.20		333.52	71.13	78.24
40 L.H., Daily Totals		$4365.20		$5621.72	$109.13	$140.54
Crew B-23A	Hr.	Daily	Hr.	Daily	Bare Costs	Incl. O&P
1 Labor Foreman (outside)	$39.60	$316.80	$64.90	$519.20	$42.60	$68.95
1 Laborer	37.60	300.80	61.65	493.20		
1 Equip. Oper. (medium)	50.60	404.80	80.30	642.40		
1 Drill Rig, Truck-Mounted		2542.00		2796.20		
1 Pickup Truck, 3/4 Ton		144.20		158.62	111.93	123.12
24 L.H., Daily Totals		$3708.60		$4609.62	$154.53	$192.07
Crew B-23B	Hr.	Daily	Hr.	Daily	Bare Costs	Incl. O&P
1 Labor Foreman (outside)	$39.60	$316.80	$64.90	$519.20	$42.60	$68.95
1 Laborer	37.60	300.80	61.65	493.20		
1 Equip. Oper. (medium)	50.60	404.80	80.30	642.40		
1 Drill Rig, Truck-Mounted		2542.00		2796.20		
1 Pickup Truck, 3/4 Ton		144.20		158.62		
1 Centr. Water Pump, 6"		346.60		381.26	126.37	139.00
24 L.H., Daily Totals		$4055.20		$4990.88	$168.97	$207.95
Crew B-24	Hr.	Daily	Hr.	Daily	Bare Costs	Incl. O&P
1 Cement Finisher	$45.00	$360.00	$71.15	$569.20	$43.18	$69.92
1 Laborer	37.60	300.80	61.65	493.20		
1 Carpenter	46.95	375.60	76.95	615.60		
24 L.H., Daily Totals		$1036.40		$1678.00	$43.18	$69.92
Crew B-25	Hr.	Daily	Hr.	Daily	Bare Costs	Incl. O&P
1 Labor Foreman (outside)	$39.60	$316.80	$64.90	$519.20	$41.33	$67.03
7 Laborers	37.60	2105.60	61.65	3452.40		
3 Equip. Oper. (medium)	50.60	1214.40	80.30	1927.20		
1 Asphalt Paver, 130 H.P.		2132.00		2345.20		
1 Tandem Roller, 10 Ton		236.80		260.48		
1 Roller, Pneum. Whl., 12 Ton		344.40		378.84	30.83	33.91
88 L.H., Daily Totals		$6350.00		$8883.32	$72.16	$100.95
Crew B-25B	Hr.	Daily	Hr.	Daily	Bare Costs	Incl. O&P
1 Labor Foreman (outside)	$39.60	$316.80	$64.90	$519.20	$42.10	$68.14
7 Laborers	37.60	2105.60	61.65	3452.40		
4 Equip. Oper. (medium)	50.60	1619.20	80.30	2569.60		
1 Asphalt Paver, 130 H.P.		2132.00		2345.20		
2 Tandem Rollers, 10 Ton		473.60		520.96		
1 Roller, Pneum. Whl., 12 Ton		344.40		378.84	30.73	33.80
96 L.H., Daily Totals		$6991.60		$9786.20	$72.83	$101.94
Crew B-25C	Hr.	Daily	Hr.	Daily	Bare Costs	Incl. O&P
1 Labor Foreman (outside)	$39.60	$316.80	$64.90	$519.20	$42.27	$68.41
3 Laborers	37.60	902.40	61.65	1479.60		
2 Equip. Oper. (medium)	50.60	809.60	80.30	1284.80		
1 Asphalt Paver, 130 H.P.		2132.00		2345.20		
1 Tandem Roller, 10 Ton		236.80		260.48	49.35	54.28
48 L.H., Daily Totals		$4397.60		$5889.28	$91.62	$122.69
Crew B-25D	Hr.	Daily	Hr.	Daily	Bare Costs	Incl. O&P
1 Labor Foreman (outside)	$39.60	$316.80	$64.90	$519.20	$42.39	$68.57
3 Laborers	37.60	902.40	61.65	1479.60		
2.125 Equip. Oper. (medium)	50.60	860.20	80.30	1365.10		
.125 Truck Driver (heavy)	40.05	40.05	64.55	64.55		
.125 Truck Tractor, 6x4, 380 H.P.		74.97		82.47		
.125 Dist. Tanker, 3000 Gallon		38.20		42.02		
1 Asphalt Paver, 130 H.P.		2132.00		2345.20		
1 Tandem Roller, 10 Ton		236.80		260.48	49.64	54.60
50 L.H., Daily Totals		$4601.43		$6158.62	$92.03	$123.17
Crew B-25E	Hr.	Daily	Hr.	Daily	Bare Costs	Incl. O&P
1 Labor Foreman (outside)	$39.60	$316.80	$64.90	$519.20	$42.50	$68.72
3 Laborers	37.60	902.40	61.65	1479.60		
2.250 Equip. Oper. (medium)	50.60	910.80	80.30	1445.40		
.25 Truck Driver (heavy)	40.05	80.10	64.55	129.10		
.25 Truck Tractor, 6x4, 380 H.P.		149.95		164.94		
.25 Dist. Tanker, 3000 Gallon		76.40		84.04		
1 Asphalt Paver, 130 H.P.		2132.00		2345.20		
1 Tandem Roller, 10 Ton		236.80		260.48	49.91	54.90
52 L.H., Daily Totals		$4805.25		$6427.97	$92.41	$123.61
Crew B-26	Hr.	Daily	Hr.	Daily	Bare Costs	Incl. O&P
1 Labor Foreman (outside)	$39.60	$316.80	$64.90	$519.20	$42.18	$68.39
6 Laborers	37.60	1804.80	61.65	2959.20		
2 Equip. Oper. (medium)	50.60	809.60	80.30	1284.80		
1 Rodman (reinf.)	52.55	420.40	85.75	686.00		
1 Cement Finisher	45.00	360.00	71.15	569.20		
1 Grader, 30,000 Lbs.		736.20		809.82		
1 Paving Mach. & Equip.		2794.00		3073.40	40.12	44.13
88 L.H., Daily Totals		$7241.80		$9901.62	$82.29	$112.52
Crew B-26A	Hr.	Daily	Hr.	Daily	Bare Costs	Incl. O&P
1 Labor Foreman (outside)	$39.60	$316.80	$64.90	$519.20	$42.18	$68.39
6 Laborers	37.60	1804.80	61.65	2959.20		
2 Equip. Oper. (medium)	50.60	809.60	80.30	1284.80		
1 Rodman (reinf.)	52.55	420.40	85.75	686.00		
1 Cement Finisher	45.00	360.00	71.15	569.20		
1 Grader, 30,000 Lbs.		736.20		809.82		
1 Paving Mach. & Equip.		2794.00		3073.40		
1 Concrete Saw		167.40		184.14	42.02	46.22
88 L.H., Daily Totals		$7409.20		$10085.76	$84.20	$114.61
Crew B-26B	Hr.	Daily	Hr.	Daily	Bare Costs	Incl. O&P
1 Labor Foreman (outside)	$39.60	$316.80	$64.90	$519.20	$42.88	$69.38
6 Laborers	37.60	1804.80	61.65	2959.20		
3 Equip. Oper. (medium)	50.60	1214.40	80.30	1927.20		
1 Rodman (reinf.)	52.55	420.40	85.75	686.00		
1 Cement Finisher	45.00	360.00	71.15	569.20		
1 Grader, 30,000 Lbs.		736.20		809.82		
1 Paving Mach. & Equip.		2794.00		3073.40		
1 Concrete Pump, 110' Boom		946.20		1040.82	46.63	51.29
96 L.H., Daily Totals		$8592.80		$11584.84	$89.51	$120.68
Crew B-26C	Hr.	Daily	Hr.	Daily	Bare Costs	Incl. O&P
1 Labor Foreman (outside)	$39.60	$316.80	$64.90	$519.20	$41.34	$67.20
6 Laborers	37.60	1804.80	61.65	2959.20		
1 Equip. Oper. (medium)	50.60	404.80	80.30	642.40		
1 Rodman (reinf.)	52.55	420.40	85.75	686.00		
1 Cement Finisher	45.00	360.00	71.15	569.20		
1 Paving Mach. & Equip.		2794.00		3073.40		
1 Concrete Saw		167.40		184.14	37.02	40.72
80 L.H., Daily Totals		$6268.20		$8633.54	$78.35	$107.92

Crews

Crew No.		Bare Costs		Incl. Subs O&P		Cost Per Labor-Hour	
Crew B-27	Hr.	Daily	Hr.	Daily	Bare Costs	Incl. O&P
1 Labor Foreman (outside)	$39.60	$316.80	$64.90	$519.20	$38.10	$62.46
3 Laborers	37.60	902.40	61.65	1479.60		
1 Berm Machine		296.80		326.48	9.28	10.20
32 L.H., Daily Totals		$1516.00		$2325.28	$47.38	$72.67
Crew B-28	Hr.	Daily	Hr.	Daily	Bare Costs	Incl. O&P
2 Carpenters	$46.95	$751.20	$76.95	$1231.20	$43.83	$71.85
1 Laborer	37.60	300.80	61.65	493.20		
24 L.H., Daily Totals		$1052.00		$1724.40	$43.83	$71.85
Crew B-29	Hr.	Daily	Hr.	Daily	Bare Costs	Incl. O&P
1 Labor Foreman (outside)	$39.60	$316.80	$64.90	$519.20	$40.99	$66.47
4 Laborers	37.60	1203.20	61.65	1972.80		
1 Equip. Oper. (crane)	51.70	413.60	82.05	656.40		
1 Equip. Oper. (oiler)	45.20	361.60	71.75	574.00		
1 Gradall, 5/8 C.Y.		882.20		970.42	15.75	17.33
56 L.H., Daily Totals		$3177.40		$4692.82	$56.74	$83.80
Crew B-30	Hr.	Daily	Hr.	Daily	Bare Costs	Incl. O&P
1 Equip. Oper. (medium)	$50.60	$404.80	$80.30	$642.40	$43.57	$69.80
2 Truck Drivers (heavy)	40.05	640.80	64.55	1032.80		
1 Hyd. Excavator, 1.5 C.Y.		1030.00		1133.00		
2 Dump Trucks, 12 C.Y., 400 H.P.		1382.00		1520.20	100.50	110.55
24 L.H., Daily Totals		$3457.60		$4328.40	$144.07	$180.35
Crew B-31	Hr.	Daily	Hr.	Daily	Bare Costs	Incl. O&P
1 Labor Foreman (outside)	$39.60	$316.80	$64.90	$519.20	$39.87	$65.36
3 Laborers	37.60	902.40	61.65	1479.60		
1 Carpenter	46.95	375.60	76.95	615.60		
1 Air Compressor, 250 cfm		201.40		221.54		
1 Sheeting Driver		5.75		6.33		
2 -50' Air Hoses, 1.5"		11.60		12.76	5.47	6.02
40 L.H., Daily Totals		$1813.55		$2855.03	$45.34	$71.38
Crew B-32	Hr.	Daily	Hr.	Daily	Bare Costs	Incl. O&P
1 Laborer	$37.60	$300.80	$61.65	$493.20	$47.35	$75.64
3 Equip. Oper. (medium)	50.60	1214.40	80.30	1927.20		
1 Grader, 30,000 Lbs.		736.20		809.82		
1 Tandem Roller, 10 Ton		236.80		260.48		
1 Dozer, 200 H.P.		1387.00		1525.70	73.75	81.13
32 L.H., Daily Totals		$3875.20		$5016.40	$121.10	$156.76
Crew B-32A	Hr.	Daily	Hr.	Daily	Bare Costs	Incl. O&P
1 Laborer	$37.60	$300.80	$61.65	$493.20	$46.27	$74.08
2 Equip. Oper. (medium)	50.60	809.60	80.30	1284.80		
1 Grader, 30,000 Lbs.		736.20		809.82		
1 Roller, Vibratory, 25 Ton		687.40		756.14	59.32	65.25
24 L.H., Daily Totals		$2534.00		$3343.96	$105.58	$139.33
Crew B-32B	Hr.	Daily	Hr.	Daily	Bare Costs	Incl. O&P
1 Laborer	$37.60	$300.80	$61.65	$493.20	$46.27	$74.08
2 Equip. Oper. (medium)	50.60	809.60	80.30	1284.80		
1 Dozer, 200 H.P.		1387.00		1525.70		
1 Roller, Vibratory, 25 Ton		687.40		756.14	86.43	95.08
24 L.H., Daily Totals		$3184.80		$4059.84	$132.70	$169.16

Crew No.		Bare Costs		Incl. Subs O&P		Cost Per Labor-Hour	
Crew B-32C	Hr.	Daily	Hr.	Daily	Bare Costs	Incl. O&P
1 Labor Foreman (outside)	$39.60	$316.80	$64.90	$519.20	$44.43	$71.52
2 Laborers	37.60	601.60	61.65	986.40		
3 Equip. Oper. (medium)	50.60	1214.40	80.30	1927.20		
1 Grader, 30,000 Lbs.		736.20		809.82		
1 Tandem Roller, 10 Ton		236.80		260.48		
1 Dozer, 200 H.P.		1387.00		1525.70	49.17	54.08
48 L.H., Daily Totals		$4492.80		$6028.80	$93.60	$125.60
Crew B-33A	Hr.	Daily	Hr.	Daily	Bare Costs	Incl. O&P
1 Equip. Oper. (medium)	$50.60	$404.80	$80.30	$642.40	$46.89	$74.97
.5 Laborer	37.60	150.40	61.65	246.60		
.25 Equip. Oper. (medium)	50.60	101.20	80.30	160.60		
1 Scraper, Towed, 7 C.Y.		114.40		125.84		
1.25 Dozers, 300 H.P.		2371.25		2608.38	177.55	195.30
14 L.H., Daily Totals		$3142.05		$3783.82	$224.43	$270.27
Crew B-33B	Hr.	Daily	Hr.	Daily	Bare Costs	Incl. O&P
1 Equip. Oper. (medium)	$50.60	$404.80	$80.30	$642.40	$46.89	$74.97
.5 Laborer	37.60	150.40	61.65	246.60		
.25 Equip. Oper. (medium)	50.60	101.20	80.30	160.60		
1 Scraper, Towed, 10 C.Y.		146.80		161.48		
1.25 Dozers, 300 H.P.		2371.25		2608.38	179.86	197.85
14 L.H., Daily Totals		$3174.45		$3819.45	$226.75	$272.82
Crew B-33C	Hr.	Daily	Hr.	Daily	Bare Costs	Incl. O&P
1 Equip. Oper. (medium)	$50.60	$404.80	$80.30	$642.40	$46.89	$74.97
.5 Laborer	37.60	150.40	61.65	246.60		
.25 Equip. Oper. (medium)	50.60	101.20	80.30	160.60		
1 Scraper, Towed, 15 C.Y.		164.80		181.28		
1.25 Dozers, 300 H.P.		2371.25		2608.38	181.15	199.26
14 L.H., Daily Totals		$3192.45		$3839.26	$228.03	$274.23
Crew B-33D	Hr.	Daily	Hr.	Daily	Bare Costs	Incl. O&P
1 Equip. Oper. (medium)	$50.60	$404.80	$80.30	$642.40	$46.89	$74.97
.5 Laborer	37.60	150.40	61.65	246.60		
.25 Equip. Oper. (medium)	50.60	101.20	80.30	160.60		
1 S.P. Scraper, 14 C.Y.		1884.00		2072.40		
.25 Dozer, 300 H.P.		474.25		521.67	168.45	185.29
14 L.H., Daily Totals		$3014.65		$3643.68	$215.33	$260.26
Crew B-33E	Hr.	Daily	Hr.	Daily	Bare Costs	Incl. O&P
1 Equip. Oper. (medium)	$50.60	$404.80	$80.30	$642.40	$46.89	$74.97
.5 Laborer	37.60	150.40	61.65	246.60		
.25 Equip. Oper. (medium)	50.60	101.20	80.30	160.60		
1 S.P. Scraper, 21 C.Y.		2688.00		2956.80		
.25 Dozer, 300 H.P.		474.25		521.67	225.88	248.46
14 L.H., Daily Totals		$3818.65		$4528.07	$272.76	$323.43
Crew B-33F	Hr.	Daily	Hr.	Daily	Bare Costs	Incl. O&P
1 Equip. Oper. (medium)	$50.60	$404.80	$80.30	$642.40	$46.89	$74.97
.5 Laborer	37.60	150.40	61.65	246.60		
.25 Equip. Oper. (medium)	50.60	101.20	80.30	160.60		
1 Elev. Scraper, 11 C.Y.		1163.00		1279.30		
.25 Dozer, 300 H.P.		474.25		521.67	116.95	128.64
14 L.H., Daily Totals		$2293.65		$2850.57	$163.83	$203.61

Crews

Crew No.	Bare Costs				Incl. Subs O&P				Cost Per Labor-Hour	

Crew B-33G	Hr.	Daily	Hr.	Daily	Bare Costs	Incl. O&P
1 Equip. Oper. (medium)	$50.60	$404.80	$80.30	$642.40	$46.89	$74.97
.5 Laborer	37.60	150.40	61.65	246.60		
.25 Equip. Oper. (medium)	50.60	101.20	80.30	160.60		
1 Elev. Scraper, 22 C.Y.		2521.00		2773.10		
.25 Dozer, 300 H.P.		474.25		521.67	213.95	235.34
14 L.H., Daily Totals		$3651.65		$4344.38	$260.83	$310.31

Crew B-33K	Hr.	Daily	Hr.	Daily	Bare Costs	Incl. O&P
1 Equipment Operator (med.)	$50.60	$404.80	$80.30	$642.40	$46.89	$74.97
.25 Equipment Operator (med.)	50.60	101.20	80.30	160.60		
.5 Laborer	37.60	150.40	61.65	246.60		
1 S.P. Scraper, 31 C.Y.		3693.00		4062.30		
.25 Dozer, 410 H.P.		601.25		661.38	306.73	337.41
14 L.H., Daily Totals		$4950.65		$5773.27	$353.62	$412.38

Crew B-34A	Hr.	Daily	Hr.	Daily	Bare Costs	Incl. O&P
1 Truck Driver (heavy)	$40.05	$320.40	$64.55	$516.40	$40.05	$64.55
1 Dump Truck, 8 C.Y., 220 H.P.		411.20		452.32	51.40	56.54
8 L.H., Daily Totals		$731.60		$968.72	$91.45	$121.09

Crew B-34B	Hr.	Daily	Hr.	Daily	Bare Costs	Incl. O&P
1 Truck Driver (heavy)	$40.05	$320.40	$64.55	$516.40	$40.05	$64.55
1 Dump Truck, 12 C.Y., 400 H.P.		691.00		760.10	86.38	95.01
8 L.H., Daily Totals		$1011.40		$1276.50	$126.43	$159.56

Crew B-34C	Hr.	Daily	Hr.	Daily	Bare Costs	Incl. O&P
1 Truck Driver (heavy)	$40.05	$320.40	$64.55	$516.40	$40.05	$64.55
1 Truck Tractor, 6x4, 380 H.P.		599.80		659.78		
1 Dump Trailer, 16.5 C.Y.		126.60		139.26	90.80	99.88
8 L.H., Daily Totals		$1046.80		$1315.44	$130.85	$164.43

Crew B-34D	Hr.	Daily	Hr.	Daily	Bare Costs	Incl. O&P
1 Truck Driver (heavy)	$40.05	$320.40	$64.55	$516.40	$40.05	$64.55
1 Truck Tractor, 6x4, 380 H.P.		599.80		659.78		
1 Dump Trailer, 20 C.Y.		141.20		155.32	92.63	101.89
8 L.H., Daily Totals		$1061.40		$1331.50	$132.68	$166.44

Crew B-34E	Hr.	Daily	Hr.	Daily	Bare Costs	Incl. O&P
1 Truck Driver (heavy)	$40.05	$320.40	$64.55	$516.40	$40.05	$64.55
1 Dump Truck, Off Hwy., 25 Ton		1351.00		1486.10	168.88	185.76
8 L.H., Daily Totals		$1671.40		$2002.50	$208.93	$250.31

Crew B-34F	Hr.	Daily	Hr.	Daily	Bare Costs	Incl. O&P
1 Truck Driver (heavy)	$40.05	$320.40	$64.55	$516.40	$40.05	$64.55
1 Dump Truck, Off Hwy., 35 Ton		1512.00		1663.20	189.00	207.90
8 L.H., Daily Totals		$1832.40		$2179.60	$229.05	$272.45

Crew B-34G	Hr.	Daily	Hr.	Daily	Bare Costs	Incl. O&P
1 Truck Driver (heavy)	$40.05	$320.40	$64.55	$516.40	$40.05	$64.55
1 Dump Truck, Off Hwy., 50 Ton		1837.00		2020.70	229.63	252.59
8 L.H., Daily Totals		$2157.40		$2537.10	$269.68	$317.14

Crew B-34H	Hr.	Daily	Hr.	Daily	Bare Costs	Incl. O&P
1 Truck Driver (heavy)	$40.05	$320.40	$64.55	$516.40	$40.05	$64.55
1 Dump Truck, Off Hwy., 65 Ton		1863.00		2049.30	232.88	256.16
8 L.H., Daily Totals		$2183.40		$2565.70	$272.93	$320.71

Crew B-34I	Hr.	Daily	Hr.	Daily	Bare Costs	Incl. O&P
1 Truck Driver (heavy)	$40.05	$320.40	$64.55	$516.40	$40.05	$64.55
1 Dump Truck, 18 C.Y., 450 H.P.		869.00		955.90	108.63	119.49
8 L.H., Daily Totals		$1189.40		$1472.30	$148.68	$184.04

Crew B-34J	Hr.	Daily	Hr.	Daily	Bare Costs	Incl. O&P
1 Truck Driver (heavy)	$40.05	$320.40	$64.55	$516.40	$40.05	$64.55
1 Dump Truck, Off Hwy., 100 Ton		2925.00		3217.50	365.63	402.19
8 L.H., Daily Totals		$3245.40		$3733.90	$405.68	$466.74

Crew B-34K	Hr.	Daily	Hr.	Daily	Bare Costs	Incl. O&P
1 Truck Driver (heavy)	$40.05	$320.40	$64.55	$516.40	$40.05	$64.55
1 Truck Tractor, 6x4, 450 H.P.		727.40		800.14		
1 Lowbed Trailer, 75 Ton		221.80		243.98	118.65	130.51
8 L.H., Daily Totals		$1269.60		$1560.52	$158.70	$195.07

Crew B-34L	Hr.	Daily	Hr.	Daily	Bare Costs	Incl. O&P
1 Equip. Oper. (light)	$48.60	$388.80	$77.15	$617.20	$48.60	$77.15
1 Flatbed Truck, Gas, 1.5 Ton		245.40		269.94	30.68	33.74
8 L.H., Daily Totals		$634.20		$887.14	$79.28	$110.89

Crew B-34M	Hr.	Daily	Hr.	Daily	Bare Costs	Incl. O&P
1 Equip. Oper. (light)	$48.60	$388.80	$77.15	$617.20	$48.60	$77.15
1 Flatbed Truck, Gas, 3 Ton		303.20		333.52	37.90	41.69
8 L.H., Daily Totals		$692.00		$950.72	$86.50	$118.84

Crew B-34N	Hr.	Daily	Hr.	Daily	Bare Costs	Incl. O&P
1 Truck Driver (heavy)	$40.05	$320.40	$64.55	$516.40	$45.33	$72.42
1 Equip. Oper. (medium)	50.60	404.80	80.30	642.40		
1 Truck Tractor, 6x4, 380 H.P.		599.80		659.78		
1 Flatbed Trailer, 40 Ton		154.00		169.40	47.11	51.82
16 L.H., Daily Totals		$1479.00		$1987.98	$92.44	$124.25

Crew B-34P	Hr.	Daily	Hr.	Daily	Bare Costs	Incl. O&P
1 Pipe Fitter	$59.75	$478.00	$93.20	$745.60	$49.82	$78.85
1 Truck Driver (light)	39.10	312.80	63.05	504.40		
1 Equip. Oper. (medium)	50.60	404.80	80.30	642.40		
1 Flatbed Truck, Gas, 3 Ton		303.20		333.52		
1 Backhoe Loader, 48 H.P.		364.20		400.62	27.81	30.59
24 L.H., Daily Totals		$1863.00		$2626.54	$77.63	$109.44

Crew B-34Q	Hr.	Daily	Hr.	Daily	Bare Costs	Incl. O&P
1 Pipe Fitter	$59.75	$478.00	$93.20	$745.60	$50.18	$79.43
1 Truck Driver (light)	39.10	312.80	63.05	504.40		
1 Equip. Oper. (crane)	51.70	413.60	82.05	656.40		
1 Flatbed Trailer, 25 Ton		113.60		124.96		
1 Dump Truck, 8 C.Y., 220 H.P.		411.20		452.32		
1 Hyd. Crane, 25 Ton		736.60		810.26	52.56	57.81
24 L.H., Daily Totals		$2465.80		$3293.94	$102.74	$137.25

Crew B-34R	Hr.	Daily	Hr.	Daily	Bare Costs	Incl. O&P
1 Pipe Fitter	$59.75	$478.00	$93.20	$745.60	$50.18	$79.43
1 Truck Driver (light)	39.10	312.80	63.05	504.40		
1 Equip. Oper. (crane)	51.70	413.60	82.05	656.40		
1 Flatbed Trailer, 25 Ton		113.60		124.96		
1 Dump Truck, 8 C.Y., 220 H.P.		411.20		452.32		
1 Hyd. Crane, 25 Ton		736.60		810.26		
1 Hyd. Excavator, 1 C.Y.		812.60		893.86	86.42	95.06
24 L.H., Daily Totals		$3278.40		$4187.80	$136.60	$174.49

Crews

Crew No.		Bare Costs		Incl. Subs O&P		Cost Per Labor-Hour	
Crew B-34S	Hr.	Daily	Hr.	Daily	Bare Costs	Incl. O&P
2 Pipe Fitters	$59.75	$956.00	$93.20	$1491.20	$52.81	$83.25
1 Truck Driver (heavy)	40.05	320.40	64.55	516.40		
1 Equip. Oper. (crane)	51.70	413.60	82.05	656.40		
1 Flatbed Trailer, 40 Ton		154.00		169.40		
1 Truck Tractor, 6x4, 380 H.P.		599.80		659.78		
1 Hyd. Crane, 80 Ton		1625.00		1787.50		
1 Hyd. Excavator, 2 C.Y.		1175.00		1292.50	111.06	122.16
32 L.H., Daily Totals		$5243.80		$6573.18	$163.87	$205.41
Crew B-34T	Hr.	Daily	Hr.	Daily	Bare Costs	Incl. O&P
2 Pipe Fitters	$59.75	$956.00	$93.20	$1491.20	$52.81	$83.25
1 Truck Driver (heavy)	40.05	320.40	64.55	516.40		
1 Equip. Oper. (crane)	51.70	413.60	82.05	656.40		
1 Flatbed Trailer, 40 Ton		154.00		169.40		
1 Truck Tractor, 6x4, 380 H.P.		599.80		659.78		
1 Hyd. Crane, 80 Ton		1625.00		1787.50	74.34	81.77
32 L.H., Daily Totals		$4068.80		$5280.68	$127.15	$165.02
Crew B-34U	Hr.	Daily	Hr.	Daily	Bare Costs	Incl. O&P
1 Truck Driver (heavy)	$40.05	$320.40	$64.55	$516.40	$44.33	$70.85
1 Equip. Oper. (light)	48.60	388.80	77.15	617.20		
1 Truck Tractor, 220 H.P.		359.60		395.56		
1 Flatbed Trailer, 25 Ton		113.60		124.96	29.57	32.53
16 L.H., Daily Totals		$1182.40		$1654.12	$73.90	$103.38
Crew B-34V	Hr.	Daily	Hr.	Daily	Bare Costs	Incl. O&P
1 Truck Driver (heavy)	$40.05	$320.40	$64.55	$516.40	$46.78	$74.58
1 Equip. Oper. (crane)	51.70	413.60	82.05	656.40		
1 Equip. Oper. (light)	48.60	388.80	77.15	617.20		
1 Truck Tractor, 6x4, 450 H.P.		727.40		800.14		
1 Equipment Trailer, 50 Ton		168.40		185.24		
1 Pickup Truck, 4 x 4, 3/4 Ton		155.60		171.16	43.81	48.19
24 L.H., Daily Totals		$2174.20		$2946.54	$90.59	$122.77
Crew B-34W	Hr.	Daily	Hr.	Daily	Bare Costs	Incl. O&P
5 Truck Drivers (heavy)	$40.05	$1602.00	$64.55	$2582.00	$43.66	$70.05
2 Equip. Opers. (crane)	51.70	827.20	82.05	1312.80		
1 Equip. Oper. (mechanic)	51.65	413.20	81.95	655.60		
1 Laborer	37.60	300.80	61.65	493.20		
4 Truck Tractor, 6x4, 380 H.P.		2399.20		2639.12		
2 Equipment Trailer, 50 Ton		336.80		370.48		
2 Flatbed Trailer, 40 Ton		308.00		338.80		
1 Pickup Truck, 4 x 4, 3/4 Ton		155.60		171.16		
1 S.P. Crane, 4x4, 20 Ton		554.60		610.06	52.14	57.36
72 L.H., Daily Totals		$6897.40		$9173.22	$95.80	$127.41
Crew B-35	Hr.	Daily	Hr.	Daily	Bare Costs	Incl. O&P
1 Labor Foreman (outside)	$39.60	$316.80	$64.90	$519.20	$46.91	$75.16
1 Skilled Worker	48.65	389.20	79.05	632.40		
1 Welder (plumber)	58.70	469.60	91.55	732.40		
1 Laborer	37.60	300.80	61.65	493.20		
1 Equip. Oper. (crane)	51.70	413.60	82.05	656.40		
1 Equip. Oper. (oiler)	45.20	361.60	71.75	574.00		
1 Welder, Electric, 300 amp		57.70		63.47		
1 Hyd. Excavator, .75 C.Y.		654.80		720.28	14.84	16.33
48 L.H., Daily Totals		$2964.10		$4391.35	$61.75	$91.49

Crew No.		Bare Costs		Incl. Subs O&P		Cost Per Labor-Hour	
Crew B-35A	Hr.	Daily	Hr.	Daily	Bare Costs	Incl. O&P
1 Labor Foreman (outside)	$39.60	$316.80	$64.90	$519.20	$45.58	$73.23
2 Laborers	37.60	601.60	61.65	986.40		
1 Skilled Worker	48.65	389.20	79.05	632.40		
1 Welder (plumber)	58.70	469.60	91.55	732.40		
1 Equip. Oper. (crane)	51.70	413.60	82.05	656.40		
1 Equip. Oper. (oiler)	45.20	361.60	71.75	574.00		
1 Welder, Gas Engine, 300 amp		145.85		160.44		
1 Crawler Crane, 75 Ton		1459.00		1604.90	28.66	31.52
56 L.H., Daily Totals		$4157.25		$5866.14	$74.24	$104.75
Crew B-36	Hr.	Daily	Hr.	Daily	Bare Costs	Incl. O&P
1 Labor Foreman (outside)	$39.60	$316.80	$64.90	$519.20	$43.20	$69.76
2 Laborers	37.60	601.60	61.65	986.40		
2 Equip. Oper. (medium)	50.60	809.60	80.30	1284.80		
1 Dozer, 200 H.P.		1387.00		1525.70		
1 Aggregate Spreader		39.60		43.56		
1 Tandem Roller, 10 Ton		236.80		260.48	41.59	45.74
40 L.H., Daily Totals		$3391.40		$4620.14	$84.78	$115.50
Crew B-36A	Hr.	Daily	Hr.	Daily	Bare Costs	Incl. O&P
1 Labor Foreman (outside)	$39.60	$316.80	$64.90	$519.20	$45.31	$72.77
2 Laborers	37.60	601.60	61.65	986.40		
4 Equip. Oper. (medium)	50.60	1619.20	80.30	2569.60		
1 Dozer, 200 H.P.		1387.00		1525.70		
1 Aggregate Spreader		39.60		43.56		
1 Tandem Roller, 10 Ton		236.80		260.48		
1 Roller, Pneum. Whl., 12 Ton		344.40		378.84	35.85	39.44
56 L.H., Daily Totals		$4545.40		$6283.78	$81.17	$112.21
Crew B-36B	Hr.	Daily	Hr.	Daily	Bare Costs	Incl. O&P
1 Labor Foreman (outside)	$39.60	$316.80	$64.90	$519.20	$44.66	$71.74
2 Laborers	37.60	601.60	61.65	986.40		
4 Equip. Oper. (medium)	50.60	1619.20	80.30	2569.60		
1 Truck Driver (heavy)	40.05	320.40	64.55	516.40		
1 Grader, 30,000 Lbs.		736.20		809.82		
1 F.E. Loader, Crl, 1.5 C.Y.		629.80		692.78		
1 Dozer, 300 H.P.		1897.00		2086.70		
1 Roller, Vibratory, 25 Ton		687.40		756.14		
1 Truck Tractor, 6x4, 450 H.P.		727.40		800.14		
1 Water Tank Trailer, 5000 Gal.		141.00		155.10	75.29	82.82
64 L.H., Daily Totals		$7676.80		$9892.28	$119.95	$154.57
Crew B-36C	Hr.	Daily	Hr.	Daily	Bare Costs	Incl. O&P
1 Labor Foreman (outside)	$39.60	$316.80	$64.90	$519.20	$46.29	$74.07
3 Equip. Oper. (medium)	50.60	1214.40	80.30	1927.20		
1 Truck Driver (heavy)	40.05	320.40	64.55	516.40		
1 Grader, 30,000 Lbs.		736.20		809.82		
1 Dozer, 300 H.P.		1897.00		2086.70		
1 Roller, Vibratory, 25 Ton		687.40		756.14		
1 Truck Tractor, 6x4, 450 H.P.		727.40		800.14		
1 Water Tank Trailer, 5000 Gal.		141.00		155.10	104.72	115.20
40 L.H., Daily Totals		$6040.60		$7570.70	$151.01	$189.27
Crew B-37	Hr.	Daily	Hr.	Daily	Bare Costs	Incl. O&P
1 Labor Foreman (outside)	$39.60	$316.80	$64.90	$519.20	$39.77	$64.78
4 Laborers	37.60	1203.20	61.65	1972.80		
1 Equip. Oper. (light)	48.60	388.80	77.15	617.20		
1 Tandem Roller, 5 Ton		157.40		173.14	3.28	3.61
48 L.H., Daily Totals		$2066.20		$3282.34	$43.05	$68.38

Crews

Crew No.	Bare Costs		Incl. Subs O&P		Cost Per Labor-Hour	
Crew B-37A	Hr.	Daily	Hr.	Daily	Bare Costs	Incl. O&P
2 Laborers	$37.60	$601.60	$61.65	$986.40	$38.10	$62.12
1 Truck Driver (light)	39.10	312.80	63.05	504.40		
1 Flatbed Truck, Gas, 1.5 Ton		245.40		269.94		
1 Tar Kettle, T.M.		132.00		145.20	15.73	17.30
24 L.H., Daily Totals		$1291.80		$1905.94	$53.83	$79.41

Crew No.	Bare Costs		Incl. Subs O&P		Cost Per Labor-Hour	
Crew B-37B	Hr.	Daily	Hr.	Daily	Bare Costs	Incl. O&P
3 Laborers	$37.60	$902.40	$61.65	$1479.60	$37.98	$62.00
1 Truck Driver (light)	39.10	312.80	63.05	504.40		
1 Flatbed Truck, Gas, 1.5 Ton		245.40		269.94		
1 Tar Kettle, T.M.		132.00		145.20	11.79	12.97
32 L.H., Daily Totals		$1592.60		$2399.14	$49.77	$74.97

Crew No.	Bare Costs		Incl. Subs O&P		Cost Per Labor-Hour	
Crew B-37C	Hr.	Daily	Hr.	Daily	Bare Costs	Incl. O&P
2 Laborers	$37.60	$601.60	$61.65	$986.40	$38.35	$62.35
2 Truck Drivers (light)	39.10	625.60	63.05	1008.80		
2 Flatbed Trucks, Gas, 1.5 Ton		490.80		539.88		
1 Tar Kettle, T.M.		132.00		145.20	19.46	21.41
32 L.H., Daily Totals		$1850.00		$2680.28	$57.81	$83.76

Crew No.	Bare Costs		Incl. Subs O&P		Cost Per Labor-Hour	
Crew B-37D	Hr.	Daily	Hr.	Daily	Bare Costs	Incl. O&P
1 Laborer	$37.60	$300.80	$61.65	$493.20	$38.35	$62.35
1 Truck Driver (light)	39.10	312.80	63.05	504.40		
1 Pickup Truck, 3/4 Ton		144.20		158.62	9.01	9.91
16 L.H., Daily Totals		$757.80		$1156.22	$47.36	$72.26

Crew No.	Bare Costs		Incl. Subs O&P		Cost Per Labor-Hour	
Crew B-37E	Hr.	Daily	Hr.	Daily	Bare Costs	Incl. O&P
3 Laborers	$37.60	$902.40	$61.65	$1479.60	$41.46	$66.93
1 Equip. Oper. (light)	48.60	388.80	77.15	617.20		
1 Equip. Oper. (medium)	50.60	404.80	80.30	642.40		
2 Truck Drivers (light)	39.10	625.60	63.05	1008.80		
4 Barrels w/ Flasher		13.60		14.96		
1 Concrete Saw		167.40		184.14		
1 Rotary Hammer Drill		23.75		26.13		
1 Hammer Drill Bit		2.35		2.59		
1 Loader, Skid Steer, 30 H.P.		173.60		190.96		
1 Conc. Hammer Attach.		114.70		126.17		
1 Vibrating Plate, Gas, 18"		36.20		39.82		
2 Flatbed Trucks, Gas, 1.5 Ton		490.80		539.88	18.26	20.08
56 L.H., Daily Totals		$3344.00		$4872.64	$59.71	$87.01

Crew No.	Bare Costs		Incl. Subs O&P		Cost Per Labor-Hour	
Crew B-37F	Hr.	Daily	Hr.	Daily	Bare Costs	Incl. O&P
3 Laborers	$37.60	$902.40	$61.65	$1479.60	$37.98	$62.00
1 Truck Driver (light)	39.10	312.80	63.05	504.40		
4 Barrels w/ Flasher		13.60		14.96		
1 Concrete Mixer, 10 C.F.		172.40		189.64		
1 Air Compressor, 60 cfm		137.40		151.14		
1 -50' Air Hose, 3/4"		3.25		3.58		
1 Spade (Chipper)		7.60		8.36		
1 Flatbed Truck, Gas, 1.5 Ton		245.40		269.94	18.11	19.93
32 L.H., Daily Totals		$1794.85		$2621.61	$56.09	$81.93

Crew No.	Bare Costs		Incl. Subs O&P		Cost Per Labor-Hour	
Crew B-37G	Hr.	Daily	Hr.	Daily	Bare Costs	Incl. O&P
1 Labor Foreman (outside)	$39.60	$316.80	$64.90	$519.20	$39.77	$64.78
4 Laborers	37.60	1203.20	61.65	1972.80		
1 Equip. Oper. (light)	48.60	388.80	77.15	617.20		
1 Berm Machine		296.80		326.48		
1 Tandem Roller, 5 Ton		157.40		173.14	9.46	10.41
48 L.H., Daily Totals		$2363.00		$3608.82	$49.23	$75.18

Crew No.	Bare Costs		Incl. Subs O&P		Cost Per Labor-Hour	
Crew B-37H	Hr.	Daily	Hr.	Daily	Bare Costs	Incl. O&P
1 Labor Foreman (outside)	$39.60	$316.80	$64.90	$519.20	$39.77	$64.78
4 Laborers	37.60	1203.20	61.65	1972.80		
1 Equip. Oper. (light)	48.60	388.80	77.15	617.20		
1 Tandem Roller, 5 Ton		157.40		173.14		
1 Flatbed Trucks, Gas, 1.5 Ton		245.40		269.94		
1 Tar Kettle, T.M.		132.00		145.20	11.14	12.26
48 L.H., Daily Totals		$2443.60		$3697.48	$50.91	$77.03

Crew No.	Bare Costs		Incl. Subs O&P		Cost Per Labor-Hour	
Crew B-37I	Hr.	Daily	Hr.	Daily	Bare Costs	Incl. O&P
3 Laborers	$37.60	$902.40	$61.65	$1479.60	$41.46	$66.93
1 Equip. Oper. (light)	48.60	388.80	77.15	617.20		
1 Equip. Oper. (medium)	50.60	404.80	80.30	642.40		
2 Truck Drivers (light)	39.10	625.60	63.05	1008.80		
4 Barrels w/ Flasher		13.60		14.96		
1 Concrete Saw		167.40		184.14		
1 Rotary Hammer Drill		23.75		26.13		
1 Hammer Drill Bit		2.35		2.59		
1 Air Compressor, 60 cfm		137.40		151.14		
1 -50' Air Hose, 3/4"		3.25		3.58		
1 Spade (Chipper)		7.60		8.36		
1 Loader, Skid Steer, 30 H.P.		173.60		190.96		
1 Conc. Hammer Attach.		114.70		126.17		
1 Concrete Mixer, 10 C.F.		172.40		189.64		
1 Vibrating Plate, Gas, 18"		36.20		39.82		
2 Flatbed Trucks, Gas, 1.5 Ton		490.80		539.88	23.98	26.38
56 L.H., Daily Totals		$3664.65		$5225.35	$65.44	$93.31

Crew No.	Bare Costs		Incl. Subs O&P		Cost Per Labor-Hour	
Crew B-37J	Hr.	Daily	Hr.	Daily	Bare Costs	Incl. O&P
1 Labor Foreman (outside)	$39.60	$316.80	$64.90	$519.20	$39.77	$64.78
4 Laborers	37.60	1203.20	61.65	1972.80		
1 Equip. Oper. (light)	48.60	388.80	77.15	617.20		
1 Air Compressor, 60 cfm		137.40		151.14		
1 -50' Air Hose, 3/4"		3.25		3.58		
2 Concrete Mixer, 10 C.F.		344.80		379.28		
2 Flatbed Trucks, Gas, 1.5 Ton		490.80		539.88		
1 Shot Blaster, 20"		214.00		235.40	24.80	27.28
48 L.H., Daily Totals		$3099.05		$4418.48	$64.56	$92.05

Crew No.	Bare Costs		Incl. Subs O&P		Cost Per Labor-Hour	
Crew B-37K	Hr.	Daily	Hr.	Daily	Bare Costs	Incl. O&P
1 Labor Foreman (outside)	$39.60	$316.80	$64.90	$519.20	$39.77	$64.78
4 Laborers	37.60	1203.20	61.65	1972.80		
1 Equip. Oper. (light)	48.60	388.80	77.15	617.20		
1 Air Compressor, 60 cfm		137.40		151.14		
1 -50' Air Hose, 3/4"		3.25		3.58		
2 Flatbed Trucks, Gas, 1.5 Ton		490.80		539.88		
1 Shot Blaster, 20"		214.00		235.40	17.61	19.37
48 L.H., Daily Totals		$2754.25		$4039.20	$57.38	$84.15

Crew No.	Bare Costs		Incl. Subs O&P		Cost Per Labor-Hour	
Crew B-38	Hr.	Daily	Hr.	Daily	Bare Costs	Incl. O&P
1 Labor Foreman (outside)	$39.60	$316.80	$64.90	$519.20	$42.80	$69.13
2 Laborers	37.60	601.60	61.65	986.40		
1 Equip. Oper. (light)	48.60	388.80	77.15	617.20		
1 Equip. Oper. (medium)	50.60	404.80	80.30	642.40		
1 Backhoe Loader, 48 H.P.		364.20		400.62		
1 Hyd. Hammer, (1200 lb.)		180.40		198.44		
1 F.E. Loader, W.M., 4 C.Y.		669.20		736.12		
1 Pvmt. Rem. Bucket		61.20		67.32	31.88	35.06
40 L.H., Daily Totals		$2987.00		$4167.70	$74.67	$104.19

Crews

Crew No.		Bare Costs		Incl. Subs O&P		Cost Per Labor-Hour	
Crew B-39	Hr.	Daily	Hr.	Daily	Bare Costs	Incl. O&P	
1 Labor Foreman (outside)	$39.60	$316.80	$64.90	$519.20	$39.77	$64.78	
4 Laborers	37.60	1203.20	61.65	1972.80			
1 Equip. Oper. (light)	48.60	388.80	77.15	617.20			
1 Air Compressor, 250 cfm		201.40		221.54			
2 Breakers, Pavement, 60 lb.		19.60		21.56			
2 -50' Air Hoses, 1.5"		11.60		12.76	4.85	5.33	
48 L.H., Daily Totals		$2141.40		$3365.06	$44.61	$70.11	
Crew B-40	Hr.	Daily	Hr.	Daily	Bare Costs	Incl. O&P	
1 Pile Driver Foreman (outside)	$48.10	$384.80	$78.55	$628.40	$47.64	$76.95	
4 Pile Drivers	46.10	1475.20	75.30	2409.60			
2 Equip. Oper. (crane)	51.70	827.20	82.05	1312.80			
1 Equip. Oper. (oiler)	45.20	361.60	71.75	574.00			
1 Crawler Crane, 40 Ton		1156.00		1271.60			
1 Vibratory Hammer & Gen.		2587.00		2845.70	58.48	64.33	
64 L.H., Daily Totals		$6791.80		$9042.10	$106.12	$141.28	
Crew B-40B	Hr.	Daily	Hr.	Daily	Bare Costs	Incl. O&P	
1 Labor Foreman (outside)	$39.60	$316.80	$64.90	$519.20	$41.55	$67.28	
3 Laborers	37.60	902.40	61.65	1479.60			
1 Equip. Oper. (crane)	51.70	413.60	82.05	656.40			
1 Equip. Oper. (oiler)	45.20	361.60	71.75	574.00			
1 Lattice Boom Crane, 40 Ton		1164.00		1280.40	24.25	26.68	
48 L.H., Daily Totals		$3158.40		$4509.60	$65.80	$93.95	
Crew B-41	Hr.	Daily	Hr.	Daily	Bare Costs	Incl. O&P	
1 Labor Foreman (outside)	$39.60	$316.80	$64.90	$519.20	$38.95	$63.63	
4 Laborers	37.60	1203.20	61.65	1972.80			
.25 Equip. Oper. (crane)	51.70	103.40	82.05	164.10			
.25 Equip. Oper. (oiler)	45.20	90.40	71.75	143.50			
.25 Crawler Crane, 40 Ton		289.00		317.90	6.57	7.22	
44 L.H., Daily Totals		$2002.80		$3117.50	$45.52	$70.85	
Crew B-42	Hr.	Daily	Hr.	Daily	Bare Costs	Incl. O&P	
1 Labor Foreman (outside)	$39.60	$316.80	$64.90	$519.20	$42.44	$70.06	
4 Laborers	37.60	1203.20	61.65	1972.80			
1 Equip. Oper. (crane)	51.70	413.60	82.05	656.40			
1 Equip. Oper. (oiler)	45.20	361.60	71.75	574.00			
1 Welder	52.65	421.20	95.15	761.20			
1 Hyd. Crane, 25 Ton		736.60		810.26			
1 Welder, Gas Engine, 300 amp		145.85		160.44			
1 Horz. Boring Csg. Mch.		474.80		522.28	21.21	23.33	
64 L.H., Daily Totals		$4073.65		$5976.57	$63.65	$93.38	
Crew B-43	Hr.	Daily	Hr.	Daily	Bare Costs	Incl. O&P	
1 Labor Foreman (outside)	$39.60	$316.80	$64.90	$519.20	$41.55	$67.28	
3 Laborers	37.60	902.40	61.65	1479.60			
1 Equip. Oper. (crane)	51.70	413.60	82.05	656.40			
1 Equip. Oper. (oiler)	45.20	361.60	71.75	574.00			
1 Drill Rig, Truck-Mounted		2542.00		2796.20	52.96	58.25	
48 L.H., Daily Totals		$4536.40		$6025.40	$94.51	$125.53	
Crew B-44	Hr.	Daily	Hr.	Daily	Bare Costs	Incl. O&P	
1 Pile Driver Foreman (outside)	$48.10	$384.80	$78.55	$628.40	$46.69	$75.69	
4 Pile Drivers	46.10	1475.20	75.30	2409.60			
2 Equip. Oper. (crane)	51.70	827.20	82.05	1312.80			
1 Laborer	37.60	300.80	61.65	493.20			
1 Crawler Crane, 40 Ton		1156.00		1271.60			
1 Lead, 60' High		74.80		82.28			
1 Hammer, Diesel, 15K ft.-lbs.		586.00		644.60	28.39	31.23	
64 L.H., Daily Totals		$4804.80		$6842.48	$75.08	$106.91	

Crew No.		Bare Costs		Incl. Subs O&P		Cost Per Labor-Hour	
Crew B-45	Hr.	Daily	Hr.	Daily	Bare Costs	Incl. O&P	
1 Equip. Oper. (medium)	$50.60	$404.80	$80.30	$642.40	$45.33	$72.42	
1 Truck Driver (heavy)	40.05	320.40	64.55	516.40			
1 Dist. Tanker, 3000 Gallon		305.60		336.16			
1 Truck Tractor, 6x4, 380 H.P.		599.80		659.78	56.59	62.25	
16 L.H., Daily Totals		$1630.60		$2154.74	$101.91	$134.67	
Crew B-46	Hr.	Daily	Hr.	Daily	Bare Costs	Incl. O&P	
1 Pile Driver Foreman (outside)	$48.10	$384.80	$78.55	$628.40	$42.18	$69.02	
2 Pile Drivers	46.10	737.60	75.30	1204.80			
3 Laborers	37.60	902.40	61.65	1479.60			
1 Chain Saw, Gas, 36" Long		45.00		49.50	.94	1.03	
48 L.H., Daily Totals		$2069.80		$3362.30	$43.12	$70.05	
Crew B-47	Hr.	Daily	Hr.	Daily	Bare Costs	Incl. O&P	
1 Blast Foreman (outside)	$39.60	$316.80	$64.90	$519.20	$41.93	$67.90	
1 Driller	37.60	300.80	61.65	493.20			
1 Equip. Oper. (light)	48.60	388.80	77.15	617.20			
1 Air Track Drill, 4"		1028.00		1130.80			
1 Air Compressor, 600 cfm		550.60		605.66			
2 -50' Air Hoses, 3"		29.80		32.78	67.02	73.72	
24 L.H., Daily Totals		$2614.80		$3398.84	$108.95	$141.62	
Crew B-47A	Hr.	Daily	Hr.	Daily	Bare Costs	Incl. O&P	
1 Drilling Foreman (outside)	$39.60	$316.80	$64.90	$519.20	$45.50	$72.90	
1 Equip. Oper. (heavy)	51.70	413.60	82.05	656.40			
1 Equip. Oper. (oiler)	45.20	361.60	71.75	574.00			
1 Air Track Drill, 5"		1243.00		1367.30	51.79	56.97	
24 L.H., Daily Totals		$2335.00		$3116.90	$97.29	$129.87	
Crew B-47C	Hr.	Daily	Hr.	Daily	Bare Costs	Incl. O&P	
1 Laborer	$37.60	$300.80	$61.65	$493.20	$43.10	$69.40	
1 Equip. Oper. (light)	48.60	388.80	77.15	617.20			
1 Air Compressor, 750 cfm		557.20		612.92			
2 -50' Air Hoses, 3"		29.80		32.78			
1 Air Track Drill, 4"		1028.00		1130.80	100.94	111.03	
16 L.H., Daily Totals		$2304.60		$2886.90	$144.04	$180.43	
Crew B-47E	Hr.	Daily	Hr.	Daily	Bare Costs	Incl. O&P	
1 Labor Foreman (outside)	$39.60	$316.80	$64.90	$519.20	$38.10	$62.46	
3 Laborers	37.60	902.40	61.65	1479.60			
1 Flatbed Truck, Gas, 3 Ton		303.20		333.52	9.47	10.42	
32 L.H., Daily Totals		$1522.40		$2332.32	$47.58	$72.89	
Crew B-47G	Hr.	Daily	Hr.	Daily	Bare Costs	Incl. O&P	
1 Labor Foreman (outside)	$39.60	$316.80	$64.90	$519.20	$40.85	$66.34	
2 Laborers	37.60	601.60	61.65	986.40			
1 Equip. Oper. (light)	48.60	388.80	77.15	617.20			
1 Air Track Drill, 4"		1028.00		1130.80			
1 Air Compressor, 600 cfm		550.60		605.66			
2 -50' Air Hoses, 3"		29.80		32.78			
1 Gunite Pump Rig		371.80		408.98	61.88	68.07	
32 L.H., Daily Totals		$3287.40		$4301.02	$102.73	$134.41	
Crew B-47H	Hr.	Daily	Hr.	Daily	Bare Costs	Incl. O&P	
1 Skilled Worker Foreman (out)	$50.65	$405.20	$82.30	$658.40	$49.15	$79.86	
3 Skilled Workers	48.65	1167.60	79.05	1897.20			
1 Flatbed Truck, Gas, 3 Ton		303.20		333.52	9.47	10.42	
32 L.H., Daily Totals		$1876.00		$2889.12	$58.63	$90.28	

Crews

Crew No.		Bare Costs		Incl. Subs O&P		Cost Per Labor-Hour	
Crew B-48	**Hr.**	**Daily**	**Hr.**	**Daily**	**Bare Costs**	**Incl. O&P**	
1 Labor Foreman (outside)	$39.60	$316.80	$64.90	$519.20	$42.56	$68.69	
3 Laborers	37.60	902.40	61.65	1479.60			
1 Equip. Oper. (crane)	51.70	413.60	82.05	656.40			
1 Equip. Oper. (oiler)	45.20	361.60	71.75	574.00			
1 Equip. Oper. (light)	48.60	388.80	77.15	617.20			
1 Centr. Water Pump, 6"		346.60		381.26			
1 -20' Suction Hose, 6"		11.50		12.65			
1 -50' Discharge Hose, 6"		6.10		6.71			
1 Drill Rig, Truck-Mounted		2542.00		2796.20	51.90	57.09	
56 L.H., Daily Totals		$5289.40		$7043.22	$94.45	$125.77	

Crew B-49	Hr.	Daily	Hr.	Daily	Bare Costs	Incl. O&P
1 Labor Foreman (outside)	$39.60	$316.80	$64.90	$519.20	$44.27	$71.38
3 Laborers	37.60	902.40	61.65	1479.60		
2 Equip. Oper. (crane)	51.70	827.20	82.05	1312.80		
2 Equip. Oper. (oilers)	45.20	723.20	71.75	1148.00		
1 Equip. Oper. (light)	48.60	388.80	77.15	617.20		
2 Pile Drivers	46.10	737.60	75.30	1204.80		
1 Hyd. Crane, 25 Ton		736.60		810.26		
1 Centr. Water Pump, 6"		346.60		381.26		
1 -20' Suction Hose, 6"		11.50		12.65		
1 -50' Discharge Hose, 6"		6.10		6.71		
1 Drill Rig, Truck-Mounted		2542.00		2796.20	41.40	45.53
88 L.H., Daily Totals		$7538.80		$10288.68	$85.67	$116.92

Crew B-50	Hr.	Daily	Hr.	Daily	Bare Costs	Incl. O&P
2 Pile Driver Foremen (outside)	$48.10	$769.60	$78.55	$1256.80	$45.30	$73.55
6 Pile Drivers	46.10	2212.80	75.30	3614.40		
2 Equip. Oper. (crane)	51.70	827.20	82.05	1312.80		
1 Equip. Oper. (oiler)	45.20	361.60	71.75	574.00		
3 Laborers	37.60	902.40	61.65	1479.60		
1 Crawler Crane, 40 Ton		1156.00		1271.60		
1 Lead, 60' High		74.80		82.28		
1 Hammer, Diesel, 15K ft.-lbs.		586.00		644.60		
1 Air Compressor, 600 cfm		550.60		605.66		
2 -50' Air Hoses, 3"		29.80		32.78		
1 Chain Saw, Gas, 36" Long		45.00		49.50	21.81	23.99
112 L.H., Daily Totals		$7515.80		$10924.02	$67.11	$97.54

Crew B-51	Hr.	Daily	Hr.	Daily	Bare Costs	Incl. O&P
1 Labor Foreman (outside)	$39.60	$316.80	$64.90	$519.20	$38.18	$62.42
4 Laborers	37.60	1203.20	61.65	1972.80		
1 Truck Driver (light)	39.10	312.80	63.05	504.40		
1 Flatbed Truck, Gas, 1.5 Ton		245.40		269.94	5.11	5.62
48 L.H., Daily Totals		$2078.20		$3266.34	$43.30	$68.05

Crew B-52	Hr.	Daily	Hr.	Daily	Bare Costs	Incl. O&P
1 Carpenter Foreman (outside)	$48.95	$391.60	$80.25	$642.00	$43.61	$70.90
1 Carpenter	46.95	375.60	76.95	615.60		
3 Laborers	37.60	902.40	61.65	1479.60		
1 Cement Finisher	45.00	360.00	71.15	569.20		
.5 Rodman (reinf.)	52.55	210.20	85.75	343.00		
.5 Equip. Oper. (medium)	50.60	202.40	80.30	321.20		
.5 Crawler Loader, 3 C.Y.		594.00		653.40	10.61	11.67
56 L.H., Daily Totals		$3036.20		$4624.00	$54.22	$82.57

Crew B-53	Hr.	Daily	Hr.	Daily	Bare Costs	Incl. O&P
1 Equip. Oper. (light)	$48.60	$388.80	$77.15	$617.20	$48.60	$77.15
1 Trencher, Chain, 12 H.P.		68.00		74.80	8.50	9.35
8 L.H., Daily Totals		$456.80		$692.00	$57.10	$86.50

Crew B-54	Hr.	Daily	Hr.	Daily	Bare Costs	Incl. O&P
1 Equip. Oper. (light)	$48.60	$388.80	$77.15	$617.20	$48.60	$77.15
1 Trencher, Chain, 40 H.P.		335.20		368.72	41.90	46.09
8 L.H., Daily Totals		$724.00		$985.92	$90.50	$123.24

Crew B-54A	Hr.	Daily	Hr.	Daily	Bare Costs	Incl. O&P
.17 Labor Foreman (outside)	$39.60	$53.86	$64.90	$88.26	$49.00	$78.06
1 Equipment Operator (med.)	50.60	404.80	80.30	642.40		
1 Wheel Trencher, 67 H.P.		1195.00		1314.50	127.67	140.44
9.36 L.H., Daily Totals		$1653.66		$2045.16	$176.67	$218.50

Crew B-54B	Hr.	Daily	Hr.	Daily	Bare Costs	Incl. O&P
.25 Labor Foreman (outside)	$39.60	$79.20	$64.90	$129.80	$48.40	$77.22
1 Equipment Operator (med.)	50.60	404.80	80.30	642.40		
1 Wheel Trencher, 150 H.P.		1890.00		2079.00	189.00	207.90
10 L.H., Daily Totals		$2374.00		$2851.20	$237.40	$285.12

Crew B-54D	Hr.	Daily	Hr.	Daily	Bare Costs	Incl. O&P
1 Laborer	$37.60	$300.80	$61.65	$493.20	$44.10	$70.97
1 Equipment Operator (med.)	50.60	404.80	80.30	642.40		
1 Rock Trencher, 6" Width		372.00		409.20	23.25	25.57
16 L.H., Daily Totals		$1077.60		$1544.80	$67.35	$96.55

Crew B-54E	Hr.	Daily	Hr.	Daily	Bare Costs	Incl. O&P
1 Laborer	$37.60	$300.80	$61.65	$493.20	$44.10	$70.97
1 Equipment Operator (med.)	50.60	404.80	80.30	642.40		
1 Rock Trencher, 18" Width		2575.00		2832.50	160.94	177.03
16 L.H., Daily Totals		$3280.60		$3968.10	$205.04	$248.01

Crew B-55	Hr.	Daily	Hr.	Daily	Bare Costs	Incl. O&P
2 Laborers	$37.60	$601.60	$61.65	$986.40	$38.10	$62.12
1 Truck Driver (light)	39.10	312.80	63.05	504.40		
1 Truck-Mounted Earth Auger		779.20		857.12		
1 Flatbed Truck, Gas, 3 Ton		303.20		333.52	45.10	49.61
24 L.H., Daily Totals		$1996.80		$2681.44	$83.20	$111.73

Crew B-56	Hr.	Daily	Hr.	Daily	Bare Costs	Incl. O&P
1 Laborer	$37.60	$300.80	$61.65	$493.20	$43.10	$69.40
1 Equip. Oper. (light)	48.60	388.80	77.15	617.20		
1 Air Track Drill, 4"		1028.00		1130.80		
1 Air Compressor, 600 cfm		550.60		605.66		
1 -50' Air Hose, 3"		14.90		16.39	99.59	109.55
16 L.H., Daily Totals		$2283.10		$2863.25	$142.69	$178.95

Crew B-57	Hr.	Daily	Hr.	Daily	Bare Costs	Incl. O&P
1 Labor Foreman (outside)	$39.60	$316.80	$64.90	$519.20	$43.38	$69.86
2 Laborers	37.60	601.60	61.65	986.40		
1 Equip. Oper. (crane)	51.70	413.60	82.05	656.40		
1 Equip. Oper. (light)	48.60	388.80	77.15	617.20		
1 Equip. Oper. (oiler)	45.20	361.60	71.75	574.00		
1 Crawler Crane, 25 Ton		1150.00		1265.00		
1 Clamshell Bucket, 1 C.Y.		47.80		52.58		
1 Centr. Water Pump, 6"		346.60		381.26		
1 -20' Suction Hose, 6"		11.50		12.65		
20 -50' Discharge Hoses, 6"		122.00		134.20	34.96	38.45
48 L.H., Daily Totals		$3760.30		$5198.89	$78.34	$108.31

Crews

Crew No.	Hr.	Daily	Hr.	Daily	Bare Costs	Incl. O&P
Crew B-58	**Hr.**	**Daily**	**Hr.**	**Daily**	**Bare Costs**	**Incl. O&P**
2 Laborers	$37.60	$601.60	$61.65	$986.40	$41.27	$66.82
1 Equip. Oper. (light)	48.60	388.80	77.15	617.20		
1 Backhoe Loader, 48 H.P.		364.20		400.62		
1 Small Helicopter, w/ Pilot		2790.00		3069.00	131.43	144.57
24 L.H., Daily Totals		$4144.60		$5073.22	$172.69	$211.38
Crew B-59	**Hr.**	**Daily**	**Hr.**	**Daily**	**Bare Costs**	**Incl. O&P**
1 Truck Driver (heavy)	$40.05	$320.40	$64.55	$516.40	$40.05	$64.55
1 Truck Tractor, 220 H.P.		359.60		395.56		
1 Water Tank Trailer, 5000 Gal.		141.00		155.10	62.58	68.83
8 L.H., Daily Totals		$821.00		$1067.06	$102.63	$133.38
Crew B-59A	**Hr.**	**Daily**	**Hr.**	**Daily**	**Bare Costs**	**Incl. O&P**
2 Laborers	$37.60	$601.60	$61.65	$986.40	$38.42	$62.62
1 Truck Driver (heavy)	40.05	320.40	64.55	516.40		
1 Water Tank Trailer, 5000 Gal.		141.00		155.10		
1 Truck Tractor, 220 H.P.		359.60		395.56	20.86	22.94
24 L.H., Daily Totals		$1422.60		$2053.46	$59.27	$85.56
Crew B-60	**Hr.**	**Daily**	**Hr.**	**Daily**	**Bare Costs**	**Incl. O&P**
1 Labor Foreman (outside)	$39.60	$316.80	$64.90	$519.20	$44.13	$70.90
2 Laborers	37.60	601.60	61.65	986.40		
1 Equip. Oper. (crane)	51.70	413.60	82.05	656.40		
2 Equip. Oper. (light)	48.60	777.60	77.15	1234.40		
1 Equip. Oper. (oiler)	45.20	361.60	71.75	574.00		
1 Crawler Crane, 40 Ton		1156.00		1271.60		
1 Lead, 60' High		74.80		82.28		
1 Hammer, Diesel, 15K ft.-lbs.		586.00		644.60		
1 Backhoe Loader, 48 H.P.		364.20		400.62	38.95	42.84
56 L.H., Daily Totals		$4652.20		$6369.50	$83.08	$113.74
Crew B-61	**Hr.**	**Daily**	**Hr.**	**Daily**	**Bare Costs**	**Incl. O&P**
1 Labor Foreman (outside)	$39.60	$316.80	$64.90	$519.20	$40.20	$65.40
3 Laborers	37.60	902.40	61.65	1479.60		
1 Equip. Oper. (light)	48.60	388.80	77.15	617.20		
1 Cement Mixer, 2 C.Y.		194.20		213.62		
1 Air Compressor, 160 cfm		156.40		172.04	8.77	9.64
40 L.H., Daily Totals		$1958.60		$3001.66	$48.97	$75.04
Crew B-62	**Hr.**	**Daily**	**Hr.**	**Daily**	**Bare Costs**	**Incl. O&P**
2 Laborers	$37.60	$601.60	$61.65	$986.40	$41.27	$66.82
1 Equip. Oper. (light)	48.60	388.80	77.15	617.20		
1 Loader, Skid Steer, 30 H.P.		173.60		190.96	7.23	7.96
24 L.H., Daily Totals		$1164.00		$1794.56	$48.50	$74.77
Crew B-62A	**Hr.**	**Daily**	**Hr.**	**Daily**	**Bare Costs**	**Incl. O&P**
2 Laborers	$37.60	$601.60	$61.65	$986.40	$41.27	$66.82
1 Equip. Oper. (light)	48.60	388.80	77.15	617.20		
1 Loader, Skid Steer, 30 H.P.		173.60		190.96		
1 Trencher Attachment		58.30		64.13	9.66	10.63
24 L.H., Daily Totals		$1222.30		$1858.69	$50.93	$77.45
Crew B-63	**Hr.**	**Daily**	**Hr.**	**Daily**	**Bare Costs**	**Incl. O&P**
4 Laborers	$37.60	$1203.20	$61.65	$1972.80	$39.80	$64.75
1 Equip. Oper. (light)	48.60	388.80	77.15	617.20		
1 Loader, Skid Steer, 30 H.P.		173.60		190.96	4.34	4.77
40 L.H., Daily Totals		$1765.60		$2780.96	$44.14	$69.52
Crew B-63B	**Hr.**	**Daily**	**Hr.**	**Daily**	**Bare Costs**	**Incl. O&P**
1 Labor Foreman (inside)	$38.10	$304.80	$62.45	$499.60	$40.48	$65.72
2 Laborers	37.60	601.60	61.65	986.40		
1 Equip. Oper. (light)	48.60	388.80	77.15	617.20		
1 Loader, Skid Steer, 78 H.P.		318.60		350.46	9.96	10.95
32 L.H., Daily Totals		$1613.80		$2453.66	$50.43	$76.68
Crew B-64	**Hr.**	**Daily**	**Hr.**	**Daily**	**Bare Costs**	**Incl. O&P**
1 Laborer	$37.60	$300.80	$61.65	$493.20	$38.35	$62.35
1 Truck Driver (light)	39.10	312.80	63.05	504.40		
1 Power Mulcher (small)		151.80		166.98		
1 Flatbed Truck, Gas, 1.5 Ton		245.40		269.94	24.82	27.31
16 L.H., Daily Totals		$1010.80		$1434.52	$63.17	$89.66
Crew B-65	**Hr.**	**Daily**	**Hr.**	**Daily**	**Bare Costs**	**Incl. O&P**
1 Laborer	$37.60	$300.80	$61.65	$493.20	$38.35	$62.35
1 Truck Driver (light)	39.10	312.80	63.05	504.40		
1 Power Mulcher (Large)		319.40		351.34		
1 Flatbed Truck, Gas, 1.5 Ton		245.40		269.94	35.30	38.83
16 L.H., Daily Totals		$1178.40		$1618.88	$73.65	$101.18
Crew B-66	**Hr.**	**Daily**	**Hr.**	**Daily**	**Bare Costs**	**Incl. O&P**
1 Equip. Oper. (light)	$48.60	$388.80	$77.15	$617.20	$48.60	$77.15
1 Loader-Backhoe, 40 H.P.		263.20		289.52	32.90	36.19
8 L.H., Daily Totals		$652.00		$906.72	$81.50	$113.34
Crew B-67	**Hr.**	**Daily**	**Hr.**	**Daily**	**Bare Costs**	**Incl. O&P**
1 Millwright	$49.15	$393.20	$77.25	$618.00	$48.88	$77.20
1 Equip. Oper. (light)	48.60	388.80	77.15	617.20		
1 Forklift, R/T, 4,000 Lb.		311.20		342.32	19.45	21.40
16 L.H., Daily Totals		$1093.20		$1577.52	$68.33	$98.59
Crew B-67B	**Hr.**	**Daily**	**Hr.**	**Daily**	**Bare Costs**	**Incl. O&P**
1 Millwright Foreman (inside)	$49.65	$397.20	$78.05	$624.40	$49.40	$77.65
1 Millwright	49.15	393.20	77.25	618.00		
16 L.H., Daily Totals		$790.40		$1242.40	$49.40	$77.65
Crew B-68	**Hr.**	**Daily**	**Hr.**	**Daily**	**Bare Costs**	**Incl. O&P**
2 Millwrights	$49.15	$786.40	$77.25	$1236.00	$48.97	$77.22
1 Equip. Oper. (light)	48.60	388.80	77.15	617.20		
1 Forklift, R/T, 4,000 Lb.		311.20		342.32	12.97	14.26
24 L.H., Daily Totals		$1486.40		$2195.52	$61.93	$91.48
Crew B-68A	**Hr.**	**Daily**	**Hr.**	**Daily**	**Bare Costs**	**Incl. O&P**
1 Millwright Foreman (inside)	$49.65	$397.20	$78.05	$624.40	$49.32	$77.52
2 Millwrights	49.15	786.40	77.25	1236.00		
1 Forklift, 8,000 Lb.		185.20		203.72	7.72	8.49
24 L.H., Daily Totals		$1368.80		$2064.12	$57.03	$86.00
Crew B-68B	**Hr.**	**Daily**	**Hr.**	**Daily**	**Bare Costs**	**Incl. O&P**
1 Millwright Foreman (inside)	$49.65	$397.20	$78.05	$624.40	$53.54	$83.58
2 Millwrights	49.15	786.40	77.25	1236.00		
2 Electricians	54.70	875.20	84.70	1355.20		
2 Plumbers	58.70	939.20	91.55	1464.80		
1 Forklift, 5,000 Lb.		331.80		364.98	5.92	6.52
56 L.H., Daily Totals		$3329.80		$5045.38	$59.46	$90.10

Crews

Crew B-68C	Hr.	Daily	Hr.	Daily	Bare Costs	Incl. O&P
1 Millwright Foreman (inside)	$49.65	$397.20	$78.05	$624.40	$53.05	$82.89
1 Millwright	49.15	393.20	77.25	618.00		
1 Electrician	54.70	437.60	84.70	677.60		
1 Plumber	58.70	469.60	91.55	732.40		
1 Forklift, 5,000 Lb.		331.80		364.98	10.37	11.41
32 L.H., Daily Totals		$2029.40		$3017.38	$63.42	$94.29

Crew B-68D	Hr.	Daily	Hr.	Daily	Bare Costs	Incl. O&P
1 Labor Foreman (inside)	$38.10	$304.80	$62.45	$499.60	$41.43	$67.08
1 Laborer	37.60	300.80	61.65	493.20		
1 Equip. Oper. (light)	48.60	388.80	77.15	617.20		
1 Forklift, 5,000 Lb.		331.80		364.98	13.82	15.21
24 L.H., Daily Totals		$1326.20		$1974.98	$55.26	$82.29

Crew B-68E	Hr.	Daily	Hr.	Daily	Bare Costs	Incl. O&P
1 Struc. Steel Foreman (inside)	$53.15	$425.20	$96.05	$768.40	$52.75	$95.33
3 Struc. Steel Workers	52.65	1263.60	95.15	2283.60		
1 Welder	52.65	421.20	95.15	761.20		
1 Forklift, 8,000 Lb.		185.20		203.72	4.63	5.09
40 L.H., Daily Totals		$2295.20		$4016.92	$57.38	$100.42

Crew B-68F	Hr.	Daily	Hr.	Daily	Bare Costs	Incl. O&P
1 Skilled Worker Foreman (out)	$50.65	$405.20	$82.30	$658.40	$49.32	$80.13
2 Skilled Workers	48.65	778.40	79.05	1264.80		
1 Forklift, 5,000 Lb.		331.80		364.98	13.82	15.21
24 L.H., Daily Totals		$1515.40		$2288.18	$63.14	$95.34

Crew B-68G	Hr.	Daily	Hr.	Daily	Bare Costs	Incl. O&P
2 Structural Steel Workers	$52.65	$842.40	$95.15	$1522.40	$52.65	$95.15
1 Forklift, 5,000 Lb.		331.80		364.98	20.74	22.81
16 L.H., Daily Totals		$1174.20		$1887.38	$73.39	$117.96

Crew B-69	Hr.	Daily	Hr.	Daily	Bare Costs	Incl. O&P
1 Labor Foreman (outside)	$39.60	$316.80	$64.90	$519.20	$41.55	$67.28
3 Laborers	37.60	902.40	61.65	1479.60		
1 Equip. Oper. (crane)	51.70	413.60	82.05	656.40		
1 Equip. Oper. (oiler)	45.20	361.60	71.75	574.00		
1 Hyd. Crane, 80 Ton		1625.00		1787.50	33.85	37.24
48 L.H., Daily Totals		$3619.40		$5016.70	$75.40	$104.51

Crew B-69A	Hr.	Daily	Hr.	Daily	Bare Costs	Incl. O&P
1 Labor Foreman (outside)	$39.60	$316.80	$64.90	$519.20	$41.33	$66.88
3 Laborers	37.60	902.40	61.65	1479.60		
1 Equip. Oper. (medium)	50.60	404.80	80.30	642.40		
1 Concrete Finisher	45.00	360.00	71.15	569.20		
1 Curb/Gutter Paver, 2-Track		997.60		1097.36	20.78	22.86
48 L.H., Daily Totals		$2981.60		$4307.76	$62.12	$89.75

Crew B-69B	Hr.	Daily	Hr.	Daily	Bare Costs	Incl. O&P
1 Labor Foreman (outside)	$39.60	$316.80	$64.90	$519.20	$41.33	$66.88
3 Laborers	37.60	902.40	61.65	1479.60		
1 Equip. Oper. (medium)	50.60	404.80	80.30	642.40		
1 Cement Finisher	45.00	360.00	71.15	569.20		
1 Curb/Gutter Paver, 4-Track		776.00		853.60	16.17	17.78
48 L.H., Daily Totals		$2760.00		$4064.00	$57.50	$84.67

Crew B-70	Hr.	Daily	Hr.	Daily	Bare Costs	Incl. O&P
1 Labor Foreman (outside)	$39.60	$316.80	$64.90	$519.20	$43.46	$70.11
3 Laborers	37.60	902.40	61.65	1479.60		
3 Equip. Oper. (medium)	50.60	1214.40	80.30	1927.20		
1 Grader, 30,000 Lbs.		736.20		809.82		
1 Ripper, Beam & 1 Shank		84.20		92.62		
1 Road Sweeper, S.P., 8' wide		662.40		728.64		
1 F.E. Loader, W.M., 1.5 C.Y.		378.40		416.24	33.24	36.56
56 L.H., Daily Totals		$4294.80		$5973.32	$76.69	$106.67

Crew B-71	Hr.	Daily	Hr.	Daily	Bare Costs	Incl. O&P
1 Labor Foreman (outside)	$39.60	$316.80	$64.90	$519.20	$43.46	$70.11
3 Laborers	37.60	902.40	61.65	1479.60		
3 Equip. Oper. (medium)	50.60	1214.40	80.30	1927.20		
1 Pvmt. Profiler, 750 H.P.		5805.00		6385.50		
1 Road Sweeper, S.P., 8' wide		662.40		728.64		
1 F.E. Loader, W.M., 1.5 C.Y.		378.40		416.24	122.25	134.47
56 L.H., Daily Totals		$9279.40		$11456.38	$165.70	$204.58

Crew B-72	Hr.	Daily	Hr.	Daily	Bare Costs	Incl. O&P
1 Labor Foreman (outside)	$39.60	$316.80	$64.90	$519.20	$44.35	$71.38
3 Laborers	37.60	902.40	61.65	1479.60		
4 Equip. Oper. (medium)	50.60	1619.20	80.30	2569.60		
1 Pvmt. Profiler, 750 H.P.		5805.00		6385.50		
1 Hammermill, 250 H.P.		1879.00		2066.90		
1 Windrow Loader		1227.00		1349.70		
1 Mix Paver 165 H.P.		2147.00		2361.70		
1 Roller, Pneum. Whl., 12 Ton		344.40		378.84	178.16	195.98
64 L.H., Daily Totals		$14240.80		$17111.04	$222.51	$267.36

Crew B-73	Hr.	Daily	Hr.	Daily	Bare Costs	Incl. O&P
1 Labor Foreman (outside)	$39.60	$316.80	$64.90	$519.20	$45.98	$73.71
2 Laborers	37.60	601.60	61.65	986.40		
5 Equip. Oper. (medium)	50.60	2024.00	80.30	3212.00		
1 Road Mixer, 310 H.P.		1929.00		2121.90		
1 Tandem Roller, 10 Ton		236.80		260.48		
1 Hammermill, 250 H.P.		1879.00		2066.90		
1 Grader, 30,000 Lbs.		736.20		809.82		
.5 F.E. Loader, W.M., 1.5 C.Y.		189.20		208.12		
.5 Truck Tractor, 220 H.P.		179.80		197.78		
.5 Water Tank Trailer, 5000 Gal.		70.50		77.55	81.57	89.73
64 L.H., Daily Totals		$8162.90		$10460.15	$127.55	$163.44

Crew B-74	Hr.	Daily	Hr.	Daily	Bare Costs	Incl. O&P
1 Labor Foreman (outside)	$39.60	$316.80	$64.90	$519.20	$44.96	$72.11
1 Laborer	37.60	300.80	61.65	493.20		
4 Equip. Oper. (medium)	50.60	1619.20	80.30	2569.60		
2 Truck Drivers (heavy)	40.05	640.80	64.55	1032.80		
1 Grader, 30,000 Lbs.		736.20		809.82		
1 Ripper, Beam & 1 Shank		84.20		92.62		
2 Stabilizers, 310 H.P.		3596.00		3955.60		
1 Flatbed Truck, Gas, 3 Ton		303.20		333.52		
1 Chem. Spreader, Towed		52.40		57.64		
1 Roller, Vibratory, 25 Ton		687.40		756.14		
1 Water Tank Trailer, 5000 Gal.		141.00		155.10		
1 Truck Tractor, 220 H.P.		359.60		395.56	93.13	102.44
64 L.H., Daily Totals		$8837.60		$11170.80	$138.09	$174.54

Crews

Crew No.	Bare Costs		Incl. Subs O&P		Cost Per Labor-Hour	
	Hr.	Daily	Hr.	Daily	Bare Costs	Incl. O&P
Crew B-75						
1 Labor Foreman (outside)	$39.60	$316.80	$64.90	$519.20	$45.66	$73.19
1 Laborer	37.60	300.80	61.65	493.20		
4 Equip. Oper. (medium)	50.60	1619.20	80.30	2569.60		
1 Truck Driver (heavy)	40.05	320.40	64.55	516.40		
1 Grader, 30,000 Lbs.		736.20		809.82		
1 Ripper, Beam & 1 Shank		84.20		92.62		
2 Stabilizers, 310 H.P.		3596.00		3955.60		
1 Dist. Tanker, 3000 Gallon		305.60		336.16		
1 Truck Tractor, 6x4, 380 H.P.		599.80		659.78		
1 Roller, Vibratory, 25 Ton		687.40		756.14	107.31	118.04
56 L.H., Daily Totals		$8566.40		$10708.52	$152.97	$191.22
Crew B-76	Hr.	Daily	Hr.	Daily	Bare Costs	Incl. O&P
1 Dock Builder Foreman (outside)	$48.10	$384.80	$78.55	$628.40	$47.47	$76.77
5 Dock Builders	46.10	1844.00	75.30	3012.00		
2 Equip. Oper. (crane)	51.70	827.20	82.05	1312.80		
1 Equip. Oper. (oiler)	45.20	361.60	71.75	574.00		
1 Crawler Crane, 50 Ton		1294.00		1423.40		
1 Barge, 400 Ton		776.00		853.60		
1 Hammer, Diesel, 15K ft.-lbs.		586.00		644.60		
1 Lead, 60' High		74.80		82.28		
1 Air Compressor, 600 cfm		550.60		605.66		
2 -50' Air Hoses, 3"		29.80		32.78	45.99	50.59
72 L.H., Daily Totals		$6728.80		$9169.52	$93.46	$127.35
Crew B-76A	Hr.	Daily	Hr.	Daily	Bare Costs	Incl. O&P
1 Labor Foreman (outside)	$39.60	$316.80	$64.90	$519.20	$40.56	$65.87
5 Laborers	37.60	1504.00	61.65	2466.00		
1 Equip. Oper. (crane)	51.70	413.60	82.05	656.40		
1 Equip. Oper. (oiler)	45.20	361.60	71.75	574.00		
1 Crawler Crane, 50 Ton		1294.00		1423.40		
1 Barge, 400 Ton		776.00		853.60	32.34	35.58
64 L.H., Daily Totals		$4666.00		$6492.60	$72.91	$101.45
Crew B-77	Hr.	Daily	Hr.	Daily	Bare Costs	Incl. O&P
1 Labor Foreman (outside)	$39.60	$316.80	$64.90	$519.20	$38.30	$62.58
3 Laborers	37.60	902.40	61.65	1479.60		
1 Truck Driver (light)	39.10	312.80	63.05	504.40		
1 Crack Cleaner, 25 H.P.		61.60		67.76		
1 Crack Filler, Trailer Mtd.		202.00		222.20		
1 Flatbed Truck, Gas, 3 Ton		303.20		333.52	14.17	15.59
40 L.H., Daily Totals		$2098.80		$3126.68	$52.47	$78.17
Crew B-78	Hr.	Daily	Hr.	Daily	Bare Costs	Incl. O&P
1 Labor Foreman (outside)	$39.60	$316.80	$64.90	$519.20	$38.18	$62.42
4 Laborers	37.60	1203.20	61.65	1972.80		
1 Truck Driver (light)	39.10	312.80	63.05	504.40		
1 Paint Striper, S.P., 40 Gallon		151.40		166.54		
1 Flatbed Truck, Gas, 3 Ton		303.20		333.52		
1 Pickup Truck, 3/4 Ton		144.20		158.62	12.48	13.72
48 L.H., Daily Totals		$2431.60		$3655.08	$50.66	$76.15
Crew B-78A	Hr.	Daily	Hr.	Daily	Bare Costs	Incl. O&P
1 Equip. Oper. (light)	$48.60	$388.80	$77.15	$617.20	$48.60	$77.15
1 Line Rem. (Metal Balls) 115 H.P.		821.60		903.76	102.70	112.97
8 L.H., Daily Totals		$1210.40		$1520.96	$151.30	$190.12

Crew No.	Bare Costs		Incl. Subs O&P		Cost Per Labor-Hour	
Crew B-78B	Hr.	Daily	Hr.	Daily	Bare Costs	Incl. O&P
2 Laborers	$37.60	$601.60	$61.65	$986.40	$38.82	$63.37
.25 Equip. Oper. (light)	48.60	97.20	77.15	154.30		
1 Pickup Truck, 3/4 Ton		144.20		158.62		
1 Line Rem.,11 H.P.,Walk Behind		65.60		72.16		
.25 Road Sweeper, S.P., 8' wide		165.60		182.16	20.86	22.94
18 L.H., Daily Totals		$1074.20		$1553.64	$59.68	$86.31
Crew B-78C	Hr.	Daily	Hr.	Daily	Bare Costs	Incl. O&P
1 Labor Foreman (outside)	$39.60	$316.80	$64.90	$519.20	$38.18	$62.42
4 Laborers	37.60	1203.20	61.65	1972.80		
1 Truck Driver (light)	39.10	312.80	63.05	504.40		
1 Paint Striper, T.M., 120 Gal.		804.60		885.06		
1 Flatbed Truck, Gas, 3 Ton		303.20		333.52		
1 Pickup Truck, 3/4 Ton		144.20		158.62	26.08	28.69
48 L.H., Daily Totals		$3084.80		$4373.60	$64.27	$91.12
Crew B-78D	Hr.	Daily	Hr.	Daily	Bare Costs	Incl. O&P
2 Labor Foremen (outside)	$39.60	$633.60	$64.90	$1038.40	$38.15	$62.44
7 Laborers	37.60	2105.60	61.65	3452.40		
1 Truck Driver (light)	39.10	312.80	63.05	504.40		
1 Paint Striper, T.M., 120 Gal.		804.60		885.06		
1 Flatbed Truck, Gas, 3 Ton		303.20		333.52		
3 Pickup Trucks, 3/4 Ton		432.60		475.86		
1 Air Compressor, 60 cfm		137.40		151.14		
1 -50' Air Hose, 3/4"		3.25		3.58		
1 Breakers, Pavement, 60 lb.		9.80		10.78	21.14	23.25
80 L.H., Daily Totals		$4742.85		$6855.14	$59.29	$85.69
Crew B-78E	Hr.	Daily	Hr.	Daily	Bare Costs	Incl. O&P
2 Labor Foremen (outside)	$39.60	$633.60	$64.90	$1038.40	$38.06	$62.31
9 Laborers	37.60	2707.20	61.65	4438.80		
1 Truck Driver (light)	39.10	312.80	63.05	504.40		
1 Paint Striper, T.M., 120 Gal.		804.60		885.06		
1 Flatbed Truck, Gas, 3 Ton		303.20		333.52		
4 Pickup Trucks, 3/4 Ton		576.80		634.48		
2 Air Compressor, 60 cfm		274.80		302.28		
2 -50' Air Hose, 3/4"		6.50		7.15		
2 Breakers, Pavement, 60 lb.		19.60		21.56	20.68	22.75
96 L.H., Daily Totals		$5639.10		$8165.65	$58.74	$85.06
Crew B-78F	Hr.	Daily	Hr.	Daily	Bare Costs	Incl. O&P
2 Labor Foremen (outside)	$39.60	$633.60	$64.90	$1038.40	$37.99	$62.21
11 Laborers	37.60	3308.80	61.65	5425.20		
1 Truck Driver (light)	39.10	312.80	63.05	504.40		
1 Paint Striper, T.M., 120 Gal.		804.60		885.06		
1 Flatbed Truck, Gas, 3 Ton		303.20		333.52		
7 Pickup Trucks, 3/4 Ton		1009.40		1110.34		
3 Air Compressor, 60 cfm		412.20		453.42		
3 -50' Air Hose, 3/4"		9.75		10.73		
3 Breakers, Pavement, 60 lb.		29.40		32.34	22.93	25.23
112 L.H., Daily Totals		$6823.75		$9793.41	$60.93	$87.44
Crew B-79	Hr.	Daily	Hr.	Daily	Bare Costs	Incl. O&P
1 Labor Foreman (outside)	$39.60	$316.80	$64.90	$519.20	$38.30	$62.58
3 Laborers	37.60	902.40	61.65	1479.60		
1 Truck Driver (light)	39.10	312.80	63.05	504.40		
1 Paint Striper, T.M., 120 Gal.		804.60		885.06		
1 Heating Kettle, 115 Gallon		81.00		89.10		
1 Flatbed Truck, Gas, 3 Ton		303.20		333.52		
2 Pickup Trucks, 3/4 Ton		288.40		317.24	36.93	40.62
40 L.H., Daily Totals		$3009.20		$4128.12	$75.23	$103.20

Crews

Crew No.		Bare Costs		Incl. Subs O&P		Cost Per Labor-Hour	
Crew B-79A	Hr.	Daily	Hr.	Daily	Bare Costs	Incl. O&P
1.5 Equip. Oper. (light)	$48.60	$583.20	$77.15	$925.80	$48.60	$77.15
.5 Line Remov. (Grinder) 115 H.P.		455.90		501.49		
1 Line Rem. (Metal Balls) 115 H.P.		821.60		903.76	106.46	117.10
12 L.H., Daily Totals		$1860.70		$2331.05	$155.06	$194.25

Crew B-79B	Hr.	Daily	Hr.	Daily	Bare Costs	Incl. O&P
1 Laborer	$37.60	$300.80	$61.65	$493.20	$37.60	$61.65
1 Set of Gases		152.00		167.20	19.00	20.90
8 L.H., Daily Totals		$452.80		$660.40	$56.60	$82.55

Crew B-79C	Hr.	Daily	Hr.	Daily	Bare Costs	Incl. O&P
1 Labor Foreman (outside)	$39.60	$316.80	$64.90	$519.20	$38.10	$62.31
5 Laborers	37.60	1504.00	61.65	2466.00		
1 Truck Driver (light)	39.10	312.80	63.05	504.40		
1 Paint Striper, T.M., 120 Gal.		804.60		885.06		
1 Heating Kettle, 115 Gallon		81.00		89.10		
1 Flatbed Truck, Gas, 3 Ton		303.20		333.52		
3 Pickup Trucks, 3/4 Ton		432.60		475.86		
1 Air Compressor, 60 cfm		137.40		151.14		
1 -50' Air Hose, 3/4"		3.25		3.58		
1 Breakers, Pavement, 60 lb.		9.80		10.78	31.64	34.80
56 L.H., Daily Totals		$3905.45		$5438.64	$69.74	$97.12

Crew B-79D	Hr.	Daily	Hr.	Daily	Bare Costs	Incl. O&P
2 Labor Foremen (outside)	$39.60	$633.60	$64.90	$1038.40	$38.29	$62.64
5 Laborers	37.60	1504.00	61.65	2466.00		
1 Truck Driver (light)	39.10	312.80	63.05	504.40		
1 Paint Striper, T.M., 120 Gal.		804.60		885.06		
1 Heating Kettle, 115 Gallon		81.00		89.10		
1 Flatbed Truck, Gas, 3 Ton		303.20		333.52		
4 Pickup Trucks, 3/4 Ton		576.80		634.48		
1 Air Compressor, 60 cfm		137.40		151.14		
1 -50' Air Hose, 3/4"		3.25		3.58		
1 Breakers, Pavement, 60 lb.		9.80		10.78	29.94	32.93
64 L.H., Daily Totals		$4366.45		$6116.45	$68.23	$95.57

Crew B-79E	Hr.	Daily	Hr.	Daily	Bare Costs	Incl. O&P
2 Labor Foremen (outside)	$39.60	$633.60	$64.90	$1038.40	$38.15	$62.44
7 Laborers	37.60	2105.60	61.65	3452.40		
1 Truck Driver (light)	39.10	312.80	63.05	504.40		
1 Paint Striper, T.M., 120 Gal.		804.60		885.06		
1 Heating Kettle, 115 Gallon		81.00		89.10		
1 Flatbed Truck, Gas, 3 Ton		303.20		333.52		
5 Pickup Trucks, 3/4 Ton		721.00		793.10		
2 Air Compressors, 60 cfm		274.80		302.28		
2 -50' Air Hoses, 3/4"		6.50		7.15		
2 Breakers, Pavement, 60 lb.		19.60		21.56	27.63	30.40
80 L.H., Daily Totals		$5262.70		$7426.97	$65.78	$92.84

Crew B-80	Hr.	Daily	Hr.	Daily	Bare Costs	Incl. O&P
1 Labor Foreman (outside)	$39.60	$316.80	$64.90	$519.20	$41.23	$66.69
1 Laborer	37.60	300.80	61.65	493.20		
1 Truck Driver (light)	39.10	312.80	63.05	504.40		
1 Equip. Oper. (light)	48.60	388.80	77.15	617.20		
1 Flatbed Truck, Gas, 3 Ton		303.20		333.52		
1 Earth Auger, Truck-Mtd.		417.20		458.92	22.51	24.76
32 L.H., Daily Totals		$2039.60		$2926.44	$63.74	$91.45

Crew B-80A	Hr.	Daily	Hr.	Daily	Bare Costs	Incl. O&P
3 Laborers	$37.60	$902.40	$61.65	$1479.60	$37.60	$61.65
1 Flatbed Truck, Gas, 3 Ton		303.20		333.52	12.63	13.90
24 L.H., Daily Totals		$1205.60		$1813.12	$50.23	$75.55

Crew B-80B	Hr.	Daily	Hr.	Daily	Bare Costs	Incl. O&P
3 Laborers	$37.60	$902.40	$61.65	$1479.60	$40.35	$65.53
1 Equip. Oper. (light)	48.60	388.80	77.15	617.20		
1 Crane, Flatbed Mounted, 3 Ton		243.80		268.18	7.62	8.38
32 L.H., Daily Totals		$1535.00		$2364.98	$47.97	$73.91

Crew B-80C	Hr.	Daily	Hr.	Daily	Bare Costs	Incl. O&P
2 Laborers	$37.60	$601.60	$61.65	$986.40	$38.10	$62.12
1 Truck Driver (light)	39.10	312.80	63.05	504.40		
1 Flatbed Truck, Gas, 1.5 Ton		245.40		269.94		
1 Manual Fence Post Auger, Gas		8.40		9.24	10.57	11.63
24 L.H., Daily Totals		$1168.20		$1769.98	$48.67	$73.75

Crew B-81	Hr.	Daily	Hr.	Daily	Bare Costs	Incl. O&P
1 Laborer	$37.60	$300.80	$61.65	$493.20	$42.75	$68.83
1 Equip. Oper. (medium)	50.60	404.80	80.30	642.40		
1 Truck Driver (heavy)	40.05	320.40	64.55	516.40		
1 Hydromulcher, T.M., 3000 Gal.		334.20		367.62		
1 Truck Tractor, 220 H.P.		359.60		395.56	28.91	31.80
24 L.H., Daily Totals		$1719.80		$2415.18	$71.66	$100.63

Crew B-81A	Hr.	Daily	Hr.	Daily	Bare Costs	Incl. O&P
1 Laborer	$37.60	$300.80	$61.65	$493.20	$38.35	$62.35
1 Truck Driver (light)	39.10	312.80	63.05	504.40		
1 Hydromulcher, T.M., 600 Gal.		127.80		140.58		
1 Flatbed Truck, Gas, 3 Ton		303.20		333.52	26.94	29.63
16 L.H., Daily Totals		$1044.60		$1471.70	$65.29	$91.98

Crew B-82	Hr.	Daily	Hr.	Daily	Bare Costs	Incl. O&P
1 Laborer	$37.60	$300.80	$61.65	$493.20	$43.10	$69.40
1 Equip. Oper. (light)	48.60	388.80	77.15	617.20		
1 Horiz. Borer, 6 H.P.		83.80		92.18	5.24	5.76
16 L.H., Daily Totals		$773.40		$1202.58	$48.34	$75.16

Crew B-82A	Hr.	Daily	Hr.	Daily	Bare Costs	Incl. O&P
2 Laborers	$37.60	$601.60	$61.65	$986.40	$43.10	$69.40
2 Equip. Opers. (light)	48.60	777.60	77.15	1234.40		
2 Dump Truck, 8 C.Y., 220 H.P.		822.40		904.64		
1 Flatbed Trailer, 25 Ton		113.60		124.96		
1 Horiz. Dir. Drill, 20k lb. Thrust		654.20		719.62		
1 Mud Trailer for HDD, 1500 Gal.		346.20		380.82		
1 Pickup Truck, 4 x 4, 3/4 Ton		155.60		171.16		
1 Flatbed trailer, 3 Ton		23.60		25.96		
1 Loader, Skid Steer, 78 H.P.		318.60		350.46	76.07	83.68
32 L.H., Daily Totals		$3813.40		$4898.42	$119.17	$153.08

Crew B-82B	Hr.	Daily	Hr.	Daily	Bare Costs	Incl. O&P
2 Laborers	$37.60	$601.60	$61.65	$986.40	$43.10	$69.40
2 Equip. Opers. (light)	48.60	777.60	77.15	1234.40		
2 Dump Truck, 8 C.Y., 220 H.P.		822.40		904.64		
1 Flatbed Trailer, 25 Ton		113.60		124.96		
1 Horiz. Dir. Drill, 30k lb. Thrust		931.20		1024.32		
1 Mud Trailer for HDD, 1500 Gal.		346.20		380.82		
1 Pickup Truck, 4 x 4, 3/4 Ton		155.60		171.16		
1 Flatbed trailer, 3 Ton		23.60		25.96		
1 Loader, Skid Steer, 78 H.P.		318.60		350.46	84.72	93.20
32 L.H., Daily Totals		$4090.40		$5203.12	$127.83	$162.60

Crews

Crew No.	Bare Costs		Incl. Subs O&P		Cost Per Labor-Hour	
Crew B-82C	Hr.	Daily	Hr.	Daily	Bare Costs	Incl. O&P
2 Laborers	$37.60	$601.60	$61.65	$986.40	$43.10	$69.40
2 Equip. Opers. (light)	48.60	777.60	77.15	1234.40		
2 Dump Truck, 8 C.Y., 220 H.P.		822.40		904.64		
1 Flatbed Trailer, 25 Ton		113.60		124.96		
1 Horiz. Dir. Drill, 50k lb. Thrust		1237.00		1360.70		
1 Mud Trailer for HDD, 1500 Gal.		346.20		380.82		
1 Pickup Truck, 4 x 4, 3/4 Ton		155.60		171.16		
1 Flatbed trailer, 3 Ton		23.60		25.96		
1 Loader, Skid Steer, 78 H.P.		318.60		350.46	94.28	103.71
32 L.H., Daily Totals		$4396.20		$5539.50	$137.38	$173.11
Crew B-82D	Hr.	Daily	Hr.	Daily	Bare Costs	Incl. O&P
1 Equip. Oper. (light)	$48.60	$388.80	$77.15	$617.20	$48.60	$77.15
1 Mud Trailer for HDD, 1500 Gal.		346.20		380.82	43.27	47.60
8 L.H., Daily Totals		$735.00		$998.02	$91.88	$124.75
Crew B-83	Hr.	Daily	Hr.	Daily	Bare Costs	Incl. O&P
1 Tugboat Captain	$50.60	$404.80	$80.30	$642.40	$44.10	$70.97
1 Tugboat Hand	37.60	300.80	61.65	493.20		
1 Tugboat, 250 H.P.		870.00		957.00	54.38	59.81
16 L.H., Daily Totals		$1575.60		$2092.60	$98.47	$130.79
Crew B-84	Hr.	Daily	Hr.	Daily	Bare Costs	Incl. O&P
1 Equip. Oper. (medium)	$50.60	$404.80	$80.30	$642.40	$50.60	$80.30
1 Rotary Mower/Tractor		370.80		407.88	46.35	50.98
8 L.H., Daily Totals		$775.60		$1050.28	$96.95	$131.29
Crew B-85	Hr.	Daily	Hr.	Daily	Bare Costs	Incl. O&P
3 Laborers	$37.60	$902.40	$61.65	$1479.60	$40.69	$65.96
1 Equip. Oper. (medium)	50.60	404.80	80.30	642.40		
1 Truck Driver (heavy)	40.05	320.40	64.55	516.40		
1 Aerial Lift Truck, 80'		634.40		697.84		
1 Brush Chipper, 12", 130 H.P.		391.40		430.54		
1 Pruning Saw, Rotary		6.65		7.32	25.81	28.39
40 L.H., Daily Totals		$2660.05		$3774.09	$66.50	$94.35
Crew B-86	Hr.	Daily	Hr.	Daily	Bare Costs	Incl. O&P
1 Equip. Oper. (medium)	$50.60	$404.80	$80.30	$642.40	$50.60	$80.30
1 Stump Chipper, S.P.		185.55		204.10	23.19	25.51
8 L.H., Daily Totals		$590.35		$846.51	$73.79	$105.81
Crew B-86A	Hr.	Daily	Hr.	Daily	Bare Costs	Incl. O&P
1 Equip. Oper. (medium)	$50.60	$404.80	$80.30	$642.40	$50.60	$80.30
1 Grader, 30,000 Lbs.		736.20		809.82	92.03	101.23
8 L.H., Daily Totals		$1141.00		$1452.22	$142.63	$181.53
Crew B-86B	Hr.	Daily	Hr.	Daily	Bare Costs	Incl. O&P
1 Equip. Oper. (medium)	$50.60	$404.80	$80.30	$642.40	$50.60	$80.30
1 Dozer, 200 H.P.		1387.00		1525.70	173.38	190.71
8 L.H., Daily Totals		$1791.80		$2168.10	$223.97	$271.01
Crew B-87	Hr.	Daily	Hr.	Daily	Bare Costs	Incl. O&P
1 Laborer	$37.60	$300.80	$61.65	$493.20	$48.00	$76.57
4 Equip. Oper. (medium)	50.60	1619.20	80.30	2569.60		
2 Feller Bunchers, 100 H.P.		1676.80		1844.48		
1 Log Chipper, 22" Tree		889.40		978.34		
1 Dozer, 105 H.P.		602.80		663.08		
1 Chain Saw, Gas, 36" Long		45.00		49.50	80.35	88.39
40 L.H., Daily Totals		$5134.00		$6598.20	$128.35	$164.96

Crew No.	Bare Costs		Incl. Subs O&P		Cost Per Labor-Hour	
Crew B-88	Hr.	Daily	Hr.	Daily	Bare Costs	Incl. O&P
1 Laborer	$37.60	$300.80	$61.65	$493.20	$48.74	$77.64
6 Equip. Oper. (medium)	50.60	2428.80	80.30	3854.40		
2 Feller Bunchers, 100 H.P.		1676.80		1844.48		
1 Log Chipper, 22" Tree		889.40		978.34		
2 Log Skidders, 50 H.P.		1822.00		2004.20		
1 Dozer, 105 H.P.		602.80		663.08		
1 Chain Saw, Gas, 36" Long		45.00		49.50	89.93	98.92
56 L.H., Daily Totals		$7765.60		$9887.20	$138.67	$176.56
Crew B-89	Hr.	Daily	Hr.	Daily	Bare Costs	Incl. O&P
1 Equip. Oper. (light)	$48.60	$388.80	$77.15	$617.20	$43.85	$70.10
1 Truck Driver (light)	39.10	312.80	63.05	504.40		
1 Flatbed Truck, Gas, 3 Ton		303.20		333.52		
1 Concrete Saw		167.40		184.14		
1 Water Tank, 65 Gal.		17.30		19.03	30.49	33.54
16 L.H., Daily Totals		$1189.50		$1658.29	$74.34	$103.64
Crew B-89A	Hr.	Daily	Hr.	Daily	Bare Costs	Incl. O&P
1 Skilled Worker	$48.65	$389.20	$79.05	$632.40	$43.13	$70.35
1 Laborer	37.60	300.80	61.65	493.20		
1 Core Drill (Large)		115.60		127.16	7.22	7.95
16 L.H., Daily Totals		$805.60		$1252.76	$50.35	$78.30
Crew B-89B	Hr.	Daily	Hr.	Daily	Bare Costs	Incl. O&P
1 Equip. Oper. (light)	$48.60	$388.80	$77.15	$617.20	$43.85	$70.10
1 Truck Driver (light)	39.10	312.80	63.05	504.40		
1 Wall Saw, Hydraulic, 10 H.P.		114.60		126.06		
1 Generator, Diesel, 100 kW		423.40		465.74		
1 Water Tank, 65 Gal.		17.30		19.03		
1 Flatbed Truck, Gas, 3 Ton		303.20		333.52	53.66	59.02
16 L.H., Daily Totals		$1560.10		$2065.95	$97.51	$129.12
Crew B-90	Hr.	Daily	Hr.	Daily	Bare Costs	Incl. O&P
1 Labor Foreman (outside)	$39.60	$316.80	$64.90	$519.20	$41.21	$66.66
3 Laborers	37.60	902.40	61.65	1479.60		
2 Equip. Oper. (light)	48.60	777.60	77.15	1234.40		
2 Truck Drivers (heavy)	40.05	640.80	64.55	1032.80		
1 Road Mixer, 310 H.P.		1929.00		2121.90		
1 Dist. Truck, 2000 Gal.		275.60		303.16	34.45	37.89
64 L.H., Daily Totals		$4842.20		$6691.06	$75.66	$104.55
Crew B-90A	Hr.	Daily	Hr.	Daily	Bare Costs	Incl. O&P
1 Labor Foreman (outside)	$39.60	$316.80	$64.90	$519.20	$45.31	$72.77
2 Laborers	37.60	601.60	61.65	986.40		
4 Equip. Oper. (medium)	50.60	1619.20	80.30	2569.60		
2 Graders, 30,000 Lbs.		1472.40		1619.64		
1 Tandem Roller, 10 Ton		236.60		260.48		
1 Roller, Pneum. Whl., 12 Ton		344.40		378.84	36.67	40.34
56 L.H., Daily Totals		$4591.20		$6334.16	$81.99	$113.11
Crew B-90B	Hr.	Daily	Hr.	Daily	Bare Costs	Incl. O&P
1 Labor Foreman (outside)	$39.60	$316.80	$64.90	$519.20	$44.43	$71.52
2 Laborers	37.60	601.60	61.65	986.40		
3 Equip. Oper. (medium)	50.60	1214.40	80.30	1927.20		
1 Roller, Pneum. Whl., 12 Ton		344.40		378.84		
1 Road Mixer, 310 H.P.		1929.00		2121.90	47.36	52.10
48 L.H., Daily Totals		$4406.20		$5933.54	$91.80	$123.62

Crews

Crew B-90C

Crew B-90C	Hr.	Daily	Hr.	Daily	Bare Costs	Incl. O&P
1 Labor Foreman (outside)	$39.60	$316.80	$64.90	$519.20	$42.00	$67.82
4 Laborers	37.60	1203.20	61.65	1972.80		
3 Equip. Oper. (medium)	50.60	1214.40	80.30	1927.20		
3 Truck Drivers (heavy)	40.05	961.20	64.55	1549.20		
3 Road Mixers, 310 H.P.		5787.00		6365.70	65.76	72.34
88 L.H., Daily Totals		$9482.60		$12334.10	$107.76	$140.16

Crew B-90D

Crew B-90D	Hr.	Daily	Hr.	Daily	Bare Costs	Incl. O&P
1 Labor Foreman (outside)	$39.60	$316.80	$64.90	$519.20	$41.32	$66.87
6 Laborers	37.60	1804.80	61.65	2959.20		
3 Equip. Oper. (medium)	50.60	1214.40	80.30	1927.20		
3 Truck Drivers (heavy)	40.05	961.20	64.55	1549.20		
3 Road Mixers, 310 H.P.		5787.00		6365.70	55.64	61.21
104 L.H., Daily Totals		$10084.20		$13320.50	$96.96	$128.08

Crew B-90E

Crew B-90E	Hr.	Daily	Hr.	Daily	Bare Costs	Incl. O&P
1 Labor Foreman (outside)	$39.60	$316.80	$64.90	$519.20	$42.43	$68.55
4 Laborers	37.60	1203.20	61.65	1972.80		
3 Equip. Oper. (medium)	50.60	1214.40	80.30	1927.20		
1 Truck Driver (heavy)	40.05	320.40	64.55	516.40		
1 Road Mixers, 310 H.P.		1929.00		2121.90	26.79	29.47
72 L.H., Daily Totals		$4983.80		$7057.50	$69.22	$98.02

Crew B-91

Crew B-91	Hr.	Daily	Hr.	Daily	Bare Costs	Incl. O&P
1 Labor Foreman (outside)	$39.60	$316.80	$64.90	$519.20	$44.66	$71.74
2 Laborers	37.60	601.60	61.65	986.40		
4 Equip. Oper. (medium)	50.60	1619.20	80.30	2569.60		
1 Truck Driver (heavy)	40.05	320.40	64.55	516.40		
1 Dist. Tanker, 3000 Gallon		305.60		336.16		
1 Truck Tractor, 6x4, 380 H.P.		599.80		659.78		
1 Aggreg. Spreader, S.P.		834.00		917.40		
1 Roller, Pneum. Whl., 12 Ton		344.40		378.84		
1 Tandem Roller, 10 Ton		236.80		260.48	36.26	39.89
64 L.H., Daily Totals		$5178.60		$7144.26	$80.92	$111.63

Crew B-91B

Crew B-91B	Hr.	Daily	Hr.	Daily	Bare Costs	Incl. O&P
1 Laborer	$37.60	$300.80	$61.65	$493.20	$44.10	$70.97
1 Equipment Oper. (med.)	50.60	404.80	80.30	642.40		
1 Road Sweeper, Vac. Assist.		986.00		1084.60	61.63	67.79
16 L.H., Daily Totals		$1691.60		$2220.20	$105.72	$138.76

Crew B-91C

Crew B-91C	Hr.	Daily	Hr.	Daily	Bare Costs	Incl. O&P
1 Laborer	$37.60	$300.80	$61.65	$493.20	$38.35	$62.35
1 Truck Driver (light)	39.10	312.80	63.05	504.40		
1 Catch Basin Cleaning Truck		573.60		630.96	35.85	39.44
16 L.H., Daily Totals		$1187.20		$1628.56	$74.20	$101.79

Crew B-91D

Crew B-91D	Hr.	Daily	Hr.	Daily	Bare Costs	Incl. O&P
1 Labor Foreman (outside)	$39.60	$316.80	$64.90	$519.20	$43.13	$69.52
5 Laborers	37.60	1504.00	61.65	2466.00		
5 Equip. Oper. (medium)	50.60	2024.00	80.30	3212.00		
2 Truck Drivers (heavy)	40.05	640.80	64.55	1032.80		
1 Aggreg. Spreader, S.P.		834.00		917.40		
2 Truck Tractor, 6x4, 380 H.P.		1199.60		1319.56		
2 Dist. Tanker, 3000 Gallon		611.20		672.32		
2 Pavement Brush, Towed		163.20		179.52		
2 Roller, Pneum. Whl., 12 Ton		688.80		757.68	33.62	36.99
104 L.H., Daily Totals		$7982.40		$11076.48	$76.75	$106.50

Crew B-92

Crew B-92	Hr.	Daily	Hr.	Daily	Bare Costs	Incl. O&P
1 Labor Foreman (outside)	$39.60	$316.80	$64.90	$519.20	$38.10	$62.46
3 Laborers	37.60	902.40	61.65	1479.60		
1 Crack Cleaner, 25 H.P.		61.60		67.76		
1 Air Compressor, 60 cfm		137.40		151.14		
1 Tar Kettle, T.M.		132.00		145.20		
1 Flatbed Truck, Gas, 3 Ton		303.20		333.52	19.82	21.80
32 L.H., Daily Totals		$1853.40		$2696.42	$57.92	$84.26

Crew B-93

Crew B-93	Hr.	Daily	Hr.	Daily	Bare Costs	Incl. O&P
1 Equip. Oper. (medium)	$50.60	$404.80	$80.30	$642.40	$50.60	$80.30
1 Feller Buncher, 100 H.P.		838.40		922.24	104.80	115.28
8 L.H., Daily Totals		$1243.20		$1564.64	$155.40	$195.58

Crew B-94A

Crew B-94A	Hr.	Daily	Hr.	Daily	Bare Costs	Incl. O&P
1 Laborer	$37.60	$300.80	$61.65	$493.20	$37.60	$61.65
1 Diaphragm Water Pump, 2"		71.00		78.10		
1 -20' Suction Hose, 2"		1.95		2.15		
2 -50' Discharge Hoses, 2"		1.80		1.98	9.34	10.28
8 L.H., Daily Totals		$375.55		$575.42	$46.94	$71.93

Crew B-94B

Crew B-94B	Hr.	Daily	Hr.	Daily	Bare Costs	Incl. O&P
1 Laborer	$37.60	$300.80	$61.65	$493.20	$37.60	$61.65
1 Diaphragm Water Pump, 4"		114.20		125.62		
1 -20' Suction Hose, 4"		3.25		3.58		
2 -50' Discharge Hoses, 4"		4.70		5.17	15.27	16.80
8 L.H., Daily Totals		$422.95		$627.57	$52.87	$78.45

Crew B-94C

Crew B-94C	Hr.	Daily	Hr.	Daily	Bare Costs	Incl. O&P
1 Laborer	$37.60	$300.80	$61.65	$493.20	$37.60	$61.65
1 Centrifugal Water Pump, 3"		78.20		86.02		
1 -20' Suction Hose, 3"		2.85		3.13		
2 -50' Discharge Hoses, 3"		3.00		3.30	10.51	11.56
8 L.H., Daily Totals		$384.85		$585.65	$48.11	$73.21

Crew B-94D

Crew B-94D	Hr.	Daily	Hr.	Daily	Bare Costs	Incl. O&P
1 Laborer	$37.60	$300.80	$61.65	$493.20	$37.60	$61.65
1 Centr. Water Pump, 6"		346.60		381.26		
1 -20' Suction Hose, 6"		11.50		12.65		
2 -50' Discharge Hoses, 6"		12.20		13.42	46.29	50.92
8 L.H., Daily Totals		$671.10		$900.53	$83.89	$112.57

Crew C-1

Crew C-1	Hr.	Daily	Hr.	Daily	Bare Costs	Incl. O&P
3 Carpenters	$46.95	$1126.80	$76.95	$1846.80	$44.61	$73.13
1 Laborer	37.60	300.80	61.65	493.20		
32 L.H., Daily Totals		$1427.60		$2340.00	$44.61	$73.13

Crew C-2

Crew C-2	Hr.	Daily	Hr.	Daily	Bare Costs	Incl. O&P
1 Carpenter Foreman (outside)	$48.95	$391.60	$80.25	$642.00	$45.73	$74.95
4 Carpenters	46.95	1502.40	76.95	2462.40		
1 Laborer	37.60	300.80	61.65	493.20		
48 L.H., Daily Totals		$2194.80		$3597.60	$45.73	$74.95

Crew C-2A

Crew C-2A	Hr.	Daily	Hr.	Daily	Bare Costs	Incl. O&P
1 Carpenter Foreman (outside)	$48.95	$391.60	$80.25	$642.00	$45.40	$73.98
3 Carpenters	46.95	1126.80	76.95	1846.80		
1 Cement Finisher	45.00	360.00	71.15	569.20		
1 Laborer	37.60	300.80	61.65	493.20		
48 L.H., Daily Totals		$2179.20		$3551.20	$45.40	$73.98

Crews

Crew No.	Bare Costs		Incl. Subs O&P		Cost Per Labor-Hour	
Crew C-3	Hr.	Daily	Hr.	Daily	Bare Costs	Incl. O&P
1 Rodman Foreman (outside)	$54.55	$436.40	$89.05	$712.40	$48.57	$79.06
4 Rodmen (reinf.)	52.55	1681.60	85.75	2744.00		
1 Equip. Oper. (light)	48.60	388.80	77.15	617.20		
2 Laborers	37.60	601.60	61.65	986.40		
3 Stressing Equipment		30.60		33.66		
.5 Grouting Equipment		81.50		89.65	1.75	1.93
64 L.H., Daily Totals		$3220.50		$5183.31	$50.32	$80.99

Crew C-4	Hr.	Daily	Hr.	Daily	Bare Costs	Incl. O&P
1 Rodman Foreman (outside)	$54.55	$436.40	$89.05	$712.40	$53.05	$86.58
3 Rodmen (reinf.)	52.55	1261.20	85.75	2058.00		
3 Stressing Equipment		30.60		33.66	.96	1.05
32 L.H., Daily Totals		$1728.20		$2804.06	$54.01	$87.63

Crew C-4A	Hr.	Daily	Hr.	Daily	Bare Costs	Incl. O&P
2 Rodmen (reinf.)	$52.55	$840.80	$85.75	$1372.00	$52.55	$85.75
4 Stressing Equipment		40.80		44.88	2.55	2.81
16 L.H., Daily Totals		$881.60		$1416.88	$55.10	$88.56

Crew C-5	Hr.	Daily	Hr.	Daily	Bare Costs	Incl. O&P
1 Rodman Foreman (outside)	$54.55	$436.40	$89.05	$712.40	$51.66	$83.69
4 Rodmen (reinf.)	52.55	1681.60	85.75	2744.00		
1 Equip. Oper. (crane)	51.70	413.60	82.05	656.40		
1 Equip. Oper. (oiler)	45.20	361.60	71.75	574.00		
1 Hyd. Crane, 25 Ton		736.60		810.26	13.15	14.47
56 L.H., Daily Totals		$3629.80		$5497.06	$64.82	$98.16

Crew C-6	Hr.	Daily	Hr.	Daily	Bare Costs	Incl. O&P
1 Labor Foreman (outside)	$39.60	$316.80	$64.90	$519.20	$39.17	$63.77
4 Laborers	37.60	1203.20	61.65	1972.80		
1 Cement Finisher	45.00	360.00	71.15	569.20		
2 Gas Engine Vibrators		62.40		68.64	1.30	1.43
48 L.H., Daily Totals		$1942.40		$3129.84	$40.47	$65.20

Crew C-7	Hr.	Daily	Hr.	Daily	Bare Costs	Incl. O&P
1 Labor Foreman (outside)	$39.60	$316.80	$64.90	$519.20	$40.93	$66.26
5 Laborers	37.60	1504.00	61.65	2466.00		
1 Cement Finisher	45.00	360.00	71.15	569.20		
1 Equip. Oper. (medium)	50.60	404.80	80.30	642.40		
1 Equip. Oper. (oiler)	45.20	361.60	71.75	574.00		
2 Gas Engine Vibrators		62.40		68.64		
1 Concrete Bucket, 1 C.Y.		23.80		26.18		
1 Hyd. Crane, 55 Ton		1128.00		1240.80	16.86	18.55
72 L.H., Daily Totals		$4161.40		$6106.42	$57.80	$84.81

Crew C-8	Hr.	Daily	Hr.	Daily	Bare Costs	Incl. O&P
1 Labor Foreman (outside)	$39.60	$316.80	$64.90	$519.20	$41.86	$67.49
3 Laborers	37.60	902.40	61.65	1479.60		
2 Cement Finishers	45.00	720.00	71.15	1138.40		
1 Equip. Oper. (medium)	50.60	404.80	80.30	642.40		
1 Concrete Pump (Small)		720.00		792.00	12.86	14.14
56 L.H., Daily Totals		$3064.00		$4571.60	$54.71	$81.64

Crew C-8A	Hr.	Daily	Hr.	Daily	Bare Costs	Incl. O&P
1 Labor Foreman (outside)	$39.60	$316.80	$64.90	$519.20	$40.40	$65.36
3 Laborers	37.60	902.40	61.65	1479.60		
2 Cement Finishers	45.00	720.00	71.15	1138.40		
48 L.H., Daily Totals		$1939.20		$3137.20	$40.40	$65.36

Crew C-8B	Hr.	Daily	Hr.	Daily	Bare Costs	Incl. O&P
1 Labor Foreman (outside)	$39.60	$316.80	$64.90	$519.20	$40.60	$66.03
3 Laborers	37.60	902.40	61.65	1479.60		
1 Equip. Oper. (medium)	50.60	404.80	80.30	642.40		
1 Vibrating Power Screed		66.30		72.93		
1 Roller, Vibratory, 25 Ton		687.40		756.14		
1 Dozer, 200 H.P.		1387.00		1525.70	53.52	58.87
40 L.H., Daily Totals		$3764.70		$4995.97	$94.12	$124.90

Crew C-8C	Hr.	Daily	Hr.	Daily	Bare Costs	Incl. O&P
1 Labor Foreman (outside)	$39.60	$316.80	$64.90	$519.20	$41.33	$66.88
3 Laborers	37.60	902.40	61.65	1479.60		
1 Cement Finisher	45.00	360.00	71.15	569.20		
1 Equip. Oper. (medium)	50.60	404.80	80.30	642.40		
1 Shotcrete Rig, 12 C.Y./hr		250.80		275.88		
1 Air Compressor, 160 cfm		156.40		172.04		
4 -50' Air Hoses, 1"		16.40		18.04		
4 -50' Air Hoses, 2"		31.00		34.10	9.47	10.42
48 L.H., Daily Totals		$2438.60		$3710.46	$50.80	$77.30

Crew C-8D	Hr.	Daily	Hr.	Daily	Bare Costs	Incl. O&P
1 Labor Foreman (outside)	$39.60	$316.80	$64.90	$519.20	$42.70	$68.71
1 Laborer	37.60	300.80	61.65	493.20		
1 Cement Finisher	45.00	360.00	71.15	569.20		
1 Equipment Oper. (light)	48.60	388.80	77.15	617.20		
1 Air Compressor, 250 cfm		201.40		221.54		
2 -50' Air Hoses, 1"		8.20		9.02	6.55	7.21
32 L.H., Daily Totals		$1576.00		$2429.36	$49.25	$75.92

Crew C-8E	Hr.	Daily	Hr.	Daily	Bare Costs	Incl. O&P
1 Labor Foreman (outside)	$39.60	$316.80	$64.90	$519.20	$41.00	$66.36
3 Laborers	37.60	902.40	61.65	1479.60		
1 Cement Finisher	45.00	360.00	71.15	569.20		
1 Equipment Oper. (light)	48.60	388.80	77.15	617.20		
1 Shotcrete Rig, 35 C.Y./hr		281.00		309.10		
1 Air Compressor, 250 cfm		201.40		221.54		
4 -50' Air Hoses, 1"		16.40		18.04		
4 -50' Air Hoses, 2"		31.00		34.10	11.04	12.14
48 L.H., Daily Totals		$2497.80		$3767.98	$52.04	$78.50

Crew C-10	Hr.	Daily	Hr.	Daily	Bare Costs	Incl. O&P
1 Laborer	$37.60	$300.80	$61.65	$493.20	$42.53	$67.98
2 Cement Finishers	45.00	720.00	71.15	1138.40		
24 L.H., Daily Totals		$1020.80		$1631.60	$42.53	$67.98

Crew C-10B	Hr.	Daily	Hr.	Daily	Bare Costs	Incl. O&P
3 Laborers	$37.60	$902.40	$61.65	$1479.60	$40.56	$65.45
2 Cement Finishers	45.00	720.00	71.15	1138.40		
1 Concrete Mixer, 10 C.F.		172.40		189.64		
2 Trowels, 48" Walk-Behind		100.40		110.44	6.82	7.50
40 L.H., Daily Totals		$1895.20		$2918.08	$47.38	$72.95

Crew C-10C	Hr.	Daily	Hr.	Daily	Bare Costs	Incl. O&P
1 Laborer	$37.60	$300.80	$61.65	$493.20	$42.53	$67.98
2 Cement Finishers	45.00	720.00	71.15	1138.40		
1 Trowel, 48" Walk-Behind		50.20		55.22	2.09	2.30
24 L.H., Daily Totals		$1071.00		$1686.82	$44.63	$70.28

Crews

Crew No.		Bare Costs	Incl. Subs O&P		Cost Per Labor-Hour	
Crew C-10D	Hr.	Daily	Hr.	Daily	Bare Costs	Incl. O&P
1 Laborer	$37.60	$300.80	$61.65	$493.20	$42.53	$67.98
2 Cement Finishers	45.00	720.00	71.15	1138.40		
1 Vibrating Power Screed		66.30		72.93		
1 Trowel, 48" Walk-Behind		50.20		55.22	4.85	5.34
24 L.H., Daily Totals		$1137.30		$1759.75	$47.39	$73.32

Crew C-10E	Hr.	Daily	Hr.	Daily	Bare Costs	Incl. O&P
1 Laborer	$37.60	$300.80	$61.65	$493.20	$42.53	$67.98
2 Cement Finishers	45.00	720.00	71.15	1138.40		
1 Vibrating Power Screed		66.30		72.93		
1 Cement Trowel, 96" Ride-On		189.00		207.90	10.64	11.70
24 L.H., Daily Totals		$1276.10		$1912.43	$53.17	$79.68

Crew C-10F	Hr.	Daily	Hr.	Daily	Bare Costs	Incl. O&P
1 Laborer	$37.60	$300.80	$61.65	$493.20	$42.53	$67.98
2 Cement Finishers	45.00	720.00	71.15	1138.40		
1 Aerial Lift Truck, 60' Boom		435.60		479.16	18.15	19.97
24 L.H., Daily Totals		$1456.40		$2110.76	$60.68	$87.95

Crew C-11	Hr.	Daily	Hr.	Daily	Bare Costs	Incl. O&P
1 Struc. Steel Foreman (outside)	$54.65	$437.20	$98.75	$790.00	$51.94	$91.49
6 Struc. Steel Workers	52.65	2527.20	95.15	4567.20		
1 Equip. Oper. (crane)	51.70	413.60	82.05	656.40		
1 Equip. Oper. (oiler)	45.20	361.60	71.75	574.00		
1 Lattice Boom Crane, 150 Ton		1813.00		1994.30	25.18	27.70
72 L.H., Daily Totals		$5552.60		$8581.90	$77.12	$119.19

Crew C-12	Hr.	Daily	Hr.	Daily	Bare Costs	Incl. O&P
1 Carpenter Foreman (outside)	$48.95	$391.60	$80.25	$642.00	$46.52	$75.80
3 Carpenters	46.95	1126.80	76.95	1846.80		
1 Laborer	37.60	300.80	61.65	493.20		
1 Equip. Oper. (crane)	51.70	413.60	82.05	656.40		
1 Hyd. Crane, 12 Ton		653.80		719.18	13.62	14.98
48 L.H., Daily Totals		$2886.60		$4357.58	$60.14	$90.78

Crew C-13	Hr.	Daily	Hr.	Daily	Bare Costs	Incl. O&P
1 Struc. Steel Worker	$52.65	$421.20	$95.15	$761.20	$50.75	$89.08
1 Welder	52.65	421.20	95.15	761.20		
1 Carpenter	46.95	375.60	76.95	615.60		
1 Welder, Gas Engine, 300 amp		145.85		160.44	6.08	6.68
24 L.H., Daily Totals		$1363.85		$2298.43	$56.83	$95.77

Crew C-14	Hr.	Daily	Hr.	Daily	Bare Costs	Incl. O&P
1 Carpenter Foreman (outside)	$48.95	$391.60	$80.25	$642.00	$46.18	$75.04
5 Carpenters	46.95	1878.00	76.95	3078.00		
4 Laborers	37.60	1203.20	61.65	1972.80		
4 Rodmen (reinf.)	52.55	1681.60	85.75	2744.00		
2 Cement Finishers	45.00	720.00	71.15	1138.40		
1 Equip. Oper. (crane)	51.70	413.60	82.05	656.40		
1 Equip. Oper. (oiler)	45.20	361.60	71.75	574.00		
1 Hyd. Crane, 80 Ton		1625.00		1787.50	11.28	12.41
144 L.H., Daily Totals		$8274.60		$12593.10	$57.46	$87.45

Crew C-14A	Hr.	Daily	Hr.	Daily	Bare Costs	Incl. O&P
1 Carpenter Foreman (outside)	$48.95	$391.60	$80.25	$642.00	$47.25	$77.17
16 Carpenters	46.95	6009.60	76.95	9849.60		
4 Rodmen (reinf.)	52.55	1681.60	85.75	2744.00		
2 Laborers	37.60	601.60	61.65	986.40		
1 Cement Finisher	45.00	360.00	71.15	569.20		
1 Equip. Oper. (medium)	50.60	404.80	80.30	642.40		
1 Gas Engine Vibrator		31.20		34.32		
1 Concrete Pump (Small)		720.00		792.00	3.76	4.13
200 L.H., Daily Totals		$10200.40		$16259.92	$51.00	$81.30

Crew C-14B	Hr.	Daily	Hr.	Daily	Bare Costs	Incl. O&P
1 Carpenter Foreman (outside)	$48.95	$391.60	$80.25	$642.00	$47.16	$76.94
16 Carpenters	46.95	6009.60	76.95	9849.60		
4 Rodmen (reinf.)	52.55	1681.60	85.75	2744.00		
2 Laborers	37.60	601.60	61.65	986.40		
2 Cement Finishers	45.00	720.00	71.15	1138.40		
1 Equip. Oper. (medium)	50.60	404.80	80.30	642.40		
1 Gas Engine Vibrator		31.20		34.32		
1 Concrete Pump (Small)		720.00		792.00	3.61	3.97
208 L.H., Daily Totals		$10560.40		$16829.12	$50.77	$80.91

Crew C-14C	Hr.	Daily	Hr.	Daily	Bare Costs	Incl. O&P
1 Carpenter Foreman (outside)	$48.95	$391.60	$80.25	$642.00	$45.08	$73.66
6 Carpenters	46.95	2253.60	76.95	3693.60		
2 Rodmen (reinf.)	52.55	840.80	85.75	1372.00		
4 Laborers	37.60	1203.20	61.65	1972.80		
1 Cement Finisher	45.00	360.00	71.15	569.20		
1 Gas Engine Vibrator		31.20		34.32	.28	.31
112 L.H., Daily Totals		$5080.40		$8283.92	$45.36	$73.96

Crew C-14D	Hr.	Daily	Hr.	Daily	Bare Costs	Incl. O&P
1 Carpenter Foreman (outside)	$48.95	$391.60	$80.25	$642.00	$46.80	$76.46
18 Carpenters	46.95	6760.80	76.95	11080.80		
2 Rodmen (reinf.)	52.55	840.80	85.75	1372.00		
2 Laborers	37.60	601.60	61.65	986.40		
1 Cement Finisher	45.00	360.00	71.15	569.20		
1 Equip. Oper. (medium)	50.60	404.80	80.30	642.40		
1 Gas Engine Vibrator		31.20		34.32		
1 Concrete Pump (Small)		720.00		792.00	3.76	4.13
200 L.H., Daily Totals		$10110.80		$16119.12	$50.55	$80.60

Crew C-14E	Hr.	Daily	Hr.	Daily	Bare Costs	Incl. O&P
1 Carpenter Foreman (outside)	$48.95	$391.60	$80.25	$642.00	$46.44	$75.75
2 Carpenters	46.95	751.20	76.95	1231.20		
4 Rodmen (reinf.)	52.55	1681.60	85.75	2744.00		
3 Laborers	37.60	902.40	61.65	1479.60		
1 Cement Finisher	45.00	360.00	71.15	569.20		
1 Gas Engine Vibrator		31.20		34.32	.35	.39
88 L.H., Daily Totals		$4118.00		$6700.32	$46.80	$76.14

Crew C-14F	Hr.	Daily	Hr.	Daily	Bare Costs	Incl. O&P
1 Labor Foreman (outside)	$39.60	$316.80	$64.90	$519.20	$42.76	$68.34
2 Laborers	37.60	601.60	61.65	986.40		
6 Cement Finishers	45.00	2160.00	71.15	3415.20		
1 Gas Engine Vibrator		31.20		34.32	.43	.48
72 L.H., Daily Totals		$3109.60		$4955.12	$43.19	$68.82

Crews

Crew No.	Bare Costs		Incl. Subs O&P		Cost Per Labor-Hour	
	Hr.	Daily	Hr.	Daily	Bare Costs	Incl. O&P
Crew C-14G						
1 Labor Foreman (outside)	$39.60	$316.80	$64.90	$519.20	$42.11	$67.54
2 Laborers	37.60	601.60	61.65	986.40		
4 Cement Finishers	45.00	1440.00	71.15	2276.80		
1 Gas Engine Vibrator		31.20		34.32	.56	.61
56 L.H., Daily Totals		$2389.60		$3816.72	$42.67	$68.16
Crew C-14H	Hr.	Daily	Hr.	Daily	Bare Costs	Incl. O&P
1 Carpenter Foreman (outside)	$48.95	$391.60	$80.25	$642.00	$46.33	$75.45
2 Carpenters	46.95	751.20	76.95	1231.20		
1 Rodman (reinf.)	52.55	420.40	85.75	686.00		
1 Laborer	37.60	300.80	61.65	493.20		
1 Cement Finisher	45.00	360.00	71.15	569.20		
1 Gas Engine Vibrator		31.20		34.32	.65	.71
48 L.H., Daily Totals		$2255.20		$3655.92	$46.98	$76.17
Crew C-14L	Hr.	Daily	Hr.	Daily	Bare Costs	Incl. O&P
1 Carpenter Foreman (outside)	$48.95	$391.60	$80.25	$642.00	$43.84	$71.64
6 Carpenters	46.95	2253.60	76.95	3693.60		
4 Laborers	37.60	1203.20	61.65	1972.80		
1 Cement Finisher	45.00	360.00	71.15	569.20		
1 Gas Engine Vibrator		31.20		34.32	.33	.36
96 L.H., Daily Totals		$4239.60		$6911.92	$44.16	$72.00
Crew C-14M	Hr.	Daily	Hr.	Daily	Bare Costs	Incl. O&P
1 Carpenter Foreman (outside)	$48.95	$391.60	$80.25	$642.00	$45.77	$74.33
2 Carpenters	46.95	751.20	76.95	1231.20		
1 Rodman (reinf.)	52.55	420.40	85.75	686.00		
2 Laborers	37.60	601.60	61.65	986.40		
1 Cement Finisher	45.00	360.00	71.15	569.20		
1 Equip. Oper. (medium)	50.60	404.80	80.30	642.40		
1 Gas Engine Vibrator		31.20		34.32		
1 Concrete Pump (Small)		720.00		792.00	11.74	12.91
64 L.H., Daily Totals		$3680.80		$5583.52	$57.51	$87.24
Crew C-15	Hr.	Daily	Hr.	Daily	Bare Costs	Incl. O&P
1 Carpenter Foreman (outside)	$48.95	$391.60	$80.25	$642.00	$44.24	$71.91
2 Carpenters	46.95	751.20	76.95	1231.20		
3 Laborers	37.60	902.40	61.65	1479.60		
2 Cement Finishers	45.00	720.00	71.15	1138.40		
1 Rodman (reinf.)	52.55	420.40	85.75	686.00		
72 L.H., Daily Totals		$3185.60		$5177.20	$44.24	$71.91
Crew C-16	Hr.	Daily	Hr.	Daily	Bare Costs	Incl. O&P
1 Labor Foreman (outside)	$39.60	$316.80	$64.90	$519.20	$41.86	$67.49
3 Laborers	37.60	902.40	61.65	1479.60		
2 Cement Finishers	45.00	720.00	71.15	1138.40		
1 Equip. Oper. (medium)	50.60	404.80	80.30	642.40		
1 Gunite Pump Rig		371.80		408.98		
2 -50' Air Hoses, 3/4"		6.50		7.15		
2 -50' Air Hoses, 2"		15.50		17.05	7.03	7.74
56 L.H., Daily Totals		$2737.80		$4212.78	$48.89	$75.23
Crew C-16A	Hr.	Daily	Hr.	Daily	Bare Costs	Incl. O&P
1 Laborer	$37.60	$300.80	$61.65	$493.20	$44.55	$71.06
2 Cement Finishers	45.00	720.00	71.15	1138.40		
1 Equip. Oper. (medium)	50.60	404.80	80.30	642.40		
1 Gunite Pump Rig		371.80		408.98		
2 -50' Air Hoses, 3/4"		6.50		7.15		
2 -50' Air Hoses, 2"		15.50		17.05		
1 Aerial Lift Truck, 60' Boom		435.60		479.16	25.92	28.51
32 L.H., Daily Totals		$2255.00		$3186.34	$70.47	$99.57
Crew C-17	Hr.	Daily	Hr.	Daily	Bare Costs	Incl. O&P
2 Skilled Worker Foremen (out)	$50.65	$810.40	$82.30	$1316.80	$49.05	$79.70
8 Skilled Workers	48.65	3113.60	79.05	5059.20		
80 L.H., Daily Totals		$3924.00		$6376.00	$49.05	$79.70
Crew C-17A	Hr.	Daily	Hr.	Daily	Bare Costs	Incl. O&P
2 Skilled Worker Foremen (out)	$50.65	$810.40	$82.30	$1316.80	$49.08	$79.73
8 Skilled Workers	48.65	3113.60	79.05	5059.20		
.125 Equip. Oper. (crane)	51.70	51.70	82.05	82.05		
.125 Hyd. Crane, 80 Ton		203.13		223.44	2.51	2.76
81 L.H., Daily Totals		$4178.82		$6681.49	$51.59	$82.49
Crew C-17B	Hr.	Daily	Hr.	Daily	Bare Costs	Incl. O&P
2 Skilled Worker Foremen (out)	$50.65	$810.40	$82.30	$1316.80	$49.11	$79.76
8 Skilled Workers	48.65	3113.60	79.05	5059.20		
.25 Equip. Oper. (crane)	51.70	103.40	82.05	164.10		
.25 Hyd. Crane, 80 Ton		406.25		446.88		
.25 Trowel, 48" Walk-Behind		12.55		13.81	5.11	5.62
82 L.H., Daily Totals		$4446.20		$7000.78	$54.22	$85.38
Crew C-17C	Hr.	Daily	Hr.	Daily	Bare Costs	Incl. O&P
2 Skilled Worker Foremen (out)	$50.65	$810.40	$82.30	$1316.80	$49.15	$79.78
8 Skilled Workers	48.65	3113.60	79.05	5059.20		
.375 Equip. Oper. (crane)	51.70	155.10	82.05	246.15		
.375 Hyd. Crane, 80 Ton		609.38		670.31	7.34	8.08
83 L.H., Daily Totals		$4688.48		$7292.46	$56.49	$87.86
Crew C-17D	Hr.	Daily	Hr.	Daily	Bare Costs	Incl. O&P
2 Skilled Worker Foremen (out)	$50.65	$810.40	$82.30	$1316.80	$49.18	$79.81
8 Skilled Workers	48.65	3113.60	79.05	5059.20		
.5 Equip. Oper. (crane)	51.70	206.80	82.05	328.20		
.5 Hyd. Crane, 80 Ton		812.50		893.75	9.67	10.64
84 L.H., Daily Totals		$4943.30		$7597.95	$58.85	$90.45
Crew C-17E	Hr.	Daily	Hr.	Daily	Bare Costs	Incl. O&P
2 Skilled Worker Foremen (out)	$50.65	$810.40	$82.30	$1316.80	$49.05	$79.70
8 Skilled Workers	48.65	3113.60	79.05	5059.20		
1 Hyd. Jack with Rods		96.50		106.15	1.21	1.33
80 L.H., Daily Totals		$4020.50		$6482.15	$50.26	$81.03
Crew C-18	Hr.	Daily	Hr.	Daily	Bare Costs	Incl. O&P
.125 Labor Foreman (outside)	$39.60	$39.60	$64.90	$64.90	$37.82	$62.01
1 Laborer	37.60	300.80	61.65	493.20		
1 Concrete Cart, 10 C.F.		60.00		66.00	6.67	7.33
9 L.H., Daily Totals		$400.40		$624.10	$44.49	$69.34
Crew C-19	Hr.	Daily	Hr.	Daily	Bare Costs	Incl. O&P
.125 Labor Foreman (outside)	$39.60	$39.60	$64.90	$64.90	$37.82	$62.01
1 Laborer	37.60	300.80	61.65	493.20		
1 Concrete Cart, 18 C.F.		99.80		109.78	11.09	12.20
9 L.H., Daily Totals		$440.20		$667.88	$48.91	$74.21
Crew C-20	Hr.	Daily	Hr.	Daily	Bare Costs	Incl. O&P
1 Labor Foreman (outside)	$39.60	$316.80	$64.90	$519.20	$40.40	$65.58
5 Laborers	37.60	1504.00	61.65	2466.00		
1 Cement Finisher	45.00	360.00	71.15	569.20		
1 Equip. Oper. (medium)	50.60	404.80	80.30	642.40		
2 Gas Engine Vibrators		62.40		68.64		
1 Concrete Pump (Small)		720.00		792.00	12.23	13.45
64 L.H., Daily Totals		$3368.00		$5057.44	$52.63	$79.02

Crews

Crew No.		Bare Costs		Incl. Subs O&P		Cost Per Labor-Hour	
Crew C-21	**Hr.**	**Daily**	**Hr.**	**Daily**	**Bare Costs**	**Incl. O&P**	
1 Labor Foreman (outside)	$39.60	$316.80	$64.90	$519.20	$40.40	$65.58	
5 Laborers	37.60	1504.00	61.65	2466.00			
1 Cement Finisher	45.00	360.00	71.15	569.20			
1 Equip. Oper. (medium)	50.60	404.80	80.30	642.40			
2 Gas Engine Vibrators		62.40		68.64			
1 Concrete Conveyer		198.80		218.68	4.08	4.49	
64 L.H., Daily Totals		$2846.80		$4484.12	$44.48	$70.06	
Crew C-22	**Hr.**	**Daily**	**Hr.**	**Daily**	**Bare Costs**	**Incl. O&P**	
1 Rodman Foreman (outside)	$54.55	$436.40	$89.05	$712.40	$52.74	$85.96	
4 Rodmen (reinf.)	52.55	1681.60	85.75	2744.00			
.125 Equip. Oper. (crane)	51.70	51.70	82.05	82.05			
.125 Equip. Oper. (oiler)	45.20	45.20	71.75	71.75			
.125 Hyd. Crane, 25 Ton		92.08		101.28	2.19	2.41	
42 L.H., Daily Totals		$2306.97		$3711.48	$54.93	$88.37	
Crew C-23	**Hr.**	**Daily**	**Hr.**	**Daily**	**Bare Costs**	**Incl. O&P**	
2 Skilled Worker Foremen (out)	$50.65	$810.40	$82.30	$1316.80	$49.01	$79.27	
6 Skilled Workers	48.65	2335.20	79.05	3794.40			
1 Equip. Oper. (crane)	51.70	413.60	82.05	656.40			
1 Equip. Oper. (oiler)	45.20	361.60	71.75	574.00			
1 Lattice Boom Crane, 90 Ton		1511.00		1662.10	18.89	20.78	
80 L.H., Daily Totals		$5431.80		$8003.70	$67.90	$100.05	
Crew C-24	**Hr.**	**Daily**	**Hr.**	**Daily**	**Bare Costs**	**Incl. O&P**	
2 Skilled Worker Foremen (out)	$50.65	$810.40	$82.30	$1316.80	$49.01	$79.27	
6 Skilled Workers	48.65	2335.20	79.05	3794.40			
1 Equip. Oper. (crane)	51.70	413.60	82.05	656.40			
1 Equip. Oper. (oiler)	45.20	361.60	71.75	574.00			
1 Lattice Boom Crane, 150 Ton		1813.00		1994.30	22.66	24.93	
80 L.H., Daily Totals		$5733.80		$8335.90	$71.67	$104.20	
Crew C-25	**Hr.**	**Daily**	**Hr.**	**Daily**	**Bare Costs**	**Incl. O&P**	
2 Rodmen (reinf.)	$52.55	$840.80	$85.75	$1372.00	$41.30	$70.03	
2 Rodmen Helpers	30.05	480.80	54.30	868.80			
32 L.H., Daily Totals		$1321.60		$2240.80	$41.30	$70.03	
Crew C-27	**Hr.**	**Daily**	**Hr.**	**Daily**	**Bare Costs**	**Incl. O&P**	
2 Cement Finishers	$45.00	$720.00	$71.15	$1138.40	$45.00	$71.15	
1 Concrete Saw		167.40		184.14	10.46	11.51	
16 L.H., Daily Totals		$887.40		$1322.54	$55.46	$82.66	
Crew C-28	**Hr.**	**Daily**	**Hr.**	**Daily**	**Bare Costs**	**Incl. O&P**	
1 Cement Finisher	$45.00	$360.00	$71.15	$569.20	$45.00	$71.15	
1 Portable Air Compressor, Gas		17.90		19.69	2.24	2.46	
8 L.H., Daily Totals		$377.90		$588.89	$47.24	$73.61	
Crew C-29	**Hr.**	**Daily**	**Hr.**	**Daily**	**Bare Costs**	**Incl. O&P**	
1 Laborer	$37.60	$300.80	$61.65	$493.20	$37.60	$61.65	
1 Pressure Washer		70.20		77.22	8.78	9.65	
8 L.H., Daily Totals		$371.00		$570.42	$46.38	$71.30	
Crew C-30	**Hr.**	**Daily**	**Hr.**	**Daily**	**Bare Costs**	**Incl. O&P**	
1 Laborer	$37.60	$300.80	$61.65	$493.20	$37.60	$61.65	
1 Concrete Mixer, 10 C.F.		172.40		189.64	21.55	23.70	
8 L.H., Daily Totals		$473.20		$682.84	$59.15	$85.36	

Crew No.		Bare Costs		Incl. Subs O&P		Cost Per Labor-Hour	
Crew C-31	**Hr.**	**Daily**	**Hr.**	**Daily**	**Bare Costs**	**Incl. O&P**	
1 Cement Finisher	$45.00	$360.00	$71.15	$569.20	$45.00	$71.15	
1 Grout Pump		371.80		408.98	46.48	51.12	
8 L.H., Daily Totals		$731.80		$978.18	$91.47	$122.27	
Crew C-32	**Hr.**	**Daily**	**Hr.**	**Daily**	**Bare Costs**	**Incl. O&P**	
1 Cement Finisher	$45.00	$360.00	$71.15	$569.20	$41.30	$66.40	
1 Laborer	37.60	300.80	61.65	493.20			
1 Crack Chaser Saw, Gas, 6 H.P.		29.20		32.12			
1 Vacuum Pick-Up System		60.90		66.99	5.63	6.19	
16 L.H., Daily Totals		$750.90		$1161.51	$46.93	$72.59	
Crew D-1	**Hr.**	**Daily**	**Hr.**	**Daily**	**Bare Costs**	**Incl. O&P**	
1 Bricklayer	$46.15	$369.20	$75.10	$600.80	$42.10	$68.50	
1 Bricklayer Helper	38.05	304.40	61.90	495.20			
16 L.H., Daily Totals		$673.60		$1096.00	$42.10	$68.50	
Crew D-2	**Hr.**	**Daily**	**Hr.**	**Daily**	**Bare Costs**	**Incl. O&P**	
3 Bricklayers	$46.15	$1107.60	$75.10	$1802.40	$43.28	$70.47	
2 Bricklayer Helpers	38.05	608.80	61.90	990.40			
.5 Carpenter	46.95	187.80	76.95	307.80			
44 L.H., Daily Totals		$1904.20		$3100.60	$43.28	$70.47	
Crew D-3	**Hr.**	**Daily**	**Hr.**	**Daily**	**Bare Costs**	**Incl. O&P**	
3 Bricklayers	$46.15	$1107.60	$75.10	$1802.40	$43.10	$70.16	
2 Bricklayer Helpers	38.05	608.80	61.90	990.40			
.25 Carpenter	46.95	93.90	76.95	153.90			
42 L.H., Daily Totals		$1810.30		$2946.70	$43.10	$70.16	
Crew D-4	**Hr.**	**Daily**	**Hr.**	**Daily**	**Bare Costs**	**Incl. O&P**	
1 Bricklayer	$46.15	$369.20	$75.10	$600.80	$42.71	$69.01	
2 Bricklayer Helpers	38.05	608.80	61.90	990.40			
1 Equip. Oper. (light)	48.60	388.80	77.15	617.20			
1 Grout Pump, 50 C.F./hr.		133.40		146.74	4.17	4.59	
32 L.H., Daily Totals		$1500.20		$2355.14	$46.88	$73.60	
Crew D-5	**Hr.**	**Daily**	**Hr.**	**Daily**	**Bare Costs**	**Incl. O&P**	
1 Bricklayer	46.15	369.20	75.10	600.80	46.15	75.10	
8 L.H., Daily Totals		$369.20		$600.80	$46.15	$75.10	
Crew D-6	**Hr.**	**Daily**	**Hr.**	**Daily**	**Bare Costs**	**Incl. O&P**	
3 Bricklayers	$46.15	$1107.60	$75.10	$1802.40	$42.29	$68.84	
3 Bricklayer Helpers	38.05	913.20	61.90	1485.60			
.25 Carpenter	46.95	93.90	76.95	153.90			
50 L.H., Daily Totals		$2114.70		$3441.90	$42.29	$68.84	
Crew D-7	**Hr.**	**Daily**	**Hr.**	**Daily**	**Bare Costs**	**Incl. O&P**	
1 Tile Layer	$42.80	$342.40	$67.60	$540.80	$38.17	$60.30	
1 Tile Layer Helper	33.55	268.40	53.00	424.00			
16 L.H., Daily Totals		$610.80		$964.80	$38.17	$60.30	
Crew D-8	**Hr.**	**Daily**	**Hr.**	**Daily**	**Bare Costs**	**Incl. O&P**	
3 Bricklayers	$46.15	$1107.60	$75.10	$1802.40	$42.91	$69.82	
2 Bricklayer Helpers	38.05	608.80	61.90	990.40			
40 L.H., Daily Totals		$1716.40		$2792.80	$42.91	$69.82	
Crew D-9	**Hr.**	**Daily**	**Hr.**	**Daily**	**Bare Costs**	**Incl. O&P**	
3 Bricklayers	$46.15	$1107.60	$75.10	$1802.40	$42.10	$68.50	
3 Bricklayer Helpers	38.05	913.20	61.90	1485.60			
48 L.H., Daily Totals		$2020.80		$3288.00	$42.10	$68.50	

Crews

Crew No.	Bare Costs		Incl. Subs O&P		Cost Per Labor-Hour	
Crew D-10	Hr.	Daily	Hr.	Daily	Bare Costs	Incl. O&P
1 Bricklayer Foreman (outside)	$48.15	$385.20	$78.35	$626.80	$46.01	$74.35
1 Bricklayer	46.15	369.20	75.10	600.80		
1 Bricklayer Helper	38.05	304.40	61.90	495.20		
1 Equip. Oper. (crane)	51.70	413.60	82.05	656.40		
1 S.P. Crane, 4x4, 12 Ton		473.40		520.74	14.79	16.27
32 L.H., Daily Totals		$1945.80		$2899.94	$60.81	$90.62
Crew D-11	Hr.	Daily	Hr.	Daily	Bare Costs	Incl. O&P
1 Bricklayer Foreman (outside)	$48.15	$385.20	$78.35	$626.80	$44.12	$71.78
1 Bricklayer	46.15	369.20	75.10	600.80		
1 Bricklayer Helper	38.05	304.40	61.90	495.20		
24 L.H., Daily Totals		$1058.80		$1722.80	$44.12	$71.78
Crew D-12	Hr.	Daily	Hr.	Daily	Bare Costs	Incl. O&P
1 Bricklayer Foreman (outside)	$48.15	$385.20	$78.35	$626.80	$42.60	$69.31
1 Bricklayer	46.15	369.20	75.10	600.80		
2 Bricklayer Helpers	38.05	608.80	61.90	990.40		
32 L.H., Daily Totals		$1363.20		$2218.00	$42.60	$69.31
Crew D-13	Hr.	Daily	Hr.	Daily	Bare Costs	Incl. O&P
1 Bricklayer Foreman (outside)	$48.15	$385.20	$78.35	$626.80	$44.84	$72.71
1 Bricklayer	46.15	369.20	75.10	600.80		
2 Bricklayer Helpers	38.05	608.80	61.90	990.40		
1 Carpenter	46.95	375.60	76.95	615.60		
1 Equip. Oper. (crane)	51.70	413.60	82.05	656.40		
1 S.P. Crane, 4x4, 12 Ton		473.40		520.74	9.86	10.85
48 L.H., Daily Totals		$2625.80		$4010.74	$54.70	$83.56
Crew E-1	Hr.	Daily	Hr.	Daily	Bare Costs	Incl. O&P
1 Welder Foreman (outside)	$54.65	$437.20	$98.75	$790.00	$51.97	$90.35
1 Welder	52.65	421.20	95.15	761.20		
1 Equip. Oper. (light)	48.60	388.80	77.15	617.20		
1 Welder, Gas Engine, 300 amp		145.85		160.44	6.08	6.68
24 L.H., Daily Totals		$1393.05		$2328.84	$58.04	$97.03
Crew E-2	Hr.	Daily	Hr.	Daily	Bare Costs	Incl. O&P
1 Struc. Steel Foreman (outside)	$54.65	$437.20	$98.75	$790.00	$51.74	$90.45
4 Struc. Steel Workers	52.65	1684.80	95.15	3044.80		
1 Equip. Oper. (crane)	51.70	413.60	82.05	656.40		
1 Equip. Oper. (oiler)	45.20	361.60	71.75	574.00		
1 Lattice Boom Crane, 90 Ton		1511.00		1662.10	26.98	29.68
56 L.H., Daily Totals		$4408.20		$6727.30	$78.72	$120.13
Crew E-3	Hr.	Daily	Hr.	Daily	Bare Costs	Incl. O&P
1 Struc. Steel Foreman (outside)	$54.65	$437.20	$98.75	$790.00	$53.32	$96.35
1 Struc. Steel Worker	52.65	421.20	95.15	761.20		
1 Welder	52.65	421.20	95.15	761.20		
1 Welder, Gas Engine, 300 amp		145.85		160.44	6.08	6.68
24 L.H., Daily Totals		$1425.45		$2472.84	$59.39	$103.03
Crew E-3A	Hr.	Daily	Hr.	Daily	Bare Costs	Incl. O&P
1 Struc. Steel Foreman (outside)	$54.65	$437.20	$98.75	$790.00	$53.32	$96.35
1 Struc. Steel Worker	52.65	421.20	95.15	761.20		
1 Welder	52.65	421.20	95.15	761.20		
1 Welder, Gas Engine, 300 amp		145.85		160.44		
1 Aerial Lift Truck, 40' Boom		301.40		331.54	18.64	20.50
24 L.H., Daily Totals		$1726.85		$2804.38	$71.95	$116.85

Crew No.	Bare Costs		Incl. Subs O&P		Cost Per Labor-Hour	
Crew E-4	Hr.	Daily	Hr.	Daily	Bare Costs	Incl. O&P
1 Struc. Steel Foreman (outside)	$54.65	$437.20	$98.75	$790.00	$53.15	$96.05
3 Struc. Steel Workers	52.65	1263.60	95.15	2283.60		
1 Welder, Gas Engine, 300 amp		145.85		160.44	4.56	5.01
32 L.H., Daily Totals		$1846.65		$3234.03	$57.71	$101.06
Crew E-5	Hr.	Daily	Hr.	Daily	Bare Costs	Incl. O&P
2 Struc. Steel Foremen (outside)	$54.65	$874.40	$98.75	$1580.00	$52.21	$92.22
5 Struc. Steel Workers	52.65	2106.00	95.15	3806.00		
1 Equip. Oper. (crane)	51.70	413.60	82.05	656.40		
1 Welder	52.65	421.20	95.15	761.20		
1 Equip. Oper. (oiler)	45.20	361.60	71.75	574.00		
1 Lattice Boom Crane, 90 Ton		1511.00		1662.10		
1 Welder, Gas Engine, 300 amp		145.85		160.44	20.71	22.78
80 L.H., Daily Totals		$5833.65		$9200.14	$72.92	$115.00
Crew E-6	Hr.	Daily	Hr.	Daily	Bare Costs	Incl. O&P
3 Struc. Steel Foremen (outside)	$54.65	$1311.60	$98.75	$2370.00	$52.25	$92.42
9 Struc. Steel Workers	52.65	3790.80	95.15	6850.80		
1 Equip. Oper. (crane)	51.70	413.60	82.05	656.40		
1 Welder	52.65	421.20	95.15	761.20		
1 Equip. Oper. (oiler)	45.20	361.60	71.75	574.00		
1 Equip. Oper. (light)	48.60	388.80	77.15	617.20		
1 Lattice Boom Crane, 90 Ton		1511.00		1662.10		
1 Welder, Gas Engine, 300 amp		145.85		160.44		
1 Air Compressor, 160 cfm		156.40		172.04		
2 Impact Wrenches		36.00		39.60	14.45	15.89
128 L.H., Daily Totals		$8536.85		$13863.78	$66.69	$108.31
Crew E-7	Hr.	Daily	Hr.	Daily	Bare Costs	Incl. O&P
1 Struc. Steel Foreman (outside)	$54.65	$437.20	$98.75	$790.00	$52.21	$92.22
4 Struc. Steel Workers	52.65	1684.80	95.15	3044.80		
1 Equip. Oper. (crane)	51.70	413.60	82.05	656.40		
1 Equip. Oper. (oiler)	45.20	361.60	71.75	574.00		
1 Welder Foreman (outside)	54.65	437.20	98.75	790.00		
2 Welders	52.65	842.40	95.15	1522.40		
1 Lattice Boom Crane, 90 Ton		1511.00		1662.10		
2 Welder, Gas Engine, 300 amp		291.70		320.87	22.53	24.79
80 L.H., Daily Totals		$5979.50		$9360.57	$74.74	$117.01
Crew E-8	Hr.	Daily	Hr.	Daily	Bare Costs	Incl. O&P
1 Struc. Steel Foreman (outside)	$54.65	$437.20	$98.75	$790.00	$52.00	$91.51
4 Struc. Steel Workers	52.65	1684.80	95.15	3044.80		
1 Welder Foreman (outside)	54.65	437.20	98.75	790.00		
4 Welders	52.65	1684.80	95.15	3044.80		
1 Equip. Oper. (crane)	51.70	413.60	82.05	656.40		
1 Equip. Oper. (oiler)	45.20	361.60	71.75	574.00		
1 Equip. Oper. (light)	48.60	388.80	77.15	617.20		
1 Lattice Boom Crane, 90 Ton		1511.00		1662.10		
4 Welder, Gas Engine, 300 amp		583.40		641.74	20.14	22.15
104 L.H., Daily Totals		$7502.40		$11821.04	$72.14	$113.66
Crew E-9	Hr.	Daily	Hr.	Daily	Bare Costs	Incl. O&P
2 Struc. Steel Foremen (outside)	$54.65	$874.40	$98.75	$1580.00	$52.25	$92.42
5 Struc. Steel Workers	52.65	2106.00	95.15	3806.00		
1 Welder Foreman (outside)	54.65	437.20	98.75	790.00		
5 Welders	52.65	2106.00	95.15	3806.00		
1 Equip. Oper. (crane)	51.70	413.60	82.05	656.40		
1 Equip. Oper. (oiler)	45.20	361.60	71.75	574.00		
1 Equip. Oper. (light)	48.60	388.80	77.15	617.20		
1 Lattice Boom Crane, 90 Ton		1511.00		1662.10		
5 Welder, Gas Engine, 300 amp		729.25		802.17	17.50	19.25
128 L.H., Daily Totals		$8927.85		$14293.88	$69.75	$111.67

Crews

Crew No.		Bare Costs		Incl. Subs O&P		Cost Per Labor-Hour	
Crew E-10	Hr.	Daily	Hr.	Daily	Bare Costs	Incl. O&P
1 Welder Foreman (outside)	$54.65	$437.20	$98.75	$790.00	$53.65	$96.95
1 Welder	52.65	421.20	95.15	761.20		
1 Welder, Gas Engine, 300 amp		145.85		160.44		
1 Flatbed Truck, Gas, 3 Ton		303.20		333.52	28.07	30.87
16 L.H., Daily Totals		$1307.45		$2045.16	$81.72	$127.82
Crew E-11	Hr.	Daily	Hr.	Daily	Bare Costs	Incl. O&P
2 Painters, Struc. Steel	$41.45	$663.20	$77.10	$1233.60	$42.27	$73.25
1 Building Laborer	37.60	300.80	61.65	493.20		
1 Equip. Oper. (light)	48.60	388.80	77.15	617.20		
1 Air Compressor, 250 cfm		201.40		221.54		
1 Sandblaster, Portable, 3 C.F.		20.40		22.44		
1 Set Sand Blasting Accessories		14.05		15.46	7.37	8.11
32 L.H., Daily Totals		$1588.65		$2603.43	$49.65	$81.36
Crew E-11A	Hr.	Daily	Hr.	Daily	Bare Costs	Incl. O&P
2 Painters, Struc. Steel	$41.45	$663.20	$77.10	$1233.60	$42.27	$73.25
1 Building Laborer	37.60	300.80	61.65	493.20		
1 Equip. Oper. (light)	48.60	388.80	77.15	617.20		
1 Air Compressor, 250 cfm		201.40		221.54		
1 Sandblaster, Portable, 3 C.F.		20.40		22.44		
1 Set Sand Blasting Accessories		14.05		15.46		
1 Aerial Lift Truck, 60' Boom		435.60		479.16	20.98	23.08
32 L.H., Daily Totals		$2024.25		$3082.59	$63.26	$96.33
Crew E-11B	Hr.	Daily	Hr.	Daily	Bare Costs	Incl. O&P
2 Painters, Struc. Steel	$41.45	$663.20	$77.10	$1233.60	$40.17	$71.95
1 Building Laborer	37.60	300.80	61.65	493.20		
2 Paint Sprayer, 8 C.F.M.		99.80		109.78		
1 Aerial Lift Truck, 60' Boom		435.60		479.16	22.31	24.54
24 L.H., Daily Totals		$1499.40		$2315.74	$62.48	$96.49
Crew E-12	Hr.	Daily	Hr.	Daily	Bare Costs	Incl. O&P
1 Welder Foreman (outside)	$54.65	$437.20	$98.75	$790.00	$51.63	$87.95
1 Equip. Oper. (light)	48.60	388.80	77.15	617.20		
1 Welder, Gas Engine, 300 amp		145.85		160.44	9.12	10.03
16 L.H., Daily Totals		$971.85		$1567.64	$60.74	$97.98
Crew E-13	Hr.	Daily	Hr.	Daily	Bare Costs	Incl. O&P
1 Welder Foreman (outside)	$54.65	$437.20	$98.75	$790.00	$52.63	$91.55
.5 Equip. Oper. (light)	48.60	194.40	77.15	308.60		
1 Welder, Gas Engine, 300 amp		145.85		160.44	12.15	13.37
12 L.H., Daily Totals		$777.45		$1259.04	$64.79	$104.92
Crew E-14	Hr.	Daily	Hr.	Daily	Bare Costs	Incl. O&P
1 Welder Foreman (outside)	$54.65	$437.20	$98.75	$790.00	$54.65	$98.75
1 Welder, Gas Engine, 300 amp		145.85		160.44	18.23	20.05
8 L.H., Daily Totals		$583.05		$950.43	$72.88	$118.80
Crew E-16	Hr.	Daily	Hr.	Daily	Bare Costs	Incl. O&P
1 Welder Foreman (outside)	$54.65	$437.20	$98.75	$790.00	$53.65	$96.95
1 Welder	52.65	421.20	95.15	761.20		
1 Welder, Gas Engine, 300 amp		145.85		160.44	9.12	10.03
16 L.H., Daily Totals		$1004.25		$1711.64	$62.77	$106.98
Crew E-17	Hr.	Daily	Hr.	Daily	Bare Costs	Incl. O&P
1 Struc. Steel Foreman (outside)	$54.65	$437.20	$98.75	$790.00	$53.65	$96.95
1 Structural Steel Worker	52.65	421.20	95.15	761.20		
16 L.H., Daily Totals		$858.40		$1551.20	$53.65	$96.95

Crew No.		Bare Costs		Incl. Subs O&P		Cost Per Labor-Hour	
Crew E-18	Hr.	Daily	Hr.	Daily	Bare Costs	Incl. O&P
1 Struc. Steel Foreman (outside)	$54.65	$437.20	$98.75	$790.00	$52.64	$92.90
3 Structural Steel Workers	52.65	1263.60	95.15	2283.60		
1 Equipment Operator (med.)	50.60	404.80	80.30	642.40		
1 Lattice Boom Crane, 20 Ton		948.90		1043.79	23.72	26.09
40 L.H., Daily Totals		$3054.50		$4759.79	$76.36	$118.99
Crew E-19	Hr.	Daily	Hr.	Daily	Bare Costs	Incl. O&P
1 Struc. Steel Foreman (outside)	$54.65	$437.20	$98.75	$790.00	$51.97	$90.35
1 Structural Steel Worker	52.65	421.20	95.15	761.20		
1 Equip. Oper. (light)	48.60	388.80	77.15	617.20		
1 Lattice Boom Crane, 20 Ton		948.90		1043.79	39.54	43.49
24 L.H., Daily Totals		$2196.10		$3212.19	$91.50	$133.84
Crew E-20	Hr.	Daily	Hr.	Daily	Bare Costs	Incl. O&P
1 Struc. Steel Foreman (outside)	$54.65	$437.20	$98.75	$790.00	$51.85	$91.04
5 Structural Steel Workers	52.65	2106.00	95.15	3806.00		
1 Equip. Oper. (crane)	51.70	413.60	82.05	656.40		
1 Equip. Oper. (oiler)	45.20	361.60	71.75	574.00		
1 Lattice Boom Crane, 40 Ton		1164.00		1280.40	18.19	20.01
64 L.H., Daily Totals		$4482.40		$7106.80	$70.04	$111.04
Crew E-22	Hr.	Daily	Hr.	Daily	Bare Costs	Incl. O&P
1 Skilled Worker Foreman (out)	$50.65	$405.20	$82.30	$658.40	$49.32	$80.13
2 Skilled Workers	48.65	778.40	79.05	1264.80		
24 L.H., Daily Totals		$1183.60		$1923.20	$49.32	$80.13
Crew E-24	Hr.	Daily	Hr.	Daily	Bare Costs	Incl. O&P
3 Structural Steel Workers	$52.65	$1263.60	$95.15	$2283.60	$52.14	$91.44
1 Equipment Operator (med.)	50.60	404.80	80.30	642.40		
1 Hyd. Crane, 25 Ton		736.60		810.26	23.02	25.25
32 L.H., Daily Totals		$2405.00		$3736.26	$75.16	$116.76
Crew E-25	Hr.	Daily	Hr.	Daily	Bare Costs	Incl. O&P
1 Welder Foreman (outside)	$54.65	$437.20	$98.75	$790.00	$54.65	$98.75
1 Cutting Torch		11.40		12.54	1.43	1.57
8 L.H., Daily Totals		$448.60		$802.54	$56.08	$100.32
Crew F-3	Hr.	Daily	Hr.	Daily	Bare Costs	Incl. O&P
4 Carpenters	$46.95	$1502.40	$76.95	$2462.40	$47.90	$77.97
1 Equip. Oper. (crane)	51.70	413.60	82.05	656.40		
1 Hyd. Crane, 12 Ton		653.80		719.18	16.34	17.98
40 L.H., Daily Totals		$2569.80		$3837.98	$64.25	$95.95
Crew F-4	Hr.	Daily	Hr.	Daily	Bare Costs	Incl. O&P
4 Carpenters	$46.95	$1502.40	$76.95	$2462.40	$47.45	$76.93
1 Equip. Oper. (crane)	51.70	413.60	82.05	656.40		
1 Equip. Oper. (oiler)	45.20	361.60	71.75	574.00		
1 Hyd. Crane, 55 Ton		1128.00		1240.80	23.50	25.85
48 L.H., Daily Totals		$3405.60		$4933.60	$70.95	$102.78
Crew F-5	Hr.	Daily	Hr.	Daily	Bare Costs	Incl. O&P
1 Carpenter Foreman (outside)	$48.95	$391.60	$80.25	$642.00	$47.45	$77.78
3 Carpenters	46.95	1126.80	76.95	1846.80		
32 L.H., Daily Totals		$1518.40		$2488.80	$47.45	$77.78

Crews

Crew No.	Bare Costs		Incl. Subs O&P		Cost Per Labor-Hour	
	Hr.	Daily	Hr.	Daily	Bare Costs	Incl. O&P
Crew F-6						
2 Carpenters	$46.95	$751.20	$76.95	$1231.20	$44.16	$71.85
2 Building Laborers	37.60	601.60	61.65	986.40		
1 Equip. Oper. (crane)	51.70	413.60	82.05	656.40		
1 Hyd. Crane, 12 Ton		653.80		719.18	16.34	17.98
40 L.H., Daily Totals		$2420.20		$3593.18	$60.51	$89.83
Crew F-7	Hr.	Daily	Hr.	Daily	Bare Costs	Incl. O&P
2 Carpenters	$46.95	$751.20	$76.95	$1231.20	$42.27	$69.30
2 Building Laborers	37.60	601.60	61.65	986.40		
32 L.H., Daily Totals		$1352.80		$2217.60	$42.27	$69.30
Crew G-1	Hr.	Daily	Hr.	Daily	Bare Costs	Incl. O&P
1 Roofer Foreman (outside)	$42.10	$336.80	$76.05	$608.40	$37.51	$67.78
4 Roofers Composition	40.10	1283.20	72.45	2318.40		
2 Roofer Helpers	30.05	480.80	54.30	868.80		
1 Application Equipment		193.00		212.30		
1 Tar Kettle/Pot		169.80		186.78		
1 Crew Truck		206.40		227.04	10.16	11.18
56 L.H., Daily Totals		$2670.00		$4421.72	$47.68	$78.96
Crew G-2	Hr.	Daily	Hr.	Daily	Bare Costs	Incl. O&P
1 Plasterer	$42.95	$343.60	$68.70	$549.60	$39.55	$63.75
1 Plasterer Helper	38.10	304.80	60.90	487.20		
1 Building Laborer	37.60	300.80	61.65	493.20		
1 Grout Pump, 50 C.F./hr.		133.40		146.74	5.56	6.11
24 L.H., Daily Totals		$1082.60		$1676.74	$45.11	$69.86
Crew G-2A	Hr.	Daily	Hr.	Daily	Bare Costs	Incl. O&P
1 Roofer Composition	$40.10	$320.80	$72.45	$579.60	$35.92	$62.80
1 Roofer Helper	30.05	240.40	54.30	434.40		
1 Building Laborer	37.60	300.80	61.65	493.20		
1 Foam Spray Rig, Trailer-Mtd.		573.25		630.58		
1 Pickup Truck, 3/4 Ton		144.20		158.62	29.89	32.88
24 L.H., Daily Totals		$1579.45		$2296.40	$65.81	$95.68
Crew G-3	Hr.	Daily	Hr.	Daily	Bare Costs	Incl. O&P
2 Sheet Metal Workers	$55.95	$895.20	$88.30	$1412.80	$46.77	$74.97
2 Building Laborers	37.60	601.60	61.65	986.40		
32 L.H., Daily Totals		$1496.80		$2399.20	$46.77	$74.97
Crew G-4	Hr.	Daily	Hr.	Daily	Bare Costs	Incl. O&P
1 Labor Foreman (outside)	$39.60	$316.80	$64.90	$519.20	$38.27	$62.73
2 Building Laborers	37.60	601.60	61.65	986.40		
1 Flatbed Truck, Gas, 1.5 Ton		245.40		269.94		
1 Air Compressor, 160 cfm		156.40		172.04	16.74	18.42
24 L.H., Daily Totals		$1320.20		$1947.58	$55.01	$81.15
Crew G-5	Hr.	Daily	Hr.	Daily	Bare Costs	Incl. O&P
1 Roofer Foreman (outside)	$42.10	$336.80	$76.05	$608.40	$36.48	$65.91
2 Roofers Composition	40.10	641.60	72.45	1159.20		
2 Roofer Helpers	30.05	480.80	54.30	868.80		
1 Application Equipment		193.00		212.30	4.83	5.31
40 L.H., Daily Totals		$1652.20		$2848.70	$41.31	$71.22
Crew G-6A	Hr.	Daily	Hr.	Daily	Bare Costs	Incl. O&P
2 Roofers Composition	$40.10	$641.60	$72.45	$1159.20	$40.10	$72.45
1 Small Compressor, Electric		12.95		14.24		
2 Pneumatic Nailers		55.30		60.83	4.27	4.69
16 L.H., Daily Totals		$709.85		$1234.28	$44.37	$77.14

Crew No.	Bare Costs		Incl. Subs O&P		Cost Per Labor-Hour	
	Hr.	Daily	Hr.	Daily	Bare Costs	Incl. O&P
Crew G-7						
1 Carpenter	$46.95	$375.60	$76.95	$615.60	$46.95	$76.95
1 Small Compressor, Electric		12.95		14.24		
1 Pneumatic Nailer		27.65		30.41	5.08	5.58
8 L.H., Daily Totals		$416.20		$660.26	$52.02	$82.53
Crew H-1	Hr.	Daily	Hr.	Daily	Bare Costs	Incl. O&P
2 Glaziers	$45.10	$721.60	$73.20	$1171.20	$48.88	$84.17
2 Struc. Steel Workers	52.65	842.40	95.15	1522.40		
32 L.H., Daily Totals		$1564.00		$2693.60	$48.88	$84.17
Crew H-2	Hr.	Daily	Hr.	Daily	Bare Costs	Incl. O&P
2 Glaziers	$45.10	$721.60	$73.20	$1171.20	$42.60	$69.35
1 Building Laborer	37.60	300.80	61.65	493.20		
24 L.H., Daily Totals		$1022.40		$1664.40	$42.60	$69.35
Crew H-3	Hr.	Daily	Hr.	Daily	Bare Costs	Incl. O&P
1 Glazier	$45.10	$360.80	$73.20	$585.60	$40.27	$65.85
1 Helper	35.45	283.60	58.50	468.00		
16 L.H., Daily Totals		$644.40		$1053.60	$40.27	$65.85
Crew H-4	Hr.	Daily	Hr.	Daily	Bare Costs	Incl. O&P
1 Carpenter	$46.95	$375.60	$76.95	$615.60	$43.90	$71.12
1 Carpenter Helper	35.45	283.60	58.50	468.00		
.5 Electrician	54.70	218.80	84.70	338.80		
20 L.H., Daily Totals		$878.00		$1422.40	$43.90	$71.12
Crew J-1	Hr.	Daily	Hr.	Daily	Bare Costs	Incl. O&P
3 Plasterers	$42.95	$1030.80	$68.70	$1648.80	$41.01	$65.58
2 Plasterer Helpers	38.10	609.60	60.90	974.40		
1 Mixing Machine, 6 C.F.		140.20		154.22	3.50	3.86
40 L.H., Daily Totals		$1780.60		$2777.42	$44.52	$69.44
Crew J-2	Hr.	Daily	Hr.	Daily	Bare Costs	Incl. O&P
3 Plasterers	$42.95	$1030.80	$68.70	$1648.80	$41.34	$65.93
2 Plasterer Helpers	38.10	609.60	60.90	974.40		
1 Lather	43.00	344.00	67.70	541.60		
1 Mixing Machine, 6 C.F.		140.20		154.22	2.92	3.21
48 L.H., Daily Totals		$2124.60		$3319.02	$44.26	$69.15
Crew J-3	Hr.	Daily	Hr.	Daily	Bare Costs	Incl. O&P
1 Terrazzo Worker	$42.95	$343.60	$67.80	$542.40	$39.23	$61.92
1 Terrazzo Helper	35.50	284.00	56.05	448.40		
1 Floor Grinder, 22" Path		115.55		127.11		
1 Terrazzo Mixer		188.20		207.02	18.98	20.88
16 L.H., Daily Totals		$931.35		$1324.93	$58.21	$82.81
Crew J-4	Hr.	Daily	Hr.	Daily	Bare Costs	Incl. O&P
2 Cement Finishers	$45.00	$720.00	$71.15	$1138.40	$42.53	$67.98
1 Laborer	37.60	300.80	61.65	493.20		
1 Floor Grinder, 22" Path		115.55		127.11		
1 Floor Edger, 7" Path		39.30		43.23		
1 Vacuum Pick-Up System		60.90		66.99	8.99	9.89
24 L.H., Daily Totals		$1236.55		$1868.93	$51.52	$77.87

Crews

Crew No.	Bare Costs		Incl. Subs O&P		Cost Per Labor-Hour	
Crew J-4A	Hr.	Daily	Hr.	Daily	Bare Costs	Incl. O&P
2 Cement Finishers	$45.00	$720.00	$71.15	$1138.40	$41.30	$66.40
2 Laborers	37.60	601.60	61.65	986.40		
1 Floor Grinder, 22" Path		115.55		127.11		
1 Floor Edger, 7" Path		39.30		43.23		
1 Vacuum Pick-Up System		60.90		66.99		
1 Floor Auto Scrubber		233.85		257.24	14.05	15.46
32 L.H., Daily Totals		$1771.20		$2619.36	$55.35	$81.86
Crew J-4B	Hr.	Daily	Hr.	Daily	Bare Costs	Incl. O&P
1 Laborer	$37.60	$300.80	$61.65	$493.20	$37.60	$61.65
1 Floor Auto Scrubber		233.85		257.24	29.23	32.15
8 L.H., Daily Totals		$534.65		$750.43	$66.83	$93.80
Crew J-6	Hr.	Daily	Hr.	Daily	Bare Costs	Incl. O&P
2 Painters	$40.35	$645.60	$64.85	$1037.60	$41.73	$67.13
1 Building Laborer	37.60	300.80	61.65	493.20		
1 Equip. Oper. (light)	48.60	388.80	77.15	617.20		
1 Air Compressor, 250 cfm		201.40		221.54		
1 Sandblaster, Portable, 3 C.F.		20.40		22.44		
1 Set Sand Blasting Accessories		14.05		15.46	7.37	8.11
32 L.H., Daily Totals		$1571.05		$2407.43	$49.10	$75.23
Crew J-7	Hr.	Daily	Hr.	Daily	Bare Costs	Incl. O&P
2 Painters	$40.35	$645.60	$64.85	$1037.60	$40.35	$64.85
1 Floor Belt Sander		14.20		15.62		
1 Floor Sanding Edger		12.60		13.86	1.68	1.84
16 L.H., Daily Totals		$672.40		$1067.08	$42.02	$66.69
Crew K-1	Hr.	Daily	Hr.	Daily	Bare Costs	Incl. O&P
1 Carpenter	$46.95	$375.60	$76.95	$615.60	$43.02	$70.00
1 Truck Driver (light)	39.10	312.80	63.05	504.40		
1 Flatbed Truck, Gas, 3 Ton		303.20		333.52	18.95	20.84
16 L.H., Daily Totals		$991.60		$1453.52	$61.98	$90.84
Crew K-2	Hr.	Daily	Hr.	Daily	Bare Costs	Incl. O&P
1 Struc. Steel Foreman (outside)	$54.65	$437.20	$98.75	$790.00	$48.80	$85.65
1 Struc. Steel Worker	52.65	421.20	95.15	761.20		
1 Truck Driver (light)	39.10	312.80	63.05	504.40		
1 Flatbed Truck, Gas, 3 Ton		303.20		333.52	12.63	13.90
24 L.H., Daily Totals		$1474.40		$2389.12	$61.43	$99.55
Crew L-1	Hr.	Daily	Hr.	Daily	Bare Costs	Incl. O&P
1 Electrician	$54.70	$437.60	$84.70	$677.60	$56.70	$88.13
1 Plumber	58.70	469.60	91.55	732.40		
16 L.H., Daily Totals		$907.20		$1410.00	$56.70	$88.13
Crew L-2	Hr.	Daily	Hr.	Daily	Bare Costs	Incl. O&P
1 Carpenter	$46.95	$375.60	$76.95	$615.60	$41.20	$67.72
1 Carpenter Helper	35.45	283.60	58.50	468.00		
16 L.H., Daily Totals		$659.20		$1083.60	$41.20	$67.72
Crew L-3	Hr.	Daily	Hr.	Daily	Bare Costs	Incl. O&P
1 Carpenter	$46.95	$375.60	$76.95	$615.60	$51.14	$81.72
.5 Electrician	54.70	218.80	84.70	338.80		
.5 Sheet Metal Worker	55.95	223.80	88.30	353.20		
16 L.H., Daily Totals		$818.20		$1307.60	$51.14	$81.72

Crew No.	Bare Costs		Incl. Subs O&P		Cost Per Labor-Hour	
Crew L-3A	Hr.	Daily	Hr.	Daily	Bare Costs	Incl. O&P
1 Carpenter Foreman (outside)	$48.95	$391.60	$80.25	$642.00	$51.28	$82.93
.5 Sheet Metal Worker	55.95	223.80	88.30	353.20		
12 L.H., Daily Totals		$615.40		$995.20	$51.28	$82.93
Crew L-4	Hr.	Daily	Hr.	Daily	Bare Costs	Incl. O&P
2 Skilled Workers	$48.65	$778.40	$79.05	$1264.80	$44.25	$72.20
1 Helper	35.45	283.60	58.50	468.00		
24 L.H., Daily Totals		$1062.00		$1732.80	$44.25	$72.20
Crew L-5	Hr.	Daily	Hr.	Daily	Bare Costs	Incl. O&P
1 Struc. Steel Foreman (outside)	$54.65	$437.20	$98.75	$790.00	$52.80	$93.79
5 Struc. Steel Workers	52.65	2106.00	95.15	3806.00		
1 Equip. Oper. (crane)	51.70	413.60	82.05	656.40		
1 Hyd. Crane, 25 Ton		736.60		810.26	13.15	14.47
56 L.H., Daily Totals		$3693.40		$6062.66	$65.95	$108.26
Crew L-5A	Hr.	Daily	Hr.	Daily	Bare Costs	Incl. O&P
1 Struc. Steel Foreman (outside)	$54.65	$437.20	$98.75	$790.00	$52.91	$92.78
2 Structural Steel Workers	52.65	842.40	95.15	1522.40		
1 Equip. Oper. (crane)	51.70	413.60	82.05	656.40		
1 S.P. Crane, 4x4, 25 Ton		599.40		659.34	18.73	20.60
32 L.H., Daily Totals		$2292.60		$3628.14	$71.64	$113.38
Crew L-5B	Hr.	Daily	Hr.	Daily	Bare Costs	Incl. O&P
1 Struc. Steel Foreman (outside)	$54.65	$437.20	$98.75	$790.00	$53.97	$88.74
2 Structural Steel Workers	52.65	842.40	95.15	1522.40		
2 Electricians	54.70	875.20	84.70	1355.20		
2 Steamfitters/Pipefitters	59.75	956.00	93.20	1491.20		
1 Equip. Oper. (crane)	51.70	413.60	82.05	656.40		
1 Equip. Oper. (oiler)	45.20	361.60	71.75	574.00		
1 Hyd. Crane, 80 Ton		1625.00		1787.50	22.57	24.83
72 L.H., Daily Totals		$5511.00		$8176.70	$76.54	$113.57
Crew L-6	Hr.	Daily	Hr.	Daily	Bare Costs	Incl. O&P
1 Plumber	$58.70	$469.60	$91.55	$732.40	$57.37	$89.27
.5 Electrician	54.70	218.80	84.70	338.80		
12 L.H., Daily Totals		$688.40		$1071.20	$57.37	$89.27
Crew L-7	Hr.	Daily	Hr.	Daily	Bare Costs	Incl. O&P
2 Carpenters	$46.95	$751.20	$76.95	$1231.20	$45.39	$73.69
1 Building Laborer	37.60	300.80	61.65	493.20		
.5 Electrician	54.70	218.80	84.70	338.80		
28 L.H., Daily Totals		$1270.80		$2063.20	$45.39	$73.69
Crew L-8	Hr.	Daily	Hr.	Daily	Bare Costs	Incl. O&P
2 Carpenters	$46.95	$751.20	$76.95	$1231.20	$49.30	$79.87
.5 Plumber	58.70	234.80	91.55	366.20		
20 L.H., Daily Totals		$986.00		$1597.40	$49.30	$79.87
Crew L-9	Hr.	Daily	Hr.	Daily	Bare Costs	Incl. O&P
1 Labor Foreman (inside)	$38.10	$304.80	$62.45	$499.60	$42.96	$71.83
2 Building Laborers	37.60	601.60	61.65	986.40		
1 Struc. Steel Worker	52.65	421.20	95.15	761.20		
.5 Electrician	54.70	218.80	84.70	338.80		
36 L.H., Daily Totals		$1546.40		$2586.00	$42.96	$71.83

Crews

Crew No.	Bare Costs		Incl. Subs O&P		Cost Per Labor-Hour	
Crew L-10	Hr.	Daily	Hr.	Daily	Bare Costs	Incl. O&P
1 Struc. Steel Foreman (outside)	$54.65	$437.20	$98.75	$790.00	$53.00	$91.98
1 Structural Steel Worker	52.65	421.20	95.15	761.20		
1 Equip. Oper. (crane)	51.70	413.60	82.05	656.40		
1 Hyd. Crane, 12 Ton		653.80		719.18	27.24	29.97
24 L.H., Daily Totals		$1925.80		$2926.78	$80.24	$121.95
Crew L-11	Hr.	Daily	Hr.	Daily	Bare Costs	Incl. O&P
2 Wreckers	$37.60	$601.60	$65.90	$1054.40	$43.88	$72.75
1 Equip. Oper. (crane)	51.70	413.60	82.05	656.40		
1 Equip. Oper. (light)	48.60	388.80	77.15	617.20		
1 Hyd. Excavator, 2.5 C.Y.		1605.00		1765.50		
1 Loader, Skid Steer, 78 H.P.		318.60		350.46	60.11	66.12
32 L.H., Daily Totals		$3327.60		$4443.96	$103.99	$138.87
Crew M-1	Hr.	Daily	Hr.	Daily	Bare Costs	Incl. O&P
3 Elevator Constructors	$76.50	$1836.00	$118.05	$2833.20	$72.67	$112.15
1 Elevator Apprentice	61.20	489.60	94.45	755.60		
5 Hand Tools		46.00		50.60	1.44	1.58
32 L.H., Daily Totals		$2371.60		$3639.40	$74.11	$113.73
Crew M-3	Hr.	Daily	Hr.	Daily	Bare Costs	Incl. O&P
1 Electrician Foreman (outside)	$56.70	$453.60	$87.75	$702.00	$57.56	$89.88
1 Common Laborer	37.60	300.80	61.65	493.20		
.25 Equipment Operator (med.)	50.60	101.20	80.30	160.60		
1 Elevator Constructor	76.50	612.00	118.05	944.40		
1 Elevator Apprentice	61.20	489.60	94.45	755.60		
.25 S.P. Crane, 4x4, 20 Ton		138.65		152.51	4.08	4.49
34 L.H., Daily Totals		$2095.85		$3208.32	$61.64	$94.36
Crew M-4	Hr.	Daily	Hr.	Daily	Bare Costs	Incl. O&P
1 Electrician Foreman (outside)	$56.70	$453.60	$87.75	$702.00	$56.94	$88.97
1 Common Laborer	37.60	300.80	61.65	493.20		
.25 Equipment Operator, Crane	51.70	103.40	82.05	164.10		
.25 Equip. Oper. (oiler)	45.20	90.40	71.75	143.50		
1 Elevator Constructor	76.50	612.00	118.05	944.40		
1 Elevator Apprentice	61.20	489.60	94.45	755.60		
.25 S.P. Crane, 4x4, 40 Ton		175.75		193.32	4.88	5.37
36 L.H., Daily Totals		$2225.55		$3396.13	$61.82	$94.34
Crew Q-1	Hr.	Daily	Hr.	Daily	Bare Costs	Incl. O&P
1 Plumber	$58.70	$469.60	$91.55	$732.40	$52.83	$82.40
1 Plumber Apprentice	46.95	375.60	73.25	586.00		
16 L.H., Daily Totals		$845.20		$1318.40	$52.83	$82.40
Crew Q-1A	Hr.	Daily	Hr.	Daily	Bare Costs	Incl. O&P
.25 Plumber Foreman (outside)	$60.70	$121.40	$94.70	$189.40	$59.10	$92.18
1 Plumber	58.70	469.60	91.55	732.40		
10 L.H., Daily Totals		$591.00		$921.80	$59.10	$92.18
Crew Q-1C	Hr.	Daily	Hr.	Daily	Bare Costs	Incl. O&P
1 Plumber	$58.70	$469.60	$91.55	$732.40	$52.08	$81.70
1 Plumber Apprentice	46.95	375.60	73.25	586.00		
1 Equip. Oper. (medium)	50.60	404.80	80.30	642.40		
1 Trencher, Chain Type, 8' D		3402.00		3742.20	141.75	155.93
24 L.H., Daily Totals		$4652.00		$5703.00	$193.83	$237.63
Crew Q-2	Hr.	Daily	Hr.	Daily	Bare Costs	Incl. O&P
2 Plumbers	$58.70	$939.20	$91.55	$1464.80	$54.78	$85.45
1 Plumber Apprentice	46.95	375.60	73.25	586.00		
24 L.H., Daily Totals		$1314.80		$2050.80	$54.78	$85.45

Crew No.	Bare Costs		Incl. Subs O&P		Cost Per Labor-Hour	
Crew Q-3	Hr.	Daily	Hr.	Daily	Bare Costs	Incl. O&P
1 Plumber Foreman (inside)	$59.20	$473.60	$92.35	$738.80	$55.89	$87.17
2 Plumbers	58.70	939.20	91.55	1464.80		
1 Plumber Apprentice	46.95	375.60	73.25	586.00		
32 L.H., Daily Totals		$1788.40		$2789.60	$55.89	$87.17
Crew Q-4	Hr.	Daily	Hr.	Daily	Bare Costs	Incl. O&P
1 Plumber Foreman (inside)	$59.20	$473.60	$92.35	$738.80	$55.89	$87.17
1 Plumber	58.70	469.60	91.55	732.40		
1 Welder (plumber)	58.70	469.60	91.55	732.40		
1 Plumber Apprentice	46.95	375.60	73.25	586.00		
1 Welder, Electric, 300 amp		57.70		63.47	1.80	1.98
32 L.H., Daily Totals		$1846.10		$2853.07	$57.69	$89.16
Crew Q-5	Hr.	Daily	Hr.	Daily	Bare Costs	Incl. O&P
1 Steamfitter	$59.75	$478.00	$93.20	$745.60	$53.77	$83.88
1 Steamfitter Apprentice	47.80	382.40	74.55	596.40		
16 L.H., Daily Totals		$860.40		$1342.00	$53.77	$83.88
Crew Q-6	Hr.	Daily	Hr.	Daily	Bare Costs	Incl. O&P
2 Steamfitters	$59.75	$956.00	$93.20	$1491.20	$55.77	$86.98
1 Steamfitter Apprentice	47.80	382.40	74.55	596.40		
24 L.H., Daily Totals		$1338.40		$2087.60	$55.77	$86.98
Crew Q-7	Hr.	Daily	Hr.	Daily	Bare Costs	Incl. O&P
1 Steamfitter Foreman (inside)	$60.25	$482.00	$94.00	$752.00	$56.89	$88.74
2 Steamfitters	59.75	956.00	93.20	1491.20		
1 Steamfitter Apprentice	47.80	382.40	74.55	596.40		
32 L.H., Daily Totals		$1820.40		$2839.60	$56.89	$88.74
Crew Q-8	Hr.	Daily	Hr.	Daily	Bare Costs	Incl. O&P
1 Steamfitter Foreman (inside)	$60.25	$482.00	$94.00	$752.00	$56.89	$88.74
1 Steamfitter	59.75	478.00	93.20	745.60		
1 Welder (steamfitter)	59.75	478.00	93.20	745.60		
1 Steamfitter Apprentice	47.80	382.40	74.55	596.40		
1 Welder, Electric, 300 amp		57.70		63.47	1.80	1.98
32 L.H., Daily Totals		$1878.10		$2903.07	$58.69	$90.72
Crew Q-9	Hr.	Daily	Hr.	Daily	Bare Costs	Incl. O&P
1 Sheet Metal Worker	$55.95	$447.60	$88.30	$706.40	$50.35	$79.45
1 Sheet Metal Apprentice	44.75	358.00	70.60	564.80		
16 L.H., Daily Totals		$805.60		$1271.20	$50.35	$79.45
Crew Q-10	Hr.	Daily	Hr.	Daily	Bare Costs	Incl. O&P
2 Sheet Metal Workers	$55.95	$895.20	$88.30	$1412.80	$52.22	$82.40
1 Sheet Metal Apprentice	44.75	358.00	70.60	564.80		
24 L.H., Daily Totals		$1253.20		$1977.60	$52.22	$82.40
Crew Q-11	Hr.	Daily	Hr.	Daily	Bare Costs	Incl. O&P
1 Sheet Metal Foreman (inside)	$56.45	$451.60	$89.10	$712.80	$53.27	$84.08
2 Sheet Metal Workers	55.95	895.20	88.30	1412.80		
1 Sheet Metal Apprentice	44.75	358.00	70.60	564.80		
32 L.H., Daily Totals		$1704.80		$2690.40	$53.27	$84.08
Crew Q-12	Hr.	Daily	Hr.	Daily	Bare Costs	Incl. O&P
1 Sprinkler Installer	$56.15	$449.20	$87.75	$702.00	$50.52	$78.97
1 Sprinkler Apprentice	44.90	359.20	70.20	561.60		
16 L.H., Daily Totals		$808.40		$1263.60	$50.52	$78.97

Crews

Crew No.	Bare Costs		Incl. Subs O&P		Cost Per Labor-Hour	
Crew Q-13	Hr.	Daily	Hr.	Daily	Bare Costs	Incl. O&P
1 Sprinkler Foreman (inside)	$56.65	$453.20	$88.55	$708.40	$53.46	$83.56
2 Sprinkler Installers	56.15	898.40	87.75	1404.00		
1 Sprinkler Apprentice	44.90	359.20	70.20	561.60		
32 L.H., Daily Totals		$1710.80		$2674.00	$53.46	$83.56
Crew Q-14	Hr.	Daily	Hr.	Daily	Bare Costs	Incl. O&P
1 Asbestos Worker	$52.35	$418.80	$84.15	$673.20	$47.13	$75.75
1 Asbestos Apprentice	41.90	335.20	67.35	538.80		
16 L.H., Daily Totals		$754.00		$1212.00	$47.13	$75.75
Crew Q-15	Hr.	Daily	Hr.	Daily	Bare Costs	Incl. O&P
1 Plumber	$58.70	$469.60	$91.55	$732.40	$52.83	$82.40
1 Plumber Apprentice	46.95	375.60	73.25	586.00		
1 Welder, Electric, 300 amp		57.70		63.47	3.61	3.97
16 L.H., Daily Totals		$902.90		$1381.87	$56.43	$86.37
Crew Q-16	Hr.	Daily	Hr.	Daily	Bare Costs	Incl. O&P
2 Plumbers	$58.70	$939.20	$91.55	$1464.80	$54.78	$85.45
1 Plumber Apprentice	46.95	375.60	73.25	586.00		
1 Welder, Electric, 300 amp		57.70		63.47	2.40	2.64
24 L.H., Daily Totals		$1372.50		$2114.27	$57.19	$88.09
Crew Q-17	Hr.	Daily	Hr.	Daily	Bare Costs	Incl. O&P
1 Steamfitter	$59.75	$478.00	$93.20	$745.60	$53.77	$83.88
1 Steamfitter Apprentice	47.80	382.40	74.55	596.40		
1 Welder, Electric, 300 amp		57.70		63.47	3.61	3.97
16 L.H., Daily Totals		$918.10		$1405.47	$57.38	$87.84
Crew Q-17A	Hr.	Daily	Hr.	Daily	Bare Costs	Incl. O&P
1 Steamfitter	$59.75	$478.00	$93.20	$745.60	$53.08	$83.27
1 Steamfitter Apprentice	47.80	382.40	74.55	596.40		
1 Equip. Oper. (crane)	51.70	413.60	82.05	656.40		
1 Hyd. Crane, 12 Ton		653.80		719.18		
1 Welder, Electric, 300 amp		57.70		63.47	29.65	32.61
24 L.H., Daily Totals		$1985.50		$2781.05	$82.73	$115.88
Crew Q-18	Hr.	Daily	Hr.	Daily	Bare Costs	Incl. O&P
2 Steamfitters	$59.75	$956.00	$93.20	$1491.20	$55.77	$86.98
1 Steamfitter Apprentice	47.80	382.40	74.55	596.40		
1 Welder, Electric, 300 amp		57.70		63.47	2.40	2.64
24 L.H., Daily Totals		$1396.10		$2151.07	$58.17	$89.63
Crew Q-19	Hr.	Daily	Hr.	Daily	Bare Costs	Incl. O&P
1 Steamfitter	$59.75	$478.00	$93.20	$745.60	$54.08	$84.15
1 Steamfitter Apprentice	47.80	382.40	74.55	596.40		
1 Electrician	54.70	437.60	84.70	677.60		
24 L.H., Daily Totals		$1298.00		$2019.60	$54.08	$84.15
Crew Q-20	Hr.	Daily	Hr.	Daily	Bare Costs	Incl. O&P
1 Sheet Metal Worker	$55.95	$447.60	$88.30	$706.40	$51.22	$80.50
1 Sheet Metal Apprentice	44.75	358.00	70.60	564.80		
.5 Electrician	54.70	218.80	84.70	338.80		
20 L.H., Daily Totals		$1024.40		$1610.00	$51.22	$80.50
Crew Q-21	Hr.	Daily	Hr.	Daily	Bare Costs	Incl. O&P
2 Steamfitters	$59.75	$956.00	$93.20	$1491.20	$55.50	$86.41
1 Steamfitter Apprentice	47.80	382.40	74.55	596.40		
1 Electrician	54.70	437.60	84.70	677.60		
32 L.H., Daily Totals		$1776.00		$2765.20	$55.50	$86.41
Crew Q-22	Hr.	Daily	Hr.	Daily	Bare Costs	Incl. O&P
1 Plumber	$58.70	$469.60	$91.55	$732.40	$52.83	$82.40
1 Plumber Apprentice	46.95	375.60	73.25	586.00		
1 Hyd. Crane, 12 Ton		653.80		719.18	40.86	44.95
16 L.H., Daily Totals		$1499.00		$2037.58	$93.69	$127.35
Crew Q-22A	Hr.	Daily	Hr.	Daily	Bare Costs	Incl. O&P
1 Plumber	$58.70	$469.60	$91.55	$732.40	$48.74	$77.13
1 Plumber Apprentice	46.95	375.60	73.25	586.00		
1 Laborer	37.60	300.80	61.65	493.20		
1 Equip. Oper. (crane)	51.70	413.60	82.05	656.40		
1 Hyd. Crane, 12 Ton		653.80		719.18	20.43	22.47
32 L.H., Daily Totals		$2213.40		$3187.18	$69.17	$99.60
Crew Q-23	Hr.	Daily	Hr.	Daily	Bare Costs	Incl. O&P
1 Plumber Foreman (outside)	$60.70	$485.60	$94.70	$757.60	$56.67	$88.85
1 Plumber	58.70	469.60	91.55	732.40		
1 Equip. Oper. (medium)	50.60	404.80	80.30	642.40		
1 Lattice Boom Crane, 20 Ton		948.90		1043.79	39.54	43.49
24 L.H., Daily Totals		$2308.90		$3176.19	$96.20	$132.34
Crew R-1	Hr.	Daily	Hr.	Daily	Bare Costs	Incl. O&P
1 Electrician Foreman	$55.20	$441.60	$85.45	$683.60	$48.37	$76.09
3 Electricians	54.70	1312.80	84.70	2032.80		
2 Helpers	35.45	567.20	58.50	936.00		
48 L.H., Daily Totals		$2321.60		$3652.40	$48.37	$76.09
Crew R-1A	Hr.	Daily	Hr.	Daily	Bare Costs	Incl. O&P
1 Electrician	$54.70	$437.60	$84.70	$677.60	$45.08	$71.60
1 Helper	35.45	283.60	58.50	468.00		
16 L.H., Daily Totals		$721.20		$1145.60	$45.08	$71.60
Crew R-2	Hr.	Daily	Hr.	Daily	Bare Costs	Incl. O&P
1 Electrician Foreman	$55.20	$441.60	$85.45	$683.60	$48.84	$76.94
3 Electricians	54.70	1312.80	84.70	2032.80		
2 Helpers	35.45	567.20	58.50	936.00		
1 Equip. Oper. (crane)	51.70	413.60	82.05	656.40		
1 S.P. Crane, 4x4, 5 Ton		275.40		302.94	4.92	5.41
56 L.H., Daily Totals		$3010.60		$4611.74	$53.76	$82.35
Crew R-3	Hr.	Daily	Hr.	Daily	Bare Costs	Incl. O&P
1 Electrician Foreman	$55.20	$441.60	$85.45	$683.60	$54.30	$84.47
1 Electrician	54.70	437.60	84.70	677.60		
.5 Equip. Oper. (crane)	51.70	206.80	82.05	328.20		
.5 S.P. Crane, 4x4, 5 Ton		137.70		151.47	6.88	7.57
20 L.H., Daily Totals		$1223.70		$1840.87	$61.19	$92.04
Crew R-4	Hr.	Daily	Hr.	Daily	Bare Costs	Incl. O&P
1 Struc. Steel Foreman (outside)	$54.65	$437.20	$98.75	$790.00	$53.46	$93.78
3 Struc. Steel Workers	52.65	1263.60	95.15	2283.60		
1 Electrician	54.70	437.60	84.70	677.60		
1 Welder, Gas Engine, 300 amp		145.85		160.44	3.65	4.01
40 L.H., Daily Totals		$2284.25		$3911.64	$57.11	$97.79

Crews

Crew No.	Bare Costs		Incl. Subs O&P		Cost Per Labor-Hour	
Crew R-5	Hr.	Daily	Hr.	Daily	Bare Costs	Incl. O&P
1 Electrician Foreman	$55.20	$441.60	$85.45	$683.60	$47.75	$75.24
4 Electrician Linemen	54.70	1750.40	84.70	2710.40		
2 Electrician Operators	54.70	875.20	84.70	1355.20		
4 Electrician Groundmen	35.45	1134.40	58.50	1872.00		
1 Crew Truck		206.40		227.04		
1 Flatbed Truck, 20,000 GVW		248.60		273.46		
1 Pickup Truck, 3/4 Ton		144.20		158.62		
.2 Hyd. Crane, 55 Ton		225.60		248.16		
.2 Hyd. Crane, 12 Ton		130.76		143.84		
.2 Earth Auger, Truck-Mtd.		83.44		91.78		
1 Tractor w/Winch		427.00		469.70	16.66	18.32
88 L.H., Daily Totals		$5667.60		$8233.80	$64.40	$93.57
Crew R-6	Hr.	Daily	Hr.	Daily	Bare Costs	Incl. O&P
1 Electrician Foreman	$55.20	$441.60	$85.45	$683.60	$47.75	$75.24
4 Electrician Linemen	54.70	1750.40	84.70	2710.40		
2 Electrician Operators	54.70	875.20	84.70	1355.20		
4 Electrician Groundmen	35.45	1134.40	58.50	1872.00		
1 Crew Truck		206.40		227.04		
1 Flatbed Truck, 20,000 GVW		248.60		273.46		
1 Pickup Truck, 3/4 Ton		144.20		158.62		
.2 Hyd. Crane, 55 Ton		225.60		248.16		
.2 Hyd. Crane, 12 Ton		130.76		143.84		
.2 Earth Auger, Truck-Mtd.		83.44		91.78		
1 Tractor w/Winch		427.00		469.70		
3 Cable Trailers		598.20		658.02		
.5 Tensioning Rig		198.40		218.24		
.5 Cable Pulling Rig		1160.00		1276.00	38.89	42.78
88 L.H., Daily Totals		$7624.20		$10386.06	$86.64	$118.02
Crew R-7	Hr.	Daily	Hr.	Daily	Bare Costs	Incl. O&P
1 Electrician Foreman	$55.20	$441.60	$85.45	$683.60	$38.74	$62.99
5 Electrician Groundmen	35.45	1418.00	58.50	2340.00		
1 Crew Truck		206.40		227.04	4.30	4.73
48 L.H., Daily Totals		$2066.00		$3250.64	$43.04	$67.72
Crew R-8	Hr.	Daily	Hr.	Daily	Bare Costs	Incl. O&P
1 Electrician Foreman	$55.20	$441.60	$85.45	$683.60	$48.37	$76.09
3 Electrician Linemen	54.70	1312.80	84.70	2032.80		
2 Electrician Groundmen	35.45	567.20	58.50	936.00		
1 Pickup Truck, 3/4 Ton		144.20		158.62		
1 Crew Truck		206.40		227.04	7.30	8.03
48 L.H., Daily Totals		$2672.20		$4038.06	$55.67	$84.13
Crew R-9	Hr.	Daily	Hr.	Daily	Bare Costs	Incl. O&P
1 Electrician Foreman	$55.20	$441.60	$85.45	$683.60	$45.14	$71.69
1 Electrician Lineman	54.70	437.60	84.70	677.60		
2 Electrician Operators	54.70	875.20	84.70	1355.20		
4 Electrician Groundmen	35.45	1134.40	58.50	1872.00		
1 Pickup Truck, 3/4 Ton		144.20		158.62		
1 Crew Truck		206.40		227.04	5.48	6.03
64 L.H., Daily Totals		$3239.40		$4974.06	$50.62	$77.72
Crew R-10	Hr.	Daily	Hr.	Daily	Bare Costs	Incl. O&P
1 Electrician Foreman	$55.20	$441.60	$85.45	$683.60	$51.58	$80.46
4 Electrician Linemen	54.70	1750.40	84.70	2710.40		
1 Electrician Groundman	35.45	283.60	58.50	468.00		
1 Crew Truck		206.40		227.04		
3 Tram Cars		412.20		453.42	12.89	14.18
48 L.H., Daily Totals		$3094.20		$4542.46	$64.46	$94.63

Crew No.	Bare Costs		Incl. Subs O&P		Cost Per Labor-Hour	
Crew R-11	Hr.	Daily	Hr.	Daily	Bare Costs	Incl. O&P
1 Electrician Foreman	$55.20	$441.60	$85.45	$683.60	$51.90	$81.14
4 Electricians	54.70	1750.40	84.70	2710.40		
1 Equip. Oper. (crane)	51.70	413.60	82.05	656.40		
1 Common Laborer	37.60	300.80	61.65	493.20		
1 Crew Truck		206.40		227.04		
1 Hyd. Crane, 12 Ton		653.80		719.18	15.36	16.90
56 L.H., Daily Totals		$3766.60		$5489.82	$67.26	$98.03
Crew R-12	Hr.	Daily	Hr.	Daily	Bare Costs	Incl. O&P
1 Carpenter Foreman (inside)	$47.45	$379.60	$77.75	$622.00	$44.45	$73.42
4 Carpenters	46.95	1502.40	76.95	2462.40		
4 Common Laborers	37.60	1203.20	61.65	1972.80		
1 Equip. Oper. (medium)	50.60	404.80	80.30	642.40		
1 Steel Worker	52.65	421.20	95.15	761.20		
1 Dozer, 200 H.P.		1387.00		1525.70		
1 Pickup Truck, 3/4 Ton		144.20		158.62	17.40	19.14
88 L.H., Daily Totals		$5442.40		$8145.12	$61.85	$92.56
Crew R-13	Hr.	Daily	Hr.	Daily	Bare Costs	Incl. O&P
1 Electrician Foreman	$55.20	$441.60	$85.45	$683.60	$52.84	$82.25
3 Electricians	54.70	1312.80	84.70	2032.80		
.25 Equip. Oper. (crane)	51.70	103.40	82.05	164.10		
1 Equipment Oiler	45.20	361.60	71.75	574.00		
.25 Hydraulic Crane, 33 Ton		187.50		206.25	4.46	4.91
42 L.H., Daily Totals		$2406.90		$3660.75	$57.31	$87.16
Crew R-15	Hr.	Daily	Hr.	Daily	Bare Costs	Incl. O&P
1 Electrician Foreman	$55.20	$441.60	$85.45	$683.60	$53.77	$83.57
4 Electricians	54.70	1750.40	84.70	2710.40		
1 Equipment Oper. (light)	48.60	388.80	77.15	617.20		
1 Aerial Lift Truck, 40' Boom		301.40		331.54	6.28	6.91
48 L.H., Daily Totals		$2882.20		$4342.74	$60.05	$90.47
Crew R-18	Hr.	Daily	Hr.	Daily	Bare Costs	Incl. O&P
.25 Electrician Foreman	$55.20	$110.40	$85.45	$170.90	$42.89	$68.63
1 Electrician	54.70	437.60	84.70	677.60		
2 Helpers	35.45	567.20	58.50	936.00		
26 L.H., Daily Totals		$1115.20		$1784.50	$42.89	$68.63
Crew R-19	Hr.	Daily	Hr.	Daily	Bare Costs	Incl. O&P
.5 Electrician Foreman	$55.20	$220.80	$85.45	$341.80	$54.80	$84.85
2 Electricians	54.70	875.20	84.70	1355.20		
20 L.H., Daily Totals		$1096.00		$1697.00	$54.80	$84.85
Crew R-21	Hr.	Daily	Hr.	Daily	Bare Costs	Incl. O&P
1 Electrician Foreman	$55.20	$441.60	$85.45	$683.60	$54.72	$84.78
3 Electricians	54.70	1312.80	84.70	2032.80		
.1 Equip. Oper. (medium)	50.60	40.48	80.30	64.24		
.1 S.P. Crane, 4x4, 25 Ton		59.94		65.93	1.83	2.01
32.8 L.H., Daily Totals		$1854.82		$2846.57	$56.55	$86.79
Crew R-22	Hr.	Daily	Hr.	Daily	Bare Costs	Incl. O&P
.66 Electrician Foreman	$55.20	$291.46	$85.45	$451.18	$46.51	$73.56
2 Electricians	54.70	875.20	84.70	1355.20		
2 Helpers	35.45	567.20	58.50	936.00		
37.28 L.H., Daily Totals		$1733.86		$2742.38	$46.51	$73.56

Crews

Crew No.	Bare Costs		Incl. Subs O & P		Cost Per Labor-Hour	
Crew R-30	Hr.	Daily	Hr.	Daily	Bare Costs	Incl. O&P
.25 Electrician Foreman (outside)	$56.70	$113.40	$87.75	$175.50	$44.33	$70.75
1 Electrician	54.70	437.60	84.70	677.60		
2 Laborers, (Semi-Skilled)	37.60	601.60	61.65	986.40		
26 L.H., Daily Totals		$1152.60		$1839.50	$44.33	$70.75
Crew R-31	Hr.	Daily	Hr.	Daily	Bare Costs	Incl. O&P
1 Electrician	$54.70	$437.60	$84.70	$677.60	$54.70	$84.70
1 Core Drill, Electric, 2.5 H.P.		55.15		60.66	6.89	7.58
8 L.H., Daily Totals		$492.75		$738.26	$61.59	$92.28

Historical Cost Indexes

The table below lists both the RSMeans® historical cost index based on Jan. 1, 1993 = 100 as well as the computed value of an index based on Jan. 1, 2015 costs. Since the Jan. 1, 2015 figure is estimated, space is left to write in the actual index figures as they become available through either the quarterly *RSMeans Construction Cost Indexes* or as printed in the *Engineering News-Record*. To compute the actual index based on Jan. 1, 2015 = 100, divide the historical cost index for a particular year by the actual Jan. 1, 2015 construction cost index. Space has been left to advance the index figures as the year progresses.

Year	Historical Cost Index Jan. 1, 1993 = 100		Current Index Based on Jan. 1, 2015 = 100		Year	Historical Cost Index Jan. 1, 1993 = 100	Current Index Based on Jan. 1, 2015 = 100		Year	Historical Cost Index Jan. 1, 1993 = 100	Current Index Based on Jan. 1, 2015 = 100	
	Est.	Actual	Est.	Actual		Actual	Est.	Actual		Actual	Est.	Actual
Oct 2015*					July 2000	120.9	58.5		July 1982	76.1	36.8	
July 2015*					1999	117.6	56.9		1981	70.0	33.9	
April 2015*					1998	115.1	55.7		1980	62.9	30.4	
Jan 2015*	206.7		100.0	100.0	1997	112.8	54.6		1979	57.8	28.0	
July 2014		204.9	99.1		1996	110.2	53.3		1978	53.5	25.9	
2013		201.2	97.3		1995	107.6	52.1		1977	49.5	23.9	
2012		194.6	94.1		1994	104.4	50.5		1976	46.9	22.7	
2011		191.2	92.5		1993	101.7	49.2		1975	44.8	21.7	
2010		183.5	88.8		1992	99.4	48.1		1974	41.4	20.0	
2009		180.1	87.1		1991	96.8	46.8		1973	37.7	18.2	
2008		180.4	87.3		1990	94.3	45.6		1972	34.8	16.8	
2007		169.4	82.0		1989	92.1	44.6		1971	32.1	15.5	
2006		162.0	78.4		1988	89.9	43.5		1970	28.7	13.9	
2005		151.6	73.3		1987	87.7	42.4		1969	26.9	13.0	
2004		143.7	69.5		1986	84.2	40.7		1968	24.9	12.0	
2003		132.0	63.9		1985	82.6	40.0		1967	23.5	11.4	
2002		128.7	62.3		1984	82.0	39.7		1966	22.7	11.0	
2001		125.1	60.5		1983	80.2	38.8		1965	21.7	10.5	

Adjustments to Costs

The "Historical Cost Index" can be used to convert national average building costs at a particular time to the approximate building costs for some other time.

Example:

Estimate and compare construction costs for different years in the same city.

To estimate the national average construction cost of a building in 1970, knowing that it cost $900,000 in 2015:

INDEX in 1970 = 28.7

INDEX in 2015 = 206.7

Time Adjustment Using the Historical Cost Indexes:

$$\frac{\text{Index for Year A}}{\text{Index for Year B}} \times \text{Cost in Year B} = \text{Cost in Year A}$$

$$\frac{\text{INDEX 1970}}{\text{INDEX 2015}} \times \text{Cost 2015} = \text{Cost 1970}$$

$$\frac{28.7}{206.7} \times \$900{,}000 = .139 \times \$900{,}000 = \$125{,}100$$

The construction cost of the building in 1970 is $125,100.

Note: The city cost indexes for Canada can be used to convert U.S. national averages to local costs in Canadian dollars.

Example:

To estimate and compare the cost of a building in Toronto, ON in 2015 with the known cost of $600,000 (US$) in New York, NY in 2015:

INDEX Toronto = 110.9

INDEX New York = 131.8

$$\frac{\text{INDEX Toronto}}{\text{INDEX New York}} \times \text{Cost New York} = \text{Cost Toronto}$$

$$\frac{110.9}{131.8} \times \$600{,}000 = .841 \times \$600{,}000 = \$504{,}600$$

The construction cost of the building in Toronto is $504,600 (CN$).

*Historical Cost Index updates and other resources are provided on the following website.
http://info.thegordiangroup.com/RSMeans.html

City Cost Indexes

How to Use the City Cost Indexes

What you should know before you begin

RSMeans City Cost Indexes (CCI) are an extremely useful tool to use when you want to compare costs from city to city and region to region.

This publication contains average construction cost indexes for 731 U.S. and Canadian cities covering over 930 three-digit zip code locations, as listed directly under each city.

Keep in mind that a City Cost Index number is a percentage ratio of a specific city's cost to the national average cost of the same item at a stated time period.

In other words, these index figures represent relative construction factors (or, if you prefer, multipliers) for Material and Installation costs, as well as the weighted average for Total In Place costs for each CSI MasterFormat division. Installation costs include both labor and equipment rental costs. When estimating equipment rental rates only, for a specific location, use 01 54 33 EQUIPMENT RENTAL COSTS in the Reference Section at the back of the book.

The 30 City Average Index is the average of 30 major U.S. cities and serves as a National Average.

Index figures for both material and installation are based on the 30 major city average of 100 and represent the cost relationship as of July 1, 2014. The index for each division is computed from representative material and labor quantities for that division. The weighted average for each city is a weighted total of the components listed above it, but does not include relative productivity between trades or cities.

As changes occur in local material prices, labor rates, and equipment rental rates (including fuel costs), the impact of these changes should be accurately measured by the change in the City Cost Index for each particular city (as compared to the 30 City Average).

Therefore, if you know (or have estimated) building costs in one city today, you can easily convert those costs to expected building costs in another city.

In addition, by using the Historical Cost Index, you can easily convert National Average building costs at a particular time to the approximate building costs for some other time. The City Cost Indexes can then be applied to calculate the costs for a particular city.

Quick Calculations

Location Adjustment Using the City Cost Indexes:

$$\frac{\text{Index for City A}}{\text{Index for City B}} \times \text{Cost in City B} = \text{Cost in City A}$$

Time Adjustment for the National Average Using the Historical Cost Index:

$$\frac{\text{Index for Year A}}{\text{Index for Year B}} \times \text{Cost in Year B} = \text{Cost in Year A}$$

Adjustment from the National Average:

$$\frac{\text{Index for City A}}{100} \times \text{National Average Cost} = \text{Cost in City A}$$

Since each of the other RSMeans publications contains many different items, any *one* item multiplied by the particular city index may give incorrect results. However, the larger the number of items compiled, the closer the results should be to actual costs for that particular city.

The City Cost Indexes for Canadian cities are calculated using Canadian material and equipment prices and labor rates, in Canadian dollars. Therefore, indexes for Canadian cities can be used to convert U.S. National Average prices to local costs in Canadian dollars.

How to use this section

1. Compare costs from city to city.

In using the RSMeans Indexes, remember that an index number is not a fixed number but a ratio: It's a percentage ratio of a building component's cost at any stated time to the National Average cost of that same component at the same time period. Put in the form of an equation:

$$\frac{\text{Specific City Cost}}{\text{National Average Cost}} \times 100 = \text{City Index Number}$$

Therefore, when making cost comparisons between cities, do not subtract one city's index number from the index number of another city and read the result as a percentage difference. Instead, divide one city's index number by that of the other city. The resulting number may then be used as a multiplier to calculate cost differences from city to city.

The formula used to find cost differences between cities for the purpose of comparison is as follows:

$$\frac{\text{City A Index}}{\text{City B Index}} \times \text{City B Cost (Known)} = \text{City A Cost (Unknown)}$$

In addition, you can use RSMeans CCI to calculate and compare costs division by division between cities using the same basic formula. (Just be sure that you're comparing similar divisions.)

2. Compare a specific city's construction costs with the National Average.

When you're studying construction location feasibility, it's advisable to compare a prospective project's cost index with an index of the National Average cost.

For example, divide the weighted average index of construction costs of a specific city by that of the 30 City Average, which = 100.

$$\frac{\text{City Index}}{100} = \% \text{ of National Average}$$

As a result, you get a ratio that indicates the relative cost of construction in that city in comparison with the National Average.

3. Convert U.S. National Average to actual costs in Canadian City.

$$\frac{\text{Index for Canadian City}}{100} \times \text{National Average Cost} = \text{Cost in Canadian City in \$ CAN}$$

4. **Adjust construction cost data based on a National Average.**
When you use a source of construction cost data which is based on a National Average (such as RSMeans cost data publications), it is necessary to adjust those costs to a specific location.

$$\frac{\text{City Index}}{100} \times \begin{array}{c}\text{"Book" Cost Based on}\\ \text{National Average Costs}\end{array} = \begin{array}{c}\text{City Cost}\\ \text{(Unknown)}\end{array}$$

5. **When applying the City Cost Indexes to demolition projects, use the appropriate division installation index.** For example, for removal of existing doors and windows, use Division 8 (Openings) index.

What you might like to know about how we developed the Indexes

The information presented in the CCI is organized according to the Construction Specifications Institute (CSI) MasterFormat 2014 classification system.

To create a reliable index, RSMeans researched the building type most often constructed in the United States and Canada. Because it was concluded that no one type of building completely represented the building construction industry, nine different types of buildings were combined to create a composite model.

The exact material, labor, and equipment quantities are based on detailed analyses of these nine building types, and then each quantity is weighted in proportion to expected usage. These various material items, labor hours, and equipment rental rates are thus combined to form a composite building representing as closely as possible the actual usage of materials, labor, and equipment used in the North American building construction industry.

The following structures were chosen to make up that composite model:

1. Factory, 1 story
2. Office, 2–4 story
3. Store, Retail
4. Town Hall, 2–3 story
5. High School, 2–3 story
6. Hospital, 4–8 story
7. Garage, Parking
8. Apartment, 1–3 story
9. Hotel/Motel, 2–3 story

For the purposes of ensuring the timeliness of the data, the components of the index for the composite model have been streamlined. They currently consist of:

- specific quantities of 66 commonly used construction materials;
- specific labor-hours for 21 building construction trades; and
- specific days of equipment rental for 6 types of construction equipment (normally used to install the 66 material items by the 21 trades.) Fuel costs and routine maintenance costs are included in the equipment cost.

A sophisticated computer program handles the updating of all costs for each city on a quarterly basis. Material and equipment price quotations are gathered quarterly from cities in the United States and Canada. These prices and the latest negotiated labor wage rates for 21 different building trades are used to compile the quarterly update of the City Cost Index.

The 30 major U.S. cities used to calculate the National Average are:

Atlanta, GA
Baltimore, MD
Boston, MA
Buffalo, NY
Chicago, IL
Cincinnati, OH
Cleveland, OH
Columbus, OH
Dallas, TX
Denver, CO
Detroit, MI
Houston, TX
Indianapolis, IN
Kansas City, MO
Los Angeles, CA

Memphis, TN
Milwaukee, WI
Minneapolis, MN
Nashville, TN
New Orleans, LA
New York, NY
Philadelphia, PA
Phoenix, AZ
Pittsburgh, PA
St. Louis, MO
San Antonio, TX
San Diego, CA
San Francisco, CA
Seattle, WA
Washington, DC

What the CCI does not indicate

The weighted average for each city is a total of the divisional components weighted to reflect typical usage, but it does not include the productivity variations between trades or cities.

In addition, the CCI does not take into consideration factors such as the following:

- managerial efficiency
- competitive conditions
- automation
- restrictive union practices
- unique local requirements
- regional variations due to specific building codes

City Cost Indexes

		UNITED STATES			ALABAMA													
	DIVISION	30 CITY AVERAGE			ANNISTON 362			BIRMINGHAM 350 - 352			BUTLER 369			DECATUR 356			DOTHAN 363		
		MAT.	INST.	TOTAL	MAT.	INST.	TOTAL	MAT.	INST.	TOTAL	MAT.	INST.	TOTAL	MAT.	INST.	TOTAL	MAT.	INST.	TOTAL
015433	CONTRACTOR EQUIPMENT		100.0	100.0		101.0	101.0		101.1	101.1		98.2	98.2		101.0	101.0		98.2	98.2
0241, 31 - 34	SITE & INFRASTRUCTURE, DEMOLITION	100.0	100.0	100.0	90.2	93.1	92.3	97.4	93.8	94.8	103.5	87.1	91.8	88.8	91.5	90.7	101.2	87.1	91.1
0310	Concrete Forming & Accessories	100.0	100.0	100.0	90.1	47.1	53.0	93.1	75.9	78.2	86.4	42.6	48.6	94.6	45.4	52.1	95.5	42.0	49.4
0320	Concrete Reinforcing	100.0	100.0	100.0	87.8	86.3	87.1	94.7	86.9	90.7	92.8	45.9	68.9	88.6	77.0	82.7	92.8	45.4	68.7
0330	Cast-in-Place Concrete	100.0	100.0	100.0	101.8	49.7	80.4	110.0	74.6	95.4	99.3	54.8	81.0	103.4	64.1	87.2	96.4	46.3	75.8
03	CONCRETE	100.0	100.0	100.0	102.1	57.0	79.9	102.2	78.4	90.5	103.0	49.3	76.6	98.3	59.3	79.1	102.3	46.0	74.7
04	MASONRY	100.0	100.0	100.0	100.1	66.8	79.3	98.8	76.0	84.6	105.9	48.6	70.2	97.3	51.3	68.6	107.1	39.7	65.1
05	METALS	100.0	100.0	100.0	104.0	91.5	100.1	104.0	94.1	101.0	102.8	76.6	94.8	106.2	86.1	100.0	102.9	75.3	94.4
06	WOOD, PLASTICS & COMPOSITES	100.0	100.0	100.0	90.5	42.7	63.7	97.4	76.1	85.4	85.0	42.4	61.1	98.9	42.9	67.6	97.4	42.4	66.6
07	THERMAL & MOISTURE PROTECTION	100.0	100.0	100.0	97.9	52.9	79.5	99.7	81.5	92.2	98.0	56.7	81.0	96.7	58.3	81.0	98.0	49.1	77.9
08	OPENINGS	100.0	100.0	100.0	98.6	50.1	87.3	103.2	76.0	96.8	98.6	44.2	85.9	106.5	50.4	93.5	98.6	44.2	86.0
0920	Plaster & Gypsum Board	100.0	100.0	100.0	90.8	41.4	57.4	94.8	75.8	81.9	87.9	41.0	56.2	95.8	41.6	59.2	98.6	41.0	59.7
0950, 0980	Ceilings & Acoustic Treatment	100.0	100.0	100.0	80.4	41.4	54.7	89.8	75.8	80.6	80.4	41.0	54.5	87.2	41.6	57.2	80.4	41.0	54.5
0960	Flooring	100.0	100.0	100.0	91.5	40.2	76.8	102.5	76.7	95.1	95.8	56.0	84.4	99.7	50.6	85.6	100.6	27.5	79.7
0970, 0990	Wall Finishes & Painting/Coating	100.0	100.0	100.0	103.7	31.5	60.1	103.9	66.5	81.3	103.7	53.3	73.3	99.6	66.8	79.8	103.7	53.3	73.3
09	FINISHES	100.0	100.0	100.0	87.0	43.0	62.8	95.4	74.8	84.1	89.9	45.7	65.5	91.9	47.7	67.6	92.5	40.1	63.6
COVERS	DIVS. 10 - 14, 25, 28, 41, 43, 44, 46	100.0	100.0	100.0	100.0	68.3	93.6	100.0	88.4	97.7	100.0	41.4	88.2	100.0	43.2	88.5	100.0	41.5	88.2
21, 22, 23	FIRE SUPPRESSION, PLUMBING & HVAC	100.0	100.0	100.0	100.0	55.4	82.0	100.0	70.4	88.1	97.3	33.3	71.5	100.0	40.8	76.1	97.3	33.2	71.5
26, 27, 3370	ELECTRICAL, COMMUNICATIONS & UTIL.	100.0	100.0	100.0	93.1	57.3	74.2	99.0	61.6	79.3	95.0	40.2	66.0	94.2	64.9	78.7	93.8	57.4	74.6
MF2014	WEIGHTED AVERAGE	100.0	100.0	100.0	98.6	61.8	82.6	100.6	76.7	90.2	98.9	49.4	77.3	99.8	57.9	81.5	99.0	49.4	77.4

		ALABAMA																
	DIVISION	EVERGREEN 364			GADSDEN 359			HUNTSVILLE 357 - 358			JASPER 355			MOBILE 365 - 366			MONTGOMERY 360 - 361		
		MAT.	INST.	TOTAL	MAT.	INST.	TOTAL	MAT.	INST.	TOTAL	MAT.	INST.	TOTAL	MAT.	INST.	TOTAL	MAT.	INST.	TOTAL
015433	CONTRACTOR EQUIPMENT		98.2	98.2		101.0	101.0		101.0	101.0		101.0	101.0		98.2	98.2		98.2	98.2
0241, 31 - 34	SITE & INFRASTRUCTURE, DEMOLITION	104.0	87.4	92.2	94.9	92.7	93.3	88.6	92.7	91.5	94.6	92.4	93.1	96.4	88.1	90.5	96.5	88.1	90.5
0310	Concrete Forming & Accessories	83.0	44.2	49.5	86.9	42.5	48.6	94.5	69.9	73.3	92.1	34.6	42.5	94.5	54.0	59.6	94.1	44.1	51.0
0320	Concrete Reinforcing	92.9	45.9	69.0	94.0	86.8	90.3	88.0	81.6	85.1	88.0	85.6	87.1	90.6	83.6	87.0	95.8	86.1	90.9
0330	Cast-in-Place Concrete	99.3	57.3	82.1	103.4	66.7	88.3	100.7	69.5	87.9	114.5	44.1	85.6	104.1	66.4	88.6	105.2	56.5	85.2
03	CONCRETE	103.3	50.9	77.5	102.9	60.8	82.2	97.0	72.9	85.2	106.6	49.4	78.5	98.9	65.3	82.4	100.2	57.9	79.4
04	MASONRY	105.9	53.2	73.0	95.6	62.4	74.9	98.8	67.7	79.4	92.9	41.2	60.7	103.7	54.2	72.9	100.0	49.7	68.6
05	METALS	102.9	76.5	94.7	104.0	93.2	100.7	106.2	90.6	101.4	104.0	90.2	99.7	105.0	91.6	100.9	104.0	91.9	100.3
06	WOOD, PLASTICS & COMPOSITES	81.3	42.4	59.5	89.3	38.7	61.0	98.9	72.6	84.2	95.9	31.6	59.9	95.9	52.8	71.8	95.3	42.4	65.6
07	THERMAL & MOISTURE PROTECTION	97.9	56.9	81.1	96.8	71.7	86.5	96.6	77.6	88.8	96.8	45.4	75.7	97.6	70.1	86.3	97.9	66.6	85.0
08	OPENINGS	98.6	44.2	85.9	102.8	49.0	90.3	106.5	67.4	97.4	102.8	48.0	90.0	102.2	58.8	92.1	103.3	53.1	91.7
0920	Plaster & Gypsum Board	87.2	41.0	56.0	88.2	37.3	53.8	95.8	72.2	79.9	92.1	30.0	50.1	95.2	51.8	65.9	94.6	41.0	58.4
0950, 0980	Ceilings & Acoustic Treatment	80.4	41.0	54.5	83.8	37.3	53.2	89.0	72.2	78.0	83.8	30.0	48.4	85.6	51.8	63.4	88.0	41.0	57.1
0960	Flooring	93.9	56.0	83.1	95.5	76.7	90.1	99.7	76.7	93.1	97.8	40.2	81.3	100.5	58.5	88.5	97.3	58.5	86.2
0970, 0990	Wall Finishes & Painting/Coating	103.7	53.3	73.3	99.6	60.7	76.1	99.6	65.9	79.3	99.6	38.6	62.8	107.3	54.3	75.3	103.3	53.3	73.1
09	FINISHES	89.2	46.8	65.9	89.4	48.7	67.0	92.3	70.7	80.4	90.5	34.4	59.6	92.7	53.5	71.1	92.8	46.1	67.0
COVERS	DIVS. 10 - 14, 25, 28, 41, 43, 44, 46	100.0	42.9	88.5	100.0	79.8	95.9	100.0	85.3	97.0	100.0	39.8	87.8	100.0	83.1	96.6	100.0	79.6	95.9
21, 22, 23	FIRE SUPPRESSION, PLUMBING & HVAC	97.3	35.6	72.4	102.0	36.3	75.5	100.0	61.6	84.5	102.0	61.4	85.6	99.9	60.7	84.1	100.0	34.2	73.4
26, 27, 3370	ELECTRICAL, COMMUNICATIONS & UTIL.	92.5	40.2	64.8	94.3	61.6	77.0	95.3	64.9	79.3	93.8	61.3	76.6	95.7	59.1	76.3	96.3	60.9	77.6
MF2014	WEIGHTED AVERAGE	98.6	50.7	77.7	99.9	60.0	82.5	99.8	72.1	87.8	100.2	57.5	81.6	99.9	65.4	84.9	99.9	57.1	81.2

		ALABAMA									ALASKA								
	DIVISION	PHENIX CITY 368			SELMA 367			TUSCALOOSA 354			ANCHORAGE 995 - 996			FAIRBANKS 997			JUNEAU 998		
		MAT.	INST.	TOTAL	MAT.	INST.	TOTAL	MAT.	INST.	TOTAL	MAT.	INST.	TOTAL	MAT.	INST.	TOTAL	MAT.	INST.	TOTAL
015433	CONTRACTOR EQUIPMENT		98.2	98.2		98.2	98.2		101.0	101.0		114.7	114.7		114.7	114.7		114.7	114.7
0241, 31 - 34	SITE & INFRASTRUCTURE, DEMOLITION	107.8	88.2	93.9	100.0	88.1	91.8	89.1	92.9	91.8	127.2	129.6	128.9	119.2	129.6	126.6	135.6	129.6	131.3
0310	Concrete Forming & Accessories	90.2	39.1	46.1	87.6	42.6	48.8	94.5	47.9	54.3	123.7	119.8	120.3	131.6	119.9	121.5	130.1	119.8	121.2
0320	Concrete Reinforcing	92.8	61.4	76.8	92.8	85.2	89.0	88.6	86.8	87.7	149.7	110.7	129.9	150.9	110.7	130.5	135.7	110.7	123.0
0330	Cast-in-Place Concrete	99.3	57.0	81.9	99.3	46.5	77.6	104.9	68.4	89.9	132.0	117.8	126.2	128.3	118.0	124.1	133.2	117.8	126.9
03	CONCRETE	106.2	51.3	79.2	101.7	53.7	78.1	99.0	63.8	81.7	138.6	116.7	127.9	123.8	116.9	120.4	136.0	116.7	126.5
04	MASONRY	105.9	41.1	65.5	109.6	38.7	65.4	97.4	65.5	77.6	184.6	125.3	147.6	189.5	125.3	149.5	176.6	125.3	144.6
05	METALS	102.8	81.9	96.4	102.8	89.3	98.7	105.3	93.2	101.6	113.6	103.4	110.5	117.6	103.5	113.3	116.3	103.4	112.3
06	WOOD, PLASTICS & COMPOSITES	90.2	35.4	59.5	87.0	42.4	62.0	98.8	44.4	68.4	129.2	118.7	123.4	136.1	118.7	126.4	129.4	118.7	123.4
07	THERMAL & MOISTURE PROTECTION	98.3	63.8	84.1	97.8	52.1	79.1	96.7	73.7	87.3	159.5	118.0	142.5	166.5	119.1	147.1	168.0	118.0	147.5
08	OPENINGS	98.6	43.5	85.8	98.6	53.1	88.0	106.5	58.8	95.4	132.7	115.6	128.7	128.9	115.7	125.8	129.8	115.6	126.5
0920	Plaster & Gypsum Board	92.2	33.9	52.8	90.1	41.0	56.9	95.8	43.1	60.2	140.9	119.1	126.2	164.8	119.1	133.9	148.1	119.1	128.5
0950, 0980	Ceilings & Acoustic Treatment	80.4	33.9	49.8	80.4	41.0	54.5	89.0	43.1	58.9	119.6	119.1	119.3	119.2	119.1	119.2	125.9	119.1	121.4
0960	Flooring	97.6	58.5	86.4	96.2	28.5	76.8	99.7	76.7	93.1	134.6	133.0	134.1	127.2	133.0	128.9	131.0	133.0	131.6
0970, 0990	Wall Finishes & Painting/Coating	103.7	53.3	73.3	103.7	53.3	73.3	99.6	45.9	67.2	136.8	116.3	124.4	133.5	121.2	126.1	132.1	116.3	122.6
09	FINISHES	91.4	42.2	64.3	90.0	40.2	62.6	92.3	51.2	69.7	134.3	122.2	127.7	132.4	122.8	127.1	134.4	122.2	127.7
COVERS	DIVS. 10 - 14, 25, 28, 41, 43, 44, 46	100.0	79.3	95.8	100.0	41.4	88.2	100.0	81.8	96.4	100.0	112.8	102.6	100.0	112.8	102.6	100.0	112.8	102.6
21, 22, 23	FIRE SUPPRESSION, PLUMBING & HVAC	97.3	34.5	72.0	97.3	33.6	71.6	100.0	34.0	73.4	100.3	105.0	102.2	100.2	108.0	103.4	100.3	105.0	102.2
26, 27, 3370	ELECTRICAL, COMMUNICATIONS & UTIL.	94.4	69.5	81.3	93.6	40.2	65.4	94.8	61.6	77.3	117.7	117.8	117.7	130.0	117.8	123.5	119.9	117.8	118.8
MF2014	WEIGHTED AVERAGE	99.5	54.6	79.9	98.7	49.9	77.4	99.8	61.2	83.0	121.1	115.6	118.7	121.0	116.4	119.0	121.3	115.6	118.8

City Cost Indexes

DIVISION		ALASKA			ARIZONA														
		KETCHIKAN			CHAMBERS			FLAGSTAFF			GLOBE			KINGMAN			MESA/TEMPE		
		999			865			860			855			864			852		
		MAT.	INST.	TOTAL	MAT.	INST.	TOTAL	MAT.	INST.	TOTAL	MAT.	INST.	TOTAL	MAT.	INST.	TOTAL	MAT.	INST.	TOTAL
015433	CONTRACTOR EQUIPMENT		114.7	114.7		93.0	93.0		93.0	93.0		92.2	92.2		93.0	93.0		92.2	92.2
0241, 31 - 34	SITE & INFRASTRUCTURE, DEMOLITION	173.8	129.6	142.4	69.6	96.1	88.4	87.3	96.3	93.7	97.3	95.6	96.1	69.6	96.3	88.6	89.0	95.8	93.8
0310	Concrete Forming & Accessories	122.9	119.8	120.2	98.8	58.1	63.7	104.4	65.1	70.5	99.0	58.1	63.7	97.0	65.1	69.5	102.2	69.3	73.9
0320	Concrete Reinforcing	117.4	110.7	114.0	97.1	85.2	91.1	97.0	85.3	91.0	108.0	85.2	96.4	97.2	85.3	91.1	108.7	85.3	96.8
0330	Cast-in-Place Concrete	259.2	117.9	201.2	90.7	73.2	83.5	90.8	73.4	83.7	94.3	72.0	85.1	90.4	73.4	83.4	95.0	72.3	85.7
03	CONCRETE	208.7	116.7	163.5	95.0	68.6	82.0	114.1	71.8	93.3	110.0	68.2	89.5	94.6	71.8	83.4	101.8	73.3	87.8
04	MASONRY	198.4	125.3	152.8	92.5	62.3	73.7	92.6	62.4	73.8	109.9	62.2	80.1	92.5	62.4	73.7	110.1	62.2	80.3
05	METALS	117.8	103.4	113.4	96.2	75.6	89.8	96.7	76.3	90.4	93.3	76.0	88.0	96.8	76.3	90.5	93.6	76.8	88.4
06	WOOD, PLASTICS & COMPOSITES	126.3	118.7	122.1	101.0	55.0	75.2	107.3	64.2	83.1	97.6	55.0	73.8	96.1	64.2	78.2	101.1	69.8	83.6
07	THERMAL & MOISTURE PROTECTION	169.3	118.0	148.3	94.6	65.0	82.5	96.2	68.1	84.7	101.9	63.0	85.9	94.6	65.6	82.7	101.2	65.1	86.4
08	OPENINGS	130.8	115.6	127.3	108.1	65.3	98.1	108.3	70.3	99.4	100.0	65.3	91.9	108.3	70.3	99.5	100.1	73.3	93.8
0920	Plaster & Gypsum Board	152.7	119.1	130.0	90.2	53.7	65.5	93.6	63.2	73.0	96.6	53.7	67.6	82.9	63.2	69.6	98.7	68.9	78.6
0950, 0980	Ceilings & Acoustic Treatment	112.7	119.1	116.9	99.7	53.7	69.5	100.5	63.2	76.0	88.9	53.7	65.8	100.5	63.2	76.0	88.9	68.9	75.8
0960	Flooring	127.2	133.0	128.9	93.3	39.5	77.9	95.4	39.7	79.5	103.5	39.5	85.2	92.1	53.7	81.1	104.7	49.5	88.9
0970, 0990	Wall Finishes & Painting/Coating	133.5	116.3	123.1	98.3	55.1	72.2	98.3	55.1	72.2	103.4	55.1	74.3	98.3	55.1	72.2	103.4	55.1	74.3
09	FINISHES	133.4	122.2	127.2	93.7	52.9	71.2	96.6	58.4	75.6	97.1	53.0	72.8	92.5	60.7	75.0	96.8	63.6	78.5
COVERS	DIVS. 10 - 14, 25, 28, 41, 43, 44, 46	100.0	112.8	102.6	100.0	82.3	96.4	100.0	83.3	96.6	100.0	82.4	96.4	100.0	83.3	96.6	100.0	84.0	96.8
21, 22, 23	FIRE SUPPRESSION, PLUMBING & HVAC	98.4	105.0	101.1	97.0	78.9	89.7	100.2	79.0	91.6	95.2	78.9	88.7	97.0	79.0	89.7	100.0	79.0	91.5
26, 27, 3370	ELECTRICAL, COMMUNICATIONS & UTIL.	129.9	117.8	123.5	104.5	70.9	86.7	103.4	61.3	81.1	97.5	61.2	78.3	104.5	61.3	81.7	94.2	61.3	76.8
MF2014	WEIGHTED AVERAGE	132.0	115.6	124.8	97.6	71.5	86.2	101.3	71.8	88.4	98.8	70.0	86.3	97.6	72.0	86.4	98.6	72.8	87.4

DIVISION		ARIZONA											ARKANSAS					
		PHOENIX			PRESCOTT			SHOW LOW			TUCSON			BATESVILLE			CAMDEN		
		850,853			863			859			856 - 857			725			717		
		MAT.	INST.	TOTAL	MAT.	INST.	TOTAL	MAT.	INST.	TOTAL	MAT.	INST.	TOTAL	MAT.	INST.	TOTAL	MAT.	INST.	TOTAL
015433	CONTRACTOR EQUIPMENT		92.7	92.7		93.0	93.0		92.2	92.2		92.2	92.2		89.1	89.1		89.1	89.1
0241, 31 - 34	SITE & INFRASTRUCTURE, DEMOLITION	89.4	96.0	94.1	76.0	96.0	90.2	99.2	95.6	96.6	85.2	95.8	92.7	77.1	84.9	82.6	78.3	84.4	82.7
0310	Concrete Forming & Accessories	103.1	68.3	73.1	100.4	52.5	59.1	106.3	69.1	74.2	102.5	68.1	72.9	85.9	44.5	50.5	83.7	30.7	38.0
0320	Concrete Reinforcing	107.0	85.4	96.0	97.0	85.2	91.0	108.7	85.2	96.8	89.9	85.3	87.5	87.4	67.5	77.3	93.6	67.4	80.3
0330	Cast-in-Place Concrete	95.1	72.4	85.8	90.7	73.1	83.5	94.3	72.1	85.2	97.8	72.3	87.3	77.4	45.3	64.2	82.6	38.3	64.4
03	CONCRETE	101.4	73.0	87.4	100.1	66.1	83.4	112.4	73.1	93.1	100.0	72.8	86.6	83.5	50.1	67.1	87.5	41.3	64.8
04	MASONRY	97.5	64.1	76.7	92.6	62.3	73.7	109.9	62.2	80.1	95.5	62.2	74.8	99.8	40.7	63.0	116.6	31.4	63.5
05	METALS	95.1	77.6	89.7	96.7	75.3	90.1	93.1	76.1	87.8	94.3	76.7	88.9	99.5	66.9	89.4	103.9	66.5	92.4
06	WOOD, PLASTICS & COMPOSITES	102.1	68.3	83.2	102.4	47.4	71.6	105.8	69.8	85.6	101.4	68.3	82.8	89.9	45.5	65.0	90.3	29.7	56.4
07	THERMAL & MOISTURE PROTECTION	100.9	67.0	87.0	95.1	64.2	82.4	102.1	65.0	86.9	102.4	64.2	86.7	98.2	43.9	76.0	94.2	35.7	70.2
08	OPENINGS	102.0	72.5	95.2	108.3	61.1	97.3	99.2	73.3	93.2	96.3	72.5	90.7	97.7	45.8	85.6	102.1	40.6	87.8
0920	Plaster & Gypsum Board	101.0	67.4	78.2	90.3	45.9	60.3	100.9	68.9	79.3	104.0	67.4	79.2	83.3	44.3	56.9	85.0	28.0	46.5
0950, 0980	Ceilings & Acoustic Treatment	96.5	67.4	77.3	98.8	45.9	64.0	88.9	68.9	75.8	89.4	67.4	75.0	83.2	44.3	57.6	83.3	28.0	47.0
0960	Flooring	104.9	51.7	89.7	94.1	39.5	78.5	106.2	45.0	88.7	95.2	44.2	80.6	96.7	59.1	85.9	99.6	39.6	82.4
0970, 0990	Wall Finishes & Painting/Coating	103.2	61.7	78.3	98.3	55.1	72.4	103.4	55.1	74.3	104.5	55.1	74.7	107.4	41.2	67.4	104.4	50.4	71.8
09	FINISHES	98.9	63.8	79.6	94.2	48.5	69.0	98.7	62.7	78.8	94.9	61.6	76.5	86.6	46.7	64.6	88.5	33.5	58.2
COVERS	DIVS. 10 - 14, 25, 28, 41, 43, 44, 46	100.0	83.9	96.7	100.0	81.4	96.3	100.0	84.0	96.8	100.0	83.9	96.7	100.0	39.5	87.8	100.0	35.6	87.0
21, 22, 23	FIRE SUPPRESSION, PLUMBING & HVAC	99.9	79.0	91.5	100.2	78.9	91.6	95.2	79.0	88.7	100.0	79.0	91.5	95.3	51.6	77.7	95.3	53.1	78.3
26, 27, 3370	ELECTRICAL, COMMUNICATIONS & UTIL.	101.0	66.7	82.9	103.0	61.2	80.9	94.6	61.2	77.0	96.5	61.3	77.9	96.5	63.5	79.0	96.8	62.1	78.5
MF2014	WEIGHTED AVERAGE	99.2	73.8	88.1	99.2	68.9	86.0	98.9	72.5	87.4	97.4	72.4	86.5	94.6	54.6	77.1	97.1	50.1	76.6

DIVISION		ARKANSAS																	
		FAYETTEVILLE			FORT SMITH			HARRISON			HOT SPRINGS			JONESBORO			LITTLE ROCK		
		727			729			726			719			724			720 - 722		
		MAT.	INST.	TOTAL	MAT.	INST.	TOTAL	MAT.	INST.	TOTAL	MAT.	INST.	TOTAL	MAT.	INST.	TOTAL	MAT.	INST.	TOTAL
015433	CONTRACTOR EQUIPMENT		89.1	89.1		89.1	89.1		89.1	89.1		89.1	89.1		108.3	108.3		89.1	89.1
0241, 31 - 34	SITE & INFRASTRUCTURE, DEMOLITION	76.4	86.3	83.5	81.8	86.7	85.3	82.1	84.9	84.1	81.4	85.8	84.5	103.2	101.1	101.7	93.1	86.8	88.6
0310	Concrete Forming & Accessories	81.2	38.5	44.4	101.3	57.3	63.4	90.6	44.8	51.1	81.1	35.7	41.9	89.6	48.2	53.9	96.4	60.0	65.0
0320	Concrete Reinforcing	87.4	71.3	79.2	88.4	72.0	80.1	87.0	63.7	75.1	91.8	67.5	79.4	84.5	70.1	77.2	93.4	68.3	80.6
0330	Cast-in-Place Concrete	77.4	49.4	65.9	88.5	70.3	81.0	85.8	43.3	68.4	84.5	38.8	65.7	84.3	55.0	72.2	87.3	70.3	80.3
03	CONCRETE	83.2	49.3	66.5	91.3	65.0	78.4	90.8	48.7	70.1	91.2	43.7	67.9	88.1	56.0	72.3	92.8	65.5	79.4
04	MASONRY	90.5	41.4	59.9	96.9	50.9	68.3	100.1	39.0	62.0	87.6	33.5	53.9	91.9	41.9	60.7	96.8	50.9	68.2
05	METALS	99.5	67.8	89.7	101.8	70.6	92.2	100.6	66.5	90.1	103.8	66.9	92.5	95.9	80.0	91.0	102.4	69.4	92.0
06	WOOD, PLASTICS & COMPOSITES	85.8	35.5	57.6	107.6	59.0	80.4	96.0	45.5	67.7	87.5	35.5	58.5	93.9	48.3	68.4	99.9	62.5	78.9
07	THERMAL & MOISTURE PROTECTION	99.0	45.3	77.0	99.4	57.5	82.2	98.5	42.6	75.6	94.4	37.9	71.2	103.3	49.1	81.1	95.7	57.8	80.2
08	OPENINGS	97.7	44.7	85.4	98.5	56.2	88.7	98.5	46.5	86.4	102.1	42.3	88.2	106.0	54.0	89.7	99.5	58.4	89.9
0920	Plaster & Gypsum Board	81.8	34.0	49.5	89.4	58.3	68.4	88.3	44.3	58.6	84.0	34.2	50.4	97.7	46.9	63.4	94.6	61.8	72.5
0950, 0980	Ceilings & Acoustic Treatment	83.2	34.0	50.9	84.8	58.3	67.4	84.8	44.3	58.2	83.3	34.2	51.0	86.2	46.9	60.3	88.0	61.8	70.8
0960	Flooring	94.0	59.1	84.0	102.9	60.8	90.9	98.8	59.1	87.5	98.5	59.1	87.2	68.9	53.4	64.5	101.0	60.8	89.5
0970, 0990	Wall Finishes & Painting/Coating	107.4	29.4	60.3	107.4	53.8	75.1	107.4	41.2	67.4	104.4	55.2	74.7	94.9	48.4	66.8	108.8	55.2	76.5
09	FINISHES	85.7	40.0	60.5	89.9	57.6	72.1	88.6	46.8	65.6	88.4	41.0	62.3	83.7	48.5	64.3	92.7	59.8	74.6
COVERS	DIVS. 10 - 14, 25, 28, 41, 43, 44, 46	100.0	48.7	89.6	100.0	78.8	95.7	100.0	49.1	89.7	100.0	36.5	87.2	100.0	45.2	88.9	100.0	79.2	95.8
21, 22, 23	FIRE SUPPRESSION, PLUMBING & HVAC	95.3	50.2	77.1	100.1	50.6	80.1	95.3	49.4	76.8	95.3	48.7	76.5	100.3	52.5	81.0	99.9	56.2	82.3
26, 27, 3370	ELECTRICAL, COMMUNICATIONS & UTIL.	90.1	51.2	69.5	93.7	64.5	78.3	95.0	37.1	64.4	99.0	67.8	82.5	100.2	63.5	80.8	100.7	70.2	84.6
MF2014	WEIGHTED AVERAGE	93.4	52.0	75.4	97.2	61.6	81.7	95.9	50.4	76.0	96.4	51.8	76.9	96.7	59.1	80.3	98.5	64.0	83.4

City Cost Indexes

		ARKANSAS													CALIFORNIA					
	DIVISION	PINE BLUFF			RUSSELLVILLE			TEXARKANA			WEST MEMPHIS			ALHAMBRA			ANAHEIM		
		716			728			718			723			917 - 918			928		
		MAT.	INST.	TOTAL	MAT.	INST.	TOTAL	MAT.	INST.	TOTAL	MAT.	INST.	TOTAL	MAT.	INST.	TOTAL	MAT.	INST.	TOTAL
015433	CONTRACTOR EQUIPMENT		89.1	89.1		89.1	89.1		89.9	89.9		108.3	108.3		100.1	100.1		100.8	100.8
0241, 31 - 34	SITE & INFRASTRUCTURE, DEMOLITION	83.7	86.8	85.9	78.3	84.9	83.0	94.9	87.1	89.3	110.5	101.1	103.8	97.9	110.9	107.1	99.4	108.3	105.7
0310	Concrete Forming & Accessories	80.7	60.0	62.8	86.6	54.7	59.1	87.3	40.1	46.6	95.5	48.5	54.9	117.1	116.4	116.5	105.4	124.2	121.6
0320	Concrete Reinforcing	93.5	68.2	80.7	88.0	67.3	77.5	93.1	67.5	80.1	84.5	70.1	77.2	106.6	113.3	110.0	93.8	113.2	103.7
0330	Cast-in-Place Concrete	84.5	70.3	78.7	81.1	45.3	66.4	92.1	43.3	72.0	88.4	55.1	74.7	95.1	120.8	105.6	92.9	123.3	105.4
03	CONCRETE	92.1	65.5	79.0	86.7	54.4	70.8	89.8	47.2	68.9	95.5	56.2	76.2	101.1	116.5	108.7	101.8	120.8	111.2
04	MASONRY	125.0	50.9	78.8	96.1	37.3	59.5	102.4	33.1	59.2	79.4	41.9	56.0	122.7	121.1	121.7	80.0	118.8	104.2
05	METALS	104.6	69.4	93.8	99.5	66.5	89.3	96.8	66.9	87.6	95.0	80.3	90.5	85.3	100.9	90.1	102.7	101.1	102.2
06	WOOD, PLASTICS & COMPOSITES	87.0	62.5	73.3	91.3	59.0	73.2	95.5	42.0	65.5	100.1	48.3	71.1	100.0	112.1	106.7	101.9	122.6	113.5
07	THERMAL & MOISTURE PROTECTION	94.5	57.8	79.5	99.2	43.9	76.5	95.2	44.5	74.4	103.7	49.1	81.3	95.1	118.1	104.5	99.4	122.0	108.7
08	OPENINGS	102.1	58.4	91.9	97.7	55.7	87.9	107.3	46.5	93.2	100.6	54.0	89.7	91.8	114.8	97.2	104.2	120.6	108.0
0920	Plaster & Gypsum Board	83.6	61.8	68.9	83.3	58.3	66.4	86.6	40.7	55.6	100.2	46.9	64.2	97.4	112.5	107.6	107.1	123.2	118.0
0950, 0980	Ceilings & Acoustic Treatment	83.3	61.8	69.2	83.2	58.3	66.8	86.6	40.7	56.4	84.3	46.9	59.7	100.2	112.5	108.3	105.0	123.2	117.0
0960	Flooring	98.2	60.8	87.5	96.2	59.1	85.6	100.4	51.0	86.3	71.0	53.4	66.0	95.4	110.2	99.6	101.9	112.0	104.8
0970, 0990	Wall Finishes & Painting/Coating	104.4	55.2	74.7	107.4	35.0	63.7	104.4	30.3	59.7	94.9	50.4	68.0	100.5	111.5	107.1	98.9	106.8	103.7
09	FINISHES	88.3	59.8	72.6	86.7	54.1	68.7	90.7	40.4	62.9	85.0	48.8	65.0	99.8	114.0	107.6	100.5	120.1	111.3
COVERS	DIVS. 10 - 14, 25, 28, 41, 43, 44, 46	100.0	79.2	95.8	100.0	41.0	88.1	100.0	32.6	86.4	100.0	45.2	88.9	100.0	110.5	102.1	100.0	112.0	102.4
21, 22, 23	FIRE SUPPRESSION, PLUMBING & HVAC	100.1	52.9	81.1	95.3	51.5	77.6	100.1	53.5	81.3	95.6	65.8	83.6	95.2	115.6	103.4	100.0	118.6	107.5
26, 27, 3370	ELECTRICAL, COMMUNICATIONS & UTIL.	97.0	70.2	82.8	93.7	44.7	67.8	98.9	36.0	65.7	101.9	65.7	82.8	118.4	120.5	119.5	92.2	105.7	99.3
MF2014	WEIGHTED AVERAGE	99.4	63.3	83.6	94.6	53.7	76.8	98.2	49.2	76.8	96.1	62.4	81.4	98.3	114.8	105.5	99.4	114.9	106.2

		CALIFORNIA																	
	DIVISION	BAKERSFIELD			BERKELEY			EUREKA			FRESNO			INGLEWOOD			LONG BEACH		
		932 - 933			947			955			936 - 938			903 - 905			906 - 908		
		MAT.	INST.	TOTAL	MAT.	INST.	TOTAL	MAT.	INST.	TOTAL	MAT.	INST.	TOTAL	MAT.	INST.	TOTAL	MAT.	INST.	TOTAL
015433	CONTRACTOR EQUIPMENT		98.8	98.8		100.3	100.3		98.5	98.5		98.8	98.8		96.5	96.5		96.5	96.5
0241, 31 - 34	SITE & INFRASTRUCTURE, DEMOLITION	100.4	106.0	104.4	118.4	107.6	110.7	110.8	103.9	105.9	103.1	105.4	104.7	89.3	103.5	99.4	95.9	103.5	101.3
0310	Concrete Forming & Accessories	104.0	124.3	121.5	115.9	149.9	145.2	114.8	133.6	131.0	104.3	134.2	130.1	111.8	115.4	114.9	106.5	115.4	114.1
0320	Concrete Reinforcing	103.4	113.2	108.4	94.5	114.4	104.6	102.5	114.1	108.4	80.6	113.7	97.5	102.7	113.3	108.1	101.9	113.3	107.7
0330	Cast-in-Place Concrete	94.0	122.6	105.7	123.4	124.8	124.0	100.8	118.0	107.9	99.4	118.9	107.4	84.5	121.4	99.7	96.2	121.4	106.5
03	CONCRETE	97.9	120.6	109.1	107.3	132.7	119.8	112.8	123.2	117.9	98.6	123.8	111.0	90.9	116.3	103.4	99.8	116.3	107.9
04	MASONRY	100.2	118.3	111.5	113.4	134.3	126.5	104.0	132.6	121.8	103.6	122.0	115.1	76.3	121.2	104.3	85.3	121.2	107.7
05	METALS	105.3	100.9	104.0	106.9	100.5	104.9	102.6	98.9	101.4	105.6	100.5	104.1	94.4	101.5	96.6	94.3	101.5	96.5
06	WOOD, PLASTICS & COMPOSITES	96.0	122.7	111.0	111.4	154.9	135.8	116.6	138.5	128.9	108.1	138.5	125.1	103.8	110.7	107.6	96.9	110.7	104.6
07	THERMAL & MOISTURE PROTECTION	98.1	116.1	105.5	105.2	137.2	118.3	103.4	119.8	110.1	88.7	116.0	99.9	98.3	118.0	106.4	98.5	118.0	106.5
08	OPENINGS	97.8	116.1	102.0	93.0	140.6	104.1	103.5	113.5	105.8	98.1	124.7	104.3	88.8	114.1	94.7	88.8	114.1	94.7
0920	Plaster & Gypsum Board	101.9	123.2	116.3	111.6	156.0	141.6	113.4	139.5	131.0	100.6	139.5	126.9	103.8	111.0	108.7	99.9	111.0	107.4
0950, 0980	Ceilings & Acoustic Treatment	96.3	123.2	114.0	106.2	156.0	139.0	110.1	139.5	129.4	92.8	139.5	123.5	100.3	111.0	107.4	100.3	111.0	107.4
0960	Flooring	106.8	110.2	107.7	108.2	127.6	113.8	106.0	116.5	109.0	113.3	133.5	119.1	104.4	110.2	106.0	101.8	110.2	104.2
0970, 0990	Wall Finishes & Painting/Coating	111.4	102.0	105.7	102.5	143.9	127.5	100.5	46.2	67.7	129.0	107.2	115.8	100.9	111.5	107.3	100.9	111.5	107.3
09	FINISHES	100.0	120.5	111.3	105.4	147.2	128.4	106.1	123.6	115.8	102.1	133.2	119.2	102.7	113.2	108.5	101.8	113.2	108.1
COVERS	DIVS. 10 - 14, 25, 28, 41, 43, 44, 46	100.0	111.9	102.4	100.0	126.4	105.3	100.0	122.2	104.5	100.0	122.2	104.5	100.0	110.4	102.1	100.0	110.4	102.1
21, 22, 23	FIRE SUPPRESSION, PLUMBING & HVAC	100.1	117.1	107.0	95.3	147.8	116.5	95.2	110.6	101.4	100.2	111.3	104.7	94.8	115.6	103.2	94.8	115.6	103.2
26, 27, 3370	ELECTRICAL, COMMUNICATIONS & UTIL.	103.6	102.5	103.1	109.4	145.0	128.2	99.1	116.7	108.4	93.8	100.8	97.5	103.8	120.5	112.6	103.5	120.5	112.5
MF2014	WEIGHTED AVERAGE	100.7	113.6	106.3	102.7	135.1	116.8	102.1	116.4	108.4	100.1	115.3	106.7	94.8	114.1	103.2	96.2	114.1	104.0

		CALIFORNIA																	
	DIVISION	LOS ANGELES			MARYSVILLE			MODESTO			MOJAVE			OAKLAND			OXNARD		
		900 - 902			959			953			935			946			930		
		MAT.	INST.	TOTAL	MAT.	INST.	TOTAL	MAT.	INST.	TOTAL	MAT.	INST.	TOTAL	MAT.	INST.	TOTAL	MAT.	INST.	TOTAL
015433	CONTRACTOR EQUIPMENT		100.1	100.1		98.5	98.5		98.5	98.5		98.8	98.8		100.3	100.3		97.7	97.7
0241, 31 - 34	SITE & INFRASTRUCTURE, DEMOLITION	96.6	107.1	104.1	107.2	105.0	105.7	102.2	105.1	104.3	96.4	106.0	103.2	124.7	107.6	112.5	103.3	104.0	103.8
0310	Concrete Forming & Accessories	108.6	124.3	122.2	104.6	134.4	130.3	100.7	134.3	129.7	116.1	112.1	112.6	104.9	149.9	143.7	107.5	124.3	122.0
0320	Concrete Reinforcing	103.6	113.4	108.6	102.3	113.8	108.2	106.2	113.8	110.0	97.3	113.1	105.3	96.7	114.4	105.7	95.5	113.1	104.4
0330	Cast-in-Place Concrete	91.2	122.0	103.9	112.6	119.2	115.3	100.8	119.2	108.4	86.0	122.4	101.0	117.0	124.8	120.2	100.1	122.7	109.4
03	CONCRETE	95.9	120.6	108.0	113.9	124.0	118.9	104.7	124.0	114.2	91.0	115.1	102.8	106.8	132.7	119.5	99.2	120.6	109.7
04	MASONRY	90.5	122.6	110.5	104.9	123.1	116.2	102.9	123.1	115.5	101.9	118.3	112.1	120.6	134.3	129.2	104.8	116.5	112.1
05	METALS	101.1	102.9	101.7	102.0	101.8	102.0	99.1	101.7	99.9	102.8	100.3	102.0	101.4	100.4	101.1	100.7	100.7	100.7
06	WOOD, PLASTICS & COMPOSITES	103.0	122.6	113.9	103.0	138.5	122.9	98.5	138.5	120.9	107.4	106.8	107.1	98.8	154.9	130.2	102.1	122.7	113.6
07	THERMAL & MOISTURE PROTECTION	97.6	121.2	107.3	102.9	121.5	110.5	102.5	118.7	109.1	94.4	110.9	101.2	103.1	137.2	117.0	97.4	119.8	106.6
08	OPENINGS	95.4	120.7	101.3	102.8	125.3	108.0	101.6	125.4	107.1	93.2	107.5	96.5	93.1	140.6	104.1	95.7	120.6	101.5
0920	Plaster & Gypsum Board	103.6	123.2	116.9	105.0	139.5	128.3	107.5	139.5	129.1	110.8	106.8	108.1	105.7	156.0	139.7	104.7	123.2	117.2
0950, 0980	Ceilings & Acoustic Treatment	110.5	123.2	118.8	109.2	139.5	129.1	105.0	139.5	127.7	94.2	106.8	102.5	108.9	156.0	139.9	96.6	123.2	114.1
0960	Flooring	102.0	112.0	104.9	102.0	117.6	106.4	102.4	112.8	105.3	112.6	110.2	111.9	103.4	127.6	110.3	104.8	112.0	106.9
0970, 0990	Wall Finishes & Painting/Coating	100.0	111.5	106.9	100.5	115.8	109.7	100.5	117.0	110.5	111.1	102.0	105.6	102.5	143.9	127.5	111.1	101.1	105.1
09	FINISHES	104.2	120.6	113.3	103.1	131.5	118.8	102.2	130.9	118.0	101.8	109.9	106.2	104.3	147.2	127.9	99.4	119.6	110.5
COVERS	DIVS. 10 - 14, 25, 28, 41, 43, 44, 46	100.0	111.9	102.4	100.0	122.2	104.5	100.0	122.2	104.5	100.0	107.8	101.6	100.0	126.3	105.3	100.0	112.2	102.5
21, 22, 23	FIRE SUPPRESSION, PLUMBING & HVAC	99.9	118.6	107.5	95.2	109.0	100.8	100.0	111.3	104.6	95.3	115.1	103.3	100.1	147.8	119.4	100.1	118.6	107.6
26, 27, 3370	ELECTRICAL, COMMUNICATIONS & UTIL.	102.0	121.3	112.2	95.7	107.4	101.9	98.2	104.0	101.3	92.0	99.9	96.2	108.5	139.3	124.8	97.7	107.3	102.8
MF2014	WEIGHTED AVERAGE	99.2	117.5	107.2	101.3	115.9	107.7	100.8	115.7	107.3	96.8	109.8	102.5	103.1	134.3	116.7	99.6	114.4	106.0

761

City Cost Indexes

		CALIFORNIA																	
	DIVISION	PALM SPRINGS 922			PALO ALTO 943			PASADENA 910 - 912			REDDING 960			RICHMOND 948			RIVERSIDE 925		
		MAT.	INST.	TOTAL	MAT.	INST.	TOTAL	MAT.	INST.	TOTAL	MAT.	INST.	TOTAL	MAT.	INST.	TOTAL	MAT.	INST.	TOTAL
015433	CONTRACTOR EQUIPMENT		99.6	99.6		100.3	100.3		100.1	100.1		98.5	98.5		100.3	100.3		99.6	99.6
0241, 31 - 34	SITE & INFRASTRUCTURE, DEMOLITION	90.9	106.3	101.8	114.4	107.6	109.6	94.9	110.9	106.3	125.1	105.0	110.8	123.8	107.6	112.3	97.9	106.3	103.9
0310	Concrete Forming & Accessories	101.9	115.3	113.4	103.0	146.3	140.3	106.2	115.3	114.1	107.8	142.7	137.9	118.7	149.7	145.5	105.8	124.2	121.7
0320	Concrete Reinforcing	107.8	113.1	110.5	94.5	114.4	104.6	107.5	113.3	110.5	118.7	113.8	116.2	94.5	114.3	104.6	104.7	113.1	109.0
0330	Cast-in-Place Concrete	88.6	123.2	102.8	104.4	124.8	112.8	90.2	120.8	102.8	118.0	119.2	118.5	120.0	124.8	122.0	96.3	123.3	107.4
03	CONCRETE	96.6	116.8	106.5	96.5	131.1	113.5	96.6	116.0	106.2	122.7	127.7	125.2	108.8	132.6	120.5	102.6	120.8	111.6
04	MASONRY	77.6	118.5	103.1	97.4	134.3	120.4	106.5	121.1	115.6	130.0	123.1	125.7	113.2	132.8	125.4	78.6	118.1	103.2
05	METALS	103.3	100.8	102.5	98.9	100.3	99.3	85.3	100.8	90.1	101.6	101.8	101.7	98.9	100.2	99.3	102.8	101.0	102.3
06	WOOD, PLASTICS & COMPOSITES	96.6	110.8	104.5	96.1	150.2	126.4	86.5	110.6	100.0	110.6	149.8	132.6	115.3	154.9	137.5	101.9	122.6	113.5
07	THERMAL & MOISTURE PROTECTION	99.1	118.7	107.2	102.7	137.5	117.0	94.9	117.4	104.1	121.6	122.1	121.8	103.3	135.9	116.7	99.6	121.1	108.4
08	OPENINGS	100.1	114.1	103.4	93.1	136.9	103.3	91.8	114.0	97.0	114.1	131.5	118.1	93.1	139.4	103.9	102.8	120.6	107.0
0920	Plaster & Gypsum Board	102.0	111.0	108.1	103.9	151.1	135.9	90.2	111.0	104.3	106.2	151.1	136.6	113.6	156.0	142.3	106.3	123.2	117.7
0950, 0980	Ceilings & Acoustic Treatment	101.7	111.0	107.8	107.1	151.1	136.0	100.2	111.0	107.3	131.5	151.1	144.4	107.1	156.0	139.3	109.2	123.2	118.4
0960	Flooring	104.6	107.2	105.4	102.4	127.6	109.6	90.8	110.2	96.3	98.0	117.6	103.6	109.9	127.6	115.0	106.0	112.0	107.7
0970, 0990	Wall Finishes & Painting/Coating	97.2	111.3	105.7	102.5	143.9	127.5	100.5	111.5	107.1	115.9	115.8	115.8	102.5	143.9	127.5	97.2	106.8	103.0
09	FINISHES	99.1	112.8	106.6	102.7	144.4	125.6	97.3	113.2	106.0	109.3	138.2	125.3	106.9	147.2	129.1	102.1	120.1	112.0
COVERS	DIVS. 10 - 14, 25, 28, 41, 43, 44, 46	100.0	110.7	102.2	100.0	125.9	105.2	100.0	110.3	102.1	100.0	123.4	104.7	100.0	126.3	105.3	100.0	112.0	102.4
21, 22, 23	FIRE SUPPRESSION, PLUMBING & HVAC	95.2	115.5	103.4	95.3	145.5	115.6	95.2	115.5	103.4	100.2	109.0	103.7	95.3	147.8	116.5	100.0	118.6	107.5
26, 27, 3370	ELECTRICAL, COMMUNICATIONS & UTIL.	95.3	105.4	100.7	108.4	150.2	130.5	115.1	120.5	118.0	99.2	107.4	103.5	109.0	132.7	121.5	91.9	105.4	99.1
MF2014	WEIGHTED AVERAGE	97.2	112.0	103.6	98.9	134.5	114.4	96.4	114.6	104.3	107.8	117.7	112.1	101.8	133.1	115.4	99.4	114.6	106.0

		CALIFORNIA																	
	DIVISION	SACRAMENTO 942,956 - 958			SALINAS 939			SAN BERNARDINO 923 - 924			SAN DIEGO 919 - 921			SAN FRANCISCO 940 - 941			SAN JOSE 951		
		MAT.	INST.	TOTAL	MAT.	INST.	TOTAL	MAT.	INST.	TOTAL	MAT.	INST.	TOTAL	MAT.	INST.	TOTAL	MAT.	INST.	TOTAL
015433	CONTRACTOR EQUIPMENT		99.9	99.9		98.8	98.8		99.6	99.6		100.1	100.1		110.7	110.7		99.3	99.3
0241, 31 - 34	SITE & INFRASTRUCTURE, DEMOLITION	100.6	113.3	109.6	116.3	105.7	108.8	77.7	106.3	98.0	101.8	104.0	103.4	127.0	113.6	117.5	133.5	100.4	109.9
0310	Concrete Forming & Accessories	103.3	137.0	132.4	111.0	137.4	133.7	109.6	115.2	114.4	105.6	113.3	112.2	104.6	150.9	144.6	107.0	149.8	143.9
0320	Concrete Reinforcing	89.4	113.8	101.8	96.0	114.2	105.2	104.7	113.0	108.9	104.2	113.1	108.7	110.1	115.0	112.6	93.5	114.5	104.1
0330	Cast-in-Place Concrete	99.0	120.3	107.7	98.9	119.6	107.4	66.6	123.2	89.8	98.9	107.8	102.6	120.0	126.4	122.6	117.1	124.2	120.0
03	CONCRETE	97.0	125.2	110.9	108.5	125.6	116.9	76.8	116.8	96.4	101.8	110.7	106.2	110.3	134.4	122.1	110.8	132.9	121.7
04	MASONRY	98.4	123.1	113.8	101.6	128.1	118.1	84.9	115.9	104.2	97.4	115.6	108.7	121.1	141.5	133.8	131.4	134.4	133.3
05	METALS	97.0	96.1	96.7	105.5	102.9	104.7	102.8	100.5	102.1	101.1	101.4	101.2	107.6	110.8	108.6	97.5	107.5	100.6
06	WOOD, PLASTICS & COMPOSITES	92.9	141.7	120.2	107.3	141.5	126.4	105.8	110.8	108.6	100.6	110.2	106.0	98.8	155.1	130.3	111.3	154.7	135.6
07	THERMAL & MOISTURE PROTECTION	110.8	121.3	115.1	95.0	126.1	107.7	98.3	117.7	106.3	99.3	107.1	102.5	104.9	141.9	120.1	98.9	139.3	115.4
08	OPENINGS	106.3	127.2	111.1	96.9	133.3	105.4	100.2	114.1	103.4	99.1	111.1	102.1	97.3	140.7	107.4	92.8	140.5	103.9
0920	Plaster & Gypsum Board	101.0	142.6	129.1	105.8	142.6	130.6	108.2	111.0	110.1	97.8	110.3	106.3	108.3	156.0	140.6	102.7	156.0	138.7
0950, 0980	Ceilings & Acoustic Treatment	107.1	142.6	130.4	94.2	142.6	126.0	105.0	111.0	109.0	107.7	110.3	109.4	117.1	156.0	142.7	103.3	156.0	138.0
0960	Flooring	102.6	117.6	106.9	107.1	121.1	111.1	107.8	110.2	108.5	98.5	112.0	102.4	103.4	127.6	110.3	95.4	127.6	104.6
0970, 0990	Wall Finishes & Painting/Coating	100.2	116.7	110.1	112.3	143.9	131.4	97.2	104.0	101.3	97.1	111.5	105.8	102.5	153.2	133.1	100.8	143.9	126.8
09	FINISHES	101.4	133.6	119.1	101.4	137.0	121.0	100.5	112.5	107.1	102.9	112.9	108.4	106.5	148.3	129.6	100.7	147.0	126.2
COVERS	DIVS. 10 - 14, 25, 28, 41, 43, 44, 46	100.0	123.0	104.6	100.0	122.6	104.6	100.0	108.1	101.6	100.0	109.7	102.0	100.0	126.8	105.4	100.0	125.9	105.2
21, 22, 23	FIRE SUPPRESSION, PLUMBING & HVAC	100.0	120.8	108.4	95.3	117.7	104.4	95.2	115.6	103.4	99.9	116.9	106.8	100.1	175.0	130.3	100.0	147.3	119.1
26, 27, 3370	ELECTRICAL, COMMUNICATIONS & UTIL.	103.5	108.7	106.2	93.2	120.9	107.8	95.3	103.4	99.6	101.2	98.5	99.8	108.6	157.8	134.6	101.1	155.9	130.1
MF2014	WEIGHTED AVERAGE	100.5	119.3	108.7	100.2	121.6	109.5	95.1	111.3	102.1	100.5	109.6	104.5	105.2	145.4	122.7	102.5	136.6	117.4

		CALIFORNIA																	
	DIVISION	SAN LUIS OBISPO 934			SAN MATEO 944			SAN RAFAEL 949			SANTA ANA 926 - 927			SANTA BARBARA 931			SANTA CRUZ 950		
		MAT.	INST.	TOTAL	MAT.	INST.	TOTAL	MAT.	INST.	TOTAL	MAT.	INST.	TOTAL	MAT.	INST.	TOTAL	MAT.	INST.	TOTAL
015433	CONTRACTOR EQUIPMENT		98.8	98.8		100.3	100.3		100.3	100.3		99.6	99.6		98.8	98.8		99.3	99.3
0241, 31 - 34	SITE & INFRASTRUCTURE, DEMOLITION	108.4	106.0	106.7	121.3	107.6	111.6	113.2	113.4	113.4	89.4	106.3	101.4	103.3	106.0	105.2	133.1	100.0	109.6
0310	Concrete Forming & Accessories	117.9	115.4	115.8	109.0	149.8	144.2	112.9	149.9	144.9	109.9	115.3	114.5	108.3	124.2	122.1	107.0	137.6	133.4
0320	Concrete Reinforcing	97.3	113.1	105.4	94.5	114.6	104.7	95.1	114.6	105.1	108.3	113.1	110.8	95.5	113.2	104.5	115.7	114.2	114.9
0330	Cast-in-Place Concrete	106.3	122.5	112.9	116.2	124.8	119.7	135.2	124.1	130.7	85.0	123.2	100.7	99.8	122.6	109.1	116.3	121.4	118.4
03	CONCRETE	106.9	116.6	111.7	105.4	132.7	118.8	126.0	132.4	129.1	94.2	116.8	105.3	99.0	120.6	109.6	113.9	126.5	120.1
04	MASONRY	103.5	118.3	112.7	112.9	137.3	128.1	93.2	137.4	120.7	74.6	118.8	102.1	102.2	119.1	112.7	135.3	128.2	130.9
05	METALS	103.5	100.6	102.6	98.7	100.6	99.3	100.1	98.8	99.7	102.9	100.8	102.2	101.2	100.9	101.1	104.5	106.3	105.0
06	WOOD, PLASTICS & COMPOSITES	109.9	110.9	110.5	104.0	154.9	132.5	101.8	154.7	131.4	107.8	110.9	109.5	102.1	122.7	113.6	111.3	141.6	128.3
07	THERMAL & MOISTURE PROTECTION	95.1	116.9	104.0	103.1	138.8	117.7	107.2	138.9	120.2	99.4	118.3	107.2	94.6	118.9	104.5	98.8	129.2	111.3
08	OPENINGS	95.1	109.7	98.5	93.1	139.4	103.8	103.8	139.3	112.1	99.5	114.1	102.9	96.6	126.0	102.4	94.0	133.4	103.2
0920	Plaster & Gypsum Board	112.2	110.0	111.4	108.9	156.0	140.8	111.5	156.0	141.6	109.6	111.0	110.6	104.7	123.2	117.2	110.1	142.6	132.0
0950, 0980	Ceilings & Acoustic Treatment	94.2	111.0	105.3	107.1	156.0	139.3	115.4	156.0	142.1	105.0	111.0	109.0	96.6	123.2	114.1	106.7	142.6	130.3
0960	Flooring	113.5	110.2	112.6	105.1	127.6	111.5	115.0	121.1	116.8	108.3	110.2	108.8	106.2	108.6	106.9	99.6	121.1	105.7
0970, 0990	Wall Finishes & Painting/Coating	111.1	101.7	105.5	102.5	143.9	127.5	98.9	142.6	125.3	97.2	106.8	103.0	111.1	101.1	105.1	101.0	143.9	126.9
09	FINISHES	103.2	112.3	108.2	104.7	147.2	128.1	107.6	145.8	128.6	102.0	112.8	107.9	100.0	119.0	110.5	103.7	137.1	122.1
COVERS	DIVS. 10 - 14, 25, 28, 41, 43, 44, 46	100.0	120.6	104.2	100.0	126.4	105.3	100.0	125.7	105.2	100.0	110.7	102.2	100.0	112.2	102.5	100.0	122.9	104.6
21, 22, 23	FIRE SUPPRESSION, PLUMBING & HVAC	95.3	115.6	103.5	95.3	142.2	114.2	95.3	169.5	125.2	95.2	114.3	102.9	100.1	118.6	107.6	100.0	117.8	107.2
26, 27, 3370	ELECTRICAL, COMMUNICATIONS & UTIL.	92.0	106.4	99.6	108.4	148.0	129.3	105.4	120.0	113.1	95.4	105.7	100.8	91.0	110.5	101.3	100.3	120.9	111.2
MF2014	WEIGHTED AVERAGE	99.4	112.0	104.9	100.9	134.6	115.6	103.5	136.7	118.0	96.9	111.8	103.4	98.9	115.1	106.0	104.5	121.8	112.0

City Cost Indexes

DIVISION		CALIFORNIA																	COLORADO		
		SANTA ROSA 954			STOCKTON 952			SUSANVILLE 961			VALLEJO 945			VAN NUYS 913 - 916			ALAMOSA 811		
		MAT.	INST.	TOTAL	MAT.	INST.	TOTAL	MAT.	INST.	TOTAL	MAT.	INST.	TOTAL	MAT.	INST.	TOTAL	MAT.	INST.	TOTAL
015433	CONTRACTOR EQUIPMENT		99.0	99.0		98.5	98.5		98.5	98.5		100.3	100.3		100.1	100.1		93.8	93.8
0241, 31 - 34	SITE & INFRASTRUCTURE, DEMOLITION	103.0	105.1	104.5	102.0	105.1	104.2	131.8	105.0	112.8	100.5	113.2	109.5	111.4	110.9	111.0	134.7	88.0	101.5
0310	Concrete Forming & Accessories	102.9	148.1	141.9	104.8	136.6	132.2	109.1	142.8	138.1	103.4	147.7	141.6	112.7	115.3	115.0	105.0	67.5	72.7
0320	Concrete Reinforcing	103.4	114.8	109.2	104.1	114.7	109.5	118.7	113.2	115.9	96.3	114.5	105.6	107.5	113.3	110.5	105.1	76.4	90.5
0330	Cast-in-Place Concrete	110.6	120.8	114.8	98.2	119.2	106.9	107.4	119.3	112.3	107.7	121.5	113.4	95.1	120.8	105.7	101.1	78.4	91.8
03	CONCRETE	113.7	130.9	122.1	103.7	125.0	114.2	125.2	127.7	126.4	102.2	130.5	116.1	110.2	116.0	113.0	113.5	73.3	93.8
04	MASONRY	103.1	137.2	124.3	102.8	123.1	115.5	127.9	112.7	118.4	71.6	133.1	109.9	122.7	121.1	121.7	126.0	71.9	92.3
05	METALS	103.2	105.0	103.8	99.3	101.8	100.0	100.6	101.1	100.7	100.1	98.1	99.5	84.4	100.8	89.5	98.2	80.2	92.7
06	WOOD, PLASTICS & COMPOSITES	98.0	154.5	129.6	104.3	141.5	125.1	112.5	149.8	133.4	90.6	154.7	126.5	95.0	110.6	103.7	98.0	67.5	80.9
07	THERMAL & MOISTURE PROTECTION	99.8	136.2	114.7	102.6	119.9	109.7	122.2	120.4	121.4	105.0	135.3	117.4	95.7	117.4	104.6	104.9	78.1	93.9
08	OPENINGS	101.0	140.4	110.2	101.6	127.1	107.5	113.9	131.5	118.0	105.6	140.5	113.7	91.6	114.0	96.8	98.3	74.1	92.7
0920	Plaster & Gypsum Board	103.9	156.0	139.2	107.5	142.6	131.2	107.1	151.1	136.9	105.8	156.0	139.8	95.3	111.0	105.9	79.0	66.3	70.4
0950, 0980	Ceilings & Acoustic Treatment	105.0	156.0	138.6	112.5	142.6	132.3	124.1	151.1	141.9	117.3	156.0	142.8	97.7	111.0	106.5	96.3	66.3	76.6
0960	Flooring	105.0	115.2	107.9	102.4	112.8	105.3	98.4	122.1	105.2	110.6	127.6	115.4	93.2	110.2	98.0	113.5	54.8	96.7
0970, 0990	Wall Finishes & Painting/Coating	97.2	142.6	124.6	100.5	117.0	110.5	115.9	115.8	115.8	99.8	142.6	125.6	100.5	111.5	107.1	114.1	39.8	69.3
09	FINISHES	101.1	143.6	124.5	103.8	132.7	119.7	108.9	139.0	125.5	104.6	145.8	127.3	99.3	113.2	107.0	102.5	61.7	80.0
COVERS	DIVS. 10 - 14, 25, 28, 41, 43, 44, 46	100.0	123.8	104.8	100.0	122.5	104.6	100.0	123.5	104.7	100.0	124.3	104.9	100.0	110.3	102.1	100.0	88.5	97.7
21, 22, 23	FIRE SUPPRESSION, PLUMBING & HVAC	95.2	167.3	124.3	100.0	111.3	104.6	95.4	109.0	100.9	100.1	127.5	111.1	95.2	115.6	103.4	95.2	72.0	85.9
26, 27, 3370	ELECTRICAL, COMMUNICATIONS & UTIL.	95.7	114.3	105.5	98.2	110.6	104.8	99.6	118.4	109.5	100.9	122.4	112.3	115.1	120.5	118.0	99.1	72.5	85.0
MF2014	WEIGHTED AVERAGE	100.8	134.7	115.6	100.9	117.2	108.0	106.8	118.1	111.7	100.1	127.2	111.9	99.1	114.6	105.9	102.2	73.7	89.8

DIVISION		COLORADO																			
		BOULDER 803			COLORADO SPRINGS 808 - 809			DENVER 800 - 802			DURANGO 813			FORT COLLINS 805			FORT MORGAN 807		
		MAT.	INST.	TOTAL	MAT.	INST.	TOTAL	MAT.	INST.	TOTAL	MAT.	INST.	TOTAL	MAT.	INST.	TOTAL	MAT.	INST.	TOTAL
015433	CONTRACTOR EQUIPMENT		97.5	97.5		95.8	95.8		100.4	100.4		93.8	93.8		97.5	97.5		97.5	97.5
0241, 31 - 34	SITE & INFRASTRUCTURE, DEMOLITION	94.8	96.4	95.9	96.9	94.7	95.3	96.1	102.6	100.7	128.5	88.0	99.7	107.1	96.0	99.2	97.7	95.9	96.4
0310	Concrete Forming & Accessories	103.1	79.9	83.1	93.6	79.4	81.3	100.0	76.0	79.3	111.4	67.7	73.7	100.7	74.7	78.3	103.6	74.9	78.9
0320	Concrete Reinforcing	98.9	76.6	87.5	98.1	80.2	89.0	98.1	80.2	89.0	105.1	76.5	90.5	99.0	76.6	87.6	99.1	76.5	87.6
0330	Cast-in-Place Concrete	101.8	80.0	92.9	104.6	88.3	97.9	99.0	82.8	92.4	116.3	78.5	100.8	114.9	78.9	100.1	99.9	78.9	91.3
03	CONCRETE	103.1	79.6	91.5	106.3	82.9	94.8	100.7	79.5	90.3	116.1	73.4	95.1	113.5	76.7	95.4	101.5	76.8	89.4
04	MASONRY	91.0	72.1	79.2	91.3	81.9	85.4	93.3	72.2	80.1	113.6	71.9	87.6	108.5	75.7	88.1	104.9	72.2	84.5
05	METALS	96.0	83.5	92.2	99.1	85.7	95.0	101.6	85.6	96.7	98.2	80.4	92.7	97.2	80.4	92.0	95.7	80.3	91.0
06	WOOD, PLASTICS & COMPOSITES	102.2	83.0	91.4	91.7	77.3	83.6	99.8	77.0	87.2	107.6	67.5	85.1	99.6	77.2	87.0	102.2	77.2	88.2
07	THERMAL & MOISTURE PROTECTION	103.1	81.7	94.3	103.9	85.6	96.4	102.4	75.8	91.5	104.9	78.1	93.9	103.5	73.5	91.2	103.0	81.2	94.1
08	OPENINGS	98.2	82.5	94.5	102.4	80.5	97.3	103.0	80.4	97.8	105.3	74.1	98.0	98.2	79.4	93.8	98.1	79.4	93.8
0920	Plaster & Gypsum Board	105.1	82.7	89.9	87.9	76.7	80.3	100.7	76.8	84.5	92.5	66.7	74.8	99.1	76.7	84.0	105.1	76.7	85.9
0950, 0980	Ceilings & Acoustic Treatment	96.9	82.7	87.5	104.4	76.7	86.2	107.3	76.8	87.2	96.3	66.3	76.6	96.9	76.7	83.6	96.9	76.7	83.6
0960	Flooring	99.7	83.8	95.2	92.0	68.2	85.2	96.7	84.8	93.3	118.2	54.8	100.0	96.6	54.8	84.6	100.1	54.8	87.1
0970, 0990	Wall Finishes & Painting/Coating	101.7	66.6	80.5	101.4	40.5	64.7	101.7	75.7	86.0	114.1	39.8	69.3	101.7	40.2	64.6	101.7	53.6	72.6
09	FINISHES	101.7	79.4	89.4	98.8	72.8	84.5	102.4	77.3	88.6	105.0	61.7	81.1	100.5	67.4	82.3	101.8	68.3	83.7
COVERS	DIVS. 10 - 14, 25, 28, 41, 43, 44, 46	100.0	89.4	97.9	100.0	92.2	98.4	100.0	88.8	97.7	100.0	88.4	97.7	100.0	88.8	97.7	100.0	88.8	97.7
21, 22, 23	FIRE SUPPRESSION, PLUMBING & HVAC	95.3	77.3	88.1	100.2	88.2	95.4	100.0	79.9	91.9	95.2	83.1	90.3	100.1	77.2	90.9	95.3	77.2	88.0
26, 27, 3370	ELECTRICAL, COMMUNICATIONS & UTIL.	97.4	84.5	90.6	100.8	82.1	90.9	102.6	83.2	92.4	98.5	69.4	83.1	97.4	84.5	90.6	97.8	84.5	90.7
MF2014	WEIGHTED AVERAGE	97.8	81.3	90.6	100.4	84.0	93.2	100.8	81.7	92.5	102.7	75.7	90.9	101.3	79.0	91.6	98.3	79.0	89.9

DIVISION		COLORADO																			
		GLENWOOD SPRINGS 816			GOLDEN 804			GRAND JUNCTION 815			GREELEY 806			MONTROSE 814			PUEBLO 810		
		MAT.	INST.	TOTAL	MAT.	INST.	TOTAL	MAT.	INST.	TOTAL	MAT.	INST.	TOTAL	MAT.	INST.	TOTAL	MAT.	INST.	TOTAL
015433	CONTRACTOR EQUIPMENT		96.7	96.7		97.5	97.5		96.7	96.7		97.5	97.5		95.2	95.2		93.8	93.8
0241, 31 - 34	SITE & INFRASTRUCTURE, DEMOLITION	143.2	95.1	109.0	107.3	96.2	99.4	128.4	94.7	104.5	94.0	95.4	95.0	137.1	91.2	104.5	120.6	91.0	99.6
0310	Concrete Forming & Accessories	102.0	75.0	78.7	96.1	74.8	77.7	110.2	74.4	79.3	98.6	78.4	81.2	101.4	74.8	78.4	107.3	79.4	83.3
0320	Concrete Reinforcing	104.0	76.5	90.0	99.1	76.4	87.6	104.3	76.3	90.1	98.9	75.1	86.8	103.9	76.4	89.9	100.6	80.2	90.2
0330	Cast-in-Place Concrete	101.1	77.9	91.6	100.0	78.9	91.3	111.9	77.1	97.6	96.1	59.1	80.9	101.1	77.8	91.5	100.4	87.8	95.2
03	CONCRETE	118.5	76.5	97.9	111.9	76.7	94.6	112.5	75.9	94.5	98.3	71.3	85.0	109.7	76.3	93.3	102.7	82.7	92.9
04	MASONRY	98.9	72.1	82.2	107.7	71.8	85.3	132.6	71.5	94.5	102.6	49.0	69.2	106.2	71.9	84.9	95.4	81.7	86.9
05	METALS	97.9	80.9	92.6	95.9	80.1	91.0	99.5	79.2	93.2	97.2	77.5	91.1	97.1	80.0	91.8	101.1	86.2	96.5
06	WOOD, PLASTICS & COMPOSITES	93.2	77.3	84.3	94.3	77.2	84.7	105.3	77.3	89.6	96.8	83.0	89.1	94.3	77.4	84.8	100.9	77.6	87.8
07	THERMAL & MOISTURE PROTECTION	104.8	79.2	94.3	103.9	75.4	92.2	104.0	68.5	89.5	102.9	66.4	87.9	105.0	79.2	94.4	103.5	83.6	95.3
08	OPENINGS	104.2	79.4	98.5	98.2	78.8	93.7	105.0	78.8	98.9	98.1	82.4	94.5	105.5	79.5	99.4	100.2	80.6	95.7
0920	Plaster & Gypsum Board	117.8	76.7	90.0	96.9	76.7	83.3	131.0	76.7	94.3	97.6	82.7	87.5	78.3	76.7	77.2	83.9	76.7	79.1
0950, 0980	Ceilings & Acoustic Treatment	95.5	76.7	83.2	96.9	76.7	83.6	95.5	76.7	83.2	96.9	82.7	87.5	96.3	76.7	83.4	104.7	76.7	86.3
0960	Flooring	112.8	50.4	95.0	94.7	54.8	83.3	117.5	54.8	99.6	95.7	54.8	84.0	115.8	45.4	95.7	114.6	84.8	106.1
0970, 0990	Wall Finishes & Painting/Coating	114.1	66.5	85.4	101.7	66.6	80.5	114.1	66.6	85.4	101.7	25.0	55.4	114.1	39.8	69.3	114.1	37.5	67.9
09	FINISHES	107.9	69.6	86.8	100.4	71.0	84.2	109.3	70.4	87.9	99.3	69.2	82.7	103.2	65.7	82.5	103.7	75.6	88.2
COVERS	DIVS. 10 - 14, 25, 28, 41, 43, 44, 46	100.0	88.8	97.7	100.0	88.8	97.7	100.0	88.8	97.7	100.0	89.4	97.9	100.0	89.1	97.8	100.0	92.7	98.5
21, 22, 23	FIRE SUPPRESSION, PLUMBING & HVAC	95.2	83.0	90.3	95.3	76.7	87.8	99.9	82.5	92.9	100.1	77.2	90.8	95.2	83.0	90.3	99.9	76.8	90.6
26, 27, 3370	ELECTRICAL, COMMUNICATIONS & UTIL.	95.7	69.4	81.8	97.8	84.5	90.7	98.2	53.8	74.7	97.4	84.5	90.6	98.2	56.4	76.1	99.1	73.0	85.3
MF2014	WEIGHTED AVERAGE	102.3	78.1	91.8	99.8	78.9	90.7	104.8	75.3	91.9	98.9	75.5	88.7	101.4	75.4	90.1	101.1	80.3	92.0

City Cost Indexes

		COLORADO			CONNECTICUT															
	DIVISION	SALIDA 812			BRIDGEPORT 066			BRISTOL 060			HARTFORD 061			MERIDEN 064			NEW BRITAIN 060		
		MAT.	INST.	TOTAL	MAT.	INST.	TOTAL	MAT.	INST.	TOTAL	MAT.	INST.	TOTAL	MAT.	INST.	TOTAL	MAT.	INST.	TOTAL
015433	CONTRACTOR EQUIPMENT		95.2	95.2		100.9	100.9		100.9	100.9		100.9	100.9		101.3	101.3		100.9	100.9
0241, 31 - 34	SITE & INFRASTRUCTURE, DEMOLITION	128.2	91.5	102.1	109.8	105.1	106.4	108.9	105.0	106.2	104.4	105.0	104.9	106.7	105.8	106.0	109.1	105.0	106.2
0310	Concrete Forming & Accessories	110.1	74.7	79.6	98.9	124.5	121.0	98.9	124.3	120.8	95.7	124.3	120.4	98.6	124.3	120.8	99.2	124.3	120.9
0320	Concrete Reinforcing	103.7	76.4	89.2	105.1	127.3	116.4	105.1	127.3	116.4	104.6	127.3	116.2	105.1	127.3	116.4	105.1	127.3	116.4
0330	Cast-in-Place Concrete	115.8	77.8	100.2	107.9	127.1	115.8	101.1	127.1	111.8	103.0	127.1	112.9	97.3	127.1	109.5	102.8	127.1	112.7
03	CONCRETE	111.2	76.3	94.1	106.6	125.7	116.0	103.4	125.6	114.3	104.0	125.6	114.6	101.6	125.6	113.4	104.2	125.6	114.7
04	MASONRY	133.9	71.9	95.3	111.8	134.0	125.6	102.9	134.0	122.3	107.7	134.0	124.1	102.5	134.0	122.1	105.2	134.0	123.1
05	METALS	96.7	79.9	91.6	100.5	125.5	108.2	100.5	125.4	108.1	105.6	125.4	111.7	97.4	125.4	106.0	96.5	125.4	105.4
06	WOOD, PLASTICS & COMPOSITES	101.9	77.4	88.2	98.8	123.4	112.6	98.8	123.4	112.6	92.3	123.4	109.7	98.8	123.4	112.6	98.8	123.4	112.6
07	THERMAL & MOISTURE PROTECTION	104.0	79.2	93.8	98.6	129.1	111.1	98.7	126.0	109.9	102.8	126.0	112.3	98.7	126.0	109.9	98.7	126.0	109.9
08	OPENINGS	98.4	79.5	94.0	101.2	132.1	108.4	101.2	132.1	108.4	100.3	132.1	107.7	103.6	132.1	110.2	101.2	132.1	108.4
0920	Plaster & Gypsum Board	78.6	76.7	77.3	101.7	123.5	116.4	101.7	123.5	116.4	97.4	123.5	115.0	103.2	123.5	116.9	101.7	123.5	116.4
0950, 0980	Ceilings & Acoustic Treatment	96.3	76.7	83.4	86.4	123.5	110.8	86.4	123.5	110.8	89.7	123.5	111.9	90.4	123.5	112.2	86.4	123.5	110.8
0960	Flooring	120.6	45.4	99.1	101.9	131.2	110.3	101.9	131.2	110.3	100.5	131.2	109.2	101.9	131.2	110.3	101.9	131.2	110.3
0970, 0990	Wall Finishes & Painting/Coating	114.1	41.7	70.4	104.9	122.9	115.8	104.9	122.9	115.8	105.9	122.9	116.2	104.9	122.9	115.8	104.9	122.9	115.8
09	FINISHES	103.6	65.9	82.8	97.2	125.0	112.6	97.3	125.0	112.6	96.3	125.0	112.1	98.4	125.0	113.1	97.3	125.0	112.6
COVERS	DIVS. 10 - 14, 25, 28, 41, 43, 44, 46	100.0	89.2	97.8	100.0	112.4	102.5	100.0	112.4	102.5	100.0	112.4	102.5	100.0	112.4	102.5	100.0	112.4	102.5
21, 22, 23	FIRE SUPPRESSION, PLUMBING & HVAC	95.2	72.0	85.8	100.0	116.4	106.6	100.0	116.4	106.6	100.0	116.4	106.6	95.2	116.4	103.8	100.0	116.4	106.6
26, 27, 3370	ELECTRICAL, COMMUNICATIONS & UTIL.	98.4	72.5	84.7	98.5	112.2	105.7	98.5	111.8	105.6	97.9	112.8	105.8	98.4	111.8	105.5	98.6	111.8	105.6
MF2014	WEIGHTED AVERAGE	101.9	75.3	90.3	101.3	120.8	109.8	100.5	120.6	109.3	101.4	120.8	109.8	98.9	120.7	108.4	100.1	120.6	109.1

		CONNECTICUT																	
	DIVISION	NEW HAVEN 065			NEW LONDON 063			NORWALK 068			STAMFORD 069			WATERBURY 067			WILLIMANTIC 062		
		MAT.	INST.	TOTAL	MAT.	INST.	TOTAL	MAT.	INST.	TOTAL	MAT.	INST.	TOTAL	MAT.	INST.	TOTAL	MAT.	INST.	TOTAL
015433	CONTRACTOR EQUIPMENT		101.3	101.3		101.3	101.3		100.9	100.9		100.9	100.9		100.9	100.9		100.9	100.9
0241, 31 - 34	SITE & INFRASTRUCTURE, DEMOLITION	108.9	105.8	106.7	101.1	105.8	104.4	109.5	105.1	106.4	110.2	105.1	106.6	109.5	105.0	106.3	109.5	105.0	106.3
0310	Concrete Forming & Accessories	98.6	124.3	120.8	98.6	124.3	120.8	98.9	124.9	121.3	98.9	124.9	121.3	98.9	124.3	120.8	98.9	124.1	120.6
0320	Concrete Reinforcing	105.1	127.3	116.4	82.4	127.3	105.2	105.1	127.4	116.5	105.1	127.4	116.5	105.1	127.3	116.4	105.1	127.2	116.4
0330	Cast-in-Place Concrete	104.5	127.1	113.8	89.1	127.1	104.7	106.2	128.6	115.4	107.9	128.6	116.4	107.9	127.1	115.8	100.8	125.8	111.0
03	CONCRETE	118.7	125.6	122.1	91.4	125.6	108.2	105.5	126.4	115.9	106.6	126.4	116.3	106.6	125.6	115.9	103.3	125.0	114.0
04	MASONRY	103.2	134.0	122.4	101.5	134.0	121.7	102.7	135.4	123.1	103.4	135.4	123.3	103.4	134.0	122.5	102.7	134.0	122.2
05	METALS	96.8	125.4	105.6	96.5	125.4	105.4	100.5	126.1	108.3	100.5	126.1	108.3	100.5	125.4	108.1	100.2	125.2	107.9
06	WOOD, PLASTICS & COMPOSITES	98.8	123.4	112.6	98.8	123.4	112.6	98.8	123.4	112.6	98.8	123.4	112.6	98.8	123.4	112.6	98.8	123.4	112.6
07	THERMAL & MOISTURE PROTECTION	98.8	126.1	110.0	98.7	126.0	109.9	98.8	129.6	111.4	98.7	129.6	111.4	98.7	126.1	110.0	98.9	124.6	109.5
08	OPENINGS	101.2	132.1	108.4	104.0	132.1	110.6	101.2	132.1	108.4	101.2	132.1	108.4	101.2	132.1	108.4	104.0	132.1	110.6
0920	Plaster & Gypsum Board	101.7	123.5	116.4	101.7	123.5	116.4	101.7	123.5	116.4	101.7	123.5	116.4	101.7	123.5	116.4	101.7	123.5	116.4
0950, 0980	Ceilings & Acoustic Treatment	86.4	123.5	110.8	84.5	123.5	110.2	86.4	123.5	110.8	86.4	123.5	110.8	86.4	123.5	110.8	84.5	123.5	110.2
0960	Flooring	101.9	131.2	110.3	101.9	131.2	110.3	101.9	131.2	110.3	101.9	131.2	110.3	101.9	131.2	110.3	101.9	133.2	110.9
0970, 0990	Wall Finishes & Painting/Coating	104.9	122.9	115.8	104.9	122.9	115.8	104.9	122.9	115.8	104.9	122.9	115.8	104.9	122.9	115.8	104.9	122.9	115.8
09	FINISHES	97.3	125.0	112.6	96.4	125.0	112.2	97.3	125.0	112.6	97.4	125.0	112.6	97.2	125.0	112.5	97.0	125.4	112.6
COVERS	DIVS. 10 - 14, 25, 28, 41, 43, 44, 46	100.0	112.4	102.5	100.0	112.4	102.5	100.0	112.6	102.6	100.0	112.6	102.6	100.0	112.4	102.5	100.0	112.4	102.5
21, 22, 23	FIRE SUPPRESSION, PLUMBING & HVAC	100.0	116.4	106.6	95.2	116.4	103.8	100.0	116.4	106.6	100.0	116.4	106.6	100.0	116.4	106.6	100.0	116.1	106.5
26, 27, 3370	ELECTRICAL, COMMUNICATIONS & UTIL.	98.4	111.8	105.5	95.3	111.8	104.0	98.5	167.2	134.8	98.5	167.2	134.8	98.0	112.2	105.5	98.5	109.6	104.4
MF2014	WEIGHTED AVERAGE	101.6	120.7	109.9	97.1	120.7	107.4	100.8	128.7	113.0	100.9	128.7	113.0	100.0	120.7	109.5	100.8	120.2	109.2

		D.C.			DELAWARE									FLORIDA					
	DIVISION	WASHINGTON 200 - 205			DOVER 199			NEWARK 197			WILMINGTON 198			DAYTONA BEACH 321			FORT LAUDERDALE 333		
		MAT.	INST.	TOTAL	MAT.	INST.	TOTAL	MAT.	INST.	TOTAL	MAT.	INST.	TOTAL	MAT.	INST.	TOTAL	MAT.	INST.	TOTAL
015433	CONTRACTOR EQUIPMENT		105.5	105.5		118.0	118.0		118.0	118.0		118.2	118.2		98.2	98.2		91.0	91.0
0241, 31 - 34	SITE & INFRASTRUCTURE, DEMOLITION	103.8	94.5	97.2	102.3	113.3	110.1	102.6	113.3	110.2	99.5	113.6	109.5	105.9	89.2	94.0	96.4	77.1	82.7
0310	Concrete Forming & Accessories	97.4	78.3	80.9	96.3	102.2	101.4	97.0	102.2	101.5	98.1	102.2	101.6	97.6	68.1	72.1	95.7	68.4	72.1
0320	Concrete Reinforcing	105.4	93.8	99.5	94.0	102.9	98.5	91.5	102.9	97.3	93.4	102.9	98.2	91.2	76.6	83.8	88.2	72.1	80.0
0330	Cast-in-Place Concrete	115.4	88.0	104.2	104.6	104.6	104.6	90.0	104.6	96.0	99.2	104.6	101.4	90.2	70.8	82.2	94.6	77.1	87.4
03	CONCRETE	107.7	85.9	97.0	102.2	104.0	103.1	94.6	104.0	99.2	99.6	104.0	101.8	91.6	71.8	81.9	94.5	73.3	84.0
04	MASONRY	98.2	80.0	86.9	107.5	98.1	101.6	103.2	98.1	100.0	108.5	98.1	102.0	99.1	65.7	78.3	102.8	68.3	81.3
05	METALS	99.9	108.6	102.6	102.0	116.7	106.5	103.5	116.7	107.5	102.0	116.7	106.5	104.7	91.1	100.5	102.4	89.6	98.2
06	WOOD, PLASTICS & COMPOSITES	95.3	76.2	84.6	94.0	101.9	98.4	95.8	101.9	99.2	92.0	101.9	97.6	96.3	69.2	81.1	84.3	67.2	74.7
07	THERMAL & MOISTURE PROTECTION	101.7	85.8	95.2	98.8	110.0	103.4	101.7	110.0	105.1	97.9	110.0	102.9	96.2	73.6	86.9	100.3	77.2	90.8
08	OPENINGS	99.9	87.9	97.1	91.3	110.9	95.9	91.6	110.9	96.1	89.9	110.9	94.8	98.6	67.4	91.3	97.5	65.5	90.1
0920	Plaster & Gypsum Board	108.0	75.5	86.0	97.1	101.8	100.3	99.5	101.8	101.1	101.7	101.8	101.8	95.6	68.7	77.4	105.4	66.6	79.2
0950, 0980	Ceilings & Acoustic Treatment	112.2	75.5	88.1	91.3	101.8	98.2	88.7	101.8	97.3	92.4	101.8	98.6	84.9	68.7	74.2	87.9	66.6	73.9
0960	Flooring	104.7	93.6	101.5	99.8	109.7	102.7	96.8	109.7	100.5	104.3	109.7	105.9	110.5	73.6	100.0	105.1	69.3	94.9
0970, 0990	Wall Finishes & Painting/Coating	108.0	82.2	92.4	97.2	106.4	102.8	97.3	106.4	102.8	95.0	106.4	101.9	104.5	74.1	86.2	98.9	70.2	81.6
09	FINISHES	100.8	80.5	89.6	96.2	103.6	100.3	95.0	103.6	99.7	98.7	103.6	101.4	96.9	69.3	81.7	94.9	68.1	80.1
COVERS	DIVS. 10 - 14, 25, 28, 41, 43, 44, 46	100.0	98.6	99.7	100.0	88.0	97.6	100.0	88.0	97.6	100.0	88.0	97.6	100.0	84.4	96.9	100.0	87.4	97.5
21, 22, 23	FIRE SUPPRESSION, PLUMBING & HVAC	100.1	92.1	96.9	100.0	117.4	107.0	100.1	117.4	107.1	100.1	117.4	107.1	99.9	76.1	90.3	100.0	66.5	86.5
26, 27, 3370	ELECTRICAL, COMMUNICATIONS & UTIL.	101.5	105.8	103.8	96.6	111.9	104.7	98.7	111.9	105.7	99.8	111.9	106.2	95.8	54.9	74.2	96.0	72.7	83.7
MF2014	WEIGHTED AVERAGE	101.1	91.9	97.1	99.3	109.2	103.6	98.8	109.2	103.3	99.3	109.2	103.6	99.0	72.9	87.6	98.6	72.6	87.2

City Cost Indexes

DIVISION		FLORIDA																	
		FORT MYERS			GAINESVILLE			JACKSONVILLE			LAKELAND			MELBOURNE			MIAMI		
		339,341			326,344			320,322			338			329			330 - 332,340		
		MAT.	INST.	TOTAL	MAT.	INST.	TOTAL	MAT.	INST.	TOTAL	MAT.	INST.	TOTAL	MAT.	INST.	TOTAL	MAT.	INST.	TOTAL
015433	CONTRACTOR EQUIPMENT		98.2	98.2		98.2	98.2		98.2	98.2		98.2	98.2		98.2	98.2		91.0	91.0
0241, 31 - 34	SITE & INFRASTRUCTURE, DEMOLITION	107.3	88.3	93.8	113.9	88.3	95.7	106.0	88.6	93.6	109.2	88.7	94.6	112.7	88.5	95.5	99.5	76.9	83.4
0310	Concrete Forming & Accessories	91.5	74.8	77.1	92.7	54.3	59.6	97.4	54.7	60.5	88.0	75.3	77.1	93.8	69.8	73.1	100.7	68.7	73.1
0320	Concrete Reinforcing	89.3	92.7	91.0	96.8	64.8	80.5	91.2	64.9	77.8	91.5	93.6	92.5	92.3	76.6	84.3	94.3	72.2	83.0
0330	Cast-in-Place Concrete	98.7	68.2	86.2	103.6	62.6	86.7	91.1	68.3	81.8	100.9	69.6	88.1	108.7	73.1	94.1	95.5	78.1	88.3
03	CONCRETE	95.1	76.9	86.2	102.6	60.7	82.0	92.1	62.8	77.7	96.8	77.8	87.5	103.0	73.3	88.4	96.1	73.7	85.1
04	MASONRY	95.8	61.9	74.7	114.0	61.2	81.1	99.1	61.2	75.5	114.1	75.8	90.2	97.5	69.8	80.2	103.1	71.9	83.7
05	METALS	104.2	96.8	102.0	103.6	85.0	97.9	103.2	85.4	97.7	104.1	97.8	102.2	113.4	91.3	106.6	102.4	88.5	98.1
06	WOOD, PLASTICS & COMPOSITES	81.1	77.4	79.0	89.9	51.8	68.5	96.3	51.8	71.4	76.5	77.4	77.0	91.5	69.2	79.0	90.6	67.2	77.5
07	THERMAL & MOISTURE PROTECTION	100.1	80.0	91.9	96.5	61.7	82.2	96.4	62.3	82.4	100.1	84.6	93.7	96.6	76.0	88.2	101.6	71.6	89.3
08	OPENINGS	98.4	74.4	92.8	96.9	52.7	86.6	98.6	55.8	88.6	98.4	75.1	93.0	97.8	70.7	91.5	99.7	65.5	91.7
0920	Plaster & Gypsum Board	101.0	77.1	84.9	91.3	50.7	63.9	95.6	50.7	65.3	97.4	77.1	83.7	91.3	68.7	76.0	103.2	66.6	78.5
0950, 0980	Ceilings & Acoustic Treatment	82.7	77.1	79.0	79.5	50.7	60.6	84.9	50.7	62.5	82.7	77.1	79.0	83.3	68.7	73.7	92.2	66.6	75.4
0960	Flooring	102.3	55.1	88.8	108.1	43.1	89.5	110.5	64.5	97.4	100.3	56.5	87.8	108.3	73.6	98.4	107.3	73.0	97.5
0970, 0990	Wall Finishes & Painting/Coating	103.7	67.6	81.9	104.5	67.6	82.3	104.5	67.6	82.3	103.7	67.6	81.9	104.5	91.9	96.9	96.0	70.2	80.5
09	FINISHES	94.4	70.7	81.3	95.3	52.5	71.7	97.0	56.7	74.8	93.5	71.0	81.1	95.9	72.3	82.9	97.1	69.2	81.7
COVERS	DIVS. 10 - 14, 25, 28, 41, 43, 44, 46	100.0	72.6	94.5	100.0	82.8	96.5	100.0	81.0	96.2	100.0	72.6	94.5	100.0	85.8	97.1	100.0	88.0	97.6
21, 22, 23	FIRE SUPPRESSION, PLUMBING & HVAC	97.4	64.3	84.1	98.8	64.1	84.8	99.9	64.2	85.5	97.4	80.8	90.7	99.9	78.2	91.2	100.0	66.4	86.4
26, 27, 3370	ELECTRICAL, COMMUNICATIONS & UTIL.	98.1	62.8	79.5	96.1	71.8	83.3	95.5	62.2	77.9	96.3	61.6	77.9	97.0	67.8	81.6	100.1	74.9	86.8
MF2014	WEIGHTED AVERAGE	98.6	72.6	87.3	100.3	66.7	85.7	98.8	66.4	84.7	99.4	77.8	90.0	101.7	76.3	90.6	99.8	73.1	88.2

DIVISION		FLORIDA																	
		ORLANDO			PANAMA CITY			PENSACOLA			SARASOTA			ST. PETERSBURG			TALLAHASSEE		
		327 - 328,347			324			325			342			337			323		
		MAT.	INST.	TOTAL	MAT.	INST.	TOTAL	MAT.	INST.	TOTAL	MAT.	INST.	TOTAL	MAT.	INST.	TOTAL	MAT.	INST.	TOTAL
015433	CONTRACTOR EQUIPMENT		98.2	98.2		98.2	98.2		98.2	98.2		98.2	98.2		98.2	98.2		98.2	98.2
0241, 31 - 34	SITE & INFRASTRUCTURE, DEMOLITION	107.6	88.4	93.9	117.6	87.4	96.1	117.7	87.9	96.5	114.0	88.4	95.8	110.9	88.1	94.7	105.1	87.7	92.7
0310	Concrete Forming & Accessories	101.4	72.0	76.1	96.7	43.9	51.1	94.6	51.8	57.7	96.0	75.0	77.9	94.9	51.0	57.1	99.9	44.1	51.8
0320	Concrete Reinforcing	96.6	73.9	85.0	95.3	72.5	83.7	97.7	73.0	85.1	92.5	93.5	93.0	91.5	86.1	88.7	98.3	64.7	81.2
0330	Cast-in-Place Concrete	112.1	70.6	95.1	95.7	56.7	79.7	118.2	65.9	96.8	106.6	69.5	91.4	102.0	64.5	86.6	97.2	56.4	80.5
03	CONCRETE	103.4	73.0	88.4	100.8	55.4	78.5	110.7	62.2	86.9	100.2	77.5	89.1	98.5	63.8	81.4	97.3	54.0	76.0
04	MASONRY	100.6	65.7	78.8	103.9	47.3	68.6	124.7	54.5	80.9	99.8	75.8	84.8	157.9	49.0	90.0	103.6	53.8	72.6
05	METALS	102.4	89.6	98.5	104.4	86.5	98.9	105.6	87.9	100.2	105.2	97.5	102.8	105.0	92.9	101.3	102.2	84.6	96.8
06	WOOD, PLASTICS & COMPOSITES	95.8	75.2	84.2	95.1	41.8	65.2	92.7	51.5	69.7	95.7	77.4	85.4	85.5	50.1	65.7	95.0	41.3	64.9
07	THERMAL & MOISTURE PROTECTION	94.9	74.3	86.4	96.7	56.6	80.2	96.6	62.3	82.5	98.1	84.6	92.5	100.3	57.8	82.8	102.6	71.5	89.9
08	OPENINGS	101.4	69.2	93.9	96.5	46.5	84.9	96.5	56.9	87.3	99.7	74.2	93.8	98.4	60.9	89.6	100.2	47.3	87.9
0920	Plaster & Gypsum Board	99.7	74.9	82.9	94.5	40.4	57.9	97.4	50.5	65.7	97.8	77.1	83.8	103.5	49.0	66.7	108.1	40.0	62.0
0950, 0980	Ceilings & Acoustic Treatment	91.4	74.9	80.5	83.3	40.4	55.1	83.3	50.5	61.7	86.2	77.1	80.3	84.5	49.0	61.2	94.1	40.0	58.5
0960	Flooring	104.1	73.6	95.4	110.1	43.0	90.9	106.1	63.0	93.7	111.1	57.9	95.9	104.1	52.2	90.1	111.1	62.0	97.0
0970, 0990	Wall Finishes & Painting/Coating	103.4	70.2	83.4	104.5	64.9	80.6	104.5	67.6	82.3	109.5	67.6	84.2	103.7	60.8	77.8	99.5	67.6	80.3
09	FINISHES	98.2	72.4	84.0	97.5	44.8	68.5	96.4	54.8	73.5	99.9	71.2	84.1	95.9	51.7	71.6	100.7	48.5	72.0
COVERS	DIVS. 10 - 14, 25, 28, 41, 43, 44, 46	100.0	85.1	97.0	100.0	46.2	89.1	100.0	46.3	89.1	100.0	72.6	94.5	100.0	55.8	91.1	100.0	65.8	93.1
21, 22, 23	FIRE SUPPRESSION, PLUMBING & HVAC	99.9	56.6	82.4	99.9	52.3	80.7	99.9	52.5	80.8	99.9	64.8	85.7	100.0	58.4	83.2	100.0	38.9	75.4
26, 27, 3370	ELECTRICAL, COMMUNICATIONS & UTIL.	97.6	60.0	77.7	94.5	59.3	75.9	98.5	55.8	75.9	97.2	61.6	78.4	96.3	61.6	77.9	103.3	60.0	80.4
MF2014	WEIGHTED AVERAGE	100.5	70.0	87.2	100.2	57.8	81.7	102.7	61.2	84.6	100.8	74.3	89.2	102.6	63.3	85.5	100.8	56.9	81.7

DIVISION		FLORIDA						GEORGIA											
		TAMPA			WEST PALM BEACH			ALBANY			ATHENS			ATLANTA			AUGUSTA		
		335 - 336,346			334,349			317,398			306			300 - 303,399			308 - 309		
		MAT.	INST.	TOTAL	MAT.	INST.	TOTAL	MAT.	INST.	TOTAL	MAT.	INST.	TOTAL	MAT.	INST.	TOTAL	MAT.	INST.	TOTAL
015433	CONTRACTOR EQUIPMENT		98.2	98.2		91.0	91.0		91.9	91.9		94.2	94.2		94.7	94.7		94.2	94.2
0241, 31 - 34	SITE & INFRASTRUCTURE, DEMOLITION	111.4	88.6	95.2	93.2	77.1	81.8	98.8	78.7	84.5	100.9	93.4	95.6	97.6	94.8	95.7	94.5	93.2	93.6
0310	Concrete Forming & Accessories	97.7	75.6	78.6	99.2	68.1	72.4	90.4	43.6	50.0	92.9	45.9	52.4	96.7	73.5	76.7	94.2	65.5	69.4
0320	Concrete Reinforcing	88.2	93.6	91.0	90.7	71.8	81.1	90.6	80.2	85.3	95.3	77.9	86.5	94.6	81.2	87.8	95.7	71.8	83.6
0330	Cast-in-Place Concrete	99.7	69.7	87.4	90.0	74.3	83.5	92.4	54.5	76.8	107.8	55.6	86.4	107.8	70.8	92.6	101.9	49.9	80.5
03	CONCRETE	97.0	77.7	87.6	91.3	72.1	81.9	93.1	55.8	74.8	104.8	56.0	80.8	102.2	74.1	88.4	97.5	61.7	79.9
04	MASONRY	103.9	75.8	86.4	102.3	66.6	80.0	102.5	49.1	69.2	81.5	53.8	64.2	95.2	66.2	77.1	95.5	43.3	63.0
05	METALS	104.0	98.1	102.2	101.1	89.3	97.4	105.0	85.5	99.0	91.9	73.3	86.2	92.8	77.0	88.0	91.6	71.6	85.5
06	WOOD, PLASTICS & COMPOSITES	89.3	77.4	82.6	89.2	67.2	76.9	85.0	38.0	58.7	91.8	40.9	63.3	96.1	75.8	84.7	93.3	70.9	80.8
07	THERMAL & MOISTURE PROTECTION	100.5	84.6	94.0	100.1	70.2	87.8	95.9	60.3	81.3	94.1	52.8	77.2	94.0	72.0	85.0	93.7	57.0	78.7
08	OPENINGS	99.7	79.2	94.9	96.8	65.5	89.5	91.1	44.2	80.2	89.5	45.7	79.3	94.7	71.9	89.4	89.5	62.2	83.1
0920	Plaster & Gypsum Board	106.1	77.1	86.5	110.2	66.6	80.7	97.5	36.5	56.3	98.7	39.4	58.7	100.9	75.4	83.6	99.8	70.4	79.9
0950, 0980	Ceilings & Acoustic Treatment	87.9	77.1	80.8	82.7	66.6	72.1	84.0	36.5	52.8	97.2	39.4	59.2	97.2	75.4	82.9	98.1	70.4	79.9
0960	Flooring	105.1	56.5	91.2	106.8	66.4	95.3	110.5	47.8	92.9	96.4	53.9	84.2	97.7	65.2	88.4	96.6	46.6	82.3
0970, 0990	Wall Finishes & Painting/Coating	103.7	67.6	81.9	98.9	70.2	81.6	105.3	52.8	73.6	106.0	46.0	69.8	106.0	85.1	93.4	106.0	46.0	69.8
09	FINISHES	97.4	71.0	82.8	94.7	67.5	79.7	98.1	43.2	67.9	95.4	45.7	68.0	95.7	73.4	83.4	95.2	60.5	76.1
COVERS	DIVS. 10 - 14, 25, 28, 41, 43, 44, 46	100.0	85.1	97.0	100.0	87.4	97.5	100.0	80.2	96.0	100.0	77.3	95.4	100.0	85.6	97.1	100.0	78.5	95.7
21, 22, 23	FIRE SUPPRESSION, PLUMBING & HVAC	100.0	80.8	92.2	97.4	62.7	83.4	99.9	68.0	87.0	95.2	69.6	84.9	99.9	70.6	88.1	100.0	68.1	84.4
26, 27, 3370	ELECTRICAL, COMMUNICATIONS & UTIL.	96.0	61.6	77.8	97.2	72.7	84.3	96.9	58.7	76.7	99.7	69.5	83.8	99.0	71.9	84.7	100.4	61.8	80.0
MF2014	WEIGHTED AVERAGE	100.1	78.4	90.6	97.4	71.1	85.9	98.5	61.3	82.3	95.5	63.9	81.7	97.6	74.4	87.5	96.3	63.6	82.1

City Cost Indexes

		GEORGIA																	
	DIVISION	COLUMBUS 318 - 319			DALTON 307			GAINESVILLE 305			MACON 310 - 312			SAVANNAH 313 - 314			STATESBORO 304		
		MAT.	INST.	TOTAL	MAT.	INST.	TOTAL	MAT.	INST.	TOTAL	MAT.	INST.	TOTAL	MAT.	INST.	TOTAL	MAT.	INST.	TOTAL
015433	CONTRACTOR EQUIPMENT		91.9	91.9		106.1	106.1		94.2	94.2		102.3	102.3		92.8	92.8		93.4	93.4
0241, 31 - 34	SITE & INFRASTRUCTURE, DEMOLITION	98.7	78.7	84.5	101.1	97.8	98.8	100.8	93.3	95.5	99.8	93.5	95.3	100.6	80.1	86.1	102.0	77.5	84.6
0310	Concrete Forming & Accessories	90.3	54.3	59.3	85.7	46.7	52.1	96.4	42.9	50.2	89.9	51.9	57.1	92.1	50.0	55.8	80.3	51.4	55.4
0320	Concrete Reinforcing	90.9	80.6	85.7	94.8	73.9	84.2	95.1	77.8	86.3	92.1	80.3	86.1	98.1	71.9	84.7	94.4	41.8	67.6
0330	Cast-in-Place Concrete	92.1	53.9	76.4	104.7	50.2	82.3	113.3	52.9	88.5	90.9	65.6	80.5	100.2	55.2	81.7	107.6	59.7	87.9
03	CONCRETE	93.0	60.4	77.0	103.9	54.6	79.7	106.7	53.7	80.6	92.6	63.3	78.2	98.0	57.3	78.0	104.2	54.0	79.6
04	MASONRY	102.6	55.2	73.1	83.8	36.7	54.4	89.5	54.2	67.5	115.9	44.5	71.4	98.8	50.9	69.0	84.7	40.6	57.2
05	METALS	104.5	86.3	98.9	92.8	82.8	89.7	91.2	72.5	85.4	100.1	85.9	95.7	101.3	82.4	95.5	96.3	72.6	89.0
06	WOOD, PLASTICS & COMPOSITES	85.0	52.1	66.6	75.2	47.5	59.7	95.9	37.9	63.4	91.8	51.2	69.0	96.0	45.9	68.0	68.5	53.9	60.3
07	THERMAL & MOISTURE PROTECTION	95.9	63.5	82.6	96.1	51.5	77.8	94.1	54.8	78.0	94.4	62.2	81.2	95.0	56.6	79.2	94.7	51.1	76.8
08	OPENINGS	91.1	59.0	83.7	90.1	49.5	80.6	89.5	40.0	77.9	90.0	52.8	81.3	94.9	48.4	84.1	91.1	40.6	79.4
0920	Plaster & Gypsum Board	97.5	51.1	66.1	86.2	46.2	59.2	100.9	36.3	57.3	103.0	50.1	67.2	95.6	44.7	61.2	86.9	52.8	63.9
0950, 0980	Ceilings & Acoustic Treatment	84.0	51.1	62.4	109.3	46.2	67.8	97.2	36.3	57.2	79.2	50.1	60.1	91.5	44.7	60.8	105.7	52.8	71.0
0960	Flooring	110.9	50.4	93.6	96.8	46.9	82.5	97.6	46.6	83.0	87.9	47.8	76.5	108.6	46.4	90.8	115.6	44.9	95.4
0970, 0990	Wall Finishes & Painting/Coating	105.3	66.5	81.9	96.3	59.8	74.2	106.0	46.0	69.8	107.7	52.8	74.6	103.8	58.3	76.3	103.5	39.2	64.7
09	FINISHES	98.0	53.5	73.5	103.0	47.1	72.2	95.9	41.6	66.0	86.7	49.9	66.4	98.2	48.9	71.0	106.9	49.0	75.0
COVERS	DIVS. 10 - 14, 25, 28, 41, 43, 44, 46	100.0	81.8	96.3	100.0	22.8	84.4	100.0	39.4	87.8	100.0	80.2	96.0	100.0	78.0	95.6	100.0	43.5	88.6
21, 22, 23	FIRE SUPPRESSION, PLUMBING & HVAC	99.9	63.4	85.2	95.2	57.0	79.8	95.2	69.5	84.8	99.9	66.7	86.6	100.0	61.8	84.6	95.7	56.7	80.0
26, 27, 3370	ELECTRICAL, COMMUNICATIONS & UTIL.	97.1	69.6	82.5	109.7	67.7	87.5	99.7	69.5	83.8	96.1	61.2	77.7	101.7	57.4	78.3	99.9	57.4	77.4
MF2014	WEIGHTED AVERAGE	98.4	65.4	84.0	97.2	59.0	80.5	96.0	61.6	81.0	97.2	64.7	83.0	99.2	60.9	82.5	97.3	55.8	79.2

		GEORGIA						HAWAII									IDAHO		
	DIVISION	VALDOSTA 316			WAYCROSS 315			HILO 967			HONOLULU 968			STATES & POSS., GUAM 969			BOISE 836 - 837		
		MAT.	INST.	TOTAL	MAT.	INST.	TOTAL	MAT.	INST.	TOTAL	MAT.	INST.	TOTAL	MAT.	INST.	TOTAL	MAT.	INST.	TOTAL
015433	CONTRACTOR EQUIPMENT		91.9	91.9		91.9	91.9		99.5	99.5		99.5	99.5		164.5	164.5		98.2	98.2
0241, 31 - 34	SITE & INFRASTRUCTURE, DEMOLITION	107.9	78.8	87.2	104.7	77.4	85.3	144.8	106.5	117.6	155.4	106.5	120.6	184.9	103.7	127.2	85.5	96.3	93.2
0310	Concrete Forming & Accessories	81.0	44.1	49.2	82.3	64.5	67.2	111.2	135.6	132.2	123.6	135.6	134.0	113.7	63.6	70.5	100.8	77.4	80.7
0320	Concrete Reinforcing	92.8	76.4	84.4	92.8	72.7	82.5	123.5	117.9	120.7	132.5	117.9	125.1	214.4	30.2	120.7	99.2	80.4	89.6
0330	Cast-in-Place Concrete	90.5	55.9	76.3	102.2	48.2	80.0	198.3	125.7	168.5	163.7	125.7	148.1	171.9	105.5	144.7	91.6	88.4	90.3
03	CONCRETE	98.1	55.8	77.3	101.2	61.4	81.6	158.1	127.7	143.2	150.8	127.7	139.5	159.9	72.3	116.9	99.1	81.9	90.7
04	MASONRY	108.9	50.5	72.5	109.7	39.3	65.8	150.5	128.3	136.7	150.6	128.3	136.7	213.9	43.5	107.7	121.4	84.2	98.2
05	METALS	104.1	84.0	97.9	103.1	77.7	95.3	108.1	107.3	107.8	120.7	107.3	116.5	139.2	76.2	119.9	102.2	80.9	95.6
06	WOOD, PLASTICS & COMPOSITES	73.4	37.6	53.4	75.0	72.4	73.5	114.5	139.2	128.3	134.4	139.2	137.1	127.1	67.4	93.7	93.1	75.9	83.4
07	THERMAL & MOISTURE PROTECTION	96.1	62.4	82.3	95.9	50.3	77.2	120.8	124.6	122.4	138.4	124.6	132.7	140.4	65.2	109.6	94.6	81.4	89.2
08	OPENINGS	87.3	42.5	76.9	87.4	57.8	80.5	111.3	132.8	116.3	121.8	132.8	124.4	116.6	54.1	102.1	99.1	71.2	92.6
0920	Plaster & Gypsum Board	90.4	36.2	53.8	90.4	72.0	78.0	109.7	140.1	130.3	151.3	140.1	143.7	219.9	55.5	108.8	92.4	75.1	80.7
0950, 0980	Ceilings & Acoustic Treatment	81.5	36.2	51.7	79.6	72.0	74.6	121.5	140.1	133.8	132.0	140.1	137.3	237.3	55.5	117.8	99.1	75.1	83.3
0960	Flooring	104.7	47.8	88.4	105.9	29.9	84.1	116.0	139.9	122.8	132.2	139.9	134.4	135.5	46.4	110.0	96.3	83.7	92.7
0970, 0990	Wall Finishes & Painting/Coating	105.3	52.2	73.2	105.3	46.0	69.5	110.4	143.8	130.6	119.4	143.8	134.1	115.9	35.9	67.6	103.2	39.5	64.7
09	FINISHES	95.6	43.6	67.0	95.2	57.4	74.4	113.7	138.6	127.4	128.4	138.6	134.0	188.8	58.9	117.2	96.0	74.8	84.3
COVERS	DIVS. 10 - 14, 25, 28, 41, 43, 44, 46	100.0	77.1	95.4	100.0	51.5	90.2	100.0	116.0	103.2	100.0	116.0	103.2	100.0	75.2	95.0	100.0	87.4	97.5
21, 22, 23	FIRE SUPPRESSION, PLUMBING & HVAC	99.9	70.0	87.9	96.9	57.1	80.8	100.2	107.7	103.2	100.3	107.7	103.3	102.6	37.4	76.3	100.0	71.8	88.6
26, 27, 3370	ELECTRICAL, COMMUNICATIONS & UTIL.	95.0	56.2	74.5	99.6	57.4	77.3	106.5	121.1	114.2	107.9	121.1	114.9	153.5	41.2	94.1	98.5	72.8	84.9
MF2014	WEIGHTED AVERAGE	98.5	61.4	82.3	98.3	59.4	81.4	114.8	120.2	117.2	119.5	120.2	119.8	136.2	58.0	102.1	100.1	78.6	90.7

		IDAHO															ILLINOIS		
	DIVISION	COEUR D'ALENE 838			IDAHO FALLS 834			LEWISTON 835			POCATELLO 832			TWIN FALLS 833			BLOOMINGTON 617		
		MAT.	INST.	TOTAL	MAT.	INST.	TOTAL	MAT.	INST.	TOTAL	MAT.	INST.	TOTAL	MAT.	INST.	TOTAL	MAT.	INST.	TOTAL
015433	CONTRACTOR EQUIPMENT		92.8	92.8		98.2	98.2		92.8	92.8		98.2	98.2		98.2	98.2		101.6	101.6
0241, 31 - 34	SITE & INFRASTRUCTURE, DEMOLITION	83.1	91.7	89.2	83.1	96.1	92.4	89.9	92.5	91.7	85.8	96.3	93.3	92.1	97.2	95.8	97.4	98.4	98.1
0310	Concrete Forming & Accessories	112.6	81.1	85.5	94.7	78.1	80.4	117.9	82.2	87.1	101.0	77.1	80.4	102.0	54.7	61.2	84.3	117.2	112.7
0320	Concrete Reinforcing	106.2	96.5	101.3	101.0	80.0	90.3	106.2	96.8	101.4	99.6	80.2	89.7	101.3	79.9	90.4	94.6	110.7	102.8
0330	Cast-in-Place Concrete	99.0	87.2	94.1	87.2	74.3	81.9	102.9	86.0	95.9	94.1	88.3	91.7	96.6	63.9	83.1	102.8	114.8	107.7
03	CONCRETE	105.7	86.1	96.1	91.3	77.2	84.4	109.3	86.2	98.0	98.3	81.7	90.1	105.8	63.3	84.9	100.1	115.2	107.5
04	MASONRY	122.9	83.9	98.6	116.5	81.0	94.4	123.4	85.9	100.0	118.0	81.1	95.3	121.6	81.1	96.4	120.3	118.3	119.1
05	METALS	96.3	86.9	93.4	110.0	79.1	100.5	95.7	88.0	93.4	110.1	80.3	100.9	110.1	78.9	100.5	97.6	113.6	102.5
06	WOOD, PLASTICS & COMPOSITES	96.8	81.0	88.0	86.8	78.8	82.3	102.4	81.0	90.4	93.1	75.9	83.4	94.2	47.2	67.9	86.4	115.4	102.6
07	THERMAL & MOISTURE PROTECTION	148.3	80.8	120.6	94.3	71.7	85.0	148.5	81.4	121.0	94.7	73.7	86.1	95.4	74.7	86.9	98.4	112.0	104.3
08	OPENINGS	117.8	73.0	107.4	102.8	68.0	94.7	117.8	75.6	108.0	99.8	66.3	92.0	102.8	48.8	90.2	94.9	104.9	97.2
0920	Plaster & Gypsum Board	161.4	80.5	106.7	78.5	78.1	78.2	162.5	80.5	107.0	80.7	75.1	76.9	82.3	45.5	57.5	91.3	115.8	107.9
0950, 0980	Ceilings & Acoustic Treatment	129.0	80.5	97.1	97.2	78.1	84.6	129.0	80.5	97.1	104.7	75.1	85.2	99.7	45.6	64.1	88.0	115.8	106.3
0960	Flooring	138.3	45.9	111.9	96.4	43.1	81.2	141.2	93.8	127.7	99.6	83.7	95.1	100.7	43.1	84.3	93.8	118.7	100.9
0970, 0990	Wall Finishes & Painting/Coating	123.8	67.8	90.0	103.3	40.3	65.3	123.8	67.8	90.0	103.1	41.1	65.7	103.3	37.3	63.4	93.8	126.4	113.5
09	FINISHES	161.8	73.3	113.0	93.3	68.7	79.7	162.9	82.8	118.8	96.6	75.0	84.7	96.7	49.7	70.8	93.3	118.9	107.4
COVERS	DIVS. 10 - 14, 25, 28, 41, 43, 44, 46	100.0	87.4	97.5	100.0	47.7	89.4	100.0	87.6	97.5	100.0	87.4	97.5	100.0	44.2	88.7	100.0	105.2	101.0
21, 22, 23	FIRE SUPPRESSION, PLUMBING & HVAC	99.5	82.6	92.7	100.9	71.7	89.1	100.7	85.6	94.6	99.9	71.8	88.6	99.9	69.3	87.6	95.1	108.8	100.7
26, 27, 3370	ELECTRICAL, COMMUNICATIONS & UTIL.	91.0	77.6	83.9	90.8	70.4	80.0	88.9	80.7	84.5	96.2	70.4	82.6	92.3	60.1	75.3	94.2	94.6	94.4
MF2014	WEIGHTED AVERAGE	107.9	82.2	96.7	99.8	74.7	88.9	108.7	84.9	98.3	101.1	77.4	90.8	102.2	67.3	87.0	97.5	109.4	102.7

City Cost Indexes

ILLINOIS

DIVISION		CARBONDALE 629			CENTRALIA 628			CHAMPAIGN 618 - 619			CHICAGO 606 - 608			DECATUR 625			EAST ST. LOUIS 620 - 622		
		MAT.	INST.	TOTAL	MAT.	INST.	TOTAL	MAT.	INST.	TOTAL	MAT.	INST.	TOTAL	MAT.	INST.	TOTAL	MAT.	INST.	TOTAL
015433	CONTRACTOR EQUIPMENT		108.0	108.0		108.0	108.0		102.4	102.4		94.3	94.3		102.4	102.4		108.0	108.0
0241, 31 - 34	SITE & INFRASTRUCTURE, DEMOLITION	98.7	99.5	99.2	99.0	100.1	99.8	106.4	99.0	101.2	107.0	95.7	99.0	93.2	99.1	97.4	101.2	99.8	100.2
0310	Concrete Forming & Accessories	90.0	109.0	106.4	91.6	112.8	109.9	90.7	115.2	111.8	96.6	156.7	148.4	91.8	114.9	111.8	87.6	114.8	111.1
0320	Concrete Reinforcing	91.2	111.6	101.6	91.2	111.8	101.7	94.6	105.1	99.9	99.0	158.1	129.1	89.3	127.1	108.6	91.1	109.9	100.6
0330	Cast-in-Place Concrete	97.2	102.5	99.4	97.7	117.3	105.7	119.2	109.5	115.2	107.5	149.4	124.7	106.1	110.5	107.9	99.3	117.0	106.6
03	CONCRETE	90.1	107.9	98.9	90.6	114.8	102.5	113.1	111.3	112.2	102.6	153.2	127.5	101.3	111.1	106.1	91.7	115.2	103.2
04	MASONRY	83.6	108.6	99.2	83.6	116.7	104.2	145.3	117.7	128.1	101.1	156.6	135.7	79.6	114.7	101.5	83.9	116.6	104.3
05	METALS	96.2	119.6	103.4	96.2	120.8	103.8	97.6	108.7	101.0	94.3	134.3	106.6	99.9	108.1	102.4	97.3	119.4	104.1
06	WOOD, PLASTICS & COMPOSITES	91.7	106.1	99.8	94.1	109.7	102.8	93.5	113.9	104.9	104.0	155.9	133.1	91.9	113.9	104.2	89.2	112.1	102.0
07	THERMAL & MOISTURE PROTECTION	97.1	101.5	98.9	97.2	111.7	103.1	99.0	113.2	104.9	98.0	145.1	117.3	103.2	109.3	105.7	97.2	110.2	102.5
08	OPENINGS	89.4	114.2	95.2	89.4	116.2	95.7	95.5	111.2	99.2	105.3	158.7	117.8	100.9	110.6	103.2	89.5	116.9	95.9
0920	Plaster & Gypsum Board	97.2	106.2	103.3	98.3	110.0	106.2	93.7	114.2	107.6	92.7	157.6	136.6	100.0	114.2	109.6	96.1	112.4	107.2
0950, 0980	Ceilings & Acoustic Treatment	91.4	106.2	101.2	91.4	110.0	103.6	88.0	114.2	105.2	99.2	157.6	137.6	98.1	114.2	108.7	91.4	112.1	105.2
0960	Flooring	122.2	120.2	121.6	123.2	115.7	121.0	96.7	120.2	103.4	96.9	148.0	111.5	109.1	116.7	111.3	121.2	115.7	119.6
0970, 0990	Wall Finishes & Painting/Coating	112.6	97.4	103.5	112.6	106.6	109.0	93.8	110.0	103.6	91.9	153.0	128.8	102.5	106.5	104.9	112.6	103.5	107.1
09	FINISHES	101.8	108.6	105.5	102.2	112.6	107.9	95.2	115.9	106.6	98.0	155.6	129.7	101.8	115.0	109.1	101.4	113.8	108.2
COVERS	DIVS. 10 - 14, 25, 28, 41, 43, 44, 46	100.0	102.6	100.5	100.0	104.1	100.8	100.0	104.4	100.9	100.0	124.5	105.0	100.0	104.3	100.9	100.0	104.4	100.9
21, 22, 23	FIRE SUPPRESSION, PLUMBING & HVAC	95.1	106.1	99.5	95.1	95.3	95.2	95.1	105.5	99.3	99.8	133.9	113.6	99.9	98.5	99.4	99.9	98.8	99.4
26, 27, 3370	ELECTRICAL, COMMUNICATIONS & UTIL.	95.4	107.7	101.9	96.8	107.7	102.6	97.4	94.9	96.1	98.3	133.9	117.1	99.5	90.4	94.7	96.4	103.5	100.1
MF2014	WEIGHTED AVERAGE	94.7	107.9	100.5	95.0	108.5	100.9	101.0	107.6	103.9	99.8	139.7	117.2	99.2	104.6	101.7	96.4	108.7	101.8

ILLINOIS

DIVISION		EFFINGHAM 624			GALESBURG 614			JOLIET 604			KANKAKEE 609			LA SALLE 613			NORTH SUBURBAN 600 - 603		
		MAT.	INST.	TOTAL	MAT.	INST.	TOTAL	MAT.	INST.	TOTAL	MAT.	INST.	TOTAL	MAT.	INST.	TOTAL	MAT.	INST.	TOTAL
015433	CONTRACTOR EQUIPMENT		102.4	102.4		101.6	101.6		92.5	92.5		92.5	92.5		101.6	101.6		92.5	92.5
0241, 31 - 34	SITE & INFRASTRUCTURE, DEMOLITION	97.6	98.9	98.5	99.9	98.3	98.8	106.9	94.9	98.4	100.5	94.6	96.3	99.2	99.2	99.2	106.1	94.9	98.1
0310	Concrete Forming & Accessories	96.3	114.0	111.6	90.6	116.6	113.1	98.4	159.6	151.2	91.8	143.2	136.2	104.6	125.3	122.4	97.8	154.0	146.3
0320	Concrete Reinforcing	92.2	99.9	96.1	94.1	110.6	102.5	99.0	150.8	125.4	99.8	147.4	124.0	94.3	144.7	119.9	99.0	156.5	128.3
0330	Cast-in-Place Concrete	105.7	108.4	106.8	105.9	106.2	106.0	107.4	146.5	123.5	100.1	132.5	113.4	105.7	122.5	112.6	107.5	141.2	121.3
03	CONCRETE	102.1	109.4	105.7	103.2	112.0	107.5	102.7	152.1	126.9	96.7	139.3	117.7	104.1	127.9	115.8	102.7	148.8	125.4
04	MASONRY	88.2	108.5	100.8	120.5	118.0	118.9	104.3	148.5	131.8	100.6	141.2	125.9	120.5	125.1	123.4	101.1	141.1	125.9
05	METALS	97.0	105.6	99.7	97.6	113.0	102.4	92.2	129.1	103.6	92.2	126.6	102.8	97.7	131.6	108.1	93.4	131.4	105.1
06	WOOD, PLASTICS & COMPOSITES	94.2	113.9	105.2	93.3	115.5	105.7	105.7	161.0	136.7	97.8	142.2	122.7	108.8	123.7	117.1	104.0	155.7	132.9
07	THERMAL & MOISTURE PROTECTION	102.7	106.6	104.3	98.5	107.6	102.2	97.8	141.6	115.8	97.0	135.7	112.9	98.7	117.8	106.5	98.2	139.3	115.1
08	OPENINGS	94.9	109.9	98.4	94.9	111.0	98.6	102.8	159.6	116.0	95.4	148.4	107.7	94.9	129.5	102.9	102.9	158.2	115.8
0920	Plaster & Gypsum Board	99.7	114.2	109.5	93.7	115.9	108.7	89.9	162.9	139.2	86.7	143.4	125.0	100.1	124.3	116.5	92.7	157.3	136.4
0950, 0980	Ceilings & Acoustic Treatment	91.4	114.2	106.4	88.0	115.9	106.3	99.2	162.9	141.0	99.2	143.4	128.3	88.0	124.3	111.9	99.2	157.3	137.4
0960	Flooring	110.1	120.2	113.0	96.6	118.7	102.9	96.5	140.3	109.1	93.7	131.9	104.6	102.8	122.6	108.5	96.9	140.3	109.3
0970, 0990	Wall Finishes & Painting/Coating	102.5	104.2	103.5	93.8	95.1	94.5	90.1	152.5	127.8	90.1	126.4	112.0	93.8	126.4	113.5	91.9	152.5	128.4
09	FINISHES	100.6	115.2	108.7	94.6	115.4	106.1	97.4	156.9	130.2	95.7	137.9	119.0	97.3	124.3	112.2	97.9	152.7	128.1
COVERS	DIVS. 10 - 14, 25, 28, 41, 43, 44, 46	100.0	70.8	94.1	100.0	105.1	101.0	100.0	124.5	105.0	100.0	120.8	104.2	100.0	102.5	100.5	100.0	122.3	104.5
21, 22, 23	FIRE SUPPRESSION, PLUMBING & HVAC	95.2	102.8	98.2	95.1	105.3	99.2	99.9	131.2	112.5	95.1	129.0	108.7	95.1	123.4	106.6	99.8	128.6	111.4
26, 27, 3370	ELECTRICAL, COMMUNICATIONS & UTIL.	97.3	107.6	102.8	95.1	86.6	90.6	97.5	137.9	118.9	92.3	137.1	116.0	92.2	137.1	116.0	97.3	128.5	113.8
MF2014	WEIGHTED AVERAGE	97.3	106.0	101.1	98.2	106.7	101.9	99.2	138.4	116.3	95.6	131.7	111.3	98.4	124.4	109.7	99.3	135.1	114.9

ILLINOIS

DIVISION		PEORIA 615 - 616			QUINCY 623			ROCK ISLAND 612			ROCKFORD 610 - 611			SOUTH SUBURBAN 605			SPRINGFIELD 626 - 627		
		MAT.	INST.	TOTAL	MAT.	INST.	TOTAL	MAT.	INST.	TOTAL	MAT.	INST.	TOTAL	MAT.	INST.	TOTAL	MAT.	INST.	TOTAL
015433	CONTRACTOR EQUIPMENT		101.6	101.6		102.4	102.4		101.6	101.6		101.6	101.6		92.5	92.5		102.4	102.4
0241, 31 - 34	SITE & INFRASTRUCTURE, DEMOLITION	100.4	98.3	98.9	96.5	98.7	98.0	98.0	97.4	97.5	99.8	99.7	99.8	106.1	94.9	98.1	99.0	99.1	99.1
0310	Concrete Forming & Accessories	93.7	117.4	114.2	94.2	111.8	109.4	92.3	104.2	102.5	98.1	131.5	126.7	97.8	154.0	146.3	92.6	115.3	112.1
0320	Concrete Reinforcing	91.7	110.8	101.4	91.8	105.3	98.7	94.1	103.7	99.0	86.7	136.9	112.2	99.0	156.5	128.3	94.2	105.1	99.7
0330	Cast-in-Place Concrete	102.7	114.1	107.3	105.9	102.9	104.7	103.6	97.3	101.0	105.2	126.4	113.8	107.5	141.2	121.3	100.8	109.2	104.2
03	CONCRETE	100.2	115.0	107.5	101.7	107.6	104.6	101.0	102.0	101.5	100.9	130.4	115.4	102.7	148.8	125.4	99.1	111.3	105.1
04	MASONRY	120.1	118.1	118.8	111.7	104.1	106.9	120.3	97.0	105.8	94.1	135.8	120.1	101.1	144.1	127.9	91.0	117.8	107.7
05	METALS	100.4	113.8	104.5	97.1	108.0	100.5	97.7	108.4	101.0	100.4	128.2	108.9	93.4	131.4	105.1	97.5	109.3	101.1
06	WOOD, PLASTICS & COMPOSITES	101.2	115.5	109.2	91.8	113.9	104.2	95.0	104.4	100.3	101.1	128.3	116.4	104.0	155.7	132.9	89.2	113.9	103.0
07	THERMAL & MOISTURE PROTECTION	99.3	112.4	104.7	102.7	103.5	103.0	98.5	98.3	98.4	101.8	129.0	113.0	98.2	139.3	115.1	104.4	112.2	107.6
08	OPENINGS	101.3	116.0	104.7	95.8	111.4	99.4	94.9	103.2	96.8	101.3	132.1	108.5	102.9	158.2	115.8	103.0	111.2	104.9
0920	Plaster & Gypsum Board	97.1	115.9	109.8	98.3	114.2	109.0	93.7	104.5	101.0	97.1	129.0	118.6	92.7	157.3	136.4	98.7	114.2	109.2
0950, 0980	Ceilings & Acoustic Treatment	93.0	115.9	108.1	91.4	114.2	106.4	88.0	104.5	98.8	93.0	129.0	116.7	99.2	157.3	137.4	102.1	114.2	110.1
0960	Flooring	99.9	118.7	105.3	109.1	108.1	108.8	97.7	105.6	100.0	99.9	122.6	106.4	96.9	140.3	109.3	113.5	107.8	111.9
0970, 0990	Wall Finishes & Painting/Coating	93.8	126.4	113.5	102.5	106.5	104.9	93.8	95.1	94.5	93.8	133.1	117.5	91.9	152.5	128.4	101.1	106.5	104.3
09	FINISHES	97.2	118.9	109.2	100.1	111.8	106.6	94.9	103.7	99.7	97.2	130.0	115.3	97.9	152.7	128.1	104.9	113.5	109.6
COVERS	DIVS. 10 - 14, 25, 28, 41, 43, 44, 46	100.0	105.2	101.0	100.0	71.2	94.2	100.0	99.0	99.8	100.0	114.3	102.9	100.0	122.3	104.5	100.0	104.5	100.9
21, 22, 23	FIRE SUPPRESSION, PLUMBING & HVAC	99.9	99.9	99.9	95.2	101.0	97.5	95.1	99.9	97.1	100.0	116.5	106.7	99.8	128.6	111.4	99.9	103.3	101.3
26, 27, 3370	ELECTRICAL, COMMUNICATIONS & UTIL.	96.1	96.7	96.4	94.7	80.8	87.3	87.3	95.5	91.6	96.4	131.1	114.7	97.3	128.5	113.8	102.4	92.0	96.9
MF2014	WEIGHTED AVERAGE	100.5	109.3	104.4	98.1	101.1	99.3	97.2	100.5	98.6	99.5	124.8	110.5	99.3	135.1	114.9	100.0	106.5	102.8

City Cost Indexes

		INDIANA																	
		ANDERSON			BLOOMINGTON			COLUMBUS			EVANSVILLE			FORT WAYNE			GARY		
DIVISION		460			474			472			476 - 477			467 - 468			463 - 464		
		MAT.	INST.	TOTAL	MAT.	INST.	TOTAL	MAT.	INST.	TOTAL	MAT.	INST.	TOTAL	MAT.	INST.	TOTAL	MAT.	INST.	TOTAL
015433	CONTRACTOR EQUIPMENT		97.0	97.0		86.5	86.5		86.5	86.5		116.0	116.0		97.0	97.0		97.0	97.0
0241, 31 - 34	SITE & INFRASTRUCTURE, DEMOLITION	93.9	96.1	95.5	85.8	94.4	91.9	82.2	94.3	90.8	91.0	123.9	114.4	94.8	96.0	95.7	94.5	99.6	98.1
0310	Concrete Forming & Accessories	97.8	81.5	83.7	101.0	80.7	83.5	95.0	78.6	80.8	94.4	82.7	84.3	96.1	75.2	78.1	97.9	116.3	113.8
0320	Concrete Reinforcing	95.8	82.8	89.2	86.8	80.9	83.8	87.2	80.9	84.0	95.1	78.1	86.4	95.8	75.5	85.5	95.8	110.8	103.5
0330	Cast-in-Place Concrete	109.4	80.2	97.4	103.6	78.2	93.1	103.1	71.2	90.0	99.0	87.9	94.4	116.3	82.8	102.5	114.3	113.9	114.1
03	CONCRETE	100.8	81.7	91.5	105.0	79.7	92.6	104.2	76.3	90.5	105.3	83.8	94.7	104.0	78.4	91.4	103.2	114.2	108.6
04	MASONRY	94.1	80.1	85.4	95.6	76.5	83.7	95.4	76.5	83.6	91.2	82.5	85.8	98.5	77.9	85.6	95.6	114.6	107.5
05	METALS	92.8	89.3	91.7	95.8	77.6	90.2	95.9	77.0	90.0	89.4	85.0	87.9	92.8	85.9	90.7	92.8	108.7	97.7
06	WOOD, PLASTICS & COMPOSITES	100.2	81.9	89.9	112.9	80.9	95.0	107.7	78.2	91.2	93.5	82.0	87.0	99.9	74.7	85.8	97.7	115.7	107.8
07	THERMAL & MOISTURE PROTECTION	107.7	75.9	94.7	95.1	78.3	88.2	94.7	78.1	87.9	99.4	83.7	93.0	107.5	77.7	95.3	106.3	108.6	107.2
08	OPENINGS	98.5	82.1	94.7	105.8	81.1	100.1	101.5	79.6	96.4	99.1	80.8	94.9	98.5	74.2	92.9	98.5	120.3	103.6
0920	Plaster & Gypsum Board	102.5	81.6	88.4	97.5	81.0	86.4	95.0	78.2	83.6	93.4	80.8	84.9	101.8	74.2	83.1	96.1	116.4	109.8
0950, 0980	Ceilings & Acoustic Treatment	87.9	81.6	83.7	80.9	81.0	81.0	80.9	78.2	79.1	85.5	80.8	82.4	87.9	74.2	78.9	87.9	116.4	106.6
0960	Flooring	100.6	85.0	96.2	105.5	75.0	96.8	100.8	75.0	93.4	100.3	80.7	94.7	100.6	78.8	94.4	100.6	123.4	107.1
0970, 0990	Wall Finishes & Painting/Coating	105.6	69.4	83.7	95.3	82.4	87.5	95.3	82.4	87.5	101.2	84.9	91.4	105.6	73.9	86.5	105.6	123.9	116.7
09	FINISHES	95.6	81.0	87.6	94.8	79.6	86.5	93.0	78.2	84.9	93.8	82.6	87.6	95.4	75.7	84.5	94.7	118.7	107.9
COVERS	DIVS. 10 - 14, 25, 28, 41, 43, 44, 46	100.0	91.8	98.3	100.0	90.8	98.1	100.0	90.5	98.1	100.0	95.9	99.2	100.0	91.7	98.3	100.0	106.1	101.2
21, 22, 23	FIRE SUPPRESSION, PLUMBING & HVAC	100.0	78.8	91.5	99.7	78.9	91.3	94.9	78.8	88.4	99.9	79.6	91.7	100.0	72.1	88.7	100.0	106.6	102.7
26, 27, 3370	ELECTRICAL, COMMUNICATIONS & UTIL.	87.0	88.4	87.7	98.9	88.0	93.2	98.1	87.9	92.7	95.1	88.3	91.5	87.7	79.2	83.2	98.7	107.6	103.4
MF2014	WEIGHTED AVERAGE	97.0	83.8	91.3	99.3	81.8	91.6	97.2	80.9	90.1	97.1	86.9	92.6	97.6	79.1	89.5	98.3	110.4	103.6

		INDIANA																	
		INDIANAPOLIS			KOKOMO			LAFAYETTE			LAWRENCEBURG			MUNCIE			NEW ALBANY		
DIVISION		461 - 462			469			479			470			473			471		
		MAT.	INST.	TOTAL	MAT.	INST.	TOTAL	MAT.	INST.	TOTAL	MAT.	INST.	TOTAL	MAT.	INST.	TOTAL	MAT.	INST.	TOTAL
015433	CONTRACTOR EQUIPMENT		93.7	93.7		97.0	97.0		86.5	86.5		104.4	104.4		95.0	95.0		94.5	94.5
0241, 31 - 34	SITE & INFRASTRUCTURE, DEMOLITION	93.5	99.3	97.7	90.3	96.1	94.4	83.1	94.3	91.0	81.2	110.4	102.0	85.6	94.6	92.0	78.2	96.8	91.4
0310	Concrete Forming & Accessories	98.6	85.6	87.4	101.2	76.5	80.0	92.6	82.6	84.0	91.5	75.4	77.6	92.6	81.0	82.6	89.5	73.3	75.5
0320	Concrete Reinforcing	99.4	83.0	91.1	86.7	81.1	83.9	86.8	82.7	84.7	86.1	74.2	80.1	96.0	82.7	89.3	87.4	77.1	82.2
0330	Cast-in-Place Concrete	101.9	86.4	95.5	108.3	82.7	97.8	103.7	81.8	94.7	97.0	74.8	87.9	108.8	78.6	96.4	100.1	74.0	89.4
03	CONCRETE	100.9	85.1	93.1	97.5	80.1	89.0	104.5	82.1	93.5	97.4	75.5	86.6	103.5	81.0	92.4	102.8	74.6	89.0
04	MASONRY	95.8	80.3	86.2	93.7	79.1	84.6	101.2	80.3	88.2	79.9	74.7	76.7	97.7	80.1	86.7	86.9	68.3	75.3
05	METALS	93.5	80.5	89.5	89.3	88.3	89.0	94.3	78.4	89.4	91.0	83.3	88.7	97.6	89.2	95.1	92.9	81.0	89.3
06	WOOD, PLASTICS & COMPOSITES	99.7	86.2	92.2	103.4	75.0	87.5	104.7	83.1	92.6	91.9	75.2	82.6	106.3	81.5	92.4	94.1	74.3	83.0
07	THERMAL & MOISTURE PROTECTION	100.7	80.8	92.5	107.3	75.7	94.3	94.7	80.6	88.9	100.1	76.7	90.5	97.7	76.8	89.2	86.8	69.6	79.7
08	OPENINGS	105.7	84.5	100.8	93.3	77.9	89.8	99.9	82.8	95.9	101.4	75.3	95.3	99.0	81.9	95.0	98.7	76.3	93.5
0920	Plaster & Gypsum Board	97.0	85.8	89.5	107.5	74.5	85.2	92.2	83.3	86.1	71.5	75.0	73.9	93.4	81.6	85.4	91.6	73.8	79.6
0950, 0980	Ceilings & Acoustic Treatment	93.8	85.8	88.6	87.9	74.5	79.1	76.7	83.3	81.0	89.7	75.0	80.0	81.7	81.6	81.6	85.5	73.8	77.8
0960	Flooring	101.4	85.0	96.7	104.6	93.3	101.3	99.7	88.8	96.6	74.1	85.0	77.2	99.9	85.0	95.6	97.9	62.8	87.9
0970, 0990	Wall Finishes & Painting/Coating	103.6	82.4	90.8	105.6	71.7	85.2	95.3	93.1	94.0	96.2	71.6	81.4	95.3	69.4	79.7	101.2	82.7	90.0
09	FINISHES	96.1	85.4	90.2	97.3	78.8	87.1	91.4	85.0	87.9	83.7	77.2	80.1	92.4	80.6	85.9	93.1	72.7	81.8
COVERS	DIVS. 10 - 14, 25, 28, 41, 43, 44, 46	100.0	93.3	98.7	100.0	91.1	98.2	100.0	91.1	98.2	100.0	43.6	88.6	100.0	90.8	98.1	100.0	43.1	88.5
21, 22, 23	FIRE SUPPRESSION, PLUMBING & HVAC	99.9	79.4	91.6	95.2	79.0	88.7	94.9	79.4	88.6	95.8	74.1	87.1	99.7	78.8	91.2	95.2	75.2	87.1
26, 27, 3370	ELECTRICAL, COMMUNICATIONS & UTIL.	101.0	88.4	94.4	91.4	78.7	84.7	97.6	82.5	89.6	93.0	73.5	82.7	90.9	79.6	84.9	93.6	75.8	84.2
MF2014	WEIGHTED AVERAGE	99.1	84.8	92.9	94.8	81.5	89.0	96.9	82.7	90.7	94.0	77.7	86.9	97.9	82.3	91.1	95.1	75.4	86.5

		INDIANA						IOWA											
		SOUTH BEND			TERRE HAUTE			WASHINGTON			BURLINGTON			CARROLL			CEDAR RAPIDS		
DIVISION		465 - 466			478			475			526			514			522 - 524		
		MAT.	INST.	TOTAL	MAT.	INST.	TOTAL	MAT.	INST.	TOTAL	MAT.	INST.	TOTAL	MAT.	INST.	TOTAL	MAT.	INST.	TOTAL
015433	CONTRACTOR EQUIPMENT		105.3	105.3		116.0	116.0		116.0	116.0		99.4	99.4		99.4	99.4		96.1	96.1
0241, 31 - 34	SITE & INFRASTRUCTURE, DEMOLITION	96.7	96.7	96.3	92.9	124.1	115.1	92.4	121.3	113.0	98.3	96.5	97.0	87.5	96.5	93.9	100.0	95.1	96.5
0310	Concrete Forming & Accessories	99.5	81.4	83.9	95.4	82.2	84.0	96.2	78.8	81.2	94.8	75.5	78.2	82.6	50.2	54.8	100.5	82.3	84.8
0320	Concrete Reinforcing	97.3	79.9	88.4	95.1	82.9	88.9	87.9	48.8	68.0	91.0	82.7	86.8	91.8	78.7	85.1	91.7	84.2	87.9
0330	Cast-in-Place Concrete	106.6	79.8	95.6	95.9	85.6	91.6	104.3	84.8	96.3	111.7	55.1	88.4	111.7	59.7	90.3	111.9	79.7	98.7
03	CONCRETE	98.3	81.8	90.2	108.4	83.7	96.2	114.3	75.3	95.1	102.6	70.8	86.9	101.3	60.2	81.1	102.7	82.3	92.7
04	MASONRY	105.7	77.8	88.3	98.9	79.6	86.9	91.3	80.8	84.8	102.3	64.9	79.0	104.2	74.1	85.4	108.2	78.0	89.4
05	METALS	92.8	99.9	95.0	90.0	87.4	89.2	84.6	68.2	79.5	88.8	93.0	90.1	88.9	87.8	88.8	91.3	92.2	91.5
06	WOOD, PLASTICS & COMPOSITES	94.8	81.6	87.4	95.7	82.1	88.1	96.1	78.9	86.5	95.1	76.6	84.8	81.6	44.9	61.1	101.9	82.6	91.1
07	THERMAL & MOISTURE PROTECTION	101.9	80.6	93.2	99.5	80.8	91.8	99.6	82.2	92.5	102.9	73.9	91.0	103.2	64.9	89.3	103.9	80.9	94.5
08	OPENINGS	96.0	80.5	92.4	99.7	82.2	95.6	96.4	67.6	89.7	94.8	70.4	89.1	99.3	53.4	88.6	100.3	81.9	96.0
0920	Plaster & Gypsum Board	90.8	81.3	84.4	93.4	81.0	85.0	93.4	77.7	82.8	102.3	75.9	84.5	97.7	43.2	60.9	106.9	82.3	90.3
0950, 0980	Ceilings & Acoustic Treatment	88.9	81.3	83.9	85.5	81.0	82.6	79.6	77.7	78.3	98.7	75.9	83.7	98.7	43.2	62.3	101.2	82.3	88.8
0960	Flooring	99.3	91.1	96.9	100.3	85.0	95.9	101.2	76.4	94.1	107.9	37.8	87.8	102.2	33.2	82.5	124.0	78.3	110.9
0970, 0990	Wall Finishes & Painting/Coating	99.2	87.1	91.9	101.2	83.1	90.3	101.2	83.4	90.5	106.3	78.5	89.5	106.3	78.7	89.6	107.6	72.2	86.3
09	FINISHES	95.4	83.9	89.1	93.8	82.8	87.7	93.1	79.6	85.6	102.5	68.9	84.0	98.7	48.3	71.0	108.4	80.6	93.1
COVERS	DIVS. 10 - 14, 25, 28, 41, 43, 44, 46	100.0	92.8	98.5	100.0	93.7	98.7	100.0	95.3	99.0	100.0	86.8	97.3	100.0	65.1	92.9	100.0	91.8	98.3
21, 22, 23	FIRE SUPPRESSION, PLUMBING & HVAC	99.9	77.1	90.7	99.9	79.6	91.7	95.2	77.6	88.1	95.4	75.2	87.3	95.4	72.6	86.2	100.2	80.5	92.2
26, 27, 3370	ELECTRICAL, COMMUNICATIONS & UTIL.	99.6	86.6	92.7	93.3	86.7	89.8	93.8	84.2	88.8	100.6	67.7	83.2	101.3	81.3	90.7	98.1	80.6	88.8
MF2014	WEIGHTED AVERAGE	98.0	84.4	92.1	97.9	86.5	92.9	95.8	81.8	89.7	97.2	75.3	87.6	97.0	71.2	85.8	100.0	83.2	92.7

City Cost Indexes

DIVISION		COUNCIL BLUFFS 515			CRESTON 508			DAVENPORT 527 - 528			DECORAH 521			DES MOINES 500 - 503,509			DUBUQUE 520		
		MAT.	INST.	TOTAL	MAT.	INST.	TOTAL	MAT.	INST.	TOTAL	MAT.	INST.	TOTAL	MAT.	INST.	TOTAL	MAT.	INST.	TOTAL
015433	CONTRACTOR EQUIPMENT		95.5	95.5		99.4	99.4		99.4	99.4		99.4	99.4		101.1	101.1		94.9	94.9
0241, 31 - 34	SITE & INFRASTRUCTURE, DEMOLITION	103.8	91.5	95.0	93.9	95.5	95.1	98.7	98.6	98.6	96.9	95.5	95.9	102.9	99.8	100.7	97.9	92.4	94.0
0310	Concrete Forming & Accessories	82.0	71.2	72.7	78.4	65.8	67.5	100.0	91.3	92.5	92.3	45.4	51.8	95.4	81.8	83.6	83.3	74.9	76.1
0320	Concrete Reinforcing	93.6	79.0	86.1	89.3	82.4	85.8	91.7	97.0	94.4	91.0	76.5	83.6	98.2	86.7	92.3	90.4	84.1	87.2
0330	Cast-in-Place Concrete	116.4	73.6	98.8	115.2	64.1	94.2	107.8	90.1	100.5	108.6	58.0	87.8	101.4	87.5	95.7	109.6	98.6	105.1
03	CONCRETE	105.0	74.3	89.9	102.9	69.3	86.4	100.7	92.5	96.7	100.5	57.0	79.1	100.0	85.4	92.8	99.4	85.5	92.6
04	MASONRY	109.8	75.0	88.1	108.1	81.9	91.7	105.3	86.8	93.7	124.0	71.4	91.2	101.0	81.2	88.7	109.2	73.9	87.2
05	METALS	96.3	89.3	94.1	93.8	90.5	92.8	91.3	102.7	94.8	89.0	85.8	88.0	99.9	96.5	98.8	89.8	92.0	90.5
06	WOOD, PLASTICS & COMPOSITES	80.4	70.9	75.1	74.1	61.6	67.1	101.9	91.0	95.8	92.2	37.1	61.3	91.8	81.1	85.8	82.1	73.5	77.3
07	THERMAL & MOISTURE PROTECTION	103.3	67.1	88.5	104.2	77.6	93.3	103.4	87.5	96.8	103.1	53.4	82.7	98.5	79.4	90.6	103.6	73.9	91.4
08	OPENINGS	99.3	75.9	93.9	108.4	63.9	98.1	100.3	90.8	98.1	97.8	48.6	86.4	102.7	86.5	98.9	99.3	79.3	94.7
0920	Plaster & Gypsum Board	97.7	70.3	79.2	93.6	60.5	71.2	106.9	90.8	96.0	101.2	35.2	56.6	90.4	80.6	83.8	97.7	72.9	81.0
0950, 0980	Ceilings & Acoustic Treatment	98.7	70.3	80.0	90.8	60.5	70.9	101.2	90.8	94.4	98.7	35.2	57.0	94.2	80.6	85.2	98.7	72.9	81.8
0960	Flooring	100.8	80.7	95.0	97.4	33.2	79.1	110.3	97.0	106.5	107.6	47.2	90.3	102.5	89.6	98.8	114.9	77.0	104.1
0970, 0990	Wall Finishes & Painting/Coating	102.2	65.8	80.2	101.0	78.7	87.5	106.3	95.5	99.8	106.3	32.9	62.0	96.2	88.6	91.6	106.7	63.8	80.8
09	FINISHES	99.4	72.1	84.4	94.2	60.2	75.4	104.3	92.7	97.9	102.2	42.6	69.4	97.1	83.7	89.7	103.7	73.5	87.1
COVERS	DIVS. 10 - 14, 25, 28, 41, 43, 44, 46	100.0	88.6	97.7	100.0	69.5	93.8	100.0	94.5	98.9	100.0	82.9	96.5	100.0	92.1	98.4	100.0	90.2	98.0
21, 22, 23	FIRE SUPPRESSION, PLUMBING & HVAC	100.2	74.2	89.7	95.2	79.7	89.0	100.2	92.8	97.2	95.4	72.9	86.3	99.9	79.7	91.7	100.2	75.3	90.1
26, 27, 3370	ELECTRICAL, COMMUNICATIONS & UTIL.	103.6	81.5	91.9	93.8	81.3	87.2	95.6	91.0	93.2	98.1	44.3	69.7	105.4	81.3	92.7	102.1	78.2	89.4
MF2014	WEIGHTED AVERAGE	100.7	78.1	90.9	98.2	77.3	89.1	99.0	93.1	96.4	98.0	64.2	83.3	100.5	85.2	93.8	99.1	80.2	90.9

DIVISION		FORT DODGE 505			MASON CITY 504			OTTUMWA 525			SHENANDOAH 516			SIBLEY 512			SIOUX CITY 510 - 511		
		MAT.	INST.	TOTAL	MAT.	INST.	TOTAL	MAT.	INST.	TOTAL	MAT.	INST.	TOTAL	MAT.	INST.	TOTAL	MAT.	INST.	TOTAL
015433	CONTRACTOR EQUIPMENT		99.4	99.4		99.4	99.4		94.9	94.9		95.5	95.5		99.4	99.4		99.4	99.4
0241, 31 - 34	SITE & INFRASTRUCTURE, DEMOLITION	102.2	94.4	96.6	102.3	95.4	97.4	98.1	90.5	92.7	102.2	90.5	93.9	108.1	94.2	98.2	109.9	95.0	99.3
0310	Concrete Forming & Accessories	78.9	45.1	49.7	83.1	45.5	50.6	90.3	73.7	75.9	83.6	56.5	60.2	84.0	37.9	44.2	100.5	66.4	71.1
0320	Concrete Reinforcing	89.3	67.2	78.1	89.2	81.8	85.4	91.0	86.1	88.5	93.6	68.4	80.7	93.6	65.3	79.2	91.7	77.9	84.7
0330	Cast-in-Place Concrete	108.2	44.2	81.9	108.2	57.5	87.4	112.4	53.7	88.3	112.5	59.8	90.9	110.3	44.7	83.4	111.0	55.5	88.2
03	CONCRETE	98.2	50.4	74.7	98.5	57.9	78.5	102.1	69.9	86.3	102.4	61.0	82.1	101.4	47.0	74.6	102.0	65.8	84.2
04	MASONRY	106.9	37.9	63.9	120.5	69.8	88.9	105.8	57.8	75.9	109.4	75.0	88.0	128.6	38.1	72.2	102.2	54.6	72.6
05	METALS	93.9	81.3	90.1	94.0	88.7	92.4	88.8	91.7	89.7	95.3	83.3	91.6	89.1	79.7	86.2	91.3	87.4	90.1
06	WOOD, PLASTICS & COMPOSITES	74.5	45.2	58.1	78.4	37.1	55.3	89.3	80.1	84.6	82.1	52.8	65.7	82.9	35.9	56.6	101.9	66.6	82.1
07	THERMAL & MOISTURE PROTECTION	103.5	57.1	84.5	103.0	63.5	86.8	103.7	64.6	87.7	102.6	65.2	87.2	102.9	48.3	80.5	103.4	64.3	87.4
08	OPENINGS	101.9	49.5	89.7	93.5	50.1	83.4	99.3	77.7	94.3	90.3	54.7	82.0	95.9	43.8	83.8	100.3	65.9	92.3
0920	Plaster & Gypsum Board	93.6	43.5	59.8	93.6	35.2	54.1	98.8	80.6	86.5	97.7	51.6	66.5	97.7	33.9	54.6	106.9	65.7	79.0
0950, 0980	Ceilings & Acoustic Treatment	90.8	43.5	59.7	90.8	35.2	54.2	98.7	80.6	86.8	98.7	51.6	67.7	98.7	33.9	56.1	101.2	65.7	77.9
0960	Flooring	98.9	47.2	84.1	100.9	47.2	85.6	117.9	50.2	98.5	101.5	34.3	82.3	103.1	34.0	83.3	110.5	53.7	94.1
0970, 0990	Wall Finishes & Painting/Coating	101.0	47.2	68.5	101.0	30.1	58.2	106.7	78.7	89.8	102.2	65.8	80.2	106.3	61.7	79.4	106.3	63.5	80.5
09	FINISHES	96.1	46.1	68.5	96.7	42.0	66.5	104.9	70.4	85.9	99.5	52.3	73.4	101.9	38.0	66.7	105.8	63.6	82.5
COVERS	DIVS. 10 - 14, 25, 28, 41, 43, 44, 46	100.0	81.7	96.3	100.0	86.5	97.3	100.0	82.4	96.4	100.0	66.1	93.2	100.0	80.8	96.1	100.0	88.2	97.6
21, 22, 23	FIRE SUPPRESSION, PLUMBING & HVAC	95.2	65.5	83.3	95.2	70.5	85.2	95.4	70.9	85.5	95.4	73.5	86.6	95.4	68.0	84.3	100.2	75.8	90.4
26, 27, 3370	ELECTRICAL, COMMUNICATIONS & UTIL.	100.3	41.9	69.4	99.4	55.6	76.3	100.4	72.0	85.4	98.1	81.5	89.3	98.1	41.4	68.1	98.1	73.0	84.8
MF2014	WEIGHTED AVERAGE	97.9	58.2	80.6	97.6	65.8	83.8	97.9	73.7	87.3	97.6	71.2	86.1	98.3	56.4	80.0	99.7	72.6	87.9

DIVISION		IOWA SPENCER 513			WATERLOO 506 - 507			KANSAS BELLEVILLE 669			COLBY 677			DODGE CITY 678			EMPORIA 668		
		MAT.	INST.	TOTAL	MAT.	INST.	TOTAL	MAT.	INST.	TOTAL	MAT.	INST.	TOTAL	MAT.	INST.	TOTAL	MAT.	INST.	TOTAL
015433	CONTRACTOR EQUIPMENT		99.4	99.4		99.4	99.4		103.4	103.4		103.4	103.4		103.4	103.4		101.6	101.6
0241, 31 - 34	SITE & INFRASTRUCTURE, DEMOLITION	108.1	92.9	97.3	107.6	95.3	98.8	111.0	95.0	99.7	112.3	95.3	100.1	114.9	94.9	100.7	103.0	92.5	95.5
0310	Concrete Forming & Accessories	90.2	37.9	45.1	94.1	55.1	60.5	96.2	54.1	59.8	99.5	58.3	64.0	93.1	58.2	63.0	87.2	65.3	68.3
0320	Concrete Reinforcing	93.6	67.1	80.1	89.9	83.9	86.8	96.6	55.1	75.5	98.9	55.2	76.7	96.5	55.1	75.4	95.3	55.7	75.2
0330	Cast-in-Place Concrete	110.3	44.7	83.4	115.8	59.2	92.6	124.2	55.7	96.1	127.1	55.8	97.8	129.3	55.5	99.0	120.1	53.6	92.8
03	CONCRETE	101.8	47.3	75.0	104.4	63.1	84.1	119.3	56.2	88.3	119.9	58.1	89.5	121.3	57.9	90.2	111.5	60.5	86.5
04	MASONRY	128.6	38.1	72.2	107.7	73.1	86.1	99.2	58.7	73.9	108.1	59.9	78.1	118.9	58.9	81.5	105.5	68.5	82.4
05	METALS	89.0	80.5	86.4	96.4	91.1	94.7	95.6	77.3	90.0	96.0	77.7	90.4	97.4	76.7	91.1	95.3	78.9	90.3
06	WOOD, PLASTICS & COMPOSITES	89.2	35.9	59.3	91.6	48.0	67.2	97.9	52.1	72.3	103.6	57.9	78.0	95.8	57.9	74.5	88.7	66.6	76.3
07	THERMAL & MOISTURE PROTECTION	103.8	48.2	81.0	103.3	71.3	90.2	94.8	60.7	80.8	99.1	61.6	83.7	99.1	60.7	83.4	93.1	75.9	86.0
08	OPENINGS	107.5	44.4	92.9	94.4	58.2	86.0	98.6	48.4	86.9	103.2	51.5	91.2	103.1	51.5	91.1	96.3	56.3	87.0
0920	Plaster & Gypsum Board	98.8	33.9	54.9	101.9	46.5	64.4	97.2	50.6	65.7	99.7	56.5	70.5	94.0	56.5	68.7	94.4	65.5	74.9
0950, 0980	Ceilings & Acoustic Treatment	98.7	33.9	56.1	94.2	46.5	62.8	86.3	50.6	62.8	84.0	56.5	65.9	84.0	56.5	65.9	86.3	65.5	72.7
0960	Flooring	105.7	34.0	85.2	106.2	64.9	94.4	103.0	39.2	84.8	103.1	39.2	84.9	99.6	39.2	82.3	98.0	36.4	80.4
0970, 0990	Wall Finishes & Painting/Coating	106.3	61.7	79.4	101.0	77.8	87.0	100.0	38.9	63.1	104.7	38.9	65.0	104.7	38.9	65.0	100.0	38.9	63.1
09	FINISHES	102.8	38.0	67.1	100.3	57.6	76.8	96.9	49.3	70.6	96.6	52.7	72.4	94.9	52.7	71.6	94.0	57.3	73.8
COVERS	DIVS. 10 - 14, 25, 28, 41, 43, 44, 46	100.0	80.8	96.1	100.0	87.8	97.5	100.0	40.8	88.0	100.0	41.4	88.2	100.0	41.4	88.2	100.0	40.5	88.0
21, 22, 23	FIRE SUPPRESSION, PLUMBING & HVAC	95.4	68.0	84.3	100.0	79.9	91.9	95.2	71.6	85.7	95.2	69.0	84.6	100.0	69.0	87.5	95.2	73.4	86.4
26, 27, 3370	ELECTRICAL, COMMUNICATIONS & UTIL.	99.9	41.4	69.0	95.8	55.7	74.6	96.5	67.7	81.3	100.0	64.5	81.2	96.8	73.8	84.6	93.9	71.4	82.0
MF2014	WEIGHTED AVERAGE	99.9	56.6	81.0	99.5	71.8	87.4	99.6	64.9	84.4	101.1	65.0	85.4	102.7	66.0	86.7	97.9	69.2	85.4

City Cost Indexes

		KANSAS																
	DIVISION	FORT SCOTT			HAYS			HUTCHINSON			INDEPENDENCE			KANSAS CITY			LIBERAL		
		667			676			675			673			660 - 662			679		
		MAT.	INST.	TOTAL	MAT.	INST.	TOTAL	MAT.	INST.	TOTAL	MAT.	INST.	TOTAL	MAT.	INST.	TOTAL	MAT.	INST.	TOTAL
015433	CONTRACTOR EQUIPMENT		102.5	102.5		103.4	103.4		103.4	103.4		103.4	103.4		99.9	99.9		103.4	103.4
0241, 31 - 34	SITE & INFRASTRUCTURE, DEMOLITION	99.8	93.4	95.2	117.7	95.2	101.7	96.2	95.0	95.4	116.0	95.1	101.1	94.4	93.1	93.5	117.3	95.0	101.5
0310	Concrete Forming & Accessories	104.2	76.8	80.6	97.2	58.3	63.7	87.9	58.2	62.2	107.9	66.3	72.0	100.9	94.2	95.1	93.5	58.2	63.0
0320	Concrete Reinforcing	94.6	93.3	94.0	96.5	55.2	75.5	96.5	55.1	75.4	95.9	62.7	79.0	91.8	97.2	94.6	97.8	55.1	76.1
0330	Cast-in-Place Concrete	111.4	54.5	87.2	100.3	55.8	82.0	92.9	53.1	76.6	129.4	53.4	98.5	95.4	97.4	96.3	100.3	53.1	80.9
03	CONCRETE	106.6	72.9	90.1	110.4	58.1	84.7	92.4	57.1	75.0	122.6	62.2	92.9	98.4	96.2	97.3	112.5	57.1	85.2
04	MASONRY	106.7	60.3	77.8	118.0	59.9	81.8	107.9	58.6	77.2	105.3	64.5	79.9	107.8	99.4	102.6	116.6	58.6	80.4
05	METALS	95.3	92.8	94.5	95.6	77.7	90.1	95.4	76.7	89.6	95.3	80.5	90.8	102.9	100.1	102.1	95.9	76.7	90.0
06	WOOD, PLASTICS & COMPOSITES	108.7	83.5	94.6	100.6	57.9	76.7	90.9	57.9	72.4	113.7	68.0	88.1	104.4	94.5	98.8	96.4	57.9	74.8
07	THERMAL & MOISTURE PROTECTION	94.0	80.8	88.6	99.4	61.6	83.9	98.0	60.6	82.7	99.1	80.5	91.5	93.7	98.0	95.5	99.5	60.6	83.6
08	OPENINGS	96.3	78.2	92.1	103.1	51.5	91.1	103.0	51.5	91.0	100.7	58.8	90.9	97.7	89.0	95.7	103.1	51.5	91.1
0920	Plaster & Gypsum Board	99.7	82.9	88.4	97.3	56.5	69.7	93.0	56.5	68.3	106.9	67.0	79.9	92.9	94.2	93.8	94.8	56.5	68.9
0950, 0980	Ceilings & Acoustic Treatment	86.3	82.9	84.1	84.0	56.5	65.9	84.0	56.5	65.9	84.0	67.0	72.8	86.3	94.2	91.5	84.0	56.5	65.9
0960	Flooring	113.7	39.4	92.5	102.0	39.2	84.0	96.8	39.2	80.3	107.3	39.2	87.8	91.4	98.7	93.5	99.8	39.2	82.5
0970, 0990	Wall Finishes & Painting/Coating	101.8	38.9	63.8	104.7	38.9	65.0	104.7	38.9	65.0	104.7	38.9	65.0	108.5	67.7	83.9	104.7	38.9	65.0
09	FINISHES	99.6	66.9	81.6	96.4	52.7	72.3	92.4	52.7	70.5	98.9	58.7	76.8	94.3	91.0	92.5	95.7	52.7	72.0
COVERS	DIVS. 10 - 14, 25, 28, 41, 43, 44, 46	100.0	46.2	89.1	100.0	41.4	88.2	100.0	41.4	88.2	100.0	42.5	88.4	100.0	61.9	92.3	100.0	41.4	88.2
21, 22, 23	FIRE SUPPRESSION, PLUMBING & HVAC	95.2	68.3	84.3	95.2	69.0	84.6	95.2	69.0	84.6	95.2	70.1	85.1	99.9	96.4	98.5	95.2	67.2	83.9
26, 27, 3370	ELECTRICAL, COMMUNICATIONS & UTIL.	93.2	71.4	81.6	98.9	67.7	82.4	93.9	67.7	80.0	96.1	71.4	83.0	98.7	97.2	97.9	96.8	73.8	84.6
MF2014	WEIGHTED AVERAGE	97.9	73.0	87.0	100.5	65.4	85.2	96.5	65.0	82.8	100.9	69.2	87.1	99.5	94.8	97.5	100.4	65.5	85.2

		KANSAS									KENTUCKY								
	DIVISION	SALINA			TOPEKA			WICHITA			ASHLAND			BOWLING GREEN			CAMPTON		
		674			664 - 666			670 - 672			411 - 412			421 - 422			413 - 414		
		MAT.	INST.	TOTAL	MAT.	INST.	TOTAL	MAT.	INST.	TOTAL	MAT.	INST.	TOTAL	MAT.	INST.	TOTAL	MAT.	INST.	TOTAL
015433	CONTRACTOR EQUIPMENT		103.4	103.4		101.6	101.6		103.4	103.4		97.3	97.3		94.5	94.5		101.1	101.1
0241, 31 - 34	SITE & INFRASTRUCTURE, DEMOLITION	105.2	95.2	98.1	98.2	91.5	93.4	101.8	93.8	96.1	112.3	86.3	93.8	78.6	97.1	91.8	87.0	98.3	95.0
0310	Concrete Forming & Accessories	89.7	54.9	59.7	98.4	40.9	48.8	95.3	51.0	57.1	87.9	105.2	102.8	85.7	83.0	83.4	89.4	82.6	83.6
0320	Concrete Reinforcing	95.9	59.4	77.3	94.9	99.5	97.2	95.3	77.3	86.1	88.8	108.7	98.9	86.2	91.3	88.8	87.0	101.8	94.5
0330	Cast-in-Place Concrete	112.3	53.7	88.2	100.4	45.5	77.9	105.3	51.7	83.3	91.1	101.1	95.2	90.4	95.5	92.5	100.8	73.6	89.6
03	CONCRETE	107.3	56.7	82.5	99.4	55.1	77.7	101.6	57.4	79.9	98.6	104.8	101.6	96.8	88.9	92.9	100.7	83.2	92.1
04	MASONRY	134.8	59.9	88.2	102.5	55.9	73.4	106.3	50.2	71.4	97.8	103.7	101.5	99.7	80.2	87.5	96.9	69.0	79.5
05	METALS	97.3	80.6	92.1	99.4	96.9	98.6	99.4	84.2	94.7	92.1	110.3	97.7	93.6	87.3	91.7	92.9	92.0	92.6
06	WOOD, PLASTICS & COMPOSITES	92.4	52.1	69.8	97.1	37.3	63.6	95.1	51.3	70.6	76.0	105.2	92.3	88.6	83.1	85.5	87.1	89.0	88.2
07	THERMAL & MOISTURE PROTECTION	98.6	61.4	83.3	96.7	67.4	84.7	97.2	56.1	80.4	91.0	97.5	93.7	86.6	80.2	84.0	99.7	67.5	86.5
08	OPENINGS	103.1	48.5	90.4	100.8	55.4	90.2	105.3	56.2	93.9	97.9	99.6	98.3	98.7	79.3	94.2	100.2	87.6	97.3
0920	Plaster & Gypsum Board	93.0	50.6	64.3	96.8	35.3	55.2	91.8	49.8	63.4	60.9	105.4	91.0	87.4	82.9	84.3	87.4	88.1	87.9
0950, 0980	Ceilings & Acoustic Treatment	84.0	50.6	62.0	92.0	35.3	54.7	88.8	49.8	63.2	80.8	105.4	97.0	85.5	82.9	83.8	85.5	88.1	87.2
0960	Flooring	98.2	61.9	87.8	103.4	73.7	94.9	105.4	63.0	93.3	79.4	102.4	86.0	95.8	86.0	93.0	98.0	41.2	81.7
0970, 0990	Wall Finishes & Painting/Coating	104.7	38.9	65.0	101.8	52.5	72.1	103.6	48.0	70.0	103.1	101.2	102.0	101.2	72.7	84.0	101.2	65.4	79.6
09	FINISHES	93.6	53.6	71.6	98.5	45.9	69.5	97.7	52.3	72.7	80.9	104.8	94.1	91.8	82.6	86.8	92.6	74.0	82.4
COVERS	DIVS. 10 - 14, 25, 28, 41, 43, 44, 46	100.0	86.0	97.2	100.0	47.2	89.3	100.0	83.0	96.6	100.0	94.1	98.8	100.0	62.3	92.4	100.0	55.8	91.1
21, 22, 23	FIRE SUPPRESSION, PLUMBING & HVAC	100.0	69.1	87.5	100.0	69.1	87.5	99.8	64.8	85.7	95.0	92.5	94.0	99.9	82.7	93.0	95.2	79.8	89.0
26, 27, 3370	ELECTRICAL, COMMUNICATIONS & UTIL.	96.6	73.8	84.5	97.5	71.8	83.9	100.9	73.8	86.5	91.4	97.3	94.5	94.0	81.5	87.4	91.4	66.1	78.0
MF2014	WEIGHTED AVERAGE	101.5	67.6	86.7	99.5	66.0	84.9	100.7	66.0	85.6	94.6	99.3	96.6	96.2	83.9	90.8	95.8	78.6	88.3

		KENTUCKY																	
	DIVISION	CORBIN			COVINGTON			ELIZABETHTOWN			FRANKFORT			HAZARD			HENDERSON		
		407 - 409			410			427			406			417 - 418			424		
		MAT.	INST.	TOTAL	MAT.	INST.	TOTAL	MAT.	INST.	TOTAL	MAT.	INST.	TOTAL	MAT.	INST.	TOTAL	MAT.	INST.	TOTAL
015433	CONTRACTOR EQUIPMENT		101.1	101.1		104.4	104.4		94.5	94.5		101.1	101.1		101.1	101.1		116.0	116.0
0241, 31 - 34	SITE & INFRASTRUCTURE, DEMOLITION	87.9	98.1	95.1	82.6	112.3	103.7	73.1	97.2	90.2	88.9	99.2	96.2	84.8	99.6	95.3	81.1	124.0	111.6
0310	Concrete Forming & Accessories	84.5	70.6	72.5	85.0	82.0	82.4	80.7	80.5	80.5	94.2	71.8	74.8	86.1	83.5	83.8	92.5	81.5	83.0
0320	Concrete Reinforcing	86.0	69.8	77.7	85.7	91.1	88.5	86.6	90.8	88.8	97.7	90.8	94.2	87.4	107.6	97.7	86.3	84.8	85.5
0330	Cast-in-Place Concrete	93.6	58.6	79.2	96.5	97.6	96.9	81.8	73.0	78.2	89.4	69.1	81.1	97.0	82.5	91.0	79.9	90.6	84.3
03	CONCRETE	93.3	66.8	80.3	98.9	89.7	94.4	88.6	80.0	84.4	93.8	74.8	84.5	97.3	87.7	92.6	93.9	85.4	89.7
04	MASONRY	94.4	61.9	74.1	111.8	101.5	105.4	83.5	75.4	78.5	94.6	79.4	85.1	94.8	71.5	80.3	103.6	92.9	96.9
05	METALS	90.4	78.2	86.6	91.0	96.9	92.8	92.8	87.2	91.1	93.0	87.8	91.4	92.9	94.2	93.3	84.4	86.9	85.1
06	WOOD, PLASTICS & COMPOSITES	76.9	75.9	76.4	85.0	71.7	77.6	83.7	82.0	82.7	91.4	68.7	78.7	84.1	89.0	86.8	91.3	79.0	84.4
07	THERMAL & MOISTURE PROTECTION	99.3	63.8	84.7	100.3	92.3	97.0	86.3	76.9	82.5	101.0	75.1	90.4	99.7	69.5	87.3	99.0	91.3	95.9
08	OPENINGS	97.9	65.6	90.4	102.4	81.7	97.6	98.7	83.2	95.1	104.5	76.7	98.0	100.6	79.9	95.8	96.8	79.9	92.9
0920	Plaster & Gypsum Board	92.3	74.6	80.3	68.8	71.4	70.6	86.7	81.7	83.3	96.5	67.2	76.7	86.7	88.1	87.6	90.2	77.8	81.8
0950, 0980	Ceilings & Acoustic Treatment	80.4	74.6	76.6	88.8	71.4	77.4	85.5	81.7	83.0	89.6	67.2	74.8	85.5	88.1	87.2	79.6	77.8	78.4
0960	Flooring	97.2	41.2	81.2	71.8	88.6	76.6	93.4	86.0	91.3	106.0	54.4	91.2	96.4	42.0	80.9	99.5	86.0	95.6
0970, 0990	Wall Finishes & Painting/Coating	107.2	49.4	72.3	96.2	83.2	88.3	101.2	76.2	86.1	109.1	79.8	91.4	101.2	65.4	79.6	101.2	96.0	98.1
09	FINISHES	90.7	63.6	75.8	82.7	82.0	82.4	90.6	80.7	85.1	97.6	68.8	81.7	91.9	74.8	82.5	91.4	83.8	87.2
COVERS	DIVS. 10 - 14, 25, 28, 41, 43, 44, 46	100.0	48.4	89.6	100.0	98.7	99.7	100.0	84.2	96.8	100.0	65.5	93.0	100.0	56.6	91.2	100.0	65.4	93.0
21, 22, 23	FIRE SUPPRESSION, PLUMBING & HVAC	95.2	71.4	85.6	95.9	92.4	94.5	95.5	80.1	89.3	100.1	82.3	92.9	95.2	80.7	89.4	95.5	77.6	88.3
26, 27, 3370	ELECTRICAL, COMMUNICATIONS & UTIL.	91.4	80.1	85.4	95.2	79.3	86.8	91.2	80.2	85.4	101.3	87.1	93.8	91.4	58.4	73.9	93.4	80.3	86.5
MF2014	WEIGHTED AVERAGE	94.0	71.8	84.3	95.9	91.4	93.9	92.8	82.0	88.1	98.1	80.7	90.5	95.2	78.7	88.0	93.7	86.1	90.4

City Cost Indexes

		KENTUCKY																
	DIVISION	LEXINGTON			LOUISVILLE			OWENSBORO			PADUCAH			PIKEVILLE			SOMERSET		
		403 - 405			400 - 402			423			420			415 - 416			425 - 426		
		MAT.	INST.	TOTAL	MAT.	INST.	TOTAL	MAT.	INST.	TOTAL	MAT.	INST.	TOTAL	MAT.	INST.	TOTAL	MAT.	INST.	TOTAL
015433	CONTRACTOR EQUIPMENT		101.1	101.1		94.5	94.5		116.0	116.0		116.0	116.0		97.3	97.3		101.1	101.1
0241, 31 - 34	SITE & INFRASTRUCTURE, DEMOLITION	90.1	100.6	97.5	84.8	97.3	93.7	91.0	124.5	114.8	83.7	123.4	111.9	123.2	85.3	96.3	77.9	98.8	92.8
0310	Concrete Forming & Accessories	96.4	73.0	76.2	95.8	82.0	83.9	90.9	85.1	85.9	88.9	79.9	81.1	96.8	88.9	90.0	87.0	75.9	77.4
0320	Concrete Reinforcing	94.4	92.2	93.3	96.8	92.6	94.7	86.3	91.8	89.1	86.8	86.2	86.5	89.3	108.4	99.0	86.6	90.8	88.7
0330	Cast-in-Place Concrete	95.8	92.7	94.5	96.4	73.9	87.2	93.2	91.1	92.4	85.2	86.5	85.7	100.1	96.7	98.7	79.9	98.3	87.5
03	CONCRETE	96.5	83.7	90.2	97.1	81.3	89.4	106.3	88.5	97.5	98.6	83.5	91.2	112.7	96.0	104.5	83.6	86.6	85.1
04	MASONRY	93.7	73.3	81.0	91.6	78.8	83.6	95.9	91.5	93.2	98.8	88.4	92.3	95.3	96.8	96.2	89.8	76.7	81.6
05	METALS	92.8	89.0	91.6	93.9	88.2	92.1	85.8	90.6	87.3	82.9	86.7	84.1	92.0	108.8	97.2	92.8	87.6	91.2
06	WOOD, PLASTICS & COMPOSITES	92.0	68.7	79.0	90.9	83.1	86.5	89.2	83.3	85.9	86.8	79.1	82.5	85.5	88.3	87.1	84.6	75.9	79.7
07	THERMAL & MOISTURE PROTECTION	99.5	84.1	93.2	96.3	78.2	88.9	99.4	92.8	96.7	99.1	78.8	90.8	91.7	79.5	86.7	99.0	71.2	87.6
08	OPENINGS	98.1	73.9	92.5	93.5	84.3	91.4	96.8	85.7	94.2	96.0	77.5	91.7	88.6	87.8	88.4	99.5	77.7	94.4
0920	Plaster & Gypsum Board	101.9	67.2	78.4	98.0	82.9	87.8	88.7	82.2	84.3	87.7	77.9	81.1	63.8	81.0	80.2	86.7	74.6	78.5
0950, 0980	Ceilings & Acoustic Treatment	84.5	67.2	73.1	88.9	82.9	84.9	79.6	82.2	81.3	79.6	77.9	78.5	80.8	88.1	85.6	85.5	74.6	78.3
0960	Flooring	101.9	70.3	92.9	98.6	86.0	95.0	98.9	86.0	95.2	97.8	59.4	86.8	83.4	102.4	88.8	96.7	41.2	80.8
0970, 0990	Wall Finishes & Painting/Coating	107.2	81.0	91.4	104.7	76.2	87.5	101.2	96.0	98.1	101.2	79.7	88.2	103.1	81.2	89.9	101.2	76.2	86.1
09	FINISHES	94.4	72.3	82.2	94.9	82.3	88.0	91.5	86.0	88.5	90.7	76.1	82.7	83.3	90.9	87.5	91.2	69.4	79.2
COVERS	DIVS. 10 - 14, 25, 28, 41, 43, 44, 46	100.0	86.5	97.3	100.0	84.9	96.9	100.0	102.9	100.6	100.0	61.8	92.3	100.0	56.3	91.2	100.0	58.5	91.6
21, 22, 23	FIRE SUPPRESSION, PLUMBING & HVAC	100.0	79.5	91.7	100.0	80.9	92.3	99.9	78.6	91.4	95.5	81.1	89.7	95.0	87.2	91.8	95.5	77.4	88.2
26, 27, 3370	ELECTRICAL, COMMUNICATIONS & UTIL.	94.2	81.5	87.5	99.4	81.5	89.9	93.5	81.4	87.1	95.8	79.9	87.4	94.4	73.4	83.3	91.8	80.2	85.6
MF2014	WEIGHTED AVERAGE	96.6	81.4	90.0	96.6	83.3	90.8	96.3	88.9	93.1	93.9	84.5	89.8	96.9	88.6	93.3	93.2	79.9	87.4

		LOUISIANA																
	DIVISION	ALEXANDRIA			BATON ROUGE			HAMMOND			LAFAYETTE			LAKE CHARLES			MONROE		
		713 - 714			707 - 708			704			705			706			712		
		MAT.	INST.	TOTAL	MAT.	INST.	TOTAL	MAT.	INST.	TOTAL	MAT.	INST.	TOTAL	MAT.	INST.	TOTAL	MAT.	INST.	TOTAL
015433	CONTRACTOR EQUIPMENT		89.9	89.9		89.5	89.5		90.1	90.1		90.1	90.1		89.5	89.5		89.9	89.9
0241, 31 - 34	SITE & INFRASTRUCTURE, DEMOLITION	101.2	87.1	91.1	104.0	87.2	92.0	102.1	87.8	91.9	103.2	88.1	92.5	103.9	87.0	91.9	101.2	87.0	91.1
0310	Concrete Forming & Accessories	82.9	43.4	48.9	97.6	60.8	65.8	78.4	46.2	50.7	96.0	54.1	59.9	96.8	56.7	62.2	82.4	43.3	48.7
0320	Concrete Reinforcing	94.9	54.3	74.2	101.1	58.9	79.6	97.9	59.3	78.3	99.3	58.9	78.7	99.3	59.1	78.9	93.8	54.0	73.6
0330	Cast-in-Place Concrete	96.2	49.9	77.2	97.2	58.8	81.4	95.8	44.0	74.5	95.3	47.9	75.8	100.3	67.3	86.8	96.2	56.8	80.0
03	CONCRETE	95.5	48.8	72.6	98.9	60.4	80.0	95.2	48.9	72.5	96.4	53.7	75.4	98.8	61.5	80.5	95.3	51.1	73.6
04	MASONRY	121.3	53.4	79.0	94.9	54.1	69.5	95.7	51.2	68.0	95.7	52.4	68.7	95.0	58.8	72.5	115.6	47.1	73.0
05	METALS	95.4	70.4	87.7	100.6	72.1	91.8	95.6	71.2	88.1	94.8	72.1	87.8	94.8	72.6	87.9	95.4	69.9	87.6
06	WOOD, PLASTICS & COMPOSITES	90.5	41.7	63.2	99.0	63.8	79.3	84.2	47.0	63.3	105.7	55.5	77.7	103.8	57.7	78.0	89.8	42.2	63.1
07	THERMAL & MOISTURE PROTECTION	95.6	62.9	82.2	96.2	65.4	83.5	96.5	61.4	82.1	97.1	63.8	83.4	96.1	66.8	84.1	95.6	60.8	81.3
08	OPENINGS	109.1	44.7	94.1	97.9	59.1	88.9	93.0	51.7	83.4	96.7	53.2	86.6	96.7	54.9	87.0	109.1	47.6	94.8
0920	Plaster & Gypsum Board	84.2	40.4	54.6	96.7	63.0	74.0	97.5	45.7	62.5	105.7	54.7	71.2	105.7	56.8	72.6	83.8	40.9	54.8
0950, 0980	Ceilings & Acoustic Treatment	84.1	40.4	55.4	91.3	63.0	72.7	94.1	45.7	62.3	92.4	54.7	67.6	93.3	56.8	69.3	84.1	40.9	55.7
0960	Flooring	98.5	65.7	89.1	101.8	65.7	91.5	96.3	65.7	87.5	104.9	65.7	93.7	104.9	72.9	95.7	98.1	57.1	86.4
0970, 0990	Wall Finishes & Painting/Coating	104.4	64.4	80.3	94.7	46.3	65.5	99.4	48.3	68.6	99.4	57.9	74.4	99.4	50.0	69.6	104.4	48.1	70.4
09	FINISHES	89.8	48.9	67.2	94.7	60.2	75.7	94.8	49.7	70.0	98.1	56.3	75.0	98.3	58.6	76.4	89.6	45.5	65.3
COVERS	DIVS. 10 - 14, 25, 28, 41, 43, 44, 46	100.0	50.0	89.9	100.0	80.8	96.1	100.0	45.4	89.0	100.0	79.4	95.8	100.0	80.3	96.0	100.0	45.7	89.0
21, 22, 23	FIRE SUPPRESSION, PLUMBING & HVAC	100.1	55.3	82.1	100.0	59.5	83.7	95.3	42.8	74.1	100.1	61.1	84.4	100.1	62.0	84.7	100.1	54.3	81.6
26, 27, 3370	ELECTRICAL, COMMUNICATIONS & UTIL.	95.7	57.9	75.7	100.8	61.6	80.1	96.7	56.0	75.2	97.9	67.4	81.8	97.4	65.0	80.3	97.7	59.6	77.6
MF2014	WEIGHTED AVERAGE	99.4	57.2	81.0	99.1	63.7	83.7	95.8	54.5	77.8	97.9	63.0	82.7	98.1	65.0	83.7	99.3	56.5	80.6

		LOUISIANA									MAINE								
	DIVISION	NEW ORLEANS			SHREVEPORT			THIBODAUX			AUGUSTA			BANGOR			BATH		
		700 - 701			710 - 711			703			043			044			045		
		MAT.	INST.	TOTAL	MAT.	INST.	TOTAL	MAT.	INST.	TOTAL	MAT.	INST.	TOTAL	MAT.	INST.	TOTAL	MAT.	INST.	TOTAL
015433	CONTRACTOR EQUIPMENT		90.4	90.4		89.9	89.9		90.1	90.1		100.9	100.9		100.9	100.9		100.9	100.9
0241, 31 - 34	SITE & INFRASTRUCTURE, DEMOLITION	103.8	91.1	94.7	105.5	87.1	92.4	104.4	88.1	92.8	91.6	101.1	98.4	93.3	101.3	99.0	91.0	101.1	98.2
0310	Concrete Forming & Accessories	95.1	65.5	69.5	98.5	44.2	51.7	90.4	64.6	68.1	95.9	94.8	94.9	93.1	96.6	96.1	88.9	94.9	94.1
0320	Concrete Reinforcing	100.2	60.1	79.8	98.3	54.4	76.0	97.9	59.1	78.2	96.1	106.7	101.5	87.7	108.1	98.1	86.8	107.9	97.5
0330	Cast-in-Place Concrete	98.0	69.4	86.3	101.1	53.6	81.6	102.9	51.6	81.8	101.5	58.8	84.0	82.2	110.6	93.9	82.2	58.9	72.7
03	CONCRETE	99.0	66.3	82.9	100.4	50.5	75.9	100.5	59.7	80.4	103.1	84.1	93.8	94.1	102.9	98.4	94.1	84.4	89.4
04	MASONRY	97.7	59.9	74.1	107.1	50.5	71.8	120.5	48.8	75.8	107.4	63.6	80.1	118.0	103.9	109.2	125.6	86.1	101.0
05	METALS	110.3	73.3	98.9	99.5	70.6	90.6	95.6	71.6	88.2	106.0	86.6	100.1	96.1	89.3	94.0	94.6	88.4	92.7
06	WOOD, PLASTICS & COMPOSITES	97.9	66.3	80.2	101.7	42.7	68.7	92.3	69.7	79.6	91.1	105.3	99.0	90.1	96.4	93.6	84.6	105.3	96.2
07	THERMAL & MOISTURE PROTECTION	94.9	68.8	84.2	95.3	59.8	80.8	96.4	61.8	82.2	100.1	65.0	85.7	97.8	84.4	92.3	97.8	71.8	87.1
08	OPENINGS	97.8	65.4	90.2	108.0	45.3	93.4	97.6	64.1	89.8	106.3	87.9	102.0	102.8	83.0	98.2	102.8	87.8	99.3
0920	Plaster & Gypsum Board	101.4	65.7	77.2	94.0	41.5	58.5	99.7	69.2	79.0	100.6	104.8	103.5	99.4	95.9	96.9	95.5	104.8	101.8
0950, 0980	Ceilings & Acoustic Treatment	98.2	65.7	76.8	88.1	41.5	57.4	94.1	69.2	77.7	97.4	104.8	102.3	86.2	95.7	92.5	84.3	104.8	97.8
0960	Flooring	105.7	65.7	94.2	101.9	61.3	90.3	102.3	44.1	85.7	103.9	55.3	90.0	97.1	114.3	102.0	95.3	55.3	83.9
0970, 0990	Wall Finishes & Painting/Coating	101.9	64.1	79.1	100.4	44.6	66.7	100.7	49.6	69.9	112.7	63.3	82.9	105.0	44.1	68.3	105.0	44.1	68.3
09	FINISHES	101.2	65.1	81.3	94.5	46.3	67.9	97.2	59.8	76.6	100.2	85.8	92.3	95.7	94.8	95.2	94.2	83.7	88.4
COVERS	DIVS. 10 - 14, 25, 28, 41, 43, 44, 46	100.0	83.1	96.6	100.0	78.1	95.6	100.0	81.4	96.2	100.0	100.5	100.1	100.0	109.8	102.0	100.0	100.5	100.1
21, 22, 23	FIRE SUPPRESSION, PLUMBING & HVAC	100.0	66.6	86.5	99.9	63.1	85.1	95.3	61.7	81.8	99.9	64.2	85.5	100.1	74.3	89.7	95.4	64.3	82.8
26, 27, 3370	ELECTRICAL, COMMUNICATIONS & UTIL.	102.1	73.3	86.9	103.5	66.6	84.0	95.3	72.3	83.7	100.9	80.1	89.9	97.2	75.8	85.9	95.2	80.5	87.2
MF2014	WEIGHTED AVERAGE	101.4	69.5	87.5	101.0	59.6	83.0	98.2	65.4	83.9	102.1	79.2	92.3	99.0	89.5	94.9	97.5	81.9	90.7

For customer support on your Commercial Renovation Cost Data, call 877.791.4977.

City Cost Indexes

DIVISION		MAINE																	
		HOULTON 047			KITTERY 039			LEWISTON 042			MACHIAS 046			PORTLAND 040 - 041			ROCKLAND 048		
		MAT.	INST.	TOTAL	MAT.	INST.	TOTAL	MAT.	INST.	TOTAL	MAT.	INST.	TOTAL	MAT.	INST.	TOTAL	MAT.	INST.	TOTAL
015433	CONTRACTOR EQUIPMENT		100.9	100.9		100.9	100.9		100.9	100.9		100.9	100.9		100.9	100.9		100.9	100.9
0241, 31 - 34	SITE & INFRASTRUCTURE, DEMOLITION	93.0	102.2	99.5	84.8	101.2	96.5	90.9	101.3	98.3	92.3	102.2	99.4	90.1	101.3	98.1	88.7	102.2	98.3
0310	Concrete Forming & Accessories	96.7	101.6	101.0	89.2	95.5	94.6	98.5	96.7	96.9	93.9	101.6	100.5	98.5	96.7	96.9	94.9	101.6	100.7
0320	Concrete Reinforcing	87.7	106.5	97.2	84.1	107.9	96.2	107.8	108.1	107.9	87.7	106.5	97.2	102.2	108.1	105.2	87.7	106.5	97.2
0330	Cast-in-Place Concrete	82.3	71.7	77.9	81.2	59.8	72.4	83.9	110.7	94.9	82.2	71.1	77.7	96.7	110.7	102.4	83.9	71.1	78.7
03	CONCRETE	95.1	91.5	93.4	89.7	85.0	87.4	94.7	102.9	98.7	94.6	91.3	93.0	101.9	102.9	102.4	92.0	91.3	91.6
04	MASONRY	101.3	76.3	85.7	113.3	87.1	97.0	101.9	103.9	103.2	101.3	76.3	85.7	112.8	103.9	107.2	95.2	76.3	83.4
05	METALS	94.8	86.6	92.3	85.5	88.6	86.4	99.5	89.3	96.4	94.8	86.6	92.3	107.9	89.3	102.2	94.7	86.5	92.2
06	WOOD, PLASTICS & COMPOSITES	94.0	105.3	100.3	89.2	105.3	98.2	96.0	96.4	96.2	91.0	105.3	99.0	93.7	96.4	95.2	91.9	105.3	99.4
07	THERMAL & MOISTURE PROTECTION	97.9	72.6	87.5	99.2	71.1	87.7	97.6	84.4	92.2	97.8	71.6	87.1	100.1	84.4	93.7	97.5	71.7	86.9
08	OPENINGS	102.9	86.1	98.7	100.0	87.8	98.7	106.2	83.0	100.8	102.9	86.1	99.0	104.5	83.0	99.5	102.8	86.1	98.9
0920	Plaster & Gypsum Board	101.6	104.8	103.8	100.0	104.8	103.3	104.9	95.7	98.7	100.1	104.8	103.3	102.5	95.7	97.9	100.1	104.8	103.3
0950, 0980	Ceilings & Acoustic Treatment	84.3	104.8	97.8	94.2	104.8	101.2	95.4	95.7	95.6	84.3	104.8	97.8	99.1	95.7	96.9	84.3	104.8	97.8
0960	Flooring	98.3	51.7	85.0	99.8	57.7	87.7	100.0	114.3	104.1	97.5	51.7	84.4	98.6	114.3	103.1	97.9	51.7	84.7
0970, 0990	Wall Finishes & Painting/Coating	105.0	137.1	124.4	96.4	38.4	61.4	105.0	44.1	68.3	105.0	137.1	124.4	102.0	44.1	67.1	105.0	137.1	124.4
09	FINISHES	95.9	97.8	97.0	96.5	83.7	89.5	98.9	94.8	96.6	95.4	97.8	96.7	98.8	94.8	96.6	95.2	97.8	96.6
COVERS	DIVS. 10 - 14, 25, 28, 41, 43, 44, 46	100.0	106.4	101.3	100.0	100.0	100.2	100.0	109.8	102.0	100.0	106.4	101.3	100.0	109.8	102.0	100.0	106.4	101.3
21, 22, 23	FIRE SUPPRESSION, PLUMBING & HVAC	95.4	73.2	86.4	95.3	76.6	87.8	100.1	74.3	89.7	95.4	73.2	86.4	100.0	74.3	89.6	95.4	73.2	86.4
26, 27, 3370	ELECTRICAL, COMMUNICATIONS & UTIL.	99.2	80.1	89.1	97.0	80.1	88.1	99.3	80.1	89.1	99.2	80.1	89.1	98.9	80.1	89.0	99.1	80.1	89.1
MF2014	WEIGHTED AVERAGE	97.2	85.6	92.1	95.2	84.7	90.6	99.6	90.1	95.5	97.1	85.5	92.0	102.0	90.1	96.8	96.3	85.5	91.6

DIVISION		MAINE			MARYLAND														
		WATERVILLE 049			ANNAPOLIS 214			BALTIMORE 210 - 212			COLLEGE PARK 207 - 208			CUMBERLAND 215			EASTON 216		
		MAT.	INST.	TOTAL	MAT.	INST.	TOTAL	MAT.	INST.	TOTAL	MAT.	INST.	TOTAL	MAT.	INST.	TOTAL	MAT.	INST.	TOTAL
015433	CONTRACTOR EQUIPMENT		100.9	100.9		99.8	99.8		103.4	103.4		105.5	105.5		99.8	99.8		99.8	99.8
0241, 31 - 34	SITE & INFRASTRUCTURE, DEMOLITION	92.8	101.1	98.7	103.0	91.5	94.9	101.5	95.6	97.3	100.5	94.5	96.2	94.3	91.8	92.5	101.4	88.6	92.3
0310	Concrete Forming & Accessories	88.4	94.8	93.9	97.2	73.6	76.8	101.4	71.1	75.3	84.3	70.2	72.2	91.4	81.0	82.4	89.2	70.2	72.8
0320	Concrete Reinforcing	87.7	106.7	97.3	100.3	80.7	90.3	105.6	80.7	92.9	105.4	78.1	91.5	86.6	72.3	79.3	86.0	79.0	82.4
0330	Cast-in-Place Concrete	82.3	58.8	72.6	110.0	75.7	95.9	107.5	76.4	94.7	116.5	77.5	100.4	91.7	85.6	89.2	101.8	48.1	79.8
03	CONCRETE	95.6	84.1	90.0	104.8	76.6	90.9	104.8	75.8	90.6	107.5	75.1	91.8	92.0	81.8	86.1	98.0	65.2	81.9
04	MASONRY	112.2	63.6	81.9	103.5	71.6	83.6	100.1	71.7	82.4	111.8	69.0	85.1	99.1	83.8	89.5	113.4	42.9	69.4
05	METALS	94.8	86.6	92.3	104.0	93.4	100.8	100.6	93.9	98.5	86.8	97.4	90.0	98.5	90.5	96.1	98.8	87.2	95.2
06	WOOD, PLASTICS & COMPOSITES	84.0	105.3	95.9	92.9	75.4	83.1	99.6	72.2	84.2	80.0	70.0	74.4	84.9	80.1	82.2	82.7	77.4	79.7
07	THERMAL & MOISTURE PROTECTION	97.9	65.0	84.4	100.2	78.8	91.4	101.9	78.6	92.3	101.9	78.7	92.4	100.2	78.9	91.5	100.3	59.0	83.4
08	OPENINGS	102.9	87.9	99.4	105.8	80.3	99.9	99.0	78.5	94.2	94.3	74.8	89.7	99.7	78.0	94.7	98.0	71.4	91.8
0920	Plaster & Gypsum Board	95.5	104.8	101.8	100.5	74.9	83.2	102.3	71.3	81.4	99.0	69.0	78.8	102.3	79.9	87.1	102.3	77.0	85.2
0950, 0980	Ceilings & Acoustic Treatment	84.3	104.8	97.8	91.3	74.9	80.6	94.0	71.3	79.1	102.2	69.0	80.4	93.3	79.8	84.4	93.3	77.0	82.6
0960	Flooring	95.0	55.3	83.6	99.8	77.2	93.3	100.7	77.2	94.0	97.8	79.7	92.6	95.5	90.8	94.2	94.7	51.7	82.4
0970, 0990	Wall Finishes & Painting/Coating	105.0	63.3	79.8	92.1	79.2	84.3	97.9	79.2	86.6	108.0	77.6	89.7	96.5	73.4	82.5	96.5	77.6	85.1
09	FINISHES	94.3	85.6	89.6	94.8	74.3	83.5	100.1	72.3	84.8	95.3	71.9	82.4	96.5	81.9	88.5	96.7	69.1	81.5
COVERS	DIVS. 10 - 14, 25, 28, 41, 43, 44, 46	100.0	100.5	100.1	100.0	86.2	97.2	100.0	86.2	97.2	100.0	85.5	97.1	100.0	90.5	98.1	100.0	75.3	95.0
21, 22, 23	FIRE SUPPRESSION, PLUMBING & HVAC	95.4	64.2	82.8	100.1	80.5	92.2	100.0	80.6	92.2	95.4	83.6	90.7	95.2	73.0	86.3	95.2	68.3	84.4
26, 27, 3370	ELECTRICAL, COMMUNICATIONS & UTIL.	99.2	80.1	89.1	99.5	91.9	95.5	102.0	91.9	96.6	101.1	99.7	100.4	98.3	81.8	89.6	97.8	65.2	80.5
MF2014	WEIGHTED AVERAGE	97.5	79.5	89.7	101.5	82.0	93.0	100.8	81.9	92.6	97.2	83.4	91.2	96.8	81.8	90.2	98.3	68.6	85.3

DIVISION		MARYLAND														MASSACHUSETTS		
		ELKTON 219			HAGERSTOWN 217			SALISBURY 218			SILVER SPRING 209			WALDORF 206			BOSTON 020 - 022, 024		
		MAT.	INST.	TOTAL	MAT.	INST.	TOTAL	MAT.	INST.	TOTAL	MAT.	INST.	TOTAL	MAT.	INST.	TOTAL	MAT.	INST.	TOTAL
015433	CONTRACTOR EQUIPMENT		99.8	99.8		99.8	99.8		99.8	99.8		98.1	98.1		98.1	98.1		106.5	106.5
0241, 31 - 34	SITE & INFRASTRUCTURE, DEMOLITION	88.4	89.5	89.2	92.6	92.3	92.4	101.3	88.6	92.3	89.0	87.6	88.0	95.4	87.3	89.7	98.9	109.1	106.2
0310	Concrete Forming & Accessories	95.4	81.0	82.9	90.3	75.1	77.2	103.9	51.7	58.9	92.8	70.4	73.5	100.2	68.7	73.0	102.7	143.8	138.1
0320	Concrete Reinforcing	86.0	105.3	95.8	86.6	72.3	79.3	86.0	63.5	74.6	104.1	77.9	90.8	104.8	78.0	91.1	108.9	155.8	132.8
0330	Cast-in-Place Concrete	82.5	75.2	79.5	87.4	85.7	86.7	101.8	46.7	79.2	119.3	79.3	102.8	133.6	76.5	110.2	104.0	151.2	123.4
03	CONCRETE	83.3	84.3	83.8	86.8	79.2	83.0	99.0	53.7	76.8	105.9	76.1	91.2	116.5	74.4	95.8	105.7	147.5	126.2
04	MASONRY	98.3	58.9	73.8	105.3	83.8	91.9	113.1	47.3	72.1	110.9	72.3	86.8	95.4	67.2	77.9	106.6	164.2	142.5
05	METALS	98.8	101.3	99.6	98.7	90.7	96.2	98.8	81.2	93.4	91.0	93.8	91.9	91.0	93.9	91.9	100.2	132.4	110.1
06	WOOD, PLASTICS & COMPOSITES	90.0	84.6	86.9	88.0	84.0	84.7	100.6	55.1	75.1	86.5	69.2	76.8	94.0	69.2	80.1	98.6	143.3	123.6
07	THERMAL & MOISTURE PROTECTION	99.9	73.5	89.1	100.7	76.8	90.5	100.6	63.7	85.4	105.4	84.9	97.0	105.8	82.7	96.4	103.4	151.9	123.3
08	OPENINGS	98.0	81.3	94.1	98.0	73.0	92.2	98.2	61.7	89.7	85.7	74.3	83.0	86.2	74.3	83.5	101.4	146.2	111.8
0920	Plaster & Gypsum Board	105.2	86.7	92.7	102.3	71.3	81.4	111.6	54.0	72.6	105.1	69.0	80.7	108.3	69.0	81.8	107.5	144.1	132.2
0950, 0980	Ceilings & Acoustic Treatment	93.3	86.7	89.0	94.3	71.3	79.2	93.3	54.0	67.5	112.2	69.0	83.8	112.2	69.0	83.8	106.1	144.1	131.1
0960	Flooring	97.0	58.5	86.0	95.2	90.8	93.9	100.7	64.7	90.4	104.0	79.7	97.2	107.7	79.0	99.5	97.9	182.7	122.2
0970, 0990	Wall Finishes & Painting/Coating	96.5	77.6	85.1	96.5	77.6	85.1	96.5	77.6	85.1	115.8	77.6	92.8	115.8	77.6	92.8	104.4	155.9	135.4
09	FINISHES	96.9	76.9	85.9	96.4	77.5	86.0	99.8	56.3	75.8	96.9	71.4	82.8	98.6	70.7	83.2	104.2	152.7	130.9
COVERS	DIVS. 10 - 14, 25, 28, 41, 43, 44, 46	100.0	59.9	91.9	100.0	89.6	97.9	100.0	38.7	87.6	100.0	84.7	96.9	100.0	83.0	96.6	100.0	119.4	103.9
21, 22, 23	FIRE SUPPRESSION, PLUMBING & HVAC	95.2	80.5	89.3	100.0	86.2	94.4	95.2	59.9	81.0	95.4	84.8	91.1	95.4	82.3	90.1	100.1	132.4	113.1
26, 27, 3370	ELECTRICAL, COMMUNICATIONS & UTIL.	99.6	91.9	95.5	98.1	81.8	89.5	96.5	67.1	81.0	98.4	99.7	99.1	95.7	99.7	97.8	100.8	135.4	119.1
MF2014	WEIGHTED AVERAGE	95.9	81.7	89.8	97.6	83.3	91.4	98.7	62.1	82.7	96.5	83.2	90.7	97.1	81.7	90.4	101.6	139.5	118.1

City Cost Indexes

DIVISION		MASSACHUSETTS																	
		BROCKTON 023			BUZZARDS BAY 025			FALL RIVER 027			FITCHBURG 014			FRAMINGHAM 017			GREENFIELD 013		
		MAT.	INST.	TOTAL	MAT.	INST.	TOTAL	MAT.	INST.	TOTAL	MAT.	INST.	TOTAL	MAT.	INST.	TOTAL	MAT.	INST.	TOTAL
015433	CONTRACTOR EQUIPMENT		102.7	102.7		102.7	102.7		103.7	103.7		100.9	100.9		101.9	101.9		100.9	100.9
0241, 31 - 34	SITE & INFRASTRUCTURE, DEMOLITION	94.5	105.7	102.4	84.8	105.6	99.6	93.5	105.8	102.2	86.3	105.6	100.0	83.0	105.3	98.9	90.0	104.3	100.1
0310	Concrete Forming & Accessories	101.6	138.7	133.6	99.2	138.5	133.1	101.6	138.8	133.6	93.8	130.8	125.7	101.2	138.8	133.6	92.1	113.2	110.3
0320	Concrete Reinforcing	106.4	155.6	131.4	85.3	160.1	123.4	106.4	160.2	133.7	83.9	151.4	118.3	83.9	155.6	120.4	87.3	124.4	106.2
0330	Cast-in-Place Concrete	97.7	151.5	119.8	81.2	151.5	110.1	94.5	152.0	118.1	85.9	150.5	112.4	85.9	148.3	111.5	88.3	131.2	105.9
03	CONCRETE	101.8	145.2	123.1	86.2	145.9	115.5	100.3	146.2	122.9	85.9	140.4	112.7	88.7	144.1	116.0	89.6	120.9	104.9
04	MASONRY	101.7	160.0	138.0	94.4	160.0	135.3	102.6	159.9	138.3	101.0	159.0	137.2	107.2	160.1	140.1	105.4	136.0	124.4
05	METALS	97.4	130.0	107.5	92.3	131.7	104.4	97.4	132.1	108.1	95.5	125.5	104.7	95.5	130.0	106.2	97.8	110.4	101.7
06	WOOD, PLASTICS & COMPOSITES	99.1	138.4	121.1	96.1	138.4	119.8	99.1	138.7	121.3	93.7	128.3	113.1	100.1	138.2	121.4	91.5	110.9	102.4
07	THERMAL & MOISTURE PROTECTION	100.9	148.7	120.5	100.1	147.0	119.4	100.8	146.3	119.5	99.2	141.5	116.5	99.3	149.1	119.7	99.2	121.9	108.5
08	OPENINGS	102.2	143.5	111.9	97.9	139.2	107.5	102.2	139.3	110.9	104.6	137.0	112.2	94.9	143.4	106.2	104.7	115.0	107.1
0920	Plaster & Gypsum Board	93.7	139.0	124.3	89.5	139.0	123.0	93.7	139.0	124.3	100.3	128.6	119.4	102.8	139.0	127.3	101.0	110.7	107.5
0950, 0980	Ceilings & Acoustic Treatment	102.2	139.0	126.4	88.2	139.0	121.6	102.2	139.0	126.4	89.0	128.6	115.0	89.0	139.0	121.9	97.4	110.7	106.1
0960	Flooring	98.6	182.7	122.7	96.7	182.7	121.3	97.7	182.7	122.0	98.1	182.7	122.3	99.7	182.7	123.5	97.3	154.0	113.5
0970, 0990	Wall Finishes & Painting/Coating	99.9	155.1	133.2	99.9	155.1	133.2	99.9	155.1	133.2	99.5	155.1	133.1	100.5	155.1	133.5	99.5	115.8	109.4
09	FINISHES	98.9	149.1	126.6	94.2	149.1	124.4	98.7	149.3	126.6	94.6	143.1	121.3	95.3	148.9	124.8	96.7	121.0	110.0
COVERS	DIVS. 10 - 14, 25, 28, 41, 43, 44, 46	100.0	117.9	103.6	100.0	117.9	103.6	100.0	118.5	103.7	100.0	109.0	101.8	100.0	117.5	103.5	100.0	104.6	100.9
21, 22, 23	FIRE SUPPRESSION, PLUMBING & HVAC	100.1	111.1	104.6	95.4	111.1	101.7	100.1	111.2	104.6	96.0	113.6	103.1	96.0	126.1	108.1	96.0	102.8	98.7
26, 27, 3370	ELECTRICAL, COMMUNICATIONS & UTIL.	99.3	100.1	99.7	96.3	100.1	98.3	99.2	100.1	99.7	100.1	105.1	102.8	96.6	128.9	113.7	100.1	98.9	99.5
MF2014	WEIGHTED AVERAGE	99.9	128.1	112.2	94.4	128.1	109.1	99.7	128.3	112.2	96.5	126.5	109.6	95.7	135.1	112.9	97.7	112.3	104.1

DIVISION		MASSACHUSETTS																	
		HYANNIS 026			LAWRENCE 019			LOWELL 018			NEW BEDFORD 027			PITTSFIELD 012			SPRINGFIELD 010 - 011		
		MAT.	INST.	TOTAL	MAT.	INST.	TOTAL	MAT.	INST.	TOTAL	MAT.	INST.	TOTAL	MAT.	INST.	TOTAL	MAT.	INST.	TOTAL
015433	CONTRACTOR EQUIPMENT		102.7	102.7		102.7	102.7		100.9	100.9		103.7	103.7		100.9	100.9		100.9	100.9
0241, 31 - 34	SITE & INFRASTRUCTURE, DEMOLITION	90.9	105.6	101.3	95.2	105.7	102.7	94.2	105.6	102.3	92.1	105.8	101.8	95.2	104.2	101.6	94.6	104.4	101.6
0310	Concrete Forming & Accessories	93.5	138.5	132.3	102.4	139.0	134.0	99.0	139.0	133.5	101.6	138.8	133.6	99.0	112.2	110.4	99.3	113.6	111.6
0320	Concrete Reinforcing	85.3	160.1	123.4	103.8	150.4	127.5	104.6	150.4	127.9	106.4	160.2	133.7	86.6	122.3	104.8	104.6	124.4	114.7
0330	Cast-in-Place Concrete	89.1	151.5	114.7	99.4	148.9	119.7	90.3	148.9	114.4	83.3	152.0	111.5	98.5	129.7	111.3	94.0	131.7	109.5
03	CONCRETE	92.4	145.9	118.7	103.7	143.5	123.2	95.1	143.3	118.8	95.0	146.2	120.2	96.2	119.5	107.6	96.9	121.2	108.8
04	MASONRY	100.6	160.0	137.6	112.3	160.7	142.5	100.2	160.0	137.5	100.6	159.9	137.6	100.9	133.2	121.0	100.5	136.9	123.2
05	METALS	93.7	131.7	105.4	98.2	128.3	107.5	98.2	126.0	106.7	97.4	132.1	108.1	98.0	109.5	101.5	100.5	110.4	103.8
06	WOOD, PLASTICS & COMPOSITES	89.3	138.4	116.8	100.6	138.4	121.8	99.8	138.4	121.4	99.1	138.7	121.3	99.8	110.9	106.0	99.8	110.9	106.0
07	THERMAL & MOISTURE PROTECTION	100.4	147.0	119.5	99.7	148.5	119.7	99.5	148.6	119.7	100.7	146.3	119.4	99.6	120.7	108.2	99.5	122.2	108.8
08	OPENINGS	98.5	139.2	108.0	99.0	142.1	109.0	105.9	142.1	114.4	102.2	139.3	110.9	105.9	114.4	107.9	105.9	115.0	108.0
0920	Plaster & Gypsum Board	85.2	139.0	121.6	105.6	139.0	128.2	105.6	139.0	128.2	93.7	139.0	124.3	105.6	110.7	109.0	105.6	110.7	109.0
0950, 0980	Ceilings & Acoustic Treatment	93.8	139.0	123.5	99.3	139.0	125.4	99.3	139.0	125.4	102.2	139.0	126.4	99.3	110.7	106.8	99.3	110.7	106.8
0960	Flooring	94.4	182.7	119.6	100.2	182.7	123.8	100.2	182.7	123.8	97.7	182.7	122.0	100.6	154.0	115.9	99.7	154.0	115.3
0970, 0990	Wall Finishes & Painting/Coating	99.9	155.1	133.2	99.6	155.1	133.1	99.5	155.1	133.1	99.9	155.1	133.2	99.5	115.8	109.4	100.9	115.8	109.9
09	FINISHES	94.4	149.1	124.6	98.8	149.1	126.5	98.7	149.1	126.5	98.6	149.3	126.5	98.8	120.3	110.6	98.7	121.2	111.1
COVERS	DIVS. 10 - 14, 25, 28, 41, 43, 44, 46	100.0	117.9	103.6	100.0	118.1	103.6	100.0	118.1	103.6	100.0	118.5	103.7	100.0	103.7	100.7	100.0	104.9	101.0
21, 22, 23	FIRE SUPPRESSION, PLUMBING & HVAC	100.1	111.1	104.6	100.1	124.8	110.0	100.1	125.7	110.4	100.1	111.2	104.6	100.1	101.4	100.6	100.1	103.3	101.4
26, 27, 3370	ELECTRICAL, COMMUNICATIONS & UTIL.	96.8	100.1	98.5	99.1	128.9	114.9	99.6	128.9	115.1	100.0	100.1	100.1	99.6	98.9	99.2	99.7	98.9	99.2
MF2014	WEIGHTED AVERAGE	97.0	128.1	110.6	100.3	134.6	115.3	99.5	134.5	114.8	99.1	128.3	111.8	99.7	111.3	104.8	100.2	112.6	105.6

DIVISION		MASSACHUSETTS			MICHIGAN														
		WORCESTER 015 - 016			ANN ARBOR 481			BATTLE CREEK 490			BAY CITY 487			DEARBORN 481			DETROIT 482		
		MAT.	INST.	TOTAL	MAT.	INST.	TOTAL	MAT.	INST.	TOTAL	MAT.	INST.	TOTAL	MAT.	INST.	TOTAL	MAT.	INST.	TOTAL
015433	CONTRACTOR EQUIPMENT		100.9	100.9		110.7	110.7		102.8	102.8		110.7	110.7		110.7	110.7		98.8	98.8
0241, 31 - 34	SITE & INFRASTRUCTURE, DEMOLITION	94.6	105.6	102.4	82.3	98.8	94.0	93.3	87.5	89.2	73.9	97.8	90.9	82.1	98.9	94.1	94.7	100.6	98.9
0310	Concrete Forming & Accessories	99.6	130.7	126.5	98.3	111.6	109.8	97.6	87.6	89.0	98.4	87.1	88.6	98.2	115.8	113.3	100.8	115.8	113.7
0320	Concrete Reinforcing	104.6	150.6	128.0	92.3	120.1	106.5	91.4	93.3	92.4	92.3	119.2	106.0	92.3	120.2	106.5	94.9	120.1	107.8
0330	Cast-in-Place Concrete	93.5	150.5	116.9	88.8	107.4	96.4	99.0	100.4	99.6	85.0	90.6	87.3	86.8	109.9	96.3	94.1	109.9	100.6
03	CONCRETE	96.6	140.2	118.0	93.5	112.0	102.6	98.1	92.5	95.4	91.7	95.2	93.4	92.6	114.7	103.5	96.6	113.6	105.0
04	MASONRY	100.0	159.0	136.8	104.4	108.4	106.9	105.2	87.2	94.0	104.0	87.0	93.4	104.3	111.6	108.9	97.9	111.6	106.5
05	METALS	101.0	125.2	108.4	94.2	118.8	101.8	98.3	87.6	95.0	94.8	115.4	101.2	94.3	119.1	101.9	94.9	102.1	97.1
06	WOOD, PLASTICS & COMPOSITES	100.3	128.3	116.0	97.5	112.8	106.0	96.5	87.2	91.3	97.5	86.5	91.3	97.5	116.8	108.3	103.0	116.8	110.7
07	THERMAL & MOISTURE PROTECTION	99.5	141.5	116.7	100.4	108.2	103.6	94.6	84.8	90.6	98.2	92.4	95.8	98.9	115.7	105.8	97.7	115.7	105.1
08	OPENINGS	105.9	136.7	113.1	99.4	110.4	101.9	96.6	82.3	93.3	99.4	93.9	98.1	99.4	112.6	102.5	99.2	112.9	102.4
0920	Plaster & Gypsum Board	105.6	128.6	121.2	103.8	112.4	109.6	93.7	82.9	86.4	103.8	85.3	91.3	103.8	116.6	112.4	100.5	116.6	111.3
0950, 0980	Ceilings & Acoustic Treatment	99.3	128.6	118.5	88.7	112.4	104.3	96.3	82.9	87.5	89.7	85.3	86.8	88.7	116.6	107.0	91.2	116.6	107.9
0960	Flooring	100.2	179.8	123.0	97.5	117.6	103.2	105.7	91.9	101.7	97.5	81.5	92.9	96.9	113.9	101.8	95.9	113.9	101.1
0970, 0990	Wall Finishes & Painting/Coating	99.5	155.1	133.1	96.2	98.7	97.7	107.9	81.9	92.2	96.2	82.0	87.6	96.2	100.8	99.0	97.3	100.8	99.4
09	FINISHES	98.7	142.5	122.9	92.8	111.1	103.1	98.0	87.6	92.3	92.6	84.8	88.3	92.6	114.3	104.5	93.9	114.3	105.1
COVERS	DIVS. 10 - 14, 25, 28, 41, 43, 44, 46	100.0	109.0	101.8	100.0	105.8	101.2	100.0	96.4	99.3	100.0	94.1	98.8	100.0	107.1	101.4	100.0	107.1	101.4
21, 22, 23	FIRE SUPPRESSION, PLUMBING & HVAC	100.1	113.6	105.5	100.0	98.8	99.5	100.0	87.4	94.9	100.0	83.6	93.4	100.0	109.2	103.7	100.0	110.2	104.1
26, 27, 3370	ELECTRICAL, COMMUNICATIONS & UTIL.	99.7	105.1	102.5	96.3	108.8	102.9	94.4	83.7	88.8	95.2	90.3	92.6	96.3	107.0	101.9	98.2	107.0	102.9
MF2014	WEIGHTED AVERAGE	100.2	126.4	111.6	97.1	107.5	101.7	98.3	87.6	93.7	96.6	91.7	94.5	97.0	111.0	103.1	97.8	109.6	102.9

773

City Cost Indexes

		MICHIGAN																
	DIVISION	FLINT 484-485			GAYLORD 497			GRAND RAPIDS 493,495			IRON MOUNTAIN 498-499			JACKSON 492			KALAMAZOO 491		
		MAT.	INST.	TOTAL	MAT.	INST.	TOTAL	MAT.	INST.	TOTAL	MAT.	INST.	TOTAL	MAT.	INST.	TOTAL	MAT.	INST.	TOTAL
015433	CONTRACTOR EQUIPMENT		110.7	110.7		104.7	104.7		102.8	102.8		93.7	93.7		104.7	104.7		102.8	102.8
0241, 31-34	SITE & INFRASTRUCTURE, DEMOLITION	71.7	98.0	90.4	88.5	85.0	86.0	93.2	87.6	89.2	96.1	93.7	94.4	109.4	87.3	93.7	93.6	87.5	89.3
0310	Concrete Forming & Accessories	101.3	89.9	91.5	95.5	77.3	79.8	97.1	84.6	86.3	87.8	85.0	85.4	92.5	86.7	87.5	97.6	87.3	88.7
0320	Concrete Reinforcing	92.3	119.6	106.2	85.2	118.6	102.2	97.8	93.2	95.5	85.0	99.8	92.5	82.8	119.4	101.4	91.4	93.4	92.4
0330	Cast-in-Place Concrete	89.4	92.4	90.6	98.7	86.1	93.5	104.1	97.6	101.4	116.8	85.4	103.9	98.6	92.2	96.0	101.0	98.8	100.1
03	CONCRETE	94.0	97.1	95.5	94.6	88.9	91.8	101.6	90.2	96.0	104.3	87.9	96.2	89.3	95.4	92.3	101.6	91.8	96.8
04	MASONRY	104.5	96.5	99.5	116.2	79.1	93.1	102.0	87.4	92.9	101.2	86.1	91.8	95.1	91.6	92.9	103.8	89.1	94.6
05	METALS	94.3	116.3	101.0	99.7	109.9	102.8	95.3	87.1	92.8	99.0	89.6	96.1	99.9	114.1	104.2	98.3	86.5	94.7
06	WOOD, PLASTICS & COMPOSITES	100.9	88.2	93.8	89.4	76.8	82.4	97.7	83.2	89.6	85.0	85.1	85.0	88.1	84.7	86.2	96.5	87.2	91.3
07	THERMAL & MOISTURE PROTECTION	98.2	96.3	97.4	93.2	74.5	85.5	97.4	76.6	88.9	96.7	81.4	90.4	92.5	92.6	92.5	94.6	85.1	90.7
08	OPENINGS	99.4	95.6	98.5	97.5	75.1	92.3	103.4	84.7	99.0	104.6	76.0	98.0	96.5	93.1	95.7	96.6	81.7	93.1
0920	Plaster & Gypsum Board	105.2	87.1	93.0	93.4	74.9	80.9	95.4	78.8	84.2	51.7	85.2	74.3	91.7	82.9	85.8	93.7	82.9	86.4
0950, 0980	Ceilings & Acoustic Treatment	88.7	87.1	87.6	95.4	74.9	81.9	104.3	78.8	87.5	94.4	85.2	88.3	95.4	82.9	87.2	96.3	82.9	87.5
0960	Flooring	97.5	95.8	97.0	97.8	92.4	96.2	105.5	85.7	99.8	122.4	93.8	114.2	96.5	82.7	92.6	105.7	82.7	99.1
0970, 0990	Wall Finishes & Painting/Coating	96.2	85.4	89.7	103.8	82.0	90.6	109.7	80.8	92.3	126.1	72.1	93.5	103.8	96.8	99.6	107.9	81.9	92.2
09	FINISHES	92.1	89.8	90.9	97.4	79.1	87.3	101.6	84.2	92.0	100.0	85.3	91.9	98.3	86.2	91.6	98.0	85.6	91.2
COVERS	DIVS. 10-14, 25, 28, 41, 43, 44, 46	100.0	95.5	99.1	100.0	88.6	97.7	100.0	96.4	99.3	100.0	87.2	97.4	100.0	100.9	100.2	100.0	96.4	99.3
21, 22, 23	FIRE SUPPRESSION, PLUMBING & HVAC	100.0	89.6	95.8	95.6	78.4	88.6	100.0	83.0	93.1	95.5	85.3	91.4	95.6	88.4	92.7	100.0	80.8	92.3
26, 27, 3370	ELECTRICAL, COMMUNICATIONS & UTIL.	96.3	96.0	96.1	92.4	79.1	85.4	99.7	79.1	88.8	98.7	84.3	91.1	96.3	108.8	102.9	94.3	81.8	87.7
MF2014	WEIGHTED AVERAGE	96.8	96.0	96.4	97.2	83.6	91.3	99.7	85.1	93.3	99.3	86.2	93.6	96.5	95.1	95.9	98.7	85.7	93.0

		MICHIGAN															MINNESOTA		
	DIVISION	LANSING 488-489			MUSKEGON 494			ROYAL OAK 480,483			SAGINAW 486			TRAVERSE CITY 496			BEMIDJI 566		
		MAT.	INST.	TOTAL	MAT.	INST.	TOTAL	MAT.	INST.	TOTAL	MAT.	INST.	TOTAL	MAT.	INST.	TOTAL	MAT.	INST.	TOTAL
015433	CONTRACTOR EQUIPMENT		110.7	110.7		102.8	102.8		96.2	96.2		110.7	110.7		93.7	93.7		98.2	98.2
0241, 31-34	SITE & INFRASTRUCTURE, DEMOLITION	95.6	98.0	97.3	91.2	87.5	88.5	86.5	98.4	95.0	74.9	97.7	91.2	82.4	93.2	90.0	95.4	97.5	96.9
0310	Concrete Forming & Accessories	98.2	89.7	90.9	98.0	84.1	86.0	94.1	112.6	110.1	98.3	87.6	89.1	87.8	75.2	76.9	86.1	89.6	89.1
0320	Concrete Reinforcing	94.3	119.4	107.1	92.1	93.3	92.7	83.7	119.9	102.1	92.3	119.2	106.0	86.3	92.5	89.5	93.7	106.6	100.2
0330	Cast-in-Place Concrete	103.5	92.1	98.8	98.6	94.8	97.1	77.8	106.3	89.5	87.7	90.5	88.9	91.2	78.2	85.9	106.0	103.0	104.8
03	CONCRETE	100.8	96.9	98.9	96.4	89.1	92.8	80.5	110.8	95.4	93.0	95.4	94.2	85.8	79.8	82.9	97.6	98.5	98.1
04	MASONRY	106.6	94.2	98.9	102.2	88.5	93.7	98.2	112.0	106.8	106.0	87.0	94.1	99.2	82.4	88.7	102.5	103.7	103.3
05	METALS	93.9	115.6	100.6	96.0	87.2	93.3	97.3	98.0	97.5	94.3	115.2	100.7	99.0	88.2	95.7	92.1	120.0	100.7
06	WOOD, PLASTICS & COMPOSITES	96.3	88.8	92.1	93.2	83.4	87.7	93.0	113.1	104.3	93.8	87.6	90.3	85.0	75.1	79.4	71.3	85.5	79.3
07	THERMAL & MOISTURE PROTECTION	98.6	89.6	94.9	93.6	78.9	87.6	96.8	111.2	102.7	99.0	92.6	96.4	95.8	71.7	85.9	104.7	95.4	100.9
08	OPENINGS	103.2	95.1	101.3	95.8	84.5	93.2	99.3	109.4	101.7	97.4	94.5	96.7	104.6	70.6	96.7	100.2	107.2	101.8
0920	Plaster & Gypsum Board	96.9	87.7	90.7	75.7	79.0	77.9	101.3	112.7	109.0	103.8	86.4	92.1	51.7	74.9	67.3	102.2	85.3	90.8
0950, 0980	Ceilings & Acoustic Treatment	94.0	87.7	89.8	97.1	79.0	85.2	88.0	112.7	104.3	88.7	86.4	87.2	94.4	74.9	81.6	122.8	85.3	98.2
0960	Flooring	106.6	85.7	100.6	104.5	85.7	99.1	94.8	113.9	100.3	97.5	81.5	92.9	122.4	94.9	114.5	104.4	120.8	109.1
0970, 0990	Wall Finishes & Painting/Coating	108.0	80.8	91.6	106.3	80.8	90.9	97.6	96.8	97.1	96.2	82.0	87.6	126.1	44.7	76.9	102.2	95.9	98.4
09	FINISHES	98.3	87.6	92.4	94.7	83.2	88.4	91.7	111.5	102.6	92.4	85.5	88.6	99.0	74.1	85.3	106.2	95.2	100.1
COVERS	DIVS. 10-14, 25, 28, 41, 43, 44, 46	100.0	100.6	100.1	100.0	96.0	99.2	100.0	103.0	100.6	100.0	94.3	98.8	100.0	84.6	96.9	100.0	96.9	99.4
21, 22, 23	FIRE SUPPRESSION, PLUMBING & HVAC	99.9	88.6	95.3	99.9	82.3	92.8	95.6	107.2	100.3	100.0	83.1	93.2	95.5	77.7	88.3	95.6	83.4	90.6
26, 27, 3370	ELECTRICAL, COMMUNICATIONS & UTIL.	98.6	93.3	95.8	94.8	79.1	86.5	98.4	104.4	101.6	93.9	90.5	92.1	94.0	79.1	86.1	104.3	104.6	104.5
MF2014	WEIGHTED AVERAGE	99.3	94.7	97.3	97.2	84.8	91.8	94.9	106.9	100.1	96.4	91.8	94.4	96.3	80.1	89.3	98.2	98.2	98.2

		MINNESOTA																	
	DIVISION	BRAINERD 564			DETROIT LAKES 565			DULUTH 556-558			MANKATO 560			MINNEAPOLIS 553-555			ROCHESTER 559		
		MAT.	INST.	TOTAL	MAT.	INST.	TOTAL	MAT.	INST.	TOTAL	MAT.	INST.	TOTAL	MAT.	INST.	TOTAL	MAT.	INST.	TOTAL
015433	CONTRACTOR EQUIPMENT		100.8	100.8		98.2	98.2		99.1	99.1		100.8	100.8		102.6	102.6		99.1	99.1
0241, 31-34	SITE & INFRASTRUCTURE, DEMOLITION	96.9	102.2	100.7	93.6	97.9	96.6	100.6	99.4	99.7	93.8	102.9	100.3	97.9	104.7	102.8	97.4	99.0	98.5
0310	Concrete Forming & Accessories	87.7	91.6	91.1	83.1	91.6	90.4	98.7	109.9	108.4	95.6	99.1	98.6	100.7	128.6	124.8	99.3	106.5	105.5
0320	Concrete Reinforcing	92.5	106.7	99.7	93.7	106.6	100.2	97.4	107.1	102.3	92.4	116.2	104.5	92.2	116.6	104.6	91.7	116.4	104.3
0330	Cast-in-Place Concrete	115.2	106.7	111.7	102.9	106.1	104.2	114.8	106.0	111.2	106.2	107.3	106.6	109.6	117.4	112.8	112.3	99.4	107.0
03	CONCRETE	101.7	100.7	101.2	95.2	100.4	97.8	106.3	108.7	107.6	97.2	106.1	101.5	103.4	122.8	112.9	102.5	106.8	104.6
04	MASONRY	126.4	112.0	117.4	126.3	110.1	116.2	107.6	117.6	113.9	114.5	110.4	112.0	109.0	126.7	120.1	104.5	114.1	110.5
05	METALS	93.2	120.0	101.5	92.1	120.0	100.6	101.7	122.4	108.0	93.1	125.5	103.0	99.1	128.9	108.3	100.4	127.3	108.7
06	WOOD, PLASTICS & COMPOSITES	89.4	85.3	87.1	68.4	85.5	78.0	98.2	109.3	104.4	98.8	96.4	97.5	106.7	127.4	118.3	103.2	105.3	104.4
07	THERMAL & MOISTURE PROTECTION	103.4	107.7	105.2	104.6	108.3	106.1	101.6	114.9	107.0	103.8	99.0	101.8	102.8	127.4	112.9	104.9	100.9	103.2
08	OPENINGS	86.9	107.1	91.6	100.2	107.2	101.8	102.5	116.5	105.8	91.5	115.5	97.1	97.1	133.6	105.6	96.4	121.5	102.2
0920	Plaster & Gypsum Board	86.9	85.3	85.8	101.1	85.3	90.5	90.2	110.0	103.6	91.5	96.8	95.1	95.8	128.6	117.9	97.6	105.8	103.2
0950, 0980	Ceilings & Acoustic Treatment	59.3	85.3	76.4	122.8	85.3	98.2	96.1	110.0	105.3	59.3	96.8	84.0	98.2	128.6	118.2	93.2	105.8	101.5
0960	Flooring	103.3	120.8	108.3	103.2	120.8	108.2	101.9	122.1	107.6	105.0	120.8	109.5	97.9	120.8	104.4	102.9	120.8	108.0
0970, 0990	Wall Finishes & Painting/Coating	96.0	95.9	95.9	102.2	95.9	98.4	101.9	110.3	107.0	108.3	104.9	106.3	98.2	126.1	115.1	97.2	104.9	101.9
09	FINISHES	89.4	96.4	93.2	105.5	96.5	100.6	98.3	112.5	106.1	91.1	103.4	97.9	98.6	127.6	114.6	97.4	109.1	103.8
COVERS	DIVS. 10-14, 25, 28, 41, 43, 44, 46	100.0	98.3	99.7	100.0	98.6	99.7	100.0	97.6	99.5	100.0	98.7	99.7	100.0	109.7	102.0	100.0	102.8	100.6
21, 22, 23	FIRE SUPPRESSION, PLUMBING & HVAC	94.8	86.1	91.3	95.6	86.0	91.7	99.8	96.5	98.4	94.8	87.7	91.9	99.9	115.6	106.2	99.9	96.2	98.4
26, 27, 3370	ELECTRICAL, COMMUNICATIONS & UTIL.	101.8	102.4	102.1	104.0	67.5	84.7	98.4	102.4	100.6	108.6	89.2	98.3	101.2	111.4	106.6	98.6	89.2	93.6
MF2014	WEIGHTED AVERAGE	96.9	100.5	98.5	98.9	95.1	97.3	101.3	107.3	103.9	97.2	101.3	99.0	100.4	120.0	108.9	99.9	104.7	102.0

City Cost Indexes

DIVISION		MINNESOTA															MISSISSIPPI		
		SAINT PAUL			ST. CLOUD			THIEF RIVER FALLS			WILLMAR			WINDOM			BILOXI		
		550 - 551			563			567			562			561			395		
		MAT.	INST.	TOTAL	MAT.	INST.	TOTAL	MAT.	INST.	TOTAL	MAT.	INST.	TOTAL	MAT.	INST.	TOTAL	MAT.	INST.	TOTAL
015433	CONTRACTOR EQUIPMENT		99.1	99.1		100.8	100.8		98.2	98.2		100.8	100.8		100.8	100.8		98.8	98.8
0241, 31 - 34	SITE & INFRASTRUCTURE, DEMOLITION	95.2	100.1	98.7	92.5	104.3	100.9	94.4	97.5	96.6	91.8	102.0	99.1	86.0	101.0	96.6	100.6	88.4	91.9
0310	Concrete Forming & Accessories	93.9	123.0	119.0	84.9	117.7	113.2	86.8	89.0	88.7	84.7	91.1	90.2	89.1	84.8	85.4	94.7	52.9	58.7
0320	Concrete Reinforcing	98.9	116.6	107.9	92.6	116.4	104.7	94.0	106.5	100.4	92.2	116.0	104.3	92.2	115.3	104.0	95.0	58.3	76.4
0330	Cast-in-Place Concrete	124.0	116.7	121.0	101.7	114.8	107.1	105.1	88.1	98.1	103.3	85.3	95.9	89.4	66.3	79.9	108.4	54.0	86.0
03	CONCRETE	111.2	120.1	115.5	92.9	116.9	104.7	96.3	93.1	94.7	92.9	94.9	93.9	83.6	85.5	84.5	101.6	55.9	79.2
04	MASONRY	111.7	126.7	121.1	111.1	119.9	116.6	102.4	104.5	103.7	115.4	112.0	113.3	125.8	92.1	104.8	102.7	44.3	66.3
05	METALS	101.1	128.6	109.6	93.9	126.8	104.0	92.2	119.4	100.6	93.0	124.5	102.7	92.9	122.4	102.0	94.9	81.7	90.8
06	WOOD, PLASTICS & COMPOSITES	93.4	119.9	108.2	86.7	114.9	102.5	72.3	85.5	79.7	86.4	85.5	85.9	90.7	82.8	86.3	99.1	56.1	75.0
07	THERMAL & MOISTURE PROTECTION	104.1	126.4	113.3	103.6	117.5	109.3	105.6	98.4	102.7	103.4	108.1	105.5	103.4	89.6	97.7	97.4	56.1	80.4
08	OPENINGS	99.8	129.5	106.7	91.5	126.7	99.7	100.2	107.2	101.8	89.0	91.9	89.7	92.7	90.4	92.2	99.6	57.7	89.8
0920	Plaster & Gypsum Board	88.9	120.9	110.6	86.9	115.8	106.4	101.8	85.3	90.7	86.9	85.6	86.0	86.9	82.8	84.1	101.7	55.2	70.3
0950, 0980	Ceilings & Acoustic Treatment	95.6	120.9	112.3	59.3	115.8	96.4	122.8	85.3	98.2	59.3	85.6	76.6	59.3	82.8	74.7	87.3	55.0	66.2
0960	Flooring	103.7	120.8	108.6	100.0	120.8	105.9	104.1	120.8	108.8	101.6	120.8	107.1	103.9	120.8	108.8	101.4	55.1	88.1
0970, 0990	Wall Finishes & Painting/Coating	105.3	126.1	117.8	108.3	126.1	119.1	102.2	95.9	98.4	102.2	95.9	98.4	102.2	104.9	103.8	99.5	43.4	65.7
09	FINISHES	97.9	123.1	111.8	88.9	119.2	105.6	106.0	95.2	100.0	88.9	94.2	91.8	89.1	91.3	90.3	93.3	52.4	70.8
COVERS	DIVS. 10 - 14, 25, 28, 41, 43, 44, 46	100.0	108.6	101.7	100.0	104.3	100.9	100.0	96.8	99.3	100.0	98.2	99.6	100.0	93.7	98.7	100.0	55.4	91.0
21, 22, 23	FIRE SUPPRESSION, PLUMBING & HVAC	99.9	113.8	105.5	99.6	113.6	105.3	95.6	83.1	90.5	94.8	106.8	99.6	94.8	84.1	90.5	100.0	49.1	79.5
26, 27, 3370	ELECTRICAL, COMMUNICATIONS & UTIL.	97.1	111.4	104.7	101.8	111.4	106.9	101.3	67.5	83.5	101.8	85.2	93.0	108.6	89.2	98.3	103.8	56.9	79.0
MF2014	WEIGHTED AVERAGE	101.4	117.9	108.6	96.9	116.0	105.2	97.8	92.4	95.4	95.5	101.3	98.0	95.9	92.3	94.3	99.2	58.1	81.3

DIVISION		MISSISSIPPI																	
		CLARKSDALE			COLUMBUS			GREENVILLE			GREENWOOD			JACKSON			LAUREL		
		386			397			387			389			390 - 392			394		
		MAT.	INST.	TOTAL	MAT.	INST.	TOTAL	MAT.	INST.	TOTAL	MAT.	INST.	TOTAL	MAT.	INST.	TOTAL	MAT.	INST.	TOTAL
015433	CONTRACTOR EQUIPMENT		98.8	98.8		98.8	98.8		98.8	98.8		98.8	98.8		98.8	98.8		98.8	98.8
0241, 31 - 34	SITE & INFRASTRUCTURE, DEMOLITION	98.3	88.2	91.1	98.9	88.4	91.5	103.9	88.4	92.9	101.0	87.9	91.7	97.6	88.4	91.1	104.4	88.3	92.9
0310	Concrete Forming & Accessories	85.0	49.2	54.2	83.3	51.2	55.6	81.7	63.7	66.1	94.2	49.2	55.4	93.6	57.5	62.5	83.3	52.9	57.1
0320	Concrete Reinforcing	99.8	41.6	70.1	101.8	42.5	71.6	100.3	60.3	79.9	99.8	52.5	75.7	100.2	56.1	77.7	102.4	39.1	70.2
0330	Cast-in-Place Concrete	105.6	53.5	84.2	110.5	58.6	89.2	108.7	55.5	86.9	113.5	53.2	88.7	98.7	56.8	81.5	108.1	55.2	86.4
03	CONCRETE	98.7	50.9	75.2	103.2	53.7	78.9	104.2	61.6	83.3	105.2	52.7	79.4	97.7	58.5	78.5	106.0	52.7	79.8
04	MASONRY	100.6	45.0	66.0	131.6	51.6	81.7	149.8	49.5	87.3	101.3	44.9	66.1	109.2	49.5	72.0	127.1	48.1	77.9
05	METALS	95.4	73.4	88.6	91.9	74.2	86.5	96.4	82.5	92.1	95.4	74.8	89.0	101.8	80.8	95.4	92.7	73.3	86.3
06	WOOD, PLASTICS & COMPOSITES	82.1	51.2	64.8	83.9	52.2	66.1	78.7	68.1	72.8	95.0	51.2	70.5	96.3	59.6	75.8	85.1	54.6	68.0
07	THERMAL & MOISTURE PROTECTION	95.7	47.2	75.8	97.3	51.5	78.5	96.0	60.0	81.2	96.1	47.1	76.0	97.5	59.2	81.8	97.4	53.1	79.3
08	OPENINGS	97.0	49.4	85.9	98.9	49.5	87.4	97.0	61.4	88.7	97.0	50.7	86.2	103.7	55.6	92.5	95.6	50.3	85.0
0920	Plaster & Gypsum Board	90.2	50.2	63.1	91.3	51.2	64.2	89.8	67.6	74.8	100.9	50.2	66.6	95.0	58.8	70.5	91.3	53.7	65.8
0950, 0980	Ceilings & Acoustic Treatment	82.9	50.2	61.4	82.0	51.2	61.8	85.7	67.6	73.8	82.9	50.2	61.4	89.7	58.8	69.4	82.0	53.7	63.4
0960	Flooring	105.7	52.8	90.6	95.5	59.3	85.1	104.1	52.8	89.4	111.1	52.8	94.4	99.3	55.1	86.7	94.2	52.8	82.4
0970, 0990	Wall Finishes & Painting/Coating	104.4	46.1	69.2	99.5	53.0	71.4	104.4	65.9	81.2	104.4	46.1	69.2	96.5	65.9	78.0	99.5	60.9	76.2
09	FINISHES	94.0	49.9	69.7	89.1	53.2	69.3	94.6	62.9	77.1	97.4	49.9	71.2	93.2	58.3	74.0	89.2	54.3	70.0
COVERS	DIVS. 10 - 14, 25, 28, 41, 43, 44, 46	100.0	55.7	91.1	100.0	56.8	91.3	100.0	58.5	91.6	100.0	55.7	91.1	100.0	57.5	91.4	100.0	37.5	87.4
21, 22, 23	FIRE SUPPRESSION, PLUMBING & HVAC	97.9	41.3	75.1	97.4	37.5	73.3	100.0	47.6	78.8	97.9	39.2	74.3	100.1	57.0	82.7	97.5	45.8	76.6
26, 27, 3370	ELECTRICAL, COMMUNICATIONS & UTIL.	97.1	46.4	70.3	100.8	49.1	73.5	97.1	62.7	78.9	97.1	43.4	68.8	105.4	62.7	82.8	102.6	56.9	78.5
MF2014	WEIGHTED AVERAGE	97.3	52.6	77.8	98.8	53.9	79.2	101.0	61.7	83.9	98.4	52.2	78.3	100.7	62.2	83.9	98.9	55.8	80.1

DIVISION		MISSISSIPPI									MISSOURI								
		MCCOMB			MERIDIAN			TUPELO			BOWLING GREEN			CAPE GIRARDEAU			CHILLICOTHE		
		396			393			388			633			637			646		
		MAT.	INST.	TOTAL	MAT.	INST.	TOTAL	MAT.	INST.	TOTAL	MAT.	INST.	TOTAL	MAT.	INST.	TOTAL	MAT.	INST.	TOTAL
015433	CONTRACTOR EQUIPMENT		98.8	98.8		98.8	98.8		98.8	98.8		106.9	106.9		106.9	106.9		101.9	101.9
0241, 31 - 34	SITE & INFRASTRUCTURE, DEMOLITION	92.2	88.0	89.2	96.2	88.7	90.9	95.6	88.1	90.3	89.6	95.6	93.9	91.3	95.4	94.2	105.2	94.6	97.7
0310	Concrete Forming & Accessories	83.3	50.9	55.4	80.4	63.3	65.7	82.2	51.2	55.5	92.3	87.3	88.0	85.2	85.1	85.1	89.0	93.0	92.5
0320	Concrete Reinforcing	103.0	41.0	71.4	101.8	56.1	78.5	97.6	52.5	74.7	101.4	102.8	102.1	102.7	88.7	95.5	96.8	106.5	101.8
0330	Cast-in-Place Concrete	96.0	53.0	78.3	102.7	59.2	84.8	105.6	54.9	84.8	91.2	84.7	88.5	90.2	86.0	88.5	103.0	85.8	95.9
03	CONCRETE	92.9	51.3	72.5	97.2	61.9	79.8	98.2	54.2	76.6	95.8	90.5	93.2	94.9	87.3	91.2	105.6	93.6	99.7
04	MASONRY	133.0	44.4	77.7	102.4	53.5	71.9	138.4	47.9	82.0	117.6	98.7	105.8	113.9	80.0	92.8	102.9	98.8	100.3
05	METALS	92.2	71.6	85.8	93.0	81.0	89.3	95.3	75.1	89.1	94.6	114.4	100.7	95.6	107.0	99.1	95.1	108.5	99.2
06	WOOD, PLASTICS & COMPOSITES	83.9	54.0	67.2	81.2	65.1	72.2	79.3	52.2	64.1	90.8	86.0	88.1	83.5	85.5	84.6	93.6	93.3	93.4
07	THERMAL & MOISTURE PROTECTION	96.9	46.4	76.2	97.1	61.6	82.5	95.7	52.9	78.2	100.3	98.9	99.7	100.2	84.2	93.6	98.3	94.5	96.8
08	OPENINGS	99.0	51.0	87.8	98.9	64.1	90.8	97.0	50.6	86.2	97.7	97.5	97.6	97.7	82.9	94.2	95.0	95.7	95.2
0920	Plaster & Gypsum Board	91.3	53.0	65.4	91.3	64.5	73.2	89.8	51.2	63.7	95.2	85.6	88.7	94.2	85.1	88.0	102.0	92.9	95.8
0950, 0980	Ceilings & Acoustic Treatment	82.0	53.0	63.0	83.9	64.5	71.2	82.9	51.2	62.0	92.6	85.6	88.0	92.6	85.1	87.6	89.1	92.9	91.6
0960	Flooring	95.5	52.8	83.3	94.2	55.1	83.0	104.4	52.8	89.6	96.4	96.8	96.5	93.3	84.0	90.6	97.3	102.5	98.8
0970, 0990	Wall Finishes & Painting/Coating	99.5	42.3	65.0	99.5	76.1	85.4	104.4	58.7	76.8	97.1	102.8	100.5	97.1	73.2	82.7	87.9	106.3	99.0
09	FINISHES	88.5	51.0	67.8	88.8	63.7	75.0	93.5	52.6	71.0	97.8	89.7	93.4	96.7	83.0	89.2	100.2	96.1	97.9
COVERS	DIVS. 10 - 14, 25, 28, 41, 43, 44, 46	100.0	59.0	91.7	100.0	59.5	91.8	100.0	56.8	91.3	100.0	88.2	97.6	100.0	93.5	98.7	100.0	88.5	97.7
21, 22, 23	FIRE SUPPRESSION, PLUMBING & HVAC	97.4	34.6	72.1	100.0	31.9	83.5	98.1	43.2	76.0	95.2	96.2	95.6	99.9	96.8	98.7	95.3	97.0	96.0
26, 27, 3370	ELECTRICAL, COMMUNICATIONS & UTIL.	99.2	49.4	72.9	102.6	57.8	78.9	96.9	49.3	71.7	96.9	82.6	89.3	96.9	101.5	99.3	92.4	78.6	85.1
MF2014	WEIGHTED AVERAGE	97.4	51.7	77.5	97.6	64.1	83.0	98.8	54.9	79.7	97.3	94.4	96.0	98.2	92.4	95.7	97.6	94.5	96.3

775

City Cost Indexes

DIVISION		MISSOURI																	
		COLUMBIA 652			FLAT RIVER 636			HANNIBAL 634			HARRISONVILLE 647			JEFFERSON CITY 650 - 651			JOPLIN 648		
		MAT.	INST.	TOTAL	MAT.	INST.	TOTAL	MAT.	INST.	TOTAL	MAT.	INST.	TOTAL	MAT.	INST.	TOTAL	MAT.	INST.	TOTAL
015433	CONTRACTOR EQUIPMENT		108.0	108.0		106.9	106.9		106.9	106.9		101.9	101.9		108.0	108.0		105.7	105.7
0241, 31 - 34	SITE & INFRASTRUCTURE, DEMOLITION	102.8	97.2	98.8	92.2	95.6	94.6	87.6	95.7	93.4	96.6	95.7	95.9	102.3	97.2	98.6	105.6	99.7	101.4
0310	Concrete Forming & Accessories	86.9	81.8	82.5	98.3	84.3	86.2	90.5	79.0	80.6	86.2	98.7	97.0	98.3	81.8	84.1	101.3	75.3	78.9
0320	Concrete Reinforcing	96.3	119.0	107.9	102.7	107.7	105.3	100.9	97.7	99.3	96.4	115.3	106.0	98.5	111.9	105.3	99.9	98.5	99.2
0330	Cast-in-Place Concrete	96.1	87.8	92.7	94.2	93.8	94.0	86.3	86.0	86.2	105.5	103.9	104.8	102.0	87.0	95.9	111.5	77.9	97.7
03	CONCRETE	93.2	92.2	92.7	98.8	93.2	96.0	92.1	86.3	89.2	101.8	104.0	102.9	98.9	90.6	94.8	105.2	81.4	93.5
04	MASONRY	154.1	90.0	114.2	114.5	81.4	93.9	109.0	101.3	104.2	97.4	103.4	101.2	112.0	90.0	98.3	96.5	85.9	89.9
05	METALS	98.4	121.2	105.4	94.5	115.6	101.0	94.6	111.9	99.9	95.6	113.2	101.0	98.0	118.3	104.2	98.2	100.5	98.9
06	WOOD, PLASTICS & COMPOSITES	91.7	78.7	84.4	99.1	82.6	89.9	88.9	73.8	80.4	89.9	97.5	94.2	100.0	78.7	88.1	106.0	74.2	88.2
07	THERMAL & MOISTURE PROTECTION	96.2	87.5	92.6	100.5	93.8	97.7	100.1	97.0	98.8	97.6	103.2	99.9	101.8	87.6	96.0	97.7	78.9	90.0
08	OPENINGS	99.5	95.0	98.4	97.7	97.2	97.6	97.7	82.3	94.1	95.1	102.8	96.9	99.1	93.1	97.7	96.2	78.5	92.1
0920	Plaster & Gypsum Board	87.2	78.0	81.0	101.3	82.1	88.3	94.5	73.0	79.9	97.3	97.3	97.3	91.3	78.0	82.3	108.2	73.2	84.5
0950, 0980	Ceilings & Acoustic Treatment	93.2	78.0	83.2	92.6	82.1	85.7	92.6	73.0	79.7	89.1	97.3	94.5	97.2	78.0	84.6	90.0	73.2	78.9
0960	Flooring	97.6	98.6	97.9	99.4	84.0	95.0	95.8	96.8	96.1	92.6	102.5	95.4	104.4	98.6	102.7	123.8	74.9	109.8
0970, 0990	Wall Finishes & Painting/Coating	105.2	77.1	88.2	97.1	78.4	85.8	97.1	102.8	100.5	92.5	103.8	99.3	102.3	77.1	87.1	87.5	74.9	79.9
09	FINISHES	92.8	83.0	87.4	99.7	82.2	90.1	97.4	83.1	89.5	97.8	99.5	98.7	97.2	83.0	89.4	107.0	75.3	89.6
COVERS	DIVS. 10 - 14, 25, 28, 41, 43, 44, 46	100.0	96.2	99.2	100.0	91.8	98.3	100.0	87.8	97.5	100.0	90.6	98.1	100.0	96.2	99.2	100.0	87.9	97.6
21, 22, 23	FIRE SUPPRESSION, PLUMBING & HVAC	99.8	96.4	98.4	95.2	97.8	96.2	95.2	97.5	96.1	95.2	100.0	97.2	99.8	98.1	99.1	100.1	72.1	88.8
26, 27, 3370	ELECTRICAL, COMMUNICATIONS & UTIL.	94.8	86.5	90.4	101.1	101.5	101.3	95.7	82.6	88.8	98.8	99.6	99.2	100.4	86.5	93.0	90.4	77.5	83.6
MF2014	WEIGHTED AVERAGE	100.2	93.9	97.5	98.1	95.0	96.8	96.2	92.4	94.6	97.1	101.5	99.1	99.9	93.7	97.2	99.5	81.7	91.8

DIVISION		MISSOURI																	
		KANSAS CITY 640 - 641			KIRKSVILLE 635			POPLAR BLUFF 639			ROLLA 654 - 655			SEDALIA 653			SIKESTON 638		
		MAT.	INST.	TOTAL	MAT.	INST.	TOTAL	MAT.	INST.	TOTAL	MAT.	INST.	TOTAL	MAT.	INST.	TOTAL	MAT.	INST.	TOTAL
015433	CONTRACTOR EQUIPMENT		103.3	103.3		99.1	99.1		101.5	101.5		108.0	108.0		99.9	99.9		101.5	101.5
0241, 31 - 34	SITE & INFRASTRUCTURE, DEMOLITION	98.9	98.2	98.4	91.8	91.4	91.5	79.1	95.4	90.7	101.5	97.1	98.3	99.0	92.8	94.6	82.4	95.4	91.7
0310	Concrete Forming & Accessories	100.1	105.4	104.6	83.2	80.6	81.0	83.3	85.0	84.7	95.0	91.9	92.3	92.4	81.4	82.9	84.3	84.9	84.9
0320	Concrete Reinforcing	94.9	119.8	107.6	101.6	98.8	100.2	104.7	94.7	99.6	96.7	94.7	95.7	95.3	106.2	100.8	104.0	98.4	101.2
0330	Cast-in-Place Concrete	103.7	106.6	104.9	94.1	83.8	89.9	72.4	86.5	78.2	98.3	96.2	97.5	102.7	83.8	94.9	77.4	86.5	81.1
03	CONCRETE	101.1	107.8	104.8	110.8	86.0	98.6	85.9	88.1	86.9	95.2	95.0	95.1	109.8	87.7	99.0	89.5	88.7	89.1
04	MASONRY	99.5	107.5	104.5	121.0	88.9	101.0	112.5	80.1	92.3	126.3	87.9	102.4	132.5	88.3	105.1	112.2	80.1	92.2
05	METALS	105.9	115.8	108.9	94.2	103.6	97.2	94.8	101.7	96.9	97.8	110.5	101.7	96.5	108.0	100.1	95.1	103.2	97.6
06	WOOD, PLASTICS & COMPOSITES	105.3	105.0	105.1	77.0	78.7	77.9	76.1	85.5	81.4	100.3	92.8	96.1	93.1	78.7	85.1	77.7	85.5	82.1
07	THERMAL & MOISTURE PROTECTION	97.8	106.8	101.5	106.8	88.2	99.2	105.2	85.9	97.3	96.4	93.4	95.1	102.5	93.1	98.7	105.3	84.5	96.8
08	OPENINGS	102.5	109.2	104.0	102.7	85.8	98.8	103.7	84.5	99.2	99.5	88.3	96.9	104.5	87.4	100.5	103.7	85.5	99.5
0920	Plaster & Gypsum Board	105.2	105.0	105.0	89.8	78.0	81.8	90.2	85.1	86.7	89.3	92.5	91.5	82.5	78.0	79.5	92.0	85.1	87.3
0950, 0980	Ceilings & Acoustic Treatment	96.7	105.0	102.1	90.9	78.0	82.4	92.6	85.1	87.6	93.2	92.5	92.7	93.2	78.0	83.2	92.6	85.1	87.6
0960	Flooring	98.0	109.0	101.2	74.4	95.8	80.5	88.3	82.0	86.5	101.2	95.8	99.7	78.8	95.8	83.7	88.7	82.0	86.8
0970, 0990	Wall Finishes & Painting/Coating	92.5	112.2	104.4	92.5	82.1	86.2	92.0	73.2	80.7	105.5	100.6	102.5	105.2	106.3	105.9	92.0	73.2	80.7
09	FINISHES	102.0	106.6	104.5	96.5	82.7	88.9	96.5	82.7	88.9	94.3	93.4	93.8	91.9	85.4	88.3	97.1	82.7	89.2
COVERS	DIVS. 10 - 14, 25, 28, 41, 43, 44, 46	100.0	100.6	100.1	100.0	87.1	97.4	100.0	93.6	98.7	100.0	89.0	97.8	100.0	90.2	98.0	100.0	94.2	98.8
21, 22, 23	FIRE SUPPRESSION, PLUMBING & HVAC	100.0	104.3	101.7	95.2	97.3	96.1	95.2	96.3	95.6	95.0	98.2	96.3	95.0	95.9	95.4	95.2	96.3	95.6
26, 27, 3370	ELECTRICAL, COMMUNICATIONS & UTIL.	100.3	100.1	100.2	95.8	82.5	88.8	96.1	101.5	99.0	93.2	86.5	89.6	94.7	128.8	112.7	95.3	101.5	98.6
MF2014	WEIGHTED AVERAGE	101.4	105.7	103.3	99.5	89.9	95.3	96.2	92.0	94.4	97.9	94.6	96.5	100.2	97.3	98.9	96.7	92.2	94.8

DIVISION		MISSOURI									MONTANA								
		SPRINGFIELD 656 - 658			ST. JOSEPH 644 - 645			ST. LOUIS 630 - 631			BILLINGS 590 - 591			BUTTE 597			GREAT FALLS 594		
		MAT.	INST.	TOTAL	MAT.	INST.	TOTAL	MAT.	INST.	TOTAL	MAT.	INST.	TOTAL	MAT.	INST.	TOTAL	MAT.	INST.	TOTAL
015433	CONTRACTOR EQUIPMENT		102.5	102.5		101.9	101.9		107.9	107.9		98.5	98.5		98.2	98.2		98.2	98.2
0241, 31 - 34	SITE & INFRASTRUCTURE, DEMOLITION	101.6	95.1	97.0	100.5	93.9	95.8	91.7	98.9	96.8	95.0	96.3	95.9	102.0	96.1	97.8	105.7	96.2	98.9
0310	Concrete Forming & Accessories	101.4	79.1	82.2	100.0	89.6	91.0	98.4	104.4	103.9	98.4	71.5	75.2	85.5	71.8	73.7	98.4	71.5	75.2
0320	Concrete Reinforcing	92.7	118.7	105.9	93.8	115.0	104.6	93.9	112.7	103.4	90.2	81.7	85.9	97.8	81.9	89.7	90.2	81.8	85.9
0330	Cast-in-Place Concrete	104.3	79.6	94.2	103.6	103.3	103.5	90.2	105.0	96.3	129.3	67.6	103.9	142.3	69.0	112.2	150.4	63.7	114.8
03	CONCRETE	105.8	87.4	96.8	100.9	99.7	100.3	94.4	107.1	100.7	109.8	72.9	91.7	113.6	73.5	94.0	119.5	71.5	96.0
04	MASONRY	102.9	89.8	94.8	99.3	96.1	97.3	95.4	111.9	105.7	128.0	73.7	94.2	123.5	76.6	94.3	128.0	75.4	95.2
05	METALS	102.8	108.8	104.7	101.9	112.4	105.2	100.3	120.2	106.4	105.8	90.0	100.9	100.1	90.1	97.1	103.2	90.0	99.2
06	WOOD, PLASTICS & COMPOSITES	101.8	77.1	88.0	106.0	87.6	95.7	99.0	103.0	101.2	98.6	70.5	82.8	85.7	70.8	77.4	99.9	70.5	83.4
07	THERMAL & MOISTURE PROTECTION	100.7	82.2	93.2	98.2	95.6	97.1	100.2	106.9	103.0	106.5	69.9	91.5	106.6	71.0	91.8	106.9	68.5	91.2
08	OPENINGS	107.2	84.3	101.9	100.6	98.6	100.1	97.6	110.9	100.7	97.3	68.7	90.6	95.6	70.0	89.6	98.5	68.0	91.4
0920	Plaster & Gypsum Board	90.4	76.3	80.9	109.7	87.0	94.4	102.2	103.1	102.8	98.0	69.8	78.9	98.7	70.2	79.4	107.3	69.8	82.0
0950, 0980	Ceilings & Acoustic Treatment	93.2	76.3	82.1	95.8	87.0	90.0	97.6	103.1	101.2	81.5	69.8	73.8	88.3	70.2	76.4	90.0	69.8	76.7
0960	Flooring	99.9	74.9	92.8	101.5	107.3	103.2	99.2	97.6	98.7	106.6	70.0	96.2	104.4	74.2	95.7	111.5	74.2	100.8
0970, 0990	Wall Finishes & Painting/Coating	99.0	74.9	84.4	87.9	86.7	87.2	97.1	106.0	102.5	105.2	86.6	94.0	103.4	51.2	71.9	103.4	86.6	93.3
09	FINISHES	96.7	78.0	86.4	103.2	91.7	96.9	100.8	103.1	102.1	96.6	72.9	83.4	97.2	69.7	82.0	101.1	73.3	85.8
COVERS	DIVS. 10 - 14, 25, 28, 41, 43, 44, 46	100.0	94.8	99.0	100.0	99.4	99.9	100.0	102.4	100.5	100.0	97.4	98.9	100.0	94.8	98.9	100.0	94.7	98.9
21, 22, 23	FIRE SUPPRESSION, PLUMBING & HVAC	99.8	74.6	89.6	100.1	89.0	95.6	100.0	104.7	103.0	100.1	74.7	89.9	100.0	75.6	90.3	100.2	70.7	88.3
26, 27, 3370	ELECTRICAL, COMMUNICATIONS & UTIL.	99.2	69.7	83.6	98.8	78.6	88.2	99.2	108.3	104.0	98.5	72.2	84.6	105.5	71.7	87.7	97.7	71.9	84.1
MF2014	WEIGHTED AVERAGE	101.7	83.7	93.8	100.6	93.5	97.5	98.7	107.8	102.7	102.7	77.1	91.5	102.6	77.3	91.6	104.1	76.2	91.9

City Cost Indexes

		MONTANA																
	DIVISION	HAVRE			HELENA			KALISPELL			MILES CITY			MISSOULA			WOLF POINT		
		595			596			599			593			598			592		
		MAT.	INST.	TOTAL	MAT.	INST.	TOTAL	MAT.	INST.	TOTAL	MAT.	INST.	TOTAL	MAT.	INST.	TOTAL	MAT.	INST.	TOTAL
015433	CONTRACTOR EQUIPMENT		98.2	98.2		98.2	98.2		98.2	98.2		98.2	98.2		98.2	98.2		98.2	98.2
0241, 31 - 34	SITE & INFRASTRUCTURE, DEMOLITION	109.2	95.8	99.7	98.5	95.8	96.6	92.2	96.0	94.9	98.5	95.4	96.3	85.1	95.8	92.7	115.5	95.6	101.3
0310	Concrete Forming & Accessories	78.4	69.0	70.3	101.5	65.2	70.2	88.7	71.4	73.8	96.8	67.6	71.6	88.7	71.4	73.8	89.2	67.9	70.8
0320	Concrete Reinforcing	98.6	81.7	90.0	102.5	81.7	91.9	100.4	83.4	91.8	98.2	81.8	89.8	99.5	83.4	91.3	99.6	81.8	90.5
0330	Cast-in-Place Concrete	153.3	60.4	115.1	115.0	62.9	93.6	123.6	62.7	98.6	135.3	60.3	104.5	104.9	64.9	88.4	151.5	60.6	114.2
03	CONCRETE	122.7	69.3	96.5	106.7	68.4	87.9	102.6	71.5	87.3	110.4	68.6	89.9	90.0	72.2	81.3	126.5	68.9	98.2
04	MASONRY	124.5	69.8	90.4	120.1	75.7	92.4	122.2	77.0	94.0	129.7	65.3	89.6	147.9	76.8	103.4	131.0	66.0	90.5
05	METALS	96.3	89.6	94.2	102.1	88.8	98.0	96.1	90.6	94.4	95.4	89.9	93.7	96.6	90.7	94.8	95.5	89.2	93.6
06	WOOD, PLASTICS & COMPOSITES	77.4	70.5	73.5	102.9	62.3	80.2	89.1	70.5	78.7	96.9	70.5	82.1	89.1	70.5	78.7	88.5	70.5	78.4
07	THERMAL & MOISTURE PROTECTION	106.6	70.0	90.8	103.9	70.7	90.3	105.9	72.8	92.3	106.2	65.7	89.6	105.5	79.1	94.7	107.1	66.0	90.3
08	OPENINGS	95.6	68.0	89.2	99.8	65.2	91.7	95.6	68.5	89.3	95.1	68.0	88.8	95.6	68.3	89.2	95.1	67.4	88.7
0920	Plaster & Gypsum Board	94.8	69.8	77.9	96.0	61.4	72.6	98.7	69.8	79.2	106.7	69.8	81.8	98.7	69.8	79.2	102.1	69.8	80.3
0950, 0980	Ceilings & Acoustic Treatment	88.3	69.8	76.2	90.8	61.4	71.5	88.3	69.8	76.2	87.5	69.8	75.9	88.3	69.8	76.2	87.5	69.8	75.9
0960	Flooring	101.8	74.2	93.9	106.7	59.1	93.1	106.2	74.2	97.1	111.3	70.2	99.5	106.2	74.2	97.1	107.6	70.2	96.9
0970, 0990	Wall Finishes & Painting/Coating	103.4	48.6	70.4	99.7	47.2	68.0	103.4	47.2	69.5	103.4	47.2	69.5	103.4	62.0	78.4	103.4	48.6	70.4
09	FINISHES	96.5	67.7	80.6	98.6	61.6	78.2	97.2	69.1	81.7	100.0	65.7	81.1	96.7	70.7	82.3	99.6	66.0	81.1
COVERS	DIVS. 10 - 14, 25, 28, 41, 43, 44, 46	100.0	70.8	94.1	100.0	94.0	98.8	100.0	93.9	98.8	100.0	90.4	98.1	100.0	93.7	98.7	100.0	90.6	98.1
21, 22, 23	FIRE SUPPRESSION, PLUMBING & HVAC	95.4	67.8	84.3	100.1	70.1	88.0	95.4	69.2	84.8	95.4	69.6	85.0	100.2	69.2	87.7	95.4	70.0	85.1
26, 27, 3370	ELECTRICAL, COMMUNICATIONS & UTIL.	97.7	71.2	83.7	104.6	71.3	87.0	102.2	69.3	84.8	97.7	76.9	86.7	103.2	67.7	84.5	97.7	76.9	86.7
MF2014	WEIGHTED AVERAGE	101.2	73.1	89.0	102.4	73.8	89.9	99.0	75.3	88.7	100.1	74.0	88.7	99.9	75.5	89.3	102.3	74.2	90.0

		NEBRASKA																
	DIVISION	ALLIANCE			COLUMBUS			GRAND ISLAND			HASTINGS			LINCOLN			MCCOOK		
		693			686			688			689			683 - 685			690		
		MAT.	INST.	TOTAL	MAT.	INST.	TOTAL	MAT.	INST.	TOTAL	MAT.	INST.	TOTAL	MAT.	INST.	TOTAL	MAT.	INST.	TOTAL
015433	CONTRACTOR EQUIPMENT		97.8	97.8		101.6	101.6		101.6	101.6		101.6	101.6		101.6	101.6		101.6	101.6
0241, 31 - 34	SITE & INFRASTRUCTURE, DEMOLITION	100.5	98.8	99.3	100.2	92.3	94.6	105.0	93.1	96.5	103.7	92.3	95.6	92.7	93.0	92.9	103.8	92.2	95.6
0310	Concrete Forming & Accessories	87.3	56.4	60.7	95.5	67.4	71.2	95.1	72.4	75.5	98.2	71.9	75.5	95.7	68.2	71.9	93.0	56.1	61.2
0320	Concrete Reinforcing	108.8	85.1	96.7	98.3	76.5	87.2	97.7	76.4	86.9	97.7	77.1	87.2	96.7	76.0	86.1	101.8	76.2	88.8
0330	Cast-in-Place Concrete	113.7	60.0	91.6	115.7	60.0	92.8	122.5	67.2	99.8	122.5	63.2	98.0	92.4	69.6	83.1	122.6	59.3	96.6
03	CONCRETE	123.9	63.5	94.2	108.8	67.3	88.4	113.7	72.1	93.3	113.9	70.6	92.6	95.6	71.0	83.5	112.9	67.9	91.8
04	MASONRY	115.6	74.1	89.7	122.7	77.8	94.7	115.4	76.1	90.9	124.9	88.0	101.9	104.8	67.2	81.4	110.7	73.9	87.8
05	METALS	101.5	78.7	94.5	96.1	84.6	92.5	97.8	86.3	94.3	98.7	85.8	94.7	99.9	85.5	95.5	96.3	83.6	92.4
06	WOOD, PLASTICS & COMPOSITES	87.0	52.2	67.5	101.2	66.7	81.9	100.4	72.2	84.6	104.1	72.2	86.3	98.4	66.7	80.7	95.7	52.2	71.4
07	THERMAL & MOISTURE PROTECTION	104.3	67.8	89.4	100.8	69.8	88.1	100.9	72.8	89.4	100.9	83.4	93.8	98.1	69.9	86.5	98.9	66.6	85.6
08	OPENINGS	94.5	58.6	86.1	95.4	66.1	88.6	95.5	69.7	89.5	95.5	69.1	89.3	107.8	63.0	97.4	95.0	56.6	86.1
0920	Plaster & Gypsum Board	82.7	50.7	61.1	90.9	65.6	73.8	90.1	71.3	77.4	91.9	71.3	78.0	96.2	65.6	75.5	93.7	50.7	64.7
0950, 0980	Ceilings & Acoustic Treatment	91.9	50.7	64.8	80.2	65.6	70.6	80.2	71.3	74.3	80.2	71.3	74.3	90.4	65.6	74.1	88.0	50.7	63.5
0960	Flooring	99.9	82.1	94.8	92.9	100.4	95.0	92.7	107.2	96.8	93.9	100.4	95.7	101.0	83.6	96.0	97.6	82.1	93.2
0970, 0990	Wall Finishes & Painting/Coating	168.5	54.8	99.9	85.5	63.7	72.4	85.5	67.6	74.7	85.5	63.7	72.4	103.8	77.9	88.1	93.8	47.7	66.0
09	FINISHES	98.4	59.5	77.0	89.9	71.9	80.0	90.0	77.1	82.9	90.6	75.2	82.1	96.9	71.7	83.0	95.6	58.6	75.2
COVERS	DIVS. 10 - 14, 25, 28, 41, 43, 44, 46	100.0	65.3	93.0	100.0	86.1	97.2	100.0	88.7	97.7	100.0	86.7	97.3	100.0	88.1	97.6	100.0	64.5	92.8
21, 22, 23	FIRE SUPPRESSION, PLUMBING & HVAC	95.3	70.5	85.3	95.2	76.3	87.4	100.0	78.0	91.1	95.2	76.0	87.5	99.9	78.0	91.0	95.1	75.9	87.4
26, 27, 3370	ELECTRICAL, COMMUNICATIONS & UTIL.	92.3	69.4	80.2	93.6	82.3	87.6	92.2	67.0	78.9	91.6	83.6	87.4	105.4	67.0	85.1	94.4	69.4	81.2
MF2014	WEIGHTED AVERAGE	101.0	70.5	87.7	98.4	77.1	89.1	100.0	77.1	90.0	99.4	79.8	90.9	100.5	74.9	89.4	98.7	71.1	86.7

		NEBRASKA												NEVADA					
	DIVISION	NORFOLK			NORTH PLATTE			OMAHA			VALENTINE			CARSON CITY			ELKO		
		687			691			680 - 681			692			897			898		
		MAT.	INST.	TOTAL	MAT.	INST.	TOTAL	MAT.	INST.	TOTAL	MAT.	INST.	TOTAL	MAT.	INST.	TOTAL	MAT.	INST.	TOTAL
015433	CONTRACTOR EQUIPMENT		92.8	92.8		101.6	101.6		92.8	92.8		95.9	95.9		98.2	98.2		98.2	98.2
0241, 31 - 34	SITE & INFRASTRUCTURE, DEMOLITION	83.6	91.5	89.2	105.2	92.3	96.0	90.0	92.1	91.5	88.6	96.5	94.2	84.5	99.0	94.8	66.9	97.6	88.8
0310	Concrete Forming & Accessories	81.7	71.1	72.5	95.4	71.5	74.8	94.3	72.4	75.4	83.7	55.8	59.6	103.3	92.8	94.2	109.2	81.3	85.2
0320	Concrete Reinforcing	98.4	66.5	82.2	101.2	77.1	88.9	99.7	76.5	87.9	101.8	66.1	83.7	103.2	118.4	111.0	102.7	113.2	108.1
0330	Cast-in-Place Concrete	116.7	62.4	94.4	122.6	64.5	98.7	101.1	75.5	90.6	108.4	57.2	87.4	100.4	85.7	94.4	95.1	80.2	89.0
03	CONCRETE	107.3	67.5	87.8	113.0	70.8	92.3	100.1	74.5	87.5	110.2	58.9	85.0	100.9	95.2	98.1	95.4	87.2	91.4
04	MASONRY	129.4	79.1	98.0	98.6	87.9	92.0	106.8	79.0	89.4	110.9	73.9	87.8	117.3	94.6	103.2	115.0	71.9	88.1
05	METALS	99.6	73.5	91.6	95.5	85.4	92.4	99.9	78.9	93.4	107.6	73.4	97.1	94.0	104.7	97.3	96.9	99.4	97.7
06	WOOD, PLASTICS & COMPOSITES	84.0	71.7	77.1	97.8	72.2	83.5	94.1	72.0	81.7	81.7	51.7	64.9	92.1	94.4	93.4	100.6	81.8	90.1
07	THERMAL & MOISTURE PROTECTION	100.8	73.6	89.6	98.9	82.8	92.3	95.5	78.7	88.6	99.6	67.6	86.5	103.5	92.3	99.0	99.9	76.1	90.1
08	OPENINGS	97.2	66.5	90.0	94.3	69.1	88.4	100.0	71.5	93.3	96.9	55.6	87.3	99.5	107.9	101.4	100.9	85.9	97.4
0920	Plaster & Gypsum Board	90.8	71.3	77.6	93.7	71.3	78.6	100.5	71.6	81.0	95.6	50.7	65.3	100.9	94.1	96.3	105.3	81.2	89.0
0950, 0980	Ceilings & Acoustic Treatment	92.0	71.3	78.4	88.0	71.3	77.0	93.0	71.6	78.9	104.9	50.7	69.3	94.0	94.1	94.1	94.4	81.2	85.7
0960	Flooring	117.4	100.4	112.5	98.5	100.4	99.1	113.4	81.9	104.4	127.4	79.8	113.8	101.9	107.9	103.6	105.4	51.1	89.9
0970, 0990	Wall Finishes & Painting/Coating	151.7	63.7	98.6	93.8	63.7	75.6	126.2	67.2	90.6	167.2	66.1	106.1	101.2	80.8	88.9	102.6	95.9	98.5
09	FINISHES	108.9	75.2	90.3	95.9	75.2	84.5	106.0	73.3	88.0	118.2	59.7	86.0	97.9	94.3	95.9	97.6	77.2	86.3
COVERS	DIVS. 10 - 14, 25, 28, 41, 43, 44, 46	100.0	85.5	97.1	100.0	66.8	93.3	100.0	87.6	97.5	100.0	62.5	92.4	100.0	89.8	97.9	100.0	62.8	92.5
21, 22, 23	FIRE SUPPRESSION, PLUMBING & HVAC	95.0	75.8	87.2	99.9	76.0	90.2	99.9	76.8	90.6	94.8	75.7	87.1	100.0	80.1	92.0	97.7	79.6	90.4
26, 27, 3370	ELECTRICAL, COMMUNICATIONS & UTIL.	92.4	82.3	87.1	92.6	77.1	84.4	99.1	82.3	90.2	89.7	86.6	88.0	103.2	96.7	99.7	100.7	96.6	98.6
MF2014	WEIGHTED AVERAGE	100.0	76.7	89.8	99.0	78.3	90.0	100.2	78.6	90.8	101.1	72.5	88.7	99.7	93.5	97.0	98.3	84.9	92.4

City Cost Indexes

DIVISION		NEVADA									NEW HAMPSHIRE								
		ELY			LAS VEGAS			RENO			CHARLESTON			CLAREMONT			CONCORD		
		893			889 - 891			894 - 895			036			037			032 - 033		
		MAT.	INST.	TOTAL	MAT.	INST.	TOTAL	MAT.	INST.	TOTAL	MAT.	INST.	TOTAL	MAT.	INST.	TOTAL	MAT.	INST.	TOTAL
015433	CONTRACTOR EQUIPMENT		98.2	98.2		98.2	98.2		98.2	98.2		100.9	100.9		100.9	100.9		100.9	100.9
0241, 31 - 34	SITE & INFRASTRUCTURE, DEMOLITION	72.2	99.1	91.3	75.4	101.0	93.6	72.2	99.0	91.3	86.9	99.6	96.0	80.9	99.6	94.2	93.7	103.9	100.9
0310	Concrete Forming & Accessories	102.3	105.7	105.2	103.5	109.6	108.8	98.7	92.7	93.5	86.8	81.1	81.9	92.6	81.1	82.7	95.5	93.6	93.8
0320	Concrete Reinforcing	101.5	111.6	106.7	93.7	119.9	107.1	96.0	119.9	108.1	84.1	93.5	88.9	84.1	93.5	88.9	95.3	94.1	94.7
0330	Cast-in-Place Concrete	102.1	105.4	103.4	98.9	107.8	102.5	108.0	85.7	98.8	97.1	70.5	86.2	89.2	70.5	81.6	112.2	90.5	103.3
03	CONCRETE	103.4	106.4	104.9	98.7	110.6	104.6	103.1	95.4	99.3	99.1	80.0	89.7	91.1	80.0	85.6	106.6	92.5	99.6
04	MASONRY	119.7	102.5	109.0	108.3	104.0	105.6	114.3	94.6	102.0	97.5	84.7	89.5	96.9	84.7	89.3	110.9	104.2	106.7
05	METALS	96.9	103.3	98.9	104.5	108.6	105.8	98.4	105.1	100.5	92.8	90.1	92.0	92.8	90.1	92.0	98.8	93.8	97.3
06	WOOD, PLASTICS & COMPOSITES	91.3	106.8	100.0	89.8	107.9	99.9	85.9	94.4	90.6	87.4	90.2	89.0	93.6	90.2	91.7	95.4	92.4	93.7
07	THERMAL & MOISTURE PROTECTION	100.3	95.7	98.4	113.8	102.6	109.2	99.9	92.3	96.8	98.8	72.3	88.0	98.7	72.3	87.9	101.9	91.5	97.6
08	OPENINGS	100.8	98.8	100.3	99.9	115.6	103.5	98.7	102.3	99.6	102.0	76.4	96.0	103.3	76.4	97.0	104.9	87.2	100.8
0920	Plaster & Gypsum Board	101.0	107.0	105.1	97.5	108.1	104.7	92.0	94.1	93.4	100.0	89.3	92.8	101.0	89.3	93.1	103.9	91.6	95.6
0950, 0980	Ceilings & Acoustic Treatment	94.4	107.0	102.7	102.7	108.1	106.3	99.4	94.1	95.9	94.2	89.3	91.0	94.2	89.3	91.0	99.9	91.6	94.4
0960	Flooring	103.0	57.2	89.9	94.6	107.9	98.4	99.9	107.9	102.2	98.4	32.1	79.4	100.6	32.1	81.0	102.7	116.9	106.8
0970, 0990	Wall Finishes & Painting/Coating	102.6	121.0	113.7	105.4	121.0	114.8	102.6	80.8	89.4	96.4	45.7	65.8	96.4	46.0	65.9	95.8	95.4	95.6
09	FINISHES	96.8	98.9	98.0	95.8	110.4	103.9	95.6	94.3	94.9	95.3	69.7	81.2	95.6	69.8	81.4	97.4	97.9	97.7
COVERS	DIVS. 10 - 14, 25, 28, 41, 43, 44, 46	100.0	63.6	92.6	100.0	104.4	100.9	100.0	89.8	97.9	100.0	92.2	98.4	100.0	92.2	98.4	100.0	104.3	100.9
21, 22, 23	FIRE SUPPRESSION, PLUMBING & HVAC	97.7	104.5	100.5	100.1	104.4	101.8	100.0	80.1	92.0	95.3	39.6	72.8	95.3	39.7	72.9	99.9	84.4	93.7
26, 27, 3370	ELECTRICAL, COMMUNICATIONS & UTIL.	101.0	109.1	105.3	105.4	119.9	113.1	101.4	96.7	98.9	98.5	52.9	74.4	98.5	52.9	74.4	99.5	83.6	91.1
MF2014	WEIGHTED AVERAGE	99.4	102.2	100.6	100.9	108.2	104.3	99.6	93.3	96.8	96.7	69.5	84.8	95.9	69.5	84.4	101.1	92.5	97.4

DIVISION		NEW HAMPSHIRE															NEW JERSEY		
		KEENE			LITTLETON			MANCHESTER			NASHUA			PORTSMOUTH			ATLANTIC CITY		
		034			035			031			030			038			082,084		
		MAT.	INST.	TOTAL	MAT.	INST.	TOTAL	MAT.	INST.	TOTAL	MAT.	INST.	TOTAL	MAT.	INST.	TOTAL	MAT.	INST.	TOTAL
015433	CONTRACTOR EQUIPMENT		100.9	100.9		100.9	100.9		100.9	100.9		100.9	100.9		100.9	100.9		99.1	99.1
0241, 31 - 34	SITE & INFRASTRUCTURE, DEMOLITION	94.0	99.9	98.2	81.0	99.9	94.4	93.6	103.9	100.9	95.7	103.9	101.5	89.2	104.0	99.7	96.4	105.2	102.7
0310	Concrete Forming & Accessories	91.3	83.0	84.2	102.9	83.0	85.8	96.9	94.0	94.4	99.3	94.0	94.8	88.2	94.3	93.4	109.0	127.5	125.0
0320	Concrete Reinforcing	84.1	93.6	88.9	84.8	93.6	89.3	103.7	94.1	98.8	105.1	94.1	99.5	84.1	94.2	89.2	76.3	118.1	97.6
0330	Cast-in-Place Concrete	97.6	72.8	87.4	87.6	72.7	81.5	110.8	113.1	111.7	92.4	113.1	100.9	87.6	113.2	98.1	86.1	133.0	105.3
03	CONCRETE	98.7	81.6	90.3	90.5	81.6	86.1	107.2	100.4	103.9	98.9	100.4	99.7	90.6	100.6	95.5	93.1	126.5	109.5
04	MASONRY	101.3	88.3	93.2	107.9	88.3	95.7	102.9	104.2	103.7	101.4	104.2	103.1	97.2	104.2	101.5	111.8	132.0	124.4
05	METALS	93.5	90.7	92.6	93.5	90.7	92.6	101.3	94.3	99.2	98.8	94.3	97.5	95.0	94.9	94.9	95.6	106.1	98.8
06	WOOD, PLASTICS & COMPOSITES	92.0	90.2	91.0	103.5	90.2	96.0	95.3	92.4	93.7	101.9	92.4	96.6	88.7	92.4	90.8	109.6	127.2	119.4
07	THERMAL & MOISTURE PROTECTION	99.2	75.3	89.4	98.8	73.8	88.5	101.5	95.0	98.8	99.6	95.0	97.7	99.3	115.0	105.7	105.5	124.2	113.4
08	OPENINGS	100.4	81.2	96.0	104.4	77.1	98.0	105.2	87.2	101.0	105.4	87.2	101.2	106.2	84.0	101.0	101.7	124.4	107.0
0920	Plaster & Gypsum Board	100.3	89.3	92.9	114.6	89.3	97.5	104.7	91.6	95.8	110.9	91.6	97.5	100.0	91.6	94.3	105.4	127.5	120.3
0950, 0980	Ceilings & Acoustic Treatment	94.2	89.3	91.0	94.2	89.3	91.0	101.7	91.6	95.0	105.1	91.6	96.2	95.1	91.6	92.8	81.3	127.5	111.6
0960	Flooring	100.2	52.6	86.6	109.9	32.1	87.6	98.6	116.9	103.8	103.7	116.9	107.4	98.5	116.9	103.8	105.4	159.9	121.0
0970, 0990	Wall Finishes & Painting/Coating	96.4	45.7	65.8	96.4	59.9	74.4	100.7	95.4	97.5	96.4	95.4	95.8	96.4	95.4	95.8	93.9	130.2	115.8
09	FINISHES	97.0	74.5	84.6	100.2	72.2	84.8	98.6	97.9	98.2	101.7	97.9	99.6	96.2	97.9	97.1	96.0	134.5	117.2
COVERS	DIVS. 10 - 14, 25, 28, 41, 43, 44, 46	100.0	93.4	98.7	100.0	98.6	99.7	100.0	104.3	100.9	100.0	104.3	100.9	100.0	104.3	100.9	100.0	111.5	102.3
21, 22, 23	FIRE SUPPRESSION, PLUMBING & HVAC	95.3	43.3	74.3	95.3	63.8	82.6	99.9	84.4	93.7	100.1	84.4	93.8	100.1	84.4	93.8	99.7	123.6	109.3
26, 27, 3370	ELECTRICAL, COMMUNICATIONS & UTIL.	98.5	63.2	79.9	99.8	55.9	76.6	99.9	83.6	91.2	101.2	83.6	91.9	99.1	83.6	90.9	91.1	138.1	116.0
MF2014	WEIGHTED AVERAGE	97.2	73.3	86.7	97.1	76.3	88.0	101.4	93.8	98.1	100.5	93.8	97.6	97.9	94.3	96.4	98.2	124.8	109.8

DIVISION		NEW JERSEY																	
		CAMDEN			DOVER			ELIZABETH			HACKENSACK			JERSEY CITY			LONG BRANCH		
		081			078			072			076			073			077		
		MAT.	INST.	TOTAL	MAT.	INST.	TOTAL	MAT.	INST.	TOTAL	MAT.	INST.	TOTAL	MAT.	INST.	TOTAL	MAT.	INST.	TOTAL
015433	CONTRACTOR EQUIPMENT		99.1	99.1		100.9	100.9		100.9	100.9		100.9	100.9		99.1	99.1		98.7	98.7
0241, 31 - 34	SITE & INFRASTRUCTURE, DEMOLITION	97.8	105.5	103.3	102.0	106.3	105.1	106.1	106.3	106.3	103.0	106.3	105.3	93.4	106.3	102.5	97.5	106.0	103.6
0310	Concrete Forming & Accessories	100.2	127.5	123.8	97.8	128.3	124.1	110.2	128.4	125.9	97.8	128.3	124.1	101.8	128.3	124.7	102.4	128.0	124.5
0320	Concrete Reinforcing	100.1	118.3	109.4	77.2	137.2	107.7	77.2	137.2	107.7	77.2	137.2	107.7	100.1	137.2	119.0	77.2	137.1	107.7
0330	Cast-in-Place Concrete	83.5	132.9	103.8	101.5	127.7	112.2	87.1	131.4	105.3	99.2	131.4	112.4	79.2	127.7	99.1	88.0	132.9	106.4
03	CONCRETE	94.0	126.5	110.0	98.8	128.7	113.5	94.5	130.1	112.0	97.0	130.0	113.2	92.1	128.6	110.0	96.2	130.2	112.9
04	MASONRY	101.3	132.0	120.4	93.0	132.5	117.6	108.7	132.5	123.5	96.9	132.5	119.1	87.0	132.5	115.4	101.3	132.0	120.5
05	METALS	101.1	102.5	101.6	93.3	116.2	100.4	94.8	116.2	101.4	93.4	116.1	100.4	98.9	113.9	103.5	93.4	113.7	99.7
06	WOOD, PLASTICS & COMPOSITES	98.4	127.2	114.5	96.2	127.2	113.5	111.7	127.2	120.4	96.2	127.2	113.5	97.1	127.3	114.0	98.2	127.1	114.4
07	THERMAL & MOISTURE PROTECTION	105.8	123.3	113.0	100.6	132.8	113.8	100.8	133.3	114.1	100.4	125.5	110.7	100.2	132.8	113.5	100.3	124.4	110.2
08	OPENINGS	104.1	124.5	108.8	105.5	127.7	110.6	103.6	127.7	109.2	102.9	127.7	108.7	101.6	127.7	107.7	97.8	127.6	104.7
0920	Plaster & Gypsum Board	101.5	127.5	119.0	100.9	127.5	118.9	107.7	127.5	121.0	100.9	127.5	118.9	104.3	127.5	120.0	102.7	127.5	119.4
0950, 0980	Ceilings & Acoustic Treatment	91.5	127.5	115.1	83.8	127.5	112.5	85.7	127.5	113.1	83.8	127.5	112.5	94.9	127.5	116.3	83.8	127.5	112.5
0960	Flooring	101.5	159.9	118.2	93.6	178.1	117.8	98.5	178.1	121.3	93.6	178.1	117.8	94.6	178.1	118.5	94.8	178.1	118.6
0970, 0990	Wall Finishes & Painting/Coating	93.9	130.2	115.8	98.4	132.5	119.0	98.4	132.5	119.0	98.4	132.5	119.0	98.5	132.5	119.0	98.5	130.2	117.6
09	FINISHES	96.4	134.5	117.4	93.7	136.9	117.5	96.9	136.9	118.9	93.5	136.9	117.4	96.6	136.9	118.8	94.5	137.5	118.2
COVERS	DIVS. 10 - 14, 25, 28, 41, 43, 44, 46	100.0	111.5	102.3	100.0	118.1	103.7	100.0	118.1	103.7	100.0	118.1	103.7	100.0	118.1	103.7	100.0	111.3	102.3
21, 22, 23	FIRE SUPPRESSION, PLUMBING & HVAC	100.0	123.6	109.5	99.7	127.4	110.9	100.1	125.5	110.3	99.7	127.4	110.9	100.1	127.4	111.1	99.7	120.8	108.2
26, 27, 3370	ELECTRICAL, COMMUNICATIONS & UTIL.	96.0	138.1	118.2	92.8	137.7	116.5	93.4	137.7	116.8	92.8	140.1	117.8	97.6	140.1	120.1	92.5	130.8	112.7
MF2014	WEIGHTED AVERAGE	99.5	124.8	110.5	97.9	127.8	110.9	98.8	127.6	111.3	97.6	128.1	110.9	97.9	127.9	111.0	97.1	126.3	109.8

City Cost Indexes

DIVISION		NEW JERSEY																	
		NEW BRUNSWICK 088 - 089			NEWARK 070 - 071			PATERSON 074 - 075			POINT PLEASANT 087			SUMMIT 079			TRENTON 085 - 086		
		MAT.	INST.	TOTAL	MAT.	INST.	TOTAL	MAT.	INST.	TOTAL	MAT.	INST.	TOTAL	MAT.	INST.	TOTAL	MAT.	INST.	TOTAL
015433	CONTRACTOR EQUIPMENT		98.7	98.7		100.9	100.9		100.9	100.9		98.7	98.7		100.9	100.9		98.7	98.7
0241, 31 - 34	SITE & INFRASTRUCTURE, DEMOLITION	109.2	106.0	107.0	107.8	106.3	106.8	105.0	106.3	105.9	110.9	106.0	107.4	103.7	106.3	105.6	95.8	106.0	103.0
0310	Concrete Forming & Accessories	103.3	128.3	124.8	97.6	128.4	124.2	99.9	128.2	124.3	97.7	127.9	123.7	100.6	128.4	124.5	98.6	127.7	123.7
0320	Concrete Reinforcing	77.2	137.2	107.7	99.8	137.2	118.8	100.1	137.2	119.0	77.2	137.1	107.7	77.2	137.2	107.7	99.8	112.4	106.2
0330	Cast-in-Place Concrete	106.3	133.2	117.4	108.8	131.5	118.1	100.8	131.4	113.4	106.3	132.8	117.2	84.3	131.4	103.6	101.8	132.8	114.5
03	CONCRETE	110.4	130.5	120.3	105.8	130.1	117.7	102.2	130.0	115.9	110.1	130.1	119.9	91.5	130.1	110.5	102.5	125.5	113.8
04	MASONRY	109.3	132.5	123.7	99.6	132.5	120.1	93.6	132.5	117.9	97.0	132.0	118.8	95.5	132.5	118.6	103.9	132.0	121.4
05	METALS	95.6	113.8	101.2	100.8	116.3	105.5	94.2	116.1	100.9	95.6	113.5	101.1	93.3	116.2	100.4	100.8	105.2	102.1
06	WOOD, PLASTICS & COMPOSITES	103.0	127.1	116.5	93.9	127.2	112.5	98.8	127.2	114.7	95.9	127.1	113.4	100.1	127.2	115.3	95.6	127.1	113.3
07	THERMAL & MOISTURE PROTECTION	106.1	132.0	116.7	101.1	133.3	114.3	100.7	125.5	110.9	106.2	124.4	113.7	101.0	133.3	114.2	105.5	124.3	113.2
08	OPENINGS	96.1	127.6	103.5	104.4	127.7	109.8	108.3	127.7	112.8	98.2	129.2	105.4	110.2	127.7	114.2	106.1	122.2	109.8
0920	Plaster & Gypsum Board	102.6	127.5	119.4	101.1	127.5	118.9	104.3	127.5	120.0	98.3	127.5	118.0	102.7	127.5	119.4	98.7	127.5	118.1
0950, 0980	Ceilings & Acoustic Treatment	81.3	127.5	111.6	97.6	127.5	117.2	94.9	127.5	116.3	81.3	127.5	111.6	83.8	127.5	112.5	93.1	127.5	115.7
0960	Flooring	103.0	178.1	124.5	95.7	178.1	119.2	94.6	178.1	118.5	100.5	159.9	117.5	94.9	178.1	118.7	102.1	172.2	122.1
0970, 0990	Wall Finishes & Painting/Coating	93.9	132.5	117.2	99.8	132.5	119.5	98.4	132.5	119.0	93.9	130.2	115.8	98.4	132.5	119.0	99.1	130.2	117.8
09	FINISHES	96.0	136.8	118.5	95.6	136.9	118.3	96.7	136.9	118.9	94.7	134.5	116.6	94.6	136.9	117.9	96.6	136.5	118.6
COVERS	DIVS. 10 - 14, 25, 28, 41, 43, 44, 46	100.0	118.0	103.6	100.0	118.1	103.7	100.0	118.1	103.7	100.0	108.8	101.8	100.0	118.1	103.7	100.0	111.3	102.3
21, 22, 23	FIRE SUPPRESSION, PLUMBING & HVAC	99.7	127.4	110.8	100.1	127.4	111.1	100.1	127.4	111.1	99.7	127.0	110.7	99.7	125.5	110.1	100.1	126.9	110.9
26, 27, 3370	ELECTRICAL, COMMUNICATIONS & UTIL.	91.8	136.8	115.6	101.2	140.1	121.8	97.6	137.7	118.8	91.1	130.8	112.1	93.4	137.7	116.6	99.3	135.4	118.4
MF2014	WEIGHTED AVERAGE	99.7	127.6	111.9	101.2	128.4	113.0	99.6	127.8	111.9	99.2	125.9	110.8	97.9	127.6	110.8	100.9	125.1	111.5

DIVISION		NEW JERSEY			NEW MEXICO														
		VINELAND 080,083			ALBUQUERQUE 870 - 872			CARRIZOZO 883			CLOVIS 881			FARMINGTON 874			GALLUP 873		
		MAT.	INST.	TOTAL	MAT.	INST.	TOTAL	MAT.	INST.	TOTAL	MAT.	INST.	TOTAL	MAT.	INST.	TOTAL	MAT.	INST.	TOTAL
015433	CONTRACTOR EQUIPMENT		99.1	99.1		110.1	110.1		110.1	110.1		110.1	110.1		110.1	110.1		110.1	110.1
0241, 31 - 34	SITE & INFRASTRUCTURE, DEMOLITION	100.7	105.5	104.1	87.6	103.3	98.7	105.8	103.3	104.0	94.1	103.3	100.6	94.1	103.3	100.6	102.1	103.3	102.9
0310	Concrete Forming & Accessories	95.2	127.6	123.1	101.7	64.6	69.7	99.2	64.6	69.3	99.2	64.5	69.2	101.7	64.6	69.7	101.7	64.6	69.7
0320	Concrete Reinforcing	76.3	116.1	96.6	98.4	68.9	83.4	107.0	68.9	87.6	108.2	68.9	88.2	107.6	68.9	87.9	103.0	68.9	85.6
0330	Cast-in-Place Concrete	92.8	133.0	109.3	98.9	70.4	87.2	95.8	70.4	85.4	95.7	70.3	85.3	99.8	70.4	87.7	93.9	70.4	84.3
03	CONCRETE	97.9	126.2	111.8	100.2	68.5	84.6	117.0	68.5	93.2	105.8	68.4	87.4	103.9	68.5	86.5	110.0	68.5	89.6
04	MASONRY	99.4	132.0	119.7	105.0	58.7	76.1	103.4	58.7	75.5	103.4	58.7	75.5	113.7	58.7	79.4	99.4	58.7	74.0
05	METALS	95.5	105.5	98.6	105.2	86.8	99.5	99.5	86.8	95.6	99.2	86.7	95.3	102.9	86.8	97.9	102.0	86.8	97.3
06	WOOD, PLASTICS & COMPOSITES	92.8	127.2	112.1	97.5	65.5	79.6	93.1	65.5	77.6	93.1	65.5	77.6	97.7	65.5	79.7	97.7	65.5	79.7
07	THERMAL & MOISTURE PROTECTION	105.7	124.2	113.3	94.8	73.5	86.0	100.3	73.5	89.3	99.2	73.5	88.7	94.9	73.5	86.2	95.9	73.5	86.7
08	OPENINGS	97.7	124.1	103.8	101.2	67.7	93.4	98.6	67.7	91.4	98.7	67.7	91.5	103.6	67.7	95.3	103.6	67.7	95.3
0920	Plaster & Gypsum Board	96.9	127.5	117.5	95.8	64.2	74.4	78.6	64.2	68.8	78.6	64.2	68.8	89.4	64.2	72.3	89.4	64.2	72.3
0950, 0980	Ceilings & Acoustic Treatment	81.3	127.5	111.6	95.5	64.2	74.9	96.3	64.2	75.2	96.3	64.2	75.2	92.6	64.2	73.9	92.6	64.2	73.9
0960	Flooring	99.6	159.9	116.9	101.9	66.0	91.6	101.2	66.0	91.2	101.2	66.0	91.2	103.7	66.0	92.9	103.7	66.0	92.9
0970, 0990	Wall Finishes & Painting/Coating	93.9	130.2	115.8	113.2	66.6	85.1	103.3	66.6	81.1	103.3	66.6	81.1	107.1	66.6	82.7	107.1	66.6	82.7
09	FINISHES	93.5	134.5	116.1	95.9	64.8	78.8	96.6	64.8	79.1	95.4	64.8	78.5	94.8	64.8	78.3	96.0	64.8	78.8
COVERS	DIVS. 10 - 14, 25, 28, 41, 43, 44, 46	100.0	111.5	102.3	100.0	82.9	96.5	100.0	82.9	96.5	100.0	82.9	96.5	100.0	82.9	96.5	100.0	82.9	96.5
21, 22, 23	FIRE SUPPRESSION, PLUMBING & HVAC	99.7	123.6	109.3	100.1	69.0	87.6	97.2	69.0	85.8	97.2	68.7	85.7	100.0	69.0	87.5	97.1	69.0	85.8
26, 27, 3370	ELECTRICAL, COMMUNICATIONS & UTIL.	91.1	138.1	116.0	91.2	71.2	80.6	92.8	71.2	81.4	90.2	71.2	80.2	88.8	71.2	79.5	88.0	71.2	79.1
MF2014	WEIGHTED AVERAGE	97.5	124.7	109.4	99.6	72.6	87.8	100.3	72.6	88.2	98.3	72.5	87.1	100.1	72.6	88.1	99.5	72.6	87.8

DIVISION		NEW MEXICO																	
		LAS CRUCES 880			LAS VEGAS 877			ROSWELL 882			SANTA FE 875			SOCORRO 878			TRUTH/CONSEQUENCES 879		
		MAT.	INST.	TOTAL	MAT.	INST.	TOTAL	MAT.	INST.	TOTAL	MAT.	INST.	TOTAL	MAT.	INST.	TOTAL	MAT.	INST.	TOTAL
015433	CONTRACTOR EQUIPMENT		86.0	86.0		110.1	110.1		110.1	110.1		110.1	110.1		110.1	110.1		86.0	86.0
0241, 31 - 34	SITE & INFRASTRUCTURE, DEMOLITION	94.0	82.9	86.1	93.6	103.3	100.5	96.3	103.3	101.3	99.8	103.3	102.0	89.8	103.3	99.4	108.3	82.9	90.3
0310	Concrete Forming & Accessories	95.7	63.4	67.8	101.7	64.6	69.7	99.2	64.6	69.3	100.4	64.6	69.5	101.7	64.6	69.7	99.3	63.4	68.3
0320	Concrete Reinforcing	104.6	68.7	86.4	104.7	68.9	86.5	108.2	68.9	88.2	103.8	68.9	86.1	106.8	68.9	87.5	100.8	68.7	84.5
0330	Cast-in-Place Concrete	90.4	62.7	79.0	97.1	70.4	86.1	95.8	70.4	85.3	105.3	70.4	91.0	95.1	70.4	85.0	104.2	62.7	87.1
03	CONCRETE	84.9	65.0	75.1	101.2	68.5	85.1	106.5	68.5	87.8	104.0	68.5	86.5	100.2	68.5	84.6	94.1	65.0	79.8
04	MASONRY	99.2	58.3	73.7	99.7	58.7	74.1	114.3	58.7	79.6	104.1	58.7	75.8	99.6	58.7	74.1	97.2	58.3	73.0
05	METALS	98.1	80.3	92.6	101.7	86.8	97.1	100.4	86.8	96.2	99.0	86.8	95.2	102.2	86.8	97.3	101.6	80.3	95.0
06	WOOD, PLASTICS & COMPOSITES	82.5	64.4	72.4	97.7	65.5	79.7	93.1	65.5	77.6	99.3	65.5	80.4	97.7	65.5	79.7	89.1	64.4	75.3
07	THERMAL & MOISTURE PROTECTION	86.0	68.6	78.8	94.5	73.5	85.9	99.3	73.5	88.7	96.9	73.5	87.3	94.5	73.5	85.9	83.3	68.6	77.2
08	OPENINGS	91.5	67.1	85.8	99.9	67.7	92.4	98.5	67.7	91.4	102.0	67.7	94.0	99.8	67.7	92.3	93.1	67.1	87.0
0920	Plaster & Gypsum Board	77.5	64.2	68.5	89.4	64.2	72.3	78.6	64.2	68.8	98.9	64.2	75.4	89.4	64.2	72.3	91.1	64.2	72.9
0950, 0980	Ceilings & Acoustic Treatment	84.2	64.2	71.0	92.6	64.2	73.9	96.3	64.2	75.2	93.0	64.2	74.0	92.6	64.2	73.9	82.4	64.2	70.4
0960	Flooring	131.2	66.0	112.5	103.7	66.0	92.9	101.2	66.0	91.2	110.8	66.0	98.0	103.7	66.0	92.9	135.3	66.0	115.4
0970, 0990	Wall Finishes & Painting/Coating	90.7	66.6	76.2	107.1	66.6	82.7	103.3	66.6	81.1	111.8	66.6	84.5	107.1	66.6	82.7	97.7	66.6	79.0
09	FINISHES	105.2	63.9	82.5	94.7	64.8	78.2	95.5	64.8	78.6	99.7	64.8	80.5	94.6	64.8	78.2	107.5	63.9	83.5
COVERS	DIVS. 10 - 14, 25, 28, 41, 43, 44, 46	100.0	80.2	96.0	100.0	82.9	96.5	100.0	82.9	96.5	100.0	82.9	96.5	100.0	82.9	96.5	100.0	80.2	96.0
21, 22, 23	FIRE SUPPRESSION, PLUMBING & HVAC	100.3	68.7	87.6	97.1	69.0	85.8	99.9	69.0	87.5	100.0	69.0	87.5	97.1	69.0	85.8	100.3	68.7	87.6
26, 27, 3370	ELECTRICAL, COMMUNICATIONS & UTIL.	92.1	71.2	81.0	90.6	71.2	80.4	91.7	71.2	80.9	102.9	71.2	86.2	88.5	71.2	79.4	92.3	71.2	81.2
MF2014	WEIGHTED AVERAGE	96.1	69.4	84.4	98.0	72.6	86.9	100.0	72.6	88.0	100.8	72.6	88.5	97.6	72.6	86.7	97.5	69.4	85.2

City Cost Indexes

		NEW MEXICO			NEW YORK													
	DIVISION	TUCUMCARI			ALBANY			BINGHAMTON			BRONX			BROOKLYN			BUFFALO		
		884			120 - 122			137 - 139			104			112			140 - 142		
		MAT.	INST.	TOTAL	MAT.	INST.	TOTAL	MAT.	INST.	TOTAL	MAT.	INST.	TOTAL	MAT.	INST.	TOTAL	MAT.	INST.	TOTAL
015433	CONTRACTOR EQUIPMENT		110.1	110.1		112.9	112.9		114.1	114.1		110.6	110.6		113.3	113.3		97.1	97.1
0241, 31 - 34	SITE & INFRASTRUCTURE, DEMOLITION	93.8	103.3	100.5	83.8	106.3	99.8	95.9	94.1	94.6	108.8	120.8	117.3	120.3	127.6	125.5	98.3	98.2	98.2
0310	Concrete Forming & Accessories	99.2	64.5	69.2	100.1	100.6	100.5	100.8	88.5	90.2	98.4	175.3	164.8	107.2	182.9	172.5	97.3	116.5	113.9
0320	Concrete Reinforcing	106.0	68.9	87.1	104.0	103.8	103.9	93.7	97.8	95.8	103.9	184.4	144.9	95.1	205.9	151.5	97.8	102.1	100.0
0330	Cast-in-Place Concrete	95.7	70.3	85.3	91.9	111.3	99.9	102.7	127.9	113.1	95.9	173.4	127.7	104.8	172.1	132.4	106.7	119.7	112.0
03	CONCRETE	105.0	68.4	87.0	99.0	105.4	102.1	95.8	105.2	100.4	96.1	174.8	134.8	107.5	181.3	143.7	103.0	114.1	108.4
04	MASONRY	114.6	58.7	79.8	101.0	112.6	108.2	109.7	127.8	121.0	92.7	177.5	145.5	120.2	177.4	155.9	105.1	121.0	115.0
05	METALS	99.2	86.7	95.3	104.1	110.3	106.0	96.3	118.9	103.3	99.9	151.6	115.8	104.4	150.1	118.4	99.6	95.4	98.3
06	WOOD, PLASTICS & COMPOSITES	93.1	65.5	77.6	98.7	97.9	98.3	104.9	84.7	93.6	94.4	174.9	139.5	106.9	185.1	150.6	99.4	116.4	108.9
07	THERMAL & MOISTURE PROTECTION	99.1	73.5	88.6	105.9	105.8	105.9	107.1	103.6	105.8	108.5	163.2	131.0	108.4	164.1	131.2	101.8	110.6	105.4
08	OPENINGS	98.5	67.7	91.3	102.6	94.0	100.6	93.0	85.0	91.1	87.9	175.5	108.3	90.4	180.0	111.2	97.7	103.9	99.1
0920	Plaster & Gypsum Board	78.6	64.2	68.8	97.2	97.6	97.5	107.4	83.8	91.4	99.2	176.9	151.7	102.8	187.6	160.2	98.3	116.7	110.7
0950, 0980	Ceilings & Acoustic Treatment	96.3	64.2	75.2	92.4	97.6	95.8	91.1	83.8	86.3	83.8	176.9	145.0	87.1	187.6	153.2	102.0	116.7	111.6
0960	Flooring	101.2	66.0	91.2	97.1	113.7	101.9	106.8	103.6	105.9	98.3	186.8	123.6	113.9	186.8	134.8	101.3	121.2	107.0
0970, 0990	Wall Finishes & Painting/Coating	103.3	66.6	81.1	103.8	94.3	98.0	93.3	98.8	96.6	102.9	157.4	135.8	123.4	157.4	143.9	98.8	112.4	107.0
09	FINISHES	95.4	64.8	78.5	94.7	101.4	98.7	94.9	91.1	92.8	94.0	175.9	139.1	108.4	181.9	148.9	100.6	117.9	110.2
COVERS	DIVS. 10 - 14, 25, 28, 41, 43, 44, 46	100.0	82.9	96.5	100.0	99.0	99.8	100.0	96.2	99.2	100.0	135.0	107.1	100.0	135.4	107.1	100.0	105.6	101.1
21, 22, 23	FIRE SUPPRESSION, PLUMBING & HVAC	97.2	68.7	85.7	100.0	102.8	101.1	100.5	90.2	96.3	100.2	165.5	126.6	99.7	165.4	126.2	100.0	96.7	98.7
26, 27, 3370	ELECTRICAL, COMMUNICATIONS & UTIL.	92.8	71.2	81.4	98.7	104.1	101.6	99.9	105.4	102.8	97.0	181.9	141.9	99.7	181.9	143.1	100.2	102.7	101.5
MF2014	WEIGHTED AVERAGE	99.0	72.5	87.4	100.1	104.7	102.1	98.5	101.4	99.8	97.7	166.1	127.5	102.8	168.5	131.4	100.3	106.2	102.9

		NEW YORK																
	DIVISION	ELMIRA			FAR ROCKAWAY			FLUSHING			GLENS FALLS			HICKSVILLE			JAMAICA		
		148 - 149			116			113			128			115,117,118			114		
		MAT.	INST.	TOTAL	MAT.	INST.	TOTAL	MAT.	INST.	TOTAL	MAT.	INST.	TOTAL	MAT.	INST.	TOTAL	MAT.	INST.	TOTAL
015433	CONTRACTOR EQUIPMENT		116.0	116.0		113.3	113.3		113.3	113.3		112.9	112.9		113.3	113.3		113.3	113.3
0241, 31 - 34	SITE & INFRASTRUCTURE, DEMOLITION	97.1	94.1	95.0	123.4	127.6	126.4	123.4	127.6	126.4	73.6	105.8	96.5	113.3	126.3	122.6	117.7	127.6	124.7
0310	Concrete Forming & Accessories	81.3	93.2	91.6	93.5	175.1	163.9	97.3	175.1	164.4	85.7	90.3	89.7	90.0	154.5	145.6	97.3	175.1	164.4
0320	Concrete Reinforcing	97.4	95.7	96.5	95.1	205.9	151.5	96.7	205.9	152.3	95.4	94.4	94.9	95.1	210.2	153.7	95.1	205.9	151.5
0330	Cast-in-Place Concrete	93.7	103.2	97.6	113.3	172.1	137.5	113.3	172.1	137.5	85.3	106.7	94.1	96.3	165.5	124.7	104.8	172.1	132.4
03	CONCRETE	90.8	98.6	94.6	113.6	177.8	145.1	114.1	177.8	145.4	88.3	97.5	92.9	99.4	167.1	132.6	106.8	177.8	141.7
04	MASONRY	103.5	102.1	102.6	124.1	177.4	157.4	118.1	177.4	155.1	100.7	105.6	103.7	114.6	166.8	147.1	122.2	177.4	156.6
05	METALS	96.7	117.3	103.0	104.4	150.1	118.4	104.4	150.1	118.4	97.7	106.1	100.3	105.8	149.1	119.2	104.4	150.1	118.4
06	WOOD, PLASTICS & COMPOSITES	85.0	92.2	89.0	90.1	174.6	137.4	94.8	174.6	139.5	87.1	87.6	87.3	86.6	152.6	123.6	94.8	174.6	139.5
07	THERMAL & MOISTURE PROTECTION	103.9	94.1	99.9	108.3	163.0	130.7	108.3	163.0	130.7	98.8	97.2	98.2	107.9	156.3	127.8	108.1	163.0	130.6
08	OPENINGS	99.4	88.9	97.0	89.0	174.4	108.8	89.0	174.4	108.8	93.3	86.2	91.7	89.0	163.4	106.3	89.0	174.4	108.8
0920	Plaster & Gypsum Board	97.8	91.7	93.7	91.6	176.9	149.2	94.1	176.9	150.0	89.1	87.0	87.7	91.2	154.2	133.8	94.1	176.9	150.0
0950, 0980	Ceilings & Acoustic Treatment	95.4	91.7	93.0	76.2	176.9	142.4	76.2	176.9	142.4	82.1	87.0	85.3	75.2	154.2	127.2	76.2	176.9	142.4
0960	Flooring	94.6	103.6	97.1	109.2	186.8	131.4	110.6	186.8	132.4	86.2	111.3	93.3	107.8	186.8	130.2	110.6	186.8	132.4
0970, 0990	Wall Finishes & Painting/Coating	98.9	90.0	93.5	123.4	157.4	143.9	123.4	157.4	143.9	101.1	87.4	92.8	123.4	157.4	143.9	123.4	157.4	143.9
09	FINISHES	95.5	94.7	95.0	103.5	175.7	143.3	104.3	175.7	143.6	86.1	93.1	90.0	102.1	159.7	133.8	103.8	175.7	143.4
COVERS	DIVS. 10 - 14, 25, 28, 41, 43, 44, 46	100.0	99.4	99.9	100.0	134.2	106.9	100.0	134.2	106.9	100.0	96.2	99.2	100.0	128.2	105.7	100.0	134.2	106.9
21, 22, 23	FIRE SUPPRESSION, PLUMBING & HVAC	95.5	91.5	93.9	95.0	165.3	123.4	95.0	165.3	123.4	95.5	95.7	95.6	99.7	154.4	121.8	95.0	165.3	123.4
26, 27, 3370	ELECTRICAL, COMMUNICATIONS & UTIL.	96.5	96.2	96.3	107.0	181.9	146.6	107.0	181.9	146.6	93.6	100.5	97.3	99.0	143.4	122.5	97.9	181.9	142.3
MF2014	WEIGHTED AVERAGE	96.7	97.4	97.0	102.6	166.8	130.6	102.5	166.8	130.5	94.0	98.6	96.0	100.9	153.3	123.7	100.8	166.8	129.6

		NEW YORK																
	DIVISION	JAMESTOWN			KINGSTON			LONG ISLAND CITY			MONTICELLO			MOUNT VERNON			NEW ROCHELLE		
		147			124			111			127			105			108		
		MAT.	INST.	TOTAL	MAT.	INST.	TOTAL	MAT.	INST.	TOTAL	MAT.	INST.	TOTAL	MAT.	INST.	TOTAL	MAT.	INST.	TOTAL
015433	CONTRACTOR EQUIPMENT		93.9	93.9		113.3	113.3		113.3	113.3		113.3	113.3		110.6	110.6		110.6	110.6
0241, 31 - 34	SITE & INFRASTRUCTURE, DEMOLITION	98.4	94.2	95.4	140.6	122.9	128.0	121.3	127.6	125.8	135.5	122.7	126.4	115.2	118.5	117.6	114.6	118.5	117.4
0310	Concrete Forming & Accessories	81.3	86.9	86.1	87.1	104.4	102.0	101.6	175.1	165.0	94.7	102.9	101.8	88.8	139.1	132.1	103.6	139.0	134.2
0320	Concrete Reinforcing	97.6	98.9	98.2	95.8	140.6	118.6	95.1	205.9	151.5	95.1	140.1	118.0	102.9	183.3	143.8	103.0	183.3	143.9
0330	Cast-in-Place Concrete	97.2	102.1	99.2	115.2	135.1	123.4	108.1	172.1	134.4	107.8	124.8	114.8	107.1	141.3	121.1	107.0	141.3	121.0
03	CONCRETE	93.8	94.3	94.0	110.8	121.5	116.0	109.9	177.8	143.3	105.5	117.2	111.2	105.4	146.9	125.8	105.0	146.9	125.6
04	MASONRY	111.8	99.9	104.4	116.0	141.3	131.8	116.9	177.4	154.6	108.6	128.7	121.1	98.1	147.3	128.8	98.1	147.3	128.8
05	METALS	94.0	92.1	93.5	106.1	117.2	109.5	104.4	150.1	118.4	106.0	116.2	109.2	99.6	135.6	110.7	99.9	135.6	110.9
06	WOOD, PLASTICS & COMPOSITES	83.5	84.0	83.8	88.4	96.4	92.9	100.9	174.6	142.2	96.1	96.4	96.3	84.6	136.2	113.5	101.4	136.2	120.9
07	THERMAL & MOISTURE PROTECTION	103.4	93.5	99.3	122.3	137.2	128.4	108.3	163.0	130.7	122.0	132.1	126.1	109.4	144.1	123.6	109.5	144.1	123.7
08	OPENINGS	99.2	85.6	96.1	95.9	117.3	100.9	89.0	174.4	108.8	91.1	117.3	97.2	87.9	148.6	102.0	88.0	148.6	102.1
0920	Plaster & Gypsum Board	88.5	83.2	84.9	91.4	96.3	94.7	98.7	176.9	151.5	92.1	96.3	94.9	94.9	137.0	123.3	106.3	137.0	127.0
0950, 0980	Ceilings & Acoustic Treatment	92.0	83.2	86.2	72.9	96.3	88.3	76.2	176.9	142.4	72.9	96.3	88.3	82.1	137.0	118.2	82.1	137.0	118.2
0960	Flooring	97.7	103.6	99.4	103.8	72.5	94.9	112.1	186.8	133.5	106.1	72.5	96.5	90.2	186.8	117.8	97.1	186.8	122.8
0970, 0990	Wall Finishes & Painting/Coating	100.2	94.2	96.6	134.4	116.6	123.8	123.4	157.4	143.9	134.4	116.6	123.8	101.1	157.4	135.1	101.1	157.4	135.1
09	FINISHES	94.6	89.9	92.0	101.0	97.1	98.8	105.2	175.7	144.0	101.4	96.3	98.6	91.2	149.2	123.2	94.6	149.2	124.7
COVERS	DIVS. 10 - 14, 25, 28, 41, 43, 44, 46	100.0	98.6	99.7	100.0	111.9	102.4	100.0	134.2	106.9	100.0	110.8	102.2	100.0	125.2	105.1	100.0	121.9	104.4
21, 22, 23	FIRE SUPPRESSION, PLUMBING & HVAC	95.4	92.9	94.4	95.4	122.3	106.3	99.7	165.3	126.2	95.4	116.7	104.0	95.5	134.1	111.1	95.5	134.1	111.1
26, 27, 3370	ELECTRICAL, COMMUNICATIONS & UTIL.	95.4	96.3	95.9	95.1	111.0	103.5	98.5	181.9	142.6	95.1	111.0	103.5	95.2	155.2	126.9	95.2	155.2	126.9
MF2014	WEIGHTED AVERAGE	96.8	92.5	94.9	102.6	118.5	109.6	102.4	166.8	130.5	101.1	115.1	107.2	97.5	141.4	116.6	97.9	141.3	116.8

City Cost Indexes

DIVISION		NEW YORK																	
		NEW YORK 100 - 102			NIAGARA FALLS 143			PLATTSBURGH 129			POUGHKEEPSIE 125 - 126			QUEENS 110			RIVERHEAD 119		
		MAT.	INST.	TOTAL	MAT.	INST.	TOTAL	MAT.	INST.	TOTAL	MAT.	INST.	TOTAL	MAT.	INST.	TOTAL	MAT.	INST.	TOTAL
015433	CONTRACTOR EQUIPMENT		111.1	111.1		93.9	93.9		98.9	98.9		113.3	113.3		113.3	113.3		113.3	113.3
0241, 31 - 34	SITE & INFRASTRUCTURE, DEMOLITION	117.3	121.5	120.2	100.5	95.5	97.0	106.1	103.0	103.9	136.5	123.5	127.2	116.6	127.6	124.4	114.5	125.9	122.6
0310	Concrete Forming & Accessories	102.6	183.3	172.2	81.3	114.4	109.8	91.1	95.4	94.9	87.1	165.0	154.3	90.1	175.1	163.4	94.6	153.4	145.3
0320	Concrete Reinforcing	109.9	210.4	161.0	96.3	102.5	99.5	99.9	102.7	101.3	95.8	141.0	118.8	96.7	205.9	152.3	96.9	184.2	141.3
0330	Cast-in-Place Concrete	107.8	177.7	136.5	100.6	120.3	108.7	104.4	104.4	104.4	111.5	137.9	122.4	99.6	172.1	129.4	97.9	164.5	125.3
03	CONCRETE	106.2	184.2	144.5	95.9	113.3	104.4	102.5	99.3	101.0	107.9	149.5	128.3	102.6	177.8	139.6	100.1	161.4	130.2
04	MASONRY	102.5	177.5	149.2	119.9	127.5	124.6	95.0	101.6	99.1	108.6	144.5	131.0	111.1	177.4	152.4	119.9	166.3	148.8
05	METALS	113.1	151.9	125.0	96.7	93.8	95.8	102.0	91.2	98.7	106.1	119.8	110.3	104.4	150.1	118.4	106.3	138.4	116.2
06	WOOD, PLASTICS & COMPOSITES	98.5	185.4	147.2	83.4	110.3	98.5	93.8	93.5	93.6	88.4	174.6	136.7	86.7	174.6	135.9	91.8	152.6	125.9
07	THERMAL & MOISTURE PROTECTION	108.6	164.5	131.5	103.5	112.9	107.4	116.3	100.5	109.8	122.3	146.8	132.3	107.9	163.0	130.5	108.8	156.1	128.2
08	OPENINGS	93.7	180.9	114.0	99.2	100.7	99.6	101.6	90.8	99.1	95.9	159.5	110.7	89.0	174.4	108.8	89.0	157.5	104.9
0920	Plaster & Gypsum Board	106.0	187.6	161.2	88.5	110.3	103.3	109.4	92.6	98.0	91.4	176.9	149.2	91.2	176.9	149.1	92.5	154.2	134.2
0950, 0980	Ceilings & Acoustic Treatment	102.2	187.6	158.4	92.0	110.3	104.1	98.1	92.6	94.5	72.9	176.9	141.2	76.2	176.9	142.4	76.1	154.2	127.5
0960	Flooring	99.5	186.8	124.5	97.7	121.2	104.4	108.9	113.7	110.3	103.8	167.2	122.0	108.3	186.8	130.7	109.2	142.7	118.8
0970, 0990	Wall Finishes & Painting/Coating	102.9	157.4	135.8	100.2	112.4	107.6	130.0	91.4	106.7	134.4	117.2	124.0	123.4	157.4	143.9	123.4	157.4	143.9
09	FINISHES	99.9	182.1	145.2	94.7	115.9	106.4	97.9	98.0	98.0	100.8	163.0	135.1	102.5	175.7	142.8	102.7	151.3	129.5
COVERS	DIVS. 10 - 14, 25, 28, 41, 43, 44, 46	100.0	136.1	107.3	100.0	107.1	101.4	100.0	98.0	99.6	100.0	121.6	104.4	100.0	134.2	106.9	100.0	127.6	105.6
21, 22, 23	FIRE SUPPRESSION, PLUMBING & HVAC	100.1	165.5	126.5	95.4	100.0	97.2	95.4	96.6	95.9	95.4	127.0	108.1	99.7	165.3	126.2	99.9	151.4	120.7
26, 27, 3370	ELECTRICAL, COMMUNICATIONS & UTIL.	104.7	181.9	145.5	94.0	100.3	97.3	91.6	90.3	90.9	95.1	119.9	108.2	99.0	181.9	142.8	100.6	133.6	118.1
MF2014	WEIGHTED AVERAGE	103.4	168.7	131.8	97.8	106.4	101.5	99.1	96.7	98.0	101.8	136.7	117.0	100.9	166.2	129.6	101.6	148.2	121.9

DIVISION		NEW YORK																	
		ROCHESTER 144 - 146			SCHENECTADY 123			STATEN ISLAND 103			SUFFERN 109			SYRACUSE 130 - 132			UTICA 133 - 135		
		MAT.	INST.	TOTAL	MAT.	INST.	TOTAL	MAT.	INST.	TOTAL	MAT.	INST.	TOTAL	MAT.	INST.	TOTAL	MAT.	INST.	TOTAL
015433	CONTRACTOR EQUIPMENT		116.7	116.7		112.9	112.9		110.6	110.6		110.6	110.6		112.9	112.9		112.9	112.9
0241, 31 - 34	SITE & INFRASTRUCTURE, DEMOLITION	85.9	109.9	102.9	84.2	106.3	99.9	119.9	120.8	120.6	111.4	116.3	114.9	94.8	105.6	102.4	73.5	104.3	95.4
0310	Concrete Forming & Accessories	98.7	98.2	98.3	103.0	100.6	100.9	88.3	183.3	170.2	96.9	135.2	129.9	99.8	91.0	92.2	101.0	87.9	89.7
0320	Concrete Reinforcing	100.1	95.7	97.9	94.4	103.8	99.2	103.9	210.4	158.1	103.0	140.9	122.3	94.7	96.4	95.6	94.7	95.7	95.2
0330	Cast-in-Place Concrete	94.2	104.4	98.4	102.9	111.3	106.4	107.1	173.5	134.4	103.8	138.2	117.9	95.4	105.9	99.7	87.3	104.7	94.4
03	CONCRETE	99.3	100.6	99.9	100.5	105.4	104.1	107.1	182.7	144.2	102.0	136.3	118.9	98.9	97.8	98.4	96.9	95.9	96.4
04	MASONRY	107.0	104.9	105.7	97.7	112.6	107.0	104.7	177.5	150.1	97.6	142.6	125.7	101.5	105.2	103.8	93.1	103.9	99.8
05	METALS	104.1	107.3	105.1	101.9	110.3	104.5	97.8	151.8	114.4	97.8	119.4	104.5	99.8	105.6	101.6	97.7	105.1	100.0
06	WOOD, PLASTICS & COMPOSITES	97.6	97.6	97.6	107.4	97.9	102.1	83.3	185.4	140.4	94.0	136.2	117.6	101.3	88.5	94.2	101.3	84.7	92.0
07	THERMAL & MOISTURE PROTECTION	103.5	101.7	102.7	100.3	105.8	102.6	108.9	164.3	131.6	109.3	142.1	122.8	102.3	97.6	100.4	90.6	97.6	93.5
08	OPENINGS	105.3	92.2	102.2	99.4	94.0	98.1	87.9	181.2	109.6	88.0	139.0	99.8	95.0	85.6	92.8	98.0	83.4	94.6
0920	Plaster & Gypsum Board	107.2	97.4	100.6	98.8	97.6	98.0	95.0	187.6	157.6	98.5	137.0	124.5	98.0	88.0	91.2	98.0	84.0	88.5
0950, 0980	Ceilings & Acoustic Treatment	100.4	97.4	98.5	88.7	97.6	94.6	83.8	187.6	152.0	82.1	137.0	118.2	91.1	88.0	89.0	91.1	84.0	86.4
0960	Flooring	93.8	113.3	99.4	92.9	113.7	98.9	94.2	186.8	120.7	93.5	185.1	119.7	94.6	102.1	96.8	92.1	102.2	95.0
0970, 0990	Wall Finishes & Painting/Coating	100.3	99.1	99.6	101.1	94.3	97.0	102.9	157.4	135.8	101.1	124.9	115.5	98.5	99.8	99.3	91.3	99.8	96.4
09	FINISHES	100.4	101.3	100.9	91.6	101.9	97.3	93.2	182.1	142.2	92.3	140.0	118.6	94.0	93.5	93.7	92.1	91.1	91.6
COVERS	DIVS. 10 - 14, 25, 28, 41, 43, 44, 46	100.0	99.8	100.0	100.0	99.0	99.8	100.0	136.1	107.3	100.0	123.6	104.8	100.0	96.5	99.3	100.0	96.2	99.2
21, 22, 23	FIRE SUPPRESSION, PLUMBING & HVAC	100.1	90.3	96.1	100.2	102.8	101.2	100.2	165.5	126.5	95.5	123.3	106.7	100.2	91.9	96.9	100.2	92.2	97.0
26, 27, 3370	ELECTRICAL, COMMUNICATIONS & UTIL.	98.7	94.6	96.6	98.1	104.1	101.3	97.0	181.9	141.9	102.8	119.9	111.8	100.0	102.2	101.1	98.1	102.2	100.3
MF2014	WEIGHTED AVERAGE	101.1	99.1	100.2	99.3	104.7	101.7	99.2	168.4	129.4	97.6	129.0	111.3	98.9	98.0	98.5	97.1	97.1	97.1

DIVISION		NEW YORK									NORTH CAROLINA								
		WATERTOWN 136			WHITE PLAINS 106			YONKERS 107			ASHEVILLE 287 - 288			CHARLOTTE 281 - 282			DURHAM 277		
		MAT.	INST.	TOTAL	MAT.	INST.	TOTAL	MAT.	INST.	TOTAL	MAT.	INST.	TOTAL	MAT.	INST.	TOTAL	MAT.	INST.	TOTAL
015433	CONTRACTOR EQUIPMENT		112.9	112.9		110.6	110.6		110.6	110.6		96.3	96.3		96.3	96.3		101.7	101.7
0241, 31 - 34	SITE & INFRASTRUCTURE, DEMOLITION	80.9	105.7	98.5	108.5	118.5	115.6	116.7	118.5	118.0	102.4	76.9	84.3	105.6	76.9	85.2	100.4	85.7	89.9
0310	Concrete Forming & Accessories	85.9	94.3	93.1	101.9	139.1	134.0	102.2	143.9	138.2	96.0	41.3	48.8	99.7	42.2	50.6	100.2	44.3	52.0
0320	Concrete Reinforcing	95.3	96.5	95.9	103.0	183.3	143.9	107.1	183.3	145.9	93.0	63.0	77.7	98.8	58.5	78.3	91.2	57.9	74.2
0330	Cast-in-Place Concrete	101.5	107.6	104.1	95.1	141.3	114.1	106.3	141.4	120.7	115.4	51.3	89.1	120.2	49.6	91.3	101.6	47.1	79.2
03	CONCRETE	109.3	99.9	104.7	95.7	146.9	120.9	105.2	149.1	126.7	106.0	50.6	78.8	108.6	49.9	79.8	98.7	49.5	74.5
04	MASONRY	94.2	107.9	102.8	97.1	147.3	128.4	102.1	147.3	130.3	93.6	43.7	62.5	101.9	51.2	70.3	86.4	37.9	56.2
05	METALS	97.8	105.6	100.2	99.3	135.6	110.5	108.9	135.7	117.1	103.1	82.3	96.7	104.0	80.9	96.9	121.4	79.9	108.6
06	WOOD, PLASTICS & COMPOSITES	83.0	92.2	88.1	99.5	136.2	120.0	99.3	142.7	123.6	97.9	40.3	65.6	103.0	41.9	68.8	96.7	44.7	67.6
07	THERMAL & MOISTURE PROTECTION	90.9	100.1	94.7	109.2	144.1	123.5	109.5	144.8	124.0	106.7	43.7	80.8	100.8	46.6	78.6	106.9	45.7	81.8
08	OPENINGS	98.0	89.4	96.0	88.0	148.6	102.1	91.1	151.7	105.2	97.1	44.1	84.8	102.5	45.0	89.1	105.8	47.1	92.2
0920	Plaster & Gypsum Board	88.7	91.7	90.8	101.3	137.0	125.4	105.6	143.6	131.3	100.1	38.3	58.3	100.0	39.9	59.4	105.9	42.8	63.3
0950, 0980	Ceilings & Acoustic Treatment	91.1	91.7	91.5	82.1	137.0	118.2	100.5	143.6	128.9	85.7	38.3	54.5	88.8	39.9	56.7	88.1	42.8	58.3
0960	Flooring	86.0	102.2	90.6	95.6	186.8	121.7	95.2	186.8	121.4	102.2	42.7	85.2	101.6	43.4	84.9	103.9	42.7	86.4
0970, 0990	Wall Finishes & Painting/Coating	91.3	96.2	94.3	101.1	157.4	135.1	101.1	157.4	135.1	113.6	40.3	69.3	113.8	49.4	74.9	105.4	37.4	64.3
09	FINISHES	89.7	95.7	93.0	92.9	149.2	123.9	98.0	153.0	128.3	95.6	40.7	65.3	95.5	42.9	66.5	96.0	42.9	66.7
COVERS	DIVS. 10 - 14, 25, 28, 41, 43, 44, 46	100.0	97.5	99.5	100.0	123.5	105.1	100.0	126.3	105.3	100.0	77.7	95.5	100.0	78.1	95.6	100.0	72.1	94.3
21, 22, 23	FIRE SUPPRESSION, PLUMBING & HVAC	100.2	85.9	94.4	100.3	134.1	114.0	100.3	134.1	114.0	100.4	53.4	81.4	100.0	54.1	81.5	100.5	52.9	81.3
26, 27, 3370	ELECTRICAL, COMMUNICATIONS & UTIL.	100.0	90.4	94.9	95.2	155.2	126.9	102.9	163.8	135.1	102.2	55.7	77.6	101.2	58.5	78.6	96.1	55.6	74.7
MF2014	WEIGHTED AVERAGE	98.5	96.3	97.5	97.6	141.4	116.7	102.5	143.7	120.2	100.8	55.3	80.9	101.9	56.8	82.2	102.9	55.3	82.1

City Cost Indexes

DIVISION		NORTH CAROLINA																	
		ELIZABETH CITY			FAYETTEVILLE			GASTONIA			GREENSBORO			HICKORY			KINSTON		
		279			283			280			270,272 - 274			286			285		
		MAT.	INST.	TOTAL	MAT.	INST.	TOTAL	MAT.	INST.	TOTAL	MAT.	INST.	TOTAL	MAT.	INST.	TOTAL	MAT.	INST.	TOTAL
015433	CONTRACTOR EQUIPMENT		106.0	106.0		101.7	101.7		96.3	96.3		101.7	101.7		101.7	101.7		101.7	101.7
0241, 31 - 34	SITE & INFRASTRUCTURE, DEMOLITION	104.6	87.4	92.4	101.7	85.6	90.3	102.3	76.9	84.2	100.3	85.7	89.9	101.2	85.5	90.1	100.3	85.6	89.8
0310	Concrete Forming & Accessories	85.3	42.5	48.4	95.6	60.3	65.2	103.1	38.7	47.5	99.9	44.4	52.1	92.0	37.1	44.6	88.2	42.1	48.4
0320	Concrete Reinforcing	89.2	45.9	67.2	96.8	58.0	77.0	93.5	56.5	74.7	90.1	58.0	73.8	93.0	56.1	74.2	92.5	45.8	68.7
0330	Cast-in-Place Concrete	101.8	47.0	79.3	121.0	48.3	91.2	112.8	52.7	88.1	100.8	47.6	79.0	115.4	48.1	87.7	111.4	44.8	84.0
03	CONCRETE	98.7	46.4	73.0	108.1	57.1	83.1	104.5	48.7	77.1	98.1	49.8	74.4	105.7	46.3	76.5	102.4	45.5	74.4
04	MASONRY	98.6	48.0	67.1	97.3	38.9	60.9	98.3	50.9	68.7	82.5	41.4	56.9	82.3	43.7	58.2	89.1	48.1	63.6
05	METALS	107.1	75.8	97.5	124.2	80.0	110.6	103.9	80.0	96.5	113.7	80.0	103.4	103.2	78.7	95.7	102.0	74.8	93.6
06	WOOD, PLASTICS & COMPOSITES	80.1	43.6	59.7	97.1	66.8	80.1	107.0	36.6	67.6	96.4	44.9	67.6	92.1	35.2	60.2	88.6	42.9	63.0
07	THERMAL & MOISTURE PROTECTION	106.2	43.3	80.4	106.3	45.9	81.5	106.9	45.2	81.6	106.7	42.8	80.5	107.0	41.7	80.2	106.8	42.8	80.6
08	OPENINGS	102.6	38.4	87.7	97.2	59.1	88.4	100.9	41.0	87.0	105.8	47.2	92.2	97.2	36.8	83.1	97.3	43.1	84.7
0920	Plaster & Gypsum Board	98.6	41.0	59.7	104.7	65.6	78.3	107.0	34.5	58.0	107.5	43.0	63.9	100.1	33.1	54.8	99.8	41.0	60.1
0950, 0980	Ceilings & Acoustic Treatment	88.1	41.0	57.1	86.5	65.6	72.7	89.0	34.5	53.2	88.1	43.0	58.4	85.7	33.1	51.1	89.0	41.0	57.4
0960	Flooring	95.7	23.2	75.0	102.4	42.7	85.3	105.4	42.7	87.4	103.9	39.7	85.6	102.1	32.8	82.3	99.6	22.6	77.6
0970, 0990	Wall Finishes & Painting/Coating	105.4	42.0	67.1	113.6	33.2	65.0	113.6	40.3	69.3	105.4	31.6	60.9	113.6	40.3	69.3	113.6	39.2	68.7
09	FINISHES	93.0	38.7	63.0	96.5	55.3	73.8	98.1	38.6	65.3	96.3	41.9	66.3	95.8	35.4	62.5	95.5	38.0	63.8
COVERS	DIVS. 10 - 14, 25, 28, 41, 43, 44, 46	100.0	79.4	95.8	100.0	74.3	94.8	100.0	77.3	95.4	100.0	75.9	95.1	100.0	77.1	95.4	100.0	71.6	94.3
21, 22, 23	FIRE SUPPRESSION, PLUMBING & HVAC	95.6	51.1	77.7	100.2	52.7	81.0	100.4	52.4	81.0	100.4	53.1	81.3	95.6	52.2	78.1	95.6	51.2	77.7
26, 27, 3370	ELECTRICAL, COMMUNICATIONS & UTIL.	95.9	34.5	63.4	101.4	50.7	74.6	101.6	57.4	78.2	95.2	55.7	74.3	99.6	57.4	77.3	99.4	46.1	71.2
MF2014	WEIGHTED AVERAGE	99.3	51.6	78.5	104.5	58.1	84.2	101.6	55.1	81.3	101.3	55.7	81.4	98.8	53.9	79.2	98.4	52.7	78.5

DIVISION		NORTH CAROLINA																NORTH DAKOTA		
		MURPHY			RALEIGH			ROCKY MOUNT			WILMINGTON			WINSTON-SALEM			BISMARCK		
		289			275 - 276			278			284			271			585		
		MAT.	INST.	TOTAL	MAT.	INST.	TOTAL	MAT.	INST.	TOTAL	MAT.	INST.	TOTAL	MAT.	INST.	TOTAL	MAT.	INST.	TOTAL
015433	CONTRACTOR EQUIPMENT		96.3	96.3		101.7	101.7		101.7	101.7		96.3	96.3		101.7	101.7		98.2	98.2
0241, 31 - 34	SITE & INFRASTRUCTURE, DEMOLITION	103.5	76.7	84.5	101.3	85.7	90.2	102.6	85.7	90.6	103.7	77.1	84.8	100.6	85.7	90.0	101.4	96.9	98.2
0310	Concrete Forming & Accessories	103.7	39.3	48.2	98.9	46.6	53.8	91.9	44.2	50.8	97.5	48.8	55.5	101.9	48.0	55.4	104.9	40.8	49.6
0320	Concrete Reinforcing	92.6	45.4	68.6	96.2	56.0	75.7	89.2	55.8	72.2	93.7	57.9	75.5	90.1	57.9	73.7	99.9	92.7	96.2
0330	Cast-in-Place Concrete	119.4	44.4	88.6	106.7	51.4	84.0	99.6	47.7	78.3	115.0	49.5	88.1	103.4	50.0	81.4	102.6	47.5	79.9
03	CONCRETE	109.2	44.1	77.2	101.7	51.7	77.1	99.5	49.3	74.9	105.9	52.4	79.6	99.4	52.2	76.2	101.5	54.2	78.3
04	MASONRY	85.3	41.2	57.8	86.6	42.2	58.9	76.7	39.5	53.5	82.8	41.7	57.2	82.7	38.6	55.2	114.6	56.4	78.3
05	METALS	100.9	74.2	92.7	105.7	79.2	97.5	106.3	78.4	97.7	102.7	79.9	95.7	110.8	79.9	101.3	103.4	88.4	98.8
06	WOOD, PLASTICS & COMPOSITES	107.8	39.7	69.7	95.1	47.5	68.5	87.4	44.7	63.5	100.1	49.0	71.5	96.4	49.4	70.1	95.3	34.7	61.4
07	THERMAL & MOISTURE PROTECTION	106.9	40.8	79.8	100.5	47.0	78.5	106.6	40.8	79.7	106.7	46.9	82.2	106.7	43.5	80.8	112.4	51.3	87.4
08	OPENINGS	97.1	38.9	83.5	104.8	47.4	91.4	101.8	41.9	87.9	97.3	51.0	86.5	105.8	49.7	92.8	108.0	50.1	94.5
0920	Plaster & Gypsum Board	106.1	37.6	59.8	99.8	45.7	63.3	100.5	42.8	61.5	102.4	47.2	65.1	107.5	47.6	67.0	98.8	33.0	54.3
0950, 0980	Ceilings & Acoustic Treatment	85.7	37.6	54.1	88.8	45.7	60.5	85.6	42.8	57.5	86.5	47.2	60.7	88.1	47.6	61.5	112.9	33.0	60.4
0960	Flooring	105.7	24.1	82.4	101.2	42.7	84.5	99.4	21.8	77.2	102.9	44.4	86.2	103.9	42.7	86.4	99.4	72.1	91.6
0970, 0990	Wall Finishes & Painting/Coating	113.6	39.4	68.8	104.2	36.9	63.5	105.4	38.5	65.0	113.6	38.7	68.4	105.4	36.7	64.0	101.9	29.4	58.1
09	FINISHES	97.7	36.1	63.7	95.5	44.7	67.5	93.9	39.3	63.8	96.3	46.8	69.0	96.3	45.7	68.4	102.7	42.8	69.7
COVERS	DIVS. 10 - 14, 25, 28, 41, 43, 44, 46	100.0	76.9	95.3	100.0	72.9	94.5	100.0	72.5	94.5	100.0	73.8	94.7	100.0	78.7	95.7	100.0	82.5	96.5
21, 22, 23	FIRE SUPPRESSION, PLUMBING & HVAC	95.6	50.9	77.6	100.0	52.5	80.8	95.6	52.7	78.3	100.4	54.9	82.0	100.4	53.3	81.4	100.1	71.9	88.7
26, 27, 3370	ELECTRICAL, COMMUNICATIONS & UTIL.	103.2	29.3	64.1	98.3	40.4	67.7	97.9	39.7	67.1	102.4	50.7	75.1	95.2	55.7	74.3	102.5	71.6	86.2
MF2014	WEIGHTED AVERAGE	99.6	48.3	77.2	100.5	54.1	80.3	98.5	52.3	78.3	100.3	55.9	81.0	101.0	56.5	81.6	103.1	66.2	87.0

DIVISION		NORTH DAKOTA																	
		DEVILS LAKE			DICKINSON			FARGO			GRAND FORKS			JAMESTOWN			MINOT		
		583			586			580 - 581			582			584			587		
		MAT.	INST.	TOTAL	MAT.	INST.	TOTAL	MAT.	INST.	TOTAL	MAT.	INST.	TOTAL	MAT.	INST.	TOTAL	MAT.	INST.	TOTAL
015433	CONTRACTOR EQUIPMENT		98.2	98.2		98.2	98.2		98.2	98.2		98.2	98.2		98.2	98.2		98.2	98.2
0241, 31 - 34	SITE & INFRASTRUCTURE, DEMOLITION	105.7	94.4	97.7	113.6	92.9	98.9	102.6	96.9	98.5	109.6	92.9	97.7	104.7	92.9	96.3	107.1	96.9	99.8
0310	Concrete Forming & Accessories	101.2	35.4	44.4	90.8	34.9	42.5	99.3	41.5	49.4	94.5	34.2	42.5	92.3	34.1	42.1	90.5	67.0	70.2
0320	Concrete Reinforcing	100.2	93.1	96.6	101.1	92.8	96.9	96.9	92.8	94.8	98.7	92.9	95.7	100.8	91.3	95.9	102.1	93.2	97.6
0330	Cast-in-Place Concrete	126.3	46.0	93.3	114.3	44.4	85.6	113.3	49.2	87.2	114.3	44.1	85.5	124.7	44.1	91.6	114.3	46.5	86.4
03	CONCRETE	110.6	51.3	81.5	109.6	50.2	80.4	105.9	55.1	80.9	106.8	49.8	78.8	109.1	48.7	79.5	105.5	65.6	85.9
04	MASONRY	121.7	64.9	86.3	124.3	60.4	84.5	114.2	56.4	78.2	115.9	64.4	83.8	135.1	32.8	71.3	114.2	65.4	83.8
05	METALS	99.9	87.4	96.0	99.8	82.5	94.5	103.0	88.7	98.6	99.8	82.1	94.4	99.8	63.7	88.7	100.1	89.3	96.8
06	WOOD, PLASTICS & COMPOSITES	95.2	32.0	59.8	83.1	32.0	54.5	92.9	35.2	60.6	87.4	32.0	56.4	87.4	32.0	55.3	82.8	70.1	75.7
07	THERMAL & MOISTURE PROTECTION	107.6	49.8	83.9	108.1	48.5	83.7	107.4	51.6	84.5	107.8	49.8	84.0	107.4	41.5	80.4	107.5	57.4	87.0
08	OPENINGS	100.7	42.7	87.2	100.7	42.7	87.2	100.5	50.3	88.9	100.7	42.7	87.2	100.7	32.2	84.8	100.9	69.3	93.5
0920	Plaster & Gypsum Board	118.4	30.2	58.8	109.1	30.2	55.8	96.6	33.5	54.0	110.5	30.2	56.2	110.2	30.2	56.1	109.1	69.4	82.3
0950, 0980	Ceilings & Acoustic Treatment	114.3	30.2	59.0	114.3	30.2	59.0	112.1	33.5	60.5	114.3	30.2	59.0	114.3	30.2	59.0	114.3	69.4	84.8
0960	Flooring	107.9	35.9	87.3	101.4	35.9	82.7	102.8	72.1	94.0	103.3	35.9	84.0	102.1	35.9	83.2	101.1	89.3	97.8
0970, 0990	Wall Finishes & Painting/Coating	104.4	22.5	54.9	104.4	31.5	60.4	101.2	70.7	82.8	104.4	28.2	58.4	104.4	22.5	54.9	104.4	27.9	58.2
09	FINISHES	108.2	32.2	66.3	105.9	33.2	65.8	103.8	47.7	72.9	106.1	32.8	65.7	105.3	32.2	65.0	105.0	67.0	84.1
COVERS	DIVS. 10 - 14, 25, 28, 41, 43, 44, 46	100.0	32.4	86.3	100.0	32.6	86.4	100.0	82.6	96.5	100.0	32.5	86.4	100.0	80.6	96.1	100.0	86.4	97.3
21, 22, 23	FIRE SUPPRESSION, PLUMBING & HVAC	95.6	74.0	86.9	95.6	66.2	83.7	100.1	77.0	90.8	100.4	33.3	73.3	95.6	35.0	71.2	100.4	60.2	84.2
26, 27, 3370	ELECTRICAL, COMMUNICATIONS & UTIL.	98.3	35.8	65.3	107.7	72.5	89.1	102.4	68.4	84.4	102.1	53.5	76.4	98.3	35.7	65.2	105.5	75.7	89.8
MF2014	WEIGHTED AVERAGE	102.0	58.6	83.1	102.8	60.9	84.6	102.7	67.7	87.4	102.7	51.9	80.4	102.1	45.2	77.3	102.7	71.3	89.0

City Cost Indexes

DIVISION		NORTH DAKOTA WILLISTON 588			OHIO AKRON 442 - 443			OHIO ATHENS 457			OHIO CANTON 446 - 447			OHIO CHILLICOTHE 456			OHIO CINCINNATI 451 - 452		
		MAT.	INST.	TOTAL	MAT.	INST.	TOTAL	MAT.	INST.	TOTAL	MAT.	INST.	TOTAL	MAT.	INST.	TOTAL	MAT.	INST.	TOTAL
015433	CONTRACTOR EQUIPMENT		98.2	98.2		94.4	94.4		90.4	90.4		94.4	94.4		99.7	99.7		99.5	99.5
0241, 31 - 34	SITE & INFRASTRUCTURE, DEMOLITION	107.5	92.9	97.1	98.1	101.6	100.6	111.0	91.6	97.2	98.2	101.0	100.2	97.0	102.6	101.0	94.1	102.4	100.0
0310	Concrete Forming & Accessories	96.3	34.9	43.3	99.0	94.3	95.0	95.0	84.3	85.8	99.0	83.7	85.8	97.4	92.3	93.0	99.3	79.2	82.0
0320	Concrete Reinforcing	103.1	92.8	97.9	99.5	92.1	95.7	93.1	87.0	90.0	99.5	75.8	87.5	90.1	80.4	85.1	95.5	79.2	87.2
0330	Cast-in-Place Concrete	114.3	44.4	85.6	94.0	98.0	95.7	111.5	96.5	105.4	94.9	95.0	95.0	101.2	99.3	100.4	93.2	92.4	92.9
03	CONCRETE	106.9	50.2	79.1	97.3	94.4	95.9	107.6	88.6	98.3	97.7	85.7	91.8	101.9	92.4	97.2	96.3	84.0	90.3
04	MASONRY	108.8	60.4	78.7	93.3	96.2	95.1	86.4	90.4	88.9	94.0	84.3	87.9	94.0	100.5	98.1	93.7	85.0	88.2
05	METALS	100.0	82.5	94.6	95.4	81.6	91.1	103.0	80.0	95.9	95.4	74.5	88.9	94.9	85.7	92.1	97.2	84.6	93.3
06	WOOD, PLASTICS & COMPOSITES	88.8	32.0	57.0	99.1	93.6	96.0	85.7	84.9	85.3	99.5	82.8	90.2	98.4	89.6	93.5	100.9	76.8	87.4
07	THERMAL & MOISTURE PROTECTION	107.7	48.5	83.5	110.8	96.5	104.9	99.4	94.8	97.5	112.0	91.4	103.5	101.2	97.4	99.7	99.2	88.7	94.9
08	OPENINGS	100.8	42.7	87.3	110.7	93.3	106.7	101.7	82.3	97.2	104.3	77.9	98.2	93.8	83.5	91.4	102.0	77.4	96.3
0920	Plaster & Gypsum Board	110.5	30.2	56.2	96.9	93.2	94.4	91.4	84.1	86.5	98.0	82.1	87.2	93.3	89.5	90.7	94.9	76.3	82.3
0950, 0980	Ceilings & Acoustic Treatment	114.3	30.2	59.0	93.0	93.2	93.1	103.3	84.1	90.7	93.0	82.1	85.8	97.9	89.5	92.4	98.7	76.3	84.0
0960	Flooring	104.1	35.9	84.6	96.9	93.4	95.9	121.3	99.0	114.9	97.1	81.7	92.7	98.3	97.7	98.1	99.3	90.7	96.8
0970, 0990	Wall Finishes & Painting/Coating	104.4	31.5	60.4	96.3	106.3	102.3	100.5	99.1	99.6	96.3	83.2	88.4	97.8	92.9	94.9	97.8	83.9	89.4
09	FINISHES	106.2	33.2	66.0	98.0	95.3	96.5	101.8	89.0	94.8	98.2	82.9	89.8	98.8	93.3	95.8	99.2	81.1	89.3
COVERS	DIVS. 10 - 14, 25, 28, 41, 43, 44, 46	100.0	32.6	86.4	100.0	98.2	99.6	100.0	52.0	90.3	100.0	95.4	99.1	100.0	92.6	98.5	100.0	89.3	97.8
21, 22, 23	FIRE SUPPRESSION, PLUMBING & HVAC	95.6	66.2	83.7	100.1	94.6	97.9	95.3	51.3	77.6	100.1	81.4	92.5	95.8	95.2	95.6	100.0	83.3	93.3
26, 27, 3370	ELECTRICAL, COMMUNICATIONS & UTIL.	102.5	72.5	86.6	99.3	93.9	96.5	95.5	98.1	96.9	98.5	92.7	95.5	95.0	85.6	90.0	93.9	79.4	86.3
MF2014	WEIGHTED AVERAGE	101.2	60.9	83.7	99.9	94.2	97.4	99.6	80.4	91.2	99.3	85.6	93.3	96.8	93.0	95.1	98.3	84.5	92.2

DIVISION		OHIO CLEVELAND 441			OHIO COLUMBUS 430 - 432			OHIO DAYTON 453 - 454			OHIO HAMILTON 450			OHIO LIMA 458			OHIO LORAIN 440		
		MAT.	INST.	TOTAL	MAT.	INST.	TOTAL	MAT.	INST.	TOTAL	MAT.	INST.	TOTAL	MAT.	INST.	TOTAL	MAT.	INST.	TOTAL
015433	CONTRACTOR EQUIPMENT		94.7	94.7		93.6	93.6		94.7	94.7		99.7	99.7		92.9	92.9		94.4	94.4
0241, 31 - 34	SITE & INFRASTRUCTURE, DEMOLITION	98.0	102.2	101.0	95.5	98.3	97.5	92.9	101.8	99.2	92.8	102.1	99.4	104.6	91.6	95.4	97.5	102.9	101.4
0310	Concrete Forming & Accessories	99.1	99.6	99.5	100.3	84.5	86.7	99.3	79.0	81.8	99.4	79.4	82.1	95.0	86.9	88.1	99.1	86.8	88.5
0320	Concrete Reinforcing	100.0	92.5	96.2	102.7	82.6	92.5	95.5	81.3	88.3	95.5	79.2	87.2	93.1	81.5	87.2	99.5	92.4	95.9
0330	Cast-in-Place Concrete	92.2	106.5	98.1	94.1	93.7	93.9	86.7	86.4	86.6	92.9	92.7	92.8	102.4	97.5	100.4	89.5	104.1	95.5
03	CONCRETE	96.5	99.8	98.2	97.9	87.1	92.6	93.3	81.8	87.6	96.2	84.2	90.3	100.4	89.3	94.9	95.2	93.2	94.2
04	MASONRY	97.8	105.6	102.7	96.6	94.0	95.0	93.2	83.5	87.1	93.5	85.5	88.5	117.7	86.4	98.2	89.9	104.6	99.1
05	METALS	96.9	84.4	93.1	96.9	80.7	91.9	96.4	78.6	91.0	96.5	84.5	92.8	103.0	81.5	96.4	96.0	82.8	91.9
06	WOOD, PLASTICS & COMPOSITES	98.2	97.1	97.6	97.1	82.3	88.8	102.2	77.1	88.1	100.9	76.8	87.4	85.6	86.3	86.0	99.1	80.9	88.9
07	THERMAL & MOISTURE PROTECTION	109.5	108.6	109.1	100.9	94.9	98.4	104.7	87.6	97.7	101.4	88.8	96.2	99.0	95.0	97.3	111.9	103.0	108.2
08	OPENINGS	100.6	95.2	99.3	102.4	80.0	97.2	101.3	78.1	95.9	98.9	77.4	93.9	101.7	79.8	96.6	104.3	86.4	100.2
0920	Plaster & Gypsum Board	96.2	96.8	96.6	93.5	81.8	85.6	94.9	76.6	82.6	94.9	76.3	82.3	91.4	85.5	87.4	96.9	80.1	85.5
0950, 0980	Ceilings & Acoustic Treatment	91.3	96.8	94.9	97.7	81.8	87.3	99.7	76.6	84.5	98.7	76.3	84.0	102.4	85.5	91.3	93.0	80.1	84.5
0960	Flooring	96.7	105.0	99.1	92.0	90.3	91.5	102.0	80.7	95.9	99.3	90.7	96.8	120.3	92.2	112.3	97.1	105.0	99.3
0970, 0990	Wall Finishes & Painting/Coating	96.3	105.1	101.6	95.5	92.9	94.0	97.8	85.0	90.1	97.8	84.7	89.9	100.5	82.7	89.8	96.3	105.1	101.6
09	FINISHES	97.5	100.9	99.4	94.1	85.9	89.6	100.2	79.3	88.7	99.1	81.4	89.3	100.8	87.2	93.3	98.0	91.1	94.2
COVERS	DIVS. 10 - 14, 25, 28, 41, 43, 44, 46	100.0	102.5	100.5	100.0	93.3	98.6	100.0	89.1	97.8	100.0	89.5	97.9	100.0	95.2	99.0	100.0	100.3	100.1
21, 22, 23	FIRE SUPPRESSION, PLUMBING & HVAC	100.0	100.5	100.2	100.1	92.0	96.8	100.9	86.6	95.2	100.6	83.6	93.7	95.3	89.0	92.8	100.1	92.1	96.9
26, 27, 3370	ELECTRICAL, COMMUNICATIONS & UTIL.	98.8	105.0	102.1	97.6	87.8	92.4	92.6	83.0	87.5	93.0	83.3	87.9	95.8	80.3	87.6	98.7	88.0	93.0
MF2014	WEIGHTED AVERAGE	99.0	100.3	99.6	98.6	89.2	94.5	98.0	84.4	92.1	97.9	85.1	92.3	100.0	86.8	94.3	98.9	93.0	96.3

DIVISION		OHIO MANSFIELD 448 - 449			OHIO MARION 433			OHIO SPRINGFIELD 455			OHIO STEUBENVILLE 439			OHIO TOLEDO 434 - 436			OHIO YOUNGSTOWN 444 - 445		
		MAT.	INST.	TOTAL	MAT.	INST.	TOTAL	MAT.	INST.	TOTAL	MAT.	INST.	TOTAL	MAT.	INST.	TOTAL	MAT.	INST.	TOTAL
015433	CONTRACTOR EQUIPMENT		94.4	94.4		93.2	93.2		94.7	94.7		98.1	98.1		96.0	96.0		94.4	94.4
0241, 31 - 34	SITE & INFRASTRUCTURE, DEMOLITION	93.9	101.5	99.3	91.8	97.1	95.6	93.2	100.5	98.4	133.1	106.9	114.5	94.8	99.1	97.8	98.0	102.0	100.8
0310	Concrete Forming & Accessories	89.1	83.7	84.5	96.7	81.2	83.3	99.3	83.5	85.7	98.3	89.4	90.7	100.3	97.1	97.6	99.0	87.5	89.1
0320	Concrete Reinforcing	90.6	76.4	83.4	94.7	82.5	88.5	95.5	81.3	88.3	92.3	85.5	88.8	102.7	85.4	93.9	99.5	85.6	92.4
0330	Cast-in-Place Concrete	87.1	94.7	90.2	85.9	92.7	88.7	89.1	86.3	87.9	93.3	94.1	93.6	94.1	100.0	96.5	93.1	96.9	94.7
03	CONCRETE	89.7	85.7	87.7	89.9	85.1	87.6	94.4	83.8	89.2	93.8	89.7	91.8	97.9	95.5	96.7	96.6	89.8	93.4
04	MASONRY	92.4	97.0	95.3	98.8	95.5	96.7	93.4	83.5	87.2	86.2	94.2	91.2	104.8	99.9	101.7	93.6	93.6	93.6
05	METALS	96.2	75.5	89.8	96.4	78.4	90.9	96.4	78.4	90.9	92.5	79.5	88.5	96.7	85.5	93.3	95.4	78.7	90.2
06	WOOD, PLASTICS & COMPOSITES	86.7	80.9	83.4	92.8	79.8	85.5	103.6	83.6	92.4	88.4	88.3	88.3	97.1	96.9	97.0	99.1	86.1	91.8
07	THERMAL & MOISTURE PROTECTION	111.3	94.4	104.4	100.5	85.2	94.2	104.6	88.2	97.9	112.8	96.4	106.1	102.5	102.7	102.6	112.1	94.3	104.8
08	OPENINGS	104.9	76.0	98.2	96.6	76.5	91.9	99.3	79.0	94.6	97.0	83.7	94.0	99.7	91.3	97.7	104.3	85.1	99.8
0920	Plaster & Gypsum Board	90.7	80.1	83.5	91.4	79.3	83.2	94.9	83.3	87.1	90.7	87.5	88.6	93.5	96.9	95.8	96.9	85.5	89.2
0950, 0980	Ceilings & Acoustic Treatment	93.9	80.1	84.8	97.7	79.3	85.6	99.7	83.3	88.9	94.8	87.5	90.0	97.7	96.9	97.2	93.0	85.5	88.1
0960	Flooring	92.5	107.5	96.8	90.9	107.5	95.6	102.0	80.7	95.9	118.6	100.4	113.4	91.2	99.7	93.6	97.1	93.9	96.2
0970, 0990	Wall Finishes & Painting/Coating	96.3	88.5	91.6	95.5	47.5	66.5	97.8	85.0	90.1	107.8	103.1	104.9	95.6	102.0	99.4	96.3	92.6	94.1
09	FINISHES	95.8	88.2	91.6	93.1	82.6	87.3	100.2	83.2	90.8	110.0	92.6	100.4	93.8	98.1	96.2	98.1	88.8	93.0
COVERS	DIVS. 10 - 14, 25, 28, 41, 43, 44, 46	100.0	96.2	99.2	100.0	54.4	90.8	100.0	89.8	97.9	100.0	95.0	99.0	100.0	98.7	99.7	100.0	96.4	99.3
21, 22, 23	FIRE SUPPRESSION, PLUMBING & HVAC	95.3	88.9	92.7	95.3	91.0	93.6	100.9	81.3	93.0	95.7	92.0	94.2	100.1	100.4	100.2	100.1	87.1	94.8
26, 27, 3370	ELECTRICAL, COMMUNICATIONS & UTIL.	96.2	79.0	87.1	91.9	79.0	85.1	92.6	87.8	90.1	86.0	114.1	101.3	97.6	104.7	101.4	98.7	86.0	92.0
MF2014	WEIGHTED AVERAGE	96.7	87.3	92.6	95.1	85.3	90.8	98.0	84.7	92.2	96.8	94.9	96.0	98.7	98.1	98.4	99.2	89.0	94.7

City Cost Indexes

DIVISION		OHIO ZANESVILLE 437-438			OKLAHOMA ARDMORE 734			OKLAHOMA CLINTON 736			OKLAHOMA DURANT 747			OKLAHOMA ENID 737			OKLAHOMA GUYMON 739		
		MAT.	INST.	TOTAL	MAT.	INST.	TOTAL	MAT.	INST.	TOTAL	MAT.	INST.	TOTAL	MAT.	INST.	TOTAL	MAT.	INST.	TOTAL
015433	CONTRACTOR EQUIPMENT		93.2	93.2		82.7	82.7		81.8	81.8		81.8	81.8		81.8	81.8		81.8	81.8
0241, 31-34	SITE & INFRASTRUCTURE, DEMOLITION	94.4	98.7	97.5	96.2	91.6	92.9	97.6	90.2	92.3	95.5	89.9	91.5	99.3	90.2	92.8	101.6	90.0	93.3
0310	Concrete Forming & Accessories	93.7	81.3	83.0	94.0	40.4	47.8	92.5	46.6	53.0	85.5	44.1	49.8	96.2	34.1	42.7	99.8	44.9	52.4
0320	Concrete Reinforcing	94.1	85.1	89.5	90.8	79.8	85.2	91.3	79.8	85.5	95.8	80.2	87.9	90.7	79.8	85.2	91.3	79.8	85.5
0330	Cast-in-Place Concrete	90.5	91.7	91.0	97.5	43.4	75.3	94.3	45.1	74.1	91.7	42.8	71.6	94.3	46.4	74.6	94.3	43.1	73.3
03	CONCRETE	93.5	85.3	89.5	95.2	49.3	72.7	94.7	52.7	74.1	92.3	50.8	71.9	95.3	47.6	71.8	98.2	51.2	75.1
04	MASONRY	95.5	85.9	89.5	103.4	57.1	74.5	129.7	57.1	84.4	95.6	62.3	74.9	110.1	57.1	77.0	105.9	53.7	73.3
05	METALS	97.4	80.4	92.1	102.5	70.1	92.5	102.6	70.1	92.6	93.5	70.7	86.5	104.0	70.0	93.6	103.2	69.8	92.9
06	WOOD, PLASTICS & COMPOSITES	88.5	79.8	83.6	97.4	37.9	64.0	96.4	46.2	68.3	88.1	43.0	62.9	100.1	29.3	60.5	104.0	46.0	71.5
07	THERMAL & MOISTURE PROTECTION	100.6	92.6	97.3	108.9	60.6	89.1	109.1	61.5	89.6	99.7	61.8	84.2	109.2	59.9	89.0	109.5	57.3	88.1
08	OPENINGS	96.6	79.3	92.5	102.9	49.0	90.3	102.9	53.3	91.4	96.3	52.0	86.0	102.9	43.9	89.1	103.0	49.2	90.5
0920	Plaster & Gypsum Board	88.2	79.3	82.2	89.8	36.6	53.8	89.4	45.2	59.6	79.4	41.9	54.1	90.5	27.8	48.1	90.7	44.9	59.8
0950, 0980	Ceilings & Acoustic Treatment	97.7	79.3	85.6	85.7	36.6	53.4	85.7	45.2	59.1	82.4	41.9	55.8	85.7	27.8	47.6	86.5	44.9	59.2
0960	Flooring	89.3	90.3	89.6	107.0	43.2	88.7	105.8	41.1	87.3	102.8	61.9	91.1	107.7	41.1	88.6	109.3	24.4	85.0
0970, 0990	Wall Finishes & Painting/Coating	95.5	92.9	94.0	104.7	51.3	72.5	104.7	51.3	72.5	103.9	51.3	72.1	104.7	51.3	72.5	104.7	33.0	61.5
09	FINISHES	92.4	83.6	87.5	92.9	40.1	63.8	92.7	44.7	66.3	90.2	46.6	66.2	93.5	34.6	61.1	94.5	38.9	63.9
COVERS	DIVS. 10-14, 25, 28, 41, 43, 44, 46	100.0	88.5	97.7	100.0	77.6	95.5	100.0	78.6	95.7	100.0	77.9	95.5	100.0	76.7	95.3	100.0	77.4	95.4
21, 22, 23	FIRE SUPPRESSION, PLUMBING & HVAC	95.3	89.9	93.1	95.6	65.8	83.6	95.6	65.9	83.6	95.5	65.4	83.4	100.3	65.8	86.4	95.6	63.8	82.7
26, 27, 3370	ELECTRICAL, COMMUNICATIONS & UTIL.	92.3	83.4	87.5	93.6	70.9	81.6	94.6	70.9	82.1	96.5	70.9	83.0	94.6	70.9	82.1	96.2	61.5	77.9
MF2014	WEIGHTED AVERAGE	95.6	86.5	91.6	98.3	61.9	82.4	99.6	63.1	83.7	95.1	63.5	81.3	100.2	60.5	82.9	99.4	59.7	82.1

DIVISION		OKLAHOMA LAWTON 735			OKLAHOMA MCALESTER 745			OKLAHOMA MIAMI 743			OKLAHOMA MUSKOGEE 744			OKLAHOMA OKLAHOMA CITY 730-731			OKLAHOMA PONCA CITY 746		
		MAT.	INST.	TOTAL	MAT.	INST.	TOTAL	MAT.	INST.	TOTAL	MAT.	INST.	TOTAL	MAT.	INST.	TOTAL	MAT.	INST.	TOTAL
015433	CONTRACTOR EQUIPMENT		82.7	82.7		81.8	81.8		89.9	89.9		89.9	89.9		83.0	83.0		81.8	81.8
0241, 31-34	SITE & INFRASTRUCTURE, DEMOLITION	95.7	91.6	92.8	89.1	90.1	89.9	90.5	87.3	88.2	90.8	87.1	88.2	95.1	92.0	92.9	95.9	90.1	91.8
0310	Concrete Forming & Accessories	99.8	44.3	52.0	83.5	43.6	49.1	96.9	66.2	70.4	101.4	33.4	42.7	97.7	57.9	63.4	92.2	46.3	52.6
0320	Concrete Reinforcing	91.0	79.8	85.3	95.5	79.8	87.5	94.0	79.9	86.8	94.9	79.3	87.0	96.2	79.9	87.9	94.9	79.9	87.2
0330	Cast-in-Place Concrete	91.2	46.4	72.8	80.4	45.2	65.9	84.3	47.5	69.2	85.3	45.8	69.1	92.9	46.9	74.0	94.2	44.7	73.8
03	CONCRETE	91.5	52.1	72.1	82.9	51.3	67.4	87.6	63.0	75.5	89.4	47.7	68.9	94.9	58.3	76.9	94.4	52.4	73.8
04	MASONRY	105.8	57.1	75.4	114.3	56.1	78.1	98.4	56.3	72.1	116.9	45.2	72.2	109.8	56.4	76.5	90.9	56.1	69.2
05	METALS	108.0	70.1	96.4	93.4	70.0	86.2	93.4	82.1	89.9	94.8	80.4	90.4	100.7	70.2	91.3	93.4	70.3	86.3
06	WOOD, PLASTICS & COMPOSITES	103.0	43.1	69.5	85.6	43.0	61.7	101.0	72.9	85.3	105.6	30.7	63.6	97.0	61.8	77.3	96.3	46.0	68.1
07	THERMAL & MOISTURE PROTECTION	108.9	61.3	89.4	99.4	60.6	83.5	99.8	64.3	85.2	99.9	48.6	78.8	101.1	62.9	85.4	99.9	66.2	86.1
08	OPENINGS	104.6	51.4	92.2	96.2	51.8	85.9	96.2	68.3	89.7	96.2	42.4	83.7	104.0	61.6	94.2	96.2	53.1	86.2
0920	Plaster & Gypsum Board	92.4	42.0	58.3	78.4	41.9	53.7	84.8	72.6	76.5	86.9	29.0	47.8	98.0	61.2	73.1	83.7	44.9	57.5
0950, 0980	Ceilings & Acoustic Treatment	93.2	42.0	59.6	82.4	41.9	55.8	82.4	72.6	75.9	90.8	29.0	50.2	96.4	61.2	73.3	82.4	44.9	57.8
0960	Flooring	109.7	41.1	90.1	101.8	41.1	84.4	108.8	61.9	95.4	111.3	39.8	90.8	106.2	41.1	87.6	105.9	41.1	87.4
0970, 0990	Wall Finishes & Painting/Coating	104.7	51.3	72.5	103.9	37.8	64.0	103.9	77.5	87.9	103.9	34.2	61.8	106.4	51.3	73.2	103.9	51.3	72.1
09	FINISHES	95.5	42.8	66.5	89.2	41.0	62.6	92.0	67.3	78.4	94.9	33.0	60.8	97.6	53.7	73.4	91.8	44.3	65.6
COVERS	DIVS. 10-14, 25, 28, 41, 43, 44, 46	100.0	78.2	95.6	100.0	77.9	95.5	100.0	81.6	96.3	100.0	76.1	95.2	100.0	80.0	96.0	100.0	78.2	95.6
21, 22, 23	FIRE SUPPRESSION, PLUMBING & HVAC	100.3	65.9	86.4	95.5	61.6	81.8	95.5	61.9	81.9	100.3	60.6	84.3	100.1	65.5	86.2	95.5	61.8	81.9
26, 27, 3370	ELECTRICAL, COMMUNICATIONS & UTIL.	96.2	70.9	82.8	94.9	64.5	78.8	96.3	64.6	79.5	94.5	52.4	72.2	101.7	70.9	85.4	94.5	70.4	81.7
MF2014	WEIGHTED AVERAGE	100.7	62.8	84.2	94.5	60.4	79.7	94.8	67.5	82.9	97.6	55.7	79.2	100.3	65.7	85.2	95.1	62.1	80.7

DIVISION		OKLAHOMA POTEAU 749			OKLAHOMA SHAWNEE 748			OKLAHOMA TULSA 740-741			OKLAHOMA WOODWARD 738			OREGON BEND 977			OREGON EUGENE 974		
		MAT.	INST.	TOTAL	MAT.	INST.	TOTAL	MAT.	INST.	TOTAL	MAT.	INST.	TOTAL	MAT.	INST.	TOTAL	MAT.	INST.	TOTAL
015433	CONTRACTOR EQUIPMENT		89.1	89.1		81.8	81.8		89.9	89.9		81.8	81.8		98.8	98.8		98.8	98.8
0241, 31-34	SITE & INFRASTRUCTURE, DEMOLITION	77.2	85.8	83.3	99.0	90.1	92.7	97.1	87.5	90.3	97.9	90.2	92.4	105.4	102.9	103.6	96.1	102.9	100.9
0310	Concrete Forming & Accessories	90.1	41.0	47.8	85.4	43.7	49.5	101.5	40.3	48.7	92.6	46.8	53.1	110.4	98.8	100.4	106.5	98.7	99.8
0320	Concrete Reinforcing	95.9	79.9	87.8	94.9	79.8	87.2	95.1	79.8	87.3	90.7	79.9	85.2	93.3	99.6	96.5	97.4	99.6	98.5
0330	Cast-in-Place Concrete	84.3	44.9	68.1	97.2	42.9	74.9	92.9	46.2	73.7	94.3	45.3	74.2	104.4	102.7	103.7	101.1	102.6	101.7
03	CONCRETE	89.9	50.9	70.7	96.0	50.6	73.7	94.7	51.0	73.2	95.0	52.8	74.3	107.1	100.0	103.6	98.8	100.0	99.4
04	MASONRY	98.7	56.2	72.2	115.6	56.1	78.5	99.3	60.0	74.8	98.6	57.1	72.7	104.1	103.3	103.6	101.1	103.3	102.5
05	METALS	93.4	81.7	89.8	93.3	69.9	86.1	98.1	81.4	93.0	102.7	70.5	92.8	91.9	96.4	93.3	92.6	96.2	93.7
06	WOOD, PLASTICS & COMPOSITES	93.0	39.1	62.8	88.0	43.0	62.8	104.8	36.6	66.6	96.5	46.0	68.2	101.7	98.4	99.9	97.6	98.4	98.1
07	THERMAL & MOISTURE PROTECTION	99.9	60.5	83.7	99.9	59.3	83.2	99.8	62.2	84.4	109.1	66.6	91.7	107.9	94.9	102.6	107.2	91.6	100.8
08	OPENINGS	96.2	49.9	85.5	96.2	51.8	85.9	97.8	47.5	86.1	102.9	53.1	91.3	96.9	102.3	98.1	97.1	102.3	98.4
0920	Plaster & Gypsum Board	82.3	37.8	52.2	79.4	41.9	54.1	86.9	35.2	51.9	89.6	44.9	59.4	107.4	98.2	101.2	106.0	98.2	100.7
0950, 0980	Ceilings & Acoustic Treatment	82.4	37.8	53.1	82.4	41.9	55.8	90.8	35.2	54.3	86.5	44.9	59.2	91.3	98.2	95.8	92.3	98.2	96.1
0960	Flooring	105.2	61.9	92.8	102.8	32.7	82.8	110.0	42.8	90.8	105.8	43.2	87.9	110.5	103.4	108.5	108.9	103.4	107.3
0970, 0990	Wall Finishes & Painting/Coating	103.9	51.3	72.1	103.9	34.1	61.8	103.9	40.5	65.6	104.7	51.3	72.5	106.4	74.8	87.3	106.4	74.8	87.3
09	FINISHES	89.9	44.3	64.8	90.4	39.2	62.2	94.7	38.9	63.9	93.0	44.9	66.5	102.7	96.8	99.5	101.2	96.8	98.8
COVERS	DIVS. 10-14, 25, 28, 41, 43, 44, 46	100.0	77.7	95.5	100.0	77.9	95.5	100.0	78.8	95.7	100.0	78.5	95.7	100.0	99.6	99.9	100.0	99.6	99.9
21, 22, 23	FIRE SUPPRESSION, PLUMBING & HVAC	95.5	61.8	81.9	95.5	65.2	83.4	100.3	63.5	85.4	95.6	65.9	83.6	95.1	100.8	97.4	99.9	100.8	100.2
26, 27, 3370	ELECTRICAL, COMMUNICATIONS & UTIL.	94.6	64.6	78.7	96.6	70.9	83.0	96.5	64.6	79.6	96.1	70.9	82.8	101.6	96.1	98.7	100.1	96.1	98.0
MF2014	WEIGHTED AVERAGE	94.3	61.4	80.0	96.5	61.8	81.4	98.1	61.6	82.2	98.3	63.4	83.1	98.9	99.4	99.1	98.6	99.3	98.9

City Cost Indexes

DIVISION		OREGON																	
		KLAMATH FALLS			MEDFORD			PENDLETON			PORTLAND			SALEM			VALE		
		976			975			978			970 - 972			973			979		
		MAT.	INST.	TOTAL	MAT.	INST.	TOTAL	MAT.	INST.	TOTAL	MAT.	INST.	TOTAL	MAT.	INST.	TOTAL	MAT.	INST.	TOTAL
015433	CONTRACTOR EQUIPMENT		98.8	98.8		98.8	98.8		96.2	96.2		98.8	98.8		98.8	98.8		96.2	96.2
0241, 31 - 34	SITE & INFRASTRUCTURE, DEMOLITION	109.2	102.9	104.7	103.4	102.9	103.0	102.5	96.4	98.2	98.5	102.9	101.6	91.9	102.9	99.7	90.5	96.4	94.7
0310	Concrete Forming & Accessories	102.9	98.6	99.2	102.0	98.6	99.1	103.3	99.0	99.6	107.7	98.9	100.1	106.0	98.8	99.8	109.9	98.7	100.2
0320	Concrete Reinforcing	93.3	99.6	96.5	94.9	99.6	97.3	92.6	99.7	96.2	98.1	99.7	98.9	103.6	99.6	101.6	90.4	99.6	95.1
0330	Cast-in-Place Concrete	104.4	102.6	103.7	104.4	102.6	103.7	105.2	103.9	104.7	103.9	102.7	103.4	97.9	102.7	99.9	82.9	103.8	91.5
03	CONCRETE	109.7	99.9	104.9	104.6	99.9	102.3	91.4	100.6	95.9	100.3	100.1	100.2	98.2	100.0	99.1	76.9	100.4	88.5
04	MASONRY	117.5	103.3	108.6	98.2	103.3	101.4	107.8	103.4	105.0	102.5	103.3	103.0	109.3	103.3	105.6	106.1	103.4	104.4
05	METALS	91.9	96.2	93.2	92.2	96.2	93.4	98.1	96.9	97.7	93.5	96.5	94.4	98.7	96.4	98.0	98.0	96.6	97.5
06	WOOD, PLASTICS & COMPOSITES	92.6	98.4	95.9	91.5	98.4	95.3	94.7	98.5	96.8	98.6	98.4	98.5	95.5	98.4	97.1	103.2	98.5	100.6
07	THERMAL & MOISTURE PROTECTION	108.2	92.9	101.9	107.8	92.9	101.7	101.1	93.5	98.0	107.1	96.7	102.9	104.9	94.9	100.8	100.5	94.4	98.0
08	OPENINGS	96.9	102.3	98.1	99.7	102.3	100.3	93.2	102.4	95.4	95.0	102.3	96.7	98.7	102.3	99.5	93.2	94.1	93.4
0920	Plaster & Gypsum Board	102.9	98.2	99.7	102.2	98.2	99.5	90.2	98.2	95.6	105.5	98.2	100.5	102.1	98.2	99.4	95.9	98.2	97.4
0950, 0980	Ceilings & Acoustic Treatment	98.9	98.2	98.4	105.4	98.2	100.7	64.4	98.2	86.6	94.2	98.2	96.8	99.3	98.2	98.5	64.4	98.2	86.6
0960	Flooring	107.6	103.4	106.4	107.1	103.4	106.1	74.7	103.4	82.9	106.4	103.4	105.5	107.2	103.4	106.1	76.7	103.4	84.4
0970, 0990	Wall Finishes & Painting/Coating	106.4	70.4	84.6	106.4	70.4	84.6	96.6	72.8	82.2	106.2	72.8	86.0	104.6	74.8	86.6	96.6	74.8	83.4
09	FINISHES	103.4	96.3	99.5	103.8	96.3	99.6	72.3	96.7	85.7	100.8	96.6	98.5	100.0	96.8	98.2	72.7	96.9	86.0
COVERS	DIVS. 10 - 14, 25, 28, 41, 43, 44, 46	100.0	99.5	99.9	100.0	99.5	99.9	100.0	90.1	98.0	100.0	99.6	99.9	100.0	99.6	99.9	100.0	99.9	100.0
21, 22, 23	FIRE SUPPRESSION, PLUMBING & HVAC	95.1	100.7	97.4	99.9	100.7	100.2	97.0	113.2	103.5	99.9	100.8	100.2	99.9	100.8	100.2	97.0	98.0	97.4
26, 27, 3370	ELECTRICAL, COMMUNICATIONS & UTIL.	100.2	80.8	89.9	103.9	80.8	91.6	92.3	97.1	94.8	100.4	103.2	101.9	108.0	96.1	101.7	92.3	97.0	94.8
MF2014	WEIGHTED AVERAGE	99.8	97.1	98.6	100.0	97.1	98.8	94.8	101.4	97.7	99.8	100.4	99.5	100.5	99.4	100.0	92.9	98.1	95.2

DIVISION		PENNSYLVANIA																	
		ALLENTOWN			ALTOONA			BEDFORD			BRADFORD			BUTLER			CHAMBERSBURG		
		181			166			155			167			160			172		
		MAT.	INST.	TOTAL	MAT.	INST.	TOTAL	MAT.	INST.	TOTAL	MAT.	INST.	TOTAL	MAT.	INST.	TOTAL	MAT.	INST.	TOTAL
015433	CONTRACTOR EQUIPMENT		112.9	112.9		112.9	112.9		109.6	109.6		112.9	112.9		112.9	112.9		112.1	112.1
0241, 31 - 34	SITE & INFRASTRUCTURE, DEMOLITION	93.6	104.5	101.3	96.8	104.4	102.2	101.4	100.1	100.5	92.4	103.5	100.3	88.1	105.8	100.7	89.2	101.2	97.8
0310	Concrete Forming & Accessories	99.2	113.0	111.1	84.2	80.8	80.9	84.1	80.8	81.3	86.4	81.9	82.5	85.7	96.0	94.6	90.2	80.0	81.4
0320	Concrete Reinforcing	94.7	108.1	101.5	91.8	102.7	97.3	93.1	80.9	86.9	93.7	103.2	98.5	92.4	109.1	100.9	91.7	103.0	97.4
0330	Cast-in-Place Concrete	86.5	104.9	94.0	96.3	86.1	92.1	106.7	70.1	91.6	92.0	92.6	92.3	85.1	96.7	89.9	91.1	69.8	82.3
03	CONCRETE	93.6	110.1	101.7	89.1	88.1	88.6	102.1	78.5	90.5	95.4	91.0	93.2	81.3	99.9	90.4	101.9	82.2	92.3
04	MASONRY	97.4	101.0	99.7	100.5	64.5	78.0	114.1	87.0	97.2	97.6	88.4	91.9	102.6	98.9	100.3	102.6	87.4	93.1
05	METALS	100.0	123.0	107.1	93.9	116.8	101.0	97.6	104.0	99.6	97.9	116.5	103.6	93.6	121.9	102.3	97.5	114.6	102.7
06	WOOD, PLASTICS & COMPOSITES	100.7	115.8	109.1	79.2	83.2	81.4	84.0	80.7	82.2	85.5	80.1	82.5	80.6	95.5	89.0	88.0	80.8	84.0
07	THERMAL & MOISTURE PROTECTION	102.3	119.1	109.2	101.4	91.0	97.1	103.0	87.7	96.7	102.3	91.7	97.9	101.1	100.7	100.9	99.3	72.5	88.3
08	OPENINGS	95.0	115.0	99.6	88.7	89.3	88.8	98.0	81.6	94.2	95.2	91.0	94.2	88.7	104.8	92.4	94.1	82.8	91.4
0920	Plaster & Gypsum Board	95.9	116.0	109.5	87.7	82.5	84.2	92.3	80.0	84.0	88.5	79.2	82.2	87.7	95.2	92.7	96.7	80.0	85.4
0950, 0980	Ceilings & Acoustic Treatment	82.8	116.0	104.6	87.0	82.5	84.0	95.0	80.0	85.1	85.4	79.2	81.3	87.9	95.2	92.7	84.0	80.0	81.4
0960	Flooring	94.6	95.9	95.0	88.2	97.5	90.9	93.1	88.2	91.7	89.4	105.3	94.0	89.0	81.0	86.7	97.9	46.2	83.1
0970, 0990	Wall Finishes & Painting/Coating	98.5	70.5	81.6	98.5	104.2	103.2	100.2	81.6	89.0	98.5	94.6	96.1	94.0	109.2	103.2	105.9	81.6	91.2
09	FINISHES	91.8	105.5	99.4	90.0	85.8	87.7	95.0	80.8	87.2	89.9	85.5	87.5	89.9	93.6	91.9	92.1	73.7	82.0
COVERS	DIVS. 10 - 14, 25, 28, 41, 43, 44, 46	100.0	105.6	101.1	100.0	95.4	99.1	100.0	98.3	99.7	100.0	99.2	99.8	100.0	102.4	100.5	100.0	96.1	99.2
21, 22, 23	FIRE SUPPRESSION, PLUMBING & HVAC	100.2	112.1	105.0	99.7	82.5	92.8	95.2	87.3	92.0	95.5	90.7	93.6	95.0	93.9	94.6	95.5	87.1	92.1
26, 27, 3370	ELECTRICAL, COMMUNICATIONS & UTIL.	99.1	98.4	98.8	89.3	112.2	101.4	94.3	112.2	103.8	92.7	112.2	103.0	89.9	111.2	101.1	89.6	86.3	87.8
MF2014	WEIGHTED AVERAGE	97.9	108.6	102.5	94.6	91.9	93.4	98.2	91.3	95.2	95.7	96.4	96.0	92.5	102.0	96.6	96.2	88.0	92.6

DIVISION		PENNSYLVANIA																	
		DOYLESTOWN			DUBOIS			ERIE			GREENSBURG			HARRISBURG			HAZLETON		
		189			158			164 - 165			156			170 - 171			182		
		MAT.	INST.	TOTAL	MAT.	INST.	TOTAL	MAT.	INST.	TOTAL	MAT.	INST.	TOTAL	MAT.	INST.	TOTAL	MAT.	INST.	TOTAL
015433	CONTRACTOR EQUIPMENT		93.9	93.9		109.6	109.6		112.9	112.9		109.6	109.6		112.1	112.1		112.9	112.9
0241, 31 - 34	SITE & INFRASTRUCTURE, DEMOLITION	106.9	90.2	95.0	106.1	100.5	102.1	93.7	105.2	101.9	97.8	102.9	101.4	89.6	103.5	99.5	86.9	104.7	99.6
0310	Concrete Forming & Accessories	83.2	128.2	122.0	83.5	84.6	84.4	98.5	88.1	89.5	90.7	95.9	95.2	100.4	87.9	89.6	80.8	88.7	87.6
0320	Concrete Reinforcing	91.6	130.9	111.6	92.6	103.5	98.1	93.7	103.3	98.6	92.6	109.1	101.0	101.4	106.2	103.9	91.9	107.1	99.6
0330	Cast-in-Place Concrete	81.7	86.7	83.8	102.9	92.9	98.8	94.7	81.9	89.4	99.0	96.3	97.9	93.2	95.8	94.3	81.7	93.1	86.4
03	CONCRETE	89.2	114.2	101.5	103.6	92.2	98.0	88.2	90.0	89.1	97.2	99.6	98.4	97.7	95.4	96.6	86.3	95.0	90.6
04	MASONRY	100.7	128.1	117.7	114.4	91.1	99.9	89.6	92.5	91.4	124.8	98.9	108.7	102.5	89.6	94.5	110.2	95.7	101.2
05	METALS	97.5	123.8	105.6	97.6	116.1	103.3	94.1	117.0	101.2	97.5	120.6	104.6	104.0	121.4	109.4	99.7	120.6	106.1
06	WOOD, PLASTICS & COMPOSITES	81.0	130.7	108.8	82.9	83.1	83.0	96.9	86.6	91.1	90.9	95.4	93.4	95.8	86.7	90.7	79.6	86.6	83.5
07	THERMAL & MOISTURE PROTECTION	100.1	131.4	112.9	103.2	96.5	100.5	101.9	92.6	98.1	102.8	100.7	102.0	102.5	110.2	105.7	101.8	106.6	103.8
08	OPENINGS	97.2	137.9	106.7	98.0	92.7	96.7	88.8	92.2	89.6	97.9	104.7	99.5	100.6	94.0	99.0	95.6	92.3	94.8
0920	Plaster & Gypsum Board	86.2	131.4	116.7	91.2	82.5	85.3	95.9	86.0	89.2	93.5	95.2	94.6	100.6	86.0	90.7	86.6	86.0	86.2
0950, 0980	Ceilings & Acoustic Treatment	82.0	131.4	114.5	95.0	82.5	86.7	82.8	86.0	84.9	94.2	95.2	94.8	92.6	86.0	88.3	83.7	86.0	85.2
0960	Flooring	79.4	134.3	95.1	92.9	105.3	96.4	93.3	91.8	92.8	96.2	68.0	88.2	103.2	91.7	99.9	87.0	94.6	89.2
0970, 0990	Wall Finishes & Painting/Coating	98.0	69.0	80.5	100.2	106.6	104.1	103.5	94.6	98.1	100.2	106.6	104.1	106.4	89.6	96.3	98.5	106.4	103.2
09	FINISHES	83.2	122.4	104.8	95.2	89.5	92.1	92.2	88.8	90.4	95.6	91.7	93.4	96.9	88.0	92.0	88.1	89.5	88.9
COVERS	DIVS. 10 - 14, 25, 28, 41, 43, 44, 46	100.0	70.6	94.1	100.0	99.1	99.8	100.0	100.8	100.2	100.0	102.2	100.4	100.0	97.4	99.5	100.0	101.1	100.2
21, 22, 23	FIRE SUPPRESSION, PLUMBING & HVAC	95.0	127.4	108.1	95.2	88.4	92.4	99.7	92.9	97.0	95.2	91.1	93.5	100.1	92.1	96.9	95.5	99.2	97.0
26, 27, 3370	ELECTRICAL, COMMUNICATIONS & UTIL.	92.1	128.1	111.1	94.9	112.2	104.0	91.1	96.7	94.0	94.9	112.2	104.1	96.8	88.3	92.3	93.6	91.7	92.6
MF2014	WEIGHTED AVERAGE	94.9	120.6	106.1	98.5	96.8	97.8	94.5	95.8	95.1	98.2	100.9	99.4	99.8	95.6	98.0	95.4	98.3	96.6

City Cost Indexes

		PENNSYLVANIA																	
DIVISION		INDIANA			JOHNSTOWN			KITTANNING			LANCASTER			LEHIGH VALLEY			MONTROSE		
		157			159			162			175 - 176			180			188		
		MAT.	INST.	TOTAL	MAT.	INST.	TOTAL	MAT.	INST.	TOTAL	MAT.	INST.	TOTAL	MAT.	INST.	TOTAL	MAT.	INST.	TOTAL
015433	CONTRACTOR EQUIPMENT		109.6	109.6		109.6	109.6		112.9	112.9		112.1	112.1		112.9	112.9		112.9	112.9
0241, 31 - 34	SITE & INFRASTRUCTURE, DEMOLITION	95.9	101.2	99.7	101.9	102.2	102.1	90.7	105.7	101.4	81.6	103.5	97.2	90.7	104.3	100.4	89.4	102.1	98.4
0310	Concrete Forming & Accessories	84.7	86.3	86.1	83.5	84.8	84.6	85.7	95.9	94.5	92.3	87.5	88.2	92.8	112.7	110.0	81.8	88.5	87.6
0320	Concrete Reinforcing	91.8	109.2	100.6	93.1	108.9	101.2	92.4	109.2	101.0	91.4	106.1	98.9	91.9	108.1	100.1	96.3	106.3	101.4
0330	Cast-in-Place Concrete	97.1	95.9	96.6	107.6	92.4	101.4	88.4	96.5	91.7	77.5	97.9	85.9	88.4	104.7	95.1	86.7	90.9	88.4
03	CONCRETE	94.7	95.2	94.9	102.9	93.2	98.2	83.8	99.7	91.6	89.8	96.0	92.8	92.6	109.8	101.1	91.3	93.7	92.5
04	MASONRY	110.1	100.1	103.9	110.9	90.6	98.3	105.4	100.1	102.1	108.2	90.1	97.0	97.3	101.1	99.6	97.3	95.1	95.9
05	METALS	97.7	120.1	104.6	97.6	118.9	104.2	93.7	121.9	102.4	97.5	120.8	104.7	99.7	122.0	106.5	98.0	114.1	102.9
06	WOOD, PLASTICS & COMPOSITES	84.8	83.1	83.8	82.9	83.1	83.0	80.6	95.5	89.0	90.8	86.7	88.5	92.2	115.8	105.4	80.4	88.4	84.9
07	THERMAL & MOISTURE PROTECTION	102.7	98.2	100.9	103.0	95.1	99.7	101.1	101.0	101.1	98.8	97.1	98.1	102.2	102.8	102.4	101.8	91.7	97.7
08	OPENINGS	98.0	94.4	97.1	98.0	90.9	96.3	88.7	101.2	91.6	94.1	98.6	95.1	95.6	115.0	100.1	92.2	93.1	92.4
0920	Plaster & Gypsum Board	92.7	82.5	85.8	91.0	82.5	85.2	87.7	95.2	92.7	98.5	86.0	90.1	89.5	110.6	107.4	87.1	87.8	87.6
0950, 0980	Ceilings & Acoustic Treatment	95.0	82.5	86.7	94.2	82.5	86.5	87.9	95.2	92.7	84.0	86.0	85.3	83.7	116.0	105.0	85.4	87.8	87.0
0960	Flooring	93.7	105.3	97.0	92.9	98.2	94.4	89.0	105.3	93.6	98.8	91.7	96.8	91.9	95.9	93.0	87.6	57.9	79.1
0970, 0990	Wall Finishes & Painting/Coating	100.2	106.6	104.1	100.2	109.2	105.6	94.0	106.6	101.6	105.9	57.1	76.5	98.5	66.4	79.1	98.5	106.4	103.2
09	FINISHES	94.8	90.0	92.1	94.6	88.8	91.4	90.1	97.4	94.1	92.0	84.5	87.9	90.2	104.1	97.9	88.9	84.7	86.6
COVERS	DIVS. 10 - 14, 25, 28, 41, 43, 44, 46	100.0	100.6	100.1	100.0	99.3	99.9	100.0	102.2	100.5	100.0	97.4	99.5	100.0	106.0	101.2	100.0	101.2	100.2
21, 22, 23	FIRE SUPPRESSION, PLUMBING & HVAC	95.2	90.8	93.4	95.2	88.7	92.5	95.0	96.6	95.8	95.5	92.3	94.2	95.5	112.0	102.2	95.5	98.4	96.7
26, 27, 3370	ELECTRICAL, COMMUNICATIONS & UTIL.	94.9	112.2	104.1	94.9	112.2	104.0	89.3	112.2	101.4	91.0	44.0	66.2	93.6	143.4	119.70	92.7	97.6	95.3
MF2014	WEIGHTED AVERAGE	97.1	99.3	98.0	98.1	97.2	97.7	92.9	103.3	97.4	95.1	89.0	92.4	95.8	114.0	103.7	94.7	96.9	95.7

		PENNSYLVANIA																	
DIVISION		NEW CASTLE			NORRISTOWN			OIL CITY			PHILADELPHIA			PITTSBURGH			POTTSVILLE		
		161			194			163			190 - 191			150 - 152			179		
		MAT.	INST.	TOTAL	MAT.	INST.	TOTAL	MAT.	INST.	TOTAL	MAT.	INST.	TOTAL	MAT.	INST.	TOTAL	MAT.	INST.	TOTAL
015433	CONTRACTOR EQUIPMENT		112.9	112.9		98.8	98.8		112.9	112.9		98.2	98.2		110.8	110.8		112.1	112.1
0241, 31 - 34	SITE & INFRASTRUCTURE, DEMOLITION	88.5	105.8	100.8	95.8	100.6	99.2	87.1	103.7	98.9	101.6	100.0	100.5	101.3	104.9	103.8	84.4	103.4	97.9
0310	Concrete Forming & Accessories	85.7	95.6	94.3	83.2	129.7	123.3	85.7	83.4	83.7	98.9	140.1	134.4	98.5	96.7	97.0	83.3	89.7	88.8
0320	Concrete Reinforcing	91.3	92.5	91.9	90.2	137.9	114.5	92.4	92.3	92.4	100.6	137.9	119.6	93.6	109.4	101.6	90.7	102.7	96.8
0330	Cast-in-Place Concrete	85.9	96.5	90.2	85.0	127.1	102.3	83.4	95.5	88.4	98.0	132.0	112.1	102.9	96.5	100.3	82.6	100.9	90.1
03	CONCRETE	81.6	96.4	88.9	89.2	130.1	109.3	80.1	90.5	85.2	99.4	136.4	117.6	100.5	100.1	100.3	93.3	97.3	95.3
04	MASONRY	101.9	97.7	99.3	110.4	124.9	119.4	101.8	95.8	98.0	96.7	130.7	117.9	103.0	102.1	102.4	102.1	92.2	95.9
05	METALS	93.7	113.6	99.8	99.2	130.0	108.7	93.7	111.5	99.2	102.0	130.2	110.7	99.0	121.3	105.9	97.7	118.8	104.2
06	WOOD, PLASTICS & COMPOSITES	80.6	95.9	89.2	79.9	130.6	108.3	80.6	80.1	80.3	98.6	141.9	122.9	100.6	95.8	97.9	80.2	88.1	84.6
07	THERMAL & MOISTURE PROTECTION	101.1	98.0	99.8	101.4	131.1	113.5	101.0	94.4	98.3	102.8	135.2	116.1	103.1	101.6	102.4	98.9	106.7	102.1
08	OPENINGS	88.7	96.6	90.5	86.9	140.1	99.2	88.7	80.3	86.7	98.6	146.2	109.7	101.8	105.0	102.5	94.1	91.8	93.6
0920	Plaster & Gypsum Board	87.7	95.6	93.0	85.6	131.4	116.5	87.7	79.2	82.0	96.6	143.1	128.0	99.8	95.6	96.9	93.5	87.5	89.5
0950, 0980	Ceilings & Acoustic Treatment	87.9	95.6	92.9	84.5	131.4	115.3	87.9	79.2	82.2	94.0	143.1	126.2	95.0	95.6	95.4	84.0	87.5	86.3
0960	Flooring	89.0	57.5	80.0	92.9	139.2	106.1	89.0	105.3	93.6	98.4	139.2	110.1	99.5	106.6	101.5	95.2	91.7	94.2
0970, 0990	Wall Finishes & Painting/Coating	94.0	109.2	103.2	98.9	147.8	128.3	94.0	106.6	101.6	99.8	154.7	133.0	100.2	119.2	111.7	105.9	106.4	106.2
09	FINISHES	90.0	94.9	89.8	90.7	133.0	114.0	89.8	87.8	88.7	98.9	141.8	122.5	97.7	100.9	99.1	90.5	91.5	90.9
COVERS	DIVS. 10 - 14, 25, 28, 41, 43, 44, 46	100.0	102.4	100.5	100.0	117.5	103.5	100.0	100.6	100.1	100.0	120.8	104.2	100.0	102.2	100.4	100.0	98.7	99.7
21, 22, 23	FIRE SUPPRESSION, PLUMBING & HVAC	95.0	93.6	94.4	95.2	127.3	108.2	95.0	92.9	94.2	100.0	131.8	112.9	99.9	100.4	100.1	95.5	98.6	96.8
26, 27, 3370	ELECTRICAL, COMMUNICATIONS & UTIL.	89.9	98.1	94.2	93.0	148.4	122.3	91.8	111.1	102.0	97.1	148.4	124.2	97.2	112.2	105.1	89.2	95.5	92.5
MF2014	WEIGHTED AVERAGE	92.5	97.9	94.8	95.0	129.5	110.0	92.4	96.9	94.4	99.7	133.6	114.5	99.9	104.6	102.0	94.9	98.5	96.5

		PENNSYLVANIA																	
DIVISION		READING			SCRANTON			STATE COLLEGE			STROUDSBURG			SUNBURY			UNIONTOWN		
		195 - 196			184 - 185			168			183			178			154		
		MAT.	INST.	TOTAL	MAT.	INST.	TOTAL	MAT.	INST.	TOTAL	MAT.	INST.	TOTAL	MAT.	INST.	TOTAL	MAT.	INST.	TOTAL
015433	CONTRACTOR EQUIPMENT		118.0	118.0		112.9	112.9		112.1	112.1		112.9	112.9		112.9	112.9		109.6	109.6
0241, 31 - 34	SITE & INFRASTRUCTURE, DEMOLITION	100.3	102.7	101.9	94.1	104.8	101.7	84.4	103.1	97.7	88.5	102.2	98.2	96.6	104.5	102.2	96.5	102.9	101.0
0310	Concrete Forming & Accessories	98.8	90.2	91.4	99.3	88.6	90.0	84.4	80.3	80.8	87.2	89.4	89.1	96.1	88.5	89.6	77.6	96.0	93.5
0320	Concrete Reinforcing	91.5	104.7	98.2	94.7	106.6	100.8	93.0	103.6	98.3	95.0	111.6	103.4	93.2	106.2	99.8	92.6	109.1	101.0
0330	Cast-in-Place Concrete	76.2	97.3	84.9	90.3	93.1	91.5	87.1	65.6	78.2	85.1	72.6	80.0	90.2	95.5	92.4	97.1	96.4	96.8
03	CONCRETE	88.2	96.7	92.4	95.4	94.9	95.1	95.5	81.0	88.4	90.1	88.9	89.5	96.7	95.5	96.1	94.4	99.7	97.0
04	MASONRY	99.5	93.0	95.4	97.7	95.7	96.4	103.1	77.9	87.4	95.1	100.4	98.4	102.3	89.7	94.5	127.0	98.9	109.5
05	METALS	99.5	121.2	106.2	102.1	120.5	107.8	97.7	117.6	103.8	99.7	115.9	104.7	97.4	120.0	104.4	97.4	120.6	104.5
06	WOOD, PLASTICS & COMPOSITES	98.5	88.1	92.6	100.7	86.2	92.6	87.6	83.2	85.1	86.4	88.4	87.5	88.9	88.1	88.5	76.7	95.4	87.2
07	THERMAL & MOISTURE PROTECTION	101.8	112.1	106.0	102.2	95.6	99.5	101.5	90.6	97.0	102.0	85.3	95.2	100.8	104.9	102.0	102.6	100.7	101.8
08	OPENINGS	91.3	98.9	93.1	95.0	91.9	94.3	92.0	89.3	91.4	95.6	88.2	93.9	94.2	92.8	93.9	97.9	104.7	99.5
0920	Plaster & Gypsum Board	96.8	87.5	90.5	98.0	85.6	89.6	89.6	82.5	84.8	88.0	87.8	87.9	92.7	87.5	89.2	89.2	95.2	93.2
0950, 0980	Ceilings & Acoustic Treatment	76.9	87.5	83.9	91.1	85.6	87.5	82.9	82.5	82.6	82.0	87.8	85.8	80.7	87.5	85.2	94.2	95.2	94.8
0960	Flooring	96.8	91.7	95.4	94.6	107.0	98.2	92.3	97.5	93.8	89.9	51.3	78.9	95.9	87.7	93.5	90.4	105.3	94.7
0970, 0990	Wall Finishes & Painting/Coating	97.3	106.4	102.8	98.5	106.4	103.2	98.5	109.2	105.0	98.5	64.6	78.0	105.9	94.3	98.9	100.2	109.2	105.6
09	FINISHES	92.0	90.8	91.3	93.9	93.1	93.4	89.4	85.8	87.4	88.9	80.0	84.0	91.3	88.7	89.9	93.2	97.5	95.6
COVERS	DIVS. 10 - 14, 25, 28, 41, 43, 44, 46	100.0	100.3	100.1	100.0	101.0	100.2	100.0	93.1	98.6	100.0	62.4	92.4	100.0	97.9	99.6	100.0	102.2	100.4
21, 22, 23	FIRE SUPPRESSION, PLUMBING & HVAC	100.1	108.8	103.7	100.2	99.2	99.8	95.5	85.9	91.6	95.5	100.0	97.3	95.5	92.0	94.1	95.2	91.1	93.5
26, 27, 3370	ELECTRICAL, COMMUNICATIONS & UTIL.	99.6	95.5	97.5	98.2	95.5	96.8	91.9	112.2	102.6	93.6	143.3	119.9	89.5	90.8	90.2	92.0	112.2	102.7
MF2014	WEIGHTED AVERAGE	97.1	102.1	99.3	98.6	99.2	98.9	95.3	92.8	94.2	95.3	101.4	97.9	95.8	95.8	95.8	97.3	101.7	99.2

City Cost Indexes

		PENNSYLVANIA																	
	DIVISION	WASHINGTON			WELLSBORO			WESTCHESTER			WILKES-BARRE			WILLIAMSPORT			YORK		
		153			169			193			186 - 187			177			173 - 174		
		MAT.	INST.	TOTAL	MAT.	INST.	TOTAL	MAT.	INST.	TOTAL	MAT.	INST.	TOTAL	MAT.	INST.	TOTAL	MAT.	INST.	TOTAL
015433	CONTRACTOR EQUIPMENT		109.6	109.6		112.9	112.9		98.8	98.8		112.9	112.9		112.9	112.9		112.1	112.1
0241, 31 - 34	SITE & INFRASTRUCTURE, DEMOLITION	96.6	102.9	101.1	95.9	102.0	100.3	101.6	97.8	98.9	86.5	104.7	99.5	87.9	103.4	98.9	85.2	103.5	98.2
0310	Concrete Forming & Accessories	84.8	96.2	94.6	85.8	86.7	86.6	89.8	128.2	122.9	89.9	88.9	89.1	92.5	61.0	65.3	86.9	88.1	87.9
0320	Concrete Reinforcing	92.6	109.3	101.1	93.0	106.2	99.7	89.3	113.6	101.7	93.7	107.1	100.6	92.5	64.6	78.3	93.2	106.2	99.8
0330	Cast-in-Place Concrete	97.1	96.4	96.8	91.2	88.9	90.3	94.1	126.4	107.4	81.7	93.0	86.4	76.4	74.2	75.5	83.0	98.1	89.2
03	CONCRETE	94.9	99.8	97.3	97.8	92.2	95.1	96.9	124.5	110.5	87.2	95.1	91.1	84.7	68.2	76.6	94.5	96.3	95.4
04	MASONRY	109.4	100.5	103.9	103.6	89.7	94.9	104.7	124.9	117.3	110.6	95.3	101.0	93.9	93.7	93.8	103.6	90.1	95.2
05	METALS	97.3	120.9	104.6	97.8	114.2	102.9	99.2	117.2	104.8	97.9	121.0	105.0	97.5	100.4	98.4	99.1	121.4	105.9
06	WOOD, PLASTICS & COMPOSITES	84.9	95.4	90.8	84.9	88.1	86.7	87.1	130.6	111.5	88.9	86.6	87.6	85.2	51.0	66.1	84.1	86.7	85.5
07	THERMAL & MOISTURE PROTECTION	102.7	101.1	102.1	102.5	89.6	97.2	101.7	129.9	113.3	101.8	106.4	103.7	99.4	102.4	100.6	99.0	110.4	103.7
08	OPENINGS	97.9	104.7	99.5	95.1	93.0	94.6	86.9	126.1	96.0	92.2	98.8	93.8	94.2	53.5	84.7	94.1	94.0	94.1
0920	Plaster & Gypsum Board	92.4	95.2	94.3	87.9	87.5	87.6	87.0	131.4	117.0	88.8	86.0	86.9	93.5	49.3	63.6	94.3	86.0	88.7
0950, 0980	Ceilings & Acoustic Treatment	94.2	95.2	94.8	82.9	87.5	85.9	84.5	131.4	115.3	85.4	86.0	85.8	84.0	49.3	61.2	83.1	86.0	85.0
0960	Flooring	93.8	105.3	97.1	89.1	50.8	78.2	95.7	139.2	108.1	90.7	94.6	91.8	94.8	49.7	81.9	96.4	91.7	95.1
0970, 0990	Wall Finishes & Painting/Coating	100.2	109.2	105.6	98.5	106.4	103.2	98.6	147.8	128.3	98.5	106.4	103.2	105.9	106.4	106.2	105.9	89.6	96.1
09	FINISHES	94.6	97.8	96.4	89.5	82.2	85.5	92.1	131.0	113.6	89.9	90.8	90.4	91.1	62.3	75.2	90.8	88.1	89.3
COVERS	DIVS. 10 - 14, 25, 28, 41, 43, 44, 46	100.0	102.2	100.4	100.0	97.9	99.6	100.0	117.3	103.5	100.0	100.8	100.2	100.0	94.6	98.9	100.0	97.6	99.5
21, 22, 23	FIRE SUPPRESSION, PLUMBING & HVAC	95.2	97.2	96.0	95.5	91.1	93.7	95.2	125.8	107.6	95.5	99.0	96.9	95.5	93.1	94.5	100.2	92.4	97.1
26, 27, 3370	ELECTRICAL, COMMUNICATIONS & UTIL.	94.3	112.2	103.8	92.7	86.5	89.5	92.9	114.9	104.5	93.6	91.7	92.6	90.0	74.6	81.8	91.0	88.3	89.5
MF2014	WEIGHTED AVERAGE	97.0	103.2	99.7	96.3	92.6	94.7	95.9	121.5	107.1	95.1	98.7	96.6	93.8	83.0	89.1	96.7	95.9	96.3

		PUERTO RICO			RHODE ISLAND						SOUTH CAROLINA					
	DIVISION	SAN JUAN			NEWPORT			PROVIDENCE			AIKEN			BEAUFORT			CHARLESTON		
		009			028			029			298			299			294		
		MAT.	INST.	TOTAL	MAT.	INST.	TOTAL	MAT.	INST.	TOTAL	MAT.	INST.	TOTAL	MAT.	INST.	TOTAL	MAT.	INST.	TOTAL
015433	CONTRACTOR EQUIPMENT		90.5	90.5		102.4	102.4		102.4	102.4		101.3	101.3		101.3	101.3		101.3	101.3
0241, 31 - 34	SITE & INFRASTRUCTURE, DEMOLITION	133.8	91.2	103.5	89.2	105.0	100.4	91.4	105.0	101.1	118.9	87.4	96.5	114.3	85.6	93.9	99.7	86.3	90.2
0310	Concrete Forming & Accessories	92.4	17.8	28.1	101.5	122.4	119.5	99.8	122.4	119.3	97.5	67.4	71.5	96.4	37.9	46.0	95.4	63.0	67.5
0320	Concrete Reinforcing	188.3	12.7	98.9	106.4	149.2	128.2	102.3	149.2	126.2	93.7	65.2	79.2	92.8	25.8	58.7	92.7	59.3	75.7
0330	Cast-in-Place Concrete	103.8	30.4	73.7	79.5	124.1	97.8	95.5	124.1	107.2	79.2	69.8	75.4	79.2	47.2	66.1	92.9	49.5	75.1
03	CONCRETE	108.7	22.2	66.2	93.2	127.5	110.0	100.4	127.5	113.7	102.7	68.9	86.1	99.9	40.9	70.9	94.4	58.9	77.0
04	MASONRY	90.1	16.9	44.5	95.4	132.5	118.5	101.3	132.5	120.7	79.6	61.4	68.3	93.9	33.3	56.1	95.1	40.9	61.4
05	METALS	118.4	34.6	92.6	97.4	125.6	106.1	103.1	125.6	110.0	101.8	83.8	96.3	101.8	68.2	91.5	103.8	80.4	96.6
06	WOOD, PLASTICS & COMPOSITES	94.4	17.5	51.3	99.0	120.9	111.3	100.1	120.9	111.7	97.0	68.6	81.1	95.3	37.8	63.1	94.0	68.4	79.7
07	THERMAL & MOISTURE PROTECTION	129.1	21.6	85.0	100.6	121.5	109.2	100.5	121.5	109.1	102.7	66.4	87.8	102.4	38.8	76.3	101.6	47.3	79.3
08	OPENINGS	152.0	15.4	120.2	102.2	128.2	108.3	107.3	128.2	112.2	99.0	64.8	91.1	99.0	36.7	84.5	102.9	63.3	93.7
0920	Plaster & Gypsum Board	159.8	14.9	61.9	92.9	120.9	111.9	94.5	120.9	112.4	104.0	67.4	79.3	107.4	35.7	59.0	109.0	67.3	80.8
0950, 0980	Ceilings & Acoustic Treatment	225.7	14.9	87.2	93.2	120.9	111.4	90.1	120.9	110.4	86.4	67.4	73.9	89.8	35.7	54.3	89.8	67.3	75.0
0960	Flooring	224.3	16.6	164.9	97.7	141.5	110.2	98.0	141.5	110.4	105.4	68.4	94.8	106.8	51.2	90.9	106.5	58.2	92.7
0970, 0990	Wall Finishes & Painting/Coating	213.5	16.8	94.8	99.9	129.2	117.6	97.7	129.2	116.7	109.5	70.3	85.8	109.5	34.7	64.3	109.5	66.9	83.8
09	FINISHES	210.6	17.6	104.2	96.6	126.9	113.3	93.7	126.9	112.0	98.3	68.0	81.6	99.5	39.8	66.6	97.8	63.5	78.9
COVERS	DIVS. 10 - 14, 25, 28, 41, 43, 44, 46	100.0	17.5	83.3	100.0	108.6	101.7	100.0	108.6	101.7	100.0	71.8	94.3	100.0	70.4	94.0	100.0	69.0	93.7
21, 22, 23	FIRE SUPPRESSION, PLUMBING & HVAC	103.3	14.0	67.3	100.1	110.6	104.3	99.9	110.6	104.2	95.7	63.1	82.5	95.7	36.5	71.8	100.5	53.7	81.6
26, 27, 3370	ELECTRICAL, COMMUNICATIONS & UTIL.	125.9	13.2	66.3	100.0	99.0	99.5	99.6	99.0	99.2	96.6	65.9	80.5	100.7	33.6	65.2	99.0	88.8	93.6
MF2014	WEIGHTED AVERAGE	122.7	24.4	79.8	98.4	117.5	106.7	100.6	117.5	108.0	98.6	69.1	85.7	99.3	44.9	75.6	99.9	65.2	84.8

		SOUTH CAROLINA															SOUTH DAKOTA		
	DIVISION	COLUMBIA			FLORENCE			GREENVILLE			ROCK HILL			SPARTANBURG			ABERDEEN		
		290 - 292			295			296			297			293			574		
		MAT.	INST.	TOTAL	MAT.	INST.	TOTAL	MAT.	INST.	TOTAL	MAT.	INST.	TOTAL	MAT.	INST.	TOTAL	MAT.	INST.	TOTAL
015433	CONTRACTOR EQUIPMENT		101.3	101.3		101.3	101.3		101.3	101.3		101.3	101.3		101.3	101.3		98.2	98.2
0241, 31 - 34	SITE & INFRASTRUCTURE, DEMOLITION	99.8	86.3	90.2	108.8	86.3	92.8	104.1	85.9	91.2	101.7	85.0	89.8	103.9	86.0	91.1	99.2	93.7	95.3
0310	Concrete Forming & Accessories	94.4	45.1	51.9	83.2	45.3	50.5	95.0	45.1	52.0	93.2	38.7	46.2	98.0	45.3	52.5	94.8	36.6	44.6
0320	Concrete Reinforcing	95.8	59.1	77.1	92.3	59.3	75.5	92.2	44.5	67.9	93.0	44.1	68.1	92.2	57.0	74.3	96.1	38.1	66.6
0330	Cast-in-Place Concrete	95.8	50.1	77.0	79.2	49.4	66.9	79.2	49.3	66.9	79.2	43.9	64.7	79.2	49.4	66.9	105.6	43.5	80.1
03	CONCRETE	96.2	51.1	74.1	94.2	51.0	73.0	93.0	48.2	71.0	90.9	43.3	67.5	93.2	50.6	72.3	102.8	40.7	72.3
04	MASONRY	91.5	37.6	57.9	79.8	40.9	55.6	77.4	40.9	54.7	101.6	34.6	59.8	79.8	40.9	55.6	115.9	55.4	78.2
05	METALS	100.9	79.6	94.3	102.6	80.0	95.6	102.6	74.5	93.9	101.8	71.2	92.4	102.6	79.1	95.4	95.9	63.6	86.0
06	WOOD, PLASTICS & COMPOSITES	98.0	44.8	68.2	79.7	44.8	60.2	93.7	44.8	66.4	92.0	38.9	62.3	98.0	44.8	68.2	99.6	36.0	64.0
07	THERMAL & MOISTURE PROTECTION	97.2	43.3	75.1	101.9	44.8	78.5	101.8	44.8	78.4	101.6	39.7	76.3	101.8	44.8	78.5	99.2	47.7	78.1
08	OPENINGS	104.6	50.5	92.0	99.1	50.5	87.8	99.0	47.0	86.9	99.0	39.4	85.2	99.0	50.0	87.6	99.1	36.7	84.6
0920	Plaster & Gypsum Board	102.3	42.9	62.2	97.1	42.9	60.5	102.6	42.9	62.3	101.9	36.8	57.9	105.5	42.9	63.2	103.7	34.3	56.8
0950, 0980	Ceilings & Acoustic Treatment	91.3	42.9	59.5	87.3	42.9	58.1	86.4	42.9	57.8	86.4	36.8	53.8	86.4	42.9	57.8	93.2	34.3	54.5
0960	Flooring	100.5	43.6	84.2	98.4	43.6	82.7	104.3	57.2	90.8	103.3	44.4	86.5	105.6	57.2	91.8	108.4	50.3	91.8
0970, 0990	Wall Finishes & Painting/Coating	106.2	66.9	82.5	109.5	66.9	83.8	109.5	66.9	83.8	109.5	39.2	67.0	109.5	66.9	83.8	102.6	37.1	63.1
09	FINISHES	95.9	46.6	68.7	94.4	47.1	68.3	96.3	49.3	70.4	95.7	39.4	64.6	97.1	49.3	70.8	100.3	38.8	66.4
COVERS	DIVS. 10 - 14, 25, 28, 41, 43, 44, 46	100.0	66.3	93.2	100.0	66.3	93.2	100.0	66.3	93.2	100.0	64.6	92.9	100.0	66.3	93.2	100.0	40.5	88.0
21, 22, 23	FIRE SUPPRESSION, PLUMBING & HVAC	100.0	53.0	81.0	100.5	53.0	81.4	100.5	53.0	81.3	95.7	44.2	74.9	100.5	53.1	81.3	100.1	39.8	75.8
26, 27, 3370	ELECTRICAL, COMMUNICATIONS & UTIL.	99.0	59.7	78.3	96.8	59.7	77.2	99.1	57.4	77.0	99.1	56.6	76.6	99.1	57.4	77.1	101.7	51.1	75.0
MF2014	WEIGHTED AVERAGE	99.3	56.5	80.6	98.2	56.9	80.2	98.3	55.8	79.8	97.8	50.3	77.1	98.6	56.7	80.3	100.5	49.6	78.3

City Cost Indexes

		SOUTH DAKOTA																	
	DIVISION	MITCHELL			MOBRIDGE			PIERRE			RAPID CITY			SIOUX FALLS			WATERTOWN		
		573			576			575			577			570 - 571			572		
		MAT.	INST.	TOTAL	MAT.	INST.	TOTAL	MAT.	INST.	TOTAL	MAT.	INST.	TOTAL	MAT.	INST.	TOTAL	MAT.	INST.	TOTAL
015433	CONTRACTOR EQUIPMENT		98.2	98.2		98.2	98.2		98.2	98.2		98.2	98.2		99.2	99.2		98.2	98.2
0241, 31 - 34	SITE & INFRASTRUCTURE, DEMOLITION	96.0	93.7	94.3	95.9	93.7	94.3	100.5	93.7	95.7	97.7	93.8	94.9	94.3	95.4	95.0	95.8	93.7	94.3
0310	Concrete Forming & Accessories	94.0	37.0	44.8	84.9	36.8	43.4	97.7	38.4	46.6	102.2	36.7	45.7	98.8	40.5	48.5	81.6	36.6	42.8
0320	Concrete Reinforcing	95.5	46.3	70.5	98.0	38.1	67.5	99.0	71.1	84.8	89.9	71.3	80.4	98.0	71.2	84.3	92.9	38.2	65.1
0330	Cast-in-Place Concrete	102.6	42.3	77.8	102.6	43.5	78.3	98.5	42.1	75.4	101.8	42.8	77.6	92.3	42.9	72.0	102.6	45.9	79.3
03	CONCRETE	100.5	41.8	71.7	100.3	40.7	71.0	99.0	47.3	73.6	99.8	46.7	73.7	95.9	48.4	72.6	99.2	41.5	70.9
04	MASONRY	103.9	53.3	72.4	113.4	55.4	77.2	117.4	53.5	77.6	113.5	56.4	77.9	105.1	53.6	73.0	141.1	57.4	88.9
05	METALS	95.0	63.8	85.4	95.0	63.6	85.3	98.4	79.1	92.4	97.8	79.7	92.2	98.2	79.3	92.4	95.0	64.1	85.5
06	WOOD, PLASTICS & COMPOSITES	98.5	36.5	63.8	87.6	36.2	58.8	104.5	36.8	66.6	103.5	33.3	64.2	96.8	39.3	64.6	83.8	36.0	57.0
07	THERMAL & MOISTURE PROTECTION	98.9	45.6	77.1	99.0	47.8	78.0	101.8	45.9	78.9	99.6	47.8	78.3	101.9	48.2	79.9	98.8	48.3	78.1
08	OPENINGS	97.7	37.3	83.6	100.4	36.3	85.5	105.7	46.7	92.0	103.1	44.8	89.6	106.5	48.1	92.9	97.7	36.5	83.4
0920	Plaster & Gypsum Board	102.1	34.8	56.7	96.2	34.6	54.5	100.1	35.1	56.2	103.2	31.6	54.8	90.7	37.7	54.9	93.9	34.3	53.6
0950, 0980	Ceilings & Acoustic Treatment	89.9	34.8	53.7	93.2	34.6	54.7	93.3	35.1	55.0	95.7	31.6	53.6	90.6	37.7	55.8	89.9	34.3	53.3
0960	Flooring	107.9	50.3	91.4	103.5	50.3	88.3	107.4	35.8	86.9	107.6	78.6	99.3	103.8	74.9	95.5	102.1	50.3	87.3
0970, 0990	Wall Finishes & Painting/Coating	102.6	40.6	65.2	102.6	41.7	65.8	105.7	44.6	68.8	102.6	44.6	67.6	101.7	44.6	67.2	102.6	37.1	63.1
09	FINISHES	99.1	39.5	66.2	97.7	39.4	65.6	100.8	37.3	65.8	100.5	43.9	69.3	97.1	46.5	69.2	96.2	38.8	64.6
COVERS	DIVS. 10 - 14, 25, 28, 41, 43, 44, 46	100.0	37.6	87.4	100.0	40.5	88.0	100.0	75.8	95.1	100.0	75.7	95.1	100.0	76.2	95.2	100.0	40.5	88.0
21, 22, 23	FIRE SUPPRESSION, PLUMBING & HVAC	95.3	39.4	72.8	95.3	39.9	73.0	100.0	64.5	85.7	100.1	65.0	85.9	100.0	38.6	75.2	95.3	39.9	73.0
26, 27, 3370	ELECTRICAL, COMMUNICATIONS & UTIL.	99.9	44.0	70.3	101.7	44.8	71.7	104.8	51.8	76.8	98.0	51.8	73.6	101.3	75.1	87.5	99.0	44.8	70.4
MF2014	WEIGHTED AVERAGE	97.8	48.4	76.3	98.5	48.8	76.8	101.7	58.4	82.8	100.4	59.5	82.6	100.0	57.7	81.5	99.0	49.1	77.2

		TENNESSEE																	
	DIVISION	CHATTANOOGA			COLUMBIA			COOKEVILLE			JACKSON			JOHNSON CITY			KNOXVILLE		
		373 - 374			384			385			383			376			377 - 379		
		MAT.	INST.	TOTAL	MAT.	INST.	TOTAL	MAT.	INST.	TOTAL	MAT.	INST.	TOTAL	MAT.	INST.	TOTAL	MAT.	INST.	TOTAL
015433	CONTRACTOR EQUIPMENT		103.8	103.8		98.2	98.2		98.2	98.2		104.5	104.5		97.7	97.7		97.7	97.7
0241, 31 - 34	SITE & INFRASTRUCTURE, DEMOLITION	104.5	97.9	99.8	89.1	86.9	87.5	94.5	86.4	88.8	97.7	96.6	97.1	111.1	86.6	93.7	90.7	87.3	88.3
0310	Concrete Forming & Accessories	97.1	56.4	62.0	82.1	62.0	64.8	82.2	35.4	41.8	89.0	44.9	51.0	83.7	38.9	45.1	95.6	61.0	65.7
0320	Concrete Reinforcing	91.2	67.4	79.1	87.4	61.2	74.1	87.4	61.2	74.1	87.4	62.0	74.4	91.7	60.3	75.7	91.2	62.3	76.5
0330	Cast-in-Place Concrete	101.3	62.7	85.5	94.2	51.6	76.7	106.7	43.5	80.7	104.2	49.3	81.6	81.6	60.2	72.8	95.2	65.7	83.1
03	CONCRETE	96.0	62.3	79.4	96.7	59.8	78.6	106.9	45.2	76.6	97.7	51.6	75.1	103.3	52.3	78.3	93.4	64.4	79.1
04	MASONRY	109.7	51.7	73.5	127.7	54.6	82.1	121.9	43.8	73.2	127.9	47.2	77.6	125.7	44.8	75.3	86.6	55.3	67.1
05	METALS	98.7	89.2	95.8	96.4	86.0	93.2	96.5	85.9	93.2	98.8	85.9	94.8	95.9	85.2	92.6	99.3	87.0	95.5
06	WOOD, PLASTICS & COMPOSITES	104.0	57.3	77.8	70.1	64.4	66.9	70.3	33.7	49.8	85.4	45.4	63.0	76.9	36.9	54.5	91.0	60.6	73.9
07	THERMAL & MOISTURE PROTECTION	99.4	59.9	83.2	93.6	59.2	79.5	94.0	45.7	74.2	96.0	54.1	78.8	94.5	52.6	77.3	92.5	61.6	79.8
08	OPENINGS	99.7	57.6	89.9	92.6	57.9	84.6	92.7	42.2	80.9	100.3	51.6	89.0	96.2	45.1	84.3	93.3	58.6	85.2
0920	Plaster & Gypsum Board	82.8	56.5	65.0	87.7	63.7	71.5	87.7	32.1	50.1	89.3	44.1	58.8	102.4	35.5	57.2	109.0	59.8	75.8
0950, 0980	Ceilings & Acoustic Treatment	96.6	56.5	70.2	77.0	63.7	68.3	77.0	32.1	47.5	86.2	44.1	58.5	93.0	35.5	55.2	93.8	59.8	71.5
0960	Flooring	99.1	58.9	87.7	89.3	20.5	69.7	89.4	58.6	80.6	87.1	40.8	73.9	96.7	40.6	80.7	101.3	55.8	88.3
0970, 0990	Wall Finishes & Painting/Coating	102.5	70.7	83.3	94.0	33.4	57.4	94.0	36.5	59.3	95.8	49.6	67.9	96.9	42.1	64.0	86.1	81.2	83.1
09	FINISHES	97.2	58.2	75.7	89.8	52.0	69.0	90.3	38.4	61.7	89.7	43.7	64.4	101.9	38.1	66.7	94.5	62.0	76.6
COVERS	DIVS. 10 - 14, 25, 28, 41, 43, 44, 46	100.0	41.8	88.3	100.0	47.9	89.5	100.0	41.0	88.1	100.0	43.0	88.5	100.0	75.5	95.1	100.0	81.7	96.3
21, 22, 23	FIRE SUPPRESSION, PLUMBING & HVAC	100.2	62.6	85.0	97.4	78.1	89.6	97.4	72.6	87.4	100.1	68.0	87.1	99.9	59.1	83.4	99.9	67.9	87.0
26, 27, 3370	ELECTRICAL, COMMUNICATIONS & UTIL.	102.3	69.2	84.8	94.2	57.4	74.7	95.9	61.6	77.8	101.0	58.5	78.5	92.5	46.3	68.1	98.1	56.0	75.8
MF2014	WEIGHTED AVERAGE	100.0	66.2	85.2	96.9	66.1	83.5	98.1	59.1	81.1	99.9	61.2	83.0	99.8	56.5	80.9	96.6	66.9	83.7

		TENNESSEE						TEXAS					
	DIVISION	MCKENZIE			MEMPHIS			NASHVILLE			ABILENE			AMARILLO			AUSTIN		
		382			375,380 - 381			370 - 372			795 - 796			790 - 791			786 - 787		
		MAT.	INST.	TOTAL	MAT.	INST.	TOTAL	MAT.	INST.	TOTAL	MAT.	INST.	TOTAL	MAT.	INST.	TOTAL	MAT.	INST.	TOTAL
015433	CONTRACTOR EQUIPMENT		98.2	98.2		102.8	102.8		103.7	103.7		89.9	89.9		89.9	89.9		89.4	89.4
0241, 31 - 34	SITE & INFRASTRUCTURE, DEMOLITION	94.3	86.5	88.8	94.1	93.7	93.8	98.9	97.5	97.9	99.4	88.7	91.8	97.3	88.6	91.1	102.4	88.4	92.4
0310	Concrete Forming & Accessories	90.0	38.1	45.2	94.3	64.9	69.0	98.9	65.8	70.3	99.0	65.5	70.1	97.9	57.1	62.7	98.6	59.1	64.5
0320	Concrete Reinforcing	87.5	61.9	74.5	99.7	68.7	83.9	96.0	67.7	81.6	92.9	52.5	72.3	99.0	51.3	74.7	96.5	48.5	72.1
0330	Cast-in-Place Concrete	104.4	57.3	85.1	94.1	61.3	80.6	90.3	65.9	80.3	96.6	63.7	83.1	94.1	63.7	81.6	93.4	66.3	82.3
03	CONCRETE	105.5	51.3	78.9	94.1	65.8	80.2	92.4	67.6	80.2	95.0	63.0	79.3	96.7	59.0	78.2	95.7	60.2	78.3
04	MASONRY	126.2	46.2	76.3	99.8	62.7	76.7	93.8	59.0	72.1	104.2	63.6	78.9	109.4	63.6	81.0	107.6	56.5	75.7
05	METALS	96.5	86.1	93.3	100.1	90.6	97.2	100.4	89.9	97.2	105.3	69.8	94.4	101.6	69.2	91.7	100.5	66.3	90.0
06	WOOD, PLASTICS & COMPOSITES	79.1	36.3	55.1	95.4	67.0	79.5	101.6	66.6	82.0	101.1	68.5	82.8	100.8	57.1	76.4	94.4	59.9	75.0
07	THERMAL & MOISTURE PROTECTION	94.0	48.7	75.4	95.5	63.4	82.4	95.7	61.5	81.7	100.0	67.7	86.7	97.9	67.3	85.3	104.2	64.8	88.1
08	OPENINGS	92.7	44.1	81.4	100.1	66.0	92.2	98.5	65.9	90.9	97.1	63.3	89.2	101.1	56.8	90.8	104.3	55.6	92.9
0920	Plaster & Gypsum Board	90.5	34.8	52.8	95.6	66.3	75.8	95.8	66.0	75.6	84.0	68.0	73.2	92.1	56.3	67.9	86.7	59.2	68.1
0950, 0980	Ceilings & Acoustic Treatment	77.0	34.8	49.2	93.9	66.3	75.8	95.5	66.0	76.1	88.9	68.0	75.2	98.0	56.3	70.6	87.9	59.2	69.0
0960	Flooring	92.1	39.3	77.0	99.4	36.1	81.3	97.7	64.2	88.2	110.9	76.5	101.1	106.5	76.5	97.9	105.7	64.1	93.8
0970, 0990	Wall Finishes & Painting/Coating	94.0	49.6	67.2	97.0	55.7	72.1	100.5	72.0	83.3	106.6	56.1	76.1	100.8	56.1	73.8	104.3	48.8	70.8
09	FINISHES	91.3	38.2	62.1	97.3	58.5	75.9	100.2	66.0	81.4	93.7	67.1	79.0	96.8	60.3	76.7	95.6	58.7	75.2
COVERS	DIVS. 10 - 14, 25, 28, 41, 43, 44, 46	100.0	27.4	85.3	100.0	82.4	96.4	100.0	82.9	96.5	100.0	82.2	96.4	100.0	68.6	93.7	100.0	80.5	96.1
21, 22, 23	FIRE SUPPRESSION, PLUMBING & HVAC	97.4	68.3	85.6	100.0	73.9	89.5	100.0	84.0	93.5	100.3	49.0	79.6	100.0	54.9	81.8	100.1	60.1	84.0
26, 27, 3370	ELECTRICAL, COMMUNICATIONS & UTIL.	95.6	62.9	78.3	101.7	65.3	82.4	97.6	63.1	79.4	99.3	48.7	72.5	101.1	64.1	81.6	96.8	63.9	79.4
MF2014	WEIGHTED AVERAGE	98.2	59.2	81.2	99.0	71.3	86.9	98.4	74.1	87.8	99.7	62.2	83.3	100.2	63.2	84.1	100.0	63.6	84.1

City Cost Indexes

TEXAS

DIVISION		BEAUMONT 776 - 777			BROWNWOOD 768			BRYAN 778			CHILDRESS 792			CORPUS CHRISTI 783 - 784			DALLAS 752 - 753		
		MAT.	INST.	TOTAL	MAT.	INST.	TOTAL	MAT.	INST.	TOTAL	MAT.	INST.	TOTAL	MAT.	INST.	TOTAL	MAT.	INST.	TOTAL
015433	CONTRACTOR EQUIPMENT		91.5	91.5		89.9	89.9		91.5	91.5		89.9	89.9		96.6	96.6		99.1	99.1
0241, 31 - 34	SITE & INFRASTRUCTURE, DEMOLITION	90.0	89.9	89.9	105.2	89.6	94.1	81.8	90.5	88.0	109.7	88.0	94.3	140.1	84.4	100.5	104.3	89.8	94.0
0310	Concrete Forming & Accessories	101.8	61.9	67.4	99.0	60.5	65.8	81.0	67.1	69.0	97.3	65.5	69.8	100.2	58.1	63.9	101.2	65.5	70.7
0320	Concrete Reinforcing	95.9	65.2	80.3	92.9	52.1	72.1	98.3	48.6	73.0	93.0	52.1	72.2	88.0	48.5	67.9	100.2	52.5	75.9
0330	Cast-in-Place Concrete	89.5	64.7	79.3	101.4	63.6	85.9	71.7	65.4	69.1	99.1	63.7	84.5	109.8	64.4	91.2	98.0	64.6	84.3
03	CONCRETE	96.3	64.2	80.5	101.2	60.7	81.3	80.9	63.7	72.5	104.0	62.9	83.8	100.1	59.9	80.4	99.0	64.1	81.9
04	MASONRY	106.8	63.9	80.0	141.8	56.6	88.7	146.3	60.4	92.7	108.5	56.6	76.2	91.0	56.6	69.6	102.4	56.7	73.9
05	METALS	95.2	75.5	89.1	99.9	69.1	90.5	94.8	70.1	87.2	102.7	69.2	92.4	94.7	77.9	89.5	100.4	80.0	94.1
06	WOOD, PLASTICS & COMPOSITES	106.1	61.8	81.3	99.5	61.9	78.4	73.5	68.5	70.7	100.4	68.5	82.6	116.0	58.4	83.7	104.1	68.6	84.2
07	THERMAL & MOISTURE PROTECTION	101.9	66.7	87.5	98.2	65.2	84.7	93.9	66.3	82.5	100.5	64.8	85.9	104.5	66.2	88.8	93.0	66.4	82.1
08	OPENINGS	94.0	62.3	86.7	94.7	59.7	86.6	97.2	61.6	88.9	94.2	63.3	87.1	109.1	54.7	96.5	99.5	63.5	91.1
0920	Plaster & Gypsum Board	102.3	61.2	74.5	87.8	61.2	69.8	90.3	68.0	75.2	83.6	68.0	73.1	98.4	57.5	70.7	96.7	68.0	77.3
0950, 0980	Ceilings & Acoustic Treatment	98.2	61.2	73.9	80.6	61.2	67.8	91.2	68.0	76.0	87.3	68.0	74.6	92.7	57.5	69.6	97.8	68.0	78.2
0960	Flooring	112.4	73.0	101.1	95.3	64.1	86.3	85.8	64.1	79.6	109.1	64.1	96.2	118.7	64.1	103.1	102.8	64.1	91.7
0970, 0990	Wall Finishes & Painting/Coating	95.1	59.4	73.6	104.2	56.1	75.1	92.5	62.2	74.2	106.6	56.1	76.1	120.1	56.5	81.7	107.0	56.1	76.3
09	FINISHES	93.7	63.4	77.0	87.7	60.7	72.8	82.3	66.2	73.4	94.0	64.6	77.8	105.9	58.7	79.9	100.0	64.7	80.5
COVERS	DIVS. 10 - 14, 25, 28, 41, 43, 44, 46	100.0	83.4	96.7	100.0	81.4	96.3	100.0	82.7	96.5	100.0	82.2	96.4	100.0	82.6	96.5	100.0	82.5	96.5
21, 22, 23	FIRE SUPPRESSION, PLUMBING & HVAC	100.2	64.1	85.6	95.4	50.0	77.1	95.4	66.5	83.8	95.5	54.9	79.1	100.2	58.9	83.5	100.0	62.4	84.9
26, 27, 3370	ELECTRICAL, COMMUNICATIONS & UTIL.	95.2	70.6	82.2	93.4	42.8	66.7	93.4	67.4	79.7	99.2	64.1	80.7	92.8	60.0	75.5	95.2	64.6	79.0
MF2014	WEIGHTED AVERAGE	97.4	68.6	84.9	98.9	59.4	81.7	94.9	68.2	83.3	99.3	64.3	84.0	100.9	63.5	84.6	99.5	67.4	85.5

TEXAS

DIVISION		DEL RIO 788			DENTON 762			EASTLAND 764			EL PASO 798 - 799,885			FORT WORTH 760 - 761			GALVESTON 775		
		MAT.	INST.	TOTAL	MAT.	INST.	TOTAL	MAT.	INST.	TOTAL	MAT.	INST.	TOTAL	MAT.	INST.	TOTAL	MAT.	INST.	TOTAL
015433	CONTRACTOR EQUIPMENT		89.4	89.4		95.8	95.8		89.9	89.9		89.9	89.9		89.9	89.9		100.5	100.5
0241, 31 - 34	SITE & INFRASTRUCTURE, DEMOLITION	121.8	88.4	98.0	105.5	81.3	88.3	108.0	87.6	93.5	101.7	87.6	91.7	102.1	88.7	92.6	107.2	88.4	93.8
0310	Concrete Forming & Accessories	96.5	57.8	63.1	106.2	65.3	70.9	99.8	65.3	70.0	97.8	63.4	68.1	98.0	65.6	70.1	89.8	64.6	68.1
0320	Concrete Reinforcing	88.7	48.4	68.2	94.3	52.1	72.8	93.1	50.9	71.6	99.2	52.0	75.2	98.4	52.4	75.0	97.8	62.4	79.8
0330	Cast-in-Place Concrete	118.2	63.2	95.6	78.5	64.6	72.8	107.2	63.6	89.3	88.4	63.7	78.3	95.0	63.7	82.1	95.2	66.4	83.4
03	CONCRETE	121.3	58.6	90.5	80.0	63.9	72.1	105.9	62.6	84.6	94.0	61.9	78.3	97.0	63.1	80.4	97.5	66.4	82.2
04	MASONRY	106.9	56.5	75.5	152.1	56.7	92.6	106.8	56.6	75.5	99.0	57.7	73.3	104.7	56.6	74.7	102.5	60.4	76.3
05	METALS	94.4	66.2	85.7	99.5	80.6	93.7	99.7	68.5	90.1	98.5	67.4	89.0	101.5	69.5	91.7	96.3	88.3	93.9
06	WOOD, PLASTICS & COMPOSITES	96.5	58.3	75.1	111.3	68.6	87.4	106.1	68.5	85.1	91.3	65.7	76.9	97.3	68.5	81.2	89.4	64.7	75.6
07	THERMAL & MOISTURE PROTECTION	101.3	65.3	86.5	96.0	66.7	84.0	98.6	65.9	85.1	99.8	64.4	85.3	96.0	65.9	83.6	93.0	67.1	82.4
08	OPENINGS	101.8	54.7	90.8	112.2	63.4	100.8	70.4	63.0	68.7	93.3	58.1	85.1	99.4	63.4	91.0	101.6	63.2	92.7
0920	Plaster & Gypsum Board	94.7	57.5	69.6	92.1	68.0	75.8	87.8	68.0	74.4	95.9	65.1	75.1	91.8	68.0	75.7	96.8	64.0	74.7
0950, 0980	Ceilings & Acoustic Treatment	89.1	57.5	68.3	84.7	68.0	73.8	80.6	68.0	72.3	91.4	65.1	74.1	90.5	68.0	75.7	94.6	64.0	74.5
0960	Flooring	100.7	64.1	90.2	90.4	64.1	82.9	121.4	64.1	105.0	110.4	64.1	97.1	117.9	64.1	102.5	100.1	64.1	89.8
0970, 0990	Wall Finishes & Painting/Coating	106.4	48.8	71.6	115.6	56.1	79.7	105.8	56.1	75.8	106.2	53.7	74.5	103.2	56.1	74.7	103.4	60.0	77.2
09	FINISHES	98.2	57.7	75.9	85.8	64.7	74.2	95.8	64.6	78.6	97.3	63.0	78.4	98.4	64.6	79.8	91.4	63.7	76.2
COVERS	DIVS. 10 - 14, 25, 28, 41, 43, 44, 46	100.0	80.5	96.1	100.0	82.5	96.5	100.0	82.2	96.4	100.0	81.0	96.2	100.0	82.3	96.4	100.0	84.5	96.9
21, 22, 23	FIRE SUPPRESSION, PLUMBING & HVAC	95.4	57.3	80.0	95.4	42.6	74.1	95.4	49.9	77.1	100.0	50.3	80.0	100.0	58.4	83.2	95.4	66.7	83.8
26, 27, 3370	ELECTRICAL, COMMUNICATIONS & UTIL.	94.9	68.8	81.1	95.9	64.6	79.3	93.3	64.5	78.1	99.1	57.4	77.0	94.7	64.5	78.8	95.0	67.9	80.7
MF2014	WEIGHTED AVERAGE	100.8	63.2	84.4	98.9	62.5	83.0	95.9	63.2	81.6	98.0	61.7	82.2	99.3	65.3	84.5	97.0	70.0	85.2

TEXAS

DIVISION		GIDDINGS 789			GREENVILLE 754			HOUSTON 770 - 772			HUNTSVILLE 773			LAREDO 780			LONGVIEW 756		
		MAT.	INST.	TOTAL	MAT.	INST.	TOTAL	MAT.	INST.	TOTAL	MAT.	INST.	TOTAL	MAT.	INST.	TOTAL	MAT.	INST.	TOTAL
015433	CONTRACTOR EQUIPMENT		89.4	89.4		96.7	96.7		100.4	100.4		91.5	91.5		89.4	89.4		91.4	91.4
0241, 31 - 34	SITE & INFRASTRUCTURE, DEMOLITION	107.8	88.4	94.0	97.9	85.4	89.0	106.0	88.2	93.4	96.6	89.9	91.8	102.1	89.4	92.3	95.8	92.3	93.3
0310	Concrete Forming & Accessories	94.1	57.9	62.9	92.3	65.3	69.0	91.8	64.9	68.4	87.0	61.6	65.2	96.6	58.0	63.3	88.4	64.9	68.1
0320	Concrete Reinforcing	89.2	45.9	67.2	100.7	52.1	76.0	97.5	61.4	79.1	98.5	50.6	74.1	88.7	48.8	68.4	99.6	52.0	75.4
0330	Cast-in-Place Concrete	100.2	63.2	85.0	89.3	63.9	78.9	92.3	66.4	81.7	98.6	64.6	84.6	84.6	63.3	75.8	103.8	62.8	86.9
03	CONCRETE	97.6	58.2	78.2	91.6	63.6	77.8	95.2	66.2	80.9	103.9	61.3	83.0	92.5	58.8	75.9	107.5	62.3	85.3
04	MASONRY	115.7	56.5	78.8	160.0	56.6	95.6	102.3	66.5	80.0	145.1	58.9	91.4	100.2	63.5	77.3	155.7	56.6	93.9
05	METALS	93.9	65.3	85.1	97.7	79.3	92.1	99.9	88.2	95.8	94.7	69.2	86.8	97.0	66.9	87.8	91.1	67.8	83.9
06	WOOD, PLASTICS & COMPOSITES	95.6	58.3	74.7	93.5	68.6	79.5	91.7	64.7	76.6	82.0	61.8	70.7	96.5	58.3	75.1	87.2	64.8	74.7
07	THERMAL & MOISTURE PROTECTION	102.0	64.5	86.6	93.0	66.0	81.9	92.6	67.8	82.4	94.9	65.9	83.0	100.2	66.5	86.4	94.5	65.2	82.5
08	OPENINGS	100.8	54.0	89.9	97.7	63.4	89.7	104.3	62.8	94.6	97.2	58.5	88.2	101.6	54.7	90.7	87.8	63.3	82.1
0920	Plaster & Gypsum Board	93.7	57.5	69.2	90.3	68.0	75.3	99.1	64.0	75.4	94.2	61.2	71.9	95.8	57.5	69.9	89.1	68.0	74.9
0950, 0980	Ceilings & Acoustic Treatment	89.1	57.5	68.3	93.6	68.0	76.8	99.6	64.0	76.2	91.2	61.2	71.5	93.3	57.5	69.8	90.3	68.0	75.7
0960	Flooring	101.1	64.1	90.5	98.6	64.1	88.7	101.4	64.1	90.6	89.4	64.1	82.1	100.5	64.1	90.1	103.5	64.1	92.2
0970, 0990	Wall Finishes & Painting/Coating	106.4	48.8	71.6	107.0	56.1	76.3	103.4	62.2	78.5	92.5	59.4	72.5	106.4	53.2	74.3	96.8	56.1	72.2
09	FINISHES	97.1	57.7	75.4	96.6	64.6	79.0	98.9	64.0	79.7	84.7	61.6	72.0	97.7	58.2	75.9	101.7	64.5	81.2
COVERS	DIVS. 10 - 14, 25, 28, 41, 43, 44, 46	100.0	80.8	96.1	100.0	82.3	96.4	100.0	84.5	96.9	100.0	81.0	96.2	100.0	80.5	96.1	100.0	82.0	96.4
21, 22, 23	FIRE SUPPRESSION, PLUMBING & HVAC	95.4	64.6	83.0	95.3	49.0	76.6	100.1	66.7	86.6	100.0	65.9	83.5	100.2	41.9	85.6	95.3	33.6	70.4
26, 27, 3370	ELECTRICAL, COMMUNICATIONS & UTIL.	91.8	63.9	77.1	92.0	64.6	77.5	96.9	67.5	81.4	93.4	68.0	80.0	95.1	60.9	77.0	92.2	55.6	72.9
MF2014	WEIGHTED AVERAGE	97.7	64.0	83.0	98.8	64.0	83.6	99.4	70.5	86.8	98.0	66.7	84.4	98.3	64.5	83.6	98.5	58.7	81.2

789

City Cost Indexes

		TEXAS																	
DIVISION		LUBBOCK 793 - 794			LUFKIN 759			MCALLEN 785			MCKINNEY 750			MIDLAND 797			ODESSA 797		
		MAT.	INST.	TOTAL	MAT.	INST.	TOTAL	MAT.	INST.	TOTAL	MAT.	INST.	TOTAL	MAT.	INST.	TOTAL	MAT.	INST.	TOTAL
015433	CONTRACTOR EQUIPMENT		98.5	98.5		91.4	91.4		96.8	96.8		96.7	96.7		98.5	98.5		89.9	89.9
0241, 31 - 34	SITE & INFRASTRUCTURE, DEMOLITION	123.8	86.1	97.0	91.2	94.0	93.1	144.1	84.4	101.7	94.6	85.3	88.0	126.7	86.1	97.8	99.7	88.0	91.4
0310	Concrete Forming & Accessories	97.9	57.2	62.8	91.7	61.7	65.9	101.2	57.5	63.5	91.3	65.3	68.8	101.9	65.4	70.5	98.9	65.4	70.0
0320	Concrete Reinforcing	94.1	52.5	72.9	101.3	64.9	82.8	88.2	48.4	67.9	100.7	50.9	75.3	95.1	52.1	73.2	92.9	52.1	72.1
0330	Cast-in-Place Concrete	96.8	64.7	83.6	92.9	64.0	81.0	119.3	64.1	96.6	83.8	63.8	75.6	102.9	64.6	87.2	96.6	63.6	83.1
03	CONCRETE	93.7	60.4	77.4	99.8	63.6	82.0	107.9	59.5	84.1	87.0	63.3	75.3	98.2	64.0	81.4	95.0	62.8	79.2
04	MASONRY	103.6	64.0	78.9	119.4	58.8	81.6	106.8	56.4	75.4	172.2	56.6	100.2	121.5	56.7	81.1	104.2	56.6	74.6
05	METALS	109.0	81.3	100.5	97.9	72.4	90.0	94.3	77.5	89.2	97.6	78.8	91.8	107.2	80.7	99.0	104.6	69.1	93.7
06	WOOD, PLASTICS & COMPOSITES	101.1	57.3	76.5	95.3	61.8	76.5	114.6	58.4	83.1	92.3	68.6	79.0	106.0	68.6	85.1	101.1	68.5	82.8
07	THERMAL & MOISTURE PROTECTION	89.3	66.2	79.8	94.3	65.6	82.5	104.8	59.8	86.3	92.8	66.0	81.8	89.5	65.4	79.7	100.0	65.9	86.0
08	OPENINGS	108.5	57.2	96.6	66.7	63.0	65.9	105.7	54.7	93.9	97.7	63.0	89.6	107.5	63.4	97.2	97.1	63.3	89.2
0920	Plaster & Gypsum Board	84.4	56.3	65.4	88.0	61.2	69.9	99.4	57.5	71.1	90.0	68.0	75.1	85.9	68.0	73.8	84.0	68.0	73.2
0950, 0980	Ceilings & Acoustic Treatment	90.6	56.3	68.1	84.4	61.2	69.1	93.3	57.5	69.8	93.6	68.0	76.8	88.1	68.0	74.9	88.9	68.0	75.2
0960	Flooring	104.4	64.1	92.8	138.2	64.1	117.0	118.1	63.8	102.6	98.2	64.1	88.4	105.7	64.1	93.8	110.9	64.1	97.5
0970, 0990	Wall Finishes & Painting/Coating	119.0	56.1	81.0	96.8	56.1	72.2	120.1	48.8	77.1	107.0	56.1	76.3	119.0	56.1	81.0	106.6	56.1	76.1
09	FINISHES	96.4	58.0	75.2	110.2	61.2	83.2	106.4	57.8	79.6	96.2	64.6	78.8	96.8	64.7	79.1	93.7	64.6	77.7
COVERS	DIVS. 10 - 14, 25, 28, 41, 43, 44, 46	100.0	81.3	96.2	100.0	80.9	96.1	100.0	80.7	96.1	100.0	82.3	96.4	100.0	81.2	96.2	100.0	80.9	96.1
21, 22, 23	FIRE SUPPRESSION, PLUMBING & HVAC	99.8	50.7	80.0	95.3	64.4	82.8	95.4	58.7	80.6	95.3	45.4	75.2	95.0	49.1	76.5	100.3	49.1	79.6
26, 27, 3370	ELECTRICAL, COMMUNICATIONS & UTIL.	97.9	64.2	80.1	93.5	70.6	81.4	92.6	36.3	62.8	92.1	64.6	77.5	97.9	64.2	80.1	99.4	64.1	80.7
MF2014	WEIGHTED AVERAGE	101.6	63.6	85.0	95.6	67.8	83.5	101.0	59.8	83.0	98.7	63.1	83.2	101.5	64.2	85.2	99.6	63.1	83.7

		TEXAS																	
DIVISION		PALESTINE 758			SAN ANGELO 769			SAN ANTONIO 781 - 782			TEMPLE 765			TEXARKANA 755			TYLER 757		
		MAT.	INST.	TOTAL	MAT.	INST.	TOTAL	MAT.	INST.	TOTAL	MAT.	INST.	TOTAL	MAT.	INST.	TOTAL	MAT.	INST.	TOTAL
015433	CONTRACTOR EQUIPMENT		91.4	91.4		89.9	89.9		91.9	91.9		89.9	89.9		91.4	91.4		91.4	91.4
0241, 31 - 34	SITE & INFRASTRUCTURE, DEMOLITION	96.6	92.6	93.8	101.5	89.7	93.1	100.9	92.0	94.5	89.8	89.1	89.3	85.9	92.0	90.2	95.1	92.4	93.2
0310	Concrete Forming & Accessories	82.7	65.4	67.8	99.3	58.7	64.3	95.0	58.1	63.2	102.8	57.8	64.0	99.0	57.4	63.1	93.6	65.4	69.3
0320	Concrete Reinforcing	98.8	52.1	75.1	92.7	52.1	72.1	93.6	49.8	71.3	92.9	49.1	70.6	98.7	52.3	75.1	99.6	52.4	75.6
0330	Cast-in-Place Concrete	84.9	63.0	75.9	95.7	64.8	83.0	85.6	65.2	77.2	78.5	63.1	72.2	85.5	62.7	76.2	101.9	63.0	85.9
03	CONCRETE	101.6	62.6	82.4	96.7	60.3	78.8	91.7	59.7	76.0	82.9	58.7	71.0	92.8	59.0	76.2	107.1	62.7	85.3
04	MASONRY	113.8	56.6	78.2	138.0	58.7	88.5	96.8	65.3	77.2	151.0	56.4	92.1	176.5	56.6	101.8	166.0	56.6	97.8
05	METALS	97.6	68.3	88.6	100.1	69.1	90.6	96.9	68.6	88.2	99.8	66.5	89.6	91.0	67.8	83.9	97.5	68.4	88.5
06	WOOD, PLASTICS & COMPOSITES	85.3	68.4	75.8	99.8	58.3	76.5	94.7	57.4	73.8	108.9	58.3	80.6	99.7	58.2	76.5	97.0	68.4	81.0
07	THERMAL & MOISTURE PROTECTION	94.8	65.3	82.7	98.0	66.0	84.9	96.4	68.2	84.8	97.7	64.8	84.2	94.1	64.2	81.9	94.6	65.3	82.6
08	OPENINGS	66.7	63.3	65.9	94.7	57.8	86.1	102.7	54.4	91.5	67.1	57.0	64.7	87.7	57.8	80.8	66.6	63.4	65.9
0920	Plaster & Gypsum Board	85.2	68.0	73.6	87.8	57.5	67.3	98.0	56.5	70.0	87.8	57.5	67.3	93.4	57.5	69.1	88.0	68.0	74.5
0950, 0980	Ceilings & Acoustic Treatment	84.4	68.0	73.6	80.6	57.5	65.4	102.7	56.5	72.3	80.6	57.5	65.4	90.3	57.5	68.7	84.4	68.0	73.6
0960	Flooring	129.9	64.1	111.1	95.3	64.1	86.4	104.1	64.1	92.6	123.2	64.1	106.3	111.1	64.1	97.6	140.1	64.1	118.4
0970, 0990	Wall Finishes & Painting/Coating	96.8	56.1	72.2	104.2	56.1	75.1	106.3	53.2	74.2	105.8	48.8	71.4	96.8	56.1	72.2	96.8	56.1	72.2
09	FINISHES	108.0	64.5	84.0	87.5	59.1	71.8	103.4	58.1	78.4	95.1	57.7	74.5	104.0	58.5	78.9	111.1	64.5	85.4
COVERS	DIVS. 10 - 14, 25, 28, 41, 43, 44, 46	100.0	82.0	96.4	100.0	81.0	96.2	100.0	81.2	96.2	100.0	81.1	96.2	100.0	80.9	96.1	100.0	82.0	96.4
21, 22, 23	FIRE SUPPRESSION, PLUMBING & HVAC	95.3	59.6	80.9	95.4	50.1	77.1	100.0	63.2	85.1	95.4	54.7	79.0	95.3	37.0	71.8	95.3	62.4	82.0
26, 27, 3370	ELECTRICAL, COMMUNICATIONS & UTIL.	89.8	55.6	71.7	97.4	44.8	69.6	96.4	60.9	77.6	94.5	59.8	76.2	93.4	60.1	75.8	92.2	55.5	72.8
MF2014	WEIGHTED AVERAGE	95.0	64.4	81.7	98.5	59.6	81.5	98.5	65.1	84.0	94.7	61.7	80.3	98.0	58.4	80.7	98.5	65.0	83.9

		TEXAS															UTAH		
DIVISION		VICTORIA 779			WACO 766 - 767			WAXAHACHIE 751			WHARTON 774			WICHITA FALLS 763			LOGAN 843		
		MAT.	INST.	TOTAL	MAT.	INST.	TOTAL	MAT.	INST.	TOTAL	MAT.	INST.	TOTAL	MAT.	INST.	TOTAL	MAT.	INST.	TOTAL
015433	CONTRACTOR EQUIPMENT		99.3	99.3		89.9	89.9		96.7	96.7		100.5	100.5		89.9	89.9		97.5	97.5
0241, 31 - 34	SITE & INFRASTRUCTURE, DEMOLITION	111.8	85.7	93.2	98.6	88.7	91.6	96.0	85.4	88.4	117.1	87.8	96.2	99.3	88.7	91.8	96.0	95.7	95.8
0310	Concrete Forming & Accessories	90.0	59.0	63.3	101.2	65.5	70.4	91.3	65.5	69.0	84.9	61.7	64.9	101.2	65.5	70.4	104.7	58.1	64.5
0320	Concrete Reinforcing	93.9	48.5	70.8	92.5	48.5	70.1	100.7	50.9	75.4	97.7	50.6	73.7	92.5	51.3	71.5	99.6	81.3	90.3
0330	Cast-in-Place Concrete	106.9	65.6	89.9	84.9	66.8	77.4	88.4	63.9	78.3	109.8	65.5	91.6	90.7	63.7	79.6	87.3	73.4	81.6
03	CONCRETE	104.9	60.9	83.3	89.7	63.3	76.7	90.6	63.4	77.2	109.1	62.5	86.2	92.4	62.8	77.9	106.0	68.3	87.5
04	MASONRY	120.0	58.9	81.9	105.2	56.6	74.9	160.7	56.6	95.8	103.6	58.9	75.8	105.7	63.6	79.5	105.8	63.2	79.3
05	METALS	94.9	80.7	90.5	102.2	67.8	91.6	97.7	79.0	92.0	96.3	81.6	91.8	102.2	69.8	92.2	102.2	79.0	95.0
06	WOOD, PLASTICS & COMPOSITES	92.5	58.4	73.4	107.1	68.5	85.5	92.3	68.6	79.0	83.1	62.0	71.3	107.1	68.5	85.5	85.3	54.9	68.3
07	THERMAL & MOISTURE PROTECTION	96.8	69.8	84.6	98.5	65.8	85.1	92.9	66.0	81.9	93.4	66.6	82.4	98.5	67.7	85.9	97.4	67.9	85.3
08	OPENINGS	101.4	56.1	90.8	77.7	62.4	74.1	97.7	63.0	89.6	101.6	58.6	91.6	77.7	63.3	74.4	94.3	56.5	85.5
0920	Plaster & Gypsum Board	93.8	57.5	69.3	88.2	68.0	74.6	90.4	68.0	75.3	92.5	61.2	71.3	88.2	68.0	74.6	78.9	53.5	61.7
0950, 0980	Ceilings & Acoustic Treatment	95.4	57.5	70.5	82.2	68.0	72.9	95.3	68.0	77.4	94.6	61.2	72.6	82.2	68.0	72.9	97.2	53.5	68.5
0960	Flooring	99.5	64.1	89.4	122.4	64.1	105.8	98.2	64.1	88.4	97.7	64.1	88.1	123.4	76.5	110.0	101.2	47.7	85.9
0970, 0990	Wall Finishes & Painting/Coating	103.6	59.4	76.9	105.7	48.8	71.4	107.0	56.1	76.3	103.4	59.4	76.8	108.5	56.1	76.8	103.3	60.7	77.5
09	FINISHES	89.4	59.6	73.0	95.8	63.8	78.1	96.8	64.6	79.1	90.8	61.7	74.8	96.3	67.1	80.2	96.2	55.3	73.6
COVERS	DIVS. 10 - 14, 25, 28, 41, 43, 44, 46	100.0	80.8	96.1	100.0	82.3	96.4	100.0	82.3	96.4	100.0	81.3	96.2	100.0	69.8	93.8	100.0	85.1	97.0
21, 22, 23	FIRE SUPPRESSION, PLUMBING & HVAC	95.4	65.9	83.5	100.2	54.7	81.8	95.3	58.5	80.5	95.4	65.0	83.1	100.2	54.2	81.6	99.9	69.5	87.6
26, 27, 3370	ELECTRICAL, COMMUNICATIONS & UTIL.	99.8	57.6	77.5	97.6	63.6	79.6	92.1	64.6	77.5	98.6	68.0	82.4	99.3	60.1	78.6	97.5	73.0	84.6
MF2014	WEIGHTED AVERAGE	99.0	65.6	84.4	96.5	64.1	82.4	98.6	65.9	84.4	98.9	67.7	85.3	97.1	64.4	82.8	99.8	70.1	86.9

City Cost Indexes

DIVISION		UTAH											VERMONT					
		OGDEN			PRICE			PROVO			SALT LAKE CITY			BELLOWS FALLS			BENNINGTON		
		842,844			845			846 - 847			840 - 841			051			052		
		MAT.	INST.	TOTAL	MAT.	INST.	TOTAL	MAT.	INST.	TOTAL	MAT.	INST.	TOTAL	MAT.	INST.	TOTAL	MAT.	INST.	TOTAL
015433	CONTRACTOR EQUIPMENT		97.5	97.5		96.6	96.6		96.6	96.6		97.5	97.5		100.9	100.9		100.9	100.9
0241, 31 - 34	SITE & INFRASTRUCTURE, DEMOLITION	84.9	95.7	92.6	93.3	93.8	93.7	92.3	94.1	93.6	84.5	95.6	92.4	88.2	100.5	96.9	87.5	100.5	96.7
0310	Concrete Forming & Accessories	104.7	58.1	64.5	107.1	48.3	56.4	106.3	58.0	64.7	107.1	58.0	64.8	98.3	85.5	87.3	95.9	106.6	105.2
0320	Concrete Reinforcing	99.2	81.3	90.1	106.8	81.1	93.7	107.7	81.3	94.3	101.5	81.3	91.2	83.2	85.7	84.5	83.2	85.7	84.5
0330	Cast-in-Place Concrete	88.7	73.4	82.4	87.4	59.5	75.9	87.4	73.3	81.6	96.9	73.3	87.2	87.0	110.7	96.7	87.0	110.7	96.7
03	CONCRETE	95.9	68.3	82.3	107.3	59.0	83.6	105.9	68.2	87.4	114.8	68.2	91.9	91.0	94.2	92.6	90.8	103.6	97.1
04	MASONRY	99.7	63.2	77.0	110.9	61.3	80.0	111.1	63.2	81.2	113.1	63.2	82.0	103.6	95.2	98.4	112.4	95.2	101.6
05	METALS	102.7	79.0	95.4	99.4	76.6	92.4	100.3	78.9	93.7	106.2	78.9	97.8	94.8	88.3	92.8	94.7	88.2	92.7
06	WOOD, PLASTICS & COMPOSITES	85.3	54.9	68.3	88.6	44.3	63.8	86.8	54.9	68.9	87.2	54.9	69.1	103.7	83.5	92.4	100.6	112.4	107.2
07	THERMAL & MOISTURE PROTECTION	96.4	67.9	84.7	99.0	59.9	82.9	99.0	67.9	86.2	103.4	67.9	88.9	97.0	82.7	91.1	97.0	81.1	90.4
08	OPENINGS	94.3	56.5	85.5	98.2	48.9	86.7	98.2	56.5	88.5	96.2	56.5	86.9	105.0	86.5	100.7	105.0	102.3	104.3
0920	Plaster & Gypsum Board	78.9	53.5	61.7	81.7	42.5	55.2	79.2	53.5	61.8	90.5	53.5	65.5	106.0	82.4	90.1	104.2	112.2	109.6
0950, 0980	Ceilings & Acoustic Treatment	97.2	53.5	68.5	97.2	42.5	61.3	97.2	53.5	68.5	91.5	53.5	66.5	91.5	82.4	85.6	91.5	112.2	105.1
0960	Flooring	99.0	47.7	84.3	102.3	35.5	83.2	102.0	47.7	86.5	103.2	47.7	87.3	100.6	104.6	101.8	99.7	104.6	101.1
0970, 0990	Wall Finishes & Painting/Coating	103.3	60.7	77.5	103.3	37.5	63.6	103.3	62.6	78.7	106.7	62.6	80.1	97.4	105.8	102.5	97.4	105.8	102.5
09	FINISHES	94.3	55.3	72.8	97.3	43.5	67.7	96.8	55.5	74.0	96.6	55.5	74.0	95.9	90.7	93.0	95.4	107.8	102.2
COVERS	DIVS. 10 - 14, 25, 28, 41, 43, 44, 46	100.0	85.1	97.0	100.0	45.8	89.0	100.0	85.0	97.0	100.0	85.0	97.0	100.0	97.6	99.5	100.0	100.7	100.1
21, 22, 23	FIRE SUPPRESSION, PLUMBING & HVAC	99.9	69.5	87.6	97.5	67.7	85.5	99.9	69.4	87.6	100.1	69.4	87.7	95.3	96.4	95.7	95.3	96.4	95.7
26, 27, 3370	ELECTRICAL, COMMUNICATIONS & UTIL.	97.9	73.0	84.7	103.5	73.0	87.4	98.4	59.8	78.0	100.5	59.8	79.0	101.5	89.3	95.1	101.5	61.2	80.2
MF2014	WEIGHTED AVERAGE	98.1	70.1	85.9	100.2	64.6	84.7	100.3	68.2	86.3	102.2	68.3	87.4	97.2	93.0	95.3	97.5	93.5	95.8

DIVISION		VERMONT																	
		BRATTLEBORO			BURLINGTON			GUILDHALL			MONTPELIER			RUTLAND			ST. JOHNSBURY		
		053			054			059			056			057			058		
		MAT.	INST.	TOTAL	MAT.	INST.	TOTAL	MAT.	INST.	TOTAL	MAT.	INST.	TOTAL	MAT.	INST.	TOTAL	MAT.	INST.	TOTAL
015433	CONTRACTOR EQUIPMENT		100.9	100.9		100.9	100.9		100.9	100.9		100.9	100.9		100.9	100.9		100.9	100.9
0241, 31 - 34	SITE & INFRASTRUCTURE, DEMOLITION	88.9	100.5	97.1	93.0	100.4	98.2	87.2	99.0	95.6	91.5	100.1	97.6	91.7	100.1	97.7	87.3	99.0	95.6
0310	Concrete Forming & Accessories	98.6	85.4	87.2	98.0	87.6	89.0	96.1	78.3	80.7	97.9	84.1	86.0	98.9	87.6	89.1	94.4	78.3	80.5
0320	Concrete Reinforcing	82.3	85.7	84.0	101.1	85.6	93.2	83.9	85.6	84.8	94.1	85.6	89.8	103.9	85.6	94.6	82.3	85.6	84.0
0330	Cast-in-Place Concrete	89.7	110.7	98.3	100.9	109.6	104.5	84.3	101.0	91.2	100.9	109.6	104.5	85.2	109.6	95.2	84.3	101.0	91.2
03	CONCRETE	93.0	94.1	93.6	101.2	94.7	98.0	88.6	87.6	88.1	100.1	93.1	96.7	94.3	94.7	94.5	88.2	87.6	87.9
04	MASONRY	111.5	95.2	101.3	115.7	93.7	102.0	111.8	78.2	90.8	109.6	93.7	99.7	93.9	93.7	93.8	138.8	78.2	101.0
05	METALS	94.7	88.2	92.7	103.5	87.5	98.6	94.8	87.3	92.5	99.6	87.4	95.9	100.6	87.4	96.5	94.8	87.3	92.5
06	WOOD, PLASTICS & COMPOSITES	104.0	83.5	92.5	99.9	88.2	93.3	100.0	83.5	90.8	97.7	83.5	89.7	104.2	88.2	95.2	94.8	83.5	88.5
07	THERMAL & MOISTURE PROTECTION	97.1	78.0	89.3	103.8	80.4	94.2	96.8	70.7	86.1	103.3	79.9	93.7	97.2	80.4	90.3	96.7	70.7	86.1
08	OPENINGS	105.0	86.5	100.7	108.4	85.0	103.0	105.0	82.4	99.7	105.8	82.4	100.3	108.4	85.0	102.9	105.0	82.4	99.7
0920	Plaster & Gypsum Board	106.0	82.4	90.1	109.7	87.2	94.5	112.4	82.4	92.2	108.9	82.4	91.0	106.6	87.2	93.5	113.8	82.4	92.6
0950, 0980	Ceilings & Acoustic Treatment	91.5	82.4	85.6	101.0	87.2	91.9	91.5	82.4	85.6	95.1	82.4	86.8	95.9	87.2	90.2	91.5	82.4	85.6
0960	Flooring	100.7	104.6	101.8	102.4	104.6	103.1	103.6	104.6	103.9	103.2	104.6	103.6	100.6	104.6	101.8	106.9	104.6	106.3
0970, 0990	Wall Finishes & Painting/Coating	97.4	105.8	102.5	102.9	85.2	92.2	97.4	85.2	90.0	101.7	85.2	91.7	97.4	85.2	90.0	97.4	85.2	90.0
09	FINISHES	96.0	90.7	93.1	99.7	90.8	94.8	97.5	84.2	90.2	97.9	84.8	90.7	97.1	90.8	93.6	98.7	84.2	90.7
COVERS	DIVS. 10 - 14, 25, 28, 41, 43, 44, 46	100.0	97.6	99.5	100.0	97.5	99.5	100.0	91.8	98.3	100.0	97.0	99.4	100.0	97.5	99.5	100.0	91.8	98.3
21, 22, 23	FIRE SUPPRESSION, PLUMBING & HVAC	95.3	96.4	95.7	99.9	71.8	88.6	95.3	64.0	82.7	95.1	71.8	85.7	100.1	71.8	88.7	95.3	64.0	82.7
26, 27, 3370	ELECTRICAL, COMMUNICATIONS & UTIL.	101.5	89.3	95.1	102.2	61.2	80.5	101.5	61.2	80.2	99.7	61.2	79.3	101.6	61.2	80.2	101.5	61.2	80.2
MF2014	WEIGHTED AVERAGE	97.8	92.8	95.6	102.4	83.6	94.2	97.4	77.8	88.8	99.5	82.8	92.2	99.7	83.6	92.7	98.6	77.8	89.6

DIVISION		VERMONT			VIRGINIA														
		WHITE RIVER JCT.			ALEXANDRIA			ARLINGTON			BRISTOL			CHARLOTTESVILLE			CULPEPER		
		050			223			222			242			229			227		
		MAT.	INST.	TOTAL	MAT.	INST.	TOTAL	MAT.	INST.	TOTAL	MAT.	INST.	TOTAL	MAT.	INST.	TOTAL	MAT.	INST.	TOTAL
015433	CONTRACTOR EQUIPMENT		100.9	100.9		103.0	103.0		101.7	101.7		101.7	101.7		106.1	106.1		101.7	101.7
0241, 31 - 34	SITE & INFRASTRUCTURE, DEMOLITION	91.5	99.2	97.0	114.3	89.8	96.9	124.5	87.7	98.3	108.8	86.1	92.7	113.5	85.1	93.3	111.8	87.6	94.6
0310	Concrete Forming & Accessories	93.1	78.9	80.8	92.2	72.8	75.5	91.2	72.0	74.7	87.1	42.5	48.6	85.4	48.7	53.7	82.5	71.4	72.9
0320	Concrete Reinforcing	83.2	85.6	84.4	83.0	87.0	85.1	93.7	84.4	88.9	93.7	68.4	80.8	93.1	72.8	82.7	93.7	84.2	88.9
0330	Cast-in-Place Concrete	89.7	101.9	94.7	105.1	81.8	95.6	102.3	80.3	93.3	101.8	48.5	79.9	105.9	55.5	85.2	104.8	69.2	90.2
03	CONCRETE	94.7	88.2	91.5	102.0	79.6	91.0	106.7	78.3	92.8	102.9	51.2	77.5	103.5	57.2	80.8	100.6	74.1	87.6
04	MASONRY	124.5	79.7	96.6	91.3	72.8	79.8	106.3	70.7	84.1	95.4	51.4	68.0	120.3	53.2	78.5	107.5	70.7	84.6
05	METALS	94.8	87.3	92.5	105.7	97.0	103.0	104.3	97.2	102.1	103.1	85.7	97.8	103.4	90.2	99.3	103.5	95.3	101.0
06	WOOD, PLASTICS & COMPOSITES	97.4	83.5	89.6	95.0	71.0	81.6	91.5	71.0	80.0	83.7	40.0	59.2	82.2	45.6	61.7	80.8	71.0	75.3
07	THERMAL & MOISTURE PROTECTION	97.1	71.4	86.6	102.7	81.5	94.0	104.7	80.6	94.8	104.2	60.1	86.1	103.8	67.7	89.0	104.0	78.1	93.4
08	OPENINGS	105.0	82.4	99.7	100.9	75.2	94.9	99.0	75.2	93.5	102.0	46.5	89.1	100.2	52.0	89.0	100.5	75.2	94.6
0920	Plaster & Gypsum Board	102.8	82.4	89.0	107.4	69.9	82.1	104.2	69.9	81.0	99.7	38.0	58.0	99.7	43.0	61.4	99.9	69.9	79.6
0950, 0980	Ceilings & Acoustic Treatment	91.5	82.4	85.6	91.4	69.9	77.3	89.8	69.9	76.7	88.9	38.0	55.4	88.9	43.0	58.8	89.8	69.9	76.7
0960	Flooring	98.6	104.6	100.4	106.0	82.9	99.4	104.5	81.9	98.0	101.0	66.5	91.1	99.7	66.5	90.2	99.7	81.9	94.6
0970, 0990	Wall Finishes & Painting/Coating	97.4	85.2	90.0	124.0	80.0	97.5	124.0	80.0	97.5	109.6	54.3	76.2	109.6	76.1	89.3	124.0	76.1	95.1
09	FINISHES	95.2	84.6	89.3	100.1	74.6	86.0	100.0	73.8	85.6	96.5	47.2	69.3	96.2	53.5	72.6	96.9	73.2	83.9
COVERS	DIVS. 10 - 14, 25, 28, 41, 43, 44, 46	100.0	92.3	98.4	100.0	88.6	97.7	100.0	87.6	97.5	100.0	67.8	93.5	100.0	79.3	95.8	100.0	87.6	97.5
21, 22, 23	FIRE SUPPRESSION, PLUMBING & HVAC	95.3	64.8	83.0	100.3	86.9	94.9	100.3	85.4	94.3	95.6	50.5	77.4	95.6	69.3	85.0	95.6	72.9	86.4
26, 27, 3370	ELECTRICAL, COMMUNICATIONS & UTIL.	101.5	61.2	80.2	97.1	102.4	99.9	94.7	99.2	97.1	96.8	35.4	64.3	96.7	71.9	83.6	99.3	99.2	99.2
MF2014	WEIGHTED AVERAGE	98.5	78.3	89.7	101.1	85.5	94.3	101.9	84.0	94.1	99.4	54.9	80.0	100.5	67.1	86.0	99.9	80.5	91.4

For customer support on your Commercial Renovation Cost Data, call 877.791.4977.

City Cost Indexes

DIVISION		VIRGINIA																	
		FAIRFAX 220 - 221			FARMVILLE 239			FREDERICKSBURG 224 - 225			GRUNDY 246			HARRISONBURG 228			LYNCHBURG 245		
		MAT.	INST.	TOTAL	MAT.	INST.	TOTAL	MAT.	INST.	TOTAL	MAT.	INST.	TOTAL	MAT.	INST.	TOTAL	MAT.	INST.	TOTAL
015433	CONTRACTOR EQUIPMENT		101.7	101.7		106.1	106.1		101.7	101.7		101.7	101.7		101.7	101.7		101.7	101.7
0241, 31 - 34	SITE & INFRASTRUCTURE, DEMOLITION	123.1	87.8	98.0	109.2	87.6	93.8	111.4	87.7	94.5	106.6	84.8	91.1	120.1	86.3	96.1	107.5	86.3	92.4
0310	Concrete Forming & Accessories	85.4	72.0	73.9	98.0	43.7	51.2	85.4	69.4	71.6	90.3	36.4	43.8	81.4	42.6	47.9	87.1	60.3	64.0
0320	Concrete Reinforcing	93.7	87.0	90.3	90.9	49.9	70.0	94.4	87.0	90.6	92.4	50.0	71.0	93.7	64.9	79.0	93.1	68.5	80.5
0330	Cast-in-Place Concrete	102.3	80.6	93.4	103.1	53.5	82.7	103.9	69.3	89.7	101.8	47.5	79.5	102.3	58.0	84.1	101.8	57.7	83.7
03	CONCRETE	106.3	78.8	92.8	101.2	49.9	76.0	100.3	73.8	87.3	101.6	44.4	73.5	104.1	53.8	79.4	101.5	62.3	82.3
04	MASONRY	106.2	71.4	84.5	103.5	44.9	66.9	106.6	70.7	84.2	96.5	47.3	65.8	104.2	50.3	70.6	112.0	53.2	75.4
05	METALS	103.6	97.0	101.5	101.0	77.1	93.6	103.5	96.2	101.3	103.1	69.9	92.9	103.4	83.9	97.4	103.3	86.3	98.1
06	WOOD, PLASTICS & COMPOSITES	83.7	71.0	76.6	96.4	44.1	67.1	83.7	68.1	75.0	86.7	35.1	57.8	79.8	39.0	56.9	83.7	62.6	71.9
07	THERMAL & MOISTURE PROTECTION	104.6	73.7	91.9	104.3	49.3	81.8	104.0	78.5	93.6	104.2	45.3	80.0	104.4	63.9	87.8	104.0	63.9	87.6
08	OPENINGS	99.0	75.2	93.5	100.7	43.0	87.3	100.2	73.6	94.0	102.0	34.4	86.3	100.5	49.4	88.6	100.5	58.8	90.8
0920	Plaster & Gypsum Board	99.9	69.9	79.6	108.6	41.5	63.2	99.9	66.9	77.6	99.7	32.9	54.5	99.7	36.9	57.3	99.7	61.3	73.7
0950, 0980	Ceilings & Acoustic Treatment	89.8	69.9	76.7	84.8	41.5	56.3	89.8	66.9	74.8	88.9	32.9	52.1	88.9	36.9	54.8	88.9	61.3	70.8
0960	Flooring	101.4	81.9	95.8	106.5	60.3	93.3	101.4	81.9	95.8	102.3	33.7	82.6	99.4	74.0	92.1	101.0	66.5	91.1
0970, 0990	Wall Finishes & Painting/Coating	124.0	80.0	97.5	112.8	38.7	68.1	124.0	76.1	95.1	109.6	35.1	64.6	124.0	54.3	81.9	109.6	54.3	76.2
09	FINISHES	98.5	74.0	85.0	98.0	46.0	69.4	97.4	71.5	83.1	96.7	35.1	62.8	97.3	47.8	70.1	96.3	60.2	76.4
COVERS	DIVS. 10 - 14, 25, 28, 41, 43, 44, 46	100.0	87.8	97.5	100.0	45.7	89.0	100.0	83.4	96.6	100.0	43.7	88.6	100.0	67.6	93.5	100.0	70.9	94.1
21, 22, 23	FIRE SUPPRESSION, PLUMBING & HVAC	95.6	85.7	91.6	95.5	43.4	74.5	95.6	85.1	91.3	95.6	62.1	82.1	95.6	66.9	84.0	95.6	69.0	84.8
26, 27, 3370	ELECTRICAL, COMMUNICATIONS & UTIL.	97.9	99.2	98.6	90.1	48.2	68.0	94.9	99.2	97.2	96.8	44.6	69.2	96.9	96.4	96.7	97.9	52.7	74.0
MF2014	WEIGHTED AVERAGE	100.7	84.0	93.4	98.7	52.4	78.5	99.5	82.7	92.1	99.3	52.5	78.9	100.2	67.2	85.8	99.9	65.6	85.0

DIVISION		VIRGINIA																	
		NEWPORT NEWS 236			NORFOLK 233 - 235			PETERSBURG 238			PORTSMOUTH 237			PULASKI 243			RICHMOND 230 - 232		
		MAT.	INST.	TOTAL	MAT.	INST.	TOTAL	MAT.	INST.	TOTAL	MAT.	INST.	TOTAL	MAT.	INST.	TOTAL	MAT.	INST.	TOTAL
015433	CONTRACTOR EQUIPMENT		106.1	106.1		106.7	106.7		106.1	106.1		106.0	106.0		101.7	101.7		106.1	106.1
0241, 31 - 34	SITE & INFRASTRUCTURE, DEMOLITION	108.2	88.8	94.4	108.9	89.8	95.3	111.7	89.0	95.6	106.7	88.6	93.8	105.9	85.5	91.4	103.6	89.0	93.2
0310	Concrete Forming & Accessories	97.3	64.1	68.7	97.9	64.3	68.9	90.6	57.2	61.8	86.4	53.1	57.7	90.3	38.7	45.8	97.6	57.2	62.8
0320	Concrete Reinforcing	90.6	72.0	81.2	94.7	72.1	83.2	90.3	73.0	81.5	90.3	72.0	81.0	92.4	64.6	78.3	97.5	73.0	85.0
0330	Cast-in-Place Concrete	100.2	66.1	86.2	105.2	66.1	89.1	106.3	53.7	84.7	99.2	66.0	85.6	101.8	48.2	79.8	98.1	53.7	79.8
03	CONCRETE	98.4	67.6	83.3	101.5	67.7	84.9	103.6	60.4	82.4	97.1	62.6	80.2	101.6	48.7	75.6	98.5	60.4	79.8
04	MASONRY	97.8	53.9	70.5	103.8	53.9	72.7	111.9	52.6	75.0	103.8	53.9	72.7	91.5	48.2	64.5	102.5	52.6	71.4
05	METALS	103.3	89.8	99.2	103.1	89.9	99.1	101.1	90.7	97.9	102.2	89.1	98.2	103.2	82.5	96.8	107.1	90.7	102.0
06	WOOD, PLASTICS & COMPOSITES	95.3	66.2	79.0	93.2	66.2	78.1	86.4	58.6	70.8	82.9	51.5	65.3	86.7	36.8	58.7	95.0	58.6	74.6
07	THERMAL & MOISTURE PROTECTION	104.3	68.4	89.6	100.9	68.4	87.5	104.3	68.3	89.5	104.3	66.8	88.9	104.2	51.1	82.4	102.2	68.3	88.3
08	OPENINGS	101.1	62.2	92.1	100.4	62.2	91.5	100.4	59.0	90.8	101.2	54.1	90.2	102.0	42.2	88.1	104.8	59.0	94.1
0920	Plaster & Gypsum Board	109.4	64.3	78.9	105.5	64.3	77.6	101.3	56.4	71.0	101.6	49.1	66.1	99.7	34.7	55.7	104.2	56.4	71.9
0950, 0980	Ceilings & Acoustic Treatment	88.2	64.3	72.5	88.9	64.3	72.7	85.7	56.4	66.4	88.2	49.1	62.5	88.9	34.7	53.3	90.6	56.4	68.1
0960	Flooring	106.5	66.5	95.1	105.6	66.5	94.4	102.5	71.2	93.6	99.5	66.5	90.0	102.3	66.5	92.0	103.4	71.2	94.2
0970, 0990	Wall Finishes & Painting/Coating	112.8	42.0	70.1	113.0	76.1	90.7	112.8	76.1	90.6	112.8	76.1	90.6	109.6	54.3	76.2	110.0	76.1	89.5
09	FINISHES	98.7	62.2	78.6	99.2	66.0	80.5	96.3	61.2	76.9	95.6	56.7	74.2	96.7	43.7	67.5	98.5	61.2	77.9
COVERS	DIVS. 10 - 14, 25, 28, 41, 43, 44, 46	100.0	82.2	96.4	100.0	82.2	96.4	100.0	79.6	95.9	100.0	72.0	94.3	100.0	44.0	88.7	100.0	79.6	95.9
21, 22, 23	FIRE SUPPRESSION, PLUMBING & HVAC	100.3	63.6	85.5	100.0	64.3	85.6	95.5	67.5	84.2	100.3	64.3	85.8	95.6	62.2	82.1	99.9	67.5	86.8
26, 27, 3370	ELECTRICAL, COMMUNICATIONS & UTIL.	92.8	63.5	77.3	95.5	58.3	75.8	93.0	71.9	81.8	91.2	58.3	73.8	96.8	54.4	74.4	96.8	71.9	83.6
MF2014	WEIGHTED AVERAGE	100.0	68.2	86.1	100.5	68.2	86.4	99.4	68.7	86.0	99.4	65.3	84.5	99.0	57.3	80.9	101.3	68.7	87.1

DIVISION		VIRGINIA									WASHINGTON								
		ROANOKE 240 - 241			STAUNTON 244			WINCHESTER 226			CLARKSTON 994			EVERETT 982			OLYMPIA 985		
		MAT.	INST.	TOTAL	MAT.	INST.	TOTAL	MAT.	INST.	TOTAL	MAT.	INST.	TOTAL	MAT.	INST.	TOTAL	MAT.	INST.	TOTAL
015433	CONTRACTOR EQUIPMENT		101.7	101.7		106.1	106.1		101.7	101.7		91.2	91.2		101.6	101.6		101.6	101.6
0241, 31 - 34	SITE & INFRASTRUCTURE, DEMOLITION	106.4	86.3	92.1	109.9	87.7	94.1	118.7	87.6	96.6	99.2	90.1	92.7	108.4	103.5	104.9	93.4	108.4	104.1
0310	Concrete Forming & Accessories	96.8	60.5	65.5	89.9	50.0	55.5	83.8	69.5	71.4	112.7	67.4	73.6	113.4	100.4	102.2	99.0	100.4	100.1
0320	Concrete Reinforcing	93.4	68.5	80.7	93.1	48.9	70.6	93.1	86.4	89.7	112.0	87.2	99.4	104.2	98.8	101.5	110.8	98.7	104.6
0330	Cast-in-Place Concrete	115.7	57.7	91.9	105.9	51.5	83.6	102.3	69.3	88.7	96.3	81.9	90.4	103.5	106.2	104.6	98.9	106.2	101.9
03	CONCRETE	105.9	62.4	84.5	102.9	52.0	77.9	103.5	73.7	88.9	107.2	76.3	92.0	98.7	101.6	100.1	98.9	101.6	100.2
04	MASONRY	97.7	53.2	70.0	106.9	51.4	72.3	101.7	59.7	75.5	102.3	81.5	89.3	110.6	100.8	104.5	103.0	99.1	100.6
05	METALS	105.5	86.5	99.7	103.4	80.5	96.3	103.5	94.7	100.8	88.3	81.6	86.2	101.8	92.4	98.9	101.2	92.1	98.4
06	WOOD, PLASTICS & COMPOSITES	95.9	62.6	77.3	86.7	49.9	66.1	82.2	68.1	74.3	113.2	63.8	85.5	111.7	100.0	105.2	91.1	100.0	96.1
07	THERMAL & MOISTURE PROTECTION	103.8	66.8	87.7	103.8	52.0	82.6	104.5	75.6	92.7	143.8	77.2	116.5	103.5	101.7	102.9	103.5	96.8	100.7
08	OPENINGS	100.9	58.8	91.1	100.5	48.1	88.3	102.1	73.7	95.5	119.0	68.7	107.3	101.3	99.8	101.0	106.9	99.9	105.3
0920	Plaster & Gypsum Board	107.4	61.3	76.2	99.7	47.5	64.4	99.9	66.9	77.6	145.7	62.5	89.5	109.4	99.9	103.0	101.3	99.9	100.4
0950, 0980	Ceilings & Acoustic Treatment	91.4	61.3	71.6	88.9	47.5	61.7	89.8	66.9	74.8	97.6	62.5	74.5	99.8	99.9	99.9	98.1	99.9	99.3
0960	Flooring	106.0	66.5	94.7	101.9	39.5	84.0	100.7	81.9	95.4	91.6	47.1	78.9	113.8	91.6	107.4	103.8	91.7	100.3
0970, 0990	Wall Finishes & Painting/Coating	109.6	54.3	76.2	109.6	34.5	64.3	124.0	87.2	101.8	100.0	58.5	75.0	95.4	91.5	93.0	99.0	91.5	94.5
09	FINISHES	99.0	61.1	78.1	96.6	46.6	69.0	97.9	72.7	84.0	110.3	61.3	83.3	104.7	97.6	100.8	98.9	97.6	98.2
COVERS	DIVS. 10 - 14, 25, 28, 41, 43, 44, 46	100.0	70.9	94.1	100.0	70.9	94.1	100.0	87.3	97.4	100.0	75.4	95.0	100.0	99.0	99.8	100.0	100.5	100.1
21, 22, 23	FIRE SUPPRESSION, PLUMBING & HVAC	100.3	64.8	86.0	95.6	58.9	80.8	95.5	87.7	92.4	95.6	83.4	90.7	100.1	99.1	99.7	100.0	99.1	99.7
26, 27, 3370	ELECTRICAL, COMMUNICATIONS & UTIL.	96.8	54.4	74.4	95.6	74.9	84.7	95.3	99.2	97.4	92.4	94.8	93.7	104.9	98.5	101.5	104.4	99.1	101.6
MF2014	WEIGHTED AVERAGE	101.5	65.1	85.6	99.7	61.8	83.2	100.0	82.2	92.2	101.5	80.2	92.2	101.6	99.6	100.7	101.1	99.4	100.4

City Cost Indexes

		\multicolumn{18}{c}{WASHINGTON}											
\multicolumn{2}{c}{DIVISION}	\multicolumn{3}{c}{RICHLAND 993}	\multicolumn{3}{c}{SEATTLE 980-981,987}	\multicolumn{3}{c}{SPOKANE 990-992}	\multicolumn{3}{c}{TACOMA 983-984}	\multicolumn{3}{c}{VANCOUVER 986}	\multicolumn{3}{c}{WENATCHEE 988}
		MAT.	INST.	TOTAL	MAT.	INST.	TOTAL	MAT.	INST.	TOTAL	MAT.	INST.	TOTAL	MAT.	INST.	TOTAL	MAT.	INST.	TOTAL
015433	CONTRACTOR EQUIPMENT		91.2	91.2		101.6	101.6		91.2	91.2		101.6	101.6		96.7	96.7		101.6	101.6
0241, 31 - 34	SITE & INFRASTRUCTURE, DEMOLITION	101.8	91.0	94.1	95.1	106.7	103.4	101.1	91.0	93.9	94.5	108.4	104.4	104.2	96.1	98.4	103.2	107.1	106.0
0310	Concrete Forming & Accessories	112.8	78.5	83.2	108.3	100.9	102.0	118.2	78.1	83.6	104.2	100.4	100.9	104.9	90.5	92.5	105.8	77.6	81.5
0320	Concrete Reinforcing	107.4	87.4	97.2	105.1	98.9	101.9	108.2	87.3	97.6	103.1	98.8	100.9	103.9	98.4	101.1	103.9	87.7	95.6
0330	Cast-in-Place Concrete	96.5	84.4	91.5	100.9	106.4	103.2	100.3	84.2	93.7	106.2	106.3	118.8	98.4	110.4	108.6	77.3	95.7
03	CONCRETE	106.8	82.2	94.7	99.6	102.1	100.8	109.1	82.0	95.8	100.4	101.6	101.0	110.7	94.6	102.8	108.4	79.6	94.2
04	MASONRY	104.3	84.3	91.8	108.2	100.9	103.6	104.9	84.3	92.0	106.2	100.9	102.9	106.7	95.6	99.8	108.6	91.4	97.9
05	METALS	88.7	84.9	87.5	102.5	94.2	99.9	90.9	84.5	89.0	103.6	92.3	100.1	101.1	91.8	98.2	101.0	86.2	96.4
06	WOOD, PLASTICS & COMPOSITES	113.4	76.1	92.5	104.9	100.0	102.2	122.4	76.1	96.5	101.0	100.0	100.4	92.9	90.2	91.4	102.5	75.8	87.5
07	THERMAL & MOISTURE PROTECTION	145.0	78.8	117.8	102.1	101.8	102.0	141.4	79.5	116.0	103.5	98.5	101.4	104.0	91.6	98.9	104.1	81.5	94.8
08	OPENINGS	121.4	71.3	109.7	107.1	99.9	100.0	121.9	71.8	110.3	102.0	99.9	101.5	98.3	92.8	97.0	101.5	71.0	94.4
0920	Plaster & Gypsum Board	145.7	75.2	98.1	103.6	99.9	101.1	137.7	75.2	95.5	107.8	99.9	102.5	105.7	90.1	95.1	110.2	74.9	86.4
0950, 0980	Ceilings & Acoustic Treatment	104.2	75.2	85.1	110.2	99.9	103.5	99.5	75.2	83.5	103.1	99.9	101.0	100.2	90.1	93.5	95.4	74.9	82.0
0960	Flooring	92.0	80.1	88.6	108.3	106.8	107.8	91.2	85.3	89.5	107.2	91.7	102.8	113.5	102.2	110.2	109.9	62.6	96.4
0970, 0990	Wall Finishes & Painting/Coating	100.0	67.6	80.4	96.3	91.5	93.4	100.2	70.9	82.5	95.4	91.5	93.0	97.6	70.8	81.4	95.4	72.6	81.6
09	FINISHES	112.0	76.7	92.6	105.0	100.6	102.6	109.8	78.1	92.3	103.4	97.6	100.2	101.5	90.6	95.5	103.8	73.5	87.1
COVERS	DIVS. 10 - 14, 25, 28, 41, 43, 44, 46	100.0	85.4	97.0	100.0	100.5	100.1	100.0	85.3	97.0	100.0	100.5	100.1	100.0	60.7	92.1	100.0	77.2	95.4
21, 22, 23	FIRE SUPPRESSION, PLUMBING & HVAC	100.5	107.6	103.3	100.0	115.1	106.1	100.4	84.3	93.9	100.2	99.2	99.8	100.3	87.0	94.9	95.4	86.7	91.9
26, 27, 3370	ELECTRICAL, COMMUNICATIONS & UTIL.	89.7	94.8	92.4	102.6	107.0	104.9	88.0	79.0	83.3	104.7	99.1	101.7	109.7	101.0	105.1	105.5	98.4	101.8
MF2014	WEIGHTED AVERAGE	103.1	89.3	97.1	101.4	104.7	102.8	103.4	82.3	94.2	101.8	99.7	100.9	102.7	92.1	98.1	101.6	86.6	95.0

		WASHINGTON			\multicolumn{15}{c}{WEST VIRGINIA}														
\multicolumn{2}{c}{DIVISION}	\multicolumn{3}{c}{YAKIMA 989}	\multicolumn{3}{c}{BECKLEY 258-259}	\multicolumn{3}{c}{BLUEFIELD 247-248}	\multicolumn{3}{c}{BUCKHANNON 262}	\multicolumn{3}{c}{CHARLESTON 250-253}	\multicolumn{3}{c}{CLARKSBURG 263-264}
		MAT.	INST.	TOTAL	MAT.	INST.	TOTAL	MAT.	INST.	TOTAL	MAT.	INST.	TOTAL	MAT.	INST.	TOTAL	MAT.	INST.	TOTAL
015433	CONTRACTOR EQUIPMENT		101.6	101.6		101.7	101.7		101.7	101.7		101.7	101.7		101.7	101.7		101.7	101.7
0241, 31 - 34	SITE & INFRASTRUCTURE, DEMOLITION	96.8	107.8	104.7	100.5	89.9	92.9	100.6	89.9	92.9	106.8	89.9	94.8	101.5	90.7	93.8	107.4	89.9	95.0
0310	Concrete Forming & Accessories	104.6	95.8	97.0	83.5	93.6	92.2	87.0	93.6	92.7	86.4	94.3	93.2	96.7	94.1	94.4	83.8	94.2	92.8
0320	Concrete Reinforcing	103.5	87.4	95.3	92.7	89.2	90.8	92.0	84.1	88.0	92.6	84.2	88.3	101.0	89.1	94.9	92.6	84.1	88.3
0330	Cast-in-Place Concrete	113.6	83.9	101.4	98.1	103.0	100.1	99.6	102.9	101.0	99.3	101.8	100.3	96.8	103.0	99.3	108.8	100.0	105.2
03	CONCRETE	105.3	89.9	97.7	96.4	97.0	97.4	95.5	96.5	100.4	95.4	98.0	99.1	96.7	97.9	104.4	94.8	99.7
04	MASONRY	99.8	83.1	89.4	98.9	98.9	98.9	93.3	98.9	96.8	104.5	98.9	101.0	95.8	97.9	97.1	108.4	98.9	102.4
05	METALS	101.6	86.3	96.9	100.9	98.1	100.0	103.4	96.3	101.2	103.6	97.1	101.6	99.6	98.9	99.4	103.6	96.9	101.5
06	WOOD, PLASTICS & COMPOSITES	101.4	100.0	100.6	83.5	93.4	89.1	85.6	93.4	90.0	84.9	93.4	89.7	98.8	93.4	95.6	81.5	93.4	88.2
07	THERMAL & MOISTURE PROTECTION	103.6	82.7	95.0	102.9	90.6	97.9	104.0	90.6	98.5	104.2	91.7	99.1	99.4	90.4	95.7	104.1	90.6	98.6
08	OPENINGS	101.5	81.0	96.7	101.3	85.8	97.7	102.5	84.6	98.3	102.5	84.6	98.4	103.6	85.8	99.4	102.5	85.1	98.5
0920	Plaster & Gypsum Board	107.4	99.9	102.4	97.0	93.0	94.3	98.9	93.0	94.9	99.3	93.0	95.1	100.3	93.0	95.4	97.2	93.0	94.4
0950, 0980	Ceilings & Acoustic Treatment	97.3	99.9	99.0	83.9	93.0	89.9	87.3	93.0	91.1	88.9	93.0	91.6	92.2	93.0	92.7	88.9	93.0	91.6
0960	Flooring	108.1	58.0	93.8	102.4	118.0	106.8	98.9	118.0	104.3	98.6	118.0	104.1	106.4	118.0	109.7	97.6	118.0	103.4
0970, 0990	Wall Finishes & Painting/Coating	95.4	70.9	80.6	107.4	91.5	97.8	109.6	94.6	100.6	109.6	94.3	100.3	106.9	95.9	100.2	109.6	94.3	100.3
09	FINISHES	102.5	86.6	93.7	94.7	98.6	96.9	94.8	98.9	97.1	95.7	98.9	97.5	99.1	98.8	99.0	95.0	98.9	97.2
COVERS	DIVS. 10 - 14, 25, 28, 41, 43, 44, 46	100.0	97.6	99.5	100.0	56.9	91.3	100.0	56.9	91.3	100.0	101.5	100.3	100.0	101.1	100.2	100.0	101.5	100.3
21, 22, 23	FIRE SUPPRESSION, PLUMBING & HVAC	100.2	107.2	103.0	95.5	93.5	94.7	95.6	92.8	94.4	95.6	95.1	95.4	100.0	93.4	97.4	95.6	95.0	95.3
26, 27, 3370	ELECTRICAL, COMMUNICATIONS & UTIL.	107.8	94.9	101.0	93.9	91.1	92.4	95.7	91.1	93.3	97.1	96.1	96.6	99.1	91.1	94.8	97.1	96.1	96.6
MF2014	WEIGHTED AVERAGE	102.0	94.1	98.6	97.9	93.3	95.9	98.3	92.9	96.0	99.6	95.5	97.8	99.9	94.8	97.6	100.1	95.4	98.1

		\multicolumn{18}{c}{WEST VIRGINIA}											
\multicolumn{2}{c}{DIVISION}	\multicolumn{3}{c}{GASSAWAY 266}	\multicolumn{3}{c}{HUNTINGTON 255-257}	\multicolumn{3}{c}{LEWISBURG 249}	\multicolumn{3}{c}{MARTINSBURG 254}	\multicolumn{3}{c}{MORGANTOWN 265}	\multicolumn{3}{c}{PARKERSBURG 261}
		MAT.	INST.	TOTAL	MAT.	INST.	TOTAL	MAT.	INST.	TOTAL	MAT.	INST.	TOTAL	MAT.	INST.	TOTAL	MAT.	INST.	TOTAL
015433	CONTRACTOR EQUIPMENT		101.7	101.7		101.7	101.7		101.7	101.7		101.7	101.7		101.7	101.7		101.7	101.7
0241, 31 - 34	SITE & INFRASTRUCTURE, DEMOLITION	104.2	89.9	94.0	105.1	90.9	95.0	116.3	89.8	97.5	104.2	90.2	94.3	101.6	90.5	93.7	109.8	90.6	96.2
0310	Concrete Forming & Accessories	85.8	92.2	91.4	95.3	95.9	95.8	84.2	93.5	92.2	83.6	82.8	82.9	84.2	93.9	92.6	88.6	90.3	90.1
0320	Concrete Reinforcing	92.6	84.0	88.3	94.1	92.9	93.5	92.6	83.9	88.2	92.7	80.6	86.6	92.6	84.1	88.3	92.0	83.9	87.9
0330	Cast-in-Place Concrete	104.0	103.5	103.7	106.9	102.5	105.1	99.7	102.9	101.0	102.8	96.7	100.3	99.3	101.6	100.2	101.5	96.4	99.4
03	CONCRETE	100.8	95.1	98.0	102.8	98.0	100.4	107.1	95.4	101.4	101.2	87.9	94.7	97.2	95.2	96.2	102.5	91.8	97.2
04	MASONRY	109.2	95.9	100.9	96.4	100.8	99.1	95.9	95.9	95.9	99.7	89.5	93.3	126.5	98.9	109.3	82.9	91.7	88.4
05	METALS	103.5	96.8	101.4	103.3	100.4	102.4	103.5	96.0	101.2	101.3	94.2	99.1	103.6	96.6	101.5	104.2	96.6	101.9
06	WOOD, PLASTICS & COMPOSITES	84.0	90.7	87.8	94.7	95.3	95.1	81.8	93.4	88.3	83.5	81.6	82.5	81.8	93.4	88.3	85.0	89.1	87.3
07	THERMAL & MOISTURE PROTECTION	104.0	90.6	98.5	103.0	90.8	98.0	104.0	89.6	98.1	103.1	79.3	93.4	104.0	90.6	98.5	103.9	89.0	97.8
08	OPENINGS	100.7	83.2	96.6	100.5	87.6	97.5	102.5	84.6	98.4	103.3	73.4	96.3	103.8	85.1	99.4	101.4	82.3	96.9
0920	Plaster & Gypsum Board	98.6	90.2	92.9	104.8	95.0	98.1	97.2	93.0	94.4	97.6	80.9	86.3	97.7	93.0	94.4	99.7	88.5	92.1
0950, 0980	Ceilings & Acoustic Treatment	88.9	90.2	89.8	86.4	95.0	92.0	88.9	93.0	91.6	86.4	80.9	82.8	88.9	93.0	91.6	88.9	88.5	88.7
0960	Flooring	98.4	118.0	104.0	109.8	121.7	113.2	97.7	118.0	103.5	102.4	118.0	106.8	97.7	118.0	103.5	101.5	118.0	106.2
0970, 0990	Wall Finishes & Painting/Coating	109.6	95.9	101.3	107.4	94.6	99.7	109.6	75.4	88.9	107.4	41.4	67.5	109.6	94.3	100.3	109.6	94.6	100.6
09	FINISHES	95.2	97.5	96.5	98.5	100.8	99.8	96.1	96.8	96.5	95.6	83.1	88.7	94.6	98.9	97.0	96.6	95.9	96.2
COVERS	DIVS. 10 - 14, 25, 28, 41, 43, 44, 46	100.0	101.2	100.2	100.0	101.8	100.4	100.0	56.9	91.3	100.0	97.7	99.5	100.0	65.8	93.1	100.0	100.5	100.1
21, 22, 23	FIRE SUPPRESSION, PLUMBING & HVAC	95.6	93.3	94.7	100.0	95.7	98.3	95.6	93.5	94.7	95.5	86.0	91.7	95.6	95.0	95.3	100.3	88.3	95.5
26, 27, 3370	ELECTRICAL, COMMUNICATIONS & UTIL.	97.1	91.1	93.9	97.8	95.3	96.5	93.0	91.1	92.0	100.1	81.8	90.4	97.3	96.1	96.7	97.2	92.7	94.8
MF2014	WEIGHTED AVERAGE	99.5	93.9	97.1	100.7	95.7	98.5	99.8	92.4	96.6	99.3	86.3	93.7	100.2	94.4	97.6	100.1	91.8	96.5

City Cost Indexes

DIVISION		WEST VIRGINIA									WISCONSIN								
		PETERSBURG 268			ROMNEY 267			WHEELING 260			BELOIT 535			EAU CLAIRE 547			GREEN BAY 541 - 543		
		MAT.	INST.	TOTAL	MAT.	INST.	TOTAL	MAT.	INST.	TOTAL	MAT.	INST.	TOTAL	MAT.	INST.	TOTAL	MAT.	INST.	TOTAL
015433	CONTRACTOR EQUIPMENT		101.7	101.7		101.7	101.7		101.7	101.7		102.3	102.3		100.4	100.4		98.2	98.2
0241, 31 - 34	SITE & INFRASTRUCTURE, DEMOLITION	100.9	90.5	93.5	103.6	90.5	94.3	110.6	90.4	96.2	94.7	108.3	104.3	96.1	102.9	101.0	99.7	98.9	99.2
0310	Concrete Forming & Accessories	87.4	91.6	91.0	83.4	91.8	90.6	90.3	93.2	92.8	100.2	100.3	100.3	97.1	99.4	99.1	105.5	98.7	99.7
0320	Concrete Reinforcing	92.0	82.5	87.2	92.6	81.0	86.7	91.4	84.0	87.7	92.3	116.0	104.3	90.4	99.2	94.9	88.7	92.4	90.6
0330	Cast-in-Place Concrete	99.3	99.1	99.2	104.0	99.1	102.0	101.5	100.1	100.9	100.6	100.9	100.7	104.0	99.2	102.0	107.6	99.7	104.3
03	CONCRETE	97.3	93.0	95.2	100.6	92.8	96.8	102.5	94.4	98.5	98.8	103.5	101.1	99.7	99.4	99.6	102.8	98.0	100.5
04	MASONRY	99.5	98.9	99.1	96.2	98.9	97.9	107.6	93.9	99.1	100.6	104.3	102.9	94.7	102.1	99.3	126.7	101.4	110.9
05	METALS	103.7	95.8	101.2	103.7	95.5	101.2	104.4	96.8	102.1	93.1	107.7	97.6	95.0	101.4	97.0	97.4	98.2	97.6
06	WOOD, PLASTICS & COMPOSITES	85.9	90.7	88.6	80.9	90.7	86.4	86.7	93.4	90.5	99.9	99.2	99.5	105.1	99.3	101.9	110.0	99.3	104.0
07	THERMAL & MOISTURE PROTECTION	104.0	86.0	96.6	104.1	83.4	95.6	104.3	90.0	98.4	102.1	96.8	99.9	102.4	86.7	95.9	104.3	86.0	96.8
08	OPENINGS	103.8	82.8	98.9	103.7	82.4	98.8	102.2	85.1	98.2	102.5	108.8	103.9	103.7	96.2	102.0	100.1	96.0	99.2
0920	Plaster & Gypsum Board	99.3	90.2	93.2	96.8	90.2	92.4	99.7	93.0	95.2	94.5	99.6	97.9	102.9	99.6	100.6	98.7	99.3	99.3
0950, 0980	Ceilings & Acoustic Treatment	88.9	90.2	89.8	88.9	90.2	89.8	88.9	93.0	91.6	88.1	99.6	95.6	90.4	99.6	96.4	81.9	99.6	93.5
0960	Flooring	99.4	118.0	104.7	97.5	118.0	103.4	102.3	118.0	106.8	99.8	118.1	105.1	91.4	114.1	97.9	110.2	114.1	111.3
0970, 0990	Wall Finishes & Painting/Coating	109.6	94.3	100.3	109.6	94.3	100.3	109.6	94.3	100.3	98.3	97.2	97.7	94.5	86.4	89.6	105.4	83.2	92.0
09	FINISHES	95.4	97.3	96.4	94.7	97.3	96.1	96.9	98.2	97.6	97.4	103.1	100.5	94.6	101.1	98.3	99.0	100.7	99.9
COVERS	DIVS. 10 - 14, 25, 28, 41, 43, 44, 46	100.0	65.5	93.0	100.0	101.2	100.2	100.0	95.4	99.1	100.0	96.6	99.3	100.0	96.8	99.3	100.0	96.5	99.3
21, 22, 23	FIRE SUPPRESSION, PLUMBING & HVAC	95.6	93.7	94.8	95.6	93.4	94.7	100.3	93.6	97.6	100.1	98.6	99.5	100.1	87.6	95.0	100.3	85.1	94.2
26, 27, 3370	ELECTRICAL, COMMUNICATIONS & UTIL.	100.5	81.8	90.6	99.8	81.8	90.3	94.4	96.1	95.3	101.0	86.7	93.4	103.8	84.9	93.8	98.2	82.0	89.6
MF2014	WEIGHTED AVERAGE	99.3	91.3	95.8	99.4	92.2	96.3	101.2	94.3	98.2	98.9	100.7	99.7	99.3	95.3	97.5	101.2	93.4	97.8

DIVISION		WISCONSIN																	
		KENOSHA 531			LA CROSSE 546			LANCASTER 538			MADISON 537			MILWAUKEE 530,532			NEW RICHMOND 540		
		MAT.	INST.	TOTAL	MAT.	INST.	TOTAL	MAT.	INST.	TOTAL	MAT.	INST.	TOTAL	MAT.	INST.	TOTAL	MAT.	INST.	TOTAL
015433	CONTRACTOR EQUIPMENT		100.2	100.2		100.4	100.4		102.3	102.3		102.3	102.3		90.5	90.5		100.8	100.8
0241, 31 - 34	SITE & INFRASTRUCTURE, DEMOLITION	100.3	105.2	103.8	90.1	102.8	99.1	93.9	108.2	104.1	92.4	108.2	103.7	94.6	98.5	97.4	94.4	103.3	100.8
0310	Concrete Forming & Accessories	108.0	100.5	101.6	84.1	99.1	97.1	99.5	97.9	98.2	102.0	100.1	100.4	104.1	118.0	116.1	92.4	100.9	99.7
0320	Concrete Reinforcing	92.1	97.9	95.0	90.1	91.5	90.9	93.5	91.3	92.3	97.9	91.6	94.7	95.7	98.2	97.0	87.8	99.0	93.5
0330	Cast-in-Place Concrete	109.4	107.3	108.5	93.5	95.9	94.5	100.0	100.9	100.4	91.1	100.8	95.1	99.8	109.6	103.8	108.3	82.5	97.7
03	CONCRETE	103.5	102.4	103.0	90.8	96.8	93.7	98.5	97.8	98.1	95.7	98.8	97.2	99.7	110.5	105.0	95.5	94.3	95.9
04	MASONRY	97.9	114.1	108.0	93.8	105.9	101.4	100.7	104.3	102.9	101.5	104.3	103.2	104.9	120.6	114.7	121.6	102.3	109.6
05	METALS	93.9	101.2	96.2	94.9	98.2	95.9	90.6	95.0	92.0	91.9	97.2	93.6	97.5	94.8	96.7	95.1	99.7	96.5
06	WOOD, PLASTICS & COMPOSITES	105.0	96.4	100.2	90.1	99.3	95.3	99.1	99.2	99.2	96.7	99.2	98.1	104.7	118.5	112.5	94.0	102.5	98.7
07	THERMAL & MOISTURE PROTECTION	102.2	108.7	104.9	101.8	90.2	97.1	101.9	82.5	93.9	102.8	105.2	103.8	100.8	114.0	106.2	103.6	92.2	98.9
08	OPENINGS	96.9	101.1	97.9	103.7	85.0	99.4	97.9	85.8	95.1	107.9	100.5	106.2	101.0	113.2	103.8	89.1	97.5	91.1
0920	Plaster & Gypsum Board	85.3	96.7	93.0	97.7	99.6	99.0	93.3	99.6	97.5	99.6	99.6	99.6	101.0	119.3	113.4	87.7	103.0	98.1
0950, 0980	Ceilings & Acoustic Treatment	88.1	96.7	93.7	89.5	99.6	96.1	83.1	99.6	93.9	94.0	99.6	97.6	95.9	119.3	111.3	58.5	103.0	87.8
0960	Flooring	118.8	110.4	116.4	85.6	114.1	93.7	99.5	110.5	102.7	103.2	110.5	105.3	98.2	114.0	102.7	103.2	114.1	106.3
0970, 0990	Wall Finishes & Painting/Coating	108.3	115.5	112.6	94.5	77.9	84.5	98.3	66.4	79.1	102.4	97.2	99.3	94.4	122.9	111.6	108.3	86.4	95.1
09	FINISHES	102.5	103.6	103.1	91.6	100.3	96.4	96.0	96.5	96.3	98.6	101.9	100.4	100.9	118.6	110.7	90.0	101.3	96.2
COVERS	DIVS. 10 - 14, 25, 28, 41, 43, 44, 46	100.0	101.1	100.2	100.0	96.7	99.3	100.0	49.2	89.8	100.0	99.7	99.9	100.0	104.6	100.9	100.0	96.1	99.2
21, 22, 23	FIRE SUPPRESSION, PLUMBING & HVAC	100.3	97.0	98.9	100.1	87.5	95.0	95.3	87.3	92.1	100.0	95.6	98.2	100.0	103.5	101.4	94.8	86.9	91.6
26, 27, 3370	ELECTRICAL, COMMUNICATIONS & UTIL.	101.5	98.7	100.1	104.2	85.1	94.1	100.8	85.1	92.5	102.1	92.1	96.8	100.8	99.5	100.1	101.9	85.1	93.0
MF2014	WEIGHTED AVERAGE	99.5	102.2	100.7	97.8	94.5	96.3	96.7	92.5	94.8	99.1	99.0	99.1	100.0	107.2	103.1	96.8	94.6	95.9

DIVISION		WISCONSIN																	
		OSHKOSH 549			PORTAGE 539			RACINE 534			RHINELANDER 545			SUPERIOR 548			WAUSAU 544		
		MAT.	INST.	TOTAL	MAT.	INST.	TOTAL	MAT.	INST.	TOTAL	MAT.	INST.	TOTAL	MAT.	INST.	TOTAL	MAT.	INST.	TOTAL
015433	CONTRACTOR EQUIPMENT		98.2	98.2		102.3	102.3		102.3	102.3		98.2	98.2		100.8	100.8		98.2	98.2
0241, 31 - 34	SITE & INFRASTRUCTURE, DEMOLITION	91.5	98.9	96.8	85.0	108.0	101.4	94.5	109.0	104.8	103.3	98.8	100.1	91.3	103.3	99.9	87.4	98.9	95.6
0310	Concrete Forming & Accessories	88.6	98.1	96.8	91.7	98.9	97.9	100.5	100.5	100.5	86.2	97.8	96.2	90.5	91.9	91.7	88.1	98.7	97.2
0320	Concrete Reinforcing	88.8	92.3	90.6	93.5	91.4	92.5	92.3	97.9	95.1	89.0	90.9	90.0	87.8	91.4	89.6	89.0	91.3	90.2
0330	Cast-in-Place Concrete	99.7	99.4	99.6	85.9	101.8	92.4	98.7	107.0	102.1	113.1	99.3	107.4	102.0	97.9	100.3	93.0	99.6	95.7
03	CONCRETE	92.4	97.6	94.9	86.4	98.5	92.4	97.9	102.3	100.1	104.7	97.2	101.0	91.8	94.2	93.0	87.3	97.8	92.5
04	MASONRY	107.9	101.4	103.8	99.6	104.3	102.5	100.6	114.0	109.0	125.3	101.4	110.4	120.8	104.5	110.6	107.4	101.4	103.6
05	METALS	95.4	96.7	95.8	91.3	96.0	92.7	94.8	101.2	96.8	95.3	96.7	95.7	96.1	98.2	96.8	95.1	97.7	95.9
06	WOOD, PLASTICS & COMPOSITES	89.8	99.3	95.1	89.2	99.2	94.8	100.2	96.4	98.1	87.3	99.3	94.0	92.3	90.1	91.0	89.2	99.3	94.9
07	THERMAL & MOISTURE PROTECTION	103.4	85.7	96.1	101.4	92.6	97.7	102.1	108.3	104.7	104.2	83.5	95.7	103.3	92.2	98.7	103.3	84.4	95.5
08	OPENINGS	95.9	95.5	95.8	98.1	92.3	96.8	102.5	101.1	102.2	96.0	87.1	93.9	88.6	91.1	89.2	96.1	95.7	96.0
0920	Plaster & Gypsum Board	86.2	99.6	95.2	86.4	99.6	95.3	94.5	96.7	96.0	86.2	99.6	95.2	87.6	90.2	89.4	86.2	99.6	95.2
0950, 0980	Ceilings & Acoustic Treatment	81.9	99.6	93.5	85.6	99.6	94.8	88.1	96.7	93.7	81.9	99.6	93.5	59.3	90.2	79.6	81.9	99.6	93.5
0960	Flooring	101.6	114.1	105.2	96.2	118.1	102.5	99.8	110.4	102.8	100.9	114.1	104.7	104.4	122.4	109.6	101.5	114.1	105.1
0970, 0990	Wall Finishes & Painting/Coating	102.2	101.4	101.7	98.3	66.4	79.1	98.3	115.5	108.7	102.2	60.4	76.9	96.0	102.8	100.1	102.2	83.2	90.7
09	FINISHES	94.0	100.9	97.8	94.1	98.6	96.6	97.4	103.6	100.8	94.8	97.0	96.0	89.3	98.6	94.4	93.7	100.7	97.5
COVERS	DIVS. 10 - 14, 25, 28, 41, 43, 44, 46	100.0	64.6	92.8	100.0	63.4	92.6	100.0	101.1	100.2	100.0	55.0	90.9	100.0	94.1	98.8	100.0	96.5	99.3
21, 22, 23	FIRE SUPPRESSION, PLUMBING & HVAC	95.6	85.7	91.5	95.3	95.3	95.3	100.1	97.0	98.8	95.6	86.7	92.0	94.8	89.4	92.6	95.6	86.4	91.9
26, 27, 3370	ELECTRICAL, COMMUNICATIONS & UTIL.	102.5	80.4	90.9	104.8	91.5	98.1	100.8	98.6	99.6	101.8	81.8	91.3	107.1	99.0	102.8	103.7	81.8	92.1
MF2014	WEIGHTED AVERAGE	96.9	91.9	94.7	95.3	96.3	95.9	99.1	102.4	100.5	99.3	91.2	95.8	96.6	96.3	96.5	96.2	93.5	95.0

City Cost Indexes

DIVISION		WYOMING																	
		CASPER			CHEYENNE			NEWCASTLE			RAWLINS			RIVERTON			ROCK SPRINGS		
		826			820			827			823			825			829 - 831		
		MAT.	INST.	TOTAL	MAT.	INST.	TOTAL	MAT.	INST.	TOTAL	MAT.	INST.	TOTAL	MAT.	INST.	TOTAL	MAT.	INST.	TOTAL
015433	CONTRACTOR EQUIPMENT		98.2	98.2		98.2	98.2		98.2	98.2		98.2	98.2		98.2	98.2		98.2	98.2
0241, 31 - 34	SITE & INFRASTRUCTURE, DEMOLITION	98.6	94.4	95.6	92.5	94.4	93.8	85.1	94.0	91.5	98.4	94.0	95.3	92.1	94.0	93.5	89.2	94.0	92.6
0310	Concrete Forming & Accessories	102.5	50.1	57.3	104.4	53.5	60.5	94.8	62.2	66.7	99.2	62.2	67.3	93.7	62.1	66.5	101.5	62.1	67.5
0320	Concrete Reinforcing	105.7	70.6	87.8	98.0	70.6	84.1	105.5	70.8	87.9	105.2	70.8	87.7	106.1	70.6	88.0	106.1	70.6	88.0
0330	Cast-in-Place Concrete	102.8	72.0	90.2	95.0	72.1	85.6	95.9	61.6	81.8	96.0	61.6	81.9	96.0	60.0	81.2	95.9	59.9	81.1
03	CONCRETE	101.7	62.1	82.3	98.6	63.7	81.4	99.0	63.9	81.8	111.8	63.9	88.3	106.7	63.3	85.4	99.5	63.2	81.7
04	MASONRY	108.2	51.9	73.1	103.7	51.9	71.4	100.3	54.8	71.9	100.3	54.8	71.9	100.3	50.5	69.3	160.4	50.4	91.8
05	METALS	99.3	72.1	90.9	101.7	72.1	92.6	97.8	72.0	89.9	97.9	72.0	89.9	98.0	71.6	89.8	98.7	71.1	90.2
06	WOOD, PLASTICS & COMPOSITES	97.2	49.7	70.6	96.5	54.3	72.9	86.4	66.9	75.5	90.4	66.9	77.2	85.3	66.9	75.0	95.4	66.9	79.4
07	THERMAL & MOISTURE PROTECTION	105.9	55.1	85.1	100.1	55.6	81.8	101.8	55.8	82.9	103.2	55.8	83.7	102.7	59.9	85.1	101.9	59.9	84.7
08	OPENINGS	108.0	53.3	95.3	108.3	55.8	96.1	112.6	61.7	100.8	112.3	61.7	100.5	112.5	61.7	100.7	113.0	61.3	101.0
0920	Plaster & Gypsum Board	97.4	48.1	64.1	89.3	52.9	64.7	85.2	65.8	72.1	85.5	65.8	72.2	85.2	65.8	72.1	97.3	65.8	76.0
0950, 0980	Ceilings & Acoustic Treatment	99.2	48.1	65.6	91.0	65.8	66.0	92.9	65.8	75.1	92.9	65.8	75.1	92.9	65.8	75.1	92.9	65.8	75.1
0960	Flooring	112.0	56.4	96.1	109.9	56.4	94.6	103.5	56.4	90.0	106.2	56.4	92.0	103.0	56.4	89.7	108.4	56.4	93.5
0970, 0990	Wall Finishes & Painting/Coating	104.7	43.1	67.5	111.2	43.1	70.0	107.1	55.4	75.9	107.1	55.4	75.9	107.1	55.4	75.9	107.1	55.4	75.9
09	FINISHES	103.2	49.6	73.6	100.0	52.3	73.7	95.0	60.8	76.1	97.2	60.8	77.1	95.7	60.8	76.5	98.1	60.8	77.5
COVERS	DIVS. 10 - 14, 25, 28, 41, 43, 44, 46	100.0	87.6	97.5	100.0	88.2	97.6	100.0	89.1	97.8	100.0	89.1	97.8	100.0	81.4	96.2	100.0	80.8	96.1
21, 22, 23	FIRE SUPPRESSION, PLUMBING & HVAC	99.9	64.1	85.4	99.9	64.1	85.4	97.4	63.4	83.7	97.4	63.4	83.7	97.4	63.4	83.7	99.8	63.3	85.1
26, 27, 3370	ELECTRICAL, COMMUNICATIONS & UTIL.	103.3	64.1	82.6	105.8	68.6	86.1	104.5	69.0	85.7	104.5	69.0	85.7	104.5	62.9	82.5	102.4	62.9	81.5
MF2014	WEIGHTED AVERAGE	102.0	63.9	85.4	101.5	65.2	85.7	99.9	66.9	85.5	101.8	66.9	86.6	101.0	65.4	85.5	103.7	65.3	87.0

DIVISION		WYOMING												CANADA					
		SHERIDAN			WHEATLAND			WORLAND			YELLOWSTONE NAT'L PA			BARRIE, ONTARIO			BATHURST, NEW BRUNSWICK		
		828			822			824			821								
		MAT.	INST.	TOTAL	MAT.	INST.	TOTAL	MAT.	INST.	TOTAL	MAT.	INST.	TOTAL	MAT.	INST.	TOTAL	MAT.	INST.	TOTAL
015433	CONTRACTOR EQUIPMENT		98.2	98.2		98.2	98.2		98.2	98.2		98.2	98.2		102.1	102.1		102.2	102.2
0241, 31 - 34	SITE & INFRASTRUCTURE, DEMOLITION	92.4	94.4	93.8	89.6	94.0	92.8	87.1	94.0	92.0	87.2	94.2	92.2	118.5	100.4	105.6	103.4	96.6	98.6
0310	Concrete Forming & Accessories	102.2	56.0	57.7	76.8	52.7	58.8	96.9	62.0	66.8	96.9	63.1	67.8	124.4	89.4	94.2	124.4	63.7	69.3
0320	Concrete Reinforcing	106.1	70.9	88.2	105.5	70.8	87.8	106.1	70.6	88.0	107.9	70.6	88.9	173.2	83.8	127.7	137.0	56.5	96.0
0330	Cast-in-Place Concrete	99.2	72.0	88.1	100.2	61.5	84.3	95.9	59.9	81.1	95.9	61.6	81.8	165.6	89.0	134.2	131.6	61.9	103.0
03	CONCRETE	106.9	62.4	85.0	103.7	59.7	82.1	99.2	63.2	81.6	99.5	64.3	82.2	150.3	88.4	119.9	124.6	62.4	94.1
04	MASONRY	100.6	56.3	73.0	100.7	54.7	72.0	100.3	50.4	69.2	100.4	53.4	71.1	171.5	98.0	125.7	160.8	64.9	101.1
05	METALS	101.6	72.4	92.6	97.8	71.7	89.8	98.0	71.5	89.9	98.6	71.5	90.3	110.2	90.2	104.0	109.0	71.5	97.5
06	WOOD, PLASTICS & COMPOSITES	97.9	50.4	71.3	88.3	54.3	69.2	88.3	66.9	76.3	88.3	66.9	76.3	119.8	88.2	102.1	101.8	64.0	80.7
07	THERMAL & MOISTURE PROTECTION	102.9	56.3	83.8	102.1	55.8	83.1	101.9	57.5	83.7	101.4	58.8	83.9	111.8	89.1	102.5	106.0	62.2	88.1
08	OPENINGS	113.2	52.7	99.1	111.1	55.8	98.3	112.8	61.7	100.9	105.6	61.7	95.4	93.9	86.5	92.2	87.9	55.9	80.5
0920	Plaster & Gypsum Board	109.4	48.8	68.5	85.2	52.8	63.3	85.2	65.8	72.1	85.4	65.8	72.1	151.1	87.7	108.2	137.6	62.7	87.0
0950, 0980	Ceilings & Acoustic Treatment	95.7	48.8	64.9	92.9	52.8	66.6	92.9	65.8	75.1	93.7	65.8	75.4	91.5	87.7	89.0	106.4	62.7	77.7
0960	Flooring	107.5	56.4	92.9	105.0	56.4	91.1	105.0	56.4	91.1	105.0	56.4	91.1	127.2	98.9	119.1	107.9	47.2	90.6
0970, 0990	Wall Finishes & Painting/Coating	109.4	43.1	69.4	107.1	34.4	63.3	107.1	55.4	75.9	107.1	55.4	75.9	116.2	89.9	100.3	117.6	51.6	77.8
09	FINISHES	102.8	50.0	73.7	95.8	51.0	71.1	95.5	60.8	76.3	95.7	61.5	76.8	115.0	91.5	102.1	111.3	59.7	82.8
COVERS	DIVS. 10 - 14, 25, 28, 41, 43, 44, 46	100.0	87.7	97.5	100.0	80.3	96.0	100.0	81.4	96.2	100.0	82.4	96.4	139.2	74.4	126.1	131.1	65.3	117.8
21, 22, 23	FIRE SUPPRESSION, PLUMBING & HVAC	97.4	64.1	83.9	97.4	63.3	83.6	97.4	63.3	83.6	97.4	64.9	84.3	101.8	99.3	100.8	102.0	69.8	89.0
26, 27, 3370	ELECTRICAL, COMMUNICATIONS & UTIL.	107.4	63.0	83.9	104.5	68.6	85.5	104.5	62.9	82.5	103.2	62.9	81.9	116.6	90.5	102.9	112.7	61.8	85.8
MF2014	WEIGHTED AVERAGE	102.6	64.3	85.9	100.4	64.3	84.7	100.0	65.3	84.9	99.3	66.3	84.9	117.7	93.0	106.9	111.5	67.3	92.2

DIVISION		CANADA																	
		BRANDON, MANITOBA			BRANTFORD, ONTARIO			BRIDGEWATER, NOVA SCOTIA			CALGARY, ALBERTA			CAP-DE-LA-MADELEINE, QUEBEC			CHARLESBOURG, QUEBEC		
		MAT.	INST.	TOTAL	MAT.	INST.	TOTAL	MAT.	INST.	TOTAL	MAT.	INST.	TOTAL	MAT.	INST.	TOTAL	MAT.	INST.	TOTAL
015433	CONTRACTOR EQUIPMENT		104.5	104.5		102.1	102.1		101.9	101.9		107.1	107.1		102.8	102.8		102.8	102.8
0241, 31 - 34	SITE & INFRASTRUCTURE, DEMOLITION	133.0	99.3	109.1	118.0	100.7	105.7	102.6	98.2	99.5	124.0	105.4	110.7	98.4	99.9	99.4	98.4	99.9	99.4
0310	Concrete Forming & Accessories	146.5	72.5	82.7	124.9	96.9	100.7	97.1	71.6	75.1	124.4	97.4	101.1	130.7	88.0	93.8	130.7	88.0	93.8
0320	Concrete Reinforcing	185.9	53.2	118.4	165.4	82.5	123.2	140.6	46.7	92.8	135.6	71.9	103.2	140.6	78.8	109.2	140.6	78.8	109.2
0330	Cast-in-Place Concrete	133.4	76.7	110.1	153.6	110.3	135.8	157.8	72.2	122.7	189.0	107.3	155.5	124.0	97.8	113.2	124.0	97.8	113.2
03	CONCRETE	145.6	71.0	109.0	141.8	98.8	120.7	139.4	67.9	104.3	159.1	96.1	128.2	124.3	89.8	107.4	124.3	89.8	107.4
04	MASONRY	233.5	67.6	130.1	168.0	102.3	127.2	163.4	71.9	106.4	196.3	91.4	130.9	163.9	87.1	116.0	163.9	87.1	116.0
05	METALS	125.6	76.8	110.6	109.1	90.9	103.5	108.2	74.2	97.7	137.0	89.3	122.3	107.4	87.6	101.3	107.4	87.6	101.3
06	WOOD, PLASTICS & COMPOSITES	154.2	73.5	109.0	123.7	95.7	108.1	92.9	71.1	80.7	102.3	97.2	99.4	135.3	88.0	108.8	135.3	88.0	108.8
07	THERMAL & MOISTURE PROTECTION	135.0	71.8	109.1	113.8	94.7	106.0	108.6	69.9	92.8	117.8	94.6	108.3	107.4	89.9	100.3	107.4	89.9	100.3
08	OPENINGS	105.5	64.8	96.1	92.4	92.7	92.5	86.1	64.9	81.1	89.9	86.2	89.1	93.6	81.1	90.7	93.6	81.1	90.7
0920	Plaster & Gypsum Board	119.3	72.2	87.4	130.1	95.5	106.7	130.7	70.0	89.7	139.4	96.6	110.5	157.0	87.4	109.9	157.0	87.4	109.9
0950, 0980	Ceilings & Acoustic Treatment	109.6	72.2	85.0	96.5	95.5	95.8	96.5	70.0	79.1	138.8	96.6	111.0	96.5	87.4	90.5	96.5	87.4	90.5
0960	Flooring	138.6	70.2	119.1	120.5	98.9	114.3	102.7	67.2	92.5	122.3	92.6	113.8	120.5	99.5	114.5	120.5	99.5	114.5
0970, 0990	Wall Finishes & Painting/Coating	126.5	58.7	85.6	118.3	98.6	106.4	118.3	64.1	85.6	127.4	112.4	118.3	118.3	91.7	102.2	118.3	91.7	102.2
09	FINISHES	122.5	71.0	94.1	112.4	97.7	104.4	106.6	70.3	86.6	125.5	98.7	110.7	115.3	90.8	101.8	115.3	90.8	101.8
COVERS	DIVS. 10 - 14, 25, 28, 41, 43, 44, 46	131.1	67.6	118.3	131.1	76.5	120.1	131.1	66.6	118.0	131.1	96.9	124.2	131.1	85.2	121.9	131.1	85.2	121.9
21, 22, 23	FIRE SUPPRESSION, PLUMBING & HVAC	102.1	84.1	94.9	102.0	102.3	102.1	102.0	84.4	94.9	101.2	92.9	97.8	102.3	91.0	97.8	102.3	91.0	97.8
26, 27, 3370	ELECTRICAL, COMMUNICATIONS & UTIL.	118.6	69.6	92.7	111.5	89.9	100.1	117.1	64.4	89.2	113.5	101.3	107.1	111.2	72.6	90.8	111.2	72.6	90.8
MF2014	WEIGHTED AVERAGE	125.2	75.8	103.7	114.9	96.9	107.1	112.9	74.6	96.2	123.5	95.7	111.4	112.4	87.6	101.6	112.4	87.6	101.6

City Cost Indexes

CANADA

DIVISION		CHARLOTTETOWN, PRINCE EDWARD ISLAND			CHICOUTIMI, QUEBEC			CORNER BROOK, NEWFOUNDLAND			CORNWALL, ONTARIO			DALHOUSIE, NEW BRUNSWICK			DARTMOUTH, NOVA SCOTIA		
		MAT.	INST.	TOTAL	MAT.	INST.	TOTAL	MAT.	INST.	TOTAL	MAT.	INST.	TOTAL	MAT.	INST.	TOTAL	MAT.	INST.	TOTAL
015433	CONTRACTOR EQUIPMENT		101.8	101.8		102.8	102.8		103.3	103.3		102.1	102.1		102.2	102.2		101.9	101.9
0241, 31 - 34	SITE & INFRASTRUCTURE, DEMOLITION	121.3	95.7	103.1	100.7	99.1	99.6	137.7	97.1	108.9	116.1	100.2	104.8	100.9	96.6	97.8	124.8	98.2	105.9
0310	Concrete Forming & Accessories	110.4	58.2	65.4	133.2	93.1	98.6	121.8	61.7	69.9	122.6	89.6	94.1	103.5	63.6	69.3	111.8	71.6	77.2
0320	Concrete Reinforcing	141.5	46.8	93.3	103.3	95.9	99.5	170.3	48.6	108.4	165.4	82.2	123.1	142.7	56.6	98.9	178.2	46.7	111.3
0330	Cast-in-Place Concrete	166.8	61.2	123.4	124.0	99.8	114.0	155.3	70.6	120.5	138.2	100.6	122.7	131.8	62.0	103.2	149.8	72.2	118.0
03	CONCRETE	147.1	57.9	103.3	114.6	95.9	105.4	182.2	63.2	123.7	134.3	92.2	113.6	132.5	62.5	98.1	162.1	67.9	115.8
04	MASONRY	178.4	62.6	106.3	162.7	95.0	120.5	229.3	64.1	126.4	166.8	94.0	121.4	160.2	64.9	100.8	244.8	71.9	137.1
05	METALS	133.9	67.6	113.5	108.8	92.2	103.7	126.0	72.9	109.7	109.1	89.6	103.1	104.4	71.8	94.3	125.9	74.2	110.0
06	WOOD, PLASTICS & COMPOSITES	97.8	57.8	75.4	137.4	93.6	112.9	128.3	60.8	90.5	122.1	89.0	103.6	100.0	64.0	80.0	116.1	71.1	90.9
07	THERMAL & MOISTURE PROTECTION	121.3	60.4	96.3	106.5	96.4	102.3	139.8	61.8	107.8	113.6	89.3	103.6	110.8	62.2	90.9	136.8	69.9	109.4
08	OPENINGS	88.2	50.0	79.3	92.2	80.6	89.5	112.4	57.1	99.5	93.6	86.2	91.9	89.1	55.9	81.4	95.3	64.9	88.2
0920	Plaster & Gypsum Board	131.6	56.3	80.7	159.4	93.2	114.6	149.0	59.3	88.4	189.4	88.6	121.3	141.2	62.7	88.2	144.6	70.0	94.2
0950, 0980	Ceilings & Acoustic Treatment	120.2	56.3	78.2	105.6	93.2	97.4	110.5	59.3	76.9	99.0	88.6	92.2	98.9	62.7	75.1	117.2	70.0	86.2
0960	Flooring	111.7	62.9	97.8	123.6	99.5	116.8	121.0	57.2	102.8	120.5	97.5	113.9	106.5	72.1	96.7	116.1	67.2	102.1
0970, 0990	Wall Finishes & Painting/Coating	125.6	42.9	75.7	117.6	106.4	110.8	126.4	61.8	87.4	118.3	91.9	102.4	120.7	51.6	79.0	126.4	64.1	88.8
09	FINISHES	117.1	57.8	84.4	118.4	96.5	106.4	122.0	60.9	88.4	121.0	91.6	104.8	110.6	64.5	85.2	120.1	70.3	92.6
COVERS	DIVS. 10 - 14, 25, 28, 41, 43, 44, 46	131.1	64.8	117.7	131.1	86.5	122.1	131.1	65.6	117.9	131.1	74.0	119.6	131.1	65.3	117.8	131.1	66.6	118.1
21, 22, 23	FIRE SUPPRESSION, PLUMBING & HVAC	102.3	63.8	86.8	102.0	92.3	98.1	102.1	71.8	89.9	102.3	100.1	101.4	102.0	69.8	89.0	102.1	84.4	95.0
26, 27, 3370	ELECTRICAL, COMMUNICATIONS & UTIL.	110.6	52.3	79.8	109.7	86.8	97.6	116.0	58.3	85.5	112.4	90.9	101.0	114.5	58.4	84.8	120.7	64.4	90.9
MF2014	WEIGHTED AVERAGE	120.0	62.8	95.1	111.4	92.9	103.3	129.6	67.6	102.6	115.0	93.3	105.5	111.9	67.5	92.5	126.1	74.6	103.7

CANADA

DIVISION		EDMONTON, ALBERTA			FORT MCMURRAY, ALBERTA			FREDERICTON, NEW BRUNSWICK			GATINEAU, QUEBEC			GRANBY, QUEBEC			HALIFAX, NOVA SCOTIA		
		MAT.	INST.	TOTAL	MAT.	INST.	TOTAL	MAT.	INST.	TOTAL	MAT.	INST.	TOTAL	MAT.	INST.	TOTAL	MAT.	INST.	TOTAL
015433	CONTRACTOR EQUIPMENT		107.1	107.1		104.4	104.4		102.2	102.2		102.8	102.8		102.8	102.8		101.9	101.9
0241, 31 - 34	SITE & INFRASTRUCTURE, DEMOLITION	144.1	105.4	116.6	123.1	101.3	107.6	105.0	96.6	99.1	98.2	99.8	99.4	98.7	99.8	99.5	109.2	98.3	101.5
0310	Concrete Forming & Accessories	123.2	97.4	100.9	122.3	92.3	96.5	120.3	64.2	71.9	130.7	87.8	93.7	130.7	87.8	93.7	108.4	78.8	82.9
0320	Concrete Reinforcing	134.2	71.9	102.5	152.9	71.8	111.7	132.8	56.7	94.1	148.7	78.8	113.2	148.7	78.8	113.2	144.3	62.7	102.7
0330	Cast-in-Place Concrete	183.9	107.3	152.4	203.9	105.1	163.3	128.3	62.0	101.1	122.3	97.8	112.2	126.4	97.7	114.6	149.4	81.9	121.7
03	CONCRETE	156.4	96.1	126.8	163.6	93.1	129.0	126.6	62.7	95.2	124.8	89.2	107.6	126.7	89.7	108.5	136.4	77.3	107.4
04	MASONRY	187.2	91.4	127.5	207.9	89.1	133.8	182.6	66.5	110.2	163.7	87.1	116.0	164.1	87.1	116.1	185.9	84.7	122.8
05	METALS	136.9	89.3	122.3	135.4	88.9	121.1	129.1	72.3	111.6	107.4	87.4	101.2	107.4	87.4	101.2	134.9	81.5	118.4
06	WOOD, PLASTICS & COMPOSITES	105.2	97.2	100.7	117.6	91.8	103.2	116.7	64.0	87.2	135.3	88.0	108.8	135.3	88.0	108.8	97.1	77.8	86.3
07	THERMAL & MOISTURE PROTECTION	122.7	94.6	111.2	121.9	91.9	109.6	116.0	63.2	94.4	107.4	89.2	100.3	107.4	88.3	99.6	119.0	79.4	102.8
08	OPENINGS	89.2	86.2	88.5	93.6	83.3	91.2	89.3	54.8	81.2	93.6	76.3	89.6	93.6	76.3	89.6	93.5	72.0	88.5
0920	Plaster & Gypsum Board	134.3	96.6	108.8	128.2	91.1	103.1	137.4	62.7	86.9	126.6	87.4	100.1	129.0	87.4	100.8	127.0	77.0	93.2
0950, 0980	Ceilings & Acoustic Treatment	139.5	96.6	111.3	104.9	91.1	95.8	116.1	62.7	81.0	96.5	87.4	90.5	96.5	87.4	90.5	118.7	77.0	91.3
0960	Flooring	122.5	92.6	114.0	120.5	92.6	112.5	116.8	75.3	104.9	120.5	99.5	114.5	120.5	99.5	114.5	109.2	88.3	103.3
0970, 0990	Wall Finishes & Painting/Coating	122.4	112.4	116.3	118.4	95.5	104.5	123.1	65.8	88.5	118.3	91.7	102.2	118.3	91.7	102.2	124.9	87.4	102.2
09	FINISHES	126.6	98.7	111.2	125.5	93.0	103.0	117.7	66.7	89.6	111.2	90.8	100.0	111.5	90.8	100.1	115.6	81.8	97.0
COVERS	DIVS. 10 - 14, 25, 28, 41, 43, 44, 46	131.1	96.9	124.2	131.1	95.2	123.9	131.1	65.3	117.8	131.1	85.2	121.9	131.1	85.2	121.9	131.1	68.5	118.5
21, 22, 23	FIRE SUPPRESSION, PLUMBING & HVAC	101.1	92.9	97.8	102.4	99.0	101.0	102.3	79.3	93.0	102.3	91.0	97.8	102.0	91.0	97.6	100.6	77.6	91.3
26, 27, 3370	ELECTRICAL, COMMUNICATIONS & UTIL.	110.6	101.3	105.7	105.8	85.4	95.1	116.1	77.1	95.5	111.2	72.6	90.8	111.8	72.6	91.1	117.0	81.2	98.1
MF2014	WEIGHTED AVERAGE	123.1	95.7	111.2	123.6	92.8	110.2	117.4	72.5	97.9	112.1	87.4	101.3	112.3	87.4	101.4	119.6	80.9	102.7

CANADA

DIVISION		HAMILTON, ONTARIO			HULL, QUEBEC			JOLIETTE, QUEBEC			KAMLOOPS, BRITISH COLUMBIA			KINGSTON, ONTARIO			KITCHENER, ONTARIO		
		MAT.	INST.	TOTAL	MAT.	INST.	TOTAL	MAT.	INST.	TOTAL	MAT.	INST.	TOTAL	MAT.	INST.	TOTAL	MAT.	INST.	TOTAL
015433	CONTRACTOR EQUIPMENT		108.9	108.9		102.8	102.8		102.8	102.8		106.1	106.1		104.4	104.4		104.3	104.3
0241, 31 - 34	SITE & INFRASTRUCTURE, DEMOLITION	115.5	112.4	113.3	98.2	99.8	99.4	98.8	99.9	99.6	120.5	103.3	108.3	116.1	104.0	107.5	99.6	104.8	103.3
0310	Concrete Forming & Accessories	120.9	94.7	98.3	130.7	87.8	93.7	130.7	88.0	93.8	123.4	90.8	95.3	122.9	89.6	94.2	113.0	87.2	90.8
0320	Concrete Reinforcing	146.4	92.0	118.7	148.7	78.8	113.2	140.6	78.8	109.2	110.3	76.2	93.2	165.4	82.2	123.1	100.8	91.9	96.3
0330	Cast-in-Place Concrete	143.5	101.0	126.1	122.3	97.8	112.2	127.4	97.8	115.2	110.3	101.6	106.7	138.2	100.6	122.7	129.0	93.6	114.4
03	CONCRETE	132.8	96.3	114.8	124.8	89.8	107.6	125.9	89.9	108.2	132.9	92.0	112.8	136.2	92.2	114.6	110.5	90.4	100.8
04	MASONRY	183.0	101.2	132.0	163.7	87.1	116.0	164.1	87.1	116.1	170.7	95.0	123.5	173.5	94.1	124.0	150.8	97.6	117.6
05	METALS	123.4	92.5	113.9	107.4	87.4	101.2	107.4	87.6	101.3	109.8	87.5	102.9	118.0	89.5	104.3	117.4	92.2	109.6
06	WOOD, PLASTICS & COMPOSITES	100.4	93.9	96.8	135.3	88.0	108.8	135.3	88.0	108.8	105.2	89.4	96.4	122.1	89.2	103.7	109.7	85.5	96.1
07	THERMAL & MOISTURE PROTECTION	118.2	95.7	109.0	107.4	89.9	100.2	107.4	89.9	100.2	123.3	86.2	108.3	113.6	90.4	104.1	108.6	92.8	102.1
08	OPENINGS	91.3	91.6	91.4	93.6	76.3	89.6	93.6	81.1	90.7	90.2	85.7	89.2	93.6	85.9	91.8	84.7	85.3	84.8
0920	Plaster & Gypsum Board	149.5	93.6	111.7	126.6	87.4	100.1	157.0	87.4	109.9	113.6	88.5	96.6	192.8	88.7	122.4	120.2	84.9	96.3
0950, 0980	Ceilings & Acoustic Treatment	122.1	93.6	103.4	96.5	87.4	90.5	96.5	87.4	90.5	96.8	88.5	91.2	112.4	88.7	96.8	98.6	84.9	89.6
0960	Flooring	123.0	102.3	117.2	120.5	99.5	114.5	120.5	99.5	114.5	119.6	55.6	101.3	120.5	97.5	113.9	110.5	102.3	108.1
0970, 0990	Wall Finishes & Painting/Coating	121.6	102.6	110.2	118.3	91.7	102.2	118.3	91.7	102.2	118.3	84.0	97.6	118.3	84.9	98.1	115.3	92.6	101.6
09	FINISHES	124.5	96.2	108.9	117.8	90.8	100.9	117.8	90.8	101.8	115.3	90.8	102.0	124.4	90.9	105.9	106.4	90.7	97.5
COVERS	DIVS. 10 - 14, 25, 28, 41, 43, 44, 46	131.1	98.8	124.6	131.1	85.2	121.9	131.1	85.2	121.9	131.1	95.6	124.0	131.1	74.0	119.6	131.1	97.0	124.2
21, 22, 23	FIRE SUPPRESSION, PLUMBING & HVAC	101.3	89.2	96.4	102.0	91.0	97.6	102.0	91.0	97.6	102.0	94.4	98.9	102.3	100.3	101.5	100.4	87.7	95.3
26, 27, 3370	ELECTRICAL, COMMUNICATIONS & UTIL.	108.3	100.5	104.2	112.9	72.6	91.6	111.8	72.6	91.1	115.1	81.9	97.5	112.4	89.6	100.3	111.5	98.0	104.4
MF2014	WEIGHTED AVERAGE	117.0	96.7	108.2	112.2	87.4	101.4	112.5	87.6	101.7	114.5	90.6	104.1	116.0	93.4	106.2	109.6	92.9	102.3

For customer support on your Commercial Renovation Cost Data, call 877.791.4977.

City Cost Indexes

		\multicolumn{18}{c}{CANADA}													
	DIVISION	LAVAL, QUEBEC			LETHBRIDGE, ALBERTA			LLOYDMINSTER, ALBERTA			LONDON, ONTARIO			MEDICINE HAT, ALBERTA			MONCTON, NEW BRUNSWICK		
		MAT.	INST.	TOTAL	MAT.	INST.	TOTAL	MAT.	INST.	TOTAL	MAT.	INST.	TOTAL	MAT.	INST.	TOTAL	MAT.	INST.	TOTAL
015433	CONTRACTOR EQUIPMENT		102.8	102.8		104.4	104.4		104.4	104.4		104.4	104.4		104.4	104.4		102.2	102.2
0241, 31 - 34	SITE & INFRASTRUCTURE, DEMOLITION	98.7	99.8	99.5	115.6	101.9	105.9	115.4	101.3	105.4	113.1	104.9	107.3	114.2	101.4	105.1	102.8	96.8	98.5
0310	Concrete Forming & Accessories	130.8	87.8	93.8	124.1	92.4	96.8	122.3	82.6	88.1	119.1	88.5	92.7	124.1	82.6	88.3	104.4	64.5	69.9
0320	Concrete Reinforcing	148.7	78.8	113.2	152.9	71.8	111.7	152.9	71.8	111.7	132.8	91.9	112.0	152.9	71.8	111.7	137.0	61.5	98.6
0330	Cast-in-Place Concrete	126.4	97.8	114.6	153.0	105.1	133.3	141.9	101.4	125.3	147.4	98.6	127.4	141.9	101.4	125.3	126.8	78.7	107.0
03	CONCRETE	126.7	89.8	108.6	139.5	93.2	116.7	134.1	87.5	111.3	132.4	92.7	112.9	134.3	87.5	111.3	122.3	69.7	96.5
04	MASONRY	164.0	87.1	116.1	181.7	89.1	124.0	163.4	82.5	113.0	182.6	98.7	130.3	163.4	82.5	113.0	160.5	65.0	101.0
05	METALS	107.4	87.4	101.3	128.8	88.9	116.5	110.0	88.8	103.4	124.9	91.8	114.7	110.0	88.7	103.4	109.0	80.4	100.2
06	WOOD, PLASTICS & COMPOSITES	135.5	88.0	109.9	121.3	91.8	104.8	117.6	82.0	97.7	108.2	86.6	96.1	121.3	82.0	99.3	101.8	64.0	80.7
07	THERMAL & MOISTURE PROTECTION	107.9	89.9	100.5	119.1	91.9	108.0	116.0	87.5	104.3	119.1	93.5	108.6	122.3	87.5	108.0	110.2	65.3	91.8
08	OPENINGS	93.6	76.3	89.6	93.6	83.3	91.2	93.6	77.9	89.9	86.4	86.6	86.4	93.6	77.9	89.9	87.9	61.0	81.6
0920	Plaster & Gypsum Board	129.3	87.4	100.9	119.1	91.1	100.2	114.4	81.0	91.8	152.4	86.1	107.6	117.2	81.0	92.8	137.6	62.7	87.0
0950, 0980	Ceilings & Acoustic Treatment	96.5	87.4	90.5	104.1	91.1	95.5	96.5	81.0	86.3	121.1	86.1	98.1	96.5	81.0	86.3	106.4	62.7	77.7
0960	Flooring	120.5	99.5	114.5	120.5	92.6	112.5	120.5	92.6	112.5	117.2	102.3	113.0	120.5	92.6	112.5	107.9	72.1	97.7
0970, 0990	Wall Finishes & Painting/Coating	118.3	91.7	102.2	118.2	104.2	109.8	118.4	81.2	96.0	122.0	99.4	108.3	118.2	81.2	95.9	117.6	51.6	77.8
09	FINISHES	111.5	90.8	100.1	112.9	94.0	102.5	110.7	84.0	96.0	120.5	91.9	104.7	110.9	84.0	96.1	111.3	64.5	85.5
COVERS	DIVS. 10 - 14, 25, 28, 41, 43, 44, 46	131.1	85.2	121.9	131.1	95.2	123.9	131.1	91.9	123.2	131.1	97.5	124.3	131.1	91.9	123.2	131.1	65.3	117.8
21, 22, 23	FIRE SUPPRESSION, PLUMBING & HVAC	100.4	91.0	96.6	102.2	95.6	99.6	102.3	95.6	99.6	101.4	86.6	95.4	102.0	92.3	98.1	102.0	71.9	89.1
26, 27, 3370	ELECTRICAL, COMMUNICATIONS & UTIL.	112.9	72.6	91.6	107.1	85.4	95.7	104.9	85.4	94.6	107.0	97.5	102.0	104.9	85.4	94.6	117.2	61.8	88.0
MF2014	WEIGHTED AVERAGE	112.1	87.4	101.3	118.4	92.2	107.0	113.4	88.9	102.8	116.4	93.3	106.3	113.8	88.2	102.5	111.8	70.1	93.6

		\multicolumn{18}{c}{CANADA}													
	DIVISION	MONTREAL, QUEBEC			MOOSE JAW, SASKATCHEWAN			NEW GLASGOW, NOVA SCOTIA			NEWCASTLE, NEW BRUNSWICK			NORTH BAY, ONTARIO			OSHAWA, ONTARIO		
		MAT.	INST.	TOTAL	MAT.	INST.	TOTAL	MAT.	INST.	TOTAL	MAT.	INST.	TOTAL	MAT.	INST.	TOTAL	MAT.	INST.	TOTAL
015433	CONTRACTOR EQUIPMENT		104.5	104.5		100.6	100.6		101.9	101.9		102.2	102.2		102.1	102.1		104.3	104.3
0241, 31 - 34	SITE & INFRASTRUCTURE, DEMOLITION	112.1	99.4	103.1	115.5	95.9	101.6	118.5	98.2	104.1	103.4	96.6	98.6	133.9	99.8	109.7	110.7	103.9	105.9
0310	Concrete Forming & Accessories	124.3	93.4	97.7	107.1	60.5	66.9	111.7	71.6	77.1	104.7	63.9	69.5	147.4	87.0	95.3	118.4	89.1	93.1
0320	Concrete Reinforcing	134.0	96.0	114.7	107.9	61.9	84.5	170.3	46.7	107.4	137.0	56.6	96.1	201.7	81.8	140.7	159.6	84.4	121.3
0330	Cast-in-Place Concrete	161.7	101.2	136.8	137.7	71.1	110.4	149.8	72.2	118.0	131.6	62.0	103.0	144.3	87.1	120.8	149.1	87.9	124.0
03	CONCRETE	139.7	96.5	118.4	117.9	65.1	92.0	160.9	67.9	115.2	124.6	62.5	94.1	163.4	86.3	125.5	135.6	88.0	112.2
04	MASONRY	167.9	95.1	122.5	162.0	64.4	101.1	228.9	71.9	131.1	160.8	64.9	101.1	235.3	89.9	144.7	153.8	94.4	116.8
05	METALS	128.0	92.7	117.1	106.3	73.9	96.3	123.6	74.2	108.4	109.0	71.8	97.5	124.6	89.2	113.7	108.6	91.4	103.3
06	WOOD, PLASTICS & COMPOSITES	114.4	93.9	102.9	102.7	59.1	78.3	116.1	71.1	90.9	101.8	64.0	80.7	157.4	87.6	118.3	116.7	88.2	100.7
07	THERMAL & MOISTURE PROTECTION	116.5	96.9	108.5	106.8	64.1	89.3	136.8	69.9	109.4	110.2	62.2	90.5	143.5	85.6	119.8	109.4	87.4	100.4
08	OPENINGS	90.1	82.4	88.3	89.3	56.8	81.7	95.3	64.9	88.2	87.9	55.9	80.5	104.2	83.7	99.4	90.4	87.9	89.8
0920	Plaster & Gypsum Board	135.1	93.2	106.8	110.7	57.8	74.9	142.7	70.0	93.6	137.6	62.7	87.0	140.3	87.1	104.4	124.7	87.7	99.7
0950, 0980	Ceilings & Acoustic Treatment	126.1	93.2	104.4	96.5	57.8	71.0	109.6	70.0	83.6	106.4	62.7	77.7	109.6	87.1	94.8	95.3	87.7	90.3
0960	Flooring	122.0	99.5	115.6	110.6	63.3	97.1	116.1	67.2	102.1	107.9	72.1	97.7	138.6	97.5	126.9	113.7	104.9	111.2
0970, 0990	Wall Finishes & Painting/Coating	120.8	106.4	112.1	118.3	66.8	87.2	126.4	64.1	88.8	117.6	51.6	77.8	126.4	91.2	105.1	115.3	106.7	110.1
09	FINISHES	119.9	96.7	107.1	107.1	61.4	81.9	118.2	70.3	91.8	111.3	64.5	85.5	125.1	89.8	105.7	107.9	93.9	100.2
COVERS	DIVS. 10 - 14, 25, 28, 41, 43, 44, 46	131.1	87.1	122.3	131.1	64.6	117.7	131.1	66.6	118.1	131.1	65.3	117.8	131.1	72.7	119.3	131.1	97.2	124.3
21, 22, 23	FIRE SUPPRESSION, PLUMBING & HVAC	101.2	92.4	97.7	102.3	76.1	91.8	102.1	84.4	95.0	102.0	69.8	89.0	102.1	98.1	100.5	100.4	100.7	100.5
26, 27, 3370	ELECTRICAL, COMMUNICATIONS & UTIL.	112.6	86.8	99.0	113.9	62.8	86.9	116.3	64.4	88.9	112.0	61.8	85.5	116.4	90.8	102.9	112.6	90.5	100.9
MF2014	WEIGHTED AVERAGE	117.8	93.2	107.1	110.7	69.5	92.7	124.1	74.6	102.5	111.6	67.9	92.5	127.3	91.1	111.5	112.3	94.3	104.4

		\multicolumn{18}{c}{CANADA}													
	DIVISION	OTTAWA, ONTARIO			OWEN SOUND, ONTARIO			PETERBOROUGH, ONTARIO			PORTAGE LA PRAIRIE, MANITOBA			PRINCE ALBERT, SASKATCHEWAN			PRINCE GEORGE, BRITISH COLUMBIA		
		MAT.	INST.	TOTAL	MAT.	INST.	TOTAL	MAT.	INST.	TOTAL	MAT.	INST.	TOTAL	MAT.	INST.	TOTAL	MAT.	INST.	TOTAL
015433	CONTRACTOR EQUIPMENT		104.3	104.3		102.1	102.1		102.1	102.1		104.5	104.5		100.6	100.6		106.1	106.1
0241, 31 - 34	SITE & INFRASTRUCTURE, DEMOLITION	108.2	104.6	105.7	118.5	100.3	105.5	118.0	100.2	105.3	116.6	99.3	104.3	110.9	96.1	100.3	124.0	103.3	109.3
0310	Concrete Forming & Accessories	121.2	91.4	95.5	124.4	85.5	90.9	124.9	88.1	93.1	124.3	72.0	79.2	107.1	60.3	66.8	112.7	85.5	89.3
0320	Concrete Reinforcing	138.9	91.9	115.0	173.2	83.8	127.7	165.4	82.3	123.1	152.9	53.2	102.2	112.7	61.8	86.8	110.3	76.8	93.2
0330	Cast-in-Place Concrete	145.6	99.9	126.8	165.6	83.1	131.7	153.6	88.7	127.0	141.9	76.1	114.9	124.8	71.0	102.7	138.2	101.6	123.2
03	CONCRETE	132.6	94.4	113.8	150.3	84.6	118.1	141.8	87.4	115.1	127.8	70.6	99.7	112.6	65.0	89.2	145.4	89.7	118.0
04	MASONRY	168.0	98.8	124.9	171.5	95.6	124.2	168.0	96.7	123.5	166.5	66.6	104.2	161.1	64.4	100.8	172.9	95.0	124.3
05	METALS	127.9	91.7	116.8	110.2	90.0	104.0	109.1	89.7	103.1	110.0	76.7	99.7	106.3	73.7	96.3	109.8	87.6	102.9
06	WOOD, PLASTICS & COMPOSITES	109.4	90.6	98.9	119.8	84.3	99.9	123.7	86.3	102.8	121.1	73.5	94.4	102.7	59.1	78.3	105.2	82.1	92.3
07	THERMAL & MOISTURE PROTECTION	124.4	93.8	111.8	111.8	86.4	101.4	113.8	91.2	104.5	107.3	71.3	92.5	106.7	63.0	88.8	117.2	86.1	104.4
08	OPENINGS	93.9	88.7	92.7	93.9	83.1	91.4	92.4	85.5	90.8	93.6	64.8	86.9	88.2	56.8	80.9	90.2	81.8	88.2
0920	Plaster & Gypsum Board	168.3	90.2	115.5	151.1	83.7	105.6	130.1	85.8	100.1	114.2	72.2	85.8	110.7	57.8	74.9	113.6	81.0	91.6
0950, 0980	Ceilings & Acoustic Treatment	125.6	90.2	102.4	91.5	83.7	86.4	96.5	85.8	89.4	96.5	72.2	80.5	96.5	57.8	71.0	96.5	81.0	86.3
0960	Flooring	112.3	97.5	108.1	127.2	98.9	119.1	120.5	97.5	113.9	120.5	70.2	106.1	110.6	63.3	97.1	116.0	76.2	104.6
0970, 0990	Wall Finishes & Painting/Coating	123.2	94.2	105.7	116.2	89.9	100.3	118.3	93.5	103.3	118.4	58.7	82.4	118.3	57.0	81.3	118.3	84.0	97.6
09	FINISHES	122.6	92.9	106.2	115.0	88.6	100.5	112.4	90.5	100.3	110.5	70.8	88.6	107.1	60.3	81.3	110.6	83.4	95.6
COVERS	DIVS. 10 - 14, 25, 28, 41, 43, 44, 46	131.1	95.7	124.0	139.2	73.2	125.9	131.1	74.2	119.6	131.1	67.2	118.2	131.1	64.6	117.7	131.1	94.8	123.8
21, 22, 23	FIRE SUPPRESSION, PLUMBING & HVAC	101.4	87.2	95.7	101.8	98.1	100.3	102.0	101.7	101.9	102.0	83.6	94.5	102.3	68.6	88.7	102.0	94.4	98.9
26, 27, 3370	ELECTRICAL, COMMUNICATIONS & UTIL.	107.3	98.7	102.8	118.2	89.6	103.1	111.5	90.4	100.4	113.6	60.1	85.3	113.9	62.8	86.9	111.7	81.9	96.0
MF2014	WEIGHTED AVERAGE	117.3	94.0	107.2	117.8	91.2	106.2	114.9	93.1	105.4	113.4	74.2	96.3	109.9	67.7	91.5	115.4	89.9	104.3

City Cost Indexes

		\multicolumn{18}{c}{CANADA}															
	DIVISION	\multicolumn{3}{c}{QUEBEC, QUEBEC}	\multicolumn{3}{c}{RED DEER, ALBERTA}	\multicolumn{3}{c}{REGINA, SASKATCHEWAN}	\multicolumn{3}{c}{RIMOUSKI, QUEBEC}	\multicolumn{3}{c}{ROUYN-NORANDA, QUEBEC}	\multicolumn{3}{c}{SAINT HYACINTHE, QUEBEC}
		MAT.	INST.	TOTAL	MAT.	INST.	TOTAL	MAT.	INST.	TOTAL	MAT.	INST.	TOTAL	MAT.	INST.	TOTAL	MAT.	INST.	TOTAL
015433	CONTRACTOR EQUIPMENT		104.9	104.9		104.4	104.4		100.6	100.6		102.8	102.8		102.8	102.8		102.8	102.8
0241, 31 - 34	SITE & INFRASTRUCTURE, DEMOLITION	110.8	99.5	102.8	114.2	101.4	105.1	124.4	98.0	105.6	98.7	99.1	99.0	98.2	99.8	99.4	98.7	99.8	99.5
0310	Concrete Forming & Accessories	125.6	93.7	98.1	139.6	82.6	90.4	117.6	90.9	94.6	130.7	93.1	98.3	130.7	87.8	93.7	130.7	87.8	93.7
0320	Concrete Reinforcing	138.3	96.0	116.8	152.9	71.8	111.7	138.0	84.8	111.0	106.0	95.9	100.9	148.7	78.8	113.2	148.7	78.8	113.2
0330	Cast-in-Place Concrete	141.8	101.6	125.2	141.9	101.4	125.3	171.5	95.7	140.4	128.6	99.8	116.8	122.3	97.8	112.2	126.4	97.8	114.6
03	CONCRETE	131.0	96.7	114.1	135.3	87.5	111.8	148.9	91.4	120.6	121.1	95.9	108.7	124.8	89.8	107.6	126.7	89.8	108.5
04	MASONRY	163.0	95.1	120.7	163.4	82.5	113.0	183.5	96.7	129.4	163.6	95.0	120.9	163.7	87.1	116.0	164.0	87.1	116.1
05	METALS	128.0	93.0	117.2	110.0	88.7	103.4	129.3	86.4	116.1	106.9	92.2	102.4	107.4	87.4	101.2	107.4	87.4	101.2
06	WOOD, PLASTICS & COMPOSITES	115.8	93.9	103.6	121.3	82.0	99.3	101.9	91.9	96.3	135.3	93.6	112.0	135.3	88.0	108.8	135.3	88.0	108.8
07	THERMAL & MOISTURE PROTECTION	118.4	97.0	109.1	132.2	87.5	113.9	123.2	86.0	107.9	107.4	96.4	102.9	107.4	89.9	100.2	107.7	89.9	100.4
08	OPENINGS	93.0	90.1	92.3	93.6	77.9	89.9	91.9	80.7	89.3	93.2	80.6	90.2	93.6	76.3	89.6	93.6	76.3	89.6
0920	Plaster & Gypsum Board	142.1	93.2	109.0	117.2	81.0	92.8	130.5	91.6	104.2	156.8	93.2	113.8	126.4	87.4	100.0	128.7	87.4	100.8
0950, 0980	Ceilings & Acoustic Treatment	126.1	93.2	104.4	96.5	81.0	86.3	125.4	91.6	103.2	95.7	93.2	94.0	95.7	87.4	90.2	95.7	87.4	90.2
0960	Flooring	125.8	99.5	118.3	123.0	92.6	114.3	120.0	65.5	104.4	120.5	99.5	114.5	120.5	99.5	114.5	120.5	99.5	114.5
0970, 0990	Wall Finishes & Painting/Coating	127.8	106.4	114.9	118.2	81.2	95.9	123.9	92.9	105.2	118.3	106.4	111.1	118.3	91.7	102.2	118.3	91.7	102.2
09	FINISHES	122.2	96.8	108.2	111.7	84.0	96.4	122.7	87.2	103.1	115.1	96.5	104.9	111.0	90.8	99.9	111.3	90.8	100.0
COVERS	DIVS. 10 - 14, 25, 28, 41, 43, 44, 46	131.1	87.3	122.3	131.1	91.9	123.2	131.1	72.4	119.3	131.1	86.5	122.1	131.1	85.2	121.9	131.1	85.2	121.9
21, 22, 23	FIRE SUPPRESSION, PLUMBING & HVAC	101.2	92.4	97.7	102.0	92.3	98.1	100.7	84.8	94.3	102.0	92.3	98.1	102.0	91.0	97.6	98.8	91.0	95.7
26, 27, 3370	ELECTRICAL, COMMUNICATIONS & UTIL.	111.4	86.8	98.4	104.9	85.4	94.6	114.4	92.4	102.8	111.8	86.3	98.6	111.8	72.6	91.1	112.5	72.6	91.4
MF2014	WEIGHTED AVERAGE	117.1	93.6	106.8	114.1	88.2	102.8	120.7	89.0	106.9	111.8	92.9	103.6	112.0	87.4	101.3	111.6	87.4	101.1

		\multicolumn{18}{c}{CANADA}
	DIVISION	\multicolumn{3}{c}{SAINT JOHN, NEW BRUNSWICK}	\multicolumn{3}{c}{SARNIA, ONTARIO}	\multicolumn{3}{c}{SASKATOON, SASKATCHEWAN}	\multicolumn{3}{c}{SAULT STE MARIE, ONTARIO}	\multicolumn{3}{c}{SHERBROOKE, QUEBEC}	\multicolumn{3}{c}{SOREL, QUEBEC}
		MAT.	INST.	TOTAL	MAT.	INST.	TOTAL	MAT.	INST.	TOTAL	MAT.	INST.	TOTAL	MAT.	INST.	TOTAL	MAT.	INST.	TOTAL
015433	CONTRACTOR EQUIPMENT		102.2	102.2		102.1	102.1		100.6	100.6		102.1	102.1		102.8	102.8		102.8	102.8
0241, 31 - 34	SITE & INFRASTRUCTURE, DEMOLITION	103.5	98.0	99.6	116.5	100.3	105.0	111.7	98.0	102.0	106.7	99.8	101.8	98.7	99.8	99.5	98.8	99.9	99.6
0310	Concrete Forming & Accessories	124.6	67.0	74.9	123.6	95.3	99.1	107.4	90.9	93.2	112.8	87.9	91.3	130.7	87.8	93.7	130.7	88.0	93.8
0320	Concrete Reinforcing	137.0	62.0	98.8	117.4	83.6	100.2	114.5	84.8	99.4	106.1	82.1	93.9	148.7	78.8	113.2	140.6	78.8	109.2
0330	Cast-in-Place Concrete	129.7	80.1	109.3	141.7	102.1	125.5	135.4	95.7	119.1	127.4	87.2	110.9	126.4	97.8	114.6	127.4	97.8	115.2
03	CONCRETE	125.2	71.4	98.8	128.6	95.4	112.3	119.6	91.4	105.7	113.0	86.8	100.2	126.7	89.8	108.5	125.9	89.8	108.2
04	MASONRY	180.8	73.7	114.0	179.1	99.5	129.5	168.8	96.7	123.8	164.4	94.4	120.8	164.1	87.1	116.1	164.1	87.1	116.1
05	METALS	108.9	81.7	100.5	109.1	90.1	103.2	104.5	86.4	99.0	108.3	90.2	102.7	107.4	87.4	101.2	107.4	87.6	101.3
06	WOOD, PLASTICS & COMPOSITES	125.7	65.4	91.9	122.6	94.7	107.0	100.5	91.9	95.7	109.6	88.3	97.6	135.3	88.0	108.8	135.3	88.0	108.8
07	THERMAL & MOISTURE PROTECTION	110.5	70.9	94.3	113.9	94.7	106.0	107.1	86.0	98.4	112.7	88.4	102.7	107.4	89.9	100.3	107.4	89.9	100.2
08	OPENINGS	87.8	60.6	81.5	94.7	89.4	93.4	88.9	80.7	87.0	86.5	85.6	86.3	93.6	76.3	89.6	93.6	81.1	90.7
0920	Plaster & Gypsum Board	152.7	64.2	92.9	150.6	94.4	112.6	123.5	91.6	101.9	119.6	87.8	98.1	128.7	87.4	100.8	156.8	87.4	109.9
0950, 0980	Ceilings & Acoustic Treatment	111.5	64.2	80.4	100.7	94.4	96.6	117.4	91.6	100.4	96.5	87.8	90.8	95.7	87.4	90.2	95.7	87.4	90.2
0960	Flooring	119.5	72.1	105.9	120.5	106.6	116.5	111.0	65.5	98.0	113.4	100.8	109.8	120.5	99.5	114.5	120.5	99.5	114.5
0970, 0990	Wall Finishes & Painting/Coating	117.6	80.2	95.0	118.3	105.7	110.7	120.7	92.9	103.9	118.3	97.9	106.0	118.3	91.7	102.2	118.3	91.7	102.2
09	FINISHES	117.9	69.0	91.0	116.1	98.9	106.6	114.3	87.2	99.3	108.1	91.5	99.0	111.3	90.8	100.0	115.1	90.8	101.7
COVERS	DIVS. 10 - 14, 25, 28, 41, 43, 44, 46	131.1	66.3	118.0	131.1	75.7	119.9	131.1	74.1	119.6	131.1	96.4	124.1	131.1	85.2	121.9	131.1	85.2	121.9
21, 22, 23	FIRE SUPPRESSION, PLUMBING & HVAC	102.0	79.8	93.0	102.0	107.8	104.3	100.5	84.8	94.2	102.0	94.8	99.1	102.3	91.0	97.8	102.0	91.0	97.6
26, 27, 3370	ELECTRICAL, COMMUNICATIONS & UTIL.	120.3	88.3	103.3	114.2	92.7	102.8	116.2	92.4	103.6	113.0	90.8	101.3	111.8	72.6	91.1	111.8	72.6	91.1
MF2014	WEIGHTED AVERAGE	114.0	77.9	98.3	114.8	97.5	107.3	111.1	89.1	101.5	110.3	92.1	102.3	112.4	87.4	101.5	112.5	87.6	101.7

		\multicolumn{18}{c}{CANADA}
	DIVISION	\multicolumn{3}{c}{ST CATHARINES, ONTARIO}	\multicolumn{3}{c}{ST JEROME, QUEBEC}	\multicolumn{3}{c}{ST JOHNS, NEWFOUNDLAND}	\multicolumn{3}{c}{SUDBURY, ONTARIO}	\multicolumn{3}{c}{SUMMERSIDE, PRINCE EDWARD ISLAND}	\multicolumn{3}{c}{SYDNEY, NOVA SCOTIA}
		MAT.	INST.	TOTAL	MAT.	INST.	TOTAL	MAT.	INST.	TOTAL	MAT.	INST.	TOTAL	MAT.	INST.	TOTAL	MAT.	INST.	TOTAL
015433	CONTRACTOR EQUIPMENT		102.1	102.1		102.8	102.8		103.3	103.3		102.1	102.1		101.8	101.8		101.9	101.9
0241, 31 - 34	SITE & INFRASTRUCTURE, DEMOLITION	100.0	101.4	101.0	98.2	99.8	99.4	123.4	99.9	106.7	100.1	100.9	100.7	128.4	95.7	105.2	114.3	98.2	102.8
0310	Concrete Forming & Accessories	110.9	94.2	96.5	130.7	87.8	93.7	123.9	84.8	90.1	107.0	89.0	91.5	112.0	58.3	65.7	111.7	71.6	77.1
0320	Concrete Reinforcing	101.7	92.0	96.7	148.7	78.8	113.2	158.9	75.5	116.4	102.5	91.3	96.8	168.1	46.8	106.4	170.3	46.7	107.4
0330	Cast-in-Place Concrete	123.2	100.6	113.9	122.3	97.8	112.2	167.5	96.4	138.3	124.2	97.0	113.0	142.3	61.2	109.0	115.6	72.2	97.8
03	CONCRETE	108.0	95.9	102.1	124.8	89.8	107.6	152.6	87.2	120.5	108.4	92.2	100.4	170.1	57.9	115.0	144.7	67.9	106.9
04	MASONRY	150.3	100.7	119.4	163.7	87.1	116.0	187.5	92.4	128.3	150.4	96.6	116.8	228.1	62.6	125.0	226.2	71.9	130.1
05	METALS	107.7	92.2	102.9	107.4	87.4	101.2	136.3	84.3	120.2	107.7	91.5	102.7	123.6	67.6	106.4	123.6	74.2	108.4
06	WOOD, PLASTICS & COMPOSITES	107.4	93.9	99.8	135.3	88.0	108.8	114.7	82.8	96.8	103.6	88.3	95.0	116.4	57.8	83.6	116.1	71.1	90.9
07	THERMAL & MOISTURE PROTECTION	108.6	96.5	103.1	107.4	89.9	100.2	127.1	91.4	112.5	108.0	91.7	101.3	136.1	61.1	105.4	136.8	69.9	109.4
08	OPENINGS	84.1	90.2	85.5	93.6	76.3	89.6	93.7	74.5	89.2	84.8	85.6	85.0	108.0	50.0	94.5	95.3	64.9	88.2
0920	Plaster & Gypsum Board	108.5	93.6	98.4	126.4	87.4	100.0	143.8	82.1	102.1	112.9	87.8	95.9	143.1	56.3	84.4	142.7	70.0	93.6
0950, 0980	Ceilings & Acoustic Treatment	95.3	93.6	94.1	95.7	87.4	90.2	120.6	82.1	95.3	90.3	87.8	88.6	109.6	56.3	74.6	109.6	70.0	83.6
0960	Flooring	109.3	98.9	106.3	120.5	99.5	114.5	116.0	59.9	99.8	107.3	100.8	105.5	116.2	62.9	100.9	116.1	67.2	102.1
0970, 0990	Wall Finishes & Painting/Coating	115.3	116.0	115.7	118.3	91.7	102.2	120.4	95.9	105.6	115.3	93.6	102.2	126.4	42.9	76.0	126.4	64.1	88.8
09	FINISHES	103.8	97.8	100.4	111.0	90.8	99.9	122.0	81.4	99.6	102.7	91.6	96.5	119.2	57.9	85.4	118.2	70.3	91.8
COVERS	DIVS. 10 - 14, 25, 28, 41, 43, 44, 46	131.1	75.2	119.8	131.1	85.2	121.9	131.1	72.6	119.3	131.1	97.1	124.3	131.1	64.8	117.7	131.1	66.6	118.1
21, 22, 23	FIRE SUPPRESSION, PLUMBING & HVAC	100.4	87.8	95.3	102.0	91.0	97.6	100.8	82.1	93.3	101.9	87.8	96.2	102.1	63.8	86.7	102.1	64.4	95.0
26, 27, 3370	ELECTRICAL, COMMUNICATIONS & UTIL.	112.9	98.8	105.5	112.4	72.6	91.3	114.4	83.2	97.9	110.3	99.2	104.4	115.4	52.3	82.0	116.3	54.4	88.9
MF2014	WEIGHTED AVERAGE	107.6	94.5	101.9	112.1	87.4	101.4	122.7	85.2	106.4	107.7	93.0	101.3	126.7	62.8	98.8	122.1	74.6	101.4

City Cost Indexes

		\multicolumn{18}{c	}{CANADA}														
\multicolumn{2}{	c	}{DIVISION}	\multicolumn{3}{c	}{THUNDER BAY, ONTARIO}	\multicolumn{3}{c	}{TIMMINS, ONTARIO}	\multicolumn{3}{c	}{TORONTO, ONTARIO}	\multicolumn{3}{c	}{TROIS RIVIERES, QUEBEC}	\multicolumn{3}{c	}{TRURO, NOVA SCOTIA}	\multicolumn{3}{c	}{VANCOUVER, BRITISH COLUMBIA}
		MAT.	INST.	TOTAL	MAT.	INST.	TOTAL	MAT.	INST.	TOTAL	MAT.	INST.	TOTAL	MAT.	INST.	TOTAL	MAT.	INST.	TOTAL
015433	CONTRACTOR EQUIPMENT		102.1	102.1		102.1	102.1		104.4	104.4		102.8	102.8		101.9	101.9		112.7	112.7
0241, 31 - 34	SITE & INFRASTRUCTURE, DEMOLITION	104.8	101.3	102.3	118.0	99.8	105.1	133.8	105.6	113.7	114.8	99.9	104.2	102.8	98.2	99.5	118.2	107.6	110.6
0310	Concrete Forming & Accessories	118.3	92.2	95.8	124.9	87.0	92.2	124.3	100.8	104.0	155.5	88.0	97.3	97.1	71.6	75.1	121.1	91.0	95.1
0320	Concrete Reinforcing	90.9	90.8	90.9	165.4	81.8	122.9	146.0	94.4	119.7	170.3	78.8	123.8	140.6	46.7	92.8	134.1	79.2	106.2
0330	Cast-in-Place Concrete	135.6	99.7	120.8	153.6	87.1	126.3	136.8	111.8	126.5	119.7	97.8	110.7	159.5	72.2	123.7	143.5	97.9	124.8
03	CONCRETE	115.8	94.5	105.3	141.8	86.3	114.5	129.7	103.1	116.7	147.4	89.8	119.1	140.2	67.9	104.7	137.1	91.4	114.6
04	MASONRY	151.0	100.6	119.6	168.0	89.9	119.3	182.0	110.9	137.7	230.7	87.1	141.2	163.6	71.9	106.5	166.7	91.4	119.7
05	METALS	107.5	91.1	102.5	109.1	89.2	103.0	132.5	93.9	120.7	122.7	87.6	111.9	108.2	74.2	97.7	139.3	90.3	124.2
06	WOOD, PLASTICS & COMPOSITES	116.7	90.9	102.2	123.7	87.6	103.5	115.1	98.9	106.0	173.0	88.0	125.4	92.9	71.1	80.7	104.8	91.5	97.4
07	THERMAL & MOISTURE PROTECTION	108.8	93.8	102.7	113.8	85.6	102.2	123.9	102.8	115.2	135.2	89.9	116.6	108.6	69.9	92.8	125.2	86.9	109.5
08	OPENINGS	83.3	88.0	84.4	92.4	83.7	90.4	90.5	96.8	92.0	105.5	81.1	99.8	86.1	64.9	81.1	88.9	86.9	88.4
0920	Plaster & Gypsum Board	139.8	90.5	106.5	130.1	87.1	101.1	133.0	98.8	109.9	173.4	87.4	115.3	130.7	70.0	89.7	129.7	90.6	103.3
0950, 0980	Ceilings & Acoustic Treatment	90.3	90.5	90.4	96.5	87.1	90.3	114.9	98.8	104.3	108.8	87.4	94.7	96.5	70.0	79.1	132.2	90.6	104.8
0960	Flooring	113.7	105.9	111.4	120.5	97.5	113.9	117.0	108.4	114.6	138.6	99.5	127.5	102.7	67.2	92.5	124.6	96.0	116.4
0970, 0990	Wall Finishes & Painting/Coating	115.3	94.6	102.8	118.3	91.2	101.9	120.4	106.7	112.1	126.4	91.7	105.4	118.3	64.1	85.6	121.7	95.6	106.0
09	FINISHES	108.5	95.1	101.1	112.4	89.8	99.9	116.9	103.0	109.2	128.7	90.8	107.8	106.6	70.3	86.6	123.8	92.7	106.6
COVERS	DIVS. 10 - 14, 25, 28, 41, 43, 44, 46	131.1	75.2	119.8	131.1	72.7	119.3	131.1	101.2	125.1	131.1	85.2	121.9	131.1	66.6	118.1	131.1	94.9	123.8
21, 22, 23	FIRE SUPPRESSION, PLUMBING & HVAC	100.4	88.0	95.4	102.0	98.1	100.4	101.0	97.3	99.5	102.1	91.0	97.7	102.0	84.4	94.9	101.2	80.5	92.8
26, 27, 3370	ELECTRICAL, COMMUNICATIONS & UTIL.	111.5	97.0	103.8	113.0	90.8	101.3	108.8	101.0	104.7	116.6	72.6	93.4	111.5	64.4	86.6	109.0	83.3	95.4
MF2014	WEIGHTED AVERAGE	108.8	93.4	102.1	115.1	91.1	104.6	118.3	101.3	110.9	124.8	87.6	108.6	112.5	74.6	96.0	119.5	89.1	106.2

		\multicolumn{18}{c	}{CANADA}														
\multicolumn{2}{	c	}{DIVISION}	\multicolumn{3}{c	}{VICTORIA, BRITISH COLUMBIA}	\multicolumn{3}{c	}{WHITEHORSE, YUKON}	\multicolumn{3}{c	}{WINDSOR, ONTARIO}	\multicolumn{3}{c	}{WINNIPEG, MANITOBA}	\multicolumn{3}{c	}{YARMOUTH, NOVA SCOTIA}	\multicolumn{3}{c	}{YELLOWKNIFE, NWT}
		MAT.	INST.	TOTAL	MAT.	INST.	TOTAL	MAT.	INST.	TOTAL	MAT.	INST.	TOTAL	MAT.	INST.	TOTAL	MAT.	INST.	TOTAL
015433	CONTRACTOR EQUIPMENT		109.3	109.3		102.1	102.1		102.1	102.1		106.5	106.5		101.9	101.9		101.9	101.9
0241, 31 - 34	SITE & INFRASTRUCTURE, DEMOLITION	123.0	107.1	111.7	142.3	96.7	109.9	95.9	101.3	99.7	118.2	100.9	105.9	118.3	98.2	104.0	154.1	101.0	116.4
0310	Concrete Forming & Accessories	112.7	90.4	93.5	129.4	59.6	69.2	118.3	90.5	94.3	128.5	69.3	77.4	111.7	71.6	77.1	132.7	80.2	87.4
0320	Concrete Reinforcing	112.0	79.1	95.2	160.8	60.0	109.5	99.6	91.9	95.7	132.9	57.9	94.7	170.3	46.7	107.4	151.4	62.7	106.2
0330	Cast-in-Place Concrete	140.3	97.2	122.6	209.3	70.3	152.2	126.2	101.9	116.2	163.9	75.0	127.4	148.3	72.2	117.0	211.2	88.1	160.6
03	CONCRETE	150.1	90.7	120.9	182.4	64.1	124.3	109.6	94.7	102.3	144.4	69.7	107.7	160.1	67.9	114.8	182.0	80.0	131.9
04	MASONRY	171.7	91.3	121.6	260.3	62.8	137.2	150.5	99.9	119.0	176.6	70.8	110.7	228.8	71.9	131.0	260.3	74.8	144.7
05	METALS	107.0	86.3	100.6	135.3	73.6	116.3	107.6	91.7	102.7	141.7	75.0	121.2	123.6	74.2	108.4	137.9	78.7	119.7
06	WOOD, PLASTICS & COMPOSITES	104.4	91.4	97.1	117.6	58.3	84.4	116.7	89.0	101.2	113.6	70.2	89.3	116.1	71.1	90.9	124.0	81.9	100.4
07	THERMAL & MOISTURE PROTECTION	116.9	86.7	104.5	139.1	62.0	107.5	108.6	93.9	102.6	120.8	71.8	100.7	136.8	69.9	109.4	151.5	78.4	121.5
08	OPENINGS	90.5	83.2	88.8	107.6	55.5	95.4	83.1	87.5	84.1	85.5	62.6	80.2	95.3	64.9	88.2	102.2	68.9	94.4
0920	Plaster & Gypsum Board	115.8	90.6	98.8	150.5	56.8	87.2	125.7	88.6	100.6	129.5	68.6	88.3	142.7	70.0	93.6	163.7	81.1	107.9
0950, 0980	Ceilings & Acoustic Treatment	98.1	90.6	93.2	144.6	56.8	86.9	90.3	88.6	89.2	136.2	68.6	91.8	109.6	70.0	83.6	138.9	81.1	100.9
0960	Flooring	116.4	76.2	104.9	141.0	61.3	118.2	113.7	103.1	110.6	117.0	75.6	105.2	116.1	67.2	102.1	135.5	92.5	123.2
0970, 0990	Wall Finishes & Painting/Coating	120.7	95.6	105.5	143.1	56.0	90.5	115.3	95.5	103.3	124.5	56.5	83.4	126.4	64.1	88.8	135.3	80.3	102.1
09	FINISHES	112.1	89.3	99.5	140.9	59.3	95.9	106.3	93.4	99.2	122.3	69.5	93.2	118.2	70.3	91.8	137.5	82.0	106.9
COVERS	DIVS. 10 - 14, 25, 28, 41, 43, 44, 46	131.1	71.5	119.1	131.1	63.9	117.6	131.1	74.7	119.7	131.1	68.7	118.5	131.1	66.6	118.1	131.1	67.1	118.2
21, 22, 23	FIRE SUPPRESSION, PLUMBING & HVAC	102.0	80.5	93.3	102.5	74.8	91.3	100.4	88.0	95.4	101.1	66.9	87.3	102.1	84.4	95.0	102.7	94.0	99.2
26, 27, 3370	ELECTRICAL, COMMUNICATIONS & UTIL.	113.1	82.7	97.0	133.5	62.2	95.8	116.0	99.0	107.0	113.3	68.0	89.4	116.3	64.4	88.9	130.5	85.2	106.5
MF2014	WEIGHTED AVERAGE	115.7	87.2	103.3	135.3	68.5	106.2	108.2	93.4	101.7	121.0	71.8	99.5	124.0	74.6	102.5	135.3	84.2	113.0

Location Factors

Costs shown in RSMeans cost data publications are based on national averages for materials and installation. To adjust these costs to a specific location, simply multiply the base cost by the factor and divide by 100 for that city. The data is arranged alphabetically by state and postal zip code numbers. For a city not listed, use the factor for a nearby city with similar economic characteristics.

STATE/ZIP	CITY	MAT.	INST.	TOTAL
ALABAMA				
350-352	Birmingham	100.6	76.7	90.2
354	Tuscaloosa	99.8	61.2	83.0
355	Jasper	100.2	57.5	81.6
356	Decatur	99.8	57.9	81.5
357-358	Huntsville	99.8	72.1	87.8
359	Gadsden	99.9	60.0	82.5
360-361	Montgomery	99.9	57.1	81.2
362	Anniston	98.6	61.8	82.6
363	Dothan	99.0	49.4	77.4
364	Evergreen	98.6	50.7	77.7
365-366	Mobile	99.9	65.4	84.9
367	Selma	98.7	49.9	77.4
368	Phenix City	99.5	54.6	79.9
369	Butler	98.9	49.4	77.3
ALASKA				
995-996	Anchorage	121.1	115.6	118.7
997	Fairbanks	121.0	116.4	119.0
998	Juneau	121.3	115.6	118.8
999	Ketchikan	132.0	115.6	124.8
ARIZONA				
850,853	Phoenix	99.2	73.8	88.1
851,852	Mesa/Tempe	98.6	72.8	87.4
855	Globe	98.8	70.0	86.3
856-857	Tucson	97.4	72.4	86.5
859	Show Low	98.9	72.5	87.4
860	Flagstaff	101.3	71.8	88.4
863	Prescott	99.2	68.9	86.0
864	Kingman	97.6	72.0	86.4
865	Chambers	97.6	71.5	86.2
ARKANSAS				
716	Pine Bluff	99.4	63.3	83.6
717	Camden	97.1	50.1	76.6
718	Texarkana	98.2	49.2	76.8
719	Hot Springs	96.4	51.8	76.9
720-722	Little Rock	98.5	64.0	83.4
723	West Memphis	96.1	62.4	81.4
724	Jonesboro	96.7	59.1	80.3
725	Batesville	94.6	54.6	77.1
726	Harrison	95.9	50.4	76.0
727	Fayetteville	93.4	52.0	75.4
728	Russellville	94.6	53.7	76.8
729	Fort Smith	97.2	61.6	81.7
CALIFORNIA				
900-902	Los Angeles	99.2	117.5	107.2
903-905	Inglewood	94.8	114.1	103.2
906-908	Long Beach	96.2	114.1	104.0
910-912	Pasadena	96.4	114.6	104.3
913-916	Van Nuys	99.1	114.6	105.9
917-918	Alhambra	98.3	114.8	105.5
919-921	San Diego	100.5	109.6	104.5
922	Palm Springs	97.2	112.0	103.6
923-924	San Bernardino	95.1	111.3	102.1
925	Riverside	99.4	114.6	106.0
926-927	Santa Ana	96.9	111.8	103.4
928	Anaheim	99.4	114.9	106.2
930	Oxnard	99.6	114.4	106.0
931	Santa Barbara	98.9	115.1	106.0
932-933	Bakersfield	100.7	113.6	106.3
934	San Luis Obispo	99.4	112.0	104.9
935	Mojave	96.8	109.8	102.5
936-938	Fresno	100.1	115.3	106.7
939	Salinas	100.2	121.6	109.5
940-941	San Francisco	105.2	145.4	122.7
942,956-958	Sacramento	100.5	119.3	108.7
943	Palo Alto	98.9	134.5	114.4
944	San Mateo	100.9	134.6	115.6
945	Vallejo	100.1	127.2	111.9
946	Oakland	103.1	134.3	116.7
947	Berkeley	102.7	135.1	116.8
948	Richmond	101.8	133.1	115.4
949	San Rafael	103.5	136.7	118.0
950	Santa Cruz	104.5	121.8	112.0

STATE/ZIP	CITY	MAT.	INST.	TOTAL
CALIFORNIA (CONT'D)				
951	San Jose	102.5	136.6	117.4
952	Stockton	100.9	117.2	108.0
953	Modesto	100.8	115.7	107.3
954	Santa Rosa	100.8	134.7	115.6
955	Eureka	102.1	116.4	108.4
959	Marysville	101.3	115.9	107.7
960	Redding	107.8	117.7	112.1
961	Susanville	106.8	118.1	111.7
COLORADO				
800-802	Denver	100.8	81.7	92.5
803	Boulder	97.8	81.3	90.6
804	Golden	99.8	78.9	90.7
805	Fort Collins	101.3	79.0	91.6
806	Greeley	98.9	75.5	88.7
807	Fort Morgan	98.3	79.0	89.9
808-809	Colorado Springs	100.4	84.0	93.2
810	Pueblo	101.1	80.3	92.0
811	Alamosa	102.2	73.7	89.8
812	Salida	101.9	75.3	90.3
813	Durango	102.7	75.7	90.9
814	Montrose	101.4	75.4	90.1
815	Grand Junction	104.8	75.3	91.9
816	Glenwood Springs	102.3	78.1	91.8
CONNECTICUT				
060	New Britain	100.1	120.6	109.1
061	Hartford	101.4	120.8	109.8
062	Willimantic	100.8	120.2	109.2
063	New London	97.1	120.7	107.4
064	Meriden	98.9	120.7	108.4
065	New Haven	101.6	120.7	109.9
066	Bridgeport	101.3	120.8	109.8
067	Waterbury	100.9	120.7	109.5
068	Norwalk	100.8	128.7	113.0
069	Stamford	100.9	128.7	113.0
D.C.				
200-205	Washington	101.1	91.9	97.1
DELAWARE				
197	Newark	98.8	109.2	103.3
198	Wilmington	99.3	109.2	103.6
199	Dover	99.3	109.2	103.6
FLORIDA				
320,322	Jacksonville	98.8	66.4	84.7
321	Daytona Beach	99.0	72.9	87.6
323	Tallahassee	100.8	56.9	81.7
324	Panama City	100.2	57.8	81.7
325	Pensacola	102.7	61.2	84.6
326,344	Gainesville	100.3	66.7	85.7
327-328,347	Orlando	100.5	70.0	87.2
329	Melbourne	101.7	76.3	90.6
330-332,340	Miami	99.8	73.1	88.2
333	Fort Lauderdale	98.6	72.6	87.2
334,349	West Palm Beach	97.4	71.1	85.9
335-336,346	Tampa	100.1	78.4	90.6
337	St. Petersburg	102.6	63.3	85.5
338	Lakeland	99.4	77.8	90.0
339,341	Fort Myers	98.6	72.6	87.3
342	Sarasota	100.8	74.3	89.2
GEORGIA				
300-303,399	Atlanta	97.6	74.4	87.5
304	Statesboro	97.3	55.8	79.2
305	Gainesville	96.0	61.6	81.0
306	Athens	95.5	63.9	81.7
307	Dalton	97.2	59.0	80.5
308-309	Augusta	96.3	63.6	82.1
310-312	Macon	97.2	64.7	83.0
313-314	Savannah	99.2	60.9	82.5
315	Waycross	98.3	59.4	81.4
316	Valdosta	98.5	61.4	82.3
317,398	Albany	98.5	61.3	82.3
318-319	Columbus	98.4	65.4	84.0

Location Factors

STATE/ZIP	CITY	MAT.	INST.	TOTAL
HAWAII				
967	Hilo	114.8	120.2	117.2
968	Honolulu	119.5	120.2	119.8
STATES & POSS.				
969	Guam	136.2	58.0	102.1
IDAHO				
832	Pocatello	101.1	77.4	90.8
833	Twin Falls	102.2	67.3	87.0
834	Idaho Falls	99.8	74.7	88.9
835	Lewiston	108.7	84.9	98.3
836-837	Boise	100.1	78.6	90.7
838	Coeur d'Alene	107.9	82.2	96.7
ILLINOIS				
600-603	North Suburban	99.3	135.1	114.9
604	Joliet	99.2	138.4	116.3
605	South Suburban	99.3	135.1	114.9
606-608	Chicago	99.8	139.7	117.2
609	Kankakee	95.6	131.7	111.3
610-611	Rockford	99.5	124.8	110.5
612	Rock Island	97.2	100.5	98.6
613	La Salle	98.4	124.4	109.7
614	Galesburg	98.2	106.7	101.9
615-616	Peoria	100.5	109.3	104.4
617	Bloomington	97.5	109.4	102.7
618-619	Champaign	101.0	107.6	103.9
620-622	East St. Louis	96.4	108.7	101.8
623	Quincy	98.1	101.0	99.3
624	Effingham	97.3	106.0	101.1
625	Decatur	99.2	104.9	101.7
626-627	Springfield	100.0	106.5	102.8
628	Centralia	95.0	108.5	100.9
629	Carbondale	94.7	107.9	100.5
INDIANA				
460	Anderson	97.0	83.8	91.3
461-462	Indianapolis	99.1	84.8	92.9
463-464	Gary	98.3	110.4	103.6
465-466	South Bend	98.0	84.4	92.1
467-468	Fort Wayne	97.6	79.1	89.5
469	Kokomo	94.8	81.5	89.0
470	Lawrenceburg	94.0	77.7	86.9
471	New Albany	95.1	75.4	86.5
472	Columbus	97.2	80.9	90.1
473	Muncie	97.9	82.3	91.1
474	Bloomington	99.3	81.8	91.6
475	Washington	95.8	81.8	89.7
476-477	Evansville	97.1	86.9	92.6
478	Terre Haute	97.9	86.5	92.9
479	Lafayette	96.9	82.7	90.7
IOWA				
500-503,509	Des Moines	100.5	85.2	93.8
504	Mason City	97.6	65.8	83.8
505	Fort Dodge	97.9	58.2	80.6
506-507	Waterloo	99.5	71.8	87.4
508	Creston	98.2	77.3	89.1
510-511	Sioux City	99.7	72.6	87.9
512	Sibley	98.3	56.4	80.0
513	Spencer	99.9	56.6	81.0
514	Carroll	97.0	71.2	85.8
515	Council Bluffs	100.7	78.1	90.9
516	Shenandoah	97.6	71.2	86.1
520	Dubuque	99.1	80.2	90.9
521	Decorah	98.0	64.2	83.3
522-524	Cedar Rapids	100.0	83.2	92.7
525	Ottumwa	97.9	73.7	87.3
526	Burlington	97.2	75.3	87.6
527-528	Davenport	99.0	93.1	96.4
KANSAS				
660-662	Kansas City	99.5	94.8	97.5
664-666	Topeka	99.5	66.0	84.9
667	Fort Scott	97.9	73.0	87.0
668	Emporia	97.9	69.2	85.4
669	Belleville	99.6	64.9	84.4
670-672	Wichita	100.7	66.0	85.6
673	Independence	100.9	69.2	87.1
674	Salina	101.5	67.6	86.7
675	Hutchinson	96.5	65.0	82.8
676	Hays	100.5	65.4	85.2
677	Colby	101.1	65.0	85.4

STATE/ZIP	CITY	MAT.	INST.	TOTAL
KANSAS (CONT'D)				
678	Dodge City	102.7	66.0	86.7
679	Liberal	100.4	65.5	85.2
KENTUCKY				
400-402	Louisville	96.6	83.3	90.8
403-405	Lexington	96.6	81.4	90.0
406	Frankfort	98.1	80.7	90.5
407-409	Corbin	94.0	71.8	84.3
410	Covington	95.9	91.4	93.9
411-412	Ashland	94.6	99.3	96.6
413-414	Campton	95.8	78.6	88.3
415-416	Pikeville	96.9	88.6	93.3
417-418	Hazard	95.2	78.7	88.0
420	Paducah	93.9	84.5	89.8
421-422	Bowling Green	96.2	83.9	90.8
423	Owensboro	96.3	88.9	93.1
424	Henderson	93.7	86.1	90.4
425-426	Somerset	93.2	79.9	87.4
427	Elizabethtown	92.8	82.0	88.1
LOUISIANA				
700-701	New Orleans	101.4	69.5	87.5
703	Thibodaux	98.2	65.4	83.9
704	Hammond	95.8	54.5	77.8
705	Lafayette	97.9	63.0	82.7
706	Lake Charles	98.1	65.0	83.7
707-708	Baton Rouge	99.1	63.7	83.7
710-711	Shreveport	101.0	59.6	83.0
712	Monroe	99.3	56.5	80.6
713-714	Alexandria	99.4	57.2	81.0
MAINE				
039	Kittery	95.2	84.7	90.6
040-041	Portland	102.0	90.1	96.8
042	Lewiston	99.6	90.1	95.5
043	Augusta	102.1	79.5	92.3
044	Bangor	99.0	89.5	94.9
045	Bath	97.5	81.9	90.7
046	Machias	97.1	85.5	92.0
047	Houlton	97.2	85.6	92.1
048	Rockland	96.3	85.5	91.6
049	Waterville	97.5	79.5	89.7
MARYLAND				
206	Waldorf	97.1	81.7	90.4
207-208	College Park	97.2	83.4	91.2
209	Silver Spring	96.5	83.2	90.7
210-212	Baltimore	100.8	81.9	92.6
214	Annapolis	101.5	82.0	93.0
215	Cumberland	96.8	81.8	90.2
216	Easton	98.3	68.6	85.3
217	Hagerstown	97.6	83.3	91.4
218	Salisbury	98.7	62.1	82.7
219	Elkton	95.9	81.9	89.8
MASSACHUSETTS				
010-011	Springfield	100.2	112.6	105.6
012	Pittsfield	99.7	111.3	104.8
013	Greenfield	97.7	112.3	104.1
014	Fitchburg	96.5	126.5	109.6
015-016	Worcester	100.2	126.4	111.6
017	Framingham	95.7	135.1	112.9
018	Lowell	99.5	134.5	114.8
019	Lawrence	100.3	134.6	115.3
020-022, 024	Boston	101.6	139.5	118.1
023	Brockton	99.9	128.1	112.2
025	Buzzards Bay	94.4	128.1	109.1
026	Hyannis	97.0	128.1	110.6
027	New Bedford	99.1	128.3	111.8
MICHIGAN				
480,483	Royal Oak	94.9	106.9	100.1
481	Ann Arbor	97.1	107.5	101.7
482	Detroit	97.8	109.6	102.9
484-485	Flint	96.8	96.0	96.4
486	Saginaw	96.4	91.8	94.4
487	Bay City	96.6	91.7	94.5
488-489	Lansing	99.3	94.7	97.3
490	Battle Creek	98.3	87.6	93.7
491	Kalamazoo	98.7	85.7	93.0
492	Jackson	96.5	95.1	95.9
493,495	Grand Rapids	99.7	85.1	93.3
494	Muskegon	97.2	84.8	91.8

801

For customer support on your Commercial Renovation Cost Data, call 877.791.4977.

Location Factors

STATE/ZIP	CITY	MAT.	INST.	TOTAL
MICHIGAN (CONT'D)				
496	Traverse City	96.3	80.1	89.3
497	Gaylord	97.2	83.6	91.3
498-499	Iron mountain	99.3	86.2	93.6
MINNESOTA				
550-551	Saint Paul	101.4	117.9	108.6
553-555	Minneapolis	100.4	120.0	108.9
556-558	Duluth	101.3	107.3	103.9
559	Rochester	99.9	104.7	102.0
560	Mankato	97.2	101.3	99.0
561	Windom	95.9	92.3	94.3
562	Willmar	95.5	101.3	98.0
563	St. Cloud	96.9	116.0	105.2
564	Brainerd	96.9	100.5	98.5
565	Detroit Lakes	98.9	95.1	97.3
566	Bemidji	98.2	98.2	98.2
567	Thief River Falls	97.8	92.4	95.4
MISSISSIPPI				
386	Clarksdale	97.3	52.6	77.8
387	Greenville	101.0	61.7	83.9
388	Tupelo	98.8	54.9	79.7
389	Greenwood	98.4	52.2	78.3
390-392	Jackson	100.7	62.2	83.9
393	Meridian	97.6	64.1	83.0
394	Laurel	98.9	55.8	80.1
395	Biloxi	99.2	58.1	81.3
396	Mccomb	97.4	51.7	77.5
397	Columbus	98.8	53.9	79.2
MISSOURI				
630-631	St. Louis	98.7	107.8	102.7
633	Bowling Green	97.3	94.4	96.0
634	Hannibal	96.2	92.4	94.6
635	Kirksville	99.5	89.9	95.3
636	Flat River	98.1	95.0	96.8
637	Cape Girardeau	98.2	92.4	95.7
638	Sikeston	96.7	92.2	94.8
639	Poplar Bluff	96.2	92.0	94.4
640-641	Kansas City	101.4	105.7	103.3
644-645	St. Joseph	100.6	93.5	97.5
646	Chillicothe	97.6	94.5	96.3
647	Harrisonville	97.1	101.5	99.1
648	Joplin	99.5	81.7	91.8
650-651	Jefferson City	99.9	93.7	97.2
652	Columbia	100.2	93.9	97.5
653	Sedalia	100.2	97.3	98.9
654-655	Rolla	97.9	94.6	96.5
656-658	Springfield	101.7	83.7	93.8
MONTANA				
590-591	Billings	102.7	77.1	91.5
592	Wolf Point	102.3	74.2	90.0
593	Miles City	100.1	74.0	88.7
594	Great Falls	104.1	76.2	91.9
595	Havre	101.2	73.1	89.0
596	Helena	102.4	73.8	89.9
597	Butte	102.6	77.3	91.6
598	Missoula	99.9	75.5	89.3
599	Kalispell	99.0	75.3	88.7
NEBRASKA				
680-681	Omaha	100.2	78.6	90.8
683-685	Lincoln	100.5	74.9	89.4
686	Columbus	98.4	77.1	89.1
687	Norfolk	100.0	76.7	89.8
688	Grand Island	100.0	77.1	90.0
689	Hastings	99.4	79.8	90.9
690	Mccook	98.7	71.1	86.7
691	North Platte	99.0	78.3	90.0
692	Valentine	101.2	72.5	88.7
693	Alliance	101.0	70.5	87.7
NEVADA				
889-891	Las Vegas	100.9	108.7	104.3
893	Ely	99.4	102.2	100.6
894-895	Reno	99.6	93.3	96.8
897	Carson City	99.7	93.5	97.0
898	Elko	98.3	84.9	92.4
NEW HAMPSHIRE				
030	Nashua	100.5	93.8	97.6
031	Manchester	101.4	93.8	98.1

STATE/ZIP	CITY	MAT.	INST.	TOTAL
NEW HAMPSHIRE (CONT'D)				
032-033	Concord	101.1	92.5	97.4
034	Keene	97.2	73.3	86.7
035	Littleton	97.1	76.3	88.0
036	Charleston	96.7	69.5	84.8
037	Claremont	95.9	69.5	84.4
038	Portsmouth	97.9	94.3	96.4
NEW JERSEY				
070-071	Newark	101.2	128.4	113.0
072	Elizabeth	98.8	127.6	111.3
073	Jersey City	97.9	127.9	111.0
074-075	Paterson	99.6	127.8	111.9
076	Hackensack	97.6	128.1	110.9
077	Long Branch	97.1	126.3	109.8
078	Dover	97.9	127.8	110.9
079	Summit	97.9	127.6	110.8
080,083	Vineland	97.5	124.7	109.4
081	Camden	99.5	124.8	110.5
082,084	Atlantic City	98.2	124.8	109.8
085-086	Trenton	100.9	125.1	111.5
087	Point Pleasant	99.2	125.9	110.8
088-089	New Brunswick	99.7	127.6	111.9
NEW MEXICO				
870-872	Albuquerque	99.6	72.6	87.8
873	Gallup	99.5	72.6	87.8
874	Farmington	100.1	72.6	88.1
875	Santa Fe	100.8	72.6	88.5
877	Las Vegas	98.0	72.6	86.9
878	Socorro	97.6	72.6	86.7
879	Truth/Consequences	97.5	69.4	85.2
880	Las Cruces	96.1	69.4	84.4
881	Clovis	98.3	72.5	87.1
882	Roswell	100.0	72.6	88.0
883	Carrizozo	100.3	72.6	88.2
884	Tucumcari	99.0	72.5	87.4
NEW YORK				
100-102	New York	103.4	168.7	131.8
103	Staten Island	99.2	168.4	129.4
104	Bronx	97.7	166.1	127.5
105	Mount Vernon	97.5	141.4	116.6
106	White Plains	97.6	141.4	116.7
107	Yonkers	102.1	143.7	120.2
108	New Rochelle	97.9	141.3	116.8
109	Suffern	97.6	129.0	111.3
110	Queens	100.9	166.8	129.6
111	Long Island City	102.4	166.8	130.5
112	Brooklyn	102.8	168.5	131.4
113	Flushing	102.5	166.8	130.5
114	Jamaica	100.8	166.8	129.6
115,117,118	Hicksville	100.9	153.3	123.7
116	Far Rockaway	102.6	166.8	130.6
119	Riverhead	101.6	148.2	121.9
120-122	Albany	100.1	104.7	102.1
123	Schenectady	99.3	104.7	101.7
124	Kingston	102.6	118.5	109.6
125-126	Poughkeepsie	101.8	136.7	117.0
127	Monticello	101.1	115.1	107.2
128	Glens Falls	94.0	98.6	96.0
129	Plattsburgh	99.1	96.7	98.0
130-132	Syracuse	98.9	98.0	98.5
133-135	Utica	97.1	97.1	97.1
136	Watertown	98.5	96.3	97.5
137-139	Binghamton	98.5	101.4	99.8
140-142	Buffalo	100.3	106.2	102.9
143	Niagara Falls	97.8	106.4	101.5
144-146	Rochester	101.1	99.1	100.2
147	Jamestown	96.8	92.5	94.9
148-149	Elmira	96.7	97.4	97.0
NORTH CAROLINA				
270,272-274	Greensboro	101.3	55.7	81.4
271	Winston-Salem	101.0	56.5	81.6
275-276	Raleigh	100.5	54.1	80.3
277	Durham	102.9	55.3	82.1
278	Rocky Mount	98.5	52.3	78.3
279	Elizabeth City	99.3	51.6	78.5
280	Gastonia	101.6	55.1	81.3
281-282	Charlotte	101.9	56.8	82.2
283	Fayetteville	104.5	58.1	84.2
284	Wilmington	100.3	55.9	81.0
285	Kinston	98.4	52.7	78.5

For customer support on your Commercial Renovation Cost Data, call 877.791.4977.

Location Factors

STATE/ZIP	CITY	MAT.	INST.	TOTAL
NORTH CAROLINA (CONT'D)				
286	Hickory	98.8	53.9	79.2
287-288	Asheville	100.8	55.3	80.9
289	Murphy	99.6	48.3	77.2
NORTH DAKOTA				
580-581	Fargo	102.7	67.6	87.4
582	Grand Forks	102.7	51.6	80.4
583	Devils Lake	102.0	58.6	83.1
584	Jamestown	102.1	45.2	77.3
585	Bismarck	103.1	66.2	87.0
586	Dickinson	102.8	60.9	84.6
587	Minot	102.7	71.3	89.0
588	Williston	101.2	60.9	83.7
OHIO				
430-432	Columbus	98.6	89.2	94.5
433	Marion	95.1	85.3	90.8
434-436	Toledo	98.7	98.1	98.4
437-438	Zanesville	95.6	86.5	91.6
439	Steubenville	96.8	94.9	96.0
440	Lorain	98.9	93.0	96.3
441	Cleveland	99.0	100.3	99.6
442-443	Akron	99.9	94.2	97.4
444-445	Youngstown	99.2	89.0	94.7
446-447	Canton	99.3	85.6	93.3
448-449	Mansfield	96.7	87.3	92.6
450	Hamilton	97.9	85.1	92.3
451-452	Cincinnati	98.3	84.5	92.2
453-454	Dayton	98.0	84.4	92.1
455	Springfield	98.0	84.7	92.2
456	Chillicothe	96.8	93.0	95.1
457	Athens	99.6	80.4	91.2
458	Lima	100.0	86.8	94.3
OKLAHOMA				
730-731	Oklahoma City	100.3	65.7	85.2
734	Ardmore	98.3	61.9	82.4
735	Lawton	100.7	62.8	84.2
736	Clinton	99.6	63.1	83.7
737	Enid	100.2	60.5	82.9
738	Woodward	98.3	63.4	83.1
739	Guymon	99.4	59.7	82.1
740-741	Tulsa	98.1	61.6	82.2
743	Miami	94.8	67.5	82.9
744	Muskogee	97.4	55.7	79.2
745	Mcalester	94.5	60.4	79.7
746	Ponca City	95.1	62.1	80.7
747	Durant	95.1	63.5	81.3
748	Shawnee	96.5	61.8	81.4
749	Poteau	94.3	61.4	80.0
OREGON				
970-972	Portland	98.8	100.4	99.5
973	Salem	100.5	99.4	100.0
974	Eugene	98.6	99.3	98.9
975	Medford	100.0	97.1	98.8
976	Klamath Falls	99.8	97.1	98.6
977	Bend	98.9	99.4	99.1
978	Pendleton	94.8	101.4	97.7
979	Vale	92.9	98.1	95.2
PENNSYLVANIA				
150-152	Pittsburgh	99.9	104.6	102.0
153	Washington	97.0	103.2	99.7
154	Uniontown	97.3	101.7	99.2
155	Bedford	98.2	91.3	95.2
156	Greensburg	98.2	100.9	99.4
157	Indiana	97.1	99.3	98.0
158	Dubois	98.5	96.8	97.8
159	Johnstown	98.1	97.2	97.7
160	Butler	92.5	102.0	96.6
161	New Castle	92.5	97.9	94.8
162	Kittanning	92.9	103.3	97.4
163	Oil City	92.4	96.9	94.4
164-165	Erie	94.5	95.8	95.1
166	Altoona	94.6	91.9	93.4
167	Bradford	95.7	96.4	96.0
168	State College	95.3	92.8	94.2
169	Wellsboro	96.3	92.6	94.7
170-171	Harrisburg	99.8	95.6	98.0
172	Chambersburg	96.2	88.0	92.6
173-174	York	96.7	95.9	96.3
175-176	Lancaster	95.1	89.0	92.4

STATE/ZIP	CITY	MAT.	INST.	TOTAL
PENNSYLVANIA (CONT'D)				
177	Williamsport	93.8	83.0	89.1
178	Sunbury	95.8	95.8	95.8
179	Pottsville	94.9	98.5	96.5
180	Lehigh Valley	95.8	114.0	103.7
181	Allentown	97.9	108.6	102.5
182	Hazleton	95.4	98.3	96.6
183	Stroudsburg	95.3	101.4	97.9
184-185	Scranton	98.6	99.2	98.9
186-187	Wilkes-Barre	95.1	98.7	96.6
188	Montrose	94.7	96.9	95.7
189	Doylestown	94.9	120.6	106.1
190-191	Philadelphia	99.7	133.6	114.5
193	Westchester	95.9	121.5	107.1
194	Norristown	95.0	129.5	110.0
195-196	Reading	97.1	102.1	99.3
PUERTO RICO				
009	San Juan	122.7	24.4	79.8
RHODE ISLAND				
028	Newport	98.4	117.5	106.7
029	Providence	100.6	117.5	108.0
SOUTH CAROLINA				
290-292	Columbia	99.3	56.5	80.6
293	Spartanburg	98.6	56.7	80.3
294	Charleston	99.9	65.2	84.8
295	Florence	98.2	56.9	80.2
296	Greenville	98.3	55.8	79.8
297	Rock Hill	97.8	50.3	77.1
298	Aiken	98.6	69.1	85.7
299	Beaufort	99.3	44.9	75.6
SOUTH DAKOTA				
570-571	Sioux Falls	100.0	57.7	81.5
572	Watertown	99.0	49.1	77.2
573	Mitchell	97.8	48.4	76.3
574	Aberdeen	100.5	49.6	78.3
575	Pierre	101.7	58.4	82.8
576	Mobridge	98.5	48.8	76.8
577	Rapid City	100.4	59.5	82.6
TENNESSEE				
370-372	Nashville	98.4	74.1	87.8
373-374	Chattanooga	100.0	66.2	85.2
375,380-381	Memphis	99.0	71.3	86.9
376	Johnson City	99.8	56.5	80.9
377-379	Knoxville	96.6	66.9	83.7
382	Mckenzie	98.2	59.2	81.2
383	Jackson	99.9	61.2	83.0
384	Columbia	96.9	66.1	83.5
385	Cookeville	98.1	59.1	81.1
TEXAS				
750	Mckinney	98.7	63.1	83.2
751	Waxahackie	98.6	65.9	84.4
752-753	Dallas	99.5	67.4	85.5
754	Greenville	98.8	64.0	83.6
755	Texarkana	98.0	58.4	80.7
756	Longview	98.5	58.7	81.2
757	Tyler	98.5	65.0	83.9
758	Palestine	95.0	64.4	81.7
759	Lufkin	95.6	67.8	83.5
760-761	Fort Worth	99.3	65.3	84.5
762	Denton	98.9	62.5	83.0
763	Wichita Falls	97.1	64.4	82.8
764	Eastland	95.9	63.2	81.6
765	Temple	94.7	61.7	80.3
766-767	Waco	96.5	64.1	82.4
768	Brownwood	98.9	59.4	81.7
769	San Angelo	98.5	59.6	81.5
770-772	Houston	99.4	70.5	86.8
773	Huntsville	98.0	66.7	84.4
774	Wharton	98.9	67.7	85.3
775	Galveston	97.0	70.0	85.2
776-777	Beaumont	97.4	68.6	84.9
778	Bryan	94.9	68.2	83.3
779	Victoria	99.0	65.6	84.4
780	Laredo	98.3	64.5	83.6
781-782	San Antonio	98.5	65.1	84.0
783-784	Corpus Christi	100.9	63.5	84.6
785	Mc Allen	101.0	59.8	83.0
786-787	Austin	100.0	63.6	84.1

For customer support on your Commercial Renovation Cost Data, call 877.791.4977.

Location Factors

STATE/ZIP	CITY	MAT.	INST.	TOTAL
TEXAS (CONT'D)				
788	Del Rio	100.8	63.2	84.4
789	Giddings	97.7	64.0	83.0
790-791	Amarillo	100.2	63.2	84.1
792	Childress	99.3	64.3	84.0
793-794	Lubbock	101.6	63.6	85.0
795-796	Abilene	99.7	62.2	83.3
797	Midland	101.5	64.2	85.2
798-799,885	El Paso	98.0	61.7	82.2
UTAH				
840-841	Salt Lake City	102.2	68.3	87.4
842,844	Ogden	98.1	70.1	85.9
843	Logan	99.8	70.1	86.9
845	Price	100.2	64.6	84.7
846-847	Provo	100.3	68.2	86.3
VERMONT				
050	White River Jct.	98.5	78.3	89.7
051	Bellows Falls	97.2	93.0	95.3
052	Bennington	97.5	93.5	95.8
053	Brattleboro	97.8	92.8	95.6
054	Burlington	102.4	83.6	94.2
056	Montpelier	99.5	82.8	92.2
057	Rutland	99.7	83.6	92.7
058	St. Johnsbury	98.6	77.8	89.6
059	Guildhall	97.4	77.8	88.8
VIRGINIA				
220-221	Fairfax	100.7	84.0	93.4
222	Arlington	101.9	84.0	94.1
223	Alexandria	101.1	85.5	94.3
224-225	Fredericksburg	99.5	82.7	92.1
226	Winchester	100.0	82.2	92.2
227	Culpeper	99.9	80.5	91.4
228	Harrisonburg	100.2	67.2	85.8
229	Charlottesville	100.5	67.1	86.0
230-232	Richmond	101.3	68.7	87.1
233-235	Norfolk	100.5	68.2	86.4
236	Newport News	100.0	68.2	86.1
237	Portsmouth	99.4	65.3	84.5
238	Petersburg	99.4	68.7	86.0
239	Farmville	98.7	52.4	78.5
240-241	Roanoke	101.5	65.1	85.6
242	Bristol	99.4	54.9	80.0
243	Pulaski	99.0	57.3	80.9
244	Staunton	99.7	61.8	83.2
245	Lynchburg	99.9	65.6	85.0
246	Grundy	99.3	52.5	78.9
WASHINGTON				
980-981,987	Seattle	101.4	104.7	102.8
982	Everett	101.6	99.6	100.7
983-984	Tacoma	101.8	99.7	100.9
985	Olympia	101.2	99.4	100.4
986	Vancouver	102.7	92.1	98.1
988	Wenatchee	101.6	86.6	95.0
989	Yakima	102.0	94.1	98.6
990-992	Spokane	103.4	82.3	94.2
993	Richland	103.1	89.3	97.1
994	Clarkston	101.5	80.2	92.2
WEST VIRGINIA				
247-248	Bluefield	98.3	92.9	96.0
249	Lewisburg	99.8	92.4	96.6
250-253	Charleston	99.9	94.8	97.6
254	Martinsburg	99.3	86.4	93.7
255-257	Huntington	100.7	95.7	98.5
258-259	Beckley	97.9	93.3	95.9
260	Wheeling	101.2	94.3	98.2
261	Parkersburg	100.1	91.8	96.5
262	Buckhannon	99.6	95.5	97.8
263-264	Clarksburg	100.1	95.4	98.1
265	Morgantown	100.2	94.4	97.6
266	Gassaway	99.5	93.9	97.1
267	Romney	99.4	92.2	96.3
268	Petersburg	99.3	91.3	95.8
WISCONSIN				
530,532	Milwaukee	100.0	107.2	103.1
531	Kenosha	99.5	102.2	100.7
534	Racine	99.1	102.4	100.5
535	Beloit	98.9	100.7	99.7
537	Madison	99.1	99.0	99.1

STATE/ZIP	CITY	MAT.	INST.	TOTAL
WISCONSIN (CONT'D)				
538	Lancaster	96.7	92.5	94.8
539	Portage	95.3	96.6	95.9
540	New Richmond	96.8	94.6	95.9
541-543	Green Bay	101.2	93.4	97.8
544	Wausau	96.2	93.5	95.0
545	Rhinelander	99.3	91.2	95.8
546	La Crosse	97.8	94.5	96.3
547	Eau Claire	99.3	95.3	97.5
548	Superior	96.6	96.3	96.5
549	Oshkosh	96.9	91.9	94.7
WYOMING				
820	Cheyenne	101.5	65.2	85.7
821	Yellowstone Nat'l Park	99.3	66.3	84.9
822	Wheatland	100.4	64.3	84.7
823	Rawlins	101.8	66.9	86.6
824	Worland	100.0	65.3	84.9
825	Riverton	101.0	65.4	85.5
826	Casper	102.0	63.9	85.4
827	Newcastle	99.9	66.9	85.5
828	Sheridan	102.6	64.3	85.9
829-831	Rock Springs	103.7	65.3	87.0
CANADIAN FACTORS (reflect Canadian currency)				
ALBERTA				
	Calgary	123.5	95.7	111.4
	Edmonton	123.1	95.7	111.2
	Fort McMurray	123.6	92.8	110.2
	Lethbridge	118.4	92.2	107.0
	Lloydminster	113.4	88.9	102.8
	Medicine Hat	113.6	88.2	102.5
	Red Deer	114.1	88.2	102.8
BRITISH COLUMBIA				
	Kamloops	114.5	90.6	104.1
	Prince George	115.4	89.9	104.3
	Vancouver	119.5	89.1	106.2
	Victoria	115.7	87.2	103.3
MANITOBA				
	Brandon	125.2	75.8	103.7
	Portage la Prairie	113.4	74.2	96.3
	Winnipeg	121.0	71.8	99.5
NEW BRUNSWICK				
	Bathurst	111.5	67.3	92.2
	Dalhousie	111.9	67.5	92.5
	Fredericton	117.4	72.5	97.9
	Moncton	111.8	70.1	93.6
	Newcastle	111.6	67.9	92.5
	St. John	114.0	77.9	98.3
NEWFOUNDLAND				
	Corner Brook	129.6	67.6	102.6
	St Johns	122.7	85.2	106.4
NORTHWEST TERRITORIES				
	Yellowknife	135.3	84.2	113.0
NOVA SCOTIA				
	Bridgewater	112.9	74.6	96.2
	Dartmouth	126.1	74.6	103.7
	Halifax	119.6	80.9	102.7
	New Glasgow	124.1	74.6	102.5
	Sydney	122.1	74.6	101.4
	Truro	112.5	74.6	96.0
	Yarmouth	124.0	74.6	102.5
ONTARIO				
	Barrie	117.7	93.0	106.9
	Brantford	114.9	96.9	107.1
	Cornwall	115.0	93.3	105.5
	Hamilton	117.0	96.7	108.2
	Kingston	116.0	93.4	106.2
	Kitchener	109.6	92.9	102.3
	London	116.4	93.3	106.3
	North Bay	127.3	91.1	111.5
	Oshawa	112.3	94.3	104.4
	Ottawa	117.3	94.0	107.2
	Owen Sound	117.8	91.2	106.2
	Peterborough	114.9	93.1	105.4
	Sarnia	114.8	97.5	107.3
	Sault Ste Marie	110.3	92.1	102.3

Location Factors

STATE/ZIP	CITY	MAT.	INST.	TOTAL
ONTARIO (CONT'D)				
	St. Catharines	107.6	94.5	101.9
	Sudbury	107.7	93.0	101.3
	Thunder Bay	108.8	93.4	102.1
	Timmins	115.1	91.1	104.6
	Toronto	118.3	101.3	110.9
	Windsor	108.2	93.4	101.7
PRINCE EDWARD ISLAND				
	Charlottetown	120.0	62.8	95.1
	Summerside	126.7	62.8	98.8
QUEBEC				
	Cap-de-la-Madeleine	112.4	87.6	101.6
	Charlesbourg	112.4	87.6	101.6
	Chicoutimi	111.4	92.9	103.3
	Gatineau	112.1	87.4	101.3
	Granby	112.3	87.4	101.4
	Hull	112.2	87.4	101.4
	Joliette	112.5	87.6	101.7
	Laval	112.1	87.4	101.3
	Montreal	117.8	93.2	107.1
	Quebec	117.1	93.6	106.8
	Rimouski	111.8	92.9	103.6
	Rouyn-Noranda	112.0	87.4	101.3
	Saint Hyacinthe	111.6	87.4	101.1
	Sherbrooke	112.4	87.4	101.5
	Sorel	112.5	87.6	101.7
	St Jerome	112.1	87.4	101.3
	Trois Rivieres	124.8	87.6	108.6
SASKATCHEWAN				
	Moose Jaw	110.7	69.5	92.7
	Prince Albert	109.9	67.7	91.5
	Regina	120.7	89.0	106.9
	Saskatoon	111.1	89.1	101.5
YUKON				
	Whitehorse	135.3	68.5	106.2

General Requirements R0111 Summary of Work

R011105-05 Tips for Accurate Estimating

1. Use pre-printed or columnar forms for orderly sequence of dimensions and locations and for recording telephone quotations.
2. Use only the front side of each paper or form except for certain pre-printed summary forms.
3. Be consistent in listing dimensions: For example, length x width x height. This helps in rechecking to ensure that, the total length of partitions is appropriate for the building area.
4. Use printed (rather than measured) dimensions where given.
5. Add up multiple printed dimensions for a single entry where possible.
6. Measure all other dimensions carefully.
7. Use each set of dimensions to calculate multiple related quantities.
8. Convert foot and inch measurements to decimal feet when listing. Memorize decimal equivalents to .01 parts of a foot (1/8" equals approximately .01').
9. Do not "round off" quantities until the final summary.
10. Mark drawings with different colors as items are taken off.
11. Keep similar items together, different items separate.
12. Identify location and drawing numbers to aid in future checking for completeness.
13. Measure or list everything on the drawings or mentioned in the specifications.
14. It may be necessary to list items not called for to make the job complete.
15. Be alert for: Notes on plans such as N.T.S. (not to scale); changes in scale throughout the drawings; reduced size drawings; discrepancies between the specifications and the drawings.
16. Develop a consistent pattern of performing an estimate. For example:
    a. Start the quantity takeoff at the lower floor and move to the next higher floor.
    b. Proceed from the main section of the building to the wings.
    c. Proceed from south to north or vice versa, clockwise or counterclockwise.
    d. Take off floor plan quantities first, elevations next, then detail drawings.
17. List all gross dimensions that can be either used again for different quantities, or used as a rough check of other quantities for verification (exterior perimeter, gross floor area, individual floor areas, etc.).
18. Utilize design symmetry or repetition (repetitive floors, repetitive wings, symmetrical design around a center line, similar room layouts, etc.). Note: Extreme caution is needed here so as not to omit or duplicate an area.
19. Do not convert units until the final total is obtained. For instance, when estimating concrete work, keep all units to the nearest cubic foot, then summarize and convert to cubic yards.
20. When figuring alternatives, it is best to total all items involved in the basic system, then total all items involved in the alternates. Therefore you work with positive numbers in all cases. When adds and deducts are used, it is often confusing whether to add or subtract a portion of an item; especially on a complicated or involved alternate.

General Requirements — R0111 Summary of Work

R011105-50 Metric Conversion Factors

Description: This table is primarily for converting customary U.S. units in the left hand column to SI metric units in the right hand column. In addition, conversion factors for some commonly encountered Canadian and non-SI metric units are included.

	If You Know		Multiply By		To Find
Length	Inches	x	25.4[a]	=	Millimeters
	Feet	x	0.3048[a]	=	Meters
	Yards	x	0.9144[a]	=	Meters
	Miles (statute)	x	1.609	=	Kilometers
Area	Square inches	x	645.2	=	Square millimeters
	Square feet	x	0.0929	=	Square meters
	Square yards	x	0.8361	=	Square meters
Volume (Capacity)	Cubic inches	x	16,387	=	Cubic millimeters
	Cubic feet	x	0.02832	=	Cubic meters
	Cubic yards	x	0.7646	=	Cubic meters
	Gallons (U.S. liquids)[b]	x	0.003785	=	Cubic meters[c]
	Gallons (Canadian liquid)[b]	x	0.004546	=	Cubic meters[c]
	Ounces (U.S. liquid)[b]	x	29.57	=	Milliliters[c,d]
	Quarts (U.S. liquid)[b]	x	0.9464	=	Liters[c,d]
	Gallons (U.S. liquid)[b]	x	3.785	=	Liters[c,d]
Force	Kilograms force[d]	x	9.807	=	Newtons
	Pounds force	x	4.448	=	Newtons
	Pounds force	x	0.4536	=	Kilograms force[d]
	Kips	x	4448	=	Newtons
	Kips	x	453.6	=	Kilograms force[d]
Pressure, Stress, Strength (Force per unit area)	Kilograms force per square centimeter[d]	x	0.09807	=	Megapascals
	Pounds force per square inch (psi)	x	0.006895	=	Megapascals
	Kips per square inch	x	6.895	=	Megapascals
	Pounds force per square inch (psi)	x	0.07031	=	Kilograms force per square centimeter[d]
	Pounds force per square foot	x	47.88	=	Pascals
	Pounds force per square foot	x	4.882	=	Kilograms force per square meter[d]
Bending Moment Or Torque	Inch-pounds force	x	0.01152	=	Meter-kilograms force[d]
	Inch-pounds force	x	0.1130	=	Newton-meters
	Foot-pounds force	x	0.1383	=	Meter-kilograms force[d]
	Foot-pounds force	x	1.356	=	Newton-meters
	Meter-kilograms force[d]	x	9.807	=	Newton-meters
Mass	Ounces (avoirdupois)	x	28.35	=	Grams
	Pounds (avoirdupois)	x	0.4536	=	Kilograms
	Tons (metric)	x	1000	=	Kilograms
	Tons, short (2000 pounds)	x	907.2	=	Kilograms
	Tons, short (2000 pounds)	x	0.9072	=	Megagrams[e]
Mass per Unit Volume	Pounds mass per cubic foot	x	16.02	=	Kilograms per cubic meter
	Pounds mass per cubic yard	x	0.5933	=	Kilograms per cubic meter
	Pounds mass per gallon (U.S. liquid)[b]	x	119.8	=	Kilograms per cubic meter
	Pounds mass per gallon (Canadian liquid)[b]	x	99.78	=	Kilograms per cubic meter
Temperature	Degrees Fahrenheit	(F-32)/1.8		=	Degrees Celsius
	Degrees Fahrenheit	(F+459.67)/1.8		=	Degrees Kelvin
	Degrees Celsius	C+273.15		=	Degrees Kelvin

[a] The factor given is exact
[b] One U.S. gallon = 0.8327 Canadian gallon
[c] 1 liter = 1000 milliliters = 1000 cubic centimeters
 1 cubic decimeter = 0.001 cubic meter
[d] Metric but not SI unit
[e] Called "tonne" in England and "metric ton" in other metric countries

General Requirements — R0111 Summary of Work

R011105-60 Weights and Measures

Measures of Length
1 Mile = 1760 Yards = 5280 Feet
1 Yard = 3 Feet = 36 inches
1 Foot = 12 Inches
1 Mil = 0.001 Inch
1 Fathom = 2 Yards = 6 Feet
1 Rod = 5.5 Yards = 16.5 Feet
1 Hand = 4 Inches
1 Span = 9 Inches
1 Micro-inch = One Millionth Inch or 0.000001 Inch
1 Micron = One Millionth Meter + 0.00003937 Inch

Surveyor's Measure
1 Mile = 8 Furlongs = 80 Chains
1 Furlong = 10 Chains = 220 Yards
1 Chain = 4 Rods = 22 Yards = 66 Feet = 100 Links
1 Link = 7.92 Inches

Square Measure
1 Square Mile = 640 Acres = 6400 Square Chains
1 Acre = 10 Square Chains = 4840 Square Yards = 43,560 Sq. Ft.
1 Square Chain = 16 Square Rods = 484 Square Yards = 4356 Sq. Ft.
1 Square Rod = 30.25 Square Yards = 272.25 Square Feet = 625 Square Lines
1 Square Yard = 9 Square Feet
1 Square Foot = 144 Square Inches
An Acre equals a Square 208.7 Feet per Side

Cubic Measure
1 Cubic Yard = 27 Cubic Feet
1 Cubic Foot = 1728 Cubic Inches
1 Cord of Wood = 4 x 4 x 8 Feet = 128 Cubic Feet
1 Perch of Masonry = 16½ x 1½ x 1 Foot = 24.75 Cubic Feet

Avoirdupois or Commercial Weight
1 Gross or Long Ton = 2240 Pounds
1 Net or Short ton = 2000 Pounds
1 Pound = 16 Ounces = 7000 Grains
1 Ounce = 16 Drachms = 437.5 Grains
1 Stone = 14 Pounds

Shipping Measure
For Measuring Internal Capacity of a Vessel:
  1 Register Ton = 100 Cubic Feet

For Measurement of Cargo:
  Approximately 40 Cubic Feet of Merchandise is considered a Shipping Ton, unless that bulk would weigh more than 2000 Pounds, in which case Freight Charge may be based upon weight.

40 Cubic Feet = 32.143 U.S. Bushels = 31.16 Imp. Bushels

Liquid Measure
1 Imperial Gallon = 1.2009 U.S. Gallon = 277.42 Cu. In.
1 Cubic Foot = 7.48 U.S. Gallons

R011110-10 Architectural Fees

Tabulated below are typical percentage fees by project size, for good professional architectural service. Fees may vary from those listed depending upon degree of design difficulty and economic conditions in any particular area.

Rates can be interpolated horizontally and vertically. Various portions of the same project requiring different rates should be adjusted proportionately. For alterations, add 50% to the fee for the first $500,000 of project cost and add 25% to the fee for project cost over $500,000.

Architectural fees tabulated below include Structural, Mechanical and Electrical Engineering Fees. They do not include the fees for special consultants such as kitchen planning, security, acoustical, interior design, etc.

Civil Engineering fees are included in the Architectural fee for project sites requiring minimal design such as city sites. However, separate Civil Engineering fees must be added when utility connections require design, drainage calculations are needed, stepped foundations are required, or provisions are required to protect adjacent wetlands.

Building Types	Total Project Size in Thousands of Dollars						
	100	250	500	1,000	5,000	10,000	50,000
Factories, garages, warehouses, repetitive housing	9.0%	8.0%	7.0%	6.2%	5.3%	4.9%	4.5%
Apartments, banks, schools, libraries, offices, municipal buildings	12.2	12.3	9.2	8.0	7.0	6.6	6.2
Churches, hospitals, homes, laboratories, museums, research	15.0	13.6	12.7	11.9	9.5	8.8	8.0
Memorials, monumental work, decorative furnishings	—	16.0	14.5	13.1	10.0	9.0	8.3

General Requirements — R0111 Summary of Work

R011110-30 Engineering Fees

Typical **Structural Engineering Fees** based on type of construction and total project size. These fees are included in Architectural Fees.

Type of Construction	Total Project Size (in thousands of dollars)			
	$500	$500-$1,000	$1,000-$5,000	Over $5000
Industrial buildings, factories & warehouses	Technical payroll times 2.0 to 2.5	1.60%	1.25%	1.00%
Hotels, apartments, offices, dormitories, hospitals, public buildings, food stores		2.00%	1.70%	1.20%
Museums, banks, churches and cathedrals		2.00%	1.75%	1.25%
Thin shells, prestressed concrete, earthquake resistive		2.00%	1.75%	1.50%
Parking ramps, auditoriums, stadiums, convention halls, hangars & boiler houses		2.50%	2.00%	1.75%
Special buildings, major alterations, underpinning & future expansion	↓	Add to above 0.5%	Add to above 0.5%	Add to above 0.5%

For complex reinforced concrete or unusually complicated structures, add 20% to 50%.

Typical **Mechanical and Electrical Engineering Fees** are based on the size of the subcontract. The fee structure for both are shown below. These fees are included in Architectural Fees.

Type of Construction	Subcontract Size							
	$25,000	$50,000	$100,000	$225,000	$350,000	$500,000	$750,000	$1,000,000
Simple structures	6.4%	5.7%	4.8%	4.5%	4.4%	4.3%	4.2%	4.1%
Intermediate structures	8.0	7.3	6.5	5.6	5.1	5.0	4.9	4.8
Complex structures	10.1	9.0	9.0	8.0	7.5	7.5	7.0	7.0

For renovations, add 15% to 25% to applicable fee.

General Requirements — R0121 Allowances

R012153-10 Repair and Remodeling

Cost figures are based on new construction utilizing the most cost-effective combination of labor, equipment and material with the work scheduled in proper sequence to allow the various trades to accomplish their work in an efficient manner.

The costs for repair and remodeling work must be modified due to the following factors that may be present in any given repair and remodeling project.

1. Equipment usage curtailment due to the physical limitations of the project, with only hand-operated equipment being used.
2. Increased requirement for shoring and bracing to hold up the building while structural changes are being made and to allow for temporary storage of construction materials on above-grade floors.
3. Material handling becomes more costly due to having to move within the confines of an enclosed building. For multi-story construction, low capacity elevators and stairwells may be the only access to the upper floors.
4. Large amount of cutting and patching and attempting to match the existing construction is required. It is often more economical to remove entire walls rather than create many new door and window openings. This sort of trade-off has to be carefully analyzed.
5. Cost of protection of completed work is increased since the usual sequence of construction usually cannot be accomplished.
6. Economies of scale usually associated with new construction may not be present. If small quantities of components must be custom fabricated due to job requirements, unit costs will naturally increase. Also, if only small work areas are available at a given time, job scheduling between trades becomes difficult and subcontractor quotations may reflect the excessive start-up and shut-down phases of the job.
7. Work may have to be done on other than normal shifts and may have to be done around an existing production facility which has to stay in production during the course of the repair and remodeling.
8. Dust and noise protection of adjoining non-construction areas can involve substantial special protection and alter usual construction methods.
9. Job may be delayed due to unexpected conditions discovered during demolition or removal. These delays ultimately increase construction costs.
10. Piping and ductwork runs may not be as simple as for new construction. Wiring may have to be snaked through walls and floors.
11. Matching "existing construction" may be impossible because materials may no longer be manufactured. Substitutions may be expensive.
12. Weather protection of existing structure requires additional temporary structures to protect building at openings.
13. On small projects, because of local conditions, it may be necessary to pay a tradesman for a minimum of four hours for a task that is completed in one hour.

All of the above areas can contribute to increased costs for a repair and remodeling project. Each of the above factors should be considered in the planning, bidding and construction stage in order to minimize the increased costs associated with repair and remodeling jobs.

General Requirements — R0129 Payment Procedures

R012909-80 Sales Tax by State

State sales tax on materials is tabulated below (5 states have no sales tax). Many states allow local jurisdictions, such as a county or city, to levy additional sales tax.

Some projects may be sales tax exempt, particularly those constructed with public funds.

State	Tax (%)	State	Tax (%)	State	Tax (%)	State	Tax (%)
Alabama	4	Illinois	6.25	Montana	0	Rhode Island	7
Alaska	0	Indiana	7	Nebraska	5.5	South Carolina	6
Arizona	5.6	Iowa	6	Nevada	6.85	South Dakota	4
Arkansas	6	Kansas	6.3	New Hampshire	0	Tennessee	7
California	7.5	Kentucky	6	New Jersey	7	Texas	6.25
Colorado	2.9	Louisiana	4	New Mexico	5.13	Utah	4.7
Connecticut	6.35	Maine	5	New York	4	Vermont	6
Delaware	0	Maryland	6	North Carolina	4.75	Virginia	5
District of Columbia	6	Massachusetts	6.25	North Dakota	5	Washington	6.5
Florida	6	Michigan	6	Ohio	5.5	West Virginia	6
Georgia	4	Minnesota	6.88	Oklahoma	4.5	Wisconsin	5
Hawaii	4	Mississippi	7	Oregon	0	Wyoming	4
Idaho	6	Missouri	4.23	Pennsylvania	6	Average	5.04 %

Sales Tax by Province (Canada)

GST - a value-added tax, which the government imposes on most goods and services provided in or imported into Canada. PST - a retail sales tax, which three of the provinces impose on the price of most goods and some services. QST - a value-added tax, similar to the federal GST, which Quebec imposes. HST - Five provinces have combined their retail sales tax with the federal GST into one harmonized tax.

Province	PST (%)	QST (%)	GST (%)	HST (%)
Alberta	0	0	5	0
British Columbia	7	0	5	0
Manitoba	8	0	5	0
New Brunswick	0	0	0	13
Newfoundland	0	0	0	13
Northwest Territories	0	0	5	0
Nova Scotia	0	0	0	15
Ontario	0	0	0	13
Prince Edward Island	0	0	0	14
Quebec	0	9.975	5	0
Saskatchewan	5	0	5	0
Yukon	0	0	5	0

R012909-85 Unemployment Taxes and Social Security Taxes

State unemployment tax rates vary not only from state to state, but also with the experience rating of the contractor. The federal unemployment tax rate is 6.0% of the first $7,000 of wages. This is reduced by a credit of up to 5.4% for timely payment to the state. The minimum federal unemployment tax is 0.6% after all credits.

Social security (FICA) for 2015 is estimated at time of publication to be 7.65% of wages up to $117,000.

General Requirements — R0131 Project Management & Coordination

R013113-40 Builder's Risk Insurance

Builder's Risk Insurance is insurance on a building during construction. Premiums are paid by the owner or the contractor. Blasting, collapse and underground insurance would raise total insurance costs above those listed. Floater policy for materials delivered to the job runs $.75 to $1.25 per $100 value. Contractor equipment insurance runs $.50 to $1.50 per $100 value. Insurance for miscellaneous tools to $1,500 value runs from $3.00 to $7.50 per $100 value.

Tabulated below are New England Builder's Risk insurance rates in dollars per $100 value for $1,000 deductible. For $25,000 deductible, rates can be reduced 13% to 34%. On contracts over $1,000,000, rates may be lower than those tabulated. Policies are written annually for the total completed value in place. For "all risk" insurance (excluding flood, earthquake and certain other perils) add $.025 to total rates below.

Coverage	Frame Construction (Class 1) Range	Frame Construction (Class 1) Average	Brick Construction (Class 4) Range	Brick Construction (Class 4) Average	Fire Resistive (Class 6) Range	Fire Resistive (Class 6) Average
Fire Insurance	$.350 to $.850	$.600	$.158 to $.189	$.174	$.052 to $.080	$.070
Extended Coverage	.115 to .200	.158	.080 to .105	.101	.081 to .105	.100
Vandalism	.012 to .016	.014	.008 to .011	.011	.008 to .011	.010
Total Annual Rate	$.477 to $1.066	$.772	$.246 to $.305	$.286	$.141 to $.196	$.180

R013113-50 General Contractor's Overhead

There are two distinct types of overhead on a construction project: Project Overhead and Main Office Overhead. Project Overhead includes those costs at a construction site not directly associated with the installation of construction materials. Examples of Project Overhead costs include the following:

1. Superintendent
2. Construction office and storage trailers
3. Temporary sanitary facilities
4. Temporary utilities
5. Security fencing
6. Photographs
7. Clean up
8. Performance and payment bonds

The above Project Overhead items are also referred to as General Requirements and therefore are estimated in Division 1. Division 1 is the first division listed in the CSI MasterFormat but it is usually the last division estimated. The sum of the costs in Divisions 1 through 49 is referred to as the sum of the direct costs.

All construction projects also include indirect costs. The primary components of indirect costs are the contractor's Main Office Overhead and profit. The amount of the Main Office Overhead expense varies depending on the the following:

1. Owner's compensation
2. Project managers and estimator's wages
3. Clerical support wages
4. Office rent and utilities
5. Corporate legal and accounting costs
6. Advertising
7. Automobile expenses
8. Association dues
9. Travel and entertainment expenses

These costs are usually calculated as a percentage of annual sales volume. This percentage can range from 35% for a small contractor doing less than $500,000 to 5% for a large contractor with sales in excess of $100 million.

General Requirements — R0131 Project Management & Coordination

R013113-60 Workers' Compensation Insurance Rates by Trade

The table below tabulates the national averages for workers' compensation insurance rates by trade and type of building. The average "Insurance Rate" is multiplied by the "% of Building Cost" for each trade. This produces the "Workers' Compensation" cost by % of total labor cost, to be added for each trade by building type to determine the weighted average workers' compensation rate for the building types analyzed.

Trade	Insurance Rate (% Labor Cost) Range	Average	% of Building Cost Office Bldgs.	Schools & Apts.	Mfg.	Workers' Compensation Office Bldgs.	Schools & Apts.	Mfg.
Excavation, Grading, etc.	3.4 % to 21.2%	9.7%	4.8%	4.9%	4.5%	0.47%	0.48%	0.44%
Piles & Foundations	6.1 to 28.3	14.3	7.1	5.2	8.7	1.02	0.74	1.24
Concrete	4.2 to 35.8	12.9	5.0	14.8	3.7	0.65	1.91	0.48
Masonry	4.9 to 36.4	13.7	6.9	7.5	1.9	0.95	1.03	0.26
Structural Steel	6.1 to 101.1	31.7	10.7	3.9	17.6	3.39	1.24	5.58
Miscellaneous & Ornamental Metals	4.3 to 31.5	12.2	2.8	4.0	3.6	0.34	0.49	0.44
Carpentry & Millwork	5.0 to 36.8	14.9	3.7	4.0	0.5	0.55	0.60	0.07
Metal or Composition Siding	6.1 to 69.2	18.4	2.3	0.3	4.3	0.42	0.06	0.79
Roofing	6.1 to 100.3	31.7	2.3	2.6	3.1	0.73	0.82	0.98
Doors & Hardware	4.1 to 36.8	11.3	0.9	1.4	0.4	0.10	0.16	0.05
Sash & Glazing	4.8 to 30.7	13.3	3.5	4.0	1.0	0.47	0.53	0.13
Lath & Plaster	3.1 to 36.2	10.9	3.3	6.9	0.8	0.36	0.75	0.09
Tile, Marble & Floors	3 to 24.8	8.9	2.6	3.0	0.5	0.23	0.27	0.04
Acoustical Ceilings	3.8 to 33.8	8.4	2.4	0.2	0.3	0.20	0.02	0.03
Painting	4.5 to 36.1	11.7	1.5	1.6	1.6	0.18	0.19	0.19
Interior Partitions	5.0 to 36.8	14.9	3.9	4.3	4.4	0.58	0.64	0.66
Miscellaneous Items	2.5 to 132.3	13.4	5.2	3.7	9.7	0.70	0.50	1.30
Elevators	1.3 to 12.8	5.3	2.1	1.1	2.2	0.11	0.06	0.12
Sprinklers	2.9 to 18.5	7.3	0.5	—	2.0	0.04	—	0.15
Plumbing	2.2 to 19.3	7.0	4.9	7.2	5.2	0.34	0.50	0.36
Heat., Vent., Air Conditioning	3.8 to 18.1	8.8	13.5	11.0	12.9	1.19	0.97	1.14
Electrical	2.4 to 13.5	5.8	10.1	8.4	11.1	0.59	0.49	0.64
Total	1.3 % to 132.3%	—	100.0%	100.0%	100.0%	13.61%	12.45%	15.18%

Overall Weighted Average 13.75%

Workers' Compensation Insurance Rates by States

The table below lists the weighted average workers' compensation base rate for each state with a factor comparing this with the national average of 13.5%.

State	Weighted Average	Factor	State	Weighted Average	Factor	State	Weighted Average	Factor
Alabama	20.6%	153	Kentucky	12.5%	93	North Dakota	8.1%	60
Alaska	12.5	93	Louisiana	19.6	145	Ohio	8.1	60
Arizona	13.8	102	Maine	11.4	84	Oklahoma	10.3	76
Arkansas	7.7	57	Maryland	13.5	100	Oregon	11.3	84
California	27.1	201	Massachusetts	11.5	85	Pennsylvania	18.3	136
Colorado	7.3	54	Michigan	13.5	100	Rhode Island	14.7	109
Connecticut	25.9	192	Minnesota	22.8	169	South Carolina	16.9	125
Delaware	14.0	104	Mississippi	14.0	104	South Dakota	14.1	104
District of Columbia	10.4	77	Missouri	13.9	103	Tennessee	12.2	90
Florida	10.8	80	Montana	8.3	61	Texas	9.3	69
Georgia	29.5	219	Nebraska	17.0	126	Utah	8.8	65
Hawaii	8.7	64	Nevada	9.2	68	Vermont	13.3	99
Idaho	10.5	78	New Hampshire	18.8	139	Virginia	8.7	64
Illinois	26.9	199	New Jersey	15.7	116	Washington	9.5	70
Indiana	5.0	37	New Mexico	15.3	113	West Virginia	7.4	55
Iowa	15.5	115	New York	19.0	141	Wisconsin	13.2	98
Kansas	9.2	68	North Carolina	17.3	128	Wyoming	6.5	48

Weighted Average for U.S. is 13.7% of payroll = 100%

Rates in the following table are the base or manual costs per $100 of payroll for workers' compensation in each state. Rates are usually applied to straight time wages only and not to premium time wages and bonuses.

The weighted average skilled worker rate for 35 trades is 13.5%. For bidding purposes, apply the full value of workers' compensation directly to total labor costs, or if labor is 38%, materials 42% and overhead and profit 20% of total cost, carry 38/80 × 13.5% = 6.4% of cost (before overhead and profit) into overhead. Rates vary not only from state to state but also with the experience rating of the contractor.

Rates are the most current available at the time of publication.

General Requirements   R0131 Project Management & Coordination

R013113-80   Performance Bond

This table shows the cost of a Performance Bond for a construction job scheduled to be completed in 12 months. Add 1% of the premium cost per month for jobs requiring more than 12 months to complete. The rates are "standard" rates offered to contractors that the bonding company considers financially sound and capable of doing the work. Preferred rates are offered by some bonding companies based upon financial strength of the contractor. Actual rates vary from contractor to contractor and from bonding company to bonding company. Contractors should prequalify through a bonding agency before submitting a bid on a contract that requires a bond.

Contract Amount	Building Construction Class B Projects	Highways & Bridges Class A New Construction	Class A-1 Highway Resurfacing
First $ 100,000 bid	$25.00 per M	$15.00 per M	$9.40 per M
Next 400,000 bid	$ 2,500 plus $15.00 per M	$ 1,500 plus $10.00 per M	$ 940 plus $7.20 per M
Next 2,000,000 bid	8,500 plus 10.00 per M	5,500 plus 7.00 per M	3,820 plus 5.00 per M
Next 2,500,000 bid	28,500 plus 7.50 per M	19,500 plus 5.50 per M	15,820 plus 4.50 per M
Next 2,500,000 bid	47,250 plus 7.00 per M	33,250 plus 5.00 per M	28,320 plus 4.50 per M
Over 7,500,000 bid	64,750 plus 6.00 per M	45,750 plus 4.50 per M	39,570 plus 4.00 per M

General Requirements   R0154 Construction Aids

R015423-10   Steel Tubular Scaffolding

On new construction, tubular scaffolding is efficient up to 60' high or five stories. Above this it is usually better to use a hung scaffolding if construction permits. Swing scaffolding operations may interfere with tenants. In this case, the tubular is more practical at all heights.

In repairing or cleaning the front of an existing building the cost of tubular scaffolding per S.F. of building front increases as the height increases above the first tier. The first tier cost is relatively high due to leveling and alignment.

The minimum efficient crew for erecting and dismantling is three workers. They can set up and remove 18 frame sections per day up to 5 stories high. For 6 to 12 stories high, a crew of four is most efficient. Use two or more on top and two on the bottom for handing up or hoisting. They can also set up and remove 18 frame sections per day. At 7' horizontal spacing, this will run about 800 S.F. per day of erecting and dismantling. Time for placing and removing planks must be added to the above. A crew of three can place and remove 72 planks per day up to 5 stories. For over 5 stories, a crew of four can place and remove 80 planks per day.

The table below shows the number of pieces required to erect tubular steel scaffolding for 1000 S.F. of building frontage. This area is made up of a scaffolding system that is 12 frames (11 bays) long by 2 frames high.

For jobs under twenty-five frames, add 50% to rental cost. Rental rates will be lower for jobs over three months duration. Large quantities for long periods can reduce rental rates by 20%.

Description of Component	Number of Pieces for 1000 S.F. of Building Front	Unit
5' Wide Standard Frame, 6'-4" High	24	Ea.
Leveling Jack & Plate	24	
Cross Brace	44	
Side Arm Bracket, 21"	12	
Guardrail Post	12	
Guardrail, 7' section	22	
Stairway Section	2	
Stairway Starter Bar	1	
Stairway Inside Handrail	2	
Stairway Outside Handrail	2	
Walk-Thru Frame Guardrail	2	

Scaffolding is often used as falsework over 15' high during construction of cast-in-place concrete beams and slabs. Two foot wide scaffolding is generally used for heavy beam construction. The span between frames depends upon the load to be carried with a maximum span of 5'.

Heavy duty shoring frames with a capacity of 10,000#/leg can be spaced up to 10' O.C. depending upon form support design and loading.

Scaffolding used as horizontal shoring requires less than half the material required with conventional shoring.

On new construction, erection is done by carpenters.

Rolling towers supporting horizontal shores can reduce labor and speed the job. For maintenance work, catwalks with spans up to 70' can be supported by the rolling towers.

General Requirements R0154 Construction Aids

R015423-20 Pump Staging

Pump staging is generally not available for rent. The table below shows the number of pieces required to erect pump staging for 2400 S.F. of building frontage. This area is made up of a pump jack system that is 3 poles (2 bays) wide by 2 poles high.

Item	Number of Pieces for 2400 S.F. of Building Front	Unit
Aluminum pole section, 24' long	6	Ea.
Aluminum splice joint, 6' long	3	
Aluminum foldable brace	3	
Aluminum pump jack	3	
Aluminum support for workbench/back safety rail	3	
Aluminum scaffold plank/workbench, 14" wide x 24' long	4	
Safety net, 22' long	2	
Aluminum plank end safety rail	2	

The cost in place for this 2400 S.F. will depend on how many uses are realized during the life of the equipment.

General Requirements — R0154 Construction Aids

R015433-10 Contractor Equipment

Rental Rates shown elsewhere in the book pertain to late model high quality machines in excellent working condition, rented from equipment dealers. Rental rates from contractors may be substantially lower than the rental rates from equipment dealers depending upon economic conditions; for older, less productive machines, reduce rates by a maximum of 15%. Any overtime must be added to the base rates. For shift work, rates are lower. Usual rule of thumb is 150% of one shift rate for two shifts; 200% for three shifts.

For periods of less than one week, operated equipment is usually more economical to rent than renting bare equipment and hiring an operator.

Costs to move equipment to a job site (mobilization) or from a job site (demobilization) are not included in rental rates, nor in any Equipment costs on any Unit Price line items or crew listings. These costs can be found elsewhere. If a piece of equipment is already at a job site, it is not appropriate to utilize mob/demob costs in an estimate again.

Rental rates vary throughout the country with larger cities generally having lower rates. Lease plans for new equipment are available for periods in excess of six months with a percentage of payments applying toward purchase.

Rental rates can also be treated as reimbursement costs for contractor-owned equipment. Owned equipment costs include depreciation, loan payments, interest, taxes, insurance, storage, and major repairs.

Monthly rental rates vary from 2% to 5% of the cost of the equipment depending on the anticipated life of the equipment and its wearing parts. Weekly rates are about 1/3 the monthly rates and daily rental rates about 1/3 the weekly rate.

The hourly operating costs for each piece of equipment include costs to the user such as fuel, oil, lubrication, normal expendables for the equipment, and a percentage of mechanic's wages chargeable to maintenance. The hourly operating costs listed do not include the operator's wages.

The daily cost for equipment used in the standard crews is figured by dividing the weekly rate by five, then adding eight times the hourly operating cost to give the total daily equipment cost, not including the operator. This figure is in the right hand column of the Equipment listings under Equipment Cost/Day.

Pile Driving rates shown for pile hammer and extractor do not include leads, crane, boiler or compressor. Vibratory pile driving requires an added field specialist during set-up and pile driving operation for the electric model. The hydraulic model requires a field specialist for set-up only. Up to 125 reuses of sheet piling are possible using vibratory drivers. For normal conditions, crane capacity for hammer type and size are as follows.

Crane Capacity	Hammer Type and Size		
	Air or Steam	Diesel	Vibratory
25 ton	to 8,750 ft.-lb.		70 H.P.
40 ton	15,000 ft.-lb.	to 32,000 ft.-lb.	170 H.P.
60 ton	25,000 ft.-lb.		300 H.P.
100 ton		112,000 ft.-lb.	

Cranes should be specified for the job by size, building and site characteristics, availability, performance characteristics, and duration of time required.

Backhoes & Shovels rent for about the same as equivalent size cranes but maintenance and operating expense is higher. Crane operators rate must be adjusted for high boom heights. Average adjustments: for 150' boom add 2% per hour; over 185', add 4% per hour; over 210', add 6% per hour; over 250', add 8% per hour and over 295', add 12% per hour.

Tower Cranes of the climbing or static type have jibs from 50' to 200' and capacities at maximum reach range from 4,000 to 14,000 pounds. Lifting capacities increase up to maximum load as the hook radius decreases.

Typical rental rates, based on purchase price are about 2% to 3% per month.

Erection and dismantling runs between 500 and 2000 labor hours. Climbing operation takes 10 labor hours per 20' climb. Crane dead time is about 5 hours per 40' climb. If crane is bolted to side of the building add cost of ties and extra mast sections. Climbing cranes have from 80' to 180' of mast while static cranes have 80' to 800' of mast.

Truck Cranes can be converted to tower cranes by using tower attachments. Mast heights over 400' have been used.

A single 100' high material **Hoist and Tower** can be erected and dismantled in about 400 labor hours; a double 100' high hoist and tower in about 600 labor hours. Erection times for additional heights are 3 and 4 labor hours per vertical foot respectively up to 150', and 4 to 5 labor hours per vertical foot over 150' high. A 40' high portable Buck hoist takes about 160 labor hours to erect and dismantle. Additional heights take 2 labor hours per vertical foot to 80' and 3 labor hours per vertical foot for the next 100'. Most material hoists do not meet local code requirements for carrying personnel.

A 150' high **Personnel Hoist** requires about 500 to 800 labor hours to erect and dismantle. Budget erection time at 5 labor hours per vertical foot for all trades. Local code requirements or labor scarcity requiring overtime can add up to 50% to any of the above erection costs.

Earthmoving Equipment: The selection of earthmoving equipment depends upon the type and quantity of material, moisture content, haul distance, haul road, time available, and equipment available. Short haul cut and fill operations may require dozers only, while another operation may require excavators, a fleet of trucks, and spreading and compaction equipment. Stockpiled material and granular material are easily excavated with front end loaders. Scrapers are most economically used with hauls between 300' and 1-1/2 miles if adequate haul roads can be maintained. Shovels are often used for blasted rock and any material where a vertical face of 8' or more can be excavated. Special conditions may dictate the use of draglines, clamshells, or backhoes. Spreading and compaction equipment must be matched to the soil characteristics, the compaction required and the rate the fill is being supplied.

General Requirements — R0154 Construction Aids

R015436-50 Mobilization

Costs to move rented construction equipment to a job site from an equipment dealer's or contractor's yard (mobilization), or to move the equipment off the job site (demobilization), are not included in the rental or operating rates, nor in the equipment cost on a unit price line or in a crew listing. These costs can be found consolidated in the Mobilization section of the data and elsewhere in particular site work sections. If a piece of equipment is already on the job site, it is not appropriate to include mob/demob costs in a new estimate that requires use of that equipment. The following table identifies approximate sizes of rented construction equipment that would be hauled on a towed trailer. Because this listing is not all-encompassing, the user can infer as to what size trailer might be required for a piece of equipment not listed.

3-ton Trailer	20-ton Trailer	40-ton Trailer	50-ton Trailer
20 H.P. Excavator	110 H.P. Excavator	200 H.P. Excavator	270 H.P. Excavator
50 H.P. Skid Steer	165 H.P. Dozer	300 H.P. Dozer	Small Crawler Crane
35 H.P. Roller	150 H.P. Roller	400 H.P. Scraper	500 H.P. Scraper
40 H.P. Trencher	Backhoe	450 H.P. Art. Dump Truck	500 H.P. Art. Dump Truck

Existing Conditions — R0241 Demolition

R024119-10 Demolition Defined

Whole Building Demolition - Demolition of the whole building with no concern for any particular building element, component, or material type being demolished. This type of demolition is accomplished with large pieces of construction equipment that break up the structure, load it into trucks and haul it to a disposal site, but disposal or dump fees are not included. Demolition of below-grade foundation elements, such as footings, foundation walls, grade beams, slabs on grade, etc., is not included. Certain mechanical equipment containing flammable liquids or ozone-depleting refrigerants, electric lighting elements, communication equipment components, and other building elements may contain hazardous waste, and must be removed, either selectively or carefully, as hazardous waste before the building can be demolished.

Foundation Demolition - Demolition of below-grade foundation footings, foundation walls, grade beams, and slabs on grade. This type of demolition is accomplished by hand or pneumatic hand tools, and does not include saw cutting, or handling, loading, hauling, or disposal of the debris.

Gutting - Removal of building interior finishes and electrical/mechanical systems down to the load-bearing and sub-floor elements of the rough building frame, with no concern for any particular building element, component, or material type being demolished. This type of demolition is accomplished by hand or pneumatic hand tools, and includes loading into trucks, but not hauling, disposal or dump fees, scaffolding, or shoring. Certain mechanical equipment containing flammable liquids or ozone-depleting refrigerants, electric lighting elements, communication equipment components, and other building elements may contain hazardous waste, and must be removed, either selectively or carefully, as hazardous waste, before the building is gutted.

Selective Demolition - Demolition of a selected building element, component, or finish, with some concern for surrounding or adjacent elements, components, or finishes (see the first Subdivision (s) at the beginning of appropriate Divisions). This type of demolition is accomplished by hand or pneumatic hand tools, and does not include handling, loading, storing, hauling, or disposal of the debris, scaffolding, or shoring. "Gutting" methods may be used in order to save time, but damage that is caused to surrounding or adjacent elements, components, or finishes may have to be repaired at a later time.

Careful Removal - Removal of a piece of service equipment, building element or component, or material type, with great concern for both the removed item and surrounding or adjacent elements, components or finishes. The purpose of careful removal may be to protect the removed item for later re-use, preserve a higher salvage value of the removed item, or replace an item while taking care to protect surrounding or adjacent elements, components, connections, or finishes from cosmetic and/or structural damage. An approximation of the time required to perform this type of removal is 1/3 to 1/2 the time it would take to install a new item of like kind (see Reference Number R220105-10). This type of removal is accomplished by hand or pneumatic hand tools, and does not include loading, hauling, or storing the removed item, scaffolding, shoring, or lifting equipment.

Cutout Demolition - Demolition of a small quantity of floor, wall, roof, or other assembly, with concern for the appearance and structural integrity of the surrounding materials. This type of demolition is accomplished by hand or pneumatic hand tools, and does not include saw cutting, handling, loading, hauling, or disposal of debris, scaffolding, or shoring.

Rubbish Handling - Work activities that involve handling, loading or hauling of debris. Generally, the cost of rubbish handling must be added to the cost of all types of demolition, with the exception of whole building demolition.

Minor Site Demolition - Demolition of site elements outside the footprint of a building. This type of demolition is accomplished by hand or pneumatic hand tools, or with larger pieces of construction equipment, and may include loading a removed item onto a truck (check the Crew for equipment used). It does not include saw cutting, hauling or disposal of debris, and, sometimes, handling or loading.

R024119-20 Dumpsters

Dumpster rental costs on construction sites are presented in two ways.

The cost per week rental includes the delivery of the dumpster; its pulling or emptying once per week, and its final removal. The assumption is made that the dumpster contractor could choose to empty a dumpster by simply bringing in an empty unit and removing the full one. These costs also include the disposal of the materials in the dumpster.

The Alternate Pricing can be used when actual planned conditions are not approximated by the weekly numbers. For example, these lines can be used when a dumpster is needed for 4 weeks and will need to be emptied 2 or 3 times per week. Conversely the Alternate Pricing lines can be used when a dumpster will be rented for several weeks or months but needs to be emptied only a few times over this period.

R024119-30 Rubbish Handling Chutes

To correctly estimate the cost of rubbish handling chute systems, the individual components must be priced separately. First choose the size of the system; a 30-inch diameter chute is quite common, but the sizes range from 18 to 36 inches in diameter. The 30-inch chute comes in a standard weight and two thinner weights. The thinner weight chutes are sometimes chosen for cost savings, but they are more easily damaged.

There are several types of major chute pieces that make up the chute system. The first component to consider is the top chute section (top intake hopper) where the material is dropped into the chute at the highest point. After determining the top chute, the intermediate chute pieces called the regular chute sections are priced. Next, the number of chute control door sections (intermediate intake hoppers) must be determined. In the more complex systems, a chute control door section is provided at each floor level. The last major component to consider is bolt down frames; these are usually provided at every other floor level.

There are a number of accessories to consider for safe operation and control. There are covers for the top chute and the chute door sections. The top chute can have a trough that allows for better loading of the chute. For the safest operation, a chute warning light system can be added that will warn the other chute intake locations not to load while another is being used. There are dust control devices that spray a water mist to keep down the dust as the debris is loaded into a Dumpster. There are special breakaway cords that are used to prevent damage to the chute if the dumpster is removed without disconnecting from the chute. There are chute liners that can be installed to protect the chute structure from physical damage from rough abrasive materials. Warning signs can be posted at each floor level that is provided with a chute control door section.

In summary, a complete rubbish handling chute system will include one top section, several intermediate regular sections, several intermediate control door (intake hopper) sections and bolt down frames at every other floor level starting with the top floor. If so desired, the system can also include covers and a light warning system for a safer operation. The bottom of the chute should always be above the Dumpster and should be tied off with a breakaway cord to the Dumpster.

Existing Conditions — R0265 Underground Storage Tank Removal

R026510-20 Underground Storage Tank Removal

Underground Storage Tank Removal can be divided into two categories: Non-Leaking and Leaking. Prior to removing an underground storage tank, tests should be made, with the proper authorities present, to determine whether a tank has been leaking or the surrounding soil has been contaminated.

To safely remove Liquid Underground Storage Tanks:
1. Excavate to the top of the tank.
2. Disconnect all piping.
3. Open all tank vents and access ports.
4. Remove all liquids and/or sludge.
5. Purge the tank with an inert gas.
6. Provide access to the inside of the tank and clean out the interior using proper personal protective equipment (PPE).
7. Excavate soil surrounding the tank using proper PPE for on-site personnel.
8. Pull and properly dispose of the tank.
9. Clean up the site of all contaminated material.
10. Install new tanks or close the excavation.

Existing Conditions — R0282 Asbestos Remediation

R028213-20 Asbestos Removal Process

Asbestos removal is accomplished by a specialty contractor who understands the federal and state regulations regarding the handling and disposal of the material. The process of asbestos removal is divided into many individual steps. An accurate estimate can be calculated only after all the steps have been priced.

The steps are generally as follows:
1. Obtain an asbestos abatement plan from an industrial hygienist.
2. Monitor the air quality in and around the removal area and along the path of travel between the removal area and transport area. This establishes the background contamination.
3. Construct a two part decontamination chamber at entrance to removal area.
4. Install a HEPA filter to create a negative pressure in the removal area.
5. Install wall, floor and ceiling protection as required by the plan, usually 2 layers of fireproof 6 mil polyethylene.
6. Industrial hygienist visually inspects work area to verify compliance with plan.
7. Provide temporary supports for conduit and piping affected by the removal process.
8. Proceed with asbestos removal and bagging process. Monitor air quality as described in Step #2. Discontinue operations when contaminate levels exceed applicable standards.
9. Document the legal disposal of materials in accordance with EPA standards.
10. Thoroughly clean removal area including all ledges, crevices and surfaces.
11. Post abatement inspection by industrial hygienist to verify plan compliance.
12. Provide a certificate from a licensed industrial hygienist attesting that contaminate levels are within acceptable standards before returning area to regular use.

Existing Conditions — R0283 Lead Remediation

R028319-60 Lead Paint Remediation Methods

Lead paint remediation can be accomplished by the following methods.
1. Abrasive blast
2. Chemical stripping
3. Power tool cleaning with vacuum collection system
4. Encapsulation
5. Remove and replace
6. Enclosure

Each of these methods has strengths and weakness depending on the specific circumstances of the project. The following is an overview of each method.

1. **Abrasive blasting** is usually accomplished with sand or recyclable metallic blast. Before work can begin, the area must be contained to ensure the blast material with lead does not escape to the atmosphere. The use of vacuum blast greatly reduces the containment requirements. Lead abatement equipment that may be associated with this work includes a negative air machine. In addition, it is necessary to have an industrial hygienist monitor the project on a continual basis. When the work is complete, the spent blast sand with lead must be disposed of as a hazardous material. If metallic shot was used, the lead is separated from the shot and disposed of as hazardous material. Worker protection includes disposable clothing and respiratory protection.

2. **Chemical stripping** requires strong chemicals be applied to the surface to remove the lead paint. Before the work can begin, the area under/adjacent to the work area must be covered to catch the chemical and removed lead. After the chemical is applied to the painted surface it is usually covered with paper. The chemical is left in place for the specified period, then the paper with lead paint is pulled or scraped off. The process may require several chemical applications. The paper with chemicals and lead paint adhered to it, plus the containment and loose scrapings collected by a HEPA (High Efficiency Particulate Air Filter) vac, must be disposed of as a hazardous material. The chemical stripping process usually requires a neutralizing agent and several wash downs after the paint is removed. Worker protection includes a neoprene or other compatible protective clothing and respiratory protection with face shield. An industrial hygienist is required intermittently during the process.

3. **Power tool cleaning** is accomplished using shrouded needle blasting guns. The shrouding with different end configurations is held up against the surface to be cleaned. The area is blasted with hardened needles and the shroud captures the lead with a HEPA vac and deposits it in a holding tank. An industrial hygienist monitors the project, protective clothing and a respirator is required until air samples prove otherwise. When the work is complete the lead must be disposed of as a hazardous material.

4. **Encapsulation** is a method that leaves the well bonded lead paint in place after the peeling paint has been removed. Before the work can begin, the area under/adjacent to the work must be covered to catch the scrapings. The scraped surface is then washed with a detergent and rinsed. The prepared surface is covered with approximately 10 mils of paint. A reinforcing fabric can also be embedded in the paint covering. The scraped paint and containment must be disposed of as a hazardous material. Workers must wear protective clothing and respirators.

5. **Remove and replace** is an effective way to remove lead paint from windows, gypsum walls and concrete masonry surfaces. The painted materials are removed and new materials are installed. Workers should wear a respirator and tyvek suit. The demolished materials must be disposed of as hazardous waste if it fails the TCLP (Toxicity Characteristic Leachate Process) test.

6. **Enclosure** is the process that permanently seals lead painted materials in place. This process has many applications such as covering lead painted drywall with new drywall, covering exterior construction with tyvek paper then residing, or covering lead painted structural members with aluminum or plastic. The seams on all enclosing materials must be securely sealed. An industrial hygienist monitors the project, and protective clothing and a respirator is required until air samples prove otherwise.

All the processes require clearance monitoring and wipe testing as required by the hygienist.

Concrete

R0311 Concrete Forming

R031113-10 Wall Form Materials

Aluminum Forms
Approximate weight is 3 lbs. per S.F.C.A. Standard widths are available from 4" to 36" with 36" most common. Standard lengths of 2', 4', 6' to 8' are available. Forms are lightweight and fewer ties are needed with the wider widths. The form face is either smooth or textured.

Metal Framed Plywood Forms
Manufacturers claim over 75 reuses of plywood and over 300 reuses of steel frames. Many specials such as corners, fillers, pilasters, etc. are available. Monthly rental is generally about 15% of purchase price for first month and 9% per month thereafter with 90% of rental applied to purchase for the first month and decreasing percentages thereafter. Aluminum framed forms cost 25% to 30% more than steel framed.

After the first month, extra days may be prorated from the monthly charge. Rental rates do not include ties, accessories, cleaning, loss of hardware or freight in and out. Approximate weight is 5 lbs. per S.F. for steel; 3 lbs. per S.F. for aluminum.

Forms can be rented with option to buy.

Plywood Forms, Job Fabricated
There are two types of plywood used for concrete forms.
1. Exterior plyform which is completely waterproof. This is face oiled to facilitate stripping. Ten reuses can be expected with this type with 25 reuses possible.
2. An overlaid type consists of a resin fiber fused to exterior plyform. No oiling is required except to facilitate cleaning. This is available in both high density (HDO) and medium density overlaid (MDO). Using HDO, 50 reuses can be expected with 200 possible.

Plyform is available in 5/8" and 3/4" thickness. High density overlaid is available in 3/8", 1/2", 5/8" and 3/4" thickness.

5/8" thick is sufficient for most building forms, while 3/4" is best on heavy construction.

Plywood Forms, Modular, Prefabricated
There are many plywood forming systems without frames. Most of these are manufactured from 1-1/8" (HDO) plywood and have some hardware attached. These are used principally for foundation walls 8' or less high. With care and maintenance, 100 reuses can be attained with decreasing quality of surface finish.

Steel Forms
Approximate weight is 6-1/2 lbs. per S.F.C.A. including accessories. Standard widths are available from 2" to 24", with 24" most common. Standard lengths are from 2' to 8', with 4' the most common. Forms are easily ganged into modular units.

Forms are usually leased for 15% of the purchase price per month prorated daily over 30 days.

Rental may be applied to sale price, and usually rental forms are bought. With careful handling and cleaning 200 to 400 reuses are possible.

Straight wall gang forms up to 12' x 20' or 8' x 30' can be fabricated. These crane handled forms usually lease for approx. 9% per month.

Individual job analysis is available from the manufacturer at no charge.

R031113-40 Forms for Reinforced Concrete

Design Economy
Avoid many sizes in proportioning beams and columns.

From story to story avoid changing column dimensions. Gain strength by adding steel or using a richer mix. If a change in size of column is necessary, vary one dimension only to minimize form alterations. Keep beams and columns the same width.

From floor to floor in a multi-story building vary beam depth, not width, as that will leave slab panel form unchanged. It is cheaper to vary the strength of a beam from floor to floor by means of steel area than by 2" changes in either width or depth.

Cost Factors
Material includes the cost of lumber, cost of rent for metal pans or forms if used, nails, form ties, form oil, bolts and accessories.

Labor includes the cost of carpenters to make up, erect, remove and repair, plus common labor to clean and move. Having carpenters remove forms minimizes repairs.

Improper alignment and condition of forms will increase finishing cost. When forms are heavily oiled, concrete surfaces must be neutralized before finishing. Special curing compounds will cause spillages to spall off in first frost. Gang forming methods will reduce costs on large projects.

Materials Used
Boards are seldom used unless their architectural finish is required. Generally, steel, fiberglass and plywood are used for contact surfaces. Labor on plywood is 10% less than with boards. The plywood is backed up with 2 x 4's at 12" to 32" O.C. Walers are generally 2 - 2 x 4's. Column forms are held together with steel yokes or bands. Shoring is with adjustable shoring or scaffolding for high ceilings.

Reuse
Floor and column forms can be reused four or possibly five times without excessive repair. Remember to allow for 10% waste on each reuse.

When modular sized wall forms are made, up to twenty uses can be expected with exterior plyform.

When forms are reused, the cost to erect, strip, clean and move will not be affected. 10% replacement of lumber should be included and about one hour of carpenter time for repairs on each reuse per 100 S.F.

The reuse cost for certain accessory items normally rented on a monthly basis will be lower than the cost for the first use.

After fifth use, new material required plus time needed for repair prevent form cost from dropping further and it may go up. Much depends on care in stripping, the number of special bays, changes in beam or column sizes and other factors.

Costs for multiple use of formwork may be developed as follows:

2 Uses	3 Uses	4 Uses
$\dfrac{\text{(1st Use + Reuse)}}{2} = \text{avg. cost/2 uses}$	$\dfrac{\text{(1st Use + 2 Reuse)}}{3} = \text{avg. cost/3 uses}$	$\dfrac{\text{(1st use + 3 Reuse)}}{4} = \text{avg. cost/4 uses}$

Concrete — R0311 Concrete Forming

R031113-60 Formwork Labor-Hours

Item	Unit	Hours Required - Fabricate	Hours Required - Erect & Strip	Hours Required - Clean & Move	Total Hours 1 Use	Multiple Use 2 Use	Multiple Use 3 Use	Multiple Use 4 Use
Beam and Girder, interior beams, 12" wide	100 S.F.	6.4	8.3	1.3	16.0	13.3	12.4	12.0
Hung from steel beams		5.8	7.7	1.3	14.8	12.4	11.6	11.2
Beam sides only, 36" high		5.8	7.2	1.3	14.3	11.9	11.1	10.7
Beam bottoms only, 24" wide		6.6	13.0	1.3	20.9	18.1	17.2	16.7
Box out for openings		9.9	10.0	1.1	21.0	16.6	15.1	14.3
Buttress forms, to 8' high		6.0	6.5	1.2	13.7	11.2	10.4	10.0
Centering, steel, 3/4" rib lath			1.0		1.0			
3/8" rib lath or slab form			0.9		0.9			
Chamfer strip or keyway	100 L.F.		1.5		1.5	1.5	1.5	1.5
Columns, fiber tube 8" diameter			20.6		20.6			
12"			21.3		21.3			
16"			22.9		22.9			
20"			23.7		23.7			
24"			24.6		24.6			
30"			25.6		25.6			
Columns, round steel, 12" diameter			22.0		22.0	22.0	22.0	22.0
16"			25.6		25.6	25.6	25.6	25.6
20"			30.5		30.5	30.5	30.5	30.5
24"			37.7		37.7	37.7	37.7	37.7
Columns, plywood 8" x 8"	100 S.F.	7.0	11.0	1.2	19.2	16.2	15.2	14.7
12" x 12"		6.0	10.5	1.2	17.7	15.2	14.4	14.0
16" x 16"		5.9	10.0	1.2	17.1	14.7	13.8	13.4
24" x 24"		5.8	9.8	1.2	16.8	14.4	13.6	13.2
Columns, steel framed plywood 8" x 8"			10.0	1.0	11.0	11.0	11.0	11.0
12" x 12"			9.3	1.0	10.3	10.3	10.3	10.3
16" x 16"			8.5	1.0	9.5	9.5	9.5	9.5
24" x 24"			7.8	1.0	8.8	8.8	8.8	8.8
Drop head forms, plywood		9.0	12.5	1.5	23.0	19.0	17.7	17.0
Coping forms		8.5	15.0	1.5	25.0	21.3	20.0	19.4
Culvert, box			14.5	4.3	18.8	18.8	18.8	18.8
Curb forms, 6" to 12" high, on grade		5.0	8.5	1.2	14.7	12.7	12.1	11.7
On elevated slabs		6.0	10.8	1.2	18.0	15.5	14.7	14.3
Edge forms to 6" high, on grade	100 L.F.	2.0	3.5	0.6	6.1	5.6	5.4	5.3
7" to 12" high	100 S.F.	2.5	5.0	1.0	8.5	7.8	7.5	7.4
Equipment foundations		10.0	18.0	2.0	30.0	25.5	24.0	23.3
Flat slabs, including drops		3.5	6.0	1.2	10.7	9.5	9.0	8.8
Hung from steel		3.0	5.5	1.2	9.7	8.7	8.4	8.2
Closed deck for domes		3.0	5.8	1.2	10.0	9.0	8.7	8.5
Open deck for pans		2.2	5.3	1.0	8.5	7.9	7.7	7.6
Footings, continuous, 12" high		3.5	3.5	1.5	8.5	7.3	6.8	6.6
Spread, 12" high		4.7	4.2	1.6	10.5	8.7	8.0	7.7
Pile caps, square or rectangular		4.5	5.0	1.5	11.0	9.3	8.7	8.4
Grade beams, 24" deep		2.5	5.3	1.2	9.0	8.3	8.0	7.9
Lintel or Sill forms		8.0	17.0	2.0	27.0	23.5	22.3	21.8
Spandrel beams, 12" wide		9.0	11.2	1.3	21.5	17.5	16.2	15.5
Stairs			25.0	4.0	29.0	29.0	29.0	29.0
Trench forms in floor		4.5	14.0	1.5	20.0	18.3	17.7	17.4
Walls, Plywood, at grade, to 8' high		5.0	6.5	1.5	13.0	11.0	9.7	9.5
8' to 16'		7.5	8.0	1.5	17.0	13.8	12.7	12.1
16' to 20'		9.0	10.0	1.5	20.5	16.5	15.2	14.5
Foundation walls, to 8' high		4.5	6.5	1.0	12.0	10.3	9.7	9.4
8' to 16' high		5.5	7.5	1.0	14.0	11.8	11.0	10.6
Retaining wall to 12' high, battered		6.0	8.5	1.5	16.0	13.5	12.7	12.3
Radial walls to 12' high, smooth		8.0	9.5	2.0	19.5	16.0	14.8	14.3
2' chords		7.0	8.0	1.5	16.5	13.5	12.5	12.0
Prefabricated modular, to 8' high		—	4.3	1.0	5.3	5.3	5.3	5.3
Steel, to 8' high		—	6.8	1.2	8.0	8.0	8.0	8.0
8' to 16' high		—	9.1	1.5	10.6	10.3	10.2	10.2
Steel framed plywood to 8' high		—	6.8	1.2	8.0	7.5	7.3	7.2
8' to 16' high		—	9.3	1.2	10.5	9.5	9.2	9.0

Concrete   R0321 Reinforcing Steel

R032110-10   Reinforcing Steel Weights and Measures

Bar Designation No.**	Nominal Weight Lb./Ft.	U.S. Customary Units			SI Units			
		Nominal Dimensions*			Nominal Weight kg/m	Nominal Dimensions*		
		Diameter in.	Cross Sectional Area, in.^{2}	Perimeter in.		Diameter mm	Cross Sectional Area, cm^2	Perimeter mm
3	.376	.375	.11	1.178	.560	9.52	.71	29.9
4	.668	.500	.20	1.571	.994	12.70	1.29	39.9
5	1.043	.625	.31	1.963	1.552	15.88	2.00	49.9
6	1.502	.750	.44	2.356	2.235	19.05	2.84	59.8
7	2.044	.875	.60	2.749	3.042	22.22	3.87	69.8
8	2.670	1.000	.79	3.142	3.973	25.40	5.10	79.8
9	3.400	1.128	1.00	3.544	5.059	28.65	6.45	90.0
10	4.303	1.270	1.27	3.990	6.403	32.26	8.19	101.4
11	5.313	1.410	1.56	4.430	7.906	35.81	10.06	112.5
14	7.650	1.693	2.25	5.320	11.384	43.00	14.52	135.1
18	13.600	2.257	4.00	7.090	20.238	57.33	25.81	180.1

* The nominal dimensions of a deformed bar are equivalent to those of a plain round bar having the same weight per foot as the deformed bar.
** Bar numbers are based on the number of eighths of an inch included in the nominal diameter of the bars.

R032110-80   Shop-Fabricated Reinforcing Steel

The material prices for reinforcing, shown in the unit cost sections of the book, are for 50 tons or more of shop-fabricated reinforcing steel and include:
1. Mill base price of reinforcing steel
2. Mill grade/size/length extras
3. Mill delivery to the fabrication shop
4. Shop storage and handling
5. Shop drafting/detailing
6. Shop shearing and bending
7. Shop listing
8. Shop delivery to the job site

Both material and installation costs can be considerably higher for small jobs consisting primarily of smaller bars, while material costs may be slightly lower for larger jobs.

Concrete — R0322 Welded Wire Fabric Reinforcing

R032205-30 Common Stock Styles of Welded Wire Fabric

This table provides some of the basic specifications, sizes, and weights of welded wire fabric used for reinforcing concrete.

	New Designation Spacing — Cross Sectional Area (in.) — (Sq. in. 100)	Old Designation Spacing — Wire Gauge (in.) — (AS & W)		Steel Area per Foot Longitudinal		Transverse		Approximate Weight per 100 S.F.	
				in.	cm	in.	cm	lbs	kg
Rolls	6 x 6 — W1.4 x W1.4	6 x 6 — 10 x 10		.028	.071	.028	.071	21	9.53
	6 x 6 — W2.0 x W2.0	6 x 6 — 8 x 8	1	.040	.102	.040	.102	29	13.15
	6 x 6 — W2.9 x W2.9	6 x 6 — 6 x 6		.058	.147	.058	.147	42	19.05
	6 x 6 — W4.0 x W4.0	6 x 6 — 4 x 4		.080	.203	.080	.203	58	26.91
	4 x 4 — W1.4 x W1.4	4 x 4 — 10 x 10		.042	.107	.042	.107	31	14.06
	4 x 4 — W2.0 x W2.0	4 x 4 — 8 x 8	1	.060	.152	.060	.152	43	19.50
	4 x 4 — W2.9 x W2.9	4 x 4 — 6 x 6		.087	.227	.087	.227	62	28.12
	4 x 4 — W4.0 x W4.0	4 x 4 — 4 x 4		.120	.305	.120	.305	85	38.56
Sheets	6 x 6 — W2.9 x W2.9	6 x 6 — 6 x 6		.058	.147	.058	.147	42	19.05
	6 x 6 — W4.0 x W4.0	6 x 6 — 4 x 4		.080	.203	.080	.203	58	26.31
	6 x 6 — W5.5 x W5.5	6 x 6 — 2 x 2	2	.110	.279	.110	.279	80	36.29
	4 x 4 — W1.4 x W1.4	4 x 4 — 4 x 4		.120	.305	.120	.305	85	38.56

NOTES: 1. Exact W—number size for 8 gauge is W2.1
2. Exact W—number size for 2 gauge is W5.4

The above table was compiled with the following excerpts from the WRI Manual of Standard Practices, 7th Edition, Copyright 2006. Reproduced with permission of the Wire Reinforcement Institute, Inc.:

1. Chapter 3, page 7, Table 1 Common Styles of Metric Wire Reinforcement (WWR) With Equivalent US Customary Units
2. Chapter 6, page 19, Table 5 Customary Units
3. Chapter 6, Page 23, Table 7 Customary Units (in.) Welded Plain Wire Reinforcement
4. Chapter 6, Page 25 Table 8 Wire Size Comparison
5. Chapter 9, Page 30, Table 9 Weight of Longitudinal Wires Weight (Mass) Estimating Tables
6. Chapter 9, Page 31, Table 9M Weight of Longitudinal Wires Weight (Mass) Estimating Tables
7. Chapter 9, Page 32, Table 10 Weight of Transverse Wires Based on 62" lengths of transverse wire (60" width plus 1" overhand each side)
8. Chapter 9, Page 33, Table 10M Weight of Transverse Wires

Concrete R0331 Structural Concrete

R033105-10 Proportionate Quantities

The tables below show both quantities per S.F. of floor areas as well as form and reinforcing quantities per C.Y. Unusual structural requirements would increase the ratios below. High strength reinforcing would reduce the steel weights. Figures are for 3000 psi concrete and 60,000 psi reinforcing unless specified otherwise.

Type of Construction	Live Load	Span	Per S.F. of Floor Area				Per C.Y. of Concrete		
			Concrete	Forms	Reinf.	Pans	Forms	Reinf.	Pans
Flat Plate	50 psf	15 Ft.	.46 C.F.	1.06 S.F.	1.71 lb.		62 S.F.	101 lb.	
		20	.63	1.02	2.40		44	104	
		25	.79	1.02	3.03		35	104	
	100	15	.46	1.04	2.14		61	126	
		20	.71	1.02	2.72		39	104	
		25	.83	1.01	3.47		33	113	
Flat Plate (waffle construction) 20" domes	50	20	.43	1.00	2.10	.84 S.F.	63	135	53 S.F.
		25	.52	1.00	2.90	.89	52	150	46
		30	.64	1.00	3.70	.87	42	155	37
	100	20	.51	1.00	2.30	.84	53	125	45
		25	.64	1.00	3.20	.83	42	135	35
		30	.76	1.00	4.40	.81	36	160	29
Waffle Construction 30" domes	50	25	.69	1.06	1.83	.68	42	72	40
		30	.74	1.06	2.39	.69	39	87	39
		35	.86	1.05	2.71	.69	33	85	39
		40	.78	1.00	4.80	.68	35	165	40
Flat Slab (two way with drop panels)	50	20	.62	1.03	2.34		45	102	
		25	.77	1.03	2.99		36	105	
		30	.95	1.03	4.09		29	116	
	100	20	.64	1.03	2.83		43	119	
		25	.79	1.03	3.88		35	133	
		30	.96	1.03	4.66		29	131	
	200	20	.73	1.03	3.03		38	112	
		25	.86	1.03	4.23		32	133	
		30	1.06	1.03	5.30		26	135	
One Way Joists 20" Pans	50	15	.36	1.04	1.40	.93	78	105	70
		20	.42	1.05	1.80	.94	67	120	60
		25	.47	1.05	2.60	.94	60	150	54
	100	15	.38	1.07	1.90	.93	77	140	66
		20	.44	1.08	2.40	.94	67	150	58
		25	.52	1.07	3.50	.94	55	185	49
One Way Joists 8" x 16" filler blocks	50	15	.34	1.06	1.80	.81 Ea.	84	145	64 Ea.
		20	.40	1.08	2.20	.82	73	145	55
		25	.46	1.07	3.20	.83	63	190	49
	100	15	.39	1.07	1.90	.81	74	130	56
		20	.46	1.09	2.80	.82	64	160	48
		25	.53	1.10	3.60	.83	56	190	42
One Way Beam & Slab	50	15	.42	1.30	1.73		84	111	
		20	.51	1.28	2.61		68	138	
		25	.64	1.25	2.78		53	117	
	100	15	.42	1.30	1.90		84	122	
		20	.54	1.35	2.69		68	154	
		25	.69	1.37	3.93		54	145	
	200	15	.44	1.31	2.24		80	137	
		20	.58	1.40	3.30		65	163	
		25	.69	1.42	4.89		53	183	
Two Way Beam & Slab	100	15	.47	1.20	2.26		69	130	
		20	.63	1.29	3.06		55	131	
		25	.83	1.33	3.79		43	123	
	200	15	.49	1.25	2.70		41	149	
		20	.66	1.32	4.04		54	165	
		25	.88	1.32	6.08		41	187	

Concrete — R0331 Structural Concrete

R033105-10 Proportionate Quantities (cont.)

4000 psi Concrete and 60,000 psi Reinforcing—Form and Reinforcing Quantities per C.Y.

Item	Size	Forms	Reinforcing	Minimum	Maximum
Columns (square tied)	10" x 10"	130 S.F.C.A.	#5 to #11	220 lbs.	875 lbs.
	12" x 12"	108	#6 to #14	200	955
	14" x 14"	92	#7 to #14	190	900
	16" x 16"	81	#6 to #14	187	1082
	18" x 18"	72	#6 to #14	170	906
	20" x 20"	65	#7 to #18	150	1080
	22" x 22"	59	#8 to #18	153	902
	24" x 24"	54	#8 to #18	164	884
	26" x 26"	50	#9 to #18	169	994
	28" x 28"	46	#9 to #18	147	864
	30" x 30"	43	#10 to #18	146	983
	32" x 32"	40	#10 to #18	175	866
	34" x 34"	38	#10 to #18	157	772
	36" x 36"	36	#10 to #18	175	852
	38" x 38"	34	#10 to #18	158	765
	40" x 40"	32	#10 to #18	143	692

Item	Size	Form	Spiral	Reinforcing	Minimum	Maximum
Columns (spirally reinforced)	12" diameter	34.5 L.F.	190 lbs.	#4 to #11	165 lbs.	1505 lb.
		34.5	190	#14 & #18	—	1100
	14"	25	170	#4 to #11	150	970
		25	170	#14 & #18	800	1000
	16"	19	160	#4 to #11	160	950
		19	160	#14 & #18	605	1080
	18"	15	150	#4 to #11	160	915
		15	150	#14 & #18	480	1075
	20"	12	130	#4 to #11	155	865
		12	130	#14 & #18	385	1020
	22"	10	125	#4 to #11	165	775
		10	125	#14 & #18	320	995
	24"	9	120	#4 to #11	195	800
		9	120	#14 & #18	290	1150
	26"	7.3	100	#4 to #11	200	729
		7.3	100	#14 & #18	235	1035
	28"	6.3	95	#4 to #11	175	700
		6.3	95	#14 & #18	200	1075
	30"	5.5	90	#4 to #11	180	670
		5.5	90	#14 & #18	175	1015
	32"	4.8	85	#4 to #11	185	615
		4.8	85	#14 & #18	155	955
	34"	4.3	80	#4 to #11	180	600
		4.3	80	#14 & #18	170	855
	36"	3.8	75	#4 to #11	165	570
		3.8	75	#14 & #18	155	865
	40"	3.0	70	#4 to #11	165	500
		3.0	70	#14 & #18	145	765

Concrete   R0331 Structural Concrete

R033105-10 Proportionate Quantities (cont.)

3000 psi Concrete and 60,000 psi Reinforcing—Form and Reinforcing Quantities per C.Y.

Item	Type	Loading	Height	C.Y./L.F.	Forms/C.Y.	Reinf./C.Y.
Retaining Walls	Cantilever	Level Backfill	4 Ft.	0.2 C.Y.	49 S.F.	35 lbs.
			8	0.5	42	45
			12	0.8	35	70
			16	1.1	32	85
			20	1.6	28	105
		Highway Surcharge	4	0.3	41	35
			8	0.5	36	55
			12	0.8	33	90
			16	1.2	30	120
			20	1.7	27	155
		Railroad Surcharge	4	0.4	28	45
			8	0.8	25	65
			12	1.3	22	90
			16	1.9	20	100
			20	2.6	18	120
	Gravity, with Vertical Face	Level Backfill	4	0.4	37	None
			7	0.6	27	
			10	1.2	20	
		Sloping Backfill	4	0.3	31	
			7	0.8	21	
			10	1.6	15	↓

		Live Load in Kips per Linear Foot							
	Span	Under 1 Kip		2 to 3 Kips		4 to 5 Kips		6 to 7 Kips	
		Forms	Reinf.	Forms	Reinf.	Forms	Reinf.	Forms	Reinf.
Beams	10 Ft.	—	—	90 S.F.	170 #	85 S.F.	175 #	75 S.F.	185 #
	16	130 S.F.	165 #	85	180	75	180	65	225
	20	110	170	75	185	62	200	51	200
	26	90	170	65	215	62	215	—	—
	30	85	175	60	200	—	—	—	—

Item	Size	Type	Forms per C.Y.	Reinforcing per C.Y.
Spread Footings	Under 1 C.Y.	1,000 psf soil	24 S.F.	44 lbs.
		5,000	24	42
		10,000	24	52
	1 C.Y. to 5 C.Y.	1,000	14	49
		5,000	14	50
		10,000	14	50
	Over 5 C.Y.	1,000	9	54
		5,000	9	52
		10,000	9	56
Pile Caps (30 Ton Concrete Piles)	Under 5 C.Y.	shallow caps	20	65
		medium	20	50
		deep	20	40
	5 C.Y. to 10 C.Y.	shallow	14	55
		medium	15	45
		deep	15	40
	10 C.Y. to 20 C.Y.	shallow	11	60
		medium	11	45
		deep	12	35
	Over 20 C.Y.	shallow	9	60
		medium	9	45
		deep	10	40

Concrete — R0331 Structural Concrete

R033105-10 Proportionate Quantities (cont.)

		3000 psi Concrete and 60,000 psi Reinforcing — Form and Reinforcing Quantities per C.Y.					
Item	Size	Pile Spacing	50 T Pile	100 T Pile	50 T Pile	100 T Pile	
Pile Caps (Steel H Piles)	Under 5 C.Y.	24" O.C.	24 S.F.	24 S.F.	75 lbs.	90 lbs.	
		30"	25	25	80	100	
		36"	24	24	80	110	
	5 C.Y. to 10 C.Y.	24"	15	15	80	110	
		30"	15	15	85	110	
		36"	15	15	75	90	
	Over 10 C.Y.	24"	13	13	85	90	
		30"	11	11	85	95	
		36"	10	10	85	90	

		8" Thick		10" Thick		12" Thick		15" Thick	
	Height	Forms	Reinf.	Forms	Reinf.	Forms	Reinf.	Forms	Reinf.
Basement Walls	7 Ft.	81 S.F.	44 lbs.	65 S.F.	45 lbs.	54 S.F.	44 lbs.	41 S.F.	43 lbs.
	8		44		45		44		43
	9		46		45		44		43
	10		57		45		44		43
	12		83		50		52		43
	14		116		65		64		51
	16				86		90		65
	18						106		70

R033105-20 Materials for One C.Y. of Concrete

This is an approximate method of figuring quantities of cement, sand and coarse aggregate for a field mix with waste allowance included.

With crushed gravel as coarse aggregate, to determine barrels of cement required, divide 10 by total mix; that is, for 1:2:4 mix, 10 divided by 7 = 1-3/7 barrels.

If the coarse aggregate is crushed stone, use 10-1/2 instead of 10 as given for gravel.

To determine tons of sand required, multiply barrels of cement by parts of sand and then by 0.2; that is, for the 1:2:4 mix, as above, 1-3/7 x 2 x .2 = .57 tons.

Tons of crushed gravel are in the same ratio to tons of sand as parts in the mix, or 4/2 x .57 = 1.14 tons.

1 bag cement = 94#
4 bags = 1 barrel

1 C.Y. sand or crushed gravel = 2700#
1 ton sand or crushed gravel = 20 C.F.

1 C.Y. crushed stone = 2575#
1 ton crushed stone = 21 C.F.

Average carload of cement is 692 bags; of sand or gravel is 56 tons.

Do not stack stored cement over 10 bags high.

R033105-70 Placing Ready-Mixed Concrete

For ground pours allow for 5% waste when figuring quantities.

Prices in the front of the book assume normal deliveries. If deliveries are made before 8 A.M. or after 5 P.M. or on Saturday afternoons add 30%. Negotiated discounts for large volumes are not included in prices in front of book.

For the lower floors without truck access, concrete may be wheeled in rubber-tired buggies, conveyer handled, crane handled or pumped. Pumping is economical if there is top steel. Conveyers are more efficient for thick slabs.

At higher floors the rubber-tired buggies may be hoisted by a hoisting tower and wheeled to location. Placement by a conveyer is limited to three floors and is best for high-volume pours. Pumped concrete is best when building has no crane access. Concrete may be pumped directly as high as thirty-six stories using special pumping techniques. Normal maximum height is about fifteen stories.

Best pumping aggregate is screened and graded bank gravel rather than crushed stone.

Pumping downward is more difficult than pumping upward. Horizontal distance from pump to pour may increase preparation time prior to pour. Placing by cranes, either mobile, climbing or tower types, continues as the most efficient method for high-rise concrete buildings.

Concrete — R0331 Structural Concrete

R033105-85 Lift Slabs

The cost advantage of the lift slab method is due to placing all concrete, reinforcing steel, inserts and electrical conduit at ground level and in reduction of formwork. Minimum economical project size is about 30,000 S.F. Slabs may be tilted for parking garage ramps.

It is now used in all types of buildings and has gone up to 22 stories high in apartment buildings. Current trend is to use post-tensioned flat plate slabs with spans from 22' to 35'. Cylindrical void forms are used when deep slabs are required. One pound of prestressing steel is about equal to seven pounds of conventional reinforcing.

To be considered cured for stressing and lifting, a slab must have attained 75% of design strength. Seven days are usually sufficient with four to five days possible if high early strength cement is used. Slabs can be stacked using two coats of a non-bonding agent to insure that slabs do not stick to each other. Lifting is done by companies specializing in this work. Lift rate is 5' to 15' per hour with an average of 10' per hour. Total areas up to 33,000 S.F. have been lifted at one time. 24 to 36 jacking columns are common. Most economical bay sizes are 24' to 28' with four to fourteen stories most efficient. Continuous design reduces reinforcing steel cost. Use of post-tensioned slabs allows larger bay sizes.

R033543-10 Polished Concrete Floors

A polished concrete floor has a glossy mirror-like appearance and is created by grinding the concrete floor with finer and finer diamond grits, similar to sanding wood, until the desired level of reflective clarity and sheen are achieved. The technical term for this type of polished concrete is bonded abrasive polished concrete. The basic piece of equipment used in the polishing process is a walk-behind planetary grinder for working large floor areas. This grinder drives diamond-impregnated abrasive discs, which progress from coarse- to fine-grit discs.

The process begins with the use of very coarse diamond segments or discs bonded in a metallic matrix. These segments are coarse enough to allow the removal of pits, blemishes, stains, and light coatings from the floor surface in preparation for final smoothing. The condition of the original concrete surface will dictate the grit coarsness of the initial grinding step which will generally end up being a three- to four-step process using ever finer grits. The purpose of this initial grinding step is to remove surface coatings and blemishes and to cut down into the cream for very fine aggregate exposure, or deeper into the fine aggregate layer just below the cream layer, or even deeper into the coarse aggregate layer. These initial grinding steps will progress up to the 100/120 grit. If wet grinding is done, a waste slurry is produced that must be removed between grit changes and disposed of properly. If dry grinding is done, a high performance vacuum will pick up the dust during grinding and collect it in bags which must be disposed of properly.

The process continues with honing the floor in a series of steps that progress from 100-grit to 400-grit diamond abrasive discs embedded in a plastic or resin matrix. At some point during, or just prior to, the honing step, one or two coats of stain or dye can be sprayed onto the surface to give color to the concrete, and two coats of densifier/hardener must be applied to the floor surface and allowed to dry. This sprayed-on densifier/hardener will penetrate about 1/8" into the concrete to make the surface harder, denser and more abrasion-resistant.

The process ends with polishing the floor surface in a series of steps that progress from resin-impregnated 800-grit (medium polish) to 1500-grit (high polish) to 3000-grit (very high polish), depending on the desired level of reflective clarity and sheen.

The Concrete Polishing Association of America (CPAA) has defined the flooring options available when processing concrete to a desired finish. The first category is aggregate exposure, the grinding of a concrete surface with bonded abrasives, in as many abrasive grits necessary, to achieve one of the following classes:

©Concrete Polishing Association of America "Glossary."

A. Cream – very little surface cut depth; little aggregate exposure

B. Fine aggregate (salt and pepper) – surface cut depth of 1/16"; fine aggregate exposure with little or no medium aggregate exposure at random locations

C. Medium aggregate – surface cut depth of 1/8"; medium aggregate exposure with little or no large aggregate exposure at random locations

D. Large aggregate – surface cut depth of 1/4"; large aggregate exposure with little or no fine aggregate exposure at random locations

The second CPAA defined category is reflective clarity and sheen, the polishing of a concrete surface with the minimum number of bonded abrasives as indicated to achieve one of the following levels:

1. Ground – flat appearance with none to very slight diffused reflection; none to very low reflective sheen; using a minimum total of 4 grit levels up 100-grit

2. Honed – matte appearance with or without slight diffused reflection; low to medium reflective sheen; using a minimum total of 5 grit levels up to 400-grit

3. Semi-polished – objects being reflected are not quite sharp and crisp but can be easily identified; medium to high reflective sheen; using a minimum total of 6 grit levels up to 800-grit

4. Highly-polished – objects being reflected are sharp and crisp as would be seen in a mirror-like reflection; high to highest reflective sheen; using a minimum total of up to 8 grit levels up to 1500-grit or 3000-grit

The CPAA defines reflective clarity as the degree of sharpness and crispness of the reflection of overhead objects when viewed 5' above and perpendicular to the floor surface; and reflective sheen as the degree of gloss reflected from a surface when viewed at least 20' from and at an angle to the floor surface. These terms are relatively subjective. The final outcome depends on the internal makeup and surface condition of the original concrete floor, the experience of the floor polishing crew, and expectations of the owner. Before the grinding, honing, and polishing work commences on the main floor area, it might be beneficial to do a mock-up panel in the same floor but in an out of the way place to demonstrate the sequence of steps with increasingly fine abrasive grits and to demonstrate the final reflective clarity and reflective sheen. This mock-up panel will be within the area of, and part of, the final work.

Concrete — R0341 Precast Structural Concrete

R034105-30 Prestressed Precast Concrete Structural Units

Type	Location	Depth	Span in Ft.		Live Load Lb. per S.F.
Double Tee	Floor	28" to 34"	60 to 80		50 to 80
	Roof	12" to 24"	30 to 50		40
	Wall	Width 8'	Up to 55' high		Wind
Multiple Tee	Roof	8" to 12"	15 to 40		40
	Floor	8" to 12"	15 to 30		100
Plank	Roof or Floor		Roof	Floor	40 for Roof
		4"	13	12	
		6"	22	18	
		8"	26	25	
		10"	33	29	100 for Floor
		12"	42	32	
Single Tee	Roof	28"	40		
		32"	80		
		36"	100		40
		48"	120		
AASHO Girder	Bridges	Type 4	100		
		5	110		Highway
		6	125		
Box Beam	Bridges	15"	40		
		27"	to		Highway
		33"	100		

The majority of precast projects today utilize double tees rather than single tees because of speed and ease of installation. As a result casting beds at manufacturing plants are normally formed for double tees. Single tee projects will therefore require an initial set up charge to be spread over the individual single tee costs.

For floors, a 2" to 3" topping is field cast over the shapes. For roofs, insulating concrete or rigid insulation is placed over the shapes.

Member lengths up to 40' are standard haul, 40' to 60' require special permits and lengths over 60' must be escorted. Over width and/or over length can add up to 100% on hauling costs.

Large heavy members may require two cranes for lifting which would increase erection costs by about 45%. An eight man crew can install 12 to 20 double tees, or 45 to 70 quad tees or planks per day.

Grouting of connections must also be included.

Several system buildings utilizing precast members are available. Heights can go up to 22 stories for apartment buildings. Optimum design ratio is 3 S.F. of surface to 1 S.F. of floor area.

Concrete — R0352 Lightweight Concrete Roof Insulation

R035216-10 Lightweight Concrete

Lightweight aggregate concrete is usually purchased ready mixed, but it can also be field mixed.

Vermiculite or Perlite comes in bags of 4 C.F. under various trade names. Weight is about 8 lbs. per C.F. For insulating roof fill use 1:6 mix. For structural deck use 1:4 mix over gypsum boards, steeltex, steel centering, etc., supported by closely spaced joists or bulb trees. For structural slabs use 1:3:2 vermiculite sand concrete over steeltex, metal lath, steel centering, etc., on joists spaced 2'-0" O.C. for maximum L.L. of 80 P.S.F. Use same mix for slab base fill over steel flooring or regular reinforced concrete slab when tile, terrazzo or other finish is to be laid over.

For slabs on grade use 1:3:2 mix when tile, etc., finish is to be laid over. If radiant heating units are installed use a 1:6 mix for a base. After coils are in place, cover with a regular granolithic finish (mix 1:3:2) to a minimum depth of 1-1/2" over top of units.

Reinforce all slabs with 6 x 6 or 10 x 10 welded wire mesh.

Masonry — R0421 Clay Unit Masonry

R042110-20 Common and Face Brick

Common building brick manufactured according to ASTM C62 and facing brick manufactured according to ASTM C216 are the two standard bricks available for general building use.

Building brick is made in three grades; SW, where high resistance to damage caused by cyclic freezing is required; MW, where moderate resistance to cyclic freezing is needed; and NW, where little resistance to cyclic freezing is needed. Facing brick is made in only the two grades SW and MW. Additionally, facing brick is available in three types; FBS, for general use; FBX, for general use where a higher degree of precision and lower permissible variation in size than FBS is needed; and FBA, for general use to produce characteristic architectural effects resulting from non-uniformity in size and texture of the units.

In figuring the material cost of brickwork, an allowance of 25% mortar waste and 3% brick breakage was included. If bricks are delivered palletized with 280 to 300 per pallet, or packaged, allow only 1-1/2% for breakage. Packaged or palletized delivery is practical when a job is big enough to have a crane or other equipment available to handle a package of brick. This is so on all industrial work but not always true on small commercial buildings.

The use of buff and gray face is increasing, and there is a continuing trend to the Norman, Roman, Jumbo and SCR brick.

Common red clay brick for backup is not used that often. Concrete block is the most usual backup material with occasional use of sand lime or cement brick. Building brick is commonly used in solid walls for strength and as a fire stop.

Brick panels built on the ground and then crane erected to the upper floors have proven to be economical. This allows the work to be done under cover and without scaffolding.

R042110-50 Brick, Block & Mortar Quantities

Type Brick	Nominal Size (incl. mortar) L H W	Modular Coursing	Number of Brick per S.F.	C.F. of Mortar per M Bricks, Waste Included 3/8" Joint	1/2" Joint	Bond Type	Description	Factor
Standard	8 x 2-2/3 x 4	3C=8"	6.75	8.1	10.3	Common	full header every fifth course	+20%
Economy	8 x 4 x 4	1C=4"	4.50	9.1	11.6		full header every sixth course	+16.7%
Engineer	8 x 3-1/5 x 4	5C=16"	5.63	8.5	10.8	English	full header every second course	+50%
Fire	9 x 2-1/2 x 4-1/2	2C=5"	6.40	550 # Fireclay	—	Flemish	alternate headers every course	+33.3%
Jumbo	12 x 4 x 6 or 8	1C=4"	3.00	22.5	29.2		every sixth course	+5.6%
Norman	12 x 2-2/3 x 4	3C=8"	4.50	11.2	14.3	Header = W x H exposed		+100%
Norwegian	12 x 3-1/5 x 4	5C=16"	3.75	11.7	14.9	Rowlock = H x W exposed		+100%
Roman	12 x 2 x 4	2C=4"	6.00	10.7	13.7	Rowlock stretcher = L x W exposed		+33.3%
SCR	12 x 2-2/3 x 6	3C=8"	4.50	21.8	28.0	Soldier = H x L exposed		—
Utility	12 x 4 x 4	1C=4"	3.00	12.3	15.7	Sailor = W x L exposed		-33.3%

Concrete Blocks Nominal Size	Approximate Weight per S.F. Standard	Lightweight	Blocks per 100 S.F.	Mortar per M block, waste included Partitions	Back up
2" x 8" x 16"	20 PSF	15 PSF	113	27 C.F.	36 C.F.
4"	30	20		41	51
6"	42	30		56	66
8"	55	38		72	82
10"	70	47		87	97
12"	85	55		102	112

Brick & Mortar Quantities
©Brick Industry Association. 2009 Feb. Technical Notes on Brick Construction 10:
    Dimensioning and Estimating Brick Masonry. Reston (VA): BIA. Table 1 Modular Brick Sizes and Table 4 Quantity Estimates for Brick Masonry.

Metals — R0505 Common Work Results for Metals

R050521-20 Welded Structural Steel

Usual weight reductions with welded design run 10% to 20% compared with bolted or riveted connections. This amounts to about the same total cost compared with bolted structures since field welding is more expensive than bolts. For normal spans of 18' to 24' figure 6 to 7 connections per ton.

Trusses — For welded trusses add 4% to weight of main members for connections. Up to 15% less steel can be expected in a welded truss compared to one that is shop bolted. Cost of erection is the same whether shop bolted or welded.

General — Typical electrodes for structural steel welding are E6010, E6011, E60T and E70T. Typical buildings vary between 2# to 8# of weld rod per ton of steel. Buildings utilizing continuous design require about three times as much welding as conventional welded structures. In estimating field erection by welding, it is best to use the average linear feet of weld per ton to arrive at the welding cost per ton. The type, size and position of the weld will have a direct bearing on the cost per linear foot. A typical field welder will deposit 1.8# to 2# of weld rod per hour manually. Using semiautomatic methods can increase production by as much as 50% to 75%.

Metals — R0512 Structural Steel Framing

R051223-10 Structural Steel

The bare material prices for structural steel, shown in the unit cost sections of the book, are for 100 tons of shop-fabricated structural steel and include:
1. Mill base price of structural steel
2. Mill scrap/grade/size/length extras
3. Mill delivery to a metals service center (warehouse)
4. Service center storage and handling
5. Service center delivery to a fabrication shop
6. Shop storage and handling
7. Shop drafting/detailing
8. Shop fabrication
9. Shop coat of primer paint
10. Shop listing
11. Shop delivery to the job site

In unit cost sections of the book that contain items for field fabrication of steel components, the bare material cost of steel includes:
1. Mill base price of structural steel
2. Mill scrap/grade/size/length extras
3. Mill delivery to a metals service center (warehouse)
4. Service center storage and handling
5. Service center delivery to the job site

R051223-15 Structural Steel Estimating for Repair and Remodeling Projects

The correct approach to estimating structural steel is dependent upon the amount of steel required for the particular project. If the project is a sizable addition to a building, the data can be used directly from the cost data book. This is not the case however if the project requires a small amount of steel, for instance to reinforce existing roof or floor structural systems. To better understand this, please refer to the unit price line for a W16x31 beam with bolted connections. Assume your project requires the reinforcement of the structural members of the roof system of a 3 story building required for the installation a new piece of HVAC equipment. The project will need 4 pieces of W16x31 that are 30' long. After pricing this 120 L.F. job using the unit prices for a W16x31, an analysis will reveal that the price is wholly inadequate.

The first problem is apparent if you examine the amount of steel that can be installed per day. The unit price line indicates 900 linear feet per day can be installed by a 5 man crew with an 90 ton crane. This productivity is correct for new construction but certainly is not for repair and remodeling work. Installation of new structural steel considers that each member is installed from the foundation to the roof of the structure in a planned and systematic manner with a crane having unrestricted access to all parts of the project. Additionally each connection is planned and detailed with full field access for fit-up and final bolting. The erection is planned and progresses such that interferences and conflicts with other structural members are minimized, if not completely eliminated. All of these assumptions are clearly not the case with a repair and remodeling job, and a significant decrease in the stated productivity will be observed.

A crane will certainly be needed to lift the members into the general area of the project but in most cases will not be able to place the beams into their final position. An opening in the existing roof may not be large enough to permit the beams to pass through and it may be necessary to bring them into the building through existing windows or doors. Moving the beams to the actual area where they will be installed may involve hand labor and the use of dollies. Finally, hoists and/or jacks may be needed for final positioning.

The connection of new members to existing can often be accomplished by field bolting with accurate field measurements and good planning but in many cases access to both sides of existing members is not possible and field welding becomes the only alternative. In addition to the cost of the actual welding, protection of existing finishes, systems and structure and fire protection must be considered.

New beams can never be installed tight to the existing decks or floors which they must support and the use of shims and tack welding becomes necessary. Additionally, further planning and cost is involved in assuring that existing loads are minimized during the installation and shimming process.

It is apparent that installation of structural steel as part of a repair and remodeling project involves more than simply installing the members and estimating the cost in the same manner as new construction. The best procedure for estimating the total cost is adequate planning and coordination of each process and activity that will be needed. Unit costs for the materials, labor and equipment can then be attached to each needed activity and a final, complete price can be determined.

Metals — R0531 Steel Decking

R053100-10 Decking Descriptions

General - All Deck Products

Steel deck is made by cold forming structural grade sheet steel into a repeating pattern of parallel ribs. The strength and stiffness of the panels are the result of the ribs and the material properties of the steel. Deck lengths can be varied to suit job conditions, but because of shipping considerations, are usually less than 40 feet. Standard deck width varies with the product used but full sheets are usually 12", 18", 24", 30", or 36". Deck is typically furnished in a standard width with the ends cut square. Any cutting for width, such as at openings or for angular fit, is done at the job site.

Deck is typically attached to the building frame with arc puddle welds, self-drilling screws, or powder or pneumatically driven pins. Sheet to sheet fastening is done with screws, button punching (crimping), or welds.

Composite Floor Deck

After installation and adequate fastening, floor deck serves several purposes. It (a) acts as a working platform, (b) stabilizes the frame, (c) serves as a concrete form for the slab, and (d) reinforces the slab to carry the design loads applied during the life of the building. Composite decks are distinguished by the presence of shear connector devices as part of the deck. These devices are designed to mechanically lock the concrete and deck together so that the concrete and the deck work together to carry subsequent floor loads. These shear connector devices can be rolled-in embossments, lugs, holes, or wires welded to the panels. The deck profile can also be used to interlock concrete and steel.

Composite deck finishes are either galvanized (zinc coated) or phosphatized/painted. Galvanized deck has a zinc coating on both the top and bottom surfaces. The phosphatized/painted deck has a bare (phosphatized) top surface that will come into contact with the concrete. This bare top surface can be expected to develop rust before the concrete is placed. The bottom side of the deck has a primer coat of paint.

Composite floor deck is normally installed so the panel ends do not overlap on the supporting beams. Shear lugs or panel profile shape often prevent a tight metal to metal fit if the panel ends overlap; the air gap caused by overlapping will prevent proper fusion with the structural steel supports when the panel end laps are shear stud welded.

Adequate end bearing of the deck must be obtained as shown on the drawings. If bearing is actually less in the field than shown on the drawings, further investigation is required.

Roof Deck

Roof deck is not designed to act compositely with other materials. Roof deck acts alone in transferring horizontal and vertical loads into the building frame. Roof deck rib openings are usually narrower than floor deck rib openings. This provides adequate support of rigid thermal insulation board.

Roof deck is typically installed to endlap approximately 2" over supports. However, it can be butted (or lapped more than 2") to solve field fit problems. Since designers frequently use the installed deck system as part of the horizontal bracing system (the deck as a diaphragm), any fastening substitution or change should be approved by the designer. Continuous perimeter support of the deck is necessary to limit edge deflection in the finished roof and may be required for diaphragm shear transfer.

Standard roof deck finishes are galvanized or primer painted. The standard factory applied paint for roof deck is a primer paint and is not intended to weather for extended periods of time. Field painting or touching up of abrasions and deterioration of the primer coat or other protective finishes is the responsibility of the contractor.

Cellular Deck

Cellular deck is made by attaching a bottom steel sheet to a roof deck or composite floor deck panel. Cellular deck can be used in the same manner as floor deck. Electrical, telephone, and data wires are easily run through the chase created between the deck panel and the bottom sheet.

When used as part of the electrical distribution system, the cellular deck must be installed so that the ribs line up and create a smooth cell transition at abutting ends. The joint that occurs at butting cell ends must be taped or otherwise sealed to prevent wet concrete from seeping into the cell. Cell interiors must be free of welding burrs, or other sharp intrusions, to prevent damage to wires.

When used as a roof deck, the bottom flat plate is usually left exposed to view. Care must be maintained during erection to keep good alignment and prevent damage.

Cellular deck is sometimes used with the flat plate on the top side to provide a flat working surface. Installation of the deck for this purpose requires special methods for attachment to the frame because the flat plate, now on the top, can prevent direct access to the deck material that is bearing on the structural steel. It may be advisable to treat the flat top surface to prevent slipping.

Cellular deck is always furnished galvanized or painted over galvanized.

Form Deck

Form deck can be any floor or roof deck product used as a concrete form. Connections to the frame are by the same methods used to anchor floor and roof deck. Welding washers are recommended when welding deck that is less than 20 gauge thickness.

Form deck is furnished galvanized, prime painted, or uncoated. Galvanized deck must be used for those roof deck systems where form deck is used to carry a lightweight insulating concrete fill.

Wood, Plastics & Comp. | R0611 Wood Framing

R061110-30 Lumber Product Material Prices

The price of forest products fluctuates widely from location to location and from season to season depending upon economic conditions. The bare material prices in the unit cost sections of the book show the National Average material prices in effect Jan. 1 of this book year. It must be noted that lumber prices in general may change significantly during the year.

Availability of certain items depends upon geographic location and must be checked prior to firm-price bidding.

Wood, Plastics & Comp. | R0616 Sheathing

R061636-20 Plywood

There are two types of plywood used in construction: interior, which is moisture-resistant but not waterproofed, and exterior, which is waterproofed.

The grade of the exterior surface of the plywood sheets is designated by the first letter: A, for smooth surface with patches allowed; B, for solid surface with patches and plugs allowed; C, which may be surface plugged or may have knot holes up to 1" wide; and D, which is used only for interior type plywood and may have knot holes up to 2-1/2" wide. "Structural Grade" is specifically designed for engineered applications such as box beams. All CC & DD grades have roof and floor spans marked on them.

Underlayment-grade plywood runs from 1/4" to 1-1/4" thick. Thicknesses 5/8" and over have optional tongue and groove joints which eliminate the need for blocking the edges. Underlayment 19/32" and over may be referred to as Sturd-i-Floor.

The price of plywood can fluctuate widely due to geographic and economic conditions.

Typical uses for various plywood grades are as follows:

AA-AD Interior — cupboards, shelving, paneling, furniture

BB Plyform — concrete form plywood

CDX — wall and roof sheathing

Structural — box beams, girders, stressed skin panels

AA-AC Exterior — fences, signs, siding, soffits, etc.

Underlayment — base for resilient floor coverings

Overlaid HDO — high density for concrete forms & highway signs

Overlaid MDO — medium density for painting, siding, soffits & signs

303 Siding — exterior siding, textured, striated, embossed, etc.

Thermal & Moist. Protec. | R0751 Built-Up Bituminous Roofing

R075113-20 Built-Up Roofing

Asphalt is available in kegs of 100 lbs. each; coal tar pitch in 560 lb. kegs. Prepared roofing felts are available in a wide range of sizes, weights and characteristics. However, the most commonly used are #15 (432 S.F. per roll, 13 lbs. per square) and #30 (216 S.F. per roll, 27 lbs. per square).

Inter-ply bitumen varies from 24 lbs. per sq. (asphalt) to 30 lbs. per sq. (coal tar) per ply, MF4@ 25%. Flood coat bitumen also varies from 60 lbs. per sq. (asphalt) to 75 lbs. per sq. (coal tar), MF4@ 25%. Expendable equipment (mops, brooms, screeds, etc.) runs about 16% of the bitumen cost. For new, inexperienced crews this factor may be much higher.

Rigid insulation board is typically applied in two layers. The first is mechanically attached to nailable decks or spot or solid mopped to non-nailable decks; the second layer is then spot or solid mopped to the first layer. Membrane application follows the insulation, except in protected membrane roofs, where the membrane goes down first and the insulation on top, followed with ballast (stone or concrete pavers). Insulation and related labor costs are NOT included in prices for built-up roofing.

Thermal & Moist. Protec. — R0752 Modified Bituminous Membrane Roofing

R075213-30 Modified Bitumen Roofing

The cost of modified bitumen roofing is highly dependent on the type of installation that is planned. Installation is based on the type of modifier used in the bitumen. The two most popular modifiers are atactic polypropylene (APP) and styrene butadiene styrene (SBS). The modifiers are added to heated bitumen during the manufacturing process to change its characteristics. A polyethylene, polyester or fiberglass reinforcing sheet is then sandwiched between layers of this bitumen. When completed, the result is a pre-assembled, built-up roof that has increased elasticity and weatherablility. Some manufacturers include a surfacing material such as ceramic or mineral granules, metal particles or sand.

The preferred method of adhering SBS-modified bitumen roofing to the substrate is with hot-mopped asphalt (much the same as built-up roofing). This installation method requires a tar kettle/pot to heat the asphalt, as well as the labor, tools and equipment necessary to distribute and spread the hot asphalt.

The alternative method for applying APP and SBS modified bitumen is as follows. A skilled installer uses a torch to melt a small pool of bitumen off the membrane. This pool must form across the entire roll for proper adhesion. The installer must unroll the roofing at a pace slow enough to melt the bitumen, but fast enough to prevent damage to the rest of the membrane.

Modified bitumen roofing provides the advantages of both built-up and single-ply roofing. Labor costs are reduced over those of built-up roofing because only a single ply is necessary. The elasticity of single-ply roofing is attained with the reinforcing sheet and polymer modifiers. Modifieds have some self-healing characteristics and because of their multi-layer construction, they offer the reliability and safety of built-up roofing.

Thermal & Moist. Protec. — R0784 Firestopping

R078413-30 Firestopping

Firestopping is the sealing of structural, mechanical, electrical and other penetrations through fire-rated assemblies. The basic components of firestop systems are safing insulation and firestop sealant on both sides of wall penetrations and the top side of floor penetrations.

Pipe penetrations are assumed to be through concrete, grout, or joint compound and can be sleeved or unsleeved. Costs for the penetrations and sleeves are not included. An annular space of 1" is assumed. Escutcheons are not included.

Metallic pipe is assumed to be copper, aluminum, cast iron or similar metallic material. Insulated metallic pipe is assumed to be covered with a thermal insulating jacket of varying thickness and materials.

Non-metallic pipe is assumed to be PVC, CPVC, FR Polypropylene or similar plastic piping material. Intumescent firestop sealant or wrap strips are included. Collars on both sides of wall penetrations and a sheet metal plate on the underside of floor penetrations are included.

Ductwork is assumed to be sheet metal, stainless steel or similar metallic material. Duct penetrations are assumed to be through concrete, grout or joint compound. Costs for penetrations and sleeves are not included. An annular space of 1/2" is assumed.

Multi-trade openings include costs for sheet metal forms, firestop mortar, wrap strips, collars and sealants as necessary.

Structural penetrations joints are assumed to be 1/2" or less. CMU walls are assumed to be within 1-1/2" of metal deck. Drywall walls are assumed to be tight to the underside of metal decking.

Metal panel, glass or curtain wall systems include a spandrel area of 5' filled with mineral wool foil-faced insulation. Fasteners and stiffeners are included.

Openings — R0813 Metal Doors

R081313-20 Steel Door Selection Guide

Standard steel doors are classified into four levels, as recommended by the Steel Door Institute in the chart below. Each of the four levels offers a range of construction models and designs, to meet architectural requirements for preference and appearance, including full flush, seamless, and stile & rail. Recommended minimum gauge requirements are also included.

For complete standard steel door construction specifications and available sizes, refer to the Steel Door Institute Technical Data Series, ANSI A250.8-98 (SDI-100), and ANSI A250.4-94 Test Procedure and Acceptance Criteria for Physical Endurance of Steel Door and Hardware Reinforcements.

Level	Model	Construction	For Full Flush or Seamless		
			Min. Gauge	Thickness (in)	Thickness (mm)
I	Standard Duty	1 Full Flush			
		2 Seamless	20	0.032	0.8
II	Heavy Duty	1 Full Flush			
		2 Seamless	18	0.042	1.0
III	Extra Heavy Duty	1 Full Flush			
		2 Seamless			
		3 *Stile & Rail	16	0.053	1.3
IV	Maximum Duty	1 Full Flush			
		2 Seamless	14	0.067	1.6

*Stiles & rails are 16 gauge; flush panels, when specified, are 18 gauge

Openings — R0851 Metal Windows

R085123-10 Steel Sash

Ironworker crew will erect 25 S.F. or 1.3 sash unit per hour, whichever is less.

Mechanic will point 30 L.F. per hour.

Painter will paint 90 S.F. per coat per hour.

Glazier production depends on light size.

Allow 1 lb. special steel sash putty per 16" x 20" light.

Openings — R0853 Plastic Windows

R085313-20 Replacement Windows

Replacement windows are typically measured per United Inch.

United Inches are calculated by rounding the width and height of the window opening up to the nearest inch, then adding the two figures.

The labor cost for replacement windows includes removal of sash, existing sash balance or weights, parting bead where necessary and installation of new window.

Debris hauling and dump fees are not included.

Openings

R0871 Door Hardware

R087110-10 Hardware Finishes

This table describes hardware finishes used throughout the industry. It also shows the base metal and the respective symbols in the three predominate systems of identification. Many of these are used in pricing descriptions in Division Eight.

US"	BMHA*	CDN	Base	Description
US P	600	CP	Steel	Primed for Painting
US 1B	601	C1B	Steel	Bright Black Japanned
US 2C	602	C2C	Steel	Zinc Plated
US 2G	603	C2G	Steel	Zinc Plated
US 3	605	C3	Brass	Bright Brass, Clear Coated
US 4	606	C4	Brass	Satin Brass, Clear Coated
US 5	609	C5	Brass	Satin Brass, Blackened, Satin Relieved, Clear Coated
US 7	610	C7	Brass	Satin Brass, Blackened, Bright Relieved, Clear Coated
US 9	611	C9	Bronze	Bright Bronze, Clear Coated
US 10	612	C10	Bronze	Satin Bronze, Clear Coated
US 10A	641	C10A	Steel	Antiqued Bronze, Oiled and Lacquered
US 10B	613	C10B	Bronze	Antiqued Bronze, Oiled
US 11	616	C11	Bronze	Satin Bronze, Blackened, Satin Relieved, Clear Coated
US 14	618	C14	Brass/Bronze	Bright Nickel Plated, Clear Coated
US 15	619	C15	Brass/Bronze	Satin Nickel, Clear Coated
US 15A	620	C15A	Brass/Bronze	Satin Nickel Plated, Blackened, Satin Relieved, Clear Coated
US 17A	621	C17A	Brass/Bronze	Nickel Plated, Blackened, Relieved, Clear Coated
US 19	622	C19	Brass/Bronze	Flat Black Coated
US 20	623	C20	Brass/Bronze	Statuary Bronze, Light
US 20A	624	C20A	Brass/Bronze	Statuary Bronze, Dark
US 26	625	C26	Brass/Bronze	Bright Chromium
US 26D	626	C26D	Brass/Bronze	Satin Chromium
US 20	627	C27	Aluminum	Satin Aluminum Clear
US 28	628	C28	Aluminum	Anodized Dull Aluminum
US 32	629	C32	Stainless Steel	Bright Stainless Steel
US 32D	630	C32D	Stainless Steel	Stainless Steel
US 3	632	C3	Steel	Bright Brass Plated, Clear Coated
US 4	633	C4	Steel	Satin Brass, Clear Coated
US 7	636	C7	Steel	Satin Brass Plated, Blackened, Bright Relieved, Clear Coated
US 9	637	C9	Steel	Bright Bronze Plated, Clear Coated
US 5	638	C5	Steel	Satin Brass Plated, Blackened, Bright Relieved, Clear Coated
US 10	639	C10	Steel	Satin Bronze Plated, Clear Coated
US 10B	640	C10B	Steel	Antique Bronze, Oiled
US 10A	641	C10A	Steel	Antiqued Bronze, Oiled and Lacquered
US 11	643	C11	Steel	Satin Bronze Plated, Blackened, Bright Relieved, Clear Coated
US 14	645	C14	Steel	Bright Nickel Plated, Clear Coated
US 15	646	C15	Steel	Satin Nickel Plated, Clear Coated
US 15A	647	C15A	Steel	Nickel Plated, Blackened, Bright Relieved, Clear Coated
US 17A	648	C17A	Steel	Nickel Plated, Blackened, Relieved, Clear Coated
US 20	649	C20	Steel	Statuary Bronze, Light
US 20A	650	C20A	Steel	Statuary Bronze, Dark
US 26	651	C26	Steel	Bright Chromium Plated
US 26D	652	C26D	Steel	Satin Chromium Plated

* - BMHA Builders Hardware Manufacturing Association
" - US Equivalent
^ - Canadian Equivalent
Japanning is imitating Asian lacquer work

Openings R0881 Glass Glazing

R088110-10 Glazing Productivity

Some glass sizes are estimated by the "united inch" (height + width). The table below shows the number of lights glazed in an eight-hour period by the crew size indicated, for glass up to 1/4" thick. Square or nearly square lights are more economical on a S.F. basis. Long slender lights will have a high S.F. installation cost. For insulated glass reduce production by 33%. For 1/2" float glass reduce production by 50%. Production time for glazing with two glaziers per day averages: 1/4" float glass 120 S.F.; 1/2" float glass 55 S.F.; 1/2" insulated glass 95 S.F.; 3/4" insulated glass 75 S.F.

Glazing Method	United Inches per Light							
	40"	60"	80"	100"	135"	165"	200"	240"
Number of Men in Crew	1	1	1	1	2	3	3	4
Industrial sash, putty	60	45	24	15	18	—	—	—
With stops, putty bed	50	36	21	12	16	8	4	3
Wood stops, rubber	40	27	15	9	11	6	3	2
Metal stops, rubber	30	24	14	9	9	6	3	2
Structural glass	10	7	4	3	—	—	—	—
Corrugated glass	12	9	7	4	4	4	3	—
Storefronts	16	15	13	11	7	6	4	4
Skylights, putty glass	60	36	21	12	16	—	—	—
Thiokol set	15	15	11	9	9	6	3	2
Vinyl set, snap on	18	18	13	12	12	7	5	4
Maximum area per light	2.8 S.F.	6.3 S.F.	11.1 S.F.	17.4 S.F.	31.6 S.F.	47 S.F.	69 S.F.	100 S.F.

Finishes — R0920 Plaster & Gypsum Board

R092000-50 Lath, Plaster and Gypsum Board

Gypsum board lath is available in 3/8" thick x 16" wide x 4' long sheets as a base material for multi-layer plaster applications. It is also available as a base for either multi-layer or veneer plaster applications in 1/2" and 5/8" thick–4' wide x 8', 10' or 12' long sheets. Fasteners are screws or blued ring shank nails for wood framing and screws for metal framing.

Metal lath is available in diamond mesh pattern with flat or self-furring profiles. Paper backing is available for applications where excessive plaster waste needs to be avoided. A slotted mesh ribbed lath should be used in areas where the span between structural supports is greater than normal. Most metal lath comes in 27" x 96" sheets. Diamond mesh weighs 1.75, 2.5 or 3.4 pounds per square yard, slotted mesh lath weighs 2.75 or 3.4 pounds per square yard. Metal lath can be nailed, screwed or tied in place.

Many **accessories** are available. Corner beads, flat reinforcing strips, casing beads, control and expansion joints, furring brackets and channels are some examples. Note that accessories are not included in plaster or stucco line items.

Plaster is defined as a material or combination of materials that when mixed with a suitable amount of water, forms a plastic mass or paste. When applied to a surface, the paste adheres to it and subsequently hardens, preserving in a rigid state the form or texture imposed during the period of elasticity.

Gypsum plaster is made from ground calcined gypsum. It is mixed with aggregates and water for use as a base coat plaster.

Vermiculite plaster is a fire-retardant plaster covering used on steel beams, concrete slabs and other heavy construction materials. Vermiculite is a group name for certain clay minerals, hydrous silicates or aluminum, magnesium and iron that have been expanded by heat.

Perlite plaster is a plaster using perlite as an aggregate instead of sand. Perlite is a volcanic glass that has been expanded by heat.

Gauging plaster is a mix of gypsum plaster and lime putty that when applied produces a quick drying finish coat.

Veneer plaster is a one or two component gypsum plaster used as a thin finish coat over special gypsum board.

Keenes cement is a white cementitious material manufactured from gypsum that has been burned at a high temperature and ground to a fine powder. Alum is added to accelerate the set. The resulting plaster is hard and strong and accepts and maintains a high polish, hence it is used as a finishing plaster.

Stucco is a Portland cement based plaster used primarily as an exterior finish.

Plaster is used on both interior and exterior surfaces. Generally it is applied in multiple-coat systems. A three-coat system uses the terms scratch, brown and finish to identify each coat. A two-coat system uses base and finish to describe each coat. Each type of plaster and application system has attributes that are chosen by the designer to best fit the intended use.

Gypsum Plaster Quantities for 100 S.Y.	2 Coat, 5/8" Thick		3 Coat, 3/4" Thick		
	Base 1:3 Mix	Finish 2:1 Mix	Scratch 1:2 Mix	Brown 1:3 Mix	Finish 2:1 Mix
Gypsum plaster	1,300 lb.		1,350 lb.	650 lb.	
Sand	1.75 C.Y.		1.85 C.Y.	1.35 C.Y.	
Finish hydrated lime		340 lb.			340 lb.
Gauging plaster		170 lb.			170 lb.

Vermiculite or Perlite Plaster Quantities for 100 S.Y.	2 Coat, 5/8" Thick		3 Coat, 3/4" Thick		
	Base	Finish	Scratch	Brown	Finish
Gypsum plaster	1,250 lb.		1,450 lb.	800 lb.	
Vermiculite or perlite	7.8 bags		8.0 bags	3.3 bags	
Finish hydrated lime		340 lb.			340 lb.
Gauging plaster		170 lb.			170 lb.

Stucco–Three-Coat System Quantities for 100 S.Y.	On Wood Frame	On Masonry
Portland cement	29 bags	21 bags
Sand	2.6 C.Y.	2.0 C.Y.
Hydrated lime	180 lb.	120 lb.

Finishes

R0929 Gypsum Board

R092910-10 Levels of Gypsum Drywall Finish

In the past, contract documents often used phrases such as "industry standard" and "workmanlike finish" to specify the expected quality of gypsum board wall and ceiling installations. The vagueness of these descriptions led to unacceptable work and disputes.

In order to resolve this problem, four major trade associations concerned with the manufacture, erection, finish and decoration of gypsum board wall and ceiling systems have developed an industry-wide *Recommended Levels of Gypsum Board Finish*.

The finish of gypsum board walls and ceilings for specific final decoration is dependent on a number of factors. A primary consideration is the location of the surface and the degree of decorative treatment desired. Painted and unpainted surfaces in warehouses and other areas where appearance is normally not critical may simply require the taping of wallboard joints and 'spotting' of fastener heads. Blemish-free, smooth, monolithic surfaces often intended for painted and decorated walls and ceilings in habitated structures, ranging from single-family dwellings through monumental buildings, require additional finishing prior to the application of the final decoration.

Other factors to be considered in determining the level of finish of the gypsum board surface are (1) the type of angle of surface illumination (both natural and artificial lighting), and (2) the paint and method of application or the type and finish of wallcovering specified as the final decoration. Critical lighting conditions, gloss paints, and thin wall coverings require a higher level of gypsum board finish than do heavily textured surfaces which are subsequently painted or surfaces which are to be decorated with heavy grade wall coverings.

The following descriptions were developed jointly by the Association of the Wall and Ceiling Industries-International (AWCI), Ceiling & Interior Systems Construction Association (CISCA), Gypsum Association (GA), and Painting and Decorating Contractors of America (PDCA) as a guide.

Level 0: Used in temporary construction or wherever the final decoration has not been determined. Unfinished. No taping, finishing or corner beads are required. Also could be used where non-predecorated panels will be used in demountable-type partitions that are to be painted as a final finish.

Level 1: Frequently used in plenum areas above ceilings, in attics, in areas where the assembly would generally be concealed, or in building service corridors and other areas not normally open to public view. Some degree of sound and smoke control is provided; in some geographic areas, this level is referred to as "fire-taping," although this level of finish does not typically meet fire-resistant assembly requirements. Where a fire resistance rating is required for the gypsum board assembly, details of construction should be in accordance with reports of fire tests of assemblies that have met the requirements of the fire rating acceptable.

All joints and interior angles shall have tape embedded in joint compound. Accessories are optional at specifier discretion in corridors and other areas with pedestrian traffic. Tape and fastener heads need not be covered with joint compound. Surface shall be free of excess joint compound. Tool marks and ridges are acceptable.

Level 2: It may be specified for standard gypsum board surfaces in garages, warehouse storage, or other similar areas where surface appearance is not of primary importance.

All joints and interior angles shall have tape embedded in joint compound and shall be immediately wiped with a joint knife or trowel, leaving a thin coating of joint compound over all joints and interior angles. Fastener heads and accessories shall be covered with a coat of joint compound. Surface shall be free of excess joint compound. Tool marks and ridges are acceptable.

Level 3: Typically used in areas that are to receive heavy texture (spray or hand applied) finishes before final painting, or where commercial-grade (heavy duty) wall coverings are to be applied as the final decoration. This level of finish should not be used where smooth painted surfaces or where lighter weight wall coverings are specified. The prepared surface shall be coated with a drywall primer prior to the application of final finishes.

All joints and interior angles shall have tape embedded in joint compound and shall be immediately wiped with a joint knife or trowel, leaving a thin coating of joint compound over all joints and interior angles. One additional coat of joint compound shall be applied over all joints and interior angles. Fastener heads and accessories shall be covered with two separate coats of joint compound. All joint compounds shall be smooth and free of tool marks and ridges. The prepared surface shall be covered with a drywall primer prior to the application of the final decoration.

Level 4: This level should be used where residential grade (light duty) wall coverings, flat paints, or light textures are to be applied. The prepared surface shall be coated with a drywall primer prior to the application of final finishes. Release agents for wall coverings are specifically formulated to minimize damage if coverings are subsequently removed.

The weight, texture, and sheen level of the wall covering material selected should be taken into consideration when specifying wall coverings over this level of drywall treatment. Joints and fasteners must be sufficiently concealed if the wall covering material is lightweight, contains limited pattern, has a glossy finish, or has any combination of these features. In critical lighting areas, flat paints applied over light textures tend to reduce joint photographing. Gloss, semi-gloss, and enamel paints are not recommended over this level of finish.

All joints and interior angles shall have tape embedded in joint compound and shall be immediately wiped with a joint knife or trowel, leaving a thin coating of joint compound over all joints and interior angles. In addition, two separate coats of joint compound shall be applied over all flat joints and one separate coat of joint compound applied over interior angles. Fastener heads and accessories shall be covered with three separate coats of joint compound. All joint compounds shall be smooth and free of tool marks and ridges. The prepared surface shall be covered with a drywall primer like Sheetrock first coat prior to the application of the final decoration.

Level 5: The highest quality finish is the most effective method to provide a uniform surface and minimize the possibility of joint photographing and of fasteners showing through the final decoration. This level of finish is required where gloss, semi-gloss, or enamel is specified; when flat joints are specified over an untextured surface; or where critical lighting conditions occur. The prepared surface shall be coated with a drywall primer prior to the application of final decoration.

All joints and interior angles shall have tape embedded in joint compound and be immediately wiped with a joint knife or trowel, leaving a thin coating of joint compound over all joints and interior angles. Two separate coats of joint compound shall be applied over all flat joints and one separate coat of joint compound applied over interior angles. Fastener heads and accessories shall be covered with three separate coats of joint compound.

A thin skim coat of joint compound shall be trowel applied to the entire surface. Excess compound is immediately troweled off, leaving a film or skim coating of compound completely covering the paper. As an alternative to a skim coat, a material manufactured especially for this purpose may be applied such as Sheetrock Tuff-Hide primer surfacer. The surface must be smooth and free of tool marks and ridges. The prepared surface shall be covered with a drywall primer prior to the application of the final decoration.

Conveying Equipment   R1420 Elevators

R142000-10   Freight Elevators

Capacities run from 2,000 lbs. to over 100,000 lbs. with 3,000 lbs. to 10,000 lbs. most common. Travel speeds are generally lower and control less intricate than on passenger elevators. Costs in the unit price sections are for hydraulic and geared elevators.

R142000-20   Elevator Selective Costs   See R142000-40 for cost development.

A. Base Unit	Passenger		Freight		Hospital	
	Hydraulic	**Electric**	**Hydraulic**	**Electric**	**Hydraulic**	**Electric**
Capacity	1,500 lb.	2,000 lb.	2,000 lb.	4,000 lb.	4,000 lb.	4,000 lb.
Speed	100 F.P.M.	200 F.P.M.	100 F.P.M.	200 F.P.M.	100 F.P.M.	200 F.P.M.
#Stops/Travel Ft.	2/12	4/40	2/20	4/40	2/20	4/40
Push Button Oper.	Yes	Yes	Yes	Yes	Yes	Yes
Telephone Box & Wire	"	"	"	"	"	"
Emergency Lighting	"	"	No	No	"	"
Cab	Plastic Lam. Walls	Plastic Lam. Walls	Painted Steel	Painted Steel	Plastic Lam. Walls	Plastic Lam. Walls
Cove Lighting	Yes	Yes	No	No	Yes	Yes
Floor	V.C.T.	V.C.T.	Wood w/Safety Treads	Wood w/Safety Treads	V.C.T.	V.C.T.
Doors, & Speedside Slide	Yes	Yes	Yes	Yes	Yes	Yes
Gates, Manual	No	No	No	No	No	No
Signals, Lighted Buttons	Car and Hall	Car and Hall	Car and Hall	Car and Hall	Car and Hall	Car and Hall
O.H. Geared Machine	N.A.	Yes	N.A.	Yes	N.A.	Yes
Variable Voltage Contr.	"	"	N.A.	"	"	"
Emergency Alarm	Yes	"	Yes	"	Yes	"
Class "A" Loading	N.A.	N.A.	"	"	N.A.	N.A.

R142000-30   Passenger Elevators

Electric elevators are used generally but hydraulic elevators can be used for lifts up to 70' and where large capacities are required. Hydraulic speeds are limited to 200 F.P.M. but cars are self leveling at the stops. On low rises, hydraulic installation runs about 15% less than standard electric types but on higher rises this installation cost advantage is reduced. Maintenance of hydraulic elevators is about the same as electric type but underground portion is not included in the maintenance contract.

In electric elevators there are several control systems available, the choice of which will be based upon elevator use, size, speed and cost criteria. The two types of drives are geared for low speeds and gearless for 450 F.P.M. and over.

The tables on the preceding pages illustrate typical installed costs of the various types of elevators available.

R142000-40   Elevator Cost Development

To price a new car or truck from the factory, you must start with the manufacturer's basic model, then add or exchange optional equipment and features. The same is true for pricing elevators.

Requirement: One-passenger elevator, five-story hydraulic, 2,500 lb. capacity, 12' floor to floor, speed 150 F.P.M., emergency power switching and maintenance contract.

Example:

Description	Adjustment
A. Base Elevator: Hydraulic Passenger, 1500 lb. Capacity, 100 fpm, 2 Stops, Standard Finish	1 Ea.
B. Capacity Adjustment (2,500 lb.)	1 Ea.
C. Excess Travel Adjustment: 48' Total Travel (4 x 12') minus 12' Base Unit Travel =	36 V.L.F.
D. Stops Adjustment: 5 Total Stops minus 2 Stops (Base Unit) =	3 Stops
E. Speed Adjustment (150 F.P.M.)	1 Ea.
F. Options:	
1. Intercom Service	1 Ea.
2. Emergency Power Switching, Automatic	1 Ea.
3. Stainless Steel Entrance Doors	5 Ea.
4. Maintenance Contract (12 Months)	1 Ea.
5. Position Indicator for main floor level (none indicated in Base Unit)	1 Ea.

Plumbing — R2201 Operation & Maintenance of Plumbing

R220102-20 Labor Adjustment Factors

Labor Adjustment Factors are provided for Divisions 21, 22, and 23 to assist the mechanical estimator account for the various complexities and special conditions of any particular project. While a single percentage has been entered on each line of Division 22 01 02.20, it should be understood that these are just suggested midpoints of ranges of values commonly used by mechanical estimators. They may be increased or decreased depending on the severity of the special conditions.

The group for "existing occupied buildings", has been the subject of requests for explanation. Actually there are two stages to this group: buildings that are existing and "finished" but unoccupied, and those that also are occupied. Buildings that are "finished" may result in higher labor costs due to the workers having to be more careful not to damage finished walls, ceilings, floors, etc. and may necessitate special protective coverings and barriers. Also corridor bends and doorways may not accommodate long pieces of pipe or larger pieces of equipment. Work above an already hung ceiling can be very time consuming. The addition of occupants may force the work to be done on premium time (nights and/or weekends), eliminate the possible use of some preferred tools such as pneumatic drivers, powder charged drivers etc. The estimator should evaluate the access to the work area and just how the work is going to be accomplished to arrive at an increase in labor costs over "normal" new construction productivity.

R220105-10 Demolition (Selective vs. Removal for Replacement)

Demolition can be divided into two basic categories.

One type of demolition involves the removal of material with no concern for its replacement. The labor-hours to estimate this work are found under "Selective Demolition" in the Fire Protection, Plumbing and HVAC Divisions. It is selective in that individual items or all the material installed as a system or trade grouping such as plumbing or heating systems are removed. This may be accomplished by the easiest way possible, such as sawing, torch cutting, or sledge hammer as well as simple unbolting.

The second type of demolition is the removal of some item for repair or replacement. This removal may involve careful draining, opening of unions, disconnecting and tagging of electrical connections, capping of pipes/ducts to prevent entry of debris or leakage of the material contained as well as transport of the item away from its in-place location to a truck/dumpster. An approximation of the time required to accomplish this type of demolition is to use half of the time indicated as necessary to install a new unit. For example; installation of a new pump might be listed as requiring 6 labor-hours so if we had to estimate the removal of the old pump we would allow an additional 3 hours for a total of 9 hours. That is, the complete replacement of a defective pump with a new pump would be estimated to take 9 labor-hours.

Plumbing — R2211 Facility Water Distribution

R221113-50 Pipe Material Considerations

1. Malleable fittings should be used for gas service.
2. Malleable fittings are used where there are stresses/strains due to expansion and vibration.
3. Cast fittings may be broken as an aid to disassembling of heating lines frozen by long use, temperature and minerals.
4. Cast iron pipe is extensively used for underground and submerged service.
5. Type M (light wall) copper tubing is available in hard temper only and is used for nonpressure and less severe applications than K and L.
6. Type L (medium wall) copper tubing, available hard or soft for interior service.
7. Type K (heavy wall) copper tubing, available in hard or soft temper for use where conditions are severe. For underground and interior service.
8. Hard drawn tubing requires fewer hangers or supports but should not be bent. Silver brazed fittings are recommended, however soft solder is normally used.
9. Type DMV (very light wall) copper tubing designed for drainage, waste and vent plus other non-critical pressure services.

Domestic/Imported Pipe and Fittings Cost

The prices shown in this publication for steel/cast iron pipe and steel, cast iron, malleable iron fittings are based on domestic production sold at the normal trade discounts. The above listed items of foreign manufacture may be available at prices of 1/3 to 1/2 those shown. Some imported items after minor machining or finishing operations are being sold as domestic to further complicate the system.

Caution: Most pipe prices in this book also include a coupling and pipe hangers which for the larger sizes can add significantly to the per foot cost and should be taken into account when comparing "book cost" with quoted supplier's cost.

Electrical — R2601 Operation & Maint. of Elect. Systems

R260105-30 Electrical Demolition (Removal for Replacement)

The purpose of this reference number is to provide a guide to users for electrical "removal for replacement" by applying the rule of thumb: 1/3 of new installation time (typical range from 20% to 50%) for removal. Remember to use reasonable judgment when applying the suggested percentage factor. For example:

Contractors have been requested to remove an existing fluorescent lighting fixture and replace with a new fixture utilizing energy saver lamps and electronic ballast:

In order to fully understand the extent of the project, contractors should visit the job site and estimate the time to perform the renovation work in accordance with applicable national, state and local regulation and codes.

The contractor may need to add extra labor hours to his estimate if he discovers unknown concealed conditions such as: contaminated asbestos ceiling, broken acoustical ceiling tile and need to repair, patch and touch-up paint all the damaged or disturbed areas, tasks normally assigned to general contractors. In addition, the owner could request that the contractors salvage the materials removed and turn over the materials to the owner or dispose of the materials to a reclamation station. The normal removal item is 0.5 labor hour for a lighting fixture and 1.5 labor-hours for new installation time. Revise the estimate times from 2 labor-hours work up to a minimum 4 labor-hours work for just fluorescent lighting fixture.

For removal of large concentrations of lighting fixtures in the same area, apply an "economy of scale" to reduce estimating labor hours.

Electrical — R2605 Common Work Results for Electrical

R260533-22 Conductors in Conduit

Table below lists maximum number of conductors for various sized conduit using THW, TW or THWN insulations.

Copper Wire Size	1/2"			3/4"			1"			1-1/4"			1-1/2"			2"			2-1/2"			3"			3-1/2"			4"		
	TW	THW	THWN	TW	THW	THWN	TW	THW	THWN	TW	THW	THWN	TW	THW	THWN	TW	THW	THWN	TW	THW	THWN	THW	THWN	THW	THWN	THW	THWN		
#14	9	6	13	15	10	24	25	16	39	44	29	69	60	40	94	99	65	154	142	93		143		192					
#12	7	4	10	12	8	18	19	13	29	35	24	51	47	32	70	78	53	114	111	76	164	117		157					
#10	5	4	6	9	6	11	15	11	18	26	19	32	36	26	44	60	43	73	85	61	104	95	160	127		163			
#8	2	1	3	4	3	5	7	5	9	12	10	16	17	13	22	28	22	36	40	32	51	49	79	66	106	85	136		
#6		1	1		2	4		4	6		7	11		10	15		16	26		23	37	36	57	48	76	62	98		
#4		1	1		1	2		3	4		5	7		7	9		12	16		17	22	27	35	36	47	47	60		
#3		1	1		1	1		2	3		4	6		6	8		10	13		15	19	23	29	31	39	40	51		
#2		1	1		1	1		2	3		4	5		5	7		9	11		13	16	20	25	27	33	34	43		
#1					1	1		1	1		3	3		4	5		6	8		9	12	14	18	19	25	25	32		
1/0					1	1		1	1		2	3		3	4		5	7		8	10	12	15	16	21	21	27		
2/0					1	1		1	1		1	2		3	3		5	6		7	8	10	13	14	17	18	22		
3/0					1	1		1	1		1	1		2	3		4	5		6	7	9	11	12	14	15	18		
4/0						1		1	1		1	1		1	2		3	4		5	6	7	9	10	12	13	15		
250 kcmil								1	1		1	1		1	1		2	3		4	4	6	7	8	10	10	12		
300								1	1		1	1		1	1		2	3		3	4	5	6	7	8	9	11		
350									1		1	1		1	1		1	2		3	3	4	5	6	7	8	9		
400											1	1		1	1		1	1		2	3	4	5	5	6	7	8		
500											1	1		1	1		1	1		1	2	3	4	4	5	6	7		
600														1	1		1	1		1	1	3	3	4	4	5	5		
700														1	1		1	1		1	1	2	3	3	4	4	5		
750														1	1		1	1		1	1	2	2	3	3	4	4		

Reprinted with permission from NFPA 70-2014, *National Electrical Code®*, Copyright © 2013, National Fire Protection Association, Quincy, MA. This reprinted material is not the complete and official position of the NFPA on the referenced subject, which is represented solely by the standard in its entirety.

Earthwork — R3123 Excavation & Fill

R312316-40 Excavating

The selection of equipment used for structural excavation and bulk excavation or for grading is determined by the following factors.
1. Quantity of material
2. Type of material
3. Depth or height of cut
4. Length of haul
5. Condition of haul road
6. Accessibility of site
7. Moisture content and dewatering requirements
8. Availability of excavating and hauling equipment

Some additional costs must be allowed for hand trimming the sides and bottom of concrete pours and other excavation below the general excavation.

Number of B.C.Y. per truck = 1.5 C.Y. bucket x 8 passes = 12 loose C.Y.

$$= 12 \times \frac{100}{118} = 10.2 \text{ B.C.Y. per truck}$$

Truck Haul Cycle:

Load truck, 8 passes	=	4 minutes
Haul distance, 1 mile	=	9 minutes
Dump time	=	2 minutes
Return, 1 mile	=	7 minutes
Spot under machine	=	1 minute
		23 minute cycle

Add the mobilization and demobilization costs to the total excavation costs. When equipment is rented for more than three days, there is often no mobilization charge by the equipment dealer. On larger jobs outside of urban areas, scrapers can move earth economically provided a dump site or fill area and adequate haul roads are available. Excavation within sheeting bracing or cofferdam bracing is usually done with a clamshell and production is low, since the clamshell may have to be guided by hand between the bracing. When excavating or filling an area enclosed with a wellpoint system, add 10% to 15% to the cost to allow for restricted access. When estimating earth excavation quantities for structures, allow work space outside the building footprint for construction of the foundation and a slope of 1:1 unless sheeting is used.

When planning excavation and fill, the following should also be considered.
1. Swell factor
2. Compaction factor
3. Moisture content
4. Density requirements

A typical example for scheduling and estimating the cost of excavation of a 15' deep basement on a dry site when the material must be hauled off the site is outlined below.

Assumptions:
1. Swell factor, 18%
2. No mobilization or demobilization
3. Allowance included for idle time and moving on job
4. No dewatering, sheeting, or bracing
5. No truck spotter or hand trimming

Fleet Haul Production per day in B.C.Y.

$$4 \text{ trucks} \times \frac{50 \text{ min. hour}}{23 \text{ min. haul cycle}} \times 8 \text{ hrs.} \times 10.2 \text{ B.C.Y.}$$

$$= 4 \times 2.2 \times 8 \times 10.2 = 718 \text{ B.C.Y./day}$$

Earthwork — R3123 Excavation & Fill

R312316-45 Excavating Equipment

The table below lists THEORETICAL hourly production in C.Y./hr. bank measure for some typical excavation equipment. Figures assume 50 minute hours, 83% job efficiency, 100% operator efficiency, 90° swing and properly sized hauling units, which must be modified for adverse digging and loading conditions. Actual production costs in the front of the book average about 50% of the theoretical values listed here.

Equipment	Soil Type	B.C.Y. Weight	% Swell	1 C.Y.	1-1/2 C.Y.	2 C.Y.	2-1/2 C.Y.	3 C.Y.	3-1/2 C.Y.	4 C.Y.
Hydraulic Excavator "Backhoe" 15' Deep Cut	Moist loam, sandy clay	3400 lb.	40%	165	195	200	275	330	385	440
	Sand and gravel	3100	18	140	170	225	240	285	330	380
	Common earth	2800	30	150	180	230	250	300	350	400
	Clay, hard, dense	3000	33	120	140	190	200	240	260	320
Power Shovel Optimum Cut (Ft.)	Moist loam, sandy clay	3400	40	170 (6.0)	245 (7.0)	295 (7.8)	335 (8.4)	385 (8.8)	435 (9.1)	475 (9.4)
	Sand and gravel	3100	18	165 (6.0)	225 (7.0)	275 (7.8)	325 (8.4)	375 (8.8)	420 (9.1)	460 (9.4)
	Common earth	2800	30	145 (7.8)	200 (9.2)	250 (10.2)	295 (11.2)	335 (12.1)	375 (13.0)	425 (13.8)
	Clay, hard, dense	3000	33	120 (9.0)	175 (10.7)	220 (12.2)	255 (13.3)	300 (14.2)	335 (15.1)	375 (16.0)
Drag Line Optimum Cut (Ft.)	Moist loam, sandy clay	3400	40	130 (6.6)	180 (7.4)	220 (8.0)	250 (8.5)	290 (9.0)	325 (9.5)	385 (10.0)
	Sand and gravel	3100	18	130 (6.6)	175 (7.4)	210 (8.0)	245 (8.5)	280 (9.0)	315 (9.5)	375 (10.0)
	Common earth	2800	30	110 (8.0)	160 (9.0)	190 (9.9)	220 (10.5)	250 (11.0)	280 (11.5)	310 (12.0)
	Clay, hard, dense	3000	33	90 (9.3)	130 (10.7)	160 (11.8)	190 (12.3)	225 (12.8)	250 (13.3)	280 (12.0)

				Wheel Loaders				Track Loaders		
				3 C.Y.	4 C.Y.	6 C.Y.	8 C.Y.	2-1/4 C.Y.	3 C.Y.	4 C.Y.
Loading Tractors	Moist loam, sandy clay	3400	40	260	340	510	690	135	180	250
	Sand and gravel	3100	18	245	320	480	650	130	170	235
	Common earth	2800	30	230	300	460	620	120	155	220
	Clay, hard, dense	3000	33	200	270	415	560	110	145	200
	Rock, well-blasted	4000	50	180	245	380	520	100	130	180

R312323-30 Compacting Backfill

Compaction of fill in embankments, around structures, in trenches, and under slabs is important to control settlement. Factors affecting compaction are:
1. Soil gradation
2. Moisture content
3. Equipment used
4. Depth of fill per lift
5. Density required

Example:

Compact granular fill around a building foundation using a 21" wide x 24" vibratory plate in 8" lifts. Operator moves at 50 F.P.M. working a 50 minute hour to develop 95% Modified Proctor Density with 4 passes.

Production Rate:

$$\frac{1.75' \text{ plate width} \times 50 \text{ F.P.M.} \times 50 \text{ min./hr.} \times .67' \text{ lift}}{27 \text{ C.F. per C.Y.}} = 108.5 \text{ C.Y./hr.}$$

Production Rate for 4 Passes:

$$\frac{108.5 \text{ C.Y.}}{4 \text{ passes}} = 27.125 \text{ C.Y./hr.} \times 8 \text{ hrs.} = 217 \text{ C.Y./day}$$

Earthwork | R3141 Shoring

R314116-40 Wood Sheet Piling

Wood sheet piling may be used for depths to 20' where there is no ground water. If moderate ground water is encountered Tongue & Groove sheeting will help to keep it out. When considerable ground water is present, steel sheeting must be used.

For estimating purposes on trench excavation, sizes are as follows:

Depth	Sheeting	Wales	Braces	B.F. per S.F.
To 8'	3 x 12's	6 x 8's, 2 line	6 x 8's, @ 10'	4.0 @ 8'
8' x 12'	3 x 12's	10 x 10's, 2 line	10 x 10's, @ 9'	5.0 average
12' to 20'	3 x 12's	12 x 12's, 3 line	12 x 12's, @ 8'	7.0 average

Sheeting to be toed in at least 2' depending upon soil conditions. A five person crew with an air compressor and sheeting driver can drive and brace 440 SF/day at 8' deep, 360 SF/day at 12' deep, and 320 SF/day at 16' deep.

For normal soils, piling can be pulled in 1/3 the time to install. Pulling difficulty increases with the time in the ground. Production can be increased by high pressure jetting.

Earthwork | R3163 Drilled Caissons

R316326-60 Caissons

The three principal types of cassions are:

(1) Belled Caissons, which except for shallow depths and poor soil conditions, are generally recommended. They provide more bearing than shaft area. Because of its conical shape, no horizontal reinforcement of the bell is required.

(2) Straight Shaft Caissons are used where relatively light loads are to be supported by caissons that rest on high value bearing strata. While the shaft is larger in diameter than for belled types this is more than offset by the saving in time and labor.

(3) Keyed Caissons are used when extremely heavy loads are to be carried. A keyed or socketed caisson transfers its load into rock by a combination of end-bearing and shear reinforcing of the shaft. The most economical shaft often consists of a steel casing, a steel wide flange core and concrete. Allowable compressive stresses of .225 f'c for concrete, 16,000 psi for the wide flange core, and 9,000 psi for the steel casing are commonly used. The usual range of shaft diameter is 18" to 84". The number of sizes specified for any one project should be limited due to the problems of casing and auger storage. When hand work is to be performed, shaft diameters should not be less than 32". When inspection of borings is required a minimum shaft diameter of 30" is recommended. Concrete caissons are intended to be poured against earth excavation so permanent forms which add to cost should not be used if the excavation is clean and the earth sufficiently impervious to prevent excessive loss of concrete.

Soil Conditions for Belling		
Good	Requires Handwork	Not Recommended
Clay	Hard Shale	Silt
Sandy Clay	Limestone	Sand
Silty Clay	Sandstone	Gravel
Clayey Silt	Weathered Mica	Igneous Rock
Hard-pan		
Soft Shale		
Decomposed Rock		

Exterior Improvements | R3292 Turf & Grasses

R329219-50 Seeding

The type of grass is determined by light, shade and moisture content of soil plus intended use. Fertilizer should be disked 4" before seeding. For steep slopes disk five tons of mulch and lay two tons of hay or straw on surface per acre after seeding. Surface mulch can be staked, lightly disked or tar emulsion sprayed. Material for mulch can be wood chips, peat moss, partially rotted hay or straw, wood fibers and sprayed emulsions. Hemp seed blankets with fertilizer are also available. For spring seeding, watering is necessary. Late fall seeding may have to be reseeded in the spring. Hydraulic seeding, power mulching, and aerial seeding can be used on large areas.

R3311 Water Utility Distribution Piping

R331113-80 Piping Designations

There are several systems currently in use to describe pipe and fittings. The following paragraphs will help to identify and clarify classifications of piping systems used for water distribution.

Piping may be classified by schedule. Piping schedules include 5S, 10S, 10, 20, 30, Standard, 40, 60, Extra Strong, 80, 100, 120, 140, 160 and Double Extra Strong. These schedules are dependent upon the pipe wall thickness. The wall thickness of a particular schedule may vary with pipe size.

Ductile iron pipe for water distribution is classified by Pressure Classes such as Class 150, 200, 250, 300 and 350. These classes are actually the rated water working pressure of the pipe in pounds per square inch (psi). The pipe in these pressure classes is designed to withstand the rated water working pressure plus a surge allowance of 100 psi.

The American Water Works Association (AWWA) provides standards for various types of **plastic pipe.** C-900 is the specification for polyvinyl chloride (PVC) piping used for water distribution in sizes ranging from 4" through 12". C-901 is the specification for polyethylene (PE) pressure pipe, tubing and fittings used for water distribution in sizes ranging from 1/2" through 3". C-905 is the specification for PVC piping sizes 14" and greater.

PVC pressure-rated pipe is identified using the standard dimensional ratio (SDR) method. This method is defined by the American Society for Testing and Materials (ASTM) Standard D 2241. This pipe is available in SDR numbers 64, 41, 32.5, 26, 21, 17, and 13.5. Pipe with an SDR of 64 will have the thinnest wall while pipe with an SDR of 13.5 will have the thickest wall. When the pressure rating (PR) of a pipe is given in psi, it is based on a line supplying water at 73 degrees F.

The National Sanitation Foundation (NSF) seal of approval is applied to products that can be used with potable water. These products have been tested to ANSI/NSF Standard 14.

Valves and strainers are classified by American National Standards Institute (ANSI) Classes. These Classes are 125, 150, 200, 250, 300, 400, 600, 900, 1500 and 2500. Within each class there is an operating pressure range dependent upon temperature. Design parameters should be compared to the appropriate material dependent, pressure-temperature rating chart for accurate valve selection.

Change Orders

Change Order Considerations

A change order is a written document, usually prepared by the design professional, and signed by the owner, the architect/engineer, and the contractor. A change order states the agreement of the parties to: an addition, deletion, or revision in the work; an adjustment in the contract sum, if any; or an adjustment in the contract time, if any. Change orders, or "extras" in the construction process occur after execution of the construction contract and impact architects/engineers, contractors, and owners.

Change orders that are properly recognized and managed can ensure orderly, professional, and profitable progress for all who are involved in the project. There are many causes for change orders and change order requests. In all cases, change orders or change order requests should be addressed promptly and in a precise and prescribed manner. The following paragraphs include information regarding change order pricing and procedures.

The Causes of Change Orders

Reasons for issuing change orders include:

- Unforeseen field conditions that require a change in the work
- Correction of design discrepancies, errors, or omissions in the contract documents
- Owner-requested changes, either by design criteria, scope of work, or project objectives
- Completion date changes for reasons unrelated to the construction process
- Changes in building code interpretations, or other public authority requirements that require a change in the work
- Changes in availability of existing or new materials and products

Procedures

Properly written contract documents must include the correct change order procedures for all parties—owners, design professionals and contractors—to follow in order to avoid costly delays and litigation.

Being "in the right" is not always a sufficient or acceptable defense. The contract provisions requiring notification and documentation must be adhered to within a defined or reasonable time frame.

The appropriate method of handling change orders is by a written proposal and acceptance by all parties involved. Prior to starting work on a project, all parties should identify their authorized agents who may sign and accept change orders, as well as any limits placed on their authority.

Time may be a critical factor when the need for a change arises. For such cases, the contractor might be directed to proceed on a "time and materials" basis, rather than wait for all paperwork to be processed—a delay that could impede progress. In this situation, the contractor must still follow the prescribed change order procedures including, but not limited to, notification and documentation.

Lack of documentation can be very costly, especially if legal judgments are to be made, and if certain field personnel are no longer available. For time and material change orders, the contractor should keep accurate daily records of all labor and material allocated to the change.

Owners or awarding authorities who do considerable and continual building construction (such as the federal government) realize the inevitability of change orders for numerous reasons, both predictable and unpredictable. As a result, the federal government, the American Institute of Architects (AIA), the Engineers Joint Contract Documents Committee (EJCDC) and other contractor, legal, and technical organizations have developed standards and procedures to be followed by all parties to achieve contract continuance and timely completion, while being financially fair to all concerned.

Pricing Change Orders

When pricing change orders, regardless of their cause, the most significant factor is when the change occurs. The need for a change may be perceived in the field or requested by the architect/engineer *before* any of the actual installation has begun, or may evolve or appear *during* construction when the item of work in question is partially installed. In the latter cases, the original sequence of construction is disrupted, along with all contiguous and supporting systems. Change orders cause the greatest impact when they occur *after* the installation has been completed and must be uncovered, or even replaced. Post-completion changes may be caused by necessary design changes, product failure, or changes in the owner's requirements that are not discovered until the building or the systems begin to function.

Specified procedures of notification and record keeping must be adhered to and enforced regardless of the stage of construction: *before, during,* or *after* installation. Some bidding documents anticipate change orders by requiring that unit prices including overhead and profit percentages—for additional as well as deductible changes—be listed. Generally these unit prices do not fully take into account the ripple effect, or impact on other trades, and should be used for general guidance only.

When pricing change orders, it is important to classify the time frame in which the change occurs. There are two basic time frames for change orders: *pre-installation change orders*, which occur before the start of construction, and *post-installation change orders*, which involve reworking after the original installation. Change orders that occur between these stages may be priced according to the extent of work completed using a combination of techniques developed for pricing *pre-* and *post-installation* changes.

Factors To Consider When Pricing Change Orders

As an estimator begins to prepare a change order, the following questions should be reviewed to determine their impact on the final price.

General

- *Is the change order work* pre-installation *or* post-installation?

  Change order work costs vary according to how much of the installation has been completed. Once workers have the project scoped in their minds, even though they have not started, it can be difficult to refocus. Consequently they may spend more than the normal amount of time understanding the change. Also, modifications to work in place, such as trimming or refitting, usually take more time than was initially estimated. The greater the amount of work in place, the more reluctant workers are to change it. Psychologically they may resent the change and as a result the rework takes longer than normal. Post-installation change order estimates must include demolition of existing work as required to accomplish the change. If the work is performed at a later time, additional obstacles, such as building finishes, may be present which must be protected. Regardless of whether the change occurs pre-installation or post-installation, attempt to isolate the identifiable factors and price them separately. For example, add shipping costs that may be required pre-installation or any demolition required post-installation. Then analyze the potential impact on productivity of psychological and/or learning curve factors and adjust the output rates accordingly. One approach is to break down the typical workday into segments and quantify the impact on each segment.

Change Order Installation Efficiency

The labor-hours expressed (for new construction) are based on average installation time, using an efficiency level. For change order situations, adjustments to this efficiency level should reflect the daily labor-hour allocation for that particular occurrence.

- *Will the change substantially delay the original completion date?*

    A significant change in the project may cause the original completion date to be extended. The extended schedule may subject the contractor to new wage rates dictated by relevant labor contracts. Project supervision and other project overhead must also be extended beyond the original completion date. The schedule extension may also put installation into a new weather season. For example, underground piping scheduled for October installation was delayed until January. As a result, frost penetrated the trench area, thereby changing the degree of difficulty of the task. Changes and delays may have a ripple effect throughout the project. This effect must be analyzed and negotiated with the owner.

- *What is the net effect of a deduct change order?*

    In most cases, change orders resulting in a deduction or credit reflect only bare costs. The contractor may retain the overhead and profit based on the original bid.

Materials

- *Will you have to pay more or less for the new material, required by the change order, than you paid for the original purchase?*

    The same material prices or discounts will usually apply to materials purchased for change orders as new construction. In some instances, however, the contractor may forfeit the advantages of competitive pricing for change orders. Consider the following example:

    A contractor purchased over $20,000 worth of fan coil units for an installation, and obtained the maximum discount. Some time later it was determined the project required an additional matching unit. The contractor has to purchase this unit from the original supplier to ensure a match. The supplier at this time may not discount the unit because of the small quantity, and the fact that he is no longer in a competitive situation. The impact of quantity on purchase can add between 0% and 25% to material prices and/or subcontractor quotes.

- *If materials have been ordered or delivered to the job site, will they be subject to a cancellation charge or restocking fee?*

    Check with the supplier to determine if ordered materials are subject to a cancellation charge. Delivered materials not used as result of a change order may be subject to a restocking fee if returned to the supplier. Common restocking charges run between 20% and 40%. Also, delivery charges to return the goods to the supplier must be added.

Labor

- *How efficient is the existing crew at the actual installation?*

    Is the same crew that performed the initial work going to do the change order? Possibly the change consists of the installation of a unit identical to one already installed; therefore, the change should take less time. Be sure to consider this potential productivity increase and modify the productivity rates accordingly.

- *If the crew size is increased, what impact will that have on supervision requirements?*

    Under most bargaining agreements or management practices, there is a point at which a working foreman is replaced by a nonworking foreman. This replacement increases project overhead by adding a nonproductive worker. If additional workers are added to accelerate the project or to perform changes while maintaining the schedule, be sure to add additional supervision time if warranted. Calculate the hours involved and the additional cost directly if possible.

- *What are the other impacts of increased crew size?*

    The larger the crew, the greater the potential for productivity to decrease. Some of the factors that cause this productivity loss are: overcrowding (producing restrictive conditions in the working space), and possibly a shortage of any special tools and equipment required. Such factors affect not only the crew working on the elements directly involved in the change order, but other crews whose movement may also be hampered. As the crew increases, check its basic composition for changes by the addition or deletion of apprentices or nonworking foreman, and quantify the potential effects of equipment shortages or other logistical factors.

- *As new crews, unfamiliar with the project, are brought onto the site, how long will it take them to become oriented to the project requirements?*

    The orientation time for a new crew to become 100% effective varies with the site and type of project. Orientation is easiest at a new construction site, and most difficult at existing, very restrictive renovation sites. The type of work also affects orientation time. When all elements of the work are exposed, such as concrete or masonry work, orientation is decreased. When the work is concealed or less visible, such as existing electrical systems, orientation takes longer. Usually orientation can be accomplished in one day or less. Costs for added orientation should be itemized and added to the total estimated cost.

- *How much actual production can be gained by working overtime?*

    Short term overtime can be used effectively to accomplish more work in a day. However, as overtime is scheduled to run beyond several weeks, studies have shown marked decreases in output. The following chart shows the effect of long term overtime on worker efficiency. If the anticipated change requires extended overtime to keep the job on schedule, these factors can be used as a guide to predict the impact on time and cost. Add project overhead, particularly supervision, that may also be incurred.

Days per Week	Hours per Day	Production Efficiency					Payroll Cost Factors	
		1 Week	2 Weeks	3 Weeks	4 Weeks	Average 4 Weeks	@ 1-1/2 Times	@ 2 Times
5	8	100%	100%	100%	100%	100%	100%	100%
	9	100	100	95	90	96.25	105.6	111.1
	10	100	95	90	85	91.25	110.0	120.0
	11	95	90	75	65	81.25	113.6	127.3
	12	90	85	70	60	76.25	116.7	133.3
6	8	100	100	95	90	96.25	108.3	116.7
	9	100	95	90	85	92.50	113.0	125.9
	10	95	90	85	80	87.50	116.7	133.3
	11	95	85	70	65	78.75	119.7	139.4
	12	90	80	65	60	73.75	122.2	144.4
7	8	100	95	85	75	88.75	114.3	128.6
	9	95	90	80	70	83.75	118.3	136.5
	10	90	85	75	65	78.75	121.4	142.9
	11	85	80	65	60	72.50	124.0	148.1
	12	85	75	60	55	68.75	126.2	152.4

Effects of Overtime

Caution: Under many labor agreements, Sundays and holidays are paid at a higher premium than the normal overtime rate.

The use of long-term overtime is counterproductive on almost any construction job; that is, the longer the period of overtime, the lower the actual production rate. Numerous studies have been conducted, and while they have resulted in slightly different numbers, all reach the same conclusion. The figure above tabulates the effects of overtime work on efficiency.

As illustrated, there can be a difference between the *actual* payroll cost per hour and the *effective* cost per hour for overtime work. This is due to the reduced production efficiency with the increase in weekly hours beyond 40. This difference between actual and effective cost results from overtime work over a prolonged period. Short-term overtime work does not result in as great a reduction in efficiency and, in such cases, effective cost may not vary significantly from the actual payroll cost. As the total hours per week are increased on a regular basis, more time is lost due to fatigue, lowered morale, and an increased accident rate.

As an example, assume a project where workers are working 6 days a week, 10 hours per day. From the figure above (based on productivity studies), the average effective productive hours over a 4-week period are:

$$0.875 \times 60 = 52.5$$

Depending upon the locale and day of week, overtime hours may be paid at time and a half or double time. For time and a half, the overall (average) *actual* payroll cost (including regular and overtime hours) is determined as follows:

$$\frac{40 \text{ reg. hrs.} + (20 \text{ overtime hrs.} \times 1.5)}{60 \text{ hrs.}} = 1.167$$

Based on 60 hours, the payroll cost per hour will be 116.7% of the normal rate at 40 hours per week. However, because the effective production (efficiency) for 60 hours is reduced to the equivalent of 52.5 hours, the effective cost of overtime is calculated as follows:

For time and a half:

$$\frac{40 \text{ reg. hrs.} + (20 \text{ overtime hrs.} \times 1.5)}{52.5 \text{ hrs.}} = 1.33$$

Installed cost will be 133% of the normal rate (for labor).

Thus, when figuring overtime, the actual cost per unit of work will be higher than the apparent overtime payroll dollar increase, due to the reduced productivity of the longer workweek. These efficiency calculations are true only for those cost factors determined by hours worked. Costs that are applied weekly or monthly, such as equipment rentals, will not be similarly affected.

Equipment

- *What equipment is required to complete the change order?*

Change orders may require extending the rental period of equipment already on the job site, or the addition of special equipment brought in to accomplish the change work. In either case, the additional rental charges and operator labor charges must be added.

Summary

The preceding considerations and others you deem appropriate should be analyzed and applied to a change order estimate. The impact of each should be quantified and listed on the estimate to form an audit trail.

Change orders that are properly identified, documented, and managed help to ensure the orderly, professional and profitable progress of the work. They also minimize potential claims or disputes at the end of the project.

Square Foot Costs

Estimating Tips

- The cost figures in this section were derived from approximately 11,000 projects contained in the RSMeans database of completed construction projects. They include the contractor's overhead and profit, but do not generally include architectural fees or land costs. The figures have been adjusted to January of the current year. New projects are added to our files each year, and outdated projects are discarded. For this reason, certain costs may not show a uniform annual progression. In no case are all subdivisions of a project listed.

- These projects were located throughout the U.S. and reflect a tremendous variation in square foot (S.F.) and cubic foot (C.F.) costs. This is due to differences, not only in labor and material costs, but also in individual owners' requirements. For instance, a bank in a large city would have different features than one in a rural area. This is true of all the different types of buildings analyzed. Therefore, caution should be exercised when using these square foot costs. For example, for courthouses, costs in the database are local courthouse costs and will not apply to the larger, more elaborate federal courthouses. As a general rule, the projects in the 1/4 column do not include any site work or equipment, while the projects in the 3/4 column may include both equipment and site work. The median figures do not generally include site work.

- None of the figures "go with" any others. All individual cost items were computed and tabulated separately. Thus, the sum of the median figures for plumbing, HVAC, and electrical will not normally total up to the total mechanical and electrical costs arrived at by separate analysis and tabulation of the projects.

- Each building was analyzed as to total and component costs and percentages. The figures were arranged in ascending order with the results tabulated as shown. The 1/4 column shows that 25% of the projects had lower costs and 75% had higher. The 3/4 column shows that 75% of the projects had lower costs and 25% had higher. The median column shows that 50% of the projects had lower costs and 50% had higher.

- There are two times when square foot costs are useful. The first is in the conceptual stage when no details are available. Then, square foot costs make a useful starting point. The second is after the bids are in and the costs can be worked back into their appropriate categories for information purposes. As soon as details become available in the project design, the square foot approach should be discontinued and the project priced as to its particular components. When more precision is required, or for estimating the replacement cost of specific buildings, the current edition of *RSMeans Square Foot Costs* should be used.

- In using the figures in this section, it is recommended that the median column be used for preliminary figures if no additional information is available. The median figures, when multiplied by the total city construction cost index figures (see City Cost Indexes) and then multiplied by the project size modifier at the end of this section, should present a fairly accurate base figure, which would then have to be adjusted in view of the estimator's experience, local economic conditions, code requirements, and the owner's particular requirements. There is no need to factor the percentage figures, as these should remain constant from city to city. All tabulations mentioning air conditioning had at least partial air conditioning.

- The editors of this book would greatly appreciate receiving cost figures on one or more of your recent projects, which would then be included in the averages for next year. All cost figures received will be kept confidential, except that they will be averaged with other similar projects to arrive at square foot cost figures for next year's book. See the last page of the book for details and the discount available for submitting one or more of your projects. ■

No part of this publication may be reproduced, stored in a retrieval system, or transmitted in any form or by any means without prior written permission of RSMeans.

50 17 | Square Foot Costs

50 17 00 | S.F. Costs

				UNIT COSTS			% OF TOTAL		
		UNIT	1/4	MEDIAN	3/4	1/4	MEDIAN	3/4	
01	0010 APARTMENTS Low Rise (1 to 3 story)	S.F.	74.50	94	125				01
	0020 Total project cost	C.F.	6.70	8.85	10.95				
	0100 Site work	S.F.	5.45	8.70	15.30	6.05%	10.55%	13.95%	
	0500 Masonry		1.47	3.62	5.95	1.54%	3.92%	6.50%	
	1500 Finishes		7.90	10.85	13.45	9.05%	10.75%	12.85%	
	1800 Equipment		2.44	3.70	5.50	2.71%	3.99%	5.95%	
	2720 Plumbing		5.80	7.45	9.50	6.65%	8.95%	10.05%	
	2770 Heating, ventilating, air conditioning		3.70	4.56	6.70	4.20%	5.60%	7.60%	
	2900 Electrical		4.33	5.75	7.80	5.20%	6.65%	8.35%	
	3100 Total: Mechanical & Electrical		15.40	19.95	24.50	16.05%	18.20%	23%	
	9000 Per apartment unit, total cost	Apt.	69,500	106,000	156,500				
	9500 Total: Mechanical & Electrical	"	13,100	20,700	27,000				
02	0010 APARTMENTS Mid Rise (4 to 7 story)	S.F.	99	119	147				02
	0020 Total project costs	C.F.	7.70	10.65	14.55				
	0100 Site work	S.F.	3.95	8	15.45	5.25%	6.70%	9.20%	
	0500 Masonry		4.13	9.05	12.40	5.10%	7.25%	10.50%	
	1500 Finishes		12.95	17.35	20.50	10.70%	13.50%	17.70%	
	1800 Equipment		2.77	4.30	5.65	2.54%	3.47%	4.31%	
	2500 Conveying equipment		2.24	2.76	3.33	2.05%	2.27%	2.69%	
	2720 Plumbing		5.80	9.30	9.85	5.70%	7.20%	8.95%	
	2900 Electrical		6.50	8.85	10.75	6.35%	7.20%	8.95%	
	3100 Total: Mechanical & Electrical		21	26	31.50	18.25%	21%	23%	
	9000 Per apartment unit, total cost	Apt.	112,000	132,000	218,500				
	9500 Total: Mechanical & Electrical	"	21,100	24,400	25,600				
03	0010 APARTMENTS High Rise (8 to 24 story)	S.F.	112	129	155				03
	0020 Total project costs	C.F.	10.90	12.65	14.95				
	0100 Site work	S.F.	4.07	6.60	9.20	2.58%	4.84%	6.15%	
	0500 Masonry		6.50	11.80	14.65	4.74%	9.65%	11.05%	
	1500 Finishes		12.45	15.55	18.35	9.75%	11.80%	13.70%	
	1800 Equipment		3.61	4.44	5.85	2.78%	3.49%	4.35%	
	2500 Conveying equipment		2.55	3.87	5.25	2.23%	2.78%	3.37%	
	2720 Plumbing		7.15	9.75	11.95	6.80%	7.20%	10.45%	
	2900 Electrical		7.70	9.75	13.15	6.45%	7.65%	8.80%	
	3100 Total: Mechanical & Electrical		23	29.50	35.50	17.95%	22.50%	24.50%	
	9000 Per apartment unit, total cost	Apt.	116,500	128,500	178,000				
	9500 Total: Mechanical & Electrical	"	23,200	28,800	30,500				
04	0010 AUDITORIUMS	S.F.	117	161	234				04
	0020 Total project costs	C.F.	7.30	10.15	14.55				
	2720 Plumbing	S.F.	6.90	10.20	12.20	5.85%	7.20%	8.70%	
	2900 Electrical		9.40	13.55	22.50	6.85%	9.50%	11.40%	
	3100 Total: Mechanical & Electrical		62.50	83	101	24.50%	27.50%	31%	
05	0010 AUTOMOTIVE SALES	S.F.	86.50	119	146				05
	0020 Total project costs	C.F.	5.70	6.80	8.85				
	2720 Plumbing	S.F.	3.93	6.85	7.45	2.89%	6.05%	6.50%	
	2770 Heating, ventilating, air conditioning		6.05	9.25	10	4.61%	10%	10.35%	
	2900 Electrical		6.95	10.90	16	7.25%	8.80%	12.15%	
	3100 Total: Mechanical & Electrical		19.35	31	37.50	17.30%	20.50%	22%	
06	0010 BANKS	S.F.	170	211	268				06
	0020 Total project costs	C.F.	12.10	16.45	21.50				
	0100 Site work	S.F.	19.40	30.50	43	7.90%	12.95%	17%	
	0500 Masonry		7.70	16.50	29.50	3.36%	6.90%	10.05%	
	1500 Finishes		15.15	23	28.50	5.85%	8.65%	11.70%	
	1800 Equipment		6.35	14	28.50	1.34%	5.55%	10.50%	
	2720 Plumbing		5.30	7.55	11.05	2.82%	3.90%	4.93%	
	2770 Heating, ventilating, air conditioning		10.05	13.45	17.90	4.86%	7.15%	8.50%	
	2900 Electrical		16.05	21.50	28	8.25%	10.20%	12.20%	
	3100 Total: Mechanical & Electrical		38.50	51.50	62	16.55%	19.45%	24%	
	3500 See also division 11 22 00								

50 17 | Square Foot Costs

50 17 00 | S.F. Costs

			UNIT	UNIT COSTS 1/4	MEDIAN	3/4	% OF TOTAL 1/4	MEDIAN	3/4
13	0010	**CHURCHES**	S.F.	114	145	190			
	0020	Total project costs	C.F.	7.05	8.90	11.75			
	1800	Equipment	S.F.	1.22	3.19	6.80	.83%	2.04%	4.30%
	2720	Plumbing		4.43	6.20	9.15	3.51%	4.96%	6.25%
	2770	Heating, ventilating, air conditioning		10.35	13.50	19.15	7.50%	10%	12%
	2900	Electrical		9.60	13.20	18	7.30%	8.80%	10.95%
	3100	Total: Mechanical & Electrical		29.50	39.50	53	18.30%	22%	25%
	3500	See also division 11 91 00							
15	0010	**CLUBS, COUNTRY**	S.F.	122	147	185			
	0020	Total project costs	C.F.	9.85	12	16.55			
	2720	Plumbing	S.F.	7.40	10.95	25	5.60%	7.90%	10%
	2900	Electrical		9.60	13.15	17.15	7%	8.95%	11%
	3100	Total: Mechanical & Electrical		51	64	67.50	19%	26.50%	29.50%
17	0010	**CLUBS, SOCIAL Fraternal**	S.F.	103	141	188			
	0020	Total project costs	C.F.	6.10	9.25	11			
	2720	Plumbing	S.F.	6.15	7.65	11.55	5.60%	6.90%	8.55%
	2770	Heating, ventilating, air conditioning		7.15	10.70	13.75	8.20%	9.25%	14.40%
	2900	Electrical		7.15	12.05	13.75	5.95%	9.30%	10.55%
	3100	Total: Mechanical & Electrical		38	41	52	21%	23%	23.50%
18	0010	**CLUBS, Y.M.C.A.**	S.F.	134	167	252			
	0020	Total project costs	C.F.	5.65	9.45	14.05			
	2720	Plumbing	S.F.	7.75	15.40	17.25	5.65%	7.60%	10.85%
	2900	Electrical		10.05	13	22.50	6.45%	7.95%	8.90%
	3100	Total: Mechanical & Electrical		41	46.50	76	20.50%	21.50%	28.50%
19	0010	**COLLEGES Classrooms & Administration**	S.F.	118	167	216			
	0020	Total project costs	C.F.	7.10	12.50	20			
	0500	Masonry	S.F.	9.30	17.15	21	5.10%	8.05%	10.50%
	2720	Plumbing		5.30	12.90	24	5.10%	6.60%	8.95%
	2900	Electrical		10.05	15.85	22	7.70%	9.85%	12%
	3100	Total: Mechanical & Electrical		42.50	58.50	77.50	23%	28%	31.50%
21	0010	**COLLEGES Science, Engineering, Laboratories**	S.F.	230	269	310			
	0020	Total project costs	C.F.	13.20	19.25	22			
	1800	Equipment	S.F.	6.25	29	31.50	2%	6.45%	12.65%
	2900	Electrical		18.95	27	41.50	7.10%	9.40%	12.10%
	3100	Total: Mechanical & Electrical		70.50	83.50	129	28.50%	31.50%	41%
	3500	See also division 11 53 00							
23	0010	**COLLEGES Student Unions**	S.F.	147	199	241			
	0020	Total project costs	C.F.	8.20	10.70	13.25			
	3100	Total: Mechanical & Electrical	S.F.	55	59.50	70.50	23.50%	26%	29%
25	0010	**COMMUNITY CENTERS**	S.F.	122	150	203			
	0020	Total project costs	C.F.	7.90	11.30	14.65			
	1800	Equipment	S.F.	2.45	4.80	7.90	1.47%	3.01%	5.30%
	2720	Plumbing		5.75	10.05	13.70	4.85%	7%	8.95%
	2770	Heating, ventilating, air conditioning		8.65	13.35	19.15	6.80%	10.35%	12.90%
	2900	Electrical		10.25	13.55	19.90	7.15%	8.90%	10.40%
	3100	Total: Mechanical & Electrical		34	43	61.50	18.80%	23%	30%
28	0010	**COURT HOUSES**	S.F.	175	206	284			
	0020	Total project costs	C.F.	13.40	16	20			
	2720	Plumbing	S.F.	8.30	11.60	13.15	5.95%	7.45%	8.20%
	2900	Electrical		18.55	21	30.50	9.05%	10.65%	12.15%
	3100	Total: Mechanical & Electrical		52	68	74.50	22.50%	26.50%	30%
30	0010	**DEPARTMENT STORES**	S.F.	64.50	87.50	110			
	0020	Total project costs	C.F.	3.46	4.48	6.10			
	2720	Plumbing	S.F.	2.01	2.54	3.85	1.82%	4.21%	5.90%
	2770	Heating, ventilating, air conditioning		4.88	9.05	13.65	8.20%	9.10%	14.80%

50 17 | Square Foot Costs

50 17 00 | S.F. Costs

			UNIT	UNIT COSTS 1/4	MEDIAN	3/4	% OF TOTAL 1/4	MEDIAN	3/4
30	2900	Electrical	S.F.	7.40	10.15	12	9.05%	12.15%	14.95%
	3100	Total: Mechanical & Electrical		13.05	16.65	29	13.20%	21.50%	50%
31	0010	**DORMITORIES Low Rise (1 to 3 story)**	S.F.	122	170	212			
	0020	Total project costs	C.F.	6.90	11.20	16.75			
	2720	Plumbing	S.F.	7.35	9.85	12.45	8.05%	9%	9.65%
	2770	Heating, ventilating, air conditioning		7.80	9.35	12.45	4.61%	8.05%	10%
	2900	Electrical		8.10	12.35	16.90	6.40%	8.65%	9.50%
	3100	Total: Mechanical & Electrical		42	45	70.50	22%	25%	27%
	9000	Per bed, total cost	Bed	52,000	57,500	123,500			
32	0010	**DORMITORIES Mid Rise (4 to 8 story)**	S.F.	150	196	242			
	0020	Total project costs	C.F.	16.55	18.20	22			
	2900	Electrical	S.F.	15.95	18.15	24.50	8.20%	10.20%	11.95%
	3100	Total: Mechanical & Electrical	"	44.50	89	90.50	25%	30.50%	35.50%
	9000	Per bed, total cost	Bed	21,400	48,800	282,500			
34	0010	**FACTORIES**	S.F.	57	85	131			
	0020	Total project costs	C.F.	3.65	5.45	9.05			
	0100	Site work	S.F.	6.50	11.85	18.75	6.95%	11.45%	17.95%
	2720	Plumbing		3.07	5.70	9.45	3.73%	6.05%	8.10%
	2770	Heating, ventilating, air conditioning		5.95	8.55	11.55	5.25%	8.45%	11.35%
	2900	Electrical		7.05	11.20	17.05	8.10%	10.50%	14.20%
	3100	Total: Mechanical & Electrical		20	32.50	41	21%	28.50%	35.50%
36	0010	**FIRE STATIONS**	S.F.	113	156	213			
	0020	Total project costs	C.F.	6.60	9.05	12.05			
	0500	Masonry	S.F.	15.40	30	38	7.50%	10.95%	15.55%
	1140	Roofing		3.68	9.95	11.35	1.90%	4.94%	5.05%
	1580	Painting		2.86	4.27	4.37	1.37%	1.57%	2.07%
	1800	Equipment		1.40	2.69	4.97	.62%	1.63%	3.42%
	2720	Plumbing		6.30	10.05	14.25	5.85%	7.35%	9.45%
	2770	Heating, ventilating, air conditioning		6.25	10.15	15.65	5.15%	7.40%	9.40%
	2900	Electrical		8.15	14.35	19.35	6.90%	8.60%	10.60%
	3100	Total: Mechanical & Electrical		41.50	53	60	18.40%	23%	27%
37	0010	**FRATERNITY HOUSES & Sorority Houses**	S.F.	113	145	199			
	0020	Total project costs	C.F.	11.25	11.70	14.10			
	2720	Plumbing	S.F.	8.55	9.75	17.90	6.80%	8%	10.85%
	2900	Electrical		7.45	16.10	19.70	6.60%	9.90%	10.65%
38	0010	**FUNERAL HOMES**	S.F.	119	162	295			
	0020	Total project costs	C.F.	12.15	13.55	26			
	2900	Electrical	S.F.	5.25	9.65	10.55	3.58%	4.44%	5.95%
39	0010	**GARAGES, COMMERCIAL (Service)**	S.F.	67.50	104	144			
	0020	Total project costs	C.F.	4.43	6.55	9.50			
	1800	Equipment	S.F.	3.80	8.55	13.30	2.21%	4.62%	6.80%
	2720	Plumbing		4.67	7.20	13.10	5.45%	7.85%	10.65%
	2730	Heating & ventilating		4.60	7.95	10.15	5.25%	6.85%	8.20%
	2900	Electrical		6.40	9.75	14.10	7.15%	9.25%	10.85%
	3100	Total: Mechanical & Electrical		12.10	27	40	12.35%	17.40%	26%
40	0010	**GARAGES, MUNICIPAL (Repair)**	S.F.	105	135	192			
	0020	Total project costs	C.F.	6.20	7.85	13.50			
	0500	Masonry	S.F.	5.20	18.20	28	4.03%	9.15%	12.50%
	2720	Plumbing		4.44	8.55	16.05	3.59%	6.70%	7.95%
	2730	Heating & ventilating		7.60	11	21	6.15%	7.45%	13.50%
	2900	Electrical		7.65	12.35	22.50	6.90%	9.40%	13%
	3100	Total: Mechanical & Electrical		34.50	56.50	68	21.50%	25.50%	35.50%
41	0010	**GARAGES, PARKING**	S.F.	39.50	56	96.50			
	0020	Total project costs	C.F.	3.62	4.92	7.15			

50 17 | Square Foot Costs

50 17 00 | S.F. Costs

				UNIT COSTS			% OF TOTAL			
			UNIT	1/4	MEDIAN	3/4	1/4	MEDIAN	3/4	
41	2720	Plumbing	S.F.	.66	1.69	2.61	1.72%	2.70%	3.85%	41
	2900	Electrical		2.12	2.80	4.08	4.52%	5.40%	6.35%	
	3100	Total: Mechanical & Electrical	↓	4.11	6.05	7.50	7%	8.90%	11.05%	
	3200									
	9000	Per car, total cost	Car	16,300	20,400	26,000				
43	0010	**GYMNASIUMS**	S.F.	108	143	196				43
	0020	Total project costs	C.F.	5.35	7.25	8.90				
	1800	Equipment	S.F.	2.55	4.78	8.60	1.76%	3.26%	6.70%	
	2720	Plumbing		5.90	7.90	10.40	4.65%	6.40%	7.75%	
	2770	Heating, ventilating, air conditioning		6.40	11.10	22.50	5.15%	9.05%	11.10%	
	2900	Electrical		8.50	11.25	14.95	6.75%	8.50%	10.30%	
	3100	Total: Mechanical & Electrical	↓	29	40.50	49.50	19.75%	23.50%	29%	
	3500	See also division 11 66 00								
46	0010	**HOSPITALS**	S.F.	206	258	355				46
	0020	Total project costs	C.F.	15.60	19.40	28				
	1800	Equipment	S.F.	5.20	10.05	17.30	.80%	2.53%	4.80%	
	2720	Plumbing		17.70	25	32	7.60%	9.10%	10.85%	
	2770	Heating, ventilating, air conditioning		26	33.50	46.50	7.80%	12.95%	16.65%	
	2900	Electrical		22.50	30.50	45.50	10%	11.75%	14.10%	
	3100	Total: Mechanical & Electrical	↓	66.50	92.50	137	28%	33.50%	37%	
	9000	Per bed or person, total cost	Bed	238,000	328,500	378,500				
	9900	See also division 11 71 00								
48	0010	**HOUSING For the Elderly**	S.F.	101	128	157				48
	0020	Total project costs	C.F.	7.20	10	12.80				
	0100	Site work	S.F.	7.05	10.95	16.05	5.05%	7.90%	12.10%	
	0500	Masonry		1.91	11.55	16.85	1.30%	6.05%	11%	
	1800	Equipment		2.45	3.37	5.35	1.88%	3.23%	4.43%	
	2510	Conveying systems		2.29	3.31	4.49	1.78%	2.20%	2.81%	
	2720	Plumbing		7.50	9.60	12.10	8.15%	9.55%	10.50%	
	2730	Heating, ventilating, air conditioning		3.86	5.45	8.15	3.30%	5.60%	7.25%	
	2900	Electrical		7.55	10.25	13.10	7.30%	8.50%	10.25%	
	3100	Total: Mechanical & Electrical	↓	26	32	41	18.10%	22.50%	29%	
	9000	Per rental unit, total cost	Unit	94,000	110,000	122,500				
	9500	Total: Mechanical & Electrical	"	21,000	24,100	28,100				
50	0010	**HOUSING Public (Low Rise)**	S.F.	85	118	154				50
	0020	Total project costs	C.F.	6.75	9.45	11.75				
	0100	Site work	S.F.	10.85	15.60	25.50	8.35%	11.75%	16.50%	
	1800	Equipment		2.31	3.77	5.75	2.26%	3.03%	4.24%	
	2720	Plumbing		6.15	8.10	10.25	7.15%	9.05%	11.60%	
	2730	Heating, ventilating, air conditioning		3.08	6	6.55	4.26%	6.05%	6.45%	
	2900	Electrical		5.15	7.65	10.65	5.10%	6.55%	8.25%	
	3100	Total: Mechanical & Electrical	↓	24.50	31.50	35	14.50%	17.55%	26.50%	
	9000	Per apartment, total cost	Apt.	93,500	106,500	133,500				
	9500	Total: Mechanical & Electrical	"	19,900	24,600	27,200				
51	0010	**ICE SKATING RINKS**	S.F.	72.50	170	187				51
	0020	Total project costs	C.F.	5.35	5.45	6.30				
	2720	Plumbing	S.F.	2.72	5.10	5.20	3.12%	3.23%	5.65%	
	2900	Electrical	↓	7.80	11.95	12.65	6.30%	10.15%	15.05%	
52	0010	**JAILS**	S.F.	189	286	370				52
	0020	Total project costs	C.F.	19.95	28	32.50				
	1800	Equipment	S.F.	8.65	25.50	43.50	2.80%	6.95%	9.80%	
	2720	Plumbing		21.50	28.50	37.50	7%	8.90%	13.35%	
	2770	Heating, ventilating, air conditioning		19.95	26.50	51.50	7.50%	9.45%	17.75%	
	2900	Electrical		23.50	31.50	40	9.40%	11.70%	15.25%	
	3100	Total: Mechanical & Electrical	↓	62	110	131	28%	31%	36%	
53	0010	**LIBRARIES**	S.F.	143	190	248				53
	0020	Total project costs	C.F.	9.55	12	15.30				

50 17 | Square Foot Costs

	50 17 00	S.F. Costs	UNIT	UNIT COSTS 1/4	MEDIAN	3/4	% OF TOTAL 1/4	MEDIAN	3/4	
53	0500	Masonry	S.F.	9.50	19.35	32.50	5.60%	7.15%	10.95%	53
	1800	Equipment		1.91	5.15	7.75	.28%	1.39%	4.07%	
	2720	Plumbing		5.15	7.30	10.20	3.38%	4.60%	5.70%	
	2770	Heating, ventilating, air conditioning		11.45	19.35	23.50	7.80%	10.95%	12.80%	
	2900	Electrical		14.40	19	25.50	8.40%	10.55%	12%	
	3100	Total: Mechanical & Electrical		47	57	67	21%	24%	28%	
54	0010	**LIVING, ASSISTED**	S.F.	129	154	180				54
	0020	Total project costs	C.F.	10.90	12.70	14.45				
	0500	Masonry	S.F.	3.75	4.53	5.55	2.36%	3.16%	3.86%	
	1800	Equipment		2.87	3.42	4.38	2.12%	2.45%	2.87%	
	2720	Plumbing		10.85	14.50	15.05	6.05%	8.15%	10.60%	
	2770	Heating, ventilating, air conditioning		12.85	13.45	14.75	7.95%	9.35%	9.70%	
	2900	Electrical		12.70	14.25	16.50	9%	9.95%	10.65%	
	3100	Total: Mechanical & Electrical		35	41	48	24%	28.50%	31.50%	
55	0010	**MEDICAL CLINICS**	S.F.	132	162	207				55
	0020	Total project costs	C.F.	9.65	12.50	16.60				
	1800	Equipment	S.F.	1.68	7.50	11.65	1.05%	2.94%	6.35%	
	2720	Plumbing		8.75	12.30	16.50	6.15%	8.40%	10.10%	
	2770	Heating, ventilating, air conditioning		10.45	13.70	20	6.65%	8.85%	11.35%	
	2900	Electrical		11.30	16.10	21	8.10%	10%	12.20%	
	3100	Total: Mechanical & Electrical		36	49	67.50	22%	27%	33.50%	
	3500	See also division 11 71 00								
57	0010	**MEDICAL OFFICES**	S.F.	124	154	190				57
	0020	Total project costs	C.F.	9.25	12.50	16.95				
	1800	Equipment	S.F.	4.09	8.10	11.35	.66%	4.87%	6.50%	
	2720	Plumbing		6.85	10.55	14.25	5.60%	6.80%	8.50%	
	2770	Heating, ventilating, air conditioning		8.25	11.95	15.75	6.10%	8%	9.70%	
	2900	Electrical		9.90	14.50	20.50	7.50%	9.80%	11.70%	
	3100	Total: Mechanical & Electrical		27	39.50	56	19.30%	22.50%	27.50%	
59	0010	**MOTELS**	S.F.	78	113	148				59
	0020	Total project costs	C.F.	6.95	9.30	15.25				
	2720	Plumbing	S.F.	7.90	10.10	12.05	9.45%	10.60%	12.55%	
	2770	Heating, ventilating, air conditioning		4.83	7.20	12.90	5.60%	5.60%	10%	
	2900	Electrical		7.40	9.35	11.65	7.45%	9.05%	10.45%	
	3100	Total: Mechanical & Electrical		22.50	31.50	54	18.50%	24%	25.50%	
	5000									
	9000	Per rental unit, total cost	Unit	39,800	75,500	82,000				
	9500	Total: Mechanical & Electrical	"	7,750	11,700	13,600				
60	0010	**NURSING HOMES**	S.F.	123	158	197				60
	0020	Total project costs	C.F.	9.65	12.05	16.45				
	1800	Equipment	S.F.	3.34	5.10	8.55	2%	3.62%	4.99%	
	2720	Plumbing		10.50	15.90	19.20	8.75%	10.10%	12.70%	
	2770	Heating, ventilating, air conditioning		11.05	16.80	22.50	9.70%	11.45%	11.80%	
	2900	Electrical		12.15	15.20	21	9.40%	10.60%	12.50%	
	3100	Total: Mechanical & Electrical		29	40.50	68	26%	28%	30%	
	9000	Per bed or person, total cost	Bed	54,500	68,000	88,000				
61	0010	**OFFICES Low Rise (1 to 4 story)**	S.F.	103	135	175				61
	0020	Total project costs	C.F.	7.35	10.15	13.40				
	0100	Site work	S.F.	8.30	14.45	21.50	5.95%	9.70%	13.55%	
	0500	Masonry		4	8.05	14.70	2.61%	5.40%	8.45%	
	1800	Equipment		.93	2.15	5.85	.57%	1.50%	3.42%	
	2720	Plumbing		3.69	5.70	8.35	3.66%	4.50%	6.10%	
	2770	Heating, ventilating, air conditioning		8.15	11.40	16.65	7.20%	10.30%	11.70%	
	2900	Electrical		8.45	12.10	17.20	7.45%	9.65%	11.40%	
	3100	Total: Mechanical & Electrical		23.50	32.50	48.50	18.20%	22%	27%	
62	0010	**OFFICES Mid Rise (5 to 10 story)**	S.F.	109	132	180				62
	0020	Total project costs	C.F.	7.75	9.90	14				

For customer support on your Commercial Renovation Cost Data, call 877.791.4977.

50 17 | Square Foot Costs

50 17 00 | S.F. Costs

			UNIT	UNIT COSTS 1/4	UNIT COSTS MEDIAN	UNIT COSTS 3/4	% OF TOTAL 1/4	% OF TOTAL MEDIAN	% OF TOTAL 3/4	
62	2720	Plumbing	S.F.	3.30	5.10	7.35	2.83%	3.74%	4.50%	62
	2770	Heating, ventilating, air conditioning		8.30	11.85	18.95	7.65%	9.40%	11%	
	2900	Electrical		8.10	10.40	15.70	6.35%	7.80%	10%	
	3100	Total: Mechanical & Electrical		21	27	51.50	18.95%	21%	27.50%	
63	0010	OFFICES High Rise (11 to 20 story)	S.F.	134	169	208				63
	0020	Total project costs	C.F.	9.40	11.75	16.85				
	2900	Electrical	S.F.	8.15	9.95	14.80	5.80%	7.85%	10.50%	
	3100	Total: Mechanical & Electrical		26.50	35.50	59.50	16.90%	23.50%	34%	
64	0010	POLICE STATIONS	S.F.	161	212	270				64
	0020	Total project costs	C.F.	12.85	15.70	21.50				
	0500	Masonry	S.F.	16.10	26.50	33.50	7.80%	9.10%	11.35%	
	1800	Equipment		2.31	11.25	17.80	.98%	3.35%	6.70%	
	2720	Plumbing		9	17.95	22.50	5.65%	6.90%	10.75%	
	2770	Heating, ventilating, air conditioning		13.20	18.70	26.50	5.85%	10.55%	11.70%	
	2900	Electrical		17.60	25.50	33	9.80%	11.70%	14.50%	
	3100	Total: Mechanical & Electrical		67	70.50	94	28.50%	31.50%	32.50%	
65	0010	POST OFFICES	S.F.	127	157	200				65
	0020	Total project costs	C.F.	7.65	9.70	11				
	2720	Plumbing	S.F.	5.15	7.10	8.95	4.24%	5.30%	5.60%	
	2770	Heating, ventilating, air conditioning		8	11.05	12.30	6.65%	7.15%	9.35%	
	2900	Electrical		10.50	14.75	17.50	7.25%	9%	11%	
	3100	Total: Mechanical & Electrical		30.50	39.50	45	16.25%	18.80%	22%	
66	0010	POWER PLANTS	S.F.	880	1,175	2,150				66
	0020	Total project costs	C.F.	24.50	53	113				
	2900	Electrical	S.F.	62.50	132	197	9.30%	12.75%	21.50%	
	8100	Total: Mechanical & Electrical		155	505	1,125	32.50%	32.50%	52.50%	
67	0010	RELIGIOUS EDUCATION	S.F.	103	137	170				67
	0020	Total project costs	C.F.	5.70	8.15	10.20				
	2720	Plumbing	S.F.	4.28	6.05	8.60	4.40%	5.30%	7.10%	
	2770	Heating, ventilating, air conditioning		10.80	12.25	17.30	10.05%	11.45%	12.35%	
	2900	Electrical		8.15	11.65	17.10	7.70%	9.05%	10.35%	
	3100	Total: Mechanical & Electrical		35.50	45	53	22%	23.50%	26%	
69	0010	RESEARCH Laboratories & Facilities	S.F.	157	222	325				69
	0020	Total project costs	C.F.	11.40	22.50	27				
	1800	Equipment	S.F.	6.80	13.40	32.50	1.22%	4.80%	9.35%	
	2720	Plumbing		11.20	19.70	31.50	6.15%	8.30%	10.80%	
	2770	Heating, ventilating, air conditioning		13.80	46.50	55	7.25%	16.50%	17.50%	
	2900	Electrical		18.45	29.50	48.50	9.45%	11.15%	14.60%	
	3100	Total: Mechanical & Electrical		59	104	147	29.50%	36%	41%	
70	0010	RESTAURANTS	S.F.	149	192	249				70
	0020	Total project costs	C.F.	12.50	16.40	21.50				
	1800	Equipment	S.F.	8.80	23.50	35.50	6.10%	13%	15.65%	
	2720	Plumbing		11.80	14.30	18.75	6.10%	8.15%	9%	
	2770	Heating, ventilating, air conditioning		14.95	21	25	9.20%	12%	12.40%	
	2900	Electrical		15.70	19.40	25	8.35%	10.55%	11.55%	
	3100	Total: Mechanical & Electrical		48.50	52	67.50	21%	25%	29.50%	
	9000	Per seat unit, total cost	Seat	5,450	7,275	8,600				
	9500	Total: Mechanical & Electrical	"	1,375	1,825	2,150				
72	0010	RETAIL STORES	S.F.	69.50	93.50	124				72
	0020	Total project costs	C.F.	4.71	6.70	9.35				
	2720	Plumbing	S.F.	2.52	4.20	7.15	3.26%	4.60%	6.80%	
	2770	Heating, ventilating, air conditioning		5.45	7.45	11.20	6.75%	8.75%	10.15%	
	2900	Electrical		6.25	8.55	12.35	7.25%	9.90%	11.60%	
	3100	Total: Mechanical & Electrical		16.65	21.50	28.50	17.05%	21%	23.50%	
74	0010	SCHOOLS Elementary	S.F.	114	141	172				74
	0020	Total project costs	C.F.	7.40	9.50	12.25				
	0500	Masonry	S.F.	8	17.30	25.50	4.89%	10.50%	14%	
	1800	Equipment		2.63	4.96	9.40	1.83%	3.13%	4.61%	

For customer support on your Commercial Renovation Cost Data, call 877.791.4977.

50 17 | Square Foot Costs

50 17 00 | S.F. Costs

			UNIT	UNIT COSTS			% OF TOTAL		
				1/4	MEDIAN	3/4	1/4	MEDIAN	3/4
74	2720	Plumbing	S.F.	6.50	9.20	12.30	5.70%	7.15%	9.35%
	2730	Heating, ventilating, air conditioning		9.80	15.55	22.50	8.15%	10.80%	14.90%
	2900	Electrical		10.70	14.25	18.20	8.45%	10.05%	11.85%
	3100	Total: Mechanical & Electrical	↓	39	47.50	60	25%	27.50%	30%
	9000	Per pupil, total cost	Ea.	13,000	19,300	43,300			
	9500	Total: Mechanical & Electrical	"	3,675	4,650	11,700			
76	0010	**SCHOOLS Junior High & Middle**	S.F.	117	145	177			
	0020	Total project costs	C.F.	7.40	9.60	10.75			
	0500	Masonry	S.F.	12.55	18.80	23	7.55%	11.10%	14.30%
	1800	Equipment		3.20	6	8.80	1.80%	3.03%	4.80%
	2720	Plumbing		6.80	8.40	10.40	5.30%	6.80%	7.25%
	2770	Heating, ventilating, air conditioning		13.60	16.55	29	8.90%	11.55%	14.20%
	2900	Electrical		11.90	14.70	18.60	8.05%	9.55%	10.60%
	3100	Total: Mechanical & Electrical	↓	37.50	48	59.50	23%	25.50%	29.50%
	9000	Per pupil, total cost	Ea.	14,800	19,400	26,100			
78	0010	**SCHOOLS Senior High**	S.F.	123	150	188			
	0020	Total project costs	C.F.	7.35	10.30	17.40			
	1800	Equipment	S.F.	3.24	7.55	10.45	1.88%	2.67%	4.30%
	2720	Plumbing		6.80	10.25	18.70	5.60%	6.90%	8.30%
	2770	Heating, ventilating, air conditioning		11.75	15.95	24.50	8.95%	11.60%	15%
	2900	Electrical		12.30	16.40	24	8.70%	10.35%	12.50%
	3100	Total: Mechanical & Electrical	↓	40.50	48	77.50	24%	26.50%	28.50%
	9000	Per pupil, total cost	Ea.	11,500	23,300	29,200			
80	0010	**SCHOOLS Vocational**	S.F.	98.50	143	177			
	0020	Total project costs	C.F.	6.10	8.80	12.15			
	0500	Masonry	S.F.	4.33	14.25	22	3.20%	4.61%	10.95%
	1800	Equipment		3.08	7.65	10.60	1.24%	3.10%	4.26%
	2720	Plumbing		6.30	9.35	13.80	5.40%	6.90%	8.55%
	2770	Heating, ventilating, air conditioning		8.80	16.40	27.50	8.60%	11.90%	14.65%
	2900	Electrical		10.25	14	19.15	8.45%	11%	13.20%
	3100	Total: Mechanical & Electrical	↓	38.50	67.50	86.50	27.50%	29.50%	31%
	9000	Per pupil, total cost	Ea.	13,700	36,700	54,500			
83	0010	**SPORTS ARENAS**	S.F.	86	115	177			
	0020	Total project costs	C.F.	4.67	8.35	10.80			
	2720	Plumbing	S.F.	4.31	7.55	15.95	4.35%	6.35%	9.40%
	2770	Heating, ventilating, air conditioning		10.75	12.70	17.60	8.80%	10.20%	13.55%
	2900	Electrical	↓	7.95	12.15	15.70	7.65%	9.75%	11.90%
85	0010	**SUPERMARKETS**	S.F.	79.50	92	108			
	0020	Total project costs	C.F.	4.42	5.35	8.10			
	2720	Plumbing	S.F.	4.44	5.60	6.50	5.40%	6%	7.45%
	2770	Heating, ventilating, air conditioning		6.55	8.65	10.55	8.60%	8.65%	9.60%
	2900	Electrical		9.30	11.35	13.50	10.40%	12.45%	13.60%
	3100	Total: Mechanical & Electrical	↓	25.50	27	36.50	23.50%	27.50%	28.50%
86	0010	**SWIMMING POOLS**	S.F.	129	300	460			
	0020	Total project costs	C.F.	10.30	12.85	14			
	2720	Plumbing	S.F.	11.90	13.60	18.25	4.80%	9.70%	20.50%
	2900	Electrical		9.75	15.65	34.50	6.05%	6.95%	7.60%
	3100	Total: Mechanical & Electrical	↓	61	80.50	106	11.15%	14.10%	23.50%
87	0010	**TELEPHONE EXCHANGES**	S.F.	160	243	315			
	0020	Total project costs	C.F.	10.65	17.05	23.50			
	2720	Plumbing	S.F.	7.20	11.15	16.30	4.52%	5.80%	6.90%
	2770	Heating, ventilating, air conditioning		16.75	33.50	42	11.80%	16.05%	18.40%
	2900	Electrical		17.40	27.50	49	10.90%	14%	17.85%
	3100	Total: Mechanical & Electrical	↓	51.50	97.50	138	29.50%	33.50%	44.50%
91	0010	**THEATERS**	S.F.	102	138	203			
	0020	Total project costs	C.F.	4.96	7.35	10.80			

50 17 | Square Foot Costs

50 17 00 | S.F. Costs

			UNIT	UNIT COSTS			% OF TOTAL			
				1/4	MEDIAN	3/4	1/4	MEDIAN	3/4	
91	2720	Plumbing	S.F.	3.34	3.88	15.85	2.92%	4.70%	6.80%	91
	2770	Heating, ventilating, air conditioning		10.45	12.65	15.65	8%	12.25%	13.40%	
	2900	Electrical		9.40	12.70	26	8.05%	9.35%	12.25%	
	3100	Total: Mechanical & Electrical	↓	24	32.50	38.50	21.50%	25.50%	27.50%	
94	0010	**TOWN HALLS City Halls & Municipal Buildings**	S.F.	112	152	199				94
	0020	Total project costs	C.F.	8.10	12.60	18.45				
	2720	Plumbing	S.F.	5	9.35	16.45	4.31%	5.95%	7.95%	
	2770	Heating, ventilating, air conditioning		9.05	17.95	26.50	7.05%	9.05%	13.45%	
	2900	Electrical		11.40	15.70	21.50	8.05%	9.45%	11.65%	
	3100	Total: Mechanical & Electrical	↓	39.50	50	76.50	22%	26.50%	31%	
97	0010	**WAREHOUSES & Storage Buildings**	S.F.	44.50	63.50	95				97
	0020	Total project costs	C.F.	2.34	3.67	6.05				
	0100	Site work	S.F.	4.62	9.15	13.80	6.05%	12.95%	19.55%	
	0500	Masonry		2.54	6.35	13.70	3.60%	7.35%	12%	
	1800	Equipment		.68	1.55	8.70	.72%	1.69%	5.55%	
	2720	Plumbing		1.49	2.68	5	2.90%	4.80%	6.55%	
	2730	Heating, ventilating, air conditioning		1.70	4.80	6.45	2.41%	5%	8.90%	
	2900	Electrical		2.65	4.98	8.20	5.15%	7.20%	10.05%	
	3100	Total: Mechanical & Electrical	↓	7.40	11.35	22.50	13.30%	18.90%	26%	
99	0010	**WAREHOUSE & OFFICES Combination**	S.F.	55	73.50	101				99
	0020	Total project costs	C.F.	2.82	4.09	6.05				
	1800	Equipment	S.F.	.95	1.84	2.74	.42%	1.19%	2.40%	
	2720	Plumbing		2.13	3.70	5.50	3.74%	4.76%	6.30%	
	2770	Heating, ventilating, air conditioning		3.36	5.25	7.35	5%	5.65%	10.05%	
	2900	Electrical		3.75	5.55	8.65	5.85%	8%	10%	
	3100	Total: Mechanical & Electrical	↓	10.45	15.85	25	14.55%	19.95%	24.50%	

Square Foot Project Size Modifier

One factor that affects the S.F. cost of a particular building is the size. In general, for buildings built to the same specifications in the same locality, the larger building will have the lower S.F. cost. This is due mainly to the decreasing contribution of the exterior walls, plus the economy of scale usually achievable in larger buildings. The area conversion scale shown below will give a factor to convert costs for the typical size building to an adjusted cost for the particular project.

The square foot base size table lists the median costs, most typical project size in our accumulated data, and the range in size of the projects.

The size factor for your project is determined by dividing your project area in S.F. by the typical project size for the particular building type. With this factor, enter the area conversion scale at the appropriate size factor and determine the appropriate cost multiplier for your building size.

Example: Determine the cost per S.F. for a 100,000 S.F. Mid-rise apartment building.

$$\frac{\text{Proposed building area} = 100{,}000 \text{ S.F.}}{\text{Typical size from below} = 50{,}000 \text{ S.F.}} = 2.00$$

Enter area conversion scale at 2.0, intersect curve, read horizontally the appropriate cost multiplier of .94. Size adjusted cost becomes .94 × $119.00 = $112.00 based on national average costs.

Note: For size factors less than .50, the cost multiplier is 1.1
For size factors greater than 3.5, the cost multiplier is .90

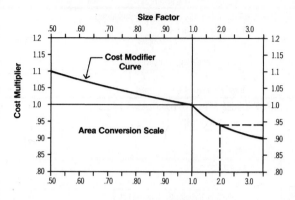

Square Foot Base Size

Building Type	Median Cost per S.F.	Typical Size Gross S.F.	Typical Range Gross S.F.	Building Type	Median Cost per S.F.	Typical Size Gross S.F.	Typical Range Gross S.F.
Apartments, Low Rise	$ 94.00	21,000	9,700 - 37,200	Jails	$ 286.00	40,000	5,500 - 145,000
Apartments, Mid Rise	119.00	50,000	32,000 - 100,000	Libraries	190.00	12,000	7,000 - 31,000
Apartments, High Rise	129.00	145,000	95,000 - 600,000	Living, Assisted	154.00	32,300	23,500 - 50,300
Auditoriums	161.00	25,000	7,600 - 39,000	Medical Clinics	162.00	7,200	4,200 - 15,700
Auto Sales	119.00	20,000	10,800 - 28,600	Medical Offices	154.00	6,000	4,000 - 15,000
Banks	211.00	4,200	2,500 - 7,500	Motels	113.00	40,000	15,800 - 120,000
Churches	145.00	17,000	2,000 - 42,000	Nursing Homes	158.00	23,000	15,000 - 37,000
Clubs, Country	147.00	6,500	4,500 - 15,000	Offices, Low Rise	135.00	20,000	5,000 - 80,000
Clubs, Social	141.00	10,000	6,000 - 13,500	Offices, Mid Rise	132.00	120,000	20,000 - 300,000
Clubs, YMCA	167.00	28,300	12,800 - 39,400	Offices, High Rise	169.00	260,000	120,000 - 800,000
Colleges (Class)	167.00	50,000	15,000 - 150,000	Police Stations	212.00	10,500	4,000 - 19,000
Colleges (Science Lab)	269.00	45,600	16,600 - 80,000	Post Offices	157.00	12,400	6,800 - 30,000
College (Student Union)	199.00	33,400	16,000 - 85,000	Power Plants	1175.00	7,500	1,000 - 20,000
Community Center	150.00	9,400	5,300 - 16,700	Religious Education	137.00	9,000	6,000 - 12,000
Court Houses	206.00	32,400	17,800 - 106,000	Research	222.00	19,000	6,300 - 45,000
Dept. Stores	87.50	90,000	44,000 - 122,000	Restaurants	192.00	4,400	2,800 - 6,000
Dormitories, Low Rise	170.00	25,000	10,000 - 95,000	Retail Stores	93.50	7,200	4,000 - 17,600
Dormitories, Mid Rise	196.00	85,000	20,000 - 200,000	Schools, Elementary	141.00	41,000	24,500 - 55,000
Factories	85.00	26,400	12,900 - 50,000	Schools, Jr. High	145.00	92,000	52,000 - 119,000
Fire Stations	156.00	5,800	4,000 - 8,700	Schools, Sr. High	150.00	101,000	50,500 - 175,000
Fraternity Houses	145.00	12,500	8,200 - 14,800	Schools, Vocational	143.00	37,000	20,500 - 82,000
Funeral Homes	162.00	10,000	4,000 - 20,000	Sports Arenas	115.00	15,000	5,000 - 40,000
Garages, Commercial	104.00	9,300	5,000 - 13,600	Supermarkets	92.00	44,000	12,000 - 60,000
Garages, Municipal	135.00	8,300	4,500 - 12,600	Swimming Pools	300.00	20,000	10,000 - 32,000
Garages, Parking	56.00	163,000	76,400 - 225,300	Telephone Exchange	243.00	4,500	1,200 - 10,600
Gymnasiums	143.00	19,200	11,600 - 41,000	Theaters	138.00	10,500	8,800 - 17,500
Hospitals	258.00	55,000	27,200 - 125,000	Town Halls	152.00	10,800	4,800 - 23,400
House (Elderly)	128.00	37,000	21,000 - 66,000	Warehouses	63.50	25,000	8,000 - 72,000
Housing (Public)	118.00	36,000	14,400 - 74,400	Warehouse & Office	73.50	25,000	8,000 - 72,000
Ice Rinks	170.00	29,000	27,200 - 33,600				

Abbreviations

A	Area Square Feet; Ampere	Brk., brk	Brick	Csc	Cosecant
AAFES	Army and Air Force Exchange Service	brkt	Bracket	C.S.F.	Hundred Square Feet
ABS	Acrylonitrile Butadiene Stryrene; Asbestos Bonded Steel	Brng.	Bearing	CSI	Construction Specifications Institute
A.C., AC	Alternating Current; Air-Conditioning; Asbestos Cement; Plywood Grade A & C	Brs.	Brass	CT	Current Transformer
		Brz.	Bronze	CTS	Copper Tube Size
		Bsn.	Basin	Cu	Copper, Cubic
		Btr.	Better	Cu. Ft.	Cubic Foot
		Btu	British Thermal Unit	cw	Continuous Wave
ACI	American Concrete Institute	BTUH	BTU per Hour	C.W.	Cool White; Cold Water
ACR	Air Conditioning Refrigeration	Bu.	bushels	Cwt.	100 Pounds
ADA	Americans with Disabilities Act	BUR	Built-up Roofing	C.W.X.	Cool White Deluxe
AD	Plywood, Grade A & D	BX	Interlocked Armored Cable	C.Y.	Cubic Yard (27 cubic feet)
Addit.	Additional	°C	degree centegrade	C.Y./Hr.	Cubic Yard per Hour
Adj.	Adjustable	c	Conductivity, Copper Sweat	Cyl.	Cylinder
af	Audio-frequency	C	Hundred; Centigrade	d	Penny (nail size)
AFUE	Annual Fuel Utilization Efficiency	C/C	Center to Center, Cedar on Cedar	D	Deep; Depth; Discharge
AGA	American Gas Association	C-C	Center to Center	Dis., Disch.	Discharge
Agg.	Aggregate	Cab	Cabinet	Db	Decibel
A.H., Ah	Ampere Hours	Cair.	Air Tool Laborer	Dbl.	Double
A hr.	Ampere-hour	Cal.	caliper	DC	Direct Current
A.H.U., AHU	Air Handling Unit	Calc	Calculated	DDC	Direct Digital Control
A.I.A.	American Institute of Architects	Cap.	Capacity	Demob.	Demobilization
AIC	Ampere Interrupting Capacity	Carp.	Carpenter	d.f.t.	Dry Film Thickness
Allow.	Allowance	C.B.	Circuit Breaker	d.f.u.	Drainage Fixture Units
alt., alt	Alternate	C.C.A.	Chromate Copper Arsenate	D.H.	Double Hung
Alum.	Aluminum	C.C.F.	Hundred Cubic Feet	DHW	Domestic Hot Water
a.m.	Ante Meridiem	cd	Candela	DI	Ductile Iron
Amp.	Ampere	cd/sf	Candela per Square Foot	Diag.	Diagonal
Anod.	Anodized	CD	Grade of Plywood Face & Back	Diam., Dia	Diameter
ANSI	American National Standards Institute	CDX	Plywood, Grade C & D, exterior glue	Distrib.	Distribution
				Div.	Division
APA	American Plywood Association	Cefi.	Cement Finisher	Dk.	Deck
Approx.	Approximate	Cem.	Cement	D.L.	Dead Load; Diesel
Apt.	Apartment	CF	Hundred Feet	DLH	Deep Long Span Bar Joist
Asb.	Asbestos	C.F.	Cubic Feet	dlx	Deluxe
A.S.B.C.	American Standard Building Code	CFM	Cubic Feet per Minute	Do.	Ditto
Asbe.	Asbestos Worker	CFRP	Carbon Fiber Reinforced Plastic	DOP	Dioctyl Phthalate Penetration Test (Air Filters)
ASCE.	American Society of Civil Engineers	c.g.	Center of Gravity		
A.S.H.R.A.E.	American Society of Heating, Refrig. & AC Engineers	CHW	Chilled Water; Commercial Hot Water	Dp., dp	Depth
		C.I., CI	Cast Iron	D.P.S.T.	Double Pole, Single Throw
ASME	American Society of Mechanical Engineers	C.I.P., CIP	Cast in Place	Dr.	Drive
		Circ.	Circuit	DR	Dimension Ratio
ASTM	American Society for Testing and Materials	C.L.	Carload Lot	Drink.	Drinking
		CL	Chain Link	D.S.	Double Strength
Attchmt.	Attachment	Clab.	Common Laborer	D.S.A.	Double Strength A Grade
Avg., Ave.	Average	Clam	Common maintenance laborer	D.S.B.	Double Strength B Grade
AWG	American Wire Gauge	C.L.F.	Hundred Linear Feet	Dty.	Duty
AWWA	American Water Works Assoc.	CLF	Current Limiting Fuse	DWV	Drain Waste Vent
Bbl.	Barrel	CLP	Cross Linked Polyethylene	DX	Deluxe White, Direct Expansion
B&B, BB	Grade B and Better; Balled & Burlapped	cm	Centimeter	dyn	Dyne
		CMP	Corr. Metal Pipe	e	Eccentricity
B&S	Bell and Spigot	CMU	Concrete Masonry Unit	E	Equipment Only; East; emissivity
B.&W.	Black and White	CN	Change Notice	Ea.	Each
b.c.c.	Body-centered Cubic	Col.	Column	EB	Encased Burial
B.C.Y.	Bank Cubic Yards	CO_2	Carbon Dioxide	Econ.	Economy
BE	Bevel End	Comb.	Combination	E.C.Y	Embankment Cubic Yards
B.F.	Board Feet	comm.	Commercial, Communication	EDP	Electronic Data Processing
Bg. cem.	Bag of Cement	Compr.	Compressor	EIFS	Exterior Insulation Finish System
BHP	Boiler Horsepower; Brake Horsepower	Conc.	Concrete	E.D.R.	Equiv. Direct Radiation
		Cont., cont	Continuous; Continued, Container	Eq.	Equation
B.I.	Black Iron	Corr.	Corrugated	EL	elevation
bidir.	bidirectional	Cos	Cosine	Elec.	Electrician; Electrical
Bit., Bitum.	Bituminous	Cot	Cotangent	Elev.	Elevator; Elevating
Bit., Conc.	Bituminous Concrete	Cov.	Cover	EMT	Electrical Metallic Conduit; Thin Wall Conduit
Bk.	Backed	C/P	Cedar on Paneling		
Bkrs.	Breakers	CPA	Control Point Adjustment	Eng.	Engine, Engineered
Bldg., bldg	Building	Cplg.	Coupling	EPDM	Ethylene Propylene Diene Monomer
Blk.	Block	CPM	Critical Path Method		
Bm.	Beam	CPVC	Chlorinated Polyvinyl Chloride	EPS	Expanded Polystyrene
Boil.	Boilermaker	C.Pr.	Hundred Pair	Eqhv.	Equip. Oper., Heavy
bpm	Blows per Minute	CRC	Cold Rolled Channel	Eqlt.	Equip. Oper., Light
BR	Bedroom	Creos.	Creosote	Eqmd.	Equip. Oper., Medium
Brg.	Bearing	Crpt.	Carpet & Linoleum Layer	Eqmm.	Equip. Oper., Master Mechanic
Brhe.	Bricklayer Helper	CRT	Cathode-ray Tube	Eqol.	Equip. Oper., Oilers
Bric.	Bricklayer	CS	Carbon Steel, Constant Shear Bar Joist	Equip.	Equipment
				ERW	Electric Resistance Welded

Abbreviations

E.S.	Energy Saver	H	High Henry	Lath.	Lather		
Est.	Estimated	HC	High Capacity	Lav.	Lavatory		
esu	Electrostatic Units	H.D., HD	Heavy Duty; High Density	lb.; #	Pound		
E.W.	Each Way	H.D.O.	High Density Overlaid	L.B., LB	Load Bearing; L Conduit Body		
EWT	Entering Water Temperature	HDPE	High density polyethelene plastic	L. & E.	Labor & Equipment		
Excav.	Excavation	Hdr.	Header	lb./hr.	Pounds per Hour		
excl	Excluding	Hdwe.	Hardware	lb./L.F.	Pounds per Linear Foot		
Exp., exp	Expansion, Exposure	H.I.D., HID	High Intensity Discharge	lbf/sq.in.	Pound-force per Square Inch		
Ext., ext	Exterior; Extension	Help.	Helper Average	L.C.L.	Less than Carload Lot		
Extru.	Extrusion	HEPA	High Efficiency Particulate Air Filter	L.C.Y.	Loose Cubic Yard		
f.	Fiber stress	Hg	Mercury	Ld.	Load		
F	Fahrenheit; Female; Fill	HIC	High Interrupting Capacity	LE	Lead Equivalent		
Fab., fab	Fabricated; fabric	HM	Hollow Metal	LED	Light Emitting Diode		
FBGS	Fiberglass	HMWPE	high molecular weight polyethylene	L.F.	Linear Foot		
F.C.	Footcandles			L.F. Nose	Linear Foot of Stair Nosing		
f.c.c.	Face-centered Cubic	HO	High Output	L.F. Rsr	Linear Foot of Stair Riser		
f'c.	Compressive Stress in Concrete; Extreme Compressive Stress	Horiz.	Horizontal	Lg.	Long; Length; Large		
F.E.	Front End	H.P., HP	Horsepower; High Pressure	L & H	Light and Heat		
FEP	Fluorinated Ethylene Propylene (Teflon)	H.P.F.	High Power Factor	LH	Long Span Bar Joist		
		Hr.	Hour	L.H.	Labor Hours		
F.G.	Flat Grain	Hrs./Day	Hours per Day	L.L., LL	Live Load		
F.H.A.	Federal Housing Administration	HSC	High Short Circuit	L.L.D.	Lamp Lumen Depreciation		
Fig.	Figure	Ht.	Height	lm	Lumen		
Fin.	Finished	Htg.	Heating	lm/sf	Lumen per Square Foot		
FIPS	Female Iron Pipe Size	Htrs.	Heaters	lm/W	Lumen per Watt		
Fixt.	Fixture	HVAC	Heating, Ventilation & Air-Conditioning	LOA	Length Over All		
FJP	Finger jointed and primed			log	Logarithm		
Fl. Oz.	Fluid Ounces	Hvy.	Heavy	L-O-L	Lateralolet		
Flr.	Floor	HW	Hot Water	long.	longitude		
FM	Frequency Modulation; Factory Mutual	Hyd.; Hydr.	Hydraulic	L.P., LP	Liquefied Petroleum; Low Pressure		
		Hz	Hertz (cycles)	L.P.F.	Low Power Factor		
Fmg.	Framing	I.	Moment of Inertia	LR	Long Radius		
FM/UL	Factory Mutual/Underwriters Labs	IBC	International Building Code	L.S.	Lump Sum		
Fdn.	Foundation	I.C.	Interrupting Capacity	Lt.	Light		
FNPT	Female National Pipe Thread	ID	Inside Diameter	Lt. Ga.	Light Gauge		
Fori.	Foreman, Inside	I.D.	Inside Dimension; Identification	L.T.L.	Less than Truckload Lot		
Foro.	Foreman, Outside	I.F.	Inside Frosted	Lt. Wt.	Lightweight		
Fount.	Fountain	I.M.C.	Intermediate Metal Conduit	L.V.	Low Voltage		
fpm	Feet per Minute	In.	Inch	M	Thousand; Material; Male; Light Wall Copper Tubing		
FPT	Female Pipe Thread	Incan.	Incandescent				
Fr	Frame	Incl.	Included; Including	M²CA	Meters Squared Contact Area		
F.R.	Fire Rating	Int.	Interior	m/hr.; M.H.	Man-hour		
FRK	Foil Reinforced Kraft	Inst.	Installation	mA	Milliampere		
FSK	Foil/scrim/kraft	Insul., insul	Insulation/Insulated	Mach.	Machine		
FRP	Fiberglass Reinforced Plastic	I.P.	Iron Pipe	Mag. Str.	Magnetic Starter		
FS	Forged Steel	I.P.S., IPS	Iron Pipe Size	Maint.	Maintenance		
FSC	Cast Body; Cast Switch Box	IPT	Iron Pipe Threaded	Marb.	Marble Setter		
Ft., ft	Foot; Feet	I.W.	Indirect Waste	Mat; Mat'l.	Material		
Ftng.	Fitting	J	Joule	Max.	Maximum		
Ftg.	Footing	J.I.C.	Joint Industrial Council	MBF	Thousand Board Feet		
Ft lb.	Foot Pound	K	Thousand; Thousand Pounds; Heavy Wall Copper Tubing, Kelvin	MBH	Thousand BTU's per hr.		
Furn.	Furniture			MC	Metal Clad Cable		
FVNR	Full Voltage Non-Reversing	K.A.H.	Thousand Amp. Hours	MCC	Motor Control Center		
FVR	Full Voltage Reversing	kcmil	Thousand Circular Mils	M.C.F.	Thousand Cubic Feet		
FXM	Female by Male	KD	Knock Down	MCFM	Thousand Cubic Feet per Minute		
Fy.	Minimum Yield Stress of Steel	K.D.A.T.	Kiln Dried After Treatment	M.C.M.	Thousand Circular Mils		
g	Gram	kg	Kilogram	MCP	Motor Circuit Protector		
G	Gauss	kG	Kilogauss	MD	Medium Duty		
Ga.	Gauge	kgf	Kilogram Force	MDF	Medium-density fibreboard		
Gal., gal.	Gallon	kHz	Kilohertz	M.D.O.	Medium Density Overlaid		
gpm, GPM	Gallon per Minute	Kip	1000 Pounds	Med.	Medium		
Galv., galv	Galvanized	KJ	Kiljoule	MF	Thousand Feet		
GC/MS	Gas Chromatograph/Mass Spectrometer	K.L.	Effective Length Factor	M.F.B.M.	Thousand Feet Board Measure		
		K.L.F.	Kips per Linear Foot	Mfg.	Manufacturing		
Gen.	General	Km	Kilometer	Mfrs.	Manufacturers		
GFI	Ground Fault Interrupter	KO	Knock Out	mg	Milligram		
GFRC	Glass Fiber Reinforced Concrete	K.S.F.	Kips per Square Foot	MGD	Million Gallons per Day		
Glaz.	Glazier	K.S.I.	Kips per Square Inch	MGPH	Thousand Gallons per Hour		
GPD	Gallons per Day	kV	Kilovolt	MH, M.H.	Manhole; Metal Halide; Man-Hour		
gpf	Gallon per flush	kVA	Kilovolt Ampere	MHz	Megahertz		
GPH	Gallons per Hour	kVAR	Kilovar (Reactance)	Mi.	Mile		
GPM	Gallons per Minute	KW	Kilowatt	MI	Malleable Iron; Mineral Insulated		
GR	Grade	KWh	Kilowatt-hour	MIPS	Male Iron Pipe Size		
Gran.	Granular	L	Labor Only; Length; Long; Medium Wall Copper Tubing	mj	Mechanical Joint		
Grnd.	Ground			m	Meter		
GVW	Gross Vehicle Weight	Lab.	Labor	mm	Millimeter		
GWB	Gypsum wall board	lat	Latitude	Mill.	Millwright		
				Min., min.	Minimum, minute		

Abbreviations

Misc.	Miscellaneous	PDCA	Painting and Decorating Contractors of America	SC	Screw Cover
ml	Milliliter, Mainline			SCFM	Standard Cubic Feet per Minute
M.L.F.	Thousand Linear Feet	P.E., PE	Professional Engineer; Porcelain Enamel; Polyethylene; Plain End	Scaf.	Scaffold
Mo.	Month			Sch., Sched.	Schedule
Mobil.	Mobilization			S.C.R.	Modular Brick
Mog.	Mogul Base	P.E.C.I.	Porcelain Enamel on Cast Iron	S.D.	Sound Deadening
MPH	Miles per Hour	Perf.	Perforated	SDR	Standard Dimension Ratio
MPT	Male Pipe Thread	PEX	Cross linked polyethylene	S.E.	Surfaced Edge
MRGWB	Moisture Resistant Gypsum Wallboard	Ph.	Phase	Sel.	Select
		P.I.	Pressure Injected	SER, SEU	Service Entrance Cable
MRT	Mile Round Trip	Pile.	Pile Driver	S.F.	Square Foot
ms	Millisecond	Pkg.	Package	S.F.C.A.	Square Foot Contact Area
M.S.F.	Thousand Square Feet	Pl.	Plate	S.F. Flr.	Square Foot of Floor
Mstz.	Mosaic & Terrazzo Worker	Plah.	Plasterer Helper	S.F.G.	Square Foot of Ground
M.S.Y.	Thousand Square Yards	Plas.	Plasterer	S.F. Hor.	Square Foot Horizontal
Mtd., mtd., mtd	Mounted	plf	Pounds Per Linear Foot	SFR	Square Feet of Radiation
Mthe.	Mosaic & Terrazzo Helper	Pluh.	Plumbers Helper	S.F. Shlf.	Square Foot of Shelf
Mtng.	Mounting	Plum.	Plumber	S4S	Surface 4 Sides
Mult.	Multi; Multiply	Ply.	Plywood	Shee.	Sheet Metal Worker
M.V.A.	Million Volt Amperes	p.m.	Post Meridiem	Sin.	Sine
M.V.A.R.	Million Volt Amperes Reactance	Pntd.	Painted	Skwk.	Skilled Worker
MV	Megavolt	Pord.	Painter, Ordinary	SL	Saran Lined
MW	Megawatt	pp	Pages	S.L.	Slimline
MXM	Male by Male	PP, PPL	Polypropylene	Sldr.	Solder
MYD	Thousand Yards	P.P.M.	Parts per Million	SLH	Super Long Span Bar Joist
N	Natural; North	Pr.	Pair	S.N.	Solid Neutral
nA	Nanoampere	P.E.S.B.	Pre-engineered Steel Building	SO	Stranded with oil resistant inside insulation
NA	Not Available; Not Applicable	Prefab.	Prefabricated		
N.B.C.	National Building Code	Prefin.	Prefinished	S-O-L	Socketolet
NC	Normally Closed	Prop.	Propelled	sp	Standpipe
NEMA	National Electrical Manufacturers Assoc.	PSF, psf	Pounds per Square Foot	S.P.	Static Pressure; Single Pole; Self-Propelled
		PSI, psi	Pounds per Square Inch		
NEHB	Bolted Circuit Breaker to 600V.	PSIG	Pounds per Square Inch Gauge	Spri.	Sprinkler Installer
NFPA	National Fire Protection Association	PSP	Plastic Sewer Pipe	spwg	Static Pressure Water Gauge
NLB	Non-Load-Bearing	Pspr.	Painter, Spray	S.P.D.T.	Single Pole, Double Throw
NM	Non-Metallic Cable	Psst.	Painter, Structural Steel	SPF	Spruce Pine Fir; Sprayed Polyurethane Foam
nm	Nanometer	P.T.	Potential Transformer		
No.	Number	P. & T.	Pressure & Temperature	S.P.S.T.	Single Pole, Single Throw
NO	Normally Open	Ptd.	Painted	SPT	Standard Pipe Thread
N.O.C.	Not Otherwise Classified	Ptns.	Partitions	Sq.	Square; 100 Square Feet
Nose.	Nosing	Pu	Ultimate Load	Sq. Hd.	Square Head
NPT	National Pipe Thread	PVC	Polyvinyl Chloride	Sq. In.	Square Inch
NQOD	Combination Plug-on/Bolt on Circuit Breaker to 240V.	Pvmt.	Pavement	S.S.	Single Strength; Stainless Steel
		PRV	Pressure Relief Valve	S.S.B.	Single Strength B Grade
N.R.C., NRC	Noise Reduction Coefficient/ Nuclear Regulator Commission	Pwr.	Power	sst, ss	Stainless Steel
		Q	Quantity Heat Flow	Sswk.	Structural Steel Worker
N.R.S.	Non Rising Stem	Qt.	Quart	Sswl.	Structural Steel Welder
ns	Nanosecond	Quan., Qty.	Quantity	St.; Stl.	Steel
nW	Nanowatt	Q.C.	Quick Coupling	STC	Sound Transmission Coefficient
OB	Opposing Blade	r	Radius of Gyration	Std.	Standard
OC	On Center	R	Resistance	Stg.	Staging
OD	Outside Diameter	R.C.P.	Reinforced Concrete Pipe	STK	Select Tight Knot
O.D.	Outside Dimension	Rect.	Rectangle	STP	Standard Temperature & Pressure
ODS	Overhead Distribution System	recpt.	receptacle	Stpi.	Steamfitter, Pipefitter
O.G.	Ogee	Reg.	Regular	Str.	Strength; Starter; Straight
O.H.	Overhead	Reinf.	Reinforced	Strd.	Stranded
O&P	Overhead and Profit	Req'd.	Required	Struct.	Structural
Oper.	Operator	Res.	Resistant	Sty.	Story
Opng.	Opening	Resi.	Residential	Subj.	Subject
Orna.	Ornamental	RF	Radio Frequency	Subs.	Subcontractors
OSB	Oriented Strand Board	RFID	Radio-frequency identification	Surf.	Surface
OS&Y	Outside Screw and Yoke	Rgh.	Rough	Sw.	Switch
OSHA	Occupational Safety and Health Act	RGS	Rigid Galvanized Steel	Swbd.	Switchboard
		RHW	Rubber, Heat & Water Resistant; Residential Hot Water	S.Y.	Square Yard
Ovhd.	Overhead			Syn.	Synthetic
OWG	Oil, Water or Gas	rms	Root Mean Square	S.Y.P.	Southern Yellow Pine
Oz.	Ounce	Rnd.	Round	Sys.	System
P.	Pole; Applied Load; Projection	Rodm.	Rodman	t.	Thickness
p.	Page	Rofc.	Roofer, Composition	T	Temperature; Ton
Pape.	Paperhanger	Rofp.	Roofer, Precast	Tan	Tangent
P.A.P.R.	Powered Air Purifying Respirator	Rohe.	Roofer Helpers (Composition)	T.C.	Terra Cotta
PAR	Parabolic Reflector	Rots.	Roofer, Tile & Slate	T & C	Threaded and Coupled
P.B., PB	Push Button	R.O.W.	Right of Way	T.D.	Temperature Difference
Pc., Pcs.	Piece, Pieces	RPM	Revolutions per Minute	Tdd	Telecommunications Device for the Deaf
P.C.	Portland Cement; Power Connector	R.S.	Rapid Start		
P.C.F.	Pounds per Cubic Foot	Rsr	Riser	T.E.M.	Transmission Electron Microscopy
PCM	Phase Contrast Microscopy	RT	Round Trip	temp	Temperature, Tempered, Temporary
		S.	Suction; Single Entrance; South	TFFN	Nylon Jacketed Wire
		SBS	Styrene Butadiere Styrene		

Abbreviations

TFE	Tetrafluoroethylene (Teflon)	U.L., UL	Underwriters Laboratory	w/	With
T. & G.	Tongue & Groove; Tar & Gravel	Uld.	unloading	W.C., WC	Water Column; Water Closet
Th., Thk.	Thick	Unfin.	Unfinished	W.F.	Wide Flange
Thn.	Thin	UPS	Uninterruptible Power Supply	W.G.	Water Gauge
Thrded	Threaded	URD	Underground Residential Distribution	Wldg.	Welding
Tilf.	Tile Layer, Floor	US	United States	W. Mile	Wire Mile
Tilh.	Tile Layer, Helper	USGBC	U.S. Green Building Council	W-O-L	Weldolet
THHN	Nylon Jacketed Wire	USP	United States Primed	W.R.	Water Resistant
THW.	Insulated Strand Wire	UTMCD	Uniform Traffic Manual For Control Devices	Wrck.	Wrecker
THWN	Nylon Jacketed Wire			W.S.P.	Water, Steam, Petroleum
T.L., TL	Truckload	UTP	Unshielded Twisted Pair	WT., Wt.	Weight
T.M.	Track Mounted	V	Volt	WWF	Welded Wire Fabric
Tot.	Total	VA	Volt Amperes	XFER	Transfer
T-O-L	Threadolet	VAT	Vinyl Asbestos Tile	XFMR	Transformer
tmpd	Tempered	V.C.T.	Vinyl Composition Tile	XHD	Extra Heavy Duty
TPO	Thermoplastic Polyolefin	VAV	Variable Air Volume	XHHW, XLPE	Cross-Linked Polyethylene Wire Insulation
T.S.	Trigger Start	VC	Veneer Core	XLP	Cross-linked Polyethylene
Tr.	Trade	VDC	Volts Direct Current	Xport	Transport
Transf.	Transformer	Vent.	Ventilation	Y	Wye
Trhv.	Truck Driver, Heavy	Vert.	Vertical	yd	Yard
Trlr	Trailer	V.F.	Vinyl Faced	yr	Year
Trlt.	Truck Driver, Light	V.G.	Vertical Grain	Δ	Delta
TTY	Teletypewriter	VHF	Very High Frequency	%	Percent
TV	Television	VHO	Very High Output	~	Approximately
T.W.	Thermoplastic Water Resistant Wire	Vib.	Vibrating	Ø	Phase; diameter
		VLF	Vertical Linear Foot	@	At
UCI	Uniform Construction Index	VOC	Volitile Organic Compound	#	Pound; Number
UF	Underground Feeder	Vol.	Volume	<	Less Than
UGND	Underground Feeder	VRP	Vinyl Reinforced Polyester	>	Greater Than
UHF	Ultra High Frequency	W	Wire; Watt; Wide; West	Z	zone
U.I.	United Inch				

Index

A

Abandon catch basin 27
Abatement asbestos 38
   equipment asbestos 37
ABC extinguisher portable 343
Abrasive aluminum oxide 48
   floor tile 298
   silicon carbide 48
   tread 298
ABS DWV pipe 413
Absorber shock 419
A/C coil furnace 452
   fan coil horizontal 665
   fan coil system 664, 665
   rooftop const volume 670
   rooftop DX 670
   rooftop DXVAV 670
   rooftop VAV 670, 671
   self-contained 672, 673
   self-contained VAV ... 672, 673
   unit elec ht/water cooled 672
   unit elec ht/water cooled
     VAV 672
   unit hot water coil water cool . 673
   unit hw coil water
     cooled VAV 673
Accelerator sprinkler system 399
Access card key 273
   control card type 273
   control fingerprint scanner ... 274
   control video camera 501
   doors and panels 238
   door basement 68
   door fire rated 238
   door floor 239
   door metal 239
   door roof 212
   door stainless steel 239
   floor 310
   flooring 310
   road and parking area 18
Accessories anchor bolt 56
   bathroom 338, 340
   door 274
   drainage 208
   drywall 296
   duct 442
   fireplace 341
   glazing 278
   gypsum board 296
   masonry 82
   roof 213
   toilet 338
Accordion door 234
Acid etch finish concrete 66
   proof floor 303
Acoustic ceiling board 299
   ceiling tile suspended 299
   tile 86
Acoustical ceiling 299, 300
   door 243
   enclosure 386
   folding door 337
   folding partition 337
   metal deck 109
   panel 337
   partition 338
   sealant 218, 295
   treatment 312
   underlayment 286
   window wall 248
Acrylic plexiglass 277
   rubber roofing 202
   sign 331
   wall coating 327

wallcovering 311
   wood block 304
Adhesive cement 306
   EPDM 181
   neoprene 181
   PVC 181
   removal floor 66
   roof 198
   wallpaper 311
Adjustable astragal 273
   frequency drives 488
Adjustment factors labor 402
Admixtures concrete 48
Aerial lift 713
   lift truck 718
   seeding 542
Agent form release 48
Aggregate exposed 530
   masonry 79
   spreader 710, 717
   stone 543
AHU ceiling concealed 462
   ceiling mounted 462
   floor standing unit 463
   floor standing unit concealed . 463
   wall mounted 463
Aids staging 17
Air cleaner electronic 448
   compressor 423, 713
   compressor portable 713
   compressor sprinkler system .. 399
   conditioner direct expansion .. 461
   conditioner fan coil 461
   cond. fan coil VRV .. 460, 462, 463
   conditioner gas heat 458
   conditioner portable 459
   conditioner receptacle 481
   conditioner removal 436
   conditioner rooftop 458
   conditioner self-contained ... 459
   conditioner window 459
   conditioner wiring 483
   conditioning computer 459
   conditioning fan 444
   conditioning ventilating .. 442-445,
                   456, 457, 459
   cooled condensing unit 456
   entraining agent 48
   filter 448
   filter mechanical media 448
   filter permanent washable .. 448
   filter roll type 448
   filter washable 448
   filtration 37
   filtration charcoal type 448
   gap fitting 418
   handler heater 453
   handler modular 453
   hose 714
   make-up unit 458
   register 447
   return grille 447
   sampling 37
   spade 714
   supply register 447
   supported structure demo ... 380
   terminal units 446
   tube system 395
   turning vane 443
   unit make-up 458
   vent roof 213
Air-compressor mobilization ... 504
Air handling unit 461
Airless sprayer 37
Air-source heat pumps 460
Alarm exit control 500

fire 500, 501
panels and devices 501
panels and devices fire 501
residential 482
sprinkler 501
valve sprinkler 398
water motor 399
All fuel chimney 449
Alteration fee 10
Alternating tread ladder 121
Alternative fastening 187
Aluminum astragal 272
   bench 540
   ceiling tile 300
   column 127, 169
   column cover 125
   commercial door 229
   conduit 680
   coping 97
   cross 366
   diffuser perforated 446
   door 222, 244
   double acting swing door .. 245
   downspout 207
   drip edge 211
   ductwork 440
   edging 544
   entrance 248
   expansion joint 211
   fence 536
   flashing 205, 421
   foil 185, 188
   frame window wall 602
   gravel stop 207
   grille 447
   gutter 209
   joist shoring 54
   light pole 705
   louver 280
   louver commercial grade ... 280
   Mansard 194
   nail 135
   oxide abrasive 48
   plank scaffold 17
   reglet 210
   roof 193
   roof panel 193
   sash 250
   screen/storm door 223
   security driveway gate 243
   service entrance cable 472
   sheet metal 207
   shingle 189
   siding 194
   siding paint 317
   siding panel 194
   sliding door 240
   steeple 347
   storefront 249
   storm door 223
   threshold 268
   threshold ramp 269
   tile 192, 298
   transom 248
   tube frame 246
   weatherstrip 272
   window 250, 599
   window demolition 222
Aluminum-framed door 248
Anchor bolt 56, 80
   bolt accessories 56
   bolt sleeve 56
   brick 81
   buck 80
   channel slot 81
   chemical 71, 100

epoxy 71, 100
expansion 100
framing 136
hollow wall 100
machinery 57
masonry 80
metal nailing 101
nailing 101
nylon nailing 101
partition 81
rafter 136
rigid 81
roof safety 17
sill 136
steel 81
stone 81
toggle bolt 101
wedge 101
Antenna TV 496
Anti-siphon device 418
Anti-slip floor treatment 323
Apartment call system 496
   hardware 263
Appliance 354
   compactor 690
   dishwasher 690
   garbage disposer 690
   plumbing 424
   range 689
   range hood 690
   refrigerator 690
   residential 354, 482
Arc fault circuit breaker 484
Arch laminated 152
   radial 152
Architectural fee 10, 808
Area clean-up 40
   window well 347
Armored cable 470
Arrow 536
Asbestos abatement 38
   abatement equipment 37
   area decontamination 40
   demolition 39
   disposal 40
   encapsulation 41
   remediation plan/method .. 37
   removal 39
   removal process 818
Ash receiver 340
Ashlar stone 95
   veneer 94
Asphalt base sheet 198
   block 533
   block floor 533
   coating 180
   curb 534
   cutting 32
   distributor 714
   felt 198, 199
   flashing 205
   flood coat 198
   grinding 525
   paver 715
   paving 531
   paving cold reused 525
   paving hot reused 525
   plant portable 717
   primer 178, 306
   roof cold-applied built-up .. 198
   roof shingle 189
   rubberized 524
   sealcoat rubberized 524
   sheathing 150
   shingle 189
   sidewalk 530

Index

surface treatment 524
Asphaltic binder 531
   concrete 531
   concrete paving 532
   cover board 186
   emulsion 524
   pavement 531
   wearing course 531
Astragal adjustable 273
   aluminum 272
   exterior moulding 272
   magnetic 272
   molding 163
   one piece 272
   overlapping 272
   rubber 272
   split 272
   steel 272
Athletic carpet indoor 309
   equipment 363
Atomizer water 38
Attachment ripper 717
Attachments skid steer 713
Attic stair 355
   ventilation fan 445
Auger earth 710
   hole 26
Auto park drain 421
Autoclave 364
   aerated concrete block 88
Automatic flush 430
   opener 263
   opener commercial 263
   sprinkler system 540
   timed thermostat 437
Automotive equipment 350
   lift 394
   spray painting booth 350
Auto-scrub machine 714
Awning canvas 345, 346
   fabric 345
   window 251, 345
   window metal-clad 253
   window vinyl clad 255

B

Backer rod 216
   rod polyethylene 56
Backerboard 181
   cementitious 293
Backfill 507, 508
   compaction 844
   dozer 507
   planting pit 541
   structural 507
   trench 505
Backflow preventer 417
   preventer solar 455
Backhoe 507, 711
   bucket 712
   extension 717
Backsplash countertop 373
Backstop basketball 363
   electric 363
Back-up block 88
Baffle roof 213
Bag disposable 38
   glove 39
Baked enamel door 227
   enamel frame 224
Balanced entrance door 248
Balcony fire escape 120
Bale hay 519
Ball check valve 404
valve bronze 403
valve plastic 404
wrecking 717
Ballast fixture 490
Ballistic door steel 243
   door wood 243
Baluster wood stair 166
Balustrade painting 324
Bamboo flooring 302
Band joist framing 114
Bank counter 353
   equipment 353
   run gravel 504
   window 353
Baptistry 366
Bar bell 363
   front 386
   grab 338
   grating tread 123
   masonry reinforcing 82
   panic 265
   restaurant 376
   touch 265
   towel 339
   tub 339
   Zee 296
Barbed wire fence 536
Barber chair 352
   equipment 352
Barge construction 720
Barometric damper 449
Barrels flasher 714
   reflectorized 714
Barricade 19, 558
   flasher 714
   tape 20
Barrier and enclosure 19
   dust 19
   parking 535
   separation 38, 43
   slipform paver 718
   weather 188
   X-ray 387
Base cabinet 369, 370
   carpet 309
   column 136
   course 531
   course drainage layers 531
   gravel 531
   masonry 92
   molding 153
   quarry tile 298
   resilient 305
   road 531
   sheet 198, 199
   sheet asphalt 198
   sink 369
   stabilizer 718
   stone 95, 531
   terrazzo 307
   vanity 371
   wall 305
   wood 153
Baseball scoreboard 364
Baseboard demolition 134
   heat 463
   heater electric 464
   register 447
Basement bulkhead stair 68
Baseplate scaffold 16
   shoring 15
Basic finish materials 284
Basketball backstop 363
   equipment 363
   scoreboard 364
Bath spa 384
steam 384
whirlpool 384
Bathroom 427
   accessories 338, 340
   exhaust fan 445
   faucet 428
   fixture 425-427
   heater & fan 445
   plumbing five fixture . . . 652, 653
   plumbing four fixture 651
   plumbing three fixture 648
   plumbing two fixture 647
Bathtub bar 339
   removal 402
   residential 427
Batt insulation 183
Bay wood window
   double hung 254
Bead blast demo 285
   casing 296
   corner 296
   parting 163
Beam & girder framing 137
   and girder formwork 49
   bond 80, 88
   bondcrete 292
   ceiling 169
   concrete 60
   concrete placement 64
   concrete placement grade 64
   decorative 169
   drywall 293
   fireproofing 213
   hanger 136
   laminated 152
   load-bearing stud wall box . . . 110
   mantel 168
   plaster 291
   precast 69
   precast concrete 68
   precast concrete tee 69
   reinforcing 58
   wood 137, 147
Bed molding 155
Bedding brick 533
   pipe 508
Bedpan cleanser 431
Belgian block 534
Bell & spigot pipe 419
   bar 363
Bench 539
   aluminum 540
   fiberglass 539
   folding 377
   greenhouse 385
   park 539
   players 540
   wood 539
   work 362
Bend EMT field 473, 474
Berm pavement 534
   road 534
Bevel siding 195
Bicycle trainer 363
Bi-fold door 228, 233
Binder asphaltic 531
Biological safety cabinet 361
Biometric identity access 274
Bi-passing closet door 234
Birch door 233, 234, 237
   molding 163
   paneling 165
   wood frame 170
Bit core drill 73
Bituminous block 533
   coating 180
concrete curbs 534
paver 715
Blade louvers movable 280
Blanket curing 68
   insulation 437
   insulation wall 183
   sound attenuation 312
Blast demo bead 285
   floor shot 285
Blaster shot 716
Blasting water 77
Bleacher telescoping 377
Blind exterior 171
   structural bolt 103
   vinyl 171
   window 368
Block asphalt 533
   autoclave concrete 88
   back-up 88
   belgian 534
   bituminous 533
Block, brick and mortar 830
Block cap 92
   concrete 88-91
   concrete bond beam 88
   concrete exterior 90
   corner 92
   decorative concrete 88
   filler 325
   floor 304
   floor wood 304
   glass 92
   glazed 92
   glazed concrete 92
   grooved 89
   ground face 89
   hexagonal face 90
   insulation 88
   insulation insert concrete 88
   lightweight 91
   lintel 91
   partition 91
   patio 95
   profile 89
   reflective 93
   scored 90
   scored split face 89
   sill 92
   slump 89
   split rib 89
   wall removal 28
   window prefabricated glass . . . 278
Blocking carpentry 137
   steel 137
   wood 137, 141
Blower insulation 715
   pneumatic tube 395
Blown-in cellulose 185
   fiberglass 185
   insulation 185
Blueboard 292
   partition 287
Bluegrass sod 542
Bluestone 94
   sill 96
   step 530
Board & batten siding 195
   batten fence 538
   ceiling 299
   control 362
   directory 331
   dock 351
   gypsum 293
   insulation 182
   insulation foam 182
   paneling 166

Index

partition gypsum 287
partition NLB gypsum 286
sheathing 150
siding wood 195
valance 370
verge 162
Boat work 720
Body harness 17
Boiler 450
   demolition 436
   electric 450
   electric steam 450
   fossil fuel 451
   gas-fired 450
   gas/oil combination 451
   hot water 450, 451
   oil-fired 451
   removal 436
   steam 450, 451
Bollards 104
   pipe 535
Bolt anchor 56, 80
   blind structural 103
   remove 100
   steel 136
Bolt-on circuit breaker 484
Bond beam 80, 88
   performance 13, 813
Bondcrete 292
Bonding agent 48
   surface 79
Bookcase 344, 370
Bookshelf 376
Boom truck 718
Booster fan 444
Booth painting 350
   restaurant 376
Bored pile 521
Borer horizontal 717
Boring and exploratory drilling ... 26
   cased 26
   horizontal 549
   machine horizontal 710
   service 549
Borosilicate pipe 432
Borrow 504
Borrowed lite 224
   lite assembled frame 225
Bosun's chair 18
Bottle storage 357
Boundary and survey marker ... 26
Bow & bay metal-clad window .. 254
   window 254
Bowl toilet 425
Bowstring truss 152
Box & wiring device 475
   beam load-bearing stud wall .. 110
   distribution 552
   electrical pull 475
   flow leveler distribution ... 552
   mail 344
   mixing 446
   pull 475
   stair 166
   storage 14
   termination 496
   trench 716
   utility 549
   VAV 446
   vent 83
Boxed headers/beams 114
   headers/beams framing 114
Boxing ring 363
Brace cross 104
Bracing 137
   framing 117

let-in 137
load-bearing stud wall 109
metal 137
metal joist 113
roof rafter 117
shoring 15
wood 137
Bracket scaffold 16
   sidewall 17
Brass hinge 269
   pipe 406
   pipe fittings 406
   screw 135
Brazed connection 473
Break glass station 501
   tie concrete 66
Breaker air gap vacuum 418
   circuit 487
   pavement 713
   vacuum 418
Brick 86, 631
   anchor 81
   bedding 533
   bedding mortar 530
Brick, block and mortar 830
Brick cart 714
   catch basin 553
   chimney simulated 340
   cleaning 77
   common 85
   common building 85
   concrete 92
   demolition 78, 284
   economy 85
   edging 544
   engineer 85
   face 85, 830
   floor 303, 533
   flooring miscellaneous 303
   manhole 553
   molding 160, 163
   oversized 84
   paving 533
   removal 27
   sidewalk 533
   sill 96
   simulated 97
   stair 93
   step 530
   structural 85
   veneer 83
   veneer demolition 78
   veneer thin 84
   vent box 83
   wall 93, 592
   wash 77
Bridge cranes 560
   sidewalk 16
Bridging 137
   framing 117
   load-bearing stud wall 109
   metal joist 113
   roof rafter 117
Broiler 358
Bronze body strainer 438
   plaque 331
   push-pull plate 268
   swing check valve 403
   valve 403
Broom cabinet 370
   finish concrete 65
   sidewalk 716
Brown coat 307
Brownstone 96
Brush chipper 711
   cutter 711, 712

Bubbler 432
Buck anchor 80
   rough 140
Bucket backhoe 712
   clamshell 711
   concrete 710
   dragline 711
   excavator 717
   pavement 717
Buggy concrete 65, 710
Builder's risk insurance 811
Building component
deconstruction 33
   deconstruction 29, 33
   demo footing & foundation .. 28
   demolition 28
   demolition selective 29
   directory 497
   greenhouse 385
   hardware 263
   insulation 185
   moving 35
   paper 188
   permit 13
   prefabricated 385
   relocation 35
   shoring 519
   sprinkler 399
   temporary 14
Built-up roof 198
   roofing 833
   roofing component 198
   roofing system 198
Bulb incandescent 492
   tee subpurlin 70
Bulk bank measure excavating .. 507
   dozer excavating 507
Bulkhead door 239
Bulkhead/cellar door 239
Bulldozer 507, 712
Bulletproof glass 279
Bullfloat concrete 710
Bullnose block 90
Bumper car 535
   dock 351
   door 267
   parking 535
   wall 266
Buncher feller 711
Burglar alarm indicating panel ... 500
Burlap curing 68
   rubbing concrete 66
Burner gas conversion 452
   gun type 452
   oil 452
   residential 452
Bush hammer 66
   hammer finish concrete 66
Butt fusion machine 714
Butterfly valve 436, 550
Butyl caulking 217
   expansion joint 210
   flashing 206
   rubber 216
   waterproofing 181
BX cable 680

C

Cab finish elevator 392
Cabinet & casework paint/coating ... 320
   base 369, 370
   broom 370
   casework 371

corner base 370
corner wall 370
current transformer 485
defibrillator 342
demolition 134
door 369, 371
electric 485
electrical hinged 475
filler 370
fire equipment 342
hardboard 369
hardware 371
hinge 372
hose rack 342
hospital 372
hotel 340
kitchen 369
laboratory 372
medicine 340
outdoor 372
oven 370
pharmacy 361
shower 340
stain 320
standard wood base 689
storage 372
strip 497
transformer 485
unit heater 669
varnish 320
wall 370, 372
Cable armored 470
   copper 470
   electric 470
   electrical 472
   jack 719
   MC 470
   mineral insulated 470
   pulling 716
   railing 126
   safety railing 107
   sheathed non-metallic 471
   sheathed Romex 471
   tensioning 716
   termination mineral insulated .. 471
   trailer 716
   UTP 496
Cables & fittings
   communication 496
Cafe door 236
Caisson concrete 521
   foundation 521
Caissons 845
Calcium chloride 48
Call system apartment 496
Canopy 346
   door 346
   entrance 346
   entry 345
   framing 103, 144
   metal 346
   patio/deck 345
   residential 345
   wood 144
Cant roof 143, 199
Cantilever needle beams 520
Canvas awning 345, 346
Cap block 92
   concrete placement pile 64
   dowel 58
   pile 62
   post 136
   service entrance 472
Car bumper 535
   tram 716
Card key access 273

Index

Carpentry finish 369
Carpet base 309
   cleaning 309
   computer room 310
   felt pad 308
   floor 309
   nylon 309
   pad 308
   pad commercial grade 308
   padding 308
   removal 284
   sheet 309
   stair 309
   tile 309
   tile removal 284
   urethane pad 308
   wool 309
Carrel 376
Carrier ceiling 290
   channel 301
Carrier/support fixture 430
Cart brick 714
   concrete 65, 710
   disposal 365
   food delivery 358
Carving stone 95
Case good metal 375
   refrigerated 356
Cased boring 26
   evaporator coils 453
Casement window 250-252, 254
   window metal-clad 253
   window vinyl 259
   window vinyl clad 598
Casework 370
   cabinet 371
   custom 369
   demolition 134
   door manufactured
      hardwood 371
   door plastic laminate 369
   frame wood 370
   ground 146
   hospital 372
   metal 372
   painting 320
   varnish 321
Cash register 351
Casing bead 296
   molding 154
Cast-in-place concrete 60, 63
   in place concrete curbs 534
   in place terrazzo 307
   iron bench 540
   iron damper 341
   iron fitting 419
   iron manhole cover 553
   iron pipe 419
   iron pipe fitting 410, 419
   iron radiator 463
   iron trap 421
   trim lock 266
Caster scaffold 16
Casting construction 124
Cast-iron column base 124
Catch basin 553
   basin brick 553
   basin cleaner 717
   basin cover remove/replace . 553
   basin precast 553
   basin removal 27
   basin vacuum 717
   door 371
Category 3 jack 496
   5 connector 496
   5 jack 496

5e jack 496
6 jack 496
Catwalk scaffold 16
Caulking 217
   polyurethane 217
   sealant 217
Cavity truss reinforcing 82
   wall grout 80
   wall insulation 184, 185
   wall reinforcing 82
Cedar closet 165
   fence 538
   paneling 166
   post 169
   roof deck 148
   roof plank 148
   shingle 190
   siding 195
Ceiling 299, 638
   acoustical 299, 300
   beam 169
   board 299
   board acoustic 299
   board fiberglass 299
   bondcrete 292
   bracing seismic 301
   carrier 290
   concealed AHU 462
   demolition 284
   diffuser 446
   drywall 294, 609
   fan 444
   finish price sheet 609, 638
   framing 138
   furring 146, 288, 289
   gypsum board 636
   insulation 183, 185
   lath 290
   luminous 300, 492
   molding 155
   mounted AHU 462
   painting 325, 326
   panels metal 299, 301
   plaster 291, 609
   price sheet 609
   register 447
   stair 355
   support 103
   suspended 289, 300
   suspended acoustical 635
   suspended plaster 637
   suspension system 301
   tile 300, 609
   wood 301
Cell equipment 366
   prison 386
Cellar door 239
   wine 357
Cellular concrete 61
   decking 109
Cellulose blown-in 185
   insulation 184
Cement adhesive 306
   color 79
   flashing 178
   grout 80, 520
   liner 548
   masonry 79
   masonry unit 88-91
   mortar 297, 298
   parging 180
   terrazzo Portland 307
   underlayment self-leveling .. 70
Cementitious backerboard 293
   waterproofing 181
   wood fiber plank 70

Central vacuum 364
Centrifugal pump 715
   type HVAC fan 444
Ceramic tile 297
   tile accessories 298
   tile countertop 373
   tile demolition 285
   tile floor 297
   tile repairs 298
   tile waterproofing membrane .. 299
   tiling 297, 299
Certification welding 13
Chain fall hoist 719
   hoist door 244
   link fence 20
   link fence paint 316
   link industrial fence 536
   saw 716
   trencher 505, 713
Chair barber 352
   bosun's 18
   lifts 394
   molding 163
   movie 361
   rail demolition 134
   restaurant 376
Chalkboard 330
   fixed 330
   freestanding 330
   liquid chalk 330
   portable 330
   wall hung 330
Chamber decontamination 38, 43
   echo 386
   infiltration 552
Channel carrier 301
   frame 224
   furring 288, 296
   grating tread 124
   metal frame 224
   siding 195
   slot 81
   slot anchor 81
Charcoal type air filter 448
Charge disposal 40
   powder 103
Check rack clothes 352
   valve ball 404
Checking bag garment 352
Checkout counter 351
   scanner 351
   supermarket 351
Checkroom equipment 351
Chemical anchor 71, 100
   cleaning masonry 77
   dry extinguisher 343
   spreader 717
   termite control 519
   toilet 716
Cherry threshold 268
Chiller direct expansion water ... 457
Chime door 487
Chimney 84
   accessories 340
   all fuel vent 449
   brick 84
   demolition 78
   fireplace 340
   foundation 60
   metal 340
   positive pressure 449
   screen 341
   simulated brick 340
   vent 449
   vent fitting 449
Chipper brush 711

   log 712
   stump 712
Chipping hammer 714
Chloride accelerator concrete .. 63
Church equipment 366
   pew 377
Chute 395
   linen 395
   refuse 395
   system 30
Circline fixture 491
Circuit breaker 487
   bolt-on 484
Circular saw 716
Circulating pump 439
   pump end suction 662, 663
Cladding sheet metal 204
Clamp water pipe ground 473
Clamshell bucket 711
Clapboard painting 317
Clay fire 96
   tile 191
   tile coping 97
Clean control joint 55
   tank 36
Cleaner catch basin 717
   crack 717
   steam 716
Cleaning & disposal equipment . 360
   brick 77
   internal pipe 548
   masonry 76
   rug 309
   up 21
Cleanout door 342
   floor 404
   pipe 404
   tee 405
Clean-up area 40
Climber mast 17
Climbing crane 718
   hydraulic jack 719
Clip plywood 136
Clock timer 482
Closer concealed 264
   door 264
   door electric automatic 264
   door electro magnetic 264
   electronic 264
   floor 264
   holder 264
Closet cedar 165
   door 228, 234
   pole 163
   rod 344
   water 425, 429
Closing coordinator door 269
Clothes check rack 352
   dryer commercial 352
CMU 88
   pointing 76
Coal tar pitch 178, 199
Coat brown 307
   glaze 325
   hook 339
   rack 371
   scratch 307
   tack 524
Coated roofing foamed 203
Coating bituminous 180
   concrete wall 327
   elastomeric 327
   flood 199
   intumescent 327
   roof 178
   rubber 182

Index

silicone	182
spray	180
trowel	180
wall	327
water repellent	182
waterproofing	180
Coffee urn	357
Coil cooling	461
Coiling doors and grille	241
Coiling-type counter door	241
Cold milling asphalt paving	525
mix paver	717
patch	525
planing	525
recycling	525
roofing	198
storage door	242
storage rooms	384
Collection concrete pipe sewage	551
sewage/drainage	553
Collector solar energy system	454
Colonial door	233, 237
wood frame	170
Coloring concrete	48
Column	169
aluminum	127, 169
base	136
base cast iron	124
bondcrete	292
brick	84
concrete	60
concrete placement	64
cover	125
cover aluminum	125
cover stainless	125
demolition	78
drywall	293
fireproof	213
formwork	49
framing	138
laminated wood	152
lath	290
plaster	291
plywood formwork	50
precast	69
precast concrete	69
reinforcing	58
removal	78
round fiber tube formwork	49
round fiberglass formwork	49
round steel formwork	49
steel framed formwork	50
structural	104
tie	81
wood	138, 147, 169
Combination device	479
storm door	233, 237
Command dog	20
Commercial automatic opener	263
dishwasher	360
door	226, 232, 239
folding partition	337
gas water heater	425
lavatory	429
mail box	344
metal door	228
refrigeration	356
service 3 phase	678
toilet accessories	338
water heater	424
water heater electric	424
Commissioning	21
Common brick	85
building brick	85
nail	135
Communicating lockset	265, 266
Communication cables & fittings	496
system	496
Compaction	517, 518
backfill	844
soil	507
structural	518
test Proctor	13
Compactor	365
earth	711
landfill	712
plate	518
residential	354
sheepsfoot	518
tamper	518
Compartment toilet	332
Component control	437
furnace	452
Composite decking	142, 175
decking woodgrained	175
door	227, 236
fabrication	175
insulation	187
metal deck	108
rafter	143
railing	175
railing encased	175
Composition floor	308
flooring	308
flooring removal	284
Compound dustproofing	48
Compressed air equipment automotive	350
Compression seals	216
Compressive strength	13
Compressor air	423, 713
reciprocating	423
reciprocating hermetic	423
refrigerant	456
Computer air conditioning	459
floor	310
room carpet	310
Concealed closer	264
Concrete	824-827
acid etch finish	66
admixture	48
asphaltic	531
beam	60
block	88-91
block autoclave	88
block back-up	88
block bond beam	88
block decorative	88
block demolition	30
block exterior	90
block foundation	90
block glazed	92
block grout	80
block insulation	185
block insulation insert	88
block interlocking	91
block lintel	91
block partition	91
block wall	577, 591
break ties	66
brick	92
broom finish	65
bucket	710
buggy	65, 710
bullfloat	710
burlap rubbing	66
bush hammer finish	66
caisson	521
cart	65, 710
cast-in-place	60, 63
cellular	61
chloride accelerator	63
coloring	48
column	60
conveyer	710
coping	97
core drilling	72
crack repair	46
crane and bucket	64
curb	534
curing	68, 533
curing compound	68
cutout	29
cylinder	13
demo	26
demolition	29, 47
direct chute	64
distribution box	552
drill	475
drilling	73
equipment rental	710
fiber reinforcing	63
finish	533
float finish	65
floor edger	710
floor grinder	710
floor patching	46
floor staining	302
form liner	53
forms	820
furring	146
granolithic finish	70
grinding	66
grout	80
hand mix	63
hand trowel finish	65
hole cutting	477
hole drilling	476
hydrodemolition	26
impact drilling	73
integral colors	48
integral topping	70
integral waterproofing	48
joist	61
lance	715
lift slab	828
lightweight	61, 830
lightweight insulating	70
machine trowel finish	65
materials	827
membrane curing	68
mixer	710
non-chloride accelerator	63
pan treads	65
panel	581
paper curing	68
paver	715
paving surface treatment	532
pile prestressed	521
pipe removal	27
placement beam	64
placement column	64
placement footing	64
placement slab	64
placement wall	65
placing	64, 827
plank	582
plantable pavers precast	533
planter	539
polishing	66
pressure grouting	520
prestressed precast	829
processing	66
pump	710
pumped	64
ready mix	63
reinforcing	58
removal	26
retarder	63
sandblast finish	66
saw	710
saw blade	71
scarify	284
septic tank	551
shingle	191
short load	63
sidewalk	530
slab	61, 580
slab conduit	477
slab cutting	71
slab X-ray	14
small quantity	63
spreader	715
stair	62
surface flood coat	47
surface repair	47
tile	191
topping	70
trowel	710
truck	710
truck holding	63
utility vault	549
vertical shoring	54
vibrator	710
volumetric mixed	63
wall	61, 576
wall coating	327
wall patching	46
water reducer	63
wheeling	65
winter	63
Concrete-filled steel pile	521
Condenser pad	465
Condensing unit air cooled	456
Conditioner portable air	459
Conductive floor	306
rubber & vinyl flooring	306
Conductor	470
& grounding	471, 472
Conductors	842
Conduit & fitting flexible	478
concrete slab	477
electrical	473
high installation	470
in slab PVC	477
in trench electrical	478
in trench steel	478
intermediate steel	474
rigid in slab	478
Cone traffic	19
Connection brazed	473
fire-department	398
motor	478
split system tubing	460
standpipe	398
utility	549
Connector category 5	496
flexible	417
gas	417
joist	136
timber	135
Construction barge	720
casting	124
management fee	10
temporary	19
time	10
Containment area lead	41
Contaminated soil	36
Contingencies	10
Continuous hinge	270
Contractor overhead	21
Control board	362
component	437
crack	524

Index

damper volume	442	
draft	449	
joint	55, 82	
joint clean	55	
joint PVC	83	
joint rubber	82	
joint sawn	55	
mirror traffic	347	
package firecycle	399	
valve heating	437	
Convection oven	358	
Cooking equipment	354, 358	
range	354	
Cooler	384	
beverage	356	
water	432	
Cooling coil	461	
tower	457	
tower fiberglass	457	
tower forced-draft	457	
tower stainless	457	
Coping	95, 97	
aluminum	97	
clay tile	97	
concrete	97	
removal	78	
terra cotta	87, 97	
Copper cable	470	
downspout	208	
drum trap	421	
DWV tubing	407	
fitting	407	
flashing	205, 206, 421	
gravel stop	207	
gutter	209	
pipe	407	
reglet	210	
roof	203	
tube fittings	408	
wall covering	310	
wire	472	
Core drill	710	
drill bit	73	
drilling concrete	72	
testing	13	
Coreboard	295	
Cork floor	306	
tile	306	
wall tile	310	
Corner base cabinet	370	
bead	296	
block	92	
guard	338	
guard metal	338	
wall cabinet	370	
Cornerboards PVC	174	
Cornice	93	
drain	423	
molding	155, 158	
painting	324	
Corridor sign	331	
Corrosion resistant fitting	433	
resistant pipe	432, 433	
resistant pipe fitting	433	
Corrosive fume resistant duct	441	
Corrugated metal pipe	553	
roof tile	191	
siding	193-195	
Cost mark-up	12	
Cot prison	376	
Counter bank	353	
checkout	351	
door	241	
door coiling-type	241	
flashing	210	
flashing aluminum	613	

flashing steel	613	
top	373	
top demolition	134	
Countertop backsplash	373	
ceramic tile	373	
engineered stone	375	
laboratory	373	
laminated	373	
laminated plastic	689	
maple	373	
marble	373	
plastic-laminate-clad	373	
postformed	373	
sink	426	
solid surface	374	
stainless steel	373	
Coupling plastic	415	
Course drainage layers base	531	
wearing	531	
Court tennis	537	
Cove base ceramic tile	297	
base terrazzo	307	
molding	155	
Cover board asphaltic	186	
board gypsum	186	
column	125	
stair tread	20	
CPVC pipe	414	
valve	404	
Crack cleaner	717	
control	524	
filler	717	
filling	524	
filling resin	47	
repair asphalt paving	527	
repair concrete	46	
repair epoxy	46	
repair resin	47	
Crane and bucket concrete	64	
bridge	560	
climbing	718	
crawler	718	
crew daily	14	
hydraulic	719	
jib	560	
mobilization	504	
overhead bridge	560	
rail	560	
tower	15, 718	
truck mounted	718	
Crawler crane	718	
drill rotary	714	
shovel	712	
Crew daily crane	14	
forklift	14	
survey	21	
Cross brace	104	
wall	366	
Crown molding	156, 163	
Crushed stone	504	
CSPE roof	200	
Cubicle detention	261	
shower	427	
toilet	335	
track & hardware	336	
Cultured stone	97	
Culvert reinforced	553	
Cupola	347	
Curb	534	
and gutter	534	
asphalt	534	
builder	710	
concrete	534	
edging	104, 534	
extruder	711	
granite	534	

inlet	534	
precast	534	
prefabricated	465	
removal	27	
roof	143	
seal	524	
slipform paver	717	
terrazzo	307	
Cured in place pipe	548	
Curing blanket	68	
compound concrete	68	
concrete	68, 533	
concrete membrane	68	
concrete paper	68	
concrete water	68	
paper	188	
Current transformer cabinet	485	
Curtain damper fire	442	
divider	364	
gymnasium divider	364	
rod	338	
stage	362	
vinyl	338	
Custom casework	369	
Cut stone curbs	534	
Cutoff pile	504	
Cutout counter	373	
demolition	29	
slab	29	
Cutter brush	711, 712	
Cutting and drilling	475	
asphalt	32	
block	373	
concrete floor saw	71	
concrete slab	71	
concrete wall saw	72	
masonry	32	
steel	101	
torch	101, 716	
wood	32	
Cylinder concrete	13	
lockset	265	
recore	265, 266	

D

Daily crane crew	14	
Dairy case	356	
Damper	442	
barometric	449	
fire curtain	442	
fireplace	341	
foundation vent	83	
multi-blade	442	
vent	449	
volume-control	442	
Darkroom door	242	
equipment	353	
sink	353	
Dead lock	265	
Deadbolt	265, 268	
Deadlocking latch	265	
Deciduous shrub	543	
Deck drain	421	
framing	141, 172	
roof	148, 150	
slab form	109	
wood	148	
Decking	832	
cellular	109	
composite	142, 175	
floor	108	
form	109	
roof	109	
woodgrained composite	175	

Deconstruction building	29, 33	
building components	33	
material handling	35	
Decontamination asbestos area	40	
chamber	38, 43	
enclosure	38, 40, 43	
equipment	38	
Decorative beam	169	
block	88	
fence	537	
Decorator device	479	
switch	479	
Deep therapy room	386	
Deep-longspan joist	107	
Defibrillator cabinet	342	
Dehumidifier	714	
Delicatessen case	356	
Delivery charge	18	
Demo air supported structures	380	
concrete	26	
footing & foundation building	28	
garden house	380	
geodesic domes	380	
greenhouses	380	
hangars	380	
lightning protection	380	
pre-engineered steel building	381	
silos	381	
sound control	381	
special purpose rooms	381	
storage tank	382	
swimming pool	382	
tension structures	383	
thermal & moisture protection	179	
X-ray/radio freq protection	383	
Demolish decking	131	
remove pavement and curb	26	
Demolition	29, 78, 133, 285, 402, 504, 817	
asbestos	39	
baseboard	134	
boiler	436	
brick	78, 284	
brick veneer	78	
cabinet	134	
casework	134	
ceiling	284	
ceramic tile	285	
chimney	78	
column	78	
concrete	29, 47	
concrete block	30	
disposal	30	
door	221	
drywall	284, 285	
ductwork	436	
electrical	468	
fencing	28	
fireplace	78	
flashing	39	
flooring	284	
framing	131	
furnace	436	
glass	222	
granite	78	
gutter	179	
gutting	32	
hammer	711	
HVAC	436	
joist	131	
lath	284	
masonry	29, 78, 285	
metal	100	
metal stud	285	
millwork	134	

870

Index

millwork & trim 134	expansion A/C 461	catch 371	overhead commercial 244
mold contaminated area 44	expansion water chiller 457	chain hoist 244	panel 232, 233
paneling 134	Directional drill horizontal 717	chime 487	paneled 228
partition 285	sign 331	cleanout 342	partition 337
pavement 26	Directory 331	closers 264, 607	passage 234, 238
plaster 284, 285	board 331	closet 228, 234, 629	patio 240
plenum 285	building 497	closing coordinator 269	plastic laminate casework 230
plumbing 402	Disappearing stairs 355	cold storage 242	plastic laminate casework 369
plywood 284, 285	Disc harrow 711	colonial 233, 237	plastic-laminate toilet 334
post 132	Discharge hose 715	combination storm 233, 237	porch 233
rafter 132	Dishwasher commercial 360	commercial 226, 232, 239	prefinished 230
railing 134	residential 354	commercial floor 239	pre-hung 236, 237
roofing 39, 179	Dispenser napkin 339	composite 227, 236	refrigerator 242
saw cutting 32	soap 339	counter 241	release 496
selective building 29	ticket 350	darkroom 242	removal 221
site 26	toilet tissue 339	darkroom revolving 242	residential 170, 228, 233
steel 100	towel 338	decorator wood 231	residential exterior 605
steel window 222	Dispensing equipment food 359	demolition 221	residential garage 244
stucco 285	Disposable bag 38	double 228	residential steel 228
terrazzo 285	Disposal asbestos 40	double acting swing 245	residential storm 223
tile 284	cart 365	dutch 227, 233	revolving 242
torch cutting 32	charges 40	dutch oven 342	rolling 241, 249
trim 134	field 551	electric automatic closer 264	roof 212
truss 134	garbage 354, 360	electro magnetic closer 264	rough buck 140
wall 285	or salvage value 33	entrance 222, 233, 237, 247	sauna 384
walls and partitions 285	waste 43	exterior 236	sectional 244
window 222	Distribution box 552	exterior pre-hung 237	sectional overhead 244
window & door 221	box concrete 552	fiberglass 236, 244	shower 340
wood 284	box flow leveler 552	fire 228, 232, 241, 501	sidelight 228, 236
wood framing 130	box HDPE 552	fire rated access 238	silencer 269
Demountable partition 336, 625	pipe water 550	fireplace 342	sill 164, 170, 268
Dentil crown molding 156	Distributor asphalt 714	flexible 246	sliding aluminum 240
Derrick crane guyed 719	Divider curtain 364	flush 230, 233	sliding glass 240
crane stiffleg 719	strip terrazzo 307	flush wood 229	sliding glass vinyl clad 240
Detection intrusion 500	Dock board 351	folding accordion 234	sliding panel 249
system 500	bumper 351	frame 223, 224	sliding patio 604
tape 549	loading 351	frame exterior wood 170	sliding vinyl clad 240
Detector explosive 500	shelter 351	frame grout 80	sliding wood 240
fire & heat 501	truck 351	frame interior 170	special 238, 243
infrared 500	Dog command 20	french 237	stain 321
metal 500	Dome 347	french exterior 236	stainless steel 237
motion 500	drain 423	garage 244	stainless steel (access) 239
smoke 501	security 347	glass 222, 240, 247, 248	steel 226, 227, 243, 835
temperature rise 501	Domed metal-framed skylight ... 262	hand carved 229	steel ballistic 243
ultrasonic motion 500	skylight 262	handle 371	steel louver 281
Detention cubicle 261	Door & window	hardboard 234	stop 247, 266, 607
equipment 366	interior paint 321, 322	hardware 263	storm 223
Detour sign 20	& window metal maintenance . 220	hardware accessories 269	swinging glass 248
Developing tank 353	& window wood	hollow metal 226	switch burglar alarm 500
Device anti-siphon 418	maintenance 220	hollow metal exterior 228	threshold 170
combination 479	accessories 274	industrial 241	torrified 235
decorator 479	accordion 234	interior flush wood 626	varnish 321
exit 264	acoustical 243	interior louvered 627	vertical lift 245
GFI 480	acoustical folding 337	interior pre-hung 238	weatherstrip 273
load management 488	aluminum 222, 244	interior solid 627	weatherstrip garage 273
panic 264	aluminum commercial 229	interior solid & louvered 627	wood 229, 232, 626
receptacle 480	aluminum double	jamb molding 163	wood ballistic 243
residential 478	acting swing 245	kick plate 271	wood double acting swing 245
wiring 475, 486, 487	aluminum-framed 248	knob 266	wood fire 231
Dewatering equipment 719	aluminum-framed entrance ... 248	knocker 269	wood panel 232
Diamond lath 290	and grilles coiling 241	labeled 227, 231	wood storm 237
plate tread 123	and panels access 238	louver 281	Doorbell system 487
Diaper station 338	and window materials 221	louvered 235, 238	Dormer gable 145
Diaphragm pump 715	balanced entrance 248	mahogany 229	Dormitory furniture 376
Dielectric union 438	bell residential 481	manufactured hardwood	Double acting swing door 245
Diesel hammer 712	bi-fold 228, 233	casework 371	hung aluminum sash window . 250
tugboat 720	bi-passing closet 234	metal 222, 228, 628	hung bay wood window 254
Diffuser ceiling 446	birch 233, 234, 237	metal access 239	hung steel sash window 251
linear 446	blind 171	metal commercial 603	hung window 255
opposed blade damper 446	bulkhead 239	metal fire 227	hung wood window 252
perforated aluminum 446	bulkhead/cellar 239	metal toilet component 333	wall pipe 554
rectangular 446	bumper 267	mirror 277	weight hinge 270
steel 447	cabinet 369, 371	molding 163	Dowel 81
T bar mount 446	cafe 236	moulded 232	cap 58
Dimmer switch 479, 487	canopy 346	opener 221, 242, 245, 263	reinforcing 58
Direct chute concrete 64	casing PVC 174	overhead 244	sleeve 58

871

For customer support on your Commercial Renovation Cost Data, call 877.791.4977.

Index

Downspout	207, 208
aluminum	207, 612
copper	208, 612
demolition	179
elbow	208
lead-coated copper	208
lead coated copper	612
steel	208
steel galvanized	612
strainer	207
Dozer	712
backfill	507
excavation	507
Draft control	449
damper vent	449
Dragline bucket	711
Drain	423
deck	421
dome	423
floor	421
main	423
roof	423
sanitary	421
Drainage accessories	208
boot roof	207
field	552
pipe	432, 433, 551, 553
trap	421
Drawer front plastic laminate	369
front wood	371
kitchen	369
pass-thru	366
security	366
track	372
wood	371
Drill concrete	475
core	710
earth	26, 521
main	550
quarry	714
rig	26
rock	26
shop	362
steel	714
tap main	550
track	714
wood	136, 539
Drilled concrete pier uncased	522
Drilling concrete	73
concrete impact	73
horizontal	549
plaster	287
rig	710
steel	102
Drinking fountain	432
fountain floor rough-in	432
fountain handicap	432
fountain wall rough-in	432
Drip edge	211
edge aluminum	211, 612
edge steel galvanized	612
Driven pile	521
Driver post	717
sheeting	714
Driveway	530, 693
gate security	243
removal	26
security gate	243
security gate opener	243
security gate solar panel	243
Drum trap copper	421
Dry fall painting	326
pipe sprinkler head	399
Drycleaner	352
Dryer commercial clothes	352
darkroom	353
hand	339
industrial	352
receptacle	481
residential	355
Dry-pipe sprinkler system	399
Dry-type transformer	483
Drywall	293, 624
accessories	296
column	293
cutout	30
demolition	284, 285
finish	839
frame	223
gypsum	293
interior surface	596
nail	135
painting	325, 326
partition	287
partition NLB	286
prefinished	294
removal	285
Dual flush flushometer	428, 430
flush valve toilet	430
Duck tarpaulin	19
Duct accessories	442
corrosive fume resistant	441
fire rated blanket	437
fitting plastic	441
flexible	443
flexible insulated	443
flexible noninsulated	443
furnace	453
heater electric	461
humidifier	465
insulation	437
liner	444
liner non-fibrous	444
mechanical	440
mixing box	446
preformed spiral	440
PVC	441
rigid plastic	441
round	440
silencer	443
thermal insulation	437
underground	555
utility	555
Ductile iron fitting	550
iron fitting mech joint	550
iron grooved-joint	412
iron pipe	550
Ductless split system	459
Ductwork	440
aluminum	440
demolition	436
fabric coated flexible	443
fabricated	440
fiberglass	441
galvanized	440
metal	440
rectangular	440
removal	436
rigid	440
Dumbbell	363
Dumbwaiter electric	390
manual	390
Dump charges	32
truck	713
Dumpsters	31, 817
Dumptruck off highway	713
Duplex receptacle	487
Dust barrier	19
collector shop	362
Dustproofing compound	48
Dutch door	227, 233
oven door	342
DWV pipe ABS	413
PVC pipe	414, 551
tubing copper	407

E

Earth auger	710
compactor	711
drill	26, 521
scraper	712
vibrator	505, 507, 518
Earthwork equipment	507
equipment rental	710
Eave vent	211
Ecclesiastical equipment	366
Echo chamber	386
Economizer shower-head water	427
Economy brick	85
Edge drip	211
Edger concrete floor	710
Edging	544
aluminum	544
curb	104, 534
Efficiency flush valve high	430
Effluent filter septic system	552
EIFS	188
Ejector pump	422
Elastomeric coating	327
liquid flooring	308
roof	202
sheet waterproofing	181
waterproofing	181
Elbow aluminum	612
copper	612
downspout	208
pipe	412
Electric backstop	363
ballast	490
baseboard heater	464
boiler	450
cabinet	485
cable	470
commercial water heater	424
duct heater	461
dumbwaiter	390
fixture	489, 491, 492
fixture exterior	492
furnace	452
generator	714
heating	464
hinge	270
lamp	492
log	341
metallic tubing	474
panelboard	484
relay	488
service	487
stair	355
switch	487, 488
traction freight elevator	390
utility	555
water cooler	432
water heater residential	424
water heater tankless	424
Electrical & telephone underground	555
cable	472
conduit	473
demolition	468, 842
field bend	473, 474
installation drilling	475
knockout	477
laboratory	361
wire	472
Electricity temporary	14
Electronic air cleaner	448
closer	264
markerboard	330
Electrostatic painting	324
Elevated conduit	470
floor	61
installation add	402
pipe add	402
slab	61
slab formwork	50
slab reinforcing	58
Elevator	390, 840
cab finishes	392
controls	393
electric traction freight	390
fee	10
freight	390
hydraulic freight	391
hydraulic passenger	392
options	393
passenger	390
residential	391
shaft wall	286
sump pump	423
Embossed print door	232
Emergency equipment laboratory	361
eyewash	431
lighting	491
lighting unit	491
shower	431
EMT	474
conduit	680
Emulsion adhesive	306
asphaltic	524
pavement	524
sprayer	714
Encapsulation asbestos	41
pipe	41
Encased railing	175
Encasing steel beam formwork	49
Enclosure acoustical	386
commercial telephone	332
decontamination	38, 40, 43
swimming pool	385
telephone	332
End grain block floor	304
Energy circulator air solar	454
system air purger solar	455
system air vent solar	455
system balancing valve solar	455
system control valve solar	455
system controller solar	454
system expansion tank solar	455
system gauge pressure solar	455
system heat exchanger solar	455
system storage tank solar	455
system vacuum relief solar	455
Engineer brick	85
Engineered stone countertop	375
Engineering fee	10, 809
Engraved panel signage	331
Entrance aluminum	248
and storefront	248
canopy	346
door	222, 233, 237, 247
door aluminum-framed	248
door fiberous glass	236
floor mat	375
frame	170
lock	268
screen	332
Entry canopy	345
EPDM adhesive	181
flashing	206
roofing	200

Index

Epoxy anchor 71, 100
   crack repair 46
   fiberglass wound pipe 433
   filled cracks 46
   floor 308
   grating 173
   grout 297, 520
   terrazzo 308
   wall coating 327
   welded wire 59
Equipment 351, 710
   athletic 363
   automotive 350
   bank 353
   barber 352
   basketball 363
   cell 366
   checkroom 351
   cleaning & disposal 360
   darkroom 353
   detention 366
   earthwork 507
   ecclesiastical 366
   exercise 363
   fire hose 398
   food preparation 357
   foundation formwork 51
   gymnasium 363
   health club 363
   insulation 437
   insurance 12
   laboratory 372
   laboratory emergency 361
   laundry 352
   loading dock 351
   lube 350
   lubrication 350
   medical 364
   medical sterilizing 364
   move 22
   moving 22
   pad 61
   parking 350
   parking collection 350
   parking control 350
   parking gate 350
   parking ticket 350
   refrigerated storage 356
   rental 710, 815, 816
   rental concrete 710
   rental earthwork 710
   rental general 713
   rental highway 717
   rental lifting 718
   rental marine 720
   rental wellpoint 719
   security 366
   security and vault 353
   shop 362
   stage 362
   theater and stage 362
   waste handling 365
Erosion control synthetic 519
Escape fire 120
Estimate electrical heating 464
Estimating 806
Evap/condens unit line sets ... 460
Evaporator coil cased 453
Evergreen shrub 543
   tree 543
Excavating bulk bank
   measure 507
   bulk dozer 507
   equipment 844
   trench 505
   utility trench 505

Excavation 505, 507, 843
   dozer 507
   footing 574
   footing or trench 574
   foundation 575
   hand 505-507
   machine 507
   planting pit 541
   septic tank 552
   structural 506
   tractor 505
   trench 505
   utility trench 692
Excavator bucket 717
   hydraulic 710
Exercise equipment 363
   ladder 363
   rope 363
   weight 363
Exhaust hood 354
   vent 447
Exhauster roof 444
Exit control alarm 500
   device 264
   light 491
   lighting 491
Expansion anchor 100
   joint 55, 210
   joint assembly 218
   joint butyl 210
   joint floor 218
   joint neoprene 210
   joint roof 210
   shield 100
   tank 438
   tank steel 438
   water chiller direct 457
Expense office 12
Explosive detection equipment .. 500
   detector 500
Exposed aggregate 530
Extension backhoe 717
   ladder 715
Exterior blind 171
   concrete block 90
   door 236
   door frame 170
   door hollow metal 228
   fixture LED 492
   fixture lighting 492
   floodlamp 492
   insulation 188
   insulation finish system .. 188
   lighting fixture 492
   molding 157
   moulding astragal 272
   paint doors & windows .. 318
   plaster 292
   pre-hung door 237
   PVC molding 174
   residential door 233
   shutter 171
   siding painting 317
   signage 331
   sprinkler 540
   surface masonry 596
   surface metal 596
   surface preparation 313
   surface wood 596
   trim 157
   trim paint 319
   wall price sheet 596
   wood door frame 170
   wood frame 170
Extinguisher chemical dry 343
   fire 343

   portable ABC 343
   standard 343
Extra work 12
Extruder curb 711
Eye wash fountain 431
Eyewash emergency 431
   safety equipment 431

F

Fabric awning 345
   flashing 205
   overlay 524
   stile 272
   waterproofing 180
   welded wire 59
Fabricated ductwork 440
Fabrication composite 175
Fabric-backed flashing 206
Face brick 85, 830
   wash fountain 431
Facial scanner unit 274
Facing stone 95
   tile structural 86
Factor 10
Fall painting dry 326
Fan air conditioning 444
   attic ventilation 445
   bathroom exhaust 445
   booster 444
   ceiling 444
   centrifugal type HVAC .. 444
   coil air conditioning 461
   house ventilation 445
   HVAC axial flow 444
   in-line 444
   kitchen exhaust 445
   paddle 482
   residential 482
   roof 444
   utility set 444
   ventilation 482
   wiring 482
Farm type siding 194
Fascia board 142
   board demolition 131
   metal 207
   PVC 174
   wood 160
Fast food equipment 357
Fastener timber 135
   wood 135
Fastening alternative 187
Faucet & fitting 428, 430, 431
   bathroom 428
   gooseneck 431
   laundry 428
   lavatory 428
   medical 431
Fee architectural 10
   engineering 10
Feller buncher 711
Felt 199
   asphalt 198, 199
   carpet pad 308
   tarred 199
   waterproofing 180
Fence 316
   aluminum 536
   and gate 536
   board batten 538
   cedar 538
   chain link 20
   chain link industrial 536
   chain link residential ... 537

   decorative 537
   fabric & accessories 537
   mesh 536
   metal 536
   misc metal 538
   picket paint 316
   plywood 20
   recycled plastic 537
   security 538
   snow 36
   steel 536, 537
   temporary 20
   tennis 537
   treated pine 538
   tubular 537
   wire 20, 538
   wood rail 538
Fencing demolition 28
Fertilizer 542
Fiber cement siding 197
   reinforcing concrete 63
   steel 60
   synthetic 60
Fiberboard insulation 186
Fiberglass area wall cap 347
   bay window 261
   bench 539
   blown-in 185
   ceiling board 299
   cooling towers 457
   door 236, 244
   ductwork 441
   floor grating 173
   fuel tank double wall ... 701, 702
   grating 173
   insulation 182, 183, 185, 437
   panel 19, 194, 313
   planter 539
   reinforced plastic panel .. 311
   single hung window 260
   slider window 260
   steeple 347
   storage tank 701
   waterproofing 181
   window 260
   wool 184
Field bend electrical 473, 474
   bend EMT 473, 474
   disposal 551
   drainage 552
   office 14
   personnel 11, 12
   seeding 542
Fieldstone 94
Fill 507, 508
   by borrow & utility bedding .. 508
   floor 61
   gravel 504
Filled crack epoxy 46
Filler block 325
   cabinet 370
   crack 717
   joint 216, 524
   strip 272
Fillet welding 102
Filling crack 524
Film equipment 353
   security 278
Filter air 448
   mechanical media 448
Filtration air 37
Fin tube radiation 463
Fine grade 542
Finish carpentry 369
   concrete 533
   floor 304, 326

Index

lime	79
nail	135
wall	310
Finishing floor concrete	65
wall concrete	66
Fir column	169
floor	304
molding	163
roof deck	148
roof plank	148
Fire & heat detectors	501
alarm	500, 501
alarm panels and devices	501
brick	96
call pullbox	501
clay	96
damage repair	315
damper curtain type	442
door	228, 232, 241, 501
door frame	224
door metal	227
door wood	231
equipment cabinet	342
escape	120
escape balcony	120
escape stairs	121
extinguisher	343
extinguisher portable	342, 343
horn	501
hose	398
hose adapter	398
hose equipment	398
hose gate valve	399
hose nozzle	398
hose rack	398
hose storage cabinet	342
hose valve	399
hydrant building	398
hydrant remove	27
panel	501
protection	342
rated blanket duct	437
rated tile	86
resistant drywall	294, 295
resistant glass	275
resistant wall	286
signal bell	501
sprinkler head	399
sprinkler system dry	675, 676
sprinkler system wet	677
Firebrick	96
Firecycle system control package	399
Fire-department connection	398
Fireplace accessories	341
box	97
built-in	341
chimney	340
damper	341
demolition	78
door	342
form	341
free standing	341
mantel	168
mantel beam	168
masonry	97
prefabricated	340, 341
Fireproofing	213
plaster	213
sprayed cementitious	213
Firestop wood	139
Firestopping	214, 834
Fitting cast iron	419
cast iron pipe	410, 419
copper	407
copper pipe	407
corrosion resistant	433
ductile iron	550
DWV pipe	414
glass	433
grooved-joint	412
grooved-joint pipe	412
malleable iron	411
malleable iron pipe	411
plastic	414
PVC	414
steel	410, 412
steel carbon pipe	412
steel pipe	412
vent chimney	449
Fixed chalkboard	330
end caisson piles	521
tackboard	330
Fixture ballast	490
bathroom	425-427
carrier/support	430
electric	489
fluorescent	489, 491
incandescent	490
incandescent vaportight	490
interior light	489
lantern	491
LED exterior	492
metal halide	490
mirror light	490
plumbing	422, 425, 439
removal	402
residential	481, 491
support handicap	430
vandalproof	490
Flagging	303
slate	534
Flange tie	81
Flasher barrels	714
barricade	714
Flashing	204-207
aluminum	205, 421, 612
asphalt	205
butyl	206
cement	178
copper	205, 206, 421, 612
counter	210
demolition	39
EPDM	206
fabric	205
fabric-backed	206
laminated sheet	206
lead	612
masonry	205
mastic-backed	206
membrane	199
neoprene	206
paperbacked	206
plastic sheet	206
polyvinyl chloride	612
PVC	206
sheet metal	205
stainless	205
steel	612
valley	189
vent	421
vent chimney	450
Flat seam sheet metal roofing	204
Flatbed truck	713, 716
truck crane	718
Flexible conduit & fitting	478
connector	417
door	246
duct	443
ductwork fabric coated	443
insulated duct	443
noninsulated duct	443
pavement repair	525
sign	331
sprinkler connector	399
Float finish concrete	65
glass	275
Floater equipment	12
Floating floor	302
pin	269
Flood coating	199
Floodlamp exterior	492
Floodlight pole mounted	492
trailer	714
tripod	714
Floor	303
access	310
acid proof	303
adhesive removal	66
asphalt block	533
brick	303, 533
carpet	309
ceramic tile	297
cleaning	21
cleanout	404
closer	264
composition	308
concrete finishing	65
conductive	306
cork	306
decking	108
door commercial	239
drain	421
elevated	61
end grain block	304
epoxy	308
expansion joint	218
fill	61
fill lightweight	61
finish	326
finish price sheet	633
framing removal	30
grating	173
grating fiberglass	173
grinding	66
hardener	48
hatch	239
honing	67
insulation	183
marble	95
mat	375
mat entrance	375
nail	135
oak	304
paint	323
paint & coating interior	323
paint removal	66
parquet	304
pedestal	310
plank	148
plate stair	120
plywood	149
polyethylene	309
quarry tile	298
receptor	422
removal	28
resilient	305, 306
rubber	305
sander	716
sealer	48
stain	323
standing unit AHU	463
standing unit concealed AHU	463
subfloor	149
terrazzo	307
tile or terrazzo base	307
tile terrazzo	307
topping	70
transition strip	308
treatment anti-slip	323
underlayment	149
varnish	324, 326
vinyl	306
vinyl sheet	305
wood	304
wood composition	304
Floor/ceiling concrete panel	581
concrete plank	582
concrete slab	580
steel joist	584
structural steel	583
Flooring	633
bamboo	302
composition	308
conductive rubber & vinyl	306
demolition	284
elastomeric liquid	308
masonry	303
miscellaneous brick	303
quartz	308
treatment	302
wood strip	304
Flue chimney metal	449
liner	84
prefab metal	449
screen	341
Fluid applied membrane air barrier	189
heat transfer	455
Fluorescent fixture	489, 491
Fluoroscopy room	386
Flush automatic	430
door	230, 233
flushometer dual	428, 430
high efficiency toilet (HET)	430
metal door	228
tube framing	246
valve	430
wood door	229, 231
Flying truss shoring	54
Foam board insulation	182
core DWV ABS pipe	414
insulation	185
roofing	203
spray rig	714
Foamed coated roofing	203
in place insulation	184
Foam-water system components	399
Fog seal	524
Foil aluminum	185, 188
metallic	185
Folder laundry	352
Folding accordion door	234
accordion partition	337
bench	377
gate	336
Food delivery cart	358
dispensing equipment	359
mixer	358
preparation equipment	357
service equipment	358
storage metal shelving	357
warmer	359
Football scoreboard	364
Footing concrete placement	64
formwork continuous	51
keyway	51
keyway formwork	51
reinforcing	58
removal	28
spread	61
Forced-draft cooling tower	457
Forest stewardship council	137

Index

Forklift 714
   crew 14
Form decking 109
   fireplace 341
   liner concrete 53
   material 820
   release agent 48
Formblock 89
Forms 820
   concrete 820
Formwork beam and girder 49
   column 49
   column plywood 50
   column round fiber tube 49
   column round fiberglass 49
   column round steel 49
   column steel framed 50
   continuous footing 51
   elevated slab 50
   encasing steel beam 49
   equipment foundation 51
   footing keyway 51
   gas station 51
   gas station island 51
   girder 49
   grade beam 52
   interior beam 49
   labor hours 821
   light base 51
   pile cap 51
   plywood 50
   radial wall 53
   retaining wall 53
   shoring concrete 54
   sign base 51
   slab blockout 52
   slab box out opening 50
   slab bulkhead 51, 52
   slab curb 51, 52
   slab edge 51, 52
   slab flat plate 50
   slab haunch 62
   slab joist dome 50
   slab joist pan 50
   slab on grade 52
   slab screed 52
   slab thickened edge 62
   slab trench 52
   slab turndown 62
   slab void 51, 52
   slab with drop panels 50
   spandrel beam 49
   spread footing 51
   steel framed plywood 53
   wall 52
   wall boxout 52
   wall brick shelf 52
   wall bulkhead 52
   wall buttress 52
   wall corbel 52
   wall lintel 53
   wall pilaster 53
   wall plywood 52
   wall prefab plywood 53
   wall steel framed 53
Fossil fuel boiler 451
Foundation caisson 521
   chimney 60
   concrete block 90
   mat 61
   mat concrete placement ... 64
   underdrain 572
   underpin 520
   vent 83
   wall 90
Fountain drinking 432

eye wash 431
face wash 431
   outdoor 383
   wash 430
   yard 383
Frame baked enamel 224
   door 223, 224
   drywall 223
   entrance 170
   exterior wood door 170
   exterior wood wall 595
   fire door 224
   labeled 224
   metal 223
   metal butt 224
   scaffold 15
   shoring 54
   steel 223
   welded 224
   window 253
   wood 370
Framing anchor 136
   band joist 114
   beam & girder 137
   boxed headers/beams 114
   bracing 117
   bridging 117
   canopy 103, 144
   ceiling 138
   columns 138
   deck 141, 172
   demolition 131
   heavy 147
   joist 139
   laminated 152
   ledger & sill 143
   lightweight 104
   lightweight angle 104
   lightweight channel 104
   load-bearing stud partition . 111
   load-bearing stud wall ... 111
   metal 596
   metal joist 115
   metal roof parapet 117
   miscellaneous wood 139
   NLB partition 288
   open web joist wood ... 139
   pipe support 105
   porch 141
   removal 130
   roof metal rafter 117
   roof metal truss 119
   roof rafters 142
   roof soffits 118
   roof truss 151
   sill & ledger 143
   sleepers 144
   slotted channel 104
   soffit & canopy 144
   timber 147
   treated lumber 144
   tube 246
   wall 144
   web stiffeners 115
   window wall 246
   wood 137, 145, 147, 597
Freestanding chalkboard 330
Freezer 356, 384
Freight elevator 390
   elevator hydraulic 391
   elevators 840
French door 237
   exterior door 236
Frequency drive adjustable .. 488
   drive variable 488
Friction pile 521, 522

Frieze PVC 174
Front end loader 507
   plastic laminate drawer ... 369
FRP panel 311
Fryer 358
Full vision glass 275
Furnace A/C coils 452
   components 452
   demolition 436
   duct 453
   electric 452
   gas-fired 452
   hot air 452
   oil-fired 452
   wall 454
Furnishing library 376
   site 377
Furniture 375
   dormitory 376
   hospital 376
   hotel 376
   move 22
   restaurant 376
Furring and lathing 287
   ceiling 146, 288, 289
   channel 288, 296
   concrete 146
   masonry 146
   metal 288
   price sheet 622
   steel 288
   wall 146, 288
   wood 146
Fusible link closer 264
Fusion machine butt 714

G

Gable dormer 145
Galley septic 552
Galvanized ductwork 440
   roof 193
   steel conduit 680
   steel reglet 210
   welded wire 59
Galvanizing lintel 105
Gap fitting air 418
Garage door 244
   door residential 244
   door weatherstrip 273
Garbage disposal 354, 360
Garden house demo 380
Garment checking bag 352
Gas boiler cast iron
   hot water 660, 661
   boiler cast iron steam ... 661
   boiler hot water 660
   boiler steam 661
   connector 417
   conversion burner 452
   fired boiler 450
   fired furnace 452
   fired space heater 454
   generator 489
   generator set 489
   log 341
   pipe 554
   space heater 668
   station formwork 51
   station island formwork . 51
   vent 449
   water heater commercial . 425
   water heater instantaneous . 425
   water heater residential . 425
   water heater tankless ... 425

Gas-fired duct heater 453
Gasket glazing 278
   joint 216
   neoprene 216
   toilet 428
Gas/oil boiler cast iron
   hot water 661
   boiler cast iron steam ... 661
   boiler hot water 661
   boiler steam 661
   combination boiler 451
Gasoline generator 489
   piping 554
Gate fence 536
   folding 336
   opener driveway security . 243
   parking 350
   security 336
   security driveway 243
   slide 536
   valve 436
General contractor's overhead . 811
   equipment rental 713
   fill 508
Generator electric 714
   emergency 489
   gas 489
   gasoline 489
   set 489
Geodesic dome demo 380
GFI receptacle 480
Girder formwork 49
   reinforcing 58
   wood 137, 148
Girt steel 106
Glass 275-277
   bead 246
   block 92
   block skylight 222
   block window prefabricated . 278
   break alarm 500
   bulletin board 330
   bulletproof 279
   demolition 222
   door 222, 240, 247, 248
   door astragal 272
   door shower 340
   door sliding 240
   door swinging 248
   fiber rod reinforcing ... 58, 59
   fire resistant 275
   fitting 433
   float 275
   full vision 275
   heat reflective 276
   insulating 276
   laminated 279
   lead 386
   low emissivity 275
   mirror 277, 339
   mosaic 298, 631
   mosaic sheet 298
   pipe 432
   pipe fitting 433
   plate 276
   reduce heat transfer . 276
   reflective 277
   sheet 276
   solar film 278
   tempered 275
   tile 277
   tinted 275
   window 276
   window wall 247
   wire 276
Glaze coat 325

Index

Glazed block 92	cover edging 699	wallboard repair 284	HDPE distribution box 552
brick 85	cover plants 543	weatherproof 151	infiltration chamber 552
concrete block 92	face block 89		Head sprinkler 399
wall coating 327	fault indicating receptacle 487	**H**	Header load-bearing stud wall .. 110
Glazing 275	rod 473		pipe wellpoint 719
accessories 278	water monitoring 13	H pile 521	wood 145
application 276	wire 473	Half round molding 163	Headers/beams boxed 114
gasket 278	Grounding 473	round vinyl clad window 255	Headrail metal toilet component . 333
plastic 277	& conductor 471, 472	round vinyl window 260	plastic-laminate toilet 334
polycarbonate 277	wire brazed 473	Halotron 343	Health club equipment 363
productivity 837	Group shower 429	Hammer bush 66	Hearth 97
Glove bag 39	wash fountain 430	chipping 714	Heat baseboard 463, 685
Glued laminated 152	Grout 80	demolition 711	electric baseboard 464
Gooseneck faucet 431	cavity wall 80	diesel 712	greenhouse 385
Gore line 536	cement 80, 520	drill rotary 714	pump 460
Grab bar 338	concrete 80	hydraulic 711, 715	pump air-source 460
Gradall 711	concrete block 80	pile 711	pump gas driven 460
Grade beam concrete	door frame 80	pile mobilization 504	pump residential 483
placement 64	epoxy 297, 520	vibratory 712	pump water heater 460
beam formwork 52	pump 716	Hammermill 717	recovery air to air 458
fine 542	topping 71	Hand carved door 229	recovery package 458
Grader motorized 711	wall 80	dryer 339	reflective film 276
Grading 514	Guard corner 338	excavation 505-507	reflective glass 276
and seeding 542	gutter 210	hole 549	temporary 49
Graffiti resistant treatment 327	house 385	scanner unit 274	transfer fluid 455
Granite 94	service 20	split shake 191	Heater & fan bathroom 445
building 94	snow 213	trowel finish concrete 65	air handler 453
curb 534	wall & corner 338	Handicap drinking fountain ... 432	floor mounted space 454
demolition 78	window 125	fixture support 430	gas residential water 425
indian 534	Guardrail scaffold 16	lever 268	gas water 425
paving block 534	temporary 20	opener 263	gas-fired duct 453
sidewalk 534	Guide rail 558	ramp 62	sauna 384
Granolithic finish concrete 70	rail removal 27	tub shower 427	space 715
Grass cloth wallpaper 311	rail timber 558	Handle door 371	unit 454
lawn 542	rail vehicle 558	latch 274	water 483
seed 542	Guide/guard rail 558	Handling waste 365	Heating 405, 450, 454
Grate precast 530	Gun type burner 452	Handrail wood 163, 168	electric 464
tree 530	Gunite 67	Hangar demo 380	estimate electrical 464
Grating area wall 347	drymix 67	Hanger beam 136	gas-fired 655
area way 347	Gutter 209	joist 136	gas-fired hot water 655
fiberglass 173	aluminum 209, 613	Hanging lintel 104	hot air 452
fiberglass floor 173	copper 209, 613	Hardboard cabinet 369	hydronic 451, 463
floor 173	demolition 179	door 234	kettle 718
stair 120	guard 210	molding 164	oil-fired 654
Gravel base 531	lead coated copper 209	overhead door 244	oil-fired hot water 654
fill 504	stainless 209	paneling 164	Heating-cooling gas forced air .. 659
stop 207	steel 209, 613	tempered 164	oil forced air 658
Grease interceptor 422	strainer 210	underlayment 149	Heavy duty shoring 15
trap 422	vinyl 209	Hardener floor 48	framing 147
Green roof membrane 192	wood 209	Hardware 263	timber 147
roof soil mixture 192	Gutting 32	apartment 263	Helicopter 719
roof system 192	demolition 32	cabinet 371	Hemlock column 169
Greenhouse 385	Guyed derrick crane 719	door 263	Hexagonal face block 90
cooling 385	Gym mat 363	finishes 836	High efficiency flush valve 430
demo 380	Gymnasium divider curtain ... 364	motel/hotel 263	efficiency toilet (HET) flush ... 430
residential 385	equipment 363	panic 264	efficiency urinal 430
Grid spike 136	floor 309	price sheet 607, 608	installation conduit 470
Griddle 358	Gypsum block demolition 30	window 263	intensity discharge lamp 492
Grille air return 447	board 293	Hardwood carrel 376	pressure fixture sodium 490
aluminum 447	board accessories 296	floor 304	pressure sodium lighting ... 490
decorative wood 168	board partition 287	grille 168	rib lath 290
painting 324	board partition NLB 286	Harness body 17	strength decorative block 89
roll up 241	board removal 285	Harrow disc 711	Highway equipment rental 717
window 256	cover board 186	Hasp 269	sign 332
Grinder concrete floor 710	drywall 293	Hat and coat strip 339	Hinge 269, 607
shop 362	fabric wallcovering 311	Hatch floor 239	brass 269
Grinding asphalt 525	lath 290	roof 212	cabinet 372
concrete 66	lath nail 135	smoke 212	continuous 270
floor 66	partition 336	Hauling 508	electric 270
Grooved block 89	partition NLB 287	cycle 508	hospital 270
joint ductile iron 412	plaster 291	Haunch slab 62	paumelle 270
joint fitting 412	restoration 284	Hay bale 519	prison 270
joint pipe 412	shaft wall 286	Hazardous waste cleanup 37	residential 269
Grooving pavement 525	sheathing 151	waste disposal 37	school 270
Ground 146	sound dampening panel ... 295	waste handling 365	security 270
clamp water pipe 473	underlayment poured 70		special 270

Index

stainless steel 269
steel 269
wide throw 270
Hip rafter 143
Hockey scoreboard 364
Hoist 560
   chain fall 719
   personnel 719
   tower 719
Holder closer 264
Holdown 136
Hole cutting electrical 477
   drilling electrical 476
Hollow core door 229, 235
   metal door 226
   metal doors and frame 237
   metal frame 223
   metal stud partition 291
   precast concrete plank 68
   wall anchor 100
Honed block 90
Honing floor 67
Hood and ventilation
   equipment 448
   exhaust 354
   range 354
Hook coat 339
   robe 339
Horizontal borer 717
   boring 549
   boring machine 710
   directional drill 717
   drilling 549
Horn fire 501
Hose adapter fire 398
   air 714
   bibb sillcock 428
   discharge 715
   equipment 398
   fire 398
   rack 398
   rack cabinet 342
   suction 715
   water 715
Hospital cabinet 372
   casework 372
   door hardware 263
   furniture 376
   hinge 270
   kitchen equipment 359
   tip pin 269
Hot air furnace 452
   air heating 452
   tub 383
   water boiler 450, 451
   water heating 450, 463
Hotel cabinet 340
   furniture 376
   lockset 265, 266
House guard 385
   telephone 496
   ventilation fan 445
Housewrap 188
Humidification equipment 385
Humidifier 465
   duct 465
   room 465
Hurricane resistant glass block .. 279
   resistant window 279
HVAC axial flow fan 444
   demolition 436
   removal 436
Hydrant building fire 398
   removal 27
   remove fire 27
   water 419

Hydrated lime 79
Hydraulic crane 719
   elevator 642
   excavator 710
   hammer 711, 715
   jack 719
   jack climbing 719
   lift 394
   seeding 542
Hydrodemolition 26
   concrete 26
Hydromulcher 717
Hydronic cabinet unit heater 669
   heating 451, 463
   space heater 668
Hypalon neoprene roofing 202

I

Ice machine 360
Icemaker 360
I-joists wood & composite 151
Impact wrench 714
Incandescent bulb 492
   exterior lamp 493
   fixture 490, 491
   interior lamp 492
Indian granite 534
Indicating panel burglar alarm .. 500
Indirect-fired make-up air unit .. 458
Indoor AHU VRV 462
   athletic carpet 309
Induction lighting 492
Industrial address system 496
   door 239, 241
   dryer 352
   lighting 489
   railing 123
   safety fixture 431
   window 251
Inert gas 36
Infiltration chamber 552
   chamber HDPE 552
Infrared broiler 358
   detector 500
Inlet curb 534
In-line fan 444
Insecticide 519
Inspection internal pipe 548
   technician 13
Installation add elevated 402
Instantaneous gas water heater . 425
   water heater 424
Insulated panels 188
   protector ADA 405
Insulating glass 276
   sheathing 149
Insulation 182, 185
   batt 183, 597
   blanket 437
   blower 715
   blown-in 185
   board 182
   building 185
   cavity wall 184
   ceiling 183, 185
   ceiling batt 609
   cellulose 184
   composite 187
   duct 437
   duct thermal 437
   equipment 437
   exterior 188
   fiberglass 182, 183, 185, 437
   finish system exterior 188

floor 183
foam 185
foamed in place 184
insert concrete block 88
isocyanurate 182
loose fill 184
masonry 185
mineral fiber 184
pipe 405
piping 405
polystyrene 182, 184
reflective 185
removal 39, 179
rigid 182
roof 186
roof deck 186
shingle 190
spray 185
sprayed-on 185
vapor barrier 188
vermiculite 185
wall 182
wall blanket 183
water heater 405
Insurance 12, 811
   builder risk 12
   equipment 12
   public liability 12
Integral colors concrete 48
   topping concrete 70
   waterproofing concrete 48
Interceptor 422
   grease 422
Interior beam formwork 49
   door frame 170
   light fixture 489
   paint door & window 321
   paint walls & ceilings 326
   pre-hung door 238
   residential door 234
   shutter 368
   wall finish price sheet 631
   wood door frame 170
Interlocking concrete block 91
Intermediate metallic conduit ... 473
Internal cleaning pipe 548
Interval timer 479
Intrusion detection 500
   system 500
Intumescent coating 327
Inverted bucket steam trap 438
Iron alloy mechanical joint pipe . 433
   body valve 436
   grate 530
Ironer laundry 352
Ironspot brick 303
Irrigation system sprinkler 540
Isocyanurate insulation 182

J

Jack cable 719
   category 3 496
   category 5 496
   category 5e 496
   category 6 496
   hydraulic 719
   pump 15
   rafter 143
   screw 519
Jackhammer 713
Jacking 549
Jib crane 560
Job condition 11
Joint assembly expansion 218

control 55, 82
expansion 55, 210
filler 216, 524
gaskets 216
push-on 550
reinforcing 82, 87
roof 211
sealant replacement 178
sealer 216, 217
Jointer shop 362
Joist concrete 61
   connector 136
   deep longspan 107
   demolition 131
   framing 139
   hanger 136
   longspan 108
   metal framing 115
   open web bar 108
   removal 131
   sister 139
   web stiffeners 115
   wood 139, 151
Jumbo brick 85
Jute mesh 519

K

Kennel fence 538
Ketone ethylene ester roofing .. 201
Kettle 358
   heating 718
   tar 716, 718
Keyless lock 265
Keyway footing 51
Kick plate 271, 607
   plate door 271
Kiln vocational 362
King brick 85
Kiosk 385
Kitchen cabinet 369
   equipment 358
   exhaust fan 445
   selective price sheet ... 689, 690
   sink 426
   sink faucet 428
   sink residential 426
   system 688
   unit commercial 355
K-lath 290
Knee action mixing valve 431
Knob door 266
Knocker door 269
Knockout electrical 477
Kraft paper 189

L

Labeled door 227, 231
   frame 224
Labor adjustment factors 402
   formwork 821
   modifiers 402
Laboratory cabinet 372
   casework metal 372
   countertop 373
   equipment 372
   safety equipment 361
   sink 361
   table 361
Ladder alternating tread 121
   exercise 363
   extension 715
   inclined metal 121

Index

reinforcing ... 82
ship ... 121
towel ... 339
vertical metal ... 121
Lag screw ... 102
Laminated beam ... 152
   countertop ... 373
   framing ... 152
   glass ... 279
   glued ... 152
   roof deck ... 148
   sheet flashing ... 206
   veneer members ... 152
   wood ... 152
Lamp electric ... 492
   high intensity discharge ... 492
   incandescent exterior ... 493
   incandescent interior ... 492
   metal halide ... 492
   post ... 126
Lampholder ... 487
Lance concrete ... 715
Landfill compactor ... 712
Landing metal pan ... 120
   stair ... 308
Landscape surface ... 309
Landscaping ... 697
Lantern fixture ... 491
Laser level ... 715
Latch deadlocking ... 265
   handle ... 274
   set ... 265, 266
Lateral arm awning retractable ... 345
Latex caulking ... 217
   underlayment ... 305
Lath ... 624
   demolition ... 284
   gypsum ... 290
   metal ... 290
Lath, plaster and gypsum board ... 838
Lath rib ... 290
Lathe shop ... 362
Lattice molding ... 163
Lauan door ... 234
Laundry equipment ... 352
   faucet ... 428
   folder ... 352
   ironer ... 352
   sink ... 427
Lavatory commercial ... 429
   faucet ... 428
   pedestal type ... 426
   removal ... 402
   residential ... 426, 429
   support ... 430
   vanity top ... 426
   wall hung ... 426
Lawn bed preparation ... 540
   establishment ... 697
   grass ... 542
   mower ... 715
   seed ... 542
Lazy Susan ... 370
Leaching field chamber ... 551
   pit ... 551
Lead coated copper downspout ... 208
   coated copper gutter ... 209
   coated downspout ... 208
   containment area ... 41
   flashing ... 205
   glass ... 386
   paint encapsulation ... 41
   paint remediation ... 41
   paint remediation methods ... 819
   paint removal ... 42
   plastic ... 386

roof ... 203
sheet ... 386
testing ... 41
Leads pile ... 711
Lean-to type greenhouse ... 385
Ledger & sill framing ... 143
Let-in bracing ... 137
Letter slot ... 344
Level laser ... 715
Leveling jack shoring ... 15
Lever handicap ... 268
Lexan ... 277
Liability insurance ... 811
Library furnishing ... 376
   shelf ... 376
Lift ... 394
   aerial ... 713
   automotive ... 394
   hydraulic ... 394
   scissor ... 713
   slab ... 62, 828
   truck aerial ... 718
   wheelchair ... 394
Lifter platform ... 351
Lifting equipment rental ... 718
Light base formwork ... 51
   exit ... 491
   fixture interior ... 489
   pole ... 705, 706
   post ... 491
   stand ... 37
   support ... 103
   temporary ... 14
   tower ... 715
Lighting ... 490-492
   darkroom ... 353
   emergency ... 491
   exit ... 491
   exterior fixture ... 492
   fixture replacement ... 468
   fluorescent ... 682
   high intensity ... 684
   high pressure sodium ... 490
   incandescent ... 683
   induction ... 492
   industrial ... 489
   metal halide ... 490
   outlet ... 481
   residential ... 481
   strip ... 489
   temporary ... 14
   track ... 491
   unit emergency ... 491
Lightning protection demo ... 380
   suppressor ... 478
Lightweight angle framing ... 104
   block ... 91
   channel framing ... 104
   concrete ... 61, 830
   floor fill ... 61
   framing ... 104
   insulating concrete ... 70
Lime ... 79
   finish ... 79
   hydrated ... 79
Limestone ... 95, 542
   coping ... 97
Line gore ... 536
   remover traffic ... 718
   sets ... 460
   sets refrigerant ... 439
Linear diffuser ... 446
Linen chutes ... 395
Liner cement ... 548
   duct ... 444
   flue ... 84

non-fibrous duct ... 444
pipe ... 548
Lint collector ... 352
Lintel ... 95
   block ... 91
   concrete block ... 91
   galvanizing ... 105
   hanging ... 104
   precast concrete ... 69
   steel ... 104
Liquid chillers screw ... 456
Lite borrowed ... 224
Load center residential ... 478
   management switch ... 488
Load-bearing stud
   partition framing ... 111
   stud wall bracing ... 109
   stud wall bridging ... 109
   stud wall framing ... 111
   stud wall header ... 110
Loader front end ... 507
   skid steer ... 713
   tractor ... 712
   vacuum ... 38
   wheeled ... 712
   windrow ... 718
Loading dock ... 351
   dock equipment ... 351
Loam ... 504, 505, 541
Lock electric release ... 265
   entrance ... 268
   keyless ... 265
   tubular ... 265
Locker metal ... 343
   steel ... 343
   wire mesh ... 343
Locking receptacle ... 481
Lockset ... 607
   communicating ... 265, 266
   cylinder ... 265
   hotel ... 265, 266
   mortise ... 266
Locomotive tunnel ... 718
Log chipper ... 712
   electric ... 341
   gas ... 341
   skidder ... 712
Longspan joist ... 108
Loose fill insulation ... 184
Louver ... 280
   aluminum ... 280
   commercial grade aluminum ... 280
   door ... 281
   galvanized steel ... 281
   midget ... 280
   operable ... 280
   PVC ... 280
   redwood ... 281
   ventilation ... 281
   wall ... 281, 282
   wood ... 168
Louvered blind ... 171
   door ... 235, 238
Lowbed trailer ... 718
Luan door hollow core ... 229
Lube equipment ... 350
Lubrication equipment ... 350
Lumber ... 137
   core paneling ... 165
   plastic ... 172
   product prices ... 833
   recycled plastic ... 172
   structural plastic ... 172
Luminaire replacement ... 468
   walkway ... 492
Luminous ceiling ... 300, 492

M

Machine auto-scrub ... 714
   excavation ... 507
   trowel finish concrete ... 65
   welding ... 717
   X-ray ... 500
Machinery anchor ... 57
Magnetic astragal ... 272
   particle test ... 13
Mahogany door ... 229
Mail box ... 344
   box call system ... 496
   box commercial ... 344
   slot ... 344
Main drain ... 423
   office expense ... 12
Maintenance door &
window metal ... 220
   door & window wood ... 220
Make-up air unit ... 458
   air unit rooftop ... 458
Mall front ... 249
Malleable iron fitting ... 411
   iron pipe fitting ... 411
Management fee construction ... 10
Manhole ... 553
   brick ... 553
   removal ... 27
Man-made soil mix ... 192
Mansard aluminum ... 194
Mantel beam ... 168
   fireplace ... 168
Manual dumbwaiter ... 390
Maple countertop ... 373
Marble ... 95
   coping ... 97
   countertop ... 373
   floor ... 95
   screen ... 336
   sill ... 96
   soffit ... 95
   stair ... 95
   synthetic ... 303
   tile ... 303
Marine equipment rental ... 720
Marker boundary and survey ... 26
Markerboard electronic ... 330
Mark-up cost ... 12
Masking ... 77
Mason scaffold ... 15
Masonry accessories ... 82
   aggregate ... 79
   anchor ... 80
   base ... 92
   brick ... 84
   cement ... 79
   cleaning ... 76, 594
   color ... 79
   cornice ... 93
   cutting ... 32
   demolition ... 29, 78, 285
   exterior surface ... 596
   fireplace ... 97
   flashing ... 205
   flooring ... 303
   furring ... 146
   insulation ... 185
   joint sealant ... 217
   nail ... 135
   pointing ... 76
   reinforcing ... 82
   removal ... 27, 78
   restoration ... 79, 594
   saw ... 716
   selective demolition ... 77

Index

sill	96
stabilization	76
step	530
toothing	76
ventilator	83
wall	88, 539, 591, 592
wall tie	80
waterproofing	80
Mass notification system	502
Mast climber	17
Mastic-backed flashing	206
Mat concrete placement foundation	64
floor	375
foundation	61
gym	363
wall	363
Material doors and window	221
handling deconstruction	35
removal/salvage	33
Materials concrete	827
MC cable	470
Meat case	356
Mechanical	841
duct	440
equipment demolition	436
fee	10
media filter	448
seeding	542
Medical equipment	364
faucet	431
sterilizing equipment	364
waste cart	365
waste disposal	365
waste sanitizer	365
Medicine cabinet	340
Membrane flashing	199
protected	202
roofing	198, 202
waterproofing	180, 524
Mercury vapor lamp	492
Mesh fence	536
radio-frequency-interference	261
stucco	290
Metal bookshelf	376
bracing	137
butt frame	224
canopy	346
case good	375
casework	372
ceiling panels	299, 301
chimney	340
corner guard	338
deck acoustical	109
deck composite	108
demolition	100
detector	500
door	222, 228, 603
door interior flush	628
door residential	228
ductwork	440
exterior surface	596
fascia	207
fence	536
fire door	228
flue chimney	449
frame	223
frame channel	224
framing parapet	117
furring	288
halide fixture	490
halide lamp	492
halide lighting	490
joist bracing	113
joist bridging	113
joist framing	115

laboratory casework	372
ladder inclined	121
lath	290
locker	343
nailing anchor	101
overhead door	244
pan ceiling	300
pan landing	120
pan stair	120
parking bumper	535
pipe removal	403
plate stair	120
rafter framing	117
roof	193
roof parapet framing	117
roof window	261
sash	251
screen	251
sheet	203
shelf	344
shelving food storage	357
shingle	190
siding	193, 194
soffit	197
stud demolition	285
stud NLB	288
support assemblies	287
threshold	268
tile	298, 631
toilet component door	333
toilet component headrail	333
toilet component panel	333
toilet component pilaster	333
toilet partition	332
truss framing	119
window	250, 251
Metal-clad double hung window	254
window bow & bay	254
window picture & sliding	254
Metal-framed skylight	262
Metallic conduit intermediate	473
foil	185
Meter socket	485
water supply	417
water supply domestic	417
Metric conversion factors	807
Microtunneling	549
Microwave detector	500
oven	354
Mill construction	147
Milling pavement	525
Millwork	153, 371
& trim demolition	134
demolition	134
Mineral fiber ceiling	299, 300
fiber insulation	184
fiberboard panel	313
insulated cable	470
roof	203
Minor site demolition	27
Mirror	277, 339
door	277
glass	277, 339
light fixture	490
plexiglass	277
wall	277
Miscellaneous painting	315, 320
Mix planting pit	541
Mixer concrete	710
food	358
mortar	710, 715
plaster	715
road	717
Mixing box	446
box constant volume	446

box duct	446
box VAV	446
valve	417, 428
Mobilization	504, 522
air compressor	504
or demobilization	18
Modification to cost	11
Modified bitumen roof	200
bitumen roofing	834
bituminous membrane SBS	200
Modifier labor	402
Modular air handler	453
office	625
office system	336
Modulating damper motorized	442
Module tub-shower	427
Moil point	714
Moisture content test	13
Mold abatement	43
abatement work area	43
contaminated area demolition	44
Molding base	153
bed	155
birch	163
brick	163
casing	154
ceiling	155
chair	163
cornice	155, 158
cove	155
crown	156
dentil crown	156
exterior	157
hardboard	164
pine	158
soffit	164
trim	163
window and door	163
wood transition	304
Money safe	353
Monitor support	103
Mop holder strip	339
roof	199
Mortar	79
admixture	80
Mortar, brick and block	830
Mortar cement	297, 298
masonry cement	79
mixer	710, 715
pigment	79
Portland cement	79
restoration	79
sand	79
thinset	298
Mortise lockset	266
Mosaic glass	298
tile	631
Motel/hotel hardware	263
Motion detector	500
Motor connection	478
support	103
Motorized grader	711
modulating damper	442
zone valve	437
Moulded door	232
Mounting board plywood	150
Movable blade louver	280
louver PVC	280
office partition	336
Move equipment	22
furniture	22
Movie screen	361
Moving building	35
equipment	22
shrub	545
structure	35

tree	545
Mower lawn	715
Muck car tunnel	718
Mud pump	710
trailer	717
Mulcher power	712
Mullion vertical	246
Multi-blade damper	442
Multizone air conditioner rooftop	458
Muntin window	256
Mylar tarpaulin	19

N

Nail	135
common	135
Nailer pneumatic	714, 715
steel	140
wood	140
Nailing anchor	101
Napkin dispenser	339
Natural fiber wall covering	310
Needle beams cantilever	520
Neoprene adhesive	181
expansion joint	210
flashing	206
gasket	216
roof	202
waterproofing	181
Newel wood stair	167
No hub pipe	419
hub pipe fitting	420
Non-chloride accelerator concrete	63
Non-destructive testing	13
Non-metallic sheathed cable	680
Non-removable pin	269
Norwegian brick	85
Nosing rubber	305
safety	305
stair	305
Nozzle fire hose	398
playpipe	398
Nut remove	100
Nylon carpet	309
nailing anchor	101

O

Oak door frame	170
floor	304
molding	163
paneling	165
threshold	170, 268
Off highway dumptruck	713
Office & storage space	14
expense	12
field	14
floor	310
overhead	12
partition	336, 625
partition movable	336
safe	353
system modular	337
trailer	14
Oil burner	452
fired boiler	451
fired furnace	452
hydraulic elevator	642
Oil-fired boiler	451
water heater	425
water heater residential	425
Olive knuckle hinge	270

Index

Omitted work 12
One piece astragal 272
One-way vent 213
Open web bar joist 108
Opener automatic 263
   door 221, 242, 245, 263
   driveway security gate 243
   handicap 263
Opening roof frame 104
Operable louver 280
Operating room equipment 364
Operation maintenance lighting . 468
Option elevator 393
Ornamental aluminum rail 127
   glass rail 127
   railing 127
   steel rail 127
   wrought iron rail 127
OSB faced panel 146
OSHA testing 40
Outdoor cabinet 372
   fountain 383
   unit VRV 460
Outlet box plastic 475
   box steel 474
   lighting 481
Oven 358
   cabinet 370
   convection 358
   microwave 354
Overhaul 32
Overhead & profit 12
   bridge cranes 560
   commercial door 244
   contractor 21, 811
   door 244, 606
   office 12
Overlapping astragal 272
Overlay fabric 524
   face door 230, 232
   pavement 524
Oversized brick 84
Overtime 12
Oxygen lance cutting 32

P

P trap 421
   trap running 421
Packaging waste 40
Pad carpet 308
   commercial grade carpet 308
   condenser 465
   equipment 61
   prefabricated 465
Padding carpet 308
Paddle fan 482
Paint & coating 313
   & coating interior floor 323
   aluminum siding 317
   chain link fence 316
   doors & windows exterior 318
   doors & windows
      interior 321, 322
   encapsulation lead 41
   exterior miscellaneous 316
   fence picket 316
   floor 323
   floor concrete 323
   floor wood 323
   interior miscellaneous 323
   remediation lead 41
   removal 42
   removal floor 66
   removal lead 42

   siding 317
   sprayer 715
   striper 717, 718
   trim exterior 319
   walls & ceilings interior .. 325, 326
   walls masonry exterior 320
Paint/coating cabinet
   & casework 320
Painted pavement markings 535
Painting 631
   balustrade 324
   booth 350
   casework 320
   ceiling 325, 326
   clapboard 317
   cornice 324
   decking 316
   drywall 325, 326
   electrostatic 324
   exterior siding 317
   grille 324
   miscellaneous 315, 320
   parking stall 536
   pavement 535
   pipe 324
   plaster 325, 326
   railing 316
   reflective 535
   shutter 316
   siding 317
   stair stringers 316
   steel siding 317
   stucco 317
   temporary road 536
   thermoplastic 536
   trellis/lattice 316
   trim 324
   truss 324
   wall 317, 325
   window 321
Pan shower 205
   slab 61
   stair metal 120
   treads concrete 65
Panel acoustical 337
   and device alarm 501
   door 232, 233
   fiberglass 19, 194, 313
   fire 501
   FRP 311
   insulated 188
   metal toilet component 333
   mineral fiberboard 313
   OSB faced 146
   plastic-laminate toilet 334
   portable 337
   prefabricated 188
   shearwall 147
   sound absorbing 313
   sound dampening gypsum ... 295
   steel roofing 193
   structural 146
   structural insulated 146
   system 165
   vision 261
   wall 97
Panelboard 484
   electric 484
   w/circuit breaker 484
Paneled door 228
   pine door 235
Paneling 164, 631
   birch 165
   board 166
   cedar 166
   cutout 30

   demolition 134
   hardboard 164
   plywood 165
   redwood 166
   wood 164
Panelized shingle 191
Panic bar 265
   device 264, 607
Paper building 188
   sheathing 188
Paperbacked flashing 206
Paperhanging 310
Paperholder 339
Parallel bar 363
Parapet metal framing 117
   wall 593
Parging cement 180
Park bench 539
Parking barrier 535
   barrier precast 535
   bumper 535
   bumper metal 535
   bumper wood 535
   collection equipment 350
   control equipment 350
   equipment 350
   gate 350
   gate equipment 350
   lot asphalt 694
   lot concrete 695
   lot paving 532
   markings pavement 536
   stall painting 536
   ticket equipment 350
Parquet floor 304
   wood 304
Particle board underlayment 149
   core door 230
Parting bead 163
Partition 336, 337
   acoustical 338
   anchor 81
   block 91
   blueboard 287
   bulletproof 353
   concrete block 91, 616
   demolition 285
   demountable 336
   door 337
   drywall 287, 619
   folding accordion 337
   folding leaf 338
   framing load bearing stud 111
   framing NLB 288
   gypsum 336
   metal stud 618, 620
   metal stud LB 620
   metal stud NLB 618
   movable office 336, 625
   NLB drywall 286
   NLB gypsum 287
   office 336
   plaster 621
   plaster & lath 621
   shower 95, 340
   steel 337
   support 103
   thin plaster 287
   tile 87
   toilet 95, 332, 334
   toilet stone 335
   wall 286
   wall NLB 286
   wood frame 140
   wood stud 617
Passage door 234, 238

Passenger elevator 390
   elevator hydraulic 392
   elevators 840
Pass-thru drawer 366
Patch core hole 13
   repair asphalt pavement 526
   roof 178
Patching concrete floor 46
   concrete wall 46
   rigid pavement 527, 528
Patio 531
   block 95
   door 240
Patio/deck canopy 345
Paumelle hinge 270
Pavement 533
   asphaltic 531
   berm 534
   breaker 713
   bucket 717
   demolition 26
   emulsion 524
   grooving 525
   markings 535
   markings painted 535
   milling 525
   overlay 524
   painting 535
   parking markings 536
   planer 718
   profiler 718
   profiling 525
   pulverization 525
   reclamation 525
   recycling 525
   repair flexible 525
   repair rigid 527
   replacement 532
   replacement rigid 528
   widener 717
Paver asphalt 715
   bituminous 715
   cold mix 717
   concrete 715
   floor 303
   roof 213
   shoulder 717
Paving asphalt 531
   asphaltic concrete 532
   block granite 534
   brick 533
   cold reused asphalt 525
   hot reused asphalt 525
   parking lot 532
   surface treatment concrete ... 532
Peastone 543
Pedestal floor 310
   type lavatory 426
Peephole 266
Pegboard 164
Performance bond 13, 813
Perlite insulation 182
   plaster 291
   sprayed 327
Permit building 13
Personal respirator 37
Personnel field 11, 12
   hoist 719
Pew church 377
   sanctuary 377
Pharmacy cabinet 361
Phone booth 332
PIB roof 201
Picket railing 122
Pickup truck 716
Pick-up vacuum 710

Index

Picture window 253
   window aluminum sash 250
   window steel sash 251
   window vinyl 260
Pier brick 84
Pigment mortar 79
Pilaster metal toilet component . . 333
   plastic-laminate toilet 334
   toilet partition 336
   wood column 169
Pile bored 521
   cap 62
   cap concrete placement 64
   cap formwork 51
   cutoff 504
   driven 521
   driving 815, 816
   friction 521, 522
   H 521
   hammer 711
   leads 711
   load test 504
   mobilization hammer 504
   prestressed 522
   sod 541
   steel 521
   steel sheet 519
   step tapered 521
   testing 504
   timber 521
   treated 521
   wood 521
   wood sheet 519
Piling sheet 519
   special costs 504
Pin powder 103
Pine door 233
   door frame 170
   fireplace mantel 168
   floor 304
   molding 158
   roof deck 148
   shelving 344
   siding 195
   stair tread 166
Pipe & fittings 404, 407,
                            408, 410-414,
                            417, 423, 428
   add elevated 402
   and fittings 841
   bedding 508
   bedding trench 506
   bollard 535
   brass 406
   cast iron 419
   cleaning internal 548
   cleanout 404
   copper 407
   corrosion resistant 433
   corrugated metal 553
   covering 405
   covering fiberglass 405
   CPVC 414
   cured in place 548
   double wall 554
   drainage 432, 433, 551, 553
   ductile iron 550
   DWV PVC 414, 551
   elbow 412
   encapsulation 41
   epoxy fiberglass wound 433
   fitting brass 406
   fitting cast iron 410, 419
   fitting copper 407
   fitting corrosion resistant 433
   fitting DWV 414

fitting glass 433
fitting grooved-joint 412
fitting malleable iron 411
fitting no hub 420
fitting plastic 414, 416
fitting soil 419
fitting steel 410, 412
fitting steel carbon 412
fitting weld steel 412
foam core DWV ABS 414
gas 554
glass 432
grooved-joint 412
inspection internal 548
insulation 405
insulation removal 39
internal cleaning 548
iron alloy mechanical joint . . . 433
liner 548
no hub 419
painting 324
plastic 413, 414
polyethylene 554
PVC 413, 551
rail aluminum 122
rail galvanized 122
rail stainless 122
rail steel 122
rail wall 122
railing 122
reinforced concrete 553
removal 27
removal metal 403
removal plastic 403
rodding 548
sewage 551, 553
sewage collection PVC 551
shock absorber 419
single hub 419
soil 419
steel 409
support framing 105
tee 413
weld joint 410
Piping designations 846
Piping gas service polyethylene . 554
   gasoline 554
   insulation 405
   storm drainage 553
Pit excavation 506
   leaching 551
Pitch coal tar 199
   emulsion tar 524
   pocket 213
Pivoted window 251
Placing concrete 64, 827
Plain tube framing 246
Plan remediation 37, 41
Planer pavement 718
   shop 362
Planing cold 525
Plank floor 148
   hollow precast concrete 68
   precast concrete roof 68
   precast concrete slab 68
   roof 148
   scaffolding 16
Plant and bulb transplanting . . . 545
   bed preparation 541
   ground cover 543
   screening 712
Planter 539
   concrete 539
   fiberglass 539
Planting 545
   price sheet 698

Plant-mix asphalt paving 531
Plaque bronze 331
Plaster 286, 624
   and drywall price sheet 624
   beam 291
   ceiling 291
   column 291
   cutout 30
   demolition 284, 285
   drilling 287
   ground 146
   gypsum 291
   mixer 715
   painting 325, 326
   partition thin 287
   perlite 291
   soffit 291
   thincoat 293
   venetian 292
   wall 291
Plasterboard 293
Plastic ball valve 404
   coupling 415
   duct fitting 441
   faced hardboard 164
   fence recycled 537
   fitting 414
   glazing 277
   laminate door 230
   lead 386
   lumber 172
   lumber structural 172
   matrix terrazzo 307
   outlet box 475
   pipe 413, 414
   pipe fitting 414, 416
   pipe removal 403
   railings 173
   sheet flashing 206
   skylight 262
   toilet compartment 335
   toilet partition 335
   valve 404
   window 258
Plastic-laminate toilet
compartments 334
   toilet components 334
   toilet door 334
   toilet headrail 334
   toilet panel 334
   toilet pilaster 334
Plastic-laminate-clad countertop . 373
Plate compactor 518
   glass 276
   heat exchanger 666
   roadway 718
   shear 136
   steel 105
   stiffener 105
   vibrating 518
   wall switch 487
Platform lifter 351
   trailer 716
Plating zinc 135
Player bench 540
Plenum demolition 285
Plexiglass 277
   acrylic 277
   mirror 277
Plow vibrator 713
Plug in tandem circuit breaker . . 484
   wall 83
Plugmold raceway 486
Plumbing 417
   appliance 424
   demolition 402

fixture 422, 425, 439
fixture removal 402
laboratory 361
system 643, 645, 647-652
Plywood 833
   clip 136
   demolition 284, 285
   fence 20
   floor 149
   formwork 50
   formwork steel framed 53
   joist 151
   mounting board 150
   paneling 165
   sheathing roof & walls 150
   shelving 344
   sidewalk 20
   siding 195
   sign 331
   soffit 164
   subfloor 149
   underlayment 149
Pneumatic nailer 714, 715
   tube system 395
Pocket door frame 170
   pitch 213
Point moil 714
   of use water heater 424
Pointing CMU 76
   masonry 76
Poisoning soil 519
Pole closet 163
Police connect panel 500
Polishing concrete 66
Polycarbonate glazing 277
Polyester room darkening shade . 368
Polyethylene backer rod 56
   floor 309
   pipe 554
   septic tank 552
   tarpaulin 19
   waterproofing 181
Polyolefin roofing thermoplastic . 201
Polypropylene siding 197
Polystyrene blind 171
   insulation 182, 184
Polysulfide caulking 217
Polyurethane caulking 217
   varnish 328
Polyvinyl soffit 164, 197
   tarpaulin 19
Polyvinyl-chloride roof 201
Porcelain tile 297
Porch door 233
   framing 141
Portable air-compressor 713
   asphalt plant 717
   chalkboard 330
   fire extinguisher 342, 343
   panel 337
   stage 362
Portland cement terrazzo 307
Positive pressure chimney 449
Post cap 136
   cedar 169
   demolition 132
   driver 717
   fence 536
   lamp 126
   light 491
   shores 54
   sign 332
   wood 138
Postal specialty 344
Postformed countertop 373
Potters wheel 362

Index

Poured gypsum underlayment ... 70
Powder actuated tool 103
   charge 103
   pin 103
Power mulcher 712
   temporary 14
   trowel 710
   wiring 487
Precast beam 69
   catch basin 553
   column 69
   concrete beam 68
   concrete column 69
   concrete lintel 69
   concrete plantable pavers . 533
   concrete roof plank 68
   concrete slab plank 68
   concrete stair 68
   concrete tee beam 69
   concrete wall 69
   concrete window sill 69
   coping 97
   curb 534
   grate 530
   members 829
   parking barrier 535
   terrazzo 307
Pre-engineered steel bldg demo . 381
Prefab metal flue 449
Prefabricated building 385
   fireplace 340
   glass block window 278
   pad 465
   panels 188
   wood stair 166
Prefinished door 230
   drywall 294
   floor 304
   hardboard paneling 164
   shelving 345
Preformed roof panel 193
   roofing & siding 193
Pre-hung door 236, 237
Preparation exterior surface ... 313
   interior surface 314
   lawn bed 540
   plant bed 541
Pressure grouting cement 520
   reducing valve water 404
   valve relief 403
   wash 314
   washer 716
Prestressed concrete piles ... 521
   pile 522
   precast concrete 829
Preventer backflow 417
Prices lumber products 833
Primer asphalt 178, 306
Prison cell 386
   cot 376
   fence 538
   hinge 270
   toilet 366
   work inside 11
Processing concrete 66
Proctor compaction test 13
Produce case 356
Product piping 554
Productivity glazing 837
Profile block 89
Profiler pavement 718
Profiling pavement 525
Project sign 21
Projected window aluminum 250
   window steel 251
   window steel sash 251

Projection screen 361
Propeller unit heaters 464
Protected membrane 202
Protection fire 342
   slope 519
   stile 272
   temporary 21
   termite 519
   winter 19, 49
   worker 38
Protector ADA insulated 405
P&T relief valve 403
Public address system 496
Pull box 475
   box electrical 475
   door 371
   plate 267
Pulverization pavement 525
Pump 550
   centrifugal 715
   circulating 439
   concrete 710
   diaphragm 715
   end suction 662
   gas driven heat 460
   grout 716
   heat 460
   in-line centrifugal 439
   jack 15
   mud 710
   sewage ejector 422
   shotcrete 710
   staging 15, 814
   submersible 423, 550, 715
   sump 355
   trash 716
   water 439, 550, 715
   wellpoint 719
Pumped concrete 64
Purlin roof 148
   steel 106
Push plate 267
Push-on joint 550
Push-pull 608
   plate 267
Putlog scaffold 16
Puttying 313
PVC adhesive 181
   conduit in slab 477
   control joint 83
   cornerboards 174
   door casing 174
   duct 441
   fascia 174
   fitting 414
   flashing 206
   frieze 174
   gravel stop 207
   louver 280
   molding exterior 174
   movable louver 280
   pipe 413, 551
   rake 174
   roof 201
   sheet 181, 309
   siding 196
   soffit 174
   trim 174
   waterstop 55

Q

Quarry drill 714
   tile 298
Quarter round molding 163

Quartz flooring 308
Quoin 95

R

Raceway 474, 476, 477
   plugmold 486
   surface 486
   wiremold 486
Rack coat 371
   hose 398
Radial arch 152
   wall formwork 53
Radiation fin tube 463
Radiator cast iron 463
Radio-frequency-interference
   mesh 261
Radiography test 13
Rafter 142
   anchor 136
   composite 143
   demolition 132
   framing metal 117
   metal bracing 117
   metal bridging 117
   wood 142
Rail aluminum pipe 122
   crane 560
   dock shelter 351
   galvanized pipe 122
   guide 558
   guide/guard 558
   ornamental aluminum 127
   ornamental glass 127
   ornamental steel 127
   ornamental wrought iron ... 127
   stainless pipe 122
   steel pipe 122
   wall pipe 122
Railing cable 126
   composite 175
   demolition 134
   encased 175
   encased composite 175
   industrial 123
   ornamental 127
   picket 122
   pipe 122
   plastic 173
   wood 163, 166, 168
   wood stair 167
Railroad tie 544
   tie step 530
   track removal 28
Raised floor 310
Rake PVC 174
   topsoil 540
   tractor 712
Rammer tamper 518
Rammer/tamper 711
Ramp handicap 62
Ranch plank floor 304
Range cooking 354
   hood 354
   receptacle 481, 487
   restaurant 358
Reach-in refrigeration 356
Ready mix concrete 63, 809, 827
Receiver ash 340
Receptacle air conditioner ... 481
   device 480
   dryer 481
   duplex 487
   GFI 480
   ground fault indicating ... 487

   locking 481
   range 481, 487
   telephone 481
   television 481
   trash 377
   waste 340
   weatherproof 481
Receptor floor 422
   shower 429
Reciprocating compressor 423
   hermetic compressor 423
   water chiller 456
Reclamation pavement 525
Recore cylinder 265, 266
Recovery package heat 458
Rectangular diffuser 446
   ductwork 440
Recycled plastic lumber 172
   rubber tire tile 375
Recycling cold 525
   pavement 525
Red bag 365
Reduce heat transfer glass ... 276
Redwood cupola 347
   louver 281
   paneling 166
   siding 195
   tub 383
   wine cellar 357
Refinish floor 304
Reflective block 93
   film heat 276
   glass 277
   insulation 185
   painting 535
   sign 331
Reflectorized barrels 714
Refrigerant compressor 456
   line sets 439
   removal 436
Refrigerated case 356
   storage equipment 356
   wine cellar 357
Refrigeration 384
   commercial 356
   reach-in 356
   residential 354
Refrigerator door 242
Refuse chute 395
Register air supply 447
   baseboard 447
   cash 351
   steel 447
Reglet aluminum 210, 613
   copper 613
   galvanized steel 210
   steel 613
Reinforced concrete pipe 553
   culvert 553
   plastic panel fiberglass .. 311
   PVC roof 201
Reinforcement welded wire 823
Reinforcing 822
   beam 58
   column 58
   concrete 58
   dowel 58
   elevated slab 58
   footing 58
   girder 58
   glass fiber rod 58, 59
   joint 82, 87
   ladder 82
   masonry 82
   sorting 58
   steel 822

Index

steel fiber 60
synthetic fiber 60
truss 82
wall 58
Relay electric 488
Release door 496
Relief valve P&T 403
valve self-closing 403
valve temperature 403
Relining sewer 548
Remediation plan 37, 41
plan/method asbestos 37
Removal air conditioner ... 436
asbestos 39
bathtub 402
block wall 28
boiler 436
catch basin 27
concrete 26
concrete pipe 27
curb 27
driveway 26
ductwork 436
fixture 402
floor 28
guide rail 27
hydrant 27
insulation 39, 179
lavatory 402
masonry 27, 78
paint 42
pipe 27
pipe insulation 39
plumbing fixture 402
railroad track 28
refrigerant 436
shingle 179
sidewalk 27
sink 402
sod 541
steel pipe 27
stone 27
tank 36
tree 545
urinal 403
utility line 27
VAT 39
water closet 403
water fountain 403
water heater 403
window 222
Removal/salvage material ... 33
Remove bolts 100
nuts 100
topsoil 540
Remove/replace catch
basin cover 553
Rendering 10
Renovation tread covers 124
Rental equipment 710
Repair asphalt pavement patch ... 526
asphalt paving crack 527
fire damage 315
slate 190
Repellent water 325
Replacement joint sealant 178
luminaire 468
pavement 532
windows 220, 835
Resaurant roof 178
Residential alarm 482
appliance 354, 482
applications 478
bathtubs 427
burner 452
canopy 345

closet door 228
device 478
dishwasher 354
door 170, 228, 233, 605
door bell 481
dryer 355
elevator 391
fan 482
fixture 481, 491
folding partition 337
gas water heater 425
greenhouse 385
gutting 32
heat pumps 483
hinge 269
kitchen sinks 426
lavatory 426, 429
lighting 481
load center 478
oil-fired water heater 425
overhead door 244
refrigeration 354
roof jack 445
service 478
service single phase 679
sink 426
smoke detector 482
stair 166
storm door 223
switch 478
transition 445
wall cap 445
wash bowl 426
washer 354
water heater 425, 483
water heater electric 424
wiring 478, 482
Resilient base 305
floor 305, 306
Resin crack filling 47
crack repair 47
Respirator 38
personal 37
Resquared shingle 190
Restaurant furniture 376
range 358
Restoration gypsum 284
masonry 79
mortar 79
window 42
Restroom plumbing
public men 645, 646
plumbing public women ... 643, 644
Retaining wall 62, 539
wall formwork 53
wall stone 539
wall timber 539
Retarder concrete 63
vapor 188
Retractable lateral arm awning ... 345
stair 355
Revolving door 242
door darkroom 242
Rib lath 290
Ribbed waterstop 55
Ridge board 143
cap 189
shingle slate 190
vent 212
Rig drill 26
Rigid anchor 81
conduit in trench 478
in slab conduit 478
insulation 182
joint sealant 218
metal-framed skylight 262

pavement patching 527, 528
pavement repair 527
pavement replacement 528
plastic duct 441
Ring boxing 363
split 136
toothed 136
Ripper attachment 717
Riser pipe wellpoint 720
rubber 305
stair 307
terrazzo 307
wood stair 167
Road base 531
berm 534
mixer 717
sign 332
sweeper 717
temporary 18
Roadway plate 718
Robe hook 339
Rock drill 26
trencher 713
Rocker switch 487
Rod backer 216
closet 344
curtain 338
ground 473
shower 338
tie 104
Roll roof 198
roofing 203
type air filter 448
up grille 241
Roller sheepsfoot 712
tandem 712
vibrating 517
vibratory 712
Rolling door 241, 249
service door 241
topsoil 540
tower scaffold 17
Romex copper 471
Roof accessories 213
accessories price sheet 612
adhesive 198
aluminum 193
baffle 213
beam 152
built-up 198
cant 143, 199
clay tile 191
coating 178
cold-applied built-up asphalt .. 198
copper 203
CSPE 200
deck 148
deck insulation 186
deck laminated 148
deck wood 148
decking 109
drain 423
drainage boot 207
elastomeric 202
exhauster 444
expansion joint 210
fan 444
fiberglass 194
fill 830
flashing vent chimney 450
frame opening 104
framing removal 30
hatch 212, 611
hatch removal 179
insulation 186
jack residential 445

joint 211
lead 203
membrane green 192
metal 193
metal rafter framing 117
metal tile 192
metal truss framing 119
mineral 203
modified bitumen 200
mop 199
nail 135
panel aluminum 193
panel preformed 193
patch 178
paver 213
paver and support 213
PIB 201
polyvinyl-chloride 201
purlin 148
PVC 201
rafter 142
rafter bracing 117
rafter bridging 117
rafter framing 142
reinforced PVC 201
resaurant 178
roll 198
safety anchor 17
sheathing 150
sheet metal 204
shingle 189
shingle asphalt 189
slate 190
soffit framing 118
soil mixture green 192
specialty prefab 207
stainless steel 204
steel 193
system green 192
thermoplastic polyolefin .. 201
tile 192
T.P.O. 201
truss 151, 152
truss wood 589, 590
vent 211
ventilator 212, 213
walkway 199
window metal 261
zinc 203
Roofing & siding preformed ... 193
built-up 609, 833
cold 198
demolition 39, 179
elastomeric 609
EPDM 200
finish price sheet 609
flat seam sheet metal 204
ketone ethylene ester 201
membrane 198, 202
modified bitumen 200
roll 203
single-ply 200
S.P.F. 203
system built-up 198
zinc copper alloy 203
Rooftop air conditioner 458
make-up air unit 458
multizone air conditioner ... 458
Room humidifier 465
Root raking loading 540
Rope exercise 363
safety line 17
Rotary crawler drill 714
hammer drill 714
Rototiller 712
Rough buck 140

Index

stone wall 94
Rough-in drinking
   fountain floor 432
      drinking fountain wall 432
      sink countertop 426
      sink raised deck 426
      sink service floor 429
      tub 427
Round duct 440
   rail fence 538
Rubber astragal 272
   base 305
   coating 182
   control joint 82
   floor 305
   floor tile 306
   nosing 305
   riser 305
   sheet 305
   stair 305
   threshold 268
   tile 306
   waterproofing 181
Rubberized asphalt 524
   asphalt sealcoat 524
Rubbish handling 30
   handling chutes 817
Rug cleaning 309
Run gravel bank 504

S

Safe office 353
Safety cabinet biological 361
   equipment laboratory 361
   fixture industrial 431
   line rope 17
   nosing 305
   railing cable 107
   shower 431
   switch 487
Salamander 716
Sales tax 11, 810
Salvage or disposal value 33
Sampling air 37
Sanctuary pew 377
Sand 79
   fill 504
   screened 79
   seal 524
Sandblast finish concrete 66
   masonry 77
Sandblasting equipment 716
Sander floor 716
Sanding 313
   floor 304
Sandstone 96
Sanitary base cove 297
   drain 421
   tee 420
Sash aluminum 250
   metal 251
   security 251
   steel 251
   wood 253
Sauna 384
   door 384
Saw blade concrete 71
   chain 716
   circular 716
   concrete 710
   cutting concrete floor 71
   cutting concrete wall 72
   cutting demolition 32
   cutting slab 71

masonry 716
shop 362
table 362
Sawn control joint 55
SBS modified bituminous
   membrane 200
Scaffold aluminum plank 17
   baseplate 16
   bracket 16
   caster 16
   catwalk 16
   frame 15
   guardrail 16
   mason 15
   putlog 16
   rolling tower 17
   specialties 16
   stairway 16
   wood plank 16
Scaffolding 813
   plank 16
   tubular 15
Scanner checkout 351
   unit facial 274
Scarify concrete 284
   subsoil 541
School door hardware 263
   equipment 363
   hinge 270
   panic device door hardware . 264
Scissor gate 336
   lift 713
Scoreboard baseball 364
   basketball 364
   football 364
   hockey 364
Scored block 90
   split face block 89
Scrape after damage 315
Scraper earth 712
Scratch coat 307
Screed, gas engine,
   8HP vibrating 710
Screen chimney 341
   entrance 332
   fence 538
   metal 251
   molding 163
   projection 361
   security 125, 251
   sight 337
   squirrel and bird 341
   urinal 336
   window 251, 257
   wood 257
Screened loam 540
   sand 79
   topsoil 541
Screening plant 712
Screen/storm door aluminum ... 223
Screw brass 135
   jack 519
   lag 102
   liquid chillers 456
   steel 135
   wood 135
Scroll water chiller 456
Seal compression 216
   curb 524
   fog 524
   pavement 524
   security 272
   slurry 525
Sealant 180
   acoustical 218, 295
   caulking 217

masonry joint 217
   replacement joint 178
   rigid joint 218
   tape 218
Sealcoat 524
Sealer floor 48
   joint 216, 217
Sealing crack asphalt paving ... 527
Seating movie 361
Sectional door 244
   overhead door 244
Security and vault equipment .. 353
   dome 347
   doors and frames 243
   drawer 366
   driveway gate 243
   driveway gate aluminum ... 243
   driveway gate steel 243
   driveway gate wood 243
   equipment 366
   fence 538
   film 278
   gate 336
   gate driveway 243
   gate opener driveway 243
   hinge 270
   sash 251
   screen 125, 251
   seal 272
   X-ray equipment 500
Seeding 542, 845
Seismic ceiling bracing 301
Selective demolition fencing 28
   demolition masonry 77
Self-closing relief valve 403
Self-contained air conditioner .. 459
Sentry dog 20
Separation barrier 38, 43
Septic galley 552
   system 551, 700
   system chamber 552
   system effluent filter 552
   tank 551
   tank concrete 551
   tank polyethylene 552
Service boring 549
   electric 487
   entrance cable aluminum .. 472
   entrance cap 472
   residential 478
   sink 429
   sink faucet 430
   station equipment 350
Sewage collection concrete pipe . 551
   collection PVC pipe 551
   ejector pump 422
   pipe 551, 553
Sewage/drainage collection ... 553
Sewer relining 548
Shade 368
   polyester room darkening .. 368
Shaft wall 286
Shake wood 191
Shear plate 136
   test 13
   wall 150
Shearwall panel 147
Sheathed nonmetallic cable .. 471
   Romex cable 471
Sheathing 150
   asphalt 150
   board 150
   gypsum 151
   insulating 149
   paper 188
   roof 150

roof & walls plywood 150
   wall 150
Sheepsfoot compactor 518
   roller 712
Sheet carpet 309
   glass 276
   glass mosaic 298
   lead 386
   metal 203
   metal aluminum 207
   metal cladding 204
   metal flashing 205
   metal roof 204
   piling 519
   steel piles 521
Sheeting driver 714
   wale 519
   wood 519, 845
Shelf bathroom 339
   library 376
   metal 344
Shellac door 321
Shell/tube heat exchanger 666
Shelter dock 351
   rail dock 351
   temporary 49
Shelving 344
   food storage metal 357
   pine 344
   plywood 344
   prefinished 345
   storage 344
   wood 344
Shield expansion 100
Shingle 189
   aluminum 189
   asphalt 189
   cedar 190
   concrete 191
   metal 190
   panelized 191
   removal 179
   roof 189
   slate 190
   steel 190
   strip 189
   wood 190
Ship ladder 121
Shock absorber 419
   absorber pipe 419
   absorbing door 221, 246
Shop drill 362
   equipment 362
Shore post 54
Shoring 519, 520
   aluminum joist 54
   baseplate 15
   bracing 15
   concrete formwork 54
   concrete vertical 54
   flying truss 54
   frames 54
   heavy duty 15
   leveling jack 15
   steel beam 54
Short load concrete 63
Shot blast floor 285
   blaster 716
Shotcrete 67
   pump 710
Shoulder paver 717
Shovel crawler 712
Shower by-pass valve 428
   cabinet 340
   cubicle 427
   door 340

Index

emergency ... 431
glass door ... 340
group ... 429
pan ... 205
partition ... 95, 340
receptor ... 429
rod ... 338
safety ... 431
spray ... 428
stall ... 427
Shower head water economizer ... 427
Shower/tub control set ... 428
Shower/tub valve spout set ... 428
Shrub and tree ... 543
broadleaf evergreen ... 543
deciduous ... 543
evergreen ... 543
moving ... 545
Shutter ... 368
exterior ... 171
interior ... 368
wood ... 171
wood interior ... 368
Sidelight ... 170
door ... 228, 236
Sidewalk ... 303, 533, 696
asphalt ... 530
brick ... 533
bridge ... 16
broom ... 716
concrete ... 530
driveway and patio ... 530
removal ... 27
temporary ... 20
Sidewall bracket ... 17
Siding aluminum ... 194
bevel ... 195
cedar ... 195
fiber cement ... 197
fiberglass ... 194
metal ... 193, 194
nail ... 135
paint ... 317
painting ... 317
panel aluminum ... 194
plywood ... 195
polypropylene ... 197
redwood ... 195
removal ... 180
stain ... 195
steel ... 195
vinyl ... 196
wood ... 195
wood board ... 195
Sign ... 21, 331
acrylic ... 331
base formwork ... 51
corridor ... 331
detour ... 20
directional ... 331
flexible ... 331
highway ... 332
plywood ... 331
post ... 332
project ... 21
reflective ... 331
road ... 332
traffic ... 331
Signage engraved panel ... 331
exterior ... 331
Signal bell fire ... 501
Silencer door ... 269
duct ... 443
Silicon carbide abrasive ... 48
Silicone ... 216
caulking ... 217
coating ... 182
water repellent ... 182
Sill ... 95
& ledger framing ... 143
anchor ... 136
block ... 92
door ... 164, 170, 268
masonry ... 96
precast concrete window ... 69
quarry tile ... 298
stone ... 94, 96
window ... 253
wood ... 143
Sillcock hose bibb ... 428
Silos demo ... 381
Silt fence ... 519
Simulated brick ... 97
stone ... 97, 98
Single hub pipe ... 419
hung aluminum sash window ... 250
hung aluminum window ... 250
zone rooftop unit ... 458
Single-ply roofing ... 200
Sink barber ... 352
base ... 369
countertop ... 426
countertop rough-in ... 426
darkroom ... 353
faucet kitchen ... 428
kitchen ... 426
laboratory ... 361
laundry ... 427
raised deck rough-in ... 426
removal ... 402
residential ... 426
service ... 429
service floor rough-in ... 429
waste treatment ... 422
Siren ... 500
Sister joist ... 139
Site demolition ... 26
furnishings ... 377
improvement ... 539, 540
preparation ... 26
Skidder log ... 712
Skid steer attachments ... 713
loader ... 713
Skylight ... 611
domed ... 262
domed metal-framed ... 262
metal-framed ... 262
plastic ... 262
removal ... 179
rigid metal-framed ... 262
roof hatch ... 611
Skyroof ... 249
Slab blockout formwork ... 52
box out opening formwork ... 50
bulkhead formwork ... 51, 52
concrete ... 61
concrete placement ... 64
curb formwork ... 51, 52
cutout ... 29
edge formwork ... 51, 52
elevated ... 61
flat plate formwork ... 50
haunch ... 62
haunch formwork ... 62
joist dome formwork ... 50
joist pan formwork ... 50
lift ... 62, 828
on grade ... 62, 573
on grade formwork ... 52
pan ... 61
saw cutting ... 71
screed formwork ... 52
stamping ... 66
textured ... 62
thickened edge ... 62
thickened edge formwork ... 62
trench formwork ... 52
turndown ... 62
turndown formwork ... 62
void formwork ... 51, 52
waffle ... 61
with drop panels formwork ... 50
Slate ... 96
flagging ... 534
repair ... 190
roof ... 190
shingle ... 190
sidewalk ... 534
sill ... 96
stair ... 96
tile ... 303
Slatwall ... 165, 311
Sleeper ... 144
framing ... 144
Sleeve anchor bolt ... 56
and tap ... 550
dowel ... 58
Slide gate ... 536
Sliding aluminum door ... 240
door ... 604
glass door ... 240
glass vinyl-clad door ... 240
mirror ... 340
panel door ... 249
vinyl-clad door ... 240
window ... 253
window aluminum ... 250
wood door ... 240
Slipform paver barrier ... 718
paver curb ... 717
Slope protection ... 519
Slot channel ... 81
letter ... 344
Slotted channel framing ... 104
Slump block ... 89
Slurry seal ... 525
seal (latex modified) ... 524
Smoke detector ... 501
hatch ... 212
vent ... 212
vent chimney ... 449
Snow fence ... 36
guard ... 213
Soap dispenser ... 339
holder ... 339
Socket meter ... 485
Sod ... 542
Sodding ... 542
Sodium high pressure fixture ... 490
low pressure fixture ... 492
Soffit ... 144
& canopy framing ... 144
drywall ... 294
marble ... 95
metal ... 197
molding ... 164
plaster ... 291
plywood ... 164
polyvinyl ... 164, 197
PVC ... 174
stucco ... 292
vent ... 211, 281
wood ... 164
Softener water ... 424
Soil compaction ... 507
decontamination ... 36
mix man-made ... 192
pipe ... 419
poisoning ... 519
tamping ... 507
test ... 13
treatment ... 519
Solar backflow preventer ... 455
energy ... 455
energy circulator air ... 454
energy system air purger ... 455
energy system air vent ... 455
energy system balancing valve ... 455
energy system collector ... 454
energy system control valve ... 455
energy system controller ... 454
energy system expansion tank ... 455
energy system gauge pressure ... 455
energy system heat exchanger ... 455
energy system storage tank ... 455
energy system thermometer ... 455
energy system vacuum relief ... 455
film glass ... 278
heating system ... 454
panel driveway security gate ... 243
system solenoid valve ... 455
Solid surface countertops ... 374
wood door ... 232
Sorting reinforcing ... 58
Sound absorbing panel ... 313
and video cable & fittings ... 496
attenuation blanket ... 312
control demo ... 381
dampening panel gypsum ... 295
movie ... 361
system ... 496
Source heat pumps water ... 461
Spa bath ... 384
Space heater ... 715
heater floor mounted ... 454
office & storage ... 14
Spade air ... 714
tree ... 713
Spandrel beam formwork ... 49
Spanish roof tile ... 191
Special door ... 238, 243
hinge ... 270
purpose rooms demo ... 381
Specialties ... 338
scaffold ... 16
telephone ... 332
S.P.F. roofing ... 203
Spike grid ... 136
unit traffic ... 351
Spinner ventilator ... 447
Spiral duct preformed ... 440
duct steel ... 440
stair ... 126, 166
Split astragal ... 272
rib block ... 89
ring ... 136
system ductless ... 459
Spotter ... 514
Spray coating ... 180
insulation ... 185
painting booth automotive ... 350
rig foam ... 714
shower ... 428
substrate ... 40
Sprayed cementitious fireproofing ... 213
Sprayed-on insulation ... 185
Sprayer airless ... 37
emulsion ... 714
paint ... 715
Spread footing ... 61, 571

Index

footing formwork 51
park soil 541
soil conditioner 541
topsoil 541
Spreader aggregate 710, 717
  chemical 717
  concrete 715
Spring bolt astragal 272
  bronze weatherstrip 272
  hinge 270
Sprinkler alarm 501
  connector flexible 399
  head 399
  irrigation system 540
  system 399, 675, 677
  system accelerator 399
  system automatic 540
  system dry-pipe 399
  underground 540
Stabilization masonry 76
Stabilizer base 718
Stacked bond block 90
Stage curtain 362
  equipment 362
  portable 362
Staging aids 17
  pump 15
  swing 18
Stain cabinet 320
  door 321
  floor 323
  lumber 152
  siding 195
  truss 324
Staining concrete floor 302
Stainless column cover 125
  cooling towers 457
  duct 440
  flashing 205
  gutter 209
  reglet 210
  screen 332
  steel cot 376
  steel countertop 373
  steel door 237
  steel doors and frame 237
  steel gravel stop 207
  steel hinge 269
  steel movable louver 280
  steel roof 204
  steel shelf 339
  steel storefront 249
Stair 630
  basement 166
  basement bulkhead 68
  brick 93
  carpet 309
  ceiling 355
  climber 394
  concrete 62
  disappearing 355
  electric 355
  fire escape 121
  floor plate 120
  grating 120
  landing 308
  marble 95
  metal pan 120
  metal plate 120
  nosing 305
  pan treads 65
  part wood 166
  precast concrete 68
  prefabricated wood 166
  railroad tie 531
  removal 133

residential 166
retractable 355
riser 307
riser vinyl 305
rubber 305
slate 96
spiral 126, 166
stringer 140, 308
stringer wood 140
temporary protection 20
tread 94, 123, 305
tread and riser 305
tread inserts 54
tread terrazzo 307
tread tile 298
wood 166
Stairlift wheelchair 394
Stairway door hardware 264
  scaffold 16
Stairwork handrails 168
Stall shower 427
  toilet 332, 334, 335
  type urinal 429
  urinal 429
Stamping slab 66
  texture 66
Standard extinguisher 343
Standing seam 203
Standpipe connection 398
Starter board & switch 484
Station diaper 338
Steam bath 384
  boiler 450, 451
  boiler electric 450
  clean masonry 77
  cleaner 716
  jacketed kettle 358
  trap 438
Steamer 359
Steel anchor 81
  astragal 272
  ballistic door 243
  beam shoring 54
  beam W-shape 105
  blocking 137
  bolt 136
  bridging 137
  conduit in slab 478
  conduit in trench 478
  conduit intermediate 474
  corner guard 338
  cutting 101
  demolition 100
  diffuser 447
  door 226, 227, 243, 835
  doors and frame stainless .. 237
  door residential 228
  downspout 208
  drill 714
  drilling 102
  edging 544
  estimating 831
  expansion tank 438
  fence 536, 537
  fiber 60
  fiber reinforcing 60
  fitting 410, 412
  flashing 205
  frame 223
  fuel tank double wall . 703, 704
  furring 288
  girts 106
  gravel stop 207
  gutter 209
  hinge 269
  joist 584

joist roof & ceiling 587
lath 290
light pole 705
lintel 104
locker 343
louver door 281
louver galvanized 281
member structural 105
movable louver stainless 280
nailer 140
outlet box 474
partition 337
pile 521
pile concrete-filled 521
piles sheet 521
pipe 409
pipe fitting 410, 412
pipe removal 27
plate 105
project structural 106
purlins 106
register 447
reinforcing 822
roof 193
roofing panel 193
sash 251, 835
screw 135
security driveway gate 243
sheet pile 519
shingle 190
siding 195
spiral duct 440
storage tank 703, 704
structural 831
stud NLB 288
tank above ground
  double wall 704
tank above ground
  single wall 704
window 250, 251
window demolition 222
Steeple 347
  aluminum 347
  tip pin 269
Step 332
  bluestone 530
  brick 530
  masonry 530
  railroad tie 530
  stone 95
  tapered pile 521
Sterilizer barber 352
Stiffener joist web 115
  plate 105
Stiffleg derrick crane 719
Stile fabric 272
  protection 272
Stone aggregate 543
  anchor 81
  ashlar 95
  base 95, 531
  cast 97
  crushed 504
  cultured 97
  curb cut 534
  fill 504
  floor 95
  ground cover 543
  paver 534
  removal 27
  retaining walls 539
  sill 94, 96
  simulated 97, 98
  step 95
  stool 96
  toilet compartment 335

tread 94
wall 94, 539, 592
Stool cap 164
  stone 96
  window 95, 96
Stop door 163
  gravel 207
  valve 428
  water supply 429
Storage bottle 357
  box 14
  cabinet 372
  metal shelving food 357
  room cold 384
  shelving 344
  tank demo 382
Storefront aluminum 249
Storm door 223
  door residential 223
  drainage manholes frame 553
  drainage piping 553
  window 257, 601
  window & door 601
  window aluminum residential . 257
Stove 342, 358
  wood burning 656
  woodburning 342
Strainer bronze body 438
  downspout 207
  gutter 210
  roof 208
  wire 208
  Y type bronze body 438
Strap hinge 270
  tie 136
Straw bale construction 385
Strength compressive 13
Stringer stair 140, 308
  stair terrazzo 307
Strip cabinet 497
  filler 272
  floor 304
  footing 61, 570
  lighting 489
  shingle 189
Striper paint 717, 718
Stripping topsoil 505
Structural backfill 507
  brick 85
  columns 104
  compaction 518
  excavation 506
  face tile 86
  facing tile 86
  fee 10
  insulated panel 146
  panel 146
  soil mixing 540
  steel 583, 831
  steel members 105
  steel projects 106
  tile 86
  welding 102
Structure moving 35
Stucco 292
  demolition 285
  mesh 290
  painting 317
Stud demolition 133
  NLB metal 288
  NLB steel 288
  NLB wall 288
  partition 140, 291
  partition framing load bearing . 111
  price sheet 622
  wall 140, 287

Index

wall blocking load-bearing ... 109
wall box beam load-bearing .. 110
wall bracing load-bearing 109
wall framing load-bearing 111
wall header load-bearing 110
wall wood 286
welded 103
Stump chipper 712
Subcontractor O&P 12
Subfloor 149
   adhesive 149
   plywood 149
   wood 149
Subflooring 634
Submersible pump 423, 550, 715
   sump pump 423
Subpurlin bulb tee 70
Suction hose 715
Sump pump 355
   pump elevator 423
   pump submersible 423
Supermarket checkout 351
   scanner 351
Supply ductile iron pipe water ... 550
Support ceiling 103
   framing pipe 105
   lavatory 430
   light 103
   monitor 103
   motor 103
   partition 103
   water closet 431
   X-ray 103
Suppressor lightning 478
Surface bonding 79
   flood coat concrete 47
   landscape 309
   preparation exterior 313
   preparation interior 314
   raceway 486
   repair concrete 47
   treatment 524
Surfacing 524
Surfactant 40
Survey crew 21
   topographic 26
Suspended acoustic ceiling tiles . 299
   ceiling 289, 300
   gypsum board ceiling 636
   space heater 668
Suspension mounted heater 454
   system ceiling 301
   system ceiling tile 610
Sweeper road 717
Swimming pool demo 382
   pool enclosure 385
Swing check valve 403
   check valve bronze 403
   clear hinge 270
   staging 18
Swing-up overhead door 244
Switch box 474
   box plastic 475
   decorator 479
   dimmer 479, 487
   electric 487, 488
   general duty 487
   load management 488
   residential 478
   rocker 487
   safety 487
   time 488
   toggle 487
Synthetic erosion control 519
   fiber reinforcing 60
   marble 303

System chamber septic 552
   components foam-water 399
   light pole 705
   septic 551
   solenoid valve solar 455
   sprinkler 399
   tube 395

T

Table laboratory 361
   saw 362
Tack coat 524
Tackboard fixed 330
Tamper 505, 714
   compactor 518
   rammer 518
Tamping soil 507
Tandem roller 712
Tank clean 36
   darkroom 353
   developing 353
   disposal 36
   expansion 438
   removal 36
   septic 551
   testing 13
   water 717
   water storage solar 455
Tankless electric water heaters .. 424
   gas water heaters 425
Tap and sleeve 550
Tape barricade 20
   detection 549
   sealant 218
   temporary 536
   tracer 549
   underground 549
Tapping crosses and sleeves ... 550
   main 550
Tar kettle 716, 718
   pitch emulsion 524
   roof 178
Tarpaulin 19
   duck 19
   Mylar 19
   polyethylene 19
   polyvinyl 19
Tarred felt 199
Tax 11
   sales 11
   social security 11
   unemployment 11
T bar mount diffuser 446
Teak floor 304
   molding 163
   paneling 165
Technician inspection 13
Tee cleanout 405
   pipe 413
   precast concrete beam 69
Telephone enclosure 332
   enclosure commercial 332
   house 496
   receptacle 481
   specialties 332
Telescoping bleacher 377
Television receptacle 481
   system 496
Temperature relief valve 403
   rise detector 501
Tempered glass 275
   hardboard 164
Tempering valve 404
   valve water 404, 417

Temporary barricade 19
   building 14
   construction 14, 19
   electricity 14
   facility 19
   fence 20
   guardrail 20
   heat 49
   light 14
   lighting 14
   power 14
   protection 21
   road 18
   road painting 536
   shelter 49
   tape 536
   toilet 716
   utilities 14
Tennis court 537
   court fences and gates 537
Tension structures demos 383
Termination box 496
   mineral insulated cable ... 471
Termite control chemical 519
   protection 519
Terne coated flashing 205
Terra cotta 87
   cotta coping 87, 97
   cotta demolition 30
   cotta tile 87
Terrazzo base 307
   cove base 307
   curb 307
   demolition 285
   epoxy 308
   floor 307
   floor tile 307
   plastic matrix 307
   precast 307
   riser 307
   venetian 307
   wainscot 308
Test load pile 504
   moisture content 13
   pile load 504
   soil 13
   ultrasonic 13
Testing 13
   lead 41
   OSHA 40
   pile 504
   tank 13
Textile wall covering 310
Texture stamping 66
Textured slab 62
Theater and stage equipment ... 362
Thermal & moisture
   protection demo 179
Thermometer solar
   energy system 455
Thermoplastic 277
   painting 536
   polyolefin roof 201
   polyolefin roofing 201
Thermostat 437
   automatic timed 437
   integral 464
   wire 483
Thickened edge slab 62
Thin brick veneer 84
   plaster partition 287
Thincoat plaster 293
Thinset ceramic tile 297
   mortar 298
Threshold 268, 608
   aluminum 268

cherry 268
door 170
oak 268
ramp aluminum 269
stone 95
walnut 268
wood 164
Thru-the-wall A/C 674
   heat pumps 674
Ticket dispenser 350
Tie adjustable wall 80
   column 81
   flange 81
   rafter 143
   railroad 544
   rod 104
   strap 136
   wall 81
   wire 81
Tile 297, 299, 305, 631
   acoustic 86
   aluminum 192, 298
   carpet 309
   ceiling 300
   ceramic 297
   clay 191
   concrete 191
   cork 306
   cork wall 310
   demolition 284
   fire rated 86
   glass 277
   grout 298
   marble 303
   metal 298
   or terrazzo base floor 307
   partition 87
   porcelain 297
   quarry 298
   recycled rubber tire 375
   roof 192
   roof metal 192
   rubber 306
   slate 303
   stainless steel 298
   stair tread 298
   structural 86
   terra cotta 87
   vinyl composition 306
   wall 297
   waterproofing memb ceramic . 299
   window sill 298
Tiling ceramic 297, 299
Tilling topsoil 541
Timber connector 135
   fastener 135
   framing 147
   guide rail 558
   heavy 147
   laminated 152
   pile 521
   retaining wall 539
Time switch 488
Timer clock 482
   interval 479
   switch ventilator 445
Tin ceiling 301
Tinted glass 275
Toaster 359
Toggle switch 487
toggle bolt ANCHOR 101
Toilet accessories 338
   accessories commercial 338
   bowl 425
   chemical 716
   compartment 332

Index

compartment plastic 335
compartment plastic-laminate . 334
compartment stone 335
dual flush valve 430
gasket 428
partition 95, 332, 334
partition removal 286
prison 366
stall 332, 334, 335
stone partition 335
temporary 716
tissue dispenser 339
Tool powder actuated 103
Toothed ring 136
Toothing masonry 76
Top demolition counter 134
vanity 374
Topographical surveys 26
Topping concrete 70
epoxy 308
floor 70
grout 71
Topsoil 504
placement and grading 542
remove 540
screened 541
stripping 505
Torch cutting 101, 716
cutting demolition 32
Torrified door 235
Touch bar 265
Towel bar 339
dispenser 338
Tower cooling 457
crane 15, 718, 815, 816
hoist 719
light 715
T.P.O. roof 201
Tracer tape 549
Track drawer 372
drill 714
lighting 491
Tractor 507
loader 712
rake 712
truck 716
Traffic cone 19
control mirror 347
line 535
line remover 718
sign 331
spike unit 351
Trailer floodlight 714
lowbed 718
mud 717
office 14
platform 716
truck 716
water 716
Trainer bicycle 363
Tram car 716
Transformer cabinet 485
dry-type 483
Transition molding wood 304
residential 445
strip floor 308
Transom aluminum 248
lite frame 224
windows 255
Transplanting plant and bulb ... 545
Trap cast iron 421
drainage 421
grease 422
inverted bucket steam 438
steam 438

Trapezoid windows 256
Trash compactor 360
pump 716
receptacle 377
Travertine 95
Tread abrasive 298
and riser stair 305
bar grating 123
channel grating 124
cover renovation 124
diamond plate 123
inserts stair 54
rubber 305
stair 94, 123, 305
stair pan 65
stone 95, 96
vinyl 305
wood stair 167
Treadmill 363
Treated lumber framing 144
pile 521
pine fence 538
Treatment acoustical 312
flooring 302
surface 524
Tree deciduous 544
evergreen 543
grate 530
guying system 545
moving 545
removal 545
spade 713
Trench backfill 505
box 716
disposal field 552
excavating 505
excavation 505
utility 505
Trencher chain 505, 713
rock 713
wheel 713
Trim demolition 134
exterior 157, 158
exterior paint 319
molding 163
painting 324
PVC 174
tile 297
wood 159
Triple weight hinge 270
Tripod floodlight 714
Trowel coating 180
concrete 710
power 710
Truck boom 718
concrete 710
crane flatbed 718
dock 351
dump 713
flatbed 713, 716
holding concrete 63
loading 506
mounted crane 718
pickup 716
tractor 716
trailer 716
vacuum 717
winch 717
Truss bowstring 152
demolition 134
framing metal 119
painting 324
plate 136
reinforcing 82
roof 151, 152

stain 324
varnish 324
Tub bar 339
hot 383
redwood 383
rough-in 427
shower handicap 427
Tube and wire kit 460
fitting copper 408
framing 246
system 395
system air 395
system pneumatic 395
Tubing connection split system 460
copper 407
electric metallic 474
Tub & shower module 427
Tubular fence 537
lock 265
scaffolding 15
Tugboat diesel 720
Tumbler holder 339
Tunnel locomotive 718
muck car 718
ventilator 718
Turndown slab 62
Turned column 169
Turning vane 443
T.V. antenna 496
inspection sewer pipeline .. 548
Two piece astragal 272

U

Ultrasonic motion detector 500
test 13
Uncased drilled concrete piers .. 522
Undereave vent 281
Underground duct 555
electrical & telephone 555
sprinkler 540
storage tank removal 818
tape 549
Underlayment 634
acoustical 286
hardboard 149
latex 305
self-leveling cement 70
Underpin foundation 520
Unemployment tax 11
Union dielectric 438
Unit air-handling 461
commercial kitchen 355
heater 454
heater propeller 464
line set evap/condens 460
Urethane foam 312
wall coating 328
Urinal 429
high efficiency 430
removal 403
screen 336
stall 429
stall type 429
wall hung 429
watersaving 429
Utility accessories 549
box 549
connection 549
duct 555
electric 555
line removal 27
set fan 444
sitework 555

temporary 14
trench 505
trench excavating 505
UTP cable 496

V

Vacuum breaker 418
breaker air gap 418
catch basin 717
central 364
cleaning 364
loader 38
pick-up 710
truck 717
wet/dry 717
Valance board 370
Valley flashing 189
rafter 143
Valve 403, 436, 455
assembly dry pipe sprinkler . 399
bronze 403
butterfly 436, 550
CPVC 404
fire hose 399
flush 430
gate 436
iron body 436
mixing 417, 428
motorized zone 437
plastic 404
relief pressure 403
shower by-pass 428
spout set shower/tub 428
sprinkler alarm 398
stop 428
swing check 403
tempering 404
water pressure 404
water supply 429
Vandalproof fixture 490
Vane air turning 443
turning 443
Vanity base 371
top 374
top lavatory 426
Vapor barrier 188
barrier sheathing 150
retarder 188
Vaportight fixture
incandescent 490
Variable frequency drives ... 488
volume damper 442
Varnish 631
cabinet 320
casework 321
door 321
floor 324, 326
polyurethane 328
truss 324
VAT removal 39
Vaulting side horse 363
VAV boxes 446
mixing box 446
VCT removal 285
Vehicle guide rails 558
Veneer ashlar 94
brick 83
core paneling 165
granite 94
member laminated 152
removal 78
Venetian plaster 292
terrazzo 307

Index

Vent box 83
   chimney 449
   chimney all fuel 449
   chimney fitting 449
   chimney flashing 450
   damper 449
   draft damper 449
   eave 211
   exhaust 447
   flashing 421
   foundation 83
   gas 449
   metal chimney 449
   one-way 213
   ridge 212
   ridge strip 281
   roof 211
   smoke 212
   soffit 211, 281
Ventilating air
  conditioning 442-445,
                     456, 457, 469
Ventilation equipment
  and hood 448
   fan 482
   louver 281
Ventilator masonry 83
   roof 212, 213
   spinner 447
   timer switch 445
   tunnel 718
Verge board 162
Vermiculite insulation 185
Vertical lift door 245
   metal ladder 121
Vibrating plate 518
   roller 517
   screed, gas engine, 8HP ... 710
Vibrator concrete 710
   earth 505, 507, 518
   plate 507
   plow 713
Vibratory hammer 712
   roller 712
Video surveillance 501
Vinyl blind 171
   casement window 259
   clad half round window 255
   clad premium window 256
   composition floor 306
   composition tile 306
   curtain 338
   faced wallboard 294
   floor 306
   gutter 209
   picture window 260
   replacement window 220
   sheet floor 305
   siding 196
   stair riser 305
   stair tread 305
   tread 305
   wall coating 328
   wallpaper 311
   window 258
   window double hung 258
   window half round 260
   window single hung 258
Vision panel 261
Vocational kiln 362
   shop equipment 362
Volume-control damper 442
VRV indoor AHU 462
   outdoor unit 460

W

Waffle slab 61
Wainscot molding 163
   quarry tile 298
   terrazzo 308
Wale sheeting 519
Walk 530
Walkway luminaire 492
   roof 199
Wall & corner guards 338
   aluminum 194
   and partition demolition .. 285
   base 305
   blocking load-bearing stud . 109
   box beams load-bearing
     stud 110
   boxout formwork 52
   bracing load-bearing stud . 109
   brick 93
   brick shelf formwork 52
   bulkhead formwork 52
   bumper 266
   buttress formwork 52
   cabinet 370, 372
   canopy 346
   cap residential 445
   ceramic tile 297
   coating 327, 631
   concrete 61
   concrete block 577, 591
   concrete finishing 66
   concrete placement 65
   corbel formwork 52
   covering textile 310
   cross 366
   cutout 29
   demolition 285
   drywall 293
   finishes 310
   forms 820
   formwork 52
   foundation 90
   framing 144
   framing load-bearing stud . 111
   framing removal 30
   furnace 454
   furring 146, 288
   grout 80
   header load-bearing stud .. 110
   hung lavatory 426
   hung urinal 429
   insulation 182
   lath 290
   lintel formwork 53
   louver 281, 282
   masonry 88, 539
   mat 363
   mirror 277
   mounted AHU 463
   NLB partition 286
   painting 317, 325
   panel 97
   paneling 165
   partition 286
   pilaster formwork 53
   plaster 291
   plug 83
   plywood formwork 52
   precast concrete 69
   prefab plywood formwork ... 53
   reinforcing 58
   retaining 62, 539
   screen 337
   shaft 286
   shear 150
   sheathing 150
   steel framed formwork 53
   stone 94, 539
   stucco 292
   stud 140, 287
   stud NLB 288
   switch plate 487
   tie 81
   tie masonry 80
   tile 297
   tile cork 310
   window 247
   wood frame 595
   wood stud 286
Wallboard repair gypsum 284
Wall covering 310, 631
   acrylic 311
   gypsum fabric 311
Wall covering natural fiber .. 310
Wallpaper 310
   grass cloth 311
   removal 311
   vinyl 311
Walnut floor 304
   threshold 268
Wardrobe 344
   wood 371
Wash bowl residential 426
   brick 77
   fountain 430
   fountain group 430
   pressure 314
Washable air filter 448
   air filter permanent 448
Washer 136
   and extractor 352
   commercial 352
   darkroom 353
   pressure 716
   residential 354
Waste cart medical 365
   cleanup hazardous 37
   disposal 43
   disposal hazardous 37
   handling 365
   handling equipment 365
   packaging 40
   receptacle 340
   treatment sink 422
Watchdog 20
Watchman service 20
Water atomizer 38
   blasting 77
   chiller direct expansion .. 457
   chiller reciprocating 456
   chiller scroll 456
   closet 425, 429
   closet removal 403
   closet support 431
   cooler 432
   cooler electric 432
   curing concrete 68
   distribution pipe 550
   fountain removal 403
   hammer arrester 419
   heater 483
   heater commercial 424
   heater gas 425
   heater gas residential 425
   heater heat pumps 460
   heater instantaneous 424
   heater insulation 405
   heater oil-fired 425
   heater point of use 424
   heater removal 403
   heater residential 425, 483
   heater residential electric . 424
   heater under the sink 424
   heater wrap kit 405
   heating hot 450, 463
   hose 715
   hydrant 419
   motor alarm 399
   pipe ground clamp 473
   pressure reducing valve ... 404
   pressure relief valve 404
   pressure valve 404
   pump 439, 550, 715
   reducer concrete 63
   repellent 325
   repellent coating 182
   repellent silicone 182
   softener 424
   source heat pumps 461
   storage solar tank 455
   supply domestic meter 417
   supply ductile iron pipe .. 550
   supply meter 417
   supply stop 429
   supply valve 429
   tank 717
   tempering valve 404, 417
   trailer 716
   well 550
Waterproofing 181
   butyl 181
   cementitious 181
   coating 180
   demolition 180
   elastomeric 181
   elastomeric sheet 181
   masonry 80
   membrane 180, 524
   neoprene 181
   rubber 181
Watersaving urinal 429
Waterstop PVC 55
   ribbed 55
Wearing course 531
Weather barrier 188
   barrier or wrap 188
Weatherproof receptacle 481
Weatherstrip 241
   aluminum 272
   door 273
   spring bronze 272
   window 274
   zinc 273
Weatherstripping 273, 608
Weathervane residential 347
Web stiffeners framing 115
Wedge anchor 101
Weight exercise 363
   lifting multi-station 363
Weights and measures 808
Weld joint pipe 410
Welded frame 224
   structural steel 831
   stud 103
   wire epoxy 59
   wire fabric 59, 823
   wire galvanized 59
Welder arc 362
Welding certification 13
   fillet 102
   machine 717
   structural 102
Well & accessories 550
   area window 347

Index

pump 550
   water 550
Wellpoint discharge pipe 719
   equipment rental 719
   header pipe 719
   pump 719
   riser pipe 720
Wet/dry vacuum 717
Wheel potters 362
   trencher 713
Wheelbarrow 717
Wheelchair lift 394
   stairlift 394
Wheeled loader 712
Whirlpool bath 384
Wide throw hinge 270
Widener pavement 717
Winch truck 717
Window 250, 262
   & door demolition 221
   air conditioner 459
   aluminum 250
   aluminum projected 250
   aluminum residential storm ... 257
   aluminum sliding 250
   and door molding 163
   awning 251, 345
   bank 353
   blind 368
   bow & bay metal-clad 254
   casement 250-252, 254
   casement wood 252
   demolition 222
   double hung 255
   double hung aluminum sash .. 250
   double hung bay wood 254
   double hung steel sash 251
   double hung vinyl 258
   double hung wood 252
   fiberglass 260
   fiberglass bay 261
   fiberglass single hung 260
   fiberglass slider 260
   frame 253
   glass 276
   grille 256
   guard 125
   half round vinyl 260
   half round vinyl-clad 255
   hardware 263
   hurricane resistant 279
   industrial 251
   metal 250, 251
   metal-clad awning 253
   metal-clad casement 253
   metal-clad double-hung 254
   muntin 256
   painting 321
   picture 253
   picture & sliding metal-clad .. 254
   pivoted 251
   plastic 258
   precast concrete sill 69
   prefabricated glass block 278
   protection film 278
   removal 222
   replacement 220
   restoration 42
   sash wood 253
   screen 251, 257
   sill 96

sill marble 95
sill tile 298
single hung aluminum 250
single hung aluminum sash ... 250
single hung vinyl 258
sliding 253
sliding wood 253
steel 250, 251
steel projected 251
steel sash picture 251
steel sash projected 251
stool 95, 96
storm 257
transom 255
trapezoid 256
trim set 164
vinyl 258
vinyl casement 259
vinyl-clad 598
vinyl-clad awning 255
vinyl-clad double hung 598
vinyl-clad half round 255
vinyl-clad premium 256
vinyl picture 260
vinyl replacement 220
wall 247, 602
wall aluminum 247
wall framing 246
weatherstrip 274
well area 347
wood 251
wood awning 251
wood bow 254
wood double hung 252
Windrow loader 718
Wine cellar 357
Winter concrete 63
   protection 19, 49
Wire copper 472
   electrical 472
   fabric welded 59
   fence 20, 538
   glass 276
   ground 473
   mesh 292
   mesh locker 343
   reinforcement 823
   strainer 208
   thermostat 483
   THW 472
   tie 81
Wiremold raceway 486
   raceway non-metallic 486
Wireway 475
Wiring air conditioner 483
   device 475, 486, 487
   device price sheet 680
   fan 482
   methods 483
   power 487
   residential 478, 482
Wood & composite I-joists 151
   awning window 251
   ballistic door 243
   base 153
   beam 137, 147
   bench 539
   block floor 304
   block floor demolition 285
   blocking 137, 141
   bow window 254

bracing 137
canopy 144
casework frame 370
ceiling 301
column 138, 147, 169
composition floor 304
cupola 347
cutting 32
deck 148
demolition 284
door 229, 232
door decorator 231
door frame exterior 170
door frame interior 170
door residential 233
double acting swing door 245
double hung window 252
drawer 371
drawer front 371
exterior surface 596
fascia 160
fastener 135
fiber ceiling 300
fiber plank cementitious 70
fiber sheathing 150
fiber underlayment 149
floor 304
floor demolition 285
folding partition 338
frame 370
frame roof & ceiling 588
frame wall 595
framing 137, 145, 147
framing demolition 130
framing miscellaneous 139
framing open web joist 139
furring 146
girder 137
gutter 209
handrail 163, 168
interior shutter 368
joist 139, 151, 586
joist floor/ceiling 586
joist open web 585
laminated 152
louver 168
nailer 140
overhead door 244
panel door 232
paneling 164
parking bumpers 535
parquet 304
partition 140
piles 521
plank scaffold 16
post 138
rafter 142
rail fence 538
railing 163, 166, 168
roof deck 148
roof truss 151
sash 253
screen 257
screw 135
security driveway gate 243
shade 368
shake 191
sheet piling 845
sheeting 519, 845
shelving 344
shingle 190

shutter 171
sidewalk 530
siding 195
siding demolition 180
sill 143
soffit 164
stair 166
stair baluster 166
stair newel 167
stair parts 166
stair railing 167
stair riser 167
stair stringer 140
stair tread 167
storm door 237
strip flooring 304
subfloor 149
threshold 164
trim 159
veneer wallpaper 310
wardrobe 371
window 251, 600
window casement 252
window demolition 222
window double hung 252
window double hung bay ... 254
window sash 253
window sliding 253
Woodburning stove 342
Wool carpet 309
   fiberglass 184
Work boat 720
   extra 12
   inside prisons 11
Worker protection 38
Workers' compensation 812
Wrecking ball 717
Wrench impact 714
Wrestling mat 363

X

X-ray barrier 387
   concrete slab 14
   equipment 500
   equipment security 500
   machine 500
   protection 386
   support 103
X-ray/radio freq
   protection demo 383

Y

Y type bronze body strainer ... 438
Yard fountains 383
Yellow pine floor 304

Z

Z bar suspension 301
Zee bar 296
Zinc copper alloy roofing 203
   divider strip 307
   plating 135
   roof 203
   terrazzo strip 307
   weatherstrip 273

Notes

Notes

Notes

Notes

Notes

Notes

Notes

Division Notes

		CREW	DAILY OUTPUT	LABOR-HOURS	UNIT	BARE COSTS				TOTAL INCL O&P
						MAT.	LABOR	EQUIP.	TOTAL	

Division Notes

		CREW	DAILY OUTPUT	LABOR-HOURS	UNIT	BARE COSTS				TOTAL INCL O&P
						MAT.	LABOR	EQUIP.	TOTAL	

Division Notes

	CREW	DAILY OUTPUT	LABOR-HOURS	UNIT	BARE COSTS				TOTAL INCL O&P
					MAT.	LABOR	EQUIP.	TOTAL	

Division Notes

	CREW	DAILY OUTPUT	LABOR-HOURS	UNIT	BARE COSTS MAT.	LABOR	EQUIP.	TOTAL	TOTAL INCL O&P

Division Notes

		CREW	DAILY OUTPUT	LABOR-HOURS	UNIT	BARE COSTS				TOTAL INCL O&P
						MAT.	LABOR	EQUIP.	TOTAL	

Division Notes

Division Notes

	CREW	DAILY OUTPUT	LABOR-HOURS	UNIT	BARE COSTS				TOTAL INCL O&P
					MAT.	LABOR	EQUIP.	TOTAL	

Other RSMeans Products & Services

RSMeans—a tradition of excellence in construction cost information and services since 1942

For more information visit the RSMeans website at www.rsmeans.com

Table of Contents
Annual Cost Guides
RSMeans Online and *CostWorks* CD Comparison Matrix
Seminars
Reference Books
Order Form

Unit prices according to the latest MasterFormat!

Book Selection Guide

The following table provides definitive information on the content of each cost data publication. The number of lines of data provided in each unit price or assemblies division, as well as the number of crews, is listed for each book. The presence of other elements such as reference tables, square foot models, equipment rental costs, historical cost indexes, and city cost indexes, is also indicated. You can use the table to help select the RSMeans book that has the quantity and type of information you most need in your work.

Unit Cost Divisions	Building Construction	Mechanical	Electrical	Commercial Renovation	Square Foot	Site Work Landsc.	Green Building	Interior	Concrete Masonry	Open Shop	Heavy Construction	Light Commercial	Facilities Construction	Plumbing	Residential
1	573	393	410	506		509	191	312	456	572	501	246	1042	403	178
2	777	279	85	733		993	178	399	213	776	734	481	1221	286	257
3	1690	340	230	1085		1471	987	355	2032	1689	1685	481	1788	316	388
4	957	21	0	922		720	179	612	1155	925	612	529	1172	0	441
5	1893	158	155	1090		841	1799	1095	720	1893	1037	979	1909	204	721
6	2452	18	18	2110		110	589	1528	281	2448	123	2140	2124	22	2660
7	1584	215	128	1623		581	757	532	524	1581	26	1317	1685	227	1037
8	2082	81	45	2609		265	1123	1753	105	2077	0	2193	2879	0	1526
9	1971	72	26	1767		313	449	2056	390	1911	15	1659	2220	54	1435
10	1026	17	10	642		215	27	842	158	1026	29	511	1118	235	224
11	1087	208	165	546		124	54	932	28	1071	0	230	1107	169	110
12	544	0	2	319		217	137	1654	14	532	0	352	1704	23	310
13	719	140	137	242		343	111	248	66	695	244	83	746	94	79
14	273	36	0	221		0	0	257	0	273	0	12	293	16	6
21	90	0	16	37		0	0	250	0	87	0	68	426	431	220
22	1157	7544	150	1190		1568	1067	838	20	1121	1677	857	7440	9347	712
23	1177	6971	580	928		158	902	776	38	1101	111	878	5187	1918	469
26	1354	454	10081	1021		793	611	1136	55	1349	562	1298	10030	399	617
27	72	0	297	34		13	0	71	0	72	39	52	279	0	22
28	96	58	136	71		0	21	78	0	99	0	40	151	44	25
31	1498	735	610	803		3217	289	7	1208	1443	3256	600	1560	660	611
32	822	53	8	886		4426	356	405	291	793	1841	420	1701	165	468
33	530	1079	538	252		2178	41	0	237	522	2159	128	1699	1286	154
34	107	0	47	4		190	0	0	31	62	193	0	128	0	0
35	18	0	0	0		327	0	0	0	18	442	0	84	0	0
41	60	0	0	32		7	0	22	0	60	30	0	67	14	0
44	75	79	0	0		0	0	0	0	0	0	0	75	75	0
46	23	16	0	0		274	261	0	0	23	264	0	33	33	0
48	12	0	25	0		0	0	25	0	0	12	17	12	0	12
Totals	24719	18967	13899	19673		19853	10154	16158	8022	24231	15597	15566	49880	16421	12682

Assem Div	Building Construction	Mechanical	Electrical	Commercial Renovation	Square Foot	Site Work Landscape	Assemblies	Green Building	Interior	Concrete Masonry	Heavy Construction	Light Commercial	Facilities Construction	Plumbing	Asm Div	Residential
A		15	0	188	150	577	598	0	0	536	571	154	24	0	1	376
B		0	0	848	2498	0	5658	56	329	1975	368	2089	174	0	2	211
C		0	0	647	926	0	1304	0	1629	146	0	816	250	0	3	588
D		1067	941	712	1859	72	2538	330	825	0	0	1345	1105	1088	4	851
E		0	0	86	260	0	300	0	5	0	0	257	5	0	5	392
F		0	0	0	114	0	114	0	0	0	0	114	3	0	6	357
G		527	447	318	249	3365	729	0	0	535	1350	205	293	677	7	307
															8	760
															9	80
															10	0
															11	0
															12	0
Totals		1609	1388	2799	6056	4014	11241	386	2788	3192	2289	4980	1854	1765		3922

Reference Section	Building Construction Costs	Mechanical	Electrical	Commercial Renovation	Square Foot	Site Work Landscape	Assem.	Green Building	Interior	Concrete Masonry	Open Shop	Heavy Construction	Light Commercial	Facilities Construction	Plumbing	Resi.
Reference Tables	yes	yes	yes	yes	no	yes	yes	yes	yes	yes	yes	yes	yes	yes	yes	yes
Models					111			25					50			28
Crews	571	571	571	550		571		571	571	571	548	571	548	550	571	548
Equipment Rental Costs	yes	yes	yes	yes		yes		yes	yes	yes	yes	yes	yes	yes	yes	yes
Historical Cost Indexes	yes	yes	yes	yes	yes	yes	yes	yes	yes	yes	yes	yes	yes	yes	yes	no
City Cost Indexes	yes	yes	yes	yes	yes	yes	yes	yes	yes	yes	yes	yes	yes	yes	yes	yes

RSMeans Building Construction Cost Data 2015

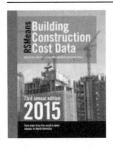

Offers you unchallenged unit price reliability in an easy-to-use format. Whether used for verifying complete, finished estimates or for periodic checks, it supplies more cost facts better and faster than any comparable source. More than 24,700 unit prices have been updated for 2015. The City Cost Indexes and Location Factors cover more than 930 areas, for indexing to any project location in North America. Order and get *RSMeans Quarterly Update Service* FREE.

$206.95 | Available Sept. 2014 | Catalog no. 60015

RSMeans Green Building Cost Data 2015

Estimate, plan, and budget the costs of green building for both new commercial construction and renovation work with this fifth edition of *RSMeans Green Building Cost Data*. More than 10,000 unit costs for a wide array of green building products plus assemblies costs. Easily identified cross references to LEED and Green Globes building rating systems criteria.

$170.95 | Available Nov. 2014 | Catalog no. 60555

RSMeans Mechanical Cost Data 2015

Total unit and systems price guidance for mechanical construction . . . materials, parts, fittings, and complete labor cost information. Includes prices for piping, heating, air conditioning, ventilation, and all related construction.

Plus new 2015 unit costs for:

- Thousands of installed HVAC/controls, sub-assemblies and assemblies
- "On-site" Location Factors for more than 930 cities and towns in the U.S. and Canada
- Crews, labor, and equipment

$203.95 | Available Oct. 2014 | Catalog no. 60025

RSMeans Facilities Construction Cost Data 2015

For the maintenance and construction of commercial, industrial, municipal, and institutional properties. Costs are shown for new and remodeling construction and are broken down into materials, labor, equipment, and overhead and profit. Special emphasis is given to sections on mechanical, electrical, furnishings, site work, building maintenance, finish work, and demolition.

More than 49,800 unit costs, plus assemblies costs and a comprehensive Reference Section are included.

$524.95 | Available Nov. 2014 | Catalog no. 60205

RSMeans Square Foot Costs 2015
Accurate and Easy-to-Use

- **Updated price information** based on nationwide figures from suppliers, estimators, labor experts, and contractors.
- Green building models
- Realistic graphics, offering true-to-life illustrations of building projects
- Extensive information on using square foot cost data, including sample estimates and alternate pricing methods

$218.95 | Available Oct. 2014 | Catalog no. 60055

RSMeans Commercial Renovation Cost Data 2015
Commercial/Multi-family Residential

Use this valuable tool to estimate commercial and multi-family residential renovation and remodeling.

Includes: **Updated costs for hundreds of unique methods, materials, and conditions that only come up in repair and remodeling, PLUS:**

- Unit costs for more than 19,600 construction components
- Installed costs for more than 2,700 assemblies
- More than 930 "on-site" localization factors for the U.S. and Canada

$166.95 | Available Oct. 2014 | Catalog no. 60045

RSMeans Electrical Cost Data 2015

Pricing information for every part of electrical cost planning. More than 13,800 unit and systems costs with design tables; clear specifications and drawings; engineering guides; illustrated estimating procedures; complete labor-hour and materials costs for better scheduling and procurement; and the latest electrical products and construction methods.

- A variety of special electrical systems, including cathodic protection
- Costs for maintenance, demolition, HVAC/mechanical, specialties, equipment, and more

$208.95 | Available Oct. 2014 | Catalog no. 60035

RSMeans Electrical Change Order Cost Data 2015

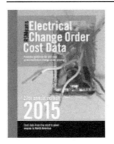

RSMeans Electrical Change Order Cost Data provides you with electrical unit prices exclusively for pricing change orders—based on the recent, direct experience of contractors and suppliers. Analyze and check your own change order estimates against the experience others have had doing the same work. It also covers productivity analysis, and change order cost justifications. With useful information for calculating the effects of change orders and dealing with their administration.

$199.95 | Available Dec. 2014 | Catalog no. 60235

RSMeans Assemblies Cost Data 2015

RSMeans Assemblies Cost Data takes the guesswork out of preliminary or conceptual estimates. Now you don't have to try to calculate the assembled cost by working up individual component costs. We've done all the work for you.

Presents detailed illustrations, descriptions, specifications, and costs for every conceivable building assembly—over 350 types in all—arranged in the easy-to-use UNIFORMAT II system. Each illustrated "assembled" cost includes a complete grouping of materials and associated installation costs, including the installing contractor's overhead and profit.

$334.95 | Available Sept. 2014 | Catalog no. 60065

RSMeans Open Shop Building Construction Cost Data 2015

The latest costs for accurate budgeting and estimating of new commercial and residential construction . . . renovation work . . . change orders . . . cost engineering.

RSMeans Open Shop "BCCD" will assist you to:

- Develop benchmark prices for change orders.
- Plug gaps in preliminary estimates and budgets.
- Estimate complex projects.
- Substantiate invoices on contracts.
- Price ADA-related renovations.

$175.95 | Available Dec. 2014 | Catalog no. 60155

For more information visit the RSMeans website at www.rsmeans.com

RSMeans Residential Cost Data 2015

Contains square foot costs for 28 basic home models with the look of today, plus hundreds of custom additions and modifications you can quote right off the page. Includes more than 3,900 costs for 89 residential systems. Complete with blank estimating forms, sample estimates, and step-by-step instructions.

Contains line items for cultured stone and brick, PVC trim, lumber, and TPO roofing.

$148.95 | Available Oct. 2014 | Catalog no. 60175

RSMeans Site Work & Landscape Cost Data 2015

Includes unit and assemblies costs for earthwork, sewerage, piped utilities, site improvements, drainage, paving, trees and shrubs, street openings/repairs, underground tanks, and more. Contains more than 60 types of assemblies costs for accurate conceptual estimates.

Includes:

- Estimating for infrastructure improvements
- Environmentally-oriented construction
- ADA-mandated handicapped access
- Hazardous waste line items

$199.95 | Available Dec. 2014 | Catalog no. 60285

RSMeans Facilities Maintenance & Repair Cost Data 2015

RSMeans Facilities Maintenance & Repair Cost Data gives you a complete system to manage and plan your facility repair and maintenance costs and budget efficiently. Guidelines for auditing a facility and developing an annual maintenance plan. Budgeting is included, along with reference tables on cost and management, and information on frequency and productivity of maintenance operations.

The only nationally recognized source of maintenance and repair costs. Developed in cooperation with the Civil Engineering Research Laboratory (CERL) of the Army Corps of Engineers.

$461.95 | Available Nov. 2014 | Catalog no. 60305

RSMeans Light Commercial Cost Data 2015

Specifically addresses the light commercial market, which is a specialized niche in the construction industry. Aids you, the owner/designer/contractor, in preparing all types of estimates—from budgets to detailed bids. Includes new advances in methods and materials.

Assemblies Section allows you to evaluate alternatives in the early stages of design/planning.

More than 15,500 unit costs ensure that you have the prices you need, when you need them.

$153.95 | Available Nov. 2014 | Catalog no. 60185

RSMeans Concrete & Masonry Cost Data 2015

Provides you with cost facts for virtually all concrete/masonry estimating needs, from complicated formwork to various sizes and face finishes of brick and block—all in great detail. The comprehensive Unit Price Section contains more than 8,000 selected entries. Also contains an Assemblies [Cost] Section, and a detailed Reference Section that supplements the cost data.

$188.95 | Available Dec. 2014 | Catalog no. 60115

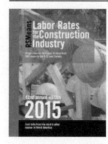

RSMeans Labor Rates for the Construction Industry 2015

Complete information for estimating labor costs, making comparisons, and negotiating wage rates by trade for more than 300 U.S. and Canadian cities. With 46 construction trades in each city, and historical wage rates included for comparison. Each city chart lists the county and is alphabetically arranged with handy visual flip tabs for quick reference.

$450.95 | Available Dec. 2014 | Catalog no. 60125

RSMeans Construction Cost Indexes 2015

What materials and labor costs will change unexpectedly this year? By how much?

- Breakdowns for 318 major cities
- National averages for 30 key cities
- Expanded five major city indexes
- Historical construction cost indexes

$359.95 per year (subscription) | Catalog no. 50145
$94.95 individual quarters | Catalog no. 60145 A,B,C,D

RSMeans Interior Cost Data 2015

Provides you with prices and guidance needed to make accurate interior work estimates. Contains costs on materials, equipment, hardware, custom installations, furnishings, and labor for new and remodel commercial and industrial interior construction, including updated information on office furnishings, and reference information.

$208.95 | Available Nov. 2014 | Catalog no. 60095

RSMeans Heavy Construction Cost Data 2015

A comprehensive guide to heavy construction costs. Includes costs for highly specialized projects such as tunnels, dams, highways, airports, and waterways. Information on labor rates, equipment, and materials costs is included. Features unit price costs, systems costs, and numerous reference tables for costs and design.

$204.95 | Available Dec. 2014 | Catalog no. 60165

RSMeans Plumbing Cost Data 2015

Comprehensive unit prices and assemblies for plumbing, irrigation systems, commercial and residential fire protection, point-of-use water heaters, and the latest approved materials. This publication and its companion, *RSMeans Mechanical Cost Data*, provide full-range cost estimating coverage for all the mechanical trades.

Contains updated costs for potable water, radiant heat systems, and high efficiency fixtures.

$206.95 | Available Oct. 2014 | Catalog no. 60215

RSMeans Online and RSMeans CostWorks CD Comparison Matrix

Both the RSMeans CostWorks CD package and the RSMeans Online estimating tool provide the same comprehensive, up-to-date, and localized cost information to improve planning, estimating, and budgeting with RSMeans—the industry's most-trusted source of construction cost data.

In addition to key differences noted in the comparison matrix below, both products let you do the following:

- Quickly locate costs in the searchable RSMeans database.
- Instantly adjust the data to your specific location for accurate costs anywhere in the U.S. and Canada.
- Create reliable, updated cost estimates in minutes.
- Manage, share, and export your estimates to MS Excel® with ease.
- Access a wide range of reference materials, including crew rates and national average labor rates.

Both the CD and Online options are available in a convenient one-year subscription!

Use the handy comparison matrix below to help you decide the best product for your cost estimating needs!

Why us?
For more than 70 years, RSMeans has provided the quality cost data construction professionals everywhere depend on to estimate with confidence, save time, improve decision-making, and increase profits.

Improve planning and decision-making
Back your estimates with complete, accurate, and up-to-date cost data for informed business decisions.
- Verify construction costs from third parties.
- Check validity of subcontractor proposals.
- Evaluate material and assembly alternatives.

Increase profits
Use RSMeans to estimate projects quickly and accurately, so you can gain an edge over your competition.
- Create accurate and competitive bids.
- Minimize the risk of cost overruns.
- Reduce variability.
- Gain control over costs.

The following matrix summarizes the key comparison points between RSMeans' two electronic cost data products.

	Question	CostWorks CD	RSMeans Online.com
1	How is the data presented?	• The data is organized by book title and sold at comparable book cost. • Several packaged bundles of five titles are available at a cost savings. • Available in a convenient one-year subscription	• The data is organized by book title and sold at comparable book cost. • Several packaged bundles of five titles are available at a cost savings. • RSMeans Online also offers a Complete Library title with all online data at a cost savings. • Available in a convenient one-year subscription
2	Where is the data stored?	With a typical install, the data is read off of your hard drive or off of the CD.	All data is stored on our secure server.
3	How are updates delivered?	Updates are available on a quarterly basis for download from the Costworks HomePort website.	Data is updated automatically on a quarterly basis, and users are notified when updates occur.
4	Where are my projects and estimates stored?	Upon installation, a "My CostWorks Projects" folder is created within the "My Documents" directory. All projects are defaulted to save here unless otherwise specified.	Estimates are stored on our secure server.
5	Does my IT team need to assist me in installing and running the program?	Yes, depending on the level of your permission. See #11 for system requirements.	While the product is entirely Web-based, you may need IT support with permissions for certain tasks. See #11 for system requirements.
6	Can users collaborate on projects/estimates?	Yes, you can copy project folders between computers.	You can create shared folders that can be viewed, copied, and edited by anyone within your account workgroup.

Comparison matrix continued

7	Can I create and use my own data with the RSMeans database?	Custom lines cannot be added to the CD database. With the Estimator tool, however, you can create and add custom cost lines to any estimate. You can also save custom lines as an Estimator template for use in creating new estimates.	You can create • a custom line within an estimate • a custom cost book consisting of custom unit (not assembly) cost lines, no more than once per quarter <u>Note</u> that custom data is searchable using the Include My Custom Data feature.
8	Can I enter my own trade/labor rates and markups?	Although you cannot adjust the trade labor rates, you can add custom markups for each report template using the Estimator tool.	For each quarterly update, you can create one custom unit cost book. You can edit the labor rates by individual trade and the crew equipment costs, and also edit the associated material and equipment markups.
9	What are my reporting options?	The Estimator tool lets you create editable custom reports and report templates. Reports can also be exported to MS Excel®.	You can generate pre-formatted estimate summaries and cost lists in both PDF and Excel formats.
10	How do I find specific cost data?	You can search by CSI line number and by keyword within a division.	You can search by CSI line number and by keyword within a division. There are several additional search options, including keyword and advanced search using special characters.
11	What are the system requirements?	• Operating system – Windows XP or later • PC requirements - 256 MB RAM (recommended), 1024 x 768 High Color Monitor, 800MHz Computer (recommended), 24X CD-ROM, 250 MB Hard Disk Space	• Resolution – Recommended resolution is 1024 x 768 or higher • Compatibility - Internet Explorer version 8.x, 9.x and Mozilla Firefox version 10 and higher • iPad - iOS 5 and higher

2015 RSMeans Seminar Schedule
Note: call for exact dates and details.

Location	Dates	Location	Dates
Seattle, WA	January and August	Jacksonville FL	September
Dallas/Ft. Worth, TX	January	Dallas, TX	September
Austin, TX	February	Charleston, SC	October
Anchorage, AK	March and September	Houston, TX	October
Las Vegas, NV	March and October	Atlanta, GA	November
Washington, DC	April and September	Baltimore, MD	November
Phoenix, AZ	April	Orlando, FL	November
Kansas City, MO	April	San Diego, CA	December
Toronto	May	San Antonio, TX	December
Denver, CO	May	Raleigh, NC	December
San Francisco, CA	June		
Bethesda, MD	June		
Columbus, GA	June	**1-800-334-3509, Press 1**	
El Segundo, CA	August		

Professional Development

eLearning Training Sessions

Learn how to use *RSMeans Online*® or *RSMeans CostWorks*® CD from the convenience of your home or office. Our eLearning training sessions let you join a training conference call and share the instructors' desktops, so you can view the presentation and step-by-step instructions on your own computer screen. The live webinars are held from 9 a.m. to 4 p.m. eastern standard time, with a one-hour break for lunch.

For these sessions, you must have a computer with high speed Internet access and a compatible Web browser. Learn more at www.rsmeansonline.com or call for a schedule: 1-800-334-3509 and press 1. Webinars are generally held on selected Wednesdays each month.

RSMeans Online®   **RSMeans CostWorks**® **CD**
$299 per person   **$299 per person**

RSMeans Online™ Training

Construction estimating is vital to the decision-making process at each state of every project. RSMeansOnline works the way you do. It's systematic, flexible and intuitive. In this one day class you will see how you can estimate any phase of any project faster and better.

Some of what you'll learn:
- Customizing RSMeansOnline
- Making the most of RSMeans "Circle Reference" numbers
- How to integrate your cost data
- Generate reports, exporting estimates to MS Excel, sharing, collaborating and more

Also available: RSMeans Online® training webinar

Facilities Construction Estimating

In this *two-day* course, professionals working in facilities management can get help with their daily challenges to establish budgets for all phases of a project.

Some of what you'll learn:
- Determining the full scope of a project
- Identifying the scope of risks and opportunities
- Creative solutions to estimating issues
- Organizing estimates for presentation and discussion
- Special techniques for repair/remodel and maintenance projects
- Negotiating project change orders

Who should attend: facility managers, engineers, contractors, facility tradespeople, planners, and project managers.

Construction Cost Estimating: Concepts and Practice

This *one-day* introductory course to improve estimating skills and effectiveness starts with the details of interpreting bid documents, ending with the summary of the estimate and bid submission.

Some of what you'll learn:
- Using the plans and specifications for creating estimates
- The takeoff process—deriving all tasks with correct quantities
- Developing pricing, using various sources; how subcontractor pricing fits in
- Summarizing the estimate to arrive at the final number
- Formulas for area and cubic measure, adding waste and adjusting productivity to specific projects
- Evaluating subcontractors' proposals and prices
- Adding Insurance and bonds
- Understanding how labor costs are calculated
- Submitting bids and proposals

Who should attend: project managers, architects, engineers, owner's representatives, contractors, and anyone who's responsible for budgeting or estimating construction projects.

Maintenance & Repair Estimating for Facilities

This *two-day* course teaches attendees how to plan, budget, and estimate the cost of ongoing and preventive maintenance and repair for existing buildings and grounds.

Some of what you'll learn:
- The most financially favorable maintenance, repair, and replacement scheduling and estimating
- Auditing and value engineering facilities
- Preventive planning and facilities upgrading
- Determining both in-house and contract-out service costs
- Annual, asset-protecting M&R plan

Who should attend: facility managers, maintenance supervisors, buildings and grounds superintendents, plant managers, planners, estimators, and others involved in facilities planning and budgeting.

Practical Project Management for Construction Professionals

In this *two-day* course, acquire the essential knowledge and develop the skills to effectively and efficiently execute the day-to-day responsibilities of the construction project manager.

Covers:
- General conditions of the construction contract
- Contract modifications: change orders and construction change directives
- Negotiations with subcontractors and vendors
- Effective writing: notification and communications
- Dispute resolution: claims and liens

Who should attend: architects, engineers, owner's representatives, project managers.

Mechanical & Electrical Estimating

This *two-day* course teaches attendees how to prepare more accurate and complete mechanical/electrical estimates, avoiding the pitfalls of omission and double-counting, while understanding the composition and rationale within the RSMeans mechanical/electrical database.

Some of what you'll learn:
- The unique way mechanical and electrical systems are interrelated
- M&E estimates—conceptual, planning, budgeting, and bidding stages
- Order of magnitude, square foot, assemblies, and unit price estimating
- Comparative cost analysis of equipment and design alternatives

Who should attend: architects, engineers, facilities managers, mechanical and electrical contractors, and others who need a highly reliable method for developing, understanding, and evaluating mechanical and electrical contracts.

RSMeans data titles are also available in online format! Go to RSMeansOnline.com for more details.

For more information visit the RSMeans website at www.rsmeans.com

Unit Price Estimating

This interactive *two-day* seminar teaches attendees how to interpret project information and process it into final, detailed estimates with the greatest accuracy level.

The most important credential an estimator can take to the job is the ability to visualize construction and estimate accurately.

Some of what you'll learn:
- Interpreting the design in terms of cost
- The most detailed, time-tested methodology for accurate pricing
- Key cost drivers—material, labor, equipment, staging, and subcontracts
- Understanding direct and indirect costs for accurate job cost accounting and change order management

Who should attend: corporate and government estimators and purchasers, architects, engineers, and others who need to produce accurate project estimates.

RSMeans CostWorks® CD Training

This one-day course helps users become more familiar with the functionality of *RSMeans CostWorks* program. Each menu, icon, screen, and function found in the program is explained in depth. Time is devoted to hands-on estimating exercises.

Some of what you'll learn:
- Searching the database using all navigation methods
- Exporting RSMeans data to your preferred spreadsheet format
- Viewing crews, assembly components, and much more
- Automatically regionalizing the database

This training session requires you to bring a laptop computer to class.

When you register for this course you will receive an outline for your laptop requirements.

Also offering web training for RSMeans CostWorks CD!

Facilities Estimating Using RSMeans CostWorks® CD

Combines hands-on skill building with best estimating practices and real-life problems. Brings you up-to-date with key concepts, and provides tips, pointers, and guidelines to save time and avoid cost oversights and errors.

Some of what you'll learn:
- Estimating process concepts
- Customizing and adapting RSMeans cost data
- Establishing scope of work to account for all known variables
- Budget estimating: when, why, and how
- Site visits: what to look for—what you can't afford to overlook
- How to estimate repair and remodeling variables

This training session requires you to bring a laptop computer to class.

Who should attend: facility managers, architects, engineers, contractors, facility tradespeople, planners, project managers and anyone involved with JOC, SABRE, or IDIQ.

Conceptual Estimating Using RSMeans CostWorks® CD

This *two-day* class uses the leading industry data and a powerful software package to develop highly accurate conceptual estimates for your construction projects. All attendees must bring a laptop computer loaded with the current year *Square Foot Costs* and the *Assemblies Cost Data* CostWorks titles.

Some of what you'll learn:
- Introduction to conceptual estimating
- Types of conceptual estimates
- Helpful hints
- Order of magnitude estimating
- Square foot estimating
- Assemblies estimating

Who should attend: architects, engineers, contractors, construction estimators, owner's representatives, and anyone looking for an electronic method for performing square foot estimating.

Assessing Scope of Work for Facility Construction Estimating

This *two-day* practical training program addresses the vital importance of understanding the SCOPE of projects in order to produce accurate cost estimates for facility repair and remodeling.

Some of what you'll learn:
- Discussions of site visits, plans/specs, record drawings of facilities, and site-specific lists
- Review of CSI divisions, including means, methods, materials, and the challenges of scoping each topic
- Exercises in SCOPE identification and SCOPE writing for accurate estimating of projects
- Hands-on exercises that require SCOPE, take-off, and pricing

Who should attend: corporate and government estimators, planners, facility managers, and others who need to produce accurate project estimates.

Unit Price Estimating Using RSMeans CostWorks® CD

Step-by-step instruction and practice problems to identify and track key cost drivers—material, labor, equipment, staging, and subcontractors—for each specific task. Learn the most detailed, time-tested methodology for accurately "pricing" these variables, their impact on each other and on total cost.

Some of what you'll learn:
- Unit price cost estimating
- Order of magnitude, square foot, and assemblies estimating
- Quantity takeoff
- Direct and indirect construction costs
- Development of contractor's bill rates
- How to use *RSMeans Building Construction Cost Data*

This training session requires you to bring a laptop computer to class.

Who should attend: architects, engineers, corporate and government estimators, facility managers, and government procurement staff.

Registration Information

Register early . . . Save up to $100! Register 30 days before the start date of a seminar and save $100 off your total fee. *Note: This discount can be applied only once per order. It cannot be applied to team discount registrations or any other special offer.*

How to register Register by phone today! The RSMeans toll-free number for making reservations is **1-800-334-3509, Press 1.**

Two-day seminar registration fee - $935. One-day *RSMeans CostWorks*® training registration fee - $375. To register by mail, complete the registration form and return, with your full fee, to: RSMeans Seminars, 700 Longwater Drive, Norwell, MA 02061.

Government pricing All federal government employees save off the regular seminar price. Other promotional discounts cannot be combined with the government discount.

Team discount program for two to four seminar registrations. Call for pricing: 1-800-334-3509, Press 1

Multiple course discounts When signing up for two or more courses, call for pricing.

Refund policy Cancellations will be accepted up to ten business days prior to the seminar start. There are no refunds for cancellations received later than ten working days prior to the first day of the seminar. A $150 processing fee will be applied for all cancellations. Written notice of cancellation is required. Substitutions can be made at any time before the session starts. **No-shows are subject to the full seminar fee.**

AACE approved courses Many seminars described and offered here have been approved for 14 hours (1.4 recertification credits) of credit by the AACE International Certification Board toward meeting the continuing education requirements for recertification as a Certified Cost Engineer/ Certified Cost Consultant.

AIA Continuing Education We are registered with the AIA Continuing Education System (AIA/CES) and are committed to developing quality learning activities in accordance with the CES criteria. Many seminars meet the AIA/CES criteria for Quality Level 2. AIA members may receive (14) learning units (LUs) for each two-day RSMeans course.

Daily course schedule The first day of each seminar session begins at 8:30 a.m. and ends at 4:30 p.m. The second day begins at 8:00 a.m. and ends at 4:00 p.m. Participants are urged to bring a hand-held calculator, since many actual problems will be worked out in each session.

Continental breakfast Your registration includes the cost of a continental breakfast, and a morning and afternoon refreshment break. These informal segments allow you to discuss topics of mutual interest with other seminar attendees. (You are free to make your own lunch and dinner arrangements.)

Hotel/transportation arrangements RSMeans arranges to hold a block of rooms at most host hotels. To take advantage of special group rates when making your reservation, be sure to mention that you are attending the RSMeans seminar. You are, of course, free to stay at the lodging place of your choice. **(Hotel reservations and transportation arrangements should be made directly by seminar attendees.)**

Important Class sizes are limited, so please register as soon as possible.

Note: Pricing subject to change.

Registration Form

ADDS-1000

Call 1-800-334-3509, Press 1 to register or FAX this form 1-800-632-6732. Visit our website: www.rsmeans.com

Please register the following people for the RSMeans construction seminars as shown here. We understand that we must make our own hotel reservations if overnight stays are necessary.

☐ Full payment of $_____ enclosed.

☐ Bill me.

Please print name of registrant(s).
(To appear on certificate of completion)

_____
_____
_____
_____

P.O. #: _____
GOVERNMENT AGENCIES MUST SUPPLY PURCHASE ORDER NUMBER OR TRAINING FORM.

Firm name_____

Address_____

City/State/Zip_____

Telephone no._____ Fax no._____

E-mail address_____

Charge registration(s) to: ☐ MasterCard ☐ VISA ☐ American Express

Account no._____ Exp. date_____

Cardholder's signature_____

Seminar name_____

Seminar City_____

Please mail check to: RSMeans Seminars, 700 Longwater Drive, Norwell, MA 02061 USA

For more information visit the RSMeans website at www.rsmeans.com

Managing the Profitable Construction Business: The Contractor's Guide to Success and Survival Strategies, 2nd Ed.
by Thomas C. Scheifler, Ph.D.

Learn from a team of construction business veterans led by Thomas C. Schleifer, who is commonly referred to as a construction business "turnaround" expert due to the number of construction companies he has rescued from financial distress. His financial acumen, combined with his practical, hands-on experience, has made him a sought-after private consultant.

$60.00 | 288 pages, hardover | Catalog no. 67370

Construction Project Safety
by John Schaufelberger, Ken-Yu Lin

This essential introduction to construction safety for construction management personnel takes a project-based approach to present potential hazards in construction and their mitigation or prevention.

Beginning with an introduction to Accident Prevention Programs and OSHA compliance requirements, the book integrates safety instruction into the building process by following a building project from site construction through interior finish. Reinforcing this applied approach are photographs, drawings, contract documentation, and an online 3D BIM model to help visualize the on-site scenarios.

$85.00 | 320 pages, hardcover | Catalog no. 67368

Project Control: Integrating Cost & Schedule in Construction
by Wayne J. Del Pico

Written by a seasoned professional in the field, *Project Control: Integrating Cost and Schedule in Construction* fills a void in the area of project control as applied in the construction industry today. It demonstrates how productivity models for an individual project are created, monitored, and controlled, and how corrective actions are implemented as deviations from the baseline occur.

$65.00 | 240 pages, softcover | Catalog no. 67366

Project Scheduling and Management for Construction, 4th Edition
by David R. Pierce

First published in 1988 by RSMeans, the new edition of *Project Scheduling and Management for Construction* has been substantially revised for both professionals and students enrolled in construction management and civil engineering programs. While retaining its emphasis on developing practical, professional-level scheduling skills, the new edition is a relatable, real world case study.

$95.00 | 272 pages, softcover | Catalog no. 67367

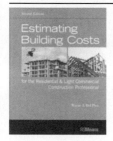

Estimating Building Costs, 2nd Edition
For the Residential & Light Commercial Construction Professional
by Wayne J. Del Pico

Explains, in detail, how to put together a reliable estimate that can be used not only for budgeting, but also for developing a schedule, managing a project, dealing with contingencies, and, ultimately, making a profit.

Completely revised and updated to reflect the CSI MasterFormat 2010 system.

$75.00 | 528 pages, softcover | Catalog no. 67343A

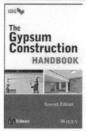

The Gypsum Construction Handbook, 7th Edition
by USG

The tried-and-true Gypsum Construction Handbook is a systematic guide to selecting and using gypsum drywall, veneer plaster, tile backers, ceilings, and conventional plaster building materials. A widely respected training text for aspiring architects and engineers, the book provides detailed product information and efficient installation methodology.

The 7th edition features updates in gypsum products, including ultralight panels, glass-mat panels, paper-faced plastic bead, and ultralightweight joint compound, and modern specialty acoustical and ceiling product guidelines. This comprehensive reference also incorporates the latest in sustainable products.

$40.00 | 576 pages, softcover | Catalog no. 67357B

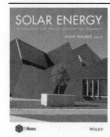

Solar Energy: Technologies & Project Delivery for Buildings
by Andy Walker

An authoritative reference on the design of solar energy systems in building projects, with applications, operating principles, and simple tools for the construction, engineering, and design professionals, the book simplifies the solar design and engineering process and provides sample documentation for the complete design of a solar energy system for buildings.

$85.00 | 320 pages, hardcover | Catalog no. 67365

RSMeans Cost Data, Student Edition

Provides a thorough introduction to cost estimating in a print and online package. Clear explanations and example-driven. The ideal reference for students and new professionals. Features include:

- Commercial and residential construction cost data in print and online formats
- Complete how-to guidance on the essentials of cost estimating
- A supplemental website with plans, problem sets, and a full sample estimate

$110.00 | 512 pages, softcover | Catalog no. 67363

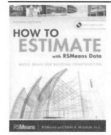

How to Estimate with Means Data & CostWorks, 4th Edition
by RSMeans and Saleh A. Mubarak, Ph.D.

This step-by-step guide takes you through all the major construction items with extensive coverage of site work, concrete and masonry, wood and metal framing, doors and windows, and other divisions. The only construction cost estimating handbook that uses the most popular source of construction data, RSMeans, this indispensible guide features access to the instructional version of *CostWorks* in electronic form, enabling you to practice techniques to solve real-world estimating problems.

$75.00 | 320 pages, softcover | Includes CostWorks CD | Catalog no. 67324C

RSMeans Estimating Handbook, 3rd Edition

Widely used in the industry for tasks ranging from routine estimates to special cost analysis projects. This handbook will help construction professionals:

- Evaluate architectural plans and specifications
- Prepare accurate quantity takeoffs
- Compare design alternatives and costs
- Perform value engineering
- Double-check estimates and quotes
- Estimate change orders

$110.00 | Over 976 pages, hardcover | Catalog No. 67276B

Value Engineering: Practical Applications
For Design, Construction, Maintenance & Operations

by Alphonse Dell'Isola, PE

A tool for immediate application—for engineers, architects, facility managers, owners, and contractors. Includes making the case for VE—the management briefing; integrating VE into planning, budgeting, and design; conducting life cycle costing; using VE methodology in design review and consultant selection; and case studies.

$90.00 | Over 427 pages, illustrated, softcover | Catalog no. 67319A

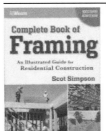

Complete Book of Framing, 2nd Edition

by Scot Simpson

This updated, easy-to-learn guide to rough carpentry and framing is written by an expert with more than thirty years of framing experience. Starting with the basics, this book begins with types of lumber, nails, and what tools are needed, followed by detailed, fully illustrated steps for framing each building element. Framer-Friendly Tips throughout the book show how to get a task done right—and more easily.

$29.95 | 368 pages, softcover | Catalog no. 67353A

Risk Management for Design and Construction

Introduces risk as a central pillar of project management and shows how a project manager can prepare to deal with uncertainty. Includes:

- Integrated cost and schedule risk analysis
- An introduction to a ready-to-use system of analyzing a project's risks and tools to proactively manage risks
- A methodology that was developed and used by the Washington State Department of Transportation
- Case studies and examples on the proper application of principles
- Combining value analysis with risk analysis

$125.00 | Over 288 pages, softcover | Catalog no. 67359

Green Home Improvement

by Daniel D. Chiras, PhD

With energy costs rising and environmental awareness increasing, people are looking to make their homes greener. This book, with 65 projects and actual costs and projected savings, helps homeowners prioritize their green improvements.

Projects range from simple water savers that cost only a few dollars, to bigger-ticket items such as HVAC systems. With color photos and cost estimates, each project compares options and describes the work involved, the benefits, and the savings.

$34.95 | 320 pages, illustrated, softcover | Catalog no. 67355

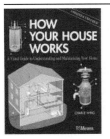

How Your House Works, 2nd Edition

by Charlie Wing

Knowledge of your home's systems helps you control repair and construction costs and makes sure the correct elements are being installed or replaced. This book uncovers the mysteries behind just about every major appliance and building element in your house. See-through, cross-section drawings in full color show you exactly how these things should be put together and how they function, including what to check if they don't work. It just might save you having to call in a professional.

$22.95 | 208 pages, softcover | Catalog no. 67351A

Life Cycle Costing for Facilities

by Alphonse Dell'Isola and Dr. Stephen Kirk

Guidance for achieving higher quality design and construction projects at lower costs! Cost-cutting efforts often sacrifice quality to yield the cheapest product. Life cycle costing enables building designers and owners to achieve both. The authors of this book show how LCC can work for a variety of projects—from roads to HVAC upgrades to different types of buildings.

$110.00 | 396 pages, hardcover | Catalog no. 67341

Construction Business Management

by Nick Ganaway

Only 43% of construction firms stay in business after four years. Make sure your company thrives with valuable guidance from a pro with 25 years of success as a commercial contractor. Find out what it takes to build all aspects of a business that is profitable, enjoyable, and enduring. With a bonus chapter on retail construction.

$60.00 | 201 pages, softcover | Catalog no. 67352

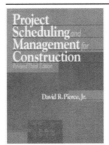

Project Scheduling and Management for Construction, 3rd Edition

by David R. Pierce, Jr.

A comprehensive, yet easy-to-follow guide to construction project scheduling and control—from vital project management principles through the latest scheduling, tracking, and controlling techniques. The author is a leading authority on scheduling, with years of field and teaching experience at leading academic institutions. Spend a few hours with this book and come away with a solid understanding of this essential management topic.

$64.95 | Over 286 pages, illustrated, hardcover | Catalog no. 67247B

The Practice of Cost Segregation Analysis

by Bruce A. Desrosiers and Wayne J. Del Pico

This expert guide walks you through the practice of cost segregation analysis, which enables property owners to defer taxes and benefit from "accelerated cost recovery" through depreciation deductions on assets that are properly identified and classified.

With a glossary of terms, sample cost segregation estimates for various building types, key information resources, and updates via a dedicated website, this book is a critical resource for anyone involved in cost segregation analysis.

$105.00 | Over 240 pages | Catalog no. 67345

Preventive Maintenance for Multi-Family Housing

by John C. Maciha

Prepared by one of the nation's leading experts on multi-family housing.

This complete PM system for apartment and condominium communities features expert guidance, checklists for buildings and grounds maintenance tasks and their frequencies, and a dedicated website featuring customizable electronic forms. A must-have for anyone involved with multi-family housing maintenance and upkeep.

$95.00 | 290 pages | Catalog no. 67346

RSMeans data titles are also available in online format! Go to RSMeansOnline.com for more details.

For more information visit the RSMeans website at www.rsmeans.com

Green Building: Project Planning & Cost Estimating, 3rd Edition

Since the widely read first edition of this book, green building has gone from a growing trend to a major force in design and construction.

This new edition has been updated with the latest in green building technologies, design concepts, standards, and costs. Full-color with all new case studies—plus a new chapter on commercial real estate.

$110.00 | 480 pages, softcover | Catalog no. 67338B

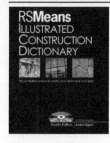
RSMeans Illustrated Construction Dictionary Unabridged 4th Edition, with CD-ROM

Long regarded as the industry's finest, *Means Illustrated Construction Dictionary* is now even better. With nearly 20,000 terms and more than 1,400 illustrations and photos, it is the clear choice for the most comprehensive and current information. The companion CD-ROM that comes with this new edition adds extra features, such as larger graphics and expanded definitions.

$110.00 | Over 880 pages, illust., hardcover | Catalog no. 67292B

Construction Supervision

Inspires excellence with proven tactics and techniques applied by thousands of construction supervisors. Leadership guidelines carve out a practical blueprint for motivating work performance and increasing productivity through effective communication. Features:

- A unique focus on field supervision and crew management
- Coverage of supervision from the foreman to the superintendent level
- An overview of technical skills whose mastery will build confidence and success for the supervisor
- A detailed view of "soft" management and communication skills

$95.00 | 464 pages, softcover | Catalog no. 67358

Building & Renovating Schools

This all-inclusive guide covers every step of the school construction process—from initial planning, needs assessment, and design, right through moving into the new facility. A must-have resource for anyone concerned with new school construction or renovation. With square foot cost models for elementary, middle, and high school facilities, and real-life case studies of recently completed school projects.

The contributors to this book—architects, construction project managers, contractors, and estimators who specialize in school construction—provide start-to-finish, expert guidance on the process.

$110.00 | Over 412 pages, hardcover | Catalog no. 67342

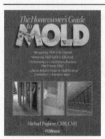
The Homeowner's Guide to Mold
By Michael Pugliese

Mold, whether caused by leaks, humidity or flooding, is a real health and financial issue—for homeowners and contractors. This full-color book explains:

- Construction and maintenance practices to prevent mold
- How to inspect for and remove mold
- Mold remediation procedures and costs
- What to do after a flood
- How to deal with insurance companies

$21.95 | 144 pages, softcover | Catalog no. 67344

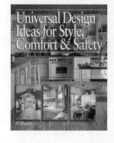
Universal Design Ideas for Style, Comfort & Safety
by RSMeans and Lexicon Consulting, Inc.

Incorporating universal design when building or remodeling helps people of any age and physical ability more fully and safely enjoy their living spaces. This book shows how universal design can be artfully blended into the most attractive homes. It discusses specialized products like adjustable countertops and chair lifts, as well as simple ways to enhance a home's safety and comfort. With color photos and expert guidance, every area of the home is covered. Includes budget estimates that give an idea how much projects will cost.

$21.95 | 160 pages, illustrated, softcover | Catalog no. 67354

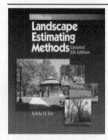
Landscape Estimating Methods, 5th Edition

Answers questions about preparing competitive landscape construction estimates, with up-to-date cost estimates and the new MasterFormat classification system. Expanded and revised to address the latest materials and methods, including new coverage on approaches to green building. Includes:

- Step-by-step explanation of the estimating process
- Sample forms and worksheets that save time and prevent errors

$75.00 | 336 pages, softcover | Catalog no. 67295C

Plumbing Estimating Methods, 3rd Edition
by Joseph J. Galeno and Sheldon T. Greene

Updated and revised! This practical guide walks you through a plumbing estimate, from basic materials and installation methods through change order analysis. *Plumbing Estimating Methods* covers residential, commercial, industrial, and medical systems, and features sample takeoff and estimate forms and detailed illustrations of systems and components.

$65.00 | 381+ pages, softcover | Catalog no. 67283B

Understanding & Negotiating Construction Contracts
by Kit Werremeyer

Take advantage of the author's 30 years' experience in small-to-large (including international) construction projects. Learn how to identify, understand, and evaluate high risk terms and conditions typically found in all construction contracts—then, negotiate to lower or eliminate the risk, improve terms of payment, and reduce exposure to claims and disputes. The author avoids "legalese" and gives real-life examples from actual projects.

$80.00 | 320 pages, softcover | Catalog no. 67350

2015 Order Form

ORDER TOLL FREE 1-800-334-3509
OR FAX 1-800-632-6732

For more information visit the RSMeans website at www.rsmeans.com

Qty.	Book no.	COST ESTIMATING BOOKS	Unit Price	Total
	60065	Assemblies Cost Data 2015	$334.95	
	60015	Building Construction Cost Data 2015	206.95	
	60045	Commercial Renovaion Cost Data 2015	166.95	
	60115	Concrete & Masonry Cost Data 2015	188.95	
	50145	Construction Cost Indexes 2015 (subscription)	359.95	
	60145A	Construction Cost Index–January 2015	94.95	
	60145B	Construction Cost Index–April 2015	94.95	
	60145C	Construction Cost Index–July 2015	94.95	
	60145D	Construction Cost Index–October 2015	94.95	
	60345	Contr. Pricing Guide: Resid. R & R Costs 2015	39.95	
	60235	Electrical Change Order Cost Data 2015	199.95	
	60035	Electrical Cost Data 2015	208.95	
	60205	Facilities Construction Cost Data 2015	524.95	
	60305	Facilities Maintenance & Repair Cost Data 2015	461.95	
	60555	Green Building Cost Data 2015	170.95	
	60165	Heavy Construction Cost Data 2015	204.95	
	60095	Interior Cost Data 2015	208.95	
	60125	Labor Rates for the Const. Industry 2015	450.95	
	60185	Light Commercial Cost Data 2015	153.95	
	60025	Mechanical Cost Data 2015	203.95	
	60155	Open Shop Building Const. Cost Data 2015	175.95	
	60215	Plumbing Cost Data 2015	206.95	
	60175	Residential Cost Data 2015	148.95	
	60285	Site Work & Landscape Cost Data 2015	199.95	
	60055	Square Foot Costs 2015	218.95	
	62014	Yardsticks for Costing (2014)	189.95	
	62015	Yardsticks for Costing (2015)	201.95	
		REFERENCE BOOKS		
	67329	Bldrs Essentials: Best Bus. Practices for Bldrs	45.00	
	67307	Bldrs Essentials: Plan Reading & Takeoff	50.00	
	67342	Building & Renovating Schools	110.00	
	67261A	The Building Prof. Guide to Contract Documents	75.00	
	67353A	Complete Book of Framing, 2nd Ed.	29.95	
	67146	Concrete Repair & Maintenance Illustrated	75.00	
	67352	Construction Business Management	60.00	
	67364	Construction Law	115.00	
	67358	Construction Supervision	95.00	
	67363	Cost Data Student Edition	110.00	
	67314	Cost Planning & Est. for Facil. Maint.	95.00	
	67230B	Electrical Estimating Methods, 3rd Ed.	75.00	
	67343A	Est. Bldg. Costs for Resi. & Lt. Comm., 2nd Ed.	75.00	
	67276B	Estimating Handbook, 3rd Ed.	110.00	
	67318	Facilities Operations & Engineering Reference	115.00	
	67338B	Green Building: Proj. Planning & Cost Est., 3rd Ed.	110.00	
	67355	Green Home Improvement	34.95	
	67357B	The Gypsum Construction Handbook	40.00	
	67148	Heavy Construction Handbook	110.00	
	67308E	Home Improvement Costs–Int. Projects, 9th Ed.	24.95	
	67309E	Home Improvement Costs–Ext. Projects, 9th Ed.	24.95	
	67344	Homeowner's Guide to Mold	21.95	
	67324C	How to Est.w/Means Data & CostWorks, 4th Ed.	75.00	

Qty.	Book no.	REFERENCE BOOKS (Cont.)	Unit Price	Total
	67351A	How Your House Works, 2nd Ed.	22.95	
	67282A	Illustrated Const. Dictionary, Condensed, 2nd Ed.	65.00	
	67292B	Illustrated Const. Dictionary, w/CD-ROM, 4th Ed.	110.00	
	67362	Illustrated Const. Dictionary, Student Ed.	60.00	
	67348	Job Order Contracting	130.00	
	67295C	Landscape Estimating Methods, 5th Ed.	75.00	
	67341	Life Cycle Costing for Facilities	110.00	
	67370	Managing the Profitable Construction Business	60.00	
	67294B	Mechanical Estimating Methods, 4th Ed.	75.00	
	67345	Practice of Cost Segregation Analysis	105.00	
	67366	Project Control: Integrating Cost & Sched.	95.00	
	67346	Preventive Maint. for Multi-Family Housing	95.00	
	67367	Project Sched. and Manag. for Constr., 4th Ed.	95.00	
	67322B	Resi. & Light Commercial Const. Stds., 3rd Ed.	65.00	
	67359	Risk Management for Design and Construction	125.00	
	67365	Solar Energy: Technology & Project Delivery	85.00	
	67145B	Sq. Ft. & Assem. Estimating Methods, 3rd Ed.	75.00	
	67321	Total Productive Facilities Management	85.00	
	67350	Understanding and Negotiating Const. Contracts	80.00	
	67303B	Unit Price Estimating Methods, 4th Ed.	75.00	
	67354	Universal Design	21.95	
	67319A	Value Engineering: Practical Applications	90.00	

MA residents add 6.25% state sales tax _____

Shipping & Handling** _____

Total (U.S. Funds)* _____

Prices are subject to change and are for U.S. delivery only. *Canadian customers may call for current prices. **Shipping & handling charges: Add 8% of total order for check and credit card payments. Add 8% of total order for invoiced orders.

Send order to: ADDV-1000

Name (please print) _____

Company _____

☐ Company
☐ Home Address _____

_____

City/State/Zip _____

Phone # _____ P.O. # _____

(Must accompany all orders being billed)

Mail to: RSMeans, 700 Longwater Drive, Norwell, MA 02061

RSMeans Project Cost Report

By filling out this report, your project data will contribute to the database that supports the RSMeans® Project Cost Square Foot Data. When you fill out this form, RSMeans will provide a $50 discount off one of the RSMeans products advertised in the preceding pages. Please complete the form including all items where you have cost data, and all the items marked (☑).

$50.00 Discount per product for each report you submit

Project Description (NEW construction only, please)

☑ Building Use (Office School . . .) _____

☑ Address (City, State) _____

_____

☑ Total Building Area (SF) _____

☑ Ground Floor (SF) _____

☑ Frame (Wood, Steel . . .) _____

☑ Exterior Wall (Brick, Tilt-up . . .) _____

☑ Basement: (check one)   ☐ Full   ☐ Partial   ☐ None

☑ Number of Stories _____

☑ Floor-to-Floor Height _____

☑ Volume (C.F.) _____

% Air Conditioned _____ Tons _____

Total Project Cost $ _____

Owner _____

Architect _____

General Contractor _____

☑ Bid Date _____

☑ Typical Bay Size _____

☑ Occupant Capacity _____

☑ Labor Force: _____ % Union   _____ % Non-Union

☑ Project Description (Circle one number in each line.)

   1. Economy   2. Average   3. Custom   4. Luxury
   1. Square   2. Rectangular 3. Irregular 4. Very Irregular

Comments _____

_____

A	☑	General Conditions	$
B	☑	Site Work	$
C	☑	Concrete	$
D	☑	Masonry	$
E	☑	Metals	$
F	☑	Wood & Plastics	$
G	☑	Thermal & Moisture Protection	$
GR		Roofing & Flashing	$
H	☑	Doors and Windows	$
J	☑	Finishes	$
JP		Painting & Wall Covering	$

K	☑	Specialties	$
L	☑	Equipment	$
M	☑	Furnishings	$
N	☑	Special Construction	$
P	☑	Conveying Systems	$
Q	☑	Mechanical	$
QP		Plumbing	$
QB		HVAC	$
R	☑	Electrical	$
S	☑	Mech./Elec. Combined	$

Please specify the RSMeans product you wish to receive.
Complete the address information.

Product Name _____

Product Number _____

Your Name _____

Title _____

Company _____

  ☐ Company
  ☐ Home   Street Address _____

City, State, Zip _____

Email Address _____

Return by mail or fax 888-492-6770.

Method of Payment:

Credit Card # _____

Expiration Date _____

Check _____

Purchase Order _____

Phone Number _____
(if you would prefer a sales call)

RSMeans
Square Foot Costs Department
700 Longwater Drive
Norwell, MA 02061
www.RSMeans.com